MW00844894

River Ecology and Management

Springer-Science+Business Media, LLC

Robert J. Naiman Robert E. Bilby
Editors

River Ecology and Management

Lessons from the Pacific Coastal Ecoregion

Sylvia Kantor
Associate and Managing Editor

With 202 Illustrations

Robert J. Naiman
School of Fisheries
University of Washington
Seattle, WA 98195
USA

Robert E. Bilby
Weyerhaeuser Company
Tacoma, WA 98477
USA

Cover: Queets River, Olympic National Park, Washington (Photo by Tim Hyatt)

Library of Congress Cataloging-in-Publication Data
River ecology and management: lessons from the Pacific coastal
ecoregion / [edited-by] Robert J. Naiman, Robert E. Bilby.
p. cm.
Includes index.
ISBN 978-0-387-95246-8 ISBN 978-1-4612-1652-0 (eBook)
DOI 10.1007/978-1-4612-1652-0
1. Stream ecology—Pacific Coast Region (North America). 2. Stream
conservation—Pacific Coast Region (North America). I. Naiman, Robert J.
II. Bilby, Robert E.
QH104.5.P32R57 1998
577.6′4′0979—dc21 97-44766

Printed on acid-free paper.

First softcover printing, 2001.

Production coordinated by Chernow Editorial Services, Inc., and managed by Terry V. Kornak; manufacturing supervised by Joe Quatela.
Typeset by Best-set Typesetter Ltd., Hong Kong.

9 8 7 6 5 4 3 2 1

ISBN 978-0-387-95246-8 SPIN 10794106

Preface

As the vast expanses of natural forests and the great populations of salmonids are harvested to support a rapidly expanding human population, the need to understand streams as ecological systems and to manage them effectively becomes increasingly urgent. The unfortunate legacy of such natural resource exploitation is well documented. For several decades the Pacific coastal ecoregion of North America has served as a natural laboratory for scientific and managerial advancements in stream ecology, and much has been learned about how to better integrate ecological processes and characteristics with a human-dominated environment. These insightful but hard-learned ecological and social lessons are the subject of this book.

Integrating land and rivers as interactive components of ecosystems and watersheds has provided the ecological sciences with important theoretical foundations. Even though scientific disciplines have begun to integrate land-based processes with streams and rivers, the institutions and processes charged with managing these systems have not done so successfully. As a result, many of the watersheds of the Pacific coastal ecoregion no longer support natural settings for environmental processes or the valuable natural resources those processes create. An important role for scientists, educators, and decision makers is to make the integration between ecology and consumptive uses more widely understood, as well as useful for effective management.

It is timely for ecologists and managers to carefully consider three questions: What will happen if stream and river policies are not underpinned by science? How can positive change be realized? How can science be better linked to decision making? Whenever these questions have been carefully considered and acted on, the results have had positive impacts on environmental quality and resource sustainability. These are some of the difficult questions addressed in this book.

The original concept for this volume arose as a result of increasing demands from three related groups: graduate students asking for a comprehensive text for classes in stream and river ecology at the

University of Washington, resource professionals seeking information on new and emerging stream management issues in the Pacific Northwest, and decision makers realizing that policies based on perception, tradition, or factors other than those founded on factual information were not helpful in resolving resource issues over the long term.

We, the editors, extensively discussed the organization and content of the book with colleagues at the University of Washington for two years and decided that an edited book would be the best approach, primarily for two reasons: (1) the volume of information available on streams and rivers had expanded exponentially since about 1980, making a coauthored volume an especially difficult challenge, and (2) regionally, the Pacific Northwest is fortunate to have an unusually talented collection of stream ecologists and managers who, over the previous decades, have provided substantial insights into their respective fields.

This book is written largely by faculty, staff, students, and colleagues affiliated with the University of Washington and the Pacific Northwest Research Station of the U.S. Forest Service. It is divided into five sections: the physical environment, the biotic environment, ecosystem processes, management, and considerations for the future. The book is designed to synthesize the enormous amount of information generated in the last two decades about streams and rivers, especially those in the Pacific coastal ecoregion. It is truly a collaborative effort with students working with faculty, corporate scientists teaming with academic scientists, and managers sharing ideas with biologists and environmentalists. This book embodies the spirit of scientists and managers working across disciplines to provide state-of-knowledge analyses about specific issues and educational opportunities while at the university and beyond.

A book of this scope requires the efforts and collaborations of scores of dedicated people, governmental agencies, and foundations. We owe a great debt of appreciation to the 64 reviewers of individual chapters, listed in the appendix, and to the 21 agencies, institutions, and private foundations that supported the research and the writing of chapters. Special appreciation is extended to the Weyerhaeuser Foundation, the Pacific Northwest Research Station of the U.S. Forest Service, and to the Simpson Company for directly supporting the compilation and publication of this volume.

Personally, we thank Peter A. Bisson, Loveday L. Conquest, Henri Décamps, Richard T. Edwards, Penelope L. Firth, Jerry F. Franklin, Catherine Gustafson, James R. Karr, John J. Magnuson, Carla Manning, Laurence Maridet, David Montgomery, Tom Quinn, Stephen C. Ralph, Jeffrey Richey, Kevin H. Rogers, James R. Sedell, Jack A. Stanford, Deanna J. Stouder, Phillippe Vervier, and numerous graduate students and technicians for willingly sharing their ideas, critiquing our approaches, and supporting our efforts over the years. We also thank Pat Owen for assistance with figures and Laura Mckee for assistance with manuscript preparation. We are especially grateful for editorial assistance provided by Daun Redfield.

Finally, we thank Sylvia Kantor for an exceptional effort in keeping the editorial process on schedule by working so effectively with the authors and editors and by embracing the most difficult tasks with grace and skill.

Robert J. Naiman
Robert E. Bilby

Contents

Preface v
Contributors xxi

1 River Ecology and Management in the Pacific Coastal Ecoregion 1
Robert J. Naiman and Robert E. Bilby

Defining Streams and Rivers 1
A Brief History of Stream Ecology 1
Recent Advances in Stream Ecology 2
Rivers as Integrators of Environmental Conditions 3
Environmental Status of Streams and Rivers 4
Providing a Sense of Place 5
The Pacific Northwest 5
The Pacific Coastal Ecoregion 7
Science and Management 8

Part I The Physical Environment

2 Channel Processes, Classification, and Response 13
David R. Montgomery and John M. Buffington

Overview 13
Introduction 13
Channel Processes 14
Conceptual Models of Channel Response 16
Examples of Channel Change 17
Sediment Supply 17
Discharge 19
Dams 19
Geomorphological Channel Classification 20
Past Classifications 20
Hierarchical Channel Classification 21
Channel Disturbance and Response Potential 31
Reach-Level Response 31

Segment-Level Response 32
External Influences 32
Debris Flow Disturbance 34
Applications for Ecosystem Analysis 36

3 Hydrology 43
Robert R. Ziemer and Thomas E. Lisle

Overview 43
Introduction 43
Hydrology of the Pacific Coastal Ecoregion 44
Runoff Processes 47
Hillslope Runoff 48
Effects of Land-Use Practices on Runoff 56
Peak Flows and Floods 56
Hydrologic Recovery from Land-Use Impacts 62

4 Stream Quality 69
Eugene B. Welch, Jean M. Jacoby, and Christopher W. May

Overview 69
Introduction 70
Acid Neutralizing Capacity, pH, and Hardness 73
Cations, Anions, and Conductivity 74
Nutrients 75
Temperature 77
Biochemical Oxygen Demand and Dissolved Oxygen 78
Metals 81
Suspended Solids 83
Water Quality and Nuisance Algae (Periphyton) 84
Nuisance Periphyton 85

Part II The Biotic Environment

5 Biotic Stream Classification 97
Robert J. Naiman

Overview 97
Introduction 98
Historical Concepts 98
Recent Concepts 99
Classification of Physical Watershed Features (a Summary) 100
Single-Scale Classification 101
Hierarchical Classification 101
Classification Coupling Biological and Physical Features 111
Vertebrate Community Classification 111
Invertebrate Community Classification 112
Plant Classification 113
An Evaluation of the Biological–Physical Approach 114
Management Based on Stream Classification 114

6 Microorganisms and Organic Matter Decomposition 120
Keller F. Suberkropp

Overview ... 120
Introduction ... 120
Heterotrophic Microorganisms ... 122
Plant Litter Decomposition ... 127
Models for Plant Litter Decomposition ... 131
Comparison of Fungal and Bacterial Activity ... 132
Fungi-Shredder Interactions ... 133
Factors Affecting Rates of Plant Litter Breakdown ... 134
Wood ... 135
Dissolved Organic Matter ... 137
Metabolism of DOM and the Microbial Loop ... 137

7 Primary Production ... 144
Michael L. Murphy

Overview ... 144
Introduction ... 144
Forms and Typical Species ... 145
Benthic Algae ... 145
Macrophytes ... 148
Phytoplankton ... 149
The Primary Production Process ... 149
Limiting Factors ... 150
Energy Flow ... 153
Distribution of Primary Production in Watersheds ... 158
Potential Response to Watershed Uses ... 160

8 Stream Macroinvertebrate Communities ... 169
Anne E. Hershey and Gary A. Lamberti

Overview ... 169
Introduction ... 169
Species Assemblages ... 170
Macroinvertebrate Taxonomic Diversity ... 170
Macroinvertebrate Life Histories ... 172
Relationship of Diversity to Physical Environment ... 175
Role of Disturbance ... 176
Functional Feeding Groups and the River Continuum Concept ... 179
Invertebrate-Mediated Processes ... 183
Detritivory ... 183
Grazing ... 184
Predator-Prey Interactions ... 185
Macroinvertebrate Drift ... 187
Secondary Production ... 189
Macroinvertebrates in Stream Food Webs ... 189
Impact of Spawning Salmon on Stream Macroinvertebrates ... 192
Effects of Land Use on Community Dynamics ... 192

9 Fish Communities 200
Gordon H. Reeves, Peter A. Bisson, and Jeffrey M. Dambacher

Overview 200
Introduction 200
Regional Diversity 201
Anadromous Life Histories 205
Population Variability 206
Watershed Scale Patterns of Diversity 208
Ecological Rules 208
Physical and Biological Processes 209
Reach Scale Diversity 214
Habitat Unit Patterns of Diversity 218
Human Impacts on Fish 220
Differential Response of Species 222
Disturbance 224
Future Management Directions 226

10 Riparian Wildlife 235
Kathryn A. Kelsey and Steven D. West

Overview 235
Introduction 235
Distribution and Abundance of Riparian Wildlife Species 236
Riparian Obligates 236
Riparian Generalists 241
Exotic Species 242
Effects of Riparian Conditions and Processes on Wildlife 243
Landscape Processes 243
Local Processes 245
Management Effects on Riparian Wildlife Communities 250
Riparian Wildlife Management Alternatives 252

Part III Ecosystem Processes

11 Dynamic Landscape Systems 261
Lee E. Benda, Daniel J. Miller, Thomas Dunne, Gordon H. Reeves, and James K. Agee

Overview 261
Introduction 261
Components of Dynamic Landscape Systems 263
Climate 263
Topography 268
Hierarchical Patterns of Channel Networks 272
Basin History.................................... 273
Dynamic Landscape Systems: Populations of Elements and Time 274
Temporal Sequencing of Storms, Fires, and Floods, and Dynamic Channel Behavior 274
Effect of Hierarchical Networks and Spatial Scale on System Properties 277

Aquatic Biology at the Landscape Systems Level 281
Applications to Watershed Science and Management 283
A Field Perspective 283
The Problem of Cumulative Effects, Natural Disturbance, and Habitat Diversity 284

12 Riparian Forests 289
Robert J. Naiman, Kevin L. Fetherston, Steven J. McKay, and Jiquan Chen

Overview 289
Introduction 290
The Physical Setting and Geomorphic Context 292
Valley Morphology 294
Hillslope Processes 294
Fluvial Processes 296
Soil Processes 296
Large Woody Debris 296
Riparian Plant Adaptations 297
Morphological and Physiological Adaptations 299
Reproductive Strategies 299
Growth Dynamics 301
Spatial Patterns of Riparian Forests: A Mosaic 302
Disturbance 302
Successional Processes 304
Case Studies 304
Riparian Forests and Ecosystem Functions 310
Riparian Forest Microclimate 310
Riparian Plant Diversity 311
Riparian Forests and Land-Use Change 311
River Regulation 311
Forest Practices 317
Lessons for Management 318

13 Function and Distribution of Large Woody Debris 324
Robert E. Bilby and Peter A. Bisson

Overview 324
Introduction 324
Abundance and Distribution of LWD in Channel Networks 325
Processes Controlling Input and Output of LWD 327
LWD Function in Stream Ecosystems 331
Influence of Land Use on LWD 338

14 Nutrient Cycles and Responses to Disturbance 347
Michael E. McClain, Robert E. Bilby, and Frank J. Triska

Overview 347
Introduction 348
The Basics of Nitrogen, Phosphorus, and Sulfur Cycling ... 349
Nutrient Spiraling 349
Natural Forms, Distributions, and Transformations 351
Input-Output Pathways and Riverine Budgets 354

Controlling Variables in Nitrogen, Phosphorus, and Sulfur Cycling 357
Hydrologic Regime 357
Temperature 359
Biological Community Composition 359
Responses to Disturbance 363
Forest Conversion and Management 363
Urbanization and Agriculture 367
Fire 367
Climate Change 367

15 Organic Matter and Trophic Dynamics 373
Peter A. Bisson and Robert E. Bilby

Overview 373
Introduction 373
Trophic Pathways 377
Autotrophic Production 377
Allochthonous Organic Matter and Heterotrophic Production 379
Organic Matter Processing 380
Organic Matter Storage and Nutrient Spiraling 384
Impacts of Human Activity 386
Human Activities and Cascading Trophic Systems 386
Loss of Riparian Vegetation 388
Loss of Salmon Carcasses 389
Why Are Some Streams More Productive Than Others? 391

16 The Hyporheic Zone 399
Richard T. Edwards

Overview 399
Introduction 399
Definition and Delineation 401
Interstitial Volume and Surface Area 404
Hydrology of Hyporheic Interactions 405
Hyporheic Zone Distribution Patterns 408
Large-Scale Geologic Factors 408
Watershed and Valley Segment Scales 409
Channel Reach Scale 410
Channel Unit Scale 410
Roughness Elements Scale 411
Spatial Scales of Management Actions 411
Temporal Scales 412
Biogeochemical Processes 412
Transient Storage 412
Nitrogen Dynamics 413
Organic Matter Utilization and the Role of Epilithon 417
Ecology and Structure of Hyporheic Invertebrate Communities 419
Controls on Community Structure 421
Food Sources and Trophic Structure 421
Epilithic Biofilms 422
Secondary Production 423

Implications for Management of Coastal Rivers of Washington 423

17 Biodiversity 430
Michael M. Pollock

Overview 430
Introduction 430
An Overview of Diversity in Riparian Corridors 431
General Theories of (Local) Diversity 432
Common Measures of Species Richness, Diversity, Evenness, and Turnover Rates 434
Describing Diversity at Multiple Spatial Scales 436
Natural Processes Influencing Biodiversity Patterns 438
Hydrologic Regimes 440
Herbivory 443
Productivity 444
Habitat Heterogeneity 444
Large-Scale Spatial Heterogeneity 446
Implications for Management 448

Part IV Management

18 Statistical Design and Analysis Considerations for Monitoring and Assessment 455
Loveday L. Conquest and Stephen C. Ralph

Overview 455
Introduction 456
Sampling Design 456
Sampling Approaches and Sampling Units 457
Replication 461
Reference Sites 461
The Issue of Scale 462
When *n* Equals One—The Argument for Case Studies 463
Choosing Parameters 464
Acquiring and Maintaining Good Data 466
Intensive and Extensive Approaches 466
Training Field Crews 466
Quality Control and Quality Assurance 467
Management of Information 467
Data Analysis 468
Parametric Procedures, Regression and Correlation, and Nonparametric Tests 468
Multivariate Procedures 470
Exploratory Data Analysis 471
Geographic Information Systems 471

19 Cumulative Watershed Effects and Watershed Analysis 476
Leslie M. Reid

Overview 476
Introduction 477

Problems in the Evaluation of Cumulative Watershed Effects 479
Technical Issues 480
Philosophical Issues 481
Sociocultural Issues 483
The Ad Hoc Approach to Cumulative Effects Evaluation 485
Standardized Methods of Cumulative Effects Analysis 488
An Index Approach: Equivalent Roaded Acres 489
A Mechanistic Impact Model: The Fish-Sediment Model 491
Professional Judgment: The California Checklist 491
Administrative Convenience Versus Technical Adequacy 492
Watershed Analysis 493
Limited Assessment with Prescriptions—Timber/Fish/Wildlife Watershed Analysis 493
Broad Assessment Without Prescriptions—Interagency Ecosystem Analysis 494
Contrasting Goals and Methods 496
Administrative Convenience Versus Technical Adequacy, Revisited 496
Tomorrow's Analyses 498

20 Rivers as Sentinels: Using the Biology of Rivers to Guide Landscape Management 502
James R. Karr

Overview 502
Introduction 503
Rivers as Sentinels 503
Biological Integrity and Cumulative Effects 505
Evolution of Biological Monitoring 506
The Index of Biological Integrity 509
Selecting IBI Metrics 510
Scoring Metrics 513
Integrating Multiple Metrics 513
What IBI Says About Streams and Watersheds 515
Detecting the Effects of Point Source Pollution 515
Identifying Multiple Sources of Degradation 515
Describing Geographic Pattern and Detecting Cause 517
Detecting Regional Variation in Human Influence 517
Detecting Change Over Time as Human Activity Changes 517
Evaluating Management Efforts 517
Statistical Power and Precision of IBI 520
A Benthic IBI for the Pacific Northwest 520
Change and Risk Assessment 523

21 Social Organizations and Institutions 529
Margaret A. Shannon

Overview 529
Introduction 530
A Drop of Water 531
Thinking Like Scientists and Managers 532
Key Concepts Defined 533

The Changing Policy Environment of Streams 538
Forming Integrative Policy Communities 540
Institutional Strategies for Collaboration 543
Typology of Organizational Decision Processes 545
Computation 546
Experimentation and Pragmatism 546
Bargaining and Advocacy with Technical Competence 547
Consensus Building and Organizational Learning 547
Future Outlook 548

22 River Law 553
Robert J. Masonis and F. Lorraine Bodi

Overview 553
Introduction 553
Sources of Law 554
Federal and State Jurisdiction 555
Laws Regulating River Systems 555
Water Quantity (In-Stream Flows) 556
Water Quality 558
Land Use 560
Biota and Habitat—Endangered Species Act 565
New Approaches 566
Linking Water Quality and Water Quantity 566
Controlling Nonpoint Source Pollution 567
Improving Environmental Protection Under the Federal Power Act 567
Future Outlook 568

23 Economic Perspectives 572
Daniel Huppert and Sylvia Kantor

Overview 572
Introduction 573
Economics and Water Resources 573
Historical Perspectives 576
Defining and Measuring Economic Value 576
Individual Values—Compensating and Equivalent Variations 577
Categories of Economic Value 579
Aggregation into "Social Value" 581
Discounting and Aggregation over Time 581
Economic Benefits and Competitive Markets 582
Measuring Nonmarket Economic Values 584
The Role of Economics in Decision Making 585
Economic Impacts of Policy Decisions 585
Impact versus Benefits 586
Economic Assessment of Water Resources 587
Economics and the Ecology of River Management 589
Forest Practices and Salmon Fishing 590
In-Stream Flow and Recreational Values 590
Sediment from Agriculture 591

Part V The Future

24 Stream and Watershed Restoration 599
Christopher A. Frissell and Stephen C. Ralph

Overview 599
Introduction 599
Defining Restoration—Scope and Scale 601
Interventions at the Microhabitat Scale 603
Large-Scale River Restoration 606
Watershed-Scale Restoration—An Example 609
Monitoring and Evaluating Restoration Projects 610
A Nested Experimental Design for Monitoring 613
Cost Accounting for Watershed Restoration 614
Watershed Restoration and Adaptive Ecosystem
Management 617
Elements of Successful Restoration and Monitoring 619

25 Nonprofit Organizations and
Watershed Management 625
Bettina von Hagen, Spencer Beebe, Peter Schoonmaker, and Erin Kellogg

Overview 625
Introduction 626
Theories of Nonprofit Formation 626
The Limits of Government 627
The Hidden Costs of Profit-Seeking 629
The Emergence of Nonprofits 629
Adapting Developing World Strategies to the Pacific
Coastal Ecoregion 631
The Role of Nonprofits in Watershed Management 632
Social and Economic Aspects of the
Pacific Coastal Ecoregion 632
Nonnprofits Building Institutional Capacity 633
Nonprofits Providing Access to Information 634
Nonprofits Restoring Degraded Watersheds 635
Nonprofits Promoting Market Incentives for
Watershed Conservation 635
The Future of Nonprofit Organizations 638

26 Watershed Management 642
Robert J. Naiman, Peter A. Bisson, Robert G. Lee, and Monica G. Turner

Overview 642
Introduction 642
Fundamental Elements of Watershed Management 643
The Natural System: Variability in Time and Space 643
A Holistic Perspective: Persistence and Invasiveness 644
Connectivity and Uncertainty 645
Human Cultures and Institutions 646

Practical Approaches for Implementing Watershed Management 646
Quantitative Analyses 646
Accepting Risk and Addressing Uncertainty 651
How Can Organizations Deal with Risk? 652
Addressing Institutional Organization and the Paradox of Scale 653
Formulating Shared Socioenvironmental Visions 654
Public Stewardship in Watershed Management 655
Fundamental Principles 658

27 Paradigms, Policies, and Prognostication about the Management of Watershed Ecosystems 662
Michael C. Healey

Overview 662
Introduction 663
Why Watersheds? 664
What Can and Cannot Be Known about Watershed Ecosystems? 666
The Process of Watershed Management 667
The Role of Science 667
Values 669
Accepting Limits 671
Achieving Goals 672
Evolving Paradigms in Watershed Management 673
Watershed Engineering 673
Environmental Assessment and Mitigation 674
Adaptive Management 675
Looking to the Future 675
Ecosystem Management 675
Setting Goals for Ecosystem Management 677
From Incrementalism to Adaptive Rationalism 678
A New Class of Problems 679

Appendix: Reviewers 683

Index 689

Contributors

James K. Agee
College of Forest Resources
University of Washington
Seattle, WA 98195
USA

Spencer Beebe
Ecotrust
Portland, OR 97209
USA

Lee E. Benda
Earth Systems Institute
Seattle, WA 98105-5832
USA

Robert E. Bilby
Weyerhaeuser Company
Tacoma, WA 98477
USA

Peter A. Bisson
USDA Forest Service
Olympia Forestry Sciences Lab
Olympia, WA 98502
USA

F. Lorraine Bodi
American Rivers Council
Seattle, WA 98122
USA

John M. Buffington
Department of Geological Sciences
University of Washington
Seattle, WA 98195
USA

Jiquan Chen
School of Forestry and Wood Products
Michigan Technological University
Houghton, MI 49931
USA

Loveday L. Conquest
School of Fisheries
University of Washington
Seattle, WA 98195
USA

Jeffrey M. Dambacher
Oregon Department of Fish and Wildlife
Corvallis, OR 97333
USA

Thomas Dunne
Graduate School of Environmental Science and Management
University of California
Santa Barbara, CA 93106
USA

Richard T. Edwards
Center for Streamside Studies
University of Washington
Seattle, WA 98195
USA

Kevin L. Fetherston
College of Forest Resources
University of Washington
Seattle, WA 98195
USA

Christopher A. Frissell
Flathead Lake Biological Station
The University of Montana
Polson, MT 59860
USA

Michael C. Healey
Westwater Research Center
University of British Columbia
Vancouver, British Columbia
V6T 1Z4
Canada

Anne E. Hershey
Department of Biology
University of Minnesota
Duluth, MN 55812-9989
USA

Daniel Huppert
School of Marine Affairs
University of Washington
Seattle, WA 98195
USA

Jean M. Jacoby
Department of Environmental Engineering
Seattle University
Seattle, WA 98122
USA

Sylvia Kantor
School of Fisheries
University of Washington
Seattle, WA 98195
USA

James R. Karr
Department of Zoology
University of Washington
Seattle, WA 98195
USA

Erin Kellogg
Ecotrust Canada
Vancouver, British Columbia
V6B 5L1
Canada

Kathryn A. Kelsey
College of Forest Resources
University of Washington
Seattle, WA 98195
USA

Gary A. Lamberti
Department of Biological Sciences
University of Notre Dame
Notre Dame, IN 46556
USA

Robert G. Lee
College of Forest Resources
University of Washington
Seattle, WA 98195
USA

Thomas E. Lisle
USDA Forest Service
Redwood Sciences Lab
Arcata, CA 95521
USA

Robert J. Masonis
American Rivers Council
Seattle, WA 98122
USA

Christopher W. May
Applied Physics Laboratory
University of Washington
Seattle, WA 98195
USA

Michael E. McClain
Facultad de Ciencias Forestates
Departamento de Manejo Forestal
Universidad Nacional Agraria La Molina
Lima
Peru

Steven J. McKay
College of Forest Resources
University of Washington
Seattle, WA 98195
USA

Daniel J. Miller
Earth Systems Institute
Seattle, WA 98105-5832
USA

David R. Montgomery
Department of Geological Sciences
University of Washington
Seattle, WA 98195
USA

Michael L. Murphy
National Marine Fisheries Service
Auke Bay Laboratory
Juneau, AK 99801-8626
USA

Robert J. Naiman
School of Fisheries
University of Washington
Seattle, WA 98195
USA

Michael M. Pollock
10,000 Years Institute
P.O. Box 2205
Seattle, WA 98111
USA

Stephen C. Ralph
US Environmental Protection Agency, Region 10
Seattle, WA 98101
USA

Gordon H. Reeves
US Forest Service
Pacific Northwest Research Station
Corvallis, OR 97331
USA

Leslie M. Reid
USDA Forest Service
Redwood Sciences Lab
Arcata, CA 95521
USA

Peter Schoonmaker
Ecotrust
Portland, OR 97209
USA

Margaret A. Shannon
Environment and Society Institute
State University of New York
Buffalo, NY 14260
USA

Keller F. Suberkropp
Department of Biology
University of Alabama
Tuscaloosa, AL 35487
USA

Frank J. Triska
US Geological Survey
Menlo Park, CA 94025-3591
USA

Monica G. Turner
Department of Zoology
University of Wisconsin
Madison, WI 53706
USA

Bettina von Hagen
Ecotrust
Portland, OR 97209
USA

Eugene B. Welch
Civil Engineering
University of Washington
Seattle, WA 98195
USA

Steven D. West
College of Forest Resources
University of Washington
Seattle, WA 98195
USA

Robert R. Ziemer
USDA Forest Service
Redwood Sciences Lab
Arcata, CA 95521
USA

1
River Ecology and Management in the Pacific Coastal Ecoregion

Robert J. Naiman and Robert E. Bilby

Streams and rivers are among the most fascinating and complex ecosystems on Earth. Touching all parts of the natural environment and nearly all aspects of human culture, they act as integrators and centers of organization within the landscape. Their roles in providing natural resources, such as fish and clean water, are well known, as are their roles in providing transportation, energy, diffusion of wastes, and recreation. What is not as well known, however, is how running waters are structured and how they function as ecological systems. For centuries, numerous human societies have exploited the natural benefits provided by running waters without understanding how these ecosystems maintain their vitality. Today, with ever increasing demands being made on streams and rivers by an exponentially increasing human population, a basic ecological understanding of the structure and dynamics of running waters is essential for formulating sound management and policy decisions.

Defining Streams and Rivers

Webster's Dictionary defines *stream* as "a flow of water," and *river* as "a natural stream of water of considerable volume flowing into an ocean, a lake etc." However, these definitions are misleading in their simplicity, perhaps reflecting the general public's level of literacy about running waters. In essence, there is no exact differentiation between a stream and a river. In the context of this book, streams are defined as relatively small volumes of water moving within a visible channel, including subsurface water moving in the same direction as the surface water, and the associated riparian vegetation. Rivers are defined in a similar manner: a relatively large volume of water moving within a visible channel, including subsurface water moving in the same direction and the associated floodplain and riparian vegetation. Both streams and rivers, as ecological systems, are highly variable over space and time, and exhibit high degrees of connectivity between systems longitudinally, laterally, and vertically (Naiman et al. 1992, 1995).

A Brief History of Stream Ecology

A number of early limnological works dealt fairly extensively with what was known at that time about the ecology of streams and rivers (Steinmann 1907, 1909, Thienemann 1925, Carpenter 1928) and up to 1965 several syntheses of a popular nature or concerned with rather narrow aspects of the subject were published (Needham 1938, Macan and Worthington 1951, Coker 1954, Hynes 1960, Illies 1961, Reid 1961, Bardach 1964). The study of streams and rivers emerged as a major area of interest, however, in 1972 with the publication of the classic treatise, *The Ecology of Running Waters* (Hynes 1972). The publication of that book, in combination with a text by Leopold et al. (1964) on the geomorphology of channels and the perspectives of Bormann and Likens (1979) on biogeochemical processes in small watersheds, provided the foundation for much of the research of the next twenty-five years.

Since the mid-1970s, a number of volumes have been published addressing major components of stream ecosystems and providing overviews of the state of knowledge. Most notably, these include syntheses on geomorphology and hydrology (Dunne and Leopold 1978, Freeze and Cherry 1979, Richards 1982, Gordon et al. 1992), riparian forests (Naiman and Décamps 1990, Malanson 1993), aquatic invertebrates (Merritt and Cummins 1984, Ward 1992), conceptual advances in structure and process (Barnes and Minshall 1983, Fontaine and Bartell 1983, Dodge 1989, Webster and Meyer 1997), the dynamics of regulated rivers (Ward and Stanford 1979, Petts 1984), textbooks (Allen 1995), methodological approaches (Hauer and Lamberti 1996), and conservation management (Boon et al. 1992).

Many of the syntheses produced in the last two decades are edited volumes, a trend that perhaps reflects the highly interdisciplinary nature of stream ecology. Effectively answering most research questions related to rivers requires understanding the fundamentals of fluvial geomorphology, hydrology, chemistry, taxonomy, and ecology. Applying the ensuing results requires understanding of the fundamentals of management, engineering, law, sociology, and economics. Given that streams and rivers are the integrators of landscape conditions and that the study of streams and rivers is highly interdisciplinary, it is easier to understand the recent (since about 1980) exponential explosion of information about stream ecology and management. There are now more than twenty professional journals and hundreds of books addressing various aspects of stream ecology. Most of what is known about streams has been generated in the last two decades, and new scientific information continues to be produced at an accelerating pace.

Although more technical information is needed about various aspects of the ecology of flowing waters, the application of existing information to the management of streams and rivers continues to be a daunting challenge. Management of streams and rivers includes legal, social, and economic considerations, as well as scientific insights. Development of the political will and the institutional infrastructure that enables management of streams and rivers to keep pace with the rapid growth in scientific information is a critical issue for the conservation and restoration of these key ecosystems.

Recent Advances in Stream Ecology

Since 1980 a nearly continuous series of theoretical and intellectual advances has fueled the growing interest in streams as ecological systems. These advances provide the conceptual foundation for the rich array of ideas addressed in the chapters that follow. Nevertheless, it is useful to identify and briefly explain several of these advances and their interrelationships to illustrate the rapid intellectual evolution taking place.

The River Continuum Concept (Vannote et al. 1980). This concept developed from the idea that river systems, from the headwaters to the mouth, present a continuous gradient of physical conditions. This gradient provides the physical template upon which the biotic communities and their associated processes developed. As a result, one would expect to observe recognizable patterns in the community structure and the input, transport, utilization, and storage of organic matter.

Nutrient Spiraling Concept (Elwood et al. 1983). The term spiraling refers to the spatially dependent cycling of nutrients and the processing (i.e., oxidation, conditioning) of organic matter in lotic ecosystems (in which there is a strong down-slope force from the water flow). In effect, the nutrient spiraling concept provides both a conceptual and quantitative framework for describing the temporal and spatial dynamics of nutrients and organic matter in flowing waters. It also allows the structural and functional aspects of the biotic communities that enhance the retention and utilization of nutrients and organic matter to be interpreted in terms of ecosystem productivity and stability.

Serial Discontinuity Concept (Ward and Stanford 1983). Few riverine systems remain free-flowing over their entire course. Rather, regulation by dams typically results in an alternating series of lentic and lotic reaches. The serial discontinuity concept attempts to attain a broad theoretical perspective on regulated rivers, similar to that proposed by the river continuum concept for unregulated rivers.

Hierarchical Controls on System Functions (Frissell et al. 1986). The relative importance of factors controlling the physical and biological components of streams changes with the spatial and temporal scales being considered. A hierarchical perspective is necessary because stream processes operate at a wide range of scales (10^{-7} to 10^{8} m spatially and 10^{-8} to 10^{7} yr temporally). This approach is an effective management tool because it classifies streams (from microhabitat to watershed scales) using relatively few variables. This important conceptual advancement addresses form or pattern of channels within each hierarchical level, as well as origins and processes of development of channels.

Ecotones (Naiman et al. 1988). There are also compelling reasons for examining the stream–river continuum as a series of discrete patches or communities with reasonably distinct boundaries, rather than a gradual gradient or a continuum. The boundary perspective is complementary to the river continuum concept, but it addresses lateral linkages (e.g., channel–riparian forest exchanges) rather than only upstream–downstream ones, and provides a better understanding of factors regulating the exchange of energy and materials between identifiable resource patches.

The Flood Pulse Concept (Junk et al. 1989). A principal driving force behind the existence, productivity, and interaction of the diverse biota in river–floodplain systems is the flood pulse. A spectrum of geomorphological and hydrological conditions produces flood pulses, which range from unpredictable to predictable and from short to long duration. Short and generally unpredictable flood pulses occur in headwater streams and in streams heavily modified by human activities, whereas long duration and generally predictable floods occur in larger rivers. The net result is that the biota and the associated system-level processes reflect the characteristics of the flood regime.

Hyporheic Dynamics (Stanford and Ward 1993). Hydrologists have long considered streams and rivers to be dynamic regions linking groundwater with water draining the Earth's surface. Ecologists, in contrast, have typically perceived streams as bounded systems, consisting of the stream bottom and the overlying water. The hyporheic zone (the subsurface region, located beneath and adjacent to streams and rivers, that exchanges water with the surface) is also an integral component of streams that significantly expands the spatial extent of lotic ecosystems. This region is now known to harbor a rich assemblage of organisms which exert a significant influence on stream metabolism, nutrient cycling, and surface-subsurface interactions.

Rivers as Integrators of Environmental Conditions

Characteristics of streams and rivers serve as integrators of broader environmental conditions because they reflect the conditions of the surrounding landscape (Naiman 1992). Activities within a watershed, whether natural or anthropogenic (human-induced), influence the most basic aspects of the hydrological cycle. Vegetation absorbs and transpires water to the atmosphere; roads channelize water to streams; urbanization and certain types of agriculture decrease soil permeability, causing an increase in erosional sheetflow; and impoundments alter the timing, frequency, and intensity of peak flows. All of these alterations, singly and in combination, directly impact the habitat distributions, the trophic structure, the physical, chemical, and biological processes (such as sediment transport, nitrogen cycling, and primary production), and the demography of the biological communities.

Biological communities in streams and rivers evolved in response to a suite of influences (including disturbances) that, within certain norms of variation, have been largely predictable over thousands of years. Alterations to this suite of influences imposes changes on the ecological processes and the biological community of the ecosystem, or more simply, ecosystem integrity (Chapter 20). Biological integrity, the capacity to support and maintain a balanced, integrated, and adaptive biological system having the full range of elements and processes expected in a region's natural habitat, reflects the environmental conditions of the surrounding landscape. Critical elements of biological integrity include genes, populations, species, and communities; critical processes include mutation, demography, biotic interactions, biogeochemical cycles, energy dynamics, and metapopulation processes. Measures of biological integrity quantify not only the relative condition of the stream but also the condition of the watershed.

Environmental Status of Streams and Rivers

The alteration of aquatic habitats associated with streams and rivers is extensive in North America (as it is elsewhere in the world), largely as a consequence of the rapid increase in human population and the associated consumption of aquatic resources and the production of waste (Gleick 1993, Naiman et al. 1995, 1998). Since about 1750 A.D., wetlands in the United States, including many associated with large rivers, already have declined by 40 to 60% (Dahl 1990), while riparian forests have been destroyed on about 70% of the rivers (Swift 1984). The Nationwide Rivers Inventory estimates a total of 5.2 million kilometers of streams in the contiguous United States, but only 2% (less than 10,000 km) have sufficiently high-quality features to be considered relatively natural rivers and thus worthy of federal protection (Benke 1990). In North America (north of Mexico), in Europe, and in the republics of the former Soviet Union, 77% of the total water discharge of the 139 largest river systems is strongly or moderately affected by fragmentation of the river channel by dams and by water regulation resulting from reservoir operation, interbasin diversion, and irrigation (Dynesius and Nilsson 1994). In effect, stream and river habitats (and their associated biotic communities) are significantly more affected by human activities than many upland habitats occupied by charismatic megafauna.

In the United States, water consumption doubled in the last 40 years. Today, industries in the United States use about 150 billion liters of water daily, and over the next several decades this consumption will increase dramatically (Powledge 1984, Pimentel et al. 1997). Not only is agricultural consumption excessive but so is household consumption, with a typical family of four requiring about 1,300 L/d (of which only about 3% is used for cooking and drinking). In contrast, the corresponding consumption rate in developing countries is only 75 L/d per person or 300 L/d per family (Powledge 1984), uncomfortably close to Gleick's (1998) estimate of a *minimum* human water requirement of 50 L/person/d (200 L/family/d). To support this level of consumption, the United States alone has built more than two million dams, 87 of which impound reservoirs of more than one million acre-feet. Reservoirs have become a significant component of the nation's hydrologic cycle because they have the capacity to store an amount of water equal to three years' annual runoff from the nation's landscapes (Graf 1993).

Water consumption coupled with impacts from pollution, overharvest of commercially valuable species, and the introduction of exotic species have seriously degraded many types of riverine systems, and populations of many riverine species have become highly fragmented and modified. For example, destruction of specific aquatic and riparian habitats has fragmented the remaining habitats, significantly influencing the movement of water, materials, and organisms. As many stream organisms require several types of habitat to support different life history stages, selective habitat destruction has reduced the viability of populations in subtle ways that have severe long-term consequences. Pacific salmon (*Oncorhynchus* spp.) provide a well-known example (NRC 1996, Stouder et al. 1996), but evidence of biological impoverishment is pervasive in riverine systems. In general, the threat to aquatic biodiversity is more serious than threats to terrestrial diversity or even the integrity of tropical rain forests. Consider that 11 to 15% of terrestrial vertebrates (birds, mammals, reptiles, amphibians) in the United States are classed as rare to extinct, while the percentages of aquatic biota similarly classed are 34% for fish, 65% for crayfish, and 75% for unionid mussels (Master 1990). Of 214 stocks of Pacific salmon south of British Columbia, 74% have a high or moderate risk of extinction (Nehlsen et al. 1991). Despite massive expenditures to improve water quality, recovery efforts were not successful in removing any of the 251 species of fish listed as rare in 1979 from the list ten years later (Williams et al. 1989). Some, however, such as the Tecopa pupfish (*Cyprinodon nevadensis calidae*), were removed because they had become extinct. In major United States river systems, commercial harvesting of fishes has declined since about 1750 A.D. by at least 80%, and in some cases, the decline has approached nearly 100% (Karr 1993). Clearly, pressures from human population growth and resource consumption will not allow streams and rivers to return to their natural state; land use change will continue. Consequently, developing predictive knowledge of how land use alterations affect aquatic resources is a major challenge for the foreseeable future.

Providing a Sense of Place

This book uses a large number of examples taken from studies in the Pacific coastal ecoregion in particular, and the Pacific Northwest in general (Figure 1.1). The vast region of the Pacific Northwest provides an extraordinary array of physical settings and examples for understanding the dynamic processes associated with streams and their management.

The Pacific Northwest

Using a watershed-based approach, the Pacific Northwest encompasses those rivers draining to the Pacific Ocean from northern California, Oregon, Washington, western British Columbia, southern Yukon, southeastern Alaska, northern Nevada, Idaho, northern Utah, northwestern Wyoming, and western Montana (Figure 1.1). The defining features of the Pacific Northwest include four of North America's largest rivers (the Columbia, Fraser, Skeena, and Stikine), as well as thousands of smaller rivers flowing to the coast. All flow into the cold but highly productive waters of the North Pacific Ocean, stretching from the Mendocino Fracture Zone off the California coast to the seamounts rising from the ocean floor off the coasts of British Columbia and Alaska (Ryan 1994). On land, the topography is dominated by three great, young mountain ranges (the Cascade, Coast, and Rocky Mountains) with many other minor, but no less spectacular, mountains interspersed between them.

Moisture-laden air masses moving seasonally from the Aleutian and Hawaiian islands across the high mountains define the general climate distribution (Ryan 1994). The west sides of the Coast and Cascade mountains receive the bulk of the precipitation as winds force the storms over the high ridges. The lower parts of many river valleys commonly receive 1 to 4m of rain annually, while 5 to 20m of snow accumulate at higher elevations. In contrast, the arid eastern sides of the mountains are typically in a rain shadow, receiving less than 0.5m of precipitation annually. The extremely complex spatial and temporal interplay between storms and mountains results in an enormous variety of terrestrial and aquatic habitats.

Many parts of the Pacific Northwest are changing rapidly in response to human population growth, increasing demands for resources, technical innovations, and evolution of the social and institutional systems. Population levels in the region have increased exponentially since 1900 (NRC 1996). Before European contact, approximately 100,000 people lived in the Pacific Northwest. The discovery of gold in California, British Columbia, and Alaska, and the opening of the Oregon Trail in the mid-nineteenth century led to a rapid influx of immigrants. The European population reached 100,000 by 1870, while the Native American population declined to about 10,000 because of diseases introduced by Europeans. By 1900 the combined population of Oregon, Washington, and Idaho alone reached one million (NRC 1996). Today 8.7 million people live in these three states, and 13 million live in the Pacific

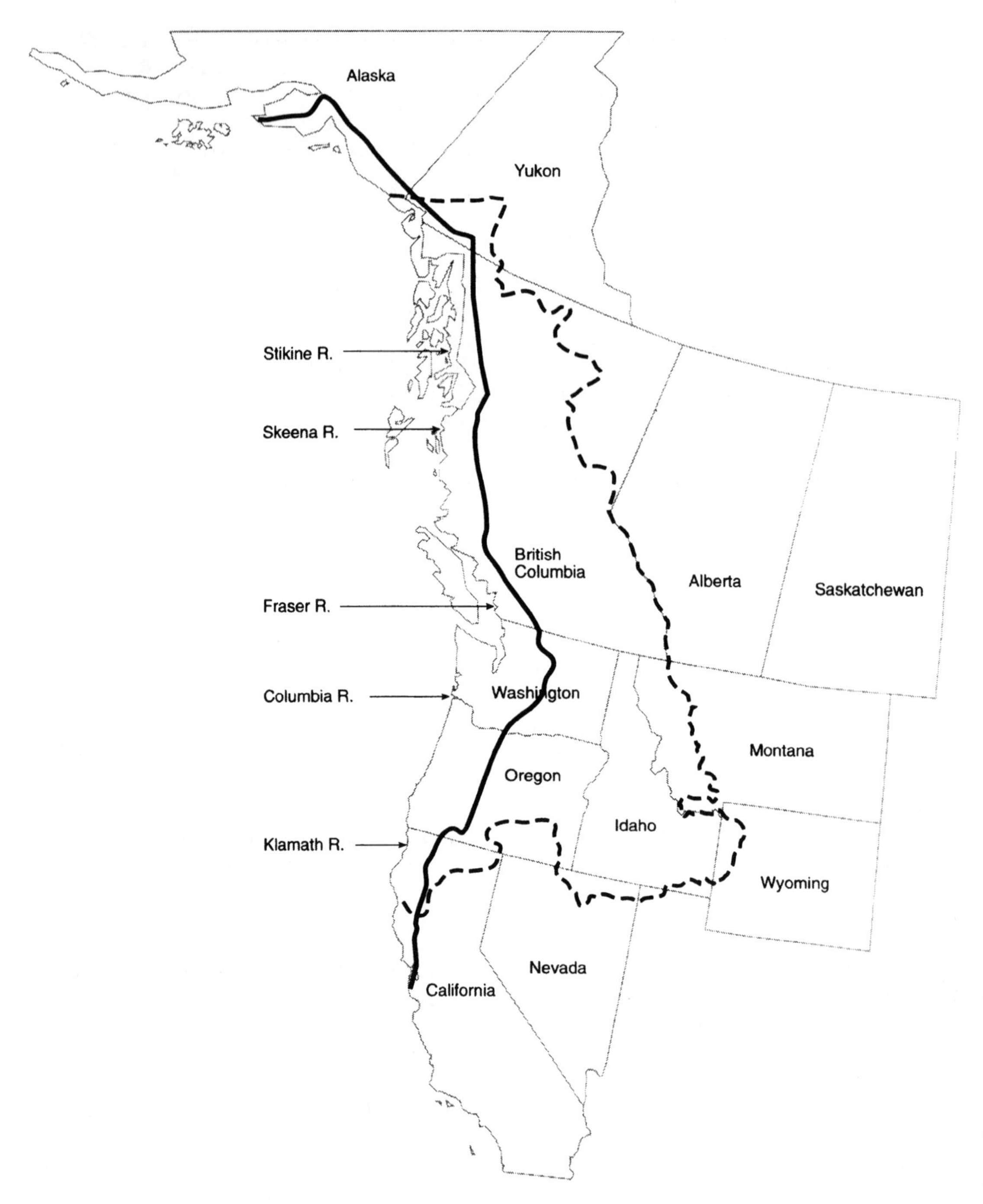

FIGURE 1.1. Watershed-based boundary of the Pacific Northwest (---) region of North America and approximate boundary of the Pacific coastal ecoregion (—) (modified from Ryan 1994, with permission).

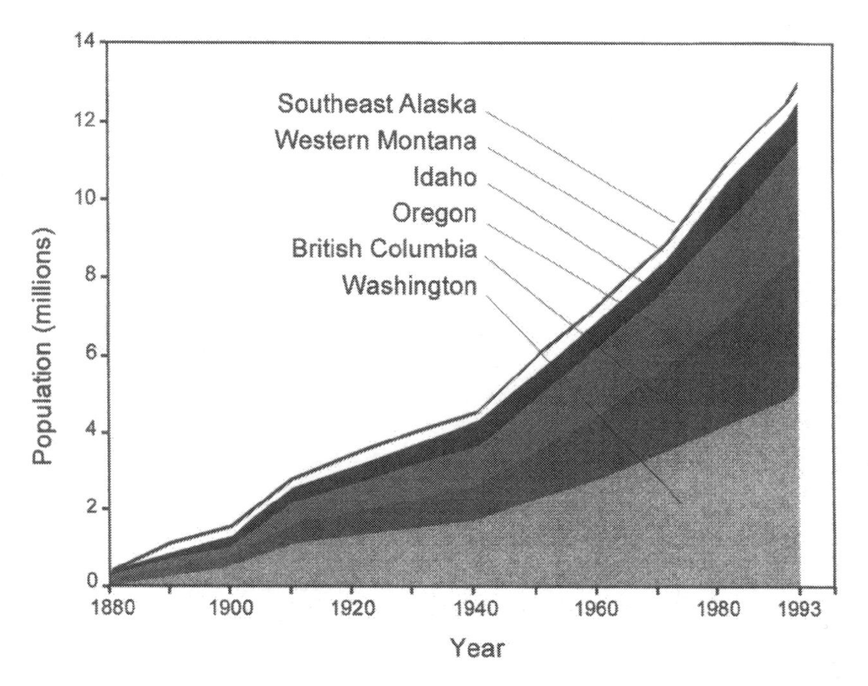

FIGURE 1.2. Human population trends for the states and provinces comprising the Pacific Northwest (from Ryan 1994, with permission).

Northwest region (Figure 1.2). The population of this region is rising faster than the average rates for Canada or the United States, primarily from immigration. At current growth rates, the population will double in the next forty years with most of that increase occurring in urban centers.

The population increase has fueled economic growth through the exploitation of the abundant and productive natural resources. Commercial exploitation of salmon began on the Columbia River in the late nineteenth century (NRC 1996). Salmon populations declined rapidly, with the most economically valuable species, chinook (*Oncorhynchus tshawytscha*) and sockeye (*O. nerka*), declining first followed by less desirable species. Current salmon populations in the Columbia River are less than 20% of estimated historical levels, while all populations south of British Columbia are at approximately 33% of their historic abundance.

Damming of streams and rivers for water, power, navigation, and flood control began in the late nineteenth century. Large numbers of impoundments have been constructed in the Pacific Northwest, some of them among the largest in the world. Nearly 3,000 dams large enough to meet federal criteria for inspection have been constructed in the Pacific Northwest south of British Columbia (NRC 1996). Land development and conversion to agricultural or commercial forest uses also has affected nearly all watersheds in the Pacific Northwest.

The Pacific Coastal Ecoregion

Embedded within the Pacific Northwest is the Pacific coastal ecoregion which extends from northern California into southeastern Alaska and east to the crests of the Cascade Mountains in the south and Coastal Mountains in the north (Figure 1.1). This region is characterized by high precipitation and a maritime climate with cool, dry summers and warm, wet winters. The Pacific coastal ecoregion encompasses an abundance of lotic ecosystems, many of which support biota of considerable economic consequence, such as Pacific salmon. It is also one of the most rapidly developing regions of North America. As a result, some of the conflicts revolving around the management of streams and rivers have achieved a high level of attention.

The Pacific coastal ecoregion supports some of the most extensive temperate rain forests in the world (Ryan 1994, Schoonmaker et al. 1997), including the world's most massive coniferous forests—the biggest and longest-lived redwoods (*Sequoia* spp.), spruces (*Picea* spp.), firs, (*Abies* spp.), and hemlocks (*Tsuga* spp.) found anywhere. These forests support thousands of identified (and many unidentified) species and supply the wood and other materials that are vital to the integrity of the streams and rivers. The wettest forests in the region are the densest forests in the world, containing more living and dead plant matter than even the tropical rain forests.

The biota of the aquatic systems draining these forests also often reach large sizes and ages, and high densities. The spawning runs of salmon are legendary. Chinook salmon may reach 55 kg, while sturgeon (*Acipenser* spp.) may live for a century and reach lengths of 6 m and a mass of 700 kg. Unfortunately, the endemic diversity of aquatic organisms in the Pacific coastal ecoregion is not as well known as it is for other ecoregions. However, this lack of information offers unique research opportunities, especially if the diversity of aquatic organisms parallels that of the terrestrial system of the Pacific coastal ecoregion.

On a global scale, the coastal temperate rain forest appears to be sufficiently homogeneous to be considered a single ecoregion for the interpretation of global biogeochemical processes and responses to climate change (Neilson and Marks 1994). The regional consistency in aquatic community structure (especially fishes) and vegetation supports this conclusion (Naiman et al. 1992). However, at smaller scales relevant to resource management and conservation, ecologically significant variations in genetic diversity, evolutionary processes, life history strategies, and other biotic characteristics are evident. These biotic characteristics are adapted to the wide variations in physical properties and disturbance histories of their respective aquatic ecosystems (Schoonmaker et al. 1997). This book characterizes the streams and rivers of the Pacific coastal ecoregion and illustrates the breadth of that heterogeneity.

The Pacific coastal ecoregion can be divided into four subregions with contrasting physiographies: southern coastal mountains, Olympic mountains, northern mainland, and northern lowlands and islands. As a consequence of the contrasting physiographies, several important latitudinal gradients are evident. First, watershed size increases from north to south. Ninety-five percent of the 608 watersheds in the northern lowlands and islands are smaller than 100 km^2, compared to only 52% in the southern coastal region. Second, the mean annual runoff from the rivers increases from 0.5 m/yr in central California to 7.4 m/yr in southeastern Alaska. Third, seasonal patterns of discharge also show significant variations with latitude (Chapter 3). Nearly 70% of the annual discharge occurs in winter in California, but only about 25% occurs in winter in British Columbia and Alaska. In California, less than 2% of the annual discharge occurs in summer, while in British Columbia and Alaska summer discharge is as high as 60%. Finally, average yearly water temperatures decrease from about 15 °C in central California to about 3 °C in Alaska (Naiman and Anderson 1997). These and other latitudinal gradients in the physical environment have significant ecological implications for population and community processes, life history strategies, and behavior of the biota.

Science and Management

This book is based on the conviction that better management is based on an understanding of fundamental ecological and related social processes. This volume focuses on river ecology and management in the Pacific coastal ecoregion because of the recent, rapid expansion in scientific knowledge about streams and rivers in the region, and because of the contentious and rapidly evolving political and social landscapes which direct the management of these systems. The interface between the science and policy of natural resource management is perhaps best illustrated by examples from this ecoregion, including the protection of riparian forests, the marbled murrelet (*Brachyramphus marmoratum*), salmon, and amphibians. Even though many of the examples in this book are from the Pacific coastal ecoregion, the fundamental ecological and management principles apply to all lotic ecosystems. The complex issues facing scientists, managers, and decision makers in this ecoregion are, in reality, no different from those in other ecoregions of North America.

Literature Cited

Allen, J.D. 1995. Stream ecology. Chapman & Hall, London, England.

Bardach, J. 1964. Downstream: A natural history of the river. Harper & Row, New York, New York, USA.

Barnes, J.R., and G.W. Minshall, eds. 1983. Stream ecology. Plenum, New York, New York, USA.

Benke, A.C. 1990. A perspective on America's vanishing streams. Journal of the North American Benthological Society **9**:77–88.

Boon, P.J., P. Calow, and G.E. Petts, eds. 1992. River conservation and management. John Wiley & Sons, Chichester, UK.

Bormann, F.H., and G.E. Likens. 1979. Pattern and process of a forested ecosystem. Springer-Verlag, New York, New York, USA.

Carpenter, K.E. 1928. Life in inland waters. Sidgwick and Jackson, London, England.

Coker, R.E. 1954. Streams, lakes, and ponds. University of North Carolina Press, Chapel Hill, North Carolina, USA.

Dahl, T.E. 1990. Wetland losses in the United States, 1780s to 1980s. US Department of the Interior, Fish and Wildlife Service, Washington, DC, USA.

Dodge, D.P., ed. 1989. Proceeding of the International Large River Symposium. Canadian Special Publication of Fisheries and Aquatic Sciences 106, Ottawa, Ontario, Canada.

Dunne, T., and L.B. Leopold. 1978. Water in environmental planning. W.H. Freeman, San Francisco, California, USA.

Dynesius, M., and C. Nilsson. 1994. Fragmentation and flow regulation of river systems in the northern third of the world. Science **266**:753–762.

Elwood, J.W., J.D. Newbold, R.V. O'Neill, and W. Van Winkle. 1983. Resource spiraling: An operational paradigm for analyzing lotic ecosystems. Pages 3–27 *in* T.D. Fontaine and S.M. Bartell, eds. Dynamics of lotic ecosystems. Ann Arbor Science, Ann Arbor, Michigan, USA.

Fontaine, T.D., and S.M. Bartell, eds. 1983. Dynamics of lotic ecosystems. Ann Arbor Science, Ann Arbor, Michigan, USA.

Freeze, R.A., and J.A. Cherry. 1979. Groundwater. Prentice-Hall, Englewood Cliffs, New Jersey, USA.

Frissell, C.A., W.J. Liss, W.J. Warren, and M.D. Hurley. 1986. A hierarchical framework for stream classification: Viewing streams in a watershed context. Environmental Management **10**: 199–214.

Gleick, P.H., ed. 1993. Water in crisis. Oxford University Press, New York, New York, USA.

Gleick, P.H. 1998. Water in crisis: Paths to sustainable water use. Ecological Applications (in press).

Gordon, N.D., T.A. Mcmahon, and B.L. Finlayson. 1992. Stream hydrology. John Wiley & Sons, Chichester, UK.

Graf, W.L. 1993. Landscapes, commodities, and ecosystems: The relationship between policy and science for American rivers. Pages 11–42 *in* Sustaining our water resources. National Academy Press, Washington, DC, USA.

Hauer, F.R., and G.A. Lamberti, eds. 1996. Methods in stream ecology. Academic Press, San Diego, California, USA.

Hynes, H.B.N. 1960. The biology of polluted waters. Liverpool University Press, Liverpool, UK.

Hynes, H.B.N. 1972. The ecology of running waters. University of Toronto Press, Toronto, Ontario, Canada.

Illies, J. 1961. Die Lebensgemeinschaft des Bergbaches. Neue Brehm-Bucherei, Wittenberg Lutherstadt, Germany.

Junk, W., P.B. Bayley, and R.E. Sparks. 1989. The flood-pulse concept in river-floodplain systems. Canadian Special Publication of Fisheries and Aquatic Sciences **106**:110–127.

Karr, J.R. 1993. Defining and assessing ecological integrity: Beyond water quality. Environmental Toxicology and Chemistry **12**:1521–1531.

Leopold, L.B., M.G. Wolman, and J.P. Miller. 1964. Fluvial processes in geomorphology. W.H. Freeman, San Francisco, California, USA.

Macan, T.T., and E.B. Worthington. 1951. Life in lakes and rivers. Collins, London, UK.

Malanson, G.P. 1993. Riparian landscapes. Cambridge University Press, Cambridge, UK.

Master, L. 1990. The imperiled status of North American aquatic animals. Biodiversity Network News **3**:1–2, 7–8.

Merritt, R.W., and K.W. Cummins, eds. 1984. An introduction to the aquatic insects of North America. Kendall/Hunt, Dubuque, Iowa, USA.

Naiman, R.J., ed. 1992. Watershed management. Springer-Verlag, New York, New York, USA.

Naiman, R.J., and E.C. Anderson. 1997. Streams and rivers: Physical and biological variability. Pages 131–148 *in* P.K. Schoonmaker, B. von Hagen, and E.C. Wolf, eds. The rain forests of home: Profile of a North American bioregion. Island Press, Washington, DC, USA.

Naiman, R.J., T.J. Beechie, L.E. Benda, D.R. Berg, P.A. Bisson, L.H. MacDonald, et al. 1992. Fundamental elements of ecologically health watersheds in the Pacific Northwest coastal ecoregion. Pages 127–188 *in* R.J. Naiman, ed. Watershed management. Springer-Verlag, New York, New York, USA.

Naiman, R.J., and H. Décamps, editors. 1990. The ecology and management of aquatic-terrestrial ecotones. UNESCO, Paris and Parthenon, Carnforth, UK.

Naiman, R.J., H. Décamps, J. Pastor, and C.A. Johnston. 1988. The potential importance of boundaries to fluvial ecosystems. Journal of the North American Benthological Society **7**:289–306.

Naiman, R.J., J.J. Magnuson, and P.L. Firth. 1998. Integrating cultural, economic, and environmental requirements for fresh water. Ecological Applications **8**:569–570.

Naiman, R.J., J.J. Magnuson, D.M. McKnight, and J.A. Stanford, eds. 1995. The freshwater imperative. Island Press, Washington, DC, USA.

NRC (National Research Council). 1996. Upstream. National Academy Press, Washington, DC, USA.

Needham, P.R. 1938. Trout streams. Comstock, Ithaca, New York, USA.

Nehlsen, W., J.E. Williams, and J.A. Lichatowich. 1991. Pacific salmon at the crossroads: Stocks at risk from California, Oregon, Idaho, and Washington. Fisheries **16**:4–21.

Neilson, R.P., and D. Marks. 1994. A global perspective of regional vegetation and hydrologic sensitivities from climate change. Journal of Vegetation Science **5**:715–730.

Petts, G.E. 1984. Impounded rivers. John Wiley & Sons, Chichester, UK.

Pimentel, D., J. Houser, E. Preiss, O. White, H. Fang, L. Mesnick, et al. 1997. Water resources: Agriculture, the environment, and society. BioScience **47**:97–106.

Powledge, F. 1984. The magnificent liquid of life. National Wildlife **22**:7–9.

Reid, G.K. 1961. Ecology of inland waters and estuaries. Reinhold, New York, New York, USA.

Richards, K. 1982. Rivers: Form and process in alluvial channels. Methuen, London, UK.

Ryan, J.C. 1994. State of the Northwest. Northwest Environmental Watch Report Number 1, Seattle, Washington, USA.

Schoonmaker, P.K., B. von Hagen, and E.C. Wolf, eds. 1997. The rain forests of home: Profile of a North American bioregion. Island Press, Washington, DC, USA.

Stanford, J.A., and J.V. Ward. 1993. An ecosystem perspective of alluvial rivers: Connectivity and the hyporheic corridor. Journal of the North American Benthological Society **12**:48–60.

Steinmann, P. 1907. Die Tierwelt der Gebirgsbache. Eine faunistisch-biologische studie. Annales de Biologie lacustre **2**:30–150.

Steinmann, P. 1909. Die neueston Arbeiten uber Bachfauna. Internationale Revue der gesamten Hydrobiologie und Hydrographie **2**:241–246.

Stouder, D.J., P.A. Bisson, and R.J. Naiman, eds. 1997. Pacific salmon and their ecosystems. Chapman & Hall, New York, New York, USA.

Swift, B.L. 1984. Status of riparian ecosystems in the United States. Water Resources Bulletin **20**:223–228.

Thienemann, A. 1925. Die Binnengewasser Mitteleuropas. Die Binnengewasser, Stuttgart, Germany.

Ward, J.V. 1992. Aquatic insect ecology. John Wiley & Sons, Chichester, UK.

Ward, J.V., and J.A. Stanford, eds. 1979. The ecology of regulated streams. Plenum, New York, New York, USA.

Ward, J.V., and J.A. Stanford. 1983. The serial discontinuity concept of lotic ecosystems. Pages 29–42 *in* T.D. Fontaine and S.M. Bartell, eds. Dynamics of lotic ecosystems. Ann Arbor Science, Ann Arbor, Michigan, USA.

Webster, J.R., and J. Meyer, eds. 1997. Stream organic matter budgets. Journal of the North American Benthological Society **16**:3–161.

Williams, J.E., J.E. Johnson, D.A. Hendrickson, S. Contreras-Balderas, J.D. Williams, M. Navarro-Mendoza, et al. 1989. Fishes of North America: Endangered, threatened, or of special concern. Fisheries **14**:2–20.

Vannote, R.L., G.W. Minshall, K.W. Cummins, J.R. Sedell, and C.E. Cushing. 1980. The river continuum concept. Canadian Journal of Fisheries and Aquatic Sciences **37**:130–137.

Part I
The Physical Environment

2
Channel Processes, Classification, and Response

David R. Montgomery and John M. Buffington

Overview

• This chapter discusses physical processes, classification, and response potential of channels in mountain drainage basins of the Pacific coastal ecoregion.

• A relatively simple set of physical processes leads to a wide variety of natural stream channels, the classification of which can guide recognition of functionally similar zones in mountain drainage basins. Different portions of mountain channel networks are dominated by different geomorphic processes and relationships between transport capacity (a function of discharge and boundary shear stress) and sediment supply (size and amount of material available for transport).

• Channel classifications use similarities of form and function to impose order on a continuum of natural stream types or morphologies. No single classification can satisfy all possible purposes or is likely to encompass all possible channel types.

• This chapter reviews geomorphological channel classifications and their use for systematizing channel morphology and physical processes for the purpose of assessing physical channel condition and response potential. Early classification systems tend to neglect the influence of woody debris or emphasize single scales of influences on channel morphology and processes. In contrast, a hierarchical approach to channel classification addresses different factors influencing channel properties over a range of spatial and temporal scales and is well suited for assessment of channel conditions and response potential in mountain drainage basins.

• The spatial distribution of reach types within a drainage basin influences the distribution of potential impacts and responses to disturbance. Alluvial channels with high transport capacities relative to sediment supply generally maintain their morphology while transmitting increased sediment loads; channels with lower ratios of transport capacity to sediment supply tend to exhibit greater morphologic response to increased sediment loads. Steep channels thereby act as sediment delivery conduits connecting zones of sediment production on hillslopes to more responsive lower-gradient channels.

• Consideration of channel bed morphology, confinement (the ratio of the width of the valley floor to the width of the bankfull channel), position in the channel network, and external influences (such as riparian vegetation and in-channel woody debris) can guide evaluation of channel condition and response potential in forested mountain drainage basins.

Introduction

Stream channels are important both as avenues of sediment transport that deliver eroded material from continents to the oceans and as environments for freshwater ecosystems that have economic and social significance (Chapter 23). Variability in sediment delivery, hydraulic discharge, and channel slope give rise to spatial and temporal variations in channel morphology and response. Over geologic time, channels respond to tectonic uplift, erosion of the landscape, and climate change. Over historical time, channels respond to changes in discharge and sediment supply from both land use and such extreme events as floods and droughts. Concern over such impacts on aquatic and riparian ecosystems, as well as human uses of fluvial systems, motivates assessment of channel change to evaluate past response to disturbance and to predict response to climate change or land use. The wide variety of channel types, adjustment of individual channels to local factors, and potential time lags between perturbation and response complicate the interpretation and prediction of changes in channel form and processes. This complexity fostered the development of classification schemes to guide identification of functionally similar portions of channel networks. This chapter discusses both the conceptual basis for understanding channel response and how channel classification can aid the study of watershed processes, assessment of channel condition, and evaluation of channel response in mountain drainage basins of the Pacific coastal ecoregion.

Channel Processes

Channels ranging in size from small ephemeral rivulets to large rivers exhibit a wide variety of morphologies, but share a number of basic processes. Over decades to centuries, channel morphology is influenced by both local and systematic downstream variation in sediment input from upslope sources (sediment supply), the ability of the channel to transmit these loads to downslope reaches (transport capacity), and the effects of vegetation on channel processes (Figure 2.1). Potential channel adjustments to altered discharge and sediment load include changes in width, depth, velocity, slope, roughness, and sediment size (Leopold and Maddock 1953).

A few key relationships describe the physics governing channel processes and illustrate controls on channel response. Conservation of energy and mass describe sediment transport and the flow of water through both the channel

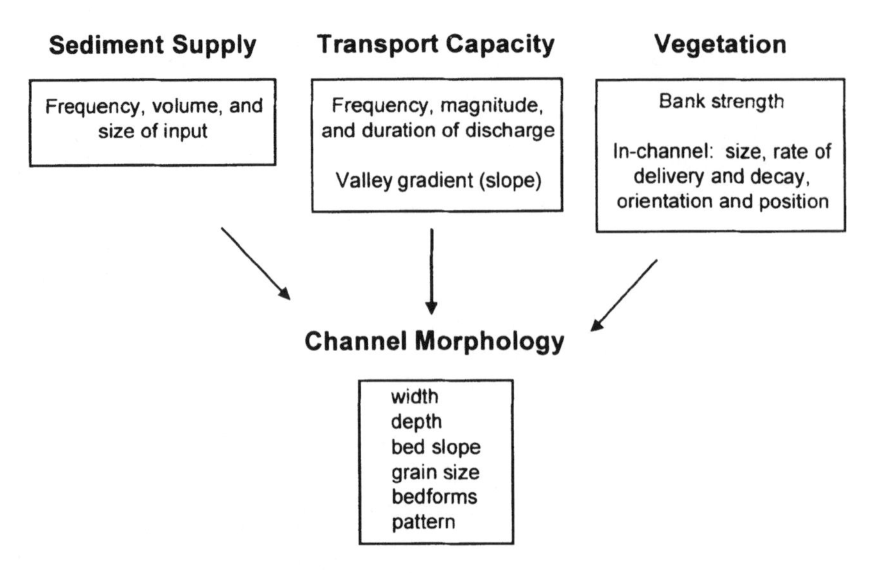

FIGURE 2.1. Decade- to century-scale influences on channel morphology include sediment supply, transport capacity, and direct and indirect effects of vegetation.

network and any point along a channel. Other relationships describe energy dissipation by channel roughness elements, the influence of boundary shear stress on sediment transport, and the geometry of the active transport zone.

Precipitation falling on a landscape moves downslope, causing erosion and maintaining channels. The frequency and magnitude of precipitation and the topographic relief onto which it falls provide the potential energy that drives these processes. Downslope movement of water converts this potential energy into kinetic energy, which is dissipated by friction and turbulence generated by the channel bed and banks. For a nonaccelerating fluid, the downstream gravitational force of the water ($mgS = \varrho ALgS$) is balanced by the shear resistance of the channel bed and banks ($\tau_0 LP$), and hence

$$\varrho ALgS = \tau_0 LP \tag{2.1}$$

where m is the mass of the water, g is gravitational acceleration, S is the water surface slope, ϱ is the fluid density, A is the channel cross-sectional area, L is the channel length, P is the wetted perimeter, and τ_0 is the total basal shear stress. Rearranging the force balance in equation (2.1) and solving for τ_0, the shear stress that flowing water exerts on the channel bed is

$$\tau_0 = \varrho gRS \tag{2.2}$$

where the hydraulic radius is $R = \frac{A}{P}$. For natural channels with a width (W) much larger than mean flow depth (D), $R \approx D$ and thus $\tau_0 \approx \varrho gDS$.

The hydraulic discharge of a channel (Q) is defined as

$$Q = WDu \tag{2.3}$$

where u is the mean flow velocity, which depends on the fluid driving force and frictional resistance of the channel. Several empirical equations relate mean flow velocity to channel resistance

$$u = \frac{D^{2/3}S^{1/2}}{n} = C\sqrt{DS} = \sqrt{\frac{8gDS}{f}} \tag{2.4}$$

where n is the Manning roughness coefficient, C is the Chezy resistance factor, and f is the Darcy–Weisbach friction factor. The total roughness of a channel reflects the rate of energy dissipation and incorporates resistance offered by bed-forming particles, bedforms, and in-channel obstructions such as large woody debris (LWD). In general, the total boundary shear stress is related to the square of velocity ($\tau_0 \propto u^2$).

The basal shear stress acting on the channel bed drives sediment transport. The fraction of the total boundary shear stress available for sediment transport, defined as the effective shear stress (τ'), depends upon the amount of energy dissipation caused by in-channel roughness other than grain resistance (e.g., bedforms, LWD, channel bends). The critical shear stress (τ_c) represents the shear stress necessary to mobilize a given grain size (d_i)

$$\tau_c = \tau_c^*(\varrho_s - \varrho)gd_i \tag{2.5}$$

where ϱ_s is the sediment density and τ_c^* is the dimensionless critical shear stress (Shields 1936). Transport of bed material occurs at effective boundary shear stresses greater than or equal to the critical shear stress ($\tau' \geq \tau_c$). In gravel- and cobble-bed channels, the bankfull stage establishes channel morphology and accomplishes most sediment transport (Wolman and Miller 1960). The frequency of the bankfull discharge varies for different channels, but commonly occurs about every 1.5 years (Williams 1978).

Modes of sediment transport include both suspension of grains within the flow (suspended load) and rolling, sliding, and saltation of grains near the channel bed (bedload). Suspended load typically accounts for the majority of transported sediment, but bedload transport dominates channel morphology. Many bedload transport equations are based on the difference between applied and critical grain shear stresses

$$Q_b = k(\tau' - \tau_c)^n \tag{2.6}$$

where Q_b is the bedload transport rate (kg/s) and k and n are empirically determined values, with n typically being greater than one. The dependence of basal shear stress on flow depth, and thus discharge, indicates that a significant change in discharge directly influences sediment transport, bed stability, and scour.

Continuity requires that differences between the sediment supply (Q_s) and the bedload transport rate be accommodated by a change in the amount of sediment stored (S_s) within a reach

$$Q_s - Q_b = \Delta S_s \tag{2.7}$$

If more sediment enters a channel reach than it can transmit, then the amount of stored sediment must increase. Continuity further requires that the thickness of the active transport layer is related to the bedload transport rate (Carling 1987) as

$$D_s = \frac{Q_b}{u_b W \varrho_s (1 - e)} \tag{2.8}$$

where D_s is the mean depth of scour, u_b is the average bedload velocity, and e is the bed porosity. Hydraulic discharge thereby controls the depth of scour defined by the thickness of the active transport layer.

Conceptual Models of Channel Response

The relatively simple set of channel processes outlined above results in a wide array of possible channel responses to changes in sediment supply, discharge, and external influences such as LWD flow obstructions. In response to changes in sediment supply or discharge, a channel may widen or deepen; change its slope through aggradation, degradation, or modified sinuosity; alter bedforms or particle size, thereby changing the frictional resistance of the bed; or alter the thickness of the active transport layer defined by the depth of channel scour. Drawing on both theory and empirical evidence, previous researchers developed conceptual models of channel response to changes in sediment load or discharge.

Gilbert (1917) hypothesized that the slope of an alluvial stream adjusts through erosion or deposition in order to transport the imposed load. Where the channel slope exceeds that necessary to transport the load, the channel incises and slope decreases, indicating a proportionality between bedload transport and channel slope

$$Q_b \propto S \tag{2.9}$$

With this reasoning, Gilbert (1917) anticipated and subsequently confirmed downstream channel incision in response to dam construction.

Lane (1955) later hypothesized that bedload discharge and sediment size adjust to hydraulic discharge and slope

$$Q_b d_i \propto QS \tag{2.10}$$

Based on this expression, Lane (1955) argued that corresponding changes in channel slope and sediment size accommodate changes in hydraulic discharge or bedload transport. Lane's expression provides a more complete model than Gilbert's, but neither expression accounts for the ability of a channel to change its basic geometry.

Schumm (1971) combined empirical relationships between discharge, bedload transport, and other descriptive morphological variables into general relationships for channel response that include channel geometry (i.e., W and D)

$$Q \propto \frac{WD\lambda}{S} \tag{2.11}$$

and

$$Q_b \propto \frac{W\lambda S}{Dp} \tag{2.12}$$

where λ is meander wavelength and p is sinuosity. Nunnally (1985) elaborated on Schumm's approach to include the median grain size of the bed surface (d_{50}).

Considering general relationships between sediment supply, discharge, and channel attributes (such as equations 2.2–2.8) additional factors can be incorporated into more comprehensive conceptual relationships for channel response in mountain drainage basins:

$$Q \propto \frac{WDQ_b D_s d_{50} n}{S_s S} \tag{2.13}$$

and

$$Q_s \propto \frac{WQ_b D_s S_s S}{Dd_{50} n} \tag{2.14}$$

Some variables in the conceptual relationships between discharge, sediment supply, and channel morphology summarized in equations (2.13) and (2.14) have thresholds of response, while others possess continuous response potential. Channel width and depth are related to discharge through equation (2.3). Discharge changes resulting in altered flow depth or velocity (2.3) have a direct influence on shear stress and hence both bedload transport (2.2, 2.5, 2.6) and depth of scour (2.8). Similarly, the control of discharge on shear stress directly affects bed surface grain size (2.5). In the absence of concurrent changes in sediment supply, increased discharge may cause higher bedload flux from a reach (2.2, 2.3, 2.6), resulting in depletion of sediment storage (2.7) or channel incision (2.3) and decreased slope; bank cutting and meander development caused by increased discharge can also decrease channel slope. Increased discharge resulting in higher ratios of transport capacity to sediment supply may be balanced by bed surface coarsening (Dietrich et al. 1989) and bedform development (Whittaker and Jaeggi 1982), both of which serve to increase resistance and stabilize the channel. In contrast, increased sediment supply can cause bed surface fining (Dietrich et al. 1989) and pool filling (Whittaker and Davies 1982), smoothing the channel bed and decreasing roughness. Bed surface fining caused by elevated sediment supply can result in greater bedload transport (2.5, 2.6) and consequently increased scour depths (2.8). Excessive sediment loading that exhausts the channel transport capacity can lead to bed aggradation, resulting in increased sediment storage (2.7) or decreased channel depth, which may, in turn, trigger alluvial bank cutting and channel widening if there are no concurrent changes in discharge (2.3); alternatively, aggradation may elevate channel slope.

Although conceptual response relationships, such as (2.13) and (2.14), allow prediction of the general direction of potential channel changes, specific responses often arise from some combination of altered discharge and sediment supply. Consequently, attribution of channel change to altered discharge or sediment supply often requires independent constraints on one of these factors. The predictions of (2.13) and (2.14) apply throughout channel networks, but the type and magnitude of response vary with site-specific channel processes and conditions. These relationships also illustrate a fundamental problem in predicting or reconstructing channel response: there are seven variables, but only two relationships—thus specific channel response is somewhat indeterminate. Fortunately, a great deal of accumulated experience relates to channel response.

Examples of Channel Change

An extensive literature on channel change highlights common responses and provides a large body of empirical evidence with which to develop and test conceptual models. Studies of channel change reveal a wide range of responses to changes in sediment supply and discharge. Increased sediment supply can induce channel widening and aggradation, decrease roughness through pool filling, and decrease bed sediment size. These responses are consistent with those predicted by (2.14). Increased discharge can cause channel widening, incision, and bed armoring, effects predicted by (2.13). Channel response to dam construction can involve a variety of effects due to changes in discharge and sediment supply, which covary.

Sediment Supply

Channel response to increased sediment supply depends on the ratio of transport capacity to the sediment supply. Significant aggradation, channel widening, bed fining, pool filling, or braiding occur where the amount of introduced sediment overwhelms the local transport capacity. Spatial variability in sediment supply may govern channel morphology in different portions of a drainage network.

Temporal variations in sediment supply also influence channel form. A classic study that illustrates progressive downstream aggradation and subsequent degradation in response to an episodic increase in sediment input is Gilbert's (1917) report on the effects of huge additions of hydraulic mining debris to rivers in the foothills of the Sierra Nevada of California from the early 1850s to the 1880s. Aggradation occurred sequentially throughout the downslope channel network as the mining debris was gradually transported through the system. Locally, channel aggradation approached 40 m by the late 1870s (Whitney 1880). Subsequent reincision of the channels was still occurring just after the turn of the century and some channels continued to respond over one hundred years after hydraulic mining ceased (James 1991).

The 1964 floods in northern California and southern Oregon also illustrate morphologic response to increased sediment supply. Channel widths doubled at some gaging stations and channel beds aggraded as much as 4 m (Kelsey 1980, Lisle 1982), except for channels with nonalluvial banks confined in narrow valleys (Lisle 1982). Kelsey (1980) estimated that a pulse of sediment originating in steep headwaters of the Van Duzen River migrated downstream at a rate of about 1 km/yr. Lisle (1982) reported that pool filling decreased channel roughness and accelerated sediment transport within aggraded reaches. Both pool frequency and the mean size of bed material also decreased in response to aggradation (Kelsey 1980). Helley and LaMarche (1973) reported increased sediment storage in large gravel bars along channel margins and described evidence for a comparable response to prehistoric floods. Channel widening on the middle fork of the Willamette River, Oregon, in response to the 1964 flood reflected increased sediment delivery from hillslopes and disturbance of riparian vegetation (Lyons and Beschta 1983). Debris flows also scoured many steep channels to bedrock (Grant et al. 1984). Over twenty years later, significant flood-delivered material remained stored in low-gradient reaches where braiding continued to rework flood deposits (Sullivan et al. 1987). Such changes in sediment storage within a channel system can persist for decades, as sediment gradually mobilizes from the reaches in which it accumulated.

The South Fork Salmon River in central Idaho presents another example of impact and recovery from significant sediment inputs. Severe storms in the early 1960s following extensive logging and road construction dramatically increased sediment loads, resulting in pool filling, burial of gravels with sand, decreased bed roughness, and fining of the channel bed (Platts et al. 1989). A coincident decline in the fish population resulted in a moratorium on logging in the watershed, which reduced the sediment supply to impacted channels. Cross sections monitored over subsequent years showed progressive reincision, as pools were reexcavated and sand was transported out of spawning gravels (Megahan et al. 1980).

An important characteristic of channel response to increased sediment loads is that different portions of a drainage network may respond differently to a single disturbance. An excellent example of spatial patterns of channel response occurred as a result of a 100-year storm in the Santa Cruz Mountains that caused widespread landsliding in January, 1982 (Ellen and Wieczorek 1988). Debris flows and high discharges scoured many of the channels with gradients steeper than 10%, resulting in major sediment delivery to lower-gradient channels (Nolan and Marron 1988). Channel response in intermediate-gradient channels was variable, with significant local aggradation associated with landslide deposition (Nolan and Marron 1988). In these channels, sand filled many pools, buried riffles, and deposited in the interstices of coarse bed material (Coats et al. 1985). Substantial aggradation and overbank deposition also occurred along steep channels from routing of landslide debris (Nolan and Marron 1988). Later that winter, subsequent flows in steep- and intermediate-gradient channels scoured much of the aggraded sediment and redistributed it downslope. Pool filling and riffle burial persisted for a longer time in lower-gradient channels (Coats et al. 1985), illustrating a strong difference in the type and persistence of channel response at different locations in the drainage network.

Changes in sediment supply also influence the character of the channel bed. For example, Perkins (1989) studied the effect of landslide-supplied sediment on channel morphology in Salmon Creek in southwestern Washington. Based on considerations of the relationship between transport capacity and sediment supply, she argued that accelerated sediment delivery increased the amount of material stored in bedforms (expanding bar volumes at the expense of pool volumes) and decreased the average grain size in the reach. In her study area, elimination of landslide-supplied sediment resulted in a long-term decrease in the amount of material stored in the bed and a greater degree of bedrock control on bed morphology. Her study illustrates how channel form and sediment storage reflect the relative balance between sediment supply and transport capacity.

The size of bed surface material also responds to changes in sediment supply. In a series of flume experiments, Jackson and Beschta (1984) showed that increasing the amount of sand in a mixed sand/gravel bed increased gravel transport and scoured previously stable gravel riffles. They also showed that the median grain size (d_{50}) of the flume bed decreased with increased sediment transport. Dietrich et al. (1989) proposed that the degree of bed surface coarsening reflects the relationship between sediment supply and transport capacity. Their flume experiments showed that decreased sediment supply resulted in surface armoring and constriction of the zone of active sediment transport. Knighton (1991) reported that channel response to large inputs of fine sediment involved both a wave of aggradation and a general fining of bed material. After passage of such a wave, the channel substrate coarsened as the bed degraded toward its original condition.

Discharge

Changes in the magnitude and frequency of channel discharge may result from alteration of either the total precipitation falling in a watershed or from changes in runoff production and routing through the channel network (Chapter 3). Climate change provides the most direct precipitation-related impact on discharge in channel networks, but opportunities to monitor the influence of climate change on channels are rare. In contrast, the impact of land management on the discharge regime and morphology of stream channels is well documented. Watershed urbanization, for example, dramatically increases peak discharges because of increased impervious area, which increases the proportion of rapid surface runoff at the expense of infiltration (Leopold 1968). Channel response to these changes typically involves channel expansion through an increase in either channel width or depth. Hammer (1972) compared relationships between drainage area and channel width for urbanized and rural drainage basins in Pennsylvania and found significant channel widening in response to increased peak flows. He also found that large impervious surfaces (such as parking lots) directly connected to the channel network (via storm sewers) enhanced channel widening. Many others have reported significant channel widening and incision as a result of urbanization in both temperate (Graf 1975, Booth 1990) and tropical catchments (Whitlow and Gregory 1989).

Changes in watershed vegetation may affect the flow regime in downstream channels through changes in water yields, summer low flows, and peak flows. Paired watershed experiments indicate that forest clearance generally increases water yields (Bosch and Hewlett 1982), but in some regions, species which revegetate a cleared forest may have higher rates of evapotranspiration and thereby reduce discharges below original levels (Harr 1983). Although they may be very important biologically, changes in low-flow conditions are generally unimportant for morphological channel response. In contrast, increases in peak runoff caused by road construction (Jones and Grant 1996) or rain on snow events in clear cut areas (Harr 1986) may significantly affect channels because of the possible change in either the frequency or magnitude of the channel-forming discharge. Channel responses to high peak flows during rain on snow events include bank erosion, channel incision, and mobilization of both bedload and LWD (Harr 1981). These effects are similar to those occurring from natural large discharge events, but a change in their frequency could impact biological systems and reach-level sediment transport.

Dams

Dam construction changes both the discharge regime and sediment supply of downstream channels, resulting in channel incision, constriction or widening, and changes in channel substrate. Many studies document channel incision and bed surface coarsening immediately downstream of dams in response to sediment impoundment (Gilbert 1917, Williams and Wolman 1984). Tributary channels also may incise in response to mainstream channel incision through upstream knickpoint propagation from their confluence. Decreased discharge below a dam may cause narrowing of the active channel width (Leopold and Maddock 1953). Tributary sediment inputs downstream of dams can cause channel aggradation (Allen et al. 1989) and accumulation of fine sediment (Wilcock et al. 1996) because of decreased transport capacity resulting from dam construction. Channel-spanning log jams can also act like dams, causing upstream fining and downstream coarsening (Rice 1994).

Collectively, the case studies presented in this section illustrate that (2.13) and (2.14) provide a reasonable conceptual framework for examining channel response. However, differences in channel form and function affect the probability of specific responses to a given perturbation. Channel classification can aid interpretation and assessment of response potential by grouping functionally similar physical environments.

Geomorphological Channel Classification

Channel classifications use similarities of form and function to impose order on a continuum of natural stream types or morphologies. A voluminous literature on channel classification attests to the wide variety of stream morphologies. Each of the channel classifications in common use has advantages and disadvantages in geological, engineering, and ecological applications (Kondolf 1995), and no single classification can satisfy all possible purposes or likely encompass all possible channel types. This chapter reviews geomorphological channel classifications and their use for systematizing channel morphology and physical processes, and for assessing physical channel condition and response potential.

Past Classifications

Early geomorphological delineations of different channel types focused on broad criteria (Powell 1875, Gilbert 1877), but recent classifications include more detailed consideration of channel pattern, bed material or mobility, sediment transport mechanisms, position within the channel network, and various combinations of slope and valley characteristics. Most geomorphological classifications are designed for large floodplain rivers, although Schumm's (1977) general delineation of erosion, transport, and deposition zones provides a conceptual framework within which to couple channel type and channel response potential throughout mountain drainage basins.

Stream order. The concept of stream order proposed by Horton (1945), and later modified by Strahler (1957), remains the most widely used channel classification. In Strahler's system, the channel segment from the head of the channel network to the first confluence constitutes a 1st-order channel. Second-order channels lie downslope of the intersection of two 1st-order channels, and so on down through the channel network. Stream order correlates with channel length and drainage area, thereby providing an indication of relative channel size and position within a channel network. Although channel ordering is a useful conceptual and organizational tool, comparisons between channel networks can prove misleading because the order assigned to a channel segment depends on the criteria used to determine where 1st-order channels begin. Representations of the extent of channels in a given watershed vary on maps of different scales, and basin topology influences the size of channels classified as a particular order. Moreover, channel networks defined from blue lines on maps, the curvature of topographic contours, or a critical gradient or drainage area can differ substantially from the network identifiable in the field (Morisawa 1957). Aside from the tautology that higher-order channels tend to be larger, there is no inherent association of channel morphology and process with stream order.

Channel bed. The nature of the channel bed provides the basis for perhaps the most fundamental geomorphological classification of stream channels. Gilbert (1914) recognized that bedrock channels have a greater transport capacity than the sediment supply, whereas alluvial channels have a transport capacity less than, or equal to, the sediment supply. Gilbert further recognized that different portions of a channel network may be composed of different channel types and patterns. Henderson (1963) later recognized two alluvial channel types based on grain size and sediment mobility characteristics: "live bed" channels that are actively mobile at most stages, and "threshold" channels that exhibit significant mobility only during high flows. These different styles of bed movement are strongly correlated with grain size, and thus with the common distinction of sand- and gravel-bed channels.

Channel patterns and processes. Several channel classifications broadly characterize general differences in channel patterns and processes. Leopold and Wolman (1957) quantitatively differentiated straight, meandering, and braided channel patterns based on relationships between slope and discharge. Schumm (1977) classified alluvial channels based on dominant modes of sediment transport (i.e., suspended, mixed, or bedload transport) and recognized three geomorphic zones within a watershed: degrading headwater channels that are the primary source of sediment and water inputs, stable mid-network channels with roughly balanced inputs and outputs, and aggrading channels low in the network characterized by extensive depositional floodplains or deltas. Mollard (1973), and later Church (1992), classified floodplain rivers into a continuum of channel patterns related to differences in discharge, slope, sediment supply, and channel stability. Kellerhals et al. (1976) used an extensive list of descriptive features to characterize large alluvial rivers in terms of stream pattern, frequency of islands, bar type, and lateral channel migration. They further emphasized that channel form and processes depend on surficial geology, the frequency and magnitude of sediment and water inputs, the relationship of a channel to its floodplain and valley walls, and the history of geologic, climatic, and anthropogenic disturbance. Church and Jones (1982) subsequently presented a classification of bar types and patterns that explicitly relates channel morphology to gradient and the volume and size of sediment supply. These fundamental distinctions can guide general interpretation of channel condition and response potential in mountain drainage basins.

A more detailed channel reach classification developed by Rosgen (1994) recognizes 7 major and 42 minor channel types based on channel pattern, entrenchment, width-to-depth ratio, sinuosity, slope, and bed material size. Although Rosgen (1994) demonstrated that his major channel types exhibit distinct roughness coefficients and hydraulic geometry relationships, the classification is not process based; a lack of any explanation of the rationale underlying Rosgen's assessment of response potential for each minor channel type emphasizes this shortcoming. In contrast, Whiting and Bradley (1993) presented a process-based classification for headwater channels that associates channel morphology with the potential for debris flow impacts, channel substrate size, and processes and rates of fluvial sediment transport. Paustian et al. (1992) provided an example of a valley-scale classification emphasizing region-specific associations with channel morphology and processes.

The classifications discussed above serve a variety of purposes, but generally are incomplete for comprehensive channel assessments in forested mountain drainage basins due to either neglect of the influence of woody debris or an emphasis on a single scale of influence on channel morphology and processes. A hierarchical channel classification approach addresses these issues and is well suited for assessment of channel conditions and response potential in mountain drainage basins.

Hierarchical Channel Classification

A hierarchical approach to channel classification addresses different factors influencing channel properties over a range of spatial and temporal scales scales (Figure 2.2) (Frissell et al. 1986). A hierarchy of spatial scales that reflects differences in processes and controls on channel morphology includes geomorphic province, watershed, valley segment, channel reach, and channel unit (Table 2.1, Figure 2.3). Each level of this spatial hierarchy provides a framework for comparing channels at increasingly finer spatial scales.

Geomorphic Provinces

Geomorphic provinces consist of regions with similar land forms that reflect comparable hydrologic, erosional, and tectonic processes over areas greater than 1,000 km^2 (Table 2.1, Figure 2.3). Major physiographic, climatic, and geological features bound geomorphic provinces and impose broad controls on channel processes. Watersheds within a geomorphic province tend to share roughly similar relief, climate, and lithologic assemblages. General controls on channel processes and morphology are reasonably similar for most large watersheds within a geomorphic province; thus, their channels are potentially comparable in terms of relationships between drainage area, discharge, sediment supply, and substrate size. Environmental conditions or the legacy of climate history may impose similar general external constraints on channels within a geomorphic province. Channels in the Olympic Mountains geomorphic province, for example, have an abundant supply of extraordinarily large logs that profoundly influence channel morphology and dynamics. Although geomorphic provinces identify broad areas likely to host comparable watersheds, the concept remains too general for predicting specific channel attributes or responses. Geomorphic provinces do, however, provide a general context for investigating and interpreting channel processes, and therefore channel response.

Watersheds

Watersheds, or drainage basins, define natural systems for routing sediment and runoff into

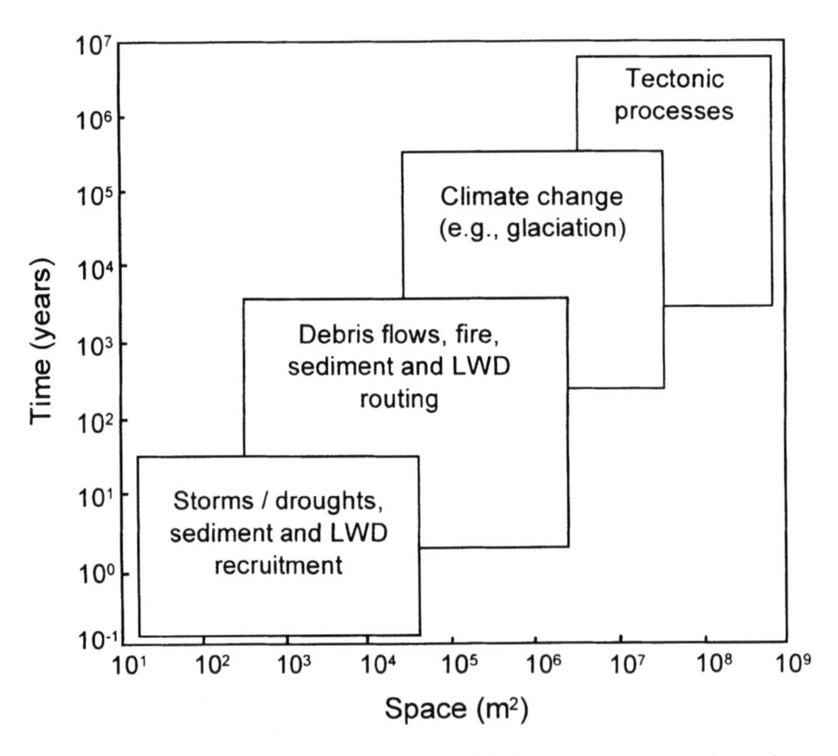

FIGURE 2.2. Range of spatial and temporal influences on stream channels.

and through channel networks (Table 2.1, Figure 2.3). While the term watershed classically refers to a drainage divide, contemporary usage also considers the drainage area upslope of any point along a channel network as a watershed. Although the appropriate scale of watershed-level classification ultimately is site specific, drainage basins of 50 to 500 km^2 provide practical units for examining the influence of watershed processes on channel morphology and disturbance regimes (Montgomery et al. 1995a). Although watershed-level classification differentiates major drainage basins within geomorphic provinces, large rivers can traverse several provinces. Classifying watersheds based on similar geologic and climatic history, lithology, and land use may highlight areas with similar controls on channel processes and identifies river systems as either well or ill suited for comparison. However, watershed level classification of channel networks neglects fundamental differences in sediment production and transport processes of finer-scale valley morphologies.

TABLE 2.1. Hierarchical levels of channel classification and associated spatial scales.

Classification level	Spatial scale
Geomorphic province	1,000 km^2
Watershed	50–500 km^2
Valley segment	10^2–10^4 m
Colluvial valleys	
Bedrock valleys	
Alluvial valleys	
Channel reaches	10^1–10^3 m
Colluvial reaches	
Bedrock reaches	
Free-formed alluvial reaches	
Cascade reaches	
Step-pool reaches	
Plane-bed reaches	
Pool-riffle reaches	
Dune-ripple reaches	
Forced alluvial reaches	
Forced step-pool	
Forced pool-riffle	
Channel units	10^0–10^1 m
Pools	
Bars	
Shallows	

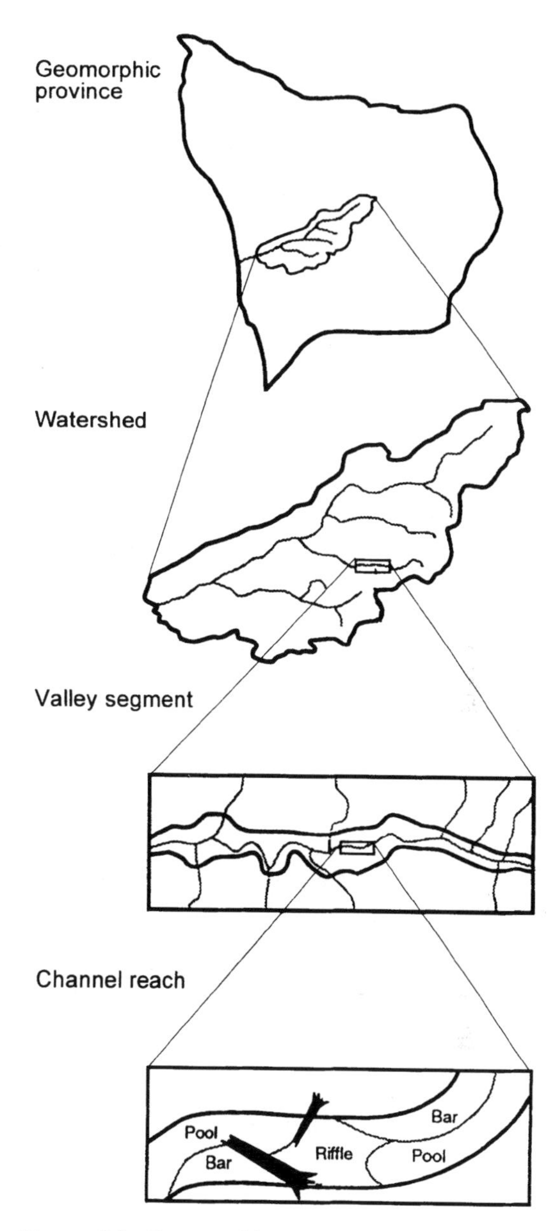

FIGURE 2.3. Geomorphic province, watershed, valley segment, channel reach, and channel unit (e.g., pools, bars, riffles) scales of classification illustrated for the Olympic Peninsula, Washington.

Valley Segments

Valley segments define portions of the drainage network exhibiting similar valley-scale morphologies and governing geomorphic processes at a scale of 10^2 to 10^4 m (Table 2.1, Figure 2.3).

In mountain drainage basins, valley segments are classified into colluvial, bedrock, and alluvial valley types based on valley fill, sediment transport processes, channel transport capacity, and sediment supply. A fourth type, estuarine valleys, are important links between terrestrial and marine environments. These divisions are similar to the valley segment classification developed by Frissell and Liss (1986) for the Oregon Coast Range. More elaborate region-specific valley segment classifications may help link channel classification with resource assessments (Paustian et al. 1992).

Colluvial valleys. Shallow and ephemeral fluvial transport in headwater valleys is relatively ineffective at transporting sediment delivered from surrounding hillslopes, resulting in accumulation of colluvial valley fills. Colluvial valley bottoms lacking evidence of a well-defined channel indicate insufficient hydraulic erosion to initiate and maintain a channel; these unchanneled valleys (regionally referred to as hollows, swales, or headwalls) extend upslope of the smallest channels in many soil-mantled environments. In steep landscapes, unchanneled valleys gradually accumulate colluvial soils transported from surrounding hillslopes and periodically deliver the stored sediment to downstream channels via debris flows. Hillslope sediment transport processes subsequently refill excavated hollows, resulting in a cycle of long-term accumulation punctuated by periodic catastrophic erosion (Dietrich and Dunne 1978). This cycle of hollow accumulation and erosion can take thousands of years (Reneau et al. 1990).

Channeled colluvial valleys downslope of hollows indicate the emergence of fluvial transport. Nevertheless, the influence of fluvial processes on colluvial valley form and incision is often secondary to transport by periodic debris flows in steep landscapes. In contrast, the maintenance of colluvial valleys in low-gradient landscapes requires incision by streams; extensive networks of colluvial valleys in low-gradient landscapes are likely a result of long-term climate change.

Bedrock valleys. Bedrock valleys typically are confined and lack significant valley fill. Narrow valley bottoms result in relatively straight channels, although deeply incised bedrock meanders may occur. Channel floors in bedrock valleys consist of either exposed bedrock or thin, patchy accumulations of alluvium. Insignificant sediment storage in bedrock valley segments indicates downstream transport of virtually all the material delivered to the channel, implying that transport capacity exceeds sediment supply over the long term.

Alluvial valleys. Channels in alluvial valleys transport and sort sediment loads supplied from upslope, but lack the transport capacity to routinely scour the valley to bedrock. Channels in alluvial valley segments may support either narrow or wide floodplains. Thick alluvial deposits in unconfined valley segments imply a long-term excess of sediment supply relative to transport capacity. Both the specific channel morphology and degree of confinement reflect the local channel slope, sediment supply, and hydraulic discharge.

Valley segment morphology is useful for distinguishing dominant sediment transport processes (fluvial versus mass wasting), inferring general long-term sediment flux characteristics (transport- versus supply-limited), and providing insight into the spatial linkages that govern watershed response to disturbance. Alluvial channels, however, exhibit a variety of morphologies, some of which appear functionally similar at the valley segment level, but which respond differently to similar perturbations in sediment load and discharge. Consequently, channel reach morphology often proves more useful than valley segment morphology for understanding channel processes and response potential.

Channel Reaches

Channel reaches exhibit similar bedforms over stretches of stream that are many channel widths in length (Table 2.1, Figure 2.3). In mountain drainage basins, channel reaches are classified into colluvial, bedrock, and alluvial reach types. These general reach types are briefly described below (for further detail see Montgomery and Buffington 1997).

Colluvial Reaches

Colluvial reaches typically occupy headwater portions of a channel network (Figure 2.4a) and occur where drainage areas are large enough to sustain a channel for the local ground slope (Montgomery and Dietrich 1988). Soil creep, tree throw, burrowing by animals, and small-scale slope instability introduce sediment into colluvial reaches. Intermittent flow reworks some portion of the accumulated material, but does not govern deposition, sorting, or transport of most valley fill because of low shear stresses (Benda 1990). Large grains, woody debris, bedrock steps, and in-channel vegetation reduce the energy available for sediment transport. Ephemeral, low discharges in colluvial reaches result in a poorly sorted bed with finer grain sizes than in downstream alluvial reaches. Episodic transport by debris flows accounts for most of the sediment transport in steep headwater channels (Swanson et al.1982).

Bedrock Reaches

Bedrock reaches exhibit little, if any, alluvial bed material or valley fill, and are generally confined by valley walls and lack floodplains (Figure 2.4b). Bedrock reaches occur on steeper slopes than alluvial reaches with similar drainage areas (Montgomery et al. 1996), an observation supporting Gilbert's (1914) hypothesis that bedrock reaches lack an alluvial bed due to a higher transport capacity than sediment supply. Steep headwater channels in mountain drainage basins may alternate through time between bedrock and colluvial morphologies in response to periodic scour by debris flows (Benda 1990). In general, bedrock reaches in low-gradient portions of a watershed reflect a high transport capacity relative to sediment supply, whereas those in steep debris-flow-prone channels also reflect recent debris flow scour.

Free-Formed Alluvial Reaches

Alluvial reaches exhibit a wide variety of bed morphologies and roughness configurations that vary with slope and position within the channel network. Montgomery and Buffington (1997) suggest that the ratio of transport capacity to sediment supply controls the roughness configurations that shape alluvial reach morphology, which can be categorized into five free-formed alluvial channel reach types: cascade, step-pool, plane-bed, pool-riffle, and dune-ripple. Transitional morphologies also occur, as this classification imposes order on a natural continuum.

Cascade reaches. Cascade reaches occur on steep slopes with high rates of energy dissipation and are characterized by longitudinally and laterally disorganized bed material, typically consisting of cobbles and boulders confined by valley walls (Figure 2.4c). Flow in cascade reaches follows a tortuous convergent and divergent path over and around individual large clasts; tumbling flow over these grains and turbulence associated with jet-and-wake flow around grains dissipates much of the mechanical energy of the flow (Peterson and Mohanty 1960, Grant et al. 1990). Large particle size relative to flow depth make the largest bed-forming material of cascade reaches mobile only during infrequent events (i.e., >25 yr, Grant et al. 1990). In contrast, rapid transport of the smaller bedload material over the more stable bed-forming clasts occurs during flows of moderate recurrence interval. Bedload transport studies demonstrate that steep, alluvial, mountain streams are typically supply limited, receiving seasonal or stochastic sediment inputs from local mass wasting events (Griffiths 1980, Whittaker 1987).

Step-pool reaches. Step-pool reaches consist of large clasts organized into discrete channel-spanning accumulations that form a series of steps separating pools containing finer material (Figure 2.4d). The stepped morphology of the bed results in alternating turbulent flow over steps and tranquil flow in pools. Channel-spanning steps provide much of the elevation drop and roughness in step-pool reaches (Whittaker and Jaeggi 1982). Step-forming clasts may be viewed as a congested zone of large grains that causes increased local flow resistance and further accumulation of large particles (Church and Jones 1982), or as

a

b

FIGURE 2.4. Photographs of reach-level channel types: (a) colluvial; (b) bedrock; (c) cascade; (d) step-pool; (e) plane-bed; (f) pool-riffle; and (g) dune-ripple.

FIGURE 2.4. *Continued.*

f

g

FIGURE 2.4. *Continued.*

macroscale antidunes (Whittaker and Jaeggi 1982). Step-pool morphologies develop during infrequent flood events and are associated with supply-limited conditions, steep gradients, coarse bed materials, and confined channels (Chin 1989). Like cascade reaches, large bed-forming material mobilizes infrequently (Grant et al. 1990), while finer pool-filling material is transported annually as bedload (Schmidt and Ergenzinger 1992). Step-pool reaches often receive episodic slugs of sediment input that travel downstream as bedload waves (Whittaker 1987).

Plane-bed reaches. Plane-bed reaches are characterized by a relatively featureless gravel/cobble bed (Figure 2.4e) that encompasses channel units (described later) often termed glides, riffles, and rapids (Bisson et al. 1982). Plane-bed reaches may be either unconfined or confined by valley walls and are distinguished from cascade reaches by the absence of tumbling flow and smaller relative roughness (ratio of the largest grain size to bankfull flow depth). Plane-bed reaches lack sufficient lateral flow convergence to develop pool-riffle morphology (discussed in next section) because of lower width-to-depth ratios and greater relative roughnesses, which may decompose lateral flow into smaller circulation cells (Ikeda 1977 and 1983). Bed surfaces are often armored in plane-bed reaches and are calculated to have a near-bankfull threshold for general mobility (Buffington 1995). Lack of depositional features, such as barforms, and typically armored bed surfaces demonstrate some supply-limited characteristics of plane-bed reaches (Dietrich et al. 1989). However, studies of armored gravel-bed channels show a general correlation of bedload transport rate and discharge during armor-breaching events (Jackson and Beschta 1982), indicating that sediment transport is not limited by supply once the bed is mobilized. Hence, plane-bed reaches represent a transition between supply- and transport-limited morphologies.

Pool-riffle reaches. Pool-riffle reaches are typically unconfined by valley walls and consist of a laterally oscillating sequence of bars, pools, and riffles (Figure 2.4f) resulting from oscillating cross-channel flow that causes flow convergence and scour on alternating banks of the channel. Concordant downstream flow divergence on the opposite side of the channel results in local sediment accumulation in discrete bars. Bedform and grain roughness provide the primary flow resistance in pool-riffle reaches. Bedforms in many pool-riffle reaches are relatively stable morphologic features, even though the material forming the bed is transported annually. Alluvial bar development requires a large width-to-depth ratio and small grains easily mobilized and aggraded by the flow (Church and Jones 1982). Pool-riffle reaches are commonly armored, exhibiting a near bankfull threshold for general bed surface mobility (Parker 1979, see also review by Buffington 1995) and a mixture of supply- and transport-limited characteristics similar to plane-bed reaches. Although the presence of depositional barforms in pool-riffle reaches suggests that they are generally more transport-limited than plane-bed reaches, the transport-limited character of both of these morphologies contrasts with the more supply-limited character of step-pool and cascade reaches.

Dune-ripple reaches. Dune-ripple reaches are unconfined, low-gradient, sand-bedded channels (Figure 2.4g). They exhibit a succession of mobile bedforms with increasing flow depth and velocity that proceeds as lower-regime plane bed, ripples, sand waves, dunes, upper-regime plane bed, and finally antidunes (Gilbert 1914, Simons et al. 1965). The primary flow resistance is provided by bedforms (Kennedy 1975), several scales of which may coexist; ripples and small dunes that climb over larger dunes as they all move down the channel. Sediment transport in dune-ripple reaches occurs at most stages, and strongly depends on discharge; as such, these reaches are transport-limited.

Forced Alluvial Reaches

External flow obstructions, such as LWD and bedrock outcrops, force local flow convergence, divergence, and sediment impoundment that respectively form pools, bars, and steps (Figure 2.5). The morphologic impact of LWD, in particular, depends on the amount, size,

FIGURE 2.5. A pool-riffle reach morphology forced by large woody debris (LWD).

orientation, and position of debris, as well as channel size (Bilby and Ward 1989, Montgomery et al. 1995b, Chapter 13, this volume) and rates of debris recruitment, transport, and decay (Bryant 1980, Murhphy and Koski 1989). In small channels, LWD is generally stable over years to decades, and individual logs can dominate channel morphology by anchoring pool and bar forms. Logs oriented transversely and located low in the flow may form steps that create local plunge pools and hydraulic jumps, buttress significant amounts of sediment, and dissipate energy otherwise available for sediment transport (Keller and Swanson 1979, Marston 1982). Single logs oriented obliquely can result in scour pools and proximal sediment storage by both upstream buttressing and downstream deposition in low-energy zones. However, in large rivers where individual pieces are mobile, debris jam formation is necessary for LWD to significantly influence channel morphology (Abbe and Montgomery 1996). When accumulated in jams, LWD can influence channel pattern and floodplain processes in large forest channels by armoring banks, developing pools, bars, and side channels, and forcing bank cutting and channel avulsions (Bryant 1980, Nakamura and Swanson 1993, Abbe and Montgomery 1996).

Flow obstructions can force specific channel morphologies on steeper slopes than is typical of analogous free-formed alluvial morphologies. In particular, LWD may force pool-riffle formation in otherwise plane-bed or bedrock reaches (Montgomery et al. 1995b, 1996). Consequently, plane-bed reaches are rare in undisturbed forested environments where LWD dominates formation of pools and bars. LWD may also force step-pool morphologies in otherwise cascade or bedrock reaches. It is important to recognize forced morphologies as distinct reach types because the interpretation of whether such obstructions govern bed morphology is crucial for understanding channel response.

Channel Units

Channel units are morphologically distinct areas that extend up to several channel widths in length and are spatially embedded within a channel reach (Figure 2.3); they are the morphologic building blocks of a reach. Channel units are classified as various types of pools, bars, and shallows (i.e., riffles, rapids, and cascades) (Bisson et al. 1982, Sullivan 1986, Church 1992). Distinctions among these units focus on topographic form, organization and areal density of clasts, local slope, flow depth and velocity, and to some extent, grain size. Different channel units have characteristic velocities and depths and provide specific habitat characteristics associated with different patterns of fish use (Chapter 9). In practice, however, definitions of these channel unit morphologies tend to overlap and vary with discharge; and channel unit classification by different observers often yields inconsistent results.

Although there is a general association of specific channel unit morphologies with reach type, prediction of channel unit properties is complicated by site-specific controls, particularly in forest environments. For example, the size, location, and orientation of individual flow obstructions control the distribution of specific pool types and their dimensions (Lisle 1986, Cherry and Beschta 1989). Furthermore, similar channel unit morphologies will have different response potentials depending on reach type and associated physical processes and conditions. For example, plunge pools in a LWD-forced pool-riffle reach may exhibit prolonged and extensive pool filling in response to increased sediment supply, while similar pools in a step-pool reach may be less responsive due to a potentially greater ratio of transport capacity to sediment supply. While the channel unit scale is biologically relevant, interpretation of the abundance, characteristics, and response potential of channel units depends upon the context imposed by reach-level channel types.

Channel Disturbance and Response Potential

Response to land use or environmental change varies for different channel types. Alluvial channels, in particular, exhibit a wide variety of potential responses. The predictive ability of conceptual models of channel response can be dramatically improved by considering the influence of reach-level channel type. Reach morphologies are associated with physical processes and environments that limit the range and magnitude of possible channel responses to changes in hydraulic discharge and sediment supply. Reach-specific response potential is further affected by external influences, such as channel confinement, riparian vegetation, and in-channel LWD. The impact of both isolated and cumulative watershed disturbance(s) on a particular reach also depends on the location of the reach within the drainage basin and the sequence of upstream reach types.

Reach-Level Response

Differences in reach morphology and physical processes result in different potential responses to similar changes in discharge or sediment supply. The specific response of a particular channel reach also reflects the intensity of disturbance or the magnitude of change, as well as the condition of the reach. While there are many possible response scenarios depending on site-specific factors, Table 2.2 illustrates typical patterns of potential response for Pacific Northwest channel reaches; direction of response is indicated by relationships (2.13) and (2.14). Changes in sediment storage dominate the response of colluvial reaches to altered sediment supply because of transport-limited conditions and low fluvial transport capacities; depending on the degree of valley fill, increased discharge can significantly change reach morphology. In contrast, bedrock, cascade, and step-pool reaches are resilient to most discharge or sediment supply perturbations because of high transport capacities and

TABLE 2.2. Interpreted reach-level channel response potential to moderate changes in sediment supply and discharge (+ = likely to change; p = possible to change; − = unlikely to change).

	Reach level morphology	Width	Depth	Roughness	Scour depth	Grain size	Slope	Sediment storage
Response	dune-ripple	+	+	+	+	−	+	+
	pool-riffle	+	+	+	+	+	+	+
	plane-bed	p	+	p	+	−	+	p
Transport	step-pool	−	p	p	p	p	p	p
	cascade	−	−	p	−	p	−	−
	bedrock	−	−	−	−	−	−	−
Source	colluvial	p	p	−	p	p	−	+

Modified from Montgomery and Buffington, 1997.

generally supply-limited conditions. Many bedrock reaches are insensitive to all but catastrophic changes in discharge and sediment load. Lateral confinement and large, relatively immobile, bed-forming clasts make channel incision or bank cutting unlikely responses to changes in sediment supply or discharge in most cascade and step-pool reaches. Potential responses in step-pool reaches include changes in bedform frequency and geometry, grain size, and pool scour depths, while only limited textural response is likely in cascade reaches. Lower gradient plane-bed, pool-riffle, and dune-ripple reaches become progressively more responsive to altered discharge and sediment supply with decreasing ratios of transport capacity to sediment supply, smaller grain sizes, and less channel confinement. Because plane-bed reaches frequently occur in both confined and unconfined valleys, they may or may not be susceptible to channel widening or changes in valley bottom sediment storage. Smaller, more mobile sediment in plane-bed and pool-riffle reaches allows potentially greater response of bed surface textures, scour depth, and slope compared to cascade and step-pool morphologies. Unconfined pool-riffle and dune-ripple reaches generally have significant potential for channel geometry response to perturbations in sediment supply and discharge. Changes in both channel and valley storage are also likely responses, as well as changes in channel roughness due to alteration of channel sinuosity and bedforms. There is less potential for textural response in dune-ripple reaches than in pool-riffle and plane-bed reaches because of smaller and more uniform grain sizes. Very high sediment loading in any unconfined reach can result in a braided morphology (Mollard 1973, Church 1992). The general progression of alluvial reach types downstream through a channel network suggests that there is a systematic downstream increase in response potential to altered sediment supply or discharge.

Segment-Level Response

Position within the network and differences between ratios of transport capacity to sediment supply allow aggregation of channel reaches into source, transport, and response segments. Source segments are headwater colluvial channels that act as transport-limited, sediment storage sites subject to intermittent debris flow scour. Transport segments are composed of morphologically resilient, supply-limited reaches (bedrock, cascade, and step-pool) that rapidly convey increased sediment inputs (Table 2.2). Response segments consist of lower-gradient, more transport-limited reaches (plane-bed, pool-riffle, and dune-ripple) in which significant morphologic adjustment occurs in response to increased sediment supply (Table 2.2).

The spatial distribution of source, transport, and response segments generally reflects the distribution of potential impacts and recovery times. Distribution of these segment types define watershed-scale patterns of sensitivity to altered discharge and sediment supply (Figure 2.6). Sediment delivered to transport segments rapidly propagates to downstream response segments, where sediment accumulates. Consequently, locations in the channel network where transport segments flow into response segments indicate places particularly susceptible to impacts from accelerated sediment supply. In this regard, the general classification of source, transport, and response segments identifies areas most susceptible to local increases in upstream sediment inputs. Because response segments are sensitive to increases in sediment supply, they are excellent sites for monitoring the effects of upstream actions and can be considered locations of critical importance in watershed monitoring programs. The relationship between channel classification and response potential provides an understanding of the linkage between upstream sediment inputs and downstream response.

External Influences

External influences on channel response include factors such as confinement, riparian vegetation, and in-channel LWD. Specific effects of these factors vary both with channel type and position within the network.

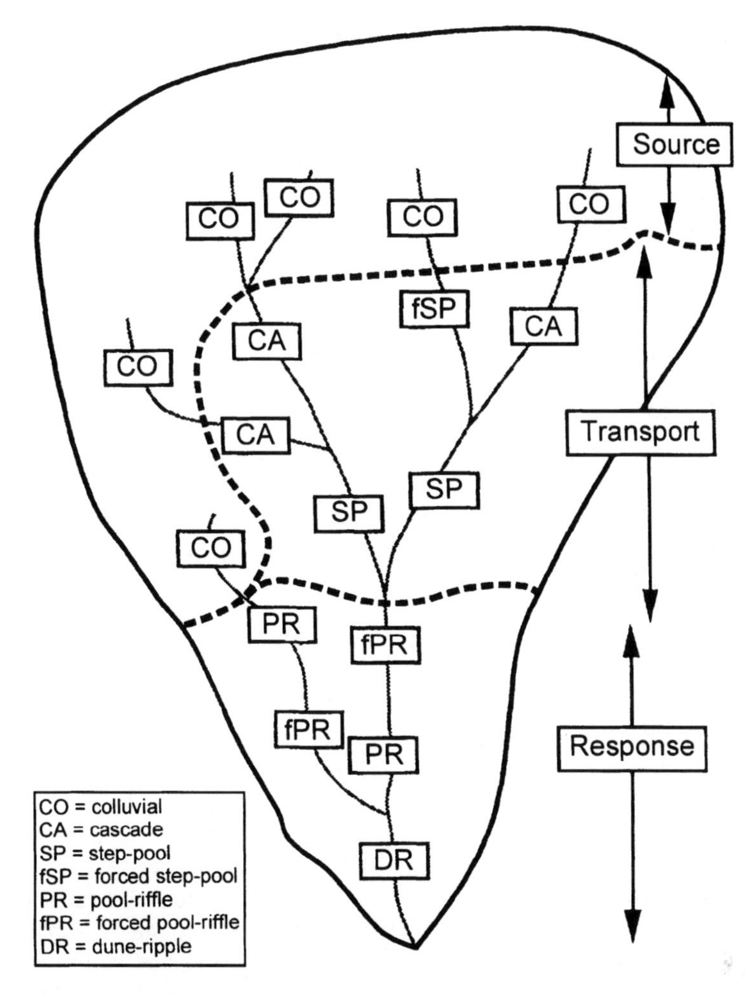

FIGURE 2.6. The spatial distribution of reach types within a drainage basin influences the distribution of potential impacts and responses to disturbance.

Confinement

Channel confinement strongly influences channel response. Channel migration and avulsion, for example, typically are rare in confined channels. The geometry of the channel above the bankfull stage also controls the response of the channel bed to high-discharge flow. Unconfined channels possess extensive floodplains across which over-bank flows spread, which limits the effect of peak discharges on channel morphology. In contrast, confined channels efficiently translate high flows into increased basal shear stress.

The degree of channel confinement may be influenced by either the long-term sediment balance of the channel or by external influences. Unconfined channels may reflect long-term alluvial aggradation and flood plain development where sediment supply exceeds transport capacity. Alternatively, unconfined channels may be controlled by tectonic boundary conditions (as in the case of alluvial fans at the base of block-faulted mountains), or reflect an inherited morphology (as in the case of underfit channels or u-shaped glacial valleys).

Isolation of unconfined channels from their floodplains can entail dramatic consequences for many biological systems. Prevention of overbank flows by dikes, or other flood control measures, may trigger channel entrenchment.

Flow diversions or regulation that prevent or decrease the frequency of floodplain inundation change both side-channel and floodplain processes. Abandonment of side channels and ponds may eliminate important aquatic habitat. Prevention of over-bank flows also stops sediment and nutrient delivery to floodplain soils, which may affect both floodplain-dwelling organisms and the long-term productivity of agricultural land.

Riparian Vegetation and Large Woody Debris

Riparian vegetation influences channel morphology and response potential by providing root strength that contributes to bank stability (Shaler 1891, Gilbert 1914), especially in relatively noncohesive alluvial deposits. The effect of root strength on channel bank stability is greatest in low-gradient, unconfined reaches where loss of bank reinforcement may result in dramatic channel widening (Smith 1976). Riparian vegetation is also an important source of roughness (Arcement and Schneider 1989) that can mitigate the erosive action of high discharges.

LWD provides significant control on the formation and physical characteristics of pools, bars, and steps, thereby influencing channel type and the potential for change in sediment storage and bedform roughness in response to altered sediment supply, discharge, or LWD loading. LWD may also decrease the potential for channel widening by armoring stream banks; alternatively it may aid bank erosion by directing flow and scour toward channel margins. Furthermore, bed surface textures and their response potential are strongly controlled by hydraulic roughness resulting from in-channel LWD and debris-forced bedforms (Buffington 1995).

Changes in amount, size, and decay rate of LWD may also affect channel processes and morphology. Alteration of channel margin vegetation may change both the age and species of wood entering the fluvial system (Murphy and Koski 1989). In small channels where LWD provides significant sediment storage, decreased supply of LWD accelerates sediment transport (Smith et al. 1993a). Channels in which LWD provides a dominant control on pool formation and sediment storage (e.g., forced pool-riffle or forced step-pool channels) are particularly sensitive to changes in the size, species, and amount of recruited LWD (Figure 2.7). Removal of LWD from forced pool-riffle reaches may alter the size and location of pools (Smith et al. 1993b) and lead to either a pool-riffle or plane-bed morphology depending on channel slope (Montgomery et al. 1995b). Similarly, loss of LWD may transform a forced step-pool reach into a step-pool or cascade reach, depending on channel slope and discharge. Where transport capacities are in extreme excess of sediment supply, forced alluvial reaches may become bedrock reaches following LWD removal (Montgomery et al. 1996).

Debris Flow Disturbance

Debris flows are primary agents of channel disturbance in mountain drainage basins. Debris flows tend to be pulsed disturbances, the effects of which vary with slope and position in the channel network. Passage of a debris flow can scour steep channels to bedrock. Deposition in lower-gradient channels typically results in local aggradation and can even obliterate the channel as a morphological feature. Recovery from debris flow impacts also differs for steep and low-gradient channels. Steep, high-energy channels (bedrock, step-pool, and cascade reaches) recover quickly from sediment deposition because of high transport capacities. In contrast, lower-gradient channels (plane-bed and pool-riffle reaches) typically take longer to recover from debris flow deposition because of their lower transport capacity. The morphology of mountain channels prone to debris flows thus reflects the time since debris flow scour, as well as position within the fluvial system. Although channel gradient generally determines the type of debris flow impacts, channel network architecture influences debris flow routing. Assessment of potential impacts of debris flow involves differentiating areas of potential debris flow initiation, scour, and deposition (Benda and Cundy 1990).

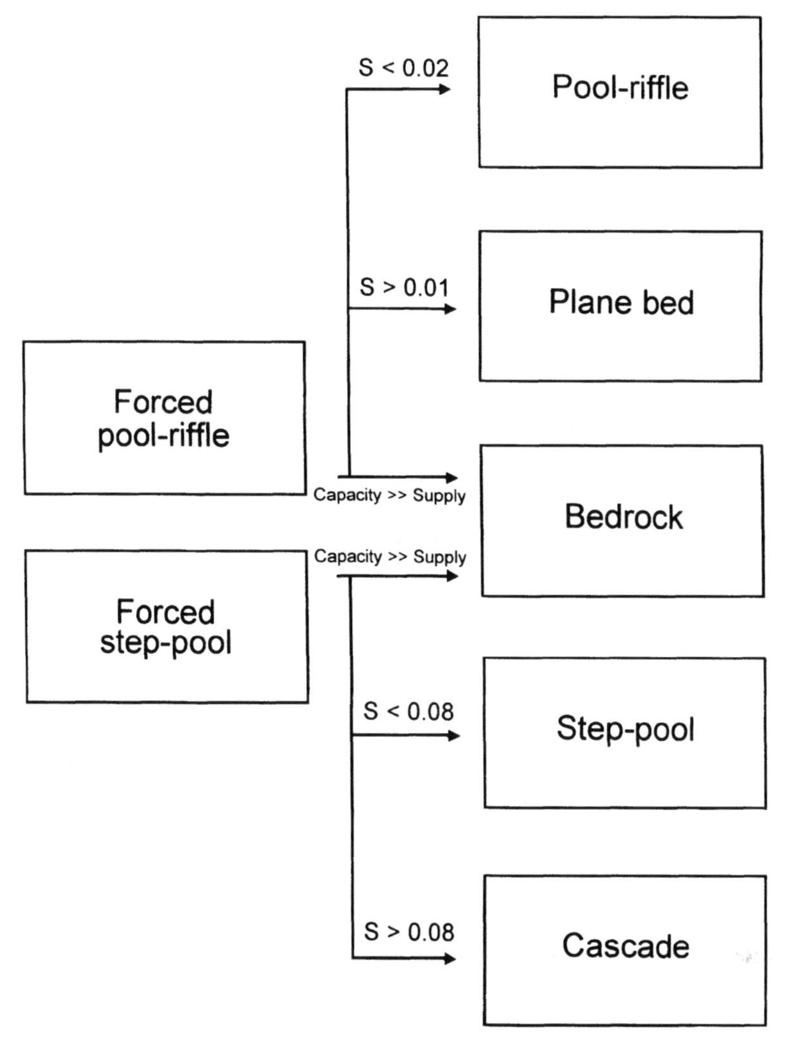

FIGURE 2.7. Potential morphologic response to LWD removal in channel reaches with forced alluvial morphology. Response depends on both channel slope (S) and the relative magnitudes of transport capacity and sediment supply.

Channel slope and tributary junction angles exert important controls on debris flow routing. Debris flows originating at the heads of long straight channels tend to scour long channel segments, and deliver sediment to downslope alluvial channels (Grant et al. 1984, Benda and Dunne 1987). Such events also may scour the base of adjacent hillslopes, hollows, and tributary channels, activating smaller failures that contribute to the sediment load imposed upon downslope channels. Debris flows originating in obliquely oriented tributaries tend to deposit at channel confluences (Grant et al. 1984, Benda and Cundy 1990). Subsequent events large enough to scour the accumulated material in the main channel can have catastrophic impacts on downstream alluvial channels. Debris flow deposition occurs when the channel slope declines to the extent that the yield strength of the flowing debris is sufficient to resist further transportation and deformation. This gradient is commonly between 5% and 10% for the range of water contents typical of debris flows (Takahashi et al. 1981, Benda and Cundy 1990), but incorporation of LWD in the leading edge of a debris flow may result in deposition on steeper slopes.

Massive inputs of sediment, such as from extensive synchronous landsliding within a basin, can set up a sediment wave that pulses through downstream channels (Figure 2.8). Passage of such waves involves local aggradation accompanied by fining of the bed. If large enough, passage of a sediment wave can change channel type; bedrock channels may become alluvial (Perkins 1989) or pool-riffle channels may become braided. Debris flow inputs can set up oscillations in channel morphology, the frequency of which varies with position in the network (Benda 1994). In historical old-growth forests of the Pacific Northwest, large stable log jams likely damped propagation of sediment waves through channel networks. Today the availability of such sediment capacitors is low in many watersheds due to stream cleaning and salvage operations, as well as harvesting large logs, from riparian forests; these large logs would otherwise serve to stabilize wood jams (Abbe and Montgomery 1996, Montgomery et al. 1996). Hence, large-scale sediment waves may be more important in today's industrial forests than in primeval forests. Channel morphology generally recovers after passage of a sediment wave at a rate that depends upon slope, confinement, sediment supply, and position in the network.

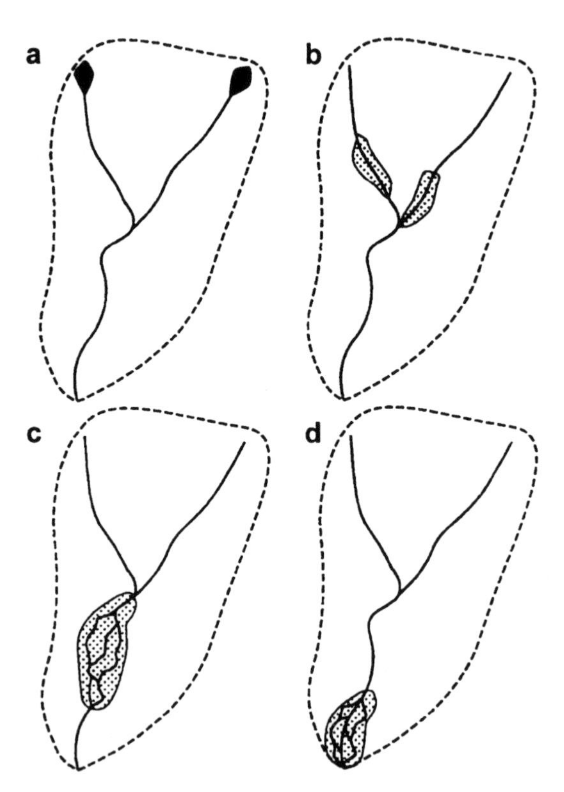

FIGURE 2.8. Sediment wave propagation through a channel network: (a) landsliding originating at the head of the channel network rapidly propagates (b) through steep headwater channels to lower-gradient reaches (c) where deposition may trigger channel braiding, which eventually (d) recovers to a single thread morphology.

Applications for Ecosystem Analysis

While the channel types discussed earlier are readily identified in the field, a method for classifying channel reaches from topographic maps or aerial photographs is useful for watershed-scale analysis and rapid assessment of reach types and response potential in mountain drainage basins. A classification that does not require visual assessment is particularly useful for designing channel assessment plans and interpreting conditions across entire watersheds. Such a classification can be based on channel gradient and confinement, two key features readily estimated from topographic maps and digital elevation models (DEMs).

Empirical association of channel type with different reach slopes provides a method for predicting reach type from topographic maps or DEMs. Frequency distributions of surveyed reach slopes for Pacific Northwest channels illustrate associations between reach type and slope gradient: pool-riffle reaches are most common on slopes less than 1%; plane-bed reaches are most common on slopes of 2% to 4%; step-pool channels are common on slopes of 4% to 8%; cascade morphologies dominate on slopes of 8% to 20%; and channels steeper than 20% typically have colluvial bed morphologies (Figure 2.9). The slope ranges for each reach type overlap, and bedrock reaches

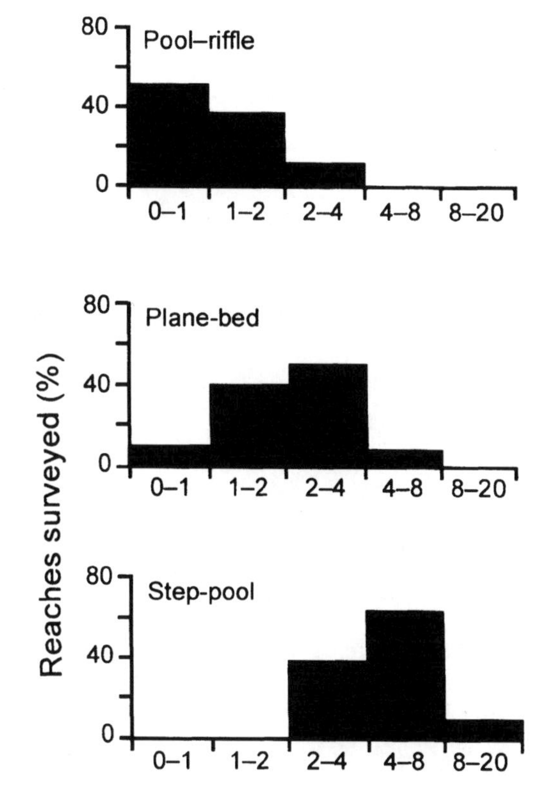

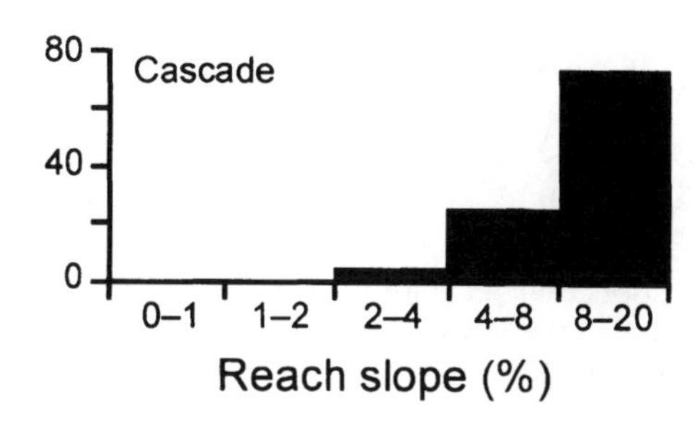

FIGURE 2.9. Frequency distributions of reach slopes for pool-riffle ($n = 48$), plane-bed ($n = 57$), step-pool ($n = 51$), and cascade ($n = 78$) morphologies surveyed in Oregon, Washington, and southeast Alaska.

are not associated with any particular slope range. Moreover, the LWD loading that controls forced reach morphologies cannot be predicted from maps or DEMs. Although predictions of channel type based on reach-average slope can provide a reasonable stratification of channel networks in mountain drainage basins of the Pacific Northwest, field observations are necessary to verify reach type, condition, and response potential.

Three simple confinement classes (totally confined, moderately confined, and unconfined) can be estimated from aerial photographs or valley widths portrayed on 7.5′ topographic maps. Although a gradient/confinement index provides a map-based stratification of channel networks useful in watershed analyses (Figure 2.10), channel confinement can be difficult to estimate from topographic maps because channel width is poorly expressed; thus, predicted slope and confinement may prove inaccurate in the field. Nevertheless, explaining discrepancies between expected conditions and those observed in the field can provide insight into local channel processes or disturbance history.

In mountain drainage basins, channel gradient and confinement provide a rough guide for identifying channels with different frequencies and magnitudes of ecologically relevant disturbance processes. Identification of headwater colluvial channels, confined alluvial channels, and unconfined alluvial channels can quickly characterize spatial patterns in the types, magnitudes, and frequencies of geomorphological disturbances that influence ecological

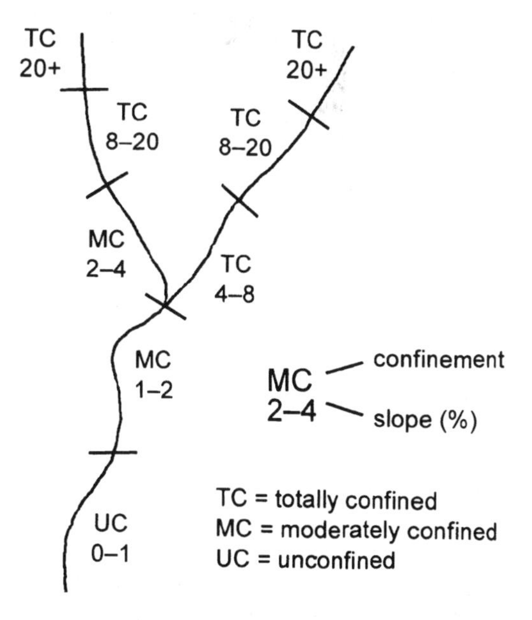

FIGURE 2.10. Application of a gradient/confinement index to channel classification.

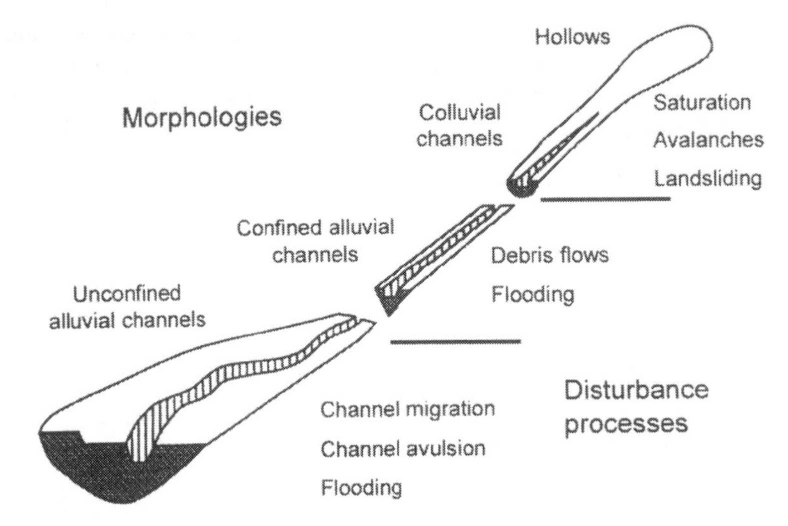

FIGURE 2.11. Differences in disturbance processes among colluvial, confined alluvial, and unconfined alluvial channels in mountain drainage basins.

organization and variability (Figure 2.11). More specific differences in disturbance processes can be evaluated based on consideration of bed morphology and disturbance history. The ecological significance of the environmental characteristics and variability associated with specific process domains depends upon the organism(s) of interest.

Channel classification cannot substitute for focused observation and clear thinking about channel processes. Channels are complex systems that need to be interpreted within their local and historical context. Classification simply provides one of a variety of tools that can be applied to particular problems—it is not a panacea. Classifications that highlight specific aspects of the linkages between channel networks and watershed processes are likely to be most useful, but careless application of any channel classification may prove misleading; no classification can substitute for an alert, intelligent, well-trained observer. Nonetheless, it is difficult to fully understand a channel reach without reference to the context defined by its bed morphology, confinement, position in the network, and disturbance history. Consideration of these factors within a spatial hierarchy can further guide interpretations of field observations and evaluation of channel conditions.

Acknowledgments. Preparation of this chapter was supported by the Washington State Timber-Fish-Wildlife agreement through grant TFW FY95-156 and by US Forest Service cooperative research agreements 93-0441 and 94-0617. The authors thank Leslie Reid and Tom Lisle for their critiques of a draft manuscript.

Literature Cited

Abbe, T.B., and D.R. Montgomery. 1996. Large woody debris jams, channel hydraulics and habitat formation in large rivers. Regulated Rivers: Research & Management **12**:201–221.

Allen, P.M., R. Hobbs, and N.D. Maier. 1989. Downstream impacts of a dam on a bedrock fluvial system. Bulletin of the Association of Engineering Geologists **26**:165–189.

Arcement, G.J., and V.R. Schneider. 1989. Guide for selecting Manning's roughness coefficients for natural channels and flood plains. United States Geological Survey Water Supply Paper 2339. US Government Printing Office, Washington, DC, USA.

Benda, L.E. 1994. Stochastic geomorphology in a humid mountain landscape. Unpublished Ph.D. dissertation. University of Washington, Seattle, Washington, USA.

Benda, L. 1990. The influence of debris flows on channels and valley floors in the Oregon Coast Range. USA Earth Surface Processes and Landforms **15**:457–466.

Benda, L., and T.W. Cundy. 1990. Predicting deposition of debris flows in mountain channels. Canadian Geotechnical Journal **27**:409–417.

Benda, L., and T. Dunne. 1987. Sediment routing by debris flows. Pages 213–223 *in* R.L. Beschta, R. Blinn, G.E. Grant, G. Ice, and F.J. Swanson, eds. Erosion and Sedimentation in the Pacific Rim. IAHS Publication 165, International Association of Hydrological Sciences, Wallingford, UK.

Bilby, R.E., and J.W. Ward. 1989. Changes in characteristics and function of woody debris with increasing size of streams in western Washington. Transactions of the American Fisheries Society **118**:368–378.

Bisson, P.A., J.L. Nielson, R.A. Palmason, and L.E. Gore. 1982. A system of naming habitat types in small streams, with examples of habitat utilization by salmonids during low stream flow. Pages 62–73 *in* N.B. Armantrout, ed. Acquisition and utilization of aquatic habitat information. American Fisheries Society, Portland, Oregon, USA.

Booth, D.B. 1990. Stream channel incision following drainage basin urbanization. Water Resources Bulletin **26**:407–417.

Bosch, J.M., and J.D. Hewlett. 1982. A review of catchment experiments to determine the effect of vegetation changes on water yield and evapotranspiration. Journal of Hydrology **55**:2–23.

Bryant, M.D. 1980. Evolution of large, organic debris after timber harvest: Maybeso creek, 1949 to 1978. USDA Forest Service General Technical Report PNW–101. Pacific Northwest Forest and Range Experiment Station, Portland, Oregon, USA.

Buffington, J.M. 1995. Effects of hydraulic roughness and sediment supply on surface textures of gravel-bedded rivers. Masters thesis. University of Washington, Seattle, Washington, USA.

Carling, P.A. 1987. Bed stability in gravel streams, with reference to stream regulation and ecology. Pages 321–347 *in* K.S. Richards, ed. River channels: Environment and process. Blackwell, Oxford, UK.

Cherry, J., and R.L. Beschta. 1989. Coarse woody debris and channel morphology: A flume study. Water Resources Bulletin **25**:1031–1036.

Chin, A. 1989. Step pools in stream channels. Progress in Physical Geography **13**:391–407.

Church, M. 1992. Channel morphology and typology. Pages 126–143 *in* P. Calow and G.E. Petts, eds. The rivers handbook. Blackwell, Oxford, UK.

Church, M., and D. Jones. 1982. Channel bars in gravel-bed rivers. Pages 291–338 *in* R.D. Hey, J.C. Bathurst, and C.R. Thorne, eds. Gravel-bed rivers: Fluvial processes, engineering and management. John Wiley & Sons, Chichester, UK.

Coats, R., L. Collins, J. Florsheim, and D. Kaufman. 1985. Channel change, sediment transport, and fish habitat in a coastal stream: Effects of an extreme event. Environmental Management **9**:35–48.

Dietrich, W.E., and T. Dunne. 1978. Sediment budget for a small catchment in mountainous terrain. Zeitschrift für Geomorphologie, Supplementband **29**:191–206.

Dietrich, W.E., J.W. Kirchner, H. Ikeda, and F. Iseya. 1989. Sediment supply and the development of the coarse surface layer in gravel-bedded rivers. Nature **340**:215–217.

Ellen, S.D., and G.F. Wieczorek. 1988. Landslides, floods, and marine effects of the storm of January 3–5, 1982, in the San Francisco Bay Region, California. United States Geological Survey Professional Paper 1434. US Government Printing Office, Washington, DC, USA.

Frissell, C.A., and W.J. Liss. 1986. Classification of stream habitat and watershed systems in south coastal Oregon and an assessment of land use impacts. Progress report prepared for the Oregon Department of Fish and Wildlife, Oak Creek Laboratory of Biology, Corvallis, Oregon, USA.

Frissell, C.A., W.J. Liss, C.E. Warren, and M.D. Hurley. 1986. A hierarchical framework for stream habitat classification: Viewing streams in a watershed context. Environmental Management **10**:199–214.

Gilbert, G.K. 1877. Report on the geology of the Henry Mountains. United States Geological and Geographical Survey, Rocky Mountain Region. US Government Printing Office, Washington, DC, USA.

——. 1914. The transportation of debris by running water. United States Geological Survey Professional Paper 86. US Government Printing Office, Washington, DC, USA.

——. 1917. Hydraulic-mining débris in the Sierra Nevada. United States Geological Survey Professional Paper 105. US Government Printing Office, Washington, DC, USA.

Graf, W.L. 1975. The impact of suburbanization on fluvial geomorphology. Water Resources Research **11**:690–692.

Grant, G.E., M.J. Crozier, and F.J. Swanson. 1984. An approach to evaluating off-site effects of timber harvest activities on channel morphology. Proceedings, Symposium on the Effects of Forest Land Use on Erosion and Slope Stability. Environment and Policy Institute, East–West

Center, University of Hawaii, Honolulu, Hawaii, USA.

Grant, G.E., F.J. Swanson, and M.G. Wolman. 1990. Pattern and origin of stepped-bed morphology in high-gradient streams, Western Cascades, Oregon. Geological Society of America Bulletin **102**:340–352.

Griffiths, G.A. 1980. Stochastic estimation of bed load yield in pool-and-riffle mountain streams. Water Resources Research **16**:931–937.

Hammer, T.R. 1972. Stream channel enlargement due to urbanization. Water Resources Research **8**:1530–1540.

Harr, R.D. 1981. Some characteristics and consequences of snowmelt during rainfall in western Oregon. Journal of Hydrology **53**:277–304.

——. 1983. Potential for augmenting water yield through forest practices in western Washington and western Oregon. Water Resources Bulletin **19**:383–393.

——. 1986. Effects of clearcutting on rain-on-snow runoff in western Oregon: A new look at old studies. Water Resources Research **22**:1095–2000.

Helley, E.J., and V.C. LaMarche, Jr. 1973. Historic flood information for northern California streams from geologic and botanical evidence. United States Geological Survey Professional Paper 485-E. US Government Printing Office, Washington, DC, USA.

Henderson, F.M. 1963. Stability of alluvial channels. Transactions of the American Society of Civil Engineers **128**:657–686.

Horton, R.E. 1945. Erosional development of streams and their drainage basins; Hydrophysical approach to quantitative morphology. Geological Society of America Bulletin **56**:275–370.

Ikeda, H. 1983. Experiments on bedload transport, bed forms, and sedimentary structures using fine gravel in the 4-meter-wide flume. Environmental Research Center Papers, Environmental Research Center, University of Tsukuba, Ibaraki, Japan.

Ikeda, S. 1977. On the origin of bars in the meandering channels. Bulletin of the Environmental Research Center, University of Tsukuba **1**:17–31.

Jackson, W.L., and R.L. Beschta. 1982. A model of two-phase bedload transport in an Oregon Coast Range stream. Earth Surface Processes and Landforms **7**:517–527.

Jackson, W.L., and R.L. Beschta. 1984. Influences of increased sand delivery on the morphology of sand and gravel channels. Water Resources Bulletin **20**:527–533.

James, L.A. 1991. Incision and morphologic evolution of an alluvial channel recovering from hydraulic mining sediment. Geological Society of America Bulletin **103**:723–736.

Jones, J.A., and G.E. Grant. 1996. Peak flow responses to clear-cutting and roads in small and large basins, western Cascades, Oregon. Water Resources Research **32**:959–974.

Keller, E.A., and F.J. Swanson. 1979. Effects of large organic material on channel form and fluvial processes. Earth Surface Processes **4**:361–380.

Kellerhals, R., M. Church, and D.I. Bray. 1976. Classification and analysis of river processes. Journal of the Hydraulics Division, American Society of Civil Engineers **102**:813–829.

Kelsey, H.M. 1980. A sediment budget and an analysis of geomorphic process in the Van Duzen River basin, north coastal California, 1941–1975. Geological Society of America Bulletin **91**:1119–1216.

Kennedy, J.F. 1975. Hydraulic relations for alluvial streams. Pages 114–154 *in* V. Vanoni, ed. Sedimentation engineering. Manual 54, American Society of Civil Engineers, New York, New York, USA.

Knighton, A.D. 1991. Channel bed adjustment along mine-affected rivers of northeast Tasmania. Geomorphology **4**:205–219.

Kondolf, G.M. 1995. Geomorphological stream channel classification in aquatic habitat restoration: Uses and limitations. Aquatic Conservation: Marine and Freshwater Ecosystems **5**:127–141.

Lane, E.W. 1955. The importance of fluvial morphology in hydraulic engineering. Proceedings of the American Society of Civil Engineers **81**:745–761.

Leopold, L.B. 1968. Hydrology for urban land planning: A guidebook on the hydrologic effects of urban land use. United States Geological Survey Circular 554. US Government Printing Office, Washington, DC, USA.

Leopold, L.B., and T. Maddock, Jr. 1953. The hydraulic geometry of stream channels and some physiographic implications. United States Geological Survey Professional Paper 252. US Government Printing Office, Washington, DC, USA.

Leopold, L.B., and M.G. Wolman. 1957. River channel patterns: Braided, meandering and straight. United States Geological Survey Professional Paper 282-B. US Government Printing Office, Washington, DC, USA.

Lisle, T.E. 1982. Effects of aggradation and degradation on riffle-pool morphology in natural gravel channels, northwestern California. Water Resources Research **18**:1643–1651.

——. 1986. Stabilization of a gravel channel by large streamside obstructions and bedrock bends, Jacoby Creek, northwestern California. Geological Society of America Bulletin **97**:999–1011.

Lyons, J.K., and R.L. Beschta. 1983. Land use, floods and channel changes: Upper Middle Fork Willamette River, Oregon (1936–1980). Water Resources Research **19**:463–471.

Marston, R.A. 1982. The geomorphic significance of log steps in forest streams. Annals of the American Association of Geographers **72**:99–108.

Megahan, W., W.S. Platts, and B. Kulesza. 1980. Riverbed improves over time: South Fork Salmon. Proceedings of the Symposium on Watershed Management, American Society of Civil Engineers, New York, New York, USA.

Mollard, J.D. 1973. Air photo interpretation of fluvial features. Proceedings of the Canadian Hydrology Symposium. National Research Council, Ottawa, Ontario, Canada.

Montgomery, D.R., and J.M. Buffington. 1997. Channel reach morphology in mountain drainage basins. Geological Society of America Bulletin **109**:596–611.

Montgomery, D.R., T.B. Abbe, N.P. Peterson, J.M. Buffington, K.M. Schmidt, and J.D. Stock. 1996. Distribution of bedrock and alluvial channels in forested mountain drainage basins. Nature **381**:587–589.

Montgomery, D.R., J.M. Buffington, R.D. Smith, K.M. Schmidt, and G. Pess. 1995b. Pool spacing in forest channels. Water Resources Research **31**:1097–1105.

Montgomery, D.R., and W.E. Dietrich. 1988. Where do channels begin? Nature **336**:232–234.

Montgomery, D.R., G.E. Grant, and K. Sullivan. 1995a. Watershed analysis as a framework for implementing ecosystem management. Water Resources Bulletin **31**:369–386.

Morisawa, M.E. 1957. Accuracy of determination of stream lengths from topographic maps. Transactions, American Geophysical Union **38**:86–88.

Murphy, M.L., and K.V. Koski. 1989. Input and depletion of woody debris in Alaska streams and implications for streamside management. North American Journal of Fisheries Management **9**:427–436.

Nakamura, F., and F.J. Swanson. 1993. Effects of coarse woody debris on morphology and sediment storage of a mountain stream system in western Oregon. Earth Surface Processes and Landforms **18**:43–61.

Nolan, K.M., and D.C. Marron. 1988. Stream channel response to the storm. Pages 245–264 *in* S.D. Ellen and G.F. Wieczorek, eds. Landslides, floods, and marine effects of the storm of January 3–5, 1982, in the San Francisco Bay Region, California. United States Geological Survey Professional Paper 1434. US Government Printing Office, Washington, DC, USA.

Nunnally, N.R. 1985. Application of fluvial relationships to planning and design of channel modifications. Environmental Management **9**:417–426.

Parker, G. 1979. Hydraulic geometry of active gravel rivers. Journal of the Hydraulics Division, American Society of Civil Engineers **105**:1185–1201.

Paustian S.J., K. Anderson, D. Blanchet, S. Brady, M. Cropley, J. Edgington, et al. 1992. A channel type users guide for the Tongass National Forest, southeast Alaska. USDA Forest Service Technical Paper R10-TP-26. Alaska Region R10.

Perkins, S.J. 1989. Interactions of landslide-supplied sediment with channel morphology in forested watersheds. Unpublished Masters thesis. University of Washington, Seattle, Washington, USA.

Peterson, D.F., and P.K. Mohanty. 1960. Flume studies of flow in steep, rough channels. Journal of the Hydraulics Division, American Society of Civil Engineers **86**:55–76.

Platts, W.S., R.J. Torquedada, M.L. McHenry, and C.K. Graham. 1989. Changes in Salmon spawning and rearing habitat from increased delivery of fine sediment to the South Fork Salmon River, Idaho. Transactions of the American Fisheries Society **118**:274–283.

Powell, J.W. 1875. Exploration of the Colorado River of the west and its tributaries. Government Printing Office, Washington, DC, USA.

Reneau, S.L., W.E. Dietrich, D.J. Donahue, A.J.T. Jull, and M. Rubin. 1990. Late Quaternary history of colluvial deposition and erosion in hollows, central California Coast Ranges. Geological Society of America Bulletin **102**:969–982.

Rice, S. 1994. Towards a model of changes in bed material texture at the drainage basin scale. Pages 159–172 *in* M.J. Kirkby, ed. Process models and theoretical geomorphology. John Wiley & Sons, Chichester, UK.

Rosgen, D.L. 1994. A classification of natural rivers. Catena **22**:169–199.

Schmidt, K.H., and P. Ergenzinger. 1992. Bedload entrainment, travel lengths, step lengths, rest periods—studied with passive (iron, magnetic) and active (radio) tracer techniques. Earth Surface Processes and Landforms **17**:147–165.

Schumm, S.A. 1971. Fluvial Geomorphology: Channel adjustment and river metamorphosis. Pages 5-1–5-22 *in* H.W. Shen, ed. River mechanics.

Volume 1. H.W. Shen, Fort Collins, Colorado, USA.

——. 1977. The fluvial system. John Wiley & Sons, New York, New York, USA.

Shaler, N.S. 1891. The origin and nature of soils. Pages 213–345 *in* J.W. Powell, ed. United States Geological Survey Twelfth Annual Report. US Government Printing Office, Washington, DC, USA.

Shields, A. 1936. Anwendung der Änlichkeits-mechanik und der Turbulenzforschung auf die Geschiebebewegung. Translated by W.P. Ott and J.C. van Uchelen. California Institute of Technology, Pasadena, California, USA.

Simons, D.B., E.V. Richardson, and C.F. Nordin. 1965. Sedimentary structures generated by flow in alluvial channels. Pages 34–52 *in* G.V. Middleton, ed. Primary sedimentary structures and their hydrodynamic interpretation. Society of Economic Paleontologists and Mineralogists, Tulsa, Oklahoma, USA.

Smith, D.G. 1976. Effect of vegetation on lateral migration of anastomosed channels of a glacier meltwater river. Geological Society of America Bulletin **87**:857–860.

Smith, R.D., R.C. Sidle, and P.E. Porter. 1993a. Effects on bedload transport of experimental removal of woody debris from a forest gravel-bed stream. Earth Surface Processes and Landforms **18**:455–468.

Smith, R.D., R.C. Sidle, P.E. Porter, and J.R. Noel. 1993b. Effects of experimental removal of woody debris on the channel morphology of a forest, gravel-bed stream. Journal of Hydrology **152**:153–178.

Strahler, A.N. 1957. Quantitative analysis of watershed geomorphology. Transactions, American Geophysical Union **38**:913–920.

Sullivan, K. 1986. Hydraulics and fish habitat in relation to channel morphology. Unpublished Ph.D. dissertation. Johns Hopkins University, Baltimore, Maryland, USA.

Sullivan, K., T.E. Lisle, C.A. Dolloff, G.E. Grant, and L.M. Reid. 1987. Stream Channels: The link between forests and fishes. Pages 39–97 *in* E.O. Salo and T.W. Cundy, eds. Streamside management: Forestry and fishery interactions. Institute of Forest Resources Contribution Number 57, University of Washington, Seattle, Washington, USA.

Swanson, F.J., R.L. Fredriksen, and F.M. McCorison. 1982. Material transfer in a western Oregon forested watershed. Pages 223–266 *in* R.L. Edmonds, ed. Analysis of coniferous forest ecosystems in the western United States. Hutchinson Ross, Stroudsburg, Pennsylvania, USA.

Takahashi, T., K. Ashida, and K. Sawai. 1981. Delineation of debris flow hazard areas. Pages 589–603 *in* T.R.H. Davies and A.J. Pearce, eds. Erosion and sediment transport in Pacific Rim steeplands. IAHS Publication 132, International Association of Hydrological Sciences, Wallingford, UK.

Whiting, P.J., and J.B. Bradley. 1993. A process-based classification for headwater streams. Earth Surface Processes and Landforms **18**:603–612.

Whitlow, J.R., and K.J. Gregory. 1989. Changes in urban stream channels in Zimbabwe. Regulated Rivers: Research and Management **4**:27–42.

Whitney, J.D. 1880. The Auriferous Gravels of the Sierra Nevada. Memoir 6. Harvard University Museum of Comparitive Zoology, Cambridge, Massachusetts, USA.

Whittaker, J.G. 1987. Sediment transport in step-pool streams. Pages 545–579 *in* C.R. Thorne, J.C. Bathurst, and R.D. Hey, eds. Sediment transport in gravel-bed rivers. John Wiley & Sons, Chichester, UK.

Whittaker, J.G., and T.R.H. Davies. 1982. Erosion and sediment transport processes in step-pool torrents. Pages 99–104 *in* D.E. Walling, ed. Recent developments in the explanation and prediction of erosion and sediment yield. IAHS Publication Number 137, International Association of Hydrological Sciences, Wallingford, UK.

Whittaker, J.G., and M.N.R. Jaeggi. 1982. Origin of step-pool systems in mountain streams. Journal of the Hydraulics Division, Proceedings of the American Society of Civil Engineers **108**:758–773.

Wilcock, P.R., G.M. Kondolf, W.V. Matthews, and A.F. Barta. 1996. Specification of sediment maintenance flows for a large gravel-bed river. Water Resources Research **32**:2911–2921.

Williams, G.P. 1978. Bank-full discharge of rivers. Water Resources Research **14**:1141–1154.

Williams, G.P., and M.G. Wolman. 1984. Downstream effects of dams on alluvial rivers. United States Geological Survey Professional Paper 1286. US Government Printing Office, Washington, DC, USA.

Wolman, M.G., and J.P. Miller. 1960. Magnitude and frequency of forces in geomorphic processes. Journal of Geology **68**:54–74.

3
Hydrology

Robert R. Ziemer and Thomas E. Lisle

Overview

- Streamflow is highly variable in mountainous areas of the Pacific coastal ecoregion. The timing and variability of streamflow is strongly influenced by form of precipitation (e.g., rainfall, snowmelt, or rain on snow).
- High variability in runoff processes limits the ability to detect and predict human-caused changes in streamflow. Changes in flow are usually associated with changes in other watershed processes that may be of equal concern. Studies of how land use affects watershed responses are thus likely to be most useful if they focus on how runoff processes are affected at the site of disturbance and how these effects, hydrologic or otherwise, are propagated downstream.
- Land use and other site factors affecting flows have less effect on major floods and in large basins than on smaller peak flows and in small basins. Land use is more likely to affect streamflow during rain on snow events, which usually produce larger floods in much of the Pacific coastal ecoregion than purely rainfall events.
- Long-term watershed experiments indicate that clear-cutting and road building influence rates and modes of runoff, but these influences are stronger for some areas, events, and seasons than for others. Logging and road building can increase areas that generate overland flow and convert subsurface flow to overland flow, thereby increasing rates and volumes of stormflow. Logging and road building can also increase runoff rates and volumes from transient snow packs during rain on snow events.
- Removal of trees, which consume water, tends to increase soil moisture and base streamflow in summer when rates of evapotranspiration are high. These summertime effects tend to disappear within several years. Effects of tree removal on soil moisture in winter are minimal because of high seasonal rainfall and reduced rates of evapotranspiration.
- The rate of recovery from land use depends on the type of land use and on the hydrologic processes that are affected.

Introduction

Streamflow is an essential variable in understanding the functioning of watersheds and associated ecosystems because it supplies the primary medium and source of energy for the movement of water, sediment, organic material, nutrients, and thermal energy. Changes in streamflow are almost invariably linked to changes in other watershed processes such as erosion, sedimentation, woody debris dynamics, and heat transfer—processes that are also important to aquatic communities and discussed in other chapters.

How forest management practices and other land uses affect hillslope runoff and streamflow has long been debated and remains controversial. The controversy is intensified by the difficulty of extrapolating results of watershed studies from one basin to the next because of the variability of hydrologic response with basin size, flow magnitude, season, climate, geology, and type and intensity of land use. Furthermore, given a certain amount of timber to harvest and the associated road systems, it is more difficult to prevent or mitigate potential hydrologic impacts than sediment impacts. The location and manner in which an area is logged or roads are built can effectively control erosion from the small areas that potentially produce disproportionately large volumes of sediment in a basin, but changes in vegetation and soil compaction may affect hydrologic response pervasively. Often such hydrologic responses are more a function of the extent of the area of logging or length of road than the methods of harvesting forests or building roads. Thus, hydrologic response and associated impacts focus the debate on *how much* timber to harvest or road to build.

This chapter provides some general information on hydrologic processes and discusses the influences of land use, particularly forestry, on runoff and streamflow in the Pacific coastal ecoregion. A more comprehensive general hydrologic background can be found in texts by Dunne and Leopold (1978), Gordon et al. (1992), Mount (1995), or Black (1996).

Hydrology of the Pacific Coastal Ecoregion

The Pacific coastal ecoregion ranges from a cool maritime climate with a rather equal seasonal distribution of precipitation in the north to a warm Mediterranean climate with dry summers and wet winters in the south. Annual precipitation and runoff generally increase from south to north (Figure 3.1a) (Naiman and Anderson 1996). The amount of precipitation is also strongly influenced by mountain ranges, which have a general north–south orientation. On the windward (western) side of the mountains, precipitation increases with elevation; whereas, on the lee (eastern) side, precipitation drops abruptly because of a pronounced rain shadow which, in extreme cases, produces desert conditions.

Seasonality of runoff is influenced by temperature as well as precipitation. At high elevations and latitudes north of about 48°N, much of the winter precipitation is stored in snowpacks. In the Olympic Range, the North Cascades of Washington, and the coastal ranges of British Columbia and southeast Alaska, melting of winter snowpacks and glaciers produces peak streamflows in spring and early summer and maintains moderate flows throughout the summer (Figure 3.1c). Glaciers are present in the Olympics and North Cascades of Washington, the high coastal ranges of British Columbia and Alaska, and on the highest volcanoes in the Cascades of Oregon and California. Meltwater runoff from glaciers commonly peaks during the warmest periods of the summer in July and August.

Runoff from snowmelt is limited by how rapidly thermal energy in the air can supply enough heat to melt snow. It takes only 1 cal to warm 1 g of water 1°C, but it takes 80 cal to melt 1 g of ice at 0°C. In addition, the low density and specific heat of air compared to that of ice (approximately 0.1% and 20%, respectively) severely limit the direct transfer of heat from air to snow. However, the transfer of latent heat from air to snow (i.e., condensation) can produce rapid snow melt. Nightly cooling often limits snowmelt runoff rates that are generated over several days. Maximum rates of snowmelt approach approximately 4 cm/day (Dunne and Leopold 1978), and resulting runoff rates tend to be less, because of mixing of runoff from areas of varying rates of snowmelt. These rates compare to runoff rates of approximately 9 cm/day during very large floods generated at least partly by rainfall in the Pacific coastal ecoregion.

The highest rates of runoff in the world, other than from dam-break floods, are generated by rainfall. However, in the Pacific coastal ecoregion, rapid snowmelt commonly accompanies the influx of warm subtropical air masses that also produce some of the highest sustained rates of rainfall, and the combination rain on snow events produce most of the largest floods. Some snowmelt during rainfall occurs every year, usually without serious consequence. However, rapid snowmelt during

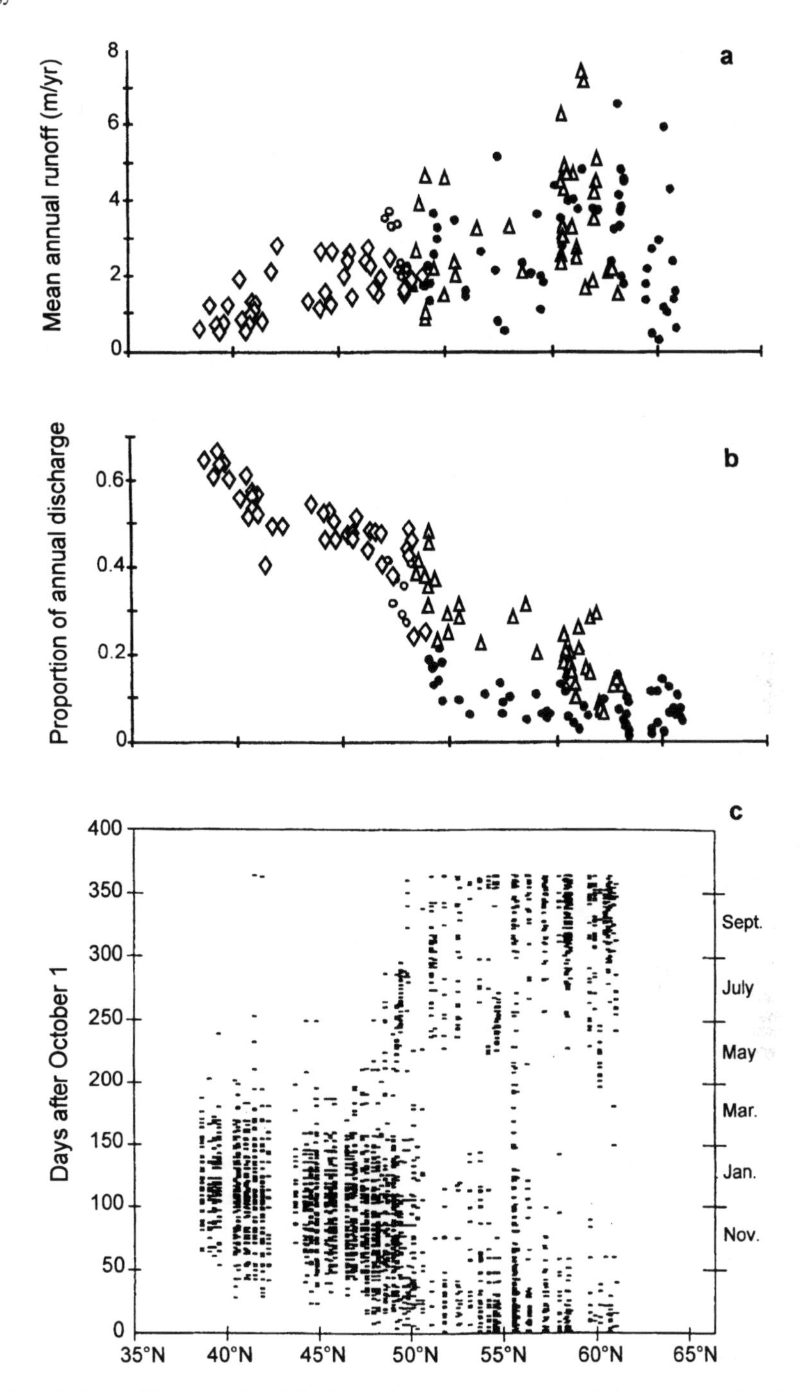

FIGURE 3.1. Variation with latitude of hydrologic characteristics for 151 rivers draining four subregions of the Pacific coastal ecoregion: (◇) southern coastal mountains, (○) Olympic Mountains, (△) northern lowlands and islands, (•) northern mainland mountains. (a) Mean annual runoff as a function of latitude. (b) Proportion of annual discharge carried during the three months of winter. (c) Temporal distribution of peak annual discharge. Each point represents the date and latitude of a maximum daily discharge for a given river in a single water year (93 rivers total having at least 10 years of record) (Naiman and Anderson 1996 with permission).

rainfall contributed to all but two of the 23 largest annual peak flows of the Willamette River at Salem, Oregon, between 1814 and 1977 (Harr 1981). In the 60-ha Watershed 2 of the H.J. Andrews Experimental Forest, major peak flows of 10 L/s/ha are five times more likely to result from rain on snow than from rain alone (Harr 1981). Although the phrase "rain on snow" implies the snow is melted directly by warm rain, the snow is primarily melted by heat transferred to the snowpack by convection and condensation of water vapor on the snowpack surface. While this distinction may seem trivial, it is important to identify the appropriate process before questions concerning the influence of land use on the magnitude of rain on snow floods can be answered. Detailed analyses of such runoff events are rare because information is almost always lacking about snow depth and density, air temperature, and form of precipitation during any given storm.

South of latitude 48°N (California, Oregon, and much of Washington), about 80% of the total annual precipitation, which ranges from 750 to 3,500 mm, falls during the six-month period between the beginning of October to the end of March. During winter, frequent frontal storms move eastward from the Pacific, typically producing relatively low-intensity precipitation (e.g., less than about 10 mm/hr for periods of 18 to 72 hours). Temperatures, regulated by latitude and elevation cause precipitation to fall as rain, snow, or a combination of the two and govern the magnitude and timing of associated peak streamflows. Patterns of streamflow discharge reflect the strong contrast in precipitation between summer and winter (Figure 3.1b). Except at high elevations, up to 70% of the annual streamflow occurs during the three months of winter (Naiman and Anderson 1996). Large streamflow peaks are generated when a winter storm is particularly strong or when several storms follow in rapid succession and produce moderately intense rainfall over a period of several days. Lack of rainfall during the summer results in low summer streamflows. Small headwater streams commonly become dry before the onset of the fall rains, except at the highest elevations where streams are fed by late-melting snowpacks or glaciers.

The proportion of annual precipitation falling as snow varies greatly with elevation and latitude. For example, in western Oregon, snow is uncommon below about 350 m. At intermediate elevations from 350 to 1,100 m, snow is intermittent, lasting only a week or two between warm periods. Above 1,100 m, one-third to three-fourths of the annual precipitation may fall as snow, which begins to accumulate in November and usually lasts until late May. Further south, in northern California, these elevational zones are about 1,000 m higher. In California, high-elevation lands occupy about 3% of the state but produce about 13% of the annual streamflow (Colman 1955). The high-elevation lands are of even greater importance in Utah, where 60% of the state's streamflow comes from the upper Uinta and Wasatch mountains which occupy only 10% of the land. In the Rocky Mountains, 85% of the annual streamflow occurs from May through July, with less than 5% occurring during the winter months (Leaf 1975).

In the southern Pacific coastal ecoregion, the general transition from rain to snow with higher elevation strongly affects the magnitude and timing of runoff events. Figure 3.2 depicts typical seasonal variations in streamflow between three streams in Washington that drain basins having different elevations. In the Wynoochee River (gaging station elevation of 12 m), streamflow reflects rainfall, which is concentrated in winter. In the Middle Fork Snoqualmie River (elevation 240 m), winter precipitation falls as either rain or snow. The largest floods occur when heavy rainfall is produced by a warm subtropical air mass that also raises the freezing level and results in a rapid melt of an existing snowpack. Smaller peak flows occur as the residual snowpack melts in late spring. In the Twisp River (elevation 850 m), winter precipitation falls as snow and produces peak flows when the snowpack melts in late spring. Consequently, the hydrology is dominated by three precipitation types: rain, rain on snow, and snow. Lower and more southern parts of the Pacific coastal ecoregion are exclusively within the rain zone, and

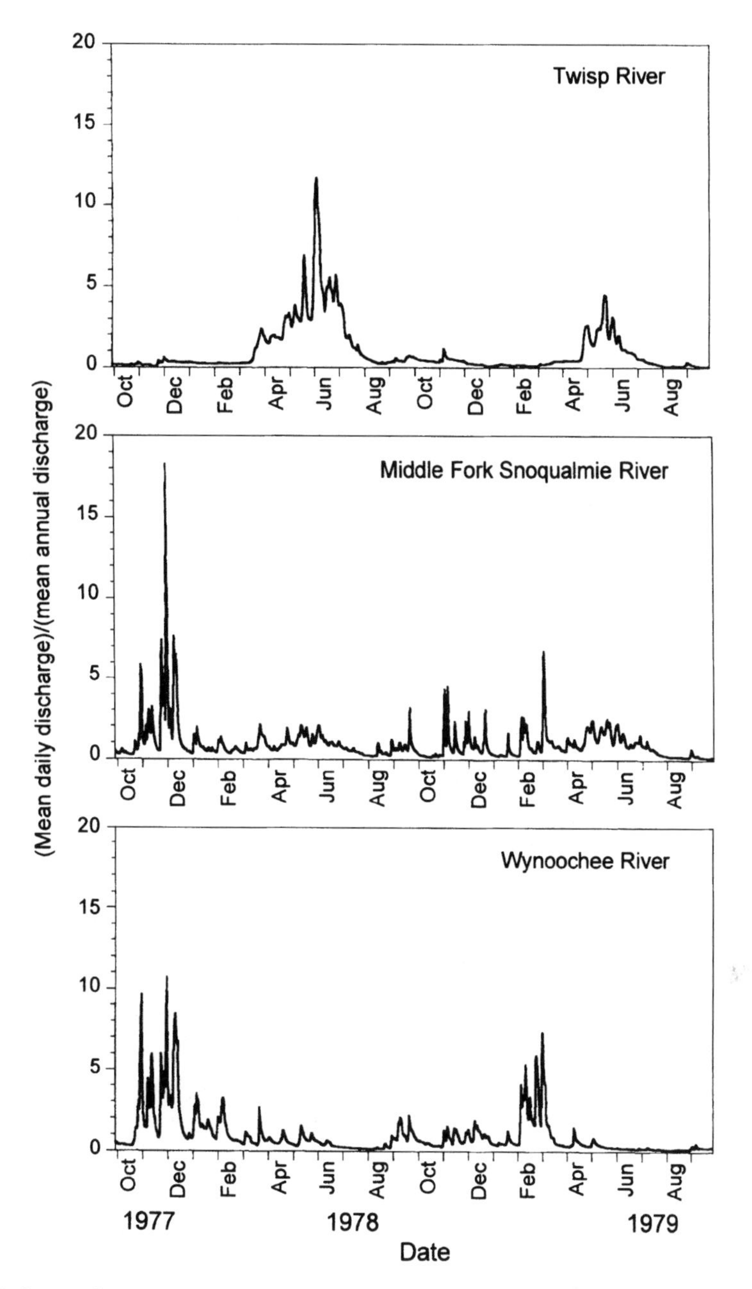

FIGURE 3.2. Hydrographs for three Washington rivers (Twisp River, drainage area = 680 km^2; Middle Fork Snoqualmie River, 430 km^2; Wynoochee River, 430 km^2) for water-years 1978 and 1979. Mean daily discharge is divided by mean annual discharge to normalize magnitudes (discharge data from USGS 1997).

northern, high-elevation areas are always within the snow zone. However, high-flow events are the result of varying combinations of rain and snow. Thus, the relationship between climate, runoff, and land use is highly complex.

Runoff Processes

Runoff processes are linked throughout a watershed from hillslopes to the mouth of the main channel. Rates of moisture movement along these pathways depend on the volume of water introduced to the system and the efficiency with which it is transported through the system. A simple mass balance equation is useful for understanding runoff processes throughout a drainage:

$$\text{Outflow} = \text{Inflow} - \text{Losses} - \text{Change in storage}$$

where outflow is runoff from a hillslope or reach of channel; inflow is rainfall or streamflow into the channel reach; losses are processes such as evapotranspiration and deep seepage; and change in storage includes, for example, pool volume and soil moisture. Therefore, the rate and capacity for filling and depleting stored water is a key element in the timing, magnitude, and duration of runoff rates in all parts of a watershed. Reduced water storage in one part of a basin leads to increased release of water downstream. For example, there is much less "storage" available for water moving over the surface of hillslopes than there is below the surface. This is in part why water flows down hillslope surfaces more than 10 times faster than it does through the soil mantle. Thus, increasing surface runoff at the expense of subsurface runoff can increase peak flows in channels downstream. Reviews of the effects of land use on the hydrology of hillslopes and channels are provided by Kirkby (1978) and Reid (1993).

Hillslope Runoff

It is useful to focus on the effects of land-use practices on hillslope runoff processes for two reasons. First, detecting and then evaluating the causes of changes in streamflow is more difficult further downstream. Second, information gained by evaluating how and where land use affects runoff processes can help determine how other processes, such as erosion and sediment delivery, are affected. Once these processes are understood, ways to prevent altered runoff and erosion may be identified. For example, whether rain or meltwater runs over the soil surface or through the soil mantle strongly influences how quickly it arrives at a stream channel and how much sediment and solutes are produced. Figure 3.3 depicts some important hydrologic pathways in the Pacific coastal ecoregion that are described below. A more comprehensive review of runoff processes is provided by Dunne and Leopold (1978) and Kirkby (1978).

Subsurface Flow

Subsurface flow accounts for nearly all of the water that is delivered to stream channels from undisturbed forested hillslopes (Harr 1977). Precipitation infiltrating the soil surface travels through the soil as either shallow subsurface flow or deep seepage that replenishes groundwater storage. Groundwater storage maintains base flows during dry periods. Water is transmitted within the soil along two different flow paths: through the soil matrix (micropores) and through macropores including root holes, soil cracks, animal burrows, and soil pipes. Subsurface flow velocities vary widely (Table 3.1). Flow in micropores is very slow (10^{-7}–10^{-6} m/hr), while flow in soil pipes (10^{-1}–10^{2} m/hr) can be as rapid as unchannelized overland flow. Normally, subsurface flow contributes little to erosion. However, if subsurface flow paths become obstructed, water can build up in the soil mantle and cause slope failure.

In many locations, large macropores or structural voids can occupy as much as 35% of the total soil volume in a forest soil (Aubertin 1971). In the coastal mountains of British Columbia, Chamberlin (1972) observed that roots made up about 50% of the upper 0.5 m

TABLE 3.1. Order of magnitude of runoff velocities by different processes.

Runoff process	Characteristic velocity (m/hr)
Subsurface	
undifferentiated	10^{-7}–10^{-4}
micropore	10^{-7}–10^{-6}
macropore	10^{-5}–10^{-4}
soil pipes	10^{-1}–10^{2}
Surface	
overland flow (unchannelized)	10^{1}–10^{2}
channel flow (gullies and stream channels)	10^{2}–10^{3}

Modified from Dunne and Leopold 1978, Kirkby 1978.

of the forest soil. When voids resulting from root decay, animal activity, and subsurface erosion by chemical (solution) and physical processes become interconnected, they form soil pipes capable of transporting subsurface water rapidly (Table 3.1) (Kirkby 1978). Where piping networks are extensive, the hydraulic conductivity of the soil matrix is of secondary importance in generating flow during storms (stormflow) (Whipkey 1965, Mosley 1979). For example, pipeflow accounted for nearly all of the stormflow from 1-ha headwater swales in northern California (Ziemer 1992) and Japan (Tsukamoto et al. 1982). In a 50-ha coastal drainage in central California, Swanson et al. (1989) reported that during a 25-year recurrence-interval storm, nearly 70% of the water was discharged through the subsurface piping network. Jones (1987) found that pipeflow was responsible for 49% of the stormflow from the Maesnant catchment in Wales.

Saturated Overland Flow

Overland flow can occur where the soil becomes fully saturated and subsurface flow emerges as *return flow* (or *exfiltration*); additional rainfall or meltwater flows over the surface as *saturated overland flow*. Soils commonly become saturated where shallow subsurface flow converges in topographic depressions or accumulates in areas of decreasing hillslope gradient. Zones of saturated overland flow are most common in valleys and swales. Zones of saturated overland flow commonly occupy small but expandable areas of drainage basins and contribute disproportionately to flows during storms; they expand during wet periods and contract during dry periods. This phenomenon is known as the *partial-area concept* of storm runoff (Betson 1964, Dunne and Leopold 1978).

In undisturbed Pacific coastal ecoregion forests, areas that generate saturated overland flow are usually confined to the base of hillslopes, near stream channels, and in swales (Figure 3.3). They can also occur where soils thin downslope over impermeable bedrock. Where hillslopes are straight, steep, and highly permeable, there is little tendency for return flow to occur. However, mechanical disturbance of areas that generate overland flow can

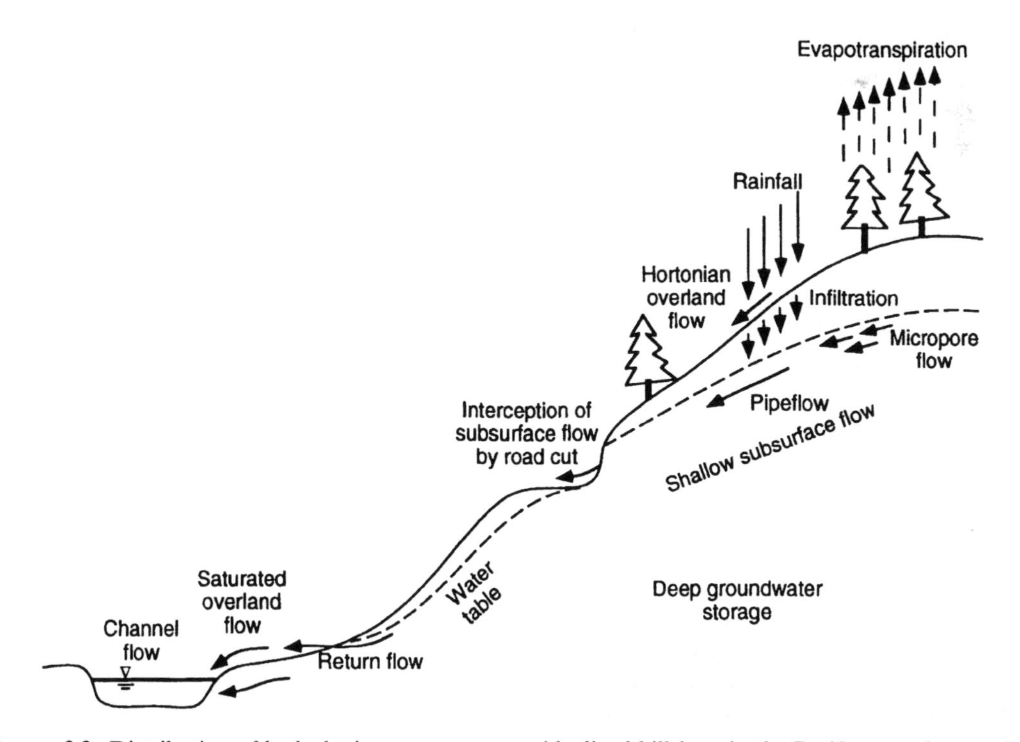

FIGURE 3.3. Distribution of hydrologic processes on an idealized hillslope in the Pacific coastal ecoregion.

create large sediment sources by decreasing soil stability where surface flow energies are high.

Hortonian Overland Flow

When water encounters the ground surface more rapidly than it can infiltrate into the soil, the excess water runs over the surface as *Hortonian overland flow* (Figure 3.3) (after Robert E. Horton, who proposed principles of surface runoff). Because infiltration rates of wetted forest soils typically exceed rainfall rates in the Pacific coastal ecoregion, Hortonian overland flow is unusual in undisturbed forests in the region. It is more likely to occur in more arid climates and agricultural lands where infiltration is less and maximum rainfall intensities are greater. Nonetheless, Hortonian overland flow can occur locally following fires in which volatilized organic molecules coat soil particles producing a water-repellent layer that prevents water from infiltrating into coarse textured soils (DeBano 1969, Beschta 1990, McNabb and Swanson 1990). These hydrophobic conditions can extend to a depth of 15 cm and persist for six or more years after the fire (Dyrness 1976, DeBano 1981).

Where and when overland flow occurs governs many of the impacts of watershed disturbances. Firstly, overland flow can travel at much greater velocities (10^1–10^2 m/hr) than undifferentiated subsurface flow (10^{-7}–10^{-4} m/hr) (Table 3.1). Therefore, increased areas of Hortonian overland flow directly contribute to streamflow peaks during storms in headwater channels, which respond to the most rapid components of runoff. Secondly, overland flow has a much greater capacity to erode and transport sediment. An increase in Hortonian overland flow in mountainous terrain is likely to be accompanied by soil loss and an increase in sediment load to streams.

Land use can increase areas of Hortonian overland flow and saturated overland flow and thereby increase hillslope erosion and stormflow magnitude in headwater channels. In rangeland, soil compaction and loss of ground cover from heavy grazing can decrease soil permeability (the soil's capacity to transmit water), and thereby increase the area and frequency of Hortonian overland flow (Horton 1933, Dunne and Leopold 1978). Hortonian overland flow is increased in urbanized areas by the expansion of impervious surfaces (e.g., roofs, streets, parking lots) (Leopold 1968, Dunne and Leopold 1978, Sauer et al. 1983). In forested lands of the Pacific coastal ecoregion, Hortonian overland flow is most commonly restricted to areas of compacted soils, such as roads, skid trails, and landings. Subsurface flow emerging in road cuts directly augments surface flow to streams through road ditches (Figure 3.3), and can contribute to erosion of road cuts and surfaces as well as road ditches (Megahan 1972). Therefore, by intercepting subsurface flow and causing Hortonian overland flow, roads can expand the channel network in a basin and thereby increase the rate of stormflow runoff (Wemple 1994). Converting subsurface flow to overland flow is especially effective in increasing stormflow volumes and rates where subsurface flow is dominated by flow through micropores, but less so where pipeflow is intercepted. When subsurface flow through the soil matrix is converted to surface flow, runoff velocity is increased by as much as five orders of magnitude (Table 3.1). But when pipes or large macropores are present, differences in runoff velocity between subsurface pipeflow and surface runoff indicate that a shift to surface runoff may result in an increase in runoff velocity of one order of magnitude or less. Mechanisms for changes in runoff processes resulting from logging and road building are conceptualized in Figure 3.4.

Although it is relatively easy to understand and detect human disturbance of runoff processes on hillslopes, evaluating downstream effects of altered runoff processes becomes increasingly difficult as the size of the basin increases. Increased surface runoff is usually easiest to detect nearest the disturbed area. One reason is that an enhanced spike in stormflow runoff attenuates downstream as the flood wave spreads out and mixes with unaffected or less-affected runoff from other parts of the basin. Also, the arrival of runoff peaks from sources upstream may or may not coincide at some point in the trunk stream. Altered

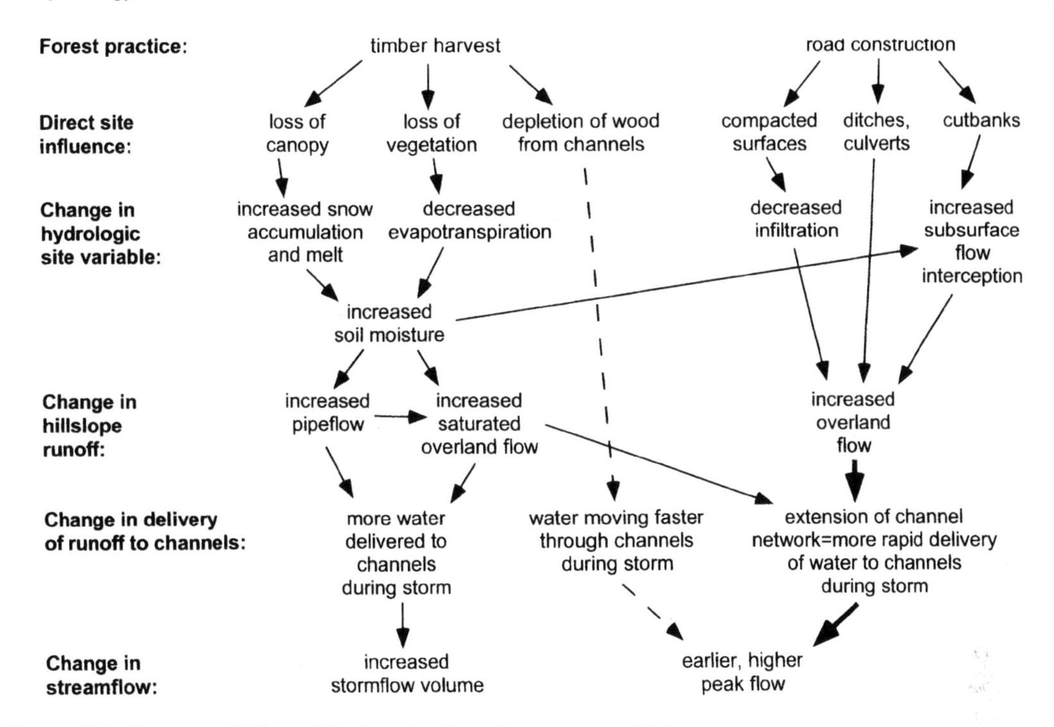

FIGURE 3.4. Conceptual hierarchical mechanisms for alteration of hillslope and channel runoff by timber harvesting and road building. The strength of linkages to higher runoff rates increases from dashed to solid to heavy lines (modified from Jones and Grant 1996).

runoff may enhance or decrease these coincidences and thus either increase or decrease downstream peak flows.

Increased erosion and sediment transport usually accompany increased runoff. For example, in the 1960s and 1970s diverted road drainage was one of the primary sources of sediment in areas of the Redwood Creek basin, California, that were logged and had roads (Figure 3.5), (Weaver et al. 1981). Plugged culverts and other failures of road drainage diverted runoff onto hillslopes where high rates of surface flow never had occurred before. The result was deep gullying—an outcome of extreme local increases in surface runoff. Eroded material was added to the already high sediment loads of the affected tributaries, but whether changes in streamflow could have been detected is debatable.

Runoff in Channels

Streamflow, in contrast to hillslope runoff, pertains only to surface flow in the channel—although subsurface flow below channels and floodplains is very important to benthic and hyporheic organisms. Differences in runoff processes between basins or regions can strongly affect the variability of peak flow in channels. For example, Pitlick (1994) compared the ratio (Q_{100}/Q_m) of the magnitude of the 100-yr flood (Q_{100}; discharge with a recurrence interval of 100 years) to the magnitude of the mean annual flood (Q_m) in five climatic zones in the western United States. In alpine areas in Colorado where snowmelt dominates flood runoff, Q_{100}/Q_m is approximately 2. In the California Coast Ranges, where large frontal storms dominate flood runoff, Q_{100}/Q_m is 3 or more, and in the Klamath Mountains, where rain on snow is more common, Q_{100}/Q_m is approximately 5.

The variation of flow during and between seasons is a key selective pressure on aquatic and riparian organisms and a primary control on channel form and process. Each season has a characteristic flow frequency which is vital to ecosystem function. Some of the seasonal

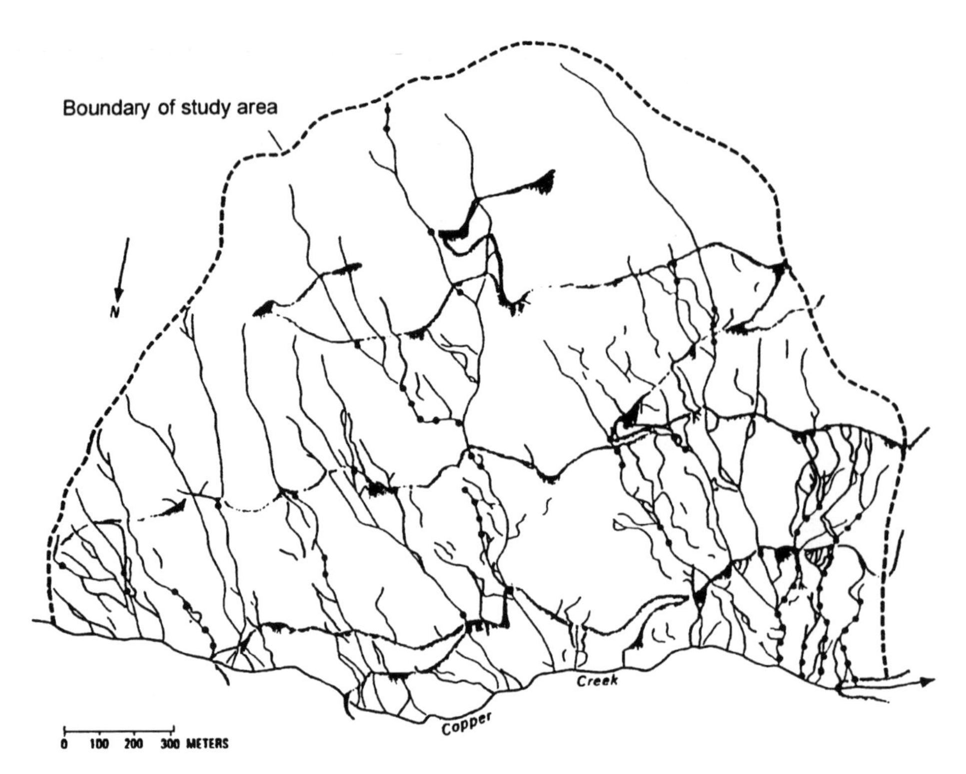

FIGURE 3.5. Gully systems that resulted from road drainage failures on a 246-ha study site in South Copper Creek sub-basin, Redwood Creek basin, California (mapped in 1979). Rills less than 0.1 m^2 in cross-sectional area are not shown. (Weaver et al. 1995).

effects of flow variability in the Pacific coastal ecoregion are summarized below:

• The timing and extent of fish spawning runs, which commonly begin in the fall, depend on flows high enough to allow fish to enter and penetrate the channel network.

• Annual floods, which commonly occur in winter and spring, distribute sediment and organic debris through the system, scour the bed, and remove newly established vegetation in the active channel. Floods can increase mortality of incubating fish embryos, and depending on incubation periods of different species and the timing of floods, they may strongly affect relative cohort populations of different species. Floods cause mortality in certain benthic invertebrates, and alter food webs, and thereby affect the entire trophic structure of aquatic communities (Wootton et al. 1996).

• Extreme flood events create new surfaces by erosion and deposition. Aquatic and riparian ecosystems in mountainous areas depend on extreme events (e.g., floods, landslides, windstorms, fire) to renew dynamic processes and maintain a mosaic of surfaces that are at various stages of evolution since the previous disturbance (Grant and Swanson 1995).

• Recessional flows in spring and early summer are occasionally punctuated by peak flows. Streamflow during this period controls the success of riparian plant seeds to germinate in channels and on streambanks and floodplains. Seeds of riparian trees commonly disperse over a time frame of a few weeks. In order to successfully germinate, seeds must be deposited high enough to avoid drowning or scour, and low enough to avoid desiccation as water tables drop with recessional stages (McBride and Strahan 1985, Lisle 1989, Segelquist et al. 1993). Seeds may be swept away or germinate, and seedlings may be drowned, desiccated, or survive depending on water stages in spring and early summer.

• Summer low flows allow sediment to settle, water to clear, and low-energy habitats to expand. Low flows also limit total aqueous living space.

Thus, the entire annual sequence of flows governs the trajectory of aquatic and riparian ecosystems. As a result, stream channels and ecosystems are in a constant state of flux within a wide range of variability, and the duration and frequency of occurrence for any given state span decades or centuries (Chapter 11). The following discussion of the relative influence of flows of different magnitude on channel form assumes relatively constant supplies of woody debris and sediment. However in nature, woody debris and sediment are supplied at widely varying rates and greatly influence the immediate and enduring effects of high-runoff events on channels and floodplains (Chapter 11).

A wide range of flows that entrain bed material and erode banks is responsible for forming stream channels. Low sediment-transporting flows occur frequently, but are too weak to move much material over a short time span. Extreme floods cause large volumes of erosion and deposition, but occur infrequently (e.g., once every several decades). A channel is a legacy of the history of flows that it has carried. If an extreme event has occurred recently, the channel retains most of the forms and dimensions created by that flood; the cumulative effects of the succeeding smaller floods are insufficient to significantly alter the channel. However, if the last extreme event occurred a decade or more ago, channel characteristics are likely to be adjusted to moderate floods that are both large enough and frequent enough to move significant volumes of material and reshape the channel. Thus, over the long term, the magnitude and frequency of a given discharge determine its effectiveness in altering the channel, and in the short term, the occurrence of a given discharge in the sequence of preceding flow events determines its role in shaping the channel's current form (Wolman and Miller 1960).

The *effective* discharge is the flow magnitude that, over a period of years, transports the most sediment or does the most work in forming the channel (Wolman and Miller 1960, Benson and Thomas 1966, Andrews 1980). Effective discharge is measured by finding the maximum product of flow frequency and sediment transport rate among equal ranges of discharge. The range of flow that transports the most sediment is commonly the one that fills the channel to the top of its banks. However, equating sediment transport to channel formation is complicated by the usual dominance of suspended sediment in the sediment load, which, because it comprises a minor fraction of bed material, may play a minor role in channel-forming processes (Wolman and Miller 1960, Wolman and Gerson 1978, Ritter 1988). The frequency of effective discharge varies widely (Nash 1994), but it commonly corresponds to bankfull discharge (the discharge that fills the channel to the top of its banks and just begins to overflow onto the floodplain) in magnitude (Andrews 1980). However, in mountainous areas of the Pacific coastal ecoregion, effective discharge can be difficult to define statistically and often exceeds bankfull discharge (Nolan et al. 1987, Grant and Wolff 1991). Moreover, many of the large bed particles that form the structure of mountain channels are moved only during rare floods (Grant 1987).

Bankfull discharge is a useful reference because it can be measured in the field, it is theoretically related to channel-forming processes, and it has a characteristic frequency (Wolman and Leopold 1956). For example, bankfull discharge is a key component of strategies to maintain channel-forming discharges in water-rights decisions. Bankfull discharge is usually equaled or exceeded, on average, every one to five years in channels that are neither aggrading nor degrading (Williams 1978). However, a consistent bankfull stage may be difficult to recognize in channels in mountainous areas of the Pacific coastal ecoregion.

Flood Routing

How do changes in hillslope runoff processes translate to changes in streamflow, and how do these propagate downstream? Downstream changes in streamflow are best understood and most easily quantified by the conservation principle introduced earlier:

Outflow = Inflow − Losses − Change in storage

To illustrate this further, imagine a reservoir (Figure 3.6). At some initial time (t = 0), the equation is balanced: there are no changes in any of the components, and the lake level is in equilibrium. At some later time (t = 1), an increase in flow at the inlet is not transmitted immediately to produce an equal increase at the outlet. Instead, the lake level rises gradually as it fills, and this rise causes a slow increase in the outflow. Later (t = 3), as the inflow decreases, the outflow exceeds inflow and decreases slowly as the lake level gradually falls. During this period of transition (t = 2), equilibrium between inflow and outflow may be achieved for a short time. The effect of storage is that the inflow hydrograph has a higher peak flow of shorter duration than the outflow hydrograph.

Although the principles outlined above are still valid, a channel differs from a reservoir in that flow is retarded, and water is thereby stored by frictional resistance along the entire channel rather than by a dam. For a section of channel, the inflow includes flow from the upstream channel plus runoff from hillslopes and tributaries entering the channel along the reach. "Reservoir" storage is provided by the channel itself along with its floodplain. Storage within the channel primarily depends on channel size and roughness: increased channel roughness causes the flow to slow and deepen; decreased roughness causes water to evacuate the channel more quickly. Simplifying channels and removing roughness elements such as riparian vegetation and large woody debris reduce channel storage of runoff and contribute to higher peak flows downstream.

Flood storage outside of the channel can vary from near zero in channels tightly constricted by valley walls to huge volumes in reaches bordered by extensive floodplains and wetlands. In mountainous areas of the Pacific coastal ecoregion, floodplains are commonly small and infrequent, and valley flats are dominated by rarely flooded terraces. Wide floodplains, on the other hand, provide a buffer to flows greater than bankfull for downstream channels. An increase in flow greatly increases the storage which dampens the increase in flow downstream. Channelization exacerbates downstream flooding by removing roughness elements and isolating a channel from its floodplain where floodwaters can be stored. One of the greatest threats to flood control, paradoxically, is confining flood waters to channels because the reduced upstream storage increases the potential for more serious flooding downstream. Increased rates of hillslope runoff or channel straightening can cause streambed erosion, and the resulting increase in channel depth can confine high flows to channels.

The exchange of flood water and sediment between channels and floodplains provides a vital link between aquatic and riparian ecosystems (Gregory et al. 1991). Flood water from the channel carries suspended sediment that settles out on floodplains and in backwaters, thus the return flow to the channel can be partially cleansed of sediment. The floor of floodplains is usually rich in organic matter that can be transported to the channel by the return flow and enrich the channel ecosystem.

Subsurface Flow in Channels and Riparian Zones

Subsurface flow in channels and floodplains performs vital ecological functions (Chapter 16). Flow in the hyporheic zone—defined by Edwards (Chapter 16) as the saturated sediments beneath and beside a river channel that contain both surface and ground water—rarely attains the discharge or velocity of channel flow, but it can become a large component of total water discharge when surface summer flows become extremely low, even in the largest rivers. Partial filling of channels with coarse sediment can cause a greater proportion of channel runoff to become subsurface flow. Subsurface discharge in Little Lost Man Creek, a small, pristine cobble-armored channel in northern coastal California, was approximately one-quarter of surface flow during the summer (Zellweger et al. 1989). Nominal subsurface flow velocities under the channel and

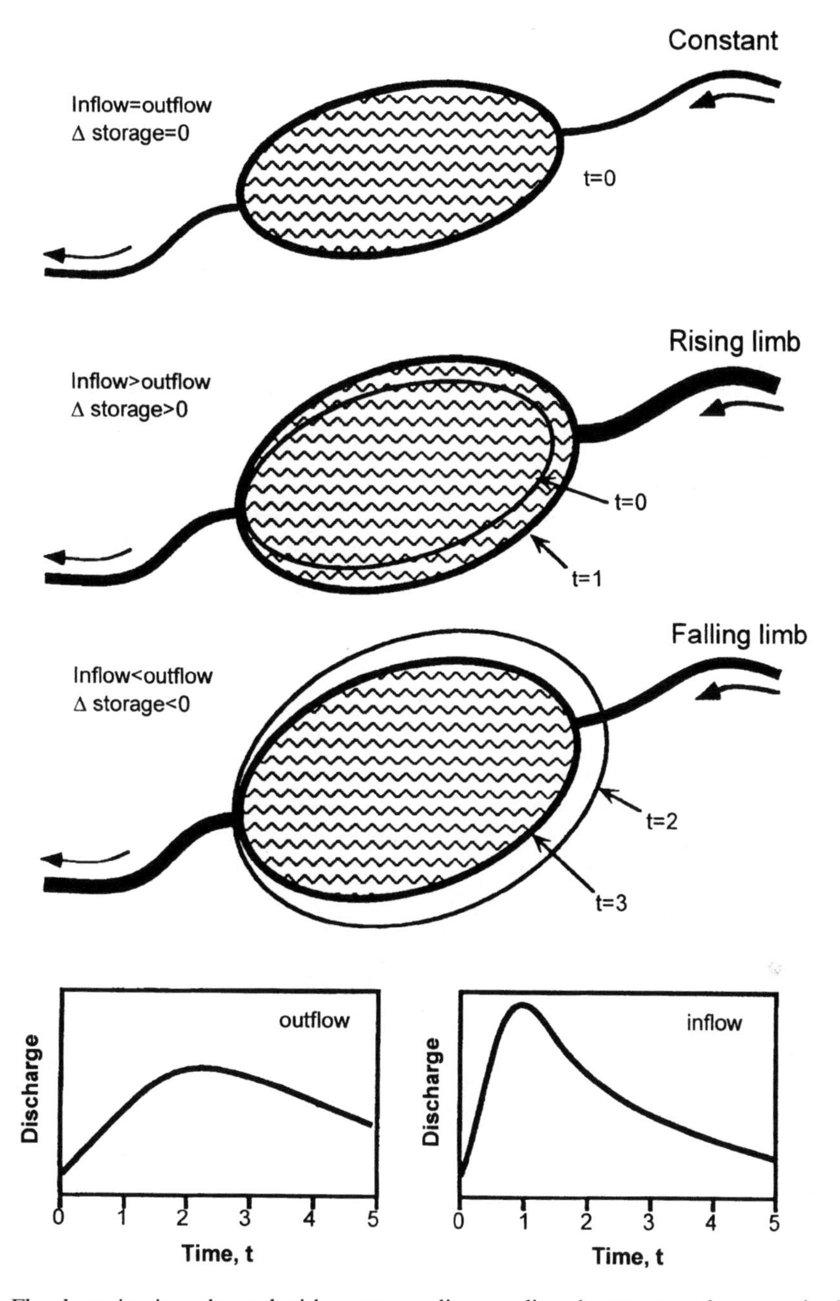

FIGURE 3.6. Flood routing in a channel with a reservoir during constant flow ($t = 0$) and the rising ($t = 1$) and falling ($t = 3$) limbs of a peak flow event. Volumes of storage in the reservoir during four time periods ($t = 0$ to 3) are represented by oval size; finer lines outline the water surface area in the reservoir during the preceding time interval. The patterned oval depicts reservoir level at corresponding inflow and outflow rates. At $t = 2$ (not shown), outflow equals inflow.

riparian corridor ranged from 0.4 to 13 m/hr (Triska et al. 1993).

Emerging subsurface flow in channels moderates more extreme seasonal water temperatures (Bilby 1984, Nielsen et al. 1994, Keller et al. 1995). For example, intergravel flow through gravel bars can emerge in still pools during low summer flow and provide cold-water refuges for salmonids when ambient stream temperatures exceed lethal limits (Nielsen et al. 1994). Other sources of cool water include surface and subsurface flow from tributaries and seeps from streambanks. Water in stratified pools in summer is commonly 3 to 9°C cooler than surface water. Similarly, seeps in off-channel habitats can provide warmer water in midwinter (Hetherington 1988).

Effects of Land-Use Practices on Runoff

Land uses such as forest practices, road building, and livestock grazing have important implications for peak flows and floods, water yield, and hydrologic recovery. The effects of land uses varies with basin size and the magnitude of flows, and recovery processes, which vary in space and time, depend on the type of disturbance and the hydrologic processes affected.

Peak Flows and Floods

Anthropogenic influences on the magnitude of floods is a recurring issue largely because of the high natural variability of flows, especially flood flows. The difficulty of detecting changes in flood size caused by land use is illustrated by Hirsch et al. (1990):

> It is not uncommon for annual floods to have a coefficient of variation (ratio of standard deviation to mean) of one or even more. Suppose the coefficient of variation of annual floods was one and the frequency distribution changed abruptly halfway through a 40-year annual flood record; in order for the change to be discernible in a statistical test with 95% power, the change in the mean would have to be at least 45%. Discriminating such a modification is further complicated by the fact that watershed change, which may modify flow, may be gradual rather than abrupt (p. 329).

Although a 45% change in flood magnitude would be undetectable in this example, it does not mean that such a change would be ecologically and culturally benign.

To address the question of whether humans influence high flows in certain basins often involves a conundrum. If an impact exists, it is costly; if an impact is assumed to exist, but really does not, it can be costly to take unnecessary preventative measures. However, determing the presence of impacts is difficult because variability in runoff makes many hydrologic impacts unpredictable and even undetectable from a statistical standpoint. How society responds to human-caused changes in floods depends on how risks that are difficult to evaluate are perceived and weighed (Chapter 26).

A debate over the beneficial influence of forests in flood protection has continued for at least a century in the United States. The arguments being made today are more moderate, but not unlike those made in the early part of the twentieth century. Concern about overexploitation of forests and the idea that forest conservation could reduce floods resulted in passage of the Weeks' Law in 1911. The Weeks' Law authorized the federal government to purchase private land to establish National Forests in the eastern United States for the protection of the watersheds of navigable streams. In the early 1900s, Chittenden (1909) stated that forest cutting alone does not result in increased runoff. During the early part of the twentieth century there were many opinions but little data to test the relationship between forests and floods. To address the varied opinions, watershed research was initiated in the 1930s in southern California (San Dimas), Arizona (Sierra Ancha), and North Carolina (Coweeta). The studies at Coweeta resulted in the first scientific evidence that conversion from a forest to a mountain farm greatly increased peak flows, but clear-cutting the forest without disturbing the forest floor did not have a major effect on peak flows (Hoover 1945). By the 1960s, there were 150 forested experimental watersheds throughout the United States. When Lull and Reinhart (1972) released their definitive paper summarizing what was known about the influence of forests and floods, about 2,000 papers had been published reporting research results about the hydrology of forested watersheds. Lull and Reinhart (1972) focused on the eastern United States. A decade later Hewlett (1982) extended the evaluation to the major forest regions of the world. Hewlett concluded, as did Chittenden (1909) and Lull and Reinhart (1972), that the effect of forest operations on the magnitude of major floods is minor in comparison with the influences of rainfall and basin storage.

Results from the Pacific coastal ecoregion are variable. Rothacher (1971, 1973) found no appreciable increase in peak flows for the largest floods as a result of clear-cutting. Paired watershed studies in the Cascades (Harr et al. 1979), Oregon Coast Range (Harr et al. 1975), and coastal northwestern California (Ziemer 1981, Wright et al. 1990) similarly suggest that the magnitude of large floods that occurred when the ground was saturated were not increased significantly by logging.

Using longer streamflow records of 34 to 55 years, Jones and Grant (1996) evaluated changes in peak flow from timber harvest and road building from a set of three small basins (0.6–1 km^2) and three pairs of large basins (60–600 km^2) in the Oregon Cascades. In the small basins, they reported that changes in small peak flows were greater than changes in large flows. In their category of "large" peaks (recurrence interval greater than 0.4 yr), flows were significantly increased in one of the two treated small basins, but the ten largest flows were apparently unaffected by treatment. They also reported that forest harvesting increased peak discharges by as much as 100% in the large basins over the past 50 years (Jones and Grant 1996). However, independent analysis of the same data set used by Jones and Grant indicated that a relationship could not be found between forest harvesting and peak discharge in the large basins (Beschta et al. 1997).

Variation with Basin Size

Effects of forest practices on storm runoff are generally more pronounced and easier to detect in small basins than in large basins. The reasons for this are both statistical and physical. The ability to detect changes in large basins is limited not only by the quality of data available, but also by the sample size of appropriate basins to study in a given hydrologic province. The scale of management units is commensurate with the area of low-order basins of 0.1 to 1 km^2, so at any time there are numerous small watersheds with very high or very low percentages of affected area that can be tested for effects on runoff. In contrast, only a small percent of the area of a large basin (drainage area $>10^2\,km^2$) is usually affected at any one time, while the rest of the basin is either pristine or recovering from past effects. Consequently, effective comparisons between treated and nontreated large basins for the same event are unlikely. The best available method to address this problem is to analyze records from small watersheds where the type and timing of land use activities can be controlled and flows can be measured accurately.

Observed effects in small basins cannot be accurately extrapolated to large basins, because processes of flood generation and routing are not represented in the same proportions. Storm peaks originating from small tributaries are lagged, damped, and desynchronized as they move downstream to contribute to flood stages in larger basins (Hewlett 1982). Stormflow response of small basins is governed primarily by hillslope processes, which are sensitive to forest practices. In contrast, stormflow response of large basins is governed primarily by the geomorphology of the channel network (Robinson et al. 1995), which is less likely to be affected by forest practices. Logging and road building commonly affect stormflow by causing the network of open-channel flow to extend upslope. This extension of the channel network is proportionately small in large basins (Beven and Wood 1993). Increases in peak runoff from hillslopes and headwaters tend to be attenuated by storage in downstream channels and floodplains. Progressing downstream, changes in channel storage of runoff (e.g., from impoundments, channel incision, widespread removal of woody debris, channelization) influence peak runoff in channels more than changes in runoff from hillslopes and headwaters.

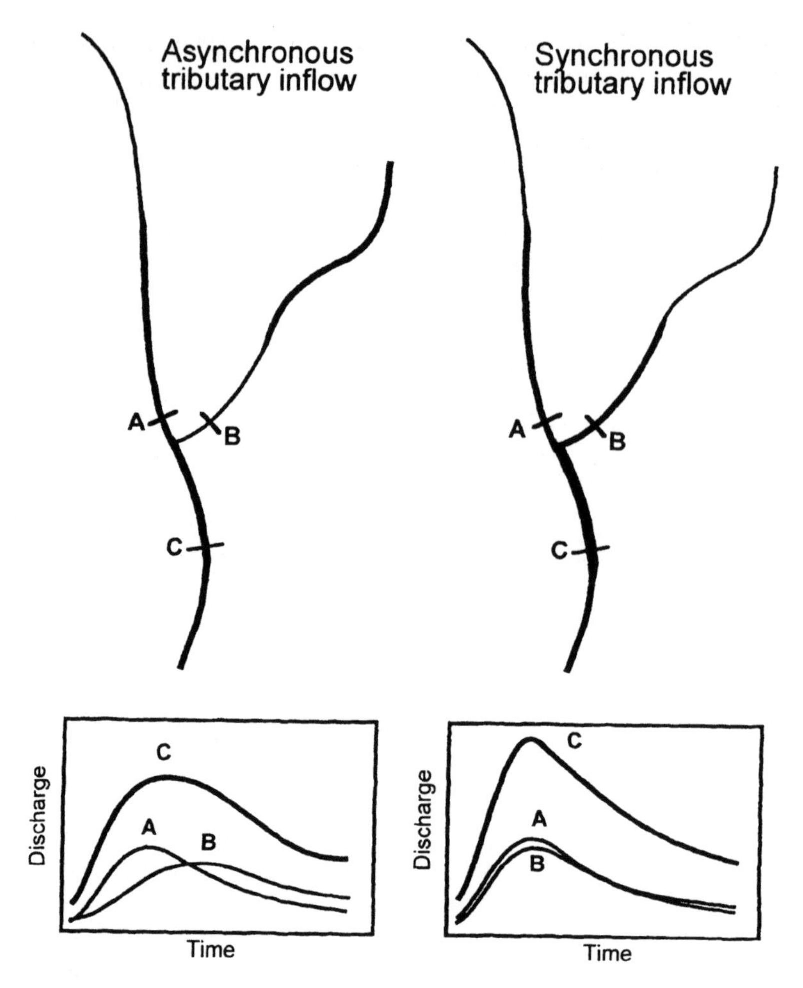

FIGURE 3.7. Effect of synchronicity of tributary hydrographs on peak discharge in the main stem. With the same volume of total flood runoff, synchronous tributary inputs create greater peak discharge in the mainstem than asynchronous inputs.

Peak discharges in mainstem channels depend in part on the synchronicity of peak discharges from tributaries. Changing the lag time between rainfall and peak runoff from tributaries as a result of hillslope disturbances without influencing flow routing in channels may or may not enhance synchronicity of the tributary inputs (Figure 3.7). Thus, no general relationship between changes in peak flows of tributaries and mainstem channels is expected. Instead, the peak flow response depends on the channel-network hydrology of each basin. However, a consistent shortening of tributary lag times tends to compress peak-flow arrival times in the mainstem and increase the probability of synchronous inputs.

Variation with Type of Precipitation

Rain on snow. There is little evidence that forest practices significantly affect large floods produced by rain. However, it is possible that clear cutting exacerbates some rain on snow floods, although the magnitude of such an effect is highly variable and difficult to measure or detect. Snow that is intercepted by a forest canopy in the Pacific coastal ecoregion is apt to melt in the canopy and reach the forest floor as melt water or as wet snow. In areas where trees have been removed, more snow accumulates on the ground. Berris (1984) reported that, in the Oregon Cascades, snow water in a clear-cut was twice that in the forest. In contrast, in colder regions where the snow is drier, wind often sweeps snowfall from large exposed openings resulting in deeper accumulations of snow in protected forests. The convective transfer of latent and sensible heat is often greater in clear cuts because wind speed at the ground is greater. Berris reported that the energy available to melt snow during a rainstorm was about 40% greater in clear-cut areas than in unlogged forests (1984).

Snow accumulation and melt. The demand for water in the west has resulted in assessments of the potential for increasing water yields by delaying delivery of melt water from the snow zone (Anderson 1963, Kattelmann et al. 1983, Ponce 1983). Management proposals have included various patterns of cutting forests (Anderson 1956), snow fences (Martinelli 1975, Tabler and Sturges 1986), intentional avalanching (Martinelli 1975), application of chemical evaporation suppressants (Slaughter 1970), and weather modification (Kattelmann and Berg 1987). Anderson (1956), for example, designed a "step and wall" forest cutting pattern of alternating cut strips and residual forest in the higher elevation snow zone to maximize snow accumulation in the openings and shade at the forest margin in order to minimize melt rate and thereby provide more water later in the summer season. Although some of these measures have technical merit, serious constraints prevent implementation. Not only do the costs of such measures generally outweigh the value of increased runoff, but also much of the high elevation land is located within National Parks, designated wilderness, or areas administratively reserved from active management. Within those few areas where manipulative land management is possible, concerns about water quality, visual impacts, wildlife, and other resource values preclude serious consideration of water yield improvement projects in the snowpack zone. In the 1950s, grand plans were developed to increase water yields from the mountains in the west (Anderson 1960, 1963), but by the 1980s none of these programs had been implemented.

Variation with Season

Effects of forest practices on streamflow vary strongly with season because of wide variations in hydrologic conditions. In much of the Pacific coastal ecoregion, summers are characterized by long, rainless periods. Since little soil moisture recharge occurs during the growing season in the west, large differences in soil moisture can develop during the summer because of differences in evapotranspiration rates between logged and unlogged watersheds. For example, a single mature pine tree in the northern Sierra Nevada depleted soil moisture to a depth of about 6 m and a distance of 12 m from the trunk (Ziemer 1968). This one tree transpired about 88 m^3 more water than a surrounding logged area. This summer transpiration loss is equivalent to about 180 mm of rain over the affected area.

With the onset of the rainy season in the fall, the soil profile begins to be recharged with moisture. In Oregon, Rothacher (1971, 1973) reported that the first storms of the fall produced streamflow peaks from a 96-ha clear-cut watershed in the H.J. Andrews Experimental Forest that ranged from 40 to 200% larger than those predicted from the prelogging relationship. In the Alsea watershed near the Oregon coast, Harris (1977) found no significant change in the mean peak flow after clear cutting a 71-ha watershed or patch cutting 25% of an adjacent 303-ha watershed. However, when Harr (1976) added an additional 30 smaller early winter runoff events to the data, average fall peak flow increased 122%. In northwestern California, Ziemer (1981) reported that selection cutting and tractor yarding of an 85-year-old second-growth redwood (*Sequoia sempervirens*) and Douglas-fir (*Pseudotsuga menziesii*) forest in the 424-ha Caspar Creek watershed increased the first streamflow peaks in the fall about 300% after logging. The effect of logging on peak flow was best explained by a variable representing the percentage of the area logged divided by the sequential storm number that began each fall. These first rains and consequent streamflow in the fall are usually small and geomorphically inconsequential in the Pacific coastal ecoregion. The large peak flows, which tend to modify stream channels and transport most of the sediment, usually occur during mid-winter after the soil moisture deficits have been satisfied in both the logged and unlogged watersheds. These large events were not significantly affected by logging in the H.J. Andrews (Rothacher 1973), Alsea (Harr 1976, Harris 1977), or Caspar Creek (Ziemer 1981) studies.

Variation with Flood Magnitude

On a regional basis, flood magnitude is usually significantly correlated with only a few variables, such as basin size and the amount and intensity of precipitation. For example, the magnitude of the largest rainfall-runoff floods from diverse areas worldwide correlates very well with basin area alone (Costa 1987). In each of five mountainous areas of the United States (Colorado alpine, Colorado foothills, Sierra Nevada, Klamath Mountains, and California Coast Ranges), the magnitude of the mean annual flood in basins with drainage areas greater than $10 km^2$ can be predicted well using only drainage area and mean annual precipitation; variables describing slope, drainage density, and percent forest cover do not significantly improve the relationship (Pitlick 1994).

Although the largest floods are most important from a flood hazard standpoint, the influence of smaller more frequent floods cannot be discounted from a channel condition or ecological standpoint. High flows occurring on average every one to five years are most important for transporting sediment and forming channels in many regions (Wolman and Miller 1960, Andrews 1980), although the less frequent large floods can have greater geomorphic effect in the Pacific coastal ecoregion, particularly in mountain channels (Nolan et al. 1987, Grant et al. 1990, Grant and Swanson 1995). Increases in the magnitude of moderate floods tend to increase sediment transport and enlarge channels either by eroding them or building higher banks. However, the response is complex and difficult to detect, because watershed delivery of sediment to channels also occurs during periods of high runoff (Chapters 2 and 11).

Different usage of the term "flood" between the general public and hydrologists can confuse public debate about effects of land use on peak streamflow. To the public, use of the term flood usually evokes the idea of a rare major discrete event that inundates and causes damage to roads, homes, businesses, or agriculture. A "normal" high streamflow event that is expected to occur each year or once every couple of years is usually not considered by laypeople to be a "flood." Human infrastructure is usually constructed to cope with such "normal" events, so property damage from these events seldom occurs. To an hydrologist, the term "flood" loosely refers to a wide range of magnitude of hydrograph peaks, including those that are contained within streambanks as well as extreme events. To avoid confusion, hydrologists should take care to state the size and frequency of the streamflow event being discussed and should exercise caution in using terminology that can be misinterpreted by the public.

There is a fundamental problem in determining whether forest practices increase size (i.e., magnitude and extent) of large floods. The problem is greater when attempting to determine whether forest practices increase the size of large floods in large river basins. First, the greater the size of the flood (or basin) being investigated, the less likely that there will be any changes caused by forest practices. Second, any such changes become harder to detect because the available sample size decreases as the size of the flood and the size of the basin increases. To evaluate changes in hydrologic response associated with land use, enough streamflow events must be observed to obtain sufficient statistical power for determining significance. This usually forces the inclusion of smaller events as "floods" to increase the number of observations. Within a 50-year record, it would be extremely fortunate to measure a 25-year streamflow event before land treatment to compare with a 25-year event after treatment. Even so, there would be little to say statistically about the events because of the small sample size. Only about five 10-year events would be expected during that 50-year record, and those events probably would be scattered throughout the record, before, during, and after treatment.

There are physical reasons why forest practices are less likely to influence large floods than small floods. While logging and road building may affect flow magnitudes by increasing the extent of more rapid surface runoff at the expense of slower subsurface runoff (discussed below), effects on runoff processes vary less with storm size as land becomes saturated (Dunne 1983) and surface runoff caused by human activity (e.g., from roads) becomes a smaller proportion of total stormflow. Moreover, as the duration of a rainfall event increases, any change in the delivery rate of runoff from hillslopes to channels resulting from forest practices becomes less important in flood magnitude.

Water Yield

Throughout much of the arid west, the lack of water during the summer growing season has been a severe constraint on agriculture, power production, urbanization, and virtually all forms of human enterprise. With the establishment of the Wagon Wheel Gap studies in Colorado in the early 1900s, serious scientific thought began to be directed toward evaluating the effect of forest manipulation on water yield. Bosch and Hewlett (1982), summarizing the results of 94 catchment experiments world wide, found extreme variation between areas, but in no case did clearing vegetation reduce water yield. In each case, clearing vegetation resulted in water yields that either remained constant or increased. In cases where water yields increased, the regrowth of vegetation following clearing returned water yields to those observed before clearing. Bosch and Hewlett concluded that the potential for increasing water yield by removing vegetation was greatest in areas having coniferous forests, less in deciduous hardwoods, and least in brush and grasslands. In addition, water yield increases following vegetation removal were greatest in high rainfall areas, and, within a given area, tended to be greater in wet years than in dry years (Ponce and Meiman 1983). These small watershed studies indicate that there is no potential for increasing water yield by manipulating vegetation in areas when precipitation is less than about 40 cm, and marginal potential when precipitation is between 40 and 50 cm (Clary 1975, Hibbert 1983).

Much of the Pacific coastal ecoregion is within a climate zone where logging might be expected to result in an increase in streamflow during the summer. For example, in a paired watershed study at Caspar Creek in northern coastal California (Figure 3.8), selective logging of 67% of the stand volume in the 484-ha South Fork watershed in the early 1970s increased the summer lowflow about 120% or about 0.3 L/s/km^2 (170 m^3 of water per day). This increase in summer flow declined with regrowth of the vegetation and returned to prelogging levels within about eight years (Keppeler and Ziemer 1990). When 12% of the 473-ha North Fork watershed was clear-cut in 1985, summer streamflow increased about 150% for one year before returning to near prelogging levels (Ziemer et al. 1996). Three to five years later, an additional 42% of the stand volume was clear cut. Summer lowflow again increased about 200% (0.6 L/s/km^2). This increased summer lowflow is anticipated to return to prelogging levels within 8 to 10 years. Similar patterns have been reported elsewhere in which water yield is observed to increase immediately after forest cutting and then return to precutting levels within a few years (Hewlett and Helvey 1970, Ursic 1986, Stednick 1996).

In the 1950s, regional proposals promised to deliver more water in water-deficient regions by clearing vegetation over large areas. Before these proposals could be implemented, not only did more detailed studies show that many of the earlier assumptions were not generally applicable, but social and environmental concerns about increased erosion, degraded aesthetic values, and habitat destruction associated with vegetation conversion began to emerge. By the mid-1980s, it had become clear that the options for increasing water yield by manipulating vegetation over large areas were quite limited (Ziemer 1987). For western Washington and Oregon, estimated sustained increases in water

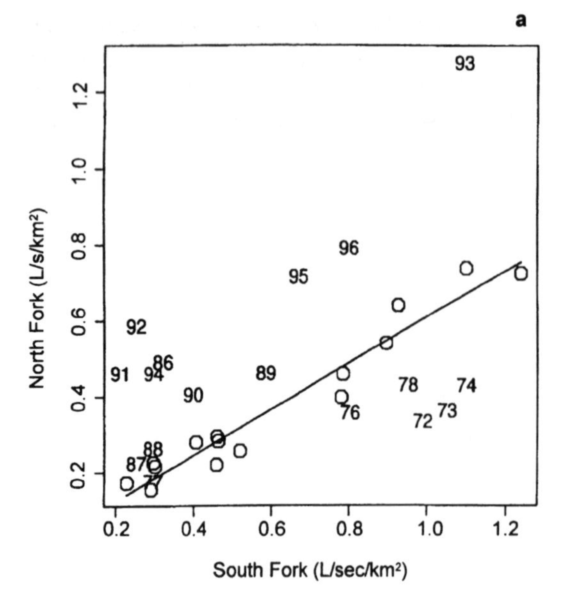

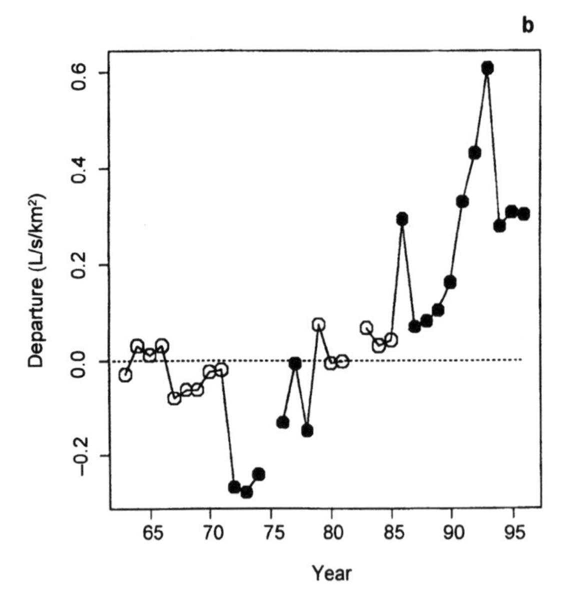

FIGURE 3.8. Relationship between minimum summer streamflow in the North and South Forks of Caspar Creek, 1963–1996. (a) The regression line is based on nondisturbance years (○); the numbers are the disturbance years. (b) Departure is the difference between observed streamflow in nondisturbance (○) and disturbance years (●) and that predicted by the nondisturbance regression (modified from Ziemer et al. 1996).

yield from most large watersheds subjected to sustained-yield forest management are at best only 3 to 6% of unchanged flows (Harr 1983).

Hydrologic Recovery from Land-Use Impacts

The period of time needed for recovery from impacts of land use is an important consideration for resource management (Figure 3.9). Recovery processes vary widely in space and time. In many cases, a return to predisturbance conditions is difficult to define or evaluate because of a lack of data, inherent variability, and an asymptotic decline in effects after a period of rapid recovery.

Recovery of hydrologic conditions following severe land use impacts depends on rates of establishment and growth of vegetation. For example, recovery following grazing is relatively rapid because grassland vegetation is small and grows and spreads rapidly enough to quickly affect runoff processes. Regrowth of vegetation and subsequent loosening of compacted soils in previously grazed areas of the Pacific Northwest can result in increased capacity for infiltration and substantially decreased overland flow velocities within a few years after removal of stock. Further, much wildland grazing occurs at high elevations where frequent freeze/thaw cycles loosen compacted soils. In the Pacific coastal ecoregion regrowth of shrubs and small trees commonly returns rates of evapotranspiration to prelogging levels within about five years (Harr 1979). However, recovery of the tree canopy to levels that restore natural rates of forest snow retention and melt rate takes several decades.

Roads are nearly permanent features on the landscape. Runoff diversion caused by roads cannot be restored by revegetation, but requires erosion or human intervention to reestablish natural drainage patterns. Abandoned roads, skid trails, and landings remain impervious to water infiltration for decades until vegetation slowly becomes established, roots begin to penetrate and break up the compacted soil, and a litter layer and soil profile develops. Road cuts continue to intercept and reroute shallow subsurface flow. Without judicious maintenance, culverts eventually plug, ditches fill, waterbars fail, and the probability of the occurrence of an hydrologic event that exceeds the road's design flows increases with time. Because road maintenance has become a low priority due to reduced funding for many agencies and landowners, risks of hydrologic or erosional impacts from roads will likely increase over time.

Removal of large trees along streams has important consequences for runoff routing in channels. Without replenishment, woody debris disappears from channels over periods of decades or longer, and the loss of channel roughness can increase channel runoff velocities and peak discharges downstream. Furthermore, loss of root strength following the removal of riparian trees may initiate accelerated channel erosion (Chapter 2).

Although water yield changes return to pretreatment levels relatively quickly, changes in the physical condition of the watershed that affect streamflow generation and routing may remain for decades. For example, following timber harvesting and site preparation in the Alsea watersheds in coastal Oregon, annual water yield increases returned to pretreatment levels within a few years, but the physical condition of the watersheds is still significantly changed 28 years after treatment (Stednick 1996).

Because land management benefits from a better understanding of the benefits and risks of watershed practices, it remains worthwhile to attempt to measure the magnitude of changes in runoff from land use practices and other influences at a variety of spatial and temporal scales and conditions, even though the outlook to accomplish this with statistical confidence is bleak. Attention to relationships between site conditions and runoff processes can provide answers where changes in river flow are elusive. The motive for investigating changes in runoff is to predict downstream impacts, not necessarily as changes in streamflow alone, but as

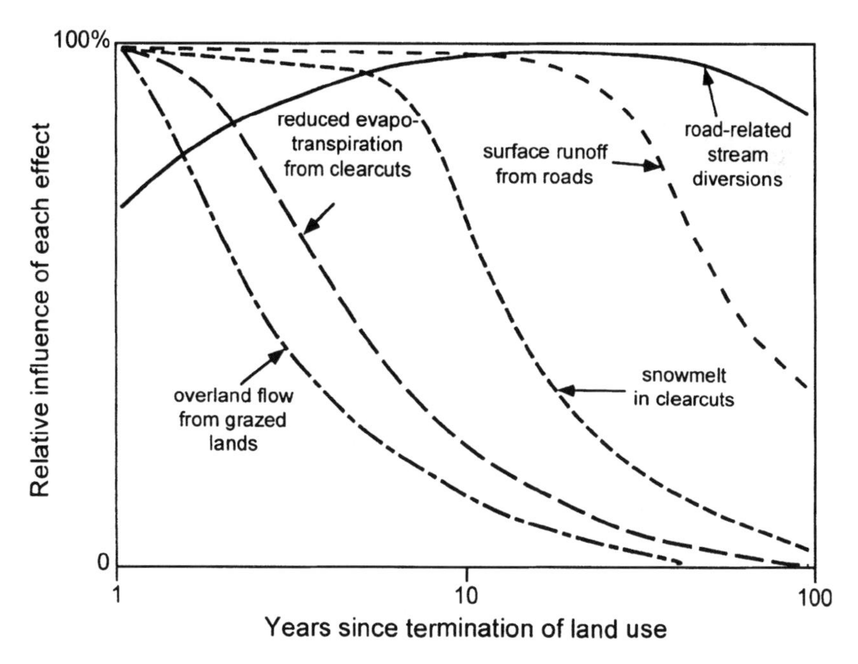

FIGURE 3.9. Characteristic temporal variation of the relative influence of various land uses upon on-site conditions after the land-use activity has terminated. Recovery of channels affected by downstream propagation of these effects is not represented. The state of recovery at any time is represented as the magnitude of the site effect relative to the maximum, which in most cases occurs just after the use is completed or terminated.

changes in the flow of all watershed products and their subsequent effects on aquatic and riparian ecosystems that often begin as a local disturbance in runoff.

Literature Cited

Anderson, H.W. 1956. Forest-cover effects on snowpack accumulation and melt, Central Sierra Snow Laboratory. American Geophysical Union Transactions **37**:307–312.

——. 1960. Research in management of snowpack watersheds for water yield control. Journal of Forestry **58**:282–284.

——. 1963. Managing California's snow zone lands for water. Research Paper PSW-6. United States Department of Agriculture, Berkeley, California, USA.

Andrews, E.D. 1980. Effective and bankfull discharges of streams in the Yampa River Basin, Colorado and Wyoming. Journal of Hydrology **46**:311–330.

Aubertin, G.M. 1971. Nature and extent of macropores in forest soils and their influence on subsurface water movement. USDA Forest Service Research Paper NE-192. Upper Darby, Pennsylvania, USA.

Benson, M.A., and D.M. Thomas. 1966. A definition of dominant discharge. Bulletin International Association of Scientific Hydrology **11**:76–80.

Berris, S.N. 1984. Comparative snow accumulation and melt during rainfall in forest and clearcut plots in western Oregon. Master's thesis. Oregon State University, Corvallis, Oregon, USA.

Beschta, R.L. 1990. Effects of fire on water quantity and quality. Pages 219–232 *in* J.D. Walstad, S.R. Radosevich, and D.V. Sandberg, eds. Natural and prescribed fire in Pacific Northwest forests. Oregon State University Press, Corvallis, Oregon, USA.

Beschta, R.L., M.R. Pyles, A.E. Skaugset, and C.G. Surfleet. 1997. Peak flow responses to clear-cutting and roads in small and large basins, western Cascades, Oregon: An alternative analysis. Unpublished report, Department of Forest Engineering, Oregon State University, Corvallis, Oregon, USA.

Betson, R.P. 1964. What is watershed runoff? Journal of Geophysical Research **69**:1541–1551.

Beven, K., and E.F. Wood. 1993. Flow routing and the hydrological response of channel networks. Pages 99–128 *in* K. Beven and M.J. Kirkby, eds. Channel network hydrology. John Wiley & Sons, New York, New York, USA.

Bilby, R.E. 1984. Characteristics and frequency of cool-water areas in a western Washington stream. Journal of Freshwater Ecology **2**:593–602.

Black, P.E. 1996. Watershed hydrology. Ann Arbor Press, Chelsea, Michigan, USA.

Bosch, J.M., and J.D. Hewlett. 1982. A review of catchment experiments to determine the effect of vegetation changes on water yield and evapotranspiration. Journal of Hydrology **55**:3–23.

Chamberlin, T.W. 1972. Interflow in the mountainous forest soils of coastal British Columbia. Pages 121–127 *in* H.O. Slaymaker and H.J. McPherson, eds. Mountain geomorphology: Geomorphological proceedings in the Canadian Cordillera. Tantalus Research, Vancouver, British Columbia, Canada.

Chittenden, H.M. 1909. Forests and reservoirs in their relation to streamflows, with particular reference to navigable rivers. American Society of Engineering Transactions **62**:248–318.

Clary, W.P. 1975. Multiple use effects of manipulating pinyon-juniper. Pages 469–477 *in* Water management, proceedings of the 1975 watershed management symposium. American Society of Civil Engineers, New York, New York, USA.

Colman, E.A. 1955. Operation wet blanket: Proposed research in snowpack management in California. USDA Forest Service, Berkeley, California, USA.

Costa, J.E. 1987. A comparison of the largest rainfall-runoff floods in the United States with those of the People's Republic of China and the world. Journal of Hydrology **96**:101–115.

DeBano, L.F. 1969. Water repellent soils: A worldwide concern in management of soil and vegetation. Agricultural Science Review **7**:11–18.

——. 1981. Water repellent soils: A state-of-the-art. USDA Forest Service General Technical Report PSW-46, Berkeley, California, USA.

Dunne, T. 1983. Relation of field studies and modeling in the prediction of storm runoff. Journal of Hydrology **65**:25–48.

Dunne, T., and L.B. Leopold. 1978. Water in environmental planning. W.H. Freeman, San Francisco, California, USA.

Dyrness, C.T. 1976. Effect of wildfire on soil wettability in the High Cascades of Oregon. USDA Forest Service Research Paper PNW-202, Portland, Oregon, USA.

Gordon, N.D., T.A. McMahon, and B.L. Finlayson. 1992. Stream hydrology, an introduction for ecologists. John Wiley & Sons, New York, New York, USA.

Grant, G.E. 1987. Assessing the effects of peak flow increases on stream channels—a rational approach. Pages 142–149 *in* R.Z. Callaham and J.J. DeVries, technical coordinators. Proceedings of the California watershed management conference, 1986 November 18–20. Wildland Resources Center Report Number 11, University of California, Berkeley, California, USA.

Grant, G.E., and F.J. Swanson. 1995. Morphology and processes of valley floors in mountain streams, western Cascades, Oregon. Pages 83–101 *in* Natural and anthropogenic influences in fluvial geomorphology. Geophysical Monograph 89, American Geophysical Union, Washington, DC, USA.

Grant, G.E., F.J. Swanson, and M.G. Wolman. 1990. Pattern and origin of stepped-bed morphology in high-gradient streams, western Cascades, Oregon. Geological Society of America Bulletin **102**:340–352.

Grant, G.E., and A.L. Wolff. 1991. Long-term patterns of sediment transport after timber harvest, western Cascade Mountains, Oregon, USA. Pages 31–40 *in* Sediment and stream water quality in a changing environment: Trends and explanation. Proceedings of the Vienna symposium. International Association of Hydrological Sciences Publication Number 203. IAHS, Exeter, UK.

Gregory, S.V., F.J. Swanson, W.A. McKee, and K.W. Cummins. 1991. An ecosystem perspective of riparian zones. BioScience **41**:540–551.

Harr, R.D. 1976. Forest practices and streamflow in western Oregon. USDA Forest Service General Technical Report PNW-49. Pacific Northwest Forest and Range Station, Portland, Oregon, USA.

Harr, R.D. 1977. Water flux in soil and subsoil on a steep forested slope. Journal of Hydrology **33**:37–58.

——. 1979. Effects of streamflow in the rain-dominated portion of the Pacific Northwest. Pages 1–45 *in* Proceedings of workshop on scheduling timber harvest for hydrologic concerns. USDA Forest Service, Portland, Oregon, USA.

——. 1981. Some characteristics and consequences of snowmelt during rainfall in western Oregon. Journal of Hydrology **53**:277–304.

——. 1983. Potential for augmenting water yield through forest practices in western Washington and western Oregon. Water Resources Bulletin **19**:383–393.

Harr, R.D., R.L. Fredriksen, J. Rothacher. 1979. Changes in streamflow following timber harvest in southwestern Oregon. USDA Forest Service Research Paper PNW-249, Portland, Oregon, USA.

Harr, R.D., W.C. Harper, J.T. Krygier, and F.S. Hsieh. 1975. Changes in storm hydrographs after road building and clear-cutting in the Oregon Coast Range. Water Resources Research **11**:436–444.

Harris, D.D. 1977. Hydrologic changes after logging in two small Oregon coastal watersheds. United States Geological Survey Water Supply Paper 2037. Washington, DC, USA.

Hetherington, E.D. 1988. Hydrology and logging in the Carnation Creek watershed—what have we learned? Pages 11–15 *in* T.W. Chamberlin, ed. Proceedings of the workshop: Applying 15 years of Carnation Creek results. Pacific Biological Station, Carnation Creek Steering Committee, Nanaimo, British Columbia, Canada.

Hewlett, J.D. 1982. Forests and floods in the light of recent investigation. Pages 543–559 *in* Canadian hydrology symposium: 82. Associate Committee on Hydrology, National Research Council of Canada, Fredricton, New Brunswick, Canada.

Hewlett, J.D., and J.D. Helvey. 1970. Effects of forest clearfelling on the storm hydrograph. Water Resources Research **6**:768–782.

Hibbert, A.R. 1983. Water yield improvement potential by vegetation management on western rangelands. Water Resources Bulletin **19**:375–381.

Hirsch, R.M., J.F. Walker, J.C. Day, and R. Kallio. 1990. The influence of man on hydrologic systems. Pages 329–359 *in* M.G. Wolman and H.C. Riggs, eds. The geology of North America. Volume O-**1**. Surface water hydrology. Geological Society of America, Boulder, Colorado, USA.

Hoover, M.D. 1945. Effect of removal of forest vegetation upon water-yields. American Geophysical Union Transactions Part **6**(1944):969–977.

Horton, R.E. 1933. The role of infiltration in the hydrologic cycle. American Geophysical Union Transactions **14**:446–460.

Jones, J.A.A. 1987. The effects of soil piping on contributing areas and erosion patterns. Earth Surface Processes and Landforms **12**:229–248.

Jones, J.A., and G.E. Grant. 1996. Peak flow responses to clear-cutting and roads in small and large basins, western Cascades, Oregon. Water Resources Research **32**:959–974.

Kattelmann, R. and N. Berg. 1987. Water yields from high-elevation basins in California. Pages 79–85 *in* R.Z. Callaham and J.J. DeVries, technical coordinators. Proceedings of the California watershed management conference, 1986 November 18–20.

Wildland Resources Center Report Number 11, University of California, Berkeley, California, USA.

Kattelmann, R.C., N.H. Berg, and J. Rector. 1983. The potential for increasing streamflow from Sierra Nevada watersheds. Water Resources Bulletin **19**:395–402.

Keller, E.A., T.D. Hofstra, and C. Moses. 1995. Summer cold pools in Redwood Creek near Orick, California, and their relation to anadromous fish habitat. Pages U1–U9 *in* K.M. Nolan, H.M. Kelsey, and D.C. Marron, eds. Geomorphic processes and aquatic habitat in the Redwood Creek basin, northwestern California. United States Geological Survey Professional Paper 1454. US Government Printing Office, Washington, DC, USA.

Keppeler, E.T. and R.R. Ziemer. 1990. Logging effects on streamflow: Water yields and summer flows at Caspar Creek in northwestern California. Water Resources Research **26**:1669–1679.

Kirkby, M.J. 1978. Hillslope hydrology. John Wiley & Sons, New York, New York, USA.

Leaf, C.F. 1975. Watershed management in the Rocky Mountain subalpine zone: The status of our knowledge. USDA Forest Service Research Paper RM-137, Fort Collins, Colorado, USA.

Leopold, L.B. 1968. Hydrology for urban land planning: A guide-book on the hydrologic effects of urban land use. United States Geological Survey Circular 554. US Government Printing Office, Washington, DC, USA.

Lisle, T.E. 1989. Channel-dynamic control on the establishment of riparian trees after large floods in northwestern California. Pages 9–13 *in* D.L. Abell, ed. Proceedings of the California riparian systems conference, 1988 September 22–24. USDA Forest Service General Technical Report PSW-110, Berkeley, California, USA.

Lull, H.W., and K.G. Reinhart. 1972. Forests and floods in the eastern United States. USDA Forest Service Research Paper NE-226, Upper Darby, Pennsylvania, USA.

Martinelli, M., Jr. 1975. Watershed yield improvement from alpine areas: The status of our knowledge. USDA Forest Service Research Paper RM-138, Fort Collins, Colorado, USA.

McBride, J.R., and J. Strahan. 1985. Establishment and survival of woody riparian species on gravel bars of an intermittent stream. American Midland Naturalist **98**:235–245.

McNabb, D.H., and F.J. Swanson. 1990. Effects of fire on soil erosion. Pages 159–176 *in* J.D. Walstad, S.R. Radosevich, and D.V. Sandberg, eds. Natural and prescribed fire in Pacific Northwest forests. Oregon State University Press, Corvallis, Oregon, USA.

Megahan, W.F. 1972. Subsurface flow interception by a logging road in mountains of central Idaho. Pages 350–356 *in* National symposium on watersheds in transition. American Water Resources Association and Colorado State University, Fort Collins, Colorado, USA.

Mosley, M.P. 1979. Streamflow generation in a forested watershed, New Zealand. Water Resources Research **15**:795–806.

Mount, J.F. 1995. California rivers and streams: The conflict between fluvial process and land use. University of California Press, Berkeley, California, USA.

Naiman, R.J., and E.C. Anderson. 1996. Ecological implications of latitudinal variation in watershed size, streamflow and water temperature in the coastal temperate rain forests of North America. Pages 131–148 *in* P.K. Schoonmaker and B. von Hagen, eds. The rain forests of home: Profile of a North American bioregion. Island Press, Washington, DC, USA.

Nash, D.B. 1994. Effective sediment-transporting discharge from magnitude-frequency analysis. Journal of Geology **102**:79–95.

Nielsen, J.L., T.E. Lisle, and V. Ozaki. 1994. Thermally stratified pools and their use by steelhead in northern California streams. Transactions of the American Fisheries Society **123**:613–626.

Nolan, K.M., T.E. Lisle, and H.M. Kelsey. 1987. Bankfull discharge and sediment transport in northwestern California. Pages 439–450 *in* R.L. Beschta, T. Blinn, G.E. Grant, F.J. Swanson, and G.G. Ice, eds. Erosion and sedimentation in the Pacific rim. International Association of Hydrological Sciences Publication Number 165. IAHS, Washington, DC, USA.

Pitlick, J. 1994. Relation between peak flows, precipitation, and physiography for five mountainous regions in the western USA. Journal of Hydrology **158**:219–240.

Ponce, S.L. 1983. The potential for water yield augmentation through forest and range management. American Water Resources Association, Bethesda, Maryland, USA.

Ponce, S.L., and J.R. Meiman. 1983. Water yield augmentation through forest and range management—issues for the future. Water Resources Bulletin **19**:415–419.

Reid, L.M. 1993. Research and cumulative effects. USDA Forest Service General Technical Report PSW-GTR-141, Berkeley, California, USA.

Ritter, D.F. 1988. Landscape analysis and the search for geomorphic unity. Geological Society of America Bulletin **100**:160–171.

Robinson, J.S., M. Sivapalan, and J.D. Snell. 1995. On the relative roles of hillslope processes, channel routing, and network geomorphology in the hydrologic response of natural catchments. Water Resources Research **31**:3089–3101.

Rothacher, J. 1971. Regimes of streamflow and their modification by logging. Pages 55–63 *in* Proceedings of the symposium on forest land use and stream environment. Oregon State University, Corvallis, Oregon, USA.

——. 1973. Does harvest in west slope Douglas-fir increase peak flow in small streams? USDA Forest Service Research Paper PNW-163, Portland, Oregon, USA.

Sauer, V.B., W.O. Thomas, Jr., V.A. Stickler, and K.V. Wilson. 1983. Flood characteristics of urban watersheds in the United States. United States Geological Survey Water Supply Paper 2207. US Government Printing Office, Washington, DC, USA.

Segelquist, C.A., M.L. Scott, and G.T. Auble. 1993. Establishment of *Populus deltoides* under simulated alluvial groundwater declines. American Midland Naturalist **130**:274–285.

Slaughter, C.W. 1970. Evaporation from snow and evaporation retardation by monomolecular films. Cold Regions Research and Engineering Laboratory Special Report 130, United States Department of the Army, Hanover, New Hampshire, USA.

Stednick, J.D. 1996. Monitoring the effects of timber harvest on annual water yield. Journal of Hydrology **176**:79–95.

Swanson, M.L., G.M. Kondolf, and P.J. Boison. 1989. An example of rapid gully initiation and extension by subsurface erosion: Coastal San Mateo County, California. Geomorphology **2**:393–403.

Tabler, R.D., and D.L. Sturges. 1986. Watershed test of a snow fence to increase streamflow: Preliminary results. Pages 53–61 *in* D.L. Kane, ed. Proceedings of the symposium: Cold regions hydrology. American Water Resources Association, Fairbanks, Alaska, USA.

Triska, F.J., J.H. Duff, and R.J. Avanzino. 1993. Patterns of hydrological exchange and nutrient transformation in the hyporheic zone of a gravel-bottom stream: Examining terrestrial-aquatic linkages. Freshwater Biology **29**:259–274.

Tsukamoto, Y., T. Ohta, and H. Noguchi. 1982. Hydrological and geomorphological studies of debris slides on forested hillslopes in Japan. Pages 187–197 *in* D.E. Walling, ed. Recent developments in the explanation and prediction of erosion and sediment yield: Proceedings of the Exeter Symposium. International Association of Hydrological Sciences Publication Number 137. IAHS, Exeter, UK.

Ursic, S.J. 1986. Forestry effects on small-stream floods. Hydrological Science and Technology: Short Papers **2**:13–15.

USGS (United States Geological Survey). 1997. Washington NWIS-W data retrieval. Online. United States Geological Survey. Available: http://h2o-nwisw.er.usgs.gov/nwis-w/WA.

Weaver, W.E., A.V. Choquette, D.K. Hagans, and J.P. Schlosser. 1981. The effects of intensive forest land-use and subsequent landscape rehabilitation on erosion rates and sediment yield in the Copper Creek drainage basin, Redwood National Park. Pages 298–312 *in* R.N. Coats, ed. Watershed rehabilitation in Redwood National Park and other Pacific coastal areas. Center for Natural Resource Studies of JMI, Davis, California, USA.

Weaver, W.E., D.K. Hagans, and J.H. Popenoe. 1995. Magnitude and causes of gully erosion in the lower Redwood Creek basin, northwestern California. Pages I1–I21 *in* K.M. Nolan, H.M. Kelsey, and D.C. Marron, eds. Geomorphic processes and aquatic habitat in the Redwood Creek basin, northwestern California. US Geological Survey Professional Paper 1454. US Government Printing Office, Washington, DC, USA.

Wemple, B.C. 1994. Hydrologic integration of forest roads with stream networks in two basins, western Cascades, Oregon. Master's thesis. Oregon State University, Corvallis, Oregon, USA.

Whipkey, R.Z. 1965. Subsurface stormflow from forested slopes. Bulletin of the International Association of Scientific Hydrology **10**:74–85.

Williams, G.P. 1978. Bank-full discharge of rivers. Water Resources Research **14**:1141–1154.

Wolman, M.G., and R. Gerson. 1978. Relative scales of time and effectiveness of climate in watershed geomorphology. Earth Surface Processes and Landforms **3**:189–208.

Wolman, M.G., and L.B. Leopold. 1956. Pages 87–109 *in* River flood plains: Some observations on their formation. United States Geological Survey Professional Paper 282-C. US Government Printing Office, Washington, DC, USA.

Wolman, M.G., and J.P. Miller. 1960. Magnitude and frequency of forces in geomorphic processes. Journal of Geology **68**:54–74.

Wootton, J.T., M.S. Parker, and M.E. Power. 1996. Effects of disturbance on river food webs. Science **273**:1558–1561.

Wright, K.A., K.H. Sendek, R.M. Rice, and R.B. Thomas. 1990. Logging effects on streamflow: Storm runoff at Caspar Creek in northwestern California. Water Resources Research **26**:1657–1667.

Zellweger, G.W., R.J. Avanzino, and K.E. Bencala. 1989. Comparison of tracer-dilution and current-meter discharge measurements in a small gravel-bed stream, Little Lost Man Creek, California. US Geological Survey Water Resources Investigation Report 89-4150. US Government Printing Office, Washington, DC, USA.

Ziemer, R.R. 1968. Soil moisture depletion patterns around scattered trees. United States Department of Agriculture Research Note PSW-166, Berkeley, California, USA.

——. 1981. Storm flow response to road building and partial cutting in small streams of northern California. Water Resources Research **17**:907–917.

——. 1987. Water yields from forests: An agnostic view. Pages 74–78 *in* R.Z. Callaham and J.J. DeVries, technical coordinators. Proceedings of the California watershed management conference, 1986 November 18–20. Wildland Resources Center Report Number 11. University of California, Berkeley, California, USA.

——. 1992. Effect of logging on subsurface pipeflow and erosion: Coastal northern California, USA. Pages 187–197 *in* D.E. Walling, T.R. Davies, and B. Hasholt, eds. Erosion, debris flows and environment in mountain regions: Proceedings of the Chendu symposium. International Association of Hydrological Sciences Publication Number 209. IAHS, Exeter, UK.

Ziemer, R.R., J. Lewis, and E.T. Keppeler. 1996. Hydrologic consequences of logging second-growth redwood watersheds. Pages 131–133 *in* J. LeBlanc, ed. Conference on coast redwood forest ecology and management, 1996 June 18–20. Humboldt State University, Arcata, California, USA.

4
Stream Quality

Eugene B. Welch, Jean M. Jacoby, and Christopher W. May

Overview

• Water quality includes physical, chemical, and biological constituents affecting a stream's physical conditions and chemical constituents. The total of these characteristics may be more accurately thought of as *stream quality*.

• Streams of the Pacific coastal ecoregion are naturally cool, clear, typically shaded and of high chemical quality, have relatively low acid neutralizing capacity, and are oligotrophic in nutrient status. As such, they and their biota are highly sensitive to nutrient enrichment, temperature increases, introduction of suspended solids, and acidic precipitation.

• High-elevation streams have less acid neutralizing capacity, hardness, total ions, nutrients, and metals than lowland waters. Increases in these characteristics in low-elevation streams are attributed to changes in soil and bedrock, as well as urban development and agriculture.

• Removal of riparian vegetation and inputs of organic material and stormwater runoff caused by agricultural activities and urbanization result in increased stream temperatures and reduced levels of dissolved oxygen (DO). The extent of DO depletion depends on the magnitude of the organic material input, the assimilation capacity of the stream (e.g., flow, volume), and the water temperature (warm water holds less oxygen than cold). Depletion of DO caused by stormwater inputs is typically not detected because it is diluted by high runoff of Pacific coastal winter storms. However, low DO in the interstitial spaces of the stream substratum (where sediment deposition decreases intragravel flows) adversely affects developing salmonid embryos and benthic invertebrates.

• Baseline concentrations of metals are usually low or undetectable in Pacific coastal streams except within urban areas or near sources of leachate water from mine tailings, old landfills, and industrial and domestic wastewater discharges.

• Pacific coastal streams are usually low in total suspended solids, except during large flooding events or in streams and rivers fed by glacial outwash. Concentrations of fine sediment are generally higher in watersheds where bedrock, soils, precipitation patterns, land use, and topography create favorable conditions for erosion. Fine sediments are typically more abundant where land-use activities such as road construction, logging, agriculture, or grazing expose soil to erosion and increase mass-wasting opportunities. Elevated levels of sediment have both lethal and sublethal effects on stream biota.

• Stream periphyton (primarily attached algae) are responsive to changes in water quality and habitat (e.g., riparian shading, light availability). In Pacific coastal streams, periphyton biomass is generally low, reflecting the low ambient nutrient content. However, periphyton respond positively to nutrient enrichment causing aesthetic, water quality, and invertebrate habitat degradation. Periphyton in small, shaded streams are usually limited by light and likely will not reach nuisance levels in response to nutrient enrichment. Periphyton are also responsive to flow conditions, proliferating during low-flow periods and decreasing when scoured by increased flows.

Introduction

Water quality is most commonly thought of as representing a water's chemical character. As such, it has a large influence on the kinds, amount, and activity of aquatic organisms present in a water body. The inorganic characteristics conventionally include hardness, defined as calcium (Ca^{2+}) plus magnesium (Mg^{2+}); alkalinity or acid neutralizing capacity (ANC), composed of bicarbonate (HCO_3^-) and carbonate (CO_3^{2-}) ions; pH, which is also a reflection of ANC; the nutrients nitrogen (N) and phosphorus (P); metals, such as copper (Cu), zinc (Zn), lead (Pb), and cadmium (Cd); and total dissolved solids (TDS), which is often estimated as conductivity (or specific conductance). The electrical conductivity and TDS in the natural waters of the region are represented primarily by the cations Ca^{2+}, Mg^{2+}, sodium (Na^+), potassium (K^+), HCO_3^-, sulfate (SO_4^{2-}), and chloride (Cl^-), with much of the Na^+ and Cl^- supplied by sea salt (Welch et al. 1986). A water's organic character often includes specific organic compounds or groups of compounds that may produce toxicity (e.g., pesticides), cause color (humic acids) or provide a substrate for heterotrophic microorganisms, which demand oxygen (Table 4.1). Other chemical constituents include dissolved oxygen (DO), biochemical oxygen demand (BOD), and chemical oxygen demand (COD).

Water quality is, however, more than just dissolved chemical constituents. This is because the principal purpose of measuring, interpreting, and managing water quality is to protect aquatic resources (Table 4.1). Therefore, assessing water quality includes physical variables such as temperature (T), total suspended solids (TSS), and solids (fine sediment) deposited in the substratum. These latter measurements, along with measurements of physical habitat such as riparian vegetation, discharge, in-stream cover, and substratum quality, are not traditionally thought of as water quality but do significantly affect aquatic life (Table 4.1).

Moreover, water quality also includes the biological constituents affecting a stream's physical conditions and chemical constituents. Periphytic algae and bacteria can create an aesthetic nuisance and alter DO or pH to such a degree that recreational uses and higher aquatic life are adversely affected. Macroinvertebrates are also an important biological attribute of the Pacific coastal ecoregion and are a valuable water quality monitoring tool (Chapters 18 and 20). Pathogenic bacteria associated with warm-blooded animal feces are detected by fecal coliform bacterial counts (FC) and gages the impact of agricultural runoff, failing septic systems, and other wastewater inputs. The total of these characteristics may be more accurately thought of as *stream quality*, the term used in this chapter.

Streams in the Pacific Northwest in general, and the Pacific coastal ecoregion in particular, range in elevation (sometimes exceeding 2,000 m), from high mountains to the sea over a relatively short distance (Naiman and Anderson 1997). Soil and bedrock types vary markedly over this elevation range, as well as among mountain ranges within the region (e.g., the coastal Cascade and Olympic mountains). Notwithstanding variation in water quality caused by such natural differences, as well as local effects of land use, the streams of the Pacific coastal ecoregion are rather unique. Because their waters are generally soft, nutrient poor, and well shaded, they are usually exceptionally clear with little or no periphytic mat development except in cases where riparian canopies are removed and wastewater or nonpoint source runoff cause nutrient enrichment. Pacific coastal streams are typically oligotrophic (Chapter 7) due largely to a combination of resistant bedrock, relatively thin or glacially compacted soil types, and high rainfall. These factors result in high dilution of ions that are slowly dissolved from bedrock and leached from soils (USGS 1993).

This chapter describes how both natural and anthropogenic factors characterize the region's

TABLE 4.1. Physiochemical water quality constituents; common symbols and acronyms, and their ecological significance.

Constituent	Acronym/symbol	Ecological significance
Acid neutralizing capacity (Alkalinity)	ANC	Determines pH and buffering capacity against pH change and biological effect.
Ash-free dry weight	AFDW	A measure of algal quantity that involves combustion of the organic material in an oven at 450°C. AFDW is usually a small percentage of the dry weight (0.5%) in many algae, although it may compose a larger portion of the dry weight of diatoms, which have silica walls.
Best management practices	BMPs	Methods, measures, or practices selected to meet nonpoint (diffuse) source control needs. BMPs include structural and nonstructural controls and operations and maintenance procedures. BMPs can be applied before, during, and after pollution-producing activities to reduce or eliminate the introduction of pollutants from diffuse sources into receiving waters.
Bicarbonate	HCO_3^-	Equals ANC if pH < 8.3 and roughly equal in equivalence to hardness ions.
Biochemical oxygen Demand	BOD	Determines oxygen-consuming potential of water due to decomposition of organic matter; can be used to predict DO.
Cadmium	Cd	Product of industrial operations; toxic to biota.
Calcium	Ca^{2+}	Primary base cation accounting for most of water ANC and hardness.
Carbon dioxide	CO_2	Gas found at about 350 ppm (0.035%) in the atmosphere. CO_2 dissolves in water to form different dissolved inorganic carbon species including carbonic acid (H_2CO_3), bicarbonate ion (HCO_3^-), and carbonate ion (CO_3^{2-}), depending on the water's pH. CO_2 is the principal form of C that is used by plants, although HCO_3^- is also used by some species. CO_2 is released by organisms during respiration.
Carbonate	CO_3^{2-}	The dominant form of dissolved inorganic carbon in natural waters at pH values > 10.3.
Chemical oxygen demand	COD	Determines total oxygen-consuming potential of water; includes BOD.
Chloride	Cl^-	The element chlorine is the most important and widely distributed of the halogen group of elements in natural water. Its dominant form in water is Cl^-, which is usually found at low concentrations relative to other anions such as SO_4^{2-} and HCO_3^-. However, lakes and streams in coastal areas can receive significant inputs of Cl^- due to atmospheric transport (rainwater in coastal areas can contain 1–20 mg Cl^-/L). Other sources include road salting and wastewater discharges.
Copper	Cu	Product of mining, transportation, and industrial operations; toxic to biota.
Dissolved oxygen	DO	Molecular oxygen dissolved in water; critical to survival of aerobic biota.
Fecal coliform bacteria	FC	Pathogen indicator; feces of warm-blooded animals; failing septic systems or sewage.
Hydrogen ion	pH	pH is the negative logarithm of the molar H^+ concentration (i.e., $-\log[H^+]$). Determined by ANC and if CO_2 is supersaturated by decomposition or under-saturated by photosynthesis; directly affects and determines solubility of substances harmful or beneficial to biota.
Intragravel dissolved oxygen	IGDO	DO present in intragravel region of streambed; affects survival to emergence of salmonid embryos; reduced by fine sediment deposition.
Lead	Pb	Most common source is leaded gasoline; toxic to biota.

(*Continued*)

Table 4.1. *Continued.*

Constituent	Acronym/symbol	Ecological significance
Magnesium	Mg^{2+}	Second most important cation accounting for remainder of hardness.
Maximum acceptable concentration	MAC	The maximum concentration above which chronic or acute toxicity is observed.
Molecular weight	MW	Molecular weight in grams of any particular compound.
Nephelometric turbidity Units	NTU	Standard measurement of turbidity involves the use of a nephelometer, which measures the intensity of scattered light from particles in a water sample. A formazin suspension of 40 NTU is used as an arbitrary standard for turbidity measurements.
Nitrate–nitrogen	NO_3^-–N	The predominant form of dissolved nitrogen in oxygenated waters. In addition to ammonium–nitrogen (NH_4^+–N), NO_3^-–N is the form used by aquatic plants for growth.
Nitrogen	N	Nutrient; can limit aquatic production in the form of ammonium and/or nitrate ion; often referred to by its constituent form (ammonium, nitrite, nitrate, or reduced nitrogen); fertilizers and waste water; can cause enrichment.
Percent total impervious area	% TIA	A standard measure of basin development or urbanization. Water does not infiltrate impervious surfaces, which include roof tops, roads, and parking lots.
Phosphorus	P	Nutrient; usually limits aquatic production in the form of phosphate ion; can be represented as soluble reactive phosphorus (SRP) or total phosphorus (TP); fertilizers, detergents, and waste water; can cause enrichment.
Potassium and sodium	K^+, Na^+	The two most abundant members of the alkali-metals group measured in natural water. Sources of these elements include igneous and sedimentary rocks.
Soluble reactive phosphorus	SRP	Dissolved inorganic phosphorus is present in water as phosphate (PO_4^{3-}) and is measured as SRP. SRP is the form of phosphorus that is available for plant uptake.
Specific conductance (conductivity)	COND	Non-specific measure of electrical resistance; proportional to quantity of cations and anions; measure of total dissolved substances.
Sulfate	SO_4^{2-}	The fully oxidized (S^{6+}) form of sulfur complexed with oxygen that is found dissolved in water. Sources of sulfate to surface waters include igneous and sedimentary rocks (in the form of reduced metallic sulfides), combustion products of fossil fuel burning (via the oxidation of SO_2, see below), and fertilizers.
Sulfur dioxide	SO_2	The dominant oxide of sulfur present in the atmosphere. SO_2 is converted to sulfuric acid (H_2SO_4) by photochemical or catalytic processes in the atmosphere, which when deposited on the land surface and in water bodies is called "acid rain."
Temperature	T	Critical to survival of cold-water organisms found in streams and rivers.
Total dissolved solids	TDS	Gravimetric measure of ions that cause conductance.
Total phosphorus	TP	Total phosphorus includes all soluble and particulate forms of inorganic and organic phosphorus. TP ranges from approximately 5 mg/L in sewage effluent.
Total suspended solids	TSS	Gravimetric measure of suspended sediment load.
Turbidity	NTU	Measure of suspended sediment using light reflectance and absorption.
Zinc	Zn	Product of urban runoff from galvanized metals, and motor oil, and tires; toxic to biota.

Hem 1989, Welch 1992, Novotny and Olem 1994, Sawyer 1994.

stream quality. In addition, specific cases illustrate how water quality affects stream periphyton and vice versa. Symbols and acronyms used are defined in Table 4.1.

Acid Neutralizing Capacity, pH, and Hardness

The acid neutralizing capacity (ANC), or alkalinity, of water is a measure of its *buffer capacity*, or resistance to a change in pH. Most of the ANC of natural waters is caused by bicarbonates, carbonates, and hydroxides (OH^-); the relative amounts of which are dependent on pH (Stumm and Morgan 1973). At pH less than 8.3 the ANC of natural waters is composed almost exclusively of HCO_3^-. At higher pH (>8.3), CO_3^{2-} increases and composes a greater portion of ANC. In Pacific coastal streams, ANC is usually quite low (<1 meq/L) and composed primarily of HCO_3^-. Water hardness is caused by divalent metallic cations, principally Ca^{2+} and Mg^{2+}, which are often associated with the ANC anions, HCO_3^- and CO_3^{2-}. Corresponding to low ANC, Pacific coastal streams are also relatively soft (<1 meq/L or 50 mg/L as calcium carbonate, $CaCO_3$).

Comparing ANC and pH levels in the Cascade and Olympic Mountains of Washington illustrates the effect of bedrock type on streams. Bedrock in the Olympic Mountains is of marine sedimentary origin, while that in the Cascades is mostly igneous (granitic), which is more resistant to weathering (WDNR 1978). As a result, ANC and pH levels in water draining from the Olympic mountains are much greater than in water from the Cascade mountains. The mean ANC of headwater lakes in the Olympics (0.40 meq/L) is ten times that of the Cascades (0.035 meq/L), and in some lakes in the Alpine Lakes Wilderness Area of the Cascades it is as low as 0.02 meq/L (1 mg/L $CaCO_3$). Average hardness in the Cascade lakes is 0.052 meq/L compared to 0.510 meq/L in the Olympic Mountain lakes (Welch et al. 1986). Streams in the Cascades often have natural contents of ionic constituents similar to the lakes from which they originate (Dethier 1979). Because Cascade lakes and streams are often completely composed of snow melt, they are extremely sensitive to acid precipitation caused by increases in sulfur dioxide (SO_2) (Welch et al. 1986). Although anthropogenic SO_4^{-2} is present in Cascade waters, the quantity is insufficient to cause significant acidification. In fact levels of SO_4^{-2} decreased following closure of a large smelter which contributed 70% of the atmospheric SO_2 emissions in the Puget Sound area (Welch et al. 1992a).

Acid neutralizing capacity, pH, and hardness naturally increase as streams flow from high to low elevation because of the increased time the water is in contact with bedrock. For example, ANC in representative western Washington waters increases from 0.40 and 0.035 meq/L observed in the headwater lakes (see above) to 0.72 and 0.20 meq/L downstream in rivers from the Olympics (Chehalis and Quinault rivers) and Cascades (Skagit, Stillaguamish, Nisqually, and Snohomish rivers), respectively (USGS 1975). Dethier (1979) observed a doubling in hardness in a Cascade Mountain creek within 8 km of the source, a headwater lake at 700 m elevation.

Streams acquire their solutes through dissolution reactions with naturally acidic precipitation. Precipitation falling on Pacific coastal watersheds has a pH of about 5 because it contains carbon dioxide (CO_2) and small amounts of sulfate (SO_4^{2-}) ion (Charlson and Rodhe 1982). Weathering (dissolution) processes can increase due to increases in the acidity of precipitation. Anthropogenic sources of SO_2 in western Washington have lowered precipitation pH below 5 in the past, but this effect has largely diminished since the closure of the region's only major smelter (Vong et al. 1988).

Acid neutralizing capacity may also increase significantly in streams and lakes that drain urbanized watersheds because more surface area of relatively erodible sources of Ca^{2+} (i.e., concrete) is exposed to naturally acidic precipitation, and sometimes, to more acidic precipitation from anthropogenic sources of strong acids (nitrate (NO_3^-) and SO_4^{2-}). Edmondson (1994) has shown that ANC in Lake Washington has nearly doubled in the past thirty years, apparently as a result of an increased ANC in tributary streams draining urbanized portions of its watershed. In contrast, ANC in the Cedar River, which drains a largely undeveloped subwatershed, has remained rather constant over the last thirty years.

There are no water quality standards for ANC and hardness because they have little or no direct effect on stream invertebrates or fish. However, ANC buffers against a change in pH, which does directly affect aquatic animals. Ecologically, an acceptable pH range is between about 6 and 9 (USGS 1993). Stream water of the Pacific coastal ecoregion is highly susceptible to pH change from acidification and photosynthesis because it is typically soft and low in ANC. Acidification from humic acids and reduced iron occurs naturally and results in lake and stream pH as low as 4.3 (Zasoski et al. 1977, Welch et al. 1986). In these cases, the acid depletes ANC. In the case of high pH caused by photosynthesis (because of high rates of periphyton productivity stimulated by nutrient enrichment), ANC is not removed. Algae simply shift the carbonate buffering system out of equilibrium with the atmosphere during the daylight hours by removing CO_2, which replaces HCO_3^- with OH^-. At night, when CO_2 is no longer depleted by photosynthesis, the system returns to equilibrium, and pH decreases (Welch 1992). This process can raise pH well above 9 during daytime, a level which violates water quality standards and may adversely affect aquatic animals. Cases where this occurs in streams of the Pacific coastal ecoregion are discussed later.

Cations, Anions, and Conductivity

The conductivity of a solution is a measure of its ability to carry an electrical current. Conductivity varies both with the number and types of ions that the solution contains. It is much easier and faster to measure specific conductivity than to analyze for specific anions and cations. For this reason, conductivity measurements are used extensively for monitoring surface waters. Conductivity provides baseline information against which changes in water quality can be detected and also is used to trace the movement of substances discharged to a water body.

Ionic content and electrical conductivity, like ANC, are also low in Pacific coastal streams. Major cations and anions, which account for nearly all the conductivity and total dissolved solids (TDS) in Pacific coastal streams, are Ca^{2+}, Mg^{2+}, Na^+, HCO_3^-, Cl^-, and SO_4^{2-}. The trends with elevation and bedrock type are similar to that for ANC, hardness and pH. Total cations plus anions in eleven Cascade and four Olympic mountain headwater lakes, where many streams originate, averaged 0.17 and 1.10 meq/L respectively, and conductivity averaged 9 and 57 µmhos/cm, respectively (Welch et al. 1986).

Total dissolved solids measures essentially the same constituents as conductivity (i.e., dissolved ions) but also includes dissolved, uncharged materials (e.g., silicic acid, organic substances). Both are nonspecific measures of dissolved solids including natural constituents as well as pollutants. In regions of high rainfall and relatively insoluble rocks, TDS concentrations in surface waters may be as low as 25 mg/L, whereas slightly saline waters may contain 1,000 to 3,000 mg/L (Hem 1970). In forested coastal streams and rivers of the Pacific Northwest, TDS concentrations are generally low. Typical TDS values range from a median TDS less than 40 mg/L in the Queets River, Washington, to a median TDS less than 100 mg/L in the Klamath River, California (Table 4.2) (USGS 1993).

Conductivity during baseflow conditions increases with urban development as indicated by basin imperviousness in the Puget Sound basin, Washington (Figure 4.1) (Olthof 1994, Bryant 1995). Because of the effects of dilution, conductivity does not increase with urban development during storm flow conditions. Thus, stream water conductivity may be an indicator of soluble substances, including nutrients and soluble metals, that enrich groundwater in urbanized areas. Conductivity also serves as an indicator of soluble reactive phosphorus (SRP) and nitrate nitrogen (NO_3-N) for periphytic algal growth (Biggs and Price 1987, Jacoby et al. 1991). Conductivity is conservative—not readily altered by biological or other processes.

TABLE 4.2. Median concentrations of total dissolved solids (TDS) and total suspended solids (TSS) in representative rivers of the Pacific coastal ecoregion for 1987–1989.

River	TDS (μg/L)	TSS (μg/L)
Skagit (WA)	35	52
Queets (WA)	37	4
Chehalis (WA)	63	5
Puyallup (WA)[a]	50	96
Nehalem (OR)	60	11
Siuslaw (OR)	50	7
South Umpqua (OR)	85	12
Rogue (OR)	80	12
Klamath (CA)	100	18
Stikine (AK)[a]	58	300
Copper (AK)[a]	94	555
Talkeetna (AK)	58	15

[a] Glacially fed rivers.
Data from USGS 1993.

As such, it represents the maximum availability (i.e., inflow concentration) of nonconservative substances such as SRP and NO_3-N (which are naturally low in streams during base flow due to uptake by periphyton) entering the stream via groundwater or hyporheic flow.

Conductivity (or TDS) can be detrimental to aquatic animals if levels reach near brackish water conditions (i.e., thousands of μmhos/cm). However, there are no specific water quality standards for ionic content, conductivity, or TDS, and these constituents do not represent major water quality problems for streams of the region.

Nutrients

Nutrients are chemicals such as carbon (C), nitrogen (N), phosphorous (P), sulfur (S), calcium, magnesium, potassium (K), iron (Fe), manganese (Mn), and sodium that are essential to the growth of living organisms. Ionic forms of these nutrients (e.g., NO_3^-, Ca^{2+}, Mg^{2+}) contribute to the conductivity of water as described above.

Nitrogen or phosphorous, and sometimes both, usually limit autotrophic production in fresh water (Welch 1992). In Pacific coastal streams and lakes, P is typically in much shorter supply than N because regional bedrock yields

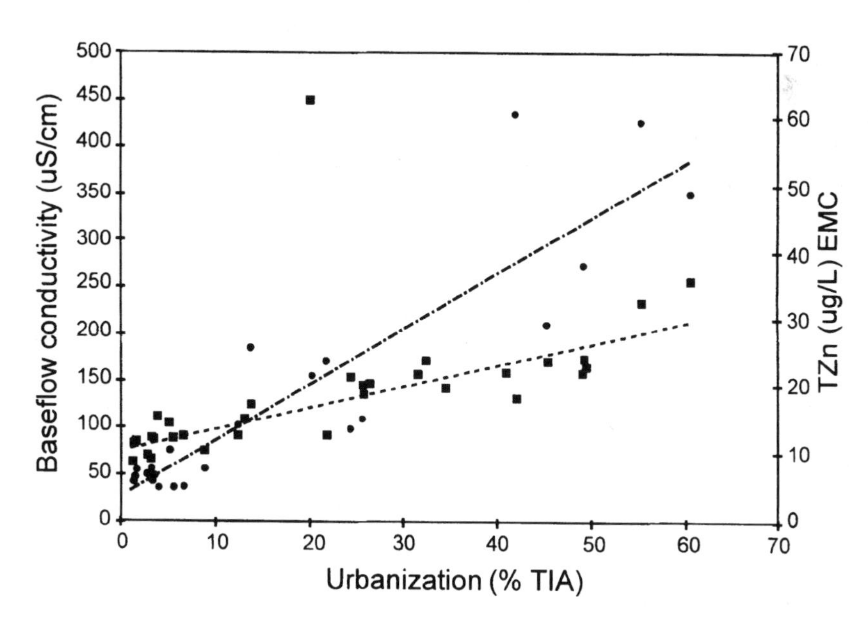

FIGURE 4.1. Relationship of urbanization (as % total impervious surface, TIA) with baseflow conductivity (■, --------, r = 0.91) and total zinc (TZn) event mean concentrations (EMC; ●, –·–·–, r = 0.91) during large storms (rainfall > 1.9 cm) in 22 Puget Sound lowland streams in Washington (modified from Bryant 1995).

TABLE 4.3. Ratios of soluble inorganic N and P in lakes and streams (winter baseflow) of Puget Sound, Washington during periods of winter nongrowth. Periphyton utilize N and P at a ratio of about 7:1 by weight.

Stream/lake	N/P Ratio
Stream	
Issaquah Creek	168
Stillaguamish River	38
South Fork Tolt River Upper	170
Thornton Creek	50
Kelsey Creek	25
Bear Creek	78
Covington Creek	104
Lake	
Sammamish	27
Washington	20
Pine	23
Long (Kitsap County)	86
Chester Morse	63

Data from Welch 1992.

low amounts of P. Ratios of N:P are typically greater than 20:1 by weight (Table 4.3). Because of the typically low concentrations of P, the region's streams are considered naturally oligotrophic and highly sensitive to enrichment. Just as ANC, hardness, and total ion content are lower in high-elevation streams and lakes than lowland waters, the same is true for N and P. Table 4.4 shows variations in N and P concentration at upstream and downstream locations in representative western Washington rivers. Increases in N and P in the Puget Sound lowlands are due partly to urbanization and agriculture in the lower portions of the watersheds. Total phosphorus (TP) increases more than soluble N because a large fraction of TP is sorbed to sediment particles (Welch 1992). Yields of N and P from developed areas are often 5 to 10 times greater than from forested areas (Reckhow and Chapra 1983). As with other water quality characteristics, TP in Puget Sound lowland streams during winter storms is related to the percentage of impervious surface area within urbanized watersheds (Bryant 1995). Stormwater runoff from urbanized areas, nonpoint source runoff from agricultural zones, and wastewater from combined sewer overflows or failing septic systems in unsewered areas are significant sources of nutrients for Pacific coastal streams. The significance of these nutrient inputs, especially P, in causing nuisance levels of periphytic algae and concomitant water quality problems is discussed later in this chapter (see also Chapters 14 and 15).

Riparian vegetation along streams also modifies cation and nutrient content in watershed runoff. Calcium and Mg^{2+} transport from uplands to a stream with 30% riparian forest in an agricultural watershed in the southeastern United States was reduced 39% and 23%, respectively by the vegetation, while N and P were reduced 68% and 30%, respectively (Lowrance et al. 1984). Gaseous loss of N via denitrification was believed to be the principal mechanism for the reduction in N transport.

As yet, there are no specific standards or criteria for P or N that pertain to nuisance levels of algae in Pacific coastal streams. However, lake-P standards or criteria on an ecoregion basis currently exist in several states, including Washington (Schlorff 1995).

TABLE 4.4. Soluble N and total P increases downstream during winter low flow in western Washington streams. Drainage areas are upstream of the sites.

Stream	Site	Drainage area (km^2)	Soluble N (μg/L)	Total P (μg/L)
Stillaguamish	Darrington (upstream)	213	335	13
	I-5 Bridge (downstream)	1,443	420	40
Nooksack	Deming (upstream)	1,544	410	30
	Brennan (downstream)	2,046	600	55
Skagit	Newhalem (upstream)	3,043	130	13
	Sedro Wooley (downstream)	7,770	273	23

Data from USGS 1975.

Although the actual establishment of nutrient (usually P) standards or criteria for lakes is a recent development, the need to control cultural eutrophication has been justified and the criteria documented for 25 years (Welch 1992). Recognizing the problem of cultural eutrophication and documenting the appropriate criteria have only begun in streams. Furthermore, a delay between recognizing problems and setting standards, similar to that for lakes, may also occur for streams. Progress has been slow because the nutrient-algae biomass relationships in streams are considerably more complicated than for lakes (Welch et al. 1988). In stream water, total nutrient concentration occurs largely in detritus rather than in the living periphytic algal mat, whereas in lakes, most of the total nutrient content in water is incorporated in phytoplankton. Thus, relationships of TP-chlorophyll *a* that apply to lakes are not as valid for streams.

Temperature

One of the most significant water quality parameter of Pacific coastal streams that affects aquatic animals is temperature. Temperature affects rates of chemical and biological processes and is critical to the survival, metabolism, reproduction, growth and behavior of salmonid fishes as well as other biota (Lantz 1970, Beschta et al. 1987, Hicks et al. 1991). The effects of temperature increases from heated effluents are often negative (because they are long term), excessive, and associated with toxic chemicals (Brown and Brazier 1972). Nevertheless, within certain limits, increases in temperature stimulate production of algae and other microorganisms, invertebrates, and fishes (Bisson et al. 1988, Hicks et al. 1991, Welch 1992). The effect may be positive where temperature increases result in summer maximums that do not exceed an organism's optimum growth (Welch 1992). During warm months (April–October), the maximum weekly temperature is determined by adding to the physiological optimum temperature (usually for growth) a factor calculated as one-third of the difference between the lethal temperature and the optimum temperature for sensitive species (US EPA 1986). For Washington state Class AA waters (extraordinary water quality), the standard is 16°C for salmonid fishes (WDOE 1988). This temperature falls within the range of the annual mean maximum stream temperatures throughout the Pacific coastal region (from about 15°C in the north to about 25°C, and sometimes higher, in the south) (Naiman and Anderson 1997).

Physical parameters that affect water temperature include: riparian vegetation, groundwater–hyporheic water interactions, tributary inflow, water depth, and air temperature (Sullivan et al. 1990). In particular, removal of riparian vegetation from timber harvesting or other land-clearing activities (urban development or agriculture) constitutes a significant alteration in stream-riparian systems.

Vegetation in the riparian zone helps regulate the microclimate of stream-riparian ecosystems (Castelle et al. 1994, Brosofske et al. 1997). Reducing vegetative cover along streams increases incident solar radiation reaching the stream. For a given rate of solar radiation, the temperature increase is directly proportional to surface area of the stream and inversely proportional to stream discharge (Sullivan et al. 1990). Increased solar energy input results in higher maximum summer temperatures and larger diurnal fluctuations, especially in small streams (Sullivan et al. 1990). On larger streams or rivers, riparian vegetation may shade only the edges and secondary channels. Nevertheless, these areas serve as important refuges for aquatic life.

Several studies indicate the importance of riparian buffers in maintaining stream water temperature and water quality (Karr and Schlosser 1977, Gregory et al. 1991, Binkley and Brown 1993, Castelle et al. 1994). In general, streams with adequate shade are cooler in summer and warmer in winter than streams lacking riparian buffers. The effectiveness of buffers decreases with increasing stream size (Brown and Brazier 1972). Small streams, with low flow rates and high width-to-depth ratios, are at greatest risk from temperature problems but are also the easiest to shade. The riparian buffer width required to maintain natural

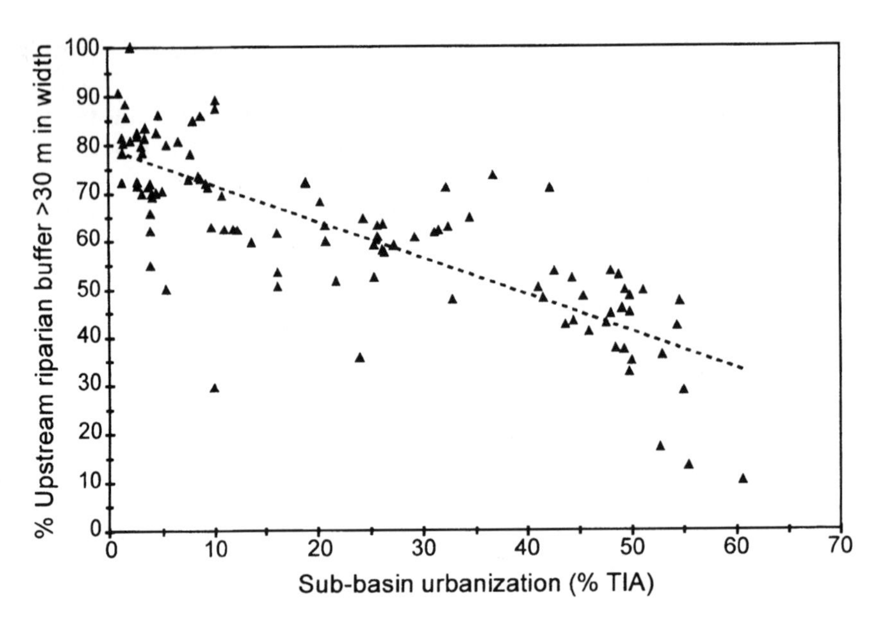

FIGURE 4.2. Relationship between percentage of riparian buffer area (width > 30 m) upstream from 120 stream segments and urbanization (% TIA) for 22 Puget Sound lowland streams in Washington (r = 0.83) (modified from May et al. 1997).

temperatures in Pacific coastal streams is a minimum of 30m, but larger (or smaller) buffers may be more appropriate depending on the management situation (Castelle et al. 1994). Brosofske et al. (1997) found that 46m was the minimum buffer width necessary to maintain a complete, unaltered riparian microclimate environment along small western Washington streams. Figure 4.2 shows a trend toward decreases in riparian buffer widths with increases in urbanized area in the Puget Sound lowlands.

Biochemical Oxygen Demand and Dissolved Oxygen

Biochemical oxygen demand is a surrogate for dissolved and particulate organic carbon (DOC and POC) that is available for utilization by microorganisms (Chapter 6). More specifically, BOD is a measure of the rate at which DO is demanded by the microbial community to aerobically oxidize organic matter. As such, BOD is usually determined over five days (BOD_5), and then used to estimate the ultimate demand (BOD_u; Figure 4.3). From the BOD_5 and exponential rate of oxygen demand, BOD_u can be calculated and used with the demand rate to predict the magnitude of DO depletion in a stream resulting from a source of organic matter input (Viessman and Welty 1985).

While BOD_5 in nonpoint source runoff is typically higher than in natural streams (i.e., 10–20 mg O_2/L versus 1–2 mg O_2/L), resulting DO depletion following storms with high runoff is not a recognized problem. This is because regional storms usually occur in winter when rates of DO demand are low and flows and DO saturation levels are high. Problems associated with low DO may occur with increased organic material input, however, within interstitial spaces in the stream substratum where intragravel flows have decreased because of filling of spaces with fine sediment. Reduction of intragravel DO, relative to water column DO, was observed in Puget Sound lowland streams as development increased (Figure 4.4). How this problem relates to sediment deposition is discussed later.

The amount of oxygen dissolved in streams is determined by the gas-absorbing capacity of

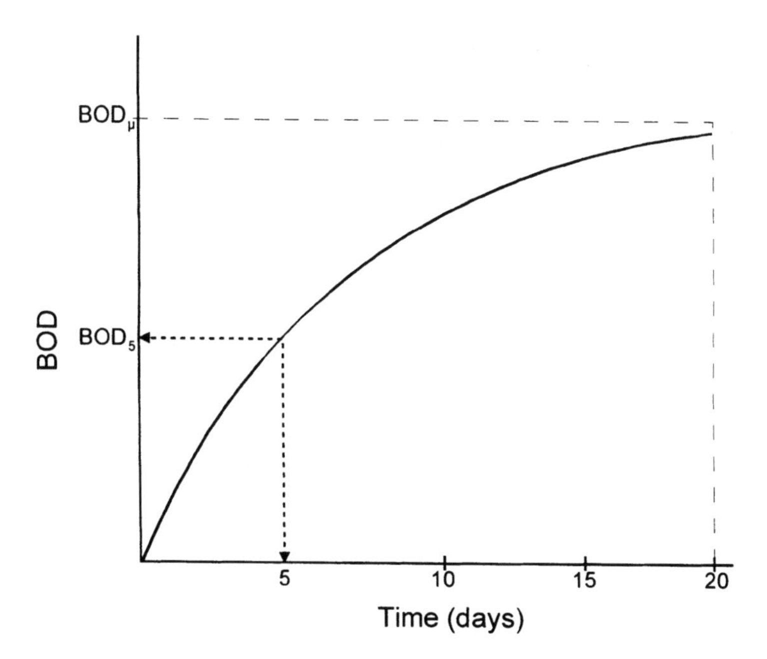

FIGURE 4.3. Biochemical oxygen demand (BOD) versus time, denoting ultimate (BOD_u) and standard 5-day BOD (BOD_5).

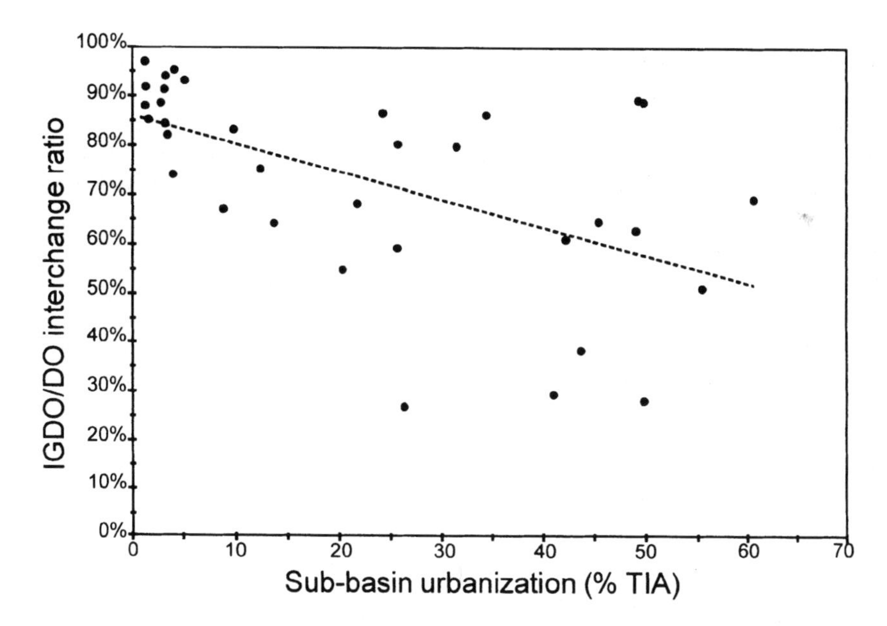

FIGURE 4.4. Relationship between ratio of intragravel dissolved oxygen (IGDO) to water column dissolved oxygen (DO) and urbanization (% TIA) in 22 Puget Sound lowland streams in Washington (modified from May et al. 1997).

the water, which is primarily dependent on temperature and turbulence and, to a lesser extent, atmospheric pressure. At a specific water temperature, the highest possible DO content is referred to as *saturation*. As temperature increases, the gas-absorbing capacity of water decreases and the corresponding DO saturation decreases. The amount of DO in stream water can be expressed as percent of saturation thus accounting for temperature dependence:

$$\% \text{ Saturation} = 100 * \left[\text{Actual DO Concentration}\left(\text{mg/L}\right)/\text{Saturation}\left(\text{mg/L}\right)\right]$$

Dissolved oxygen concentrations vary depending on the physical, chemical, and biological conditions of individual streams. In general, DO concentrations vary spatially and change over time, both on a diurnal and seasonal basis. Reaches with high photosynthetic activity become supersaturated during daylight hours, but DO can decrease during the night as a result of respiration. Chapra and DiToro (1991) reported that diurnal variations in DO are particularly important in small streams (high substratum to volume ratio) with high primary production. In turbulent, well-mixed cascades or riffles, DO is normally near saturation and may even become supersaturated. High concentrations of decomposing organic matter, low flows, or elevated temperatures can reduce DO levels in other stream reaches (including off-channel pools and wetlands). Inflows of deoxygenated groundwater can create pockets of low DO concentration until mixed with oxygenated surface waters.

Forested streams and rivers are typically at or near oxygen saturation because of generally turbulent conditions and low primary production (Naiman et al. 1992, Chapter 7 this volume). For example, the median DO concentration in four coastal rivers of Oregon (Nehalem, Siuslaw, South Umpqua, and Rogue) was greater than 10mg/L during water-years 1987–1989 (USGS 1993). Anthropogenic activities (e.g., urbanization, agricultural and forestry practices, waste discharge), however, have a detrimental impact on stream temperature, DO concentration, and aquatic biota dependent on natural conditions. Low concentrations of DO sharply reduce habitat suitability for fish and invertebrates (Beschta et al. 1987). In addition to reducing DO concentrations, higher-than-normal water temperatures alter the fish species composition, favoring warm-water species over salmonids (Reeves et al. 1987, Chapter 9 this volume). Higher-than-normal water temperature in winter can be harmful to salmonids by causing more rapid egg development and emergence. Emerging fry are then subject to late-winter and early-spring high flows and may suffer increased mortality or washout (Holtby 1988). Higher summer temperatures also lead to higher juvenile mortality caused by thermal metabolic stress, competition for cool-water refugia, and reduced DO levels (Bisson et al. 1988). Reduced concentrations of DO adversely affect the swimming performance of both juvenile and adult salmonids, even causing a halt to migration at lower levels (Bjorn and Reiser 1991).

Agricultural activities and urbanization result in increased stream temperatures and reduced DO levels as a result of the clearing of riparian vegetation, inputs of organic pollutants, and stormwater runoff (Karr and Schlosser 1977, Smith and Eilers 1978, Greb and Graczyk 1993). In most cases, the effects of agricultural land use alter DO more than urbanization (Karr and Schlosser 1977, Williamson 1985, Olthof 1994). However, sewage inputs from combined sewer overflows during major storms and sediment inputs from stormwater runoff reduce primary production through shading and inhibit reaeration (Alonso et al. 1975).

Forest practices also significantly impact stream temperatures and DO concentrations. In some cases, clear-cutting and slash burning the watersheds of small streams resulted in DO decreases of more than 40% lasting up to three years after logging (Ringler and Hall 1975). Such decreases mainly result from the loss of shading from riparian vegetation and the influx of large amounts of organic matter to the channel, which increase BOD. Organic inputs also produce nuisance biomass levels of filamentous bacteria such as *Sphaerotilus*, a bacterium commonly known as "sewage fungus" (Curtis and

Harrington 1971). This problem usually occurs downstream from sources of low molecular weight DOC. Relatively low BOD_5 concentrations (3–5 mg/L) have been shown to produce extensive mats of *Sphaerotilus* if the DOC is largely low molecular weight compounds such as sugars (Warren et al. 1964), dairy waste (Quinn and McFarlane 1989), landfill leachate (Brende 1975), or pulp mill waste (Ormerod et al. 1966). However, naturally complex DOC sources such as leaf leachate usually do not result in mats of filamentous bacteria (Cummins et al. 1972).

TABLE 4.5. Total metal concentrations in representative Puget Sound lowland streams following winter storms.

Stream	% TIA[a]	Total Copper (μg/L)	Total Zinc (μg/L)	Total Lead (μg/L)
Thornton	61	13	71	3
Kelsey	47	9	40	ND[b]
Juanita	45	6	66	ND
Percival	22	3	33	ND
Little Bear	14	6	25	ND
Big Bear	11	10	5	ND
Covington	8	ND	5	ND
Rock	3	ND	9	ND

[a] % TIA = % total impervious area.
[b] ND = undetectable or below detection limit.
Data from Bryant 1995.

Metals

Except in urbanized areas and in sources of leachate water, such as mine tailings, old landfills, or industrial and domestic discharges, baseline concentrations of metals are usually low or undetectable in Pacific coastal streams. Elevation or bedrock type, generally do not influence metal concentrations. However, atmospheric deposition and lower pH occasionally result in atypical situations with higher metal concentrations upstream than downstream (Dethier 1979). Metals are usually reported as total acid extractable, however, this measure is unrelated to the fraction available for animal or plant uptake, or *bioavailability*. Concentrations of the more common toxic metals (copper, lead, and zinc) are relatively low in representative Puget Sound lowland streams (Table 4.5). Other toxic metals, such as mercury, cadmium, and nickel, may occur in ecologically significant concentrations and are usually associated with industrial wastewater. Aluminum, another toxic metal, is associated with effects of acid precipitation as well as industrial discharges.

To determine the significance of metal concentrations for aquatic animals, levels are often compared to published water quality criteria (US EPA 1986). Maximum acceptable concentrations (MACs) are developed from continuous, flow-through bioassays in which metals are measured as the free ion. Sometimes filterable (dissolved) metal is reported and may include organically complexed metals, which are usually biologically unavailable. However, compounds with low molecular weight (MW) are biologically more available than those with high molecular weight (Davies 1986). In the case of Cu, toxic effects are clearly related to the free ion (Cu^{2+}), which is measured with a specific ion electrode (Buckley 1993).

Another factor that affects the toxicity of metals to aquatic organisms is water hardness (primarily the result of Ca^{2+} and Mg^{2+}), which reduces the toxic effect of a metal through competitive binding at the gill surface. As a result, as hardness increases, concentrations at which metals are toxic also increase (Welch 1992). MACs are typically adjusted for hardness according to bioassay-determined models (WDOE 1988), however, as of 1996 there has been no adjustment for the effective free ion fraction of the reported total acid extractable concentration. MACs for the more commonly occurring toxic metals in Pacific coastal streams are generally low (Table 4.6) compared to other regions. Toxic effects at concentrations above such low MACs indicate that animals in the low-hardness Pacific coastal streams are susceptible to metal toxicity resulting from human activities in the watershed.

Metal concentrations are typically higher in stormwater runoff from urbanized areas. A principal source of two of the three most common toxic metals, copper and zinc, is the corrosion of car parts and pipes. Until the early 1970s

TABLE 4.6. Maximum acceptable concentrations (MACs) for metals in low hardness (30 mg/L as $CaCO_3$) waters.

Metal	Acute (1-hour mean) (μg/L)	Chronic (4-day mean) (μg/L)
Copper (Cu)	5.7	4.2
Zinc (Zn)	42.0	37.0
Cadmium (Cd)	1.0	0.4

Data from US EPA 1986.

the source of the third, lead, was primarily leaded gasoline. Levels of metal in runoff from non-urbanized areas are typically about 100 times less than runoff from urbanized basins (Welch 1992). Data from several Puget Sound lowland streams show that while total Zn concentration increases with urbanization (Figure 4.1), total Cu, Zn, and Pb are frequently undetectable (Table 4.5). However, even when present in streams with highly developed watersheds, in-stream concentrations do not approach levels typical of undiluted stormwater (Bryant 1995). The toxic significance of metal concentrations is often difficult to interpret, partly because the fraction that is biologically available is usually unknown. Also, synergism among several metals as well as with other water quality variables, such as low DO, high temperature, and low pH, suggests that toxicity may be occurring at less than the respective MACs. Thus, the metal's bioavailability as well as the synergistic factors above must be considered to determine if the water is toxic.

The total fractions of metals (i.e., particulate, dissolved) often occur at levels that can adversely affect aquatic organisms in streams that drain developed watersheds, yet there is little documented evidence that this occurs (Welch 1992). For example, note that the total Cu and Zn concentrations from streams in the more developed watersheds in Table 4.5 are equal to or greater than the MACs in Table 4.6. Although the dissolved (filterable) fractions (averaged 55 ± 27% [SD] of total for Zn) are mostly less than the individual MACs, the additive effect of both metals should be sufficient to cause toxicity (see below).

Industrial sites present similar scenarios where metal concentrations are high enough to cause toxicity, yet there is no observable effect on aquatic organisms. For example, concentrations of Cu, Zn, and Cd were higher in lower Mill Creek, Washington downstream from a hazardous waste site than at an upstream location (Table 4.7). Cleanup of the site during the late 1980s resulted in substantial reductions (>50%) in these metal concentrations. An additive model based on MAC criteria (US EPA 1986) with adjustment for hardness (WDOE 1988), was used to estimate the total toxicity present from the three metals (Table 4.7) (Welch 1992) and indicated that mortality should have occurred before and after cleanup alike. However, no mortalities were ever reported, even though anadromous cutthroat trout (*Oncorhynchus clarki*) and coho salmon (*O. kisutch*)passed through the affected area. Assuming there were, in fact, no mortalities, there are at least two possible explanations: (1) much of the reported total metal was biologically unavailable (complexed with high MW organic compounds, such as humic acids), or (2) migratory salmonids only passing through the "toxic" region of the stream were not exposed

TABLE 4.7. Metal concentrations in Mill Creek upstream and downstream of the Western Processing site in June 1984*.

Metal	Upstream (in μg/L)	Downstream (in μg/L)
Cadmium (Cd)	<0.1	14
Copper (Cu)	<1.0	29
Zinc (Zn)	11	1,016

* Acute toxic units (TU) for the metal concentrations at the downstream site were calculated using EPA criteria from Table 4.6 and an additive model:

$$TU = \frac{C_1}{MAC_1} + \frac{C_2}{MAC_2} + \frac{C_3}{MAC_3}$$

$$= \frac{Cd}{MAC, Cd} + \frac{Cu}{MAC, Cu} + \frac{Zn}{MAC, Zn}$$

$$= \frac{14}{1} + \frac{29}{5.7} + \frac{1016}{42} = 43.3$$

Data from WDOE 1988.

enough to be adversely affected, and juveniles did not rear in the reach due to poor physical habitat and water quality. The low abundance and diversity of invertebrates in the region (primarily pollution-tolerant oligochaetes and chironomids) indicate poor-quality habitat.

Suspended Solids

Suspended material in water is typically measured as turbidity, which measures the scattering of light by particles, or as total suspended solids (TSS), which is a gravimetric determination of the mass of solids. Turbidity is expressed in nephelometric turbidity units (NTU) whereas TSS is expressed in mg/L. Because a wider variety of particles (including the colloids responsible for color) contribute to turbidity, the two measurements may not always be directly related. Water-quality standards are based on turbidity because it is easier and quicker to measure than TSS.

Natural Pacific coastal streams are usually low in turbidity except during large flooding events (Adams and Beschta 1980). Streams and rivers fed by glacial outwash, however, often have naturally high sediment loads (Table 4.2). The concentration of fine sediment in stream channels varies considerably throughout the region depending on natural and anthropogenic factors. The amount of fine sediment (nominally, particles $<0.85\,mm$ diameter) present as suspended load or deposited on the streambed depends on the nature and magnitude of erosion processes in the watershed as well as on in-stream morphology and hydrologic regime (Chapter 2). Erosion processes include sheet and gully erosion, streambank erosion, and mass-wasting events. The latter two processes are the most significant erosion processes in natural Pacific coastal streams (Swanson et al. 1987).

Geomorphic characteristics of the stream channel significantly influence suspended sediment transport and storage. Stream gradient, in-stream structure (organic debris), streambank vegetation, and underlying geology are the prime determinants. Levels of fine sediment (measured as turbidity or TSS) are generally higher in watersheds where bedrock, soils, precipitation patterns, and topography create favorable conditions for erosion. Fine sediments are typically more abundant where land-use activities such as road construction, logging, urban development, agriculture, or grazing expose soil to erosion and increase opportunities for mass-wasting (Everest et al. 1987, Swanson et al. 1987, Hicks et al. 1991, Platts 1991). Studies by Booth (1991) indicate that most sediment input into urban streams results from in-stream sources, primarily bank erosion caused by excessive streamflows thus illustrating the importance of hydrologic variables.

Changes in sediment load upset the equilibrium between sediment entering the stream and sediment transported through the channel. Major disruptions to water quality can occur when sediment delivery substantially exceeds the transport capacity of the stream (Madej 1978). The amount of sediment deposited and in-stream turbidity increase dramatically in such situations. Urban development, especially on steep and unstable slopes, can increase the load of suspended sediments enough for effects to be observed within an entire basin. Byron and Goldman (1989) found a strong correlation between the annual average TSS concentration and the proportion of watershed development on slopes greater than 9%. Others have also confirmed the relationship of increasing human impacts and the increase in in-stream turbidity and TSS, most notably during storm events (Williamson 1985, Ryan 1991, Charbonneau and Kondolf 1993, Olthof 1994). The effects of timber harvest and roads for logging on sediment production is also well documented (Ryan 1991, Binkley and Brown 1993, Quinn and Peterson 1994).

Elevated levels of TSS may have both lethal and sublethal effects on stream biota. Salmon (*Oncorhynchus* spp.) display alterations in migrations, reduced feeding and growth, and even direct mortality in response to elevated levels of TSS (Noggle 1978, Sigler et al. 1984, Bjorn and Reiser 1991). Sublethal effects have been observed at levels as low as a few hundred mg/L and lethal effects at concentrations in excess of 1000 mg/L. Moreover, deposition of fine sediment on the streambed causes reduced survival of salmon eggs and alevins, and reduces biomass and taxa richness of invertebrates (Chapman 1988, Welch 1992, Quinn and Peterson 1994). Reduction of intragravel flow and a resultant decrease in intragravel dissolved oxygen (IGDO) levels has a significant impact on incubating salmonid embryos (Chapman 1988).

The United States EPA (1986) recommends a maximum turbidity of 25 NTU to protect designated beneficial uses in fresh waters of the United States. Most state regulations for turbidity or TSS are based on a background or baseline level (Lloyd 1987). Expected values for turbidity in natural streams are difficult to specify because turbidity depends on discharge and varies regionally (Naiman et al. 1992, Naiman and Anderson 1997). Continuous or event monitoring is usually required to detect elevated levels of turbidity caused by specific land-use activities. Sediment problems often can be prevented most efficiently by controlling erosion and runoff. In most cases, retention of forested buffer strips and the use of current best management practice (BMPs) prevent unacceptable changes in water quality due to sediment (Binkley and Brown 1993). BMPs have effectively reduced adverse effects of nonpoint source pollution (including sediment) and other land uses such as forest practices, road building, and construction (Schueler 1987, Binkley and Brown 1993, Horner et al. 1994, Quinn and Peterson 1994).

An important water-quality problem associated with sedimentation and turbidity is pollutant contamination of sediment particles. This is especially so in urban environments where runoff contains trace metals, organic compounds, and industrial chemicals (Williamson 1985). Toxic pollutants may also occur in runoff from agricultural or silvicultural areas where herbicides, fertilizers, and pesticides are applied. Most aquatic sediments are highly susceptible to pollutant uptake (Pitt et al. 1995) through several processes: dissolved organic compounds and metals sorb to sediment particles before they are deposited, metal ions exchange with calcium, magnesium, and other minerals in sediments, and precipitation reactions occur (Horner et al. 1994). Once in the sediment, pollutants generally do not decompose or change form (Pitt et al. 1995). Pollutants as metals (Zn, Pb, and Cu) and organic compounds (hydrocarbons, herbicides, pesticides, and industrial chemicals) tend to resist biodegradation and, consequently, progressively accumulate (Horner et al. 1994). Eventually such accumulations in sediment may increase several orders of magnitude greater than that in the overlying water column, thereby exceeding concentrations that cause both chronic and acute toxicity (Chapman 1973, Pagenkopf 1983, Davies 1986, Hoiland and Rabe 1992, Kiffney and Clements 1993, Clements 1994). Most aquatic organisms (including salmonid alevins and macroinvertebrates) are vulnerable to contaminated sediment because at some point in their life cycle they spend a portion of time on or near the streambed. Bioavailability, however, is the key factor responsible for biological reactions to sediment-bound toxic chemicals. Specific effects depend on other water constituents, sediment composition (particle size and mineral and organic content), and temperature. Unfortunately, mechanisms and environmental variables controlling bioavailability and toxicity remain poorly known (Pitt et al. 1995).

Water Quality and Nuisance Algae (Periphyton)

In Pacific coastal streams, periphyton assemblages (attached micro- and macroorganisms) are the stream communities most responsive to changes in water quality and habitat, and are composed mainly of primary producers (algae). In contrast to planktonic algae (the main primary producers in standing waters) which are suspended in the water column, periphyton are attached to the surface of rocks and other substrata. Periphyton may respond positively to nutrient enrichment (Stockner and Shortreed 1978, Horner et al. 1983 and 1990, Bothwell 1988 and 1989, Welch et al. 1989) causing aesthetic, water quality, and invertebrate habitat degradation (Biggs 1985, Quinn and McFarlane 1989). Periphyton are also affected by flow conditions, often proliferating during low-flow periods and decreasing when scoured by increased flows and when grazed by invertebrates, which are an important variable controlling periphyton biomass (Lamberti and Moore 1984, Jacoby 1987, Lamberti et al. 1987, Chapter 7 this volume).

Periphyton biomass is generally low, reflecting the low ambient nutrient content of Pacific coastal streams (Chapter 7). However, nuisance periphyton have been observed in streams receiving nutrients from point or nonpoint sources (Welch et al. 1988, Tanner and Anderson 1996). The potential for nuisance periphyton to proliferate is greater during low flows and in lighted reaches of small streams or in nutrient-enriched larger (more open) rivers. Periphyton in small, shaded streams are usually limited by light and are not likely to reach nuisance levels in response to nutrient enrichment (Purcell 1994). On the other hand, dense mats of the filamentous green algae *Ulothrix* are sometimes observed during spring in oligotrophic streams of the Pacific coastal ecoregion where it is not expected. Invertebrate grazing in streams is a confounding factor that has prevented definition of a relationship between limiting nutrient concentration and algal biomass analogous to those relationships defined for lakes (Gregory 1983, Welch et al. 1992). Relationships between periphyton and controlling factors such as nutrients, flow, and grazing invertebrates, as well as the effects of periphyton on water and stream quality are explored further in the following section.

Nuisance Periphyton

Nuisance periphyton usually occur as dense mats or long, oscillating strands of green filamentous algae, which result primarily from inorganic nutrient (N or P) enrichment. Filamentous green algae typically associated with nuisance conditions include *Cladophora*, *Stigeoclonium*, *Ulothrix*, and *Spirogyra* (Welch 1992). Blue-green algae, notorious for causing water quality problems in lakes, are not common nuisances in streams of the Pacific coastal ecoregion. However, the filamentous blue-green algae *Phormidium* attained nuisance levels in laboratory channels in Seattle (Walton et al. 1995) suggesting that under appropriate conditions blue-green algae can proliferate in streams or rivers. In nutrient-poor or shaded streams, diatoms typically dominate the periphyton community forming a thin, slippery mat on the surface of rocks. Representative genera include *Navicula*, *Gomphonema*, *Diatoma*, *Cymbella*, *Cocconeis*, and *Synedra* (Welch 1992). Diatoms may also create nuisance conditions due to nutrient enrichment in which case they form dense brownish mats that appear flocculent or filamentous.

Nuisance conditions are more readily determined subjectively than quantitatively. Recognition of nuisance periphyton is often instantaneous and related to the intended uses of the stream. For example, anglers instantly recognize impaired fishing quality in a stream where long, stringy filaments cover the bottom substrata. Other undesirable effects caused by dense periphyton biomass include the following (Welch 1992):

- Oxygen depletion caused by periphyton decomposition and respiration
- pH elevation caused by high rates of photosynthesis
- Production of undesirable tastes and odors in water supplies and fish flesh
- Impediment of water movement
- Obstruction of water intake structures
- Restriction of intragravel water flow and oxygen replenishment
- Impairment of recreation and aesthetic enjoyment
- Degradation of benthic invertebrate habitat.

The adverse effects of dense periphyton mats on water quality and aesthetic conditions are exemplified in the South Umpqua River in southwestern Oregon (Anderson et al. 1994). Along much of the 100 km of river studied, photosynthetic and metabolic activities of abundant algal growths (primarily *Cladophora*, *Ulothrix*, and *Spirogyra*) resulted in pH as high as 9.3 and DO as low as 1 mg/L. Furthermore, aesthetic enjoyment was impaired by strong odors and unsightly filamentous mats. These conditions are attributed to discharge to the main stem and a tributary from six wastewater

a

b

FIGURE 4.5. Nuisance biomass levels of *Cladophora* in the Umpqua River, Oregon (Photos: Chauncey Anderson, USGS, Portland, Oregon). Examples of mat formation are shown in (a) and (b).

treatment plants. Total phosphorus concentrations can exceed 200 μg/L and much of it is dissolved and readily available for algal uptake. Periphyton biomass routinely exceeded 200 g ash-free dry weight (AFDW)/m^2 during summer, with a high of 770 mg AFDW/m^2 measured in September 1991 near Roseberg (Figure 4.5) (Tanner and Anderson 1996). Similar nuisance levels of filamentous green algae (mostly *Cladophora*) also occurred over a 150 km stretch of the Clark Fork River, Montana (Watson 1990).

Results from experiments in laboratory channels and literature surveys suggest that nuisance periphyton biomass may be represented by concentrations greater than 100–150 mg chlorophyll *a* (chl *a*)/m^2 (Horner et al. 1983). In six western Washington streams percent coverage of the stream bed with periphyton (directly related to biomass) supports this nuisance threshold (Welch et al. 1988). At biomass levels greater than 100 mg chl *a*/m^2, filamentous species tend to dominate and noticeably affect the stream's aesthetic quality. However, in this case, this biomass density has not proven to deplete dissolved oxygen or suppress benthic invertebrates. Furthermore, periphyton biomass could not be related to limiting nutrient concentration (i.e., SRP) in the ambient water (Welch et al. 1988) as had been observed in the laboratory channels (Horner et al. 1983). A clear relationship between periphyton biomass and the ambient concentration of limiting nutrients is undetermined as yet, although recent studies have enhanced understanding of the interactive factors that influence the amount and type of periphyton ultimately attained in streams.

Critical Nutrient Concentration

The stimulatory effect of inorganic nutrient (particularly P) enrichment on periphyton biomass has been demonstrated in numerous studies (Stockner and Shortreed 1978, Newbold et al. 1981, Horner et al. 1983, Perrin et al. 1987, Bothwell 1989, Horner et al. 1990). In the naturally oligotrophic streams of the region, nutrient enrichment has profound effects on both periphyton biomass and composition. Perrin et al. (1987) increased periphyton biomass (from <10 to >100 mg chl *a*/m^2) by increasing SRP (from 1 to 20 μg/L) in a British Columbia stream. Filamentous green algae (*Ulothrix* and *Spirogyra*) replaced diatoms in enriched areas during the late summer low-flow period, suggesting the importance of not only increased nutrients but also low flows (and therefore little scouring), and possibly low grazing to filamentous growth. In another oligotrophic British Columbia system, periphytic diatoms in streamside channels increased ten times with SRP enrichment (from <1 to 9 μg/L) (Stockner and Shortreed 1978). Bothwell (1985, 1989) found that maximum biomass of diatoms (150 mg chl *a*/m^2) in flow-through channels (Thompson River system, British Columbia) was attained at SRP concentration of 25 μg/L, although growth rate was saturated at low SRP levels (1–2 μg/L).

Studies in laboratory channels at the University of Washington (Seattle) have focused in part on the level of nutrient enrichment needed to produce nuisance biomass levels of periphyton. Filamentous green and blue-green algae attained biomass levels exceeding 500 mg chl *a*/m^2 at enrichment levels of about 10 μg SRP/L and higher (Horner et al. 1983). In these same channels, Walton et al. (1995) produced a maximum biomass of diatoms and blue-green algae (*Phormidium*) of almost 1,000 mg chl *a*/m^2 in three weeks at an in-channel SRP concentration of nearly 15 to 20 μg/L. Periphyton composed primarily of the filamentous green algae, *Mougeotia*, required SRP of approximately 7 μg/L to attain maximum biomass (350 m chl *a*/m^2) in two weeks (Horner et al. 1990).

These studies indicate that a maximum biomass-limiting concentration of P for filamentous periphyton in Pacific coastal streams may range from 7 to 20 μg SRP/L (probably as an annual mean). However, the relationship is not linear; such a linear relationship between ambient nutrient concentrations and periphyton biomass has not been observed in natural streams either in the Pacific Northwest (Welch et al. 1988) or worldwide (Jones et al. 1984, Welch et al. 1988, Kjeldsen 1994). Biggs and Close (1989) did find a slightly stronger relationship ($r^2 = 0.53$) than previous work by correlating mean annual periphyton biomass with average SRP content over a 13-month period in New Zealand streams.

Welch (1992) suggests that direct relationships between limiting nutrients and periphyton biomass in streams are difficult to define because of difficulties in determining available nutrient concentrations, the failure of total nutrient measurements in stream water to include and represent a limitation of the living periphytic biomass, and the importance of other factors (e.g., grazing, scouring) in controlling periphyton biomass. The previously cited research indicates that stream periphyton biomass is determined by the soluble fraction of limiting nutrients in water passing over the substrata. Thus, high periphyton biomass is often associated with low soluble nutrient concentrations because the periphyton mat has already extracted it from the water. Total nutrient measured in stream water contains sloughed detritus from substrata, but not the living biomass in that substrata. Whereas in lakes, total nutrient concentration in the water includes both available nutrients and those nutrients already converted to biomass. This accounts for the strong relationship between total limiting nutrients (most of which is in algal cells) and algal biomass in lakes (Dillon and Rigler 1974). Determining available nutrient concentrations in streams is further complicated by inputs from diffuse groundwater sources (Chapter 16). Lower than expected periphyton biomass in the presence of high nutrient concentrations also occurs and may be related to grazing by invertebrates or scouring by high flows.

Grazing

Grazing by benthic herbivores is an important determinant of periphyton biomass and composition (see reviews by Gregory 1983, Lamberti and Moore 1984, and Chapter 7 this volume). Grazers effectively reduce biomass of large filamentous algal species that form nuisance mats in streams. In particular, large grazers such as the caddisfly larvae (*Dicosmoecus gilvipes*), which are common in the lower elevations of the Northwest, efficiently remove both periphytic diatoms as well as filamentous algae from rock substrata (Jacoby 1987, Lamberti et al. 1987, Feminella et al. 1989, De Nicola et al. 1990, Walton 1995). At natural densities caddisflies reduced periphyton biomass (mostly diatoms with some filamentous green algae) by 80% in an unenriched Cascade Mountain foothill stream (Jacoby 1987). In the laboratory channels, caddisflies reduced filamentous blue-green algae and diatoms to 7% of the ungrazed level in unenriched treatments and 12% in enriched (10 μg SRP/L) treatments (Walton et al. 1995). Smaller grazers such as mayflies are apparently less efficient at removing filamentous forms perhaps because of physical constraints (e.g., mouth parts too small, smothering of substrata) (Jacoby 1987, Hill and Knight 1988, De Nicola et al. 1990).

Observations in western Washington streams indicate that high densities of invertebrate grazers are associated with low periphyton biomass, even in nutrient-enriched waters (Welch et al. 1988). Similarly, an inverse relationship between invertebrate grazers and periphyton biomass was demonstrated in seven New Zealand streams (Welch et al. 1992a). In these streams, which received point-source enrichment, grazing was more important than nutrients in determining periphyton biomass.

The effects of nutrients and herbivory on stream periphyton are complex and interactive (Stewart 1987, Steinman et al. 1991, Walton et al. 1995). Grazing by caddisflies controlled periphyton biomass in laboratory channels in spite of enrichment (Figure 4.6) (Walton et al. 1995). Size of grazers appears to be important; larger grazers most effectively controlled nuisance periphyton. The large grazing minnow, *Campostoma anomalum*, reduced periphyton biomass in natural stream pools in Oklahoma even when nutrients were added (Stewart 1987).

Hydrologic Factors

The primary hydrologic factors that influence periphyton accrual are flood events (frequency and intensity), current velocity, and substrata stability (Biggs 1995). Floods are a major reset mechanism, initiating new cycles of periphyton accrual, community succession, and biomass sloughing (Fisher et al. 1982, Biggs 1995). In many cases, hydrologic factors influence periphyton biomass at least as much as nutrients and grazing. In Pacific coastal streams, hydrological influences are usually seasonal, with increased flooding and current velocity, and decreased substrate stability occurring primarily during late fall and winter storm events. During dry summers, streamflow is low and periphyton accrual (particularly filamentous forms) is enhanced by reduced current velocities and high substratum stability. Dense periphyton mats in the Umpqua River, Oregon, attain nuisance levels during such summer low-flow conditions when water temperature and light levels are high. During summer, less dilution and greater retention of nutrient inputs from wastewater and ground water also enhance periphyton development.

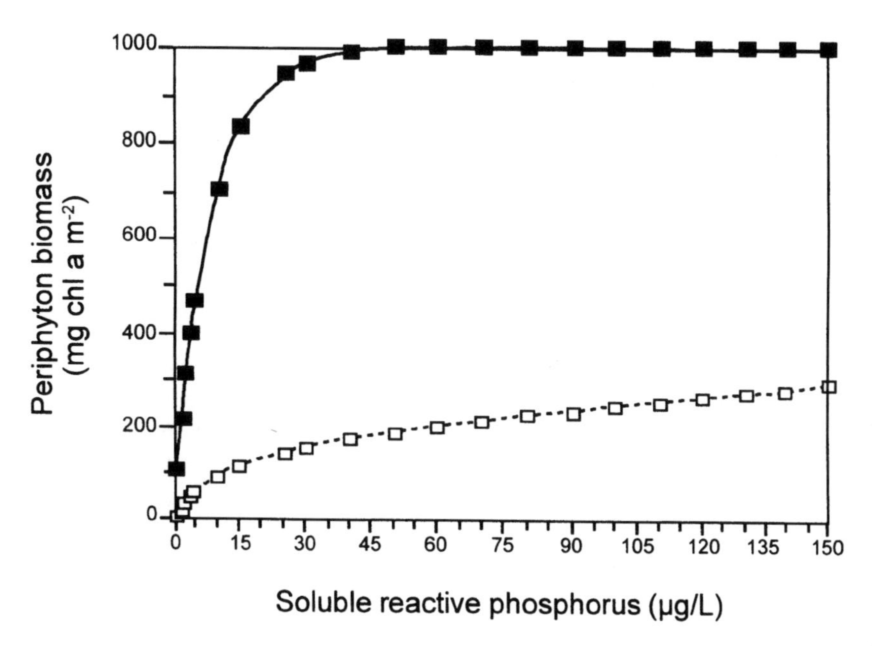

FIGURE 4.6. Model prediction of periphyton biomass with (--□--) and without (—■—) grazing by caddisfly (*Dicosmoecus gilvipes*) based on channel experiments (caddisfly density ~100/m^2, light intensity 194 μE/m^2·s, velocity 20 cm/s and SRP concentrations of 2, 6, and 10 μg/L) (Walton et al. 1995).

High current velocity experienced during flood events increases frictional shear stress and thus scours periphyton from the substrata. At current velocities above approximately 50 cm/s, loss rates of periphyton increase, substantially reducing biomass (Horner and Welch 1981). When velocity was constant during mat development in laboratory channels, periphyton losses were low even at velocities of 70 cm/s. Such low losses can be attributed to adaptation since losses increased dramatically when current velocity increased from 20 to 60 cm/s (Horner et al. 1983, 1990), suggesting that velocity change is more important than absolute velocity. Furthermore, suspended material (e.g., glacial flour, soil particles) in natural systems may enhance scouring (Horner et al. 1990). Increased sediment transport and current velocity during fall and winter flood events likely plays a role in reduction of nuisance periphyton mats that have accumulated during low-flow summer conditions in the Pacific Northwest.

There is evidence of interactions between nuisance periphyton biomass, invertebrate grazing, and scouring by floods. For example, Power (1992) found that *Cladophora* reached high biomass levels followed by senescence and detachment from substrata during winter flooding and scour in unregulated rivers in northern California. In regulated rivers, invertebrate grazing maintained persistent, low standing crops of *Cladophora*. Power (1992) suggests that chronically grazed *Cladophora* does not build up sufficient biomass to become self-limited. This example illustrates the complexity of interactive factors that can influence the occurrence of nuisance periphyton.

Management Implications

Knowledge of the factors that control filamentous periphyton is needed to predict the potential effects of nutrient enrichment in streams. Where grazing potential is high, increased nutrient inputs may not produce nuisance periphytic biomass levels. On the other hand, nutrient enrichment in the absence of grazers may result in nuisance periphyton and associated water and habitat quality degradation. It follows that protection of stream grazers through habitat enhancement (e.g., stable substrata, moderate velocity, reduced flood flows and sedimentation, adequate refuges, cool water temperatures, good water quality) may become a practical periphyton control strategy in cases where nutrient reduction is inadequate or impractical.

Literature Cited

Adams, J.N., and R.L. Beschta. 1980. Gravel bed composition in Oregon coastal streams. Canadian Journal of Fisheries and Aquatic Sciences **37**:1514–1521.

Alonso, C.V., J.R. McHenry, and J.C. Hong. 1975. The influence of suspended sediment on the reaeration of uniform streams. Water Research **9**:695–700.

Anderson, C.W., D.Q. Tanner, and D.B. Lee. 1994. Water-quality data for the South Umpqua River basin, Oregon, 1990–1992. U.S. Geological Survey Open-File Report 94–40. Portland, Oregon, USA.

Beschta, R.L., R.E. Bilby, G.W. Brown, L.B. Holtby, and T.D. Hofstra. 1987. Stream temperature and aquatic habitat: Fisheries and forestry interactions. Pages 191–232 *in* E.O. Salo and T.W. Cundy, eds. Streamside management: Forestry and fisheries interactions. Institute of Forest Resources Contribution Number 59, University of Washington, Seattle, Washington, USA.

Biggs, B. 1985. Algae, a blooming nuisance in rivers. Soil and Water **21**:27–31.

Biggs, B.J.F. 1995. The contribution of flood disturbance, catchment geology and land use to the habitat template of periphyton in stream ecosystems. Freshwater Biology **33**:419–438.

Biggs, B.J.F., and M.E. Close. 1989. Periphyton biomass dynamics in gravel bed rivers: The relative effects of flows and nutrients. Freshwater Biology **22**:209–231.

Biggs, B.J.F., and G.M. Price. 1987. A survey of filamentous algal proliferations in New Zealand rivers. New Zealand Journal Marine and Freshwater Research **21**:175–191.

Binkley, D., and T.C. Brown. 1993. Forest practices as nonpoint sources of pollution in North America. Water Resources Bulletin **29**:729–740.

Bisson, P.A., J.L. Nielsen, and J.W. Ward. 1988. Summer production of coho salmon stocked in Mount St. Helens streams 3–6 years after the 1980 eruption. Transactions of the American Fisheries Society **117**:322–335.

Bjorn, T.C., and D.W. Reiser. 1991. Habitat requirements of salmonids in streams. Pages 83–138 *in* W.R. Meehan, ed. Influences of forest and rangeland management on salmonid fishes and their habitats. American Fisheries Society Special Publication Number 19, Bethesda, Maryland, USA.

Booth, D.B. 1991. Urbanization and the natural drainage system. Northwest Environmental Journal **7**:93–118.

Bothwell, M.L. 1985. Phosphorus limitation of lotic periphyton growth rates: An intersite comparison using continuous-flow throughs (Thompson River system; British Columbia). Limnology and Oceanography **30**:527–542.

——. 1988. Growth rate responses of lotic periphytic diatoms to experimental phosphorus enrichment: The influence of temperature and light. Canadian Journal of Fisheries and Aquatic Sciences **45**:261–270.

——. 1989. Phosphorus-limited growth dynamics of lotic periphytic diatom communities: Areal biomass and cellular growth rate responses. Canadian Journal of Fisheries and Aquatic Sciences **46**:1293–1301.

Brende, H.B. 1975. Effects of sanitary landfill leachate in Issaquah Creek. Masters thesis. Department of Civil Engineering, University of Washington, Seattle, Washington, USA.

Brosofske, K.D., J. Chen, R.J. Naiman, and J.F. Franklin. 1997. Harvesting effects on microclimatic gradients from small streams to uplands in western Washington. Ecological Applications **7**:1188–1200.

Brown, G.W., and J.R. Brazier. 1972. Controlling thermal pollution in small streams. US Environmental Protection Agency Report EPA R2-72-083. Prepared for Office of Research and Monitoring, US Environmental Protection Agency, Washington, DC, USA.

Bryant, J. 1995. The effects of urbanization on water quality in Puget Sound lowland streams. Masters thesis. Department of Civil Engineering, University of Washington, Seattle, Washington, USA.

Buckley, J.A. 1993. Biologically available copper in wastewater: Estimation by the response of the aquatic macrophyte *Lemna minor* (Duckweed). Ph.D. dissertation. Department of Civil Engineering, University of Washington, Seattle, Washington, USA.

Byron, E.R., and C.R. Goldman. 1989. Land-use and water quality in tributary streams of Lake Tahoe, California–Nevada. Journal of Environmental Quality **18**:84–88.

Castelle, A.J., A.W. Johnson, and C. Conolly. 1994. Wetland and stream buffer size requirements—a review. Journal of Environmental Quality **23**:878–882.

Chapman, D.W. 1988. Critical review of variables used to define effects of fines in redds of large salmonids. Transactions of the American Fisheries Society **117**:1–21.

Chapman, G. 1973. Effects of heavy metals on fish. Oregon State University Water Resources Research Institute (OSU-WRI) Report SEMN-WR-D16. 73, Corvallis, Oregon, USA.

Chapra, S.C., and D.M. DiToro. 1991. Delta method for estimating primary production, respiration, and reaeration in streams. Journal of Environmental Engineering **117**:640–655.

Charbonneau, R., and G. Kondolf. 1993. Land-use change in California, USA: Nonpoint source water quality impacts. Environmental Management **17**:453–460.

Charlson, R.J., and H. Rodhe. 1982. Factors controlling the acidity of natural rainwater. Nature **295**:683–685.

Clements, W. 1994. Benthic invertebrate community responses to heavy metals in the upper Arkansas River basin, Colorado. Journal of the North American Benthological Society **13**:30–44.

Cummins, K.W., J.J. Klug, R.G. Wetzel, R.C. Peterson, K.F. Superkropp, A.B. Manny, et al. 1972. Organic enrichment with leaf leachate in experimental lotic ecosystems. BioScience **22**:719–722.

Curtis, E.J., and D.W. Harrington. 1971. The occurrence of sewage fungus in rivers of the United Kingdom. Water Research **5**:281–290.

Davies, P. 1986. Toxicology and chemistry of metals in urban runoff. Pages 60–78 *in* Urbonas and Roesner, eds. Urban runoff quality-impact and quality enhancement technology. American Society of Civil Engineering (ASCE), New York, New York, USA.

De Nicola, D.M., C.D. McIntire, G.A. Lamberti, S.V. Gregory, and L.R. Ashkenas. 1990. Temporal patterns of grazer-periphyton interactions in laboratory streams. Freshwater Biology **23**:475–489.

Dethier, D.P. 1979. Atmospheric contributions to stream water chemistry in the North Cascade Range, Washington. Water Resources Research **15**:787–794.

Dillon, P.J., and F.H. Rigler. 1974. The phosphorus-chlorophyll relationship in lakes. Limnology and Oceanography **19**:767–773.

Edmondson, W.T. 1994. Sixty years of Lake Washington: A curriculum vitae. Lake and Reservoir Management **10**:75–84.

Everest, F.H., R.L. Beschta, J.C. Scrivener, K.V. Koski, J.R. Sedell, and C.J. Cederholm. 1987. Fine sediment and salmonid production: A paradox. Pages 98–142 *in* E.O. Salo and T.W. Cundy, eds. Streamside management: Forestry and fisheries interactions. Institute of Forest Resources Contribution Number 59, University of Washington, Seattle, Washington, USA.

Feminella, J.W., M.E. Power, and V.H. Resh. 1989. Periphyton responses to grazing and riparian canopy in three northern California coastal streams. Freshwater Biology **22**:445–457.

Fisher, S.G., L.J. Gray, N.B.S. Grimm, and D.E. Busch. 1982. Temporal succession in a desert stream ecosystem following flash flooding. Ecological Monographs **52**:903–1010.

Greb, S.R., and D.J. Graczyk. 1993. Dissolved oxygen characteristics of streams. Water Science and Technology **28**:575–581.

Gregory, S.V. 1983. Plant-herbivore interactions in stream systems. Pages 157–189 *in* J.R. Barnes and G.W. Minshall, eds. Stream ecology: Application and testing of general ecology. Plenum Press, New York, New York, USA.

Gregory, S.V., F.J. Swanson, W.A. McKee, and K.W. Cummins. 1991. An ecosystem perspective of riparian zones: Focus on links between land and water. BioScience **41**:540–551.

Hem, J.D. 1970. Study and interpretation of the chemical characteristics of natural water, second edition. United States Geological Survey Water-Supply Paper 1473. US Government Printing Office, Washington, DC, USA.

Hem, J.D. 1989. Study and interpretation of the chemical characteristics of natural water. Second edition. United States Geological Survey Water-Supply Paper 1473. US Government Printing Office, Washington, DC, USA.

Hicks, B.J., J.D. Hall, P.A. Bisson, and J.R. Sedell. 1991. Responses of salmonids to habitat changes. Pages 438–517 *in* W.R. Meehan, ed. Influences of forest and rangeland management on salmonid fishes and their habitats. American Fisheries Society Special Publication Number 19, Bethesda, Maryland, USA.

Hill, W.R., and A.W. Knight. 1988. Concurrent grazing effects of two stream insects on periphyton. Limnology and Oceanography **33**:15–26.

Hoiland, W., and F. Rabe. 1992. Effects of increasing zinc levels and habitat degradation on macroinvertebrate communities in three north Idaho streams. Journal of Freshwater Ecology **7**:373–380.

Holtby, L.B. 1988. Effects of logging on stream temperatures in Carnation Creek, B.C. and associated impacts on coho salmon. Canadian Journal of Fisheries and Aquatic Sciences **45**:502–515.

Horner, R.R., J.J. Skupien, E.H. Livingston, and H.E. Shaver. 1994. Fundamentals of urban runoff management: Technical and institutional issues. Terrene Institute and United States Environmental Protection Agency, Washington, DC, USA.

Horner, R.R., and E.B. Welch. 1981. Stream periphyton development in relation to current velocity and nutrients. Canadian Journal of Fisheries and Aquatic Science **38**:449–457.

Horner, R.R., E.B. Welch, M.R. Seeley, and J.M. Jacoby. 1990. Responses of periphyton to changes in current velocity, suspended sediment and phosphorus concentration. Freshwater Biology **24**:215–232.

Horner, R.R., E.B. Welch, and R.B. Veenstra. 1983. Development of nuisance periphytic algae in laboratory streams in relation to enrichment and velocity. Pages 121–131 *in* R.G. Wetzel, ed. Periphyton of freshwater ecosystems. Dr. W. Junk Publishers, The Hague, The Netherlands.

Jacoby, J.M. 1987. Alterations in periphyton characteristics due to grazing in a Cascade foothills stream. Freshwater Biology **18**:495–508.

Jacoby, J.M., D.D. Bouchard, and C.R. Patmont. 1991. Response of periphyton to nutrient enrichment in Lake Chelan, WA. Lake and Reservoir Management **7**:33–43.

Jones, J.R., M.M. Smart, and J.N. Burroughs. 1984. Factors related to algal biomass in Missouri Ozark streams. Internationale Vereinigung für Theoretische und Angewandte Limnologie **22**:1867–1875.

Karr, J.R., and I.J. Schlosser. 1977. Impact of nearstream vegetation and stream morphology on water quality and stream biota. Ecological Research Series EPA 600-3-77-097. United States Environmental Protection Agency, Environmental Research Laboratory, Office of Research and Development, Athens, Georgia, USA.

Kiffney, P., and W. Clements. 1993. Bioaccumulation of heavy metals by benthic invertebrates at the Arkansas River, Colorado. Environmental Toxicology and Chemistry **12**:1507–1517.

Kjeldsen, K. 1994. The relationship between phosphorus and peak biomass of benthic algae in small lowland streams. Internationale Vereinigung für Theoretische und Angewandte Limnologie **25**:1530–1533.

Lamberti, G.A., L.R. Ashkenas, S.V. Gregory, and A.D. Steinman. 1987. Effects of three herbivores on periphyton communities in laboratory streams. Journal of the North American Benthological Society **6**:92–104.

Lamberti, G.A., and J.W. Moore. 1984. Aquatic insects as primary consumers. Pages 164–195 *in* V.H. Resh and D.M. Rosenberg, eds. The Ecology of Aquatic Insects. Praeger, New York, New York, USA.

Lantz, R.L. 1970. Influence of water temperature on fish survival, growth, and behavior. Pages 182–193 *in* J. Morris, ed. Proceedings of a Symposium on Forest Land Uses and the Stream Environment at Oregon State University, Corvallis, Oregon, USA.

Lloyd, D.S. 1987. Turbidity as a water quality standard for salmonid habitats in Alaska. North American Journal of Fisheries Management **7**:34–45.

Lowrance, R., R. Todd, J. Fail, Jr., O. Henrickson, Jr., R. Leonard, and L. Asmussen. 1984. Riparian forests as nutrient filters in agricultural watersheds. BioScience **34**:374–377.

Madej, M.A. 1978. Response of a stream channel to an increase in sediment load. Masters thesis. Department of Geology, University of Washington, Seattle, Washington, USA.

May, C.W., E.B. Welch, R.R. Horner, J.R. Karr, and B.W. Mar. 1997. Quality indices for urbanization effects in Puget Sound lowland streams. Department of Civil Engineering, University of Washington Water Resources Series Technical Report, No. 154. Seattle, Washington, USA.

Naiman, R.J., and E.C. Anderson. 1997. Streams and rivers: Their physical and biological variability. Pages 131–148 *in* P.K. Schoonmaker, B. von Hagen, and E.C. Wolf. The rainforests of home: Profile of a North American bioregion. Island Press, Washington, DC, USA.

Naiman, R.J., T.J. Beechie, L.E. Benda, D.R. Berg, P.A. Bisson, L.H. MacDonald, et al. 1992. Fundamental elements of ecologically healthy watersheds in the Pacific Northwest coastal ecoregion. Pages 127–188 *in* R.J Naiman, ed. Watershed management: Balancing sustainability and environmental change. Springer-Verlag, New York, New York, USA.

Newbold, J.D., J.W. Elwood, R.V. O'Neill, and W. Van Winkle. 1981. Measuring nutrient spiraling in streams. Canadian Journal of Fisheries and Aquatic Sciences **38**:860–863.

Noggle, C.C. 1978. Behavioral, physiological, and lethal effects of suspended sediments on juvenile salmonids. Masters thesis. School of Fisheries, University of Washington, Seattle, Washington, USA.

Novotny, V., and H. Olem. 1994. Water quality: Prevention, identification, and management of diffuse pollution. Van Nostrand Reinhold, New York, New York, USA.

Olthof, J. 1994. Puget lowland stream habitat and relations to basin urbanization. Masters thesis. Department of Civil Engineering, University of Washington, Seattle, Washington, USA.

Ormerod, J.D., B. Grynne, and K.S. Ormerod. 1966. Chemical and physical factors involved in heterotrophic growth response to organic pollution. Internationale Vereinigung für Theoretische und Angewandte Limnologie **16**:906–910.

Pagenkopf, G. 1983. Gill surface interaction model for trace-metal toxicity to fishes: Role of complexation, pH, and water hardness. Environmental Science and Technology **17**:342–347.

Perrin, C.J., M.L. Bothwell, and P.A. Slaney. 1987. Experimental enrichment of a coastal stream in British Columbia: Effects of organic and inorganic additions on autotrophic periphyton production. Canadian Journal of Fisheries and Aquatic Sciences **44**:1247–1256.

Pitt, R., R. Field, M. Lalor, and M. Brown. 1995. Urban stormwater toxic pollutants: Assessment, sources, and treatability. Water Environment Research **67**:260–275.

Platts, W.S. 1991. Effects of livestock grazing on salmonid habitat. Pages 389–424 *in* W.R. Meehan, ed. Influences of forest and rangeland management on salmonid fishes and their habitats. American Fisheries Society Special Publication Number 19, Bethesda, Maryland, USA.

Power, M.E. 1992. Hydrologic and trophic controls of seasonal algal blooms in northern California rivers. Archiv für Hydrobiologie **125**:385–410.

Purcell, M. 1994. Factors regulating periphyton in an urbanizing environment. Masters thesis. Department of Civil Engineering, University of Washington, Seattle, Washington, USA.

Quinn, J.M., and P.N. McFarlane. 1989. Epilithon and dissolved oxygen depletion in the Manawatu River, New Zealand. Water Research **23**:825–832.

Quinn, T.P., and N.P. Peterson. 1994. The effect of forest practices on fish populations. TFW-F4-94-001. Washington Department of Natural Resources, Timber-Fish-Wildlife Program, Olympia, Washington, USA.

Reckhow, K.H., and S.C. Chapra. 1983. Engineering approaches for lake management. Volume 1. Butterworths, Boston, Massachusetts, USA.

Reeves, G.H., F.H. Everest, and J.D. Hall. 1987. Interactions between the redside shiner and steelhead trout in western Oregon: The influence of water temperature. Canadian Journal of Fisheries and Aquatic Sciences **44**:1603–1613.

Ringler, N.H., and J.D. Hall. 1975. Effects of logging on water temperature and dissolved oxygen in spawning beds. Transactions of the American Fisheries Society **104**:111–121.

Ryan, P.A. 1991. Environmental effects of sediment on New Zealand streams: A review. New Zealand Journal of Marine and Freshwater Research **25**:207–221.

Sawyer, C.N., P.L. McCarty, and G.F. Parkin. 1994. Chemistry for environmental engineers, 4th edition. McGraw-Hill, New York, New York, USA.

Schlorff, E. 1995. Nutrient criteria: Review and analysis for Washington state lakes. Second Draft—November 1995. Washington State Department of Ecology, Water Quality Program, Olympia, Washington, USA.

Schueler, T.R. 1987. Controlling urban runoff: A practical manual for planning and designing urban best management practices. Metro Washington, DC, Council of Governments Publication #87703. Prepared for Metropolitan Washington Water Resources Planning Board, Washington, DC, USA.

Sigler, J.W., T.C. Bjorn, and F.H. Everest. 1984. Effects of chronic turbidity on density and growth of steelhead and coho salmonids. Transactions of the American Fisheries Society **113**:142–150.

Smith, R., and R. Eilers. 1978. Effects of stormwater on stream dissolved oxygen. Journal of the Environmental Engineering Division, American Society of Civil Engineers **104**:549–559.

Steinman, A.D., P.J. Mulholland, and D.B. Kirschetel. 1991. Interactive effects of nutrient reduction and herbivory on biomass, taxonomic structure and P uptake in lotic periphyton communities. Canadian Journal of Fisheries and Aquatic Sciences **48**:1951–1959.

Stewart, A.J. 1987. Responses of stream algae to grazing minnows and nutrients: A field test for interactions. Oecologia **71**:1–7.

Stockner, J.G., and K.S. Shortreed. 1978. Enhancement of autotrophic production by nutrient addition in a coastal rain-forest stream on Vancouver Island. Journal of Fisheries Research Board of Canada **35**:28–34.

Stumm, W., and J.J. Morgan. 1973. Aquatic Chemistry. Wiley-Interscience, New York, New York, USA.

Sullivan, K., J. Tooley, K. Doughty, J. Caldwell, and P. Knudsen. 1990. Evaluation of prediction models and characterization of stream temperature regimes in Washington. TFW-WQ3-90-006. Washington State Department of Natural Resources, Olympia, Washington, USA.

Swanson, F.J., L.E. Brenda, S.H. Duncan, G.E. Grant, W.F. Megahan, L.M. Reid, et al. 1987. Mass failures and other processes of sediment production in Pacific Northwest landscapes. Pages 9–38 *in* E.O. Salo and T.W. Cundy, eds. Streamside management: Forestry and fisheries interactions. Institute of Forest Resources Contribution Number 59, University of Washington, Seattle, Washington, USA.

Tanner, D.Q., and C.W. Anderson. 1996. Assessment of water quality, nutrients, algal productivity, and management alternatives for low-flow conditions, South Umpqua River basin, Oregon, 1990–1992. US Geological Survey Water Resources Investigations Report 96-4082, Portland, Oregon USA.

US EPA (United States Environmental Protection Agency). 1986. Quality criteria for water 1986. EPA 440-5-86-001. United States Environmental Protection Agency, Office of Water Regulations and Standards, Washington, DC, USA.

USGS (United States Geological Survey). 1975. Water resources data for Washington. Part 2. United States Geological Survey, Tacoma, Washington, USA.

——. 1993. National Water Summary 1990–91. USGS Water Supply Paper #2400, Washington, DC, USA.

Viessman, W., Jr., and C. Welty. 1985. Water management, technology and institutions. Harper & Row, San Francisco, California, USA.

Vong, R.J., L. Moseholm, D.S. Covert, P.D. Sampson, J.F. O'Loughlin, M.N. Stevenson, et al. 1988. Changes in rainwater acidity associated with closure of a copper smelter. Journal of Geophysics Research **93**:7169–7179.

Walton, S.P., E.B. Welch, and R.R. Horner. 1995. Stream periphyton response to grazing and changes in phosphorus concentration. Hydrobiologia **302**:31–46.

Warren, C.E., J.H. Wales, G.E. Davis, and P. Doudoroff. 1964. Trout production in an experimental stream enriched with sucrose. Journal of Wildlife Management **28**:617–660.

Watson, V. 1990. Control of algal standing crop by P and N in the Clark Fork River. Proceedings of the Clark Fork Symposium, University of Montana, Missoula, Montana, USA.

WDOE (Washington Department of Ecology). 1988. Water quality criteria. Department of Ecology, State of Washington, Olympia, Washington, USA.

WDNR (Washington Department of Natural Resources). 1978. Geology of Washington. WDNR Publication Number 12, Olympia, Washington, USA.

Welch, E.B. 1992. Ecological effects of wastewater, 2nd edition. Chapman & Hall, London, UK.

Welch, E.B., R.R. Horner, and C.R. Patmont. 1989. Prediction of nuisance periphytic biomass: A management approach. Water Research **23**:401–405.

Welch, E.B., J.M. Jacoby, R.R. Horner, and M.R. Seeley. 1988. Nuisance biomass levels of periphytic algae in streams. Hydrobiologia **157**:161–168.

Welch, E.B., J.M. Quinn, and C.W. Hickey. 1992a. Periphyton biomass related to point-source nutrient enrichment in seven New Zealand streams. Water Resources Research **26**:669–675.

Welch, E.B., D.E. Spyridakis, K.B. Easthouse, and T.J. Smayda. 1992b. Response to a smelter closure in Cascade Mountain lakes. Water, Air and Soil Pollution **61**:325–338.

Welch, E.B., D.E. Spyridakis, and T. Smayda. 1986. Temporal variability in acid sensitive high elevation lakes. Water, Air and Soil Pollution **31**:35–44.

Williamson, R.B. 1985. Urban stormwater quality. New Zealand Journal of Marine and Freshwater Research **19**:413–427.

Zasoski, R.J., H. Dawson, and L. Lestelle. 1977. Impacts of forest management practices on aquatic environment—Phase III. Select Water Resources Abstracts 1077–1088. Quinault Resource Development Project, Taholah, Washington, USA.

Part II
The Biotic Environment

5
Biotic Stream Classification*

Robert J. Naiman

Overview

• Stream classification implies that sets of observations or characteristics can be organized into meaningful groups based on measures of similarity or difference. Although this chapter focuses on biotic classification systems and management applications, it briefly reviews selected physical classification systems discussed by Montgomery and Buffington (Chapter 2) that are important for understanding biotic patterns.

• Early attempts at whole river classifications were generally unsuccessful because of the biophysical variability inherent over large spatial scales. At a smaller spatial scale, classification of river segments by stream order, linkage number, and drainage density has been useful because the scale is more appropriate for understanding patterns of biotic zonation either by using fish or invertebrates as indicators of segment types or by delineating zones where ecological processes (such as primary production or detrital dynamics) are occurring.

• Recent concepts emphasize multidisciplinary bases for classification related to increasingly small spatial units and hierarchical rankings of linkages between the geologic and climatic settings, stream habitat features, and biota. These are divided into ultimate (large spatial and long temporal scales) and proximate (small and short scales) controls on system characteristics. Classifications of physical features use either a single scale system (e.g., stream order, linkage number, or ecoregion) or a hierarchical system which nests characteristic features over a variety of spatial and temporal scales.

• Rosgen's (1994) classification system is used to illustrate the classification of present-day channels relative to their geologic setting and related fluvial processes such as water hydraulics and material transport.

• Most classification systems coupling biological and physical features employ vertebrate (mostly fish) and invertebrate (mostly insect) distributions. Occasionally, riparian and aquatic vegetation are utilized. Overall, these approaches usually sacrifice precision for generality. The usefulness of biologically based systems in stream management is diminished because of the intensive efforts demanded to measure and monitor community characteristics. However, the need for such a system remains great because of the significance of the biological community to regional ecological integrity.

• The best approach to stream classification depends on the scope and the nature of the question being asked. However, in general, the system chosen should have the ability to encompass broad spatial and temporal scales, to integrate structural and functional characteristics under different disturbance regimes, to convey information about mechanisms

*This chapter is an updated version of Naiman et al. 1992

controlling stream features, and to accomplish the goal at low cost and at a high level of uniform understanding among resource managers.

Introduction

Classification systems have been used for centuries to organize information about ecological systems. Yet the classification of fluvial systems remains a difficult topic because running waters exhibit such dynamic changes with time and space. Each stream possesses a set of characteristics (e.g., morphology, hydrology, productivity, and so forth) which change in response to the local climate, geology, and disturbance regime.

The term "classification" implies that sets of observations or characteristics can be organized into meaningful groups based on measures of similarity or difference (Gauch 1982, Hawkins et al. 1993). Implicitly, relatively distinct boundaries exist and may be identified by a set of discrete variables. However, the classification of streams is complicated by both longitudinal and lateral linkages, by changes that occur in the physical features over time, and by boundaries between apparent patches that are often indistinct (Naiman et al. 1992, Rosgen 1994). For example, geomorphic and ecological characteristics of streams vary spatially from the headwaters to the sea (Langbein and Leopold 1966, Vannote et al. 1980), as well as temporally in response to disturbance patterns (Bisson et al. 1982, Wissmar and Swanson 1990).

Stream classification is essential for understanding the distribution of ecological patterns within drainage networks and for developing management strategies that are responsive to the ecological patterns. This chapter reviews the principles of stream classification through an analysis of conceptual approaches previously used to develop several contrasting schemes. Historic and extant classification systems, based on a variety of spatial scales (from microhabitats to ecoregions) incorporate several combinations of physical and biological components that are important to riverine systems. Chapter 2 provides a detailed discussion of some of the more useful physically based classifications while this chapter focuses on biotic classification systems and management applications. Nevertheless, it is prudent to briefly summarize the physical classification systems discussed by Montgomery and Buffington (Chapter 2) which are important for understanding biotic patterns.

Historical Concepts

The history of stream classification is reviewed comprehensively by Wasson (1989), Naiman et al. (1992), and Rosgen (1994). The dominant conceptual themes of the early efforts range from biological to physical features over spatial scales of a few meters to hundreds of square kilometers. At the larger scale, one of the original whole-river schemes was developed for New Jersey (USA) rivers (Davis 1890). Davis classified streams as young, mature, or old on the basis of observed erosion patterns. Later, Shelford (1911) attempted to produce a biological classification scheme for whole rivers in Michigan (USA) based on his idea of succession. However, because of longitudinal differences in physical and biological characteristics, whole-stream classification has been of little use.

In contrast, basin-wide classification systems based on drainage network characteristics such as stream order, linkage number (total number of 1st-order streams), and drainage density (Horton 1945, Strahler 1957) have proven useful, but too simplistic, for elucidating biotic patterns within drainage basins. Many became important tools—if only locally adapted—during the early part of the century. Early classifications generally were based on perceived patterns of biotic zonation using species of fish or invertebrates as indicators of segment types (Carpenter 1928, Ricker 1934, Huet 1954). In addition, numerous specialists of certain orders of stream invertebrates (e.g., Plecoptera, Ephemeroptera, Trichoptera) also utilized their data to propose organizational patterns (Macan 1961, Illies and Botosaneanu 1963). These early attempts at stream classification recognized that biotic zonation patterns generally were correlated with gradient or other abiotic features such as temperature or water chemistry, although Huet (1954) also recognized the importance of larger spatial scales by incorporating valley form. Later, classifications using stream order and linkage number were successful in describing patterns of ecological processes, such as primary production and detrital dynamics (Minshall et al. 1983, Naiman et al. 1987). In spite of widespread recognition of distinct biotic zones along rivers, there were many early critics because key physical parameters change gradually along the stream continuum (e.g., slope and width), and the biological characteristics change in a similar manner.

There are two general limitations to these historic systems. First, the reliance on species as indicators of ecological zones means that the biotic zonation schemes are only valid in basins with similar zoogeographic, geologic, and climatic histories. Despite relating physical factors to biotic patterns, these schemes failed to construct a conceptual framework for stream classification that could transcend regions. Second, for both physical and biological zonation systems there were no features relating geologic and climatic processes, which regulate the physical features of streams, to the classification system. Therefore, these efforts were ineffective at relating watershed-scale processes to dynamic changes in channel features (Naiman et al. 1992). However, the application of landscape ecology concepts (such as patches, boundaries, and connectivity) to rivers is now becoming a useful approach to overcome some of this difficulty (Décamps 1984, Ward and Stanford 1987, Rosgen 1994).

Recent Concepts

Ideally, a classification system should be based on a hierarchical ranking of linkages between the geologic and climatic settings, the stream habitat features and the biota (Hawkins et al. 1993). These—the geomorphic and climatic processes that shape the abiotic and biotic features of streams—provide a conceptual and practical foundation for understanding the structure and processes of fluvial systems (Chapter 2, Chapter 11). Furthermore, an understanding of process allows streams to be viewed in a larger spatial and temporal perspective, and to infer the direction and magnitude of potential changes due to natural and human disturbances. A stream classification system, based on patterns and processes and how they are expressed at different temporal and spatial scales, is the basis for successful management (Rosgen 1994).

Conceptually, individual stream classification units can be thought of as an integrated collection of *ultimate* and *proximate* controls on system characteristics. These terms generally correspond to higher and lower levels of a hierarchical ranking of controlling factors. Ultimate controls refer to a set of geologic factors that act over large areas ($>1\,km^2$), are stable over long time scales ($>10^4\,yr$), and dictate the range of conditions possible in a drainage network. These include physical characteristics, such as regional geology and climate, and biotic characteristics, such as zoogeography (Moyle and Li 1979, Briggs et al. 1990). Proximate controls refer to local geomorphic and biotic processes important at small scales ($<10^2\,m^2$), which can change stream characteristics over relatively short time periods ($<10^4\,yr$). Proximate controls are constrained by ultimate controls and include such physical processes as discharge, temperature, hillslope erosion, channel migration, and sediment transport; and the biotic processes of reproduction, competition, disease, and predation—all of which may be influenced by an equally diverse array of human impacts. Within this conceptual framework, management strategies to effectively maintain important physical and ecological structures may be tailored to local conditions.

Frissell et al. (1986) discuss the topic of ultimate and proximate controls utilizing ideas from hierarchy theory to construct a continuum of habitat sensitivity to disturbance and recovery time (Figure 5.1). In their scenario, microhabitats are most sensitive to disturbance and watersheds are least sensitive. Furthermore,

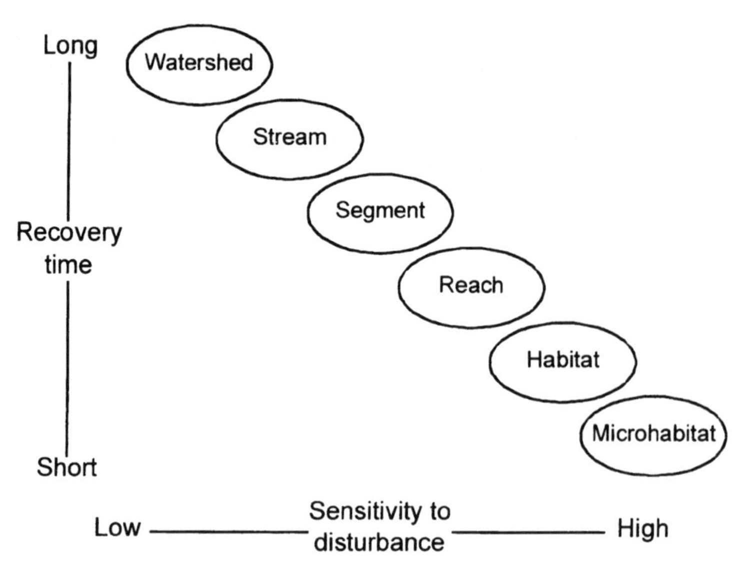

FIGURE 5.1. Relationship between recovery time and sensitivity to disturbance for different spatial scales associated with stream systems (After Frissell et al. 1986 and from Naiman et al. 1992. Reproduced with permission). See Table 5.1 for definition of spatial scales.

individual events that affect smaller-scale habitat characteristics generally do not affect larger-scale system characteristics (however, collectively they can have an impact), whereas large-scale disturbances directly influence smaller-scale features of streams. For example, on a small spatial scale, deposition at one site may be accompanied by scouring at another site nearby, and the reach or segment does not appear to change significantly. In contrast, a large-scale disturbance (such as a debris flow) is initiated at the segment level and reflected in all lower levels of the hierarchy (reach, habitat, microhabitat). On a temporal scale, siltation of microhabitats may disturb the biotic community over the short term. However, if the disturbance is of limited scope and intensity, the system may recover quickly to predisturbance levels.

Tailoring management strategies to stream types implies that the classification system includes the physical and biotic characteristics of the stream, as well as the disturbance regime creating and maintaining those characteristics. Successful classification systems are able to categorize the types and frequencies of disturbance that may impact the stream and predict adjustments in the physical and biotic characteristics.

Classification of Physical Watershed Features (a Summary)

River classification based on geomorphic characteristics came into prominence in the 1940s (Horton 1945, Leopold and Wolman 1957). This approach became important to fisheries biologists and land managers because geomorphic patterns are strongly linked to patterns of species distribution and abundance (Huet 1954, Bisson et al. 1988, Morin and Naiman 1990). Almost all classification schemes based on physical habitat features have been founded on the perception that stream units (i.e., segment, reach, habitat) are discrete and can therefore be delineated. However, that is not always the case. Subtle gradations between segment types, reach types, and habitat types are common. Fortunately, dramatic and abrupt physical changes in stream width, depth, and velocity also are found (Frissell et al. 1986, Kellerhals and Church 1989). For example, Frissell et al. (1986) defined longitudinal boundaries of segment types by easily measured tributary junctions, major waterfalls, or other structural discontinuities, while reaches were defined less clearly by changes in channel gradient. Ultimately, however, the scope of the issue, or nature of the question being considered, should determine the appropriate scale(s) of resolution.

Single-Scale Classification

Observed patterns in drainage networks led to the development of the stream order concept (Horton 1945, Strahler 1957). Within geographic regions, this system has been correlated with physical and biotic features of streams (for example, see Minshall et al. 1983, Naiman et al. 1987). However, it is much less reliable at predicting patterns and behavior of stream characteristics across regions, or at microscales within regions. For example, major differences in stream size (Minshall et al. 1983) and response to disturbance (Resh et al. 1988) can be encountered for streams of the same order between regions because of variability in geology and hydrology. More importantly, stream order by itself provides little information on processes controlling longitudinal and lateral patterns, and therefore makes predictions of response to both natural and human disturbance imprecise. In spite of its almost universal usage in the United States, the value of the stream order classification scheme is only as an indicator of relative biotic and stream segment characteristics and position within a given drainage network. When properly used, however, it can be valuable as an *accounting* tool in categorizing biological and physical data.

Other more recent approaches to classification include large-scale schemes developed for their potential usefulness to regional water resource and fisheries managers. Bailey (1978) defined 11 ecoregions that delineated large areas ($>10^3\,km^2$) of the United States based on climate, physiography, and vegetation. These were chosen because they were thought to be important in stratifying in-channel features (Rohm et al. 1987). The system has now been tested successfully in at least three areas (the upper Midwest, Arkansas, and Oregon, USA) with respect to chemical characteristics and fish species distribution (Larsen et al. 1986, Whittier et al. 1988). Rohm et al. (1987) were able to categorize fish assemblage, physical habitat (e.g., percentage riffle, pool) and water chemistry (e.g., alkalinity, conductivity) patterns into six ecoregions (Omernik 1987). The ecoregion concept is effective for grouping of streams where large-scale resolution is required (e.g., Rohm et al. 1987).

Hierarchical Classification

Recent stream classification systems in North America have been based on a hierarchical perspective linking large regional scales (ecoregions) with small microhabitat scales (Table 5.1). This approach is especially useful since stream processes occur at scales spanning 16 orders of magnitude (10^{-7}–10^{8} m spatially and 10^{-8}–10^{7} yr temporally; Minshall 1988). Several classification systems have been developed using nested landscape or channel features (Warren 1979, Frissell et al. 1986, Cupp 1989, Hawkins et al. 1993, Rosgen 1994, Chapter 2 this volume). The value of hierarchical stream classification is greatest when broadly applied (e.g., global, national, regional scales; Frissell et al. 1986). However, the approach is flexible enough to be modified for subregional purposes. Furthermore, it is important to understand the relative roles of controlling factors in determining the long-term and short-term characteristics of streams since the relative importance of the factors changes with spatial and temporal scales. Finally, a hierarchical approach requires fewer variables at any one level for classification. Within most geographic regions, managers and scientists need only one or two spatial and temporal scales to classify streams (Table 5.2).

One of the first hierarchical classification systems was developed by Warren (1979). He described 11 levels ranging from regional ($>10^2\,km^2$) to microhabitat ($<1\,m^2$) defined largely by four variables (substrate, climate, water chemistry, and biota). Warren did not propose a concrete classification system, but his

TABLE 5.1. Some events or processes controlling stream habitat on different spatio-temporal scales.

System level	Linear spatial scale[a] (m)	Evolutionary events[b]	Developmental processes[c]	Time scale of continuous potential persistence[a] (yr)
Stream	10^3	Tectonic uplift subsidence; catastrophic volcanism; sea level changes; glaciation; climate shifts	Planation; denudation; drainage network development	10^6–10^5
Segment	10^2	Minor glaciation, volcanism; earthquakes; very large landslides; alluvial or colluvial valley infilling	Migration of tributary junctions and bedrock nick-points; channel floor incision; development of new 1st-order channels	10^4–10^3
Reach	10^1	Debris torrents; landslides; log input or washout; channel shifts, cutoffs; channelization, diversion or damming by humans	Aggradation/degradation associated with large sediment-storing structures; bank erosion; riparian vegetation succession	10^2–10^1
Habitat or Channel Unit	10^0	Input or washout of wood, boulders etc.; small bank failures; flood scour or deposition; thalweg shifts; numerous human activities	Small-scale lateral or elevational changes in bedforms; minor bedload resorting	10^1–10^0
Microhabitat	10^{-1}	Annual sediment and organic matter transport; scour of stationary substrates; seasonal macrophyte growth and cropping	Seasonal depth and velocity changes; accumulation of fines; microbial breakdown of organics; periphyton growth	10^0–10^{-1}

[a] Space and time scales indicated are approximate for a 2nd- or 3rd-order mountain stream. Caution is advised in using absolute spatial scales for the hierarchy. Depending on the specific situation, for example, a channel reach may be tens to hundreds of meters long while a habitat unit may be less than one meter to several meters long. Perhaps a better spatial index that preserves geomorphic similitude is scaling by channel width (Chapter 2 this volume) because there is no absolute association of channel size with stream order.
[b] Evolutionary events change potential capacity; that is, extrinsic forces that create and destroy systems at that scale.
[c] Developmental processes are intrinsic, progressive changes following a system's genesis in an evolutionary event.
From Frissell et al. 1986 with permission.

TABLE 5.2. Habitat spatial boundaries, conforming with the temporal scales of Table 5.1.

System level	Capacity time scale[a] (yr)	Vertical boundaries[b]	Longitudinal boundaries[c]	Lateral boundaries[d]	Linear spatial scale[a] (m)
Stream	10^6–10^5	Total initial basin relief; sea level or other base level	Drainage divides and sea coast, or chosen catchment area	Drainage divides; bedrock faults, joints controlling ridge valley development	10^3
Segment	10^4–10^3	Bedrock elevation; tributary junction or falls elevation	Tributary junctions; major falls, bedrock lithological or structural discontinuities	Valley sideslopes or bedrock outcrops lateral migration	10^2
Reach	10^2–10^1	Bedrock surface; relief of major sediment-storing structures	Slope breaks; structures capable of withstanding <50 year flood	Mean annual flood channel; mid-channel bars; other flow-splitting obstructions	10^1
Habitat or Channel Unit	10^1–10^0	Depth of bedload subject to transport in <10-year flood; top of water surface	Water surface and bed profile slope breaks; location of genetic structures	Same as longitudinal	10^0
Microhabitat	10^0–10^{-1}	Depth to particles immovable in mean annual flood; water surface	Zones of differing substrate type, size, arrangement; water depth and velocity		10^{-1}

[a] Scaled to approximate a 2nd- or 3rd-order mountain stream. See cautions in Table 5.1, footnote *a*.
[b] Vertical dimension refers to upper and lower surfaces.
[c] Longitudinal dimension refers to upstream–downstream extent.
[d] Lateral dimension refers to cross-channel or equivalent horizontal extent.
From Frissell et al. 1986 with permission.

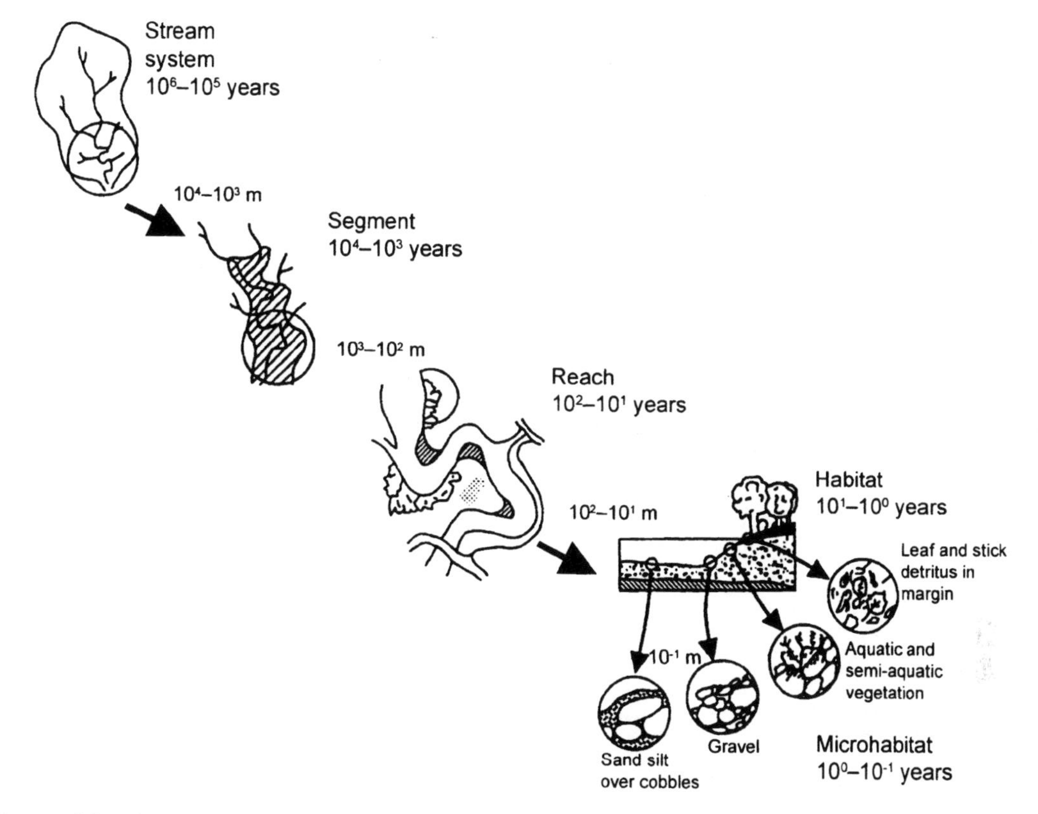

FIGURE 5.2. Hierarchical organizations of a stream system and its habitat subsystems. Linear spatial scale, approximated to 4th- to 6th-order mountain stream, is indicated as well as the temporal duration of existing channel features (modified from Frissell et al. 1986 with permission).

contribution to the conceptual evolution of stream classification is worth noting because he presented an explicit theoretical structure for a complex hierarchical system. He stressed the importance of assessing the potential of a stream (i.e., all possible developmental states and performances that a system may exhibit while still maintaining its integrity as a coherent unit) rather than its current condition. Evaluating potential states for a system assists in distinguishing natural variability from human disturbance.

Frissell et al. (1986) extended Warren's approach by incorporating spatially nested levels of resolution (e.g., watershed, stream, valley segment, reach, habitat unit (e.g., pool/riffle), and microhabitat) (Table 5.2, Figure 5.2). An important conceptual advancement, this system addressed form or pattern within each hierarchical level, as well as origins and processes of development. Nawa et al. (1990) used this approach to show that both fish species composition and the sensitivity of channels to disturbance varied between different valley segment types.

Other significant hierarchical classification systems for broad scales of resolution include that of Lotspeich (1980) and Brussock et al. (1985). Brussock et al. (1985) developed a hierarchical system for large rivers based primarily on predictable patterns of variation in channel form. Channel form is an important parameter because it overlays in-channel features (e.g., relief, lithology, and discharge) controlling the physical state of the stream (e.g., temperature, depth, substrate, and velocity) which, in turn, influences the character of biotic resources. Variations in channel form are believed to be related to lithology, gradient, and climate (state factors), as they act on substrate particle size,

TABLE 5.3. Valley bottom and sideslope geomorphic characteristics used to identify the five valley segment types in Figure 5.3.

Valley segment type[a]	Average Channel gradient[b]	Sideslope gradient[c]	Valley bottom width[d]	Channel patterns	Stream order[e]	Landform and geomorphic features
F2 Alluviated lowlands	≤1%	>5%	>5X	Unconstrained; highly sinuous	Any	Wide floodplains typically formed by present or historic large rivers within flat to gently rolling lowland landforms; sloughs, oxbows, and abandoned channels commonly associated with mainstream rivers
V1 V-shaped, moderate-gradient bottom	2–6%	30–70%	<2X	Constrained	≥2	Deeply incised channels with steep competent sideslopes; very common in uplifted mountainous topography; less commonly associated with marine or glacial outwash terraces in lowlands and foothills
V4 Alluviated mountain valley	1–4%	Channel adjacent slopes <10%; increase to ≥30%+	2–4X	Unconstrained; high sinuosity with braids and side-channels common	2–5	Deeply incised channels with relatively wide floodplains; distinguished as "alluvial flats" in otherwise steeply dissected mountainous terrain
U4 Active glacial out-wash valley	1–7%	Initially <5%, increasing to >60%	<4X	Unconstrained; highly sinuous and braided	1–3	Stream corridors directly below active alpine glaciers; channel braiding and shifting common; active channel nearly as wide as valley bottom
H3 Very high-gradient valley wall/head-water	11%+	>60%	<2X	Constrained; stair–stepped	1–2	Small channels moderately entrenched into high-gradient mountain slopes or headwater basins; bedrock exposures and outcrops common; localized alluvial/colluvial terrace deposition

[a] Valley segment type names include alphanumeric mapping codes in italic (from Cupp, 1989a, b).
[b] Valley bottom gradient is measured in lengths of *ca.* 300 m or more.
[c] Sideslope gradient characterizes the hillslopes within 1,000 horizontal and *ca.* 100 m vertical distance from the active channel.
[d] Valley bottom width is a ratio of the valley bottom width to active channel width.
[e] Stream order defined by Strahler (1957).

bed load, and competence. Examining streams throughout the United States, they described seven regions based on differences in state factors. They related channel form to community structure and confirmed L.B. Leopold's assertion that stream channel form can be predicted along the length of the river within geographic regions (Leopold et al. 1964).

In the Pacific Northwest, three hierarchical classification systems are widely used in resource management (Cupp 1989, Hawkins et al. 1993, Rosgen 1994) while the system described by Montgomery and Buffington (Chapter 2) is gaining acceptance. Cupp (1989) adapted the hierarchical concept of Frissell et al. (1986) to small forested streams in Washington using eight hierarchical levels ranging from ecoregion to microhabitat. Valley segments are distinguished by average channel gradient and valley form (Table 5.3, Figure 5.3). Initial field tests show that stream segment types are correlated with habitat (Beechie and Sibley 1990). Hawkins et al. (1993) refined the Bisson et al. (1982) system of salmonid habitat classification by first identifying which physical characteristics were needed to describe specific channel units and then ranking their importance as descriptive features useful in defining and discriminating among different types of channel units. They recommend a three-level hierarchy

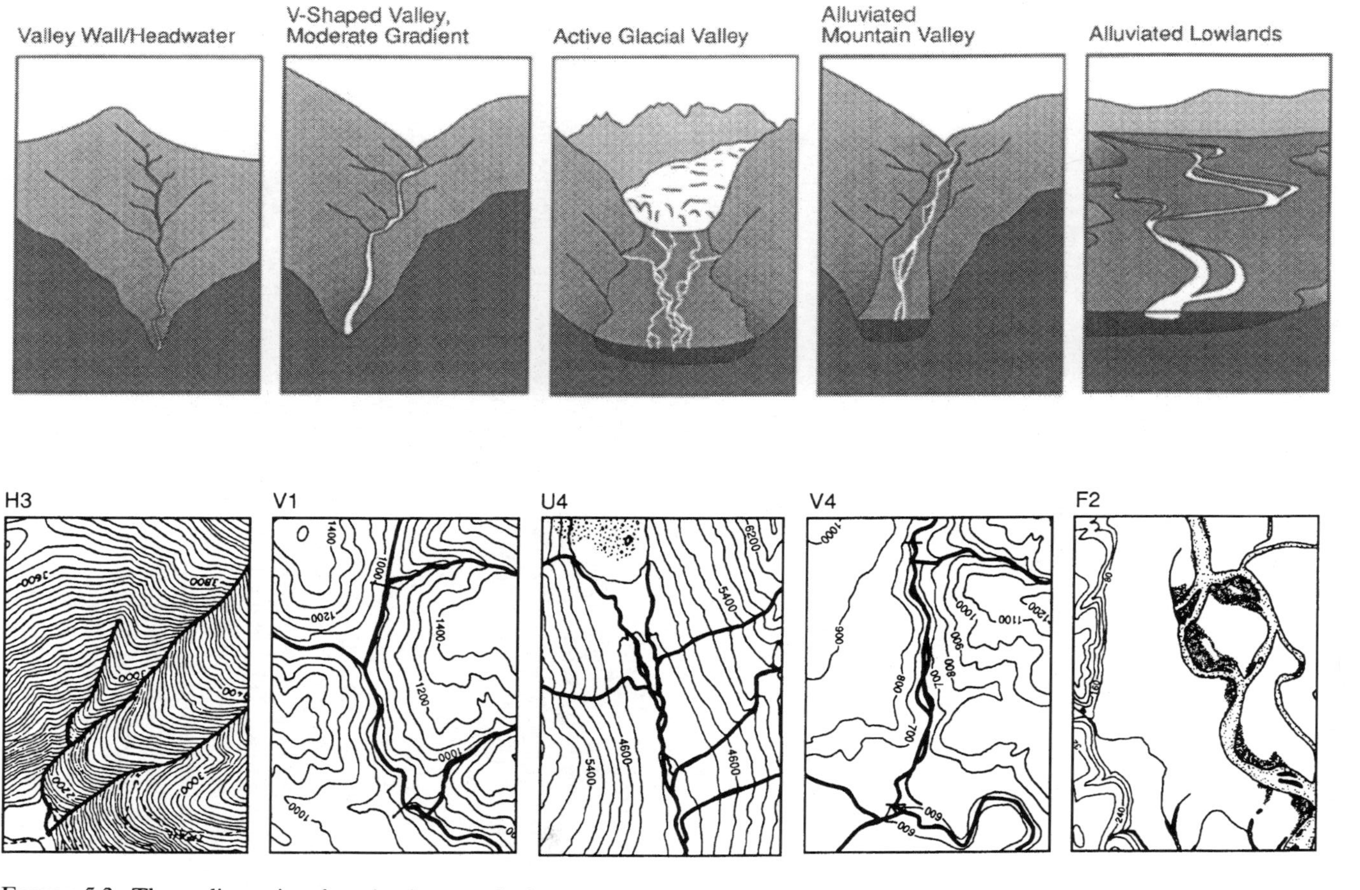

FIGURE 5.3. Three-dimensional projections made from topographic maps assist in determining segment type. See Table 5.3 for physical characteristics of each stream type (Naiman et al. 1992 reproduced with permission).

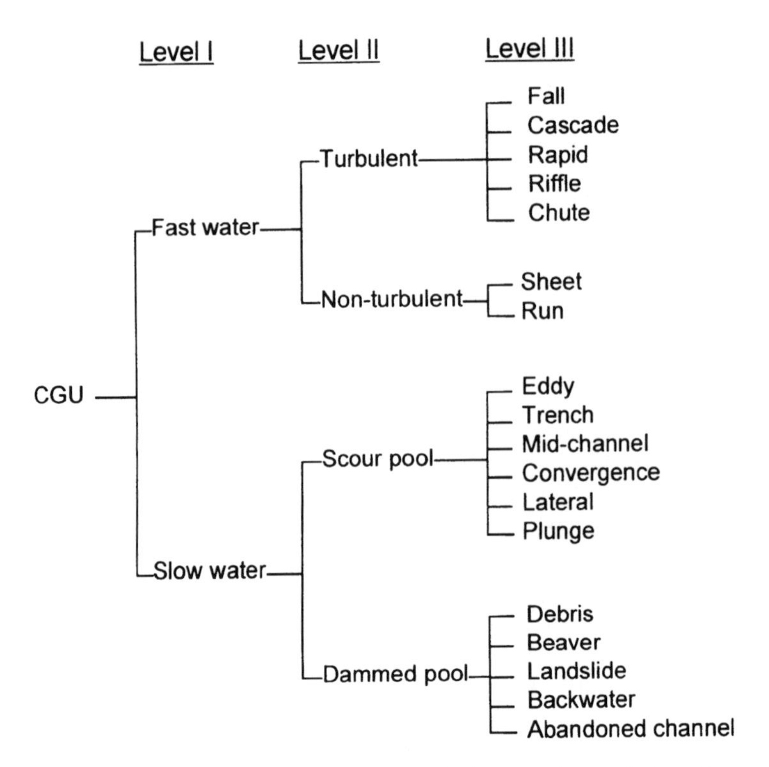

FIGURE 5.4. Similarity dendrogram illustrating how channel geomorphic units (CGU) can be classified with increasing levels of resolution. Three levels of resolution are shown that can be used to distinguish classes (Hawkins et al. 1993 with permission).

(Figure 5.4): (A) At the coarsest level of resolution are pool and riffles; (B) at base flows riffles are either turbulent or not, and pools are created by either scour or material deposition in the channel; (C) the fast- and slow-water classes are further subdivided based on other physical criteria related to specific fish-habitat considerations. Fishes and other stream organisms appear to distinguish among these habitats at one or more levels of the hierarchy but, unfortunately, there are few published data available for empirical tests of the system. Rosgen (1994) developed a classification system based on geomorphic and in-channel characteristics, including channel gradient, sinuosity, width-to-depth ratio, bed material, entrenchment, channel confinement, soil erodibility, and stability. It also includes subcategories that may change over short temporal scales and are characterized by riparian vegetation, channel width, organic debris, flow regime, meander patterns, depositional features, and sediment supply. Rosgen's stream-type classification system has been used widely for site-specific riparian forest and fisheries management, and for predicting geomorphic and hydrologic processes.

Rosgen's (1994) classification system requires further explanation because of its wide use. It is based on a morphological arrangement of the aforementioned stream characteristics organized into relatively homogeneous stream types. He correctly assumes that contemporary channel morphology is governed by physical laws resulting in observable stream features and fluvial processes (such as water hydraulics and transported materials). A change in any one of the fluvial (i.e., physical) processes causes channel adjustments which lead to changes in other fluvial processes, resulting in new channel features.

Rosgen (1994) recognizes four hierarchical aspects to classification: (I) broad geomorphological characterization, (II) morphological description of the channel, (III) stream condition, and (IV) verification (Table 5.4). The first two aspects address the character of the channel, forming the basis of his system, and are discussed below. Aspect III addresses the state of the stream further describing existing conditions that influence the response of channels to imposed change and provides specific information for prediction. Aspect IV addresses verification of reach-specific information on channel processes which is used to evaluate predictions. Interested readers are referred to the original publication for additional information on Aspects III and IV.

Geomorphologic Characterization (Aspect I). The purpose of Aspect I is to provide a broad characterization that integrates the landform and fluvial features of valley morphology with channel relief, pattern, shape, and dimension. Aspect I combines the influences of climate, depositional history, and vegetative life zones on channel morphology. Generalized categories of stream types are initially delineated using descriptions of dominant slope range, valley and channel cross-sections, and plan-view patterns (Figure 5.5 and Table 5.5).

The longitudinal profile serves to identify slope categories for stream reaches. For example, streams of type Aa+ have channel gradients greater than 10% with frequently spaced, vertical drop scour-pools (Figure 5.5 and Table 5.5). The cross-sectional profile also can be inferred at this broad level as well as information concerning floodplains,

TABLE 5.4. Hierarchy of river inventories.

Level of detail	Inventory description	Information required	Objectives
I	Broad geomorphological characterization	Landform; lithology; soils; climate; depositional history; basin relief; valley morphology; river profile morphology, general river pattern	To describe generalized fluvial features using remote sensing and existing inventories of geology, landform evolution, valley morphology, depositional history, and associated river slopes; relief and patterns used for generalized categories of major stream types and associated interpretations
II	Morphological description (channel types)	Channel patterns; entrenchment ratio; width-to-depth ratio; sinuosity; channel material; slope	To delineate homogeneous stream types that describe specific slopes, channel materials, dimensions, and patterns from "reference reach" measurements; provides a more detailed level of interpretation and extrapolation than Level I
III	Stream "state" of condition	Riparian vegetation; depositional patterns; meander patterns; confinement features; fish habitat indices; flow regime; river size category; debris occurrence; channel stability index; bank erodibility	To further describe existing conditions that influence the response of channels to imposed change and provide specific information for prediction methodologies (such as stream bank erosion calculations); provides for very detailed descriptions and associated prediction/ interpretation
IV	Verification	Involves direct measurements and observations of sediment transport, bank erosion rates, aggradation/degradation processes, hydraulic geometry, biological data such as fish biomass, aquatic insects, riparian vegetation evaluations, etc.	Provides reach-specific information on channel processes; used to evaluate prediction methodologies; to provide sediment, hydraulic, and biological information related to specific stream types; and to evaluate effectiveness of mitigation and impact assessments for activities by stream type

Modified from Rosgen 1994 with kind permission of Elsevier Science-NL, Sara Burgerhartstrant 25, 1,055 KV Amsterdam, The Netherlands.

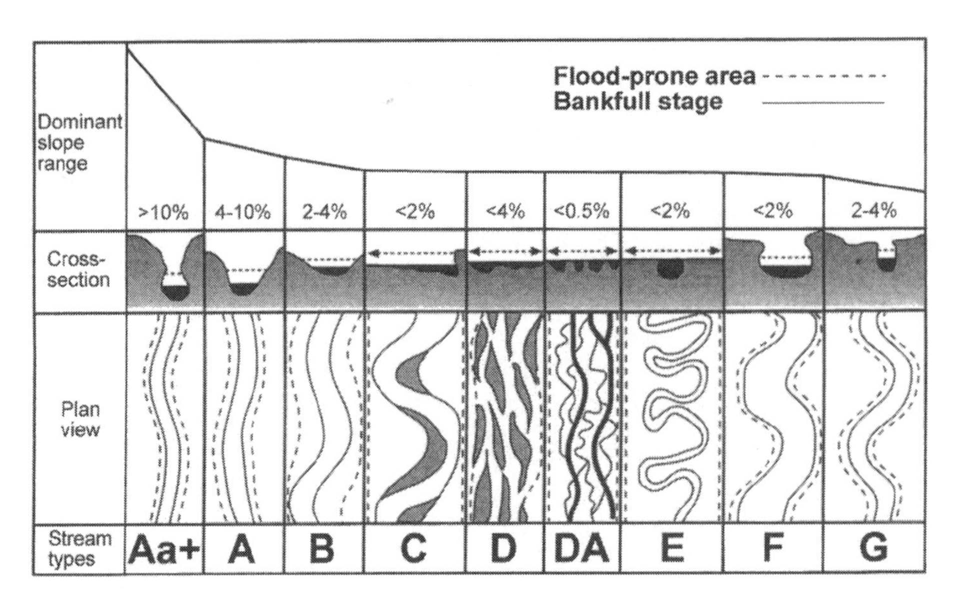

FIGURE 5.5. Longitudinal, cross-sectional, and plan views of major stream types (reprinted from Rosgen 1994 with kind permission of Elsevier Science-NL, Sara Burgerhartstrmt 25, 1055 KV Amsterdam, The Netherlands).

terraces, structural control features, confinement, entrenchment, and valley versus channel dimensions. For example, the type A streams are narrow, deep, confined, and entrenched while the width of the channel and the valley are similar (Figure 5.5 and Table 5.5). The plan view morphology is simply the pattern of the river from above. For example, type A streams are relatively straight while type C streams are meandering (Figure 5.5).

Morphological Description (Aspect II). After streams are separated into the major categories of A through G (Figure 5.5 and Table 5.5), Aspect II is applied separating them into discreet slope ranges and dominant substrate particle sizes. This results in 42 subcategories of stream types (Figure 5.6). In reality, however, there is a normal range of values for each criterion and this important observation is incorporated into Rosgen's classification system. This aspect recognizes and describes a morphological continuum within and among stream types. The continuum is applied where values outside the normal range are encountered but do not warrant a unique stream type. For example, selected channel slopes in Figure 5.6 are sorted by subcategories of: a+ (>10%), a (4–10%), b (2–3.9%), c (<2%), and c− (<0.01%).

The emphasis on channel materials is equally important, as they are critical not only for sediment transport and hydraulic influences (such as channel roughness) but also for the modification of the river's form, plan, and profile. Interpretation of biological function and stability also require this information. Using the "pebble count" method of Wolman (1954), with a few modifications for large bank materials and sand (Rosgen 1994), the particle size distribution of channel materials can be determined easily in the field.

Although the classification systems developed by Cupp (1989) and Rosgen (1985, 1994) are both based on geomorphic and geologic landscape features, they illustrate two fundamentally different approaches in classification. Rosgen's system is based on present stream characteristics (e.g., channel width, sinuosity). Cupp's system is based on the presumed potential states of the stream (i.e., all possible natural states that may occur in stream features within given segment types). Therefore, Rosgen's method is responsive to the effects of natural and human-induced disturbance as manifested

TABLE 5.5. Summary of delineative criteria for broad-level classification.

Stream type	General description	Entrenchment ratio	Width-to-depth ratio	Sinuosity	Slope	Landform/soils/features
Aa+	Very steep, deeply entrenched, debris transport streams	<1.4	<12	1.0–1.1	>10%	Very high relief; erosional, bedrock, or depositional features; debris flow potential; deeply entrenched streams; vertical steps with deep scour pools; waterfalls
A	Steep, entrenched, cascading step-pool streams; high energy/ debris transport associated with depositional soils; very stable if bedrock or boulder dominated channel	<1.4	<12	1.0–1.2	4–10%	High relief; erosional or depositional and bedrock forms; entrenched and confined streams with cascading reaches; frequently spaced, deep pools in associated step-pool bed morphology
B	Moderately entrenched, moderate-gradient, riffle dominated channel with infrequently spaced pools; very stable plan and profile; stable banks	1.4–2.2	>12	>1.2	2–3.9%	Moderate relief, colluvial deposition and/or residual soils; moderate entrenchment and width-to-depth ratio; narrow,gently sloping valleys; rapids predominate with occasional pools
C	Low-gradient, meandering, point-bar, riffle-pool, alluvial channels with broad, well-defined floodplains	>2.2	>12	>1.4	<2%	Broad valleys with terraces in association with floodplains and alluvial soils; slightly entrenched with well-defined meandering channel; riffle-pool bed morphology
D	Braided channel with longitudinal and transverse bars; very wide channel with eroding banks	n/a	>40	n/a	<4%	Broad valleys with alluvial and colluvial fans; glacial debris and depositional features; active lateral adjustment with abundance of sediment supply
DA	Anastomosing (multiple channels) narrow and deep with expansive well-vegetated floodplain and associated wetlands; very gentle relief with highly variable sinuosities; stable streambanks	>4.0	<40	variable	<0.05%	Broad, low-gradient valleys with fine alluvium and/or lacustrine soils; anastomosed (multiple channel) geologic control creating fine deposition with well-vegetated bars that are laterally stable with broad wetland floodplains
E	Low-gradient, meandering riffle-pool stream with low width-to-depth ratio and little deposition; very efficient and stable; high meander width ratio	>2.2	<12	>1.5	<2%	Broad valley/meadow; alluvial materials with floodplain; highly sinuous with stable, well-vegetated banks; riffle-pool morphology with very low width-to-depth ratio
F	Entrenched meandering riffle-pool channel on low gradients with high width-to-depth ratio	<1.4	>12	>1.4	<2%	Entrenched in highly weathered material; gentle gradients, with a high width-to-depth ratio; meandering, laterally unstable with bank-erosion rates; riffle-pool morphology
G	Entrenched "gully" step-pool and low width-to-depth ratio on moderate gradients	<1.4	<12	>1.2	2–3.9%	Gully, step-pool morphology with moderate slopes and low width-to-depth ratio; narrow valleys or deeply incised in alluvial or colluvial materials, i.e. fans or deltas; unstable, with grade control problems and high bank erosion rates

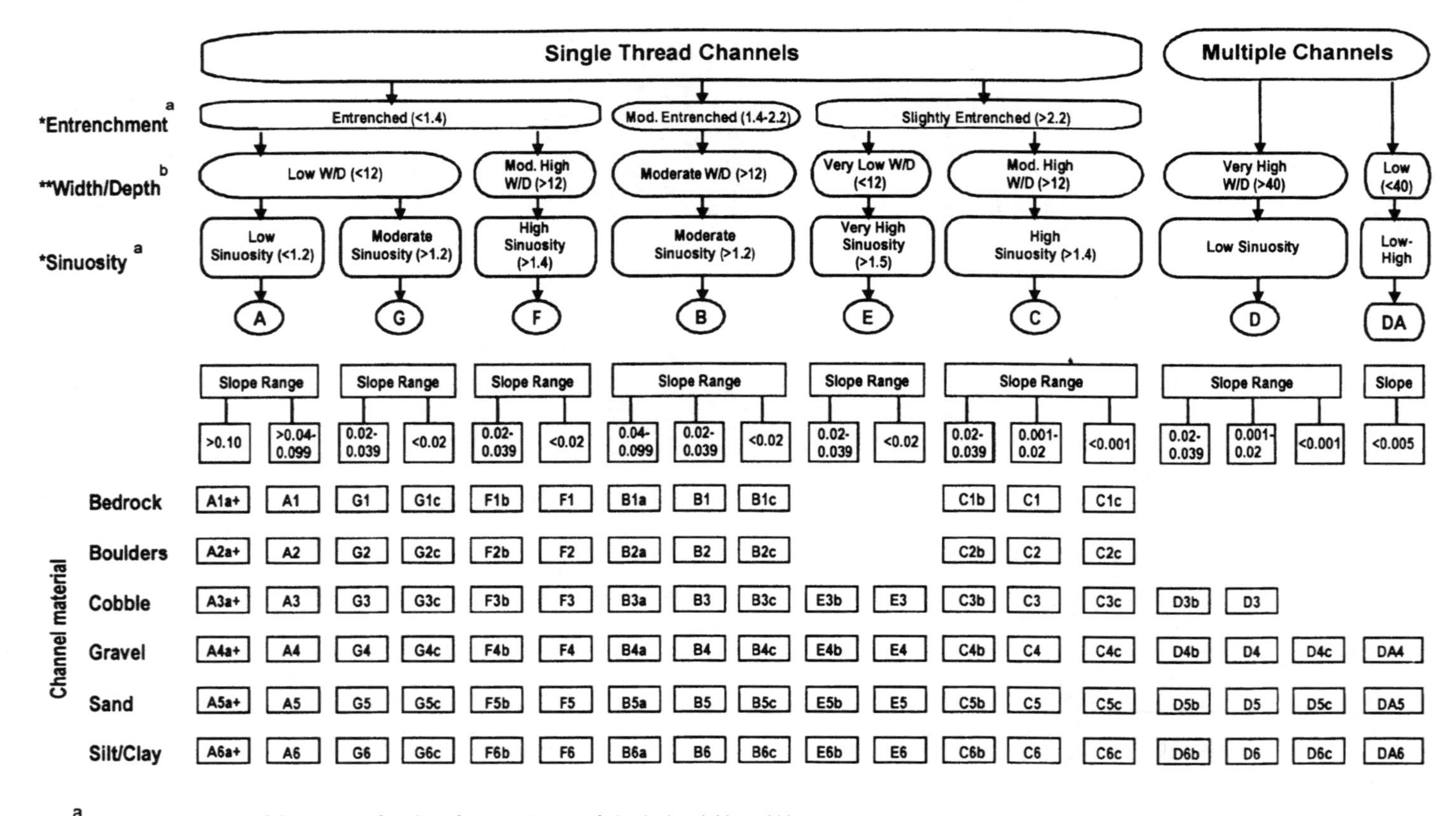

[a] Values can vary by ±0.2 units as a function of the continuum of physical variables within stream reaches.
[b] Values can vary by ±2.0 units as a function of the continuum of physical variables within stream reaches.

FIGURE 5.6. Key to classification of natural rivers (reprinted from Rosgen 1994 with kind permission of Elsevier Science-NL, Sara Burgerhartstrmt 25, 1055 KV Amsterdam, The Netherlands).

by variations in width-to-depth ratio or changes in riparian vegetation, whereas Cupp's approach is responsive to disturbances only at the large segment scale, or to severe small-scale disturbances, such as debris flows or hillslope failures, that cause channel features to deviate outside some predicted range or alter the mean state of the system.

There is disagreement about whether the principal units of classification should be temporally stable (e.g., valley segment) or dynamic (e.g., stream types). Arguments for temporal stability suggest that a reach, once classified, is of little management value if it changes naturally over the time scale of land-use practices (Frissell et al. 1986). In contrast, a dynamic classification based on smaller, evolving units provides a more accurate description of present conditions in the reach (e.g., active channel width, riffle/pool ratio). Both perspectives may be useful for management assessments depending on the specific objectives of the assessment.

Classification Coupling Biological and Physical Features

Coupling biotic resources with the physical features of streams has practical value for both science and management. Existing systems have been based on patterns of species distribution, community structure, and biotic function. Biotic communities serve as integrators of ecological conditions expressed over different time and space scales and, therefore, can be sensitive indicators of environmental vitality. Most classification systems have been based on fish (e.g., Huet 1954, Karr 1981) or invertebrate assemblages (e.g., Illies and Botosaneanu 1963, Cummins 1974, Wright et al. 1984). However, several recent systems have been based on patterns of riparian vegetation (Harris 1988) and aquatic plants (Holmes 1989). In general, all biotic classification schemes assume a predictable relationship between stream biota and geomorphic and hydrologic controlling factors acting on the system.

Vertebrate Community Classification

Fish have formed the basis for stream classification systems for several biological and political reasons. Hawkes (1975) argued that fish probably best reflect the general ecological conditions of rivers because they are presumed to be at the top of the aquatic food chain and, therefore, are integrative of the condition of the entire environmental system. In addition, because many commercial, recreational, and endangered fish species inhabit rivers, there has been continued need to categorize and manage their habitat. Fisheries managers and scientists have the growing responsibility of identifying fish community associations, their ecological requirements, and designing suitable ways of maintaining their integrity in the face of continued habitat deterioration (Naiman et al. 1995, Stouder et al. 1996). Despite the merits of this type of classification, there are limitations that often impede widespread application.

In general, models coupling biological and physical features usually sacrifice precision for generality and assume that fish populations are limited by habitat rather than intra- or interspecific competition, extrinsic factors (e.g., fishing mortality or disease), or natural disturbance (Bisson et al. 1982, Fausch et al. 1988). Although there is inherent value in using sensitive fish species in stream habitat models, individual species often show high yearly variability in production independent of physical habitat conditions (Hall and Knight 1981). In contrast, the entire fish community may provide a more accurate indication of habitat conditions, especially if community parameters are more stable over time than population parameters, and relate predictably to habitat features (e.g., complexity, size) and habitat change (Gorman and Karr 1978, Berkman and Rabeni 1987, Hughes et al. 1987) (Table 5.4).

Ultimately, zoogeographic factors restrict the geographic scope of classification schemes based on the structure of fish assemblages. However, spatial variability in physical and biotic factors shaping community dynamics also can limit geographic scope. Environmental disturbance regimes vary with climate and geology (Poff and Ward 1989). In streams where seasonal flow patterns are predictable, communities may be persistent and resilient (Moyle and Vondracek 1985). However, in streams with highly variable and unpredictable flow patterns, communities can exhibit sharp temporal fluctuations in structure (Matthews 1982), and this is especially so in the Pacific coastal ecoregion where so many of the fish populations are anadromous (Chapter 9 this volume, Stouder et al. 1996). Anadromous fish often out-compete resident species and, where spawning salmonid populations are large, excavation of the streambed during spawning significantly alters the invertebrate community (which also is used in classification, see below). Furthermore, within climatic regions, the influence of floods can vary depending on channel form and substrate (Resh et al. 1988).

Biotic factors further compound species-habitat relationships. In stream segments where competition and predation are important factors, fluctuations in physicochemical conditions can alter the intensity and direction of competitive and predator-prey interactions (Fraser and Cerri 1982, Reeves et al. 1987, Chapter 9 this volume). Further, variability in productivity between streams may also contribute to wide ranging diversity patterns (Bunn and Davies 1990, Morin and Naiman 1990, Chapter 17 this volume). Moreover, there are various human activities which produce major alterations in fish community composition (e.g., species introductions, chemical pollution, harvest) without altering physical habitat structure.

Many recent investigations have examined spatial patterns, both within and between streams of functional characteristics in fish communities (Gorman and Karr 1978, Moyle and Li 1979, Schlosser 1982 and 1987, Berkman and Rabeni 1987). Moyle and Li (1979) speculated that, while species composition may often be unstable, there may be stability in trophic structure in given habitat settings. Furthermore, Schlosser (1987) hypothesized that there is a predictable longitudinal pattern in characteristics of fish communities (e.g., trophic diversity, demography, seasonal stability) in warm-water streams.

The literature on the ecology of stream fish communities is replete with empirical support for stochastic and deterministic structure (Grossman et al. 1982, Moyle and Vondracek 1985, Matthews 1986) as well as strong (Gorman and Karr 1978) and weak species-habitat relationships (Schlosser 1982). Such disparity is ultimately a consequence of the physicochemical features of the drainage basin (i.e., geology, climate), channel (i.e., substrate, depth-to-width ratio), and habitats (i.e., depth, velocity, large organic debris). These are the primary determinants of the physical template influencing the life history attributes, population dynamics, and community structure and function of stream fishes.

For both scientific and management purposes, it is particularly important to characterize community patterns and controlling processes under different physical conditions. Zalewski and Naiman (1985) speculated that the relative importance of biotic and abiotic controls over fish community characteristics varies along a continuum from upstream to the mouth. Poff and Ward (1989) described a conceptual model relating factors of community regulation to characteristics of the flow regime. Site-specific management can be applied on the basis of understanding community patterns and controlling processes.

Invertebrate Community Classification

Classification schemes based on patterns in benthic invertebrate community structure also have been important tools. Hawkes (1975) discussed the value of developing biotic classification schemes which couple the macroinvertebrate distribution with physicochemical stream features. Macroinvertebrates are good indicators of both short- and long-term change, as well as local and large-scale disturbances because they exhibit diverse life history strategies (Minshall 1988). However, factors limiting the utility of fish classification (e.g., zoogeography, disturbance regimes, biotic interactions, and productivity) also restrict the utility of invertebrate-based classification systems by altering species-habitat relationships.

In Britain, for example, invertebrate assemblages form the basis for the classification of unpolluted rivers and are used to develop procedures for predicting faunal assemblages at given sites from a small set of physicochemical variables (Wright et al. 1989). This particular invertebrate classification system was developed following an intensive biological and physicochemical survey of rivers. Environmental variables measured are those suspected of playing major roles in determining the distribution of the invertebrate fauna, and those which are altered by chemical and thermal pollution and regulation of river discharge regimes (Armitage 1984). As a predictive model it is valuable in detecting environmental stress and identifying species-rich communities, both important elements in stream management (Wright et al. 1989). This type of predictive system, coupling the invertebrate classification scheme to environmental variables, is largely successful because it employs a small set of variables regulating invertebrate distribution which change with direct impacts on water quality. However, the ability to link this approach to larger landscape features of the watershed (e.g., hierarchical classification) is diminished because the human-induced alterations are considered to be on water quality and in-channel (on-site) physical features rather than larger-scale changes. Other potential drawbacks to this approach are the influence of larger-scale geologic features on water quality changes (Armitage 1984) and the demand for exhaustive field monitoring of invertebrate assemblages or in-channel physicochemical variables to establish such a system.

Another approach is based on the functional attributes of invertebrates (Chapter 8). For two decades there has been an emphasis on organizing species into ecologically meaningful trophic guilds and elucidating changes in the functional role of assemblages along the length of rivers (Cummins 1974). This approach takes advantage of changes in trophic diversity that occur naturally along the longitudinal profile of rivers (Chapter 15). Like fish community classification schemes, the value of classifying streams by invertebrate functional groups is the independence from taxonomic structure, which enables comparisons of different basins over larger regions. Minshall (1988), however, argued against relying on such a trophic group classification, using evidence of Hawkins et al. (1983), who were unable to find shifts in functional groups in habitats degraded by logging. An added drawback to such a classification is the difficulty of categorizing diverse species into realistic functional feeding groups for all life-history stages.

Plant Classification

Various classification systems based on riparian vegetation patterns have also been developed (Harris 1988, Swanson et al. 1988, Baker 1989). This has considerable potential for stream management because riparian forests are active boundaries at the interface between upland and aquatic systems, and therefore may be sensitive indicators of environmental change (Naiman and Décamps 1990, Chapter 12 this volume).

The fundamental classification unit of riparian zones is the community type. This is defined either by present vegetative composition or potential climax vegetation (Swanson et al. 1988). Inferences are drawn regarding environmental gradients and successional relationships between community types. Stratification of community types is based on overstory or understory vegetation. The understory (herbs and shrubs), because of its higher turnover rate, is a better indicator of current soil and hydrologic conditions, whereas the canopy is a better integrator of longer temporal patterns. As with other biotic classification systems, the most valuable riparian classification schemes center on relationships to physical factors associated with the river environment.

Many authors have addressed the need for ranking riparian zones with respect to conservation value or ecological potential (Slater et al. 1987, Harris 1988, Swanson et al. 1988, Baker 1989, Gregory et al. 1991, Gurnell et al. 1994). Slater et al. (1987) used species richness, rarity, and frequency-of-occurrence to formulate the conservation value of different stream segments. Although these biotic variables were independent of taxonomic structure, the value of this classification system was diminished by the absence of a relationship between riparian habitat variables and the aquatic biota.

Harris (1988) classified riparian vegetation (i.e., species composition) in relation to six geomorphic valley types in the Sierra Nevada mountains of California. Incorporating concepts from landscape ecology and hierarchical relationships of different landscape elements, he limited classification units to the stream segment scale and addressed the importance of larger-scale factors in determining smaller-scale patterns. His geomorphic-vegetation units differed in their sensitivity to management, yet were useful for purposes of resource inventory, detailed ecological studies, and prediction of human-induced alterations. Although Harris suggested several reasons for the stream segment-vegetation relationships, processes governing the observed patterns could not be determined. Nonetheless, the classification system developed by Harris was an important step forward in coupling different landscape processes to biotic resources and in attempting to predict the sensitivity of stream segments to disturbance.

Another approach undertaken in the late 1970s in Britain classifies rivers from the distribution of aquatic plant assemblages (Holmes 1989). The basis for this system is that plants integrate short- and long-term conditions in the river, and that they play an important role in the ecology of stream fish and invertebrates as food and shelter. This approach requires an extensive survey of rivers, including a complete documentation of plant species diversity and habitat variables. A computer-aided classification system is essential to stratify rivers and river segments hierarchically. This approach has been successful largely because rivers in Britain do not exhibit the strong longitudinal shifts in physical features seen in western North America (Holmes 1989).

An Evaluation of the Biological–Physical Approach

The usefulness of biologically based stream classification systems in stream management may be diminished because such approaches demand, at least initially, intensive efforts to measure and monitor community characteristics (Chapter 18). This is especially true for invertebrates, somewhat less so for fish and vegetation. Furthermore, species-habitat relationships are often confounded by such factors as zoogeography, disturbance, biotic interactions, and productivity. If biotic classification systems are to have broad application, they must be related to physical features of the watershed in order to make inferences on the effects of land-use changes. In this regard Harris (1988) comes closest to accomplishing this objective. Yet disturbances to different watershed elements (e.g., habitat, riparian zone, hillslope) can produce similar impacts on the stream biota. In the absence of information on the cause of stream degradation and the linkage between the physical and biotic components of the system, it remains difficult to gauge the recovery potential of stream biota.

Management Based on Stream Classification

Although the number and diversity of specific stream classification systems are large, there appears to be a consensus developing on the fundamental attributes of an enduring classification system. These attributes relate to the ability to encompass broad spatial and temporal scales, to integrate structural and functional characteristics under various disturbance regimes, to convey information about underlying mechanisms controlling in-stream features, and to accomplish this at low cost and at a high level of uniform understanding among resource managers (Naiman et al. 1992, Hawkins et al. 1993). No existing classification system adequately meets all of the model attributes. Even though the concepts of Frissell et al. (1986), Cupp (1989), Hawkins et al. (1993), and Rosgen (1994) are regarded as important intellectual advancements, they do not provide the level of understanding of channel processes needed to predict channel responses to specific types of watershed disturbances (such as debris flows or hillslope failures). Consequently, specific links between physical and biological processes within these classification systems remain poorly defined. Physical, process-based approaches that offer more promise in meeting these attributes, discussed by Montgomery and Buffington (Chapter 2 this volume and 1997), are currently being used to address a wide variety of stream-related management issues in the Pacific Northwest. Nevertheless, the generally narrow perspective provided by all existing classification systems limits their effectiveness. For example, all current stream classifications for regulating forest practices in the Pacific coastal ecoregion rely simply on the presence or absence of salmonids coupled with some index of stream size (e.g., channel width, mean annual discharge, or stream order). The relative degree of regulatory protection decreases with decreasing stream size and is virtually nonexistent in streams without salmon or trout. Generally, there is no explicit consideration of the underlying geomorphic context or the potential response of the channel segment to disturbance. A similar case could be made for the classification schemes applied by the water quality regulatory agencies which classify streams based on a comparison of physiochemical characteristics with a region-wide set of "desired" criteria. Such classifications are narrowly focused only on the properties of the system which fall under the legal jurisdiction of the regulatory agency.

Despite these caveats, the hierarchical classification system has been useful in making resource managers in the Pacific Northwest aware of the diversity of stream types and the need for a variety of management prescriptions for habitat protection and conservation. This is especially important in a region with approximately forty subcategories of stream segments, and where nearly 80% of the ancient forests have been cut in the last century to sustain a US $9.0 billion /yr forest products industry employing more than 60,000 people. The evolving stream classification system currently used as part of the Washington Forest Practices Regulations (Chapter 2), for example, allows resource managers and scientists to consider, in some cases, alternative forestry practices (e.g., silvicultural techniques, cutting patterns) that are tailored to specific stream and valley bottom configurations rather than using narrowly defined techniques and regulations applied across a few stream sizes and types. Simple prescriptive management, such as riparian zones of fixed width, is less effective than management techniques adapted to local topography and natural disturbance regimes.

This has been effectively demonstrated by Benda et al. (1992), who showed how the zonation of geomorphic surfaces in a 260 km^2 montane valley could be used to focus attention on streams where salmonid habitat value was highest. The valley was stratified at a large scale (>50 km^2) by geologic structure and associated geomorphology, and at a smaller scale (<10 km^2) by older lacustrine clay terraces and the more recent floodplain of the main river. Additionally they quantified differences in the habitat characteristics (channel width, large organic debris, and spawning gravel) of streams on the various geomorphic surfaces. The valley was then partitioned into areas of high and low risk based on the physical habitat characteristics of the streams.

This is only one example of an emerging perspective for streams and riparian zones which uses classification as a basis for designing new approaches for resource management. The placement of logging access roads, decisions on when, where, and how much tree harvest should occur, and development of silvicultural restoration techniques and of system models all require adherence to stream type. The most effective stream and riparian models include aquatic and terrestrial disturbance regimes, unique species mixtures, spatial and temporal heterogeneity, and microclimate gradients—all of which vary by stream type. Further, emerging silvicultural techniques for riparian tree species account for genetic vitality, stand development, and system complexity—factors that are specific to stream types (Berg 1995).

Even though the search for an ideal classification system is not complete, the fundamental principles of an ideal system are reasonably well articulated. However, it will be necessary for resource managers to adapt guiding principles using an adaptive management approach for specific situations (Holling 1978, Chapter 27 this volume). The task is difficult and requires a holistic, long-term perspective, but once in place, it provides a solid foundation for making resource decisions that affect the environmental quality of streams for decades.

Acknowledgments. I thank John M. Buffington, Robert E. Bilby, and two anonymous reviewers for insightful comments and suggestions, which substantially improved the content. Special thanks are due to D. L. Lonzarich, S. Ralph, and T. Beechie who initially stimulated me to learn more about this important topic.

Literature Cited

Armitage, P.D. 1984. Environmental changes induced by stream regulation and their effect on lotic macroinvertebrate communities. Pages 139–165 *in* A. Lillehammer and S.J. Saltveit, ed. Regulated rivers. Oslo Universitetsforlaget, Norway.

Bailey, R.G. 1978. Description of the ecoregions of the United States. USDA Forest Service, Intermountain Region, Ogden, Utah, USA.

Baker, W.L. 1989. Classification of the riparian vegetation of the montane and subalpine zones in western Colorado. Great Basin Naturalist **9**:214–228.

Beechie, T.J., and T.H. Sibley. 1990. Evaluation of the TFW stream classification system on the South Fork Stillaguamish streams. The State of Washington Water Research Center, Project Number A-164-WASH, Seattle, Washington, USA.

Benda, L., T.J. Beechie, A. Johnson, and R.C. Wissmar. 1992. The geomorphic structure of salmonid habitats in a recently deglaciated river basin, Washington State. Canadian Journal of Fisheries and Aquatic Sciences **49**:1246–1256.

Berg, D.R. 1995. Riparian silvicultural design and assessment in the Pacific Northwest Cascade Mountains, USA. Ecological Applications **5**:87–96.

Berkman, H.E., and C.F. Rabeni. 1987. Effect of siltation on stream fish communities. Environmental Biology of Fishes **18**:285–294.

Bisson, P.A., J.L. Nielsen, R.A. Palmason, and L.E. Grove. 1982. A system of naming habitat types in small streams, with examples of habitat utilization by salmonids during low streamflow. Pages 62–73 *in* N.B. Armantrout, ed. Acquisition and utilization of aquatic habitat inventory information. American Fisheries Society Western Division, Bethesda, Maryland, USA.

Bisson, P.A., K. Sullivan, and J.L. Nielson. 1988. Channel hydraulics, habitat use, and body form of juvenile coho salmon, steelhead, and cutthroat trout in streams. Transactions of the American Fisheries Society **117**:262–273.

Briggs, B.J.F., M.J. Duncan, I.G. Jowett, J.M. Quinn, C.W. Hickey, R.J. Davies-Collwy, et al. 1990. Ecological characterization, classification, and modeling of New Zealand rivers: An introduction and synthesis. New Zealand Journal of Marine and Freshwater Research **24**:277–304.

Brussock, P.P., A.V. Brown, and J.C. Dixon. 1985. Channel form and stream ecosystem models. Water Resources Bulletin **21**:859–866.

Bunn, S.E., and P.M. Davies. 1990. Why is the stream fauna of south-western Australia so impoverished? Hydrobiologia **194**:169–176.

Carpenter, K.E. 1928. Life in inland waters. Macmillan, New York, New York, USA.

Cummins, K.W. 1974. Structure and function of stream ecosystems. BioScience **24**:631–641.

Cupp, C.E. 1989a. Identifying spatial variability of stream characteristics through classification. Masters thesis. University of Washington, Seattle, Washington, USA.

——. 1989b. Stream corridor classification for forested lands of Washington. Washington Forest Protection Association, Olympia, Washington, USA.

Davis, W.M. 1890. The rivers of northern New Jersey, with note on the classification of rivers in general. National Geographic Magazine **2**:82–110.

Décamps, H. 1984. Towards a landscape ecology of river valleys. Pages 163–178 *in* J.H. Cooley and F.G. Golley, eds. Trends in ecological research for the 1980s. Plenum, New York, New York, USA.

Fausch, K.D., C.L. Hawkes, and M.G. Parsons. 1988. Models that predict standing crop of stream fish from habitat variables: 1950–85. USDA Forest Service General Technical Report PNW-GTR 213, Pacific Northwest Field Station, Portland, Oregon, USA.

Fraser, D.F., and R.D. Cerri. 1982. Experimental evaluation of predator–prey relationships in a patchy environment: Consequences for habitat use in minnows. Ecology **63**:307–313.

Frissell, C.A., W.J. Liss, C.E. Warren, and M.D. Hurley. 1986. A hierarchical framework for stream classification: Viewing streams in a watershed context. Environmental Management **10**:199–214.

Gauch, H.G. 1982. Multivariate analysis in community ecology. Cambridge University Press, Cambridge, UK.

Gorman, O.T., and J.R. Karr. 1978. Habitat structure and stream fish communities. Ecology **59**:507–515.

Gregory, S.V., F.J. Swanson, W.A. McKee, and K.W. Cummins. 1991. An ecosystem perspective of riparian zones. BioScience **41**:540–551.

Grossman, G.D., P.B. Moyle, and J.O. Whittaker. 1982. Stochasticity in structural and functional characteristics in an Indiana stream fish assemblage: A test of community theory. American Naturalist **120**:423–454.

Gurnell, A.M., P. Angold, and K.J. Gregory. 1994. Classification of river corridors: Issues to be addressed in developing an operational methodology. Aquatic Conservation: Marine and Freshwater Ecosystems **4**:219–231.

Hall, J.D., and N.J. Knight. 1981. Natural variation in abundance of salmonid populations in streams and its implication for design of impact studies. Report Number EPA-600/S3-81-021, United States Environmental Protection Agency, Corvallis, Oregon, USA.

Harris, R.R. 1988. Associations between stream valley geomorphology and riparian vegetation as a basis for landscape analysis in eastern Sierra Nevada, California, USA. Environmental Management **12**:219–228.

Hawkes H.A. 1975. River zonation and classification. Pages 312–374 *in* B.A. Whitton, ed. River ecology. Blackwell, London, UK.

Hawkins, C.P., J.L. Kershner, P.A. Bisson, M.D. Brgant, L.M. Decker, and S.V. Gregory, et al. 1993. A hierarchical approach to classifying stream habitat features. Fisheries **18**:3–12.

Holling, C.S., ed. 1978. Adaptive environmental assessment and management. John Wiley & Sons, New York, New York, USA.

Holmes, N.T.H. 1989. British rivers: A working classification. British Wildlife **1**:20–36.

Horton, R.E. 1945. Erosional development of streams and their drainage basins: Hydrophysical approach to quantitative morphology. Geological Society of America Bulletin **56**:275–370.

Huet, M. 1954. Biologie, profils en long et en travers des eaux courants. Bulletin Français de Pisciculture **175**:41–53.

Hughes, R.M., R.M. Rexstad, and E. Bond. 1987. The relationship of aquatic ecoregions, river basins, and physiographic provinces to the ichthygeographic regions in Oregon. Copeia **1987**:423–432.

Illies, J., and L. Botosaneanu. 1963. Problèmes et méthodes de la classification et de la zonation écologique des eaux courantes considerées surtout du point de vue faunistique. Mitteilungen der Internationalen Vereinigung für Theoretische and Angewandte Limnologie **12**:1–57.

Karr, J.R. 1981. Assessment of biotic integrity using fish communities. Fisheries **6**:21–26.

Kellerhals R., and M. Church. 1989. The morphology of large rivers: Characterization and management. Pages 31–48 *in* D.P. Dodge, ed. Proceedings of the International Large River Symposium. Canadian Special Publications of Fisheries and Aquatic Sciences 106, Ottawa, Ontario, Canada.

Langbein, W.B., and L.B. Leopold. 1966. River meanders—theory of minimum variance. Professional Paper. United States Geological Survey Report Number 422H, Washington, DC, USA.

Larsen, D.P., J. Omernik, R.M. Hughes, D.R. Dudley, C.M. Rohm, T.R. Whittie, et al. 1986. The correspondence between spatial patterns in fish assemblages in Ohio streams and aquatic ecoregions. Environmental Management **10**:815–828.

Leopold, L.B., and M.G. Wolman. 1957. River channel patterns: Braided, meandering and straight. United States Geological Survey Professional Paper 282-B, Washington, DC, USA.

Leopold, L.B., M.G. Wolman, and J.P. Miller. 1964. Fluvial processes in geomorphology, W.H. Freeman, San Francisco, California, USA.

Lotspeich, F.B. 1980. Watersheds as the basic ecosystem: This conceptual framework provides a basis for a natural classification system. Natural Resources Bulletin **16**:581–586.

Macan, T.T. 1961. A review of running waters. Verhandlungen der Internationalen Vereinigung für Theoretische und Angewandte Limnologie **14**:587–602.

Matthews, W.J. 1982. Small fish community structure in Ozark streams: Structured assemble patterns or random abundance of species. American Midland Naturalist **107**:42–54.

——. 1986. Fish community structure in a temperate stream: Stability, persistence and a catastrophic flood. Copeia 388–397.

Minshall, G.W. 1988. Stream ecosystem theory: A global perspective. Journal of the North American Benthological Society **7**:263–288.

Minshall, G.W., R.C. Petersen, K.W. Cummins, T.H. Bott, J.R. Sedell, C.E. Cushing, et al. 1983. Interbiome comparison of stream ecosystem dynamics. Ecological Monographs **53**:1–25.

Montgomery, D.R., and J.M. Buffington. 1997. Channel reach morphology in mountain drainage basins. Geological Society of America Bulletin **109**:596–611.

Morin, R., and R.J. Naiman. 1990. The relationship of stream order to fish community dynamics in boreal forest watersheds. Polskie Archiwum Hydrobiologii **37**:135–150.

Moyle, P.B., and H.W. Li. 1979. Community ecology in predator–prey relationships in warmwater streams. Pages 171–181 *in* R.H. Stroud and H.E. Clepper, eds. Predator-prey systems in fisheries management. Sport Fishing Institute, Washington, DC, USA.

Moyle, P.B., and B. Vondracek. 1985. Persistence and structure of the fish assemblage in a small California stream. Ecology **66**:1–13.

Naiman, R.J., and H. Décamps, eds. 1990. The ecology and management of aquatic–terrestrial ecotones. UNESCO, Paris, France, and Parthenon, Carnforth, UK.

Naiman, R.J., D.G. Lonzarich, T.J. Beechie, and S.C. Ralph. 1992. General principles of classification and the assessment of conservation potential in rivers. Pages 93–123 *in* P. Boon, P. Calow, and G. Petts, eds. River conservation and management. John Wiley and Sons, Chichester, UK.

Naiman, R.J., J.M. Magnuson, D.M. McKnight, and J.A. Stanford, eds. 1995. The freshwater imperative. Island Press, Washington, DC, USA.

Naiman, R.J., J.M. Melillo, M.A. Lock, T.E. Ford, and S.R. Reice. 1987. Longitudinal gradients of ecosystem processes and community structure in a subarctic river continuum. Ecology **68**:1139–1156.

Nawa, R.K., C.A. Frissell, and W.J. Liss. 1990. Life history and persistence of anadromous fish stocks in relation to stream habitats and watershed classification. Annual Progress Report, Oregon Department of Fish and Wildlife, Portland, Oregon, USA.

Omernik, J.M. 1987. Ecoregions of the conterminous United States. Annals of the Association of American Geographers **77**:118–125.

Poff, N.L., and J.V. Ward. 1989. Implications of stream flow variability and predictability for lotic community structure: A regional analysis for streamflow patterns. Canadian Journal of Fisheries and Aquatic Sciences **48**:1805–1818.

Reeves, G.H., F.H. Everest, and J.D. Hall. 1987. Interactions between the redside shiner (*Richardsonius balteatus*) and the steelhead trout (*Salmo gairdneri*) in western Oregon: Influence of water temperature. Canadian Journal of Fisheries and Aquatic Sciences **44**:1603–1613.

Resh, V.H., A.V. Brown, A.P. Covich, M.E. Gurtz, H.W. Li, G.W. Minshall, et al. 1988. The role of disturbance in stream ecology. Journal of the North American Benthological Society **7**:433–455.

Ricker, W.E. 1934. An ecological classification of certain Ontario streams. Publications of the Academy of Natural Sciences of Philadelphia **101**:277–341.

Rohm, C.M., J.W. Giese, and C.G. Bennett. 1987. Evaluation of an aquatic ecoregion classification of streams in Arkansas. Journal of Freshwater Ecology **4**:127–139.

Rosgen, D.L. 1985. A stream classification system. Pages 91–95 *in* R.R. Johnson, C.D. Zeibell, D.R. Patton, P.F. Pfolliott, and R.H. Hamre, eds. Riparian ecosystems and their management: Reconciling conflicting uses. USDA Forest Service General Technical Report M-120, Rocky Mountain Forest and Range Experimental Station, Fort Collins, Colorado, USA.

——. 1994. A classification of natural rivers. Catena **22**:169–199.

Schlosser, I.J. 1982. Fish community structure and function along two habitat gradients in a headwater stream. Ecological Monographs **52**:396–414.

——. 1987. A conceptual framework for fish communities in small warmwater streams. Pages 17–24 *in* W.J. Matthews and D.C. Heins, eds. Community and evolutionary ecology of North American stream fishes. University of Oklahoma Press, Norman, Oklahoma, USA.

Shelford, V.E. 1911. Ecological succession I. Stream fishes and method of physiographic analysis. Biological Bulletin **21**:9–35.

Slater, F.M., P. Curry, and C. Chadwell. 1987. A practical approach to the examination of the conservation status of vegetation in river corridors in Wales. Biological Conservation **43**:259–263.

Stouder, D., P.A. Bisson, and R.J. Naiman, eds. 1996. Pacific salmon and their ecosystems. Chapman & Hall, New York, New York, USA.

Strahler, A.N. 1957. Quantitative analysis of watershed geomorphology. American Geophysical Union Transactions **38**:913–920.

Swanson, S., R. Miles, S. Leonard, and K. Genz. 1988. Classifying rangeland riparian areas: The Nevada task force approach. Journal of Soil and Water Conservation **43**:259–263.

Vannote, R.L., G.W. Minshall, K.W. Cummins and J.R. Sedell. 1980. The river continuum concept. Canadian Journal of Fisheries and Aquatic Sciences **37**:130–137.

Ward, J.V., and J.A. Stanford. 1987. The ecology of regulated streams: Past accomplishments and directions for future research. Pages 391–409 *in* J.F. Craig and J.B. Kemper, eds. Regulated streams: Advances in ecology. Plenum, New York, New York, USA.

Warren, C.E. 1979. Toward classification and rationale for watershed management and stream protection. United States Environmental Protection Agency Report Number EPA-600/3-79-059, Corvallis, Oregon, USA.

Wasson, J.G. 1989. Eléments pour une typologie fontionelle des eaux courants: 1. Revue critique de quelques approches existantes. Bulletin d'Ecologie **20**:109–127.

Whittier, T.R., R.M. Hughes, and D.P. Larsen. 1988. Correspondence between ecoregions and spatial patterns in stream ecosystems in Oregon. Canadian Journal of Fisheries and Aquatic Sciences **25**:1264–1278.

Wissmar, R.C., and F.J. Swanson. 1990. Landscape disturbances and lotic ecotones. Pages 65–89 *in* R.J. Naiman and H. Décamps, eds. The ecology and management of aquatic-terrestrial ecotones. UNESCO, Paris, France, and Parthenon, Carnforth, UK.

Wolman, M.G. 1954. A method of sampling coarse river-bed material. Transactions of the American Geophysical Union **35**:951–956.

Wright, J.F., D. Moss, P.D. Armitage, and M.T. Furse. 1984. A preliminary classification of running water sites in Great Britain based on macro-invertebrate species and the prediction of community type using environmental data. Freshwater Biology **14**:221–256.

Wright, J.F., P.D. Armitage, M.T. Furse, and D. Moss. 1989. Prediction of invertebrate communities using stream measurements. Regulated Rivers: Research and Management **4**:147–155.

Zalewski, M., and R.J. Naiman. 1985. The regulation of fish communities by a continuum of abiotic-biotic factors. Pages 3–9 *in* J.S. Alabaster, ed. Habitat modification and freshwater fisheries. Butterworths, London, UK.

6
Microorganisms and Organic Matter Decomposition

Keller F. Suberkropp

Overview

- Forested stream ecosystems are generally considered heterotrophic. They depend on inputs of allochthonous organic matter from the riparian zone for most of their energy. This organic matter is the base of the detrital food web in which heterotrophic microorganisms (fungi, bacteria, and protozoa) decompose and transform organic matter into energy available to organisms at higher trophic levels.
- Both dissolved organic matter (DOM) and particulate organic matter (POM) are decomposed in streams. The decomposition, or breakdown, of POM such as leaf litter and wood occurs as a result of leaching, microbial mineralization, and fragmentation. The predominant microorganisms associated with the breakdown of leaf litter are fungi, in particular, aquatic hyphomycetes, an ecological group of fungi that are adapted to flowing waters. These fungi produce extracellular enzymes that soften (macerate) leaf litter. Fungal species vary in their palatability to invertebrate detritivores. The growth and activity of palatable fungi make leaf litter a better food source for these animals.
- Biomass and production of fungi predominate over that of bacteria on coarse particulate organic matter (CPOM) such as leaves. As the particle size of the organic matter decreases, biomass and production of bacteria increases as that of fungi declines.
- Fungi growing on leaf detritus differ in their palatability and food quality to leaf-shredding invertebrates (shredders). The improvement in food quality of leaf detritus as a result of fungal colonization appears to be the result of increased concentrations of fungal biomass and fungal degradative enzymes. These enzymes convert leaf tissue into a more assimilable food source and may continue to act in the digestive tract of the shredder.
- Rates of leaf breakdown are affected both by internal factors, such as litter quality (nitrogen, lignin, tannin content) and external environmental factors (nutrient concentrations in stream water, temperature, presence of shredders, pH, current velocity, and oxygen availability).
- Wood typically breaks down at slower rates than leaf litter. Although less is known about the organisms and processes involved in wood decay in streams, microbial activity on woody substrates is significant.
- Decomposition of fine particulate and dissolved organic matter is dominated by bacteria. The flow of energy from dissolved organic matter through bacteria to protozoa and filter feeding invertebrates represents the *microbial loop*—a trophic pathway that plays an important role in nutrient regeneration in forested streams.

Introduction

Heterotrophic microorganisms (fungi, bacteria, and protozoa) in streams and rivers use decaying organic matter as a source of carbon (C)

TABLE 6.1. Major size classes of organic matter found in streams.

Class	Size	Type of substance
Coarse particulate organic matter (CPOM)	>1 mm	Leaves, needles, twigs, branches, fruits, woody boles
Fine particulate organic matter (FPOM)	0.5 μm–1 mm	Degraded CPOM, microbial cells, animal parts, soil organic matter
Dissolved organic matter (DOM)	<0.5 μm	Leachates, exudates, soluble and colloidal organic matter in ground water and runoff

Modified from Cummins 1974.

and energy. Sources and types of this organic matter vary. A significant amount of the organic matter in streams is allochthonous (originates in the surrounding terrestrial environment) (Chapter 15), and enters the stream as detritus—a complex of decaying organic matter and its associated microbiota. In addition, primary production by algae and macrophytes in streams (Chapter 7) gives rise to autochthonous (originating within the stream) organic matter and, if not consumed directly, may become detrital organic matter.

Organic matter can be classified according to solubility and size (Cummins 1974). Dissolved organic matter (DOM) is functionally defined as that organic matter that passes through filters with pore sizes from 0.2 to 0.5 μm (Table 6.1). Particulate organic matter (POM) is divided into coarse particulate organic matter (CPOM) and fine particulate organic matter (FPOM) (Table 6.1). Both classes can be further subdivided into additional size classes, for example, ultrafine particulate organic matter (UPOM, 0.5–50 μm), depending on the questions being addressed (Minshall et al. 1983). All types of organic matter may be reported as particulate or dissolved organic carbon (POC, DOC) which typically comprises 45 to 50% of the dry weight of organic matter.

In addition to being used by microorganisms, some organic matter is used by stream macroorganisms for energy and nutrients, some is transported through the system, and some is stored. Energy (organic matter) budgets constructed for forested streams indicate the relative importance of different sources and processes to the stream ecosystem (Fisher and Likens 1973, Hornick et al. 1981, Triska et al. 1982, Naiman et al. 1986, Calow 1992). An organic matter budget for a small, high gradient, forested stream in the Pacific Northwest (Table 6.2) indicates the importance of allochthonous inputs from the riparian vegetation (litterfall and lateral movement) in comparison with autochthonous production (algae and moss photosynthesis) (Triska et al. 1982). In this case, outputs of organic matter are primarily attributed to downstream transport of both DOM and POM, and to respiration by microorganisms associated with detritus. In headwater forested streams such as this, the riparian forest canopy provides shade which limits the amount

TABLE 6.2. An organic matter budget for a small forested stream (Watershed 10, a 10.2-ha catchment in the H.J. Andrews Experimental Forest, Oregon) during a wet year.

Inputs	Total 790 kg/yr	Standing crop	Total 10,585 kg	Outputs	Total 756 kg/yr
Litterfall	22%	Large CPOM (woody boles)	82%	POM transport	32%
Lateral movement	43%	Small CPOM (twigs, leaves)	15%	Microbial respiration	24%
Throughfall	7%				
		FPOM	3%		
DOM transport	26%			DOM transport	41%
Algae & moss photosynthesis	3%	Algae & moss	0.3%	Algae & moss respiration	2%

Modified from Triska et al. 1982.

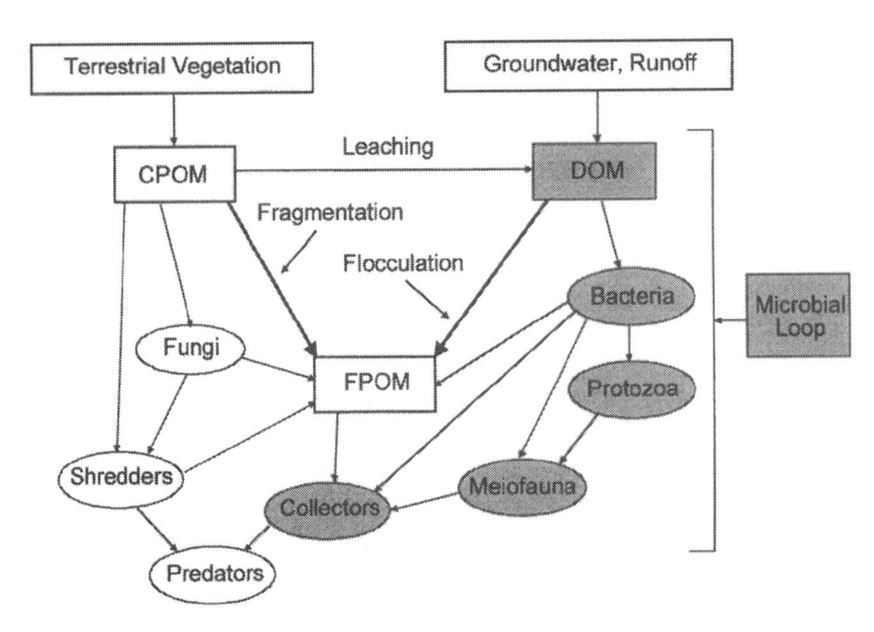

FIGURE 6.1. A detrital food web characteristic of a forested stream. Major components of the microbial loop are shaded.

of autochthonous production and contributes large inputs of organic matter in the form of leaves, branches and DOM of various types. Such allochthonous sources of organic matter must enter into detrital food webs to be used by stream organisms. Decomposers (i.e., heterotrophic bacteria and fungi) are the key organisms that use this organic matter. These microorganisms grow at the expense of organic matter and convert at least a portion into microbial biomass. Microbial biomass plus modified detritus serves as a food source for detritivores (both invertebrate and vertebrate) that are common in streams.

A typical detrital food web found in such headwater streams is summarized in Figure 6.1. This food web shows two major pathways for the use of organic matter in streams. CPOM such as leaf litter is decomposed primarily by fungi, consumed by shredders and, as a result, converted into carbon dioxide, DOM, and FPOM. Dissolved organic matter is decomposed primarily by bacteria. This transfers energy throughout the microbial loop in which bacteria are consumed by protozoa, meiofauna (e.g., copepods, nematodes) or filtering collectors (e.g., blackflies) and eventually by higher trophic levels.

This chapter introduces the ecology of the heterotrophic microorganisms (fungi, bacteria, protozoa) which are at the base of the detrital food webs of streams. Decomposition, the driving process of detrital food webs, is described by discussing the role of microorganisms and the factors that affect rates of breakdown of terrestrial leaf litter (a type of CPOM). Finally, sources and decomposition of DOM and its role in the microbial loop of stream ecosystems are characterized.

Heterotrophic Microorganisms

Fungi. Fungi play a number of roles in stream ecosystems, ranging from decomposers to parasites and pathogens of aquatic plants and animals. Even though fungi have been detected in a variety of habitats, the ecology of fungi found in most stream habitats (sediments, rock surfaces) is poorly understood. Most of the ecological information currently available focuses on fungi involved in the colonization and breakdown of leaf litter after it enters a stream. Before leaves are shed from riparian vegetation they are typically colonized by a variety of terrestrial fungi, so that leaves entering a stream or river already contain a number of fungal inhabitants. Little is known about the activity of terrestrial fungi once leaves enter the water. However, evidence suggests that if temperatures are low, as is typically the case during autumn in temperate climates, many of these fungi are not very active (Bärlocher and Kendrick 1974, Suberkropp and Klug 1976). In addition, there is little if any dispersal of these fungi underwater. Once leaves enter a stream, they are rapidly colonized by stream fungi. These fungi belong to an ecological group known as aquatic hyphomycetes (also known as Ingoldian fungi) and consist of asexual forms (anamorphic species) of higher fungi (mainly Ascomycota with some Basidiomycota). The sexual stage (teleomorph) of most aquatic hyphomycetes has not been described.

Since the initial discovery of aquatic hyphomycetes growing on leaves in streams (Ingold 1942), many studies have examined the ecology of these fungi and their role in leaf decomposition. Recent reviews have examined the biology and ecology of aquatic hyphomycetes (Bärlocher 1992, Suberkropp 1992a), their interactions with invertebrates (Bärlocher 1985, Suberkropp 1992b), and their role in decomposition (Chamier 1985, Maltby 1992a, Gessner et al. 1997).

Aquatic hyphomycetes, commonly found growing on leaves and wood in flowing waters worldwide, are successful in their role as decomposers as a result of both morphological and physiological adaptations that allow them to colonize and use plant litter in flowing waters. In many cases, aquatic hyphomycetes produce spores only when submerged in water (an unusual property for higher fungi). Because of their characteristic tetraradiate or sigmoid shapes (Figure 6.2) and the mucilage produced at the ends of the arms (Read et al. 1992), spores of aquatic hyphomycetes attach more efficiently and wash off substrata less readily than fungal spores of other shapes (Webster 1959, Webster and Davey 1984). Aquatic hyphomycetes are also adapted for growth at low temperatures common in temperate climates during autumn leaf-fall. Aquatic hyphomycetes commonly found on leaves in temperate streams during autumn and winter grow optimally at 15 to 20 °C but can grow at 1 °C (Suberkropp 1984). In contrast, species common in summer (or in tropical streams) grow optimally at temperatures as high as 25 to 30 °C, but typically do not grow at temperatures below 5 °C. Most aquatic hyphomycete species also are capable of producing a variety of extracellular enzymes that can degrade major structural plant polysaccharides, including cellulose, pectin, and xylan (Suberkropp and Klug 1980, Chamier 1985, Suberkropp 1992a), making them well adapted to obtaining carbon and energy from plant litter.

In laboratory microcosms, individual species of aquatic hyphomycetes create changes in leaves that are very similar to the changes that occur to leaves in streams (Suberkropp 1991), suggesting that these fungi are major mediators of leaf breakdown in these ecosystems. During leaf breakdown, aquatic hyphomycetes allocate a great deal of their energy to reproduction (production of spores) (Bärlocher 1982a, Suberkropp 1991, Gessner and Chauvet 1994). For example, in two fungi grown on leaf litter in the laboratory, spore production accounts for 46% and 81% of the total production, and the weight of the spores produced accounts for about 10% of the loss in leaf organic matter (Suberkropp 1991). In streams, the weight of the spores produced during leaf breakdown accounted for 1 to 4% of the loss in leaf organic matter (Findlay and Arsuffi 1989, Suberkropp 1991, Baldy et al. 1995). In temperate streams, aquatic hyphomycete spore concentrations in the water reach maxima shortly after peak leaf inputs in the autumn (Iqbal and Webster 1973). During autumn, spore concentrations in the water can increase to levels that are 10 to 100 times the concentrations that occur at other times of the year (Iqbal and Webster 1973, Thomas et al. 1989). Because the spores of many aquatic hyphomycetes are morphologically distinct, the species composition of these fungi in a stream can be determined from the spores present in the water (Iqbal and Webster 1973) or accumulated in foam (Chauvet 1991).

One of the problems in studying the ecology of litter-decaying fungi is that they grow within the litter they are degrading. Consequently, it is difficult to determine the amount (biomass) of

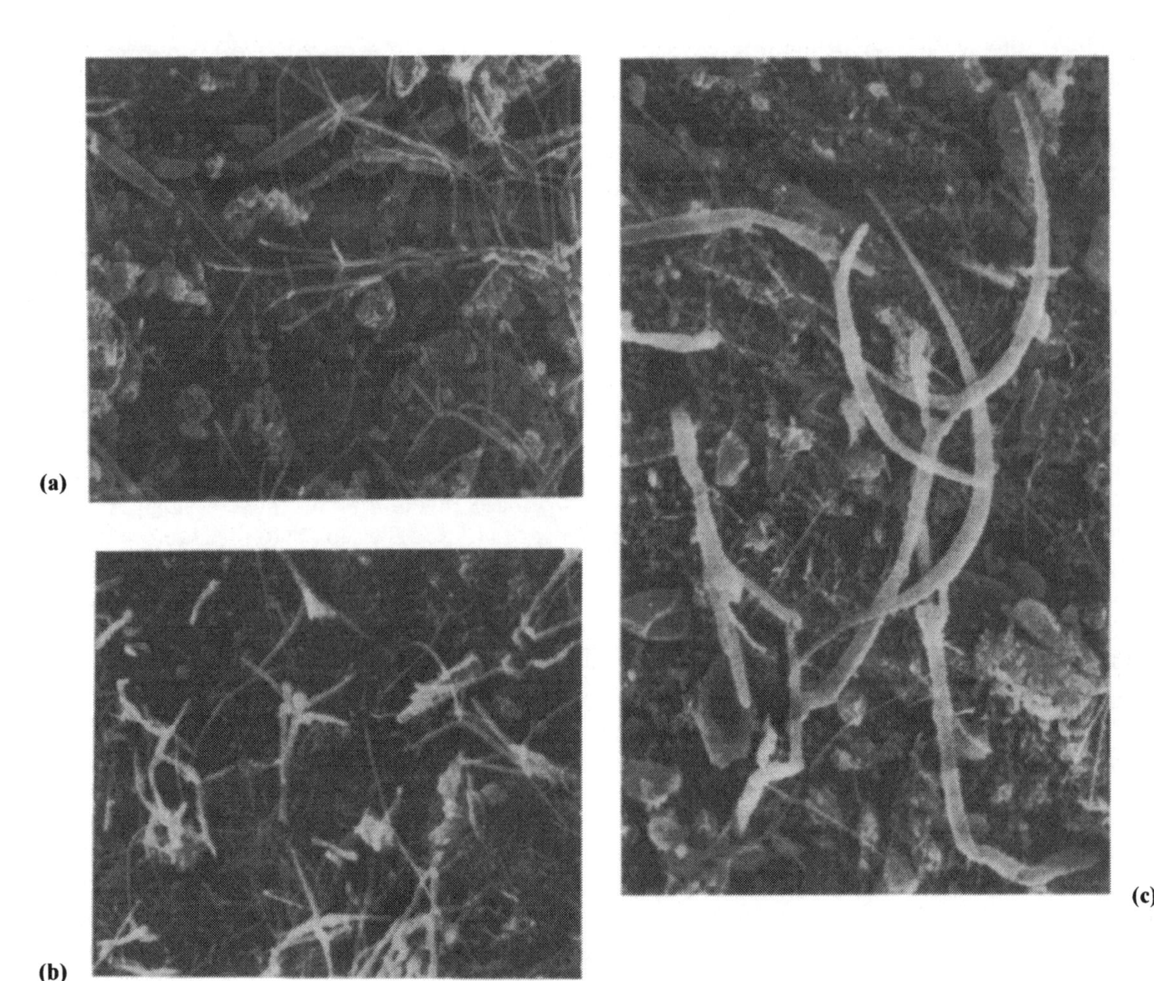

FIGURE 6.2. Scanning electron micrographs of spores of three aquatic hyphomycetes growing on decomposing leaf litter: (a) *Alatospora acuminata*, tetraradiate spores, (b) *Tetracladium marchalianum*, tetraradiate spores, (c) *Lunulospora curvula*, sigmoid spores.

fungi associated with litter. However, recent advances in methods allow the involvement of fungi associated with leaf litter to be evaluated quantitatively during decomposition. Ergosterol, a sterol associated with membranes of higher fungi, is thought to degrade rapidly after fungal death (Gessner and Newell 1997). It has been used to estimate fungal biomass in a variety of environments (Newell 1992) including streams (Gessner and Chauvet 1994, Suberkropp 1995). Although the amount of fungal biomass associated with leaves can vary depending on state of breakdown (Figure 6.3) and environmental conditions, maximum values of fungal biomass based on ergosterol content range from 4 to 18% of the weight of leaf detritus. Unfortunately, there are several groups of fungi and fungal-like organisms (e.g., Chytridiomycota and Oomycota) common in aquatic habitats such as streams that do not contain ergosterol. Therefore, measuring concentrations of ergosterol may underestimate the biomass of fungal-like organisms in certain habitats. For leaves colonized in streams, however, ergosterol appears to be a good measure of fungal biomass, since it correlates well with other measures of microbial biomass, such as ATP concentrations (Figure 6.3) (Golladay and Sinsabaugh 1991, Suberkropp et al. 1993).

Rates at which acetate is incorporated into ergosterol (Newell and Fallon 1991) have been used to determine fungal production on leaves in streams (Suberkropp 1995, Gessner and Chauvet 1997). Initial studies using this method indicate that fungal production is greatest following submersion (2–8 weeks depending on temperature and other characteristics of the water) and then declines (but remains at significant levels) throughout the remainder of leaf breakdown. This method shows great promise for estimating the annual production of leaf-decaying fungi within a stream since instantaneous growth rates can be determined from subsamples of naturally occurring litter. Estimates from three small (2nd order) forested streams in the southeastern United States indicate that the production of leaf-decaying fungi is highest during autumn to early winter when the concentration of leaves in the stream is at its maximum. Annual production of leaf-decaying fungi in these streams ranges from 25 to 34 g/m^2 (Suberkropp 1997, B.R. Methvin and K. Suberkropp, unpublished data).

Bacteria. Bacteria (Figure 6.4) are found in all stream habitats including the water column (McDowell 1984, Edwards 1987), sediments (Bott and Kaplan 1985, Findlay et al. 1986), rock surfaces (epilithon) (Lock et al. 1984), and detritus (Suberkropp and Klug 1976, Findlay and Arsuffi 1989, Baldy et al. 1995). Some bacteria are pathogens of aquatic animals and cause diseases in fish populations in the Pacific Northwest (Li et al. 1987). Little is known about the species of bacteria in streams except those that affect public health or are capable of unusual metabolic transformations. Although some studies have isolated and determined types of bacteria present in several different habitats (Suberkropp and Klug 1976, Baker and Farr 1977), knowledge of bacteria species diversity in streams is limited because only a small fraction of the total bacteria (as determined from direct microscopic counts) can be grown on laboratory media (Maltby 1992a).

Nonetheless, a number of studies have examined the role of bacteria in biofilm formation and function. Biofilms are the complex of microorganisms and organic matter attached to the surfaces of rocks, wood, plants, sediment particles and detritus (Lock 1993). When surfaces are exposed to light they develop biofilms

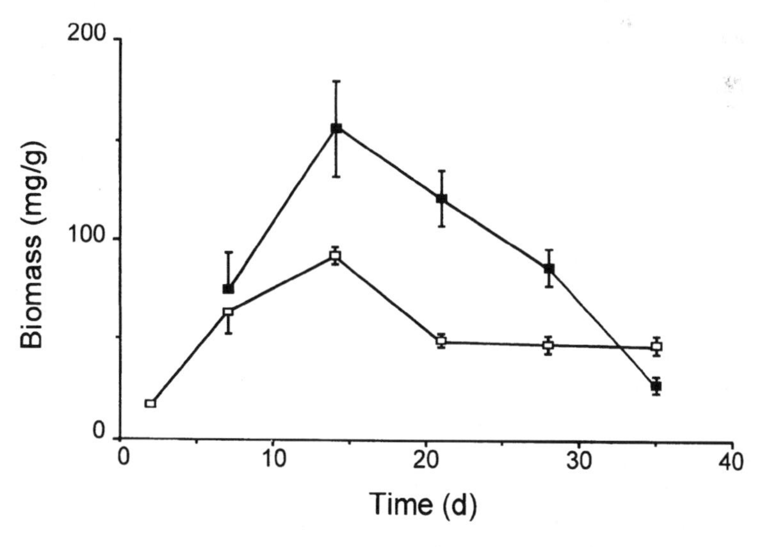

FIGURE 6.3. Changes in microbial biomass determined from ATP (□——□) and ergosterol (■——■) concentrations associated with yellow poplar leaves (*Liriodendron tulipifera*) as they are broken down in a stream. To convert ATP and ergosterol concentrations to biomass, factors of 1.75 mg ATP/g microbial biomass and 5.5 mg ergosterol/g fungal biomass were used (Suberkropp et al. 1993, with permission). Error bars indicate ± SE.

FIGURE 6.4. A scanning electron micrograph of a microcolony of rod shaped bacteria occurring on a decomposing leaf taken from a stream (Suberkropp and Klug 1974 with permission).

with both autotrophic (algae) and heterotrophic (bacteria, protozoa) microorganisms. The algae associated with these biofilms are referred to as periphyton (Chapter 7). In biofilms that form on the surfaces of rocks, microorganisms and their enzymes and metabolic products are held together and to the surface by extracellular polysaccharides (Lock et al. 1984). This polysaccharide matrix is important in stabilizing and attaching extracellular enzymes (Lock 1993) as well as providing a primary carbon reserve for biofilm bacteria when DOM concentrations are low (Freeman and Lock 1995).

The bacterial community has been characterized in several different stream habitats by determining total cell numbers (or biomass) or by estimating bacterial production from rates of incorporation of radioactive substrates (e.g., [^{3}H]thymidine into DNA or [^{14}C]acetate or [^{32}P]PO_4 into phospholipids). Concentrations of bacteria suspended in the water range widely (10^7–10^{10}/L) with highest densities found in blackwater rivers (rivers rich in humic acids and low in nutrient concentrations) where DOM concentrations are high (Edwards 1987). In several streams, bacterial concentrations increase during periods of high discharge (Baker and Farr 1977, McDowell 1984, Edwards 1987). Evidence suggests that such increases are caused by scouring of stream sediments and flooding of the surrounding terrestrial environment.

In streams, biofilms are thought to have the highest densities of bacteria (Geesey et al. 1978), while sediments typically exhibit the greatest amount of bacterial biomass and production (Edwards et al. 1990). A number of factors, including temperature, DOM concentrations, and inorganic nutrients, were correlated with bacterial production in sediments of a forested stream in the northeastern United States (Bott and Kaplan 1985). Total bacterial production in the silt and sand sediments was 26.4 g/C/m^2/yr, about 40% of the annual algal productivity of a wooded site in this stream (Bott and Kaplan 1985). In a comprehensive study of the bacterial activity associated with four different habitats in a blackwater river in the southeastern United States, Edwards et al. (1990) found that the greatest bacterial biomass and production was associated with the sandy sediments of the main channel. However, bacterial production in the river was not great enough to account for the biomass present, particularly during periods of high water, suggesting that a significant portion of the bacteria was carried into the water from the floodplain.

Protozoa. Little is known about the activities and significance of protozoa (Figure 6.5) in flowing waters. Densities of ciliates and flagellates are generally higher in stream sediments than on rock or leaf surfaces (Bott and Kaplan 1989). Densities of both ciliates and flagellates can vary over ten times in sediments of different streams (Baldock and Sleigh 1988, Bott and Kaplan 1989). Carlough and Meyer (1989) found high densities of protozoa in the water column of a blackwater river. These protozoa can filter an average of 16% of the bacte-

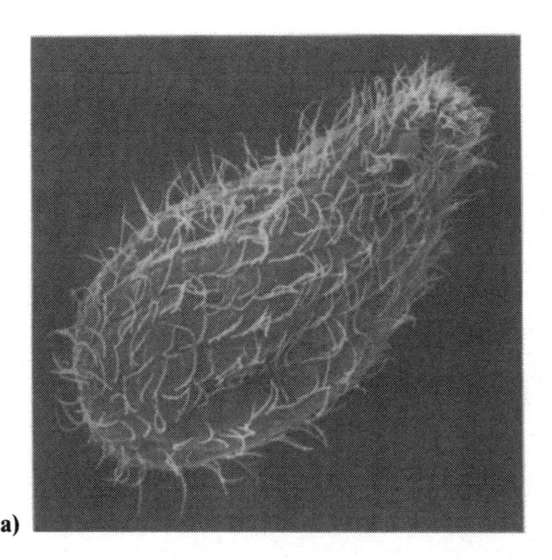

(a)

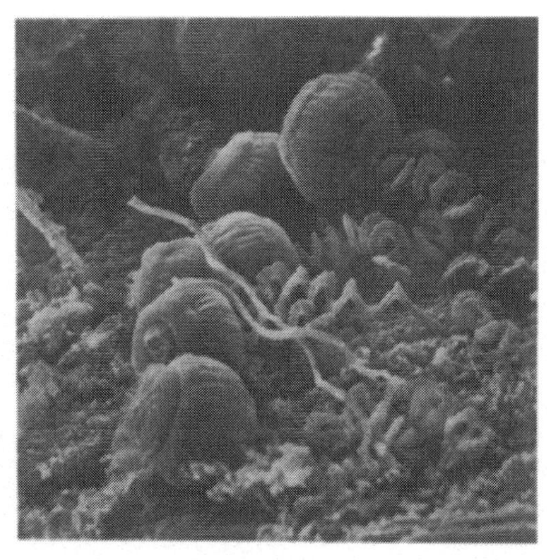

(b)

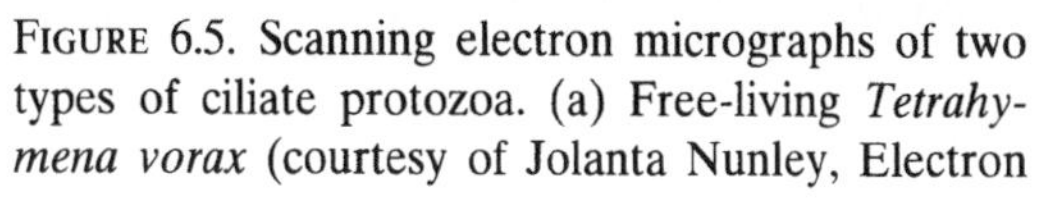

FIGURE 6.5. Scanning electron micrographs of two types of ciliate protozoa. (a) Free-living *Tetrahymena vorax* (courtesy of Jolanta Nunley, Electron Microscopy Laboratory, Biological Sciences, University of Alabama); (b) sessile *Vorticella* spp. (Suberkropp and Klug 1974 with permission).

ria from the water column each day suggesting that protozoa may play a significant role in transferring energy from bacteria to higher trophic levels (i.e., in the microbial loop) (Carlough and Meyer 1991). In both clear water and blackwater streams, flagellates dominate the protozoan biomass and grazing (Bott and Kaplan 1989, Carlough and Meyer 1989). Much more information concerning the biomass, production, and feeding rates of protozoa from a variety of environments is needed before their role in detrital food webs is fully understood.

Plant Litter Decomposition

The decomposition of plant litter, particularly leaf litter, has been the subject of numerous studies in flowing waters (reviewed by Webster and Benfield 1986, Boulton and Boon 1991, Maltby 1992b). During decomposition in streams, leaf litter is subjected to leaching and fragmentation by abrasion from the current. This action converts significant amounts of leaf litter to other forms of organic matter (DOM and FPOM) that are carried downstream and decomposed elsewhere. Since observed losses of organic matter are not completely due to mineralization (i.e., conversion to CO_2), it is questionable whether such losses should be referred to as *decomposition* (Webster and Benfield 1986, Boulton and Boon 1991). In other environments, where losses to physical forces (particularly fragmentation) are significant, the term *breakdown* describes loss of litter weight (Hanlon 1982), and *decomposition* refers primarily to mineralization. In streams, loss of organic matter from plant litter is called *breakdown* (Webster and Benfield 1986) or *processing* (Petersen and Cummins 1974) to reflect the fact that a variety of processes (both biotic and physical) are involved in causing losses in organic matter. Rates of breakdown are typically determined by placing known amounts of leaf litter in packs or mesh bags (Figure 6.6) in a stream and periodically determining the amount of organic matter remaining.

Plant litter breakdown in streams occurs in three phases: leaching, microbial colonization which modifies and mineralizes litter, and fragmentation by physical forces and invertebrate feeding (Figure 6.7) (Cummins 1974, Webster and Benfield 1986, Boulton and Boon 1991). These processes occur simultaneously although

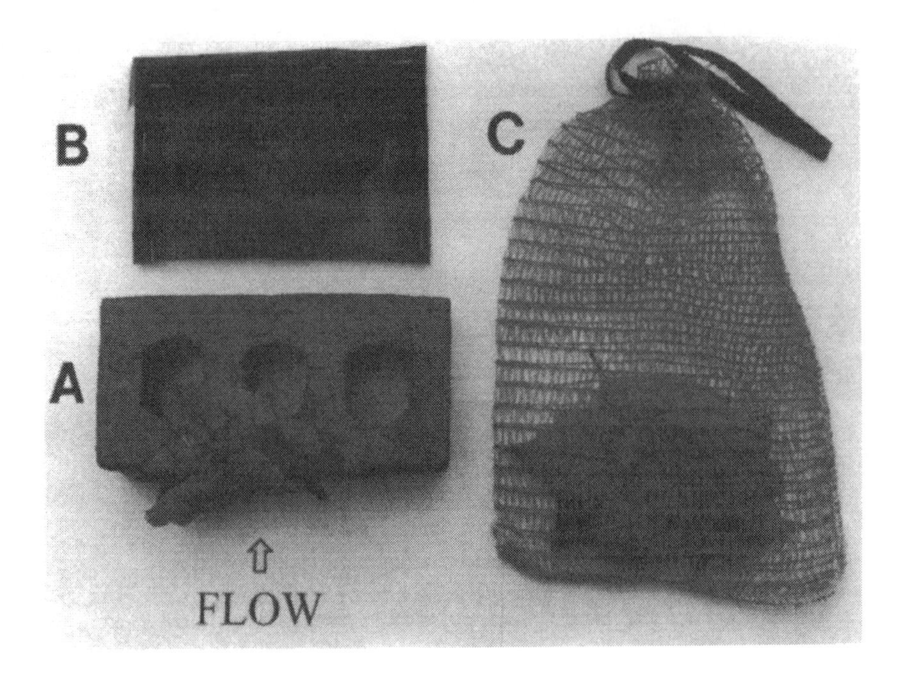

FIGURE 6.6. Examples of leaf pack and litter bags used to follow weight loss of plant litter: (A) leaf pack tied to brick which is positioned in a stream so that it is perpendicular to the flow of the water as indicated, thereby simulating natural accumulations of leaves; (B) fine-meshed litter bag; (C) coarse-meshed litter bag.

the relative contribution of each may vary (Figure 6.7). Microbial colonization has also been referred to as *conditioning*. This term reflects the observation that the growth and partial degradation of leaf litter by microorganisms improves the quality of detritus as a food source for stream detritivores. When leaves become softened and more fragile later in the

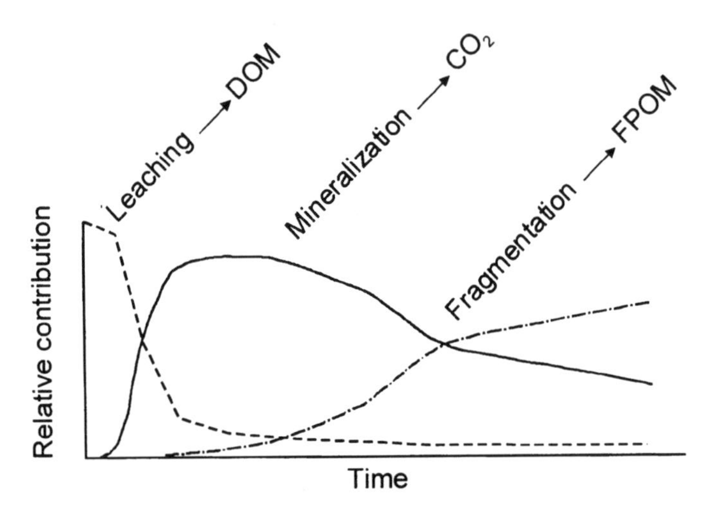

FIGURE 6.7. Representation of the relative contribution of leaching, mineralization, and fragmentation to weight loss throughout the course of breakdown of a hypothetical leaf in a stream. The relative contribution of each of these processes may vary depending on leaf species and environmental conditions.

breakdown process, fragmentation then increases losses.

Dry leaves entering streams lose a significant fraction of their dry weight to leaching during the first several days of immersion (Figure 6.8). Such losses range from 10 to 30% of the initial dry weight (Petersen and Cummins 1974). For fresh leaves (i.e., naturally shed, but not dry) initial leaching losses are much less (Figure 6.8), primarily because leaf cell membranes are still intact and soluble materials are retained for longer periods of time (Gessner and Schwoerbel 1989). Leaching can continue throughout leaf breakdown, because components of leaves which are solubilized by microbial enzymatic activity can be lost as DOM. Such losses may become significant as demonstrated by Findlay and Arsuffi (1989) who found that 3 to 25% of the loss from leaves during breakdown may occur as DOM.

After entering the water, leaves are colonized by stream microorganisms that begin to use leaf organic molecules for energy and production of microbial biomass. During leaf breakdown (Figure 6.9a), microbial respiration increases to a maximum (Figure 6.9b) and degradation by microbes mineralizes a portion of the leaf organic matter (Triska et al. 1975, Suberkropp 1991). Microbial biomass also increases to a maximum (Figure 6.9c). Microbial cells are released from leaf detritus as FPOM through sloughing and sporulation (Findlay and Arsuffi 1989). Microorganisms, particularly the aquatic hyphomycetes, produce a variety of extracellular enzymes that degrade leaf structural polysaccharides (Chamier 1985, Suberkropp 1992a). Pectin degrading enzymes soften and macerate leaf tissue (Suberkropp and Klug 1980, Chamier 1985) by degrading the middle lamella of the plant cell wall, the portion of the cell wall responsible for maintaining tissue integrity in plants. As leaf tissue becomes softer, it is more easily fragmented by the physical forces of the water current. In addition, as microbial biomass increases and the leaf decomposes, invertebrate detritivores may begin feeding and thus contribute to the conversion of leaf detritus to FPOM. Overall, 30 to 50% of

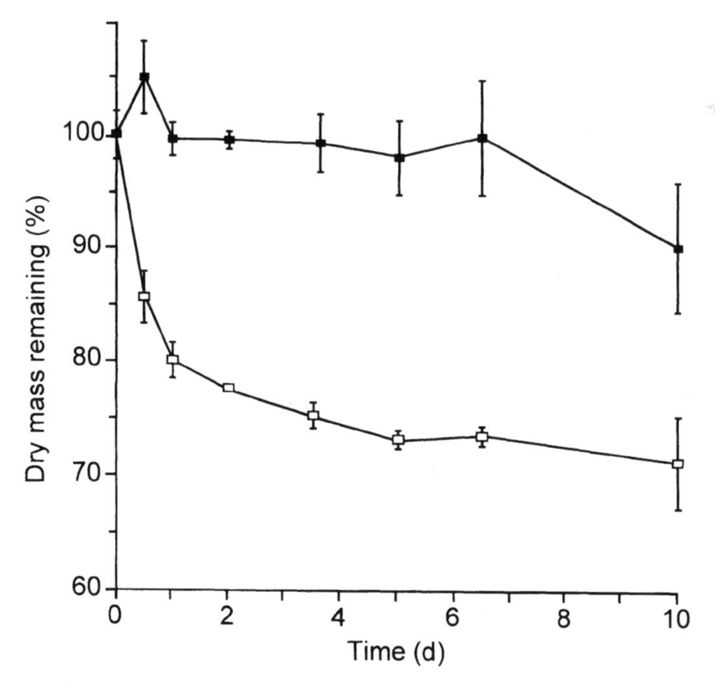

FIGURE 6.8. Leaching rates of alder (*Alnus glutinosa*) leaves that are fresh (non-dried; ■——■) or dry (air-dried; □——□) prior to placing in water (Gessner and Schwoerbel 1989 with permission). Error bars indicate ± SD.

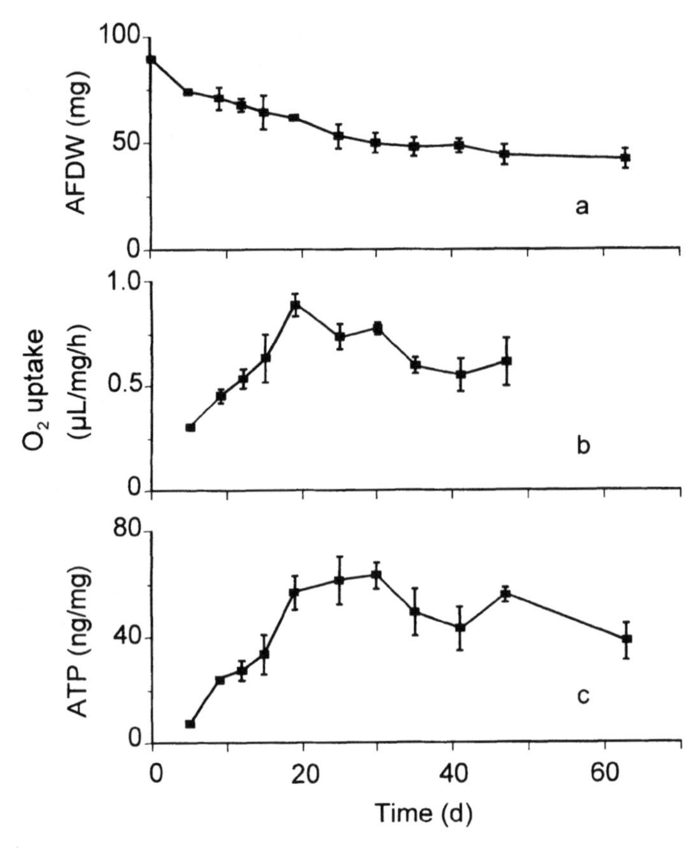

FIGURE 6.9. Changes in leaf weight, rates of microbial respiration, and microbial biomass during breakdown of yellow poplar (*Liriodendron tulipifera*) leaves in a headwater stream. (a) remaining ash-free dry weight (AFDW); (b) rates of microbial respiration (O_2 uptake) associated with leaf material; (c) concentrations of microbial biomass (as ATP) associated with leaf material (Suberkropp 1991 with permission). Error bars indicate ± SE.

leaf organic matter is mineralized to CO_2 by microbial decomposers (Elwood et al. 1981, Sinsabaugh and Findlay 1995), but significant amounts are also converted to FPOM and DOM by a variety of biotic and physical processes.

Naturally shed leaf litter is generally low in nitrogen (N) and phosphorus (P) and thus has high C:N and C:P ratios. During breakdown, the concentrations and absolute amounts of both of these nutrients increase (Suberkropp et al. 1976, Melillo et al. 1984, Chauvet 1987, Gessner and Chauvet 1994), a process called *immobilization*. Two proposed mechanisms for N immobilization are accumulation of N in microbial biomass and accumulation of N in complexes formed between partially degraded organic matter and microbial polymers (Melillo et al. 1984). Early estimates of microbial biomass associated with leaves in streams (Iversen 1973) suggested concentrations were not high enough to account for the amount of N immobilized. Therefore, the first mechanism, incorporation of N into microorganisms, was discounted as the primary mechanism for nitrogen immobilization. The second mechanism for N immobilization relies on observations that microbial modification of plant polymers (e.g., polyphenols, lignin) makes them reactive. Consequently, breakdown products may condense with carbohydrates and nitrogen-rich compounds (proteins, peptides) in a humification process (Melillo et al. 1984). This depolymerization-recondensation cycle results in the formation of progressively more recalcitrant nitrogen-containing compounds that accumulate in detritus thereby increasing its N content (Suberkropp et al. 1976, Melillo et al. 1984). The relative contribution of these processes to nitrogen immobilization are not known and are likely to vary depending on type of litter and environmental conditions. However, incorporation of N into microbial biomass may be more important than previously thought, since improved methods indicate that the concentrations of microbial biomass associated with decaying leaves are significantly greater than suggested by earlier studies (see Newell 1992, Gessner et al. 1997).

Models for Plant Litter Decomposition

The most commonly used model to characterize rates of plant litter breakdown in streams is the negative exponential model:

$$W_t/W_0 = e^{-kt}$$

where W_t is the ash-free dry weight (AFDW) remaining after time, t in days, W_0 is the initial AFDW, k is the breakdown rate, and e is the base of the natural logarithm. This model was originally developed for decomposition of organic matter in terrestrial environments (Jenny et al. 1949, Olson 1963). It was used to describe the breakdown of leaf litter in stream systems by Petersen and Cummins (1974), who proposed that leaf species exhibit a continuum of breakdown rates that are arbitrarily assigned to one of three "processing ranges" (fast, medium, and slow, Table 6.3). Subsequent research indicates that the rate of breakdown is affected by a number of factors, and a single leaf species may exhibit a range of breakdown rates depending on environmental conditions. The breakdown rate, k, can be determined by linear regression of log transformed weight data or by application of nonlinear regression methods. Chauvet (1987) argues that nonlinear regression methods based on only one coefficient are preferable because they do not distort the estimation of that coefficient as do regressions based on logarithmic transformation.

Since its introduction, the negative exponential model has been the subject of a number of criticisms; other mathematical equations may describe the data more precisely, the fit is sometimes poor and variables that affect breakdown rates (e.g., temperature) are not present in the model (for a more complete discussion of models of organic matter breakdown, see Wieder and Lang 1982, Webster and Benfield 1986, Boulton and Boon 1991). Temperature is taken into account by modifications of the negative exponential model in which degree days substitute for days as the time component (Minshall et al. 1983), or by a two variable model in which breakdown rate is a function of temperature (Hanson et al. 1984). Additional modifications of the negative exponential model divide leaf components into two (labile and refractory) (see Wieder and Lang 1982) or more fractions (Minderman 1968) and determine the overall breakdown rate by summing

TABLE 6.3. Ranges of breakdown rates (k) for leaves in the three groups (fast, medium, slow) proposed by Petersen and Cummins (1974). Examples of leaves from plant species common in the riparian zones of streams in the Pacific Northwest that generally fall within these groups.

	Breakdown rates k(/day)		
Leaf species	Fast >0.010	Medium 0.005–0.010	Slow <0.005
Red alder (*Alnus rubra*)	0.012–0.017		
Vine maple (*Acer circinatum*)		0.007–0.020	
Bigleaf maple (*Acer macrophyllum*)			0.002–0.011
Douglas-fir and western hemlock (*Pseudotsuga menziesii and Tsuga heterophylla*)			0.002–0.013

Modified from Sedell et al. 1975.

individual rates. Although these and other models may fit the data better or provide more realism, the breakdown rate, k, calculated with the negative exponential model provides a single number that often describes the process of leaf breakdown in streams fairly accurately (Webster and Benfield 1986). Consequently, it has remained the most commonly used parameter to compare litter breakdown under different circumstances.

Comparison of Fungal and Bacterial Activity

Evidence suggests that the growth and activity of fungi predominate over that of bacteria during the breakdown of leaf litter in streams. Direct comparisons indicate that fungal biomass associated with leaves in streams ranges from 3 to 100 times that of bacteria (Findlay and Arsuffi 1989, Baldy et al. 1995, Weyers and Suberkropp 1996). Patterns of biomass accumulation of these two types of microorganisms also differ (Figure 6.10). Fungal biomass typically peaks relatively early, whereas bacterial biomass increases and levels off or continues to increase throughout breakdown. The pattern exhibited by fungal biomass more closely follows changes in total microbial biomass and respiration associated with leaves (Suberkropp 1991).

Fungi are not the predominant microorganism on all types of POM in streams. The concentration of fungal biomass (as measured by ergosterol concentrations) associated with POM in three streams was correlated with particle size of the detritus (Sinsabaugh et al. 1992). Fungal biomass was highest on particles larger than 4 mm in diameter and was undetectable in some of the smaller size fractions (<250 μm diameter). Similarly, in the Hudson River estuary (Figure 6.11), both fungal biomass and production were positively correlated with particle size, but the opposite trend was observed for bacteria (Sinsabaugh and Findlay 1995). Fungal production (determined from differences in biomass) was 5 to 600 times greater than bacterial production (determined from rates of incorporation of [^{3}H]thymidine into DNA) on particles greater than 4 mm,

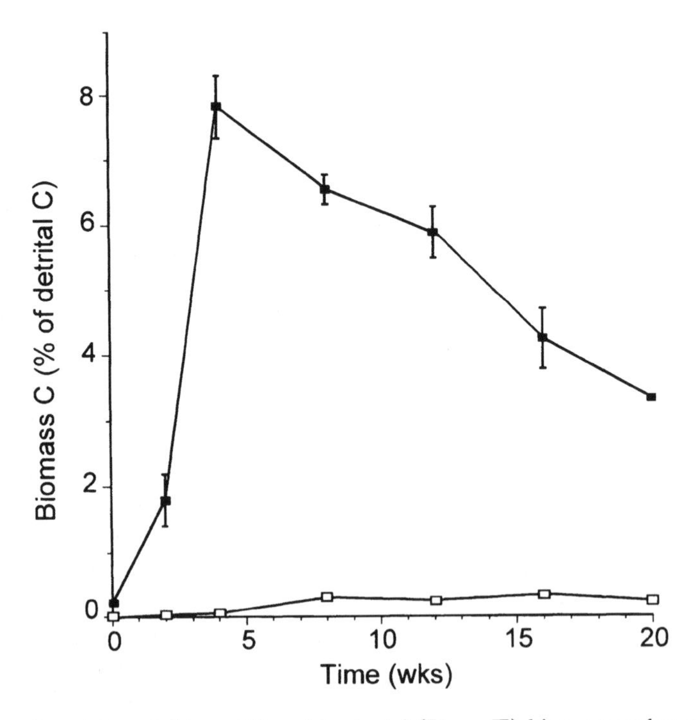

FIGURE 6.10. Dynamics of fungal (■——■) and bacterial (□——□) biomass carbon (C) associated with willow (*Salix alba*) leaves during breakdown (Baldy et al. 1995 with permission). Error bars indicate ± SE.

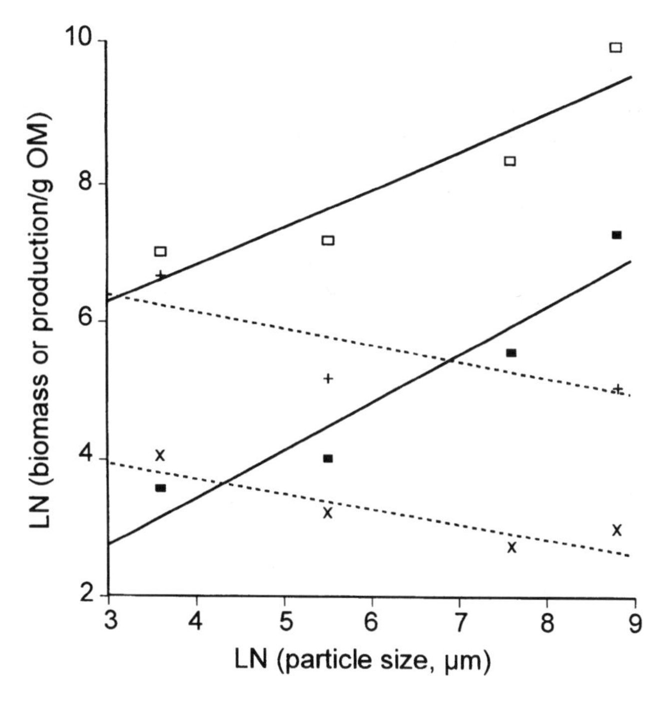

FIGURE 6.11. The distribution of microbial biomass carbon (μgC) and production (μgC/day) per g organic matter (OM) in relation to detrital particle size in surface sediments of the Hudson River estuary. Ln is the natural logarithm. Solid lines represent fungi; dashed lines represent bacteria. Fungal productivity (μgC/d/gOM), ■; fungal biomass (μgC/gOM), □; bacterial productivity (μgC/d/gOM), ×; bacterial biomass (μgC/gOM), + (Sinsabaugh and Findlay 1995 with permission).

whereas bacterial production was 2 to 44 times greater than fungal production on particles less than 63 μm. Using rates of incorporation of radioactive substrates ([^{14}C]acetate incorporation into ergosterol for fungi, [^{3}H]leucine incorporation into protein for bacteria), direct comparisons of fungal and bacterial production associated with leaves indicate that fungi production generally exceeds that of bacteria 0.6 to 88 times (Weyers and Suberkropp 1996). It appears that the ability of fungi to penetrate CPOM by extending their hyphae during growth gives them a competitive advantage over bacteria on large particles such as leaves, whereas on smaller particles with greater surface-area-to-volume ratios, the ability to colonize surfaces gives bacteria a competitive advantage.

Fungi-Shredder Interactions

Invertebrate detritivores that feed on leaf detritus belong to the functional feeding group of stream invertebrates called shredders (Cummins 1974, Chapter 8 this volume) The interactions between leaf-decaying fungi and shredders in streams are dynamic and exhibit a greater level of complexity than many predator–prey interactions because the leaf on which the fungi are growing can also be used as a resource by the shredder (Bärlocher 1985, Suberkropp 1992b). Fungal colonization of leaf material affects both the palatability and food quality of leaf detritus. When shredders are offered choices of leaves colonized by different fungi, they preferentially feed on certain species and avoid others (Bärlocher and Kendrick 1973a, Suberkropp et al. 1983, Graça et al. 1993). Caddisfly larvae can discriminate among fungal species that have colonized different patches within the same leaf (Figure 6.12) suggesting that these shredders have the ability to feed selectively on fungi as they occur on leaves in streams (Arsuffi and Suberkropp 1985). Palatability of leaf detritus increases after a period of fungal growth and modification of the leaf

FIGURE 6.12. Selective feeding by caddisfly larvae on patches within an aspen leaf that had been colonized by different fungi. Four leaf disks that had been colonized by different fungi were attached to four quadrants of the leaf for five days. During this period, the fungi on the disks grew into the leaf. The disks were then removed and caddisfly larvae were allowed to feed on the leaf. Note that one patch (lower left) colonized by one fungal species was skeletonized by larval feeding. Two other patches (upper and lower right) colonized by different fungal species were sampled but not completely skeletonized and one patch (upper left) was not skeletonized at all. See Arsuffi and Suberkropp (1985).

(Arsuffi and Suberkropp 1984). Since many fungal species cause similar changes when growing on leaves (e.g., in concentrations of fungal biomass, nitrogen content, softening of leaf, enzymology), the basis for palatability differences are difficult to determine (Bärlocher and Kendrick 1973a, Suberkropp et al. 1983). It is possible that fungi produce specific chemicals that act as feeding stimulants or deterrents. For example, lipids such as those extracted from fungi appear to act as feeding stimulants for late instar caddisfly larvae (Cargille et al. 1985). However, differences in lipid composition have not been linked to differences in palatability of fungi.

The growth and survival of shredders varies considerably depending on the fungal species in their diet, suggesting that fungi also differ in quality as food sources (Bärlocher and Kendrick 1973b, Rossi and Fano 1979, Arsuffi and Suberkropp 1986). Bärlocher (1985) suggests that several factors are responsible for the increased the food quality attributed to fungal colonization of leaves. First, increased food quality is attributed to the fungal biomass associated with leaves. Though fungal biomass accounts for less than 20% of the detritus consumed, it is assimilated more efficiently by shredders than leaf tissue (Bärlocher and Kendrick 1975) and, consequently, may account for a significant amount of the detritus that is assimilated. In addition, fungi improve the food quality of leaves by producing extracellular enzymes that act on leaf polysaccharides and convert them into products that are digestible by shredders (Bärlocher 1982b). For those shredders that maintain favorable conditions in their digestive tracts, fungal enzymes can remain active after ingestion (acquired enzymes), thereby contributing to the digestive capabilities of the organism (Bärlocher 1982b). Available evidence indicates that the relative contribution of the leaf and its associated microbiota to the food quality of the detritus varies and is affected by the stage of colonization, species of fungi growing on the leaf, and digestive adaptations of the shredder.

Factors Affecting Rates of Plant Litter Breakdown

Rates of litter breakdown are affected both by internal factors, such as characteristics of the leaves or litter quality, and external environmental factors (Melillo et al. 1984, Webster and Benfield 1986). Leaves of different tree species exhibit inherent differences in breakdown rates; for example, alder leaves are broken down more rapidly than conifer needles placed in the same stream (Table 6.3). Deciduous leaves and needles from different species of plants exhibit a continuum of breakdown rates (Petersen and Cummins 1974, Webster and Benfield 1986). Factors that contribute to such differences among leaf species include nutrients—higher concentrations of N in leaf tissue may lead to faster breakdown; fiber content—higher fiber content (particularly lignin) results in slower breakdown; and chemical inhibitors—presence of compounds (higher concentrations) such as tannins, waxes, cutins, or compounds specific to particular tree species can reduce breakdown rates (Webster and Benfield 1986). These litter quality characteristics undoubtedly interact to influence the rate of breakdown. For example, Melillo et al. (1982) suggested that the ratio of lignin to nitrogen was a better indicator of leaf quality than either alone. Such factors influence breakdown rates primarily by their effects on microbial (fungal) activity. In a study of the breakdown of seven leaf species in a stream, Gessner and Chauvet (1994) found that breakdown rates were negatively correlated with initial lignin concentration, but were not significantly correlated with initial N or P contents of the leaves. Breakdown rates and lignin concentrations were significantly correlated with maximum fungal biomass, mycelial production, and maximum sporulation rates. This correlation suggests that lignin concentrations determine carbon availability for fungi associated with decomposing litter (Gessner and Chauvet 1994), and that rates of leaf breakdown are controlled by carbon limitation of the fungi (Gessner et al. 1997). Since N and P contents of the leaves were not correlated with leaf breakdown rate or fungal activity, it is likely that these nutrients were available in sufficient amount from the litter, water, or both in this stream.

A number of external environmental factors also affect rates of litter breakdown, and in some cases, these factors may override litter quality. For example, Suberkropp and Chauvet (1995) found that yellow poplar leaves (*Liriodendron tulipifera*), which typically break down quickly, placed in eight streams exhibited differences in breakdown rates as large as those observed for leaf species on either end of the breakdown continuum (Petersen and Cummins 1974). These differences were attributed to differences in water chemistry, particularly nitrate concentrations, in the streams. Other studies also indicate that in some streams breakdown rates can be stimulated by increased concentrations of nutrients (N, P) (Elwood et al. 1981, Meyer and Johnson 1983).

In addition to nutrient concentrations of the water, other environmental factors affecting breakdown rates include the presence of invertebrate detritivores (shredders) and temperature (Webster and Benfield 1986). Shredders can increase rates of litter breakdown through their feeding activities; leaf breakdown is more rapid in streams with high densities of shredders (Webster and Benfield 1986, Stewart 1992). Removal of shredders from streams with high shredder densities greatly reduces breakdown rates of litter (Wallace et al. 1982). Temperature affects microbial mineralization directly during leaf breakdown. In laboratory studies and in studies comparing breakdown rates seasonally or in streams of one region that differ in temperature, breakdown rates are greater in environments with higher temperatures (Webster and Benfield 1986). However, when examined across latitudinal gradients, differences in temperature do not always account for the variation in breakdown rates (Irons et al. 1994). Irons et al. (1994) suggest that this apparent lack of influence of temperature is due to changes in the relative contribution of shredders and decomposing organisms to leaf breakdown (Figure 6.13). In streams at high latitudes, shredder densities are higher, and shredder feeding appears to contribute more to weight loss than microbial mineralization. This leads to higher rates of breakdown than would be predicted based on the low temperatures of the stream. In lower latitudes, invertebrate feeding is less important than microbial activity in determining breakdown rates (Irons et al. 1994). These examples point to the complex nature of leaf breakdown in streams and suggest that interactions between internal and external factors that control leaf breakdown rates can be difficult to predict.

Wood

Considerable amounts of woody debris can accumulate in streams (Table 6.2) (Triska et al. 1982). Wood, particularly large boles and branches, has a long residence time and plays important structural roles in stream habitats (Harmon et al. 1986, Chapter 13 this volume). Wood forms debris dams that retain POM

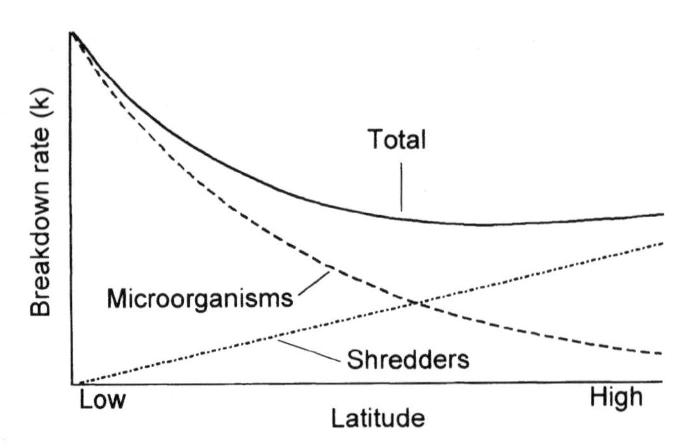

FIGURE 6.13. Conceptual model of the relative contribution of microorganisms and shredders to total breakdown rates of leaf litter in streams at different latitudes (Irons et al. 1994 with permission).

(Bilby 1981) and provide habitat for aquatic organisms (Anderson and Sedell 1979). Wood has also proven to be a potentially important site for nitrogen fixation in streams of the Pacific Northwest (Buckley and Triska 1978).

While there have been relatively few studies of wood decomposition in streams, they indicate that wood typically breaks down more slowly than leaf litter (Melillo et al. 1983, Harmon et al. 1986). This is likely because wood has a higher lignin and lower nitrogen content than leaves (Melillo et al. 1983). In addition, when large pieces of wood enter streams and become waterlogged, the availability of oxygen in woody tissues is reduced (Hedges et al. 1985). This is particularly important for wood breakdown since lignin degradation is an oxygen-requiring process that is primarily mediated by certain groups of aerobic fungi (Kirk and Farrell 1987). Consequently, most lignocellulolytic activity associated with large pieces of wood is on the outer surface (Aumen et al. 1983).

Wood breakdown is generally affected by the same type of environmental factors that affect leaf litter breakdown. For example, the mineralization of lignocellulose prepared from wood is stimulated by additions of N and P to the water (Aumen et al. 1985). Melillo et al. (1983) found that wood of five tree species broke down faster in a 1st-order stream than in larger streams and suggested that this could be due to higher concentrations of P in the 1st-order stream.

Although wood is broken down more slowly than leaf litter in streams, it supports significant microbial activity. When expressed in terms of surface area rather than weight, small pieces of wood exhibit higher rates of respiration and enzyme activities than leaves placed in a stream for comparable times (Golladay and Sinsabaugh 1991, Tank et al. 1993). One reason for this may be that a more stable biofilm forms on wood than on leaves since surfaces of submerged wood persist for longer times. The organisms in such biofilms may depend, in part, on DOM in stream water and autochthonous production (Sinsabaugh et al. 1991, Lock 1993). In addition, differences in breakdown rates between leaves and wood may relate to the relative contributions of fragmentation and mineralization to weight loss of these materials. The high amount of fragmentation of leaves during the later stages of breakdown may magnify the differences in breakdown rates between leaves and woody detritus.

There is currently some uncertainty concerning the types of microorganisms involved in wood decay in streams. Unlike decaying leaves, there have been no comparisons of bacterial and fungal biomass or production associated with woody debris. Observations with the scanning electron microscope indicate that bacteria and actinomycetes are more abundant than fungi on large submerged logs (Aumen et al. 1983, Harmon et al. 1986). However, several studies have noted a rich community of aquatic hyphomycetes and other Ascomycota occurring on wood submerged in flowing water (Willoughby and Archer 1973, Shearer and Webster 1991, Shearer 1992). In addition, concentrations of ergosterol (an index of fungal biomass) are higher (per cm^2) in pieces of decaying wood than in decaying leaves in the same river (Golladay and Sinsabaugh 1991).

Dissolved Organic Matter

Dissolved organic matter (DOM) present in flowing waters originates from a variety of sources including riparian and floodplain soil organic matter, groundwater entering streams through the hyporheic zone (Chapter 16), throughfall or leachates from terrestrial vegetation, and exudates and leachates from both autotrophic and heterotrophic stream organisms (Kaplan and Newbold 1993). DOM is generally the dominant form of organic matter in transport; ratios of DOM:POM in flowing waters average 6:1 but can reach much higher values (Naiman 1982, Wetzel 1984). DOM includes an extremely diverse group of substances and is not well characterized for many systems. Composition of DOM may vary from stream to stream as well as seasonally.

In general, DOM is divided into humic and nonhumic substances (Wetzel 1983). Nonhumic substances are generally more labile and consist of organic molecules (both monomeric and polymeric) derived primarily from compounds such as carbohydrates, organic acids, proteins and peptides. These compounds typically exist at low concentrations because they are rapidly taken up and used by microorganisms. Humic substances consist of colored, polyelectrolytic organic acids that exhibit a wide range in molecular size (molecular weights of 500–5000 or higher) and chemical composition (Thurman 1985). Operational definitions of humic substances differ depending on whether they are being isolated from soils or water. In soils, humic substances are extracted in a solution of sodium hydroxide (0.1 M NaOH), whereas in natural waters, they are isolated on XAD resins (nonionic macroporous resins), weak-base ion-exchange resins, or a comparable procedure (Thurman 1985). In spite of these differences in definition, humic substances from soil and water appear to have similar properties. Humic substances are divided into humic acids and fulvic acids based on their solubility in acid (Thurman 1985). Humic acids precipitate in acid (pH 1–2) whereas fulvic acids are soluble in both acid and base. Fulvic acids are generally of lower molecular weight and have a higher content of oxygen-containing functional groups than humic acids. Some large humic acids are colloidal in nature, but both soluble and colloidal material may form complexes with organic and inorganic materials (Wetzel 1983).

Humic substances are formed from plant and animal polymers (particularly those with aromatic ring structures such as lignin or tannins) that have been partially degraded by microorganisms and ultraviolet oxidation followed by polymerization (Thurman 1985). Humic substances typically contain a variety of aromatic rings and aliphatic sidechains as well as variable amounts of nitrogen and other nutrients. Because of the complex chemical arrangements and different types of bonds between constituents, humic substances are generally more resistant to decay than nonhumic substances and, consequently, can account for the majority of DOM in flowing waters. Thurman (1985) suggests the composition of the DOM in an average river contains 40% fulvic acids, 10% humic acids, 30% hydrophilic acids (or hydrophilic humic substances, a group of poorly characterized substances that do not bind to XAD resins at acid pH), and 20% simple compounds such as carbohydrates, carboxylic acids, amino acids, and hydrocarbons.

Metabolism of DOM and the Microbial Loop

Bacteria appear to be the primary microorganisms to use DOM in streams, although there have been few studies of the use of DOM by other organisms. DOM can also be removed from solution by physical adsorption. Dahm (1981) found that leachate from red alder leaves was rapidly removed (97% in 48h) from solution in stream microcosms. For this type of DOM, microbial uptake was slower but removed more of the leachate (77%) than physical adsorption (20%). The proportion of DOM present in rivers that can be degraded by bacteria ranges from less than 1% for intermediate molecular weight fractions to over 50% for low molecular weight fractions (Meyer et al. 1987). The bioavailability of DOM to bacteria (i.e., the amount of bacterial growth per unit of DOM) exhibits a wide range (1–100μg bacterial C per mg DOC, Meyer 1994) depending on the source of the water. DOM in transport may also be a source of carbon and energy for attached bacteria. Bott et al. (1984) examined this question by moving sediment communities that had become adapted to the DOM concentration at one site in a stream to a second site that was exposed to lower DOM concentrations. After the transfer, bacterial numbers and biomass declined by 53 to 55% suggesting that DOM in transport supported a sizable portion of the bacteria in sediment. However, in other forested streams, community respiration associated with sediments is correlated with the amount of particulate organic matter in the sediment and not with concentrations of DOM in the water column (Hedin 1990).

Several studies have suggested that bacteria become adapted or acclimatized to the types of DOM present in a stream. For example, DOM from grass leachate, as opposed to oak leaf leachate, supported growth of bacteria suspended in the water from an upstream grassland site of a prairie stream. In a downstream wooded site which had been exposed to both grass and oak leaves, bacterial growth was supported by both types of leachate (McArthur and Marzoff 1986). Kaplan and Bott (1983, 1985) found that bacterial biomass and activity in the surface sediments of a forested stream increased more rapidly (within 2–5h) after addition of DOM from sources present in the landscape than after the addition DOM from foreign sources (a minimum of 48h). During use of DOM in these studies, low molecular weight compounds, particularly carbohydrates were preferentially metabolized (Kaplan and Bott 1983).

The metabolism of DOM by bacteria leads into the microbial loop of streams (Meyer 1990 and 1994). The microbial loop was originally described for marine planktonic food webs in which organic matter, rather than passing directly from primary producers to higher trophic levels, is channeled to bacteria which are consumed by protozoa and then by organisms at higher trophic levels (Pomeroy 1974). A microbial loop (see Figure 6.1) has also been described for freshwater streams and rivers (Meyer 1990, 1994). In streams, bacteria in the water column and attached to surfaces metabolize DOM which is primarily of terrestrial origin rather than from autochthonous primary production (Meyer 1990, 1994). Bacteria are, in turn, consumed by a variety of organisms ranging in size from microflagellates and meiofauna (copepods, nematodes) to filtering insect larvae (blackflies). Few have studied the impact of feeding by meiofauna on bacteria in streams; however, Borchardt and Bott (1995) found that meiofauna were generally present in sediments at densities that were too low to ingest more than a small fraction (0.03–0.08%) of bacterial production. In contrast, Perlmutter and Meyer (1991) determined that meiofauna could ingest up to 22% of the daily bacteria production associated with leaf litter. Since there are greater numbers of macroscopic consumers, such as larvae of blackflies and midges that feed directly on bacteria, there are potentially fewer trophic transfers in streams than in, for example, marine food webs (Meyer 1990, 1994). Consequently, more energy can be transferred to the higher trophic levels in stream food webs than in food webs with greater numbers of transfers (Meyer 1990). Regardless of how much energy passes to higher trophic levels, the pathway in which energy from DOM flows through bacteria to protozoa is essential for ecosystem stability, particularly with regard to nutrient regeneration (Wetzel 1992). Further research is needed on the microbial loop in streams to determine its role in energy and nutrient transfers (Meyer 1990).

Organisms in forested streams are closely linked to the riparian vegetation (Cummins et al. 1989), and changes in the amount or species composition of this vegetation can alter stream biotic diversity and function. Likewise, changes in the stream environment that affect the breakdown of organic matter may affect higher trophic levels that depend on this material for energy. Although research reveals that a number of factors affect specific processes in the decomposer subsystem of streams, few large-scale manipulations have focused on microorganisms or the processes they mediate. Consequently, not enough is currently known to accurately predict the effects of perturbations to the decomposer subsystem on the functioning of the stream ecosystem, particularly over long time scales. For example, it is known that higher temperatures and increased concentrations of nutrients such as N and P speed up rates of leaf breakdown. However, it is unknown whether increased temperatures or greater nutrient inputs to a stream would increase the rates of leaf breakdown enough so that, later in the year, leaf detritus would be unavailable as food needed by invertebrate detritivores to complete their life cycles. Such effects, although difficult to document, have the potential to alter food webs that are central to stream ecosystem function.

Literature Cited

Anderson, N.H., and J.R. Sedell. 1979. Detritus processing by macroinvertebrates in stream ecosystems. Annual Review of Entomology **24**:351–377.

Arsuffi, T.L., and K. Suberkropp. 1984. Leaf processing capabilities of aquatic hyphomycetes: Interspecific differences and influence on shredder feeding preferences. Oikos **42**:144–154.

Arsuffi, T.L., and K. Suberkropp. 1985. Selective feeding by caddisfly (Trichoptera) detritivores on leaves with fungal colonized patches. Oikos **45**:50–58.

Arsuffi, T.L., and K. Suberkropp. 1986. Growth of two stream caddisflies (Trichoptera) on leaves colonized by different fungal species. Journal of the North American Benthological Society **5**:297–305.

Aumen, N.G., P.J. Bottomley, and S.V. Gregory. 1985. Impact of nitrogen and phosphorus on [^{14}C]lignocellulose decomposition by stream wood microflora. Applied and Environmental Microbiology **49**:1113–1118.

Aumen, N.G., P.J. Bottomley, G.M. Ward, and S.V. Gregory. 1983. Microbial decomposition of wood in streams: Distribution of microflora and factors affecting [^{14}C]lignocellulose mineralization. Applied and Environmental Microbiology **46**:1409–1416.

Baker, J.H., and I.S. Farr. 1977. Origins, characterization and dynamics of suspended bacteria in two chalk streams. Archiv für Hydrobiologie **80**:308–326.

Baldock, B.M., and M.A. Sleigh. 1988. The ecology of benthic protozoa in rivers: Seasonal variation in numerical abundance in fine sediments. Archiv für Hydrobiologie **111**:409–421.

Baldy, V., M.O. Gessner, and E. Chauvet. 1995. Bacteria, fungi and the breakdown of leaf litter in a large river. Oikos **74**:93–102.

Bärlocher, F. 1982a. Conidium production from leaves and needles in four streams. Canadian Journal of Botany **60**:1487–1494.

——. 1982b. The contribution of fungal enzymes to the digestion of leaves by *Gammarus fossarum* Koch (Amphipoda). Oecologia **52**:1–4.

——. 1985. The role of fungi in the nutrition of stream invertebrates. Botanical Journal of the Linnean Society **91**:83–94.

——. 1992. The ecology of aquatic hyphomycetes. Springer-Verlag, Berlin, Germany.

Bärlocher, F., and B. Kendrick. 1973a. Fungi and food preferences of *Gammarus pseudolimnaeus*. Archiv für Hydrobiologie **72**:501–516.

Bärlocher, F., and B. Kendrick. 1973b. Fungi in the diet of *Gammarus pseudolimnaeus* (Amphipoda). Oikos **24**:295–300.

Bärlocher, F., and B. Kendrick. 1974. Dynamics of the fungal population on leaves in a stream. Journal of Ecology **62**:761–791.

Bärlocher, F., and B. Kendrick. 1975. Assimilation efficiency of *Gammarus pseudolimnaeus* (Amphipoda) feeding on fungal mycelium or autumn-shed leaves. Oikos **26**:55–59.

Bilby, R.E. 1981. Role of organic debris dams in regulating the export of dissolved and particulate matter from a forested watershed. Ecology **62**:1234–1243.

Borchardt, M.A., and T.L. Bott. 1995. Meiofaunal grazing of bacteria and algae in a Piedmont stream. Journal of the North American Benthological Society **14**:278–298.

Bott, T.L., and L.A. Kaplan. 1985. Bacterial biomass, metabolic state, and activity in stream sediments: Relation to environmental variables and multiple assay comparisons. Applied and Environmental Microbiology **50**:508–522.

Bott, T.L., and L.A. Kaplan. 1989. Densities of benthic protozoa and nematodes in a Piedmont stream. Journal of the North American Benthological Society **8**:187–196.

Bott, T.L., L.A. Kaplan, and F.T. Kuserk. 1984. Benthic bacterial biomass supported by streamwater dissolved organic matter. Microbial Ecology **10**:335–344.

Boulton, A.J., and P.I. Boon. 1991. A review of methodology used to measure leaf litter decomposition in lotic environments: Time to turn over an old leaf? Australian Journal of Marine and Freshwater Research **42**:1–43.

Buckley, B.M., and F.J. Triska. 1978. Presence and ecological role of nitrogen-fixing bacteria associated with wood decay in streams. Verhandlungen der Internationalen Vereinigung für Theoretische und Angewandte Limnologie **20**:1333–1339.

Calow, P. 1992. Energy budgets. Pages 370–378 *in* P. Calow and G.E. Petts, eds. The rivers handbook: Hydrological and ecological principles. Volume 1. Blackwell Scientific, London, UK.

Cargille, A.S., K.W. Cummins, B.J. Hanson, and R.R. Lowry. 1985. The role of lipids as feeding stimulants for shredding aquatic insects. Freshwater Biology **15**:455–464.

Carlough, L.A., and J.L. Meyer. 1989. Protozoans in two southeastern blackwater rivers and their importance to trophic transfer. Limnology and Oceanography **34**:163–177.

Carlough, L.A., and J.L. Meyer. 1991. Bactivory by sestonic protists in a southeastern blackwater river. Limnology and Oceanography **36**:873–883.

Chamier, A.-C. 1985. Cell-wall-degrading enzymes of aquatic hyphomycetes: A review. Botanical Journal of the Linnean Society **91**:67–81.

Chauvet, E. 1987. Changes in the chemical composition of alder, poplar and willow leaves during decomposition in a river. Hydrobiologia **148**:35–44.

——. 1991. Aquatic hyphomycete distribution in southwestern France. Journal of Biogeography **18**:699–706.

Cummins, K.W. 1974. Structure and function of stream ecosystems. BioScience **24**:631–641.

Cummins, K.W., M.A. Wilzbach, D.M. Gates, J.B. Perry, and W.B. Taliaferro. 1989. Shredders and riparian vegetation. BioScience **39**:24–30.

Dahm, C.N. 1981. Pathways and mechanisms for removal of dissolved organic carbon from leaf leachate in streams. Canadian Journal of Fisheries and Aquatic Sciences **38**:68–76.

Edwards, R.T. 1987. Sestonic bacteria as a food source for filtering invertebrates in two southeastern blackwater rivers. Limnology and Oceanography **32**:221–234.

Edwards, R.T., J.L. Meyer, and S.E.G. Findlay. 1990. The relative contribution of benthic and suspended bacteria to system biomass, production, and metabolism in a low-gradient blackwater river. Journal of the North American Benthological Society **9**:216–228.

Elwood, J.W., J.D. Newbold, A.F. Trimble, and R.W. Stark. 1981. The limiting role of phosphorus in a woodland stream ecosystem: Effects of P enrichment on leaf decomposition and primary producers. Ecology **62**:146–158.

Findlay, S.E.G., and T.L. Arsuffi. 1989. Microbial growth and detritus transformations during decomposition of leaf litter in a stream. Freshwater Biology **21**:261–269.

Findlay, S., J.L. Meyer, and R. Risley. 1986. Benthic bacterial biomass and production in two blackwater rivers. Canadian Journal of Fisheries and Aquatic Sciences **43**:1271–1276.

Fisher, S.G., and G.E. Likens. 1973. Energy flow in Bear Brook, New Hampshire: An integrative approach to stream ecosystem metabolism. Ecological Monographs **43**:421–439.

Freeman, C., and M.A. Lock. 1995. The biofilm polysaccharide matrix: A buffer against changing organic substrate supply? Limnology and Oceanography **40**:273–278.

Geesey, G.G., R. Mutch, J.W. Costerton, and R.B. Green. 1978. Sessile bacteria: An important component of the microbial population in small mountain streams. Limnology and Oceanography **23**:1214–1223.

Gessner, M.O., and E. Chauvet. 1994. Importance of stream microfungi in controlling breakdown rates of leaf litter. Ecology **75**:1807–1817.

Gessner, M.O., and E. Chauvet. 1997. Growth and production of aquatic hyphomycetes in decomposing leaf litter. Limnology and Oceanography **42**:496–505.

Gessner, M.O., and S.Y. Newell. 1997. Bulk quantitative methods for the examination of eukaryotic organoosmotrophs in plant litter. Pages 295–308 *in* C.J. Hurst, G.R. Knudsen, M.J. McInerney, L.D. Stetzenbach, and M.V. Walter, eds. Manual of environmental microbiology. American Society for Microbiology, Washington, DC, USA.

Gessner, M.O., and J. Schwoerbel. 1989. Leaching kinetics of fresh leaf-litter with implications for the current concept of leaf-processing in streams. Archiv für Hydrobiologie **115**:81–90.

Gessner, M.O., K. Suberkropp, and E. Chauvet. 1997. Decomposition of plant litter by fungi: Marine and freshwater ecosystems. Pages 303–322 *in* D.T. Wicklow and B. Söderström, eds. The

Mycota: Environmental and microbial relationships. Volume IV. Springer Verlag, Berlin, Germany.

Golladay, S.W., and R.L. Sinsabaugh. 1991. Biofilm development on leaf and wood surfaces in a boreal river. Freshwater Biology **25**:437–450.

Graça, M.A.S., L. Maltby, and P. Calow. 1993. Importance of fungi in the diet of *Gammarus pulex* and *Asellus aquaticus* I: feeding strategies. Oecologia **93**:139–144.

Hanlon, R.D.G. 1982. The breakdown and decomposition of allochthonous and autochthonous plant litter in an oligotrophic lake (Llyn Frongoch). Hydrobiologia **88**:281–288.

Hanson, B.J., K.W. Cummins, J.R. Barnes, and M.W. Carter. 1984. Leaf litter processing in aquatic systems: A two variable model. Hydrobiologia **111**:21–29.

Harmon, M.E., J.F. Franklin, F.J. Swanson, P. Sollins, S.V. Gregory, J.D. Lattin, et al. 1986. Ecology of coarse woody debris in temperate ecosystems. Advances in Ecological Research **15**:133–302.

Hedges, J.I., G.L. Cowie, J.R. Ertel, R.J. Barbour, and P.G. Hatcher. 1985. Degradation of carbohydrates and lignins in buried woods. Geochimica et Cosmochimica Acta **49**:701–711.

Hedin, L.O. 1990. Factors controlling sediment community respiration in woodland stream ecosystems. Oikos **57**:94–105.

Hornick, L.E., J.R. Webster, and E.F. Benfield. 1981. Periphyton production in an Appalachian Mountain trout stream. American Midland Naturalist **106**:22–36.

Iqbal, S.H., and J. Webster. 1973. Aquatic hyphomycete spora of the River Exe and its tributaries. Transactions of the British Mycological Society **62**:331–346.

Ingold, C.T. 1942. Aquatic hyphomycetes of decaying alder leaves. Transactions of the British Mycological Society **25**:339–417.

Irons, J.G., M.W. Oswood, R.J. Stout, and C.M. Pringle. 1994. Latitudinal patterns in leaf litter breakdown: Is temperature really important? Freshwater Biology **32**:401–411.

Iversen, T.M. 1973. Decomposition of autumn-shed beech leaves in a spring brook and its significance for the fauna. Archiv für Hydrobiologie **72**:305–312.

Jenny, H., S.P. Gessel, and F.T. Bingham. 1949. Comparative study of decomposition rates of organic matter in temperate and tropical regions. Soil Science **68**:419–432.

Kaplan, L.A., and T.L. Bott. 1983. Microbial heterotrophic utilization of dissolved organic matter in a piedmont stream. Freshwater Biology **13**:363–377.

Kaplan, L.A., and T.L. Bott. 1985. Acclimation of stream-bed heterotrophic microflora: Metabolic responses to dissolved organic matter. Freshwater Biology **15**:479–492.

Kaplan, L.A., and J.D. Newbold. 1993. Biogeochemistry of dissolved organic carbon entering streams. Pages 139–165 *in* T.E. Ford, ed. Aquatic microbiology: An ecological approach. Blackwell Scientific, Oxford, UK.

Kirk, T.K., and R.L. Farrell. 1987. Enzymatic "combustion": The microbial degradation of lignin. Annual Review of Microbiology **41**:465–505.

Li, H.W., C.B. Schreck, C.E. Bond, and E. Rexstad. 1987. Factors influencing changes in fish assemblages of Pacific northwest streams. Pages 193–202 *in* W.J. Matthews and D.C. Heins, eds. Community and evolutionary ecology of North American stream fishes. University of Oklahoma Press, Norman, Oklahoma, USA.

Lock, M.A. 1993. Attached microbial communities in rivers. Pages 113–138 *in* T.E. Ford, ed. Aquatic microbiology: An ecological approach. Blackwell Scientific, Oxford, UK.

Lock, M.A., R.R. Wallace, J.W. Costerton, R.M. Ventullo, and S.E. Charlton. 1984. River epilithon: Toward a structural-functional model. Oikos **42**:10–22.

Maltby, L. 1992a. Heterotrophic microbes. Pages 165–194 *in* P. Calow and G.E. Petts, eds. The rivers handbook: Hydrological and ecological principles. Volume 1. Blackwell Scientific, London, UK.

——. 1992b. Detritus processing. Pages 331–353 *in* P. Calow and G.E. Petts, eds. The rivers handbook: Hydrological and ecological principles. Volume 1. Blackwell Scientific, London, UK.

McArthur, J.V., and G.R. Marzolf. 1986. Interactions of the bacterial assemblages of a prairie stream with dissolved organic carbon from riparian vegetation. Hydrobiologia **134**:193–199.

McDowell, W.H. 1984. Temporal changes in numbers of suspended bacteria in a small woodland stream. Verhandlungen der Internationalen Vereinigung für Theoretische und Angewandte Limnologie **22**:1920–1925.

Melillo, J.M., J.D. Aber, and J.F. Muratore. 1982. Nitrogen and lignin control of hardwood leaf litter decomposition dynamics. Ecology **63**:621–626.

Melillo, J.M., R.J. Naiman, J.D. Aber, and K.N. Eshleman. 1983. The influence of substrate quality and stream size on wood decomposition dynamics. Oecologia **58**:281–285.

Melillo, J.M., R.J. Naiman, J.D. Aber, and A.E. Linkins. 1984. Factors controlling mass loss and nitrogen dynamics of plant litter decaying in northern streams. Bulletin of Marine Science **35**:341–356.

Meyer, J.L. 1990. A blackwater perspective on riverine ecosystems. BioScience **40**:643–651.

——. 1994. The microbial loop in flowing waters. Microbial Ecology **28**:195–199.

Meyer, J.L., R.T. Edwards, and R. Risley. 1987. Bacterial growth on dissolved organic carbon from a blackwater river. Microbial Ecology **13**:13–29.

Meyer, J.L., and C. Johnson. 1983. The influence of elevated nitrate concentration on rate of leaf decomposition in a stream. Freshwater Biology **13**:177–183.

Minderman, G. 1968. Addition, decomposition, and accumulation of organic matter in forests. Journal of Ecology **56**:355–362.

Minshall, G.W., R.C. Petersen, K.W. Cummins, T.L. Bott, J.R. Sedell, C.E. Cushing, et al. 1983. Interbiome comparison of stream ecosystem dynamics. Ecological Monographs **53**:1–25.

Naiman, R.J. 1982. Characteristics of sediment and organic carbon export from pristine boreal forest watersheds. Canadian Journal of Fisheries and Aquatic Sciences **39**:1699–1718.

Naiman, R.J., J.M. Melillo, and J.E. Hobbie. 1986. Ecosystem alteration of boreal forest streams by beaver (*Castor canadensis*). Ecology **67**:1254–1269.

Newell, S.Y. 1992. Estimating fungal biomass and productivity in decomposing litter. Pages 521–561 *in* G.C. Carroll and D.T. Wicklow, eds. The fungal community: Its organization and role in the ecosystem, second edition. Marcel Dekker, New York, New York, USA.

Newell, S.Y., and R.D. Fallon. 1991. Toward a method for measuring instantaneous fungal growth rates in field samples. Ecology **72**:1547–1559.

Olson, J.S. 1963. Energy storage and the balance of producers and decomposers in ecological systems. Ecology **44**:322–332.

Perlmutter, D.G., and J.L. Meyer. 1991. The impact of a stream-dwelling harpacticoid copepod upon detritally associated bacteria. Ecology **72**:2170–2180.

Petersen, R.C., and K.W. Cummins. 1974. Leaf processing in a woodland stream. Freshwater Biology **4**:343–368.

Pomeroy, L.R. 1974. The ocean's food web, a changing paradigm. BioScience **24**:499–504.

Read, S.J., S.T. Moss, and E.B.G. Jones. 1992. Attachment and germination of conidia. Pages 135–151 *in* F. Bärlocher, ed. The ecology of aquatic hyphomycetes. Springer-Verlag, Berlin, Germany.

Rossi, L., and A.E. Fano. 1979. Role of fungi in the trophic niche of the congeneric detritivorous *Asellus aquaticus* and *A. coxalis* (Isopoda). Oikos **32**:380–385.

Sedell, J.R., R.J. Triska, and N.S. Triska. 1975. The processing of conifer and hardwood leave in to coniferous forest streams: I. Weight loss and associated invertebrates. Verhandlungen der Internationalen für Theoretische und Angewandte Limnologie **19**:1617–1627.

Shearer, C.A. 1992. The role of woody debris. Pages 77–98 *in* F. Bärlocher, ed. The ecology of aquatic hyphomycetes. Springer-Verlag, Berlin, Germany.

Shearer, C.A., and J. Webster. 1991. Aquatic hyphomycete communities in the river Teign. IV. Twig colonization. Mycological Research **95**:413–420.

Sinsabaugh, R.L., and S. Findlay. 1995. Microbial production, enzyme activity, and carbon turnover in surface sediments of the Hudson River estuary. Microbial Ecology **30**:127–141.

Sinsabaugh, R.L., S.W. Golladay, and A.E. Linkins. 1991. Comparison of epilithic and epixylic biofilm development in a boreal river. Freshwater Biology **25**:179–187.

Sinsabaugh, R.L., T. Weiland, and A.E. Linkins. 1992. Enzymic and molecular analysis of microbial communities associated with lotic particulate organic matter. Freshwater Biology **28**:393–404.

Stewart, B.A. 1992. The effect of invertebrates on leaf decomposition rates in two small woodland streams in southern Africa. Archiv für Hydrobiologie **124**:19–33.

Suberkropp, K. 1984. The effect of temperature on the seasonal occurrence of aquatic hyphomycetes. Transactions of the British Mycological Society **82**:53–62.

——. 1991. Relationships between the growth and sporulation of aquatic hyphomycetes on decomposing leaf litter. Mycological Research **95**:843–850.

——. 1992a. Aquatic hyphomycete communities. Pages 729–747 *in* G.C. Carroll and D.T. Wicklow, eds. The fungal community: Its organization and role in the ecosystem, second edition. Marcel Dekker, New York, New York, USA.

——. 1992b. Interaction with invertebrates. Pages 118–134 *in* F. Bärlocher, ed. The ecology of aquatic hyphomycetes. Springer-Verlag, Berlin, Germany.

——. 1995. The influence of nutrients on fungal growth, productivity, and sporulation during leaf breakdown in streams. Canadian Journal of Botany **73** (Supplement **1**):S1361–S1369.

——. 1997. Annual production of leaf-decaying fungi in a woodland stream. Freshwater Biology **38**:169–178.

Suberkropp, K., T.L. Arsuffi, and J.P. Anderson. 1983. Comparison of the growth, enzymatic activity and palatability of aquatic hyphomycetes grown on leaf litter. Applied and Environmental Microbiology **46**:237–244.

Suberkropp, K., and E. Chauvet. 1995. Regulation of leaf breakdown by fungi in streams: Influences of water chemistry. Ecology **76**:1433–1445.

Suberkropp, K., M.O. Gessner, and E. Chauvet. 1993. Comparison of ATP and ergosterol as indicators of fungal biomass associated with decomposing leaves in streams. Applied and Environmental Microbiology **59**:3367–3372.

Suberkropp, K., G.L. Godshalk, and M.J. Klug. 1976. Changes in the chemical composition of leaves during processing in a woodland stream. Ecology **57**:720–727.

Suberkropp, K., and M.J. Klug. 1974. Decomposition of deciduous leaf litter in a woodland stream. I. A scanning electron microscope study. Microbial Ecology **1**:96–103.

Suberkropp, K., and M.J. Klug. 1976. Fungi and bacteria associated with leaves during processing in a woodland stream. Ecology **57**:707–719.

Suberkropp, K., and M.J. Klug. 1980. The maceration of deciduous leaf litter by aquatic hyphomycetes. Canadian Journal of Botany **58**:1025–1031.

Tank, J.L., J.R. Webster, and E.F. Benfield. 1993. Microbial respiration on decaying leaves and sticks in a southern Appalachian stream. Journal of the North American Benthological Society **12**:394–405.

Thomas, K., G.A. Chilvers, and R.H. Norris. 1989. Seasonal occurrence of conidia of aquatic hyphomycetes (fungi) in Lees Creek, Australian Capital Territory. Australian Journal of Marine and Freshwater Research **40**:11–23.

Thurman, E.M. 1985. Organic geochemistry of natural waters. Marinus Nijhoff/Dr. W. Junk, Dordrecht, The Netherlands.

Triska, F.J., J.R. Sedell, and B. Buckley. 1975. The processing of conifer and hardwood leaves in two coniferous forest streams: II Biochemical and nutrient changes. Verhandlungen der Internationalen Vereinigung für Theoretische und Angewandte Limnologie **19**:1628–1639.

Triska, F.J., J.R. Sedell, and S.V. Gregory. 1982. Coniferous forest streams. Pages 292–332 *in* R.L. Edmonds, ed. Analysis of coniferous forest ecosystems in the western United States. Hutchinson Ross, Stroudsburg, Pennsylvania, USA.

Wallace, J.B., J.R. Webster, and T.F. Cuffney. 1982. Stream detritus dynamics: Regulation by invertebrate consumers. Oecologia **53**:197–200.

Webster, J. 1959. Experiments with spores of aquatic hyphomycetes: I. Sedimentation, and impaction on smooth surfaces. Annals of Botany, N. S. **23**:595–611.

Webster, J., and R.A. Davey. 1984. Sigmoidal conidial shape in aquatic fungi. Transactions of the British Mycological Society **83**:43–52.

Webster, J.R., and E.F. Benfield. 1986. Vascular plant breakdown in freshwater ecosystems. Annual Review of Ecology and Systematics **17**:567–594.

Wetzel, R.G. 1983. Limnology, second edition. Saunders College Publishing, Philadelphia, Pennsylvania, USA.

——. 1984. Detrital dissolved and particulate organic carbon functions in aquatic ecosystems. Bulletin of Marine Science **35**:503–509.

——. 1992. Gradient-dominated ecosystems: Sources and regulatory functions of dissolved organic matter in freshwater ecosystems. Hydrobiologia **229**:181–198.

Weyers, H.S., and K. Suberkropp. 1996. Fungal and bacterial production during the breakdown of yellow poplar leaves in 2 streams. Journal of the North American Benthological Society **15**:408–420.

Wieder, R.K., and G.E. Lang. 1982. A critique of the analytical methods used in examining decomposition data obtained from litter bags. Ecology **63**:1636–1642.

Willoughby, L.G., and J.F. Archer. 1973. The fungal spora of a freshwater stream and its colonization pattern on wood. Freshwater Biology **3**:219–239.

7 Primary Production

Michael L. Murphy

Overview

• Aquatic primary producers—benthic algae, macrophytes, and phytoplankton—play a key role in the trophic support of stream ecosystems.

• Primary production depends on many factors, including light, nutrients, temperature, streamflow, and herbivores. Light is often limiting in small forest streams, which may receive less than 5% of full sunlight. With increased light, photosynthesis can increase up to a level of light saturation, about 20 to 60% of full sunlight, depending on species and acclimation conditions. Phosphorus and nitrogen are often limiting in more open streams.

• Grazing of live tissues and collecting of autochthonous detritus (i.e., from in-stream primary production) are the main avenues of energy flow from primary producers to consumers. At higher trophic levels, salmonids feed mostly on invertebrate drift produced through autochthonous pathways regardless of the input of allochthonous organic matter (i.e., produced outside the stream).

• Within a watershed, production is typically low in the headwaters because of shade from forest canopy, and high in more open, mid-order streams and rivers. In a large (9th-order) watershed, 1st- to 3rd-order streams may produce only 10 to 20% of the total annual gross primary production, despite having over 80% of total stream length.

• Watershed uses can have varied effects on aquatic primary production. Removing riparian canopy and increasing nutrients will often increase productivity leading to greater production of invertebrates and fish.

• Effects of watershed uses, however, should be viewed from a long-term perspective and in the context of associated changes in physical habitat. Maintaining healthy stream habitats and biodiversity can be accomplished by managing for naturally functioning streams with diverse energy sources.

Introduction

Aquatic primary production is a basic source of energy for stream ecosystems. Production by green plants and algae is called *primary production* because it constitutes the only significant energy gateway into the earth ecosystem. Algae and aquatic plants, together with allochthonous organic matter (i.e., produced outside a stream), provide trophic support for invertebrates, fish, and other animals that make up the diverse communities in running waters. Past ideas about stream ecosystems often downplayed the significance of autochthonous (i.e., produced in stream) primary production and emphasized the dependence on allochthonous energy sources, particularly litter from streamside vegetation (Minshall 1978). More recently, ecologists have pointed out the importance of autochthonous primary production, not only in open, nonforested sites, but even in small, forested streams usually perceived as depending on forest litter as the primary energy source (Bilby and Bisson 1992, Chapter 15 this volume).

Aquatic primary production is sometimes underrated because of the small amount of algae and plants typically present in a stream compared to the much larger amount of stored allochthonous organic matter. However, a small algal biomass can support a much larger biomass of consumers because of rapid turnover (in hours to days) (McIntire 1973, Lamberti et al. 1989, see Hershey and Lamberti this volume). In contrast, turnover of allochthonous detritus is much slower, measured in years or decades (Naiman 1983). Autochthonous organic matter is also more nutritious than allochthonous matter (Anderson and Cummins 1979), and high aquatic primary production, occurring seasonally, can significantly enrich the detritus pool for invertebrate consumers (Hawkins et al. 1982).

Primary production has an important role in both stream productivity and in mediating impacts of watershed uses. Land uses affect primary production by altering riparian vegetation, nutrients, and other habitat features and have consequences at a stream's highest trophic levels. Increased primary production after a disturbance sometimes completely masks otherwise detrimental effects of decreased habitat quality (Hawkins et al. 1983). In order to conserve or restore stream productivity and biodiversity, it is important to understand how land uses affect aquatic primary production.

This chapter describes primary production in streams—the typical forms of algae and plants, environmental factors and energy gateways into the food web, distribution in watersheds, and effects of watershed uses.

Forms and Typical Species

Aquatic primary producers in streams and rivers occur in three common forms: benthic algae, macrophytes, and phytoplankton. Benthic algae are attached to the stream bottom and submerged debris. Macrophytes include angiosperms rooted in the stream bottom, mosses, and other bryophytes. Phytoplankton consists of algae suspended or freely floating in water. The relative abundance of these three types differs with stream size, gradient, and exposure to sunlight.

Benthic Algae

Benthic algae covering stream substrates have two general forms: microscopic, unicellular algae (especially diatoms) that form thin layers on stream substrates (Figures 7.1a and 7.1b); and macroalgae (green, blue–green and red algae) that grow as filaments, sheets, or mats (Figures 7.2a and 7.2b). Benthic algae typically occur with bacteria and fungi in an assemblage called periphyton, which usually also includes inorganic sediments and organic matter as an integral part of this complex association. Macroalgae such as *Ulothrix* form the primary matrix of algal mats, giving the mat its basic structure, but diatoms (e.g., *Achnanthes*) are also an important component (Figure 7.3). The macroalgae often dominate in low-gradient, open streams, whereas diatoms usually dominate in higher-gradient, shaded streams (Minshall 1978). Streams frequently contain a variety of surfaces and microhabitats for different kinds of algal assemblages. Even the backs of snails are seasonally important for certain specialized algae (Stock and Ward 1991).

The various growth forms of periphytic algae are related to their function in primary production, nutrient uptake, and protection from grazing (Steinman et al. 1992). Growth forms include branched or unbranched filamentous forms (e.g., *Draparnaldia* and *Oedogonium*); branched filaments in a gelatinous or mucilaginous matrix (e.g., *Batrachospermum*); plate-like colonies of cells in a gelatinous matrix (*Tetraspora*); and prostrate forms (e.g., *Stigeoclonium* basal cells and several diatoms). Carbon fixation and nutrient uptake are directly related to the alga's surface-to-volume ratio (Steinman et al. 1992) and are highest in branched filamentous species and lowest in gelatinous species. Gelatinous filamentous algae are least vulnerable to grazers, whereas loosely attached diatoms and unbranched filaments are most vulnerable (Lamberti and Resh 1983, Steinman et al. 1992).

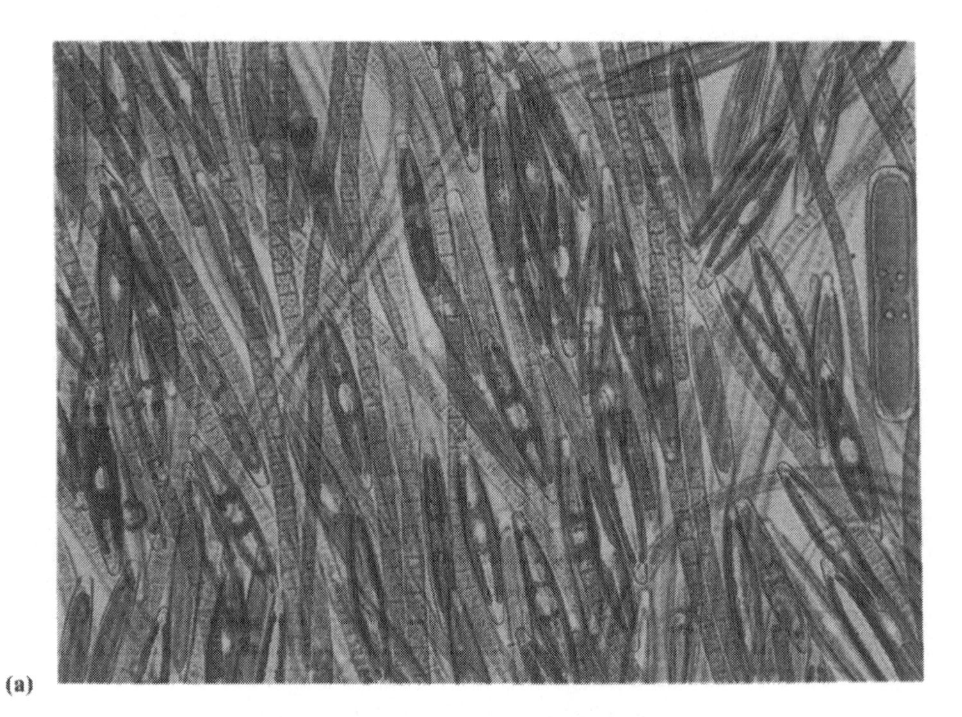
(a)

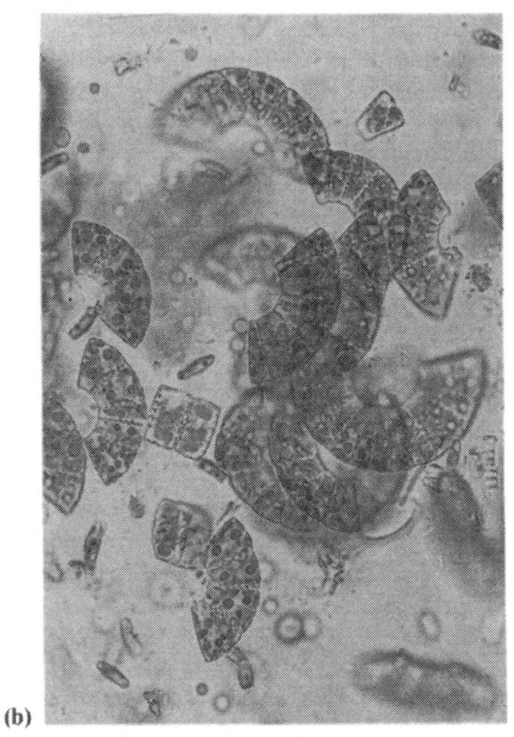
(b)

FIGURE 7.1. (a) The cigar-shaped diatom *Navicula* is a large genus with many common species occupying a broad range of habitats in fresh and salt water. (b) The benthic diatom *Meridion* forms fan-shaped colonies. It is common in headwater streams, sometimes occurring as a brown scum on the stream bottom. (Photos: Canter-Lund and Lund 1995 with permission).

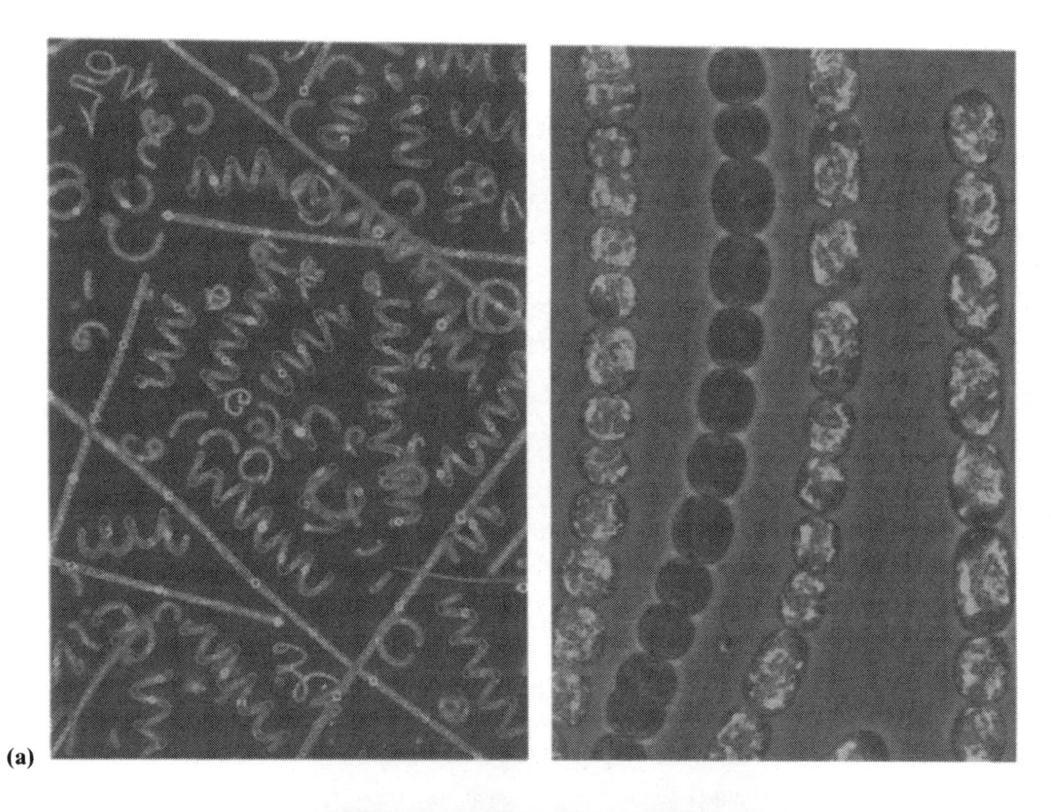

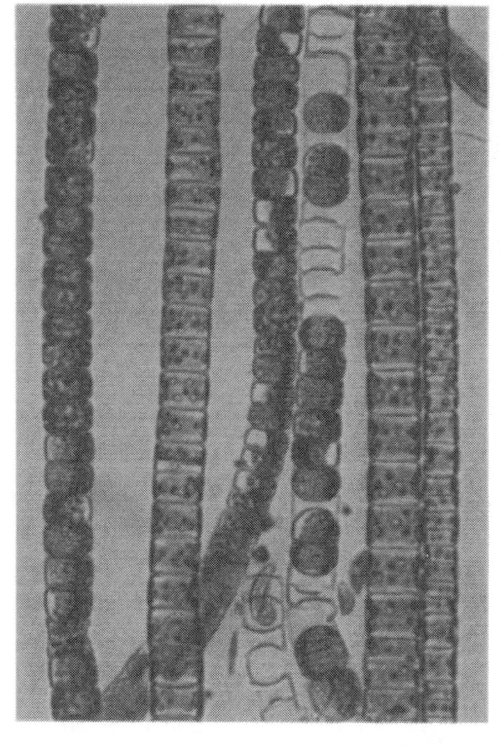

FIGURE 7.2. (a) *Anabaena* is one of the blue-green algae that are widespread in fresh water and capable of nitrogen fixation. (b) The filamentous green alga *Ulothrix* often forms dense mats during blooms in spring and summer (Photos: Canter-Lund and Lund 1995 with permission).

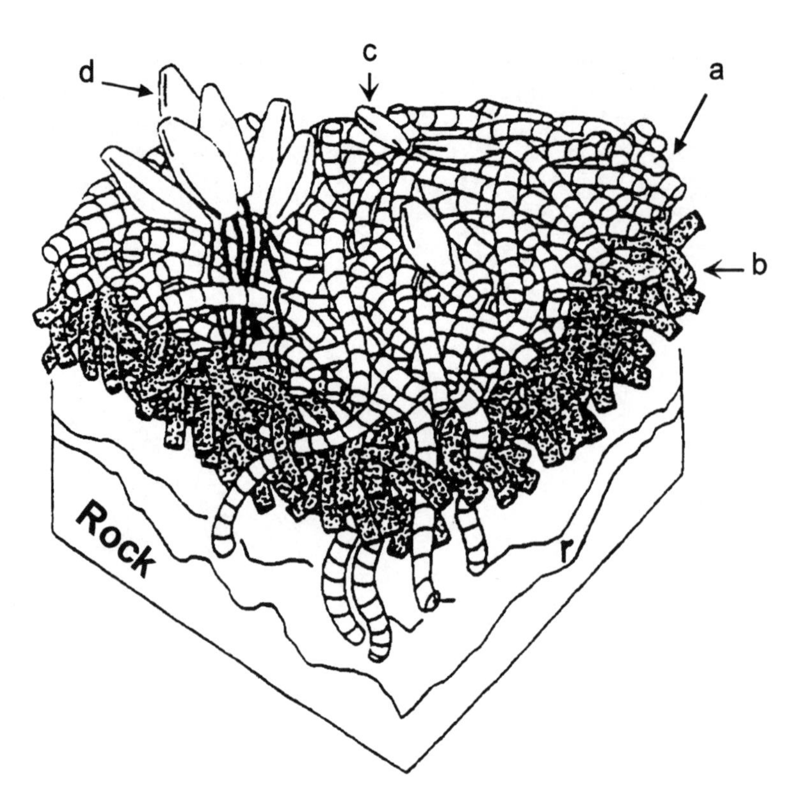

FIGURE 7.3. Drawing of an Oscillatoria mat on rock substrate, showing upper metabolically active filaments (a), lower dead algal filaments (b), loosely attached diatoms (c), and stalked diatoms (d) (Stock and Ward 1991 with permission).

Dominant species and growth forms usually change seasonally in response to changes in light, temperature, and streamflow. Seasonal changes are usually gradual. Major species tend to wax and wane in a succession of broad peaks, with long periods of overlap (Wehr 1981). Benthic algae often consist of permanent taxa that are always present, with certain taxa being seasonally dominant (Walter 1984). Filamentous green algae (e.g., *Ulothrix*) often dominate in spring and early summer, giving way to other green and blue-green algae (e.g., *Oedogonium*, *Pectonema*, *Phormidium*) in early autumn, and then to mainly diatoms in winter (Wehr 1981, Walter 1984). Overlying the gradual seasonal succession are rapid changes resulting from sudden increases in discharge during freshets. Species susceptible to export accumulate during periods of low streamflow but are exported during the next storm (Rounick and Gregory 1981, Steinman and McIntire 1990).

Macrophytes

Principal macrophytes include the angiosperms which have differentiated roots, leaves, and vascular tissue. They have four main growth habits: emergent plants rooted below water with aerial leaves (e.g., *Pontederia*); floating attached plants with submerged roots (e.g., *Nymphaea*); floating unattached plants (e.g., *Lemna*); and rooted submerged plants (e.g., *Potamogeton*) (Riemer 1984). A river may be lined by emergents along shore, have floating plants in protected water, and have submerged plants midstream. Vascular macrophytes are most abundant in low-gradient streams with open canopy (Fisher and Carpenter 1976).

Besides angiosperms, bryophytes (mosses and liverworts) can be important in both densely shaded headwater streams (Steinman and Boston 1993) and large rivers (Naiman and Sedell 1980, Naiman 1983). Characteristic genera include the moss *Fontinalis* and the leafy liverwort *Porella*. Often the same bryophyte species that occur submerged in streams also grow terrestrially along shore (Glime and Vitt 1987). Mosses have several characteristic growth forms, including clumps with long, free-floating filaments, and in fast water, short filaments that appear to be sheared by flowing water and bedload (Brusven et al. 1990). Mosses are perennial, taking several years to accumulate (Naiman and Sedell 1980), and they are sensitive to substrate instability (Englund 1991). Hence, they are often most abundant in fast water on stable boulders and bedrock (Suren 1991, Steinman and Boston 1993).

Phytoplankton

True riverine phytoplankton, or "potamoplankton," are normally restricted to slow-flowing rivers and sloughs and do not usually occur in small streams. Nearly all suspended algae in small streams are detached and drifting cells of benthic algae (Swanson and Bachmann 1976), called "tychoplankton" (Reid and Wood 1976). Typical potamoplankton of slow-flowing rivers includes Centric diatoms (e.g., *Stephanodiscus*) and small green algae (e.g., *Scenedesmus*). Seasonally, diatoms often bloom in spring as river discharge decreases followed by mixed blooms of green algae and diatoms in summer (Garnier et al. 1995).

For potamoplankton to develop in a river, the retention time must be long enough so that population growth by cell division outweighs the loss by dilution and discharge. At a given site, potamoplankton increase as river discharge declines, whereas suspension of detached cells from benthic algae increases as discharge goes up (Jones and Barrington 1985). These two opposing trends determine the composition and total density of algae suspended in a river.

The Primary Production Process

Driving primary production is the energy of photons from the sun captured as chemical bond energy by a plant's chlorophyll and other photosynthetic pigments (Figure 7.4) (Jorgensen 1977). Photosynthesis takes carbon, usually from carbon dioxide (CO_2) or bicarbonate, and hydrogen from water to fix carbohydrate and release oxygen. The plant uses some of the photosynthetic products for maintenance and either stores the rest or uses it for growth. These products subsequently may be used by other organisms in the food web.

Analogous to the first law of thermodynamics, the energy equation for primary production is

$$GPP = NPP + R$$

where GPP is gross primary production, NPP is net primary production, and R is respiration (McNaughton and Wolf 1973). The net production NPP represents the excess energy above the plant's maintenance costs which accumulates as biomass. Net production is what becomes available to other organisms through the food web.

The relationship between biomass and production determines the turnover rate at which

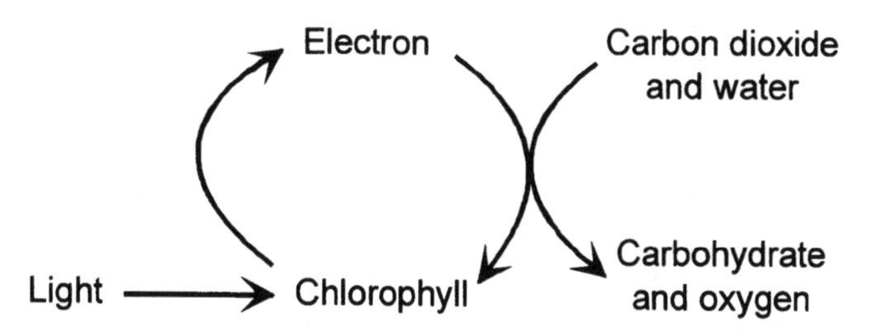

FIGURE 7.4. The basic photosynthetic process of green plants and algae, in which photons excite electrons in the chlorophyll molecule, and the resultant energy is used for carbon fixation (modified from McNaughton and Wolf 1973).

the standing biomass is replaced by new biomass production. Standing biomass itself is not necessarily a good measure of potential production because a small biomass with a high turnover rate may support a great abundance of consumers. Periphytic diatoms, for example, may be kept at a low standing biomass by stream scour and invertebrate grazing (Minshall 1978), but because they reproduce rapidly, their production is often many times greater than their standing biomass (McIntire 1973).

Limiting Factors

The rate of primary production in streams is a function of many factors, including availability of light and nutrients, temperature, herbivores, physical characteristics of the stream, and periodic disturbances (e.g., high streamflow). These factors interact with one another, and different ones predominate in different situations. Limiting factors also depend on whether production is expressed per unit algal biomass or unit stream area. In a given stream, for example, biomass-specific production may be limited by light, whereas areal-specific production could be limited by low algal biomass due to grazing (Steinman 1992).

The most common limiting factors for primary production in streams are sunlight and nutrients (Gregory et al. 1987). Low light is usually the first limiting factor in small streams (Hill and Harvey 1990), and low nutrients limit production at higher light levels (Lowe et al. 1986, Perrin et al. 1987, Hill and Knight 1988). Many diatoms, however, are adapted to low light, and areal-specific production may actually be limited by nutrients and grazers even in small shaded streams (Steinman 1992).

Sunlight

Most aquatic plants and algae are adapted to low light intensity (Riemer 1984). The ability to photosynthesize efficiently under low light is advantageous to submerged plants. The critical light level at which respiration equals photosynthesis is known as the compensation point, below which plants and algae would eventually starve to death because they respire food faster than they produce it.

With increased light, photosynthesis can increase up to the level of light saturation. At higher light levels, photosynthesis may actually decline because of photooxidation of enzymes and chlorophyll inactivation (Jorgensen 1977). The saturation level for many algae and vascular macrophytes is about 30 to 60% of full sunlight (McIntire 1973, Riemer 1984). The saturation level, however, varies depending on species and their acclimation conditions. The light intensity of saturation is positively correlated with ambient light conditions and is in the range of about 100 to 400 μmol quanta m^2/s in 2nd- to 4th-order streams (Boston and Hill 1991). Light saturation for shade-adapted benthic algae from small streams is approximately 20% of full sunlight (Gregory et al. 1987). The important consequence is that more sunlight will not increase primary production if photosynthesis is already light-saturated.

In small woodland streams, dim light under the dense forest canopy usually limits primary production (Gregory et al. 1987, Hill and Knight 1988, Feminella et al. 1989, Hill and Harvey 1990). Light at the surface of forested headwater streams often is less than 5% of full sunlight (Gregory et al. 1987). If light increases by removing riparian vegetation or as streams get larger, algal biomass and production can increase with adequate nutrients.

Nutrients

With sufficient light, primary production usually then becomes limited by nutrients. The mechanism of nutrient limitation in streams is different from that in lakes, where biomass can increase until a nutrient is exhausted (Russell-Hunter 1970). Streams receive a continual supply of nutrients from upstream, but nutrient uptake is limited by the rate of diffusion through the boundary layer around algal cells (Elwood et al. 1981, Klotz 1985). Under this diffusion limitation, nutrient uptake and cell reproduction may be nutrient-limited, even though nutrient supply is continuous. Large rivers, however, may behave like lakes if retention time is long enough for potamoplankton to deplete available nutrients (Garnier et al. 1995).

Availability of carbon is usually sufficient in streams because of water turbulence and high CO_2 solubility (Wetzel 1975, see Chapter 14 this volume). In some situations, however, free CO_2 in equilibrium with the atmosphere may be inadequate for high rates of photosynthesis. Ability to use bicarbonate instead of CO_2 is an adaptive advantage, particularly for many submerged angiosperms (Wetzel 1975). Aquatic bryophytes, however, may be limited by low CO_2 in alkaline streams because, unlike most other aquatic plants, they utilize only free CO_2 and can not use bicarbonate (Wetzel 1975, Glime and Vitt 1987). In highly calcareous waters above pH 8, free CO_2 concentration is negligible, and carbon species are dominated by bicarbonate and carbonate. Thus, mosses are generally restricted to soft waters of lower pH and abundant CO_2 (Wetzel 1975).

Silica is a required mineral for diatoms but not for most other algae, and diatom cell division stops without it (Werner 1977). The spring diatom bloom of phytoplankton in many lakes and some large rivers may deplete silica to limiting concentrations, after which other algae develop in a silica-limited environment (Wetzel 1975, Garnier et al. 1995).

Phosphorus (P) and nitrogen (N) are the two elements that most frequently limit aquatic primary production and are most commonly implicated in eutrophication from nonpoint source pollution (Riemer 1984). Both nutrients can limit algal growth in unpolluted waters, and adding either or both can increase productivity (Stockner and Shortreed 1978). If light and other factors (such as micronutrients, Patrick 1978), are sufficient, P often will be the first nutrient to become limiting (Wetzel 1975) because, in most fresh waters, P is an order of magnitude less abundant than N (Wetzel 1975).

The major sources of P in unpolluted waters are dust in precipitation and the slow weathering of rock (Wetzel 1975). The P content of groundwater is generally low, even where rocks and soils have high P content, because P-containing minerals are highly insoluble and strongly retained by biological and chemical processes in soil (Wetzel 1975, Mulholland 1992).

Phosphorus is present in natural waters usually in extremely small amounts in many different forms: dissolved as inorganic orthophosphate, suspended as organic colloids, adsorbed onto particulate organic and inorganic sediment, and contained in organic matter (Wetzel 1975). Soluble reactive phosphorus, consisting of ionic orthophosphates, is the only significant form available to plants and algae and constitutes less than 5% of the total phosphorus in most natural waters (Wetzel 1975). Dissolved P is usually in equilibrium with the amount bound to sediments, whose exchange capacity is increased by organic and inorganic colloids. Phosphorus is most available at a slightly acidic pH of 6 to 7. At a lower pH, it combines readily with aluminum, iron, and manganese (Tate et al. 1995), and at a higher pH, it becomes associated with calcium as apatite and phosphate minerals (Wetzel 1975).

Retention of P in streams is mostly a biological process regulated by algae, bacteria, and fungi. Aquatic microflora are capable of rapid uptake of significant amounts of P, as much as 95% within 100m of stream length (Gregory 1978, Tate et al. 1995). Physical sorption generally accounts for less than 20% of P retention on stream substrates (Gregory 1978, Mulholland 1992) but can be much higher, such as the high sorption by iron oxide in acid streams (Tate et al. 1995). Streams have prolonged periods of net in-stream retention of nutrients punctuated by large net losses during storms (Meyer and Likens 1979). In-stream processes tend to transform nutrients that are transported downstream from inorganic forms to dissolved or particulate organic forms (Meyer and Likens 1979, Mulholland 1992).

The reservoir of N in the atmosphere as free gas (N_2) cannot be used by plants and most algae until N-fixing bacteria or blue–green algae transform it into nitrate or ammonia (Wetzel 1975). Nitrogen occurs in fresh waters in numerous forms: dissolved molecular N_2, organic compounds, ammonia, nitrite, and nitrate. Sources of N include N-fixation, precipitation, surface runoff, and groundwater. Losses of N occur as stream outflow, denitrification of nitrate to N_2 by bacteria, and deposition in sediments (Wetzel 1975). Unlike P, N inorganic ions are highly soluble in water and readily leach out of soils and into streams. Concentration of N in stream water often follows a diel pattern, lowest in mid-afternoon and highest after dark, due to varying uptake by benthic algae (Triska et al. 1983).

Ammonia is generated by heterotrophic bacteria as the primary end-product of decomposition of organic matter, either directly from proteins or from other nitrogenous organic compounds (Wetzel 1975). Ammonia in water is present primarily as NH_4^+ and undissociated NH_4OH, which can be toxic. Although ammonia would be a good source of nitrogen for plants and algae, and many can use it at alkaline pH values, most plants and algae grow better with nitrate as their nitrogen source.

Whether N or P limit production depends on their relative and absolute abundance. The optimal N:P ratio for primary production is about 15:1 molar ratio (Elwood et al. 1981). A lower ratio indicates N is limiting; a higher ratio indicates P is limiting. Although a high N:P ratio indicates potential P limitation, this actually depends on the absolute P concentration (Bothwell 1985). Phosphorus limitation begins at a very low concentration ($<4\,\mu g\,PO_4/L$), and even a small increase above that level can produce a large amount of benthic algal biomass (Bothwell 1989). Growth of benthic algae in typical streams may be P-saturated at less than $1\,\mu g PO_4/L$ because of low temperature ($<15\,°C$) and low algal biomass (Bothwell 1985, Bothwell 1988). More P (up to $25\,\mu g\,PO_4/L$) may be needed to saturate photosynthesis at higher temperatures and higher algal biomass (Horner et al. 1983).

The optimum N:P ratio also depends on the particular algal species (Rhee 1978). A low N:P ratio (N limitation) favors growth of N-fixing algae, such as the blue–green *Nostoc* and the diatom *Epithemia* which has symbiotic blue–green inclusions (Fairchild and Lowe 1984). Adding P to a river can favor certain taxa, particularly the N-fixers.

Nutrient levels in unpolluted streams vary geographically, depending on geologic parent material (Chapter 4 this volume). Streams with volcanic bedrock generally have low N and relatively high P, and thus are likely to be N-limited, whereas streams with glacial or granitic geology are more likely to be P-limited (Gregory et al. 1987). Streams with sedimentary bedrock, such as dolomite, tend to have the most P (Golterman 1975) and be N-limited. Northern California streams are often poor in N with a low N:P ratio, and are N-limited when canopy is removed (Hill and Knight 1988). In contrast, addition of P to an Alaska tundra river increased primary production because P concentration was very low ($<4\,\mu g PO_4/L$) compared to N (Peterson et al. 1985).

Besides the limiting effects of light and nutrients, other controlling, lethal, and accessory factors (Fry 1947) also influence primary production. Controlling factors, of which temperature is most important, regulate metabolism without actually entering the process. Temperature has little effect on rate of photosynthesis at low light intensity, but under high light, temperature directly affects photosynthesis by essentially raising the level of light saturation (Jorgensen 1977). Turbidity acts as an accessory factor by clouding the water (Lloyd et al. 1987). Streambed scour and invertebrate grazing are lethal factors that reduce areal-specific primary production by keeping algal biomass low (Hill et al. 1992).

Grazing

Natural densities of snails, caddisflies, isopods, minnows, and other grazers can limit primary production by cropping benthic algae to a low biomass (Elwood and Nelson 1972, Gregory 1983, Lamberti and Resh 1983, McAuliffe 1984, Murphy 1984, Power et al. 1988, see Chapter 8 this volume). Effects of grazers on benthic algae are most evident during periods of low streamflow when grazers are concentrated (Hill and Knight 1987, Feminella et al. 1989).

Although intensive grazing can limit areal-specific primary production, moderate grazing often increases biomass-specific production by changing algal structure (i.e., species composition, physiognomy, age, and chlorophyll content) (Lamberti and Resh 1983, Hill and Knight 1987) and by enhancing nutrient supply by excretion (McCormick and Stevenson 1991). Biomass-specific production is highest when biomass is low because of self-limitation within the algal assemblage (Hill et al. 1992). Ungrazed benthic algae often change from a diatom film to a dense turf of filamentous green algae in which self-shading and reduced circulation cause light and nutrient limitation in lower layers. Growth is then confined to a thin upper layer, whereas lower layers only respire. Grazers thus help to keep benthic algal communities in a productive, early-successional state characterized by low biomass and rapid turnover (Lamberti and Resh 1983, Jacoby 1987).

Grazing affects benthic algal structure because of differential vulnerability of algal species and physical disturbance while feeding. Upper-layer diatoms that are loosely attached to the substrate are most susceptible to grazers, whereas small, adnate diatoms are more resistant (Lamberti and Resh 1983, Hart 1985). Heavy grazing converts benthic algae from filamentous types with diverse overstory to less-diverse, closely attached diatoms (Jacoby 1987). Grazers also affect the benthic algal assemblage by disturbing substrate surfaces (Sumner and McIntire 1982), which reduces abundance of loose-layer diatoms (Hill and Knight 1987).

Coevolution of algae with grazers has resulted in grazer-resistant algal species that depend on grazers to control their competitors. The characteristics that help certain algae resist grazers—toughness, unpalatability, mucilage, or protective growth form—also slow their growth and make them more susceptible to being overgrown by faster-growing, less-resistant species (Power et al. 1988).

Effects of grazing on benthic algae depend on the grazer species (Lamberti et al. 1987). The caddisfly *Dicosmoecus*, for example, possesses robust mandibles that can remove most algae except for basal holdfast structures. The snail *Juga* has a fine-toothed radula that is effective at removing diatoms and filamentous algae. Mayflies, having delicate brush-like mouthparts effective for harvesting diatoms rather than long filaments, tend to have little effect on benthic algal structure (Jacoby 1987).

Energy Flow

Energy flow from aquatic primary producers involves several output processes to various consumer compartments (Figure 7.5). Aquatic net primary production is the source of autochthonous organic matter in a given reach. It is also the source of part of the allochthonous

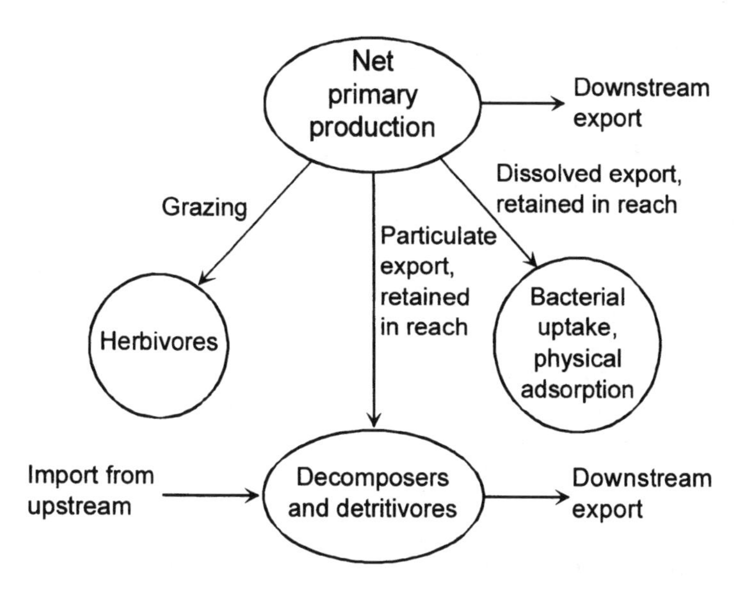

FIGURE 7.5. Energy-flow pathways for net primary production in a stream reach.

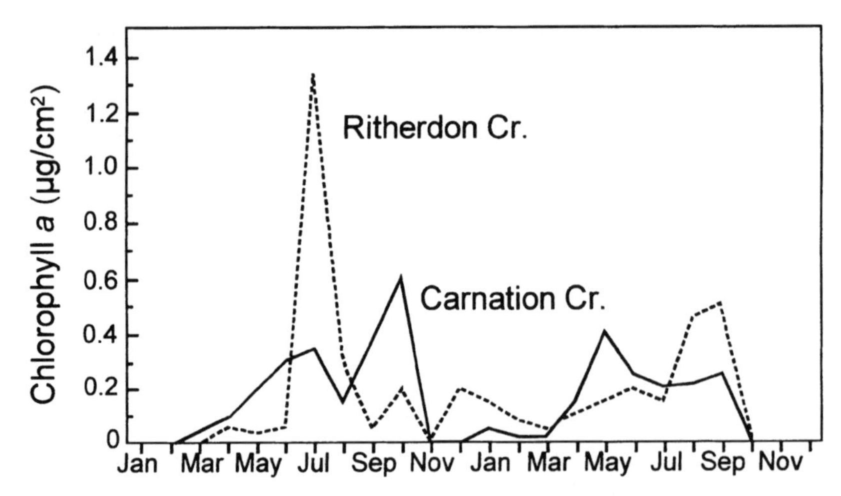

FIGURE 7.6. Seasonal growth of benthic algae, as measured by amount of chlorophyll *a*, in Carnation Creek (old-growth forest) and Ritherdon Creek (clear-cut without buffer) on Vancouver Island, British Columbia. Carnation Creek sites received 5 to 47% of available light; Ritherdon Creek received 72% of available light. (modified from Stockner and Shortreed 1976. Reprinted by permission of Kluwer Academic Publishers).

input, the part produced upstream by algae and aquatic plants and transported into the reach via streamflow. Energy flows out of net primary production through excretion of dissolved organic matter, grazing, and decomposition of particulate organic matter. Of these, the principal avenues of energy flow from primary producers are through direct grazing of living tissues and collecting of autochthonous detritus.

Different forms of aquatic plants and algae vary in the seasonal patterns of production and mortality that influence timing of energy flow and relative importance of grazing and detrital pathways. Benthic algae often show two seasonal peaks: in spring before vegetation leafs out and in autumn after leaves have fallen (Figure 7.6) (Minshall 1978, Sumner and Fisher 1979). The decline later in autumn is due to declining light and temperature and scour by seasonal storms (Rounick and Gregory 1981). Macrophyte biomass and production are highest in mid-summer and decline sharply in fall (Figure 7.7). Macrophytes are not grazed extensively, and biomass accumulates until plants die in late summer and autumn (Mann 1975, Minshall 1978). Macrophyte detritus decomposes quickly (50% weight loss the first week; Anderson and Sedell 1979), and most decomposition occurs near the site of production (Fisher and Carpenter 1976).

The Grazing Pathway

The amount of net primary production consumed by grazers is frequently considered negligible (e.g., Stockner and Shortreed 1976, Bothwell 1989), but in many common situations, grazing is an important pathway of energy flow. Abundance of grazers depends on stream size, as predicted by river continuum theory. For example, in western Oregon, invertebrates that scrape benthic algae make up only 1 to 12% of total invertebrates in 1st-order streams, but almost 25% in 3rd- to 7th-order streams (Figure 7.8) (Hawkins and Sedell 1981). This relative abundance indicates the importance of grazing in energy flow in the different stream orders.

In general, benthic diatoms provide the most nutritious food source for stream herbivores. Diatoms tend to have higher nutritive quality (Patrick 1978) and are more readily assimilated than other algae (Lamberti et al. 1989). Stream invertebrates most commonly reject filamentous and gelatinous algae (Gregory

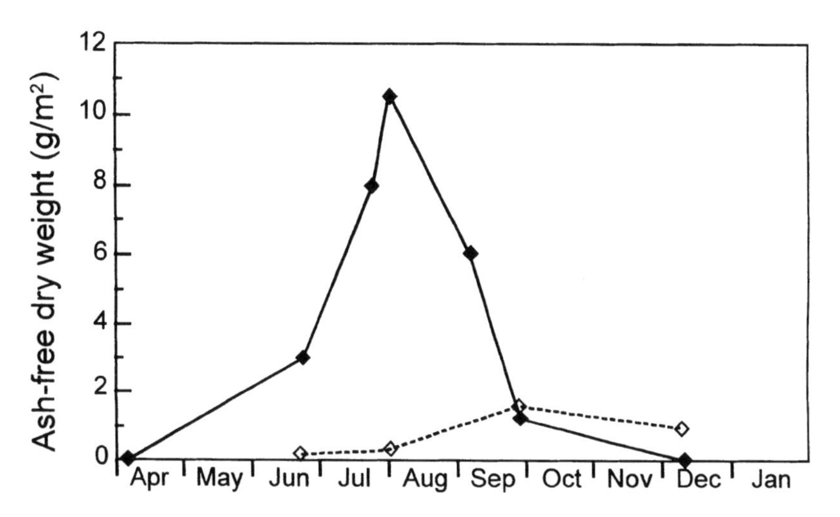

FIGURE 7.7. Living biomass (◆—◆) and downstream export (◇---◇) of vascular macrophytes (*Potamogeton*, *Callitriche*, and others) in the Fort River, Massachusetts (modified from Fisher and Carpenter 1976 with permission).

1983, Steinman et al. 1992). Live macrophytes generally are not consumed because of their tough cell walls, lignified structures, and low N content (Gregory 1983).

Consumption of living benthic algae by grazers varies depending on season, streamflow, and other factors. Although invertebrate grazers potentially have profound effects on benthic algae, they are usually prevented from having major effects because of periodic disturbances from storm events that significantly reduce their abundance (Steinman and McIntire 1990). Both plants and animals are regulated to some degree by catastrophic events. The major "reset mechanism" in lotic ecosystems, high streamflow, occurs several times per year (Naiman et al. 1992), a frequency similar to algal generation (Steinman and McIntire 1990). The marked spatial heterogeneity in current, light, and substrate of streams

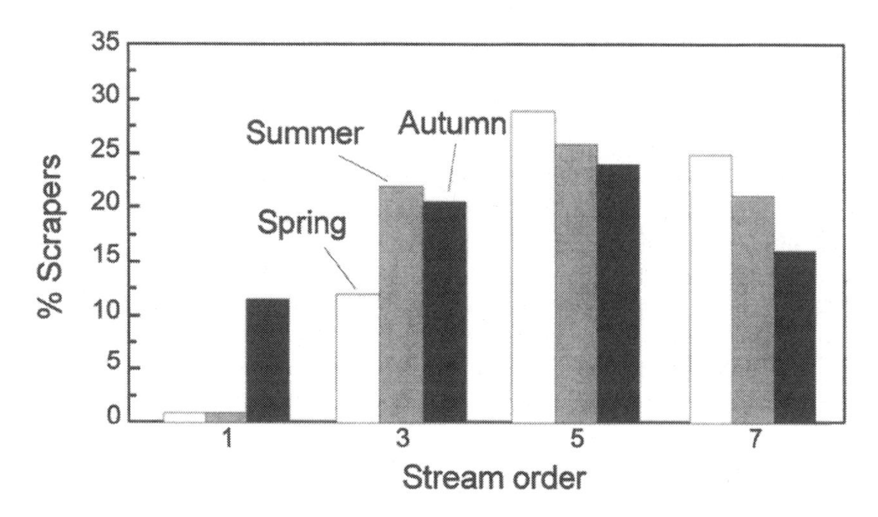

FIGURE 7.8. Relative abundance of the scraper functional group of macroinvertebrates in spring, summer, and autumn in relation to stream order in the McKenzie River basin, Oregon (modified from Hawkins and Sedell 1981 with permission).

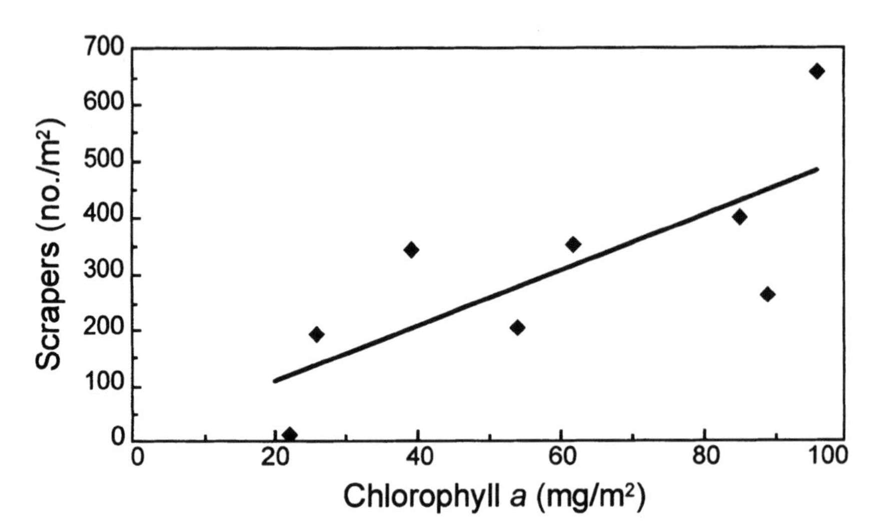

FIGURE 7.9. Density of the scraper functional group of macroinvertebrates in relation to amount of benthic algal chlorophyll *a* on rock surfaces (modified from Hawkins and Sedell 1981 with permission).

can mask grazing effects even when grazer density is high (Hill and Knight 1987). Such fluctuations and heterogeneity prevent grazers from attaining densities that reduce benthic algae (Hill and Knight 1987). In streams with stable flow or during extended dry periods when streamflow and wetted area slowly diminish, grazers can become concentrated enough to significantly impact their food resource.

In return, aquatic plants and algae have important consequences on grazer populations (Lamberti and Resh 1983, McAuliffe 1984). Availability of benthic algae regulates the distribution, abundance, and growth of scrapers (Figure 7.9) (Hawkins and Sedell 1981, Gregory 1983). Grazing invertebrates generally compete exploitatively for food (Hart 1987), and competition is most common during periods of summer low streamflow (Hill and Knight 1987). Grazer populations may be food-limited at sites where primary production is severely restricted by dim light (Lamberti et al. 1989) and during periods of stable or low streamflow, even if other factors limit them during other times (Hill and Knight 1987, Hart and Robinson 1990).

The Detrital Pathway

Although grazing is important at times in many streams, much of the biomass produced by aquatic primary producers is typically utilized through detrital pathways after the live or dead algae are exported from the production site (Minshall 1978, Lamberti et al. 1989, Scrimgeour et al. 1991). For example, diatoms that detach from the streambed are strained from the water by filter feeders like the net-spinning caddisfly *Hydropsyche* (Fuller and Mackay 1981) or are collected after they settle by deposit feeders like the chironomid *Paratendipes* (Hawkins and Sedell 1981). Besides exporting particulate detritus, algae also excrete dissolved organic matter that is then taken up by bacteria, a process which is an important energy pathway (Peterson et al. 1985).

Export of benthic algae from production areas to detrital pools results from sloughing, dislodgement by grazers, and scour by flowing water and sediment. Sloughing occurs when underlying layers of algal mats die and pieces float away (Naiman 1976). Dislodgement results when grazers disturb benthic algae while feeding. It can exceed consumption, particularly when algal biomass is high (Lamberti et al. 1989, Scrimgeour et al. 1991). Scour during periodic freshets causes major export of benthic algae, especially in autumn and winter (Rounick and Gregory 1981). Filamentous algae are susceptible to export if currents exceed 50 cm/s (Horner and Welch 1981), whereas prostrate forms are more resistant to export (Steinman and McIntire 1990). Redd-digging salmon can also cause extensive scour and export (Walter 1984).

To contribute energy to a stream's food web, organic matter from primary production must first be retained in the stream channel so that it can be processed. Export and retention, therefore, largely determine the contribution that aquatic primary producers make to a stream ecosystem. Organic matter and nutrients in stream ecosystems are repeatedly transported, retained, metabolized, and exported in a cycling process called *spiraling* (Newbold et al. 1983, see Chapter 15 this volume). Spiraling describes the coupled processes of cycling and downstream transport of nutrients and organic matter in streams (Mulholland et al. 1983, Newbold et al. 1983). Spiral length represents the average downstream distance traveled in completing one cycle. Streams with short spirals have high retention capacity and efficiently utilize organic matter and nutrients. Both stream channel obstructions and invertebrate collectors help retain particulate autochthonous organic matter.

Although greatly outweighed by allochthonous organic matter of terrestrial origin, autochthonous detritus is important in secondary production, even in heavily forested streams with abundant allochthonous inputs (Bilby and Bisson 1992). Most of the organic matter present in streams is usually of poor nutritional quality (e.g., high carbon-to-nitrogen [C:N] ratio), and large differences in quantity of detritus represent negligible differences in actual food availability (Hawkins et al. 1982). Benthic algae are more nutritious than allochthonous organic matter because of their low C:N ratio and high protein content (Anderson and Cummins 1979). Live algae growing on detritus (Naiman 1983) also augment food quality. Abundance of invertebrates does not correlate well with total amount of organic detritus because invertebrates are usually more influenced by food quality than quantity (Ward and Cummins 1979, Hawkins et al. 1982). However, invertebrate abundance does correlate with the chlorophyll content or respiration rate of detritus (Figure 7.10) (Hawkins and Sedell 1981, Hawkins et al. 1982). Thus, algal detritus is important because it enriches the detritus pool, providing a richer substrate for microbial colonization and invertebrate consumption.

Fish and other predators also benefit from the energy flow from autochthonous organic matter. By the nature of their drift-feeding behavior, juvenile salmonids focus on food from autochthonous pathways. Invertebrates that

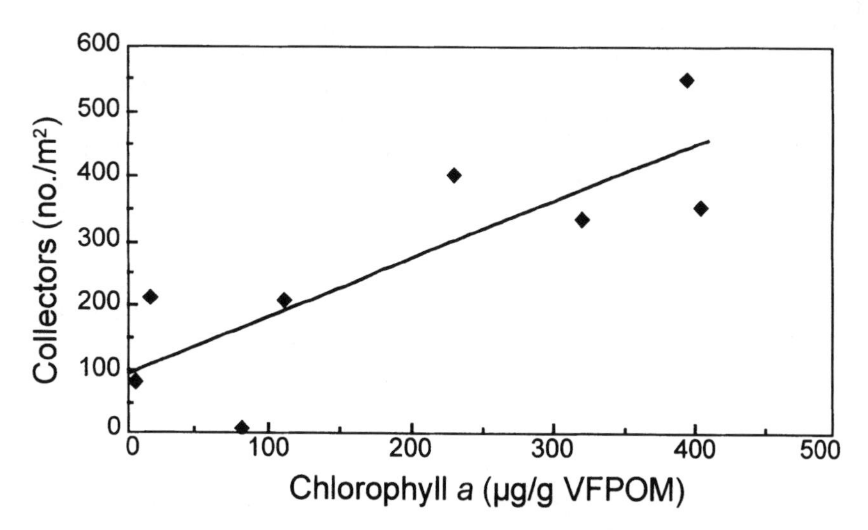

FIGURE 7.10. Density of the collector functional group of macroinvertebrates in relation to amount of chlorophyll *a* in very fine particulate organic matter (VFPOM) (modified from Hawkins and Sedell 1981 with permission).

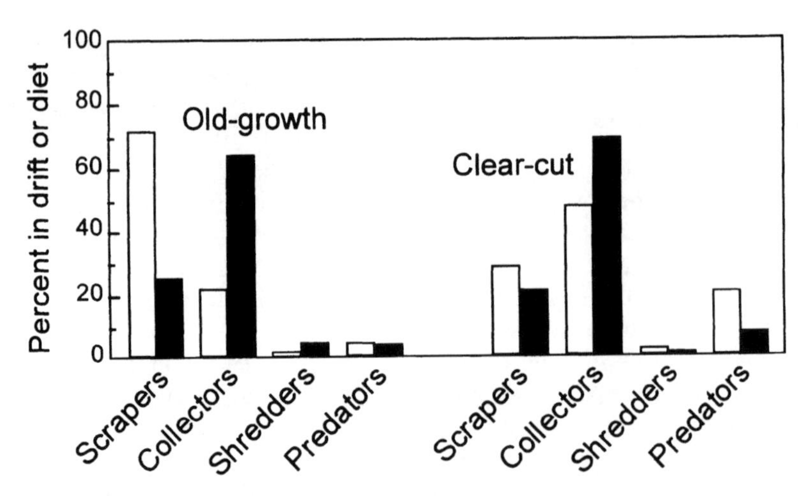

FIGURE 7.11. Percentages of invertebrate functional groups in drift (clear) and coho salmon diet (shaded) in old-growth and clear-cut sections of streams in Washington (modified from Bilby and Bisson 1992, reproduced with permission of the Minister of Supply and Services Canada 1996).

utilize autochthonous production belong mainly to the scraper and collector-gatherer functional groups (Cummins 1974), the groups most frequently eaten by salmonids (Figure 7.11) (Hawkins et al. 1983, Murphy and Meehan 1991, Bilby and Bisson 1992, see Chapter 9, this volume). Insects in these groups (e.g., Baetidae, Chironomidae, Simuliidae) are generally small-bodied and multivoltine, have rapid turnover, and are prone to drift (Waters 1969). In contrast, shredders of allochthonous organic matter (e.g., Limnephilidae) are uncommon in salmonid diets (Figure 7.11) (Jenkins et al. 1970, Griffith 1974). They tend to have only one generation per year, and their advanced instars are often large and armored with wood, shell, or stone cases so that they do not often drift and are hard for fish to ingest. Although the contributions to the diet from both terrestrial and adult aquatic insects are important (e.g., Bjornn et al. 1992), most aquatic insects eaten by juvenile salmon are primarily supported by autochthonous organic matter (Bilby and Bisson 1992). Thus, autochthonous pathways are of overriding importance in the trophic support of juvenile salmonids regardless of the amount of allochthonous material entering the stream.

Distribution of Primary Production in Watersheds

Aquatic primary production changes predictably in response to trends in geomorphology and fluvial processes as streams get larger. These trends in primary production and other ecosystem structures and functions are embodied in the concept of the river continuum (Vannote et al. 1980, Minshall et al. 1983). The river continuum portrays entire drainage networks as integrated ecosystems that change systematically from headwaters to river mouth.

In a typical forested river continuum, primary production is low in the headwaters, high in mid-order streams and swift rivers, and low again in large, sluggish rivers (Vannote et al. 1980). In small headwater streams, primary production is severely restricted by shade from forest canopy and geomorphic features. Mid-order streams have gaps in the canopy that allow in more sunlight, and benthic algal production increases (Naiman 1983). Finally, in large, sluggish rivers, shade from riparian trees and geomorphic features is insignificant, but water depth and turbidity restrict light penetration, and autotrophs are largely reduced to

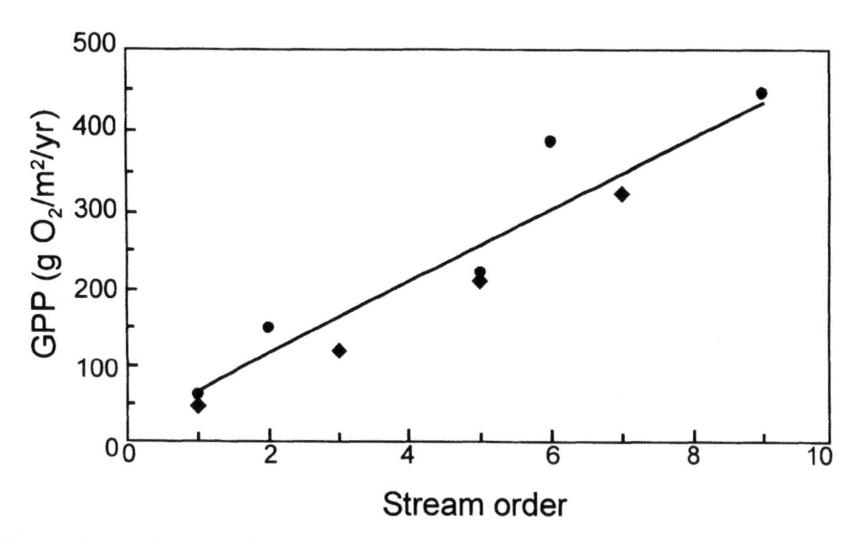

FIGURE 7.12. Total annual gross primary production (GPP) by benthic algae, mosses, and macrophytes in relation to stream order in watersheds in Oregon ◆ and Quebec ●. Production in lower-order streams is mostly by benthic algae; production in higher-order streams is augmented by moss and macrophytes (data from Naiman and Sedell 1980 and Naiman 1983). The regression equation is $Y = 55.7X^{0.93}$; $R^2 = 0.93$.

phytoplankton and macrophytes near shore (Naiman and Sedell 1980). Production, however, may continue to increase in a downstream direction at least up to the size of a moderately large river (Figure 7.12), if not too sluggish and turbid (Naiman and Sedell 1980, Naiman 1983).

Most primary production within a watershed occurs in mid-order and larger streams. The relationship between area-specific primary production and stream order (Figure 7.12) combined with data on surface area of each order give the distribution of primary production within a watershed (Naiman and Sedell 1981, Naiman 1983). For example, in a 9th-order river basin in Quebec, Canada, 1st- to 3rd-order streams contain 23% of the basin's total stream surface area but yield only 16% of the total annual gross production of benthic algae, despite representing 87% of the total stream length (Table 7.1). In contrast, streams larger than 6th order are only 2% of total stream

TABLE 7.1. Distribution of gross primary production (GPP) of benthic algae in the Moisie River basin, Quebec.

Stream order	Stream length (km)	Stream width (m)	Surface area (km^2)	Annual GPP per m^2 (g/O_2/m^2)	Total annual GPP (tonnes O_2)	(%)
1	16,142	0.3	4.8	42.6	204	1.2
2	8,249	2.1	17.3	52.0	900	5.3
3	3,842	6.7	25.7	61.4	1,578	9.2
4	1,879	15.6	29.3	70.9	2,077	12.2
5	1,072	30.1	32.3	80.3	2,594	15.2
6	471	51.3	24.2	89.8	2,173	12.7
7	340	80.6	27.4	99.2	2,718	15.9
8	292	119.1	34.8	108.6	3,779	22.1
9	54	168.2	9.1	118.1	1,075	6.3
Total	32,341		204.9		17,098	100.0

Hydrology data, Naiman 1983; production data, Naiman and Sedell 1981, with permission.

length, but account for 35% of total surface area. Because of this large area and open canopy, these large rivers account for 44% of annual gross production of benthic algae for the watershed.

Potential Response to Watershed Uses

Watershed uses have profound multiple effects on aquatic primary production and other functions of stream ecosystems. Timber harvest, livestock grazing, agriculture, urban development, and other activities affect aquatic primary production by altering riparian vegetation, streamflow, sediment, channel structure, and other watershed features. These changes alter the light and nutrient regimes, physical habitat for aquatic plants and algae, and spiraling of autochthonous organic matter through the system.

The most obvious effect on aquatic primary production from land uses that alter riparian vegetation is the result of increased sunlight reaching the stream during early seral stages after disturbance. Canopy removal often causes increased primary production (Lowe et al. 1986) and increased energy flow through the food web leading to greater production of invertebrates and fish (Murphy and Meehan 1991, Bilby and Bisson 1992). Nutrient enrichment of naturally more open, mid-order streams also increases growth and abundance of algae, invertebrates, and fish (Hershey et al. 1988, Johnston et al. 1990). Effects of canopy removal depend on stream size. Larger streams are naturally more open; thus canopy removal has a smaller effect on primary production.

Strong linkages exist between riparian vegetation, light reaching the stream, primary production, microbial respiration, invertebrate production, and ultimately vertebrate production. Increased benthic algal production in open streams is available to higher trophic levels both as live algae and as detritus. The algae enrich the detritus as a substrate for microbes, resulting in increased respiration (Figure 7.13). The abundant microbes and associated live algae improve food quality, encouraging greater invertebrate production (Behmer and Hawkins 1986), which is further increased by a shift in species composition to multivoltine collector-gatherers with fast turnover (Gregory et al. 1987). Finally, increased availability of invertebrate prey leads to greater biomass and production of fish (Murphy et al. 1981, Johnston et al. 1990, Bilby and Bisson 1992).

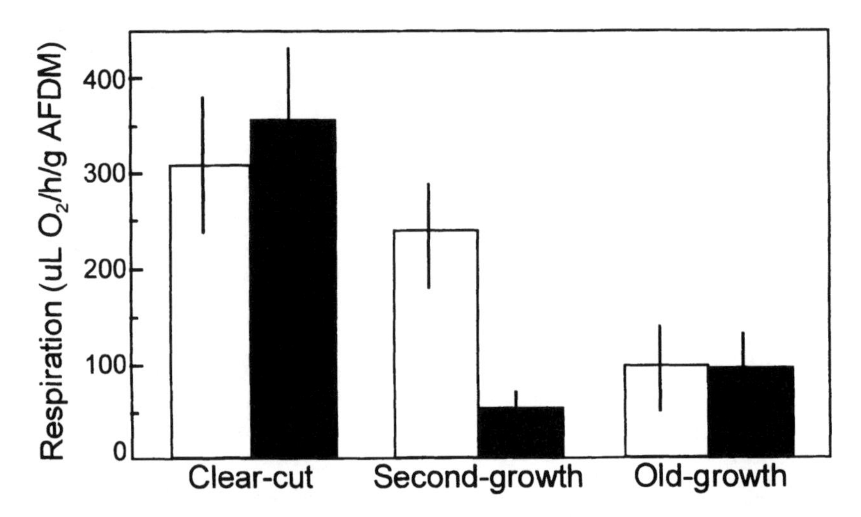

FIGURE 7.13. Respiration rate per *g* ash-free dry weight (AFDM) of fine particulate organic matter from streambeds in clear-cut, second-growth, and old-growth sites with high gradient (clear) or low gradient (shaded) in western Oregon. Vertical bars are ±1 standard error of the mean (modified from Murphy et al. 1981 with permission).

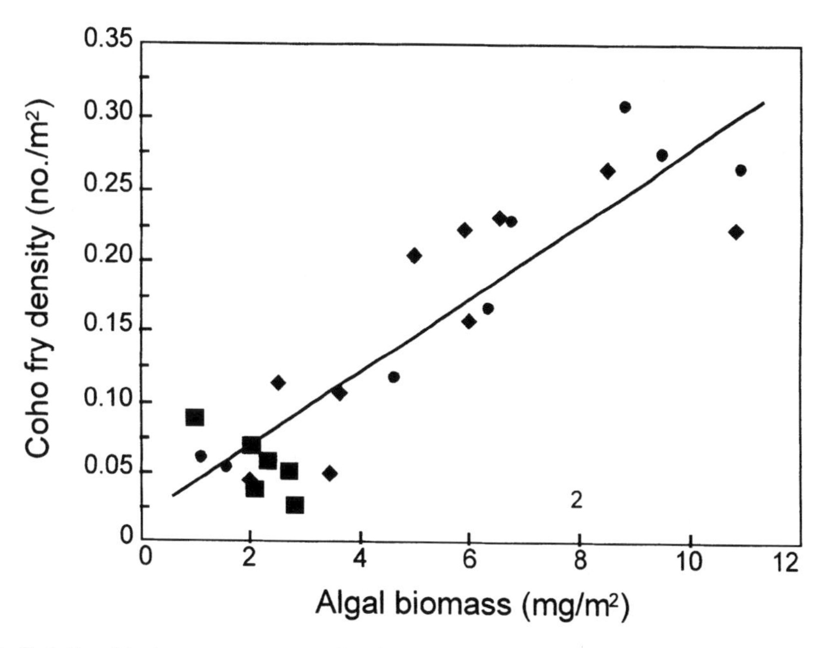

FIGURE 7.14. Relationship between summer density of coho salmon fry and benthic algal biomass (ash-free dry matter) in old-growth ■, buffered ◆, and clear-cut ● reaches of streams in southeast Alaska (modified from Murphy et al. 1984 with permission).

Besides increasing primary production, canopy removal also usually reduces inputs of allochthonous organic matter from streamside vegetation (Duncan and Brusven 1985a and 1985b, Gregory et al. 1987, Bilby and Bisson 1992). Increased food quality of detritus derived from autochthonous sources, however, can more than offset the decline in allochthonous sources (Bilby and Bisson 1992).

Higher food quality in open streams mainly benefits certain functional feeding groups: grazers, collectors, and predators (Murphy et al. 1981, Hawkins et al. 1982). Because invertebrates do not respond equally to canopy removal, species diversity may decline (Newbold et al. 1980). Reduced diversity, however, may be a result of increased dominance of certain grazer and collector-gatherers (e.g., *Baetis*, *Nemoura*, and Chironomidae) that consume algae and algal detritus rather than from reduced species richness (Newbold et al. 1980). Within the functional groups that benefit from increased primary production, such as collector-gatherers and invertebrate predators, canopy removal can increase species diversity (Murphy and Hall 1981, Hawkins et al. 1982).

Salmonids and other vertebrates often increase in abundance and growth after canopy removal because of greater food availability (Murphy and Meehan 1991). Where food is limiting and other habitat and population factors are suitable, density of coho salmon fry in summer is directly related to the abundance of algae (Figure 7.14). Evidence suggests higher fry density results from smaller feeding territories (Dill et al. 1981) because of an increase in invertebrate prey (Murphy et al. 1981, Hawkins et al. 1983). In a comparison of energy flow in forested and clear-cut sites (Bilby and Bisson 1992), total fish production was greater in the clear-cut than in the old-growth site, despite the old-growth site receiving five times more terrestrial organic matter. The increase in vertebrate production occurred mainly in spring and early summer, coinciding with the production cycle of benthic algae (Figure 7.15). A similar ratio between the clear-cut and old-growth sites (CC:OG ratio) for both fish and algal production (Table 7.2) indicated that primary production was mainly responsible for supporting fish populations in both sites.

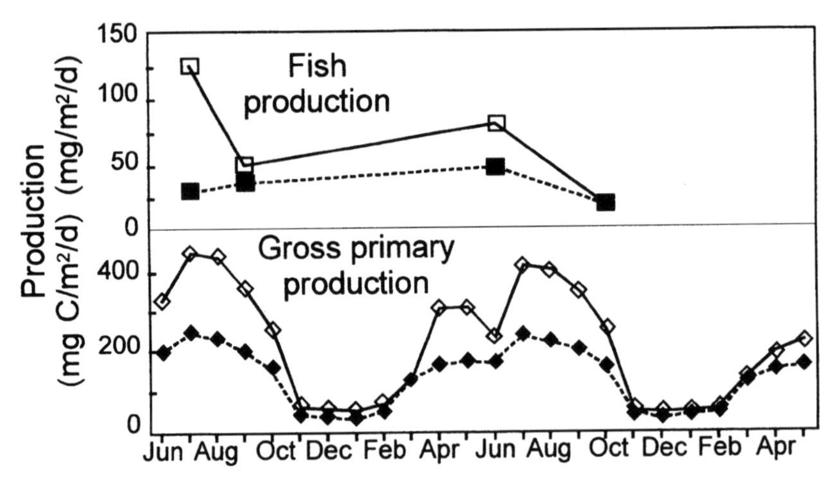

FIGURE 7.15. Seasonal production of benthic algae and fish (coho salmon, cutthroat trout, and shorthead sculpin) in clear-cut □ and old-growth ■ reaches of streams (data from Bilby and Bisson 1992, reproduced with permission of the Minister of Supply and Services Canada 1996).

The overall effects of watershed uses result from the interaction of changes in both trophic processes and physical habitat (Murphy and Meehan 1991). For example, increased aquatic primary production after canopy removal can temporarily mask detrimental effects on physical habitat, such as increased accumulation of fine sediment (Hawkins et al. 1983). In a comparison of paired shaded and open stream reaches with different sediment levels (Hawkins et al. 1983), densities of invertebrates and vertebrates decreased with increased fine sediment in shaded sites, but densities increased independently of sediment in open sites (Figure 7.16). Thus, increased primary production from removing riparian canopy apparently mitigates and sometimes completely masks otherwise detrimental effects associated with decreased habitat quality. As the canopy closes, however, effects of decreased habitat quality may become more evident.

Furthermore, species that are not food-limited do not necessarily benefit from increased primary production. Increased density of salmonid fry in summer, for example, may be nullified by decreased woody debris where winter habitat is the population's limiting factor (Murphy et al. 1986). Studies on coho salmon show increased salmon production after canopy removal, whereas other species may decrease with associated habitat degradation (Hicks et al. 1991, Reeves et al. 1993).

Over the long term, increases in primary production in early seral stages may be outweighed by longer-lasting reductions as a result of increased shade in later seral stages (Murphy and Hall 1981, Sedell and Swanson 1984). Second-growth hardwoods and young conifers produce

TABLE 7.2. Comparison of mean daily input of terrestrial organic matter, net production of benthic algae, and fish production from spring to early autumn in two streams in clear-cut (CC) and old-growth (OG) forest in Washington.

Variable	Clear-cut ($mg/m^2/d$)	Old-growth ($mg/m^2/d$)	CC:OG Ratio
Terrestrial input	164	851	0.2
Net algal production	482	301	1.6
Fish production			
Oncorhynchus kisutch	34	18	1.9
O. clarki	8.2	7.4	1.1
Total salmonids	42	25	1.7
Cottus confusus	4.5	2.5	1.8
Total (all species)	47	28	1.7

Modified from Bilby and Bisson 1992, reproduced with permission of the Minister of Supply and Services Canada 1996.

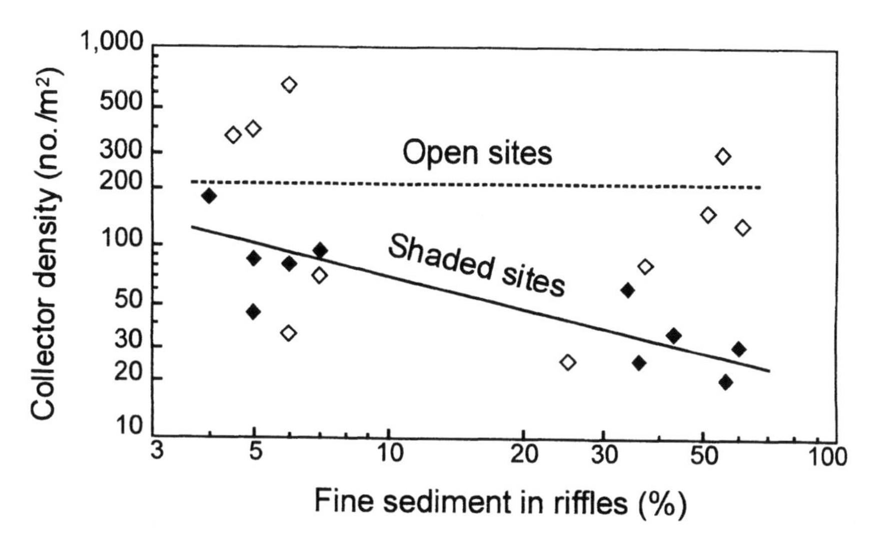

FIGURE 7.16. Relationships between density of the collector functional group of macroinvertebrates and percent fine sediment (<1 mm) in riffles in open ◇ and shaded ◆ reaches of streams in western Oregon and California. The Y-axis is in log scale; the X-axis is in arcsin scale (modified from Hawkins et al. 1983, reproduced with permission of the Minister of Supply and Services Canada 1996).

a denser canopy and lack the canopy gaps common in old-growth forest (Sedell and Swanson 1984, Bjornn et al. 1992). Streams in dense second growth in western Oregon, for example, have lower trout biomass and lower insect species richness than in adjacent old-growth (Murphy and Hall 1981). Besides reduced primary production, second-growth streams often lack important habitat features for salmonids, particularly pools and large woody debris (Bisson et al. 1987). Thus, effects of watershed uses on aquatic primary production need to be viewed from a long-term perspective and in the context of other changes in physical habitat. The focus should expand beyond initial changes, effects in one season, or effects on a single species.

Aquatic primary production plays a key role in the trophic support of stream ecosystems. Opening the canopy surrounding a stream and increasing nutrients often results in a more productive stream for invertebrates and fish. However, the desirability and integrity of stream systems is not solely determined by productivity (Karr and Dudley 1981). Stream organisms are adapted to complex environments, and although an environmental change may increase production, it may also cause undesirable shifts in community structure and loss of many taxa typical of unmodified streams (Lemly 1982, Hawkins et al. 1983). While enhancing productivity is an important management goal, managing for naturally functioning watersheds and stream ecosystems, with diverse energy sources and full complements of appropriate functional groups, is probably a better approach to maintaining healthy stream habitats and biodiversity.

Literature Cited

Anderson, N.H., and K.W. Cummins. 1979. The influence of diet on the life histories of aquatic insects. Journal of the Fisheries Research Board of Canada **36**:335–342.

Anderson, N.H., and J.R. Sedell. 1979. Detritus processing by macroinvertebrates in stream ecosystems. Annual Review of Entomology **24**:351–377.

Behmer, D.J., and C.P. Hawkins. 1986. Effects of overhead canopy on macroinvertebrate production in a Utah stream. Freshwater Biology **16**:287–300.

Bilby, R.E., and P.A. Bisson. 1992. Allochthonous versus autochthonous organic matter contributions to the trophic support of fish populations in clear-cut and old-growth forested streams.

Canadian Journal of Fisheries and Aquatic Sciences **49**:540–551.

Bisson, P.A., R.E. Bilby, M.D. Bryant, C.A. Dolloff, G.B. Grette, R.A. House, et al. 1987. Large woody debris in forested streams in the Pacific Northwest: Past, present, and future. Pages 143–190 *in* E.O. Salo and T.W. Cundy, eds. Streamside management: Forestry and fishery interactions. Institute of Forest Resources Contribution Number 57, University of Washington, Seattle, Washington, USA.

Bjornn, T.C., M.A. Brusven, N.J. Hetrick, R.M. Keith, and W.R. Meehan. 1992. Effects of canopy alterations in second-growth forest riparian zones on bioenergetic processes and responses of juvenile salmonids to cover in small southeast Alaska streams. USDA Forest Service Technical Report 92-7, Pacific Northwest Research Station, Forest Sciences Laboratory, Juneau, Alaska, USA.

Boston, H.L., and W.R. Hill. 1991. Photosynthesis-light relations of stream periphyton communities. Limnology and Oceanography **36**:644–656.

Bothwell, M.L. 1985. Phosphorus limitation of lotic periphyton growth rates: An intersite comparison using continuous-flow troughs (Thompson River system, British Columbia). Limnology and Oceanography **30**:527–542.

——. 1988. Growth rate responses of lotic periphytic diatoms to experimental phosphorus enrichment: The influence of temperature and light. Canadian Journal of Fisheries and Aquatic Sciences **45**:261–270.

——. 1989. Phosphorus-limited growth dynamics of lotic periphyton diatom communities: Areal biomass and cellular growth rate responses. Canadian Journal of Fisheries and Aquatic Sciences **46**:1293–1301.

Brusven, M.A., W.R. Meehan, and R.C. Biggam. 1990. The role of aquatic moss on community composition and drift of fish-food organisms. Hydrobiologia **196**:39–50.

Cummins, K.W. 1974. Structure and function of stream ecosystems. BioScience **24**:631–641.

Dill, L.M., R.C. Ydenberg, and A.H.G. Fraser. 1981. Food abundance and territory size in juvenile coho salmon (*Oncorhynchus kisutch*). Canadian Journal of Zoology **59**:1801–1809.

Duncan, W.F.A., and M.A. Brusven. 1985a. Energy dynamics of three low-order southeast Alaska streams: Autochthonous production. Journal of Freshwater Ecology **3**:155–166.

Duncan, W.F.A., and M.A. Brusven. 1985b. Energy dynamics of three low-order southeast Alaska streams: Allochthonous processes. Journal of Freshwater Ecology **3**:233–248.

Elwood, J.W., and D.J. Nelson. 1972. Periphyton production and grazing rates in a stream measured with a ^{32}P material balance method. Oikos **23**:295–303.

Elwood, J.W., J.D. Newbold, A.F. Trimble, and R.W. Stark. 1981. The limiting role of phosphorus in a woodland stream ecosystem: Effects of P enrichment on leaf decomposition and primary producers. Ecology **62**:146–158.

Englund, G. 1991. Effects of disturbance on stream moss and invertebrate community structure. Journal of the North American Benthological Society **10**:143–153.

Fairchild, G.W., and R.L. Lowe. 1984. Artificial substrates which release nutrients: Effects on periphyton and invertebrate succession. Hydrobiologia **114**:29–37.

Feminella, J.W., M.E. Power, and V.H. Resh. 1989. Periphyton responses to invertebrate grazing and riparian canopy in three northern California coastal streams. Freshwater Biology **22**:445–457.

Fisher, S.G., and S.R. Carpenter. 1976. Ecosystem and macrophyte primary production of the Fort River, Massachusetts. Hydrobiologia **49**:175–187.

Fry, F.E.J. 1947. Effects of the environment on animal activity. Publications of the Ontario Fisheries Research Laboratory 68. University of Toronto Press, Toronto, Ontario, Canada.

Fuller, R.L., and R.J. Mackay. 1981. Effects of food quality on the growth of three *Hydropsyche* species (Trichoptera: Hydropsychidae). Canadian Journal of Zoology **59**:1133–1140.

Garnier, J., G. Billen, and M. Coste. 1995. Seasonal succession of diatoms and Chlorophyceae in the drainage network of the Seine River: Observations and modeling. Limnology and Oceanography **40**:750–765.

Glime, J.M., and D.H. Vitt. 1987. A comparison of bryophyte species diversity and niche structure of montane streams and stream banks. Canadian Journal of Botany **65**:1824–1837.

Golterman, H.L. 1975. Chemistry. Pages 39–80 *in* B.A. Whitton, ed. River ecology. University of California Press, Berkeley, California, USA.

Gregory, S.V. 1978. Phosphorus dynamics on organic and inorganic substrates in streams. Verhandlungen der Internationalen Vereinigung für Theoretische und Angewandte Limnologie **20**:1340–1346.

——. 1983. Plant-herbivory interactions in stream systems. Pages 159–189 *in* J.R. Barnes and G.W. Minshall, eds. Stream ecology: Application and testing of general ecological theory. Plenum, New York, New York, USA.

Gregory, S.V., G.A. Lamberti, D.C. Erman, K.V. Koski, M.L. Murphy, and J.R. Sedell. 1987. Influence of forest practices on aquatic production. Pages 233–255 *in* E.O. Salo and T.W. Cundy, eds. Streamside management: Forestry and fishery interactions. Institute of Forest Resources Contribution Number 57, University of Washington, Seattle, Washington, USA.

Griffith, J.S.J. 1974. Utilization of invertebrate drift by brook trout (*Salvelinus fontinalis*) and cutthroat trout (*Salmo clarki*) in small streams in Idaho. Transactions of the American Fisheries Society **103**:440–447.

Hart, D.D. 1985. Grazing insects mediate algal interactions in a stream benthic community. Oikos **44**:40–46.

——. 1987. Experimental studies of exploitative competition in a grazing stream insect. Oecologia **73**:41–47.

Hart, D.D., and C.T. Robinson. 1990. Resource limitation in a stream community: Phosphorus enrichment effects on periphyton and grazers. Ecology **71**:1494–1502.

Hawkins, C.P., M.L. Murphy, and N.H. Anderson. 1982. Effects of canopy, substrate composition, and gradient on the structure of macroinvertebrate communities in Cascade Range streams of Oregon. Ecology **63**:1840–1856.

Hawkins, C.P., M.L. Murphy, N.H. Anderson, and M.A. Wilzbach. 1983. Density of fish and salamanders in relation to riparian canopy and physical habitat in streams of the northwestern United States. Canadian Journal of Fisheries and Aquatic Sciences **40**:1173–1185.

Hawkins, C.P., and J.R. Sedell. 1981. Longitudinal and seasonal changes in functional organization of macroinvertebrate communities in four Oregon streams. Ecology **62**:387–397.

Hershey, A.E., A.L. Hiltner, M.A.J. Hullar, M.C. Miller, J.R. Vestal, M.A. Lock, et al. 1988. Nutrient influence on a stream grazer: *Orthocladius* microcommunities respond to nutrient input. Ecology **69**:1383–1392.

Hicks, B.J., J.D. Hall, P.A. Bisson, and J.R. Sedell. 1991. Responses of salmonids to habitat changes. Pages 483–518 *in* W.R. Meehan, ed. Influences of forest and rangeland management on salmonid fishes and their habitats. Special Publication 19, American Fisheries Society, Bethesda, Maryland, USA.

Hill, W.R., H.L. Boston, and A.D. Steinman. 1992. Grazers and nutrients simultaneously limit lotic primary productivity. Canadian Journal of Fisheries and Aquatic Sciences **49**:504–512.

Hill, W.R., and B.C. Harvey. 1990. Periphyton responses to higher trophic levels and light in a shaded stream. Canadian Journal of Fisheries and Aquatic Sciences **47**:2307–2314.

Hill, W.R., and A.W. Knight. 1987. Experimental analysis of the grazing interaction between a mayfly and stream algae. Ecology **68**:1955–1965.

Hill, W.R., and A.W. Knight. 1988. Nutrient and light limitation of algae in two northern California streams. Journal of Phycology **24**:125–132.

Horner, R.R., and E.B. Welch. 1981. Stream periphyton development in relation to current velocity and nutrients. Canadian Journal of Fisheries and Aquatic Sciences **38**:449–457.

Horner, R.R., E.B. Welch, and R.B. Veenstra. 1983. Development of nuisance periphytic algae in laboratory streams in relation to enrichment and velocity. Pages 121–134 *in* R.G. Wetzel, ed. Periphyton of freshwater ecosystems. Dr. W. Junk Publishers, The Hague, The Netherlands.

Jacoby, J.M. 1987. Alterations in periphyton characteristics due to grazing in a Cascade foothill stream. Freshwater Biology **18**:495–508.

Jenkins, T.M.J., C.R. Feldmeth, and G.V. Elliott. 1970. Feeding of rainbow trout (*Salmo gairdneri*) in relation to abundance of drifting invertebrates in a mountain stream. Journal of the Fisheries Research Board of Canada **27**:2356–2361.

Johnston, N.T., C.J. Perrin, P.A. Slaney, and B.R. Ward. 1990. Increased juvenile salmonid growth by whole-river fertilization. Canadian Journal of Fisheries and Aquatic Sciences **47**:862–872.

Jones, R.I., and R.J. Barrington. 1985. A study of the suspended algae in the River Derwent, Derbyshire, UK. Hydrobiologia **128**:255–264.

Jorgensen, E.G. 1977. Photosynthesis. Pages 150–168 *in* D. Werner, ed. The biology of diatoms. University of California Press, Berkeley, California, USA.

Karr, J.R., and D.R. Dudley. 1981. Ecological perspective on water quality goals. Environmental Management **5**:55–68.

Klotz, R.L. 1985. Factors controlling phosphorus limitation in stream sediments. Limnology and Oceanography **30**:543–553.

Lamberti, G.A., L.R. Ashkenas, S.V. Gregory, and A.D. Steinman. 1987. Effects of three herbivores on periphyton communities in laboratory streams. Journal of the North American Benthological Society **6**:92–104.

Lamberti, G.A., S.V. Ashkenas, S.V. Gregory, A.D. Steinman, and C.D. McIntire. 1989. Productive capacity of periphyton as a determinant of plant-animal interactions in streams. Ecology **70**:1840–1856.

Lamberti, G.A., and V.H. Resh. 1983. Stream periphyton and insect herbivores: An experimental study of grazing by a caddisfly population. Ecology **64**:1124–1135.

Lemly, A.D. 1982. Modification of benthic insect communities in polluted streams: Combined effects of sedimentation and nutrient enrichment. Hydrobiologia **87**:229–245.

Lloyd, D.S., J.P. Koenings, and J.D. LaPerriere. 1987. Effects of turbidity in fresh waters of Alaska. North American Journal of Fisheries Management **7**:18–33.

Lowe, R.L., S.W. Golladay, and J.R. Webster. 1986. Periphyton response to nutrient manipulation in streams draining clearcut and forested watersheds. Journal of the North American Benthological Society **5**:221–229.

Mann, K.H. 1975. Patterns of energy flow. Pages 248–263 *in* B.A. Whitton, ed. River ecology. University of California Press, Berkeley, California, USA.

McAuliffe, J.R. 1984. Resource depression by a stream herbivore: Effects on distributions and abundances of other grazers. Oikos **42**:327–333.

McCormick, P.V., and R.J. Stevenson. 1991. Grazer control of nutrient availability in the periphyton. Oecologia **86**:287–291.

McIntire, C.D. 1973. Periphyton dynamics in laboratory streams: A simulation model and its implications. Ecological Monographs **43**:399–420.

McNaughton, S.J., and L.L. Wolf. 1973. General ecology. Holt, Rinehart and Winston, New York, New York, USA.

Meyer, J.L., and G.E. Likens. 1979. Transport and transformation of phosphorus in a forest stream ecosystem. Ecology **60**:1255–1269.

Minshall, G.W. 1978. Autotrophy in stream ecosystems. BioScience **28**:767–771.

Minshall, G.W., R.C. Petersen, K.W. Cummins, T.L. Bott, J.R. Sedell, C.E. Cushing, et al. 1983. Interbiome comparison of stream ecosystem dynamics. Ecological Monographs **53**:1–25.

Mulholland, P.J. 1992. Regulation of nutrient concentrations in a temperate forest stream: Roles of upland, riparian, and instream processes. Limnology and Oceanography **37**:1512–1526.

Mulholland, P.J., J.D. Newbold, J.W. Elwood, and C.L. Hom. 1983. The effect of grazing intensity on phosphorus spiraling in autotrophic streams. Oecologia **53**:358–366.

Murphy, M.L. 1984. Primary production and grazing in freshwater and intertidal reaches of a coastal stream, southeast Alaska. Limnology and Oceanography **29**:805–815.

Murphy, M.L., and J.D. Hall. 1981. Varied effects of clearcut logging on predators and their habitat in small streams of the Cascade Mountains, Oregon. Canadian Journal of Fisheries and Aquatic Sciences **38**:137–145.

Murphy, M.L., C.P. Hawkins, and N.H. Anderson. 1981. Effects of canopy modification and accumulated sediment on stream communities. Transactions of the American Fisheries Society **110**:469–478.

Murphy, M.L., J. Heifetz, S.W. Johnson, K.V. Koski, and J.F. Thedinga. 1986. Effects of clear-cut logging with and without buffer strips on juvenile salmonids in Alaskan streams. Canadian Journal of Fisheries and Aquatic Sciences **43**:1521–1533.

Murphy, M.L., K.V. Koski, J. Heifetz, S.W. Johnson, D. Kirchoffer, and J.F. Thedinga. 1984. Role of large organic debris as winter habitat for juvenile salmonids in Alaska streams. Proceedings of the annual conference of the Western Association of Fish and Wildlife Agencies **64**:251–262.

Murphy, M.L., and W.R. Meehan. 1991. Stream ecosystems. Pages 17–46 *in* W.R. Meehan, ed. Influences of forest and rangeland management on salmonid fishes and their habitats. Special Publication 19, American Fisheries Society, Bethesda, Maryland, USA.

Naiman, R.J. 1976. Primary production, standing stock, and export of organic matter in a Mohave Desert thermal stream. Limnology and Oceanography **21**:60–73.

——. 1983. The annual pattern and spatial distribution of aquatic oxygen metabolism in boreal forest watersheds. Ecological Monographs **53**:73–94.

Naiman, R.J., T.J. Beechie, L.E. Benda, D.R. Berg, P.A. Bisson, L.H. MacDonald, et al. 1992. Fundamental elements of ecologically healthy watersheds in the Pacific Northwest coastal ecoregion. Pages 127–188 *in* R.J. Naiman, ed. Watershed management: Balancing sustainability and environmental change. Springer-Verlag, New York, New York, USA.

Naiman, R.J., and J.R. Sedell. 1980. Relationships between metabolic parameters and stream order in Oregon. Canadian Journal of Fisheries and Aquatic Sciences **37**:834–847.

Naiman, R.J., and J.R. Sedell. 1981. Stream ecosystem research in a watershed perspective. Verhandlungen der Internationalen Vereinigung für Theoretische und Angewandte Limnologie **21**:804–811.

Newbold, J.D., J.W. Elwood, R.V. O'Neill, and A.L. Sheldon. 1983. Phosphorus dynamics in a woodland stream ecosystem: A study of nutrient spiraling. Ecology **64**:1249–1265.

Newbold, J.D., D.C. Erman, and K.B. Roby. 1980. Effects of logging on macroinvertebrates in streams with and without buffer strips. Canadian Journal of Fisheries and Aquatic Sciences **37**:1076–1085.

Patrick, R. 1978. Effects of trace metals in the aquatic ecosystem. American Scientist **66**:185–191.

Perrin, C.J., M.L. Bothwell, and P.A. Slaney. 1987. Experimental enrichment of a coastal stream in British Columbia: Effects of organic and inorganic additions on autotrophic periphyton production. Canadian Journal of Fisheries and Aquatic Sciences **44**:1247–1256.

Peterson, B.J., J.E. Hobbie, M.A. Lock, T.E. Ford, J.R. Vestal, V.L. McKinley, et al. 1985. Transformation of a tundra river from heterotrophy to autotrophy by addition of phosphorus. Science **22**:1383–1386.

Power, M.E., A.J. Stewart, and W.J. Matthews. 1988. Grazer control of algae in an Ozark Mountain stream: Effects of short-term exclusion. Ecology **69**:1894–1898.

Reeves, G.H., F.H. Everest, and J.R. Sedell. 1993. Diversity of juvenile anadromous salmonid assemblages in coastal Oregon basins with different levels of timber harvest. Transactions of the American Fisheries Society **122**:309–317.

Reid, G.K., and R.D. Wood. 1976. Ecology of inland waters and estuaries. Van Nostrand, New York, New York, USA.

Rhee, G.Y. 1978. Effects of N:P atomic ratios and nitrate limitation on algal growth, cell composition, and nitrate uptake. Limnology and Oceanography **23**:10–25.

Riemer, D.N. 1984. Introduction to freshwater vegetation. AVI, Westport, Connecticut, USA.

Rounick, J.S., and S.V. Gregory. 1981. Temporal changes in periphyton standing crop during an unusually dry winter in streams of the Western Cascades, Oregon. Hydrobiologia **83**:197–205.

Russell-Hunter, W.D. 1970. Aquatic productivity: An introduction to some basic aspects of biological oceanography and limnology. The Macmillan Company, London, UK.

Scrimgeour, G.J., J.M. Culp, M.L. Bothwell, F.J. Wrona, and M.H. McKee. 1991. Mechanisms of algal patch depletion: Importance of consumptive and non-consumptive losses in mayfly-diatom systems. Oecologia **85**:343–348.

Sedell, J.R., and F.J. Swanson. 1984. Ecological characteristics of streams in old-growth forests of the Pacific Northwest. Pages 9–16 *in* W.R. Meehan, T.R. Merrell, Jr. and T.A. Hanley, eds. Fish and wildlife relationships in old-growth forests. American Institute of Fishery Research Biologists, Morehead, North Carolina, USA.

Steinman, A.D. 1992. Does an increase in irradiance influence periphyton in a heavily-grazed woodland stream? Oecologia **91**:163–170.

Steinman, A.D., and H.L. Boston. 1993. The ecological role of aquatic bryophytes in a heterotrophic woodland stream. Journal of the North American Benthological Society **12**:17–26.

Steinman, A.D., and C.D. McIntire. 1990. Recovery of lotic periphyton communities after disturbance. Environmental Management **14**:589–604.

Steinman, A.D., P.J. Mulholland, and W.R. Hill. 1992. Functional responses associated with growth form in stream algae. Journal of the North American Benthological Society **11**:229–243.

Stock, M.S., and A.K. Ward. 1991. Blue-green algal mats in a small stream. Journal of Phycology **27**:692–698.

Stockner, J.G., and K.R.S. Shortreed. 1976. Autotrophic production in Carnation Creek, a coastal rainforest stream on Vancouver Island, British Columbia. Journal of the Fisheries Research Board of Canada **33**:1553–1563.

Stockner, J.G., and K.R.S. Shortreed. 1978. Enhancement of autotrophic production by nutrient addition in a coastal rainforest stream on Vancouver Island. Journal of the Fisheries Research Board of Canada **35**:28–34.

Sumner, W.T., and S.G. Fisher. 1979. Periphyton production in Fort River, Massachusetts. Freshwater Biology **9**:205–212.

Sumner, W.T., and C.D. McIntire. 1982. Grazer-periphyton interactions in laboratory streams. Archiv für Hydrobiologie **93**:135–157.

Suren, A.M. 1991. Bryophytes as invertebrate habitat in two New Zealand alpine streams. Freshwater Biology **26**:399–418.

Swanson, C.D., and R.W. Bachmann. 1976. A model of algal exports in some Iowa streams. Ecology **57**:1076–1080.

Tate, C.M., R.E. Broshears, and D.M. McKnight. 1995. Phosphate dynamics in an acidic mountain stream: Interactions involving algal uptake, sorption by iron oxide, and photoreduction. Limnology and Oceanography **40**:938–946.

Triska, F.J., V.C. Kennedy, R.J. Avanzino, and B.N. Reilly. 1983. Effect of simulated canopy cover on regulation of nitrate uptake and primary production by natural periphyton assemblages. Pages 129–159 *in* T.D. Fontaine and S.M. Bartell, eds. Dynamics of lotic ecosystems. Ann Arbor Science, Ann Arbor, Michigan, USA.

Vannote, R.L., G.W. Minshall, K.W. Cummins, J.R. Sedell, and C.E. Cushing. 1980. The river continuum concept. Canadian Journal of Fisheries and Aquatic Sciences **37**:130–137.

Walter, R.A. 1984. A stream ecosystem in an old-growth forest in southeast Alaska: Part II: Structure and dynamics of the periphyton community. Pages 57–69 *in* W.R. Meehan, T.R. Merrell, Jr. and T.A. Hanley, eds. Fish and wildlife relationships in old-growth forests. American Institute of Fishery Research Biologists, Morehead City, North Carolina, USA.

Ward, G.M., and K.W. Cummins. 1979. Effects of food quality on growth of a stream detritivore, *Paratendipes albimanus* (Meigen) (Diptera: Chironomidae). Ecology **60**:57–64.

Waters, T.F. 1969. The turnover ratio in production ecology of freshwater invertebrates. American Naturalist **103**:173–185.

Wehr, J.D. 1981. Analysis of seasonal succession of attached algae in a mountain stream, the North Alouette River, British Columbia. Canadian Journal of Botany **59**:1465–1474.

Werner, D. 1977. Silicate metabolism. Pages 110–149 *in* D. Werner, ed. The biology of diatoms. University of California Press, Berkeley, California, USA.

Wetzel, R.G. 1975. Limnology. W.B. Saunders, Philadelphia, Pennsylvania, USA.

8
Stream Macroinvertebrate Communities

Anne E. Hershey and Gary A. Lamberti

Overview

• Macroinvertebrates are found in all streams and play crucial roles in organic matter dynamics and trophic energy transfer in stream ecosystems.

• The Pacific Northwest has a high diversity of macroinvertebrates, particularly arthropods, because of the large variety of stream habitats that have not been heavily impacted by human activity.

• Studies of macroinvertebrates in Pacific Northwest streams have shown that riparian canopy has strong influences on community structure, as predicted by the river continuum concept (RCC). Small, heavily shaded streams in Oregon's Cascade Mountains tend to have a predominance of leaf shredders and predators. Midsized, less shaded streams tend to have high numbers of algal scrapers and fine-particle collectors. Large rivers contain similar numbers of predators, scrapers, and collectors.

• Disturbance of Pacific Northwest streams by floods and debris flows greatly reduces local macroinvertebrate density and diversity, but recolonization generally occurs within weeks (floods) or months (debris flows). Drift, oviposition, and migration from the hyporheic zone all contribute to macroinvertebrate recovery.

• Channel geomorphology has a major influence on macroinvertebrate production. Broad, alluviated channels support the highest macroinvertebrate production, whereas narrow, sediment-depleted channels have the lowest production.

• Macroinvertebrates are intimately involved in organic matter processing and trophic transfer in streams; thus, stream ecologists and watershed managers need a broad understanding of macroinvertebrate community structure and dynamics.

• Recent landscape-level research has emphasized the importance of watershed conditions, stream geomorphology, and woody debris in determining characteristics of macroinvertebrate communities.

Introduction

Macroinvertebrates play significant roles in stream ecosystems. As a group, macroinvertebrates are the primary food source for most stream fishes. Their taxonomic, habitat, and life-history diversity ensures that an array of food types are available to many fish species over the entire annual cycle. They also conduct the less apparent but no less important work of decomposing leaf litter and small particles of organic debris on the stream bottom or in the water column, and of grazing stream algae, fungi, and bacteria. Considerable information is available on invertebrate responses to a variety of environmental conditions, and thus invertebrates may be used as indicators of stream conditions. In the Pacific Northwest, stream macroinvertebrates have been studied from many perspectives, including population ecology, taxonomic diversity, community structure, food relations, and their role in stream ecosystem function. In one Oregon drainage, stream macroinvertebrates were used to test a fundamental theory in stream ecology, the River Continuum Concept. However, most ecological literature on stream macroinvertebrates from the Pacific coastal ecoregion, and elsewhere, is relatively recent, that is, from the past 20 years. This reflects the current interest in the topic, but also demonstrates that knowledge in the discipline is necessarily limited.

This chapter presents the general diversity and life histories of macroinvertebrates in Pacific coastal streams; describes the role of macroinvertebrates in the processing of organic matter and as members of stream food webs; explains some unique attributes of stream macroinvertebrates, such as drift dynamics; and discusses the effects of land use on macroinvertebrate community structure and dynamics.

Species Assemblages

The very high diversity of macroinvertebrates found in streams reflects a variety of ecological and evolutionary processes. The species assemblage within a stream include macroinvertebrates from several phyla, each with species that have diversified in response to the variability in the physical environment, the variety of microhabitats present, the seasonality of resources in the stream, and interspecific interactions within the assemblage. Within each phylum, different life history constraints govern responses to these physical and biological selective pressures.

Macroinvertebrate Taxonomic Diversity

Stream macroinvertebrates include representatives from several major invertebrate phyla, including Porifera, Cnidaria, Platyhelminthes, Nematoda, Annelida, Mollusca, and Arthro-

TABLE 8.1. Common macroinvertebrates in a mountain stream of the Pacific Northwest. The symbol "?" indicates there is uncertainty about the life history of the indicated taxon.

Taxon	Functional feeding group	Life history*
Ephemeroptera (mayflies)		
Ameletus spp.	Scraper, collector-gatherer	univoltine
Baetis spp.	Collector-gatherer	multivoltine
Cinygmula reticulata	Scraper, collector-gatherer	univoltine
Drunella doddsi	Predator	univoltine
Ephemerella spp.	Collector-gatherer, scraper	univoltine
Heptagenia spp.	Scraper, collector-gatherer	univoltine ?
Paraleptophlebia spp.	Collector-gatherer	univoltine
Tricorythodes minutus	Collector-gatherer	univoltine
Odonata (damselflies, dragonflies)		
Argia spp.	Predator	univoltine ?
Octogomphus specularis	Predator	semivoltine ?
Plecoptera (stoneflies)		
Calineuria californica	Predator	merovoltine
Isoperla spp.	Predator	univoltine
Pteronarcys spp.	Shredder	merovoltine
Sweltsa spp.	Predator	semivoltine
Zapada cinctipes	Shredder	univoltine
Megaloptera (dodsonflies, alderflies)		
Orohermes spp.	Predator	semivoltine ?
Sialis spp.	Predator	univoltine
Trichoptera (caddisflies)		
Agapetus occidentis	Scraper, collector-gatherer	univoltine
Anagapetus bernea	Scraper	univoltine

TABLE 8.1. *Continued*

Taxon	Functional feeding group	Life history*
Arctopsyche grandis	Collector-filterer	univoltine
Cheumatopsyche spp.	Collector-filterer	univoltine
Dicosmoecus gilvipes	Scraper, shredder, predator	univoltine
Eocosmoecus spp.	Shredder	semivoltine
Glossosoma pentium	Scraper	bivoltine
Helicopsyche borealis	Scraper	univoltine
Heteroplectron californicum	Shredder	semivoltine
Hydropsyche spp.	Collector-Filterer	univoltine
Lepidostoma spp.	Shredder	univoltine
Micrasema spp.	Shredder, collector-gatherer	univoltine ?
Neophylax spp.	Scraper	univoltine
Onocosmoecus unicolor	Shredder	univoltine
Rhyacophila spp.	Predator	univoltine ?
Coleoptera (beetles)		
Acneus spp.	Scraper	univoltine ?
Lara avara	Shredder	merovoltine
Optioservus spp.	Scraper, collector-gatherer	semivoltine ?
Diptera (true flies)		
Antocha spp.	Collector-gatherer	bivoltine ?
Blepharicera spp.	Scraper	univoltine
Chironomidae	all	various
Deuterophlebia spp.	Scraper	univoltine
Lipsothrix spp.	Shredder	merovoltine
Simulium spp.	Collector-filterer	multivoltine
Tabanus spp.	Predator	univoltine
Tipula spp.	Shredder, collector-gatherer	univoltine ?
Gastropoda (snails)		
Juga spp.	Scraper	long-lived
Decapoda (crayfish)		
Pacifastacus spp.	Scraper, shredder, predator	long-lived
Amphipoda (amphipods)		
Gammarus spp.	Shredder, collector-gatherer	univoltine ?

* See text for explanation of voltinism.
Taxa and life history taken mainly from Wallace and Anderson (1996); functional feeding group taken from Merritt and Cummins (1996a).

poda. However, Annelida, Mollusca, and especially Arthropoda are by far the best represented (Table 8.1) and comprise most of the faunal biomass. Among arthropods, Insecta is the most important group. Among insects, the orders Ephemeroptera (mayflies), Odonata (damselflies and dragonflies), Plecoptera (stoneflies), and Trichoptera (caddisflies) are strictly aquatic in their immature stages. Collembola (springtails), Neuroptera (especially suborder Megaloptera—alderflies, dobsonflies, fishflies, hellgrammites), Hemiptera (suborder Heteroptera—bugs), Coleoptera (beetles), Diptera (flies, especially Suborder Nematocera but also some Brachycera), and occasionally Lepidoptera (moths and butterflies) have many aquatic representatives that are prominent members of stream communities. Many species of Coleoptera and Hemiptera have adult as well as larval stages that are aquatic. Noninsect arthropods, including amphipod, isopod, and decapod crustaceans, are also conspicuous components of stream communities. Mollusca, including several families of Gastropoda (snails and limpets) and Bivalvia (clams and mussels), are common in many streams. Annelida, represented by a few families of Oligochaeta (worms) and

Hirudinea (leeches), are common in some stream habitats. *Hydra* is a reasonably common freshwater representative of the Cnidaria that is found in some streams, and several species of Turbellaria (flatworms), the free-living members of the phylum Platyhelminthes, also are found in streams.

A list of the major subdivisions of macroinvertebrate groups that occupy Pacific coastal streams does not do justice to their species diversity. A relatively modest sampling effort in a pristine stream may yield well over a hundred species of macroinvertebrates, including many that are difficult or impossible to identify in the immature stages. Although this high diversity fascinates systematists, it also baffles ecologists who seek to understand the factors that control population and community structure and ecosystem function. Thus, stream macroinvertebrates have been categorized into a smaller number of groups based on their functional role in streams. These are known as *functional feeding groups*; this system is discussed later in the chapter.

Macroinvertebrate Life Histories

Stream macroinvertebrates also show tremendous diversity in their life history patterns. This includes variation in life cycle length, developmental strategies, and seasonality of the various life history stages. Such variety serves to separate taxa seasonally, thus providing a temporally changing composition to stream communities.

The frequency with which an organism completes its life cycle is referred to as *voltinism*. Depending on environmental constraints, especially temperature and nutrition, as well as evolutionary factors, stream invertebrates may be univoltine (one generation per year), multivoltine (more than one generation per year; e.g., bivoltine = two per year, trivoltine = three per year), semivoltine (two-year life cycle), or merovoltine (three or more-year life cycle). Commonly, larger invertebrates have longer life cycles than small species. For example, in the Kuparuk River in arctic Alaska, the relatively small *Baetis* mayflies are univoltine, but the larger *Brachycentrus* caddisfly has a three-year life cycle (Hershey et al. 1997). In Pacific coastal streams, life cycles range from several months to several years, but most aquatic insects are univoltine (Table 8.1).

In the Pacific coastal ecoregion, most aquatic insects have life cycles with synchronized development (Figure 8.1). Environmental factors, especially temperature and photoperiod (day length), as well as evolutionary factors serve to synchronize invertebrate life cycles (Butler 1984). For example, many insects overwinter in the egg stage and hatch in the spring (Figure 8.1, right). Others overwinter as larvae, emerging in the spring when the appropriate temperature or photoperiod cues are present (Figure 8.1, left). Although there may be several evolutionary as well as environmental mechanisms contributing to synchronous development, and contributing factors vary among species (Butler 1984), the result is that within a species, individuals in a *cohort*, or age class, emerge during a very short interval (Figure 8.1), enhancing the probability of mating success. Because species within any stream exhibit a range of life histories, the mix of species in the stream community continually changes through the annual cycle.

Differences in developmental strategies are most evident between insect and noninsect macroinvertebrates, although important differences also occur within the class Insecta. Most noninsect macroinvertebrates spend their entire life cycle in the stream, whereas insects have an aerial adult phase and reproduction

FIGURE 8.1. Examples of life cycle synchrony of stream macroinvertebrates in response to temperature and photoperiod cues. In the Pacific Northwest, many species overwinter as larvae (left). Larval growth and maturation occurs at approximately the same rate for the entire cohort, such that pupation and emergence will also be synchronized. Other species overwinter in the egg stage, then begin larval development in the spring (right). Synchronous development patterns increase the probability that adults will find mates, thus enhancing reproductive success. Larval instars indicated as: F = final instar, F-1 = penultimate instar, etc. P = pupae (open bars) (modified from Anderson and Bourne 1974).

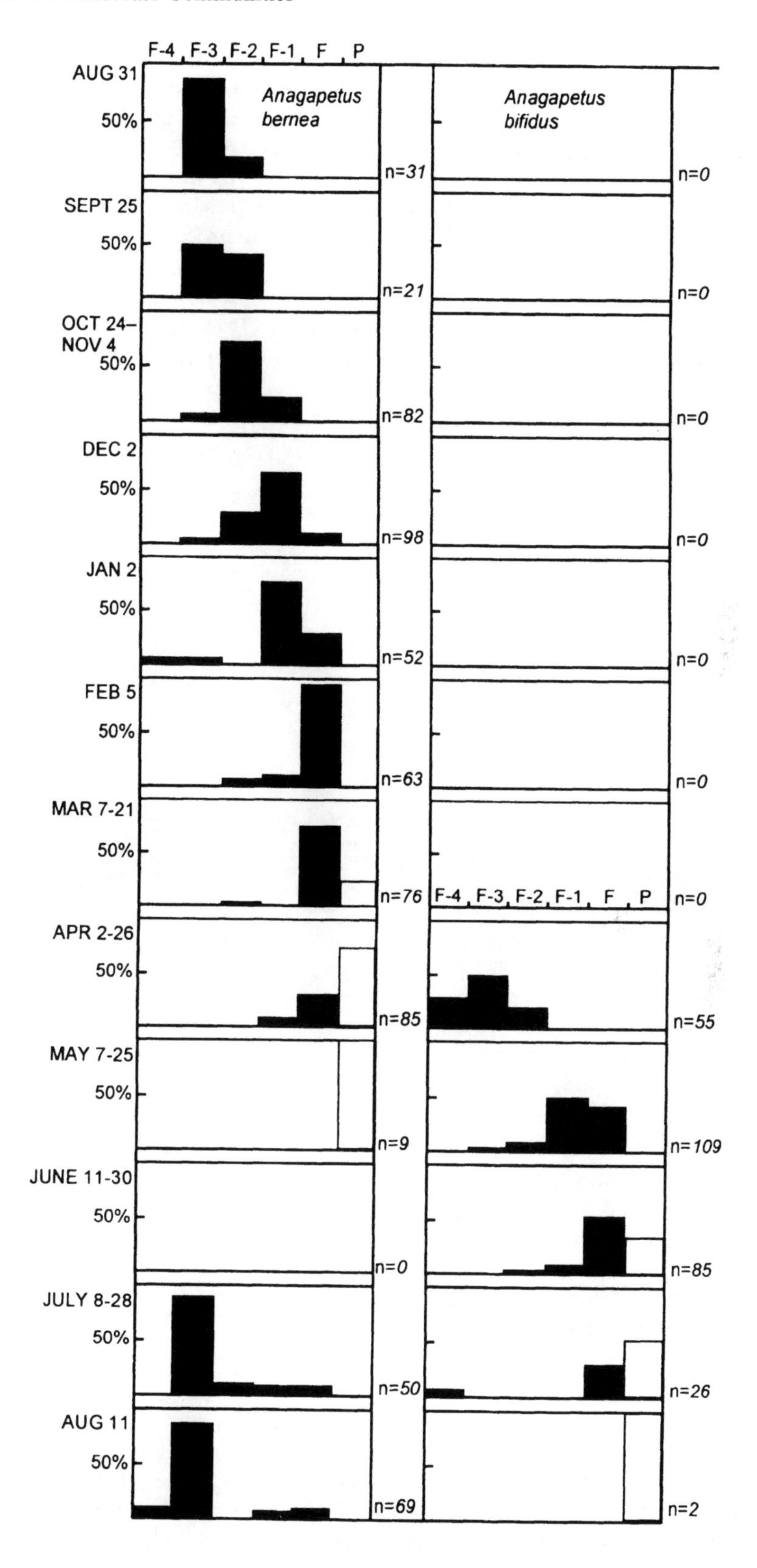

F-4
F-3
F-2
F-1
F
P
AUG 31
SEPT 25
OCT 24–NOV 4
DEC 2
JAN 2
FEB 5
MAR 7-21
APR 2-26
MAY 7-25
JUNE 11-30
JULY 8-28
AUG 11
50%
Anagapetus bernea
Anagapetus bifidus
n=31
n=21
n=82
n=98
n=52
n=63
n=76
n=85
n=9
n=0
n=50
n=69
n=0
n=55
n=109
n=85
n=26
n=2

occurs outside of the aquatic environment. Thus, for many insects, the adult phase is important to dispersal, permitting upstream oviposition, or oviposition between watersheds.

From a developmental perspective (Figure 8.2), most insects are categorized as either *hemimetabolous* or *holometabolous*. However, Collembola are considered *ametabolous*, having a primitive developmental pattern that leads some entomologists to exclude it from the Insecta (Borror et al. 1989). Primarily hemimetabolous orders, which (in streams) include Ephemeroptera, Odonata, Hemiptera, and Plecoptera, are characterized by gill breathing immature forms that may differ considerably from adults but are generally insectlike (Figure 8.2). After hatching, the immature stages are referred to as nymphs or larvae. Newly hatched larvae are said to be in the first instar, and larval development proceeds through a series of instars. The number of instars varies at the family level, but may also depend on environmental conditions (Borror et al. 1989). Most stream insects, however, are holometabolous, and undergo complete metamorphosis (Figure 8.2). The immature instars are referred to as larvae and do not resemble the adults. Although there

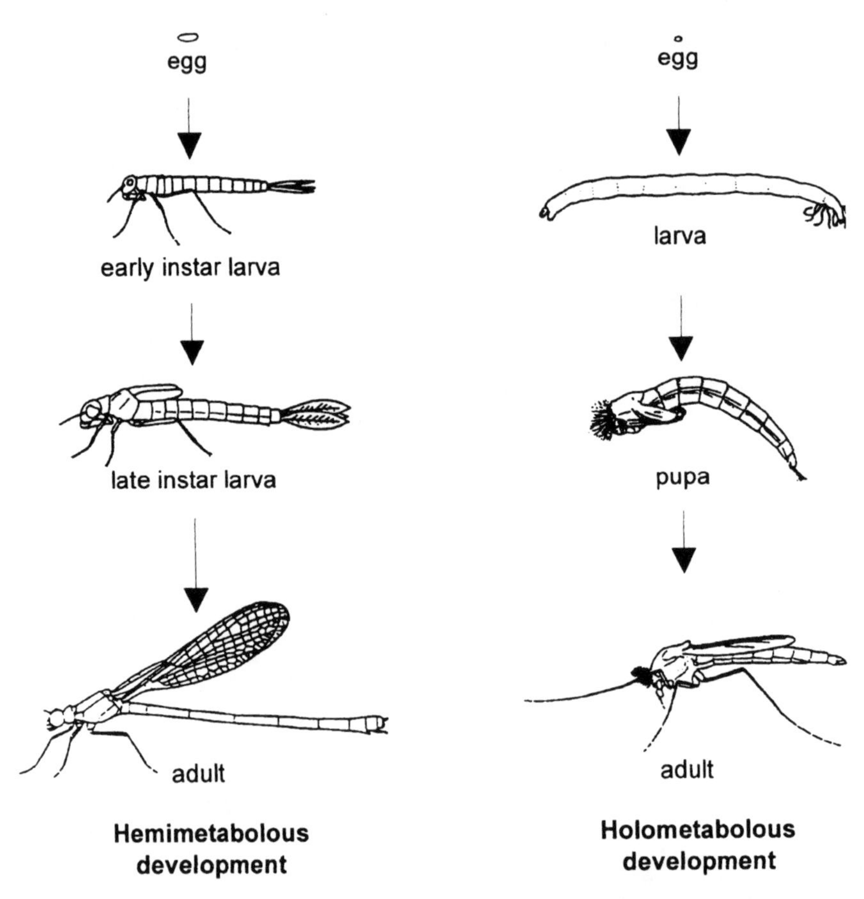

FIGURE 8.2. The two major forms of insect development seen in stream macroinvertebrates. In gradual metamorphosis (left), or *hemimetabolous* development, immature forms are insectlike. Adult characteristics develop gradually through the larval instars, and wings inflate in the final molt to the adult form. Insects demonstrating complete metamorphosis (right), or *holometabolous* development, have several larval instars that that are less insectlike. Adult characteristics develop as a result of radical restructuring of the anatomy and physiology during the pupal stage (McCafferty 1981 © Jones and Bartlett Publishers. Reprinted with permission).

is considerable variation in larval appearance, many larval forms are somewhat wormlike, grublike, or caterpillarlike. Development of adult characteristics occurs in a protected habitat or cocoon during a pupal stage that follows the last larval instar. The insect does not feed during this stage, and appears to be quiescent. However, metabolic activity is very high as larval tissues are reorganized to form the morphologically and physiologically distinct adult form.

Relationship of Diversity to Physical Environment

Many aspects of the physical stream environment affect the composition and abundance of stream macroinvertebrates. These factors include substratum, current velocity, water temperature, dissolved oxygen concentration, and water chemistry. Water quality sets broad limits to the types of macroinvertebrates that may be present. Temperature imposes fundamental constraints on animal physiology; some macroinvertebrates prefer a narrow range of temperature, usually cool water (cold *stenotherms*), whereas others can tolerate a broader range of temperatures (*eurytherms*). Most streams of the coastal and inland mountains of the Pacific Northwest remain cool year-round, and thus have a stenothermal cool-water fauna, but many streams of the interior basin get quite warm seasonally, and thus have a eurythermal warm-water fauna (Li et al. 1994, Tait et al. 1994). Dissolved oxygen concentrations generally are near saturation in most streams and thus rarely limit macroinvertebrate populations. However, oxygen "sags" can occur in river sections with low turbulence (and therefore low reaeration), especially at night when photosynthesis does not occur but producers and consumers still use oxygen, or where municipal, industrial, or agricultural pollutants produce high oxygen demand downstream of their input. In these cases, macroinvertebrates that can tolerate low oxygen conditions are prevalent (e.g., many dipterans and oligochaetes). Other aspects of water chemistry can restrict the presence of macroinvertebrates that require specific ions, such as gastropods and bivalves that need calcium for shell formation (Chapter 4).

At the local habitat scale, substratum and current velocity are probably the most important factors determining the types of macroinvertebrate taxa present. The stream substratum has obvious importance because the vast majority of stream macroinvertebrates spend most of their lives attached to substrata. Lotic substrata can be broadly divided into inorganic substrata (geologic material ranging from silt to boulders) and organic substrata (organic material ranging from fine particles to logs). The particle size of inorganic matter has a large influence on macroinvertebrate community structure. Coarser bed materials (e.g., gravel, cobble, boulders) generally provide more interstitial habitat for macroinvertebrates than fine sediments (e.g., sand, silt). For example, water penny beetles (family Psephenidae), hellgrammite larvae (Megaloptera: Corydalidae), and perlid stoneflies (Plecoptera) frequently are found in interstitial spaces on the undersides of rocks. Many case-building caddisflies (Trichoptera) pupate in dense aggregations on protected faces of cobbles and boulders where interstitial water flow carries the dissolved oxygen needed for metamorphosis (Resh et al. 1981). A lower diversity of invertebrates is typically found in fine sediments because the tight packing of particles restricts physical habitat and the trapping of detritus and can limit the availability of oxygen (Allan 1995). However, taxa adapted to such habitats (e.g., chironomid midges, oligochaete worms, certain mayflies) may be abundant. For example, Soluk (1985) recorded densities of chironomid midges exceeding 80,000/m^2 in the sand substratum of an Alberta river. Organic substrata such as large woody debris or leaf packs often are "hot spots" for invertebrate activity because they provide both substratum and nutritional resources. In a survey of Oregon streams, Dudley and Anderson (1982) found 52 invertebrate taxa that were closely associated with wood, and another 129 taxa that were associated facultatively with woody debris.

Flow affects many aspects of macroinvertebrate biology, including body form, food acquisition, and movement. Taxa found in riffles have morphological adaptations that enable them to resist being swept away by the current. Examples include species that are streamlined (e.g., *Baetis* mayflies), dorsoventrally compressed (e.g., heptageniid mayflies and psephenid beetles), possess suctorial disks (e.g., blephariceridflies), use rock "ballast" in their cases to resist the flow (e.g., *Glossosoma* caddisflies), use silk to anchor themselves (e.g., black fly larvae), or are small enough (e.g., chironomid midges) to fit almost entirely within the *boundary layer*, a thin layer of slow-moving water at the substrate surface that occurs even in rapid-flow river reaches.

Many invertebrates take advantage of swift flow by allowing current to transport food items to them. Examples include caddisflies of the family Hydropsychidae that construct silken nets to capture suspended food particles and black flies (Simuliidae) that filter food with specialized mouthparts (Figure 8.3). Some invertebrates prefer slow current velocities, such as in pools or along stream edges, where organic matter is deposited and thus can be "gathered" and ingested (e.g., phantom crane flies-members of the Ptychopteridae). These taxa often are more wormlike than swift-water forms. The unidirectional flow of streams ensures that the overall direction of in-stream macroinvertebrate movement is downstream, of which a large part is insect *drift* (the downstream transport of organisms in the current—discussed later in this chapter).

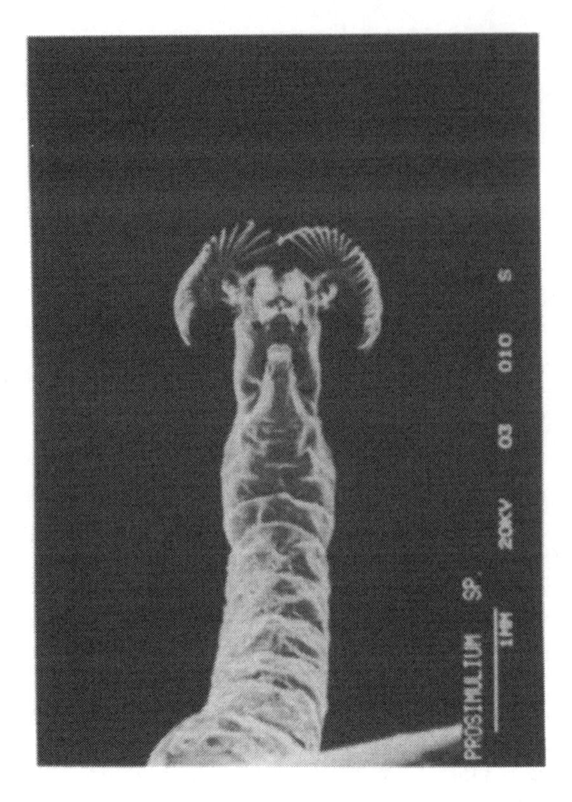

FIGURE 8.3. The larval black fly is a common filter feeder in riffle habitats of Pacific Northwest streams. The fan-shaped objects, or cephalic fans, are a modification of the mouthparts that collect fine particulate organic matter (FPOM) from the water column. The larva orients itself in the current by forming a silk pad on the rock, then attaches itself to the pad using tiny hooks on the posterior portion of its abdomen (Photo: D.A. Craig).

Variation in flow (floods to desiccation) is the major cause of natural disturbance in streams and is responsible for large, usually temporary, reductions in macroinvertebrate abundance and diversity (Lamberti et al. 1991). The *hyporheic* (interstitial) zone and the undersurfaces of rocks provide refugia during flood events (Figure 8.4) (Chapter 16). During drought, the deep hyporheic zone is especially important, but is usually accessible only to smaller invertebrates (typically about 1–5 mm). The spatial extent of the hyporheic zone varies with local geomorphology. In bedrock-dominated reaches, it is virtually nonexistent, whereas in the Flathead River, Montana the hyporheic zone may extend up to 3 km laterally from the river, supporting relatively large stoneflies several millimeters in length (Stanford and Ward 1988). Finally, recolonization by egg-laying adults or by individuals drifting from upstream (Table 8.2) serves to reintroduce taxa that become locally extinct during drought, flood, or other disturbance events (Fisher et al. 1982).

Role of Disturbance

A *disturbance* is defined as a discrete event that disrupts population, community, or ecosystem structure, often by changing resource abundance or the physical environment (Sousa 1984,

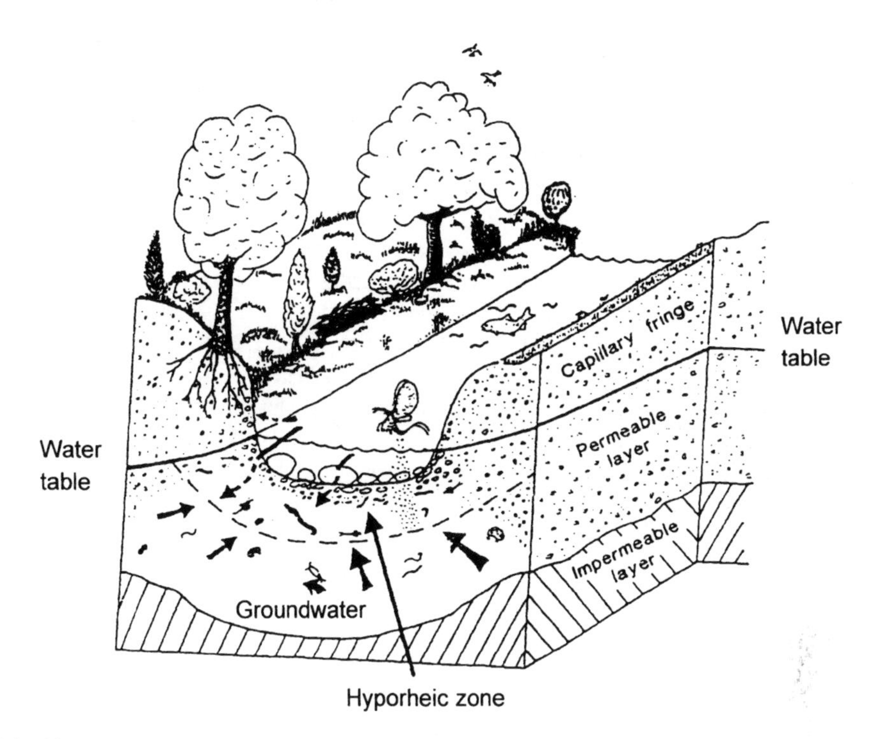

FIGURE 8.4. Channel cross-section during summer low-flow, showing position of hyporheic zone relative to surface water and groundwater (modified from Williams 1993 with permission).

TABLE 8.2. Mechanisms of macroinvertebrate recolonization of stream reaches.

Mechanism	Time scales	Example	References
Downstream drift	Nocturnal, constant, episodic; dispersal stages of some invertebrates	Many macroinvertebrates	Waters 1972
Drift from tributaries	Nocturnal, constant, episodic; dispersal stages of some invertebrates	Many macroinvertebrates	Cairns 1990
Upstream flight	Seasonal, depending on emergence period	*Baetis* mayflies	Hershey et al. 1993
Flight from other watersheds	Seasonal, depending on emergence period	Many macroinvertebrates, *Baetis* mayflies	Wallace et al. 1986, Schmidt et al. 1995
Upstream swimming	Seasonal	*Leptophlebia cupida*, *Paraleptophlebia* mayflies	Hayden and Clifford 1974, Bird and Hynes 1981
Upstream crawling	Daily	*Dicosmoecus* caddisflies, *Juga* snails	Hart and Resh 1980, L.M. Blair and G.A. Lamberti, personal communication
Movement from hyporheic refuge	Episodic, seasonal	Many early instar macroinvertebrates; overwintering stages of many macroinvertebrates in cold climates; many residents of intermittent streams	Sedell et al. 1990, Delucchi 1989, Hershey unpublished data

Resh et al. 1988). Effects of various types of disturbance on stream macroinvertebrate communities have been studied from many perspectives, including toxicants entering the stream, anthropogenic modifications of the channel, scour due to high discharge, drought, overexploitation of native fish species, and introduction of exotic species. The intermediate disturbance hypothesis (Figure 8.5), as modified for streams (Ward and Stanford 1983), predicts that biotic diversity will be greatest in communities subjected to intermediate levels of disturbance. At low levels of disturbance, competitive interactions will result in lower diversity because of exclusion of species (McAuliffe 1984). Such communities often are dominated by large, relatively long-lived caddisflies or snails (Figure 8.5) (Lamberti and Resh 1983, McAuliffe 1984). On the other extreme, high disturbance also will result in lower diversity because of exclusion of poor colonists or long-lived species. These communities tend to be dominated by small, short-lived species such as chironomids and baetid mayflies (Figure 8.5) (Fisher and Gray 1983).

The types and frequency of both natural and anthropogenic disturbances vary regionally as a result of climatic, geomorphic, and land-use conditions. In the Pacific coastal ecoregion, streams in different regions have quite different hydrologies, leading to different timing and magnitude of flood flows (Naiman and Anderson 1996) that scour bed materials and deplete macroinvertebrate populations (Chapter 3). For example, streams in maritime coastal areas can experience floods any time between November and April. In contrast, Cascade Mountain streams show peak flows during

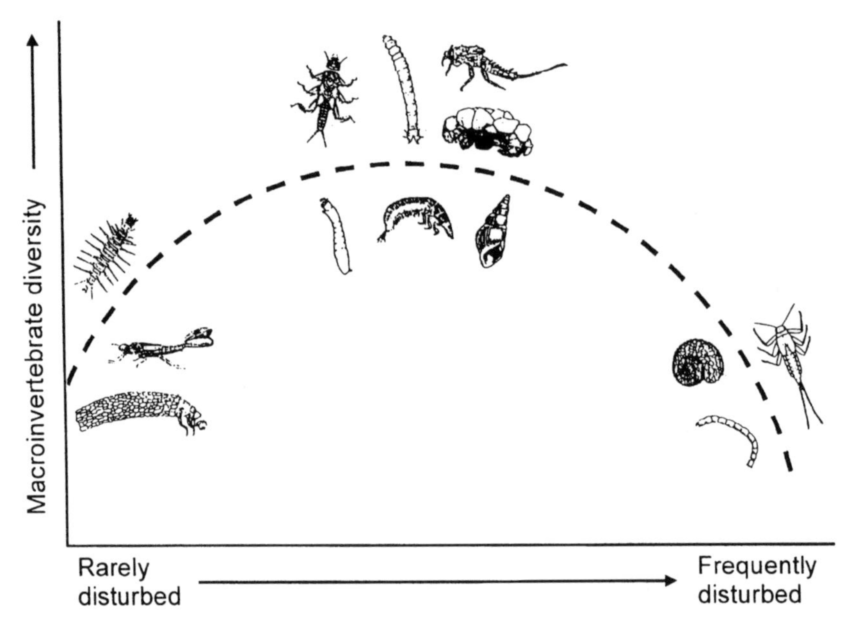

FIGURE 8.5. The intermediate disturbance hypothesis predicts the relationship between macroinvertebrate species diversity and frequency of disturbance (e.g., floods) in streams of the Pacific Northwest (redrawn from Ward and Stanford 1983 after Connell 1978), with examples of macroinvertebrates that can be found under different conditions. Rarely disturbed streams often contain large, long-lived taxa such as (top to bottom): *Neohermes* hellgrammite, *Argia* damselfly, and *Dicosmoecus* caddisfly. Streams with intermediate levels of disturbance usually contain a diverse fauna including (clockwise from upper left): *Calineuria* stonefly, *Tipula* cranefly, *Drunella* mayfly, *Glossosoma* caddisfly, *Juga* snail, *Hydropsyche* caddisfly, and *Simulium* blackfly. Streams that are disturbed frequently may contain small-bodied taxa with short life cycles or high reproductive rates such as (clockwise from upper left): *Helicopsyche* caddisflies, *Baetis* mayflies, and chironomid midges.

fall rains and not again until snowmelt in late spring. Interior streams may have peak flows only during spring snowmelt. However, a major disturbance in the Pacific coastal streams is the occasional "rain-on-snow" events that "reset" the stream biota as a result of extreme high flows, deep scour of streambed sediments, and even initiation of debris flows down stream channels (Lamberti et al. 1991). Recent such events in the Oregon Cascades occurred in December 1964, February 1986, and February 1996.

Stream macroinvertebrates have evolved various mechanisms to deal with such flow variation, including life history adjustments to minimize the presence of vulnerable stages during peak flows, behavioral movement into protected hyporheic and lateral habitats during floods, and high reproductive rates to compensate for losses. Despite these adaptations, severe flood disturbance can still deplete the macroinvertebrate fauna. Macroinvertebrates recolonize streams from several sources: egg laying by adults, usually aerial in the case of insects; drift from upstream areas; and movement along the bed from upstream, downstream, hyporheic, or lateral habitats (Table 8.2) (Smock 1996).

Flood impacts are exacerbated if they are accompanied by *debris flows*, mass movements of sediment and debris through stream channels initiated by landslides (Swanson et al. 1987). In the Pacific coastal ecoregion, clear-cutting of forested land has increased the frequency and severity of landslides and debris flows, especially on steep slopes with shallow soils (Swanson et al. 1987, NRC 1996). Lamberti et al. (1991) studied the response and recovery of a Cascade Mountain (Oregon) stream to a flood and debris flow triggered by the February 1986 rain on snow event (see above). They found that over 99% of the macroinvertebrates were removed by the debris flow, but that a flood alone in an upstream reach reduced macroinvertebrates by 90% compared to postrecovery levels. In both the flood-impacted and debris flow–impacted reaches, macroinvertebrate species richness and total abundance recovered within one year of the disturbance. In fact, after two years, the debris-flow reach had about twice the invertebrate density as the upstream reach, which Lamberti et al. (1991) attributed to higher benthic primary production after the debris flow removed the riparian canopy. The eruption of Mt. St. Helens, Washington, eliminated the biota from entire drainage systems and drastically changed stream channels, yet in the first year after the eruption, 98 species of macroinvertebrates were found in one devastated stream (Anderson 1992). Five years after the eruption, 141 species were found in that stream, although nearly half of these were the ubiquitous chironomid midges, and the fauna continued to change in composition over time, suggesting an unstable community. Both of these examples attest to the remarkable resilience of stream macroinvertebrates to disturbance. These episodic events, however, contrast with chronic anthropogenic disturbance, such as inputs of pollutants or increased temperature because of removal of riparian vegetation. Chronic disturbances tend to preclude recovery to predisturbance conditions (Cairns 1990).

Functional Feeding Groups and the River Continuum Concept

A major breakthrough in studies of consumer-mediated processes in stream ecosystems was the development of the functional feeding group concept for stream consumers (Cummins 1974, Cummins and Klug 1979, Cummins and Merritt 1996, Merritt and Cummins 1996a). The functional feeding group approach categorizes stream consumers into functional rather than taxonomic groups (Figure 8.6). Thus, rather than dozens or hundreds of consumers to be studied, there are a small number of groups of organisms which can be studied collectively from the perspective of their function in the stream ecosystem. This approach categorizes consumers based on their mechanisms for obtaining food and the particle sizes of the food obtained, and is not based on what they eat, *per se*.

The major functional feeding groups are: (1) *scrapers* (grazers), which consume algae and associated material; (2) *shredders*, which consume leaf litter or other coarse particulate or-

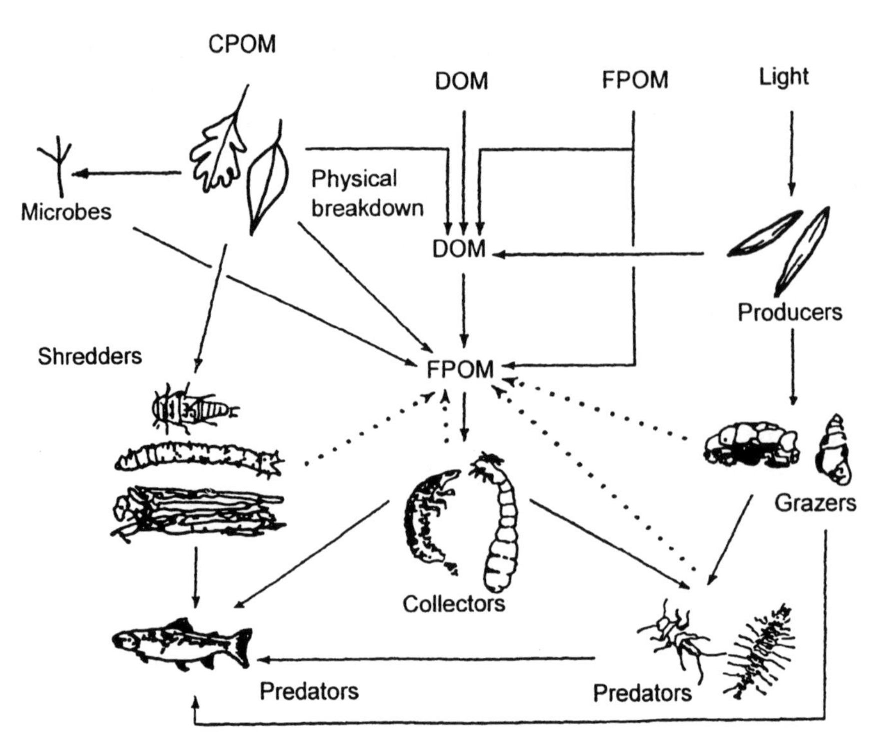

FIGURE 8.6. Simplified food web for a forest stream in the Pacific Northwest, showing major energy inputs and use by different macroinvertebrate functional feeding groups (modified from Allan 1995 after Cummins 1973 with permission). Macroinvertebrate examples are: Shredders, *Peltoperla* stonefly, *Tipula* cranefly, *Hydatophylax* caddisfly; Collectors, *Hydropsyche* caddisfly, *Simulium* blackfly; Grazers, *Glossosoma* caddisfly, *Juga* snail; Predators, *Calineuria* stonefly, *Orohermes* hellgrammite. Broken lines indicate fate of fecal matter.

ganic matter (CPOM; organic particles >1 mm in diameter) including wood; (3) *collector-gatherers*, which collect fine particulate organic matter (FPOM; organic particles <1 mm and >0.45 μm) from the stream bottom; (4) *collector-filterers*, which collect FPOM from the water column using a variety of filtering devices; and (5) *predators*, which feed on other consumers (Figure 8.6). Because each consumer species need not be studied individually to unravel major components of organic matter processing (see below), the functional feeding group approach greatly simplifies the study of stream ecosystems. It also provides a strong basis for comparative studies of streams; biomass and productivity of consumer groups can be compared between streams, whereas it is much more difficult (and often less informative) to make such comparisons on a species by species basis.

Although development of this approach has been an important catalyst for the development of other major paradigms of stream ecology, the functional feeding group concept is not without limitations. Assignment of individual organisms from stream samples to a functional feeding group requires identification of the organism at least to the family level, and more often to the genus level. This is a time-consuming task, although a short-cut approach effective for many taxa is provided by Merritt and Cummins (1996b) as a pictorial key. However, food selection may vary even at the species level according to habitat or food availability. Thus, published functional group designations are not always reliable. In fact, a very large number of stream organisms are known to fall into more than one functional feeding group or may change functional feeding group during development (Merritt and

Cummins 1996a), and many macroinvertebrates are omnivorous to some extent (Lamberti 1996). Despite these shortcomings, the functional feeding group concept serves as a useful, if somewhat imperfect, means to evaluate consumer functions in a given stream and to compare these functions between streams.

The functional feeding groups have been incorporated into a major theoretical advancement in stream ecology, the River Continuum Concept (Figure 8.7) (Vannote et al. 1980). The River Continuum Concept predicts that small, heavily shaded streams will have high inputs of riparian detritus (allochthonous CPOM) and thus shredders will dominate the macroinvertebrate community. Collectors also will be abundant because high-quality FPOM will be produced as CPOM is fragmented. In medium-sized streams, light inputs increase and thus benthic algal production will increase. In response, shredders are replaced by scrapers, while collectors remain abundant. In large

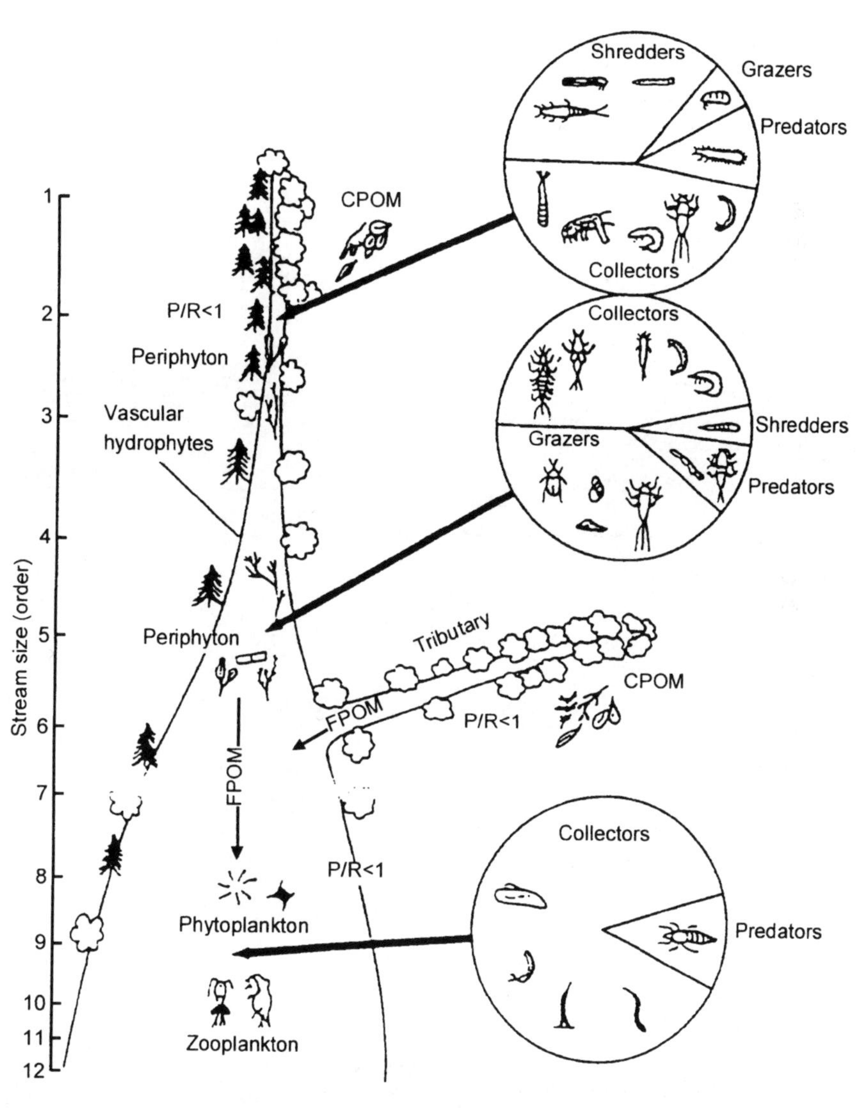

FIGURE 8.7. Predictions of the relative abundance of macroinvertebrate functional feeding groups in different stream orders according to the River Continuum Concept (modified from Allan 1995 after Vannote et al. 1980 with permission).

rivers, benthic algal production and direct riparian inputs decline. Food resources for macroinvertebrates are dominated by FPOM in suspension; collectors dominate the macroinvertebrate community. In all streams, predators comprise a small but fairly stable proportion of the fauna. These predictions were tested by Hawkins and Sedell (1981) for the McKenzie River drainage in Oregon, which consisted of streams ranging in size from 1st- to 7th-order. Hawkins and Sedell (1981) found that functional feeding group distributions were consistent with predictions of the River Continuum Concept (Figure 8.8). These longitudinal shifts were more pronounced than seasonal fluctuations in community structure in the individual Oregon streams. Other studies of functional feeding groups have shown differing degrees of conformance to the River Continuum Concept (Winterbourn et al. 1981, Minshall et al. 1983). Some groups, such as shredders, predictably decline with increasing stream size whereas other groups, such as scrapers, do not fit tightly with River Continuum Concept expectations. Hawkins et al. (1982) found that riparian canopy most highly influenced functional feeding group composition in six small streams in Oregon, but that absence of canopy increased the abundance of all functional feeding groups possibly because both autochthonous (within the reach) and allochthonous (from outside the reach) inputs were received. Hence, consumers responded to the abundance and composition of the available food resources.

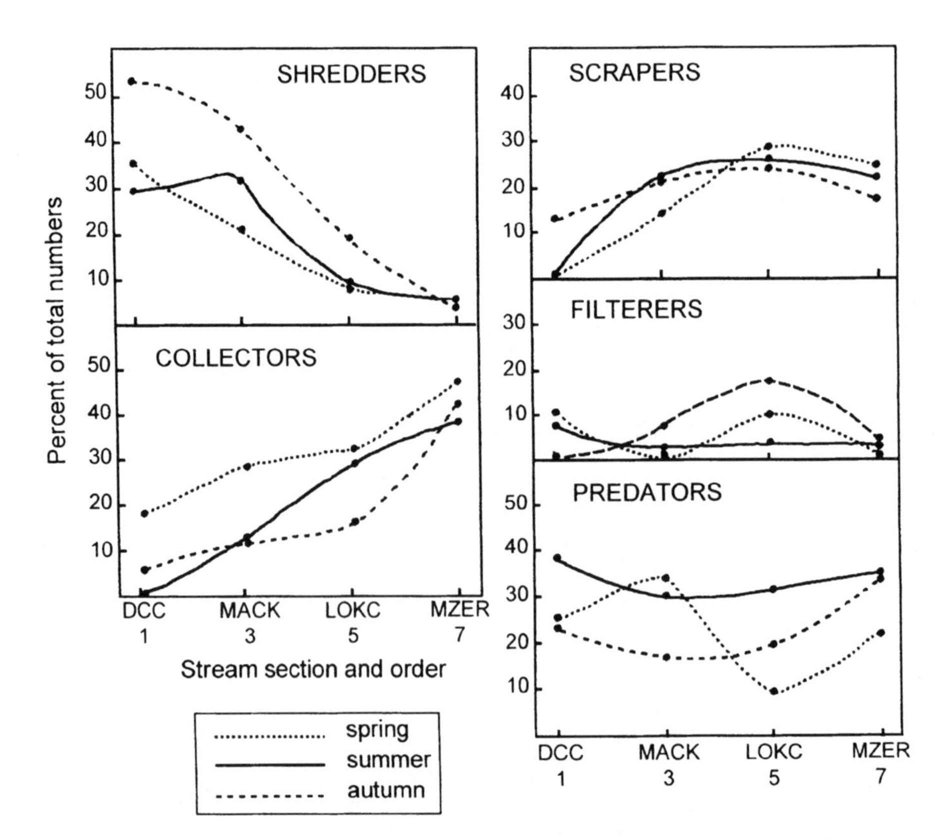

FIGURE 8.8. Longitudinal changes in relative abundance of macroinvertebrate functional feeding groups at different seasons within an Oregon river system (Hawkins and Sedell 1981 with permission). Stream size increases along the *x*-axis; DCC = Devilsclub Creek, MACK = Mack Creek, LOKC = Lookout Creek, MZER = McKenzie River. Note the general conformance to predictions of the River Continuum Concept (Figure 8.7).

Invertebrate-Mediated Processes

Macroinvertebrates are the major consumers in streams, and thus are intimately involved in the flows of matter and energy that occur in these ecosystems. Specific relationships between invertebrates and stream resources that are of particular interest include detritivory, grazing, and predation. Processes that integrate one or more of these specific relationships include downstream drift and secondary production.

Detritivory

Stream ecosystems are strongly influenced by the input of terrestrial organic matter from adjoining ecosystems. Much of this detritus enters as a seasonal pulse of leaves, which is then processed in the stream by microbial and macroinvertebrate communities, and serves as the major energy source for many stream consumers. This seasonality interacts with invertebrate life histories such that detrivores are often seasonally abundant in streams when their food resource also is plentiful. The relative importance of the detritivore food chain compared to the grazer food chain is a matter of some debate (Minshall 1978), but varies in individual streams as a function of stream order and riparian conditions (Hawkins and Sedell 1981, Cummins et al. 1989, Gregory et al. 1991).

Leaves that enter during leaf drop, as well as decomposing macrophytes, together comprise most of the nonwoody CPOM in streams and serve as the major food resource for stream shredders (Figure 8.6). Microbes, which colonize the detritus, form an integral part of the CPOM complex, and are nutritionally important to shredders (Anderson and Sedell 1979, Chapter 6 this volume). Dominant shredder taxa vary according to stream type and biogeography. In Pacific coastal streams, shredders are found in nearly all of the aquatic insect orders (Table 8.1). Some of the more obvious shredders are *Pteronarcys* stoneflies, *Tipula* craneflies, *Lepidostoma* caddisflies, and amphipods. Low-order streams often have macroinvertebrate communities composed of 30–50% shredders (Figure 8.8) (Hawkins and Sedell 1981) by virtue of their closed canopy and high input of CPOM per unit of stream surface (Vannote et al. 1980). Experimental reduction of shredders in a stream using an insecticide resulted in lower CPOM decomposition (20–40% that of a reference stream) and 5- to 15-fold lower FPOM transport (Wallace et al. 1982). Thus, macroinvertebrate shredders are critical to organic matter processing in low-order streams.

Although formation of FPOM from CPOM by shredders is an important process, FPOM is also formed by a variety of other mechanisms. All macroinvertebrate consumers produce feces, which are an important component of FPOM (Shepard and Minshall 1981, Hershey et al. 1996a). Bacteria use dissolved organic matter (DOM), thereby incorporating it into bacterial biomass, which is particulate (Meyer 1990). Bacteria produce extracellular materials, which are sloughed into the FPOM pool (Wotton 1988). FPOM is also produced by flocculation of DOM due to a variety of physical and chemical conditions (Dahm 1981, Petersen 1986, Wotton 1988, Ward et al. 1994). The mass of FPOM usually exceeds that of nonwoody CPOM by about an order of magnitude (Wallace and Grubaugh 1996). Collector macroinvertebrates (collector-gatherers and collector-filterers) that use this resource are important components of the macroinvertebrate community in all streams (Figures 8.7 and 8.8) (Vannote et al. 1980, Hawkins and Sedell 1981). However, because FPOM is formed by many processes, it varies widely in quality; some FPOM is highly labile, but much is refractory (Newbold et al. 1983). The variety of collectors that use FPOM is also very high.

DOM is generally considered the exclusive resource of stream microbes, which assimilate labile components of DOM into microbial biomass (Lock et al. 1984, Meyer 1990). However, larval black flies also ingest DOM (by a currently unknown mechanism), a process that is important to the size and quality of the FPOM pool downstream of black fly aggregations (Hershey et al. 1996). Although direct DOM

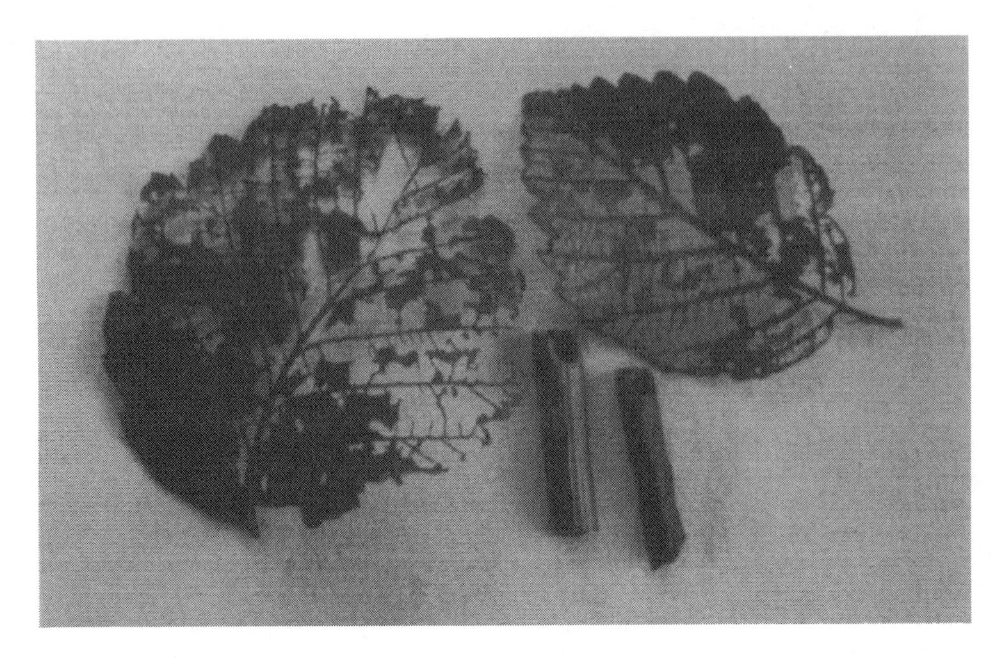

FIGURE 8.9. Wood cases of the caddisfly *Heteroplectron californicum* next to shredded alder leaves. Note opening chewed by caddisflies at upper end of twigs; larva resides inside the twig within a silk-lined tunnel (Photo: Jon Speir, reproduced from Anderson 1976 with permission).

ingestion by stream macroinvertebrates is not thoroughly understood, its importance is likely limited to a few taxonomic groups.

Large woody debris (LWD) often forms a large pool of detritus in stream ecosystems, especially in the Pacific coastal ecoregion (Anderson and Sedell 1979, Swanson et al. 1982, Maser and Sedell 1994, Chapter 13 this volume). Wood may cover over 25% of the beds of small streams in old-growth forests (Anderson and Sedell 1979). However, only very specialized macroinvertebrates use this material as food, and these likely do not ingest quantities that are significant to wood decomposition (Wallace and Anderson 1996). LWD is important as macroinvertebrate habitat, and as a mechanism for retaining CPOM and FPOM resources that are used by macroinvertebrates. In addition, LWD is a major factor defining pool and riffle habitat within a stream, which determines many aspects of macroinvertebrate community structure. Wood itself has low nutritional value, but the microorganisms that coat submerged wood (e.g., algae, bacteria, fungi, protozoans) provide a rich food resource for macroinvertebrates. A few invertebrates ("wood gougers," a special group of shredders) are adapted to feed directly on the wood and include the cranefly larva, *Lipsothrix,* and the elmid beetle, *Lara avara* (Wallace and Anderson 1996). These taxa bore into and feed strictly on wood and have very long life cycles (3–6 years) as a result of the low nutritive value of their food. Examining a piece of water-logged wood will reveal the tunnels of these borers and often the insects themselves. Wood is also used by some larval caddisflies for case construction. *Heteroplecton californicum* caddisflies, which may be found in shallow pools, hollow out twigs to construct portable cases (Figure 8.9) (Anderson and Sedell 1979). When the larvae move, their cases appear as moving twigs, and the larvae themselves are not visible.

Grazing

Lotic primary producers include algae, bryophytes (mosses and liverworts), and rooted macrophytes (Chapter 7). Benthic algae frequently are the dominant plants in Pacific coastal streams, and thus support most of the grazing macroinvertebrates. However, mosses and macrophytes can be important locally. Mosses, because of their fibrous texture and long-lived nature, provide excellent habitat for a variety of stream invertebrates from chironomids to caddisflies (Suren and Winterbourn 1992). However, mosses may be more important for entrapping food particles than serving as food themselves (Glime and Clemons 1972, Suren and Winterbourn 1992). Rooted macrophytes become more important in the finer sediments of larger rivers and are exploited both as habitat and as substrate for periphytic growth. Macrophytes are less often eaten by macroinvertebrates while alive; after death, macrophytes are incorporated into the detritus food web (Minshall 1978).

Benthic algae grow on virtually any stream substratum exposed to light, including the surfaces of benthic macroinvertebrates. Common stream algae include diatoms (Chrysophyta), green algae (Chlorophyta), and blue–green algae (Cyanophyta) (Chapter 7). Macroinvertebrate grazers span many taxonomic groups, but insects and gastropods are conspicuous grazers in many streams. In streams of the Pacific coastal ecoregion, river snails (Pleuroceridae), larval caddisflies (Trichoptera), and mayflies (Ephemeroptera) appear to be particularly important grazers and have the ability to individually or collectively exert large effects on the abundance, productivity, and community structure of benthic algae (Lamberti and Resh 1983, Hawkins and Furnish 1987, Jacoby 1987, Lamberti et al. 1987, 1989). For example, well over 90% of the plant production can be harvested by the assemblage of macroinvertebrate grazers (Gregory 1983, Feminella and Hawkins 1995, Steinman 1996). *Juga silicula* snails and *Dicosmoecus gilvipes* caddisflies are very common herbivores in streams of the Pacific coastal ecoregion (Li and Gregory 1989, Furnish 1990, Tait et al. 1994), often dominating the grazer biomass because of their large size. Lamberti et al. (1987, 1989) found that each of these species at typical densities could harvest about 60% of benthic algal production, and that much of the remaining algal biomass was dislodged and exported by their grazing activities. Furthermore, these large grazers have negative effects on other benthic invertebrates, which they "bulldoze" and thereby displace as they harvest periphyton (Hawkins and Furnish 1987, Lamberti et al. 1992).

Many streams in the Pacific coastal ecoregion have a low standing crop of benthic algae, sometimes barely perceptible to the naked eye. In comparison with obvious pools of dead organic matter (e.g., leaves, wood, FPOM, DOM), it is tempting to dismiss the significance of grazing. However, a specific standing crop of algae can support 10 to 20 times higher biomass of herbivores because of high algal turnover rates (McIntire 1973), sometimes referred to as an "inverted trophic pyramid." For example, Mayer and Likens (1987) found that an abundant caddisfly (*Neophylax aniqua*) consumed mostly periphyton in a small, heavily shaded stream whose energy base was previously thought to be almost totally detrital based (Fisher and Likens 1973).

Despite high use of periphyton, there are few examples of host-specific associations between herbivores and benthic algae, for reasons that are still unclear (Gregory 1983). In streams of the Pacific coastal ecoregion and elsewhere, a mutualistic relationship may exist between the chironomid midge *Cricotopus nostocicola* and the blue-green alga *Nostoc parmelioides* (Brock 1960). A first-instar midge larva finds and enters a small, globular colony of *Nostoc* and begins feeding on *Nostoc* cells, in the process changing the colony morphology from the globular form to an ear-like form about one centimeter in diameter. These dark-green colonies often are quite obvious on rocks, and the single larva can be seen clearly within the colony (Figure 8.10). The midge grows and pupates within the colony, finally emerging as an adult. This association also may benefit the colony, which has a higher photosynthetic rate when the fly larva is present (Ward et al. 1985, Dodds 1989). Amazingly, the midge will reattach the colony using silk produced by the larva if it becomes dislodged from the substratum during a disturbance (Brock 1960, Dodds and Marra 1989), which doubtless benefits both the colony and the larva.

Predator-Prey Interactions

A large variety of macroinvertebrates function as predators in stream ecosystems. Although predators are found in nearly all of the major

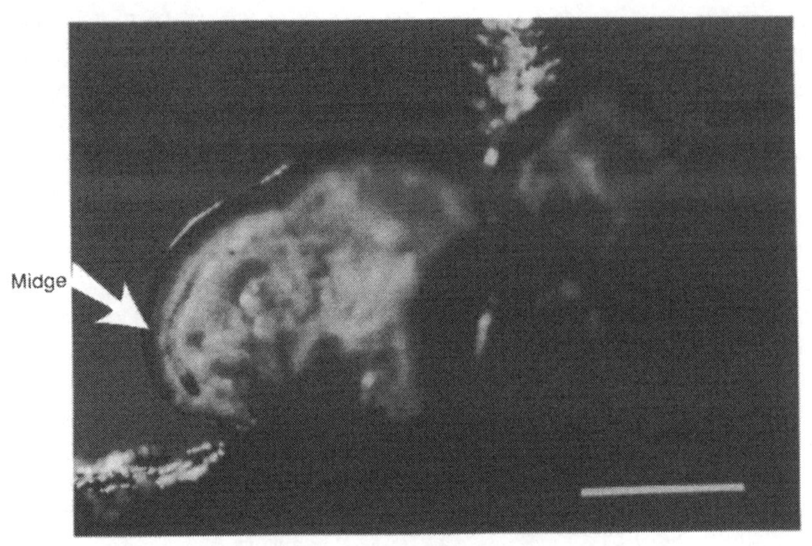

FIGURE 8.10. Colonies of the blue-green alga *Nostoc parmelioides* containing the midge larva *Cricotopus nostocicola*. Upper panel: *Nostoc* "ear" form with midge larva in left side of colony; Lower panel: *Nostoc* colonies attached to a rock. Bar = 5 mm (Photos: Louis Nelson, reproduced from Ward et al. 1985 with permission).

groups of aquatic macroinvertebrates, a few groups, notably the Odonata, aquatic Hemiptera, and Megaloptera, are strictly predators, whereas other groups are more heterogeneous in their trophic affiliations, but may have important subdivisions that are predators. For example, among the Plecoptera, most members of the families Perlidae, Perlodidae, and Chloroperlidae are primarily predators (Merritt and Cummins 1996b). In the Pacific coastal ecoregion, common lotic predators include the stonefly *Calineuria californica*, the robust mayfly *Drunnella doddsi*, the hellgrammite *Orohermes*, and the crayfish *Pacifastacus leniusculus* (Table 8.1).

Predators are often categorized according to their method of acquiring and subduing their prey (Peckarsky 1982). Many predators, such as stoneflies, actively search for their prey, whereas others, such as dragonfly larvae, are better known as ambush or "sit-and-wait" predators. Once the prey is captured, the method of ingestion also is variable, but predators typically either engulf their prey intact, or feed on fluid, using specialized (haustellate) mouthparts for piercing prey and sucking partially digested body tissues in fluid form. Liquidation of body tissues is accomplished via digestive enzymes injected into the prey through the ejection channel of their mouthparts (Borror et al. 1989). Clearly the relative sizes of predators and prey, as well as the mode of feeding used by the predator, are important mechanical constraints governing predator-prey interactions. Engulfing predators are limited to prey small enough to be subdued and swallowed intact (Hershey 1987), whereas piercing predators have access to a broader variety of prey sizes that are large enough to be handled and to accommodate the piercing mouthparts.

The role of predators in controlling the distribution and abundance of prey is a central question in ecology, and thus has also received considerable attention in stream ecology. Stream macroinvertebrates have been studied from two major perspectives: the potential for their communities to be structured by fish predation, and as predators interacting with and possibly controlling other macroinvertebrates.

Experimental studies of fish predation have not always led to the same conclusions. In a comprehensive experimental removal of trout from a rocky mountain stream, Allan (1982) found no difference in macroinvertebrate density or diversity between removal and reference sections of the stream. However, Power (1990) has shown that predatory fish in California's Eel River control predatory invertebrates (especially damselflies) which in turn control the abundance of larval chironomids, affecting both algal biomass and the physical appearance of the algal mat in the river. The general importance of such trophic cascades in stream ecosystems is not fully understood, but it is clear that fish predation does not control macroinvertebrate abundance in all river systems. Because of the threat of fish predation, however, many invertebrates feed more actively on rock surfaces at night, and are more prone to drift at night in search of better food patches (Kohler 1985).

The role of predatory macroinvertebrates in streams has been studied using a combination of gut analyses, laboratory stream microcosms, and in-stream experiments. Numerous studies of gut contents have yielded a wealth of information on which invertebrate species function as predators and which species or groups serve as their major prey. However, experimental studies have provided considerable insight into their role in structuring stream communities. Although notable exceptions exist (Power 1990), invertebrate predators probably do not control the abundance of their prey in most stream ecosystems, because stream communities are very open to immigration and emigration of both predators and prey (Cooper et al. 1990). However, behavioral avoidance of predators by prey may be very important in structuring communities. For example, predatory stoneflies affect mayfly distributions because mayflies will drift to avoid contact with them (Peckarsky and Dodson 1980). When contact occurs, they further minimize the threat of predation by projecting their cerci forward in a scorpion posture which renders them more difficult for the stonefly to handle (Peckarsky 1980). In stream habitats where freshwater polyps (*Hydra*) are very abundant, long hairs on *Cricotopus sylvestris* midge larvae greatly decrease the risk of predation relative to the short haired species, *C. bicinctus* (Figure 8.11), or *C. sylvestris* individuals with experimentally shortened hairs (Hershey and Dodson 1987). Although predator–prey interactions are diverse, and many specific interactions remain to be elucidated, these examples illustrate that macroinvertebrate predators alter prey distributions, prey behavior, and likely have been important in selecting for morphological adaptations of prey for predator defense.

Macroinvertebrate Drift

Drift is the downstream transport of organisms in the current (usually expressed as

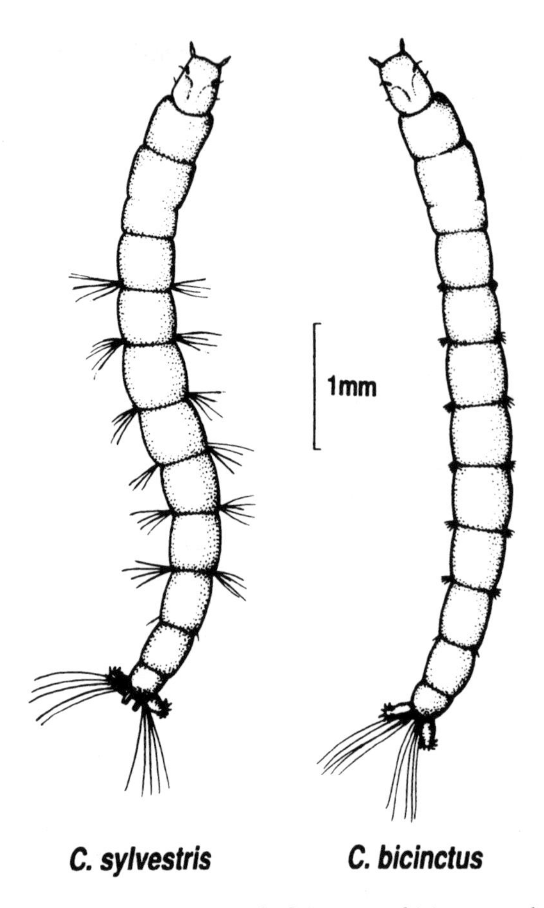

FIGURE 8.11. Larvae of *Cricotopus bicinctus* and *C. sylvestris* (Diptera:Chironomidae) illustrating the long hairs on *C. sylvestris*. The long hairs provide *C. sylvestris* with protection against *Hydra*, a freshwater cnidarian that preys on chironomids and other aquatic invertebrates (from Hershey and Dodson 1987 with permission).

number/m^3). Some macroinvertebrates are more inclined to drift than others, but drifting macroinvertebrates are temporary residents of the water column. At any given time, the proportion of fauna in the drift, compared to that on the stream bottom, is very small (0.01–0.5%; Ulfstrand 1968, Waters 1972).

Experimental studies of drifting invertebrates often have focused on mayflies, especially *Baetis* spp., because they are among the most common of drifters and are relatively easy to identify. Estimates of drift distance vary widely. Most experimental studies in small to mid-sized streams have shown that insects drift only a short distance (a few meters or less) during any drift event (Allan and Fiefarek 1989, Wilzbach and Cummins 1989). However, Benke et al. (1986) suggested that insects drifted between snags in higher-order blackwater rivers, a drift distance on the scale of kilometers.

Drift is characterized as either passive or active. Macroinvertebrates may drift passively or inadvertently if they are dislodged, such as might occur during a high discharge event. However, most experimental studies have suggested that most drift is active, or a behavioral response to some stimulus. This can occur catastrophically due to chemical spills, or on a smaller scale in response to depleted food resources or the presence of a predator. It was once thought that drift was a density dependent response, such that drifting insects represented a portion of the population that exceeded the carrying capacity of the stream (Waters 1972). However, empirical studies of drift have yielded little support for density dependent behavior (Hinterleitner-Anderson et al. 1992).

In most streams, drift densities are much higher at night than during daylight hours. Because fish feed heavily on drifting invertebrates, nighttime drift behavior provides invertebrates with more protection from fish predation. Because many macroinvertebrates forage more on the upper surfaces of rocks at night, it was once thought that dislodgement was simply more likely at night. However, Kohler (1985) found that well-fed *Baetis* foraged only at night on tops of stones while starved nymphs foraged during both day and night but drifted similarly to fed nymphs. Thus, activity during foraging was not related to drift (i.e., they were not accidentally dislodged) and drift entry was active.

Drift clearly is an important mechanism for invertebrate dispersal (Table 8.2), especially for early larval instars. It is also an important means for recolonizing reaches that are impacted by natural or anthropogenic disturbance (Cairns 1990). However, if there were no upstream movement of macroinvertebrates (Table 8.2), downstream drift could result in a net displacement of populations, net depletion of upstream reaches, and loss of invertebrate mediated ecosystem functions in depleted areas. The colonization cycle proposed by Müller (1954) states that insects displaced downstream as larvae later fly upstream as adults to oviposit, thereby repopulating depleted areas. Upstream flight has been observed for some species (Madsen et al. 1973) and quantified for a *Baetis* population in arctic Alaska (Hershey et al. 1993).

Secondary Production

Secondary production is the amount of animal biomass that is produced in a given area over a given time period, usually expressed in g/m^2/yr. For an animal cohort, this measurement integrates both individual growth rate and survivorship of individuals within the cohort (Benke 1984). Within a food web, secondary production at each trophic level represents the amount of biomass that is available to be consumed by the next trophic level. At the ecosystem level, an estimate of secondary production is important as a currency to express energy or material flow through the consumer components of the ecosystem. Secondary production is often used to examine energy flow through different functional feeding groups as a mechanism for evaluating the relative roles of these different groups of consumers.

From a management perspective, secondary production often is used to assess the capacity of a stream reach to support fish. However, very often there does not appear to be enough invertebrate production to support the observed fish production, a problem referred to as Allen's paradox (Allen 1951). This may be due to underestimates of the production-to-biomass ratio (P/B) for individual species (Benke 1996), or perhaps failure to consider the production of chironomid midges (Berg and Hellenthal 1992). Recently, Huryn (1996) has shown that trout can consume a very large proportion of the benthic macroinvertebrates in a stream.

Although conceptually very useful, secondary production is time consuming and expensive to measure. The varied techniques for measuring secondary production, reviewed by Benke (1984, 1996), involve sampling efforts that quantify consumer standing stocks at several points in the growing season, at least a trophic level understanding of the various consumers, and some knowledge of life histories of the dominant consumer components (Table 8.3). However, in the process of measurement, considerable insight is gained into the stream fauna, which often sheds light on which consumers play key roles in the various stream ecosystem processes.

Macroinvertebrates in Stream Food Webs

Benthic macroinvertebrates provide the critical linkage in energy flow from microbial to vertebrate populations in stream ecosystems. Food web analyses are used to understand these linkages, and thereby integrate organic matter processing with community interactions. The goals of food web studies for a particular stream are to identify organic matter sources for the various consumers and to elucidate the trophic structure of the web. In most streams, either three or four trophic levels are present including: (1) primary producers and detritus; (2) primary consumers, including detritivores (shredders and collectors) and grazers (scrapers); (3) secondary consumers (invertebrate predators); and (4) tertiary consumers (vertebrate predators that consume invertebrate predators). These designations are oversimplifications because omnivory is common in streams, such that many consumers occupy more than one trophic level (Lamberti 1996). The food web in any particular stream reflects the combination of factors altering resource and invertebrate abundance and distribution. Thus, wide variation in food web structure occurs among streams. Gut content studies are often used to delineate trophic structure. However, some items are difficult or impossible to recognize, while others may be overlooked because of rarity or temporal variability.

Recently, stable isotopes have been used to study food-web relationships in streams (Fry

TABLE 8.3. Methods for measuring secondary production of stream macroinvertebrates.

Method	Data needed	Comments	Other references
Allen curve	Density (no./m^2) for several dates spanning development of a cohort; mean individual biomass for each date (often obtained from length *X* weight regressions [Smock 1980]).	Graphical presentation of density versus mean individual biomass. The area under the curve corresponds to the secondary production of a population. This method was developed prior to the widespread availability of electronic calculators and computers.	Allen 1951
Increment summation	Density (no./m^2) for several dates spanning development of a cohort; mean individual biomass for each date (often obtained from length *X* weight regressions [Smock 1980]).	Production (P) is calculated as the summation of incremental measurements of the product of mean density ($\bar{N}$) and change in mean individual biomass (ΔW) between two sampling dates: $P = \Sigma\bar{N}\Delta W$.	Waters 1997; Waters and Crawford 1973
Summation removal	Density (no./m^2) for several dates spanning development of a cohort; mean individual biomass for each date (often obtained from length *X* weight regressions [Smock 1980]).	Production lost, or the product of the change in density and the mean individual mass, is summed over the production interval: $P = \Sigma\bar{W}\Delta N$.	Waters 1977
Instantaneous growth	Growth rate; mean biomass	Production is the product of growth rate and mean biomass, where growth rate (g) is determined in laboratory experiments, or from field samples of distinct cohorts, and biomass (B) is determined from field samples: $P = gB$. Growth is temperature and size-dependent, so accuracy of this method depends on making measurements under the appropriate conditions.	Ross and Wallace 1981; Hauer and Benke 1987
Size frequency	Mean annual density; size distribution of individuals throughout year; mean biomass for each size category; cohort production interval (mean development time from hatching to emergence).	This method is widely used and does not depend on the presence of distinct cohorts in field populations. The mean size-frequency distribution, which is skewed toward smaller size categories due to mortality, represents an average cohort. For populations in which distinct cohorts are known, this method gives the same result as the increment summation method.	Benke 1979

Modified from Benke 1984, 1996.

and Sherr 1984, Minagawa and Wada 1984, Peterson and Fry 1987, Fry 1991, Peterson et al. 1993, Schuldt and Hershey 1995, Hershey and Peterson 1996). One useful aspect of carbon stable isotopes is that the various organic matter sources for stream consumers often have different relative abundances of the isotopes ^{13}C and ^{12}C. The difference between the ratio ^{13}C to ^{12}C in a consumer and a standard (or *del* (δ) value, expressed as parts per thousand, or per mil) can be used to infer the consumer's food source because the relative abundances of these isotopes change only slightly as organic matter is processed by various consumers. In contrast, the stable isotopes of nitrogen, ^{15}N and ^{14}N, change as food is processed by consumers because the various metabolic processes use, or *fractionate*, these isotopes differently. The net result is that with each trophic level, consumers become enriched (by about 3 per mil) in ^{15}N relative to ^{14}N, as ^{15}N is retained in the tissues and ^{14}N is excreted. Thus consumers in a stream food web can generally be assigned to a trophic level even if their precise food resource is not known (Fry 1991, Peterson et al. 1993, Hershey and Peterson 1996).

When organic matter sources in a stream are isotopically distinct, a combination of ^{13}C and ^{15}N analyses can yield considerable insight into a stream food web. Such an analysis was performed for the dominant consumers and their food resources of Lookout Creek in Oregon (Figure 8.12). In Lookout Creek, the $\delta^{15}N$ values revealed that macroinvertebrates were trophically distinct from two predatory vertebrates studied, whereas the $\delta^{13}C$ values showed that consumers relied more heavily on stream algae than on terrestrial detritus. The macroinvertebrate scrapers studied (the snail *Juga* and the caddisfly *Dicosmoecus*) appeared to consume algae, whereas the shredding caddisfly *Heteroplectron* apparently ate terrestrial detritus. The stonefly *Calineuria*, a presumed predator, occupied a trophic level close to primary consumers and thus may have

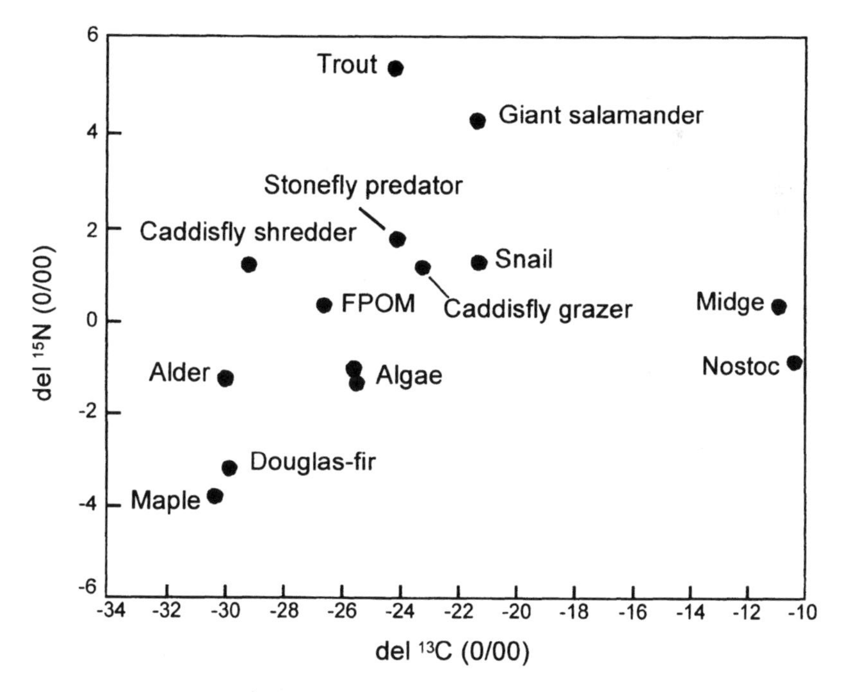

FIGURE 8.12. Stable isotope cross-plot of selected members of Lookout Creek, Oregon, food web (modified from Fry 1991 with permission). Organisms horizontally positioned over other sources (i.e., have similar del ^{13}C) are inferred to consume those materials. Organisms displaced vertically 2–4 del ^{15}N units upwards are inferred to belong to the next higher trophic level in the food web.

consumed some plant matter along with small invertebrates that were not measured. The endosymbiotic midge larva *Cricotopus* (described earlier) was positioned precisely over its algal host *Nostoc*. Cutthroat trout (*Oncorhynchus clarki*) and Pacific giant salamanders (*Dicamptodon ensatus*) consumed invertebrates, as shown by their direct vertical positioning over the ^{13}C of their food and a 3–4 per mil difference in ^{15}N (Figure 8.12). Gut contents of trout revealed a variety of invertebrates whereas salamanders had exclusively snails in their guts (R. Wildman, Oregon State University, personal communication).

Impact of Spawning Salmon on Stream Macroinvertebrates

A conspicuous feature of Pacific coastal streams is spawning salmon (*Oncorhynchus* spp.). Salmon are a major food resource for native peoples of the region and have considerable commercial and recreational value. Fisheries managers are very concerned with overexploitation of salmon and with protection of genetic diversity in the populations. Protection of stream habitats used by spawning fishes and rehabilitation of stream reaches where spawning habitat has been destroyed (see below) are important management objectives.

Spawning salmon are an important resource for stream macroinvertebrate communities. Pacific salmon die after spawning, and the carcasses provide nutrients used by stream algae (Brickell and Goering 1970, Mathisen et al. 1988), which are then fed upon by grazers (Richey et al. 1975, Kline et al. 1990). Salmon carcasses also serve as a high quality source of CPOM for stream shredders (Kline et al. 1990). Use of stable isotope technology has provided quantitative data on the importance of salmon carcasses to the stream food web. Salmon returning to streams to spawn have an isotopic signature that reflects their position in the marine food web. They are enriched in ^{15}N, reflecting their predation on marine fishes, and enriched in ^{13}C, reflecting their dependence on marine sources of carbon. By comparing the isotopic signature of algae and invertebrates in reaches not used by salmon to reaches where salmon spawn, the proportion of marine derived N (from salmon) in a consumer group can be traced. Reliance on C from salmon can be detected but is more difficult to quantify because it is not conserved in the food web (Kline et al. 1990). For example, using this approach, Kline et al. (1990) showed that at least 90% of the N in periphyton from Sashin Creek, in southeastern Alaska, was derived from spawning salmon or salmon eggs. Herbivores contained 50 to 100% salmon-derived N. The quantitative importance of salmon-derived nutrients varies with size of the salmon run relative to discharge and other sources of nutrients (Schuldt and Hershey 1995), but salmon carcasses and eggs are clearly integral to the invertebrate food webs of Pacific coastal streams.

Effects of Land Use on Community Dynamics

Hynes (1975) first articulated the importance of employing a landscape perspective for understanding stream ecosystem structure and function. The realization of this linkage between the stream and its terrestrial setting was a critical development in stream ecosystem theory (Minshall et al. 1985). This relationship is well appreciated qualitatively, but poorly understood quantitatively. Local geomorphology, vegetation, and climate determine the quality and quantity of organic matter entering a stream. These processes and disturbance history at a site determine the spatial and temporal patterns of riparian zone characteristics that dictate the physical characteristics of the stream habitat (Gregory et al. 1991). Thus, riparian processes and natural disturbances, or disturbances due to changes in land use, alter macroinvertebrate habitats in the stream channel. These disturbances operate over temporal scales ranging from geologic alterations of landform to ecological succession of the riparian forest, as well as over spatial scales spanning watersheds to local patchiness of algal growth on individual rocks within the channel (Gregory et al. 1991).

Macroinvertebrate communities reflect effects of land use within the watershed because changes in land use alter stream habitat and water quality (Gregory et al. 1987, Merritt and Lawson 1992). As discussed above, macroinvertebrate communities are sensitive to predominant sources of organic matter, dictated largely by stream order, local geomorphology, and riparian vegetation (Vannote et al. 1980, Naiman et al. 1987, Naiman and Anderson 1996). Biogeography (Resh 1992, Hershey et al. 1995) and water quality factors (Richards and Minshall 1992, Lamberti and Berg 1995) also constrain macroinvertebrate communities. The types of alterations depend on the nature and intensity of the land-use change (Cairns 1990), but local geomorphology modifies some of the effects of land-use change in addition to constraining land cover and dictating what land uses might occur in a watershed (McIntire and Colby 1978, Naiman et al. 1993).

One of the most conspicuous features of stream channel geomorphology in the Pacific coastal ecoregion is LWD contributed by the riparian forest (Van Sickle and Gregory 1990, Chapter 13 this volume). In streams this material serves many functions, and many of these functions affect macroinvertebrate populations. Macroinvertebrates readily colonize LWD, using it as both a stable substratum and, in some cases, a food resource (Anderson and Sedell 1979). LWD alters the physical characteristics of the channel by reducing bed shear stress during high-discharge events, thereby stabilizing the substratum (Naiman and Anderson 1996). Thus, factors that alter input of LWD from the watershed have a strong indirect effect on stream macroinvertebrates because of the close association between macroinvertebrate community structure and disturbance events (Figure 8.5). Stream fish populations, which also interact with the macroinvertebrate community, generally increase with LWD volume in response to both habitat and invertebrate food resources (Bisson et al. 1987).

Land use changes, such as logging, that alter LWD have large impacts on macroinvertebrate communities. In the Pacific coastal ecoregion, logging has greatly altered the size and amount of LWD that enters stream channels (Sullivan et al. 1987). Removal of riparian trees changes inputs of LWD for centuries (Maser and Sedell 1994), but recent changes in forest practices now offer some protection to the riparian zone from excessive harvest (Gregory and Ashkenas 1990). Streams that lack LWD sources due to past logging are candidates for rehabilitation, which now emphasizes the input of logs in natural volumes and arrangements to improve stream habitat and to augment the food base (Chapter 24).

In addition to removing LWD and thereby altering detrital inputs and channel structure, logging also removes vegetative cover within the watershed. This results in increased sedimentation into the stream; reduced infiltration, which decreases ground water recharge; and increased runoff, which increases the magnitude of stream discharge change following storm events (Gregory et al. 1991, Naiman and Anderson 1996). Although the Pacific coastal ecoregion is not heavily populated compared to some regions, other land uses, including urban, agricultural, and industrial development may follow timber harvest. Road construction is usually associated with all types of human land use. Road construction often destabilizes hillsides, which leads to landslides and sometimes debris flows in stream channels downslope (Lamberti et al. 1991, Swanson et al. 1987, NRC 1996); both of these events constitute a major disturbance to the stream ecosystem (see above). These changes in land use dramatically alter watershed hydrology and hence stream hydrology. As discussed above, macroinvertebrate communities are strongly affected by stream hydrology, but are also sensitive to the various forms of pollution that result from many land uses (Richards et al. 1993). Thus, macroinvertebrate communities are intimately linked to land use in the watershed, as well as to local conditions in the stream channel.

Literature Cited

Allan, J.D. 1982. The effects of reduction in trout density on the invertebrate community of a mountain stream. Ecology **63**:1444–1455.

——. 1995. Stream ecology. Chapman & Hall, London, UK.

Allan, J.D., and B.P. Feifarek. 1989. Distances traveled by drifting mayfly nymphs: Factors influencing return to the substrate. Journal of the North American Benthological Society **8**:322–330.

Allen, K.R. 1951. The Horokiwi Stream. A study of a trout population. New Zealand Department of Fisheries Bulletin **10**:1–238.

Anderson, N.H. 1976. The distribution and biology of the Oregon Trichoptera. Oregon State University Agricultural Experiment Station Technical Bulletin 134. Oregon State University, Corvallis, Oregon, USA.

——. 1992. Influence of disturbance on insect communities in Pacific Northwest streams. Hydrobiologia **248**:79–92.

Anderson, N.H., and J.R. Bourne. 1974. Bionomics of three species of glossosomatid caddisflies (Trichoptera: Glossosomatidae) in Oregon. Canadian Journal of Zoology **52**:405–411.

Anderson, N.H., and J.R. Sedell. 1979. Detritus processing by macroinvertebrates in stream ecosystems. Annual Review of Entomology **24**:351–377.

Benke, A.C. 1979. A modification of the Hynes method for estimating secondary production with particular significance for multivoltine populations. Limnology and Oceanography **24**:168–174.

——. 1984. Secondary production of aquatic insects. Pages 289–322 *in* V.H. Resh and D.M. Rosenberg, eds. The ecology of aquatic insects. Praeger, New York, New York, USA.

——. 1996. Secondary production of macroinvertebrates. Pages 557–578 *in* F.R. Hauer and G.A. Lamberti, eds. Methods in stream ecology. Academic Press, San Diego, California, USA.

Benke, A.C., R.J. Hunter, and F.K. Parrish. 1986. Invertebrate drift dynamics in a subtropical blackwater river. Journal of the North American Benthological Society **5**:173–190.

Berg, M.B., and R.A. Hellenthal. 1992. Role of Chironomidae in energy flow of a lotic ecosystem. Netherlands Journal of Aquatic Ecology **26**:471–476.

Bilby, R.E., and J.W. Ward. 1989. Changes in characteristics and function of woody debris with increasing stream size in western Washington. Transactions of the American Fisheries Society **118**:368–378.

Bird, G.A., and H.B.N. Hynes. 1981. Movement of immature insects in a lotic habitat. Hydrobiologia **77**:103–112.

Bisson, P.A., R.E. Bilby, M.D. Bryant, C.A. Dolloff, G.B. Brette, R.A. House, et al. 1987. Large woody debris in forested streams in the Pacific Northwest: Past, present, and future. Pages 143–190 *in* E.O. Salo and T.W. Cundy, eds. Streamside management: Forestry and fishery interactions. Institute of Forest Resources Contribution Number 57, University of Washington, Seattle, Washington, USA.

Borror, D.J., C.A. Triplehorn, and N.F. Johnson. 1989. An introduction to the study of insects, 6th edition. Saunders College Publishing, Philadelphia, Pennsylvania, USA.

Brickell, D.C., and J.J. Goering. 1970. Chemical effects of salmon decomposition on aquatic ecosystems. Pages 125–138 *in* R.S. Murphey, ed. First international symposium of water pollution control in cold climates. US Government Printing Office, Washington, DC, USA.

Brock, E.M. 1960. Mutualism between the midge *Cricotopus* and the alga *Nostoc*. Ecology **41**:474–483.

Butler, M.G. 1984. Life histories of aquatic insects. Pages 24–55 *in* V.H. Resh and D.M. Rosenberg, eds. The ecology of aquatic insects. Praeger, New York, New York, USA.

Cairns, J., Jr. 1990. Lack of theoretical basis for predicting rate and pathways of recovery. Environmental Management **14**:517–526.

Connell, J.H. 1978. Diversity in tropical rain forests and coral reefs. Science **199**:1302–1310.

Cooper, S.D., S.J. Walde, and B.L. Peckarsky. 1990. Prey exchange rates and the impact of predation in streams. Ecology **71**:1503–1514.

Cummins, K.W. 1973. Trophic relations of aquatic insects. Annual Review of Entomology **18**:183–206.

——. 1974. Structure and function of stream ecosystems. BioScience **24**:631–641.

Cummins, K.W., and M.J. Klug. 1979. Feeding ecology of stream invertebrates. Annual Review of Ecology and Systematics **10**:147–172.

Cummins, K.W., and R.W. Merritt. 1996. Ecology and distribution of aquatic insects. Pages 74–86 *in* R.W. Merritt and K.W. Cummins, eds. An introduction to the aquatic insects of North America, 3rd edition. Kendall/Hunt, Dubuque, Iowa, USA.

Cummins, K.W., M.A. Wilzbach, D.M. Gates, J.B. Perry, and W.B. Taliaferro. 1989. Shredders and riparian vegetation. BioScience **39**:24–30.

Dahm, C.N. 1981. Pathways and mechanisms for removal of dissolved organic carbon from leaf leachate in streams. Canadian Journal of Fisheries and Aquatic Science **38**:68–76.

Delucchi, C.M. 1989. Movement patterns of invertebrates in temporary and permanent streams. Oecologia **78**:199–207.

Dodds, W.K. 1989. Photosynthesis of two morphologies of *Nostoc parmelioides* (Cyanobacteria) as related to current velocities and diffusion patterns. Journal of Phycology **25**:258–262.

Dodds, W.K., and J.L. Marra. 1989. Behaviors of the midge, *Cricotopus* (Diptera: Chironomidae) related to mutualism with *Nostoc parmelioides* (Cyanobacteria). Aquatic Insects **11**:201–208.

Dudley, T.L., and N.H. Anderson. 1982. A survey of invertebrates associated with wood debris in aquatic habitats. Melanderia **39**:1–21.

Feminella, J.W., and C.P. Hawkins. 1995. Interactions between stream herbivores and periphyton: A quantitative analysis of past experiments. Journal of the North American Benthological Society **14**:465–509.

Fisher, S.G., and L.J. Gray. 1983. Secondary production and organic matter processing by collector macroinvertebrates in a desert stream. Ecology **64**:1217–1224.

Fisher, S.G., L.J. Gray, N.B Grimm, and D.E. Busch. 1982. Temporal succession in a desert stream ecosystem following flash flooding. Ecological Monographs **52**:93–110.

Fisher, S.G., and G.E. Likens. 1973. Energy flow in Bear Brook, New Hampshire: An integrative approach to stream ecosystem metabolism. Ecological Monographs **43**:421–439.

Fry, B. 1991. Stable isotope diagrams of freshwater food webs. Ecology **72**:2293–2297.

Fry, B., and E. Sherr. 1984. $\delta^{13}C$ measurements as indicators of carbon flow in marine and freshwater ecosystems. Contributions in Marine Science **27**:13–47.

Furnish, J.L. 1990. Factors affecting the growth, production, and distribution of the stream snail *Juga silicula* Gould. Dissertation. Oregon State University, Corvallis, Oregon, USA.

Glime, J.M., and R.M. Clemons. 1972. Species diversity of stream insects on *Fontinalis* spp. compared to diversity on artificial substrates. Ecology **53**:458–464.

Gregory, S.V. 1983. Plant-herbivore interactions in stream systems. Pages 157–189 *in* J.R. Barnes and G.W. Minshall, eds. Stream ecology. Plenum Press, New York, New York, USA.

Gregory, S.V., and L. Ashkenas. 1990. Riparian management guide. Willamette National Forest. USDA Forest Service, Pacific Northwest Region, Corvallis, Oregon, USA.

Gregory, S.V., G.A. Lamberti, D.C. Erman, K.V. Koski, M.L. Murphy, and J.R. Sedell. 1987. Influence of forest practices on aquatic production. Pages 233–255 *in* E.O. Salo and T.W. Cundy, eds. Streamside management: Forestry and fishery interactions. Institute of Forest Resources Contribution Number 57, University of Washington, Seattle, Washington, USA.

Gregory, S.V., F.J. Swanson, W.A. McKee, and K.W. Cummins. 1991. An ecosystem perspective of riparian zones. BioScience **41**:540–552.

Hart, D.D., and V.H. Resh. 1980. Movement patterns and foraging ecology of a stream caddisfly larva. Canadian Journal of Zoology **58**:1174–1185.

Hauer, F.R., and A.C. Benke. 1987. Influence of temperature and river hydrograph on black fly growth rates in a subtropical blackwater river. Journal of the North American Benthological Society **6**:251–261.

Hawkins, C.P., and J.K. Furnish. 1987. Are snails important competitors in stream ecosystems? Oikos **49**:209–220.

Hawkins, C.P., M.L. Murphy, and N.H. Anderson. 1982. Effects of canopy, substrate composition, and gradient on the structure of macroinvertebrate communities in Cascade Range streams of Oregon. Ecology **63**:1840–1856.

Hawkins, C.P., and J.R. Sedell. 1981. Longitudinal and seasonal changes in functional organization of macroinvertebrate communities in four Oregon streams. Ecology **62**:387–397.

Hayden, W., and H.F. Clifford. 1974. Seasonal movements of the mayfly *Leptophlebia cupida* (Say) in a brown-water stream of Alberta, Canada. American Midland Naturalist **91**:90–102.

Hershey, A.E. 1987. Tubes and foraging behavior in larval Chironomidae: Implications for predator avoidance. Oecologia **73**:236–241.

Hershey, A.E., W.B. Bowden, L.A. Deegan, J.E. Hobbie, B.J. Peterson, G.W. Kipphut, et al. 1997. The Kuparuk River: A long-term study of biological and chemical processes in an Arctic river. Pages 107–129 *in* A.M. Milner and M.W. Oswood, eds. Freshwaters of Alaska. Springer-Verlag, New York, New York, USA.

Hershey, A.E., and S.I. Dodson. 1987. Predator avoidance by *Cricotopus*: Cyclomorphosis and the importance of being big and hairy. Ecology **68**:913–920.

Hershey, A.E., R.W. Merritt, and M.C. Miller. 1995. Trophic dynamics, diversity, and life history features of black flies (Diptera: Simuliidae) in arctic Alaskan streams. Pages 283–295 *in* S.F. Chapin and C. Kroner, eds. Arctic and alpine biodiversity. Springer-Verlag, New York, New York, USA.

Hershey, A.E., R.W. Merritt, M.C. Miller, and J.S. McCrea. 1996. Organic matter processing by larval black flies in a temperate woodland stream. Oikos **75**:524–532.

Hershey, A.E., J. Pastor, B.J. Peterson, and G.W. Kling. 1993. Stable isotopes resolve the drift paradox for *Baetis* mayflies in an arctic river. Ecology **74**:2315–2325.

Hershey, A.E., and B.J. Peterson. 1996. Stream food webs. Pages 511–530 *in* F.R. Hauer and G.A. Lamberti, eds. Methods in stream ecology. Academic Press, San Diego, California, USA.

Hinterleitner-Anderson, D., A.E. Hershey, and J.A. Schuldt. 1992. The effects of river fertilization on mayfly (*Baetis* sp.) drift patterns and population density in an arctic river. Hydrobiologia **240**:247–258.

Huryn, A.D. 1996. An appraisal of the Allen paradox in a New Zealand trout stream. Limnology and Oceanography **41**:243–352.

Hynes, H.B.N. 1975. The stream and its valley. Internationale verein für Theoretische und Angewandte Limnologie Verhandlungen **19**:1–15.

Jacoby, J.M. 1987. Alterations in periphyton characteristics due to grazing in a Cascade foothill stream. Freshwater Biology **18**:495–508.

Kline, T.C., J.J. Goering, O.A. Mathisen, and P.H. Poe. 1990. Recycling of elements transported upstream by runs of Pacific salmon: 1. ^{15}N and ^{13}C evidence in Sashin Creek, southeastern Alaska. Canadian Journal of Fisheries and Aquatic Sciences **47**:136–144.

Kohler, S.L. 1985. Identification of stream drift mechanisms: An experimental and observational approach. Ecology **66**:1749–1761.

Lamberti, G.A. 1996. The role of periphyton in benthic food webs. Pages 533–572 *in* R.J. Stevenson, M.L. Bothwell, and R.L. Lowe, eds. Algal ecology: Freshwater benthic ecosystems. Academic Press, San Diego, California, USA.

Lamberti, G.A., L.R. Ashkenas, S.V. Gregory, and A.D. Steinman. 1987. Effects of three herbivores on periphyton communities in laboratory streams. Journal of the North American Benthological Society **6**:92–104.

Lamberti, G.A., and M.B. Berg. 1995. Invertebrates and other benthic features as indicators of environmental change in Juday Creek, Indiana. Natural Areas Journal **15**:249–258.

Lamberti, G.A., S.V. Gregory, L.R. Ashkenas, A.D. Steinman, and C.D. McIntire. 1989. Productive capacity of periphyton as a determinant of plant-herbivore interactions in streams. Ecology **70**:1840–1856.

Lamberti, G.A., S.V. Gregory, L.R. Ashkenas, R.C. Wildman, and K.M.S. Moore. 1991. Stream ecosystem recovery following a catastrophic debris flow. Canadian Journal of Fisheries and Aquatic Sciences **48**:196–208.

Lamberti, G.A., S.V. Gregory, C.P. Hawkins, R.C. Wildman, L.R. Ashkenas, and D.M. DeNicola. 1992. Plant-herbivore interactions in streams near Mount St. Helens. Freshwater Biology **27**:237–247.

Lamberti, G.A., and V.H. Resh. 1983. Stream periphyton and insect herbivores: An experimental study of grazing by a caddisfly population. Ecology **64**:1124–1135.

Li, H.W., G.A. Lamberti, T.N. Pearsons, C.K. Tait, J.L. Li, and J.C. Buckhouse. 1994. Cumulative effects of riparian disturbances along high desert trout streams of the John Day basin, Oregon. Transactions of the American Fisheries Society **123**:627–640.

Li, J.L., and S.V. Gregory. 1989. Behavioral changes in the herbivorous caddisfly *Dicosmoecus gilvipes* (Limnephilidae). Journal of the North American Benthological Society **8**:250–259.

Lock, M.A., R.R. Wallace, J.W. Costerton, R.M. Ventullo, and S.E. Charlton. 1984. River epilithon: Toward a structural-functional model. Oikos **42**:10–22.

Madsen, B.L., J. Bengtson, and I. Butz. 1973. Observations on upstream migration by imagines of some Plecoptera and Ephemeroptera. Limnology and Oceanography **18**:678–681.

Maser, C. and J.R. Sedell. 1994. From the forest to the sea. St. Lucia Press, Delray Beach, Florida, USA.

Mathisen, O.A., P.L. Parker, J.J. Goering, T.C. Kline, P.H. Poe, and R.S. Scalan. 1988. Recycling of marine elements transported into freshwater by anadromous salmon. Verhandlungen der Internationalen Vereinigung für Theoretische und Angewandte Limnologie **23**:2249–2258.

Mayer, M.S., and G.E. Likens. 1987. The importance of algae in a shaded headwater stream as food for an abundant caddisfly (Trichoptera). Journal of the North American Benthological Society **6**:262–269.

McAuliffe, J.R. 1984. Competition for space, disturbance, and the structure of a benthic stream community. Ecology **65**:894–908.

McCafferty, W.P. 1981. Aquatic entomology: The fisherman's and ecologists' illustrated guide to insects and their relatives. Jones and Bartlett, Boston, Massachusetts, USA.

McIntire, C.D. 1973. Periphyton dynamics in laboratory streams: A simulation model and its implications. Ecological Monographs **43**:399–420.

McIntire, C.D., and J.A. Colby. 1978. A hierarchical model of lotic ecosystems. Ecological Monographs **48**:167–190.

Merritt, R.W., and K.W. Cummins, eds. 1996a. An introduction to the aquatic insects of North America, 3rd edition. Kendall/Hunt, Dubuque, Iowa, USA.

Merritt, R.W., and K.W. Cummins. 1996b. Trophic relations of macroinvertebrates. Pages 453–474 *in* F.R. Hauer and G.A. Lamberti, eds. Methods in stream ecology. Academic Press, San Diego, California, USA.

Merritt, R.W., and D.L. Lawson. 1992. The role of macroinvertebrates in stream-floodplain dynamics. Hydrobiologia **248**:65–77.

Meyer, J.L. 1990. A blackwater perspective on riverine ecosystems. BioScience **40**:643–651.

Minagawa, M., and E. Wada. 1984. Stepwise enrichment of ^{15}N along food chains: Further evidence and the relation between δ^{15}N and animal age. Geochimica et Cosmochimica Acta **48**:1135–1140.

Minshall, G.W. 1978. Autotrophy in stream ecosystems. BioScience **28**:767–771.

Minshall, G.W., K.W. Cummins, R.C. Petersen, C.E. Cushing, D.A. Bruns, J.R. Sedell, et al. 1985. Developments in stream ecosystem theory. Canadian Journal of Fisheries and Aquatic Sciences **42**:1045–1055.

Minshall, G.W., R.C. Petersen, K.W. Cummins, T.L. Bott, J.R. Sedell, C.E. Cushing, et al. 1983. Interbiome comparison of stream ecosystem dynamics. Ecological Monographs **53**:1–25.

Müller, K. 1954. Investigations on the organic drift in North Swedish streams. Institute of Freshwater Research, Drottingholm, Report **34**:133–148.

Naiman, R.J., and E.C. Anderson. 1996. Streams and rivers of the coastal temperate rain forest of North America: Physical and biological variability. Pages 131–148 *in* P.K. Schoonmaker and B. von Hagen, eds. The rain forests of home: An exploration of people and place. Island Press, Washington, DC, USA.

Naiman, R.J., T.J. Beechie, L.E. Benda, D.R. Berg, P.A. Bisson, L.H. MacDonald, et al. 1993. Fundamental elements of ecologically healthy watersheds in the Pacific Northwest coastal ecoregion. Pages 127–188 *in* R.J. Naiman, ed. Watershed management. Springer-Verlag, New York, New York, USA.

Naiman, R.J., J.M. Melillo, M.A. Lock, T.E. Ford, and S.R. Reice. 1987. Longitudinal patterns of ecosystem processes and community structure in a subarctic river continuum. Ecology **68**:1138–1156.

Newbold, J.D., J.W. Elwood, R.V. O'Neill, and A.L. Sheldon. 1983. Phosphorus dynamics in a woodland stream ecosystem: A study of nutrient spiraling. Ecology **64**:1249–1265.

NRC (National Research Council). 1996. Upstream. National Academy Press, Washington, DC, USA.

Peckarsky, B.L. 1980. Predator–prey interactions between stoneflies and mayflies: Behavioral observations. Ecology **61**:932–943.

——. 1982. Aquatic insect predator–prey relations. BioScience **32**:261–266.

Peckarsky, B.L., and S.I. Dodson. 1980. Do stonefly predators influence benthic distributions in streams? Ecology **61**:1275–1282.

Peterson, B.J., L. Deegan, J. Helfrich, J. Hobbie, M. Hullar, B. Moller, et al. 1993. Biological responses of a tundra river to fertilization. Ecology **74**:653–672.

Peterson, B.J., and B. Fry. 1987. Stable isotopes in ecosystem studies. Annual Review of Ecology and Systematics **18**:293–320.

Petersen, R.C. 1986. In situ particle generation in a southern Swedish stream. Limnology and Oceanography **31**:432–437.

Power, M.E. 1990. Effects of fish in river food webs. Science **250**:411–415.

Resh, V.H. 1992. Year-to-year changes in the age structure of a caddisfly population following loss and recovery of a springbrook habitat. Ecography **15**:314–317.

Resh, V.H., A.V. Brown, A.P. Covich, M.E. Gurtz, H.W. Li, G.W. Minshall, et al. 1988. The role of disturbance in stream ecology. Journal of the North American Benthological Society **7**:433–455.

Resh, V.H., T.S. Flynn, G.A. Lamberti, E.P. McElravy, K.L. Sorg, and J.R. Wood. 1981. Responses of the sericostomatid caddisfly *Gumaga nigricula* (McL.) to environmental disruption. Series Entomologica **20**:311–318.

Richards, C., G.E. Host, and J.W. Arthur. 1993. Identification of predominant environmental factors structuring stream macroinvertebrate communities within a large agricultural catchment. Freshwater Biology **29**:285–294.

Richards, C., and G.W. Minshall. 1992. Spatial and temporal trends in stream macroinvertebrate species assemblages: The influence of watershed disturbance. Hydrobiologia **241**:173–184.

Richey, J.E., M.A. Perkins, and C.R. Goldman. 1975. Effects of Kokanee salmon (*Oncorhynchus nerka*) decomposition on the ecology of a subalpine stream. Journal of the Fisheries Research Board of Canada **32**:817–820.

Ross, D.H., and J.B. Wallace. 1981. Production of *Brachycentrus spinae* Ross (Trichoptera: Brachycentridae) and its role in seston dynamics of a southern Appalachian stream (USA). Holarctic Ecology **6**:270–284.

Schmidt, S.K., J.M. Huges, and S.E. Bunn. 1995. Gene flow among conspecific populations of *Baetis* sp. (Ephemeroptera): Adult flight and larval drift. Journal of the North American Benthological Society **14**:147–157.

Schuldt, J.A., and A.E. Hershey. 1995. Impact of salmon carcass decomposition on Lake Superior tributary streams. Journal of the North American Benthological Society **14**:259–268.

Sedell, J.R., G.H. Reeves, F.R. Hauer, J.A. Standford, and C.P. Hawkins. 1990. Role of refugia in recovery from disturbances: Modern fragmentation and disconnected river systems. Environmental Management **14**:711–724.

Shepard, R.B., and G.W. Minshall. 1981. Nutritional value of lotic insect feces compared with allochthonous materials. Archive für Hydrobiologica **90**:467–488.

Smock, L.A. 1996. Macroinvertebrate movements: Drift, colonization, and emergence. Pages 371–390 *in* F.R. Hauer and G.A. Lamberti, eds. Methods in stream ecology. Academic Press, San Diego, California, USA.

Soluk, D.A. 1985. Macroinvertebrate abundance and production of psammophilous Chironomidae in shifting sand areas of a lowland river. Canadian Journal of Fisheries and Aquatic Sciences **42**: 1296–1302.

Sousa, W.P. 1984. The role of disturbance in natural communities. Annual Review of Ecology and Systematics **15**:353–391.

Stanford, J.A., and J.V. Ward. 1988. The hyporheic habitat of river ecosystems. Nature **335**:64–66.

Steinman, A.D. 1996. Effects of grazers on freshwater benthic algae. Pages 341–373 *in* R.J. Stevenson, M.L. Bothwell, and R.L. Lowe, eds. Algal ecology: Freshwater benthic ecosystems. Academic Press, San Diego, California, USA.

Sullivan, K., T.E. Lisle, C.A. Dolloff, G.E. Grant, and L.M. Reid. 1987. Stream channels: The link between forests and fishes. Pages 39–97 *in* E.O. Salo and T.W. Cundy, eds. Streamside management: Forestry and fishery interactions. Institute of Forest Resources Contribution Number 57, University of Washington, Seattle, Washington, USA.

Suren, A.M., and M.J. Winterbourn. 1992. The influence of periphyton, detritus and shelter on invertebrate colonization of aquatic bryophytes. Freshwater Biology **27**:327–339.

Swanson, F.J., L.E. Benda, S.H. Duncan, G.E. Grant, W.F. Megahan, L.M. Reid, and R.R. Zeimer. 1987. Mass failures and other processes of sediment production in Pacific Northwest forest landscapes. Pages 9–38 *in* E.O. Salo and T.W. Cundy, eds. Streamside management: Forestry and fishery interactions. Institute of Forest Resources Contribution Number 57, University of Washington, Seattle, Washington, USA.

Swanson, F.J., S.V. Gregory, J.R. Sedell, and A.G. Campbell. 1982. Land-water interactions: The riparian zone. Pages 267–291 *in* R.L. Edmonds, ed. Analysis of coniferous forest ecosystems in the western United States. US/IBP Synthesis Series 14. Hutchinson and Ross, Stroudsburg, Pennsylvania, USA.

Tait, C.K., J.L. Li, G.A. Lamberti, T.N. Pearsons, and H.W. Li. 1994. Relationships between riparian cover and the community structure of high desert streams. Journal of the North American Benthological Society **13**:45–56.

Ulfstrand, S. 1968. Benthic animals in Lapland streams. Oikos Supplement **10**:1–120.

Vannote, R.L., G.W. Minshall, K.W. Cummins, J.R. Sedell, and C.E. Cushing. 1980. The river continuum concept. Canadian Journal of Fisheries and Aquatic Sciences **37**:130–137.

Van Sickle, J., and S.V. Gregory. 1990. Modeling inputs of coarse woody debris to streams from falling trees. Canadian Journal of Forest Research **20**:1593–1601.

Wallace, J.B., and N.H. Anderson. 1996. Habitat, life history, and behavioral adaptations of aquatic insects. Pages 41–73 *in* R.W. Merritt and K.W. Cummins, eds. An introduction to the aquatic insects of North America, 3rd edition. Kendall/Hunt, Dubuque, Iowa, USA.

Wallace, J.B., and J.W. Grubaugh. 1996. Transport and storage of FPOM. Pages 191–215 *in* F.R. Hauer and G.A. Lamberti, eds. Methods in stream ecology. Academic Press, San Diego, California, USA.

Wallace, J.B., D.S. Vogel, and T.F. Cuffney. 1986. Recovery of a headwater stream from an insecticide-induced community disturbance. Journal of the North American Benthological Society **5**:115–126.

Wallace, J.B., J.R. Webster, and T.F. Cuffney. 1982. Stream detritus dynamics: Regulation by invertebrate consumers. Oecologia **53**:197–200.

Ward, A.K., C.N. Dahm, and K.W. Cummins. 1985. *Nostoc* (Cyanophyta) productivity in Oregon stream ecosystems: Invertebrate influences and differences between morphological types. Journal of Phycology **21**:223–227.

Ward, G.M., A.K. Ward, C.N. Dahm, and N.G. Aumen. 1994. Origin and formation of organic and inorganic particles in aquatic systems. Pages 27–56 *in* R.S. Wotton, ed. The biology of particles in aquatic systems, 2nd edition. Lewis, Boca Raton, Florida, USA.

Ward, J.V., and J.A. Stanford. 1983. The intermediate disturbance hypothesis: An explanation for biotic diversity patterns in lotic ecosystems. Pages 347–356 *in* T.D. Fontaine and S.M. Bartell, eds. Dynamics of lotic ecosystems. Ann Arbor Science, Ann Arbor, Michigan, USA.

Waters, T.F. 1972. The drift of stream insects. Annual Review of Entomology **17**:253–272.

——. 1977. Secondary production in inland waters. Advances in Ecological Research **10**:91–164.

Waters, T.F., and G.W. Crawford. 1973. Annual production of a stream mayfly population: A comparison of methods. Limnology and Oceanography **18**:286–296.

Williams, D.D. 1993. Nutrient and flow vector dynamics at the hyporheic/groundwater interface and their effects on the interstitial fauna. Hydrobiologia **251**:185–198.

Wilzbach, M.A., and K.W. Cummins. 1989. An assessment of short-term depletion of stream macroinvertebrate benthos by drift. Hydrobiologia **185**:29–39.

Winterbourn, M.J., J.S. Rounick, and B. Cowie. 1981. Are New Zealand stream ecosystems really different? New Zealand Journal of Marine and Freshwater Research **15**:321–328.

Wotton, R.S. 1988. Dissolved organic material and trophic dynamics. BioScience **38**:172–177.

9
Fish Communities

Gordon H. Reeves, Peter A. Bisson, and Jeffrey M. Dambacher

Overview

• Biological communities are best understood in terms of community dynamics and functional organization because of the difficulty in delineating communities as well-defined units. The structure and composition of fish communities in Pacific coastal stream systems vary with spatial scale from watershed to habitat units. The relative influence of biotic and abiotic factors has been the subject of much debate but generally varies according to spatial scale.

• Compared to other parts of North America, stream fish communities in the Pacific coastal ecoregion are relatively depauperate. A major reason for this is the large influence of geologic history, primarily extensive tectonic activity and glaciation. Although the species richness of the region is low, there is great diversity in life history and morphological types within and among populations.

• At the watershed scale, species richness generally increases downstream. This pattern follows the species-area relationship. Species gradually disappear and additional ones appear (are added) along the stream network. Physical factors strongly influence community structure and composition in headwater areas, whereas, biological factors are more important in lower portions of the watershed.

• At the reach scale, fish communities are generally more diverse in unconstrained reaches than in constrained reaches. These differences are attributable to differences in productivity and habitat diversity between reach types.

• Biological factors and habitat complexity are major influences on fish communities at the habitat unit level. Generally the more diverse the habitat unit, the greater the community diversity. The influence of biotic factors, primarily competition, varies with the species present in the habitat. In some situations, competition for space or food restricts species presence. In other cases, partitioning of habitat and food resources allows more species to inhabit a particular unit.

• The diversity of stream fish communities throughout much of the Pacific coastal ecoregion has declined as a result of human activities. The general pattern of response is changes in relative abundance rather than loss of species. Some species respond positively while others decline in abundance.

• Integration of macroscale physical and biological processes into a broader spatial and temporal framework assists in understanding the dynamics of stream fish communities as well as recovering and maintaining native stream fish communities in the Pacific coastal ecoregion.

Introduction

Ecological communities are currently viewed as dynamic entities whose members vary in space and time. A community consists of coadapted species with similar geographic distributions, but most species are not in obligatory associations with each other because populations of species tend to change along environmental gradients (Whittaker 1962). Consequently, it is difficult to identify communities as well-defined units. Thus, community ecology now emphasizes the dynamics and functional organization of communities rather than the classification of communities into discrete units or types (Underwood 1986, Krebs 1994).

The organization of natural communities is influenced by environmental factors and biotic interactions (Sousa 1984, Schlosser 1987, Menge and Olson 1990, Rosenzweig 1994). Factors such as habitat heterogeneity, frequency and magnitude of physical disturbance, and life history attributes of organisms influence the structure and composition of communities, with the relative importance of physical and biological processes generally changing along environmental gradients (Connell 1975 and 1978, Karr and Freemark 1983 and 1985, Wiens 1984). Community composition is also influenced by historical processes, such as speciation and dispersal (Holt 1993).

The relative influence of physical and biological factors on the structure and composition of stream fish communities has been the subject of much debate. The "stochastic school" maintains that physical and temporal factors are dominant influences (Grossman et al. 1982). This view argues that species abundance varies and is largely determined by differential responses to unpredictable environmental changes. In contrast, the "deterministic" school argues that interactive biological mechanisms, such as competition and predation, are the primary determinants of community structure and composition (Yant et al. 1984). Power and Matthews (1983), Moyle and Vondracek (1985), and Schlosser and Ebel (1989) cite examples of streams in different part of the United States where biological interactions exert a strong influence on fish communities. Schoener (1987) noted that, as a group, stream fishes are unusual in the extent to which communities are influenced by both physical and biological factors. Schlosser (1987) argues that, at least for fish communities in small warmwater streams, the relative importance of physical and biological processes varies with location within the stream network.

The ecology of many salmonid fishes in the Pacific coastal ecoregion has been studied extensively but relatively little is known about nonsalmonids or the organization of stream fish communities. Several studies have examined interactions between species, again primarily salmonids (Hartman 1965, Everest and Chapman 1972, Bisson et al. 1988). Li et al. (1987) reviewed factors influencing communities of stream fish in the Pacific Northwest, which includes the Pacific coastal ecoregion, but could not reach firm conclusions about how the communities were structured. One reason for this is that factors influencing the structure and composition of stream fish communities vary with spatial scale (Table 9.1). This chapter examines the structure and composition of fish communities at different spatial scales—regional, watershed, reach and habitat unit. For an explanation of terms and a general review of the concept of biodiversity, see Chapter 17.

Regional Diversity

Relative to other areas of North America, the Pacific Northwest (including the Pacific coastal ecoregion) has comparatively fewer species of fish. The western United States has about one half of the families and one quarter of the fish species found in the eastern United States (Smith 1981, Minckley et al. 1986). Mahon (1984) estimated that in western areas there are 5 to 10 native fish species in a $10{,}000\,km^2$ watershed compared to 50 species in a similar sized watershed in Ontario, Canada.

Geological history has been a major influence on the fish fauna of the Pacific coastal ecoregion (Miller 1959). The ecoregion and surrounding areas have been more extensively affected by tectonic activity and glaciation than other areas of North America (McPhail and Lindsey 1986, Minckley et al. 1986). During the last glacial period, ice sheets covered much of Alaska, British Columbia, and northern Washington reaching their maximum extent about 15,000 years ago (Figure 9.1).

TABLE 9.1. Hierarchical organization of factors regulating fish species diversity in Pacific coastal streams.

Hierarchical level	Spatial scale (km^2)	Physical events and processes	Biological events and processes	Environmental gradients
Region	10^8–10^6	Tectonism, volcanism, glaciation, sea level fluctuations	Invasion of oceanic taxa, speciation	Ocean/freshwater productivity, aridity, temperature
Basin	10^5–10^3	Stream capture, formation of falls, climatic change	Dispersion and isolation, subspecific endemism, stock differentiation, range expansion and contraction	Elevation, stream order/size, temperature
Reach	10^2–10^0	Alluvial/colluvial valley formation, fluvial disturbance	Species refugia and recolonization	Stream gradient, constraint, habitat diversity, productivity
Habitat unit	10^{-1}–10^{-4}	Sediment and organic debris storage and transport	Competition, predation	Habitat diversity, complexity

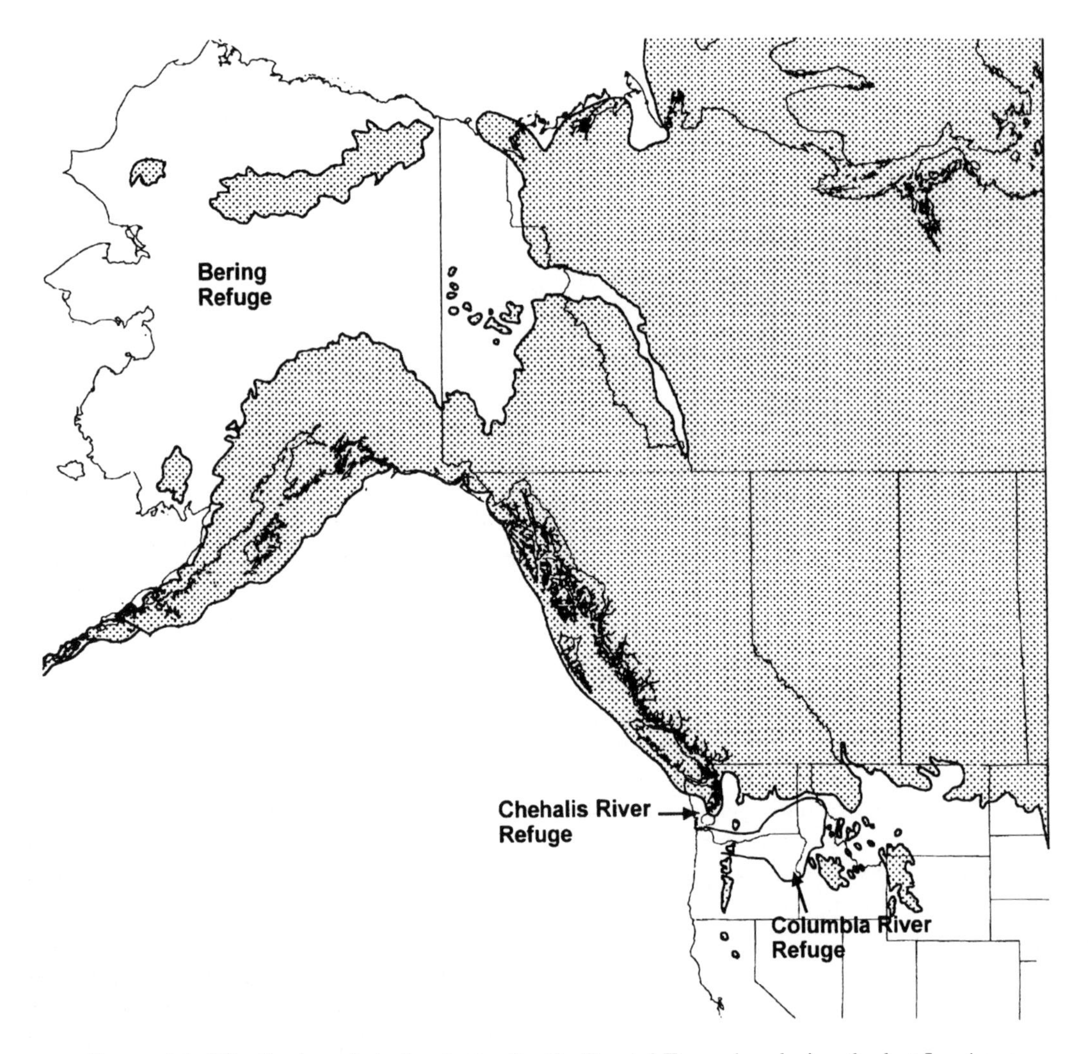

FIGURE 9.1. Distribution of glaciers in the Pacific Coastal Ecoregion during the last Ice Age.

Continental glaciers reached south into western Washington near what is now the southern tip of Puget Sound and along the Cascade Mountain Range into Oregon. At this time, three major refugia for fishes were present: the Bering refuge to the north, and refugia in the Chehalis and Columbia River basins in Washington (McPhail and Lindsey 1986). As the glaciers gradually retreated these three areas served as major sources of colonists for drainages in northern Washington and British Columbia. However, the relatively short duration of postglacial colonization (about 10,000 years) has contributed to a paucity of species in northern portions of the Pacific coastal ecoregion (McPhail and Lindsey 1986). The ability of many closely related species or stocks to hybridize when brought together (Behnke 1992, Smith et al. 1995) suggests that the relatively short time period following the last glaciation has been insufficient for evolutionary processes to produce a native fauna that rivals in species richness the much more diverse faunas of eastern United States and other parts of the world.

Modern fishes of the Pacific coastal ecoregion have been influenced by tectonic activities such as earthquakes, volcanism, and uplift resulting from the collision of continental plates. Primarily these activities separate formerly continuous habitats or join formerly isolated ones (Minckley et al. 1986). Changes in local conditions caused by tectonic activity, such as formation of mountains, alter environmental conditions and thus influence species abundance and distribution.

The history and specific effects of tectonic activity on fishes in the Pacific coastal ecoregion is too complex to present in detail. Suffice it to say, tectonic activity is responsible for the large number of endemic (i.e., species with local distributions) and the limited number of species in some river systems (Minckley et al. 1986). For example, the Klamath River in southern Oregon consists of discrete upper and lower areas that are separated by Klamath Falls. The upper area, historically connected to the Sacramento River system (California), was isolated from the Sacramento River in the late Miocene period (i.e., about 10 million years ago) after tectonic activity. However, the two systems share common species that are not found in the lower Klamath River (Minckley et al. 1986). Fishes in the lower river resemble those in the Rogue River and other coastal systems. Another example of the effect of tectonic activities is seen in coastal Oregon streams. Much of coastal Oregon was under water in the Paleocene (i.e., 54–65 million years ago). Many streams, particularly smaller ones, had no connection with interior systems. Limited reinvasion of these systems by marine-tolerant species, such as salmonids (*Salmonidae*), lampreys (*Petromyzontidae*) and sculpins (*Cottidae*) occurred after ocean levels receded. Minckley et al. (1986) and McPhail and Lindsey (1986) provide descriptions on this topic.

As a group, fish in the Pacific coastal ecoregion and western North America have different morphologies and life histories compared to species from eastern North America (Table 9.2). A greater proportion of the fauna in the Pacific coastal ecoregion and western North America are larger (>30 cm standard length [SL]), longer lived, older at age of first reproduction, and have longer reproduction spans and higher absolute fecundity than the fauna from eastern North America (Miller 1959, Moyle 1976). For example, in two major Pacific coastal ecoregion river systems (the Klamath and Columbia) large fishes represent 64% and 41% of the fauna, respectively (Moyle and Herbold 1987). Small fishes comprise 29% and 22% of the fauna, respectively. The fish fauna of eastern North America is dominated by smaller (<10 cm SL), early reproducing fishes (Mahon 1984). In two eastern rivers, large fish represent 18% and 27% and small fish 41% and 55% of the fauna (Moyle and Herbold 1987).

Body size and life history features are adaptations to regional environmental conditions. Smaller size and earlier age at first reproduction is advantageous in environments where conditions may be more stable and adult survivorship is variable but reproductive success is high (Miller 1979, Mann et al. 1984). Larger body size, delayed age at first reproduction, and high relative fecundity and adult survivorship (but low reproductive success in any given year) are advantageous traits in

TABLE 9.2. General characteristics of fish faunas of the Mississippi River drainage and western North America.

Characteristic	Mississippi drainage	Western North America
Species richness		
Sample richness	High (10–30 species)	Low (<10)
Cumulative richness	High; increases rapidly in small streams, then levels out	Low; increases as stream size increases
Life history		
Bionomics	Short lives, early maturation, low fecundity, predominately short reproductive spans	Long lives (2+ years), late maturation, high fecundity, predominately long reproduction spans
Parental care	Brood hiders and guarders	Little parental care
Migration	Limited movements for most species; spawning migrations for some large species	Spawning migrations common, but occurrence variable
Trophic specialization	Relatively unspecialized invertebrate feeders ommon	Specialists common

Modified from Moyle and Herbold 1987 © University of Oklahoma Press.

variable environments (Moyle and Herbold 1987).

The freshwater fish fauna of the Pacific Northwest is dominated by lampreys, salmon, trout and chars (both anadromous and nonanadromous), minnows such as shiners, dace, chubs, and squawfish (Cyprinidae), suckers (Catostomidae), and sculpins. Stickleback (Gasterosteidae) are also common in many systems. Species richness is much greater in southern portions of the Pacific coastal ecoregion (i.e., Oregon and northern California), which were not glaciated during the last major glacial period. Fish communities in river basins in the northern portions are dominated by euryhaline (i.e., salt tolerant) species capable of dispersal through salt water. Endemism is fairly common in many of the isolated drainages of this area (Smith 1981, Minckley et al. 1986), particularly among lampreys, salmonids, cyprinids, and cottids (Moyle and Herbold 1987). In the Columbia Zoogeographic Province, 58% of the native fishes are endemic, while 37% of the fishes of the Klamath Zoogeographic Province (California and Oregon) are endemic (Moyle and Cech 1982). Most species with widespread distributions are capable of dispersal through the ocean or through fluvial connections between river basins (Miller 1959). Stream capture (action of a river acquiring a headwater of a second river as a result of differential erosion rates between the two systems, or a change in landscape features resulting from tectonic activity) facilitates dispersal for species dwelling in adjacent headwaters (Bond 1963, Smith 1981).

Despite this paucity of species, native fishes are integral components of, and exert strong influences on, all trophic levels of ecosystems in the Pacific coastal ecoregion (Willson and Halupka 1995). Both terrestrial and aquatic larger animals occupying several trophic levels, (e.g., raccoons, dippers, deer and grizzly bears) use anadromous salmonids (*Oncorhynchus* spp.) as food sources during some part of the year in streams throughout the Pacific coastal ecoregion (Cederholm et al. 1989, Willson and Halupka 1995) (Table 9.3). Returning adults also provide nutrients for a suite of trophic levels (Bilby et al. 1996). Salmonid-derived nutrients, ^{15}N and ^{13}C, are incorporated into stream biota by consumption of eggs, carcasses, and fry, and sorption onto dissolved organic matter released by decomposing carcasses. Juvenile salmonids in streams with larger returning adult runs experience faster growth rates than do juveniles in streams with smaller runs. Decomposing carcasses also provide important sources of nitrogen for riparian vegetation. Additionally, because of predictable migra-

TABLE 9.3. Mammals and birds known to consume salmon carcasses in the Pacific coastal ecosystem.

Mammals	Birds
Beaver (*Castor canadensis*)	Bald eagle (*Haliaeetus leucocephalus*)
Black bear (*Ursus americanus*)	Chickadees (*Parus* spp.)
Blacktail deer (*Odocoileus hemionus*)	Crow (*Corvus* spp.)
Bobcat (*Lynx rufus*)	Dipper (*Cinclus mexicanus*)
Cougar (*Felis concolor*)	Fox sparrow (*Passerella iliaca*)
Coyote (*Canis Latrans*)	Gray jay (*Perisoreus canadensis*)
Deer mouse (*Peromyscus maniculatus*)	Gulls (*Larus* spp.)
Douglas squirrel (*Tamiasciuris douglasii*)	Hairy woodpecker (*Picoides villosus*)
Elk (*Cervus elaphus*)	Hermit thrush (*Catharus guttatus*)
Flying squirrel (*Glaucomys sabrinus*)	Kingfisher (*Megaceryle alcyon*)
Grizzly bear (*U. artos*)	Kinglets (*Regulus* spp.)
Masked shrew (*S. cinereus*)	Merlin (*Falco columbarius*)
Mink (*M. vison*)	Nuthatches (*Sitta* spp.)
Mole (*Scapanus* spp.)	Pine siskin (*Carduelis pinus*)
Mountain beaver (*Aplodontia rufa*)	Pygmy owl (*Glaucidium gnoma*)
Otter (*Lutra canadensis*)	Raven (*C. corax*)
Raccoon (*Procyon lotor*)	Red-tailed hawk (*Buteo jamaicensis*)
Skunk (*Mephitis mephitis*)	Robin (*Turdus migratorius*)
Wandering shrew (*S. vagrans*)	Ruffed grouse (*Bonasa umbellus*)
Water shrew (*Sorex palustris*)	Sapsucker (*Sphyrapicus varius*)
Weasel (*Mustela* spp.)	Song sparrow (*Melospiza melodia*)
	Stellers jay (*Cyanocitta stelleri*)
	Varied thrush (*Ixoreus naevius*)
	Winter wren (*Troglodytes roglodytes*)

From Cederholm et al. 1989, Willson and Halupka 1995.

tions, high food quality, and high commercial, recreational, and aesthetic value, many species are important components of human cultural, social, and economic systems throughout the region (Stouder et al. 1996).

Anadromous Life Histories

Anadromous fishes comprise approximately 25% of the freshwater fishes in many river systems of the Pacific coastal ecoregion. Anadromous species begin life in freshwater, move to the marine environment to grow and mature, and then return to freshwater to reproduce. Species with anadromous life histories include some lampreys, sturgeons (*Acipenser transmontanus, A. medirostris*), salmon and trout, and candlefish (*Thaleichthys pacificus*). The predominance of fish with anadromous life histories increases from the equator to the poles in the northern hemisphere because of differences in the relative production of freshwater and marine habitats (Gross et al. 1988). Northern marine environments, where fish with anadromous life histories prevail, are more productive than freshwater environments. In more southern environments where production of freshwater systems is greater, fish with catadromous life histories prevail (i.e., begin life in the marine environment, move to freshwater to grow and mature and return the marine environment to reproduce). Fish with catadromous life histories are not found in the Pacific coastal ecoregion.

Large annual fluctuations in stream habitat conditions also contribute to the predominance of anadromous species in the Pacific coastal ecoregion. Flows vary by several orders of magnitude annually (10–1,000 m^3/sec) and many streams experience wide fluctuations in water temperature. Fish must survive until they are able to reproduce, and then they must reproduce successfully to be successful in such an environment. Successful reproduction is more likely in fish that are highly fecund and have relatively large bodies capable of burying eggs deeper in substrate.

Population Variability

What the Pacific coastal ecoregion lacks in species diversity, is at least partially compensated in variability within and among local watershed populations. Fish populations vary phenotypically and genetically to a substantial degree. Species or populations can exhibit multiple life history patterns. For example, chinook salmon (*O. tshawytscha*) exhibit two types of general life histories. The stream type rear in fresh water for a year or more as juveniles and may enter fresh water many months prior to spawning as adults, whereas the ocean type rear in streams for only a few months as juveniles before migrating to sea and spawn shortly after reentering freshwater as adults. When both types of chinook are present in rivers, the two are often temporally and spatially isolated in terms of reproduction. Stream-type chinook often use headwater areas for spawning and rearing, and spawn in the early fall. Ocean types spawn in the late fall and rear lower in the drainage system (Healey 1991). Chinook salmon may exhibit multiple life histories even within a single population. Reimers (1973) identified five distinct life history variations within a population of ocean-type chinook salmon in Oregon's Sixes River. These variations centered around the length of time spent in headwater tributaries, the river mainstem, and the estuary, and were adaptations to highly unpredictable rearing conditions in the Sixes River system and in the ocean. Thus, for species with extended freshwater spawning and rearing periods (such as chinook and coho salmon, steelhead, and sea-run cutthroat trout) different life history patterns are favored as environmental conditions change from year to year (Healey 1991, Healey and Prince 1995, Stouder et al. 1996). It is likely that life history polymorphism within populations has both genetic and environmental components.

Genetic and phenotypic variation, within and between stream fish populations, indicates that although each population may contain most of the genetic material for the species, phenotype expressed as adaptation to local conditions is remarkably plastic (Table 9.4) (Healey and Prince 1995). The genetic and environmental basis of this variability is complex, but the great majority of evidence points to strong selective tendencies for local adaptation.

Spawning populations of anadromous salmon exhibit highly specific local adaptations for a number of traits such as the complicated

TABLE 9.4. Percentage of genotypic and phenotypic variation of anadromous salmonids within populations, among populations within regions, and among regions.

	Variation (%)		
Species and characteristic	Within populations	Among populations	Among regions
Chinook			
Allozyme variation[a]	87.7	4.6	7.7
Allozyme variation[b]	94.1	3.3	2.6
Pink[c]			
Fry length	30.3	41.4	28.2
Adult weight	13.3	1.5	85.2
Chum[c]			
Fecundity	7.9	5.1	87.0
Age at maturity	14.0	0.5	85.5

[a] Utter et al. (1989).
[b] Gharrett et al. (1987).
[c] Groot and Margolis (1991).
From Healey and Prince 1995.

homing behavior, temperature adjustments, unique local mating behavior, and adjustments of smolts to local feeding conditions. These adaptations are most likely to be quantitative characteristics that are dependent on the effects of many genes, each of which has only a small effect individually (polygenes)—it would likely be difficult to "replace" a local population with transplants from nonlocal populations. Therefore, the more complex the life cycle, the more difficult it would be to replace a local population (NRC 1996).

In fact, attempts to establish Pacific salmon elsewhere often have been unsuccessful. Yet, in a few instances, populations have been successfully transplanted and adaptive radiation has occurred quickly. A dramatic example of the ability of anadromous salmonids to adapt to local conditions is demonstrated by the phenotypic divergence of chinook salmon in several rivers on the South Island of New Zealand, where several distinct life history patterns developed after introduction of this species from a single Sacramento River donor population about 90 years ago (Quinn and Unwin 1993). The degree of phenotypic differentiation that developed over a span of about 20 generations may be among the highest known for aquatic vertebrates (Miller 1961) and indicates that under certain conditions significant evolutionary divergence can occur in a matter of decades (Healey and Prince 1995).

The extent to which populations of fishes other than anadromous salmonids have developed unique genetic, morphological, and life history characteristics in Pacific Northwest rivers is not nearly as well known. Non-anadromous species may not have differentiated into "stocks" to the degree exhibited by Pacific salmon but available evidence suggests that significant variation among populations is the rule rather than the exception, particularly for those species with widespread distributions. Examples of non-anadromous taxa possessing locally variable populations include lampreys (*Lampetra* spp.; Kan 1975), speckled dace (*Rhinichthys osculus*; Zirges 1973, Sada et al. 1995), longnose dace (*R. cataractae*; Bisson and Reimers 1977), suckers (*Catostomus* and other genera; Smith 1966), sculpins (*Cottus* spp; McAllister and Lindsey 1961, Bond 1963), and threespine sticklebacks (*Gasterosteus aculeatus*; Miller and Hubbs 1969, Hagen and Gilbertson 1972).

Virtually every study comparing variability among populations of Pacific coastal freshwater fishes over relatively large areas reports significant variation in some morphological feature. For example, there are strong differences in body size, morphology and egg size of chum salmon (*O. keta*) in British Columbia rivers (Beacham and Murray 1987). Individuals of populations from larger rivers have larger heads, thicker caudal peduncles, and larger fins than those from smaller rivers, and early spawning populations have older fish, larger eggs, and later emerging fry than later spawning groups. These characteristics appear to be adaptations to local conditions. Fish in larger rivers need to move larger substrates and deposit eggs deeper than fish spawning in smaller streams.

Local populations of the same species may also exhibit variation in morphologies and life histories. For example, anadromous forms of the threespine stickleback are generally larger and have higher gill raker, fin ray, and lateral plate counts than resident (i.e., nonanadromous) forms (Bell 1984). In coastal areas of northern California, morphological differences of threespine stickleback are not as well defined (Snyder and Dingle 1989), but there are differences in life history features of the two types. Anadromous forms are older and larger when they first reproduce and more fecund than resident forms. Juvenile coho salmon (*O. kisutch*) from a lake and inlet stream in British Columbia exhibit differences in behavior and morphology (Swain and Holtby 1989). The stream fish are more aggressive than lake-rearing individuals; as juveniles stream fish are territorial. Lake fish have more posteriorly placed pectoral fins, shallower bodies, and smaller, less colorful dorsal and anal fins than stream fish. Their diminished aggression and accompanying body features are an adaptation to schooling behavior in open waters.

Reasons for such variability are complex. Some variation may be the result of founder effects (establishment of a new population by a few original founders which carry only a small fraction of the total genetic variation of the parental population); some may be caused by local selective pressures coupled with prolonged geographical isolation. But whatever the reasons, morphological, physiological, and life history characteristics of fishes in this region can change fairly rapidly. Protecting this local variability in morphological, physiological, and life history characteristics among and within local populations is at the heart of applying the "evolutionarily significant unit" (populations of anadromous salmonids that share common genetic and/or life history features and are a unit of consideration for the Endangered Species Act listing) approach to the definition of distinct population segments under the Endangered Species Act (Waples 1991) and is strongly endorsed by the National Research Council (NRC 1996).

Watershed Scale Patterns of Diversity

Generally, the diversity of stream fish communities increases from headwaters to lower portions of river basins. That is, within a drainage basin the number of species increases in higher-order channels (i.e., larger streams). This pattern has been observed in the United States in the midwest (Horwitz 1978, Schlosser 1987), in the south (Boschung 1987), and in the east (Sheldon 1968). However, it has not been well established in the Pacific coastal ecoregion, primarily because most studies of stream fishes in the region have focused on salmonids rather than other species, nor have the longitudinal distribution of stream fishes been considered. Nonetheless, Li et al. (1987) described a generalized distribution of fish (salmonids and other species) in river systems of the Pacific coastal ecoregion (Figure 9.2) which follows the diversity pattern described above. In addition, when the number of species (and collection locations) is plotted relative to elevation (a surrogate for stream size and basin area) for four different river systems in Oregon and there is a significant ($p < 0.05$) negative correlation between numbers of species collected and elevation (Figure 9.3). These examples suggest that the pattern of increasing species richness with stream order holds in the Pacific coastal ecoregion.

Ecological Rules

Strictly from a species richness perspective, the general pattern of increasing species diversity moving down a watershed fits the *species-area relationship* of community ecology. Simply stated, this concept, which has been established for more than a century, predicts that the number of species increases with the area considered (Williams 1964). Rosenzweig (1994) cited studies on a suite of organisms where this relationship holds and characterized it as the "most supported rule in ecology." However, the species-area relationship may break down at large-spatial scales, such as regions, when there are large differences in productivity across an area (Levin 1974). Caution should be used in interpreting species-area relationships at smaller spatial scales, such as watersheds, because species numbers can be influenced by factors such as rates of disturbance and competition, and regional processes such as speciation and biogeographical dispersal (Ricklefs 1987, Caswell and Cohen 1993).

Another possible explanation for the observed patterns in fish distribution at the watershed scale is that diversity is related to productivity (Connell and Orias 1964, McArthur and Pianka 1966, Currie 1991, Huston 1994). Diversity is generally greatest at intermediate levels of productivity. It declines at lower and higher levels of productivity (Rosenzweig and Abramsky 1993, see also Chapter 17). The reason for this pattern is the subject of much debate in ecology. Rosenzweig and Abramsky (1993) believe that the most plausible explanation is that resource and habitat heterogeneity is greatest at intermediate levels of productivity. Increased heterogeneity allows for greater diversity. The productivity rule is most applicable at large spatial scales ($10^6 km^2$) (Wright et al. 1993). It also depends on taxonomic scale; the relationship is

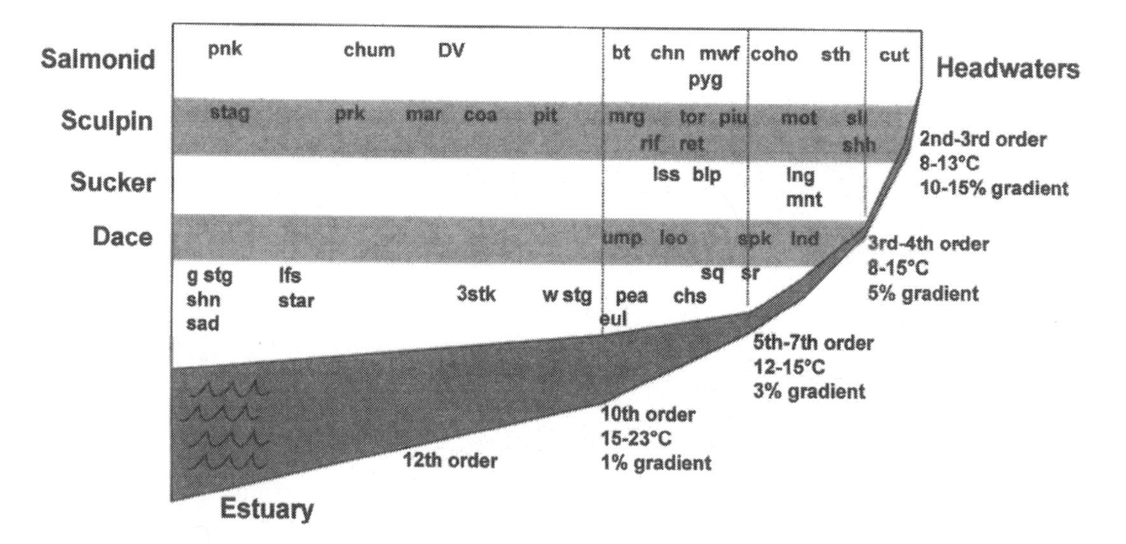

FIGURE 9.2. Generalized distribution of native fishes throughout watersheds along a continuum of gradient, stream order, and water temperature (from Li et al. 1987 © University of Oklahoma Press, with permission).

bt = bull trout
chn = chinook salmon
cut = cutthroat trout
eul = eulechon
lfs = longfin smelt
lss = largescale sculpin
mot = mottled sculpin
pea = peamouth
pink = pink salmon
ret = reticulate sculpin
shh = shorthead sculpin
spk = speckled dace
stag = staghorn sculpin
3 stk = threespined stickleback
w stg = white sturgeon

blp = bridgelip sucker
chs = chiselmouth
coho = coho salmon
g stg = green sturgeon
lng = longnose sucker
mar = marbled sculpin
mrg = margined sculpin
pit = pit sculpin
prk = prickly sculpin
rif = riffle sculpin
shin = shiner perch
sq = squawfish
star = starry flounder
tor = torrent sculpin

chum = chum salmon
coa = coast range sculpin
DV = Dolly Varden
leo = leopard dace
lnd = longnose dace
mnt = mountain sucker
mwf = mountain whitefish
piu = piute sculpin
pyg =pygmy whitefish
sad = saddleback gunnel
sli = slimy sculpin
sr = sandroller
sth = steelhead trout
ump = Umpqua dace

weaker at lower taxonomic levels (i.e., family and genus) (Wright et al. 1993). Rosenzweig and Abramsky (1993) and Rosenzweig (1994) reviewed the applicability of this rule and concluded that it generally held in terrestrial systems but was less applicable in aquatic systems. The reason for this is not known.

Further, the River Continuum Concept (Vannote et al. 1980, Chapter 1), which predicts changes in biological and physical attributes in stream systems moving from headwaters to larger rivers (Table 9.5), implies that the productivity rule is less applicable because other aspects of aquatic systems also influence diversity. In other words, the pattern of downstream change in biotic communities (e.g., downstream increase of fish species) is influenced, not only by changes in energy input and processing but also by changes in physical features such as width, depth, flow velocity and volume, and temperature.

Physical and Biological Processes

Within a watershed, biological (biotic: e.g., competition and predation) and physical (abiotic: e.g., flow, pool depth, etc.) processes strongly influence the structure and composition of fish communities. The relative influence of physical and biological processes varies

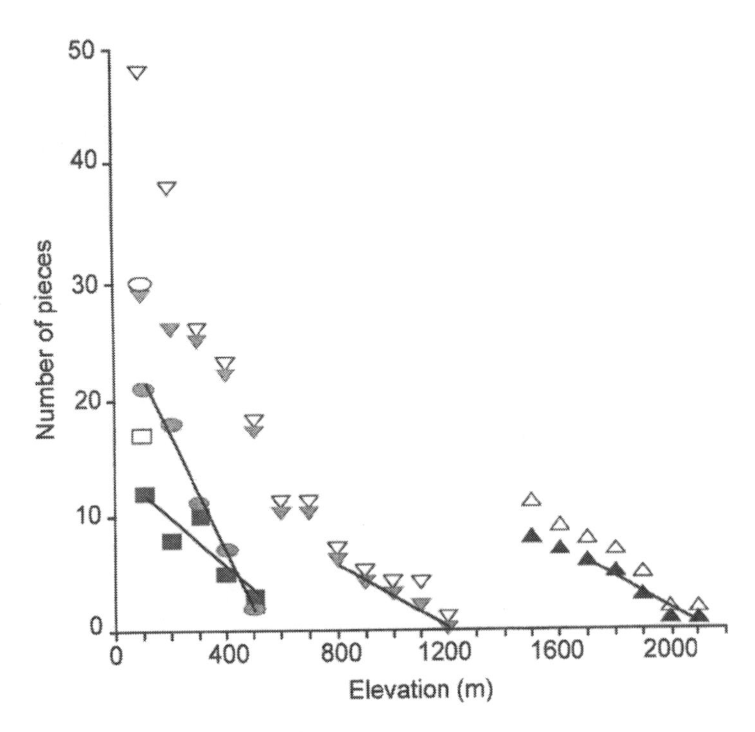

FIGURE 9.3. Relationship between basin elevation and number of fish species for four Oregon watersheds. (North Coast = ■; South Coast = ●; Willamette = ▼; Goose Lake = ▲). Solid symbols are native fishes and open symbols are introduced species.

along the stream network (Figure 9.4). Schlosser (1987) proposed a conceptual framework to explain how fish communities in warmwater streams are structured based on the relationship between changes in habitat heterogeneity (i.e., the diversity of habitat types and the complexity of habitat conditions), and the strength and type of biotic processes in different parts of the watershed (Figure 9.5). For example, according to Schlosser's framework, habitat heterogeneity is low in headwaters (although annual physical conditions, such as flow and pool depth, are more variable), primarily because deep pools are lacking and competi-

TABLE 9.5. Summary of basin features predicted by the River Continuum Concept.

	Location		
Feature	Headwaters (1st–3rd order)	Medium-sized streams (4th–6th order)	Large rivers (>6th order)
Energy Source	Allochthonous detritus	Autochthonous primary production	Autochthonous primary production
Production-respiration ratio	<1	>1	>1
Riparian vegetation influence	Strong	Moderate	Weak, localized
Benthic macroinvertebrate Functional feeding group	Shredders Collectors	Collectors Grazers	Collectors
Fish	Insectivores	Piscivores, Insectivores	Piscivores, Insectivores, Planktivores

Vannote et al. 1980.

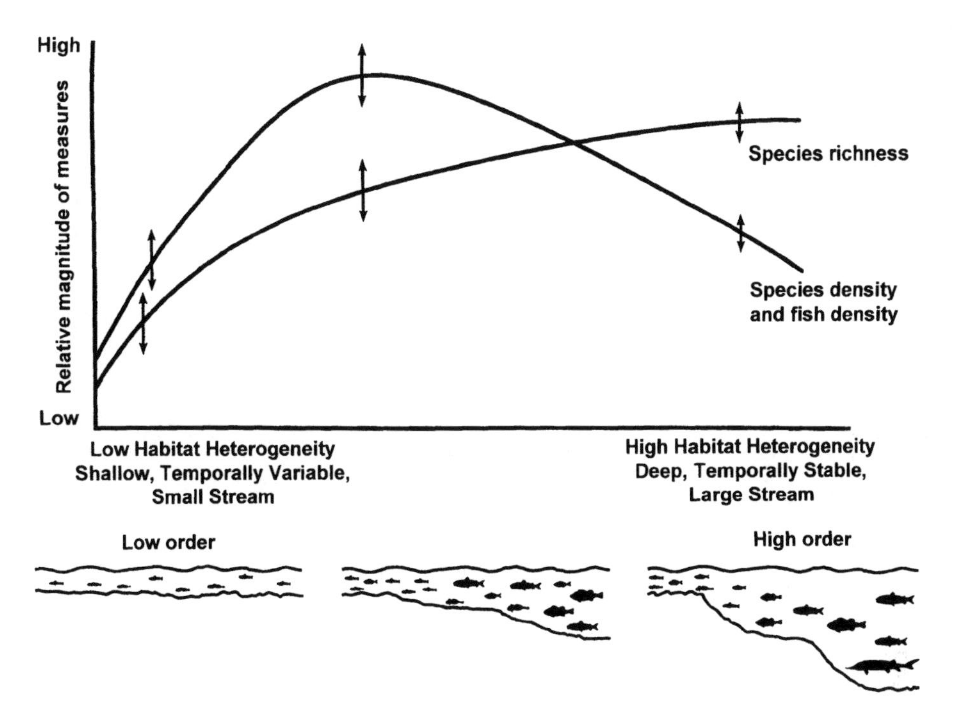

FIGURE 9.4. Hypothetical pattern of fish community attributes along a gradient of increasing habitat heterogeneity and pool development in a small warmwater stream. The patterns of species composition and size structure of fishes in the community are shown at the bottom of the figure. Arrows indicate the relative temporal variability in the parameters (from Schlosser 1987 © University of Oklahoma Press with permission).

tion is the dominant biotic process. Consequently, the number of species is low and trophic relationships are simple compared to other parts of the system. Fish communities in these streams demonstrate high variability in numbers and relative abundance as a result of the large variability in conditions and lack of refugia (Schlosser 1987). Headwater fishes are able to persist, in part, because of the ability to recolonize areas quickly following a disturbance. These fishes have high fecundity rates and mobile juveniles. Schlosser's (1987) framework further predicts that habitat heterogeneity and community diversity continues to increase as stream size and order increase (Figure 9.4). In larger rivers, the relative influence of biotic processes may be greater than in other parts of the watershed (Schlosser 1987). For example, piscivores (i.e., fish that feed predominately on other fish) are a dominant component of the community and presumably are responsible, in part, for the decline of smaller headwater species. The structure and composition of fish communities are more stable than communities in headwater streams because of decreased variability in environmental conditions in larger streams.

It is not clear how applicable Schlosser's (1987) framework is to fish communities in Pacific coastal ecoregion streams, particularly with regard to the relative importance of biotic and abiotic factors. Although studies of the relative importance of biological and physical processes to fish community structure in Pacific coastal streams are lacking, the pattern of distribution presented by Li et al. (1987) appears to follow the pattern (Figure 9.2) (for at least a large portion of the basin) described by Schlosser (1987) (Figure 9.5) and Vannote et al. (1980) (Table 9.5). However, in large rivers (i.e. >10th order), numbers of native fish species appear to decline, possibly because fewer niches are available in these more homogenous habitats.

Fish communities in Pacific coastal ecoregion headwater streams tend to be relatively less diverse compared to lower parts of the stream network. Typical communities have a high

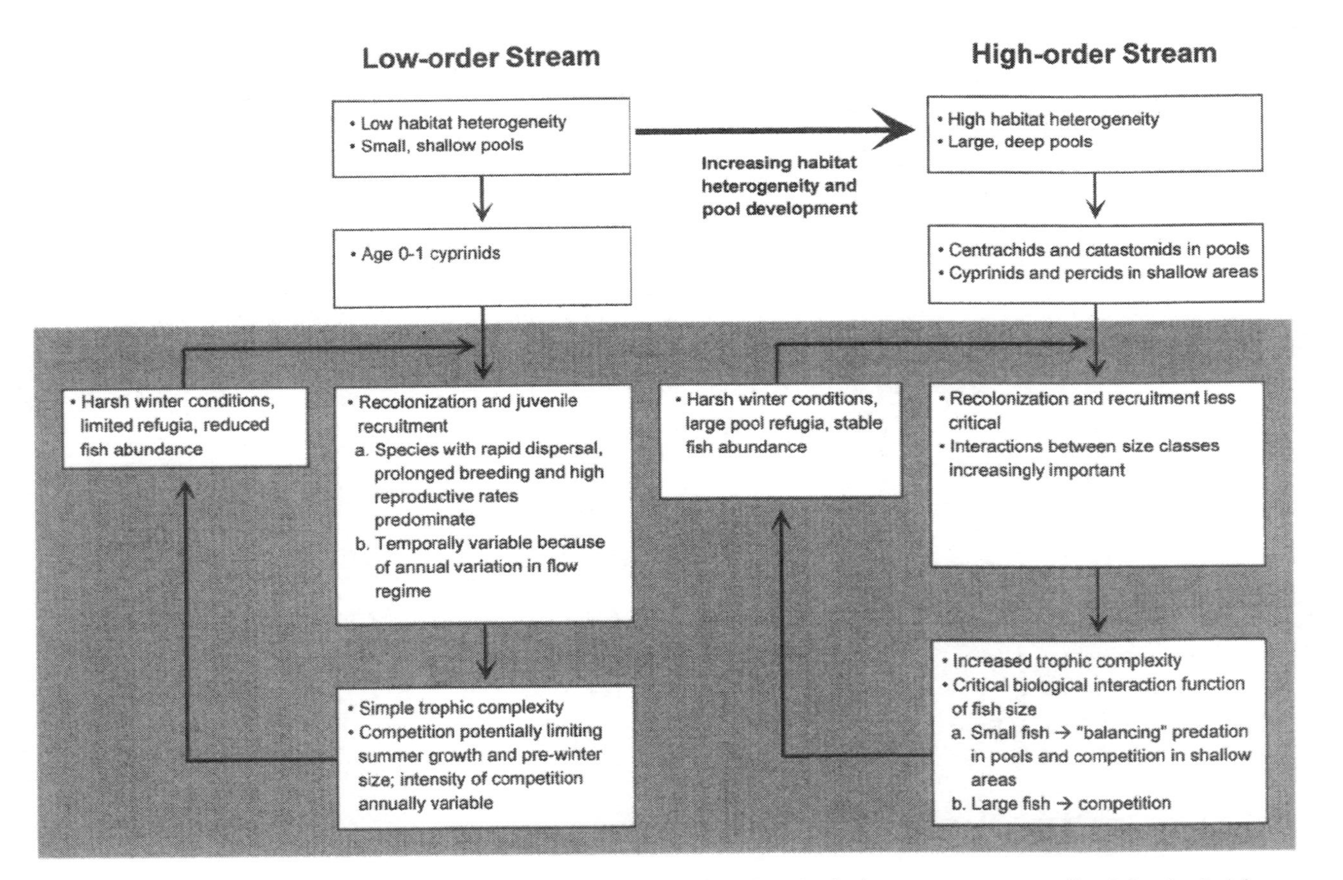

FIGURE 9.5. Conceptual framework of processes determining fish community structure along a gradient of habitat heterogeneity and pool development in a small warmwater stream. Critical ecological processes are outlined in shaded boxes (from Schlosser 1987 © University of Oklahoma Press, with permission).

degree of trophic and microhabitat segregation (Moyle and Herbold 1987) and contain one or two species of lamprey, trout, dace, and sculpin—fishes with very different morphologies and behaviors (Moyle and Cech 1982). Salmonids have very agile, torpedo-shaped bodies that allow them to occupy foraging positions in the water column and swim in rapid bursts (Moyle and Cech 1982) (Figure 9.6a). They actively prey on benthic macroinvertebrates, terrestrial insects, and fishes. Dace are active bottom dwellers with small subterminal mouths and feed on benthic organisms (Figure 9.6b). Sculpins are bottom

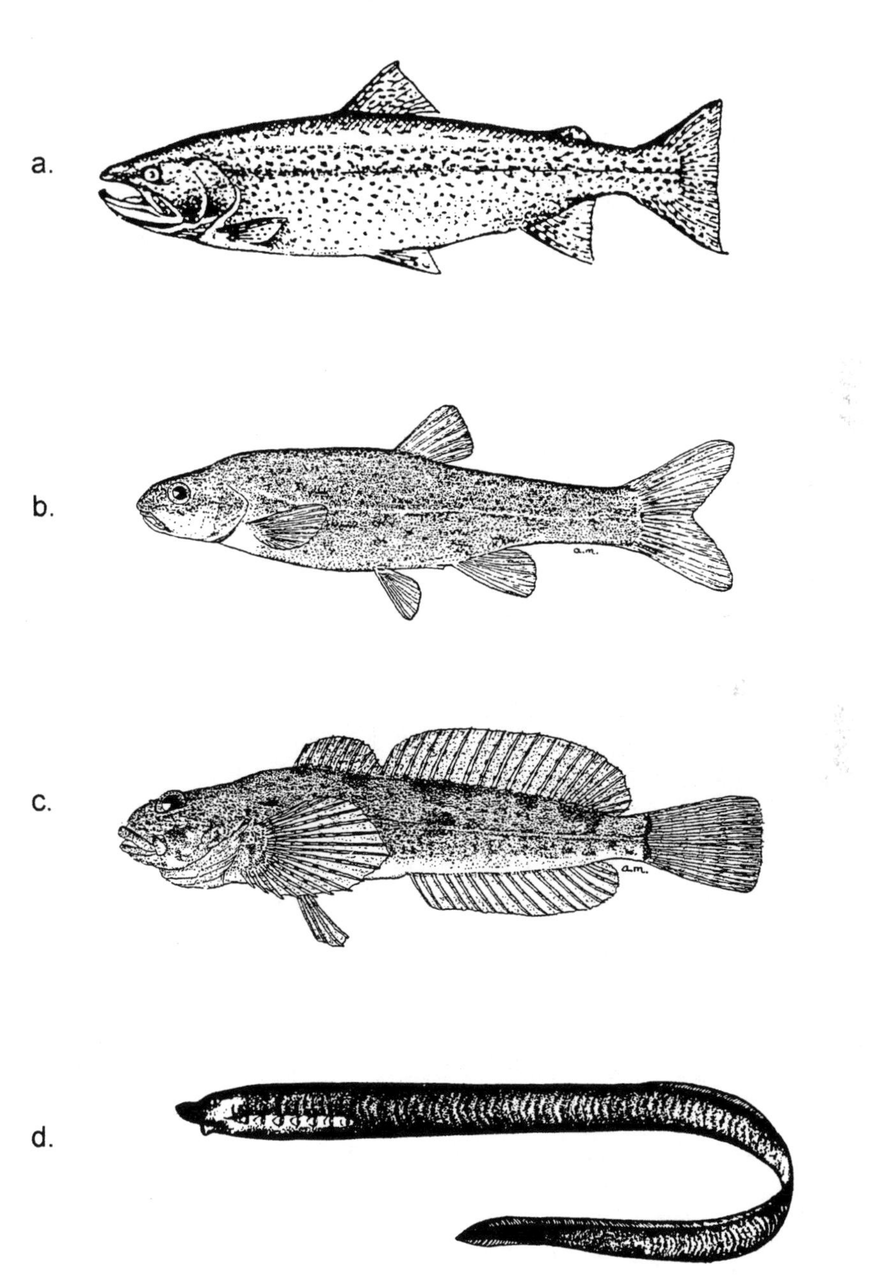

FIGURE 9.6. Fishes (a = cutthroat trout, b = speckled dace, c = riffle sculpin, and d = western brook lamprey) found in headwater streams in the Pacific coastal ecoregion.

dwellers with dorsoventrally flattened bodies and large pectoral fins and mouths; they ambush their prey (benthic macroinvertebrates and fish) (Figure 9.6c). Lamprey ammocoetes (i.e., immature forms) have vermiform bodies that can burrow in stream substrates (Figure 9.6d). As a result of these differences in morphology and behavior, biological factors, (particularly competition) have little influence on the community structure and composition (Zalewski and Naiman 1985). Thus, physical factors related to the dynamic nature of these streams, low habitat heterogeneity, and restricted temperature regimes are the most important influences on fish community structure, particularly in headwater streams.

Schlosser (1987) predicts that more diverse communities develop moving downstream (Figure 9.5). Environmental conditions tend to be less variable than in headwaters. Habitat heterogeneity and numbers of larger, deeper pools increases in downstream, high-order channels. Many headwater species also may be present downstream. However, the age classes of the species present differ between the two areas; headwaters primarily contain juveniles and downstream areas primarily contain adults (Figure 9.5). Predation, particularly on smaller individuals, and competition are the dominant processes influencing community structure in downstream areas.

Differences in temporal and localized distribution of fish allow similar species to coexist in the different parts of the drainage network. Where the distribution of congeneric species such as anadromous salmonids overlap, differences in fine-scale distributions and timing of activities limit interactions for particular resources. For example, in smaller watersheds (i.e., 4th–5th order streams, as in Fish Creek, Oregon) coho salmon, winter and summer steelhead trout, and spring chinook salmon use the same parts of the basin for spawning and early rearing (Figure 9.7). However, chinook and coho salmon use larger size gravels for spawning (1.3–10.0 cm diameter) than steelhead trout (0.6–10.0 cm) (Bjornn and Reiser 1991). Chinook salmon and steelhead trout generally use deeper areas ($\geq$24 cm) for spawning than coho salmon ($\geq$18 cm). Further, these species spawn at different times which results in the temporal segregation of juveniles when habitat requirements overlap. Chinook salmon spawn from mid-September to mid-November, coho salmon from mid-October to mid-January, summer steelhead trout from December to May, and winter steelhead trout from mid-March to late-June (Figure 9.8). The temporal and fine-scale partitioning of habitats allows similar species to use particular habitats and thus increases species richness and diversity.

Differences in the rates of species additions observed in Figure 9.3 can be attributed to various factors; for example, variations in basin morphology and water temperature. Stream gradient and size most strongly mediate the rate of species additions in the North and South Coast streams, whereas water temperature is a stronger influence in higher elevation headwater streams. Numbers of fish species increase at a significantly higher rate (2.1 and 4.9 species/100 m drop in elevation, respectively; $p < 0.01$) in the low elevation headwaters in the North and South Coast streams in Oregon, than in the higher elevation headwaters in the Willamette River (1.4 species/100 m drop in elevation) and Goose Lake watershed (1.4 species/100 m drop in elevation) (Figure 9.3). The influence of cold water is more extensive in the higher elevation streams and may limit additions of species preferring warmer water. Another factor that affects the rate of species additions is the influence of different zoogeographic provinces, or life zones. For example, the high rates of species additions in the South Coast streams is likely due to the fact that the area encompasses three different zoogeographic provinces.

Reach Scale Diversity

The structure and composition of aquatic biotic communities also varies with reach-level characteristics. A reach is an integrated series of geomorphic units sharing a common landform pattern (Grant et al. 1990). Reaches are influenced by variation in channel slope, local side slopes, valley floor width, riparian vegetation,

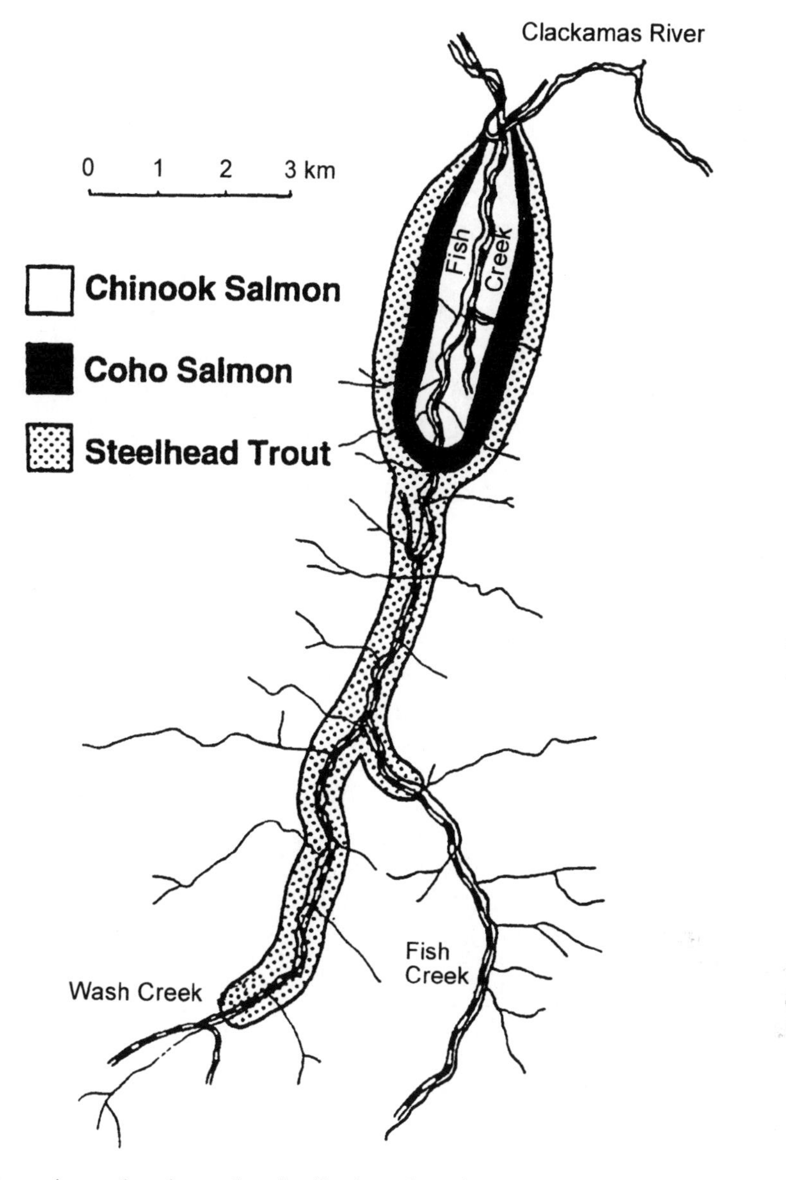

FIGURE 9.7. Spawning and early rearing distribution of anadromous salmonids in Fish Creek, Oregon.

and bank material (Frissell et al. 1986). Gregory et al. (1989) classified reaches as constrained (active channel to valley floor width ratio <2) and unconstrained (active channel to valley floor width ratio >2) (Figure 9.9a and 9.9b).

In several small coastal Oregon streams, salmonid community composition varies between these two reach types (Reeves, unpublished data). Trout (age 1+) are generally the dominant component of the community in constrained reaches (Figure 9.10a). In unconstrained reaches salmon (coho and chinook) and trout (age 1+) are more evenly distributed in terms of relative abundance (Figure 9.10b). In the Elk River, Oregon, unconstrained reaches contain approximately 15% of the total available habitat but accounted for 30% of the estimated juvenile anadromous salmonids. In southwestern Washington streams, Cupp

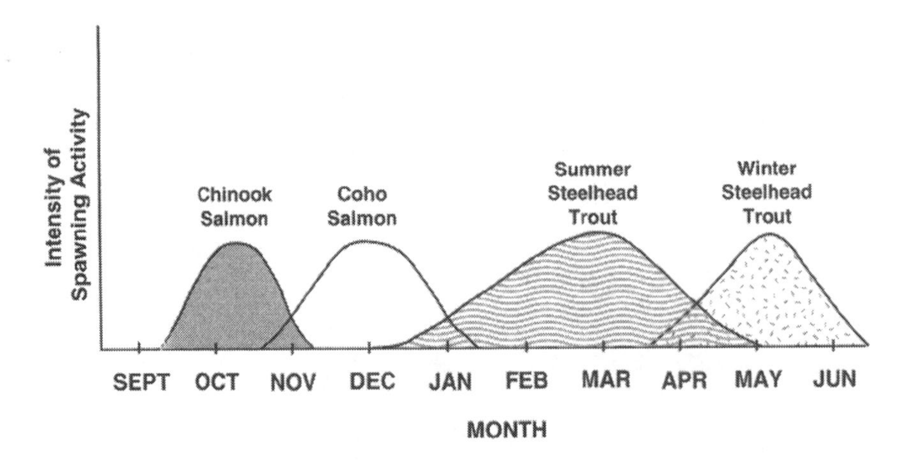

FIGURE 9.8. Spawning intensity and duration of anadromous salmonids in Fish Creek, Oregon.

(1989) found the greatest abundance of salmonids in lower elevation, lower gradient stream reaches with wide valleys. In unconstrained reaches of the McKenzie River headwaters in Oregon densities of cutthroat and rainbow trout are more than twice those in constrained reaches (Gregory et al. 1989).

This pattern, however, may not apply to lower portions of large watersheds. The relationship between reach features and salmonid communities varies with location in the stream network in Drift Creek, Oregon (drainage area 140 km^2) (Schwartz 1990). In the mainstem of Drift Creek there is no clear pattern of commu-

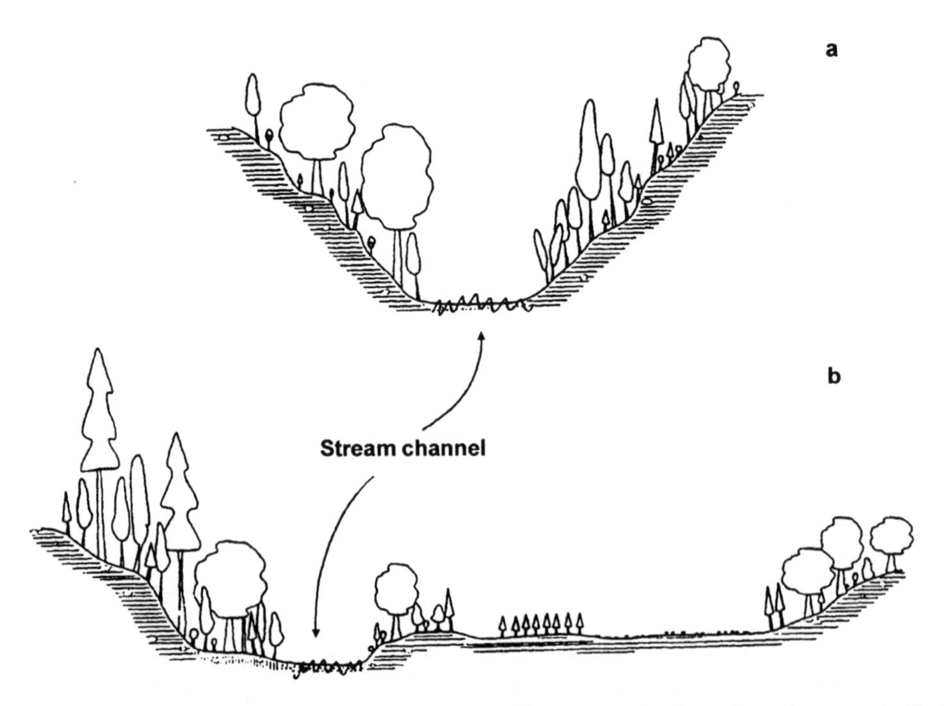

FIGURE 9.9. Generalized diagram of (a) constrained and (b) unconstrained reaches of streams in the Pacific coastal ecoregion.

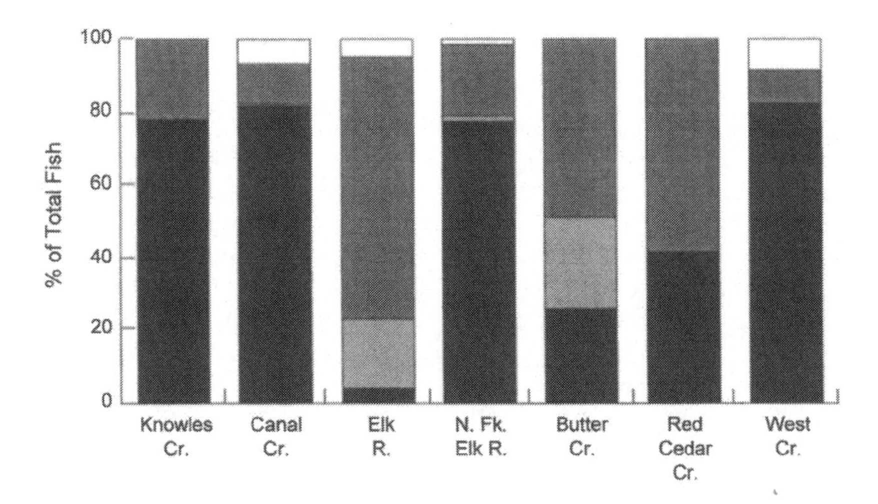

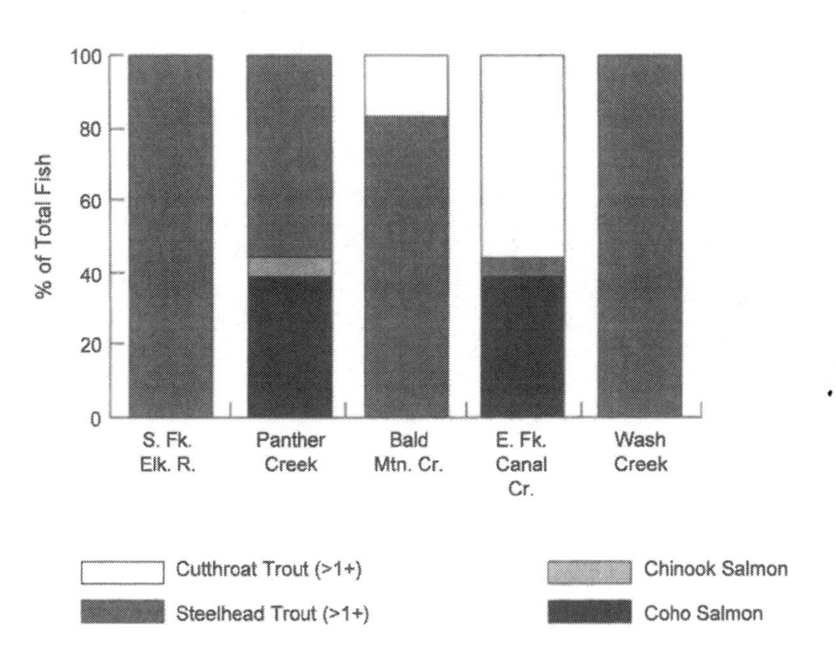

FIGURE 9.10. Composition of juvenile anadromous salmonid assemblages in (a) constrained and (b) unconstrained Oregon coastal streams.

nity structure with regard to reach types. Communities of juvenile anadromous salmonids in unconstrained and constrained reaches contained salmon and trout in about the same relative abundance. In tributaries, however, higher gradient constrained reaches are dominated by trout; whereas, lower gradient, unconstrained reaches contain salmon and trout.

Several factors explain the differences in the salmonid communities in the different reach types. Unconstrained reaches contain a greater diversity of habitat types than constrained reaches (Gregory et al. 1989, Schwartz 1990). Constrained reaches are typically dominated by fast water habitat and offer little refugia from high flow events, whereas, unconstrained reaches contain slow water (e.g., pools and side channels) and fast water (e.g., riffles) habitat and provide lateral refuges during floods.

Fish community diversity and production are positively correlated at the reach scale. Primary and secondary production are generally greater in unconstrained reaches than in constrained reaches. For example, primary production and densities of benthic macroinvertebrates are higher in unconstrained reaches of Elk River than in constrained reaches (Zucker 1993). Unconstrained reaches are also sites of greater hyporheic zone exchange (Grimm and Fisher 1984, Triska et al. 1989, Chapter 16) and greater hydraulic and particulate organic matter retention (Lamberti et al. 1989), factors which contribute to higher production and, therefore, a greater potential for diversity.

Habitat Unit Patterns of Diversity

Diversity of fish species is directly related to habitat unit features. There are two primary types of habitat units: riffles, which are topographic high points in the bed profile and are composed of coarser sediments, and pools which are low points with finer substrates (Richards 1982, O'Neill and Abrahams 1987). At base flows, riffles are shallow and have a steep water-surface gradient with rapid flow. In contrast, pools are deeper and generally have a gentle surface slope with slower flow (O'Neill and Abrahams 1987, Richards 1978).

Riffles and pools do not always have clear boundaries, but they are distinct ecological habitats. Fish inhabiting them differ markedly in taxonomic composition and morphological, physiological, and behavioral traits. Riffle dwellers (e.g., sculpins, dace, and age 0+ trout) are bottom-oriented fish, possessing large pectoral fins to help maintain position. Some lack an air bladder or can adjust the air in the swim bladder to reduce buoyancy. Riffle dwellers are solitary or part of small loose knit groups. Pool dwellers (e.g., coho salmon), often found in aggregations, are more active swimmers with more dorsally-ventrally compressed bodies and smaller fins.

Within a habitat unit, structural features, substrate, flow velocity, and pool depth influence biotic diversity (Sheldon 1968, Evans and Noble 1979, Angermeier 1987). Increased complexity resulting from the combination of these factors creates a greater array of microhabitats. Complexity can mediate competition between species. Structural complexity provides protection from predators, alters foraging efficiency (Wilzbach 1985), and influences social interactions (Fausch and White 1981, Glova 1986). In a Washington stream, Lonzarich and Quinn (1995) observed a general increase in species diversity with increasing complexity of pools and different responses of species to habitat features. Numbers of juvenile coho salmon, steelhead trout (age 1+), and cutthroat trout were directly correlated with depth; however, sculpin (*C. aleuticus*) did not respond to changes in habitat features. A similar pattern for salmonids and habitat features has been observed in a small coastal Oregon stream (D.H. Olson, personal communication, USDA Forest Service, Pacific Northwest Research Station, Corvallis, Oregon). In that stream, salmonid diversity (Shannon–Weiner Index, see Chapter 17) increased with maximum pool depth, pool surface area, and volume of wood. Community diversity increases during the summer with increasing levels of habitat complexity (S. Feith and S.V. Gregory, personal communication, Department of Fisheries and Wildlife, Oregon State University, Corvallis, Oregon). In another small stream in western Oregon, biomass (an indicator of abundance) of speckled dace, sculpins (*C. perplexus* and *C. gulosus*), and juvenile cutthroat trout also increased with increasing levels of habitat complexity. However, biomass of coho salmon showed no response to the changes in habitat complexity even though habitat complexity may influence coho density at other seasons (Nickelson et al. 1992, Quinn and Peterson 1996) and life history stages (McMahon and Holtby 1992).

Complexity within habitat units also influences the diversity of fish assemblages (Gorman and Karr 1978, Schlosser 1982, Angermeier and Karr 1984). Reduced habitat complexity, resulting primarily from the loss of large wood, in part, explains differences in juvenile anadromous salmonid communities in coastal Oregon streams with varying levels of timber harvest activity (Reeves et al. 1993). Communities in streams with reduced habitat complexity are less diverse than those with higher habitat complexity. In addition, interactions between coho salmon and steelhead and cutthroat trout may be altered as a result of habitat simplification. Further, changes in microhabitat features favors some species but decreases suitability for others. For example, densities of fish decrease in southeastern Alaska and mid-western streams when habitat structure is removed or simplified (Dolloff 1986, Elliott 1986, Berkman and Rabeni 1987).

Location of a habitat unit in the stream network also influences community diversity within the unit. As discussed previously, diversity of stream fish communities tends to increase downstream (Figures 9.2 and 9.3). A major reason for this is that pool depth also increases creating, among other things, a greater array of microhabitats which results in the addition of species (Sheldon 1968, Schlosser 1987). However, at present the role of habitat unit size, complexity, and location is only understood for small streams.

The influence of biological factors, such as competition and predation, on the structure and composition of fish communities in habitat units of streams is largely unknown at present. Studies that have examined interactions between species have focused on salmonids. Surprisingly, virtually nothing is known about interactions between salmonids and nonsalmonids, or among nonsalmonids, in the Pacific coastal ecoregion. Researchers in other regions have found that the influence of competition on community diversity varies depending on local conditions (Fausch and White 1981, Matthews 1982, Schlosser and Toth 1984, Moyle and Vondracek 1985, Grossman and Freeman 1987, Dolloff and Reeves 1990, Grossman and Boule 1991).

Two forms of interactions have been identified for stream fishes, interactive and selective segregation (Nilsson 1967). In *interactive segregation*, species are capable of using the same niche but one species is dominant and precludes the subordinate species from preferred habitats. The dominant species is generally more aggressive or more efficient at exploiting a particular resource. Thus, the subordinant species will move into the preferred habitat only when the dominant species is absent. Habitat use by juvenile coho salmon and trout (age 1+) is influenced, in part, by interactions among these species. Coho salmon are aggressive and preclude steelhead (Hartman 1965) and cutthroat trout (Glova 1978) from the head of pools, where food resources are highest. Steelhead trout dominate cutthroat trout and generally preclude them from habitats in larger stream systems (i.e., $>15\,km^2$) in British Columbia (Hartman and Gill 1968). Similar patterns of segregation have been observed between rainbow and cutthroat trout (Nilsson and Northcote 1981) and cutthroat and Dolly Varden (Andrusak and Northcote 1971, Hindar et al. 1988) in lakes. Reeves et al. (1987) found that habitat use by redside shiner and juvenile steelhead trout was determined by interactive segregation. Under cool water temperatures (i.e., $<20\,°C$), steelhead trout precluded shiners from riffles, where food is most abundant, by aggressively driving the shiners away. Shiners formed loose aggregations in pools in the presence of trout. When trout were absence, shiners moved to riffles.

In contrast to interactive segregation, *selective segregation* involves differential use of available resources by each species (Nilsson 1967). Each species uses resources not "selected" by the other, resulting in neutral interactions among groups. Differences in resource use can arise from differences in instinctive behavior or body morphology. For example, selective segregation between juvenile steelhead trout and chinook salmon reduces interactions for habitat and food in Idaho streams (Everest and Chapman 1972). The fish use similar habitats at any given size but, because of differences in life history features, interactions for space are minimal. In this case, because chinook salmon spawn in the fall and steelhead trout in the spring, the species are different in size; chinook salmon tend to be larger because of earlier time of emergence. Similar patterns occur between juvenile coho and chinook salmon in British Columbia (Lister and Genoe 1970). In small Alaskan streams, selective segregation results in habitat partitioning between juvenile coho salmon and Dolly Varden (Dolloff and Reeves 1990). Each species occupies similar habitats when alone and in the presence of the other species. Coho salmon occupy mid-water positions that are defended from other fish. Dolly Varden are more closely associated with the stream bottom and are seldom territorial. Differences in life history features, such as those outlined above, as well as differences in morphology, behavior, and physiology that lead to selective segregation, are genetically encoded over time (Nilsson 1967).

Human Impacts on Fish

The structure and composition of many native fish communities are modified by anthropogenic activities throughout the vast majority of Pacific coastal ecoregion. Several species of native fishes are extinct and many others are in need of special management considerations because of low or declining numbers (Williams et al. 1989, Nehlsen et al. 1991, Frissell 1993). More than 314 stocks of Pacific salmon are considered at moderate to high risk of extinction in coastal Washington, Oregon, and northern California (Forest Ecosystem Management Assessment Team 1993). In the same area, several resident fish, including bull trout (*Salvelinus confluentus*), Oregon chub (*Oregonichthys crameri*), and Olympic mudminnow (*Novumbra hubbsi*) are considered at risk because of low or declining populations (Williams et al. 1989). Slaney et al. (1996) found that more that 700 stocks of anadromous salmonids in British Columbia and the Yukon have a moderate to high risk of extinction. An additional 230 need special management consideration and 142 are already extinct. These stocks represent about 10% of the total estimated number of stocks found in that area. Habitat alteration and introduction of nonnative fishes are the most frequently cited factors associated with the losses and declines (Li et al. 1987, Hicks et al. 1991, Bisson et al. 1992).

Physical habitats in rivers of all sizes throughout the Pacific coastal ecoregion have been simplified by human activities (Hicks et al. 1991). Large river systems (e.g., the Willamette River in Oregon) have been extensively channelized and diked for flood control and transportation (Figure 9.11) (Sedell and Froggatt 1984). Consequently, secondary channels, backwaters, and oxbows, which are important habitats for many juvenile fishes have been lost. McIntosh et al. (1994) reported that large pools, which are important habitats for many species and age-classes of fish, declined more than 50% in larger streams in western Oregon and other parts of the Pacific Northwest in the last 50 years. Structural elements such as large wood and boulders, which create habitat and complexity, also have been reduced in many systems by various activities, including timber harvest, urbanization, agriculture, and livestock grazing (Bisson et al. 1992).

It is difficult to generalize about the response of fish communities to habitat alteration because responses vary with the individual situations. Some species may respond positively to environmental changes and others negatively. The result of this differential response is generally a decrease in diversity because of changes in relative evenness rather than loss of species richness. In the Willamette River, Oregon, channelization has contributed, in part, to the decline of Oregon chub because of loss of off-channel habitats (Li et al. 1987). However, other species such as prickly sculpin (*C. asper*), redside shiner, northern squawfish (*Ptychocheilus oregonensis*), and chiselmouth (*Arocheilus alutaceus*) have increased because of increased availability of food in and on rock revetments (Hjort et al. 1984, Li et al. 1987). The structure and composition of fish communities of a pristine and a channelized urban stream in the Puget Sound area of Washington were similar in terms of species richness but evenness was greater in the pristine stream (Scott et al. 1986) (Figure 9.12). Bisson and Sedell (1984) and Hicks (1990) noted similar

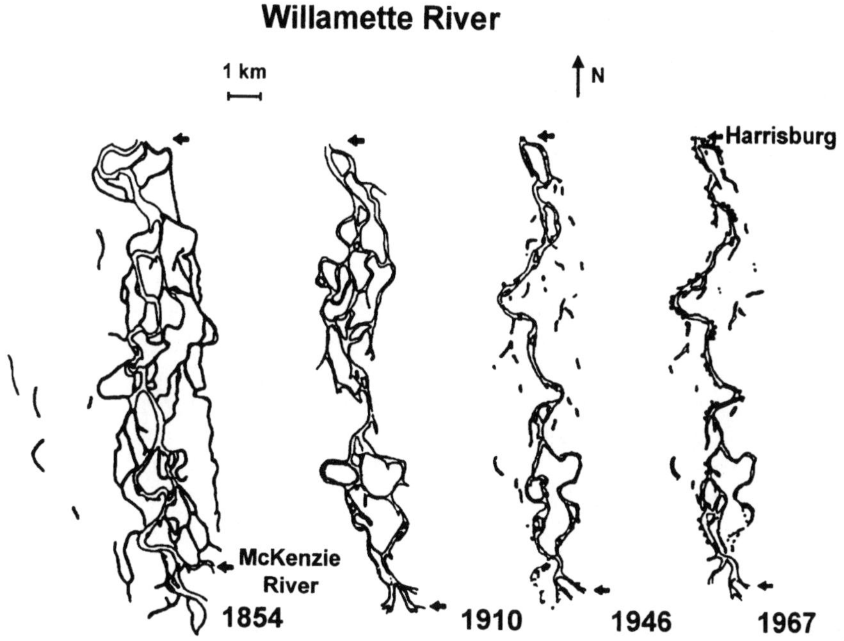

Willamette River

1 km

N

Harrisburg

McKenzie River

1854 1910 1946 1967

FIGURE 9.11. Changes in the floodplain and channel of the Willamette River, Oregon over 113 years as a result of channelization and diking (from Sedell and Froggatt 1984 with permission).

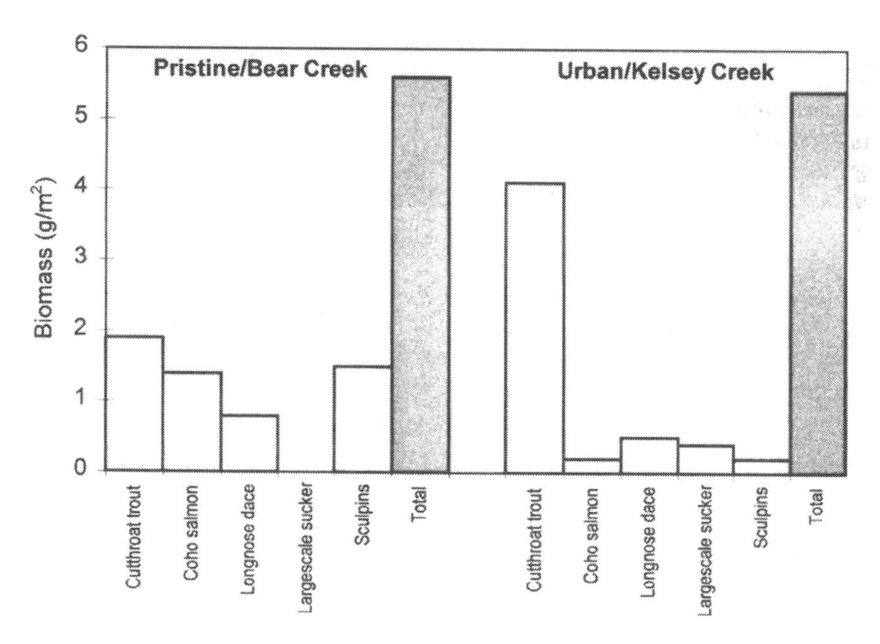

FIGURE 9.12. Composition of fish communities in a pristine stream (Bear Creek) and an urban stream (Kelsey Creek), Puget Sound, Washington (from Scott et al. 1986 with permission).

responses in communities of juvenile anadromous salmonids in streams in Washington and Oregon, respectively, that had been impacted by timber harvest. The diversity of assemblages of juvenile salmonids in coastal Oregon streams was greater in streams where less than 25% of the basin was subjected to timber harvest and associated activities compared to streams where more than 25% of the basin was harvested (Reeves et al. 1993). Again, the decrease in diversity resulted from changes in relative abundance rather than loss of species. Coho salmon generally increased in number whereas cutthroat trout declined. Similar responses to timber harvest have been observed in British Columbia (Hartman 1988, Holtby 1988, Scrivener and Brownlee 1989) and Oklahoma (Rutherford et al. 1987).

Differential Response of Species

Changes in community diversity resulting from the differential response to alterations just described are a consequence of specific environmental changes. For example, water temperatures mediate competitive interactions between redside shiner and juvenile steelhead trout (Reeves et al. 1987). At temperatures of 19 to 22°C, shiners displaced trout through exploitative competition (i.e., more efficiently obtaining food), whereas, trout are dominant at temperatures of 12–15°C because of interference competition (i.e., preventing access to food by defending territories). Dambacher (1991) attributed the distribution pattern of trout and shiners in Steamboat Creek, Oregon, to the changes in competitive interactions associated with changes in water temperature and reach gradient. Shiners dominate in low gradient, warm reaches, and steelhead dominate in cool reaches and warm reaches with high gradients. Water temperature mediates interactions for space and food between shiners and juvenile chinook salmon in the Wenatchee River, Washington (Hillman 1991) as well as competition for space between the riffle sculpin (*C. gulosus*) and the speckled dace in a California stream (Baltz et al. 1982).

Alteration of environmental conditions may reduce habitat suitability for some species but increases it for others. Though freshwater fish exist over a wide range of conditions, a narrow range is generally most favorable for a species (Larkin 1956). Relative abundances of the species in the community shift when conditions change; those favored by the new conditions increase and vice versa. This pattern prevails in juvenile anadromous salmonid assemblages in streams in coastal Oregon altered as a result of timber harvest activities (Reeves et al. 1993). Numbers of coho salmon decline in steeper gradient systems but numbers of trout (age 1+) decline in lower gradient systems. Loss of structure (primarily large woody debris) in higher-gradient streams results in the loss of slow-water habitat which is favorable to coho salmon. Coho salmon are better suited to slow water because their bodies are dorsally–ventrally compressed and their relatively large fins allow better maneuverability. Lack of pools and high-velocity areas associated with large woody debris in lower-gradient streams results in conditions less suitable to trout, particularly cutthroat. Trout are more cylindrical and thus, better suited for fast water (Bisson et al. 1988). Rutherford et al. (1987) attributed changes in fish communities in Oklahoma streams altered by timber harvest to the tolerance of species for environmental extremes; more tolerant species were able to maintain themselves better than less tolerant species.

Many exotic fishes (i.e., nonnative) have become established in many river systems in the Pacific coastal ecoregion as a result of changes in environmental conditions associated with anthropogenic activities. Consequently, some native communities have been altered as a result of competition with, or predation from, various exotic species. In the Columbia River, for instance, dams increase water temperatures, create large areas of slow water, and decrease the amount of riverine environment. Before the introduction (intentional and accidental) of exotic species, native predators (e.g., burbot [*Lota lota*], bull trout, cutthroat trout), which all have small mouths, foraged primarily on smaller fish species and age-classes (Figure 9.13a) (Li et al. 1987). Many exotic predators (e.g., smallmouth bass [*Micropterus dolomieui*], channel catfish [*Italurus*]) have larger jaws and

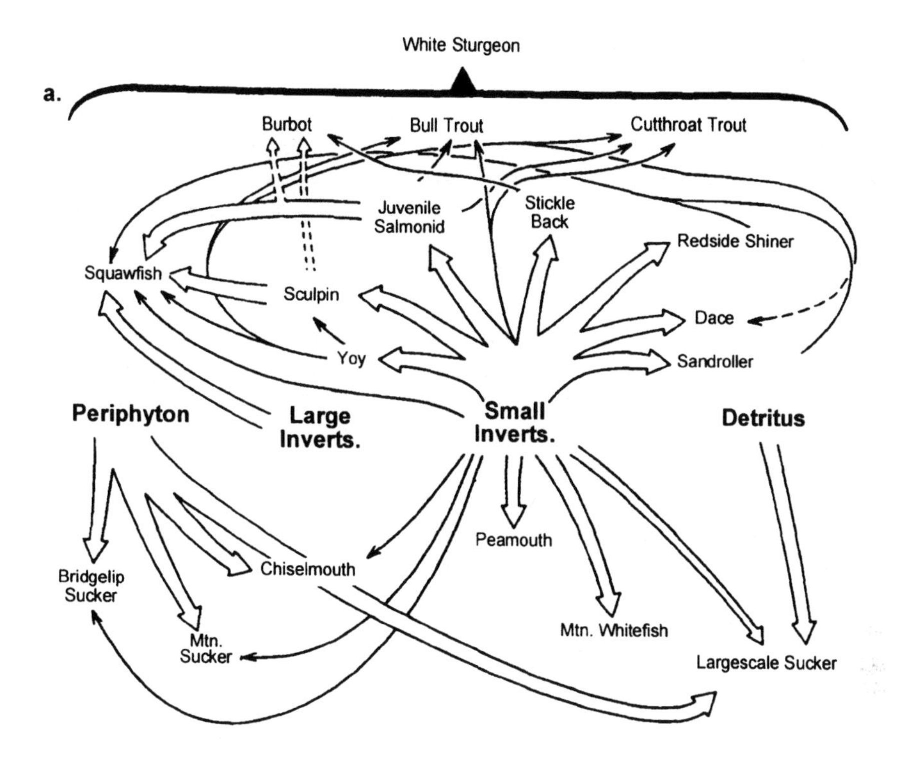

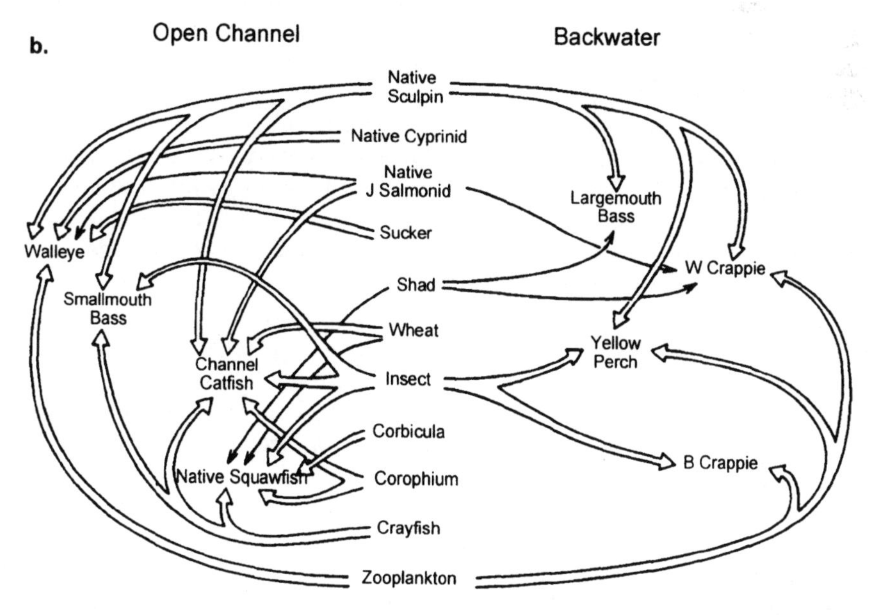

FIGURE 9.13. Hypothetical food webs of the middle and lower Columbia River (a) before 1800 and (b) following major changes resulting from human activity, ca. 1888–1983 (from Li et al. 1987 © University of Oklahoma Press with permission).

prey on larger individuals previously not affected by native predators. Also, unlike native predators, some exotic predators such as walleye (*Stizostedion vitreum*) are schooling nocturnal predators. Native salmonids (primarily smolts or immature individuals moving to the ocean) migrate at night to avoid predation by native fishes. However, walleye and other exotic fishes are efficient nocturnal feeders thereby reducing many native fish populations (Wydoski and Whitney 1979, Maule and Horton 1984) while becoming dominant components of the community (Figure 9.13b).

Competition with introduced species also causes a decline in native fishes. Wide-scale introduction of brook trout (*S. fontinalis*) has been implicated in the decline of native bull trout (Ratliff and Howell 1992). Brook trout and bull trout use similar food (Wallis 1948) and habitat (Dambacher et al. 1992) but brook trout reproduce at an earlier age and can numerically overwhelm bull trout. They also reduce the reproductive capacity of bull trout through hybridization (Leary et al. 1993). In addition, interactions with introduced competitors and predators, in conjunction with habitat loss and alteration, has contributed to the decline of Oregon chub in the Willamette River, Oregon (Pearsons 1989).

Disturbance

Disturbance strongly influences the structure and composition of biotic communities (Krebs 1994). Periodic disturbance creates conditions that allow some species to persist and that prevent competitively superior species from dominating communities, particularly at local scales (such as reaches or habitat units). The response of communities to disturbance depends, among other things, on the duration, intensity, and frequency of the disturbance. Yount and Neimi (1990) modified the definition of disturbance established by Bender et al. (1984) as "press" or "pulse." A *pulse disturbance*, such as flood or wildfire in an undisturbed system, allows an ecosystem to remain within a normal range of conditions or domains and to recover conditions that were present prior to disturbance. A *press disturbance* forces an ecosystem to a different set of conditions or domains. Yount and Neimi (1990) considered many anthropogenic activities to be press disturbances, for example, the construction of dams on the Columbia River. Dams alter the system from flowing to standing water and reduce variation in flow below the dams. These and other changes caused by dams have resulted in pronounced changes in the structure and composition of the fish community (Figure 9.13a and 9.13b) (Li et al. 1987). Many stream biota cannot recover from the effects of anthropogenic disturbances because, lacking an analogue in the natural disturbance regime, they may not have evolved the appropriate characteristics for recovery (Gurtz and Wallace 1984). Changes in environmental conditions are often too drastic and rapid for native fishes to adapt.

Modification of the type, frequency, or magnitude of natural disturbances alters fish communities (White and Pickett 1985, Hobbs and Huenneke 1992). Changes in the disturbance regime are manifested in extirpation of some species, increases in species favored by postdisturbance habitats, and invasion of exotics (Levin 1974, Harrison and Quinn 1989, Hansen and Urban 1992). Alteration of annual flooding on the Colorado River has contributed to the loss of many habitats and environmental conditions, and ultimately the decline of many unique native fishes (Tyus 1987, 1991). Additionally, dam construction has altered water temperatures. Spring water temperatures are lower because colder water is released from the bottom of the reservoirs. The effect is reduced reproductive success and decline of such native species as the razorback sucker (*Xyrauchen texanus*).

Streams throughout the Pacific coastal ecoregion operate differently from natural systems because of human disturbances (Li et al. 1987). Yet, the response of native fish communities in the Pacific coastal ecoregion to alteration of disturbance regimes, perhaps with the exception of those impacted by large dams, is poorly understood or appreciated at present. Nonetheless, the alteration of natural disturbance regimes has contributed to the widespread decline of anadromous salmonids (Reeves et al. 1995). Landscapes in the Pacific

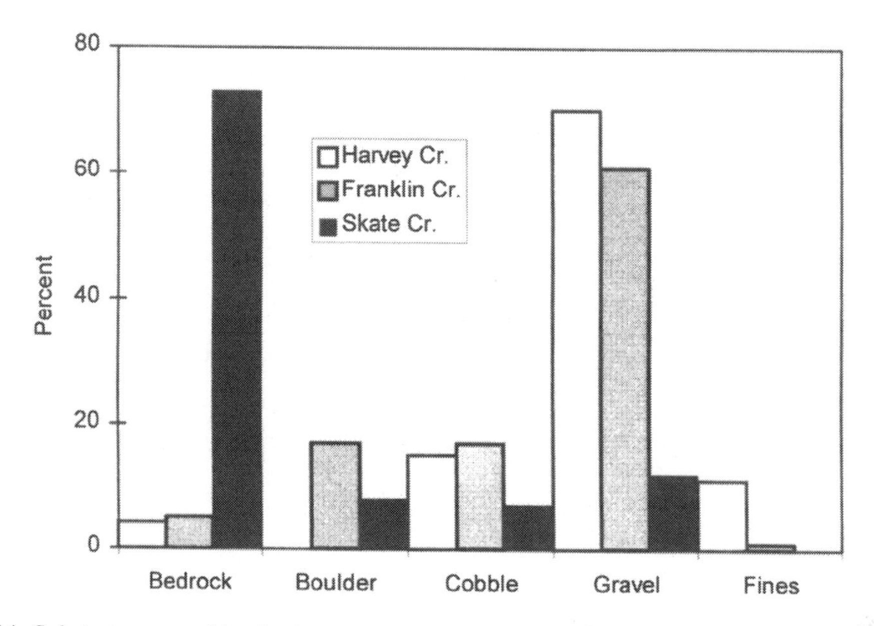

FIGURE 9.14. Substrate composition in three streams of the central Oregon Coast Range at different times since the last major natural disturbance. Time since the last major natural disturbance was 90–100 years for Harvey Creek, 160–180 years for Franklin Creek, and more than 3000 years for Skate Creek (from Reeves et al. 1995 with permission).

coastal ecoregion historically were dynamic in space and time (Naiman et al. 1992, Chapter 11). In the central coast of Oregon, for example, periodic catastrophic wildfire and hillslope failure occurred on average every 250 to 300 years (Benda 1994, Chapter 11) and were followed by periodic extirpations of local populations (Reeves et al. 1995). Over time, disturbed areas developed different conditions as a result of changes in the amount of sediment (Figure 9.14) and wood in the stream channel (Table 9.6). The landscape was a mosaic of habitat conditions in streams, similar to variation in successional states observed in terrestrial systems. The relative abundances of species changed over time (Table 9.6). Coho salmon were present at all times while trout were most abundant at intermediate times (160–180 yrs) from disturbance, when habitat conditions were most diverse.

Anadromous salmonids have evolved strategies (e.g., straying of adults and high fecundity) to persist in the dynamic landscapes and streams systems of the Pacific coastal ecoregion (Naiman et al. 1992, Reeves et al. 1995). Straying adults (which are relatively fecund) as well

TABLE 9.6. Composition of juvenile anadromous salmonids and mean number of pieces of large wood (mean diameter >0.3 m and lenght >3 m) per 100 m in three streams in the central Oregon Coast Range at different times since last major natural disturbance.

Stream	Mean number of pieces of large wood/100 m	Years since last major disturbance	Mean percent of estimated total numbers		
			Coho Salmon (Age 0)	Steelhead Trout (Age 1+)	Cutthroat Trout (Age 1+)
Harvey Creek	7.9	90–100	98.0	1.0	1.0
Franklin Creek	12.3	160–180	85.0	12.5	2.3
Skate Creek	23.5	>300	100.0	0.0	0.0

Modified from Reeves et al. 1995.

TABLE 9.7. Features of natural and human disturbance regimes.

	Natural disturbance	Human disturbance
Magnitude	High	Low-Moderate
Frequency	Low	High
Area Impacted	Small	Large
Degree to which Natural Processes Retained	Strong	Weak
Legacy of Disturbance	Complex	Simple

as moving juveniles can colonize newly disturbed areas. But human activities, such as timber harvest, agriculture, and urbanization, have altered the historical natural disturbance regime (Table 9.7). Affected landscapes are less heterogeneous than the natural systems and disturbed more frequently and over larger areas. As a result, the range and type of conditions in aquatic systems are much simpler and thus do not support the most diverse communities (Reeves et al. 1995).

The differential response of native fishes to anthropogenic disturbances and the increasing dominance of exotic species in Pacific coastal ecoregion rivers and streams suggests that press disturbances have become a major influence on fish communities and aquatic ecosystems. It is unlikely that native fish communities will recover unless the role of natural disturbance in structuring these communities is better regulated or mimicked (Reeves et al. 1995).

Future Management Directions

Although managing for biodiversity is currently the objective of many management plans in the Pacific coastal ecoregion, it is not exactly clear what biodiversity means. Diversity has two components, species richness (i.e., numbers) and relative abundance (see Chapter 17 for more detail) and management plans for each component may be very different.

A focus on species richness without concern for relative abundance of other species results in development of conditions more favorable for the species of interest at the expense of other species. Each species has a specific range of conditions over which it performs best (Larkin 1956). It is unlikely that this range overlaps exactly for any two species. Thus, creating optimal conditions for a particular species may result in conditions less favorable for other species. Consequently, what are perceived to be relatively small changes in environmental conditions can reduce the abundance of nontarget species and have significant effects on the structure and composition of stream fish communities. Focusing on relative abundance requires management plans which develop a range of conditions suitable for a suite of species. This may mean that maximum abundance or production of any particular species is not reached. Clearly, it is important to identify the type of diversity desired before developing management plans.

Until recently, habitat management decisions were premised on the belief that habitats could be manipulated with technology to benefit fish (Sedell et al. 1997). This perspective has changed, partly because of the awareness of the continued trends toward widespread habitat simplification. Management of fish communities has shifted towards an ecosystem approach (Frissell et al. 1997, Lichatowich 1997). Although the definition of ecosystem management is still evolving, aspects related to fish communities include consideration of natural disturbance regimes, landscape contexts, and temporal and spatial dimensions matched to the dynamics of native fish populations (Bisson et al. 1997). Integration of macro-scale physical and biological processes into a broader spatial and temporal ecosystem framework is necessary to explain the dynamics of stream fish communities (Power et al. 1988) and to recover and maintain native fish communities in the Pacific coastal ecoregion.

Literature Cited

Andrusak, H., and T.G. Northcote. 1971. Segregation between adult cutthroat trout (*Salmo clarki*) and Dolly Varden (*Salvelinus malma*) in small coastal British Columbia lakes. Journal of the Fisheries Research Board of Canada **28**:1259–1268.

Angermeier, P.L. 1987. Spatiotemporal variation in habitat selection by fishes in small Illinois streams. Pages 52–60 *in* W.J. Matthews and D.C. Heins, eds. Community and evolutionary ecology in North American stream fishes. University of Oklahoma Press, Norman, Oklahoma, USA.

Angermeier, P.L., and J.R. Karr. 1984. Relationship between woody debris and fish habitat in small warmwater streams. Transactions of the American Fisheries Society **113**:716–726.

Baltz, D.M., P.B. Moyle, and N.J. Knight. 1982. Competitive interactions between benthic fishes, riffle sculpin, *Cottus gulosus*, and speckled dace, *Rhinichthys osculus*. Canadian Journal of Fisheries and Aquatic Sciences **39**:1502–1511.

Beacham, T.D., and C.B. Murray. 1987. Adaptive variation in body size, age, morphology, egg size, and developmental biology of chum salmon (*Oncorhynchus keta*) in British Columbia. Canadian Journal of Fisheries and Aquatic Sciences **44**:244–261.

Behnke, R.J. 1992. Native trout of western North America. American Fisheries Society Monograph 6, Washington, DC, USA.

Bell, M.A.1984. Evolutionary phenetics and genetics. Pages 431–528 *in* B.J. Turner, ed. Evolutionary genetics of fishes. Plenum, New York, New York, USA.

Benda, L.E. 1994. Stochastic geomorphology in a humid mountain landscape. Ph.D. dissertation. University of Washington, Seattle, Washington, USA.

Bender, E.A., T.J. Case, and M.E. Gilpin. 1984. Perturbation experiments in community ecology: Theory and practice. Ecology **65**:1–13.

Berkman, H., and C.F. Rabeni. 1987. Effect of siltation on stream fish communities. Environmental Biology of Fishes **18**:285–294.

Bilby, R.E., B.R. Fransen, and P.A. Bisson. 1996. Incorporation of nitrogen and carbon from spawning coho salmon into the trophic system of small streams: Evidence from stable isotopes. Canadian Journal of Fisheries and Aquatic Sciences **53**:164–173.

Bisson, P.A., G.H. Reeves, R.E. Bilby, and R.J. Naiman. 1997. Watershed management and Pacific salmon: Desired future conditions. Pages 447–474 *in* D.J. Stouder, P.A. Bisson, and R.J. Naiman, eds. Pacific salmon and their ecosystems: Status and future options. Chapman & Hall, New York, New York, USA.

Bisson, P.A., and P.E. Reimers. 1977. Geographic variation among Pacific Northwest populations of the longnose dace (*Rhinichthys cataractae*). Copeia **1977**:518–522.

Bisson, P.A., and J.R. Sedell. 1984. Salmonid populations in streams in clearcut vs. old-growth forests of western Washington. Pages 121–129 *in* W.R. Meehan, T.R. Merrell, Jr., and T.A. Hanely, eds. Fish and wildlife relationships in old-growth forests: Proceedings of a symposium. American Institute of Fishery Biologists, Morehead City, North Carolina, USA.

Bisson, P.A., K. Sullivan, and J.L. Nielsen. 1988. Channel hydraulics, habitat use, and body form of juvenile coho salmon, steelhead trout, and cutthroat trout. Transactions of the American Fisheries Society **117**:262–273.

Bisson, P.A., T.P. Quinn, G.H. Reeves, and S.V. Gregory. 1992. Best management practices, cumulative effects, and long-term trends in fish abundance in Pacific Northwest river systems. Pages 189–232. *in* R.J. Naiman, ed. Watershed management: Balancing sustainability and environmental change. Springer-Verlag, New York.

Bjornn, T.C., and D.W. Reiser. 1991. Habitat requirements of salmonids in streams. Pages 83–138. *in* W.R. Meehan, ed. Influence of forest and rangeland management on salmonid fishes and their habitats. American Fisheries Society Special Publication 19, Bethesda, Maryland, USA.

Bond, C.E. 1963. Distribution and ecology of freshwater sculpins, genus *Cottus*, in Oregon. Ph.D. dissertation. University of Michigan, Ann Arbor, Michigan, USA.

Boschung, H. 1987. Physical factors and the distribution and abundance of fishes of the Upper Tombigbee River system of Alabama and Mississippi, with emphasis on the Tennessee–Tombigbee Waterway. Pages 178–183 *in* W.J. Matthews and D.C. Heins, eds. Community and evolutionary ecology in North American stream fishes. University of Oklahoma Press, Norman, Oklahoma, USA.

Caswell, H., and J.E. Cohen. 1993. Local and regional regulation of species-area relationships: A patch-occupancy model. Pages 99–107 *in* R.E. Ricklefs and D. Schulter, eds. Species diversity in ecological communities. University of Chicago Press, Chicago, Illinois, USA.

Cederholm, C.J., D.B. Houston, D.L. Cole, and W.J. Scarlett. 1989. Fate of coho salmon (*Oncorhynchus kisutch*) carcasses in spawning streams. Canadian Journal of Fisheries and Aquatic Sciences **46**:1347–1355.

Connell, J.H. 1975. Some mechanisms producing structure in natural communities. Pages 460–490 *in* M. Cody and J.M. Diamond, ed. Ecology and evolution of communities. Harvard University Press, Cambridge, Massachusetts, USA.

——. 1978. Diversity in tropical rain forests and coral reefs. Science **199**:1302–1310.

Connell, J.H., and E. Orias. 1964. The ecological regulation of species diversity. American Naturalist **98**:399–414.

Cupp, C.E. 1989. Identifying spatial variability of stream characteristics through classification. Masters thesis. University of Washington, Seattle, Washington, USA.

Currie, D.J. 1991. Energy and large-scale patterns of animal- and plant-species richness. American Naturalist **137**:27–49.

Dambacher, J.M. 1991. Distribution, abundance, and emigration of juvenile steelhead (*Oncorhynchus mykiss*) and analysis of stream habitat in the Steamboat Creek basin, Oregon. M.Sc. thesis, Oregon State University, Corvallis, Oregon, USA.

Dambacher, J.M., M.W. Buktenica, and G.L. Larson. 1992. Distribution, abundance, and habitat utilization of bull and brook trout in Sun Creek, Crater Lake National Park, Oregon. Pages 30–36 *in* P.J. Howell and D.V. Buchanan, eds. Proceedings of the Gearhart Mountain bull trout workshop. Oregon Chapter of the American Fisheries Society, Corvallis, Oregon, USA.

Dolloff, C.A. 1986. Effects of stream cleaning on juvenile coho salmon and Dolly Varden in southeast Alaska. Transactions of the American Fisheries Society **115**:743–755.

Dolloff, C.A., and G.H. Reeves. 1990. Microhabitat partitioning among stream-dwelling juvenile coho salmon, *Oncorhynchus kisutch*, and Dolly Varden, *Salvelinus malma*. Canadian Journal of Fisheries and Aquatic Sciences **47**:2297–2306.

Elliot, S.T. 1986. Reductions of Dolly Varden and macrobenthos after removal of logging debris. Transactions of the American Fisheries Society **115**:392–400.

Evans, J.W., and R.L. Noble. 1979. The longitudinal distribution of fishes in an east Texas stream. American Midland Naturalist **101**:333–343.

Everest, F.H., and D.W. Chapman. 1972. Habitat selection and spatial interaction by juvenile chinook salmon and steelhead trout in two Idaho streams. Journal of the Fisheries Research Board of Canada **29**:91–100.

Fausch, K.D., and R.J. White. 1981. Competition between brook trout (*Salvelinus fontinalis*) and brown trout (*Salmo trutta*) for position in a Michigan stream. Canadian Journal of Fisheries and Aquatic Sciences **38**:1220–1227.

Forest Ecosystem Management Assessment Team. 1993. Forest ecosystem management: An ecological, economic, and social assessment. Report of the Forest Ecosystem Management Assessment Team. USDA Forest Service, Portland, Oregon, USA.

Frissell, C.A. 1993. Topology of extinction and endangerment of native fishes in the Pacific Northwest and California (USA). Conservation Biology **7**:342–354.

Frissell, C.A., W.J. Liss, R.E. Greswell, R.K. Nawa, and J.L. Ebersole. 1997. A resource in crisis: Changing the measure of salmon measurement. Pages 411–444 *in* D.J. Stouder, P.A. Bisson, and R.J. Naiman, eds. Pacific salmon and their ecosystems: Status and future options. Chapman & Hall, New York, New York, USA.

Frissell, C.A., W.J. Liss, C.E. Warren, and M.D. Hurley. 1986. A hierarchical framework for stream habitat classification: Viewing streams in a watershed context. Environmental Management **10**:199–214.

Gharrett, A.J., S.M. Shirley, and G.R. Tromble. 1987. Genetic relationships among populations of Alaskan chinook salmon (*Oncorhynchus tshawytscha*). Canadian Journal of Fisheries and Aquatic Sciences **44**:765–774.

Glova, G.J. 1978. Pattern and mechanism of resource partitioning between stream populations of juvenile coho salmon (*Oncorhynchus kisutch*) and coastal cutthroat trout (*Salmo clarki clarki*). Ph.D. dissertation. University of British Columbia, Vancouver, British Columbia, Canada.

——. 1986. Interaction for food and space between experimental populations of juvenile coho salmon (*Oncorhynchus kisutch*) and coastal cutthroat trout (*Salmo clarki*) in a laboratory stream. Hydrobiologia **131**:155–168.

Gorman, O.T., and J.R. Karr. 1978. Habitat structure and stream fish communities. Ecology **59**: 507–515.

Grant, G.E., F.J. Swanson, and M.G. Wolman. 1990. Pattern and origin of stepped-bed morphology in high-gradient streams, western Cascades, Oregon. Geological Society of America Bulletin **102**:340–352.

Gregory, S.V., G.A. Lamberti, and K.M.S. Moore. 1989. Influence of valley floor land forms on stream ecosystems. Pages 3–8 *in* D.L. Abell, ed. Proceedings of the California riparian systems conference: Protection, management, and restoration for the 1990s. Proceedings on 22–24 September, 1988. Davis, California. USDA. Forest Service General Technical Report PSW-110,

Pacific Southwest Forest and Range Experiment Station, Berkeley, California, USA.

Grimm, N.B., and S.G. Fisher. 1984. Exchange between interstitial and surface water: Implications for stream metabolism and nutrient cycling. Hydrobiologia **111**:219–228.

Groot, C., and L. Margolis, eds. 1991. Pacific salmon life histories. University of British Columbia Press, Vancouver, British Columbia, Canada.

Gross, M.R., R.M. Coleman, and R.M. McDowell. 1988. Aquatic productivity and the evolution of diadromous fish migration. Science **239**:1291–1293.

Grossman, G.D., and V. Boule. 1991. Effects of roseyside dace (*Clinostomus funduloides*) on microhabitat use of rainbow trout (*Oncorhynchus mykiss*). Canadian Journal of Fisheries and Aquatic Sciences **48**:1235–1243.

Grossman, G.D., and M.C. Freeman. 1987. Microhabitat use in a stream fish assemblage. Journal of Zoology **212**:151–176.

Grossman, G.D., P.B. Moyle, and J.O. Whitaker, Jr. 1982. Stochasticity in structural and functional characteristics of an Indiana stream fish assemblage: A test of community theory. American Naturalist **120**:423–454.

Gurtz, M.E., and J.B. Wallace. 1984. Substrate-mediated response of invertebrates to disturbance. Ecology **65**:1556–1569.

Hagen, D.W., and L.G. Gilbertson. 1972. Geographic variation and environmental selection in *Gasterosteus aculeatus* in the Pacific Northwest, America. Evolution **26**:32–51.

Hansen, A.J., and D.L. Urban. 1992. Avian response to landscape patterns: The role of life histories. Landscape Ecology **76**:163–180.

Harrison, S., and J.F. Quinn. 1989. Correlated environments and the persistence of metapopulations. Oikos **56**:292–298.

Hartman, G.H. 1965. The role of behavior in the ecology and interaction of underyearling coho salmon (*Oncorhynchus kisutch*) and steelhead trout (*Salmo gairdneri*). Journal of the Fisheries Research Board of Canada **22**:1035–1081.

——. 1988. Preliminary comments on results of studies of trout biology and logging impacts in Carnation Creek. Pages 175–180 *in* T.W. Chamberlain, ed. Proceedings of the workshop applying 15 years of Carnation Creek results. Pacific Biological Station, Carnation Creek Steering Committee, Nanaimo, British Columbia, Canada.

Hartman, G.H., and C.A. Gill. 1968. Distributions of juvenile steelhead and cutthroat trout (*Salmo gairdneri* and *S. clarki*) within streams in southwestern British Columbia. Journal of the Fisheries Research Board of Canada **22**:33–48.

Healey, M.C. 1991. Life history of chinook salmon (*Oncorhynchus tshwaytscha*). Pages 311–394 *in* C. Groot and L. Margolis, eds. Pacific salmon life histories. University of British Columbia Press, Vancouver, British Columbia, Canada.

Healey, M.C., and A. Prince. 1995. Scales of variation in life history tactics of Pacific salmon and the conservation of phenotype and genotype. Pages 176–184 *in* J. Nielsen, ed. Evolution and the aquatic ecosystem: Defining unique units in population conservation. American Fisheries Society Symposium 17, Bethesda, Maryland, USA.

Hicks, B.J. 1990. The influence of geology and timber harvest on channel geomorphology and salmonid populations in Oregon Coast Range streams. Ph.D. dissertation. Oregon State University, Corvallis, Oregon, USA.

Hicks, B.J., J.D. Hall, P.A. Bisson, and J.R. Sedell. 1991. Response of salmonid populations to habitat changes caused by timber harvest. Pages 483–518 *in* W.R. Meehan, ed. Influence of forest and rangeland management on salmonid fishes and their habitats. American Fisheries Society Special Publication 19, Bethesda, Maryland, USA.

Hillman, T. 1991. The effect of temperature on the spatial interaction of juvenile chinook salmon and the redside shiner and their morphological differences. Ph.D. dissertation. Idaho State University, Pocatello, Idaho, USA.

Hindar, K., B. Jonsson, J.H. Andrew, and T.G. Northcote. 1988. Resource utilization of sympatric and experimentally allopatric cutthroat trout and Dolly Varden charr. Oecologia **74**:481–491.

Hjort, R.C., P.L. Hulett, L.D. LaBolle, and H.W. Li. 1984. Fish and invertebrates of revetments and other habitats in the Willamette River, Oregon. United States Army Engineers EWQOS Technical Report E-84-9, Waterways Experiment Station, Portland, Oregon, USA.

Hobbs, R.J., and L.F. Huenneke. 1992. Disturbance, diversity, and invasion: Implications for conservation. Conservation Biology **6**:324–337.

Holt, R.D. 1993. Ecology at the mesoscale: The influence of regional processes on local communities. Pages 77–89 *in* R.E. Ricklefs and D. Schulter, eds. Species diversity in ecological communities. University of Chicago Press, Chicago, Illinois, USA.

Holtby, L.B. 1988. Effects of logging on stream temperatures in Carnation Creek, British Columbia, and associated impacts on coho salmon (*Oncorhynchus kisutch*). Canadian Journal of Fisheries and Aquatic Sciences **45**:502–515.

Horwitz, R.J. 1978. Temporal variability patterns and the distributional patterns of stream fishes. Ecological Monographs **48**:307–321.

Huston, M.A. 1994. Letters to Science: Biological diversity and agriculture. Science **265**:458–459.

Kan, T.T. 1975. Systematics, variation and distribution of lampreys of the genus *Lampetra* in Oregon. Ph.D. dissertation. Oregon State University, Corvallis, Oregon, USA.

Karr, J.R., and K.E. Freemark. 1983. Habitat selection and environmental gradients: Dynamics in the "stable" tropics. Ecology **64**:1481–1494.

——. 1985. Disturbance and vertebrates: An integrative perspective. Pages 153–168 *in* S.T.A. Pickett and P.S. White, eds. Natural disturbance: The patch dynamic perspective. Academic Press, New York, New York, USA.

Krebs, C.J. 1994. Ecology: The experimental analysis of distribution and abundance, fourth edition. HarperCollins, New York, New York, USA.

Larkin, P.A. 1956. Interspecific competition and population control in freshwater fish. Journal of the Fisheries Research Board of Canada **13**:327–342.

Lamberti, G.A., S.V. Gregory, L.R. Ashkenas, R.C. Wildman, and C.D. McIntire. 1989. Influence of channel geomorphology on retention of dissolved and particulate matter in a Cascade Mountain stream. Pages 33–39 *in* D.L. Abell, ed. Proceedings of the California riparian systems conference: Protection, management, and restoration for the 1990s. Proceedings on 22–24 September, 1988. Davis, California. USDA Forest Service General Technical Report PSW-110, Pacific Southwest Forest and Range Experiment Station, Berkeley, California, USA.

Leary, R.F., F.W. Allendorf, and S.H. Forbes. 1993. Conservation genetics of bull trout in the Columbia and Klamath River drainages. Conservation Biology **7**:856–865.

Levin, S.A. 1974. Dispersion and population interactions. American Naturalist **104**:413–423.

Li, H.W., C.B. Schreck, C.E. Bond, and E. Rexstad. 1987. Factors influencing changes in fish assemblages of Pacific streams. Pages 193–202 *in* W.J. Matthews and D.C. Heins, eds. Community and evolutionary ecology in North American stream fishes. University of Oklahoma Press, Norman, Oklahoma, USA.

Lichatowich, J. 1997. Evaluating salmon management institutions: The importance of performance measurements, temporal scales, and production cycles. Pages 69–87 *in* D.J. Stouder, P.A. Bisson, and R.J. Naiman, eds. Pacific salmon and their ecosystems: Status and future options. Chapman & Hall, New York, New York, USA.

Lister, D.B., and H.S. Genoe. 1970. Stream habitat utilization by cohabitating underyearling chinook salmon (*Oncorhynchus tshawytscha*) and coho salmon (*O. kisutch*) salmon in the Big Quilicum River, British Columbia. Journal of the Fisheries Research Board of Canada **27**:1215–1224.

Lonzarich, D.G. 1994. Stream fish communities in Washington: Pattern and processes. Ph.D. dissertation. University of Washington, Seattle, Washington, USA.

Lonzarich, D.G., and T.P. Quinn. 1995. Experimental evidence for the effect of depth and structure on the distribution, growth, and survival of stream fishes. Canadian Journal of Zoology **73**:2223–2230.

McAllister, D.E., and C.C. Lindsey. 1961. Systematics of the freshwater sculpins (*Cottus*) of British Columbia. Bulletin of the National Museum of Canada **172**:66–89.

McArthur, R.H. and E.R. Pianca. 1966. On the optimal use of a patchy environment. American Naturalist **100**:603–609.

McIntosh, B.M., J.R. Sedell, J.E. Smith, R.C. Wissmar, S.E. Clarke, and G.H. Reeves. 1994. Historical changes in fish habitat for selected river basins of eastern Oregon and Washington. Northwest Science **68**(special issue):36–53.

McMahon, T.E., and L.B. Holtby. 1992. Behavior, habitat use, and movements of coho salmon (*Oncorhynchus kisutch*) during seaward migration. Canadian Journal of Fisheries and Aquatic Sciences **49**:1478–1485.

McPhail, J.D., and C.C. Lindsey. 1986. Zoogeography of the freshwater fishes of Cascadia (the Columbia systems and rivers north to the Stikine). Pages 615–637 *in* C.H. Hocutt and E.O. Wiley, eds. The zoogeography of North American freshwater fishes. John Wiley & Sons, New York, New York, USA.

Mahon, R. 1984. Divergent structure in fish taxocenes of north temperate streams. Canadian Journal of Fisheries and Aquatic Sciences **41**:330–350.

Mann, R.H.K., C.A. Mills, and D.T. Crisp. 1984. Geographical variation in life history tactics of some species of freshwater fish. Pages 171–186 *in* R. Wooten, ed. Reproduction. Academic Press, New York, New York, USA.

Matthews, W.J. 1982. Small fish community structure in Ozark streams: Structured assembly patterns or random abundances. American Midland Naturalist **107**:42–54.

Maule, A.G., and H.F. Horton. 1984. Feeding ecology of walleye, *Stizostedion vitreum vitreum*, in the mid-Columbia River, with emphasis on the interactions between walleye and juvenile anadromous salmonids. Fisheries Bulletin **82**:411–418.

Menge, B.A. and O.M. Olson. 1990. Role of scale and environmental factors in regulating community structure. Trends in Ecology and Evolution **5**:52–57.

Miller, P. 1979. Adaptations and implications of small size in teleosts. Symposium of the Zoological Society of London **4**:263–306.

Miller, R.R. 1959. Origin and affinities of the freshwater fish fauna of western North America. Pages 187–222 *in* C.L. Hubbs, ed. Zoogeography. American Association for the Advancement of Science Publication 51. Washington, DC, USA.

——. 1961. Speciation rates of some freshwater fishes of western North America. Pages 537–560 *in* Vertebrate speciation. University of Texas Press, Austin, Texas, USA.

Miller, R.R., and C.L. Hubbs. 1969. Systematics of *Gasterosteus aculeatus* with particular reference to intergradation and introgression along the Pacific coast of North America: A commentary on a recent contribution. Copeia **1969**:52–69.

Minckley, W.L., D.A. Hendrickson, and C.E. Bond. 1986. Geography of western North American freshwater fishes: Description and relationship to intracontinental tectonism. Pages 519–614 *in* C.H. Hocutt and E.O. Wiley, eds. The zoogeography of North American freshwater fishes. John Wiley & Sons, New York, New York, USA.

Moyle, P.B. 1976. Inland fishes of California. University of California Press, Berkeley, California, USA.

Moyle, P.B., and J.J. Cech, Jr. 1982. Fishes: An introduction to ichthyology. Prentice-Hall, Englewood Cliffs, New Jersey, USA.

Moyle, P.B., and B. Herbold. 1987. Life history patterns and community structure in stream fishes of western North America: Comparisons with eastern North America and Europe. Pages 25–32 *in* W.J. Matthews and D.C. Heins, eds. Community and evolutionary ecology in North American stream fishes. University of Oklahoma Press, Norman, Oklahoma, USA.

Moyle, P.B. and B. Vondracek. 1985. Persistence and structure of the fish assemblage in a small California stream. Ecology **66**:1–13.

Naiman, R.J., T.J. Beechie, L.E. Benda, D.R. Berg, P.A. Bisson, L.H. McDonald, et al. 1992. Fundamental elements of ecologically healthy watersheds in the Pacific Northwest coastal ecoregion. Pages 127–188 *in* R.J. Naiman, ed. Watershed management: Balancing sustainability and environmental change. Springer-Verlag, New York, New York, USA.

Nehlsen, W., J.E. Williams, and J.E. Lichatowich. 1991. Pacific salmon at the crossroads: Stocks at risk from California, Oregon, Idaho, and Washington. Fisheries **16**(2):4–21.

Nickelson, T.E., J.D. Rogers, S.L. Johnson, and M.F. Solazzi. 1992. Seasonal changes in habitat use by juvenile coho salmon (*Oncorhynchus kisutch*) in Oregon coastal streams. Canadian Journal of Fisheries and Aquatic Sciences **49**:783–789.

Nilsson, N.A. 1967. Interactive segregation between fish species. Pages 295–313 *in* S.D. Gerking, ed. The biological basis of freshwater fish production. Blackwell Scientific, Oxford, UK.

Nilsson, N.A., and T.G. Northcote. 1981. Rainbow trout (*Salmo gairdneri*) and cutthroat trout (*S. clarki*) interactions in coastal British Columbia lakes. Canadian Journal of Fisheries and Aquatic Sciences **38**:1228–1246.

NRC (National Research Council). 1996. Upstream: Salmon and society in the Pacific Northwest. National Academy Press, Washington, DC, USA.

O'Neil, M.P., and A.D. Abrahams. 1987. Objective identification of pools and riffles. Water Resources Research **20**:921–926.

Pearsons, T.N. 1989. Ecology and decline of a rare western minnow: The Oregon chub (*Oregonichthys crameri*). Masters thesis. Oregon State University, Corvallis, Oregon, USA.

Poff, N.L., and J.D. Allen. 1995. Functional organization of stream fish assemblages in relationship to hydrological variability. Ecology **76**:606–627.

Power, M.E., and W.J. Matthews. 1983. Algae-grazing minnows (*Campostoma anomalum*), piscivorous bass (*Micropterus* spp.), and the distribution of attached algae in a small prairie-margin stream. Oecologia **60**:328–332.

Power, M.E., R.J. Stout, C.E. Cushing, P.P. Harper, F.R. Hauer, W.J. Matthews, et al. 1988. Biotic and abiotic controls in river and stream communities. Journal of the North American Benthological Society **7**:503–524.

Quinn, T.P., and N.P. Peterson. 1996. The influence of habitat complexity and fish size on over-winter survival and growth of individually marked juvenile coho salmon (*Oncorhynchus kisutch*) in Big Beef Creek, Washington. Canadian Journal of Fisheries and Aquatic Sciences **53**:1555–1564.

Quinn, T.P., and M.J. Unwin. 1993. Variation in life history patterns among New Zealand chinook salmon (*Oncorhynchus tshawytscha*) populations.

Canadian Journal of Fisheries and Aquatic Sciences **50**:1414–1421.

Ratliff, D.E. and P.J. Howell. 1992. The status of bull trout populations in Oregon. Pages 10–17 *in* P.J. Howell and D.V. Buchanan, eds. Proceedings of the Gearhart Mountain bull trout workshop. Oregon Chapter of the American Fisheries Society, Corvallis, Oregon, USA.

Reeves, G.H., L.E. Benda, K.M. Burnett, P.A. Bisson, and J.R. Sedell. 1995. A disturbance-based ecosystem approach to maintaining and restoring freshwater habitats of evolutionarily significant units of anadromous salmonids in the Pacific Northwest. Pages 334–349 *in* J. Nielsen, ed. Evolution and the aquatic ecosystem: Defining unique units in population conservation. American Fisheries Society Symposium 17, Bethesda, Maryland, USA.

Reeves, G.H., F.H. Everest, and J.D. Hall. 1987. Interactions between the redside shiner (*Richardsonius balteatus*) and the steelhead trout (*Salmo gairdneri*) in western Oregon: The influence of water temperature. Canadian Journal of Fisheries and Aquatic Sciences **43**:1521–1533.

Reeves, G.H., F.H. Everest, and J.R. Sedell. 1993. Diversity of juvenile anadromous salmonid assemblages in coastal Oregon basins with different levels of timber harvest. Transactions of the American Fisheries Society **122**:309–317.

Reimers, P.E. 1973. The length of residence of juvenile fall chinook salmon in Sixes River, Oregon. Volume 4. Research Reports of the Fish Commission of Oregon, Portland, Oregon, USA.

Richards, K. 1982. Rivers: Form and process in alluvial channels. Methuen, New York, New York, USA.

Richards, K.S. 1978. Simulation of flow geometry in a riffle-pool stream. Earth Surface Processes **3**:345–354.

Ricklefs, R.E. 1987. Community diversity: Relative roles of local and regional processes. Science **235**:167–171.

Rosenzweig, M.L. 1994. Species diversity in space and time. Cambridge University Press, New York, New York, USA.

Rosenzweig, M.L., and Z. Abramsky. 1993. How are diversity and productivity related. Pages 52–65 *in* R.E. Ricklefs and D. Schluter, eds. Species diversity in ecological communities. University of Chicago Press, Chicago, Illinois, USA.

Rutherford, D.A., A.A. Echelle, and O.E. Maughan. 1987. Changes of the fauna of the Little River drainage, southeastern Oklahoma, 1948–1951 to 1981–1982: A test of the hypothesis of environmental degradation. Pages 178–183 *in* W.J. Matthews and D.C. Heins, eds. Community and evolutionary ecology in North American stream fishes. University of Oklahoma Press, Norman, Oklahoma, USA.

Sada, D.W., H.B. Britten, and P.F. Brussard. 1995. Desert aquatic ecosystems and the genetic and morphological diversity of Death Valley system speckled dace. Pages 350–359 *in* J. Nielsen, ed. Evolution and the aquatic ecosystem: Defining unique units in population conservation. American Fisheries Society Symposium 17, Bethesda, Maryland, USA.

Schlosser, I.J. 1982. Fish community structure and function along low habitat gradients in a headwater stream. Ecological Monographs **52**:393–414.

——. 1987. A conceptual framework for fish communities in small warmwater streams. Pages 17–24 *in* W.J. Matthews and D.C. Heins, eds. Community and evolutionary ecology in North American stream fishes. University of Oklahoma Press, Norman, Oklahoma, USA.

Schlosser, I.J., and K.K. Ebel. 1989. Effects of flow regime and cyprinid predation on a headwater stream. Ecological Monographs **59**:41–57.

Schlosser, I.J., and L.A. Toth. 1984. Niche relationships and population ecology of rainbow (*Etheostoma caeruleum*) and fantail (*E. flabellare*) darters in temporally variable environments. Oikos **42**:229–238.

Schoener, T.W. 1987. Axes of controversy in community ecology. Pages 8–16 *in* W.J. Matthews and D.C. Heins, eds. Community and evolutionary ecology in North American stream fishes. University of Oklahoma Press, Norman, Oklahoma, USA.

Schwartz, J.S. 1990. Influence of geomorphology and land use on distribution and abundance of salmonids in a coastal Oregon basin. M.Sc. thesis, Oregon State University, Corvallis, Oregon, USA.

Scott, J.B., C.R. Steward, and Q.J. Stober. 1986. Effects of urban development on fish population dynamics in Kelsey Creek, Washington. Transactions of the American Fisheries Society **115**:555–567.

Scrivener, J.C., and M.J. Brownlee. 1989. Effects of forest harvesting on gravel quality and incubation survival of chum salmon (*Oncorhynchus keta*) and coho salmon (*Oncorhynchus kisutch*) in Carnation Creek, British Columbia. Canadian Journal of Fisheries and Aquatic Sciences **46**:681–696.

Sedell, J.R., and J.L. Froggatt. 1984. Importance of streamside forests to large rivers: The isolation of the Willamette River, Oregon, USA, from its floodplain by snagging and streamside forest removal. Internationale Vereinigung für Theoretishe und Angewandte Limnologie Verhandlungen **22**:1828–1834.

Sedell, J.R., G.H. Reeves, and P.A. Bisson. 1997. Habitat policy for salmon in the Pacific Northwest. Pages 375–388 *in* D.J. Stouder, P.A. Bisson, and R.J. Naiman, eds. Pacific salmon and their ecosystems: Status and future options. Chapman & Hall, New York, New York, USA.

Sheldon, A.L. 1968. Species diversity and longitudinal succession in stream fishes. Ecology **49**:193–198.

Slaney, T.L., K.D. Hyatt, T.G. Northcote, and R.J. Fielden. 1996. Status of anadromous salmon and trout in British Columbia and Yukon. Fisheries **20**(10):20–35.

Smith, G.R. 1966. Distribution and evolution of the North American catostomid fishes of the subgenus *Pantosteus*, genus *Catostomus*. Miscellaneous publications of the University of Michigan Museum of Zoology **129**:1–132.

Smith, G.R. 1981. Late Cenozoic freshwater fishes of North America. Annual Review of Ecology and Systematics **12**:163–193.

Smith, G.R., J. Rosenfield, and J. Porterfield. 1995. Processes of origin and criteria for preservation of fish species. Pages 44–57 *in* J. Nielsen, ed. Evolution and the aquatic ecosystem: Defining unique units in population conservation. American Fisheries Society Symposium 17, Bethesda, Maryland, USA.

Snyder, R.S., and H. Dingle. 1989. Adaptive, genetically based differences in life history between estuary and freshwater threespine sticklebacks (*Gasterosteus aculeatus*). Canadian Journal of Zoology **67**:2448–2454.

Sousa, W.P. 1984. The role of disturbance in natural communities. Annual Review of Ecology and Systematics **15**:353–391.

Stouder, D.J., P.A. Bisson, and R.J. Naiman. 1996. Where are we? Resources at the brink. Pages 1–11 *in* D.J. Stouder, P.A. Bisson, and R.J. Naiman, eds. Pacific salmon and their ecosystems: Status and future options. Chapman & Hall, New York, New York, USA.

Swain, D.P., and L.B. Holtby. 1989. Differences in morphology and behavior between juvenile coho salmon (Oncorhynchus kisutch) rearing in a lake and in its tributary stream. Canadian Journal of Fisheries and Aquatic Sciences **46**:1406–1414.

Triska, F.J., V.C. Kennedy, R.J. Avanzino, G.W. Zellweger, and K.E. Bencala. 1989. Retention and transport of nutrients in a third-order stream in northwestern California: Hyporheic processes. Ecology **70**:1893–1905.

Tyus, H.M. 1987. Distribution, reproduction, and habitat use of the razorback sucker in the Green River, Utah, 1979–1986. Transactions of the American Fisheries Society **116**:111–116.

——. 1991. Ecology and management of the Colorado squawfish. Pages 379–402 *in* W.L. Minkley and J.E. Deacons, eds. Battle against extinction: Native fish management in the American west. The University of Arizona Press, Tucson, Arizona, USA.

Underwood, A.J. 1986. What is a community? Pages 351–368 *in* D.M. Raup and D. Jablonski, eds. Patterns and processes in the history of life. Springer-Verlag, New York, USA.

Utter, F., G. Milner, G. Stahl, and D. Teel. 1989. Genetic population structure of chinook salmon (*Oncorhynchus tshawytscha*) in the Pacific Northwest. Fisheries Bulletin **87**:239–264.

Vannote, R.L., G.W. Minshall, K.W. Cummins, J.R. Sedell, and C.E. Cushing. 1980. The river continuum concept. Canadian Journal of Fisheries and Aquatic Sciences **37**:130–137.

Wallis, O.L. 1948. Trout studies and a stream survey of Crater Lake National Park, Oregon. M.Sc. thesis, Oregon State University, Corvallis, Oregon, USA.

Waples, R.S. 1991. Definition of "species" under the Endangered Species Act: Application to Pacific salmon. National Oceanic and Atmospheric Administration Technical Memorandum NMFS F/NWC-194, United States Department of Commerce, National Oceanic and Atmospheric Administration, National Marine Fisheries Service, Washington, DC, USA.

White, P.S., and S.T.A. Pickett. 1985. Natural disturbance and patch dynamics: An introduction. Pages 3–13 *in* S.T.A. Pickett and P.S. White, eds. The ecology of natural disturbance and patch dynamics. Academic Press, Orlando, Florida, USA.

Whittaker, 1962. Classification of natural communities. Botanical Review **28**:1–239.

Wiens, J.A. 1984. On understanding a non-equilibrium world: Myth and reality in community patterns and processes. Pages 439–457 *in* D.R. Strong, Jr., D. Simberloff, L.G. Abele, and A.B. Thistle, eds. Ecological communities: Conceptual issues and the evidence. Princeton University Press, Princeton, New Jersey, USA.

Williams, C.B. 1964. Patterns in the balance of nature. Academic Press, London, UK.

Williams, J.E., J.E. Johnson, D.A. Henderson, S. Conteras-Balderas, J.D. Williams, M. Navarro-

Mendoza, D.E. McAllister, and J.E. Deacon. 1989. Fishes of North America endangered, threatened, or of special concern. Fisheries **14**(6):2–21.

Willson, M.F., and K.C. Halupka. 1995. Anadromous fish as keystone species in vertebrate communities. Conservation Biology **9**:489–497.

Wilzbach, M.A. 1985. Relative roles of food abundance and cover determining the habitat distribution of stream dwelling cutthroat trout (*Salmo clarki*). Canadian Journal of Fisheries and Aquatic Sciences **42**:1668–1672.

Wright, D.H., D.J. Currie, and B.A. Maurer. 1993. Energy supply and patterns of species richness on local and regional patterns. Pages 66–74 *in* R.E. Ricklefs and D. Schluter, eds. Species diversity in ecological communities. University of Chicago Press, Chicago, Illinois, USA.

Wydoski, R.S., and R.R. Whitney. 1979. Inland fishes of Washington. University of Washington Press, Seattle, Washington, USA.

Yant, P.R., J.R. Karr, and P.L. Angermeier. 1984. Stochasticity in stream fish communities: An alternative interpretation. American Naturalist **124**:573–582.

Yount, J.D., and G.J. Niemi. 1990. Recovery of lotic communities and ecosystems from disturbance—a narrative review of case studies. Environmental Management **14**:547–570.

Zalewski, M., and R.J. Naiman. 1985. The regulation of riverine fish communities by a continuum of abiotic–biotic factors. Pages 3–9 *in* J.S. Alabaster, ed. Habitat modifications and freshwater fisheries. Butterworth, London, UK.

Zirges, M.H. 1973. Morphological and meristic characteristics of ten populations of blackside dace, *Rhinichthys osculus nubilus* (Girard) from western Oregon. Masters thesis. Oregon State University, Corvallis, Oregon, USA.

Zucker, S.J. 1993. Influence of channel constraint on primary production, periphyton biomass, and macroinvertebrate biomass in streams of the Oregon Coast Range. Masters thesis. Oregon State University, Corvallis, Oregon, USA.

10
Riparian Wildlife

Kathryn A. Kelsey and Stephen D. West

Overview

• Contrasts between riparian and adjacent upland microclimates, vegetation, and animal communities are less dramatic in the relatively moist Pacific coastal ecoregion than in more arid regions of the United States.

• Approximately 29% of wildlife species found in riparian forests in the Pacific coastal ecoregion are riparian obligates (species that depend upon riparian and aquatic resources and experience severe population declines when riparian forests are removed). Riparian communities also support riparian generalists (species that use both riparian and upland habitats) and exotic species.

• The composition of riparian wildlife communities is influenced by three important continua occurring at both local and landscape levels. First, riparian obligates depend on certain habitat characteristics associated with stream order. Large streams and rivers support riparian communities greatly different from those associated with small streams. Second, wildlife species respond to habitat characteristics associated with forest successional stage. Third, the type, frequency, duration, and severity of natural or human induced disturbances contribute to the unique characteristics of individual stands or forest harvest units.

• Effective management options for wildlife, in terms of attaining stated objectives at low cost, will be sensitive to conditions of basin geomorphology, riparian forest successional stage and disturbance.

• Although many human induced disturbances (dams, flood-control devices, alterations to streambanks) change riparian zones dramatically and permanently, timber harvest practices can be designed to minimize impacts on wildlife communities. Buffer strips along small forested streams protect in-stream habitat for salmonids and provide wildlife habitat. Testing alternative management strategies for riparian buffer zones is important for maximizing suitable wildlife habitat and minimizing buffer strip failure because of windthrow.

• The future of this region's wildlife communities depends in part on the ability to design and implement cost-effective forest harvest regulations for riparian zones that provide suitable habitat for the long-term persistence of native species.

Introduction

Much of the understanding of wildlife associations with riparian habitat is based on research conducted in arid and semiarid regions where riparian zones provide habitat for communities with greater species richness and abundance than the surrounding uplands (see Stevens et al. 1977, Odum 1979, Thomas et al. 1979, Szaro 1980, Tubbs 1980, Malanson 1993). Contrasts between riparian and adjacent upland microclimates, vegetation, and animal communities are much less dramatic in the relatively moist Pacific coastal ecoregion. Consequently, wildlife communities of Pacific coastal riparian zones are not strikingly more diverse than those of adjacent upland areas. Pacific coastal riparian zones do, however, provide three important functions for wildlife communities. First, the presence of riparian habitat in a landscape increases wildlife species diversity by providing habitat for obligate riparian species, for species that seek edge habitats, and for species associated with early successional plant communities. Second, the riparian zones provide refugia for upland forest species when upland habitat experiences a major disturbance (e.g., clearcutting, fire). Finally, riparian zones function as topographic landmarks to visually cue species during migration or dispersal.

This chapter describes riparian wildlife community organization, processes, and management in the Pacific coastal ecoregion. It provides a basic understanding of the role riparian zones play in providing wildlife habitat and of the wildlife issues arising from selected land use practices. A description of the current distribution and abundance of riparian wildlife is followed by discussion of ecological conditions and processes that influence wildlife communities in riparian zones. Finally, management issues concerning riparian wildlife are presented, highlighting alternative strategies for forest land management. Most of the discussion excludes wildlife communities along large rivers because economic development has extensively altered these systems. Emphasis on mid-sized and small stream systems serves to underscore their integration with timberland management issues.

Distribution and Abundance of Riparian Wildlife Species

The wildlife community within riparian zones is composed of obligate and generalist riparian species as well as introduced species. Use of riparian habitat varies, depending on the life history characteristics of individual species and the availability of suitable habitat elsewhere, in some cases rendering the distinction between obligate and generalist discretionary. Riparian obligates are species considered so highly dependent on riparian and aquatic resources that they would disappear with the loss of this habitat from a drainage basin. Amphibian species dominate the list of riparian obligates when the proportion of obligates is compared to the total number of species present (Table 10.1). Species that utilize both riparian and upland forests are considered riparian generalists. The discussion of riparian obligates and generalists is limited to native species. A later section on introduced species briefly describes influences of nonnative species on riparian wildlife communities.

Riparian Obligates

Amphibians. Amphibians are a diverse class of vertebrates primarily adapted to moist habitats. Their water-permeable skin and intolerance of high temperatures limit the activity of many species to periods of precipitation and to moist forested habitats (Stebbins and Cohen 1995). Body temperatures reflect environmental temperatures and are regulated behaviorally by moving to cooler or warmer areas. In a cold location, body functions slow allowing amphibians to survive for months without food. During periods of dry weather, most species estivate, becoming inactive for varying lengths of time. Of the 30 amphibian species found in the Pacific coastal ecoregion, 60% require aquatic habitat for reproduction (Table 10.1). Three genera breed primarily in lotic waters: giant salamanders (*Dicamptodon* spp.), torrent salamanders (*Rhyacotriton* spp.), and the tailed frog (*Ascaphus truei*; Leonard et al. 1993). Five genera breed in lentic waters or behind beaver dams in stream systems: mole salamanders (*Ambystoma* spp.), newts (*Taricha* spp.), the western toad (*Bufo boreas*), the Pacific tree frog (*Hyla regilla*; for nomenclature see Cocroft 1994), and true frogs (*Rana* spp.; Leonard et al. 1993). Long-toed salamander (*Ambystoma macrodactylum*), newt, Pacific tree frog and wood frog (*Rana sylvatica*) larvae, however, develop in less than two months (Nussbaum et al. 1983), enabling them to successfully reproduce in temporary pools that may not be associated with riparian zones. Individuals of both giant salamanders and mole salamanders

TABLE 10.1. Numbers, by taxonomic class, of native riparian obligate and upland species compared to the total number of native species in the Pacific coastal ecoregion. The proportion of riparian obligates to all species is shown in the "Riparian obligate (%)" column.

	Riparian obligates	Upland specialists	All species	Riparian obligate (%)
Amphibians	18	7	30	60
Reptiles	3	12	19	16
Birds	78	93	231	34
Mammals	13	31	107	12
Total	112	143	387	29

exhibit paedomorphosis (attain sexual maturity without metamorphosis) and are restricted to aquatic habitat (Nussbaum et al. 1983) (Figure 10.1). One species of lungless salamander, Dunn's salamander (*Plethodon dunni*), is also considered a riparian obligate even though it may never enter the water. It is most often found along stream banks, and prefers moister habitat than other western plethodontids (Dumas 1956, Nussbaum et al. 1983, Stebbins 1985, Leonard et al. 1993).

Reptiles. The cool, moist riparian forests of the northwestern Pacific coast do not provide optimal habitat for many reptilian species adapted to warmer, drier environments. Exceptions are the western pond turtle (*Clemmys marmorata*) and the painted turtle (*Chrysemys picta*) that spend most of their lives within riparian habitat or in the water (Nussbaum et al. 1983, Brown et al. 1995). The Pacific coast aquatic garter snake (*Thamnophis atratus*; formerly the western aquatic garter snake, *T. couchii*, Lawson and Dessauer 1979, Rossman and Stewart 1987, Collins 1990, Brown et al. 1995) is the most aquatic snake in the Pacific Northwest. It rarely ventures far from water, requiring aquatic resources for food and cover.

Birds. Obligate riparian bird species require aquatic or riparian habitat for either breeding, nesting, roosting, or feeding, although many species do not require aquatic or riparian

FIGURE 10.1. Paedomorphic Pacific giant salamander from western Washington Cascade Mountains (Photo: S. West).

habitat at all times. Of the approximately 230 species found throughout the Pacific coastal ecoregion, one-third are classified as riparian obligates (Table 10.1) including species belonging to the following groups: loons, grebes, cormorants, ducks, geese, hawks and falcons, herons, cranes, rails, coots, kingfishers, and some of the passerine birds. The most frequently observed riparian obligates are the loons, grebes, cormorants, herons, ducks, and geese. Resident species are found on or near water year round.

Many bird species nest within riparian zones because of the close proximity to aquatic food resources and the availability of nest sites and materials (Figure 10.2). Rearing young birds demands tremendous energy; these energy expenditures can be reduced by nesting near food resources. A brood of three young osprey (*Pandion haliaetus*) and one adult, for example, requires more than 1 kg of fish daily (Van Daele and Van Daele 1982). Nests are found in large trees or on raised platforms along lakes and rivers. The seasonality of riparian food resources influences the times when obligates are observed in riparian zones. In spring and summer, the emergence of adult flying insects from aquatic nymphs and pupae provides abundant food resources for insectivorous birds and the young of other birds (e.g., geese and ducks). During winter months, bald eagle (*Haliaeetus leucocephalus*) feed on salmon carcasses in many river systems. Along Alaska's Chiklat River, 3,000 to 4,000 eagles gather during the salmon run (Ehrlich et al. 1988).

The distribution of riparian birds varies with the size and gradient of streams (Figure 10.3). In a comparison of bird communities along large and small rivers on the Olympic Peninsula in Washington, Lock (1991) found that large rivers supported greater bird abundance, species richness, and species diversity than small rivers where bird communities resembled upland communities. Wide, low-gradient rivers provide habitat for waterfowl, heron, osprey, and other large-bodied birds. In contrast, forest streams in the mountains are narrow with steep gradients and do not provide habitat suitable for large-bodied riparian obligates such as waterfowl. The American dipper (*Cinclus mexicanus*), a small-bodied passerine, is well adapted to conditions common to cascading, interior forest streams. The dipper has strong legs, an unusually large preen gland, and ocular nictitating membranes that enable it to exploit under water habitat and prey on aquatic invertebrates and fish (Terres 1980).

Mammals. Mammalian riparian obligates can be discussed using taxonomic groups with similar ecological characteristics (e.g., body size, locomotion). For example, small mammals, such as insectivores, rodents, and hares, provide an important prey base for carnivorous animals. Small mammals differ from other mammals in their body size, relatively short life spans, high reproductive rates, and limited movement. Because of the relatively limited potential for long distance movement, riparian obligates are found year round in riparian habitats. Such small mammals include water shrew (*Sorex palustris*; Bailey 1936, Conaway 1952, Anthony et al. 1987), marsh shrew (*S. bendirii*; Pattie 1973, Hooven and Black 1976, Anthony et al. 1987), water vole (*Microtus richardsoni*; Bailey 1936, Hooven and Black 1976, Ludwig 1984, Doyle 1985, Anthony et al. 1987), longtail vole (*M. longicaudus*; Maser et al. 1981, McComb et al. 1993a and b), and northern bog lemming (*Synaptomys borealis*; Layser and Burke 1973, Wilson et al. 1980). Muskrat (*Ondatra zibethica*; Willner et al. 1980) and beaver (*Castor canadensi*; Hill 1982) are also restricted to riparian sites.

Most captures of bats in a given landscape are made over stream courses and ponds because bats drink each day, are known to feed on emerging aquatic insects, and most importantly, can be caught within restricted riparian flyways (Figure 10.4). Despite this association with riparian areas, there are no widely recognized riparian obligate species of bats in the Pacific Northwest coastal ecoregion. This is in part because several bat species can use varied habitat but perhaps just as well from a general lack of information on habitat-use patterns within this region (Christy and West 1991). Only recently have studies been initiated to compare bat activity in riparian and upland

FIGURE 10.2. (a) Epicormic branches on red alder (*Alnus rubra*) tree trunk in Gifford Pinchot National Forest, Washington (Photo by S. Pearson); (b) Pacific slope flycatcher at nest constructed in epicormic branches of red alder tree within riparian forest in western Washington (Photo: P. Gibert).

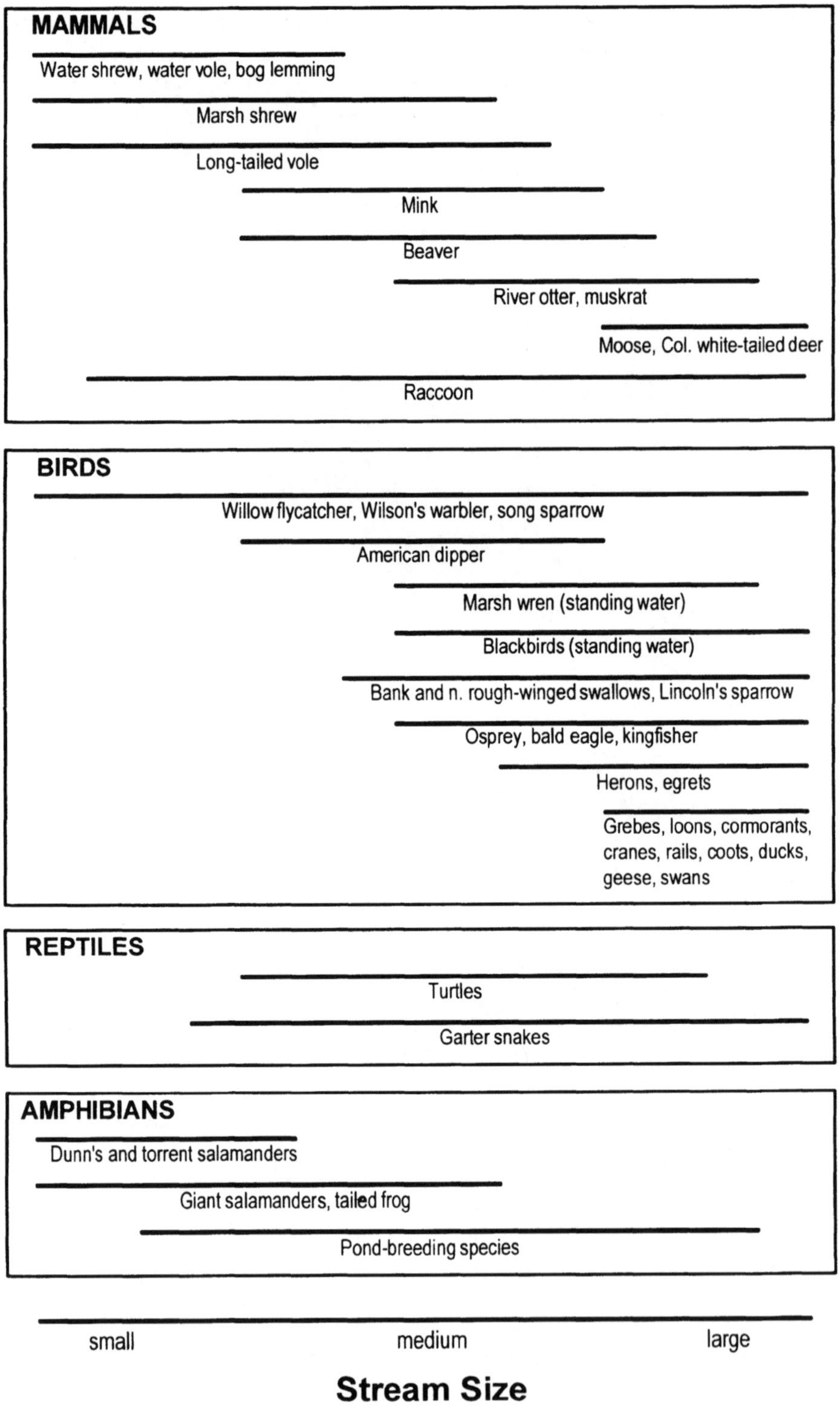

MAMMALS

Water shrew, water vole, bog lemming

Marsh shrew

Long-tailed vole

Mink

Beaver

River otter, muskrat

Moose, Col. white-tailed deer

Raccoon

BIRDS

Willow flycatcher, Wilson's warbler, song sparrow

American dipper

Marsh wren (standing water)

Blackbirds (standing water)

Bank and n. rough-winged swallows, Lincoln's sparrow

Osprey, bald eagle, kingfisher

Herons, egrets

Grebes, loons, cormorants, cranes, rails, coots, ducks, geese, swans

REPTILES

Turtles

Garter snakes

AMPHIBIANS

Dunn's and torrent salamanders

Giant salamanders, tailed frog

Pond-breeding species

small medium large

Stream Size

FIGURE 10.3. Distribution of selected native riparian obligates as a function of stream size.

FIGURE 10.4. Harp trap used to capture bats flying over a second order stream in western Washington (Photo: K. Kelsey).

habitats. With the use of ultrasonic detectors (Fenton 1988), large differences in activity between riparian habitats and their associated uplands, as indexed by echolocation call frequencies, point to a strong preference for riparian habitats (Cross 1988, Thomas 1988). The higher frequency of calls indicative of foraging (feeding buzzes) indicates that riparian zones are being used preferentially for feeding. In ongoing research along small forested streams in western Washington, more than 99% of all calls are of small, slow-flying, and highly maneuverable species of the genus *Myotis* (West unpublished data). Whether bats are day roosting (vs. night roosting) within riparian forests is a current topic of research with implications for timber harvest regulations. Early evidence from eastern Washington forests suggests that bats may opt for roosts with warmer microclimates provided by snags and trees located in the uplands (Campbell et al. 1996, Frazier 1997). If ongoing studies corroborate this pattern, forest management practices may have to be adapted to provide healthy riparian zones for feeding and upland snags (of an acceptable morphology) for roosting.

Most carnivores have large home ranges and use both riparian and upland habitats. Exceptions include river otter (*Lutra canadensis*) and mink (*Mustela vison*), both of which are considered semiaquatic riparian obligates (Dalquest 1948). Riparian habitat along streams and lakes provides dens, breeding areas, and easy access to fish, their primary prey. Mink also feed on birds, small mammals, frogs, and crayfish. Both species are excellent swimmers.

Ungulates find a diversity of herbaceous plants and deciduous shrubs typical of early successional plant communities in riparian zones. Although most ungulates are considered riparian generalists, two species are considered riparian obligates. Columbian white-tailed deer (*Odocoileus virginianus leucurus*) are restricted to riparian zones along the Columbia River. Their diet is limited to high quality food because of a small rumen volume to body weight ratio (Dublin 1980). To meet their nutritional needs, Columbian white-tailed deer must limit amounts of grasses and increase amounts of browse and forbs (Hanley 1982, Hofmann 1988), which are available year round in the riparian zones of large rivers. Moose (*Alces alces*) also rely on riparian and aquatic vegetation throughout the year (Ingles 1965). During winter months, riparian willow stands provide forage and adjacent conifer forests are used for cover (Coady 1982).

Riparian Generalists

Species considered riparian generalists use upland as well as riparian habitat. Generalists use riparian habitat because it provides early successional and deciduous plant species, particularly ground cover, in contrast to adjacent forests with suppressed understory plant communities. Following a disturbance in the upland forest, riparian habitat often provides a refuge with structural characteristics of an older forest within a young forest matrix.

Amphibians. Amphibian generalists include species that do not require aquatic habitat for breeding, such as lungless salamanders (*Plethodon vandykei*, *P. vehiculum*, *Aneides flavipunctatus*). These salamanders are found in humid places under rocks, talus, and decaying wood regardless of the proximity to water.

Reptiles. Reptiles use riparian zones more frequently when adjacent uplands are dry with reduced canopy cover. Within the Pacific coastal ecoregion, reptilian diversity is much greater in northern California and Oregon than in Washington, British Columbia, or southeast Alaska (Stebbins 1985). Active reptiles require higher body temperatures than amphibians and, as ectotherms, their body temperatures are limited by environmental conditions. Garter snakes (*Thamnophis* spp.), the sharp-tailed snake (*Contia tenuis*), the ringneck snake (*Diadophis punctatus*), and the rubber boa (*Charina bottae*) are all considered riparian generalists (Brown et al. 1995). These reptiles find increased cover sites and food resources along the terrestrial-aquatic ecotone. In addition, species that are good swimmers (e.g., rubber boa and garter snakes) use aquatic habitat to escape from terrestrial predators.

Birds. Roughly one-quarter of the region's bird species are riparian generalists, foraging, nesting, and roosting in both riparian and upland habitats. In uplands where the forest canopy obstructs sunlight and understory vegetation is sparse, riparian zones provide nesting habitat and cover for ground and shrub species, as well as flowers and fruit for nectivores and frugivores. Western birds associated with deciduous trees, such as Pacific slope flycatcher (*Empidonax difficilis*) and Swainson's thrush (*Catharus ustulatus*), are commonly observed in riparian alder forests but are not limited to these areas.

Mammals. Most species of small mammals, including mice, voles, woodrats, and jumping mice, can be considered riparian generalists. The moist, often rocky and exposed soils of riparian zones appear to be important for vagrant shrew (*Sorex vagrans*; Terry 1981), coast mole (*Scapanus orarius*; Hartman and Yates 1985), and Townsend's vole (*Microtus townsendii*; Bailey 1936). Hare and mountain beaver (*Aplodontia rufa*) can also be considered riparian generalists. Marmots, ground squirrels, chipmunks, and tree squirrels generally favor upland areas which provide more visibility for these species which, except for the northern flying squirrel (*Glaucomys sabrinus*), are active during the day.

Most carnivores find drinking water, food resources, and prey in both upland and riparian zones. High prey densities in riparian zones during summer droughts attract carnivores such as bobcat (*Relis rufus*; Sweeney 1978, Raedeke et al. 1988) and ermine (*Mustela erminea*; Fitzgerald 1977, Simms 1979, Doyle 1990). Raccoon (*Procyon lotor*), also closely associated with riparian zones, have naked feet and long toes which make them well adapted for walking in shallow water and on muddy stream bottoms, as well as for climbing trees (Dalquest 1948). Raccoon patrol shorelines in search of fish, amphibians, insects, mollusks, and crayfish.

All ungulates, except the mountain goat (*Oreamnos americanus*), rely seasonally on riparian zones for drinking water, high quality forage, and thermal cover. According to studies by Witmer and deCalesta (1983), in the spring, female Roosevelt elk (*Cerrus elaphus roosevelti*) in the Oregon Coast range use riparian forest more than expected when pregnancy and lactation increase nutritional needs and calving limits mobility. Witmer and deCalesta (1983) also found that in summer when air temperatures are high and succulent forage is sparse, elk use riparian forest on north-facing slopes more than expected. Ungulates also use mineral deposits, such as natural salt licks along eroded river banks (Butler 1995).

Exotic Species

The introduction of non-native species into new environments results in at least one of four possible outcomes: the new environment cannot meet the needs of the species and the population dies out; the species finds the new environment quite favorable, and the population multiplies quickly and outcompetes native species; the population, having reached a high density, degrades the habitat, driving out native species; the species thrives with no apparent harm to native species. Introductions where habitats cannot support a non-native species frequently are unnoticed. Examples of the second scenario are introductions of the bullfrog (*Rana catesbeiana*), turtles (especially the snapping turtle [*Chelydra serpentina*], and the red-eared slider [*Trachemys scripta*]), the European starling (*Sturnus vulgarus*), old world rats and mice (Family: Muridae), and nutria (*Myocastor coypus*). Evidence of the negative impact of each of these introduced species on native wildlife is well documented (e.g., Hayes and Jennings 1986, Brown 1994, Ingold 1994). Horses, cattle, and other livestock are examples of introduced animals that force out native species by degrading habitat (Hall 1988, Fleischner 1994). These animals use riparian zones for forage, drinking water and shade during hot days. Livestock grazing and trampling often leads to extensive modification of riparian vegetation, degradation of streambanks, and increased rates of erosion (Fleischner 1994). Although lacking for this region, an example of the fourth outcome is the introduction of certain game birds (e.g., chukar, *Alectoris chukar*) to regions east of the Cascade Mountain crest.

Effects of Riparian Conditions and Processes on Wildlife

Composition of riparian wildlife communities varies over space and time. Processes of ecological succession and the disturbance regimes that initiate them influence the composition of wildlife communities at both local and landscape scales (Table 10.2). Although the distinction between small and large scale is useful for discussion purposes, ecological conditions exist along a continuum and change as a result of natural and human-induced processes. Here a distinction is made between the impact of natural versus human-induced disturbances on wildlife communities. Human-induced disturbances are discussed later.

Landscape Processes

Although wildlife communities experience environmental changes at the local level, land-

TABLE 10.2. Effects of landscape and local characteristics on riparian wildlife communities.

Level	Process/characteristic	Effects on riparian wildlife
Landscape	Basin geomorphology	Extent and structure of riparian habitat
		Primary production within stream
	Landscape connectivity	Distribution of nutrients
		Transport of material
		Downstream effects of wildlife species
		Movement of animals along riparian corridors
	Animal influences	Successional stage of riparian habitat
		Extent and structure of riparian habitat
	Disturbances	
	Natural	Successional stage of riparian habitat
		Increase in complexity of habitat structure
	Human-induced	Successional stage of riparian habitat
		Widespread extirpation of species
		Introduction of nonnative species
		Increase in homogeneity of habitat structure
Local	Successional stage of riparian vegetation	Availability of food, cover, and breeding habitat
	Contrast between riparian and upland vegetation	Uniqueness of wildlife community
	Animal interactions	Trophic relationships
		Symbiotic relationships
		Availability of food, cover, and breeding habitat

scapes display great spatial and temporal variability which affect local conditions. Landscape processes are described here in terms of events which affect spatial variability and include basin geomorphology, animal influences, and natural disturbances. An understanding of landscape-level patterns helps answer questions concerning animal movement and dispersal, source-sink phenomena, gene flow, and responses to potential dispersal barriers.

Geomorphology. The intrinsic characteristics which influence wildlife distribution are shaped by stream conditions and processes (Figure 10.3). Low-order streams (generally 1st–3rd) found in upper basins tend to have high gradients, steep side slopes, nearly continuous forest canopy, narrow strips of riparian vegetation, and production based on allochthonous inputs to the stream (Naiman et al. 1992, see Chapters 2, 7, and 15 this volume). Riparian wildlife communities in upper basins are distinguished by high densities of stream-breeding amphibians, absence of fish, and few riparian obligate bird species. Lower-basin rivers (generally 6th order and greater) have low gradients, extensive flood plains, open canopies, wide riparian zones, and productive autotrophic communities. Riparian zones along these large streams and rivers provide habitat for ungulates, reptiles, raptors, heron, kingfisher, waterfowl, and river otter. Mid-order streams function as transition areas between small streams and large rivers where the overlap of species distributions creates different communities. For example, greater rates of garter snake predation on larval and paedomorphic giant salamander occur along mid-sized streams where garter snake are more abundant than along smaller streams (Lind and Welsh 1994).

Connectivity. In general, streams provide a fundamental connection between high and low elevations within a landscape. Geomorphological processes upstream exert a tremendous influence on downstream reaches (Chapter 2); however, the influence of upstream wildlife communities on downstream communities has rarely been examined. One study, however, designed to examine downstream effects of American dipper predation found that invertebrates did not drift downstream to escape predation (Jenkins and Ormerod 1996). Though data are sparse, evidence suggests that the presence of beaver dams influences downstream conditions and possibly wildlife as well. Certainly, catastrophic failure of beaver dams results in extreme impacts to downstream reaches. Whether dam failure is the result of high precipitation rates, beaver breaching their own dams, or otter burrowing through dams, flooding can damage downstream beaver dams, riparian vegetation, and streamside amphibian and small mammal communities (Butler 1995).

Ecologists have theorized that some species use riparian zones as conduits for movement and dispersal across the landscape. It is clear that riparian obligates move along riparian zones but it is much less certain that other species commonly do so. Aerial photos of the Kabetogama Peninsula in northern Minnesota from 1940 to 1986 show the progressive development of beaver impoundments as they move along connecting rivers and streams (Naiman et al. 1988). Migrant passerine birds use riparian habitat more often than non-riparian habitat as stopover points during migration (Stevens et al. 1977). Evidence suggests riparian zones in low to mid-elevations are used as travel corridors more than the smaller, high elevation streams because the lower areas provide more habitat for movement and procurement of food and cover, and they are more visible within the landscape.

Animal influences. The presence of certain species within a drainage basin also influences spatial variability. For example, beaver activity dramatically alters stream and riparian environments. When the work of many beaver within one basin is examined collectively, the habitat and hydrologic changes take on greater significance. The diversity and extent of habitats directly impacted by beaver depend on the population density. A study distinguishing eight categories of wetland vegetation on the Kabetogama Peninsula, Minnesota, over 46 years shows a complex pattern of changing vegetation and habitat as the number of beaver impoundments and, consequently, the spatial extent of open water and saturated soils increased (Naiman et al. 1988). The spatial and temporal variation in vegetation types demonstrates the existence of numerous successional courses rather than the simplest successional paradigm of forested stream, to pond, to meadow, and back to forested stream.

Disturbance. At any point in time, wildlife communities reflect a dynamic assemblage of species interacting with each other and the environment. Perturbations affecting these complex relationships alter the community, initiating a process of readjustment. For example, riparian forest avian communities change abruptly when upland forests are harvested (Stauffer and Best 1980, Triquet et al. 1990, Thompson et al. 1992, Croonquist and Brooks 1993, West unpublished data). When riparian buffer strips are left along streams, changes in the bird community and species interactions within riparian habitat occur. Many nest failures within riparian buffer strips in western Washington result from predation by corvid species (e.g., ravens, jays) and other predators common to forest edge habitat (West unpublished data). If reproductive success continuously decreases until the replacement forest canopy closes, then populations of forest avian species are certain to decline (Yahner 1988).

The frequency, duration, and severity of natural disturbances directly affect riparian habitat, thereby influencing associated wildlife communities. High water and wind disturbance help preserve conditions necessary for early successional plant communities by impeding the growth of trees at the stream edge. Fire, on the other hand, occurs less frequently in coastal forests and tends to have a lesser impact on riparian than on upland vegetation (Agee 1988). Increased humidity, higher fuel moisture, and more deciduous vegetation, all characteristic of riparian zones, help to reduce fire damage. Wildlife species finding refuge from fire within riparian zones provide a source population for recolonization of adjacent burned areas. Extensive landslides through stream channels (debris flows) occur infrequently yet greatly influence stream morphology, substrate characteristics, streamside landforms, and riparian vegetation (Swanson et al. 1987, Chapters 2 and 11 this volume). Although there is a lack of quantitative evidence indicating how landslides affect wildlife communities, general responses relate to the severity and areal extent of habitat alteration. Removal of streambed and bank materials reduces habitat for salamanders, reptiles, and burrowing small mammals and reduces vegetation cover and forage for birds and ungulates.

Native species have evolved with natural disturbances and possess strategies which allow them to persist through the associated changes. For example, the northwestern salamander (*Ambystoma gracile*) and Pacific tree frog recolonized areas within the blast zone of Mt. St. Helens, Washington—areas that offer very little suitable terrestrial habitat. Two years after the eruption, several northwestern, Cascade torrent (*Rhyacotriton cascadae*) and Van Dyke's salamanders were observed in the blast zone where they apparently survived the intense heat of the blast under 3 m of snow (Zalisko and Sites 1989).

Tailed frog tadpoles possess morphological adaptations that allow them to overwinter in streams where discharge rates can peak as much as ten times the low summer flow rates. The tadpole uses its mouth as a suction cup and holds onto the underside of rocks to avoid being carried downstream (Gradwell 1971). Wildlife species' adaptations to flooding that deposits sediment, rocks, and wood in low gradient areas are common. During summer along large rivers, numerous birds and mammals find perches, food, and cover among piles of woody debris at the river's edge (Steel 1993). Debris piles that are used during one summer may be destroyed by flooding the following winter and new piles formed in different places, contributing to spatial and temporal variability within the riparian corridor.

Local Processes

The landscape-level processes described above influence local conditions within riparian communities which, in turn, influence species distribution and abundance (Table 10.2). At the local level, wildlife communities respond to the successional stage of the plant community, the interface of upland and riparian habitats, and the presence of certain animals.

Succession. Wildlife community composition is influenced by the structure and composition of vegetation associated with different forest successional stages (Figure 10.5). Early successional plant communities, present in many riparian zones, consist largely of grasses, forbs, shrubs, and deciduous trees (see Chapter 12) and provide food, nesting and cover for wildlife. Frequent flooding and the accompanying scouring and deposition of streambed materials keep much of the forest in early successional communities. In presettlement times, when old-growth forests dominated the region's uplands, a large proportion of the landscape's early successional forest was associated with riparian zones. In the future, riparian landscapes may be largely associated with mid- to late-successional forest strips because of the impacts of reducing or eliminating flood events and establishing persistent riparian buffers in the areas managed for timber resources.

Wildlife use of riparian zones varies with forest composition and structure. During the dry, hot days of summer, riparian forests offer shade, drinking water, and herbs that have not yet dried out. The abundant flowers, fruits, and seeds in riparian zones attract nectivores, such as the rufous hummingbird (*Selasphorus rufus*), frugivores, such as the cedar waxwing (*Bombycilla cedrorum*) and granivores. Many birds are associated with the high foliage density of deciduous areas (Stauffer and Best 1980, Morrison and Meslow 1983, Manuwal 1986, Martin 1988). This association has been attributed to foliage effects on food abundance (MacArthur and MacArthur 1961, Karr and Roth 1971, Willson 1974), thermal environment (Calder 1973, Walsberg 1981), nest site availability (MacKenzie et al. 1982), and nest predation rates (Nolan 1978, Best and Stauffer 1980, Martin 1988). The complex habitat structures present in deciduous riparian zones are of great importance to some bird species when the adjacent uplands are devoid of understory and mid-story vegetation, a common condition in managed forests, 20 to 40 years old, that have not been thinned. Wilson's warbler (*Wilsonia pusilla*) and western tanager (*Piranga ludoviciana*), although absent from dense forest stands with little understory, are common in older and younger stands with understory and mid-story vegetation (West, unpublished data). Several species of small mammals are common in the early successional vegetation of riparian zones, particularly members of the genus *Microtus*. Increases in their populations attract raptors and other predators, contributing to the diversity of riparian wildlife communities. In areas where second- and third-growth managed forests dominate the landscape and understories are suppressed, riparian forests and clearcuts 6 to 15 years old may provide the only palatable winter forage for deer and elk (Schroer et al. 1988).

Upland and riparian habitat. Much of the interest in riparian wildlife communities developed from research in arid regions where upland habitat conditions are strikingly different from riparian conditions (see Smith 1977, Stevens et al. 1977, Best et al. 1979, Odum 1979, Stauffer and Best 1980, Szaro 1980, Tubbs 1980, Knopf 1985). The Pacific coastal ecoregion presents an entirely different situation in which similarities between riparian habitat along small and mid-sized streams and adjacent upland habitat are stronger than in arid and semiarid regions. Riparian vegetation extends an average of 2 to 10 m away from very small streams, depending on stream size and valley morphology (Carlson 1991). While both riparian and upland areas are forested, compositional and structural differences do exist. The primary difference is in the ratio of deciduous to evergreen vegetation. Riparian zones are often dominated by deciduous trees, shrubs, and forbs, creating habitat that is more dense in the under- and mid-stories (Doyle 1990, Lock 1991,

FIGURE 10.5. Distribution of selected mammals, birds, reptiles, and amphibians according to forest successional stage (thick line = primary habitat, thin line = secondary habitat, dashed line = gap in distribution). Definitions of mature and old-growth forest follow Spies and Franklin (1991).

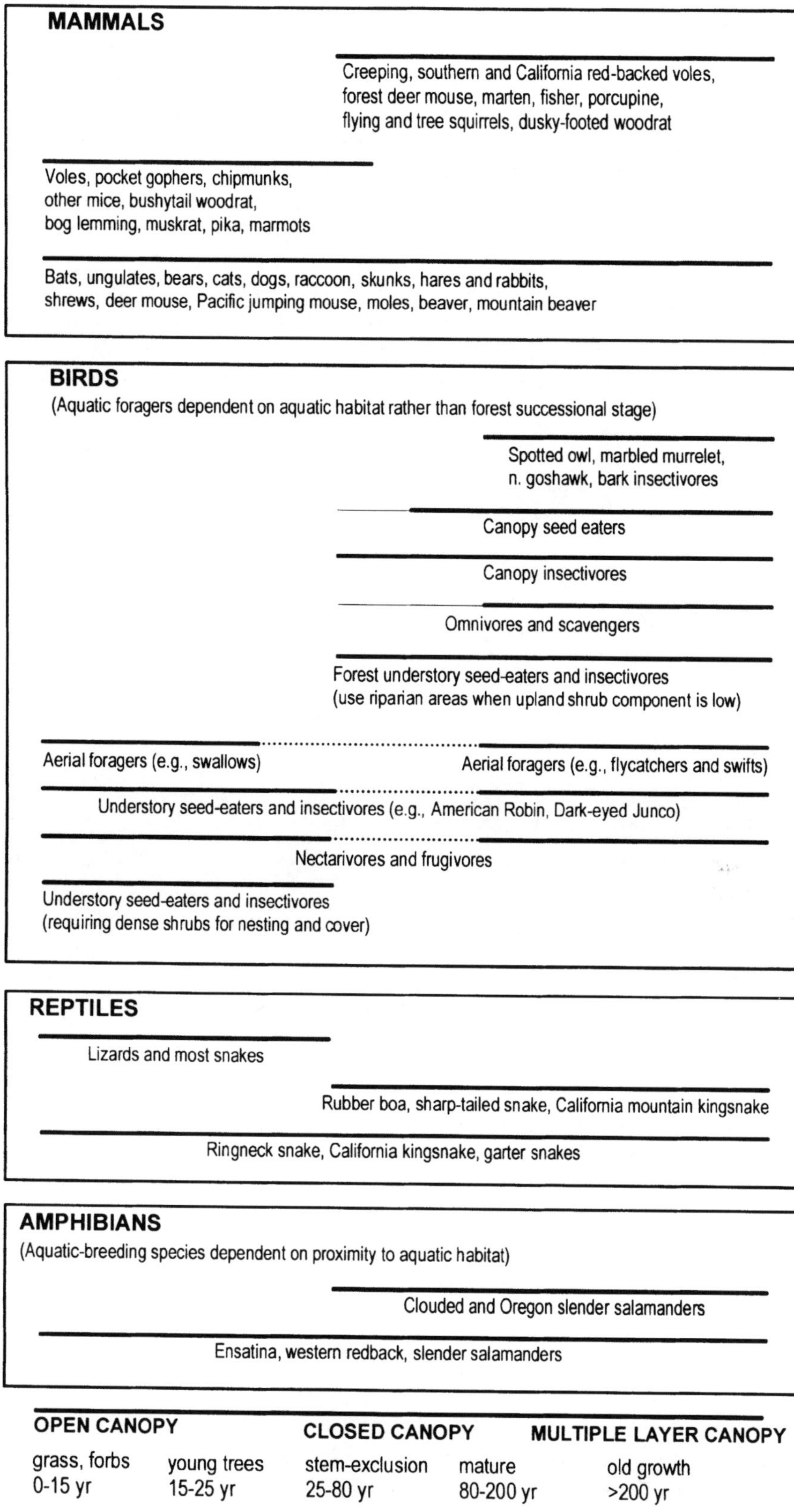

MAMMALS

Creeping, southern and California red-backed voles, forest deer mouse, marten, fisher, porcupine, flying and tree squirrels, dusky-footed woodrat

Voles, pocket gophers, chipmunks, other mice, bushytail woodrat, bog lemming, muskrat, pika, marmots

Bats, ungulates, bears, cats, dogs, raccoon, skunks, hares and rabbits, shrews, deer mouse, Pacific jumping mouse, moles, beaver, mountain beaver

BIRDS

(Aquatic foragers dependent on aquatic habitat rather than forest successional stage)

Spotted owl, marbled murrelet, n. goshawk, bark insectivores

Canopy seed eaters

Canopy insectivores

Omnivores and scavengers

Forest understory seed-eaters and insectivores (use riparian areas when upland shrub component is low)

Aerial foragers (e.g., swallows)

Aerial foragers (e.g., flycatchers and swifts)

Understory seed-eaters and insectivores (e.g., American Robin, Dark-eyed Junco)

Nectarivores and frugivores

Understory seed-eaters and insectivores (requiring dense shrubs for nesting and cover)

REPTILES

Lizards and most snakes

Rubber boa, sharp-tailed snake, California mountain kingsnake

Ringneck snake, California kingsnake, garter snakes

AMPHIBIANS

(Aquatic-breeding species dependent on proximity to aquatic habitat)

Clouded and Oregon slender salamanders

Ensatina, western redback, slender salamanders

OPEN CANOPY		CLOSED CANOPY	MULTIPLE LAYER CANOPY	
grass, forbs 0-15 yr	young trees 15-25 yr	stem-exclusion 25-80 yr	mature 80-200 yr	old growth >200 yr

McGarigal and McComb 1992, West unpublished data) while upland areas are dominated by conifers, evergreen shrubs (Doyle 1990, McGarigal and McComb 1992, West unpublished data), and snags (McGarigal and McComb 1992), characteristics to which birds and bats respond. Ground cover characteristics influence the presence and abundance of amphibians and small mammals active on the forest floor. Some ground cover variables, such as amounts of downed wood (Doyle 1990, McComb et al. 1993a, 1993b), bare soil, leaf litter, lichen (Doyle 1990), and grass (McGarigal and McComb 1992), do not differ significantly between riparian and adjacent upland areas. Other variables, such as amounts of forbs (McGarigal and McComb 1992), herbs (Doyle 1990, McGarigal and McComb 1992), and rocks (Doyle 1990), are greater in riparian than upland areas.

Birds respond to structural habitat features in the forest understory, mid-story, and canopy that provide nesting, perching, and foraging habitat. Consequently, species composition and abundance of birds differ between riparian and upland areas (Table 10.3). Results of bird

TABLE 10.3. Summary of studies comparing amphibian, reptile, bird, and small mammal communities between riparian (R) and upland (U) habitats in the Pacific Northwest coastal ecoregion. Unique species category represents the number of species observed only within riparian or within upland habitat

Taxon	Region	Number of study sites	Forest age (years)	Unique species		Total species	[a]Significant species		Reference
				Riparian	Upland		R > U	U > R	
Amphibians	Oregon coast	6	120–140	1	1	9	0	1	McComb et al. 1993b
	Oregon coast	3	[b]40–50	1	1	8	0	1	McComb et al. 1993a,c
	Washington Cascades	18	40–60	3	1	13	2	2	West unpublished data
Reptiles	Oregon coast	3	[b]40–50	1	0	1	0	0	McComb et al. 1993a,c
Birds	Oregon coast	6	120–140	3	11	55	1	3	McComb et al. 1993b
	Washington Cascades	18	40–60	4	7	32	2	0	West unpublished data
Small Mammals	[c]S. Oregon Cascades	1	>250	2	2	10	0	0	Cross 1985
	[d]S. Oregon Cascades	1	unknown	1	2	7	0	0	Cross 1985
	S. Oregon Cascades	1	4	2	1	9	2	0	Cross 1985
	S. Oregon Cascades	4	>250	5	0	8	2	0	Cross 1985
	S. Oregon Cascades	1	>250	3	0	10	2	0	Cross 1985
	Oregon Cascades	4	100–250	6	3	19	6	2	Doyle 1990
	Oregon coast	6	120–140	2	0	18	4	4	McComb et al. 1993b
	Oregon coast	3	[b]40–50	7	4	24	0	0	McComb et al. 1993a,c
	Washington Cascades	18	40–60	2	0	15	4	2	West unpublished data

[a] Number of species with captures significantly greater in eiher riparian or upland habitat ($p < 0.05$).
[b] Alder forest.
[c] Upland forest compared with riparian buffer strip averaging 67 m along one side of stream.
[d] Upland forest compared with riparian buffer strip averaging 13 m along one side of stream.

community studies in western Oregon indicate significantly greater species richness and abundance along upland transects, 400m from second- or third-order streams, than along riparian transects (McGarigal and McComb 1992). One-third of the avian species were observed only in upland habitat while only 9% were unique to riparian zones. The greater number of snags and large trees in upland areas appears to have a more significant influence on avian species distribution than the riparian deciduous vegetation. Evidence suggests that riparian zones do not provide high quality habitat for cavity nesters during the spring breeding season.

McComb et al. (1993a, 1993b) found that amphibian species diversity, equitability, richness, and abundance did not differ between riparian and upland transects. In these studies, stream-breeding species in Oregon, adult tailed frog and Pacific giant salamander, were captured more frequently in riparian zones than in uplands, but the difference was not significant. In Washington, Pacific giant salamanders were captured significantly more frequently in riparian zones than in uplands 100m from the stream, and adult tailed frogs were captured approximately 1.5 times more frequently in uplands than riparian zones (West unpublished data). Even though both species are dependent on in-stream habitat for breeding and larval development, adults move away from streams outside of the breeding season. Dispersal and migration distances are unknown at this time. Capture rates of pond-breeding species were low suggesting that study sites were not located near breeding ponds. The roughskin newt, northwestern salamander, and red-legged frog (*Rana aurora*) were captured in similar numbers on riparian and upland transects (McComb et al. 1993a, 1993b; West unpublished data). Captures of terrestrial-breeding salamanders varied by species. The *Ensatina* salamander was captured in significantly greater numbers in upland areas, while western redback salamander captures were not significantly different along riparian and upland transects. Despite low sample sizes, McComb et al. (1993a, 1993b) and West (unpublished data) found Dunn's salamander more abundant in riparian zones.

Comparisons of small mammal communities indicate significantly greater species richness in the Oregon Cascades riparian zones (Doyle 1990) and greater species diversity and equitability in the Oregon Coast Mountains (McComb et al. 1993b) and Washington Cascades riparian zones (West unpublished data) when compared to adjacent uplands. Differences in small mammal communities between adjacent habitats vary as a function of contrast in plant community structure and species composition. Slight differences are found between riparian zones and uplands in alder forests (McComb et al. 1993a), somewhat more contrast in coniferous forests (Doyle 1990, McComb et al. 1993b, West unpublished data), and strong differences in coniferous forests in southwestern Oregon (Cross 1985). In the Cross study, upland areas are mostly devoid of ground cover and very few shrubs are present, yielding results that show species richness and diversity to be similar to that of arid regions where riparian and upland habitats are distinct.

The similarity of bird, amphibian, and small mammal communities between upland and riparian habitat suggests that adjacent uplands contribute significantly to the diversity of wildlife communities in western Washington and Oregon riparian forests. Due to the slight contrast between upland and riparian habitat, many species appear to recognize the transition as continuous habitat rather than two distinct habitats. The data presented in Table 10.3 highlight the similarity of riparian and adjacent upslope wildlife communities, while exposing the need for research at the northern and southern ends of the Pacific coastal ecoregion.

Animal influences. While habitat characteristics influence animal communities, animals also affect habitats. An excellent example is dam building activities of beaver which result in removal of trees adjacent to streams and water impoundment, thus creating and maintaining wetlands (Finley 1937, Naiman and Melillo 1984, Naiman et al. 1994, Butler 1995). These actions not only reset the ecological development of the riparian forest, they modify habitat to the point of creating an entirely new environment. Beaver activities alter stream morphology and patterns of discharge, decrease current velocity, increase retention times of sediment and organic matter, and expand areas of flooded soil (Naiman et al. 1988, Butler 1995, Pollock et al. 1995). As beaver remove trees, insolation increases followed by increased rates of primary production (see Chapter 7). Lotic invertebrate communities are replaced by lentic communities. In fact, the total density and biomass of invertebrates in beaver ponds is two to five times greater than in stream riffles prior to damming (McDowell and Naiman 1986).

The interaction of beaver activity and hydrologic processes influences the successional status of riparian zones. Beaver ponds may return to stream habitat or fill with sediment to become meadows, wetlands, or bogs depending on a range of factors, including topography, soil characteristics, existing vegetation, fire, and herbivory (Naiman and Melillo 1984, Naiman et al. 1988). The modification of riparian habitat by beaver also influences wildlife community composition. Pond-breeding salamanders and frogs utilize beaver ponds for breeding, egg deposition, and larval development in areas where they did not utilize the previous stream habitat. Conversely, the distribution of stream-breeding amphibians is interrupted as they avoid impoundments. Reptiles, such as the western pond turtle, make use of sunny areas and warm temperatures. Avian and small mammal communities shift from forest associated species to those commonly found in grassy meadows and along forest edges. Bats commonly forage for insects over ponds and soon discover new ponds. Although thermal and protective cover for ungulates is generally reduced in areas inhabited by beaver, such losses are offset by increases in high quality forage. Given the striking reduction in beaver populations from trapping (Dalquest 1948) and from habitat loss as agricultural, industrial, and residential areas expand, it is likely that only a fraction of the habitat and wildlife diversity that was present historically exists today. Pond-breeding amphibian populations, in particular, are probably much lower now and more unevenly distributed than when beaver impoundments were prevalent elements of drainage networks.

Herbivory also may exert a profound effect on riparian community structure. Studies using exclosures designed to exclude elk and deer from foraging areas reveal significant differences in the vegetation community structure inside and outside of the exclosures (Hanley and Taber 1980, Woodward et al. 1994). Herbivory and trampling reduce growth of shrub species (e.g., *Rubus spectabilis*, *Sambucus racemosa*) and stimulate growth of ground-covering herbaceous species. Such changes reduce availability of fruit, nectar, and understory vegetation often used by nesting birds.

Management Effects on Riparian Wildlife Communities

Riparian management can be considered the regulation of human-induced disturbances within riparian zones. Goals range from the protection of wildlife habitat to the maintenance of right-of-ways and the development of roads. In contrast to recurrent patterns of disturbance and recovery found under natural disturbance regimes, human-induced disturbances often have unidirectional effects on riparian communities. Urban and agricultural development converts riparian habitat to other purposes. Building and farming activities within or adjacent to riparian zones often involve channeling and confining rivers. These anthropogenic disturbances result in the loss of riparian and wetland habitats, permanent alteration of hydrologic regimes, and dramatic changes in wildlife communities. Natural stream processes, such as erosion and deposition of materials, are altered and streambed characteristics change. Edge species frequently found in urban areas and agricultural fields replace forest riparian species. Construction of large hydroelectric dams results in the permanent loss of thousands of hectares of riparian and upland habitat. Although new riparian zones form along reservoir edges, stream processes are interrupted and soils, flood regimes, vegetation, and wildlife communities are altered. Dams interrupt habitat connectivity for aquatic animals, prevent periodic downstream flooding that deposits sediment and suspended organic materials, and alter water levels (Chapter 12).

Many types of riparian management have so completely altered riparian zones that intervention is extremely difficult and expensive. However, timber harvest practices along streams and rivers can be designed to minimize impacts on wildlife communities. In some respects, timber harvest mimics the effects of natural disturbances on a landscape. Trees are lost and forest succession is reinitiated by the disturbance. However, standing live trees, snags, and downed logs that have tremendous value as components of wildlife habitat and often remain following natural disturbances are removed during timber harvest (Maser et al. 1988, Franklin 1992). Recent experimental and regulatory efforts to mitigate effects of timber harvest on wildlife require leaving a certain number of trees, snags, and downed logs for every hectare of forest cut (Washington State Forest Practices Board 1988, Stofel 1993).

While the primary means of protecting riparian and aquatic areas from the impacts of timber harvest in adjacent upslope areas is to leave trees along both sides of a stream or river, questions arise concerning the configuration of these buffer strips and their effectiveness for providing wildlife habitat. Buffer strips are intended to provide shade, thus preventing dramatic changes in stream water temperatures; maintain bank stability by reducing or eliminating stream bank disturbances and preventing erosion; reduce runoff of fertilizers, herbicides, and pesticides into streams; act as barriers to logging debris; provide continuous inputs of wood and other terrestrial organic matter to streams, essential habitat components for aquatic life; and provide habitat for wildlife.

A basic question of buffer strip function is how do wildlife use riparian habitat within a buffer strip along small and mid-size streams? Washington and Oregon researchers have implemented an experimental approach by observing wildlife (amphibians, breeding birds, bats, small mammals) in riparian and upland habitats both within buffer strips and adjacent clearcuts before harvest and after harvest (West unpublished data, McComb personal communication). A similar study in Washington examined responses of tailed frog tadpoles and Pacific giant salamander larvae to buffer strips on perennial, nonfish-bearing streams (Kelsey 1995). Results indicate that clearcutting, even when combined with buffer strips, has a detrimental effect on tadpoles. These results raise questions about the effectiveness of buffer strips in protecting wildlife from impacts of timber harvest.

Windthrow of standing trees bordered by clearcuts is not a new problem. Foresters are often reluctant to leave trees in areas with a history of blowdown. Effects of blowdown along streams can be undesirable when uprooted trees destabilize the streambank and expose large areas of loose soil prone to erosion during rainy periods. Three of the six intended functions of buffer strips listed above are compromised when blowdown occurs: shade is reduced (unless tree crowns fall over the stream); bank stability is diminished and sediment input to the stream is increased; and long-term sources for in-stream woody debris are reduced. While a downed tree does provide wildlife habitat and cover for small mammals and amphibians at the point where it contacts the ground or streambank, both downed and standing trees are necessary to maintain habitat for riparian wildlife communities. Although windthrow of trees occurs naturally in forests, buffer strips increase the probability of blowdown by suddenly exposing trees to winds that previously were attenuated by the surrounding forest (Steinblums et al. 1984). It is difficult at present to determine precisely where buffer strips are more likely to fail, but eventually such predictions will be needed if buffer strips are to be successfully included in riparian wildlife management plans. This is especially so if several harvest rotations are anticipated because initial establishment of riparian buffer zones is the most difficult. The challenge is to retain standing trees in the buffer until young trees in the surrounding harvest area reach sufficient size to reduce wind velocity at the buffer margin. After trees within the buffer acquire wind-firmness, subsequent harvests are less demanding in terms of manipulations needed to avoid windthrow.

Riparian Wildlife Management Alternatives

Riparian management alternatives are most effectively developed within a continuum of forest harvest intensity bounded at one extreme by the removal of all trees (clear-cutting) and at the other by the retention of all trees (no timber harvest). Along the continuum, forest managers may decide to leave a certain percentage of trees in one of many possible configurations. The percentage removed represents a position along the continuum. For example, a 10m buffer strip left along a stream flowing through a 35ha clearcut falls closer to the clearcut end than the retention end of the continuum. Wildlife responses to management scenarios vary along the continuum, but not necessarily uniformly, and the existence of response thresholds is likely. Expected responses are based on knowledge of the life history characteristics of individual species and information from field experiments and surveys.

Numerous alternative riparian management strategies can be designed by varying buffer widths and the proportion of trees removed. To preserve preharvest microclimate conditions within riparian zones, buffer strips of 100 to 150m may be left. Although wind velocities remain elevated up to 150 to 185m from a forest or clearcut edge, temperature, humidity, light intensity, and soil moisture reflect interior forest levels at 100m from the clearcut (Chen et al. 1995). Such large buffers increase the likelihood of providing habitat for wildlife species (such as the tailed frog, Pacific slope flycatcher, and southern red-backed vole) associated with interior forest conditions.

Without the threat of windthrow, minimal buffer widths (e.g., 10m on both sides of a small stream) appear to provide adequate stream protection for amphibians and habitat structure for wildlife associated with early successional forests (Figure 10.6a). As the probability of windthrow increases, greater protection for riparian trees is needed. This can be accomplished by increasing the width of buffer strips and establishing areas of selective harvest along the buffer (Figure 10.6b). By gradually decreasing the density of trees through selective cutting, the landowner derives economic benefits while at the same time protecting the stream and riparian corridor. In areas of selective logging, retention of shrubs, particularly berry-producing shrubs (e.g. *Rubus* spp., *Vaccinium* spp., *Ribes* spp.), snags, and damaged trees is an important consideration. These features provide food, cover, and breeding resources for various wildlife species. The actual number of trees removed should be determined on a site-by-site basis to accommodate operational constraints and site topography.

Beyond considerations of buffer width and tree density, a variety of options can be investigated. Topping trees prone to windthrow and leaving standing trunks approximately 6 to 8m high (Figure 10.6c) creates buffer strips that are less subject to blowdown and provides wildlife habitat for woodpeckers, cavity nesting birds, and arboreal small mammals. These fabricated snags continue to function as long-term sources of wood for stream and riparian zones. Outside of the band of topped trees, a rather narrow buffer strip, similar to Figure 10.6a, can be left to shade the stream. Another option is to whorl-prune trees, removing alternate whorls of branches to break up wind currents and to decrease wind interception on buffer strip edges (Agee 1995).

Abandoning the idea of continuous buffer strips in some areas would allow more extensive protection in other areas. Leaving islands of trees along stream reaches particularly sensitive to impacts from timber operations preserves riparian microclimate conditions and provides habitat for riparian and upland species (Figure 10.6d). If streamside shrubs and ground cover are left intact and trees, slash, or other logging debris are prevented from entering the stream, then timber can be harvested to the stream's edge upstream and downstream of the buffer island. This alternative provides two primary benefits. First, wildlife with limited mobility and species particularly sensitive to forest harvest are protected from habitat changes resulting from timber harvest. Second, land owners have the flexibility to provide

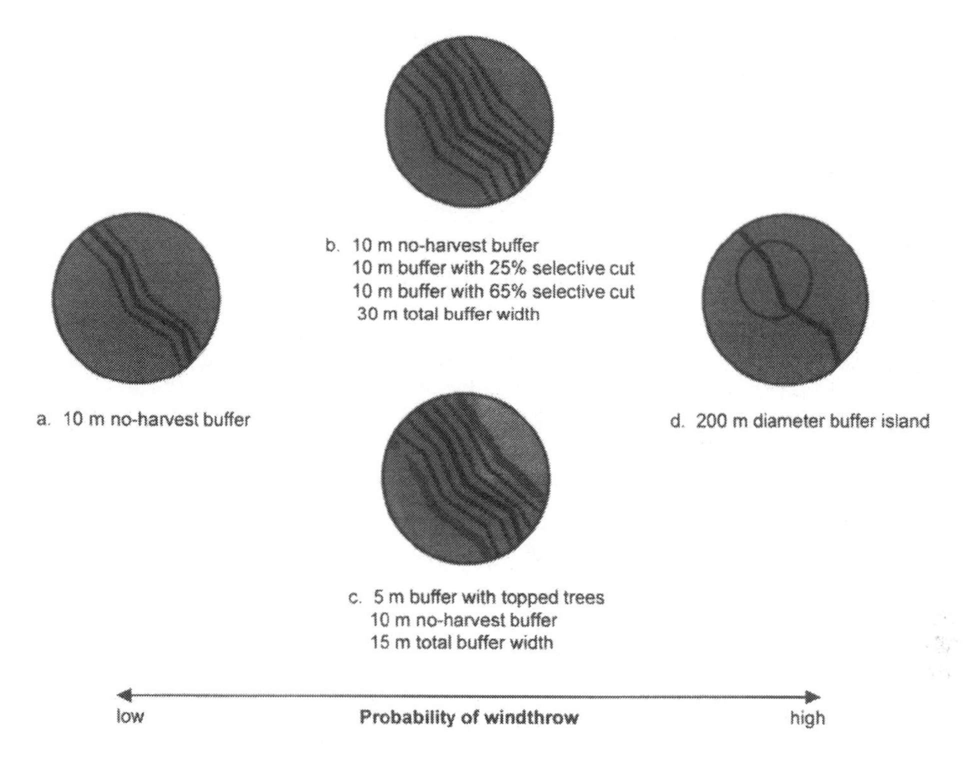

FIGURE 10.6. Representation of alternative forest harvest management strategies designed to optimize wildlife habitat and reduce incidence of windthrow of riparian trees along first to fourth order streams. (a) Stream shown with 10m, no harvest buffer strip of trees on each side of stream; appropriate when probability of windthrow is low; (b) Stream shown with 30m buffer strip of trees on each side of stream: the first 10m from the stream is a no harvest zone, 10 to 20m from stream is a 25% selective cut zone, 20 to 30m from stream is a 60% selective cut zone. This buffer configuration is suggested in areas with a moderate risk of windthrow; (c) Stream shown with 15m buffer strip. Trees within 5m of stream are topped to heights of 6 to 8m. Area 5 to 15m from stream is a selective harvest area to preserve shade over stream. Suggested in areas with moderate probability of windthrow; (d) Stream shown with 200m diameter buffer island intended for use along headwater streams with high probabilities of windthrow, unique riparian communities, or unstable soils.

extensive protection of stream and riparian habitat in unstable areas, or areas with high riparian obligate species diversity, while harvesting trees to the stream edge in other, less sensitive areas. Variations of this alternative include modifications to the size of the buffer island, the degree of selective cutting within the buffer, and the treatment of upstream and downstream areas.

Broad management prescriptions are less effective than management based on local conditions. In effect, no single management option optimizes environmental and timber management concerns for all wildlife situations. Fixed prescriptions for the design of buffer strips derived from "average" conditions result in excessive retention of trees in some areas and insufficient protection in others. Successful applications of management options consider wildlife goals at each site. For example, forest practices regulations designed specifically to protect the torrent salamander only need to be implemented in areas of their known distribution and not throughout western Washington and Oregon. Only localized protection of small tributaries is necessary to preserve their breeding habitat.

Another element of successful management decisions regarding riparian zones is consideration of the entire drainage basin rather than just the individual harvest unit. This broad view of habitat management is facilitated by the use of Geographic Information Systems (GIS) to produce detailed maps of large areas. However, basins are divided into small, independently managed regions by ownership boundaries which streams, as well as wildlife, do not recognize. Attempts by one landowner to reduce logging impacts on streams are often negated by a second owner who implements minimum regulations (Hamilton 1996). Research to monitor wildlife communities and assess wildlife responses to timber harvest practices are hampered by multiple land ownership. Cooperative efforts are difficult to arrange, often bring delays as decisions and approval for protocols occur at many administrative levels (see Chapter 21), and must not violate anti-trust laws. There are no easy solutions to these difficulties except to stress the importance of communication and cooperation between landowners, regulators, and biologists (Chapters 12 and 27). As demands on natural resources increase with a growing human population, the ability to successfully coordinate wildlife management and research efforts becomes even more important.

Riparian wildlife communities in the Pacific coastal ecoregion may not be the primary drivers of design criteria for riparian management because of their similarities to upland wildlife communities. Nonetheless, strategies to provide wildlife habitat must include both riparian and upland forests due to the fluidity of habitat use between the two areas. Further, it is important to recognize that riparian management practices that make sense intuitively often do not function as intended. Therefore, wildlife monitoring projects must accompany the implementation of alternative management options to assure that new strategies meet their goals.

Literature Cited

Agee, J.K. 1988. Successional dynamics in riparian zones. Pages 31–43 *in* K.J. Raedeke, ed. Streamside management: Riparian wildlife and forestry interactions. Institute of Forest Resources Contribution Number 59, University of Washington, Seattle, Washington, USA.

——. 1995. Management of greenbelts and forest remnants in urban forest landscapes. Pages 128–138 *in* G.A. Bradley, ed. Urban forest landscapes: Integrating multidisciplinary perspectives. University of Washington Press, Seattle, Washington, USA.

Anthony, R.G., E.D. Forsman, G.A. Green, G. Witmer, and S.K. Nelson. 1987. Small mammal populations in riparian zones of different aged coniferous forests. Murrelet **68**:94–102.

Bailey, V. 1936. The mammals and life zones of Oregon. North American Fauna **55**:1–416.

Best, L.B., D.F. Stauffer, and A.R. Geier. 1979. Evaluating the effects of habitat alteration on birds and small mammals occupying riparian communities. Pages 117–124 *in* R.R. Johnson and J.F. McCormick, technical coordinators. Strategies for protection and management of floodplain wetlands and other riparian ecosystems. USDA Forest Service General Technical Report WO-12, Fort Collins, Colorado, USA.

Best, L.B., and D.F. Stauffer. 1980. Factors affecting nesting success in riparian bird communities. Condor **82**:149–158.

Brown, B.T. 1994. Rates of brood parasitism by brown-headed cowbirds on riparian passerines in Arizona. Journal of Field Ornithology **65**:160–168.

Brown, H.A., R.B. Bury, D.M. Darda, L.V. Diller, C.R. Peterson, and R.M. Storm. 1995. Reptiles of Washington and Oregon. Seattle Audubon Society, Seattle, Washington, USA.

Butler, D.R. 1995. Zoogeomorphology. Cambridge University Press, Cambridge, UK.

Calder, W.A. 1973. Microhabitat selection during nesting of hummingbirds in the Rocky Mountains. Ecology **54**:129–134.

Campbell, L.A., J.G. Hallett, and M.A. O'Connell. 1996. Conservation of bats in managed forests: Use of roosts by *Lasionycteris noctivagans*. Journal of Mammalogy **77**:976–984.

Carlson, A. 1991. Characterization of riparian management zones and upland management areas with respect to wildlife habitat, 1988–1990 cumulative report. Timber/Fish/Wildlife Report WLI 91-001. Washington Department of Wildlife, Olympia, Washington, USA.

Chen, J., J.F. Franklin, and T.A. Spies. 1995. Growing-season microclimatic gradients from clearcut edges into old-growth Douglas-fir forests. Ecological Applications **5**:74–86.

Christy, R.E., and S.D. West. 1991. Biology of bats in Douglas-fir forests. USDA Forest Service General

Technical Report PNW-308, Pacific Northwest Research Station, Portland, Oregon, USA.

Coady, J.W. 1982. Moose. Pages 902–922 *in* J. Chapman and G. Feldhamer, eds. Wild mammals of North America. The Johns Hopkins University Press, Baltimore, Maryland, USA.

Cocroft, R.B. 1994. A cladistic analysis of chorus frog phylogeny (Hylidae: *Pseudacris*). Herpetologica **50**:420–437.

Collins, J.T. 1990. Standard common and current scientific names for North American amphibians and reptiles. Herpetological Circular Number 19. Society for the Study of Amphibians and Reptiles, Lawrence, Kansas, USA.

Conaway, C.H. 1952. Life history of the water shrew (*Sorex palustris navigator*). American Midland Naturalist **48**:219–248.

Croonquist, M.J., and R.P. Brooks. 1993. Effects of habitat disturbance on bird communities in riparian corridors. Journal of Soil and Water Conservation **48**:65–70.

Cross, S.P. 1985. Responses of small mammals to forest riparian perturbations. Pages 269–275 in R.R. Johnson, ed. Riparian ecosystems and their management: Reconciling conflicting uses. Proceedings of the First North American Riparian Conference. USDA Forest Service General Technical Report RM-120, Fort Collins, Colorado, USA.

——. 1988. Riparian systems and small mammals and bats. Pages 93–112 *in* K.J. Raedeke, ed. Streamside management: Riparian wildlife and forestry interactions. Institute of Forest Resources Contribution Number 59, University of Washington, Seattle, Washington, USA.

Dalquest, W.W. 1948. Mammals of Washington. Volume 2, Museum of Natural History. University of Kansas Publications, Lawrence, Kansas, USA.

Doyle, A.T. 1985. Small mammal micro- and macrohabitat selection in streamside ecosystems (Oregon, Cascade range). Ph.D. dissertation. Oregon State University, Corvallis, Oregon, USA.

——. 1990. Use of riparian and upland habitats by small mammals. Journal of Mammalogy **71**:14–23.

Dublin, H.T. 1980. Relating deer diets to forage quality and quantity: The Columbian white-tailed deer (*Odocoileus virginianus leucurus*). Master's thesis. University of Washington, Seattle, Washington, USA.

Dumas, P.C. 1956. The ecological relations of sympatry in *Plethodon dunni* and *Plethodon vehiculum*. Ecology **37**:485–495.

Ehrlich, P.R., D.S. Dobkin, and D. Wheye. 1988. The birder's handbook: A field guide to the natural history of North American birds. Simon & Schuster, New York, New York, USA.

Fenton, M.B. 1988. Detecting, recording, and analyzing vocalizations of bats. Pages 91–104 *in* T.H. Kunz, ed. Ecological and behavioral methods for the study of bats. Smithsonian Institution, Washington, DC, USA.

Finley, W. 1937. The beaver: Conserver of soil and water. Transactions of the North American Wildlife Conference **2**:295–297.

Fitzgerald, B.M. 1977. Weasel predation on a cyclic population of the montane vole, *Microtus montanus*, in California. Journal of Animal Ecology **46**:367–397.

Fleischner, T.L. 1994. Ecological costs of livestock grazing in western North America. Conservation Biology **8**:629–644.

Franklin, J.F. 1992. Scientific basis for new perspectives in forests and streams. Pages 25–72 *in* R.J. Naiman, ed. Watershed management: Balancing sustainability and environmental change. Springer-Verlag, New York, New York, USA.

Frazier, M.W. 1997. Roost site characteristics of the long-legged Myotis (*Myotis volans*) in the Teanaway River Valley of Washington. Master's thesis. University of Washington, Seattle, Washington, USA.

Gradwell, N. 1971. *Ascaphus* tadpole: Experiments on the suction and gill irrigation mechanisms. Canadian Journal of Zoology **49**:307–332.

Hall, F.C. 1988. Characterization of riparian systems. Pages 7–12 *in* K. Raedeke, ed. Streamside management: Riparian wildlife and forestry interactions. Institute of Forest Resources Contribution Number 59, University of Washington, Seattle, Washington, USA.

Hamilton, D. 1996. The role of private forests in ecosystem management: Policy issues, themes and recommendations. Professional paper, College of Forest Resources, University of Washington, Seattle, Washington, USA.

Hanley, T.A. 1982. The nutritional basis for food selection by ungulates. Journal of Range Management **35**:146–151.

Hanley, T.A., and R.D. Taber. 1980. Selective plant species inhibition by elk and deer in three conifer communities in western Washington. Forest Science **26**:97–107.

Hartman, G.D., and T.L. Yates. 1985. *Scapanus orarius*. Mammalian Species **253**:1–5.

Hayes, M.P., and M.R. Jennings. 1986. Decline of ranid frog species in western North America: Are bullfrogs (*Rana catesbeiana*) responsible? Journal of Herpetology **20**:490–509.

Hill, E.H. 1982. Beaver (*Castor canadensis*). Pages 256–281 *in* J.A. Chapman and G.A. Feldhamer, eds. Wild mammals of North America. Johns Hopkins University Press, Baltimore, Maryland, USA.

Hofmann, R.R. 1988. Morphophysiological evolutionary adaptations of the ruminant digestive system. Pages 1–20 *in* A. Dobson and M.J. Dobson, eds. Aspects of digestive physiology in ruminants: Proceedings of Physiological Sciences. Comstock, Ithaca, New York, USA.

Hooven, E.F., and G.C. Black. 1976. Effects of some clearcutting practices on small mammal populations in western Oregon. Northwest Science **50**:189–208.

Ingles, L.G. 1965. Mammals of the Pacific states. Stanford University Press, Stanford, California, USA.

Ingold, D.J. 1994. Influence of nest-site competition between European starlings and woodpeckers. Wilson Bulletin **106**:227–241.

Jenkins, R.K.B., and S.J. Ormerod. 1996. The influence of a river bird, the dipper (*Cinclus cinclus*), on the behaviour and drift of its invertebrate prey. Freshwater Biology **35**:45–56.

Karr, J.R., and R.R. Roth. 1971. Vegetation structure and avian diversity in several New World areas. American Naturalist **105**:423–435.

Kelsey, K.A. 1995. Responses of headwater stream amphibians to forest practices in western Washington. Ph.D. dissertation. University of Washington, Seattle, Washington, USA.

Knopf, F.L. 1985. Significance of riparian vegetation to breeding birds across an altitudinal cline. Pages 105–111 *in* R.R. Johnson, C.D. Ziebell, D.R. Patton, P.F. Ffolliott, and R.H. Hamre, technical coordinators. Riparian ecosystems and their management: Reconciling conflicting uses. USDA Forest Service General Technical Report RM-120, Fort Collins, Colorado, USA.

Lawson, R., and H.C. Dessauer. 1979. Biochemical genetics and systematics of garter snakes of the *Thamnophis elegans-couchii-ordinoides* complex. Occasional Papers of the Museum of Zoology at Louisiana State University **56**:1–24.

Layser, E.F., and T.E. Burke. 1973. The northern bog lemming and its unique habitat in northeastern Washington. Murrelet **54**:7–8.

Leonard, W.P., H.A. Brown, L.L.C. Jones, K.R. McAllister, and R.M. Storm. 1993. Amphibians of Washington and Oregon. Seattle Audubon Society, Seattle, Washington, USA.

Lind, A.J., and H.H. Welsh, Jr. 1994. Ontogenetic changes in foraging behaviour and habitat use by the Oregon garter snake, *Thamnophis atratus hydrophilus*. Animal Behavior **48**:1261–1273.

Lock, P. 1991. Old growth riparian birds of the Olympic Peninsula: Effects of stream size on community structure. Master's thesis. University of Washington, Seattle, Washington, USA.

Ludwig, D.R. 1984. *Microtus richardsoni*. Mammalian Species **223**:1–6.

MacArthur, R.H., and J.W. MacArthur. 1961. On bird species diversity. Ecology **42**:594–598.

MacKenzie, D.I., S.G. Sealy, and G.D. Sutherland. 1982. Nest-site characteristics of the avian community of the dune-ridge forest, Delta Marsh, Manitoba: A multivariate approach. Canadian Journal of Zoology **60**:2212–2223.

Malanson, G.P. 1993. Riparian landscapes. Cambridge University Press, Cambridge, UK.

Manuwal, D.A. 1986. Characteristics of bird assemblages along linear riparian zones in western Montana. Murrelet **67**:10–18.

Martin, T.E. 1988. Habitat and area effects on forest bird assemblages: Is nest predation an influence? Ecology **69**:74–84.

Maser, C., B.R. Mate, J.F. Franklin, and C.T. Dyrness. 1981. Natural history of some Oregon coast mammals. USDA Forest Service General Technical Report PNW-GTR-133, Pacific Northwest Forest and Range Experiment Station, Portland, Oregon, USA.

Maser, C., R.F. Tarrant, J.M. Trappe, and J.F. Franklin, eds. 1988. From the forest to the sea: A story of fallen trees. USDA Forest Service General Technical Report PNW-GTR-229, Pacific Northwest Forest and Range Experiment Station, Portland, Oregon, USA.

McComb, W.C., C.L. Chambers, and M. Newton. 1993a. Small mammal and amphibian communities and habitat associations in red alder stands, central Oregon Coast Range. Northwest Science **67**:181–188.

McComb, W.C., K. McGarigal, and R.G. Anthony. 1993b. Small mammal and amphibian abundance in streamside and upslope habitats of mature Douglas-fir stands, western Oregon. Northwest Science **67**:7–14.

McDowell, D.M., and R.J. Naiman. 1986. Structure and function of a benthic invertebrate stream community as influenced by beaver *Castor canadensis*. Oecologia **68**:481–489.

McGarigal, K., and W.C. McComb. 1992. Streamside versus upslope breeding bird communities in the central Oregon Coast Range. Journal of Wildlife Management **56**:10–23.

Morrison, M.L., and E.C. Meslow. 1983. Avifauna associated with early growth vegetation on clearcuts in the Oregon Coast Ranges. USDA Forest Service Research Paper PNW-305, Pacific Northwest Forest and Range Experiment Station, Portland, Oregon, USA.

Naiman, R.J., T.J. Beechie, L.E. Benda, D.R. Berg, P.A. Bisson, L.H. MacDonald, et al. 1992. Fundamental elements of ecologically healthy watersheds in the Pacific Northwest coastal ecoregion. Pages 127–188 *in* R.J. Naiman, ed. Watershed management: Balancing sustainability and environmental change. Springer-Verlag, New York, New York, USA.

Naiman, R.J., C.A. Johnston, and J.C. Kelley. 1988. Alteration of North American streams by beaver. BioScience **38**:753–762.

Naiman, R.J., and J.M. Melillo. 1984. Nitrogen budget of a subarctic stream altered by beaver (*Castor canadensis*). Oecologia (Berlin) **62**:150–155.

Naiman, R.J., G. Pinay, C.A. Johnston, and J. Pastor. 1994. Beaver influences on the long term biogeochemical characteristics of boreal forest drainage networks. Ecology **75**:905–921.

Nolan, V., Jr. 1978. The ecology and behavior of the Prairie Warbler, *Dendroica discolor.* Ornithological Monographs Number 26, Lawrence, Kansas, USA.

Nussbaum, R.A., E.D. Brodie, Jr., and R.M. Storm. 1983. Amphibians and reptiles of the Pacific Northwest. University Press of Idaho, Moscow, Idaho, USA.

Odum, E.P. 1979. Ecological importance of the riparian zone. Pages 2–4 *in* R.R. Johnson and J.F. McCormick, technical coordinators. Strategies for protection and management of floodplain wetlands and other riparian ecosystems. USDA Forest Service General Technical Report WO-GTR-12, Fort Collins, Colorado, USA.

Pattie, D. 1973. *Sorex bendirii.* Mammalian Species **27**:1–2.

Pollock, M.M., R.J. Naiman, H.E. Erickson, C.A. Johnston, J. Pastor, and G. Pinay. 1995. Beaver as engineers: Influences on biotic and abiotic characteristics of drainage basins. Pages 117–126 *in* C.G. Jones and J.G. Lawton, eds. Linking species and ecosystems. Chapman & Hall, New York, New York, USA.

Raedeke, K.J., R.D. Taber, and D.K. Paige. 1988. Ecology of large mammals in riparian systems of Pacific Northwest forests. Pages 113–132 *in* K.J. Raedeke, ed. Streamside management: Riparian wildlife and forestry interactions. Institute of Forest Resources Contribution Number 59, University of Washington, Seattle, Washington, USA.

Rossman, D.A., and G.R. Stewart. 1987. Taxonomic reevaluation of *Thamnophis couchii* (Serpentes: Colubridae). Occasional Papers of the Museum of Zoology at Louisiana State University 63–25, Baton Rouge, Louisiana, USA.

Schroer, G.L., G. Witmer, and E.E. Starkey. 1988. Effects of land use practices on Roosevelt elk winter ranges, western Washington and Oregon. Pages 68–75 *in* Proceedings of the 1988 western states and provinces elk workshop, July 13–15, 1988, Wenatchee, Washington. Washington Department of Wildlife, Olympia, Washington, USA.

Simms, D.A. 1979. North American weasels: Resource utilization and distribution. Canadian Journal of Zoology **57**:504–520.

Smith, K.G. 1977. Distribution of summer birds along a forest moisture gradient in an Ozark watershed. Ecology **58**:810–819.

Spies, T.A., and J.F. Franklin. 1991. The structure of natural young, mature and old-growth Douglas-fir forests in Washington and Oregon. Pages 90–109 *in* L.F. Ruggiero, K.B. Aubry, A.B. Carey, and M.H. Huff, technical coordinators. Wildlife and vegetation of unmanaged Douglas-fir forests. USDA Forest Service General Technical Report PNW-GTR-285, Pacific Northwest Research Station, Portland, Oregon, USA.

Stauffer, D.F., and L.B. Best. 1980. Habitat selection by birds of riparian communities: Evaluating effects of habitat alterations. Journal of Wildlife Management **51**:616–621.

Stebbins, R.C. 1985. Western reptiles and amphibians. Houghton Mifflin, Boston, Massachusetts, USA.

Stebbins, R.C., and N.W. Cohen. 1995. A natural history of amphibians. Princeton University Press, Princeton, New Jersey, USA.

Steel, E.A. 1993. Woody debris piles: Habitat for birds and small mammals in the riparian zone. Master's thesis. University of Washington, Seattle, Washington, USA.

Steinblums, I.J., H.A. Froehlich, and J.K. Lyons. 1984. Designing stable buffer strips for stream protection. Journal of Forestry 82:49–52.

Stevens, L.E., B.T. Brown, J.M. Simpson, and R.R. Johnson. 1977. The importance of riparian habitat to migrating birds. Pages 156–164 *in* R.R. Johnson and D.A. Jones, technical coordinators. Importance, preservation, and management of riparian habitat: A symposium. USDA Forest Service General Technical Report RM-43, Fort Collins, Colorado, USA.

Stofel, J.L. 1993. Evaluating wildlife responses to alternative silvicultural practices. Master's thesis. University of Washington, Seattle, Washington, USA.

Swanson, F.J., L.E. Benda, S.H. Duncan, G.E. Grant, W.F. Megahan, L.M. Reid, et al. 1987. Mass failures and other processes of sediment production in Pacific Northwest forest landscapes. Pages 9–38 *in* E.O. Salo and T.W. Cundy, eds. Streamside management: Forestry and fishery interactions. Institute of Forest Resources Contribution Number 57, University of Washington, Seattle, Washington, USA.

Sweeney, S.J. 1978. Diet, reproduction, and population structure of bobcat (*Lynx rufus fasciatus*) in western Washington. Master's thesis. University of Washington, Seattle, Washington, USA.

Szaro, R.C. 1980. Factors influencing bird populations in southwestern riparian forests. Pages 403–418 *in* R.M. DeGraaf, technical coordinator. Management of western forests and grasslands for nongame birds. USDA Forest Service General Technical Report INT-86, Ogden, Utah, USA.

Terres, J.K. 1980. The Audubon Society encyclopedia of North American birds. Alfred A. Knopf, New York, New York, USA.

Terry, C.J. 1981. Habitat differentiation among three species of *Sorex* and *Neurotrichus gibbsii* in Washington. American Midland Naturalist **106**: 119–125.

Thomas, D.W. 1988. The distribution of bats in different ages of Douglas-fir forests. Journal of Wildlife Management **52**:619–626.

Thomas, J.W., C. Maser, and J.E. Rodieck. 1979. Riparian zones. Pages 40–47 *in* J.W. Thomas, ed. Wildlife habitats in managed forests: The Blue Mountains of Oregon and Washington. USDA Forest Service Agricultural Handbook Number 553. USDI Bureau of Land Management Wildlife Management Institute, Washington, DC, USA.

Thompson, F.R. III, W.D. Dijak, T.G. Kulowiec, and D.A. Hamilton. 1992. Breeding bird populations in Missouri Ozark forests with and without clearcutting. Journal of Wildlife Management **56**:23–30.

Triquet, A.M., G.A. McPeek, and W.C. McComb. 1990. Songbird diversity in clearcuts with and without a riparian buffer strip. Journal of Soil and Water Conservation **45**:500–503.

Tubbs, A.A. 1980. Riparian bird communities of the Great Plains. Pages 419–433 *in* R.M. DeGraaf, technical coordinator. Management of western forests and grasslands for nongame birds. USDA Forest Service General Technical Report INT-86, Ogden, Utah, USA.

Van Daele, L.J., and H.A. Van Daele. 1982. Factors affecting the productivity of ospreys nesting in west-central Idaho. Condor **84**:292–299.

Walsberg, G.E. 1981. Nest-site selection and the radiative environment of the warbling vireo. Condor **83**:86–88.

Washington State Forest Practices Board. 1988. Washington state forest practices, rules, and regulations. Olympia, Washington, USA.

Willner, G.R., G.A. Feldhamer, E.E. Zucker, and J.A. Chapman. 1980. *Ondatra zibethicus*. Mammalian Species **141**:1–8.

Willson, M.F. 1974. Avian community organization and habitat structure. Ecology **55**:1017–1029.

Wilson, C., J.D. Reichel, and R.E. Johnson. 1980. New records for the northern bog lemming in Washington. Murrelet **61**:104–106.

Witmer, G.W., and D.S. de Calesta. 1983. Habitat use by female Roosevelt elk in the Oregon Coast Range. Journal of Wildlife Management **47**:933–939.

Woodward, A., E.G. Schreiner, D.B. Houston, and B.B. Moorhead. 1994. Ungulate-forest relationships in Olympic National Park: Retrospective exclosure studies. Northwest Science **68**:97–110.

Zalisko, E.J., and R.W. Sites. 1989. Salamander occurrence within Mt. St. Helens blast zone. Herp Review **20**:84–85.

Part III
Ecosystem Processes

11
Dynamic Landscape Systems

Lee E. Benda, Daniel J. Miller, Thomas Dunne, Gordon H. Reeves, and James K. Agee

Overview

• Dynamic landscape processes influence the supply, storage and transport of water, sediment, and wood, thereby shaping many aspects of riparian and aquatic habitats. These processes comprise the disturbance regime of a watershed.

• The study of natural disturbance (and cumulative effects) in riverine and riparian areas requires a fundamental shift in focus from individual landscape elements (such as a forest, a hillslope, and a stream reach) over short timescales (years) to populations of landscape elements over long time scales (decades to centuries). The study of landscapes as a system expands the focus from predictions about exact future states to predictions about the relationships between large-scale properties of landscapes (i.e., climate, topography, and channel networks) and the long-term behavior of aquatic systems.

• Temporal patterns of landscape behavior are best described by frequency distributions which estimate the probability of a specific event occurring. Likewise, describing spatial patterns amongst a population of landscape elements in any year requires proportioning their characteristics amongst the range of all possible environmental conditions, and this also is best described by frequency distributions.

• Characteristics of landscapes that vary naturally over time can be described by four components: (1) climate, which drives environmental variability; (2) topography, which comprises a population of diverse hillslopes that creates spatial variability in the sediment and wood supplied to channels; (3) channel networks, which govern how sediment and wood are routed through a population of linked stream reaches and unevenly redistributed in time and space; and (4) basin history, which effects the volume of sediment and wood stored on hillslopes and in stream channels, and which influences how sediment and woody debris are redistributed during storms, fires, wind, and floods.

• The study of landscapes as systems, focusing on the collective behavior of populations of landscape elements over time, provides the necessary framework for investigating natural disturbance and cumulative effects. The field application of this framework provides insights into how channel and riparian morphologies are related to the recent environmental history of a watershed.

Introduction

Powerful climatic and geomorphic processes shape the landscape of the Pacific coastal ecoregion. Climatic conditions produce wildfires and windstorms that modify large tracts of forests, enabling new species to contribute to a diversity of forest ages and structures. Fires, in particular, controlled the age distribution of natural forests prior to fire suppression throughout most of the mid- and southern parts of the Pacific coastal ecoregion (Teensma 1987,

Morrison and Swanson 1990, Agee 1993). Fires and storms trigger geomorphic processes such as bank erosion, surface erosion and gullying, and shallow and deep landslides; processes which control the supply of sediment and wood to streams (Dietrich and Dunne 1978, Swanson 1981, Swanston 1991). Once in stream channels, sediment and wood are transported episodically and redistributed unevenly through the channel network by floods.

Collectively, these climatic and geomorphic processes comprise the disturbance regime of a watershed. The term disturbance refers to a disruption in an environment that leads to a biological response (Pickett and White 1985). Fully understanding the role of disturbance in shaping aquatic ecosystems requires estimating its regime frequencies, magnitudes, and spatial distributions of landscape processes. Likewise, cumulative effects (because they involve a history of human activities dispersed in time and space) can beviewed as a modification of a regime—a shift in frequency, magnitude, and spatical distribution of processes.

Disturbance is embodied in the temporal behavior of a single (or set of interacting) landscape element(s), such as forests, hillsides, and streams over decades to centuries. However, in any year, the history of a dynamic climate or the history of disturbance is represented by the environmental condition of a population of landscape elements (e.g., hundreds to thousands of forest stands, hillsides, or stream reaches). Therefore, the study of disturbance fundamentally involves changes over time and populations of landscape elements.

Natural disturbance is of great interest to researchers and natural resource managers because dynamic (temporal) aspects of landscapes are an inherent characteristic of ecosystems in the region (Swanson et al. 1988), and because natural disturbance can be contrasted with human impacts or disturbances to reveal the long-term consequences of resource management. The study of disturbance in landscapes of the Pacific coastal ecoregion has focused primarily on processes in terrestrial environments such as revegetation following volcanism (Franklin 1990), fires (Teensma 1987, Morrison and Swanson 1990), wind (Borman et al. 1995), snow avalanches (Hemstrom and Franklin 1982), and rockfalls (Oliver 1981). Although some of the disturbances influencing aquatic systems have been recognized, they have not been well quantified (Everest and Meehan 1981, Sedell and Swanson 1984, Minshall et al. 1985, Frissell et al. 1986, Resh et al. 1988, Naiman et al. 1992). As a consequence, descriptions of streams in the context of their watersheds have emphasized spatial determinism (which include classification systems of channel morphology and stream biota) (Vannote et al. 1980, Frissell et al. 1986, Rosgen 1995, Montgomery and Buffington 1997, Chapters 2 and 5). Given the importance of disturbance in aquatic systems, why has disturbance been difficult to define?

Most quantitative theories of landscape processes, and their derivative predictive models, address the behavior of a single landscape process (such as fire, flooding, sediment transport, and slope stability), or a single landscape element (such as an individual forest stand, hillslope, or stream reach), over short time scales; for example, responses to a single fire, storm, or flood. Disturbance regimes (and cumulative effects which can be viewed as an alteration of a regime) in aquatic systems remain unquantified largely because theories and models designed to predict numerical solutions about exact future states of single processes at small scales (e.g., a few years) are inappropriate for understanding behavior of populations of processes occurring at larger scales (e.g., decades to centuries). Limitations in data and computing power, and the unpredictability of the weather, further confound applications of data-intensive, small-scale theories and models to the problem of predicting long-term ecosystem behavior.

Understanding the consequences of disturbance (or cumulative effects) in aquatic systems requires a fundamental shift in focus from individuals to populations (of landscape elements) and from short- to long-time scales. For example, the behavior of a population of landscape elements (such as all point sources of sediment and wood in a watershed), and the interaction of that population with other populations (such as a set of linked stream reaches comprising a whole network) involves routing materials over time between highly variable hillslope sources and stream reaches. The outcome of the collective behavior of such populations, represented, for example, in the characteristic frequencies and magnitudes of sediment transport and storage in a channel network, reflect a system property of a landscape. The term *system*, in this context, refers to the interacting group of landscape elements (i.e., topography, vegetation, soils, fires, rainstorms, channel geometry, and so on) that give rise to such long-term patterns of behavior. Hence, the study of aquatic disturbance (or cumulative effects) lies within the domain of the study of landscapes as systems

The study of landscapes as systems is different than the study of single landscape elements over short time scales. Viewing the behavior of a forest, a hillslope, or a stream channel over decades to centuries requires replacing predictions about exact future states with more general predictions of long-term patterns of behavior. This demands a degree of simplification in scientific analysis and description, referred to as *coarse graining* (Gell-Mann 1996), and the use of estimated probability distributions to overcome lack of scientific understanding and data, computing limitations, and uncertainty about future climate. Such an approach can be used to parlay empirical knowledge and theory available at smaller spatial and temporal scales to produce new understanding of landscape behavior (theories, models, and hypotheses) at larger scales.

Long-term patterns of landscape-behavior, including disturbance regimes, are best described in terms of frequency distributions from which estimates of probability of occurrence can be made. Likewise, describing the environmental condition of a population of similar landscape elements in any year (governed by climate history and land use) requires proportioning characteristics among the range of all possible environmental conditions. This also is best described by frequency distributions, and in this chapter aspects of long-term patterns of landscape behavior, or natural disturbance are described in terms of those statistical measures. A new system-scale framework is described using field data and simulation models to describe aspects of long-term behavior or natural disturbance in the Oregon Coast Range and in southwestern Washington.

Components of Dynamic Landscape Systems

The study of landscapes as systems (in terms of populations of upland or riparian forests, hillsides, and stream channels that vary naturally over time) focuses on four basic components: (1) climate, which drives environmental variability and emphasizes the importance of time in a landscape systems perspective; (2) topography, which represents a population of diverse hillslopes responsible for the spatial variability of sediment and wood sources; (3) channel networks, which govern how sediment and wood delivered from interactions between climate and topography are transported downstream through a population of linked stream reaches; and (4) basin history, which governs the volume of sediment and wood stored on hillslopes and in stream channels in any year, and thereby influences how sediment and woody debris are redistributed during future storms, fires, and floods.

Climate

In Washington, Oregon, and northern California the contemporary climate has been in place for the past several thousand years resulting in relatively stable vegetative communities (Heusser 1977, Leopold et al. 1982, Brubaker 1991). Nevertheless, smaller-scale variations in climate (decades to centuries), such as the neoglaciation of the seventeenth and eighteenth centuries, have influenced erosion and channel morphology in mountain areas (Church 1983).

The following discussion of climate concentrates on the effects of precipitation, flooding, and fire. Although not discussed here, windstorms are also important controls on the age distribution of forests along coastal areas and in northern parts of the region (southeast Alaska), and windthrow may be an important modifier of soil chemistry and vegetation productivity (Borman et al. 1995).

Precipitation

Precipitation (rain and snow), as the source of groundwater and stream flow, is the primary natural driver of change in Pacific Northwest forest and stream environments. Long-lasting, intense rainstorms saturate soils and trigger shallow landslides and debris flows, even under forest canopies (Figure 11.1) (Pierson 1977, Dietrich and Dunne 1978, Hogan et al. 1995). High seasonal rainfall accelerates movement of deep-seated landslides and earth flows. When it is not raining, forests become dry and the potential for wildfire increases (Agee 1993). Temporal variation in precipitation creates seasonal

FIGURE 11.1. Aerial photograph taken in 1939 of a 4th-order basin in the western Olympic Mountains, Washington. The storm of record (1934) triggered numerous debris flows (indicated by arrows) that deposited into the main channel. Orme (1990) and Hogan et al. (1995) estimated the frequency of such concentrated landsliding based on precipitation alone to be between 40 and 70 years in the northwestern part of the region.

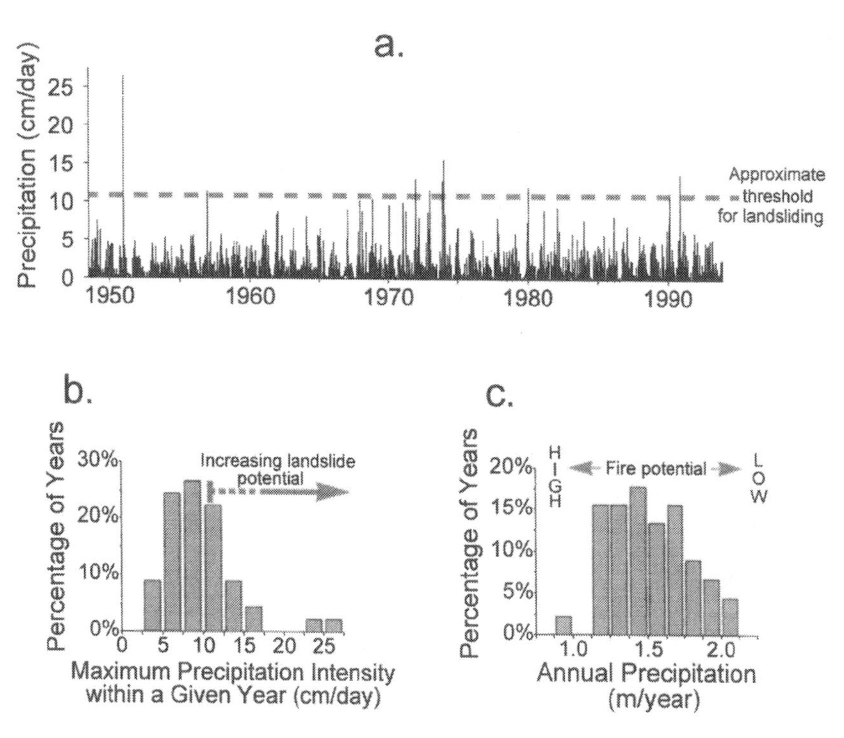

FIGURE 11.2. (a) A time series records the sequence of rain (and snow) storms over 47 years at the Randle Ranger Station in southwestern Washington. The sequence of rainstorms initiates temporal variability in the landscape by triggering landslides, generating floods, and creating drought conditions conducive to wildfire. The rainfall threshold for triggering landslides is estimated from an empirical relationship among rainfall intensity, duration, and landslide occurrence (Caine 1980). (b) A frequency distribution of daily maximum precipitation indicates the proportion of time that landslides can be generated (approximately 33% of the time or once every three years). (c) The frequency distribution of annual total rainfall indicates the amount of time that wildfires may be triggered due to moisture deficits (<1.0 m/yr occurs less than 3% of the time).

changes in stream flows and annual fluctuations in flood peaks, in sediment delivery to stream channels, and in fire potential.

A sequence of daily precipitation measured over 47 years for a site in southwestern Washington (Figure 11.2a) indicates that rainfall varies widely (0 to 25 cm in one day). Despite this variability, these data are useful for evaluating the influence of local climate on channel and floodplain morphologies. The frequency of storms that trigger landslides, for example, is indicated by a threshold estimated from an empirical relationship among rainfall intensity and duration, and landslide occurrence (Caine 1980). Sometimes several landslide-triggering storms occur in one year or in consecutive years; and sometimes nearly a decade passes with no large storms at all.

The role of precipitation in landslide or fire generation is best viewed through a frequency distribution. By condensing data in the time series of precipitation (Figure 11.2a) to a distribution of values, information on the temporal sequence of events is lost but insight as to how often an event of a given magnitude occurs is gained (Figures 11.2b and 11.2c). For example, the frequency of storms that trigger landslides is about once every three years (e.g., approximately 33% probability in any year, Figure 11.2b) and provides information on how often landslide debris might be expected to enter channels in that landscape. Likewise, the potential for large, intense fires increases dramatically when the annual precipitation total is less than 1.0m, but such low values of annual precipitation occur only about 3% of the time (Figure 11.2c). Hence, fire-producing droughts are much rarer than landslide-triggering storms.

Floods

Ultimately streambeds, banks, and riparian floodplains are shaped by the temporal sequence of flows and sediment loads carried through the channel. Although sediment may arrive directly from adjacent slopes, most material found in the streambed and floodplain has been carried some distance from upstream. The size and frequency of floods determine the capacity of a stream to move sediment into or out of a channel and to overflow its banks. Thus, the relationship between floods (potential sediment transport capacity) and sediment supply (the history of erosion) ultimately controls the ability of floods to modify channels and floodplains, and to create channel refuge habitats, such as wood jams and side channels (Chapter 2).

The temporal variation of flows typical of Pacific coastal streams is illustrated by the time series in Figure 11.3a and the associated frequency distribution of discharge measured at a gaging station in southwestern Washington (discharge varies from 2 to 200 m^3/s) (Figure 11.3b). Approximate thresholds for bed mobilization (i.e., bedload sediment transport) and overbank flows are determined from substrate size, channel width, and bank height for the gauged channel with a drainage area of 200 km^2 (Figure 11.3a). Frequency of channel-forming events can be defined by representing the time

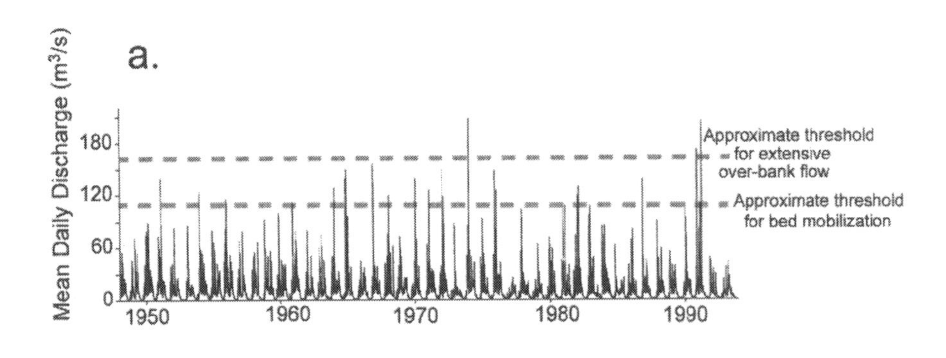

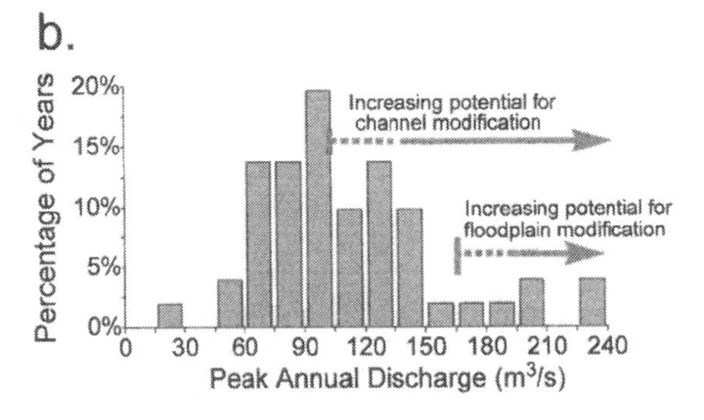

FIGURE 11.3. (a) A time series of channel discharge dictated by the sequence of rainfall (and snowfall) over previous days or weeks. In this case, the extent of channel and floodplain modification varies greatly from year to year because only large flows (>100 m^3/sec) are capable of moving sediment from or onto the channel bed, eroding banks, and inundating floodplains. (b) Frequency distribution of floods derived from time series of stream discharge in Figure 11.3a. Frequency of bedload transport is about 30% while overbank flows occur 10–15% of the time.

series of flows as a frequency distribution. For example, floods transport bedload approximately once every three years (e.g., 33% of winter floods) (Figure 11.3b). The opportunity for floodplain modification is much less, only about 7% (once about every 14 yrs). This indicates that the frequency of floods is sufficient to redistribute the sediment that originates from frequent landslides.

Floods are also important for transporting and redistributing wood as well as creating log jams. Floods large enough to transport wood may occur frequently, but transport of wood depends on the wood supply as well as piece length and slope (Nakamura and Swanson 1993). Therefore, the transport of woody debris by stream flow (and the development of debris jams) strongly depends on the temporal sequence of flooding magnitude and wood supply, the same way sediment transport depends on the temporal sequence of flood magnitude and erosion.

Fire

Periodic droughts, strong east winds, and the massive buildup of woody fuels create conditions favorable for fire, particularly in the southern and middle Pacific coastal ecoregion (Agee 1993). The frequency and magnitude (intensity and size) of fires strongly influence the distribution of forest ages, the recruitment of wood to streams, and the frequency and magnitude of erosion.

Fire influences erosion by destroying vegetation and by sharply decreasing the rate at which water infiltrates into soil. Roots of trees and shrubs reinforce soils laterally and bind thin soil to partially fractured bedrock. As a consequence, landslides and debris flows are more likely to occur after fires destroy vegetation and when root strength is lowest (O'Loughlin 1972). Thus, forest mortality by fire increases the potential for landslides and debris flows (Figure 11.4) (Swanson 1981, Benda and Dunne 1997). Fires in drier areas with hydrophobic soils can lead to widespread surface erosion, gullying, and the release of large quantities of sediment to the valley floor (Klock and Helvey 1976).

Variation in the frequency of fires across the Pacific coastal ecoregion is controlled by the frequency of ignitions (i.e., by lightning), the frequency and severity of droughts, the availability of combustible organic material, and topography. In addition, ignitions by Native Americans may have been an important factor in certain parts of the region (Schoonmaker et al. 1995). Low-frequency, high-intensity fires associated with high tree mortality are referred to as stand-replacing fires (Agee 1993). The average time interval between stand-replacing fires (the mean fire recurrence interval or fire cycle) at a specific location in a landscape has been estimated in various parts of the Pacific coastal ecoregion and is about 400 years for cedar (*Thuja plicata*)/spruce (*Picea sitchensis*)/hemlock (*Tsuga heterophylla*) forests of the Olympic Peninsula, Washington (Agee 1993); 200 to 300 years for the Douglas-fir (*Psuedotsuga mensiezii*)/western hemlock (*Tsuga heterophylla*) forests of the central Oregon Coast Range (Teensma et al. 1991, Long 1995); 150 to 200 years for Douglas-fir/western hemlock forests in the western Cascade Mountains of Washington and Oregon (Teensma 1987, Morrison and Swanson 1990), and 80 to 100 years for lodgepole pine (*Pinus contorta*) forests in southwestern Oregon (Gara et al. 1985).

Variation in fire frequency affects the frequency and location of erosion and wood input to streams (i.e., by determining rates of mortality and stand age) and therefore the disturbance regime of the channel. Different parts of the landscape burn at different frequencies, depending on the topography. North-facing slopes are typically moister and cooler than south-facing slopes and are therefore less susceptible to stand-replacing fires, particularly in the wetter areas of the ecoregion. Ridges and low-order valleys (e.g., 1st- and 2nd-order) are more susceptible to fire than nearby larger valley floors because fires tend to burn upslope. A fire simulation model, using fire probabilities obtained from field data in a 500 km^2 area in southwest Washington, illustrates how different topographic positions result in different fire regimes (Figure 11.5). Although fire recurrence intervals in the landscape of southwestern

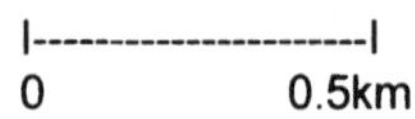

FIGURE 11.4. Several large and intense fires between 1933 and 1939 in the north-central Oregon Coast Range (the Tillamook fires) caused widespread forest death and created a forest of standing, dead trees in a portion of the Kilchus River basin. Aerial photographs (1954, 1:12,000-scale) reveal a concentration of shallow landslides (small arrows) in the burned area. Some of the failures triggered debris flows in 1st- and 2nd-order channels (large arrow) and deposited sediment directly into a 4th-order channel (Benda and Dunne 1997a).

Washington averages 300 years, fire intervals of 150 years occur on ridges and in low-order stream valleys (e.g., 1st- and 2nd-order), and fire intervals can exceed 500 years for wide and low-gradient valley floors.

Topography

The second component in the study of landscapes as dynamic systems is the diverse population of topographic patterns in a watershed which creates a discontinuous and spatially variable supply of sediment and wood to channel networks. Interactions between climate and topography are fundamental components shaping landscapes. In the Pacific coastal ecoregion intense rain and wind result in erosion and flooding that form stream and valley floor morphologies, especially after widespread fire. Heterogeneous topography coupled with a dynamic climate results in a supply of sediment and wood to streams that varies temporally as well as spatially. Although this chapter focuses on contemporary climate, it also recognizes that present erosion rates are influenced by past climate processes.

Sediment Supply

In terms of a landscape system, topography is represented as a population of diverse hillslopes and hence a population of diverse erosion sources. Two areas (the Oregon Coast Range and southern Cascade Mountains of Washington) illustrate how diverse topography leads to spatially variable sediment supply (Figure 11.6a and 11.6b). Two types of mass movement occur in the highly dissected, low-relief topography of the Oregon Coast Range: one is shallow landslides from steep, unchanneled swales (referred to as bedrock hollows) and the other is debris flows from 1st- and 2nd-order channels spaced along both sides of the main river at irregular intervals (Figure 11.6a). Landslides or debris flows emanating from any one site are rare, but mass wasting at the watershed scale (containing thousands of landslide sites) is a common occurrence (illustrated later in this chapter).

A population of landslides and debris flows over the entire channel system has a large and persistent influence on the valley floor environment. The immediate short-term effects of mass wasting are readily apparent as the large influx of sediment and wood overwhelms the capacity of the Oregon Coast Range stream to transport material downstream (deposited sediment buries the channel), the average size of channel substrate decreases, and riparian vegetation is destroyed. Over time the stream gradually removes much of the deposited material, larger particles (boulders) are exposed and new plant communities are established (Chapter 12). Persistent, long-term morphologic effects include creation of log jams, development of terraces, accumulation of boulders, and construction of debris fans that force the channel against the opposite valley wall (Benda 1990). The density and location of these

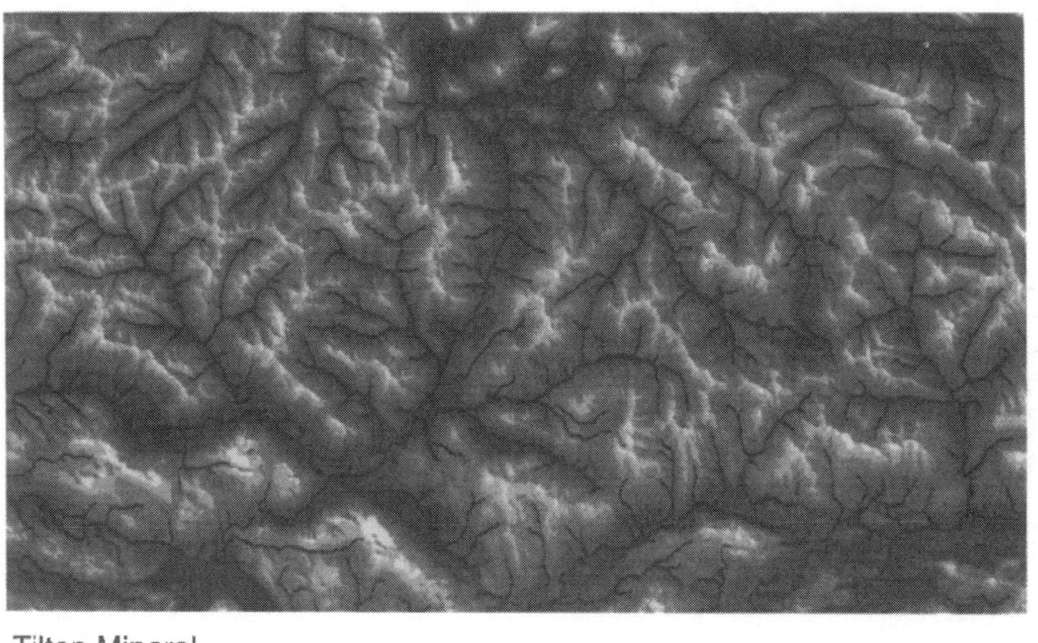

FIGURE 11.5. Topographic controls on fire. Stand-age maps obtained from 1939 aerial photography over a 1000 km^2 area in southwest Washington indicate correlations between forest age and slope aspect (southwest facing slopes burn most frequently) and local relief (ridge tops burn more frequently than valley floors). These empirical relationships were used to construct this map which shows the relative likelihood of fire over time as a function of topography. Values are normalized so that the average over the entire area equals unity. In this area, fire cycles are predicted to be less than 200 years on ridge tops and greater than 500 years on valley floors, with a mean of 280 years. Channel network is indicated by black lines and ridges and south facing slopes are indicated by lighter shading.

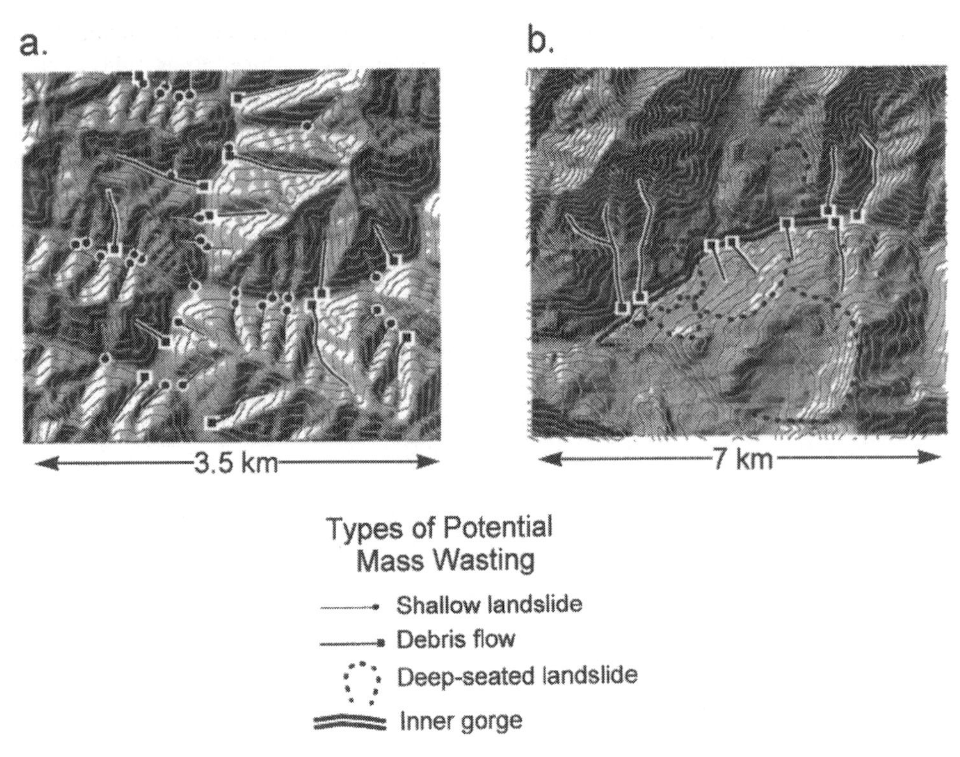

FIGURE 11.6. (a) High density of shallow landslides and debris flows characterizes the highly dissected terrain of the Oregon Coast Range. (b) In contrast, pervasive deep-seated landslides have modified the topography in the southern Cascade Mountains of Washington, resulting in a lower density of debris flows and a higher density of small shallow landslides along inner gorges. Regional variation in the processes and spatial patterns of erosion leads to regional differences in the frequency and magnitude of sediment and wood supply to streams.

deposits throughout the channel system (Figure 11.6a) dictates the type, diversity, and distribution of many of the channel and riparian environments within the watersheds.

The southern Cascade Mountains illustrate a different and more complex topography, one smoothed by extensive deep-seated landslides (Figure 11.6b). Compared to the Oregon Coast Range, a greater diversity of mass wasting processes is active, including deep-seated bedrock landslides at a variety of spatial scales and pervasive shallow landslides in inner gorges. The spatial density of shallow landslides adjacent to streams is greater than in the Oregon Coast Range partly because steep inner gorges have formed at the toes of large deep-seated landslides. In contrast, the spatial density of debris flows is lower, in part, because deep-seated landslides have lowered midslope gradients. Deep-seated landslides likewise have altered channel and valley floor morphology, creating local low-gradient areas upstream of slides and local steepening at the downstream slide edges. Channels are confined within inner gorges where deep-seated landslides have created narrow valley floors.

Together, topography and climate create a pattern of sediment supply to stream channels that is variable in space and time. The status of sediment storage in any particular channel depends on the population of all upstream sediment sources from the numerous hillslopes.

Wood Supply

The supply of woody debris to streams is also influenced by climate and topography. The size of trees that fall into streams depends on tree age which is controlled, in part, by the frequency of fires, windstorms, or erosion. For example, fires generally do not consume entire trees and the majority of dead trees fall within several decades after a fire (Agee 1993) leading to large fluctuations in the size and number of trees recruited to small channels.

Concentrating on fire in this example, the change in in-stream wood volume over time differs for two different fire regimes (Figure 11.7a). Stand-replacing fires every 500 years

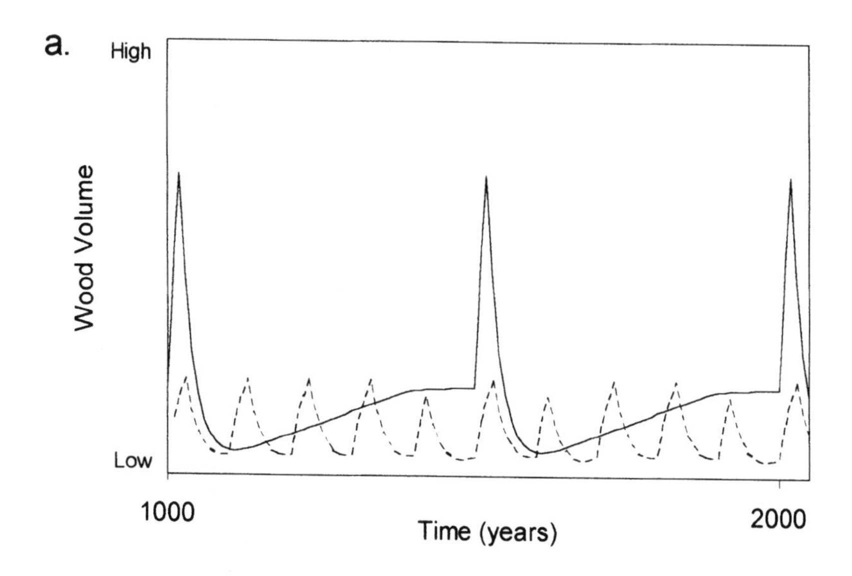

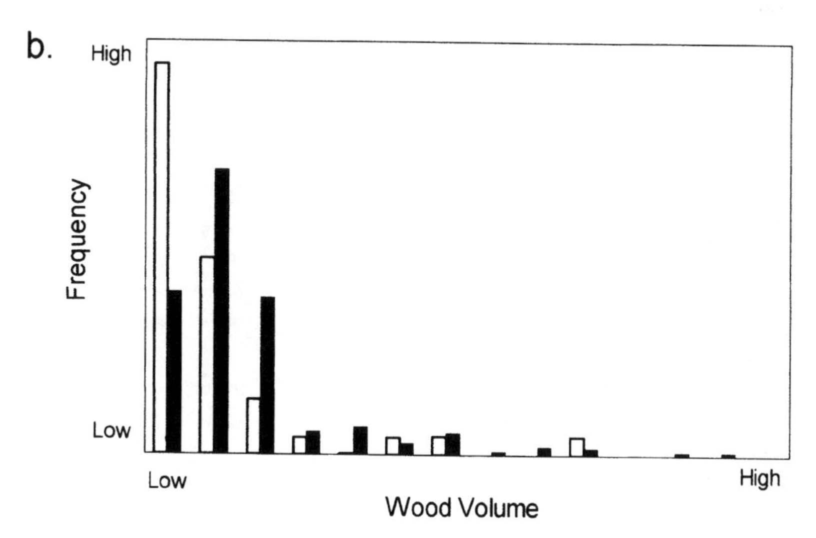

FIGURE 11.7. (a) The changing in-stream volume of wood for two different fire regimes, and (b) their accompanying frequency distributions predicted by a simulation model. The 500-year fire cycle (——; solid shading) yields abrupt and large increases in wood supply after a fire, and almost all of the time large volumes of wood are present in the channel. In contrast, the 100-year fire cycle (-----; no shading) results in lower volumes of wood, and a lower range of variability in wood loading.

(similar to the fire frequency in the coastal rain forest of Washington) leads to abrupt and large increases in wood supply as trees killed by fire fall to the ground less than 50 years later. After this initial large input of wood, wood supply slowly increases from stand mortality as standing biomass increases. In contrast, more frequent fires every 100 years (similar to drier forests of eastern Washington and Oregon) result in a smaller punctuated supply of wood because large amounts of biomass do not accumulate. The difference between the two simulated fire regimes is apparent in the frequency distributions (Figure 11.7b). The distribution of wood volume for the 100-year fire cycle indicates the dominance of low-wood supply conditions. But perhaps more importantly, the figure shows that lower volumes of wood are common. In contrast, the frequency distribution for the 500-year fire cycle shows significantly higher wood volumes for large portions of the time, and very low volumes of wood are rare. The predicted frequency distributions from these two fire regimes allow consideration of how in-stream volumes of wood change as one moves from coastal rain forests to drier interior forests.

The variability in fire frequency for different parts of the Pacific coastal ecoregion also contributes to a spatially variable wood supply at the watershed scale. Different parts of watersheds burn at different frequencies (Figure 11.5), with small, steep valleys having a higher fire frequency compared to wider valley floors.

Other processes influencing input of woody debris to streams include bank erosion, land slides and debris flows. In large streams woody debris is commonly recruited by bank erosion, which can occur every several years (Keller and Swanson 1979). Landslides and debris flows (and snow avalanches), with or without fires, are also important sources of woody debris (Swanson and Lienkaemper 1978, Chapter 13, this volume). In the Queen Charlotte Islands, British Columbia, landslides under forest canopies, with a recurrence interval of approximately 40 years, result in large numbers of debris jams (Hogan et al. 1995). Although landslides and debris flows may only contribute minor volumes of woody debris because of their low frequencies compared to decay rates, during periods of disturbance wood loading by mass wasting may overwhelm all other sources of wood.

The ability of woody debris to affect channel morphology depends on stream size (Bilby and Ward 1989). Even large rivers can be greatly influenced by wood (Sedell and Frogatt 1984, chapters 12 and 13, this volume). Trees falling into a stream reach significantly modify stream morphology in several ways (Chapters 2 and 13). For instance, scouring of the channel bed around the downed trees modifies aquatic habitats as, pools, side channels, and sediment storage areas are formed (Keller and Swanson 1979, Sullivan et al. 1987). In addition, log jams create small terrace-like deposits along valley floors (Kochel et al. 1987) and stabilize portions of floodplains allowing development of mature conifers (Abbe and Montgomery 1996). Woody debris jams temporarily store significant amounts of sediment (Megahan 1982), although eventually, decaying jams release stored sediment. High sediment supply in concert with debris jams can lead to channel avulsions and the isolation of jams from active channels.

Hierarchical Patterns of Channel Networks

The third component of landscapes viewed as systems is the population of linked stream reaches arrayed in a hierarchical channel network. Two aspects of hierarchical channel networks are considered: how fluctuations in sediment supply (and transport) are mixed and modulated downstream through the network and how different sediment transport regimes are abruptly joined at tributary confluences.

Fluctuations in Sediment Supply and Channel Morphology

Recall that the first two components of a dynamic landscape, climate and diverse topography (viewed as a population of erosion and wood sources), result in a punctuated supply of sediment and wood to channels. The sudden deposition of sediment into a channel previously carrying little sediment may create temporary zones with greater sediment volume and higher sediment transport rates that can either remain stationary (such as point bars) or migrate downstream in the form of a sediment wave (Beschta 1978, Church 1983, Benda and Dunne 1997). Dimensions of recognizable sediment waves range between several hundred to several thousand meters in length and from one half to four meters in height (Church 1983, Pickup et al. 1983, Madej and Ozaki 1996). Sediment waves also may disperse rapidly in steeper, confined channels, such as in canyons. Channel beds that do not undergo major fluctuations in sediment supply may exhibit relative stability (i.e., no sediment waves and little scour and fill).

Within the channel networks sediment waves (or an accumulation of sediment) may merge downstream because the velocity of bedload particles (annual transport distance) generally decreases downstream. Accumulations of sediment or large sediment waves have formed in mainstem channels from erosion in numerous tributary subbasins (Jacobson 1995, Madej and Ozaki 1996). In addition, sediment waves may merge and travel serially downstream when a climatic disturbance (e.g., storm or fire) occurs over a large area of a watershed and releases sediment in numerous adjacent tributaries that all join to become a single larger channel (Benda 1994).

If fluctuations in sediment supply are of sufficient magnitude, a cycle of scour and fill creates floodplains and terraces (Gilbert 1917, Hack and Goodlett 1960, Griffiths 1979, Pickup et al. 1983). This cycle, thus, can lead to formation of new riparian areas and may reset the age of riparian vegetation. Fluctuations in sediment supply, in conjunction with floods, also creates side channels on newly formed terraces and slough channels in low-gradient floodplains, that provide critical refuge habitats.

Fluctuations in sediment supply also lead to changes in channel morphology. Transient increases in sediment supply in the form of waves lead to temporary increases in the widths of channels and floodplains (Beschta 1984, James 1991), fining of the substrate (Coates and Collins 1984, Roberts and Church 1986), braided channels (Roberts and Church 1986), decreasing pools in conjunction with increasing riffles (Madej and Ozaki 1996), and death of riparian forests (Janda et al. 1975). Passage of sediment waves on a variety of scales leads to frequent bed filling and scouring which creates a channel environment unfavorable for fish spawning (Nawa et al. 1989). However, sediment waves also cause bank erosion and tree fall which forms pools and creates sediment storage areas, including spawnable riffles.

Tributary Confluences and Discontinuities in Transport Regimes

Different sediment transport regimes (i.e., process, frequency, magnitude, and particle sizes) are abruptly joined at tributary confluences. For example, a discontinuity in transport process often occurs where 1st- and 2nd-order channels, that are prone to debris flows, join higher-order channels. Debris flows transport large volumes and sizes of sediment and wood into lower-gradient channels. As a consequence, such confluences may be characterized by boulder deposits, fans, terraces and debris jams (Benda 1990, Grant and Swanson 1995, Hogan et al. 1995). Discontinuities in transport regimes also occur at confluences of large streams. An increase in the probability of sedimentation at, and immediately downstream of, confluences results from a higher proportion of bedload compared to the main river (Benda 1994).

Basin History

Basin history is the fourth component influencing landscape characteristics. The temporal sequence of past climatic and erosional events (that has either supplied or removed sediment and wood from hillsides or channels) influences how sediment and wood are redistributed during future storms, fires, and floods.

The history of soil moisture over a winter period (i.e., numerous storms) affects the rate of stream flow and soil stability (Pierson 1977) which affects stream power and the rate of sediment transport and the volume of sediment supply (through bank erosion and landslides) during any individual storm. Each storm alters the distribution of sediment, so that the volume and location of sediment stored in a channel reflects the sequence of storms over several decades to a century. Hence, the temporal ordering of storms and associated erosion in a basin greatly influences channel response and controls cycles of sedimentation and erosion of valley floors and channels (Beven 1981).

The amount of sediment stored on hillslopes that may ultimately become incorporated into landslides and debris flows also is affected by basin history. The timing of a landslide or debris flow is influenced by the rate of sediment accumulation since the time the site last failed (Sidle 1987, Dunne 1991). Hence, longer periods between landslide-triggering events at the watershed scale (e.g., involving a population of landslide sites) will likely result in a greater volume of sediment and wood released to a channel network.

Glaciation strongly influences both local and landscape-scale geomorphic processes. Late Pleistocene glaciation (e.g., 12,000–15,000 yrs ago) created large valley floor reservoirs of outwash sand and gravel, and lacustrine deposits of silt and clay in some of the northern areas of the Pacific coastal ecoregion. These glacial deposits initiated new mechanisms of erosion (i.e., deep-seated landslides in glacial sediments) and increased erosion rates for thousands of years after the glacier retreated (Church and Ryder 1972, Benda et al. 1992). In the 17th and 18th centuries smaller alpine glacial events in some areas, such as British Columbia, caused shorter-term changes in sediment supply and sediment transport (Church 1983).

Dynamic Landscape Systems: Populations of Elements and Time

Studying the system properties of landscapes focuses on relationships among the major attributes of climate, topography, networks, and basin history and, hence, the space-time structure of material fluxes and aquatic morphologies. The remaining portion of this chapter describes some of these relationships, including how effects of scale and channel networks cause shapes of frequency distributions to evolve downstream. Biological aspects of landscape systems and general applications to land use are also presented.

Observed records (e.g., rainfall in Figure 11.2a) are typically too short to fully characterize relatively rare events such as fires, storms, and landslides. Empirical distributions can be combined with numerical models designed to simplify certain small-scale hillslope and channel processes to characterize behavior over longer time periods. Use of such coarse-grained models allows problems of increasing complexity to be addressed, leading to testable predictions (hypotheses) on landscape-scale patterns of behavior (Benda and Dunne 1997).

Temporal Sequencing of Storms, Fires, and Floods, and Dynamic Channel Behavior

A large flood may not cause major channel changes in the absence of sediment to transport. However, in the presence of large quantities of sediment originating from recent erosion, the same flood may cause large modifications to channels and floodplains. The key to understanding the dynamic behavior of channels is the short-term (years to decades) synchronicity of climatic events that produce erosion and flooding. For example, a sequence of stand-replacing fires followed within a decade or two by large storms may lead to concentrated landslides because this is the period of low root strength prior to revegetation (Figure 11.4). In drier areas, where fires create soils prone to surface erosion, a fire-storm sequence may result in massive influx of sediment to channels and valley floors.

Using an example from a computer simulation of the steep, highly dissected central Oregon Coast Range where shallow landslides and debris flows are dominant erosion processes, the sequence of rainstorms (Figure 11.8a) and fires (Figure 11.8b) governs the

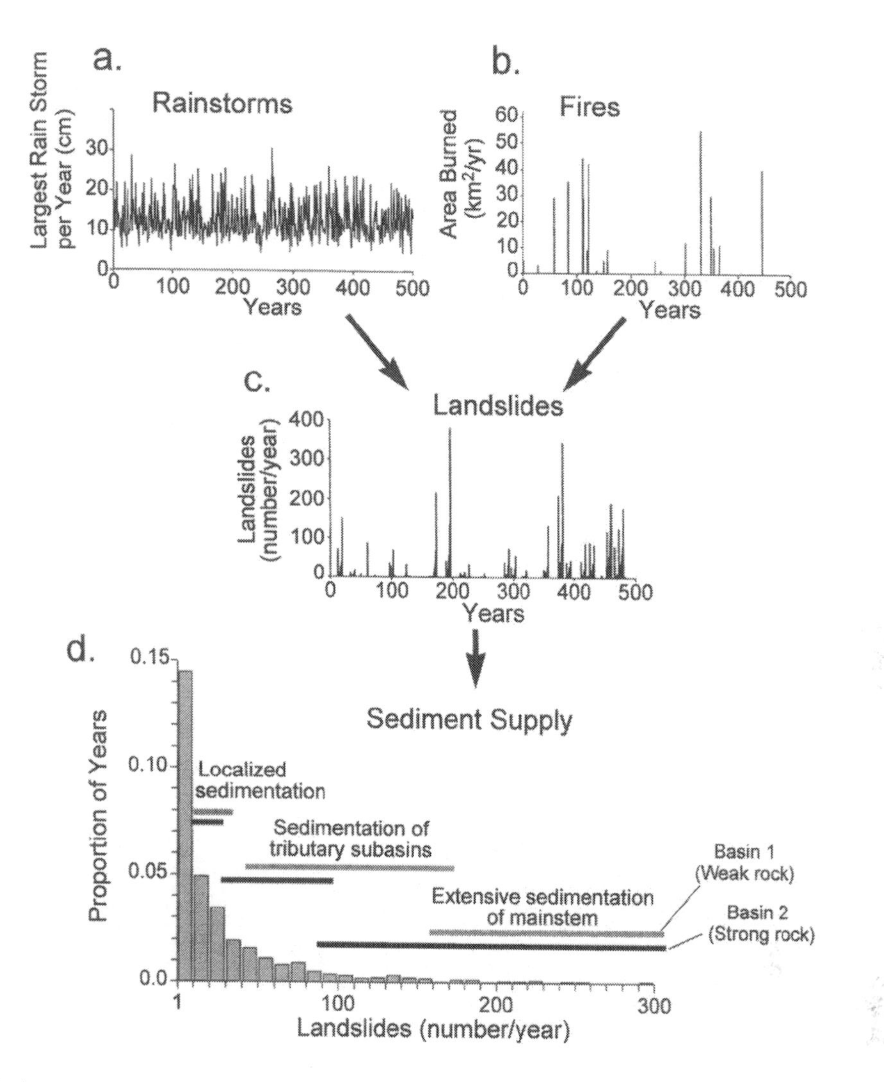

FIGURE 11.8. A sequence of rainstorms (a) and fires (b) generates a sequence of landslides within a basin (c) which results in an intermittent sequence of sediment delivery to the channel system. The time sequence of sediment supply is represented as a distribution (d) indicating how likely various magnitudes of sedimentation occur. Channel response depends on the size distribution and durability of the sediment delivered.

sequence of erosional events (Figure 11.8c). The time series of sediment supply is represented as a frequency distribution (Figure 11.8d) to predict the long-term pattern of sediment input to a channel network.

Considering the timing of sediment supply to a network (Figure 11.8c) in the context of the timing of large floods is necessary to anticipate the ability of erosion or floods to modify channels and floodplains, and ultimately aquatic habitats. By considering the frequency distribution of sediment supply (Figure 11.9a) in concert with the frequency distribution of floods (Figure 11.9b), the frequency distribution of channel and floodplain changes can be conceptually portrayed (Figure 11.9c). This distribution illustrates how often certain channel and floodplain changes are likely to occur and therefore provides insight on how the channel and floodplain will evolve over time and space.

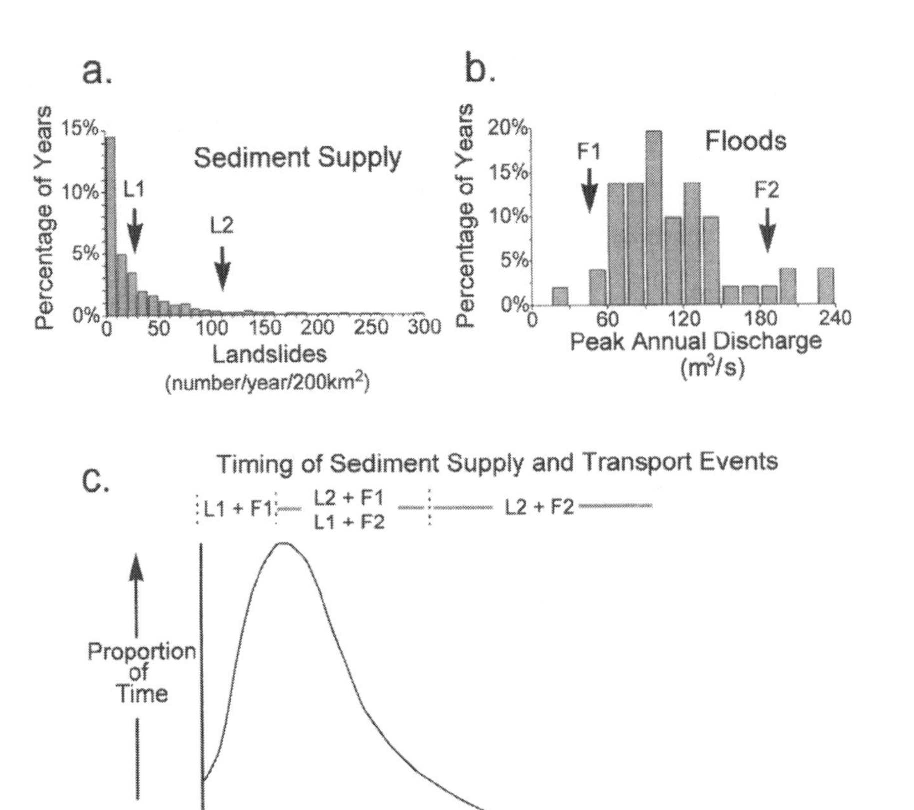

FIGURE 11.9. The extent to which floods affect channel and floodplain characteristics depends on the volume of sediment and the size of the flood (i.e., temporal sequencing). Hence, the frequency with which impacts of a given magnitude occur depends on the frequency distribution of sediment input to the channel system (a), the frequency distribution of flood flows (b), the persistence of sediment in the channel, and (c) the relative timing between these events. Relative timing between flooding (F) and sediment supply from landslides (L) is indicated by L1 + F1 and so on.

An example from Idaho illustrates this concept. In the summer of 1995, a stand-replacing fire with an estimated recurrence interval of approximately 150 years was followed very shortly thereafter by an extreme rainstorm triggering widespread surface erosion and gullying, accompanied by floods and fluvial mobilization of sediment in channels. The result was widespread sedimentation throughout an entire network of channels, which caused channel avulsions and braiding, floodplain deposition, and the creation of new terrace and floodplain surfaces (Figure 11.10). In the frequency distribution of channel changes (Figure 11.9c), this type of event would be located in the right tail indicating how often new terraces, floodplains, and channels (e.g., aquatic and riparian habitats) are formed along the valley floor. In the absence of the fire, a flood of similar magnitude likely would be located in the left hand tail of the distribution in Figure 11.9c, resulting in only minor changes to channels and floodplains. The influence of sediment supply on the geomorphic impacts of floods also has been documented in British Columbia (Church 1983), northern California (Madej and Ozaki 1996), and New Zealand (Beschta 1984).

Effect of Hierarchical Networks and Spatial Scale on System Properties

Viewing topography as a population of hillslopes and sediment sources means that the number of sediment sources and the number of linked stream reaches increases with the spatial scale of the basin and with distance downstream. In addition, frequency of climatic perturbations, such as fires and storms, also increases as basin area increases. As a consequence, the system properties of landscapes, as represented in frequency distributions of environmental conditions, evolves with scale and distance downstream through a network.

Topography as a Population of Hillslope Sediment Sources

The effects of increasing basin area on the dynamic behavior of channels can be illustrated with a simulation model. The model is applied to a 6th-order 200 km^2 basin in the Oregon Coast Range where erosion is dominated by debris flows (e.g., Figure 11.6a) following fires and storms (Benda and Dunne 1997). Episodes of concentrated landslides occur within burned areas during rainstorms for a decade or so after a stand-replacing fire, with fewer landslides at other times and places. The frequency of concentrated landslides predicted within a 3 km^2 3rd-order basin is about once every 200 years with 2 to 13 landslides per fire episode (approximately the 20 yr period following a fire with reduced rooting strength) (Figure 11.11a). As the basin area increases to 25 km^2, the frequency of landslides increases to once every 50 to 100 years with 5 to 165 landslides per fire or storm episode (Figure 11.11b). Finally, at a basin area of 200 km^2, landslides are relatively common with about 10 landslides occurring every few years (Figure 11.11c). Figure 11.11a–c show that the likelihood of landslides increases with basin size because a larger drainage area encompasses a greater number of potential landslide sites and the potential for a greater number of fires and storms. Thus, the frequency and magnitude of sediment supply to channels increases downstream in a channel

FIGURE 11.10. Gully erosion after a fire caused channel and valley floor sedimentation in a basin in Idaho. Such events correspond to the right hand tail in the distribution shown in Fig. 11.9c.

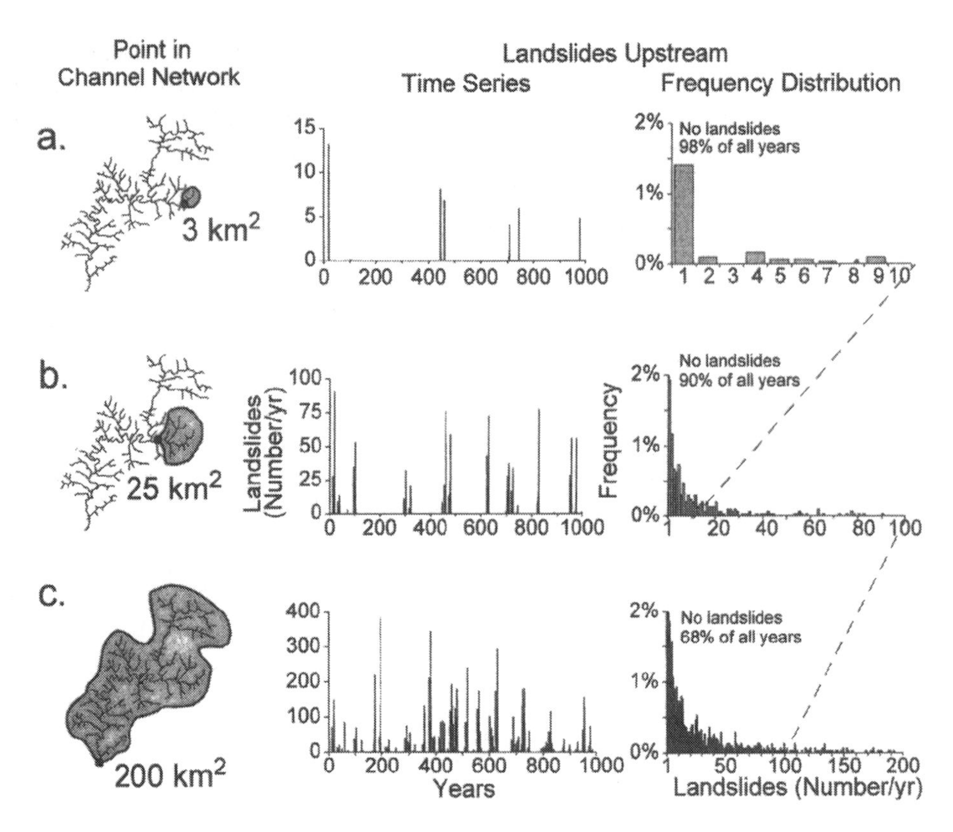

FIGURE 11.11. A 1000-year simulation of landslides in a 200-km² basin in the Oregon Coast Range indicates that the likelihood of landslides and associated sediment and wood delivery to the channel system increases with increasing basin area. (a) A 3-km² headwater basin experiences infrequent landslides. (b) A 25-km² tributary subbasin contains a greater total number of potential landslide sites and is more likely to include a fire: hence, landslides occur more frequently and in greater numbers within this larger area. (c) Numerous subbasins and a vary large number of potential landslide sites are contained within the entire 200-km² basin. Landslides occur somewhere within the basin a third of all years. On rare occasions, when large fires are followed by intense storms, well over 100 landslides can occur within the basin in a single year.

network, and has important consequences for the dynamic behavior of sediment transport, channel sediment storage, and channel morphology.

Channel Networks as a Population of Linked Stream Reaches

The simulation model also can be used to illustrate how variations in the supply of sediment from landslides and debris flows over long time periods (Figure 11.11a–c) interact with a converging hierarchical channel network (Benda and Dunn 1997). Again, the model is applied to a 200 km² basin in the Oregon Coast Range.

The model predicts an inherent imbalance between sediment supply and transport in 3rd-order channels resulting in centuries of sediment poor conditions interrupted by decades of sediment rich conditions after fires and major storms (Figure 11.12a). In the Oregon Coast Range, sediment poor conditions are characterized by bedrock channels with accumulations of boulders next to debris flow fans and bedrock outcrops. Woody debris, in single pieces and jams, creates local areas of sediment storage. Sediment rich conditions, in contrast, are characterized by gravel and cobble bed channels with buried boulders; some wood may be buried under gravel. The frequency in fluctuations in channel-stored sediment increases with basin area and corresponding stream size (Figure 11.12a–c) because of an increasing number of landslide and debris flow source areas and increasing probability of fires (i.e., fire occurrence increases with basin area). However, the magnitude of sediment supply fluctuations (i.e., represented as sediment depth in Figure 11.12a–c) decreases downstream in the Oregon Coast Range because of mechanical breakdown of weak sandstone bedload and widening channels.

The many small channels in the branching, hierarchical network (represented as $3\,km^2$ basins in Figure 11.12a) deliver small, disparate

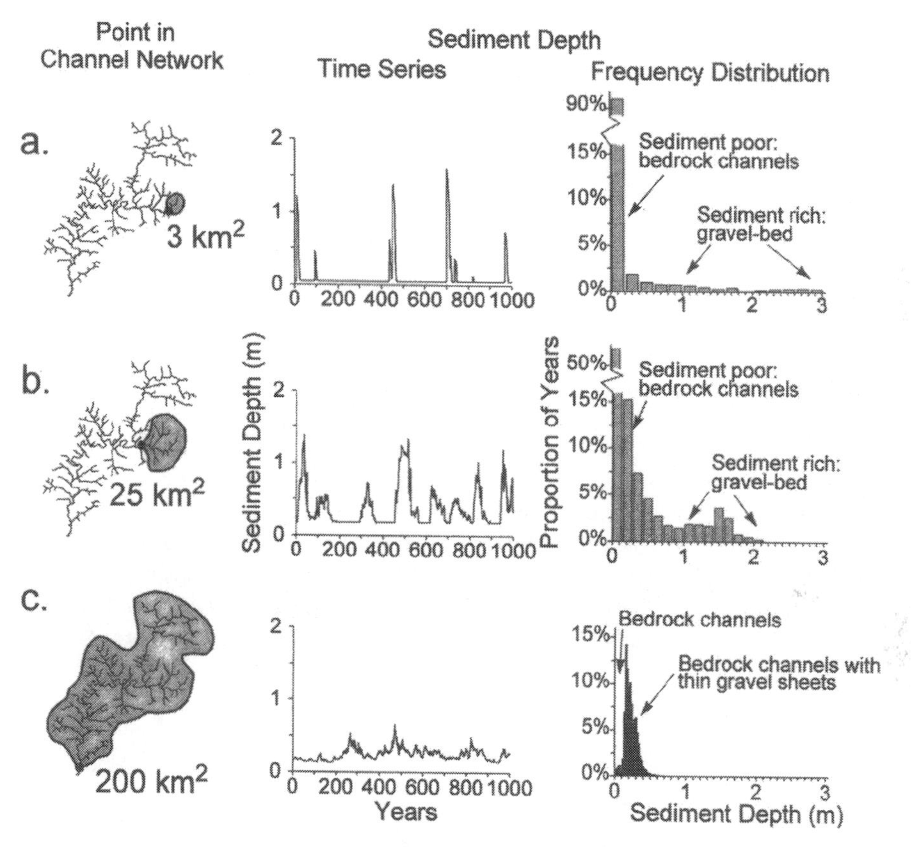

FIGURE 11.12. Sediment depth as a function of time and position in the channel network corresponding to the time series and frequency distributions of landslides shown in Figure 11.11. The volume of sediment found within the channel depends on the number, frequency, and distance to upstream sediment sources. (a) For a small channel draining a 3-km^2 headwater basin, upstream sediment inputs from landslides are infrequent and the channel is generally sediment poor, although periods of deep burial can occur. (b) A larger channel draining 25-km^2 has a greater frequency of upstream sediment inputs so that the channel here contains sediment of appreciable depths (>0.2 m) a larger proportion of the time. Since the channel is larger and sediment must, on average, be carried a longer distance through the channel to this point, sediment depths rarely exceed 1 m. (c) A great number of upstream sediment sources are integrated at the mouth of the 200-km^2 basin, resulting in a relatively constant volume of sediment at this point. In all cases, the time series of sediment depths show fluctuations at several time scales: long-term fluctuations are dictated by the frequency of large fires, shorter term fluctuations are a consequence of annual variability in flood peaks. The depth of sediment in large part determines channel-bed morphology.

sediment pulses to larger channels while decreasing stream power in the main channel promotes accumulation of the pulses (Figure 11.12b). The simulation model predicts that 10 to 30% of bed material is transported in sediment waves (0.5m thick and sufficient to change channel morphology) depending on stream size, with the greatest proportion carried as waves in the center of the network. In a drainage area of 200km^2, the sediment supply is relatively continuous and the magnitude of sediment transport variations is diminished, although small, biologically relevant fluctuations in channel-stored sediment continue (Figure 11.12c).

The right-skewed distribution of sediment storage at 3km^2 indicates that the range of variability is high but that the most probable condition is one of low sediment supply (Figure 11.12a). The shift of the distribution towards the right at 25km^2 indicates a reduction in the range of variability but an increase in likelihood of sediment in storage as many more hillslope sediment sources are integrated (Figure 11.12b). The more symmetrical distribution at 200km^2 (Figure 11.12c) indicates that variation in sediment depth is more evenly distributed about the mean, although low sediment storage is the most probable condition because of rapid breakdown of weak sandstone bedload and wide channels. Hence, the branching, hierarchical network results in an evolution of sediment distributions downstream.

Classification of Landscape System Behavior

The system behavior of a landscape, embodied in the form of frequency distributions, can be classified to describe how frequencies and magnitudes in the supply and storage of sediment and woody debris (or the range and magnitude of variability in these materials) vary through a network. Such a dynamic or disturbance-based classification system explicitly links channel behavior to basin climate and to the behavior (in time and space) of the upstream population of sources of sediment and wood.

The model predictions of frequencies and magnitudes of sediment supply for the 200km^2 basin in the Oregon Coast Range (Figure 11.11a–c) and sediment routing (Figure 11.12a–c) are used to classify the entire 3rd- through 6th-order network (Table 11.1). A similar classification system based on wood loading and routing also could be developed but is not included in this example.

The morphological consequences of the fluctuations in sediment supply (Table 11.1) are controlled partly by certain valley and channel characteristics that vary spatially through the channel network, but that are constant in time.

TABLE 11.1. Classification of a 200km^2 channel network in the Oregon Coast Range based on the frequency and depth of sediment deposition causing channel aggradation. The effects of tributary confluences are not included. Low-frequency (fewer than once every two centuries), high-magnitude (thickness greater than 1.0m) events occur in low-order channels; high-frequency (every five to ten years), low-magnitude (0.3m) events occur in high-order channels. The central part of the network (drainage areas of 20–50km^2) has the highest probability of encountering significantly aggraded channels.

		Channel aggradation				
Stream order	Drainage area (km^2)	Frequency	Magnitude Avg/Max (m)	Duration Avg/Max (yrs)	Channel length Avg/Max (km)	Substrate
Low	0.5–1.0	<0.05	0.6/2.5	2/20	0.2/2	Bedrock/boulders
↓	1–10	0.20	0.5/2.5	5/80	0.5/10	Gravel/bedrock
	10–35	0.25	0.5/2.0	5/90	1/12	Gravel/bedrock
	35–70	0.20	0.4/1.5	10/120	1/8	Bedrock/boulders
High	70–200	0.10	0.3/0.5	1/80	0.2/4	Bedrock/boulders

Adapted from Benda and Dunne (1997).

For example, the response of a channel segment to variations in sediment supply depends greatly on channel gradient and confinement. Spatial patterns of channel gradient and confinement have been used to stratify channel networks into broad zones of general channel types (e.g., step-pool channels with steep, bedrock floors covered with boulders or flatter meandering pool-riffle channels) (Rosgen 1995, Montgomery and Buffington 1997, Chapters 2 and 5, this volume). Spatially deterministic classification schemes predict a general morphological state and aid in estimating the sensitivity of channel reaches to change, such as fluctuations in sediment supply. A landscape systems-level classification system estimates the frequency, duration, and magnitude of likely changes. In addition, because the systems-level classification scheme is based on the entire frequency distribution (Figures 11.8 and 11.9), insights into how the formation of floodplains and terraces vary in the network also are gained.

Aquatic Biology at the Landscape Systems Level

Aquatic habitat is often viewed at the habitat unit scale (e.g., individual pools, riffles). Although this is an important scale of observation, understanding landscapes as systems requires that the focus on individual units shift to a focus on the population of units and the behavior of the population over time as dictated by the dynamic interactions of climate, topography, and channel networks. Hence, the contextual focus is on process linkages and time, rather than the state of habitat at any one site in a year.

A landscape systems perspective requires that the focus of climatic and geomorphic processes (such as storms, fires, floods, and landslides) be expanded beyond terms of environmental damage. Poor environmental conditions associated with these events can be short lived. Dynamic landscape processes also leave long-term legacies in the form of habitat development in channels, floodplains, and valley floors. Hence, dynamic processes must be viewed as both an environmental risk but also as a mechanism creating habitat, including refugia. In addition, aquatic organisms living in a dynamic environment have evolved a suite of adaptations for survival (Chapters 9, 10, 12, and 17).

Environmental Risk

Many environmental risks arise naturally from interactions between climate and geomorphology. For example, in the absence of side channels and woody debris jams (which act as habitat refugia), annual winter floods with high velocity flows flush juvenile and adult fish downstream, resulting in high mortality rates. Bedload transport during floods causes bed scour and fill (often linked with fluctuating sediment volume) that excavates or buries redds (areas within stream gravel beds where fish have deposited eggs) (Nawa et al. 1989). In contrast, low flows prevent access to spawning areas decreasing available spawning area for adult anadromous fish as well as increasing temperatures which also may lead to reduced survival (Reiser and Bjornn 1979). Erosion delivers fine sediment that fills inter-gravel spaces and thus reduces oxygen flux, which causes fish eggs to suffocate (Everest et al. 1987). In addition, mass wasting buries instream aquatic habitat (Everest and Meehan 1981).

Habitat Development Including Refugia

Processes creating environmental risks also create reach-scale babitats, including refugia. Wind, bank erosion, landslides, and fire all supply woody debris to the channel, where it accumulates in jams forming areas with low water velocities (including deep pools). These low velocity areas around debris jams provide refuge for adult and juvenile fish during winter floods (Dolloff 1983, Sullivan et al. 1987). In addition, boulders deposited in channels by mass wasting are used by fish as low-velocity refugia (Everest and Meehan 1981).

Large floods, often in conjunction with high sediment supplies and woody debris jams, cause temporary channel braiding which creates side channels (Keller and Swanson 1979), both in mountain valley floors and along lower-gradient floodplains (Church 1983). Debris jams formed during floods also contribute to the formation of side channels and mid-channel islands (Abbe and Montgomery 1996). Side channels, which may persist for decades or longer, are utilized by fish escaping the high flows and velocity of the active channel during individual storms or during entire winters (Scarlett and Cederholm 1984, Sullivan et al. 1987).

At the reach scale, habitat refugia act in the context of the environmental risks just described. However, refugia also occur at a variety of spatial scales including subbasins or watersheds (Sedell et al. 1990). Relatively infrequent but large magnitude events (e.g., fires, large storms) trigger widespread erosion and an increase in sediment supply which decreases both vertical and lateral channel stability (Figure 11.10). Thus, refugia at the reach scale may not be available or may require time to develop, and the environmental risk to fish may be high enough to result in local extirpations (Reeves et al. 1995). Nearby basins untouched by fires or storms may have more stable channels and pose less environmental risk and, thus, function as watershed-scale refugia. Fish may escape to less hostile subbasins of a watershed and, eventually, some may recolonize newly forming habitats in the previously impacted subbasins. Recall, however, that large quantities of sediment and wood moving into channels during low frequency, high magnitude storms or fires, and which initially created an environmental risk, may evolve over decades to centuries into channel habitats.

Because processes creating environmental risks also create habitats, the regime of landscape processes (frequency, magnitude, composition, and spatial distribution) is central to understanding this duality. Frequency distributions of floods (Figure 11.3), erosion (Figure 11.11), and sediment transport (Figure 11.12) help to quantify changes in channel environments (e.g., Figure 11.9c) and therefore provide a measure to understand how environmental risk and refugia are related. For example, flows that transport bedload occur approximately 70% of the time in 5th-order channels in southwestern Washington (Figure 11.3b) which suggest that flow velocities capable of scouring redds and flushing fish downstream (e.g., an environmental risk) occur an average of once every year. However, the same frequency distribution also reveals that major floods capable of bank erosion and tree recruitment (i.e., the formation of habitat refugia) occur about 10 to 15% of the time (once every 5 to 7 yrs). Hence, consideration of cumulative effects at the system level should seek to understand how the relative proportions of environmental risk versus habitat development have changed due to land uses.

Biological Adaptations

The ability of fish to move from risk prone areas to low velocity refugia during floods, and to utilize changing habitat conditions during storms, is based on a suite of behavioral adaptations. These adaptations include juvenile and adult movements to avoid or utilize changing habitat conditions over days (i.e., a single storm) to years at the reach to watershed scales (Quinn 1984, Reeves et al. 1995). In addition, fish encountering potentially hostile conditions during spawning tend to lay a large numbers of eggs (Heard 1991), an adaptation to potentially high rates of egg loss caused by scour (which varies from year to year).

Dynamic Fish Habitat and Community Structure

The variations in channel sediment storage predicted by the simulation model in the Oregon Coast Range (Figure 11.12a–c) cause variations in fish habitat. Fires and large storms occurring asynchronously across the Oregon Coast Range create a spatial diversity of channel morphologies (in several adjacent basins with channel segments of similar gradient and drainage area). For example, in Knowles Creek basin (5th-order, 30 km^2), an absence of recent wildfires has contributed to a low sediment supply, the channel is dominated by bedrock, and pools are shallow. Harvey Creek basin, burned by a forest-replacing wildfire in the late 1800s, contains large volumes of sediment, is dominated by a gravel substrate, contains deep pools that dry up in the summer. Franklin Creek contains an intermediate volume of sediment and has the greatest diversity of substrates and pool depths.

Each of the three different channel morphologies contain different fish communities (Reeves et al. 1995). Knowles Creek has only a single species, coho salmon (*O. kisutch*). Franklin Creek, with a greater diversity of substrate sizes, accommodates a greater diversity of species: 85% coho salmon, 13% steelhead trout (*O. mykiss*), and 2% cutthrout trout (*O. clarki*). Harvey Creek is also dominated by coho salmon, with only incidental portions (1% each) of steelhead and cutthrout trout. Although the relative differences are not great, there is a significant correlation between species diversity and sediment supply.

Linking the field data on channel morphology, fish community structure, and estimates of sediment depth to the probability distribution of predicted sediment depth previously discussed (Figure 11.12b) indicates how often each of the fish habitat conditions are likely to occur, either in a particular stream reach over time or across many stream reaches of similar channel slope and drainage area at a single time. The predicted frequency distribution of sediment volume (Figure 11.12b) suggests that the low sediment supply of Knowles Creek (dominated by coho salmon) is the most-commonly occurring environmental state in the central Oregon Coast Range (about 70% of the time). The effects of large organic debris were not considered and may include create. Locally increased habitat and fish diversity. The frequency distribution in Figure 11.12b also suggests that the sediment rich state of Harvey Creek would be fairly infrequent, occurring perhaps less than 10% of the time. In addition, the intermediate sediment storage state of Franklin Creek with a higher fish diversity, is predicted to occur about 15 to 20% of the time.

The comparison of field studies of habitats and fish communities with model predictions of long-term channel behavior (Figure 11.12b) provides insights into natural disturbance or habitat diversity. This comparison suggests that under natural conditions in the central Oregon Coast Range, with it's low fire frequency (about 300 yrs), habitat diversity is relatively low. Model predictions coupled with field observations also suggest that if the frequency of disturbance is slightly higher (a fire cycle of 150 to 200 yrs, for example), the higher sediment (and wood) supply could potentially increase habitat diversity.

Applications to Watershed Science and Management

On close examination stream channels are extremely complex. Valley-floor conditions vary reach by reach and are governed by the vagaries of local topography and the history of mass wasting. Bed texture and slope vary over the scale of meters and are altered by every log that falls into the channel. Sediment transport, even under steady flow, varies across channels and from moment to moment. The sediment under one's feet may have originated from a fire several hundred years ago 20 kilometers upstream or it may have originated from nearby bank erosion during the last flood. Hence, stream channels contain information over all length and time scales, providing a classic definition of a complex system (Bak 1994).

A Field Perspective

A basic rule of streams, particularly at small scales, is spatial variability and, if one returns to a certain stream reach year after year, temporal variability. Anticipating spatial and temporal variability in stream morphology is exceedingly difficult because of the complex environmental interactions. Compounding this problem is the uncertainty about recent basin history that may be responsible for present channel morphology.

Although interpreting the origin and significance of small-scale channel and habitat features is important, the study of landscapes as systems provides other insights into channel morphology and the watersheds in which they are embedded. A systems perspective helps define the range and magnitude of variability that may be encountered in the field (i.e., through the use of frequency distributions) and, furthermore, some of the morphological consequences of that variability (e.g., long-term legacies in the form of terrace formation). Moreover, frequency and probability distributions contain information on how the major landscape properties of climate, topography, channel networks, and basin history are linked to channel changes. Interpreting channel conditions using frequency distributions means interpreting the environmental status of the watershed and its potential for providing sediment and wood.

Applying the landscape systems perspective to watersheds requires measurements and monitoring that seek to quantify the range of natural characteristics in channels, floodplains, terraces, fans, and forests. In dealing with a population of channel reaches the complexities of an individual reach, or the variation that occurs from reach to reach, becomes less important. The systems approach focuses instead on the range and relative abundance of certain channel attributes found within the entire population of reaches. Furthermore, for a population of channel features, the systems perspective endeavors to link how the shapes of the frequency distributions are controlled by interactions among climate, topography, and channel network structure, focusing mostly on the recent basin history. A landscape perspective represents a more coarse-grained approach to stream or watershed interpretation, similar to the coarse grained approach of making simplifications when applying quantitative theories and models of small-scale processes to larger-scale behavior through the use of simulation models.

The Problem of Cumulative Effects, Natural Disturbance, and Habitat Diversity

Concerns over how human activities, including fishing, forestry, and agriculture impact fish populations in the Pacific coastal ecoregion are often condensed into three broad concepts: cumulative effects, natural disturbance, and habitat diversity. Cumulative effects in streams is the accumulation and manifestation of human activities dispersed in space and time at any point in a channel network (Chapter 19). Information on natural disturbance provides an important baseline from which to interpret the ecological significance of cumulative effects or a long history of human impacts. Habitat diversity is related to natural disturbance through fires, storms, and floods occurring asynchronously that create a mosaic of habitats in watersheds or across landscapes in any year.

Cumulative effects, natural disturbance, and habitat diversity have two important things in common. First, they all pertain to the system behavior of a landscape. That is, they pertain to the behavior (or the condition in any year) of a population of landscape elements, such as upland and riparian forest stands, erosion sources, channel and valley floor environments, and fish stocks, over periods relevant to the governing physical processes (commonly decades to centuries). Second, they all remain poorly defined despite significant scientific efforts to address them in the past. For example, a credible scientific method for measuring or predicting cumulative effects in streams is lacking because of the difficulty in applying existing quantitative theories and models (such as for sediment transport and slope stability) that focus on small-scale landscape behavior (e.g., a hillslope or stream reach over a few years) to populations of hillslopes and stream reaches which have evolved over long time periods. Further, the aquatic or riparian natural disturbance regime has never been fully defined for any landscape for the same reason.

The inability to make significant progress in defining natural disturbance regimes in terrestrial or aquatic systems, or in measuring or predicting cumulative effects, largely is due to an absence of scientific theory explaining landscape behavior in terms of a population of landscape elements over appropriate temporal and spatial scales. Moreover, the availability of theories addressing landscape processes that act at small spatial and temporal scales often form the scientific basis for evaluating many broader environmental issues precisely because of the absence of system-scale theory. Unfortunately, the specific models cannot successfully address system-scale questions of cumulative effects, natural disturbance, and habitat diversity. The landscape systems framework described in this chapter could potentially be expanded into a methodology for evaluating and predicting cumulative effects because it deals with populations of landscape elements over time as well as the condition of populations of elements in any year.

Literature Cited

Abbe, T.B., and D.R. Montgomery. 1996. Large woody debris jams, channel hydraulics and habitat formation in large rivers. Regulated Rivers: Research and Management **12**:201–222.

Agee, J.K. 1993. Fire ecology of Pacific Northwest forests. Island Press, Washington DC, USA.

Agee, J.K., and M.H. Huff. 1987. Fuel succession in a western hemlock/Douglas-fir forest. Canadian Journal of Forest Research **17**:697–704.

Bak, P. 1994. Self-organized criticality: A holistic view of nature. *In*: G. Cowan, D. Pines, and D. Metzer, eds. Complexity, metaphor, models, and reality. SFI Studies in the Sciences of Complexity, Proceedings. Vol. XIX. Addison-Welsey, Reading, Massachusetts, USA.

Benda, L. 1988. Debris flows in the Oregon Coast Range. Master's thesis. Department of Geological Sciences, University of Washington, Seattle, Washington, USA.

——. 1990. The influence of debris flows on channels and valley floors of the Oregon Coast Range, USA Earth Surface Processes and Landforms **15**:457–466.

——. 1994. Stochastic geomorphology in a humid mountain landscape. Ph.D. dissertation. Department of Geological Sciences, University of Washington, Seattle, Washington, USA.

Benda, L., T.J. Beechie, A. Johnson, and R.C. Wissmar. 1992. The geomorphic structure of salmonid habitats in a recently deglaciated river basin, Washington state. Canadian Journal of Fisheries and Aquatic Sciences **49-6**:1246–1256.

Benda, L., and T. Dunne. 1997a. Stochastic forcing of sediment supply to the channel network from landsliding and debris flow. Water Resources Research **33**:2849–2863.

Benda, L., and T. Dunne. 1997b. Stochastic forcing of sediment routing and storage in channel networks, Water Resources Research, **33**:2865–2880.

Beschta, R.L. 1978. Long term patterns of sediment production following road construction and logging in the Oregon Coast Range. Water Resources Research **14**:1011–1016.

——. 1984. River channel response to accelerated mass soil erosion. Pages 155–164 *in* Symposium on Effects of forest land use on erosion and slope stability. Environment and policy institute, East-West Center, University of Hawaii, Honolulu, Hawaii, USA.

Beven, K. 1981 The effect of ordering on the geomorphic effectiveness of hydrological events. Pages 510–526 *in* T.R.H. Davies and A.J. Pearce, eds. Erosion and sediment transport in Pacific Rim Steeplands, Proceedings of the Christchurch Symposium, January 1981. International Association of Hydrological Sciences, Washington, DC, USA.

Bilby, R.E., and J.W. Ward. 1989. Changes in characteristics and function of woody debris with increasing size of streams in western Washington. Transactions of the American Fisheries Society **118**:368–378.

Borman, B.T., H. Spaltenstein, M.H. McClellan, F.C. Ugolini, K. Cromack, Jr., and S.M. Nay. 1995. Rapid soil development after windthrow disturbance in pristine forests. Journal of Ecology **83**:747–757.

Brubaker, L.B. 1991. Climate change and the origin of old-growth Douglas-fir forests in the Puget Sound lowland. Pages 17–24 *in* L.F. Ruggiero, K.B. Aubry, and M.H. Brookes, eds. Wildlife and vegetation of unmapped Douglas-fir forests. USDA Forest Service General Technical Report PNW-GTR-285, Pacific Northwest Research Station, Portland, Oregon, USA.

Caine, N. 1980. Rainfall intensity—duration control of shallow landslides and debris flows. Geografika Annaler **62A**:23–27.

Church, M. 1983. Pattern of instability in a wandering gravel bed channel. Pages 169–180 *in* International association of sedimentologists special publication 6.

Church, M., and J.M. Ryder. 1972. Paraglacial sedimentation: A consideration of fluvial processes conditioned by glaciation. Geological Society of America Bulletin **83**:3059–3072.

Coats, R., and L. Collins. 1984. Streamside landsliding and channel change in a suburban forested watershed: Effects of an extreme event. Pages 155–164 *in* symposium on effects of forest land

Use on Erosion and Slope Stability. Environment and Policy Institute, East-West Center, University of Hawaii, Honolulu, Hawaii, USA.

Dietrich, W.E., and T. Dunne. 1978. Sediment budget for a small catchment in mountainous terrain. Zietschrift für Geomorphologie, Neue Folge Supplemetband **29**:191–206.

Dolloff, C.A. 1983. The relationships of woody debris to juvenile salmonid production and microhabitat selection in small southeast Alaska streams. Ph.D. thesis. Montana State University, Bozeman, Montana, USA.

Dunne, T. 1991. Stochastic aspects of the relations between climate, hydrology and landform evolution. Transactions Japanese Geomorphological Union **12**:1–24.

Everest, F.H., and W.R. Meehan. 1981. Forest management and anadromous fish habitat productivity. Transactions of the North American Wildlife and Natural Resources Conference **46**:521–530.

Everest, F.H., R.L. Beschta, J.C. Scrivener, K.V. Koski, J.R. Sedell, and C.J. Cederholm. 1987. Fine sediment and salmonid production: A paradox. Pages 98–142 *in* E.O. Salo and T.W. Cundy, eds. Streamside management: Forestry and fishery interactions. Institute of Forest Resources Contribution Number 57, University of Washington, Seattle, Washington, USA.

Franklin, J.F. 1990. Biological legacies: A critical management concept from M. St. Helens. Pages 216–219 *in* Transactions of the Fifty-fifth North American Wildlife and Natural Resources Conference. Wildlife Management Institute, Washington, DC, USA.

Frissell, C.A., W.J. Liss, C.E. Warren, and M.D. Hurley. 1986. A hierarchical framework for stream habitat classification: Viewing streams in a watershed context. Environmental Management **10**:199–214.

Gara, R.I., W.R. Littke, J.K. Agee, D.R. Geiszler, J.D. Stuart, and C.H. Driver. 1985. Influence of fires, fungi, and mountain pine beetles on development of a lodgepole pine forest in south-central Oregon. Pages 153–162 *in* D.M. Baumgartner, ed. Lodgepole pine: The species and its management. Cooperative Extension, Washington State University, Pullman, Washington, USA.

Gell-Mann, M. 1994 The quark and the jaguar: Adventures in the simple and the complex. W.H. Freeman and Company, New York, USA.

Gilbert, G.K. 1917. Hydraulic-mining debris in the Sierra Nevada. United States Geological Survey Professional Paper 105.

Gomez, B., R.L. Naff, and D.W. Hubbell. 1989. Temporal variations in bedload transport rates associated with the migration of bedforms. Earth Surface Processes and Landforms **14**:135–156.

Grant, G.E., and F.J. Swanson. 1995. Morphology and processes of valley floors in mountain streams, Western Cascades, Oregon. Pages 83–111 *in* J.E. Costa et al., eds. Natural and anthropogenic influences in fluvial geomorphology. Geophysical Monograph 89, American Geophysical Union, Washington DC, USA.

Griffiths, G.A. 1979. Recent sedimentation history of the Waimakariri River, New Zealand. Journal of Hydrology (New Zealand) **18**:6–28.

Hack, J.T., and J.C. Goodlett. 1960. Geomorphology and forest ecology of a mountain region in the Central Appalachians. United States Geological Survey Professional Paper 347.

Heard, W.R. 1991. Life history of pink salmon (*Oncorhynchus gorbuscha*). Pages 121–230 *in* C. Groot and L. Margolis, eds. Pacific salmon life histories. University of British Columbia Press, Vancouver, British Columbia, Canada.

Hemstrom, M.A., and J.F. Franklin. 1982. Fire and other disturbances of the forests in Mt. Rainier National Park. Quaternary Research **18**:32–51.

Heusser, C.J. 1977. Quaternary palynology of the Pacific slope of Washington. Quaternary Research **8**:282–306.

Hogan D.L., S.A. Bird, and M.A. Haussan. 1995. Spatial and temporal evolution of small coastal gravel-bed streams: The influence of forest management on channel morphology and fish habitats. 4th International Workshop on Gravel-bed Rivers, August 20–26, Gold Bar, Washington, USA.

Jacobson, R.B. Spatial controls on patterns of land-use induced stream disturbance at the drainage-basin-scale—an example from gravel-bed-streams of the Ozark Plateaus. American Geophysical Union EOS (abstract) H52c-6:151.

James, L.A. 1991. Incision and morphologic evolution of an alluvial channel recovering from hydraulic mining sediment. Geological Society America Bulletin **103**:723–736.

Janda, R.J., K.M. Nolan, D.R. Harden, and S.M. Colman. 1975. Watershed conditions in the drainage basin of Redwood Creek, Humboldt County, California, as of 1973. United States Geological Survey Open-File Report 75-568, Menlo Park, California, USA.

Keller E.A., and F.J. Swanson. 1979. Effects of large organic material on channel form and fluvial processes. Earth Surface Processes **4**:361–380.

Klock, G.O., and J.D. Helvey. 1976. Debris flows following wildfire in north central Washington. Pages 91–98 *in* Proceedings Third Federal Interagency Sedimentation Conference, Denver, Colorado, USA.

Kochel, R.C., D.F. Ritter, and J. Miller. 1987. Role of tree dams in the construction of pseudo-terraces and variable geomorphic response to floods in Little River Valley, Virginia. Geology **15**:718–721.

Leopold, E.B., R. Nickmann, J.I. Hedges, and J.R. Ertel. 1982. Pollen and lignin records of the late Quaternary vegetation, Lake Washington. Science **218**:1305–1307.

Long, C.J. 1995. Fire history of the central Coast Range, Oregon: 9000 year record from Little Lake. Master's thesis. University of Oregon, Eugene, Oregon, USA.

Madej, M.A., and V. Ozaki. 1996. Channel response to sediment wave propagation and movement, Redwood Creek, California, USA. Earth Surface Process and Landforms **21**:911–927.

Megahan, W.F. 1982. Channel sediment storage behind obstructions in forested drainage basins draining the granitic bedrock of the Idaho Batholith. Pages 114–121 *in* Sediment budgets and routing in forested drainage basins. USDA Forest Service General Technical Report PNW-GTR-141, Pacific Northwest Forest and Range Experiment Station, Portland, Oregon, USA.

Minshall, G.W., K.W. Cummins, R.C. Petersen, C.E. Cushing, D.A. Bruns, J.R. Sedell, et al. 1985. Developments in stream ecosystem theory. Canadian Journal of Fisheries and Aquatic Science **42**:1045–1055.

Montgomery, D.R., and J.M. Buffington. 1993. Channel classification, prediction of channel response, and assessment of channel condition. Washington State Timber Fish and Wildlife Technical Report TFW-SH10-93-002, University of Washington, Seattle, Washington, USA.

Morrison, P., and F. Swanson. 1990. Fire history in two forest ecosystems of the central western Cascades of Oregon. USDA Forest Service General Technical Report PNW-254, Pacific Northwest Forest and Range Experiment Station, Portland, Oregon, USA.

Nakamura, F., and F.J. Swanson. 1993. Effects of coarse woody debris on morphology and sediment storage of a mountain stream system in western Oregon. Earth Surface Processes and Landforms **18**:43–61.

Naiman, R.J., T.J. Beechie, L.E. Benda, D.R. Berg, P.A. Bisson, L.H. McDonald, et al. 1992. Fundamental elements of ecologically healthy watersheds in the Pacific Northwest coastal ecoregion. Pages 127–188 *in* R.J. Naiman, ed. Watershed management: Balancing sustainability and environmental change. Springer-Verlag, New York, New York, USA.

Nawa, R.K., C.A. Frissell, and W.J. Liss. 1989. Life history and persistence of anadromous fish stocks in relation to stream habitat and watershed classification. Annual progress report. Oak Creek Laboratory of Biology, Department of Fish and Wildlife, Oregon State University, Corvallis, Oregon, USA.

O'Loughlin, C.L. 1972. The stability of steepland forest soils in the Coast Mountains, southwest British Columbia. Ph.D. thesis. Department of Geology, University of British Columbia, Vancouver, British Columbia, Canada.

Oliver, C.D. 1981. Forest development in North America following major disturbances. Forest Ecology and Management **3**:153–68.

Orme, A.R. 1990. Recurrence of debris production under coniferous forest, Cascade foothills, northwest United States. Pages 67–84 *in* J.B. Thornes, ed. Vegetation and erosion. John Wiley & Sons, New York, New York, USA.

Pickett, S.T.A., and P.S. White, eds. 1985. The ecology of natural disturbance and patch dynamics. Academic Press, New York, New York, USA.

Pickup, G., R.J. Higgins, and I. Grant. 1983. Modeling sediment transport as a moving wave: The transfer and deposition of mining waste. Journal of Hydrology **60**:281–301.

Pierson, T.C. 1977. Factors controlling debris-flow initiation on forested hillslopes in the Oregon Coast Range. Ph.D. dissertation. Department of Geological Sciences, University of Washington, Seattle, Washington, USA.

Quinn, T.F. 1984. Homing and straying in Pacific salmon. Pages 357–362 *in* J.D. McCleave, G.P. Arnold, J.J. Dodson, and W.H. Neil, eds. Mechanisms of migration in fish. Plenum, New York, New York, USA.

Reeves, G.H., L.E. Benda, K.M. Burnett, P.A. Bisson, and J.R. Sedell. 1995. A disturbance-based ecosystem approach to maintaining and restoring freshwater habitats of evolutionarily significant units of anadromous salmonids in the Pacific Northwest. *In* J.L. Nielson and D.A. Powers, eds. Evolution and the aquatic ecosystem: Defining unique units in population conservation. American Fisheries Society Symposium 17, Bethesda, Maryland, USA.

Reiser, D.W., and T.C. Bjornn. 1979. Influence of forest and rangeland management on anadromous fish habitat in the western United States and Canada. Part 1: Habitat requirements of anadromous salmonids. USDA Forest Service General Technical Report PNW-96 Pacific Northwest Forest and Range Experiment Station, Portland, Oregon, USA.

Resh, V.H., A.V. Brown, A.P. Covich, M.E. Gurtz, H.W. Li, G.W. Minshall, et al. 1988. The role of disturbance in stream ecology. Journal of the North American Benthological Society **7**:433–455.

Roberts, R.G., and M. Church. 1986. The sediment budget in several disturbed watersheds, Queen Charlotte Ranges, British Columbia. Canadian Journal of Forest Research **16**:1092–1106.

Rosgen, D.L. 1995. A stream classification system. Pages 91–95 *in* R.R. Johnson, C.D. Zeibell, D.R. Patton, P.F. Folliott, and R.H. Hamre, eds. Riparian ecosystems and their management: Reconciling conflicting uses. USDA Forest Service General Technical Report RM-120.

Roth, G., F. Siccardi, and R. Rosso. 1989. Hydrodynamic description of the erosional development of drainage patterns. Water Resources Research **25**:319–332.

Scarlett, W.S., and C.J. Cederholm. 1984. Juvenile coho salmon fall-winter utilization of two small tributaries in the Clearwater River, Washington. Pages 227–242 *in* J.M. Walton and D.B. Houston, eds. Proceedings of the Olympic Wild Fish Conference March 23–25, 1983. Peninsula College, Fisheries Technology Program, Port Angeles, Washington, USA.

Schoonmaker, P.K., B. von Hagen, and E.C. Wolf, eds. 1997. The rain forests of home: Profile of a North American bioregion. Island Press, Washington, DC, USA.

Sedell, J.R., and J.L. Froggatt. 1984. Importance of streamside forests to large rivers: The isolation of the Willamette River, Oregon, USA from its floodplain by snagging and streamside forest removal. Internationale Vereinigung für theoretische und angewandte Limnologie Verhandlungen **22**:1828–1834.

Sedell, J.R., G.H. Reeves, F.R. Hauer, J.A. Standford, and C.P. Hawkins. 1990. Role of refugia in recovery from disturbances: Modern fragmented and disconnected river systems. Environmental Management **14**:711–724.

Sedell, J.R., and F.J. Swanson. 1984. Ecological characteristics of streams in old-growth forests of the Pacific Northwest. Pages 9–16 *in* W.R. Meehan, T.R. Merrell, Jr., and T.A. Hanley, eds. Proceedings of Symposium Fish and Wildlife Relationships in Old Growth Forests. American Institute of Fishery Research Biologists, Asheville, North Carolina, USA.

Sidle, R.C. 1987. A dynamic model of slope stability in zero-order basins. Proceedings from Symposium on the Erosion and Sedimentation in the Pacific Rim. International Association of Hydrologic Science **165**:101–110.

Sullivan, K., T.E. Lisle, C.A. Dolloff, G.E. Grant, and L.M. Reid. 1987. Stream channels: The link between forests and fishes. Pages 39–97 *in* E.O. Salo and T.W. Cundy, eds. Streamside management: Forestry and fishery interactions. Institute of Forest Resources Contribution Number 57, University of Washington, Seattle, Washington, USA.

Swanson, F.J. 1981. Fire and geomorphic processes. USDA Forest Service General Technical Report WO-26:401–420, Pacific Northwest Range and Experiment Station, Portland, Oregon, USA.

Swanson, F.J., T.K. Kratz, N. Caine, and R.G. Woodmansee. 1988. Landform effects on ecosystem patterns and processes. BioScience **38**:92–98.

Swanson, F.J., and G.W. Lienkaemper. 1978. Physical consequences of large organic debris in Pacific Northwest streams. USDA Forest Service General Technical Report, PNW-69, Pacific Northwest Forest and Range Experiment Station, Portland, Oregon, USA.

Swanston, D.N. 1991. Natural processes. American Fisheries Society Special Publication **19**:139–179.

Teensma, P.D.A. 1987. Fire history and fire regimes of the central western Cascades of Oregon. Ph.D. dissertation. University of Oregon, Eugene, Oregon, USA.

Teensma, P.D.A., J.T. Rienstra, and M.A. Yeiter. 1991. Preliminary reconstruction and analysis of change in forest stand age classes of the Oregon Coast Range from 1850 to 1940. USDI Bureau of Land Management Technical Note T/N OR-9. Portland, Oregon, USA.

Vannote, R.L., G.W. Minshall, K.W. Cummins, J.R. Sedell, and C.E. Cushing. 1980. The river continuum concept. Canadian Journal Fisheries and Aquatic Sciences **37**:130–137.

12
Riparian Forests

Robert J. Naiman, Kevin L. Fetherston, Steven J. McKay, and Jiquan Chen

Overview

• The term *riparian* refers to the biotic communities and the environment on the shores of streams, rivers, ponds, lakes, and some wetlands. In this chapter, the phrase *riparian forest* refers to vegetation directly adjacent to rivers and streams. The riparian forest extends laterally from the active channel to the uplands, thereby including active floodplains and the immediately adjacent terraces.

• Riparian forests of the Pacific coastal ecoregion are floristically and structurally the most diverse vegetation of the region. As complex ecological systems localized at the land–water ecotone, riparian forests maintain generally high levels of beta and gamma diversity, exhibit high rates of nutrient cycling and productivity, provide specialized ecological functions (e.g., maintain water quality), and exert strong influences on adjacent ecological systems by modifying the flow of materials and information (i.e., sound, visual communication) across the landscape. Alterations to riparian forests have effects that go far beyond the specific activity, with the ecological consequences felt throughout the entire river corridor.

• The spatial extent of the riparian forest varies longitudinally and laterally throughout the drainage network. The extent is a function of valley morphology, physical processes (such as climate and hydrology), and vegetative legacies and life history strategies. The basis for understanding riparian forest dynamics, therefore, is the watershed's geomorphology and the physical processes occurring therein.

• Plants of the riparian forest have numerous morphological, physiological and reproductive adaptations for life in variable environments. Specific adaptations include those related to flooding, sediment deposition, physical abrasion and stem breakage. For example, some riparian plants tree species establish and grow upon large woody debris (LWD), while other species require the mineral soils of floodplains, and even others need saturated or flooded soils.

• Three closely related processes and interactions determine patterns of riparian forest development and structure: responses to disturbances, soil dynamics, and biological characteristics related to succession. Several case studies illustrate the variety of successional patterns in natural settings.

• Complex, interactive relationships between riparian forest structure and ecosystem function impart vitality, not only to the forest, but also to the stream environment. In channels, LWD partially controls the routing of water and sediment which creates and maintains a variety of habitats. Roots improve stream bank stability and sequester nutrients. Floodplain vegetation captures and holds particles while the forest canopy controls the light regime as well as provides seasonal nourishment for organisms in the form of leaves, needles, and wood. Riparian influences on microclimate and plant diversity are determined by forest structure at several hierarchical spatiotemporal scales.

• Hydrologic alteration caused by river regulation and forest management alters a variety of processes necessary for the establishment and maintenance of riparian vegetation. Riparian forests are affected not only by alteration of the hydrologic regime but also by human-induced changes in the transport of materials. The site-specific ecological consequences of these two changes remain difficult to predict but include alterations in species composition and distribution and changes in sediment moisture retention and soil biogeochemical cycles.

Introduction

Riparian forests are one of the biosphere's most complex ecological systems but also one of the most important for maintaining the vitality of the landscape and its rivers (Naiman and Décamps 1990, 1997). Riparian forests are highly variable in space and time, reflecting the inherent physical heterogeneity of the drainage basin, the processes shaping its morphology, and the character of the biotic community (Figure 12.1). Riparian forests are the products of interactions among biophysical factors (whether they are occurring in the present or in the past), while the forests have strong feedback influences as well on geological structure and the processes by which they are shaped.

Riparius is a Latin word meaning "of or belonging to the bank of a river." The term *riparian* has generally replaced the Latin *riparius* and refers to biotic communities living on the shores of streams, rivers, ponds, lakes, and some wetlands. Riparian zones, as interfaces between terrestrial and aquatic systems, encompass sharp gradients in environmental and community processes (Naiman et al. 1993, Naiman and Décamps 1997). Riparian zones are an unusually diverse mosaic of landforms, communities, and environments within the larger landscape. For example, they serve as a framework for establishing the organization, diversity, and dynamics of communities associated with aquatic ecosystems (Gregory et al. 1991, Décamps 1996). Often it is difficult to precisely delineate the spatial extent of the riparian zone because its physical heterogeneity is expressed in an array of plant life history strategies and successional patterns, while

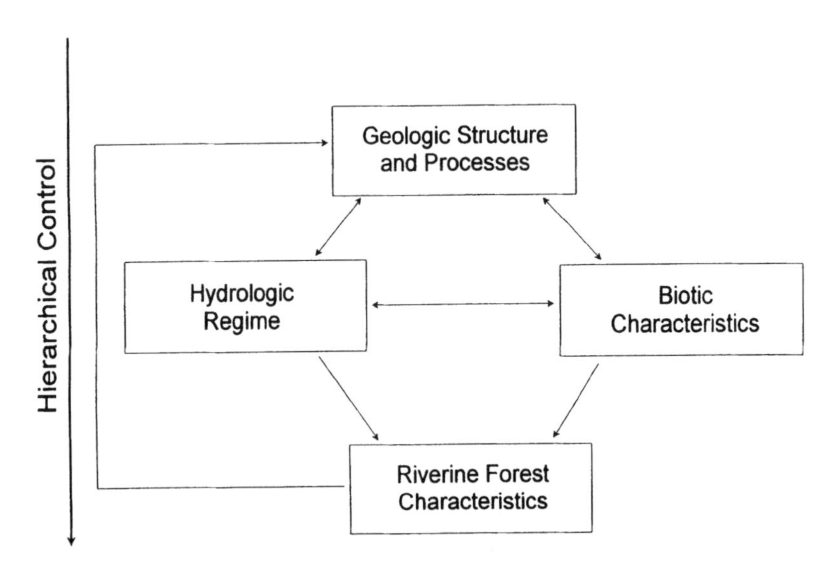

FIGURE 12.1. Overview of fundamental interactions between geologic structure, hydrologic regime, biotic characteristics, and the character of the riparian forest. The relative level of hierarchical control on riparian features is indicated by the arrow on the left.

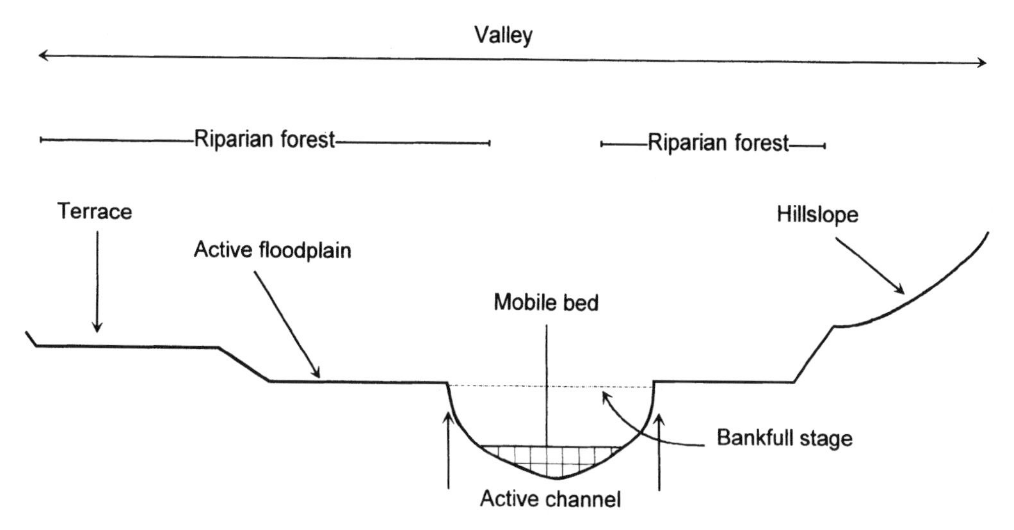

FIGURE 12.2. Local topographic landforms found within a typical alluvial river valley.

the functional attributes depend not only on community composition but also on the environmental setting.

In this chapter, the phrase *riparian forest* refers to floodplain vegetation or vegetation directly adjacent to rivers and streams. The riparian forest extends laterally from the active channel to include the active floodplain and terraces (Figure 12.2). Forests contributing organic matter such as leaves, branches, and large woody debris directly to the active channel or the floodplain are included in this definition (Gregory et al. 1991).

Riparian forests of the Pacific coastal ecoregion are floristically and structurally the most diverse vegetation of the region. This diversity is expressed in relatively high levels of species richness, plant biomass, and structural complexity. For example, riparian forest biomass is among the greatest of any ecosystem on earth, and riparian zones in the southern range of the Pacific coastal ecoregion contain the tallest tree in the world, the coast redwood (*Sequoia sempervirens*). Spatially, riparian forests are heterogeneous mosaics of vegetation patches possessing a wide variety of species (in various stages of succession) arranged in linear patterns corresponding to valley bottoms.

The climate, vegetation associations, and physiography of the Pacific coastal ecoregion have been described by Franklin and Dyrness (1973), Bailey (1980), Pojar et al. (1987), and Alaback (1995). These temperate rain forests have maritime climates with over 1,400 mm annual precipitation and cool temperatures year around (Alaback 1995). North American rain forests are divided into four principle zones: (1) the subpolar rain forest, (2) the permanently humid rain forest, (3) the seasonal rain forest, and (4) the warm temperate rain forest (Table 12.1). Fog is a key climatic element and fog-drip interception is important in defining the geographic limit of the most productive forests in the Pacific coastal ecoregion (Alaback 1995). Cool, foggy conditions are optimal for the growth of the native coniferous species (Waring and Franklin 1979).

Riparian forests of the Pacific coastal ecoregion range from alpine forests along steep, constrained, headwater channels (Figure 12.3a) to floodplain forests along unconfined, low-gradient, alluvial channels (Figure 12.3b). Characteristic patterns of riparian forest responses to climate and disturbance include long, narrow patches of low-stature slide alder (*Alnus sinuata*) parallel to steep gradient avalanche channels (Figure 12.3a); fan-shaped deciduous forest patches overlying alluvial and debris-flow deposits emptying from confined steep valley tributaries to higher order valley floors; and multiple-aged valley forest patches

TABLE 12.1. Climatic regions of the coastal Pacific Northwest rain forests.

Zone	Latitude	Summer rainfall	Snow Level	Mean annual temperature (°C)	Riverine fire occurence
Subpolar rainforest (SP)	59–60° North	>20% annual	Sea level snow is common	4	N/A[a]
Perhumid rainforest (PH)	50–58° North	>10% annual	Snow transient in winter	7	N/A
Seasonal rainforest (S)	43–50° North	<10% annual	Snow transient in winter	10	Fire rare in riverine forests
Warm temperate rainforest (WT)	38–43° North	<5% annual	N/A	12	Extended droughts, fire occurs any time

[a] N/A = Not applicable.
After Alaback 1995.

of coniferous islands set within a matrix of deciduous forest dominated by black cottonwood (*Populus trichocarpa*) and willow (*Salix* spp.) (Figure 12.3b). The resulting mosaic of riparian patches, often dominated by deciduous species, generally is set within a valley dominated by coniferous species.

This mosaic of plant communities and attendant structural and species diversity develops in response to varied disturbances (Table 12.2). Disturbances to riparian vegetation include avalanches, debris flows, floods, ice scour, fire, wind, herbivory, drought, plant disease, and insect infestations. The prevalence of any disturbance type is a function of climate, plant species composition, hillslope stability, and position within the drainage network (Swanson et al. 1988, Agee 1993).

Although riparian vegetation patterns are initiated by physical processes, it is the interaction of species-specific life history strategies (i.e., reproductive mechanisms and physiological adaptations) with physical processes that result in the characteristic patterns of vegetation colonization, establishment, and succession (Menges and Waller 1983, Oliver et al. 1985, Walker et al. 1986). Riparian forests throughout the Pacific coastal ecoregion, as well as temperate montane regions world wide, exhibit characteristic vegetative patterns in response to similar climate, physiography, plant genera distribution, and disturbance regimes. A classic example is willow. *Salix* spp. includes many pioneer species that are found in areas characterized by high frequency disturbances such as repeated scour and deposition.

The characteristic vegetative patterns and the driving biophysical processes influencing the biotic patterns are the subjects of this chapter. Riparian forests are particularly instructive for demonstrating linkages between conceptual advancements and practical applications in ecology. This chapter summarizes several important aspects of the physical setting, discusses life history adaptations to the physical environment, examines the spatial patterns of riparian communities within drainage networks (including several case studies), and concludes by describing biotic responses to river regulation and forest management.

The Physical Setting and Geomorphic Context

The dynamics of the riparian forest varies longitudinally and laterally throughout a drainage network as a function of valley morphology, physical processes, vegetative legacies, and life history strategies. The identification of watershed geomorphology and physical processes, thus, forms the basis for understanding the spatial extent of the riparian forest. Given the importance of geomorphic processes in shaping riparian forests, as well as the role forests play in structuring channel morphology (see later), it is important to briefly review fundamental concepts related to valley morphology, hillslope processes, fluvial processes, soil

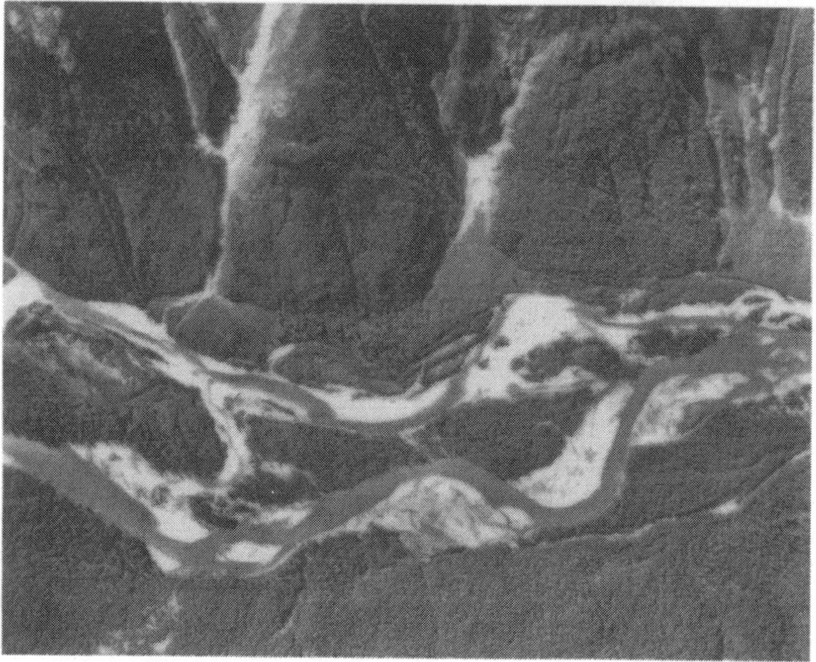

FIGURE 12.3. (a) Avalanche channels on steep mountain slopes in the Queets River valley, Washington (photo: K.L. Fetherston). (b) Alluvial channels of the Kitlope River, British Columbia.

TABLE 12.2. Characteristics of disturbance regimes.

Descriptor	Definition
Type	The agent of disturbance: flood, fire, wind, debris-flow, etc.
Frequency	Mean number of events per time period, expressed in several ways: Probability—Decimal fraction of events per year Return interval—The inverse of probability; years between events Rotation/cycle—Time needed to disturb an area equal in size to the study area
Predictability	A scaled inverse function of variation in frequency
Extent	Area disturbed per time period or event
Magnitude	Described as intensity (physical force, such as energy released per unit area and time for a fire, or windspeed for a hurricane), or severity (a measure of the effect on the organism or ecosystem)
Timing	The seasonality of the disturbance, linked to differential susceptibility of organisms to damage based on phenology

Adapted from White and Pickett 1985, as cited in Agee 1993.

processes, and large woody debris. For further development of these concepts, see Chapters 2, 11, and 13.

Valley Morphology

Valley landforms are physical templates for vegetation colonization, establishment, and development (Hupp 1982, Grant 1986, Grant and Swanson 1995). The valley floor in mountainous drainage networks may be delineated into discrete geomorphic surfaces formed by fluvial or hillslope processes (Grant and Swanson 1995). The dominance of fluvial or hillslope processes is a function of channel network position. For example, landforms along streams with steep gradients are largely created by hillslope processes (such as avalanches and debris flows), whereas the influence of fluvial processes (such as meandering) increases as channels become larger and gradient decreases (Montgomery and Buffington 1993).

Extensive plant colonization occurs annually over the active channel (which includes the floodplain) (Figure 12.2). The active channel, defined by the horizontal limit of continuous riparian vegetation, is flooded at least annually (Church 1992). The floodplain in alluvial rivers is the relatively flat area between the active channel and the adjacent terrace or hillslope constructed by the river in the present climate and overflowed at times of high discharge (Dunne and Leopold 1978). Inundated annually to biannually at times of high discharge, the water directly affects the growing conditions of floodplain vegetation. Riparian vegetation able to persist through floods reduces future fluvial disturbance (Harris 1987).

Hillslope Processes

Hillslope processes include avalanches, debris flows, and landslides (Table 12.3). These originate external to the floodplain or channel and influence the organization, development, and composition of riparian forests in high-gradient, montane river networks.

Avalanches in high-gradient alpine channels result in characteristically long narrow vegetation patches parallel to the channel (Figure 12.3a) (Cushman 1981). The high-frequency but low-magnitude impact of avalanches maintains the vegetation in a state of early succession. Shrubs and trees associated with these channels have life history adaptations (e.g., stem flexibility, asexual reproduction) enabling them to survive in this harsh environment.

Landslides in montane river networks both disturb riparian vegetation and constrain the extent of floodplain forest within the valley (Grant and Swanson 1995). Landslides may reduce the extent of riparian vegetation by burial or increase it by damming channels and thereby increasing the length of shoreline available for colonization.

Debris flows are the primary disturbance of riparian vegetation in many steep montane streams. Debris flows initiate along low-order

TABLE 12.3. Selected vegetation disturbance types and characteristics in the coastal Pacific Northwest.

Type	Zone[1]	Return interval	Magnitude[2]	Network position	Comments	References
Avalanche	SP, PH, S	Variable	1	Initiation > 22°[a] gradient	Absence of trees along avalanche channel	[a]Cushman (1981)
Debris-flow	SP, PH, S, WT	500 yrs[a] 750–1,500 yrs[b]	1, 2, 3	Initiation > 26°[c] scour > 6°[a,c,d]	Dominates morphology and disturbance frequency of steep mountain channels[c]	[a]Benda and Dunne (1987) [b]Swanson et al., (1982) [c]Campbell (1975) [d]Reneau and Dietrich (1987)
Flood	SP, PH, S, WT	Approximately 1–2 yrs[a]	1, 2, 3	Entire channel network	Disturbance by erosion or sediment burial	[a]Williams (1978)
Fire	S, WT	Variable 33–50 yrs[a] 500–600 yrs[b]	1,2	Entire channel network	Coast redwood (*Sequoia sempervirens*) has a moderate severity fire regime[a]	[a]Agee (1993) [b]Veirs (1980)
Ice	SP, PH	Annual	1	Variable, entire channel network	Little quantitative data available	
Wind	SP, PH, S, WT	Variable	1	Throughout the channel network	Little quantitative data available	
Herbivory	SP, PH, S, WT	Annual	1	<8° channel gradient	Beaver (*Castor canadensis*), ungulates (elk [*Cervus canadensis*], moose [*Alces alces*], deer [*Odocoileus* spp.])	

1) See Table 1 for definitions of zones.
2) MAGNITUDE: 1 = vegetation disturbance (partial to complete); 2 = vegetation disturbance and upper soil profile; 3 = complete removal of vegetation and soil.

channels with gradients greater than 50% (Montgomery and Buffington 1993). Debris flows begin when a landslide enters a channel or with the failure of saturated soils adjacent to the channel. The debris flow may move rapidly downslope to high-order channels (Reneau and Dietrich 1987, Montgomery 1991) increasing in size as the channel bed and riparian vegetation are entrained. The accumulated materials travel in-channel until the debris encounters a downstream reach with a gradient low enough to cause deposition (Montgomery and Buffington 1993).

The response of riparian vegetation to debris flow disturbance varies with valley configuration (Grant and Swanson 1995). Long linear strips of deciduous vegetation associated with high-gradient, low-order channels are evidence of channel scour within confined valleys. Vegetated, debris flow depositional fans occur where steep confined channels empty into low-gradient valleys forming a characteristic landform feature in many montane river valleys (Figure 12.3b).

Fluvial Processes

Fluvial processes influencing riparian vegetation in alluvial valleys include stream power, basal shear stress, channel migration, and sediment deposition (Richards 1982). The characteristic vegetative pattern of most low-gradient valleys is maintained by the frequency and magnitude of fluvial disturbances.

The amount of force exerted on the channel bed and vegetation growing in the active channel and floodplain during a flood (i.e., basal shear stress) is a product of fluid density, gravitational acceleration, flow depth, and water surface slope. Vegetation located in the active channel must resist the shear stress of flowing water and has a greater probability of being physically disturbed with increases in channel slope or discharge. The active channel is the harshest environment (physically) for vegetation establishment because it has the highest frequency of flooding of any area in the river valley. In addition to shear stress during floods, plants colonizing the active channel or floodplain are exposed to abrasion and burial from bedload and large wood movement, and to physical removal by erosion.

Soil Processes

Soil development within alluvial environments is highly variable. Frequent erosional and depositional disturbances from flooding create a complex mosaic of soil conditions in the active floodplain that fundamentally influences vegetation colonization and establishment (Oliver and Larson 1996). Well-drained soil or recently deposited mineral alluvium may be found adjacent to very poorly drained organic soils in abandoned high-flow channels or hillslope seeps. This heterogeneity in soil conditions is a major factor in maintaining high plant diversity.

Large Woody Debris

Although produced biologically, large woody debris (LWD) plays a significant geomorphic role in the evolution of channel morphology. Furthermore, it plays a significant, and only recently recognized, ecological role in the colonization and establishment of coniferous species on the floodplain, the formation of forested islands that coalesce into the larger floodplain forest, and the spatial organization of floodplain forests in alluvial valleys (Fetherston et al. 1995, Chapter 13).

LWD accumulates in structurally distinctive debris jams in alluvial channels (Abbe and Montgomery 1996). The jams and the accumulated sediment provide sites on the active floodplain where vegetation colonization and forest island establishment can occur. If undisturbed, forest islands grow and coalesce into the forested floodplain (Figure 12.4). Riparian features created by debris jams may be very long-lived. LWD buried in alluvium can persist up to 500 years in redwood-dominated alluvial valleys (Keller and Tally 1979) and up to 17,000 years in pine forests in Tasmania, Australia (Nanson et al. 1995). Preliminary data from LWD in the Queets River, Washington, have dated exposed LWD jams to over 300 years (Abbe and Montgomery 1996), and buried wood appears to be at least 10 times older. The persistence of LWD in the active floodplain for centuries is evidence of the importance of intact riparian forests to the integrity of channel and floodplain processes.

In addition, LWD on active channel, floodplain, and terrace surfaces provides sites for colonization and establishment of coniferous and deciduous trees (Figure 12.4). Harmon and Franklin (1989) noted the function of LWD as nursery logs for tree species, such as western hemlock (*Tsuga heterophylla*) and Sitka spruce (*Picea sitchensis*), in coastal forests of western Oregon and Washington. For example, over 90% of both western hemlock and Sitka spruce seedling recruitment occurs on large woody debris on terraces of the South Fork Hoh River, Olympic National Park, Washington (McKee et al. 1982). The implication of these findings, at least in the Olympic rain forest, is that LWD is necessary for the long-term maintenance of the coniferous component of the riparian forest. In contrast, the reduction in basal shear stress immediately downstream of LWD provides sites for seedling germination of deciduous species.

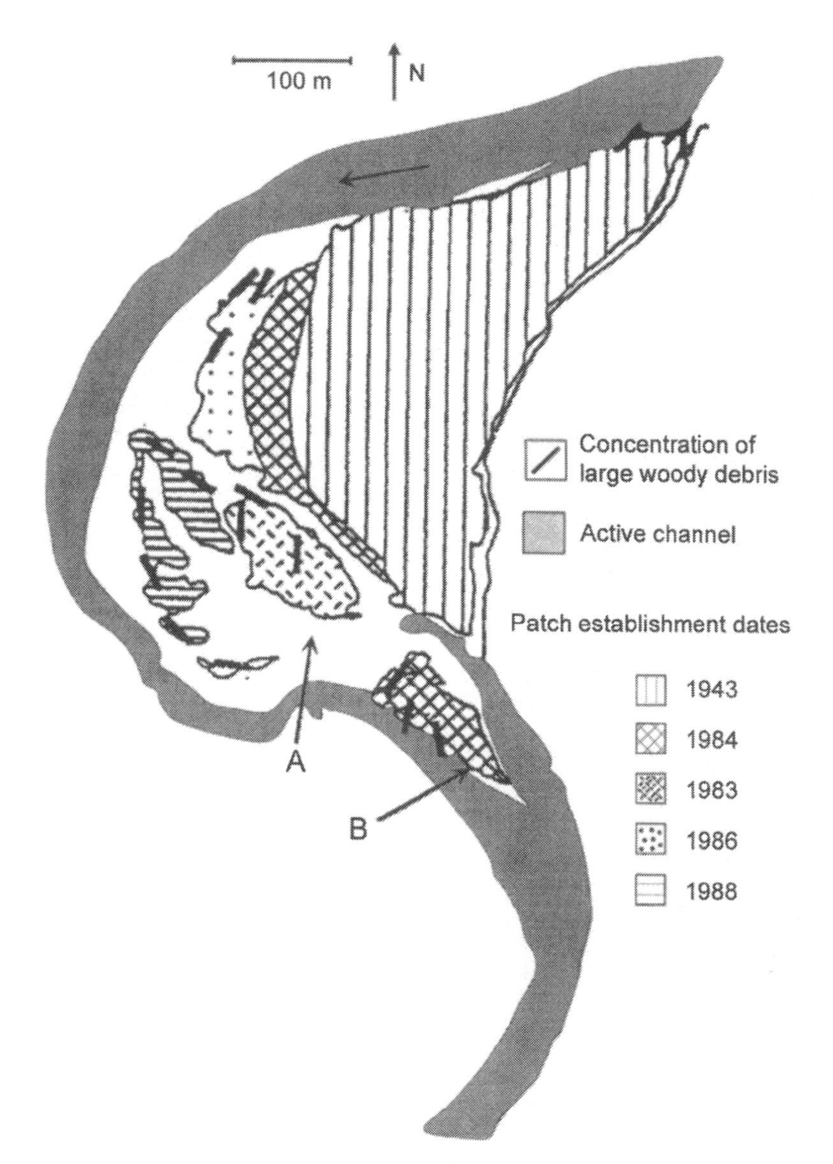

FIGURE 12.4. Riparian forest patches associated with an unconfined alluvial channel in the Queets River, Washington. Forest islands *A* and *B* are growing ($A = 767\,m^2/yr$; $B = 546\,m^2/yr$) behind accumulations of LWD. If undisturbed, these patches will coalesce to form a fully vegetated floodplain (from Fetherston et al. 1995 with kind permission from Elsevier Science-NL, Sara Burgerhartstraat 25, 1055 KV Amsterdam, The Netherlands).

Riparian Plant Adaptations

Plants of the riparian forest have numerous morphological, physiological, and reproductive adaptations which suit them for life in high-energy and wet environments (Brinson 1990, Malanson 1993, Mitsch and Gosselink 1993, Naiman and Décamps 1997). Adaptations by riparian plants include tree species growing upon LWD, species establishing upon mineral soils of the floodplain surface, and obligate hydrophytes growing in saturated or flooded soils. In addition, many riparian plants are adapted to cope with flooding, sediment deposition, and physical abrasion and stem breakage.

The active channel and floodplain are harsh environments for the colonization and establishment of plants. Even though vegetation may colonize a variety of sites, it only establishes successfully on "safe sites." Safe sites have suitable conditions for germination (water and oxygen), are refuges from herbivores and floods, and have environmental conditions compatible with the life history requirements of plants which eventually survive (Harper 1977).

The classification of plants into guilds of similar life history strategies is useful for understanding riparian forest succession and species distribution. Functional adaptations of plants fall into four broad categories (Table 12.4) (Grime 1979, Noble and Slatyer 1980, Agee 1993):

- Invader—produces large numbers of wind- and water-disseminated propagules that colonize alluvial substrates.
- Endurer—resprouts after breakage or burial of either the stem or roots from floods or after being partially eaten (e.g., herbivory).
- Resister—withstands flooding for weeks during the growing season, moderate fires, or disease epidemics.

TABLE 12.4. Relative adaptation to primary disturbance types of major riverine tree and shrub species in the coastal Pacific Northwest.

	Adaptation				
Species	Zone[a]	Flood[b]	Avalanche	Fire	Debris-flow
Sitka spruce (*Picea sitchensis*)	SP, PH, S, WT	Avoider/resister	N/A[c]	Avoider	Avoider
Black cottonwood (*Populus trichocarpa*)	SP, PH, S, WT	Invader/resister	N/A	Invader	Resister/endurer
Red alder (*Alnus rubra*)	SP, PH, S, WT	Invader/resister	Resister	Invader	Endurer
Sitka alder (*A. sitchensis*)	SP, PH, S, WT	Invader/endurer/resister	Invader/endurer/resister	Invader/endurer	Invader
Sitka willow (*Salix sitchensis*)	SP, PH, S, WT	Invader/endurer/resister	Invader/endurer/resister	Invader/endurer	Invader/endurer/resister
Scouler's willow (*S. scouleriana*)	SP, PH, S, WT	Invader/endurer/resister	Invader/endurer/resister Resister	Invader/endurer Resister	Invader/endurer/resister Resister
Douglas-fir (*Pseudotsuga menziesii*)	PH, S, WT	Avoider	Resister	Resister	Avoider
Western hemlock (*Tsuga heterophylla*)	SP, PH, S, WT	Avoider	Resister	Avoider	Avoider
Western redcedar (*Thuja plicata*)	PH, S, WT	Avoider	Resister	Avoider	Avoider
Grand fir (*Abies grandis*)	PH, S, WT	Aoider	Resister	Avoider	Avoider
Coast redwood (*Sequoia sempervirens*)	WT	Resister/endurer	N/A	Endurer	Endurer

[a] Climatic Zones are: SP = Subpolar Rain Forest; PH = Perhumid Rain Forest; S = Seasonal Rain Forest; WT = Warm Temperate Rainforest.
[b] Flooding impact is sand/gravel deposition around tree base.
[c] N/A = not within avalanche zone.
Adapted and modified from Agee 1993.

- Avoider—lacks adaptations to specific disturbance types; individuals germinating in an unfavorable habitat do not survive.

Categorizing some of the major riparian tree species in the Pacific coastal ecoregion illustrates the variety, as well as some of the similarities of life history strategies among riparian plants. For example, Sitka willow (*Salix sitchensis*) and Scouler's willow (*S. scouleriana*) are ubiquitous throughout the region as pioneer plants adapted to several types of disturbance (Table 12.4). Seeds of these species germinate and establish in the postfire landscape, individuals resprout following light intensity fires that have not destroyed the root system, and adventitious roots appear when the stems are fragmented by floods, debris flows or herbivory. These adaptive characters make them well suited as invaders, endurers, or resisters depending on local conditions. In contrast, Sitka spruce is more restricted, colonizing elevated LWD and mineral substrates on the floodplain (Harmon and Franklin 1989, McKee et al. 1982). Once established, Sitka spruce is resistant to both flooding and sediment deposition. However, it is fire sensitive and, in response to that type of disturbance, it is classified as an avoider (Table 12.4). Note that those trees most often thought of as riparian species (black cottonwood, Sitka spruce, alder, willow, coast redwood) are also species categorized as resisters or endurers to specific riparian disturbances.

Morphological and Physiological Adaptations

Numerous morphological adaptations of plants (e.g., adventitious roots, stem buttressing, root and stem flexibility) are responses to anoxia in the soil, unstable substrate conditions, or reproductive requirements. Because metabolic depletion of oxygen in the rhizosphere (i.e., rooting zone) occurs rapidly when soils are flooded, vascular plants have evolved several mechanisms and structural adaptations to anoxic environments (Mitsch and Gosselink 1993). These include air spaces (aerenchyma) in the roots and stems, which allow the diffusion of oxygen from aerial portions of the plant to the roots (making their rhizospheres aerobic), and adventitious roots and "knee roots" (pneumatophores) that grow above the anoxic zone enabling oxygen absorption by the plant. Note, however, that flooding of well-drained alluvium during the cool, dormant season does not lead to the classic physiological plant adaptations to anoxic conditions found in flooded wetlands.

The development of aerenchyma is mediated by increased levels of the hormone ethylene, which is synthesized when soil conditions turn anoxic (Kozlowski et al. 1991, Mitsch and Gosselink 1993). Root and stem aerenchyma are common among ethylene-producing species of the families Cyperaceae (sedges) and Juncaceae (rushes), which are found in floodplains with poor soil drainage. Related to ethylene synthesis is a process known as rhizosphere oxygenation. It occurs in wetland plants where oxygen moves from roots to the adjacent soil creating a minute oxidized zone. This process is believed to effectively mediate the toxic effects of soluble reduced ions found in anoxic soil, such as manganese (Mitsch and Gosselink 1993). The second structural adaptation to anaerobic conditions, adventitious roots and pneumatophores, occurs in a number of floodplain trees (e.g., black cottonwood, willow, red alder [*Alnus rubra*], and coast redwood) (Figure 12.5).

Flooding also mechanically disturbs plants through erosion of the soil surface and abrasion by transported sediment and LWD. Stem flexibility among woody species (i.e., willow, alder, vine maple [*Acer circinatum*], and cottonwood) growing in the active channel and floodplain allows these species to accommodate high levels of shear stress accompanying seasonal floods and avalanches. Major floods and avalanches in the Pacific coastal ecoregion occur during periods when deciduous trees are without leaves, reducing the amount of drag on the plant stem.

Reproductive Strategies

Plant life history strategies include a suite of coadapted characteristics that enhance repro-

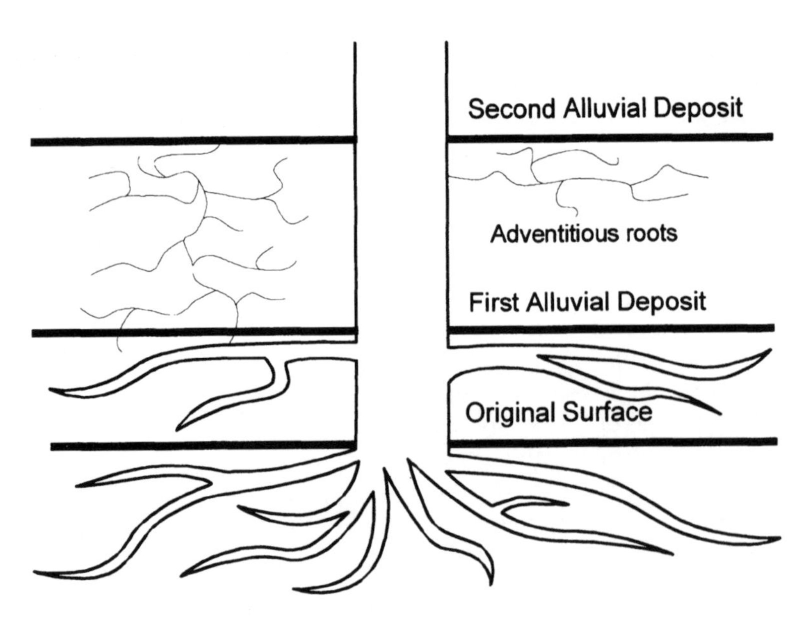

FIGURE 12.5. The roots systems of coast redwood adapt to periodic siltation by developing new adventitious roots higher on the trunk, at the level of sediment deposition (Reprinted with permission from Stone and Vasey 1968 © 1968 American Association for the Advancement of Science).

ductive success in specific environmental conditions (Barbour et al. 1987). Some primary strategies include balancing sexual versus asexual reproduction, optimizing seed size, timing of dormancy, mode of seed dispersal, and seed longevity. Examples are provided below for seed dispersal and asexual reproduction.

A unique adaptation to the riparian environment is a strategy in which seed dispersal coincides with the seasonal retreat of floodwaters when moist seedbeds are available for successful germination and colonization. Many species rely on transport of seeds and vegetative fragments by flowing water, a phenomenon known as hydrochory, for dispersal to new sites. Schneider and Sharitz (1988), working in a South Carolina swamp forest, found that bald cypress (*Taxodium distichum*) and water tupelo (*Nyssa aquatica*) rely on water more than wind to disperse seeds away from parental trees. Similarly, Thébaud and Debussche (1991) studied the invasion of flowering ash (*Fraxinus ornus*), typically a wind-dispersed species, along a river in southern France. They found that dispersal downstream is much more extensive than in other directions, suggesting water transports seeds during autumn flooding. In Sweden, Johansson et al. (1996) found a positive relationship between diaspore floating capacity (length of time a seed will float) and the frequency of species in the riparian vegetation. Nevertheless, hydrochory is only one means for dispersal. Dispersal by animals (zoochory) and by wind (anemochory) may be even more important but few empirical data exist for comparison.

In addition to sexual reproduction by seeds, many riparian plant species reproduce vegetatively (asexually). Notable examples include redwood, willow, poplar (*Populus* spp.), and ash. Multiple sprouts can result from burial during floods (Sigafoos 1964, Everett 1968, Zinke 1981) and abrasion during floods can stimulate stump sprouts (Bégin and Payette 1991, Tardif and Bergeron 1992). Riparian poplars (cottonwood), for example, exhibit the trait of branch abscission, or cladoptosis, in which healthy branch tips are shed, develop adventitious roots on moist soils, and grow into new but genetically identical trees (Figure 12.6) (Galloway and Worrall 1979, Dewit and Reid 1992).

a

b

FIGURE 12.6. (a) Cottonwood (*Populus* spp.) occasionally shed branch tips in the spring which, upon coming into contact with a moist surface, develop adventitious roots and grow into genetically identical trees. The process is known as cladoptosis. (photo: R.F. Stettler). (b) Broken or cut branches also form adventitious roots upon contact with moist surfaces forming suckers at several points (photo: S. McKay).

Growth Dynamics

The growth dynamics of trees—density, basal area, biomass, and production—provide integrated measurements of vitality in specific environments. Unfortunately, comprehensive studies of the growth of riparian trees are few. Nevertheless, the available data suggest that riparian tree communities attain high densities, basal area, and biomass while also attaining high production rates.

The few observations of density and basal area suggest considerable variability on different geomorphic surfaces and at different successional stages. The density of riparian trees in the Pacific coastal region reported to date ranges from 42 to 793 stems/ha ($\bar{x}$ = 221 stems/ha) with no apparent trend associated with stand age or latitude (Table 12.5). Younger stands, however, tend to be dominated by red alder, whereas older stands are dominated by Sitka spruce or western hemlock, especially in the northern latitudes. Basal area ranges from 29 to 83 m^2/ha ($\bar{x}$ = 58 m^2/ha) with Sitka spruce dominating the older and more northern stands. Cottonwood and bigleaf maple (*Acer macrophyllum*) are of intermediate importance in several of the older stands and red alder is a surprisingly minor component in all but the very youngest stands.

Available data from other regions suggest that riparian forests in the southeastern United States and the humid tropics tend toward greater stem density and basal area than those in more arid regions and more northern latitudes (Brinson 1990). Although values vary within regions, largely because of stand age, the variation within regions is usually less than an order of magnitude. In general, the basal area of riparian forests is as great as or greater than that of upland forests.

Data on biomass and production of riparian forests in the Pacific coastal ecoregion, are few and from limited areas. In other regions of North America, however, the aboveground biomass of riparian forests ranges between 100 and 300 t/ha with few exceptions (Brinson 1990). Leaves represent 1 to 10% of the total. Belowground biomass varies from 12 to 190 t/ha, but much of this variation may be the result of sampling methods and site-specific conditions. In general, however, belowground biomass tends to be less than aboveground biomass but ranges from 5 to 120%. Published values for aboveground riparian production ranges from 6.5 to 21.4 t/ha/yr; litter fall averages 47% of the annual production. The limited data do not reveal latitudinal or successional gradients. Rates of belowground production have received virtually no attention in riparian forests. Overall, most evidence supports the concept that riparian forests have relatively high rates of production in comparison with upland forests. Further, the data suggest that water or nutrient limitations on production are often ameliorated in the riparian environment.

Spatial Patterns of Riparian Forests: A Mosaic

Disturbance regimes, soil characteristics (discussed earlier), and biological processes related to succession, three closely related and strongly interactive factors, determine patterns of riparian forest development. This section briefly reviews major forms of riparian disturbance, discusses successional processes, and illustrates the strong interactions with a series of case studies.

Disturbance

Disturbances commonly influencing riparian zones in the Pacific coastal ecoregion include fire, wind, and fluvial transport processes such as flooding and ice scour. As mentioned at the beginning of this chapter, and discussed in the chapters on geomorphology (Chapter 2) and hydrology (Chapter 3) disturbance types and regimes vary systematically throughout the region and throughout the river network. Riparian vegetation is largely an expression of how these physical forces shape vegetative communities within and among catchments.

Fire

Fire is rare in Sitka spruce riparian forests of the northern Pacific coastal ecoregion, but plays a significant role in coastal redwood forests at the southern boundary of the region. As a result, Sitka spruce and other associated riparian tree species have life history strategies to avoid fires (Table 12.4). For example, western hemlock, spruce, and western redcedar (*Thuja*

TABLE 12.5. Riparian forest structure, composition, age, and disturbance history.

Site	Species	Density (#/ha)	Basal area (m^2/ha)	Age	Disturbance
Kadashan River, Southeast Alaska (Pollock et al. 1998)	Red alder (*Alnus rubra*)	28.0	2.9	>250 yrs	Flooding, erosion
	Western hemlock (*Tsuga heterophylla*)	60.0	2.0		
	Sitka spruce (*Picea sitchensis*)	104.0	61.7		
	All species	192.0	66.6		
Kitlope River, British Columbia (Fetherston, unpublished data) forested island	Red alder	14.3	4.2	<150 yrs	Channel migration, flooding
	Black cottonwood (*Populus trichocarpa*)	6.8	5.1		
	Western hemlock	14.3	9.1		
	Western redcedar (*Thuja plicata*)	14.3	6.8		
	Sitka spruce	64.2	33.5		
	All species	113.9	58.7		
South Fork Hoh River, Washington (McKee et al. 1982) lower terrace	Sitka spruce	33.1	52.8	205 yrs	Channel migration
	Western hemlock	24.7	10.3	258 yrs	
	Western redcedar	—	—		
	Douglas-fir (*Pseudotsuga menziesii*)	—	—		
	Red alder	5.3	1.9		
	Bigleaf maple (*Acer macrophyllum*)	0.7	0.4		
	Pacific silver fir (*Abies amabalis*)	—	—		
	All species	64	66.3		
Upper terrace	Sitka spruce	57.8	52.8	220 yrs	Channel migration
	Western hemlock	79.9	15.8	266 yrs	
	Western redcedar	2.1	1.0		
	Douglas-fir	1.7	2.9		
	All species	142.0	81.8		
Hoh River—Main Stem, Alder Flat (Fonda 1974)	Red alder	763.0	36.4	65–75 yrs	Flooding
	Sitka spruce	21.7	0.8		
	Black cottonwood	7.9	1.6		
	All species	792.6	38.8		
First terrace	Sitka spruce	100.8	19.8	370–430 yrs	No data
	Bigleaf maple	81.0	11.6		
	Black cottonwood	65.2	14.8		
	Red alder	10.9	1.2		
	Douglas-fir	5.9	1.4		
	Western hemlock	4.0	1.6		
	All species	267.8	50.4		
Second terrace	Bigleaf maple	34.6	27.0	No data	Landslides
	Sitka spruce	7.4	2.0		
	All species	42.0	29.0		
Central Oregon Coast Range, Flynn Creek (Poage and Spies 1996)	Douglas-fir	51.6	34.96	>145 yrs	Partial burn, wind, herbivory, flooding, pathogens, hillslope process
	Sitka spruce	0.4	0.02		
	Western redcedar	0.4	0.10		
	Western hemlock	0.0	0.0		
	Bigleaf maple	0.0	0.0		
	Red alder	80.4	12.67		
	All species	185.2	82.83		
Trout Creek	Douglas-fir	48.0	33.87	>145 yrs	Partial burn, wind, herbivory, flooding, pathogens, mass movement
	Sitka spruce	0.0	0.0		
	Western redcedar	8.0	1.43		
	Western hemlock	102.0	11.49		
	Bigleaf maple	16.5	2.41		
	Red alder	16.0	1.78		
	All species	190.5	50.98		

plicata) are shallow rooted, thin barked, and shade tolerant (Agee 1993). In contrast, coast redwood is well adapted to all but the most extreme fires by virtue of its stature and thick, fire resistant bark (Table 12.4).

Wind

Wind is considered a major factor in the development of coastal riparian forests (Agee 1993, Oliver and Larson 1996). Anecdotal reports indicate that windthrow and wind-breakage of dominant trees are associated with strong winter storms that frequent the Pacific Northwest coast (Boyce 1929, Oswald 1968, Ruth and Harris 1979). Significant windstorms along the Olympic Peninsula, Washington have occurred in 1921, 1962, 1979, and 1993. However, the regional effects of wind at the forest stand and landscape scales largely remain unknown even though Orr (1963) calculated that $2.6 \times 10^7 m^3$ of wood was blown down in the 1962 wind storm alone.

Floods and Ice

Unregulated rivers in the Pacific coastal ecoregion exhibit fairly predictable hydrologic regimes which vary latitudinally with the varying influence of snowpack and with watershed dimensions (Naiman and Anderson 1997, Chapter 3). The amount and timing of transported materials also varies considerably from river to river and strongly influences riparian vegetation (Naiman and Décamps 1997). Additionally, ice scour following the thaw of frozen rivers (primarily in the northern portion of the region) removes vegetation from the active channel and floodplain, creating zones of annual disturbance.

Successional Processes

Forest development or succession can be described as occurring in four stages (Figure 12.7) (Oliver and Larson 1996):

1. *Stand initiation stage.* Plants colonize sites following a disturbance or formation of a new depositional landform. The stand initiation stage is the developmental period before the growing space is fully occupied.

2. *Stem exclusion stage.* After all the available growing space is occupied, species with a competitive advantage in size or growth expand into the space utilized by other plants, eliminating them from the community. New plants, for the most part, are excluded from colonizing. A predictable vertical sorting of individuals (e.g., vertical stratification by height) in even-aged stands occurs during this stage with some individuals growing faster than others.

3. *Understory initiation stage.* Mortality of large trees in the overstory initiates development of an understory. Understory initiation is characterized by the invasion of shade tolerant herbs, shrubs, and trees. These can be the same species as those present during the stand initiation stage but they grow slowly, creating a stand with multiple canopy layers.

4. *Old growth stage.* Individual trees die as a forest stand ages, opening up canopy space which can be occupied by advanced regeneration in the understory. When the trees which invaded immediately following the initial disturbance die, the stand enters into an old growth condition. This is an autogenic process whereby trees regenerate and grow without the influence of external disturbances. Note, however, that wind (which is an allogenic processes) also may be important in shaping the forest in many cases. Riparian forests are described as old growth if they have structural characteristics such as large, living old trees; large, dead standing trees; massive fallen logs; relatively open canopies with foliage in many layers; and a diverse understory. Structurally, old growth riparian forests have the greatest horizontal and vertical variation of all the successional stages, with both large and small trees growing in separate and intermixed patches (Alaback 1982a, 1982b).

Case Studies

The presence of distinct riparian forest patterns provides a clear illustration of how physical processes affect vegetation. These patterns are created by interactions between physical

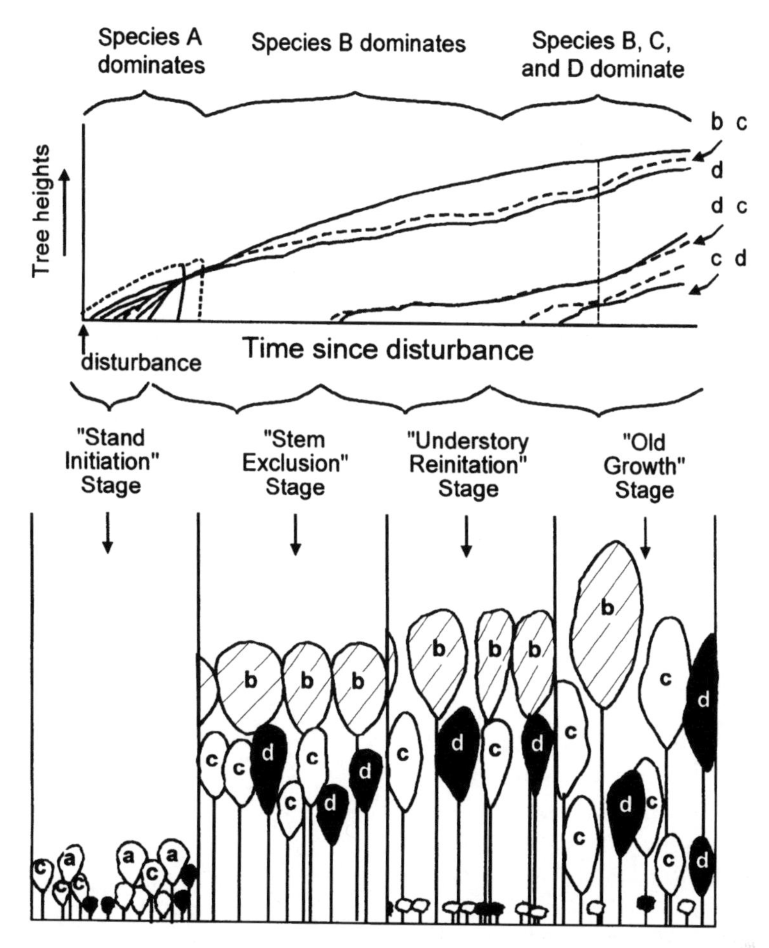

FIGURE 12.7. Four schematic stages of forest development following major disturbances. All trees forming the riparian forest establish soon after the disturbance (stem initiation); however, the dominant tree type changes as stem numbers decrease and vertical stratification of species progresses (stem exclusion, understory reinitiation, and old growth stages). The height attained and the time lapsed during each stage vary with species, disturbance and site (from Oliver 1981 with kind permission of Elsevier Science-NL, Sara Burgerhartstraat 25, 1055 KV Amsterdam, The Netherlands).

processes and plant life history strategies. Case studies illustrate the diversity of geomorphic and climatic settings, as well as the similarity of riparian successional processes in the Pacific coastal ecoregion. The case studies presented here illustrate examples of several primary types of disturbance of the region: glacial retreat, avalanche, debris-flow, flood, and fire.

Glacial Retreat

Glacially fed rivers are common from Washington north. The rapid retreat of glaciers following the Little Ice Age (1550–1850) resulted in identifiable sequences of vegetation succession along the retreating glacial terminus. Glacier Bay, Alaska, traditionally has been the focus of studies of succession following glacial retreat (Cooper 1923, Reiners et al. 1971, Sidle and Milner 1989). Following glacial retreat, till and outwash surfaces are initially colonized by mosses followed by fireweed (*Epilobium latifolium*) and scattered sedges (*Carex* spp.). The principal woody pioneer species is yellow mountain ariens (*Dryas drummondii*) with scattered mountain alder (*Alnus incana*), black cottonwood and willow. Approximately 100 years after glacial retreat an even-aged forest of Sitka spruce develops (Sidle and Milner 1989). Stands of alder, cottonwood and willow occupy stream banks with an understory of Sitka spruce. In stands older than 100 years, wood accumulates in piles on the floodplain and stream banks, stabilizing gravel bars and providing new sites for vegetation colonization and establishment within the active channel and floodplain (Sidle and Milner 1989).

Avalanches

Snow avalanches occur annually throughout much of the steep-gradient, high-altitude zones of the Pacific coastal ecoregion. Avalanches frequently follow headwater stream channels, impacting riparian vegetation. The low growing and flexible woody species found in riparian avalanche zones are capable of rapid growth and vegetative reproduction, enabling them to withstand the force of sliding snow as well as to respond rapidly to repeated disturbance (Cushman 1981). Thus, the hallmarks of riparian vegetation in avalanche channels are a low-stature physiognomy, flexible stems, rapid growth in saturated soils, and asexual reproduction.

In Washington's Cascade Mountains, the composition of the plant community in riparian avalanche zones is determined primarily by elevation. However, the physiognomy of the vegetation and the community structure are driven by avalanche frequency (Figure 12.8) (Cushman 1981). Woody species comprise 70% of the cover and herbs, particularly bracken fern (*Pteridium aquilinum*) and several species of sedges are significant components of the vegetation. Characteristic long linear patches of different vegetation types parallel the steep-gradient, high-elevation, avalanche-dominated streams (Figure 12.3a). Undisturbed vegetation adjacent to avalanche tracks above 1,350m in elevation is dominated by mountain hemlock (*Tsuga mertensiana*) while between 1,350 and 1,550m there is a continuous forest of mountain hemlock, yellow cedar (*Chamaecyparis nootkatensis*) and subalpine fir (*Abies lasiocarpa*). Other shrubs and vines typically associated with riparian avalanche tracks below 1,350m elevation include vine maple, slide alder, salmonberry (*Rubus spectabilis*), false azalea (*Menziesia ferruginea*), blueberry and huckleberry (*Vaccinium* spp.), red elderberry (*Sambucus racemosa*), and hardhack (*Spiraea douglasii*). Above 1,350m elevation the typical shrubs and vines are devil's club (*Oplopanax horridum*), stink currant (*Ribes bracteosum*), thimbleberry (*Rubus parviflorus*), and various species associated with heather meadow (*Phyllodoce*), lush meadow, and krummholz.

Glacial Retreat, Avalanches, Floods, Rock Slides

In one of the most comprehensive studies to date of physical disturbance processes and riparian forest succession, Oliver et al. (1985) reconstructed the vegetation history of the recently deglaciated Nooksack Cirque in the North Cascade Range of Washington. The study area is located in the Pacific silver fir (*Abies amabilis*) (600–1,300m) and the mountain hemlock zones (>1,300m) (Franklin and Dyrness 1973). Advances and retreats of the Nooksack glacier, snow avalanches, rock slides, mass movements, and channel migration created a complex forest mosaic (Figure 12.9). Chronic avalanche disturbance resulted in extensive areas of slide alder and vine maple with living stems pointed downhill. Following glacial retreat, Sitka spruce initiated succession by emerging from shrub thickets; similar patterns were reported by Cooper (1923, 1931) and Sidle and Milner (1989) for Glacier Bay, Alaska, where moss, fireweed, and shrubs initially colonized new substrates.

Debris Flows and Floods

Debris flows and floods create and maintain local landforms upon which riparian vegetation develops. In a comprehensive study of 21 sites in mature to old growth riparian forests of the western Cascade Mountains in Washington, Rot (1995) found distinct plant communities

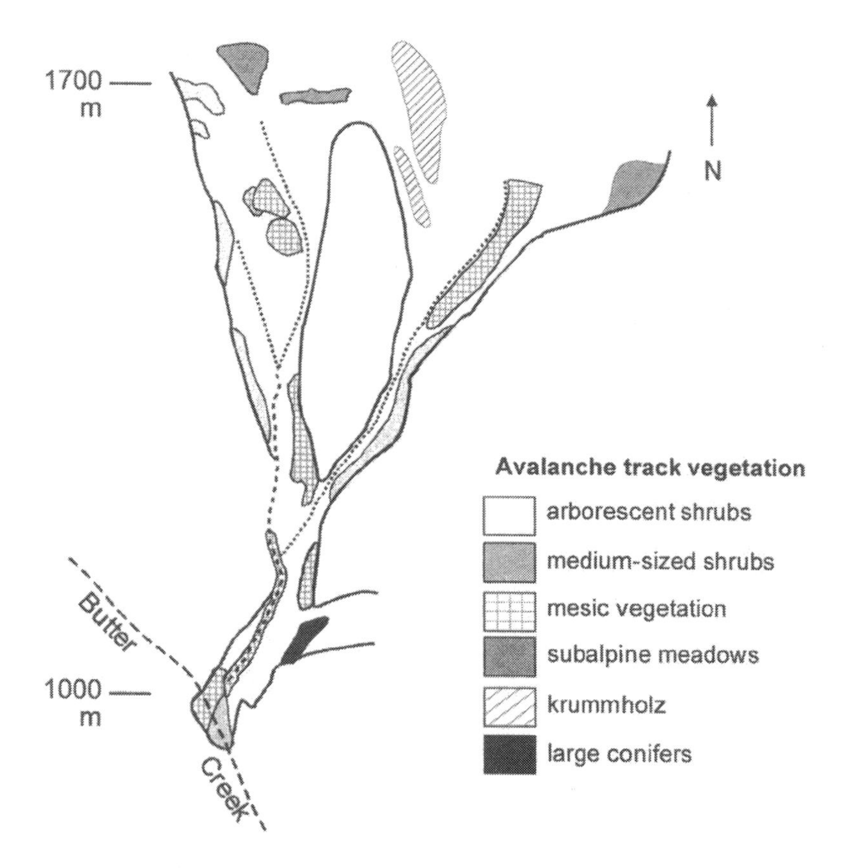

FIGURE 12.8. Avalanche track vegetation mosaic (Butter Creek, Mount Rainer National Park; after Cushman 1981). Physiognomic vegetation types include: (1) arborescent shrubs (*Acer circinatum*, *Alnus sinuata*, *Sambucus racemosa*). (2) medium sized shrubs (*Rubus* spp., *Vaccinium* spp., *Menziesii ferruginea*, *Spiraea douglasii*, *Spiraea densiflora*). (3) mesic vegetation (*Oplopanax horridum*, *Pteridium aquilinum*, *Ribes bracteosum*, *Rubus parviflorus*, *Rubus spectabalis*, and *Salix* spp.) (4) subalpine meadows (*Aster ledophyllus*, *Carex spectabilis*, *Lupinus latifolius*, *Valeriana sitchensis*). (5) krummholz, stunted specimens of *Abies lasiocarpa*, *Chamaecyparis nootatensis*, and *Tsuga mertensiana* grow on protected sites. (6) large conifers are found outside the avalanche tracks.

occupying different riparian landforms. Subdividing the riparian zone into four landform types (floodplain, low terrace, high terrace, and hillslope) he found that both successional processes and fluvial disturbances control community composition. Floodplains associated with 3rd- and 4th-order streams were dominated by deciduous species, especially red alder (62% of stems), while conifers dominated the overstory of other landforms. Pacific silver fir and western hemlock were most common on low terraces (77% of stems), while Douglas-fir (*Psuedotsuga menziesii*) and western hemlock dominated high terraces (72% of stems). Hillslopes supported mostly western hemlock with some Douglas-fir. Discriminate component analyses suggest that vegetation on floodplains and low terraces is strongly influenced by moisture and temperature gradients, whereas vegetation on high terraces and hillslopes is influenced by biotic factors (probably herbivory, competition, or disease). Further, propagation from buried woody plants and seedlings also can contribute to revegetation because debris flows and floods may bury vegetation (Gecy and Wilson 1990).

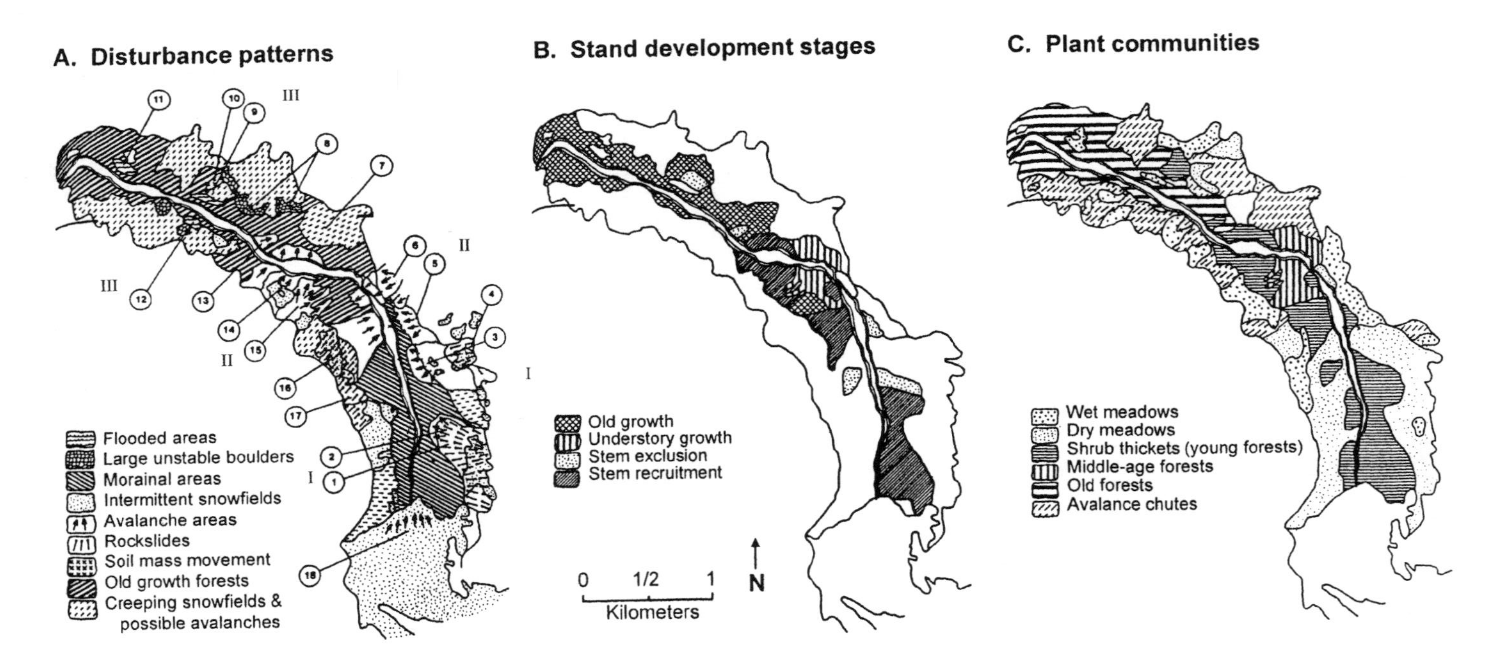

FIGURE 12.9. Specific disturbance types often result in similar patterns of vegetation composition and structure as shown in a mid-elevation valley (Nooksack Cirque area) in the north Cascades of Washington (after Oliver et al. 1980, 1985). (a) Disturbance pattern types. Primary disturbance areas: I = glacial activity within the last 200 years; II = primary disturbances (possibly glacier) occured about 600 years B.P.; III = no obvious evidence of a major disturbance. Secondary disturbances: 1, 4, 16, and 17 chronic rockslides; 2, 3, 5, 6, 7, and 15, chronic snow avalanches; 8, unstable rocks; 9, soil mass movement from ca. 1830–1880; 10, chronic soil mass movement; 11, chronic flooded area; 12, chronic beaver pond flooding; 13 and 14, avalanches from ca. 1875–1885; 18, ice calving from glacier (chronic). (b) Patterns of forest stand by development stage. (c) Patterns of forest stands based on predominant species type (Oliver and Larsen 1990. Reprinted by permission of John Wiley & Sons, Inc.).

Floods

Riparian forests in alluvial valleys initiate following the erosion of existing landforms and the deposition of new ones. The Tanana River floodplain, even though it lies in the Alaskan interior, is useful for understanding vegetation development on floodplains of the Pacific coastal ecoregion (Walker et al. 1986). The pattern of floodplain primary succession results from interactions between stochastic flood events and life history traits of the dominant species of willow, alder, poplar, and spruce (Figure 12.10). The timing, intensity, and extent of flooding determines the pattern of vegetative colonization (Walker et al. 1986). Following colonization, the life history traits of species occupying a site determine patterns of successional replacement. A similar pattern of vegetation colonization and establishment has been identified for several rivers in the Pacific coastal ecoregion (Cooper 1923, Fonda 1974, Hawk and Zobel 1974, Sidle and Milner 1989, Fetherston et al. 1995). Even though climate and flood regimes vary greatly among the above rivers, similar processes dictate the development of the riparian forest community.

Fire

Fire plays an important role in riparian forest development in temperate to warm climate zones. In contrast, fire is rare in coastal riparian forests from the Olympic Mountains north to

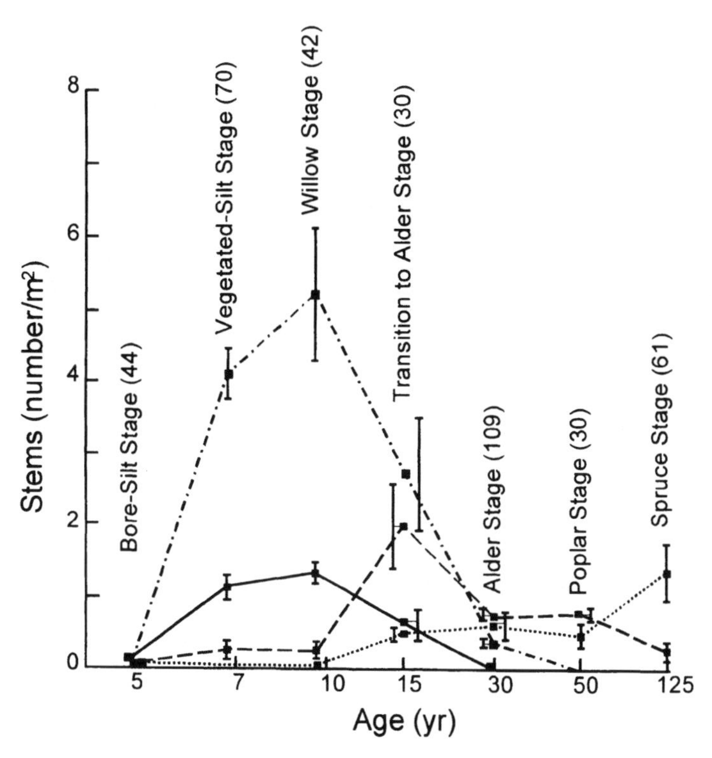

FIGURE 12.10. Successional dynamics of riparian vegetation of the Tanana River floodplain, Alaska. Stem densities of willow (——), alder (-----), poplar (-•-•-•), and spruce (••••••) in seven successional stages (mean ± SE). All stems over 2 m tall and less than 50 mm dbh are recorded. Numbers of quadrats sampled (1 × 5 m) in each successional stage are in parentheses. Although the density of popular is substantial in the early willow stage, the basal area (not shown) of willow is greater (i.e., all the popular are very young). Approximate age of each stage is shown across the bottom of the figure (from Walker et al. 1986 with permission).

Alaska. These latter forests are not fireproof but usually have a greater relative humidity and fuel moisture content, as well as a greater percentage of deciduous vegetation, which is less flammable than coniferous vegetation (Agee 1988).

Coast redwood forests are disturbed by both flooding and fire. Fires rarely kill mature redwood but kill hemlock and grand fir (*Abies grandis*) in the understory. Tan oak (*Lithocarpus densiflorus*) and California bay (*Umbellularia californica*) persist after fire by sprouting vegetatively (Agee 1993). The depositional environment of floodplains and terraces, coupled with the elimination or reduction of understory competition by fire, results in growing conditions that support coast redwood—the largest trees in the world.

Riparian Forests and Ecosystem Functions

Processes such as biogeochemical cycles and primary production are functions that characterize an ecosystem (Brinson and Reinhardt 1996). The riparian forest influences ecosystem function not only within the riparian area but also greatly affects the stream environment (Table 12.6). In the channel, LWD controls the routing of water and sediment which determines channel form. Roots improve stream bank stability and sequester nutrients. Vegetation in the floodplain captures and holds particles while the forest canopy controls the light regime and provides seasonal nourishment in the form of leaves, needles, and fine wood (Naiman and Décamps 1997).

Perhaps three of the most important functions of the riparian forest are related to microclimate, biodiversity, and biogeochemical cycles. Many aspects of biogeochemical cycles and biodiversity are addressed by McClain et al. (Chapter 14), Edwards (Chapter 16), and Pollock (Chapter 17). However, specific patterns of local microclimate and riparian plant diversity require additional consideration.

Riparian Forest Microclimate

Riparian forests exert strong controls on the microclimate of streams, but again, comprehensive studies of the riparian forest microclimate itself are surprisingly few (Gregory et al. 1991, Naiman et al. 1992). Available data suggest strong differences between upland and riparian microclimates. Stream water temperatures are highly correlated with riparian soil temperatures and there are microclimatic gradients in the riparian forest for air, soil and surface temperatures, and relative humidity, but not for short-wave solar radiation or wind velocity (Brosofske et al. 1997).

TABLE 12.6. Riparian forest ecosystem functions.

Site	Structure	Function
In-channel	Large woody debris recruited from hillslope and floodplain forests	Control routing of water and sediment Control aquatic habitat dynamics—pools, riffles, cover Provides wildlife habitat Source of scour pools—fish resting, hiding, and feeding zones Contributes to formation of forest islands and floodplains
Stream banks	Roots	Increase bank stability Create overhanging bank cover Nutrient uptake from hyperheos and streamwater
Floodplain	Stems and low-lying canopy	Retard movement of sediment, water and transported woody debris
Aboveground/above channel	Canopy and stems	Shade control of temperature and stream primary productivity Source of large and fine plant detritus Provides wildlife habitat
Riverine forest	Corridor	Facilicates movement of fish and wildlife

Microclimatic data collected along a 180-m line transect from the active channel to the uplands of a small (>2 m wide) Cascade Mountain stream suggest that the stream environment (active channel) clearly differed from riparian and upland forest environments (Figure 12.11) (Brosofske et al. 1997). The stream significantly influenced air temperatures in the riparian forest for up to 60 m on either side of the channel in summer (Figure 12.11a), either through direct cooling or by supplying water for evaporative cooling by vegetation. Average air temperatures for the riparian zone, adjacent forest, and an upland clearcut were 19 °C, 20.5 °C, and 24.8 °C, respectively, over 12 days. Gradual changes in soil temperature and relative humidity from the stream to the upland were also detected (Figure 12.11b and 12.11c). No clear difference was identified for wind velocities (Figure 12.11d). The abrupt microclimatic change from the active channel to a closed canopy forest interior is suspected to significantly affect plant community composition.

In addition, dense forest canopies often restrict the understory light regime to less than 2% of total solar radiation, but this does not seem to affect seedling densities (MacDougall and Kellman 1992). However, the differential effects of the reduced light regime on seedling survivorship over the longer term is not known.

Finally, riparian forests, especially in warmer climates and seasons, reduce stream discharge through evapotranspiration; red alder stands decrease flow substantially more than conifer stands (Hicks et al. 1991). Reduced discharge decreases the extent of inchannel habitat for the fauna, and this may become critical in certain situations. Overall, the ecological consequences of many of these apparently important microclimatic gradients and processes remain to be discovered and quantified.

Riparian Plant Diversity

A decrease in disturbance frequency and productivity from the valley floor to the hillslopes, in combination with the microclimate gradients, influences plant diversity (e.g., plant species richness) within river valleys in the Pacific coastal ecoregion (Gregory et al. 1991, Pollock et al. 1998, Chapter 17). This has been documented in several studies. For example, plant species diversity decreased from the active channel to the hillslope along streams on the west slope of Oregon's Cascade Mountains, three streams in California's Sierra Nevada Mountains, and seven stream buffers in western Washington (Figure 12.12). The peak in plant diversity found in the active channel and floodplain is associated with the large variety of vegetative age classes and landforms created by erosion and depositional processes. Furthermore, large wood accumulations on the floodplain are significant sites for colonization by coniferous species within an otherwise deciduous tree matrix, as well as sites for the retention of numerous plant propagules in a favorable microclimate (McKee et al. 1982, Fetherston et al. 1995). Large wood not only adds to the diversity of plant habitat but also significantly increases the structural complexity of floodplains (Abbe and Montgomery 1996, Chapter 13).

Riparian Forests and Land-Use Change

Land-use change in the Pacific coastal ecoregion, like other regions in North America, has been extensive in recent decades. Forest harvest, conversions to agriculture and urban centers, water withdrawals, erosion, as well as a variety of other alterations have profoundly impacted stream channels and associated riparian vegetation. Two types of landuse impacts (e.g., river regulation and forest management) are discussed below as basic illustrations of how changes in hydrology and sediment transport regimes alter riparian plant communities.

River Regulation

Anthropogenic regulation of river flow alters a variety of hydrologic processes necessary for

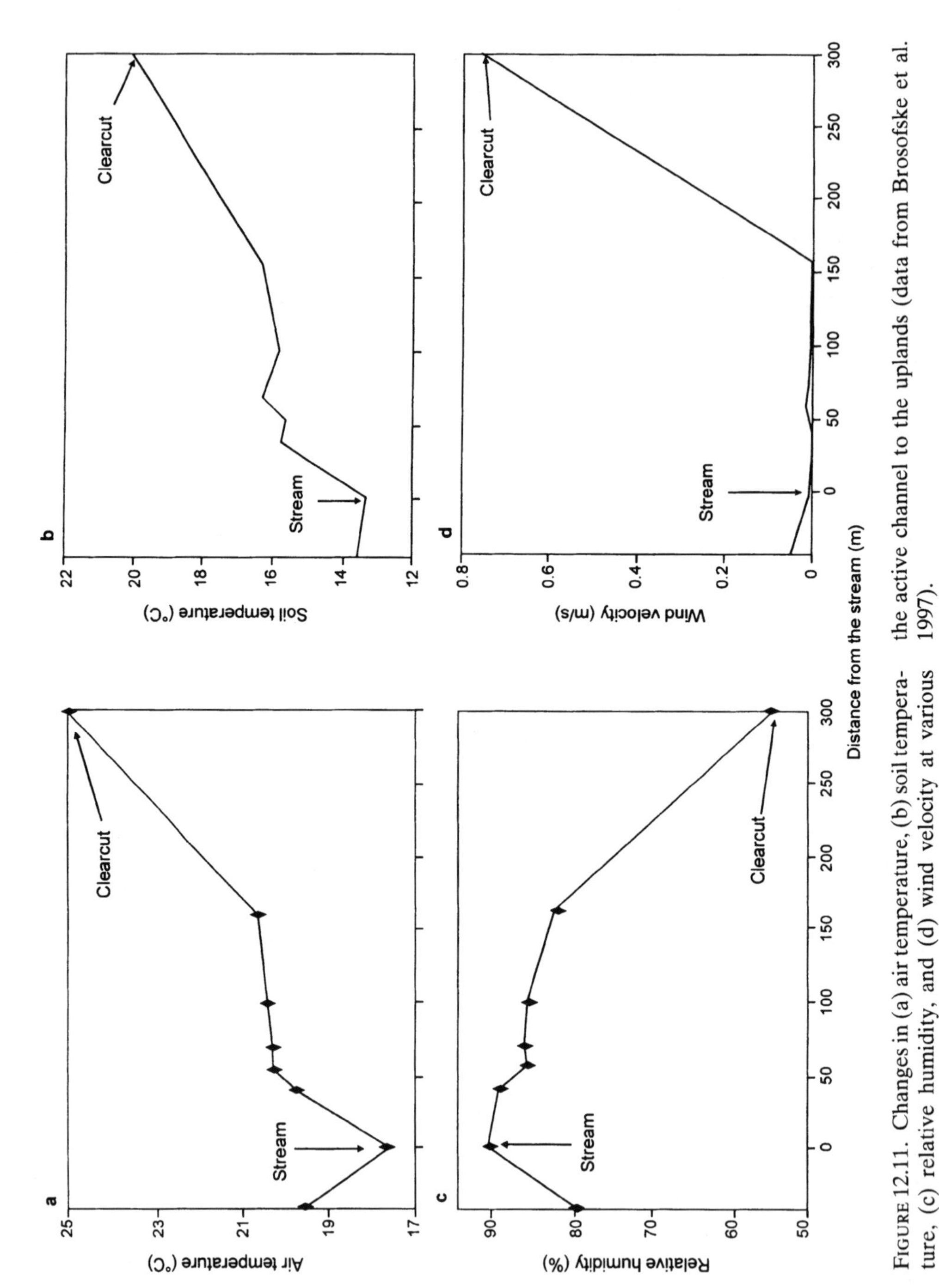

FIGURE 12.11. Changes in (a) air temperature, (b) soil temperature, (c) relative humidity, and (d) wind velocity at various distances along a small Cascade Mountain stream channel from the active channel to the uplands (data from Brosofske et al. 1997).

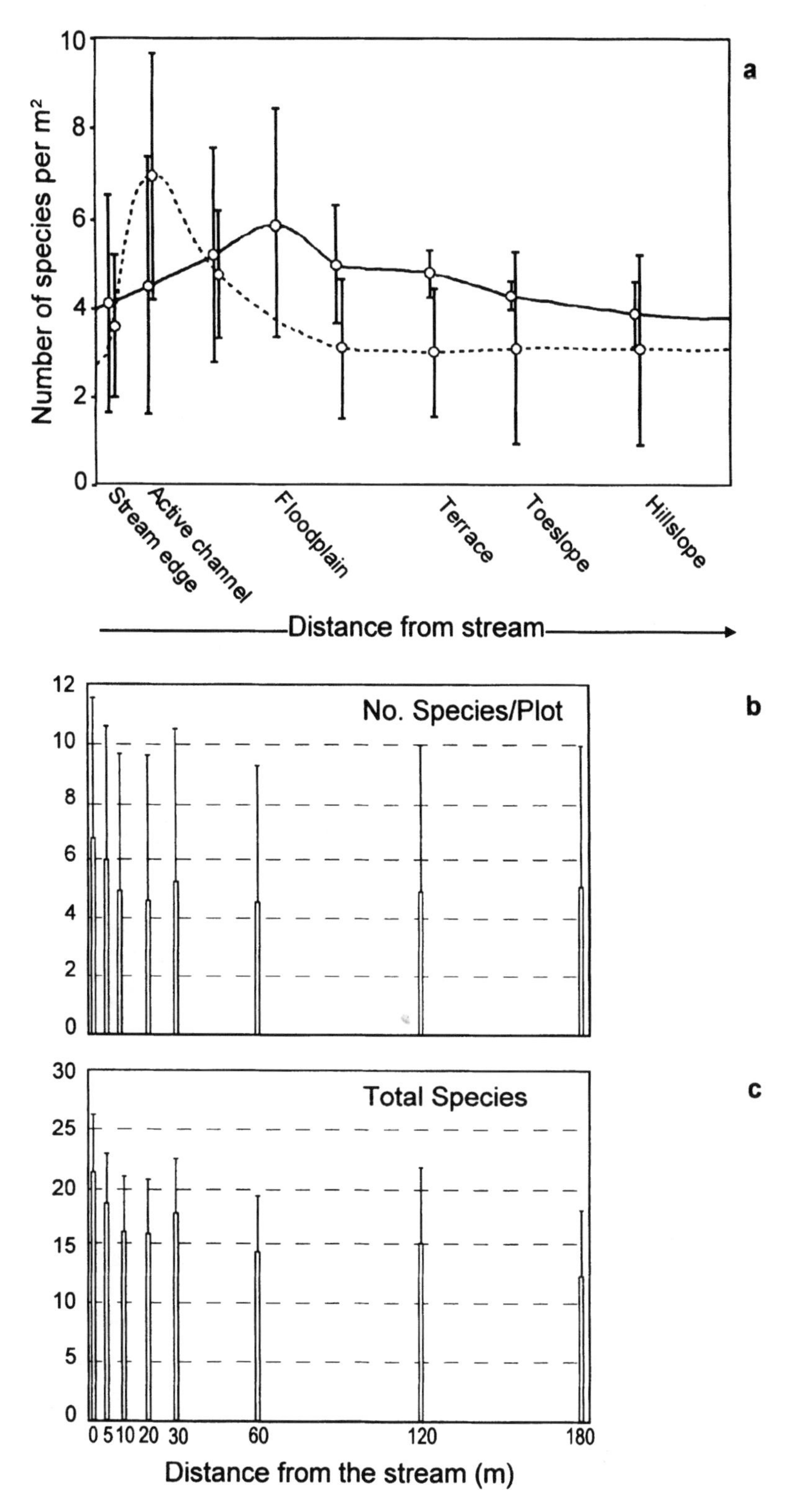

FIGURE 12.12. Gradients of species richness along lateral transects from the stream channel to upper hillslopes for streams on the west slope of the Cascade Mountains, Oregon and Washington (——) and Sierra Nevada, California (-----) (from Gregory et al. 1991 © American Institute of Biological Sciences and Chen et al. unpublished data). (a) Richness is expressed as number of species in 1 m^2 plots sampled on longitudinal, 1 × 5 m belt transects. Points represent means for more than 45 transects per surface; bars are SD. (b) Plant species richness is greatest within 10 m of the stream as indicated by the number of species per plot as well as (c) the total number of species.

the establishment and maintenance of vegetation (Petts et al. 1992, Dynesius and Nilsson 1994, Naiman et al. 1995). Even though studies in Pacific coastal riparian forests are not common, useful information about the effects of river regulation may be drawn from other regions of the northern hemisphere due to the many shared riparian genera (e.g., *Salix*, *Populus*, *Alnus*, *Fraxinus*, and *Acer*).

Although widespread dam construction in the Pacific coastal ecoregion largely has lagged behind that of the arid interior east of the Cascade Mountains, dam construction was well underway by the 1950s (International Commission on Large Dams 1964). The earliest dams were relatively small, but subsequent dams rapidly increased in size, with reservoir capacities exceeding 100 million m^3 in the Eel, Klamath, Skokomish, Baker, Skagit, Powell, Stave, and Coquitlan watersheds by 1930, and the Columbia River by 1943. Dams change transport of materials and alter the hydrologic regime, disrupting riparian forest species composition and distribution, sediment moisture retention and soil biogeochemistry.

Material Transport

Material transport is most obviously altered when dams directly act as barriers to moving materials. Large volumes of sediment accumulate behind dams, shortening the life span of hydroelectric dams or reducing the effectiveness of flood-control dams. The reduced water velocity behind the dam results in the settling out of finer sediments that would otherwise continue downstream. This reduction in transport reduces sediment deposition downstream, in turn reducing the rate of downstream channel migration in unconstrained reaches. In fact, immediately below reservoirs, the sediment load may be so reduced that channel downcutting occurs as the river's competence (capacity) for carrying suspended materials returns to equilibrium.

It is well known that in natural rivers, newly deposited sediments provide terrestrial habitat within floodplains and river meanders create new habitat for terrestrial plant establishment. However, with less material transport these processes occur less frequently and pioneer species often decline. For example, reduced rates of sedimentation and meander migration are thought to be partially responsible for declining riparian cottonwood populations along the Milk River of southern Alberta and northern Montana, and along the Missouri River (Johnson et al. 1976, Bradley and Smith 1986). Johnson (1992) found that below the Missouri River's Garrison Dam, a pre-dam condition of net deposition, albeit slight, gave way to a post-dam condition of net sediment erosion and channel widening. Simultaneously, the area occupied by young, pioneer forests (primarily cottonwood and willow) decreased from 47 to 6%.

Besides reducing sediment volume, reservoirs change the nature of the transported sediments. Sediment size tends to decrease, and the chemical composition can change, with specific minerals settling out differentially according to mass. Changes in sediment size alter the suite of plant species which can successfully occupy a site. Increased sediment size downstream from dams increases roughness due to increased hydraulic conductivity, reducing or eliminating some riparian species.

Changes in the chemical composition also affect the species composition. For example, geochemistry and sediment texture are important factors for segregating closely related tree species (Schnitzler et al. 1992). In many systems, regular deposition of nutrient-rich sediment is necessary for forest maintenance because the soils are poorly developed or because they experience rapid leaching associated with high water tables. A decreased nutrient supply, induced by flood controls, can result in decreased productivity (Schnitzler 1994) as opposed to increased radial growth which may occur following alluvial sediment deposition by floods (Stone and Vasey 1968, Zinke 1981). Further, periodic deposition, which eliminates potentially competing species, such as Douglas fir, grand fir, and tan oak (Stone and Vasey 1968), normally does not occur along regulated rivers, thus these species can become established in large numbers in the former floodplain.

Impoundment of rivers behind reservoirs essentially severs the river continuum, reducing transport of organic matter and sediments (Stanford et al. 1996). Thus, number of plant propagules are reduced below the dams. In addition, the retention of fluvially transported LWD above the dam may reduce the number of appropriate sites for plant establishment. For example, Schneider and Sharitz (1988) found that seeds tended to concentrate against logs, trees, and other obstructions, while Fetherston et al. (1995) found that large woody debris stabilizes river banks and islands, creating new habitat.

Altered Hydrologic Regimes

The control of flow variability is the principal purpose for building dams, regardless of whether the dams are for flood control, hydroelectric power, or water supply. Changes in flow variability affect the timing of high and low flow events, the magnitude of flows, and the duration and frequency of floods. Although these features are related, each has a different impact on riparian forest establishment and succession (Junk et al. 1989).

Changes in the timing of flow events are often subtle, yet the effects can be dramatic. Many riparian plant species depend upon specific conditions at specific times for successful seedling establishment, and changes in flow patterns often shift the timing of those conditions. Interactions between timing of seed dispersal and fluctuating water levels is a major factor determining zonation of willow and poplar species (Van Splunder et al. 1995). Willow and cottonwood require moist, unvegetated substrates for germination, and such conditions are common immediately following spring flooding as water levels gradually decrease, which is also the period during which willow and cottonwood seeds disperse. However, subsequent flooding will drown or uproot new seedlings. The precise combination of conditions for survival cannot be met every year at any particular site but these conditions occur often enough for establishment (Wilson 1970, Noble 1979, Stromberg et al. 1991). By changing the timing of the decline in water levels, seed dispersal may occur too early or too late for successful seedling establishment and other species may establish instead. Even minor changes in timing can have an impact, as coexisting species often have slightly different dispersal timings. For example, coexisting, closely related willow species disperse at slightly different times (up to a month), but with some overlap, and thus establish at slightly different heights along the shorelines (Niiyama 1990). Minor changes differentially favor one species over another, changing the relative importance of a species. Nilsson et al. (1991), however, found that the total number of riparian plant species was essentially the same for two parallel Swedish rivers, one regulated and the other unregulated. The regulated river, however, had a preponderance of uncommon species and few common species relative to the unregulated river.

Timing of flooding also affects the survival of species. Perennial riparian species can withstand periodic inundation to various degrees. The timing of inundation, however, may be critical. Some species are more vulnerable to water-induced stress during the growing season than during the dormant season, and shifts in the timing of inundation increases the mortality rates of those species. Specific differences in germination timing may be partially responsible for differential flood survival (Streng et al. 1989). Further, it has been demonstrated experimentally that summer flooding has a much greater impact than winter flooding on six tree species from the Pacific Northwest (Minore 1968).

Changes in the magnitude of river flows are usually the most obvious effects of dams, influencing riparian forests in several unique ways. The reduction of maximum flow volumes reduces the area of floodplain directly affected by flooding. This often results in a reduction of riparian vegetation as less flood-tolerant upland species extend further into the floodplains. Decreased peak flows have been credited with the expansion of invasive salt cedar or tamarisk (*Tamarix pentandra*), at the expense of Fremont cottonwood (*Populus fremontii*), along rivers in the arid southwestern United States (Irvine and West 1979, Stromberg et al. 1993).

Reducing flood frequency appears to reduce the frequency of asexual reproduction by some species of willow and cottonwood, thus altering the genetic structure of downstream populations (Barnes 1985, Tardif and Bergeron 1992, Rood et al. 1994). Furthermore, flood resistance varies dramatically even within populations of the generally flood-tolerant black cottonwood, a trait that is possibly determined genetically (Smit 1988). If so, then reduced selection for this trait could result from decreased flood frequency.

Changes in minimum flow also are critical for survival. Decreased flow results in increased drought stress and thus mortality to seedlings and mature plants. Cottonwood, for example, is highly vulnerable to drought stress; in fact, cottonwood display the highest vulnerability to drought-induced xylem cavitation ever measured in North American trees (Tyree et al. 1994). Decreased summer flows correlate with decreased water potential levels in riparian poplar, willow, and birch (*Betula* spp.). This effect is particularly strong among juveniles, suggesting that juveniles may selectively experience higher mortality rates because of increased water stress (Smith et al. 1991). Reily and Johnson (1982) documented reductions in radial growth among relatively shallow-rooted riparian trees, citing a lowering of the water table from dam construction as the probable cause, and Rood and Heinze-Milne (1989) and Rood et al. (1995) cite similar causes for the widespread decline of cottonwood communities in southern Alberta. In contrast, increased minimum water flows appear to reduce drought stress and increase vegetation survival.

The duration of river flows are often altered by dams as well, thereby influencing riparian forests directly. Extended inundation of forests increases injury or mortality in nearly all species or age classes (Hosner 1958 and 1960, Minore 1968, Harrington 1987). Flooding duration is an especially important factor in shoreline vegetation species richness and abundance (Robertson et al. 1978, Nilsson and Keddy 1988). Barnes (1985, 1991) found the number of riverine island tree species increased with local elevation relative to the wetted channel. In contrast to the normally gradual decrease of flows following spring flooding in natural rivers, the decrease in water level may be accelerated in regulated rivers. This is particularly important for small, newly established seedlings that have not developed extensive root systems, yet must acquire water while living on substrates with a low water retention capacity (i.e., cobble bars). As a result, successful seedling establishment is greatly reduced for riparian plants such as willow and cottonwood.

Changes in the frequency of flood events have dramatic impacts on population age structures and successional processes. Increased flooding frequency can eliminate entire age classes (Streng et al. 1989), while decreased frequency can reduce establishment (Reily and Johnson 1982, Bradley and Smith 1986, Schneider and Sharitz 1988, Stromberg et al. 1991). In both cases, the result is a simplification of age structures. Furthermore, the specific nature of both processes result in changes to the relative proportions of species within a site, in turn influencing subsequent competitive and successional processes, potentially changing the nature of the community (Reily and Johnson 1982, Stromberg et al. 1991).

Although many river features modified by dams are related to transport and hydrological regimes, numerous riparian forest processes are affected. Lethal and sublethal impacts can occur at all life stages and ages, and nutrient dynamics, decomposition, and moisture retention are also fundamentally modified. As drought and flooding change, so do population dynamics and genetic vigor. Further complications include interactions between environmental components, making analysis and prediction of the effects more difficult (Stanford et al. 1996).

The most probable scenario for riparian forests downstream from reservoirs is for continued simplification. The spatial extent of riparian forests tends to continue to decrease as channel migration, sediment deposition, and flooding are reduced. Reduced flood frequencies and shifts in the timing of flow events result in less frequent establishment of riparian species. Riparian species with competitive flood tolerant advantages become less important under such conditions and upland species begin to encroach upon the floodplain forest (Décamps et al. 1988). As a result, the extent of riparian forests will likely be reduced to narrow ribbons along the flood plain.

Forest Practices

In addition to direct alteration of the hydrological and sediment transport regimes by dams, disturbances (e.g., direct harvest, road construction, channelization, bridging) caused by harvesting of timber in riparian or lowland forests are equally profound and extensive (Figure 12.13) (Meehan 1991). Direct effects of overstory tree removal include alteration of energy fluxes and microclimate (e.g., increase in incoming short-wave solar radiation during the day, outgoing long-wave radiation at night, and wind velocity; Chen et al. 1993); changes in material budgets (e.g., direct input of precipitation, water loss through evaporation, nutrient loss through leaching); reduction in input of organic materials (e.g., leaves, seeds, logs) to the stream; and loss of movement corridors and critical habitats for plants and wildlife. These direct changes cause further indirect impacts, such as alteration of water quality (e.g., temperature, pH, turbidity, oxygen, pollutants, and so forth), stream flow, number and composition of species in the stream and riparian areas, ground water table, and physical stability of the stream channel and surrounding land (MacDonald et al. 1991). In general, forest practices have the strongest influence on riparian zones when they alter normal regional streamflow at very high or very low flows (Chamberlin et al. 1991). These are the flows

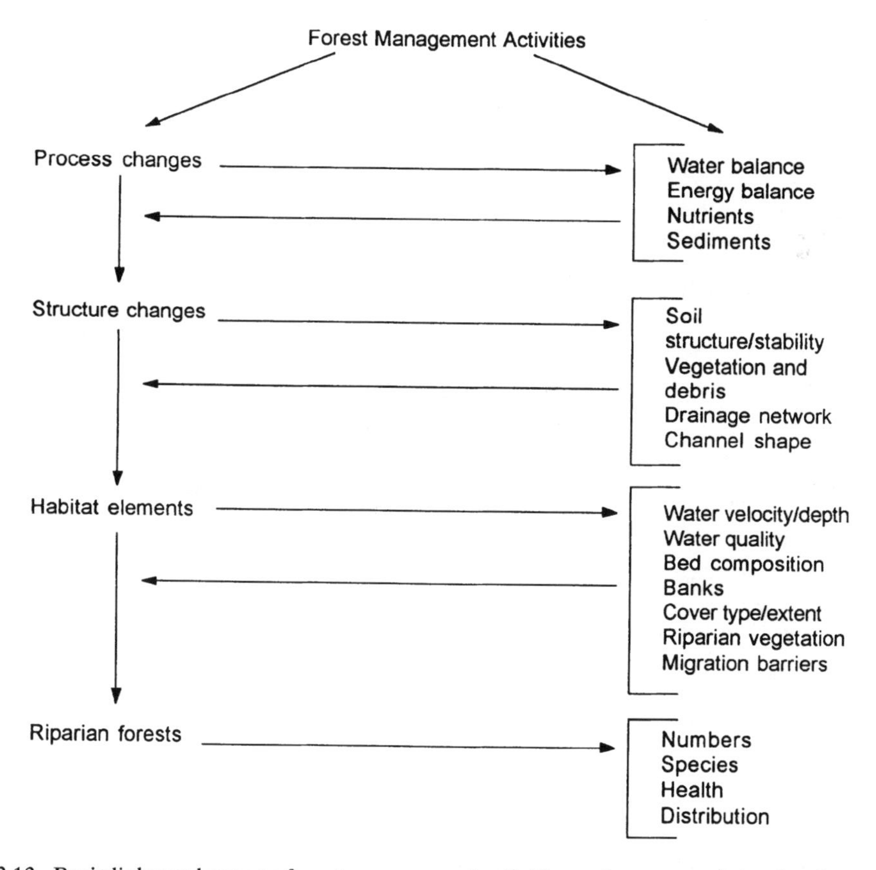

FIGURE 12.13. Basic linkages between forest management activities and stream and riparian forest characteristics (modified from Chamberlin et al. 1991).

with the greatest impacts on plants whether caused by drought, inundation, abrasion, sedimentation, or erosion.

Riparian forests are sensitive to silvicultural treatments in nearby upland forests as well. Brosofske et al. (1997) found that timber harvesting in uplands near small streams in western Washington caused significant changes in riparian microclimate, even when riparian buffer strips were preserved. They also found stream water temperature linearly correlated with upland soil temperature, probably due to significant ground water input to the streams from the surrounding landscape. Forest removal produces higher soil temperatures at the harvest site and indirectly causes an increase in stream water temperature. Forest harvesting in uplands also creates a large amount of edge area (if riparian forests are preserved during the harvest). Previous studies of microclimate across forest-clearcut edges in the Pacific coastal ecoregion have suggested that edge effects can sometimes extend more than 240 m into the forest from the harvest boundary (Chen et al. 1995). Consequently, these edge influences may contribute to significant alteration of the riparian environment and related ecosystem processes.

Lessons for Management

As complex ecological systems localized at the land-water ecotone, riparian forests maintain generally high levels of biodiversity (Chapter 17), exhibit high rates of nutrient cycling and productivity, provide specialized ecological functions (e.g., water quality), and exert strong influences on adjacent ecological systems by modifying the flow of materials and information across the landscape. Alterations to these important systems have ecological effects that go far beyond the specific activity, with the consequences reverberating throughout the entire river corridor. Effective management of riparian forests requires a deep understanding of complex systems and a willingness to acknowledge their key role in maintaining environmental vitality at local as well as larger scales.

Acknowledgments. Research support from the Andrew W. Mellon Foundation, the Weyerhaeuser Foundation, and The Pacific Northwest Research Station of the U.S. Forest Service is gratefully acknowledged. The authors thank R.E. Bilby and two anonymous reviewers for offering insightful comments which greatly improved the content.

Literature Cited

Abbe, T.E., and D.R. Montgomery. 1996. Large woody debris jams, channel hydraulics, and habitat formation in large rivers. Regulated Rivers **12**:201–221.

Agee, J.K. 1988. Successional dynamics in forest riparian zones. Pages 31–43 *in* K.J. Raedeke, ed. Streamside management: Riparian wildlife and forestry interactions. Contribution Number 59. Institute of Forest Resources, University of Washington, Seattle, Washington, USA.

——. 1993. Fire ecology of the Pacific Northwest forests. Island Press, Washington, DC, USA.

Alaback, P.B. 1982a. Dynamics of understory biomass in Sitka spruce-western hemlock forests of Southeast Alaska. Ecology **63**:1932–1948.

——. 1982b. Forest community structural changes during secondary succession in Southeast Alaska. Pages 70–79 *in* J.E. Means, ed. Forest Succession and Stand Development Research in the Northwest: Proceedings of the Symposium. Oregon State University, Forest Research Laboratory, Corvallis, Oregon, USA.

——. 1995. Biodiversity patterns in relation to climate: The coastal temperate rain forests of North America. Pages 105–133 *in* R. Lawford, P. Alaback, and E.R. Fuentes, eds. High latitude rain forests and associated ecosystems of the west coast of the Americas: Climate, hydrology, ecology and conservation. Ecological Studies **116**, Springer-Verlag, New York, New York, USA.

Bailey, R.G. 1980. Description of the ecoregions of the United States. US Department of Agriculture, Miscellaneous Publication Number 1391, Washington, DC, USA.

Barbour, M.G., J.H. Burk, and W.A. Pitts. 1987. Terrestrial Plant Ecology. Benjamin/Cummings, Menlo Park, California, USA.

Barnes, W.J. 1985. Population dynamics of woody plants on a river island. Canadian Journal of Botany **63**:647–655.

——. 1991. Tree populations on the islands of the lower Chippewa River in Wisconsin. Bulletin of the Torrey Botanical Club **118**:424–431.

Bégin, Y., and S. Payette. 1991. Population structure of lakeshore willows and ice-push events in subarctic Québec, Canada. Holarctic Ecology **14**:9–17.

Benda, L., and T. Dunne. 1987. Sediment routing by debris flows. Pages 213–223 *in* R.L. Beschta, R. Blinn, G.E. Grant, G. Ice, and F.J. Swanson, eds. Erosion and sedimentation in the Pacific Rim. International Association of Hydrological Sciences Publication 165. Corvallis, Oregon, USA.

Boyce, J.S. 1929. Deterioration of wind-thrown timber on the Olympic Peninsula, Washington. US Department of Agriculture Technical Bulletin Number 104. Washington, DC, USA.

Bradley, C.E., and D.G. Smith. 1986. Plains cottonwood recruitment and survival on a prairie meandering river floodplain, Milk River, southern Alberta and northern Montana. Canadian Journal of Botany **64**:1433–1442.

Brinson, M.M. 1990. Riparian forests. Pages 87–141 *in* A.E. Lugo, M.M. Brinson, and S. Brown, eds. Forested wetlands. Elsevier Scientific Publishers, Amsterdam, The Netherlands.

Brinson, M.M., and R. Reinhardt. 1996. The role of reference wetlands in functional assessment and mitigation. Ecological Applications **6**:69–76.

Brosofske, K.D., J. Chen, R.J. Naiman, and J.F. Franklin. 1997. Effects of harvesting on microclimate from small streams to uplands in western Washington. Ecological Applications **7**:1188–1200.

Campbell, R.H. 1975. Soil slips, debris flows and rainstorms in the Santa Monica Mountains and vicinity, southern California. US Geological Survey Professional Paper 851.

Chamberlin, T.W., R.D. Harr, and F.H. Everest. 1991. Timber harvesting, silviculture, and watershed processes. Pages 181–205 *in* W.R. Meehan, ed. Influences of forest and rangeland management on salmonid fishes and their habitats. American Fisheries Society Special Publication 19, Bethesda, Maryland, USA.

Chen, J., J.F. Franklin, and T.A. Spies. 1993. Contrasting microclimates among clearcut, edge, and interior of old-growth Douglas-fir forest. Agricultural and Forest Meteorology **63**:219–237.

Chen, J., J.F. Franklin, and T.A. Spies. 1995. Growing-season microclimatic gradients from clearcut edges into old-growth Douglas-fir forests. Ecological Applications **5**:74–86.

Church, M. 1992. Channel morphology and typology. Pages 126–143 *in* P. Calow and G.E. Petts, eds. The rivers handbook, Volume I. Blackwell Scientific Publications, Oxford, UK.

Cooper, W.S. 1923. The recent ecological history of Glacier Bay, Alaska. II. The present vegetation cycle. Ecology **4**:223–246.

——. 1931. A third expedition to Glacier Bay, Alaska. Ecology **12**:61–95.

Cushman, M.J. 1981. The influence of recurrent snow avalanches on vegetation patterns in the Washington Cascades. Ph.D. dissertation. University of Washington, Seattle, Washington, USA.

Décamps H. 1996. The renewal of floodplain forests along rivers: A landscape perspective. Verhandlungen der Internationalen Vereinigung für Theoretische und Angewandte Limnologie **26**:35–59.

Décamps H., M. Fortuné, F. Gazelle, and G. Pautou. 1988. Historical influence of man on the riparian dynamics of a fluvial landscape. Landscape Ecology **1**:163–173.

Dewit, L., and D.M. Reid. 1992. Branch abscission in balsam poplar (*Populus balsamifera*): Characterization of the phenomenon and the influence of wind. International Journal of Plant Science **153**:556–564.

Dunne, T., and L.B. Leopold. 1978. Water in Environmental Planning. W.H. Freeman, New York, New York, USA.

Dynesius, M., and C. Nilsson. 1994. Fragmentation and flow regulation of river systems in the northern third of the world. Science **266**:753–762.

Everett, B.L. 1968. Use of the cottonwood in an investigation of the recent history of a flood plain. American Journal of Science **266**:417–439.

Fetherston, K.L., R.J. Naiman, and R.E. Bilby. 1995. Large woody debris, physical process, and riparian forest succession in montane river networks of the Pacific Northwest. Geomorphology **13**:133–144.

Fonda, R.W. 1974. Forest succession in relation to river terrace succession in Olympic National Park, Washington. Ecology **55**:927–942.

Franklin, J.F., and C.T. Dyrness. 1973. Natural vegetation of Oregon and Washington, USDA Forest Service General Technical Report PNW-8, Pacific Northwest Forest and Range Experiment Station, Portland, Oregon, USA.

Galloway, G., and J. Worrall. 1979. Cladoptosis: A reproductive strategy in black cottonwood. Canadian Journal of Forest Research **9**:122–125.

Gecy, J.L., and M.V. Wilson. 1990. Initial establishment of riparian vegetation after disturbance by debris flows in Oregon. American Midland Naturalist. **123**:282–291.

Grant, G.E. 1986. Downstream effects of timber harvest activities on the channel and valley floor morphology of western Cascade streams. Ph.D. dissertation. Johns Hopkins University, Baltimore, Maryland, USA.

Grant, G.E., and F.J. Swanson. 1995. Morphology and processes of valley floors in mountain streams, western Cascades, Oregon. Pages 83–101 *in* J.E. Costa, A.J. Miller, K.W. Potter, and P.R. Wilcock, eds. Natural and anthropogenic influences in fluvial geomorphology. Geophysical Monograph 89, American Geophysical Union, Washington, DC, USA.

Gregory, S.V., F.J. Swanson, W.A. McKee, and K.W. Cummins. 1991. An ecosystem perspective of riparian zones. BioScience **41**:540–551.

Grime, J.P. 1979. Plant Strategies and Vegetation Processes. John Wiley & Sons, New York, New York, USA.

Harmon, M.E., and J.F. Franklin. 1989. Tree seedlings on logs in *Picea-Tsuga* forests of Oregon and Washington. Ecology **70**:48–59.

Harper J.L. 1977. Population Biology of Plants. Academic Press, New York, New York, USA.

Harrington, C.A. 1987. Responses of red alder and black cottonwood seedlings to flooding. Physiologia Plantarum **69**:35–48.

Harris, R.R. 1987. Occurrence of vegetation on geomorphic surfaces in the active floodplain of a California alluvial stream. American Midland Naturalist **118**:393–405.

Hawk, G.M., and D.B. Zobell. 1974. Forest succession on alluvial landforms of the McKenzie River Valley, Oregon. Northwest Science **48**:245–265.

Hicks, B.J., R.L. Beschta, and R.D. Harr. 1991. Long-term changes in stream flow following logging in western Oregon and associated fisheries implications. Water Resources Bulletin **27**:217–226.

Hosner, J.F. 1958. The effects of complete inundation upon seedlings of six bottomland tree species. Ecology **39**:371–373.

——. 1960. Relative tolerance to complete inundation of fourteen bottomland tree species. Forest Science **6**:246–251.

Hupp, C.R. 1982. Stream-grade variation and riparian-forest ecology along Passage Creek, Virginia. Bulletin of the Torrey Botanical Club. **109**:488–499.

International Commission on Large Dams. 1964. World Register of Dams. Paris, France.

Irvine, J.R., and N.E. West. 1979. Riparian tree species distribution and succession along the lower Escalante River, Utah. Southwestern Naturalist **24**:331–346.

Johansson, M., C. Nilsson, and E. Nilsson. 1996. Do rivers function as corridors for plant dispersal? Journal of Vegetation Science **7**:593–598.

Johnson, W.C. 1992. Dams and riparian forests: Case study from the upper Missouri River. Rivers **3**:229–242.

Johnson, W.C., R.L. Burgess, and W.R. Keammerer. 1976. Forest overstory vegetation and environment on the Missouri River floodplain in North Dakota. Ecological Monographs **46**:59–84.

Junk, W.J., P.B. Bailey, and R.E. Sparks. 1989. The flood pulse concept in river-floodplain systems. Pages 110–127 *in* D.P. Dodge, ed. Proceedings of the International Large River Symposium. Canadian Special Publication in Fisheries and Aquatic Science 106, Ottawa, Ontario, Canada.

Keller, E.A., and T. Tally. 1979. Effects of large organic debris on channel form and fluvial processes in the coastal redwood environment. Pages 169–197 *in* D.D. Rhodes and G.P. Williams, eds. Adjustments of the Fluvial System. Proceedings of the 10th Annual Geomorphology Symposia, Binghamton, New York. Kendall/Hunt, Dubuque, Iowa, USA.

Kozlowski, T.T., P.J. Kramer, and S.G. Pallardy. 1991. The Physiological Ecology of Woody Plants. Academic Press, New York, New York, USA.

MacDonald, L.H., A. Smart, R.C. Wissmar. 1991. Monitoring guidelines to evaluate effects of forestry activities on streams in the Pacific Northwest and Alaska. US Environmental Protection Agency, Technical Report EPA/910/9-91-001, Seattle, Washington, USA.

MacDougall, A., and M. Kellman. 1992. The understory light regime and patterns of tree seedlings in tropical riparian forest patches. Journal of Biogeography **19**:667–675.

Malanson, G.P. 1993. Riparian Landscapes. Cambridge University Press, Cambridge, UK.

McKee, A., G. Laroi, and J.F. Franklin. 1982. Structure, composition, and reproductive behavior of terrace forests, South Fork Hoh River, Olympic National Park. Pages 22–29 *in* E.E. Starkey, J.F. Franklin, and J.W. Matthews, eds. Ecological research in National Parks of the Pacific Northwest. National Park Cooperative Studies Unit, Corvallis, Oregon, USA.

Meehan, W.R., ed. 1991. Influences of Forest and Rangeland Management on Salmonid Fishes and Their Habitats. American Fisheries Society Special Publication 19, Bethesda, Maryland, USA.

Menges, E.S., and D.M. Waller. 1983. Plant strategies in relation to elevation and light in floodplain herbs. The American Naturalist **122**:454–473.

Minore, D. 1968. Effects of artificial flooding on seedling survival and growth of six northwestern tree species. USDA Forest Service Research Note PNW-92, Portland, Oregon, USA.

Mitsch, W.J., and J.G. Gosselink. 1993. Wetlands. Van Nostrand Reinhold, New York, New York, USA.

Montgomery, D.R. 1991. Channel initiation and landscape evolution. Ph.D. dissertation. University of California, Berkeley, California, USA.

Montgomery, D.R., and J.M. Buffington. 1993. Channel classification, prediction of channel response, and assessment of channel condition. Report TFW-SH10-93-002. Washington Department of Natural Resources, Olympia, Washington, USA.

Naiman, R.J., and E.C. Anderson. 1997. Streams and rivers: Their physical and biological variability. Pages 131–148 *in* P.K. Schoonmaker, B. von Hagen, and E.C. Wolf, eds. The rain forests of home: Profile of a North American bioregion. Island Press, Washington, DC, USA.

Naiman, R.J., T.J. Beechie, L.E. Benda, D.R. Berg, P.A. Bisson, L.H. MacDonald, et al. 1992. Fundamental elements of ecologically healthy watersheds in the Pacific Northwest coastal ecoregion. Pages 127–188 *in* R.J. Naiman, ed. Watershed management, Springer-Verlag, New York, New York, USA.

Naiman, R.J., and H. Décamps. 1997. The ecology of interfaces—riparian zones. Annual Review of Ecology and Systematics **28**:621–658.

Naiman, R.J., and H. Décamps. 1990. Aquatic-terrestrial ecotones: Summary and recommendations. Pages 295–301 *in* R.J. Naiman and H. Décamps, eds. Ecology and management of aquatic-terrestrial ecotones. UNESCO, Paris, and Parthenon Publishing Group, Carnforth, UK.

Naiman R.J., H. Décamps, and M. Pollock. 1993. The role of riparian corridors in maintaining regional biodiversity. Ecological Applications **3**:209–212.

Naiman, R.J., J.J. Magnuson, D.M. McKnight, and J.A. Stanford, eds. 1995. The freshwater imperative: A research agenda. Island Press, Washington, DC, USA.

Nanson, G.C., M. Barbetti, and G. Taylor. 1995. River stabilization due to changing climate and vegetation during the late Quaternary in western Tasmania, Australia. Geomorphology **13**:145–158.

Niiyama, K. 1990. The role of seed dispersal and seedling traits in colonization and coexistence of *Salix* species in a seasonally flooded habitat. Ecological Research **5**:317–331.

Nilsson, C., A. Ekblad, M. Gardfjell, and B. Carlberg. 1991. Long-term effects of river regulation on river margin vegetation. Journal of Applied Ecology **28**:963–987.

Nilsson, C., and P.A. Keddy. 1988. Predictability of change in shoreline vegetation in a hydroelectric reservoir, northern Sweden. Canadian Journal of Fisheries and Aquatic Science **45**:1896–1904.

Noble, I.R., and R.O. Slatyer. 1980. The use of vital attributes to predict successional changes in plant communities subject to recurrent disturbances. Vegetatio **43**:5–21.

Noble, M.G. 1979. The origin of *Populus deltoides* and *Salix* interior zones on point bars along the Minnesota River. American Midland Naturalist **102**:59–67.

Oliver, C.D. 1981. Forest development in North America following major disturbances. Forest Ecology and Management **3**:153–168.

Oliver, C.D., A.B. Adams, A.R. Weisbrod, J. Dragovon, R.J. Zasoski, and K. Bardo. 1980. Nooksack Cirque natural history final report. Unpublished report, College of Forest Resources, University of Washington, Seattle, Washington, USA.

Oliver, C.D., A.B. Adams, and R.J. Zasoski. 1985. Disturbance patterns and forest development in a recently deglaciated valley in the northwestern Cascade Range of Washington, USA Canadian Journal of Forest Research **15**:221–232.

Oliver, C.D., and B.C. Larson. 1996. Forest Stand Dynamics. John Wiley & Sons, New York, New York, USA.

Orr, P.W. 1963. Wind-thrown timber survey in the Pacific Northwest, 1962. USDA Forest Service, Pacific Northwest Region, Portland, Oregon, USA.

Oswald, D.D. 1968. The timber resources of Humboldt County, California. USDA Forest Service Resource Bulletin PNW-26, Portland, Oregon, USA.

Petts, G.E., A.R.G. Large, M.T. Greenwood, and M.A. Bickerton. 1992. Floodplain assessment for restoration and conservation: Linking hydrogeomorphology and ecology. Pages 217–234 *in* P.A. Carling and G.E. Petts, eds. Lowland Floodplain Rivers. John Wiley & Sons, Chichester, UK.

Poage, N.J., and T.A. Spies. 1996. A tale of two unmanaged riparian forests. Coastal Oregon

Productivity Experiment (COPE) Report 9, Corvallis, Oregon, USA.

Pojar, J., K. Klinka, and D.V. Meidinger. 1987. Biogeoclimatic ecosystem classification in British Columbia. Forest Ecology and Management **22**:119–154.

Pollock, M.M., R.J. Naiman, and T.A. Hanley. 1998. Plant species richness in riparian wetlands—A test of biodiversity theory. Ecology **79**:94–105.

Reily, P.W., and W.C. Johnson. 1982. The effects of altered hydrologic regime on tree growth along the Missouri River in North Dakota. Canadian Journal of Botany **60**:2410–2423.

Reiners, W.A., I.A. Worley, and D.B. Lawrence. 1971. Plant diversity in a chronosequence at Glacier Bay, Alaska. Ecology **52**:55–69.

Reneau, S.L., and W.E. Dietrich. 1987. Size and location of colluvial landslides in a steep forested landscape. Pages 39–48 *in* R.L. Beschta, T. Blinn, G.E. Grant, G.G. Ice, and F.J. Swanson, eds. Erosion and sedimentation in the Pacific Rim. International Association of Hydrological Sciences Publication 165, Corvallis, Oregon, USA.

Richards, K. 1982. Rivers: Form and Process in Alluvial Channels. Methuen, London, UK.

Robertson, P.A., G.T. Weaver, and J.A. Cavanaugh. 1978. Vegetation and tree species patterns near the northern terminus of the southern floodplain forest. Ecological Monographs **48**:249–267.

Rood, S.B., and S. Heinze-Milne. 1989. Abrupt downstream forest decline following river damming in southern Alberta. Canadian Journal of Botany **67**:1744–1749.

Rood, S.B., C. Hillman, T. Sanche, and J.M. Mahoney. 1994. Clonal reproduction of riparian cottonwoods in southern Alberta. Canadian Journal of Botany **72**:1766–1774.

Rood, S.B., J.M. Mahoney, D.E. Reid, and L. Zilm. 1995. Instream flows and the decline of riparian cottonwoods along the St. Mary River, Alberta, Canada. Canadian Journal of Botany **73**:1250–1260.

Rot, B.W. 1995. The interaction of valley constraint, riparian landform, and riparian plant community size and age upon channel configuration of small streams of the western Cascade Mountains, Washington. Masters thesis, College of Forest Resources, University of Washington, Seattle, Washington, USA.

Ruth, R.H., and A.S. Harris. 1979. Management of western hemlock-Sitka spruce forests for timber production. USDA Forest Service General Technical Report PNW-88, Portland, Oregon, USA.

Schneider, R.L., and R.R. Sharitz. 1988. Hydrochory and regeneration in a bald cypress-water tupelo swamp forest. Ecology **69**:1055–1063.

Schnitzler, A. 1994. European alluvial hardwood forests of large floodplains. Journal of Biogeography **21**:605–623.

Schnitzler, A., R. Carbiener, and M. Trémolières. 1992. Ecological segregation between closely related species in the flooded forests of the upper Rhine plain. New Phytologist **121**:293–301.

Sidle, R.C., and A.M. Milner. 1989. Stream development in Glacier Bay National Park, Alaska, USA Arctic and Alpine Research **21**:350–363.

Sigafoos, R.S. 1964. Botanical evidence of floods and flood plain deposition. US Geological Survey Professional Paper 485-A, Washington, DC, USA.

Smit, B.A. 1988. Selection of flood-resistant and susceptible seedlings of *Populus trichocarpa* Torr & Gray. Canadian Journal of Forest Research **18**:271–275.

Smith, S.D., A.B. Wellington, J.L. Nachlinger, and C.A. Fox. 1991. Functional responses of riparian vegetation to streamflow diversion in the eastern Sierra Nevada. Ecological Applications **1**:89–97.

Stanford, J.A., J.V. Ward, W.J. Liss, C.A. Frissell, R.N. Williams, J.A. Lichatowich, et al. 1996. A general protocol for the restoration of regulated rivers. Regulated Rivers: Research and Management **12**:391–413.

Streng, D.R., J.S. Glitzenstein, and P.A. Harcombe. 1989. Woody seedling dynamics in an east Texas floodplain forest. Ecological Monographs **59**:177–204.

Stone, E.C., and R.B. Vasey. 1968. Preservation of coastal redwood on alluvial flats. Science **159**:157–161.

Stromberg, J.C., D.T. Patten, and B.D. Richter. 1991. Flood flows and dynamics of Sonoran riparian forests. Rivers **2**:221–235.

Stromberg, J.C., B.D. Richter, D.T. Patten, and L.G. Wolden. 1993. Response of a Sonoran riparian forest to a 10-year return flood. Great Basin Naturalist **53**:118–130.

Swanson, F.J., T.K. Kratz, N. Caine, and R.G. Woodmansee. 1988. Landform effects on ecological processes and features. BioScience **38**:92–98.

Swanson, F.J., R.L. Fredrickson, and F.M. McCorison. 1982. Material transfer in a western Oregon forested watershed. Pages 233–266 *in* R.L. Edmonds, ed. Analysis of Coniferous Forest Ecosystems in the Western United States. Hutchinson/Ross, Stroudsburg, Pennsylvania, USA.

Tardif, J., and Y. Bergeron. 1992. Analyse écologique des peuplements de frêne noir (*Fraxinus nigra*) des rives du lac Duparquet, nord-ouest du Québec. Canadian Journal of Botany **70**:2294–2302.

Thébaud, C., and M. Debussche. 1991. Rapid invasion of *Fraxinus ornus* L. along the Hérault River system in southern France: The importance of seed dispersal by water. Journal of Biogeography **18**:7–12.

Tyree, M.T., K.J. Kolb, S.B. Rood, and S. Patiño. 1994. Vulnerability to drought-induced cavitation of riparian cottonwoods in Alberta: A possible factor in the decline of the ecosystem? Tree Physiology **14**:455–466.

Van Splunder, I., H. Coops, L.A.C.J. Voesenek, and C.W.P.M. Blom. 1995. Establishment of alluvial forest species in floodplains: The role of dispersal timing, germination characteristics and water level fluctuations. Acta Botanica Neerlandica **44**:269–278.

Veirs, S.D. 1980. The influence of fire in coast redwood forests. Pages 93–95 *in* M.A. Stokes and J.H. Dietrich, eds. Proceedings of the Fire History Workshop. USDA Forest Service, General Technical Report, Rocky Mountain Forest and Range Experiment Station-81, Fort Collins, Colorado, USA.

Walker, L.R., J.C. Zasada, and F. Stuart Chapin III. 1986. The role of life history processes in primary succession on an Alaskan floodplain. Ecology **67**:1243–1253.

Waring, R., and J.F. Franklin. 1979. Evergreen coniferous forests of the Pacific Northwest. Science **204**:1380–1386.

Williams, G.P. 1978. Bank-full discharge of rivers. Water Resources Research **14**:1141–1154.

Wilson, R.E. 1970. Succession in stands of *Populus deltoides* along the Missouri River in southeastern South Dakota. American Midland Naturalist **83**:330–342.

Zinke, P.J. 1981. Floods, sedimentation, and alluvial soil formation as dynamic processes maintaining superlative redwood groves. Pages 26–49 *in* R.N. Coats, ed. Proceedings of a Symposium on Watershed Rehabilitation in Redwood National Park and other Pacific Coastal Areas. Center for Natural Resource Studies, JMI, Sacramento, California, USA.

13
Function and Distribution of Large Woody Debris

Robert E. Bilby and Peter A. Bisson

Overview

- Wood is more abundant in streams in the Pacific coastal ecoregion than in streams anywhere else in North America. Abundance of large woody debris (LWD) decreases with increasing channel size. Size of large woody debris pieces increases with channel size.
- Input of large wood to stream channels occurs as a result of chronic bank cutting, windthrow, and stem suppression. Catastrophic occurrences, such as debris torrents, floods, and fires, can deposit large quantities of wood in channels in short periods of time.
- Large woody debris is removed from stream channels by leaching, microbial decomposition, fragmentation by invertebrates, physical fragmentation, and downstream transport. The relative importance of these processes varies with stream size.
- Large woody debris is a primary determinant of channel form in small streams, creating pools and waterfalls and affecting channel width and depth. Wood has less effect on channel form in larger streams.
- The presence of large woody debris facilitates deposition of sediment and the accumulation of finer organic matter. Dramatic increases in sediment and organic matter export occur immediately following removal or disturbance of LWD.
- Particulate organic matter accumulated by large woody debris is an important food source for many stream-dwelling invertebrates. Addition of wood to channels causes increased abundance of macroinvertebrates and changes species composition.
- Pools formed by large woody debris in streams are an important habitat for many species of stream fishes. Fish also use large woody debris as a source of cover.
- Sediment accumulated by woody debris provides a substrate for establishment of early-successional plant species. Large woody debris in riparian areas provides an important germination site for several conifer species.
- The quantity of woody debris in channels in the Pacific coastal ecoregion has decreased over time as a result of various land use practices including removal of wood from rivers for navigation and fish passage, splash damming, and clearing of riparian trees.

Introduction

Stream and river ecosystems are intricately interconnected physically, chemically, and biologically with the terrestrial ecosystems through which they flow. In the forested landscape of the Pacific coastal ecoregion, one of the most obvious indications of this connection is the great abundance of large woody debris (LWD) deposited in stream channels. LWD has a wide variety of influences on lotic ecosystems, dictating channel form, providing sites for storage of organic matter and sediment, and modifying the movement and transformation of nutrients (Bisson et al. 1987). The influence of wood on the structural and functional characteristics of streams affects the biological community (Bisson et al. 1987, Maser and Sedell 1994) including dynamics of riparian forest succession (Fetherston et al. 1995). Large wood on the riparian forest floor provides habitat for many species of wildlife (Bartels et al. 1985, Steel 1993). This chapter examines the spatial and temporal variability in LWD distribution and abundance through drainage networks, the processes of wood delivery and elimination, the influence of large wood on stream ecosystems, and the effect of land-use practices on LWD.

Abundance and Distribution of LWD in Channel Networks

Definitions of the size of LWD vary according to the objectives of a particular study. For example, investigators examining the contribution of LWD to the total organic matter load of a stream often include relatively fine material (as small as 2.5 cm in diameter) in their definition (Harmon et al. 1986). However, investigators examining the influence of wood on channel morphology often employ a much larger minimum size (usually 10 cm in diameter and 2 m in length) in their definition (Bilby and Ward 1989, Maser and Sedell 1994).

The use of multiple definitions of LWD makes comparisons of wood abundance among studies difficult. Nonetheless, research in the Pacific coastal ecoregion reveals general patterns of wood abundance (Table 13.1). Redwood (*Sequoia sempervirens*) forests of northern California exhibit the highest wood biomass with levels for some stream reaches as high as 180 kg/m^2, a value considerably greater than reported from any other region (Harmon et al. 1986). LWD biomass decreases in a northerly direction, with lowest levels observed in the sitka spruce forests of southeast Alaska. However, LWD abundance throughout the Pacific coastal ecoregion is much greater than that measured in other forested areas of North America (Harmon et al. 1986).

Amount and distribution of LWD in stream channels is strongly influenced by channel size (Figures 13.1 and 13.2) (Swanson et al. 1982, Bilby and Ward 1989). Small channels tend to contain abundant LWD that is distributed randomly. Wood is more easily transported in larger channels, leading to a reduction in the amount and aggregation of the remaining pieces. The size of these wood aggregations increases in a downstream direction while their frequency decreases (Swanson et al. 1982, Bisson et al. 1987). Despite the general tendency for wood abundance to decrease in larger channels, accumulations of LWD in large rivers

TABLE 13.1. Biomass of large woody debris in small streams (channel width <10 m) flowing through undisturbed, mature forests of the Pacific Coastal Ecoregion. Values from other areas in North America are shown for comparison.

Location	Primary tree species	Number of reaches inventoried	Average channel width (m)	LWD biomass (kg/m^2)
Northern Rocky Mountains, Idaho	Pine (*Pinus* spp.)	3	4.4	2.2
White Mountains, New Hampshire	Red spruce (*Picea rubens*), Eastern hemlock (*Tsuga canadensis*)	2	4.2	2.2
Northern Rocky Mountains, Idaho	Engelmann spruce (*Picea engelmanii*)	2	3.0	2.8
Smoky Mountains, Tennessee	Mixed hardwoods	5	5.1	5.0
Smoky Mountains, Tennessee	Red spruce, balsam fir (*Abies balsamea*)	2	4.8	7.2
Southeast Alaska	Sitka spruce (*Picea sitchensis*), western hemlock (*Tsuga heterophylla*)	4	3.5	6.6
Coastal British Columbia	Sitka spruce, western hemlock	5	—	31.6
Cascade Mountains, Oregon	Douglas-fir (*Pseudotsuga menziesii*)	24	3.5	34.7
Northern California	Coast redwood (*Sequoia sempervirens*)	8	6.8	74.2

Calculated from information presented in Harmon et al. (1986).

occasionally reach an enormous size. In the ninteenth century a wood jam on the Red River in Louisiana was estimated at 300 km in length (Lobeck 1939), and accumulations of several kilometers in length were reported on rivers along the Pacific coast of North America (Sedell and Luchessa 1982).

LWD amount and distribution also is affected by the density and species composition of the riparian forest. Tree density in riparian forests is positively related to LWD amount in streams in eastern Washington (Bilby and Wasserman 1989). LWD produced by conifers, which tends to be larger than that produced by hardwoods, is less likely to flush downstream, and significantly lower decay rates increase the longevity of LWD in the system (Harmon et al. 1986). Thus, streams flowing through mature stands of conifer in the Pacific Northwest tend to contain larger amounts of wood with larger average piece size than channels located in younger forests, which often are dominated by

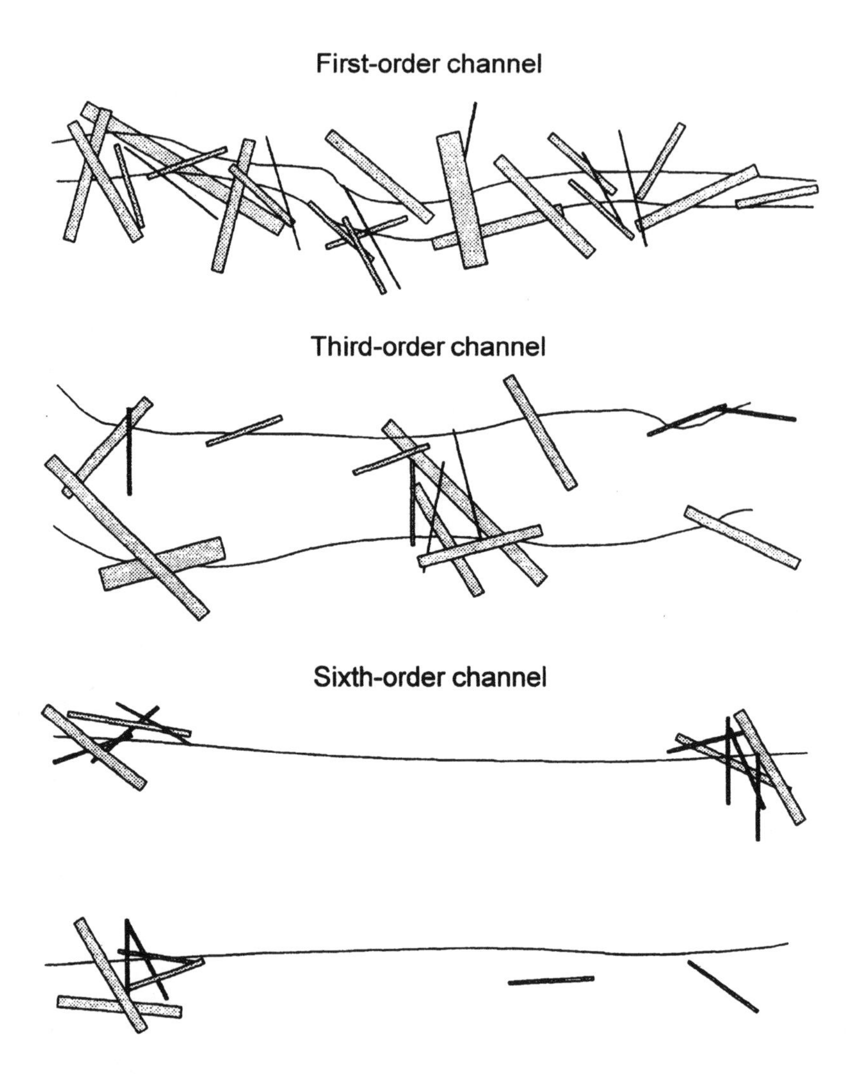

FIGURE 13.1. Typical distribution of LWD in channels of various size. Aggregation of wood increases with channel size and total wood abundance decreases. This figure is based on maps of LWD in the McKenzie River basin, Oregon (Swanson et al. 1982).

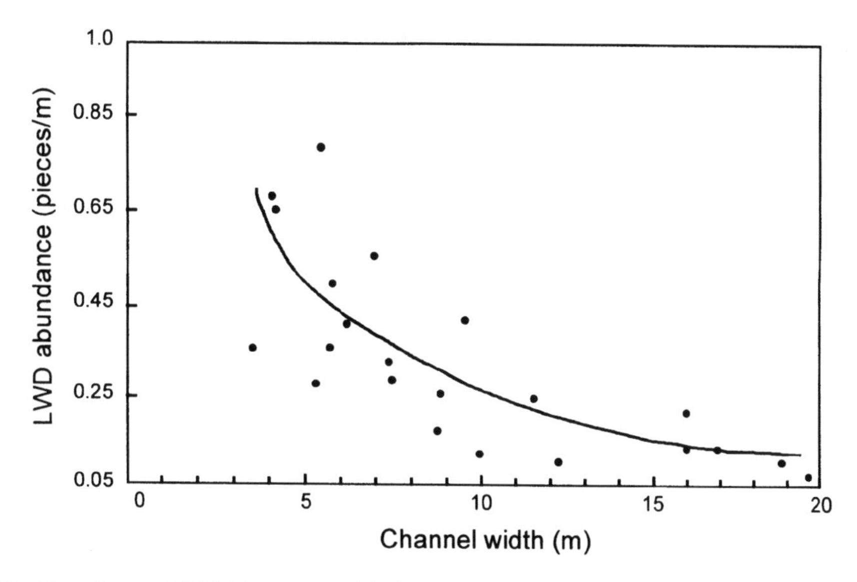

FIGURE 13.2. Abundance of LWD increases with decreasing channel size in old-growth forests in southwestern Washington (modified from Bilby and Ward 1989).

smaller, hardwood species (Grette 1985, Bilby and Ward 1991).

Channel type also is related to the abundance of LWD. Confined channels with boulder or bedrock substrate contain only about half the number of pieces of wood found in similarly sized, unconfined reaches with finer substrate (Bilby and Wasserman 1989). This difference is likely the result of an increased rate of LWD input to unconfined channels from bank cutting (Murphy and Koski 1989) and to the greater capacity for transporting wood downstream during high flow in the higher-energy, confined channels. Susceptibility of a watershed to catastrophic events, such as windstorms or landslides, also may significantly impact the amount of LWD in some stream channels (Keller and Swanson 1979, Bisson et al. 1987).

Stream size plays a major role in determining the size of LWD pieces retained in stream channels. Generally, the average size (diameter, length, or volume) of LWD in a stream channel increases with stream size (Figure 13.3) (Bilby and Ward 1989). The increase in LWD size is caused by the greater capacity for large channels to transport wood. Smaller pieces of wood are selectively flushed from larger channels leaving only larger pieces and causing a decrease in wood amount but an increase in average piece size.

Processes Controlling Input and Output of LWD

Input processes. Both chronic and episodic processes are responsible for delivering LWD to streams (Figure 13.4) (Keller and Swanson 1979, Bisson et al. 1987). Chronic mechanisms include the regular introduction of wood as a result of tree mortality or gradual bank undercutting. These processes tend to add small amounts of wood at frequent intervals. Much of the LWD in unconfined channels is introduced by undercutting of trees on the bank (Grette, 1985, Murphy and Koski, 1989), whereas windthrow is the principal mechanism of wood delivery to channels with confined, erosion-resistant banks (Lienkaemper and Swanson 1987).

The rate at which these chronic input processes deliver LWD to a channel varies as a

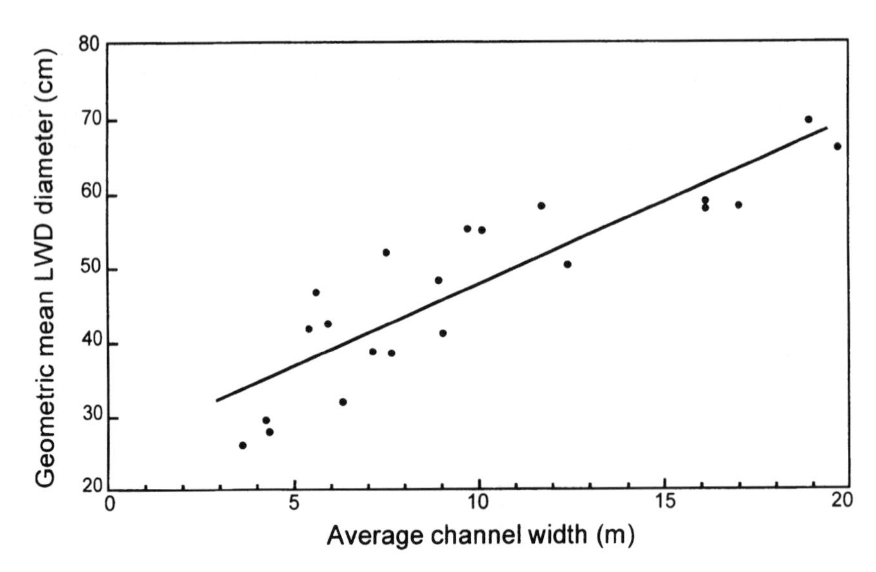

FIGURE 13.3. Relationship between channel width and geometric mean diameter of pieces of LWD for channels in old-growth forests in southwestern Washington (modified from Bilby and Ward 1989).

function of successional stage of the riparian stand. Red alder (*Alnus rubra*), a common early-successional species in riparian areas in the Pacific coastal ecoregion, has a relatively short life span and begins to die and contribute LWD to the channel approximately 60 years after stand establishment (Grette 1985). Shade-tolerant conifers, such as western redcedar

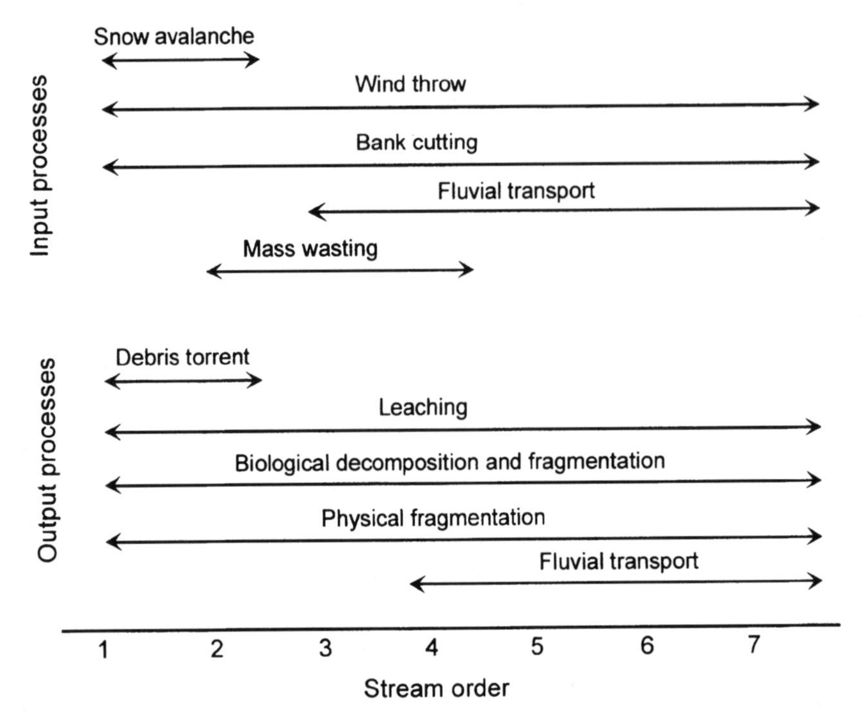

FIGURE 13.4. Processes of wood input and output to channels in the Pacific coastal ecoregion. Different processes predominate in different sized channels (modified from Keller and Swanson 1979).

(*Thuja plicata*) or western hemlock (*Tsuga heterophylla*), establish in the alder understory, then occupy the site and contribute wood to the channel as a result of stem suppression (mortality due to competition). Evidence suggests that stem suppression is the major process contributing LWD to stream channels in the western Cascade Mountains of Washington in stands up to 300 years old (Rot 1995). In stands older than this, mortality of dominant trees due to disease and windthrow is the primary process delivering wood to channels.

The length of time it takes (following a vegetation-removing disturbance) for a riparian area to produce woody debris large enough to remain in the channel varies with stream size. In large channels, input of LWD takes longer to resume and the rate of LWD accrual is slower following a severe disturbance (Bilby and Ward 1991). For example, in 3rd-order channels on the Olympic Peninsula, measurable contributions of wood from disturbed riparian areas did not occur until 60 years after harvest (Grette 1985). Bilby and Wasserman (1989) suggest that streamside vegetation in southwestern Washington must be at least 70 years old to provide stable material to streams more than 15 m wide.

Episodic input events, including catastrophic windthrow, fire, or severe floods, occur infrequently but can add massive amounts of wood to the channel network in a very short period of time. Landslides and debris torrents can transport huge amounts of wood from hillslopes and headwater tributary channels to downstream reaches (Keller and Swanson 1979). Input by this mechanism, however, is restricted to lower-order channels in steep terrain, where landslides are common (Figure 13.4) (Swanson et al. 1987). Infrequent, severe windstorms in the Pacific coastal ecoregion have been responsible for leveling very large areas of forest (Harmon et al. 1986). A single windstorm on October 12, 1962, blew down $2.6 \times 10^7 m^3$ of wood (Orr 1963). Fire occurrence varies as a function of aspect, elevation, and other factors. However, fires recur in most Pacific coastal forests at intervals of 200 to 1,000 years, often resulting in the delivery of large amounts of LWD to stream channels (Agee 1988). Very severe floods also add large amounts of wood to channels through accelerated bank cutting and transport of wood stored on the floodplain into the channel (Keller and Swanson 1979). Input of LWD by flooding tends to be particularly prevalent in large channels with extensive floodplains.

The area from which LWD is supplied to the channel varies as a function of the species composition and age of the riparian vegetation, topography of the streamside area, characteristics of the channel, and direction of the prevailing wind (Steinblums et al. 1984, Grette 1985, Murphy and Koski 1989, McDade et al. 1990). Both empirical and theoretical analyses of the probability of input of LWD to a channel as a function of distance from the streambank have been developed (Murphy and Koski 1989, McDade et al. 1990, Robison and Beschta 1990b, Van Sickle and Gregory 1990, Lorenzen et al. 1994). In general, these analyses suggest that the primary zone of input is equivalent to the height of the tallest trees growing along the stream. The probability of a tree within the riparian zone entering the stream when it falls decreases with distance from the channel edge and varies due to differences in tree height, a function of stand age and tree species (Figure 13.5) (McDade et al. 1990, Van Sickle and Gregory 1990). In general, 70 to 90% of the input of LWD occurs within 30 m of the channel edge. However, trees growing anywhere on the floodplain of an unconfined reach may ultimately be captured by the stream due to lateral migration of the channel across the valley bottom.

Output processes. Leaching, fragmentation, microbial decay, invertebrate consumption, and fluvial transport all contribute to the ultimate demise of a piece of wood in a stream (Figure 13.4) (Keller and Swanson 1979). However, because wood is resistant to solution, leaching, the gradual dissolution of the wood by water, plays a minor role in LWD decomposition (Harmon et al. 1986).

Fragmentation, the physical breakdown of the wood by the force of flowing water, is one of the principal means of wood degradation in streams (Aumen 1985). The process of fragmentation is accelerated by microbial decay,

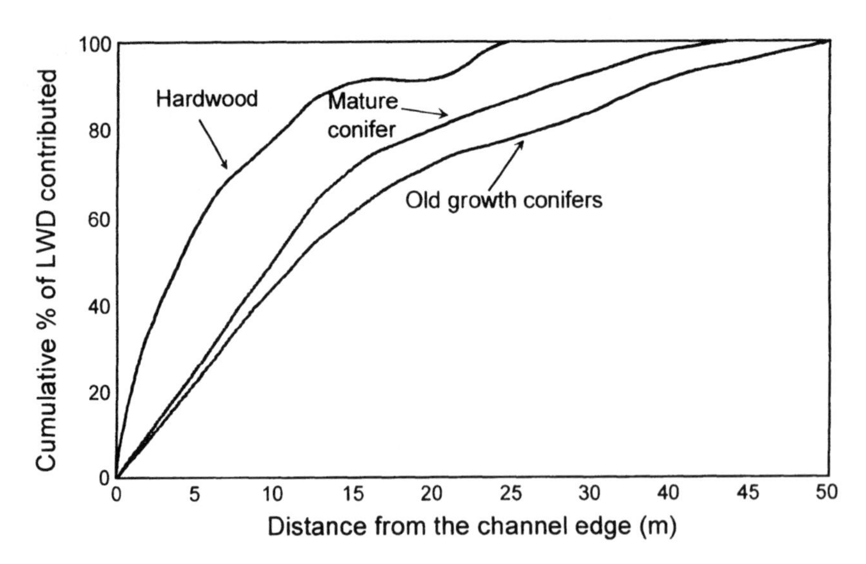

FIGURE 13.5. The cumulative proportion of LWD delivered to a stream as a function of distance from the channel edge. Differences in tree height between hardwood trees, mature conifers and old-growth conifers are indicated by the three curves (modified from McDade et al. 1990).

which weakens the wood. Microbial decay of wood in streams is conducted primarily by bacteria (Crawford and Sutherland 1979) and occurs in a thin layer on the wood surface (Aumen 1985). Decomposition by fungi, the primary organisms responsible for wood decay in terrestrial environments, is limited in streams because of the high water content and low oxygen levels within the wood (Savory 1954). Invertebrates that feed on wood accelerate the decomposition process by fragmenting the wood and providing fresh surfaces for microbial colonization. However, these highly specialized invertebrates generally consume relatively small amounts of wood and have a minor impact on decomposition rate (Wallace and Anderson 1996).

The importance of downstream transport as a process of wood removal from streams varies with channel size. In small, steep headwater tributaries, debris torrents can remove wood from long stretches of stream channels, depositing this material in a large accumulation at the terminus of the torrent (Swanson et al. 1976). Downstream transport of wood during periods of elevated flow is a process which is most common in larger channels (Keller and Swanson 1979).

LWD in Pacific coastal streams can persist for a very long time (Franklin et al. 1981). A study using dendrochronological methods identified pieces of wood that had remained for up to 108 years in streams in the Cascade Mountains of Oregon (Swanson and Lienkaemper 1978). Size and species of logs (mean piece volume = $1.6\,m^3$) in streams on Washington's Olympic Peninsula decayed at the rate of 1%/yr, with large pieces persisting longer than small pieces (Grette 1985). In southeast Alaska, pieces 10 to 30cm in diameter never persisted for more than 110 years, whereas pieces greater than 60cm in diameter persisted up to 226 years (Murphy and Koski 1989). The rate of LWD depletion in this study was higher in tightly constrained, high-energy channels than it was in low-gradient systems with broad floodplains. A study in the Cascade Mountains of Oregon indicated that western redcedar decayed most slowly, followed in order by Douglas-fir (*Pseudotsuga menziesii*), western hemlock, and red alder (Swanson and Lienkaemper 1978). Rates of microbial respiration on LWD in stream channels suggest that total decomposition requires from 5 to 200 years, depending on piece size (Anderson et al. 1978).

LWD Function in Stream Ecosystems

Channel form. Large wood has a major impact on channel form in 1st- through 4th-order streams in the Pacific coastal ecoregion. Wood tends to increase average channel width and increase variability in width (Zimmerman et al. 1967, Trotter 1990). In several small streams in northern California, channel width near large accumulations of LWD were 27 to 124% greater than average channel width of the reach (Keller and Swanson 1979). LWD also forms and stabilizes gravel bars and other depositional sites (Lisle 1986, Abee and Montgomery 1995, Fetherston et al. 1995), forms waterfalls (Heede 1972), creates pools (Robison and Beschta 1990a) and influences channel meandering and bank stability (Swanson and Lienkaemper 1978, Cherry and Beschta 1989). Although LWD tends to influence the channel form of larger systems less, occasional large LWD accumulations may increase channel width, create bars and other depositional features, and encourage the development of meander cutoffs (Keller and Swanson 1979).

Wood is often the primary agent forming pools in plane-bedded and step-pool channels and plays a role in forming or modifying pools in other channel types as well (Montgomery and Buffington 1993, definition of channel types in Chapter 2 of this volume). In small, high-gradient, stepped-pool channels, wood forms a waterfall by obstructing flow and creates a plunge pool. Obstructions other than wood may form waterfalls in these small channels, but wood is most often responsible (Keller and Swanson 1979, Bilby and Ward 1991). LWD forms pools by concentrating flow and scouring the bed in larger, lower-gradient, plane-bedded channels. In such channels, pools are infrequent when LWD is rare. Over 80% of the pools in a small stream in southwest Washington are associated with wood (Bilby 1984). Similarly, 80% of the pools in a series of small streams in the Idaho panhandle are associated with wood (Sedell et al. 1985), and 86% of the pools in a northern California stream are associated with large roughness elements, the majority of which are LWD (Lisle and Kelsey 1982). The proportion of stream surface area occupied by pools range from 4 to 11% for several small streams in British Columbia with little LWD; pools in nearby reaches with abundant wood occupy from 27 to 45% of the surface area (Fausch and Northcote 1992).

The relative importance of LWD in pool formation decreases with increasing channel size and decreasing gradient (Bilby and Ward 1989, Montgomery and Buffington 1993, Chapter 2 this volume). Pool formation is primarily dictated by dynamics of flow in channels exhibiting a pool-riffle morphology, rather than by channel obstructions like large woody debris. However, channel-spanning accumulations of wood occasionally form even in large rivers and can create lake-like conditions upstream (Sedell and Luchessa 1982). In addition, LWD forms pools along the channel margins or in secondary channels of large rivers, which can provide important habitat for some species of fish (Bisson et al. 1987).

Pieces of LWD associated with large accumulations of finer organic matter (e.g., twigs, needles, leaves) are more likely to form pools than individual pieces. Bilby and Ward (1991) found that 77% of the pieces of LWD with fine debris accumulations greater than $0.5\,m^3$ formed pools compared with 26% of LWD pieces with accumulations less than $0.5\,m^3$.

LWD influences pool size as well as pool frequency. The deepest pools tend to be associated with large roughness elements, like LWD (Lisle and Kelsey 1982). Average pool depth decreased following experimental removal of wood from several stream reaches in the area impacted by the 1980 eruption of Mt. St. Helens (Lisle 1995). Depth and sinuosity in several reaches of a small British Columbia stream with abundant LWD was greater than in nearby reaches which lacked wood (Fausch and Northcote 1992). Surface area of LWD-formed pools is positively correlated with the size of piece of wood, or wood accumulation, forming that pool (Bilby and Ward 1989). The effect of piece size on pool surface area can be quite dramatic. For example, based on the relationships in Bilby and Ward (1989), a piece of LWD

30 cm in diameter and 5 m in length in a stream with a channel width of 8 m would produce a pool with a surface area of 1.4 m^2. A piece of LWD 60 cm in diameter and 5 m in length would produce a pool with a surface area of 3.9 m^2.

Wood also affects channel form through the creation of waterfalls. Waterfalls form plunge pools and influence sediment transport in streams. The greater the proportion of the drop in elevation of a stream caused by waterfalls, the less efficient the system is at moving sediment downstream (Heede 1972). The proportion of channel drop accounted for by summing the heights of waterfalls caused by LWD ranged from 30 to 80% in streams in the western Oregon Cascades (Keller and Swanson 1979) and 6% in a stream in the Oregon Coast Range (Marston 1982). The proportion of elevation drop caused by LWD decreased with increasing stream size for 22 stream reaches in areas of old-growth forest in western Washington, ranging from greater than 15% in channels less than 10 m wide to less than 5% for channels 10 to 20 m wide (Bilby and Ward 1989). In channels wider than 20 m, waterfalls formed by LWD are very rare.

Movement of particulate matter. LWD controls routing of sediment and particulate organic matter through channel networks by creating areas of low flow velocity and shear stress where this material can be stored. The primary method by which LWD decreases shear stress in small, high-gradient streams is through the formation of step-pools that produce a depositional site upstream from the waterfall and along the margins of the plunge pool (Heede 1972, Montgomery and Buffington 1993, Chapter 2 this volume). In larger systems, areas of reduced shear stress form downstream from the wood accumulation or between the LWD and the stream bank (Keller and Swanson 1979, Lisle 1986).

Depositional sites associated with LWD tend to be small but frequent in small streams, In channels less than 7 m wide flowing through old-growth forest in western Washington, 39% of the LWD pieces were associated with sites of sediment deposition (Bilby and Ward 1989). The frequency with which LWD formed depositional sites decreased with increasing stream size, with 26% of the pieces accumulating sediment in channels 7 to 10 m wide and 19% in channels over 10 m wide. The proportion of the channel covered by sediment associated with LWD decreased with increasing stream size as well. Depositional sites formed by LWD covered 19% of the streambed in channels 5 m wide, decreasing to 3% in channels 15 m wide. The decrease was due to a reduction in the amount of LWD and in the proportion of LWD pieces forming depositional areas.

The average size of depositional sites increases with channel size (Bilby and Ward 1989). This is, in part, the result of the steep banks, high gradient, and step-pool morphology of small streams, which tend to limit the size of depositional areas relative to those formed downstream from LWD accumulations in channels with a pool-riffle or plane-bed morphology (Montgomery and Buffington 1993). In addition, larger pieces or accumulations of LWD create larger depositional areas, and average piece size and frequency of large aggregations of LWD increases with channel size (Bilby and Ward 1991).

LWD may form very large mid-channel or channel-margin gravel bars in large channels (Abbe and Montgomery 1995, Fetherston et al. 1995, Chapter 12 this volume). These gravel bars frequently increase in size over time as additional LWD is accumulated at the upstream edge of the bar. Establishment of vegetation further stabilizes the bar and encourages additional deposition. Ultimately, a vegetative community becomes established that is oldest at the upper middle point of the bar and decreases in age towards the margins.

The influence of LWD on sediment routing can be demonstrated by determining the proportion of stored sediment associated with wood in a stream reach and by measuring sediment transport before and after LWD removal. Wood was responsible for the storage of 49% of the sediment in seven, small, Idaho watersheds (Megahan 1982) and 87% of the sediment in the channel of a small stream in New Hampshire (Bilby 1981). Removal of wood from a 250 m reach of a stream in the Oregon

Coast Range released 5250 m^3 of sediment (Beschta 1979). The winter following the removal of redwood LWD from a 100 m reach of a northern California stream 60% of the stored sediment was exported (MacDonald and Keller 1983). Experimental removal of LWD from a 175-m reach of a 2nd-order channel in New Hampshire led to a doubling in the rate of particulate matter export from the entire watershed the following year (Bilby 1981).

Wood in streams is responsible for storing large amounts of particulate organic matter, such as leaves, needles, or twigs. Naiman and Sedell (1979) found that amount of particulate organic matter (<10 cm) was strongly related to abundance of LWD in reaches of the McKenzie River watershed in western Oregon (Figure 13.6). Trotter (1990) found that the presence of LWD in three stream reaches more than doubled the amount of stored coarse particulate organic matter. LWD was responsible for storing 75% of the organic matter in 1st-order channels and 58% in 2nd-order channels in the White Mountains of New Hampshire (Bilby and Likens 1980). Removal of LWD from a 175 m-reach of 2nd-order channel increased export of coarse particulate organic matter (>1 mm) 138% and fine particulate organic matter (<1 mm) 632% (Bilby and Likens 1980), whereas, addition of LWD to three stream reaches in the southern Appalachian Mountains led to an increase in stored particulate organic matter from 88 to 1,568 g/m^2 (Wallace et al. 1995). In the absence of LWD, much of the terrestrial organic matter entering streams is flushed rapidly downstream (Naiman and Sedell 1980). Reduced quantities of particulate organic matter decrease the productivity and change the composition of the macroinvertebrate community (Wallace et al. 1995, Chapter 8 this volume).

An important source of particulate organic matter in Pacific coastal ecoregion streams are Pacific salmon (*Oncorhynchus* spp.), which die after spawning. Monitoring of several hundred, tagged coho salmon (*O. kisutch*) carcasses in a number of western Washington streams revealed that 60% were retained by LWD (Cederholm et al. 1989). Materials transported to freshwater by spawning salmon makes a substantial contribution to the productivity of the typically, oligotrophic systems in this region.

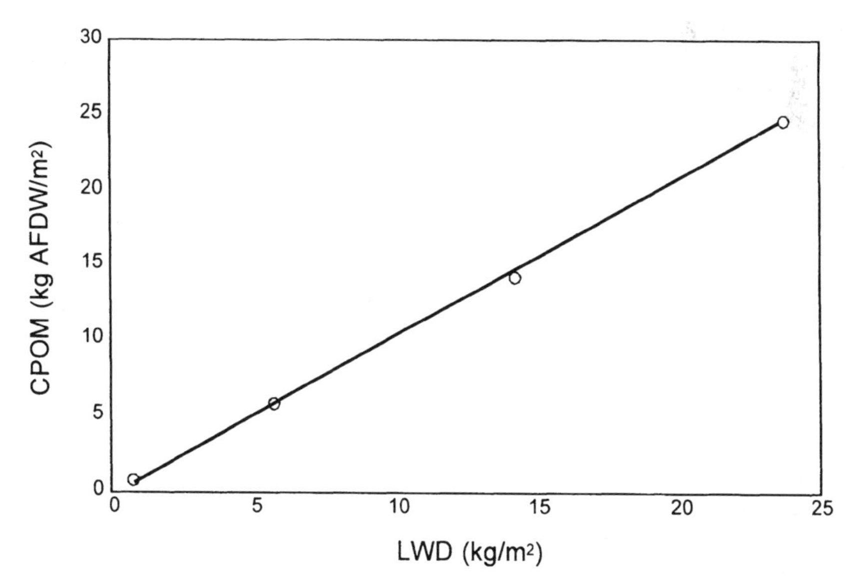

FIGURE 13.6. The relationship between large woody debris (LWD) and coarse particulate organic matter (CPOM) in channels within the McKenzie River watershed in Oregon. LWD is defined as larger than 10 cm and CPOM as smaller than 10 cm. Data from Naiman and Sedell (1979). Regression equation: CPOM Amount = 1.04 (LWD Amount) − 0.13; $r^2 = 0.99$.

Kline et al. (1990 1994) found that spawning pink salmon (*O. gorbuscha*) and sockeye salmon (*O. nerka*) were the source of most nitrogen (N) in the tissues of invertebrates and juvenile fishes in two Alaskan watersheds. Twenty-five to 40% of the N and carbon (C) in the tissues of juvenile salmonid fishes in a small stream in western Washington were derived from spawning coho salmon (Bilby et al. 1996). Salmon carcasses also are an important food resource for many species of wildlife (Cederholm et al. 1989). Availability of this material is reduced in stream reaches containing little LWD.

By controlling the rate of particulate matter transport in channels, LWD influences the rate of movement of nutrients through drainage systems. Export of elements contained in particulate matter increased as much as 88-fold following removal of wood from a short stretch of stream channel in New Hampshire (Table 13.2) (Bilby 1981).

Transport of dissolved matter, however, is less impacted by LWD. Dissolved organic carbon concentration increased after removal of the wood from a stream reach in New Hampshire, but the concentration of 10 ions did not change (Bilby 1981). Nonetheless, LWD can influence the movement of dissolved nutrients by affecting the rate at which these materials are removed from stream water. Addition of logs to a stream in North Carolina slightly decreased uptake rate for ammonium while nitrate uptake rate increased (Wallace et al. 1995). However, the uptake rate of phosphate was not affected by the addition of logs.

Water quality. Low oxygen concentrations in stream water as a result of decomposition of LWD seldom occurs in streams because wood has a very low surface area to volume ratio that minimizes the area available for microbial activity. In addition, wood is composed of material relatively resistant to decomposition, thus the rate of oxygen consumption by decomposing organisms on wood is low (Harmon et al. 1986), and the rapid, turbulent flow of the water helps facilitate reoxygenation from the atmosphere (Ice 1978). However, introduction of large amounts of woody debris to beaver ponds or to slowly flowing streams during logging can cause decreases in oxygen concentration below levels needed to support salmonid fishes (Hall and Lantz 1969).

TABLE 13.2. The estimated export of various elements from a 175-m reach of a 2nd-order stream in the White Mountains of New Hampshire. Relationships between dissolved and particulate matter export and discharge were developed before and after removal of LWD. Values presented represent estimates derived by applying the relationships to daily discharges during a single year.

Element	Export with LWD (kg/y)	Export without LWD (kg/y)	Increase without LWD (%)
Si (silicon)	710	2,350	231
Al (aluminum)	84.7	554	554
Fe (iron)	32.5	213	555
Ca (calcium)	275	331	20
Na (sodium)	152	225	48
K (potassium)	63.3	212	235
Mg (magnesium)	66.5	110	65
Mn (manganese)	1.1	7.0	536
P (phosphorus)	1.1	5.3	382
S (sulfur)	389	392	0.8
C (carbon)	791	1,940	145
N (nitrogen)	57.5	72.7	26
Total Export	5,510	13,440	144

Modified from Bilby 1981.

Three common tree species in the Pacific coastal ecoregion, western redcedar, western hemlock, and sitka spruce (*Picea sitchensis*), produce leachate toxic to aquatic organisms (Buchanan et al. 1976). The toxicity of leachates from wood is exacerbated by their tendency to reduce streamwater pH (Allee and Smith 1974). Under most conditions, however, leaching of materials from wood occurs at a very slow rate, keeping concentrations well below toxic levels (Bisson et al. 1987). Impacts have been noted only where large quantities of woody material are introduced into small, slowly flowing streams during logging. Leachate from LWD appears to be more toxic to salmonid fishes than aquatic insects (Peters et al. 1976) with eggs and fry particularly susceptible (Buchanan et al. 1976).

Macroinvertebrates. LWD is directly utilized by macroinvertebrates in streams as substrate and a source of food (Table 13.3). In addition, the role wood plays in accumulating organic matter and sediment creates habitats favored by certain types of aquatic invertebrates that may be rare elsewhere in the channel. Over 50 taxa of macroinvertebrates in five orders are closely associated with wood (Dudley and Anderson 1982).

Coarse particulate organic matter is the primary food source for shredding macroinvertebrates, and these organisms process coarse material into finer particles that are used by collector/gather macroinvertebrates (Merritt and Cummins 1978). As indicated in the previous section, coarse particulate organic matter availability is greatly reduced in the absence of LWD. Productivity, abundance, and biomass of macroinvertebrates tends to be greatest in areas of high particulate organic matter availability (Gurtz and Wallace 1984, Huryn and Wallace 1987, Smock et al. 1989, Richardson 1991). This has been demonstrated by the addition of LWD to a southern Appalachian stream which resulted in a 24-fold increase in invertebrates and a 2.1-fold increase in biomass at the location where wood was placed (Figure 13.7) (Wallace et al. 1995).

LWD influences invertebrate community composition as well as overall abundance. Wallace et al. (1995), found that addition of LWD to a small stream caused increases in certain taxa and functional feeding groups (Diptera, especially chironomids, and noninsect invertebrates) and decreases in others (Ephemeroptera) (Figure 13.7). Biomass of collectors and predators increased while that of scrapers and filterers decreased. Reductions in scrapers was attributed to a lack of suitable substrate caused by deposition of fine organic matter and sediment near the wood. Reduced availability of suspended organic matter caused by decreased current velocities near the added LWD accounted for declines in filterers.

TABLE 13.3. Various uses of LWD by orders of freshwater insects.

Type of wood utilization	Orders of invertebrates
Ingestion	Coleoptera, Plecoptera, Trichoptera, Diptera
Oviposition site	Trichoptera, Hemiptera, Diptera
Burrowing during rearing	Coleoptera, Ephemeroptera
Attachment to surface during rearing	Coleoptera, Diptera, Trichoptera, Ephemeroptera
Pupation site	Trichoptera, Diptera

Summarized from information presented by Harmon et al. (1986).

There are a number of invertebrates that live or feed on wood and the microflora it supports. Several genera of elmid beetles are known to ingest wood (White 1982); *Lara avara* apparently feeds only on this material (Anderson et al. 1978). The stable substrate provided by the wood is important for species which cannot tolerate a frequently shifting bottom, such as filter-feeding insects (Cudney and Wallace 1980). A majority of the insect productivity may be associated with LWD in these soft-bottomed systems. Biomass of insects on wood in a southeastern United States stream was 5 to 10 times that on the sand of the streambed (Benke et al. 1984).

Marine invertebrates also use wood that has been transported to estuaries and the ocean by streams and rivers. LWD in estuaries and the ocean provides a stable substrate for attachment for numerous sessile invertebrates and serves as a food source for shipworms and other wood-boring marine organisms (Maser and Sedell 1994). Experimental addition of wood to an estuary resulted in increased densities of three crustaceans: grass shrimp (*Palaemonetes pugio*), blue crab (*Callinectes sapidus*), and mud crab (*Rhithropanopeus harrisii*) (Everett and Ruiz 1993).

Fish. Much of the work on the role of woody debris as habitat and cover for fish has focused on salmonids in Pacific Northwest streams (Sedell et al. 1984, Murphy et al. 1985, Fausch and Northcote 1992). Juvenile coho salmon and older age classes of cutthroat trout (*O. clarki*) and steelhead (*O. mykiss*) prefer the pool habitat created by LWD over faster-water habitat types (Bisson et al. 1982). Pools provide a location where fish can maintain their position with a minimum of effort, yet food items

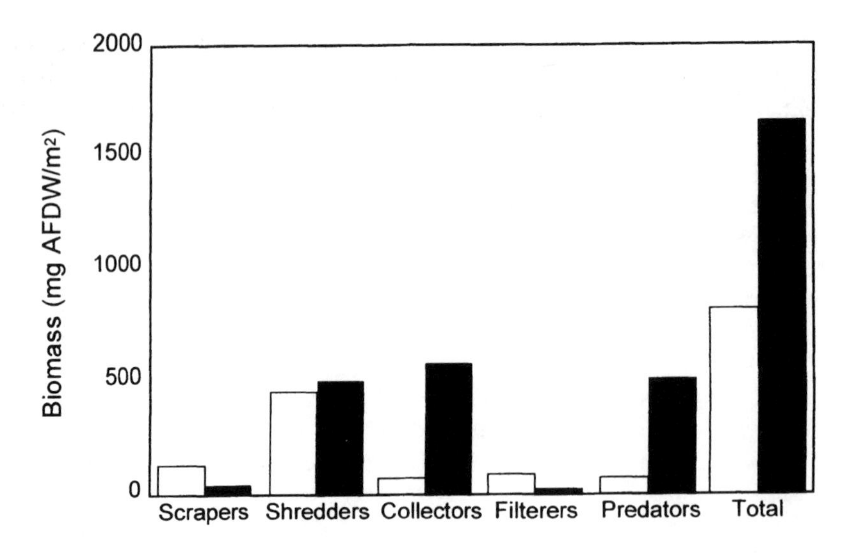

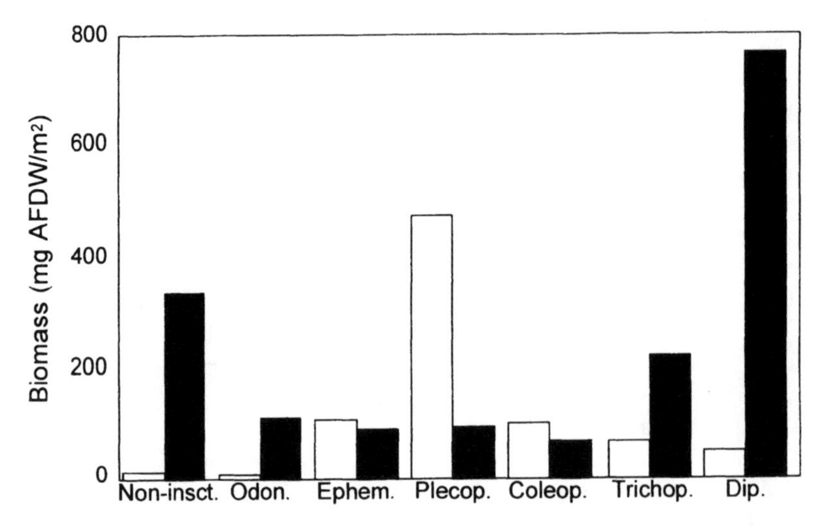

FIGURE 13.7. Biomass of invertebrates at channel cross sections with LWD ■ and without LWD □ in a southern Appalachian stream. The upper panel segregates organisms by functional category, the lower panel by taxonomic category. Taxonomic abbreviations are: Non-insct. = invertebrates other than insects; Odon. = Odonata; Ephem. = Ephemeroptera; Plecop. = Plecoptera; Coleop. = Coleoptera; Trichop. = Trichoptera; Dip. = Diptera (data from Wallace et al. 1995).

carried by the current are abundant (Dill et al. 1981, Fausch 1984). In faster water, food availability may be high but the metabolic cost of maintaining position can negate this advantage.

Fish populations are typically larger in streams with plenty of LWD than in systems with little wood. Stream reaches with large amounts of wood in southern British Columbia supported standing stocks of juvenile coho and cutthroat trout five times higher than reaches in the same system with little wood (Figure 13.8) (Fausch and Northcote 1992). Comparison of winter population levels of juvenile coho in 54 stream reaches in southeast Alaska revealed that average coho salmon density in streams with wood volume less than $50 m^3$ per 30-m stream section was only 25% the average density in streams with greater wood volumes (Murphy et al. 1985). Decreases in fish abundance have been documented following wood removal from channels throughout the Pacific Northwest (Lestelle 1978, Bryant 1983, Dolloff 1986, Elliott 1986). Deliberate additions of LWD to streams resulted in increased abundance of juvenile salmonids in streams on the coast of Oregon (House and Boehne 1986) and British Columbia (Ward and Slaney 1979). An increase in the abundance of adult coho salmon was attributed to increased wood in a coastal Oregon stream, presumably due to improved survival of the juvenile fish (Crispin et al. 1993). In high-gradient streams on the Oregon coast, loss of wood and associated pools led to a decrease in coho salmon (Reeves et al. 1993). Loss of wood in low-gradient channels led to a decrease in the abundance of cutthroat trout due to a decrease in the number of deep pools preferred by older cutthroat and a reduction in structurally complex habitat that would enable juvenile trout to compete successfully with the larger, juvenile coho salmon.

More complex wood structures, such as rootwads or accumulations of multiple pieces, tend to attract more fish than single logs (Sedell et al. 1984). In Kloiya Creek, British Columbia, 99% of the coho salmon fry and 83% of the steelhead parr were associated with rootwads placed in the mid-channel area of the stream, where cover had previously been scarce (Shirvell 1990). Examination of the propensity for juvenile coho salmon to leave experimental channels with varying conditions of shade, flow velocity, and woody cover indicated that woody cover was the most important factor in promoting continued residence during high flows (McMahon and Hartman 1989).

Large accumulations of wood may block the passage of anadromous fishes. For many years this was perceived as a serious problem and wood was removed from channels in order to prevent blockages (Merrell 1951). However, many LWD accumulations which may be blocks at low flows are passable at higher discharges. In addition, these blocks normally occur in steeper channels where habitat available for spawning and rearing by anadromous fish is often limited. It is estimated that, historically, only 5 to 20% of available anadromous fish habitat was inaccessible because of natural blockages formed by LWD (Sedell et al. 1984).

Some marine fishes also are attracted to LWD. Addition of wood to an estuary caused increased abundance of several species of fishes (Everett and Ruiz 1993). The wood created conditions which increased invertebrate populations, the primary food source for the fish,

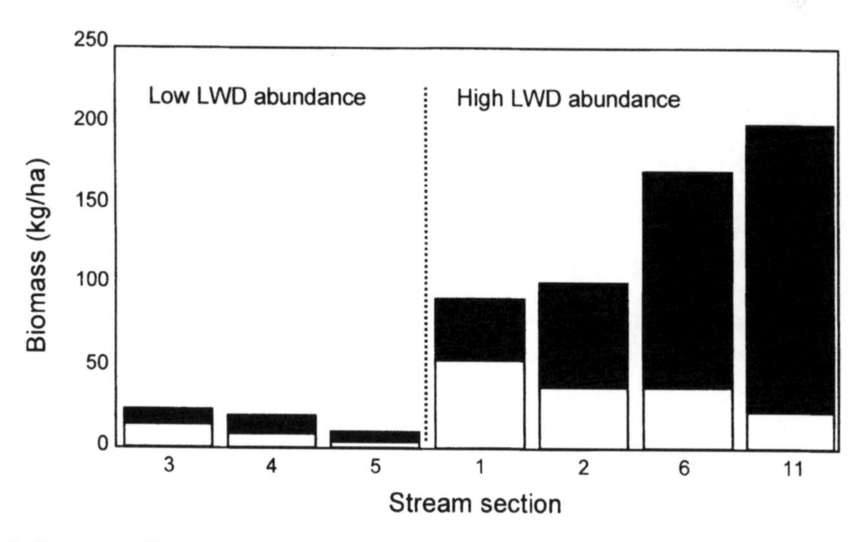

FIGURE 13.8. Biomass of juvenile coho salmon □ and cutthroat trout ■ in southern British Columbia stream reaches with little LWD and abundant LWD (data from Fausch and Northcote 1992).

and provided cover. Floating wood is utilized by pelagic, ocean fishes for cover and shade (Maser and Sedell 1994), and smaller fishes and other organisms attracted to wood are eaten by larger fish. This association is appreciated by commercial fishermen. Purse-seining beneath driftwood is a common method of fishing for tuna (Thunnidae).

Riparian successional processes. The highly diverse vegetation supported by riparian areas in the Pacific Northwest (Campbell and Franklin 1979, Pollock 1995) is the product of small-scale spatial heterogeneity of environmental conditions and frequent disturbance. The physical characteristics and vegetation of the streamside area differ from those upslope as a result of interactions with the stream, including frequent inundation, saturated soils, and physical disturbance of the streamside vegetation due to flood flows, mass soil movements or ice damage (Agee 1988). LWD also contributes to the spatial heterogeneity in riparian areas. Woody material in the channel or within the riparian area creates sites where sediment and organic matter transported by the stream collects. These depositional sites provide locations for the establishment of pioneer plant species, commencing the successional process (Fetherston et al. 1995). Ultimately, patches of vegetation established by this process may be removed during a disturbance event, delivering LWD to the channel and reinitiating the process. Even in old-growth forests, a substantial proportion of the riparian area supports vegetation normally associated with early successional conditions (Rot 1995).

The role of LWD in creating depositional sites suitable for occupation by plants is influenced by topography of the site, size of the channel, and the types of disturbance to which the channel is susceptible (Fetherston et al. 1995). Smaller streams in mountainous terrain have steep channel gradients and stream banks, which limit the size of depositional sites created by LWD (Bilby and Ward 1989). These channels also are more prone to avalanches and debris flows, events which often remove vegetation and much of the soil from riparian areas. As a result, the vegetative communities established on these sites tend to be short-lived. In contrast, the lack of confinement, low gradients, and extensive floodplains associated with larger rivers provide conditions where LWD accumulations can form large depositional areas which are less frequently influenced by disturbance events which remove vegetation.

Some common conifer species in Pacific Northwest forests germinate on decomposing, downed logs, commonly referred to as nurse logs (Harmon et al. 1986). Eighty percent of the conifer regeneration in riparian areas in the Oregon Coast Range occurs on woody debris (Thomas et al. 1993). More than 90% of the conifer regeneration occurred on decomposing logs on floodplain terraces adjacent to the Hoh River in Washington (McKee et al. 1984). Nurse logs elevate the seedlings, which reduces competition with other plants and provides lower soil moisture, enabling establishment of species which cannot tolerate extended periods of soil saturation.

Influence of Land Use on LWD

A variety of land use practices employed in the Pacific coastal ecoregion over the last century have altered the amount and characteristics of large wood in streams. Removal of wood accumulations in large rivers began in the nineteenth century to improve navigation and enable logs from upstream forests to be floated to downstream mills (Sedell and Luchessa 1982). The practice of splash damming was developed to move logs to watercourses large enough to allow them to be floated to the mills. This practice entailed the construction of dams on relatively small streams. Cut timber was placed in the pond formed by the dam and in the channel below the dam. The impounded water was then released, carrying the logs downstream on the crest of the resultant flood. Over 70 splash dams were operated in streams draining to Grays Harbor and Willapa Bay in southwestern Washington during the early part of the twentieth century (Figure 13.9) (Wendler and Deschamps 1955), and the practice was common throughout the Pacific coastal ecoregion (Sedell and Luchessa 1982). Trans-

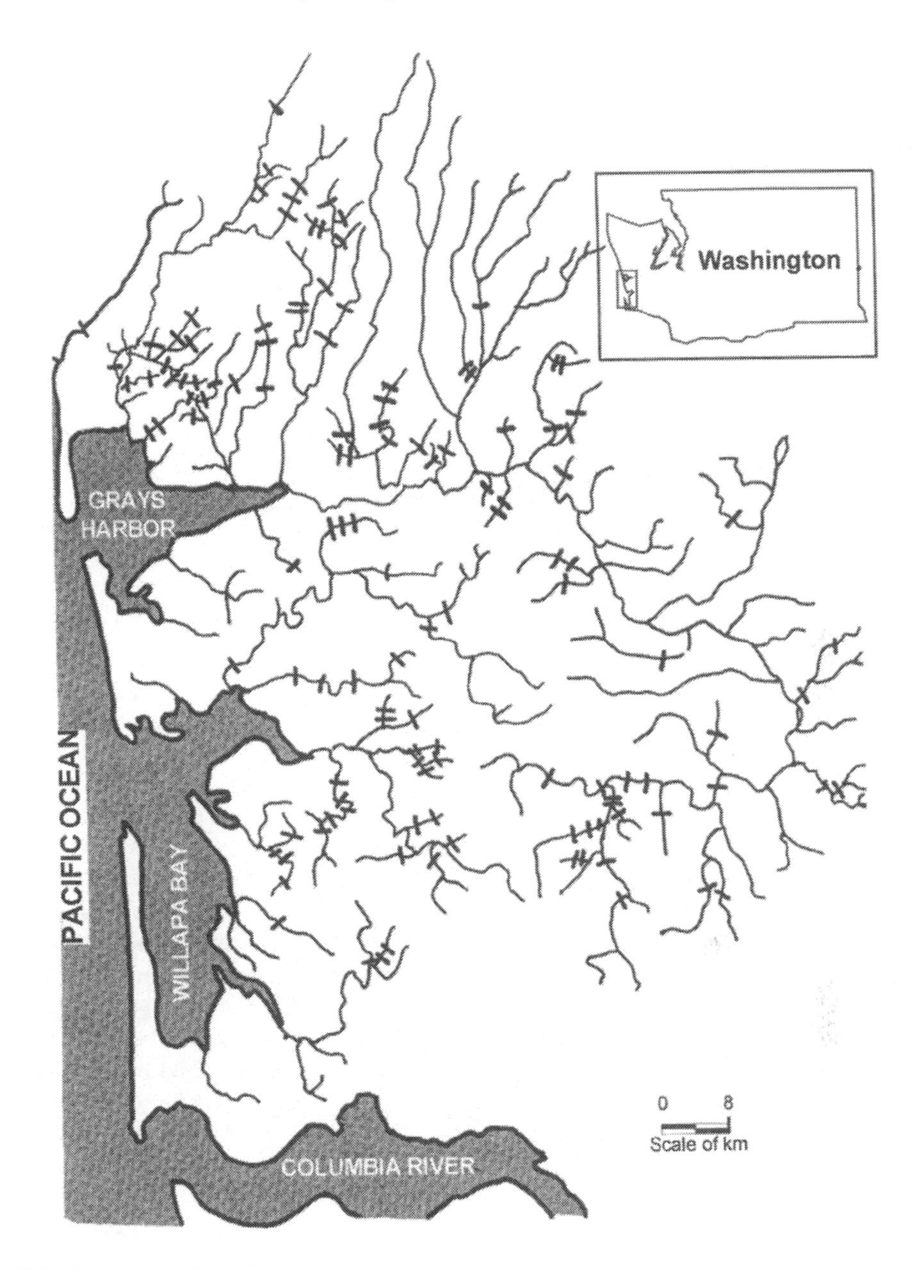

FIGURE 13.9. Locations of splash dams within the Grays Harbor and Willapa Bay drainages in southwestern Washington. Most of these dams operated during the early part of the twentieth century (from Wendler and Deschamps 1955 with permission).

port of logs by splash damming greatly reduced the abundance of LWD in streams where this practice was employed (Sedell and Luchessa 1982).

Until the 1970s, LWD was viewed as an impediment to the upstream migration of anadromous fishes and was deliberately removed to improve passage (Merrell 1951). Stream cleaning was pursued aggressively throughout the Northwest from shortly after World War II until the early 1970s (Narver 1971, Hall and Baker 1982). In many cases, wood which did not prevent upstream access by anadromous fish was removed (Bisson et al. 1987). Even in cases where access was improved, detrimental impacts were associated with this practice, including reduced channel stability and the release of large amounts of stored sediment which damaged fish habitat downstream (Hall and Baker 1982, Bilby 1984).

Removal of trees from riparian areas as a result of logging, agriculture, or development activities decreases LWD by removing the future source of wood input to the channel (Swanson and Lienkaemper 1978, Likens and Bilby 1982). Debris flows, which also remove trees from riparian areas in steep terrain, increase in response to certain land-management activities including road construction and logging (Kauffman 1987, Swanson et al. 1987, Hartman and Scrivener 1990). Reduction in wood input to the channel leads to a gradual decrease in LWD over time as residual material decomposes but is not replaced (Swanson and Lienkaemper 1978, Grette 1985, Bisson et al. 1987, Bilby and Ward 1991). In some cases, especially when removal of riparian trees is followed by commercial salvage of wood from the stream or removal of LWD to accelerate water flow, the rate of decrease in LWD after removal of riparian vegetation can be very rapid (Bilby and Ward 1991).

Logging also alters the characteristics and distribution of LWD. Ralph et al. (1994) reported a decrease in the average diameter of LWD pieces in watersheds subjected to moderate or intensive levels of logging activity. More than 60% of the LWD pieces in stream reaches flowing through old-growth forest were over 50cm in diameter but only 40 to 45% were this large in reaches in watersheds managed for wood production. In addition, a much larger proportion of the LWD in the logged basins was located along channel margins above the level of water during summer (Figure 13.10). These changes in LWD size and distribution caused a reduction in pool frequency and depth as well as a corresponding increase in fast-water habitats.

Regulations regarding the treatment of stream channels and riparian areas during forest management activities are in place throughout most of the Pacific coastal ecoregion (Chapter 22). The most recent revisions have incorporated considerations for retaining LWD in streams and providing a future supply from the riparian area. Approaches which are currently being applied to address concerns for LWD include establishing buffer strips along the stream in which no harvest is permitted (e.g., Alaska) or that require leaving a specific number of trees per length of channel along the stream (e.g., Washington), or specify a minimum basal area of trees that must be retained along the stream (e.g., Oregon). The efficacy of these various approaches has yet to be determined. Few regulations governing agricultural practices or development in streamside areas address LWD.

The combined effect of various land-use practices over the last century has changed the species composition and age structure of riparian forests in much of the Pacific coastal ecoregion (Booth 1991, Carlson 1991, Franklin 1992). In undisturbed Pacific Northwest watersheds, patches of riparian vegetation, in various stages of recovery from disturbance, form a linear, mosaic pattern (Naiman et al. 1992). The complex assemblage of riparian vegetation types is dictated by upslope erosional processes, the frequency of disturbance, and the topographic and edaphic characteristics of the site. The amount, size, and species of LWD in channels reflects the condition of the adjacent riparian area (Rot 1995). It has been estimated that 60 to 70% of Pacific Northwest forests were in late successional condition (>200 years old) prior to extensive timber harvest in the region (Franklin and Spies 1984, Booth 1991). In contrast, recent surveys of the riparian vegetation on commercial forest land in western Washington indicate that the majority of riparian areas are in an early successional condition (<60 years) many with an overstory of hardwoods—primarily the pioneer species red alder (Carlson 1991).

Reestablishment of the diverse vegetation reflective of the interaction between riparian areas and natural disturbance processes is required in order to reverse the trend of decreasing LWD in Pacific coastal watersheds. A key

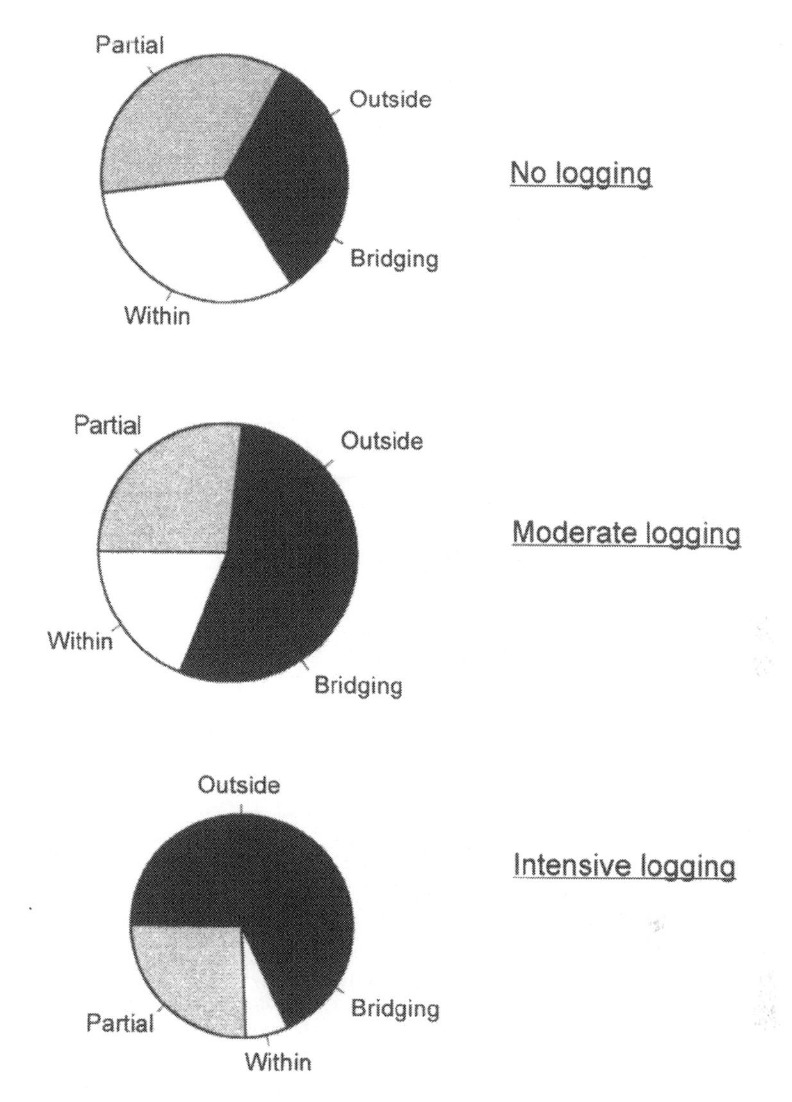

FIGURE 13.10. Distribution of LWD in the stream channels of watersheds subjected to varying levels of logging intensity. *Outside* indicates wood located outside of the low-flow wetted channel, *partial* indicates pieces with a portion of their length in the water, *within* indicates pieces completely in the wetted channel, and *bridging* indicates pieces suspended over the low-flow channel (modified from Ralph et al. 1994).

challenge to achieving this objective is developing a management approach that recognizes and encompasses all the processes responsible for the delivery of LWD to stream channels. The approach also must address management-related alterations in the type, frequency, and severity of disturbances that impact riparian and aquatic systems. Results from such a management approach may take decades to centuries before substantial increases in LWD in streams are achieved. Therefore, successful management approaches will include an adaptive monitoring process which will enable periodic assessment of progress against objectives and allow for corresponding revisions to management plans.

Literature Cited

Abee, T.B., and D.R. Montgomery. 1996. Large woody debris jams, channel hydraulics, and habitat formation in large rivers. Regulated Rivers **12**:201–222.

Agee, J.K. 1988. Successional dynamics of forest riparian zones. Pages 31–43 *in* K.J. Raedeke, ed. Streamside management: Riparian wildlife and forestry interactions. Institute of Forest Resources Contribution Number 59, University of Washington, Seattle, Washington, USA.

Allee, B.J., and M.J. Smith. 1974. Impact of forest management practices on the aquatic environment. Final Report. Quinault Resource Development Program C-4370. Tahola, Washington, USA.

Anderson, N.H., J.R. Sedell, L.M. Roberts, and F.J. Triska. 1978. The role of aquatic invertebrates in processing wood debris in coniferous forest streams. American Midland Naturalist **100**:64–82.

Aumen, N.G. 1985. Characterization of lignocellulose decomposition in stream wood samples using ^{14}C and ^{15}N techniques. Ph.D. dissertation. Oregon State University, Corvallis, Oregon, USA.

Bartels, R., J.D. Dell, R.L. Knight, and G. Schaefer. 1985. Dead and down woody material. Pages 172–186 *in* E.R. Brown, ed. Management of wildlife and fish habitats in forests of western Oregon and Washington. USDA Forest Service Publication Number R6-F&WL-192-1985, Pacific Northwest Region, Portland, Oregon, USA.

Benke, A.C., T.C. Van Arsdall, D.M. Gillespie, and F.K. Parrish. 1984. Invertebrate productivity in a subtropical blackwater river: The importance of habitat and life history. Ecological Monographs **54**:25–63.

Beschta, R.L. 1979. Debris removal and its effect on sedimentation in an Oregon Coast Range stream. Northwest Science **53**:71–77.

Bilby, R.E. 1981. Role of organic debris dams in regulating the export of dissolved and particulate matter from a forested watershed. Ecology **62**:1234–1243.

——. 1984. Post-logging removal of woody debris affects stream channel stability. Journal of Forestry **82**:609–613.

Bilby, R.E., B.R. Fransen, and P.A. Bisson. 1996. Incorporation of nitrogen and carbon from spawning coho salmon into the trophic systems of small streams: Evidence from stable isotopes. Canadian Journal of Fisheries and Aquatic Science **53**:164–173.

Bilby, R.E., and G.E. Likens. 1980. Importance of organic debris dams in the structure and function of stream ecosystems. Ecology **61**:1107–1113.

Bilby, R.E., and J.W. Ward. 1989. Changes in characteristics and function of woody debris with increasing size of streams in western Washington. Transactions of the American Fisheries Society **118**:368–378.

Bilby, R.E., and J.W. Ward. 1991. Characteristics and function of large woody debris in streams draining old-growth, clear-cut, and second-growth forests in southwestern Washington. Canadian Journal of Fisheries and Aquatic Sciences **48**:2499–2508.

Bilby, R.E., and L.J. Wasserman. 1989. Forest practices and riparian management in Washington state: Data based regulation development. Pages 87–94 *in* R.E. Gresswell, B.A. Barton, and J.L. Kershner, eds. Practical approaches to riparian resource management. United States Bureau of Land Management, Billings, Montana, USA.

Bisson, P.A., R.E. Bilby, M.D. Bryant, C.A. Dolloff, G.B. Grette, R.A. House, et al. 1987. Large woody debris in forested streams in the Pacific Northwest: Past, present, and future. Pages 143–190 *in* E.O. Salo and T.W. Cundy, eds. Streamside management: Forestry and fishery interactions. Institute of Forest Resources Contribution Number 57, University of Washington, Seattle, Washington, USA.

Bisson, P.A., J.L. Nielson, R.A. Palmasson, and L.E. Grove. 1982. A system of naming habitat types in small streams with examples of habitat utilization by salmonids during low stream flow. *In* N.B. Armantrout, ed. Acquisition and utilization of aquatic habitat inventory information. American Fisheries Society, Western Division, Bethesda, Maryland, USA.

Booth, D.E. 1991. Estimating prelogging old-growth in the Pacific Northwest. Journal of Forestry **89**:25–29.

Bryant, M.D. 1983. The role and management of woody debris in west coast salmonid nursery streams. North American Journal of Fisheries Management **3**:322–330.

Buchanan, D.V., P.S. Tate, and J.R. Moring. 1976. Acute toxicities of spruce and hemlock bark extracts to some estuarine organisms in southeastern Alaska. Journal of the Fisheries Research Board Canada **33**:1188–1192.

Campbell, A.G., and J.F. Franklin. 1979. Riparian vegetation in Oregon's western Cascade Mountains: Composition, biomass, and autumn phenology. Coniferous Forest Biome Bulletin Number 14. University of Washington, Seattle, Washington, USA.

Carlson, A. 1991. Characterization of riparian management zones and upland management areas with respect to wildlife habitat. Washington Timber/Fish/Wildlife Report Number T/F/W-WLI-91-001. Washington Department of Natural Resources, Olympia, Washington, USA.

Cederholm, C.J., D.B. Houston, D.L. Cole, and W.J. Scarlett. 1989. Fate of coho salmon (*Oncorhynchus kisutch*) carcasses in spawning streams. Canadian Journal of Fisheries and Aquatic Science **46**:1347–1355.

Cherry, J., and R.L. Beschta. 1989. Coarse woody debris and channel morphology: A flume study. Water Resources Bulletin **25**:1031–1036.

Crawford, D.L., and J.B. Sutherland. 1979. The role of actinomycetes in the decomposition of lignocellulose. Developments in Industrial Microbiology **20**:143–151.

Crispin, V., R. House, and D. Roberts. 1993. Change in instream habitat, large woody debris, and salmon habitat after the restructuring of a coastal Oregon stream. North American Journal of Fisheries Management **13**:96–102.

Cudney, M.D., and J.B. Wallace. 1980. Life cycles, microdistribution and production dynamics of six species of net-spinning caddisflies in a large southeastern (USA) river. Holarctic Ecology **3**:169–182.

Dill, L.M., R.C. Ydenberg, and A.H.G. Fraser. 1981. Food abundance and territory size in juvenile coho salmon (*Oncorhynchus kisutch*). Canadian Journal of Zoology **59**:1801–1809.

Dolloff, C.A. 1986. Effects of stream cleaning on juvenile coho salmon and Dolly Varden in southeast Alaska. Transactions of the American Fisheries Society **115**:743–755.

Dudley, T.L., and N.H. Anderson. 1982. A survey of invertebrates associated with wood debris in aquatic habitats. Melanderia **39**:1–21.

Elliott, S.T. 1986. Reduction of a Dolly Varden population and macrobenthos after removal of logging debris. Transactions of the American Fisheries Society **115**:392–400.

Everett, R.A., and G.M. Ruiz. 1993. Coarse woody debris as a refuge from predation in aquatic communities: An experimental test. Oecologia **93**:475–486.

Fausch, K.D. 1984. Profitable stream positions for salmonids: Relating specific growth rate to net energy gain. Canadian Journal of Zoology **62**:441–451.

Fausch, K.D., and T.G. Northcote. 1992. Large woody debris and salmonid habitat in a small coastal British Columbia stream. Canadian Journal of Fisheries and Aquatic Science **49**:682–693.

Fetherston, K.L., R.J. Naiman, and R.E. Bilby. 1995. Large woody debris, physical process, and riparian forest development in montane river networks of the Pacific Northwest. Geomorphology **13**:133–144.

Franklin, J.F. 1992. Scientific basis for new perspectives in forests and streams. Pages 25–72 *in* R.J. Naiman, ed. Watershed management: Balancing sustainability and environmental change. Springer-Verlag, New York, New York, USA.

Franklin, J.F., K. Cromack, Jr., W. Denison, A. McKee, C. Maser, J. Sedell, et al. 1981. Ecological characteristics of old-growth Douglas-fir forests. USDA Forest Service General Technical Report PNW-118. Pacific Northwest Forest and Range Experiment Station, Portland, Oregon, USA.

Franklin, J.F., and T.A. Spies. 1984. Characteristics of old-growth Douglas-fir forests. Pages 328–334 *in* New forests for a changing world: 1983 National Convention Proceedings. Society of American Foresters, Bethesda, Maryland, USA.

Grette, G.B. 1985. The abundance and role of large organic debris in juvenile salmonid habitat in streams in second growth and unlogged forests. Master's thesis, University of Washington, Seattle, Washington, USA.

Gurtz, M.E., and J.B. Wallace. 1984. Substrate mediated response of stream invertebrates to disturbance. Ecology **65**:1556–1569.

Hall, J.D., and C.O. Baker. 1982. Rehabilitating and enhancing stream habitat. Part 1: Review and evaluation. USDA Forest Service General Technical Report PNW-138. Pacific Northwest Forest and Range Experiment Station, Portland, Oregon, USA.

Hall, J.D., and R.L. Lantz. 1969. Effects of logging on the habitat of coho salmon and cutthroat trout in coastal streams. Pages 335–375 *in* T.G. Northcote, ed. Symposium on salmon and trout in streams. H.R. MacMillan Lectures in Fisheries. University of British Columbia, Vancouver, British Columbia, Canada.

Harmon, M.E., J.F. Franklin, F.J. Swanson, P. Sollins, S.V. Gregory, J.D. Lattin, et al. 1986. Ecology of coarse woody debris in temperate ecosystems. Advances in Ecological Research **15**:133–302.

Hartman, G.F., and J.C. Scrivener. 1990. Impacts of forestry practices on a coastal stream ecosystem, Carnation Creek, British Columbia. Canadian Bulletin of Fisheries and Aquatic Science **223**:1–148.

Heede, B.H. 1972. Influences of a forest on the hydraulic geometry of two mountain streams. Water Resources Bulletin **8**:523–530.

House, R.A., and P.L. Boehne. 1986. Effects of instream structure on salmonid habitat and populations in Tobe Creek, Oregon. North American Journal of Fisheries Management **6**:38–46.

Huryn, A.D., and J.B. Wallace. 1987. Local geomorphology as a determinant of macrofaunal production in a mountain stream. Ecology **68**:1932–1942.

Ice, G.G. 1978. Reaeration in a turbulent stream system. Ph.D. dissertation. Oregon State University, Corvallis, Oregon, USA.

Kauffman, P.R. 1987. Channel morphology and hydraulic characteristics of torrent-impacted streams in the Oregon Coast Range, USA. Ph.D. dissertation. Oregon State University, Corvallis, Oregon, USA.

Keller, E.A., and F.J. Swanson. 1979. Effects of large organic material on channel form and fluvial processes. Earth Surface Processes **4**:361–380.

Kline, T.C. Jr., J.J. Goering, O.A. Mathisen, P.H. Poe, and P.L. Parker. 1990. Recycling of elements transported upstream by runs of Pacific salmon: I. δ15N and δ13C evidence in Sashin Creek, southeastern Alaska. Canadian Journal of Fisheries and Aquatic Science **47**:136–144.

Kline, T.C. Jr., J.J. Goering, O.A. Mathisen, P.H. Poe, P.L. Parker, and R.S. Scanlan. 1994. Recycling of elements transported upstream by runs of Pacific salmon: II. δ15N and δ13C evidence in the Kvichak River watershed, Bristol Bay, southwestern Alaska. Canadian Journal of Fisheries and Aquatic Science **50**:2350–2365.

Lestelle, L.C. 1978. The effects of forest debris removal on a population of resident cutthroat trout in a small, headwater stream. Master's thesis. University of Washington, Seattle, Washington, USA.

Lienkaemper, G.W., and F.J. Swanson. 1987. Dynamics of large woody debris in streams in old-growth Douglas-fir forests. Canadian Journal of Forestry Research **17**:150–156.

Likens, G.E., and R.E. Bilby. 1982. Development, maintenance and role of organic debris dams in New England streams. Pages 122–128 *in* F.J. Swanson, R.J. Janda, T. Dunne, and D.N. Swanston, eds. Sediment budgets and routing in forested drainage basins. USDA Forest Service Research Paper PNW-141. Pacific Northwest Forest and Range Experiment Station, Portland, Oregon, USA.

Lisle, T.E. 1986. Stabilization of a gravel channel by large streamside obstructions and bedrock bends, Jacoby Creek, northwestern California. Geological Society of America Bulletin **97**:999–1011.

——. 1995. Effects of coarse woody debris and its removal on a channel affected by the 1980 eruption of Mt. St. Helens, Washington. Water Resources Research **31**:1791–1808.

Lisle, T.E., and H.M. Kelsey. 1982. Effects of large roughness elements on the thalweg course and pool spacing. Pages 134–135 *in* L.B. Leopold, ed. American geomorphological field group field trip guidebook. Leopold, Berkeley, California, USA.

Lobeck, A.K. 1939. Geomorphology. McGraw-Hill, New York, New York, USA.

Lorenzen, T., C. Andrus, and J. Runyon. 1994. The Oregon forest practices act water protection rules: Scientific and policy considerations. Oregon Department of Forestry, Salem, Oregon, USA.

MacDonald, A., and E.A. Keller. 1983. Large organic debris and anadromous fish habitat in the coastal redwood environment: The hydrologic system. Technical Completion Report OWRT Project B-213-CAL. Water Resources Center, University of California, Davis, California, USA.

Marston, R.A. 1982. The geomorphic significance of log steps in forest streams. Annual of the Association of American Geologists **72**:99–108.

Maser, C., and J.R. Sedell. 1994. From the forest to the sea: The ecology of wood in streams, rivers, estuaries, and oceans. St. Lucie Press, Delray Beach, Florida, USA.

McDade, M.H., F.J. Swanson, W.A. McKee, J.F. Franklin, and J. Van Sickle. 1990. Source distances for coarse woody debris entering small streams in western Oregon and Washington. Canadian Journal of Forest Research **20**:326–330.

McKee, A., G. LeRoi, and J.F. Franklin. 1984. Structure, composition and reproductive behavior of terrace forests, South Fork Hoh River, Olympic National Park. Pages 22–29 *in* E.E. Starkey, J.F. Franklin, and J.W. Mathews, eds. Ecological research in national parks of the Pacific Northwest. National Park Service Cooperative Study Unit, Corvallis, Oregon, USA.

McMahon, T.E., and G.F. Hartman. 1989. Influence of cover complexity and current velocity on winter habitat use by juvenile coho salmon

(*Oncorhynchus kisutch*). Canadian Journal of Fisheries and Aquatic Science **46**:1551–1557.

Megahan, W.F. 1982. Channel sediment storage behind obstructions in forested drainage basins draining the granitic bedrock of the Idaho batholith. Pages 114–121 *in* F.J. Swanson, R.J. Janda, T. Dunne, and D.N. Swanston, eds. Sediment budgets and routing in forested drainage basins. USDA Forest Service General Technical Report PNW-141. Pacific Northwest Forest and Range Experiment Station, Portland, Oregon, USA.

Merrell, T.R. 1951. Stream improvement as conducted in Oregon on the Clatskanie River and tributaries. Fish Commission, Oregon Research Briefs **3**:41–47.

Merritt, R.W., and K.W. Cummins. 1978. An introduction to the aquatic insects of North America. Kendall/Hunt, Dubuque, Iowa, USA.

Montgomery, D.R., and J.M. Buffington. 1993. Channel classification, prediction of channel response, and assessment of channel condition. Draft report to the Sediment, Hydrology, and Mass Wasting Committee of the Washington State Timber/Fish/Wildlife Agreement. Department of Geological Sciences and Quaternary Research Center, University of Washington, Seattle, Washington, USA.

Murphy, M.L., and K.V. Koski. 1989. Input and depletion of woody debris in Alaska streams and implications for streamside management. North American Journal of Fisheries Management **9**:427–436.

Murphy, M.L., K.V. Koski, J. Heifetz, S.W. Johnson, D. Kirchofer, and J.F. Thedinga. 1985. Role of large organic debris as winter habitat for juvenile salmonids in Alaska streams. Proceedings, Western Association of Fish and Wildlife Agencies **1984**:251–262.

Naiman, R.J., T.J. Beechie, L.E. Benda, P.A. Bisson, L.H. MacDonald, M.D. O'Connor, et al. 1992. Fundamental elements of ecologically healthy watersheds in the Pacific Northwest coastal ecoregion. Pages 127–188 *in* R.J. Naiman, ed. Watershed management: Balancing sustainability and environmental change. Springer-Verlag, New York, New York, USA.

Naiman, R.J., and J.R. Sedell. 1979. Benthic organic matter as a function of stream order in Oregon. Archiv für Hydrobiologie **87**:404–422.

Naiman, R.J., and J.R. Sedell. 1980. Relationships between metabolic parameters and stream order in Oregon. Canadian Journal of Fisheries and Aquatic Science **37**:834–847.

Narver, D.W. 1971. Effects of logging debris on fish production. Pages 100–111 *in* J.T. Krygier and J.D. Hall, eds. Proceedings of a Symposium on Forest Land Uses and the Stream Environment. Oregon State University, Corvallis, Oregon, USA.

Orr, P.W. 1963. Windthrown timber survey in the Pacific Northwest, 1962. USDA Forest Service, Pacific Northwest Region, Portland, Oregon, USA.

Peters, G.B., H.J. Dawson, B.F. Hrutfiord, and R.R. Whitney. 1976. Aqueous leachate from western redcedar: Effects on some aquatic organisms. Journal of the Fisheries Research Board (Canada) **33**:2703–2709.

Pollock, M.M. 1995. Patterns of plant diversity in southeast Alaskan wetlands—Can they be explained by biodiversity theory? Ph.D. dissertation. University of Washington, Seattle, Washington, USA.

Ralph, S.C., G.C. Poole, L.L. Conquest, and R.J. Naiman. 1994. Stream channel morphology and woody debris in logged and unlogged basins of western Washington. Canadian Journal of Fisheries and Aquatic Science **51**:37–51.

Reeves, G.H., F.H. Everest, and J.R. Sedell. 1993. Diversity of juvenile anadromous salmonid assemblages in coastal Oregon basins with different levels of timber harvest. Transactions of the American Fisheries Society **122**:309–317.

Richardson, J.S. 1991. Seasonal food limitation of detritivores in a montane stream: An experimental test. Ecology **72**:873–887.

Robison, E.G., and R.L. Beschta. 1990a. Characteristics of coarse woody debris for several coastal streams of southeast Alaska, USA. Canadian Journal of Fisheries and Aquatic Science **47**:1684–1693.

Robison, E.G., and R.L. Beschta. 1990b. Identifying trees in riparian areas that can provide coarse woody debris to streams. Forest Science **36**:790–801.

Rot, B.W. 1995. The interaction of valley constraint, riparian landform, and riparian plant community size and age upon channel configuration of small streams of the western Cascade Mountains, Washington. Master's thesis. University of Washington, Seattle, Washington, USA.

Savory, J.G. 1954. Breakdown of timber by ascomycetes and fungi imperfecti. Annals of Applied Biology **41**:336–347.

Sedell, J.R., and K.J. Luchessa. 1982. Using the historical record as an aid to salmonid habitat enhancement. Pages 210–223 *in* N.B. Armantrout, ed. Acquisition and utilization of aquatic habitat inventory information. Western Division, American Fisheries Society, Bethesda, Maryland, USA.

Sedell, J.R., F.J. Swanson, and S.V. Gregory. 1985. Evaluating fish response to woody debris. *In* T.J. Hassler, ed. Proceedings of the Pacific Northwest stream habitat workshop. Humboldt State University, Arcata, California, USA.

Sedell, J.R., J.E. Yuska, and R.W. Speaker. 1984. Habitats and salmonid distribution in pristine, sediment-rich river valley systems: South Fork Hoh and Queets River, Olympic National Park. Pages 33–46 *in* W.R. Meehan, T.R. Merrell, Jr., and T.A. Hanley, eds. Fish and wildlife relationships in old-growth forests. American Institute of Fishery Research Biologists, Juneau, Alaska, USA.

Shirvell, C.S. 1990. Role of instream rootwads as juvenile coho salmon (*Oncorhynchus kisutch*) and steelhead trout (*O. mykiss*) cover habitat under varying streamflows. Canadian Journal of Fisheries and Aquatic Science **47**:852–861.

Smock, L.A., G.M. Metzler, and J.E. Gladden. 1989. The role of organic debris dams in the structuring and functioning of low-gradient headwater streams. Ecology **70**:764–775.

Steel, E.A. 1993. Woody debris piles: Habitat for birds and small mammals in the riparian zone. Master's thesis. University of Washington, Seattle, Washington, USA.

Steinblums, I., H.A. Froelich, and J.K. Lyons. 1984. Designing stable buffer strips for stream protection. Journal of Forestry **82**:49–52.

Swanson, F.J., L.E. Benda, S.H. Duncan, G.E. Grant, W.F. Megahan, L.M. Reid, et al. 1987. Mass failures and other processes of sediment production in Pacific Northwest forest landscapes. *In* E.O. Salo and T.W. Cundy, eds. Streamside management: Forestry and fishery interactions. Contribution Number 57. Institute of Forest Resources, University of Washington, Seattle, Washington, USA.

Swanson, F.J., S.V. Gregory, J.R. Sedell, and A.G. Campbell. 1982. Land-water interactions: The riparian zone. Pages 267–291 *in* R.L. Edmonds, ed. Analysis of coniferous forest ecosystems in the western United States. Hutchinson Ross, Stroudsburg, Pennsylvania, USA.

Swanson, F.J., and G.W. Lienkaemper. 1978. Physical consequences of large organic debris in Pacific Northwest streams. USDA Forest Service General Technical Report PNW-69. Pacific Northwest Forest and Range Experiment Station, Portland, Oregon, USA.

Swanson, F.J., G.W. Lienkaemper, and J.R. Sedell. 1976. History, physical effects, and management implications of large organic debris in western Oregon streams. USDA Forest Service General Technical Report PNW-56. Pacific Northwest Forest and Range Experiment Station, Portland, Oregon, USA.

Thomas, J.W., M.G. Raphael, R.G. Anthony, E.D. Forsman, A.G. Gunderson, R.S. Holthavsen, et al. 1993. Viability assessments and management considerations for species associated with late-successional and old-growth forests of the Pacific Northwest: The report of the scientific analysis team. USDA Forest Service, Portland, Oregon, USA.

Trotter, E.H. 1990. Woody debris, forest-stream succession, and catchment geomorphology. Journal of the North American Benthological Society **9**:141–156.

Van Sickle, J., and S.V. Gregory. 1990. Modeling inputs of large woody debris to streams from falling trees. Canadian Journal of Forest Research **20**:1593–1601.

Wallace, J.B., J.R. Webster, and J.L. Meyer. 1995. Influence of log additions on physical and biotic characteristics of a mountain stream. Canadian Journal of Fisheries and Aquatic Science **52**:2120–2137.

Ward, B.R., and P.A. Slaney. 1979. Evaluation of in-stream enhancement structures for the production of juvenile steelhead trout and coho salmon in the Keogh River: Progress 1977 and 1978. Fisheries Technical Circular 45. British Columbia Fish and Wildlife Branch Victoria, British Columbia, Canada.

Wendler, H.O., and G. Deschamps. 1955. Logging dams on coastal Washington streams. Fisheries Research Papers **1**:27–38. Washington Department of Fisheries, Olympia, Washington, USA.

White, D.S. 1982. Elmidae. Pages 10.99–10.110 *in* A.R. Brigham, W.U. Brigham, and A. Gnilka, eds. Aquatic insects and oligochaetes of North and South Carolina. Midwest Aquatic Enterprises, Mahomet, Illinois, USA.

Zimmerman, R.C., J.C. Goodlet, and G.H. Comer. 1967. The influence of vegetation on channel forms of small streams. Pages 255–275 *in* Symposium on river morphology. International Association of Scientific Hydrology Publication Number 75.

14

Nutrient Cycles and Respon Disturbance

Michael E. McClain, Robert E. Bilby, and Frank J. Tri

Overview

• This chapter examines the cycling of nitrogen (N), phosphorus (P), and sulfur (S) in stream and river corridors of the Pacific coastal ecoregion under natural and managed conditions.

• In its inorganic form, nitrogen occurs primarily as nitrate (NO_3^-) and ammonium (NH_4^+). Microbially mediated transformations between these species generally move in the direction NH_4^+ to NO_3^- through the process of nitrification. Under anoxic conditions, NO_3^- may be further transformed through microbially mediated denitrification into molecules of nitrogen (N_2) and nitrous oxide (N_2O). Both NO_3^- and NH_4^+ are assimilated by organisms into amino acids and other organic molecules.

• Sulfur occurs as sulfate (SO_4^{2-}) and sulfide (S^{2-}) in its inorganic form, but only SO_4^{2-} is available for assimilation by organisms. As with N, the occurrence of SO_4^{2-} versus S^{2-} depends on redox conditions and microbially mediated reactions, with S^{2-} dominating reduced environments and SO_4^{2-} dominating oxidized environments.

• Phosphorus is unique relative to nitrogen and sulfur in that it has no redox chemistry. It remains in the ($P\text{–}O_4$) state in both organic and inorganic forms. In its dissolved inorganic form, it occurs as orthophosphate (PO_4^{3-}), whereas in its organic form it occurs as $P\text{–}O_4$ bound into larger organic molecules. Phosphorus is also important to organisms in the inorganic form of apatite ($Ca_5[PO_4]_3F$), which forms an essential part of teeth and bones.

• Because these elements are essential nutrients for living organisms, their natural distributions are strongly influenced by biological, as well as chemical and physical processes. Forest plants and microbes are instrumental in the initial uptake of nitrogen from atmospheric sources, and phosphorus and sulfur from geologic sources, and forest to stream transfers are the primary pathways by which these elements are input to river corridors.

• Once in the river corridor, these elements follow a convoluted downstream course, which includes cycling between organic and inorganic forms, chemical species transformations in response to changing redox conditions, sorptive partitioning onto particulate surfaces, movement into and out of streamside soils, and periods of immobilization and storage. The specific course of these reactions and physical trajectories varies mainly as a function of hydrologic regime, temperature, and biological community composition.

• Concentrations and fluxes of nitrogen, phosphorus, and sulfur change in response to changing stream discharge. Fluxes increase with increasing discharge, but concentrations are less predictable and correlations may be positive or negative, depending on the season. Hydrologic exchange between the channel and its hyporheic zone significantly impacts nitrogen, phosphorus, and sulfur cycling, owing to the intensification of biological and sorptive processes in streamside soils and sediments and ary storage of these elements in ater.

perature influences elemental distri- mainly by controlling rates of microbial essing.

• The composition of the biological community strongly controls cycling reactions by regulating nutrient uptake and regeneration rates. Pacific salmon (*Oncorhynchus* spp.) act as a net source of marine derived nutrients to the river system, whereas beaver (*Castor canadensis*) strongly influence cycling reactions by modifying the physical structure and hydrologic characteristics of the corridor.

• Natural and anthropogenic disturbances disrupt nutrient cycles by altering the nature of controlling processes (i.e., hydrologic regime, temperature, and biological community).

• Direct logging of river corridors for silvicultural and agricultural purposes profoundly impacts nutrient cycling, although the severity of the impact depends on the logging technique used. In general, dissolved and particulate nutrient concentrations tend to temporarily increase following logging, with NO_3^- showing the greatest increase among the measured parameters.

• Forest fertilization using urea (NH_3–N) may temporarily increase stream nitrogen concentrations by nearly two orders of magnitude, but such concentration spikes are significantly reduced when unfertilized buffer strips are left along stream channels.

• Urbanization also tends to increase nutrient inputs to river corridors while reducing the capacity of the riverine biological community to uptake excess nutrients.

• Fire strongly disrupts nutrient cycles in river corridors, but the net effects on nutrient budgets are difficult to predict. Large amounts of nitrogen may be volatilized as ammonia (NH_3) during the fire, but rainfall on ash tends to flush increased concentrations into the soil and stream systems.

• Climate change is now an acknowledged phenomenon which is predicted to intensify during the next century. Its impacts on riverine nutrient cycles will come mainly through altered hydrologic regimes, temperatures, and biological communities. Among the few specific predictions are decreased food quality of leaf litter because of decreased leaf nitrogen content and increased soil organic matter decomposition rates with concomitant decreases in stream inputs.

Introduction

Rivers are fundamental components of regional and global biogeochemical cycles, acting as both transport pathways and sites of elemental transformations. Coastal rivers of the Pacific Northwest form a particularly dynamic link between the region's highly productive temperate forests and nearby marine ecosystems. Nutrients derived from forests are accumulated in rivers and transferred downstream to estuarine systems. This transfer, however, is highly dynamic as transported nutrients participate in a variety of chemical and biological interactions. These interactions are dominated by sequential processes of biological uptake, remineralization, and microbially mediated redox transformations. An atom of a nutrient species may pass through the sequence of uptake and mineralization many times during its downstream journey, repeatedly cycling between organic and inorganic molecular species.

Three elements of key importance to the structural and physiological requirements of the biota of the river corridor are nitrogen (N), phosphorus (P), and sulfur (S). Under natural conditions, riverine cycles of nitrogen, phosphorus, and sulfur are intricately woven into the overall ecological balance of coastal forest ecosystems. Together, climate and geology produce the soils and moisture regimes that support the forests of the region. The forests, in turn, regulate the initial uptake of nitrogen, phosphorus, and sulfur from atmospheric and geologic sources. These elements are then cycled repeatedly through the forest's trophic system, the components of which are adapted to conserve and retain nutrients. Biological processes exert a primary control on the natural distributions of nitrogen, phosphorus, and sulfur in the recycling nutrient pool, but other abiotic processes such as adsorption to the surfaces of soil minerals also serve to retain nutrients.

The efficiency of nutrient retention in the forest ecosystem is not 100 percent, however, and a small quantity of nitrogen, phosphorus, and sulfur is continually input to streams and rivers. The bulk of these inputs occur through hydrologic flowpaths such as groundwater baseflow, storm flow, and canopy throughfall; biological pathways such as direct litterfall and lateral movement from the adjacent forest floor; and inorganic material flowpaths such as landslides and bank erosion.

A natural disturbance regime is part of healthy forest and river ecosystems (Naiman et al. 1992). Disturbances such as landslides and wildfires promote a more resilient and productive system by introducing heterogeneity. The key to maintaining disturbance as a healthy rather than detrimental influence is to separate events in space and time so as not to overwhelm an ecosystem's natural capacity to recover. In the heavily managed watersheds of the coastal ecoregion, disturbance has become more frequent and intense. Where watersheds have become urbanized, unnatural pressures are more or less continuous. Scientists and managers alike are faced with the challenge of understanding the beneficial aspects of disturbance on nutrient cycles and developing methods of assessing and minimizing the detrimental effects.

This chapter explores the cycling of nitrogen, phosphorus, and sulfur in rivers of the Pacific coastal ecoregion at several levels. It begins by describing the forms and concentrations of nitrogen, phosphorus, and sulfur encountered in riverine systems, the fundamental reactions involved in their cycles, and the principal input and output pathways determining their budgets. Next, it examines the principal biological and environmental variables influencing the direction and rate of cycling processes. Finally, some attention is devoted to the effects of natural and anthropogenic disturbances in temperate forest watersheds, where disturbance amounts to a sudden change in one or more of the variables controlling the course and rate of cycling. Responses of nutrient cycling in Pacific coastal watersheds to disturbances such as deforestation for silvicultural and agricultural purposes, urbanization, and fire are considered. The potential responses of these cycles to changes in global climate are also considered. The goal is to clarify the relationships between disturbances and underlying cycling processes, thereby conveying a more process-based understanding of the consequences of watershed management decisions.

The Basics of Nitrogen, Phosphorus, and Sulfur Cycling

Webster's Seventh New Collegiate Dictionary defines cycling as "a course or series of events or phenomena that recur regularly and usually lead back to the starting point" (Figure 14.1a). *Biogeochemical cycling*, as used in ecosystem sciences, generally refers more broadly to a whole set of processes involving the element of interest, including physical transport and storage, changes in form (organic vs inorganic, solute vs particulate), and chemical transformations in elemental speciation (oxidation state, complexation) and association (adsorption). Cycling in the literal sense does occur, however, if the "starting point" is considered a chemical form rather than physical point.

Nutrient Spiraling

In riverine ecosystems, transport is clearly an important component of all biogeochemical cycles, but because of the unidirectional nature of river flow, the completion of a cycle does not return an atom to its physical starting point. Instead, the atom is displaced some distance downstream, thereby stretching the visualized circular cycle into the shape of a spiral (Figure 14.1b). The concept of nutrient spiraling was developed in the late 1970s and early 1980s (Webster and Patten 1979, Newbold et al. 1982, Elwood et al. 1983) and remains a useful framework in which to examine the cycling of nutrients in rivers. According to the concept, one idealized *cycle* is completed when a nutrient atom has, in sequence, been taken up by an organism from a dissolved available state, passed through the food chain, and returned to a dissolved available state for reutilization

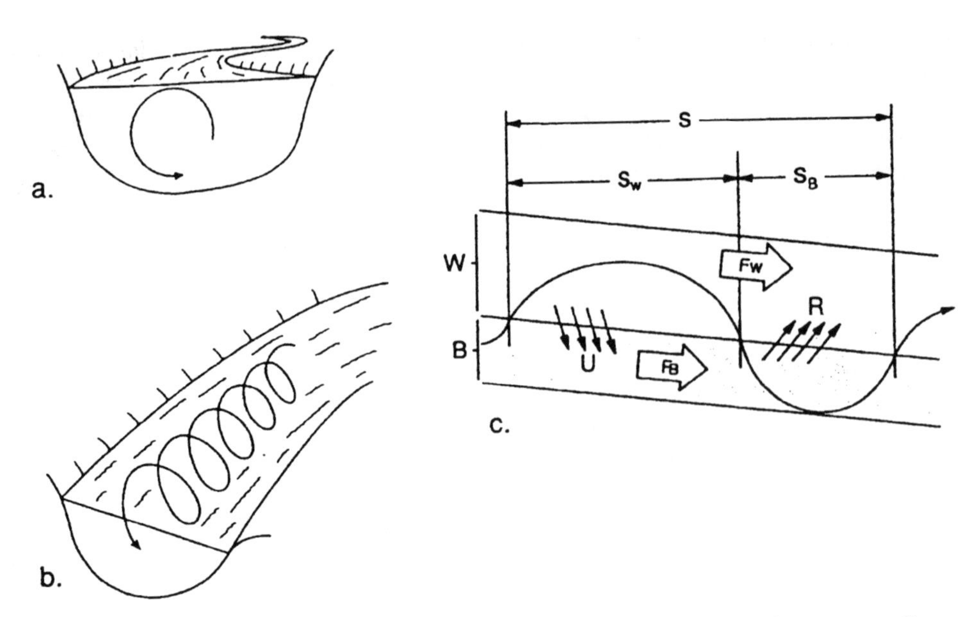

FIGURE 14.1. Schematic representations of nutrient spiraling. (a) Simple cycling with no consideration of transport. (b) Combined cycling and transport, with resulting spiral form. (c) Representation of spiraling between inorganic (dissolved in water column, W) and organic (bound in organisms or organic detritus, B) compartments. *Spiraling length* (S) is the total downstream distance over which a cycle is completed; S_W and S_B are the downstream distances over which nutrients are transported in the inorganic and organic forms, respectively. F_W and F_B represent the downstream fluxes in each compartment per unit width of channel, and U and R represent areal rates of nutrient uptake and regeneration (from Newbold 1992. Reprinted by permission of Blackwell Science, Inc.)

(Newbold et al. 1982). Two particularly important qualities of the concept are that it explicitly considers the longitudinal connectedness of river corridors, and it provides a simple quantitative framework in which to evaluate system variables such as productivity, remineralization, and nutrient limitation.

Spiraling length (S), the longitudinal distance associated with one cycle, is the central term in the concept. For any nutrient (n), S_n is represented by the basic equation

$$S_n = Vt_c,$$

where V is the average downstream velocity of the nutrient atom and t_c is the average time required for the nutrient atom to complete one cycle. The uptake (U) and regeneration (R) of nutrients by the biotic community of the stream are critical terms in determining spiraling lengths (Figure 14.1c). Under ideal conditions, when $U = R$, S may be expressed by as a simple function of a nutrient's biotic uptake rate (U, in mass/area/time) and its total downstream flux per unit width (F_t, in mass/length/time) in the inorganic and organic forms:

$$S = \dagger \frac{F_t}{U} \tag{14.1}$$

For a more detailed quantitative development of the concept and a discussion of its applicability to questions of nutrient limitation and nutrient retention, the reader is referred to Newbold (1992). The general concept of nutrient spiraling is also useful because it can be effectively incorporated into larger riverine ecosystem theories such as the river continuum concept (Vannote et al. 1980), serial discontinuity concept (Ward and Stanford 1983), patch dynamics concept (Pringle et al. 1988), flood pulse concept (Junk et al. 1989), and riverine productivity model (Thorp and Delong 1994).

Natural Forms, Distributions, and Transformations

Nitrogen, phosphorus, and sulfur occur in a variety of chemical forms which alternate as cycling reactions proceed (Table 14.1; Figure 14.2). In keeping with the idealized *cycle* presented in the previous paragraph, the first step in the cycle is biological uptake (*assimilation*) of these elements in their dissolved available states. In the cases of nitrogen and sulfur, assimilation is almost always accompanied by a decrease in the oxidation state of the element. The main exception is the assimilation of ammonium (NH_4^+), which involves no redox reactions. Phosphorus occurs in only one oxidation state (+5), so redox reactions do not directly influence its cycling.

Nitrogen is available to plants primarily in the form of nitrate (NO_3^-) and ammonium (NH_4^+) (Figure 14.2a). Certain plants that contain symbiotic nitrogen-fixing bacteria (*Rhizobium*) in their roots are able to utilize gaseous nitrogen (N_2) directly. Red alder (*Alnus rubra*) is one such plant in the Northwest that is particularly important. Nitrogen-fixing bacteria also have been identified in association with decaying wood in streams (Buckley and Triska 1978). Triska et al. (1989a) measured the uptake of NO_3^- injected into the stream channel of Little Lost Man Creek in northern California. Over the course of their 10-day injection experiment, approximately 19% of injected NO_3^- was taken up by the stream's biota, which amounted to an average uptake rate of 26 mg/m^2/d. Uptake rates were greatest during daylight hours but continued through the night.

Phosphorus is available to plants in the form of orthophosphate (PO_4^{3-}), which is also the only inorganic form in which phosphorus occurs in appreciable amounts (Figure 14.2b). Gregory (1978) examined isotopically labeled (^{32}P) PO_4^{3-} uptake by primary producers in Mack Creek, Oregon, and found epilithic algae to exhibit higher PO_4^{3-} uptake *rates* than either filamentous algae or riparian vegetation. He noted, however, that because of its larger standing crop, riparian vegetation may assimilate larger amounts of PO_4^{3-}.

Sulfur is available in the form of sulfate (SO_4^{2-}), which is the predominant dissolved inorganic form of sulfur in oxygenated river waters (Figure 14.2c). Rates of SO_4^{2-} uptake have not been measured in Pacific coastal rivers. Animals, of course, can only acquire these elements by ingesting them in an already organic form. This may occur several times as the organically bound element makes its way through the food chain.

Once incorporated by plants and animals, each of these elements serves vital metabolic and structural functions. Nitrogen is a component of all amino acids and therefore proteins. In this form it contributes to many functions, including structural support, movement, and defense against foreign substances. As a component of enzymes it catalyzes chemical reactions in cells. Nitrogen also occurs in alkaloids and urea $CO(NH_2)_2$. Organic forms of phosphorus include such important molecules as nucleic acids (RNA and DNA), adenosine triphosphate (ATP), and phospholipids (components of cell membranes). As a component of bones and teeth, phosphorus occurs with calcium (Ca) as the mineral apatite

TABLE 14.1. Species of nitrogen (N), phosphorus (P), and sulfur (S) encountered in river corridors.

Aqueous	Gas	Mineral	Organic
NO_2^- (nitrite), NO_3^- (nitrate)	N_2, (nitrogen), N_2O (nitrous oxide)	—	$CO{-}HN_2$ (amines)
NH_4^+ (ammonium)	NO (nitrogen oxide), NH_3 (ammonia)	—	$CH_2{-}NH_2$ (amides)
PO_4^{3-} (phosphate)	—	$Ca_5(PO_4)_3F$ (apatite)	$P{-}O_4$
SO_4^{2-} (sulfate),	H_2S (hydrogen sulfide)	FeS (pyrite)	$CH_2{-}SH$ (thiol)
HS^- (hydrogen sulfide)			
S^{2-} (sulfide)	—	—	$CH_3{-}S{-}CH_3$ (sulfide)
			–SS– (disulfide)

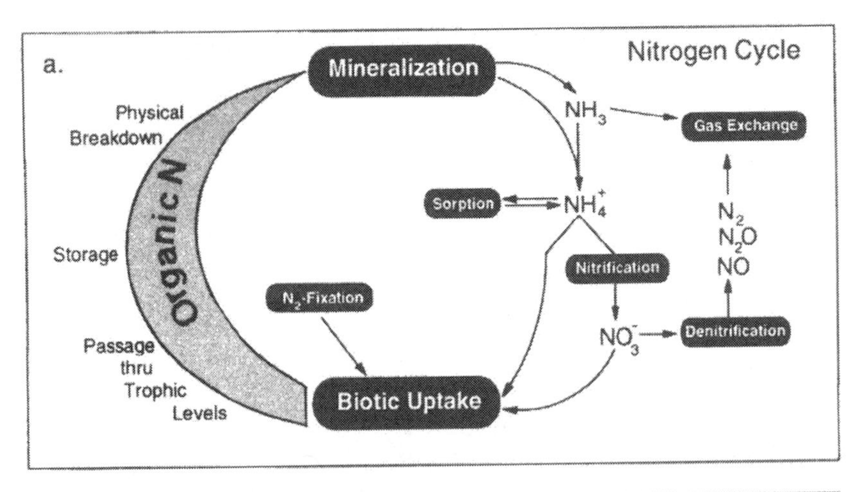

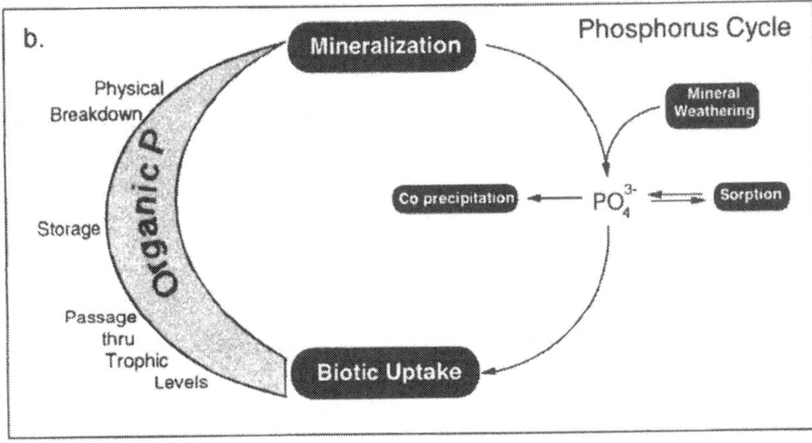

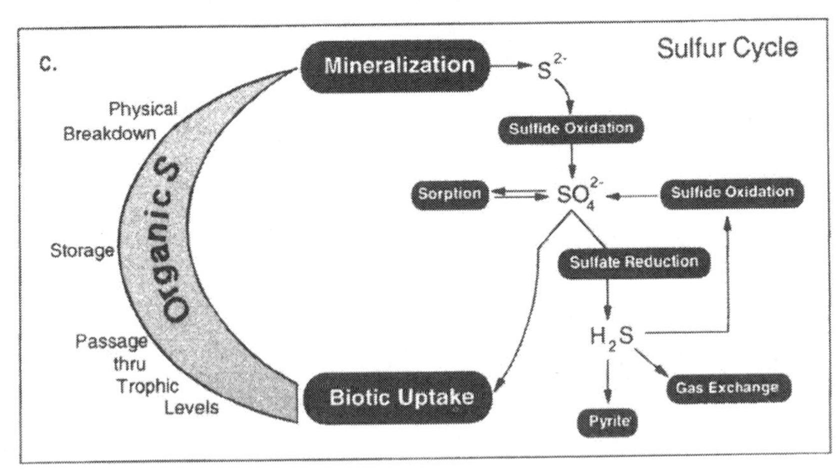

FIGURE 14.2. (a) nitrogen, (b) phosphorus, and (c) sulfur cycling. Element species are shown along with processes determining their concentrations and compositions.

($Ca_5[PO_4]_3F$). Sulfur is less abundant than nitrogen and phosphorus in living organisms, but nevertheless it serves essential functions as a component of two amino acids (cysteine and methionine) and several other less abundant molecules. Occurrence in living tissue represents the intersection point in the cycles of nitrogen, phosphorus, and sulfur. In the organic form, these elements co-occur in more or less constant proportions, roughly 35:2:1 in plants (Campbell 1990).

At the time of death, decomposition and *mineralization* reactions combine to transform nitrogen, phosphorus, and sulfur back into inorganic forms. These reactions rely on the action of heterotrophic bacteria and fungi who utilize reduced carbon as an energy source and liberate organically bound nitrogen, phosphorus, and sulfur in the process. Nitrogen emerges from these reactions in the form of ammonia (NH_3) and NH_4^+ (Figure 14.2a). Although some of the NH_3 may volatilize (Freney et al. 1983), in the acidic waters common to the Pacific coastal ecoregion, most rapidly reacts with free protons (H^+) to produce NH_4^+. At this point, remineralized nitrogen is available for reutilization by organisms and the idealized *cycle* is effectively complete.

Various other energy-producing reactions may occur prior to nitrogen reassimilation. In anaerobic waters, NH_4^+ remains the stable form of dissolved inorganic nitrogen, but in aerobic river waters its concentrations are generally low. In Pacific coastal rivers and streams, natural concentrations of NH_4^+ are generally less than 1 mM, but in oxygen depleted hyporheic waters adjacent to the stream concentrations may exceed 20 mM. Another factor contributing to low ambient concentrations of NH_4^+ is its adsorption to cation-exchange sites on streambed sediments and adjacent hyporheic soils. At Little Lost Man Creek, California, Triska et al. (1994) reported concentrations of exchangeable NH_4^+ ranging from 10 meq/100g sediment in the stream channel to 115 meq/100g soil in the hyporheic zone 18 m inland of the channel.

Under aerobic conditions, NH_4^+ is oxidized to NO_3^- in a two-step *nitrification* process which may be represented as follows:

$$NH_4^+ + \tfrac{3}{2}O_2 \Rightarrow NO_2^-(\text{nitrite}) + H_2O + 2H^+ \quad (14.2)$$

$$NO_2^- + \tfrac{1}{2}O_2 \Rightarrow NO_3^- \quad (14.3)$$

These reactions are carried out by autotrophic bacteria of the genera *Nitrosomonas* and *Nitrobacter*, respectively. There are several possible intermediate compounds, the most important of which is nitrous oxide (N_2O), a greenhouse gas that may be lost from the ecosystem by gas exchange. Nitrate is relatively unreactive with organic or mineral surfaces. Thus, upon entering aqueous environments it moves freely downstream until taken up once more by the biota or transformed through redox reactions. In Pacific coastal rivers, natural concentrations of NO_3^- range from near 0 to 30 mM.

If NO_3^- is transported into an anoxic zone, such as generally occurs in streambed sediments, the hyporheic zone, or some micro-zone within a particle of degrading organic matter, it may be reduced by *denitrification* to N_2. The gases NO (nitrogen oxide) and N_2O are also byproducts of denitrification. These gases are generally lost from the ecosystem, thereby balancing the N_2 input via nitrogen-fixation (Jaffe 1992). The hyporheic zone is a particularly important site of denitrification reactions (Triska et al. 1993). In the hyporheic zone of Little Lost Man Creek, Duff and Triska (1990) reported increasing rates of denitrification with increasing distance from the stream channel and decreasing amounts of dissolved oxygen.

Phosphorus released from decomposing organic matter reenters the aquatic ecosystem as PO_4^{3-} (Figure 14.2b). Natural concentrations of PO_4^{3-} in Pacific coastal streams and rivers are generally less than 1 mM (see Chapter 4). At neutral or acidic pH, PO_4^{3-} is generally bonded to one or two hydrogen (H) atoms in the forms of phosphoric acids, HPO_4^{2-} and $H_2PO_4^{2-}$. Like the first NH_4^+ to emerge from the organic form, this PO_4^{3-} is immediately available for reutilization and the idealized cycle is complete. However, again several reactions may occur before reassimilation. Concentrations remain low in natural riverine systems due to strong *adsorptive* reactions with iron (FeO_x) and aluminum oxides (AlO_x) and clay minerals. These adsorption reactions, along with *coprecipitation* reactions with Fe III, Ca, and Al, make PO_4^{3-} biologically unavailable to organisms. Because of adsorption, coprecipitation, and biotic assimilation, total phosphorus concentrations in river corridors are generally dominated by particulate forms of the element. This is especially true where phosphate minerals such as apatite occur.

Sulfur liberated from decomposing organic matter reenters the inorganic form as sulfide (S^{2-}), but is quickly oxidized to SO_4^{2-} (*sulfide oxidation*) and is again available for reutilization (Figure 14.2c). The behavior of SO_4^{2-} at this stage is similar to that of NO_3^-, in that it too is rather inert and moves more or less freely with the flowing water, although not as freely as NO_3^-. Sulfate has not been analyzed in many Pacific coastal rivers but, where data are available, concentrations range from 10 to 30 mM (see Chapter 4). When SO_4^{2-} is transported into an anaerobic zone, it is reduced (*sulfate reduction*) by the action of heterotrophic bacteria to hydrogen sulfide (H_2S). The H_2S produced may then react with Fe^{2+} to form the mineral pyrite (FeS), or it may be transported back into an aerobic zone where it reacts spontaneously with oxygen and is oxidized back to SO_4^{2-} (Chen and Morris 1972).

As breakdown and mineralization reactions proceed, a fraction of the organically bound nitrogen, phosphorus, and sulfur persists in the refractory residual pool of particulate organic matter (POM). Coarse POM (>63 μm) is often identifiable as fragments of plant litter, but fine POM is composed largely of molecularly unidentifiable and refractory organic matter termed *humin* (Hatcher and Spiker 1988). Fine POM remineralizes over long periods (months to years) and is subject to physical processes of transport and storage, as well as abiotic sorptive reactions with other organic matter or mineral surfaces.

The great variety of transformations involving nitrogen, phosphorus, and sulfur adds considerable complexity to the idealized representation of cycling within the nutrient spiraling concept (Figure 14.2a–c). Moreover, rivers are open systems, where nitrogen, phosphorus, and sulfur are continually exchanged with adjoining terrestrial and atmospheric systems and output to the ocean. These input and output fluxes are examined more closely in the following section.

Input-Output Pathways and Riverine Budgets

Nitrogen, phosphorus, and sulfur enter riverine ecosystems via several hydrological, geological, and biological flowpaths (Figure 14.3). Fluxes along these flowpaths are reported as mass per area of river surface per unit time (e.g., kg/m^2/s) or mass per distance of river length per time (kg/m/s). Total fluxes to the river are calculated by multiplying these values by the total area or length of the river, respectively. Similarly, exports from the river may be reported in these units when considering lateral flowpaths like gas exchange, but exports associated with river discharge are reported simply as mass per time (e.g., kg/s). Groundwater baseflow is a continuous hydrologic flowpath, while stormflow and canopy throughfall are episodic and strongly seasonal. Geologic flowpaths, such as erosion, also are episodic and largely associated with seasonal precipitation patterns. The biological pathways of direct litterfall from the overhanging forest canopy and lateral movement from the adjacent forest floor are seasonally variable but continuous throughout the year. All input flowpaths are active along the entire length of river systems Their relative importance, however, may vary in response to changing geomorphology, riparian plant communities, and channel size. Output pathways are less numerous than input pathways. Clearly the dominant pathway is river discharge to estuaries, but nitrogen and sulfur are also lost along the river's length via gas exchange (N_2, N_2O, NO, H_2S) with the atmosphere (Figure 14.3).

A technique which has proven particularly informative in investigating ecosystem nutrient cycles is the construction of input–output budgets. Budgets are calculated by subtracting the sum of output fluxes from the sum of input fluxes. Differences in the fluxes reflect net ecosystem processes such as nutrient accumulation

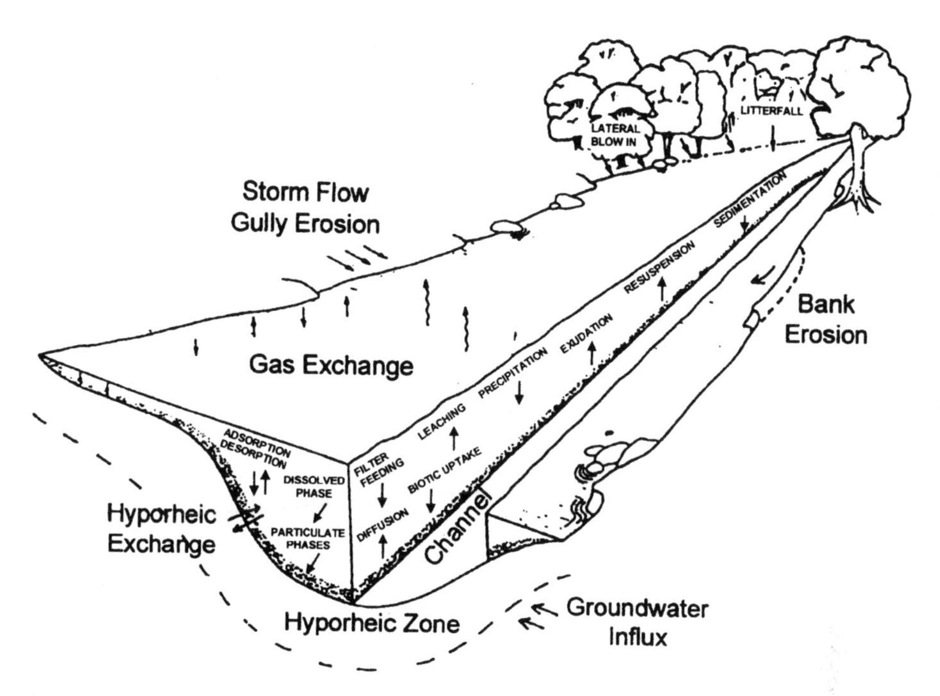

FIGURE 14.3. Physical, chemical, and biological pathways and processes impacting nitrogen (N), phosphorus (P), and sulfur (S) cycling in Pacific coastal streams and rivers (modified from Meyer et al. 1988 with permission).

or depletion. Nutrient budgets are most often constructed at the scale of the entire watershed, where precipitation represents the dominant input and stream and groundwater discharges the dominant outputs (Likens et al. 1977). Budgets such as this have been constructed for nitrogen, phosphorus, sulfur and other elements in several Pacific coastal watersheds (Fredriksen et al. 1975, Scrivener 1975, Feller and Kimmins 1979, Larson 1979).

Unfortunately, budgets for the streams themselves are rarely determined. One exception is the work of Triska et al. (1984), who constructed such a budget for nitrogen in a small stream in the western Cascades of Oregon (H.J. Andrews Experimental Forest, Watershed 10). By considering input–output flux pathways and nitrogen forms individually, these researchers succeeded in developing a detailed stream budget which also shed light on internal ecosystem processes. For the two years of data they reported, 74% of nitrogen input to the stream was in the dissolved form through groundwater baseflow and canopy throughfall, 21% of input occurred through litterfall and lateral movement, and the remaining 5% of input occurred through N_2 fixation within the stream channel (Table 14.2). Overall, more than 90% of the nitrogen input to the stream was organic and derived from biotic sources in the surrounding forest. Inputs of dissolved inorganic nitrogen (NO_3^-, NO_2^-, NH_4^+) accounted for less than 5% of nitrogen inputs. Total nitrogen inputs exceeded total outputs by 34%, indicating significant retention of nitrogen within the stream ecosystem during the period of the study. The greatest degree of retention and processing was in the pool of nitrogen held in leaf and needle litter. Only 13% of nitrogen input as leaf and needle litter exited the watershed in a recognizable form, indicating that 87% was either mineralized, leached into dissolved form, or broken down into unrecognizable fine POM. Of the nitrogen input as wood, only 40% was output from the reach. The remaining 60% was retained in the stream, mostly in storage, but some fraction also contributed to the growing pool of fine POM. Fine POM is more easily transported in the stream system, and although it is more refractory, it

TABLE 14.2. Nitrogen budget for Watershed 10 at the H.J. Andrews Experimental Forest.

Pathway	Form	g/m^2	% Total
Nitrogen inputs			
Groundwater baseflow	NO_3^-–N (nitrate)	0.50	3
	DON (dissolved organic nitrogen)	10.56	69
Throughfall	Organic	0.30	2
Litterfall	Organic	1.35	9
Lateral movement	Mixed	1.78	12
In-stream N_2-fixation	Organic	0.76	5
	Total	15.25	100
Nitrogen outputs			
Stream discharge	NO_3^-–N (nitrate)	0.43	4
	DON (dissolved organic nitrogen)	8.38	74
	LPON (large particulate organic nitrogen)	0.87	7
	FPON (fine particulate organic nitrogen)	1.66	15
Denitrification	N_2 (nitrogen), N_2O (nitrus oxide)	?	?
	Total	11.36	100

From Triska et al. 1984.

constitutes an important input of nitrogen and other nutrients to downstream reaches. Magnitudes of nitrogen loss via gas exchange were not quantified, but they were assumed to be minimal (Triska et al. 1984).

Although comprehensive stream budgets have not been constructed for phosphorus and sulfur in the Pacific coastal ecoregion, within the organic forms, phosphorus and sulfur budgets are expected to be similar to that of nitrogen. Organic forms input via litterfall and lateral movement are largely retained in the stream channel. It may be, however, that fluxes of inorganic phosphorus and sulfur are proportionally more important than those for N. This prediction stems from two important points; phosphorus and sulfur occur in organic matter in smaller concentrations than nitrogen, and, whereas the ultimate source of nitrogen is the atmosphere and all sources to the stream depend on plant-derived nitrogen, both phosphorus and sulfur have geological sources, which release these elements through mineral weathering.

Table 14.3 summarizes the forms of nitrogen, phosphorus, and sulfur encountered in riverine

TABLE 14.3. Summary of reactions impacting nitrogen (N), phosphorus (P), and sulfur (S) cycling.

Process/reaction	Reactant(s)	Product(s)	Mediator
Assimilation	NO_3^- (nitrate), NH_4^+ (ammonium) PO_4^{3-} (phosphate), SO_4^{2-} (sulfate)	Organic N, P, S	Plants, fungi, bacteria
Mineralization	Organic N, P, S	NH_3 (ammonia), NH_4^+, PO_4^{3-}.S^{2-} (sulfide)	Bacteria, fungi
Nitrification	NH_4^+	NO_2^- (nitrite), NO_3^-	Bacteria
Denitrification	NO_3^-	N_2, N_2O (nitrous oxide), NO (nitrogen oxide)	Bacteria
Sulfate reduction	SO_4^{2-}	H_2S (hydrogen sulfide)	Bacteria
Sulfide oxidation	S^{2-}	SO_4^{2-}	Bacteria
Adsorption	NH_4^+, PO_4^{3-} Organic N, P, S	Sorbed forms	Abiotic
Coprecipitation	PO_4^{3-}, S^{2-}	Amorphous Fe (iron) and A1 (aluminum) oxides, FeS (iron sulfide)	Abiotic

systems, the fundamental reactions involved in their biogeochemical cycles, and the principal input and output pathways determining their budgets. The environmental variables determining the course and rate of these cycles are treated next.

Controlling Variables in Nitrogen, Phosphorus, and Sulfur Cycling

The natural distributions and cycling of nitrogen, phosphorus, and sulfur in riverine ecosystems are determined by the balance of inputs, internal processing, and outputs. These factors, in turn, respond to an interrelated suite of environmental variables including the hydrologic regime, temperature (air and water), and the composition and activity of the biological community. In the Pacific coastal ecoregion, relatively strong gradients in these variables occur with changing altitude, latitude, and season.

Hydrologic Regime

Hydrologic regimes permeate every facet of nitrogen, phosphorus, and sulfur cycling, because of the primary role of water as a transport medium and solvent. Water also reduces gas diffusion rates, thereby promoting anoxia in organic-rich bed sediments or hyporheic soils where oxygen consumption exceeds oxygen influx rates. While biological processes mainly vary seasonally and diurnally, hydrologic regimes vary on the scale of individual storm events. Sudden increases in river discharge (freshets) wash in large amounts of organic and inorganic debris, flush out areas of stagnant surface water and groundwater, and generally stir up the system. The flushing action of freshets may even aerate previously anoxic zones. Precipitation feeding the increased discharge washes the tree canopy and transfers the leached material to the river by direct throughfall or overland flow pathways. During the course of a freshet, concentrations of suspended matter in the river swell and both dissolved and particulate loads increase (Bilby and Bisson 1992). Less dramatic seasonal patterns in hydrologic regime also occur. Increased baseflow during the winter rainy season produces wetlands and wider channels. The degree of seasonality in discharge and the intensity of freshets varies geographically with latitude. Thus, multi-temporal variations in hydrologic regime are superimposed on geographic variations, producing complex patterns of variability (see Chapter 3).

It is not easy to predict how dissolved inorganic nitrogen, phosphorus, and sulfur concentrations will respond to increasing discharge. In order for concentrations to rise, storm waters must dissolve more nutrients (per liter) than baseflow waters. This is the case during autumn freshets when new rains dissolve nutrients that have accumulated in the forest canopy, forest floor, and upper soil horizons during the preceding dry season. Later in the rainy season, however, after the forest has been repeatedly washed by winter rains, the discharge relationship breaks down or becomes negative (representing dilution by storm water).

Clear correlations between discharge and nitrogen, phosphorus, and sulfur concentrations in undisturbed Pacific coastal rivers are rare. Positive correlations between discharge and concentrations of NO_3^- (Scrivener, 1975), SO_4^{2-} (Scrivener 1975, Feller and Kimmins 1979), and PO_4^{3-} (Fredriksen 1975, Gall 1986) have been reported, but they are not consistent in time or between studies. Scrivener (1975) reported positive correlations between NO_3^- and discharge only during autumn freshets; later in the autumn and winter the relationship became negative (Figure 14.4). Similarly, both Scrivener (1975) and Feller and Kimmins (1979) reported positive correlations between discharge and SO_4^{2-} only on select occasions; otherwise correlations were unclear or negative. No data are available concerning correlations between discharge and dissolved organic forms of these elements.

Regardless of the significance of concentration-discharge relationships, overall fluxes of dissolved and particulate inorganic and organic

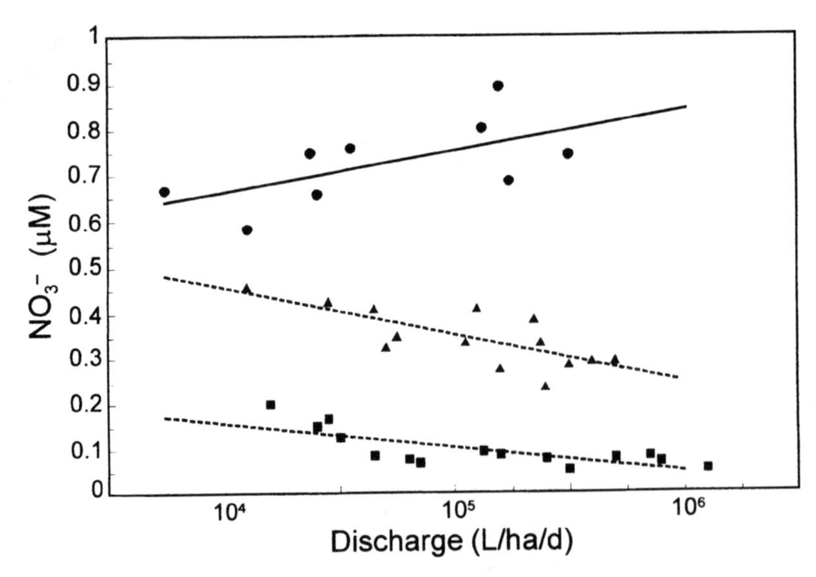

FIGURE 14.4. Nitrate (NO_3^-) concentrations as a function of discharge during three freshets at Carnation Creek, British Columbia, Canada (●; October 12–14, 1973, ▲; October 18–20, 1973, and ■; February 5–14, 1974). Trends illustrate the positive correlation between [NO_3^-] and discharge at the beginning of the rainy season and the subsequent negative correlation. The positive correlation is thought to result from the flushing of accumulated NO_3^- from plants and soils, whereas the later negative trend results from dilution of baseflow stream water by low-NO_3^- storm runoff (modified from Scrivener 1982).

nitrogen, phosphorus, and sulfur tend to increase with increasing flow. High flows create higher energy environments in rivers which suspend and transport previously deposited sediment and organic particulate material (Chapter 2). These materials, along with their associated nutrients, are carried downstream until energy levels decrease and they are again deposited. When plant litter and soils eroded from the adjacent hillslopes are added to this in-stream derived load, increases in downstream fluxes of particulate matter may become substantial.

Hydrological characteristics determine the dimensions of the hyporheic zone, the area beneath and adjacent to the channel where stream water mixes with upland groundwater. Many chemical and biological processes which play a key role in controlling the nutrient dynamics of stream systems occur in the hyporheic zone (Chapter 16). The size of the zone within which these processes may operate depends on the interaction between exchange flow out of the river channel and groundwater inflow from uplands. Upland groundwater influxes are controlled by upland water table gradients and the permeability of upland soils. Similarly, exchange between a river and its hyporheic zone is strongly dependent upon channel hydraulics (i.e., pressure gradients) and the grain size of channel bed sediments. In general, irregularities in stream channel shape and coarse bed sediments promote exchange. During tracer injection experiments at Little Lost Man Creek in northern California, Triska et al. (1989b) found hyporheic water 10 m from the stream channel to contain 47 to 76% stream water, with the remaining water derived from upland groundwater inputs. Triska et al. (1989b) estimated nominal travel times ranging from 5 to 19 days for stream derived water to reach the 10-m distance. This illustrates the role of the hyporheic zone as a transient storage point for stream water during downstream transport, but from a nutrient standpoint, the hyporheic zone may very well be a final sink for NO_3^- and SO_4^{2-} which may be reduced to N_2 and H_2S, respectively, and released to the atmosphere. Slow water movement through the hyporheic zone, coupled with biotic uptake by riparian vegetation (Gregory 1978), greatly reduces nutrient spiraling lengths relative to those exhibited by atoms of the same nutrient in the open water of the channel.

Temperature

Temperature is a fundamental variable in many of the processes involved in nutrient cycling. It varies daily and seasonally as well as along latitudinal and altitudinal geographic gradients. Naiman and Anderson (1996) reported mean annual water temperatures which decreased from 15°C in central California to about 3°C in south-central Alaska. Above 48°N latitude, mean annual minimum temperatures drop to below 0°C, while at the southern extreme of the ecoregion mean annual maximum temperatures exceed 25°C. The diurnal range in temperatures decreases with increasing latitude.

Changing temperatures impact the biotic and abiotic processes that govern nutrient movement through riverine systems. As temperatures increase, so does microbial activity. This leads to higher rates of plant litter processing and higher rates of redox reactions. These reactions increase rates of both nutrient mineralization and uptake, depending upon the reaction considered. Plant metabolism is also accelerated by increasing temperatures, thereby increasing nutrient uptake. The metabolic rate of invertebrates, fishes and other cold-blooded stream animals increases with temperature, necessitating increased rates of food intake, and thus nutrient assimilation, and increased elimination of waste products. In cold-water streams (≤1–2°C), an interesting condition may develop where bacterial activity is shut down but algal production is not (J. Stockner, personal communication).

Biological Community Composition

The biological community occupying the river corridor plays a key role in nearly every aspect of nitrogen, phosphorus, and sulfur cycling. It comprises not only channel communities and processes, but also communities and processes in riparian zones, floodplains, and even the river's estuary. Plant litter from riparian and aquatic plants are major sources of nutrients to the river. Plants also remove and temporarily store an appreciable mass of these nutrients through uptake from channel and hyporheic waters (Chapter 12). Microbial communities mediate chemical transformations of nitrogen, phosphorus, and sulfur including dissimilatory-transformations and redox reactions as well as assimilatory uptake and biosynthesis (Meyer 1994). Riverine invertebrate and fish communities ingest various forms of organic matter and associated nutrients (Chapters 8, 9, and 15). These organisms subsequently release nutrients in waste products and as they decompose after death.

Susceptibility of litter to microbial decomposition and consumption by invertebrates varies by plant species, with important consequences for the availability of elements to higher trophic levels (Chapter 6). Input vectors for plant nutrients include direct litterfall (primarily autumn) as well as more stochastically driven lateral transport, lateral movement, and flooding of adjacent riparian zones. Herbaceous vegetation provides the most labile nutrient source to channel communities due to generally higher content of nitrogen, phosphorus, and sulfur, and lower content of metabolically refractory structural tissues such as cellulose and lignin. The degree of direct nutrient availability for biosynthesis is termed "quality." Litter quality generally proceeds from annual herbaceous to deciduous to coniferous to woody debris. A natural, intact, riparian corridor provides a variety of litter types, which decompose at various rates and are introduced to the channel by a variety of seasonal and stochastic vectors. As a result, a significant mass of nutrients is almost always available in various states of decay to the channel community.

Nutrients such as nitrogen, assimilated by microbial decomposers, can actually increase the nutrient capital of decomposing litter as refractory structural components such as cellulose and lignin are partially digested. The result is a lowering of the carbon to nitrogen (C:N) ratio in decomposing litter which improves food quality for litter consuming invertebrates known as shredders. This improvement in food quality is termed *conditioning*. The conditioning process is especially important in Pacific coastal streams where much of the litter is refractory. Anderson and Grafius (1975) demonstrated the importance of conditioning time on decay and consumption of red alder litter and Sedell et al. (1975) reported similar findings on needle litter in Oregon streams. Triska and Buckley (1978) reported increases in nitrogen capital during decomposition of Douglas-fir (*Psuedotsuga menziesii*) litter from six Oregon coastal and Cascade streams. They also reported that invertebrate biomass associated with litter packs tended to be greatest on litter with the lowest C:N ratio.

Nutrient uptake from channel water supports primary production (photosynthesis) by both benthic (periphyton) and planktonic algae (Chapter 7). In oligotrophic waters primary production can result in large diel shifts in nutrient concentration under low-flow conditions. The magnitude of diel shifts depends on stream order because channel width determines canopy cover, and heavy shading inhibits photosynthesis. Gregory (1979) reported virtually no NO_3^- uptake on a 1st-order stream in the Oregon Cascades but a greater than 80% NO_3^- decrease from midnight concentration in a 5th-order stream (Figure 14.5). Thus primary production by benthic algae can be a significant mechanism for dissolved nutrient retention. Nutrient transformation into tissue also produces an important food resource to grazer invertebrates, which are themselves a major link in the lotic food chain. Egesta from grazer and shredder invertebrates constitutes effective breakdown and repackaging of nutrients in particulate organic forms which are more transportable in the river current. Conversely, uptake of dissolved nutrients by algae or decomposers promotes retention. The result is a balance between transport and retention which directly determines the structure and function of riverine communities.

Symbiotic relationships between microbes and higher plants such as exists with nitrogen fixation can have important consequences for nutrient input into Pacific coastal streams and rivers. As mentioned previously, red alder roots often contain nodules infected with an actinomycete-like endophyte which facilitates nitrogen fixation. In the Coast Range of Washington, Oregon, and northern California, alder often invades following clear-cutting. Because alder lives long, it can make significant long-term contributions to the nitrogen capital of

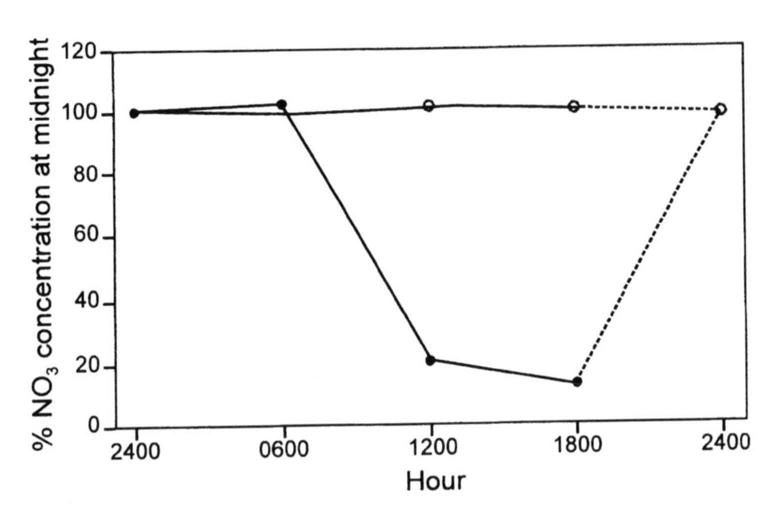

FIGURE 14.5. Percentage change from midnight concentration of nitrate in a summer-derived sample from 1st-order (○; Watershed 10) and 5th-order (●; Lookout Creek) streams from the H.J. Andrews Experimental Forest (modified from Gregory 1979)

affected watersheds and ultimately to the streams draining them.

Between 1975 and 1985 the role of nitrogen fixation by alder was examined as a potential substitute for urea fertilization (Atkinson and Hamilton 1978, Miller and Murray 1979). The average annual accumulation of nitrogen by fixation varies by site, stand age, and duration of significant fixation. In the Cedar River watershed in Washington, annual nitrogen accumulation in a 38-year old stand was estimated at 85.3 kg/ha/yr (Cole et al. 1978). Fixation at sites in the Coast Range of Oregon vary from 11 to 45 kg/ha/yr (Berg and Doerksen 1975), although rates as high as 320 kg/ha/yr (Newton et al. 1968) have been reported. Nitrogen accretion does not occur indefinitely. Cromack et al. (1979) estimated long-term accretion after 500 years was no greater than that of 15 years because fixation slows as nitrogen accumulates in soils, while denitrification and transport to the watershed's stream gradually balance fixation and atmospheric inputs.

Transport of nitrogen from fixation-enriched hillslopes can be a long-term source of both particulate and dissolved nitrogen to the adjacent aquatic environment. Direct litter input by alder can be an enhanced source of nitrogen to channel microbial and invertebrate communities since it typically has a higher nitrogen content than most other litter species. The initial nitrogen content of red alder litter typically exceeds 2%, whereas that of other common litter species like bigleaf maple (*Acer macrophyllum*), vine maple (*Acer circinatum*), and needle litter of Douglas-fir, and western hemlock (*Tsuga heterophylla*) is approximately 0.5% (Triska et al. 1975). Alder dominated riparian areas can also promote elevated dissolved nitrogen concentrations in stream water. Goldman (1961) noted that alder-lined (*Alnus tenuifolia*) springs on the east side of Castle Lake, California, averaged 10 times the NO_3^- concentration of west shore springs which lacked alder. He reported that the enhanced nitrogen input increased primary production in Castle Lake. Triska et al. (1994) analyzed for interstitial dissolved inorganic nitrogen and NH_4^+ sorption in riparian sediments beneath the water table at Little Lost Man Creek, California. Their study reach included a 25-year old, alder standing clear cut along the stream's west bank and old-growth forest along the east bank. Nitrate concentrations were typically far higher in riparian groundwater beneath the clear-cut bank than beneath the old-growth. Concentrations were highest during the late fall-to-spring rainy season when transport of infiltrated precipitation is high. At summer the base-flow NO_3^- concentration of stream water was significantly lower upstream of the alder-dominated clear-cut (Triska et al. 1995), and increased dramatically within short reaches along the clear-cut (Triska et al. 1989a).

Both beaver (*Castor canadensis*) and Pacific salmon (*Oncorhynchus* spp.) influence the biogeochemistry of watersheds in the Pacific coastal ecoregion. The abundance of these animals, however, is much reduced from historical levels throughout much of the region (Johnson and Chance 1974, Nehlsen et al. 1991), reducing their influence on the chemical processes of streams and rivers. Nevertheless, investigations on the impact of beaver and salmon on nutrient dynamics in systems where these animals are abundant indicates they can have a very significant effect (Brickell and Goring 1970, Naiman et al. 1988).

Five species of Pacific salmon die after spawning once. As a result, these anadromous fishes transport nutrients and organic matter from the north Pacific Ocean to streams and rivers of the coastal ecoregion. The species utilize different parts of the watershed for spawning and spawn at different times of the year, thereby distributing the input of nutrients provided by the carcasses both temporally and spatially.

Spawning sockeye (*Oncorhynchus nerka*) and pink (*Oncorhynchus gorbuscha*) salmon affect stream water chemistry. NO_3^- concentration increased four-fold and organic nitrogen concentration increased three-fold for a month following spawning by pink salmon (Brickell and Goring 1970). In Lake Dalnee, British Columbia, 26% of the annual phosphorus input is provided by spawning sockeye salmon in the lake's inlet stream (Krokhin 1968). Richey et al.

(1975) found that spawning kokanee salmon (landlocked sockeye salmon) contributed 44 kg/yr of phosphorus to a small tributary of Lake Tahoe, California. Increased phosphorus concentrations in the stream water and a consequent algal bloom were attributed to the spawning fish.

There appear to be several pathways by which nutrients released by decomposing salmon may be assimilated into the stream ecosystem. Richey et al. (1975) found that autotrophic production utilized much of the phosphorus released during decomposition of kokanee salmon carcasses. Bilby et al. (1996) examined the contribution of spawning coho salmon (*Oncorhynchus kisutch*) to a small stream system in western Washington using nitrogen and carbon stable isotopes. They found that up to 30% of the nitrogen and 39% of the carbon in the stream biota was contributed by the spawning salmon (Table 14.4). Autotrophic uptake accounted for very little assimilation of the nitrogen or carbon released during decomposition of the carcasses. Sorption onto the streambed substrate of dissolved organic matter released by the rotting carcasses, heterotrophic uptake, and direct consumption of the carcasses by insects and juvenile fishes inhabiting the stream were the primary pathways of incorporation.

Table 14.4. Percent of nitrogen (N) and carbon (C) contributed by spawning salmon to riparian vegetation and aquatic biota of a small stream in western Washington. Approximately 350 coho salmon (*Oncorhyncus kisutch*) per km of channel length spawned at the study site. Proportions of marine-derived nutrients were determined using stable isotope analysis.

Sample type	% Marine N	% Marine C
Riparian foliage	17.5	0
Epilithic organic matter	20.7	25.2
Grazers	24.8	29.2
Shredders	23.8	0
Collector–gatherers	14.4	29.4
Invertebrate predators	10.9	27.5
Age 0 cutthroat trout (*Oncorhynchus clarki*)	18.5	23.4
Age 1 and 2 cutthroat trout	25.6	24.8
Age 0 coho salmon (*O. kisutch*)	30.6	39.5

Data taken from Bilby et al. (1996).

Table 14.5. Historical and present levels of nutrients delivered from streams in the Willapa Bay drainage of Washington by spawning salmon (*Oncorhynchus* spp.). Historical values assume no harvest of returning fish

	Historical	Present
	(kg/km of stream length)	
Biomass	1,110	86
Nitrogen	111	9.0
Phosphorus	4.0	0.3

Historically, many Pacific coastal watersheds supported large numbers of spawning salmon, often exceeding 100,000 fish in larger systems. Pacific salmon range in size from approximately 2 kg for pink salmon, the smallest species, to greater than 20 kg for chinook salmon (*Oncorhynchus tshawytscha*). Therefore, the contribution of nitrogen and phosphorus from the fish was substantial (Table 14.5). The reduction in the numbers of naturally spawning salmon in many watersheds of the Pacific coastal ecoregion has led to a corresponding decrease in the delivery of nutrients and organic matter and a reduction in the productivity of these stream ecosystems. Many watersheds currently support salmon runs of less than 10% their historical size (Nehlsen et al. 1991). Reduction in the number of spawning fish not only reduces the number of juvenile fish to repopulate available habitat but also decreases the productivity of the stream, reducing its capacity to support future generations of salmon.

Beaver influence biogeochemical cycles of streams by changing the hydrologic regime and altering patterns and rates of organic matter and sediment transport. Impoundments created by beaver flood forest soils, creating anoxic conditions (Naiman et al. 1994). These conditions produce an increase in the amount of reduced nitrogen and conditions conducive to the release of nitrogen through denitrifica-

TABLE 14.6. Increase in the total amount of various nutrients in a 2563-ha area in Minnesota influenced by beaver (*Castor canadensis*). Few beaver were present in the watershed in the 1920s. The initial nutrient values assume all soils in the watershed exhibited levels similar to those currently displayed by non-flooded forest soils in the area.

Nutrient	% Increase after recovery of beaver population
Total N	72
NO_3^-–N	208
NH_4^+–N	295
Total P	43
SO_4^{2-}	82

Data from Naiman et al. (1994).

tion (Naiman et al. 1988, Triska et al. 1994). The impoundments also collect and retain large amounts of organic matter. Organic matter sources include the inundated forest, material carried into the pond by the stream and transported to the pond from the upslope forest by the beaver. Naiman et al. (1988) reported a three-fold increase in organic matter standing stock following impoundment of a stream by beaver. The nutrients contained in the organic matter increase the overall abundance of nutrients in systems where beaver are active (Table 14.6).

Responses to Disturbance

Healthy riverine ecosystems are balanced systems, where inputs, outputs, and chemical transformations maintain nutrient concentrations within ranges appropriate for organisms living in the river. The preceding sections examined the current understanding of the biogeochemical cycling of nitrogen, phosphorus, and sulfur in Pacific coastal rivers, and although much remains to be learned, several fundamental linkages between cycling processes and other biophysical characteristics of river and forest systems have been elucidated. Change is clearly a fundamental component of the biogeochemistry of riverine ecosystems. These changes often cycle at various time scales (e.g., diurnal, seasonal, or interannual). However, infrequent severe disturbances also play a key role in maintaining the long-term diversity and productivity of riverine ecosystems (Naiman et al. 1992).

In many steep, headwater stream segments, debris flows are an important pathway by which sediment and large woody debris are transported to and down streams (Bisson et al. 1987). This is despite the fact that their recurrence frequency is estimated at once every 500 years (Swanson et al. 1982). Wildfires are another catastrophic event that causes widespread disruption of watershed processes in the short-term, but on longer time scales is beneficial to forest and stream ecosystems. Wildfires tend to be rare in the Pacific coastal ecoregion due to the wet climate. Hemstrom and Franklin (1982) estimate an average recurrence interval of once every 450 years. Although these events often result in mortality among organisms, over longer time scales they deliver much of the larger structural materials, such as large wood, to channels (Naiman et al. 1992). These large materials provide a great deal of structural heterogeneity to the stream, creating depositional sites for organic matter, sediments, and associated nutrients (Bilby 1981).

In many of the heavily managed and urbanized watersheds of the Pacific coastal ecoregion, natural disturbance regimes have been overridden by anthropogenic effects. This section examines these effects and their impacts on natural biogeochemical cycles.

Forest Conversion and Management

Forest conversion to managed silviculture is the most widespread type of anthropogenic disturbance impacting rivers of the coastal ecoregion. Trees accumulate nutrients from primary atmospheric and geologic sources, supply organic nutrients to rivers, and remove nutrients from hyporheic and riparian groundwater. Their canopies control water temperature by providing shade and fallen trees significantly affect channel morphology. The roots of trees dampen the hydrograph of streams and rivers by absorbing infiltrating groundwater, and they stabilize colluvium on hillslopes, thereby decreasing rates of erosion. Removal of trees along a river corridor disrupts each of these processes. Macroinvertebrate and fish communities may also be altered by logging activities (Newbold et al. 1980, Hartman and Scrivener 1990). Consequently, the biological drivers of nutrient cycling are affected at all trophic levels. Owing to the tight linkages between ecosystem processes in watersheds, disturbances at the streamside (and even on the hillslope) propagate throughout the system.

In one of the first studies of its kind in the Pacific coastal ecoregion, Brown et al. (1973) investigated the effects of logging on concentrations of NO_3^- and PO_4^{3-} in two streams of the Oregon Coast range. In the Deer Creek watershed, the forest was "patch cut," with 25% of the forest cut in three logging units; slash was burned in one of the three patches. Concentrations of NO_3^- and PO_4^{3-} had no significant response to this treatment. In the Needle Branch watershed, where the forest was fully clear cut and the slash burned, mean annual concentrations of NO_3^- in the stream increased 240% (from 13 to 31mM) and persisted at elevated concentrations for six years. PO_4^{3-} concentrations remained unchanged in both watersheds following logging. In a similar study, Fredriksen et al. (1975) reported data for three watersheds in the Oregon Cascades. Their results differed slightly from those of Brown et al. (1973) in that 25% patch cuts at Fox Creek and the South Umpqua Experimental Forest produced measurable increases in NO_3^- concentrations, and 100% clear-cutting at the H.J. Andrews Experimental Forest produced increased PO_4^{3-} concentrations in the first year following the cut. The pattern of increased NO_3^- and unchanged PO_4^{3-} concentrations was also reported at Carnation Creek on Vancouver Island (Scrivener 1982) and in other watersheds of the H.J. Andrews Experimental Forest (Martin and Harr 1989).

Differences in the responses of NO_3^- and PO_4^{3-} concentrations may stem in large part from the different surface reactivity of these species. Both are released in large amounts by burning and decomposing slash, and both remain in inorganic forms more readily due to the decrease in uptake by trees. But PO_4^{3-} appears to be effectively immobilized in the system by adsorption to mineral surfaces, a process that does not affect NO_3^-. Data are not available concerning responses of stream concentrations of SO_4^{2-} or dissolved organic forms of nitrogen, phosphorus, or sulfur following logging, but one might expect temporary increases in these species as well. It should be noted that NO_3^- concentrations after logging have not been measured in excess of drinking water standards (700mM). Moreover, elevated nutrient concentrations and water temperatures at Carnation Creek produced no discernible increases in stream periphyton productivity (Shortreed and Stockner 1982).

Large increases in suspended sediment loads are a common response to logging activities (Figure 14.6). The increases are attributed in part to the removal of trees, but road construction appears to play an even larger role (Brown and Krygier 1971, Beschta 1978). In the four Oregon watersheds discussed in the preceding paragraph, maximum suspended sediment concentrations increased by as much as two orders of magnitude following logging (Fredriksen et al. 1975). The delivery to streams of nutrients bound to soil particles must also increase as a result of logging. However, no studies in the Pacific coastal ecoregion have specifically addressed these inputs. If the sediment generated by logging has the same nutrient content as sediment generated under undisturbed conditions, nutrient input by this pathway would increase in proportion with the increase in sediment delivery. However, the sources of sediment after logging are more diverse than those in an undisturbed situation and nutrient content may be different. The only certainty is that as suspended sediment loads increase, so will fluxes of particulate nitrogen, phosphorus, and sulfur. Revegetation is the most effective means of reducing increased sediment loads (Fredriksen et al. 1975).

Cutting of the riparian forest decreases nutrient inputs from direct litterfall. Input of litter to a stream flowing through a recently clear-cut area in western Washington averaged about $16 g/m^2/yr$ ash-free dry weight (AFDW), whereas input to a nearby stream in an old-

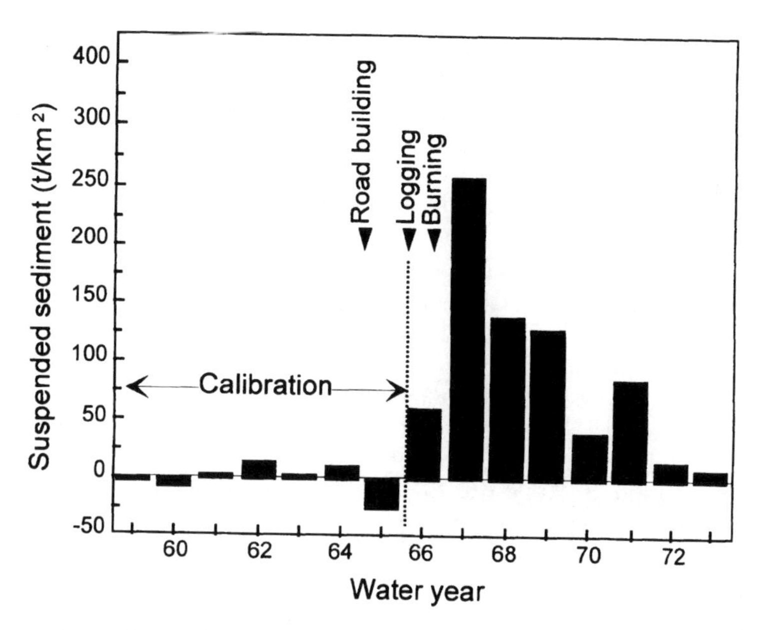

FIGURE 14.6. Suspended sediment yield in the Needle Branch watershed, western Oregon, before and after road building, 82% clear-cut logging, and burning (modified from Beschta 1978).

growth forest received about 230 g/m^2/yr AFDW (Bilby and Bisson 1992).

Overall fluxes of organic nutrients probably increase temporarily in streams following logging. Increased nutrient inputs due to higher amounts of sediment and dissolved nutrients delivered to the channel generally outweigh any decreased input of nutrients in litterfall. Outputs increase due to greater runoff, during both summer and winter (Harr et al. 1982, Harr 1986), and accelerated flushing of stored particulate matter from the streambed due to elimination of depositional sites as woody debris decomposes (Bilby 1981). Returning to the nutrient spiraling concept and equation (1), increased fluxes (F_t) from logging will increase spiraling lengths (S), unless there are commensurate increases in nutrient uptake (U).

The degree of disturbance associated with deforestation is greatly dependent upon the method of logging employed and the geomorphology of the watershed. During the first half of this century little consideration was given to the impact of logging on stream systems, as economic pressures were the primary determinant in the design of management practices. More recently, however, as knowledge of ecosystem functioning has improved and societal pressures for ecosystem protection have come into play, forest management has adopted new techniques which substantially reduce the degree of disturbance to river corridors. This trend continues today, as the latest management techniques aim to further minimize the degree of disturbance within the river corridors as well as on the surrounding hillslopes (Franklin 1992).

The effects of forest management are felt not only at the time of tree cutting. Continued management through replanting and regrowth also alters natural nutrient cycles. Nitrogen is most often the limiting nutrient in coniferous forests of the Pacific coastal ecoregion (Weetman et al. 1992). Consequently, for decades, foresters have applied urea pellets (~200 kg N/ha) to replanted stands every 5 to 10 years in order to enhance tree growth. While unfertilized buffers are established along larger streams, urea is often introduced directly into smaller systems. In the Mohun drainage basin on northern Vancouver Island, Perrin et al. (1984) compared the responses of stream nitrogen concentrations following forest fertilization in catchments with and without buffer strips. Nitrogen in streams without buffer strips peaked at concentrations which were 20 times greater than those in streams with buffer strips (Figure 14.7). Both systems, however, had nitrogen concentrations in excess of those in unfertilized catchments. Bisson et al. (1992) reported nearly instantaneous increases in the concentration of dissolved organic nitrogen (DON) in stream water from two fertilized watersheds from western Washington (Louse and Ludwig Creeks). In Louse Creek, DON concentrations increased from less than 50 mM to greater than 4,000 mM. Less dramatic, but equally rapid increases in DON were reported in two small tributaries to Lens Creek on southern Vancouver Island (Hetherington 1985). DON concentrations in both of these systems returned to near pre-fertilization levels within 3 to 10 days. By comparison, concentrations of inorganic nitrogen (NH_3 and NO_3^-) increased and remained elevated for weeks in the case of ammonia and up to a year or more in the case of nitrate. In another study conducted in two southeastern Alaskan streams where only NH_3 and NO_3^- were monitored following fertilization, stream concentrations responded in a similar fashion (Meehan et al. 1975).

Herbicides are often applied to reduce vegetation competing with young conifer trees in newly planted stands. Measures to protect streams from deleterious effects of herbicide contamination are built into procedures of application and, in general, these measures are quite effective (MacDonald et al. 1991), although herbicide residues are commonly detected in streams for a few days following application (Fredriksen et al. 1975). Nutrient concentrations in streamwater draining areas treated with herbicide often increase. Applications of the herbicide glyphosate (Roundup®) resulted in a doubling of PO_4^{3-} concentrations for one to two years (Hartman and Scrivener, 1990). Concentrations of NO_3^- also increased, but only following large storm events. Hartman and Scrivener (1990) attribute the increase in PO_4^{3-} concentrations to leaching of

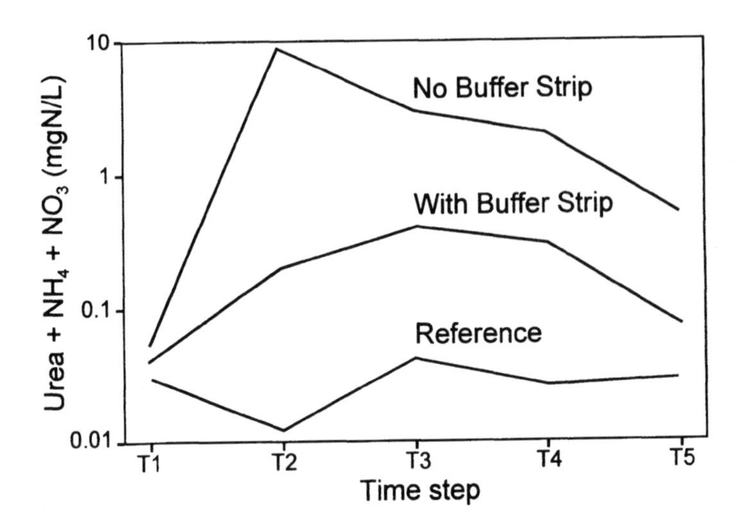

FIGURE 14.7. Mean concentrations of combined dissolved nitrogen (N) versus time in the Mohun drainage system of British Columbia, Canada, following application of urea fertilizer in two catchments. Stream dissolved nitrogen concentrations in catchments where a 50 m buffer strip was left unfertilized remained significantly less than stream dissolved nitrogen concentrations in catchments not containing a 50 m buffer strip that was left unfertilized, but concentrations were still higher than in the reference catchments. Time steps are defined as follows: T1 encompasses an 8 month period prior to fertilization, T2 encompasses the first 3 days following fertilization, T3 encompasses days 4–18 following fertilization, T4 encompasses days 19–59 following fertilization, and T5 encompasses day 60 through the completion of the study (modified from Perrin et al. 1984).

the P-rich deciduous vegetation killed by the treatment.

Urbanization and Agriculture

Converting forest land to agriculture or urban/suburban development dramatically alters natural nutrient cycles. In the Pacific coastal ecoregion, deforestation is the first step in land conversion, but instead of replanting trees, the forest is permanently removed. Furthermore, the channels of streams and rivers flowing through these areas are often engineered to guard against flooding and erosion, permanently destroying many of their natural geomorphologic features. In addition to the input pathways described earlier, sewage and industrial effluent from urban areas and fertilizers from agricultural fields and residential landscaping become major input pathways, raising nutrient concentrations to many times their natural concentrations. In response to the high nutrient levels and the associated high concentrations of organic matter, autotrophic and heterotrophic microbial communities bloom, and natural species compositions are severely disrupted. The consequences of all of these effects to water quality may be so severe that human health is directly threatened, especially given the presence of other toxins in these effluents. See Chapter 4 for a more detailed discussion of water quality.

Fire

The occurrence of wildfires is one disturbance that has actually decreased as a result of management programs in Pacific coastal watersheds. The good and bad points of this reduction are a topic of considerable debate, and the effects on nutrient cycles are not clear. When fire sweeps through a watershed, large amounts of organically bound nutrients are mineralized and enter the soil column. In the case of nitrogen, however, NH_3 volatilization during the fire may result in significant nitrogen losses from the ecosystem. Grier (1975) reported nitrogen losses of 97% from the forest floor and 33% from the A1 horizon of the mineral soil following a fire in the Entiat valley of central Washington. The total loss amounted to 39% of the nitrogen stock in the ecosystem. These losses appear to have occurred almost entirely via volatilization, given that waters leaching the remaining ash layer contained undetectable amounts of nitrogen (Grier 1975). Volatilization losses of PO_4^{3-} and SO_4^{2-} are likely to be less important.

The net effect of severe fires is not completely unlike that of logging and slash burning. The forest is largely destroyed and large amounts of non-volatile nutrients (including base cations) are input to the soil. Destruction of the vegetation results in decreased rates of nutrient uptake, as well as decreased inputs of nutrients to streams from riparian vegetation. Woody debris within the streams may also be burned. Hillslope instability is possible, although not to the extent associated with improperly built roads. Likewise, the potential for forest regeneration is also very high, as seedlings profit from the mostly increased levels of available nutrients.

Climate Change

The future course of global climate change is still far from clear, but there now appears to be general agreement that global atmospheric temperatures are rising and will continue to rise into the next century (Titus and Narayanan 1995, IPCC 1996). The most extreme ecosystem responses to climate change are expected in marginal or stressed systems, but the net effects are likely to be felt to some extent in all living systems. Possible impacts include changing patterns of rainfall (and therefore runoff), shifts in growing season temperatures, and changes in soil moisture levels. Increased fires as well as altered vegetation patterns are a concern (Risser 1992).

In an analysis not limited to the Northwest, Meyer and Pulliam (1992) considered various potential impacts of climate change to stream systems. Rising carbon dioxide (CO_2) and anticipated rising temperatures may lead to changes in the food quality of litter as leaf nitrogen contents decrease. Increased plant metabolism may deplete nutrients in soil and groundwater, and enhanced organic matter decomposition in warmer soils could reduce inputs of dissolved organic nutrient forms to streams. Overpeck et al. (1990) reasoned that increasing temperatures will alter the natural disturbance regimes of forests and streams. In southeastern Alaska, increasing air temperatures would probably lead to decreased water temperatures, as streams receive more runoff from melting glaciers (Oswood et al. 1992). Thus, there are many possible responses of nutrient cycles to a changing climate, but until our understanding of these responses improves, or until verifiable evidence of changed cycles appears, discussions will continue to be qualified by words like *may*, *could*, and *might*.

Acknowledgments. The authors thank John Stockner, Robert Beschta, and Robert Naiman for their careful reviews of the manuscript.

Literature Cited

Anderson, N.H., and E. Grafius. 1975. Utilization and processing of allochthonous materials by stream Trichoptera. Verhandlungen Internationale Vereinigung für Theoretische und Angewandte Limnologie **19**:3082–3088.

Atkinson, W.A., and W.I. Hamilton. 1978. The value of red alder as a source of nitrogen in Douglas-fir/alder mixed stands. Pages 337–351 *in* D.G. Briggs, D.S. DeBell, and W.A. Atkinson, eds. Utilization and management of alder. USDA Forest Service General Technical Report PNW 70, Pacific Northwest Forest and Range Experiment Station, Portland, Oregon, USA.

Berg, A., and A. Doerksen. 1975. Natural fertilization of a heavily thinned Douglas-fir stand by understory red alder. Forest Research Laboratory Research Note Number 56, Oregon State University, Corvallis, Oregon, USA.

Beschta, R.L. 1978. Long-term patterns of sediment production following road construction and logging in the Oregon coast range. Water Resources Research **14**:1011–1016.

Bilby, R.E. 1981. Role of organic debris dams in regulating the export of dissolved and particulate matter from a forested watershed. Ecology **62**:1234–1243.

Bilby, R.E., and P.A. Bisson. 1992. Allochthonous versus autochthonous organic matter contributions to the trophic support of fish populations in clear-cut and old-growth forested streams. Canadian Journal of Fisheries and Aquatic Sciences **49**:540–551.

Bilby, R.E., B.R. Fransen, and P.A. Bisson. 1996. Incorporation of nitrogen and carbon from spawning coho salmon into the trophic system of small streams: Evidence from stable isotopes. Canadian Journal of Fisheries and Aquatic Sciences **53**:164–173.

Bisson, P.A., G.G. Ice, C.J. Perrin, and R.E. Bilby. 1992. Effects of forest fertilization on water quality and aquatic resources in the Douglas-fir region. Pages 179–193 *in* H.N. Chappell, G.F. Weetman, and R.E. Miller, eds. Improving nutrition and growth of western forests. College of Forest Resources, University of Washington, Seattle, Washington, USA.

Bisson, P.A., R.E. Bilby, M.D. Bryant, C.A. Dollof, G.B. Grette, R.A. House, et al. 1987. Large woody debris in forested streams in the Pacific Nothwest: Past, present, and future. Pages 143–190 *in* E.O. Salo and T.W. Cundy, eds. Streamside management: Forestry and fishery interactions. Contribution Number 57. Institute of Forest Resources, University of Washington, Seattle, Washington, USA.

Brickell, D.C., and J.J. Goering. 1970. Chemical effects of salmon decomposition on aquatic ecosystems. Pages 125–138 *in* R.S. Murphy, ed. First international symposium on water pollution control in cold climates. US Government Printing Office, Washington, DC, USA.

Brown, G.W., A.R. Gahler, and R.B. Marston. 1973. Nutrient losses after clear-cut logging and slash burning in the Oregon coast range. Water Resources Research **9**:1450–1453.

Brown, G.W., and J.T. Krygier. 1971. Clear-cut logging and sediment production in the Oregon coast range. Water Resources Research **7**:1189–1198.

Buckley, B.M., and F.J. Triska. 1978. Presence and ecological role of nitrogen-fixing bacteria associated with wood decay in streams. Verhandlungen Internationale Vereinigung für Theoretische und Angewandte Limnologie **20**:1333–1339.

Campbell, N.A. 1990. Biology. Benjamin/Cummings, New York, New York, USA.

Chen, K.Y., and J.C. Morris. 1972. Kinetics of oxidation of aqueous sulfide by O_2. Environmental Science and Technology **6**:529–537.

Cole, D.W., S.P. Gessel, and J. Turner. 1978. Comparative mineral cycling in red alder and Douglas-fir. Pages 327–336 *in* D.G. Briggs, D.S. DeBell, and W.A. Atkinson, eds. Utilization and management of alder. USDA Forest Service General Technical Report PNW 70, Pacific Northwest Forest and Range Experiment Station, Portland, Oregon, USA.

Cromack, K., C.C. Delwiche, and D.H. McNabb. 1979. Prospects and problems of nitrogen management using symbiotic nitrogen fixers. Pages 210–223 *in* J.C. Gordon, C.T. Wheelen, and D.A. Perry, eds. Symbiotic Nitrogen Fixation in the Management of Temperate Forests: Proceedings of a Workshop. Forest Sciences Laboratory, Oregon State University, Corvallis, Oregon, USA.

Duff, J.H., and F.J. Triska. 1990. Denitrification in sediments from the hyporheic zone adjacent to a small forested stream. Canadian Journal of Fisheries and Aquatic Sciences **47**:1140–1147.

Elwood, J.W., J.D. Newbold, R.V. O'Neill, and W. Van Winkle. 1983. Resource spiraling: An operational paradigm for analyzing lotic ecosystems. Pages 3–27 *in* T.D. Fontaine and S.M. Bartell, eds. Dynamics of lotic ecosystems. Ann Arbor Science (Butterworth), Ann Arbor, Michigan, USA.

Feller, M.C., and J.P. Kimmins. 1979. Chemical characteristics of small streams near Haney in southwestern British Columbia. Water Resources Research **15**:247–258.

Franklin, J.F. 1992. Scientific basis for new perspectives in forests and streams. Pages 25–72 *in* R.J. Naiman, ed. Watershed management. Springer-Verlag, New York, New York, USA.

Fredriksen, R.L. 1975. Nitrogen, phosphorus, and particulate matter budgets of five coniferous forest ecosystems in the western Cascades range, Oregon. Ph.D. dissertation. Oregon State University, Corvallis, Oregon, USA.

Fredriksen, R.L., D.G. Moore, and L.A. Norris. 1975. The impact of timber harvest, fertilization, and herbicide treatment on streamwater quality in western Oregon and Washington. Pages 283–313 *in* B. Bernier and C.H. Winget, eds. Forest soils and forest land management. Laval University Press, Quebec, Canada.

Freney, J.R., J.R. Simpson, and O.T. Denmead. 1983. Volatilization of ammonia. Pages 1–32 *in* J.R. Freney and J.R. Simpson, eds. Gaseous loss of nitrogen from plant-soil systems. Martinus Nijhoff, The Hague, The Netherlands.

Gall, B.F. 1986. A multivariate comparison of the physical and chemical characteristics of streams draining three adjacent watersheds in northwestern Washington. Master's thesis. Western Washington University, Bellingham, Washington, USA.

Goldman, C.R. 1961. The contribution of alder trees *Alnus tenuifolia* to the primary production of Castle Lake, California. Ecology **42**:282–288.

Gregory, S.V. 1978. Phosphorus dynamics on organic and inorganic substrates in streams. Verhandlungen Internationale Vereinigung für Theoretische und Angewandte Limnologie **20**:1340–1346.

——. 1979. Primary production in Pacific Northwest streams. Ph.D. dissertation. Oregon State University, Corvallis, Oregon, USA.

Grier, C.C. 1975. Wildfire effects on nutrient distribution and leaching in a coniferous ecosystem. Canadian Journal of Forest Research **5**:599–607.

Harr, R.D. 1986. Effects of clearcutting on rain-on-snow runoff in western Oregon: A new look at old studies. Water Resources Research **22**:1095–1100.

Harr, R.D., A. Levno, and R. Mersereau. 1982. Streamflow changes after logging 130-year-old Douglas fir in two small watersheds. Water Resources Research **18**:637–644.

Hartman, G.F., and J.C. Scrivener. 1990. Impacts of forestry practices on a coastal stream ecosystem, Carnation Creek, British Columbia. Canadian Bulletin of Fisheries and Aquatic Sciences **223**: 136–148.

Hatcher, P.G., and E.C. Spiker. 1988. Selective degradation of plant biomolecules. Pages 59–74 *in* F.H. Frimmel and R. Christman, eds. Humic substances and their role in the environment. John Wiley & Sons, New York, New York, USA.

Hemstrom, M.A., and J.F. Franklin. 1982. Fire and other disturbances of the forests in Mt. Rainier National Park. Quaternary Research **18**:32–51.

Hetherington, E.D. 1985. Streamflow nitrogen loss following forest fertilization in a southern Vancouver Island watershed. Canadian Journal of Forest Research **15**:34–41.

IPCC (Intergovernmental Panel on Climate Change). 1996. Climate Change 1995: Impacts, adaptations and mitigation of climate change: Scientific–technical analyses. Contribution of Working Group II to the Second Assessment Report of the Intergovernmental Panel on Climate Change. Cambridge University Press, New York, New York, USA.

Jaffe, D.A. 1992. The nitrogen cycle. Pages 263–284 *in* S.S. Butcher, R.J. Charlson, G.H. Orians, and G.V. Wolfe, eds. Global biogeochemical cycles. Academic Press, San Diego, California, USA.

Johnson, D.R., and D.H. Chance. 1974. Presettlement overharvest of upper Columbia River beaver populations. Canadian Journal of Zoology **52**:1519–1521.

Junk, W.J., P.B. Baley, and R.E. Sparks. 1989. The flood pulse concept in river-floodplain systems. Pages 110–127 *in* D.P. Dodge, ed. Proceedings of the International Large River Symposium. Canadian Special Publication in Fisheries and Aquatic Sciences **106**.

Krokhin, E.M. 1968. Effect of size of escapement of sockeye salmon spawners on the phosphate content of a nursery lake. Journal of the Fisheries Research Board of Canada **57**:31–39.

Larson, A.G. 1979. Origin of the chemical composition of undisturbed forested streams, western Olympic peninsula, Washington State. Ph.D. dissertation. University of Washington, Seattle, Washington, USA.

Likens, G.E., F.H. Bormann, R.S. Pierce, J.S. Eaton, and N.M. Johnson. 1977. Biogeochemistry of a forested watershed. Springer-Verlag, New York, New York, USA.

MacDonald, L.H., A.W. Smart, and R.C. Wissmar. 1991. Monitoring guidelines to evaluate effects of forestry activities on streams in the Pacific Northwest and Alaska. EPA 910/9-91-001. US Environmental Protection Agency, Region 10, Seattle, Washington, USA.

Martin, C.W., and R.D. Harr. 1989. Logging of mature Douglas-fir in western Oregon has little effect on nutrient output budgets. Canadian Journal of Forest Research **19**:35–43.

Meehan, W.R., F.B. Lotspeich, and E.W. Mueller. 1975. Effects of forest fertilization on two southeast Alaska streams. Journal of Environmental Quality **4**:50–55.

Meyer, J.L. 1994. The microbial loop in flowing waters. Microbial Ecology **28**:195–198.

Meyer, J.L., W.H. McDowell, T.L. Bott, J.W. Elwood, C. Ishizaki, J.M. Melack, et al. 1988. Elemental dynamics in streams. Journal of the North American Benthological Society **7**:410–432

Meyer, J.L., and W.M. Pulliam. 1992. Modification of terrestrial-aquatic interactions by a changing climate. Pages 177–191 *in* P. Firth and S.G. Fisher, eds. Global climate change and freshwater ecosystems. Springer-Verlag, New York, New York, USA.

Miller, R.E., and M.D. Murray. 1979. Fertilizer versus red alder for adding nitrogen to Douglas-fir forests of the Pacific Northwest. Pages 356–373 *in* J.C. Gordon, C.T. Wheelen, and D.A. Perry, eds. Symbiotic nitrogen fixation in the management of temperate forests: Proceedings of a workshop. Forest Sciences Laboratory, Oregon State University, Corvallis, Oregon, USA.

Naiman, R.J., and E.C. Anderson. 1996. Ecological implications of latitudinal variations in watershed size, streamflow, and water temperature in the coastal temperate rain forest of North America. Pages 131–148 *in* P. Schoonmaker and B. von Hagen, eds. The rain forests of home: Profile of a North American bioregion. Island Press, Washington, DC, USA.

Naiman, R.J., T.J. Beechie, L.E. Benda, D.R. Berg, P.A. Bisson, L.H. MacDonald, et al. 1992. Fundamental elements of ecologically healthy watersheds in the Pacific Northwest coastal ecoregion. Pages 127–188 *in* R.J. Naiman, ed. Watershed management. Springer-Verlag, New York, New York, USA.

Naiman, R.J., C.A. Johnston, and J.C. Kelley. 1988. Alteration of North American streams by beaver. BioScience **38**:753–762.

Naiman, R.J., G. Pinay, C.A. Johnston, and J. Pastor. 1994. Beaver influences on the long-term biogeochemical characteristics of boreal forest drainage networks. Ecology **75**:905–921.

Nehlsen, W., J.E. Williams, and J.A. Lichatowich. 1991. Pacific salmon at the crossroads: Stocks at risk from California, Oregon, Idaho and Washington. Fisheries **16**:4–21.

Newbold, J.D. 1992. Cycles and spirals of nutrients. Pages 379–408 *in* P. Calow and G.E. Petts, eds. The rivers handbook: Hydrological and ecological principles. Blackwell Scientific, Boston, Massachusetts, USA.

Newbold, J.D., D.C. Erman, and K.B. Roby. 1980. Effects of logging on macroinvertebrates in streams with and without buffer strips. Canadian Journal of Fisheries and Aquatic Sciences **37**:1076–1085.

Newbold, J.D., R.V. O'Neill, J.W. Elwood, and W. Van Winkle. 1982. Nutrient spiraling in streams: Implications for nutrient limitations and invertebrate activity. The American Naturalist **120**:628–652.

Newton, M., B.A. El Hassan, and J. Zavitkovski. 1968. Role of red alder in western Oregon forest succession. Pages 73–84 *in* J.M. Trappe, J.F. Franklin, R.F. Tarrant and G.H. Hansen, eds. The biology of alder: Proceeding of a symposium. Pacific Northwest Forest and Range Experimental Station, Portland, Oregon, USA.

Oswood, M.W., A.M. Milner, and J.G. Irons III. 1992. Climate change and Alaskan rivers and streams. Pages 192–210 *in* P. Firth and S.G. Fisher, eds. Global climate change and freshwater ecosystems. Springer-Verlag, New York, New York USA.

Overpeck, J.T., D. Rind, and R. Goldberg. 1990. Climate-induced changes in forest disturbance and vegetation. Nature **343**:51–53.

Perrin, C.J., K.S. Shortreed, and J.G. Stockner. 1984. An integration of forest and lake fertilization: Transport and transformations of fertilizer elements. Canadian Journal of Fisheries and Aquatic Sciences **41**:253–262.

Pringle, C.M., R.J. Naiman, G. Bretschko, J.R. Karr, M.W. Oswood, J.R. Webskr, et al. 1988. Patch dynamics in lotic ecosystems: The stream as a mosaic. Journal of the North American Benthological Society **7**:503–524.

Richey, J.E., M.A. Perkins, and C.R. Goldman. 1975. Effects of kokanee salmon (*Oncorhynchus nerka*) decomposition on the ecology of a subalpine stream. Journal of the Fisheries Research Board of Canada **32**:817–820.

Risser, P.G. 1992. Impacts on ecosystems of global environmental changes in Pacific Northwest watersheds. Pages 12–24 *in* R.J. Naiman, ed. Watershed management. Springer-Verlag, New York, New York, USA.

Scrivener, J.C. 1975. Water, water chemistry, and hydrochemical balance of dissolved ions in Carnation Creek watershed, Vancouver Island, July 1971–May 1974. Fisheries and Marine Service Technical Report 564.

——. 1982. Logging impacts on the concentration patterns of dissolved ions in Carnation Creek, British Columbia. Pages 64–80 *in* G.F. Hartman, ed. Proceedings of the Carnation Creek Workshop: A ten year review. Pacific Biological Station, Nanaimo, British Columbia, Canada.

Sedell, J.R., F.J. Triska, and B.M. Buckley. 1975. The processing of coniferous and hardwood leaves in two coniferous forest streams. Verhandlungen Internationale Vereinigung für Theoretische und Angewandte Limnologie **19**:617–627.

Shortreed, K.S., and J.G. Stockner. 1982. Impact of logging on periphyton biomass and species composition in Carnation Creek: A coastal rain forest stream on Vancouver Island, British Columbia. Pages 197–209 *in* G.F. Hartman, ed. Proceedings of the Carnation Creek Workshop: A ten year review. Pacific Biological Station, Nanaimo, British Columbia, Canada.

Swanson, F.J., S.V. Gregory, J.R. Sedell, and A.G. Campbell. 1982. Land-water interactions: The riparian zone. Pages 267–291 *in* R.L. Edmonds, ed. Analysis of coniferous forest ecosystems in the western United States. Hutchinson Ross, Stroudsburg, Pennsylvania, USA.

Thorp, J.H., and M.D. Delong. 1994. The riverine productivity model: An heuristic view of carbon sources and organic processing in large river systems. Oikos **70**:305–308.

Titus, J.G., and V.K. Narayanan. 1995. The probability of sea level rise. EPA 230-R-95-008. US Environmental Protection Agency, Office of Policy, Planning, and Evalxuation, Climate Change Division, Adaptation Branch, Washington, DC, USA.

Triska, F.J., J.R. Sedell, and B.M. Buckley. 1975. The processing of conifer and hardwood leaves in two coniferous forest streams. II. Biogeochemical and nutrient changes. Verhandlungen Internationale Vereinigung für Theoretische und Angewandte Limnologie **19**:1628–1639.

Triska, F.J., and B.M. Buckley. 1978. Patterns of nitrogen uptake and loss in relation to litter disappearance and associated invertebrate biomass in six streams of the Pacific Northwest, USA. Verhandlungen Internationale Vereinigung für Theoretische und Angewandte Limnologie **20**:1324–1332.

Triska, F.J., V.C. Kennedy, R.J. Avanzino, G.W. Zellweger, and K.E. Bencala. 1989a. Retention and transport of nutrients in a third-order stream in northwestern California: Channel processes. Ecology **70**:1877–1892.

Triska, F.J., V.C. Kennedy, R.J. Avanzino, G.W. Zellweger, and K.E. Bencala. 1989b. Retention and transport of nutrients in a third-order stream in northwestern California: Hyporheic processes. Ecology **70**:1893–1905.

Triska, F.J., J.R. Sedell, K. Cromack, Jr., S.V. Gregory, and F.M. McCorison. 1984. Nitrogen budget for a small coniferous forest stream. Ecological Monographs **54**:119–140.

Triska, F.J., J.H. Duff, and R.J. Avanzino. 1993. The role of water exchange between a stream channel and its hyporheic zone in nitrogen cycling at the terrestrial-aquatic interface. Hydrobiologia **251**:167–184.

Triska, F.J., A.P. Jackman, J.H. Duff, and R.J. Avanzino. 1994. Ammonium sorption to channel and riparian sediments: A transient storage pool for dissolved inorganic nitrogen. Biogeochemistry **26**:67–83.

Triska, F.J., V.C. Kennedy, R.J. Avanzino, and K.C. Stanley. 1995. Inorganic nitrogen uptake and regeneration in a small stream at summer low flow: Long-term clear cutting and short-term storm related impacts. Pages V1–V13 *in* K.M. Nolan, H.M. Kelsey, and D.C. Marron, eds. Geomorphic processes and aquatic habitat in the Redwood Creek Basin, northwestern California.

United States Geological Survey Professional Paper 1454. US Government Printing Office, Denver, Colorado, USA.

Vannote, R.L., G.W. Minshall, K.W. Cummins, J.R. Sedell, and C.E. Cushing. 1980. The river continuum concept. Canadian Journal of Fisheries and Aquatic Sciences **37**:130–137.

Ward, J.V., and J.A. Stanford. 1983. The serial discontinuity concept of lotic ecosystems. Pages 29–42 *in* T.D. Fontaine and S.M. Bartell, eds. Dynamics of lotic ecosystems. Ann Arbor Science, Ann Arbor, Michigan, USA.

Webster, J.R., and B.C. Patten. 1979. Effects of watershed perturbations on stream potassium and calcium dynamics. Ecological Monographs **49**:51–72.

Weetman, G.F., E.R.G. McWilliams, and W.A. Thompson. 1992. Nutrient management of Douglas-fir and western hemlock stands: The issues. Pages 17–27 *in* H.N. Chappell, G.F. Weetman and R.E. Miller, eds. Improving nutrition and growth of western forests. College of Forest Resources, University of Washington, Seattle, Washington, USA.

15
Organic Matter and Trophic Dynamics

Peter A. Bisson and Robert E. Bilby

Overview

- This chapter examines the sources of nutrients and organic matter in streams, the mechanisms by which organic materials are physically and biologically transformed as they pass from small to progressively larger channels, and the trophic processes by which organic matter supports consumer organisms.
- Organic matter is incorporated into stream ecosystems through two general pathways: autotrophy, the production of new plant material through photosynthesis; and heterotrophy, the assimilation of organic matter by consumers.
- Material derived from photosynthesis can come from within the stream (autochthonous production) or from terrestrial sources (allochthonous production). Depending on the type of input, organic matter is utilized by a variety of consumer groups whose mode of feeding is adapted to different physical states of organic materials.
- Inputs of coarse particulate organic matter (CPOM) to headwater streams are transformed through processes of microbial decomposition, consumption by macroinvertebrates, and physical abrasion to fine particles and dissolved organic matter, which are utilized by aquatic communities downstream.
- The relative abundance of different types of producer and consumer organisms changes from small headwater streams to large rivers in response to the availability and character of organic matter inputs, as well as storage and transport processes.
- Using salmonid fishes as a focus of the discussion, differences in productivity between streams can often be explained by variations in the types and availability of food resources. For populations of anadromous salmonids in the Pacific coastal ecoregion, productivity is strongly influenced by seasonal shifts in organic matter origin. Autotrophic processes dominate in spring and summer, whereas heterotrophic processes dominate in autumn and winter.
- Human activities depriving streams of nutrients and organic matter or reducing the capacity of aquatic communities to store and process these materials (e.g., removal of streamside vegetation, loss of coarse woody debris, and reduction of salmon carcasses) often lead to changes in the trophic system that ultimately impair salmonid productivity. Anthropogenic nutrient enrichment may enhance trophic processes that support the production of undesirable organisms.

Introduction

Just as the characteristics of flowing water—velocity, turbulence, temperature, sediment, and solute content—influence the distribution and abundance of plants and animals in streams, so too the movements and transformations of organic matter influence the composition and productivity of different members of the aquatic community. Whether the goal is protecting invertebrates and salamanders in a small stream or maximizing the yield of commercially valuable fishes in a large river, knowledge of both physical processes and trophic dynamics is needed for effective management. This chapter reviews the sources and transformations of organic matter in flowing water ecosystems and examines some impacts of human activities on organic matter pathways supporting the production of important aquatic resources. Although processes discussed are shared by river systems throughout the world, many examples are drawn from western North America, where much of the research on trophic dynamics has examined factors contributing to the support of salmonid fishes.

The following information is meant to provide a very general overview of organic matter and trophic dynamics in streams of the Pacific coastal ecoregion. Two important references contain more detailed and comprehensive information about organic matter and trophic dynamics in rivers. The first, H.B.N. Hynes' *The Ecology of Running Waters* (Hynes 1970), is a treatise of exceptional depth and breadth summarizing much of what was known about stream ecology up to that time. It is still an invaluable reference for any aquatic scientist. The second and more recent book, J.D. Allan's *Stream Ecology: Structure and Function of Running Waters* (Allan 1995), describes many newer advances in knowledge of lotic ecosystems, for example, the paradigm that riverine communities are structured from headwaters to mouth by a changing pattern of organic matter sources and nutrient transformations. Together, these two works provide a wealth of information and references.

Organic matter and nutrient transformations in streams are often discussed in the context of aquatic productivity. Warren (1971) defined the concept of productivity as the capacity of an ecosystem to generate a product of interest. Productivity is measured by the survival and growth of members of a population under a certain set of conditions; the population's actual performance—its production—is limited by properties of the ecosystem in which its members reside. The nature and timing of organic matter and nutrient inputs usually exert a powerful influence on the capacity of streams and rivers to produce different species. Measurement of ecosystem productivity and measurement of ecosystem "health" or "integrity," using parameters such as species composition, may be quite different. It is possible for species diversity to be low in streams that are highly productive for certain taxa. Conversely, species diversity may be relatively high in unproductive streams because of a variety of physical and biological factors, as well as time lags for recovery and colonization of competitively dominant organisms after a disturbance (Patrick 1975, Power 1990, Wootton 1996).

Movements and transformations of organic matter influence the productive capacity of rivers for salmonids. Fishes are frequently taken as indicators of stream productivity and aquatic health, but other organisms often dominate secondary production and may be far more sensitive indicators of aquatic integrity (Karr 1991, Chapter 20, this volume). Nevertheless, fish production has generally been studied more widely than production of other aquatic taxa. A comparison of salmonid fishes in rivers throughout the world (Table 15.1) reveals that annual production (the product of average population biomass and mean individual growth rates over a year expressed as new tissue produced per unit of area) varies greatly from stream to stream. Production of salmonids is relatively low in most streams world wide, but a few streams are very productive (Figure 15.1), although not as productive as streams supporting warm-water species where annual production can exceed 100 g/m^2/yr (Naiman 1976).

Many factors influence productivity. Physical characteristics of the stream environment affect a population's abundance, hence its production. Food availability also influences productivity (Chapman 1966); the 10-fold difference in salmonid production between the most and least productive sites in Table 15.1 would be difficult to explain strictly on the basis of differences in physical habitat among the rivers. In many instances, large differences in the capacity of streams to produce salmonid fishes, or other consumer organisms, are strongly related to availability of food resources (Le Cren 1969, Warren 1971, Gregory et al. 1987).

TABLE 15.1. Production estimates of salmonid populations in streams rank
approximately highest to lowest.

Species location	Production (g/m²/yr)	
Brown trout (*Salmo trutta*)		
Horokiwi Stream, New Zealand	54.7[1]	Alle
Black Brows Beck, England	12.2–33.9	Ellic
Candover Brook, England	13.4–28.7	Man
Tadnoll Brook, England	20.0	Man
Granslev a, Denmark	12.6–25.7	Mor
Bisballe baek, Denmark	18.8	Mor
Duhonw Stream, England	10.5–19.8	Miln
Orred baek, Denmark	10.0–18.6	Mortensen (1977b)
Docken's Water, England	12.1	Mann (1971)
River Tarrant, England	12.0	Mann (1971)
Walla Brook, England	11.6	Horton et al. (1968)
Black Brows Beck, England	10.0	Le Cren (1969)
Shelligan Burn, Scotland	6.7–12.7	Egglishaw and Shackley (1977)
Hinaki Stream, New Zealand	8.9	Hopkins (1971)
Hinau Stream, New Zealand	8.5	Hopkins (1971)
Chwefru Stream, England	6.9–9.3	Milner et al. (1978)
Bere Stream, England	2.6–12.9	Mann (1971)
Kingswell Beck, England	7.4	Le Cren (1969)
Valley Creek, Minnesota, USA	<0.1–13.2	Waters (1983)
Marteg Stream, England	5.9–6.0	Milner et al. (1978)
Loucka Creek, Czechoslovakia	1.7–8.6	Libosvarsky (1968)
Hall Beck, England	5.2	Le Cren (1969)
Nether Hearth Sike, England	5.0	Le Cren (1969)
Devils Brook, England	4.8	Mann (1971)
Bidno Stream, England	3.2–6.0	Milner et al. (1978)
Blackhoof River, Minnesota, USA	4.4	Waters et al. (1990)
Nunn Creek, Colorado, USA	3.4–5.2	Scarnecchia and Bergersen (1987)
Appleworth Beck, England	3.0	Le Cren (1969)
Cow Creek, Colorado, USA	1.4–1.8[2]	Scarnecchia and Bergersen (1987)
Normandale Creek, Ontario, Canada	1.4[3]	Gordon & MacCrimmon (1982)
Atlantic salmon—*Salmo salar*		
Shelligan Burn, Scotland	5.5–12.6	Egglishaw and Shackley (1977)
Duhonw Stream, England	6.3–11.0	Milner et al. (1978)
Bere Stream, England	7.2	Mann (1971)
Chwefru Stream, England	5.7–5.9	Milner et al. (1978)
Marteg Stream, England	2.4–6.9	Milner et al. (1978)
Bidno Stream, England	1.9–6.9	Milner et al. (1978)
Catamaran Brook, New Brunswick, Canada	2.1–5.1	Randall and Paim (1982)
Little River, New Brunswick, Canada	2.2–2.7	Randall and Paim (1982)
Meo River, Quebec, Canada	0.5	Leclerc and Power (1980)
Chinook salmon (*Oncorhynchus tshawytscha*)		
Lemhi River, Idaho, USA	3.0	Goodnight and Bjornn (1971)
Orwell Brook, New York, USA	0.6[4]	Johnson (1980)
Coho salmon (*Oncorhynchus kisutch*)		
Schultz Creek, Washington, USA	5.7–21.6[4]	Bisson et al. (1988)
Flynn Creek, Oregon, USA	6.0–16.2[5]	Chapman (1965)
Deer Creek, Oregon, USA	5.4–12.7[5]	Chapman (1965)
Needle Branch, Oregon, USA	4.9–11.5[5]	Chapman (1965)
Hoffstadt Creek, Washington, USA	2.7–10.3[4]	Bisson et al. (1988)
Herrington Creek, Washington, USA	2.3–9.7[4]	Bisson et al. (1988)
Orwell Brook, New York, USA	2.7[4]	Johnson (1980)
Tye Creek, Alaska, USA	0.6–2.0[4]	Dolloff (1986)
Toad Creek, Alaska, USA	0.6–1.0[4]	Dolloff (1986)
Bush Creek, British Columbia, Canada	0.9	Tripp and McCart (1983)

(*Continued*)

d

	Production (g/m²/yr)	Reference
k, British Columbia,	0.6	Tripp and McCart (1983)
ndale Creek, Ontario, Canada	0.5[3]	Gordon and MacCrimmon (1982)
ow trout (*Oncorhynchus mykiss*)		
othwell's Creek, Ontario, Canada	13.2	Alexander and MacCrimmon (1974)
Big Springs Creek, Idaho, USA	10.4	Goodnight and Bjornn (1971)
Blackhoof River, Minnesota, USA	9.6	Waters et al. (1990)
Normandale Creek, Ontario, Canada	5.6[3]	Gordon and MacCrimmon (1982)
Dale 1 Creek, Colorado, USA	4.5–5.4[2]	Scarnecchia and Bergersen (1987)
Dale 2 Creek, Colorado, USA	3.0–3.8[2]	Scarnecchia and Bergersen (1987)
Lemhi River, Idaho, USA	2.4	Goodnight and Bjornn (1971)
Valley Creek, Minnesota, USA	<0.1–4.5	Waters (1983)
Orwell Brook, New York, USA	0.9[4]	Johnson (1980)
Cutthroat trout (*Oncorhynchus clarki*)		
Little Green Creek, Colorado, USA	2.2–3.6	Scarnecchia and Bergersen (1987)
Roaring Creek, Colorado, USA	2.3–3.3	Scarnecchia and Bergersen (1987)
Right Hand Fork, Colorado, USA	1.5–3.6	Scarnecchia and Bergersen (1987)
Upper Deschutes River, Washington, USA	1.1–2.1[4]	Bilby and Bisson (1987)
West Fork Creek, Washington, USA	0.4–2.5[4]	Bilby and Bisson (1987)
Bush Creek, British Columbia, Canada	0.4–2.1	Tripp and McCart (1983)
Berry Creek, Oregon, USA	0.5	Nickelson (1974)
Banon Creek, British Columbia, Canada	0.1–0.8	Tripp and McCart (1983)
Brook trout (*Salvelinus fontinalis*)		
Big Spring, Pennsylvania, USA	30.0	Cooper and Scherer (1967)
McCreavy Creek, Colorado, USA	12.6–18.4	Scarnecchia and Bergersen (1987)
Lawrence Creek, Wisconsin, USA	9.3–10.6	Hunt (1966)
Valley Creek, Minnesota, USA	2.5–16.7	Waters (1983)
Kaikhosru Creek, Quebec, Canada	6.1–6.6	O'Connor and Power (1976)
Larry's Creek, Pennsylvania, USA	5.8	Cooper and Scherer (1967)
Caribou River, Minnesota, USA	5.8	Waters et al. (1990)
Cow Creek, Colorado, USA	4.9–6.5[2]	Scarnecchia and Bergersen (1987)
Porcupine Creek, Colorado, USA	4.8–4.9	Scarnecchia and Bergersen (1987)
Dale 2 Creek, Colorado, USA	4.4–5.7[2]	Scarnecchia and Bergersen (1987)
Indian Creek, Colorado, USA	4.4–5.2	Scarnecchia and Bergersen (1987)
Dale 1 Creek, Colorado, USA	4.0–4.9[2]	Scarnecchia and Bergersen (1987)
Gallienne Creek, Quebec, Canada	3.7–3.9	O'Connor and Power (1976)
Sherry Creek, Quebec, Canada	1.4–4.1	O'Connor and Power (1976)
Me 'o River, Quebec, Canada	1.9	O'Connor and Power (1976)
Davis Creek, Colorado, USA	1.7–1.9	Scarnecchia and Bergersen (1987)
Tchinicaman Creek, Quebec, Canada	1.5–16	O'Connor and Power (1976)
Guys Run, Virginia, USA	0.5–1.9	Neves and Pardue (1983)
Dolly Varden (*Salvelinus malma*)		
Toad Creek, Alaska, USA	0.3–0.8[4,6]	Dolloff (1983)
Knob Creek, Alaska, USA	0.1–0.8[4,6]	Dolloff (1983)
Aha Creek, Alaska, USA	0.3–0.5[4,6]	Dolloff (1983)
Tye Creek, Alaska, USA	0.1–0.9[4,6]	Dolloff (1983)

[1] Many scientists believe this figure to have been an overestimate. Le Cren (1969) suggested an alternative estimate of 38 g/m²/yr while Chapman (1966) suggested 45 g/m²/yr.

[2] Biomass ratios of different species from Scarnecchia and Bergersen (1987: Table 7, p. 322) were used to estimate each species' contribution to total annual production (Table 5, p. 321).

[3] Brown trout and rainbow trout production were estimated from Figure 4 (p. 466) of Gordon and MacCrimmon (1982).

[4] The interval over which production was estimated was less than 1 year, but included the late spring and summer growth period when much of the annual production takes place.

[5] The production interval was 14 months for each cohort.

[6] The estimate did not include production of age 0+ fish.

From Hanson and Waters 1974 and Chapman 1978.

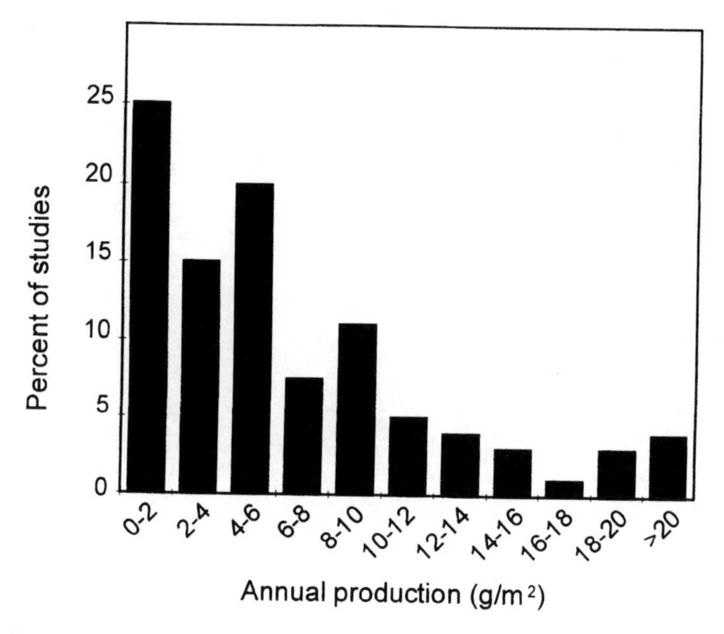

FIGURE 15.1. Distribution of estimates of annual production of salmonid fishes in streams of the world (based on 92 studies from Table 15.1 [Hanson and Waters 1974, Chapman 1978]).

Food availability as a factor limiting the production of stream-dwelling fishes is often overlooked by habitat managers. Most projects designed to improve the production of salmonids involve manipulating elements of the physical habitat such as logs or boulders, placing structures to prevent streambank erosion, or planting trees and shrubs to increase shade (NRC 1992 and 1996, Hunt 1993). Although these projects may remedy habitat damage caused by human activities, they usually do not address the food web contribution to production. Given the potential gains in productivity that can accompany increased food availability, consideration of the trophic system is critical to managing aquatic resources in rivers. Ignoring this important component imposes significant constraints on the effectiveness of restoration (Gregory et al. 1987, NRC 1992).

Trophic Pathways

Autochthonous and allochthonous, two major sources of organic matter, fuel lotic ecosystems. Autochthonous organic matter is generated by autotrophic members of the aquatic community (Table 15.2) that synthesize organic matter from inorganic materials in the presence of sunlight through the process of photosynthesis (Minshall 1978). On the other hand, organic inputs to lotic ecosystems that originate from terrestrial plants and animals, or organic materials of marine origin, are termed allochthonous inputs (Table 15.3). The relative abundance of these types of organic matter plays a key role in structuring the biotic community of the stream (Vannote et al. 1980) and influences nutrient dynamics of the system (Newbold et al. 1981 and 1982).

Autotrophic Production

Important autotrophs in rivers include vascular plants (principally aquatic angiosperms), bryophytes (mosses and liverworts), periphyton and phytoplankton (green and red algae), some bacteria (principally blue-green algae), and protists (diatoms, yellow–brown algae, and euglenoids). Periphyton refers to algae growing on the stream bottom; phytoplankton refers to algae suspended in the water column. The

TABLE 15.2. Major types of autotrophs and heterotrophs in streams.

Autotrophs		Heterotrophs	
Group	Growth forms	Group	Feeding types
Macrophytes (flowering plants, mosses and liverworts)	Emergent Floating-leaved Free-floating Submergent Crustose	*Microheterotrophs* (bacteria, fungi, protozoans, and micrometazoans)	Microbial decomposers Suspension feeders Grazers Predators Shredders
Periphyton (diatoms, green algae, blue–green algae, euglenoids, yellow–brown algae, red algae)	Prostrate Stalked/short filamentous Filamentous Filamentous with epiphytes Gelatinous	*Macroinvertebrates* (all larger invertebrate metazoans)	Gougers Suspension feeders/filterers Collector-gatherers Grazers Predators
Phytoplankton (algae, protists, and cyanobacteria)	Sloughed periphyton True phytoplankton	*Aquatic vertebrates* (fishes and amphibians)	Herbivore-detritivore Benthic invertebrate feeder Surface/drift feeder Generalized invertevore Planktivore Omnivore Piscivore Parasite

From Berkman and Rabeni 1987; Allan 1995, Merritt and Cummins 1996a.

former category dominates primary production in swiftly flowing streams while phytoplankton often dominates in slow moving, large rivers (Vannote et al. 1980). A more complete discussion of the structure and function of autotrophs in flowing waters is found in Chapter 7.

Through photosynthesis, autotrophic production creates new plant material within the stream itself. Part of this material is consumed by herbivorous animals, including a wide variety of micro- and macroinvertebrates, fishes, and larval amphibians. Some of the plant material becomes senescent and dies, after which it may settle into the stream substrate and become part of the detrital pool, or it can be carried downstream by the current. In either event, dead plants serve as a food source for detritivorous animals as well as bacteria and fungi, which form important links in heterotrophic organic matter pathways.

TABLE 15.3. Size classes and examples of different categories of organic matter.

Category	Abbreviation	Size range (μm)	Example
Dissolved organic matter	DOM	<0.45	Leaf leachate
Fine particulate organic matter:	FPOM	>0.45–<1,000	
Ultrafine	UPOM	0.45–25	Bacteria
Very fine	VPOM	25–45	Pollen
Fine	FPOM	45–100	Macroinvertebrate feces
Small	SPOM	100–250	Algal detritus
Medium–large	MPOM	250–1,000	Very small leaf fragments
Coarse particulate organic matter	CPOM	>1,000	Needles; dead aquatic animals

Allochthonous Organic Matter and Heterotrophic Production

Heterotrophic organisms derive energy from organic matter, living or dead, much of which may ultimately derive from outside the stream (Table 15.3). Important heterotrophs in flowing waters include large and small organisms: microbes (bacteria and fungi), protozoans and micrometazoans that consume mostly dead organic matter, and animals (macroinvertebrates and vertebrates) that feed on living or dead organic matter—the most conspicuous of which are usually insects (Table 15.4) (Chapter 8). In heavily forested headwater streams, heterotrophic production is supported primarily by terrestrial organic matter, while in open river channels heterotrophic production can be supported by autochthonous detrital pathways (Vannote et al. 1980).

Inputs of allochthonous material are highly seasonal. Dissolved organic matter (DOM) (Table 15.3) enters river networks throughout the year but may be particularly significant during early fall rains or in periods of spring snow melt (Triska et al. 1984, Naiman et al. 1992). Fine particulate organic matter (FPOM) enters rivers in great quantities during flooding (Welcomme 1979). Coarse particulate organic matter (CPOM) varies widely in size, from needles and leaves entering rivers during autumn to large woody debris entering stream channels during major storms (Naiman and Sedell 1979). Carcasses of anadromous

TABLE 15.4. Functional groups of aquatic insects, their food and feeding mechanisms, examples of taxa, and typical food size.

Functional group	Subdivision of functional group		Example taxa	Food particle size (μm)
	Dominant food	Feeding mechanism		
Shredders	Living vascular hydrophyte plant tissue	Herbivores: chewers and miners of live macrophytes	Trichoptera: Phryganeidae, Leptoceridae	>1,000
	Decomposing vascular plant tissue and wood; coarse particulate organic matter	Detritivores: chewers, wood borers, and gougers	Diptera: Tipulidae, Chironomidae	>1,000
Collectors	Decomposing fine particulate organic matter	Detritivores: filterers or suspension feeders	Trichoptera: Hydropsychidae Diptera: Simuliidae	<1,000
	Decomposing fine particulate organic matter	Detritivores: gatherers or deposit feeders (includes surface film feeders)	Ephemeroptera: Ephemeridae Diptera: Chironomidae	<1,000
Scrapers	Periphyton: attached algae and associated material	Herbivores: grazing scrapers of mineral and organic surfaces	Trichoptera: Glossosomatidae Coleoptera: Psephenidae Ephemeroptera: Heptageniidae	<1,000
Predators (engulfers)	Living animal tissue	Carnivores: attack prey, pierce tissues and cells, and suck fluids	Hemiptera: Belostomatidae	>1,000
	Living animal tissue	Carnivores: ingest whole animals or parts	Plecoptera: Perlidae Odonata	>1,000

From Merritt and Cummins 1996a, 1996b.

fishes, a major source of allochthonous material to many coastal watersheds in the Pacific coastal ecoregion, become available during adult spawning periods (Kline et al. 1990 and 1993, Bilby et al. 1996) which may be very short in some rivers or nearly year-round in others.

Organic Matter Processing

A generalized model of a riverine trophic pathways (Figure 15.2) shows some important pathways of organic matter as it is consumed, stored, and excreted by members of different functional groups. The flow of materials through different pathways is strongly influenced by position along the stream network (Vannote et al. 1980). In a heavily shaded 1st-order stream in the western Cascade Mountains of Oregon, Triska et al. (1982) found the organic budget dominated by inputs of terrestrial litter (mostly wood fragments and detritus >10 cm diameter) which comprised over 80% of the organic standing crop (Table 15.5). Detritus entered the stream as litterfall from overhanging vegetation and through lateral movement along the forest floor. Microbial respiration accounted for much of the community outputs, but a considerable amount of organic material was exported from the stream as small particles or dissolved organic matter. Primary production was very low under the dense forest canopy of this headwater stream. Community respiration (microbial plus aquatic plant plus macroinvertebrate respiration) far exceeded photosynthesis by algae and moss (Table 15.5).

Although much of the allochthonous organic material in headwater streams enters as CPOM, it is converted through a variety of mechanisms (Figure 15.3) to FPOM. Microbial colonization takes place rapidly after CPOM enters the stream, forming a layer of bacteria and fungi on its surface. As this layer forms, it also includes a polysaccharide matrix inhabited by protozoans, micrometazoans, and early instars of macroinvertebrates such as midges and copepods. Microbial decomposers consume DOM leached from CPOM, and it is this rich layer of microbes, small animals, and organic

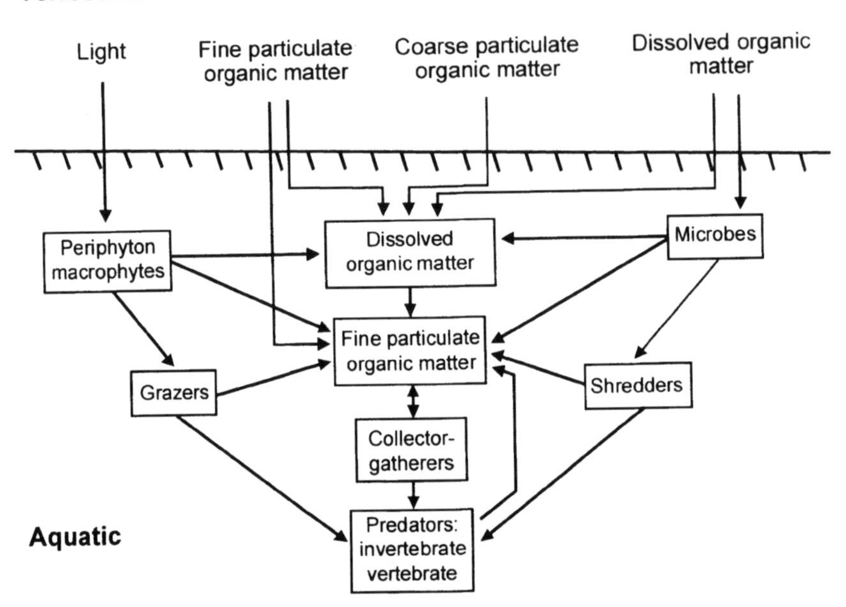

FIGURE 15.2. Generalized food web for low- to mid-order forested streams (from Hildrew et al. 1987, Allan 1995).

TABLE 15.5. Annual organic matter budget for a 1st–order stream in the H.J. Andrews Experimental Forest of western Oregon from 1973–1974.

Inputs			Standing crop			Outputs		
Category	kg/y	%	Category	kg/y	%	Category	kg/y	%
Litterfall	161–170	22–33	Large detritus (>10 cm)	8,692–8,698	82–83	Particulate organic	37–245	11–32
Throughfall	41–57	7–8	Small detritus (10 cm to 1 mm)	1,382–1,535	13–15	Microbial respiration	183–186	24–55
Lateral movement	200–333	41–43	Fine particulate organic matter (FPOM) (<1 mm to 75 μm)	87	1	Macroinvertebrate respiration	2	0.3–1
Gross primary production:			Ultrafine detritus (<75 μm to 45 μm)	233	2	Primary producer respiration:		
Algae	<1	0.1–0.2				Algae	<1	0.1–0.3
Moss	23	3–5				Moss	15	2–4
Dissolved organic matter	64–206	13–26	Primary producer biomass	31	0.3	Dissolved organic matter	96–310	28–41
			Macroinvertebrates	0.8–1	<0.1			
Total	490–790		Total	10,432–10.585		Total	337–756	

From Triska et al. 1982.

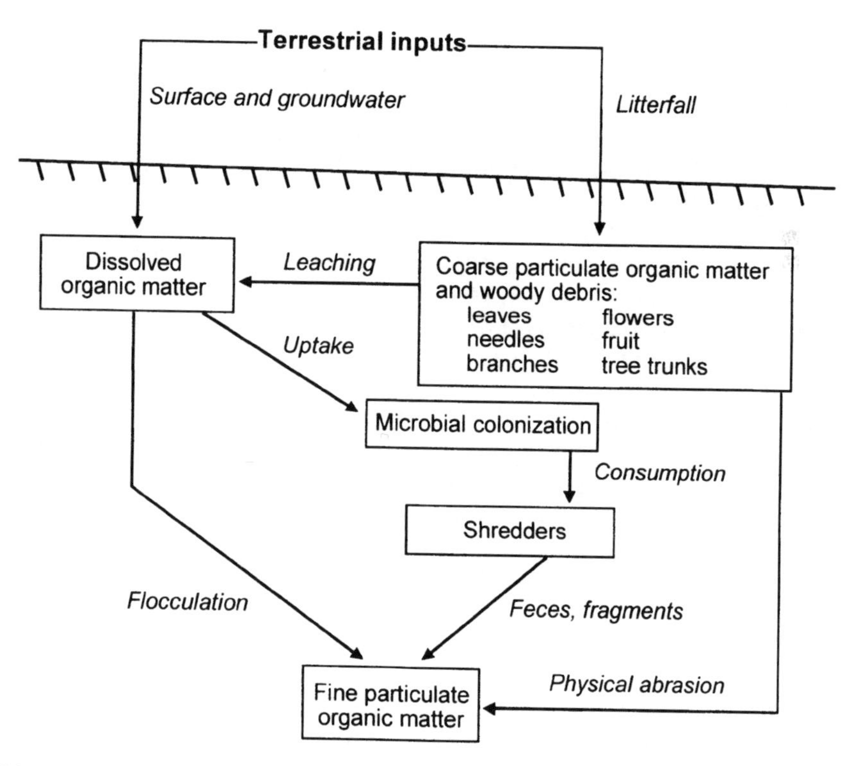

FIGURE 15.3. Processes through which coarse particulate organic matter (CPOM) and dissolved organic matter (DOM) are converted to fine particulate organic matter (FPOM) (from Cummins and Klug 1979, Allan 1995).

substances that provides the main food source to macroinvertebrates specialized for feeding on terrestrial foliage and woody debris (shredders). Dead aquatic macrophytes can also be processed in this way. In the digestive tracts of shredder macroinvertebrates, CPOM is converted to FPOM, which is excreted. Shredders actually derive most of their food value from the microbial coating on CPOM, not from the plant material itself (Anderson and Sedell 1979). In addition to conversion of CPOM to FPOM by macroinvertebrates, FPOM can be formed by the flocculation of DOM and by the physical breakdown of larger particles as they are carried by the stream or pulverized by bedload movement (Figure 15.3).

As headwater streams come together to form larger channels, the relative importance of allochthonous and autochthonous inputs changes significantly (Figure 15.4). In a comparison of community structure and organic matter pathways among 1st-, 3rd-, 5th-, and 7th-order streams in the McKenzie River system of Oregon, Naiman and Sedell (1980) found that autotrophic production exceeded heterotrophic consumption (i.e., the ratio of gross photosynthesis to community respiration, P/R, was greater than one) in both 5th- and 7th-order streams during all seasons. Respiration exceeded photosynthesis in the 3rd-order stream only during winter, but on an annual basis the 3rd-order stream was predominantly heterotrophic (P/R < 1) because it contained a very large pool of detritus accompanied by considerable microbial respiration. The balance between allochthonous and autochthonous pathways gradually shifted from a strongly heterotrophic-based stream community in the headwaters to an autotrophic-based community in the larger streams. For the McKenzie River, this shift from aquatic communities in which P/R is less than one to those in which P/R exceeds one occurred in channels of approximately 3rd- to 5th-order (Hawkins and Sedell 1981).

The hypothesis that the relative balance of photosynthesis and respiration shifts downstream in a predictable pattern in response to changing organic matter sources and increasing light reaching the channel is an important part of the river continuum concept (RCC, Vannote et al. 1980). This hypothesis suggests that the species and functional composition of plant, invertebrate, and vertebrate communities changes according to the physical template of the stream and its valley as well as its sources of organic matter. According to this hypothesis, small headwater streams are dominated by large-bodied shredder and collector-gatherer macroinvertebrates (i.e., those invertebrates specialized for consuming CPOM of terrestrial origin). Mid-order streams, only partially shaded, have a much more significant component of grazing organisms specialized for feeding on periphyton. Large rivers are dominated by three distinct trophic communities: detritivores feeding on fine particulate organic matter that had been consumed and excreted by organisms upstream, phytoplankton and zooplankton, and aquatic macrophytes and their associated fauna along the riverine-streams edge. The role of floodplains as an important source of fine particulate organic matter for larger rivers was recognized (Minshall et al. 1985) after the original formulation of the river continuum concept by Vannote et al. (1980).

Overall, the river continuum concept predicts that low-order streams in forested landscapes should have a very low P/R ratio and a high ratio of coarse to fine particulate organic matter (CPOM/FPOM), reflecting heavily shaded conditions and inputs of coarse material from the surrounding riparian zone. In mid-order streams, the model predicts that P/R should reach a peak and CPOM/FPOM should decline, reflecting an abundance of periphyton and conversion of CPOM to FPOM in the headwaters. In high-order streams, CPOM/FPOM should be very low and P/R should decline somewhat because of heterotrophic processing of floodplain-derived detritus. The river continuum concept predicts that biological diversity should be greatest in mid-order streams because of the wide variety of organic inputs.

An examination of the composition of animal communities from different parts of the McKenzie River system in Oregon generally supports this hypothesis (Hawkins and Sedell

1981). Approximately 75% of the densities of macroinvertebrates in the 1st-order tributary were shredders and the rest were mostly predators (Figure 15.4). In the 3rd-order stream, shredders were still a dominant functional group but grazers and collectors assumed greater importance. Grazers and collector-gatherers became the dominant feeding guilds of the 5th-order stream while shredders were reduced to less than 10% of the densities. In the 7th-order stream, collector–gatherers assumed dominance. Interestingly, in all stream sizes predators comprised about one-fourth of the total standing crop, although the composition shifted from macroinvertebrate and amphibian dominated predator species in headwater streams to communities dominated by predatory fishes in the larger channels.

A number of investigators have tested the river continuum concept in both temperate and tropical streams, although tests of the concept in high-order river basins are relatively few because so many large rivers have been dammed. Minshall et al. (1985, 1992) tested predictions of the RCC model in Idaho's Salmon River, a free-flowing 8th-order drainage system. They found that communities fit expectations in some respects: shredders dominated CPOM-

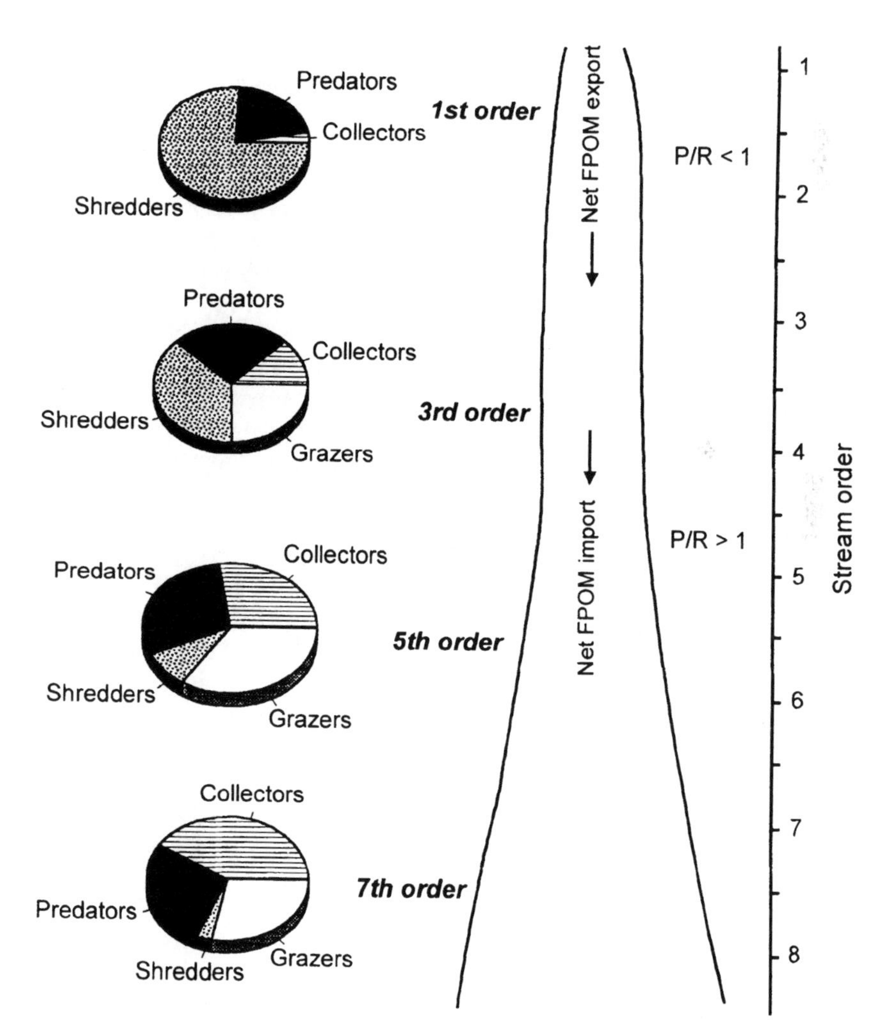

FIGURE 15.4. Changes in the relative importance of allochthonous and autochthonous inputs and the abundance of different functional groups of aquatic organisms in the McKenzie River system, Oregon, from 1st- to 7th-order stream channels (data from Hawkins and Sedell 1981).

rich headwater channels, periphyton photosynthesis was greatest in mid-order reaches, and FPOM and heterotrophic consumption by collector–gatherers dominated the largest streams. Some functional groups, however, did not conform to predictions. This was attributed to anthropogenic disturbances in the Salmon River basin which reduced CPOM inputs in the headwaters and increased sediment caused by hillslope erosion. In New Zealand streams, Winterbourne et al. (1981) were unable to associate community composition with functional group proportions suggested by RCC. They suggested that a hydrologic regime prone to frequent disturbance could change communities in ways that were not easily predictable, but there has been some debate over whether Winterbourne et al. (1981) used the concept correctly. Overall, the river continuum concept has been useful in describing general changes in organic matter inputs and processing from small headwater streams to large floodplain rivers, but there appear to be many deviations from specific predictions of the model in response to local conditions (Statzner and Higler 1986, Statzner et al. 1988, Allan 1995).

Organic Matter Storage and Nutrient Spiraling

Without processes for storing organic matter in stream channels, much of the material would be rapidly carried downstream without being incorporated into the aquatic community. Storage mechanisms are somewhat different for each of the principal organic matter classes: DOM, FPOM, and CPOM. DOM can be stored in ponded water created by wetlands and beaver dams (Naiman and Melillo 1984) or in hyporheic water deep within the substrate of the stream network (Stanford and Ward 1992). These "reservoirs" of DOM are important nutrient sources for the stream and the release of DOM from them often depends on flow (Triska et al. 1989, 1990).

CPOM enters stream channels seasonally or during large disturbances such as landslides, fires, or floods. The ability of channels to retain this material is strongly influenced by their fluvial-geomorphic character (Gregory et al. 1991). CPOM storage is often greatest in low-gradient, alluvial river systems and least in tightly constrained, bedrock-dominated channels (Sedell and Swanson 1984, Montgomery and Buffington 1993). Large woody debris plays an essential role in the retention of both CPOM and FPOM (Bilby 1981, Harmon et al. 1986, Chapter 13, this volume). As a rule, hydraulically complex streams store more CPOM than streams with simplified channels (D'Angelo et al. 1993).

FPOM can be stored in interstitial spaces in coarse sediments of the streambed, in backwaters formed by eddies, and adjacent to large structural elements of the channel such as large woody debris and boulders (Cushing et al. 1993). Storage capacity for FPOM is largely a function of stream power, a hydrological measure of the capacity of the stream to do "work" defined as a function of the shear stress at the bed and the mean velocity of the stream in cross section (Gordon et al. 1992). The greater the stream power, the greater the tendency to suspend and transport FPOM, which is negatively buoyant but nevertheless has a low specific gravity and is easily carried by the current. Characteristics of the channel that either decrease shear stress or reduce the average velocity of the flowing water will increase FPOM storage capacity. Turbulence plays a significant role in keeping FPOM in suspension; highly turbulent streams have greater FPOM transport capacities.

Organic matter and inorganic nutrients cycle back and forth between biologically incorporated, stored, and fluvially transported states as they are taken up and subsequently released (Figure 15.5). The term nutrient spiraling has been applied to the path traced by nutrients as they are assimilated by living organisms, returned to the stream by decomposition, respiration or excretion, and eventually reincorporated into the aquatic community further downstream (Webster and Patten 1979). Procedures for calculating the rate at which nutrients travel downstream are outlined in Webster and Ehrman (1996). Soluble nutrients are termed conservative if their concentration does not change through biotic or abiotic uptake processes (e.g., chloride, which is usually

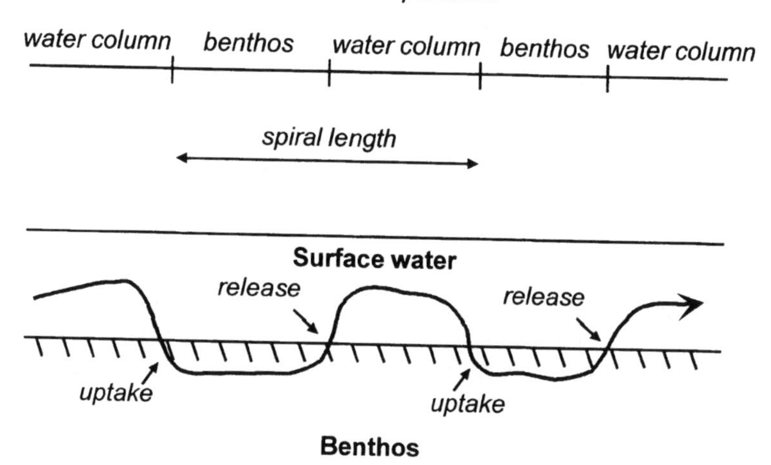

FIGURE 15.5. Two-dimensional nutrient spiraling diagram showing cyclic uptake of nutrients by aquatic biota and subsequent release by remineralization to the water column. The spiral length is the longitudinal distance traveled by a nutrient atom (e.g., nitrogen or phosphorus) over one complete cycle of uptake and release (from Newbold 1992, Allan 1995).

present in streams far in excess of biological demand), or nonconservative if in-stream processes do alter their concentration (e.g., nitrogen and phosphorus). Models of solute movement can be very complex and require computers to solve systems of differential equations that describe variable stream morphology, groundwater and tributary inputs, transient storage, abiotic processes (adsorption, desorption, precipitation, and dissolution), and biotic exchanges (heterotrophic uptake, plant uptake, leaching, and mineralization). As they are recycled between immobilized (benthic) and remobilized (streamwater) states, nonconservative, biologically important nutrients trace a spiral (Figure 5) whose characteristics can be described by the equation:

$$S = SW + SB \tag{15.1}$$

where S represents spiraling length, SW represents uptake length, and SB represents turnover length. Spiraling length is the linear distance along the stream traveled by a nutrient atom while completing a cycle from biotic to abiotic and back to biotic form. The uptake length is the distance traveled in abiotic form dissolved in the water column, while turnover length is the distance traveled in biotic form before being remobilized and returned to the water column. Uptake length is the inverse of uptake rate:

$$SW = 1/kC \tag{15.2}$$

where kC represents the rate of biological uptake of the nutrient atom in its abiotic form (Webster and Ehrman 1996).

Transient storage of nutrients in streams with highly retentive characteristics results in relatively short distances traveled downstream at each phase of the cycle of incorporation and release. Conversely, streams with poor nutrient and organic matter retention characteristics possess longer "spirals." Newbold et al. (1981, 1982) constructed a model of the distance traveled by nutrients within the water column and alternately stored in plant and animal components of the ecosystem as they moved downstream. Other investigators have found that a variety of abiotic controls (precipitation and sorption, seasonal hydrologic regimes, severe storms, patterns of subsurface flow) and biotic processes (autotrophic production, sequestration in long-lived or decomposition-resistant organisms, spawning of anadromous fishes, nitrogen fixation and denitrification) govern the rate of nutrient spiraling (Stream Solute Workshop 1990, Webster and Ehrman 1996). Both standing stock and productivity of members of aquatic communities, including salmonids, are strongly influenced by nutrient dynamics (Mundie 1969, Gregory et al. 1987).

Impacts of Human Activity

It is impossible to describe all of the changes in the trophic system of rivers resulting from anthropogenic disturbances; the reader is once again referred to more comprehensive treatises (Naiman and Decamps 1990, Naiman 1992, National Research Council 1992). Stanford and Ward (1992) summarized several types of human disturbance that disrupt ecological linkages between terrestrial and aquatic ecosystems, alter the concentration of nutrients or toxic substances, or strongly alter aquatic food webs (Table 15.6). Activities resulting in impounded water and regulated streamflows disrupt the downstream movement of dissolved and particulate organic matter, prevent or inhibit the exchange of sediment and organic matter between the river and its floodplain, reduce the exchange of water between the stream and hyporheic flows, and create sediment and nutrient sinks in the drainage. Water pollution alters the transformations (flux rates) of materials within aquatic food webs by reducing or eliminating some members of the aquatic community, accelerates the rate of eutrophication, or results in sedimentation that reduces organic matter storage within the substrate. Introductions of exotic species frequently alter trophic pathways in aquatic food webs and change the biomass and production of different trophic groups (Li et al. 1987).

Human Activities and Cascading Trophic Systems

Effects of species introductions, faunal changes caused by pollution, or harvest of fishes and invertebrates may generate changes throughout a stream's trophic system. Organisms near

TABLE 15.6. Categories, examples, and effects of some common types of anthropogenic disturbances that alter the trophic dynamics of lotic ecosystems.

Category	Examples	Effects
Stream regulation	Dams, water diversions, dredging, diking, revetments	Lotic reaches replaced by reservoirs: loss of upstream–downstream continuity, migration barriers, flood and nutrient sink, stimulates biophysical constancy in downstream environments Channel reconfiguration and simplification: loss of lateral connectivity, removal of large woody debris, isolation of riparian and hyporheic components of floodplains Diversion of water to other river basins: loss of particulate organic matter, dewatering of stream channels, introduction of exotic species, increases in pollutant concentrations
Water pollution	Point source discharges of industrial wastes, nonpoint source runoff from agriculture, urban, highway, and forestry development, airborne pollutants	Deposition of airborne pollutants: eutrophication, acidification Deposition of waterborne pollutants: toxicity, eutrophication Accelerated erosion: sedimentation of stream substrate, eutrophication
Food web manipulations	Introduction of exotic plants and animals, harvest of fishes and invertebrates, aquaculture	Harvest of fishes and invertebrates, aquaculture and hatcheries: biomass and production shifts, loss of marine- or lake-derived nutrients from anadromous or adfluvial species Introduction of exotic species: species displacement, cascading trophic effects

From Stanford and Ward 1992.

the apex of lotic food webs, usually vertebrates or large-bodied invertebrates, can significantly impact the abundance of their prey. If prey are herbivorous, the influence of predation by a carnivore on herbivores can influence the standing crop of algae in the stream. These "top-down" changes are called *trophic cascades*. In streams where aquatic communities are structured by density-dependent processes (competition, predation), experiments by Power and her associates (Power 1984, Power et al. 1985, Power 1990, 1992) have shown that changes in predator populations influence the abundance of grazers, which in turn regulate the biomass of periphyton. For example, Power (1990) manipulated the summer food web of pools in a northern California river in which the community consisted of a relatively large predator (juvenile steelhead trout [*Oncorhynchus mykiss*]) that consumed other fishes and large predatory invertebrates, two small fish species (roach [*Hesperoleucas symmetricus*] and stickleback [*Gasterosteus aculeatus*]) that fed primarily on small invertebrates, predatory invertebrates (larval damselflies [Lestidae]) that also consumed small invertebrates, and a periphyton community dominated by a filamentous green alga (*Cladophora*), diatoms, and a blue-green alga (*Nostoc*). When fish were present, damselflies and other small predators were limited and midge (Orthocladiinae) populations flourished, grazing the periphyton community to a thin layer on the boulder-cobble substrate. When fish were experimentally removed, invertebrate predators increased, grazing invertebrates declined, and periphyton grew into a thick mat covering the stream bottom.

The strength and importance of cascading trophic interactions such as those described by Power (1990) appear to differ greatly within and among streams and it is possible that in the majority of cases predators do not control the structure and abundance of lower trophic levels. Differences in predator–prey linkages, disturbance regimes, and recruitment of colonizing organisms from upstream sources can obscure cascading trophic interactions (Allan 1995). In a review of experimental predation studies, Cooper et al. (1990) concluded that the ability of predators to structure lower trophic levels declines with increasing rates of prey recruitment. Hawkins and Furnish (1987) found that grazing can be dominated by herbivores such as snails (*Juga*) that are generally not eaten by vertebrate predators and are thus not affected by predator densities. Physical features of stream channels also influence the strength of trophic cascade effects. For example, Power (1992) was unable to detect strong top-down predator effects on the structure of communities in gravel riffles, which were more complex and may have provided more spatial refugia than the pool community in the earlier study (Power 1990). Feminella and Hawkins (1995) concluded that most studies have shown no or only weak effects of stream predators on herbivores, but often strong effects of herbivores on periphyton. On the other hand, "bottom up" controls based on food resource limitation (reviewed in Allan 1995) appear to exert a strong control over the food web structure of many streams.

Additionally, a stream's disturbance regime may affect the characteristics of its food web. Wootton et al. (1996) presented experimental evidence that scouring floods in northern California strongly influenced the abundance of predator-resistant grazers (a caddisfly [*Dicosmoecus*]). When this large, armored benthic grazer was abundant, algal biomass was lower and densities of predator-susceptible grazers and their predators was likewise reduced. Wootton et al. (1996) found that streams with regulated flows were often dominated by predator-resistant grazers, whereas streams with a natural flow regime, including periodic flooding, possessed more predator-susceptible grazers and thus more insectivorous fishes.

Human-induced changes in the species composition of stream communities can have dramatic and unanticipated trophic effects, with changes occurring at every trophic level from primary producers to top level consumers (Ross 1991). In many instances the deliberate introduction of a large carnivore has radically altered the structure of lake communities (e.g., the release of Nile perch [*Tilapia nilotica*] into rift lakes of central Africa) (Kaufman 1992). In the Pacific coastal ecoregion, the most extensive changes to riverine ecosystems caused by species introductions have occurred in large, impounded rivers such as the Columbia River and the Sacramento River, where large numbers of exotic species have transformed the food web and largely displaced the native fish fauna from much of the middle and lower river (Moyle et al. 1986, Li et al. 1987).

Loss of Riparian Vegetation

Loss of riparian vegetation, especially trees, has been one of the most pervasive changes to riverine ecosystems resulting from human activities. From the smallest headwater streams to the largest floodplain rivers, riparian vegetation influences organic matter inputs and trophic pathways. Removal of riparian vegetation eliminates a major source of terrestrial organic matter, reduces shade to small streams causing a shift in the autotrophic community, and often leads to streambank erosion that increases fine sediment in the channel. All of these changes affect trophic processes in rivers (Gregory et al. 1991). Loss of riparian trees also reduces the recruitment of large woody debris, which plays an important role in storing sediment and FPOM, provides habitat for aquatic organisms, mediates channel topography, and itself forms an important source of organic matter (Harmon et al. 1986, Ward and Aumen 1986, Bisson et al. 1987, Chapter 13, this volume).

Timber harvesting in riparian zones has been a common cause of riparian alteration in western North America. Overall, removal of trees from riparian zones in forested landscapes has had negative consequences for stream-dwelling salmonids (Hicks et al. 1991) but increases in summer biomass and production of headwater salmonid populations after logging have been reported from enough sites in the Pacific Northwest (e.g., Murphy and Hall 1981, Hawkins et al. 1983, Bisson and Sedell 1984, Bilby and Bisson 1987, Holtby and Scrivener 1989) to suggest that trophic pathways supporting salmonids have actually been enhanced by timber harvest in these headwater areas.

A comparison of the organic matter inputs to paired old-growth forested and clear-cut watersheds (Bilby and Bisson 1992) demonstrated that headwater salmonid populations relied primarily on autotrophically based food in summer (Table 15.7). In this study, approximately 90% of the food items consumed by juvenile coho salmon (*Oncorhynchus kisutch*) in both the clear-cut and old-growth streams were organisms supported by autochthonous organic matter sources. Especially prominent in the diets of coho and cutthroat trout (*O. clarki*) were larval midges and baetid mayflies, two invertebrate groups that ingest algae and algal based detritus. Although total organic matter inputs to the stream in the old-growth forest were approximately two times greater than at the clear-cut site, production of fishes in the old-growth forested stream was less than one half that observed at the clear-cut site. Ratios of summer production of two salmonids and a sculpin between the clear-cut and old-growth sites were remarkably similar to the ratios of autochthonous inputs at these streams (Table 15.7), strongly supporting the hypothesis that the fish community was supported by autotrophic food pathways in summer regardless of the condition of the forest canopy.

Similar increases in summer populations of salmonids in the Pacific coastal ecoregion have been reported after losses of riparian vegetation caused by other land uses (e.g., channelization, livestock management) (Chapman and Knudsen 1980). Increased salmonid productivity has apparently been related to greater amounts of light reaching stream channels, stimulating photosynthesis and boosting secondary production of heterotrophs dependent on periphyton. Yet, juvenile salmonids do not always benefit from enhanced autotrophic production following vegetative canopy removal (Gregory et al. 1987). Elevated stream temperature accompanying increased sunlight may favor species better adapted to warmer waters if such species are present in the system. For example, Reeves et al. (1987) found that stream warming after logging in riparian zones led to increased populations of redside shiners (*Richardsonius balteatus*), which displaced juvenile steelhead. In warmer streams, shiners became more aggressive and outcompeted steelhead for pre-

TABLE 15.7. Average daily organic matter inputs from allochthonous and autochthonous sources in two 4th-order streams, a clear-cut and an old-growth forested watershed in the Cascade Mountains of Washington; the production of fishes inhabiting these sites from late spring through early autumn; and the ratios of organic inputs and fish production at the two sites.

	Clear–cut ($mg/m^2/d$)	Old–growth ($mg/m^2/d$)	Ratio (clear-cut/old-growth)
Annual organic matter inputs			
Allochthonous sources	164	851	0.19
Autochthonous sources	482	301	1.60
Fish production			
Coho salmon (*Oncorhynchus kisutch*)	34	18	1.89
Cutthroat trout (*O. clarki clarki*)	8.2	7.4	1.10
Shorthead sculpin (*Cottus confusus*)	4.5	2.5	1.82

From Bilby and Bisson 1992.

ferred foraging locations. Salmonid production in small forested streams of the Pacific coastal ecoregion generally tends to increase following removal of streamside vegetation at northern latitudes or high altitudes (Dolloff 1983, Holtby and Scrivener 1989). In the southern part of the region most of the enhanced autotrophic production has benefited cyprinid fishes and other species preferring warmer waters (Reeves et al. 1987, R.E. Bilby and B.R. Fransen, Weyerhaeuser Company, Tacoma, Washington, USA, personal communication).

Loss of Salmon Carcasses

Another pervasive effect of human activity in many of the river basins draining to the Pacific Ocean has been a large reduction in the numbers of anadromous salmonids naturally spawning in their natal rivers (Nehlsen et al. 1991). There are a multitude of reasons for declines in the abundance of spawning salmon, including natural as well as anthropogenic causes, but in terms of impact on the trophic system of streams, loss of carcasses has been one of the most significant (NRC 1996). Salmon carcasses appear to influence food pathways supporting rearing juveniles and other stream-dwelling fishes in several ways. First, mineralized nutrients from carcass decomposition stimulate autotrophic production (Richey et al. 1975, Kline et al. 1990, 1993). Second, organic leachates from carcass tissues stimulate microbial uptake and enhance the heterotrophic food web (Bilby et al. 1996). Third, carcasses are consumed by macroinvertebrates, which in turn may become prey for fishes (Piorkowski 1995). Finally, carcasses, eggs, and alevins are eaten directly by fish. The latter pathway has received relatively little attention, but recent evidence has shown that direct consumption of carcasses, eggs, and alevins provide an extremely valuable food resource for juvenile salmonids during a period when other food resources are scarce (Figure 15.6).

Aside from providing trophic support to salmonids, salmon carcasses serve as vectors of marine-derived nutrients that benefit many other organisms in watersheds. Carcasses are a seasonally important resource for a variety of terrestrial scavengers (Cederholm et al. 1989). Nutrients from carcasses deposited on streambanks and riverine terraces by floods and scavenger activities fertilize riparian vegetation (Bilby et al. 1996). High flows carry carcasses as well as nutrients from their decomposition to lower rivers and estuaries, where they may enhance both autotrophic and heterotrophic production.

To what extent carcass deposition in streams has been reduced relative to predevelopment levels is not well known in most areas. Long-term records of spawning counts are available from only a few streams, and historical runs are usually reconstructed from catch statistics (Bisson et al. 1992). However, in watersheds where fishery policies have emphasized hatch-

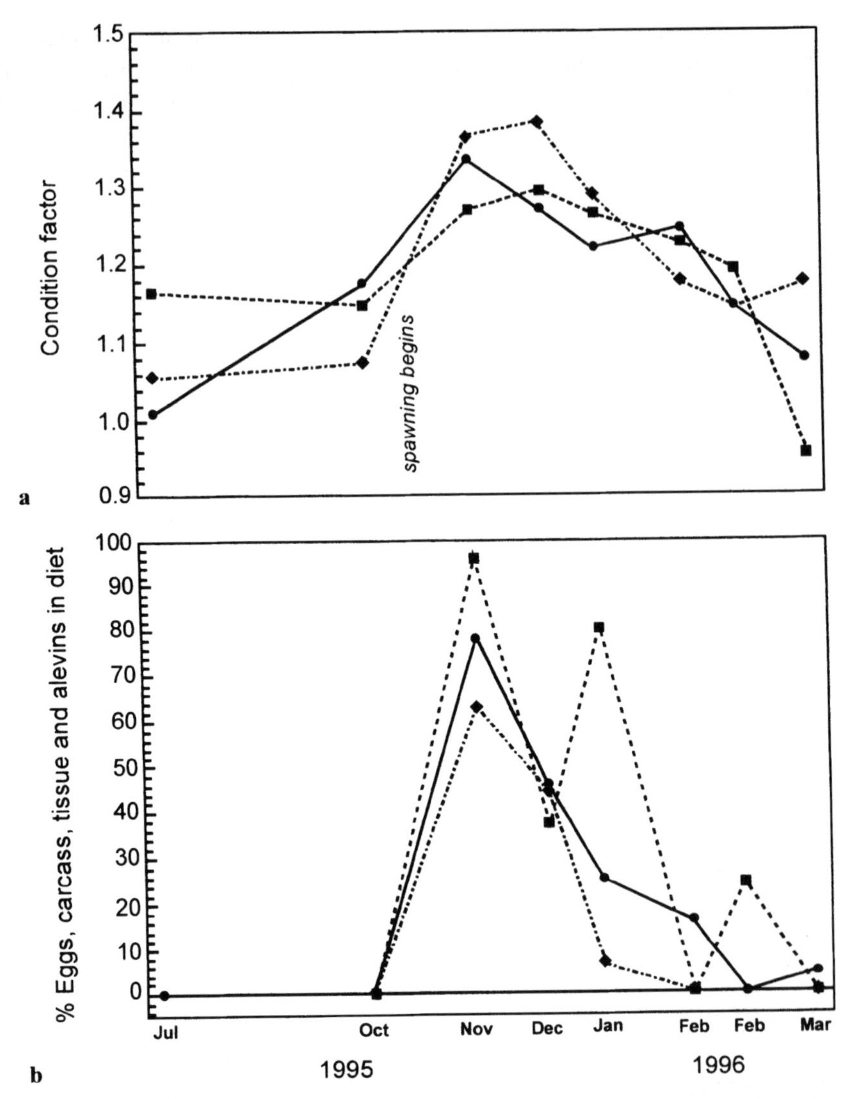

FIGURE 15.6. Influence of spawning by adult coho salmon on a) condition factor (a commonly used measure of overall fish condition or "fatness") and b) diet of juvenile salmonids (steelhead [age 0]-●-; steelhead [age 1]-■-; and coho [age 0]-◆-) in a small tributary of the Willapa River, Washington (unpublished data of B.R. Fransen, J. Walter, and R.E. Bilby, Weyerhaeuser Company, Tacoma, Washington, USA).

ery production and permitted high fishing rates, numbers of salmon spawning naturally in rivers and their tributaries are typically quite low (Nickelson et al. 1992, Washington Department of Fisheries et al. 1993). In one coastal basin in southwestern Washington state, naturally spawning salmon are estimated to be approximately one tenth of the numbers that spawned in the system earlier in the twentieth century (NRC 1996). This reduction has deprived the basin of almost 3,000 metric tons (wet weight) of salmon carcass biomass each year. Similar reductions are likely to have occurred elsewhere. For example, estimates of predevelopment salmon runs in the Columbia River range from about 8 to 16 million fish annually (Chapman 1986) but current runs total less than two million adults, and most of these are hatchery-produced fish that will either be caught in fisheries or return to the hatchery of their origin. Very few spawn naturally in the Columbia River basin, relative to the salmon runs that occurred there in the ninteenth century. The significance of this loss for the trophic systems of streams and rivers is seriously underappreciated.

Why Are Some Streams More Productive Than Others?

Consider, again, the question of productivity—the capacity of a stream to produce a product of interest. The chapter began with a comparison of the production of salmonids in rivers throughout the world (Table 15.1). Most streams are relatively unproductive for salmonids; only a few demonstrate high levels of productivity (Figure 15.1). What makes these streams productive, while so many others are not? One hypothesis is that they possess superior physical habitat, but a careful examination of descriptions of study sites reveals that many unproductive streams contain abundant pools and cover, two features often emphasized in assessment of salmonid habitat quality. Some of the most productive streams (e.g., British chalk streams and some spring-fed creeks) tend to have very fine-grained substrates that would be considered distinctly suboptimal for salmonids. Additionally, some of the coldest sites are quite unproductive (e.g., those in Alaska, the coastal rain forests of the Pacific Northwest, the Rocky Mountains of western North America, the boreal forests of eastern Canada). Physical characteristics such as cold water, channel morphology, and coarse substrates, so often identified as key limiting factors in lotic environments, do not appear to be primarily responsible for regulating the productivity of salmon and trout in streams. This is not to discount the importance of habitat, rather to point out that other factors have a powerful influence on productivity.

The most productive streams appear to contain abundant food. Although most of the studies in Table 15.1 have not quantitatively assessed food availability, productive sites tend to possess hard waters with relatively high inorganic nutrient concentrations; moderate temperatures, especially in spring-fed streams where temperatures are buffered by groundwater inputs year round; relatively low vegetative canopy coverage allowing ample sunlight to reach the streams; and abundant macrophytes and mosses, or dense growths of filamentous algae. These conditions are indicative of high levels of autotrophic production, which may be the most important trophic pathway for aquatic macroinvertebrates preferred by salmonids during spring and summer (Bilby and Bisson 1992). Although there are exceptions, production of salmonids is often more strongly influenced by high growth rates than by dense populations (Warren 1971). Growth rates are a function of food availability, metabolic costs of obtaining and processing food, and density-dependent interactions including competition and predation (Fausch 1984). High growth rates require abundant food resources (Dill et al. 1981), even when physical habitat and water quality are favorable for growth. Population density is mediated by habitat quality but growth rates can be low when densities are relatively high, even in high-quality habitat (Chapman 1966, Mason 1976, Bilby and Bisson 1992, Fransen et al. 1993). For stream-dwelling salmonids, these observations suggest that food availability may be one of the most important factors controlling production.

Further evidence for the importance of food is provided by studies of experimental stream enrichment. Virtually every attempt to increase production by adding inorganic nutrients or organic matter to rivers in the Pacific coastal ecoregion has resulted in increased salmonid production. The most striking increases have come from placing food organisms directly in streams (Mason 1976) or attempting to increase autotrophic production (Johnston et al. 1990, Slaney and Ward 1993). Salmonid production also has been enhanced by increasing heterotrophic pathways (Warren et al. 1964, Mundie et al. 1983, Perrin et al. 1987), but results have not been as obvious. Of course, these studies did not attempt to add so much organic matter or nutrients that the streams suffered from obvious effects of excessive enrichment characteristic of polluted waters.

Chapman (1966) hypothesized that food would limit production of Pacific coastal salmonids during conditions of summer low flow, and available rearing space would limit production during winter when fish need refuge from high flows. Mundie (1974) speculated that heterotrophic food pathways were generally the most important food resources for juvenile salmon and suggested that it might be possible to create artificial channels in which food organisms available to salmonids could be deliberately enhanced. Based on recent findings, the conclusions of these two seminal papers with regard to the trophic dynamics of Pacific coastal rivers containing anadromous salmonids can be updated. First, available evidence suggests that food availability may limit the productive capacity of streams for salmonids throughout the year, including autumn and winter. Second, autotrophic production is the principal source of trophic support for salmonids during spring and summer, but heterotrophic food pathways dominate during autumn and winter when streams receive inputs of leaves and salmon carcasses. In the spring, as photoperiod lengthens, scouring freshets subside, and stream temperatures warm, salmonids begin to feed primarily on organisms supported by autochthonous production.

If salmonids in Pacific coastal rivers rely primarily on alternating autotrophic and heterotrophic food pathways throughout the year, natural resource managers must consider both sources of organic matter when formulating land use and fishery policies. To attempt to return all streams to a densely forested, heavily shaded condition, for example, would negatively impact autochthonous production, although returning riparian zones to dense, mature conifer forest on a broad regional scale is unlikely to happen in the foreseeable future. On the other hand, benefits of openings in the riparian canopy, such as openings created by natural disturbances, should be acknowledged and included in the planning process for any watershed. In addition, the role of salmon carcasses must be expanded to include their trophic contribution to the stream ecosystem. To limit the numbers of returning adults strictly to those needed to adequately populate available habitat in the stream network with fry is to misunderstand the notion of carrying capacity and to guarantee that autumn and winter food webs will be impaired. As relationships between salmonid productivity and food pathways become better understood, natural resource managers will be in a much better position to determine what rivers can and cannot produce.

Acknowledgments. The authors gratefully acknowledge the support of the USDA Forest Service and the Weyerhaeuser Company during the preparation of this paper. Charles P. Hawkins and Robert J. Naiman offered helpful suggestions for improvement of the manuscript. Brian R. Fransen generously provided unpublished data on juvenile salmonid utilization of carcasses in the Willapa River system.

Literature Cited

Alexander D.R., and H.R. MacCrimmon. 1974. Production and movement of juvenile rainbow trout (*Salmo gairdneri*) in a headwater of Bothwell's Creek, Georgian Bay, Canada. Journal of the Fisheries Research Board of Canada **31**:117–121.

Allan, J.D. 1995. Stream ecology: Structure and function of running waters. Chapman & Hall, London, UK.

Allen, K.R. 1951. The Horokiwi Stream: A study of a trout population. New Zealand Department of Fisheries Bulletin **10**:1–238.

Anderson, N.H., and J.R. Sedell. 1979. Detritus processing by macroinvertebrates in stream ecosystems. Annual Review of Entomology **24**: 351–377.

Berkman, H.E., and C.F. Rabeni. 1987. Effect of siltation on stream fish communities. Environmental Biology of Fishes **18**:285–294.

Bilby, R.E. 1981. Role of organic debris dams in regulating the export of dissolved and particulate matter from a forested watershed. Ecology **62**:1234–1243.

Bilby, R.E., and P.A. Bisson. 1987. Emigration and production of hatchery coho salmon (*Oncorhynchus kisutch*) stocked in streams draining an old-growth and a clear-cut watershed. Canadian

Journal of Fisheries and Aquatic Sciences **44**:1397–1407.

Bilby, R.E., and P.A. Bisson. 1992. Allochthonous versus autochthonous organic matter contributions to the trophic support of fish populations in clear-cut and old-growth forested streams. Canadian Journal of Fisheries and Aquatic Sciences **49**:540–551.

Bilby, R.E., B.R. Fransen, and P.A. Bisson. 1996. Incorporation of nitrogen and carbon from spawning coho salmon into the trophic system of small streams: Evidence from stable isotopes. Canadian Journal of Fisheries and Aquatic Sciences **53**:164–173.

Bisson, P.A., R.E. Bilby, M.D. Bryant, C.A. Dolloff, G.B. Grette, R.A. House, et al. 1987. Large woody debris in forested streams in the Pacific Northwest: Past, present, and future. Pages 143–190 *in* E.O. Salo and T.W. Cundy, eds. Streamside management: Forestry and fishery interactions. Institute of Forest Resources Contribution Number 57, University of Washington, Seattle, Washington, USA.

Bisson, P.A., J.L. Nielsen, and J.W. Ward. 1988. Summer production of coho salmon stocked in M. St. Helens streams 3–6 years after the 1980 eruption. Transactions of the American Fisheries Society **117**:322–335.

Bisson, P.A., T.P. Quinn, G.H. Reeves, and S.V. Gregory. 1992. Best management practices, cumulative effects, and long-term trends in fish abundance in Pacific Northwest river systems. Pages 189–232 *in* R.J. Naiman, ed. Watershed management: Balancing sustainability and environmental change. Springer-Verlag, New York, New York, USA.

Bisson, P.A., and J.R. Sedell. 1984. Salmonid populations in streams in clearcut vs. old-growth forests of western Washington. Pages 121–129 *in* W.R. Meehan, T.R. Merrell, Jr., and T.A. Hanley, eds. Fish and wildlife relationships in old-growth forests. American Institute of Fisheries Research Biologists, Juneau, Alaska, USA.

Cederholm, C.J., D.B. Houston, D.L. Cole, and W.J. Scarlett. 1989. Fate of coho salmon (*Oncorhynchus kisutch*) carcasses in spawning streams. Canadian Journal of Fisheries and Aquatic Sciences **46**:1347–1355.

Chapman, D.W. 1965. Net production of juvenile coho salmon in three Oregon streams. Transactions of the American Fisheries Society **94**:40–52.

——. 1966. Food and space as regulators of salmonid populations in streams. American Naturalist **100**:345–357.

——. 1978. Production in fish populations. Pages 5–25 *in* S.D. Gerking, ed. Ecology of freshwater fish production. Blackwell Scientific, Oxford, UK.

——. 1986. Salmon and steelhead abundance in the Columbia River in the nineteenth century. Transactions of the American Fisheries Society **115**:662–670.

Chapman, D.W., and E. Knudsen. 1980. Channelization and livestock impacts on salmonid habitat and biomass in western Washington. Transactions of the American Fisheries Society **109**:357–363.

Cooper, E.L., and R.C. Scherer. 1967. Annual production of brook trout (*Salvelinus fontinalis*) in fertile and infertile streams of Pennsylvania. Proceedings of the Pennsylvania Academy of Sciences **41**:65–70.

Cooper, S.D., S.J. Walde, and B.L. Peckarsky. 1990. Prey exchange rates and the impact of predators on prey populations in streams. Ecology **71**:1503–1514.

Cummins, K.W., and M.J. Klug. 1979. Feeding ecology of stream invertebrates. Annual Review of Ecology and Systematics **10**:147–172.

Cushing, C.E., G.W. Minshall, and J.D. Newbold. 1993. Transport dynamics of fine particulate organic matter in two Idaho streams. Limnology and Oceanography **38**:1101–1115.

D'Angelo, D.J., J.R. Webster, S.V. Gregory, and J.L. Meyer. 1993. Transient storage in Appalachian and Cascade mountain streams as related to hydraulic parameters. Journal of the North American Benthological Society **12**:223–235.

Dill, L.M., R.C. Ydenberg, and A.H.G. Fraser. 1981. Food abundance and territory size in juvenile coho salmon (*Oncorhynchus kisutch*). Canadian Journal of Zoology **59**:1801–1809.

Dolloff, C.A. 1983. The relationships of wood debris to juvenile salmonid production and microhabitat selection in small southeast Alaska streams. Ph.D. thesis, Montana State University, Bozeman, Montana, USA.

——. 1986. Effects of stream cleaning on juvenile coho salmon and Dolly Varden in southeast Alaska. Transactions of the American Fisheries Society **115**:743–755.

Egglishaw, H.J., and P.E. Shackley. 1977. Growth, survival and production of juvenile salmon and trout in a Scottish stream, 1966–1975. Journal of Fish Biology **11**:647–672.

Elliott, J.M. 1985. Population dynamics of migratory trout, *Salmo trutta*, in a Lake District stream, 1966–83, and their implications for fisheries management. Journal of Fish Biology **27**(Supplement A):35–43.

Fausch, K.D. 1984. Profitable stream positions for salmonids: Relating specific growth rate to net energy gain. Canadian Journal of Zoology **62**:441–451.

Feminella, J.W., and C.P. Hawkins. 1995. Interactions between stream herbivores and periphyton: A quantitative analysis of past experiments. Journal of the North American Benthological Society **14**:465–509.

Fransen, B.R., P.A. Bisson, R.E. Bilby, and J.W. Ward. 1993. Physical and biological constraints on summer rearing of juvenile coho salmon (*Oncorhynchus kisutch*) in small western Washington streams. Pages 271–288 *in* L. Berg and P. Delaney, eds. Proceedings of a workshop on coho salmon, held May 26–28, 1992, at Nanaimo, British Columbia. Department of Fisheries and Oceans, Vancouver, British Columbia, Canada.

Goodnight, W.H., and T.C. Bjornn. 1971. Fish production in two Idaho streams. Transactions of the American Fisheries Society **100**:769–780.

Gordon, D.J., and H.R. MacCrimmon. 1982. Juvenile salmonid production in a Lake Erie nursery stream. Journal of Fish Biology **21**:455–473.

Gordon, N.D., T.A. McMahon, and B.L. Finlayson. 1992. Stream hydrology: An introduction for ecologists. John Wiley & Sons, Chichester, UK.

Gregory, S.V., G.A. Lamberti, D.C. Erman, K.V. Koski, M.L. Murphy, and J.R. Sedell. 1987. Influence of forest practices on aquatic production. Pages 233–255 *in* E.O. Salo and T.W. Cundy, eds. Streamside management: Forestry and fishery interactions. Institute of Forest Resources Contribution Number 57, University of Washington, Seattle, Washington, USA.

Gregory, S.V., F.J. Swanson, and W.A. McKee. 1991. An ecosystem perspective of riparian zones. BioScience **40**:540–551.

Hanson, D.L., and T.F. Waters. 1974. Recovery of standing crop and production rate of a brook trout population in a flood-damaged stream. Transactions of the American Fisheries Society **103**:431–439.

Harmon, M.E., J.F. Franklin, F.J. Swanson, P. Sollins, S.V. Gregory, J.D. Lattin, et al. 1986. Ecology of coarse woody debris in temperate ecosystems. Advances in Ecological Research **15**:133–302.

Hawkins, C.P., and J.K. Furnish. 1987. Are snails important competitors in stream ecosystems? Oikos **49**:209–220.

Hawkins, C.P., M.L. Murphy, N.H. Anderson, and M.A. Wilzbach. 1983. Density of fish and salamanders in relation to riparian canopy and physical habitat in streams of the northwestern United States. Canadian Journal of Fisheries and Aquatic Sciences **40**:1173–1185.

Hawkins, C.P., and J.R. Sedell. 1981. Longitudinal and seasonal changes in functional organization of macroinvertebrate communities in four Oregon streams. Ecology **62**:387–397.

Hicks, B.J., J.D. Hall, P.A. Bisson, and J.R. Sedell. 1991. Response of salmonids to habitat changes. American Fisheries Society Special Publication **19**:83–138.

Hildrew, A.G., M.K. Dobson, A. Groom, et al. 1987. Flow and retention in the ecology of stream invertebrates. Verhandlungen Internationale Vereinigung für Theoretishe und Angewandte Limnologie **24**:1742–1747.

Holtby, L.B., and J.C. Scrivener. 1989. Observed and simulated effects of climatic variability, clear-cut logging and fishing on the numbers of chum salmon (*Oncorhynchus keta*) and coho salmon (*O. kisutch*) returning to Carnation Creek, British Columbia. Pages 62–81 *in* C.D. Levings, L.B. Holtby, and M.A. Henderson, eds. Proceedings of the national workshop on effects of habitat alterations on salmonid stocks. Canadian Special Publication of Fisheries and Aquatic Sciences 105, Ottawa, Ontario, Canada.

Hopkins, C.L. 1971. Production of fish in two small streams in the North Island of New Zealand. New Zealand Journal of Marine and Freshwater Research **5**:280–290.

Horton, P.A., R.G. Bailey, and S.I. Wilsdon. 1968. A comparative study of the bionomics of the salmonids of three Devon streams. Archiv für Hydrobiologie **2**:187–204.

Hunt, R.L. 1966. Production and angler harvest of wild brook trout in Lawrence Creek, Wisconsin. Technical Bulletin 35, Wisconsin Conservation Department, Milwaukee, Wisconsin, USA.

——. 1993. Trout stream therapy. University of Wisconsin Press, Madison, Wisconsin, USA.

Hynes, H.B.N. 1970. The ecology of running waters. University of Toronto Press, Toronto, Ontario, Canada.

Johnson, J.H. 1980. Production and growth of subyearling coho salmon, *Oncorhynchus kisutch*, chinook salmon, *Oncorhynchus tshawytscha*, and steelhead, *Salmo gairdneri*, in Orwell Brook, tributary of Salmon River, New York. NOAA Fishery Bulletin **78**:549–554.

Johnston, N.T., C.J. Perrin, P.A. Slaney, and B.R. Ward. 1990. Increased juvenile salmonid growth by whole-river fertilization. Canadian Journal of Fisheries and Aquatic Sciences **47**:862–872.

Karr, J.R. 1991. Biological integrity: A long-neglected aspect of water resource management. Ecological Applications **1**:66–84.

Kaufman, L. 1992. Catastrophic change in species-rich freshwater ecosystems. BioScience **42**:846–858.

Kline, T.C., Jr., J.J. Goering, O.A. Mathisen, P.H. Poe, and P.L. Parker. 1990. Recycling of elements transported upstream by runs of Pacific salmon: I. δ15N and δ13C evidence in Sashin Creek, southeastern Alaska. Canadian Journal of Fisheries and Aquatic Sciences **47**:136–144.

Kline, T.C., Jr., J.J. Goering, O.A. Mathisen, P.H. Poe, P.L. Parker, and R.S. Scanlan. 1993. Recycling of elements transported upstream by runs of Pacific salmon: II. δ15N and δ13C evidence in the Kvichak River watershed, Bristol Bay, southwestern Alaska. Canadian Journal of Fisheries and Aquatic Sciences **50**:2350–2365.

Le Cren, E.D. 1969. Estimates of fish populations and production in small streams of England. Pages 269–280 *in* T.G. Northcote, ed. Symposium on salmon and trout in streams. H.R. MacMillan Lectures in Fisheries, University of British Columbia, Vancouver, British Columbia, Canada.

Leclerc, J., and G. Power. 1980. Production of brook charr and ouananiche in a large rapid, tributary of Caniapiscau River, northern Quebec. Environmental Biology of Fishes **5**:27–32.

Li, H.W., C.B. Schreck, C.E. Bond, and E. Rexstad. 1987. Factors influencing changes in fish assemblages of Pacific Northwest streams. Pages 193–202 *in* W.J. Matthews and D.C. Heins, eds. Community and evolutionary ecology of North American stream fishes. University of Oklahoma Press, Norman, Oklahoma, USA.

Libosvarsky, J. 1968. A study of brown trout population (*Salmo trutta* morpha *fario* L.) in Loucka Creek (Czechoslovakia). Acta Sciences Natural Brno **2**:1–56.

Mann, R.H.K. 1971. The populations, growth, and production of fish in four small streams in southern England. Journal of Animal Ecology **40**:155–190.

Mann, R.H.K., J.H. Blackburn, and W.R.C. Beaumont. 1989. The ecology of brown trout *Salmo trutta* in English chalk streams. Freshwater Biology **21**:57–70.

Mason, J.C. 1976. Response of underyearling coho salmon to supplemental feeding in a natural stream. Journal of Wildlife Management **40**:775–788.

Merritt, R.W. and K.W. Cummins, eds. 1996a. An introduction to the aquatic insects of North America, 3rd edition. Kendall/Hunt, Dubuque, Iowa, USA.

Merritt, R.W., and K.W. Cummins. 1996b. Trophic relations of macroinvertebrates. Pages 453–474 *in* F.R. Hauer and G.A. Lamberti, eds. Methods in stream ecology. Academic Press, San Diego, California, USA.

Milner, N.J., A.S. Gee, and R.J. Hemsworth. 1978. The production of brown trout, *Salmo trutta*, in tributaries of the Upper Wye, Wales. Journal of Fish Biology **13**:599–612.

Minshall, G.W. 1978. Autotrophy in stream ecosystems. BioScience **28**:767–771.

Minshall, G.W., K.W. Cummins, R.C. Petersen, C.E. Cushing, D.A. Burns, J.R. Sedell, and R.L. Vannote. 1985. Developments in stream ecosystem theory. Canadian Journal of Fisheries and Aquatic Sciences **42**:1045–1055.

Minshall, G.W., R.C. Peterson, T.L. Bott, et al. 1992. Stream ecosystem dynamics of the Salmon River, Idaho: An 8th-order system. Journal of the North American Benthological Society **11**:111–137.

Montgomery, D.R., and J.M. Buffington. 1993. Channel classification, prediction of channel response, and assessment of channel condition. Report TFW-SH10-93-002. Washington State Timber/Fish/Wildlife Agreement, Department of Natural Resources, Olympia, Washington, USA.

Mortensen, E. 1977a. Population, survival, growth and production of trout *Salmo trutta* in a small Danish stream. Oikos **28**:9–15.

——. 1977b. The population dynamics of young trout (*Salmo trutta* L.) in a Danish brook. Journal of Fish Biology **10**:23–33.

——. 1982. Production of trout, *Salmo trutta*, in a Danish stream. Environmental Biology of Fishes **7**:349–356.

Moyle, P.B., H.W. Li, and B.A. Barton. 1986. The Frankenstein effect: Impact of introduced fishes on native fishes in North America. Pages 415–426 *in* R.H. Stroud, ed. Fish culture in fisheries management. American Fisheries Society, Bethesda, Maryland, USA.

Mundie, J.H. 1969. Ecological implications of the diet of juvenile coho in streams. Pages 135–152 *in* T.G. Northcote, ed. Symposium on salmon and trout in streams. H.R. MacMillan Lectures in Fisheries, University of British Columbia, Vancouver, British Columbia, Canada.

Mundie, J.H. 1974. Optimization of the salmonid nursery stream. Journal of the Fisheries Research Board of Canada **31**:1827–1837.

Murphy, M.L., and J.D. Hall. 1981. Varied effects of clear-cut logging on predators and their habitat in small streams of the Cascade Mountains, Oregon. Canadian Journal of Fisheries and Aquatic Sciences **38**:137–145.

Mundie, J.H., S.M. McKinnel, and R.E. Traber. 1983. Responses of stream zoobenthos to enrichment of gravel substrates with cereal grain and soybean. Canadian Journal of Fisheries and Aquatic Sciences **40**:1702–1712.

Naiman, R.J. 1976. Primary production, standing stock, and export of organic matter in a Mojave Desert thermal stream. Limnology and Oceanography **21**:60–73.

——, ed. 1992. Watershed management: Balancing sustainability and environmental change. Springer-Verlag, New York, New York, USA.

Naiman, R.J., T.J. Beechie, L.E. Benda, P.A. Bisson, L.H. MacDonald, M.D. O'Connor, et al. C. Oliver, P. Olson, and E.A. Steel. 1992. Fundamental elements of ecologically healthy watersheds in the Pacific Northwest coastal ecoregion. Pages 127–188 *in* R.J. Naiman, ed. Watershed management: Balancing sustainability and environmental change. Springer-Verlag, New York, New York, USA.

Naiman, R.J., and H. Decamps, eds. 1990. The ecology and management of aquatic-terrestrial ecotones. UNESCO, Paris, and Parthenon, Carnforth, UK.

Naiman, R.J., and J.M. Melillo. 1984. Nitrogen budget of a subarctic stream altered by beaver (*Castor canadensis*). Oecologia (Berlin) **62**:150–155.

Naiman, R.J., and J.R. Sedell. 1979. Characteristics of particulate organic matter transported by some Cascade mountain streams. Journal of the Fisheries Research Board of Canada **36**:17–31.

Naiman, R.J., and J.R. Sedell. 1980. Relationships between metabolic parameters and stream order in Oregon. Canadian Journal of Fisheries and Aquatic Science **37**:834–847.

Nehlsen, W., J.E. Williams, and J.A. Lichatowich. 1991. Pacific salmon at the crossroads: Stocks at risk from California, Oregon, Idaho and Washington. Fisheries **16**:4–21.

Neves, R.J., and G.B. Pardue. 1983. Abundance and production of fishes in a small Appalachian stream. Transactions of the American Fisheries Society **112**:21–26.

Newbold, J.D. 1992. Cycles and spirals of nutrients. Pages 379–408 *in* P. Calow and G.E. Petts, eds. The rivers handbook. Volume One: Hydrological and ecological principles. Blackwell Scientific, Oxford, UK.

Newbold, J.D., J.W. Elwood, R.V. O'Neill, and W. Van Winkle. 1981. Measuring nutrient spiraling in streams. Canadian Journal of Fisheries and Aquatic Sciences **38**:860–863.

Newbold, J.D., P.J. Mulholland, J.W. Elwood, and R.V. O'Neill. 1982. Organic spiraling in stream ecosystems. Oikos **38**:266–272.

Nickelson, T.E. 1974. Population dynamics of coastal cutthroat trout in an experimental stream. Master's thesis. Oregon State University, Corvallis, Oregon, USA.

Nickelson, T.E., J.W. Nicholas, A.M. McGie, R.B. Lindsay, D.L. Bottom, R.J. Kaiser, et al. 1992. Status of anadromous salmonids in Oregon coastal basins. Available from the Research and Development Section, Oregon Department of Fisheries and Wildlife, Corvallis, Oregon, USA.

NRC (National Research Council). 1992. Restoration of aquatic ecosystems. National Academy Press, Washington, DC, USA.

——. 1996. Upstream: Salmon and society in the Pacific Northwest. National Academy Press, Washington, DC, USA.

O'Connor, J.F., and G. Power. 1976. Production by brook trout (*Salvelinus fontinalis*) in four streams in the Matamek watershed, Quebec. Journal of the Fisheries Research Board of Canada **33**:6–18.

Patrick, R. 1975. Stream communities. Pages 445–459 *in* M.L. Cody and J.M. Diamond, eds. Ecology and evolution of communities. Belknap Press of Harvard University, Cambridge, Massachusetts, USA.

Perrin, C.J., M.L. Bothwell, and P.A. Slaney. 1987. Experimental enrichment of a coastal stream in British Columbia: Effects of organic and inorganic additions on autotrophic periphyton production. Canadian Journal of Fisheries and Aquatic Sciences **44**:1247–1256.

Piorkowski, R.J. 1995. Ecological effects of spawning salmon on several south-central Alaskan streams. Ph.D. thesis. University of Alaska, Fairbanks, Alaska, USA.

Power, M.E. 1984. Habitat quality and the distribution of algae-grazing catfish in a Panamanian stream. Journal of Animal Ecology **53**:357–374.

——. 1990. Effects of fish in river food webs. Science **250**:811–814.

——. 1992. Habitat heterogeneity and the functional significance of fish in river food webs. Ecology **73**:1675–1688.

Power, M.E., W.J. Matthews, and A.J. Stewart. 1985. Grazing minnows, piscivorous bass, and stream algae: Dynamics of a strong interaction. Ecology **66**:1448–1456.

Randall, R.G., and U. Paim. 1982. Growth, biomass, and production of juvenile Atlantic salmon (*Salmo salar* L.) in two Miramichi River, New Brunswick, tributary streams. Canadian Journal of Zoology **60**:1647–1659.

Reeves, G.H., F.H. Everest, and J.D. Hall. 1987. Interactions between the redside shiner (*Richardsonius balteatus*) and the steelhead trout (*Salmo gairdneri*) in western Oregon: The influence of water temperature. Canadian Journal of Fisheries and Aquatic Sciences **44**:1602–1613.

Richey, J.E., M.A. Perkins, and C.R. Goldman. 1975. Effects of kokanee salmon (*Oncorhynchus nerka*) decomposition on the ecology of a subalpine stream. Journal of the Fisheries Research Board of Canada **32**:817–820.

Ross, S.T. 1991. Mechanisms structuring stream fish assemblages: Are there lessons from introduced species? Environmental Biology of Fishes **30**:359–368.

Scarnecchia, D.L., and E.P. Bergersen. 1987. Trout production and standing crop in Colorado's small streams, as related to environmental features. North American Journal of Fisheries Management **7**:315–330.

Sedell, J.R., and F.J. Swanson. 1984. Ecological characteristics of streams in old-growth forests of the Pacific Northwest. Pages 9–16 *in* W.R. Meehan, T.R. Merrell, Jr., and T.A. Hanley, eds. Fish and wildlife relationships in old-growth forests. American Institute of Fishery Research Biologists, Juneau, Alaska, USA.

Slaney, P.A., and B.R. Ward. 1993. Experimental fertilization of nutrient deficient streams in British Columbia. Pages 128–141 *in* G. Shooner and S. Asselin, eds. Le developpement du saumon Atlantique au Quebec: Connaitre les regles du jeu pour reussir. Colloque international de la Federation quebecoise pour le saumon Atlantique. Quebec, Decembre 1992. Collection *Salmo salar* Number 1.

Stanford, J.A., and J.V. Ward. 1992. Management of aquatic resources in large catchments: Recognizing interactions between ecosystem connectivity and environmental disturbance. Pages 91–124 *in* R.B. Naiman, ed. Watershed management: Balancing sustainability and environmental change. Springer-Verlag, New York, New York, USA.

Statzner, B., J.A. Gore, and V.H. Resh. 1988. Hydraulic stream ecology: Observed patterns and potential applications. Journal of the North American Benthological Society **7**:307–360.

Statzner, B., and B. Higler. 1986. Stream hydraulics as a major determinant of benthic invertebrate zonation patterns. Freshwater Biology **16**:127–139.

Stream Solute Workshop. 1990. Concepts and methods for assessing solute dynamics in stream ecosystems. Journal of the North American Benthological Society **9**:95–119.

Tripp, D., and P. McCart. 1983. Effects of different coho stocking strategies on coho and cutthroat trout production in isolated headwater streams. Canadian Technical Report of Fisheries and Aquatic Sciences 1212, Canada Department of Fisheries and Oceans, Ottawa, Ontario, Canada.

Triska, F.J., V.C. Kennedy, R.J. Avazino, G.W. Zellweger, and K.E. Bencala. 1989. Retention and transport of nutrients in a third order stream: Hyporheic processes. Ecology **70**:1893–1905.

Triska, F.J., V.C. Kennedy, R.J. Avanzino, G.W. Zellweger, and K.E. Bencala. 1990. *In situ* retention-transport response to nitrate loading and storm discharge in a third-order stream. Journal of the North American Benthological Society **9**:229–239.

Triska, F.J., J.R. Sedell, K. Cromack, Jr., S.V. Gregory, and F.M. McCorison. 1984. Nitrogen budget for a small coniferous forest stream. Ecological Monographs **54**:119–140.

Triska, F.J., J.R. Sedell, and S.V. Gregory. 1982. Coniferous forest streams. Pages 292–332 *in* R.L. Edmonds, ed. Analysis of coniferous forest ecosystems in the western United States. United States International Biological Program Synthesis Series 14, Hutchinson Ross, Stroudsburg, Pennsylvania, USA.

Vannote, R.L., G.W. Minshall, K.W. Cummins, J.R. Sedell, and C.E. Cushing. 1980. The river continuum concept. Canadian Journal of Fisheries and Aquatic Sciences **37**:130–137.

Ward, G.M., and N.G. Aumen. 1986. Woody debris as a source of fine particulate organic matter in coniferous forest stream ecosystems. Canadian Journal of Fisheries and Aquatic Sciences **43**: 1635–1642.

Warren, C.E. 1971. Biology and water pollution control. W.B. Saunders, Philadelphia, Pennsylvania, USA.

Warren, C.E., J.H. Wales, G.E. Davis, and P. Doudoroff. 1964. Trout production in an experimental stream enriched with sucrose. Journal of Wildlife Management **28**:617–660.

Washington Department of Fisheries, Washington Department of Wildlife, and Western Washington

Treaty Indian Tribes. 1993. 1992 Washington State salmon and steelhead stock inventory. Available from the Washington Department of Fisheries and Wildlife, Olympia, Washington, USA.

Waters, T.F. 1983. Replacement of brook trout by brown trout over 15 years in a Minnesota stream: Production and abundance. Transactions of the American Fisheries Society **112**:137–146.

Waters, T.F., M.T. Doherty, and C.C. Krueger. 1990. Annual production and production: Biomass ratios for three species of stream trout in Lake Superior tributaries. Transactions of the American Fisheries Society **119**:470–474.

Webster, J.R., and T.P. Ehrman. 1996. Solute dynamics. Pages 145–160 *in* F.R. Hauer and G.A. Lamberti, eds. Methods in stream ecology. Academic Press, San Diego, California, USA.

Webster, J.R., and B.C. Patten. 1979. Effects of watershed perturbation on stream potassium and calcium dynamics. Ecological Monographs **49**:51–72.

Welcomme, R.L. 1979. Fisheries ecology of floodplain rivers. Longman, London, UK.

Winterbourne, M.J., J.S. Rounick, and B. Cowie. 1981. Are New Zealand ecosystems really different? New Zealand Journal of Marine and Freshwater Research **15**:321–328.

Wootton, J.T., M.S. Parker, and M.E. Power. 1996. Effects of disturbance on river food webs. Science **273**:1558–1561.

16
The Hyporheic Zone

Richard T. Edwards

Overview

- The *hyporheic zone* is the volume of saturated sediment beneath and beside streams and rivers where ground water and surface water mix.
- Hyporheic zones in alluvial rivers are one of the dominant links between the riparian forest and the stream channel. The porous, hydraulically conductive sediments characteristic of alluvial rivers of the Pacific coastal ecoregion indicate the presence of extensive hyporheic zones.
- Hyporheic zones are hotspots of biological diversity that contain intensive physical and chemical gradients.
- Hyporheic zone processes can dominate surface water quality. Rivers with extensive hyporheic zones retain and process nutrients with greater efficiency than rivers without. Organic matter elimination can be two times greater in rivers with intact hyporheic zones.
- Upwelling nutrients from hyporheic zones influence primary production within surface communities and accelerate the recovery of surface production from floods and other disturbances.
- Effective stream management and restoration requires defining system boundaries to include all major ecosystem structural and functional attributes. Lack of consideration of hyporheic processes when predicting ecosystem responses to natural and anthropogenic stresses is a significant source of uncertainty in current management planning.

Introduction

Effective stream management requires that ecosystem boundaries be defined to capture all critical ecosystem linkages, so that key processes determining ecosystem structure and function are considered. Boundaries defining rivers have been expanded vertically and horizontally as understanding of the nature and strength of river-watershed linkages has increased (Bencala 1993). Deeper sediments, periodically inundated floodplains, and wetlands are all recognized to lie within stream boundaries. In the last decade, a previously unrecognized habitat, the *hyporheic zone*, has been recognized as a critical component of many streams and rivers, but it is still poorly understood. Defining stream ecosystem boundaries by the extent of hyporheic zone distributions ensures that critical ecosystem processes are included in management decisions (Triska et al. 1989a).

The hyporheic zone comprises saturated sediments beneath and beside a river channel containing some proportion of water from the surface channel (Figure 16.1). In this chapter, *ground water* refers to subsurface water that has not yet entered a surface flow channel, and *surface water* refers to water that has entered the stream channel directly, as rainfall or surface runoff, or indirectly as ground water. The mixing of these two water masses, which often differ significantly, stimulates biological activity. Thus, the hyporheic zone is an ecotone between ground water and surface water, blending properties of both water sources (Table 16.1) (Gibert 1991, Sabater and Vila 1992).

The potential importance of hyporheic zones to river ecosystems stems from biological and chemical activity and strong physicochemical and biological gradients that occur within them. Hyporheic biogeochemical processes strongly influence surface water quality (Bencala 1984). Streams with extensive hyporheic zones retain and process solutes more efficiently than those without (Valett et al. 1996). Further, decomposition in hyporheic habitats may double the ability of streams to eliminate organic wastes (Pusch 1996). Hyporheic habitats contain a diverse and abundant fauna (Williams and Hynes 1974) often dominating the biological productivity of rivers (Smock et al. 1992). Extensive hyporheic zones may serve as a refuge for stream biota, buffering them from disturbances in discharge and food supply. Seasonal and postdisturbance patterns of productivity differ from streams with reduced hyporheic flows (Valett et al. 1994).

Although hyporheic studies in Pacific coastal ecoregion streams are few, studies in similar Pacific Northwest rivers suggest that hyporheic zones are widespread and important. For example, hyporheic fauna are abundant 2 to 3 km from the nearest surface water in the Flathead River, Montana (Stanford and Gaufin 1974, Stanford and Ward 1993). Hyporheic zone nitrogen (N) cycling influences total ecosystem nitrogen flux in Little Lost Man Creek, northern California, (Triska et al. 1989a, Triska et al. 1993b). Rivers draining the Western

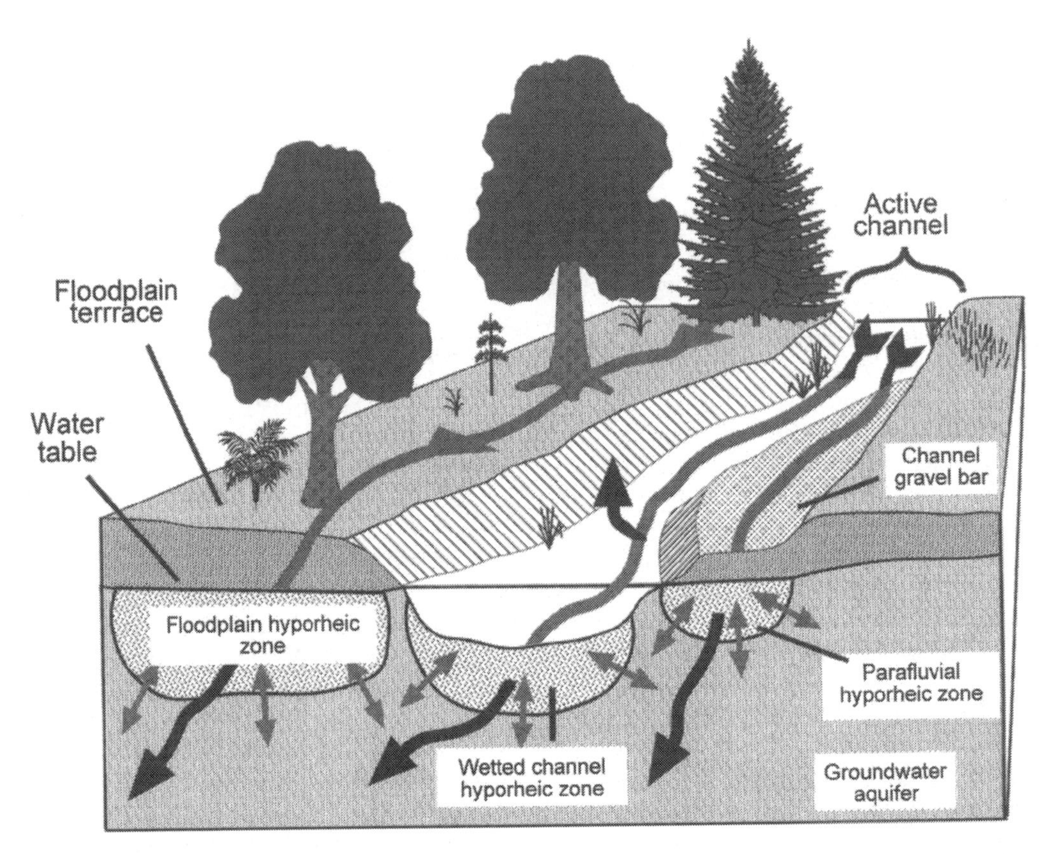

FIGURE 16.1. Location of three major types of hyporheic zones beneath and adjacent to a river. Single ended arrows symbolize subsurface flowpaths. Double ended arrows indicate mixing with adjacent water masses. Although the three zones are shown as distinct regions, in any catchment, they are connected to varying degrees depending upon sediment and hydraulic head characteristics.

TABLE 16.1. Common physical and chemical characteristics of surface and groundwater in unpolluted drainage basins.

Characteristic	Groundwater	Surface water
Light	No	Yes
Physical disturbance	Low	High
Temperature	Stable	Variable
Chemical composition	Generally low	Variable to high
DOM content	Generally high	Variable
Oxygen content	Often depleted	Often near saturation
Current velocities	Very low	Relatively high
Contact with sediment	Very high	Low

Cascades contain hyporheic zones that are sites of intense processing of carbon (C) and nitrogen (Vervier and Naiman 1992, Wondzell and Swanson 1996b). Accumulations of high-porosity alluvial gravel in many Pacific coastal streams suggest the existence of hyporheic zones comparable to those of the Flathead River, Little Lost Man Creek, and rivers draining the Western Cascades.

Upland management practices and land uses can significantly alter the input and distribution of sediment and water, thereby impacting hyporheic zone processes. Management practices that alter the hyporheic zone and thus also impact stream water quality, productivity, and diversity are often overlooked when only the surface stream channel is considered. Understanding the role of this hidden ecotone is perhaps one of the last frontiers of stream management in the Pacific coastal ecoregion.

Definition and Delineation

The hyporheic zone is defined on physical or biological bases depending upon the interest of the researcher or the research methods available at a given site. The most general definition: the saturated interstitial areas beneath the stream bed and into the stream banks that contain some proportion of channel water or that have been altered by channel water infiltration (White 1993), emphasizes the presence of surface water as a key property of hyporheic zones. Triska et al. (1989a) quantify the definition by setting a minimum surface water content of 10% as a threshold for defining the extent of hyporheic waters. Defining the hyporheic zone by water source is favored by researchers interested in biogeochemical processes. A more biological approach, however, defines the hyporheic zone as the area where hyporheic fauna occur (Stanford and Ward 1988). Hyporheic fauna, known as the *hyporheos*, are often distinguished by life history characteristics or adaptations to life within sediment interstices.

Boulton et al. (1992) further categorized hyporheic zones into subzones or biotopes based on distinctive communities in zones of different physical, chemical, biological, or biogeochemical characteristics (Figure 16.1). According to this system the volume beneath the wetted channel surface in close contact biologically and hydraulically is referred to as the hyporheic zone or shallow hyporheic zone, the volume beneath dry bars within the active channel boundaries is the parafluvial zone, and the hyporheic zone outside the bankfull boundaries beneath the adjacent riparian zone is the floodplain hyporheic zone. Hyporheic terminology remains ambiguous because of the varying orientation of investigators and the recent burst of growth in this new field.

Although the hyporheic zone is conceptually simple to define, it is considerably harder to delineate in practice. Consequently, it is often defined and delineated operationally. The most direct approach to determining the extent and location of hyporheic zones in the field is to use conservative tracers (Triska et al. 1989a). This method introduces a nonreactive material, which is detectable in low concentration, into the surface channel where it mixes with stream water as it moves downstream. Surface water labeled with tracer infiltrates the hyporheic zone and is detected in sampling wells placed in the saturated zone adjacent to and beneath the channel. The proportion of surface water within each well can be calculated by comparing tracer concentrations in surface water with those in nearby wells (Figure 16.2). The extent of surface water penetration can be determined by mapping the position and percentage of surface water composition (Triska et al. 1989a). Although this method is the most direct, it is impractical in large rivers because the tracer becomes so diluted it cannot be detected.

The parafluvial zone can be delineated by tracking the movement of fluorescent dyes through lateral channel bars in trenches dug perpendicular to the expected pathways (Holmes et al. 1994b). This technique can determine the spatial extent of the hyporheic zone as well as the position and shape of individual flow paths because hyporheic flows are often focused in paleochannels, or underflow paths (paths of higher hydraulic conductivity) (Harvey and Bencala 1993, Stanford et al. 1994). It is only useful, however, in streams with fine-grained sediments that are easily trenched.

Intrusion of surface water into hyporheic zones can also be inferred by monitoring concentrations of naturally occurring solutes or characteristics such as pH, electrical conductivity, various ions, or temperature (White et al. 1987, Fortner and White 1988, Stanford and Ward 1988, Hendricks and White 1991, Evans et al. 1995). Where surface and ground water are measurably different in one of these variables, hyporheic flows can be inferred from contour plots of the variable (Figure 16.3).

In larger rivers, where dilution makes tracer techniques impractical, most delineation tech-

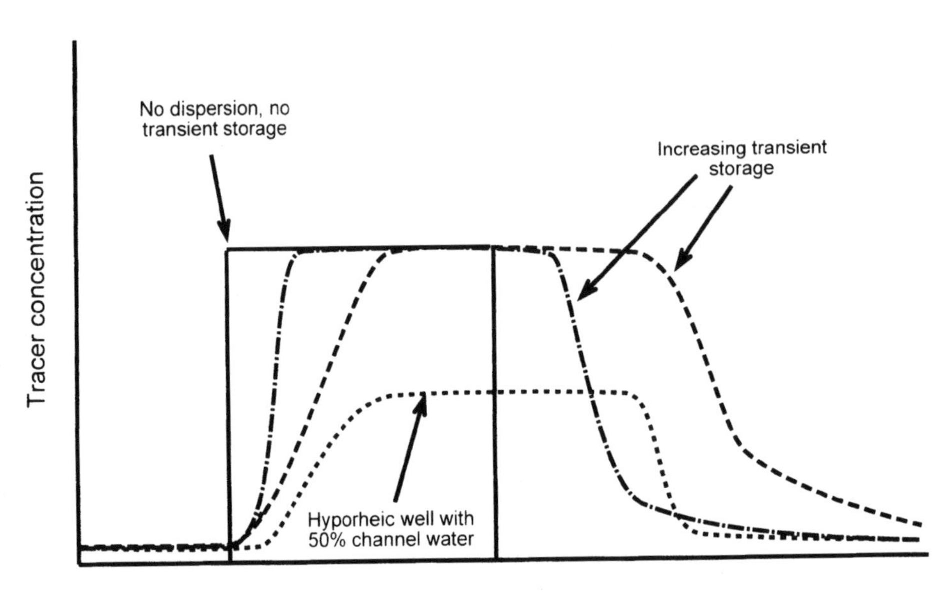

FIGURE 16.2. Time series plots of concentrations of a conservative tracer added at constant rate upstream. Transient storage creates a delay in reaching steady state and a more gradual rise and fall in concentration. With no dispersion or transient storage (——) the tracer concentration plot would be rectangular. As the amount of transient storage increases from moderate (-·-) to high (---), the time to reach the plateau increases and the return to baseline concentrations is slowed. The bottom curve illustrates the tracer concentrations in a hyporheic well with 50% surface water and 50% groundwater (----).

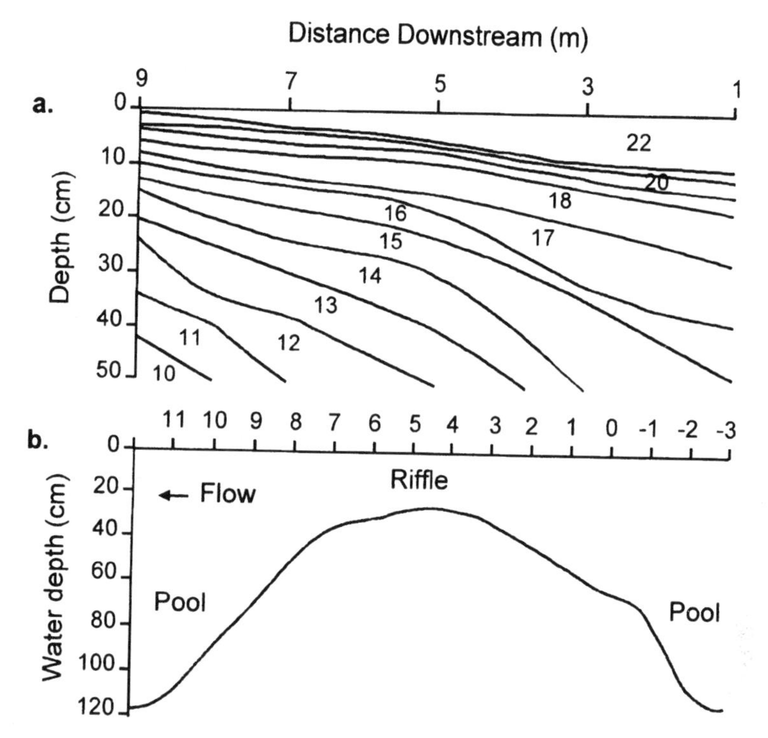

FIGURE 16.3. (a) Contour diagram of temperature (°C) within a hyporheic zone through a longitudinal transect of a riffle channel unit. Surface water downwelling into a riffle is indicated by warmer surface temperature contours intruding into sediment. The 0 reference for the top diagram is the sediment surface. (b) Water depths along the riffle transect, illustrate the relationship between sediment topography and temperature contours (from White et al. 1987, with permission).

niques are indirect and inductive. For example, close hydraulic connectivity, which suggests the existence of a hyporheic zone, can be demonstrated by monitoring water heights within wells and the adjacent surface channel (Figure 16.4) (Stanford and Ward 1993). When temporal patterns of water heights in wells and surface water coincide, a hydraulic connection is inferred.

The hyporheic zone is also commonly delineated by mapping the occurrence of hyporheic fauna in water pumped from wells (Williams 1989). Organisms found in subsurface waters can be classified by morphology or taxonomy according to their known affinity for ground water, hyporheic, or surface water habitats (Gibert et al. 1994). The usefulness of this approach, however, is limited by knowledge of the taxonomy and life histories of new species because descriptions of hyporheic fauna are limited. It is not clear how strongly the distribution of various interstitial taxa correlates with hydraulically defined boundaries; therefore, biological and physical approaches to delineating hyporheic zones do not necessarily identify the same boundaries. However, this approach is more direct when biological characteristics of the hyporheic zone are the primary focus.

All delineation techniques require installation of wells (which are expensive to install) in the presumed area of hyporheic water movement to withdraw water or organisms for analysis. Therefore, determining hyporheic zone boundaries is expensive and resolution is low because of the limited number of wells that can be installed. Consequently, knowledge of hyporheic zone boundaries remains limited except in a few intensively studied streams.

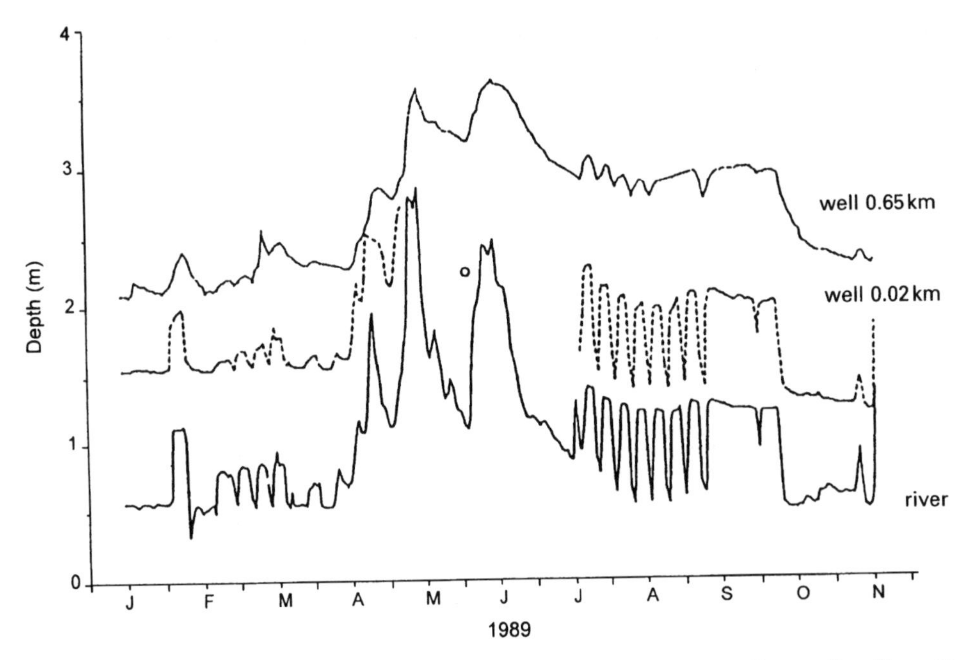

FIGURE 16.4. Hydrographs from Flathead River and adjacent floodplain wells at different distances from river. Close correspondence in water height responses indicates hydraulic coupling (from Stanford and Ward 1993, with permission).

Interstitial Volume and Surface Area

Hyporheic zones are important to river ecosystems in part because of their relatively large interstitial volume and surface area (compared with the overlying stream). These characteristics are important determinants of the types of organisms that exist in the hyporheic zone (Williams and Hynes 1974). Hyporheic interstitial volume (habitat volume) can be estimated from sediment porosity—the proportion of a given sediment volume occupied by water. Porosity, which is determined by particle size, shape, and packing density, ranges from 25% to 70% in sediments common in Pacific coastal ecoregion streams (Table 16.2). Thus, the interstitial volume associated with accumulated sediments provides a potentially large habitat for hyporheic biota. For example, a stream channel 20 cm deep flowing over 100 cm of sediment could have 2.5 times as much habitat volume (per unit channel area) within the hyporheic zone as that in the overlying channel. Triska et al. (1989) estimated that the interstitial water volume in Little Lost Man Creek, California, was at least as great as the entire surface water volume. In floodplain rivers, where lateral hyporheic zones are large compared with the surface channel area, interstitial volume differences are even greater. Stanford and Ward (1988) estimated hyporheic habitat

TABLE 16.2. Porosity and hydraulic conductivity (K = cm/s) for sediments found in Pacific coastal streams.

Sediment type	Porosity (%)	Hydraulic conductivity (cm/s)
Gravel	25–40	10^{-1}–10^{2}
Sand	25–50	10^{-4}–10^{0}
Silty sand	—	10^{-5}–10^{-1}
Silt	35–50	10^{-7}–10^{-3}
Glacial till	—	10^{-10}–10^{-4}
Clay	40–70	10^{-10}–10^{-7}

Modified from Freeze and Cherry 1979.

volumes to be 2,400 times greater than channel habitat volume along a floodplain reach of the Flathead River, Montana.

Comparisons of the relative sediment surface areas are even more dramatic because the combined surface area of particles (which varies with particle size) in a volume of sediment is quite high for particles 1–100 mm in diameter (Figure 16.5). For example, the surface area available for colonization by organisms such as bacteria, fungi, and protozoans is at least 2,000 times greater in the hyporheic zone as on surface sediments in a stream 20 cm deep with a bed composed of 2 mm diameter sand. The organic matter and associated community of microorganisms on the surface of sediment particles (known as epilithon) are biochemically reactive and are responsible for much of the uptake and transformation of dissolved chemicals and organic matter (Bärlocher and Murdoch 1989, Fischer et al. 1996, Chapter 6). Such large surface area ratios suggest a much greater capacity for water quality changes within reaches dominated by hyporheic zones.

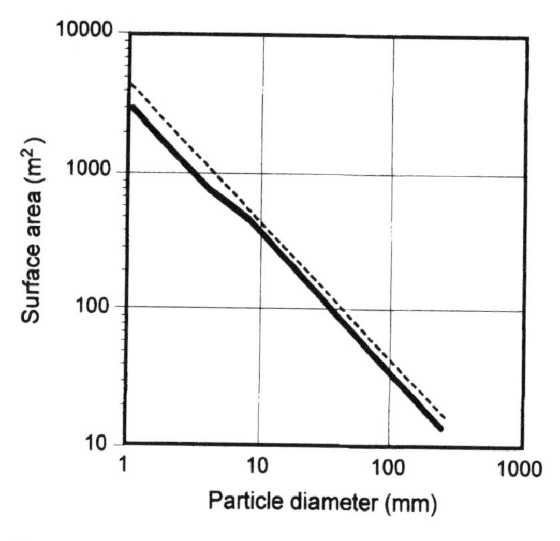

FIGURE 16.5. Approximate total saturated surface area within a cubic meter of sediment for a range of sediment diameters. Upper and lower lines indicate values for the naturally-occurring range of sediment porosities (------ low porosity, —— high porosity).

Hydrology of Hyporheic Interactions

To understand how natural and anthropogenic disturbances alter the extent and distribution of hyporheic zones, it is necessary to understand factors controlling the penetration of surface water into porous sediments. Flow through a saturated semipermeable medium can be described by Darcy's law as:

$$v = -K^{*}\Delta h/\Delta l$$

where v is the specific discharge of water through the sediment (m/s), $-K$ is a proportionality constant known as the *hydraulic conductivity* (m/s), Δh is the difference in *hydraulic head* at two points (m), and Δl is the distance between the two points (m) (Freeze and Cherry 1979).

Therefore, the physical conditions required for hyporheic flows are sediments with adequate hydraulic conductivity and hydraulic head differentials across the conductive medium. Hydraulic conductivity is a measure of the resistance to flow and hydraulic head is a measure of the energy per unit mass of water available to overcome resistance within a porous medium. Hydraulic conductivity can be measured *in situ* (Bouwer 1989, Cedergren 1989) or estimated from the size distribution of sediment particles (Freeze and Cherry 1979). Sediment characteristics such as porosity, particle shape and size, particle size distribution, and packing density control hydraulic conductivity, which varies over 12 orders of magnitude (10^{-10}–10^{2}) for sediments in Pacific coastal streams (Table 16.2). Except for clays, which are highly porous but poorly conductive, bed materials with high porosity are also highly conductive. The high porosity of alluvium in the Pacific coastal ecoregion creates large interstitial volumes and high flow velocities—optimal traits for hyporheic habitat.

Hydraulic head, is a combination of water elevation and water pressure. Water flows from higher to lower head regions reflecting the loss of energy associated with flow. Hydraulic head is measured with a piezometer which, in its simplest form, is an open tube with one end buried in the sediment and the other end open to the atmosphere (Figure 16.6). Water rises in the tube in response to the combined components of the hydraulic head; therefore, the water's height within the tube is a measure of the hydraulic head at the bottom of the tube. Head measurements are expressed as the elevation (m) of the water above an arbitrary datum.

Head measurements are useful for predicting water movements at a site. Where water moves vertically from channel to sediment, the water level in a piezometer inserted into the bed is lower than the stream water surface (Figure 16.6a). Where water is upwelling from sediment to surface, the water level in the peizometer is higher than that of the overlying surface water (Figure 16.6c). Where no vertical movements occur, the piezometer water level is the same height as the overlying channel water surface (Figure 16.6b). Vertical movements of hyporheic water can also be detected where the channel surface is dry by comparing heads in nested piezometers installed at different depths. Dividing the difference in head between two vertical points by the distance between them yields the vertical hydraulic gradient (VHG), which is a measure of the potential for flow between the two points.

By mapping heads across a piezometer or well grid, inferences can be made about the direction of subsurface flow (Figure 16.7). Because water flows from higher to lower heads, it moves at right angles across head contour lines. Head contour diagrams suggest potential horizontal flow directions, but the pattern and velocity of actual flow varies with the hydraulic conductivity of the sediments.

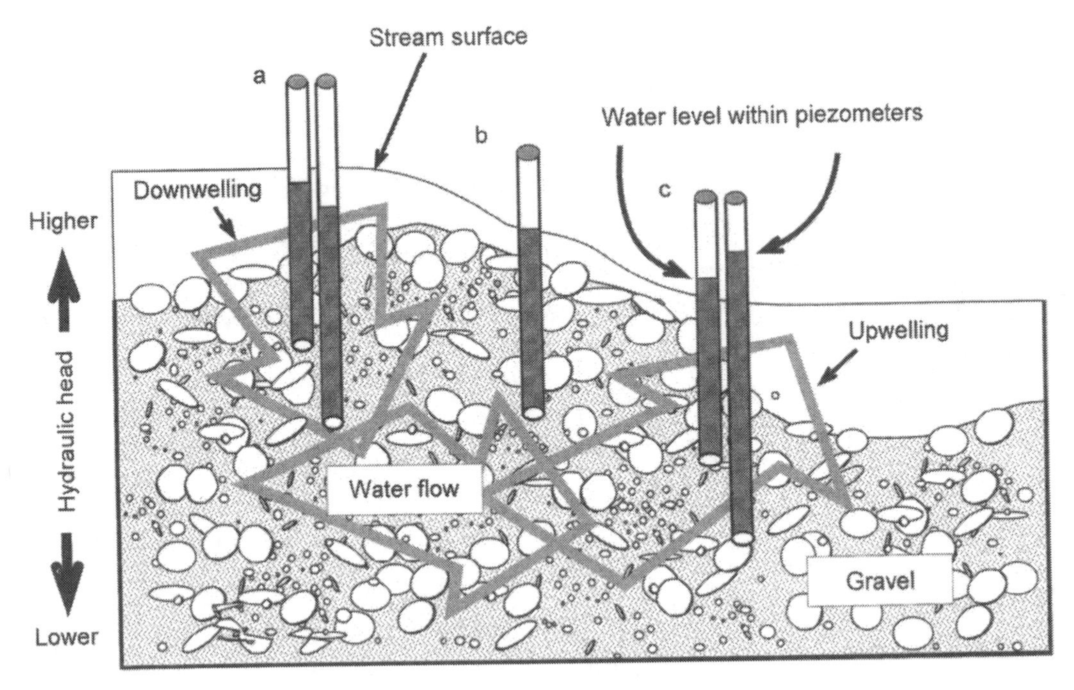

FIGURE 16.6. Piezometers are used to measure hydraulic head and detect potential or actual vertical water movements. Water flows from higher to lower head areas. Within the wetted channel, movement into the sediments can be detected using one or more piezometers. (a) Where downwelling occurs, the stream surface head is greater than that in a piezometer. (b) Where no vertical water movement occurs, the piezometer head is equal to the water surface head. (c) Where water is upwelling back into the overlying surface water, the head within the piezometer is greater than the steam water surface. Where surface water is absent, such as parafluvial bars or floodplains, vertical water movements can be detected as head differences among two or more nested piezometers at a site.

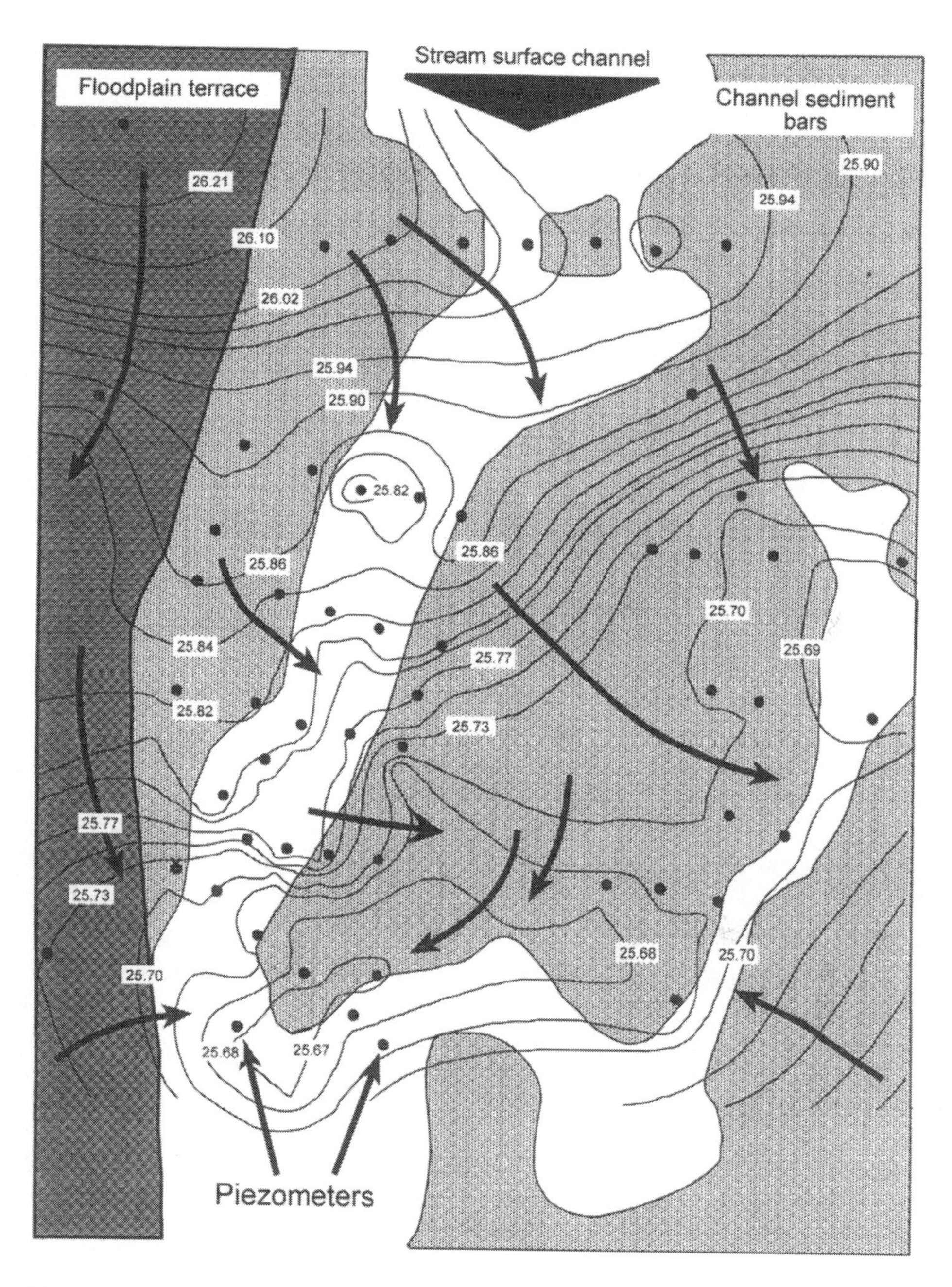

FIGURE 16.7. Hydraulic head contours (m) at a depth of 50 cm into the sediment within a floodplain back channel in the Queets River, Washington. Dots indicated location of piezometers. Potential water flow (indicated by arrows) is at right angles to the contour lines.

Additional information on hydraulic conductivity is needed to confirm that flow actually occurs.

Although the hyporheic zone is defined as the interface between surface channel water and ground water, little is known about the effects of ground water in hyporheic zone processes. The groundwater component, which varies greatly in source and hydrologic behavior, has important, but poorly understood, implications for hyporheic dynamics. The simple model of Darcian flow adequately explains the flow patterns of water within saturated sediments. However, the vertical and lateral boundaries of the hyporheic zone are also influenced by the volume of groundwater inputs (Olson 1995).

Groundwater inputs are influenced by saturated and unsaturated flows driven by the interaction between topography and soil characteristics such as saturation, storage volume, and conductivity (Winter 1978, Olson 1995). From the perspective of hyporheic fauna and biogeochemical processes, such complexity means that the exact boundaries of the hyporheic zone are highly variable in space and time. How much the physicochemical nature of the resulting mix of ground water and surface water reflects the history of the variable sources of the groundwater component is unknown. From a management perspective, the complexity of groundwater contributions suggests that upland land uses outside the riparian corridor influence the total composition of hyporheic water sources and consequent processes.

Hyporheic Zone Distribution Patterns

The distribution of hyporheic zones in a drainage basin varies systematically with the physical processes that control sediment routing and discharge. The interaction of geomorphic and hydrologic processes determines the location, volume, and shape of sediment accumulations in the channel network, the shape and particle size distribution of the substratum, and the magnitude and location of hydraulic head differentials necessary to drive water through a porous medium.

Because hyporheic processes influence surface water chemistry, hyporheic distribution patterns may create observable basin-wide patterns in surface water chemistry where the degree of hydraulic coupling and size of hyporheic zones varies longitudinally. Understanding the potential impact of disturbance on hyporheic processes requires an understanding of how the set of conditions operating at each scale respond (Table 16.3, Figure 16.8). Because channel morphology sets the physical template, the process-based channel classification approach introduced in Chapter 2 forms a useful foundation to understanding the implications of physical controls on hyporheic zone distributions.

Large-Scale Geologic Factors

Streams of different geologic composition often exhibit broad differences in the extent and nature of hyporheic interactions. These differences have important effects on river processes. For example, differences in bedrock geology among three headwater streams in New Mexico resulted in predictable differences in dissolved oxygen concentration and nitrogen retention that were attributed to variations in hydraulic conductivity of alluvial sediments derived from parent materials of sandstone, volcanic tuff, and granite (Valett et al. 1996). Although basin-wide trends in each watershed were not examined, this study demonstrates that bedrock geology can impose large-scale constraints on potential hyporheic interactions by

TABLE 16.3. Common head-forming elements, flowpath length, and residence times as a function of spatial scale. Flowpath residence time is the average travel time along the flowpath.

Scale	Flowpath length (m)	Flowpath residence time	Head-forming elements
Roughness element	10^{-2}–10^{-1}	Minutes	Boulders, sand waves, vegetation, fish nests, beaver dams, logs
Channel unit	10^{-1}–10^{1}	Minutes to hours	Pools, riffles, bars
Reach	10^{1}–10^{2}	Hours to days	Reach morphology
Basin	$>10^{3}$	Weeks to months	Valley shape, channel constraint

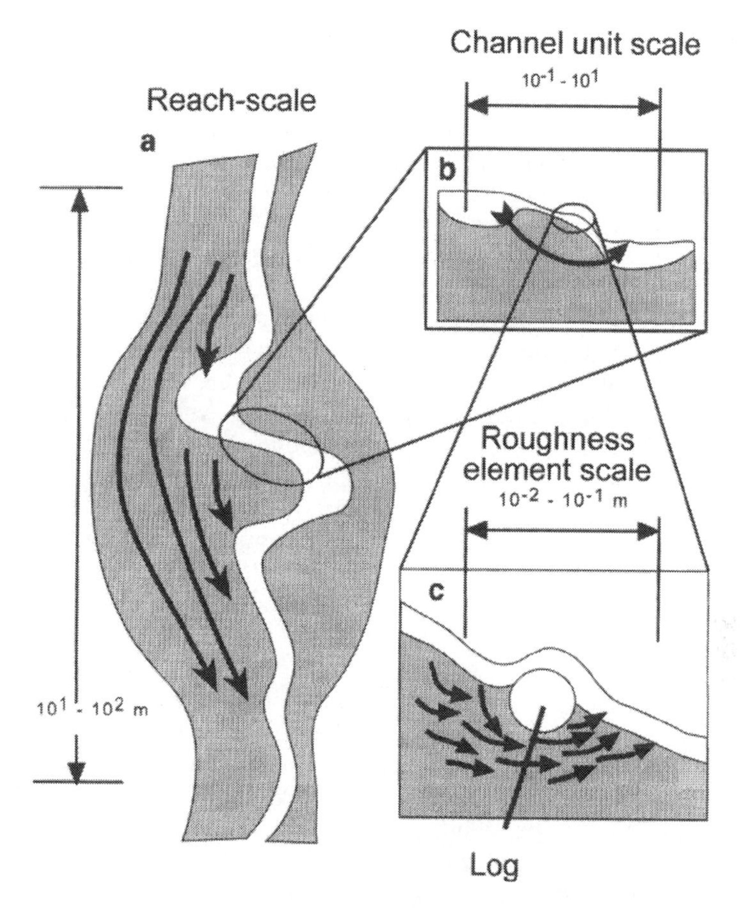

FIGURE 16.8. Nested scales of head-forming elements and resulting hyporheic flowpaths at the (a) reach, (b) channel unit, and (c) roughness element scale. Arrows indicate hyporheic flowpaths. Flows generated at one scale may override the influence of flows at larger or smaller scales.

controlling the size and texture of channel sediments. Bedrock geology influences hydraulic conductivity of sediment at the geomorphic province, watershed, or valley segment scale depending upon regional geologic conditions.

Watershed and Valley Segment Scales

The volume of sediment in a valley segment varies with the balance between sediment input and transport. As channels widen and gradients decrease downstream, increased accumulations of sediment create a potentially larger potential hyporheic zone. The hydraulic conductivity of sediment accumulations varies with the characteristics of the parent material and the amount of sorting that has occurred during and after deposition. The potential for hydraulic head gradients decreases in the lower basins as stream slope decreases. White (1993) suggests that systematic changes in channel slope, the ratio of channel volume to hyporheic volume, and direction and proportional volume of hyporheic flows occur predictably throughout a drainage basin, however, empirical evidence is lacking.

Changes in channel constraint superimpose basin-scale patterns on hyporheic zone distributions (Stanford and Ward 1988). For example, the distribution pattern of hyporheic habitats where constrained and unconstrained valley segments alternate resembles a giant string of pearls (Stanford and Ward 1988). A large volume of hyporheic downwelling occurs where the channel changes from constrained to unconstrained. At the downstream end of the unconstrained reach, water upwells into the surface above the constrained reach (Stanford and Ward 1993). This phenomenon can induce subsurface flows many kilometers long that drive water through lateral riparian floodplains (Stanford and Ward 1993), creating a hyporheic zone up to 3 km wide and 10 m deep. Such flows are often focused in paleochannels, sites of former channel meanders. Although documented only in the very large Flathead River in Montana, similar but smaller patterns of valley constraint occur in Pacific coastal rivers.

Channel Reach Scale

Evidence suggests that differences in hyporheic processing among colluvial, bedrock, and alluvial reaches are significant.

Colluvial reaches—Colluvial reaches contain poorly sorted finer sediments. Despite large hydraulic heads created by steep channel beds, hyporheic interactions may be limited in colluvial reaches because hydraulic conductivity within the poorly sorted sediments is lower. Consequently, hyporheic zones in colluvial reaches likely exhibit characteristics similar to those in reaches with sediments derived from siltstone and sandstone as described by Valett et al. 1996.

Bedrock reaches—In bedrock reaches hyporheic zones are limited to small pockets of sediment. Although the impact of hyporheic processes in such reaches is small, pockets of alluvium may serve as important refugia for invertebrates during floods.

Alluvial reaches—The development of hyporheic zones is most significant where large volumes of porous sediments accumulate. Although Montgomery and Buffington (Chapter 2) describe five forms of alluvial channel reach types, comparative hyporheic studies of reach types are lacking. Most reach-scale studies have focused on step-pool or pool-riffle reaches. Nested within those scales are studies of flow through channel unit features such as bars and riffles or individual roughness elements. Where porous alluvial sediments dominate, the primary control of surface water penetration in a reach is the hydraulic head differential created by channel topography.

Reach-scale topography can create extensive flows parallel to the wetted channel (Triska et al. 1990, Triska et al. 1993a) or adjacent floodplain (Wondzell and Swanson 1996a). Flow paths can travel through relic channels (paleochannels), sometimes even passing underneath the wetted channel (Triska et al. 1989a). At this scale, surface water can interact extensively with ground water entering the reach. Position along the flow path influences the relative proportions of surface and ground water, and residence times (time required to flow through the hyporheic zone) are sufficient (hours to days) for the development of significant oxygen gradients in the flow path. The volume of hyporheic flow at the reach scale can equal the volume of flow entering as net groundwater input, a fact which emphasizes the importance of hyporheic exchanges in increasing contact between channel water and substratum (Figure 16.9) (Harvey and Bencala 1993).

Channel Unit Scale

In pool-riffle reaches, head differences occur at breaks in water surface elevation between channel unit features. Water tends to downwell at pool tails and riffle heads, and upwell in riffle tails and pool heads (Figure 16.6). In meandering channels, differences in water elevation across channel bars create flows that short circuit the surface pathways. These flows occur over distances from one to several meters, emerging at the base of the same riffle or the downstream side of a bar. At this scale water can penetrate sediments quickly, establishing a rapid equilibrium with surface water (Munn and Meyer 1988). Flow paths are relatively short with residence times measured in minutes or hours. Bed topography at this scale is significantly correlated to biologically significant variables such as community respiration rate, organic matter distributions, and carbon turnover rates (Pusch 1996).

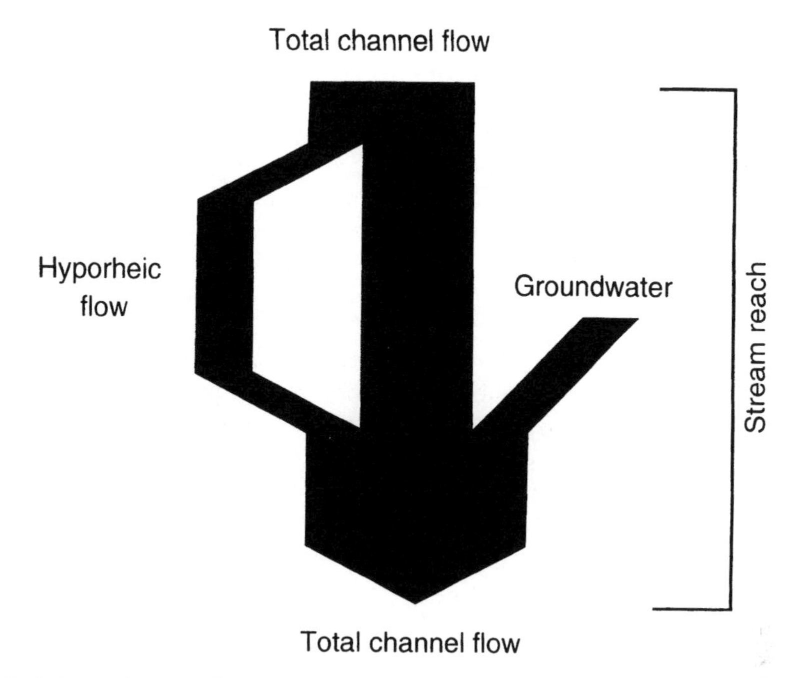

FIGURE 16.9. Relative volume of hyporheic return flow and groundwater input along a gaining reach of a montane stream. The width of the arrow is proportional to the volume of flow (modified from Harvey and Bencala 1993).

Roughness Elements Scale

The flow of water over channel roughness elements such as boulders, sand waves, clumps of macrophytes, lamprey nests, beaver dams, and logs creates upstream head elevations and downstream head depressions that cause flow under and around the object (Savant et al. 1987, White 1990). Flows of this scale are localized with short flow paths (10's of centimeters) and short residence times (minutes). An important small-scale head forming process in Pacific coastal rivers is fish spawning (Vaux 1968). The redd structure created by spawning salmon (*Oncorhynchus* spp.) creates local head differences that force water into the sediment thus enhancing the flow of oxygenated water which is important for egg development.

Spatial Scale of Management Actions

Management practices occur at different scales from site-specific alterations in wood input to basin-wide changes in sediment transport or discharge. Actions at different scales have different consequences for hyporheic zones depending upon their effects on channel morphology and hydraulic heads. For example, alterations in wood inputs change the number of pieces of buried wood, thus, affecting the number and distribution of short hyporheic flow paths. Such alterations also affect larger-scale hyporheic flows by changing sediment storage in the reach. Basin-wide changes in sediment supply and retention impact hyporheic zones at a variety of scales by influencing channel morphology and the total volume of saturated sediment. These responses underscore the need to incorporate an understanding of hyporheic processes in management plans.

Flows at different scales are interdependent. Where shorter flow paths are nested within longer flows, each may influence the other. When the length and residence times of hyporheic flows are altered, key biophysical variables such as oxygen concentration, water chemistry, and quality and quantity of food supply to hyporheic organisms change. To predict responses of the hyporheic ecosystem to manipulations and disturbances at different scales, an understanding of how physical and biologic processes interact to modify the hyporheic environment is required.

Temporal Scales

The extent of the hyporheic zone, patterns of downwelling and upwelling, and the volume and velocity of hyporheic flows vary with season and hydrologic condition. For example, in a small stream in Colorado, low-flow hyporheic flow paths became groundwater input sites during snow melt (Harvey and Bencala 1993). In New Mexico, upwelling and downwelling patterns reversed following floods, or with seasonal changes in baseflow (Valett 1993) and the proportion of surface water entering hyporheic zones changed with discharge (Valett et al. 1996). Therefore, anthropogenic alterations to river discharge (e.g., dams, irrigation withdrawals) change the extent, volume, and timing of hyporheic exchanges.

Natural and anthropogenic processes affecting rates of sediment input and transport can also impose large-scale temporal patterns on hyporheic zone distributions. For example, steep gradients and unstable hillslopes of Pacific coastal basins create episodic inputs of colluvium (Chapter 2). The subsequent redistribution of this material creates sediment waves that move downstream over periods ranging from years to decades. A transient hyporheic zone moves with the sediment waves creating temporal variations in basin-wide distributions of hyporheic processes that override local controls.

Biogeochemical Processes

The pattern of influence of hyporheic zones on surface water throughout the watershed is determined by the interaction among the physical factors controlling the distribution of hyporheic zones, the volume of flow through them, and the biochemical processes occurring within them. The influence of hyporheic biogeochemical processes on stream water quality stems from the combination of an enormous, highly reactive surface area and long periods of sediment and water contact. Removal of solutes at sediment surfaces occurs by abiotic sorption to non-living surfaces and by biotic uptake mediated by epilithic biofilms (Broshears et al. 1993). Both processes occur quickly relative to the residence time of water within the hyporheic zone, producing a rapid reduction in solute concentration within the water column (Bencala et al. 1984). Two conditions are necessary for hyporheic processing to significantly influence river chemistry. Water must enter the hyporheic zone in sufficient quantity and stay there long enough, and process rates must be high enough to effect chemistry changes as water passes through the hyporheic zone.

Transient Storage

The removal of solutes from advecting water is enhanced by the phenomenon known as *transient storage* (Runkel and Bencala 1995). Transient storage refers to the temporary residence of water moving downstream within more slowly moving pools. This water enters the hyporheic zone, mixes with water there, then reenters the surface channel. Tracer injections combined with mathematical modeling are used to estimate the magnitude of transient storage (Bencala et al. 1993, Broshears et al. 1993, Runkel and Bencala 1995).

Tracers are injected into the surface channel for an extended period of time (8 hr–5 d) until the tracer-labeled water mixing with hyporheic pools reaches equilibrium. At this point, downstream tracer concentrations plateau (Figure 16.2). Transient storage is revealed (in time-series plots) as a dampening of the rate of rise and fall of concentrations. When the storage volume is large relative to stream discharge, the concentration curve trails off more gradually and takes longer to reach preinjection baseline concentrations. Mathematical transient storage models simulating tracer behavior in a stream are used to estimate the relative size of transient storage pools (Bencala et al. 1993, D'Angelo et al. 1993).

Solute uptake within the substratum can be reversible or irreversible (Bencala et al. 1983). Physicochemical sorption of certain ions is influenced by reversible equilibrium processes. Clay minerals or oxides of metals such as iron (Fe) or aluminum (Al) serve as sites of abiotic sorption, and sorption of inorganic phosphorus (P) or ammonium (NH_4) onto clays or iron oxides (FeO) act as removal mechanisms (Triska et al. 1994). Reversible sorption releases material back into the water column over a range of intervals, producing lags in downstream transport that increase the residence time of material in a stream reach. Experimental tracer injections in streams receiving acidic drainage from mines confirm the importance of transient storage and sediment sorption in influencing solute concentrations and the downstream flux of metals (Bencala et al. 1984, Kuwabarra et al. 1984).

Biological uptake can irreversibly remove material from the water, or reintroduce it in a different form at a later time. Permanent removal of carbon or nitrogen occurs when cellular metabolism converts an organic or inorganic molecule to a gaseous phase, which is then lost through diffusion into the atmosphere (Chapter 14). Key processes leading to permanent removal through gas loss are respiration, which converts organic carbon into carbon dioxide (CO_2) and methane (CH_4), and denitrification, which converts nitrate to nitroten gas ($NO_3 \rightarrow NO_2 \rightarrow NO \rightarrow N_2O \rightarrow N_2$) (nitrate, nitrite, nitric oxide, nitrous oxide, nitrogen gas). All elements incorporated into living biomass also can be removed from a stream by migration of the organism.

Transient hydraulic storage combined with reversible and irreversible uptake increases the availability of energy and materials to biota by retarding the downstream loss of biologically labile materials. The net result is decreased nutrient spiraling lengths (Chapter 14), increased retention, and reduced downstream solute loading (Valett et al. 1996).

Nitrogen Dynamics

To understand biogeochemical patterns within hyporheic zones, it is necessary to know how physical and hydrological properties interact with chemical and biological processes. The specific spatial pattern and efficiency of solute transformation and removal vary according to the properties of each individual constituent. Detailed descriptions of all solute reactions (Stumm 1970, Lynch 1988) are beyond the scope of this chapter, but a discussion of hyporheic nitrogen cycling synthesizes basic concepts.

Nitrogen, important because of its role in cellular metabolism and plant growth, limits primary production in many Pacific coastal streams (Thut and Haydu 1971, Gregory et al. 1987). In streams impacted by urban or agricultural runoff, excessive nitrogen inputs create low water quality conditions, or eutrophication, in downstream impoundments. Riparian zones are important because they effectively remove fertilizer nitrogen from water entering streams and prevent contamination through uptake by riparian vegetation and denitrification by bacteria (Peterjohn and Correll 1984, Lowrance and Pionke 1989, Lowrance 1992).

Forests in Pacific coastal riparian zones, however, may act both as nitrogen sources and sinks. A bacterium (*Frankia alni*) growing in the roots of Red alder (*Alnus rubra*), a common tree species in Pacific coastal riparian zones, fixes large amounts of nitrogen in excess of the tree's growth needs (Cole et al. 1990). This interaction results in large pools of organic nitrogen in the associated riparian soils (Figure 16.10) (Luken and Fonda 1983). Although the fate of nitrogen is largely unknown, increased hyporheic nitrogen concentrations and hyporheic-stream nitrogen transfers have been reported from two Pacific Northwest streams (Triska et al. 1989b, Triska et al. 1990, Wondzell and Swanson 1996b). Duff and Triska (1990) suggest that soil nitrogen may also be lost via denitrification at other sites within the same hyporheic zone.

Whether the hyporheic zone acts as a nitrogen source or sink is determined by the balance of physical, chemical, and biological properties operating within a site, which ultimately determines the oxygen concentration. Water entering the hyporheic zone from the surface channel carries oxygen and labile organic

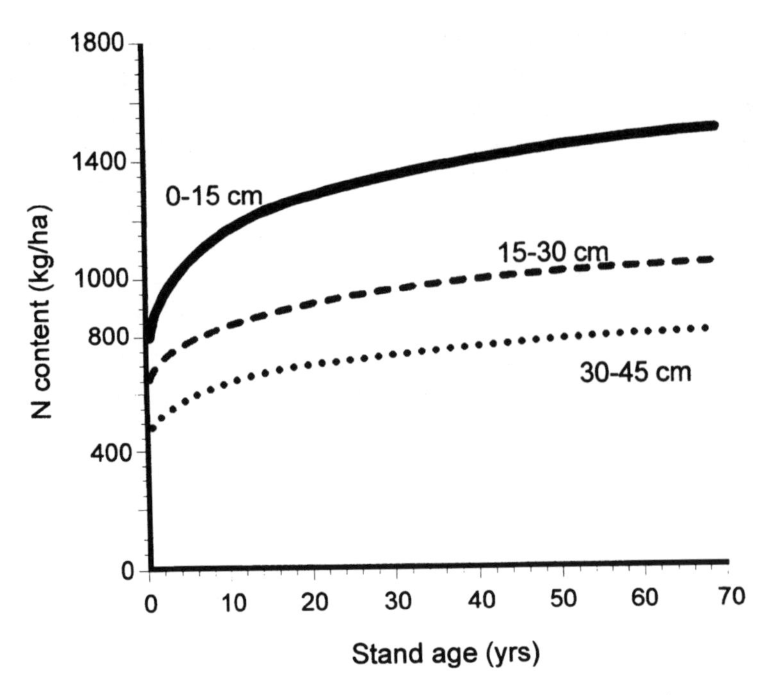

FIGURE 16.10. Total soil nitrogen concentrations at three depths in soils from a chronosequence in the Hoh River, Washington (data from Luken and Fonda 1983).

matter. As water moves through the hyporheic zone, decomposition of the organic matter consumes oxygen, creating oxygen gradients along the flow path (Findlay et al. 1993, Holmes et al. 1994, Findlay and Sobczak 1996). The extent of the oxygen gradient is determined by the interplay between flow path length, water velocity, the ratio of surface to ground water, and the amount and quality of organic matter. Oxygen concentrations control the type of biochemical transformations occurring and the chemical nature of the end products (Chapter 14) and thereby determine whether a point is a net source or sink of nitrogen.

An example of simultaneous opposing pathways of nitrogen cycling within the hyporheic zone of a single reach is well documented in a study of Little Lost Man Creek, California, one of the few rivers in the Pacific coastal ecoregion where hyporheic biogeochemical processes have been extensively researched (Triska et al. 1989a and b, Duff and Triska 1990, Triska et al. 1990, Triska et al. 1993a and b). Triska and others examined influences on nitrogen dynamics within the hyporheic zone using laboratory studies and field data from coinjections of conservative tracers and various forms of inorganic nitrogen. Sampling wells along the study reach differed widely in distance from and degree of coupling to the surface channel water. These differences determined the form of hyporheic nitrogen and whether locations were dominated by nitrification or denitrification (Table 16.4).

The surface water content of sampling wells in Little Lost Man Creek ranged from 47 to 100%, while oxygen concentrations in the wells varied more widely (0–100% saturation). Wells near the stream had greater proportions of surface water, higher oxygen concentrations, and lower ammonium concentrations than more remote wells. Higher than predicted concentrations of nitrate in wells near the stream suggest inputs of nitrogen came from decomposing organic matter. Sediment incubations demonstrated that nitrification potential within the high-oxygen wells near the channel could increase concentrations of nitrate through oxidation of ammonium. Field injections of ammonium increased nitrate concentrations in the wells, confirming that nitrification occurred near the high-oxygen wells.

Wells farther from the channel contained lower proportions of advected channel water compared to wells closer to the channel. Oxygen and nitrate concentrations were also lower, and ammonium concentrations were higher. Sediment incubations confirmed the presence of denitrification potential within these low-oxygen wells. Field injections of nitrate and acetylene (HC≡CH), which inhibits the final step in denitrification, resulted in increased concentrations of N_2O, confirming that denitrification was occurring *in situ*. The source of the nitrogen and organic matter supporting denitrification was nitrogen-rich ground water entering the hyporheic zone from a riparian bank dominated by alder.

Similar trends along well-oxygenated hyporheic flow paths also occurred in Sycamore Creek, Arizona where nitrate consistently increased as oxygen decreased (Holmes et al. 1994b). Nitrification assays on sediments indicated, and coinjections of ammonium and chloride (Cl) confirmed, that nitrification rates were highest in the upstream sections of the flow path near the downwelling water source (Holmes et al. 1994b).

Spatial patterns of respiration suggest that nitrification is driven by microbial respiration of advected organic matter (Jones et al. 1995) and subsequent nitrification of released ammonium. Decreased respiration further along the flow path confirms that labile inputs are rapidly consumed in the upper part of the flow path, a result confirmed by other studies (Findlay et al. 1993, Marmonier et al. 1995). However, in contrast to the studies by Triska and others, denitrification in Sycamore Creek was greatest in the well-oxygenated upper sections of the parafluvial flow paths (Holmes et al. 1996). This was attributed to the greater supply of nitrate created by the high nitrification rates in those areas. Denitrification in well-oxygenated zones was likely within anaerobic microzones that were not detected in the bulk water samples.

A generalized diagram of biogeochemical trends along a well-oxygenated hyporheic flow path is illustrated in Figure 16.11. Mineralization is rapid relative to flow velocity creating an enriched source of nitrate and other nutrients in water that reenters the channel. Increased hyporheic zone nutrient concentrations and an apparent stimulation of primary production in the surface channel at sites of upwelling nutrients are reported for a variety of systems (Ward 1989, Coleman and Dahm 1990, Valett et al. 1990, Valett et al. 1994).

In rivers of the Pacific coastal ecoregion, where primary production is limited by the availability of nitrogen, the input of nitrogen from hyporheic zones significantly influences primary and secondary production and invertebrate grazing in the stream. Data from the Queets River, Washington, suggest that hyporheic transfers of alder-derived nitrogen may be even more important in these large alluvial floodplain rivers than in smaller streams because of the dominance of alder in floodplain forests. Floodplain reaches of these rivers are composed of a mosaic of riparian forest patches of various age and vegetative composition inhabiting a highly dissected floodplain through

TABLE 16.4. Conditions controlling the dominant form of inorganic nitrogen and dominant nitrogen transformations within hyporheic zones.

Dominant form Dominant transformation	NO_3 mineralization, nitrification	NH_4 denitrification
Condition	high hydraulic conductivities	low hydraulic conductivities
	head of flowpath	mid to tail of flowpath
	short flowpath	long flowpath
	adequate oxygen supply	reduced oxygen supply
	adequate organic nitrogen supply	adequate nitrate supply
	adequate supply of labile organic carbon	adequate supply of labile organic carbon

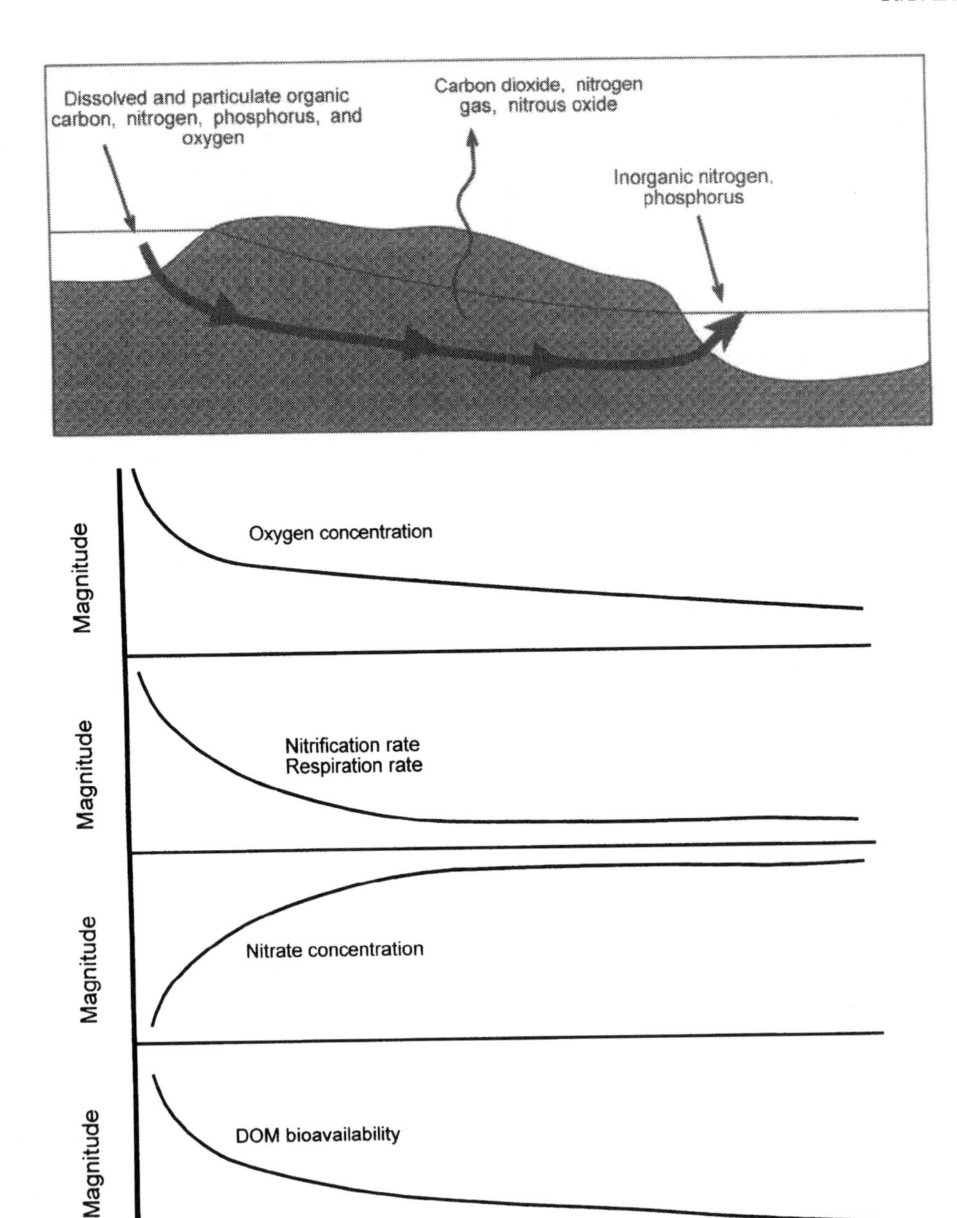

FIGURE 16.11. Generalized trends in key variables along a well-oxygenated parafluvial flowpath dominated by surface water. Downwelling water brings oxygen and organic matter to the epilithic community. Utilization of organic matter reduces oxygen concentrations and releases inorganic forms of nitrogen and phosphorus. Nitrification of ammonia released by metabolism further reduces oxygen concentrations. As the labile organic matter is degraded, the remaining material is less biologically available. As a consequence, all transformation rates decrease downstream along the flow path.

which high-flow backchannels and floodplain tributaries meander (Chapter 12). The accumulation of nitrogen during the successional development of landforms colonized by alder creates a patchy mosaic of potential nitrogen inputs.

The interaction between patchy distributions of soil nutrients and variable hyporheic flows beneath the floodplain forest creates a spatially heterogeneous pattern of hyporheic nutrient flux upwelling into the channel. Tracer studies in a floodplain backchannel in the Queets River, Washington, indicate that hyporheic pathways interact with riparian soils in a complex manner. Groundwater flows from floodplain terraces may pass beneath the active channel before upwelling into the wetted channel or mixing with other subsurface waters. Shallow hyporheic flows respond within 10 hours to additions of surface tracers, whereas, deeper, longer hyporheic flow paths show increasing levels of tracers long after shallow wells have returned to background levels (~80h).

Nutrients supplied from the substratum have different effects on algal diversity and community structure than nutrients imported with surface water (Pringle 1990). Hyporheic nutrient inputs seeping up into nutrient-poor backchannel habitats profoundly affect the distribution of primary production and the community structure of epilithic algae in alluvial floodplain habitats in these systems. Distinct seasonal patterns of subsurface nitrogen concentration in the Queets River suggest that nitrogen from alder terraces influences subsurface nitrogen concentrations in adjacent, nonvegetated zones and channels (Figure 16.12). Epilithic algae are concentrated at sites of upwelling hyporheic water where standing stocks of epilithic chlorophyll in backchannels are seven times greater than in downwelling zones.

Organic Matter Utilization and the Role of Epilithon

Organic matter decomposition and oxygen consumption in hyporheic zones are largely mediated by benthic microbes. Organic material carried in downwelling water or deposited during floods provides a food source for a community of microorganisms living on the sediment surfaces known as *epilithon*. This epilithic community, which is ubiquitous on surface sediments, is composed of layers of bacteria, fungi, protozoans, meiofauna, and an associated polysaccharide matrix (Karlstrom 1978, Lock et al. 1984). Epilithic bacteria within this community rapidly take up and metabolize dissolved organic matter (DOM, organic material <0.45 μm) (Bott et al. 1984).

Epilithon is likely a dominant component of the hyporheic community because of the enormous surface area of hyporheic sediment. Although studies are few, there is evidence that hyporheic epilithon is common (Bärlocher and Murdoch 1989, Fischer et al. 1996, Pusch 1996). Bacteria within deep sediments often dominate total sediment bacterial biomass (Meyer 1988). Bacteria and protozoans readily colonized rocks incubated in hyporheic wells at several locations in the Flathead River, Montana (Stanford et al. 1994). Cell densities on the rock surfaces were two orders of magnitude greater than those in the hyporheic water. In the hyporheic zone of the Steina River, a mountain stream in Germany, hyporheic bacterial biomass equaled the combined biomass of hyporheic and surface fauna (Fischer et al. 1996). Bacterial production was ten times that of the fauna and equal to that of surface algae.

Respiration by epilithon is important both in creating oxygen gradients within hyporheic flow paths and influencing total ecosystem metabolism. Respiration within the hyporheic zone is a major fraction of total river metabolism (Grimm and Fisher 1984, Edwards and Meyer 1987, Pusch and Schwoerbel 1994). Although it is difficult to separate microbial respiration from that of larger organisms, evidence suggests microbes are responsible for most oxygen consumption. Hyporheic oxygen and DOM concentration patterns in Wappingera Creek, New York, corresponded to hydraulic head gradients, indicating rapid epilithic utilization of imported organic matter along the flow path (Findlay et al. 1993). Fifty percent of the DOM entering the hyporheic zone was available to sediment bacteria. Further along the flow paths, DOM availability and oxygen consumption were lower, indicating that the organic matter (driving respiration) entered with the advecting water. Similar patterns are documented for DOM and oxygen within a gravel bar

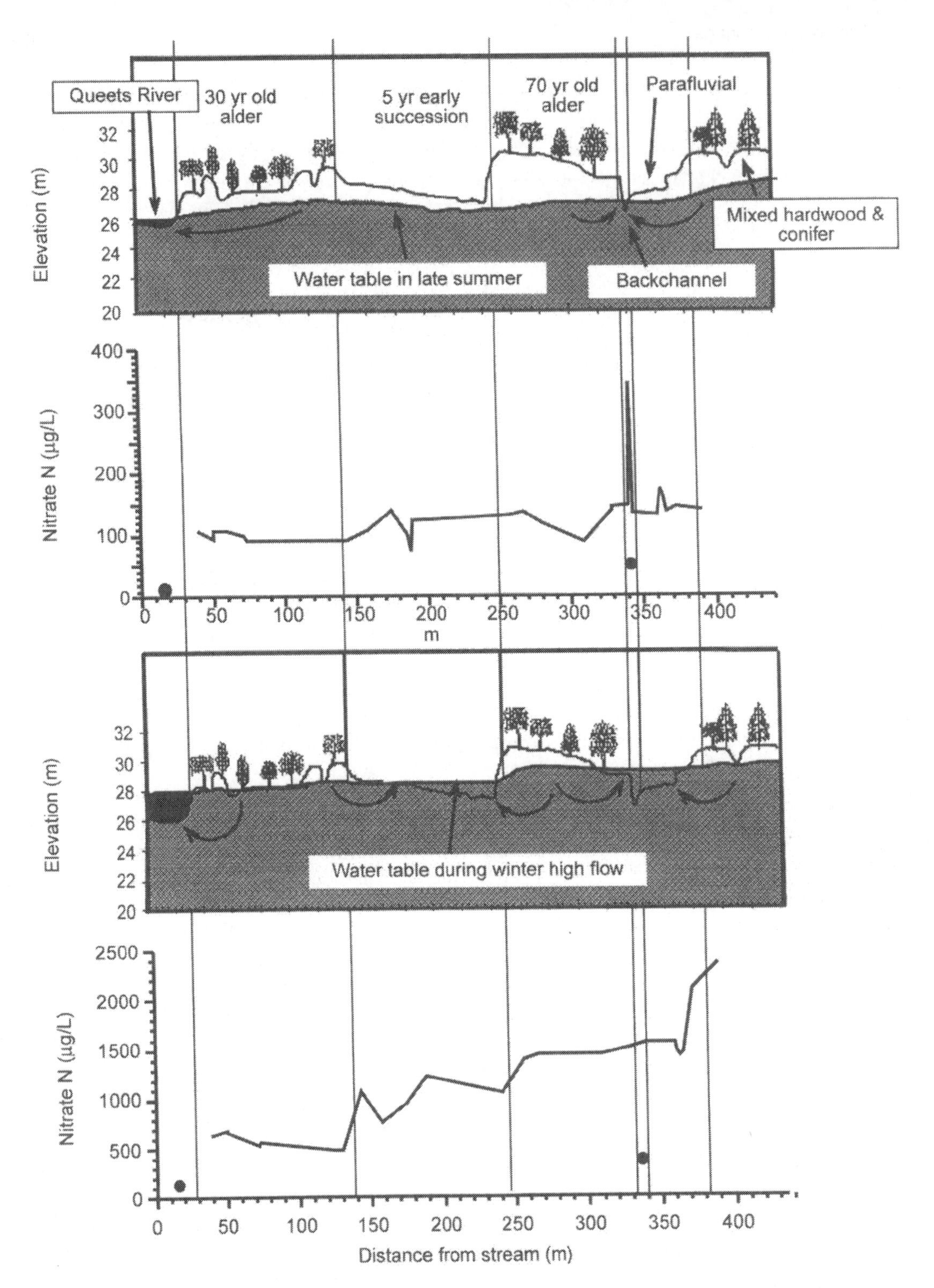

FIGURE 16.12. Subsurface water table levels and nitrate concentrations along an alluvial floodplain transect in the Queets River, Washington (● surface water concentrations, ——subsurface well concentrations) during summer and winter. Nitrate concentrations are higher during winter high water periods when the water table is in contact with the shallow soils. Water surface elevations indicate that water and nitrogen are moving from the 70-year-old alder terrace to adjacent surface channels. In both seasons, surface water concentrations within the backchannel are elevated compared with mainchannel Queets River values, indicating that subsurface water from the adjacent terraces is supplying added nitrate.

in a western Washington river (Vervier and Naiman 1992).

In parafluvial bars in Sycamore Creek, Arizona, oxygen and DOM concentrations decreased consistently along a series of parafluvial flow paths (Holmes et al. 1994b). Respiration rates were greatest in downwelling zones (Jones et al. 1995) and the greatest oxygen losses occurred in the first 5 to 10 meters, suggesting that DOM availability decreased along the flow paths (Holmes et al. 1994b). In Steina Creek, Germany, 96% of hyporheic respiration was attributable to microbes. Respiration rates were highly correlated with the amount of particulate organic matter loosely associated with gravel in the sediment pores (Pusch and Schwoerbel 1994, Fischer et al. 1996).

In addition to nitrogen cycling, oxygen gradients may also be important for carbon cycling pathways. Generation and loss of methane by anaerobic hyporheic metabolism may be an important pathway for carbon removal from streams (Dahm 1984, Jones et al. 1994). It has been suggested that anaerobic areas within stream sediments are common and occur in microzones surrounded by oxygenated water (Dahm et al. 1991, Valett et al. 1996). Further, surface upwellings of labile products of anaerobic metabolism may be a significant, though unevenly distributed, food resource for aerobic biofilms.

Ecology and Structure of Hyporheic Invertebrate Communities

Hyporheic zones vary from primarily surface water to almost pure ground water. Biota inhabiting the hyporheic zone, the hyporheos, exhibit various adaptations to this range. Animals dwelling in surface waters or channel bottom sediments are often referred to as epigean fauna, whereas those living in subsurface habitats are called hypogean fauna. Numerous terms, reflecting great variation in life history and behavior of aquatic fauna, have been used to describe the affinity of various taxa of the hypogean fauna for life within interstitial habitats. Gibert (1994), however, developed a system that classifies taxa by life history and affinity for interstitial habitats and thus, eliminates confusion over vague or conflicting terms (Figure 16.13).

The hyporheos near the open channel comprises a mixture of organisms known as stygophiles. Organisms that are common in surface sediments (occasional hyporheos) or have epigean adult life stages (amphibites) often dominate. Occasional hyporheos use the hyporheic zone seasonally or during early life history stages (Williams and Hynes 1974). Amphibites, for example rarely occur in or near the surface as larvae but migrate to the surface to metamorphose, emerge, mate, and lay eggs (Stanford and Ward 1993). Their eggs wash into the hyporheic zone or larvae migrate back into the hyporheic zone after hatching. Organisms normally found only in the subsurface (permanent hyporheos) may also be present near the open channel (Coleman and Hynes 1970, Williams and Hynes 1974, Williams 1989). Stygoxen species may also occur, for example, while seeking refuge during spates or droughts (Williams and Hynes 1974).

Further from the stream surface, the hyporheic community is dominated by stygophiles and stygobites, a collection of species which usually are restricted to subsurface habitats. This group includes copepods, ostracods, isopods, and amphipods, species which do not need to reach the surface to reproduce. As the proportion of surface water decreases, the proportion of organisms that are highly specialized for hypogean life (stygobites) increases (Bretschko 1992).

Interstitial fauna exhibit unique traits and adaptations to life in sediment pores. Long, slender, and flexible bodies facilitate movement through interstices, but small, blunt, hard bodies also help force their way through (Williams 1984). Obligate groundwater species (stygobites) are blind or lack pigmentation (Boulton et al. 1991). Some organisms simply are very small. The suitability of different shapes and sizes of organisms for penetrating different types of sediment interstices may explain the correlation between species

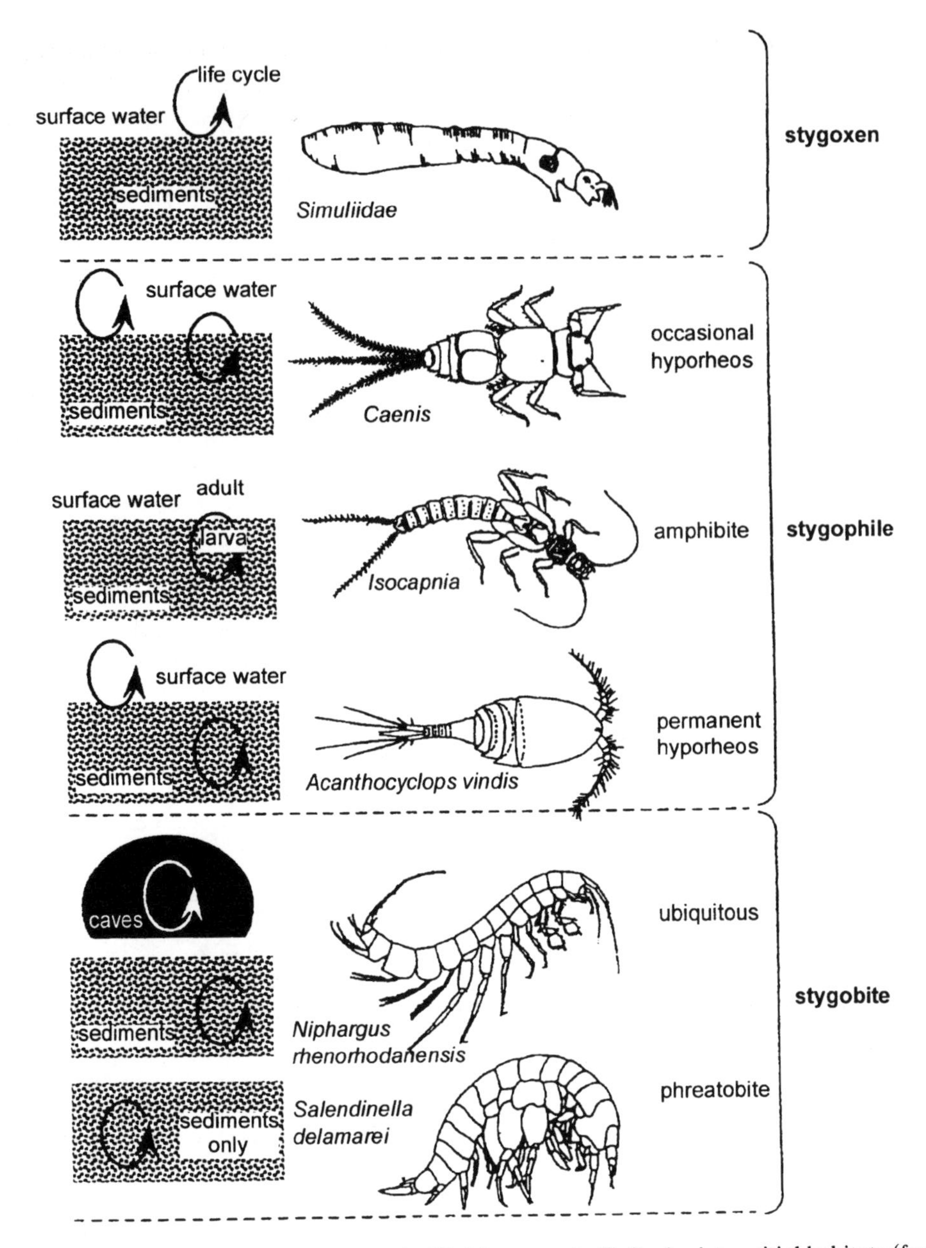

FIGURE 16.13. Hyporheic fauna classification by life history and affinity for interstitial habitats (from Gibert et al. 1994, with permission).

abundance and porosity reported in some systems (Maridet et al. 1992).

Current understanding of the abundance, distribution, community structure, productivity, and trophic structure of subsurface communities is limited and generally biased toward hyporheic zones less than two meters beneath the wetted channel. Understanding of most community variables is limited by the difficulty of sampling and analyzing more remote hyporheos. Sediment cores provide a picture of hyporheic communities but provide quantitative data only in shallow, relatively fine-grained sediments. In coarser substrata, freeze coring also provides quantitative data (Williams and Hynes 1974, Maridet et al. 1992) but is restricted to depths less than two meters. In coarse-grained, deep sediments characteristic of many of the Pacific coastal streams, sampling hyporheos requires pumping wells to collect organisms entrained in the water. Notwithstanding these limits, a picture of an abundant, diverse, and fascinating fauna emerges from the available data.

Boulton et al. (1992) identified 56 hyporheic taxa in a desert stream. Stanford and Ward (1993) found communities of up to 165 species in the hyporheic zone of Montana rivers. Although the diversity of hyporheic fauna in Pacific coastal streams is unknown, Stanford and Gaufin (1974) found adults of common amphibites throughout the Pacific Northwest. Preliminary samples collected from wells in a floodplain backchannel of the Queets River, Washington yielded at least 15 hyporheic taxa, including rotifers, crustaceans, annelids, and insect larvae. Although these samples represent only one location, hyporheic invertebrates were found in wells throughout the reach.

Evidence suggests that benthic organisms may find refuge from surface perturbations in the hyporheic zone (Williams and Hynes 1974, Marchant 1988, Grimm et al. 1991, Griffith and Perry 1993). As a refuge the hyporheic zone may decrease the impact of floods and droughts and serve as a source of colonizers for habitats defaunated by disturbance. Recolonization studies in experimentally defaunated sediments indicate that movement of organisms from hyporheic habitats can be a significant source of recruits (Williams and Hynes 1976, Boulton et al. 1991). However, a study by Palmer et al. (1992) reported little evidence for a hyporheic refuge for meiofauna in a sandy-bottomed river in Virginia. Further, Smock et al. (1994) reported that the hyporheic zone of a sandy-bottomed coastal plain river was ineffective as refuge during drought-induced dewatering because of anoxic conditions. The importance of the hyporheic zone as a refuge is clearly not fully understood.

Controls on Community Structure

Studies of community structure and distribution confirm that the hyporheos is highly variable in space and time reflecting the heterogeneity of interstitial habitats (Marmonier 1991, Kowarc 1992, Schmid 1993). The relative proportions of different faunal types vary in response to changes in flow, chemistry, or food availability (Williams 1989, Marmonier 1991). In areas of downwelling, surface organisms occur deeper in the sediments whereas, in areas of upwelling, stygobites (sub-surface organisms) occur closer to the surface (Dole-Olivier and Marmonier 1992). Many investigators have found strong correlations between community structure and physico-chemical variables (Palmer 1990, Chafiq et al. 1992, Creuzé des Châtelliers et al. 1990, Sterba et al. 1992, Stanley and Boulton 1993), whereas, others have found none (Williams 1989). In some studies both trends occur within the same river at different locations (Dole-Olivier and Marmonier 1992).

Factors that may influence the distribution of fauna in hyporheic zones include temperature, flow direction and velocity, location along flow path, concentration of various chemical constituents, geomorphology, porosity, grain size, abundance of fine particulate organic matter, residence time, and oxygen concentration. Oxygen is particularly important because it is required by most organisms and is strongly correlated with physical factors affecting flow (Danielopol 1989). Along the Rhine River and below the terminus of an Australian desert stream, hyporheic species richness was inversely proportional to oxygen concentration (Creuzé des Châtelliers et al. 1992, Cooling and Boulton 1993). However, Strayer (1994) suggests oxygen may have relatively little influence on distributions of stygobite species that are adapted to very low oxygen concentrations. Many of the physicochemical variables that affect the distribution of organisms are highly correlated. Most are directly related to porosity and hydraulic conductivity, which may be the fundamental controls on hyporheic community structure Strayer (1994).

Food Sources and Trophic Structure

The hyporheic food web depends on organic matter imported from outside its boundaries (allochthonous material), because there is no light to support photosynthesis within the hyporheic zone. Organic matter enters in particulate or dissolved forms which are influenced by different sets of variables. Inputs of organic matter are continuous (e.g., when material is entrained within the downwelling surface water) or episodic (e.g., when scour and fill events deposit material during floods). The location and size of the hyporheic zone affects the timing and nature of organic inputs.

Hyporheic zones within the active channel receive coarse particulate organic matter (CPOM > 1 mm) during episodic events such as floods or channel movements (Herbst 1980, Jones et al. 1995). In the Ogeechee River, Georgia, leaves and wood are deposited deep within the hyporheic sediments on the falling limb of floods (Meyer 1988). These inputs are a major source of food for hyporheic bacteria, whose productivity rates correlate directly with organic matter content. Episodic inputs of this type occur at frequencies proportional to the recurrence interval of channel-forming events such as bankfull floods. In rivers with natural hydrologic regimes, CPOM inputs recharge hyporheic carbon pools every few years.

Floodplain hyporheic zones receive few episodic CPOM inputs over short time intervals; therefore, most material percolates from overlying soils or is entrained within the ground water or advecting surface water. However, in alluvial reaches of Pacific coastal streams, buried wood from previous channel migrations is often abundant and can last for centuries. Large amounts of slowly decomposing buried wood provide a continuous source of food to hyporheic fauna despite the low return frequencies of fresh inputs. Wood can enter the food web directly through fungal or bacterial decomposition or through direct consumption by hyporheic macroinvertebrates. Dissolved organic matter (DOM) leached from buried wood may also be a significant localized food source. Although much is known about the community of decomposers on logs on the forest floor, very little is known of the corresponding community on logs buried within the hyporheic zone.

Fine particulate organic matter (FPOM < 1 mm) can enter the hyporheic zone with advecting water and be transported along flow paths, thus providing a continual source of new detritus (Boulton et al. 1991). The maximum size of particles in transport and the distance traveled is a function of the porosity, length and velocity of the hyporheic flow paths. For short flow paths close to the advecting water source, FPOM import may be a major source of food throughout the flow path. For long flow paths characteristic of large floodplain hyporheic zones, the supply of FPOM declines with distance along the flow paths.

Dissolved organic matter entering with downwelling water is quickly absorbed or consumed by microbes, creating gradients in DOM quality along the flow path. Where flow paths are short or fast, DOM quality may not vary with position, but flow paths longer than a few meters tend to have gradients in DOM quality that impact the abundance and productivity of epilithic microbes. Labile, DOM entering with ground water may alter this pattern.

In alluvial rivers of the Pacific coastal ecoregion, flow paths greater than 10 meters tend to occur beneath large river floodplains overlain by productive riparian forests. Here, the downward transport of DOM leached from overlying soils may be more important to hyporheic food webs than material entrained within advecting surface water. Movement of soil organic matter down into the hyporheic zone is influenced by the overlying vegetation structure, soil decomposition dynamics, and chemical partitioning during water movement. Organic matter supply may be seasonal where overlying vegetation is deciduous. Thus, hyporheic communities, may be sensitive to changes in riparian forest vegetation resulting from successional dynamics, floods, or riparian land use (e.g., silviculture, agriculture).

Epilithic Biofilms

Epilithic biofilms are a potentially important food source for higher trophic levels. Bärlocher and Murdoch (1989) found that artificial substrates placed within the hyporheic zone quickly became covered with epilithic biofilms. When treated with digestive enzymes from hyporheic organisms the biofilm released sugars and amino acids, high-quality foods for epilithic grazers. The amount of organic matter released by acid hydrolysis of the epilithon was five times the amount attributable to the living biomass of bacteria and fungi, suggesting the presence of an additional source of organic matter in the biofilm.

This additional organic matter likely comes from the polysaccharide matrix that comprises the nonliving component of epilithon. This reactive layer may be responsible for much of the immediate uptake of DOM by epilithon. Rapid abiotic sorption of DOM onto the biofilm matrix has been detected by several investigators (Rutherford and Hynes 1987, Freeman and Lock 1995, Findlay and Sobczak 1996). Subsequent uptake of sorbed DOM by epilithic microbes allows growth during periods when imported DOM supplies are reduced in quantity or quality. Freeman and Lock (1995) suggest that physical sorption followed by biotic uptake buffers biofilm organisms from perturbations in food supply and increases community resilience. Consumption of the matrix by invertebrate scrapers also directly transfers DOM to higher trophic levels, bypassing metabolic losses attributed to transfer of DOM through the microbial food loop (Pomeroy 1974).

Secondary Production

Only one study to date has quantified secondary production within the hyporheic zone. Smock et al. (1992) found that 65% of total channel invertebrate production occurred within the hyporheic zone of a sandy-bottomed stream in Virginia. Production was dominated by small collector–gatherers (e.g., chironomids, ceratopogonids, early instar caddisflies, mayflies, beetles). Higher production in the hyporheic zone was attributed to the greater stability of hyporheic sediments compared with surface. Production decreased as depth increased and oxygen concentrations decreased.

The fate of invertebrate production in the hyporheic zone is poorly known. Jackson and Fisher (1986) reported a significant return of stream production to riparian zone insectivores by emerging insect adults, but did not distinguish hyporheic from benthic production. Return of amphibite biomass (Figure 16.13) during adult life stages may transfer material from the hyporheic zone to the surface riparian zone. Emergent hyporheic stoneflies are abundant in the drift of Montana rivers and are episodically important in diets of stream consumers (Stanford et al. 1994). Consumption of shallow hyporheos by fish and invertebrate predators in channels is also likely. Hyporheic fauna that migrate toward the surface are subject to predation by vertebrate and invertebrate predators moving into the shallow hyporheic zone in search of prey.

Implications for Management of Coastal Rivers of Washington

Hyporheic processes profoundly influence the integrity and productivity of stream ecosystems in the Pacific coastal ecoregion. Water quality in most of these streams is excellent (Chapter 4) but as land uses change, maintaining such quality requires an understanding of the hidden hyporheic habitat. Changes to the hyporheic zone occur through actions that affect the extent of the hyporheic zone, the strength of its connection to surface waters, or the way it interacts with riparian soils (Table 16.5). Management activities that alter sediment inputs or hydrology impact hydraulic conductivity or hydraulic head patterns, thereby decreasing the available hyporheic habitat and the strength of its connection to surface water. Channelization and diking also alter subsurface flow patterns and result in similar effects.

Decreased surface water quality may have unanticipated effects on hyporheic processes, further accelerating overall water quality degradation. For example, increased biological oxygen demand (BOD) as a result of organic loading (e.g., sewage inputs) can decrease oxygen concentrations in the surface water. When the oxygen content of water advecting into the

Table 16.5. Potential direct and indirect effects of various management activities on hyporheic processes.

Activity	Direct ecosystem response	Indirect hyporheic response
Dams	Reduced maximum discharge, altered flood frequency and timing, reduced sediment transport, altered temperature regime	Reduced subsurface flows, reduced extent of hyporheic zone, lower oxygen concentrations, less fine sediment flushing, reduced interstitial space, lower DOM and POM inputs, reduced secondary production
Forestry	Decreased input of large woody debris, increased coarse and fine sediment input, altered riparian vegetation	Changes in distribution and volume of hyporheic zone, altered riparian soil chemistry, altered riparian nutrient inputs, changes in stream primary production
Agriculture	Elimination of riparian vegetation, groundwater withdrawal, fertilizer applications, pesticide inputs, diking, channelization	Alterations of riparian soil organic matter and nutrient stocks, elimination of riparian habitat, reductions in hyporheic flows, reversals of subsurface flowpaths, elimination hyporheic zones, reduction in invertebrate production and diversity
Urbanization	Changes in hydrology, increased fine sediment inputs, increased organic loading, toxic material inputs, increased flood magnitudes, channel incisement, reduced riparian zone	Elimination of hyporheic zone, anaerobic conditions, reduction or elimination of hyporheic fauna, shift to undesirable fauna, reduced biodiversity and production

hyporheic zone is lowered, the boundaries of aerobic and anaerobic biogeochemical processes change with subsurface patterns of oxygen. Shifts in the pattern of hyporheic nutrient retention and transformation caused by changes in the redox environment and organic carbon supply impact solute retention and transformation efficiency. Ecosystem productivity and diversity in surface and subsurface habitats are likely altered.

Many management issues in Pacific coastal streams center around the impacts of land use on salmon populations. Salmon utilize the hyporheic zone during egg development and after hatching (Vaux 1968). As a result, spawning behavior of some salmon populations is influenced by subsurface ground water and hyporheic flows. For example, in the Kamchatcha River, Russia, 30 to 40 % of chum salmon (*O. keta*) spawn in areas of upwelling hyporheic water (Leman 1993). Research on how hyporheic flows at channel unit or reach scales affect conditions in salmon redds is lacking, but interactions between nested scales of subsurface flows suggests that larger scale hyporheic flows might override the flows induced by redd geometry.

Salmon habitat productivity is linked to riparian characteristics that control the input of light, nutrients, and organic matter (Gregory et al. 1987). Changes in pattern of riparian vegetation created by urbanization, agriculture, or riparian silviculture influence floodplain habitat productivity by altering the pattern of soil nutrients and subsurface linkages. In large alluvial reaches of Pacific coastal streams, riparian floodplain habitats are important salmon rearing areas (Sedell et al. 1984, Cederholm and Reid 1987, Hartman and Brown 1987, Swales and Levings 1987). Enhanced subsurface nitrogen inputs via hyporheic exchanges may be a critical factor supporting biological productivity in these habitats.

An understanding of hyporheic habitats and processes is necessary for making decisions about the management of streams and their associated riparian zones (FEMAT 1993). Because Pacific coastal rivers are most often characterized by alluvial gravel beds with relatively high gradients and fast currents, hyporheic zones are extensive and hyporheic processes will strongly influence water chemistry, nutrient retention, ecosystem productivity, and biodiversity. Management and restoration programs aimed at achieving sustainable resource management must recognize the importance of hyporheic habitats to succeed.

Acknowledgments. The author would like to thank Bob Bilby, Robert J. Naiman, Maury Valett, Stacey Poulson, and an anonymous reviewer for their helpful reviews of the manuscript. Sylvia Kantor assisted with Figures 16.3, 16.4, 16.5, and 16.6.

Literature Cited

Bärlocher, F., and J.H. Murdoch. 1989. Hyporheic biofilms—a potential food source for interstitial animals. Hydrobiolgia **184**:61–67.

Bencala, K.E. 1984. Interactions of solutes and streambed sediment 2. A dynamic analysis of coupled hydrologic and chemical processes that determine solute transport. Water Resources Research **20**:1804–1814.

——. 1993. A perspective on stream-catchment connections. Journal of the North American Benthological Society **12**:44–47.

Bencala, K.E., J.H. Duff, J.W. Harvey, A.P. Jackman, and F.J. Triska. 1993. Modeling within the stream-catchment continuum. Pages 163–187 *in* A.J. Jakeman, M.B. Beck, and M.J. McAleer, eds. Modeling change in environmental systems. John Wiley & Sons, New York, New York, USA.

Bencala, K.E., A.P. Jackman, V.C. Kennedy, R.J. Avanzino, and G.W. Zellweger. 1983. Kinetic analysis of strontium and potassium sorption onto sand and gravel in a natural channel. Water Resources Research **19**:725–731.

Bencala, K.E., V.C. Kennedy, G.W. Zellweger, A.P. Jackman, and R.J. Avanzino. 1984. Interactions of solutes and streambed sediment 1. An experimental analysis of cation and anion transport in a mountain stream. Water Resources Research **20**:1797–1803.

Bott, T.L., L.A. Kaplan, and F.T. Kuserk. 1984. Benthic bacterial biomass supported by streamwater dissolved organic matter. Microbial Ecology **10**:335–344.

Boulton, A.J., S.E. Stibbe, N.B. Grimm, and S.G. Fisher. 1991. Invertebrate recolonization of small patches of defaunated hyporheic sediments in a Sonoran desert stream. Freshwater Biology **26**:267–277.

Boulton, A.J., H.M. Valett, and S.G. Fisher. 1992. Spatial distribution and taxonomic composition of the hyporheos of several Sonoran desert (Arizona, USA) streams. Archiv für Hydrobiologie **125**:37–61.

Bouwer, H. 1989. The Bouwer and Rice slug test-An update. Groundwater **27**:304–309.

Bretschko, G. 1992. Differentiation between epigeic and hypogeic fauna in gravel streams. Regulated Rivers: Research & Management **7**:17–22.

Broshears, R.E., K.E. Bencala, B.A. Kimball, and D.M. McKnight. 1993. Tracer-dilution experiments and solute-transport simulations for a mountain stream, St. Kevin Gulch, Colorado. US Geological Survey Water-Resources Investigations.

Cedergren, H.R. 1989. Seepage, drainage, and flow nets. John Wiley & Sons, New York, New York, USA.

Cederholm, C.J., and L.M. Reid. 1987. Impact of forest management on coho salmon (*Oncorhynchus kisutch*) populations in the Clearwater River, Washington. *In* E.O. Salo and T.W. Cundy, eds. Streamside management: Forestry and fishery interactions. University of Washington Press, Seattle, Washington, USA.

Chafiq, M., J. Gibert, P. Marmonier, M.J. Dole Olivier, and J. Juget. 1992. Spring ecotone and gradient study of interstitial fauna along two floodplain tributaries of the River Rhone, France. Regulated Rivers: Research & Management **7**:103–115.

Cole, D.W., J. Compton, H.V. Miegroet, and P. Homann. 1990. Changes in soil properties and site productivity caused by red alder. Water, Air and Soil Pollution **54**:231–246.

Coleman, M.J., and H.B.N. Hynes. 1970. The vertical distribution of the invertebrate fauna in the bed of a stream. Limnology and Oceanography **15**:31–40.

Coleman, R.L., and C.N. Dahm. 1990. Stream geomorphology: Effects on periphyton standing crop and primary production. Journal of the North American Benthological Society **9**:293–302.

Cooling, M.P., and A.J. Boulton. 1993. Aspects of the hyporheic zone below the terminus of a South Australian arid zone stream. Australian Journal of Marine and Freshwater Research **44**:411–426.

Creuzé des Châtelliers, M., P. Marmonier, O.M.J. Dole, and E. Castella. 1992. Structure of interstitial assemblages in a regulated channel of the River Rhine (France). Regulated Rivers: Research & Management **7**:23–30.

Creuzé des Châtelliers, M., and J. Reygrobellet. 1990. Interactions between geomorphological processes, benthic and hyporheic communities: First results on a bypassed canal of the French Upper Rhone River. Regulated Rivers: Research & Management **5**:139–158.

D'Angelo, D.J., J.R. Webster, S.V. Gregory, and J.L. Meyer. 1993. Transient storage in Appalachian and Cascade mountain streams as related to hydraulic characteristics. Journal of the North American Benthological Society **12**:223–235.

Dahm, C.N. 1984. Uptake of dissolved organic carbon in mountain streams. Verhandlungen der Internationalen Vereinigung für Theorestische und Angewandte Limnologie **22**:1842–1846.

Dahm, C.N., D.L. Carr, and R.L. Coleman. 1991. Anaerobic carbon cycling in stream ecosystems. Verhandlungen der Internationalen Vereinigung für Theorestische und Angewandte Limnologie **24**:1600–1604.

Danielopol, D.L. 1989. Groundwater fauna associated with riverine aquifers. Journal of the North American Benthological Society **8**:18–35.

Dole-Olivier, M.J., and P. Marmonier. 1992. Patch distribution of interstitial communities: Prevailing factors. Freshwater Biology **21**:177–191.

Duff, J.H., and F.J. Triska. 1990. Denitrification in sediments from the hyporheic zone adjacent to a small forested stream. Canadian Journal of Fisheries and Aquatic Sciences **47**:1140–1147.

Edwards, R.T., and J.L. Meyer. 1987. Metabolism of a sub-tropical low gradient blackwater river. Freshwater Biology **17**:251–263.

Evans, E.C., M.T. Greenwood, and G.E. Petts. 1995. Thermal profiles within river beds. Hydrologic Process **9**:19–25.

FEMAT (Forest Ecosystem Management Assessment Team). 1993. Forest ecosystem management: An ecological, economic, and social assessment. USDA Forest Service, US Department of Commerce, National Oceanic and Atmospheric Administration, National Marine Fisheries Service, USDI Bureau of Land Management, Fish and Wildlife Service, National Park Service, Environmental Protection Agency. Portland, Oregon, USA.

Findlay, S., and W. Sobczak. 1996. Variability in removal of dissolved organic carbon in hyporheic sediments. Journal of the North American Benthological Society **15**:35–41.

Findlay, S., D. Strayer, C. Goumbala, and K. Gould. 1993. Metabolism of streamwater dissolved organic carbon in the shallow hyporheic zone. Limnology and Oceanography **38**:1493–1499.

Fischer, H., M. Pusch, and J. Schwoerbel. 1996. Spatial distribution and respiration of bacteria in stream-bed sediments. Archiv für Hydrobiologie **137**:281–300.

Fortner, S.L., and D.S. White. 1988. Interstitial water patterns: A factor influencing the distributions of some lotic aquatic vascular macrophytes. Aquatic Botany **31**:1–12.

Freeman, C., and M.A. Lock. 1995. The biofilm polysaccharide matrix: A buffer against changing organic substrate supply? Limnology and Oceanography **40**:273–278.

Freeze, R.A., and J.A. Cherry. 1979. Groundwater. Prentice Hall, Englewood Cliffs, New Jersey, USA.

Gibert, J. 1991. Groundwater systems and their boundaries: Conceptual framework and prospects in groundwater ecology. Verhandlungen der Internationalen Vereinigung für Theorestische und Angewandte Limnologie **24**:1605–1608.

Gibert, J., J.A. Stanford, M.-J. Dole-Oliver, and J.V. Ward. 1994. Basic attributes of groundwater ecosystems and prospects for research. Pages 8–40 *in* J. Gibert, D.L. Danielpol, and J.A. Stanford, eds. Groundwater ecology. Academic Press, San Diego, California, USA.

Gregory, S.V., G.A. Lamberti, D.C. Erman, K.V. Koski, M.L. Murphy, and J.R. Sedell. 1987. Influence of forest practices on aquatic production. *In* E.O. Salo and T.W. Cundy, eds. Streamside management: Forestry and fishery interactions. Institute of Forest Resources, University of Washington. Seattle, Washington, USA.

Griffith, M.B., and S.A. Perry. 1993. The distribution of macroinvertebrates in the hyporheic zone of two small Appalachian headwater streams. Archives für Hydrobiologia **126**:373–384.

Grimm, N.B., and S.G. Fisher. 1984. Exchange between interstitial and surface water: Implications for streams metabolism and nutrient cycling. Hydrobiologia **111**:219–228.

Grimm, N.B., H.M. Valett, E.H. Stanley, and S.G. Fischer. 1991. Contribution of the hyporheic zone to stability of an arid land stream. Verhandlungen der Internationalen Vereinigung für Theorestische und Angewandte Limnologie **20**:1595–1599.

Hartman, G.F., and T.G. Brown. 1987. Use of small, temporary, floodplain tributaries by juvenile salmonids in a west coast rain-forest drainage basin, Carnation Creek, British Columbia. Canadian Journal of Fisheries and Aquatic Sciences **44**:262–270.

Harvey, J.W., and K.E. Bencala. 1993. The effect of streambed topography on surface–subsurface water exchange in mountain catchments. Water Resources Research **29**:89–98.

Hendricks, S.P., and D.S. White. 1991. Physicochemical patterns within a hyporheic zone of a northern Michigan river, with comments on surface water patterns. Canadian Journal of Fisheries and Aquatic Sciences **48**:1645–1654.

Herbst, G.N. 1980. Effects of burial on food value and consumption of leaf detritus by aquatic invertebrates in a lowland forest stream. Oikos **35**:411–424.

Holmes, R.M., S.G. Fisher, and N.B. Grimm. 1994. Nitrogen dynamics along parafluvial flow paths: Importance to the stream ecosystem. Second International Conference on Groundwater Ecology, Atlanta, GA. American Water Resources Association, Herndon, Virginia, USA.

Holmes, R.M., S.G. Fisher, and N.B. Grimm. 1994b. Parafluvial nitrogen dynamics in a desert stream ecosystem. Journal of the North American Benthological Society **13**:468–478.

Holmes, R.M., J. Jeremy, B. Jones, S.G. Fisher, and N.B. Grimm. 1996. Denitrification in a nitrogen-limited stream ecosystem. Biogeochemistry **33**:125–146.

Jackson, J.K., and S.G. Fisher. 1986. Secondary production, emergence, and export of aquatic insects of a Sonoran Desert stream. Ecology **67**:629–638.

Jones, J.B., Jr., R.M. Holmes, S.G. Fisher, N.B. Grimm, and D.M. Greene. 1994. Methanogenesis in Sonoran desert stream ecosystems. Biogeochemistry **31**:155–173.

Jones, J.B., Jr., S.G. Fisher, and N.B. Grimm. 1995. Vertical hydrologic exchange and ecosystem metabolism in a Sonoran desert stream. Ecology **76**:942–952.

Karlstrom, U. 1978. Role of the organic layer on stones in detrital metabolism in streams. Verhandlungen der Internationalen Vereinigung für Theorestische und Angewandte Limnologie **20**:1463–1470.

Kowarc, K.V. 1992. Depth distribution and mobility of a harpacticoid copepod within the bed sediment of an alpine brook. Regulated Rivers: Research & Management **7**:57–63.

Kuwabarra, J.S., H.V. Leland, and K.E. Bencala. 1984. Copper transport along a Sierra Nevada stream. Journal of Environmental Engineering **110**:646–655.

Leman, V.N. 1993. Spawning sites of Chum salmon, *Oncorhynchus keta*: Microhydrological regime and viability of progeny in redds (Kamchatka River Basin). Journal of Ichthyology **33**:104–117.

Lock, M.A., R.R. Wallace, J.W. Costerton, R.M. Ventullo, and S.E. Charlton. 1984. River epilithon: Toward a structural-functional model. Oikos **42**:10–22.

Lowrance, R. 1992. Groundwater nitrate and denitrification in a coastal plain riparian forest. Journal of Environmental Quality **21**:401–405.

Lowrance, R.R., and H.B. Pionke. 1989. Transformations and movement of nitrate in aquifer systems. *In* R.F. Follett, ed. Nitrogen management and groundwater protection. Elsevier, New York, New York, USA.

Luken, J.O., and R.W. Fonda. 1983. Nitrogen accumulation in a chronosequence of red alder communities along the Hoh River, Olympic National Park, Washington. Canadian Journal of Forest Research **13**:1228–1237.

Lynch, J.M., and J.E. Hobbie, eds. 1988. Microorganisms in action: Concepts and applications in microbial ecology. Blackwell Scientific Publications, Boston, Massachusetts, USA.

Marchant, R. 1988. Vertical distribution of benthic invertebrates in the bed of the Thomson River, Victoria. Australian Journal of Marine and Freshwater Research **39**:775–784.

Maridet, L., J.G. Wasson, and M. Philippe. 1992. Vertical distribution of fauna in the bed sediment of three running water sites: Influence of physical and trophic factors. Regulated Rivers: Research & Management **7**:45–55.

Marmonier, P. 1991. Effect of alluvial shift on the spatial distribution of interstitial fauna. Verhandlungen der Internationalen Vereinigung für Theorestische und Angewandte Limnologie **24**:1613–1616.

Marmonier, P., D. Fontvieille, J. Gibert, and V. Vanek. 1995. Distribution of dissolved organic carbon and bacteria at the interface between the Rhône River and its alluvial aquifer. Journal of the North American Benthological Society **14**:382–392.

Meyer, J.L. 1988. Benthic bacterial biomass and production in a blackwater river. Proceedings of the International Association of Theoretical and Applied Limnology **23**:1832–1838.

Montgomery, D.R., and J.M. Buffington. 1997. Channel reach morphology in mountain drainage basins. Geological Society of America Bulletin. **109**:596–611.

Munn, N.L., and J.L. Meyer. 1988. Rapid flow through the sediments of a headwater stream in the southern Appalachians. Freshwater Biology **20**:235–240.

Olson, P.L. 1995. Shallow subsurface flow systems in a montane terrace-floodplain landscape: Sauk

River, North Cascades, Washington. Ph.d dissertation, College of Forest Resources, University of Washington, Seattle, Washington, USA.

Palmer, M.A. 1990. Temporal and spatial dynamics of meiofauna within the hyporheic zone of Goose Creek, Virginia. Journal of the North American Benthological Society **9**:17–25.

Palmer, M.A., A.E. Bely, and K.E. Berg. 1992. Response of invertebrates to lotic disturbance: A test of the hyporheic refuge hypothesis. Oecologia **89**:182–194.

Peterjohn, W.T., and D.L. Correll. 1984. Nutrient dynamics in an agricultural watershed: Observation of a riparian forest. Ecology **65**:1466–1475.

Pomeroy, L.R. 1974. The ocean's food web, a changing paradigm. BioScience. **24**:499–504.

Pringle, C.M. 1990. Nutrient spatial heterogeneity: Effects on community structure, physiognomy, and diversity of stream algae. Ecology **71**:905–920.

Pusch, M. 1996. The metabolism of organic matter in the hyporheic zone of a mountain stream, and its spatial distribution. Hydrobiologia **323**:107–118.

Pusch, M., and J. Schwoerbel. 1994. Community respiration in hyporheic sediments of a mountain stream (Steina, Black Forest). Archiv für Hydrobiologie **130**:35–52.

Runkel, R.L., and K.E. Bencala. 1995. Transport of reacting solutes in rivers and streams. Pages 137–164 *in* V.P. Singh, ed. Environmental hydrology. US Government.

Rutherford, J.E., and H.B.N. Hynes. 1987. Dissolved organic carbon in streams and groundwater. Hydrobiologia **154**:33–48.

Sabater, F., and P.B. Vila. 1992. The hyporheic zone considered as an ecotone. Pages 35–43 *in* Joandomenec Ros and Narcis Prat, eds. Homage to Ramon Margalef; or, why there is such pleasure in studying nature. University De Barcelona, Barcelona, Spain.

Savant, S.A., D.D. Rieble, and L.J. Thibodeaux. 1987. Convective transport within stable river sediments. Water Resources Research **23**:1763–1768.

Schmid, P.E. 1993. Random patch dynamics of larval Chironomidae (Diptera) in the bed sediments of a gravel stream. Freshwater Biology **30**:239–255.

Sedell, J.R., J.E. Yuska, and R.W. Speaker. 1984. Habitats and salmonid distribution in pristine, sediment-rich river valley systems: S. Fork Hoh and Queets River, Olympic National Park. Fish and wildlife relationships in old-growth forests. American Institute of Fishery Research Biologists, Juneau, Alaska, USA.

Smock, L.A., J.E. Gladden, J.L. Riekenberg, L.C. Smith, and C.R. Black. 1992. Lotic macroinvertebrate production in three dimensions: Channel surface, hyporheic, and floodplain environments. Ecology **73**:876–886.

Smock, L.A., L.C. Smith, J.B. Jones, Jr., and S.M. Hooper. 1994. Effects of drought and a hurricane on a coastal headwater stream. Archiv für Hydrobiologie **131**:25–38.

Stanford, J.A., and A.R. Gaufin. 1974. Hyporheic communities of two Montana rivers. Science **185**:700–702.

Stanford, J.A., and J.V. Ward. 1988. The hyporheic habitat of river ecosystems. Nature **335**:64–66.

Stanford, J.A., and J.V. Ward. 1993. An ecosystem perspective of alluvial rivers: Connectivity and the hyporheic corridor. Journal of the North American Benthological Society **12**:48–60.

Stanford, J.A., J.V. Ward, and B.K. Ellis. 1994. Ecology of the alluvial aquifers of the Flathead River, Montana. Pages 367–390 *in* J. Gibert, D.L. Danielopol, and J.A. Stanford, eds. Groundwater ecology. Academic Press, New York, New York, USA.

Stanley, E.H., and A.J. Boulton. 1993. Hydrology and the distribution of hyporheos: Perspectives from a mesic river and a desert stream. Journal of the North American Benthological Society **12**:79–83.

Sterba, O., V. Uvira, P. Mathur, and M. Rulik. 1992. Variations of the hyporheic zone through a riffle in the R. Morava, Czechoslovakia. Regulated Rivers: Research & Management **7**:31–43.

Strayer, D.L. 1994. Limits to biological distributions in groundwater. Pages 287–310 *in* J. Gibert, D.L. Danielpol, and J.A. Stanford, eds. Groundwater ecology. Academic Press, San Diego, California, USA.

Stumm, W., and J.J. Morgan. 1970. Aquatic chemistry. Wiley-Interscience, New York, New York, USA.

Swales, S., and C.D. Levings. 1987. Role of off-channel ponds in the life of coho salmon (Oncorhynchus kisutch) and other juvenile salmonids in the Coldwater River, British Columbia. Canadian Journal of Fisheries and Aquatic Sciences **46**:232–242.

Thut, R.N., and E.P. Haydu. 1971. Effects of forest chemicals on aquatic life in surface waters. Pages 159–171 *in* J.T. Krygier and J.D. Hall, eds.

Forest land uses and the stream environment. Oregon State University Press. Corvallis, Oregon, USA.

Triska, F.J., J.H. Duff, and R.J. Avanzino. 1990. Influence of exchange flow between the channel and hyporheic zone on nitrate production in a small mountain stream. Canadian Journal of Fisheries and Aquatic Sciences **47**:2099–2111.

Triska, F.J., J.H. Duff, and R.J. Avanzino. 1993a. Patterns of hydrological exchange and nutrient transformation in the hyporheic zone of a gravel-bottom stream: Examining terrestrial-aquatic linkages. Freshwater Biology **29**:259–274.

Triska, F.J., J.H. Duff, and R.J. Avanzino. 1993b. The role of water exchange between a stream channel and its hyporheic zone in nitrogen cycling at the terrestrial-aquatic interface. Hydrobiologia **251**:167–184.

Triska, F.J., V.C. Kennedy, R.J. Avanzino, and K.C. Stanley. 1994. Ammonium sorption to channel and riparian sediments: A transient storage pool for dissolved inorganic nitrogen. Biogeochemistry **26**:67–83.

Triska, F.J., V.C. Kennedy, R.J. Avanzino, G.W. Zellweger, and K.E. Bencala. 1989a. Retention and transport of nutrients in a third-order stream: Channel processes. Ecology **70**:1877–1892.

Triska, F.J., V.C. Kennedy, R.J. Avanzino, G.W. Zellweger, and K.E. Bencala. 1989b. Retention and transport of nutrients in a third–order stream in northwestern California: Hyporheic processes. Ecology **70**:1894–1905.

Triska, F.J., V.C. Kennedy, R.J. Avanzino, G.W. Zellweger, and K.E. Bencala. 1990b. In situ retention-transport response to nitrate loading and storm discharge in a third-order stream. Journal of the North American Benthological Society **9**:229–239.

Valett, H.M. 1993. Surface hyporheic interactions in a Sonoran desert stream: Hydrologic exchange and diel periodicity. Hydrobiologia **259**:133–144.

Valett, H.M., S.G. Fisher, N.B. Grimm, and P. Camill. 1994. Vertical hydrologic exchange and ecological stability of a desert stream ecosystem. Ecology **75**:548–560.

Valett, H.M., S.G. Fisher, and E.H. Stanley. 1990. Physical and chemical characteristics of the hyporheic zone of a Sonoran desert stream. Journal of the North American Benthological Society **9**:201–215.

Valett, H.M., J.A. Morrice, C.N. Dahm, and M.E. Campana. 1996. Parent lithology, surface-groundwater exchange, and nitrate retention in head water streams. Limnology and Oceanography **41**:333–345.

Vaux, W.G. 1968. Intragravel flow and interchange of water in a streambed. Fisheries Bulletin **66**:479–489.

Vervier, P., and R.J. Naiman. 1992. Spatial and temporal fluctuations of dissolved organic carbon in subsurface flow of the Stillaguamish River. Archiv für Hydrobiologie **123**:401–412.

Ward, J.V. 1989. The four-dimensional nature of lotic ecosystems. Journal of the North American Benthological Society **8**:2–8.

White, D.S. 1990. Biological relationships to convective flow patterns within stream beds. Hydrobiologia **196**:149–158.

——. 1993. Perspectives on defining and delineating hyporheic zones. Journal of the North American Benthological Society **12**:61–69.

White, D.S., C. Elzinga, and S. Hendricks. 1987. Temperature patterns within the hyporheic zone of a northern Michigan river. Journal of the North American Benthological Society **6**:85–91.

Williams, D. 1989. Towards a biological and chemical definition of the hyporheic zone in two Canadian rivers. Freshwater Biology **22**:189–208.

Williams, D.D. 1984. The hyporheic zone as a habitat for aquatic insects and associated arthropods. Pages 430–455 *in* V.H. Resh and D.M. Rosenberg, eds. The ecology of aquatic insects. Praeger, New York, New York, USA.

Williams, D.D., and H.B.N. Hynes. 1974. The occurrence of benthos deep in the substratum of a stream. Freshwater Biology **4**:233–256.

Williams, D.D., and H.B.N. Hynes. 1976. The recolonization mechanisms of stream benthos. Oikos **27**:265–272.

Winter, T.C. 1978. Numerical simulation of steady state three-dimensional groundwater flow near lakes. Water Resources Research **14**:245–254.

Wondzell, S.M., and F.J. Swanson. 1996a. Seasonal and storm dynamics of the hyporheic zone of a 4th-order mountain stream. I: Hydrologic processes. Journal of the North American Benthological Society **15**:3–19.

Wondzell, S.M., and F.J. Swanson. 1996b. Seasonal and storm dynamics of the hyporheic zone of a 4th-order mountain stream. I: Nutrient cycling. Journal of the North American Benthological Society **15**:20–34.

17
Biodiversity

Michael M. Pollock

Overview

• Riparian corridors are the most biologically diverse component of Pacific coastal watersheds. These areas are utilized by unusually high numbers of plants, wildlife, aquatic invertebrates, and most freshwater and anadromous fishes. However, for many taxa (e.g., most invertebrates) the number of species utilizing riparian corridors is poorly documented.

• Factors affecting diversity levels in riparian corridors include disturbances (floods, spatial heterogeneity created as a result of floods, debris flows, lateral migration of rivers, and the activity of beaver), productivity, herbivory, and, on the watershed scale, variation in local climate as streams flow from high to low elevations. Collectively, these forces create a mosaic of nonequilibrium habitats existing at various spatial and temporal scales that allow a large number of species to coexist.

• The influence of these factors on diversity patterns is explained in terms of existing biodiversity theory. The usefulness of common diversity indices for describing the multiscale nature of diversity in riparian corridors is also assessed.

Introduction

Why are some riparian and stream communities more diverse than others, and what are the reasons for such patterns? These issues are particularly relevant to the study of riparian corridors because they are unusually diverse systems. The factors that influence diversity can be divided into two basic categories: broad physical or geographic controls that affect the total number of species that occur in a region, and local factors that determine how many species in the regional species pool can coexist in a given area or community.

Broad geographic and physical controls influencing diversity patterns between regions are discussed only briefly because this book focuses on one ecoregion. Energy availability, which roughly corresponds to latitude, is a broad physical factor that has been most strongly correlated with regional patterns of species richness in North America (Currie and Paquin 1987, Currie 1991). Unfortunately, the reasons for this are not at all clear, and the relationship between large-scale energy and large-scale diversity patterns does not hold at finer scales (Huston 1994a). For example, within the Pacific coastal states and provinces, the number of species of vascular plants declines as one moves northward (decreasing energy availability), but no such trends have been observed for freshwater fishes or freshwater benthic invertebrates (Huston 1994b). Other factors within this region such as variation in climatic conditions, glaciation, volcanic activity, and other geologic events probably have more to do with large-scale variation in diversity than available solar energy.

Local factors influencing diversity are similar in many regions and in many different types of aquatic and terrestrial ecosystems. Such factors include disturbance regimes, predation, herbivory, spatial heterogeneity, and productivity. This chapter largely is concerned with how local factors produce the wide spectrum of diversity in both the terrestrial and aquatic communities within riparian corridors of the Pacific coastal ecoregion (sensu Naiman et al. 1992).

An Overview of Diversity in Riparian Corridors

Natural riparian corridors are the most diverse, dynamic, and complex biophysical habitats on the terrestrial portion of the Earth (Naiman et al. 1993). Riparian corridors, which encompass both terrestrial and aquatic systems, contain steep environmental gradients and a diversity of biological communities. Because riparian corridors contain a diverse mosaic of landforms, communities, and environmental conditions, they also support a diversity of species. The riparian corridor encompasses the stream channel and that portion of the terrestrial landscape where vegetation may be influenced by elevated water tables during floods (Naiman et al. 1993). The riparian corridor is frequently disturbed by floods and debris flows, creating a shifting mosaic of landforms. Consequently, the species richness of riparian vegetation varies considerably in space and time, and this variation has important influences on the richness and composition of both riparian and in-stream biota. Riparian vegetation regulates light and temperature regimes, provides nourishment to aquatic (and terrestrial) biota, and acts as a source of large woody debris (LWD), which in turn influences sediment routing, channel morphology, and other aspects of in-stream habitat (Swanson and Lienkaemper 1978, Naiman et al. 1992, Chapter 13). Riparian vegetation also regulates the flow of water and nutrients from uplands to the stream (Peterjohn and Correll 1984) and provides habitat for an unusually high number of birds and mammals (Raedeke 1988). As such, the maintenance or restoration of diversity may largely be determined by the dynamics and structure of vegetation in riparian corridors.

Although the riparian corridor has been recognized for its high levels of biodiversity, it is still not known how many species are present in most watersheds. No studies are available to suggest what percentage of all life in the Pacific coastal ecoregion utilizes riparian corridors, but available evidence suggests that for some groups of species the numbers are quite high. Biodiversity in riparian corridors is best documented for plants. Studies of plant diversity in Pacific coastal riparian corridors mirror findings throughout the world that these areas have high levels of diversity (Nilsson 1986, Salo et al. 1986, Kalliola and Puhakka 1988, Junk 1989, Nilsson et al. 1989, Tabacchi et al. 1990, Kalliola et al. 1992, Décamps and Tabacchi 1994). In the Pacific coastal ecoregion, two surveys have demonstrated that riparian corridors are highly diverse relative to the surrounding uplands (Gregory et al. 1991, Pollock et al. 1998). For example, 74% of all plant species within the 145-km^2 Kadashan watershed in southeast Alaska were found in riparian corridors. In Washington, a high percentage of wildlife also utilizes riparian corridors (Raedeke 1988). Of the approximately 480 species of wildlife in Washington, 291 (60%) occur in wooded riparian habitats, and 68 species of mammals, birds, amphibians, and reptiles require riparian ecosystems to satisfy a vital habitat need during all or part of the year. Another 103 species are more numerous in riparian ecosystems, or use them more heavily than upland habitats (Raedeke 1988). These numbers are particularly impressive given that riparian corridors usually account less than 10% of the total area of most watersheds.

In addition to plants and wildlife, most Pacific coastal ecoregion freshwater and anadromous fish utilize streams for some part of their life cycle (Li et al. 1987). Also, an unknown number of freshwater algae, bryophytes, insects and other invertebrates, bacteria, fungi, and viruses utilize these streams. Numbers of such freshwater species in the Pacific coastal ecoregion are not available and, in general, the taxonomy and diversity of these taxa are poorly characterized for most regions of the world.

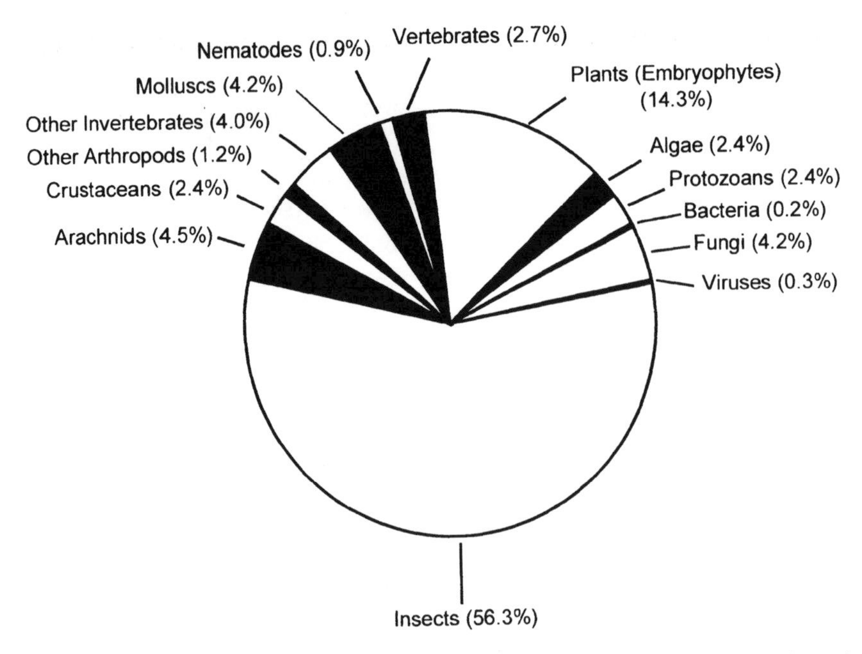

FIGURE 17.1. Proportion of major groups of organisms in terms of species described to date (estimated total is about 1.7 million). Groups included are those considered likely to contain in excess of 100,000 species when as yet undescribed species are taken into account, along with vertebrates for comparison (modified from Groombridge 1992. © World Conservation Monitoring Centre).

This is unfortunate, because surveys suggest that most species, at least on global and continental scales, are insects and other invertebrates (Figure 17.1) (Groombridge 1992). Vascular plants and vertebrates account for only 17% of all species identified to date, and consist of only 2.8% of all estimated species (Groombridge 1992). Although these data apply to the global scale, they probably reflect the situation in the Pacific coastal ecoregion. Taxa making up the majority of species in riparian corridors have been poorly characterized. For example, Pennak (1989) observed that there were an estimated 11,000 species of freshwater invertebrates (excluding protozoa and parasites) in the United States, whereas ten years earlier, the figure was estimated at 10,000 species. The majority of these groups are currently undergoing taxonomic refinement. Such rapid changes in the known species pool and constant taxonomic realignments suggests that there is much to learn about the cladistics (evolutionary relationships) and species diversity of freshwater invertebrates.

Historically, most studies exploring the mechanisms controlling biodiversity have examined plants and vertebrates (see Huston 1994a for a review). Consequently, most theories, hypotheses, and mechanisms attempting to explain biodiversity patterns have only been applied to these organisms. Although this chapter uses biodiversity theory to explain patterns of species richness of organisms in riparian corridors, it remains to be seen how appropriate these major biodiversity theories are for explaining patterns of diversity among the more primitive plants and animals (e.g., periphyton and invertebrates).

General Theories of (Local) Diversity

Two general theories of biodiversity explain local variation in species richness between communities, both of which have the potential to at least partially explain biodiversity patterns in riparian corridors. These are the *dynamic equilibrium model* (Huston 1979, 1994b) and the

TABLE 17.1. Relationship between productivity, disturbance, and diversity, as predicted by the dynamic equilibrium model.

Productivity	Disturbance frequency	Diversity	Explanation
High	High	High	Competition and disturbances are in dynamic equilibrium
High	Low	Low	Competitive exclusion eliminates less competitive species
Low	Low	High	Competition and disturbances are in dynamic equilibrium
Low	High	Low	Frequent disturbances eliminate slowly maturing species

From Huston 1979, 1994b.

theory of resource competition in a heterogeneous environment (Tilman 1982).

Huston's (1979, 1994b) dynamic equilibrium model makes important predictions about site patterns of species diversity based on the interaction of productivity and disturbance. The dynamic equilibrium model suggests that there are predictable relationships between productivity, disturbance and diversity, and explains the effect of the interaction of disturbance and productivity on patterns of species richness among competing species (Table 17.1, Figure 17.2).

Although the model describes a continuum in productivity–disturbance–diversity space, it can be reduced to four basic states to illustrate the interacting effects of productivity and disturbance on diversity: (1) In highly productive, undisturbed systems, diversity tends to be low because dominant species eliminate nondominant species through the process of competitive exclusion. If the community is pro-

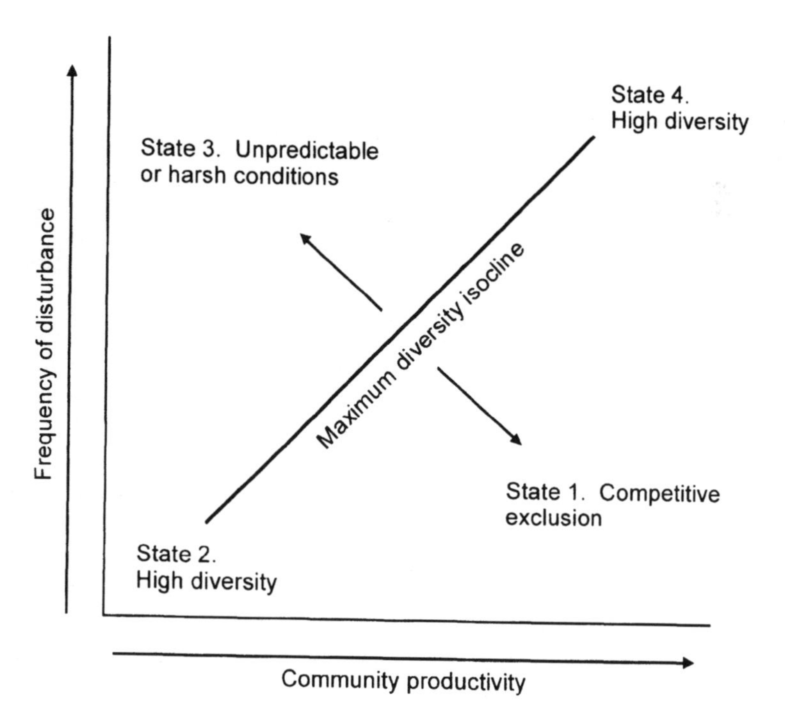

FIGURE 17.2. The dynamic equilibrium model. A dynamic equilibrium exists between disturbance and productivity. Highest diversity (States 2 and 4) occurs along the isocline, where there is a balance between competitive exclusion by dominant species (which keeps diversity low, State 1) and disturbance (i.e., unpredictable or harsh conditions that prevent species survival, State 3) (modified from Huston 1994b).

ductive but disturbances are more frequent, then community diversity can increase because competitive dominants are killed or their growth is slowed by the disturbance, and they cannot competitively exclude other species. (2) Conversely, if a site is unproductive but disturbances are infrequent, then diversity will be high because the rate of competitive exclusion is slow. (3) Unproductive, frequently disturbed sites tend to be low in diversity because frequent disturbances eliminate slow growing species before they have a chance to reproduce. (4) Productive, frequently disturbed sites tend to be high in diversity because growth rates are high, but disturbances are frequent enough that competitive exclusion does not occur.

In contrast, the theory of resource competition in a heterogeneous environment predicts that species diversity is controlled primarily by resource heterogeneity and productivity, and the influence of disturbance is not emphasized (Table 17.2) (Tilman 1982). Tilman (1982) makes several predictions about relationships between plant diversity, resource richness, and spatial heterogeneity. First, species richness is expected to be highest in relatively resource-poor habitats, but should decline rapidly in extremely resource-poor habitats and decline slowly as resource richness increases. This results in a unimodal resource-richness-species richness curve (Figure 17.3)—similar to previously published models such as Grime's (1973) productivity-diversity model. Second, communities with the greatest diversity should have many codominant species, whereas more resource-rich, lower-diversity communities should be dominated by a few species, with most species being rare. Finally, for a given level of resource-richness, increased spatial heterogeneity should lead to increased species richness, with the most marked effects in resource-poor habitats. Together the models of Tilman and Huston suggest that potential productivity, disturbance, and spatial heterogeneity are key factors controlling local patterns of diversity, a suggestion supported by numerous field and experimental observations and other theoretical models (Rosenzweig 1971, Horn and MacArthur 1972, Grenney et al. 1973, Grime 1973, Levin and Paine 1974, Cody 1975, Al-Mufti et al. 1977, Connell 1978, Sousa 1979, Schlosser 1982, Tonn and Magnuson 1982, Brown and Gibson 1983, Tilman 1983, Ward and Stanford 1983, Robinson and Minshall 1986, Abrams 1988, Pringle 1990, Brown and Brussock 1991, Currie 1991, Scarsbrook and Townsend 1993).

TABLE 17.2. Relationship between productivity, spatial heterogeniety, and diversity as predicted by the theory of resource competition in a heterogenous environment (Tilman 1982). Increased spatial heterogeneity of resources increases diversity at any average level of productivity, and effects are most pronounced in resource-poor habitats.

Productivity	Diversity	Explanation
High	Low	Competitive exclusion eliminates less competitive species
Intermediate	High	Conditions are suitable for growth and competition is not severe, resulting in many species co-dominating
Low	Low	Unproductive conditions prevent most species from thriving

Common Measures of Species Richness, Diversity, Evenness, and Turnover Rates

The terms *species richness*, *diversity*, and *evenness* are used to describe the number or relative abundance of a group of species in a particular area. Species richness simply refers to the total number of species in a given area, regardless of their relative abundances. Species evenness refers to the evenness in species abundances. If all species are equally abundant, then evenness is maximized. Traditionally, ecologists have used the term diversity synonymously with the Shannon–Wiener diversity index, which measures both the number and evenness of species. The Shannon–Wiener diversity index is calculated by the equation:

$$H' = -\sum_{i=1}^{n} p_i \log p_i \quad (1)$$

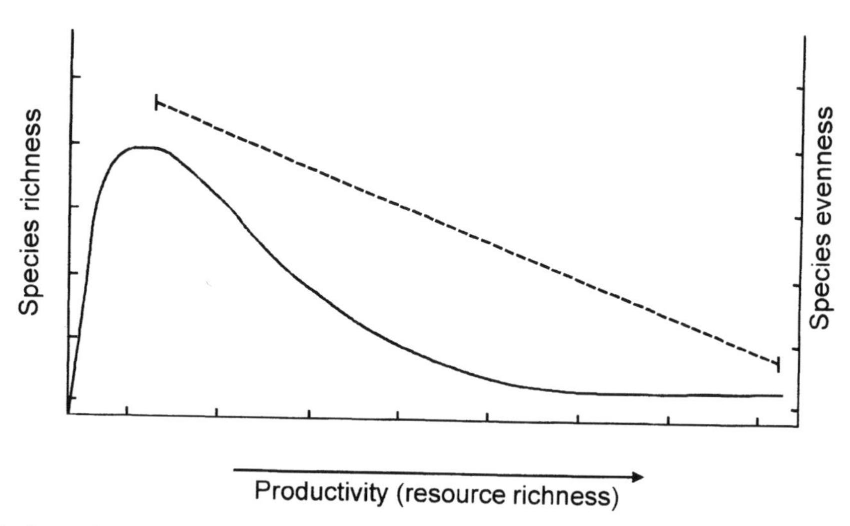

FIGURE 17.3. Several tenets of the theory of resource competition in a heterogeneous environment (Tilman 1982). Species richness (—) rises rapidly to a peak as productivity increases to relatively low levels, then slowly declines as productivity continues to increase to high levels. Species evenness (----) is highest in moderately unproductive habitats where diversity is high, then decreases as productivity increases.

where H' = the Shannon–Wiener diversity index, p_i = the proportional abundance of the ith species, and n = the species richness of the sample (Pielou 1975).

This index is used to account for differences in abundances, as well as the total number of species in a community. For instance, a community of birds with ten individuals of ten different species is considered more diverse (H' = 1.00) than a community with 91 individuals of one species and one individual each of the remaining nine species (H' = 0.22). Species richness does not account for these differences because the number of species is ten in both communities (Table 17.3, community 1 versus 2). In spite of the popularity of the Shannon–Wiener diversity index, it is not always an appropriate measure of species diversity for many taxa, particularly plants, because many rare species and few dominant species often occur in a community (Whittaker 1972, 1977). Further, a large number of rare species has little affect on the diversity index because of the mathematical nature of the Shannon–Wiener diversity index (Table 17.3, community 2 versus 3). Thus, communities with large differences in the number of species have only small differences in diversity indices. Additionally, communities with a high number of unevenly distributed species can have a lower diversity index than communities with considerably fewer, but more evenly distributed species (Table 17.3, community 1 versus 3).

The Shannon–Wiener diversity index has been used primarily by vertebrate ecologists, where the abundance of a species can be counted as the number of individuals (as opposed to biomass) and where more equitable distribution of species is expected. However, for plants and probably most other taxa, typical community abundance distributions fit a negative exponential curve (i.e., there are a few abundant species, and most species are relatively uncommon). In these situations, the Shannon–Wiener diversity index provides little comparative information between communities. Additionally, information is lost because the Shannon–Wiener index combines two pieces of data, the number of species and their abundances, into one number. In most cases, it is probably more straightforward to use the easily interpretable measure of species richness to understand the diversity of a community. To also understand species evenness, an evenness measure such as Pielou's equitability index (Table 17.3) (Pielou 1975), which measures

TABLE 17.3. Comparison of species richness, Shannon–Wiener's diversity index, and Pielou's equitability index between three hypothetical communites (*pi* = proportional abundance of the *i*th species; *H′* = Shannon–Weiner diversity index; *SR* = Species Richness; *J′* = Pielou's equitability index).

Community 1				Community 2				Community 3			
spp	abundance	*pi*	*pi**log*pi*	spp	abundance	*pi*	*pi**log*pi*	spp	abundance	*pi*	*pi**log*pi*
A	10.00	0.10	−0.10	A	90.00	0.90	−0.04	A	90.00	0.90	−0.041
B	10.00	0.10	−0.10	B	1.00	0.01	−0.02	B	1.00	0.01	−0.020
C	10.00	0.10	−0.10	C	1.00	0.01	−0.02	C	1.00	0.01	−0.020
D	10.00	0.10	−0.10	D	1.00	0.01	−0.02	D	1.00	0.01	−0.020
E	10.00	0.10	−0.10	E	1.00	0.01	−0.02	E	1.00	0.01	−0.020
F	10.00	0.10	−0.10	F	1.00	0.01	−0.02	F	1.00	0.01	−0.020
G	10.00	0.10	−0.10	G	1.00	0.01	−0.02	G	1.00	0.01	−0.020
H	10.00	0.10	−0.10	H	1.00	0.01	−0.02	H	1.00	0.01	−0.020
I	10.00	0.10	−0.10	I	1.00	0.01	−0.02	I	1.00	0.01	−0.020
J	10.00	0.10	−0.10	J	1.00	0.01	−0.02	J	1.00	0.01	−0.020
		H′	1.00			H′	0.122	K	0.10	0.00	−0.003
		SR	10			SR	10	L	0.10	0.00	−0.003
		J′	1.00			J′	0.22	M	0.10	0.00	−0.003
								N	0.10	0.00	−0.003
								O	0.10	0.00	−0.003
								P	0.10	0.00	−0.003
								Q	0.10	0.00	−0.003
								R	0.10	0.00	−0.003
								S	0.10	0.00	−0.003
								T	0.10	0.00	−0.003
										H′	0.25
										SR	20
										J′	0.19

only the evenness of a community and says nothing about species richness, is more appropriate. Pielou's (1975) equitability index is a simple index derived by dividing the Shannon index by the $\log_{10}$ of the number of species in the community:

$$J' = \left(-\sum_{i=1}^{n} p_i \log p_i\right) \Big/ \log(n) \qquad (2)$$

where J' = Pielou's equitability index, p_i = the proportional abundance of the *i*th species, and n = the species richness of the sample. When evenness is at a maximum (i.e., all species are equally abundant) J' = 1 (Pielou 1975).

Currently, the term diversity is more generally used to mean species richness. The term "Shannon–Wiener diversity" or "Shannon diversity" should be used when referring to diversity measures using that index. In this chapter, the term diversity is used interchangeably with the term species richness.

Species turnover rate refers to the change in species composition between sites. A wide variety of methods have been developed to measure such turnover rates, however, unlike measures to inventory species, no one particular technique to measuring turnover rates has emerged. Among ecologists, for example, species turnover rate may refer to the change in species composition along a transect (Whittaker 1972) or to turnover rates along an ordination axis (Cody 1986). Excellent discussions on techniques for measuring species turnover rates and species diversity are presented by Pielou (1975) and Magurran (1988).

Describing Diversity at Multiple Spatial Scales

Seven components of biodiversity are used to describe patterns at different spatial scales, which are applicable to any taxa or groups of taxa. The seven components are known as microhabitat (point or internal-alpha) diversity, between-microhabitat (pattern or internal-beta) diversity, alpha diversity, beta diversity, gamma diversity, delta diversity and epsilon diversity (Table 17.4). The number of species found in a microhabitat (e.g., a quadrat) is considered point diversity, while the rate of change in species composition between microhabitat in a community is between-microhabitat diversity. The number of species found in a habitat patch is called alpha diversity while the rate of species turnover between different types of habitat patches is called beta diversity. The term gamma diversity describes diversity levels over large areas, but has been used to describe two different types of coarse scale diversity patterns, leading to some confusion over the term. Whittaker (1972, 1977) defined the term as the total number of species found in all the different types of habitat along an environmental gradient or coenocline. In this sense, gamma diversity is simply the total number of species found in all communities sampled. Cody (1986), however, used the term to mean the turnover rate of species in a particular habitat type over a large geographical area, or what he called, "the rate at which additional species are encountered as geographic replacements within a habitat type in different localities." This latter definition allows gamma-diversity to be used as a measure of endemism. Cody's definition of gamma diversity corresponds to what Whittaker (1977) called delta diversity. Whittaker referred to delta diversity as the change or turnover of species between geographic areas or along climatic gradients. Finally, epsilon diversity refers to the total number of species observed in a region or in the total number of gamma-scale landscapes sampled.

Point, alpha, gamma, and epsilon diversity are considered *inventory* diversity measures because they tabulate the number of species in a given area. Pattern, beta, and delta diversity are considered *differentiation* diversity measures because they measure the rate of change

TABLE 17.4. Seven types of species diversity.

Diversity type	Inventory or differentiation diversity	Scale	Riparian plant example	Aquatic invertebrate example
Internal Alpha (point)	Inventory	Diversity of a microhabitat or sample within a homogenous community	1 × 1 m quadrat	A stone
Internal Beta (pattern)	Differentiation	Change between microhabitats within a homogenous community	Changes between quadrats	Changes between stones
Alpha (within-habitat)	Inventory	Community diversity	Floodplain forest, channel shelf, depositional bar, etc.	Riffle, pool, glide, cascade, etc.
Beta (between-habitat)	Differentiation	Change between communities	Changes between habitat types or between different communities of the same habitat type in a watershed	
Gamma (landscape)	Inventory	Total diversity of all sampled communities in a geographic area	All species in sampled communities within a watershed	
Delta (geographic differentiation)	Differentiation	Change between geographic areas or along climatic gradients	Changes in all species between watersheds or between a specific habitat in different watersheds	
Epsilon (regional)	Inventory	Total diversity of a region	All sampled species in many watersheds	

Adapted from Whittaker 1972, 1977.

in the number of species between areas. Although these measures were initially designed to measure the diversity of plants at different spatial scales from quadrats to large geographic regions, there is no specific spatial scale attached to these definitions. In reality, most studies of diversity only examine several scales at most, and alpha, beta and gamma diversity are the terms most commonly used, regardless of the spatial scales involved.

Field observations indicate that the predicted pattern of plant diversity as a function of scale in Pacific coastal riparian corridors is hypothetically as follows: alpha and gamma diversity are high, beta diversity is moderate, and delta diversity is low (Figure 17.4) (Pollock, unpublished data). This expected pattern of diversity reflects the fact that in riparian corridors a number of relatively small habitat patches in close proximity have high numbers of species (so alpha diversity is high). However, many of these species have wide ecological amplitudes, and species composition between habitat patches overlaps considerably, lowering beta diversity. Gamma diversity is high because there is a wide array of habitat types. Species turnover rates may be low, but there are enough different habitats that species continue to accumulate. Delta diversity (e.g., changes in species composition between watersheds) is low, because there is very little endemism among riparian plants in this region. Finally, over large areas, species diversity slowly increases as the size of the area increases, primarily because of changes in climate. Therefore epsilon diversity in riparian corridors is low if the area only includes climatically similar watersheds, and increases as the area sampled includes a wider array of climatic conditions (Figure 17.4). Few studies have compared the rate of increase in species in riparian corridors relative to uplands at such large scales. Most studies lending support to the idea that riparian corridors are floristically diverse have largely examined alpha or gamma diversity (Nilsson 1986, Kalliola and Puhakka 1988, Nilsson et al. 1989, Décamps and Tabacchi 1993, Pollock et al. 1998), while little or no research has examined beta, delta, or epsilon diversity.

Natural Processes Influencing Biodiversity Patterns

The high diversity of Pacific coastal riparian corridors is related to disturbances caused by floods, spatial heterogeneity created as a result of other geomorphological events such as debris flows and the lateral migration of rivers, LWD input and beaver activity, productivity, grazing, and, at the watershed scale, variation in local climate as streams flow from high to low elevations (Table 17.5). Collectively, these forces create a mosaic of nonequilibrium habitats of various ages which allow a large number of species to coexist.

Flood frequency and spatial heterogeneity (in the form of topographical variation) are particularly good predictors of plant species richness (Pollock et al. 1998). Floods destroy habitat patches and create new patches as well as alter rates of competitive displacement in patches that are not completely destroyed. Topographical heterogeneity creates spatial variation in the frequency and duration of flooding. Such differences are exploited by different species of plants, allowing them to exist in close proximity. Variation in site productivity, in combination with disturbances, results in a variety of riparian communities with different suites of species, while the effects of animal activity (whether grazing by ungulates or dam-building by beaver), also increase site diversity. Finally, temperature and precipitation are major determinants of the distribution of plant species. Riparian corridors pass through most of the altitudinal variation within a watershed, and thus, encompass a range of temperatures and precipitation suitable to the needs of many plants in the regional flora.

The following section examines possible influences on biodiversity in the Pacific coastal ecoregion, emphasizing factors controlling plant diversity because of the tremendous influence that riparian vegetation has on both the morphology and biology of streams. Hydrologic regimes (flooding), animal influences, productivity, and spatial heterogeneity appear to exert the strongest influence on diversity

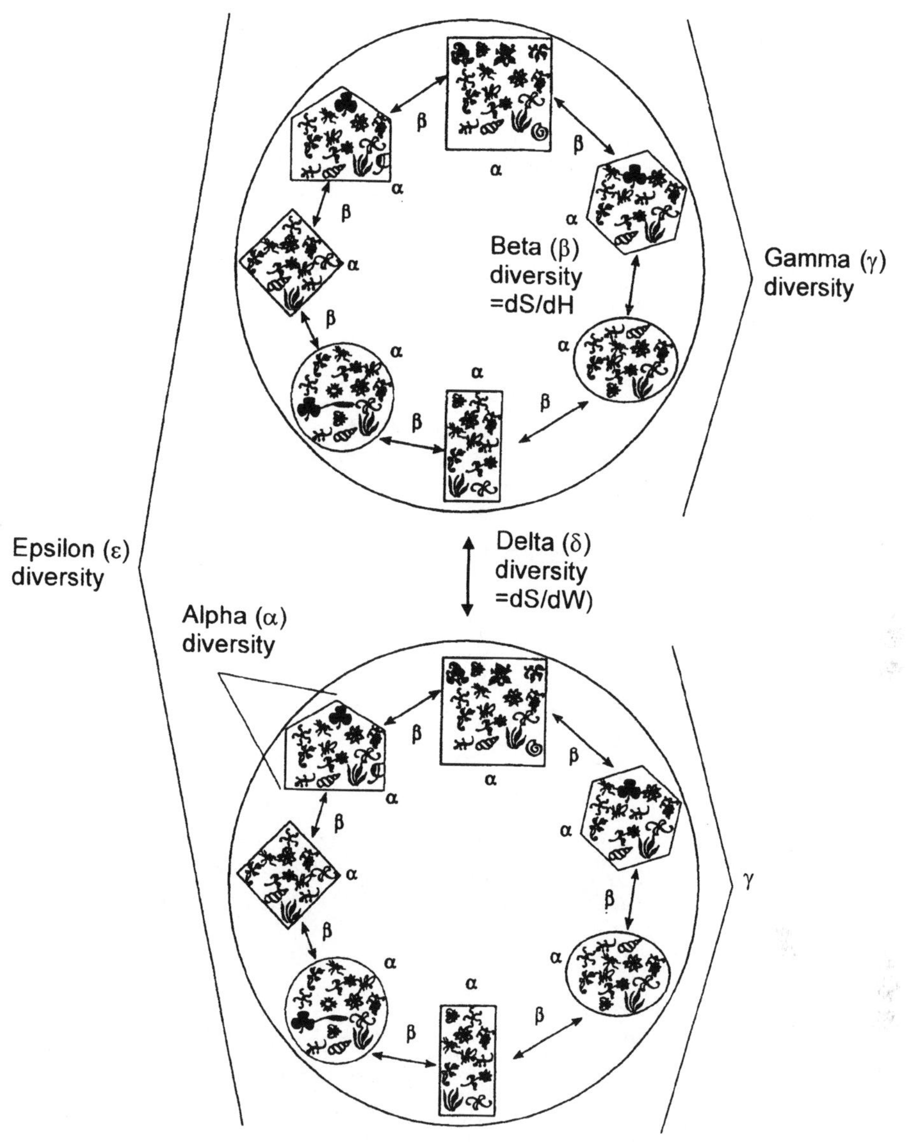

FIGURE 17.4. Hypothetical model of biodiversity at different scales in riparian corridors. Large circles represent watersheds, geometric figures within the circles represent habitat types, symbols within each habitat type represent individual species and arrows represent species turnover. Alpha diversity is generally high, and the change in species with respect to habitat types (dS/dH = beta diversity) is moderately high. Subsequently, gamma diversity at the scale of the watershed is high, but changes in species between watersheds (dS/dW = delta diversity) is low. Epsilon diversity, the total number of species in many watersheds, is not much higher than gamma diversity because differences in species composition between watersheds are not large.

patterns in riparian corridors. Unfortunately, the factors controlling in-stream diversity have not been well studied as compared to riparian plants, but available evidence suggests that some of the processes controlling in-stream diversity are similar, at least for invertebrates and periphyton (Reice et al. 1990). A discussion of fish diversity in Pacific coastal

TABLE 17.5. Major natural processes that affect species diversity in riparian corridors.

Process	Primary type(s) of diversity affected
Nondestructive flooding	Alpha
Spatial heterogeneity created by destructive flooding (e.g., lateral river migration, avulsions), debris flows, LWD input, and beaver activity	Beta, gamma
Productivity	Alpha
Herbivory (aquatic and terrestrial)	Alpha
Variation in climate caused by changes in altitude	Beta, gamma

rivers is presented by Reeves (Chapter 9), and wildlife is discussed by Kelsey and West (Chapter 10).

Hydrologic Regimes

The frequency and power (sensu Leopold et al. 1964) of floods are probably the most important determinant of species diversity within riparian corridors. Floods periodically create and destroy habitat patches, as well as modifying existing ones. The destruction of old patches (e.g., floodplains) creates new types of habitats (e.g., gravel bars), allowing new species and novel combinations of species to exist. By disturbing, but not completely destroying older riparian habitats, floods slow down the process of competitive exclusion and allow noncompetitive plant species to survive for longer periods (as predicted by the intermediate disturbance hypothesis [Connell 1978]), thereby enhancing alpha diversity. Available evidence suggests that plant communities of Pacific coastal riparian corridors exhibit a consistent pattern of diversity (Gregory et al. 1991, Pollock et al. 1998). In general, plant diversity is low on depositional bars near the stream, rapidly increases on active channel shelves, then decreases again in floodplain forests, and further decreases in terrace and upland communities (Figures 17.5 and 17.6).

Plant diversity is low on depositional bars near the river, where the unstable substrate and frequent scouring by floods generally favors ruderal plants that can complete their life cycle between major floods (usually within one

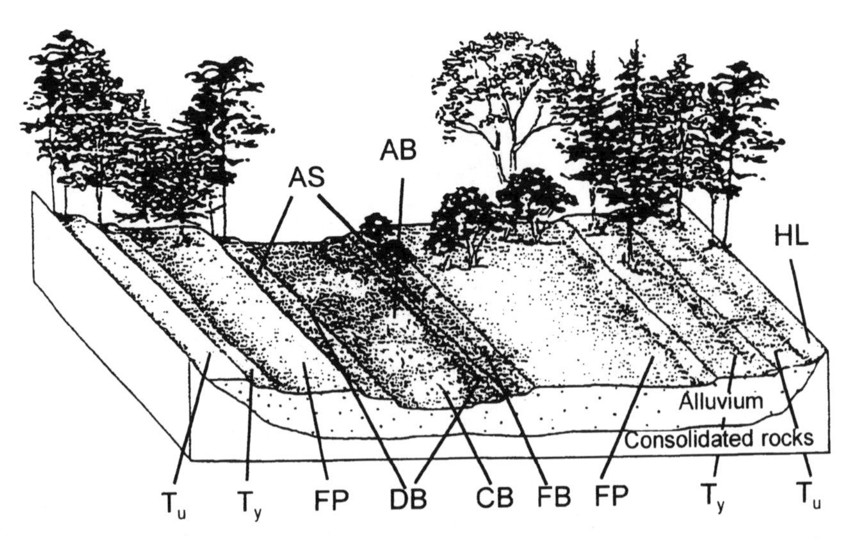

FIGURE 17.5. Fluvial landforms showing channel bed (CB), depositional bar (DB), active-channel bank (AB) active-channel shelf (AS), floodplain bank (FB), floodplain (FP), low terrace (Tl), upper terrace (Tu), and hillslope (HL) (from Hupp and Osterkamp 1985 and Hupp 1988).

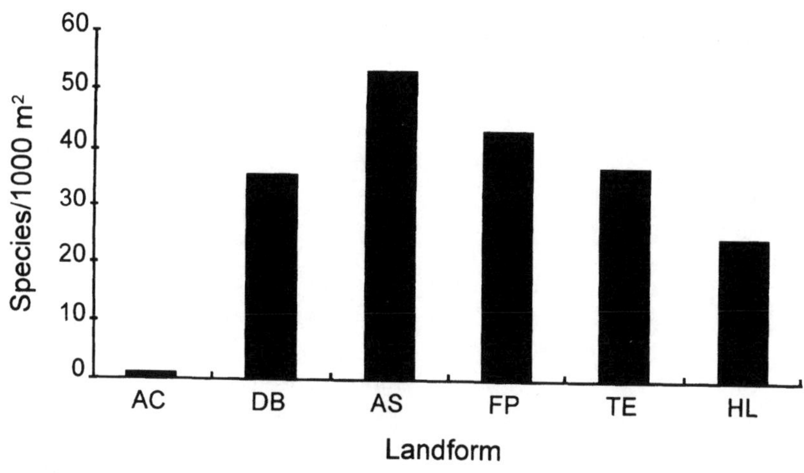

FIGURE 17.6. Plant species richness (species/ 1,000 m^2) varies as a function of landform: active channel (AC), depositional bar (DB), active-channel shelf (AS), floodplain (FP), terrace (TE), and hillslope (HL). Species richness is highest on active channel shelves. Frequently flooded (e.g. active channels and depositional bars) and nonflooded (e.g. terraces and hillslopes) have relatively low numbers of species. Definition of landforms follow Leopold et al. (1964) and Hupp (1988).

growing season). However, this habitat is also stressful even in the absence of floods. The porous nature of the typically gravel deposits precludes storage of much water, making plants susceptible to drought during the summer. Additionally, summer surface temperatures on gravel bars may be considerably higher than in adjacent communities (approximately 50–60 °C, R.J. Naiman, University of Washington, personal communication), further stressing plants trying to colonize these areas. On active channel shelves (Hupp and Osterkamp 1985, Hupp 1988) flood frequencies are lower, allowing woody vegetation such as willow (*Salix* spp.), cottonwood (*Populus* spp.), alder (*Alnus rubra*), conifers (e.g., Sitka spruce [*Picea sitchensis*]), stink currant (*Ribes bracteosum*) and salmonberry (*Rubus spectabilis*) to become established. These sites are relatively narrow transitional zones between depositional bars and floodplains, and the frequency of flooding varies considerably as a function of elevation relative to the channel. Diversity can be quite high. Plants other than ruderals and other early successional species can become established, but frequent flooding still reduces the growth of competitive dominants, thereby preventing them from eliminating other species. Floods also occasionally open up patches within active channel shelves by removing or killing vegetation. These open areas can then be recolonized by ruderals. As a result, the flooding regime on active channel shelves allows for a high number of species to coexist.

Communities on well-established floodplains are usually less diverse than active channel shelves. Floods are infrequent on floodplains, on average occurring every few years (Leopold et al. 1964, Dunne and Leopold 1978). Floods tend to occur in the late fall or winter, when trees are not actively growing, thus reducing the potential of floods to inflict stress upon the dominant species. Well developed floodplain forests dominated by conifers often have a dense canopy. This allows a shade-tolerant understory to develop that is in many ways floristically similar to upland forests. However, these understories also usually include a number of species such as ladyfern (*Athyrium filix-femina*) and skunk cabbage (*Lysichitum americanum*), that normally grow in wet microsites on the floodplain. On a larger scale, such as an entire drainage, floodplains are species rich (high gamma diversity) because in addition to forested communities, they contain a number of rarer, non-forested habitat types such as emergent wetlands, beaver ponds, and oxbow lakes. On forested terraces or uplands, floods and saturated soils are rare or nonexistent, and diversity further decreases. Thus, the general overall plant diversity pattern in riparian communities supports Connell's (1978) idea that both undisturbed and highly disturbed communities do not support high numbers of species.

Flood frequency, or variation in water levels, also allows competing species to coexist within streams. McAuliffe (1984) demonstrated that variations in discharge prevent the complete monopolization of space by a competitively dominant aquatic invertebrate. In a small stream coming from a lake, the sessile caddisfly larva *Leucotrichia pictipes* dominates on large stones in deep water. *Leucotrichia* grows slowly, completing just one generation per year, but where it is abundant, it competitively excludes other sessile organisms such as the moth larva *Parargyractis confusalis*, midge larva species of the genera *Rheotanytarsus*, and *Eukiefferiella*, and several mobile insects including species from the genera *Baetis*, *Glossoma*, and *Simulium*. However, in shallow sections of the stream, changes in discharge result in periodic dessication, and in deeper waters, small stones are frequently overturned by high flows in the course of the year. Both of these habitats are not conducive to the survival of slow growing competitors such as *Leucotrichia*. Benthic substrates disturbed by dessication have few *Leucotrichia* but are dominated by *Eukiefferiella* , which are able to complete most of their life cycle (from hatching to emergence) in less than one month. Small stones that are frequently overturned are dominated by short-lived sessile insects such as *Rheotanytarsus* or by mobile species such as *Baetis*. These data suggest that in the absence of floods that overturn smaller stones, or in the absence of fluctuating water levels, *Leucotrichia* would dominate even more of the stream channel, and other species would be competitively excluded or severely reduced in abundance.

In another example, Hemphill and Cooper (1983) demonstrated that the scouring action of winter floods allows *Simulium virgatum*, an aquatic blackfly with a short life span, to coexist with a competitively dominant species, *Hydropsyche oslari* (caddisfly). After a severe flood, the simuliid rapidly colonize scoured areas, then are slowly replaced by *H. oslari*. This suggests that in the absence of regular floods, competitive exclusion would take place (*H. oslari* eliminating *S. virgatum*). Conversely if floods are extremely frequent, only the opportunistic *S. virgatum* can complete its life cycle between floods, and the slower growing *H. oslari* can be eliminated.

It has also been demonstrated that severe disturbances caused by changes in discharge reduce diversity among benthic invertebrates when compared to streams that experienced less frequent changes in discharge (Siegfried and Knight 1977, Ward and Stanford 1983). In general, if severe flooding is frequent, benthic invertebrate communities are dominated by pioneer species (e.g. species able to rapidly complete their life cycles such as simulliids and chironomids), whereas slower growing competitive dominants are reduced in number.

The diversity of benthic invertebrates also reflects the severity of natural floods and human-induced hydrological disturbances in a pattern that supports the intermediate disturbance hypothesis (Stanford and Ward 1983, Ward and Stanford 1983). Severely disturbed sites such as a stream below an acid mine drainage or a hydroelectric dam, which experience extreme diel fluctuations in discharge are relatively low in diversity, presumably because only pioneer species survive. Sites such as cold springs or springbrooks, where discharge is relatively constant and severe flooding rare, also have low species diversity, presumably because of competitive exclusion. Sites on streams that experience moderate disturbances from natural floods are the most diverse, because environmental conditions are neither constant nor extreme.

The role of flood disturbances in promoting coexistence among periphytic species (algae, fungi, bacteria and protozoa) is less clear. Evidence of distinct phases of periphyton colonization following a disturbance exists, but it is not clear if competitive dominants eliminate other species in the absence of future disturbances. Periphyton succession begins with the development of an organic matrix and a bacterial flora, followed by algae and protozoa in several seral stages (Steinman and McIntire 1990). Although this suggests that competitive hierarchies exist, it does not predict what happens to species diversity in the absence of disturbance. Other evidence suggests that species diversity may actually decline with repeated disturbances (Luttenton and Rada 1986). Such decline appears to be the result of reduced structural complexity of the periphyton assemblage. Many periphyton species embed themselves in the matrix created by the dominant species, usually filamentous algae. When the dominant algae are removed by disturbance, many other species disappear. Prostrate species such as diatoms benefit from the removal of filamentous algae, but this does not appear to compensate for the loss of species caused by structural simplification of the system.

Herbivory

In the Pacific coastal ecoregion, herbivorous ungulates such as elk (*Cervus elaphus*) are most likely to affect diversity patterns in riparian vegetation, while a number of invertebrate aquatic grazers affect the composition of periphyton communities. The effect of grazing on species richness has been well documented in a number of systems. One example of herbivore effects is in the riparian corridors of Olympic National Park, Washington. Extensive grazing by elk in the rich alluvial rain forests has reduced the abundance of understory dominant shrubs such as salmonberry and huckleberry (*Vaccinium* spp.). Grazing is so extensive in many places that no vascular plant species are abundant, but a diverse flora is maintained because the elk prevent any one species from dominating (Woodward et al. 1994). In particular, removal of the dense shrub layer allows more light to penetrate to the herbaceous layer, allowing numerous forbs and ferns to become established. The ability of grazers to increase the diversity of plants has been documented in many other systems as well (Darwin 1859, Tansley and Adamson 1925, Summerhayes 1941, Harper 1969, Grime 1973 and 1979, Lubchenko 1978). In each case, grazing reduces the abundance of competitive dominants and allows noncompetitive species of small stature to coexist.

Herbivory increases diversity when the most dominant plants in a community are removed. However, if herbivore densities are high, then the effect of herbivory can result in fewer species in an area because preferred species are eaten (resulting in local extirpation). Experimental studies support this idea (Lubchenko 1978, Hart 1985, Jacoby 1987). For example, Jacoby (1987) demonstrated that the caddisfly *Discomoecus gilvipes* reduces periphyton (Shannon) diversity in a Cascade Mountain stream by preferentially feeding on green filamentous algae (*Ulothrix*, *Stigeoclonium*, *Spirogyra*) and overstory and stalked forms of diatoms (*Synedra* and *Gomphonema*). While all algal species are consumed by the caddisfly, overstory and filamentous forms are almost entirely removed from the community. Interestingly, in the control sites, the absence of grazing does not lead to the competitive elimination of any species from the community. Periphyton communities may differ from terrestrial vascular plant communities in that competitive exclusion does not occur. Although not well studied, it appears that early periphyton colonists persist in the community even in the presence of competitive dominants such as filamentous algae (Hart 1985, Jacoby 1987, Steinman et al. 1987, Steinman and McIntire 1990). If succession in algal communities involves only the addition of new species but not loss of the early colonizers, then many aspects of diversity theory (which was largely developed based on the characteristics of terrestrial plants) may not be applicable to stream periphyton. In general, periphyton community dynamics in lotic systems and the influence of herbivory on these dynamics are not well understood (Fisher 1983, Gregory 1983). Short generation times of algal species, the strong influence of seasonal changes (Busch 1978) and the persistence of algal propagules on substrata (Wehr 1981) complicate the use of traditional theories to explain the structure of these communities (Steinman et al. 1987).

In general, understanding changes in diversity along periphyton successional trajectories lags far behind the understanding of temporal diversity patterns in vascular plants. It is too early to say if relationships between disturbance and diversity observed for other taxa are applicable to periphyton. Only careful observations and experiments will tell if diversity patterns in periphyton communities can be predicted by existing theory.

Productivity

Although productivity has been directly correlated to diversity patterns in many systems, it is now generally accepted that relationships between diversity and productivity are dramatically influenced by disturbance regimes. The relationship between productivity and community diversity can best be understood if the disturbance regime is also taken into consideration, as illustrated by the dynamic equilibrium model of species diversity (Figure 17.2) (Huston 1979 and 1994b). This model has been tested in Pacific coastal riparian communities (Pollock et al. 1998). Species-rich communities had intermediate levels of disturbance and intermediate flood frequencies. Unproductive, frequently flooded sites were species-poor, and productive, undisturbed sites were also species-poor. Furthermore, field observations suggest that the unproductive, frequently flooded sites were dominated by r-selected taxa, such as *Epilobium* and *Agrostis* spp. whereas the productive, undisturbed sites were dominated by comparatively slower growing shrubs and clonal herbaceous plants (K-selected species) such as *Carex aquatilis*, *Calamagrostis canadensis*, *Vaccinium ovalifolium* and *Oplopanax horridus*, in accord with the predictions of the dynamic equilibrium model. Unproductive, infrequently flooded sites were species-poor, also in agreement with the model. Presumably in these latter sites, disturbances are so infrequent that exclusion occurs in spite of low rates of competitive displacement.

Less empirical evidence is available about in-stream relationships between diversity and productivity. However, one study of periphyton suggests that the relationship between productivity and diversity is predicted by the dynamic equilibrium model (Yount 1956). His study of relationships between diversity and diatom productivity demonstrates that productive sites are less diverse than unproductive sites. Placing a series of microscope slides in high light (productive) and low light (unproductive) stream environments, Yount observed that both productive and unproductive sites initially have similar numbers of diatom species, but that the productive sites soon become dominated by high numbers of relatively few taxa. The unproductive sites maintain high number of species in relatively low abundances for over 100 days, then diversity begins to drop as diatom densities reach higher levels. Yount's (1956) study suggests that competitive exclusion rapidly reduces species diversity at the more productive sites, while slowly reducing diversity at the less productive sites.

Organic enrichment from sewage effluents dramatically increases the productivity of periphyton communities, and provides an indication of what happens to diversity in a highly productive system. Protozoan species richness appears to increase in response to moderate levels of organic enrichment, then decline at high levels of enrichment as certain species (e.g., *Carchesium*, *Urocentrum turbo*, and *Vorticella microstoma*) dominate (Hynes 1963, Henebry and Cairns 1980). The long-term response to enrichment (in the absence of disturbance) in diatom communities, appears as a decrease in both species richness and evenness (Patrick 1969).

Habitat Heterogeneity

Habitat heterogeneity is an important determinant of species diversity for both riparian and in-stream communities. Riparian plants respond to variation in environmental parameters such as light availability, substrate, and degree of soil saturation. The distribution of riparian flora is a function of elevation above the stream channel. Elevation approximates the frequency and duration that a site is flooded. In some cases vegetation has been used as an indicator of flood frequency (Teversham and Slaymaker 1970, Hupp and Osterkamp 1985). Many plants are sensitive to soil saturation, others are tolerant of flooded soils, and some require varying degrees of soil saturation in order to survive (Hook and Crawford 1978, Kozlowski 1984, Crawford 1987, Kozlowski et al. 1991). Thus, a wide range of topographical heterogeneity at a site that enhances the range of soil saturation conditions (i.e., from infrequently flooded to permanently flooded) results in high species diversity. Variation in light conditions and substrate type (e.g., gravel versus organic versus clay) also probably contributes to the floristic diversity of Pacific coastal riparian communities.

In-stream habitat heterogeneity is also important for maintaining diverse communities. Benthic invertebrates segregate themselves primarily along substrate particle size (and velocity) gradients, substrate type, temperature, and water quality gradients (Ward 1992 and references cited therein). For example, Minshall et al. (1985) examined changes in benthic invertebrate diversity in a Pacific Northwest watershed as a function of stream size. Examining a range of streams from 1st to 8th order (mean discharge ranged from 0.04–336 m^3/s), they found the highest numbers of species in mid-size streams (where mean discharge is approximately 1 m^3/s). The high species diversity in these streams is attributed to the high habitat heterogeneity created by variation in diel and annual temperature regimes. Both headwater streams and large rivers are characterized by more stable temperature regimes (Figure 17.7), whereas mid-size streams have much larger variation in daily and annual temperature regimes. Maximum temperature variation occurs in mid-size streams because of an increase in solar radiation from the opening of the riparian forest, relatively shallow water, and moderate discharges. In headwater streams, the canopy is generally closed, reducing temperature fluctuations, whereas in large streams, the large volume of water buffers streams against large temperature fluctuations even though the riparian canopy does little to regulate stream temperatures. Similar relationships between variation in temperature

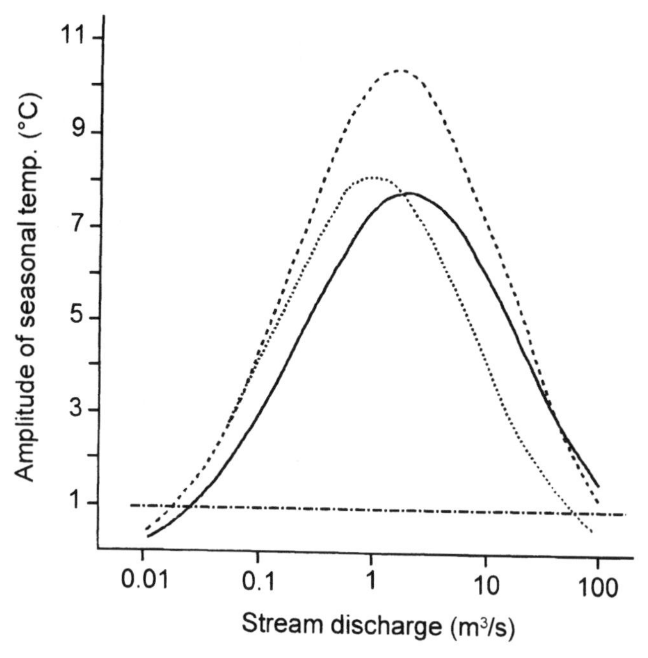

FIGURE 17.7. Amplitude of seasonal temperature plotted against the log of average stream discharge (m^3/s). Winter (-·-·), spring (······), summer (-----), autumn (—). Benthic invertebrate diversity is highest in streams with moderate discharge, where temperature variations are the greatest. Curves were fitted by linear regression using a lognormal model. Plotted points are actual values (from Minshall et al. 1985).

and benthic invertebrate diversity are expected in other systems.

On a smaller scale Minshall et al. (1985) found that spatially heterogeneous habitats have more species than relatively homogenous habitats. In this case, small areas of headwater streams (e.g., 10 rocks in close proximity) have higher numbers of species than larger streams with larger homogenous patches (e.g., pools and riffles). Headwater habitat differences occur on finer spatial scales where pools of slow water occur next to patches of faster water (e.g., chutes, rapids, riffles, and glides) all within a few meters. Conversely, in larger downstream reaches a single habitat type such as a pool, riffle, or glide might extend for tens to hundreds of meters. Thus, headwater streams have high numbers of species in relatively small areas; whereas larger streams require a greater area to accumulate similar numbers of species. This illustrates the fact that within streams, as in other systems, comparisons of diversity are only meaningful when the spatial scales of observation are taken into account.

Pringle (1990) demonstrated experimentally *in situ* that nutrient spatial heterogeneity is an important mechanism for maintaining species diversity in periphytic algae. Nutrient spatial heterogeneity included spatial variation in concentrations of the same nutrient, different nutrients, and different forms of the same nutrient. The most striking differences are found between nutrient poor and nutrient rich systems. In nutrient poor systems, diversity is higher because the effective spatial heterogeneity is higher, while in nutrient rich systems, diversity is relatively low because the effective nutrient spatial heterogeneity is also low (sensu Tilman 1982). Pringle's results suggest that periphyton diversity is influenced by the spatial heterogeneity of nutrients in both the water column and the substratum. Spatial variation in light energy, current velocity, and microtopographic relief also control the distribution, and thus the diversity of periphyton (Steinman and McIntire 1986 and 1990, Nicola and McIntire 1990).

Habitat heterogeneity created by the activities of insects can also tremendously influence periphyton diversity. For example, the tube building activities of chironomids provide a unique substrate that is characterized by a diverse flora (Pringle 1985). Periphyton colonizing chironomid tubes are distinct from surrounding substrates, and are often considerably more diverse. The tube-building chironomids increase microscale heterogeneity by creating a structurally complex and stable habitat in an unstable (sand-dominated) environment, thereby providing refugia for upperstory diatom taxa from grazing by Baetid mayflies (Pringle 1985).

Large-Scale Spatial Heterogeneity

Biodiversity in riparian corridors is also affected by spatial heterogeneity at large scales where severe flooding and other powerful fluvial processes act to create and destroy distinct geomorphic landforms supporting different plant communities. Riparian corridors are unique in that no other ecosystem experiences geomorphic disturbances on such a regular and consistent basis. Lateral channel migration is the primary mechanism shifting one landform type to another (e.g., from floodplain to depositional bar). The result is the creation of a dynamic mosaic of plant communities bordering the banks of streams and rivers. This unique type of disturbance regime is intimately linked to the high diversity in these systems. The wide variety of habitats found in riparian environments are a direct result of these highly destructive (and constructive) disturbances.

In addition to lateral migration of channels, spatially distinct mesoscale landforms are created and maintained by other fluvial processes such as flooding, avulsions, debris flows, sediment deposition, and beaver activity (Swanson et al. 1988, Naiman et al. 1992, 1993). Such landforms include terraces, floodplains, active-channel shelves, depositional bars, abandoned channels, overflow channels, backwaters, oxbow lakes, beaver ponds, meander scrolls, alluvial fans, and mid-channel islands (Leopold et al. 1964, Dunne and Leopold 1978, Swanson et al. 1988, Gregory et al. 1991, Mitsch and Gosselink 1993, Naiman et al. 1993, Chapter 11). This results in a complex mosaic of wetland plant associations reflecting the hydrogeomorphic conditions of each landform (Hupp and Osterkamp 1985). Species diversity is high both within patches and among patches (Kalliola and Puhakka 1988, Gregory et al. 1991).

Overall species diversity is high in riparian corridors because of the great heterogeneity of habitats. Patch diversity means that niches are spatially separated, a condition which helps minimize competition by physically separating species. Such heterogeneity does not facilitate the coexistence of species per se (alpha or point diversity), but rather acts to create a number of environmentally distinct patches where different species dominate (beta diversity). This contrasts to the previously discussed mechanism (see hydrology section) whereby less destructive flooding enhances diversity within a patch by slowing the process of competitive exclusion. Such contrasts illustrate how floods of differing magnitudes influence diversity in completely different ways. Less destructive floods slow down the rate of competitive exclusion within a habitat patch, thereby increasing diversity within a patch, whereas less common but more powerful floods create (and destroy) habitat patches, thereby increasing the diversity of habitat types in the riparian corridor.

For example, avulsions (sudden changes in channel location or form) create new channels and result in the abandonment of old ones, which may remain dry or fill with water, creating backwaters or overflow channels. Beaver dams create side channels and floodplain tributaries, while debris flows and floods create alluvial fans. All these patch types are relatively uncommon in comparison with habitats resulting from channel meandering. However, the uncommon patch types contribute disproportionately to the diversity of riparian corridors because they contain many species not found in the common patch types of depositional bars, active-channel shelves, and floodplain forests (Chapter 12) and are often quite floristically diverse. A key role that fluvial processes play in enhancing riparian diversity is in the creation of these rarer habitat patches. However, very little is known about rates of formation, frequency, distribution, or longevity. The characteristics of disturbance regimes maintaining spatial heterogeneity of these patch types in the riparian environment at a level which maximizes overall diversity is unclear.

Beaver ponds are a particularly important habitat type because they contain unusually high number of species from a number of different taxa. Beaver increase both the diversity of habitats and the diversity of species within a particular site. For example, in a southeast Alaskan watershed, the vegetation of beaver-influenced riparian communities is more diverse (1.4–2.7 times) than the vegetation in other riparian communities (Pollock et al. 1994). Additionally, beaver create open, flooded emergent meadows, a type of community that is relatively rare in the riparian corridor. As a result, a number of plant species are found in beaver-influenced communities that are rare or nonexistent elsewhere in the riparian corridor. Johnston and Naiman (1990) documented the existence of 32 different wetland types created by beaver over a 250 km^2 area in Minnesota, demonstrating that these animals significantly increase habitat diversity in a watershed. For example, beaver impound water and create low-velocity, mud-bottom reaches. In streams where such a habitat type is rare or nonexistent, beaver increase benthic diversity by adding this new habitat type to the system. The periphyton and benthic fauna of beaver ponds is quite different from free-flowing streams (McDowell and Naiman 1986, Naiman et al. 1988, Coleman and Dahm 1990, White 1990, Clifford et al. 1993). Beaver-created habitat also influences the composition of birds and mammals using the riparian corridor. Although beaver activity tends to increase the biomass of small mammals, whether or not it increases the diversity of this group of organisms is not clear (Medin and Clary 1990, 1991).

Few studies have examined patch dynamics of the more common habitat types within riparian corridors (e.g., point bar succession), but those that have suggest that patch turnover rates are of a frequency which allows development of a wide range of patch ages. Everitt (1968) observed that the age frequency distribution of forest patches in a meandering section of the Little Missouri river floodplain fits a negative exponential curve, that is young stands are much more common than old stands (see Johnson 1994). Agee (1988) provides data which also shows that the patch age frequency of floodplain communities in meander bends of the North Fork Flathead River in Montana also fit a negative exponential curve. In this example, the mean patch age is 225 years, while the median age is about 150 years. Thus, 63% of the patches are below the mean age, again demonstrating that the distribution of patch ages creates ample habitat for early and mid successional species, while relatively fewer sites are occupied by late successional species. This indicates that the disturbance regime driven by floods and lateral channel migration in floodplains creates a patch age distribution capable of supporting a mix of early, mid, and late successional species. Further, this age distribution suggests that few short-lived species are eliminated by competitive exclusion, and few slow growing species are eliminated by frequent mass mortality. Thus, a wide spectrum of life history strategies occur within a riparian floodplain. Other studies of rivers in western North America also have demonstrated that riparian corridors contain a mix of patch types of different ages which are dominated by different suites of species (Teversham and Slaymaker 1970, Fonda 1974, Nanson and Beach 1977, Walker et al. 1986).

Implications for Management

In riparian corridors, local patterns of diversity are determined primarily by the interacting influences of site productivity, flood-based disturbances, and spatial heterogeneity. Diversity is maintained by two primary mechanisms: (1) processes such as mild floods and herbivory which reduce the rate or periodically retard the process of competitive exclusion and (2) destructive floods, other fluvial processes, and activities, such as dam building by beaver, that maintain a variety of habitat types or niches. The first mechanism allows species to coexist in the same proximal space, that is, where they compete for resources (alpha diversity). The second mechanism physically separates species by creating physically distinct habitats, each of which are preferable to different species (beta diversity). Because of the wide variety of habitat types and the high within-habitat diversity, gamma diversity is also high (see Figure 17.4). Site productivity also influences alpha diversity, primarily by determining the rate of competitive exclusion. Productivity and disturbances that retard the rate of competitive exclusion interact to determine the level of diversity at a given site.

Floods act as both a disturbance to slow the rate of competitive exclusion and as an agent increasing spatial heterogeneity of riparian corridors. In riparian communities, nondestructive floods (i.e., floods that do not remove physical substrate from a community) slow the rate of competitive exclusion, whereas destructive floods (i.e., those that remove physical substrate) tend to increase the large-scale spatial heterogeneity of riparian communities. In the stream channel itself, floods act to sort sediment, scour pools and entrain material from the banks (such as LWD), all of which help to increase the spatial heterogeneity of habitats within the stream. The scouring action of floods also disturbs periphyton and benthic invertebrate communities, thus serving to slow the process of competitive exclusion to the extent which it exists in aquatic communities.

The ability of floods to act both as an agent maintaining a spatially heterogeneous environment as well as slowing rates of competitive exclusion is probably the single most important factor accounting for the unusually high levels of diversity in riparian corridors throughout the world. Influences such as herbivory, the activity of beaver, and geomorphic disturbances such as debris flows are of secondary importance. Much of the habitat diversity associated with riparian corridors is found on unconstrained, low gradient valleys, where fluvial processes act to create the largest variety of landforms. The key to maintaining diverse riparian corridors is to allow rivers to create and destroy habitats at natural rates. From a management viewpoint, this means allowing rivers to flood, meander, and move so as to create a dynamic mosaic of physically complex habitat patches supporting an abundance of life forms.

Literature Cited

Abrams, P.A. 1988. Resource productivity–consumer species diversity: Simple models of competition in spatially heterogeneous environments. Ecology **69**:1418–1433.

Agee, J.K. 1988. Successional dynamics in riparian forests. Pages 31–43 *in* K.J. Raedeke, ed. Streamside management: Riparian wildlife and forestry interactions. Institute of Forest Resources, Contribution Number 59, University of Washington, Seattle, Washington, USA.

Al-Mufti, M.M., C.L. Sydes, S.B. Furness, J.P. Grime, and J.R. Band. 1977. A quantitative analysis of shoot phenology and dominance in herbaceous vegetation. Journal of Ecology **65**: 759–791.

Brown, A.V., and P.P. Brussock. 1991. Comparisons of benthic invertebrates between riffles and pools. Hydrobiologia **220**:99–108.

Brown, J.H., and A.C. Gibson. 1983. Biogeography. Mosby, St. Louis, Missouri, USA.

Busch, D.E. 1978. Successional changes associated with benthic assemblages in experimental streams. Ph.D. dissertation. Oregon State University, Corvallis, Oregon, USA.

Clifford, H.F., G.M. Wiley, and R.J. Casey. 1993. Macroinvertebrates of a beaver-altered boreal stream of Alberta, Canada, with special reference to the fauna on the dams. Canadian Journal of Zoology **71**:1439–1447.

Cody, M.L. 1975. Towards a theory of continental species diversities. Pages 214–257 *in* M.L. Cody and J.M. Diamond, eds. Ecology and evolution of communities. Belknap, Cambridge, Massachusetts, USA.

Cody, M.L. 1986. Diversity, rarity and conservation in Mediterranean-climate regions. Pages 122–152 *in* M.E. Soulé, ed. Conservation biology, the science of scarcity and diversity. Sinauer, Sunderland, Massachusetts, USA.

Coleman, R.L., and C.N. Dahm. 1990. Stream geomorphology: Effects on periphyton standing crop and primary production. Journal of the North American Benthological Society **9**:293–302.

Connell, J.H. 1978. Diversity in tropical rain forests and coral reefs. Science **199**:1302–1310.

Crawford, R.M.M., ed. 1987. Plant life in aquatic and amphibious environments. Blackwell Scientific, Oxford, UK.

Currie, D.J. 1991. Energy and large-scale patterns of animal and plant species richness. American Naturalist **137**:27–49.

Currie, D.J., and V. Paquin. 1987. Large-scale biogeographical patterns of species richness or trees. Nature **329**:326–327.

Darwin, C. 1859. On the origin of species. 1964, reprinted edition. Harvard University Press, Cambridge, Massachusetts, USA.

Décamps, H., and E. Tabacchi. 1994. Species richness in riparian vegetation along river margins. Pages 1–20 *in* P.S. Giller, A.G. Hildrew, and D.G. Rafaelli, eds. Aquatic ecology: Scale, pattern and process. Blackwell Scientific, London, UK.

Dunne, T., and L.B. Leopold. 1978. Water in environmental planning. W.H. Freeman, New York, New York, USA.

Everitt, B.L. 1968. Use of the cottonwood in an investigation of the recent history of a floodplain. American Journal of Science **266**:417–439.

Fisher, S.G. 1983. Succesion in streams. Pages 7–27 *in* J.R. Barnes and G.W. Minshall, eds. Stream ecology: Application and testing of general theory. Plenum, New York, New York, USA.

Fonda, R.W. 1974. Forest succession in relation to river terrace development in Olympic National Park, Washington, USA. Ecological Monographs **55**:927–942.

Gregory, S.V. 1983. Plant-herbiovre interactions in stream systems. Pages 157–190 *in* J.R. Barnes and G.W. Minshall, eds. Stream ecology: Application and testing of general theory. Plenum, New York, New York, USA.

Gregory, S.V., F.J. Swanson, W.A. McKee, and K.W. Cummins. 1991. An ecosystem perspective of riparian zones. BioScience **41**:540–551.

Grenney, W.J., D.A. Bella, and H.C.J. Curl. 1973. A theoretical approach to interspecific competition in phytoplankton communities. American Naturalist **107**:405–425.

Grime, J.P. 1973. Competitive exclusion in herbaceous vegetation. Nature **242**:344–347.

——. 1979. Plant strategies and vegetative processes. John Wiley & Sons, Chichester, UK.

Groombridge, B., ed. 1992. Global biodiversity, status of the earth's living resources. Chapman & Hall, London, UK.

Harper, J.L. 1969. The role of predation in vegetation diversity. Brookhaven Symposium on Biology **22**:48–62.

Hart, D.D. 1985. Grazing insects mediate algal interactions in a stream benthic community. Oikos **44**:40–46.

Hemphill, N., and S.D. Cooper. 1983. The effect of physical disturbance on the relative abundances of two filter-feeding insects in a small stream. Oecologia **58**:378–383.

Henebry, M.S., and J. Cairns, Jr. 1980. Monitoring of stream pollution using protozoan communities on artificial substrates. Transactions of the American Microscopic Society **99**:151–160.

Hook, D.D., and R.M.M. Crawford, eds. 1978. Plant life in anaerobic environments. Ann Arbor Science, Ann Arbor, Michigan, USA.

Horn, H.S., and R.H. MacArthur. 1972. Competition among fugitive species in a harlequin environment. Ecology **53**:749–752.

Hupp, C.R. 1988. Plant ecological aspects of flood geomorphology and paleoflood history. Pages 335–356 *in* V.R. Baker, R.C. Kochel, and P.C. Patton, eds. Flood geomorphology. John Wiley & Sons, New York, New York, USA.

Hupp, C.R., and W.R. Osterkamp. 1985. Bottomland vegetation along Passage Creek, Virginia, in relation to fluvial landforms. Ecology **66**:670–681.

Huston, M.A. 1979. A general hypothesis of species diversity. American Naturalist **113**:81–101.

——. 1994a. Biological diversity: The coexistence of species on changing landscapes. Cambridge University Press, Cambridge, UK.

——. 1994b. Biological diversity and agriculture. Science **265**:458–459.

Hynes, H.B.N. 1963. The biology of polluted waters. University of Liverpool Press, Liverpool, UK.

Jacoby, J.M. 1987. Alterations in periphyton characteristics due to grazing in a Cascade foothill stream. Freshwater Biology **18**:495–508.

Johnson, W.C. 1994. Woodland expansion in the Platte River, Nebraska: Patterns and causes. Ecological Monographs **64**:45–84.

Johnston, C.A., and R.J. Naiman. 1990. Aquatic patch creation in relation to beaver population trends. Ecology **71**:1617–1621.

Junk, W. 1989. Flood tolerance and tree distribution in the central Amazon floodplains. Pages 47–64 *in* L.B. Holme-Nielson, I.C. Nielson, and H. Basley, eds. Tropical forests: Botanical dynamics, speciation, and diversity. Academic Press, Orlando, Florida, USA.

Kalliola, R., and M. Puhakka. 1988. River dynamics and vegetation mosaicism: A case study of the River Kamajohka, northernmost Finland. Journal of Biogeography **15**:703–719.

Kalliola, R., J. Salo, M. Puhakka, M. Rajasilta, T. Haeme, R.J. Neller et al. 1992. Upper Amazon channel migration. Implications for vegetation perturbance and succession using bitemporal Landsat MSS images. Naturwissenschaften **79**:75–79.

Kozlowski, T.T., ed. 1984. Flooding and plant growth. Academic Press, Orlando, Florida, USA.

Kozlowski, T.T., P.J. Kramer, and S.G. Pallardy. 1991. The physiological ecology of woody plants. Academic Press, New York, New York, USA.

Leopold, L.B., M.G. Wolman, and J.P. Miller. 1964. Fluvial processes in geomorphology. W.H. Freeman, San Francisco, California, USA.

Levin, S.A., and R.T. Paine. 1974. Disturbance, patch formation, and community structure. Proceedings of the National Academy of Science, USA **71**:2744–2747.

Li, H.W., C.B. Shreck, C.E. Bond, and E. Rexstad. 1987. Factors influencing changes in fish assemblages of Pacific Northwest streams. Pages 193–202 *in* W.J. Matthews and D.C. Heins, eds. Community and evolutionary ecology of North American fishes. University of Oklahoma Press, Oklahoma City, Oklahoma, USA.

Lubchenko, J. 1978. Plant species diversity in a marine intertidal community: Importance of herbivore food preference and algal competitive abilities. American Naturalist **112**:23–39.

Luttenton, M.R., and R.G. Rada. 1986. Effects of disturbance on epiphytic community architecture. Journal of Phycology **22**:320–326.

Magurran, A.E. 1988. Ecological diversity and its measurement. Princeton University Press, Princeton, New Jersey, USA.

McAuliffe, J.R. 1984. Competition for space, disturbance, and the structure of a benthic stream community. Ecology **65**:894–908.

McDowell, D.M., and R.J. Naiman. 1986. Structure and function of a benthic invertebrate stream community as influenced by beaver (*Castor canadensis*). Oecologia **68**:481–489.

Medin, D.E., and W.P. Clary. 1990. Bird populations in and adjacent to a beaver pond ecosystem in Idaho. USDA Forest Service Research Paper INT-432, Intermountain Research Station, Ogden, Utah, USA.

Medin, D.E., and W.P. Clary. 1991. Small mammals of a beaver pond ecosystem and adjacent riparian habitat in Idaho. USDA Forest Service Research Paper INT-445, Intermountain Research Station, Ogden, Utah, USA.

Minshall, G.W., R.C. Petersen, Jr., and C.F. Nimz. 1985. Species richness in streams of different size from the same drainage basin. American Naturalist **125**:16–38.

Mitsch, W.J., and J.G. Gosselink. 1993. Wetlands. Van Nostrand Reinhold, New York, New York, USA.

Naiman, R.J., T.J. Beechie, L.E. Benda, D.R. Derg, P.A. Bisson, L.H. MacDonald, M.D. O'Connor, P.L. Olson, and E.A. Steel. 1992. Fundamental elements of ecologically healthy watersheds in the Pacific Northwest coastal ecoregion. Pages 127–188 *in* R.J. Naiman, ed. Watershed management. Springer-Verlag, New York, New York, USA.

Naiman, R.J., H. Décamps, and M. Pollock. 1993. The role of riparian corridors in maintaining regional biodiversity. Ecological Applications **3**:209–212.

Naiman, R.J., C.A. Johnston, and J.C. Kelley. 1988. Alteration of North American streams by beaver. BioScience **38**:753–761.

Nanson, G.C., and H.F. Beach. 1977. Forest succession and sedimentation on a meandering river floodplain, northeast British Columbia, Canada. Journal of Biogeography **4**:229–251.

Nicola, D.M., and C.D. McIntire. 1990. Effects of substrate relief on the distribution of periphyton in laboratory streams. 1. Hydrology. Journal of Phycology **26**:624–633.

Nilsson, C. 1986. Changes in plant community composition along two rivers in northern Sweden. Canadian Journal of Botany **64**:589–592.

Nilsson, C., G. Grelsson, M. Johansson, and U. Sperens. 1989. Patterns of plant species richness along riverbanks. Ecology **70**:77–84.

Patrick, R. 1969. Diatom communities. Pages 151–164 *in* J.J. Cairns, ed. The structure and function of fresh-water microbial communities. Virginia Polytechnic Institute and State University Press, Blackburg, Virginia, USA.

Pennak, R.W. 1989. Freshwater invertebrates of the United States. John Wiley & Sons, New York, New York, USA.

Peterjohn, W.T., and D.L. Correll. 1984. Nutrient dynamics in an agricultural watershed: Observations on the role of a riparian forest. Ecology **65**:1466–1475.

Pielou, E.C. 1975. Ecological diversity. John Wiley & Sons, New York, New York, USA.

Pollock, M.M., R.J. Naiman, H.E. Erickson, C.A. Johnston, J. Pastor, and G. Pinay. 1994. Beaver as engineers: Influences on biotic and abiotic characteristics of drainage basins. Pages 117–126 *in* C.G. Jones and J.H. Lawton, eds. Linking species to ecosystems. Chapman & Hall, New York, New York, USA.

Pollock M.M., R.J. Naiman, and T.A. Hanley. 1998. Plant species richness in forested and emergent wetlands—A test of biodiversity theory. Ecology **79**:94–105.

Pringle, C.M. 1985. Effects of chironomid (Insecta: Diptera) tube-building activities on stream diatom communities. Journal of Phycology **21**:185–194.

——. 1990. Nutrient spatial heterogeneity: Effects on community structure, physiognomy, and diversity of stream algae. Ecology **71**:905–920.

Raedeke, K.J. 1988. Introduction. Pages xiii–xv *in* Streamside management: Riparian wildlife and forestry interactions. K.J. Raedeke, ed. Institute of Forest Resources Contribution Number 59, University of Washington, Seattle, Washington, USA.

Reice, S.R., R.C. Wissmar, and R.J. Naiman. 1990. Disturbance regimes, resilience and recovery of animal communities and habitats in lotic ecosystems. Environmental Management **14**:647–659.

Robinson, C.T., and G.W. Minshall. 1986. Effects of disturbance frequency on stream benthic community structure in relation to canopy cover and season. Journal of the North American Benthological Society **5**:237–248.

Rosenzweig, M.L. 1971. Paradox of enrichment: Destabilization of exploitation in ecological time. Science **171**:385–387.

Salo, J., R. Kalliola, I. Haekkinen, Y. Maekinen, P. Niemelae, M. Puhakka et al. 1986. River dynamics and the diversity of Amazon lowland forest. Nature **322**:254–258.

Scarsbrook, M.R., and C.R. Townsend. 1993. Stream community structure in relation to spatial and temporal variation: A habitat template study of two contrasting New Zealand streams. Freshwater Biology **29**:395–410.

Schlosser, I.J. 1982. Fish community structure and function along two habitat gradients in a headwater stream. Ecological Monographs **52**: 395–414.

Siegfried, C.A., and A.W. Knight. 1977. The effects of washout in a Sierra foothills stream. American Midland Naturalist **98**:200–207.

Sousa, W.P. 1979. Disturbance in marine intertidal boulder fields: The nonequilibrium maintenance of species diversity. Ecology **60**:1225–1239.

Stanford, J.A., and J.V. Ward. 1983. Insect species diversity as a function of environmental variability and disturbance in stream systems. Pages 265–278. *In* J.R. Barnes and G.W. Minshall, eds. Stream ecology: Application and testing of general ecological theory. Ann Arbor Science, Ann Arbor, Michigan, USA.

Steinman, A.D., and C.D. McIntire. 1986. Effects of current velocity and light energy on the structure of periphyton assemblages in laboratory streams. Journal of Phycology **22**:352–361.

Steinman, A.D., and C.D. McIntire. 1990. Recovery of lotic periphyton communities after disturbance. Environmental Management **14**:589–604.

Steinman, A.D., C.D. McIntire, S.V. Gregory, G.A. Lamberti, and L.R. Ashkenas. 1987. Effects of herbivore type and density on taxonomic structure and physiognomy of algal assemblages in laboratory streams. Journal of the North American Benthological Society **6**:175–188.

Summerhayes, V.S. 1941. The effect of voles (*Microtus agrestis*) on vegetation. Journal of Ecology **29**:14–48.

Swanson, F.J., T.K. Kratz, N. Caine, and R.G. Woodmansee. 1988. Landform effects on ecosystem patterns and processes. BioScience **38**:92–98.

Swanson, F.J., and G.W. Lienkaemper. 1978. Physical consequences of large organic debris in Pacific Northwest streams. USDA Forest Service Research Paper GTR-69, Pacific Northwest Research Station, Corvallis, Oregon, USA.

Tabacchi, E., A.M. Planty-Tabacchi, and H. Décamps. 1990. Continuity and discontinuity of the riparian vegetation along a fluvial corridor. Landscape Ecology **5**:92–98.

Tansley, A.G., and R.S. Adamson. 1925. Studies of the vegetation of the English chalk. III. The chalk grasslands of the Hampshire-Sussex border. Journal of Ecology **13**:117–223.

Teversham, J.M., and J. Slaymaker. 1970. Vegetation composition in relation to flood frequency in Lillooet River Valley, British Columbia. Catena (Cremlingen-Destedt, Germany) **3**:191–201.

Tilman, D. 1982. Resource competition and community structure. Princeton University Press, Princeton, New Jersey, USA.

——. 1983. Plant succession and gopher disturbance along an experimental gradient. Oecologia **60**: 285–292.

Tonn, W.M., and J.J. Magnuson. 1982. Patterns in the species composition and richness of fish assemblages in northern Wisconsin lakes. Ecology **63**:1149–1166.

Walker, L.R., J.C. Zasada, and F.S. Chapin. 1986. The role of life history processses in primary succession on an Alaskan floodplain. Ecology **67**:1508–1523.

Ward, J.V. 1992. Aquatic insect ecology. 1. Biology and habitat. John Wiley & Sons, New York, New York, USA.

Ward, J.V., and J.A. Stanford. 1983. The intermediate disturbance hypothesis: An explanation for biotic diversity patterns in lotic ecosystems. Pages 347–355 *in* T.D. Fontaine III and S.M. Bartell, eds. Dynamics of lotic ecosystems. Ann Arbor Science, Ann Arbor, Michigan, USA.

Wehr, J.P. 1981. Analysis of seasonal succession of attached algae in a mountain stream, the North Alouette River, British Columbia. Canadian Journal of Botany **59**:1465–1474.

White, D.S. 1990. Biological relationships to convective flow patterns within stream beds. Hydrobiologia **196**:149–158.

Whittaker, R.H. 1972. Evolution and measurement of species diversity. Taxon **21**:213–251.

——. 1977. Evolution of species diversity in land communities. Evolutionary Biology **10**:1–67.

Woodward, A., E.G. Shreiner, D.B. Houston, and B.B. Moorhead. 1994. Ungulate-forest relationships in Olympic National Park: Retrospective exclosure studies. Northwest Science **68**:96–110.

Yount, J.L. 1956. Factors that control species number in Silver Springs, Florida. Limnology and Oceanography 286–295.

Part IV
Management

18
Statistical Design and Analysis Considerations for Monitoring and Assessment

Loveday L. Conquest and Stephen C. Ralph

Overview

- The use of statistical methods is an important element of monitoring and assessing ecological responses of streams and rivers.
- Clearly defined study objectives determine the resulting sampling design and ultimately the types of data analysis approaches; there is no "one size fits all" sampling approach.
- Investigators may need to alter traditional statistical designs to adapt to field conditions; hence designs for stream studies may not meet strict definitions of traditional experimental designs. Advantages and disadvantages of whatever sampling design is chosen must be understood.
- True statistical replicates can be difficult to find in stream studies; care must be taken to avoid unintentional pseudoreplication. The concept of regional reference sites addresses the issue of appropriate "control" replicates; thorough knowledge of an area is necessary when considering candidates for such sites.
- Case studies (an *n* of one) are useful for providing knowledge about underlying processes; they do not provide a basis for actual statistical inference.
- Statistical inference can be made only at the same scale of the units at which sampling occurred. Process models or hierarchical frameworks allow some extension of conclusions to different levels of resolution.
- Parameter selection is important. Parameters must be linked to the change that is viewed as significant and must relate directly to the resources of concern; resource variability must not be overwhelmed by confounding natural factors of scale or time.
- To maximize data value, data quality should be known, data type and quality should be consistent and comparable, and data should be accessible to potential users beyond those who conduct the initial study. Quality assurance and quality control in field methods are important for acquiring reliable data.
- In conducting statistical tests, investigators are usually interested in whether there is enough of an effect to warrant a particular action by resource managers. Information on an approximate estimate of an effect may be better than simply a statement of statistical significance.
- Multivariate procedures are useful for analyzing samples with multiple, correlated responses; they are often used to create classification schemes for stream and watershed-related units. Geographic Information Systems (GIS) organize and display multivariate spatial information by linking together different types of data gathered at different scales and pertaining to different shapes.
- Incorrect statistical analyses on high quality data can always be redone; no application of complicated statistical techniques will salvage data from a poor study design or one that is not implemented well. The usefulness of data analysis depends upon measuring meaningful responses at an appropriate scale and using a good sampling design.

Introduction

An important component of the monitoring and assessment process is the use of statistical methods, covering everything from the broad design aspects of a project to the final particulars of the data analysis and presentation of results. By its very nature, monitoring of streams and rivers involves sampling, and sampling and measurement inevitably involve the use of statistics to answer posed questions (MacDonald et al. 1991). At the same time, it is unwise to view field study results from streams and rivers as if they were taken from laboratory experiments under carefully controlled conditions. The availability of computer software may tempt one to engage in blind applications of various statistical techniques without careful thought as to the quality of the data, issues of replication, or the inherent variability of the sampling units.

This chapter addresses how to turn objectives for monitoring or assessment programs into proper statistical design, and from there how to proceed to proper data acquisition, information management, and data analysis. Figure 18.1 illustrates the process of a study to assess the effects of logging practices on stream channel morphology and woody debris (Ralph et al. 1994). The more focused objective (the specific purpose of the study) is to compare the physical in-stream habitat and channel condition under three types of timber harvest practice, while accounting for variation in stream channel morphology. Clear objectives at the beginning of the study determine many of the characteristics of the resulting sampling design. Sampling units should be chosen with care, and the issue of pseudoreplication (i.e., treatments are not replicated or replicates are not statistically independent) must be addressed regarding possible effects on data analysis and interpretation. A valid study includes regional reference sites, the closest one can come to an experimental control. And when the objective is to obtain long-term information on underlying biological and physical stream processes, a case study (an n of one) may be in order.

In choosing what responses to measure, one must pay particular attention to the study objectives. As an example, Figure 18.1 shows the relationship between the response variables on channel morphology and habitat characteristics, the sampling units (stream), and the study objective. Responses should relate to the resource of concern, reflect the effects of disturbance activities on stream and hillslope processes, and not be overwhelmed by natural variability. Acquiring and managing good data are also important; quality control and quality assurance procedures need to be developed and implemented. There are numerous methods to choose from regarding data analysis, but not every study has to include a hypothesis test, and there is much to be learned from exploratory techniques made possible by advances in computing (statistical tools include both hypothesis tests and exploratory techniques). Even so, availability of highly technical tools is no excuse for neglecting careful sampling design, parameter selection, and field execution of any monitoring or assessment project.

Finally, when conducting studies on streams and rivers, collaboration among experts representing different areas of knowledge is essential. For example, unless the statistician understands the limitations of translating classical sampling or experimental design theory to the natural world, one can end up with sampling designs that are statistically correct (and even elegant) but ecologically nonsensical. It is also usually necessary for the stream ecologist or forest scientist to understand some general principles of statistical design and how they affect the data analysis in order to plan monitoring designs that yield reliable results. A truly effective monitoring or assessment study includes scientists with different expertise working in tandem.

Sampling Design

Freshwater aquatic systems have been studied using a large variety of sampling designs. A general sampling plan is determined by a number of criteria, a crucial one of which is the

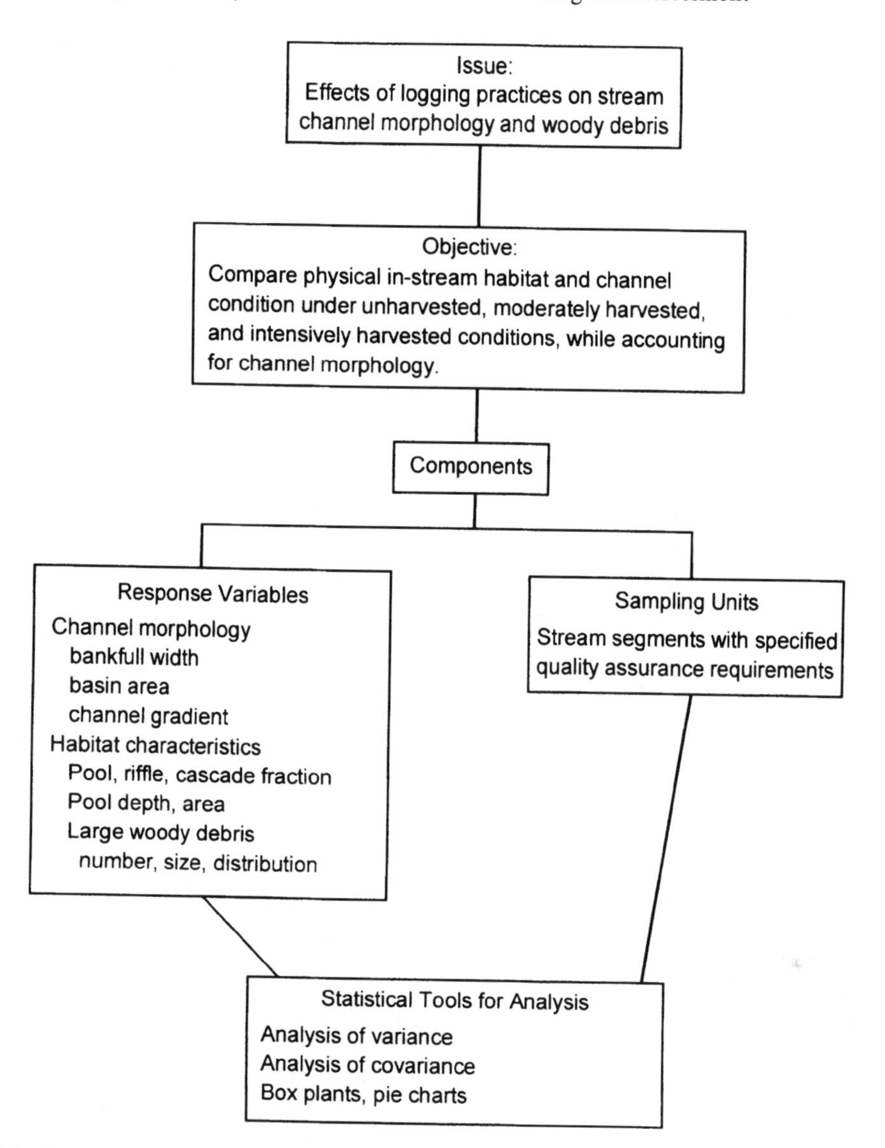

FIGURE 18.1. Process for a study investigating effects of commercial timber harvest on in-stream habitat (based on Ralph et al. 1994).

overall objective of the experiment or study. Is the purpose of the study some type of treatment comparison (e.g., disturbed versus undisturbed areas); is it a habitat study; is it to track a number of ecological and environmental variables over time to assess change in general conditions? Different objectives can lead to different sampling approaches. There is no "one size fits all" design, and investigators may be disappointed to discover that a design that is optimal for achieving one particular objective of a project with multiple objectives is far from optimal for achieving another.

Sampling Approaches and Sampling Units

Table 18.1 lists commonly used approaches for sampling streams and rivers. The simplest approach is simple random sampling, where no previous knowledge about inherent variability

TABLE 18.1. Sampling approaches for streams and rivers.

Sampling approach	Advantages	Disadvantages	Example
Simple random sampling	No special assumptions required	Hard to implement in field; if response is patchy, variances can be high; uniform coverage unlikely	Gilbert (1987); estimating pollution concentration in stream sediments
Stratified random sampling	Accounts for patchiness; obtains lower variance; obtains separate estimates for each stratum; easy to compare overall; cost allocation straight forward	Strata must be well defined before sampling; requires prior information on heterogeneity to define strata	Ralph et al. (1991); stratification of stream channel segments by landform and geological features
Systematic sampling	Easier to do in field than random sampling; provides uniform coverage of target population	Unsuspected periodicity can bias estimates; can be difficult to obtain good standard errors	Hankin and Reeves (1988); use of systematic sampling and visual estimation methods for total habitat area and total fish abundance in small streams
Two-stage sampling	Natural use in selecting stream segments and then estimating fish abundance or other habitat characteristic from segments; does not require segments to be of equal length	Using stream segments of equal length does not use full power of this procedure	Hankin (1984); application of two-stage sampling in stream sections of unequal sizes to estimate fish abundance

is required. A disadvantage to this approach is that resulting variances for estimated parameters may be quite high when sampling patchy phenomena and when uniform coverage of the target population is unlikely. Field implementation of simple random sampling (without considerations of field conditions) can also be expensive and time consuming. Thus sampling designs often incorporate stratified sampling, where a heterogeneous area is divided (stratified) into more homogeneous subareas, and the subareas are sampled at different densities to account for differences in variability. For example, random sampling in rivers according to habitat type is inevitably stratified random sampling (Norris et al. 1994). Because the strata need to be well defined before sampling, previous knowledge on inherent variability is required-thus the need for good pilot studies. Systematic sampling (e.g., sampling a stream every 10 km) provides uniform coverage and is usually easier to implement in the field. However, systematic sampling assumes that the measured responses are randomly distributed and exhibit no periodicity; otherwise, biased estimators can result. Also, obtaining good standard errors for estimates under systematic sampling can be difficult. Lastly, multistage (usually two-stage) sampling is used whenever stream segments are first selected, than sampled to estimate fish abundance or describe other stream characteristics. This approach does not require stream segments to be of equal length, as long as they are sampled proportionate to their size (Hankin 1984).

Designs for sampling stream characteristics tend to focus on such sampling units as kilometers of a stream, stream order, habitat units, or specific sites. For example, stream reach was designated as the sampling unit for describing fish community structure as part of the United States National Water Quality Assessment Program (Meador et al. 1993). Geomorphological characteristics such as meander wavelength were used to define the length of a given stream reach. In addition to ensuring the collection of a representative sample of the fish community, reach length was kept to a reasonable distance for sampling. Ralph et al. (1994) (see Figure 18.1) used channel morphology and habitat characteristics of stream segments to compare unharvested forest areas to moderately and intensively logged basins. To have meaningful comparisons, it was also necessary that sites cover a broad geographic range and be stratified by basin area and channel gradient, because these factors were directly related to the outcome variables.

A sampling design is often developed to make some type of comparison (e.g., comparing a disturbed to an undisturbed site). The simplest approach in this case involves monitoring a pair of similar sites (i.e., as identical as possible), one of which is subjected to some type of disturbance. However, a single pair is essentially an unreplicated design (one untreated site, one treated); data from other pairs are necessary for making inferences about the disturbance (Hurlbert 1984, MacDonald et al. 1991). The use of replicated pairs enables real statistical evaluation.

Luchetti et al. (1987) describe a monitoring project on the Snohomish River, Washington, with sites upstream and downstream from a major agricultural area. The pairs vary considerably in terms of the size of the stream being monitored and the amount of agricultural activity; five pairs are on the mainstem, and ten pairs are on tributary reaches. Sampling also is stratified by climatic condition. The design of this monitoring plan allows investigators to document the magnitude and impact of commercial agriculture on water quality.

The replicated paired-site approach does not meet the definition of a traditional experimental design (MacDonald et al. 1991). Experimental units such as sites, streams, or rivers are almost never randomly allocated to treatments like timber harvest, road building, agricultural disturbance, or the installation of a power plant. Rather, investigators must search for pairs of comparable sites, where one member of each pair is subjected to the disturbance of interest, and the other member is not. Upstream–downstream comparisons usually regard the upstream site as the untreated site and the downstream site as the treated site. Because of the nonrandom assignment of sites, Stewart-Oaten et al. (1992) recommend the inclusion of untreated pairs (i.e., both the upstream and downstream sites are untreated) in their Before-After-Control-Impact-Paired (BACIP) design. Then the upstream–downstream differences can be compared between untreated pairs and pairs where the disturbance effect is included. The BACIP design thus accounts for the natural variation that is unrelated to the disturbance between upstream and downstream sites. Alternatively, a calibration period can be used to establish the relationship between upstream and downstream untreated sites; a change in the relationship following disturbance activities indicates disturbance effects over and above site differences.

An overall sampling approach is sometimes a combination of separate sampling designs. An example is the U.S. Environmental Protection Agency's (EPA) Environmental Monitoring and Assessment Program (EMAP) design to sample rivers and estuaries in the Virginian Province (from the Chesapeake Bay to Cape Cod) (Weisberg et al. 1992). This sampling design, which assesses ecological condition, was created by assembling a list representing all estuarine systems greater than 2.6km . The estuaries on this list were then stratified as small estuaries and tidal rivers, large tidal rivers, and large estuaries. For the large estuaries stratum, the general EMAP areal grid (a regular hexagonal grid with a random orientation) was applied to select 54 sampling sites as the center points of hexagons 18km apart. For large tidal rivers, systematic sampling was used. The mouth of the river provided the starting point and the first transect was located randomly between 0 and 25km upstream. Additional transects were located 25km from the first. Sampling sites were then selected randomly along each transect to yield a total of 25 sampling sites per river. The 137 small estuarine systems were stratified geographically by listing them from north to south into groups of four, one system being selected at random from each group. This ensured that the 32 sampled systems were dispersed geographically. Figure 18.2 displays the resulting sample sites from this approach. Samples are collected on one-quarter of the sampling sites each year, returning to any particular site four years later. Through this sampling plan, each year a

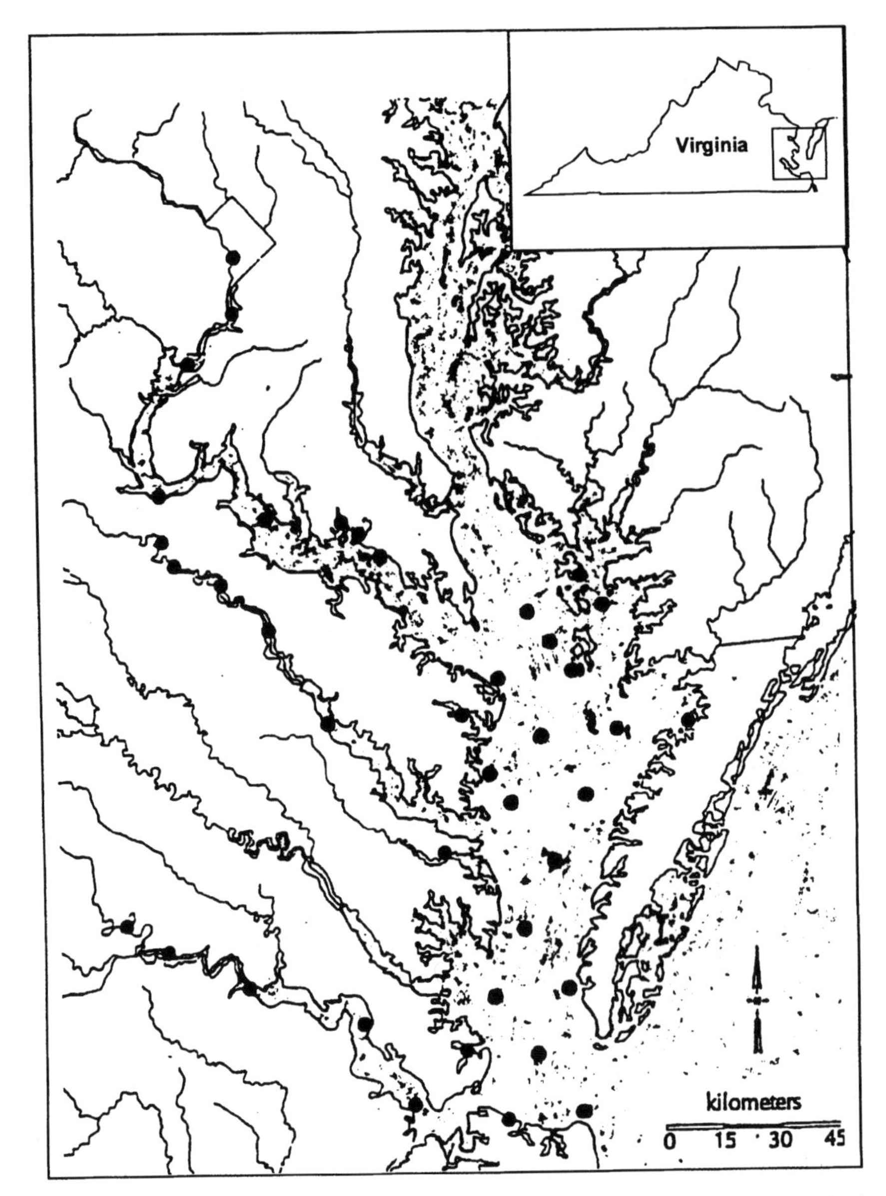

FIGURE 18.2. Base sampling stations from the EMAP sampling design for rivers and estuaries in the Virginian Province Source, Norfolk, Virginia (from Weisberg et al. 1992).

different set of sites yields areal information on the same region.

Studies of streams and rivers nearly always involve sampling over time and space. The area covered in a study may be large enough so that spatial heterogeneity is a concern; or the temporal aspect of the sampling may be such that temporally correlated data points become an issue, along with accounting for seasonal effects. Therefore, sampling must be designed at both the correct spatial scale and temporal scale in order to detect the effects of interest. A short-term disturbance requires sampling on a time scale small enough to detect its effects, whereas effects of a chronic disturbance requires a longer time scale of measurement (Underwood 1994). Monitoring large areas in time and space may make sampling all sites at every sampling time difficult; therefore some kind of alternating or rotating design, where some sites remain unsampled for a period of time, may be warranted. Urquhart et al. (1993) compared two types of sampling designs for long-term assessment of condition. One type visited a panel of sites over consecutive sampling times, with partial replacement of sites each time; this is often known as a "rotating panel" (Skalski 1990). Such a design can be augmented by adding one or more sites that are sampled each time, thus providing complete temporal coverage for some sites. The other type, the "serially alternating design" (which can also be augmented with sites sampled every time), returned to the same site every *k*th sampling time. (As described above, the U.S. EPA's EMAP design is a serially alternating design, with sets of sites visited once every four years.) Statistical models incorporating linear trend, site effects, temporal effects, and temporal correlations were used to measure the resulting precision for estimating trend in the two types of design. The finding that the serially alternating design is always as good as, and often better than, the rotating panel design in terms of precision is one reason for the choice of the EMAP design; the results also yielded numerically specific guidelines to achieve certain statistical efficiencies for combinations of temporal and spatial sampling (e.g., specified precision).

Replication

Although scientists (statistical and otherwise) have pondered about what constitutes a replicate (see Steel and Torrie 1980, pp. 124–125, for an admonishment about the distinction between sampling and experimental units), it was Hurlbert (1984) who coined the term "pseudoreplication" to denote the situation where either treatments are not replicated (though samples may be) or replicates are not statistically independent. Studies on streams and rivers present real challenges to achieving true statistical replicates. True statistical replicates are independent; often this independence is inferred by simply knowing that the response on one sampling unit does not physically influence the response on another sampling unit. Because a stream has a series of connections from the uplands to the mouth, sometimes this is inferred by the sheer distance between sampling units (e.g., stream segments from entirely different basins). If independence cannot be established, it may be wise to assess spatial or temporal patterns to see whether observations are truly uncorrelated (e.g., regional precipitation patterns could result in correlated flow measurements). If substantial correlation exists, the data analysis should take this into account, usually by incorporating correlation terms into variance estimates. Gilbert (1987) provides formulas for sample size calculation and error bounds (confidence intervals) when data are correlated (e.g., among upstream–downstream measurements). Conquest (1993) uses these formulas in examples for resource monitoring. Urquhart et al. (1993), in comparing the rotating panel to the serially alternating design, use models that assume temporal correlation. Millard et al. (1985) incorporate spatial and temporal correlation in statistical models to detect ecological change.

Reference Sites

Although an upstream–downstream approach may be useful for assessing an unreplicated point disturbance (as in Stewart-Oaten et al. 1992), this is not always true for broader comparisons (e.g., comparing different types of landscape disturbance, or comparing different intensities of timber harvest). As Hughes (1985) points out, if an entire stream is heavily disturbed, an upstream–downstream approach for selecting reference sites is no longer appropriate. This issue has given rise to the concept of regional reference sites based on similarities in watershed characteristics (e.g., land-surface form, soil, natural vegetation, land use) as an integral part of the design aspect for monitoring streams and rivers (Hughes 1985, Hughes et al. 1986). The regional reference site takes into account that a pristine, completely undisturbed site is unrealistic and that regional variation does indeed exist—hence the origin of the term, *regional reference site*.

Thorough knowledge of an area is necessary when considering candidates for regional reference sites. Maps of a region, aerial photographs, and information from remote sensing data can be used to compile an initial list of candidate sites; at some point, streams must be walked to assess their suitability as a reference sites. Relatively lower levels of human disturbance (e.g., as in wilderness areas, wildlife refuges, and parks), similarity in stream size and stream channel characteristics, and zoogeographic patterns should all be considered in choosing reference sites (Hughes et al. 1986). Ralph et al. (1994), in a study comparing logged and unlogged basins in Washington, relied upon wilderness sites to find stream segments unaffected by logging or road construction.

The general process for choosing reference sites involves the issue of spatial variability. Large-scale environmental patterns are difficult to assess without some subdivision of a heterogeneous landscape into more homogeneous subareas (Green 1979). Such a process inevitably leads to the development of classification units. One way to achieve classification is to use clearly identified variables and logical rules to subdivide units in an hierarchical fashion (Conquest et al. 1994). For example, ecoregions were devised to provide a regional framework for aquatic ecosystem management (Hughes and Larsen 1988). Omernik and Gallant (1986) mapped ecoregions of the Pacific Northwest and indeed for most of the United States (Omernik 1987). At the stream scale, emphasis on geomorphic templates and the relationship of a stream to its watershed across a wide range of spatial scales are prominent in the development of recent stream classification systems (Lotspeich and Platts 1982, Rosgen 1985, Frissell et al. 1986, Naiman et al. 1992, Montgomery and Buffington 1993, Chapters 2 and 4, this volume).

When the importance of a set of response variables is understood and those variables have been measured on a large number of available sites, it may be possible to use multivariate, data-intensive, statistical approaches to develop classification units. *Cluster analysis* is a multivariate statistical technique to create groupings of units. While mathematical criteria (minimizing distances within groups, maximizing distances between groups) drive the statistical process, the final groupings must make biological or physical sense to be useful (Conquest et al. 1994). Huang and Ferng (1990a, 1990b) used cluster analysis to classify 78 watershed units into different watershed zones. Richardson and Healey (1994) used cluster analysis to classify 250 unpolluted sites from the headwaters to the mainstream in the Fraser River Basin Benthic Monitoring Program. The major advantage of cluster analysis is that many response variables can be used in a classification scheme with the optimization done via computer according to well-established statistical rules. The major disadvantage is that data must be taken from a large number of sites.

The Issue of Scale

A real difficulty with much stream monitoring and assessment is that often the scale at which the field sampling actually occurs (e.g., at a particular site) differs from the desired scale for making inferences (e.g., a watershed). An inference made within a limited geographic setting may become much more tenuous when attempting to make inferences for larger-scale regions. For example, Riemann and McIntyre (1996) claim that monitoring only a few index populations of bull trout (*Salvelinus confluentus*) does not represent dynamics of the larger regional populations. A hierarchical framework of different scales allows data to be integrated from diverse sources and at different levels of resolution, while allowing a user to select the level of resolution most appropriate to the objectives (Conquest et al. 1994). Some hierarchical classification schemes (e.g., Omernik and Gallant 1986, Omernik 1987, Hughes and Larsen 1988) have been used to incorporate landscape features at different scales of resolution. Frissell et al. (1986) described a hierarchical approach to construct a continuum of habitat sensitivity to disturbance and recovery time (Figure 18.3). In this concept, microhabitats within streams are the most sensitive to disturbance and watersheds

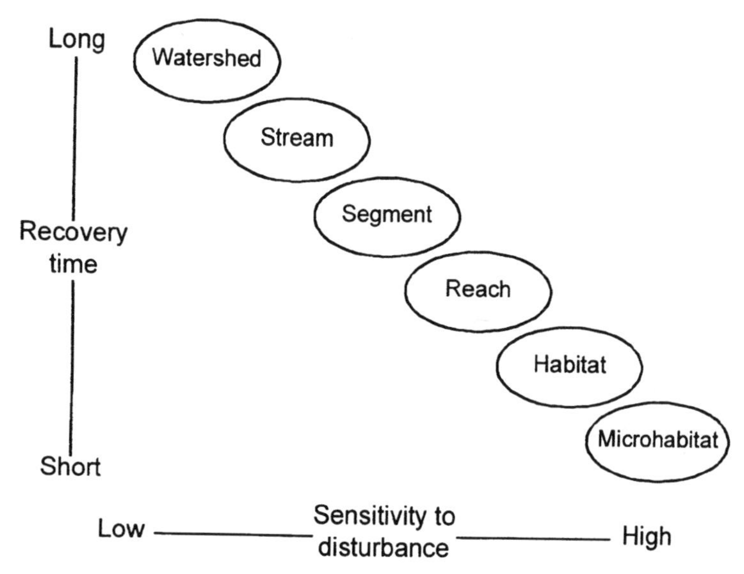

FIGURE 18.3. Relationship between recovery time and sensitivity to disturbance for different spatial scales associated with stream systems (from Naiman et al. 1992. Reproduced with permission).

the least. However, physical and biological recovery from catastrophic or cumulative effects may take a long time in large rivers, while recovery at the microhabitat scale may be short, depending upon the nature, extent and persistence of the disturbance (Rosgen 1994). Under conditions where the effect is chronic and widespread, such as in highly urbanized streams, the microhabitat features may be largely eliminated.

In addressing the scale of inference, an essential question to ask is: from what population or universe have experimental units been taken? If all the experimental units come from the same stream, then the inference extends to that stream only and not "all streams of that type," because only one stream has been observed. Similarly, although it might be tempting to extend the inference from a number of streams sampled in a single watershed to all watersheds of that type, the statistical inference is limited to that particular watershed. Statements about all watersheds of this type rely upon knowledge of physical or biological processes, knowledge that allows one to generalize without statistical sampling. One way to obtain process knowledge is via a case study.

When *n* Equals One—The Argument for Case Studies

Elementary statistics courses emphasize that to have a credible experiment or observational study, more than one statistical replicate is required. When studying landscapes, watersheds, or streams, just what constitutes a replicate may not be immediately clear. Some of the studies that Hurlbert (1984) cites as having pseudoreplication (treating subsampling as real replication, or assuming there are more replicates than really exist) include case studies, where a single lake, stream, or watershed has been studied in detail. One might conclude that because a case study by definition constitutes a sample size of one, no inference of any kind is possible. Just what is the argument for a case study, particularly one that may involve very long-term field monitoring? The motivation behind case studies is different from that behind statistical inference, in which samples from a defined population are necessary to make inferences from samples to the population. The primary purpose of a case study is to reveal information about fundamental underlying biological, physical, and ecological processes in a system. Armed with basic knowledge about processes, investigators can then formulate hypotheses and devise other studies, with real statistical replicates, to test those hypotheses. As an example, the National Science Foundation's Long-Term Ecological Research areas (LTERs) have been chosen as an entire group of case studies in different ecoregions of the United States, yielding knowledge about long-term ecological processes, rather than as a sample of "statistical replicates." Naturally, case studies also have limitations in that some features cannot be varied to investigate process changes.

Hartman and Scrivener (1990) presented an example of a before-and-after case study that assesses the effects of forestry practices in a coastal stream ecosystem (Carnation Creek) in British Columbia over 17 years. A case study like this adds to the general understanding of the effect of silvicultural activities on features of stream condition, such as large organic debris, temperature and stream flow, diversity of stream habitats, and composition of spawning gravel. The authors concluded that because managing unstable streams is difficult, it is more important to understand how the variability in physical conditions and fish population characteristics is induced by logging and how to avoid it than to determine the average impact of logging activities on a fish stock (Hartman and Scrivener 1990). While other stream systems are expected to have different characteristics, there is much in terms of the process information (based on accepted stream ecology theory) that carries over from Carnation Creek to other coastal stream ecosystems. Another long-term case study, which has been monitored since 1955 by the U.S. Forest Service, takes place in a 3076-ha area of the Hubbard Brook Experimental Forest in New Hampshire (Likens and Bormann 1995). Similar and adjacent watershed-ecosystems in this study have allowed scientists to conduct experimental studies of deforestation, using reference and experimental sites (Bormann and Likens 1979).

One thing that a case study (or sequence of case studies) does not yield is a regional estimate of ecological condition with stated levels of uncertainty. If regional estimates are desired, then a probability-based sampling scheme is required. The U.S. EPA's EMAP sampling design that uses a regular spatial sampling grid with a random start is set up to achieve regional estimates of ecological condition with specified levels of uncertainty. For example, it allows EPA to estimate the percentage of eutrophic lakes in the northeastern region of the United States, with error bounds around the estimate (Urquhart 1994). Much of the discussion over the usefulness of the EMAP design concerns whether scientists place more emphasis on information about underlying ecological processes or on estimates of condition for large regions. Those favoring the probability-based sampling design emphasize the need for regional estimates with associated statistical confidence intervals; those favoring careful choice of certain types of sites for long-term monitoring emphasize the need for understanding underlying processes.

Choosing Parameters

The overarching purpose of environmental monitoring programs is the detection and documentation of ecologically significant changes in response to disturbance. In various forms, chemical contamination, physical changes, and ecological measurements have been proposed as environmental monitors (Ausmus 1984). Ecosystem level monitoring and assessment programs require procedures that can assess influences of natural and human-induced disturbances contributing to cumulative impacts and that relate to the various processes that determine how and where these effects are manifested (Wissmar 1993). To be meaningful, the selected attribute or parameter must be linked to the change that is viewed as significant. This linkage is most effectively established through a clear articulation of specific objectives that can be translated into testable hypotheses through rigorous experimental design. For example, asking a general question about the adverse effects of logging in a particular watershed gives no indication as to whether the concern is directed toward overall biological integrity or the presence of sportfish.

More specific questions are needed to determine what combination of biological, physical, and chemical responses are chosen for study (MacDonald et al. 1991).

Developing a monitoring plan requires understanding of the variability of selected parameters over time and throughout the spatial area of concern. This inherent variability determines if change can be detected, and if that change has any ecological significance. Norris et al. (1994) consider the influence of variability on statistical power to detect change and the use of variance-stabilizing transformations. Typically, however, investigators do not have previous knowledge of this variability before initiating a monitoring plan. Therefore, a recommended approach includes selecting sites for monitoring stations for both long-term baseline data (trends in key parameters) and for reference conditions, to understand the range or variability in chosen parameters under conditions relatively undisturbed by humans.

For a parameter to be useful it should satisfy these criteria: (1) the parameter must change in a measurable way in response to the effects of land-use activities and their consequences on stream and hillslope processes; (2) the parameter must relate directly to the resources of concern (i.e., designated beneficial uses; MacDonald and Smart 1993); and (3) the parameter's variability must be limited and not likely to be overwhelmed by confounding natural factors of scale or time. Uncertainty about what to measure and lack of knowledge about the linkage between cause and effect manifested in the chosen parameters can lead to a tendency to collect data on a broad suite of variables without clearly stating questions to be answered. When the inherent variability of a particular parameter is not known, the risk of making a wrong conclusion from the results are all the more probable (Conquest et al. 1994). One way to address this problem is to select a parameter for measurement at a number of sites where development or human alteration is relatively minimal or absent (reference situations), and from this parameter establish an expected range of values for the parameter (Green 1989).

MacDonald et al. (1991) and MacDonald and Smart (1993) discuss the effects of water quality parameters on the major designated uses of water (domestic and agricultural water supply, hydropower, recreation, biological integrity, warm- and cold-water fishes) from forested watersheds in the Pacific Northwest and Alaska. These parameters include water column responses (e.g., temperature, pH, conductivity, dissolved oxygen, nitrogen, phosphorus), flow (peak flows, low flows, water yield), sediment (suspended, turbidity, bedload), channel characteristics (e.g., bank stability, large woody debris, bed material, channel width/depth ratio), riparian characteristics (riparian canopy opening and riparian vegetation), and aquatic organisms (bacteria, algae, invertebrates, fish). The authors also discuss the sensitivity of the aforementioned water quality monitoring parameters to forest management activities (harvesting, road building and maintenance, applications of fertilizers, herbicides, and pesticides) along with mining, grazing, and recreation. For example, road building and maintenance can have large effects on peak flow, sediment, channel characteristics, invertebrates, and fish. Fertilizer application affects the amounts of nitrogen and phosphorus in a stream. Herbicide application affects riparian canopy opening and vegetation, and aquatic organisms. Therefore, part of the process of moving from original broad monitoring objectives to deciding exactly which responses to measure involves determining which parameters relate to specific uses (e.g., dissolved oxygen is directly related to biological integrity in a stream) and determining the sensitivity of parameters to various management activities (e.g., the presence of grazing and fertilizer use affects a stream's level of dissolved oxygen). Ralph et al. (1991) include both channel geometry variables (e.g., channel bankfull width and depth, riffle gradient, slope failure area, sediment particle size, gravel embeddedness) and habitat variables (e.g., discharge, residual pool depth, woody debris characteristics, habitat unit characteristics, canopy closure, adjacent land use, riparian vegetation type) to assess effects of changes in flow and sediment at both the stream channel segment and habitat unit scales. The authors also provide details on the standardized field methods used to record the responses.

One limitation of the existing regulatory framework for nonpoint source pollution is that many traditional water quality regulatory criteria (e.g., chemical parameters) are either insensitive or largely irrelevant to land use effects (such as timber harvest) that more typically manifest themselves in altering physical rather than chemical processes (Chapter 4). However, land use practices are known to alter numerous and ecologically significant water resource parameters (e.g., riparian canopy opening) for which few or no specific standards or criteria presently exist (MacDonald and Smart 1993).

Acquiring and Maintaining Good Data

Data acquisition and maintenance constitute an important part of any monitoring plan. Much attention is typically given to data acquisition, because decisions about measurement techniques, field crews, sampling locations, and sampling times are involved. Data maintenance is of less intrinsic interest to environmental managers; however, one of the most challenging aspects of conducting an extensive monitoring program is the task of creating and maintaining an orderly flow of information from the field, to the analysts, and then to the decision makers (Conquest et al. 1994).

Intensive and Extensive Approaches

Typically, the methods employed to conduct assessment and monitoring differ depending on the scope and scale of the objectives. Intensive monitoring may employ field techniques that are designed to address specific data needs at limited sites, and thus may be more time intensive and quantitative in nature. Examples include establishing permanent channel reference sites, where a series of cross-section transects and longitudinal profiles are taken using survey techniques with high accuracy and precision (Madej 1978, Olson-Rutz and Marlow 1992, Harrelson et al. 1994). In extensive monitoring, field methods may also be used to quickly characterize conditions across multiple sites (Platts et al. 1983, 1987). Methods involving subjective assessments are inherently prone to observer bias, although they do provide a quick and synoptic overview of stream and channel conditions. In practice, both intensive and extensive field methods should play a prominent role in assessment and monitoring programs (Bryant 1995). Intensive field methods help validate the assumptions derived from initial interpretation of extensive field data, and as such greatly aid the process of refining these techniques. Judicial use of a suite of methods and assessment techniques, employed at strategic intervals, promotes understanding of the stream and its catchment in context.

Training Field Crews

Once the response variables are decided on and measurement techniques are standardized, field methods must be calibrated to ensure observations and measurements that are as consistent, accurate, and precise as possible. Considerable resources may have to be devoted to this part of any monitoring project. A field manual written especially for a project can describe in detail the procedures and rationale for employing the various methods used in obtaining measurements (Conquest et al. 1994). Field staff who understand why a particular method is being used and how it contributes to the data will gather the data more effectively than staff who are simply following directions without understanding why. The field manual also can be of immense use to future researchers who are using project data years after the project's completion. In addition to the field manual of procedures, training sessions for field crews are necessary. Staff should receive training in the various measurement methods for recording data, especially under diverse field conditions. For a long-term project with turnover in field staff, this training is absolutely necessary; even seasoned field crews benefit from refresher courses. Though not always practical, Conquest et al. (1994) recommend the employment of field crews on a long-term basis (instead of sequences of temporary hires and layoffs) to improve data recording quality and to avoid excessive costs for training new crews at the beginning of each field season.

Quality Control and Quality Assurance

Quality control is necessary to ensure that methods are applied properly. This may involve two or more field teams measuring the same things independently for later comparison. It is important that reported changes are not artifacts of measurement error or observer bias. For example, Bott et al. (1978) compared different methods for measuring primary productivity and community respiration in streams to assess measurement bias. The Timber-Fish-Wildlife Ambient Monitoring Project (Washington state) investigated the consistency with which dimensions of habitat units could be visually estimated (Conquest et al. 1991). Because enough conditions existed that the reliability of subsampling data might be questioned, it was recommended that all habitat units in a stream segment be measured instead of only a subsample. These examples illustrate the type of problems that can occur as a consequence of the inherent difficulties of field training and quality assurance in conducting a large-scale project under challenging field conditions. A comprehensive field training program also includes quality assurance relating to parameters involving an element of subjectivity.

Management of Information

Research information is of little or no value unless it is accessible to those who use it (Stafford 1993). Shampine (1993) notes three general characteristics of monitoring data to maximize data value: the data quality should be known, data type and quality should be consistent and comparable, and data should be accessible to potential users beyond those who conduct the initial study. Beyond the objectives of a given study, a data set also should be available for other uses. The creation of the Forest Science Data Bank (FSDB) at Oregon State University (Stafford et al. 1984) is a good example of data compilation and documentation that ensures usability by researchers other than those originally involved with a particular project. This process includes documenting enough information on older studies to ensure their future value, along with establishing standard procedures for acquiring new data sets with full documentation. A master file is continually updated and includes information on both data structure and study methodology, in addition to key words, descriptive narrative information, references, contact persons, computer programs used, and data file organization. The availability of a series of standard forms ensures uniform documentation and a set of standard information for all projects. Over the years the FSDB has been integrated into the general research process of all forestry-related projects in Oregon, from the initial stages of sampling design, through the field research, to the information retrieval stage (Stafford et al. 1988). The flexibility of the FSDB allows documentation of long-term collaborative research, joint studies, and comparative analyses. Recent developments include the evolution in hardware from mainframe computers to microcomputers, the evolution of software, personal computer links, and the use of local area networks to access the Internet and the World Wide Web, which considerably enhances data availability to multiple users. Currently, interfaces with Geographic Information Systems (GIS) and graphic capabilities to display multivariate relationships and complex interdependencies are emerging as well.

Because numerous studies in a region lead to large quantities of information scattered in various data sets, Fyles et al. (1995) advocate a system to convey the reliability of the information along with the information itself as a way of better serving remote users (both in space and time). *Metadata*, or data about data, give information about a data set—the who (e.g., bibliography), what (responses measured), where (locations and how they were selected), how (methods used to generate the data), and other summary information, such as indicators of data quality. For example, a data quality score can indicate whether a data set is suspect as a result of recognized weaknesses, whether a data set is internally consistent, and how easily a data set can be compared with others.

Data Analysis

The usefulness of data analysis depends in large part on measuring meaningful responses, at an appropriate scale, using a good sampling design. The sampling design can determine what types of analyses are possible, and both are driven by the objectives of the monitoring project. Stafford (1985) presents a statistical primer showing fundamental aspects of several different experimental designs and their use in forestry. When statisticians are clearly informed about study objectives, they are better poised to make clear recommendations about the data analysis. There are several groups of statistical methods used in analyzing, monitoring and assessing data (Table 18.2). Table 18.2 is not intended to be a comprehensive list of every statistical technique ever devised; however, it does list commonly used classes of methods and tests for river data. The discussion below begins with the most frequently used parametric, normality-based tests.

Parametric Procedures, Regression and Correlation, and Nonparametric Tests

Parametric methods are based upon the assumption that responses behave according to the normal (bell-shaped) probability distribution. Methods such as *t*-tests and analysis of variance are used to demonstrate whether or not observed differences between group means are statistically significant. Much material under the topic of statistical experimental design has to do with devising designs that assign experimental units to different treatment groups in a manner that allows for comparisons of desired treatments while accounting for other major sources of variation and using random sampling or random assignment to invoke certain probability laws.

In testing something as fundamental as the equality of two population means or variances, in many cases investigators are not really interested in exact equality versus inequality; rather they are interested in whether there is enough of a difference to warrant a particular action by resource managers. Such procedures lie more in the realm of estimating effects and assessing their ecological importance. Hypothesis testing and confidence interval estimation are complementary procedures; historically the emphasis has fallen more on the former than the latter. In many cases, a resource manager makes better decisions with information on an approximate estimate of an effect rather than simply a statement of statistical significance (Stewart-Oaten et al. 1992).

Investigating the association of two or more variables uses regression or correlation techniques. *Correlation analysis* is used to investigate the association between two variables, whereas *regression analysis* is used when a particular variable can definitely be labeled as the response variable andwhen other variables are seen as predictor variables. Hankin and Reeves (1988) use regression analysis relating measured habitat area to visual observer estimates, thus allowing a correction for bias in observer estimates of habitat areas; they also use regression to relate electrofishing estimates of fish abundance to diver counts. Keleher and Rahel (1996) develop a linear regression model relating air temperature to latitude and elevation to assess the effects of potential climate changes on salmonid fish habitat in Wyoming streams. *Analysis of covariance* is a technique that incorporates both elements of analysis of variance (comparison of group means) and regression. Ralph et al. (1994) use both analysis of variance and analysis of covariance (see Figure 18.1) to assess the effects of three different types of logging intensities (low, medium, high) on the abundance of large woody debris (LWD) in a stream segment, while taking into account known relationships between LWD and such stream characteristics as bankfull width and basin area.

Nonparametric or distribution-free tests are used when the response variable does not follow a normal distribution. Many nonparametric

TABLE 18.2. Commonly used statistical tools for stream and river data.

Statistical tests/tools	Use	Assumption	Remarks
Parametric tests	Estimation or comparison of means and variances	Normal distribution, equal variances, independent observations	Most powerful to detect change when assumptions are met
t-tests, confidence intervals			
Analysis of variance (ANOVA)			
Tests on variances			
Regression and correlation	Linear association or prediction	Independent sampling points, normal distribution	More advanced techniques available for correlated sampling points
Nonparametric tests	Comparison of means without relying on normality	Independent observations, equal shapes of the distributions	Not as powerful as parametric tests, particularly for small sample sizes
Mann–Whitney test			
Wilcoxon rank test (paired data)			
Kruskal–Wallis ANOVA			
Bootstrap techniques	Confidence intervals and parameter estimation (distribution-free)	Independent observations	Computer-intensive techniques
Multivariate techniques	Groupings of sampling units or responses	Normal distribution for any testing	Useful in classification schemes
Cluster analysis			
Principal component analysis			
Factor anlaysis			
Discriminant analysis			
Multivariate permutation procedures	Nonparametric, multivariate group comparisons	Independent samples	Mann-Whitney test and Kruskal-Wallis test are special cases
Exploratory data analysis (EDA)	Data display only, no testing	None	Very useful for data visualization
Histograms			
Box plots			
Bar/pie charts			
Geographic information system (GIS)	Management and display of spatial data	All data require spatial coordinate information from maps, aerial photos, or ground-truthing	Can integrate data from a variety of sources and translate statistical results to maps

tests assume that the observations are independent, so the user is not really free of all the underlying assumptions, just some of them. Nonetheless, in view of the prevalence of nonnormality in ecological data, distribution-free methods are quite useful (Potvin and Roff 1993). The most common nonparametric tests are based on the ranks of the data; thus data need only be specified as to an ordering, rather than actual measurements. Well-known nonparametric tests include the Mann–Whitney test (the nonparametric equivalent of the two-sample t-test), the Wilcoxon signed rank test for paired data (nonparametric equivalent of the paired t-test), and the Kruskal–Wallis test (equivalent to one-way analysis of variance) to compare three or more groups. These and other tests are described in many statistics texts (e.g., Sokal and Rohlf 1995, Zar 1996). Instead of treating nonparametric procedures separately, Zar (1996) applies each nonparametric test to the same set of data used to illustrate its parametric equivalent. Sokal and Rohlf (1995) include a key to statistical methods at the front and back of their book, guiding the reader to the proper test depending upon the number of response variables, the number of samples and other relevant characteristics (e.g., is there pairing or blocking going on), and whether the data are continuous, ranked, or attribute.

In addition to the classical nonparametric tests based upon ranks, computer-intensive, distribution-free methods are available that rely upon either random reshuffling of the data (permutation tests) or resampling many times from the original data (bootstrap techniques). When the sampling distribution for the data is unknown and cannot be derived, bootstrap techniques may be especially useful for parameter estimation and confidence intervals. Fore et al. (1994) use bootstrap resampling to investigate statistical properties of an index of biological integrity (IBI) (Karr 1981 and 1991, Karr et al. 1986) used to evaluate water resources. The statistical models for Karr's IBI include effects such as measurement error, variability of fish assemblages through time, and location. Bootstrap results validate the use of standard analysis of variance techniques; power analyses show the IBI distinguishing between five and six categories of biotic integrity.

Multivariate Procedures

Multivariate statistical methods generally refer to the analysis of vectors of responses (i.e., each sampling unit yields a vector of two or more different responses rather than a single response). Multivariate procedures, including principal components analysis, cluster analysis, and discriminant analysis, allow the user to summarize otherwise redundant information that accumulates when a variety of correlated responses is measured on the same unit. When any kind of formal testing is involved, multivariate normality of the data is essential. Afifi and Clarke (1984) offer a good introduction to basic multivariate techniques for the inexperienced, including annotated computer output from several statistical software programs. Goff (1986) illustrates the use of a principal components analysis, multivariate analysis of variance, discriminant analysis, and multiple regression to assess the effects of thirteen physical and biological factors (in the nesting environment) on reproductive success for smallmouth bass in Lake Erie. Nonparametric multivariate procedures are also available. Clarke (1993) assesses changes in community structure using species assemblages. Zimmerman et al. (1985) extend the concept of comparing groups based on ranks to a procedure for comparing groups based on several simultaneous responses.

Factor analysis and *principal components analysis* are related multivariate techniques used to reduce a long list of responses carrying redundant information to a much shorter list of uncorrelated components, where each component is some linear combination of the original variables. For example, Huang and Ferng (1990a) used principal components analysis to reduce 27 watershed management variables to a shorter list of six main factors. *Cluster analysis* creates groupings of experimental units (rather than variables) using either hierarchical or nonhierarchical approaches. Huang and Ferng (1990a, 1990b) used nonhierarchical clustering to partition 78 watersheds into zones of similar basin morphometry and assimilative capacity. Researchers must take care that cluster analysis results have some biological or physical meaning, because the final clustering pattern produced by the computer reflects optimization of certain statistical criteria and nothing more (Conquest et al. 1994). Patterns may also change with changes in the clustering criteria, such as using a different distance or similarity measure (Afifi and Clarke 1984).

Discriminant analysis is a multivariate procedure that allows the user to decide which responses do the best job of separating two or more previously defined groups. Bailey (1984) used discriminant analysis to show hydrologic differences between two ecoregions; Cupp

(1989) used it to confirm a rule-based system defining nine valley segment types in Washington's Gifford Pinchot National Forest. Multivariate techniques are often used in a descriptive manner because ecological data rarely meet the requirement of multivariate normality for hypothesis tests. As such, multivariate techniques are helpful for pointing out that certain groups do differ according to certain responses measured without placing too much emphasis on the exact results of the associated statistical tests.

Exploratory Data Analysis

Exploratory data analysis (EDA) refers to the process of data examination simply to explore existing patterns without prior hypotheses to test. This is distinguished from *confirmatory data analysis*, which usually does involve some parameter estimation or hypothesis testing. Exploratory data analysis is more of a philosophical approach to data analysis rather than a specific tool or set of tools (Porter 1993). Tukey (1977) emphasizes the four underlying themes of exploratory data analysis: revelation, residuals, reexpression, and resistance. Revelation emphasizes data display to reveal patterns and relationships within the data; residuals are what remain after removing from the total variation in the data the variation explained by a proposed statistical model (many statistical models are based upon the residuals being independent, normally distributed with homogeneous variance); reexpression deals with transformations of the data; resistance refers to procedures that dampen the effects of outliers on the data (such as the practice of using medians instead of means).

Standard EDA techniques for simple data display include barcharts and piecharts, histograms, probability plots, scatterplots, boxplots, and stem-and-leaf plots. Ralph et al. (1994) used barcharts, piecharts, scatterplots, and boxplots to display stream characteristics. More computer-intensive techniques, called "animated EDA," are usually used to investigate characteristics of multivariate data. They include three-dimensional rotating scatterplots and three-dimensional surface plots with rotation, which allow the user to investigate more than three dimensions of the data at a time. These techniques use the fact that complex interrelationships often show up best graphically, especially when the data are viewed in more than three dimensions.

Geographic Information Systems

Different disciplines may use different standard sampling regimes as part of the same monitoring effort. Using different sampling protocols may require viewing the data using a variety of scales and resolutions; thus, creating the need for systems which can store and present data from different scales. Geographic Information Systems (GIS) have been devised to do just this. A GIS uses a hierarchical structure to link together different types of data gathered at different scales and pertaining to different shapes. For example, combining data from streams and lakes means having to combine data from shapes that are, respectively, linear and areal in nature. Although not a panacea, GISs do offer a meaningful, information-rich way of data summarization and presentation. As an example, Keleher and Rahel (1996) used a GIS to assess the potential impact of global warming on salmonid populations. The authors used data on fish distributions and maximum summer air temperatures to identify upper thermal limits for salmonid fishes in Wyoming streams. Following regression modeling relating air temperature to latitude and elevation, the GIS translates such information into maps of current and future potential distributions of cold water trout with increasing temperatures, at the same time computing potential habitat loss.

A GIS database is essentially a collection of layers of information, such that a particular layer occurs at a given scale. The notion of scale may be thought of as the smallest distance over which change can be measured. To create maps and images at various scales and to be able to link different scales together, a complex set of aggregation rules exists based on a hierarchical structure, with an index system for discrete features. A GIS database, like any other database, is not immune to errors. Misclassification errors may occur in classifying various units, as a result of either observer or instrument error. Digitizing errors may occur in creation of the database. Nevertheless, increasing use of Geographic Information Systems, and the scale linkages they provide, is likely to result in better inferential connections at the various levels of a watershed.

In regard to statistical aspects of monitoring, investigators need to pay careful attention to the creation and implementation of the study design. Although incorrect data analyses or graphs from a good design can always be redone, no amount of complicated statistical techniques can salvage the data from an ill-conceived design or one that is not well implemented. Careful thinking about study objectives, and ensuring that the design (including the sampling approach and choice of responses to measure) meets those objectives are among the best steps investigators can take to achieve successful monitoring outcomes.

Literature Cited

Afifi, A.A., and V. Clarke. 1984. Computer-aided multivariate analysis. Lifetime Learning, Wadsworth, Belmont, California, USA.

Ausmus, B.S. 1984. An argument for ecosystem level monitoring. Environmental Monitoring and Assessment **4**:275–293.

Bailey, R.G. 1984. Testing an ecosystem regionalization. Environmental Mangement **19**:239–248.

Bormann, L.H., and G.E. Likens. 1979. Pattern and process of a forested ecosystem. Springer-Verlag, New York, New York, USA.

Bott, T.L., J.T. Brock, C.E. Cushing, S.V. Gregory, D. King, and R.C. Petersen. 1978. A comparison of methods for measuring primary productivity and community respiration in streams. Hydrobiologia **60**:3–12.

Bryant, M.D. 1995. Pulsed monitoring for watershed and stream restoration. Fisheries **20**:6–13.

Clarke, K.R. 1993. Nonparametric multivariate analyses of changes in community structure. Australian Journal of Ecology **18**:117–143.

Conquest, L.L. 1993. Statistical approaches to environmental monitoring: Did we teach the wrong things? Environmental Monitoring and Assessment **26**:107–124.

Conquest, L.L., R.L. Burr, R.F. Donnelly, J.B. Chavarria, and V.F. Gallucci. 1996. Sampling methods for stock asessment for small-scale fisheries in developing countries. Pages 179–225 *in* V.F. Gallucci, S.B. Saila, D.J. Gustafson, and B.J. Rothschild, eds. Stock assessment: Quantitative methods and applications for small-scale fisheries. CRC, Boca Raton, Florida, USA.

Conquest, L.L., T.P. Cardoso, K.D. Seidel, and S.C. Ralph. 1991. Using visual estimation in a watershed monitoring study. Page C1–3 *in* S. Ralph, T. Cardoso, L. Conquest, R. Naiman, eds. Status and trends of instream habitat in forested lands of Washington: The Timber-Fish-Wildlife ambient monitoring project, 1989–1991 biennial progress report. TFW-AM9-91-002. Timber, Fish, & Wildlife, Olympia, Washington, USA.

Conquest, L.L., S.C. Ralph, and R.J. Naiman. 1994. Implementation of large-scale stream monitoring efforts: Sampling design and data analysis issues. Pages 69–90 *in* S. Loeb, ed. Biological monitoring of freshwater ecosystems. Lewis, Chelsea, Michigan, USA.

Cupp, C.E. 1989. Stream corridor classification for forested lands of Washington. Report for Washington Forest Protection Association, Olympia, Washington, USA.

Fore, L., J. Karr, and L. Conquest. 1994. Statistical properties of an index of biological integrity used to evaluate water resources. Canadian Journal of Fisheries & Aquatic Sciences **51**:1077–1087.

Frissell, C.A. 1992. Cumulative effects of land use on salmon habitat in S.W. Oregon coastal streams. Ph.D. dissertation. Oregon State University, Corvallis, Oregon, USA.

Frissell, C.A., W.J. Liss, C.E. Warren, and M.D. Hurley. 1986. A hierarchical framework for stream habitat classification: Viewing streams in a watershed context. Environmental Management **10**:199–214.

Fyles, T.M., P.R. West, and B.A. King. 1995. A meta-data approach to information management in the Georgia basin. Puget Sound Research '95 Proceedings. Puget Sound Water Authority, Olympia, Washington, USA.

Gilbert, R.O. 1987. Statistical methods for environmental pollution monitoring. Van Nostrand Reinhold, New York, New York, USA.

Goff, G.P. 1986. Reproductive success of male smallmouth bass in Long Point Bay, Lake Erie. Transactions of the American Fisheries Society **115**:415–423.

Green, R.H. 1979. Sampling design and statistical methods for environmental biologists. John Wiley & Sons, New York, New York, USA.

——. 1989. Power analysis and practical strategies for environmental monitoring. Environmental Research **50**:195–205.

Hankin, D.G. 1984. Multistage sampling designs in fisheries research: Applications in small streams. Canadian Journal of Fisheries & Aquatic Sciences **41**:1575–1591.

Hankin, D.G., and G.H. Reeves. 1988. Estimating total fish abundance and total habitat area in small streams based on visual estimation methods. Canadian Journal of Fisheries & Aquatic Sciences **45**:834–844.

Harrelson, C.C., C.L. Rawlins, and J.P. Potyondy. 1994. Stream channel reference sites: An illustrated guide to field techniques. USDA Forest Service General Technical Report GTR-RM-245. Rocky Mountain Forest and Range Experiment Station, Fort Collins, Colorado, USA.

Hartman, G.G., and J.C. Scrivener. 1990. Impacts of forestry practices on a coastal stream ecosystem, Carnation Creek, British Columbia. Canadian Bulletin of Fisheries & Aquatic Sciences **223**:148.

Huang, S.L., and J.J. Ferng. 1990a. Applied land classification for surface water quality management: I. Watershed classification. Journal of Environmental Management **31**:107–126.

Huang, S.L., and J.J. Ferng. 1990b. Applied land classification for surface water quality management: II. Land process classification. Journal of Environmental Management **31**:127–141.

Hughes, R.M. 1985. Use of watershed characteristics to select control streams for estimating effects of metal mining wastes on extensively disturbed streams. Environmental Management **9**:253–262.

Hughes, R.M., and D.P. Larsen. 1988. Ecoregions: An approach to surface water protection. Journal of the Water Pollution Control Federation **60**:486–493.

Hughes, R.M., D.P. Larsen, and J.M. Omernik. 1986. Regional reference sites: A method for assessing stream potentials. Environmental Management **10**:629–635.

Hurlbert, S.H. 1984. Pseudoreplication and the design of ecological field experiments. Ecological Monographs **54**(2):187–211.

Karr, J.R. 1981. Assessment of biotic integrity using fish communities. Fisheries **6**:21–27.

——. 1991. Biological integrity: A long-neglected aspect of water resource management. Ecological Applications **1**:66–84.

Karr, J.R., K.D. Fausch, P.L. Angermeier, P.R. Yant, and I.J. Schlosser. 1986. Assessing biological integrity in running waters: A method and its rationale. Special Publication Number 5, Illinois Natural History Survey, Champaign, Illinois, USA.

Keleher, C.J., and F.J. Rahel. 1996. Thermal limits to salmonid distributions in the Rocky Mountain region and potential habitat loss due to global warming: A geographic information system (GIS) approach. Transactions of the American Fisheries Society **125**:1–13.

Likens, G.E., and F.H. Bormann. 1995. Biogeochemistry of a forested ecosystem, 2nd edition. Springer-Verlag, New York, New York, USA.

Lotspeich, F., and W. Platts. 1982. An integrated land-aquatic classification system. American Journal of Fisheries Management **2**:138–149.

Luchetti, G.K., K. Thornburgh, and C. Rawson. 1987. A water quality monitoring program for major agricultural areas of the Snohomish River basin. Report for Tulalip Fisheries Department Water Quality Laboratory, Marysville, Washington, USA.

MacDonald, L.H., and A.W. Smart. 1993. Beyond the guidelines: Practical lessons for monitoring. Environmental Monitoring and Assessment **26**:203–218.

MacDonald, L.H., A.W. Smart, and R.C. Wissmar. 1991. Monitoring guidelines to evaluate effects of forestry activities on streams in the Pacific Northwest and Alaska. EPA 910/9-91-001. US Environmental Protection Agency Region 10, Seattle, Washington, USA.

Madej, M.A. 1978. Response of a stream channel to an increase in sediment load. Master's thesis. Department of Geology, University of Washington, Seattle, Washington, USA.

Meador, M.R., T.F. Cuffney, and M.E. Gurtz. 1993. Methods for sampling fish communities as part of the National Water-Quality Assessment Program. United States Geological Survey Open-file Report 93–104. United States Geological Survey Water Resources Division, Raleigh, North Carolina, USA.

Millard, S.P., J.R. Yearsley, and D.P. Lettenmaier. 1985. Space–time correlation and its effects on methods for detecting aquatic ecological change. Canadian Journal of Fisheries and Aquatic Sciences **42**:1391–1400.

Montgomery, D.R., and J.M. Buffington. 1993. Channel classification, prediction of channel response and assessment of channel condition. Report #TFW-SH10-93-002 for the SHAMW committee of the Washington State Timber/Fish/Wildlife Agreement. University of Washington, Seattle, Washington, USA.

Naiman, R.J., D.G. Lonzarich, T.J. Beechie, and S.C. Ralph. 1992. General principles of classification and the assessment of conservation potential in rivers. Pages 93–124 *in* P.J. Boon, P. Carlo, and G.E. Petts, eds. River conservation and management. John Wiley & Sons, Chichester, UK.

Norris, R.H., E.P. McElravy, and V.H. Resh. 1994. The sampling problem. Pages 282–306 *in* P. Calow and G.E. Petts, eds. The rivers handbook: Hydrological and ecological principles. Blackwell Scientific, London, UK.

Olson-Rutz, K.M., and C.B. Marlow. 1992. Analysis and interpretation of stream channel cross-sectional data. North American Journal of Fisheries Management **12**:55–61.

Omernik, J.M. 1987. Ecoregions of the coterminous United States. Annals of the Association of American Geographers **77**:118–125.

Omernik, J.M., and A.L. Gallant. 1986. Ecoregions of the Pacific Northwest (with map). EPA/600/3-96/033. United States Environmental Protection Agency, Environmental Research Laboratory, Corvallis, Oregon, USA.

Platts, W.S., C. Armour, G.D. Booth, M. Bryant, J.L. Bufford, P. Cuplin, et al. 1987. Methods for evaluating riparian habitats with applications to management. USDA Forest Service General Technical Report INT-211. USDA Forest Service, Ogden, Utah, USA.

Platts, W.S., W.F. Megahan, and G.W. Minshall. 1983. Methods for evaluating stream, riparian, and biotic conditions. USDA Forest Service General Technical Report INT-183. USDA Forest Service, Ogden, Utah, USA.

Porter, M.L. 1993. Exploratory data analysis uncovers unexpected relationships. Personal Engineering and Instrumentation 21–28.

Potvin, C., and D.A. Roff. 1993. Distribution-free and robust statistical methods: Viable alternatives to parametric statistics? Ecology **74**:1617–1628.

Ralph, S.T., G.P. Cardoso, L. Conquest, and R. Naiman. 1991. Status and trends of instream habitat in forested lands of Washington: The Timber-Fish-Wildlife ambient monitoring project, 1989–1991 biennial progress report. TFW-AM9-91-002. Timber, Fish, & Wildlife, Olympia, Washington, USA.

Ralph, S.T., G.C. Poole, L.L. Conquest, and R.J. Naiman. 1994. Stream channel morphology and woody debris in logged and unlogged basins of western Washington. Canadian Journal of Fisheries & Aquatic Sciences **51**:37–51.

Richardson, J.S., and M.C. Healey. 1994. Autocorrelations in ecosystems across space, time and disciplines: Case of the Fraser River [abstract]. *In* Sampling Design for Aquatic Network Systems workshop, 1995 September 11–14, Timberline, Oregon, USA.

Riemann, B.E., and J.D. McIntyre. 1996. Spatial and temporal variability in bull trout redd counts. North American Journal of Fisheries Management **16**:132–141.

Rosgen, D.L. 1994. A classification of natural rivers. Catena **22**:169–199.

——. 1985. A stream classification. *In* R.R. Johnson et al., eds. Riparian ecosystems and their management: Reconciling conflicting uses. First North American Riparian Conference, Tucson, Arizona, USA.

Shampine, W.J. 1993. Quality assurance and quality control in monitoring programs. Environmental Monitoring and Assessment **26**:143–151.

Skalski, J. 1990. A design for long-term status and trends. Journal of Environmental Management **30**:139–144.

Sokal, R.R., and F.J. Rohlf. 1995. Biometry. W.H. Freeman, New York, New York, USA.

Stafford, S.G. 1985. A statistics primer for foresters. Journal of Forestry **83**:148–157.

——. 1993. Data, data everywhere but not a byte to read: Managing monitoring information. Environmental Monitoring and Assessment **26**:125–141.

Stafford, S.G., P.B. Alaback, G.J. Koerper, and M.W. Klopsch. 1984. Creation of a forest data bank. Journal of Forestry **82**:432–433.

Stafford, S.G., G. Spycher, and M.W. Klopsch. 1988. Evolution of the forest science data bank. Journal of Forestry **86**:50–51.

Steel, R.G., and J.H. Torrie. 1980. Principles and procedures of statistics: A biometrical approach. McGraw-Hill, New York, New York, USA.

Stewart-Oaten, A., J.R. Bence, and C.W. Osenberg. 1992. Assessing effects of unreplicated perturbations: No simple solutions. Ecology **73**:1396–1404.

Tukey, J.W. 1977. Exploratory data analysis. Addison-Wesley, Reading, Massachusetts, USA.

Underwood, A.J. 1994. Spatial and temporal problems with monitoring. Pages 101–123 *in* P. Calow and G.E. Petts, eds. The rivers handbook, hydrological and ecological principles. Volume 2. Oxford, Blackwell Scientific, London, UK.

Urquhart, N.S. 1994. Anatomy of sampling studies of ecological responses through time [abstract]. *In* Sampling Design for Aquatic Network Systems workshop; 1995 September 11–14, Timberline, Oregon, USA.

Urquhart, N.S., W.S. Overton, and D.S. Birkes. 1993. Comparing sampling designs for monitoring ecological status and trends: Impact of temporal patterns. Pages 71–86 *in* V. Barrett and K.F. Turkman, eds. Statistics for the environment. John Wiley & Sons, New York, New York, USA.

Weisberg, S.B., J.B. Frithsen, A.F. Holland, J.F. Paul, K.J. Scott, J.K. Summers, et al. 1992. EMAP-Estuaries Virginian Province 1990 demonstration project report EPA 600/R-92/100. United States Environmental Protection Agency, Environmental Research Laboratory, Narragansett, Rhode Island, USA.

Wissmar, R.C. 1993. The need for long-term stream monitoring programs in forest ecosystems of the Pacific Northwest. Environmental Monitoring & Assessment **26**:219–234.

Zar, J.H. 1996. Biostatistical analysis, 3rd edition. Prentice-Hall, Englewood Cliffs, New Jersey, USA.

Zimmerman, G.M., H. Goetz, and P.W. Mielke, Jr. 1985. Use of an improved statiscal method for group comparisons to study effects of prairie fire. Ecology **66**:606–611.

19
Cumulative Watershed Effects and Watershed Analysis

Leslie M. Reid

Overview

• Cumulative watershed effects are environmental changes that are affected by more than one land-use activity and that are influenced by processes involving the generation or transport of water. Almost all environmental changes are cumulative effects, and almost all land-use activities contribute to cumulative effects.

• An understanding of cumulative watershed effects is necessary if land-use activities and restoration projects are to be designed that accomplish their intended objectives. Cumulative effects first must be evaluated to decide what actions are appropriate. The likely direct and indirect effects of the planned actions must then be assessed.

• Technical issues that complicate analysis of cumulative effects include the large spatial and temporal scales involved, the wide variety of processes and interactions that influence cumulative effects, and the lengthy lag-times that often separate a land-use activity and the landscape's response to that activity.

• Analysis strategies contain implicit assumptions about the role of humans on the landscape, the limits of responsibility, and how the natural world functions. Controversy over methods often revolves around philosophical differences concerning these assumptions.

• Cumulative effects analysis requires a nontraditional approach to information: patterns are usually more important than details; an interdisciplinary focus is more useful than multiple monodisciplinary foci; and a large area is more than the sum of its parts, so it must be evaluated as a unit.

• Ad hoc methods for evaluating cumulative effects are well developed and have been widely used for nearly a century. An ad hoc method is designed to address a particular kind of problem in a particular place and often cannot be applied elsewhere without modification.

• Standardized analysis methods were developed in the 1970s and 1980s to fulfill requirements of the National Environmental Policy Act, but most of these methods lack technical credibility or are limited in the kinds of problems they can address or the areas in which they can be applied. Examples include use of index values, mechanistic models, and checklists for applying expert judgment.

• More recently, methods of watershed analysis were developed to provide the background information needed for evaluating cumulative effects.

• The watershed analysis method employed in Washington state uses an understanding of past environmental changes to develop prescriptions for land-use practices, but it does not assess the likely cumulative effects of future activities.

• The ecosystem analysis method used on federal lands in much of the Pacific Northwest provides background information about ecosystem and landscape interactions that can be used for later cumulative effects assessments when projects are being planned.

Introduction

In 1969 the U.S. Congress formally recognized that even if each land-use project is allowed to produce only a small environmental impact, enough small impacts can accumulate to have a large effect. This realization took the form of a requirement of the National Environmental Policy Act (NEPA) that the cumulative impacts of proposed actions be evaluated. The Council on Environmental Quality (CEQ) defined a *cumulative impact* (or *cumulative effect*) as:

> "the impact on the environment which results from the incremental impact of the action when added to other past, present, and reasonably foreseeable future actions regardless of what agency (federal or non-federal) or person undertakes such other actions. Cumulative impacts can result from individually minor but collectively significant actions taking place over a period of time." (CEQ Guidelines, 40CFR 1508.7, issued 23 April 1971)

In practice, virtually every environmental impact is influenced by multiple land-use activities and is, therefore, a cumulative impact. Similarly, almost any change caused by a land-use activity contributes to a cumulative impact because it affects something that is also affected by other land-use changes. A cumulative impact thus is not a new type of impact, but is a new way of looking at the impacts that people have always confronted.

The CEQ definition is important because it implies that the severity of an impact must be judged from the point of view of the impacted party. Before this, the potential impact of a project could be evaluated in isolation. If a project was not going to produce very much sediment, for example, it probably was not going to have much of an impact. But now the context for the project would need to be examined. Even if a project did not itself produce enough sediment to fill in a reservoir, the incremental effect of the project could be significant if previous activities had already imperiled the reservoir. Or if the project was only one of a series of such projects, then its incremental addition could increase the severity of the projects' combined impact. In short, if the reservoir was filling in, then an impact was occurring, and any contribution to that impact would increase the severity of the impact.

A *cumulative watershed effect* is a special type of cumulative effect that is influenced by processes that involve the generation or transport of water. Clogging of spawning gravels by sediment from eroding road surfaces is considered a cumulative watershed impact, as is decreased woody debris in streams caused by upstream changes in forest composition. In each case, the change is influenced by water flowing through a watershed.

Cumulative effects are often perceived as a hazy concept that is important only because the law requires that they be analyzed before land-use plans are accepted. However, a basic understanding of cumulative effects is useful for many applications and is particularly necessary for designing projects or land-use strategies that are sustainable through time.

As an example, consider the problems described in Box 19.1. The first of these represents a relatively low-budget restoration effort on a local scale, while the second involves the design of a multimillion dollar program to be administered over a large region. Despite the difference in spatial and economic scales, the same kinds of information are needed to address both of these problems:

a. What areas are important for fish, and why?
b. Where has habitat been impaired?
c. What aspects of the habitat have changed?
d. What caused those changes?
e. What is the relative importance of the various habitat changes to fish?
f. What is the present trend of changes in the system?
g. Which changes are reversible?
h. What is the expected effectiveness of potential remedies?
i. What are the effects of those remedies on other land uses and ecosystem components?
j. What are the relative costs of the potential remedies over the long term?

Answers to these questions will provide most of the information needed for a decision about what actions in what areas will have the biggest return for the least effort. In both cases it is necessary to know what conditions were like in

Box 19.1. What Would You Do?

Problem 1: You are the biologist responsible for deciding how to spend the $5,000 raised to improve anadromous fish habitat in the 250-km^2 Wilrick Creek system. What do you need to know to make the best decisions? List the types of information you would need.

Problem 2: You are the biologist responsible for deciding how to spend $2,200,000 earmarked by your agency for enhancing anadromous fish habitat in western Oregon. What do you need to know to make the best decisions? Again, list the information needed.

the past (*a*, *b*, *c*), how they have changed (*b*, *c*, *d*, *e*), and what they will be like in the future (*f*, *g*, *h*, *i*, *j*). These are the same questions that must be answered to evaluate cumulative effects. For both of the problems represented by Box 19.1, cumulative watershed effects need to be evaluated twice: once to determine what needs to be done and once to identify the effects of a proposed solution.

Now consider the projects described in Box 19.2. In each of these cases, project outcomes did not meet project objectives—something went wrong. These examples are typical of the kinds of misjudgments that result when people fail to look beyond the symptoms of a problem to evaluate its broader context. In Case 1, the resource specialists knew that they had a big problem—fish were dying—and they had a good supply of dynamite on hand for solving fish-habitat problems. They simply were not aware that the waterfall had its own constituency of interest groups; they had been trapped by monodisciplinary assumptions and failed to consider the broader context for their

Box 19.2. Less-Than-Perfect Projects

Case 1: Fisheries biologists were about to dynamite a small cascade on a popular recreation river to decrease fishing pressure on migrating steelhead: fish would hold for awhile in the plungepool, thus increasing their vulnerability to anglers. The popular tourist vista and major draw for whitewater recreationists was spared because kayakers heard of the plans and protested.

Case 2: Thousands of dollars were spent to build structures on a floodplain to promote deposition and thus initiate revegetation of the floodplain. Because the structures were built in the path of a migrating meander, they were washed away during the first flood.

Case 3: Tens of thousands of dollars were spent to plant young alders across the face of an immense landslide to protect downstream salmon habitat. The slide failed again during the next storm, removing the alders.

Case 4: Hundreds of thousands of dollars have gone into carrying out detailed habitat inventories of stream channels in an area. After nearly a decade of inventory work, the first examination of the accumulating data showed that most of the variables being measured were not correlated with habitat use.

Case 5: Millions of dollars were spent to obliterate roads in the lower third of a watershed to prevent further aggradation of the channel. However, most of the sediment contributing to aggradation comes from upstream of the treated area.

proposed solution. Question *i* of the list (*What effect will those remedies have on other land uses and ecosystem components?*) had been overlooked.

Case 2 is more sophisticated. The vegetation experts saw the link between the physical process of deposition and their goal of revegetation, but their analysis was limited to conditions at the site of interest. They did not consider the broader spatial or temporal context for those conditions and so did not recognize the inevitability of the meander's migration. Questions *f* and *h* (*What is the present trend of changes? What is the expected effectiveness of remedies?*) had been ignored.

Case 3 introduces an additional facet of the context problem. Here, an ineffectual solution to the landslide problem was instituted, ignoring questions *f*, *g*, and *h* (*What is the present trend of changes? Which changes are reversible? What is the expected effectiveness of remedies?*). In addition, sediment was assumed to be a big problem because the landslide scar was large and visible. The actual importance of this sediment source, relative to others in the system, was not addressed before the solution was implemented. Question *e* (*What is the relative importance of the various changes?*) had been neglected.

The inventory in Case 4 was an attempt to provide information that specialists would need to restore fish habitat. Unfortunately, design of a reasonable inventory was impossible under the circumstances. Not enough was known about what variables were important in the area (questions *a* and *e*: *What areas are important for fish, and why? What is the relative importance of the various habitat changes to fish?*) and geomorphologists had not been consulted about what patterns and scales of variability to expect (questions *c*, *d*, and *f*: *What aspects of the habitat have changed? What caused those changes? What is the present trend of changes in the system?*). The mistake was not recognized sooner because the success of the project was measured by the length of channel inventoried rather than by the knowledge gained, so there was little motivation to turn numbers into knowledge. No one had been given the responsibility for looking at the results.

In Case 5, evaluation of the causes for change (question *d*) would have revealed that much of the source for the aggradation problem was upstream, and evaluation of question *h* (*What is the expected effectiveness of remedies?*) would have shown that the proposed solution would have little effect on the problem.

Each of these five efforts suffered from inattention to the context of the identified problem. Context also has been overlooked where restoration projects are designed to enhance summer habitat but the major constraint to fish survival is degraded winter habitat (Nickelson et al. 1992), where habitat improvement structures are built that cannot survive typical winter flows (Frissell and Nawa 1992), and when limited restoration money is spent to make small improvements in a few large sediment sources rather than to ensure that a thousand small sediment sources do not become large sources. These examples all involve aquatic habitat restoration, but parallels can be found in any aspect of wildland resource management.

Each of the projects described in Box 19.2 was designed to redress a cumulative watershed impact, but none of the projects incorporated an adequate analysis of the problem. Had these problems been evaluated to understand how the original impacts occurred and to identify the watershed-scale context for the impacts, most of these failures could have been avoided. Thus, an understanding of cumulative effects is important not just because it is an administrative requirement, but because it is essential for designing successful projects.

Problems in the Evaluation of Cumulative Watershed Effects

The concept behind cumulative watershed effects is a simple one: environmental impacts are influenced by multiple factors. However, this simple concept makes the evaluation of potential land-use impacts a difficult task. Not only is the problem technically difficult, but it is further complicated by lack of a common understanding of philosophical aspects of the problem and by analytical constraints imposed by societal value systems.

Technical Issues

Much of the difficulty of cumulative effects analysis arises from the large number of diverse biological and physical processes that influence most land-use impacts. Interactions between processes make it possible for impacts to accumulate through both space and time, they ensure that no single field of expertise can adequately address the cumulative effects problem, and they introduce time lags in the expression of impacts.

Consider the example described in Sidebar 19.3. In this case, increased stream temperature was identified as a cumulative watershed impact in the Pilot Creek watershed of northwestern California (USDA Forest Service 1994). Changes in riparian vegetation, channel form, and hydrologic regime are all suspected of contributing to the impact, but the relative importance of these changes is unknown.

Land-use activities influencing the temperature regime in Pilot Creek include logging, roadbuilding, and fire control. These activities have occurred for a long time in many parts of the watershed and have contributed to environmental changes that accumulated through time. Stream temperatures probably changed as vegetation began to change in the 1850s, and further temperature changes accompanied riparian logging a hundred years later and

Box 19.3. Causes for Increased Temperatures in Pilot Creek

Pilot Creek drains an 80-km^2 watershed in the Coast Ranges of northwestern California. The stream has provided good trout fishing in the past, but recent surveys suggest that summer water temperatures along the main channel are higher than they would be under natural conditions. Evaluation of past and present conditions in the watershed disclosed several kinds of changes likely to have influenced water temperatures and identified potential causes for those changes (USDA Forest Service 1994):

1. Riparian forest cover has decreased (increasing solar heating)
   - Debris flows originating in logged areas destroyed vegetation.
   - Aggrading and widening channels destroyed vegetation.
   - The 1964 flood destroyed vegetation.
   - Logging along the main channel removed the original forest cover.
   - 7% of the watershed burned in 1987.
   - A possible decrease in summer flows may have stressed riparian vegetation.
2. The channel has aggraded and widened (making more water surface available to be heated)
   - Debris flows contributed sediment.
   - The 1964 storm produced sediment.
   - Long-term sources such as road surfaces and earthflows contribute sediment.
   - Reduced root cohesion in logged riparian areas accelerated bank erosion.
3. Summer flows may be lower than in the past (allowing the entire water column to warm more quickly)
   - Recently there have been more than 5 years of drought.
   - Reduction of fire frequency (since the end of burning by Native Americans in about 1850) has allowed conifers to encroach on grasslands, increasing summer evapotranspiration.
   - Aggradation in the channel forces part of the flow underground.
   - Broader channels increase rates of evaporation.
   - Decreased riparian cover increases evaporation.

upslope logging even more recently. But effects also accumulated through space. Upslope logging, the vegetation change, and logging of downstream riparian zones occurred in different parts of the watershed, yet they all affected temperatures in the same downstream channels.

Not only were several different activities involved, but multiple mechanisms for change were important: temperatures were affected by changes in hydrology, sedimentation, and vegetation. In addition, each of these mechanisms influenced the others. A change in erosion rate, for example, led to channel aggradation, which decreased surface flows by creating a porous substrate. Aggradation also contributed to the destruction of riparian vegetation. Because of the variety of these interacting changes, no single expert was capable of evaluating the issue. It took a fisheries biologist to recognize that temperature changes were important in the watershed; an anthropologist to identify long-term changes in fish abundance; an archeologist to recognize that vegetation had changed; a plant ecologist and a soil scientist to identify the extent of the vegetation change; and a geologist to evaluate changes in erosion rates and channel form.

Even with progressive changes occurring on the hillslopes, Pilot Creek did not change much until the floods of 1955 and 1964. The cumulative impacts of a hundred years of changing land use became visible only after two major storms had occurred. The storms would have happened in any case, but the changes wrought by the storms under disturbed conditions were very different from those that would have appeared under natural conditions. Indeed, large storms of the late 1800s, although similar in character, resulted in much less dramatic changes in the region (Harden 1995). A recent drought provides a similar example. Channel widening had reduced the water depth in the Mad River downstream of Pilot Creek, but normal water years still provided deep enough flow for chinook salmon (*Oncorhynchus tshawytscha*) to migrate upstream. During the drought, however, fish were blocked by shallow reaches (Ken Gallagher, Mad River Hatchery, Arcata, California, personal communication). The original channel form would have provided sufficient flow depth to allow passage even during droughts. Altered conditions that were tolerable during normal years thus were intolerable during the drought, and it became clear that conditions must be maintained at a level that provides a margin of safety even under the most severe stresses.

Time lags occur in the expression of cumulative watershed effects even when a large-magnitude event is not needed to disclose those effects. For example, logging-related landslides might not occur until several years after logging, when roots are sufficiently rotted to destabilize the slope (Sidle 1985). Even then, it takes a long time for gravel to be transported down a channel, so sediment from the slides might not accumulate at sites downstream until decades later (Madej and Ozaki 1996).

These characteristics of cumulative watershed effects mean that analysts must evaluate much larger spatial and temporal scales than they have been accustomed to in the past. Impact analysis must take into account the influence of rare events, which are difficult to observe and may not even have occurred during the period of record for an area. Analysis also must be interdisciplinary. Each of these requirements represents an excursion into the least-understood aspects of the natural sciences.

Philosophical Issues

Three other problems confronting cumulative effects analysts are more philosophical in nature: the standard of comparison appropriate for assessing the importance of an environmental change is rarely evident; there is no inherent limit to the distance downstream over which impacts might occur; and there can be no generalizable measure of impact severity. These problems touch upon deeply held beliefs about the role of humans on the landscape, the limits of responsibility, and how the natural world functions.

Cumulative impact assessments are intended to evaluate environmental change, but change can only be recognized and measured in relation to an unchanged condition. Thus, the "natural" condition of a watershed often needs to be defined. Non-Native Americans usually consider the conditions that met the European explorers to be "natural." However, a sophisticated program of land management that included the extensive use of fire predated European exploration (Lewis 1993). In the case of Pilot Creek, the pre-Euro–American vegetation and hydrologic regimes (and their resulting influences on aquatic habitat) reflected centuries of intentional burning by Native Americans. Thus, "natural" is an ambiguous term—does it include some land-use effects but not others? Throughout this paper the term "natural" refers to the conditions under which the native flora and fauna were evolving at the time of European contact.

The need to identify natural conditions is particularly strong when a cumulative effects analysis is used to identify goals for restoration projects or for sustainable wildland management. Such goals are often designed using the concept of "natural range of variability" (Fullmer 1994). With this approach, the range of conditions that occurred naturally (the maximum and minimum channel widths, for example) are adopted as the bounds for acceptable conditions. The underlying idea is attractive: try to make future conditions look like conditions of the past. However, in most areas past conditions are not well enough known to define maximum and minimum values for most variables. Further, a system can become incapable of supporting its natural ecosystem even when conditions remain within the specified bounds. The Mississippi River flood of 1993 was well within the natural range of variability for that system, for example, but a yearly recurrence of such flows would create riparian and aquatic ecosystems very different from those encountered by European explorers. Thus, what is required as a goal is not simply that conditions remain within a tolerable range, but that the system reassume the temporal and spatial distribution of conditions that originally sustained it (Bisson et al. 1997, Frissell et al. 1997). A distribution, of course, is even more difficult to define than a range.

In other cases, restoration and wildland management goals are defined according to the perceived needs of target species, such as anadromous fish. For example, riparian management objectives defined for federal lands in the Pacific Northwest east of the Cascades (USDA and USDI 1994a) comprise a list of acceptable threshold values for six physical channel variables, including pool frequency and width-to-depth ratio. However, real streams in natural settings do not adhere to averages, and reaches that are themselves inhospitable to salmonids may contribute to the maintenance of salmonid populations downstream (G. Reeves, USDA Forest Service, Corvallis, Oregon, personal communication). Similarly, landslides that seem to devastate channels over the short term may be the mechanism by which long-term habitat quality is maintained (Reeves et al. 1995).

Furthermore, adoption of channel design specifications for the benefit of a desired species may harm other components of the aquatic ecosystem. Although silty streambeds are considered detrimental to salmonids, Pacific lamprey (*Lampetra tridentata*) require fine-grained substrates for rearing (Moyle 1976). Similarly, modification of channels to suit the presumed needs of steelhead (*Oncorhynchus mykiss*) can reduce habitat quality for the foothill yellow-legged frog (*Rana boylii*; Fuller and Lind 1992). If all streams were "restored" to conform to the USDA and USDI (1994a) specifications, the biological integrity of the overall aquatic ecosystem thus would be compromised. In addition, such targets rarely reflect the range of conditions actually provided by natural habitats. For example, the USDA and USDI (1994a) riparian management objectives specify that width-to-depth ratios of less than 10 are desired for all channels, even though these values are not characteristic of many natural channels important to salmonids, as demonstrated by Rosgen's generalized descriptions of stream types (Rosgen 1994).

Recently, however, restoration and management goals have occasionally been defined using a more realistic and tractable approach. If land is managed to re-create the distribution of processes (such as landsliding or treefall) that was present naturally, then the processes themselves will reestablish the temporal and spatial array of natural conditions (such as turbidity or woody-debris loads). Managing for processes is easier and likely to be more successful than managing for conditions because processes are more directly influenced by land-use activities than are the conditions which those processes control. In other words, if reestablishment of natural channel conditions is desired, then the system can be managed to ensure that the processes that affect channel conditions—the production and transport of water, sediment, and organic material—are not altered by management activities. This is the rationale used in the Northwest Forest Plan for the establishment of buffer zones along stream channels (FEMAT 1993, USDA and USDI 1994b). Working against this approach is the tendency to redefine goals to make oversight easier: it is hard to tell from inspections if processes are being maintained appropriately, while it is easy to determine if the width-to-depth ratio is less than 10.

The second philosophical problem is that the potential for accumulating impacts does not end at the mouth of a watershed. In the case of the Pilot Creek watershed, few people have heard of the place, and environmental changes there would go largely unnoticed. However, the intakes that supply Mad River water to most Humboldt County residents are located 50 km downstream from the mouth of Pilot Creek. What happens in Pilot Creek takes on regional importance because of its downstream effect on 80,000 people. The fact that the perceived significance of a change depends on its context implies that there is no inherent limit to the distance downstream over which potential impacts must be considered. Does this mean that an impact analysis for a 40-ha timber sale on Mount Shasta should evaluate its potential effects on shrimp in San Francisco Bay, 400 km downstream? The San Francisco Bay shrimp industry might still exist if such broad-scale connections had been considered in the past.

The third philosophical problem is that there can be no generally applicable measure of impact severity. Temperatures in Pilot Creek have probably increased, and anadromous salmonids are stressed by high temperatures (Bjornn and Reiser 1991). Therefore, according to traditional reasoning, increased temperatures in Pilot Creek are bad. But did the temperature changes in Pilot Creek actually have a negative effect on salmonid populations? Or were they small enough to be irrelevant? Or did they increase primary productivity to a point that benefited the overall system? Or did they merely compensate for earlier cooling due to reduced area of grasslands? The importance of an environmental change can be interpreted only by examining the influences of that change in the particular context in which it occurred. The same magnitude of change may prove beneficial in one setting and harmful in another.

These philosophical issues strongly influence the strategies used for cumulative effects analysis. Any framework for analysis must incorporate approaches for defining reference conditions, identifying the area over which impacts may be relevant, and evaluating the actual importance of changes. Once the approaches are selected, they become fundamental building blocks of the analysis method and cannot be altered without modifying the entire analysis strategy. But because selection of these approaches is based on the formulator's world view, these decisions often become magnets for controversy among those with differing world views. Unlike technical problems, philosophical problems cannot be resolved by amassing facts; they must be addressed by reconciling or selecting among different people's views of the world.

Sociocultural Issues

Efforts to analyze cumulative watershed effects are also complicated by the expectations and traditions of the Euro-American sociocultural system. Because of the extraordinary variety of processes and interactions that can influence an environmental impact, evaluation of cumulative watershed effects requires a different analytical approach than that ordinarily taught to specialists. The education of a specialist in Western society is usually a rigorous journey toward an increased knowledge of detail in an increasingly restricted field of study. Specialists tend to focus on cataloging details rather than on integrating them into a broader understanding of large systems, and they often assume that it is someone else's responsibility to do the integration once the "specialized" work has been completed. Understanding cumulative watershed effects requires an entirely different approach.

First, the cumulative effects analyst must place higher importance on the understanding of general patterns than on the collection of precise data. At the scale of 500-km^2 watersheds, collection of detailed information is often counterproductive. For example, the anomalous relationship between soil and vegetation maps for the Pilot Creek watershed was understood only after distracting details were removed (Figure 19.1): the halos of forested grassland soils around the shrinking remnants of grassland record the earlier extent of the grasslands.

Second, the analyst must strive for understanding of interactions between components of the environment rather than for detailed understanding of isolated components. Cumulative effects result from interactions between environmental changes; they cannot be understood without understanding those interactions, and an understanding of the interactions cannot be gained simply by understanding the components in isolation. Because Western science has traditionally focused on the central subject areas of defined disciplines rather than on their boundaries, it is difficult for specialists to realize that an interdisciplinary

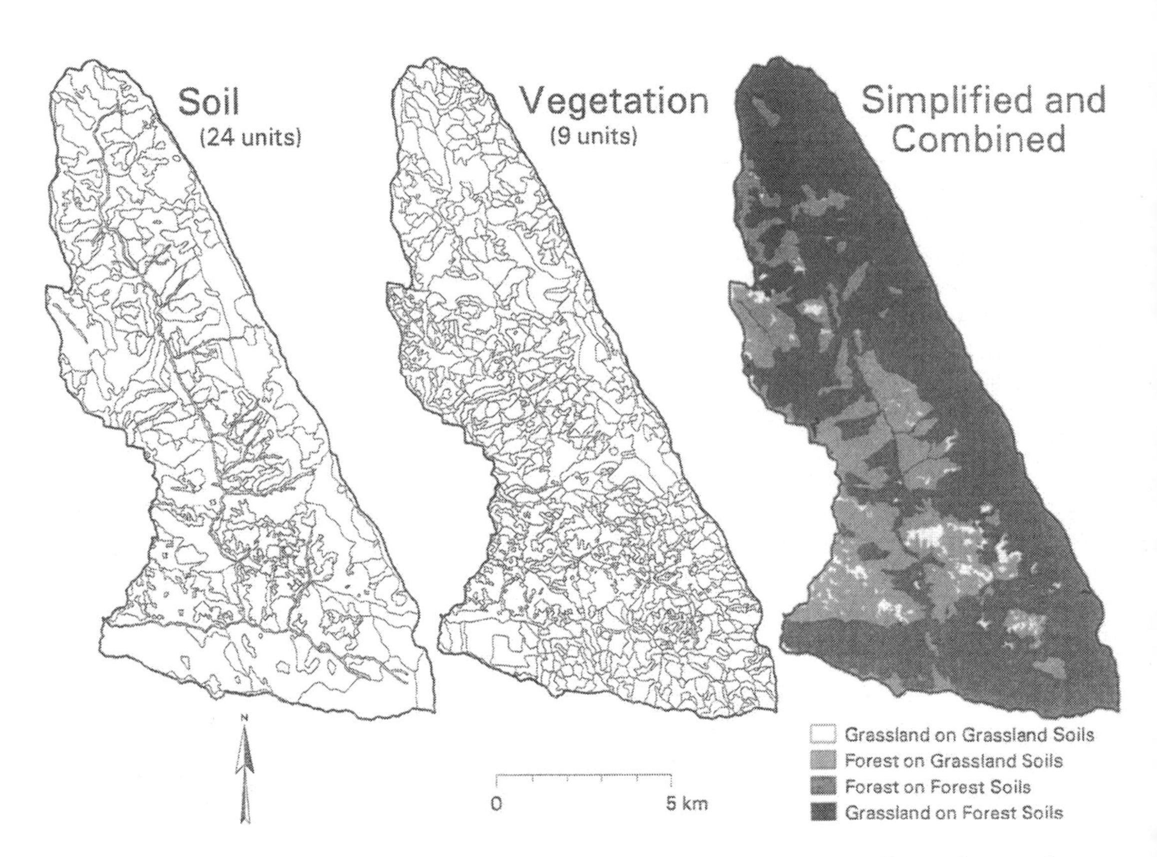

FIGURE 19.1. Soil, vegetation, and combined maps for the Pilot Creek watershed. Twenty-four soil units and nine vegetation units are outlined on the soil and vegetation maps, respectively. These sets of units were each reduced to two general categories (forest and grassland soils for the soil map; forest and grassland for the vegetation map) and superimposed to reveal the pattern of forest encroachment shown on the combined map.

understanding cannot be derived from a detailed understanding of each individual component (Holling 1993). Consider, though, that demarcations between disciplines are cultural artifacts; a different cultural outlook could easily define the interactions as the core areas of disciplines. If "*B*" is the problem, it makes little sense to limit one's attention to "*A*" and "*C*".

Third, the analyst must address each potential impact at the scale required by that impact. In particular, an understanding of small watersheds cannot be scaled up to explain the behavior of a large watershed. For example, a climatic regime characterized by intense, localized thunderstorms produces rare but large changes in the channel that drains any particular small watershed. In a large channel that drains a thousand of these tributary watersheds, the same pattern of storms produces frequent, low-magnitude changes. Science in the past has concentrated on the workings of small watersheds because that is the scale most tractable for experimental watershed studies. The result is a tendency for analysts to divide a large watershed into subareas of a familiar size and to evaluate those, thus ignoring the problems introduced by the larger scale. Unfortunately, it is at the larger, more poorly understood scales that cumulative watershed effects affect most people.

Finally, because so many influences and interactions are involved in the expression of an environmental impact, an understanding of cumulative watershed effects is most often based on qualitative descriptions and order-of-magnitude estimates. The problem cannot be reduced to the deterministic stimulus-and-response models beloved by Western science, and even stochastically based models are limited in their applicability. This must be viewed not as an affliction to be cured, but as an indication that different approaches are necessary when addressing problems whose very nature arises from their complexity. Uncertainty is inherent in the field of cumulative watershed effects.

The conflict between traditional scientific values and the needs of cumulative effects analysis is particularly evident in the treatment of measurement precision: the level of precision necessary to understand cumulative effects is often lower than that considered acceptable by traditional science. For example, knowing exactly how much sediment is eroded from a road contributes little to an understanding a watershed's cumulative impacts. Instead, what is relevant is knowing whether roads in the watershed produce a lot more sediment than grazing, whether a particular kind of road produces a lot more sediment than other kinds of roads, or whether sediment is even a problem in the area.

Because the skills and approach needed for cumulative effects analysis are not those fostered by traditional approaches to science, many would-be analysts find it difficult to forgo their accustomed approaches and adopt those most useful for the problem. Any task becomes difficult if one is equipped with the wrong tools.

The Ad Hoc Approach to Cumulative Effects Evaluation

Despite the complexity of the problem, the questions that must be answered during an evaluation of cumulative effects—what was the past like, how did changes occur, and what will the future be like—are questions that need to be answered for many applications, whether or not the words "cumulative effects" are used. Not surprisingly, they are questions that have been answered routinely for decades.

Until requirements for cumulative effects evaluations were mandated by NEPA, most evaluations of environmental change focused on specific problems in specific areas. Usually, the approach used was ad hoc: people provided an answer relevant to a particular problem using whatever techniques were best suited to that problem. One of the best examples of an ad hoc cumulative effects assessment was carried out more than 80 years ago by G.K. Gilbert (Box 19.4; Gilbert 1917).

Ad hoc methods have been applied to many problems that involve off-site cumulative watershed effects: Will changes caused by Hurricane Iniki aggravate future floods? Can Castleford's water supply be increased by

Box 19.4. G.K. Gilbert and the Fate of San Francisco Bay

Hydraulic mining in the Sierra Nevada had introduced vast volumes of sediment into California rivers, and officials in San Francisco worried that the sediment would eventually shut down the Port of San Francisco. They asked Grove Karl Gilbert to figure out when and by how much the Port would be affected.

Gilbert's first step was to identify the possible mechanisms for damage. Preliminary calculations quickly allowed him to discard all mechanisms except shoaling of the bay-mouth bar because of decreased tidal flow as mining-related sediments aggrade in other parts of San Francisco Bay.

Gilbert then identified the questions that needed to be answered to solve the problem: How much of the excess sediment contributed by mining will reach the bay? When will it arrive? How much of an effect will it have on tidal flow? How much of an effect will other future activities, such as marsh reclamation, have on tidal flow? How much will a decrease in tidal flow reduce the transport capacity of currents across the bay-mouth bar? Reconnaissance-level field work provided the information he needed to compare the volume of mining sediment to background sediment input rates; channel cross-sections disclosed the location of stored and mobile sediment and its rate of transport to the bay; and hydraulic calculations allowed estimation of sediment transport by tidal currents given various scenarios of sedimentation and coastal development.

From this information, Gilbert concluded that the effects of mining-related sedimentation would be small compared to the effects from development of tidal marshes. The CEQ's definition of cumulative impacts is foreshadowed by Gilbert's description: "Every acre of reclaimed tide marsh implies a fractional reduction of the tidal current in the Golden Gate. For any individual acre the fraction is minute, but the acres of tide marsh are many, and if all shall be reclaimed the effect at the Golden Gate will not be minute" (Gilbert 1917).

clear-cutting the reservoir's catchment? Why are wells in West Valley drying up? How should the $5,000 raised to improve fish habitat in Wilrick Creek be spent? Each of these problems concerns a specific environmental change in a specific area, and the intent of each inquiry is well defined. The methods used vary with every application because the methods selected are those best suited to the particular questions asked.

The most important phase of an ad hoc cumulative effects analysis is usually the initial step of identifying the question to be answered. The analyst usually must delve into the motivations of those desiring the analysis to determine exactly how the results are intended to be used. With this information, the analyst can select the appropriate scope and level of precision needed for the results. Sometimes only qualitative information is needed, and order-of-magnitude or relative values are sufficient to solve many problems. For example, an analysis of the potential effects of a hurricane on future flooding required only order-of-magnitude estimates of hurricane-related erosion to identify areas where channel aggradation could become a problem (Reid and Smith 1992). Where erosion from the storm was an order of magnitude less than the average annual erosion rate, potential aggradation could be ignored (Table 19.1).

Because ad hoc analysis of cumulative environmental changes is usually done on a small scale, investigators from few disciplines are usually involved, so the full scope of a problem may not be immediately evident. Thus, a second step in solving ad hoc problems is to identify the variety of factors that may influence the problem. Without this effort, proposed solutions often fail to meet their objectives. For example, fisheries biologists asked to restore fish habitat in a watershed generally inventory in-stream conditions and modify the unsatisfactory reaches. Solutions often improve local conditions until the underlying problems reappear to destroy the modifications (Frissell and Nawa 1992). Given the same problem, geomorphologists and hydrologists usually identify the underlying causes for change, such as an increase in landslide frequency, and then design solutions that reverse those causes, such as improving road drainage structures or removing roads (Spreiter et al. 1996). These solutions lead to permanent habitat recovery over the long term, but the fish may already be extinct by the time they take effect. What is needed is a melding of the two viewpoints, combined with input from other disciplines that provide insight into riparian vegetation changes, fire frequencies, and so on.

The third step is usually one of triage. Just enough information is gathered to determine which influences are small, and can be ignored; which are big, and thus require little precision in their evaluation; and which must be examined in more detail to determine their importance.

Once the major foci of the investigation have been identified, the problem can be simplified by making generalizations about the area to be examined. Subareas that are internally uniform are identified, and each is characterized or evaluated as a whole. Common criteria for stratification include climate, geology, topography, and vegetation, though others have been used for specific applications. Using this approach, stream-temperature regimes or habitat use might be described for forested granite watersheds, forested basalt watersheds, and chaparral granite watersheds. An investigation designed to address a different problem in the same area is likely to find a different classification more useful; it might distinguish between lands above and below the transitional snow line, for example, or among small, medium, and large channels.

The remainder of the investigation usually focuses on answering the three major questions

TABLE 19.1. An order-of-magnitude sediment budget for sediment contributed by Hurricane Iniki to watersheds and hydrologic zones on the island of Kauai, Hawaii.

	Increased sediment input from hurricane			
Watershed or zone	Sheet erosion	Landslides	Uprooting	Total
1. Wainiha	++	+++	+	+++
2. Lumahai	++	+++	+	+++
3. Waioli	–	–	+	+
4. Hanalei	–	–	+	+
5. Kalihiwai	–	–	+	+
6. Kilauea	–	–	+	+
7. Anahola	–	+	+	++
8. Kapaa	–	+	+	+
9. Wailua	–	+	+	++
10. Hanamaulu	–	–	+	+
11. Huleia	–	–	+	+
12. Waikomo	–	–	+	+
13. Lawai	–	–	+	+
14. Wahiawa	–	–	+	+
15. Hanapepe	++	++	+	++
16. Canyon zone	–	–	+	+
17. Waimea	+	++	+	++
18. Na Pali zone	++	++	+	++

Expected annual sediment inputs are on the order of 1,000 t-km^{-2}-yr^{-1} (– = $<$1 t-km^{-2}-yr^{-1}; + = 1–10 t-km^{-2}-yr^{-1}; ++ = 10–100 t-km^{-2}-yr^{-1}; +++ = 100–1,000 t-km^{-2}-yr^{-1}).

Adapted from Reid and Smith (1992).

for each land class: what did the past look like? how did changes occur? and what will the future look like? Depending on the type of information available, different approaches might be taken. Investigators usually use some combination of direct observation of processes, evidence of past process activity, historical records, information from analogous sites, modeling, and reasoning from a basic understanding of the biological, physical, and socioeconomic processes that affect the issue. Monitoring usually requires an intractably long duration to produce interpretable results and so is rarely useful during analysis. After the analysis is complete, however, monitoring is often used to evaluate the effectiveness of whatever solutions are implemented (Chapter 18).

If there are good records of past conditions, then past and present can be compared to define the changes. Aerial photographs and early maps provide the best record of the past in many remote areas, and sequences of images can document changes in channel form and reveal associations between channel changes and land-use activities (Ryan and Grant 1991). Interviews with long-term residents often provide useful information about past biological and physical conditions. If records do not exist, past conditions can often be deduced from an understanding of the mechanisms by which changes occur. The history of land use is first examined to identify direct changes to topography, vegetation, hydrology, fauna, and soils. An understanding of physical and biological processes is then used to infer how these direct changes (e.g. altered forest composition on hillslopes) influenced the conditions of interest (e.g. the size, amount, and stability of large woody debris in the main channel).

Downstream impacts usually involve changes in the transport of water or sediment through channels, so methods for describing such changes are useful. Sediment and water budgets describe the inputs, storage, and outflows of sediment or water in a system and are widely used for such applications. The precision, scope, and focus of such budgets are readily modified to suit the needs of each application (Reid and Dunne 1996).

Future conditions are controlled by three types of influences. First, the future will reflect trends set by past and present conditions, and some future conditions can be estimated by extrapolating these trends. Second, "time bombs" set by past changes will appear suddenly when conditions are appropriate. If logging roads were designed to withstand 25-year-recurrence-interval storms, for example, an unprecedented level of damage is likely the first time a larger storm occurs. Third, the future will reflect the results of future activities, and these usually must be inferred from socioeconomic projections of changing land-use patterns.

Because each ad hoc application is different, no analytical technique is applicable to all investigations. However, one problem is shared by most: some essential information is usually missing. For example, it might be necessary to know an erosion rate where no measurements exist. In such cases, a best estimate for the missing value often is made using information from similar areas, and the potential significance of an error is evaluated by calculating the effects of particular levels of over- and underestimation. Results also might be presented as a range between maximum and minimum likely values estimated using information from elsewhere.

Standardized Methods of Cumulative Effects Analysis

In 1969 the primary reason for evaluating cumulative watershed effects changed: NEPA now required environmental impact evaluations before federal land-use plans or permits would be approved. Most earlier evaluations needed to consider only a particular kind of impact in a limited area, but now the potential for all types of impacts had to be evaluated over large areas administered by the Forest Service and Bureau of Land Management. State legislation soon followed, and states now had to oversee thousands of private analyses. The ad hoc approach no longer was feasible because each evaluation would require too much expert review to assure the validity of

TABLE 19.2. Extent to which standardized cumulative effects evaluation methods provide the characteristics desired of such methods.

Goal	Cumulative effects evaluation method					
	Ad hoc	ERA	Fish-sediment model	CDF	TFW	Interagency
Administrative preferences						
Consistent set of methods used	no	yes	yes	yes	yes*	no
Results in a number or yes/no	yes*	yes	yes	yes	no	no
Completed in less than a month	yes*	yes	yes	yes	no	no
Can be done by non-experts	no	yes	yes	yes	no	no
Compliance judged by procedure	no	yes	yes	no	no	no
Reproducible results	yes*	yes	yes	no	yes	yes*
Accepted by peers	yes*	no	yes	no	yes	yes*
Can be used anywhere	yes	no**	no	yes	no	yes
Antidote to litigation	yes	yes	yes	no	yes	***
Technical requirements						
Methods tested and validated	yes*	no	no	no	***	***
Evaluates any impact	yes	no	no	yes	no	yes
Evaluates any mechanism	yes	no	no	yes	no	yes
Evaluates temporal accumulations	yes	no	yes	yes	yes	yes
Evaluates spatial accumulations	yes	yes	yes	yes	yes	yes
Evaluates distant off-site impacts	yes	no	no	no	no**	yes
Based on valid assumptions	yes*	no	yes*	yes*	yes*	yes*
Best available methods allowed	yes	no	yes	yes*	yes	yes

* Varies with the application.
** Has the potential but not as generally applied.
*** Is being or will soon be tested.

analysis techniques. Oversight had not been a problem before because most analyses directly benefited those who paid for the work. Now, however, analyses were done primarily to suit the needs of a third party: the regulatory agencies.

With the change in the scope and intent of analysis came a change in what a desirable analysis method should be like. Impact analysis had to be consistently applicable to facilitate administrative oversight; it had to produce numerical or yes-or-no results to allow consistent decision making; and it had to be accomplishable over a short period. Methods had to be tightly constrained and consistent so they could be carried out by nonexperts. The validity of an analysis had to be judged on procedural grounds: analyses would be accepted if they were carried out according to the agreed-on recipe. Whoever carried out the method in a particular area would have to arrive at the same answer as anyone else.

On this basis, states and land-management agencies began developing their own official procedures for cumulative effects analysis. The approaches adopted range from calculating indices of land-use intensity to using data-intensive mechanistic models; and from following a "cookbook" that requires no expertise to using professional judgment to fill out a checklist. Three examples illustrate the variety of approaches taken and the short-comings entailed by each (Table 19.2).

An Index Approach: Equivalent Roaded Acres

The method designed for National Forests in California is based on an assumption that the potential for cumulative watershed effects increases with land-use intensity in a watershed (USDA Forest Service 1988). Analysts inventory the areas affected by past activities and apply coefficients to adjust the effects of each activity to the same scale and to adjust for the extent of recovery. The scores for each activity are then summed to calculate the total score for a watershed, and this total is compared to the "threshold of concern." In watersheds where scores approach the threshold, further analysis identifies the condition of the watershed, and projects might be planned to decrease the score (e.g., road removal), or projects might be deferred until natural recovery has lowered the score to an allowable level. The intensity of all activities is described in terms of the area of road surface that would provide the same effect, the "equivalent roaded acre" (ERA). Most watersheds analyzed are smaller than 200 km^2.

Administratively, the ERA method is convenient and useful. Coefficients for defining the "road-likeness" of an activity and for describing the rate at which an activity site recovers can be defined for large areas, and subsequent analyses require little fieldwork. Little expertise is required to tally road lengths and logged areas; the method produces a numerical score that can be compared directly to the threshold of concern; and the procedure is fast, reproducible, documented, and relatively consistent.

Unfortunately, the method contains flaws that undermine its technical adequacy. For example, coefficients for recovery refer to recovery at the site of land use but not to recovery from the off-site impacts of that activity. Thus, a logged area is considered recovered when it is revegetated, even though sediment from logging-related landslides may still be present in downstream channels. All else being equal, calculation of ERAs would show that subsequent logging on the site would have no potential cumulative impact, yet the resulting influx of sediment would add to that already stored in the system and so would create a cumulative impact.

In addition, because only one set of coefficients is used to describe each activity, the method implicitly assumes either that only one mechanism for impact is possible in an area or that some composite variable is meaningful. However, a tractor-logged slope might be very much like a road in terms of sediment production but very different in terms of hydrologic change. If both hydrologic and sediment impacts are of concern, then two different sets of coefficients should be used to compare roads and logging. A third problem is that the scope of concern is limited to the watershed being managed, so an analysis for Pilot Creek would calculate ERAs only for the Pilot Creek watershed. Such an analysis could not evaluate Pilot Creek's contribution to impacts on the municipal water supply downstream. Thus the method is inherently incapable of evaluating the downstream cumulative watershed effects that are of most concern to most people.

Other problems arise from the manner in which the ERA method is applied. For example, the procedure requires calibration of many coefficients for each area, and such calibration ordinarily would be based on a lot of monitoring data for each activity. Monitoring is expensive and time consuming, however, so the necessary coefficients are usually estimated using professional judgment. This approach might not present a problem if the results of activities were then monitored to test the predictions, or if the entire program were tested by statistical comparisons between prediction and reality, but no such monitoring has been carried out.

Furthermore, those applying the method rarely specify the impact mechanisms important in the area of application. Instead of identifying hydrologic change, for example, as the major mechanism and using a measure of hydrologic change to define coefficients, most analysts estimate coefficients using a visual estimate of channel disturbance. The variety of mechanisms and activities that influence channel form makes an algebraic solution for coefficients impossible, even if conditions were to be carefully monitored. Further, even if coefficients could be defined, they would be relevant only for the smallest watersheds. Large watersheds require a different set of coefficients than small watersheds because large channels respond differently than small ones, but too many activities have already occurred in large watersheds to allow coefficients to be back-calculated.

These considerations suggest that the apparent simplicity of the index approach is deceptive. To use such an approach appropriately requires that different indices be defined for different areas to account for differences in impacts, impact mechanisms, and recovery trajectories. Appropriate use also requires a tremendous monitoring effort both to calibrate the method for the variety of conditions present and to test the validity of the results.

A Mechanistic Impact Model: The Fish-Sediment Model

In central Idaho, the U.S. Forest Service took a different approach to evaluating cumulative effects. Here, decades of collaborative efforts between researchers and resource specialists had provided data that defined relationships between particular land-use activities and rates of in-stream sedimentation (Cline et al. 1981), and between in-stream sedimentation and salmonid populations (Stowell et al. 1983). When the need for a cumulative effects method arose, the necessary information was on hand to construct a model that could be applied directly to the cumulative impact of most concern in the area: the impact of sediment on salmonids.

The resulting Fish-Sediment Model first calculates the sediment input expected from a planned land-use activity using calibrated coefficients that vary with locale and kind of activity. The effects of coexisting and prior activities are also calculated, and the results for previous activities are modified according to the expected recovery rate of the downstream channel. Results then allow an estimate of future channel conditions, from which the expected level of impact to fish can be calculated. Because the model takes into account the recovery rate of the impact instead of that of the driving variables, it allows evaluation of temporal accumulations of impact. Analysis focuses on changes to moderate-sized, low-gradient, alluvial channels.

Like the ERA approach, the Fish-Sediment Model is administratively convenient. Methods are standardized; results provide a numerical prediction of impact; the required level of expertise is not high; most work can be done in an office once the coefficients are calibrated; and given the same information, all users will produce the same answer.

Unlike the ERA approach, the Fish-Sediment method is based on a wealth of data that demonstrates the connection between specific land-use activities and a specific impact. Thus, the method is widely accepted as the best method available for predicting embeddedness of spawning gravel in steep granitic terrain of the northern Rocky Mountains where logging, roads, and fire are the major influences. However, the method cannot be applied to other areas where other impact mechanisms apply, other impacts are important, or the necessary data are missing. The method also cannot be used in channels larger than those for which it was calibrated, and this restriction limits the spatial scope of analyses. A similar approach could be developed for other impacts in other areas, but development would require the same painstaking accumulation of data that was needed to construct the model for central Idaho.

Professional Judgment: The California Checklist

The California Department of Forestry and Fire Protection (CDF) has a very different role than the U.S. Forest Service. The CDF manages little land, so it rarely has to evaluate cumulative effects. Instead, like similar resource agencies in other states, the CDF is primarily a regulatory agency with responsibilities for designing regulations for activities related to logging on private lands, reviewing applications for land-use plans, and inspecting activities for compliance with regulations. The CDF thus was responsible for designing a cumulative effects evaluation method for use by private timberland owners in California.

The CDF method (CDF 1998) consists of a checklist that leads the user through an evaluation of the impacts that may be important in an area and of the influences that may contribute to those impacts. The user is responsible for selecting appropriate methods to address each point. The report consists of yes-or-no answers to questions, about whether cumulative impacts are present in the area and whether the proposed project is likely to cause or add to significant cumulative impacts, accompanied by a narrative explanation of the answers. The report is sent to the state for approval as part of a timber harvest plan. State agencies that regulate or manage resources that might be affected by the timber harvest review the plan and decide whether it is adequate. Analyses usually consider impacts within an area of approximately 10 to 40 km^2, but a larger area may be evaluated if the analyst considers it necessary.

The CDF method has the advantage of being flexible. The method permits evaluation of whatever impacts are important in each area, and it allows consideration of the temporal and spatial scales that are relevant to those impacts. It also allows use of any analytical techniques that are applicable to the identified problems. This method also has several administrative advantages: it can be used consistently throughout the state, it provides a yes-or-no answer, it can be completed quickly, and it requires little specialized expertise beyond that of a state-certified professional forester. The analysis is considered adequate if the report seems reasonable to the reviewers.

From the point of view of the regulatory agencies, however, the method has the weakness that review must be based on the content of the narrative report rather than on whether the procedure was followed. This requirement places heavy responsibility on reviewers to detect faulty reasoning and inadequate methods, but examination of approved reports shows that reviewers have overlooked some fundamental errors. One report explained that the presence of riparian vegetation downstream meant that all logging-related sediment would be filtered out before it could contribute to a cumulative impact, though such reasoning would lead to the obviously absurd conclusion that no channel with riparian vegetation carries suspended sediment. Another report limited its scope to the project area, arguing that impacts would not occur downstream if they did not appear at the site of the activity. This argument obviously conflicts with the definition of cumulative watershed effects, yet it was not challenged in review.

The CDF method could be effective because of its flexibility, but its credibility depends on the diligence and expertise of those reviewing each report. Heavy reliance on expert oversight is necessary for methods based on professional judgment.

Administrative Convenience Versus Technical Adequacy

The three standardized analysis methods described above differ in their strengths and weaknesses (Table 19.2), but together they reveal a pattern that is shared by a variety of other institutionally based methods (Reid 1993). First, each standardized method is convenient from the point of view of the institutions responsible for the analyses. In contrast, the ad hoc approach would be difficult to carry out in an institutional setting because it depends on expertise both for analysis and oversight. Administrative convenience is important for institutional methods because whatever method is adopted must be accomplishable.

Second, the most technically credible methods are those based on considerable amounts of data and having the most restricted areas of applicability. Both the Fish-Sediment Model and individual ad hoc analyses are restricted in scope because the methods used are designed for a particular kind of problem in a particular place.

Third, a lack of technical validity has not stopped methods from being used, indicating that the utility of an analysis is not necessarily based on its results. Standardized analysis methods did not exist until regulatory mandates made them necessary. If the regulators accept the results of an analysis, the analysis has fulfilled its primary objective; the *results* of the analysis do not aid the institutions in accomplishing their goals as much as the *completion* of an acceptable analysis does. The institutions' ideal method thus is one that is easy to carry out and is acceptable to the regulatory agencies. More credible methods would have required too much expertise, too much data, and too much time.

The people who designed the first generation of cumulative effects methods were not blind to the need for scientific validity, but they were also very aware of the constraints imposed by their institutional contexts. Until pressure on the institutions increased to the point that the institutions allowed those contexts to be broadened, the first-generation methods were the best possible. It was largely through examining the successes and shortcomings of these early analysis efforts that a new approach to the evaluation of cumulative effects could be designed.

Watershed Analysis

The credibility of the standardized methods began to waiver under public scrutiny. A panel of experts criticized the ERA approach for not addressing cumulative effects directly; the Fish-Sediment Model could not be applied to other regions because it does not consider the variety of impacts and mechanisms important elsewhere; and the CDF checklist did not protect approved logging plans from lawsuits concerning impacts downstream of the mandated assessment area. In each case, the validity or applicability of the method was challenged. Meanwhile, an entirely different approach to impact evaluation was being designed in Washington State: *watershed analysis*.

Watershed analysis is not a stand-alone method for evaluating the cumulative effects of a proposed project. Instead, it is a formalization of the procedure used by ad hoc cumulative impact evaluations to develop the locale-specific understanding necessary to evaluate project impacts in the future. The Washington method combines this background analysis with a procedure for using the results to plan land-use activities that are intended to avoid cumulative impacts.

Limited Assessment With Prescriptions—Timber/Fish/Wildlife Watershed Analysis

Under the threat of increased litigation in the late 1980s, an agreement was crafted to protect public resources while allowing logging to continue on private and state lands in Washington State. This effort was known as the Timber/Fish/Wildlife Agreement, or TFW. Fundamental to the agreement was the realization that any evaluation procedure had to produce credible results, because people could not be persuaded to modify their land-use activities unless they knew that the reasons for doing so were valid. People would once again be analyzing cumulative effects for their own benefit, so validity was again an important goal. But the motivations once again had changed. Validity was important not because a mistake costs money, as was the case with the ad hoc analyses, but because the foundation of the agreement would collapse if results were not valid. None of the participants wanted to embark on the cycle of litigation that the agreement was designed to avoid.

The method that TFW developed consists of two parts—resource assessment and management prescription—that together are referred to as watershed analysis. The resource assessment describes fisheries impacts, their causes, and land-use activities influencing those impacts in different parts of a 40- to 200-km^2 watershed. Assessment teams usually include a fisheries biologist, a hydrologist, a riparian ecologist, and experts on landslides, surface erosion, and capital improvements. Other specialists may be brought in to work on questions that the assessment team cannot answer. All those interested in the watershed may participate, but reports are prepared by certified analysts who have received official training in the procedure.

A manual (Washington Forest Practices Board 1995) provides recommended methods for carrying out analyses, but assessment teams may use other methods if they are appropriate for the problems encountered. The manual includes modules that describe how to evaluate landslides, surface erosion, hydrologic change, riparian function, channel function, fish habitat, and public works. Each module was designed to suit the conditions and impact mechanisms commonly found in Washington, and each is directed toward evaluating impacts on anadromous fish and capital improvements. Results from the assessment modules provide the information needed to answer a series of "synthesis questions," which explore the mechanisms of impact, and to identify areas sensitive to particular kinds of change.

The resource assessment provides the objective information needed for the second, more subjective phase of the TFW method, in which management practices are prescribed by a "field managers team" composed of foresters, engineers, fisheries specialists, and hydrologists. Prescriptions are made for lands of all ownerships within the watershed, so small landowners in the area need not invest their own time or resources in environmental planning. Activities are not proscribed, but prescriptions for particular areas may define widths of riparian buffers, types of unstable ground that must be protected, specifications for road construction and maintenance, and similar requirements. Participation in the program is voluntary, but the motivation to participate is strong because adherence to the prescriptions absolves landowners of the need for further environmental evaluations. Prescriptions are reviewed by relevant state agencies and by all interested parties, and the long-term effects of implemented prescriptions are intended to be monitored.

The TFW method does not fit the profile of the administrative ideal established for earlier cumulative effects methods (Table 19.2) because it achieves the same ends in a very different way. By design, the method is acceptable to all, and each analysis is carefully scrutinized by the disparate groups that participate in it.

A weakness of the approach is its focus on impacts to aquatic habitat. Even the original intent to include analysis of wildlife needs (the W in TFW) was postponed when participants realized that achieving consensus on fisheries issues was a big enough challenge (Dr. Kate Sullivan, Weyerheauser Co., personal communication). The method also suffers from a reliance on untested analytical techniques. Procedures for calculating road-surface erosion, peak-flow changes from clear-cutting, and sediment yields from undisturbed watersheds have not yet been demonstrated to provide valid results. Revisions of the manual are planned and future versions are likely to correct some of these problems.

The TFW approach does not evaluate cumulative effects in the traditional sense; it does not describe the likely cumulative impacts of a particular project or land-use strategy. Instead, the approach uses an understanding of past cumulative effects to produce a set of land-use prescriptions. Because the impacts of those prescriptions are not themselves evaluated, analyses may overlook potential future cumulative impacts. For example, if the road density in a watershed is low and has not caused an identified impact, then prescriptions for future roads do not necessarily consider the impacts from increased road density.

Broad Assessment Without Prescriptions—Interagency Ecosystem Analysis

Following President William Clinton's "Timber Summit" of 1993, federal agencies worked together to design an approach to federal land management in the Pacific Northwest that would ensure the sustainability of natural ecosystems and rural economies and would satisfy requirements for lifting a federal court injunction on logging of federal lands in the area (FEMAT 1993). Under the resulting Northwest Forest Plan (USDA and USDI 1994b), large tracts of federal land would be set aside as forest reserves, and land-use activities in other areas would be tailored to suit the needs and capabilities of those areas. Central to the strategy was the use of watershed analysis to identify constraints and opportunities in each area. Watershed analysis originally had been intended to focus on aquatic ecosystems, but its role was quickly expanded to include analysis of terrestrial ecosystems and human communities, and the procedure was renamed "ecosystem analysis at the watershed scale" (REO 1995a). Ecosystem analysis was intended to be part of larger-scale analysis efforts that considered entire river basins and the region as a whole (FEMAT 1993).

Interagency ecosystem analysis is designed to produce integrated descriptions of the influence of physical, biological, and socioeconomic processes on environmental impacts in watersheds of 50 to 500 km^2. Watersheds were selected as the analysis unit to facilitate evaluation of impacts on the aquatic and riparian environment and because they are readily identifiable features. Other issues are not as conveniently evaluated on a watershed basis, but it was realized that a fundamental knowledge of processes can be applied to understand landscape patterns within any arbitrary set of boundaries. In effect, each issue is to be evaluated using information from whatever scales are relevant to that issue, but the implications of the results are to be highlighted for the watershed in question.

The primary task of ecosystem analysis is to reorganize existing information so that the connections between different components of the system become evident and the relevant interdisciplinary context for each issue can be examined (USDA and USDI 1994b); these connections are to be understood before management actions are planned. In particular, analysis is intended to

- Identify what issues in the watershed are important to what constituencies.
- Facilitate interagency communication.
- Identify physical, biological, and socioeconomic interactions in the watershed.
- Understand the watershed's role in a larger spatial context.
- Identify information that will be needed in the future for making management decisions.
- Explain to the public how the ecosystem and landscape function in the watershed.

Originally, analysis was not expected to provide prescriptions or recommendations for land-use activities, as these would require subjective weighting of conflicting values and desires. Instead, analysis was to provide an objective basis of understanding and cooperation that would contribute to any future land-use decisions. However, agency land managers later modified the procedure to include limited recommendations.

Analysis teams usually consist of federal fisheries biologists, wildlife biologists, hydrologists, and specialists in social sciences, vegetation, and earth sciences. Because of the variety of issues and landscapes that ecosystem analysis must address, no method is applicable everywhere, so particular techniques are not specified. Instead, the analysis manual (REO 1995a) describes a six-step strategy for analysis (Table 19.3). A supplement to the manual provides examples of techniques that might be useful in different areas (REO 1995b).

TABLE 19.3. Steps of federal interagency ecosystem analysis at the watershed scale.

1. Characterize the watershed and show its relation to the larger area.
2. Identify issues of concern and the questions that will need to be answered.
3. Describe existing conditions.
4. Describe the conditions that existed before Euro-American disturbances.
5. Describe how the changes in condition have come about.
6. Recommend management strategies and identify information needs and monitoring goals.

REO (1995a).

An on-going review of completed analyses (Reid, unpublished data) indicates that many are approaching the primary goal for analysis: the resource specialists preparing the analyses are becoming more familiar with issues, processes, and interactions in the landscapes they help manage. The level of understanding produced by many of the analyses could have prevented failure in the cases described by Box 19.2. However, many of the reports provide data compendia rather than integrated analyses, and inconsistencies between adjacent chapters are common. Such problems could be remedied by providing training in analysis strategies, putting more effort into interdisciplinary analysis, and soliciting outside technical review.

Interagency ecosystem analysis does not produce a traditional cumulative impact report, but it provides the background information needed by later cumulative effects evaluations. Ecosystem analysis describes the kinds of impacts likely to be important for particular land uses and ecosystem components in different parts of the watershed. Later, project-level impact evaluations identify the changes likely from a proposed project and interpret them in light of the ecosystem analysis results. Most of the work necessary for a cumulative impact assessment thus is done by ecosystem analysis, and project-level impact evaluations need only describe the effects of the proposed project in the context of the larger watershed, basin, and region.

Contrasting Goals and Methods

The TFW and interagency analysis methods share the underlying philosophy that effective management decisions require an understanding of how a system works, but the two approaches differ markedly in procedure and scope. Examination of the reasons for these differences illustrates the importance of carefully defining objectives before selecting strategies or methods.

The most fundamental difference between the two methods is in the application of results. The two-part TFW procedure is designed to produce a prescription for management practices, while interagency analyses are, for the most part, limited to providing the objective information upon which later prescriptions can be based. This limited application for interagency analyses was adopted because of the realization that any prescription requires subjective consideration of divergent desires and is thus a political decision subject to the decision-making procedures set forth by NEPA. In contrast, the TFW approach can make subjective decisions because the differing political interests have already agreed that consensus is less painful than litigation.

Other differences arise because federal land managers have more options than are usually available to a consortium of private interests. On federal lands, activities can be scheduled over decades on large tracts of the landscape, and particular areas can be designated for particular uses or values. The greater flexibility in land-use options on federal lands means that there is less need for up-front comprehensive prescriptions. Instead, prescriptions are made when a project is planned so that they reflect the condition of the watershed at that time.

In addition, federal management is directed toward a variety of resources, whereas past industrial management tended to focus on a single resource. On federal lands, the project in question might be timber harvest, construction of a campground, or any of a number of other activities. Appropriate prescriptions differ according to the kind of activity, the nature of surrounding activities, and the condition of the ecosystem at the time, so comprehensive prescriptions would be difficult to design before activities are proposed. Results of interagency ecosystem analysis thus provide the background of understanding necessary to better carry out all aspects of land management rather than focusing on prescriptions for a particular land use.

The wide range of issues that federal land managers must consider is also reflected in the broader scope of the interagency approach, and the diverse character of federal lands ensures that no single set of recommended techniques would be appropriate everywhere. In contrast, TFW's limited focus on impacts to anadromous salmonids and capital improvements in Washington State allowed specification of analysis techniques that are well suited to those issues in that area.

Because of their differing contexts, the TFW analysis approach would not suit the needs of federal land managers, just as the interagency approach would not be appropriate for TFW. In essence, watershed analysis is not a solution, but a tool that can come in many different shapes and sizes. Which shape and size is most useful—or whether any version is useful—depends on the type of problem being addressed and the objectives of those addressing it.

Administrative Convenience Versus Technical Adequacy, Revisited

Watershed analysis is a modification of ad hoc environmental impact analysis to fit the needs of an institutional setting. Problems with existing implementations of watershed analysis reflect the compromises necessary when the context for an approach changes. The attributes required to make the approach valid can restrict the method's ability to satisfy administrative needs, and vice versa (Table 19.2). For example, to ensure credibility, the interagency analysis method requires a higher level of expertise than federal agencies would have desired; to ensure feasibility, many analyses lack participation by certain kinds of experts (e.g., sociologists and geomorphologists) not available in those institutions. In the case of TFW, a full solution would require analysis of wildlife issues, but this aspect of the problem was postponed to ensure feasibility.

Each of the potential applications identified for watershed analysis results (Table 19.4) carries a different level of difficulty, uncertainty, political significance, and subjectivity, so each requires a different level of oversight and review. Applications which involve technically difficult problems or which have inherently uncertain results require technical review to establish the credibility of the analysis, while those intended merely to compile information need less oversight.

Applications having high levels of political or economic significance must also be examined to identify analyst biases and implicit philosophical assumptions. "Political significance" is used here in a very broad sense to indicate situations in which a particular group might benefit from a particular result. Thus, identification of a specific inventory need has political significance because it might mean that funding is taken from one group of specialists in an agency or corporation and given to another.

Applications that involve subjectivity require even more careful review. Subjectivity is introduced when applications require value judgments, and any choice between options that carry different levels of benefit or damage to different interests requires a value judgment. Such judgments are necessary for many applications and are the fundamental purpose for land management planning, but there is often a temptation to ensure that a particular outcome prevails when stakes are high (Bella 1997). It is in this situation that the design and implementation of an unbiased procedure for review and oversight become especially important. Not only must investigator biases be understood and the technical foundation for a decision be examined, but the rationale for the value-based decisions must be carefully scrutinized. Watershed analysis procedures that are restricted to objective outcomes thus require less review and oversight and are less likely to be controversial than those which produce recommendations or prescriptions for land management. The TFW approach succeeds in producing widely accepted land-use prescriptions because its

TABLE 19.4. Potential applications for watershed analysis and the extent to which each application is characterized by various attributes.

	Attributes of application					Are analysis results applied in this way?	
Potential application	Difficulty	Uncertainty	Political significance	Subjectivity	Need for review	TFW	Interagency
Identify available information	l	l	l	l	l	yes	yes
Compile relevant information	l	l	l	l	l	yes	yes
Identify inventory needs	m	l	m	l	m	no	varies
Identify monitoring needs	m	l	m	l	m	yes	yes
Describe past conditions	m	m	m	l	m	yes	yes
Understand existing impacts	m	m	m	l	m	yes	yes
Predict future conditions	h	m	h	l	h	no	varies
Identify effects of activity type	m	m	h	l	h	yes	yes
Identify effects of planned activity	m	m	h	l	h	no	no
Identify desired conditions	vh	m	vh	h	vh	no	varies
Recommend suitable activities	m	m	h	h	h	no	yes
Design land-use prescriptions	vh	m	vh	vh	vh	yes	no

The final two columns indicate whether the application is included in the Timber/Fish/Wildlife (TFW) and Interagency Ecosystem Analysis approaches (l = low, m = medium, h = high, vh = very high).

implementation is overseen by the full range of concerned interest groups.

Tomorrow's Analyses

The TFW and federal interagency approaches to watershed analysis are the best developed and most widely implemented at this point, but both have flaws. Neither represents a final stage in the development of methods for evaluating cumulative effects, and both will contribute to further development of analysis strategies and tactics as they themselves are further revised.

Efforts to improve cumulative impact analysis methods are currently progressing in two complementary directions. First, the need continues for ad hoc evaluations, and analysis techniques are continually being developed to support such efforts. Second, the drive toward development of a formal procedure for screening impacts over large areas also continues. Idaho has developed a general method for watershed analysis (Idaho Department of Lands 1995), California and Oregon are considering methods, and there is interest in developing an analysis strategy for use on federal lands throughout the United States.

Any effort to design a new general method for cumulative impact analysis will require making preliminary decisions about the intended use of the results, the range of topics to be considered, the spatial scale of analysis, and the level of oversight and review to be sought. Of these, a very specific definition of the intent and goals of analysis is most important because it strongly influences the remaining decisions. Any general method must be capable of considering the variety of impacts that people are likely to care about, and it must do so in a useful and credible way (Table 19.5). Unless the administrative unit over which analyses are to be carried out is small and uniform, these requirements imply that a single set of recommended methods is untenable. The landscape is too diverse for a single set of techniques to be relevant, useful, and valid everywhere.

TABLE 19.5. Desired characteristics of a generally applicable watershed analysis method.

1. Fits the particular needs of the agency or organization instituting it.
2. Evaluates any potentially important impacts.
3. Evaluates impacts at any point downstream.
4. Evaluates impacts accumulating through both time and space.
5. Evaluates the influence of any expected kind of land-use activity.
6. Evaluates any lands within the analysis area.
7. Uses the best available analysis methods for each aspect of the analysis.
8. Incorporates new information as understanding grows.
9. Can be done for a reasonable cost over a reasonable length of time.
10. Produces a readable and useable product.
11. Is credible and widely accepted.

Ad hoc and standardized analysis methods are independent of one another, although particular techniques developed for one might be applicable to the other. What is shared by both kinds of analysis is a philosophy that calls upon impact analysts to look at the world from a point of view in which

- General patterns are more important than details.
- Qualitative understanding is more important than precise numbers.
- Change is more important than stasis, which has never existed.
- Understanding of process and interaction is more important than description of condition.
- The issue defines the inquiry.
- Extreme conditions are usually more important than average conditions.
- Uncertainty is certain.

Development of this mindset is the biggest problem confronting effective analysis. Those who wish to evaluate cumulative impacts must be capable of departing from their accustomed ways of approaching a problem. In particular, analysts must be able to work in an interdisciplinary format; they must be able to see connections and patterns across very large areas; and they must be flexible enough to step away from a limited set of standardized procedures.

These traits can be encouraged through careful design of an analysis strategy to focus on problems or issues rather than on disciplines, to address types of sites rather than specific sites, and to specify that the team as a whole plans and prioritizes tasks rather than leaving the responsibility to individual members (Reid 1996).

Development of the necessary skills could also be encouraged by changes in educational curricula. Students who have participated on problem-oriented interdisciplinary teams are better able to recognize and solve interdisciplinary problems and to communicate effectively with those in other fields.

Changes are also needed in institutional infrastructures. Institutional budgets are too often designated by discipline, so that an increase in the wildlife budget implies a decrease in the hydrology budget. This framework creates competition among disciplines rather than promoting the cooperation that is increasingly needed. Reorganization of work groups by problem (e.g., "the cumulative effects unit") rather than by discipline (e.g., "the fisheries unit") would facilitate communication and understanding between disciplines.

Meanwhile, some fundamental research questions have been left largely unanswered. Relatively little is known about problems that sit on the boundaries between disciplines, and methods for aggregating the results of small-scale studies to address large-scale questions are poorly developed. There is a strong need for studies of all types that focus on understanding patterns and processes over large areas or long time scales.

Over the past several decades, the scale and complexity of questions that natural resource specialists are being asked to address has grown considerably. This trend can be seen as intimidating, and those who prefer designing precise answers to well-defined, well-controlled questions are not confident working in this setting. On the other hand, those who are intrigued by complex systems, who enjoy finding creative solutions to unprecedented problems, and who see beauty in the connections between disparate influences are finding today's challenges to be the most rewarding of their careers.

Literature Cited

Bella, D.A. 1996. Organizational systems and the burden of proof. Pages 617–638 *in* D.J. Stouder, P.A. Bisson, and R.J. Naiman, eds. Pacific salmon and their ecosystems: Status and future options. Chapman & Hall, New York, New York, USA.

Bisson, P.A., G.H. Reeves, R.E. Bilby, and R.J. Naiman. 1997. Watershed management and pacific salmon: Desired future conditions. Pages 447–474 *in* D.J. Stouder, P.A. Bisson, and R.J. Naiman, eds. Pacific salmon and their ecosystems: Status and future options. Chapman & Hall, New York, New York, USA.

Bjornn, T.C., and D.W. Reiser. 1991. Habitat requirements of salmonids in streams. Pages 83–138 *in* W.R. Meehan, ed. Influences of forest and rangeland management on salmonid fishes and their habitats. American Fisheries Society Special Publication 19. American Fisheries Society, Bethesda, Maryland, USA.

CDF (California Department of Forestry and Fire Protection). 1998. California forest practice rules. California Department of Forestry and Fire Protection, Sacramento, California, USA.

Cline, R., G. Cole, W. Megahan, R. Patten, and J. Potyondy. 1981. Guide for predicting sediment yields from forested watersheds. US Department of Agriculture, Forest Service Northern Region and Intermountain Region.

FEMAT (Forest Ecosystem Management Assessment Team). 1993. Forest ecosystem management: An ecological, economic, and social assessment. US Government Printing Office, Washington, DC, USA.

Frissell, C.A., W.J. Liss, R.E. Gresswell, R.K. Nawa, and J.L Ebersole. 1997. A resource in crisis: Changing the measure of salmon management. Pages 411–444 *in* D.J. Stouder, P.A. Bisson, and R.J. Naiman, eds. Pacific salmon and their ecosystems: Status and future options. Chapman & Hall, New York, New York, USA.

Frissell, C.A., and R.K. Nawa. 1992. Incidence and causes of physical failure of artificial fish habitat structures in streams of western Oregon and Washington. North American Journal of Fisheries Management **12**:182–197.

Fuller, D.D., and A.J. Lind. 1992. Implications of fish habitat improvement structures for other stream vertebrates. Pages 96–104 *in* R.R. Harris, D.C. Erman, and H.M. Kerner, eds. Proceedings of the symposium on biodiversity of northwestern California. Wildland Resources Center Report 29. University of California, Berkeley, USA.

Fullmer, D.G. 1994. Sustainability and ecosystem management: An analysis of the concept. Appendix I-A *in* Draft Region 5 Ecosystem Management Guidebook, Volume 2: Appendices. USDA Forest Service Pacific Southwest Region, San Francisco, California, USA.

Gilbert, G.K. 1917. Hydraulic-mining debris in the Sierra Nevada. US Geological Survey Professional Paper 105. US Government Printing Office, Washington, DC, USA.

Harden, D.R. 1995. A comparison of flood-producing storms and their impacts in northwestern California. Pages D1–D9 *in* K.M. Nolan, H.M. Kelsey, and D.C. Marron, eds. Geomorphic processes and aquatic habitat in the Redwood Creek Basin, northwestern California. US Geological Survey Professional Paper 1454. US Government Printing Office, Washington, DC, USA.

Holling, C.S. 1993. Investing in research for sustainability. Ecological Applications **3**:552–555.

Idaho Department of Lands. 1995. Forest practices cumulative watershed effects process for Idaho. Idaho Department of Lands, Boise, Idaho, USA.

Lewis, H.T. 1993. Patterns of Indian burning in California: Ecology and ethnohistory. Pages 55–116 *in* T.C. Blackburn and K. Anderson, eds. Before the wilderness: Environmental management by native Californians. Ballena Press, Menlo Park, California, USA.

Madej, M.A., and V. Ozaki. 1996. Channel response to sediment wave propagation and movement, Redwood Creek, California, USA. Earth Surface Processes and Landforms **21**:911–927.

Moyle, P.B. 1976. Inland fishes of California. University of California Press, Berkeley, California, USA.

Nickelson, T.E., M.F. Solazzi, S.L. Johnson, and J.D. Rodgers. 1992. Effectiveness of selected stream improvement techniques to create suitable summer and winter rearing habitat for juvenile coho salmon in Oregon coastal streams. Canadian Journal of Fisheries and Aquatic Sciences **49**:790–794.

Reeves, G.H., L.E. Benda, K.M. Burnett, P.A. Bisson, and J.R. Sedell. 1995. A disturbance-based ecosystem approach to maintaining and restoring freshwater habitats of evolutionarily significant units of anadromous salmonids in the Pacific Northwest. Pages 334–349 *in* J.L. Nielsen, ed. Evolution and the aquatic ecosystem: Defining unique units in population conservation. American Fisheries Society Symposium 17. American Fisheries Society, Bethesda, Maryland, USA.

REO (Regional Ecosystem Office). 1995a. Ecosystem analysis at the watershed scale, version 2.2. US Government Printing Office, Portland, Oregon, USA.

REO (Regional Ecosystem Office). 1995b. Ecosystem analysis at the watershed scale section II: Analysis methods and techniques, version 2.2. US Government Printing Office, Portland, Oregon, USA.

Reid, L.M. 1993. Research and cumulative watershed effects. USDA Forest Service General Technical Report PSW-GTR-141. Pacific Southwest Research Station, Albany, California, USA.

Reid, L.M. 1996. Enabling interdisciplinary analysis. Pages 624–626 *in* Proceedings: Watershed '96. Prepared by Tetra Tech, Inc., under EPA contract no. 68-C3-0303.

Reid, L.M., and T. Dunne. 1996. Rapid evaluation of sediment budgets. Catena Verlag GMBH, Reiskirchen, Germany.

Reid, L.M., and C.W. Smith. 1992. The effects of Hurricane Iniki on flood hazard on Kauai. Report to the Hawaii Department of Land and Natural Resources, Honolulu, Hawaii, USA.

Rosgen, D.L. 1994. A classification of natural rivers. Catena **22**:169–199.

Ryan, S.E., and G.E. Grant. 1991. Downstream effects of timber harvesting on channel morphology in Elk River Basin, Oregon. Journal of Environmental Quality **20**:60–72.

Sidle, R.C. 1985. Factors influencing the stability of slopes. Pages 17–25 *in* D. Swanston, ed. Proceedings of a workshop on slope stability: Problems and solutions in forest management. USDA Forest Service General Technical Report PNW-180. Pacific Northwest Forest and Range Experiment Station, Portland, Oregon, USA.

Spreiter, T.A., J.F. Franke, and D.L. Steensen. 1996. Disturbed lands restoration: The redwood experience. Pages 140–146 *in* J. LeBlanc, ed. Proceedings of the conference on coast redwood forest ecology and management. June 18–20, 1996, Humboldt State University, Arcata, California, USA.

Stowell, R., A. Espinosa, T.C. Bjornn, W.S. Platts, D.C. Burns, and J.S. Irving. 1983. Guide to predicting salmonid response to sediment yields in Idaho Batholith watersheds. US Department of Agriculture, Forest Service Northern Region and Intermountain Region.

USDA and USDI (United States Department of Agriculture and United States Department of the Interior). 1994a. Environmental assessment for the implementation of interim strategies for managing anadromous fish-producing watersheds in eastern Oregon and Washington, Idaho, and portions of California. United States Department of Agriculture and United States Department of the Interior, Washington, DC, USA.

USDA and USDI (United States Department of Agriculture and United States Department of the Interior). 1994b. Record of decision for amendments to Forest Service and Bureau of Land Management planning documents within the range of the northern spotted owl; Standards and guidelines for management of habitat for late-successional and old-growth forest related species within the range of the northern spotted owl. US Government Printing Office, Portland, Oregon, USA.

USDA Forest Service. 1988. Cumulative off-site watershed effects analysis. In: USDA Forest Service Region 5 Soil and Water Conservation Handbook. FSH 2509.22. US Department of Agriculture Forest Service, San Francisco, California, USA.

USDA Forest Service. 1994. Pilot Creek watershed analysis. USDA Forest Service, Six Rivers National Forest, Eureka, California, USA.

Washington Forest Practices Board. 1995. Standard methodology for conducting watershed analysis under Chapter 222-22 WAC. Version 3.0. Department of Natural Resources Forest Practices Division, Olympia, Washington, USA.

20 Rivers as Sentinels: Using the Biology of Rivers to Guide Landscape Management

James R. Karr

Overview

• Humans alter the surface of Earth in ways, on scales, and at frequencies unprecedented in recent history. Resource and environmental managers must identify and minimize the effects of changes that have negative consequence for human society.

• Because rivers integrate all that happens in their landscapes, their condition, especially their biological condition, reveals much about the consequences of human actions.

• The condition of rivers in the Pacific Northwest indicates that much of the region's rich natural capital has been spent.

• Existing laws do not adequately protect rivers because they are at odds with the physical connectedness of water, and, worse, they commonly ignore the biological components of aquatic ecosystems.

• Human actions jeopardize the biological integrity of water resources by altering physical habitat, modifying seasonal flow of water, changing the food base of the system, changing interactions within the stream biota, and polluting water with chemical contaminants.

• Conventional monitoring and evaluation studies, such as tracking chemical pollution or population sizes of target species, are inadequate to protect overall river condition in part because they are conceptually narrow, in part because they are not well suited for distinguishing variation caused by natural events from variation caused by human actions.

• Biological monitoring in the twentieth century began with a restricted focus (organic pollution, toxic chemicals) but is shifting to a more integrative approach that evaluates the condition of aquatic biota from diverse perspectives.

• Integrative, multimetric biological indexes used to develop biological standards (criteria) are more comprehensive and robust than chemical standards, and they are effective at diagnosing degradation, defining its cause(s), and suggesting treatments to halt or reverse damage.

• Multimetric biological monitoring is a central feature of water resource assessments throughout the United States (48 states have or are developing multimetric approaches), and it has been used on all continents but Antarctica.

• The index of biological integrity (IBI) is one multimetric approach used to examine the influence of humans on fish, invertebrate, and algal assemblages.

• IBI has substantial statistical power to detect the effects (point and nonpoint pollution, physical habitat alteration, flow alteration, and complex cumulative impacts) of diverse human actions (agriculture, livestock grazing, logging, recreation, and urbanization) on water resources.

• IBI can be used to define spatial and temporal patterns in water resource conditions and to evaluate the effects of management efforts.

• A benthic invertebrate index of biological integrity (B-IBI) proposed for use in the Pacific Northwest includes ten metrics: total number of taxa; number of mayfly, stonefly, caddisfly, long-lived, intolerant, and clinger taxa; proportion of individuals belonging to tolerant taxa and to predatory taxa; and percent dominance of the three most abundant taxa.

• Rivers are sentinels: they give early warning of the risks human activities engender. Society can no longer afford to ignore these risks or behave as if they did not exist.

Introduction

Environmental change is a reality, and it is continuous. Change on Earth is driven by wind and water; geological forces; astronomical events; and the work of microorganisms, plants, and animals. During the past two centuries, human activities have become the principal driver of change on Earth. Human influences are massive, they are incessant, and they are global. For the first time, a biological agent—a single species—rivals geophysical forces in shaping the surface of the Earth.

Human-caused change may be positive, neutral, or negative. The challenge faced by environmental scientists is to distinguish among these three alternatives by detecting and interpreting the causes and consequences of change, especially those that alter living systems. Resource and environmental managers want to detect and treat changes that have negative consequences; at the same time, they want to avoid wasting resources—treating changes that do not have negative consequences.

Rivers as Sentinels

Rivers and streams serve as a continent's circulatory system, and the study of those rivers, like the study of blood, can diagnose the health not only of the rivers themselves but of their landscapes (Sioli 1975). Changes anywhere on the landscape are likely to influence rivers. People change the landscape as they harvest forests. They compromise river health when they build dams to generate power or control flooding, or when they mine lands for minerals. They degrade rivers when they construct industrial parks to manufacture goods, shopping centers to sell those goods, homes to shelter families, and farms and tree plantations to supply food and fiber. They degrade rivers when they construct transportation corridors to bring people and goods together.

Before these activities altered Pacific Northwest rivers and landscapes, humans prospered for thousands of years on the region's wealth of forest, river, and marine resources. When European colonists arrived about 200 years ago, they forever altered the interactions between regional natural systems and human populations. Now demand for water in the Northwest exceeds supply; forests and fish populations are decimated; soils are eroded; and marine resources are damaged and depleted. These and other changes in the health of Northwest landscapes indicate that modern stewardship of regional resources has fallen short. The region's rich natural capital has been squandered.

Pacific Northwest rivers, and especially their living inhabitants, show signs of degradation. For example, plummeting populations of anadromous salmonids (Nehlsen et al. 1991) and other, nonanadromous native fishes follow directly from dams, excessive water withdrawals for irrigation, and other destruction of habitat; fish population declines are the consequence of overharvest by sport and commercial fishers and of the effects of invading exotic species. In fact, fish often continued to decline even when society tried to protect native fish stocks. To offset fishery losses from overharvest and habitat destruction, stocking streams with fish raised in hatcheries began in the 1800s (Bowen 1970), but for more than 100 years, it has been obvious that fish stocking exacerbates the decline of wild salmonid populations (White et al. 1995).

Although it is naive to assume that the changes people cause can be stopped altogether, it is equally naive to suggest that these changes can proceed without consequences. Laws, regulations, and management programs have proliferated to reduce human damage to river systems, but too often, society's interests in water resources still are not protected. American legal doctrines such as "prior appropriation" in the West and the Clean Water Act are simply at odds with the physical connectedness of water; worse, they completely ignore the biological components of aquatic ecosystems (Karr 1990 and 1995a, Johnson and Paschal 1995, Chapter 22, this volume). Water in rivers is connected in four dimensions: upstream and downstream (longitudinally); across channels and hyporheic and groundwater zones (vertically); on the surface from uplands through riparian corridors to the channels (horizontally); and through the water cycle from clouds to precipitation to surface and then ground water (temporally).

Existing laws and their implementing regulations treat water narrowly—as if surface and groundwater were not connected, as if point and nonpoint sources of pollution could be treated in isolation. The laws undervalue the immense diversity of goods and services supplied by aquatic ecosystems. Improvements in the laws, and in their implementing regulations, are urgently needed if, as the Clean Water Act mandates, society is "to restore and maintain the physical, chemical, and biological integrity of the nation's waters."

For 25 years, this mandate was largely ignored in water policy (Karr and Dudley 1981, US EPA 1990, Karr 1991). Three approaches to the use and management of water resources kept the focus narrow, incomplete, and inadequate:

Water was viewed as a fluid for human use. Too many water resource professionals saw "the forms of life in a river [as] purely incidental, compared with the main task of a river, which is to conduct water runoff from an area toward the oceans" (Einstein 1972).

Pollution was the only threat to water resources, and dilution was the solution. People managed for "water quality" (degrees of chemical contamination). In 1965 an Illinois water official observed, "Regardless of how one may feel about the discharge of waste products into surface waters, it is accepted as a universal practice and . . . a legitimate use of stream waters" (Evans 1965). Surface waters existed to receive the discharge of human society.

Only a few aquatic species "counted" as important to human society. Society sought to maximize sport or commercial harvest of selected species. Production (larger harvests of fish or shellfish) became the goal, and technological fixes like hatcheries became the means to supplement falling wild populations (Meffe 1992). Fish ladders helped migrating adults pass upstream over dams, but no provisions were made for helping young fish go around the dams as they migrated downstream toward the ocean. Many biologists removed large woody debris from stream channels to make passage easier, never mind that fish had been passing such barriers for centuries, or that the wood actually created fish habitat (Maser and Sedell 1994).

Although these three philosophies have not been abandoned, a growing number of water resource professionals recognize their inadequacies. For example, neither of the first two attitudes give any value to the life forms associated with aquatic ecosystems. Yet, citizen watershed teams and government agencies now recognize the importance of rivers and their biota and are struggling to establish long-term monitoring and assessment programs. Weber (1981) argued that the only way to determine the biological integrity of aquatic assemblages is to examine their properties in the field and to compare them with assemblages expected in the absence of degradation caused by human actions. Increasingly, water managers are being called upon to evaluate the biological effects of their management decisions, for no other aspect of a river gives a more integrative perspective about the condition of a river than its biota.

For decades biologists have intuitively recognized the connection between river conditions and biota, but have usually focused on pollution tolerance or on population sizes of a few target species. The pollution-tolerance approach concentrated on the ability of individuals to survive stress from pollution, especially organic contamination (Kolkwitz and Marsson 1908). Even water resource biologists rarely looked at phenomena other than chemical contamination. Fishery managers, on the other hand, tracked the population sizes of species with recreational or commercial importance.

Agencies charged with reducing pollution (departments of environmental quality) or enhancing populations (departments of fish and game) developed independently but rarely interacted, even though they managed the same water body. Each organization's programs were carried out as if scientists knew how organisms would respond to their management prescriptions; monitoring to see whether a program was actually effective occurred rarely or not at all. Failure to evaluate management decisions allowed wasteful, even destructive, practices to continue for decades (Maser and Sedell 1994, White et al. 1995).

In short, water resource managers either ignored biological systems or implemented policy with only narrow conceptions of biological conditions and their importance to human society. Reductionist viewpoints dominated water management; legal and regulatory programs avoided most biological issues and contexts; precise biological goals were not well developed or defined; field methods to measure biological condition were not standardized; formal processes to evaluate and express biological condition were not established; links between field measurements and enforceable goals were weak; and approaches to measuring biological condition were not cost effective (Karr 1991).

Now, approaches that are more comprehensive have been developed and are being adopted by state and federal agencies. Forty-two states now use multimetric biological assessments of biological condition and six states are developing biological assessment approaches, whereas only three states used multimetric biological approaches in 1989 (US EPA 1996a). At last, efforts to monitor the biological integrity of water resources as mandated by the Clean Water Act 25 years ago are emerging (Karr 1991, Davis and Simon 1995, US EPA 1996a,b).

Biological Integrity and Cumulative Effects

The biota that evolves and maintains itself in a region possesses biological integrity—the capacity to support and maintain a balanced, integrated, and adaptive biological system having the full range of elements and processes expected in a region's natural habitat (Karr and Dudley 1981, Angermeier and Karr 1994, Karr 1996). Critical elements of biological integrity comprise genes, populations, species, and assemblages; critical processes encompass mutation, demography, biotic interactions, biogeochemical cycles, energy dynamics, and metapopulation processes. Together these elements and processes create a regional complex of living organisms and act over a variety of spatial and temporal scales. Living systems thus are embedded in dynamic evolutionary and biogeographic contexts (Karr 1996).

Human impacts on the biological integrity of water resources are complex and cumulative. When humans modify landscapes or stream channels, changes in biological integrity are likely (Table 20.1). Human actions jeopardize the biological integrity of water resources by altering one or more of five principal factors: physical habitat, seasonal flow of water, the food base of the system, interactions within the stream biota, and chemical contamination (Karr et al. 1986, Karr 1991). These features are connected physically, chemically, and biologically. Furthermore, the biota of streams evolved in the presence of, and now depends on, variability in time and connectivity throughout landscapes (Meyer et al. 1988, Pringle et al. 1988, Naiman 1992). Attempts to protect water resources must take account of all these factors and the cumulative effects that human actions impose on them (Chapter 19).

Systems possessing biological integrity can withstand, or recover rapidly from, most natural disturbance (Rapport et al. 1985, Karr et al. 1986). But biological integrity declines if the natural disturbance regime is altered by a type, intensity, or frequency of perturbation that lies outside the biota's adaptive experience, especially if those disturbances become incessant. Urbanization, for example, compromises the biological integrity of streams by severing the connections among segments of a watershed and by altering hydrology, water quality, energy sources, habitat structure, and biotic interactions.

TABLE 20.1. Five primary classes of water resource attributes and components altered by the cumulative effects of human activity, with examples of degradation in Northwest watersheds.

Attribute	Components	Degradation
Water quality	Temperature; turbidity, dissolved oxygen; acidity; alkalinity; heavy metals, toxic substances; organic and inorganic chemicals	Increased temperature Oxygen depletion Chemical contaminants
Habitat structure	Substrate type; water depth and current speed; spatial and temporal complexity of physical habitat	Sedimentation and loss of spawning gravel Lack of large woody debris Destruction of riparian vegetation and banks Lack of deep pools Altered distribution of constrained and unconstrained channel reaches
Flow regime	Water volume; timing of flows	Altered flows limiting survival of salmon and other aquatic organisms at various phases in their life cycles
Food (energy) source	Type, amount, and size of organic particles entering stream; seasonal pattern of energy availability	Altered supply of organic material from riparian corridor Reduced or unavailable nutrients from the carcasses of adult salmon after spawning
Biotic interactions	Competition; predation; disease; parasitism; mutualism	Increased predation on young by native and exotic species Overharvest by sport and commercial fishers

Modified from Karr 1995a.

The most practical and cost-effective approach to determine if human actions are degrading biological integrity is systematic biological monitoring and assessment (Davis and Simon 1995, Karr and Chu 1998). Such monitoring provides both numeric and narrative descriptions of resource condition, which can be compared among watersheds, across a single watershed, and over time (Karr 1991), and does so at costs that are often less than that of complex chemical monitoring. For example, costs (per evaluation) for ambient monitoring (Yoder and Rankin 1995) are low (benthic invertebrates, US$824; fish, US$740) compared to chemical and physical water quality (US$1,653) and bioassays (US$3,573–$18,318).

Evolution of Biological Monitoring

Ecological systems, especially their biological components, are exceedingly complicated and have been studied from many perspectives. Theoretical ecologists try to understand natural variation within assemblages of organisms over space and time, along with the evolutionary and thermodynamic principles that mediate this variation. They ask questions such as, Why does the number of species vary from place to place on the surface of the Earth? What regulates the size of animal and plant populations? How do global biogeochemical cycles regulate ecosystem structure and function? For the most part, theoretical ecologists work in natural systems subjected to relatively little influence from human actions.

Applied ecologists, on the other hand, try to understand natural variation as well as how natural systems respond to stresses imposed by human society. They ask questions such as, What do we measure to understand responses to stress? How do we interpret the results? How do we distinguish natural variation from human-induced change? What are the likely consequences of the changes we see? How do we tell citizens, policymakers, and political leaders what is happening and how to fix it?

Biological monitoring has evolved rapidly during the twentieth century as knowledge of aquatic ecosystems has changed and human-imposed stresses have become more complex and pervasive. Early water quality specialists developed biotic indexes that were sensitive to organic pollution (sewage) and sedimentation (Kolkwitz and Marsson 1908), an approach that continues in modern biotic indexes (Chutter 1972, Hilsenhoff 1982, Lenat 1988, 1993). As toxic chemicals became more pervasive, water managers recognized the limitations of early biotic indexes and began to screen for the biological effects of synthetic as well as natural chemical contaminants. Biologists exposed fish or invertebrates experimentally to contaminants and documented toxicological dose-response curves. They observed that for a given body size very low doses of a chemical contaminant led to little or no response, but as dose increased response also increased. The goal was to establish quantitative chemical criteria to use in water quality standards, criteria that would protect human health or populations of desirable aquatic species.

But just as biotic indexes measure mostly the effects of organic pollution, chemical criteria based on toxicology generally address only a small number of chemical contaminants. Chemical criteria based on dose-response curves for single toxicants do not incorporate synergistic or other interactions of multiple chemicals in the environment. Further, the toxicological approach ignores other human impacts on aquatic biota, such as habitat alteration.

At the same time, biologists responsible for sport and commercial fisheries built hatcheries (White et al. 1995), "enhanced" habitat by first removing, and later adding, large woody debris (Maser and Sedell 1994), and introduced exotic species as alternatives to native fisheries (Courtenay and Moyle 1992, Allan and Flecker 1993). These population enhancement efforts are now recognized not only as inadequate but also, in many circumstances, as damaging to regional biological resources.

In a recent review, Fausch et al. (1990) identified four major measurement approaches commonly used to detect and understand the effects of human actions on aquatic organisms: indicator taxa or guilds; species richness, diversity, and evenness; multivariate statistics; and multimetric indexes such as the index of biotic integrity (IBI) (Karr et al. 1986, Davis and Simon 1995, Karr and Chu 1998).

The most common approach is to track human-induced change in abundances (or population size or density) of indicator taxa or guilds. Managers and scientists conducting field assessments of environmental impacts must isolate the effect of interest from "noise" caused by natural spatial and temporal variation (Osenberg et al. 1994). Compared with physical or chemical parameters, past biological studies have not accomplished this task well. Data from long-term studies of marine invertebrates (Osenberg et al. 1994), for example, show that temporal variability in population-based parameters (e.g., densities of organisms) show about three times the variability of individual-based parameters (e.g., size and body condition of individuals) and nearly four times the variability of chemical–physical parameters (e.g., water temperature, sediment quality, water column elements). Repeated sampling of standard water quality parameters in streams and rivers typically yields coefficients of variation (CV) of 20 to 25%, whereas CVs for standard measures of population size are much larger at 50 to 300% in standard sampling designs (J.R. Karr, personal observation).

Early field assessments used "control-impact" (CI) or "before-after" (BA) sampling designs, which often do not detect or reveal patterns relevant for water resource management. To overcome this problem, Green (1979) proposed a single design, "before-after-control-impact" (BACI), combining the two to separate the effect of human activity from other sources of variability in space or time. Because the BACI design confounds interactions between time and location, however, still other statistical approaches were proposed: "before-after-control-impact paired series" (BACIPS; Stewart-Oaten et al. 1986) and "beyond BACI" (Underwood 1991, 1994). Understanding the interaction of variability and magnitude of effect is also critical for determining statistical power in planning and interpreting environmental impacts (Osenberg et al. 1994). The growing sophistication of these designs demonstrates the effort to improve field assessment protocols. But the complexity of these designs goes beyond the planning, sample size, and analytical capability of most routine monitoring efforts; the narrow biological context limits the ability of managers to detect other relevant biological signals. Schmitt and Osenberg (1996) provide an excellent review of these sampling designs and their use.

During the 1960s, ecological research favored measures of species diversity to evaluate biological communities. More than 25 years ago, however, Hurlbert (1971) raised concerns about the statistical properties of diversity indexes, and others have since questioned their biological properties (Wolda 1981, Fausch et al. 1990). Diversity indexes often respond erratically to systematic changes in assemblages; they are often inconsistent and dependent on initial conditions—all of which leads to ambiguous interpretations (Wolda 1981, Boyle et al. 1990). These measures were nevertheless advocated for use in water management (Wilhm and Dorris 1968) and used in a few cases (e.g., to set Florida water quality standards, although Florida is now moving away from use of diversity indexes to multimetric evaluations; Barbour et al. 1996). Few scientists and managers recommend species diversity indexes today, largely because biologically more reasonable and statistically more reliable approaches are available. Unfortunately, however, these early diversity indexes left a negative semantic legacy that surfaces whenever the word index appears. Many concerns expressed by Suter (1993) are examples of that legacy.

Many researchers advocate multivariate statistical approaches because they ostensibly provide an objective way to explore variation in biological data. Principal components analysis (PCA), the most common procedure (James and McCulloch 1990), and other ranking techniques attempt to extract maximum statistical variance from variance–covariance matrices, usually across species or sites (Ludwig and Reynolds 1988). These approaches assume that describing the maximum variation will identify the most meaningful indication of biological condition. Ordination, another multivariate procedure, was developed for data that follow a multivariate normal distribution (Fore et al. 1996), a rare pattern in data from biological monitoring.

Although multivariate methods seem to provide an objective view of data, they often lose valuable information. Many users of multivariate methods focus on presence and absence of taxa or numbers of animals present, rather than knowledge of natural history or organismal responses to human actions; they emphasize statistical (variance-covariance) rather than biologically relevant patterns and their consequences; and they exclude rare species from analyses in an effort to avoid the effect of zeros in a data matrix. For example, one participant in a workshop on study design and data analysis in benthic macroinvertebrate assessments concluded that rare species simply add "noise to the community structure signal and . . . little information to the data analysis, and therefore, rare species should be excluded from the data matrix. We recommend excluding all taxa that contribute less than 1% of the total number or occur at less than 10% of the sites" (Reynoldson and Rosenberg 1996; also see Norris 1995). Such exclusion represents a substantial loss of important biological information.

Without objective statistical tests to determine what is different—and the biological consequences of those differences—multivariate analyses rarely give results that go beyond common-sense knowledge (Karr and Martin 1981, Fore et al. 1996, Stewart-Oaten 1996). Gotelli and Graves (1996:137) even suggest that "multivariate analysis has been greatly abused by ecologists." They further note that the practice of "drawing polygons (or amoebas) around groups of [points], and interpreting the results often amounts to ecological palmistry. Ad hoc 'explanations' often are based on the original untransformed variables, so that the multivariate transformation offers no more insight than the original variables did."

Gotelli and Graves conclude that patterns in multivariate analysis should be compared against a properly formulated null model. Comparing the results of PCA analysis of real data with similar matrices of random numbers, however, shows that the percent of variation described may be similar for both, especially for the second and higher principal components; loadings of original variables on principal axes are often as high for random numbers as for real data; and matrix size is an important determinant of the amount of variation extracted (Karr and Martin 1981). Multivariate techniques failed to discern known deterministic relationships in one study (Armstrong 1967), and in another study they manufactured relationships in data sets containing no such relationships (Rexstad et al. 1988).

Used cautiously, multivariate statistics can point to patterns when little is known about the underlying natural history of a biota (Gerritsen 1995). But an approach that actively and explicitly uses knowledge about streams and landscapes, invertebrates and fish, and the effects of humans on those places and organisms is preferable.

In addition to biological knowledge, multimetric indexes also use knowledge from earlier monitoring approaches, integrating that information into a single method while avoiding theoretically flawed indicators (e.g., species diversity indexes). They are also wider in scope (Davis 1995, Simon and Lyons 1995). The set of metrics (or biological attributes) incorporated into a multimetric index integrates information from ecosystem, community, population, and individual levels (Karr 1991, Barbour et al. 1995, Gerritsen 1995). Multimetric indexes are generally dominated by metrics of taxa richness (number of taxa) because structural changes, such as shifts among taxa, generally occur at lower levels of stress than do changes in ecosystem processes (Karr et al. 1986, Schindler 1987 and 1990, Howarth 1991, Karr 1991). But the most effective and integrative multimetric indexes explicitly embrace several concepts including taxa richness, indicator taxa or guilds (e.g., tolerant and intolerant), health of individual organisms, and assessment of processes (e.g., as reflected by trophic structure or reproductive biology of the sampled assemblage).

Like the multimetric indexes used to track national economies (Mitchell and Burns 1938), multimetric biological indexes measure many dimensions of complex ecological systems. Multimetric economic indexes assess economic health against a standard fiscal period; indexes of biological integrity assess the biological well-being of sites against a regional "baseline condition" reflecting the relative absence of human influence.

Stream ecologists recognize that stream segments (or habitat patches)—even those in the same region—may not share the same evolutionary and biogeographic history, and thus the expectations of biological condition are not the same at all places. Thus classifying system types (e.g., stream size or stream type) for a given region is essential in defining baseline conditions within the region; the conditions reflecting the absence of human influence in one patch may be quite different from baseline conditions in another. Again, the goal is to understand and isolate, through sampling design and analytical procedures, patterns that derive from natural variation in environments. Too many existing studies confound these patterns with human-induced variation, making interpretation of biological signals difficult or impossible.

The Index of Biological Integrity

The index of biological integrity (Karr 1981) was the first comprehensive multimetric index applied to assess biological condition in running waters. When used properly, IBI detects degradation of living systems, diagnoses the likely causes of degradation, identifies management actions that can halt or reverse degradation, and monitors living systems to find out if management efforts to restore degraded sites are successful. The foundation for IBI was established in a project in Allen County, Indiana, that began in 1973, soon after passage of the 1972 Clean Water Act (PL 92-500). Most project participants concentrated on chemical benchmarks; some, including U.S. Environmental Protection Agency officials, even denied the relevance of a biological perspective.

Because the chemical approach was inadequate, resource condition was assessed based on the resident fish assemblage. The resultant index for midwestern streams, or as adapted for use in the Willamette River, Oregon (Table 20.2), incorporated metrics reflecting species

TABLE 20.2. Metrics used to assess fish communties in the midwestern United States (modified from Karr 1981 and Fausch et al. 1984) and the Willamette River, Oregon. Modified from Hughes and Gammon 1987.

Category	Midwestern IBI metric	Oregon IBI metric
Species composition and abundance	Total number of native fish species	Same
	Number of darter species	Number of sculpin species
	Number of sunfish species	Number of minnow species
	Number of sucker species	Same
	Number of intolerant species	Same
	Proportion of green sunfish individuals	Proportion of common carp individuals
Trophic composition	Proportion of omnivorous individuals	Same
	Proportion of insectivorous cyprinid individuals	Proportion of insectivorous individuals
	Proportion of top carnivorous individuals	Proportion of catchable salmonid individuals (>20 cm)
Fish abundance and condition	Number of individuals in sample	Same
	Proportion of hybrid individuals	Proportion of individuals from introduced species
	Proportion of individuals with disease, tumors, fin damage, or skeletal anomalies	Same
	None	Total fish biomass

richness, indicator taxa, trophic guilds, fish abundance, presence of exotic species, and condition of individual fish. Efforts to adapt this fish IBI to benthic invertebrates often proposed similar metrics. Those metrics were carefully evaluated before protocols were published for the invertebrate community index (ICI; Ohio EPA 1988, DeShon 1995) and the benthic index of biological integrity (B-IBI; Kerans and Karr 1994, Fore et al. 1996) but not for rapid bioassessment protocol III (RBP III; Plafkin et al. 1989).

Successful application of multimetric biological indexes requires four tasks: selecting measurable attributes that provide reliable and relevant signals about the effects of human activities; developing sampling protocols and designs that ensure that those attributes are measured accurately; defining analytical procedures to extract and understand relevant patterns in the sample data; and communicating those results to policymakers and to society so that all stakeholder communities can contribute to the development of environmental policy. Failure of multimetric indexes generally stems from inadequate attention to one or another of these four activities.

Although an infinite variety of biological attributes can be measured, a much smaller number of attributes provide useful information about the impact of human activities on local and regional biological systems. Some attributes vary little or not at all (e.g., the number of scales on the lateral line of a fish); others vary substantially (e.g., weight, which varies with age or environmental context). Variation may be natural or anthropogenic. Natural variation derives from temporal (diurnal, seasonal, annual) and spatial (stream size, stream channel type) sources. Thus the first step in IBI development is to define how humans influence aquatic ecosystems (Table 20.1) (Karr et al. 1986, Karr 1991, Karr and Chu 1998) and what biological attributes can be measured with precision to provide reliable information about biological condition.

Selecting IBI Metrics

Before toxicologists could establish chemical criteria, they documented the response of organisms to the presence of specific contaminants in experimental conditions and established dose-response curves for those contaminants acting on the same organisms (Figure 20.1a). Similarly, selecting a metric for inclusion in a multimetric index such as IBI requires establishing an empirical relationship

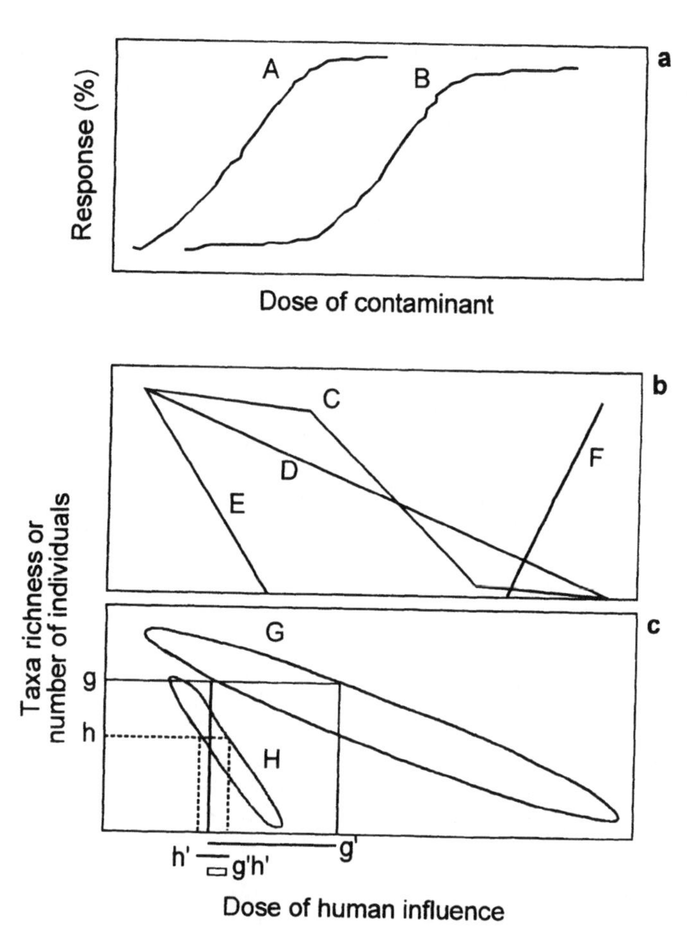

FIGURE 20.1. Generalized dose-response curves for toxicology (upper panel) and cumulative ecological effects of human activity (middle and lower panels). (a) Curves A and B are dose-response curves for two chemical contaminants acting on the same organisms. (b) Curves C through F are cumulative ecological dose-responses for several biological attributes along a gradient of human influence (e.g., urbanization in a study watershed): predator taxa richness (C), total taxa richness (D), intolerant taxa richness (E), and % of individuals in tolerant taxa (F). (c) Because natural variation in the range of human impact is associated with a given value of the biological attribute on the *y*-axis, single metrics (ellipses G and H) only generally define the level of human impact (e.g., projection of range of value g and h [lines g′ and h′] onto the *x*-axis). Simultaneous use of multiple metrics [g′h′] narrows the identified range of actual degradation on the *x*-axis.

between the attribute proposed for measurement and a gradient of human influence. Unlike the contaminant-specific dose-response curves of toxicology, the ecological dose-response curves used to define metrics for an IBI reflect specific human activities that exist in a set of study watersheds. Curves C through F (Figure 20.1b) are cumulative ecological dose-responses for several biological attributes along a gradient of human influence (e.g., urbanization in a study watershed): predator taxa richness (C), total taxa richness (D), intolerant taxa richness (E), and % of individuals in tolerant taxa (F).

Study watersheds may be carefully selected to involve a gradient of only one human activity (e.g., area logged or grazing intensity), or they may reflect the cumulative impacts of many human activities (e.g., chemical contaminants, varied land uses, riparian condition). These ecological dose-response curves are, in essence, empirically derived descriptors of the responses of biological systems to human-induced stress. Because natural variation in the range of human impact is associated with a given value of the biological attribute on the y-axis, a single metric only generally defines the level of human impact (Figure 20.1c). Simultaneous use of multiple metrics narrows the identified degree of biotic integrity (or amount of degradation) on the x-axis.

Regardless of whether fish, invertebrates, or other taxa are used, the search for a small set of biological attributes that reliably signal resource condition along gradients of human influence yields the same basic list of attributes (Miller et al. 1988, Karr 1991, Barbour et al. 1995, Davis and Simon 1995). With only minor tuning, the list can be adapted to specific regions (Miller et al. 1988). For example, taxa richness of darters (Percidae) in midwestern North America is replaced by taxa richness of sculpins (Cottidae) in the Pacific Northwest; both metrics reflect the presence or absence of riffle-dwelling benthic insectivores. In the only application of the fish IBI to the Pacific Northwest, Hughes and Gammon (1987) used half the midwestern IBI metrics (total number of native fish, sucker, and intolerant species; proportion omnivores; and abundance) and modified others (number of darters replaced by sculpins, sunfish (Centrarchidae) by minnow (Cyprinidae) species; proportion of green sunfish (*Lepomis cyanellus* Rafinesque) replaced by carp (*Cyprinus carpio* Linnaeus), top carnivores by catchable salmonids (>20 cm), and hybrids by introduced species).

Six recent studies of stream invertebrates in three regions (the Tennessee River valley [Kerans and Karr 1994], Pacific Northwest [Kleindl 1995, Fore et al. 1996, Patterson 1996, Karr and Chu 1998], west-central Japan [Rossano 1995, 1996]) suggest 10 metrics, out of about 30 routinely tested, as reliable indicators of human impact in diverse circumstances (Table 20.3). Further, these metrics are robust regardless of sampling method (e.g., Surber samplers, Hess samplers, or kicknets). Metric selection is based on informal, especially graphical, methods rather than formal statistical hypothesis testing (Fausch et al. 1984, Fore et al. 1996). Formal methods can "obscure the information in the data rather than clarify it" because they are interpreted as what should be

TABLE 20.3. Four groups of metrics selected for inclusion in a Pacific Northwest benthic invertebrate index of biological integrity (B-IBI) based on six studies of invertebrate responses to human activities.

Metric	Predicted response	Geographic area					
		TENN	SWOR	SEOR	PUSD	JAPN	GTNP
Taxa richness and composition							
Total number of taxa	Decrease	+	+	+	+	+	+
Number of Ephemeroptera taxa	Decrease	+	+	−	+	+	+
Number of Plecoptera taxa	Decrease	+	+	+	+	−	+
Number of Trichoptera taxa	Decrease	+	+	+	+	+	+
Number of long-lived taxa	Decrease	0	+	+	+	0	
Tolerance							
Number of intolerant taxa	Decrease	+	+	+	+	+	+
% of individuals in tolerant taxa	Increase	+	+	−	+	+	+
Feeding ecology							
% of predator individuals	Decrease	+	−	+	+	−	+
Number of clinger taxa	Decrease	0	0	0	+	+	0
Population attributes							
% dominance (2 or 3 taxa)	Increase	+	+	−	−	−	+

A (+) indicates that the metric varied systematically along a gradient of human impact for that data set; (−) indicates that the metric did not vary systematically; (0) indicates that the metric was not tested for that data set.
Sources: Tennessee (TENN), Kerans and Karr 1994; southwestern Oregon (SWOR), Fore et al. 1996; southeastern Oregon (SEOR); Puget Sound (PUSD), Kleindl 1995; Japan (JAPN), Rossano 1995; Grand Teton National Park (GTNP), Patterson 1996.

done rather than as judgments about what can be done (Stewart-Oaten 1995).

In the Northwest, best results are obtained from sampling protocols with three replicate samples taken from a single riffle during September (Fore et al. 1996). Choice of a standard sampling time avoids the need to interpret variation because of seasonal patterns in arthropod abundances and distributions. Surber samples provide an excellent foundation for biological assessment, but kicknet samples may also be used. The important point is that samples must be collected and processed with carefully defined standard procedures; samples from different sites or times should not be mixed. Results improve if each sample replicate contains at least 400 fish (Fore et al. 1994) or 500 invertebrates, all of which are counted.

Scoring Metrics

Different biological contexts for different metrics produce different quantitative ranges for each; relative abundances may vary from 0 to 100%, and taxa richness may vary from zero to a few taxa (stoneflies) or 0 to 40 (total taxa richness). Thus, before metric results can be combined into a single integrative index, the data must be converted into a common scoring base. Whether invertebrates or fish are sampled, each site must be compared with a reference standard for each metric. By the convention established in the original IBI (Karr 1981), a site receives a score of 5 if a metric value lies at or near the value expected at a site minimally altered by humans, 3 if moderately degraded, and 1 if severely degraded.

Defining scoring criteria requires some understanding of normal biological responses to human activity. In the case of species (or taxa) richness, one expects a consistent decline (or increase) in number of species present as human influence increases. Taxa richness is generally trisected (Figure 20.2a, b): if 27 species are expected at an undisturbed site, for example, the score is 5 if 19 to 27 species are present, 3 with 10 to 18 species, and 1 if 0 to 9 species in a standard sample. More complex relationships are often scored according to the structure of the relevant dose-response curve (Figure 20.2c,d) (Karr 1992, Kerans and Karr 1994, Fore et al. 1996). Because species richness for most taxa varies with stream size, lines defining maximum expected species richness across a gradient of stream size is appropriate in each sampling region (Figure 20.3) (Fausch et al. 1984).

Integrating Multiple Metrics

Comparisons among sites in space and time are easier if both narrative and numeric expressions of resource conditions exist. The challenge is to combine the substantial information contained in each individual metric into an easily interpreted and communicated quantitative expression about the relative condition of sites within a region or about the same site over time. The quantitative expression (IBI) is the sum of the converted metric scores (5, 3, or 1). Detailed information about individual sites, including the status of individual metrics, is not lost because the components of the multimetric index can still be examined individually in both quantitative and narrative form (Simon and Lyons 1995). Both empirical and statistical evaluations of variation in IBIs across time at numerous sites suggests that variation at sites with little or no change in human influence over time is 8 to 10% (Fore et al. 1994).

Managers can use the IBI, or a narrative description of each biological attribute of a site, to communicate with policymakers or citizens. The existence of numeric and narrative descriptions makes it possible to compare sites, thereby permitting more effective analysis of biotic condition. Policymakers can then establish priorities for protection or restoration. Because metric responses along gradients are essentially dose-response curves, managers can also predict the effects of certain proposed actions (e.g., when evaluating permit requests). Knowing general patterns of biological response makes it possible to predict the consequences of alternative actions, thereby leading to better management decisions. Such knowledge also makes it possible to predict degradation that might be expected following increased human activities in a watershed or, alternatively, the expected improvement after certain activities have stopped.

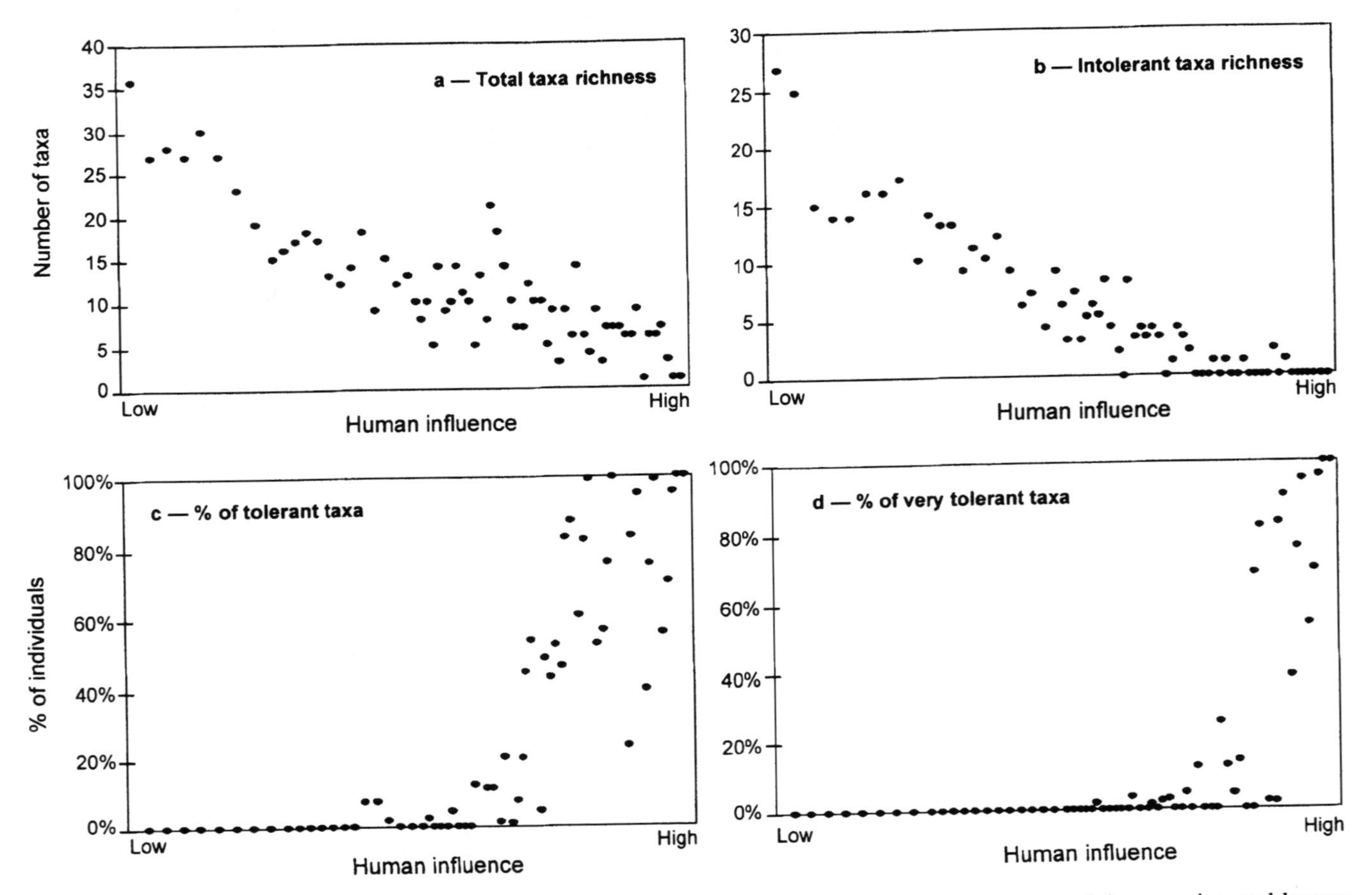

FIGURE 20.2. Ecological dose-response patterns across a gradient of human influence for four attributes of invertebrate assemblages. The human influence gradient integrates type and amount of effluent; presence of dams, weirs, and levees; and the condition of the riparian corridor for 65 streams in west-central Japan (modified from Rossano 1995).

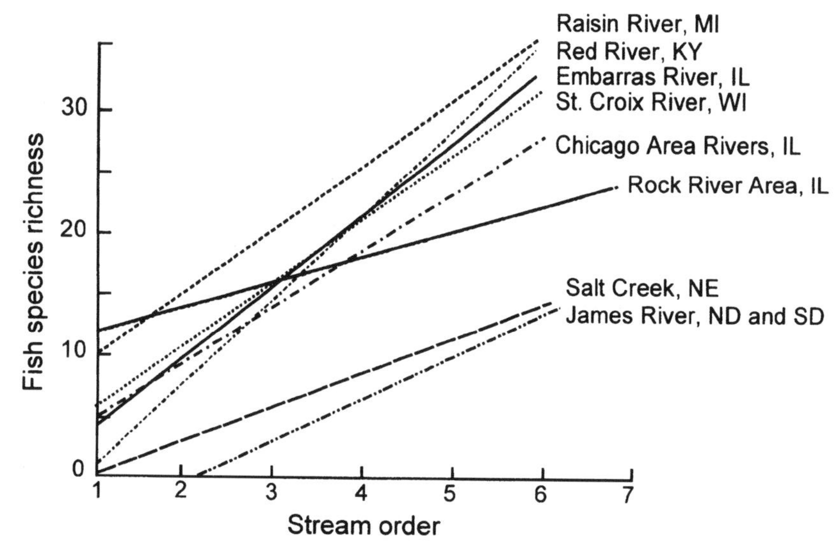

FIGURE 20.3. Lines of maximum species richness (fish) as a function of stream size (expressed as stream order) based on historical data from midwestern streams (from Fausch et al. 1984). These lines show that species richness varies predictably across regions and stream orders, generally in two groups: forested regions in the eastern Midwest (upper 6 lines) and Great Plains (lower 2 lines).

When used correctly, the IBI approach costs no more, and often less, than conventional chemical screening; it is sensitive to the full array of human impacts on living systems; and it can provide sophisticated and interpretable data in a short time.

What IBI Says About Streams and Watersheds

IBIs have now been defined for a number of taxa (fish, insects, algae, birds), habitat types (streams, wetlands, estuaries), and geographic regions (all continents but Antarctica). As the following examples illustrate, IBIs can inform a variety of assessment and management contexts.

Detecting the Effects of Point Source Pollution

IBI successfully detected variation in biological conditions (aquatic invertebrates) along the length of a stream in the North Fork Holston River in Virginia and Tennessee (Figure 20.4) (Kerans and Karr 1994). Four years of data on benthic invertebrates yielded high B-IBIs upstream of a sludge pond. Immediately below the sludge pond, B-IBI dropped sharply, then increased slowly downstream from the pond. Although B-IBIs varied over the years, rankings of sites along the stream remained strikingly consistent across the four-year sampling period.

Identifying Multiple Sources of Degradation

Most streams are influenced by several human activities. One important test of IBI is its ability to detect different human impacts. Big Ditch is a 3rd-order stream in a largely agricultural watershed (90% row crops) in east-central Illinois (Figure 20.5) (Karr et al. 1987). The regional landscape is degraded, and so is the stream channel, which is no more than a homogeneous raceway with sand, gravel, and rock substrates and no riparian vegetation. A municipal sewage treatment plant just above sampling station 2 pours 75 million liters per day of wastewater into the stream. The fish IBI was moderate (fair) above, and low (poor) below this effluent

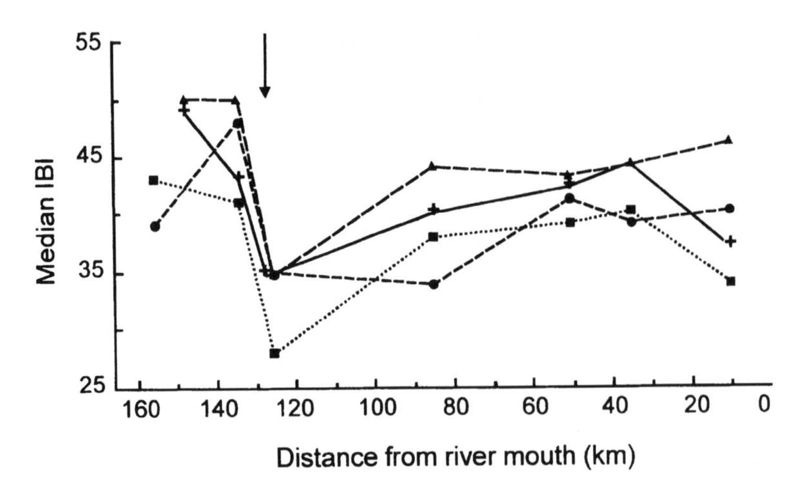

FIGURE 20.4. Median B-IBI in riffles of the North Fork Holston River in Virginia and Tennessee from 1973 to 1976 (-●- 73, ···■··· 74, —+— 75, —▲— 76). Arrow indicates location of a streamside sludge pond (from Kerans and Karr 1994, with permission).

source. IBI increased slowly along the stream from stations 2 to 4 (recently channelized) and rose sharply at stations 5 and 6 (not recently channelized). Point-source inputs, time since channelization, and intensive regional land use interacted to produce a complex longitudinal pattern of biological condition along this single stream; IBI detected this pattern.

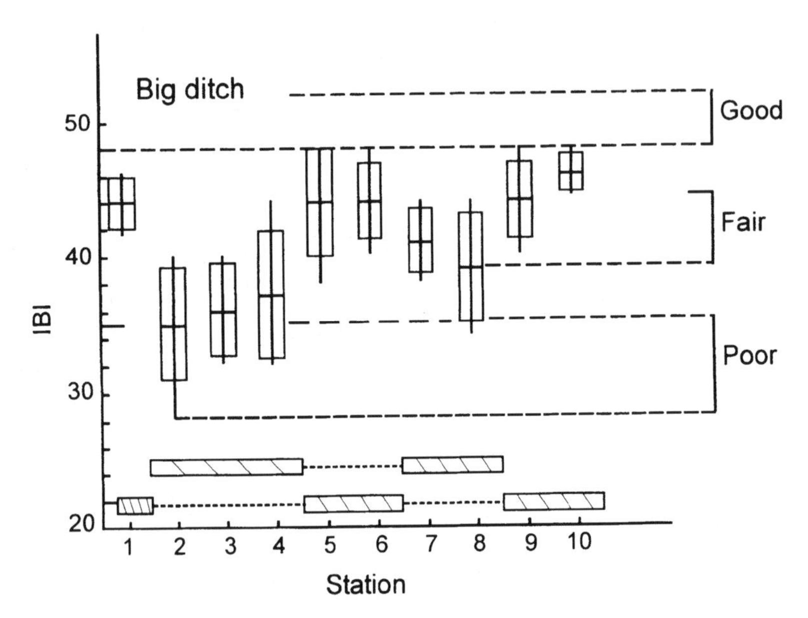

FIGURE 20.5. Fish IBI for ten sampling stations during 1978–1980 on Big Ditch, a channelized, 3rd-order stream in east-central Illinois (from Karr et al. 1986, with permission). Hatched bars indicate groups of sites for which means are statistically indistinguishable ($p < 0.05$, Student-Newman-Keuls test). Vertical line through each station bar shows the range; the bar itself extends one standard deviation above and below the mean (horizontal line).

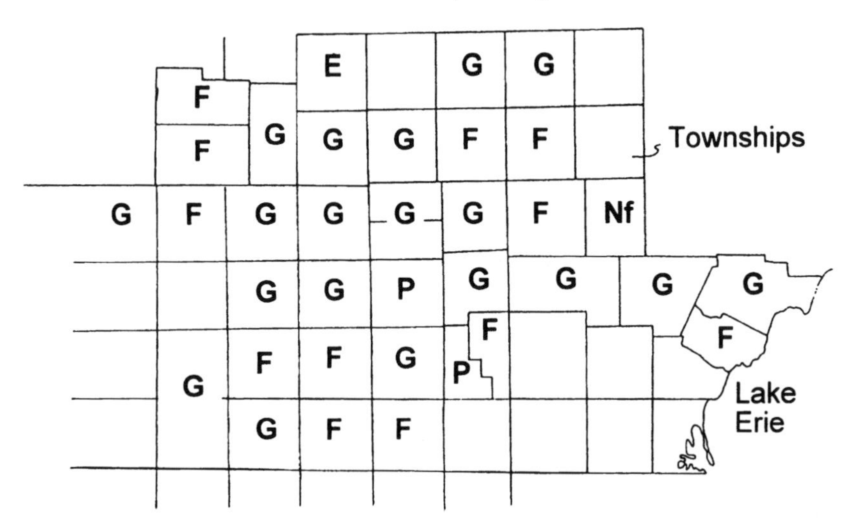

FIGURE 20.6. Integrity classes as revealed by IBI for each township in the Raisin River Watershed, Michigan (from Fausch et al. 1984). Areas with low IBIs are associated with larger towns, extensive agricultural areas, and feedlots. IBIs range as follows: excellent (E), 56–60; good (G), 48–52; fair (F), 40–44; poor (P), 28–34; very poor (VP), 12–22; no fish (NF), no IBI.

Describing Geographic Pattern and Detecting Cause

Sample sites within the Raisin River, Michigan, show different levels of biological integrity among the townships within the watershed (Figure 20.6) (Karr et al. 1986). Areas with low IBI (fish) are associated with larger towns, extensive agricultural areas, and feed lots. Geographic analysis informs managers where degradation is worst, diagnoses the causes of degradation, and suggests where regulatory actions or incentive programs are required to improve water resource condition.

Detecting Regional Variation in Human Influence

Degrees of degradation from human actions in different regions can be detected with IBI (Figure 20.7). IBIs (fish) were high for five sites in Arkansas selected as representing the best conditions in each major region of the state (Karr et al. 1986). IBI ranked 92% of sites in the Red River, Kentucky, which runs through the Daniel Boone Wilderness, as good or better (Fausch et al. 1984). Sites sampled in the Raisin River, Michigan, and the seven-county area around Chicago, Illinois, ranged from excellent to severely degraded (no fish). Average IBIs for the heavily urbanized Chicago area were well below those of the Raisin River.

Detecting Change Over Time as Human Activity Changes

A small woodlot in Wertz Drain, Allen County, Indiana, had a high fish IBI, reflecting good habitat quality (sinuous channel, well-developed pools and riffles, trees shading the channel in a good riparian corridor) (Karr et al. 1987). Upstream of the woodlot poorly executed bank stabilization in 1976 resulted in the delivery of substantial sediment downstream, which filled pools and degraded habitat within the woodlot. The resident fish community deteriorated, then recovered slowly over the next five years (Figure 20.8).

Evaluating Management Efforts

Management decisions driven solely by chemical criteria may not benefit biological integrity.

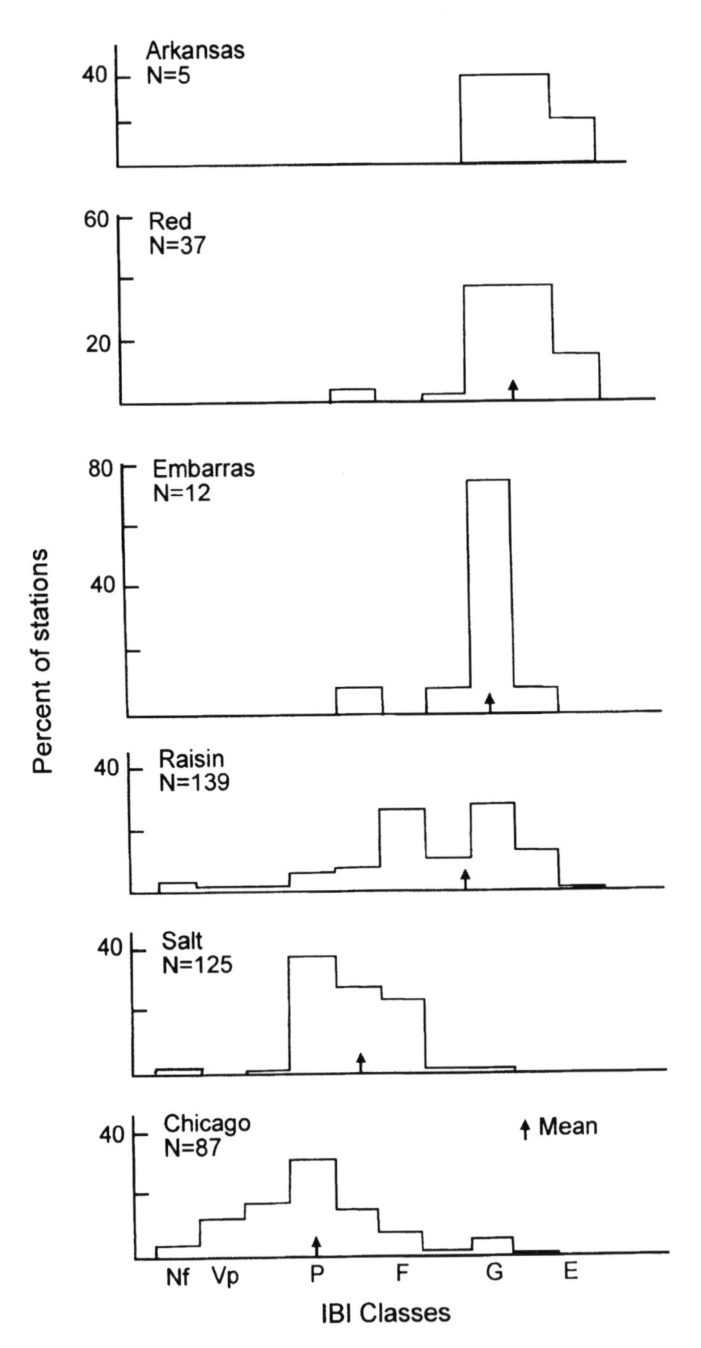

FIGURE 20.7. Distribution of sites by IBI classes in six midwestern regions or watersheds (modified from Fausch et al. 1986). Integrity classes as defined in Figure 20.6.

For example, downstream from a wastewater treatment plant on Copper Slough in east-central Illinois discharging effluent that had undergone standard secondary treatment with chlorination, fish IBIs were significantly lower than upstream (Figure 20.9) (Karr et al. 1985). When chlorine was not present in discharge water, however, upstream and downstream

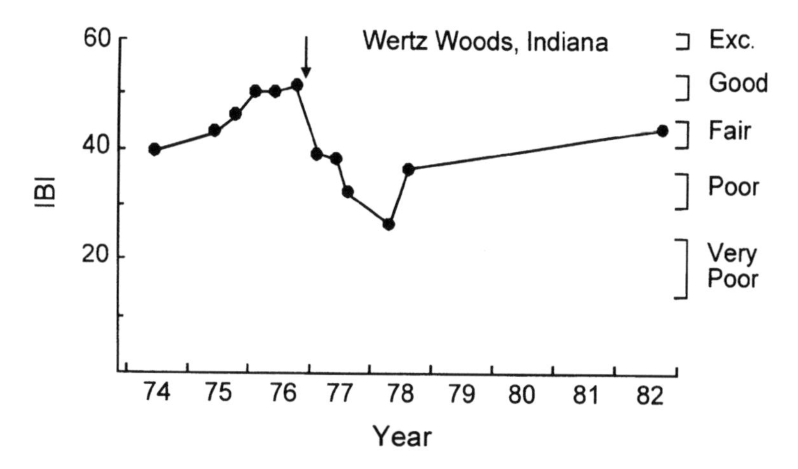

FIGURE 20.8. Changes in IBI over time in Wertz Drain at Wertz Woods, Allen County, Indiana (from Karr et al. 1986, with permission). Sedimentation in Wertz Woods from upstream channel restoration in spring, 1977 (arrow) caused deterioration of both habitat quality and the resident fish community. IBIs clearly indicate this decline followed by a slow improvement.

sites did not differ statistically. Finally, installing expensive tertiary treatment for nitrates, mandated by chemical water quality standards, did not improve the biological status of the stream. In an Ohio study, fish IBI increased substantially after significant reductions in pollutants (e.g., oxygen-demanding wastes, suspended solids, ammonia, some nutrients, and in some cases heavy metals and other toxins) in the effluent from two wastewater treatment

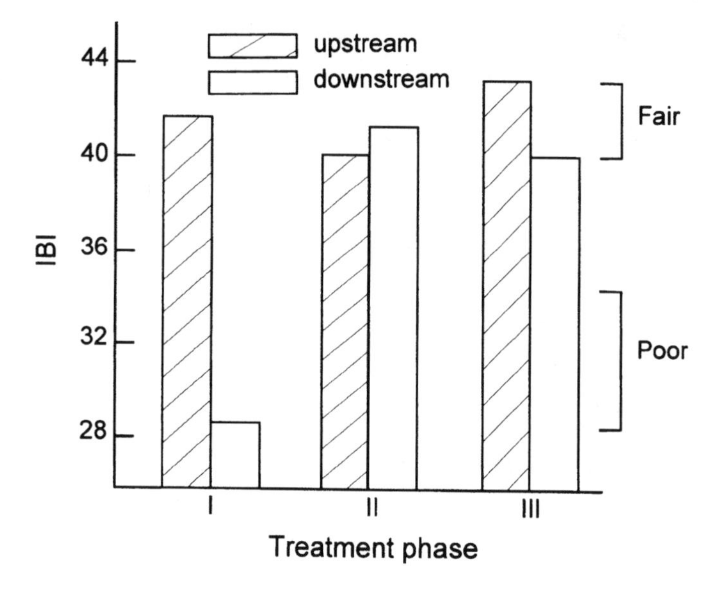

FIGURE 20.9. IBIs for sampling stations upstream and downstream of wastewater treatment effluent in Copper Slough, east-central Illinois. Phase I, standard secondary treatment ($p < 0.001$); phase II, secondary treatment without chlorination (ns); phase III, secondary treatment without chlorination but with tertiary denitrification (ns).

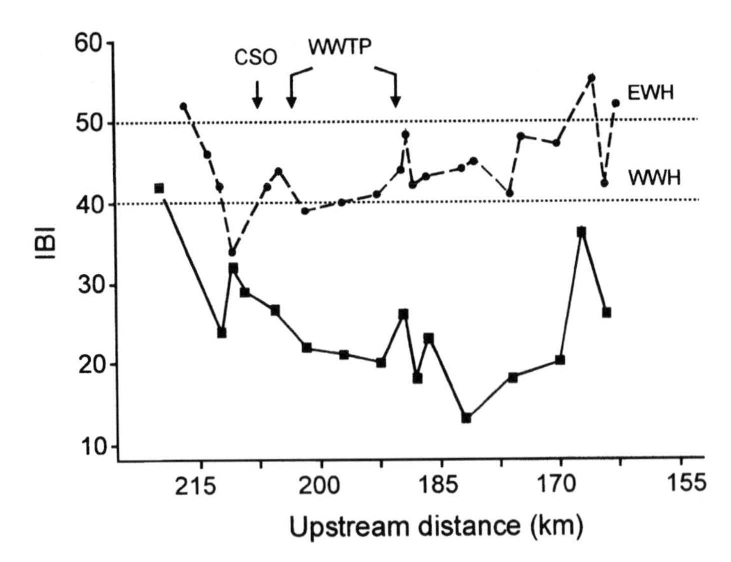

FIGURE 20.10. Longitudinal trend in IBI for the Scioto River, Ohio, at Columbus, 1979 (–■–) and 1991 (–●–) (data from Yoder and Rankin 1995). Arrows indicate locations of wastewater treatment plants (WWTP) and combined sewer overflow (CSO). Dotted lines indicate numeric biological criteria based on IBI used in enforcement of water quality standards for two levels of designated use under the federal Clean Water Act (excellent warmwater habitat [EWH] and warmwater habitat [WWH]).

plants on the Scioto River (Figure 20.10) (Yoder and Rankin 1995).

Statistical Power and Precision of IBI

IBI is versatile because it supports the use of standard analysis techniques such as ANOVA or *t*-tests designed to evaluate specific hypotheses. Statistical power analysis (Figure 20.11), based on fish data collected by Ohio EPA, demonstrates that IBI can detect six distinct categories of resource conditions (Fore et al. 1994). IBI is thus an effective monitoring tool, both to communicate qualitative biological condition and to provide quantitative assessments for use in legal or regulatory contexts.

These examples demonstrate that biological assessment of stream health can be applied in diverse contexts. In contrast, failure to incorporate direct biological evaluations into water resource or fishery management (as in the case of tertiary sewage treatment and chlorine laden effluent) can waste financial resources as well as degrade biological resources.

A Benthic IBI for the Pacific Northwest

Tests of the IBI concept in recent years demonstrate that a benthic invertebrate IBI (B-IBI) provides an effective assessment of river condition in varied circumstances (Ohio EPA 1988, Karr and Kerans 1991, Kerans and Karr 1994, DeShon 1995, Fore et al. 1996, Karr 1997). Macroinvertebrates, such as insects, crustaceans, mollusks, and worms, lend themselves to biomonitoring in the Pacific Northwest. Northwest streams have few fish species (low species richness) but many kinds of macroinvertebrates (Fore et al. 1996). Macroinvertebrate assemblages are ubiquitous, abundant, relatively easy to sample, and encompass taxa with differential responses to a broad spectrum of human activities. Because the life cycles of some benthic invertebrates extend several years, they are better integrators of past human influences than water chemistry.

Until recently, few systematic studies have been done on the biota of Pacific Northwest streams, especially with regard to human effects on those systems. In contrast, the impacts of landscape alteration on hydrological processes and water quality are well known (Booth 1991), as are the general impacts of human activity on salmon populations (Bottom 1996) and the effects of urbanization on Lake Washington, Washington State (Edmundson 1991). The shifts in algal assemblages of Lake Washington involved an easily defined culprit: nutrient enrichment caused by growing quantities of human waste released into the lake. The solution was simple and specific. King County created a regional sewage management agency to control the problem; the improvement in the lake was rapid and dramatic (Edmundson 1991). But without detailed biological studies of streams, managers and scientists cannot always so clearly convey the full consequences of poorly planned development; nor can they reverse degradation.

The first systematic stream study examined urbanization, the largest threat to streams in the Puget Sound lowlands. A two-year study of 24 Puget Sound streams was launched in 1994 under a grant from the Centennial Clean Water Program of the Washington Department of Ecology. Benthic invertebrate samples (three replicates) were collected in each stream from a riffle segment chosen on the basis of cobble size, cover, flow, and slope.

Nine attributes (mayfly, stonefly, caddisfly, intolerant, long-lived, and total taxa richness; relative abundance of planaria and amphipods, tolerant taxa, and predator taxa) of stream invertebrates changed systematically along a gradient of human influence (measured as percent impervious area). Aerial and satellite images were analyzed to determine the area within each watershed in each land use. Imperviousness, which is a measure of a landscape's ability to absorb rainfall, varies from 2% for natural forest, 5% for clear-cut areas, 20% for medium-density single-family housing (3–7 units/ha) to 60% for light-industrial and 90% for commercial development (Law 1994). Percent impervious area is directly and linearly related to biological integrity; no threshold of degradation stands out (Figure 20.12). Invertebrate assemblages varied more with impervious area (91% variation among sites), than among replicate samples from a single site (9% variation among replicates). Some biological attributes declined as human influence increases (e.g.,

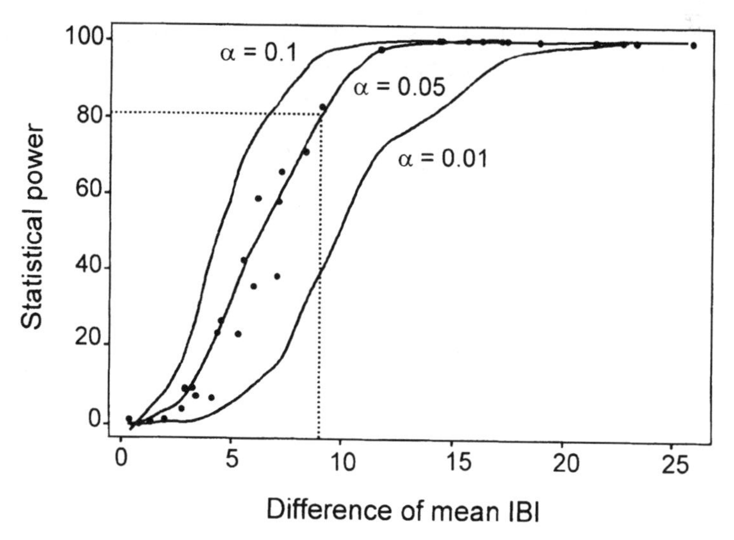

FIGURE 20.11. Statistical power analysis curves for IBI estimated from nine locations sampled three times (from Fore et al. 1994, with permission). Actual points are shown only for $\alpha = 0.05$; other values of α are shown as smoothed lines. For 80% power (dotted line), IBI can reliably detect a difference in scores of about 8 points at an level of $\alpha = 0.05$. IBIs can range from 12 to 60, thus IBI can detect six overlapping categories of biotic integrity.

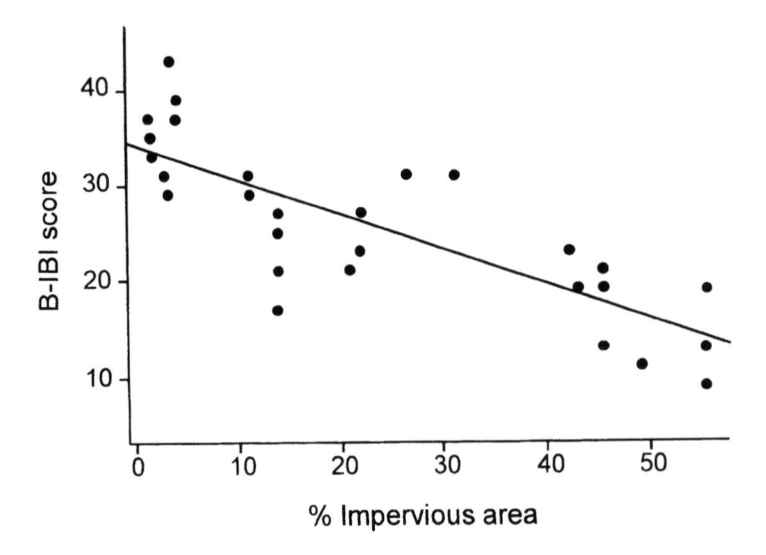

FIGURE 20.12. Relationship between B-IBI and % impervious area for streams sampled in the Puget Sound lowlands in September 1994.

total taxa richness and richness of mayflies, stoneflies, caddisflies, and intolerant taxa); other attributes increased (e.g., number of tolerant taxa such as worms).

The least degraded streams in this study (e.g., Rock, Griffin, and Big Anderson Creeks) had B-IBIs ranging from 35 to 45 (out of 45); the most degraded streams (Des Moines, Juanita, Kelsey, and Thornton Creeks) had B-IBIs below 15. B-IBIs for streams sampled in both years changed very little from one year to the next. None of 13 sites sampled in both years changed by more than four IBI points, while 10 changed by 2 or less.

B-IBI can also be related to the health of salmonid populations. For sites with fish data, a B-IBI above 35 had ratios of juvenile salmon (*Oncorhynchus* spp.) to juvenile cutthroat trout (*Oncorhynchus clarki*) greater than 5; sites scoring below 35 B-IBI had salmon-to-cutthroat ratios of less than 2 (Horner et al. 1996). In short, macroinvertebrate assemblages are ideal for discerning biological impacts of logging (Fore et al. 1996), livestock grazing, recreation (Patterson 1996), and urbanization (Kleindl 1995, Rossano 1995, 1996).

Regardless of locations, four broad classes of metrics (taxa richness and composition, tolerant and intolerant taxa, feeding ecology, and population attributes) reflect gradients of human influence and successfully distinguish least- and most-disturbed sites when benthic invertebrates are used (Fore et al. 1996). Generally, trophic metrics are less reliable for invertebrates than for fishes. Because the characteristics of invertebrate assemblages vary among regions, as do quality of data sets and knowledge of regional faunas, the number of attributes with demonstrated dose-response curves also varies: 13 metrics were used in Tennessee (Kerans and Karr 1994), 10 in logged landscapes of southwest Oregon (Fore et al. 1996), 9 in lowland streams in Puget Sound (Kleindl 1995), 10 in Grand Teton National Park (Patterson 1996), and 11 in a study of 115 streams in west-central Japan (Rossano 1995, 1996).

Integrating these studies, a regionally appropriate IBI with 10 metrics in four groups is recommended: (A) taxa richness and composition: total number of taxa, number of mayfly, stonefly, caddisfly, and long-lived taxa; (B) taxa tolerance: number of intolerant taxa and number of clinger taxa, proportion of individuals belonging to tolerant taxa; (C) trophic or feeding ecology: proportion of individuals that are predators; and (D) population attributes:

dominance (percent) of the three most abundant taxa (Table 20.3).

In sum, when the multimetric IBI approach is used to assess invertebrate responses along gradients of human influence, the results are strikingly consistent in four geographic areas (Tennessee, Grand Teton Mountains, Pacific Northwest, and Japan) and for a number of major human impacts (agriculture, grazing, logging, mining, recreation, and urbanization). In addition to conveying information about site condition, IBI, along with other knowledge of sampled watersheds, can speed identification of the causes of degradation and thus suggest management guidelines to prevent further degradation or reverse past damage.

In the Puget Sound study, for example, the B-IBI expected from general watershed conditions for one stream (Coal Creek) was higher than the B-IBI observed. This discrepancy led to the discovery of biological effects remaining from an old coal mine in the watershed. The B-IBI from a number of reference sites identified by Wyoming state agencies was also lower than expected. A number of these sites, assumed to be minimally influenced by human actions, were then discovered to be biologically degraded (K. King, Wyoming DEQ, personal communication). In this case, the existing methods of defining reference sites turned out to be inadequate (Patterson 1996).

As 15 years of experience shows, water resource protection improves when biological indexes, conventional water chemistry, and hydrologic analyses complement one another; the results are more cost effective and better safeguard biological resources (Davis and Simon 1995).

Change and Risk Assessment

Risks exist when the health or well-being of human society is threatened by natural events or by the consequences of what humans themselves have done (Mazaika et al. 1995, Commission on Risk Assessment and Risk Management 1996). People may or may not be able to prepare for natural disasters, but people can avoid the risks ensuing from their own actions. Unfortunately, widespread evidence of environmental degradation indicates that people have not effectively assessed and protected themselves from human-induced ecological risks. Instead, people behave as if they were immune to such risks; they assume that ecological risks do not exist or that the effects of human actions can be reversed when a crisis develops (Karr 1995b).

Ecological risk assessment must be broader than conventional assessment of risks to human health. Its focus must not be exclusively human but must incorporate the integrity of all living systems. Protecting living systems from chemical contamination alone—the focus of conventional risk assessment—is not enough. Integrity encompasses all species, the health of individuals, and the status of regional landscapes. It is these landscapes that supply clean air and water, produce food and fiber, and offer recreation and other amenities that make up the quality of life. Biological integrity includes both the elements of living systems and the processes that generate and maintain those systems (Karr 1993, 1996). Systems with biological integrity maintain the natural capital assets (goods and services) needed to sustain human society (Prugh et al. 1995).

In some respects, the term *ecological risk assessment* obscures the real issue—*biological* risks. The laws of physics and chemistry are immutable, but living, biological systems and their supporting elements and processes can be lost. Nonhuman living systems are bellwethers of conditions that threaten humans; they are also vital to human society. Ecological risk assessment must directly remedy the threat to human society caused by depletion of Earth's living systems.

Multimetric indexes are important because they effectively accomplish the mandate defined by the EPA Science Advisory Board (SAB 1990): "Attach as much importance to reducing ecological risk as is attached to reducing human health risk." Going beyond the human health consequences of chemical contaminants requires explicit recognition that people can no longer behave as if ecological risks did not exist (Karr 1996). Society must improve its understanding and measurement of ecological risks. Multimetric biological assessment, such as that provided by IBI, identifies the condition of living systems, which is the best primary endpoint to assess environmental quality. Systematic monitoring of biological resource conditions helps communities keep the natural assets that support them. It is also a major step toward restoring degraded systems—reversing the trend toward resource damage and depletion that has prevailed during the twentieth century.

Acknowledgments. Over the last 24 years, the U.S. Environmental Protection Agency funded much of the author's work in stream and watershed ecology, which led to and fostered the development of the index of biological integrity. The Washington Department of Ecology supported the study of lowland Puget Sound streams, and additional work leading to this paper was funded in part by the Consortium for Risk Evaluation with Stakeholder Participation (CRESP), established under U.S. Department of Energy Cooperative Agreement #DE-FC01-95EW55084.S. The author thanks Ellen W. Chu and several anonymous reviewers for their time and help in making this a better manuscript.

Literature Cited

Allan, J.D., and A.S. Flecker. 1993. Biological diversity conservation in running waters. BioScience **43**:32–43.

Angermeier, P.L., and J.R. Karr. 1994. Biological integrity versus biological diversity as policy directives. BioScience **44**:690–697.

Armstrong, J.S. 1967. Derivation of theory by means of factor analysis, or Tom Swift and his electric factor analysis machine. American Statistician **21**:17–21.

Barbour, M.T., J.B. Stribling, and J.R. Karr. 1995. Multimetric approach for establishing biocriteria and measuring biological condition. Pages 63–77 *in* W.S. Davis and T.P. Simon, eds. Biological assessment and criteria: Tools for water resource planning and decision making. Lewis, Boca Raton, Florida, USA.

Barbour, M.T., J. Gerritsen, G.E. Griffith, R. Frydenborg, E. McCarron, J.S. White, et al. 1996. A framework for biological criteria for Florida streams using benthic macroinvertebrates. Journal of the North American Benthological Society **15**:185–211.

Booth, D.B. 1991. Urbanization and the natural drainage system: Impacts, solutions, and prognoses. Northwest Environmental Journal **7**:93–118.

Bottom, D.L. 1996. To till the water: A history of ideas of fisheries conservation. *In* D.J. Stouder, P.A. Bisson, and R.J. Naiman, eds. Pacific salmon and their ecosystems: Status and future options. Chapman & Hall, New York, New York, USA.

Bowen, J.T. 1970. A history of fish culture as related to the development of fishery programs. American Fisheries Society Special Publication **7**:71–93.

Boyle, T.P., G.M. Smillie, J.C. Anderson, and D.R. Beason. 1990. A sensitivity analysis of nine diversity and seven similarity indices. Research Journal Water Pollution Control Federation **62**:749–762.

Chutter, F.M. 1972. An empirical biotic index of the quality of water in South African streams and rivers. Water Research **6**:19–30.

Commission on Risk Assessment and Risk Management. 1996. Risk assessment and risk management in regulatory decision making. Draft report. Commission on Risk Assessment and Risk Management, Washington, DC, USA.

Courtenay, W.R., Jr., and P.B. Moyle. 1992. Crimes against biodiversity: The lasting legacy of fish introductions. *In* Transactions 57th North American Wildlife and Natural Resources Conference. Wildlife Management Institute, Washington, DC, USA.

Davis, W.S. 1995. Biological assessment and criteria: Building on the past. Pages 15–29 *in* W.S. Davis and T.P. Simon, eds. 1995. Biological assessment and criteria: Tools for water resource planning and decision making. Lewis, Boca Raton, Florida, USA.

Davis, W.S., and T.P. Simon, eds. 1995. Biological assessment and criteria: Tools for water resource planning and decision making. Lewis, Boca Raton, Florida, USA.

DeShon, J.E. 1995. Development and application of the invertebrate community index (ICI). Pages 217–243 *in* W.S. Davis and T.P. Simon, eds. Biological assessment and criteria: Tools for water resource planning and decision making. Lewis, Boca Raton, Florida, USA.

Edmundson, W.T. 1991. The uses of ecology: Lake Washington and beyond. University of Washington Press, Seattle, Washington, USA.

Einstein, H.A. 1972. Sedimentation (suspended solids). Pages 309–318 *in* R.T. Oglesby, C.A. Carlson, and J. McCann, eds. River ecology and man. Academic Press, New York, New York, USA.

Evans, R. 1965. Industrial wastes and water supplies. Journal of the American Water Works Association **57**:625–628.

Fausch, K.D., J.R. Karr, and P.R. Yant. 1984. Regional application of an index of biotic integrity based on stream fish communities. Transactions American Fisheries Society **113**:39–55.

Fausch, K.D., J. Lyons, P.L. Angermeier, and J.R. Karr. 1990. Fish communities as indicators of environmental degradation. American Fisheries Society Symposium **8**:123–144.

Fore, L.S., J.R. Karr, and L.L. Conquest. 1994. Statistical properties of an index of biological integrity used to evaluate water resources. Canadian Journal Fisheries and Aquatic Sciences **51**:1077–1087.

Fore, L.S., J.R. Karr, and R.W. Wisseman. 1996. Assessing invertebrate responses to human activities: Evaluating alternative approaches. Journal of the North American Benthological Society **15**:212–231.

Gerritsen, J. 1995. Additive biological indices for resource management. Journal of the North American Benthological Society **14**:451–457.

Gotelli, N.J., and G.R. Graves. 1996. Null models in ecology. Smithsonian Institution Press, Washington, DC, USA.

Green, R.H. 1979. Sampling design and statistical methods for environmental biologists. John Wiley and Sons, New York, New York, USA.

Hilsenhoff, W.L. 1982. Using a biotic index to evaluate water quality in streams. Department of Natural Resources Technical Bulletin Number 132, Madison Wisconsin, USA.

Horner, R.R., D.B. Booth, A. Azous, and C.W. May. 1996. Watershed determinants of freshwater ecosystem character and functioning: Results of 10 years of research in the Puget Sound region. *In* L. Roesner, ed. Effects of watershed development and management on aquatic ecosystems. Engineering Foundation, New York, New York, USA.

Howarth, R.W. 1991. Comparative responses of aquatic ecosystems to toxic chemical stress. Pages 169–195 *in* J. Cole, G. Lovett, and S. Findlay, eds. Comparative analyses of ecosystems: Patterns, mechanisms, and theories. Springer-Verlag, New York, New York, USA.

Hughes, R.M., and J.R. Gammon. 1987. Longitudinal changes in fish assemblages and water quality in the Willamette River, Oregon. Transactions of the American Fisheries Society **116**:196–209.

Hurlbert, S.H. 1971. The nonconcept of species diversity: A critique and alternative parameters. Ecology **52**:577–586.

James, F.C., and C.E. McCulloch. 1990. Multivariate analysis in ecology and systematics: Panacea or Pandora's box? Annual Review of Ecology and Systematics **21**:129–166.

Johnson, R.W., and R. Paschal. 1995. The limits of prior appropriation. Illahee **11**:40–50.

Karr, J.R. 1981. Assessment of biotic integrity using fish communities. Fisheries **6**(6):21–27.

——. 1990. Biological integrity and the goal of environmental legislation: Lessons for conservation biology. Conservation Biology **4**:244–250.

——. 1991. Biological integrity: A long-neglected aspect of water resource management. Ecological Applications **1**:66–84.

——. 1992. Measuring biological integrity: Lessons from streams. Pages 83–104 *in* S. Woodley, J. Kay, and G. Francis, eds. Ecological integrity and the management of ecosystems. St. Lucie, Delray Beach, Florida, USA.

——. 1993. Defining and assessing ecological integrity. Environmental Toxicology and Chemistry **12**:1521–1531.

——. 1995a. Clean water is not enough. Illahee **11**:51–59.

——. 1995b. Risk assessment: We need more than an ecological veneer. Human and Ecological Risk Assessment **1**:436–442.

——. 1996. Ecological integrity and ecological health are not the same. Pages 100–113 *in* P. Schulze, ed. Engineering within ecological constraints. National Academy, Washington, DC, USA.

Karr, J.R., and D.R. Dudley. 1981. Ecological perspective on water quality goals. Environmental Management **5**:55–68.

Karr, J.R., K.D. Fausch, P.L. Angermeier, P.R. Yant, and I.J. Schlosser. 1986. Assessment of biological integrity in running water: A method and its rationale. Illinois Natural History Survey Special Publication Number 5, Champaign, Illinois, USA.

Karr, J.R., R.C. Heidinger, and E.H. Helmer. 1985. Sensitivity of the index of biotic integrity to changes in chlorine and ammonia levels from wastewater treatment facilities. Journal of the Water Pollution Control Federation **57**:912–915.

Karr, J.R., and B.L. Kerans. 1991. Components of biological integrity: Their definition and use in development of an integrative IBI. Pages 1–16 *in* T.P. Simon and W.S. Davis, eds. Environmental indicators: Measurement and assessment endpoints. Proceedings of the 1991 Midwest Pollution Control Biologists' Meeting. EPA 905/R-92/003. US Environmental Protection Agency, Chicago, Illinois, USA.

Karr, J.R., and T.E. Martin. 1981. Random numbers and principal components: Further searches for the unicorn. Pages 20–24 *in* D. Capen, ed. The use of multivariate statistics in studies of wildlife habitat. USDA Forest Service General Technical Report RM-87.

Karr, J.R., P.R. Yant, K.D. Fausch, and I.J. Schlosser. 1987. Spatial and temporal variability of the index of biotic integrity in three midwestern streams. Transactions of the American Fisheries Society **116**:1–11.

Karr, J.R., and E.W. Chu. 1998. Restoring life in running waters: Better biological monitoring. Island Press, Washington, DC. In press.

Kerans, B.L., and J.R. Karr. 1994. A benthic index of biotic integrity (B-IBI) for rivers of the Tennessee Valley. Ecological Applications **4**:768–785.

Kleindl, W.J. 1995. A benthic index of biotic integrity for Puget Sound lowland streams, Washington, USA. Master's thesis. University of Washington, Seattle, Washington, USA.

Kolkwitz, R., and M. Marsson. 1908. Okologie der pflanzlichen saprobien. Berichte der Deutschen botanischen Gesellschaft **26a**:505–519. Translated 1967. Ecology of plant saprobia. Pages 47–52 *in* L.E. Kemp, W.M. Ingram, and K.M. Mackenthum, eds. Biology of water pollution. Federal Water Pollution Control Administration, Washington, DC, USA.

Law, A.W. 1994. The effects of watershed urbanization on stream ecosystem integrity. Master's thesis. University of Washington, Seattle, Washington, USA.

Lenat, D. 1988. Water quality assessment of streams using a qualitative collection method for benthic macroinvertebrates. Journal of the North American Benthological Society **7**:222–233.

——. 1993. A biotic index for the southeastern United States: Derivation and list of tolerance values, with criteria for assigning water quality ratings. Journal of the North American Benthological Society **12**:279–290.

Ludwig, J.A., and J.F. Reynolds. 1988. Statistical ecology. John Wiley & Sons, New York, New York, USA.

Maser, C., and J.R. Sedell. 1994. From the forest to the sea: The ecology of wood in streams, rivers, estuaries, and oceans. St. Lucie, Delray Beach, Florida, USA.

Mazaika, R., R.T. Lackey, and S.L. Friant, eds. 1995. Ecological risk assessment: Use, abuse, and alternatives. Human and Ecological Risk Assessment **1**:337–458.

Meffe, G.K. 1992. Techno-arrogance and halfway technologies: Salmon hatcheries on the Pacific coast of North America. Conservation Biology **6**:350–354.

Meyer, J.L., W.H. McDowell, T.L. Bott, J.W. Elwood, C. Ishizaki, J.M. Melack, et al. 1988. Elemental dynamics in streams. Journal of the North American Benthological Society **7**:410–432.

Miller, D.L., P.M. Leonard, R.M. Hughes, J.R. Karr, P.B. Moyle, L.H. Schrader, et al. 1988. Regional applications of an index of biotic integrity for use in water resource management. Fisheries **13**(5):12–20.

Mitchell, W.C., and A.F. Burns. 1938. Statistical indicators of cyclical events. National Bureau of Economic Research, New York, New York, USA.

Naiman, R.J., ed. 1992. Watershed management: Balancing sustainability and environmental change. Springer-Verlag, New York, New York, USA.

Nehlsen, W., J.E. Williams, and J.A. Lichatowich. 1991. Pacific salmon at the crossroads: Stocks of salmon at risk from California, Oregon, Idaho, and Washington. Fisheries **16**(2):4–21.

Norris, R.H. 1995. Biological monitoring: The dilemma of data analysis. Journal of the North American Benthological Society **14**:440–450.

Ohio EPA (Environmental Protection Agency). 1988. Biological criteria for the protection of aquatic life. Volumes 1–3. Ecological Assessment Section, Division of Water Quality Monitoring and Assessment, Ohio EPA, Columbus, Ohio, USA.

Osenberg, C.W., R.J. Schmitt, S.J. Holbrook, K.E. Abu-Saba, and A.R. Flegal. 1994. Detection of environmental impacts: Natural variability, effect size, and power analysis. Ecological Applications **4**:16–30.

Patterson, A.J. 1996. The effect of recreation on biotic integrity of small streams in Grand Teton National Park. Master's thesis. University of Washington, Seattle, Washington, USA.

Plafkin, J.L., M.T. Barbour, K. Porter, S. Gross, and R.M. Hughes. 1989. Rapid bioassessment protocols for use in streams and rivers: Benthic macroinvertebrates and fish. EPA/440/4-89/001. US Environmental Protection Agency, Washington, DC, USA.

Pringle, C.M., R.J. Naiman, G. Bretschko, J.R. Karr, M. Oswood, J. Webster, et al. 1988. Patch dynamics in lotic systems: The stream as a mosaic. Journal of the North American Benthological Society **7**:503–524.

Prugh, T., R. Costanza, J.H. Cumberland, H. Daly, R. Goodland, and R.B. Norgaard. 1995. Natural capital and human economic survival. ISEE, Solomons, Maryland, USA.

Rapport, D.J., H.A. Regier, and T.C. Hutchinson. 1985. Ecosystem behavior under stress. American Naturalist **125**:617–640.

Rexstad, E.A., D.D. Miller, C.H. Flather, E.M. Anderson, J.W. Hupp, and D.R. Anderson. 1988. Questionable multivariate statistical inference in wildlife habitat and community studies. Journal of Wildlife Management **52**:794–798.

Reynoldson, T.B., and D.M. Rosenberg. 1996. Sampling strategies and practical considerations in building reference data bases for prediction of invertebrate community structure. Pages 1–31 *in* R.C. Bailey, R.H. Norris, and T.B. Reynoldson, eds. Study design and data analysis in benthic macroinvertebrate assessments of freshwater ecosystems using a reference site approach. Ninth Annual Technical Information Workshop, June 4, 1996. 44th Annual Meeting of the North American Benthological Society, Kalispell, Montana, USA.

Rossano, E.M. 1995. Development of an index of biological integrity for Japanese streams (IBI-J). Master's thesis. University of Washington, Seattle, Washington, USA.

——. 1996. Diagnosis of stream environments with index of biotic integrity. (In Japanese and English.) Museum of Lakes and Streams, Sankaido, Tokyo, Japan.

SAB (Science Advisory Board). 1990. Reducing risk: Setting priorities and strategies for environmental protection. SAB-EC-90-021. US Environmental Protection Agency, Washington, DC, USA.

Schindler, D.W. 1987. Determining ecosystem responses to anthropogenic stress. Canadian Journal Fisheries and Aquatic Sciences **44**(Supplement 1):6–25.

——. 1990. Experimental perturbations of whole lakes as tests of hypotheses concerning ecosystem structure and function. Oikos **57**:25–41.

Schmitt, R.J., and C.W. Osenberg, eds. 1996. Detecting ecological impacts: Concepts and applications in coastal habitats. Academic Press, San Diego, California, USA.

Simon, T.P., and J. Lyons. 1995. Application of the index of biotic integrity to evaluate water resource integrity in freshwater ecosystems. Pages 245–262 *in* W.S. Davis and T.P. Simon, eds. 1995. Biological assessment and criteria: Tools for water resource planning and decision making. Lewis, Boca Raton, Florida, USA.

Sioli, H. 1975. Tropical rivers as expressions of their terrestrial environments. Pages 275–288 in F.B. Golley and E. Medina, eds. Tropical ecological systems: Trends in terrestrial and aquatic research. Springer-Verlag, New York, New York, USA.

Stewart-Oaten, A. 1995. Rules and judgments in statistics: Three examples. Ecology **76**:2001–2009.

——. 1996. Goals in environmental monitoring. Pages 17–28 *in* R.J. Schmitt and C.W. Osenberg, eds. 1996. Detecting ecological impacts: Concepts and applications in coastal habitats. Academic Press, San Diego, California, USA.

Stewart-Oaten, A., W.W. Murdoch, and K.R. Parker. 1986. Environmental impact assessment: "Pseudoreplication" in time? Ecology **67**:929–940.

Suter, G.W., II. 1993. A critique of ecosystem health concepts and indexes. Environmental Toxicology and Chemistry **12**:1533–1539.

Underwood, A.J. 1991. Beyond BACI: Experimental designs for detecting human environmental impacts on temporal variations in natural populations. Australian Journal of Marine and Freshwater Research **42**:569–587.

——. 1994. On beyond BACI: Sampling designs that might reliably detect environmental disturbances. Ecological Applications **4**:3–15.

US EPA (US Environmental Protection Agency). 1990. Biological criteria: National program guidance for surface waters. EPA 440-5-90-004. US EPA, Office of Water Regulations and Standards, Washington, DC, USA.

——. 1996a. Summary of state biological assessment programs for streams and rivers. EPA 230-R-96-007. US EPA, Office of Policy, Planning, and Evaluation, Washington, DC, USA.

——. 1996b. Biological assessment methods, biocriteria, and biological indicators: Bibliography of selected technical, policy, and regulatory literature. EPA 230-B-96-001. US EPA, Office of Policy, Planning, and Evaluation, Washington, DC, USA.

Weber, C.I. 1981. Evaluation of the effects of effluents on aquatic life in receiving waters—an overview. Pages 3–13 *in* J.M. Bates and C.I. Weber, eds. American Society for Testing and Materials Special Publication Number 730, Philadelphia, Pennsylvania, USA.

White, R.J., J.R. Karr, and W. Nehlsen. 1995. Better roles for fish stocking in aquatic resource management. American Fisheries Society Symposium **15**:527–547.

Wilhm, J.L., and T.C. Dorris. 1968. Biological parameters for water quality criteria. BioScience **18**:477–481.

Wolda, H. 1981. Similarity indices, sample size, and diversity. Oecologia **50**:296–302.

Yoder, C.O., and E.T. Rankin. 1995. Biological response signatures and the area of degradation value: New tools for interpreting multimetric data. Pages 263–286 *in* W.S. Davis and T.P. Simon, eds. 1995. Biological assessment and criteria: Tools for water resource planning and decision making. Lewis, Boca Raton, Florida, USA.

21
Social Organizations and Institutions

Margaret A. Shannon

Overview

• Understanding society is fundamental to stream science and management because people establish institutional and organizational boundaries based on streams and rivers. As a result, social organizations and institutions create a complex tapestry of relationships shaping and constraining stream processes. Because science and policy are negotiated, who is involved in the process and the institutional and organizational context are critical elements of stream science and stream management.

• Institutions, organizations, and policy communities are key concepts for understanding the social institutions and organizations that shape stream ecology and management. Loosely organized networks of policy communities are the mechanism whereby innovations occur and are translated into both new institutional arrangements and new organizational forms.

• Historically, policy communities organized around specific resources. Thus agencies, constituents, and legislators were locked into intertwined institutional relationships commonly referred to as "the iron triangle." Today, pressures from both ecosystem science and ecosystem management are forcing changes in resource institutions and demanding new organizational forms based on integration, dialogue, and participation. Working collaboratively, developing common approaches to shared problems, creating new organizational and administrative mechanisms for implementing new policies, and integrating science and policy are all elements of the new environment of stream management.

• The norms of interaction between people and organizations vary systematically with the forms of their interaction, thereby forming a typology of policy community cultures. Natural resource policy communities based on specific resources (e.g., timber, water, wildlife), tend to have a mediative culture in which the various stakeholders and organizations seek to maximize their individual interests and values, while in part accommodating the interests and values of others. However, a policy community with a collaborative culture is one in which members understand that they share a common fate and assume responsibility for one another.

• The openness of the institutional environment and the skills of the policy entrepreneurs are critical to developing effective collaboration across jurisdictional and organizational boundaries. Building collaborative policy communities that can lead to new, integrative institutional arrangements requires an organizational strategy to increase complexity by creating new relationships. When interests are sufficiently expanded, it is possible to identify common ground.

• When the culture of policy communities is collaborative its norms are based on personal relationships and civic friendship. Trust and learning occurs when organizations embrace experimental and tentative solutions to problems.

• A typology of organizational decisions based on purposes and objectives in combination with the available technology is helpful for understanding how organizations approach decisions. Computation, experimentation and pragmatism, bargaining and advocacy with technical competence, and consensus building and organizational learning illustrate a range of decision-making processes common to the Pacific coastal ecoregion.

Introduction

At a time when our society is pulled as far apart as possible by the centrifugal force of individualism, water teaches us interdependence and cooperation. . . . Water is the last, best hope we may have to remind us . . . we do in fact "all live downstream" (Baril 1995).

More than two millennia ago, the Greek philosopher Heraclitus wrote, "you can never step in the same river twice," an idea which resonates immediately with stream ecologists. The most fundamental notions of stream ecology stem from an understanding of the universe as a place of constant change, unfolding relationships, emergent structures, repetitive patterns, and complexity seeming to border on chaos. As ecologists often repeat, "ecosystems not only are more complex than we think, they are more complex than we can think."

Yet, in their daily life and professional careers, stream ecologists also create institutional stability and organizational predictability by defending disciplinary boundaries, establishing predictable career paths, and organizing research projects often lasting decades. As scientists, stream ecologists seek to systematically construct a body of lasting knowledge that can be applied across time and place.

Streams provide an excellent vantage point from which to grasp patterns of change and stability. Water moves; it flows; it connects; it changes form. Like water, social organizations and institutions connect, shape, transform, adapt and change form. Stream ecologists and managers intuitively recognize the irony of using streams and rivers as jurisdictional boundaries, for they connect more than they divide. However, the proclivity of people to establish institutional and organizational boundaries based on streams and rivers makes understanding society fundamental to stream science and management. Social organizations and institutions create a complex tapestry of relationships shaping and constraining stream processes.

Fear of the accumulation of power in any one jurisdiction or organization results in fragmentation of power and authority across numerous jurisdictional, administrative, and organizational entities in the modern (liberal) state (Meidinger, 1998). As a result, public policy regarding trees, streams, fish, wildlife, birds, vertebrates, nonvertebrates, plants, terrestrial habitat, aquatic habitat, riparian habitat, and so on is fragmented across numerous jurisdictions, often with conflicting authorities. Efforts to create a more holistic and integrated ecosystem-based policy must first overcome these jurisdictional and organizational boundaries. This requires collaboration across jurisdictions, among multiple organizations and stakeholders, and about multiple resources and problems.

As society seeks better policies for protecting and managing stream resources, scientists and managers need to understand the institutional and organizational environment of streams so that they can contribute to improved collaboration (Shannon 1992c). This chapter begins by describing the social context of a typical stream in the Pacific coastal ecoregion, a description that could fit anywhere by simply renaming the subdivisions of jurisdiction and authority. Portions of interviews with several participants in recent bioregional science and policy assessments provide insight into how managers and scientists use their understanding of streams to shape new policies and management opportunities. These interviews clearly link the processes of science with the possibilities for policy and management. They make it very clear that streams are as much social products as ecological entities.

These examples and definitions of key concepts provide the background for an integrative analysis of institutional strategies, organizational decision-making strategies, and the culture of relationships among those shaping the policies and management of streams and rivers. While analytically distinguishable, what people do is clearly interrelated with what is possible, who is involved, where they are working, how they define the problem, and who has the power or authority to do something about it. These interdependencies are woven through the examples presented throughout this analysis.

Working collaboratively, developing common approaches to shared problems, creating new organizational and administrative mechanisms for implementing new policies, and integrating science and policy are all elements of the new environment of stream management. This chapter illustrates how these new ways of working together fit with current organizations and institutions, and what it may take to realize the promise of improved policies regarding the streams and rivers in the Pacific coastal ecoregion and around the world.

A Drop of Water

Think of a drop of water that falls one winter as a snowflake on top of Mt. Olympus in the center of Olympic National Park, Washington. When spring arrives and the snowpack melts, it enters a rivulet flowing into a stream along with many other drops forming a rushing torrent. Eventually, this stream crosses the boundary of Olympic National Park into the Olympic National Forest, where it is subject to the laws, policies, and traditions of the National Forest System. After gathering more strength, the stream flows off federal lands and onto State of Washington Experimental Forest lands. Next, it crosses onto corporately owned private industrial forest land. Then the stream flows by homes and towns, where land ownership is fragmented among numerous individuals and municipalities. Just before it enters Puget Sound, it passes through a Native American Reservation. In this river, the drop passes by a town, by houses and factories, under bridges, over a viaduct, through a culvert, and swirls beneath the docks into Puget Sound. The drop merges with the waters of Puget Sound and flows out toward the Pacific Ocean and then out of sight.

All along the way, the drop of water moved through the jurisdictions of different institutions governing the use of the stream (Figure 21.1). Sometimes the institutions shaped individual rights to use property-in-land or property-in-water for specific purposes. Sometimes, like property, the institutions were jurisdictional and gave specific authorities to various agencies or owners to directly manage natural resources or to regulate their use by others. When the drop entered the ocean, it was lost in the myriad, overlapping, crisscrossing national and international jurisdictions that govern the ocean waters. Table 21.1 is a partial list of typical federal, state, and tribal management and regulatory agencies that may influence the path of water.

From the perspective of the water, as long as the natural flow of the stream continues unimpeded, the interwoven human institutions serve to maintain necessary channel and stream processes. However, when the flow or nature of the stream is changed or diverted for human purposes, then these intersecting human institutions and organizations change the stream and its environment. For example, in the case of the drop of water, its natural course continued relatively unimpeded, but if there had been a diversion of the water from the stream into an irrigation storage dam, then the reduced water in the channel might be insufficient to maintain the channel processes essential for the habitat requirements of associated aquatic and riparian species.

Or, if people continue to build more houses, use more drinking or yard water from the stream, and put more waste products into it, the flow would be reduced and the quality of water changed. If the uses of the forested areas change, resulting in fewer trees, more roads, more people fishing, boating, and trampling along the river's edge, then the land around the stream is changed, affecting not only the quality of the water (temperature, turbidity, channel processes) but also the capacity of the riparian functions to sustain the plants and animals living near the stream. The more that social life shapes the course of the stream, the more important understanding social organizations and institutions becomes for effective stream science or management.

Thinking Like Scientists and Managers

No matter who they work for or what they study, stream scientists and managers will find themselves working with others. How do managers and scientists working on streams think about this complex social environment? How do they make sense of the complex tapestry of agencies, authorities, landowners, citizens and others in their daily work? Selections from interviews with stream scientists and managers give some insight into these questions (Boxes 21.1 and 21.2). These interviews were done as part of a study (done under contract for the USDA Forest Service) of eleven bioregional science-policy assessments to develop working guidelines for the relationships between scientists and managers.

Listening to these key players, it is clear that both science and policy are negotiated, and thus who is involved is crucial to what is known as well as what is done. For any stream, what is known and what is done depends on who is involved, what questions are important to them, and what options are viewed as organizationally viable by different stakeholders. Thus, institutional and organizational context is an essential element for stream science and stream management.

Key Concepts Defined

Three key concepts—*institutions, organization*, and *policy community*—must be defined to provide a clear conceptual basis for describing and

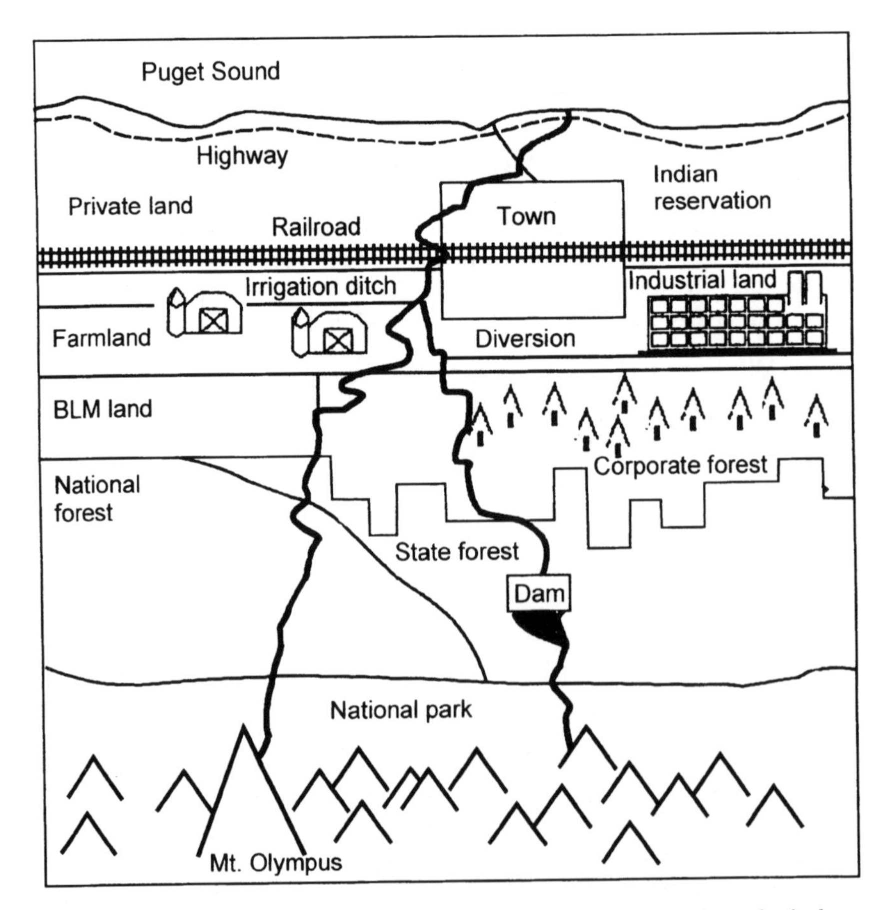

FIGURE 21.1. Schematic map of land ownership and jurisdiction for a hypothetical stream.

TABLE 21.1. Partial list of federal, state, and tribal management and regulatory agencies.

Regulated use	Agency
Land use	U.S. Department of Agriculture
	U.S. Forest Service
	Natural Resource Conservation Service
	U.S. Department of Interior
	U.S. Park Service
	Fish and Wildlife Service
	Bureau of Land Management
	Bureau of Indian Affairs
	U.S. Department of Commerce
	National Marine Fisheries Service
	State of Washington
	Department of Natural Resources
Water use	State of Washington
	Department of Ecology
	Federal Energy Regulatory Commission
	Local Irrigation Cooperatives
	Municipal Water Departments
Water quality	State of Washington
	Department of Ecology
	U.S. Environmental Protection Agency
	Puget Sound Water Quality Authority
Aquatic habitat	State of Washington
	Department of Fish and Wildlife
	Native American tribes
	U.S. Department of Agriculture
	Forest Service
	U.S. Department of Interior
	Fish and Wildlife Service
Fisheries	State of Washington
	Department of Fish and Wildlife
	Native American tribes
	Pacific Fisheries Commission
	U.S. Department of Interior
	Fish and Wildlife Service

discussing social institutions and organizations. Research suggests that innovations occur as a result of enterprising individuals working within loosely organized networks of policy communities. Once new ideas, concepts, and approaches occur and are repeated, these innovations are translated into both new institutional arrangements and new organizational forms.

Institutions—Institutions can be defined as persistent patterns of relationships among people and organizations that transcend physical boundaries and tend to be stable over relatively long periods of time. Institutions provide templates for patterns of social organization and the capacity to produce and reproduce social relationships, organizations, and cultural life ways.

Just as a stream channel shapes the flow of water, social institutions shape social life into organized patterns. At one level, institutions define broad cultural ways of social and economic life (e.g., hunting and gathering, agriculture, or industrialization). Evolution of these macrolevel institutions works at a time scale approaching that of climate change or fluctuations in rainfall and temperature (White 1980, 1991 and 1995, Cronon 1983 and 1991, Perlin 1989).

In general, institutions are intergenerational, and thus their rules and norms affect intergenerational equity, provide for the transfer of property rights across generations, shape the forms of organization characteristic of different eras, and overall, maintain a way of life shared by members of a given society. Generally, institutional change is slow but steady in an evolutionary sense. Occasionally, in the face of dramatic shifts in conditions (e.g., the overrunning of Native Americans inhabiting the Pacific Northwest by European settlers in the 1800s), institutional change can be disjunctive as a new social order is created. Viewed from the scale of human generations, institutions provide a relatively consistent template for society in different times and places (Meidinger, 1998).

Human institutions are multilayered, overlapping, and integral to one another across all scales, just as a stream reach is embedded within a watershed and within the larger hydrologic cycles of the globe. In the same way that the hydrologic cycle sends the drops of water to reproduce streams, so social institutions guide individual actions to produce and reproduce organizations day after day, year after year, generation after generation, until society changes and evolves. Then, some organizations disappear and new ones form, reflecting new institutional arrangements (Powell and DiMaggio 1991).

Organization—Organization is the mobilization of people and resources to accomplish desired purposes. Organizations have an identi-

Box 21.1. Water Rights Adjudication Processes: Federal Reserve Water Right Claims in State Courts. Confidential interview from the case on federal in-stream water right claims in Colorado for the U.S. Forest Service Science-Policy project done in 1996.

Stream Scientist

Interviewer: What is meant by channel maintenance?

Stream Scientist: It's the amount of flow required within a channel system to actually maintain the integrity of the system, or its channel geometry. Maintaining channel geometry means that the channels don't shrink from reduced capacity to move sediment and then become more prone to frequent flooding.

Interviewer: When people say in-stream flow, what are they talking about?

Stream Scientist: I guess when I think of it, it would be the developing policy and procedures for determining how best to determine the amount of water and then manage that water to protect the wide variety of things, including the channels themselves and the fisheries values, recreation values, scenic values, whatever.

Interviewer: Why did a water rights adjudication process begin?

Stream Scientist: A power company was not getting all the water to which it had legal title, because upstream users were using some of the water that should have gone though its turbines and generated electricity and revenues. So the shareholders sued the power company, and as the case unfolded, the state courts decided that they would adjudicate the whole river basin. I don't think they understood at that time what they were getting into.

Interviewer: Who do you work with on this research?

Stream Scientist: I guess if you include all the people dealing with adjudications in some way, its probably 100 to 200 people. Probably 30% researchers and 70% managers of resources specifically related with a given adjudication.

Interviewer: How are research needs identified?

Stream Scientist: There was an interagency symposium dealing with in-stream flows. It was mainly to get the agencies to mutually discuss what their problems were, what types of information were available, and who was doing the type of research that might apply to other agencies. It was more of an exchange to get to know the other agencies, their problems, and how they're trying to solve things. At the end of the symposium, we had a half-day workshop that dealt with where to go from here, how to get the agencies to prioritize what we think are important needs to get research off the ground or to exchange information.

Interviewer: Are lawyers involved?

Stream Scientist: Agency lawyers are involved in prioritizing what needs to be done and the urgency with which it needs to be done. OGC (Office of General Counsel, USDA) and (Department of) Justice attorneys are involved in the different adjudications. . . . I would guess each individual adjudication and its involvement with non-federal agencies or the public probably varies a lot. Some of the states are willing to negotiate with the federal government on certain types of claims, and then there is more public and state interaction.

Box 21.2. PACFISH: An Interim Strategy for the Management of Anadromous Fish in the Pacific Northwest. Confidential interviews from the analysis of the Pacific Salmon Fisheries assessment for the US Forest Service Science-Policy project, 1996

Fisheries Scientist

Interviewer: What led to PACFISH?

Fisheries Scientist: An article was published saying that more than 100 stocks of salmon and steelhead were already extinct, over 200 were at a high risk of extinction if current trends continued, another 100 stocks were at moderate risk of extinction, and only a few were secure. That was the red flag. Federal agencies managed a significant proportion of the land with anadromous stream habitat and they recognized that they would have to play a significant role in the protection and recovery of these stocks.

Interviewer: How did you get involved with PACFISH?

Fisheries Scientist: I'm a fisheries scientist. I've worked over 20 years monitoring anadromous fish all across the Pacific Northwest. When the need for a science assessment on salmon arose, I was asked to work on it. The idea was to do an assessment, look at the status of the critters, and then put together a strategy for management. This would be a cooperative effort within the Forest Service, and it eventually included the Bureau of Land Management, as well as other land managing and regulatory agencies.

Interviewer: How was the scientific work developed?

Fisheries Scientist: A lot of the research had already been done regarding status of stocks and status of habitat. In the Columbia River Basin a survey had been done about 50 years ago that had a well-documented sampling design and good quantitative data. As a result, PNW (Pacific Northwest) Research Station scientists resurveyed the exact areas and made direct comparisons across 50 years of time. That was very useful and very enlightening in establishing changes in habitat over five decades. It really showed the degree of degradation of habitat broadly across the Pacific Northwest over that period of time. A lot of it was associated with forestry and grazing activities on federal lands as well as with other human impacts on the landscape. This study put current habitat conditions in context and clearly showed that even the best habitat today was considerably degraded compared to 50 years ago.

Interviewer: Was the assessment a catalogue of habitat conditions in different watersheds?

Fisheries Scientist: Yes, that's part of it. Once there was a good picture of the condition of habitats in the Columbia Basin and other parts of the Northwest, then the issue became what sort of strategy would we develop to protect the remaining good habitats and try to restore or enhance damaged habitats. That became the two-pronged approach of the PACFISH Assessment.

Interviewer: Scientists didn't necessarily stay out of the protection issues at that point?

Fisheries Scientist: No, we didn't. We basically put together the strategy and then it was refined with managers and so on. But our little team of scientists created the components of that strategy. Those were things like defining what good habitat is. That is, how you would describe it quantitatively: how many pools per mile, what

Continued

Box 21.2. *Continued*

water depth, what water temperature, and things like that. We worked on the standards and guides that went along with PACFISH, although that's kind of a management thing. The managers were heavily involved in that, too.

Aquatic Scientist

Aquatic Scientist: Yeah, PACFISH is even, in my opinion, getting more successful in that more people are saying: "hey, we've got to something different than business as usual." What PACFISH has done is broadened it from outside the BLM (Bureau of Land Management) and the Forest Service, and on their review teams they had people from Fish and Wildlife Service, the National Marine Fisheries Service, and the Environmental Protection Agency going around with them to these different districts, examining projects and talking about the success of the implementation of these guidelines.

Interviewer: Why do you think these large-scale assessments will continue?

Aquatic Scientist: I think in fact to get some clarity as to species viability analysis, or cumulative effects, and put these together with the social issues and other ecological issues, you have to look broader than a single town. Get the owls, fish, or other organisms in the picture, it becomes much more of a regional pattern, and when you add in the economics and social stuff, its more like a Greater Yellowstone Ecosystem idea. When ecological issues are wide ranging, outside the bounds of one region or one forest, then these assessments need to be done. The political and administrative boundaries of the Forest Service are inadequate to deal with the social and ecological realities. And yet, we are still stuck with these administrative boundaries that are where the legal line in the sand is drawn.

Land Manager

Interviewer: How did managers participate in the PACFISH assessment?

Land Manager: The new perspective that emerged from this assessment was that what we were saying was not what we were doing, and we had to make some changes. The current policies were not adequate as clearly evidenced by the comparisons to conditions 50 years ago. So, we recognized the importance of having field people involved in crafting the solution. The field team had a very fluid membership. Some meetings we would have 100 people with representatives from each of the regions. Depending on the purpose of the meeting, we would have specialists, too. They would say, "ok, how do we get more wood in the water, because we've removed too much wood from the streams over the years." The silviculturist would suggest something; the hydrologists would talk about the upper reaches of the stream, and we would talk about sedimentation.

So the standards and guides moved forward slowly until we had general consensus by the field people that this was what was needed. Sometimes we invited line managers to these meetings, and that served us well because even though we had no decision-making power as a group we knew we were influencing what decisions would be made. So this partnership of scientists and field managers provided the credibility we needed to develop a credible, implementable strategy. Very congenial relationships developed and a lot of trust was created as we argued things out over the months. That congeniality and respect for one another's roles did a lot to overcome the resistance to the PACFISH strategy when it was implemented as interim direction by modifying all the federal land management plans in

Continued

Box 21.2. *Continued*

the Pacific Northwest, except Alaska which had to develop a separate analysis.

Decision Maker

Interviewer: Sounds like the science and policy process was very closely intertwined in this case?

Decision Maker: Right. I'll be real candid with you. I think when we began PACFISH there was probably not the best level of trust between scientists and managers. The process we went through together in PACFISH developed a lot of trust. There was skepticism at the beginning that the problem really was this big and needed such a big policy change over such a large region. I think it finally became clear that based on the science that we had and the information that we had, people said "heck yes, we got to do this." I think that's where everybody came out. So through the whole process, I think a lot of skepticism went away, and folks said, "hey, this is real stuff, we got good science here and we got to get something done."

Interviewer: Who was involved in the ongoing consultations in PACFISH?

Decision Maker: We met with all the tribes and tribal governments in the Columbia River Basin. We met with all kinds of different groups and talked to them about what we were doing. We had pretty wide involvement with a lot of different people and other scientists looking at it at the same time. We were involved with a lot of people, talking to them about the situation to see what they knew and what they thought about it. Some of the tribes have tremendous fisheries biologists and we talked to those folks. They offered us advice. We had pretty good coverage, I think.

fiable structure of roles and responsibilities. Some organizations are more capable than others of self-renewal and self-creation over time, making them resilient in the face of a changing environment (Hage and Aiken 1970, Kickert 1993). Organization is necessary when individuals cannot accomplish their goals alone but must work with others to achieve them. It is easiest to think of organizations as populated by individuals and social groups joined together to achieve a common goal (e.g., to manage public lands, to build houses, bridges or dams, or to provide drinking water).

Organizations allow people to act across time and space, within varying degrees of social complexity, and with the constant awareness that every action creates a reaction that reverberates through both the social and biophysical environments. It is clear that any one spot on a stream is embedded within a myriad of relationships, and thus connected to numerous institutions and organizations.

Policy community—A policy community is composed of a diverse network of public and private organizations generally associated with the formation and implementation of policy in a given resource area (e.g., streams, water, wildlife). Policy communities are interactive networks of alliances around common interests. They create bridges across institutions by linking multiple organizations. Policy communities have distinctive cultures because each one links a distinct set of agencies, jurisdictions, people, and places.

Collaborative policy communities are based upon the kind of norms found in face-to-face relationships—friendship. Friendship in this case has a very special meaning. It is civic friendship, or the capacity to empathize with another's situation and take it into account when making decisions which affect others and the common good (Stanley 1983). Aristotle used this idea of friendship in his concept of citizen. This is not the friendship of common usage, but the concept of a critical friend who forces questioning of a course of action based upon concern for the other (Forester 1996). Civic friendship is based upon the recognition of a shared problem and a mutual interest in solving it (Shannon 1992a, b, c). To better understand this idea, consider the words of Katherine Baril, Water Resources Coordinator, Jefferson County, Washington.

"The [Puget Sound Water Quality] Authority also adopted watershed planning guidelines which required that all diverse interests in the watershed come together—first to prioritize local drainage, and then to form local watershed planning councils. Those first meetings were often very intense and disagreeable as farmers, environmentalists, tribal fishermen, and teachers met for the first time. Now, in my watershed, we've learned a new rule of engagement called the 'Take your enemy to lunch' rule. It developed because we've learned that the best thing you can do is take a person you can't stand and get to know them better. We learned to put enemies together in car pools or bus rides for three-hour watershed tours. We asked that people take off those caps which identify their employers or result in people stereotyping their interests based on their hats. No, we want to get to know each other as individuals with interests as diverse as our watershed.... We learned that it is valid time to build understanding, respect and relation ships.... This doesn't happen because agencies do watershed planning, it happens when people do" (Baril 1995:130).

Trust is the word heard most often when people describe why something worked well (Shannon 1987b, 1996). Lack of trust is consistently first on the list when efforts fail to produce desired results. Building trust requires both personal commitments to do so and organizational forms which are based upon interaction and dialogue (Friedman 1987, Heckscher and Donnellon 1994, Forester 1996). Such organizational forms are evolving slowly, but progress, as illustrated by Katherine Baril's words, is evident in the management of streams and rivers.

The Changing Policy Environment of Streams

Historically, natural resource policies grew from the nation-building efforts of the nineteenth century. Policies embodied the claims of "Manifest Destiny": America was foreordained to encompass the entire continent and committed its natural resources to building a strong economy and new society in the process (White 1991). As a result, natural resource policy communities became classic examples of the "iron law of oligarchy," for they locked agencies, constituents, and legislators into intertwined institutional relationships (Michels 1962). In *Crossing the Next Meridian* (1992), Charles Wilkinson evocatively named these power blocks the "lords of yesterday." Quoting a 1951 comment to a reporter by an engineer in the Bureau of Reclamation, Daniel Worster (1993:135) captured the Bureau's view of rivers as tools for development, "we enjoy pushing rivers around." These "lords of yesterday" operated the engines of industrializing America (Hurst 1956), and natural resources were simply fuel (Worster 1993, Hirt 1994).

In 1969, ecologist Eugene Odum declared that "the landscape is not just a supply depot but is also the *oikos*—the home—in which we must live" (Worster 1993:161). With this phrase, he gave voice to a major shift in social values (Caldwell 1970a). Arguments for the importance of resource values based not on direct economic utility, but on existence value have ancient roots (Worster 1977). These arguments led to the reforming of natural resource policies and laws in the 1960s and early 1970s. Reforms included greater attention to multiple uses (resource uses were to be considered on an equal basis and a desired mix would be based on a broadly defined concept of the public interest) as well as a shift of policy priorities away from resources as fuel for the engines of industry and toward protection of environmental quality and the sustainability of ecosystems (Hirt 1994). State environmental and resource policies followed suit, and thus new ecosystem-based approaches to resource management on private as well as public lands are now common practice (John 1994).

Policy communities are those loose networks of individuals and organizations which form an alliance to shape public policy. Since scientific inquiry creates theories and concepts which can shape, change, inform, and direct policy, scientists are participants in policy communities. While scientists are always part of these policy alliances, their participation becomes controversial when new theories and concepts necessitate changes in policy and management

(Shannon et al. 1997). Thus, the policy changes in the 1960s and 1970s were based upon new concepts from systems theory in which components of systems could not be extracted without affecting the rest of the system. It was this core concept of whole systems that fundamentally challenged the principles of viewing resources as extractable components and land as merely the garden producing resources for human consumption.

The ideas of general systems theory affected not only concepts of ecological systems, but also models of political and social systems. By the 1970s, systems approaches were basic theory in nearly every field of scientific inquiry. Thus, it is not surprising that the 1970s laws and policies applied these concepts in reforming policy, and also included requirements for multidisciplinary analysis and public participation (across all areas of public policy). These requirements for inclusiveness and openness were intended to break the lock of the "iron triangles" over public policy (Schattschneider 1960). In natural resource policy, these "triangles" proved exceptionally strong and capable of resisting demands for change (Caldwell et al. 1994). This intractability led to the intense political struggles of the last fifteen years across the country as both agencies and landowners found themselves finally forced, usually by court decrees, to change their management policies and practices. Central to the debates about policy change in the Pacific coastal ecoregion were streams, rivers, and aquatic systems.

When considering streams and rivers, probably no scientific concept has had more effect on management and policy than "connectivity." Put simply, the idea of connectivity linked aquatic systems with riparian and upland systems so as to create one interdependent system of interlocking components. As a result, connectivity challenged the jurisdictional divisions of authority across agencies, governments, laws, and property ownership. It demanded a more integrative policy and a more inclusive decision process across as well as within organizations and policy communities. Suddenly, the legal basis for challenging practices of timber management, for example, could rest on the maintenance of stream systems and the management of upland forest areas had to provide the maintenance requirements of aquatic systems (Caldwell et al. 1994, Sedell et al. 1994). The policy processes of the last decade in the Pacific coastal ecoregion center around these issues. Policy continues to evolve and to transform the actions on public and private lands in order to adequately protect the long-term sustainability of stream systems (New York Academy of Sciences 1995).

It is in this way that ecosystem science and management are demanding more integrative organizational forms (Meidinger 1997). Old forms of organization were jurisdictionally fragmented (e.g., irrigation and dam-building were in different agencies), bureaucratically organized, resource or land use specific, and strongly linked to specific resource outputs. New organizational forms create networks of linkages across jurisdictions, form loosely coupled internal structures around interdisciplinary task groups, and are self organizing to allow continuous adaptation of management and policy to new values and knowledge.

Both ecosystem science and ecosystem management demand new organizational forms based on integration, dialog, and participation. Such organizational forms require new institutional templates to create and recreate them over time. Thus, the increasing interdependence between spheres of resource policy and management are creating strong pressures to build cross-linked institutional strategies—everyone is downstream sometime. More than ever, the policy environment is characterized by being multiproblem, multiissue, multiinterest, multiconstituency, multiorganization, multiinstitution, multijurisdiction, multiauthority, multination. Effective action, or any action at all, across these fragmented relationships requires strategies for forming new institutions (Thompson and Schwartz 1990).

In summary, policy communities once locked in "iron triangles" are becoming more inclusive and participatory as the shifts in social values are first reflected in new laws, and then in court decisions (Caldwell et al. 1994, Meidinger 1997). These definitive shifts in policy are creating new organizational forms and new decision-making processes as agencies and landowners work to take into consideration more than just the direct economic gains from resource use and production. Needed to facilitate cross-jurisdictional coordination and interorganizational collaboration and social learning are policy communities that can create new opportunities to work across resources, jurisdictions, and organizations, and thereby create the capacity for integrative policy solutions (Meidinger 1987, Shannon 1991).

What makes policy communities capable of innovation and change, however, are individuals (Wildavsky 1987, Burns and Stalker 1994). John Forester (1989, 1996) shows clearly that individuals capable of engaging others in open dialogue are those who can straddle the roles of mediator (a neutral position relative to the issue or outcome) and negotiator (an advocacy position relative to the issue or outcome). Within policy communities, the "negotiator" works to coalesce interest, attention and concern in developing strong policy alliances. Simultaneously, the "mediator" actively listens to all points of view, and thereby builds trust across all factions of the community. When individuals can effectively play both roles, they possess the form of leadership which is manifested as "teacher and facilitator" because they inspire others to behave differently and to build new relationships (Dewey 1927). As a result of this relative openness of policy communities, change in either institutions or organizations often begins with these new relationships and new expectations from within policy communities.

Forming Integrative Policy Communities

There was a time when the adage "a stream cleans itself every ten miles" was a sufficient policy to govern the relationships between downstream and upstream users. Fewer people, fewer noxious chemicals, and fewer long-term effects of human action meant that little coordination among owners and users was necessary for everyone to be satisfied. Such times are largely past.

Policy communities not only comprise organizations with "formal authority" (authority created from above) but also ad hoc organizations (e.g., the Willapa Alliance in southwestern Washington) in which "legitimate authority" is given by the people involved (Benveniste 1989). Sometimes several agencies at one or more levels of government must work together. Riparian property owners are inextricably linked to each other as well as to public agencies. Thus, jurisdictional authority over a stream is held by many different people, organizations, and institutions.

How will multiple individuals and organizations working within different institutional environments come together, develop mutually acceptable policies, and carry them out? The jurisdictions listed in Table 21.1 demonstrate that different agencies or owners have authority over specific aspects of streams: the land (public and private), water quality, water quantity, riparian habitat, aquatic habitat, water use and allocation, and so on. Federal agencies, created by Congress, are given a specific delegation of executive authority, and this fragmentation of authority continues at state and local levels. This history drives numerous recent efforts to develop integrated policies for streams, including water quantity and quality as well as riparian management and channel processes.

A policy community includes those individuals and organizations associated with shaping the working solutions to social problems that shape the institutions (Wildavsky 1987). Thus, a primary purpose of bioregional science-policy assessments like PACFISH (Box 21.2) is to create an integrative policy community by providing opportunities for personal relationships to grow and thereby build both trust and implementation capacity.

Shaped by people and time, policy communities have different cultures which overlap each reach of a stream. While cultures persist beyond individual lives and soon are considered a part of social reality, individuals create, maintain, and change culture through their personal relationships to one another (Becker 1982, Geertz 1973). An issue termed "contentious" generally describes the quality of the relationships among individuals and organizations as angry and uncivil. Yet, another place with similar issues and similar institutional and organizational settings may be characterized by a collaborative culture through the efforts of individuals to build and maintain civil, cooperative, working relationships.

Thus, not all policy communities are alike—who initiates dialogue, where people meet, who is involved, what they discuss, and how they understand one another all vary and shape the relationships and culture of the policy community (Yaffee and Wondolleck 1997). Following is a description of a typology of policy community cultures that can be applied generally to any policy community based upon its features.

As Figure 21.2 illustrates, the norms of interaction between people and organizations vary systematically with the forms of their interaction, thereby forming a typology of policy community cultures (Shannon 1987a, b). For example, when the quality of interaction is highly formalized (as in court proceedings or arbitration) and the norms of interaction are highly impersonal (e.g., plaintiff and defendant communicate only via legal representation in court), the culture is characterized by efforts to maintain autonomy while exercising power. Individuals interact through their roles, in their formal capacities, and as represented by legal counsel. Reliance upon legal challenges, legislative actions, or media advertisements are examples of such efforts to affect what others can do. One effective role of lawsuits in policy disputes is to pull the policy community culture toward an autonomous culture in an effort to overcome the power of entrenched oligarchies or charismatic individuals.

If the only interaction between individuals, organizations, or agencies occurs in a court of law, or formalized hearings, integrative policies are less likely because individuals and organizations are acting on the basis of formal roles, rules, and procedures. Often policy communities built around an autonomous culture gridlock because of the social distance between stakeholders. However, during science-policy assessments, impersonal relationships can grow into personal relationships, thus transforming the culture of the policy community. Once power has shifted making new policy options possible, stakeholders discover they can shape new policies by drawing on personal relationships such as those illustrated in the PACFISH example (Box 21.2). This openness of policy communities to the efforts by individuals to create personal relationships and to work together is why change in institutions and organizations tends to originate in policy communities.

Although ecological components of streams (the fish, water, stream bed, and land) are integrally related, natural resource policy

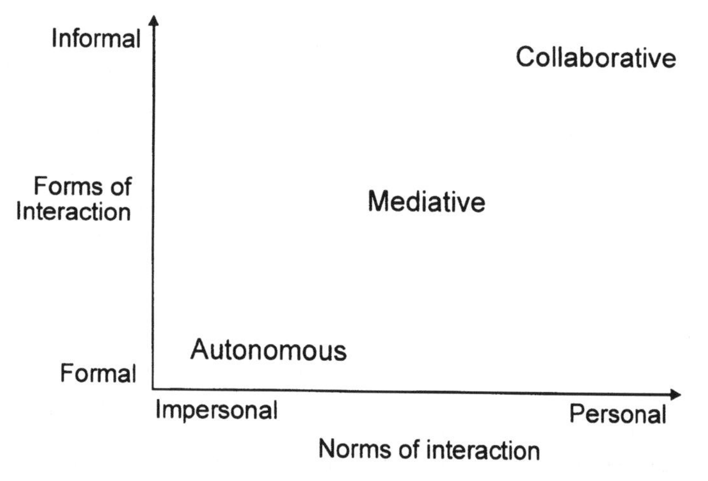

FIGURE 21.2. Typology of cultures in policy communities based on the norms and forms of interaction among individuals and organizations.

communities typically have a mediative culture because the various stakeholders seek to maximize their individual interests and values. A mediative culture typifies legislative politics and formal public participation in land-use planning processes. Individuals and groups develop and take pride in maintaining long-term relationships with one another. They accept that sometimes they will be "winners" and sometimes "losers," yet their consent arises from commitment to the fairness of the process, or a sense of procedural justice. The underlying culture of these policy communities is that entry and access into the process is open and available to all. Thus, outcomes are considered fair if all affected interests are represented and governed by the same rules. Not surprising, however, in practice a policy community with a mediative culture tends to result in policies favored by the interest(s) holding greater power (usually wealth), claim to exclusive jurisdiction, specialized knowledge, or other political resources. Policy uncertainty occurs when the outcomes are no longer clearly determined by these forms of political power.

Whether decision makers consider habitat quality when making decisions regarding how streams are used and managed illustrates the shifts in political power occurring in the Pacific Northwest. State agencies are responsible for the allocation of water rights and rights to capture fish and wildlife, but they have little jurisdiction over habitat relationships (ELI 1977). State natural resource and forestry agencies have sufficient authority over land use to require minimum levels of riparian protection on privately held land (Chapter 22). When the potential for extinction threatens a species, the federal Endangered Species Act strengthens government authority if an area is designated as critical habitat. Agencies managing federal lands have some provisions for consideration of stream and habitat consequences of management activities, the strongest being the current regulation of the USDA Forest Service regarding maintenance of viable, well-distributed wildlife populations (36 CFR Part 219). But until recently, the Forest Service, pulled by constituents, Congress, and professional training, gave little weight to considerations of stream quality or habitat maintenance. Rather, protection of stream quality was counted as a cost to producing commercial wood products, and efforts to protect streams only reduced revenues to the U.S. Treasury. Entrenched policy relationships combined with the professional training of staff, as well as the values and culture of the Forest Service, all contributed to the stability of these policies (Steen 1976, Clary 1986, Wilkinson 1992, Hirt 1994).

For policy to embrace efforts to protect streams, including retaining large, valuable trees and creating wide buffer strips along streams (where land-use activities are limited), political and economic power had to shift. The Clinton Administration embraced a new cultural norm requiring agencies to work together, speak with one voice, and develop collaborative partnerships with each other (FEMAT 1993, Gore 1995). In addition, federal court decisions determined that previous federal agency policies were inadequate in meeting the requirements of law and regulations, and demanded change before allowing federal timber management to continue in the Pacific Northwest (Caldwell et al. 1994). These events realigned power relationships and agencies were forced to develop new policies.

Whereas a mediative culture assumes that the public interest is created as individual interests are satisfied, and stakeholders consent to proposed decisions (Landy and Plotkin 1982), a collaborative culture assumes that the search for the common good binds people together as citizens. Citizens are obligated to join together in deliberating about the common good and shaping the destiny of the political community (Mansbridge 1980 and 1990, Kemmis 1990, Sandel 1996). Through such dialog and public deliberation, citizens shape their work to achieve the common good (Reich 1985).

The public philosophy of a collaborative policy community is one of belonging, a shared concern for the whole, and a moral bond with the community whose fate is at stake (Sandel 1996). The words of the scientists and managers involved in PACFISH (Box 21.2) indicate that the definitive nature of several scientific studies led to the (albeit reluctant) recognition that the salmon crisis was a shared problem, and no one could escape responsibility. As a result, PACFISH developed a strongly collaborative culture as scientists from many agencies worked closely with resource managers and specialists. The PACFISH strategies they created included a system of "key watersheds" and "riparian protection buffers" to protect streams on federal lands from the negative effects of timber harvest and other management activities (Sedell et al. 1994). The PACFISH strategies already provided interim direction across federal lands in the Pacific Northwest in 1993 when President Clinton asked the Forest Ecosystem Management Assessment Team (FEMAT) to develop an integrative policy for the protection of old growth forests and associated species, including anadromous fish. The Integrated Ecological Assessment of the Columbia River Basin had also developed new management and policy guidelines based on the PACFISH approach.

In addition to formally authorized policy processes like PACFISH, numerous civic associations also provide leadership by bringing together the agencies, land owners, municipalities, nongovernmental organizations, and citizens needed to fully deliberate the use and management of streams and rivers (John 1994). One organizational form of these groups is the *watershed association*. Watershed associations are growing, in part because states are reorganizing their policy processes along the boundaries of watersheds. Collectively, watershed associations are changing how agencies and others do business by creating a new culture based upon collaboration (Lee and Stankey 1992, Shannon 1992c). Watershed associations become integrative, collaborative policy communities as they engage in public deliberation (Reich 1985). The deliberative capacity of these relationships is remarkable when built on respect for difference and the value of diversity as demonstrated by the "take your enemy to lunch rule" described earlier by Baril.

While a collaborative culture may seem like either a hopeless dream or remnant of some past ideal time, management of streams and rivers proves otherwise. A policy community with a collaborative culture means that members accept their common fate and agree to assume responsibility for one another. Collaborative cultures often characterize relationships worked out over centuries and manifested in commonly understood meanings. The very institutional framework of Western water law which is expressed as "use it or lose it," and "first in time and first in right," make a collaborative culture difficult to realize but, as the watershed associations demonstrate, not impossible (Baril 1995).

What lessons can be drawn from understanding the cultures of policy communities associated with streams? First, streams are characterized by numerous, overlapping, inconsistent, and strong cultures associated with the individuals, organizations, and institutions they connect. Second, a collaborative culture is necessary to build institutional strategies that facilitate the kind of deliberation regarding multiple problems from multiple standpoints required to develop integrative policies. Third, a collaborative policy community evolves from specific efforts to work together, sometimes on small problems, like the watershed association example, and other times, on large ones, like PACFISH. Collaboration is not an abstract quality; it describes the effort needed to create, nurture, and sustain strong relationships of trust. Finally, policy communities are dynamic and closely reflect the norms and values of their participants.

Institutional Strategies for Collaboration

The role of institutions is to provide templates for reproducing patterns of relationships and organizations over time (Powell and DiMaggio 1991). However, as discussed earlier, innovations result from new relationships and new forms of interaction. Thus, individuals can greatly expand the integrative and deliberative capacity of institutions simply by talking to each other, seeking to understand multiple perspectives, and searching for ways to expand the problem to encompass enough issues to provide options for solutions. New relationships provide opportunities for innovations to become institutionalized (Burns and Stalker 1994). Thus, enhancing complexity by building new relationships is the organizational strategy necessary for building collaborative policy communities which can in turn lead to new, integrative institutional arrangements.

The Washington State Department of Natural Resources 1992 Forest Plan embraced landscape management as the means to implement new policy guidelines. As a part of a the state's effort to implement this landscape management approach, a group of scientists was convened to consider the ecological, managerial, economic, institutional, organizational, and land ownership strategies necessary for its implementation.

In an effort to provide an institutional, organizational, and social assessment, Shannon et al. (1995) examined a dozen examples where cross-boundary institutional arrangements were formally authorized, informally constituted, or simply devised ad hoc. The study design began with a literature review, developed a conceptual framework for understanding landscape institutions, proposed an ideal institutional arrangement as an evaluative benchmark, and briefly analyzed a dozen case studies of existing institutional strategies at varying geographic scales (Shannon et al. 1995).

In brief, the findings indicate that effective institutional strategies vary according to the geographic scope of the problem and the administrative capacity of the agencies (Shannon et al. 1995). First, in situations where the stakeholders are diverse, distant, and in direct competition with one another for access to resources, formal authorization of a "council" with specified authority is necessary (e.g., Northwest Power Planning Council, Tennessee Valley Authority, Great Lakes International Joint Commission). In such situations, enforcement is essential in order to achieve integrated action across multiple jurisdictions, across several states or nations, and with significant differences in local conditions. These cases are a reminder that the fragmentation of authority institutionalized into American government is both a centerpiece of the liberal state and a serious impediment to joint action (Meidinger 1997).

Second, when stakeholders are closer (geographical scale is smaller), are more similar to one another, and have opportunities for joint gains, then more informal approaches to landscape institutions can work (e.g., Timber/Fish/Wildlife Agreement, Washington, and Nisqually Resource Management Council, Washington). These informal, network organizations link across existing organizations, similar to the PACFISH assessment. They do not have direct coercive authority to force compliance, but rather build credibility for a course of action through face-to-face relationships and provide the substantive policy guidance that different organizations adopt in concert.

Third, when stakeholders share a localized place that includes a common community and there is no external organization that can singlehandedly implement broad-scale policy, then ad hoc organizations that provide a forum for agencies and others to work together may coalesce (e.g., Applegate Partnership, Oregon and Western Oregon Special Forest Products Council). Although watershed partnerships are a classic example of an ad hoc organization, they also illustrate the importance of shared social space (necessary for personal relationships) for forming the basis of legitimate authority (Forester 1992). This common institutional form is not new. Local civic associations have historically taken responsibility for the care and protection of streams. Indeed, it was this civic institution which provided the template for the formal institutional strategy of the New England River Basin Commissions.

In summary, these institutional strategies are associated with different cultures based on the nature of the relationships within the policy community and the norms governing those relationships. In general, the more personal the relationships, the more localized and informal the institutional strategy. Such strategies typically have a collaborative culture and the capacity for deliberation based upon both technical information and local knowledge. As institutional strategies affect larger bioregions with greater diversity of places and stakeholders, they tend to require more formalized coordination as found in procedures and rules. The more formalized they become, the more their culture tends to be mediative as representatives negotiate the best deal for themselves. When either the claims are highly rule based (e.g., conflicts among property claims) or the geographic distance between stakeholders precludes effective personal relationships, then the institutional strategies become more procedurally rigid and dependent on authority. In these instances, the culture is characterized by the autonomy of the stakeholders, which is maintained by separated institutional spheres (e.g., conflicts between water right claims are resolved in courts).

These interdependent relationships among institutional strategies and policy community cultures suggest that organizations make decisions differently when they participate in one culture compared to another, or are located within different institutional strategies (Wildavsky 1987). Indeed, while organizations nominally have particular styles of decision making, in practice how they make decisions is strongly affected by the context within which they work (Shannon 1987a, b). The next section explores a typology of decision-making approaches.

Typology of Organizational Decision Processes

Social institutions do not treat streams and rivers as wholes, but divide them into thousands of fragments, each forming the core of an organized set of relationships. Clearly, the institutions and organizations related to streams and rivers are not neatly arrayed on simple dimensions of scale and authority, but are woven into a remarkable pattern with an intricate and elusive design.

This image raises the question of how decisions and choices are made. What forms do decision processes take in these different, highly diverse organizations? The typology in Figure 21.3 provides a guide to understanding how organizations approach decisions based on their purposes and objectives in combination with the available *technology* (the combination of knowledge and capacity to use it) (Shannon 1987a, b). Organizations can be understood as acting on clear purposes when either there is a single overriding objective or there is a single organization with simple jurisdiction. Purposes are ambiguous when there are many, different objectives to be simultaneously achieved or there are several diverse and differentially affected organizations with potential jurisdiction. Similarly, technology is known when both the science is understood and the capacity to use it is well established in management routines. Technology is uncertain when either the scientific or managerial knowledge is rudimentary or the organizational policies to manage according to available knowledge are inadequate.

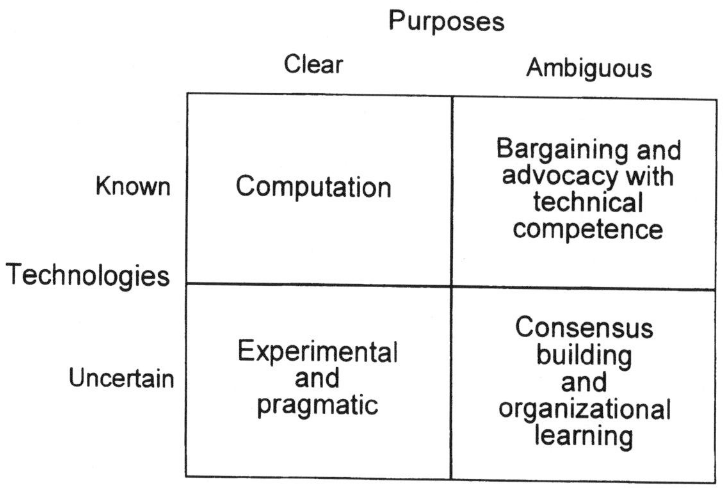

Technologies \ Purposes	Clear	Ambiguous
Known	Computation	Bargaining and advocacy with technical competence
Uncertain	Experimental and pragmatic	Consensus building and organizational learning

FIGURE 21.3. Typology of organizational decision processes based on the clarity of objectives and the relative ability to implement a course of action.

To understand these relationships consider, for example, what is happening when managers argue over the certainty of the knowledge regarding the kinds of riparian zone protection needed for stream habitat qualities. When they debate whether buffers of 8m, 30m, or the distance of two average tree lengths are sufficient to protect stream processes, are managers more concerned about the affects on other resource values and programs, such as timber harvest or livestock grazing, than on the habitat qualities of the stream? If so, then decisions must be negotiated by organizational program managers and scientific knowledge may become an impediment to protecting program commitments. However, if managers are uncertain as to the applicability of watershed scale analysis to a particular reach of a stream, then working in partnership with scientists to employ an adaptive management approach in which different buffer widths are tested against a set of criteria is ideal. Both decision strategies require an organizational commitment to risk taking and innovation, rather than the obstinate application of standardized management prescriptions (Daniels and Walker 1996).

Computation

When purposes are clear and technology is known, decisions can be made by computation. The computational form of decision process fits best in a bureaucracy working within an autonomous culture and a formal institutional arrangement (Reich 1985). By viewing decisions as the neutral application of rules, bureaucracies treat such choices as nonpolitical, meaning they are not choices among values (Cortner and Shannon 1993). When organizations insulate their decision processes from politics when values are at stake, the decisions err toward formalism; they do not respond to political differences when they should (Shannon 1987a, 1991).

Most federal agency manuals and procedures assume purposes are clear and technology is known. Partly, this assumption reflects a bureaucratic approach of developing a single rule and applying it to many places regardless of local variations (Kaufman 1960). Federal agency procedures in general assume that unambiguous cost-benefit ratios can be determined and that administrators simply choose the array of goods and services that has the best ratio of benefits to costs. Since 1982, Forest Service regulations require that cost effective options are selected from the array of alternative decisions, and that this is a straightforward matter of adding up all the values and subtracting all the costs (36 CFR Part 219).

It is possible to account for this simplifying approach in terms of the cultures of policy communities. First, bureaucratic organizations are designed to simplify the decision process by predetermining the decisions given the definition of the objective (Kaufman 1960). In the case of the Forest Service, the definition of forest management was long simplified to maximizing the production of commercial wood products. Second, the agency mission and policies were formalized into rules which, in the case of the Forest Service, gave little weight to the negative effects of timber harvest on streams and aquatic habitat. Such management routines allowed decisions to sell timber to appear as clearly the best option, as if unaffected by political influence inside or outside of the agency (Steen 1976, Clary 1986).

Experimentation and Pragmatism

When purposes are fairly clear or when the jurisdiction over the decision is largely within one organization, but the knowledge or management experience for how to achieve such purposes is uncertain, traditional methods of experimentation or scientific adaptive management can be applied. The objective may be stated quite simply—protect channel maintenance processes—but knowledge about channel maintenance processes and how management can achieve this goal are uncertain. For example, would ensuring bankfill at critical times of the year or allowing high water levels to maintain channel size and vegetation within the stream bed be sufficient to maintain channel processes? If so, what constitutes bankfill, and when are the critical periods for different streams? In this situation, decisions are stated as hypotheses about what is expected to happen as a result of management actions. Together scientists and managers develop a series of hypotheses about how the channel will respond to different conditions, and they monitor the actual outcomes over time. This type of decision process still presumes that decisions are the province of experts and require technical knowledge, but it acknowledges that they must negotiate solutions based upon insufficient information. This is often called scientific adaptive management (Lee 1993, McLain and Lee 1996).

Bargaining and Advocacy with Technical Competence

Decision processes require coordination among organizations and often exemplify a mediative culture based on bargaining and advocacy amongst the stakeholders over technical issues. In such cases, an organization provides the political forum for debate by allowing experts to interrogate one another and to test their ideas in practice. The policy of multiple use management encourages such processes. When interdisciplinary teams work effectively, they negotiate problem definitions and solutions based on their expertise and experience, and they pursue agreement on a reasonable course of action in the face of multiple (often conflicting) objectives (James 1971).

Recall, however, that organizational efforts to treat debates about values (purposes) as arguments about methods (technology) err toward formalism, following conventional management approaches when change is required. To say that technologies are known does not necessarily mean that they are uncontested. Indeed, the very purpose of policy standards and guidelines is to provide guidance for decisions when there are contested goals (GAO 1997). The danger is that "means" (technology) become defined as desired "ends" (purposes).

Clear-cutting is an example of a means that became an end. Originally, clear-cutting of forests in the Pacific Northwest served many purposes. Highly valued commercial tree species needed open areas and sunlight to regenerate. Roads were nonexistent and expensive to build. By building a short piece of road, clear-cutting a patch of trees, and then moving on, the patchwork of harvested areas would both pay for the construction of the transportation infrastructure and create the conditions for a sustainable flow of wood products. Even in the face of other, conflicting purposes, the political strength of this technology overwhelmed other possibilities. The clear-cutting controversy has lasted over three decades, and its resolution is far from clear (Caldwell et al. 1994). The National Forest Management Act of 1976 requires using the "optimum" method of harvest, and the law, as well as the regulations, specifies the criteria for determining what constitutes optimum. Nonetheless, in spite of its known shortcomings and deleterious effects on streams and aquatic habitats, large-scale clear-cutting remains a common method of timber harvest across the Pacific Northwest.

The clear-cutting controversy illustrates the importance of public deliberation about values when purposes are multiple. Even though public opposition to clearcutting was growing as a result of changing social values, forestry organizations continued to treat the debate as a challenge to their technical expertise. Arguments defending clear-cutting lost sight of it as simply a management practice designed to achieve the social goal of producing wood fiber. The more its defenders forgot that clear-cutting was a management technique which could be critically evaluated and modified while still achieving the goal of wood fiber production, the more acrimonious the public debate became as organizations and individuals concerned about other purposes, such as stream quality, argued for different management approaches (e.g., stream buffers with limited or no harvesting, smaller patch sizes or partial harvest approaches).

Consensus Building and Organizational Learning

A fourth type of decision process acknowledges the organizational and institutional complexity of natural resource management. Often such complexity is hidden by the application of bureaucratic rules and routines. Analytical methods of decision making assume political complexity away in the application of cost-benefit analysis. Addressing decisions and impacts only from the perspective of one organization's jurisdiction, authority, and technology assumes institutional complexity away. Ignoring the social values implicit in alternative technologies assumes social complexity away.

Attending to organizational and institutional complexity necessarily means that purposes are multiple, overlapping and conflicting. It also means that technology is necessarily uncertain, if only because knowledge and technique are contested. Decision processes with complexity rest on consensus building and organizational learning. The concept of consensus building rests on the importance of building a collaborative culture wherein all the stakeholders consent to sharing a problem and the need to act to solve it.

The idea of organizational learning places organizations in an adaptive stance relative to problems, solutions, and other organizations (Friedman 1987, Forester 1995, 1996). This means that an organization acts less as an advocate for its favored technologies and more as a participant deliberating the nature of the problem and negotiating a potential solution. A collaborative culture urges organizations to cope with change by acting within the situation, with respect to purposes and objectives, and with regard to one another (Shannon and Antypas 1996).

Learning is an outcome of deliberation. Learning means that individuals or organizations change what they do and how they do it. Only through organizational learning processes can institutional changes be realized. A collaborative culture increases the deliberative capacity of organizations (Shannon 1987a,b) and supports adaptive management within them (McLain and Lee 1996).

Forester (1995:25) reminds us that "deliberative encounters are there to be created within conflictual institutions, between organizations and actors who face uncertain consequences and ambiguous mandates, between citizens, politicians, developers, and other planners with agendas of their own." It is particularly important for stream scientists and managers to realize this because although streams are often treated as institutional and jurisdictional boundaries, stream scientists and managers and citizens realize that streams form natural linkages and can foster the deliberation and collaboration on a shared problem (Hoover and Shannon 1995).

Future Outlook

While there are reasons to be optimistic about processes of institutional change and the opportunities for individuals to work together in collaborative relationships, the Pacific Northwest also illustrates the difficulty of doing so. The Northwest Forest Plan (in its second year of implementation) embraced the organizational approach that allows field staff to work with broad-based standards and guides for stream protection and develop site-specific policy. Yet, as scientists in the region note, it is commonplace that even the best watershed analysis finds reluctant support from field specialists when it leads to new management approaches. Why? A possible conclusion is that institutions have not changed enough to allow organizations to embrace uncertainty and complexity of new management strategies. Fear of organizational retribution, fear of political challenge, and fear of downgraded performance ratings hold in place old patterns of routine and old management prescriptions. Yet, the solution may lie in sustained interaction among the entire policy community to build consensus around new approaches, to try them and see if they work, and to foster organizational learning by changing routine decisions and ways of working together.

Change requires individuals to engage in hard work as they build trust with each other, trust within their organization, trust among organizations that must work together on a shared problem, and finally trust within the institutional environment that the new policies will lead to improved outcomes. In addition to the natural slowness of change, policy change is problematic. While the institutional environment must change to prevent organizations and policy from reverting to traditional forms, the source of change is innovation through personal relationships formed in policy communities. Thus, the institutional environment holds the templates for reproducing organizations, policies, management activities, and careers. In order for new organizational forms, new policy solutions, new management approaches, and new personal futures to emerge, policy communities must first produce innovations, and then move those innovations through organizational decision processes into new organizational approaches. It is through the building of strong networks of "friendship" that individual managers and scientists can create through their own action the policy communities necessary to open institutions to change and allow new polices to work within the multiple organizations related to any one stream.

The policy processes in the Pacific Northwest have indeed developed strong new management approaches for the protection and restoration of streams across the region. The examples of new institutional strategies emerging here and around the country give some insight of the changes in progress. Scientists and managers working in the twenty-first century will find a very different organizational and institutional environment than those of the twentieth century. While the outlines of the new organizations are just emerging, the opportunities are in place for collaboration among individuals and organizations to realize the promise of these new policies and institutional strategies for better stream policies in the future.

Acknowledgments. Material presented in this chapter was developed in part by research funding from the USDA Forest Service Pacific Northwest Research Station through cooperative research, along with a research contract to conduct interviews with participants in bioregional science-policy assessments to better understand their organization and management. Computer assistance from Chris Meidinger, substantive comments of Errol Meidinger, Amanda Graham, and members of the Ph.D. Research Seminar of the Department of Public Administration, Maxwell School of Citizenship and Public Affairs, Syracuse University are gratefully acknowledged. The anonymous comments of the reviewers and editors of this volume were extremely helpful in suggesting useful and important revisions, and in confirming the approach of this analysis.

Literature Cited

Baril, K.L. 1995. The Puget Sound Water Quality Initiative: A case study in using the tools IV. Proceedings of the 4th National Watershed Conference, "Opening the toolbox: Strategies for successful watershed management" held in Charleston, West Virginia on May 21–24, 1995.

Becker, H. 1982. Culture: A sociological view. The Yale Review **71**:513–527.

Benveniste, G. 1989. Mastering the politics of planning. Jossey-Bass, San Francisco, California, USA.

Burns, T., and G.M. Stalker. 1994. The management of innovation. Oxford University Press, Oxford, UK.

Caldwell, L.K. 1970a. Environment: A challenge for modern society. Natural History Press, New York, New York, USA.

——. 1970b. The ecosystem as a criterion for public lands policy. Natural Resources Journal **10**:203–220.

Caldwell, L.K., C.F. Wilkinson, and M.A. Shannon. 1994. Making ecosystem policy: Three decades of change. Journal of Forestry **92**(4):7–10.

Clary, D.A. 1986. Timber and the Forest Service. University Press of Kansas, Lawrence, Kansas, USA.

Cortner, H.J., and M.A. Shannon. 1993. Embedding public participation in its political context. Journal of Forestry **91**(7):14–16.

Cronon, W. 1983. Changes in the land: Indians, colonists, and the ecology of New England. Hill and Wang, New York, New York, USA.

——. 1991. Nature's metropolis: Chicago and the Great West. W. W. Norton, New York, New York, USA.

Daniels, S.E., and G.B. Walker. 1996. Collaborative learning: Improving public deliberation in ecosystem-based management. Environmental Impact Assessment Review **16**:71–102.

Dewey, J. 1927. The public and its problems. The Swallow Press, Chicago, Illinois, USA.

ELI (Environmental Law Institute). 1977. The evolution of national wildlife law. Prepared for the Council on Environmental Quality. US Government Printing Office, Washington, DC, USA.

FEMAT (Forest Ecosystem Management Assessment Team). 1993. Forest ecosystem management: An ecological, economic, and social assessment. USDA Forest Service, US Department of Commerce, National Oceanic and Atmospheric Administration, National Marine Fisheries Service, USDI Bureau of Land Management, Fish and Wildlife Service, National Park Service, Environmental Protection Agency. Portland, Oregon, USA.

Forester, J. 1989. Planning in the face of conflict: Negotiation and mediation strategies in local land use regulation. (Originally published in the APA Journal, Summer 1987.) Reprinted *in* J. Forester, ed. Planning in the face of power. University of California Press, Berkeley, California, USA.

——. 1992. Envisioning the politics of public-sector dispute resolution. Studies in Law, Politics, and Society **12**:247–286.

——. 1995. Democratic deliberation and the promise of planning. 1995 Lefrak Lecture, University of Maryland, College Park, Maryland, USA.

——. 1996. Beyond dialogue to transformative learning: How deliberative rituals encourage political judgment in community planning processes. Pages 295–333 *in* S. Esquith, ed. Democratic dialogues: Theories and practices, Poznan studies in the philosophy of the sciences and the humanities, University of Poznan, Poland. Rodopi, Alanta, Georgia, USA.

Friedman, J. 1987. Planning in the public domain: From knowledge to action. Princeton University Press, Princeton, New Jersey, USA.

GAO. 1997. Forest Service decision-making: Greater clarity needed on mission priorities. Statement of Barry T. Hill, Associate Director, before the Subcommittee on Forests and Public Lands Management, Subcommittee on Energy and Natural Resources, US Senate, Tuesday, February 25, 1997. GAO/T-RCED-97-81.

Geertz, C. 1973. The interpretation of cultures. Basic Books, New York, New York, USA.

Gore, A. 1995. Common sense government: Works better and costs less. Random House, New York, New York, USA.

Hage, J., and A. Aiken. 1970. Social change in complex organizations. Random House, New York, New York, USA.

Heckscher, C., and A. Donnellon. 1994. The post-bureaucratic organizations. Sage, Thousand Oaks, California, USA.

Hirt, P.W. 1994. A conspiracy of optimism: Management of the national forests since World War II. University of Nebraska Press, Lincoln, Nebraska, USA.

Hoover, A., and M.A. Shannon. 1995. Coordination of wildlife corridor policies across jurisdictional boundaries: The need for institutional corridors. Greenway Issue of Landscape and Urban Planning **33**:433–459.

Hurst, J.W. 1956. Law and the conditions of freedom in the nineteenth century United States. University of Wisconsin Press, Madison, Wisconsin, USA.

James, L.D. 1971. Remedial flood plain management as the focus for an experiment in interdisciplinary team research. Environmental Resources Center (ERC-0771), Georgia Institute of Technology, Atlanta, Georgia, USA.

John, D. 1994. Civic environmentalism. Congressional Quarterly Press, Washington, DC, USA.

Kaufman, H. 1960. The forest ranger, a study in administrative behavior. The Johns Hopkins Press, Baltimore, Maryland, USA.

Kemmis, D. 1990. Community and the politics of place. University of Oklahoma Press, Norman, Oklahoma, USA.

Kickert, W.J.M. 1993. Autopoiesis and the science of (public) administration: Essence, sense and nonsense. Organization Studies **14**:261–278.

Landy, M.K., and H.A. Plotkin. 1982. Limits of the market metaphor. Society. May/June issue.

Lee, K. 1993. Compass and gyroscope: Integrating science and politics for the environment. Island Press, Covelo, California, USA.

Lee, R.G., and G.R. Stankey. 1992. Evaluating institutional arrangements for regulating large watersheds and river basins. Pages 30–37 *in* P.W. Adams and W.A. Atkinson, compilers. Watershed resources: Balancing environmental, social, political and economic factors in large basins. Conference Proceedings. Forest Engineering Department, Oregon State University, Corvallis, Oregon, USA.

Mansbridge, J.J. 1980. Beyond adversary democracy. Basic Books, New York, New York, USA.

——, ed. 1990. Beyond self-interest. University of Chicago Press, Chicago, Illinois, USA.

McLain, R., and R. Lee. 1996. Adaptive management: Promises and pitfalls. Journal of Environmental Management **20**:437–448.

Meidinger, E.E. 1987. Regulatory culture. Law and Policy **9**:355–386.

——. 1997. Organizational and legal challenges for ecosystem management. Pages 361–379 *in* K.A. Kohm and J.F. Franklin, eds. Creating a forestry for the 21st century: The science of ecosystem management. Island Press, Covelo, California, USA.

——. 1998. Laws and institutions in cross-boundary stewardship. *In* R.L. Knight and P.B. Landres, eds. Stewardship across boundaries. Island Press, Covelo, California, USA.

Michels, R. 1962. Political parties: A sociological study of the oligarchical tendencies of modern democracy. Collier, New York, New York, USA. (Originally published by The Free Press, Glencoe, Illinois, 1915.)

New York Academy of Sciences. 1995 (August). Science and endangered species preservation: Rethinking the environmental policy process: A sustainable forests exchange. Special Report, Science Policy Program of the New York Academy of Sciences, New York, New York, USA.

Perlin, J. 1989. A forest journey: The role of wood in the development of civilization. W.W. Norton, New York, New York, USA.

Powell, W.W., and P.J. DiMaggio, eds. 1991. The new institutionalism in organizational analysis. University of Chicago Press, Chicago, Illinois, USA.

Reich, R. 1985. Public administration and public deliberation: An interpretive essay. The Yale Law Journal **94**:1617–1641.

Sandel, M.J. 1996. Democracy's discontent: America in search of a public philosophy. Belknap Press of Harvard University Press, Cambridge, Massachusetts, USA.

Schattschneider, E.E. 1960. The semisovereign people. The Dryden Press, Hinsdale, Illinois, USA.

Sedell, J.R., G.H. Reeves, and K.M. Burnett. 1994. Development and evaluation of aquatic conservation strategies. Journal of Forestry **92**(4):28–31.

Shannon, M.A. 1987a. Forest planning: Learning with people. Pages 233–252 *in* M.L. Miller, R.P. Gale, and P. Brown, eds. Social science in natural resource management systems. Westview, Boulder, Colorado, USA.

——. 1987b. Building trust: Forest plans as social contracts. Pages 229–240 *in* R.G. Lee, D.R. Field, and W.R. Burch, eds. Community and forestry. Westview, Boulder, Colorado, USA.

——. 1991. Resource managers as policy entrepreneurs. Journal of Forestry **89**(6):27–30.

——. 1992a. Community governance: An enduring institution of democracy. Senate Report, Symposium on Multiple Use and Sustained Yield: Changing Philosophies for Federal Land Management. Congressional Research Service, Library of Congress, Washington, DC, USA.

——. 1992b. Building public decisions: Learning through planning. Pages 227–338 *in* Forest Service planning: Accommodating uses, producing outputs, and sustaining ecosystems. Volume II: Contractor's Documents. Office of Technology Assessment OTA-F-505. US Government Printing Office, Washington, DC, USA.

——. 1992c. Achieving a common approach to problems and directions. Pages 38–44 *in* P.W. Adams and W.A. Atkinson, compilers. Watershed resources: Balancing environmental, social, political and economic factors in large basins. Conference proceedings. Forest Engineering Department, Oregon State University, Corvallis, Oregon, USA.

Shannon, M.A., and A.R. Antypas. 1996. Civic science is democracy in action. Northwest Science Forum **70**(1):66–69.

Shannon, M.A., and A.R. Antypas. 1997. Open institutions: Uncertainty and ambiguity in 21st century forestry. Pages 437–445 *in* K.A. Kohm and J.F. Franklin, eds. Creating a forestry for the 21st century: The science of ecosystem management. Island Press, Covelo, California, USA.

Shannon, M.A., A.C. Graham, A.R. Antypas, T. Steger, C. Hobart, and G. Smith. In progress. Organizing to learn and learning to organize: A comparative organizational and institutional assessment of the ten adaptive management areas in the Pacific Northwest. USDA Forest Service Research Report, Pacific Northwest Research Station, Seattle, Washington, USA.

Shannon, M.A., E. Meidinger, and R. Clark. 1996. Science advocacy is inevitable: Deal with it. Proceedings, Society of American Foresters Annual Meeting.

Shannon, M.A., G. Smith, and C. Robinson. 1995. Institutional strategies for landscape management. Landscape Management Project, Department of Natural Resources, Washington State, USA.

Stanley, M. 1983. The mystery of the commons: On the indispensability of civic rhetoric. Social Research **50**:851–883.

Steen, H. 1976. The US Forest Service: A history. University of Washington Press, Seattle, Washington, USA.

Thompson, M., and M. Schwartz. 1990. Divided we stand: Redefining politics, technology, and social choice. University of Pennsylvania Press, Philadelphia, Pennsylvania, USA.

White, R. 1980. Land use, environment and social change: The shaping of Island County, Washington. University of Washington Press, Seattle, Washington, USA.

——. 1991. "It's your misfortune and none of my own:" A history of the American West. University of Oklahoma Press, Norman, Oklahoma, USA.

——. 1995. The organic machine: The remaking of the Columbia River. Hill and Wang, New York, New York, USA.

Wildavsky, A. 1987. Choosing preferences by constructing institutions: A cultural theory of preference formation. American Political Science Review **81**:3–21.

Wilkinson, C.F. 1992. Crossing the next meridian: Land, water, and the future of the West. Island Press, Washington, DC, USA.

Worster, D. 1977. Nature's economy: A history of ecological ideas. Cambridge University Press, Cambridge, England.

——. 1993. The wealth of nature: Environmental history and the ecological imagination. Oxford University Press, New York, New York, USA.

Yaffee, S.L., and J.M. Wondolleck. 1997. Building bridges across agency boundaries. Pages 381–396 *in* K.A. Kohm and J.F. Franklin, eds. Creating a forestry for the 21st century: The science of ecosystem management. Island Press, Covelo, California, USA.

22
River Law

Robert J. Masonis and F. Lorraine Bodi

Overview

- Management of rivers and streams in the Pacific coastal ecoregion is directed by various federal and state laws. Those laws can be substantive (i.e., requiring the attainment of a specific resource condition) or procedural (i.e., requiring consideration of environmental impacts in resource management decisions).
- Substantive river laws range from laws designed to protect entire river corridors, to laws that focus on protecting specific components of riverine systems (e.g., biota, water quality), to laws that regulate specific human activities in the watershed (e.g., logging, hydropower).
- There are federal and state laws protecting river corridors deemed to be of high value in their undeveloped state.
- The diversion and use of river water is primarily governed by state law, with some federal regulation.
- Water quality is regulated under the federal Clean Water Act, which allows states and qualified Native American tribes to establish water quality standards that are at least as stringent as federal water quality standards.
- Wetland development is regulated under Section 404 of the Clean Water Act.
- Forestry-related activities on federal lands are regulated under federal law (i.e., National Forest Management Act, Federal Land Policy and Management Act). State forest practices laws regulate such activity on state and private lands.
- Hydropower development is governed primarily under the Federal Power Act, although a state water right and a state certification that the project will meet state water quality standards is required for federal licensing.
- The Endangered Species Act, a federal law, is the chief law protecting biota in riverine systems.
- The numerous listings and proposed listings of anadromous fish under the Endangered Species Act signals that the existing legal regime needs improvement. Existing laws could be used to improve river protection, including limiting water withdrawals to protect water quality; controlling nonpoint source pollution; and giving greater consideration to natural resource values in the regulation of hydropower development.
- Collaborative, consensus-driven management of rivers and streams is the current trend, but the success of these processes in protecting rivers will likely require defined, enforceable legal standards.

Introduction

Ecologists often describe the physical landscape as a complex mosaic of patches with different structural and functional characteristics. Similarly, the legal "landscape" regulating the management of river systems in the Pacific coastal ecoregion can be described as a mosaic of federal and state/provincial laws and regulations. This complex web of laws directs and

TABLE 22.1. Primary laws regulating river ecosystem management in British Columbia.

Law and citation	Subject regulated
Navigable Waters Protection Act R.S.C. 1985, c. N-22	Construction in and along navigable waters
Fisheries Act R.S.C. 1985, c. F-14	Fish habitat and water quality
Waste Management Act S.B.C. 1982, c. 41	Water quality
Water Act R.S.B.C. 1979, c. 429	Water rights
Water Protection Act S.B.C. 1995, c. 34	Water diversion and transfers
Forest Practices Code Act S.B.C. 1994, c. 41	Logging and associated road building
Forest Act R.S.B.C. 1979, c. 140	Logging and associated road building
Park Act R.S.B.C. 1979, c. 309	River corridor protection in provincial parks
National Parks Act R.S.C. 1985, c. N-14	River corridor protection in national parks

regulates the numerous land and water uses that affect river and stream health. A sound understanding of the legal framework is, therefore, essential for effective river and stream resource management. This chapter addresses only federal and state laws in the United States. A list of the principal laws regulating rivers in British Columbia is provided in Table 22.1.

It is beyond the scope of this chapter to fully describe and explain all of the myriad laws that control, either directly or indirectly, human activities that affect rivers. Rather, this chapter provides a general overview of the federal and state laws that are of paramount importance in the Pacific coastal ecoregion. Different sources of law (statutory, regulatory, and common) are described, federal and state jurisdiction are explained, and principal laws regulating land use and water development activities are summarized. Finally, suggestions for improving the application of existing laws are explored.

Sources of Law

At both the federal and state levels, river laws originate in several different ways. They can be embodied in the federal or a state constitution. For example, the Alaska constitution states that fish, wildlife, and waters in their natural state are reserved to the people of Alaska for common use.[1] Laws created through court decisions are referred to as common law. The prior appropriation doctrine and the public trust doctrine, discussed later in this chapter, are examples of common law legal rules. Most often, however, river laws are products of legislation (statutes) passed by federal or state legislatures. Once enacted, statutes are codified in the respective federal or state statutory compilations.

Statutes typically set forth general requirements and obligations that apply to specific actors, such as federal and state agencies or private citizens, but often do not prescribe specific standards or procedures required for compliance. Consequently, statutes often authorize the federal or state agency charged with implementing and enforcing the statute to develop and issue regulations that provide specific standards, criteria, and procedures. Regulations implementing statutes are legally binding and enforceable.

For example, Section 404 of the federal Clean Water Act prohibits the discharge of dredged or fill material into the navigable waters of the United States unless the United

States Army Corps of Engineers issues a permit authorizing such action.[2] The statute does not, however, identify the specific criteria that the Army Corps must consider in determining whether a Section 404 permit should be issued. The specific criteria are set forth in regulations issued by the Army Corps under Section 404.[3] In the first instance, federal and state agencies typically apply laws embodied in statutes and regulations to specific facts. Final agency determinations can be reviewed judicially, but courts often defer to an administrative agency's expertise.

Federal and State Jurisdiction

In many watersheds, federal, state, and local laws regulate land and water use. Whether federal and/or state law applies to a specific resource issue is a question of jurisdiction. Generally, federal law applies where either the resource or activity in question is located on federal land (e.g., logging on national forests), or where the resource or activity affects interstate commerce (e.g., the regulation of hydroelectric development on navigable rivers). Resources and activities located on state or private land are generally within state jurisdiction (e.g., logging on private or state lands; allocation of water rights), although actions on state and private property may be subject to additional federal regulation if necessary to effectuate the purpose of a federal law (e.g., protection of a federally recognized endangered or threatened species).

Some resources and resource uses are subject to both federal and state regulation. This dual regulation, known as *concurrent jurisdiction*, is constitutionally based. The federal government has regulatory authority under various clauses of the U.S. Constitution, primarily the Commerce Clause that empowers Congress to govern interstate commerce[4] (Johnson and Paschal 1995). Similarly, the states' sovereign right to regulate water derives from the Tenth Amendment to the U.S. Constitution, which reserves for the states powers not expressly reserved for the federal government (Trelease 1971).

Water is a prime example of a resource over which the federal government and the individual states have concurrent jurisdiction. The respective regulatory rights of the federal and state governments over water are discussed later in this chapter. At times this arranged marriage of federal and state law is less than blissful. Where there is conflict, federal law overrides state law under the Supremacy Clause of the U.S. Constitution.[5] Thus, federal law may limit or nullify otherwise legitimate state powers over water.[6] This is commonly referred to as federal preemption of state law. In other areas, such as water quality regulation, the federal government delegates its regulatory authority to the individual states subject to federal oversight.

Statutes determine which entity within the federal or a state government has regulatory authority over a resource. For example, in Washington state, the Department of Natural Resources is vested with the authority to regulate logging on nonfederal (state and private) lands, and the Department of Ecology is authorized to regulate both water allocation and water quality.

Laws Regulating River Systems

Laws regulating rivers and adjacent land areas are quite varied. This chapter focuses primarily on substantive laws (i.e., those requiring either the attainment of specific resource conditions or the use of specific protective measures related to land use and water development projects). Omitted, but nonetheless important, are predominantly procedural laws, such as the National Environmental Policy Act, the Coastal Zone Management Act, state environmental policy acts, and state growth management laws, which require consideration of environmental impacts in resource planning and development decisions.

The substantive laws covered here are by no means the only ones relevant to river system management. They were selected based on their importance to the regulation of human activities in and along Pacific Northwest coastal rivers, and on their direct influence on the five fundamental components of river systems: geomorphology, hydrology, water quality, riparian zones, and in-stream habitat. These laws are presented as enacted. The implementation of these laws is a separate but equally important issue. Without adequate implementation, even the most well conceived and drafted law will not attain the desired outcome.

Water Quantity (In-Stream Flows)

Water is the most fundamental component of a river system. Without sufficient water to maintain the ecological processes essential to river health, the eventual demise of a river, or a river reach, is guaranteed. Irrigation diversions of rivers and streams in the arid interior West have resulted in the complete dewatering of some major rivers. For example, the Snake River often runs dry during the irrigation season at Milner Dam in south-central Idaho. In contrast, the typically copius precipitation and numerous water sources of the Pacific coastal ecoregion limit the need to divert water from river channels. Nevertheless, the burgeoning population of the Pacific Northwest is increasing the demand for water, and the diminution of in-stream flows—for hydroelectric power, agricultural uses, and industrial, municipal, and residential water supply—is an increasing threat to river health.

State Law

Water rights statutes—The right to take water from a river for consumptive use is governed primarily by state law. A water right is *usufructory*, which means that the right is limited to the use of water; the water itself remains a publicly owned resource.[7] Alaska, California, Oregon, and Washington have enacted state water rights statutes which prescribe the process by which water rights are granted. Those statutes codify the common law doctrine of *prior appropriation* which evolved during the settlement of the West in the nineteenth century.

The prior appropriation doctrine was first formally recognized by the California Supreme Court in 1855, and establishes a system for allocating water based on seniority of water use: early water users have rights superior to subsequent users (Johnson and Paschal 1995). Appropriative water rights are recognized as property interests protected under the U.S. Constitution and can be sold or leased (Bell and Johnson 1991). This use-based system of water rights contrasts with the riparian doctrine prevalent in the eastern and midwestern states, where water rights are tied to ownership of land adjacent to the water source and cannot be severed from the land. Today, water rights in California, Oregon, and Washington are primarily based on the prior appropriation doctrine, although each state recognizes some riparian rights. Alaska water rights are determined exclusively under the doctrine of prior appropriation (Reisner and Bates 1990).

A water right generally does not expire, but is conditional. It is limited to the amount of water necessary to serve a recognized beneficial use and can be reduced or surrendered if the water is not used for that designated purpose. A water right can also be lost if it is not used or is improperly used. A water right is abandoned if use is intentionally discontinued for a period of time (usually several years) designated in state law. A right can also be forfeited if use is not continued for a period of time, even if there is no intent to relinquish the water right (Reisner and Bates 1990). Moreover, a water right carries with it an obligation not to *waste* water through inefficient methods of diversion and use (Blumm and Schwartz 1995). Although legally enforceable, these aspects of the prior appropriation doctrine are rarely enforced (Reisner and Bates 1990). The result is often unauthorized water diversions and inefficient diversion methods.

State water rights statutes require that a person seeking to divert and use water from a river apply for and obtain a water right permit from the designated state agency (the Alaska Department of Natural Resources, the California State Water Resources Control Board, the Oregon Water Resources Department, or the Washington Department of Ecology). The State of Washington's statute governing water rights permitting is typical. It provides that a water right permit will be granted if the following criteria are met: (1) there is water available for appropriation, (2) the appropriation is for a beneficial use, (3) the appropriation will not impair existing rights, and (4) the appropriation will not be detrimental to the public welfare or public interest.[8]

Ecosystem components dependent upon in-stream flows, such as biota and water quality, can be protected in several ways under state water allocation statutes. Natural resources are part of the public interest in rivers. For example, in Washington state the Department of Ecology must "consider the total environmental and ecological factors to the fullest" in making its public interest determination.[9] As the laws are written, natural resources should, therefore, be considered in the permitting process, and should provide a basis for denying or conditioning new water rights permits as needed to maintain in-stream flows (Shupe and MacDonnell 1993). The fact that many of the rivers in the Pacific coastal ecoregion are overappropriated is evidence that the permitting process does not adequately protect in-stream flows. There is evidence, however, that change is afoot. In July, 1996, the Washington Pollution Control Hearing Board (PCHB)—the state water court—in a consolidated appeal upheld the Department of Ecology's decisions denying 140 water right applications based primarily on inadequate stream flows.[10] The PCHB's decision is particularly noteworthy because many of the applications were for ground water rights, not surface water rights. The ground water applications were rejected because the water sought was found to be in "hydraulic continuity" with streams with low flows.

A state also can reserve in-stream water rights through the water right permitting process to protect a river's ecological health. This is a relatively recent development because, historically, only diversionary uses of water were considered "beneficial." Water remaining in the channel was viewed as wasted (Reisner and Bates 1990). Today, all Pacific Northwest states recognize the protection of environmental values as a beneficial use of water for which a water right may be reserved.[11]

Alaska, California, Oregon, and Washington have established statutorily defined procedures for obtaining in-stream flow rights. Entities entitled to obtain in-stream flow rights vary by state. For example, in Oregon the Department of Fish and Wildlife and the Department of Environmental Quality are authorized to apply to the Water Resources Commission for in-stream flow rights.[12] In contrast, Alaska allows any local, state, or federal government agency or any private person or organization to apply for an in-stream flow right (although private parties cannot possess in-stream flow rights, only diversionary rights). Even if in-stream flow rights are obtained, however, they may not result in higher flows because of late priority dates. Senior appropriative water rights take precedence.

Public trust doctrine—A state's authority to issue water rights is limited, at least theoretically, by its concomitant responsibility to protect public interests in water. The *public trust doctrine* is a state law construct rooted in ancient common law (Sax 1970). It was initially embraced by the U.S. Supreme Court during the nineteenth century as requiring states to protect the public interest in fishing, navigation, and commerce on navigable waters.[13] Today, the doctrine has been cut free from the mooring of those traditional public trust interests, and is now increasingly used to protect additional public interests in water, such as environmental quality, recreation, and aesthetics (Johnson and Paschal 1995).

All four Pacific coastal states recognize the public trust doctrine, although its geographic and substantive scope varies (Beck 1991). California recognizes public trust interests in certain nonnavigable as well as navigable waters,[14] and has added ecological and recreation values to the list of interests protected under the doctrine.[15] California is the only state in the region which has expressly recognized that the public trust interest in ecologically based in-stream flows can supersede conflicting private appropriative rights.[16] Washington recognizes public trust interests in yet undefined recreation and ecological values[17] (Bodi 1989), but has not extended the doctrine beyond navigable waters.[18] Alaska and Oregon also recognize a public interest in in-stream flows, although the precise scope of public trust protection remains undefined (Blumm and Schwartz 1995).

Federal Law

Reserved water rights—In-stream flows are largely, but not exclusively, within the regulatory province of the states. Public lands set aside by the federal government as reserves for specific purposes, such as tribal reservations, national parks, national forests, and national fish and wildlife refuges, carry with them an entitlement to the amount of water necessary to fulfill the purpose of the reservation (MacDonnell and Rice 1993). *Federal reserved water rights,* as they are commonly known, are a product of common law brought about by the need to carry out congressional intent embodied in statutes, treaties, and executive orders that do not expressly provide an entitlement to water. Federal reserved water rights are independent of state water rights and have as a priority date the date that the reservation was established, unless the reservation was created with the intent of preserving preexisting uses (Blumm 1992).

Federal reserved water rights have been construed narrowly by the courts, presumably to limit conflicts with state water rights. The right is limited to the minimum necessary to fulfill the primary purpose of the reservation.[19] Whether the protection of fish and wildlife or some other reason for maintaining in-stream flows (e.g., recreation) is a primary purpose of a federal reservation depends on the type of reservation and on the congressional pronouncements regarding its designation. (For a detailed discussion of the federal reserved water rights for specific types of reservations, see Blumm 1992.) For example, the U.S. Supreme Court determined that the maintenance of in-stream flows for fish, wildlife, and recreation was not a primary purpose in creating national forests, and thus not a basis for reserved water rights.[20] In contrast, the establishment of national parks appears to create a reserved water right for in-stream flows sufficient to support fish and wildlife because the preservation of such values is an express purpose of the park system (Blumm 1992).

Regarding tribal reservations, federal courts look at the means by which the tribe in question supports itself to determine the primary purpose of the reservation and the amount of water in the implied water right. Thus, a tribe that depends upon a fishery for its subsistence is entitled to the amount of water necessary to maintain the fishery.[21] A tribal reserved water right to maintain a fishery is not limited to reservation land, and may be asserted to protect off-reservation fishing sites[22] (Beck 1991, Blumm 1992). Tribal reserved water rights have been effectively asserted to restrict a variety of water development activities, such as dam construction, dam operation, and irrigation withdrawals (Blumm 1992).

Regulatory rights—In addition to reserved water rights, the federal government possesses federal regulatory water rights to the extent necessary to implement federal laws. Regulatory water rights are independent of federal land ownership and supersede conflicting state water rights (Blumm 1992). The primary federal statutes that create regulatory water rights are the Clean Water Act, the Endangered Species Act, and the Federal Power Act. These regulatory rights are discussed in greater detail later in this chapter.

Water Quality

Clean Water Act

In contrast with the laws regulating water allocation, which with the few exceptions discussed above are largely determined by the individual states, water quality regulation is an amalgam of federal and state law. The federal Clean Water Act (CWA),[23] passed by Congress in 1972 and amended several times, is the principal law regulating water quality. It establishes water quality as an issue of national concern subject to both federal and state regulation. Its express purpose is to "restore and maintain the chemical, physical, and biological integrity of the Nation's waters."[24] Though the act explicitly preserves for the states the right to allocate water,[25] the federal government can circumscribe that right as needed to effectuate the CWA's purpose of protecting water quality (Blumm 1992).

The CWA weaves a web of federal and state law to achieve that ambitious objective. The statutory provisions most directly related to water quality include Section 303, which requires states to adopt water quality standards; Section 301, which prohibits the "discharge" of "pollutants" from a "point source" unless a permit is issued under Sections 402 or 404 of the Clean Water Act; and Section 319, which requires states to control "nonpoint source" pollution through management plans and best management practices (BMPs). Section 404, which regulates the use of wetlands, is discussed in the Land Use section later in this chapter.

State water quality standards—Section 303 of the CWA requires all states to develop water quality standards for waters within their boundaries.[26] Eligible Native American tribes may establish water quality standards applicable to reservation waters.[27] Water quality standards consist of three basic components: (1) designated uses of the waterway (e.g., salmonid spawning, agriculture, primary and secondary contact recreation), (2) numerical or narrative criteria to support the designated uses, and (3) a general prohibition against further degradation of existing water uses (Ransel 1995).

State standards can be more stringent than those established under federal law and must be revised every three years and approved by the Environmental Protection Agency, the federal agency responsible for enforcing the CWA. Water quality standards are enforceable against both private and public actors, including federal land managers.[28] The means for maintaining compliance with state water quality standards include controlling point and nonpoint source pollution. Pollution is defined broadly as "the man-induced alteration of the chemical, physical, biological, and radiological integrity of the water."[29]

Activities requiring a federal license are not exempt from state water quality standards. According to Section 401 of the CWA, a federal license or permit for activities that may result in "discharge" to state waters requires state certification that the action will not violate state water quality standards.[30] Examples include licenses to construct and operate hydroelectric dams under the Federal Power Act, wetlands permits, and permits for point source discharges.

Traditionally, states imposed Section 401 conditions only on the specific chemical or numeric water quality criteria component of state water quality standards. That tradition was recently broken when the U.S. Supreme Court affirmed the State of Washington's legal right to impose conditions necessary to attain designated uses of a waterbody and to implement the state's antidegradation policy[31] (Ransel 1995). The significance of this ruling is discussed in greater detail in the New Approaches section of this chapter.

Point source controls—A point source is defined as "any discernible, confined and discrete conveyance," which includes pipes, ditches, and boats.[32] The CWA controls point source pollution by prohibiting the discharge of pollutants into surface waters unless the discharger first obtains a National Pollution Discharge Elimination System (NPDES) permit from the EPA or from a state or Native American tribe with an approved permitting program. Point source permits are conditioned upon the discharger meeting specific numeric technology-based effluent limitations established by the EPA. Citizens are expressly permitted to seek judicial enforcement of effluent standards in NPDES permits.[33] If compliance with established effluent limitations is insufficient to meet state water quality standards, Section 302 mandates that NPDES permits include more stringent effluent limitations.[34]

Nonpoint source controls—Nonpoint sources of pollution include runoff from agriculture, logging, and mining operations. Section 319 of the CWA is the primary statutory mechanism for controlling nonpoint source pollution.[35] Prior to the enactment of Section 319 of the CWA in 1987, nonpoint source pollution was addressed through Section 208 of the CWA,[36] which requires states to develop area-wide waste treatment management plans that include a process for identifying nonpoint sources and establishing control measures. Section 208 was ineffective at preventing nonpoint source pollution, and the EPA eventually stopped funding the development of Section 208 planning (Percival et al. 1992). Section 319 requires states to prepare and submit to the EPA assessment reports identifying waters within the state for which point source controls are inadequate to meet water quality standards, and the significant sources of pollution for each listed water.[37] Each state must then develop a comprehensive management program to control nonpoint source pollution, which identifies BMPs that will be used to control nonpoint source pollution,[38] and a schedule for implementation.[39] States are directed to develop such programs on a watershed basis.[40]

Both Section 319 assessment reports and nonpoint management programs must be submitted to the EPA for approval. States with approved nonpoint programs are eligible for financial assistance to implement their programs,[41] but are not required to adopt a regulatory program. Some have argued that the voluntary nature of Section 319, coupled with technical and political obstacles to establishing effective regulatory programs, has limited its effectiveness (Mandelker 1989, Whitman 1991).

The Coastal Zone Amendment Reauthorization Act of 1990 (CZARA Amendments), which amended the Coastal Zone Management Act (CZMA), provides another mechanism for reducing nonpoint source pollution. The CZARA Amendments established a coastal nonpoint pollution control program requiring states with federally approved coastal zone management programs to develop programs to protect coastal watersheds from nonpoint source pollution.[42] In contrast with nonpoint programs under the Clean Water Act, which are largely voluntary, nonpoint programs under the CZARA Amendments must contain "enforceable policies and mechanisms to implement" nonpoint source control programs.

Total Maximum Daily Loads—Where point and nonpoint source controls fail to attain state water quality standards, Section 303 of the CWA requires the establishment of Total Maximum Daily Loads (TMDLs) for the pollutant(s) causing the degradation.[43] TMDLs establish the maximum amount of daily pollutant loadings from all sources. Biennially, states are required to submit to EPA a list of water quality impaired waters for which TMDLs must be promulgated. The Pacific coastal states have been slow to implement TMDLs for water quality impaired waters, although recently the threat of litigation has focused more attention on TMDL development. For example, in 1987 the state of Oregon entered a consent decree requiring development of TMDLs for phosphorous and ammonia for ten Oregon rivers, including the Tualatin.

Land Use

Nonpoint source pollution is just one of the ways in which land use activities threaten the ecological integrity of rivers within the Pacific coastal ecoregion. Land-use activities such as river channelization for flood control, logging and development in riparian and upland areas adjacent to the river channel, and dam construction can alter ecological processes essential to river health. This section focuses on some of the primary federal and state laws regulating land use within the ecoregion's watersheds.

River Corridor Protection

Both federal and state statutes provide special protection for specific rivers and river segments based on their high noncommodity environmental values. In 1968 Congress passed the National Wild and Scenic Rivers Act (WSRA),[44] a federal statute under which entire rivers or river segments and adjacent lands can be set aside and protected from development. Following the federal government's lead, all four states in the ecoregion enacted similar statutes establishing a system for protecting intrastate waters from development.[45] British Columbia has also taken steps recently to protect significant rivers in the province. On May 2, 1995, the provincial government announced the creation of the British Columbia Heritage Rivers System, which is intended to increase public awareness of the natural, cultural, and recreational values of rivers, and improve river stewardship (British Columbia's Heritage Rivers Board 1995). Discussion in this section is limited to the United States WSRA.

The WSRA was spawned by recognition that unrestricted river development would be inimical to "vital national conservation purposes."[46] Those purposes include preserving scenic and aesthetic values; protecting fish and wildlife; preserving geologic, cultural, and historic assets; promoting recreational uses; protecting water quality; and preserving the free-flowing character of rivers for present and future generations.[47] Rivers or river segments with "outstandingly remarkable values" are eligible for protection under the WSRA. Rivers selected for protection are classified as *wild, scenic,* or *recreational.* Wild and scenic rivers must be unimpounded and essentially unpolluted, the only difference being that limited shoreline development and road access is permitted along scenic rivers.[48] Recreational rivers may have been impounded or diverted in the past, may have some shoreline development, and are readily accessible.[49]

Rivers protected under the WSRA are managed by the federal agency with jurisdiction over the land containing or adjoining the rivers. The federal agency must develop a management plan to protect the exceptional values for which the river was selected.[50] Both the Departments of Interior (Park Service) and Agriculture (Forest Service) have established guidelines to assist in the management of designated rivers.[51] The Forest Service and Bureau of Land Management (BLM) also have adopted internal policies implementing the WSRA (USDA 1987, USDI 1992). Particularly noteworthy for the Pacific coastal ecoregion is the Forest Service policy prohibiting logging in wild river corridors except to clear trails, protect hikers, or control fire. In addition to river-specific management prescriptions established by federal agencies, the WSRA generally proscribes specific types of activities along designated rivers, including new dam construction authorized by the Federal Energy Regulatory Commission.[52]

Land-use restrictions under the WSRA apply only to federal lands. Private rights to use federal land within a protected river corridor cannot be limited if they predate the Act.[53] The Departments of Interior and Commerce may acquire (through purchase, exchange, or donation) non-federal lands within the boundaries of designated national wild and scenic river systems.[54] Cooperative management agreements with states and their political subdivisions are also encouraged to preserve a designated river's outstandingly remarkable values.[55]

Finally, a designated river carries with it a reservation of water necessary to preserve the river's outstandingly remarkable values[56] (Beck 1991). This water reservation, like all federal reserved water rights, has the potential to limit conflicting state water rights (Frost 1993).

Wetlands Regulation

Wetlands serve various functions vital to the health of river ecosystems, including maintaining water quality by fixing and recycling nutrients and filtering pollutants, buffering flood surges by dissipating and absorbing high flows, and providing habitat for both aquatic and terrestrial species (roughly 35 percent of endangered and threatened species are wetland dependent) (Office of Technology Assessment 1984). Unfortunately, the value of wetlands was largely unrecognized until the 1960s. Consequently, it is estimated that over half of the pre-European settlement wetlands in the United States have been eliminated. Today, despite significant improvements in wetland protection, a net loss of wetlands continues (Dahl and Johnson 1991).

Growing concern for the loss of wetlands provided the impetus for Section 404 of the Clean Water Act (CWA),[57] which precludes the discharge of "dredged and fill" material into navigable waters of the United States without a permit issued by the U.S. Army Corps of Engineers.[58] Permits can be either project specific or general, the latter covering activities on a state, regional, or national level which are deemed to have minimal adverse environmental effects.[59]

Wetlands are broadly defined in the regulations promulgated by the U.S. Army Corps of Engineers as those areas with sufficient water content in the soil to support a prevalence of vegetation adapted for saturated soil conditions.[60] Thus, three primary criteria for wetlands delineation are soils, hydrology, and vegetation. Wetlands meeting these criteria are often referred to as *jurisdictional wetlands.* There has been a concerted effort in recent years to narrow the legal definition of wetlands to allow for greater development. This is a primary issue in the current debate over reauthorization of the Clean Water Act. The geographic scope of the permitting authority of the U.S. Army Corps of Engineers under Section 404 is similarly broad, encompassing all wetlands—interstate and intrastate—that could affect interstate commerce.[61] For example, an isolated, intrastate wetland falls within Section 404 if it is used by migratory and endangered birds.[62]

Section 404 permits are required for activities that involve a discharge of "dredged and fill" material, with the intent of altering the hydrology (i.e., converting the wetland to dry land by raising the bottom elevation).[63] Discharges of other than dredged and fill material, though not covered under Section 404, may require a discharge permit under another section of the CWA. The draining of wetlands, because it does not involve a discharge, is not covered by Section 404.

Certain activities are expressly exempted from Section 404 permitting, including several of particular significance to the Pacific coastal ecoregion—"normal" farming and silviculture activities, and the construction and maintenance of farm and forest roads where BMPs minimize environmental impacts.[64] The exemptions do not apply, however, to otherwise exempt activities that would convert a wetland to a new use that would impair the wetland's natural hydrology.[65] This "recapture" provision in the CWA limits the scope of the exemptions.

The U.S. Army Corps of Engineers cannot issue a Section 404 permit unless it determines that the permit would be in the public interest, a process that entails weighing the benefits and detriments that would flow from the proposed action.[66] Factors considered include economics, fish and wildlife values, floodplain values, water quality, and property ownership.[67]

In addition to its own public interest review, the U.S. Army Corps of Engineers is required to deny permits that do not comport with the EPA's Section 404(b) guidelines.[68] The EPA guidelines preclude discharges that would have an unacceptable impact (either individual or cumulative) on the aquatic ecosystem.[69] The guidelines expressly prohibit discharges for which there are "practicable alternatives," discharges that would cause or contribute to violations of state water quality standards, and discharges that would jeopardize endangered or threatened species (including habitat destruction).[70] They also prohibit discharges that would cause or contribute to the "significant degradation of waters of the United States."[71] Finally, discharges cannot be approved without "appropriate and practical" mitigation measures.[72] Although the EPA has the authority to reject a Section 404 permit issued by the Corps in violation of the guidelines,[73] it has exercised that authority only 11 times since 1972 (US EPA 1995).

Forest Practices Regulation

Unsound logging practices have profoundly affected coastal rivers in the Pacific Northwest in several ways, including increasing sediment loads and reducing large woody debris inputs (Doppelt 1993). These deleterious impacts have garnered much attention in the last decade due primarily to the decline of salmon and steelhead populations in many coastal rivers. Several federal and state laws are intended to regulate logging and associated road building to reduce adverse impacts on river health. At the federal level, the National Forest Management Act (NFMA)[74] and the Federal Land Policy and Management Act (FLPMA)[75] regulate forest practices on national forest and Bureau of Land Management lands, respectively. "The Northwest Forest Plan," a land management plan adopted by the Departments of Agriculture and Interior to protect the habitat of the northern spotted owl and other old-growth dependent species, also restricts logging on federal lands in coastal watersheds in California, Oregon, and Washington (USFS 1993).

Two federal laws regulating logging apply only to federal lands in Alaska. The Tongass Timber Reform Act[76] establishes additional protections for the Tongass National Forest, the largest national forest in the United States with a land area approximately the size of West Virginia. The Alaska National Interest Lands Claim Act[77] limits logging, mining, and other land-use activities on federal lands to minimize the impact to subsistence uses of natural resources (e.g., hunting and fishing). State forest practices laws in all four Pacific coastal states primarily regulate logging and roading on state and private lands. Management prescriptions vary by state and are typically established by regulation. State forest practices acts and the National Forest Management Act (NFMA) are discussed in greater detail below.

National Forest Management Act (NFMA)—Although the Bureau of Land Management manages some federal land in the Pacific coastal ecoregion, most is managed by the U.S. Forest Service. The Multiple Use Sustained Yield Act (MUSY),[78] passed in 1960, established a multiple use mandate to guide the management of national forests. Watershed protection, fish and wildlife, and timber production are among the legitimate uses recognized under MUSY. To provide greater direction to the Forest Service regarding how to plan to meet the "multiple use" mandate, Congress passed NFMA in 1976. NFMA requires the Forest Service to develop land management plans for each national forest, and specifies substantive and procedural planning objectives and guidelines.

NFMA directly addresses watershed protection within national forests. It precludes logging unless "soil, slope or other watershed condition[s]" will not be "irreversibly damaged," and requires protective measures wherever logging is "likely to seriously and adversely affect water conditions or fish habitat."[79] The Forest Service regulations implementing those statutory directives establish a riparian buffer zone of approximately 30 m (or to the extent of riparian vegetation domination), within which activities that would "seriously and adversely affect water conditions or fish habitat" are prohibited.[80] In addition, they require the use of protection and mitigation measures identified in Forest Service handbooks which are either regional in scope or tailored to specific physiographic or climatic provinces.[81]

The regulations implementing NFMA also require the protection of sufficient habitat to maintain "minimum viable" vertebrate fish and wildlife populations.[82] What constitutes a minimum viable population of a species depends upon the species in question and how "population" is defined. Critics of the Forest Service's implementation of the viable populations requirement contend that the focus has been on maintaining fish species as opposed to genetically distinct stocks, thereby allowing individual stock extirpation and the preservation of only mimimal amounts of aquatic and riparian habitat (Doppelt 1993).

Federal Land Planning and Management Act (FLPMA)—Like NFMA, FLPMA requires that BLM lands be managed on a multiple-use basis.[83] The protection of water resource values and fish and wildlife habitat are listed as management objectives.[84] Moreover, the BLM must give priority to protecting areas of critical environmental concern.[85]

The regulations implementing FLPMA's multiple-use mandate are quite broad and aspirational in nature. In contrast with the NFMA regulation requiring the maintenance of "minimum viable populations" of vertebrates, the FLPMA regulation addressing the protection of fish and wildlife habitat is more vague, requiring only that BLM lands be managed to ensure a "sustained yield of fish and wildlife" and public access to fish and wildlife resources.[86] The regulation covering watershed protection states that BLM lands must be managed "to minimize soil erosion, siltation, and other destructive consequences of uncontrolled water flows; and to maintain and improve storage, yield, quality and quantity of surface and subsurface waters."[87] These regulations give the BLM broad management discretion, and do not establish any specific criteria or standards for determining whether the objectives are attained.

State Forest Practices Acts (SFPAs)—The Pacific coastal states have statutes governing forest practices on state and private lands.[88] Administration and enforcement of the state forest practices acts (SFPAs) is entrusted to the Alaska Department of Environmental Conservation, the California Department of Forestry and Fire Protection, the Oregon Department of Forestry, and the Washington Department of Natural Resources. The SFPAs establish the authority of each state to regulate forest practices, articulate state timber management policy, and authorize the responsible state agency to promulgate regulations implementing the statute. SFPA regulations typically enumerate specific management standards and prescriptions applicable to timber harvesting and associated activities such as road building. Though the specific regulations vary by state, the type of regulations do not. All four states have SFPA regulations designed to protect vegetation in riparian areas and water quality.

The definition of riparian zone differs by state, and generally differs among stream classes (e.g., fish bearing or nonfish bearing). Within the defined riparian zone, regulations typically prescribe timber harvest restrictions, road-building and equipment-use restrictions, slash removal, and stream-crossing restrictions. SFPA regulations typically require the use of BMPs, both within and outside the riparian zone, to meet state water quality standards. For example, the Oregon Forest Practices Act directs the state forestry board to establish BMPs to limit nonpoint source pollution.[89]

Hydropower Regulation

Mountainous, high-gradient terrain and numerous rivers have made the Pacific Northwest a prime area for hydroelectric development. Consequently, hydroelectric development has substantially altered the hydrology of many Pacific coastal rivers. The Federal Power Act (FPA)[90] regulates most hydroelectric development in the Pacific Northwest, covering all non-federal hydropower projects on navigable waters of the United States, on federal lands, or on nonnavigable waters subject to Commerce Clause jurisdiction[91] (Beck 1991). Hydroelectric development not regulated under the FPA includes "federal" projects, such as the dams on the Columbia-Snake system operated by the U.S. Army Corps of Engineers, that were authorized under different federal statutes, and projects on some nonnavigable waterways, which are regulated by the states. Oregon is the only state in the Pacific Northwest with a comprehensive hydropower licensing statute.[92] The FPA was enacted in 1920 and amended several times since. The latest amendments, embodied in the Electric Consumers Protection Act of 1986, are intended to strengthen the FPA's protection of nonpower values in light of the failure of the Federal Energy Regulatory Commission (FERC), the independent federal agency responsible for implementing the statute, to adequately safeguard fish, wildlife, water quality, and other public uses of the nation's rivers.[93]

The FPA authorizes FERC to issue hydropower licenses for up to 50 years if the license is deemed to be in the public interest. In determining whether licensing a project is in the public interest, FERC must examine all relevant issues[94] and must give equal consideration to power and nonpower values of the resource. Nonpower values include "the protection, mitigation of damage to, and enhancement of fish and wildlife (including related spawning grounds and habitat), the protection of recreational opportunities, and the preservation of other aspects of environmental quality."[95] The public interest inquiry is the same regardless of whether it is a license to construct a project or a license to operate an existing project.[96]

Moreover, the FPA mandates license terms that provide "adequate" and "equitable" protection, mitigation, and enhancement measures for fish and wildlife and their habitat.[97] Fish and wildlife mitigation measures must be based on recommendations by the National Marine Fisheries Service (NMFS), the U.S. Fish and Wildlife Service (USFWS), and state fish and wildlife agencies. Recommendations can be rejected only if they are not supported by substantial evidence or are inconsistent with the FPA. NMFS and USFWS also have the authority to mandate fish passage facilities at FERC-licensed dams.[98]

Hydropower development proposed on federal reserved lands (e.g., national forests) is subject to two additional requirements. First, FERC cannot license a project that would be inconsistent with the reservation's purpose. Second, the federal land management agency in charge of the reservation can prescribe mandatory license conditions deemed necessary to protect and carry out the purpose of the reservation.[99] A federal agency's conditioning authority cannot be exercised arbitrarily; it must be reasonably related to the protection of the reservation.[100]

In contrast with the Clean Water Act, which expressly acknowledges the primary right of states to regulate water quality, the Federal Power Act is an exercise of the federal government's broad Commerce Clause power and preempts most state laws regulating hydropower development within FERC's jurisdiction.[101] The Supreme Court has interpreted the FPA as preserving only the states' right to regulate hydropower development to the extent necessary to protect existing state proprietary interests in water.[102] What constitutes a state proprietary interest "saved" by the FPA is unclear, though the Supreme Court has determined that, if not embodied in a separate water right, state-imposed minimum in-stream flows as conditions on a water right at a FERC-licensed project are not proprietary and are therefore preempted by the FPA.[103]

The FPA's broad preemptive sweep does not extend to federal laws such as the Clean Water Act. Consequently, a FERC license is not, alone, sufficient to build and/or operate a hydroelectric project. Pursuant to Section 401 of the CWA, a FERC licensee must first obtain from the responsible state agency certification that the project will be consistent with state water quality standards—designated uses, numeric and narrative criteria, and the antidegradation policy. Thus, for example, a state can condition project construction or operation on maintaining a minimum in-stream flow if that flow is necessary to meet a designated use of the river, such as salmon spawning,[104] and FERC can not reject a condition included in a state's Section 401 certification. Construction of new hydroelectric projects also requires a dredge and fill permit from the Army Corps of Engineers under Section 404 of the Clean Water Act and a state water right.

Biota and Habitat—Endangered Species Act

In contrast with the land-use laws discussed previously, which are designed to ensure that specific anthropogenic activities do not lead to ecological damage, the Endangered Species Act (ESA)[105] is crisis driven, coming into play only after a species is at risk of extinction. It is, in essence, a safety net. ESA protections apply only to species listed as either *endangered* (i.e., in danger of extinction) or *threatened* (i.e., likely to become an endangered species within the foreseeable future).[106]

The ESA is designed to conserve endangered species through the protection of the ecosystems upon which they depend, as well as the species themselves.[107] The ESA protects the ecosystem or habitat requirements of a listed species primarily through designating *critical habitat*, prohibiting critical habitat destruction and adverse modification, and prohibiting the adverse modification of habitat (even if not designated as critical) if it would result in a "taking" of a listed species (Yagerman 1990).

The ESA generally requires the designation of critical habitat at the time of species listing.[108] Critical habitat includes those geographic areas within the species' current range that contain the physical or biological features essential to species conservation and that require special management considerations or protections. It may also encompass areas outside the species' current geographic range that are "essential for conservation of the species."[109] The full range of habitat needs are to be considered in the designation process, including space, food, water, light, nutrients, cover or shelter, sites for reproduction, and historic species distribution.[110] Critical habitat must be determined based on the best scientific data available and on economic considerations.[111]

Habitat protection is effectuated through two means. First, Section 7 of the ESA requires federal agencies to consult with the U.S. Fish and Wildlife Service (USFWS) or the National

Marine Fisheries Service (NMFS), whichever agency is responsible, prior to undertaking action that may affect a listed species.[112] Formal consultation is required whenever there is a threshold determination that an agency might adversely affect a listed species.[113] That determination is made either through a biological assessment,[114] which is required for all agency actions involving a "major construction activity,"[115] or through informal consultation[116] (Kilbourne 1991).

Formal consultation culminates in a biological opinion by USFWS or NMFS.[117] If the biological opinion results in a determination that the proposed action would "jeopardize" the species (i.e., appreciably reduce the likelihood of both survival and recovery of a listed species in the wild)[118] or diminish its critical habitat, the action cannot proceed, and USFWS or NMFS must attempt to identify "reasonable and prudent" alternatives that would be consistent with the ESA.[119] Approval of a proposed action can also be conditioned upon specific mitigation measures.[120]

Second, the ESA precludes the "taking" of a listed species by private citizens as well as government agencies.[121] "Harming" a listed species is defined as a taking.[122] Federal courts have construed "harm" broadly to include the adverse modification of designated critical habitat,[123] even if that habitat is privately owned. Habitat modification that would constitute a taking can proceed only if an incidental take permit is first obtained. The requirements for obtaining an incidental take permit are stringent, including the preparation of habitat conservation plans (HCPs) designed to minimize adverse impacts and ensure that the taking will not jeopardize the species.[124] Thus, for example, the ESA could preclude state water right holders from withdrawing water beyond the point at which the decrease in flow would harm a listed species.[125]

As many anadromous fish stocks in the Pacific Northwest continue to decline, the ESA is likely to play a greater role in ecosystem management within the Pacific coastal ecoregion. Its presence is already evident in the protections now in place to conserve Snake River salmon and steelhead, and ESA listings of stocks in coastal rivers appear imminent. Of particular note is its applicability to state and private lands, which has engendered an often acrimonious debate over ESA reform.

New Approaches

Anadromous fish listings under the Endangered Species Act signal that existing river laws are currently not providing adequate protection for Pacific coastal riverine ecosystems. This failure is attributable to various causes including anachronistic state water allocation laws enacted before the complexity and connectedness of riverine systems was well understood, the lack of consistent, coordinated management objectives across jurisdictional boundaries (see Chapter 21), and the failure to fully implement and enforce existing laws. Though there are calls for new river laws, such as a comprehensive federal law regulating riverine ecosystems regardless of land ownership, the following discussion introduces three key areas where reform is needed within the existing legal regime.

Linking Water Quality and Water Quantity

A rudimentary understanding of river ecology is sufficient to understand the inextricable link between water quantity and water quality (see Chapter 5). For many years, however, that linkage was not recognized in the law; water quantity and water quality were viewed as separate issues. That artificial distinction was finally jettisoned in a seminal 1994 U.S. Supreme Court decision, *Washington Department of Ecology v. PUD No. 1 of Jefferson County*, in which water quantity was deemed to be an integral part of water quality.[126] The Court reasoned that the definition of pollution in the Clean Water Act encompasses the physical and biological integrity of water, as well as chemical integrity.[127]

The legal recognition that sufficient water quantity is essential to water quality has potentially wide-ranging ramifications. First, river development activities that would deplete flows can now be conditioned on maintaining flow levels necessary to achieve both the numeric and narrative criteria and designated use components of the applicable water quality standards. For example, in *Jefferson County*, the Court held that the Washington State Department of Ecology, through its Section 401 certification authority, could condition licensing of a hydroelectric project on the maintenance of a minimum flow in the bypass reach to protect "salmonid migration, rearing, spawning, and harvesting," designated uses of the Dosewallips River. The Court also said that the flow requirement was justified to comply with the antidegradation provision of Washington's water quality standards.

The Court's rulings also suggest that rivers with flows insufficient to support designated uses can be listed as "water quality limited" under Section 303(d) of the Clean Water Act. Indeed, following on the heels of the *Jefferson County* decision, Washington state has listed approximately 13 stream segments west of the Cascades as water quality limited because of insufficient flows attributable to water withdrawals. In short, the linkage of water quality and quantity in *Jefferson County* was transformational, opening a number of potential avenues for improving water quality through in-stream flow regulation.

Controlling Nonpoint Source Pollution

Controlling nonpoint source pollution has proven a much more formidable challenge than controlling point source pollution due to several obstacles, including lack of Best Management Practices implementation and monitoring, and the voluntary nature of many state nonpoint source control programs. Approximately 100% of sediment, 82% of nitrogen, and 84% of phosphorous reaching the nation's surface waters from anthropogenic activities originate from nonpoint sources (Conservation Foundation 1987). More than two decades after the Clean Water Act was enacted, many rivers and streams in the Pacific coastal ecoregion are water quality impaired, due largely to nonpoint source pollution. For example, in Washington state, approximately 86% of water bodies assessed for water quality by the Department of Ecology are not supporting one or more of their designated uses (Washington State Department of Ecology 1995).

One avenue for reducing nonpoint source loads is the Clean Water Act's requirement that states develop and implement Total Maximum Daily Loads (TMDLs) for water quality impaired rivers.[128] To date, despite the mandatory nature of the TMDL requirement, few TMDLs have been established due to technical, informational, and political obstacles (Hildreth 1994). In recent years, states have stepped up their TMDL development due in large part to litigation to force EPA to require states to develop TMDLs.[129] If the impediments can be overcome, TMDLs could be a significant step toward improving water quality.

However, establishing pollutant load limits will not, alone, reduce nonpoint source pollution. Effective strategies and management practices must be developed and implemented to reduce nonpoint loads. Existing nonpoint source programs have relied heavily on voluntary control measures with only limited success. To achieve TMDLs, mandatory load-reduction measures, timelines for implementing such measures, and the availability of significant sanctions for non-compliance will likely be needed (Gould 1990).

Improving Environmental Protection Under the Federal Power Act

Many of the hydroelectric dams on Pacific coastal rivers were licensed and constructed before the ecological damage caused by dam construction and operation was well understood. Consequently, hydropower projects were often licensed without adequate protection for riverine biota or due regard for maintaining the ecological functions that depend upon a natural hydrologic regime. Fortunately, the Federal Power Act limits hydropower licenses to 50-year terms and requires the Federal Energy Regulatory Commission (FERC) to evaluate anew, giving "equal consideration" to developmental (i.e., power generation and flood control) and nondevelopmental values (i.e., fish and wildlife), whether an existing project should be relicensed, and if so, under what conditions.

Despite the great potential to strike a balance between environmental and power values through the relicensing process, that balance has not been achieved due in large part to FERC's policy of assessing the environmental impacts of hydropower projects using current river conditions (i.e., the river with the project) as the baseline for analysis. Consequently, past and continuing environmental damage caused by project construction and operation during the original license term is essentially written off, and the relicensing decision is made based on the current degraded environment.

FERC's use of a current condition baseline has drawn sharp criticism, and will likely be challenged in the courts. The position that FERC must assess all project impacts in its relicensing decisions appears to be consistent with the statutory language and case law interpreting the FPA as requiring FERC to look at projects anew considering all issues relevant to the public interest.[130] Moreover, it is consonant with the fundamental ecological tenet that the goal of river restoration should be to reestablish, to the extent possible, the natural hydrologic regimes that prevailed prior to significant human manipulation of riverine systems.

Future Outlook

The state and federal laws discussed in this chapter form the nucleus of the laws shaping riverine system management in the Pacific coastal ecoregion. Though these laws possess great potential for protecting and restoring the ecological health and integrity of coastal rivers, much of that protection and restoration potential remains unrealized, and, consequently, many river systems continue to be degraded. Insufficient funding of resource agencies, lack of political will, inadequate understanding of the ecological and economic importance of healthy rivers, the complexity of riverine systems, and the strength of economic and property interests promoting watershed development have coalesced to keep the promise of river protection unfulfilled.

Although there is an ecological need for better implementation and enforcement of existing laws, there is also a movement within both the federal and state legislatures to reduce the level of environmental regulation primarily because it is viewed by some as impeding economic growth and impinging private property rights. The attempt to curtail environmental regulation manifests in efforts to amend state and federal laws, to diminish environmental protections, and to reduce the budgets of agencies charged with implementing and enforcing such statutes. Given the political impediments to legal reform at the national and state levels, focus has shifted to collaborative, locally based solutions to river management issues. Local watershed councils are an example of this approach, where representatives from all affected interest groups work with federal and state agencies to develop watershed-specific management plans. Collaborative, consensus-driven watershed management appears to be the wave of the future, although its long-term effectiveness has yet to be proven. Collaborative processes should be coupled with clearly defined, legally enforceable standards to protect our rivers and streams.

Acknowledgments. The authors would like to thank Michael Blumm and David Avren for their contributions.

Literature Cited

Beck, R.E., ed. 1991. Waters and water rights. Michie, Charlottesville, Virginia, USA.

Bell, C.D., and N.K. Johnson. 1991. State water laws and federal water users: The history of conflict, the prospects for accommodation. Environmental Law **21**:1–88.

Blumm, M.C. 1992. Unconventional waters: The quiet revolution in federal and tribal minimum streamflows. Ecology Law Quarterly **19**:445–480.

Blumm, M.C., and T. Schwartz. 1995. Mono Lake and the evolving public trust in Western water. Arizona Law Review **37**:701–738.

Bodi, F.L. 1989. The public trust doctrine in the state of Washington: Does it make any difference to the public? Environmental Law **19**:645–650.

British Columbia Heritage Rivers Board. 1995. British Columbia's heritage rivers: Inaugural candidates for a provincial system. British Columbia Heritage Rivers Board, Vancouver, British Columbia, Canada.

Conservation Foundation. 1987. State of the environment: A view toward the nineties. Conservation Foundation, Washington, DC, USA.

Dahl, T.E., and C.E. Johnson. 1991. Status and trends of wetlands in the coterminus United States, mid-1970s to mid-1980s. US Department of the Interior, Fish and Wildlife Service, Washington, DC, USA.

Doppelt, B. 1993. Entering the watershed: A new approach to save America's river ecosystems. Island Press, Washington, DC, USA.

Frost, P.M.K. 1993. Protecting and enhancing wild and scenic rivers in the West. Idaho Law Review **29**:313–360.

Gould, G.A. 1990. Agriculture, nonpoint source pollution, and federal law. University of California Davis Law Review **23**:461–498.

Hildreth, R. 1994. Instream flow protection—vain hope or possibility? Chapter F *in* The Sinking Creek decision: Water rights in the 21st century. University of Washington School of Law, Seattle, Washington, USA.

Johnson, R., and R. Paschal. 1995. The limits of prior appropriation. Illahee **11**:40–50.

Kilbourne, J.C. 1991. The ESA under the microscope: A closeup look from a litigator's perspective. Environmental Law **21**:499–586.

MacDonnell, L., and T. Rice. 1993. The federal role in in-place protection. Pages 5-1–5-38 *in* L. MacDonnell and T. Rice, eds. Instream flow protection in the west, revised edition. University of Colorado School of Law, Boulder, Colorado, USA.

Mandelker, D.R. 1989. Controlling nonpoint source water pollution: Can it be done? Chicago-Kent Law Review **65**:479–502.

Office of Technology Assessment. 1984. Wetlands: Their use and regulation. Publication Number OTA-0-206. Office of Technology Assessment, Washington, DC, USA.

Percival, R. et al. 1992. Environmental regulation: Law, science, and policy. Little, Brown, Boston, Massachusetts, USA.

Ransel, K.P. 1995. The sleeping giant awakens: PUD No. 1 of Jefferson City. v. Washington Department of Ecology. Environmental Law **25**:255–284.

Reisner, M., and S. Bates. 1990. Overtapped oasis: Reform or revolution for Western water. Island Press, Washington, DC, USA.

Sax, J.L. 1970. The Public Trust Doctrine in natural resource law: Effective judicial intervention. Michigan Law Review **68**:471, 475.

Shupe, S., and L. MacDonnell. 1993. Recognizing the value of in-place uses of water in the West: An introduction to the laws, strategies, and issues. Pages 1-1–1-22 *in* L. MacDonnell and T. Rice, eds. Instream flow protection in the West, revised edition. University of Colorado School of Law, Boulder, Colorado, USA.

Trelease, F.J. 1971. Federal–state relations in water law. National Water Commission Study Number 5. National Water Commission, Arlington, Virginia, USA.

USDA (United States Department of Agriculture). 1987. United States Forest Service land and resource management planning handbook. United States Department of Agriculture, Washington, DC, USA.

USDI (United States Department of Interior). 1992. Manual for wild and scenic rivers. US Department of the Interior, Washington, DC, USA.

US EPA (United States Environmental Protection Agency). 1995. Wetlands fact sheets. Publication Number EPA 843-F-95-001e. US Environmental Protection Agency, Washington, DC, USA.

USFS (United States Forest Service). 1993. Record of decision for amendments to Forest Service and Bureau of Land Management planning documents within the range of the Northern Spotted Owl. 1v. USDA Forest Service and USDI Bureau of Land Management, Washington, DC, USA.

Washington State Department of Ecology. 1995. Washington state water quality assessment report [305(b)]. Publication Number WQ-95-65a. Washington Department of Ecology, Olympia, Washington, USA.

Whitman, R. 1991. Clean water or multiple use? Best management practices for water quality control in the national forests. Ecology Law Quarterly **16**:909–966.

Yagerman, K. 1990. Protecting critical habitat under the Federal Endangered Species Act. Environmental Law **20**:811–856.

Endnotes

1. Alaska Const. art. VIII, § 3.
2. 33 U.S.C. § 1344 (West 1986).
3. 33 C.F.R. Pt. 325 (1995).

4. U.S. Const., art. I, § 8, cl. 3.
5. U.S. Const. art. VI, cl. 2.
6. *Ray v. Atlantic Richfield Co.*, 435 U.S. 151 (1978).
7. *Crook v. Hewitt*, 4 Wash. 749, 31 P. 28 (1892).
8. Wash. Rev. Code Ann. §§ 90.03.290 and 90.44.060 (1992 & 1996 Supp.).
9. *Stempel v. Dept. of Water Resources*, 82 Wash. 2d 109, 117 (1973).
10. *In the Matter of Appeals from Water Rights Decisions of the Department of Ecology. Order on Motions for Summary Judgement*, PCHB Nos. 968-96181 (nonsequential) July 16, 1996.
11. Alaska Stat. § 46.15.145 (1995 Supp.) ("protection of fish and wildlife habitat, migration, and propagation . . . and water quality purposes"); Cal. Water Code § 1707 (1996 Supp.) ("preserving or enhancing wetlands habitat, fish and wildlife resources, or recreation in or on the water"); Or. Rev. Stat. Ann. § 537.336 (1988) ("conservation, maintenance and enhancement of aquatic and fish life, wildlife, and fish and wildlife habitat"); Wash. Rev. Code Ann. § 90.22.010 (1992 & 1996 Supp.) ("protecting fish, game, birds or other wildlife resources, or recreational or aesthetic values of said public waters whenever it appears to be in the public interest").
12. Or. Rev. Stat. § 536.325 (1988).
13. *Illinois Central Railroad Co. v. Illinois*, 146 U.S. 387 (1892).
14. *National Audubon Soc'y v. Superior Court*, 33 Cal. 3d 419, 437 *cert. denied*, 464 U.S. 977 (1983) (holding that public trust doctrine applies to non-navigable tributaries that affect navigable waters).
15. *Marks v. Whitney*, 6 Cal. 3d 251, 259 (1971).
16. *See United States v. State Water Resources Control Board*, 182 Cal. App. 3d 82, at 118–120 (1986).
17. *See Wilbour v. Gallagher*, 77 Wa. 2d 306 (1969).
18. *See Rettkowski v. Washington*, 858 P.2d 232, 239 (Wash. 1993).
19. *United States v. New Mexico*, 438 U.S. 696, 700 (1978) (reserved rights limited to the amounts without which "the purposes of the reservation would be entirely defeated").
20. *Id.* at 707–08.
21. *See, e.g., Colville Confederated Tribes v. Walton*, 647 F.2d 42 (9th Cir. 1981), *reh'g denied*, 752 F.2d 397 (9th Cir. 1985); *cert. denied*, 475 U.S. 1010 (1986).
22. *United States v. Washington*, 384 F. Supp. 312, 332–33, 359–82 (W.D. Wash. 1974).
23. Federal Water Pollution Control Act of 1972, 33 U.S.C. § 1251 *et seq.* (West 1986 & 1996 Supp.).
24. 33 U.S.C. § 1251(a).
25. *Id.* § 1251(a).
26. 33 U.S.C. § 1311(b)(1)(C).
27. *Id.* § 1377(e).
28. *See Marble Mountain Audubon Society v. Rice*, 914 F.2d 179, 182 (9th Cir. 1990) (Forest Service required to comply with all state water quality standards).
29. 33 U.S.C. § 1362(19).
30. *Id.* § 1341. Qualified Indian tribes may also establish a 401 certification program. *Id.* at § 1377(e).
31. *PUD No. 1 of Jefferson County v. Washington Dep't of Ecology*, 114 S.Ct. 1900, 1910–12 (1994).
32. 33 U.S.C. § 1362(14).
33. *Id.* § 1365(a)(1).
34. 33 U.S.C. § 1312(a).
35. *Id.* § 1329.
36. *Id.* § 1288(b).
37. 33 U.S.C. § 1329(a)(1).
38. *Id.* § 1329(b)(2)(A).
39. *Id.* § 1329(b)(2)(C).
40. *Id.* § 1329(b)(4).
41. *Id.* § 1329(h).
42. 16 U.S.C. § 1455b (West 1996 Supp.).
43. 33 U.S.C. § 1313(d).
44. 16 U.S.C. § 1271–87b (West 1986 & 1996 Supp.).
45. Alaska Stat. §§ 42.23.400–.510 (1995 Supp.); Cal. Pub. Res. Code §§ 5093.50–.69 (1984 & 1996 Supp.); Or. Rev. Stat. § 390.815 (1988 & 1994 Supp.); Wash. Rev. Code Ann. §§ 79.72.010–.900 (1992 & 1996 Supp.).
46. 16 U.S.C. § 1271.
47. *Id.* §§ 1271, 1273(b), 1282(a), 1283(c).
48. *Id.* § 1273(b)(1)&(b)(2).
49. *Id.* § 1273(b)(3).
50. *Id.* § 1283.
51. 47 Fed. Reg. 39,459 (September 7, 1982).
52. 16 U.S.C. § 1278(a).
53. *Id.* § 1281(a).
54. *Id.* § 1277(a), (e), & (f).
55. *Id.* § 1281(e).
56. *Id.* § 1284(c).
57. 33 U.S.C. § 1344 (West 1986 & 1996 Supp.).
58. *Id.* § 1344(a).
59. *Id.* § 1344(e).
60. 33 C.F.R. § 328.3(b)(1995).

61. 33 C.F.R. § 328.3(a).
62. *United States v. Malibu Beach. Inc.*, 711 F. Supp. 1301, 1312 (D. N.Y. 1989).
63. 33 C.F.R. § 323.2(k).
64. 33 U.S.C. § 1344(f)(1)(A),(E).
65. *Id.* § 1344(f)(2).
66. 33 C.F.R. § 320.4(a).
67. *Id.*
68. 40 C.F.R. § 230.
69. *Id.* § 230.1(c).
70. *Id.* § 230.10(b).
71. *Id.* § 230.10(c).
72. *Id.* § 230.10(d).
73. 33 U.S.C. § 1344(c).
74. 16 U.S.C. §§ 1600 *et seq.* (West 1986 & 1996 Supp).
75. 43 U.S.C. § 1701 *et seq.* (West 1986 & 1996 Supp).
76. 16 U.S.C. § 539d,e (West 1996 Supp).
77. 16 U.S.C. § 3100 *et seq.* (West 1986 & 1996 Supp.).
78. 16 U.S.C. § 528 (West 1986).
79. 16 U.S.C. § 1604(g)(E).
80. 36 C.F.R. § 219.27(e) (1995).
81. *Id.* § 219.27(f).
82. *Id.* § 1604(3)(E)(iii).
83. 43 U.S.C. § 1701(a)(7).
84. *Id.* § 1701(a)(8).
85. *Id.* § 1701(c)(3).
86. 43 C.F.R. § 1725.3–3(b).
87. *Id.* § 1725.3–3(h).
88. Alaska Stat. § 41.17.010 *et seq.* (1995 Supp.); Cal. Pub. Res. Code § 4511 *et seq.* (1984 & 1996 Supp.); Or. Rev. Stat. Ann. § 527.610 *et seq.* (1988 & 1994 Supp.); Wash. Rev. Code Ann. § 76.09.010 *et seq.* (1992 & 1996 Supp.).
89. Or. Rev. Stat. Ann. § 527.765.
90. 16 U.S.C. § 791 *et seq.* (1986 & 1996 Supp.).
91. *Id.* § 817(1).
92. Or. Rev. Stat. § 543.010–543.620.
93. *See* H.R. Rep. No. 507, 99th Cong., 2d Sess. 30, *reprinted in* 1986 U.S. Code Cong. & Admin. News 2496, 2503–05.
94. *Confederated Tribes and Bands of the Yakima Indian Nation v. FERC*, 746 F.2d 466, 471 (9th Cir. 1984) (citing *Udall v. FPC*, 387 U.S. 428, 450 (1967)), *cert. denied*, 471 U.S. 1116 (1985).
95. 16 U.S.C. § 803(a).
96. *Yakima*, 746 F.2d at 470.
97. 16 U.S.C. § 803(j).
98. *Id.* § 811.
99. *Id.* § 797(e).
100. *Escondido Mutual Water Co. v. La Jolla Band of Mission Indians*, 466 U.S. 765, 776–79 (1984).
101. *California v. FERC*, 495 U.S. 490 (1990); *First Iowa Hydro-Electric Power Coop. v. Federal Power Commission*, 328 U.S. 152 (1946).
102. *California v. FERC*, 495 U.S. at 494–95.
103. *Id.*
104. *Washington Dept. of Ecology v. PUD NO. 1 of Jefferson County*, 114 S. Ct. 1900 (1994).
105. 16 U.S.C. § 1531 *et seq.* (West 1986 & 1996 Supp.).
106. *Id.* § 1533(a)(1).
107. *Id.* § 1531(b).
108. 16 U.S.C. § 1533(a)(3)(A).
109. *Id.* § 1532(5)(a).
110. 50 C.F.R. § 424.12(b) (1995).
111. 16 U.S.C. § 1533(b)(2).
112. *Id.* § 1536(c)(1).
113. 50 C.F.R. § 402.14(b).
114. 16 U.S.C. § 1536(c)(1).
115. *Id.* § 1536(c); 50 C.F.R. § 402.12(b)(1).
116. 50 C.F.R. § 402.13.
117. 16 U.S.C. § 1536(a),(b).
118. 50 C.F.R. § 402.02.
119. 16 U.S.C. § 1536(a)(2), (b)(3)(A).
120. *Id.* § 1536(b)(4).
121. *Id.* § 1538(a)(1).
122. *Id.* § 1532(19).
123. *Palila v. Hawaii Dept. of Land and Natural Resources*, 639 F.2d 495, 497–98 (9th Cir. 1981).
124. 16 U.S.C. § 1539(a)(2).
125. *E.g.*, *Riverside Irrigation District v. Andrews*, 758 F.2d 508, 514 (10th Cir. 1985).
126. *PUD No. 1 of Jefferson County v. Washington Department of Ecology*, 114 S.Ct. 1900, 1912–13 (1994).
127. *Id.*
128. 33 U.S.C. § 1313(d).
129. *E.g.*, *Alaska Ctr. for the Env't v. Reilly*, 762 F. Supp. 1422, 1425 (W.D. Wash. 1991) (Alaska had not submitted any TMDLs to EPA in a decade).
130. *E.g.*, *Confederated Tribes and Bands of the Yakima Indian Nation v. FERC*, 746 F.2d 466, 471 (9th Cir. 1984), *cert. denied*, 471 U.S. 1116 (1985).

23 Economic Perspectives

Daniel Huppert and Sylvia Kantor

Overview

• River management is bounded by the functioning of both the river ecosystem and the socioeconomic system. Society faces the economic problem of allocating scarce river ecosystem resources to satisfy numerous economic demands. Economic analysis of benefits and costs helps to assess the human consequences of these allocation decisions.

• Before the 1960s economic studies focused on benefits and costs of river development—flood control, navigation projects, hydroelectric dams, and water diversion for irrigation and domestic water supply.

• Federal agencies are required to apply the well-developed benefit-cost procedures in planning for water resources projects. Retrospective case studies by Haveman (1972) indicate that benefits are often overestimated, and costs underestimated, in water project planning.

• More recently, environmental economics research has emphasized the measurement of nonmarket values such as recreation, endangered fish species, and in-stream flow. Nonmarket values are derived from river ecosystems directly (e.g., river fishing, boating) or indirectly (e.g., lower flood losses due to preservation of natural wetlands). Nonmarket values are divided into *use* values and *nonuse* or existence values (e.g., the value of protecting endangered fish species that do not contribute to any fishery).

• Economists provide decision-makers with cost-effectiveness, benefit-cost, or economic impact analysis. Cost-effectiveness analysis shows, for example, the costs of alternative means of improving water quality or of increasing salmon survival through hydroelectric systems. Benefit-cost analysis compares the benefits of, for example, preserving stream flows and fish stocks with the cost of reduced irrigated agriculture, hydroelectricity, and flood control. Regional income and employment impacts indicate, for example, the magnitude of local, short-term economic changes resulting from the decline of a river fishery or the construction of a new irrigation project.

• Ecological–economic research goes beyond estimating values derived from river ecosystems to causal modeling of linkages between economic consequences, river management actions, and ecosystem functions. Examples include forest/salmon fishery management in Oregon and Idaho (Loomis 1988, 1989), policy concerning agricultural sediment production in the Tom Beall watershed of Washington (Brusven et al. 1995), and assessment of agricultural diversion versus steelhead fishing in Oregon's John Day River (Johnson and Adams 1988). Comprehensive models of this sort are rare and difficult to validate.

• Few full-scale ecological–economic models for river management have been developed. The ecological–economic analyses reviewed in this chapter rely upon very simplified models that assume, for example, that fish stocks increase directly with stream flow, or that sedimentation rates affect fish and water quality in easily understood ways.

• Uncertainties in both economic estimates and ecological responses to management action make the use of economic models in river management a challenging but, nonetheless, important field.

Introduction

Rivers have always been important to the economy of the Pacific coastal ecoregion. Native Americans used rivers for transportation and for fishing. River flood plains are extensively developed for urban and agricultural use. Many rivers are diverted to irrigate crops, dredged to maintain shipping channels, and impounded to reduce floods, generate electricity, and provide flatwater recreation. Because rivers are components of complex, interacting systems, river developments often impinge upon economic activities and ecological functions dependent upon natural river conditions. Planners in the Columbia River find that expanding naturally spawning populations of salmon (*Oncorhynchus* spp.) may mean sacrificing some hydroelectric power generation (Lee 1993). Greater agricultural production may impinge on both salmon and hydropower. Improved river habitats for fish and wildlife may mean less timber production. In managing rivers society faces the classic economic problem of allocating scarce resources among social needs. The scarce resource is not just the water, but encompasses the whole river ecosystem. In recent decades, economists have recognized undeveloped rivers as scarce and valuable natural assets that provide recreational and ecological services. Crucial economic choices involve balancing the social values of off-stream water uses (i.e., municipal and industrial water supply, irrigated agriculture) and in-stream uses (i.e., hydroelectric power, navigation, recreation, and natural ecosystem maintenance).

The human population in the Pacific Northwest has been growing at twice the national growth rate. If this trend continues, the expanding regional economy will put added pressure on increasingly scarce natural environments and resources. Reconciling economic expansion with sustained river-dependent wildlife, fish, water quality, and river recreation will be a major challenge. The facilities and structures of economic development often block fish migration, alter river flow regimes, raise water temperatures, rearrange the deposition of sediments, and entrain and kill juvenile fish. In developed landscapes, creeks and tributaries receive point and non-point source pollution from mines, mills, roads, farms, and parking lots. The long-term alteration of river ecosystems by these kinds of structural changes has contributed to the notable decline in salmon populations of the Pacific coastal ecoregion. The decline in salmon is also attributed to overfishing, ocean conditions, and hatchery practices (NRC 1995). Hence, society faces choices in managing rivers requiring economic tradeoffs, and deciding "how much is enough" for salmon, wildlife, recreation, electrical power, agriculture, and economic growth.

As depicted in Figure 23.1, the economics of river management is tied to the functioning of the river ecosystem and to the socioeconomic system. The socioeconomic system affects the ecosystem through outputs of pollution and sediment, and through structures and activities that directly change the river morphology and hydrology. The ecosystem responds to these effects in complex ways that alter the economic outputs obtained from the ecosystem. Economic outputs of the ecosystem include fish, hydropower, water supplies, scenic amenities, and recreation. The allocation of water and other river system components among competing uses, the protection of fish and wildlife, and the regulation of industry have complex social, economic, and legal consequences. Economic analysis of river ecosystems focuses on the quantities and values of the full array of economic benefits enjoyed by residents of the region.

Economics and Water Resources

One of the basic tasks of economics is to explain and understand the function of decentralized, competitive markets in which consumers and producers engage in voluntary exchanges

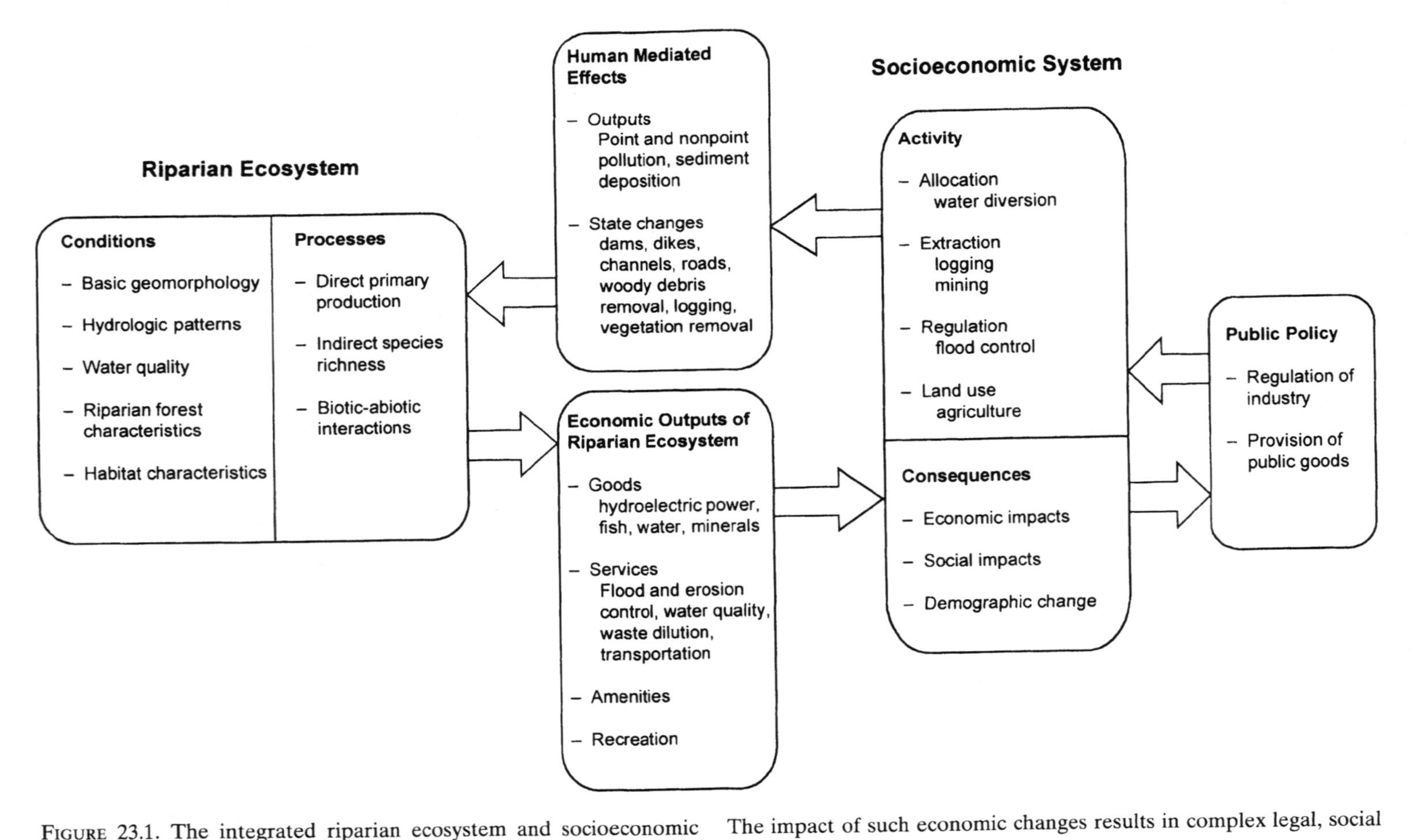

FIGURE 23.1. The integrated riparian ecosystem and socioeconomic system. Socioeconomic activities affect riparian ecosystem processes and conditions which then alter economic outputs from the ecosystem. The impact of such economic changes results in complex legal, social and economic consequences.

guided largely by individual self interest. The grandfather of modern economics, Adam Smith, explained that unregulated, competitive activity could yield socially desirable results. The individual producer, acting alone, with only self interest as a guide, would be lead "as if by an invisible hand" to supply the goods and services most needed by consumers. In modern economics terminology, economies organized by competitive allocation of *private* goods in markets are said to be *economically efficient*, which means that consumer preferences are satisfied as much as possible given the resources available. Just as importantly, economics provides a rigorous understanding of how and why markets provide too few *public* goods, and why markets do not adequately control the external costs associated with pollution and indirect effects on ecosystems.

By definition, pure public goods are nonexclusive which means nobody can be excluded from enjoying or consuming them. Public goods are also "nonrival in consumption," which means that one person's consumption or enjoyment is not affected by another's. Examples of public goods that are nonrival in consumption include clean air, visual amenities of rivers, and bird watching. Because people cannot be excluded from consumption by a price or entry fee, private firms find it unprofitable to supply public goods. While pure public goods are uncommon, there are many quasipublic goods that are somewhat nonexclusive or nonrival. For example, many people can enjoy viewing riparian wildlife without significantly impacting the resource, but it is possible to exclude people from close observation of wildlife by fencing riparian land. Further, some quasipublic goods are depleted by economic use (e.g., overhunting riparian wildlife). Because private firms can recover costs for some uses of riparian land (e.g., raising livestock, vacation housing, private campgrounds), the market economy is expected to supply these goods. The unregulated market economy, however, does not provide adequate amounts of public goods.

Further, river ecosystems are intensely affected by such externalities as agricultural runoff, and industrial and municipal effluent. Miners benefit from dredging streambeds for gold, but the dredging can suspend sediments and restructure gravel beds and pools that are important to fish. The mining activity imposes an external cost on the fishery and other river users, and the miners do not pay compensation for the reduced value of fishing and water quality as a result of dredging. The unregulated private market will supply too much gold and too few natural river conditions. Some of the economic externalities could, in principle, be controlled through more comprehensive private negotiation and contracting. For example, the fishing and the mining industries could agree on levels and timing of dredging. Absent social conventions and legal support for private contracting between all affected river users, the unregulated external costs and poorly supplied public goods reflect *market failure*. One response to market failure is collective decision making and government regulation. (Kahn 1995, Chapter 27 this volume).

While collective action can balance public and private use of rivers, governments are not necessarily well suited for the job. An agency exercising authority to optimize an ecosystem (balancing fish habitat, trout fishing, river boating, mining, and pollutant levels) is also a bureaucratic organization striving for internal objectives that often differ from broader social objectives. (Chapter 21) Success in a government agency is often associated with larger budgets or a larger workforce, and this drives the incentive structures and behavior of office directors and department heads (Niskanen 1971). Further, government agency directors serve political leaders whose outlook may be limited by election cycles. Political systems often have shorter time horizons than do ecosystem processes. Agency stewards of ecosystems may be compelled to satisfy relatively short-term demands rather than longer term economic objectives. Some analysts claim that placing important natural resources and ecosystems in the hands of politically driven bureaucracies can lead to *government failure* (Anderson 1983). Properly quantified and interpreted, economic studies of river management can provide an independent check on both market driven and bureaucratic decisions.

Historical Perspectives

Economic studies of rivers initially focused on flood control and navigation projects. The Flood Control Act of 1936 directed federal agencies (particularly the Army Corps of Engineers and the Bureau of Reclamation) to assess the benefits and costs of proposed projects. Natural rivers were viewed mainly as raw resources to be developed, and river development meant dredging for inland waterways, diking to reduce local flooding, and diversion for irrigating lands and domestic water supply. By the 1960s hydroelectric dams became a major focus for federal river projects. Because the costs of federal projects are born by all taxpayers while most benefits accrue locally, regional interests frequently campaign for river projects even when the local benefits are small relative to federal costs. Congress requires that federal agencies use economics to screen projects before seeking appropriations. This procedural requirement does not, of course, prevent the funding of economically indefensible projects for other social or political reasons.

Benefit-cost analysis, the principal economic assessment method, developed rapidly in response to federal agency needs after World War II. The Federal Inter-Agency River Basin Committee, formed in 1946 by agencies concerned with water-resource problems, appointed a Subcommittee on Benefits and Costs which produced its final report in 1950. Officially titled "Proposed Practices of Economics Analysis of River Basin Projects," but widely known as the Green Book, that report set forth a complete set of principles for water project evaluation. Otto Eckstein's classic *Water-Resource Development: The Economics of Project Evaluation* (1958) reviewed and summarized the project evaluation practices of major agencies and set forth the state-of-the-art for economic assessment in four categories of river development: flood control, navigation, irrigation, and electric power. These categories reflect a view of rivers as assemblages of independent economic outputs, most of which are marketable goods (e.g., electricity, barging of freight, irrigation water). Eckstein barely mentions recreation benefits from water projects, and does not acknowledge the costs of development to nonmarket goods or the indirect effects of river development occurring through ecosystem responses.

During the 1960s and 1970s, intensified interest in the economic value of nonmarket goods and environmental conditions led to a broader concept of nonmarket economic benefits, including outdoor recreation and scenic amenities (Krutilla and Fisher 1975). Values associated with pristine natural rivers, with instream flows, and with maintenance of wildlife species in natural environments are now given more attention. Rivers and other features of the natural landscape have *existence value* to people who simply appreciate the beauty of a functioning natural ecosystem. The Green Book was expanded and updated in the Water Resources Council's Principles and Standards for Planning Water and Related Land Resources (Water Resources Council 1973, pg. 29778). The Principles and Standards were replaced in 1983 by the Economic and Environmental Principles and Guidelines for Water and Related Land Resources Implementation Studies (U.S. Water Resources Council 1983). Since 1983 there has been an outpouring of research on economic values for nonmarket goods and services, much of which pertains to fishing, boating, water quality, and scenic values.

Defining and Measuring Economic Value

Two basic premises of economic welfare theory are that the purpose of economic activity is to increase the material well-being of individuals in society, and that each individual is the best judge of his or her own welfare. An individual's well-being depends on the market goods and services consumed, on goods and services provided by government programs, and on a wide variety of nonmarket services flowing from natural ecosystems. A change in river ecosystem triggers changes in many related goods and services. Hence, a full economic valuation must incorporate the values that individuals place on all aspects of the ecosystem. An economic assessment of river management begins with a prediction of ecosystem changes due to the management measures; it then quantifies the influence of those ecosystem changes on economic goods and services; and it ends with a valuation of the economic changes. Thus, the analysis flows from management actions to ecosystem conditions, to changes in economic services, to individual economic values. The economist identifies the important economic service flows, estimates their magnitude and timing, and then estimates their effects on individual welfare.

Individual Values—Compensating and Equivalent Variations

Quantifying economic value of a change in river ecosystem functioning requires a yardstick—or *numeraire*. Economists use the income of individuals as a measure of their ability to purchase welfare-enhancing goods and services. Income is the amount of purchasing power accruing periodically through wages, salaries, professional fees, interest on loans, rents on property, and profits of enterprises. Income does not measure the individual's mental state or level of happiness. Increased income permits the individual to increase consumption of whatever mix of goods and services is desired. A decrease in income, or an increase in price at a given income level, causes a decrease in individual economic welfare. Similarly, a decrease in a public good or ecosystem service, such as water quality or wildlife, decreases individual welfare. If a person must pay for an improvement in ecosystem services (through taxes or access fees), the person's welfare may be improved or diminished depending on whether the improvement is worth the payment. Similarly, if a person is paid compensation for a degradation in ecosystems, her welfare could be improved or diminished, depending upon whether the compensation covers the value of lost services. In principle, it should be possible to find a level of payment or compensation that leaves her economic welfare unchanged as the ecosystem services increase or decrease. The decrease or increase in income that is just sufficient to compensate for the ecosystem change (called the *compensating variation*) is the standard economic measure of individual value.

The maximum amount the individual would be willing to pay (WTP) for an increase in, say, river fishing opportunities is just the amount that could be deducted from his income in exchange for improvement while leaving him as well off as before. If required to pay any more than this, he would refuse the offer. The minimum amount he would be willing to accept in compensation (WTA) for a deterioration in fishing is the increased income that leaves him as well off with the deterioration. This welfare measure is hypothetical; it does not depend upon whether the payment is actually made (in the case of a welfare improving change) or the compensation is actually paid (in the case of a welfare decreasing change). Whether payments should actually be made, or compensation be awarded, is not determined by economic reasoning.

The individual economic value assigned to any economic good depends upon the baseline from which the value is measured. If the baseline is taken to be the status quo before the ecosystem change, the individual value is the WTP for improvements to welfare or the WTA for losses. This approach to valuation is called the *compensating variation*. On the other hand, if the condition after the ecosystem change is taken as the baseline, the individual value could be measured as the WTA to forego a benefit or the WTP to avoid a loss. This approach to valuation is called the *equivalent variation*. The equivalent variation approach could be appropriate, for example, if the *status quo* is deficient in delivering on past contracts or fails to satisfy an established right—the individual has a right to an ecosystem state that reflects past commitments or makes up for past wrongs. As summarized in Table 23.1, the four possible measures of economic value depend on whether the change is positive or negative and on what baseline is chosen.

If WTA and WTP for a given change differ little, it does not matter much which is used to measure the economic value of a change in river ecosystem. Further, because any attempt

TABLE 23.1. Four possible measures of individual economic value for a river ecosystem change.

	Compensating variation Equals the income adjustment that would leave a person as well off with as without the change.	Equivalent variation Equals the income adjustment that would leave a person as well off without as with the change.
Economic welfare gain Increase in goods and services (e.g., increased salmon runs, increased crops from irrigated agriculture).	WTP Maximum amount an individual would be willing to pay to obtain the beneficial change.	WTA Minimum amount an individual would be willing to forego for the beneficial change.
Economic welfare loss Decrease in goods and services (e.g., loss of whitewater recreation due to damming of river, increase in electricity price).	WTA Minimum amount an individual would accept in compensation for experiencing the loss.	WTP Maximum amount an individual would be willing to pay to avoid the loss.

WTP = Willingness to pay, WTA = Willingness to accept compensation.

to measure the welfare change will involve statistical errors, small differences between the two measures are not practically measurable. In fact, Robert Willig (1976) claimed exactly this—that under normal circumstances the differences between the two measures of value would be overshadowed by the errors of estimation. He concluded that either measure of value is a good approximation for the other. However, Hanneman (1991) shows that the difference between WTA and WTP can be substantial, especially if the valued good has no comparable substitutes among other market or nonmarket goods and services. If a fishery is to be eliminated by a hydropower dam or by effluent from a gold mine, the compensation demanded by former fishers may approximate the amount they would pay to prevent the elimination if that fishery were one among a number of similar fisheries. If the fishery at risk is the only one available, or if it has some unique social significance, the WTA may greatly exceed the WTP. In this case, an economic analyst cannot avoid the question of which baseline to choose for valuation, whether to use equivalent or compensating variation in assigning a value.

Some analysts state flatly that the selection of WTA or WTP should be based upon property rights (Freeman 1993). If person *A* proposes to engage in an action that diminishes the value enjoyed by person *B*, the question of how to "value" that diminution in value should be answered, according to this view, as follows: If *B* has a right to the status quo, then *A* must compensate *B*, and the relevant value is *B*'s willingness to accept compensation for the loss (WTA). In contrast, if *A* has a right to engage in the action that diminishes *B*'s value, then *B* must pay *A* to forego the action, and the relevant value is *B*'s willingness to pay (WTP) to prevent the loss. The simplicity of this solution, and its clear connection to principles of property and nuisance law, have led many to accept it.

Most river management and development activities involve a complex tangle of conflicting and uncertain rights and responsibilities. It is often difficult in practice to know whose rights are to be taken as the baseline for selection of values. For example, the Northwest Power Planning Act of 1980 attempts to "protect, mitigate and enhance fish and wildlife" while also requiring an "adequate, efficient, economical and reliable power supply" (Hildreth 1994). Congress ratified treaties with Native American tribes that commit the U.S. government to provide salmon harvest opportunities in the Columbia River basin, and it passed legislation authorizing the construction of dams to generate electricity, control floods, improve navigation on the river, and provide recreation. States allocate and recognize rights to divert water for irrigation and other purposes that clearly conflict with maintenance of instream flows, natural river ecosystem uses and salmon survival (Huffman 1983). Examination by economists or ecologists of legislation, treaties, commitments, and entitlements in the river will not resolve the assignment of property rights to specific resources or ecosystem. In the face of ambiguous property rights, some economists suggest that both WTP and WTA values be displayed for decision makers (Zerbe and Dively 1994). Another option would be to assume that no one has property rights to river ecosystems. If everyone has to pay for benefits received from the river, WTP values would be used for every benefit and cost component.

Categories of Economic Value

Environmental economists have defined a bewildering array of values during the past two decades. Some people classify values based upon how ecosystem functions lead to economic value. For example, *direct values* result from direct use of the ecosystem, as in recreational fishing, boating, or nature viewing. *Indirect values* result from a chain of events, such as improved water quality which leads to better fishing or swimming conditions or decreased human health problems. The value of improved water quality is derived from the direct values of fishing, swimming, and better health. Another distinction is between *market values,* which are associated with goods and services supplied by private firms and sold at a price in markets, and *nonmarket values*, which are derived from public goods like environmental amenities. In principle, tracking of market prices and quantities through time enables economists to estimate market demand and supply curves which facilitate the valuation of market goods. The market essentially generates evidence of value that can be interpreted. Nonmarket goods, such as most outdoor recreation and nature observation, generate different types of evidence that cannot be evaluated directly from a market model.

Finally, a whole taxonomy of values has been developed by environmental economists to focus on the behavioral and motivational sources of value (Figure 23.2). *Use value* is a broad category associated with using the environment or ecosystem, such as outdoor recreational activity, irrigated farming, electricity, and wildlife observation. Subcategories of use value identify *consumptive values* as those that literally consume part of the environment (e.g., fishing) or less obviously lead to a deterioration in the quality of the environment (e.g., use of a river to transport and disperse effluent). Examples of *nonconsumptives uses* are wildlife viewing, nature appreciation, and hiking. Nonconsumptive uses are often associated with public goods, because public goods are by definition not consumed by use. Further, people may be willing to pay to assure that river environments are available for future use—a simple extension of evaluation over time. However, given uncertainty that river conditions will exist in the future, people may be willing to pay a premium, over and above the value of future use, to assure that the river is available to use in the future. This *option value* is like an excess insurance payment—an amount over and above the expected insurable damage which the individual pays to reduce risk.

Nonuse value covers a broad category of human motivations that are unrelated to any tangible use of the environment. A classic example is the value people place on knowing that endangered fish are protected, even though they never expect to actually see that species of fish. The value that many people place on preserving spectacular natural features like the Grand Canyon, even though they do not plan to visit the place, is a nonuse value. Often called *existence value* or, more recently, *passive use value,* nonuse value can be associated with particular human motives. Holding existence value for the altruistic motive of bequeathing environmental assets to future generations are said to reflect *bequest value.* Existence values based on an obligation to the resource itself (stewardship) are said to reflect *preservation* or *intrinsic value.* The development and explication of these value categories is largely a means of understanding and explaining why people place value on environmental features and conditions. The categories themselves do not reflect separate values that

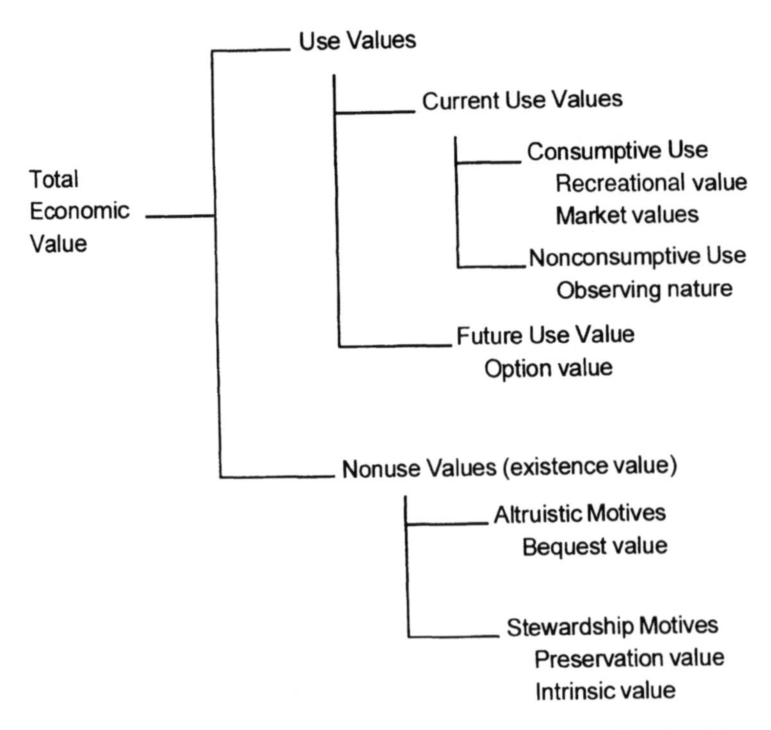

FIGURE 23.2. A taxonomy of behavioral and motivational sources of economic value. Use values plus nonuse values equal total economic value.

can be individually estimated and summed up to get a total value. As discussed below, the methods for estimating environmental values usually do not distinguish among the various motives for holding value.

The meaning and scope of nonuse values in law and in practice have been debated vigorously. To estimate economic damages from the Exxon Valdez oil spill in 1989, for example, researchers used a nationwide survey to elicit the magnitude of individual existence values lost as a result of the spill. Loomis (1996) mounted a similar nationwide survey to estimate the value of restoring the Elwha River in Washington State. Critics of such estimates claim that people unfamiliar with the specific site before and after the accident could not reasonably hold a legitimate value for the loss. Others claim that values expressed for the specific object of these surveys actually represent more inclusive values for resources in a broad class. For example, the expressed value for a specific salmon run might represent the value for salmon runs in general. Some fundamental philosophical issues of this sort remain unresolved at present. Although most economists now recognize that nonuse values can be important for environmental assets, the scope and importance of those values is largely an empirical question that can be answered only through more extensive research.

Economic research on rivers encompasses values of all types. Values have been estimated for both in-stream uses and diversions of water. In-stream use values are associated with whitewater rafting, canoeing, power boating, fishing, bird watching, shipping, visual amenities, power generation, and enjoyment of pristine conditions. Diversions and impoundments of rivers are largely for flood control, municipal water supply, irrigation of agricultural lands, and hydropower generation. Where one use of the river diminishes the value of another use, the lost value is an opportunity cost. Hence, almost every river project will entail a number of positive values and opportunity costs, all of which must be addressed in an economic assessment.

Aggregation into "Social Value"

As a guide to collective decisions, individual values are typically aggregated to reflect the net value to the society as a whole. This can be done by estimating values for a representative sample of the public or by measuring the aggregate value as reflected in aggregate consumer behavior. The *net social value* equals the sum of positive values to individuals who gain and negative values to people who lose. Restoring streamside vegetation may improve wildlife habitat, thus increasing values associated with hunting, fishing, and wildlife watching. But, that restoration may require reduced agriculture or restricted urban land use, thus decreasing the values to farmers and land owners. The aggregate value simply combines the positive and negative effects.

Aggregating individual values into a single indicator of social benefit necessarily raises the ethically charged issue of wealth distribution. The summing up process treats a dollar of benefits to anyone, regardless of circumstances, as equivalent to a dollar of benefits to anyone else. Is an incremental dollar of benefits to river boaters really equivalent to another dollar of costs to local farmers? In planning hydropower plants, should an increment of public benefit from lower electricity rates be treated as equal to an increment of cost to a Native American fishery? By summing up the benefits without special consideration of the relative merit of particular subgroups, the net economic value criterion ignores a number of social issues. Most people do hold ethical views about the fairness of particular shifts in economic welfare among interest groups, and public decision processes do consider the distribution of economic benefits along with the magnitude of those benefits. For example, the fulfillment of past commitments, such as treaties, binding contracts, property rights, and sometimes even political promises serve as principles of "fairness." Hence, economic net social benefit cannot function as an all-purpose economic criterion for policy analysis. It needs to be considered alongside information revealing the structure and distribution of the resulting benefits and costs.

Discounting and Aggregation over Time

Changes in river ecosystems are dynamic processes that work out over years, if not decades. Consequently, the benefits and costs flowing out over many time periods need to be considered simultaneously. One way of doing this is to simply add up the net benefits over a standard length of time. This procedure implicitly assumes that a dollar of benefits delivered in one year has the same importance as a dollar of benefits in any other year. This assumption, however, neglects the fact that people do, for a variety of reasons, discount values expected to occur in the future. To the extent that people place lower value on more distant benefits (that is, they express a positive time rate of preference), the discounting of future economic values in an economic valuation is consistent with the use of human preferences as an indicator of value. Markets for loans and investments reveal the rates at which individuals trade off present for future consumption; and the market interest rate, which functions as the price in such markets is affected by both supply (quantity of funds people offer to lend) and demand (quantity of funds demanded). Because it establishes the rate at which present income can be traded for future income, the interest rate indicates the "time preference" of individuals participating in the market. The interest rate also reflects the opportunity cost of scarce capital. That is, the transfer of \$1,000 from a use that yields a rate of return of 5%, to a use that does not generate a return involves the sacrifice of \$50/yr in income. The use of funds is said to entail an opportunity cost of 5%. Using discounted value of future returns as an investment criteria assures that capital is allocated to those projects which generate enough income to cover the opportunity cost of capital. In other words, discounting is needed for economic efficiency.

The net present value (*NPV*) is calculated using the following formula:

$$NPV = \sum_{t=1}^{t=N} \left(\frac{B_t - C_t}{(1+d)^t} \right),$$

where B_t and C_t represent the aggregate benefits and costs in year t, d represents the annual discount rate, and N is the "time horizon" or length of planning period. The higher the discount rate, the less weight is placed upon net benefits occurring further in the future. When benefits and costs are correctly reckoned in real, inflation-corrected monetary units, most economists support the use of discount rates in the range of 1% to 4% (Freeman 1993). When the levels of benefits and costs are uncertain, a risk averse decision maker might wish to use a higher discount rate.

Discounting of future net benefits raises concerns for sustainability and economic equity across generations of people. Some ecologists disparage discounting as unethical or as an expression of a selfish or myopic attitude toward the future. A hydroelectric dam may provide valuable electricity for 75 years while leaving future generations with accumulated sediment behind the dam, species extinction, and communities dependent upon cheap power. A dollar of benefit discounted at 5% for 75 years has a present value of only $0.026. Hence, the net present value reflects mainly the 75 years of benefits and only slightly the subsequent costs. If the project trades early benefits for later costs, a decision based solely on an NPV could treat future generations unfairly. Norgaard and Howarth (1991), however, note that conservationists face a dilemma. Conservationists typically prefer lower discount rates because lower rates favor the management of slow growing trees, the protection of biodiversity, and the conservation of exhaustible resources. Yet conservationists also argue for the use of high discount rates when higher rates make projects with deleterious environmental consequences appear uneconomic. For example, conservationists join with the economics profession to argue that water development projects should be evaluated using discount rates based on market rates of interest rather than the lower interest rate on tax-exempt government bonds. Conservationists want to be able to select the discount rate strategically depending on what favors conservation interests.

Norgaard and Howarth recommend that we distinguish between economic efficiency criteria and intergenerational equity concerns. Economic efficiency is not a criterion for equitable wealth distribution, neither within a generation nor between generations. A transfer of economic wealth from one group to another has to be justified by moral standards, not by economic efficiency. Such transfers would be carried out under an appropriate social program, not by the operation of competitive markets. Still, if conservation of ecosystem assets for future generations (e.g., preserving riparian habitat, protecting endangered species) is justified on equity grounds, the means of implementing such a wealth transfer could be evaluated using efficiency criteria. An inefficient transfer would entail higher costs than necessary to the current generation.

Using an arbitrary, low discount rate in NPV calculations will fail to screen out projects that generate low net benefits in the future. If capital invested in the economy generates a net return of 5%, then a dollar invested today could yield $38.83 of value to people living 75 years from today. Using low discount rates in project selection will permit projects that generate benefits less than $38.83 in 75 years, possibly leaving future generations with less wealth. Further, projects generating low rates of return are no more likely to be ecologically sensible than projects generating higher rates of return. A balanced approach establishes an explicit intergenerational equity criterion as a constraint, and then chooses specific investments in natural assets and manufactured assets using NPV analysis that meet the equity constraint. The use of the NPV criterion assures that projects selected yield adequate net rates of return.

Economic Benefits and Competitive Markets

The equilibrium price in a competitive market is that price which equates the quantity of goods demanded with the quantity supplied. In Figure 23.3, the price P_1 clears the market with Supply 1; while price P_2 clears the market with Supply 2. The amount that consumers are willing to pay for additional units of a good, as represented by the height of the demand, de-

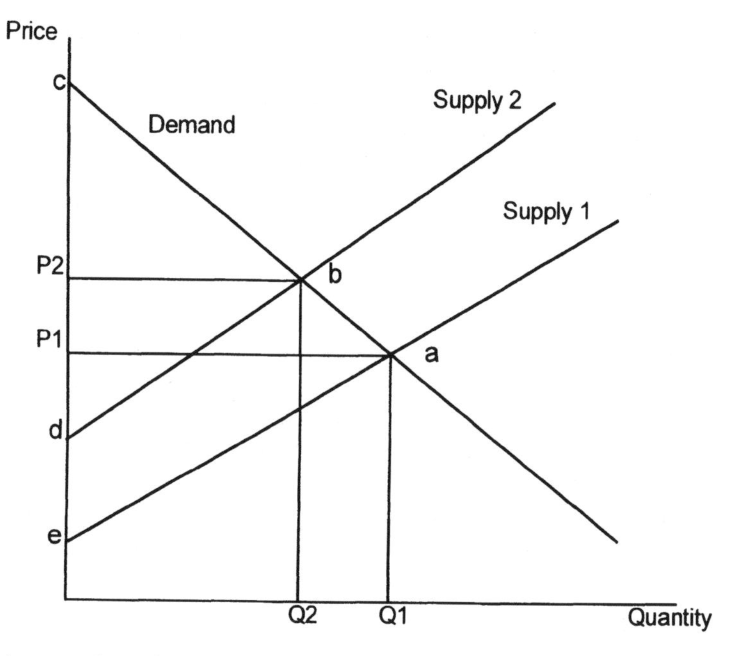

FIGURE 23.3. Market supply and demand diagram showing the effect of a shift in supply on price and quantity.

clines as consumer preferences are increasingly satisfied. The sum of the marginal values to consumers, the area under the demand curve, is the consumers' willingness to pay for a given quantity. The marginal cost of increasing supply is represented by the height of the supply curve (the supply price). The upward slope on the supply curve means that because of increasing costs, sellers will increase the supply only when offered a higher price. At the equilibrium price, the marginal cost equals the marginal value of an additional unit of the good. Hence, the marginal *net* value of the good at the competitive equilibrium is zero. If the quantity supplied were restricted below the quantity demanded at the equilibrium price, consumers would be willing to pay more than the current marginal cost for additional units. The net economic value per unit declines continuously as the quantity is increased up to the market equilibrium. Expanding quantity supplied beyond the market equilibrium would generate negative net marginal value. In this sense, the market equilibrium is the optimum quantity, at least for purely private goods sold in competitive markets.

A common measure of economic value for market goods is the *consumer surplus* which is the maximum amount that consumers could be induced to pay for the market equilibrium quantity minus the amount that they actually do pay. For Supply 1, the amount actually paid for Q_1 is equal to $P_1 x Q_1$ (Figure 23.3). The consumer surplus equals the area of the triangle P_1ac. A shift to supply curve 2 causes an increase in market price to P_2, and a reduction in quantity supplied and demanded to Q_2. The supply reduction decreases the consumer surplus to the area of triangle P_2bc. As measured by consumer surplus, consumers are always better off with greater supply and lower price.

A parallel concept, *producer surplus,* represents the amount that producers are paid in the market minus the minimum amount they would have to be paid to induce them to supply the market equilibrium quantity. Because supply price equals the marginal cost of producing and selling the good in question, the area under Supply 1 in Figure 23.3 is the total cost of production. Since sales revenue just equals the amount consumers pay ($P_1 x Q_1$), the producer surplus is the area over the supply curve and below the price, the area of triangular P_lae in Figure 23.3. The producer surplus is reflected in profits to business enterprises and rents earned by owners of land or other scarce natural resources (e.g., water, forests, minerals). Higher price and greater quantity sold always increases the producer surplus with upward sloping supply curves.

Consumer surplus as measured by market demand curves is related to the basic concepts of individual value—WTP and WTA. An individual's consumer surplus lies somewhere between the consumers' willingness to pay for a good and willingness to accept compensation for having to do without the good. As noted above, WTP and WTA values for a change in quantity are similar if the change considered does not entail a major alteration in the individual's income or does not affect an important good that has no substitutes. So long as the WTP and WTA are not far apart, consumer surplus is an good indicator of the economic value to consumers. Market-based consumer and producer surpluses can be used to measure value of commercial fish taken from the river, because these market values account for net economic benefits to the fishing fleet and to fish consumers. If there were markets for water and if there were no external costs of water diversion, water diverted from rivers could be evaluated using consumer and producer surpluses. Unfortunately for the market model, there are few competitive water markets of any significance, and even where there are partial markets (e.g., the water banks in Idaho and California), external costs imposed on third parties make the market price an unreliable indicator of marginal cost.

Measuring Nonmarket Economic Values

As summarized in Freeman (1993), the methods for estimating values can be categorized based upon two characteristics—whether they use observed or hypothetical behavior, and whether the value is revealed directly or inferred indirectly from the data (Table 23.2). The first distinction depends on whether the estimate uses observed behavior of people in their everyday economic life (prices paid, fishing trips taken), or responses to hypothetical questions (e.g., "Would you be willing to pay $10 per year for cleaner water?"). The second characteristic is whether the monetary values come directly from data or whether they must be inferred through an indirect interpretation of data based upon a model of individual behavior and choice.

Each valuation method involves a combination of these two characteristics. *Direct observed methods* utilize observed behavior of individuals who make purchases of the good in question. While this may seem an impossibility for nonmarket goods, some researchers have experimented with simulated markets—markets operated specifically to determine what people will pay for a good. For example, Bishop and Heberlein (1979) describe a simulated market for goose hunting in Wisconsin which offered actual payments for individual hunting permits. The market is simulated only in the sense that the market would not exist

TABLE 23.2. Methods for estimating economic values.

	Data type	Data source
Direct	Direct observed behavior	Direct hypothetical questions
	Competitive market price	Contingent valuation (e.g., bidding games and willingness to pay (WTP) questions)
	Simulated markets	
Indirect	Indirect observed behavior	Indirect hypothetical questions
	Travel cost recreation models	Contingent ranking
	Avoidance expenditures	Contingent activity
	Referendum voting	Contingent referendum

Modified from Mitchell and Carson 1989.

outside the study itself. *Direct hypothetical methods*, often called *contingent valuation*, use direct valuations provided by individuals in response to hypothetical questions (Mitchell and Carson 1989). Typically, the questions must be couched in terms of a logically plausible circumstance, and the answers are contingent upon the information presented as well as upon the structure of the question. *Indirect observed methods* use observed behavior that is related to an environmental or other value in the context of an economic model. For example, if recreationists rationally choose recreation sites and activities based upon the characteristics and costs of sites available, observations of recreational trip-taking behavior combined with data on trip characteristics and costs can be used to estimate travel cost demand curves. The estimated demand curve for a fishing site predicts the number of trips or anglers as a function of access cost for the site, and the area under that demand curve represents the consumer surplus for the site. The *indirect hypothetical method* is a variant of the contingent valuation approach that substitutes questions about behavior under hypothetical circumstances for direct questions about value. For example, individuals are asked whether they would continue to fish in a particular stream if a fee of $10 is charged, if the stream is contaminated by hydrocarbons, or if big trout are stocked in the stream. The hypothetical behavior is then used to estimate behavioral models (like a recreational demand for stream use) that indirectly generate estimated values.

The Role of Economics in Decision Making

Economics is often used to evaluate the consequences of decisions or to assist in the selection of new projects. Some decisions are *cost sensitive*—the decision of whether to build a dam or rehabilitate a riparian habitat may depend on whether the anticipated cost is greater than some threshold (e.g., available funds). Cost may be the only relevant economic factor in such a decision. In other circumstances, the decision involves selection of a means to achieve a given end. Whether to construct flood control structures from earth or concrete may depend on which structural material achieves a given level of flood protection at least cost. Whether to generate electricity through hydroelectric dam or combined-cycle gas combustion turbines may depend on which delivers the lower cost per megawatt hour. In these cases, a ranking of alternatives based upon cost per unit output is called *cost-effectiveness analysis*. For complex river system changes, a cost effectiveness approach examines a variety of means to achieve a goal like doubling the size of the fish population (Paulsen and Wernstadt 1995). When projects are selected for net economic contribution to social welfare, benefit-cost analysis is the appropriate technique.

Economic Impacts of Policy Decisions

Public policy debates rarely focus on economic benefits as defined here. Instead, people focus on the *economic impacts* of policy choices or the number of jobs created. Economic impact analysis predicts gross changes in incomes, market value of goods and services (often called production), and employment levels in a regional economy resulting from policy choices. Construction of navigation channels, hydropower dams, attractive facilities for tourists and visitors, or environmental regulations that affect the pattern of economic development all have economic impacts. Popular methods of economic impact assessment typically rely on simple linear economic models (called input–output models) that assume constant prices and constant costs per unit of output. In a simplified model, a rise or fall in any economic sector generates a proportional change in that sector's purchases from other sectors, a proportional change in employment and household earnings from that sector, and a secondary proportional change in regional incomes and employment affecting the whole economy. By accepting the simplification of linearity, these models make possible a simultaneous examination of dozens of overlapping and competing economic activities.

Three stages of economic impact are often identified. These are primary (or direct) impacts, secondary (or indirect) impacts, and induced impacts. Primary impact is the increase in value of final goods and services (or of income) in sectors directly affected. Secondary impact is the increase in value of final goods produced (or income) in sectors linked through purchases or sales to the directly affected sector. Induced impact is the broad effect on the local economy due to increased incomes and spending by households directly and indirectly affected. To give a concrete example, Radtke and Davis (1994) estimate that incomes for Columbia River gill net fishermen are 36% to 55% of gross sales revenue, with a midpoint estimate of 45.5%. For an increase of $100 in fish sales, the direct income impact is a $45.50 increase in income to fishermen. The remaining portion of gross sales revenue—the $54.50 cost of fishing—represents purchases of fuel, gear, and insurance from other firms. Some of these fishery inputs are purchased in the local economy, creating indirect increased incomes among suppliers to the fishing sector—this is called a backward linkage. If 50% of fishing cost is for goods and services produced locally, and if these suppliers generate $0.60 of direct local income (via wage payments) per dollar of sales, each $100 dollars of fishery output has an indirect income impact of $100 × 54.5% × 50% × 0.60 = $16.35 in the fishery support sector. If the salmon are utilized by local fish processers, indirect income impacts will be created through these forward linkages as well. Finally, because increased incomes in fishing and related households will encourage increased consumer spending, expanded consumer expenditures have an induced economic impact throughout the retail sector of the economy. The full impact includes the direct fishing incomes, indirect incomes in associated sectors, and general income impacts due to increased consumer spending in the regional economy.

Impacts Versus Benefits

A simple example illustrates the important difference between net economic benefits and economic impacts. Suppose construction of a hydroelectric dam increases employment by 1,000 full-time jobs for two years, and generates a local economic income impact of $2 million. To those local businesses, real estate agents, and politicians whose careers or incomes are enhanced by the project, this would seem to be a favorable outcome. The economic impacts would be positive in this sense. On the other hand, a benefit-cost analysis would seek to assess the benefits of the dam relative to the benefits of a natural river, and it would treat the incomes paid to local workers as a cost. And the value of the hydroelectric power would be equal to the amount consumers are willing to pay for it over and above what it costs to produce. Hence, a project generating large local economic income impacts could easily be rated negative using benefit-cost analysis which focuses on the net benefits of hydropower and flow regulation over natural river flows, riparian habitats and whitewater recreation.

Similarly, the economic value of sport fishing on a stream is the amount that participants are willing to pay for fishing there over and above the costs they actually incur. From the perspective of a community dependent upon expenditures by sport fishers, however, the economic value of fishing may be of little interest. Community leaders are concerned about the regional income generated by local businesses catering to fishers (e.g., bait and tackle shops, restaurants, motels). The prosperity and stability of a community may be adversely affected by loss of recreational business. If sport fishing shifts to other locations because of declining fish abundance or closure of access roads, increased fishing expenditures in the new locations will cause off-setting positive economic impacts. One community's or one economic sector's negative impact is linked to another community's positive impact. It is inconsistent to focus only on the local negative impacts when a natural ecosystem is protected (e.g. when a dam is not built), while ignoring the positive impacts that may occur broadly and over time. A narrow focus on local impacts often systematically overstates the negative aspects of ecosystem protection (Power 1996). While the benefit-cost analysis systematically explores all sources of value to all identifiable groups over time, impact analysis highlights the local gains and losses in income and employment.

Economic Assessment of Water Resources

A number of specific economic assessment procedures have become standard in the water resources planning field. Some of these are reviewed below.

Flood control benefits—The main benefit of flood control is reduced property damage at high flood stages. The expected damage is calculated by multiplying probability of achieving a given flood stage by amount of property damage caused at that stage. As noted by Haveman (1972), the benefit of a given flood control project is the reduction in expected loss of life and property damage. The fact that the presence of flood control structures may encourage building in the floodplain, which ultimately increases the damage per flood event, contributes to the difficulty in assessing the benefits. The trade off between more frequent floods with low damage and less frequent flooding with catastrophic damage is difficult to appraise. As the dynamic effects of flood control structures and their linkage to ecosystem services have become better understood, emphasis has shifted from flood control to the broader concept of floodplain management (Dzurik 1990).

River Navigation—The U.S. Army Corps of Engineers has long developed rivers to provide barging and ship traffic to areas otherwise open only to truck or railroad. The value of providing river navigation can be gauged by the reduction in costs of transport compared with alternative means (Haveman 1972). Water transportation is generally lower cost than highways or railroads in terms of fuel consumption and land use. Many areas of the arid west would not have been developed at all, or would have developed at slower pace, without cheap river transportation. For example, opening the Lower Snake River to navigation created significant regional economic impacts (Peterson 1995). From a national perspective, however, it is unclear whether such federal support for regional transportation generates sufficient benefits to justify the project costs. A proposal to draw down reservoirs behind the Lower Snake River dams to increase downstream salmon migration rates, would disrupt river navigation and raises the question of economic benefit (Huppert and Fluharty 1996). Martin et al. (1992) found the costs of the river closures to be relatively small, given the available alternative modes of transportation, and the fact that shippers can adjust grain storage and shipping location decisions during the drawdown. While this conclusion does not demonstrate that costs of the navigation projects are unjustified by associated benefits, it does raise concerns about the project's economic assessment.

Agricultural water supply—The benefits to agriculture from water supplied by rivers include production of crops in newly irrigated arid lands and of high-valued, high water consuming crops (e.g., irrigated pasturage, legumes, potatoes, and sugar beets) instead of low-value, low water consuming crops (e.g., dry pasturage). Benefits are typically measured as the increased net farm income plus increased regional incomes due to forward and backward linkages. Many federal irrigation projects provide subsidies to agricultural firms. The expansion in crops from irrigated farms can depress market prices, causing a reduction in crop production elsewhere from unsubsidized farms. Wahl (1989) estimates that irrigators repay only an average of 14% of costs on U.S. Bureau of Reclamation projects. Further, when new lands are brought into production, increased erosion and sedimentation may impose significant costs on other users through reduced water quality. Also, diversion of water for agriculture decreases instream flows which may diminish the habitat for aquatic organisms, such as trout and salmon, that are an alternative source of economic value. The combination of excess construction costs and negative indirect effects can bring into question the existence of positive economic benefits for some federal irrigation projects. This does not mean that all water diversions for agriculture are economically suspect, but it does raise concerns about the appropriate allocation of water between agriculture and in-stream uses.

Municipal water supply—Water supply projects are developed to assure the quantity and quality (e.g., safety, alsence of contaminants) of water for domestic use. These projects are not a frequent target of economic analysis because of the high priority of domestic water supply.

Hydropower—Electrical energy is easily generated in many Pacific Northwest streams due to rapid elevation changes, high flow volumes, and many suitable locations for hydroelectric dams. In the Columbia River basin, because river runoff peaks naturally in the spring as snow pack melts, and regional energy demand for space heating peaks in the late fall and winter, shaping hydropower generation to meet demand involves construction of large storage reservoirs and regulation of flows (Northwest Power Planning Council 1996). Because water power competes with energy from other sources, mainly nuclear and fossil fuel-fired thermal energy plants, the value of hydropower is often reckoned via the *replacement cost method*, which assigns a value equal to the cost of providing power from the next best source. In 1995 the wholesale replacement cost of hydropower was in the range of $0.022 to $0.0263 per kWh (Huppert and Fluharty 1996). The amount of power produced by a given flow of water depends upon the depth of the reservoir and water pressure created (i.e., the head), and the efficiency of the turbines. Using the head at each federal and Idaho Power Company dam on the Snake and Columbia Rivers, and a power generation rate of 0.87 kWh per foot of head per acre–foot of water, Hamilton and Whittlesey (1996) calculated that each acre–foot of water released from Upper Snake River would generate 1,621 kWh of electricity. This translates into a market value of between $35.66 and $42.63 per acre–foot of water flowing from the Upper Snake River.

Most major hydropower facilities are multipurpose projects yielding reservoir recreation, flood control, and navigation as well as electrical power. Haveman (1972) raises the conceptual problem of assigning construction and operating costs to specific project outputs, but federal agencies do allocate project costs among authorized purposes based upon standard formulas. For example, about 87% of construction costs are allocated to hydropower dams operated by the U.S. Army Corps of Engineers in the Columbia River basin (Bonneville Power Administration 1996). Based upon megawatts of power produced at these dams annually, the average cost per kWh is $0.0065, well below the current market value of the power. To extend this partial analysis into a full multipurpose project assessment, other project benefits need to be evaluated along with operating costs and external costs imposed on fish, wildlife, and other beneficiaries of natural instream flows.

Recreation—Fishing, camping, and boating on rivers, lakes and reservoirs are major components of outdoor recreation activity in the Pacific coastal ecoregion. Travel-cost models and contingent value models are frequently used to measure the value of river-based recreation. Both approaches to recreational valuation involve data collection, statistical modeling, and economic interpretations that call for specialized methods the description of which is beyond the scope of this chapter. Freeman (1993) provides an up-to-date summary of methods. A study by Johnson et al. (1994) reviews estimates of recreational fishing values for anadromous fish species in the Pacific Northwest. They found values per salmon fishing day ranging from $18.07 to $80.92 and values per steelhead fishing day ranging from $30.87 to $159.12. The estimated values vary across studies due to differences in research methodology, to random variation in sample statistics, and to real differences in the underlying values. The accumulated studies of recreational fishing demonstrate clearly that recreational fishing values depend upon a number of key variables, including distance of fishing sites from residential locations, scenic characteristics of sites, season of the year, and rate of catch for the most desired species.

Multi-purpose project assessment—Nearly every river project or management plan, by design or circumstance, will positively or negatively impact numerous economic goods and services. Overall, a multipurpose project assessment must account for the economic value of goods and services that increase in magnitude (benefits), the economic value of goods and services that decrease in magnitude (costs), plus the direct costs of project planning and implementation. Howe's (1987) retrospective assessment of the Colorado-Big Thompson project, a federal project that provides an average of 230,000 acre–feet of water to a semiarid region of northeastern Colorado, provides a good example. Water is diverted from the Colorado River on the west side of the State and pumped to an elevation where it flows down a tunnel to the eastern side. Hydroelectric power is produced as the water descends, and the water is eventually delivered through natural streams and channels for distribution by the Northern Colorado Water Conservancy District. Appraising all past and prospective future costs and benefits to agriculture, recreation, and hydropower, Howe quantifies the important regional net benefits and national net benefits.

After adjusting all dollar amounts to 1960 dollars and discounting each year's cost to the year 1960, the present value of direct project costs for construction, maintenance, and operations through 1980 is $493.2 million, which is substantially higher than original agency cost estimates. Additional economic costs of reduced flow in the Colorado River are incurred as a result of increased salinity in the river, reduced water supplies in the Colorado River basin, and reduced opportunities for recreation. These additional costs, again through 1980, have a present value in 1960 of $57.5 million. Howe shows that only $107.9 million (19.6%) of the national costs of $550.7 million was imposed on the project region through repayment of construction costs to the federal government, sharing of operating and maintenance costs, and payments for U.S. Bureau of Reclamation wholesale electric power. Finally, prospective costs extending out indefinitely in the future are added to the costs incurred through 1980 to yield a net present value of costs (in 1960 dollars) of $591.8 million in the nation, but only $117.5 million in Colorado (Table 23.3).

TABLE 23.3. Present value of Colorado-Big Thompson project benefits and costs (in millions of 1960 dollars), using a discount rate of 5%.

	Benefits	Costs	Net benefits	Benefit-cost ratio
National	$354.8	$591.8	$−237.0	0.60
Regional	$1,305.3	$117.5	$1,187.8	11.11

Modified from Howe 1987.

Howe assessed the net increase in value of agricultural production, hydroelectric energy, and recreation relative to what would likely have occurred without the project. The national agricultural benefits include increased household incomes because of the conversion from dryland to irrigated cropland in Colorado, minus associated losses experienced outside the project area due to curtailed incomes in competing farming areas. The value added to raw farm products by food processors (i.e., difference between crop price and wholesale price of products) is also included in the estimated national economic benefit. In assessing regional benefits of the project, the increase in household incomes is not adjusted to reflect offsetting decreased incomes in other regions, and incomes associated with food processing firms that re-locate to Colorado in response to the development are counted as regional benefits. Because of this and other income transfers, the net benefits to Colorado are nearly four times the national benefits. The present value of national net benefits is *negative* $237 million, while the regional net benefit is *positive* $1,187.8 million. While river projects often fail to satisfy national benefit-cost criteria, they obtain political support because beneficiaries are concentrated and politically mobilized while the losers are widely scattered and less actively engaged.

Economics and the Ecology of River Management

While the economic assessment of water resources typically consolidates multiple benefits and costs for particular projects, a broader view of economic-ecological systems (as depicted in Figure 23.1) suggests the need to incorporate a wider range of economic consequences that are linked through time by ecosystem functions and conditions. In this broader view, policy analysis would simultaneously consider all the economic services provided by rivers and river projects along with water quality/contamination issues and long-term biological sustainability. Implementation of this approach requires quantifiable predictions of ecosystem response to human interventions, even though the responses are complex and variable over time. Experience suggests that river and landscape ecologists are often unable to provide the predictions demanded by sophisticated ecological-economic models. This section reviews a few examples that illustrate the potential and current limitations of such models.

Forest Practices and Salmon Fishing

In studies of the Siuslaw National Forest in Oregon and the Porcupine-Hyalite wilderness area of the Gallatin National Forest in Montana, Loomis (1988, 1989) used bioeconomic models to assess the effects of timber harvest strategy on fishing values. The Siuslaw forest model linked specific incremental affects of logging activity to economic benefits from salmon fishing. Loomis combined a habitat model, which relates watershed characteristics to carrying capacity of fish habitat, and a Fish Habitat Index model (Heller et al. 1983) for the Siuslaw national forest. The model included stream conditions that affect salmon and steelhead at several life stages, including sediment increase, temperature increase, debris index, and debris torrent index. Each of these stream conditions is affected by logging.

Loomis' model incorporated the link between logging, stream condition, and fish production to evaluate the impacts of five timber harvesting strategies, each of which was evaluated by amount of timber harvested, acres of forest managed for timber, catchable salmon, and catchable steelhead. Commercial salmon harvests were valued at prices paid to fishermen (exvessel price). Recreational harvests were assigned a value equal to the marginal consumer surplus derived from a multisite, travel cost demand model. Timber values were derived from the relevant U.S. Forest Service plans. Loomis found that a shift from the baseline Forest Service logging plan to a complete stoppage of logging would increase the value of the fishery from $1.55 million to $1.67 million over a thirty year period. Because timber from the Siuslaw forest has a high commercial value, the increased fish value of $170 thousand would not compensate for the loss of timber value. Hence, this model did not provide an economic rationale for reduced logging. Additional research on nonfishing recreation and existence values might reverse the ranking of logging over forest preservation.

In the Porcupine-Hyalite wilderness, Loomis (1989) considered two management options: (1) establish wilderness protection over 145,000 acres in the area, and (2) leave 70,000 acres in near natural condition, allow intensive grazing on 21,354 acres, and allow timber harvesting on about 36,000 acres of private forest. As in the Siuslaw study, timber harvesting and associated road construction harms fisheries through increased sedimentation of streams, resulting in lower standing stocks of trout and reduced angler catch rates. Loomis estimated the value of recreational fishing in the area using travel cost demand equations, and he determined the effect of reduced logging in a manner similar to that used in the Siuslaw study. In this area of the Gallatin National Forest, he found that timber harvest had less value than the fishing that would be lost under the logging alternative—a result that would justify selection of the wilderness option.

In-Stream Flow and Recreational Values

As recreational and existence values of natural river ecosystems have increased in importance, a wide variety of economic studies have focused on the linkage between in-stream flow and recreational values. Shelby et al. (1992) reviewed 31 studies that quantify the effects of flow on recreational activities and economic values. The recreational characteristics affected by flow include catch rates in sport fishing, sound levels, visual quality, and suitability for rafting, canoeing, kayaking, wading, camping, and hiking. Nearly every recreational river activity is affected directly by flow rates or indirectly by the influence of flow on riparian vegetation, water quality, and fish stocks. The link between recreation and river flow is based on statistical relationships between observed flow and user activities or responses to questions. In several studies, the flow rates are controlled to establish experimentally the effect of flow on recreation. In some cases the study objective is simply to establish minimum or maximum flow needs; in other cases the objective is to quantify the effect of flow rate on recreational value. EA Engineering, Science, and Technology (1991) controlled flows below a dam on the McKenzie River in Oregon at three levels to determine flow suitability for drift boats, rafts, canoes, and kayaks. That study and others found that different flow regimes are suitable for different activities. Regardless of flow, trip satisfaction was largely determined by boater skill, experience, and expectations.

Johnson and Adams (1988) studied the effects of stream flow on the economic benefits of steelhead recreational fishing in order to assess the merits of alternative water allocations in the John Day River. Where fish production is impaired by hydroelectric projects, improved fish production is often legislatively or judicially mandated (see Chapter 22). Increased streamflow is a means of compensating for losses in fish production, but at the same time may conflict with off-stream uses such as irrigated agriculture. To assess the economic value of in-stream flow allocations, Johnson and Adams combined a steelhead production model with an on-site contingent valuation analysis.

The analysis proceeded as follows (Figure 23.4). First, to estimate the effect of flow on steelhead fishing on the John Day River, the relationship between streamflow and fish productivity was quantified using a fish production model. Second, a contingent value survey was used to establish the relationship between steelhead catch rates and angler's WTP for steelhead fishing. The combined result of these methods indicated that an additional acre–foot of water in the production of recreational steelhead fishing is worth $2.40 to anglers. This value is then compared with the value of an additional acre–foot of water for offstream uses. In this case agricultural irrigation value is estimated at $10 to $24 an acre–foot. Because the benefit of diverting an acre–foot to agriculture exceeds the value of an acre–foot used for in-stream flow, a reallocation of in-stream flow to recreational fishery production would seem economically inefficient.

Nonetheless, Johnson and Adams warned that this value comparison can be misleading for several reasons. The value of water allocation to fishery production is highly sensitive to the spatial and temporal pattern in each competing use. In some situations, water used for recreation may be used downstream by agriculture. In addition, fishery production depends on a combination of factors including in-stream flow, riparian habitat, downstream passage conditions, water quality, and other ecosystem characteristics. Johnson and Adams' fish production model assumed that all factors other than in-stream flow remained constant. With habitat improvements in the John Day River, the fish production increase could be much greater than predicted. The variation both temporally and spatially of riparian ecosystem characteristics poses a challenge to effective bio-economic analysis.

Sediment from Agriculture

Brusven et al. (1995) developed an economic policy model to assess the impact of different farming practices on sustainable farming, erosion, and stream health in the Tom Beall watershed, Idaho. They used farm incomes, erosion rates, and associated damage values as an economic indicator, benthic macroinvertebrates as an indicator of stream health, and agricultural nonpoint source pollution as an indicator for water quality. If conventional tillage is replaced with a reduced tillage program, which involves payments from a government program, the average erosion rate decreases by 43% (10.9 t/ha/yr), farm income increases an average of $23.81 per hectare on-site, and erosion damage costs decrease an average of $3.28 per hectare

(Table 23.4). Although the total benefit per hectare of $27.09 is short of the $63.72 per hectare farm program cost (Table 23.4), benefits from improved water quality and reduced sedimentation (hence lower dredging costs) in the reservoir behind Lower Granite Reservoir should be considered. Brusven et al. (1995) suggest that an alternative system of maintaining farm income and erosion control would likely be more efficient than the current system of price supports and conservation compliance (e.g., reduced tillage). Further, they propose that sustainable management practices ought to include a balance of maintaining farm profitability, minimizing nonpoint source pollution, and dredging sediment build-up behind the dams.

Although this model addresses only a few components of a very complex system, it does examine how stream health is affected by land use practices such as urbanization, industrialization, and farming. In this case, farming practices were examined by assessing the affects of sediment deposition which impairs dams and affects stream health as well as nonpoint source pollution which impacts water use, and increases the cost of water treatment, flood protection, navigation channels, irrigation systems and reser-

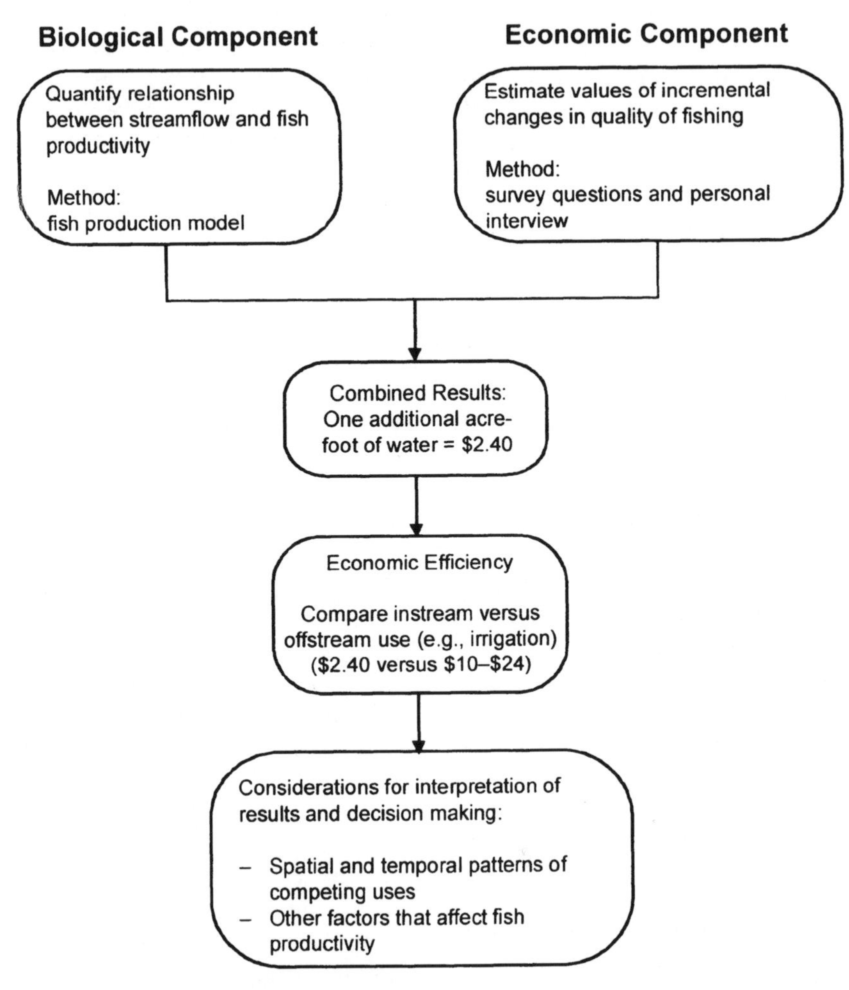

FIGURE 23.4. Schematic diagram of Johnson and Adams (1988) methodology for John Day River instream flow assessment. A fish production model and a contingent value survey were used to establish a value to anglers for an additional acre–foot of water which is then compared to the value of an additional acre–foot of water for offstream uses.

TABLE 23.4. Results of profit-maximizing policy models for farms in the Tom Beall watershed, Idaho.

	Average erosion rate (ton/ha/yr)	Average farm income ($/ha/yr)	Average on-site erosion damage ($/ha/yr)	Farm program costs ($/ha/yr)
Conventional tillage (no government program)	25.7	−15.56	10.79	0
Reduced tillage (government program)	14.8	8.52	7.51	63.72
Difference	−10.9	24.08	−3.28	63.72

Modified from Brusven et al. 1995.

voir storing capacities. Nonpoint source pollution also degrades fish habitat which in turn decreases the value of commercial and recreational fisheries. This is further complicated by the impact dams have on stream ecosystem characteristics such as flows, temperature, channel geometry, sediment deposition, water quality, nutrient cycling, and food web interactions. However, the model described by Brusven et al. does not include analysis of how these characteristics are affected by dams. Such an omission may be indicative of the complexity involved in more holistic approaches to river resource economics.

While traditional economic studies of rivers have focused on rather narrow and short-term projects, the economic–ecological approach calls for assessment of complex interactions and longer-term effects. From this broader perspective, the economic contributions of river ecosystems extend beyond the usual project outputs of navigation, irrigation, water supply, hydroelectric power, and recreation. Economists have expanded their concepts of economic benefits to incorporate many non-market goods and services provided by river ecosystems. Intense competition for use of scarce river resources, in the presence of public goods and pervasive externalities, makes the valuation of river resources a crucial input to collective decision making in the Pacific coastal ecoregion.

If properly conceived and implemented, economic studies can account for benefits and costs of changes in river ecosystems, and they can contribute to the social processes that guide river management decisions. But these economic assessments are useful only when management measures generate predictable changes in river ecosystems which, in turn, cause predictable changes in economic welfare. Where ecological science cannot support such predictions, it is difficult to ground the economics on a realistic assessment of biophysical consequences. In such situations, decision makers are often confronted by purveyors of ideological fervor rather than rigorous analysis. Adequate economics accounting in river management requires a long term partnership between social and natural sciences. Research must focus on aspects of river systems that generate significant economic values and on connections between ecosystem characteristics and functions that most clearly link short-term actions to longer term economic consequences.

Literature Cited

Anderson, T.L. 1983. Water crisis: Ending the policy drought. Johns Hopkins University Press, Baltimore, Maryland, USA.

Bishop, R.C., and T.A. Heberlein. 1979. Measuring values of extramarket goods: Are indirect measures biased? American Journal of Agricultural Economics **61**:926–930.

Bonneville Power Administration. 1996. 1995 Annual report: The benefits of power. Bonneville Power Administration, Portland, Oregon, USA.

Brusven, M.A., D.J. Walker, and R.C. Biggam. 1995. Ecological–economic assessment of a sediment-producing stream behind Lower Granite Dam on the Lower Snake River, USA. Regulated Rivers **10**:373–387.

Dzurik, A.A. 1990. Water resources planning. Rowman and Littlefeld, Savage, Maryland, USA.

EA Engineering, Science, and Technology. 1991

Boating and fishing suitability of the lower McKenzie River, Oregon. EA Engineering, Science, and Technology, Lafayette, California, USA.

Eckstein, O. 1958. Water-resource development: The economics of project evaluation. Harvard University Press, Cambridge, Massachusetts, USA.

Freeman III, A.M. 1993. The measurement of environmental and resource values. Resources for the Future, Washington, DC, USA.

Hamilton, J.R., and N.K. Whittlesey. 1996. Cost of using water from the Snake River Basin to augment flows for endangered species. Presented April 1996 at Sustainable Fisheries Conference, Victoria, British Columbia, Canada.

Hanneman, M. 1991. Willingness to pay and willngness to accept, how much can they differ? American Economics Review **81**:635–647.

Haveman, R.H. 1972. The economic performance of public investments: An ex post evaluation of water resources investments. Resources for the Future, Washington, DC, USA.

Heller, D., J. Maxwell, and M. Parsons. 1983. Modeling the effects of forest management on salmonid habitat. USDA Forest Service, Siuslaw National Forest, Corvallis, Oregon, USA.

Hildreth, R. 1994. Legal aspects of Columbia-Snake River salmon recovery. Ocean and Coastal Law Memo 41. School of Law, University of Oregon, Eugene, Oregon, USA.

Howe, C.W. 1987. Project benefits and costs from national and regional viewpoints: Methodological issues and case study of the Colorado-Big Thompson project. Natural Resources Journal **27**:1–20.

Huffman, J. 1983. Instream water use: Public and private alternatives. Pages 249–282 *in* T. Anderson, ed. Water rights: Scarce resource allocation, bureaucracy, and the environment. The Pacific Institute for Public Policy Research, San Francisco, California, USA.

Huppert, D.D., and D.L. Fluharty. 1996. Economics of Snake River salmon recovery. A Report to the National Marine Fisheries Service. University of Washington, School of Marine Affairs. Seattle, Washington, USA.

Johnson, N., and R.M. Adams. 1988. Benefits of increased streamflow: The case of the John Day River steelhead fishery. Water Resources Research **24**:1839–1846.

Johnson, R.L., H.D. Radtke, S.W. Davis, and R.W. Berrens. 1994. Economic values and impacts of anadromous sportfishing in Oregon coastal rivers: Assessment of available information. Prepared for The Center for the Study of the Environment, Santa Barbara, California, USA.

Kahn, J.R. 1995. The economic approach to environmental and natural resources. Harcourt Brace & Company, Orlando, Florida, USA.

Krutilla, J.V., and A.C. Fisher. 1975. The economics of natural environments: Studies in the valuation of commodity and amenity resources. Johns Hopkins University Press, Baltimore, USA.

Lee, K.N. 1993. Compass and gyroscope: Integrating science and politics for the environment. Island Press, Washington, DC, USA.

Loomis, J.B. 1988. The bioeconomic effects of timber harvesting on recreational and commercial salmon and steelhead fishing: A case study of the Siuslaw National Forest. Marine Resource Economics **5**:43–60.

Loomis, J.B. 1989. A bioeconomic approach to estimating the economic effects of watershed disturbance on recreational and commercial fisheries. Journal of Soil and Water Conservation **44**:83–87.

Loomis, J.B. 1996. Measuring the economic benefits of removing dams and restoring the Elwha River: Results of a contingent valuation survey. Water Resources Research **32**:441–447.

Martin, M., J.E. Hamilton, and K. Casavant. 1992. Implications of a drawdown of the Snake-Columbia River on barge transportation. Water Resources Bulletin **25**:673–680.

Mitchell, R.C., and R.T. Carson. 1989. Using surveys to value public goods: The contingent valuation method. Johns Hopkins University Press, Baltimore, Maryland, USA.

Niskanen, W. 1971. Bureaucracy and representative government. Aldine and Atherton, Chicago, Illinois, USA.

Norgaard, R.B., and R.B. Howarth. 1991. Sustainability and discounting the future. Pages 88–101 *in* Robert Costanza, ed. Ecological economics: The science and management of sustainability. Columbia University Press, New York, New York, USA.

Northwest Power Planning Council. 1996. Northwest power in transition: Opportunities and risks. Draft Fourth Northwest Conservation and Electric Plan. NPPC 96-5. Portland, Oregon, USA.

NRC (National Research Council). 1996. Upstream: Salmon and society in the Pacific Northwest. National Academy Press, Washington, DC, USA.

Paulsen, C.M., and K. Wernstadt. 1995. Cost-effectiveness analysis for complex managed hydrosystems: An application to the Columbia River Basin. Journal of Environmental Economics and Management **28**:388–400.

Peterson, K.C. 1995. River of life, channel of death: Fish and dams on the Lower Snake. Confluence Press, Lewiston, Idaho, USA.

Power, T. 1996. Lost landscapes and failed economics. Island Press, Washington, DC, USA.

Radtke, H., and Davis, S. 1994. Some estimates of the asset value of the Columbia River gillnet fishery based on present value calculations and gillnetters' perceptions. Report prepared for Salmon for All, Astoria, Oregon, USA.

Shelby, B., T.C. Brown, and J.G. Taylor. 1992. Streamflow and recreation. USDA Forest Service, Rocky Mountain Forest and Range Experiment Station. General Technical Report RM-209. Fort Collins, Colorado, USA.

US Water Resources Council. 1983. Economic and environmental principles for water and related land resources implementation studies. US Government Printing Office, Washington, DC, USA.

Water Resources Council. 1973. Principles and standards for planning water and related land resources. Page 29778 *in* 38 Federal Register. US Government Printing Office, Washington, DC, USA.

Wahl, R.W. 1989. Markets for federal water: Subsidies, property rights, and the Bureau of Reclamation. Resources for the Future, Washington, DC, USA.

Willig, R.D. 1976. Consumer's surplus without apology. American Economic Review **66**:589–597.

Zerbe, R.O., and D.D. Dively. 1994. Benefit–cost analysis in theory and practice. Harper Collins, New York, New York, USA.

Part V
The Future

24
Stream and Watershed Restoration

Christopher A. Frissell and Stephen C. Ralph

Overview

• Restoration is the process of returning a river or watershed to a condition that relaxes human constraints on the development of natural patterns of diversity. Restoration does not create a single, stable state but enables the system to express a range of conditions dictated by the biological and physical characteristics of the watershed and its natural disturbance regime.

• Most restoration efforts to date have focused on the alteration of physical habitat characteristics at small spatial scales, most often the placement of logs, rocks, or wire gabions in a channel to create pools or collect gravel. The effect of such efforts on the production and survival of the target fish species is uncertain.

• Relatively few projects have attempted restoration at the reach scale. However, this approach may be well suited for severely degraded stream reaches, although accurate documentation of the effectiveness of the approach is not available.

• Restoration of an entire watershed is very rarely attempted. However, addressing restoration from this broad spatial perspective is often necessary to relax human constraints on system function. A well-designed and evaluated watershed restoration project conducted in Redwood National Park illustrates the potential effectiveness of a comprehensive approach.

• Restoration efforts are constrained by a lack of a clear understanding of how human activities have altered the processes at work within a watershed. In large part, this deficiency is the result of the failure to include monitoring as an integral part of restoration projects. Evaluation and monitoring may pay large dividends in terms of developing a full understanding of which approaches to restoration work and which do not. Monitoring also may enable the identification of small adjustments in a program that greatly increase effectiveness or reduce costs.

Introduction

In the Pacific Northwest, decades of benign neglect, denial, and inertia in aquatic resource management have only recently yielded to the reluctant realization that a once substantial heritage of aquatic resources may become a mere relic of their historical potential to provide economic, cultural, and spiritual sustenance. The challenge to reverse these declines is formidable. The inevitability of Endangered Species Act (ESA) listings that mandate comprehensive salmon recovery plans for depleted stocks provides political imperative to promote restoration of aquatic communities. The growing number of Habitat Conservation Plans (HCPs) agreements negotiated between federal agencies and private landowners to address future aquatic habitat and fish species needs under the ESA provide an opportunity to evaluate watershed-scale recovery processes and timeliness in managed landscapes. Similarly, recent federal court decisions under the Clean Water Act (CWA) could require state water quality agencies and the Environmental Protection Agency to design and implement water resource recovery plans to restore imperiled water courses (presently thousands of water bodies fail to conform to established water quality criteria) that currently are impaired in their capacity to support aquatic resources. Growing social and political interest has the potential to greatly accelerate the adoption of restoration as a solution. This chapter presents a perspective on the present capability of aquatic restoration to address the challenge of protecting and restoring the integrity of Pacific coastal rivers and streams. The chapter discusses how holistic restoration might reverse declining trends.

In the Pacific coastal ecoregion, restoration of freshwater ecosystems is undergoing fundamental changes. In the last decade especially, fish biologists and stream ecologists have benefited greatly from the techniques and knowledge of physical scientists working in the related fields of fluvial geomorphology, sediment transport, channel hydraulics, and hydrology. The assessment techniques common to these disciplines, some of which were developed decades ago, have aided in assessing hillslope and channel processes that control input and routing of sediment and water through a stream system (see discussion in Chapters 2, 3, and 11).

From a long-standing emphasis on the small-scale, site-specific addition of artificial structures to favor individual species, restoration is graduating to an increasing emphasis on regional and watershed-scale reestablishment of the biophysical processes and structures that promote natural ecosystem recovery (Minns et al. 1996). This transition necessitates thinking across larger scales of space and time. It promises a future not of endless bounty, but at least of more effective conservation of biological resources, with the potential of net gains in the productivity and resilience of some ecosystems and species.

Unfortunately, existing institutional, political, and educational constraints tend to hinder wider acceptance and participation in such a conceptual transition (Frissell et al. 1997). If aquatic ecosystem restoration devolves into thinking simplistically about habitat alteration, the result could be economically and biologically costly failure on a scale larger than ever before. Successful negotiation of this transition requires a thoughtful and cautious approach that avoids blind faith in generic prescriptions of dubious benefit, acknowledges inherent uncertainty in biological outcomes, and embraces monitoring and evaluation as integral elements of any restoration project (Minns et al. 1996).

Although much of the current enthusiasm for watershed restoration is based on laudable objectives, success often is limited by a fundamental misunderstanding of the cause and effect relationship of processes operating at a scale that can overwhelm limited, site specific restoration efforts. The massive scale of the task and the need for strategic and focused solutions does not become clear until the magnitude and extent of environmental change that has occurred in the Pacific coastal ecoregion over the past century (Bisson et al. 1992) is fully appreciated. The effects of these changes have recently become all too obvious, as underscored by the depleted status of once abundant Northwest native stocks of salmon and trout (Nehlsen et al. 1991, Frissell 1993). The specific causes of these outcomes, however, are not so easily identified.

Fundamental concepts about and the need for integrated, interdisciplinary approaches to stream and watershed restoration have gained wide acceptance in aquatic ecology only in the last decade or so (Gore 1985, Heede and Rinne 1990). Although the evaluation of stream habitat projects is not a new concept (Ehlers 1956, Hunt 1971, Gard 1972), accurate documentation of the ecological success or failure of contemporary stream and watershed restoration efforts has only recently begun to appear in print (Williams et al. 1997). Many such reports are after-the-fact evaluations of small-scale projects at a limited number of locations (Beschta et al. 1994, Parry and Seaman 1994, BioWest 1995). However, a truly sustained watershed-scale effort to evaluate a fish habitat restoration program took place in Fish Creek in the Clackamas Basin in Oregon (Everest et al. 1987, Reeves et al. 1991, Reeves et al. 1997). These evaluations provide a valuable primer on how to judge success of restoration efforts and emphasize the importance of clearly defining the physical and biological rationales and expected outcomes from restoration treatments.

In the face of past failure to achieve restoration goals, scientists and managers must reconcile the need to balance the political and social imperative to "do something" with the need to be judicious with their actions and accountable for the outcomes. Choosing where and how to invest increasingly scarce public resources to ensure the best chance of a beneficial and sustainable outcome requires wisdom and creativity. The potential effectiveness of restoration actions depends upon how the rates and patterns of processes that control the character or expression of aquatic communities have changed. Different land uses affect these processes in different ways and to differing degrees across the diverse landscape of the Pacific coastal ecoregion. Sorting out the history of the interaction of processes and land use is the first step in the path to successful restoration (Cairns 1989, Beechie et al. 1994, Sear 1994, Frissell and Bayles 1996, Frissell et al. 1996, Stanford et al. 1996).

In some situations, for example removing a road culvert to restore access for fish to miles of historical spawning and rearing habitats, the "fix" is obvious. Similarly, where the river experiences seasonally excessive water withdrawals, restoration of some level of in-stream flow regimes may allow at least partial recovery of the historical aquatic habitat potential. In many other cases, however, the solution is much more elusive. Several investigators provide excellent summaries identifying individual strengths and weaknesses of the host of techniques that are commonly used for alteration and rehabilitation of stream habitats in the Pacific coastal ecoregion (Wesche 1985, Reeves et al. 1991, Parry and Seaman 1994).

This chapter focuses on the ecological and ecosystem management contexts of restoration activities, and emphasizes the need to strategically reconfigure and rescale concepts and approaches to stream ecosystem restoration. Several case studies that provide relatively positive examples of restoration projects at a range of scales are examined. Discussion of the central practical importance of monitoring and evaluation emphasizes the need for professional humility, controlled experimentation, and critical evaluation to improve future ecological stream and watershed restoration efforts.

Defining Restoration—Scope and Scale

Restoration is the act of returning a river or watershed (or assisting its recovery) to a condition in which it can function ecologically in a self-sustaining way, more nearly resembling its former function prior to human-induced disturbance (Cairns 1989, Bisson et al. 1992, Sear 1994). Taking a dynamic, coevolutionary view of streams and watersheds, Ebersole et al. (1997) and Frissell et al. (1996) define restoration as the act of relaxing human constraints on the development of natural patterns of diversity. In this view, a restored ecosystem does not necessarily return to a single ideal and stable state (i.e. pristine) but is free to express a range of natural successional trajectories and states, as constrained by the historical biological and physical capacity of its encompassing environment. The principal effect of human disturbance is to alter or suppress key successional stages, thereby eliminating certain desirable characteristics of diversity that the ecosystem would otherwise include. Most such capacity to develop system diversity is retained in the biota and the suite of processes that shape habitats and can be expressed if specific human constraints are relaxed (Regier et al. 1989, Stanford et al. 1996, Ebersole et al. 1997). This definition implies idealy that restoration measures should not focus on directly recreating natural structures or states, but on identifying and reestablishing the conditions under which natural states create themselves. The focus is on ecosystem processes and patterns at larger scales, within which local habitats and individual organisms are embedded (Frissell et al. 1986, Naiman et al. 1992, Kondolf and Larson 1995). However, population extinctions, introductions or invasions of nonindigenous species, and major changes in geological features and processes can permanently alter the capacity of the watershed system to recover former states, precluding full restoration. Moreover, extensive human occupation of ecosystems frequently leads to permanent loss of ecosystem developmental capacity and diversity, and in a growing number of cases, these losses have to be accepted as permanent constraints on restoration (Sear 1994, Frissell et al. 1997, Stanford et al. 1996).

The National Research Council (NRC 1992) and others (Regier et al. 1989, Sear 1994, Kondolf and Larson 1995, Kondolf et al. 1996, Stanford et al. 1996) stress the importance of taking a systems approach to river restoration, that is, understanding and working in harmony with the dynamic physical forces (processes) associated with flowing water to restore natural or normative patterns of hydrological and ecosystem processes (i.e., fluvial restoration). In a general sense, the unintended consequence of watershed-scale land use and resource extraction in the Pacific coastal ecoregion is typically expressed in changes to the fundamental driving forces of watershed processes: changes to the flow regime, to the input and routing of sediment and large woody debris, and in the functional capacity of riparian areas. Without reestablishment of the dynamic equilibrium of natural biological and physical processes, recovery of the biotic community to its full productive potential may never occur. This is especially true in rivers where sediment aggradation, channel widening and consequent decrease in depth have created aquatic environments that provide poor habitat to meet the life history requirements of native fish, aquatic insects and amphibians. Once this complex ecosystem is disrupted, the time frame necessary for natural recovery of meander length, amplitude, radius of curvature, bankfull width and width-to-depth ratio, and other physical features is highly uncertain. Although trends toward channel recovery from aggraded states are documented (Lisle 1981 and 1982), reestablishment of mature floodplain and riparian forests is necessary for full recovery of most Pacific coastal streams, and may take centuries (Bisson et al. 1992, FEMAT 1993). The requisite experience to reliably estimate the time required for recovery of dynamic equilibrium to watershed processes is simply lacking.

Scientists and managers must take care not to mistake an apparent trend toward a recovering condition as evidence that natural recovery has been achieved or even fully initiated (Espinosa et al. 1997). Until recovery trends are manifested in some self-sustaining, relatively naturally functioning condition, restoration or recovery has not truly occurred. For example, Platts et al. (1989) found that the channel bed conditions in the South Fork Salmon River in Idaho partially recovered during the 20 years following a massive influx of sediment, but then stabilized at an incompletely recovered state. Platts et al. suggest that in its present state of arrested recovery, the river remains highly sensitive to even very small anthropogenic increases in erosion and sediment delivery. In this condition, extensive areas of the river no longer provide substrate conditions suitable to support spawning habitat required by chinook salmon (*Oncorhynchus tshawytscha*). Time, careful protection, and perhaps additional erosion control measures are necessary to create the conditions that could allow South Fork Salmon River ecosystem to eventually recover enough to provide the full complement of historical chinook salmon habitats.

Opportunities to assist in the recovery of watershed-scale processes and ecosystem dynamics may be seriously limited without a clear understanding of the basin wide and historical context of a proposed project (or array of projects) (Reeves et al. 1991, Frissell and Nawa 1992, Wissmar et al. 1994, Minns et al. 1996). The need for integrated, watershed-scale restoration programs is increasingly recognized (FEMAT 1993, Williams et al. 1997) and some resources are becoming available for such efforts. However, large-scale restoration projects are being implemented at a pace that continues to vastly outdistance the ability to accurately evaluate their results and effectively apply such knowledge to future projects. New projects fail to receive the potential benefit of knowledge gained from adequate analysis of past mistakes and successes because resources committed to monitoring and evaluation are limited.

The following section describes a number of restoration activities that demonstrate varying approaches to solve common problems affecting water resource and habitat integrity. Examples include those applied at the habitat unit scale (i.e., microhabitat); the reach scale (linked habitat units encompassing as few as several pool-riffle sequences to several kilometers of stream and floodplain systems); to an example applied at the whole watershed scale. This latter example illustrates a model comprehensive approach to address multiple factors operating over large areas and over many decades.

Interventions at the Microhabitat Scale

Until the mid 1990s, agency-sponsored stream "improvement" and "enhancement" programs in the Pacific coastal ecoregion almost exclusively emphasized restoration at the microhabitat or pool and riffle scale (see Chapter 5 for discussion of classification levels). Placement of log weirs (Figure 24.1), wire gabions, and other in-stream structures to create individual pool riffle habitat units were favored over such other techniques as riparian tree planting and off-channel pond development. Although in-stream habitat structures are clearly well suited for artificially altering the structure of pools and riffles in certain kinds of stream reaches (Wesche 1985, House and Boehne 1986, Reeves et al. 1991), their effectiveness in increasing the production and survival of fish has always been and remains uncertain (Carufel 1964, Platts and Nelson 1985, Hamilton 1989, Reeves et al. 1991, Reeves et al. 1997). Moreover, in some situations, such projects have unintended adverse effects on native fish and wildlife species (Rinne and Turner 1991, Fuller and Lind 1992), and it is probable that many other negative side effects occur but have not been noticed or documented because evaluations have focused on a narrow set of habitat parameters or target fish species.

FIGURE 24.1. Typical log weir structure placed in a channel by the U.S. Forest Service in an attempt to enhance production of a small run of native summer steelhead (*Oncorhynchus mykiss*) in Trout Creek, a tributary of the Wind River in the Cascades Range of south-western Washington (Photo: R. Nawa, Oregon State University).

One intervention commonly employed at the microhabitat scale is treatment or replacement of streambed gravel to provide improved spawning habitat for salmonid fishes (Reeves et al. 1991). Such projects often fail, because the problem (poor quality or absent spawning gravel) is caused by a persistent source (e.g., chronically accelerated input of fine sediments, changes in flow regime and channel form that preclude storage of appropriate-sized particles on the bed). For example, in a massive and costly project in the Merced River in California, reconstructed gravel riffles were destroyed during a peak stream flow with a return period of just 1.5 years (Kondolf et al. 1996). The project designers failed to consider the full scope of effects on sediment and flow regimes caused by dam regulation and other human modifications, including long-term alteration of channel configuration and hydraulics of the reach where the project took place (Kondolf et al. 1996).

Localized interventions are most effective where the principal cause of ecosystem damage is a local alteration, such as the historical forestry practice of removing trees and downed woody debris from a stream channel and riparian area. A relatively well-documented and partially successful project involving habitat modification at the pool and riffle occurred in the North Fork of Porter Creek in western Washington (Cederholm et al. 1997). This low-gradient (2%) stream with a mean bankfull channel width of about 10 m supports both resident and anadromous salmonids. Like many similar forested streams in the Pacific coastal ecoregion, this one has reduced levels of instream woody debris as a consequence of past forestry practices. Restoration efforts focused on creating pool habitat and cover through the reintroduction of large woody debris. A 1,500 m section of the creek was divided into three study sections, a control (no treatment), an "engineered" treatment with intensive and carefully designed placement of structures and anchoring of debris, and a "logger's choice" treatment where woody debris was placed (as was practical in the field), with limited preplacement design and little or no anchoring.

After completion, the levels of large woody debris increased 10 times in the engineered section and 2.7 times in the logger's choice reach of North Fork Porter Creek (Cederholm et al. 1997). The surface area of pools increased significantly in both treated reaches, while it declined somewhat in the control reach. Subsequent increases in juvenile coho salmon (*O. kisutch*) wintering in both treated habitat reaches (judged against preproject levels from 1989–1994 survey data) were observed. Coho smolt production (measured as counts of outmigrating individuals) also increased significantly after large woody debris additions, but declined in the control section. Lack of juvenile steelhead (*O. mykiss*) production or utilization in the treated sites, suggests that habitats associated with the input of woody debris were more useful to coho than steelhead.

Costs of treatment were US$164 per lineal stream meter for the engineered reach and US$13 for the logger's choice reach. However, a cost efficiency analysis projected over the anticipated project life, estimated costs adjusted per coho smolt produced in the two treatments to be about equal (US$13–US$15 per smolt), largely because woody debris placed in the engineered treatment was considered more persistent (expected to last ca 25 yr) than that in the logger's choice treatment (expected to last ca 5 yrs). Even a casual inspection and extrapolation of these costs suggest that projects of this type are quite expensive, and the aggregate cost over a significant fraction of many thousands of kilometers of such streams across the Pacific coastal ecoregion is prohibitive. Moreover, the benefits of such projects at best are likely to accrue only to a subset of the species of interest.

Several projects that create, enlarge, or excavate low-flow connections of floodplain ponds with main channels along major alluvial river segments have increased the survival and production of juvenile salmonids, especially coho salmon, in the Pacific coastal ecoregion (Cederholm et al. 1988). These projects typically constitute field interventions defined at the scale of a single pool, riffle, or off-channel habitat. Although they are conceived and designed at the microhabitat or habitat scale, the biological and physical effects (both beneficial and adverse) of such in-stream projects can potentially propagate to influence a larger area of the reach within which they are nested. Peterson (1982) investigated the role of off-channel ponds in the overwintering survival of coho juveniles in tributaries of the Clearwater River, Washington. To increase the availability of winter habitat, explosives were used to create pools in Paradise Creek, an existing 10-km long floodplain springbrook (i.e., wall-based channel) of the Clearwater River (Cederholm et al. 1988). The short stream reach chosen for blasting the pools was a mud bottom section that meandered through a sedge marsh wetland complex. A 1.5-m high dam was built at the downstream outlet to control water levels in the ponds. Monitoring of fish movement and survival occurred for two years prior to and two years after pond construction using a two-way fish trap constructed to allow capture of fish moving into and out of the ponds. Winter survival was estimated from recapture of a subsample of freeze-branded juveniles. The average overwinter survival (i.e., from one year's winter to the next spring) of marked juvenile coho that had immigrated into Paradise Creek increased from 11% of juveniles surviving before, to 56% after pond construction. Based on an extensive smolt trapping study in the main river (Cederholm et al. 1988), the investigators estimated that restoration efforts accounted for a 2.8% (1,968 smolts) increase in the total annual smolt yield from the Clearwater basin for the two years evaluated. The total project cost was less than US$10,000.

Although the development of floodplain ponds holds promise for improved survival of coho salmon and perhaps other fishes, some important considerations may limit this technique. Based on verbal accounts from the managers of projects in Washington, Oregon, and British Columbia, to remain effective such projects often require a high level of diligent monitoring and maintenance, especially of connecting channels and control structures. Moreover, seasonally dense aggregations of juvenile fish in small ponds typically attract concentrations of avian, terrestrial, and piscine predators. Finally, floodplain ponds may afford important breeding and rearing habitats for native frogs and salamanders, especially where fish access is limited (C. Frissell and B. Cavallo, Flathead Lake Biological Station, The University of Montana, Polson, Montana, unpublished data). Amphibian eggs and larvae are highly vulnerable to fish predation and to physical disturbance, such as gravel excavation or alteration of water levels associated with project construction and management. Thus, alterations of floodplain habitats intended to benefit fish may adversely affect amphibians and other native aquatic animals and plants, many of which are themselves regionally declining and threatened.

High failure rates and escalating costs of maintenance have been reported for many in-stream structure projects and other site-specific interventions in the western United States (Ehlers 1956, Carufel 1964, Platts and Nelson 1985, Rinne and Turner 1991, Frissell and Nawa 1992, Beschta et al. 1994, Kondolf et al. 1996) and elsewhere (Sear 1994). Such projects are consistently unsuccessful in watersheds where high erosion and sedimentation rates, high peak flows, or other watershed alterations or stresses are pervasive (Hamilton 1989, Frissell and Nawa 1992). Where the physical integrity of these structures fail, they can cause significant collateral damage to downstream habitat and non-target biota (Frissell and Nawa 1992). The cumulative ecological costs of such failures and the maintenance burdens incurred are rarely recognized by those completing the projects.

In general, treating large-scale land-use related problems—such as deforestation, accelerated erosion, and altered hydrologic and sediment regimes—with small-scale interventions (such as importation of gravel for spawning or construction of in-stream structural devices) is ineffective and costly (Frissell and Nawa 1992, Doppelt et al. 1993, Beschta et al. 1994, Kondolf et al. 1996, Poole et al. 1997). To be effective, restoration interventions must be scaled appropriately to the cause and consequences of ecosystem damage (Sear 1994,

Kondolf and Larson 1995, Kondolf et al. 1996). Moreover, the causes of the problem must be controlled before the damage itself can be effectively repaired (Platts and Rinne 1985, Frissell and Nawa 1992, Sear 1994, Stanford et al. 1996, Reeves et al. 1997).

Larger-Scale River Restoration

Cumulative ecological alterations of streams and rivers occur most often at the larger spatial scales of stream reaches, valley segments, or entire drainage basins. Because these large-scale changes can seriously limit the recovery potential of anadromous fishes like salmon, restoration efforts have been scaled up to address these problems. The following examples are but a few of these initial efforts—but the lessons learned from them will help such efforts in the future.

An example of a restoration effort to directly treat a highly altered stream–riparian–floodplain system at the scale of several contiguous kilometers took place on the Blanco River in southwestern Colorado (Rosgen 1988, NRC 1992). In the course of the three-year river and floodplain reconstruction project, a stable 20-m wide channel (with a large pool to riffle ratio) was excavated to replace nearly 4.34 km of braided river channel, which had widened to nearly 121 m and contained few pools to support resident trout (Figure 24.2a,b).

Before restoration efforts, the affected reach had been channelized by the U.S. Army Corps

(a)

(b)

FIGURE 24.2. (a) Aerial views of the Blanco River showing the highly braided, unstable channel form resulting from channel changes after channel straightening and flood dike construction but before restoration efforts. This channel form reflects the D4 type in Rosgen's classification scheme. (b) Aerial views of more stable single thread channel form (C4) following substantial reconstruction of historical channel meander pattern, width-to-depth cross-section, channel slope, and reconnection with historical floodplain system (from Rosgen 1996).

of Engineers (COE) to protect adjacent grazing lands from erosion following a flood in 1970. After the flood, portions of the river were straightened, widened and entrenched within a levee network so that what had been a stepped, low-flow channel, terrace, and floodplain, was converted to a wide, flat-bottomed, trapezoid-shaped channel in cross-section. The resulting loss of the historical meander pattern (as seen in plan view) decreased the length of the stream and increased local channel slope, leading initially to degrading of bed elevation through down and headward erosion until local channel slope reached equilibrium. Enlarging the width-to-depth ratio by channelization reduced the sediment transport capacity of the channel, which resulted in deposition of bed material (gravel and cobble) in the reach after flooding. With build up of gravel deposits in its center, the channel subsequently took on a highly unstable, braided character while active lateral migration and bank erosion once again threatened agricultural use of the adjacent lands. Subsequent floods and increased shear stress (a function of bed slope and channel shape) along channel margins led to erosion of the constructed levees, increased additional sediment input, and bed aggradation. The river's value for resident fish production declined further because the now shallow, wide river was exposed to severe freezing in winter and elevated summer temperatures, and had lost much of its former structural habitat complexity.

The primary goal of the Blanco River restoration project was to stabilize the channel in a configuration that resembled its natural state of dynamic equilibrium, so that it could handle floods and provide in-stream habitat for fish, even under low flow conditions. Rosgen's approach was to rebuild the channel geometry to interact naturally with its floodplain and historical river valley. This involved narrowing the channel and reestablishing the historical floodplain to accommodate overbank flows, thus reducing shear stress along the channel margins. In selecting design criteria, project managers first located similar streams in the vicinity that were undisturbed and found that their dimensions and channel patterns were consistent with a particular channel type in the Rosgen stream classification scheme (Rosgen 1994, Chapter 5, this volume). The channel characteristics of these nearby, more intact streams served as the design criteria or template for the reconstruction of the 4.34-km reach of the Blanco River. To determine the river's previous character, physical dimensions were calibrated locally by evaluating a time series of aerial photos of the braided reach, and using a stable downstream reach as the model for reconstruction. Sources and volumes of sediment input, both upstream and from bank erosion, were evaluated to ensure that the restored channel would have the appropriate transport capacity. This evaluation helped project managers target the causes of the river's instability and deliberately tailor solutions to match the processes that controlled them. The final design was based on existing flow and channel relationships inferred from Rosgen's predictive model for comparable channels (NRC 1992).

Treatments on the Blanco River project included carefully surveying the site and shaping the desired channel pattern with bulldozers and scrapers. Natural materials (e.g., layers of logs and root wads held in place by boulders) were set in place to protect the newly excavated outside meander bends from erosion. Rock weirs were placed selectively within the channel to focus the flow and scour local pool habitat. A key element of this project was revegetation of riparian zones with mature plant materials salvaged from adjacent sites on the floodplain to simulate vegetation processes that occur naturally after a large flood. These plantings greatly accelerated the stabilization of banks with root masses and vegetation cover.

As a result of this project, a more natural looking, stable river channel developed, deep pools returned, bank erosion decreased, and in-stream habitats and fish production reportedly increased (Figure 24.2a,b) (NRC 1992). Costs for the Blanco River restoration work were about US$30 per lineal foot of stream; the total cost for the project (born by the landowner) was about US$400,000. From this investment, the landowner recovered nearly 68.8 hectares of agricultural land in the floodplain, at a cost about equal to the market rate for similar lands in the valley.

Although the available information suggests the Blanco River project was successful relative to the preexisting physical and biological conditions, documentation of the success of the project is not available in the scientific literature. This lack of documentation is an unfortunate but common shortcoming that prevents the claims of success for many restoration projects from being critically evaluated (Hamilton 1989, Reeves et al. 1991, Frissell and Nawa 1992), and hinders broader recognition of truly successful projects that could be used as models for other projects.

Off-Channel Habitat Restoration at the Reach Scale

Rivers that were systematically dredged during gold placer mining activities are numerous in western North America, and because of the severity of channel and floodplain alteration and habitat loss (Wissmar et al. 1994) these rivers are often targeted for habitat restoration efforts (Richards et al. 1992). Dredging in river reaches during placer mining changes the magnitude of the channel, and the general character of the river valley imposes both constraints and important (but limited) opportunities for designing restoration features. An advantage of projects in dredged or otherwise highly altered rivers is that the risk of adverse impacts from failed treatments or unanticipated effects generally is relatively low (because existing habitat values and biodiversity are usually quite low). Nevertheless, care should be taken to avoid further negative impacts.

In the case of the Yankee Fork of the Salmon River, Idaho, a reach targeted for restoration was, prior to dredging, important historical spawning and rearing for now severely depleted native chinook salmon (Richards et al. 1992). As part of the extensive efforts of the Northwest Power Planning Council and others to recover this threatened population of salmon, a series of off-channel juvenile rearing habitats were excavated within the dredge and settling ponds isolated from but adjacent to the existing channel. This project focused on controlling flow and establishing stable surface–water linkages from the ponds to the river.

It was not practical to observe the reaction of native chinook juveniles to these newly created habitats because the number of returning adult chinook salmon was and continues to be so low. Thus, managers evaluated the Yankee Fork project by releasing 60,000 juvenile chinook salmon of hatchery origin into two series of ponds, and tracking their movements over time. Observed densities of juvenile hatchery chinook in the constructed habitats were higher than expected. However, Richards et al. (1992) caution that such a novel and perhaps somewhat unnatural rearing environment (and the artificial method of fish introduction) could confound natural behavior and habitat selection in ways that have unexpected consequences for evaluation studies, and for survival or growth of wild fish in the altered stream system. Conclusive evaluation of the biological success of such a project requires a much longer period and should include tracking salmon through a complete generation to the returning adults. The second generation of adult returns should reflect any improvement in the productive capacity of the system; however, improvement may not be evident until 6 to 10 years (or more) after completion of the habitat treatments.

Similar projects involving excavation or reconnection of off-channel ponds along rivers in the Pacific Northwest appear to have sustained high densities of juvenile fishes for two or more salmon generations. For example, in Fish Creek, Oregon, reconnected off-channel ponds have clearly benefited coho salmon far more than numerous, more costly in-channel treatments (Everest et al. 1987, Reeves et al. 1991). However, other comparable projects in the same region are unsuccessful in sustaining natural production, for reasons that are unclear. Again, published evaluations spanning the time scale of two or more salmon generations are generally lacking. In Washington, the relatively indiscriminate and often poorly documented release of fish from state, federal, and tribal hatcheries severely confounds the relationship between locally observed fish production and local habitat conditions. Although reconnected off-channel ponds have been successful and cost efficient, keeping these ponds functional requires constant vigilance and maintenance (Dave Heller, USDA Forest Service, Pacific Northwest Region, Portland, Oregon, personal communication). This suggests such treatments may be less satisfactory in remote areas lacking ready access for monitoring and maintenance.

Watershed-Scale Restoration—An Example

A truly watershed-scale restoration project took place in Redwood Creek Basin, Redwood National Park, California. For more than a decade, experimental treatments have focused on upslope erosion sources which are the principal and continuing cause of problems in the stream channels (i.e., persistent bed aggradation and channel instability) (Weaver et al. 1987). This well-conceived and remarkably well-documented program is a preeminent model for future, large-scale watershed restoration programs (Weaver and Hagans 1995, Ziemer 1997).

The restoration program in Redwood Creek Basin did not begin as an effort focused on fish, but rather the broader goal was generally to restore natural watershed and stream channel processes. This goal sought protection of multiple natural resources inside the park boundaries, not the least of which was riparian old-growth redwood trees threatened by aggradation and lateral erosion of Redwood Creek itself. The program began with a barrage of qualitative and quantitative assessments conducted by physical and biological scientists (Kelsey et al. 1981), and field restoration efforts started almost simultaneously. Acknowledging the experimental nature of such projects, teams of physical and biological scientists designed and executed an extensive monitoring and evaluation effort to examine the effectiveness of each restoration method in the context of watershed-scale and site-specific processes. Most projects were geared toward reducing potential sediment sources to streams (Figure 24.3), but ancillary projects for special purposes also were conducted (e.g., modifying and removing logging debris jams that block fish passage). The watershed restoration program included detailed mapping and inventory to identify problem sites, development and interdisciplinary review of prescriptions for site treatments, implementation of field treatments, and monitoring and evaluation to adjust methods and evaluate their effectiveness in reducing overall erosion rates (Weaver et al. 1987). Some field treatments such as road ripping, changing orientation of road surface drainage, construction of cross-road drains and ditches, excavation of road fill at stream crossings, removal of unstable fill along roads and landings, and placement of rock armor in newly excavated channels involved using heavy equipment (Figure 24.4). Other treatments relied on hand labor, including constructing check dams; hand placing rock armor, flumes, water ladders, contour trenches, wooded terraces, and gravel catchers; and planting bank protection features such as living willow wattles and stem cuttings, mulching and seeding, and transplanting several species of container grown plants (Weaver et al. 1987). Different experimental areas received different combinations of treatments. Table 24.1 summarizes some of the many treatments used in Redwood Creek Basin and includes costs and assessed benefits (in terms of reduction of sediment inputs to streams).

In response to the monitoring and evaluation program, over time the program emphasized the use of heavy equipment, particularly at stream crossings and on unstable slopes where the potential for delivery of substantial volumes of sediment to streams could be averted (Weaver et al. 1987). Hand labor treatments, although locally effective in reversing surface and gully erosion processes, proved less effective in reducing sediment delivery problems in the basin overall. In most circumstances if sufficient care was taken during equipment operations, park staff found they could rely largely on natural revegetation. Several valuable manuals providing evaluation and guidance for watershed restoration methods have emerged from the Redwood National Park program (Weaver et al. 1987, Spreiter 1992) and from subsequent efforts of

FIGURE 24.3. Rills and gullies beginning to form on the surface of a newly constructed logging road on private timber lands above the Pistol River in southwestern Oregon. Left unattended, such poorly constructed or badly maintained roads are potential major sediment sources to streams, and likely initiation sites of stream diversions, debris flows, and landslides that can have serious and long-lasting adverse effects on stream habitat. Selective obliteration or structural rehabilitation of the thousands of kilometers of such roads residing on forest, range, and crop lands is a major restoration challenge and opportunity across the Pacific coastal ecoregion. The greatest gains from such work are preventative; the most effective treatments must be applied before a large storm triggers a major episode of erosion and sediment delivery from the road to stream channels (Photo credit: Christopher A Frissell, Oregon State University).

specialists from this program applying their earned expertise to other areas of northern California and elsewhere (Weaver and Hagans 1994).

Nonetheless, the Redwood Creek Basin example also illustrates the tragedy of cumulative effects of watershed disturbance. Although millions of public dollars were invested in restoration of the lower two-thirds of the basin, the downstream impacts from logging on unprotected, private timberlands in the headwaters of the basin threaten the benefits of erosion control efforts within the borders of Redwood National Park (Kelsey et al. 1981, Hagans et al. 1986, Hagans and Weaver 1987).

Monitoring and Evaluating Restoration Projects

Although evaluation of the outcome of restoration actions is essential for improving and documenting their effectiveness, long-term evaluation (in effect, monitoring) of restoration actions usually is left out of the overall planning and implementation cost equation. In most cases, the enthusiastic rush to address perceived problems leaves little time for planning and implementing follow-up actions to learn what did and did not work, and measures of success are often based on inappropriate

FIGURE 24.4. Excavation of a stream crossing during a road obliteration project in Redwood National Park. Road fill is removed to expose the original stream channel and side slope contour, thus restoring natural drainage patterns, and excavated material is transported by dump truck to a waste site away from streams and potentially unstable hillslopes (Photo credit: Christopher A Frissell, Oregon State University).

criteria (Minns et al. 1996). For example, historically, evaluation of the "success" of some in-stream habitat structures on federal forest lands in the Pacific coastal ecoregion was based on whether the structures survived the first year's winter storm events, rather than whether they provided any measurable ecological benefits to fish. Unfortunately, many expensive projects did not even pass the first test (Frissell and Nawa 1992). Further, past allocations of annual restoration funding to U.S. Forest Service ranger districts have been based more on the number of projects built the previous year (i.e., whether previous short-term targets were met), rather than on the demonstrated effectiveness of restoration efforts. When local specialists have managed to find resources to evaluate projects, the results have often been ambiguous because of design or data limitations (House 1996).

Committing to a long-term program of monitoring is the most practical and effective way to assess a restoration program and document the recovery of ecosystems. Although most restoration projects are hampered by a lack of baseline and reference data, the use of such data in a well-designed, systematic monitoring program provides the best opportunity to document overall success or failure of a program (NRC 1992). Existing conditions can be characterized through collection of pretreatment, baseline data, which provides a partial basis for comparison of the treatment effects following the application of restoration activities. Reference data is especially important because it provides a measure of site potential, or a sense of what level of recovery is reasonable and desirable given the larger context of the stream or basin. Reference data may come from a reach of the river targeted for restoration or from a reach in a different river with similar characteristics of channel morphology and basin hydrology. Assessing the unrestored condition of a river allows investigators to gauge the effects of restoration efforts, and to better understand natural rates of recovery (Cairns 1989). Bryant (1995) describes a well-reasoned approach to long term monitoring for tracking the outcome

TABLE 24.1. Selected watershed restoration techniques and measures of their cost and benefits (in 1987 US dollars) in Redwood National Park.

Restoration method	Unit	Unit cost	Estimated unit benefit	Unit cost per benefit
Excavation of road stream crossings using heavy equipment, with end hauling of excavated fill	L-1-5 Road in Bond Creek unit and M-7-5-1 Road in Bridge Creek unit (11 crossings)	$1,213–11,597 per crossing	195–1,500 m^3 potential sediment removed per crossing	$3.90–9.45 per m^3 of potential sediment removed (does not include prevented gully erosion)
Excavation of skid trail stream crossings, with limited end hauling	Skid trails in Bond Creek unit (10 crossings)	$80–$1,345 per crossing	5–155 m^3 potential sediment removed per crossing	$4.45–19.00 per m^3 of potential sediment removed (does not include prevented gully erosion)
Outsloping and cross-draining major haul roads	L-1-5 Road in Bond Creek unit and M-7-5-1 Road in Bridge Creek unit (11 crossings)	$7,600–$71,000 per road mile	Uncertain; if untreated, only a limited portion of mobilized sediment would enter streams	Uncertain (but small compared to stream crossing removal)
Partial removal of in-stream logging debris, excavation and endhauling of aggraded alluvium and adjacent unstable hillslope	Bridge Creek	$10,250 (including riprapping of exposed streambanks)	1,300 m^3 of sediment and potential sediment removed	$7.88 per m^3 of potential sediment removed
Outsloping and ripping landing fill material	Bridge Creek and Copper Creek units (total 12 landings treated)	$1,960–8,200 per landing (high end reflects larger landings and longer end haul to dispose of fill)	Uncertain; if untreated, only a limited portion of mobilized sediment would enter streams	Uncertain; probably greater than road surface outsloping, less than stream crossing treatments
Hand construction of checkdams in stream channels and gullies	Bond Creek, Bridge Creek, and Copper Creek units	$19.52–47.60 per structure or treated site; not quantified by channel length	Benefits limited by malfunction and short life span of many structures, pre-project erosion	Uncertain; probably small since most erosion occurred prior to treatment
Hand labor: mulching and revegetation using grass seed, stem cuttings, wattles, and transplants	Bond Creek, Bridge Creek, and Copper Creek units	3–28% of total project costs; not quantified by area	Uncertain but small; natural revegetation was rapid except where hindered by grass seed or mulch	Uncertain, possibly negative when natural revegetation is hindered

of restoration efforts. This strategy of pulsed monitoring combines extensive long-term surveys repeated over long intervals (10–15 years) interspersed with intensive short-term (3–5 years) studies that focus on specific questions. Ideally, pretreatment baseline studies and parallel studies at reference sites are included. Bryant's approach addresses broad watershed-scale conditions while examining treatment effects and outcomes at the most ecologically significant stream reaches.

Of course, fundamental principles of the design and implementation of evaluation and monitoring programs must be understood whether they are applied on a basin-wide or limited project scale. Ecologists or technicians involved with the design and placement of such habitat structures may lack engineering and other technical support to define the problem behind the need for the project and to design and implement solutions that take into account the watershed context of such restoration activities (Reeves et al. 1991, Frissell and Nawa 1992, Beschta et al. 1994, Sear 1994, Kondolf et al. 1996). MacDonald et al. (1991) describe many fundamental principles of monitoring water quality and in-stream conditions in situations where timber harvesting is the dominant land use. Like any environmental study, effective restoration efforts must identify the purpose, questions, hypotheses, models, sampling designs, statistical analyses, tests of hypotheses, and the interpretation and presentation of results (Green 1984, Kershner 1997).

Designing restoration projects requires explicit definition of the setting and problem to be corrected, which determines the nature of the objectives and the hypotheses that form the core of the evaluation and monitoring approach. Until and unless this critical but often neglected step is taken, subsequent monitoring is not likely to provide the desired information (Minns et al. 1996). MacDonald et al. (1991) and Platts et al. (1987) address some aspects of selecting monitoring methods and designing sampling plans, but this subject continues to warrant further research and development (Kondolf 1995, Poole et al. 1997). No standard checklist, blueprint, or catalogue exists for monitoring and evaluation that can be applied to all river and stream restoration programs and projects. (A more thorough discussion of monitoring and assessment is presented in Chapter 18.)

A number of methods are commonly applied to quantify physical channel conditions before and following restoration activities (Platts et al. 1983 and 1987, Olson-Rutz and Marlow 1992, Meador et al. 1993, Nawa and Frissell 1993, Harrelson et al. 1994). However, many habitat parameters routinely applied in stream surveys have not been fully evaluated as to their usefulness or sensitivity in detecting meaningful changes (Peterson et al. 1992, Poole et al. 1997). Other techniques that focus on biotic assemblages have been successfully applied to such questions (Roth et al. 1996), but only recently to Pacific coastal rivers, and widespread implementation awaits testing and further regional development (Karr et al. 1986, Miller et al. 1988, Pflakin et al. 1989, Imhoff et al. 1996).

Monitoring is about detecting if, when, and how change occurs, and what change itself means in ecological terms. Change reflects natural processes that are an essential part of natural history and evolution. Maintaining natural ecological values of river systems becomes increasingly difficult when the rate of change induced by human activities accelerates, the overall magnitude and persistence of the effects increases, and the spatial distribution of human activities affects growing portions of the landscape (Doppelt et al. 1993, Frissell and Bayles 1996, Minns et al. 1996). Effective assessment and subsequent monitoring provide the major link between science and management by directly determining whether objectives are met (e.g. whether habitat complexity is maintained, whether loss of riparian vegetation and bank erosion remain within expected natural patterns) (Naiman et al. 1992, Stanford and Poole 1996). Such assessment is especially critical when there is uncertainty about whether management objectives (including investing public dollars in restoration measures) will actually address the underlying causes of the apparent conditions targeted for correction. Such uncertainty is the rule for the complex task of restoring streams that suffer from multiple impacts across entire watersheds.

A Nested Experimental Design for Monitoring

Ideally, restoration projects should be approached on an experimental basis. The effects of a restoration action should be measured against the performance of a comparable, untreated system that can serve as a reasonable basis to judge performance. Studies that attempt to relate habitat conditions to specific species responses, such as the abundance of returning adult migratory salmonids, are plagued by factors beyond the control of experimental design (Lichatowich and Cramer 1979), including ocean or riverine survival, fisheries harvest as a mortality factor, and related conditions acting independent of early life history of juvenile salmon. Yet, documenting desired biological responses is viewed as the bottom line for a restoration program. Although spatial and temporal factors prohibit strict experimental controls in field studies, the ability to demonstrate a biological response to a restoration project in the face of natural variation depends strongly on careful selection and monitoring of untreated control or reference systems (Minns et al. 1996).

The most effective monitoring designs should include designated functional controls nested at a series of spatial scales (Frissell et al. 1986, Minns et al. 1996, Poole et al. 1997), from habitat units at the scale of pools and riffles (nested within a reach in which other units have been treated) through reaches and valley segments, up to and including whole watersheds if possible. Many of the possible consequences of scale discussed earlier can be explicitly evaluated with such a hierarchical approach (Poizat and Pont 1997). A nested design allows quantitative assessment of possible regional (off-site) or indirect ecological effects of site-specific habitat alterations.

For example, it is often hypothesized but rarely substantiated that reach- or stream-scale juvenile fish recruitment increases following localized improvements in survival or growth, or that the concentration of adult fish increases where habitat complexity expanded with placement of artificial structures (Gowan and Fausch 1996). On the other hand, it is equally plausible that many projects may simply attract and concentrate fish that would otherwise be dispersed over a larger area, with no net effect (or possible adverse effect) on overall production. A spatially nested design allows these various hypotheses about biological responses to be tested (Figure 24.5). If restoration treatments have local effects only, fish abundance should increase in treated habitat units, with no response (positive or negative) in untreated habitat units. If both local and regional benefits accrue, fish abundance should increase in both treated habitats and at least some nearby untreated control habitats in restored streams. If concentration without compensation occurs, abundance should increase in treated habitats with a corresponding decrease in untreated habitats. Untreated streams should show no consistent or comparable trend either at individual habitat unit or reach scales.

Effective evaluation using such a hierarchically nested design must be conducted over a series of posttreatment years. The power of the analysis is potentially increased with prolonged pretreatment monitoring to assess interannual correlation among habitat units within streams and among treatment and control streams. Interpretation of biological mechanisms underlying the results is critical to understanding the significance and generality of such an experiment. Such understanding can be improved markedly by marking individuals to assess the movements and growth of animals among habitats in both the treated and control streams (Gowan and Fausch 1996).

Application of a similar spatially nested and hierarchical design to a physical monitoring program also addresses questions about possible spatial diffusion of physical effects (Frissell et al. 1986, Sear 1994), such as whether installation of in-stream structures changes the thermal regime of an entire reach, or whether local control of bank erosion results in desired changes in the sediment dynamics or structure of downstream habitats (and the distance from the treatment site that any benefits measurably accrue). Such a hierarchical approach to evaluation of stream restoration projects has been applied rarely anywhere (Imhoff et al. 1996), and perhaps never in the Pacific coastal ecoregion.

Cost Accounting for Watershed Restoration

In an effort to integrate socioeconomic information into regional planning for watershed restoration and salmon recovery, Fluharty et al. (1996) developed the Habitat Restoration Cost Estimation Model (HRCEM) for the National Marine Fisheries Service (NMFS). As part of their Pacific Northwest regional strategy to address threatened and endangered stocks of Pacific salmon, NMFS is interested in the costs

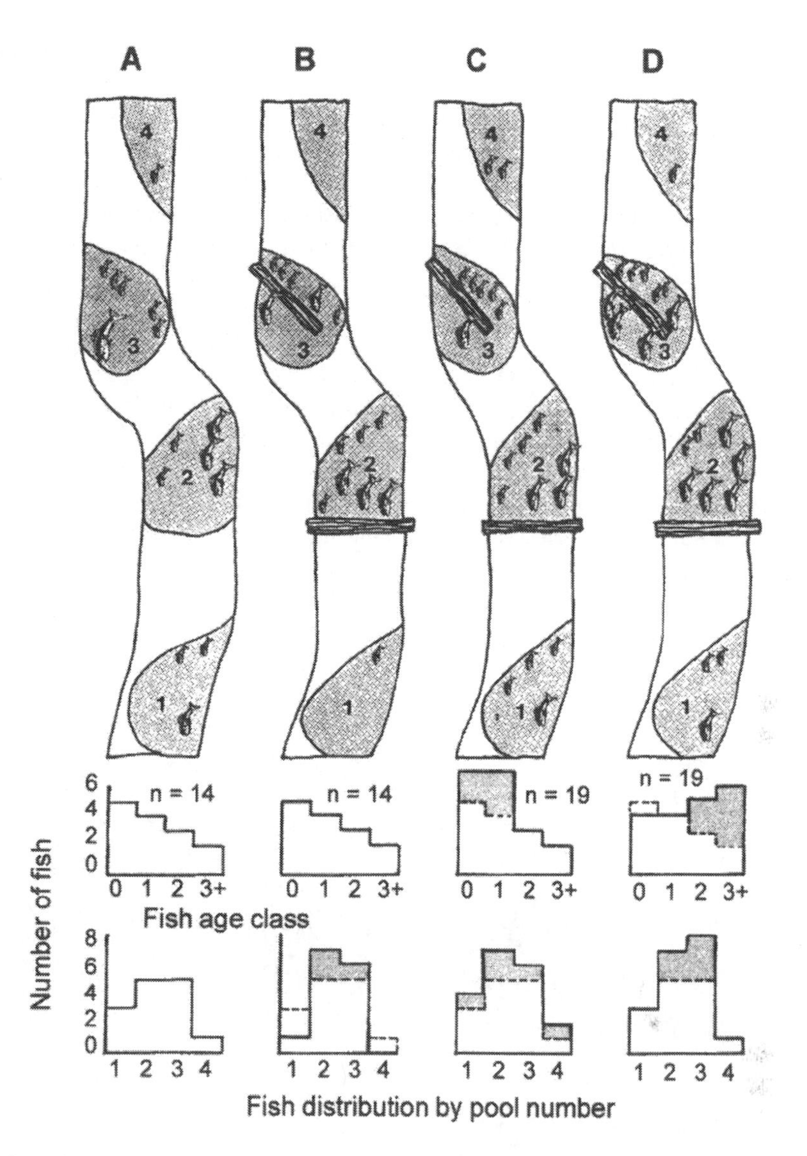

FIGURE 24.5. Greatly simplified depiction showing four hypothetical scenarios of fish response to alteration of stream habitats at the scales of reach and pool habitats (i.e., habitat unit scale). Four pools are numbered in downstream sequence. The pretreatment baseline or control is shown in panel A; treatment (panels B–D) is installation of a log weir in pool 2 and lateral deflector in pool 3. Inset graphs depict hypothetical population age structure and spatial distribution under each response scenario (relative to the baseline, dashed bars) two years after treatment. Under scenario B, apparent increases in fishes in treated pools are offset by decreases in untreated pools (which would not be detected if untreated pools were not monitored). In scenario C, populations are supplemented by increased survival and juvenile recruitment within the reach. In scenario D, populations are supplemented by increased retention of older immigrants originating outside the reach. Even though a local positive response is observed in the treated habitat units in all cases, only scenario C represents clear evidence of net increased production attributable to the treatment at a reach or larger scale. Scenario D might imply a net positive response, assuming retention of immigrants in the treated reach opens space for fishes in other reaches that would otherwise have been lost to the population through competitive displacement (Gowan and Fausch 1996) or cannibalism. Moreover, even under a scenario like C it is possible that increased juvenile recruitment will be nullified by survival bottlenecks at later life stages (Reeves et al. 1991); to ascertain whether gains in juvenile survival for this species (assumed to mature at age two or three) translate into increased adult populations requires monitoring through three or more years posttreatment.

and distribution of efforts for watershed restoration. HRCEM combines several information sources (including hydrologic data from the U.S. Geological Survey, Rivers Information Systems for California, Idaho, Oregon and Washington, and USDA National Resources Inventory and Major Land Resource Area databases) into an integrated data system based on river reach and watershed acre units at the 1:250,000 scale. The resulting information is used to develop estimates of the nature and extent of administratively adopted land management practices meant to deter erosion, water quality, and other impacts of agriculture and other land uses (i.e., Best Management Practices) or to determine what rehabilitation activities would promote aquatic habitat recovery or protection. Together with estimates of costs for various methods of river alteration and habitat rehabilitation, the HRCEM method results in a regionally aggregated estimate of project installation, operation, and maintenance costs annualized over a 20-year period. The model can be used to evaluate scenarios using different levels of management effort and methods. Among the aggregated "restoration" measures included in HRCEM are streambank and shoreline protection (rip-rap and revetments), fish stream improvement measures (in-stream structures), livestock fencing along riparian areas (Figure 24.6), and revegetation and tree planting in critical areas. Costs for each measure (on a per stream miles basis) are then applied to estimates of anadromous habi-

FIGURE 24.6. Three-year-old fenced cattle exclosure constructed along Elk Creek, a tributary of Joseph Creek in northeastern Oregon. A joint venture of Bonneville Power Administration, Oregon Department of Fish and Wildlife, and the private landowner from whom the exclosed plot was leased, this approach attempts to address causes, rather than symptoms of stream ecosystem deterioration. However, note that during a flood of approximately 5–10 year return period, the stream waters are well outside the fenced area (at left). The small sedge wetland in the middle foreground, also outside the exclosure, occupies an old channel swale, and is probably connected to the present channel through groundwater flow paths. In such biologically important alluvial valleys, channel migration and the presence of broad floodplains and fluvial wetland complexes challenge the conventional notion of protection through exclusion of activities within narrow streamside buffers. Long-term restoration of such streams depends on recovery of the ecological integrity (woody vegetation, soil structure, and natural channel and groundwater dynamics) of the entire valley floor (Photo credit: Christopher A Frissell, Oregon State University).

tat (e.g., miles of spawning habitat) within any particular basin.

Fluharty et al. (1996) suggest the HRCEM can further the use of economic data in comprehensive watershed management planning to help establish priorities for rehabilitation efforts and to identify the hidden costs of failing to provide adequate protection of aquatic resources now, in lieu of dubious restoration in the future. They caution, however, that assumptions in the model regarding the costs for various management practices need to be validated because the estimated costs varied widely between watersheds throughout California, Idaho, Oregon, and Washington.

This ambitious approach has additional limitations that require further attention or improvement. Presently, the model does not incorporate estimates of actual fish production benefits or restoration success that might accrue from the various restoration activities. Although this is understandable given the difficulty of predicting the success of such projects, unless the likely effectiveness of restoration projects is somehow evaluated in ecological terms, economic assessments of any kind may remain of limited value. Many of the kinds of projects incorporated in the Fluharty model and similar efforts to date (including many in-stream fish habitat structures and tree planting in riparian areas) are questionable in terms of their actual net benefits, in large part because details of project implementation were not readily available at the scale that would allow such assessment. Traditional stream bank stabilization projects using large rock or concrete structures for example, should be assumed to have negative net value for native fishes, their habitats, and aquatic ecosystem integrity in the long term. As directed by the NMFS, the HRCEM looked at scenarios with hundreds of miles of such treatments in coastal river basins, ostensibly for the purpose of salmon recovery. Possibly such scenarios reflect the present well-intentioned ambitions of certain management agencies and commercial interests, but they may be misguided ecologically if they are not critically evaluated in terms of the ecological context in which they will occur.

Fluharty et al. (1996) recognize that the most effective economic assessments of restoration projects focus not simply on the costs of applying a smorgasbord of remedial actions, but on a fuller accounting of the costs and benefits of redirecting human activities on the landscape in configurations that facilitate natural recovery of aquatic ecosystems (Chapter 23). Such evaluations may eventually show that substantial recovery is possible in many ecosystems with little regional net economic cost and only modest capital investment (Doppelt et al. 1993, Stanford et al. 1996).

Numerous watershed-scale restoration programs are now underway in the Pacific coastal ecoregion, (e.g., Big Quilcene River, Washington; Grande Ronde River, Oregon; Upper Clark Fork River, western Montana), as a result of identification by FEMAT (1993) and related regional assessments, but few if any of these programs show evidence of an institutional commitment to long-term, integrated monitoring and evaluation of outcomes that is equal to the environmental and technological challenges posed. In developing effective restoration programs, citizens' involvement councils and funds spent on retrospective watershed analyses or quantified engineering specifications cannot substitute for a carefully designed and controlled, adequately staffed, extensive monitoring and evaluation program.

Watershed Restoration and Adaptive Ecosystem Management

Determining the success of watershed or river restoration depends on the nature and the temporal and spatial extent of the degradation. For example, it is a relatively simple matter to evaluate the effectiveness of reducing pollution if the damage is caused by the discrete point discharge of factory or sewage effluents into a stream. The problem (low dissolved oxygen, elevated temperatures, elevated nutrient levels) can be easily defined and the cause-effect relationship easily established with existing knowledge. The response of the biological community following effluent reduction is relatively simple to predict, assuming the effects are primarily local in scale.

Predicting the outcome of restoration efforts at the scale of a whole watershed, stream network, or large river segment, however, is highly problematic due in large part to the diffuse, persistent, and time-dependent nature of cumulative effects operating within a basin (Montgomery 1995, Kershner 1997), and the problem that as a science, ecology still largely lacks (and may always lack) a robust predictive capability for specific sites and cases (Bella and Overton 1972, Cairns 1989, Ludwig et al. 1993). These limitations present an imperative for ecologists to link their descriptive skills with the better developed predictive capabilities of physical scientists to better understand and predict the outcome of physical processes acting on biological systems in a given context (Sear 1994, Stanford et al. 1996, Stanford and Poole 1996). One promising avenue in this respect is *adaptive management*, a management system designed to provide the information necessary for defensible, timely tests of the assumptions upon which interim decisions are based, so that informed "mid-course correction" of management programs can take place (Walters and Hilborn 1978, Walters and Holling 1990). Adaptive management is not a new idea, but government agencies recently have promoted it as a chief premise of ecosystem management (FEMAT 1993, Dombeck 1996, Thomas 1996). As difficult as implementing a sound adaptive management plan for a large ecosystem seems to be (Halbert 1993, Walters et al. 1993), adaptive management is essentially nothing more than avoiding past mistakes on a large scale, and learning from both successes and failures.

Unfortunately, adaptive management and environmental monitoring are often prescribed by managers as a "general tonic" that allows environment-damaging activities to proceed, as long as they are somehow monitored. Further, reliable predictive capability about linkages between specific watersheds, aquatic habitats, and biota remains limited. Large variation in the rate and pattern of natural processes and disturbance events often triggers changes in aquatic ecosystems that are neither predictable nor easily interpreted at particular locales on time scales of years to decades. Therefore, conservative assumptions are necessary to avoid unanticipated, irreversible consequences of management actions (Bella and Overton 1972, Ludwig et al. 1993, Montgomery 1995, Frissell and Bayles 1996). This fundamental precautionary principle includes deliberately avoiding the assumption that habitat restoration methods alone will suffice to induce full physical and biotic recovery of streams, a politically appealing presumption that is often employed as a conventional rationale for proceeding with human alteration of the landscape (Doppelt et al. 1993).

A corollary axiom of adaptive ecosystem management could be that distributing human activities and risks over the landscape would ensure that strategically selected, significantly large areas remain relatively free of the multiple threats posed by human activities. What this means is that the first step in an ecologically effective restoration program is a deliberate assessment of all management actions to ensure that regulatory systems are rigorous enough to minimize the future need for restoration. In other words, avoid making the same mistakes everywhere. For example, landscape alteration increases the need to establish and maintain relatively natural "safe havens" (Li et al. 1995) or large-scale refugia (Sedell et al. 1990, Frissell and Bayles 1996, Frissell 1997) to conserve representative aquatic habitats and biota at key points across the landscape (Ebersole et al. 1997). This axiom applies equally well to restoration, in which the recovery of specific habitats may be increasingly critical for maintaining key species or biotic resources on the landscape, as it does to the activities that alter landscapes and generate the need for restoration in the first place. Such a strategic, scientific assessment of priorities and reallocation of management impacts and restoration efforts is one stated goal of state and federal procedures for watershed analysis (FEMAT 1993, Montgomery et al. 1995, Chapter 19), but few if any existing examples of watershed analysis to date come close to meeting this goal (Keeton 1995,

Weaver and Hagans 1995, Frissell and Bayles 1996, Collins and Pess 1997a, b).

Elements of Successful Restoration and Monitoring

In terms of improving interdisciplinary working knowledge of successful restoration techniques, benefits of evaluation and monitoring may be vastly more important than the particular biological benefits directly accrued from any single project. Monitoring is also invaluable for identifying small adjustments that can greatly increase the ecological efficacy and reduce the costs of a project or program, as exemplified in the monitoring program of Redwood National Park's large watershed restoration program. Resources devoted to monitoring and evaluation early in the program revealed that many early methods were relatively ineffective in the face of large-scale alteration of watershed processes. The results allowed managers to target resources to other methods that were more clearly cost-effective for addressing the overriding ecological objectives, including sediment reduction and restoration of slope and drainage network stability (Weaver et al. 1987).

To be successful, a substantial portion of the cost of even the most routine restoration projects must be dedicated to monitoring and evaluation (including pre- and postproject studies) (Minns et al. 1996). In fact, projects of a highly experimental nature may merit allocation of a predominant share of total project costs for monitoring and evaluation. The requirement for specific monitoring plans cannot be excluded from funds for planning. New institutional mechanisms are needed to establish site-dedicated, carryover funding to ensure long-term, postproject monitoring. Moreover, agencies must cooperatively develop institutional arrangements and fiscal support for independent evaluation and quality assurance of data and program monitoring. A regional, independent scientific panel or commission to direct the design and implementation of stream and watershed monitoring criteria and programs may be appropriate. It would, however, require a significant discretionary budget for a separate, competitive research grants program to develop innovative monitoring methods. The scientific oversight panel should be engaged in evaluating the efficacy and timeliness of management program responses to monitoring results (Stanford and Poole 1996). The oversight process must include provision that allows agencies and citizens access to monitoring data. Effective adaptive and cooperative ecosystem management requires closely integrating monitoring impacts of ongoing development activities and restoration programs (Minns et al. 1996).

The most fundamental challenge facing successful restoration of aquatic systems is to establish a clear understanding of the cause and effect relationships between the physical processes at work within a watershed, how the expression of these processes (rate, magnitude, and distribution) has been altered by human activities, and what short- and long-term restoration strategies best address such factors. Such strategies must consider the inherent constraints dictated by the characteristics of the watershed, the legacy of natural disturbances, and the distribution, magnitude, and persistence of management-induced changes. The examples in this chapter illustrate both the promise and problems of restoration efforts at the microhabitat, reach, and watershed scales. From the simple to the highly engineered approach, the inherent and collateral costs of recovery techniques and the spatial extent of the problem suggests traditional approaches need to be fundamentally rethought.

Literature Cited

Beechie, T., E. Beamer, and L. Wasserman. 1994. Estimating coho salmon rearing habitat and smolt production losses in a large river basin, and implications for habitat restoration. North American Journal of Fisheries Management **14**:797–811.

Bella, D.A., and W.S. Overton. 1972. Environmental planning and ecological possibilities. Journal of the Sanitary Engineering Division, American Society of Civil Engineers **98**(SA3):579–592.

Beschta, R.L., W.S. Platts, J.B. Kauffman, and M.T. Hill. 1994. Artificial stream restoration: Money well spent or an expensive failure? *In* Proceedings of a symposium on environmental restoration.

The Universities Council on Water Resources, Montana State University, Bozeman, Montana, USA.

BioWest. 1995. Stream habitat improvement evaluation project. Final report prepared for Utah Reclamation Mitigation and Conservation Commission and US Department of Interior, Salt Lake City, Utah, USA.

Bisson, P.A., T.P. Quinn, G.H. Reeves, and S.V. Gregory. 1992. Best management practices, cumulative effects, and long-term trends in fish abundance in Pacific Northwest river systems. Pages 189–232 *in* R.J. Naiman, ed. Watershed management: Balancing sustainability and environmental change. Springer-Verlag, New York, New York, USA.

Bryant, M.D. 1995. Pulsed monitoring for watershed and stream restoration. Fisheries **20**(11):6–13.

Cairns, J., Jr. 1989. Restoring damaged ecosystems: Is predisturbance condition a viable option? The Environmental Professional **11**:152–159.

Carufel, L.H. 1964. Statewide fisheries investigations: Post-development evaluation (investigations). Federal aid to fisheries restoration project completion report, project number F-7–R-7. Arizona Game and Fish Department, Phoenix, Arizona, USA.

Cederholm, C.J., R.E. Bilby, P.A. Bisson, B.R. Fransen, W.J. Scarlett, J.W. Ward, et al. 1997. Response of juvenile salmon and steelhead to placement of large woody debris in a coastal Washington stream. North American Journal of Fisheries Management **17**:947–963.

Cederholm, C.J., W.J. Scarlett, and N.P. Peterson. 1988. Low-cost enhancement technique for winter habitat of juvenile coho salmon. North American Journal of Fisheries Management **8**:438–441.

Collins, B.D., and G.R. Pess. 1997a. Critique of Washington state's Watershed Analysis Program. Journal of American Water Resources Association **33**:997–1010.

Collins, B.D., and G.R. Pess. 1997b. Evaluation of forest practice prescriptions from Washington state's Watershed Analysis Program. Journal of American Water Resources Association **33**:969–996.

Dombeck, M.P. 1996. Thinking like a mountain: BLM's approach to ecosystem management. Ecological Applications **6**:699–702.

Doppelt, B., M. Scurlock, C. Frissell, and J. Karr. 1993. Entering the watershed: A new approach to save America's river ecosystems. Island Press, Washington, DC, USA.

Dunne, T., and L.B. Leopold. 1978. Water in environmental planning. W.H. Freeman, New York, New York, USA.

Ebersole, J.L., W.J. Liss, and C.A. Frissell. 1997. Restoration of stream habitats in the western United States: Restoration as re-expression of habitat capacity. Environmental Management **21**:1–14.

Ehlers, R. 1956. An evaluation of stream improvement devices constructed eighteen years ago. California Fish and Game **42**:203–217.

Espinosa, F.A. Jr., J.J. Rhodes, and D.A. McCullough. 1997. The failure of existing plans to protect salmon habitat in the Clearwater National Forest in Idaho. Journal of Environmental Management **49**:205–230.

Everest, F.H., G.H. Reeves, J.R. Sedell, D.B. Hohler, and T. Cain. 1987. The effects of habitat enhancement on steelhead trout and coho salmon smolt production, habitat utilization, and habitat availability in Fish Creek, Oregon, 1983–86. Project 84-11. 1986 Annual Report for the Bonneville Power Administration, Division of Fish and Wildlife, Portland, Oregon, USA.

FEMAT (Forest Ecosystem Management Assessment Team). 1993. Forest ecosystem management: An ecological, economic and social assessment. Report of the Forest Ecosystem Management and Assessment Team (a federal interagency consortium), Portland, Oregon, USA.

Fluharty, D., E.G. Doyle, D. Huppert, and B. Amjoun. 1996. The economic costs of critical habitat designation for coho salmon and steelhead populations listed under the ESA. A report to the National Marine Fisheries Service. School of Marine Affairs, University of Washington, Seattle, Washington, USA.

Frissell, C.A. 1993. Topology of extinction and endangerment of native fishes in the Pacific Northwest and California (U.S.A.). Conservation Biology **7**:342–354.

Frissell, C.A. 1997. Ecological principles. Pages 96–115 *in* J.E. Williams, C.A. Wood, and M.P. Dombeck, eds. Watershed restoration: Principles and practices. American Fisheries Society, Bethesda, Maryland, USA.

Frissell, C.A., W.J. Liss, C.E. Warren, and M.D. Hurley. 1986. A hierarchical approach to stream habitat classification: Viewing streams in a watershed context. Environmental Management **10**:199–214.

Frissell, C.A., and D. Bayles. 1996. Ecosystem management and the conservation of aquatic biodiversity and ecological integrity. Journal of the American Water Resources Association **32**:229–240.

Frissell, C.A., and R.K. Nawa. 1992. Incidence and causes of physical failure of artificial habitat structures in streams of western Oregon and Washington. North American Journal of Fisheries Management **12**:182–197.

Frissell, C.A., W.J. Liss, R.K. Nawa, R.E. Gresswell, and J.L. Ebersole. 1997. Measuring the failure of salmon management. Pages 411–444 *in* D.J. Stouder, P.A. Bisson, and R.J. Naiman, eds. Pacific salmon and their ecosystems: Status and future options. Chapman & Hall, New York, New York, USA.

Fuller, D.D., and A.J. Lind. 1992. Implications of fish habitat improvement structures for other stream vertebrates. Pages 96–104 *in* H.M. Kerner, ed. Proceedings of the symposium on biodiversity of northwestern California, October 28–30, Santa Rosa, California. Report 29. Wildland Resources Center, Division of Agriculture and Natural Resources, University of California, Berkeley, California, USA.

Gard, R. 1972. Persistence of check dams in a trout stream. Journal of Wildlife Management **36**:1363–1367.

Gore, J.A., ed. 1985. The restoration of rivers and streams—theories and experiences. Butterworth, Stoneham, Massachusetts, USA.

Gowan, C., and K.D. Fausch. 1996. Long-term demographic responses of trout populations to habitat manipulations in six Colorado streams. Ecological Applications **6**:931–946.

Green, R.H. 1984. Statistical and nonstatistical considerations for environmental monitoring studies. Environmental Monitoring and Assessment **4**:293–301.

Hagans, D.K., and W.E. Weaver. 1987. Magnitude, causes, and basin response to fluvial erosion, Redwood Creek Basin, northern California. Pages 419–428 *in* R.L. Beschta, T. Blinn, G.E. Grant, F.J. Swanson, and G.G. Ice, eds. Erosion and sedimentation on the Pacific rim. International Association of Hydrological Sciences Publication 165, IAHS, Institute of Hydrology, Walingsford, Oxfordshire, UK.

Hagans, D., W.E. Weaver, and M.A. Madej. 1986. Long-term off-site effects of logging and erosion in the Redwood Creek Basin, northern California. National Council for Air and Stream Improvement Technical Bulletin **490**:38–66.

Halbert, C.L. 1993. How adaptive is adaptive management? Implementing adaptive management in Washington State and British Columbia. Reviews in Fisheries Science **1**:261–283.

Hamilton, J.B. 1989. Response of juvenile steelhead to instream deflectors in a high-gradient stream. Pages 149–157 *in* R.E. Gresswell, B.A. Barton, and J.L. Kershner, eds. Practical approaches to riparian resources management. American Fisheries Society, Montana Chapter, Bethesda, Maryland, USA.

Harrelson, C.C., C.L. Rawlins, and J.P. Potyondy. 1994. Stream channel reference sites: An illustrated guide to field technique. USDA Forest Service General Technical Report GTR RM-245, Rocky Mountain Forest and Range Experiment Station, Fort Collins, Colorado, USA.

Heede, B.H., and J.N. Rinne. 1990. Hydrodynamic and fluvial morphologic processes: Implications for fisheries management and research. North American Journal of Fisheries Management **10**:249–268.

House, R.L. 1996. An evaluation of stream restoration structures in a coastal Oregon stream, 1981–1993. North American Journal of Fisheries Management **16**:272–281.

House, R.A., and P.L. Boehne. 1986. Effects of instream structures on salmonid habitat and populations in Tobe Creek, Oregon. North American Journal of Fisheries and Aquatic Sciences **6**:38–46.

Hunt, R.L. 1971. Responses of a brook trout population to habitat development in Lawrence Creek. Wisconsin Department of Natural Resources Technical Bulletin 82, Madison, Wisconsin, USA.

Imhoff, J.G., J. Fitzgibbon, and W.K. Annable. 1996. A hierarchical evaluation system for characterizing watershed ecosystems for fish habitat. Canadian Journal of Fisheries and Aquatic Sciences **53**(Supplement 1):312–326.

Karr, J.R., K.D. Fausch, P.L. Angermeier, P.R. Yant, and I.J. Schlosser. 1986. Assessing biological integrity in running waters: A method and its rationale. Special Publication Number 5, Illinois Natural History Survey, Champaign, Illinois, USA.

Keeton, B. 1995. A critique of federal watershed analysis. Wild Fish and Forests, Spring 1995:4–6. Pacific Northwest Regional Office of The Wilderness Society, Seattle, Washington, USA.

Kelsey, H., M.A. Madej, J. Pitlick, M. Coghlan, D. Best, R. Belding, et al. 1981. Sediment sources and sediment transport in the Redwood Creek Basin: A progress report. Redwood National Park Technical Report 3, National Park Service, Arcata, California, USA.

Kershner, J.L. 1997. Monitoring and adaptive management. Pages 116–131 *in* J.L. Williams, C.A. Wood, and M.P. Dombeck, eds. Watershed restoration: Principles and practices. American Fisheries Society, Bethesda, MD.

Kondolf, G.M. 1995. Geomorphological stream channel classification in aquatic habitat restoration: Uses and limitations. Aquatic Conservation: Freshwater and Marine Systems **5**:127–141.

Kondolf, G.M., and M. Larson. 1995. Historical channel analysis and its application to riparian and aquatic habitat restoration. Aquatic Conservation: Freshwater and Marine Systems **5**:109–126.

Kondolf, G.M., J.C. Vick, and T.M. Ramirez. 1996. Salmon spawning habitat rehabilitation on the Merced River, California: An evaluation of project planning and performance. Transactions of the American Fisheries Society **125**:899–912.

Li, H.W., K. Currens, D. Bottom, S. Clarke, J. Dambacher, C. Frissell, et al. 1995. Safe havens: Refuges and evolutionarily significant units. American Fisheries Society Symposium **17**:371–380.

Lichatowich, J., and S. Cramer. 1979. Parameter selection and sample sizes in studies of anadromous salmonids. Information Report Series, Fisheries Number 80-1. Oregon Department of Fish and Wildlife, Portland, Oregon, USA.

Lisle, T.E. 1981. Channel recovery from recent large floods in north coastal California: Rates and processes. Pages 153–160 *in* R.L. Coats, ed. Watershed rehabilitation in Redwood National Park and other Pacific coastal areas. Center for Natural Resource Studies, JMI and National Park Service, Arcata, California, USA.

——. 1982. Effects of aggradation and degradation on riffle-pool morphology in natural gravel channels, northwestern California. Water Resources Research **18**:1643–1651.

Ludwig, D., R. Hilborn, and C. Walters. 1993. Uncertainty, resource exploitation, and conservation: Lessons from history. Science **260**:17–36.

MacDonald, L.H., A.W. Smart, and R.C. Wissmar. 1991. Monitoring guidelines to evaluate effects of forestry activities on streams in the Pacific Northwest and Alaska. EPA/910/9-91-001. US Environmental Protection Agency Region 10, Seattle, Washington, USA.

Meador, M.R., T.F. Cuffney, and M.E. Gurtz. 1993. Methods for sampling fish communities as part of the national Water-Quality Assessment Program. Open-file Report 93-104. United States Geological Survey, Water Resources Division, Raleigh, North Carolina, USA.

Miller, D.L., et al. 1988. Regional applications of an index of biotic integrity for use in water resource management. Fisheries **13**:12–20.

Minns, C.K., J.R.M. Kelso, and R.G. Randall. 1996. Detecting the response of fish to habitat alterations in freshwater ecosystems. Canadian Journal of Fisheries and Aquatic Sciences **53**(Supplement 1):403–414.

Montgomery, D.R. 1995. Input- and output-oriented approaches to implementing ecosystem management. Environmental Management **19**:183–188.

Montgomery, D.R., G.E. Grant, and K. Sullivan. 1995. Watershed analysis as a framework for implementing ecosystem management. Water Resources Bulletin **31**:369–386.

Naiman, R.J., T.J. Beechie, L.E. Benda, D.R. Berg, P.A. Bisson, L.H. MacDonald, et al. 1992. Fundamental elements of ecologically healthy watersheds in the Pacific Northwest coastal ecoregion. Pages 127–188 *in* R.J. Naiman, ed. Watershed management: Balancing sustainability and environmental change. Springer-Verlag, New York, New York, USA.

Nawa, R.K., and C.A. Frissell. 1993. Measuring scour and fill of gravel streambeds with scour chains and sliding-bead monitors. North American Journal of Fisheries Management **13**:634–639.

Nehlsen, W., J.E. Williams, and J.A. Lichatowich. 1991. Salmon at the crossroads: Stocks at risk from California, Oregon, Idaho, and Washington. Fisheries **16**(2):4.

NRC (National Research Council). 1992. Restoration of aquatic ecosystems: Science, technology, and public policy. National Academy Press, Washington, DC, USA.

Olson-Rutz, K.M., and C.B. Marlow. 1992. Analysis and interpretation of stream channel cross-sectional data. North American Journal of Fisheries Management **12**:55–61.

Parry, B.L., C.M. Rozen, and G.A. Seaman. 1994. Restoration and enhancement of aquatic habitats in Alaska: Case study reports, policy guidance, and recommendations. Technical Report 94-3, Habitat Restoration Division, Alaska Department of Fish and Game, Juneau, Alaska, USA.

Peterson, N.P. 1982. Immigration of juvenile coho salmon (*Oncorhynchus kisutch*) into riverine ponds. Canadian Journal of Fisheries and Aquatic Sciences **39**:1308–1310.

Peterson, N., A. Hendry, and T.P. Quinn. 1992. Assessment of cumulative effects on salmonid habitat: Some suggested parameters and target conditions. Report number TFW-F3-92-001 prepared for the Washington Department of Natural

Resources, Timber, Fish, and Wildlife, Olympia, Washington, USA.

Pflakin, J.L., M.T. Barbour, K.D. Porter, S.K. Gross, and R.M. Hughes. 1989. Rapid bioassessment protocols for use in streams and rivers: Benthic macroinvertebrates and fish. Document Number EPA/444/4-89-001. US Environmental Protection Agency, Office of Water, Assessment and Watershed Protection Division, Washington, DC, USA.

Platts, W.S., C. Armour, G.D. Booth, M. Bryant, J.L. Bufford, P. Cuplin, et al. 1987. Methods for evaluating riparian habitats with applications to management. USDA Forest Service General Technical Report INT-221, Intermountain Research Station, Ogden, Utah, USA.

Platts, W.S., W.F. Megahan, and G.W. Minshall. 1983. Methods for evaluating stream, riparian, and biotic conditions. USDA Forest Service General Technical Report INT-183, Intermountain Research Station, Ogden, Utah, USA.

Platts, W.S., and R.L. Nelson. 1985. Stream habitat and fisheries response to livestock grazing and instream improvement structures, Big Creek, Utah. Journal of Soil and Water Conservation **40**:374–379.

Platts, W.S., and J.N. Rinne. 1985. Riparian and stream enhancement management and research in the Rocky Mountains. North American Journal of Fisheries Management **5**:115–125.

Platts, W.S., R.J. Torquemada, M.L. McHenry, and C.K. Graham. 1989. Changes in salmon spawning and rearing habitat from increased delivery of fine sediment to the South Fork Salmon River, Idaho. Transactions of the American Fisheries Society, **118**:629–645.

Poizat, G., and D. Pont. 1997. Multi-scale species-habitat relationships: Juvenile fish in a large river section. Freshwater Biology **36**:611–622.

Poole G.C., C.A. Frissell, and S.C. Ralph. 1997. In-stream habitat unit classification: Inadequacies for monitoring and some consequences for management. Journal of the American Water Resources Association **33**:879–896.

Reeves, G.H., D.B. Houer, B.E. Hansen, F.H. Everest, J.R. Sedell, T.C. Hickmon, et al. 1997. Fish habitat, restoration in the Pacific Northwest: Fish Creek of Oregon. Pages 335–359 *in* J.E. Williams, C.A. Wood, and M.P. Donbeck, eds. Watershed restoration: Principles and practices. American Fisheries Society, Bethesda, MD.

Reeves, G.H., J.D. Hall, T.D. Roelofs, T.L. Hickman, and C.O. Baker. 1991. Rehabilitating and modifying stream habitats. Pages 519–557 *in* W.R. Meehan, ed. Influences of forest and rangeland management on salmonid fishes and their habitats. American Fisheries Society Special Publication 19, Bethesda, Maryland, USA.

Regier, H.A., R.L. Welcomme, R.J. Steedman, and H.F. Henderson. 1989. Rehabilitation of degraded river ecosystems. Pages 86–97 *in* D.P. Dodge, ed. Proceedings of the International Large River Symposium. Canadian Special Publication in Fisheries and Aquatic Sciences 106.

Richards, C., P.J. Cernera, M.P. Ramey, and D.W. Reiser. 1992. Development of off-channel habitats for use by juvenile chinook salmon. North American Journal of Fisheries Management **12**:721–727.

Rinne, J.N., and P.R. Turner. 1991. Reclamation and alteration as management techniques, and a review of methodology in stream reclamation. Pages 219–244 *in* W.L. Minckley and J.E. Deacon, eds. Battle against extinction: Native fish management in the American west. University of Arizona Press, Tucson, Arizona, USA.

Rosgen, D.L. 1988. Conversion of a braided river pattern to meandering: A landmark restoration project. Presented at the California Riparian Systems Conference, September 22–24, Davis, California, USA.

——. 1994. A classification of natural rivers. Catena **22**:169–199.

——. 1996. Applied River Morphology. Wildland Hydrology, Pagosa Springs, Colorado, USA.

Roth, N.E., J.D. Allan, and D.L. Erickson. 1996. Landscape influences on stream biotic integrity assessed at multiple spatial scales. Landscape Ecology **11**:141–156.

Sear, D.A. 1994. River restoration and geomorphology. Aquatic Conservation: Freshwater and Marine Ecosystems **4**:169–177.

Sedell, J.R., G.H. Reeves, F.R. Hauer, J.A. Stanford, and C.P. Hawkins. 1990. Role of refugia in modern fragmented and disconnected river systems. Environmental Management **14**:711–724.

Spreiter, T. 1992. Redwood National Park watershed restoration manual. Redwood National Park, Watershed Restoration Program, Orick, California, USA.

Stanford, J.A., and G.C. Poole. 1996. A protocol for ecosystem management. Ecological Applications **6**:741–744.

Stanford, J.A., J.V. Ward, W.J. Liss, C.A. Frissell, R.N. Williams, J.A. Lichatowich, et al. 1996. A general protocol for restoration of regulated rivers. Regulated Rivers: Research and Management **12**:391–413.

Thomas, J.W. 1996. Forest Service perspective on ecosystem management. Ecological Applications **6**:703–705.

Walters, C., R.D. Goruk, and D. Radford. 1993. Rivers Inlet sockeye salmon: An experiment in adaptive management. North American Journal of Fisheries Management **13**:253–262.

Walters, C.J., and C.S. Holling. 1990. Large-scale management experiments: Learning by doing. Ecology **71**:2060–2068.

Walters, C.J., and R. Hilborn. 1978. Ecological optimization and adaptive management. Annual Review of Ecology and Systematics **1978**:157–188.

Weaver, W., and D. Hagans. 1995. Protecting and restoring salmonid watersheds from sediment inputs. Wild Fish and Forests, Spring 1995:6–7. Pacific Northwest Regional Office of The Wilderness Society, Seattle, Washington, USA.

Weaver, W.E., and D.K. Hagans. 1994. Handbook for farm and ranch roads: A guide for planning, designing, constructing, reconstructing, maintaining, and closing wildland roads. Mendocino County Resource Conservation District, Ukiah, California, USA.

Weaver, W.E., M.M. Hektner, D.K. Hagans, L.J. Reed, R.A. Sonneville, and G.J. Bundros. 1987. An evaluation of experimental rehabilitation work, Redwood National Park. Technical Report 19, United States National Park Service, Redwood National Park, Arcata, California, USA.

Wesche, T.A. 1985. Stream channel modifications and reclamation structures to enhance fish habitat. Pages 103–163 *in* J.A. Gore, ed. The restoration of rivers and streams: Theory and experience. Butterworth, Stoneham, Massachusetts, USA.

Williams, J.E., C.A. Wood, and M.P. Dombeck, eds. 1997. Watershed restoration: Principles and practices. American Fisheries Society, Bethesda, MD.

Wissmar, R.C., J.E. Smith, B.A. McIntosh, H.W. Li, G.H. Reeves, and J.R. Sedell. 1994. A history of resource use and disturbance in riverine basins of eastern Oregon and Washington (early 1800s–1900s). Northwest Science **68**:1–35.

Ziemer, R.R. 1997. Temporal and spatial scales. Pages 80–95 *in* J.E. Williams, C.A. Wood, and M.P. Dombeck, eds. Watershed restoration: Principles and practices. American Fisheries Society, Bethesda, MD.

25 Nonprofit Organizations and Watershed Management

Bettina von Hagen, Spencer Beebe, Peter Schoonmaker, and Erin Kellogg

Overview

- New economic and environmental challenges in watershed management require organizations that are as rich and complex as the problems they are designed to address. The growing nonprofit sector plays a vital role in the evolving institutional landscape of the Pacific coastal ecoregion.
- Two theories on the formation of nonprofit organizations prevail: they arise as a response to government failure, or they provide a needed and desired alternative to the private for-profit sector.
- As currently configured the government is not well adapted to successful management of natural resources. It is constrained by short political cycles and artificial political boundaries and it divides ecosystems into components (e.g., air, water, soil, and fish) and fails to recognize the critical links among them.
- The for-profit private sector has made natural resources decisions that are based incomplete information, do not factor environmental costs (externalities) into the market, and have had disastrous environmental consequences (e.g., the introduction of exotic species).
- The limitations of government and private business have led to dramatic growth in the nonprofit sector in the United States in the last three decades. Between 1972 and 1982, employment in the nonprofit sector grew twice as fast as that in the public and private sectors.
- Nongovernmental organizations (NGOs), the nonprofit equivalent in developing countries, have also flourished in the last few decades and offer valuable lessons to the Pacific coastal ecoregion. Conservation is seldom addressed without considering its effect on development options and taking into account human needs. As a result, credit, marketing, and technical assistance to encourage appropriate, conservation-based development is an important activity of NGOs in developing countries.
- Nonprofit organizations serve four key functions in watershed management: building institutional capacity, providing access to information, restoring degraded watersheds, and promoting market incentives for conservation.
- Institutions that have traditionally sustained small communities in the Pacific coastal ecoregion, increasingly find themselves unable to adapt to changing economic and environmental circumstances. Many of these communities are creating new, nonprofit organizations to provide the leadership needed to chart a more promising course for the future. Nonprofit organizations provide a critical link between professional scientists and local communities, and help community members access information and build the knowledge base needed for long-term community and ecosystem vitality.
- Throughout the Pacific coastal ecoregion, nonprofit organizations have been particularly successful in restoring degraded watersheds, either by developing regional restoration

strategies or by physically restoring individual streams and watersheds.

• Several nonprofit organizations in the Pacific coastal ecoregion are using innovative strategies, such as environmental banking and the creation of markets for environmental services (e.g., water purification, carbon storage) to encourage watershed conservation and restoration.

• Nonprofit organizations combine several of the advantages of the private for-profit and public sectors, and have proven effective in promoting watershed conservation and restoration in the region. The most effective and enduring strategies for watershed restoration and management emerge when the lines between profit, nonprofit, public, and private sectors blur. Nonprofit organizations in the Pacific coastal ecoregion have pioneered these kinds of collaborations, and will play an increasing role in catalyzing the adaptive institutional landscape that is needed to address watershed management in the region.

Introduction

In policy and management circles and in the increasingly acerbic debate between liberals and conservatives, the relative merits of public and private resource management have long been discussed. Conservatives point to bureaucratic entanglements, economic inefficiency, and mismanagement in the public sector as a rationale for privatization. Liberals point to corporate abuses and the inequity of wealth distribution as arguments for a strong public sector. Often ignored in the debate is the flourishing of a third approach—the private, nonprofit sector. Nonprofits can use a private, entrepreneurial approach to the production of public goods, giving rise to flexibility and adaptability, rapid decision making, and more effective collaboration among disciplines (Livernash 1992). These qualities are particularly important in managing human activities in watersheds and in other natural resource arenas, where knowledge is limited and complexity and uncertainty are especially prevalent.

Lee (1992) concluded, "What we do know with certainty is that sustainable watershed management begins by building ecologically effective human organizations." Sustainable management requires the simultaneous consideration of economic, social and ecological dimensions, which is only possible through broad bottom-up participation guided by extensive local knowledge. Large organizations, whether public or private, tend to erect barriers to the flow of information that lead to suboptimal decision making (Lee 1992, Chapter 21, this volume). In addition, large organizations seek economies of scale. They favor a strategy which seeks uniformity and common denominators across resources, rather than a strategy which seeks the highest value for each resource used. This former approach requires less investment in information and unique skills and a greater investment in systems which maximize uniformity and economic efficiency. As a consequence, highly skilled approaches, which often require subjective judgment and creativity and add economic value to resources, are not favored in either the production of private or public goods by large organizations.

This chapter addresses the role of small, private, nonprofit organizations in the management of watersheds, and the streams and rivers they comprise. It begins by reviewing theories of nonprofit formation and considering their applicability in watersheds of the Pacific coastal ecoregion. The emergence of nongovernmental organizations (NGOs) in the developing world and the increasing adoption of "southern" (developing-country) NGO approaches by "northern" (developed world) nonprofits is explored. Finally, profiles of some emerging nonprofit organizations in the Pacific coastal ecoregion and predictions about the future institutional landscape are presented.

Theories of Nonprofit Formation

Why do nonprofits exist at all in democratic societies with stable governments and well-functioning markets? There are two primary theories: nonprofits arise as a response to government failure (Weisbrod 1988), and/or nonprofits provide a needed and desired alternative to the private for-profit sector (Hansmann 1986).

The public sector's primary role is generally perceived as a supplier of public goods, defined as goods which are fully accessible to all and in which one person's consumption does not diminish the amount available for others (Tietenberg 1992). In theory, the political process reveals demand for public goods such as education, biodiversity, clean water, and national defense. The government then has the power to tax and regulate, to assure the production of social goods, and to limit the production of social "ills," or socially detrimental goods (Meyer 1992). However, political officials may be much more concerned with reelection and personal benefit seeking than with the social good. In addition, the politically determined level of provision of a public good may leave some citizens unsatisfied (Meyer 1992). Finally, government provision of goods is a function of relatively short political cycles, leading to an inconsistent and potentially unreliable provision of services. Unsatisfied consumers may thus turn to the nonprofit sector for adequate provision of the public good they desire. Nonprofits can also correct monopolies of power caused by the alliance of government and the moneyed elite by providing a countervailing force, organizing disadvantaged citizens, and providing diversity of opinion (Ware 1989, Brinkerhoff and Goldsmith 1992).

Those who view nonprofit formation as a response to market failure argue that for-profit firms have a clear incentive to meet consumer demands at minimal cost. Although this ensures efficiency, it also provides an incentive to sacrifice quality when the quality of output is hard to measure, such as the restoration of watersheds, or when the potential for "contract failure" (e.g., the failure to deliver promised goods in the future) exists (Hansmann 1986). In addition, for-profit firms tend to serve the more lucrative part of the market and to externalize costs. Because nonprofits may not by definition distribute profits, they are viewed as more trustworthy in providing goods in which the quality is not readily apparent. Finally, nonprofits play a significant role in serving underdeveloped markets and markets that produce both public and private goods. For-profit firms tend to invest less in markets where a large part of the output is for public benefit, as they cannot capture the full benefits of their investment. For example, long-rotation, diverse forests produce not only timber for private benefit but also wildlife habitat and water storage and purification, among other public goods, and are chronically undersupplied by for-profit firms.

The government and market failure theories described above are essentially demand-side models, suggesting nonprofits are formed because of consumer demand. Rose-Ackerman (1986) argues that supply-side forces are also at work, for example in the case of motivated non-profit entrepreneurs. Entrepreneurs create new organizations and have the capacity to transfer their personal values and motivations to the new organization, which are sometimes better reflected in a nonprofit rather than a for-profit structure (Young 1986). These values shape the organization's outputs and clientele and further diversify consumers' and investors' options (Rose-Ackerman 1986).

The Limits of Government

Society has come to understand, somewhat belatedly, that policies and programs that address individual natural resources (such as streams or watersheds) within arbitrarily established political boundaries do not always serve it well. Air, water, soils, climate, microorganisms, vegetation, fish, and wildlife exist in complex relationships which resist simplified and fragmented political approaches. President Nixon noted this interdependence in an address to Congress in 1970, "Despite its complexity, for pollution control purposes the environment must be perceived as a single interrelated system . . . A single source may pollute the air with smoke and chemicals, the land with solid wastes, and a river or lake with chemicals and other wastes. Control of the air pollution may produce more solid wastes, which then pollutes the land or water" (Nixon 1970). Nevertheless, the Environmental Protection Agency (EPA), created in the same year, had and continues to have separate programs for air, water, and waste. Subsequent legislation has continued to focus on individual programs rather than comprehensive management (John 1994). Individual state, county and municipal programs have generally mirrored this fragmented approach. The fragmentation is not only thematic but also geographic. Agencies responsible for managing the use of natural resources are organized around political boundaries, which are seldom relevant to the natural resources being considered.

Effective policies and programs generally require time and patience—both of which the political process does not necessarily foster. With each political cycle, commitments are reexamined and wide swings in program orientation and funding can and do occur. Thus with much fanfare, the Oregon legislature launched the Watershed Health Program in 1993, an innovative program that catalyzed the formation of local watershed councils and provided over US$12 million in funding and technical assistance for watershed restoration projects (Oregon House Bill 2215, Oregon Senate Bill 81). In the next legislative session in 1995, the program was severely reduced to about 20% of the prior session's funding largely because of the challenges of political cycles, bureaucracy, and political entanglements (Gregory, Daily Courier, Grants Pass, Oregon, August 6, 1994; Roler, Daily Courier, Grants Pass, Oregon, August 9, 1994; Soscia 1996). Successful environmental initiatives are rarely achievable in a two- or even four- or six-year cycle, and the cycle of premature birth and early death of programs is highly inefficient.

The current salmon crisis in the Pacific Northwest highlights the critical nature and the absurdity of this thematic, temporal, and geographic fragmentation. Anadromous Pacific salmon (*Oncorhynchus* spp.) spawn in fresh water, migrate to the sea, and return to their natal streams to reproduce, and, for most species, to die. For some populations, such as those in central Idaho, this can mean a one-way migration of more than 1,500 km, crossing many geographic and political boundaries in the process (National Research Council 1995). Pacific salmon populations are declining throughout most of their North American range, victims of a complex array of forces, including the effects of logging, farming, grazing, overfishing, urbanization, dams, industrial activities, hatcheries and ocean conditions (Nehlsen and Lichatowich 1996). Prospects for salmon recovery are dim: "unless agencies cooperate more effectively, salmon populations are unlikely to recover . . . one problem is that current institutions and the boundaries of their jurisdictions usually do not match the spatial, temporal, or functional scales of the salmon problem" (National Research Council 1995).

Adaptive management has emerged as one of the public sector's more innovative responses to environmental decision making. Recent environmental crises, from the massive flooding of Pacific Northwest rivers to the precipitous decline in the biotic integrity of stream systems (Karr 1992), underscore the difficulty of predicting and controlling ecosystem responses, despite good science and engineering. "Ecosystems are not only more complex than we think, but more complex than we *can* think" Egler (1977), and the implications of this truth are finally becoming clear. Adaptive management recognizes that since human understanding of nature is imperfect, human interactions with nature should be experimental and monitored (Lee 1993). Adaptive management calls for accelerated learning through an interdisciplinary team that posits explicit hypotheses about how systems function and tests these hypotheses through systematic monitoring and evaluation. Although compelling in theory, practice has been more difficult, in part because public agencies interpret experimental failure not as an opportunity to learn, but as a manager's inability to perform (Chapter 21, this volume, Lee 1992).

Thus government, as currently configured, is not well adapted to successful management of natural resource issues. Where long time scales are critical, the public sector is hostage to short political cycles. Where watersheds and bioregions are the natural and most effective decision-making unit, governments are constrained to artificial political boundaries. While ecosystems are interdependent and complex, political administration generally divides ecosystems into components—air, water, soil, fish, wildlife—and fails to recognize the critical links among them. Even explicit decision-making systems, such as adaptive management, which recognize the role of uncertainty, complexity and learning in "managing" ecosystems, seem incongruent with the rewards and appetite for risk-taking and innovation in the public sector.

The Hidden Costs of Profit-Seeking

Although conservation has been largely the domain of the public sector in the United States, development has been the province of the private sector. Economic development is undertaken to produce primarily private goods, which are traded for the mutual benefit of the buyer and seller. Prevailing theory holds that buyers and sellers of goods and services, acting individually, create a collective marketplace in which goods are allocated so efficiently in response to supply and demand that it appears that an invisible hand is at work. The theory of perfect, efficient markets rests on some key premises. First, efficient markets are based on the presumption that both the buyer and the seller have complete and timely information on the goods or services being traded. History is replete with examples of natural resource decisions, made on the basis of incomplete information, which had disastrous, unforeseen consequences. For example, the Mysis (*Mysis* sp.) shrimp introduced into Flathead Lake, Montana, as a food source for Kokanee salmon (*O. nerka*) instead competed with young salmon for zooplankton. The resulting decline in salmon then caused the local bald eagle (*Haliaeetus leucocephlus*) population to decline (Spencer et al. 1991). Similar though less disastrous decisions in the Pacific coastal ecoregion include the fertilization of coastal lakes in British Columbia resulting in increases in stickleback (*Gasterosteus aculeatus*) which outcompeted the target sockeye salmon populations (*O. nerka*) (Webb et al. 1993).

Second, efficient market theory presumes that the buyer and the seller bear all the costs and receive all the benefits of the traded goods. In decisions concerning watersheds in particular, which operate as complex, interdependent ecosystems, this condition is rarely met. Instead, a farmer's decision to irrigate heavily may affect not only the purchasers of his or her produce, but also the fishing industry due to reduced survival of salmon smolts. A decision by a wood products company to log its property may result in costs to a downstream city, which must now filter its water due to increased siltation. The costs of suburban development are borne not only by the buyers of houses in the development, but also by displaced wildlife and by existing residents who experience increased congestion. In order for efficient, mutually beneficial allocation of goods to occur, all these so-called "external" costs, or "externalities," would have to be paid by the buyer or the seller. If they are not, efficient market theory suggests that goods which contain high external costs (costs not paid by the buyer or the seller) will be over-supplied, resulting in high social and environmental costs.

The Emergence of Nonprofits

Clearly, both the public and the for-profit private sector are limited in making decisions and allocating goods. These limitations, magnified by the recognition that environmental limits are increasingly inescapable, have given rise to an explosive growth in the private, nonprofit sector. Alternatively called the voluntary, third, intermediate, or independent sector, the nonprofit sector is made up of private organizations in which profits legally may not be distributed and income is exempt from taxes (Weisbrod 1988). In the United States, nonprofits are divided into noncharitable organizations, which pursue activities that primarily benefit their own members, and charitable organizations, which serve religious, charitable, scientific, literary, or educational purposes (Ben-Ner and Van Hoomissen 1993). Charitable organizations are governed by the Internal Revenue Code section 501(c)(3), which establishes operating and organizational criteria that qualifying nonprofits must meet, in addition to providing a public benefit. Charitable organizations which meet the 501(c)(3) criteria qualify for

TABLE 25.1. Growth of United States nonprofit sector by number of applications to the IRS for tax exempt status.

Year	Number of applications
1956	5,000 to 7,000
1965	13,000
1985	45,000

Based on Weisbrod 1988.

TABLE 25.2. Growth of United States nonprofit sector by financial assets.

Year	Assets in billions
1975	$280
1987	$500

Based on O'Neill 1989.

additional benefits including the tax deductibility of donations, subsidized interest rates on borrowing through the issuance of tax-free bonds, and lower postal rates. It is difficult to track the actual number and types of nonprofit organizations in the United States, but it is clear that this sector has grown dramatically (Table 25.1 and Table 25.2). Based on tax status figures, the number of nonprofit 501(c)(3) organizations in the United States has nearly tripled, from 137,487 in 1968 to 366,071 in 1985 (Ware 1989). By another estimate, the number of philanthropic nonprofit organizations climbed beyond one million in 1986 (Hodgkinson and Weitzman 1986). Between 1972 and 1982 employment in the nonprofit sector grew twice as fast as it did in the public and private sectors and reached about seven percent of the total workforce by 1986 (Hodgkinson and Weitzman 1986, O'Neill 1989). In Canada the nonprofit, or charitable society sector, has not grown as dramatically as it has in the United States; most likely because the Canadian government provides more social services than that of the United States government.

Nonprofit organizations combine the advantages, and some of the disadvantages, of both the public and private sectors, while offering some unique benefits and constraints (Table 25.3). First, like private for-profits, nonprofits are privately organized and relatively independent of political influence and cycles. (Al-

TABLE 25.3. Relative advantages and disadvantages of the private for-profit, public, and nonprofit sectors in the United States.

Organization	Advantage	Disadvantage
Private for-profit	• Is privately organized • Operates independently of political influence and cycles • Is innovative, flexible, and responsive • Tends to be efficient	• Decisions are overshadowed by private and monetary gain • Externalizes social and environmental costs of production in order to capture the full benefits of investments
Public	• Can use Adaptive Management approach • Has no incentive to shift cost for private gain • Has power to tax and regulate	• Is fragmented • Operates within artificial/political boundaries • Is constrained by short political cycles • Lacks incentives to minimize cost • Tends to be inefficient • Lacks initiative for risk-taking
Nonprofit	• Lacks incentive to shift cost for private gain • Lacks legal or cultural barriers for forming alliances with public, corporate, and individual interests • Serves underdeveloped markets • Operates relatively independent of political influence and cycles • Can be innovative, flexible, and responsive • Uses volunteer labor	• Is subject to the "free rider" problem • Cannot force/impose action • Lacks incentive to minimize cost • Tends to be inefficient • Lacks accountability

though if non-profits arise in response to a government failure that is addressed by a subsequent administration, or they depend on government funding, then they will be affected.) Nonprofits are also free of the centralized bureaucracy of government, leading to innovation, flexibility, and responsiveness (Livernash 1992). In addition, nonprofits benefit from volunteer labor, which can be significant—in 1985, the nonprofit sector garnered 80% of all volunteer labor totaling 5.4 million full-time workers (Weisbrod 1988). Since nonprofits, like the public sector, do not distribute profits, they have less incentive to shift costs for private gain (Hansmann 1986). However, also like the public sector, the lack of a profit-maximizing objective can translate into a lack of incentive to minimize costs and lead to inefficiencies (Meyer 1992). This tendency towards inefficiency is generally less pronounced in the nonprofit sector than in the public sector, as the desire for continuing support of donors can provide almost as powerful an incentive as profit distribution. Finally, relative to the public sector, nonprofits do not have the power to tax or to regulate, leading to the "free rider" problem in which consumers may desire and benefit from a service, such as clean water, but are not obliged to pay for it.

Some would view the inability of nonprofits to force or impose action as severely limiting. In the environmental arena in particular, "command and control" approaches have been the preferred approach since the early 1970s and have been effective with many environmental problems. However, a new era of environmental policy is emerging, one that recognizes the interdependence of species and habitats, the complexity of ecosystems, the lack of understanding of even fundamental processes, and the difficulty of predicting change. In this new environment, nonregulatory tools, such as information and incentives, complement traditional approaches and can be more effective than sanctions (John 1994). John (1994) calls this approach "civic environmentalism," a collaborative, integrative approach in which there is bargaining among a diverse set of participants resulting in voluntary action.

Adapting Developing World Strategies to the Pacific Coastal Ecoregion

In the developing world, voluntary organizations are generally referred to as nongovernmental organizations (NGOs), rather than nonprofits, because they emerged largely as an alternative to government programs and resources. In addition, the tax status of NGOs varies from country to country, so the definition of NGOs as privately established and governed organizations which are supported by voluntary contributions for a stated philanthropic purpose is more universally applicable (Meyer 1993). Livernash (1992) defines NGOs in terms of their common visions and beliefs: the reduction of poverty, sustainable development, local participation and control, and bottom-up approaches.

Much of the explosion in NGO formation over the past two decades is attributed to growing interest and availability of funds in developed countries for sustainable development and resource conservation. Nonetheless there has been an equally significant explosion in locally created indigenous organizations that are not reliant on external funding. As a result of some of the inherent advantages of NGOs, such as the ability to innovate and act quickly because of their smaller size and independence, as well as the fact that many developing-country governments were in default or potential default of "northern" debt, donor funds have been channeled to NGOs. Structural adjustment mandates following renegotiation of developing world debt has further curtailed public sector development activities, as did the growing infatuation with privatization (Meyer 1992, Meyer 1993). These forces are reflected in government policy in the United States In 1981, Congress passed legislation to expand development assistance funding through NGOs, resulting in substantial government funding to United States-based NGOs for developing-country programs (Meyer 1993). In Britain, 15% of the foreign aid budget of 1.7 billion British pounds is now channeled through NGOs (Brett 1993). The Organization for Economic Cooperation and Development

(OECD) countries contributed $2.5 billion of official development aid through NGOs in 1992 (Meyer 1993).

The dynamics and evolution of NGOs in the developing world are relevant and potentially instructive to nonprofit activity in watersheds in the United States and Canada. First, conservation is seldom addressed by developing world NGOs without considering its effect on development options and alternatives. The seemingly obvious premise that successful conservation and development are mutually dependent is adopted out of necessity. With relatively higher population density, higher poverty, and a greater direct reliance on natural ecosystems for food, fiber, and fuel, protected area strategies will not succeed unless human needs are met. While the context is different in the United States, the premise remains the same—successful long-term conservation must take into account human needs, especially the needs of local communities in and around the conservation areas.

Second, southern (developing-country) NGOs have played a role in providing credit, marketing and technical assistance to communities in areas with underdeveloped markets to spur appropriate, conservation-based development. Rural, coastal watersheds in the United States share many of these same barriers to development, including lack of access to information, transportation barriers, poor access to credit, and concentration of land and resource ownership in the hands of large, external interests. Adapting southern NGO strategies and developing local responses to address these market failures is an increasing focus of nonprofits in developed countries, especially in the Pacific coastal ecoregion.

The Role of Nonprofits in Watershed Management

In the following section, some of the emerging nonprofit organizations working in the Pacific coastal ecoregion of the United States and Canada are profiled, and the role of these nonprofits in addressing the limitations of the public and private for-profit sectors is examined. Nonprofit organizations serve four key functions in watershed management: building institutional capacity, providing access to information, restoring degraded watersheds, and promoting market incentives for conservation. Described first, however, is the socioeconomic climate of the Pacific coastal ecoregion that gave rise to these nonprofits.

Social and Economic Aspects of the Pacific Coastal Ecoregion

The abundant natural resources of the Pacific coastal ecoregion first attracted Native Americans, then Euro-American explorers and settlers, and finally corporate extractive interests and more recently, recreationists (Schoonmaker et al. 1996).One of the bitter ironies of the Pacific coastal ecoregion is the contrast between its rich natural productivity and diversity and the condition of many of its human communities. In general, Pacific coastal forest communities tend to experience higher unemployment, lower income and higher rates of emigration of youth than communities outside the region (Radtke et al. 1996). In part, this simply reflects larger trends. Greater scale is now needed to attract and retain many businesses and decision-making is increasingly centered in urban areas. However, it also reflects the fact that ownership of resources in the region has been highly concentrated in the hands of a few outside owners. For example, a recent study of land ownership in Oregon coastal counties revealed that the largest ten land owners owned 76% of total private forest land (Table 25.4) (Willer 1995). This ownership pattern is repeated throughout coastal California, Oregon, and Washington, while in British Columbia and Alaska virtual monopolies have been granted to a handful of transnational timber companies. Although some outside owners are responsive to community concerns, they are nevertheless primarily responsible to outside shareholders who generally do not have long-term commitments to the region. This lack of long-term committment is reflected in a

TABLE 25.4. Percent of forest land owned by the ten largest private forest land owners in counties of coastal Oregon.

County	Total area (hectares)	Portion owned by large landowners	
		Area	Percentage
Clatsop	110,106	103,208	94%
Tillamook	74,643	59,586	80%
Lincoln	135,569	113,321	84%
W. Lane	106,058	66,599	63%
W. Douglas	200,190	166,603	82%
Coos	190,959	129,774	68%
Benton	72,685	48,919	67%
Polk	79,629	65,575	82%
Yamhill	54,021	29,675	55%
Washington	68,700	40,623	59%
Columbia	117,153	90,508	77%
Totals	1,209,714	914,391	75.5%

 Based on Willer 1995.

variety of ways. For example, a study of forest-dependent communities in California found that outside ownership was correlated with higher levels of poverty (Kusel and Fortmann 1991). It is also reflected in an economic structure geared more towards resource extraction for export, with low value retained locally from the extracted goods, rather than an economy structured for long-term community vitality and ecosystem health.

Nonprofits Building Institutional Capacity

The overarching role of nonprofit organizations in watershed management is that of building, or at a minimum enriching, the strength and diversity of local institutions. Small community-based organizations like The Willapa Alliance in southwestern Washington, the Institute for Watershed Arts and Sciences on the Oregon coast, the Mattole Restoration Council in California, and many of the watershed councils that are forming up and down the coast provide the needed structure and forum for addressing related environmental and economic problems. From schools to hospitals, from the Lions Club to economic development councils, the institutions that have traditionally sustained small communities along the coast of California, Oregon, Washington, British Columbia, and Alaska find themselves increasingly unable to meet the demands of changing economic and environmental circumstances. Recognizing that government does not have the resources to solve their problems, these communities are looking inward, and creating new, nonprofit organizations that are able to bring the community together to set goals and chart a more promising course for the future, and more importantly, to provide the leadership to affect the changes that are critical to this course. These organizations have a board and staff (whether paid or unpaid) and the basic institutional structure to form partnerships with private and public sectors in order to capture resources and carry out programs.

The Willapa Alliance, in southwestern Washington provides a good example of an organization created by community members to facilitate the long-term conservation and appropriate development of the Willapa watershed. Founded in 1992, the Alliance is composed of members throughout the 275,000-ha watershed—oyster growers, fishers, farmers, ranchers, forest land owners, physicians, and other business owners, as well as tribal representatives—and has raised close to a million dollars for programs in natural resource management, science and information, and civic involvement. The Alliance recently convened almost one hundred community leaders in Pacific County (which roughly mirrors the Willapa watershed boundary) to discuss critical watershed issues, and collectively assess and prioritize the economic, social, and environmental issues the community would need to address as a whole in the coming years in order to ensure community well-being.

Building the basic local institutional capacity in watersheds along the Pacific coast of North America is fundamental to delivering the other services of nonprofit organizations—providing access to information, restoring degraded watersheds, and promoting market incentives for watershed conservation.

Nonprofits Providing Access to Information

Nonprofit organizations play an increasingly important role in conducting critical research and providing access to information. In many ways NGOs provide a link between professional scientists and local communities. The distinction between the local amateur and the professional scientist is actually fairly recent. Less than two hundred years ago the amateur scientist was the rule, often organized into local scientific associations; the professional was the exception.

The industrial revolution and emergence of graduate degree programs began a trend toward professionalism and specialization in science, resulting in a tacit contract between society and professional scientists. Society supported scientists to make visible gains in theory and applications through private businesses, philanthropic organizations, and tenured appointments at colleges and universities.

This contract changed radically when the public sector emerged as a major financial supporter of basic and applied research. During the middle of the twentieth century, scientists were entrusted to set research priorities and allocate funding primarily through the peer-review process, and partially through the advisory role of scientific societies and ad hoc advisory boards. Today society is calling for more direct accountability from the science community; the contract is being renegotiated (Shannon and Antypas 1996).

Scientists have always been motivated by an ultimate goal of addressing society's needs, but proximate goals of advancing theory, publishing, and securing funding sometimes obscure the former. One result is that the needs of rural communities and their watersheds, where much natural resource science is conducted, can be misinterpreted or ignored. Worse yet, the information accumulated by local naturalists—fishers, farmers, foresters—can be underutilized, in part because some of it is anecdotal or untestable. This is illustrated in Willapa Bay, where agency and university scientists monitor and study certain phenomena—oyster condition, salmon escapement—often with the help of local volunteers, but not other conditions such as plankton and beach drift, which are monitored solely by local residents (Sayce 1996). One avenue for addressing this challenge is the Willapa Science Advisory Group, composed of local and regional scientists, which provides a forum for communication, priority setting and partnership development.

The integration of professional scientific pursuits with local knowledge accumulated by amateur scientists represents one of the most promising avenues for synthesizing knowledge and fulfilling society's current expectations from the scientific community. Nonprofit organizations represent a promising vehicle for traveling this new path.

Interrain Pacific (formerly Pacific GIS), a Portland-based nonprofit, was formed to foster understanding by explicitly combining scientific and local knowledge and by making this knowledge widely accessible. Geographic information systems (GIS) have been used to organize, analyze, and display information derived from both public agencies and local communities (Backus 1996). For example, Interrain Pacific is working with the Ahousaht Band Council on Vancouver Island, British Columbia, to develop a village-owned geographic information system designed to give the Ahousaht the capacity to better manage its fisheries. The system draws on resource information at the ministries as well as historical knowledge, far outstripping government knowledge of the fisheries resources of the Ahousaht territory, and strengthening the Ahousaht's voice in managing the resources in their territory (Backus 1996).

The Prince William Sound Science Center in Cordova, Alaska was formed by local fishers and scientists concerned about increasing demands on the Sound's natural resources. A need for an independent local source of credible scientific information was not being provided by the public sector. The Exxon Valdez oil spill in 1989 changed the direction of the fledgling Science Center from baseline research to exploring the changes in the Sound attributable to the oil spill. In 1992, a dramatic decline of pink salmon (*O. gorbuscha*) and herring (*Clupea harengus pallasi*), important commercial species, led to a blockade by fishers and the subsequent creation of a fisheries research planning group. The Science Center helped organize this planning group, involving scientists and non-scientists, to identify the key questions about the biological and physical processes of Prince William Sound. The presence of an independent, community-based research center allowed a credible research plan to emerge which had broad support from the local and the professional scientific community (Bird 1996).

Nonprofits Restoring Degraded Watersheds

Restoration of degraded watersheds and declining fisheries has also absorbed the attention of nonprofit organizations. The role of nonprofits has ranged from developing regional restoration strategies to physically restoring individual streams. For example, the Willapa Alliance recently released the first documents on the Willapa fisheries recovery strategy process (Figure 25.1). These reports were the culmination of a year-long effort of local fishers, corporate interests, local government and state agencies that the Willapa Alliance convened and guided from initial discussions to plan formulation. While newer and smaller than most of the group's membership, the Alliance was successful because of its nonprofit attributes—independence, flexibility, and the capacity to innovate.

Nonprofits have also been successful in spurring the restoration of individual rivers and salmon runs. In the Mattole watershed in northern California, residents have taken action to reverse the decline of the river's salmon runs. The Mattole Salmon Group began by raising fish in hatchboxes (in this remote 787.3-km^2 basin), and soon expanded to understanding how the watershed works. The Mattole Restoration Council, formed in 1986, began by creating maps comparing the watershed's old-growth cover in 1948 and 1988 and changes in channel configuration from 1942 to 1992 (Figure 25.2), and went on to complete a watershed assessment in 1989 and a recovery plan for the lower basin in 1995 (Mattole Restoration Council 1989, Mattole Restoration Council 1995). These comprehensive documents formed the basis for projects including habitat enhancement in streams, landslide stabilization, and reforestation (Zuckerman 1996).

Nonprofits Promoting Market Incentives for Watershed Conservation

There is a need and an opportunity for organizations to create and restore markets for long-term community stewardship. Two nonprofits, ShoreTrust Trading Group, in Ilwaco, Washington, and Pacific Forest Trust, in Boonville, California, address this need.

Shorebank Enterprise Pacific (formerly ShoreTrust Trading Group) resulted from a partnership between Ecotrust, a Portland-based nonprofit which fosters conservation-based development in the region, and Shorebank Corporation, a bank-holding company which attempts to revitalize disinvested communities. Shorebank Enterprise Pacific provides marketing and technical assistance and credit to local businesses which are socially and environmentally restorative. Table 25.5 summarizes principles and business practices of conservation-based development put forth by the Shorebank Enterprise Pacific. Shorebank Enterprise Pacific is the first component of a permanent development institution to bring much-needed entrepreneurial resources—information, credit, product brokering, venture capital—to the region. The goal of the initiative is to restore market forces and conditions so that local entrepreneurs, who have a long-term interest in the welfare of their ecosystem, may participate in an economy which supports their concerns (Hauth 1996, von Hagen and Kellogg 1996).

Pacific Forest Trust was created to restore, enhance and preserve forest lands in the coastal Pacific Northwest by removing the barriers that inhibit landowners' commitment to long-term forest stewardship. The challenge, in this region and beyond, is that the economic returns from forests have been focused almost exclusively on timber. Timber-focused management strategies often lead to a loss of the vast array of other products and environmental services which natural forests provide, such as edible mush-

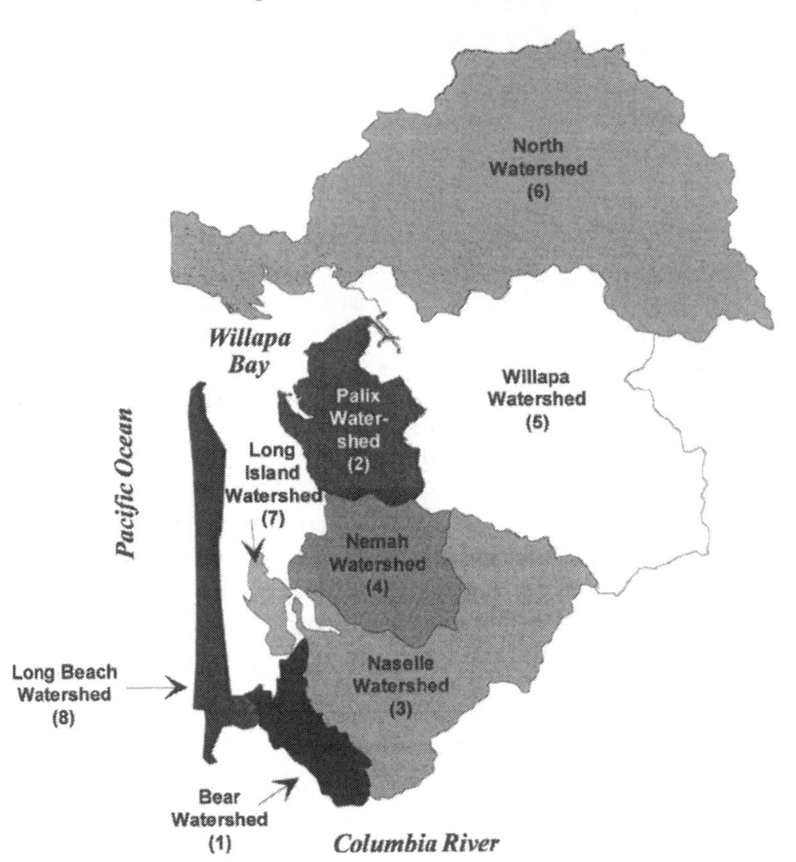

FIGURE 25.1. The Willapa Fisheries Recovery Team was convened by the nonprofit Willapa Alliance to develop a strategy for sustaining Willapa fish populations. The group devised a matrix for prioritizing subbasins in the Willapa watershed for restoration efforts (priority ranking from 1 to 8 is indicated in parentheses). The team determined the ranking using geographic information systems (GIS) analyses of salmon population status and ecological quality and risk which incorporated factors such as sedimentation, stream flows and temperature, spawning gravel abundance, water quality, mass wasting potential, riparian vegetation, and road to area ratios (map created by Interrain Pacific, Portland, Oregon).

rooms, medicinal plants, biodiversity, water purification, habitat, and carbon storage. Pacific Forest Trust attempts to create markets for these environmental services, so that landowners can be financially rewarded for their stewardship (Best 1996).

For example, recognizing the capacity of well-managed Pacific Northwest forests to store carbon and thus diminish the threat of global warming, Pacific Forest Trust markets the carbon storage capacity of private forest land. This capacity is marketed to carbon producers, such as utilities, which need to offset the carbon dioxide they release under EPA regulations. Computer models developed by Pacific Forest Trust suggest that landowners who manage for longer rotations (80–120 years) and use selective cutting can store significantly more carbon than those who manage on the 40 to 60 year cycles typical of current industrial forest

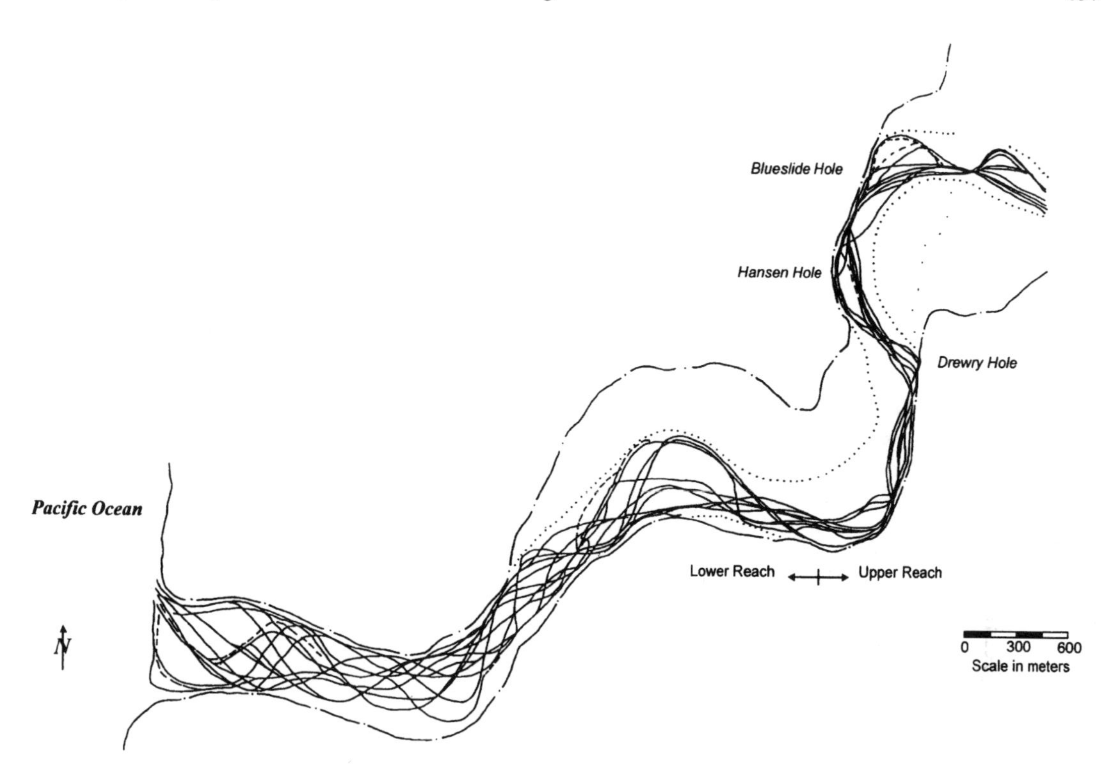

FIGURE 25.2. The Mattole Restoration Council in northern California has been carefully documenting changes in the watershed, such as the channel configuration shifts that occurred between 1942 and 1992 depicted here, in order to better understand and restore salmonid populations in the Mattole River. Key to line symbols: —— centerline of low-flow channel; - - - - prominent overflow channel; –·– valley wall [boundary between hillslope and valley bottom]; ······ meander belt; · · · · meander belt before levee constructions.

TABLE 25.5. Conservation-based development principles and business practices.

Principle	Business practice
Utilize and reduce waste.	• Do not allow emissions of waste to exceed the assimilative capacity of the ecosystem.
Save energy.	• Use appropriate energy sources and technology. • Apply the highest possible standards of energy efficiency.
Add value locally.	• Process products locally, rather than exporting unprocessed raw material.
Enhance human and biological productivity and diversity.	• Strive to balance the ethnic, gender, and age composition of employees, board members and investors. • Mobilize entrepreneurial energy of residents with modest financial resources. • Offer education and training opportunities. • Maximize production efficiency in preference to increasing use of raw materials. • Aim to restore native plant and animal species and communities. • Practice land and water management guided by natural patterns of disturbance. • Harvest resources only at a rate below natural regeneration. • Strive to reduce pollution, sedimentation, soil erosion, and the use of persistent chemicals.
Improve public and private community institutions and civic processes.	• Practice corporate citizenship that strengthens local government, schools, and community organizations. • Support local businesses and financial institutions.

Shorebank Enterprise Pacific, Ilwaco, Washington 1996a.

practices (Best 1996). Managing forests for longer rotations leaves more large woody debris in stream channels, a condition which is important for stream ecosystem processes (Chapter 13).

The Future of Nonprofit Organizations

Nonprofits are successful because they can make long-term commitments, they generally have no legal or cultural barriers restricting alliances with public, corporate or individual interests, and they can adapt their structure and governance to best accomplish the restoration strategy. When the State of Oregon attempted to address the drastic decline of salmon runs, it was largely through the formation of nonprofit watershed councils (Department of Water Resources 1993). Local nonprofit organizations offer one of the best hopes for enduring solutions which are relevant to local conditions and supported by local people. Nonprofit organizations that focus on watersheds and rivers range from the local to the international, and vary widely in terms of the issues they address. Flyfishing clubs, trout protection societies, river protection groups, and general environmental groups often engage in collaborative programs that cross jurisdictional boundaries and better reflect the ecological contours of the issue they address. Collaboration may be especially important for small nonprofits that need to pool resources, on big river systems like the Columbia where there are numerous and competing interests upstream and downstream of each other, and in watersheds facing common problems.

At the same time, nonprofits are not a panacea to society's ills in general or watershed health in particular. The nonprofit structure has its limitations, including loss of innovation as operations scale up, a tendency for potential contributors to "free ride" (receive benefits without contributing since nonprofits cannot assess taxes for providing public goods), a potential lack of incentives for efficiency, and a potential lack of accountability (Livernash 1992, Meyer 1992).

Nevertheless, nonprofits are a part of the solution which also includes more responsible corporate behavior, a more creative and efficient government, and an increased role for individual initiative and responsibility. Effective solutions emerge when the lines between profit, nonprofit, public and private blur, and organizations and partnerships emerge which adopt or combine the best attributes of each.

What are the appropriate roles and relationship between the nonprofit and the public sector? The general sense is that the nonprofit sector and the public sector are not substitutes for each other, but complements (Annis 1987, Korten 1987, Rudi Frantz 1987, Brett 1993, John 1994). Organizational capacity of government seems to set the stage for nonprofit proliferation. Where the state is strong, democratic, and prosperous, nonprofit organizations tend to be more numerous and important (Annis 1987). At the same time, the style of government can influence the success and relevance of nonprofit organizations. Korten (1987) suggests the optimal role of government is the creation of the human and institutional will and capacity to develop resources sustainably and equitably, rather than its current focus on transferring financial resources. In this people-centered approach to development, the role of government is to create a supportive national policy environment, while nonprofits play a key role as independent yet interlinked organizations through which people can define and pursue their individual and collective interests. Watershed councils organized at the state and provincial level have the potential to function in this manner, pursuing local projects but supporting common policy goals.

The general literature reveals less about potential relationships between the profit and nonprofit sectors. The literature focuses more on the unfair competition nonprofits provide in the provision of private goods (Ware 1989, Schiff and Weisbrod 1993) than on avenues of potential collaboration. Yet the last decade has seen increased "nonprofit behavior," in the form of environmental and social policies and programs by for-profit firms, largely in response to expanding green consumerism and socially conscious investing. For example, Collins Pine, a forest products company, sought out third party certification for its 38,000-ha Collins-Almanor Forest and is currently marketing wood products that are produced sustainably. Other forest products companies participate actively in environmental restoration and watershed councils. At the same time, organizations with conservation and social objectives are beginning to adopt for-profit institutional structures to capture bottom-line financial accountability and efficiency and to access capital markets, while remaining values-driven through socially and environmentally conscious investors, target beneficiaries, and staffs.

For example, the San Francisco based EcoTimber was created as a for-profit distributor of environmentally certified hardwoods to promote sustainable forest management (Grant 1995). Its for-profit status encourages entrepreneurial behavior and efficiency and allows it to compete for business along with other for-profit distributors, yet its mission is also environmental and social. The mission is upheld by its investors, which include foundations and conservation organizations. ShoreBank Pacific, the partnership between Ecotrust and Shorebank Corporation to provide entrepreneurial resources, is another example. A for-profit bank holding company is the umbrella for an array of for-profit and nonprofit organizations which promote environmental management. Again, the for-profit elements provide access to capital markets and market discipline; the investors ensure continued focus on environmental and social objectives (Ecotrust and Shorebank Corporation 1996).

The environmental and development challenges facing watersheds at all levels are large and daunting. It is clear that many of these challenges require unprecedented coordination and collaboration at regional, national, and international scales. It seems equally clear that essential building blocks for large-scale cooperation are local capacity, knowledge, and initiative—not as sufficient conditions, but as necessary ones. Nonprofit organizations play a key role in nurturing and organizing this local capacity within watersheds through organizing information, reforming policy, accessing resources, protecting natural areas, and providing a countervailing force to large corporate interests. However, to be effective at larger scales nonprofits need a strong and effective government—both to provide a policy framework and to catalyze and support their development and carry out many routine functions and regulations. Since conservation and development are inextricably intertwined, the private for-profit sector is also critically important to the health and vitality of watersheds. Here, profit incentives need to be balanced through ownership by social and environmental interests, through internalizing social and environmental costs, through increased local ownership and accountability, and through increased partnerships with the nonprofit and public sectors. To meet the complex and interdependent social, political, economic and ecological challenges facing society requires institutions which are as rich, complex and intertwined as the problems they are designed to address. Nonprofits will play an increasing role in catalyzing and nurturing this adaptive institutional landscape.

Acknowledgments. Many thanks to Elizabeth Coleman for research and invaluable help with the tables and figures, to Edward Wolf and Nancy Hauth for their comments and insights, and to Sarah O'Connor for her assistance preparing the manuscript.

Literature Cited

Annis, S. 1987. Can small-scale development be large-scale policy? The case of Latin America. World Development **15**(Supplement):129–134.

Backus, E. 1996. Validating "vernacular" knowledge: The Ahousaht GIS project, Vancouver Island, British Columbia. Pages 299–301 *in* P.K. Schoonmaker, B. von Hagen, and E.C. Wolf, eds. The rain forests of home: Profile of a North American bioregion. Island Press, Covelo, California, USA.

Ben-Ner, A., and B. Gui. 1993. Introduction *in* A. Ben-Ner and B. Gui, eds. The nonprofit sector in the mixed economy. The University of Michigan Press, Ann Arbor, Michigan, USA.

Ben-Ner, A., and T. Van Hoomissen. 1993. Nonprofit organizations in the mixed economy: A demand and supply analysis in the nonprofit sector in the mixed economy. Pages 27–58 *in* A. Ben-Ner and B. Gui, eds. The nonprofit sector in the mixed economy. The University of Michigan Press, Ann Arbor, Michigan, USA.

Best, C. 1996. The Pacific Forest Trust. Pages 197–201 *in* P.K. Schoonmaker, B. von Hagen, and E.C. Wolf, eds. The rain forests of home: Profile of a North American bioregion. Island Press, Covelo, California, USA.

Bird, N. 1996. The Prince William Sound Science Center. Pages 209–211 *in* P. K. Schoonmaker, B. von Hagen, and E. C. Wolf, eds. The rain forests of home: Profile of a North American bioregion. Island Press, Covelo, California, USA.

Brett, E.A. 1993. Voluntary agencies as development organizations: Theorizing the problem of efficiency and accountability. Development and Change **24**:269–303.

Brinkerhoff, D.W., and A.A. Goldsmith. 1992. Promoting the sustainability of development institutions: A framework for strategy. World Development **20**:369–383.

Department of Water Resources. 1993. Oregon watershed health/salmon recovery proposal. Department of Water Resources, State of Oregon, Salem, Oregon, USA.

Ecotrust and Shorebank Corporation. 1996. ShoreTrust, the first environmental bancorporation: A regional development strategy. Implementation of Phase III, ShoreTrust Bank Draft. Ecotrust, Portland, Oregon, USA.

Egler, F. 1977. The nature of vegetation: Its management and mismanagement. Avon Forest, Norfolk, Connecticut, USA.

Frantz, T.R. 1987. The role of NGOs in the strengthening of civil society. World Development **15**(Supplement):121–127.

Grant, J. 1995. Looking for good wood: Encouraging sustainable forestry. Woodwork, April, 1995:70–74.

Hansmann, H.B. 1986. The role of nonprofit enterprise. Pages 57–84 *in* S. Rose-Ackerman, ed. The economics of nonprofit institutions: Studies in structure and policy. Oxford University Press. New York, New York, USA.

Hauth, N. 1996. ShoreTrust trading group. Pages 354–357 *in* P. K. Schoonmaker, B. von Hagen, and E.C. Wolf, eds. The rain forests of home: Profile of a North American bioregion. Island Press, Covelo, California, USA.

Hodgkinson, V., and M.S. Weitzman. 1986. Dimension of the independent sector: A statistical profile. 2nd edition. Independent Sector, Washington, DC, USA.

John, D. 1994. Civic environmentalism: Alternatives to regulation in states and communities. Congressional Quarterly, Washington DC, USA.

Karr, J. 1992. Ecological integrity: Protecting Earth's life support systems. Pages 223–238 *in* R. Costanza, B.G. Norton, and B.D. Haskell. Ecosystem health: New goals for environmental management. Island Press, Washington DC, USA.

Korten, D.C. 1987. Third generation NGO strategies: A key to people-centered development. World Development **15**(Supplement):145–159.

Kusel, J., and L. Fortmann. 1991. Well-being in forest-dependent communities. Department of Forestry and Resource Management, University of California, Berkeley, California, USA.

Lee, K.N. 1993. Compass and gyroscope: Integrating science and politics for the environment. Island Press, Washington DC, USA.

Lee, R.G. 1992. Ecologically effective social organization as a requirement for sustaining watershed ecosystems. Pages 73–90 *in* R.J. Naiman, ed. Watershed management: Balancing sustainability and environmental change. Springer-Verlag, New York, New York, USA.

Livernash, R. 1992. The growing influence of NGOs in the developing world. Environment **34**:13–20; 42–43.

Mattole Restoration Council. 1995. Dynamics of recovery: A plan to enhance the Mattole estuary. Mattole Restoration Council, Petrolia, California, USA.

Mattole Restoration Council. 1989. Elements of recovery: An inventory of upslope sources of sedimentation in the Mattole River watershed. Mattole Watershed Council, Petrolia, California, USA.

Meyer, C. 1992. A step back as donors shift institution building from the public to the "private" sector. World Development **20**:1115–1126.

——. 1993. Opportunism and NGOs: Entrepreneurship and green north–south transfers. World Development **23**:1277–1289.

National Research Council. 1995. Upstream: Salmon and society in the Pacific Northwest. National Academy Press, Washington DC, USA.

Nehlsen, W., and J. Lichatowich. 1996. Pacific salmon: Life histories, diversity, productivity. Pages 213–226 *in* P.K. Schoonmaker, B. von Hagen, and E.C. Wolf, eds. The rain forests of home: Profile of a North American bioregion. Island Press, Covelo, California, USA.

Nixon, R. 1970. Message of the President relative to reorganization plans no. 3 and 4 of 1970 in Environmental quality: The first annual report of the

Council on Environmental Quality. US Government Printing Office, Washington DC, USA.

O'Neill, M. 1989. The third America: The emergence of the nonprofit sector in the United States. Jossey-Bass, San Francisco, California, USA.

Radtke, H., S. Davis, R. Johnson, and K. Lindberg. 1996. Economic and demographic transition on the Oregon coast. Pages 329–348 *in* P.K. Schoonmaker, B. von Hagen, and E.C. Wolf, eds. The rain forests of home: Profile of a North American bioregion. Island Press, Covelo, California, USA.

Rose-Ackerman, S. 1986. Introduction. Pages 3–17 *in* S. Rose-Ackerman, ed. The economics of nonprofit institutions: Studies in structure and policy. Oxford University Press, New York, New York, USA.

Sayce, K. 1996. Local science in Willapa Bay, Washington. Pages 301–304 *in* P. K. Schoonmaker, B. von Hagen, and E.C. Wolf, eds. The rain forests of home: Profile of a North American bioregion. Island Press, Covelo, California, USA.

Schiff, J., and B. Weisbrod. 1993. Competition between for-profit and nonprofit organizations in commercial markets. Pages 127–148 *in* A. Ben-Ner, and B. Gui, eds. The nonprofit sector in the mixed economy. The University of Michigan Press, Ann Arbor, Michigan, USA.

Schoonmaker, P.K., B. von Hagen, and E.C. Wolf, eds. 1996. The rain forests of home: Profile of a North American bioregion. Island Press, Covelo, California, USA.

Shannon, M. and A. Antypas. 1996. Civic science is democracy in action. Northwest Science **70**:66–69.

Soscia, M.L. 1996. Oregon's watershed health program. Pages 304–306 *in* P. K. Schoonmaker, B. von Hagen, and E.C. Wolf, eds. The rain forests of home: Profile of a North American bioregion. Island Press, Covelo, California, USA.

Spencer, C.N., B.R. McClelland, and J.A. Stanford. 1991. Shrimp stocking, salmon collapse and eagle displacement: Cascading interactions in the food web of a large aquatic ecosystem. BioScience **41**:14–21.

Tietenberg, T. 1992. Environmental and natural resource economics. HarperCollins, New York, New York, USA.

von Hagen, B., and E. Kellogg. 1996. Entrepreneurs and ecosystems: Building sustainable economies. Northwest Report **19**:10–15.

Ware, A. 1989. Between profit and state: Intermediate organizations in Britain and the United States. Princeton University Press, Princeton, New Jersey, USA.

Webb, T., J. Korman, and D.R. Marmorek. 1993. Kennedy Lake Workshop Report. Environmental and Social Systems Analysts, Vancouver, British Columbia, Canada.

Weisbrod, B.A. 1988. The nonprofit economy. Harvard University Press, Cambridge, Massachusetts, USA.

Willer, C. 1995. Gated lands: A report on the ownership of Oregon's private coast range forests. Coast Range Association, Newport, Oregon, USA.

Young, D.R. 1986. Entrepreneurship and the behavior of nonprofit organizations: Elements of a theory. Pages 161–184 *in* S. Rose-Ackerman, ed. The economics of nonprofit institutions: Studies in structure and policy. Oxford University Press, New York, New York, USA.

Zuckerman, S. 1996. Restoration as practice in the Mattole. Pages 91–94 *in* P.K. Schoonmaker, B. von Hagen, and E.C. Wolf, eds. The rain forests of home: Profile of a North American bioregion. Island Press, Covelo, California, USA.

26 Watershed Management[1]

Robert J. Naiman, Peter A. Bisson, Robert G. Lee, and Monica G. Turner

Overview

• Management at the watershed scale is a major challenge facing present and future generations. Watershed management requires integrating scientific knowledge of ecological relationships within a complex framework of cultural values and traditions to provide socio-environmental integrity. This implies that socioenvironmental integrity can operate for the long term and over large spatial scales—especially within hydrologically identifiable boundaries.

• Development of a watershed management perspective incorporates variability in time and space, takes a holistic approach toward the persistence of ecological features, treats human cultures and institutions as inherent features, and addresses system connectivity and uncertainty.

• Several approaches are presented for implementing watershed management that relate to public stewardship (monitoring and education), accepting and dealing with risk, addressing uncertainty, formulating a shared vision, quantitatively analyzing socio-environmental conditions, and structuring institutional organization.

• Although there is no set methodology for achieving effective watershed management, fundamental principles related to cooperation, balance, fairness, integration, trust, responsibility, communication, and adaptability are essential for guiding the process.

Introduction

Fresh water, and freshwater ecosystems, are the most basic components of watershed management (Naiman et al. 1995a,b). Freshwater issues, more than ever, embody the complexity that characterizes natural resource management. Changes in human demography, resource consumption, cultural values, institutional processes, technological applications, and information all contribute to the increasing complexity. Despite attempts to manage change, changes continue to occur and the consequences remain difficult to predict at scales commensurate with the changes themselves (Naiman 1992, Lee 1993). Understanding the abilities and limits of freshwater ecosystems to respond to human-generated pressures is central to long-term social stability as well as ecological vitality. Yet, even though human actions and cultural values drive the environmental issues, few holistic approaches for watershed management offer effective resolution.

In the current debate over the scope of watershed (and ecosystem) management (Grumbine 1994, U.S. MAB 1994,

[1] This chapter is an expanded version of the original chapter in Kohm, K.A., and J.F. Franklin (editors). 1997. Creating a forestry for the 21st century: The science of ecosystem management. Island Press, Washington, DC.

Montgomery et al. 1995), it is widely recognized that there are significant technical and cultural constraints to effective implementation. These constraints are related to such important issues as identifying appropriate spatial and temporal scales, monitoring and assessment, developing an adaptive management process, and developing cultural values and philosophies that allow watershed management to be successful (Levin 1993, Grumbine 1994, Harwell et al. 1996). Nonetheless, the ability of a rapidly increasing human population to dramatically impact local, regional, and global ecosystems makes it essential to incorporate an ecological perspective into watershed management if there is to be a healthy resource base for future generations.

The first part of this chapter suggests several features which are fundamental to contemporary watershed management. The second part then presents several practical approaches for implementing effective watershed management programs.

Fundamental Elements of Watershed Management

Initially, it is important to recognize that there are four watershed-scale features which provide the foundation for effective management: variability in time and space, persistence and invasiveness of species, system connectivity and uncertainties, and the role of human cultures and institutions. These features are closely related to specific goals frequently endorsed as being fundamental to ecosystem management (Grumbine 1994, U.S. MAB 1994; Table 26.1).

The Natural System: Variability in Time and Space

Natural processes (i.e., climate, soil formation, geological disturbances, and so forth) structure the diversity, productivity, and availability of natural resources on which human societies depend. The challenge is to understand how naturally variable systems operate and to predict the environmental consequences of human activities in these systems (Naiman et al. 1995a,b).

The vitality of natural ecosystems is created and maintained by substantial variation in time and space (Reice et al. 1990, Reice 1994). Natural systems are constantly changing in a complex mosaic of time periods and spatial dimensions (Turner 1990). For example, the ecological characteristics of riparian forests are structured by a complex array of dynamic and spatially variable hydrological processes that erode and deposit materials, deliver nutrients, and remove waste products (Gregory et al. 1991; Figure 26.1). Variability in time and space

TABLE 26.1 Principles for management at the watershed scale.

- Use an ecological approach that would recover and maintain the biological diversity, ecological function, and defining characteristics of natural ecosystems.
- Recognize that humans are part of ecosystems—they shape and are shaped by the natural systems; the sustainability of ecological and societal systems are mutually dependent.
- Adopt a management approach that recognizes ecosystems and institutions are characteristically heterogeneous in time and space.
- Integrate sustained economic and community activity into the management of ecosystems.
- Develop a shared vision of desired human and environmental conditions.
- Provide for ecosystem governance at appropriate ecological and institutional scales.
- Use adaptive management as the mechanism for achieving both desired outcomes and new understandings regarding ecosystem conditions.
- Integrate the best science available into the decision-making process, while continuing scientific research to reduce uncertainties.
- Implement ecosystem management principles through coordinated government and non-government plans and activities.

Modified from U.S. MAB 1994.

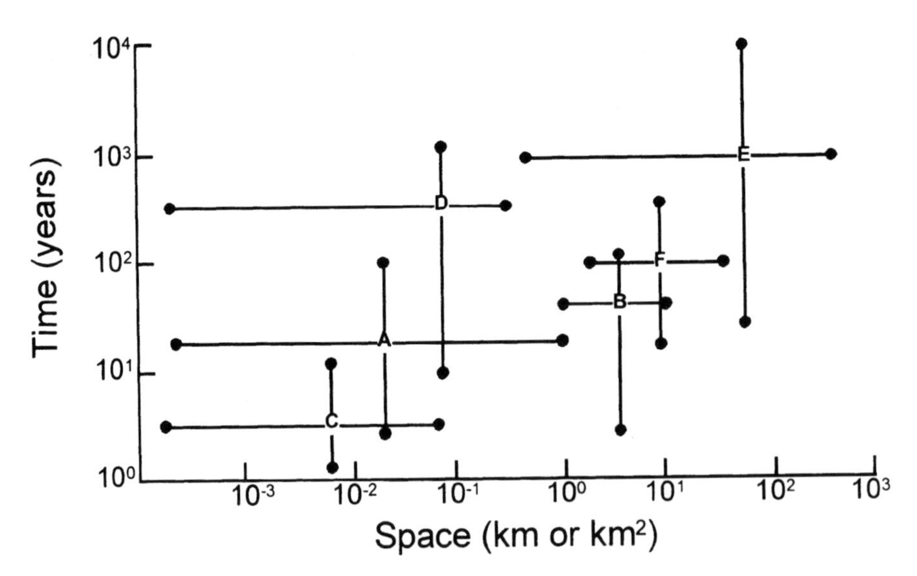

FIGURE 26.1. Illustration of the diversity of spatial and temporal scales influencing the creation and maintenance of riparian forests in the coastal temperate rainforest of North America. Colonization surfaces created by flooding (A), colonization surfaces created by debris flow (B), seedling germination and establishment (C), longevity and size of species patches (D), perisistence and movement of dead wood in channel (E), and impact of herbivores (F).

results in the biological diversity and productivity characteristically found in riparian environments (Fetherston et al. 1995, Naiman and Décamps 1997). A key managerial challenge is balancing human needs with variations in physical and chemical characteristics so that significant declines or losses of species and ecological attributes (i.e., biodiversity, productivity, resilience) do not occur.

A Holistic Perspective: Persistence and Invasiveness

The persistence of ecological attributes for the long term (i.e., decades to centuries) requires maintaining a naturally variable environmental regime as well as isolation from invading organisms that can alter the natural regime. When the natural environmental regime is altered, adjustments occur within the ecosystem (i.e., relative abundance of species or biogeochemical processes) producing new combinations of biophysical environments susceptible to the invasion of exotic organisms and the establishment of non-native ecological processes and structures (Drake et al. 1989). Understanding and quantifying persistence and invasiveness of species (and their ecological processes) are important for watershed management because these components are sensitive to change, integrate change over broad spatial and temporal scales, and can be used as measures of change. There is often a cultural identity with many species, and, in addition, the ecological processes are essential for sustaining human populations (Botkin 1990). There are a variety of quantitative approaches and technical tools that already exist for analyzing persistence and invasiveness at the watershed scale, and many other techniques are in the design and testing stages (see later). Existing techniques include new approaches to statistical analyses, patch and boundary analyses, modeling cumulative effects, indices of biotic integrity, and knowledge-based land-use analysis systems (Karr 1991, Risser 1993, Fortin and Drapeu 1995, Turner et al. 1996). These techniques are especially useful tools for setting goals related to desired future conditions and for preliminary examinations of the long term effects of new or anticipated institutional regulations and policy (Turner et al. 1996, Wear et al. 1996).

Connectivity and Uncertainty

The goal of watershed management is to let all components of the human and nonhuman communities exist in a relative but dynamic state of balance (Naiman 1992, U.S. MAB 1994). This goal explicitly recognizes strong connections between the social and environmental components at multiple scales. This means managing for connectivity between components, as well as managing the components themselves. For example, consideration must be given to water, fish, soils, forests, education, resource extraction, and cultural values, as well as to the strong interactions which occur between them (Stanford and Ward 1992).

Unfortunately, quantitative approaches for managing connectivity are not well formulated. There remains considerable uncertainty among scientists and decision makers as to how to proceed, while the magnitude of current socioenvironmental issues requires decisions now. This means accepting risk since actions cannot wait until all the information is available. How can this be accomplished at the watershed scale? There is no definitive answer or one right way. However, from a wide range of empirical studies from many scientific disciplines, it is known that major advances often come at the interfaces between human, natural, and management sciences (Figure 26.2). Following is a discussion of approaches used by small organizations addressing risk, groups helping to define social and environmental viewpoints for future conditions, and researchers and managers struggling to monitor and assess change at regional scales.

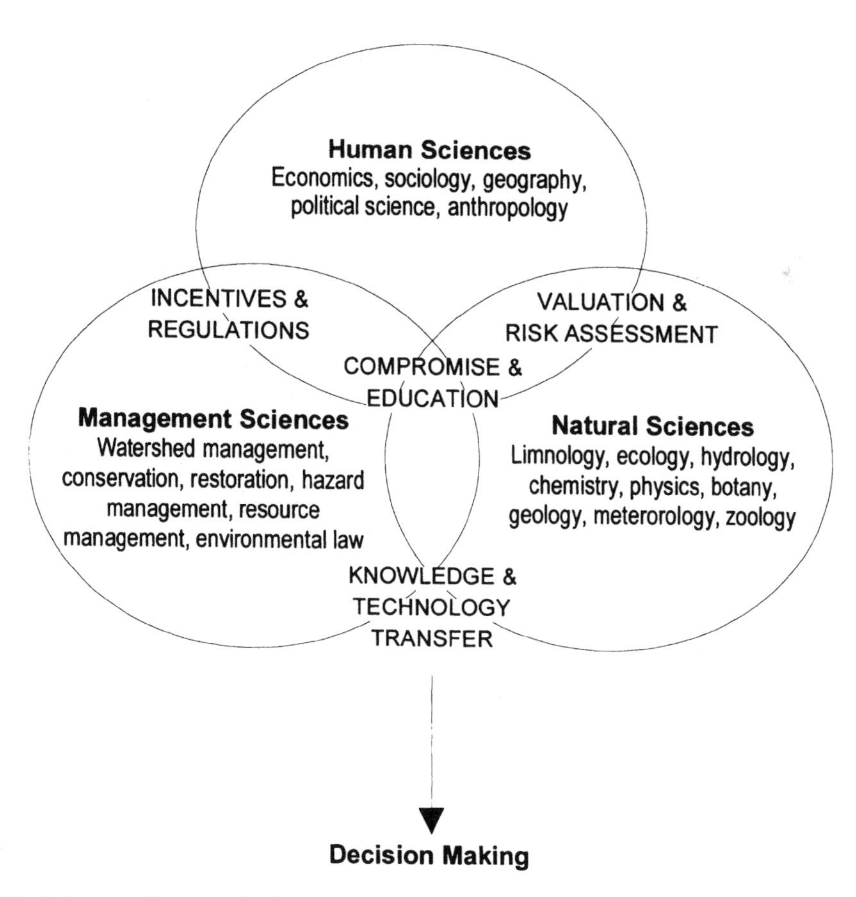

FIGURE 26.2. Advances in watershed management come at the interfaces between natural, human, and management sciences (from Naiman et al. 1995a, with permission).

Human Cultures and Institutions

In human-dominated watersheds, the land mosaic (i.e., patches and boundaries) is created by a mixture of cultural practices, traditions, myths, and institutions (Lee et al. 1992, Décamps et al. 1998). The spatial extent and temporal duration of each patch and boundary type are ultimately determined by laws, regulations, taxation, technologies, cultural values and beliefs, and traditional land use practices that pertain to the utilization of natural resources (Turner 1990).

Developing an integrated socioenvironmental system means confronting and resolving important issues related to social and ecological literacy, the role and accommodation of changing cultural values, the increasing migration of peoples away from traditional homelands and cultures (i.e., cultural mixing), balancing consumption rates and population growth, weathering political change, and establishing knowledge-based cooperative institutions (Lee 1993). These issues are closely interrelated and cannot be resolved separately. How to implement an integrated program that addresses these and related issues may not be immediately apparent since each watershed has a unique set of issues to resolve. There are, however, basic principles and practical approaches to guide the development of effective watershed management.

Practical Approaches for Implementing Watershed Management

Quantitative Analyses

Attempts to manage watersheds with more than one demand on the principal resources have been ineffective for the most part. Well-known examples include the Columbia River, the Sacramento-San Joaquin rivers, and the Colorado River watersheds. An inability to identify appropriate spatial and temporal scales for management, cumulative effects from multiple users, conflicting management goals, lack of accepted statistical or realistic modeling approaches, and a dearth of indices for evaluating a dynamic socioenvironmental system all contributed to the difficulties (Lee 1993, Volkman and Lee 1994). Fortunately, as public awareness of watershed-level issues has improved, so has the array of quantitative approaches for assessing complex issues that have several causes and competitive solutions. Watershed analysis techniques, quantitative measures, assessing risk with integrated socioenvironmental models, and development of socioenvironmental indices are but a few of the empirical approaches available. Table 26.2 summarizes the available empirical approaches and their advantages and disadvantages. While the availability of quantitative tools may improve the ability to address watershed management issues, past failures cannot be totally attributed to the absence of such tools. Moreover, quantitative tools alone will not solve current or future problems.

Watershed Analysis

Quantitative approaches to document the status and dynamics of entire watersheds are still in the early stages of development (Montgomery et al. 1995). Most of the techniques developed have been concentrated in western states heavily impacted by forest management (Table 26.2) but there are notable exceptions such as South Florida (Harwell et al. 1996). The intent of watershed analysis is to provide a scientifically based understanding of the environmental processes and their interactions occurring within a watershed (U.S. Government 1994, Washington Forest Practices Board 1994). This understanding, which focuses on specific issues, values, and uses within the watershed, is essential for making sound management decisions. The fundamental steps involved in watershed analysis and some of the basic products to be expected from the process are summarized in Table 26.3. Protecting beneficial uses, such as those identified by state and federal environmental laws (e.g., Clean Water Act and Endangered Species Act), is a fundamental objective for watershed analysis. Watershed analysis encompasses the entire watershed because of the strong fluvial linkages among

TABLE 26.2. Summary of empirical approaches for watershed management, applicability, and advantages and disadvantages of selected approaches and relevant case studies.

Approach	Description	Applicability	Advantages	Disadvantages	Case studies
1. Watershed analysis	Provides a scientifically based understanding of environmental processes and their interactions	Largely limited to forested watersheds of 50–500 km^2, although it can be adapted to other situations	Provides a spatially explicit description of resources, hazards, environmental variation, and potentials, as well as potential conflicts over resource use; is adaptable to new technological methodologies	Requires highly trained, inter-disciplinary teams, familiar with the terrain; assumes demonstrable linkages between physical patches and biological processes	Washington Forest Practices Board, 1994; Montgomery et al., 1995
2. Quantitative measures	Inventory of the abundance and spatial arrangement of vegetation land cover, or habitat characteristics	All watersheds	Provides a resource inventory for establishing spatial and temporal trends; takes advantage of existing GIS databases; acts to centralize storage of information; requires personnel with only moderate levels of training	Often requires a substantial investment in database development; data availability is often incomplete; requires long-term monitoring and analyses to be useful	Turner and Gardner, 1991; Turner et al., 1995, 1996
3. Integrated socioenvironmental models	Models explicitly combining the social, economic, and environmental factors influencing watershed characteristics	Still in an experimental stage; best applied to watersheds with few, direct human influences on resources	Allows a holistic (and more realistic) perspective to be developed where human activities and values are a central component of the ecosystem; allows a wide range of social choices to be evaluated	Database development is expensive and time consuming; essential data are often incomplete; requires a moderate-to-high level of technical expertise	Le Maitre et al. 1993; Warwick et al. 1993; Berry et al. 1996; Wear et al. 1996
4. Indices of socioenvironmental conditions	Components contributing to the long-term vitality of an social–economic–environmental system	Watersheds with a significant human population	Provides a regular report to the citizens; improves literacy about watershed-scale issues; develops stewardship for the long-term; easily maintained	Requires regular monitoring and analysis of data, some of which may be difficult to obtain	Willapa Alliance and Ecotrust 1995

TABLE 26.3. Fundamental steps and basic products expected from watershed analysis.

Steps
1. Identify issues, describe desired conditions, and formulate key questions.
2. Identify key processes, functions, and conditions.
3. Stratify the watershed.
4. Assemble analytic information needed to address the key questions.
5. Describe past and current conditions.
6. Describe condition trends and predict effects of future land management.
7. Integrate, interpret, and present findings.
8. Manage, monitor, and revise information.
Products
1. A description of the watershed including its natural and cultural features.
2. A description of the beneficial uses and values associated with the watershed and,
3. when supporting data allow, statements about compliance with water quality standards.
4. A description of the distribution, type, and relative importance of environmental process.
5. A description of the watershed's present condition relative to it's associated values and uses.
6. A map of interim conservation areas.

headwater areas, valley floors, and downstream users.

Watershed analysis requires a high level of expertise (Montgomery et al. 1995). However, earlier attempts such as the California checklist for cumulative effects required little technical expertise, were largely ineffective, and are no longer used (Chapter 19). The current watershed analysis procedure (Table 26.3) is designed to be carried out by an interdisciplinary team of resource professionals who are already experts in their fields, and who are familiar with the area to be evaluated. Different methods apply to different areas, and teams must use their professional judgment to select or design appropriate methods. Watershed analysis is also an iterative and evolving process. Analytical methods improve or are replaced as experience and knowledge grow.

The results of watershed analysis may include a description of resource needs, capabilities, and opportunities; the range of natural variation; spatially explicit information that will facilitate environmental and cumulative effects analyses for the National Environmental Policy Act (NEPA) regulations; and a description of processes and functions operating within the watershed (Montgomery et al. 1995; Table 26.3). Watershed analysis also identifies potentially conflicting objectives and uses within watersheds. However, watershed analysis is not a decision-making process per se; it is a process that derives information to assist in decision making. Watershed analysis is, nevertheless, a substantial advancement over management approaches used in the past because it brings factual information to the decision-making process.

Watershed analysis assumes there are demonstrable linkages between physical patches and biological processes and that human values and perspectives do not change. These are flawed assumptions that contradict the later discussion on risk. Despite the promise watershed analysis brings to management, there are impediments to its application. Local and regional political influences, nonbinding agreements, the lack of long-term accountability for institutions, decision makers, and land managers, and the avoidance of an interactive synthesis of information are potentially fatal flaws in the concept. Further, to date there has been no scientific validation of the approach, which was developed primarily by physical scientists, and there is little understanding of how biological attributes (i.e., community composition and so forth) modify physical–biological relationships.

Despite these flaws, watershed analysis is now a part of the regulatory framework for managing state and privately owned commercial forests in Washington (Washington Forest Practices Board 1994). The Washington Department of Natural Resources, the agency charged with implementing watershed analysis, has identified a number of forested subbasins (5th–6th order river systems) termed Watershed Analysis Units (WAUs) within which watershed analysis forms a basis for local forest practice decisions. The first WAU to be analyzed was the Tolt River drainage, located in the Puget Sound basin of western Washington.

The Tolt River watershed includes mixed ownership dominated by private forest land, but the drainage also includes a reservoir that supplies water to Seattle. Because the Tolt River contains valuable fishery resources (salmon, trout and steelhead) as well as an important drinking water supply, many interest groups participated in the watershed analysis process.

The Tolt watershed analysis procedure identified areas where salmonid habitat features such as stream temperature and large woody debris abundance were degraded, as well as areas where delivery of sediment to streams would be likely from unpaved logging roads and geologically unstable slopes. Prescriptions for preventing or mitigating these problems (Tolt Watershed Analysis Prescriptions 1993) were developed by a team that included six foresters representing the Washington Department of Natural Resources and the Weyerhaeuser Company, a forest road engineer, a tree physiologist, an environmental analyst from the Washington Department of Ecology, two aquatic biologists from the Tulalip Indian tribe, and a forest hydrologist. The prescriptions for future forestry operations are not voluntary; the land owners must comply or be subject to civil and criminal prosecution.

Over 40 people officially participated in the Tolt watershed analysis in addition to the 12 members of the prescription team. The five-month process itself was at times contentious. This was perhaps to be expected given the diversity of interests. Nonetheless, members of the watershed analysis team generally agreed that the process of working together was at least as important as the process of using available data to guide management decisions.

Quantitative Measures

Watersheds can be characterized by a variety of quantitative measures when digital data are available. Most simply, the total area and proportion of the watershed occupied by each cover type (i.e., vegetation or habitat) can be identified and its area and perimeter recorded. Analyses of the total number of patches, arithmetic mean patch size, standard deviation of mean patch size, size of the largest patch, weighted average patch size, amount of interior habitat, total edge, and mean patch shape are easily computed (Table 26.2). In addition to metrics describing individual cover types, edges between habitats, which are sensitive measures of habitat fragmentation, can be tabulated as the length of edge between each pair of land cover classes (e.g., forest-grassy, forest-unvegetated, grassy-unvegetated) or as edge-to-area ratios.

While the development of quantitative measures of watershed condition has proceeded rapidly, empirical studies that test for significant relationships between watershed metrics and ecological condition (e.g., presence or abundance of species, water quality) are still few in number (Johnston et al. 1990). There is a clear need to identify the most important watershed metrics to monitor as well as the levels beyond which socioenvironmental conditions change significantly. In addition, it is essential to be aware of the assumptions and constraints that are implicit in the metrics. For example, the selection of the land cover categories to be used in the analysis in part determines the results, and the spatial scale of the data—both the total extent of the area and the resolution, or grid cell size—can strongly influence the numerical results (Turner et al. 1989a,b).

Integrated Socioenvironmental Models

The risk of undesirable future conditions can be assessed by using integrated socioenvironmental models to explore alternative land management scenarios (Le Maitre et al. 1993, Warwick et al. 1993, Flamm and Turner, 1994a, b). An example of such a model is the Land-Use Change and Analysis System (LUCAS) (Berry et al. 1996, Wear et al. 1996). LUCAS is a spatial simulation model at the watershed scale in which the probability of land being converted from one land cover type to another depends on a variety of social, economic, and ecological factors (Figure 26.3). Conditional transition probabilities are estimated empirically by comparing land cover at different times (e.g., from decade to decade) and then used to simulate potential future conditions (Turner et al. 1996).

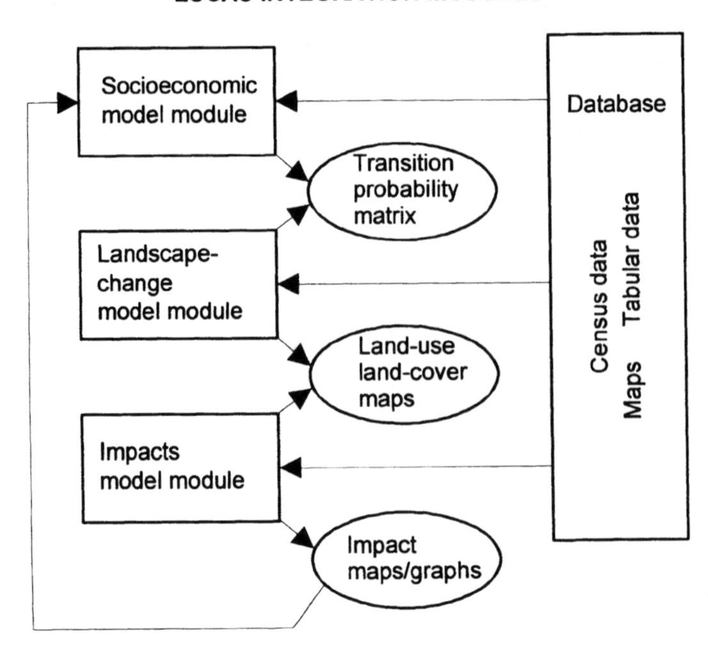

FIGURE 26.3. Integration of social, economic, and environmental aspects of watershed management can be accomplished with the use of the Land Use Change Analysis System (LUCAS), a modeling environment (Berry et al. 1996, with permission © 1996 IEEE).

Simulations begin with an initial map of land cover, and equations are used to generate a transition probability for each grid cell in the watershed map based on ownership type, elevation, slope, aspect, distance to road, distance to market, and population density (Flamm and Turner 1994a, b, Wear et al. 1996). An integrated modeling approach permits the effects of a wide range of alternatives to be evaluated. For example, the effects of residential development in different locations within the basin or the effects of moving a large parcel of land into or out of intensive timber production can be examined. Linking projected land cover maps with effects on ecological indicators (such as species persistence or water quality) allows the potential long-term implications of alternative human decisions to be compared.

Indices of Socioenvironmental Conditions

Methods for separately examining the status and trends of environmental, social, and economic factors are well established (Finstenbusch and Wolf 1981, Burch and DeLuca 1984, Karr 1991). However, watershed management requires integrated socioenvironmental indices that provide a holistic understanding of watershed condition (Table 26.2). In a broad sense, a socioenvironmental index is a report to citizens, resource users, and government agencies on the vitality of a space they hold in common. Ideally, a socioenvironmental index should provide usable information on the important aspects of a watershed's environment, economy, and communities.

Components of a socioenvironmental index are chosen to reflect the unique characteristics of the watershed. For example, in the Willapa Bay watershed of western Washington, shellfish aquaculture, timber production, fishing, and agriculture are important in maintaining the local economy and culture. The Willapa Alliance, a consortium of concerned citizens and resource users, has developed an index based on indicators of environmental quality, economic vitality, and community health (Table 26.4). Environmental quality is indicated by oyster condition that reflects water quality, by changes in vegetation cover which reflects terrestrial condition, by escapement of wild and hatchery salmon (*Oncorhynchus* spp.) which reflect ecological conditions in the rivers and streams, and by counts of wetland and riparian birds which reflect habitat condition. Economic conditions are gauged by annual finfish, shellfish, and timber harvests (i.e., resource production), income distribution, local unemployment rates, and bank loans per capita (i.e., credit and local investment). Community health is measured by the percentage of healthy birthweight babies, high school graduation rates, and voter turnout in county elections (i.e., participation). Each category has alternate candidates that can be used in developing the socioenvironmental index. Adult literacy rates could replace high school graduation rates, vegetative biodiversity could replace bird abundance, and so forth. Nevertheless, the point is that citizens and resource users within the watershed have an integrated socioeconomic index that can be used to develop a holistic understanding about the watershed and to create the stewardship necessary for long-term sustainability (The Willapa Alliance and Ecotrust 1995).

TABLE 26.4. Socioenvironmental index developed for the Willapa River watershed, Washington.

Natural resource-based industries	Shellfish harvest, timber production, fishing, and agriculture
Socioenvironmental index	
Environmental quality	Oyster condition
	Vegetation cover
	Salmon escapement
	Bird abundance
Economic conditions	Finfish, shellfish, and timber harvest
	Income distribution
	Unemployment rates
	Bank loans per capita
Community health	Birthweight of healthy babies
	Rates high school graduation
	Voter turnout

The Willapa Alliance and Ecotrust 1995.

Accepting Risk and Addressing Uncertainty

Attempts to make decisions by identifying, evaluating, and formulating management strategies for risks associated with watershed management need to include a broad socio-environmental perspective, recognition of spatial scales ranging from sites to global ecological and social processes, and an explicit consideration of the temporal transfers of risks, especially those involving decisions that may transfer risks to future generations or impose unacceptable rates of change on current generations.

Risk is a product of the probability of a negative (unwanted) event and the consequences of that event (Campbell 1969, Slovic 1984, Suter 1992). Estimating risks involves identification and evaluation (Schrader-Frechette 1991). Risk management is also part of the process of subjecting unwanted events to rational interpretations because it involves selection of the most effective and efficient means of reducing harm. A comprehensive discussion of risk assessment is beyond the scope of this chapter, but is provided by Fava et al. (1991), Bartell (1992), Harwell and Gentile (1992), and U.S. EPA (1993).

Competing risks are often a basic cause for conflict. Decisions about allocation of land or resources involve judgments about the extent to which the risk should be placed on environmental features or on human well-being. Scientists are poorly prepared for making these judgments, because decisions on how to balance risks involve political and ethical considerations for which scientists are seldom adequately trained and generally unauthorized. Scientists are often best at identifying problems, not resolving them (Ludwig et al. 1993).

The scientific role is to identify what may be at risk, discover what causes risks to increase or decrease (risk attribution), estimate the relative importance of causes (risk assessment), use knowledge to suggest options for reducing risks (risk management), and design effective monitoring programs that facilitate learning. Scientists are most useful to the decision-making process if risk attribution is framed by hypoth-

Null hypothesis is:	Accepted	Rejected
True	Correct decision	Type I error
False	Type II error	Correct decision

FIGURE 26.4. Type I and Type II errors in decision making arise according to inherent philosophies and training of specific disciplines (modified from Sokal and Rohlf 1981).

eses linking causes to effects (i.e., using *If-Then* statements).

Risk assessment requires explicit statements about the degree of confidence scientists and managers have in cause-and-effect relationships. Statements involving statistical confidence limits are generally required but rarely available for large systems. However, they may be approximated by the "best judgment" of knowledgeable experts. Opportunities for risk management improve with knowledge of cause-and-effect relationships, often permitting scientists to suggest management strategies that sufficiently mimic natural structure or processes to allow resource use without imposing unacceptable risks on species, natural processes, or society.

Unfortunately, it remains difficult to frame decisions based on the allocation of risk because scientists are trained to avoid making Type I errors (rejecting a null hypothesis when it is true) while paying less attention to Type II errors (failing to reject the null hypothesis when it is false; Figure 26.4). Scientists are generally concerned about avoiding false-positives because the professional mission is to contribute accurate information for building an understanding of fundamental processes (Schrader-Frechette 1991). When contributing to risk assessment, it is far more important to avoid overlooking important cause-and-effect relationships that are true (avoid Type II error) which could, for example, result in loss of a species or the disintegration of a human community. The appropriate scientific role is, of course, to pay attention to both types of error.

Addressing Type I and Type II errors becomes especially complex when there are competing risks. Concern about losing a species or causing unnecessary human suffering often leads to placing emphasis on avoiding a Type II errors while minimizing Type I errors. For example, mistakenly accepting the hypothesis that sustaining coho salmon (*Oncorhynchus kisutch*) requires preservation of all remaining ancient forests and unregulated rivers, could cause unnecessary human suffering and economic losses. Similarly, mistakenly accepting the hypothesis that timber harvest reductions will cause a large increase in rural poverty could place the survival of certain animal species at risk. Hence, extra effort needs to be paid to minimizing both Type I and Type II errors through better interdisciplinary communication. It is a natural tendency to ignore risks that are not well understood. This is especially true if the implications of research conclusions allow scientists to externalize Type II errors on subjects understood better by other disciplines.

How Can Organizations Deal with Risk?

Formal organizations, especially those that have persisted for a long time, tend to conserve their mission and structure (Selznick 1966, Meyer and Scott 1983). Organizational mission and structure are threatened by an uncertain environment, especially one that imposes risks. For example, natural disturbances such as fires, insect outbreaks, floods, and windstorms often exceed the capacity of land management organizations to cope with the disruption. Similarly, the emergence of new and powerful political clients bring risks of diminished political support and stability.

Organizations tend to limit actions that will threaten organization structure and mission, even when those actions might better position the organization to deal with emerging risks (Meyer and Scott 1983, Bella 1987). These tendencies are related to the statistical concepts of Type I and Type II error discussed previously (Figure 26.4). Organizations, like individual scientists, are usually concerned with minimizing the risk of making a Type I error. Likewise, organization members are worried about false-positives (asserting an effect where none exists) because they do not want to needlessly upset organizational relationships and purposes upon which they depend for security, rewards, and identity (Schrader-Frechette 1991).

Consumers or clients depending on the organization for service or protection are, by contrast, concerned with minimizing Type II errors. They are more concerned about the social acceptability or environmental safety of technologies or organization practices. Hence, they do not want to allow an organization to increase the risk of public injuries or losses (Schrader-Frechette 1991).

As in statistics, there is an inverse relationship between these two types of error—these two sources of risk. Customers or the public might be hurt by decreasing risk to the organization, and the organization might be hurt by decreasing the risk to customers or public. These inverse relationships are well illustrated in land management organizations responsible for implementing watershed management plans.

Federal land management organizations are placed at risk by dramatically changing management practices (such as timber harvest, fishing or grazing) to avoid the risk of losing species valued by society. Normal routines are disrupted, changes in work force are implemented, insecurity becomes contagious, and the identity provided by organizational culture becomes confused by drastic institutional change. The very existence of an organization can be threatened by sudden change in organization and mission. A frantic search for new mission, or paradigm, generally accompanies these periods of unrest and insecurity, with stability returning if the organization is successful in finding a new mission and a structure for implementation. Watershed management and ecosystem management are now part of such a search for a new mission by many of the large, material resource-based industries and the provincial, state, and federal agencies in Canada and the United States.

Adaptive organizations avoid risks accompanying such turmoil by continually striving to maintain a balance between Type I and Type II errors. Large industrial organizations, such as those responsible for managing forested lands, have sought to manage internal and market or regulatory risks that accompany increasing uncertainty by reorganizing their structure and diversifying their mission. Horizontal, team-based management with accountability to attain performance objectives is replacing the centralized and homogeneous mission characteristic of traditional, bureaucratic command-and-control structures (Reich 1991). This often includes product diversification and relative independence of production units. Greater flexibility and efficiency are gained by down-sizing salaried staff and contracting for services such as roadbuilding, resource extraction, inventory, and environmental assessment. These innovations have enabled large formal organizations to capture some of the advantages of flexibility characteristic of market systems that have a large number of independent producers.

Large public organizations seldom have the opportunity to develop decentralized structures suitable for adapting to an uncertain environment. Government, as opposed to markets, centralizes power and defines organizational mission and structure from the top down. One of the greatest challenges (and opportunities) facing public land management organizations in North America is to develop permissible ways of diversifying organizational mission and developing a structure capable of responding to diverse and dynamic social, political, and geographic concerns and conditions. Successful implementation of watershed management may depend on how well public organizations meet these challenges.

Addressing Institutional Organization and the Paradox of Scale

Watershed-scale activities must ultimately be integrated across a larger region, including the landscapes that make up that region. The challenge of integration involves a paradox of scale: in some cases large-scale (regional) ecological systems can be most effectively regulated by small-scale (local) social organizations (Lee and Stankey 1992). Since peoples' interests, commitments, and knowledge are generally localized, bottom-up approaches that aggregate the local initiatives of citizens may be the most likely to succeed in achieving regional goals (Dryzek 1987). Experience throughout the world has shown that regional ecological stability is more likely to be achieved by permitting greater variability in land use practices and, within limits of critical biological thresholds, allowing and encouraging localized fluctuations in management practices (Korten 1987, Ostrom 1990, Wheatley 1993).

The paradox of scale is a general principle found to apply to systems as diverse as business organizations, chemical and physical processes, and ecological systems (Wheatley 1993). Stability in larger-scale processes arises when the smaller-scale processes are allowed freedom to operate. This is illustrated by business organizations when individual and small-group initiatives respond to prices or other incentives by developing resources within the limits set by larger-scale organizations.

Maintenance of local initiatives, commitments, and knowledge also helps promote sustainability by insulating the management of ecological processes from the political cycles that affect large-scale organizations. National- and state- or provincial-level policies for regulating ecological systems are generally affected by the policies and preferences of political elites currently in power. Since political elites cycle in and out of power, especially in democratic systems, top-down control becomes a source of substantial instability. Hence, ecological regulations that rely on top-down control become highly unstable, and result in levels of unpredictability that discourage local initiatives requiring long-term commitments and investments of time and money. Sustainable watershed management requires the social and political stability often associated with local initiatives (also see Firey 1960).

The continuity of commitments and knowledge embodied in small-scale local organizations also helps foster effective adaptive management (Lee 1993, Pinkerton 1993). Political cycles in top-down administrations make it difficult to sustain long-term data-gathering and monitoring. But even more difficult for large-scale organizations are commitments to take experimental actions for purposes of monitoring results. When commitments to experiments are made, they often lack the diversity of trials necessary for eliminating multiple rival hypotheses about the operation of complex systems. Local initiatives, when supported by the generalized commitment of large-scale organizations, can be far more effective in implementing adaptive management (Lee 1993). Experimental practices are insulated from the influence of political elites, fostering commitments to undertake experiments, and ensuring that a diversity of trials will be put in place (McLain and Lee 1996).

Formulating Shared Socioenvironmental Visions

The paradox in many environmental and social approaches is that they have not proven to be effective over the long-term (>10 yr). The evidence is clear: loss of species, destruction of habitat, declining productivity, unstable social systems, and the disintegration of cultures are occurring on regional to global scales. How might these trends be reversed? Can it be accomplished by accepting risk at individual to institutional scales? One approach is to develop a shared socioenvironmental vision of future conditions (Figure 26.5). In an ideal sense, this may prove to be a nearly impossible task, although the process of trying to identify socioenvironmental endpoints for the short- and long-term is an exercise that aids communication and acts as an effective form of education about the diverse cultural beliefs and values embedded in a watershed. For example, environmental endpoints may be related to the extent and condition of riparian forests, to acceptable levels of water quality and aquatic habitat, or the persistence of viable populations of ecologically or culturally valuable plants and

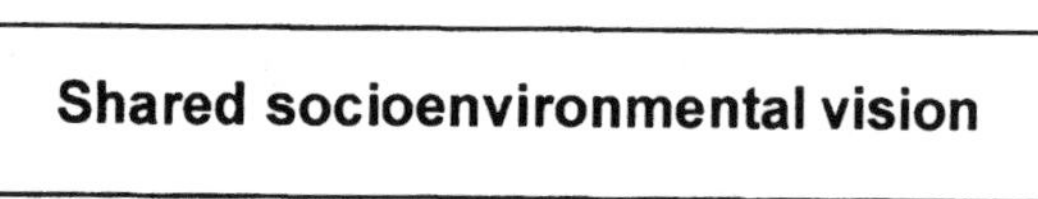
Identify criteria
• Environmental endpoints
– Riparian forest condition
– Species persistence
– Water and habitat quality
• Social endpoints
– Literacy
– Adaptive institutions
– Partnerships
– Stewardship and responsibility
Personal responsibility and stewardship
• Long-term commitments by leadership
• Empowerment of citizens
• Communication of vision
• Education about value of vision
• Active monitoring
• Continued learning
Shared socioenvironmental vision

FIGURE 26.5. Components of a shared socioenviromental vision for watersheds.

animals. Social endpoints may relate to the level of literacy about the structure and functioning of the socioenvironmental system, the development of flexible (or adaptive) institutions that are able to respond to new and as yet unforeseen issues, the formulation of unique partnerships between private industry, citizens, academia, tribes, and government, and the realization of levels of personal stewardship and responsibility that allow for the long-term maintenance of a balanced socioenvironmental system.

Successful examples of the development of shared socioenvironmental visions, and the various methods used to attain those visions, can be found for British Columbia (Fraser River), Florida (Kissismee River), New England (Connecticut River), Northern California (Metolius River), western Washington (Willapa Bay), and many other watersheds where citizens share a common concern about their future. However, there are two fundamental aspects inherent in the successful attempts: long-term commitment by the citizens who initially provide much of the vision and leadership, and empowerment of citizens with the responsibility for their own future (Lee 1993).

Public Stewardship in Watershed Management

Concerned and educated citizens are fundamental to watershed management. They represent an essential reservoir of human resources whose involvement can benefit management organizations and increase the overall level of awareness of socioenvironmental conditions. Watershed management requires thoughtful stewardship that cannot be attained solely by government regulations or technical specialists. Citizens can play an important role in monitoring socioenvironmental conditions, but they require continuing education to keep abreast of scientific and cultural advances.

Monitoring

Public involvement in coordinated monitoring activities instills a sense of ownership. Citizens taking an active interest in changes within the watershed provide inputs to decision makers based on firsthand, objective observations. This is a learning opportunity for those setting policy as well as those seeking to influence watershed management decisions.

There are unique advantages to including citizens in monitoring programs. First, with limited institutional budgets and staff availability, funds for collecting information about watershed features are usually directed to sites believed to be severely degraded. Many watersheds in need of monitoring are ignored unless volunteer efforts are undertaken. Thus, monitoring by volunteer groups or networks of individuals provides valuable information on watershed condition that may not be high on political priority lists. Second, public involvement in monitoring projects helps ensure data continuity. Staff turnover and job transfers within public agencies and large landowner organizations often occur at rates less than the duration of monitoring programs, resulting in discontinuities in data collection or undocumented changes in techniques. Local citizens working with public and private organizations fill gaps inevitably created when monitoring staffs change, and provide insight to new staff members that might not be otherwise obtained within existing organizational structures. Third, the sheer numbers of citizens available to assist with monitoring make it possible to conduct large-scale adaptive management experiments that would otherwise be impossible with limited agency or landowner resources. However, there are some disadvantages which include the challenge of maintaining interest over long periods of time and potential inconsistency in data collection (Ralph et al. 1994).

Understanding changes in watershed conditions requires distinguishing between localized and large-scale effects, assessing system responses that separate human-related impacts from uncontrolled environmental factors, and having institutional agreements that provide for decades-long measurements (Walters and Holling 1990). These requirements generally go beyond the capabilities of individual organizations; thus, cooperative monitoring programs must become the rule instead of the exception.

Expectations of the abilities of citizens to take samples and perform routine scientific tests must be tempered by a general lack of advanced technical training. Therefore, monitoring tasks need to focus, in most cases, on measurements readily understandable and not requiring specialized skills. Often this precludes the collection of biological samples. However, there are a number of monitoring activities well within the abilities of average citizens, including:

Photographs—The importance of time-series photographs cannot be overstated. Some of the most valuable information about historical conditions is derived from photographs, particularly those where locations can be clearly identified. In addition, reference photopoints within watersheds are helpful in tracking long-term trends in vegetative structure and stream conditions. Reference photopoints can also be used to display the effects of seasonal changes and large disturbances such as floods and fires. Important photographs often exist in family albums or businesses, and public involvement can bring these historical records to light.

Water Samples—Long-term trends in water quality require regularly scheduled sampling, but the number of sites that can be routinely monitored by agencies is limited by the availability of automated sampling equipment and staff time. For example, the U.S. Geological Survey (USGS) monitored water quality parameters in many watersheds after passage of federal water laws in the 1960s and 1970s, but was forced to abandon many sites in the late 1970s when funding for monitoring programs expired. A network of water sampling locations established within a watershed, with periodic samples obtained by local volunteers, is an especially effective means of monitoring easily observed parameters such as suspended sediment. Monitoring programs can be coordinated by appropriate regulatory agencies, which would supply sample bottles and instructions for handling and process samples. Likewise, maximum–minimum thermometers placed throughout the watershed and checked at regular intervals by citizen's groups or individual landowners provide an indication of changes in temperature fluctuations over time. Easily measured parameters such as sediment and temperature have immediate, significant effects on aquatic ecosystems. They also provide important information about erosion and upstream riparian conditions.

Habitat Measurements—Stream morphology is an integrative measurement of overall watershed condition, and pools are very sensitive to change. Pools are important habitat for certain types of aquatic organisms, including many fish species. Citizen participation in simple habitat measures such as counting the number of large pools increases the area of a watershed for which inventory information is available. Sportsman's clubs and conservation organizations (including adopt-a-stream groups) are especially suited to this type of project. For example, in the Willapa Basin seasonally unemployed fisherman collect fish habitat information throughout the watershed. Training is provided by the Willapa Alliance, who also oversee and coordinate the field effort and compile and analyze the data. Funding is provided by a federal program.

Riparian Forest Surveys—Riparian forests are critical to watershed health, yet insufficient attention is paid to their condition. Riparian plots where surveyors periodically identify and count the number of trees within the boundaries, measure changes in species composition and growth, and note causes of mortality provide integrated long-term information about watershed characteristics. Plots do not have to be revisited every year, as long as their locations are well documented; they can be resurveyed by the same group or rotated among several groups over longer periods. Information generated by these surveys is useful for verifying remote sensing data, providing riparian vegetation information for watershed analysis, and teaching citizens about the dynamic nature of watershed processes.

Socioeconomic Conditions—Socioeconomic conditions are already well monitored, but data are seldom given for conditions within watershed boundaries. Annually based socioeconomic indices may include such factors as annual capital investment, resource exports, unemployment, high school literacy, healthy births and so forth, which provide essential information about human conditions in a larger community including a watershed (Table 26.4). Commitments to maintain and enhance the biological and physical conditions of watersheds are most likely to arise when local economies and societies are healthy and the population is well informed.

Public Outreach

Effective watershed management requires that scientists and managers provide knowledge about watershed processes and management techniques to citizens on a regular basis. Although citizens and local groups usually act with good intentions, they do not always have the benefit of current professional insights into human and environmental processes. The result may be that restoration and enhancement projects fail to achieve their objectives, or worse, that they actually impair socio-environmental functions. For example, stream cleaning projects have largely been discontinued by public agencies, but are occasionally sponsored by citizen groups. When asked why these projects have been undertaken, many citizens continue to be unaware of the ecological functions of woody debris or to regard these functions as secondary in importance to the need to provide unimpeded fish passage (even woody debris removal diminishes fish production in the long term).

How can the educational and scientific communities maintain socioenvironmental literacy and instill a sense of stewardship among citizens? First, scientists need to explain the importance of watershed connectivity, the role of natural disturbances in maintaining productivity, the need to view watershed management in terms of large landscape units, and how social and environmental components work as an integrated system. Agricultural and forestry extension services, where citizens turn for advice from local specialists familiar with the region, serve as models for the establishment of an integrated watershed extension service. Watershed extension specialists, serving as local sources of the latest information, can act as liaisons between small and large landowners, natural resource consumers, and management agencies.

Second, colleges and universities must do more to educate citizens about important watershed management issues. Although educational institutions sponsor many meetings, the presentations are often too technical for citizens. Weekend or evening workshops aimed at communicating applied watershed science to a general audience are needed to facilitate increased public understanding of management options. Workshops featuring a combination of university faculty, research scientists, resource managers, citizens, and environmental policy makers are essential if we are to develop effective watershed management based on an integrated socioenvironmental perspective.

Fundamental Principles

Watershed management is an ongoing experiment guided by fundamental principles and a common vision of the future, and utilizes a multitude of approaches to achieve an integrated and balanced socioenvironmental system (Naiman 1992, Lee 1993). There is no universal methodology for achieving effective watershed management. However, fundamental principles related to cooperation, balance, fairness, integration, communication, and adaptability can help guide the process:

1. Recognize that watershed management demands unparalleled cooperation among citizens, industry, governmental agencies, private institutions, and academic organizations. In most situations, the complexity of information processing and the scope of socioenvironmental change exceeds the capacity of any single group to manage a watershed effectively.
2. Balance technical solutions (e.g., fish hatcheries, waste management, and so forth) to specific human-generated problems with the wide-scale maintenance of appropriate environmental components that provide similar ecological services.
3. Minimize decisions based only on conceptualization and perception; data-driven policy and management decisions need to become the standard for resolving issues.
4. Apply regulations guiding the structure and behavior of the socioenvironmental system evenly and fairly throughout the watershed. For example, basic regulations (such as riparian protection and chemical applications) should not differ across forestry, agricultural, and urban areas but should encourage citizen initiatives and landowner incentives that result in greater protection and reduced chemical applications.
5. Accept human activities as fundamental elements of the watershed along with the structure and dynamics of the environmental components. Both have inherent rights to exist for the long term.

These principles, when combined with approaches outlined in this chapter, provide only the initial steps in achieving effective watershed management. Cultural values, social behavior, and environmental characteristics will continue to evolve. Unfortunately, a critical evaluation of the approaches for watershed management outlined here will not be possible for several decades. Will the evaluation be positive? If so, it will be because citizens, regulators, educators, and industries shared a common long-term vision and adapted to change by implementing appropriate approaches to meet that vision.

Literature Cited

Bartell, S., R.H. Gardner, and R.V. O'Neil. 1992. Ecological risk estimations. Lewis, Chelsea, Michigan, USA.

Bella, D.A. 1987. Organizations and systematic distortion of information. Journal of Professional Issues in Engineering **113**:360–370.

Berry, M., R. Flamm, B. Hazen, and R. MacIntyre. 1996. LUCAS: A system for modeling land-use change. IEEE Computational Science and Engineering **3**:24–35.

Botkin, D.B. 1990. Discordant harmonies: A new ecology for the 21st century. Oxford University Press, New York, New York, USA.

Burch, W.R., and D.R. DeLuca. 1984. Measuring the social impact of natural resource policies. University of New Mexico Press, Albuquerque, New Mexico, USA.

Campbell, D.T. 1969. Reforms as experiments. American Psychologist **24**:409–429.

Décamps, H., R.J. Naiman, and J.C. Lefeuvre. 1998. Landscape ecology and regional development. *In* M. Hadley, ed. Integrating conservation, development, and research. UNESCO, Paris, France. (In press).

Drake, J.A., H.A. Mooney, R. diCastri, R.H. Groves, F.J. Kruger, M. Rejmánek, et al. eds. 1989. Biological invasions: A global perspective. SCOPE 37. John Wiley & Sons, New York, New York, USA.

Dryzek, J.S. 1987. Rational ecology: Environment and political choice. Basil Blackwell, Oxford, UK.

Fava, J., L. Barnthouse, J. Falco, M. Harwell, and K. Reckhow. 1991. Peer review workshop report on a framework for ecological risk assessment. US EPA Risk Assessment Forum, Washington, DC, USA.

Fetherston, K.L., R.J. Naiman, and R.E. Bilby. 1995. Large woody debris, physical process, and riparian forest development in montane river networks of the Pacific Northwest. Geomorphology **13**:133–144.

Finstenbusch, K., and C.P. Wolf. 1981. Methodology of social impact assessment. Hutchinson and Ross, Stroudsburg, Pennsylvania, USA.

Firey, W.A. 1960. Man, mind, and land: A theory of resource use. Free Press, Glencoe, Illinois, USA.

Flamm, R.O., and M.G. Turner. 1994a. Multidisciplinary modeling and GIS for landscape management. Pages 201–212 *in* V.A. Sample, ed. Forest ecosystem management at the landscape level: The role of remote sensing and integrated GIS in resource management planning, analysis and decision making. Island Press, Washington, DC and Covelo, California, USA.

Flamm, R.O., and M.G. Turner. 1994b. Alternative model formulations of a stochastic model of landscape change. Landscape Ecology **9**:37–46.

Fortin, M.J. and P. Drapeu. 1995. Delineation of ecological boundaries: Comparison of approaches and significance tests. Oikos: A Journal of Ecology **72**:323.

Gregory, S.V., F.J. Swanson, and W.A. McKee. 1991. An ecosystem perspective of riparian zones. BioScience **40**:540–551.

Grumbine, R.E. 1994. What is ecosystem management? Conservation Biology **8**:27–38.

Harwell, M.A., and J. Gentile. 1992. Report on the ecological risk assessment guidelines strategic planning workshop. US EPA Risk Assessment Forum, Washington, DC, USA.

Harwell, M., J.F. Long, A. Bartuska, J.H. Gentile, C.C. Harwell, V. Meyers, and J.C. Ogden. 1996. Ecosystem management to achieve ecological sustainability: The case of south Florida. Environmental Management **20**:497–522.

Johnston, C.A., N.E. Detenbeck, and G.J. Niemi. 1990. The cumulative effect of wetlands on stream water quality and quantity: A landscape approach. Biogeochemistry **10**:105–141.

Karr, J.R. 1991. Biological integrity: A long-neglected aspect of water resource management. Ecological Applications **1**:66–84.

Korten, D. 1987. Community management: Asian experience and perspectives. Kumaian Press, West Hartford, Connecticut, USA.

Le Maitre, D.C., B.W. van Wilgen, and D.M. Richardson. 1993. A computer system for catchment management: Background, concepts, and development. Journal of Environmental Management **39**:121–142.

Lee, K.N. 1993. Compass and gyroscope: Integrating science and politics for the environment. Island Press, Washington, DC and Covelo, California, USA.

Lee, R.G., R.O. Flamm, M.G. Turner, C. Bledsoe, P. Chandler, C.M. DeFerrari, et al. 1992. Integrating sustainable development and environmental vitality: A landscape ecology approach. Pages 499–521 *in* R.J. Naiman, ed. Watershed management: Balancing sustainability and environmental change. Springer-Verlag, New York, New York, USA.

Lee, R.G., and G.S. Stankey. 1992. Major issues associated with managing watershed resources. Pages 30–37 *in* P.W. Adams and W.A. Atkinson, eds. Balancing environmental, social political, and economic factors in managing watershed resources. Oregon State University, Corvallis, Oregon, USA.

Levin, S.A., ed. 1993. Science and sustainability. Ecological Applications **3**:550–589.

Ludwig, D., R. Hilborn, and C. Walters. 1993. Uncertainty, resource exploitation, and conservation: Lessons from history. Science **260**:17–36.

McLain, R.J., and R.J. Lee. 1996. Adaptive management: Promises and pitfalls. Journal of Environmental Management **20**:437–448.

Meyer, J.W., and R.W. Scott, eds. 1983. Organizational environments: Ritual and rationality. Sage, Beverly Hills, California, USA.

Montgomery, D.R., G.E. Grant, and K. Sullivan. 1995. Watershed analysis as a framework for implementing ecosystem management. Water Resources Bulletin **31**:1–18.

Naiman, R.J., ed. 1992. Watershed management: Balancing sustainability and environmental change. Springer-Verlag, New York, New York, USA.

Naiman, R.J., and H. Décamps. 1997. The ecology of interfaces – Riparian zones. Annual Review of Ecology and Systematics **28**:621–658.

Naiman, R.J., J.J. Magnuson, D.M. McKnight, and J.A. Stanford. 1995a. The freshwater imperative: A research agenda. Island Press, Washington, DC, USA.

Naiman, R.J., J.J. Magnuson, D.M. McKnight, J.A. Stanford, and J.R. Karr. 1995b. Freshwater ecosystems and management: A national initiative. Science **270**:584–585.

Ostrom, E. 1990. Governing the commons: The evolution of institutions for collective action. Cambridge University Press, New York, New York, USA.

Pinkerton, E.W. 1993. Co-management efforts as social movements. Alternatives **19**:34–38.

Ralph, S.C., G.C. Poole, L.L. Conquest, and R.J. Naiman. 1994. Stream channel condition and in-stream habitat in logged and unlogged basins of western Washington. Canadian Journal of Fisheries and Aquatic Sciences **51**:37–51.

Reice, S.R. 1994. Nonequilibrium determinants of biological community structure. American Scientist **82**:424–435.

Reice S.R., R.C. Wissmar, and R.J. Naiman. 1990. Disturbance regimes, resilience, and recovery of animal communities and habitats in lotic ecosystems. Environmental Management **14**:647–659.

Reich, R. 1991. The work of nations: Preparing ourselves for 21st century capitalism. AA Knopf, New York, New York, USA.

Risser, P.G. 1993. Ecotones. Ecological Applications **3**:367–368.

Schrader-Frechette, K.S. 1991. Risk and rationality: Philosophical foundations for popular reforms. University of California Press, Berkeley, California, USA.

Selznick, P. 1966. TVA and the grassroots: A study on the sociology of formal organizations. Harper & Row. New York, New York, USA.

Slovic P. 1984. Low-probability/high consequence risk analysis and the public. Pages 505–508 *in* R.A. Waller and V.T. Covello, eds. Low-probability high-consequence risk analysis. Plenum, New York, New York, USA.

Sokal, R.R., and R.J. Rohlf. 1981. Biometry. W.H. Freeman & Co., San Francisco, California, USA.

Stanford, J.A., and J.V. Ward. 1992. Management of aquatic resources in large catchments: Recognizing interactions between ecosystem connectivity and environmental disturbance. Pages 91–124 *in* R.J. Naiman, ed. Watershed management: Balancing sustainability and environmental change. Springer-Verlag, New York, New York, USA.

Suter, II G.W. 1992. Ecological risk assessment. Lewis, Boca Raton, Louisiana, USA.

Tolt Watershed Analysis Prescriptions. 1993. Weyerhaeuser, Tacoma, Washington USA.

Turner, M.G. 1990. Spatial and temporal analysis of landscape patterns. Landscape Ecology **4**:21–30.

Turner, M.G., V.H. Dale, and R.H. Gardner. 1989a. Predicting across scales: Theory development and testing. Landscape Ecology **3**:245–252.

Turner, M.G., R.V. O'Neill, R.H. Gardner and B.T. Milne. 1989b. Effects of changing spatial scale on the analysis of landscape pattern. Landscape Ecology **3**:153–162.

Turner, M.G. and R.H. Gardner, eds. 1991. Quantitative methods in landscape ecology. Springer-Verlag, New York, New York, USA.

Turner, M.G., G.J. Arthaud, R.T. Engstrom, S.J. Hejl, J. Liu, S. Loeb, et al. 1995. Usefulness of spatially explicit animal models in land management. Ecological Applications **5**:12–16.

Turner, M.G., D.N. Wear, and R.O. Flamm. 1996. Land ownership and land-cover change in the southern Appalachian Highlands and the Olympic Peninsula. Ecological Applications **6**:1150–1172.

US EPA. 1993. A framework for ecological risk assessments. US EPA Risk Assessment Forum, Washington, DC, USA.

US Government. 1994. A federal agency guide for pilot watershed analysis. Version 1.2, Pacific Northwest Research Station, US Forest Service, Portland, Oregon, USA.

US MAB. 1994. Isle au Haut principles: Ecosystem management and the case of south Florida. Man and the Biosphere Program, US Department of State, Washington, DC, USA.

Volkman, J.M., and K.N. Lee. 1994. The owl and Minerva: Ecosystem lessons from the Columbia. Journal of Forestry **92**:48–52.

Walters, C.J., and C.S. Holling. 1990. Large-scale management experiments and learning by doing. Ecology **71**:2060–2068.

Warwick, C.J., J.D. Mumford, and G.A. Norton. 1993. Environmental management expert systems.

Journal of Environmental Management. **39**:251–270.

Washington Forest Practices Board. 1994. Standard methodology for conducting watershed analysis. Version 2.1, Washington Department of Natural Resources, Olympia, Washington, USA.

Wear, D.N., M.G. Turner, and R.O. Flamm. 1996. Ecosystem management with multiple owners: Landscape dynamics in southern Appalachian watershed. Ecological Applications **6**:1173–1188.

The Willapa Alliance and Ecotrust. 1995. Willapa indicators for a sustainable community. Unpublished report. South Bend, Washington, USA.

Wheatley, M.J. 1993. Leadership and the new Science: Learning about organization from an orderly universe. Barrett-Koehler, San Francisco, California, USA.

27
Paradigms, Policies, and Prognostication about the Management of Watershed Ecosystems

Michael C. Healey

Overview

• In the Pacific Northwest, as elsewhere, the adverse consequences of historical approaches to water and river basin management are forcing a reappraisal of management policy and philosophy. Managing from an ecosystem perspective is touted as the way to ensure maintenance of ecological integrity and biodiversity in the future. This chapter describes some of the obstacles to realizing the goal of watershed ecosystem management and how they might be surmounted.

• A fundamental problem with ecosystem management is in defining ecosystems that are sufficiently autonomous and about which there is sufficient understanding of causal mechanisms to attempt management. Watershed ecosystems appear to have these necessary characteristics.

• Watersheds also constitute natural geographic units on which management institutions can be based. At present, however, few jurisdictional boundaries correspond with watersheds, and the jurisdictional boundaries of the various institutions that should be included in a program of ecosystem management do not correspond with watersheds either. Such correspondence is a necessary condition for ecosystem management.

• Although scientific understanding of ecosystems is important, such understanding alone is not sufficient. Furthermore, ecosystems are complex and nonlinear. Their behavior defies simple quantitative prediction. As a consequence, ecosystem management must depend heavily on judgment under uncertainty and on linking science with environmental policy through adaptive experimentation.

• The process of watershed management can be constructed around four key questions: (1) what is the nature of the ecosystem; (2) what are society's desired ecosystem conditions; (3) what is feasible; and (4) how can desired conditions be achieved? These questions provide a basis for defining sets of human activities that are ecologically sustainable, economically viable, and socially acceptable. This integration of environment, economy, and society is at the heart of ecosystem management.

• The natural and social sciences are capable of providing a sufficient answer only to the first question. The remaining questions have more to do with human values and the tradeoffs that must be made in achieving social consensus. Nevertheless, ecological science can help to inform debates about values and the policies and management programs to achieve particular ecological objectives.

• A number of watershed management experiments are in progress in the Pacific Northwest but none incorporates all of four essential principles of ecosystem management: (1) organized around naturally bounded ecosystem units; (2) based on a mix of environmentally sustainable human activities; (3) focused on ends rather than means; and (4) employs adaptive experimentation and learning.

• For two centuries social evolution in the western world has been driven by the resource development paradigm supported by reductionist science and engineering. Recently, however, problems have begun to emerge (e.g., global warming, declining human male sperm counts) that defy solution through traditional disciplinary analysis and technique. Such problems appear to demand more holistic and integrative approaches to managing human relations with the natural world. In the Pacific Northwest there is a clear opportunity to explore such approaches through experiments in watershed management.

Introduction

The Pacific Northwest, generously endowed with water, seems to be insulated from the impending global water crisis (Gleick 1993; Vidal, *Vancouver Sun*, 12 August 1995). Indeed, residents of the region are more likely to curse an excess of water than a scarcity, and approaches to the management of rivers and streams have been anything but cautious. Nevertheless, together with the rest of the world, the Pacific Northwest is becoming increasingly caught up in a global ecological revolution that will force a new perspective on water resources and their management. In fact, this new perspective is already emerging as the costs of humankind's historical disregard for the ecological role of water become apparent (Lee 1993, Reisner 1993, Black 1995) and as powerful interests seek ways to redistribute the region's perceived wealth (Bocking 1987, Windsor 1992).

The adverse ecological consequences of traditional approaches to river basin management are readily apparent. The waters of the Sacramento and San Joaquin Rivers in California are overcommitted to the extent that the flow of water is sometimes reversed in the San Joaquin Delta. Once world-renowned salmon (*Oncorhynchus* spp.) populations in these rivers have been reduced to remnants in danger of extinction, and changes in a myriad of less famous species testify to lost ecological integrity (Brown 1986, Black 1995). So many dams have been constructed on the Columbia River and its tributaries that this once turbulent waterway is now reduced to a series of pools. Upstream migration of adult salmon is seriously impeded, and the downstream migration of their offspring is significantly delayed. Accumulations of squawfish in the reservoirs and below the spillways of the dams deplete the numbers of migrating salmon smolts (Lee 1993). Hatcheries that were intended to compensate for the lost salmon have not replaced wild salmon production and, in fact, may have exacerbated the problems of wild stock conservation (Hilborn 1992). Damming and diversion of waters from a tributary of the upper Fraser River, British Columbia's most productive salmon river, are believed to have imperiled valuable runs of sockeye and chinook salmon. Technological measures to mitigate the effects of this diversion have been repudiated (BCUC 1994). In the lower Fraser, diking, channelization, and a cocktail of industrial, agricultural and domestic effluents have contributed to a 70% loss of wetland habitat, dramatically altered carbon budgets for the river, and caused a broad spectrum of pathological abnormalities among resident fishes (Healey 1994, Healey and Richardson 1996). Hydropower development on the Peace River has resulted in subtle changes in the hydrology of the Peace-Athabasca delta in northern Alberta. These changes appear to have significantly and permanently altered successional processes in the delta, a United Nations World Heritage ecosystem (Healey 1994).

As these dramatic examples testify, approaches to river management during the last century have brought about a myriad of undesirable consequences that are now greatly regretted and must be redressed (Lee 1993, Canada 1995, NRBS 1995, US Fish and Wildlife Service 1995). In large measure, these undesirable side effects of historical river management derived from a lack of understanding of rivers and river basins as dynamic ecosystems. However, they also derived from the assumption that it was right for humans to take command of these ecosystems and that our clever technology gave us mastery over the laws of nature. These attitudes are still present as evidenced by the recent resurrection of North American Water and Power Alliance (NAWAPA)

(Windsor 1992). NAWAPA is a monumental proposal to replumb western North America and divert its northward flowing and many of its westward flowing rivers south to where, according to NAWAPA's proponents, the people are and the water is needed. No one should presume that the economic, ecological, and ethical absurdity of NAWAPA will guarantee its rejection. The history of western water development is a legacy of such schemes (Reisner 1993).

Yet, the world is changing. Ecologically sensitive approaches to development are now debated and promoted both nationally (CCREM 1987, Federal Interagency Floodplain Management Task Force 1992) and internationally (WCED 1987, United Nations 1993). Ecosystem management, sustainable development, ecological integrity, and biodiversity are the new buzzwords of environmental management. There is widespread anticipation and expectation that humankind is on the threshold of a new way of relating to the natural world—a way that is cognizant of both humankind's complete dependence on and fundamental ignorance of the natural world (Costanza 1991)—a way that links environment, economy, and society in an "ecosystem approach" to management. Despite enthusiasm for the ecosystem approach in recent resource management literature, however, the means of implementing this new relationship remains elusive. This chapter presents some of the obstacles to realizing the goal of ecosystem management of rivers and their catchments and some ways of imagining how these obstacles might be overcome.

Why Watersheds?

The first and perhaps most significant obstacle is the general lack of consensus about what the phrase *ecosystem management* means. Presumably ecosystem management involves a focus on ecology and ecosystems, but this is not as helpful as one might hope. The concept of the ecosystem is itself rather fuzzy (Simberloff 1980, Worster 1990). One important problem with defining an ecosystem is determining boundaries that are objective and establish a meaningful degree of autonomy for the "system" inside. Without such boundaries the ecosystem can be criticized as simply an arbitrary patch on the global landscape. Another important problem is defining a set of causal relationships that have broad ecological meaning and are also subject to human manipulation and adjustment. Without such relationships the management of ecosystems is impossible. Although by no means a perfect solution, using watersheds as the ecological units for structuring the concept of ecosystem management helps resolve these issues.

First of all, water is the integrating resource in all ecosystems. All organic life depends on water and all living entities are either continuously bathed in water or have developed elaborate packaging designed to conserve their internal water. Through the hydrological cycle, water is continually cycled among great reservoirs in the oceans, in the atmosphere, in the soil, and on the surface of the earth, creating a communication network that links organisms, materials and processes in every part of the biosphere (Figure 27.1). On land, watersheds (or catchments) are the natural geographic units that capture water from the atmospheric reservoir and cycle it either back to the atmosphere or to the soil, surface water, and oceanic reservoirs. Each catchment has its own geological, morphological, hydrological, and biological characteristics. Stream and river basins, thus, form independent, relatively autonomous and easily identifiable "ecosystem" units within the overall context of the hydrological cycle. They fulfill the fuzzy definition of an ecosystem reasonably well.

Furthermore, larger catchments are comprised of smaller catchments and stream segments that form a nested hierarchy of ecosystem units (Frissell et al. 1986). This makes possible the logical subdivision of larger catchments into sets of smaller catchments with varying degrees of ecological uniformity. The nested hierarchy of smaller catchments within larger makes sense both ecologically and intuitively. Each catchment within the nested hierarchy can be characterized as a mosaic of

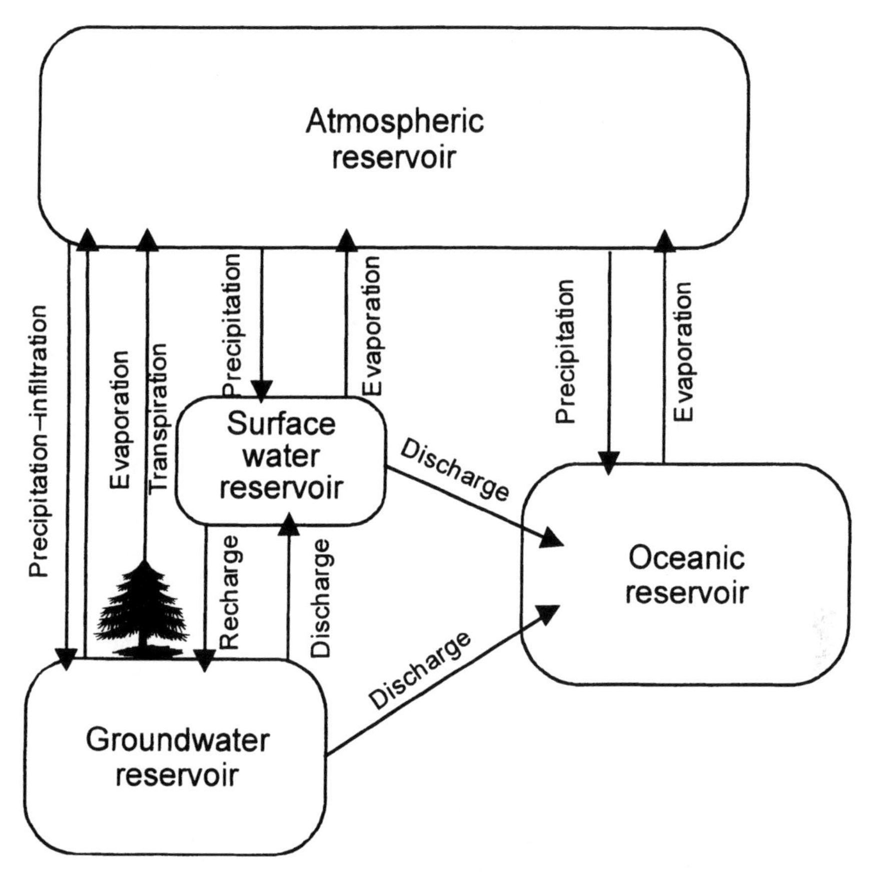

FIGURE 27.1. The hydrological cycle showing the major pathways by which water moves among reservoirs in the atmosphere, soil (groundwater), rivers and lakes (surface water), and ocean.

habitat or community types at several geographic scales and levels of detail. Many of the basic ecological processes of production, energy flow, and nutrient cycling in these watershed communities are understood. Throughout a watershed, rivers make important contributions to the linkages among habitat types and to the long-term evolution of land forms and habitats. Rivers are also often the most sensitive indicators of human-induced changes to an ecosystem. Thus, there is probably sufficient ecological understanding of watersheds to plan and implement ecosystem management with rivers and, perhaps, related groundwater basins as sensitive indicators of ecosystem change.

Watersheds not only make sense ecologically, they also constitute a natural geographic unit on which management institutions can be based. This is a feature that is particularly important for linking science with policy. The dendritic pattern of rivers is so familiar that the concept of watersheds within watersheds is one with which nonspecialists can immediately identify. The familiarity of the watershed concept provides a basis of common understanding that helps ecologists explain to policy makers more complex topics of ecosystem management, such as ecological integrity.

Finally, using watersheds as units for ecosystem management allows for a logical emphasis on the linkages between land and water (Table 27.1). These linkages have generally been ignored in the design of resource legislation and management agencies. The legislative and institutional separation of the land and waters, which has been mirrored in heroic feats of engineering to make the separation a reality, is pos-

TABLE 27.1. Examples of the impacts of the land on the river ecosystem and of the impacts of the river on the terrestrial ecosystem illustrating the interrelation between the two systems.

Impact of land on river ecosystem	Impact of river on terrestrial ecosystem
Local relief determines gradient	Carries surplus water away from terrestrial ecosystem
Local geology influences water chemistry, sediment load	Carries sediment and organic debris from upstream to downstream locations
Local groundwater regime determines base flow	Contributes to groundwater recharge in certain locations
Catchment vegetation influences water supply, storm hydrographs	Transports sediment and nutrients out of channel and onto flood plain during freshet
Riparian vegetation influences channel form, evolution, migration	Creates intermittent disturbance regime in flood plain affecting vegetational succession
Riparian vegetation influences stream temperature, light available for photosynthesis	Contributes to the creation of new land in lacustrine and estuarine deltas
Riparian vegetation provides organic carbon to river ecosystem	Creates new habitat for colonization by riparian species such as poplar
Riparian vegetation provides insect food to stream fishes	Provides food and water sources for numerous species that are primarily terrestrial

sibly the clearest indication of the ecological naiveté of traditional approaches to river basin management.

What Can and Cannot Be Known about Watershed Ecosystems?

Our daily confrontation with the ecological naiveté of traditional river and watershed management seems to belie the scholarly contents of this and other recent publications on river ecology and management (Gore and Petts 1989, Calow and Petts 1992, Naiman 1992). The preceding chapters clearly demonstrate the accumulating wealth of technical information about rivers and their catchments. Such knowledge is not simply an encyclopedia of unconnected facts. A growing list of integrating concepts—the river continuum concept (Vannote et al. 1980), the flood pulse concept (Junk et al. 1989), the serial discontinuity concept (Ward and Stanford 1983), the riverine productivity concept (Thorp 1994)—provides structure to the facts and a rich intellectual framework for speculating about the response of catchments to human activity. Yet, this wealth of knowledge about rivers has not paved the way to ecosystem management. The key to ecosystem management may lie in further research and study. This argument is particularly appealing to those who see ecosystem and environmental management primarily as a technical problem. Paradoxically, however, the problem posed by ecosystems and by watersheds as particular examples of ecosystems is at once both a technical problem and a problem not resolvable by technological means.

This apparent paradox arises from two important and possibly interrelated features of ecosystems. The first of these is that ecosystems are "medium number" systems (O'Neill et al. 1986). That is to say, ecosystems are made up of a moderate number (a few hundred to a few thousand) of interacting subsystems. It is virtually impossible to predict the future states of such a system when it is disturbed. The number of interactions is too large for straightforward analytical solution (as with the behavior of planets in a solar system) and too small to be smoothed out through some emergent law of averages (as the behavior of molecules in a vessel of gas is averaged in the gas laws). Attempting to resolve the behavior of ecosystems through the study of their interacting subsystems is rather like trying to discover the gas laws by studying the behavior of individual molecules.

The second feature is that ecosystems display patterns suggestive of chaotic behavior (Schaffer 1985). Whether ecosystem behavior is truly chaotic remains to be resolved. Nevertheless, ecosystems are characterized by "surprise" events on a wide range of time and space scales (Holling 1987 and 1992, Healey 1990, Costanza et al. 1993). Rivers are the embodiment of dynamic hydrological forces operating within a heterogeneous and complex physical matrix (their "catchment"). They are particularly likely to deliver dramatic surprises, including floods, abrupt channel shifts, and debris torrents, all of which have associated ecological consequences.

This "unknowable" character of rivers and river basins is part of their fascination as ecosystems. But their "unknowableness" also means it is not possible to predict their behavior the way that the behavior of structural materials in a bridge or the airfoil of a jet plane can be predicted. Fortunately, this does not mean that the goal of ecosystem management must be abandoned. What it does mean is that approaches to the management of ecosystems must differ from approaches to the management of traffic on highways or the exploitation of individual fish populations. In the latter two instances, management is based on simple analytic models that predict quantities (e.g., vehicles, fish) that can be accommodated or harvested in a specified period of time. Such quantitative statements about ecosystem behavior may never be possible. Questions about the quantitative behavior of ecosystems are typically of the sort that Weinberg (1972) termed "transscientific." These are questions that can be framed in the language of science but cannot be answered by the traditional means of science. A familiar example of such a transscientific question about river ecosystems is: "How much can a river's hydrology be altered without endangering its ecological integrity?" This question is at the heart of the ongoing debate about in-stream flows for fish and other aquatic life. Notwithstanding increasingly elaborate attempts to provide a technical solution (Walder 1996), the question is not soluble by traditional reductionist science. It is not soluble because the solution demands an orderliness and a consistency in the behavior of riverine ecosystems that does not exist. Such questions can only be answered in terms of relative risk to ecological integrity with different models or approaches often giving different results (Figure 27.2).

The Process of Watershed Management

The fact that watershed ecosystems are "unknowable" and "unpredictable" in the ways that would allow management to be based primarily on technical understanding does not mean that natural scientists have been wasting their time. The level of technical understanding achieved so far about the behavior of rivers and river basins is crucial to any effective program of ecosystem management. By revealing what can and cannot be said about the behavior of rivers, science has opened the way to a workable process for ecosystem management. However, it is a process that will be much more dependent on judgment under uncertainty and on linking science with environmental policy and management than is traditional.

The process of watershed management can be constructed around four key questions (Figure 27.3): (1) what is the nature of the ecosystem; (2) what are society's desired ecosystem conditions; (3) what is feasible; and (4) how can desired conditions be achieved? These questions are seldom explicitly stated as a basis for deciding issues of resource and environmental management but they are implicit in every management action. Technical information (from both the natural or the social sciences) can provide a sufficient answer to the first question. However, technical information is not sufficient and is necessary only in varying degrees for answering the remaining three.

The Role of Science

The scientific study of watershed ecosystems makes three important contributions toward answering the first question (Weinberg 1972). The first of these is a technical description of the elements that make up the ecosystem and their functional interrelationships. This is not a casual description; rather, it is a description that has been subjected to the rigorous process of quality control that characterizes the scientific method. The description is, therefore, a credible one. It is this kind of description that most of this book is about. Such a description is a

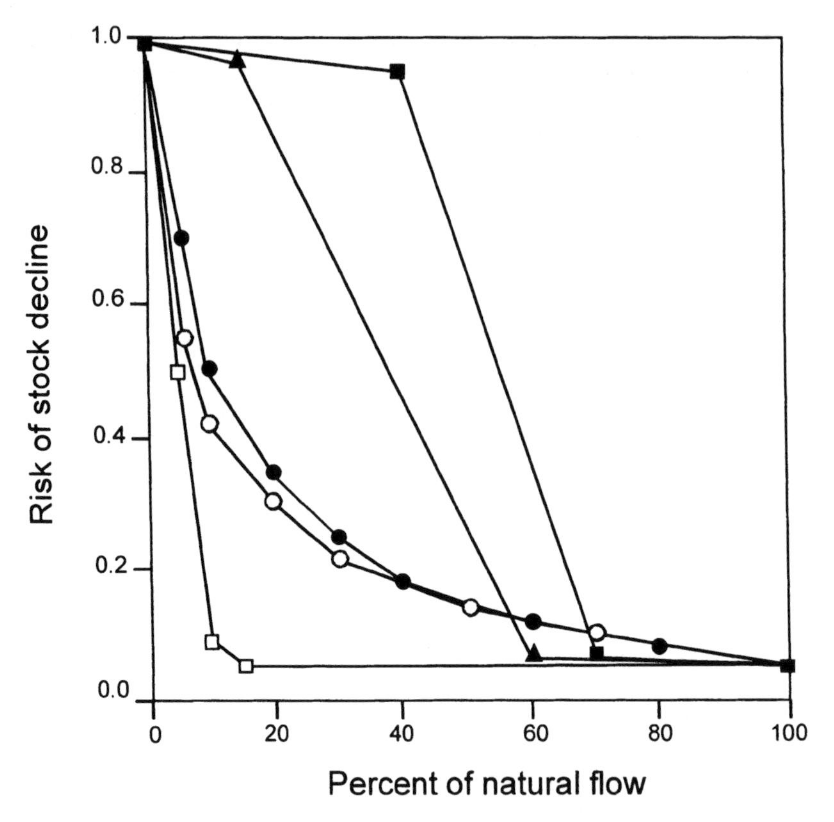

FIGURE 27.2. Relative risk to chinook salmon conservation from changes in river flow in the Nechako River, B.C. Each risk curve is based on a different model linking fish habitat to river discharge (—■— = case study analysis [Burt and Mundie 1982]; —▲— = Tennant [1976]; —●— = length of wetted side channels [Russell et al. 1983]; —○— = wetted width [Swift 1976]; —□— = IFIM [Bovee 1982]). The variation in the curves illustrates the uncertainty in the relationship between flow and fish conservation.

necessary but not a sufficient basis for ecosystem management.

Second, science provides a codified language that allows informed and informative debate about the ecosystem in question. This is not a trivial contribution. The existence of an agreed terminology and rules for its use is the basis of communication. A recurrent source of complaint and confusion in the debate about ecosystem management, for example, has been over the meaning and application of such terms as "ecological integrity" and "biodiversity" (Angermeir and Karr 1994). Confusion and debate about core concepts such as these show clearly that the overall concept of ecosystem management is still in its early formative stages.

The third contribution of science is the paradigm of the ecosystem. The emergence of this paradigm is crucial. Without it there could be no conception of or prospect of ecosystem management. Furthermore, the concept of the ecosystem has gone beyond serving as a focus for "normal science" in ecology (Kuhn 1970) to providing a plausible world view and a basis for political theorizing (Paehlke 1989). The popularization of the ecosystem concept, spearheaded by the environmental movement in North America, is bringing about a wholesale change in the way society thinks about its relationship with the natural world. The significance of this shift in public consciousness is evidenced by the fact that a little more than a generation ago "ecology" was not in the political lexicon, whereas now prominent politicians write books on the subject (Gore 1992). It is hard to imagine a clearer example of how basic science has influenced public perception and public policy.

Values

The second question, "What are society's desired ecosystem conditions?" is fundamentally about values. Values drive the way humankind interacts with the environment. This question, therefore, highlights the importance of humankind in ecosystem management. Traditionally, ecosystem science has not considered humankind as a part of the ecosystem but rather as an external agent whose activities impact the ecosystem. In ecosystem management, however, it is important that humans be considered an integral part of the ecosystem. To do otherwise is to perpetuate the myth that humankind is separate from and independent of natural ecosystems. Excluding humans also distracts from the importance of human values in decisions about ecosystem management.

A clash of values is usually at the core of conflicts over land and water use. Several recent conflicts over river basin management in western Canada have revealed the intensity of the disagreement over values. A proposal to dam the Oldman River in southwestern Alberta, for example, raised a storm of protest and pitted irrigation farmers against environmentalists, Native Canadians, and other segments of the community (Table 27.2). Forty kilometers of highly productive trout stream and its associated riparian poplar forest as well as numerous archeological and Native Canadian spiritual sites were to be sacrificed, at public expense, in the construction of an irrigation reservoir. The lost trout and wildlife habitat was to be replaced by enhancement of rivers and coulees upstream from the reservoir while archeological sites were to be "salvaged" by

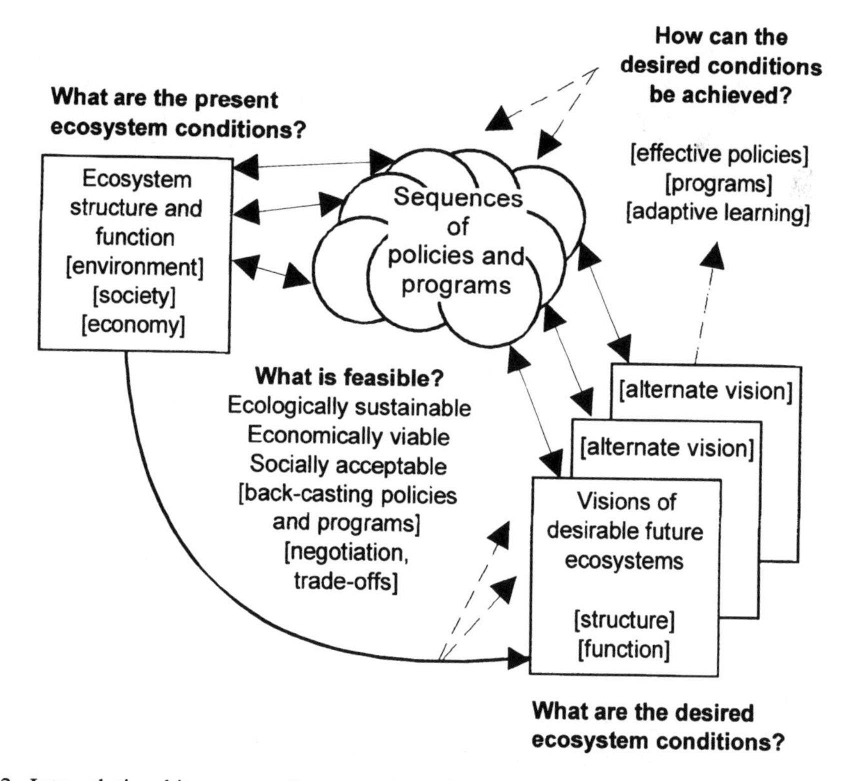

FIGURE 27.3. Interrelationships among four questions that are central to developing policies and programs for watershed ecosystem management.

TABLE 27.2. Values in conflict over the construction of the Oldman River Dam.

Values	Proponent's view	Opponent's view
Economic	Irrigation agriculture is the economic life blood of the region. The dam has a high benefit-cost ratio (>1.6) even from a national perspective.	Irrigation agriculture is sustained only through massive subsidy. The dam has a low benefit-cost ratio (<1.0) even from a regional perspective.
Environmental	The extensive program of fishery and wildlife habitat enhancement and new recreation facilities more than compensates for modest losses in natural habitats.	The project destroyed or endangered rare plant communities, bull trout, 40 km of productive trout stream, and regional biodiversity. The mitigation is short term and inadequate.
Heritage	The project led to the discovery and salvage of archeological materials that will keep anthropologists busy for a generation.	The project flooded irreplaceable archeological sites critical to Peigan Indian and Canadian heritage.
Peigan Indian	The project will provide water for irrigation agriculture on the Peigan reserve and enhance their economic opportunities.	The project destroyed Peigan historic and spiritual sites. Traditional lifestyles were endangered.
Social	The project removes barriers to social and economic growth caused by limited water.	Multigenerational farming families were evicted from their land within the reservoir. The Peigan were treated as second class citizens.

removing as many artifacts as possible. Despite the program of environmental mitigation, opponents of the project believed that components of the ecosystem that they valued highly (e.g., wild trout and charr, wintering habitat for deer, riparian poplar forests, rare plant communities, Native Canadian spiritual and archeological sites) would be either irreversibly destroyed or put at unacceptable risk.

The reservoir was intended primarily to assist a small number of irrigation farmers to grow crops, the sale of which must also be subsidized at public expense. Irrigation farming in southern Alberta is not economically viable in its own right. Yet the farmers are proud. They think of themselves as independent, beholden to no man, and are opposed to government interference in their lives. The farmers and their supporters believed that construction of the reservoir was necessary to sustain and enhance aspects of the ecosystem that they valued (e.g., the ability of its soils to produce crops and to generate economic wealth).

Although the clash of values among environmentalist, aboriginal, irrigation farmer, and other segments of society was particularly evident in this dispute, debate about values was not prominent in the various hearings that preceded and were concurrent with construction of the dam (Healey and Hennessey 1996). It appears to be a peculiarity of Western society that it abjures serious debate about values. Instead, debates about values are disguised as debates about science or technology. In the case of the Oldman River Dam, volumes of technical information were mustered by both sides to demonstrate that the reservoir was or was not economically justified as a public project, that impact analyses had or had not identified all the significant environmental consequences, and that proposed wildlife and fisheries mitigation was or was not technologically feasible (DOE 1992).

Expressions of values, although relevant, were neither necessary nor sufficient for decisions about the project. Questions about whose values should prevail were sometimes raised but seldom pursued. In the end, it was the politically better-organized irrigation farmers whose values prevailed. The decision to build the Oldman Dam was not a decision based on objective analysis. It was a decision confirming a particular vision of how best to manage the ecosystem of the Oldman River watershed. The vision espoused by the irrigation farmers, of an orderly and productive landscape, is rooted in the agrarian paradigm of taming and civilizing the wilderness with its attendant mythology about farmers and farming. Furthermore, the irrigation farmers argued that they were more environmentalists than any of the "tree hugging urbanites" who opposed the dam. They lived daily with the vagaries of nature, working the land; they left crop in the fields for the white tailed deer and prairie chicken; they were the stewards of the ponds and wetlands where the waterfowl bred; and it was they who would suffer first if the ecosystem was degraded. In their testimony the irrigation farmers argued that farming was simply applied ecology and that their activities improved rather than degraded the ecosystem.

The story of the Oldman Dam has played out many times in a variety of venues in western North America (Reisner 1993). Each has been a testimony to the power of the prevailing paradigmatic world view and the values inherent in that view. Each has also been a testimony to the power of scientific and technical information in the service of a particular set of values. The fact that experts on both sides of a watershed management dispute are able to put forward plausible ecological models to support their positions is also a testimony to the fact that there are no absolutes in ecology. The best available scientific information is open to a variety of interpretations. Which interpretation is most likely to influence policy and social decisions depends greatly on the prevailing societal values and how they are organized politically.

The answer to the question, "What are society's desired ecosystem conditions?" provides the foundation on which programs of resource management are built. The answer cannot be provided by science. Nevertheless, the science of ecology can help to inform debates about desirable future conditions and the policies and management programs to achieve them. Ecological science can also help speed the way to fuller acceptance of the ecosystem paradigm by revealing the contradictions, the traps and the dangers inherent in other paradigms. It is important to remember, however, that there are many ecologically acceptable visions of the future. Part of the challenge of ecosystem management will be to search among the alternatives for those that are most socially acceptable and economically viable (Figure 27.3).

Accepting Limits

A critical aspect of that search involves the third question, "What is feasible?" This question implies the existence of limits and constraints. Although there are important technical elements to the question, its ultimate answer rests heavily on tradeoffs among competing interests and values (Figure 27.3). The issue of feasibility raises the specter of winners and losers and of having to apply restraint. This question tends to get brushed aside as unworthy of humans in their role as the techno-bureaucratic regulators of the natural world. The idea of limits is particularly unpalatable in a social climate that eschews the very notion of limitation and in which winning is the only important outcome.

Society often tallies the tangible benefits and costs of its actions, but usually only as a means to maximize benefits to particular interests. In the imaginary world of free buyers and sellers, everyone ends up optimally satisfied. And when the physical size of the economy is growing rapidly, as it has been in the western industrialized world for several generations, every round of transactions makes most people "better off," creating the illusion that there are no limits (Daly and Cobb 1994). This paradigm of "no limits" is well exemplified by the way that the tradeoff between the capacity of rivers to produce salmon and their capacity to generate hydropower or irrigation water was resolved in the Pacific Northwest. Since "either–or" was socially unacceptable, the proponents of dams and reservoirs embraced the notion of "both–and." They called for both dams to store water and hatcheries or other fish-generating technology to replace the lost natural productivity. Even where the tradeoff was primarily a result of imbalance between natural productivity and the demand for fish to harvest, technology and "making the pie bigger" were seen as the solution. The present legacy of unproductive hatcheries and devastated salmon runs has been the result (Nehlson et al. 1991, Hilborn 1992, Black 1995).

If society is to adopt the ecosystem paradigm as a basis for management, it will be necessary also to accept the notion of limits. This may, in fact, be the greatest challenge of ecosystem management. It is a challenge that was effectively sidestepped by the World Commission on Environment and Development (WCED 1987) which "solved" the problem of low material living standards in undeveloped countries by projecting a several-fold increase in economic activity. As yet, the idea of limits has no place in the rhetoric or platform of any prominent political organization. The fallacy of limitless economic growth has and is being challenged on a number of fronts (Rees 1990, Daly 1991). Alternative points of view, however, have only a tenuous toehold in contemporary social and political institutions.

In exploring the issue of feasibility it is helpful to work backward from the desired future state of the ecosystem to its present condition by asking what chain of policies would have to be established to take us from where we are now to where we want to be (Figure 27.3). This "backcasting" has proven very helpful in exploring policy scenarios and the linkages between policy and programs in other complex social decisions (Robinson 1988). It also helps uncover inherent conflicts and contradictions in many visions of the future. By a process of iteration through the questions—what is the nature of the ecosystem, what are society's desired ecosystem conditions, and what is feasible—ecologically sound yet achievable future conditions can be determined.

Achieving Goals

The final question, "How can desired conditions be achieved?" implies the joining of policy and programs in a happy synergy. Policies establish broad social objectives while programs are the specific activities undertaken to move toward those objectives. In the case of dams and fish hatcheries, for example, the policy was to maximize the benefits to society through multiple use of rivers, and the programs were the construction and operation of dams, fishways, hatcheries, hydroelectric turbines, irrigation canals, and so on. Until recently, it seemed that this union was a problem-solving process. It is now being recognized as a problem-expanding process. Every human intervention in the natural world creates a multitude of adjustments, many of which necessitate further interventions, which themselves create adjustments requiring interventions.

The idea that policies and programs provide solutions is a legacy from industrial engineering in which technology often appears to provide durable solutions to problems. For example, once the physics was sufficiently well understood, with a few notable exceptions, bridges and buildings did not fall down, dams held back water, and engines did not blow up. But, in the case of these physical systems, the feedbacks from the natural world were relatively simple, and the man-made physical system (e.g., the bridge or the engine) was sufficiently "autonomous" that its behavior could be predicted and managed. In other instances, the new problems created by a particular solution were simply ignored, or became someone else's problem, and did not appear in the assessment of the effectiveness of a particular solution. Thus, changes in land-use patterns and ecosystem dynamics were not part of the equation when dikes were constructed to prevent floods in major rivers of the Pacific Northwest and elsewhere. Yet, the overall cost of flood damage has not declined as a result of the extensive diking programs, whereas the ecological production dynamics of the diked rivers has changed dramatically (Federal Interagency Flood Plain Management Task Force 1992, Healey and Richardson 1996). The apparent security provided by dikes encouraged construction on floodplains so that more structures were there to be damaged when the inevitable flood overtopped the dikes. Meanwhile, dikes denied the river access to its flood plain during normal seasonal freshets while human uses of the floodplain greatly altered floodplain plant communities. Both processes have changed the nature and the availability of organic carbon to the river channel with ambiguous and unmonitored consequences for the riverine ecosystem.

Many of the consequences of the technological approach to river basin management were not anticipated at the time of their implementation. Nor is it likely that they could have been anticipated. Achieving ecosystem management is, therefore, much more than a matter of finding the right policy and the right technique. It is also a matter of accepting and even embracing uncertainty. Ecosystems are dynamic and ever changing, and unanticipated changes in the system are inevitable.

To deal with uncertainty, the successful watershed ecosystem manager will have a strong focus on ends and on the definition of "what are society's desired ecosystem conditions" and will continually adapt means to nudge the system toward those ends. The successful ecosystem manager will also know that the lower order processes with which sectoral management institutions are chiefly concerned (e.g., forest harvest, water supply) derive from and are integrated by higher order processes (e.g., rates and pathways of energy flow, the hydrologic cycle). Ecosystem management will be cognizant of and responsive to these nested levels of ecological integration. The successful ecosystem manager will not expect rigid compliance of the ecosystem with human expectation but will expect human adaptation to some vagaries of nature, particularly where these are important to ecological sustainability. Examples of this kind of adaptation include policies and zoning regulations that prohibit most kinds of construction on flood plains and the relocation of dikes away from the river bank to permit a significant amount of natural channel migration by the river and the seasonal flooding of at least a portion of the flood plain. The successful ecosystem manager will know that every ecosystem is different and that management will have to adapt to local circumstances. Further, the successful ecosystem manager will know that every human alteration of the ecosystem is an experiment and will treat each of them as such (Lee 1993).

In short, watershed ecosystem management requires new ways of thinking about policies and programs and more flexible and responsive institutional arrangements. These new ways of thinking must include a closer linking of ecosystem science with policy and program design. Such links will include a focus on the interaction between policies, programs, and ecosystem change through appropriate monitoring (Lee 1993). The more flexible and responsive institutional arrangements must include a role for adaptive experimentation and the adjustment of operating rules in response to new information (Healey and Hennessey 1994). The implications of this new way of thinking for program evaluation are particularly important. The issue will not be whether programs are delivered effectively (a concern with "means") but whether the ecosystem is responding in the expected way (a concern with "ends"). Monitoring ecosystem responses will, therefore, be an integral part of policy and program implementation (Chapter 18).

Evolving Paradigms in Watershed Management

Watershed Engineering

The ways of thinking about and managing watershed ecosystems outlined above run strongly counter to the way things have been done in the past. Until very recently water was accorded value only in a narrow utilitarian sense. Watercourses that were not dammed, diverted, pumped or polluted were regarded as "flowing wasted to the sea" (Bocking 1987). Under this doctrine and in the name of water resource development, the West embarked on a program of ecosystem restructuring of an unprecedented scale. Rivers were there to be bent (or straightened) to the human will. Managers were confident that any unanticipated side effects could be dealt with through further technological innovation. In other resource sectors (e.g., forestry, fisheries, mining) the policy was the same. This era, which lasted for a brief 100 years in the Pacific Northwest, produced some of the most impressive engineering achievements (e.g., the Shasta Dam and other high dams on the Sacramento/San Joaquin, Trinity, Columbia, Snake, Nechako and Peace Rivers) and also some of the greatest present day ecological headaches (e.g., salmonid population declines, unprecedented soil erosion, drying of the Peace-Athabasca delta). The terms *watershed scale management* and *integrated resource management* were born and embraced during this era. But they were largely restricted to mean the combining of diverse disciplinary expertise to satisfy a particular sectoral objective, with a clear focus on economic and industrial development.

The mounting intersectoral problems resulting from this approach were at first met with technological ad hoc solutions. Where forest removal led to siltation and degradation of river channels, technology was used to contain or limit erosion and stabilize or rehabilitate river channels. Where dams destroyed fish habitat, new habitat was constructed elsewhere or fish were raised in hatcheries (Black 1995). The pinnacle of this ad hoc approach was reached only in the past two decades with extensive and elaborate programs to mitigate the adverse effects of large-scale river development. In Canada this pinnacle is well illustrated by mitigation schemes included as part of the Oldman Dam project in Alberta and the Kemano Completion project in British Columbia (DOE 1992, BCUC 1994). An equally illustrative example in the United States is the Northwest Power Act of 1980, which mandated the rehabilitation of salmon runs in the Columbia River (Lee 1993).

In the case of the Oldman Dam, flooded trout habitat was to be compensated by constructing weirs and groins to improve trout habitat in river channels upstream from the reservoir. The loss of valley bottom winter habitat for mule deer was to be compensated by creating patches of forest habitat in coulees adjacent to the reservoir (DOE 1992). In the case of Kemano Completion, dramatically reduced flows in the Nechako River were to be mitigated by an extensive program of physical habitat enhancement in the reduced river and by temperature control using deep reservoir water (BCUC 1994). Both the Oldman Dam and Kemano Completion were vigorously opposed by environmental organizations and Native Canadians on the basis that the technological fixes were unworkable and the projects themselves morally wrong. The environmental lobby forced cancelation of a government/industry agreement over salmon conservation under reduced flows in the Nechako River, although future river flows remain uncertain. The Oldman Dam project proceeded, but with a much greater emphasis than was originally planned on monitoring and adapting mitigation technology to changing river conditions. The Northwest Power Act of 1980 was intended to reverse decades-long declines in Columbia River salmon. In its implementation, the Act has seen a steady emphasis on achieving salmonid rehabilitation within the context of long-established operating regimes for the dams. To date this approach has achieved little measurable success. Advocates for the fishery resource call for dramatically changed operating regimes for the dams and even for the decommissioning of dams to conserve dwindling salmon stocks.

Each of these projects reveals the high costs, the technological uncertainty, and the ecological complexity of artificial measures to sustain desired elements of ecosystem function in the face of large structural changes in the ecosystem. A principle lesson from these and similar projects is the inadequacy, except in a few instances, of technological substitutes for natural ecological processes. Many segments of society paid a high price for the failed promise of technology to substitute for natural ecosystem processes.

Environmental Assessment and Mitigation

As the ecological and social costs of big budget, big project development began to emerge, alienation grew in the communities most adversely affected. Communities and governments both began to question the belief that all development is good. North American society demanded a broader consideration of the environmental and social implications of proposed developments. As a result, beginning in the 1970s, procedures for environmental and social impact assessment of large development projects were adopted. These procedures are still evolving in both the United States and Canada. While the results may be dissatisfying, the requirement for environmental and social impact assessment has brought about dramatic changes to the way that major projects are designed, constructed, and operated (Reuss

1992). Unfortunately, the introduction of environmental assessments does not appear to have resulted in the cancellation of many large projects, nor have impact assessments and the mitigation they fostered prevented adverse environmental consequences of large projects. Society continues to alter the environment dramatically without understanding what the long-term ecological consequences might be (Turner et al. 1990, McMichael 1993).

In recent years, communities have become further disillusioned with the institutions of environmental management and with centralized government decision making. They are demanding a direct and more meaningful voice in decisions that affect their well being. At present, several experiments in cooperative, community-based watershed management are underway. These range in geographic scale and in the degree of autonomy and authority accorded community-based groups. Large-scale watershed projects involved in basin-wide planning and integration of management activities include the Northwest Power Planning Council (Lee 1993), the Fraser River Basin Management Board (FBMB 1994), and the Northern River Basins Study Board (NRBS 1995). Small-scale local projects, such as the Salmon River Watershed Management Partnership in British Columbia and the Nisqually River Management Project in Washington State, directly involve community groups in data collection and stream rehabilitation.

A common feature of these cooperative, community-based experiments is that they bring a broad range of interests to the table and seek to negotiate acceptable compromise across a number of contentious social, economic, and environmental issues. The extent to which these experiments will succeed in integrating community concerns into watershed management processes is still uncertain. An emerging issue is the fatigue experienced by citizen volunteers in participating in the endless rounds of meetings that such negotiations demand. However, these experiments have begun the process of engaging communities and concerned citizens in developing new and more holistic visions of watershed management policies.

Adaptive Management

Although the recent experiments in cooperative management and decision making are exciting, they continue to be rooted in the existing sectoral institutional structure and generally lack the flexibility and authority to effect real change. They also tend to approach ecosystem management as though it is primarily a matter of taking a broader range of issues and interests into account. The fundamental change in thinking needed and described in this chapter is not yet apparent (e.g., that technology is seldom a satisfactory substitute for ecosystem processes and that ecosystem management needs a strong focus on ends as opposed to means). However, some important elements of an ecosystem approach are emerging from these experiments. For example, at the urging of Kai Lee (1993), the Northwest Power Planning Council adopted an *adaptive management* approach to project planning and implementation. This means that the council has taken the important step of acknowledging uncertainty and treating projects and programs as experiments from which it can learn. In Canada, the Northern River Basins Study Board, a committee made up primarily of local citizens, was given the responsibility and the budgetary authority to administer a major watershed study in Northern Alberta and to advise the provincial and federal governments about ecosystem management based on the results of the study. This was the first time in Canada that government had relinquished financial control to such a locally constituted board. The Board developed a strong sense of ownership of the study and enthusiasm for retaining local control of development processes once the project is complete (NRBS 1996). Although none of the present experiments could be considered a fully fledged example of ecosystem management, they are the nests from which workable elements of ecosystem management will emerge.

Looking to the Future

Ecosystem Management

This chapter has suggested that ecosystem management has not yet been achieved,

TABLE 27.3. Description of four principles of ecosystem management.

Principle	Description
Ecosystem management is organized around naturally bounded ecosystem units.	Few administrative boundaries coincide with either natural ecosystem units or with those of other administrative units. The integration of activities that ecosystem management demands requires that administrative boundaries match ecosystem boundaries.
Ecosystem management is based on a mix of environmentally sustainable human activities.	Under the paradigm of ecosystem management, approaches to development will involve first asking what kinds of activities are ecologically sustainable and then asking which of those will most benefit individuals or the community.
Ecosystem management focuses on ends rather than means.	Traditional approaches to management of resources has been heavily oriented toward means. As long as human activities were a relatively minor ecological force, a focus on means was not unreasonable. Now, however, the emphasis must shift to a greater concern with ends.
Ecosystem management is adaptive.	Because ecosystems are not predictable, it is necessary to incorporate active programs of experimentation, learning, and policy review into the policy implementation process.

although trends in resource management are leading in that direction. If so, then it is important to ask how one might recognize when ecosystem management has been achieved. Since an agreed definition of ecosystem management has yet to be established, knowing when it has been achieved it is not a simple matter. Despite the uncertainty, four principles, summarized in Table 27.3, will help guide true ecosystem management. The principles provide a set of criteria by which ecosystem management can be recognized.

Ecosystem Management Is Organized Around Naturally Bounded Ecosystem Units

Few administrative boundaries presently coincide with natural ecosystem units, and in fewer instances still do the management units defined by different regulatory agencies coincide (Rueggeberg and Dorcey 1991). It will be impossible to achieve the integration of activities that ecosystem management demands as long as different institutions continue to administer their programs on unrelated geographic units. Watersheds offer an attractive option on which to structure such integration, although other natural units are also feasible (e.g., bio-geo-climatic zones or ecoregions and ecoprovinces).

Ecosystem Management Is Based on a Mix of Environmentally Sustainable Human Activities

Although long-term sustainable yield is usually an objective of renewable resources management, effective ways of integrating this objective across a mix of renewable resources, let alone ways to incorporate the use of nonrenewable and nondivisible resources, have yet to be discovered. The issue is not simply one of integration, however. In choosing activities to incorporate into the sustainable mix, society must recognize the limited capacity of ecosystems to support human activities. Under the paradigm of ecosystem management, development must involve first asking what kinds of activities are ecologically sustainable and then asking which of those will most benefit individuals or the community. In contrast, the present approach first asks what activities are most economically beneficial and then tries to make those activities environmentally acceptable. This is not to say that social and economic considerations are not important. Successful ecosystem management must include not only a sustainable environment but also a viable economy and a nurturing society (Figure 27.3). However, neither human economy or human society can persist in a dysfunctional ecosystem. Present approaches to watershed management have not embraced this principle.

Ecosystem Management Focuses on Ends Rather Than Means

Traditional approaches to resource management have been heavily oriented toward means. Such a focus is motivated by a desire for fair and equitable treatment of all parties and on a perceived need to limit the discretion of regulatory bodies. As long as human activities were a relatively minor ecological force, a focus on means was not unreasonable. Now, however, human activity is the dominant ecological force on the planet. Subordinating ends to means is no longer an affordable luxury. If we are to avoid Hobbes' political Leviathan (Oakeshott 1962), however, a focus on ends must also accompany a significant decentralization of political and economic power. Society is experimenting with various forms of political decentralization in resource management, some examples of which are outlined above. Further decentralization and community ownership of many aspects of environmental decision making will accompany ecosystem management. However, global society is also in the midst of dramatic centralization—a consolidation of corporate economic power. The growing asymmetry between political and corporate power bases does not bode well for ecosystem management in the short term.

Ecosystem Management Is Adaptive

Because ecosystems are not well behaved, it is impossible in many instances to predict the consequences of human activities. Nevertheless, policies and programs are put into place with the expectation that they will do good rather than ill. However, policies and programs generally are not subject to any routine practice of monitoring and evaluation, making assessment of their consequences difficult (Lee 1993). Ignoring the information value of ecosystem responses to human activities has resulted in a huge opportunity cost and probably slowed the evolution of management practices by many years. As Lee (1993) has argued, information has value not only as an initiator of action but also as a product of action. Ignoring or delaying the opportunity to learn from ecosystem responses to policy and program initiatives hampers any attempts at ecosystem management. In ecosystem management, however, monitoring is not the passive accumulation of information. It is a dynamic feedback process in which policies generate programs which are tested through adaptive experimentation and monitoring. The results of monitoring generate reappraisal of policies and programs and their adjustment or replacement as required. It is important to note that this is a fundamentally different process than the usual program review and evaluation that takes place in government. Program review and evaluation emphasizes program delivery and budgetary accountability. Adaptive management emphasizes the outcome of programs in relation to specified ecosystem objectives and how to adjust programs and policies to changes in the ecosystem.

Setting Goals for Ecosystem Management

Ecosystems are complex, nonlinear systems (Costanza et al. 1993), and we cannot predict their future states with any certainty. However, a range of more or less desirable future states for any watershed can be imagined (Figure 27.3). In the past, society has avoided any pretense at collectively defining desirable future states of the ecosystems of which it is a part. To do so seems to run counter to the belief that in a free society individuals should not be shackled to any one else's vision of the future. However, societies have given up much individual freedom in the past to achieve collective benefits. Political rhetoric typically involves an appeal to collective social or economic visions of the future. The image of a free society in which everyone prospers is, after all, a vision of a future to which many aspire. It is not such a radical step to include a vision of desired ecosystem structure and function in the planning of policies and programs. In fact, cherished economic and social visions may be unattainable without a corresponding and appropriate ecological vision.

Most scientists probably have some cognitive picture of the desired ecosystem in mind when they think about the management of watershed ecosystems. Other sectors of society also have an ecological vision in mind when they contribute to community plans and other forward looking exercises (Arduino and Healey 1995). It is time to get these private, individual visions out in the open where they can be discussed and debated. Then the painful job of stripping away the utopian aspects of individual visions and developing a workable shared ecological vision can begin.

From Incrementalism to Adaptive Rationalism

Thinking narrowly and acting locally seems to be a part of the institutional culture of western society—as does fearing losses more than valuing gains. As a consequence, decisions are characteristically remedial and incremental (Lindblom 1959) and not based on a very holistic vision of the future. The incremental, remedial approach, however, is a trap into which society endlessly retreats from a myriad of problems. The alternative, deliberately attempting to advance toward a desired future, has yet to come to pass. Lindblom (1959) argued that this "rational" alternative is unattainable in part because society is incapable of imagining, let alone analyzing, all the alternatives of its actions. The future is too uncertain for any complete analysis of how even a small change will affect society. The only "safe" option for the beleaguered decision maker is to make small adjustments within familiar territory and hope that others will adapt to his or her mistakes. Lindblom (1959) termed this decision-making behavior "muddling through."

Although Lindblom's (1959) analysis correctly characterizes the traditional behavior of institutional decision makers, his criticism of a more rational and analytic approach to decisions has an important weakness. It derives from his assumption that a rational approach will only be effective if it is able to encompass all possible consequences of a decision. Since no decision maker possesses such a godlike insight, Lindblom rejected rational methods. A less ambitious goal of simply improving the quality of decisions, however, makes it possible to envision a valuable role for rational approaches. This is particularly true if rational analysis is included as a component of adaptive management. As Lee (1993) has argued, complex ecological systems behave too uncertainly to be managed under traditional sectoral regulation. Each policy initiative and management program must be viewed as an experiment with broad implications across a range of sectors and from which society can learn about the behavior of the system. By properly designing, implementing, and monitoring policy and program initiatives, society can develop the tools and the sensitivity to manage human-ecosystem interrelationships.

The adaptive approach allows the problem-solving power of the scientific method to be brought to bear on environmental problems in much the way that science is used to advance medicine (Healey and Hennessey 1994). But, to be most effective, adaptive management should begin with a careful and rational analysis of policy options and their implications. This will allow major outcomes of policy or programs to be forecast and ensure that monitoring is designed to maximize learning. Adaptive management is the hatchway from which to escape the inadequate confines of incremental remedial management. Together, rational analysis and adaptive management are the key to implementing ecosystem management.

Although the adaptive approach is critical to ecosystem management, escaping incrementalism will not be easy or painless. Just as Kuhn (1970) has argued that the history of science is characterized by long periods of "normal science" punctuated by unpredictable "revolutions," incrementalism is the "normal management" that characterizes the long periods between great "policy revolutions." Such a policy revolution in environmental management may be imminent. When it happens, however, the revolution will likely initiate a period of great controversy and struggle during which environmental conditions will worsen and environmental science will be reviled as a failure before society adjusts to the principles of ecosystem management. This is because humankind faces some looming environmental problems of a new kind. Problems to which traditional institutions and science cannot easily adapt.

A New Class of Problems

For the past two centuries western social evolution has been driven by the resource-development paradigm. Within this paradigm, reductionist science and engineering flourished as the source of innovation and technique to sustain an ever growing material standard of living in industrialized countries. During most of this period the human economy was a small fraction of the global ecosystem. It was possible for society to reap the benefits of resource exploitation, economic and intellectual specialization, and sectoral rivalry while the costs were absorbed by the global ecosystem with little apparent harm. The problems addressed by science and technology were generally of two types: those that were generated from within a discipline and those assigned to a discipline by managers and policy makers. Problems of the first sort were characteristically well bounded and soluble within the established paradigms of the discipline. Problems of the second sort were either immediately soluble within the discipline or were "translated" into problems that were soluble (Bocking 1995).

Recently, however, a new class of problems has begun to emerge. These are problems that are thrust upon both policy makers and scientists, problems for which there are no known disciplinary, interdisciplinary, or institutional solutions within normal science and management, and problems that cannot be translated into normal science and engineering. Highly publicized examples include increasing atmospheric CO_2, tropical deforestation, thinning of the stratospheric ozone, world population growth, widespread desertification, and declining sperm counts and genital malformation in human males (McMichael 1993). Sadler (1990) has termed these "wicked problems." They are, in large measure, a consequence of overusing the absorptive and regenerative capacity of the biosphere. Dealing with these problems demands a restructuring of humankind's whole relationship with the natural world. Managing within the ecosystem paradigm is one way of restructuring this relationship. However, that such a restructuring is desirable, let alone necessary or imperative, has yet to be generally accepted. And, when faced with overwhelming problems, the tendency of humankind is to retrench and cling to old institutions rather than expand and experiment with new ideas (Healey 1990). This is why things will likely get worse before they get better. Furthermore, a revolution in the way society relates to the natural world is not assured. Human history is a graveyard of civilizations that were unable to come to terms with the limitations imposed by their environment (Pointing 1991).

Historical precedent notwithstanding, there is cause for optimism about the longer term future. As noted above, experiments with elements of ecosystem management are underway and debates about new ways of conceiving the relationship of humankind with the natural world are occurring (e.g. Brown 1981, WCED 1987, Pinkerton 1989, Turner et al. 1990, Costanza 1991, McMichael 1993). Although North Americans cannot ignore their responsibilities as members of the global commons, North America in general and the Pacific Northwest in particular is in a favorable position with regard to the balance of population to resources. People of the Pacific Northwest have the opportunity to explore new ideas and new institutional arrangements without jeopardizing the very fabric of their society. Perhaps the impending global crisis should be regarded as an opportunity to establish this region as both the breeding ground and the proving ground for radical approaches to the management of watershed ecosystems that may become the normal management of the future. The rich endowment of diverse watersheds, of streams and rivers and the strong tradition of ecosystem science empower citizens of the Pacific Northwest to take this opportunity.

Acknowledgments. This paper was prepared while the author was on sabbatical in the Political Science Department at the University of Rhode Island. The author is indebted to Gerry

Tyler of Political Science and Steve Olson of the Coastal Resources Center for providing support and facilities for his sabbatical. He is also indebted to his colleague, Tim Hennessey, for the many debates and discussions that contributed to the ideas in this paper and for commenting on a draft version. Michael Black, John Williams, and several anonymous referees also provided helpful comments. The research on which the paper is based was supported by an eco-research grant from the Tri-Council Secretariat of the Canadian Granting Councils and by a Fulbright Fellowship to the author.

Literature Cited

Angermeir, P.L., and J.R. Karr. 1994. Biological integrity versus biological diversity as policy directives: Protecting biotic resources. BioScience **44**:690–697.

Arduino, S., and M.C. Healey. 1995. Analysis of objectives and policy options of communities of the lower Fraser Valley based on official community plans. Working Paper, Basin Ecosystem Study, Sustainable Development Research Institute, University of British Columbia, Vancouver, British Columbia, Canada.

Black, M. 1995. Tragic remedies: A century of failed fishery policy on California's Sacramento River. Pacific Historical Review **64**:37–70.

BCUC (British Columbia Utilities Commission). 1994. Kemano completion project review. Report and recommendations to the Lieutenant Governor in Council. British Columbia Utilities Commission, Vancouver, British Columbia, Canada.

Bocking, R. 1987. Canadian water, a commodity for export? Pages 105–135 *in* M. Healey and R. Wallace, eds. Canadian aquatic resources. Canadian Bulletin of Fisheries and Aquatic Science 215. Department of Fisheries and Oceans, Ottawa, Ontario, Canada.

Bocking, S. 1995. Aquatic science in Canada: A case study of research in the Mackenzie basin. Royal Society of Canada, Ottawa, Ontario, Canada.

Bovee, K.D. 1982. A guide to stream habitat analysis using the instream flow incremental methodology. Instream flow information paper 12. FWS/OBS-82/86. USDI Fish and Wildlife Service, Office of Biological Services.

Brown, L.R. 1981. Building a sustainable society. W.W. Norton, New York, New York, USA.

Brown, R.L. 1986. Annual report of the Interagency Ecological Studies Program for the Sacramento–San Joaquin estuary for 1984. California Department of Water Resources, Sacramento, California, USA.

Burt, D.W., and J.H. Mundie. 1986. Case histories of regulated stream flow and its effect on salmonid populations. Canadian Technical Report of Fisheries and Aquatic Science 1477.

Calow, P., and G.E. Petts. 1992. The rivers handbook, Volume 1. Blackwell Scientific, Oxford, UK.

Canada. 1995. The Fraser River Action Plan. 1994–1995 progress report. Departments of Environment and Fisheries and Oceans, Ottawa, Ontario, Canada.

Canadian Council of Resource and Environment Ministers. 1987. Report of the National Task Force on Environment and Economy. Ottawa, Ontario, Canada.

Costanza, R. 1991. Ecological economics: The science and management of sustainability. Columbia University Press, New York, New York, USA.

Costanza, R., L.Wainger, C. Folke, and K-G. Maler. 1993. Modeling complex ecological and economic systems. BioScience **43**:545–555.

Daly, H.E. 1991. Elements of environmental macroeconomics. Pages 32–46 *in* R. Costanza, ed. Ecological economics: The science and management of sustainability. Columbia University Press, New York, New York, USA.

Daly, H.E., and J. Cobb. 1994. For the common good: Redirecting the economy towards community, the environment, and a sustainable future. Revised edition. Beacon Press, Boston, Massachusetts, USA.

DOE (Department of Environment). 1992. Final report of the Environmental Assessment Review Panel for the Oldman Dam Project. Environment Canada, Ottawa, Ontario, Canada.

FBMB (Fraser River Basin Management Board). 1994. Annual report. Fraser River Basin Management Board. Vancouver, British Columbia, Canada.

Federal Interagency Floodplain Management Task Force. 1992. Floodplain management in the United States: An assessment report. Volume 1. Summary Report, Contract Number TV-72105A. Prepared by Natural Hazards Research and Applications Information Center, University of Colorado, Boulder, Colorado, USA.

Frissell, C.A., W.J. Liss, C.E. Warren, and M.D. Hurley. 1986. A hierarchical framework for stream habitat classification: Viewing streams in a watershed context. Environmental Management **10**:199–214.

Gleick, P. 1993. Water in crisis: A guide to the world's fresh water resources. Oxford University Press, New York, New York, USA.

Gore, A. 1992. Earth in the balance: Ecology and the human spirit. Houghton Mifflin, Boston, Massachusetts, USA.

Gore, J.A., and G.E. Petts. 1989. Alternatives in regulated river management. CRC Press, Boca Raton, Florida, USA.

Healey, M.C. 1990. The implications of climate change for fisheries management policy. Transactions of the American Fisheries Society **119**:366–373.

——. 1994. The effects of dams and dikes on fish habitat in two Canadian river deltas. Pages 385–398 *in* W.J. Mitsch, ed. Global wetlands old world and new. Elsevier, Amsterdam, The Netherlands.

Healey, M.C., and T. Hennessey. 1996. The role of science in decisions about water resources management in the Nechako River, BC, and the Oldman River, AB, Canada. International Association for Impact Assessment, IAIA '96, Symposium proceedings.

Healey, M.C., and T.M. Hennessey. 1994. The utilization of scientific information in the management of estuarine ecosystems. Ocean and Coastal Management **23**:167–191.

Healey, M.C., and J. Richardson. 1996. Changes in the productivity base and fish populations of the lower Fraser River associated with historical changes in human occupation. Proceedings of the International Large Rivers Symposium June 1995, Krems, Austria. Archiv für Hydrobiologie, Supplement **113**:279–290.

Hilborn, R. 1992. Hatcheries and the future of salmon in the northwest. Fisheries **17**:5–8.

Holling, C.S. 1987. Simplifying the complex: The paradigms of ecological function and structure. European Journal of Operations Research **30**:139–146.

——. 1992. Cross-scale morphology, geometry and dynamics of ecosystems. Ecological Monographs **62**:447–502.

Junk, W.J., P.H. Bayley, and R.F. Sparks. 1989. The flood pulse concept in river floodplain systems. Pages 110–127 *in* D.P. Dodge ed. Proceedings of the International Large River Symposium. Canadian Special Publication of Fisheries and Aquatic Science 106. Department of Fisheries and Oceans, Ottawa, Ontario, Canada.

Kuhn, T. 1970. The structure of scientific revolutions. University of Chicago Press, Chicago, Illinois, USA.

Lee, Kai. 1993. Compass and gyroscope. Island Press, Washington, DC, USA.

Lindblom, C. 1959. The "science" of muddling through. Public Administration Review **19**:79–88.

McMichael, A.J. 1993. Planetary overload: Global environmental change and the health of the human species. Cambridge University Press, Cambridge, UK.

Naiman, R.J. ed. 1992. Watershed management: Balancing sustainability and environmental change. Springer-Verlag. New York, New York, USA.

Nehlsen, W., J.E. Williams, and J.A. Lichatowich. 1991. Pacific salmon at the crossroads: Stocks at risk from California, Oregon, Idaho and Washington. Fisheries **16**:4–21.

NRBS (Northern River Basins Study). 1995. Annual report for the 1994–95 fiscal year. Northern River Basins Study, Edmonton, Alberta, Canada.

——. 1996. Northern River Basins Study Report to the Ministers. Northern River Basins Study, Edmonton, Alberta, Canada.

Oakeshott, M., ed. 1962. Thomas Hobbes Leviathan. Collier Books, New York, New York, USA.

O'Neill, R.V., D.L. DeAngelis, J.B. Waide, and T.F.H. Allen. 1986. A hierarchical concept of ecosystems. Princeton University Press, Princeton, New Jersey, USA.

Paehlke, R.C. 1989. Environmentalism and the future of progressive politics. Yale University Press, New Haven, Connecticut, USA.

Pinkerton, E. 1989. Cooperative management of local fisheries. University of British Columbia Press, Vancouver, British Columbia, Canada.

Pointing, C. 1991. A green history of the world. Sinclair Stevenson, London, UK.

Rees, W. 1990. The ecology of sustainable development. The Ecologist **20**:18–23.

Reisner, M. 1993. Cadillac desert. Revised and updated edition. Penguin Books, New York, New York, USA.

Reuss, M. 1992. Coping with uncertainty: Social scientists, engineers, and federal water resources planning. Natural Resources Journal **32**:101–135.

Robinson, J.B. 1988. Unlearning and backcasting: Rethinking some of the questions we ask about the future. Technological Forecasting and Social Change **33**:325–338.

Rueggeberg, H.I., and A.H.J. Dorcey. 1991. Governance of aquatic resources in the Fraser River Basin. Pages 201–240 *in* A.H.J. Dorcey and J.Griggs, eds. Water in sustainable development:

Exploring our common future in the Fraser River Basin. Westwater Research Centre, University of British Columbia, Vancouver, British Columbia, Canada.

Russell, L.R., K.R. Conlin, O.K. Johansen, and U. Orr. 1983. Chinook salmon studies in the Nechako River: 1980, 1981, 1982. Canadian Manuscript Report of Fisheries and Aquatic Science 1728.

Sadler, B. 1990. An evaluation of the Beaufort Sea environmental assessment panel review. Environmental Assessment and Review Office, Ottawa, Ontario, Canada.

Schaffer, W.M. 1985. Order and chaos in ecological systems. Ecology **66**:93–106.

Simberloff, D. 1980. A succession of paradigms in ecology: Essentialism to materialism and probabilism. Synthese **43**:3–39.

Swift, C.H. 1976. Estimation of stream discharges preferred by steelhead trout for spawning and rearing in western Washington. United States Geological Survey Open File Report 75–155.

Tennant, D. 1976. Instream flow regimes for fish, wildlife, recreation and related environmental resources. Pages 359–373 *in* J.F. Ovsborn and C.H. Allman, eds. Proceedings of the symposium and specialty conference on instream flow needs. Volume II. American Fisheries Society, Bethesda, Maryland, USA.

Thorp, J.H. 1994. The riverine productivity model: An heuristic view of carbon sources and organic processing in large river ecosystems. Oikos **70**:305–308.

Turner II, B.L., W.C. Clark, R.W. Kates, J.F. Richards, J.T. Mathews, W.B. Meyer. 1990. The earth as transformed by human action. Cambridge University Press, Cambridge, England.

United Nations. 1993. Agenda 21: Program of action for sustainable development. United Nations, New York, New York, USA.

US Fish and Wildlife Service. 1995. Working paper: Habitat restoration actions to double natural production of anadromous fish in the central valley of California. 3 Volumes. Prepared for the US Fish and Wildlife Service under the direction of the Anadromous Fish Restoration Program Core Group, Stockton, California, USA.

Vannote, R.L., G. Minshall, K.W. Cummins, J.R. Sedell, and C.E. Cushing. 1980. The river continuum concept. Canadian Journal of Fisheries and Aquatic Science **37**:130–137.

Walder, G. 1996. Proceedings of the Northern River Basins Study Instream Flow Needs Workshop. Report Number 66. Northern River Basins Study, Edmonton, Alberta, Canada.

Ward, J.V., and J.A. Stanford. 1983. The serial discontinuity concept of lotic ecosystems. Pages 29–42 *in* T.D. Fontain and S.M. Bartell, eds. Dynamics of lotic ecosystems. Ann Arbor Science Publications, Ann Arbor, Michigan, USA.

WCED (World Commission on Environment and Development). 1987. Our common future. Report of the World Commission on Environment and Development (Gro Harlem Bruntland Chair). Oxford University Press, Oxford, UK.

Weinberg, A. 1972. Science and trans-science. Minerva **10**:209–222.

Windsor, J. 1992. Water export, should Canada's water be for sale. Canadian Water Resources Association, Cambridge, Ontario, Canada.

Worster, D. 1990. The ecology of order and chaos. Environmental History Review **14**:1–18.

Appendix

Reviewers

Richard M. Adams
Department of Agricultural and Resource Economics
Oregon State University
Corvallis, OR 97331-3601
USA

James K. Agee
College of Forest Resources
University of Washington
Seattle, WA 98195
USA

Norm Anderson
Oregon State University
Department of Entomology
Corvallis, OR 97331
USA

Kelly Austin
Weyerhaeuser Company
Tacoma, WA 98477
USA

Katherine Baril
Washington State University Cooperative Extension
WSU Community Learning Center, Jefferson County
Port Hadlock, WA 98331
USA

Bob Beschta
Forest Engineering
Oregon State University
Corvallis, OR 97331-5704
USA

Michael Blumm, Esq.
Lewis and Clark Law School
Portland, OR 97219
USA

Susan Bolton
College of Forest Resources
University of Washington
Seattle, WA, 98195
USA

Max Bothwell
National Hydrology Research Institute
Saskatoon, Saskatchewan S7N 3H5
Canada

Tom Bott
Stroud Water Research Center
Avondale, PA 19311
USA

Gardner Brown
Department of Economics
University of Washington
Seattle, WA 98195
USA

John M. Buffington
Department of Geological Sciences
University of Washington
Seattle, WA 98195
USA

David R. Butler
Department of Geography
University of North Carolina
Chapel Hill, NC 27599-3220
USA

Mike Church
Department of Geography
University of British Columbia
Vancouver, British Columbia V6T 1W5
Canada

Hanna J. Cortner
School of Renewable Natural Resources
University of Arizona
Tucson, AZ 85721
USA

Don Erman
Centers for Water and Wildlife Resources
University of California at Davis
Davis, CA 95616
USA

Stuart Findlay
Institute for Ecosystem Studies
Mary Flagler Cary Arboretum
Millbrook, NY 12545
USA

Penny Firth
National Science Foundation
Division of Environmental Biology
Arlington, VA 22230
USA

Stuart Fisher
Zoology Department
Arizona State University
Tempe, AZ 85287-1501
USA

Alexander Flecker
Cornell University
Division of Biological Sciences
Section of Ecology and Suystematics
Ithaca, NY 14853-2701
USA

Steve Golladay
Joseph W. Jones Ecological Center
Newton, GA 31770
USA

James Gore
Director, Environmental Science Graduate Program
Columbus State University
Columbus, GA 31907-5645
USA

Nancy Grimm
Department of Zoology
Arizona State University
Tempe, AZ 85287-1501
USA

Tom Hanley
USDA Forest Service
PNW Station
Juneau, AK 99801-8545
USA

Andy Hansen
Department of Biology
Montana State University
Bozeman, MT 59717-0346
USA

Mark Harwell
University of Miami
Miami, FL 33149
USA

F. Richard Hauer
Flathead Lake Biological Station
The University of Montana
Polson, MT 59860-9659
USA

Clifford Hupp
US Geological Survey
National Center
Reston, VA 22092
USA

Daniel D. Huppert
School of Marine Affairs
University of Washington
Seattle, WA 98195
USA

Jeff Kershner
Department of Fish and Wildlife
Utah State University
Logan, UT 84322-5210
USA

Fran Korten
The Ford Foundation
New York, NY 10017
USA

Thomas E. Lisle
USDA Forest Service
PSW Research Station
Redwood Sciences Laboratory
Arcata, CA 95521
USA

Stanford Loeb
Department of Systematics and Ecology
University of Kansas
Lawrence, KS 66045
USA

John B. Loomis
Agriculture & Resource Economics
Colorado State University
Fort Collins, CO 80523
USA

Donald Ludwig
Surrey, British Columbia V4A 6Z1
Canada

Michael E. McClain
Department of Oceanography
University of Washington
Seattle, WA 98195
USA

J. Donald McPhail
The University of British Columbia
Vancouver, British Columbia V6T 1Z1
Canada

Walter Megahan
Port Townsend, WA 98368
USA

Richard Merritt
Department of Entomology
Michigan State University
East Lansing, MI 48824-111
USA

Wayne Minshall
Department of Biological Sciences
Idaho State University
Pocatello, ID 83209
USA

David R. Montgomery
Department of Geological Sciences
University of Washington
Seattle, WA 98195
USA

Linn Montgomery
Department of Biological Sciences
Northern Arizona University
Flagstaff, AZ 86011-5640
USA

Dennis Newbold
Stroud Water Research Center
Avondale, PA 19311
USA

Christer Nilsson
Department of Ecological Botany
Umea University
S 901 87 Umea
Sweden

Tony Olsen
Environmental Protection Agency
Corvallis, OR 97333
USA

Cassie Phillips
WWC
Weyerhaeuser Company
Tacoma, WA 98477
USA

Evelyn Pinkerton
School of Resource and Environmental Management
Simon Fraser University
Burnaby, British Columbia V5A 1S6
Canada

Tom Quinn
School of Fisheries
University of Washington
Seattle, WA 98195
USA

Leslie M. Reid
USDA Forest Service
Redwood Sciences Laboratory
Arcata, CA 95521
USA

Leonard Smock
Virginia Commonwealth University
Richmond, VA 23284-2012
USA

Susan Stafford
Department of Forest Science
Oregon State University
Corvallis, OR 97331-5705
USA

John Stednick
Department of Earth Resources
Colorado State University
Fort Collins, CO 80523
USA

Jan Stevenson
Department of Biology
Center of Environmental Sciences
University of Louisville
Louisville, KY 40292
USA

John Stockner
Canada Fisheries and Oceans
West Vancouver Laboratory
West Vancouver, British Columbia V7V 1N6
Canada

Vicky Sturtevant
Department of Sociology
Southern Oregon State College
Ashland, OR 97520
USA

Kate Sullivan
Weyerhaeuser Company
Tacoma, WA 98477
USA

Frank Triska
Water Resources Division
US Geological Survey
Menlo Park, CA 94025
USA

H.M. (Maury) Valett
Department of Biology
University of New Mexico
Albuquerque, NM 87131
USA

Bettina von Hagen
Ecotrust
Portland, OR 97209
USA

Bruce Wallace
Department of Entomology
University of Georgia
Athens, GA 30602-2603
USA

Ray J. White
Edmonds, WA 98020
USA

Richard White
Department of History
University of Washington
Seattle, WA 98195
USA

Jack E. Williams
USDA Forest Service
Intermountain Research Station
Boise, ID 83702
USA

John Williams
Hydrology and Water Resources Planning
Davis, CA 95616
USA

Index

A

Abrasion, 298, 299, 300, 318
Acetylene, 415
Acid neutralizing capacity, 70, 71, 73–74
Actinomycetes, 136–137
Active channel, 291, 294, 296, 401
Adaptations, 419
  biological, 282
  functional, 298
  morphological, 176
  riparian plant, 297–299
    growth dynamics, 301–302
Adsorption, physical 137–139
Adventitious roots, 299, 300, 301
Aerenchyma, 299
Agencies, regulatory, 533
Aggradation, 18
Aggregation, 581
Agricultural diversion, 572
Agriculture, 284, 311, 349, 424, 502, 515, 523, 572, 591
Alder (*Alnus glutinosa*), 129, 134, 309, 441. *See also* Red Alder; Slide Alder
Alderflies. *See* Megaloptera
Alder terrace, 417
Alevins, 84
Algae, 121, 126, 169, 184, 185, 377, 378, 379, 380
  diversity of, 417
  See also Benthic algae; Blue-green algae; Epilithic algae; Filamentous green algae; Periphytic algae; Planktonic algae
Allen's paradox, 189, 190
Allochthonous organic matter, 120, 121, 144–145, 147
Aluminum, 405, 413
Alluvial rivers, 21, 399, 422
Alluvium, 299
Amenities, scenic, 576
American dipper (*Cinclus mexicanus*), 238
Ammonium, 347, 413, 414
Amphibians, 236–242, 248, 378, 383, 602, 604
Amphibites, 419
Amphipoda (amphipods) (*Gammarus* spp.), 171, 183
Anadromous fish, 112, 324, 337, 340, 430, 431
  management of, 535–537
Anadromous salmonids, 82, 215, 216, 220, 224, 373, 503
Anaerobic microzones, 415
Anchoring, 604
Andrews Experimental Forest, 46, 59, 381
Anemochory, 299
Angiosperms, 148, 377, 148, 155
  Lemna, 148
  Nymphaea, 148
  Pontederia, 148
  Potamogeton, 148, 155
Anions, 74–75
Annelida, 170, 171
Annual transport distance, 273
Anthropogenic activities, 220, 222, 224. *See also* Human activities
Apatite, 347, 351
Appropriation, prior, 556
Aquatic garter snake (*Thamnophis atratus*), 237
Arthropoda, 165, 170–171. *See also* Odonata, Plecoptera, and Trichoptera
Ascomycota, 123, 137
Army Corps of Engineers, 587
Ash, flowering (*Fraxinus ornus*), 300
Ash-free dry weight, 72
  in plant litter decomposition, 131
  in primary production, 160
Assemblages, 502, 504, 505, 506, 508, 509, 521, 522
  algae, 521
  aquatic plant, 111, 114
  species, 170
Assimilation, 351
Autumn, 389, 392
Avalanches, 262, 292, 293, 294, 295, 305, 306
Avoider, 298, 299
Avulsion, 275, 440, 446

B

BACIP. *See* Before-After-Control-Impact Paired Design
Backchannels, floodplain, 416, 417
Bacteria,
  and dissolved organic matter, 183
  and hyporheic zone, 405, 413, 417, 422

and LWD, 330
and organic matter
decomposition of, 120, 121, 122, 125–126, 132–133, 136–137, 138
dissolution of, 183
and trophic dynamics, 378–379
types of
fecal coliform, 71
heterotrophic, 120, 122, 126
pathogenic, 70, 122, 125
Bald eagle (*Haliaeetus leucocephalus*), 238
Balsam fir (*Abiers balsamea*), 325
Banks, 262
Bank undercutting, 328
Bars, 21, 29–30, 30–31, 401, 406, 410
Basal area, 301, 302
Basidiomycota, 123
Basin history, 261, 273, 274
Basin morphology, 208, 210, 214
Bats, 238, 241
Beaver (*Castor canadensi*), 238, 242, 348, 361, 362, 363, 447
and dam building activities, 249–250, 411
Bedload, 15, 266
Bedrock geology, 408, 409
Bed surface coarsening, 17
Bed surface fining, 17
Beetles. *See* Coleoptera
Before-after-control-impact-paired (BACIP) design, 459, 507
Before-after-control-impact (BACI), 507
Before-after (BA) sampling design, 507
Benefit-cost analysis, 572
Benthic algae, 144, 145–148, 153–154, 182, 185, 368
Best management practices, 71, 84, 559, 561
Big Ditch stream, 516
Bigleaf maple (*Acer macrophyllum*), 131, 361
Bioassays, 81, 506
Bioavailability, 81, 84
Biochemical oxygen demand, 70, 71, 78–81
Biochemical transformations, 414
Biodiversity, 310, 318, 430, 432, 446, 644, 651, 662, 664, 668
and natural processes, 438–448
Biofilm, formation of, 125, 136
Biogeographical dispersal, 208
Biological integrity, 4, 502, 505–506, 509, 517, 521, 523
Biological oxygen demand, 423
Biological uptake, 413
Biomagnification, 04
Biomass, 291, 302, 301, 335, 336, 413, 417
microbial, 122, 124, 129
plant, 291
Biota, stream, 419
Biotopes, 407
Birds, 237–242, 248
Bivalvia (clams and mussels), 171
Blackflies, 122, 138
Blackwater rivers, 188
Blanco River Project, 606–608
Blueberry, 306
Blue-green algae, 87, 185, 147, 148, 152, 185, 186, 192, 387
Anabaena, 147
Nostoc, 152, 185, 186, 192, 387
Oedogonium, 148
Pectonema, 148
Phormidium, 87, 148
Boating, 572, 576
Bobcat (*Relis rufus*), 242
Body size, 203, 507
Boundary layer, 176
Branch abscission, 300
Breakdown rates, 127, 131, 134
Breeding, 605
Broad geomorphic characterization, 106, 107
delineative criteria for, summary of, 109
Brook trout (*Salvelinus fontinalis*), 376
Brown trout (*Salmo trutta*), 375
Bryophytes, 148, 184, 377
See also Liverworts; Mosses
Budgets
carbon, 663
energy, 121
nutrient, 355, 356
riverine, 354–357
Buffer capacity, 73, 84
Buffer island, 253
Buffers, riparian, 77–78, 235, 245
Bull frog (*Rana catesbeiana*), 243
Bureau of Reclamation, 576
Business, private, 575, 579, 625
Butterflies. *See* Lepidoptera
Buttressing, stand, 299

C

Caddisflies. *See Trichoptera*
California Bay (*Umbellularia californica*), 310
California Department of Forestry and Fire Protection, 491–492
Candlefish (*Thaleichthys pacificus*), 205
Canopy, forest, 388
Carbon, 410, 413
Carbon cycling, 419
Carbon dioxide, 367, 413
Carcasses, 192, 204, 205, 334, 373, 389–391
Carnation Creek, Vancouver Island, 154, 364
Carnivores, 238
Cascade Mountains, 5, 270
ANC and pH levels, 73
and flood and debris flow, reaction to, 179
Cascades, trophic, 387
Catchment, 664, 666, 667
Cation-exchange sites, adsorption to, 353
Cations, 70, 73, 74–75
Cedar (*Thuja plicata*), 267, 306, 328–329
See also Western red cedar; Yellow cedar
Cellulose, 123
Channel bed, 14, 15, 17, 21
Channels
braided, 17, 21
changes, examples of, 17–20
confinement of, 33
disturbance in, 31–36
erosion of, 48
form of, 103–104, 331–332
hydraulics if, 600
length of, 15
morphology of, 272–273, 284, 331, 408

networks of, 261–264, 272, 278–280
physical condition of, 20
processes in, 14–16
size of, 324 ,325, 332
slope of, 13, 16, 17
typography of, 410
types of, 327, 331
alluvial, 13
confined, 327
plane-bedded, 331
relic, 410
step-pool, 331
unconfined, 327
wetted, 400, 401, 406, 410, 417
Channel unit, 30–31
Channel unit scale, 408, 409, 410
Chemical contamination, 504, 505, 507, 523
Chemical oxygen demand, 70, 71
Chezy resistance factor, 15
Chiklat River, Alaska, 238
Chinook salmon (*Oncorhynchus tshawytscha*), 7, 8, 206, 214, 220, 362, 375, 602, 668
Chipmunks, 242
Chironomid (*Paratendipes*), 156, 178, 185
Chlorine, 518, 520
Chlorophyll, 417
Chukar (*Alectoris chukar*), 243
Ciliates, 126
Civic friendship, 529
Cladoptosis, 300, 301
Clams. *See* Bivalvia
Classification
basin-wide, 98–99
biological-physical classification approaches, evaluation of, 114
biotic, 97–98
historical concepts, 98–99
recent concepts, 99–100
channels, 14, 20–31
process-based channel, 408
and riparian vegetation, 113–114
hierarchical, 101–111, 113
Michigan rivers, biological classification scheme for, 98
natural rivers, classification of, 110
network, 280
plant, 113–114
salmonid habitat, 103–104
Sierra Nevada Mountains, California, classification of, 114
single-scale, 101
stream, 111, 114–116
units of, 98–99
Clay minerals, 413
Clean Water Act, 503, 504, 505, 509, 553, 558, 559, 561, 565, 600
Clear-cutting, 57, 58–59
Climate, 44, 61, 261–268, 291
change in, and nutrient cycling, 367
Cnidaria, 170, 172
Coarse graining, 284
Coastal Mountains, 5
Coastal Zone Management Act, 560
Coast mole (*Scarpanus orarius*), 242
Coast redwood (*Sequoia sempervirens*), 291, 325
Coefficient of variation, 507
Cohort, 172, 189
Coho salmon (*Oncorhynchus kisutch*), 82, 207, 214, 218, 222, 333, 335, 336, 362, 375, 388, 389, 604
Colembola (springtails), 171, 174
Coleoptera (beetles)
Acneus spp., 171
Lara avara, 171, 185
Optioservus spp., 171
Collaborative culture, 529
Collector-filterers, 170–171, 180
Collector-gatherers, 158, 161, 170–171, 180, 378, 380, 423
Collectors, 335, 379
Colluvium, 412
Colonization, 292, 294, 296, 298, 309
microbial, 127
Colorado River, and human influences, 224
Columbia River, 5
dams on, 224
and hypothetical food webs, 223
Columbia Zoogeographic Province, 204
Charitable organizations, 629
Command and control approach, 630
Common law, 554
Communities, 113
composition of, 201, 291, 625
ecology of, 201, 208
respiration of, 380
structure of, 111, 282, 283, 421
types of
biological, 3–4, 359
biotic, 289
early successional, 236, 241
ecological, 208–209
hyporheic invertebrate, 419–423
macroinvertebrate, 169, 193
policy, 529, 532, 537–538, 539, 540–543
Compensation, 577
Competition, 201, 219, 222, 387
Concurrent jurisdiction, 555
Condensation, 44
Conditioning, 128, 359–360
Conditions
anaerobic, 299
anoxic, 299, 421, 424
baseline, 509
redox, 347, 424
Conductivity, 70, 74–75
electrical, 402
hydraulic, 49, 62–63, 405, 406, 407, 408, 409
Confidence interval estimation, 469
Confinement, 33–34
classes of, 37
Conifers, 134, 291, 307, 311, 338, 441
Connectivity, 244, 250
Conservation, 113–114, 625, 626, 629, 632
Conservation-based development, 625, 632, 635, 637
Consumers, 179, 189, 191, 582, 583
Consumption
invertebrate, 324
oxygen, 334

water, 4
Contingent variation, 585
Control impact, 507
Contour plots, 402
Cooperative management, 675
Coots, 238
Copepods, 122
Copper Slough, Illinois, 518, 519
Coprecipitation, 354
Cormorants, 238
Corridors
riparian, 243, 430, 431, 438, 504, 514, 517
diversity in, 431–438
river, 247, 248, 250, 265
Cost-effectiveness analysis, 572, 585
Cottonwood (*Populus trichocarpa*), 292, 301, 305, 315, 441
Council on Environmental Quality (CEQ), 477
Coweeta, North Carolina, 56
Crab
blue (*Callinectes sapidus*), 335
mud (*Rhithropanopeus harrisii*), 335
Cranes, 238
Crayfish. *See* Decapoda
Crustaceans, 520
and LWD, 335
Cumulative effects, 261, 262, 284, 476, 477, 479, 485, 489, 505, 511, 610, 646, 648
Council on Environmental Quality (CEQ) definition of, 477
Cumulative impact, 477, 485
Currant, stink (*Ribes bracteosum*), 306, 441
Current velocity, 175
Cutthroat trout (*Oncorhynchus clarki*), 82, 376, 388, 389
Cycles
biogeochemical, 290, 310, 349, 362, 505
hydrological, 4
nutrient, 347–349
Cypress, bald (*Taxodium distichum*), 300

D

Dace, 207, 213, 218, 222
Dams
Dams, 19–20, 224, 503, 504, 514
construction of, 16, 19–20, 314
hydroelectric, 572, 585, 586
See also Beaver
Damselflies. *See* Odonata
Daniel Boone Wilderness, 517
Darcy's law, 405
Darcy-Weisbach friction factor, 15
Data analysis, 468–471
Debris
flow of, 34–36, 179, 264, 273, 291, 294–295, 299, 305, 306–307
torrents of, 328
Decapoda (crayfish) (*Pacifastacus* spp.), 172, 186
Decapod crustaceans, 171
Decay, microbial, 330
Decision making, 530, 537, 545
and economics, 585
Decomposition, 122, 123, 124, 127, 353, 400, 414, 417, 422
hyporheic, 422
microbial, 373, 378
Decry, 330
Deep seepage, 48
Deer Creek Watershed, 364
Degradation, sources of, 502, 503, 504, 509, 511, 512, 513, 515–516, 517, 521, 523
Demography, 505
Dentrification, 362, 413
Depositional fans, 295
Detrital dynamics, 99
Detrital food web. *See* Food webs
Detrital pathway, 379
Detritus, 120, 121, 122, 125, 129, 132, 139, 183, 189
autochthonous, 144, 154
Detrivores, 120, 122, 128, 129, 133, 189
Detrivory, 183–184
Developing countries, 625, 631
Devil's club (*Oplopanax horridum*), 306
Diatoms, 87, 145, 185, 387
Epithemia, 153
Gomphonema, 443
Meridion, 146
Navicula, 146
Synedra, 443
Diel, 360, 445
Differential response of species, 222–224
Diking, 424, 576
Diptera (true flies)
Antocha spp., 171
Blepharicera spp., 171
Chironomidae, 171
Deuterophlebia spp., 171
Lipsothrix spp., 171, 184
Simulium spp., 171, 178, 180
Tabanus spp., 171
Tipula spp., 171, 178, 180, 183
Discharge
bankfull, 53
effective, 53
hydraulic, 13–14, 15, 16
stream, 364
Discounting, 581, 582
Disease, 298, 307
Dissimilatory transformations, 359
Dissolved organic carbon, 121
Dissolved organic nitrogen, 366
Dissolved oxygen, 69, 70, 71, 78–81, 175, 408
Distribution
frequency, 261–275, 279, 284
invertebrate, 113
LWD, 325–327, 340, 341
normal, 508
probability, 263, 283
species, 111
wealth, 581, 582
Distribution-free tests, 468, 470
Disturbance, 99–100, 176–179, 224, 289, 290, 292, 302, 308, 399, 401, 408, 503, 505
human-induced versus natural, 236, 243, 245, 250, 251
and nutrient cycling
and climate change, 367–368
forest conversion, 363–367
and urbanization and agriculture, 367
regime, 261, 262, 263, 386, 431, 446
types of
natural, 261, 262, 263, 284, 599

press, 224
pulse, 224
Diversity, 401, 507, 508, 601
indexes of, 508
Dodsonflies. *See* Megaloptera
Dolly Varden trout (*Salvelinus malma*), 220, 376
Dominance, 523
Dose-response curve, 507, 510, 513, 522
Douglas fir (*Pseudotsuga menziesii*), 59, 131, 267, 307, 325, 330, 360
Downwelling, 406, 409, 412
Dragonflies. *See* Odonata
Drainage network, 98, 294
Dredging, 608
Drift, 176, 187–188, 423
dynamics of, 170
macroinvertebrate, 176, 187–189
Drought, 267, 292, 316, 421
Ducks, 238
Dunn's salamander (*Plethodon dunni*), 237
Dynamics, trophic, 374

E

Eastern hemlock (*Tsuga canadensis*), 325
Ecological-economic analysis, 572
Ecological integrity, 667
Ecological patterns, 98
Ecologists
applied, 506
theoretical, 506
Economic analysis, 572
Economic assessment, 577, 587
Economic benefits, 582
Economic efficiency, 575, 581
Economic impact analysis, 572
Economics, 573, 589
and water resources, 573–576
Economic welfare theory, 571
Ecoregion, 101, 430, 553
Ecosystem management, 619, 642–643, 662, 664, 665, 666, 667, 668, 669, 671, 672, 675–678
elements of, 643–646
connectivity and uncertainty, 645
holistic perspective, 644
human activities, 646
natural processes, 643–644
principles of, 643
Ecosystems
analysis of, interagency, 494–496
behavior of, long-term, 261
boundaries of, 399
conditions of, 577
functions of, 572, 579
types of
adaptive, 617, 618, 628, 629, 630, 675, 677, 678
aquatic, 290
lotic, 374, 377
river, 572
Ecotones, 3, 399, 401
Ecotrust, 635, 639
Egesta, 360
Elevation, 44, 306, 316
Elk (*Cervus elaphus*), 443
Elk Creek, Oregon, 616
Endangered Species Act, 553, 565–566, 599
Endurer, 298, 299
Energy, 121, 126, 379
Energy dynamics, 505
Engelmann spruce (*Picea engelmanii*), 325
Enhancement, 603
Environmental change, 502, 503
Environmental economics, 572, 579
Environmental Protection Agency (EPA), 559
Environmental risk, 282
Ephemeroptera (mayflies), 153, 174, 175, 177, 178, 185
Ameletus spp., 170
Baetus spp., 170, 172, 176, 177, 178, 188, 189
Cinygmula reticulata, 170
Drunella doddsi, 170, 178, 186
Ephemerella spp. 170
Heptagenia spp., 170
Paraleptophlebia spp., 170
Tricorythodes minutus, 170
Epilithic algae, 351, 417
Epilithic biofilms, 412, 422
Epilithon, 405, 417
Equilibrium, dynamic, 433, 444, 602
Equilibrium price, 582, 583
Equivalent Road Acres, 489, 490
Equivalent variation, 577
Ergosterol, 124, 125, 132, 133, 137
Ermine (*Mustela erminea*), 242
Erosion, 48, 262, 283, 299, 318, 591, 605, 613
patterns of, 98
surface, 262
types of
gully, 83
sheet, 83
streambank, 83
Establishment, 296, 298, 309, 314, 316
Estivate, 236
Estuaries, 335
Ethylene, 299
European colonists, 503
Eurytherms, 175
Eutrophication, 77, 151, 386, 413
Evaluation, 599, 600, 603
Evaporative cooling, 311
Evapotranspiration, 48, 311
Evenness, 507
Exclusions, competitive, 433, 442
Exotic fishes, 220, 222, 224
Exploratory data analysis, 469, 471
Externalities, 625, 629
Extracellular enzymes, 120, 123, 126, 129, 134

F

Falcons, 238
False azalea (*Menziesia ferruginea*), 306
Farms, 573
Fauna
epigean, 419
hypogean, 400
hyporheic, 401, 408, 422
Fecundity, 203
Federal agencies, 572
Federal Energy Regulatory Commission (FERC), 561, 564, 567
Federal Land Planning and Management Act (FLPMA), 553, 563
Federal Power Act (FPA), 559, 564, 567–568
Federal reserved water rights, 558
FEMAT, 494–495, 602, 617

Fern, bracken (*Pteridium awuilinum*), 306
Fertilization, 348–365
Fertilizer nitrogen, 413
Field crews, training of, 466–467
Filamentous green algae, 85–87, 153, 251
Cladophora, 85, 86, 89, 387, 443
Mougeotia, 87
Spirogyra, 85, 87, 443
Stigeoclonium, 443
Ulothrix, 85, 87, 145, 147, 148, 443
Filterers, 335. *See also* Collector-filterers
Fir, 306, 310, 325. *See also* Balsam fir; Grand fir; Pacific silver fir; Subalpine fir
Fire, 267–268, 261–282, 302–304, 309–310, 329, 348
and recurrence interval, 267
and simulation model, 267
stand-replacing, 277
Fireweed (*Epilobium latifolium*), 305
First instar, 174
Fish community, 111–112, 200–201, 273, 282, 284, 324, 334, 337, 503, 504, 507, 509, 510, 512, 513, 519, 520, 522, 572
diversity of, 208, 211, 214
regional, 201–205
species, 202
habitat, 282, 283
life history attributes, 203
morphologies of, 200, 203
population variability in, 206–208
spawining, 411
Fisheries Act, 554
Fish-sediment model, 491
Fish stocks, 572
Flagellates, 126, 127
Flathead River, 176, 404
Flatworms. *See* Turbellaria
Flood Control Act, 576
Floodplain, 51, 52, 53, 273, 275, 277, 289, 310, 311, 315, 399, 401, 404, 410, 416, 417
forests in, 317
and hyporheic zone, 400, 401
Flood pulse concept, 3, 250, 666
Flood routing, 53–54
Floods, 54, 56–57, 60, 338, 572, 587, 261, 262, 273, 274, 304, 306, 309, 338, 421, 572, 587
Flood storage, 59
Flow
alteration of, 502, 517, 518
path of, 408, 417
length of, 408
regime of 604
types of
instream, 576, 590
intergravel, 56, 84
overbank, 607
overland, 49–50, 50–51
peak, 51, 56, 604, 605
return flow, 49
subsurface, 48–49, 54, 56
velocity of, 15
Fluid density, 15
Flumes, 609
Fluvial geomorphology, 600
Food base, 502, 505
Food chain, aquatic, 111
Food loop, microbial, 120, 122, 127, 137–139
Food supply, 374, 375, 400, 411
Food webs, 180, 189–192, 386
Columbia River, hypothetical, 223
detrital, 120, 122
generalized, 380
and hyporheic zone, 421, 422
Forest Act, 554
Forest
age of, 261
distribution of, 261
invertebrate, 324
oxygen, 334
water, 4
practices, 115, 284, 317, 590
Forest Practices Code Act, 554
Forest Ecosystem Management Assessment Team. *See* FEMAT
Forests
deciduous, 292
rain, 8, 291
temperate, 8
riparian, 289–294
and ecosystem functions, 310–311
and forest practices, 317–318
and land-use change, 311–317
and life history, 289, 294, 298, 299, 305
microclimate of, 310–311
mosaic of, 302
forest patches, 422
physical setting of, 294–297
spatial patterns of, 302–310
and succession, 340, 328, 338
Forest Science Data Bank (FSDB), 467
Formation, pool, 331
Founder effects, 208
Fragmentation, 127, 136, 329–330
Franklin Creek, and disturbances, 225, 282–283
Fraser River, 5
Fremont cottonwood (*Populus fremontii*), 315
Frequency, pool, 340
Freshet, 357
Freshwater polyps, 187
Friction, 15
Frissell, C.A., 99–100, 101
Frogs. *See* Bull frogs; Pacific tree frogs; Red-legged frogs; Tailed frogs; True frogs; Wood frogs Fulvic acid, 137
Functional groups, 172, 179–181, 335, 380, 384
Fungal biomass, estimates of, 124, 132, 134
Fungi
and hyporheic zone, 405, 417, 422
and organic matter, decomposition of, 120, 121, 122–125, 132, 133, 378, 380
See also Hyphomycetes, aquatic

G

Game birds, 243
Garter snakes (*Thamnophis* spp.), 242. *See also* Aquatic garter snake;

Western aquatic garter snake
Gastropoda (snails and limpets) (*Juga*) spp., 172, 178, 191
Geese, 238
Gene flow, 244
Genes, 505
Genetic variation, 206
Geographic information systems, 254, 471–471, 636, 634
Geologic composition, 408
Geomorphic characteristics, 100
Geomorphic province, 22, 408
Geomorphology, 245
Giant salamander (*Dicamptodon* spp.), 236
Gifford Pinchot National Forest, Washington, 239
Glacial retreat, 305–306
Glaciation, 200, 201–203
Glaciers, 46
Global change, 367
Goods and services, 576
  economic, 577
  monexclusive, 575
  private, 583
  public, 574, 579, 627
  quasipublic, 575
Government management efforts
  failure of, 575, 625, 626, 631
  limitations of, 627–629
Gradient/confinement index, 37
Gradients
  hydraulic, vertical, 406
  latitudinal, 8, 135
  oxygen, 414, 417, 419
  physicochemical, 400
Grand fir (*Abies grandis*), 310
Granite, 408
Grass shrimp (*Palaemonetes pugio*), 335
Gravel, 274, 599, 609
  alluvial, 424
Gravel bars, 332
Gravitational acceleration, 15
Grazers
  and hyporheic zone, 423
  as a limiting factor in primary production, 145, 152–153, 155–156, 179–180
  and trophic dynamics, 382
Grazing, 88, 144, 169, 184–185
  invertebrate, 415
  livestock, 502, 522, 616
Grazing minnow (*Campostoma anomalum*), 88
Grebes, 238
Green consumerism, 638
Ground squirrel, 242
Groundwater, 399, 400, 402, 408, 411
  and interactions with hyporheic water, 77
  obligate, 419
  storage of, 48
Guilds, 5607, 509, 510
Gullying, 275

H

Habitat Restoration Cost Estimation Model, 614
Habitats, 335, 340
  alteration of, 220, 222, 507
  complexity of, 200, 218–219
  conservation plans for, 566, 599
  destruction of, 4–5
  diversity of, 283, 284
  heterogeneity of, 201, 208, 210, 211, 212
  partitioning of, 220
  patches, 437, 438, 440, 447
  sensitivity to disturbance, continuum of, 99–100
  structure of, 505
    instream, 603
  types of
    aquatic, 261, 281, 599
    critical, 565
    off-channel, 603, 608
    physics, 502, 505
    rearing, 605
    riparian, 246, 248–249
    winter, 605
  unit patterns of diversity, 102, 103, 218–220
    scale, 615
  volume of, 404
Hand labor, 609
Hardhack (*Spiraea douglasii*), 306
Hardness, 70, 73
Hardwoods, 326, 327
Hares, 238, 242
Harvey Creek, and disturbances, 225
Hatchery, 573, 608
Hawks, 238
Heather meadow (*Phyllodoce*), 306
Heminetabolous, 174
Hemiptera, 174
Hemlock (*Tsuga* spp.), 8, 267. *See also* Eastern hemlock; Western hemlock Herbicides, 366
Herbivory, 295, 430, 431, 443
Herons, 238
Heteroptera, 171
Hierarchical approach, 614
Hierarchical control on system functions, 3
Hillslope stability, 292
Hirudinea (leeches), 172
Holometabolous, 174
Huckleberry (*Vaccinium* spp.), 306, 443
Hudson River estuary, 132, 133
Human activities, 108, 112, 113, 284, 507, 510, 511, 513, 515, 521, 599, 613, 617
  fish, impact on, 220–221
  impacts of, 505, 507, 510, 511, 512, 515, 523
  regional variations in, 517
  and trophic dynamics, 373, 374, 386–388
Human population, 4, 5, 7
Humic acids, 70, 137
Humin, 354
Hunting, 581
Hydraulic head, 400, 405, 406, 409, 417
Hydraulic storage, transient, 402
Hydraulics, 604
Hydroelectric power, 315
Hydrologic factors, and stream quality, 88–89
Hydrology, 43–44, 193, 505, 573, 600
Hydrophilic acid, 137
Hydrophytes, obligate, 297
  Odonata (damselflies, dragonflies), 174, 186
  Argia spp., 170, 178
  Octogomphus specularis, 170
Hydropower, 564–565, 573, 588
Hyphomycetes, aquatic, 120, 122, 129, 137
Hyporheos, 401, 419, 421

I

Ice scour, 292, 302, 304
Immobilization, 130, 131, 385
Imperviousness, 521
Implementation, 610
Impoundment, 315, 362, 363, 413
Income, 577, 584
Incrementalism, 678
Increment summation, 190
Index of biological integrity, 502, 507, 509–515, 516, 517, 518, 520, 521, 524, 644
Indicator taxa, 507, 509, 510
Individual economic welfare, 577
Inflow, 48, 54
Ingoldian fungi. *See* Hyphomycetes, aquatic
Insectivores, 238
Insects, 171–172, 520
  infestations of, 292
  and LWD, uses of, 335
  and trophic dynamics, 379
Instantaneous growth, 190
Institutions, 529–532, 533
  and strategies for collaboration, 543–545
Instream water uses, 573
Interactions
  biotic, 505
  fungi-shredder, 133–134
  predator-prey, 185–187
Intergenerational equity, 582
Interstitial spaces, 175, 384
Interstitial surface area, 401
Interstitial volume, 404, 405
  hyporheic, 404
Institutional capacity, 625
Interrain Pacific, 634, 636
Intervention, 605
Intragravel dissolved oxygen, 71, 78
Invader, 298, 299
Invasiveness, 643, 644
Invertebrates, 170–171, 330, 334, 335, 507, 509, 512, 513, 514, 515, 521, 522
  benthic
    community structure, 112–113
    index of biological integrity (B-IBI), 502, 510, 515, 520–523
  and functional groups, 113
  and life cycles, 172
  stream, 98
Iron, 413
Iron oxide, 413
Irrigation, 572, 576
Isopods, 153, 171

J

Jays, 245
Juvenile salmonids, 214, 217, 218, 219, 220, 222, 225, 334, 337

K

Kabetogama Peninsula, Minnesota, 244
Kauai, Hawaii, 487
Kingfishers, 238
Klamath River, 203
Klamath Zoogeographic Province, 204
Knee roots, 299
Knowles Creek basin, 282
Krummholz, 306, 307
Kuparuk River, 172

L

Ladders, 609
Ladyfern (*Athyrium filix-femina*), 441
Lamprey, 203, 204, 205, 207, 211, 214
Landforms, successional development of, 416
Land mosaic, 646
Landscape
  behavior, classification of, 263–281
  ecology of, 99
  elements, 261, 262, 263
    population of, 261, 262, 274
  institutional, 625, 626, 639
  processes, 243–244
    animal influences, 244–245
    connectivity, 244
    disturbance, 245
    geomorphology, 244
  systems, 261–263, 274
    components of, 265–274
    elements of, 274–283
Landslides, 263–265, 274, 275, 294, 295, 327, 329, 340
Land use, 56, 62–64, 170, 192–194, 408, 424, 605
  changes in, 311
  and Federal Land Planning and Management Act, 563
  and LWD, influences on, 338–341
  and National Forest Management Act, 563
  and river corridor, protection of, 560–561
  and State Forest Practices Act, 563–564
Land-Use Change and Analysis System (LUCAS), 649, 650
Large organic debris, 464
Large woody debris. *See* LWD
Larvae, 172, 174, 184
Latent heat, 59
Law, 503, 504
Leachate, 137–138, 334
Leaching, 127, 129, 328, 329
Leaf litter,
  and nutrient cycling, 348
  and organic matter, decomposition of, 120, 122, 124, 127, 130, 132, 136
  and stream macroinvertebrate communities, 169
  See also Plant litter decomposition Leeches. *See* Hirudinea
Lepidoptera (moths and butterflies), 171
Levels, trophic, 144, 204
Life histories
  anadromous, 205–206
  catadromous, 205
  macroinvertebrate, 170, 172–175, 173, 179
Light, 401
Lignin, 120, 134, 136, 137
Limpets. *See* Gastropoda
Linkages, 98, 99
Lipids, 134
Litter quality characteristics, 120
Litterfall, 121
Little Lost Man Creek, California, 351, 414–415
Liverworts (*Porella*,) 148
Loading, organic, 423
Local capacity, 633
Local knowledge, 626, 634

Lodgepole pine (*Pinus contorta*), 267
Logging, 3, 59–60, 341, 348, 364, 365, 411, 502, 522, 590
Longtail voles (*Microtus longicaudus*), 238
Long-Term Ecological Research, 464
Long-toed salamander (*Ambystoma macrodactylum*), 236
Lookout Creek, Oregon, 191
Loons, 238
Lungless salamander (*Plethodon vandykei, Plethodon vehiculum, Aneides flavipunctatus*), 242
Lush meadow, 306
LWD, 14, 34, 175, 184, 193, 222, 225, 289, 296–297, 331, 324, 332, 340, 438, 604, 649, 659
  abundance of, 324, 327, 333, 339
  decomposition of, 329, 334
  function of
    and channel form, 331–332
    and fish, 335–338
    and macroinvertebrates, 334–335
    and particulate matter, movement of, 332–335
    and riparian successional processes, 338
    and water quality, 334
  input of, 324, 327–329
  output of, 327, 329–330
  size of, 324, 325, 327

M

Macroinvertebrates, 112, 169–170, 334, 335, 378, 379, 380, 508, 520, 522
Macroinvertebrate taxonomic diversity, 170–172
Macroorganisms, 121
Macrophytes, 121, 144, 145, 148–149, 378, 411
Macropores, 48
Mammals, 238, 242, 248
Management, 98, 106, 220, 226, 388, 503, 504, 505, 509, 515, 517–520, 521, 523, 572
  and biodiversity, 448–449
  and hyporheic zone, 399, 401, 408, 411, 424
  and LWD, 341
  and riparian forests, 318
  and riparian wildlife, 250–254
  watershed, 283–285
Manning roughness coefficient, 15
Maple. *See* Bigleaf maple; Vine maple
Marbled murrelet, 8
Markets, 575, 625, 626, 629, 635
  competitive, 573, 582
  equilibrium of, 583
  failure of, 575
  and incentives, 625, 629, 633, 635
  value of, 579, 585
Marmots, 242
Marsh shrew (*Sorex bendirii*), 238
Mattole Restoration Council, 633, 635, 637
Maximum acceptable concentrations, 72, 81–82
Mayflies. *See* Ephemeroptera
McKenzie River, 155, 182, 216, 382
Meadows, subalpine, 307
Meander wave length, 16
Mediative culture, 529
Megaloptera (dodsonflies, alderflies)
  Orohermes spp., 170, 186
  Sialis spp., 170
Meiofauna, 122, 138, 417, 421
Merovoltine, 170–171, 172
Metabolism, 419
Metals, concentrations of, 81–83
Metamorphosis, 174
Methane, 413
Methods
  economic values, estimation of, 584
  hypothetical, 584, 585
  nonparametric, 470
  observed, 584
  parametric, 470
  pebble count, 108
Metrics, 502, 509, 510, 511, 512, 513, 522
Microbes, 379, 417, 422, 423
Microclimate, 289
Microflagellates, 138
Microflora, 335
Microhabitat, 99–100, 102, 103
  scale, 101, 603
Micrometazoans, 378, 379
Microorganisms, 121, 129, 130
  decomposition of, 120–122
  types of
    autotrophic, 126
    heterotrophic, 120, 121, 122–127
Micropores, 48
Middle Fork Snoqualmie River, 46
Midges, 138, 175, 178, 185, 186, 187, 188, 189, 192, 387
Mill Creek, 82
Mineralization, 127, 128, 135, 136, 352, 353, 359, 415
Minerals, 503
Mines, 573, 575
Mink (*Mustela vison*), 241
Minnows, 153
Mississippi River drainage, and fish faunas of, 204
Models, 572
  bioeconomic, 591
  channel response, 14, 16–17
  digital elevation, 36
  economic, 573
  habitat, stream, 111
  integrated socioenvironmental, 646, 647, 649
  plant litter decomposition, 131–132
  simulation, 277, 283
  spatially deterministic, 281
Mohun drainage basin, Vancouver Island, 365, 366
Moisie River basin, Quebec, 159
Mole salamander (*Ambystoma* spp.), 236
Mollusca, 170, 171
Mollusks, 520
Monitoring, 502, 506–509, 524, 599, 600, 610, 613, 619
  ambient, 506
  pulsed, 612
Moose (*Alces alces*), 241
Mosses, 148, 149, 184
Moths. *See* Lepidoptera

Mountain alder (*Dryas drummondii*), 305
Mountain beaver (*Aplodontia rufia*), 242
Mountain goat (*Oreamnos americanus*), 242
Multimetric biological indexes, 502, 510, 523
Multiple Use Sustained Yield Act, 563
Multipurpose project assessment, 588
Multivariate statistical analysis, 462, 470, 502, 507, 509
Multivoltine, 170–171, 172
Municipal water supply, 588
Muskrat (*Ondatra zibethica*), 238
Mussels. *See* Bivalvia
Mutation, 505

N

National Environmental Policy Act, 476, 477, 488, 648
National Forest Management Act, 553, 563
National Marine Fisheries Service, 564, 614
National Parks Act, 554
National Wild and Scenic Rivers Act, 560
Nationwide Rivers Inventory, 4
Native fishes, 209, 224–225, 226
Navigable Waters Protection Act, 554
Navigation, 572, 576
Nectivores, 246
Needle Branch watershed, 364, 365
Nematoda, 122, 170
Neoglaciation, 263
Nested design, 613
Neuroptera, 171. *See also* Metaloptera
New Jersey rivers, whole-river scheme, 98
Newts (*Taricha* spp.), 249, 236
Nitrate, 413, 415, 418
Nitric oxide, 413
Nitrification, 347, 353, 356
Nitrite, 413
Nitrogen, 120, 130, 131, 135, 136, 139, 151–152
  cycling of, 344, 353, 365, 413, 419
    hyporheic, 400
  dynamics 0f, 413–419
  retention of, 408
  sources of, 413
Nitrogen oxide, 347, 351, 353, 356
Nitrogen sinks, 413
Nitrous oxide, 413
Nonconsumptive uses, 579
Nongovernmental organizations, 625, 626, 627, 631, 632, 633, 634
Nonmarket economic benefits, 584
Nonnative fishes. *See* Exotic fishes
Nonprofit organizations, 624, 625, 635, 637, 638, 639
  degraded watersheds, restoration of, 635
  emergence of, 629–631
  formation of, 626–627
    theories concerning, 625
  future of, 638–639
  information, providing access to, 634–635
  watershed conservation, promotion of market incentives for, 635–638
Nonrival in consumption, 575
Nonsalmonids, 219
North American Water and Power Alliance, 663, 664
Northern bog lemming (*Synaptomys borealis*), 238
Northern flying squirrel (*Glaucomys sabrinus*), 242
North Fork Holston River, 516
Northwestern salamander (*Ambystoma gracile*), 245, 249
Northwest Power Act, 674
Northwest Power Planning Act, 578
Nuisance algae. *See* Periphyton
Numeraire, 577
Nurse logs, 338
Nutria (*Myocastor coypus*), 243
Nutrients, 75–77, 138–139, 334, 373, 374, 415, 417
  concentration of, critical, 87–88
  limitations of, 350
  nonconservative, 385
  in primary production, 150–152
  sources of, 384
  and transformations, 374
  types of
    inorganic, 350
    soil, 416
  and uptake, 359, 360
Nymphs, 174, 188

O

Oak leaves, 138
Obligates, riparian, 236–241
Off-stream water uses, 573
Oldman River Dam, 670
Old world mice (Muridae), 243
Oligochaeta (worms), 171, 175, 520, 522
Olympic Mountains, 22, 264
  ANC and pH levels, 73
Olympic Peninsula, Washington, 238
Oregon Coast Range, 270, 274
Organic carbon, 413
Organic matter
  class found in streams, 121
  decomposition of, 120–122, 131
  flocculation of, 381
  and hyporheic zone, 410, 414, 416, 417
  processing of, 380
  storage of, 333, 373, 384–386
  and trophic dynamics, 373, 374, 377, 379
  types of
    autochthonous, 121, 122, 145, 153
    coarse particulate, 120, 333, 335, 373, 380, 422
    detrital, 121
    dissolved, 74, 78, 120, 121, 122, 137–139, 373, 378, 379, 380, 417, 421, 422
    fine, 324, 335
    fine particulate, 121, 380, 422
    labile, 413, 416
    particulate, 78, 120, 121, 354, 355

terrestrial , 388
Organisms
aquatic, 281
consumer, 373
Organization for Economic Cooperation and Development, 632
Organizations, 529–533
ad hoc, 544
charitable, 629
decision processes, typology of, 545–548
advocacy, 547
bargaining, 547
computation, 546
consensus building, 547–548
experimentation, 546–547
organizational learning, 545, 547
pragmatism, 546–547
Orthophosphate, 347, 351
Osprey (*Pandion haliaetus*), 238
Outflow, 48, 54
Overharvest, 503
Oviposition, 174
Oxidation, 351
Oxides, 413

P

PACFISH, 535–537, 540–543
Pacific coastal region, 5, 7–8, 10–11
Pacific Forest Trust, 635, 636
Pacific Northwest coastal ecoregion, 5–7, 235
Pacific salmon, 4, 333, 348, 361
Pacific silver fir (*Abies amabilis*), 306
Pacific slope flycatcher (*Empidonax difficilis*), 242
Pacific tree frog (*Hyla regilla*), 236
Paedomorphic, 237
Painted turtle (*Chrysemys picta*), 237
Paleochannels, 402, 410
Parasites, 122
Park Act, 554
Partial-area concept, 49
Particle size, 108, 273
Particulate matter, 332–335
Passerines, 238
Patch and boundary analyses, 646
Patch dynamics concept, 350
Pathways, trophic, 380, 388
Pectin, 123
Pelagic ocean fishes, 338
Perch, Nile (*Tilapia nilotica*), 387
Periphytic algae, 70, 76
Batrachospermum 145
Draparnaldia, 145
Oedogonium, 145
Stigeoclonium, 145
Tetraspora 145
Periphytic mat, 70
Periphyton, 69–70, 84–90, 126, 185, 377, 378, 379
Persistence, 653, 654
pH, 73, 353, 402
Phantom crane flies. *See* Ptychopteridae
Phenotypic variation, 206, 207
Phosophoric acid, 353
Phosphorus, 130, 139, 151–152
cycling of, 347, 349, 353, 354
and hyporheic zone, 413
soluble reactive, 72, 74
Photoperiod, 172
Photosynthesis, 144, 149, 373, 377, 380
Phytoplankton, 144, 145, 149, 377, 378
Scenedesmus,
Stephanodiscus, 149
Piezometer, 405, 406
Pilot Creek watershed, 480, 484
Pine (*Pinus* spp.), 267, 325
Pink salmon (*Oncorhynchus gorbuscha*), 334
Piscivores, 211
Pistol River, Oregon, 609
Planktonic algae, 84, 360
Plant diversity, riparian, 311
Plant litter decomposition, 127–137
fungal and bacterial activity, comparison of, 132–133
fungi-shredder interactions, 133–134
and negative exponential model, 131–132
rates of, factors affecting, 134–135
and wood, 135–137
Platyhelminthes, 170, 172
Plecoptera (stoneflies), 174, 175, 186
Calineuria californica, 170, 180, 184, 186, 191
Isoperla spp., 170
Pteronarcys spp., 170, 183
Sweltsa spp., 170
Zapada cinctipes, 170
Pneumatophores, 299
Policy decisions, economic impacts of, 572, 585-586
Pollution
nonpoint source, 84, 151, 502
point source, effects of, 515,
water, 386
Polymerization, 130–131
Polysaccharide matrix, 380, 417, 422
Polysaccharides, degradation of, 123, 129, 134
Ponds, 609
Pools, 17–18, 29–30, 30–31, 218, 273–283, 408, 599, 603
size of, 331
volume of, 48
Poplar (*Populus* spp.), 300, 307
Population, 502, 503, 504, 506, 521, 522
density of, 391
Porifera, 170
Porosity, 404, 420, 422
Potamoplankton. *See* Phytoplankton
Precipitation, 264–266, 291
annual, 46
Predation, 201, 214, 219, 383, 387, 391
Predators, 170–171, 180, 335, 379, 422
Preservation, 579
Principal components analysis, 508
Private sector, 625, 626, 629, 630
Processes
biogeochemical, 407, 408, 412, 424
biophysical, 600
fluvial, 294, 296
geomorphic, 261, 262
hillslope, 294–296
invertebrate-mediated, 183–189
metapopulation, 505
soil, 296

sorptive, 367
Processing, 127, 131
Procession, microbial, 348
Producers, 189, 583
Production
agricultural sediment production, 572
allochthonous, 373
autochthonous, 373
autrotrophic, 75, 362, 399–379
fish, 607
heterotroplc, 379
primary, 99, 144–145, 149–158, 399
aquatic, 144–145, 153–158
energy equation for, 149
limiting factors, 150–153
and nutrient cycling, 360
secondary, 189
and hyporheic zone, 422
measuring, methods for, 190
Production-to-biomass ratio, 189
Productivity, 112, 374, 391–392, 430, 431, 444, 643, 644
riverine, 250, 666
rule, 208
Profit-seeking, costs of, 625, 627, 629
Protocols, 508, 510
Protozoa, 120, 121, 122, 125–126, 126–127, 378, 379, 405, 417
free-living *Tetrahymena vorax*, 127
sessile *Vorticella spp.*, 127
Psephenidae (water penny beetles), 175
Psuedoreptication, 463
Ptychopteridae (phantom crane flies), 176
Public interest, 564
Public trust doctrine, 557
Puget Sound, 76, 78, 81, 82, 220, 221, 522

Q

Quality assurance, 467
Quality control, 467
Queets River, Washington, 418, 421

R

Raccoon (*Procyon lotor*), 242
Rails, 238
Rainbow trout (*Oncorhynchus mykiss*), 376, 387
Rainfall, 48
Rain on snow, 44, 46, 58–59, 179
Raisin River, Michigan, 517
Rats, old world, 243
Ravens 245
Reaches, 100, 102, 103
scale, 599, 608
channel, 410
diversity, 214–218, 222, 284
types of
alluvia, 25, 29–30, 422
bedrock, 25, 410
cascade, 25
channel, 25, 48
colluvial, 25, 410
dune-ripple, 29
plane-bed, 29
pool-riffle, 29, 103, 410
step pool, 25, 29
Reach-level response, 31–32
Reaeration, 80
Recovery, 600, 602, 617
hydraulic, 62–64
Recreation, 502, 522
Red alder (*Alnus rubra*), 131, 137–138, 239, 328, 351
Redds, 282, 411, 424
Red-earred slider (*Trachemys scripta*), 243
Red elderberry (*Sambucus racemosa*), 306
Red-legged frog (*Rana aurora*), 249
Red River, Kentucky, 517
Redside shiners (*Richardsonius balteatus*), 388
Red spruce (*Picea rubens*), 325
Reduction, 362
Redwood National Park, 51, 610–612
Redwood (*Sequoia* spp.), 8, 59, 291, 300, 325
Refuge, 400, 421
Refugia, 203, 281–282, 618
Regime
hydraulic, 348
hydrologic, 290, 304, 315–317
Regional net benefits, 589
Regulations, 554, 563, 588
and forest practices, 562–563
and wetlands, 561–562
and rivers, 311–314, 558–567
Remineralization, 350
Replication, 461
Reproduction
asexual, 300
sexual, 300
Reptiles, 237, 242, 248
Reservoirs, 314, 315, 588
and storage, 54
Resource competition theory, 434
Resources, allocation of, 572
Respiration, 132, 136, 380, 413, 417, 419
microbial, 129
Response
hydraulic, 62–63
hydrologic, 44
Restoration, scope and scale, 399, 424, 601–612, 619
and adaptive ecosystem management, 617–618
cost accounting for, 614–617
evaluation of, 613
of larger-scale rivers, 606–608
of off-channel habitats, 608–609
at microhabitat scale, 603–6606
monitoring of, 612–613
design for, 613–614
success, elements of, 618–619
of watershed-scale projects, 609–612
Revegetation, 607
Rhizosphere, 299
Riffles, 218, 403
dwellers, 218
scale, 603
Ringneck snake (*Diadophis punctatus*), 242
Risk assessment, ecological, 523, 608
Ritherdon Creek, 153
River continuum concept, 2, 3, 181–182, 209, 350, 381, 666
River law, 553–555
sources of, 554–555
federal jurisdiction, 555
state jurisdiction, 555
River management, economics and ecology of, 589–593
River otter (*Lutra canadensis*), 241
Rivers, definition of, 1

Roach (*Hesperoleucas symmetricus*), 387
Roads, 56, 193, 573
Rocky Mountains, 5
Rodents, 238
Roosevelt elk (*Cervus elaphus roosevelti*), 242
Rooted macrophytes, 184
Roots, 413
  strength of, 267
Rootwads, 337
Rosgen, D.L., 106–111
Roughness, 14, 17
  element scale, 409, 411
Rubber boa (*Charina bottae*), 242
Ruderal plants, 440
Runoff, 44, 47–56, 51–53, 193, 413
  hillslope, 48–56
  nonpoint source, 78
  subsurface, 48–49
  surface, 50–51

S

Sacramento River, 207
Salamanders, 192, 249. *See also* Dunn's salamander; Giant salamander; Long-toed salamander; Lungless salamander; Mole salamander; Northwestern salamander; Torrent salamander
Salmon (*Oncorhynchus* spp.), 83, 192, 334, 337, 362, 424, 573, 572, 590, 599, 614. *See also* Chinook salmon; Coho salmon; Cutthroad trout; Pacific salmon; Pink salmon; Rainbow trout; Sockeye salmon Salmonberry (*Rubus spectabilis*), 306, 441, 443
Salmon Creek, 18–19
Salmonids, 77, 80, 115, 203, 204, 213, 218, 219, 375–376. *See also* Anadromous salmonids
Sampling design, 507, 508, 509
Sandstone, 408
Sand waves, 408, 411
San Francisco Bay, 486
Santa Cruz Mountains, 18
Saturated sediment, 411
Saturation, 80
Scale, 261, 262, 263, 283, 463, 468
  paradox of, 654
  temporal, 102, 412
Scioto River, Ohio, 520
Scoring metrics, 513
Scour, 16, 273
Scrapers, 154, 155, 158, 170–171, 179–180, 182, 335, 379
Sculpin, 203, 207, 213–214, 218, 389
Sedges, scattered (*Carex* spp.), 305
Sediment, 122, 261–273, 591, 507, 519, 602, 605
  cores of, 420
  delivery of, 13. 48, 275
  density of, 15
  porosity of, 405
  routine of, 332
  supply of, 14, 268–270, 277
  waves of, 36, 273, 412
Seeds, dispersal of, 300
Seedling establishment, 315, 316
Segregation
  interactive, 219
  microhabitat, 213, 219
  selective, 219–220
  trophic, 213
Semivoltine, 170–171, 172
Sentinels, 503–505
Serial discontinuity concept, 3
Sewage, 367, 423, 515, 520, 521
Sewage fungus (*Sphaerotilus*), 80, 81
Shallows, 30–31
Sharp-tailed snake (*Contia tenuis*), 242
Shear stress, 332, 607
  basal, 15, 296
  critical, 15
ShoreTrust Trading Group, 635
Shorthead sculpin (*Cottus confusus*), 389
Shredders, 120, 122, 133–134, 135, 136, 158, 170–171, 179–180, 183, 378, 379, 382
Shrews. *See* Marsh shrew; Vagrant shrew; Water shrew
Shrubs
  arborescent, 307
  medium-sized, 307
Sites
  clear-cut, 153, 161, 162
  depositional, 332, 338
  industrial, impact of, 82–83
  old-growth, 154, 161, 162
Sitka spruce (*Picea sitchensis*), 296, 325
Sixes River, Oregon, 206
Skate Creek, and disturbances, 225
Sinuosity, 16, 21
Size frequency, 190
Skeena River, 5
Skunk cabbage (*Lysichitum americanum*), 441
Slash burning, 367–368
Slide alder (*Alnus sinuata*), 291
Smolt, 604, 605
Snail (*Juga*), 152, 185, 387. *See also* Gastropoda
Snapping turtle (*Chelydra serpentina*), 243
Snow, 58–59
Snowmelt, 44, 46, 59
Snowpack, 46
Sockeye salmon (*Oncorhynchus nerka*), 7, 334, 361–362
Soil, 137
  matrix, 48
  moisture, 48
  pipes, 49
Solutes, 400
  transformation of, 413
Sorption
  abiotic, 413, 422
  irreversible, 413
  reversible, 413
Source-silk phenomena, 244
South Fork Salmon River, 18
South Island of New Zealand, 207
South Umpqua Experimental Forest, 364
Spatial heterogeneity/homogeneity, 430, 434, 446, 448
Spatial scale, 98, 99, 103, 208, 599, 614
  hierarchy of, 21–22
  multiple, 436–438
Spawning, 273, 282

Species, 502, 503, 504, 505, 506, 507, 508, 509, 512, 523
composition of, 340
diversity of, 238, 249, 432, 434
equitability of, 249
evenness of, 434
richness of, 200, 204, 226, 235, 238, 249, 291, 313, 316, 430–432, 434, 443, 507, 513, 515, 520
types of
congeneric, 214
corvid, 245
early successional species 324
endemic, 203, 204
exotic, 235, 242, 245, 386, 503, 507, 509
nonindigenous, 602
obligate, 235
pioneer, 305, 340
riparian wildlife, 236
Species-area relationship (of community ecology), 200, 208
Spiraling, 157
length of, 365, 385, 413
nutrient, 2, 349–350, 384–386
Springtails. *See* Collembola
Spruce (*Picea* spp.), 8, 267, 307. *See also* Engelmann spruce; Red spruce; Sitka spruce Squirrels. *See* Ground squirrel; Northern flying squirrel; Tree squirrel
Stable isotopes, 189–190, 191, 192
Stages
bankfull, 15
old-growth, 304
pupal, 175
stand exclusion, 304
stand initiation, 304
successional, 601
understory initiation, 304
Starling, European (*Sturnus vulgarus*), 243
State forest practices acts (SFPAs), 563, 564
Statistical design and analysis, 455–463
case studies, 455, 463–464
data acquisition, 466–468
hypothesis testing, 456, 464
sampling design, 456–463
Statistical information, management of, 467–468
Statistical power, 502, 507, 520
Statistical variance, 508
Stenotherms, 175
Stewardship, 580
Stickleback (*Gasterosteus aculeatus*), 387
Stikine River, 5
Stochastic school, 201
Stocks, 599
Stoneflies. *See* Plecoptera
Storage
mechanisms for, 384
transient, 385, 402, 412–413
water, 48
Storm, 261, 262, 274
Stream, 103
Streambed, 262
Streamflow, 46, 48, 50, 51
Stream order concept, 20, 98, 101
Streams, 1, 244, 538–540
capture of, 204
condition of, 106–107
ecology of,
morphology of, 657
quality of, 69, 70
segments of, 509
units of, 100
Structure, 291, 311
artificial, 603
deterministic, 112
stochastic, 112
trophic, 420, 421
vertebrate community, 11–112
Sturgeon (*Acipenser* spp.), 8, 205
Stygobites, 419, 420
Stygophiles, 419, 420
Stygoxen, 419
Subalpine fir (*Abies lasiocarpa*), 306
Substances
humic, 137
nonhumic, 137
Substrate, 175, 273, 283
inorganic, 175
lotic, 175
organic, 175
Succession, 246, 291, 292, 298, 302, 304, 305, 306, 315
Sulfate, 354
Sulfide, 356
Sulfur, cycling of, 347, 349, 359
Sunlight, as a limiting factor in primary production, 150
Surplus, 583
Sustainable development, 631, 664
Swainson's thrush (*Catharus ustulatus*), 242
Systems, 261, 262, 263, 283
oligotrophic, 69, 70. 76, 333, 360
trophic, cascading, 386

T

Tailed frog (*Ascaphus truei*), 236
Tamarisk (*Tamarix pentandra*), 315
Tan oak (*Lithocarpus densiflorus*), 310
Taxa richness, 509, 511, 512, 513, 521, 522
Taxa tolerance, 522
Tecopa pupfish (*Cyprinodon nevadensis calidae*), 5
Tectonic activity, 200, 203
Teleomorph, 123
Temperature, 291, 373
and fish communities, 206, 214, 222
and hyporheic zone, 401, 402
lethal, 77
and macroinvertebrate communities, 172
and nutrient cycling, 359–363
and organic matter, decomposition of, 123, 123, 126, 131, 135, 139
physiological, optimum, 77
and stream quality, 77–78
water, 175, 209, 214
Terraces, 273, 284
Theory, 284
Thimbleberry (*Rubus parviflorus*), 306
Timber/Fish/Wildlife watershed analysis, 493–494
Timber harvesting, 318
Topography, 261–272, 274, 277, 281, 283
Torrent salamander (*Rhyacotriton* spp.), 236, 245
Total boundary slicer stress 02

Total Maximum Daily Loads, 560
Townsend's voles (*Microtus townsendii*), 242
Toxicology, 502, 507, 511
Tracers, 402, 417
  conservative, 401, 402
Transformations, 347, 354, 363
Transport, 14
  fluvial, 328
  material, 106, 314–315
  sediment, 262, 266, 273, 600
Travel cost, 585, 588
Trees, 267
Tree squirrel, 242
Tributary, 77, 273
Trichoptera (caddisflies), 175, 178, 185
  Agapetus occidentis, 170
  Anagapetus bernea, 170
  Arctopsyche grandis, 171
  Brachycentrus, 172
  Cheumatopsyche spp., 171
  Dicosmoecus gilvipes, 88, 133, 134, 152, 153, 171, 177, 178, 185, 191, 387, 443
  Eocosmoecus spp., 171
  Glossosoma pentium, 171, 176, 178, 180
  Helicopsyche borealis, 171
  Heteroplectron californicum, 171, 184. 191
  Hydatophylax, 180
  Hydropsyche spp., 156, 171, 176, 178, 180
  Lepidostoma spp., 171, 183
  Micrasema spp., 171
  Neophylax spp., 171, 185
  Onocosmoecus unicolor, 171
  Rhyacophila spp., 171
Trout *Salvelinus* spp., 205, 213, 218, 219, 220, 222, 224, 335, 600, 606. *See also* Brook trout; Dolly Varden trout True flies. *See* Diptera
True frog (*Rana* spp.), 236
Turbellaria (flatworms), 172
Turbidity, 72, 83, 84
Turbulence, 373
Turtles. *See* Painted turtle; Snapping turtle; Western pond turtle
  Twisp River, 46
Type I/Type errors, 652, 653

U
Ungulates, 238
United States Fish and Wildlife Service, 564, 565
Univoltine, 170–171, 172
Uplands, 235, 241, 242, 246, 248, 249, 289, 312, 318, 504
Upwelling, 406, 412
Urbanization, 367, 424, 502, 511, 521, 522, 523
Urea, 366

V
Vagrant shrew (*Sorex vagrans*), 242
Valley
  morphology of, 292
  types of
    alluvial, 24
    bedrock, 25
    colluvial, 24
Valley segments, 23–24, 103, 104, 112
  scale, 409
  types of,
    constrained valley, 409
    unconstrained valley, 409
Values
  bequest, 579
  direct, 579
  economic, measurement of, 576–585
    aggregation, 581–582
    categories of, 579–580
    impacts versus benefits, 586
    individual values, 577, 584
    nonmarket values, measurement of, 572, 579
  existence, 572, 571, 579
  intrinsic, 579
  net present, 581
  net social, 581
  nonuse, 572, 579
  option, 579
  recreational, 588, 590
  social, 573, 581
  use, 579, 572
Van Duzen River, 18
Variability
  spatial, 261, 263
  temporal, 265
Variation
  compensating, 577
  natural, 506, 509, 510, 511, 512
  temporal, 507
Vascular plants, 377
Vegetation
  mesic, 307
  overstory, 113
  patterns of, 292, 296
  riparian, 34, 111, 114, 204, 340
    and hyporheic zone, 413, 424
    loss of, 388–389
    and organic matter, decomposition of, 121, 122, 137, 193
    and primary production, 160, 161
  understory, 113
Velocity, water, 414
Vertebrates, 379
Vine maple (*Acer circinatum*), 131, 361
Volcanic tuff, 408
Volcanism, 262
Voles. *See* Longtail voles; Townsend's voles; Water voles Voltinism, 172

W
Wagon Wheel Gap studies, 61
Washington state
  coastal rivers, 423–425
  western streams, 76
Waste Management Act, 554
Water Act, 554
Water
  chemistry of, 175, 238, 411
  diversion of, 572
  elevation of, 405
  hardness of, 70, 73, 81
  hydraulics of, 106
  management of, 505, 507, 508
  and pressure, 405
Waterfalls, 324, 332
Water penny beetles. *See* Psephenidae
Water Protection Act, 554

Water quality, 68, 74, 84–90, 334, 402, 405, 423, 572, 576, 600
  parameters of, 507
  physiochemical constituents, 71–72
  regulation of, 558–560, 566–567
    Clean Water Act, 558–559
    nonpoint source controls, 559–560
    point source controls, 559
    Total Maximum Daily Loads, 560
Water quantity, 556–558, 556–567
  federal law, 558
  public trust, doctrine of, 557–558
  regulatory rights, 558
  statutes concerning, 556–557
Water resources, economic assessment of, 573, 574, 587–589
  agricultural water supply, 587–588
  hydropower, 588
  multipurpose project assessment, 588–589
  municipal water supply, 588
  recreation, 588
  river navigation, 587
Water rights, adjudication of, 534
Waters
  lentic, 250
  lotic, 250
Watershed effects, cumulative, 476–479
  analysis, standardized methods of, 488–489
    patterns among, 492–493
  evaluation, ad hoc approach to, 485–488
  evaluation, problems in, 479–485
    philosophical issues, 481–483
    sociocultural issues, 483–485
Watershed engineering, 673–674
Watersheds, 22–23, 103, 114, 261, 262, 283, 599, 617
  analysis of, 472, 476, 493–499, 617, 646, 647, 648
    Watershed Analysis Units (WAUs), 648
  conservation of, and nonprofit organizations, 635–638
  management of, 662, 664, 663, 665, 666, 667, 671, 676, 677, 678, 679
    paradigms, evolution of, 673–675
    and paradox of scale, 653–654
    and public stewardship, 655–658
    quantitative analysis of, 646–650
    risk, management of, 652–653
    socioenvironmental visions, formulation of shared, 654–655
    and uncertainty, 651–652
    and nonprofit organizations, 625, 626, 632–638
  physical features, classification of, 100–111
  in primary production, distribution of, 144–145, 158–160
    responses to, 160–163
  restoration of degraded by nonprofit organizztions, 635
Watershed-scale restoration, 600, 602, 609, 614, 617
Watershed scales, 208–214, 409, 599
Water shrew (*Sorex palustris*), 238
Water tupelo (*Nyssa aquatica*), 300
Water voles (*Microtus richardsoni*), 238
Weeks' Law, 56
Welfare, 576
Wells, sampling of, 421
Wenatchee River, 222
Wertz Drain, Allen County, Indiana, 517, 519
Western aquatic garter snake (*Thamnophis couchi*), 237
Western hemlock (*Tsuga heterophylla*), 131, 267, 296, 306, 325, 329, 361
Western pond turtle (*Clemmys marmorata*), 237
Western red cedar (*thuja plicata*), 328–329
Western tanager (*Piranga ludoviciana*), 246
Western toad (*Bugo boreas*), 236
Wetland, 553, 561
White-tailed deer, 241
Width-to depth ratio, 21, 77, 602
Wildlife, 573, 575
  riparian, 235–236
    effects of conditions on, 243–250
    management, effects on, 250–254
Willamette River, 220, 221, 510
Willapa Alliance, 633, 635
Willing to accept, 577
Willing to pay, 577
Willow (*Salix alba*), 132, 292, 300, 305, 441
Wilson's warbler (*Wilsonia pusilla*), 246
Wind, 261, 281, 304
Windthrow, 304, 327, 328, 329, 331
Wood, 175, 184, 261, 262, 263, 273, 324, 329, 330, 340, 411, 422
  decay of, 120, 136–137
  delivery of, 325
  supply of, 270–272
Wood frog (*Rana sylvatica*), 236
Woody debris, 53, 135–137
  as habitat, 324, 333, 335
Worms. *See* Oligochaeta
Wynoochee River, 46

X

Xylan, 123

Y

Yellow cedar (*Chamaecyparis nootkatensis*), 306
Yellow mountain ariens (*Dryas drummondii*), 305
Yellow poplar leaves (*Liriodendron tulipifera*), 125, 130, 135
Yield, water, 59, 61–62

Z

Zonation patterns, biotic, 98–99
Zones
  ecological, 99
  groundwater, 504
  hyporheic, 3, 54, 169, 176, 177, 170, 353, 355, 358, 359, 399, 400–402, 407, 408, 412, 421, 504
    distribution of, 399–412
    hydrology of, 405
    shallow, 417
  parafluvial, 400, 401, 402
  riparian, 113, 115, 407, 413, 556, 564
Zoochory, 300
Zoogeographic provinces, 204
Zoogeography, 99, 111–112

MW00844903

Encyclopedia of WORLD GEOGRAPHY

Volume II

H – R

R.W. McCOLL, Ph.D.

GENERAL EDITOR

Facts On File, Inc.

Encyclopedia of World Geography

Published by Facts On File, Inc.

Facts On File, Inc.
132 West 31st Street
New York NY 1001

Library of Congress Cataloging-in-Publication Data

Encyclopedia of world geography / R.W. McColl, general editor.— 1st ed.
p. cm.
Includes bibliographical references and index.
ISBN 0-8160-5786-9 (hardcover : alk. paper)
1. Geography—Encyclopedias, Grades 9 and up. I. McColl, R. W.

G133.E483 2005
910'.3—dc22 2005006435

Facts On File books are available at special discounts when purchased in bulk quantities for businesses, associations, institutions, or sales promotions. Please call our Special Sales Department in New York at (212) 967-8800 or (800) 322-8755.

You can find Facts On File on the World Wide Web at http://www.factsonfile.com

GOLSON BOOKS, LTD.

Geoff Golson, President and Editor
Robert W. McColl, Ph.D., General Editor, Encyclopedia of World Geography
Richard W. Dawson, Ph.D., Associate Editor, Encylopedia of World Geography
Kevin Hanek, Design Director
Laurie Rogers, Copyeditor and Proofreader
Gail Liss, Indexer

PHOTO CREDITS

Photo Disc, Inc.: Pages 65, 83, 100, 113, 130, 131, 136, 137, 166, 168, 171, 199, 243, 266, 278, 303, 335, 337, 372, 406, 438, 460, 484, 485, 493, 545, 552, 553, 577, 590, 601, 627, 633, 649, 679, 806, 689, 710, 736, 771, 772, 821, 828, 843, 850, 851, 872, 887, 905, 934, 935, 963.Pages vii: Perry-Castañeda Library, University of Texas at Austin; 3: www.acongagua.org.uk; 11: www.pitt.edu; 21: Steven Allison, www.stanford.edu; 22: http://theopenline.cc; 39: NASA; 45: www.planetek.it; 91: www.cruisevents.com; 197: www.wallpaperdave.com; 216: U.S. Navy; 257: Sarun Charumilind, http://studentweb.med.harvard.edu; 262: Library of Congress; 265: www.zoutenzoewaterparels.com; 299: www.paconserve.org; 302: www.coastalmanagement.com; 313: www.waxvisual.com; 327: National Park Service; 371: www.anders.com; 446: NOAA; 507: USGS; 515: www.kathy loperevents.com; 524: U.S. Marines; 57: Library of Congress; 606: NASA; 775: http://euro rivercruises.com; 809: NASA; 812: Blackwell Publishing; 908: NOAA; 951: Library of Congress; 966: http://environment.cornell.edu; 991: www.blogula-rasa.com; 1005: www.gnu.org.

Printed in the United States of America

EB GB 10 9 8 7 6 5 4 3 2 1

This book is printed on acid-free paper.

Contents

Volume II
Entries H – R 403

Resource Guide 1011
Glossary 1015
Appendix A: World Rankings 1029
Appendix B: World Atlas 1109
Index 1153

Encyclopedia of

WORLD GEOGRAPHY

Volume II

H – R

Hainan Island

THE SECOND-largest island (after TAIWAN) in CHINA with an area of 13,104 square mi (33,940 square km), Hainan Island is the most southerly province located in the South China Sea. The island is situated south of Guangdong province and west of VIETNAM. The topographic structure of the island resembles a staircase, with towering mountains in the middle and descending hills, plateaus, and plains toward the periphery.

Hainan Island enjoys a tropical maritime climate with abundant rainfall and year-round sunshine. The annual average temperature is about 77 degrees F (25 degrees C) and the average rainfall is 63 in (160 cm). Winter and spring are considered the dry seasons, while summer and autumn are wet. Good sunlight, heat and water allow a rich variety of crops like rice, tropical fruits, and coconut to be cultivated on the island. It is rich in mineral resources such as salt, natural gas, iron ore, and oil.

But the island is also exposed to tropical storms and typhoons that hit the southeastern coast from July to September. Although geologically stable, it has a history of volcanic eruptions and mild earthquakes. Diseases ranging from malaria to mosquito-borne viruses plague the island.

Hainan Island is home to over 4,200 plant species and some 560 animal species, making it a destination for researchers. But deforestation could well threaten species extinction. The island is also a growing destination for tourists. Mainland Chinese and foreigners mostly flock to the southern coastal city of Sanya for surf and sun. In addition, the island has played host to international events like the Miss World Contest and Boao Forum for Asia.

Historically, the Li ethnic peoples were the first to inhabit the island 3,000 years ago. Other ethnic groups include the Miao and Hui peoples. The Han Chinese form the majority group. Backward and remote, the island was a place for exiles. In the early 20th century, many Hainanese emigrated abroad in search of a better life. Today, about 2 million Hainan Chinese are scattered in 53 countries worldwide.

In 1988, the island was made a special economic zone to boost foreign investment. The gross domestic product has increased tremendously, but economic development has gone through some boom-and-bust cycles. While rubber and agriculture are traditional industries, tourism, petrochemicals, biopharmaceuticals and machinery and electronic industries are expanding.

BIBLIOGRAPHY. China Internet Information Center, www.china.org.cn (September 2004); China Business Guide, www.chinaknowledge.com (September, 2004); Peter Hotez, *The Other Hainan* (Albert B. Sabin Vaccine Institute, 2001);

"China's Hainan Island at a Glance with China-U.S. Plane," Associated Press (April 2, 2001).

BENNY TEH CHENG GUAN
KANAZAWA UNIVERSITY, JAPAN

Haiti

Map Page 1137 **Area** 27,750 square mi (44,658 square km) **Population** 7,527,817 **Capital** Port-au-Prince **Highest Point** 8,792 ft (2,680 m) **Lowest Point** 0 m **GDP per capita** $1,400 **Primary Natural Resources** bauxite, copper, calcium carbonate, gold.

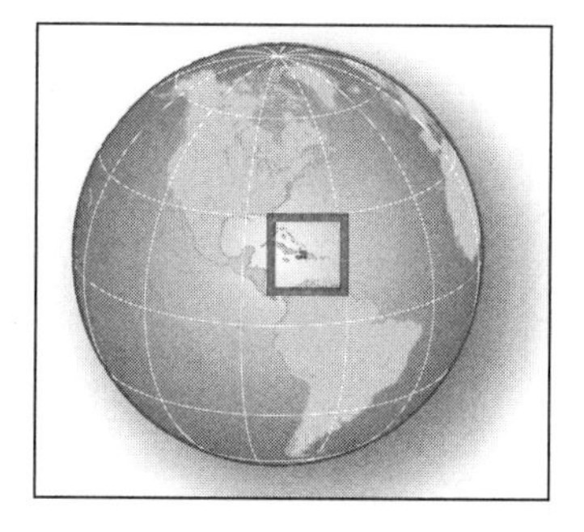

HAITI IS LOCATED on the western third of the island of Hispaniola in the CARIBBEAN SEA. The country is mainly mountainous, with 60 percent of the land on gradients of at least 20 percent. The mountain ranges of Massif de la Hotte, the Massif de la Selle, and the Chaine du Bonet encompass much of the land. Two large peninsulas are present in the country. The peninsulas are separated by the Golfe de la Gonave.

The fertile river valleys and deep forests, which once provided beautiful sights, have diminished to a few. Haiti has suffered from deforestation, which has left only three percent of the land untouched. Much of the nutrient-rich topsoil has been washed away into the surrounding waters, which has destroyed some of the marine life. Because of the distinct range of elevations, the country has numerous plant species: 5,000 are spread throughout the land, including 600 fern and 300 orchid species. These plants thrive in the hot and humid climate throughout the year, with temperatures ranging from 65 degrees F (20 degree C) to 85 degrees F (30 degrees C).

The history of Haiti is full of turbulence. After almost 200 years of rule, SPAIN ceded the western portion of Hispaniola to FRANCE in 1697. This area named Saint Dominique soon became one of the wealthiest nations in the Caribbean. Sugar plantations arose throughout the nation, and slaves were brought over in large numbers.

In 1791, a slave revolt, led by Toussaint Louverture, broke out across the colony. Louverture named himself lieutenant governor of the state in 1796. However, in October 1801, after a victorious campaign in Europe, Napoleon ordered an expeditionary force to Haiti to reclaim its lost colony. By November 1803, the last of Napoleon's forces were routed and on January 1, 1804, the war-ravaged colony was soon declared independent of French rule and renamed Haiti.

In 1844, Santo Domingo (soon renamed the Dominican Republic) declared its independence from the struggling country. Haiti seemed to be spiraling into obscurity in the world stage. From 1843 to 1915, 22 dictators ruled the country. In 1915, the UNITED STATES invaded Haiti hoping to create stability in the country. By 1937, the American troops left the country, with the belief that the country was a modernized and thriving.

For the next century, the Haitian government continued to struggle. During World War II, increased world market prices increased Haitian trade and greater exports. However, after the fighting ended, the country was still mired in poverty. François Duvalier was elected president of the country in 1957. Throughout his power, Duvalier struggled with the Catholic Church and citizens fleeing the country. He died in 1971, but his son, Jean-Claude Duvalier, took over the position. The Haitian economy continued to suffer and its external public debt increased to a staggering $366 million in 1980. The Haitian people continued to flee.

During January 1986, urban resistance overspread the country, led in particular by Jean-Bertrand Aristide. On February 7, 1986, Duvalier left Haiti, and General Henri Namphy, with American support, took control of the country. The Haitian government continued to promise the United States that elections would occur in November 1987. Three years later, Aristide won the presidential election. He took office in early 1991 but was overthrown by a violent coup.

During this three-year period, a military de facto regime ruled the country and refused to return to a constitutional government. The United Nations ultimately passed a resolution and sent in a multinational force to restore order in the country. Before the American-led force actually stepped into the country, a deal was brokered, which the Haitian de facto government peacefully accepted. Aristide was returned to the country, where he became president and the constitutional government was restored under the watchful eye of the United Nations.

For the next 10 years, the international optimism surrounding Aristide slowly diminished. He became embroiled in a variety of corruptive practices and many questioned the fairness of the presidential elections. Opposition parties rose and by early 2003, violence

from all political sides began to flare up once again. In February 2004, Aristide fled Haiti, and afterward, U.S. Marines arrived along with 3,600 other international peacekeepers, to stabilize the situation. Haiti once again is trying to reestablish itself from its troubled past.

BIBLIOGRAPHY. Charles E. Cobb, Jr., "Haiti, Against All Odds," *National Geographic* (November 1987); Paul Farmer, *The Uses of Haiti* (Common Courage Press, 1994); *World Factbook* (CIA, 2004); Paisley Dodds, "U.S. Ambassador Says Haiti's Aristide Was Sad and Passive," *San Francisco Gate* (April 13, 2004); U.S. Department of State, Bureau of Western Hemisphere Affairs, "Background Note: Haiti," www.state.gov (April 2004).

GAVIN WILK
INDEPENDENT SCHOLAR

hammada

HAMMADA IS AN Arabic word used to describe "desert pavement." The deserts of the world that have sand dunes capture the imagination, but many of the world's desert areas are bleak, stony deserts. Usually, the stony deserts are level plains that are virtually devoid of vegetation and also have very little, if any, soil. These deserts can have a relatively smooth, rocky surface that is hard, like road paving. Hammadas form in regions where the soil is either saline or alkaline, with little ability to absorb water, and winds are strong.

Desert paving or hammada (North Africa) is called by a number of names in the different hot and cold deserts of the world. The names are banada, desert crust, deflation armor, desert armor (North America), desert pavement, GOBI (CHINA), gibber (AUSTRALIA), lag gravel or reg (Africa, if the fine material remains), serir (Africa, if no fine material remains) and stone pavement.

Hammada describes a dark, stony desert surface without sand or vegetation. The dark color is "desert varnish" caused by wind-borne clay particles that carry bacteria living on them. The particles coat the rocks of the pavement coloring them with a dark sheen.

Desert regions covered with hammada have a surface that is fairly smooth. The wind has blown away most of the sand, soil, and dust. A layer of rocks on bedrock has formed that is so tightly packed that it forms a solid surface. There may be a layer of soil or dust under the hammada, but it can form directly on top of clean bedrock. The bare rocks that form hammada surfaces are usually relatively small and fit as tightly as a mosaic. Earth scientists describe these rock fragments as either primary or secondary. The primary stones are usually coarse, while the fine secondary material comes from the disintegration of the larger primary material.

There are several geomorphic theories explaining how hammada forms. One theory is that the pavement is a lag deposit. This theory says that the rock fragments of the pavement are what remains after the wind has blown away (deflated) all the small, fine-grained sand and dust. A second theory argues that moving water has deflated all of the fine material, leaving the hammada material behind to form the desert pavement.

A third theory sees the cycles of wetting and drying of the material as the cause. The exact nature of the process is not fully understood. Some claylike soils expand when wet, which forces stones to move upward. Another theory says that salt acts like water does when it expands or contracts in the freezing or melting process to create a stony surface. Another theory is that hammadas are caused by chilling and heating, as mechanisms that push stones to the surface of the desert and keep them there. Another theory is that several physical processes move the fine particles of sand and dust down between the stones of the hammada surface.

Hammada has been found on ALLUVIAL FANS, in dry WADIS, in terraces, on plateaus, on plains, and on bedrock. It is believed that hammada is also on the surface of the windswept planet Mars. Since hammada forms in such varied deserts conditions, it may be that all the theories are correct explanations.

The surfaces of many hammadas are hard enough for planes to use as landing strips. Vehicles can travel across the more solid variety, but excessive use can cause dust to rise from particles below the hammada surface. In the Atacama desert of Peru, the ancient Nazca people used hammada to make enormous drawings.

Tank battle maneuvers and tank battles in North Africa, KUWAIT, and IRAQ have disturbed large areas of hammada, causing dust and sand dunes to form elsewhere. The damage to these deserts will take many years for nature to repair.

BIBLIOGRAPHY. A. D. Abrams and A. J. Parsons, *Geomorphology of Desert Environments* (Chapman & Hall, 1994); R.U. Cooke and Andrew Warren, *Geomorphology in*

Deserts (University College Press, 1993); E.S. Hills, ed., *Arid Lands: A Geographical Appraisal* (Methuen & Co., 1996); L.D. McFadden, S.G. Wells, and M.J. Jercinovich, "Influences of Eolian and Pedogenic Processes on the Origin and Development of Desert Pavements," *Geology* (v.15, 1987); Sid Perkins, "Thin Skin," *Science News* (v.165/1, 2004); Douglas V. Prose and Howard G. Wilshire, "The Lasting Effects of Tank Maneuvers on Desert Soils and Intershurb Flora," U.S. Geological Survey (Open-File Report 00-512); D. Sharon, "On the Nature of Hammadas in Israel," *Zeitung Für Geomorph* (v.6, 1962); S.H. William and J.R. Zimbelman, "Desert Pavement Evolution: An Example of the Role of Sheetflood," *Journal of Geology* (v.102, 1994).

ANDREW J. WASKEY
DALTON STATE COLLEGE

Hawaii

HAWAII IS a U.S. state located in the central PACIFIC OCEAN near the EQUATOR and consists of eight major islands: Hawaii, Maui, Kahoolawe, Lanai, Molokai, Oahu, Kauai, and Niihau; 129 smaller islands are also part of Hawaii. Hawaii is not the equivalent of the Hawaiian Islands or the Hawaiian Chain, which includes other islands, such as Midway, that are not part of the state of Hawaii. At its greatest expanse, Hawaii extends for 1,523 mi (2,600 km); it has a total land mass of 6,423 square mi (16,729 square km). The highest point is Mauna Kea at more than 13,796 feet (4,205 m) and the lowest point is at sea level. The capital city is Honolulu on the island of Oahu. The population of Hawaii is 1,211,537 (2000 census), and its largest city is Honolulu (876,156). The state population is evenly divided by gender and has a density of 188.6 persons per square mile. The median age in 2000 was 36.2 years.

The high cliffs and rough coasts of Hawaii are a result of the volcanic origin of the Pacific Ocean islands.

The island of Hawaii is the largest of the islands and has two active volcanoes, Mauna Loa and Kilauea; it supports extensive agricultural activity. Maui is called the Valley Island because of the dominance of two volcanic mountains; while a haven for tourists, Maui also has sustained valuable production of sugar. Kahoolawe is a small, uninhabited island near Maui. Lanai is the center of the pineapple business. Molokai, also near Maui, is an island of canyons and mountains; its central region is fertile for crops. Oahu has two mountain ranges, the Koolau Range in the east and the Waianae Range in the west. While supporting an expanding metropolitan Honolulu, Oahu is an agricultural center with sugar and pineapple plantations. Kauai, known as the Garden Island, is rugged; its Mount Waialeale has the greatest rainfall (460 in or 1,168 cm) per year on Earth. Niihau is rather arid; it is privately owned by the Robinson family of Kauai.

Hawaii's gross state product in 2000 was $39.1 billion. The economy is based on tourism ($10.9 billion), U.S. military defense spending ($4.4 billion), and the sugar and pineapple businesses ($276.1 million.) The Hawaiian state government is interested in diversifying the economy and has invested in developing science and technology, film and television production, sports, oceanic research and development, education, and floral and specialty food products. In trade, Hawaii transacted $3.31 billion in merchandise in 2000 and had exports totaling $407.7 million. Imported petroleum products provide more than 90 percent of Hawaii's energy requirements (2000).

Originally settled by Polynesians more than 1,500 years ago, Captain James Cook was the first European to visit Hawaii (1778). During the 19th century the native Hawaiian population (estimated about 300,000) was decimated by diseases imported by European and American adventurers, businessmen, and missionaries.

By 1900, the native population had been reduced to about 70,000. During the same time, American immigrants overthrew the native government and moved Hawaii to annexation by the UNITED STATES. American expansion in Hawaii and the PHILIPPINES was denounced as imperialism; for the most part, American culture absorbed Hawaiian culture, although recent years have witnessed a resurgent native Hawaii culture. Americans viewed Hawaii as strategically and economically significant; the American navy centered its Pacific headquarters in Hawaii, and American businesses reaped profits from sugar, pineapple, and other fruit products. The attack on Pearl Harbor and the subsequent war against Japan led to an expanded American presence in the Pacific.

During the second half of the 20th century, the United States relinquished its naval bases in the Philippines and became dependent upon Hawaii as the focal point for its Pacific fleet. Further, the naval observatories on Hawaii are important astronomical centers for research and satellite management.

BIBLIOGRAPHY. Evelyn Colbert, *The Pacific Islands: Paths to the Present* (Westview Press, 1997); Joseph Feher, *Hawaii: A Pictorial History* (Bishop Museum Press, 1969); Neal R. Peirce, *The Pacific States of America: People, Politics, and Power in the Five Pacific Basin States* (W.W. Norton, 1972); Marshall Sahlins, *Islands of History* (University of Chicago Press, 1985).

WILLIAM T. WALKER, PH.D.
CHESTNUT HILL COLLEGE

Heard and McDonald Islands

TWO OF THE MOST isolated spots on the globe, the Heard and McDonald Islands have only been visited a few times, and much remains to be discovered about them. The islands are located at the bottom of the world, where the INDIAN OCEAN meets the Antarctic seas, and have recently been declared a United Nations UNESCO World Heritage site due to their pristine natural habitat, nearly untouched by human hands.

The first sighting of the larger island, Heard, is attributed to the British captain Peter Kemp in 1833, but also to the American captain John J. Heard, in 1853. Much smaller McDonald island, 27 mi (43.5 km) to the west, was spotted a year later by British captain William McDonald. Hunters, in the 1850s to 1870s, nearly depleted the islands of their population of varieties of seals (notably elephant seal and fur seal), king penguins and whales, but these populations have now mostly returned to their former numbers. To further protect this wildlife, the government of AUSTRALIA established the Heard Island and McDonald Islands Marine Reserve in 2002. Thirty-four species of bird come here to breed, including penguins, albatrosses, and giant petrels. Human visitors are tightly restricted to occasional scientific expeditions.

The islands, about 1,000 mi (1,700 km) off the coast of ANTARCTICA, are the most volcanically active in the sub-Antarctic region. They are rises on the Kerguélen Plateau, which also includes the Îles Kerguélens, about 434 mi (700 km) to the northwest (territory of FRANCE). Neither of the islands has much vegetation, though kelp is abundant offshore. Rain or snow and extensive cloud cover keep the islands very wet most of the year and are accompanied by nearly constant high winds. Heard Island is a circular volcanic cone, dominated by the Big Ben Massif topped by Mawson Peak, the only active volcano in Australian territory. A mountainous headland extends 6.2 mi (10 km) to the northwest (Laurens Peninsula), connected by a narrow ridge little more than 330 ft (100 m) wide. Heard is surrounded by numerous outlying rocks, islets and reefs. A few kilometers offshore are the Shag Islands.

About 80 percent of Heard is glaciated, with ice up to 495 ft (150 m) deep. Ice cliffs make up much of the coast, adding to the island's inaccessibility. The McDonald Island group is made up of one volcanic peak and several smaller rocky islets (Flat Island, Meyer Rock).

Much smaller than Heard, McDonald has been landed on only twice in its history, and very little is known about it. In fact, the size of McDonald seems to have recently doubled—from .39 square mi (1 square km) to an estimated 1 square mi (2.45 square km)—through a volcanic eruption sometime between 1997 and 2001, when changes were detected on a satellite photograph. It is believed that this eruption is continuing, radically altering the shape of McDonald Island.

BIBLIOGRAPHY. *World Factbook* (CIA, 2003); Australian Antarctic Division, www-new.aad.gov.au (May 2004); World Conservation Monitoring Centre, www.wcmc.org.uk (May 2004); Smithsonian Institution's Global Volcanism Program, www.volcano.si.edu (May 2004).

JONATHAN SPANGLER
SMITHSONIAN INSTITUTION

heartland

THE HEARTLAND THEORY was developed by Scottish geographer, Sir Halford J. MACKINDER. He read his paper "The Geographical Pivot of History" before the Royal Geographical Society in London in 1904, and soon afterward published his views on the influence of geography on politics in the *Geographical Journal.*

Mackinder's argument introduced into POLITICAL GEOGRAPHY the view that the globe could develop into two power blocs. The center of world power lay in the "heartland." This was roughly the territory east of the DANUBE RIVER to the URAL MOUNTAINS. Mackinder claimed that whoever unified and dominated Central and Eastern Europe would be able to control the three continents of the "World Island"—Europe, Asia, and Africa. His hypothesis, which became famous, was: "Who rules Eastern Europe commands the Heartland. Who rules the Heartland commands the World Island. Who rules the World Island commands the world." The other power bloc, because of what he viewed as the relative decline of sea power, was the "maritime lands."

Mackinder had rejected the theory of sea power as the key to world domination proposed in *The Influence of Sea Power upon History* by Captain Alfred T. Mahan of the U.S. Navy. Mahan developed a number of factors of state power, such as geographical position, extent of territory, population, and national and governmental character to explain state power. Mahan had used his study to propose policies for the American government to follow. These included annexing the Hawaiian Islands, controlling the CARIBBEAN SEA, and building a canal across PANAMA. Theodore Roosevelt was to use these ideas as the basis of his foreign policy.

In contrast, Mackinder claimed that land-based power was increasing in scope because of improvements in communications, transportation, and armaments. The heartland power would be the ultimate land power that would be impossible to expel from its natural fortress. He believed that there was a Eurasian core that would be impregnable to naval power. As the center of a great fertile landmass, it would be difficult to penetrate with sea power and easily defended with land forces.

Mackinder's views were read and discussed by many people immediately. His hypothesis soon became one of the most intensely debated ideas in the history of geography. Mackinder's views were eagerly accepted by a number of German politicians. In particular by Karl Haushofer, who with others adopted a *geopolitik* program. Their views influenced Adolf Hitler, who used them as a rationale for conquering Europe.

Among those who disagreed with Mackinder's theory was N.J. Spykman. He argued that Mackinder had exaggerated the Heartland's potential for power. The real power was not in the "pivot," as Mackinder had termed the heartland, but in what Mackinder had called the "inner crescent" or the "rimland" in Spykman's terminology. Spykman's hypothesis became, "He who controls the Rimland rules Eurasia; Who rules Eurasia controls the destinies of the world."

Missing from the heartland theories of Mackinder and Spykman was air power. In 1943 Mackinder dismissed air power as a technological change that would make a difference in the balance of power. However, in 1950 A.P. de Seversky argued that the key to American survival and supremacy was air power.

Another critic of Mackinder was A.R. Hall. In 1955, Hall argued that Mackinder had ignored the existence of another "heartland," namely the Anglo-American alliance.

In 1956, D.W. Meinig proposed changes to Mackinder's heartland theory and to the rimland theory. His criticism included changes to the definitions and criteria of Mackinder's theory. In 1964, D.J.M. Hooson extended the heartland theory by identifying the "core" areas, such as the industrial centers.

In the last quarter of the 20th century, two major groups debated and extended Mackinder's ideas. The first group was composed of specialists in strategic studies. One important representative of this approach was Colin S. Grey, who published *The Geopolitics of the Nuclear Era* (1979). The other group was composed of scholars in traditional POLITICAL GEOGRAPHY. These included authors such as Geoffrey Parker (*Western Geopolitical Thought in the Twentieth Century* 1985), among others.

After World War II, the Soviet Union effectively controlled the heartland. Mackinder, however, had proposed ideas of an Atlantic community in 1924 that developed into the NORTH ATLANTIC TREATY ORGANIZATION (NATO). It addition the Cold War policy of containment of communism was, in effect, a siege of "fortress" heartland. Global politics of the 20th century were deeply influenced by Mackinder's geopolitical vision. The War on Terror is also being waged with a geopolitical eye on the Middle Eastern "rimland."

BIBLIOGRAPHY. Harm J. de Blij, *Systematic Political Geography* (Wiley, 1973); S.B. Cohen, *Geography and Politics in a World Divided* (Random House, 1963); Brian W.

Blouet, *Halford Mackinder: A Biography* (Texas A&M University Press, 1987); Saul Bernard Cohen, *Geopolitics of the World System* (Rowman & Littlefield, 2003); A.R. Hall, "Mackinder and the Course of Events," *Annals of the American Association of Geographers* (v.45/2, 1955); D.J.M. Hooson, *A New Soviet Heartland?* (Van Nostrand, 1964); James Trapier Lowe, *Geopolitics and War: Mackinder's Philosophy of Power* (University Press of America, 1981); Halford J. Mackinder, "The Geographical Pivot of History," *Geographical Journal* (v.23, 1904); D.W. Minig, "Heartland and Rimland in Eurasian History," *Western Political Quarterly* (v.9, 1956); W.H. Parker, *Mackinder: Geography as an Aid to Statecraft* (Clarendon Press, 1982); A.P. de Seversky, *Air Power: Key to Survival* (Simon and Schuster, 1950); G. Sloan, "Sir Halford J. Mackinder: The Heartland Theory Then and Now," *Journal of Strategic Studies* (v.22, June-September 1999); Francis P. Sempa, *Geopolitics: From the Cold War to the 21st Century* (Transaction Publishers, 2002); N.J. Spykman, *The Geography of the Peace* (Harcourt, Brace, 1944); P.J. Taylor, "From Heartland to Hegemony: Changing the World in Political-Geography," *Geoforum* (v.25/4, 1994).

Andrew J. Waskey
Dalton State College

hemisphere

THE WORD *hemisphere* means "half a sphere." In geography, the term refers to half the Earth, and the enclosing boundary line of a hemisphere is a great circle. A space traveler viewing the Earth from a great distance will see only half the earth, a hemispheric perspective. This is true because the Earth is a sphere and only one side of the Earth can be seen in any one particular view.

Our only look at the entirety of the Earth's surface is through world maps. The map view of the world is important for that reason despite the fact that every map 1) is a scaled-down replica of the Earth, 2) presents a generalized view of the Earth's surface, and 3) is a distortion of the Earth's surface to varying degrees. There are an infinite number of possible hemispheric views of the Earth. When an observer looks directly at any particular point on the Earth's surface, a hemisphere is defined. However, there are a number of hemispheres that have special importance for the discriminating viewer. For instance, if the equator is the great circle enclosing a hemisphere, the viewer will be seeing either the Northern Hemisphere or the Southern Hemisphere. In both cases, the center of the hemispheric view will be one of the poles. Viewing the Southern Hemisphere would show vast expanses of the southern oceans, interrupted by the southern extremes of South America and Africa, about half of AUSTRALIA, all of NEW ZEALAND, and a few scattered islands in the PACIFIC OCEAN. In the center of the view would be the geographical South Pole and the continent of ANTARCTICA, a true landmass covered by heavy sheets of ice. For example, South Pole station sits on ice that measures 8,000 ft (2,438 m) in thickness. The presence of a preponderance of water in the Southern Hemisphere has a great impact of the climate in this region. Since large water bodies do not heat up and cool as quickly as comparably sized land areas, the annual changes in temperature are much lower.

The view of the Northern Hemisphere is distinctly different. At the center of the view is the geographical North Pole, a point that is impossible to permanently mark on the surface because of the constant movement of the ice on the Arctic Ocean. South of the pole are the northern regions of the great landmasses of the northern hemisphere. Particularly imposing in this regard is the longitudinal sweep of Eurasia extending over 180 degrees of longitude. RUSSIA alone boasts of having 11 time zones. The northernmost reaches of North America are also prominent in this view, its land area bracketed by the ATLANTIC and PACIFIC oceans. The climatic impact of the presence of large land areas in the Northern Hemisphere is profound. The variability in annual temperatures is extreme in this region. It is not uncommon for areas within the continents of Eurasia and North America to have average winter temperature near 0 degrees F (-17.7 degrees C) and summer averages in the 70 and 80 degrees F (21 to 26 degrees C) range, temperature ranges not found in the southern hemisphere.

Familiar to everyone are the Eastern and Western Hemispheres. The two are separated by a great circle comprising two longitude lines, both running from pole to pole in opposite directions. The longitude line designated 0 degrees passes through Greenwich, England, and serves as one of the dividers between the two hemispheres. Its counterpart passes through the Pacific Ocean 180 degrees from the prime meridian. Hence, it is identified as longitude 180 degrees. An observer standing at any location along either of these lines is on the boundary between the eastern and western hemispheres. The prime meridian is the conventional starting point for assigning degree values to longitude lines.

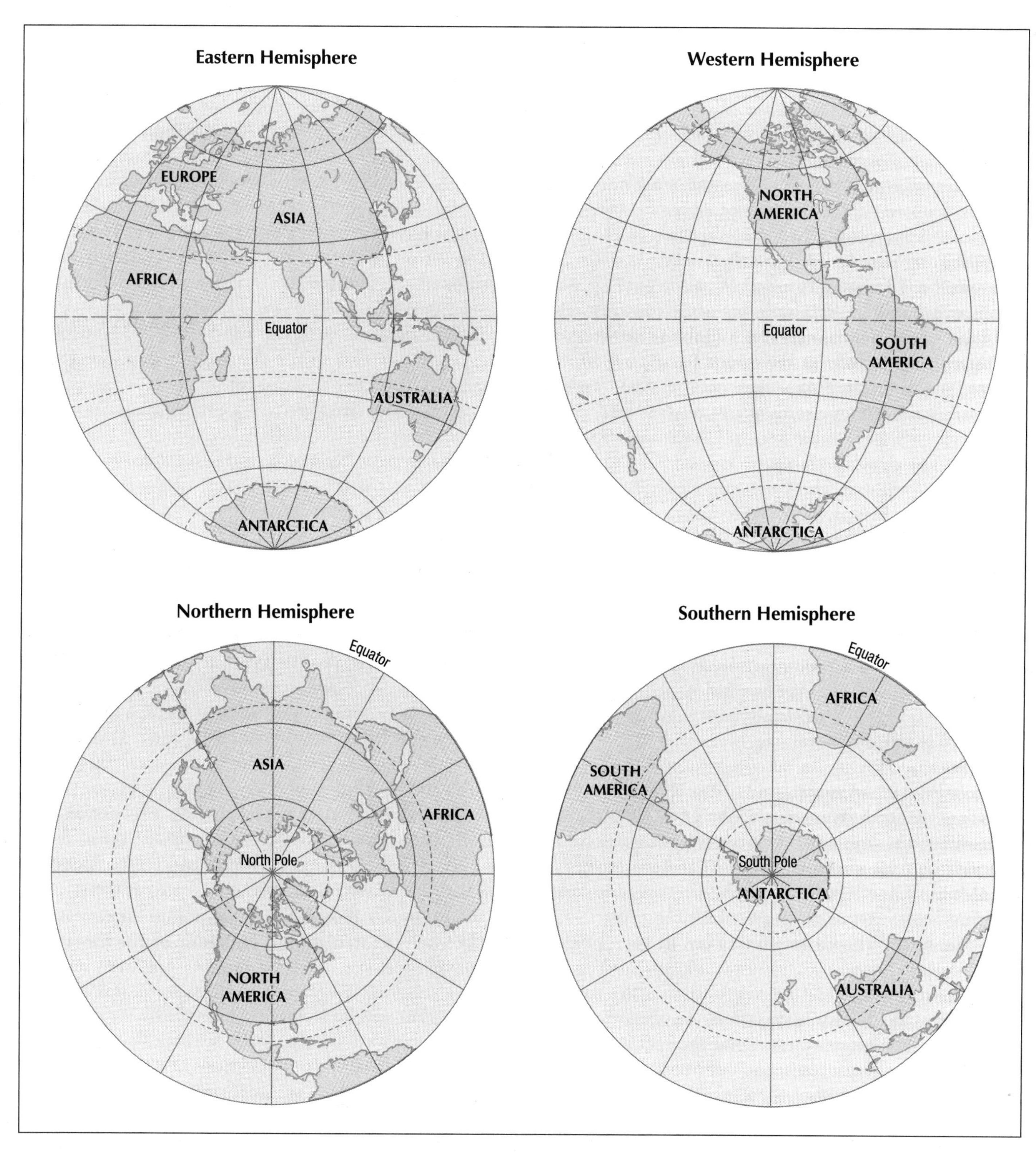

The area west of the prime meridian to longitude 180 degrees is the Western Hemisphere. Conversely, the area to the east of the prime meridian is the Eastern Hemisphere. The longitude lines in both hemispheres are measured in degrees from the prime meridian. Time zones are also directly related to the 360 degrees of longitude. The time of day increases east of the prime meridian one hour for every 15 degrees of longitude. On the other hand, the time of day decreases west of the prime meridian one hour for every 15 degrees of longitude. The longitude line at 180 degrees is identified as the INTERNATIONAL DATE LINE. A traveler cross-

ing the International Date Line in an easterly direction and entering the Western Hemisphere would then be in the previous day. On the other hand, travelers going from the Western Hemisphere into the Eastern Hemisphere would find themselves in the next day. Although 180 degrees longitude designates the International Date Line, the line is displaced to the west at one point in order to include the Aleutians and keep all of Alaska in the same day. Similarly, the line is displaced to the east in the southern Pacific Ocean to include the Cook Islands in the same day as Australia and New Zealand.

Two additional hemispheric views illustrate the land and water hemispheres. If a globe is oriented so that the Cook Islands in the South Pacific are in the center, the water hemisphere may be seen. This is the hemisphere that has the maximum amount of water in view. In addition to a few islands in the South Pacific, the only land areas visible are portions of Australia, Antarctica, and South America. All of New Zealand is seen. The water hemisphere perspective clearly illustrates the enormous extent of the Pacific Ocean.

If the globe is oriented with the country of Turkey in the center, then the land hemisphere comes into view. In this perspective the viewer is seeing the hemisphere that has the maximum land area. The entire continent of Africa is present as well as South Asia, Europe, and a large part of the Asian landmass to the east. In addition, the eastern portion of North America is in view.

BIBLIOGRAPHY. John Campbell, *Map Use and Analysis* (McGraw-Hill, 2001); Arthur Getis, Judith Getis, and Jerome D. Fellmann, *Introduction to Geography* (McGraw-Hill, 2004); Phillip C. Muehrcke, *Map Use: Reading, Analysis, Interpretation* (JP Publications, 2001).

GERALD R. PITZL, PH.D.
MACALESTER COLLEGE

Himalayas

THE HIMALAYAN mountain region, located between INDIA and TIBET, has the world's highest peaks. It stretches from the INDUS RIVER in the west to the Brahamaputra in the east and has a length of 1,500 mi (2,414 km) and a width from 100 to 150 mi (161 to 241 km). Northwest of the Indus, the region of mountain ranges that extends to a junction with the Hindu Kush, south of the Pamir range, is known as Trans-Himalaya. Thus, the Himalayas represent the southern face of the great central elevated region—the plateau of Tibet—the northern face of which is buttressed by the Kunlun.

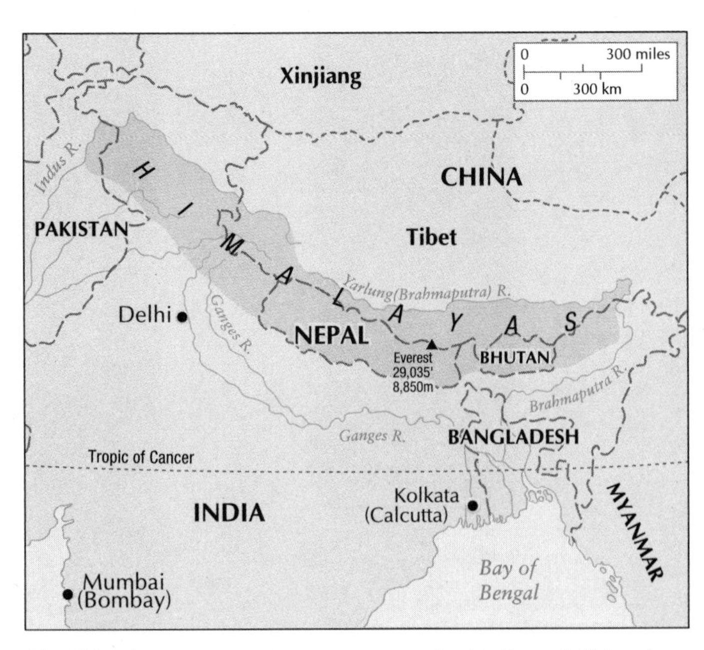

The Himalayan mountain range, separating India and China, has the highest peaks in the world.

The physiography of the Himalayan mountain system can be classified into three parallel longitudinal zones:

The Great Himalayas. The main ranges, which lie in the north, rise above the snow line and have an average elevation of 20,000 ft (6,096 m) above sea level. They include the highest peaks of EVEREST (29,035 ft or 8,850 m), K2 (Godwin Austen) (28,251 ft or 8,611 m) and Kangchenjunga (28,168 ft or 8,586 m).

The Lesser Himalayas. The middle ranges, which are closely related to and lie south of the Great Himalayas, form an intricate mountain system with an average height of 12,000 to 15,000 ft (3,657 m to 4572 m) above sea level.

The Outer Himalayas. These comprise the Siwalik and other ranges, which lie between the Lesser Himalayas and the plains and have an average height of 3,000 to 4,000 ft (914 to 1,219 m) above sea level.

The above classification is a useful generalization but does not represent the peculiar and complex features of the Himalayan system. These include:

The Great Northern Watershed. On the north and northwest of Kashmir is the great water divide, which separates the Indus drainage area from that of the Yarkand and other rivers of Chinese Turkistan

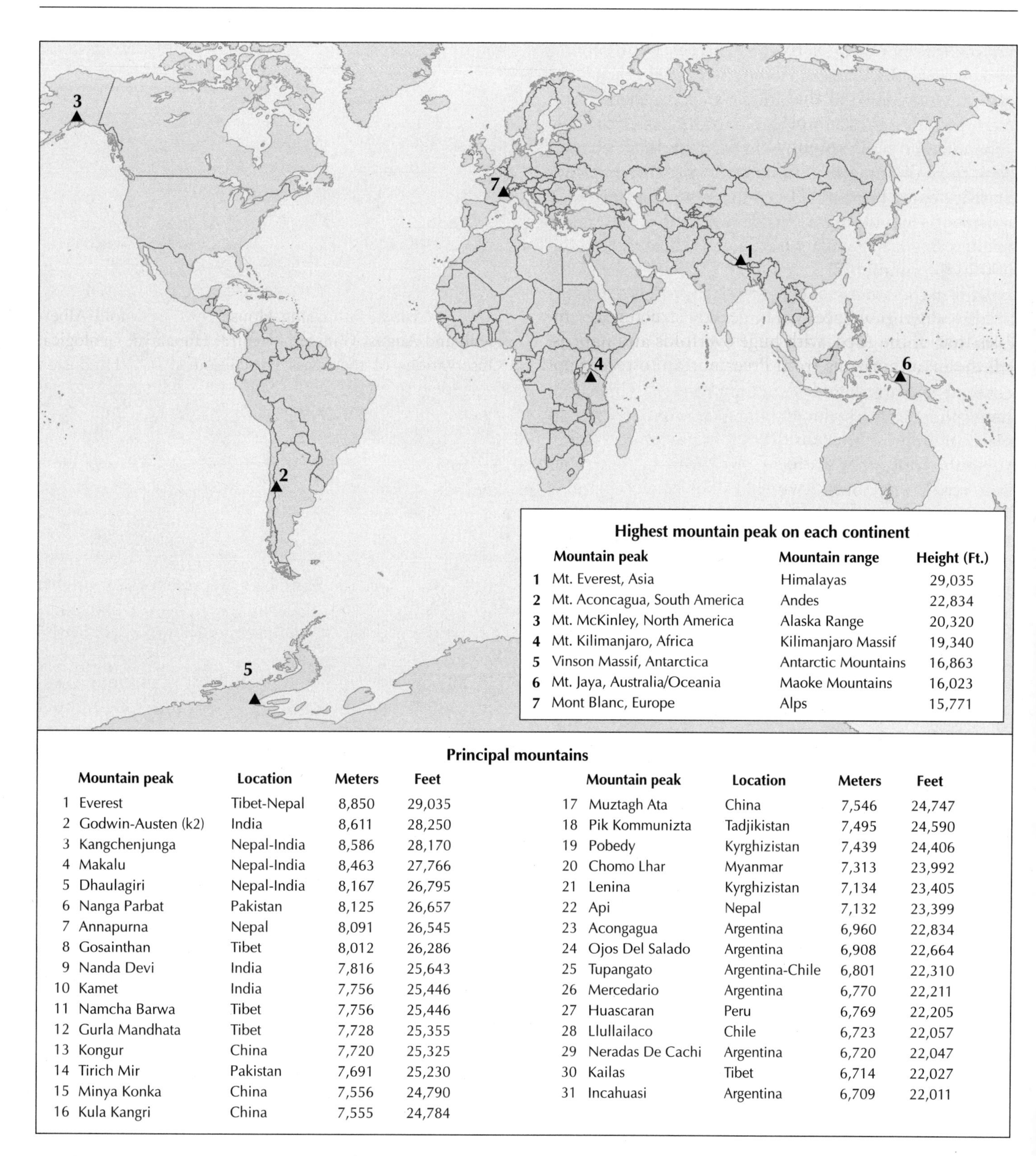

Highest mountain peak on each continent

	Mountain peak	Mountain range	Height (Ft.)
1	Mt. Everest, Asia	Himalayas	29,035
2	Mt. Aconcagua, South America	Andes	22,834
3	Mt. McKinley, North America	Alaska Range	20,320
4	Mt. Kilimanjaro, Africa	Kilimanjaro Massif	19,340
5	Vinson Massif, Antarctica	Antarctic Mountains	16,863
6	Mt. Jaya, Australia/Oceania	Maoke Mountains	16,023
7	Mont Blanc, Europe	Alps	15,771

Principal mountains

	Mountain peak	Location	Meters	Feet
1	Everest	Tibet-Nepal	8,850	29,035
2	Godwin-Austen (k2)	India	8,611	28,250
3	Kangchenjunga	Nepal-India	8,586	28,170
4	Makalu	Nepal-India	8,463	27,766
5	Dhaulagiri	Nepal-India	8,167	26,795
6	Nanga Parbat	Pakistan	8,125	26,657
7	Annapurna	Nepal	8,091	26,545
8	Gosainthan	Tibet	8,012	26,286
9	Nanda Devi	India	7,816	25,643
10	Kamet	India	7,756	25,446
11	Namcha Barwa	Tibet	7,756	25,446
12	Gurla Mandhata	Tibet	7,728	25,355
13	Kongur	China	7,720	25,325
14	Tirich Mir	Pakistan	7,691	25,230
15	Minya Konka	China	7,556	24,790
16	Kula Kangri	China	7,555	24,784
17	Muztagh Ata	China	7,546	24,747
18	Pik Kommunizta	Tadjikistan	7,495	24,590
19	Pobedy	Kyrghizistan	7,439	24,406
20	Chomo Lhar	Myanmar	7,313	23,992
21	Lenina	Kyrghizistan	7,134	23,405
22	Api	Nepal	7,132	23,399
23	Acongagua	Argentina	6,960	22,834
24	Ojos Del Salado	Argentina	6,908	22,664
25	Tupangato	Argentina-Chile	6,801	22,310
26	Mercedario	Argentina	6,770	22,211
27	Huascaran	Peru	6,769	22,205
28	Llullailaco	Chile	6,723	22,057
29	Neradas De Cachi	Argentina	6,720	22,047
30	Kailas	Tibet	6,714	22,027
31	Incahuasi	Argentina	6,709	22,011

(Sinkiang). It is not the great Himalayas but the Muztagh range which, with the Karakoram mountains, trends southward, forming a continuous mountain barrier and the true water divide west of the Tibetan plateau.

Eastern Tibet. The Tibetan plateau, or Chang, breaks up at about the meridian of latitude 92 degrees E, to the east of which the affluent of the Tsangpo (the Dihang and subsequently the Brahamaputra) drain from wild, rugged mountain slopes. In this region are

the sources of all major rivers of China and Myanmar (Burma).

It is now proved that Mount Everest, which appears from the Tibetan plateau as a single dominating peak, has no rival among Himalayan altitudes and is definitely the world's highest mountain. Everest was climbed by Sir Edmund Hillary and Tenzing Norgay in 1953. In Asia, there are 94 peaks exceeding 24,000 ft (7315 m); all but two are in the Himalayas or Karakoram.

Much of the Himalayan area is still very imperfectly known geologically. The general structure resembles that of the Alps, with huge overfolds and nappes; all the main horizons from Precambrian to recent appear to be represented. A very large number of rock groups have been distinguished, described, and given local names. It is certain that during Mesozoic times, the Himalayan area was occupied by the great geosyncline, which coincided with the Tethys Sea or ocean basin. The sediments laid down in the Tibetan section of this great basin constitute the Tibetan zone, in which fossiliferous beds of Paleozoic and Mesozoic ages differ entirely in facies from those farther south. The second or Himalayan zone, which comprises the Great and Lesser Himalayas, is composed chiefly of metamorphic rocks and sediments that are generally unfossiliferous. It is believed that the elevation of this central axis took place mainly in Eocene-Oligocene times and that during this phase the important nummulitic limestones were deposited in a series of basins, notably in Ladakh. The main orogeny would seem to have resulted from the northward movement of the ancient block that is now seen in peninsular India and that underlies the Indo-Gangetic plain. Continued movement in Miocene times folded the nummulitic limestones; the final phase of the mountain building came in post-Pliocene times and has scarcely yet ceased—as the Assam earthquake bears witness—and folded intensely the Pliocene Siwalik sediments of the southern flank of the Outer Himalayas.

The uplift of the Himalayas was a gradual process protracted over a very long period and had a very marked effect upon the scenery, the topography, and the river system. The last is not consequent upon the structure, but the principle rivers were of an age anterior to the tertiary earth movements and the drainage is spoken of as antecedent. During the slow process of uplift, folding, and faulting, the rivers were able to keep, for the most part, to their original courses, although their erosive power was increased because of increased gradients.

BIBLIOGRAPHY. Sir Edmund Hillary, *Schoolhouse in the Clouds* (Doubleday, 1964); Ian Cameron, *Mountains of the Gods* (Century, 1984); Christina Noble, *Over the High Passes* (Prentice Hall, 1987); Mike Harding, *Footloose in the Himalayas* (M. Joseph, 1989); D.N. Wadia, *Geology of India* (Macmillan, 1919); R.W. Hingston, *A Naturalist in Himalaya* (H. F. & G. Witherby, 1920); Chetan Singh, *Natural Premises: Ecology and Peasant Life in the Western Himalaya* (Oxford University Press, 1998); Edward Percy Stebbing, *Stalks in the Himalaya: Jottings of a Sportsman-Naturalist* (John Lane, 1912); S.K. Chadha, ed., *Himalayan Ecology* (Ashish Publishing House, 1989); Arnold Albert Heim and August Gansser, "Central Himalaya: Geological Observations of the Swiss Expedition 1936" (Hindustan Publishing Company, 1975); William F. Ruddiman, *Tectonic Uplift and Climatic Change* (Plenum Press, 1997); D.D. Maithani, ed., *Central Himalaya: Ecology, Environmental Resources and Development* (Daya Publishing House, 1991); Michael Su, *Geomorphology and Global Tectonics* (Wiley, 2000).

JITENDRA UTTAM
JAWAHARLAL NEHRU UNIVERSITY, INDIA

Hindu Kush

HINDU KUSH (or Hindukush) is the main mountain range in the Central Asian state of AFGHANISTAN. Hindu Kush is the westernmost extension of the Pamir mountain, the Karakoram (K-2), and the western HIMALAYAS.

The origin of the term *Hindu Kush* (which translates as "Hindu Killer" or "Killer of Hindus") is a point of controversy among scholars. Three possibilities have been put forward in this context. One, the mountains concerned are tribute to Indian slaves who perished in the difficult mountainous terrain while being transported from India to slave markets of Central Asia. Two, the name is merely a corruption of Hindu Koh, the pre-Islamic name of the mountains that divided Hindu-populated southern Afghanistan from non-Hindu northern Afghanistan. The third possibility is that the name is a posited Avestan appellation meaning "water mountain."

The mountains of the Hindu Kush system diminish in height as they stretch in a westward direction. Toward the middle near Afghanistan's capital city Kabul, they extend from 14,763 to 19,685 ft (4,500 to 6,000 m); in the west, they attain heights of 11,482 to 13,123

ft (3,500 to 4,000 m). The average altitude of the Hindu Kush is about 14,700 ft (4,500 m). Th Hindu Kush mountain range stretches about 600 mi (966 km) laterally, and its median north-south measurement is about 149 mi (240 km). Only about 373 mi (600 km) of the Hindu Kush system is called the Hindu Kush mountains. The rest of the system consists of numerous smaller mountain ranges, including the Salang, Koh-e-Baba, Koh-e-Paghman, Spin Ghar, Sian Koh, Suleiman, Selseleh-e-band-e-Turkistan, and Koh-e-Khwaja Mohammad. The western Safid Koh, Doshakh and the Siah Band are known as the Paropamisus by scholars of Central Asia. Three rivers flow from the Hindu Kush mountain range, namely the Helmand River, the Hari Rud, and the Kabul River that also provide water to major cities and regions of Afghanistan.

Huge caravans pass through high passes (*kotal*) transecting the mountains. The most important mountain pass in the Hindu Kush range is the Kotal-e-Salang, which links Kabul and points south to northern part of Afghanistan. Before the Salang road was constructed, the most famous passes in the Hindu Kush region were the Khyber Pass (3,369 ft or 1,027 m) and the Kotal-e-Lataband (8,199 ft or 2,499 m). The roads through the Salang and Tang-e-Gharu passes are critical strategic routes between Afghanistan and RUSSIA and other old Soviet republics.

The Hindu Kush is sparsely populated and inhabitants subsist year-round on livestock and crops. The Kalash people are one of the main inhabitant groups of Hindu Kush and claim to be descendants of Alexander the Great. They have their own distinctive laws, religion, and culture.

BIBLIOGRAPHY. Alberto Cacopardo and Augusto Cacopardo, *Gates of Peristan; History, Religion, and Society in the Hindu Kush* (Instituto Italiano per Africa Oriente, 2001); Caroe Olaf, *The Pathans* (Macmillan, 1965); Tyler Frazer, *Afghanistan* (Oxford University Press, 1967); Joseph Greenberg, *Indo-European and Its Closest Relatives* (Stanford University Press, 2000).

MOHAMMED BADRUL ALAM
MIYAZAKI INTERNATIONAL COLLEGE, JAPAN

hinterland

FROM THE GERMAN *umland*, a hinterland is inland territory behind and bordering a town on a coast or river, or the backcountry extending from an inland town. In both cases, the hinterland generally falls under the legal and economic jurisdiction of the same state to which the city belongs. The hinterland supplies a city with food, fuel, raw materials and, in the modern period, labor. It also serves as a market for its manufactured goods and services. The flow of goods and services between the hinterland and city is facilitated by a communication infrastructure of roads, canals, and bridges.

The development of the hinterland is connected with the rise of cities, appearing first with ancient Mesopotamian ceremonial and administrative centers, whose population specialized in the maintenance of ritual, on the one hand, and trade or redistribution, on the other hand. Because the center's population was occupied primarily with non-food producing activities, it depended entirely for its food and fuel on its agricultural hinterland. The size and prosperity of a ceremonial and administrative city, therefore, depended on the agricultural productivity of its countryside, and tributary relationships were developed to control its resources.

A city that also engaged in long-distance trade, such as the polis of classical Greece (5th to 4th centuries B.C.E.), was less dependent on its hinterland if food could be obtained from farther afield. With expanding trade and a rise in population, the polis outstripped its hinterland's capacity for food production. The hinterland of classical Athens, Attica, was suited to the cultivation of olives and grapes, used in the production of olive oil and wine, which then were exchanged for grain from the Greek BLACK SEA colonies. Similarly, Quanzhou, a metropolitan port city in Southeast CHINA's Quannan region, served as one of the main trading centers in the South China Sea and Southeast Asia from 1000 to 1400. With Quanzhou's rapid expansion in the 11th and 12th centuries, its agriculturally weak backcountry, the Fujian hinterland, made the transition from food cultivation to that of cash crops such as lichee, sugarcane, and cotton and the manufacture of ceramics and metal products.

This commoditized city-hinterland relationship therefore served to include Greek and Chinese farmers on marginal arable lands in the international market. Additionally, premodern trading centers that lacked even agriculturally poor hinterlands, such as Timbuktu in West Africa or many of the early medieval coastal trading sites in northern Europe, either declined or were abandoned altogether when trade routes shifted away from these centers.

Even an agriculturally rich hinterland benefited from its relationship with the city. For example, the introduction of more animals to provision estates and pull the recently invented carriages of the elite in 17th- and 18th-century PARIS meant that more fodder had to be brought to the city. Hinterland peasants responded by planting fields with more oats and meadows with forage crops, delivering these products to the city and carting away the manure from urban stalls, which they would use in turn to fertilize nitrogen-poor fields. Paris basin hinterland farmers thus took advantage of the opportunities offered by economic changes in their core region and thus generally fared better than the outland French peasantry.

MODERN ADAPTATIONS

In the 19th and 20th centuries, metropolitan hinterlands encountered further adaptations with the growth of long-distance transportation networks, communication, industrialization and banking. As with classical Athens, modern LONDON's south English hinterland is poorly equipped to feed the metropolis, but as a major financial center, London is able to import its food from distant mainland markets. Like Attica, London's lowland backcountry takes part in the long-distance market, using the revenues from its industries to purchase imports.

Third-world hinterlands, in contrast, suffer from slow technological development and poverty, resulting in the migration of the hinterland population to metropolises such as KOLKATA (CALCUTTA), which cannot support such a large population. A final example of a metropolitan hinterland is designated by the U.S. Census Bureau's term the Metropolitan Statistical Area (MSA), comprising three zones: the city, bounded by its corporate limits; an urbanized area contiguous to the city; and a nonurbanized area economically connected to the metropolis.

BIBLIOGRAPHY. M.M. Austin and P. Vidal-Naquet, *Economic and Social History of Ancient Greece: An Introduction* (University of California Press, 1977); Philip T. Hoffman, *Growth in a Traditional Society: The French Countryside, 1450–1815* (Princeton University Press, 1996); Richard Pearson, Li Min, and Li Guo, "Port City and Hinterlands: Archaeological Perspectives on Quanzhou and Its Overseas Trade," *The Emporium of the World: Maritime Quanzhou, 1000–1400* (Brill, 2001).

HEIDI M. SHERMAN
UNIVERSITY OF MINNESOTA, TWIN CITIES

historical geography

GEOGRAPHERS SEEK TO understand the world by examining spatial relationships. The types of questions they might ask are: Why are things located where they are? How are places different from each other? How are places like each other? How are places interconnected with each other? How do people affect their natural environment and how does the natural environment affect people? In many instances, the answers to these questions are related directly to what the world is like today.

For historical geographers, these questions are adapted to consider the role of time. For example, a historical geographer might ask questions such as: How did people, things, and landscape elements come to be located where they are? How did a place come to be like other places? How did it develop differently from other places? How have people been affected by the natural environment? How have they altered the environment as well? In short, historical geography might be described as the study of past places.

Some of the earliest attempts at what might be considered historical geography are rooted in ancient GREECE. Although typically identified as a historian, Herodotus has often been regarded by geographers as one of their own. Based upon his own extensive travels and keen observations, Herodotus developed a sophisticated understanding of how the processes of physical geography played out over extended periods of time, resulting in what was his contemporary landscape.

While individual travel and exploration aided in the development of the historical geographies of Herodotus, the expansion of Islam further developed the field of historical geography. By the mid-8th century C.E., religious conquests had brought northern Africa and the Iberian Peninsula under Islamic control. What transpired was an exchange of ideas between East and West. At the same time, Muslim concepts such as the use of the decimal system made their way into Europe, and Greek and Roman texts were translated into Arabic for the first time. Like the Greeks, Muslim scholars such as Al-Biruni incorporated the role of time within the processes of physical geography. In his study of India, Al-Biruni attempted to explain the formation and distribution of alluvial deposits, predating the development of similar ideas in Europe by centuries. Considered perhaps the most significant historical geographer of the medieval Muslim world, Ibn-Khaldun has been cited as the first to explicitly link the physical environment to human activity and culture

through time—thereby establishing the human-environment connection so crucial to the broader field of geography as well.

Nineteenth-century European historical geographers continued to study the relationship between humans and the natural environment with respect to time, but were furthered in their research by theoretical developments in the biological and social sciences. In 1859, Charles Darwin introduced the notion of natural descent in his historic volume, *The Origin of Species.* Drawing in part upon Thomas Malthus's ideas concerning population growth and the limitations of the natural environment, Darwin concluded that environments were capable of supporting a limited number of organisms and only those organisms best biologically suited to an environment would be able to remain in that environment and successfully reproduce. Those less well suited to the environment would ultimately face extinction because of competition from better adapted organisms.

SOCIAL DARWINISM

Although Darwin himself did not specifically include humans within his understanding of evolutionary processes, other scholars like English philosopher Herbert Spencer would. Called social Darwinism, the work of Spencer (who coined the phrase "survival of the fittest") drew parallels between human and animal societies and their relationships to the natural environment. Over time, humans had adapted to their natural environments.

But as historical geography would suggest, natural environments were always changing as well and humans would once again have to adapt or face extinction. Social Darwinism expanded these ideas to include not only the natural environment, but also the social environment. Only those people best adapted to the natural environment and how society had altered and organized itself would thrive and prosper while those not who were unsuccessful in their struggle for survival would eventually die out with time.

Evolutionary ideas like those of Darwin and Spencer were eventually incorporated into the work of historical geographers. German geographer Friedrich Ratzel has often been described as having been greatly influenced by the notion that human societies, like organisms, struggle to survive in specific natural environments. Ratzel suggested that states develop throughout history within the particular constraints of natural environmental resources, including space itself. When the population of a state begins to exceed or meet the capacity of these resources, it must expand its boundaries or be destined to go into a stage of decline and be threatened by more expansive and powerful states. Such expansionist theories of how the history of states was affected by the geography of a territory marked an advancement not only in historical geography, but likewise in the study of migration and geopolitics.

HEARTLAND THEORY

Originally trained in the natural sciences like Ratzel, English geographer Halford J. MACKINDER also tied human culture and political organization through time to the environment. Through his heartland theory, Mackinder suggested that world history could be explained through an examination of the resources and accessibility of various natural environments. Prehistoric humans, he claimed migrated out of the "heartland" to other parts of the globe. Control of the strategic heartland then, had and would continue to dictate who held dominion over the rest of the world.

Just as Ratzel and Mackinder discussed historic diffusion of population and its impact on global politics past and present, European contributions to the field of historical geography diffused across the ATLANTIC OCEAN to the UNITED STATES as well. The influence of Ratzel is especially evident in the work of one of his American students: Ellen Churchill Semple. Although she served as the first female president of the Association of American Geographers in 1921, Semple is most typically remembered for her role as an environmental determinist.

Environmental determinism suggested that the cultural characteristics of a human society would largely evolve according to the physical geography of the group's home territory. As a result, some groups of people in the world were declared predisposed to immorality and laziness as a result of long-term exposure to tropical climatic conditions or because they lived in along mountain passes. Conversely, other groups, most notably those of Northern Europe, were deemed more successful due to the manner in which their physical environment challenged them and contributed to admirable cultural traits. Environmental determinism allowed for the integration of evolutionary history, culture, and physical geography.

Although much of Semple's work falls within the classification of environmental determinism, it also at times backs away from being absolutely deterministic. Semple suggested that while environmental conditions may influence people's actions and livelihoods, they cannot explicitly control people. Environmental condi-

tions may restrict or enable certain ways of life, but ultimately people make their own choices as to how to react to these situations. According to Semple, people did, however, tend to react to specific environments in predictable ways that could be described in terms of spatial patterns of behaviors and corresponding environments.

A contemporary of Semple, geologist Albert Perry Brigham, also began his career in historical geography by examining the role of environmental influences upon society. Later in his career, however, Brigham criticized fellow geologists and geographers who studied this link between humans and the environment, but failed to approach the matter scientifically. He criticized environmental determinists for creating broad generalizations concerning the effect of the environment upon culture that were not based upon scientific evidence but were rather descriptive accounts. Specifically, Brigham rejected their attempts to explain human society based solely upon climatic influences. The natural environment, he explained, was far too complex for geographers to be able to isolate singular environmental characteristics as the historical root cause of race, character, or culture.

Human adjustment to the physical environment through time continued as a primary area of focus after World War I with the work of Harlan H. Barrows. Through the development of the subfield of human ecology, Barrows redirected the emphasis of environmental determinists upon physical environmental controls toward a human-centered approach. Human ecology sought to understand the impact of humans upon the earth through time rather than merely the impact of the earth upon humans.

CHOROLOGY

An even stronger rejection of environmental determinism in American geography was led by Carl O. SAUER in the 1920s. For Sauer, the primary purpose of geography should be chorology, or the study of areas. Rather than constrain geographers within the limits of environmental influences, geography should study places in terms of regular characteristics that tied them together. These regularities could then be analyzed to understand how areas differ as well as how they are interconnected.

For Sauer, one important way of classifying areas was the cultural landscape. Drawing upon both human ecology and chorology, geographers like Sauer examined how humans transformed the physical environment of the Earth to make it suitable for their own needs. What remained was an imprint of human habitation: the cultural landscape. At times, practitioners of the cultural landscape approach to geography went so far as to challenge environmental determinist traditions by suggesting that human decision making was often a much greater determinant of location than were such physical geography components as climate, soils, and relief.

Chorology and the accompanying cultural landscape studies allowed for the explicit inclusion of history into geography. Cultural landscapes could be examined with regard to which people had lived in them and created them. From this trend emerged the work of geographers such as Derwent Whittlesey and his notions of sequent occupance. According to Whittlesey, a group of people that occupies a place leaves its cultural imprint upon the physical environment based upon its way of life. With the passage of time, preexisting groups must readapt the cultural landscape to meet their changing needs. Similarly, new groups to a location may also alter the preexisting cultural landscape to coincide with their society's needs and desires. What emerges in both instances is a succession of imprints upon the cultural landscape, each one in essence adding another layer to what the contemporary geographer sees. By working backward through time, a historical geographer could uncover each layer and come to understand how the past contributed to the creation of current cultural landscapes.

METHODS AND RESOURCES

As the work of the early 20th-century scholars suggests, a key strength of historical geography lies within its ability to incorporate the notion of change. Historical geographers incorporate elements of the past to understand the present and future geographies. So how do historical geographers approach their work? What kinds of resources do they use?

Among the resources most used by historical geographers are archival documents, historic records of the past. Often these sources are stored by the government in national, state, and local archives; public and university libraries; as well as government offices. Examples of the type of information available in these locations include the original Federal Manuscript Census forms filled out over a century ago by hand; property ownership and land use records; natural resource inventories; and some of the primary tools of the trade for geographers—maps and photographs.

Maps and photographs have proven especially important for historical geographers who study the cul-

tural landscape. But equally important for them is the use of fieldwork that includes studying the landscapes as they are today. By looking at the material culture, or built environment, of a cultural landscape, the historical geographer may trace the elements back through history to understand what took place in the past.

For example, by studying the types of houses built or field patterns used, a historical geographer might be able to tie those elements back to another place in the world and in doing so uncover historic migration patterns. With that in mind, they might also attempt to relate the appearance of the cultural landscape to political, economic, or religious belief systems present in a society.

Fieldwork is also crucial for those historical geographers specializing in the transformation of the physical environment. After going into the field to take current measurements of physical phenomena such as river sediment discharge or inventory soil types, they can then compare those new measurements to those of the same places in past times. This process allows physical geographers to understand and chart historical change in the environment. Like cultural geographers, physical geographers may also combine fieldwork and archival documents in their work. Old maps, photographs, and even personal accounts of past environments can be used to fill in the gaps between time periods as well.

HISTORICAL GEOGRAPHY TODAY

While the basic idea of historical geography as the study of past places remains today, the scale of those places and spatial relationships has grown considerably in more recent decades. One of the major areas of interest in historical geography today examines the process of globalization. Globalization describes how different places in the world are becoming more and more interconnected with each other every day. Many historical geographers are now interested in the history of geopolitics and the changes in world economies. By examining these systems at the largest scale possible, the global scale, historical geographers seek not only to understand how politics and economics impacted societies around the world in the past, but also to predict what new alliances and conflicts may arise in our future.

Migration studies are still vital to understanding the historical processes of globalization. In some instances, historical geographers examine migration patterns with the goal of trying to understand how countries such as the United States expanded as well as how its settlers learned to adapt to new environmental challenges. While these historical geographers may focus on the progress of settlers and new nations, others may examine the issue of migration from the standpoint of the preexisting, or indigenous, population. Migration and settlement do not occur in empty space, but rather, in the case of colonialism, they affect the people who were already living in those places for centuries beforehand.

By understanding the changes that occurred with the development of colonialism centuries ago, historical geographers can hope to better explain the often difficult conflicts that emerge with the collapse of colonial empires in the 20th century.

This reexamination of colonialism from the perspective of native populations has also drawn attention to other groups whose stories have often been left untold. One area of historical geography that has grown substantially in more recent decades is one that focuses on feminist historical geography. Throughout history, women and men have often held different amounts of political, economic, and social power. Feminist historical geographers attempt to shed light on how these inequalities affected the way that women experienced historic events and places in dramatically different ways from men.

Modern technology has certainly aided in the practice of historical geography in more recent decades. With the development of cheaper and faster computing systems and the Internet, access to historic geographic data has grown substantially. Now, with only the click of a computer mouse, historical geographers can locate data sources on the Internet once available only on paper in archives. This turn to the digital world has also been reflected in the increasing use of geographic information systems (GIS) in historical geography. GIS often serves as a means for efficiently storing, retrieving, and analyzing historical spatial data that would have previously occupied thousands of printed volumes and taken thousands of hours to examine individually by hand. Understanding the past has certainly been aided by present technology.

Despite the centuries that have elapsed since the chronicles of Herodutus and Ibn-Khaldun, their work nonetheless still reflects ideas common to that of more contemporary geographers, even those using GIS. The goal of historical geography is not simply to provide a single snapshot image of past places, but rather, like these historic scholars, to come to understand how places of the past are related to those of today. Historical geography does not try to discover ultimate causes

or a single origin; rather, it recognizes that geographical relationships occur within a continuum of time—past, present, and future. Historical geographers attempt to reconstruct past places in the interest of understanding contemporary places and their potential impacts upon future ones.

BIBLIOGRAPHY. Alan R.H. Baker, *Geography and History: Bridging the Divide* (Cambridge University Press, 2003); Andrew H. Clark, "Historical Geography," *American Geography: Inventory and Prospect*, Preston E. James and Clarence F. Jones, eds. (Syracuse University Press, 1954); Craig E. Colten et al., "Historical Geography," *Geography at the Dawn of the 21st Century*, Gary L. Gaile and Cort J. Willmott, eds. (Oxford University Press, 2003); H.C. Darby, "On the Relations of Geography and History," *Transactions of the Institute of British Geographers* (v.19/1, 1953). Carville Earle, et al., "Historical Geography," *Geography in America*, Gary L. Gaile and Cort J. Willmott, eds. (Merrill Publishing Company, 1989); Anne Kelly Knowles, *Past Time, Past Place: GIS for History* (ESRI Press, 2002); David N. Livingstone, *The Geographical Tradition: Episodes in the History of a Contested Enterprise* (Blackwell, 1994); Geoffrey J. Martin and Preston E. James, *All Possible Worlds: A History of Geographical Ideas* (Wiley, 1993).

TONI ALEXANDER
KANSAS STATE UNIVERSITY

Hokkaidô Island

NORTHERNMOST OF the four main islands of JAPAN, Hokkaidô is surrounded by the Sea of Okhotsk to the north and east, the Sea of Japan to the west, and the PACIFIC OCEAN to the south. Its northern location and frigid ocean currents from the Sea of Okhotsk make Hokkaidô's climate considerably colder and drier than the rest of Japan; the island is known for its long winters. Hokkaidô is about the size of IRELAND, and almost square in shape, roughly 186 mi (300 km) on each side.

Japan and RUSSIA dispute territorial jurisdiction over four small islands to the north of Hokkaidô (Etorofu, Kunashiri, Habomai, and Shikotan). Japan had exercised control over these islands until the former Soviet Union occupied them at the end of World War II. (Jurisdiction over the much larger island of Sakhalin, or Karafuto as the Japanese call it, is split between Japan and Russia.) Hokkaidô was only fully incorporated into Japan in modern times. Until the 19th century, most of Ezo (the previous name for Hokkaidô) was a wilderness, home to the indigenous Ainu people. Today, the population of Ainu in Hokkaidô has dwindled to about 16,000 from death and assimilation. Many place names in Hokkaidô are in the Ainu language, including the prefecture's capital, Sapporo, and cities such as Kushiro and Muroran.

Japan's Meiji government renamed the island Hokkaidô in 1869, set national borders encompassing the entire island as part of Japan, and encouraged settlers to move north. Yet Hokkaidô remains sparsely populated by Japanese standards. Its population of about 5.7 million is less than 5 percent of Japan's overall population. On the other hand, Hokkaidô's area of 30,315 square mi (78,515 square km) makes it the second-largest island in Japan (after Honshû). The large area of Hokkaidô combined with its relatively small number of inhabitants yields a population density of 186 persons per square mi (72 persons per square km), only about one-fifth the national average.

The forests of Hokkaidô produce lumber and forest products, and the island is the largest source of coal in Japan, followed by southernmost major island of Kyûshû. The waters around Hokkaidô are rich sources of fish (including the local specialty of crab). Agriculture is another important part of the economy, and Hokkaidô provides much of Japan's food, including potatoes, onions, beans, wheat, and milk and other dairy products. Although the climate is far less suited to wet paddy field agriculture than lands to the south, new seed types permit the growing of rice, the single largest crop in Hokkaidô. In contrast to most of Japan, Hokkaidô has many dispersed farmsteads, with much larger farms, and employs far more draft animals. Parts of Hokkaidô resemble New England, with broad fields, rolling hills, and grazing cattle.

Tourism is another important part of the economy in Hokkaidô. The island is home to several large national parks, with clear deep lakes, snowy mountains, and unspoiled forests. The first Winter Olympics to be held outside Europe or the United States were hosted by Sapporo in 1972. Every February since 1950, Sapporo has also held an annual Snow Festival, featuring ice sculptures and snow statues; the festival attracted 2 million visitors in 2003.

Development of Hokkaidô has lagged behind the rest of Japan, in part because of the isolation of this northern territory and its separation from the main island of Honshû by long distances and the Tsugaru Strait. In 1988, the Seikan Tunnel linked Hokkaidô

with Honshû by underground rail, and the Bullet Train (Shinkansen) now has its northern terminus in Hokkaidô. Yet because of the distances involved, air travel remains a common route to Hokkaidô.

BIBLIOGRAPHY. Martin Collcutt, Marius Jansen, and Isao Kumakura, *Cultural Atlas of Japan* (Facts On File, 1988); "Hokkaido," *Kodansha Encyclopedia of Japan* (Kodansha Ltd., 1983); Kayano Shigeru, *Our Land Was a Forest: An Ainu Memoir* (Westview Press, 1994); Glenn T. Trewartha, *Japan: A Geography* (University of Wisconsin Press, 1965).

LAWRENCE FOURAKER, PH.D.
ST. JOHN FISHER COLLEGE

homeland security

AS DEFINED BY the executive branch of the U.S. government, homeland security is "a concerted national effort to prevent terrorist attacks within the United States, reduce America's vulnerability to terrorism, and minimize the damage and recover from attacks that do occur" (July 2002 National Strategy for Homeland Security, issued by President George W. Bush).

In more simple terms, the primary end of homeland security is to secure the nation's borders. This is not to say, however, that the concept of homeland security is well defined or clear at all times. Much debate has occurred, particularly around the establishment of the Department of Homeland Security within the U.S. government, over the exact definition of homeland security and what types of both domestic and international issues are contained within such a concept. Further, the ultimate goal of homeland security is not always clear and may change over time, including at any one point in time prevention of terrorist attacks and/or minimization of harm that may occur as a result of such attacks. Both concepts seem to fall within the overall goal of "homeland security," but each has different organizational and resource requirements, and each impacts multiple governmental agencies and offices in a different manner.

Although the concept of homeland security has taken on a new sense of immediacy and familiarity, the idea of providing for homeland security or national security has a historical foundation within the UNITED STATES dating as far back (as an identifiable concept) as the Federalist papers. As noted by Harold Relyea, both John Jay and Alexander Hamilton wrote about "national security" in the Federalist papers, Hamilton suggesting that during war, "the energy of the Executive is the bulwark of the national security." As Relyea observes, however, the concept of national security truly took shape during the period between World War I and World War II, when the idea of a need for a unified, national interest in securing the nation and a national approach to foreign policy was blossoming.

The concept of homeland security has become most well known within the context of the post-September 11 landscape of the American government. Following the terrorist events, on October 8, 2001, through an executive order, Bush established the Office of Homeland Security to carry out this national strategy, appointing former Pennsylvania Governor Tom Ridge as assistant to the president for Homeland Security. In the immediate aftermath of September 11, the concept of homeland security was a symbolic, idealistic belief in the need to provide security for the nation. Over time, however, the concept took shape in the form of concrete policy decisions issued by Congress and the president himself. Ultimately, in June 2002, Bush sought to make homeland security a permanent, established part of the federal government, proposing the creation of both a Department of Homeland Security and an executive office of the president devoted to certain aspects of homeland security, namely the National Office for Combating Terrorism, the most extensive reorganization of the federal government since the Harry Truman administration of the 1940s.

In July 2002, Bush released a National Strategy for Homeland Security outlining his vision for securing homeland security and the role that the new department was to play. This strategy outlines the definition for homeland security provided above and identifies four foundations intended to help maintain homeland security: law, science and technology, information sharing and systems, and international cooperation. Each area or foundation is given a role in providing homeland security in the post-September 11 arena. As these foundations indicate, providing for homeland security impacts more than simply the federal-level agencies. It is a highly decentralized mission, which impacts both the federal and the state government, from the national down to the local level. By recent estimates, nearly 70 federal agencies alone are involved in some aspect of homeland security, and one can imagine hundreds of state and local agencies and offices likewise impacted.

BIBLIOGRAPHY. Sharon Caudle, "Homeland Security: A Challenging Environment," *The Public Manager* (v.32,

Spring 2003); Ivo H. Daalder and I. M. Destler, "Behind America's Front Lines: Organizing to Protect Homeland Security," *Brookings Review* (v.20, Summer 2002); Eric Larson, and John F. Peters, *Preparing the U.S. Army for Homeland Security: Concepts, Issues, and Options* (Rand, 2001); Harold C. Relyea, "Organizing for Homeland Security," *Presidential Studies Quarterly* (v.33, September); D.R. Yergen, *Shattered Peace: The Origins of the Cold War and the National Security State* (Houghton Mifflin, 1978).

AMY WILSON
UNIVERSITY OF WASHINGTON

Honduras

Map Page 1136 **Area** 43,278 square mi (112,090 square km) **Population** 6,669,789 **Capital** Tegucigalpa **Highest Point** 9,416 ft (2,870 m) **Lowest Point** 0 m **GDP per capita** $2,600 **Primary Natural Resources** timber, gold, silver, copper, lead, zinc.

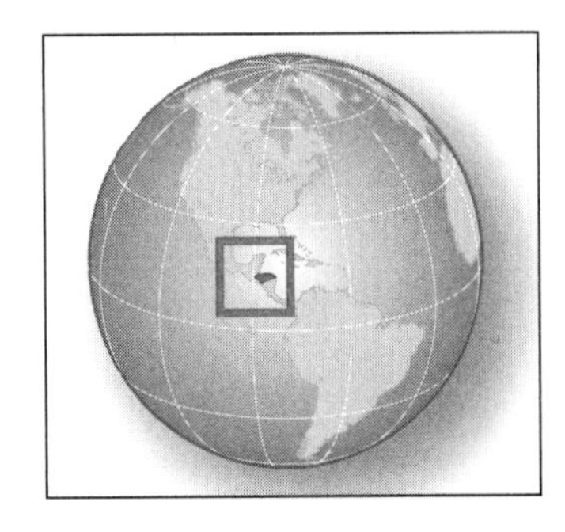

Honduras is located right in the heart of Central America, sharing borders with EL SALVADOR, NICARAGUA, and GUATEMALA. Most of the people are considered mestizos, or of a mixed heritage of white and Amerindian. The country's area includes the three Caribbean islands ceded to the country by the UNITED STATES in 1971, Roatan, Utila, and Guaranja. Honduras's topography is varied, with mostly highlands through the central part of the country, flanked by coastal, tropical coasts. Currently, 7.7 percent of the country is preserved by the government, and includes tropical rainforest and fragile coastland. Education at the primary level is free and compulsory; the literacy rate for both men and women is about 30 percent. Healthcare is very poor, and conditions are some of the worst in the Western Hemisphere. The life expectancy is about 67 years old.

Like its neighbors, Honduras possesses an enormous amount of natural resources in its rainforests and coastline. One of the most precious of these resources, and also the source of notorious conflicts with Nicaragua, is the Miskito coast of the southern portion of the Honduran Caribbean coast. This preserved forest, which is home to the Miskito indigenous group, is under partial jurisdiction of the Honduran government. Home to thousands of fish and bird species, this untouched "biogem" presents an opportunity for the country to develop sustainable tourism in its lesser developed regions.

Another natural resource of Honduras, and likewise a source of contention with its neighbors, is the Gulf of Fonseca, which is located on the western coast. El Salvador and Nicaragua also share a border with this gulf, creating a dynamic of competition and culpability for aquaculture and pollution. Shrimp farming has proliferated in the gulf, creating environmental degradation for native wild species as well as tremendous coastal development. Since three countries share the gulf, conflicts often arise when fishing harvests are diminished; addressing problems of over-fishing and water pollution to alleviate small harvests becomes a game of finger pointing.

Political life in Honduras has been tenuous and complex. Original settlements of the warlike Lencas and Jicaques were taken over by Spanish colonizing forces in 1524. Despite declaring its independence in 1838, Honduras's government has still suffered economic and political forces, such as America's United Fruit Company, that lie outside its boundaries. United Fruit Company had a long presence in the Caribbean and was notorious for its enslaving conditions of Honduran workers whose lands were monopolized and whose livelihoods were at the mercy of the huge corporation. The company has since left the country and been broken up, but present-day fruit distributors, like Dole and Del Monte, provide hauntingly familiar dynamics between their mass production and local communities.

BIBLIOGRAPHY. *Worldmark Encyclopedia of the Nations: Americas* (Gale Publishing, 1998); Barry Turner, ed., *The Statesman's Yearbook: Politics, Cultures and Economies of the World* (Macmillan, 2003).

LINDSAY HOWER JORDAN
AMERICAN UNIVERSITY

Hong Kong

HONG KONG IS A FORMER British colony that returned to the People's Republic of CHINA on June 30, 1997, and is now officially the Hong Kong Special Administrative Region. Hong Kong consists of the island of Hong Kong and about 200 islands and the New Ter-

ritories that are on Kowloon Peninsula on the mainland. Hong Kong has a tropical monsoon climate. Its winters are cool and humid. Rain falls from spring through the summer, and autumn is warm and sunny.

Hong Kong covers an area of 421 square mi (1,092 square km) and has a population of 7,394,170 (2003). Its highest point is Tai Mo Shan at 3,142 ft (958 m) and its lowest point is the South China Sea at 0 m. The gross domestic product per capita is $26,000 (2002).

Since the end of British rule, Hong Kong has functioned under the "one country, two systems" relationship with China. According to the Joint Declaration in 1984 reached by Britain and China, Hong Kong, as a special administrative region of China, would have autonomy in its internal affairs and its economic system for 50 years after the end of colonial rule. Hong Kong is a limited democracy that is governed according to the Basic Law approved in 1990 by the National People's Congress. The chief executive is the head of government who is advised by an executive council. Laws are made by the Legislative Council, which is a unicameral body of 30 seats indirectly elected, 24 directly elected seats, and 6 chosen by an 800-member election committee.

Hong Kong began as a small fishing community and a den for smugglers that was of little significance in its early history. Hong Kong was part of a trading network that centered around Canton in the 18th and early 19th centuries. Great Britain became interested in Hong Kong as a commercial base and as a colony for its deepwater harbor. In 1841, the island of Hong Kong was occupied by British forces during the Opium War. By the Treaty of Nanjing in 1842, China ceded the island of Hong Kong to Britain "in perpetuity." In 1860, after the Second Opium War, Britain obtained a 99-year lease on the Kowloon Peninsula, which became the New Territories. Throughout the 19th century, the development of Hong Kong from a barren island to a major Asian port was made possible by cooperation between British administrators and a Chinese business elite.

In 1912, Hong Kong experienced a major influx of refugees who escaped the turmoil in China following the overthrow of the Qin Dynasty, increasing the population from 600,000 in 1920 to 1.6 million in 1941. On December 7, 1941, Japan attacked Hong Kong and occupied it for more than three years, reducing the population to 600,000. With World War II behind them, the people of Hong Kong expressed a desire to participate more in the political process. After the communist takeover in China in 1949, many Hong Kong residents were worried about their future. Hong Kong was again inundated with about 700,000 refugees who escaped from the civil war on the mainland. However, there came to be a quiet understanding between Mao Zedong and the British that China would maintain the status quo because of trade benefits Hong Kong provided and the $5,000 in rent that British paid for the lease in the New Territories.

Between 1971 and 1982, Hong Kong underwent major reforms in housing, education, police, health care, and infrastructure through the leadership of Governor Sir Murray MacLehose. Beginning in 1982, British Prime Minister Margaret Thatcher met with Chinese Premier Deng Xiaoping to discuss Hong Kong's future. In 1984, both countries issued the Joint Declaration specifying Hong Kong's return to the People's Republic of China in 1997 as a special administrative region. The suppression of student protesters at Tiananmen Square caused alarm among many residents of Hong Kong about their relationship with BEIJING. In 1990, the Basic Law, which serves as Hong Kong's constitution was passed by the National People's Congress.

Hong Kong's population is generally homogeneous; 95 percent is Chinese and the remainder is mostly Indian and European. Cantonese Chinese and English serve as the official languages for the territory.

Hong Kong has one of the most developed economies in the world. From its beginnings as a free port, Hong Kong relies almost exclusively on international trade. Having little natural resources, Hong Kong must import its food and raw materials. Hong Kong has extensive investment trade and investment ties with China.

BIBLIOGRAPHY. John Flowerdew, *The Final Years of British Hong Kong: The Discourse of Colonial Withdrawal* (Macmillan, 1998); Tak-Wing Ngo, ed., *Hong Kong's History: State and Society under Colonial Rule* (Routledge, 1999); *World Factbook* (CIA, 20043).

DINO E. BUENVIAJE
UNIVERSITY OF CALIFORNIA, RIVERSIDE

Horn of Africa

THE PENINSULA OF NORTHEAST Africa is called the Horn of Africa. It lies opposite of the southern Arabian Peninsula. This area is also known as the Somali

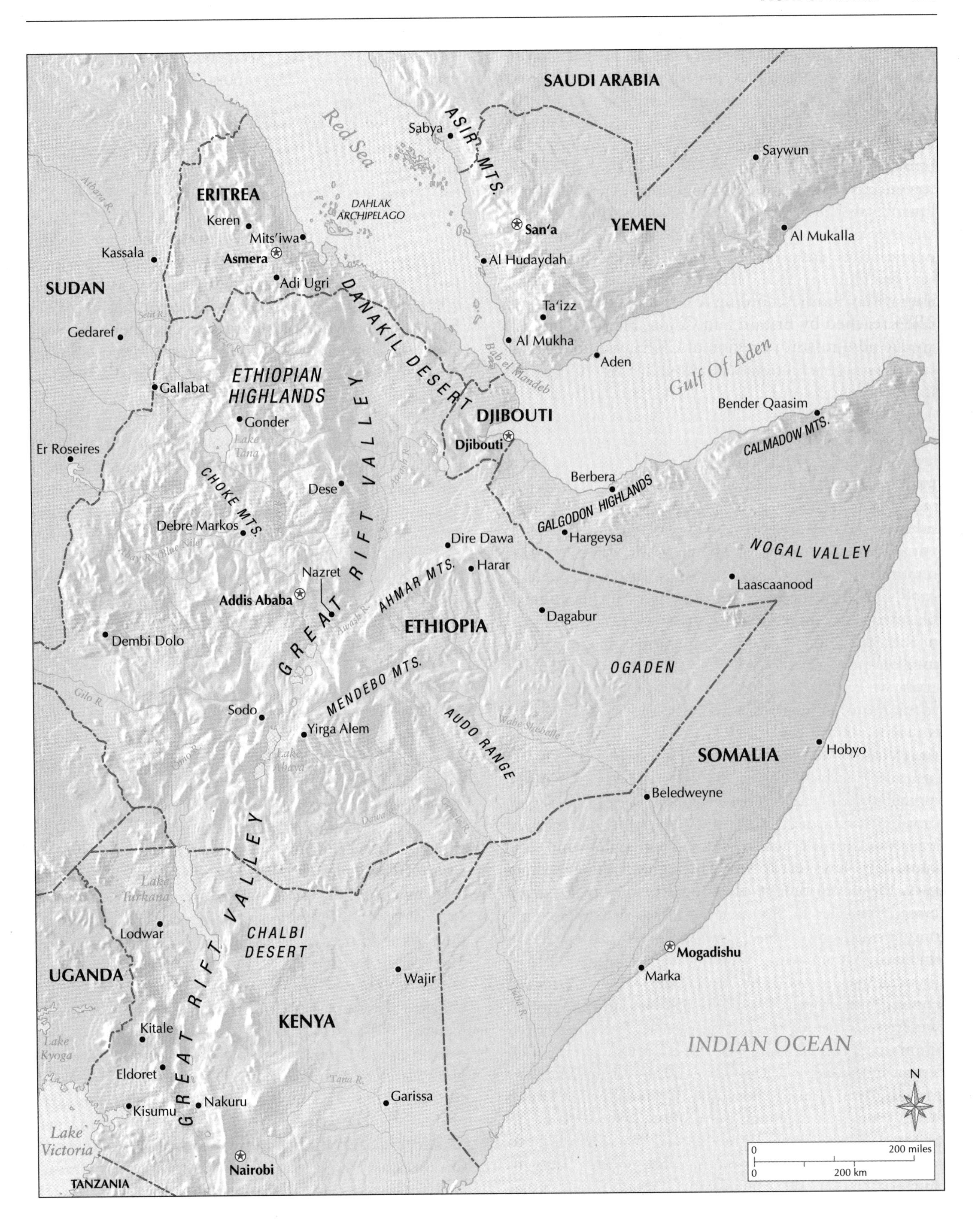

SAUDI ARABIA
Red Sea
Sabya
ASIR MTS.
Saywun
ERITREA
DAHLAK ARCHIPELAGO
Keren
Mits'iwa
San'a
YEMEN
Al Mukalla
Kassala
Asmera
Al Hudaydah
SUDAN
Adi Ugri
DANAKIL DESERT
Ta'izz
Gedaref
Al Mukha
Aden
Bab el Mandeb
Gulf Of Aden
Gallabat
ETHIOPIAN HIGHLANDS
DJIBOUTI
Bender Qaasim
Gonder
Lake Tana
Djibouti
CALMADOW MTS.
Er Roseires
CHOKE MTS.
Dese
Berbera
GALGODON HIGHLANDS
Debre Markos
Dire Dawa
Hargeysa
NOGAL VALLEY
Nazret
Harar
AHMAR MTS.
Laascaanood
Addis Ababa
ETHIOPIA
Dagabur
Dembi Dolo
GREAT RIFT VALLEY
OGADEN
MENDEBO MTS.
Sodo
Yirga Alem
AUDO RANGE
Lake Abaya
SOMALIA
Hobyo
Beledweyne
Lake Turkana
Lodwar
CHALBI DESERT
Mogadishu
Marka
UGANDA
Wajir
KENYA
Kitale
Lake Kyoga
INDIAN OCEAN
Eldoret
N
Kisumu
Nakuru
Garissa
Lake Victoria
Nairobi
0 200 miles
0 200 km
TANZANIA

Peninsula because within it lies the countries of SOMALIA and eastern ETHIOPIA. It is the easternmost extension of the African continent separating the Gulf of Aden from the INDIAN OCEAN. It is sometimes also used as a name for the entire region of countries in northeast Africa including ERITREA, DJIBOUTI, Ethiopia, and Somalia and sometimes even including parts of SUDAN and KENYA. The Horn of Africa also alludes to the shape of this area because it sticks out like that of a rhinoceros horn from the continent of Africa.

The Horn of Africa is made up of a wedge of land that is cut north to south by two great geographical features: the NILE RIVER Valley and the Great RIFT VALLEY. Between these two features are high plateaus and rugged volcanic mountains. West of the White Nile River spans the great and vast SAHARA DESERT. The Great Rift Valley rises to just over 1 mi (1.6 km) above sea level in central Ethiopia and then drops well below sea level in the Danakil (or Dallol) Depression. Some places in the depression are over 328 ft (100 m) below sea level making it one of the lowest places on Earth not covered in water.

It is known as one of the most inhospitable places on Earth because the landscape is sand and volcanic rock. Much of the area is still active volcanically and tends to have intermittent earthquakes. It is also the hottest place on earth with temperatures that can reach up to 145 degrees F (63 degrees C) in the sun. From the coast of Eritrea, the Danakil Depression drops even further into the depths of the Red Sea.

The coastal lowlands along the Gulf of Aden and the Red Sea are mainly desert, as is most of Eritrea and northern Sudan. Southern Sudan is covered in savanna. Much of Ethiopia is covered in highlands that were once thickly forested. Over millennia, this area suffered from deforestation but there are still some small isolated areas of tropical forest. Much of this area could be used for agriculture but has yet to be truly developed. All of the countries of the Horn have petroleum, natural gas, gold, silver, copper, and iron ore but not much of this has really been developed either.

The main reason this area is important is that the Horn of Africa commands the Red Sea and the northwestern portion of the Indian Ocean. There are many good ports on both sides of the Horn. Ships traveling in the area headed for the suez canal and the Mediterranean from Asia, the Persian Gulf, and East Africa all have to navigate through the narrow waterway at BAB EL MANDEB, the entrance to the Red Sea. Thus, its importance is not only geographical but political as well. The people of the Horn subsist in vastly different styles ranging from those that are hunter-gatherers and fishermen to nomads and farmers to industrial workers and more urban modern professions.

Most of the people of the Horn are farmers who still use draft animals to turn the land. Per capita income, literacy, and life expectancy are among the lowest in the world on the Horn. There are over 200 languages and dialects spoken over the Horn. There are even more ethnic groups that are further broken down into tribes and clans many of these who have deeply divided loyalties. There has almost always been conflict on the Horn for some reason or another, whether if be for political, ethnic, economic, or religious purposes and it does not look like that will change any time soon.

BIBLIOGRAPHY. Paul B. Henze, *The Horn of Africa: From War to Peace* (St. Martin's Press, 1991); James E. Dougherty, *The Horn of Africa: A Map of Political-Strategic Conflict* (Institute for Foreign Policy Analysis, 1982); Gunther Schlee "Redrawing the Map of the Horn: The Politics of Difference," *Africa* (Oxford University Press, 2003).

CHRISTY A. DONALDSON
MONTANA STATE UNIVERSITY

horst

HORST IS THE German word for a high-nesting area for a bird, so the term infers a high elevation. Early geomorphologists fittingly applied the German word to a fault block mountain range. The faulting and vertical displacement is due to tension stresses in the crust. Tension causes crustal rock to spread apart and break to form steeply inclined normal faults.

The fault lines are normal to the direction of spreading and parallel to each other at the surface, but their planes dip away from each other beneath the surface. A horst is left standing either by sinking of the crust on either side of a pair of normal faults or by physical lifting of a crustal block between the faults. Geomorphologists often use the German word *graben* (which means "trench" or "grave") for the low-lying block between two horsts.

Horsts have either symmetrical or asymmetrical profiles. Equal rates of vertical movement along the parallel faults produce a symmetrical profile. Horsts of this sort may have flatlike tops, but they are more likely in various stages of erosion. Horsts have asym-

metric (tilted) profiles when the vertical movement is along only one of the bounding faults. The resulting mountain has a steep escarpment on one side and a gentle back slope on the opposite side. The Sierra Nevada of CALIFORNIA are a classic example of a asymmetrical fault block mountains; their steep eastern escarpment faces NEVADA and the gentler western back slope faces California's Central Valley.

Horsts (ranges) in combination with grabbens (basins) form a basin-and-range topography; Nevada is in the middle of this terrain type, which extends into IDAHO, OREGON, California, UTAH, ARIZONA, and NEW MEXICO as well as northern MEXICO. Countless horsts interrupt the landscape of the region. For example, California's Owens Valley (a graben) separates the Sierra Nevada from the White Mountains (also a horst). Another example is the Panamint Mountains and Armagosa Mountains; they align on either side of Death Valley. Countless horsts occur in other parts of the world, wherever normal faulting has occurred. The Vosges Mountains of France, the Harz Mountains of GERMANY, the island country of TAIWAN, and several highland ranges of KENYA and ETHIOPIA in eastern Africa are horsts. The SINAI PENINSULA between water-filled grabben of the Gulfs of Suez and Aqaba is another prominent example.

BIBLIOGRAPHY. H.F. Garner, *The Origin of Landscapes: A Synthesis of Geomorphology* (Oxford University Press, 1974); Michael A. Summerfeld, *Global Geomorphology: An Introduction to the Study of Landforms* (Addison-Wesley, 1996); F.K. Lehner and James Lajos Urai, *Aspects of Tectonic Faulting* (Springer-Verlag, 2000).

RICHARD A. CROOKER
KUTZTOWN UNIVERSITY

Huang (Yellow) River

THE YELLOW RIVER, called the Huáng Hé by the Chinese, is the second longest river in CHINA reaching 5,500 km (3,000 mi), second only to the CHANGJIANG (Yangzi River). The regions to the north and south of the Huang River are called, respectively, the Hebei (north) and Henan (south). The chief tributaries of the Huang River are the Wei and Fen Rivers. The Huang River originates in the Yekuzonglie basin at an elevation of 4,500 m (13,500 ft) in the Bayan Kera Mountains and flows in an easterly direction through nine regions: Qinghai, Sichuan, Gansu, Ningxia, Inner Mongolia, Shanxi, Shaanxi, Henan, and Shandong, where it finally reaches the Bohai Sea. The Huang River has made at least five major direction changes from 602 B.C.E. to today. It is believed that the most devastating direction change occurred in 1194, killing hundreds of thousands of Chinese and destroying the economy. The latest direction change occurred from 1855 to 1897, resulting in the current easterly flow.

The Huang River gets its name from its color caused by the sediment carried by the flow. Large-scale erosion occurring in the northern regions continues to be a problem for the Chinese government, as it increases the danger level of the river through buildup of sediment farther downstream. In fact, the Huang River remains the most dangerous river in China.

The Huang River basin is where Chinese civilization originates. Some historians believe this is because of a unique land structure found only here and no where else in the world. The land structure is called Tai Yuan, meaning a table land on a plateau. What is left on the Tai Yuan gives researchers an idea of the landscape prior to it being carved up by the river. Researchers explain that these land structures were ideal for agricultural development, community activities, and unrestricted movement because it is believed that the current Tai Yuans were once much closer together, if not a single unit.

A large problem associated with the Huang River is the continuous flooding, due in large part to the heavy sediment that is carried off by erosion and left to rest in the lower regions, causing the flow to slow and preventing the river from reaching the Bohai Sea. The river, on an annual basis, fails to reach the sea one-third of the time, which is largely attributed to an increase in ciphering off of water by families, factories, and farmers and, of course, the large amount of sediment left behind in the lower regions from erosion in the upper regions. In 1955, the Upper and Middle Huang River Administration Bureau was created under the Huang River Conservancy Commission to initiate soil conservancy and sediment reduction.

Silt deposition has created large problems for certain areas of the Huang River. In some areas the river bed is elevated 60 to 70 ft (18 to 21 m) above the surrounding towns. Since the 2nd century B.C.E., the Huang River has broken the levees some 1,500 times, inundating the surrounding towns. For example, in 1642 the levees broke, killing some 300,000 people. And again in 1938, the Chinese government decided to flood the levees in an attempt to stop the Japanese in-

vaders, resulting in the deaths of 900,000 Japanese and Chinese combined.

Currently, the Chinese government has undertaken a project that rivals no other water project in the world: to transport water from the overflowing Changjiang in the south to the Huang River in the north. Three canals will deliver the water to the drier regions at a distance of 750 mi (1,250 km). A conservative cost estimate for this project is $30 billion; however, that can easily reach up to $60 billion or higher as the routes that the government wants to take are dangerous and, in some cases, require almost impossible engineering feats.

Work has begun on two of the south-north canals and engineers from the east are working on an ancient waterway called the GRAND CANAL. These two canals are expected to serve a population of 20 million, a population that must ration water until the project is built. The most expensive canal is the third, which is planned to deliver water from the higher regions of the Changjiang to the higher regions of the Huang through rugged mountains. This water will provide farmers of the most important agricultural regions the resources to adequately produce corn, sorghum, winter wheat, vegetables, and cotton.

Finally, it must also be noted that the canal project put forward by the Chinese government is highly controversial because of the nature of the project, the amount of money needed to finance it, and the impact it will have on the surrounding environment.

BIBLIOGRAPHY. K. Huus, "The Yellow River's Desperate Plight," www.msnbc.com/news (May 2004); University of Massachusetts, Dartmouth, "Sediment as Resources," www.cis.umassd.edu May 2004); University of Massachusetts, "Why the Yellow River Basin Is Called the Cradle of Chinese Civilization?" www.cis.umassd.edu (May 2004); University of Massachusetts, Dartmouth, "The Core Project," www.cis.umassd.edu (May 2004).

ARTHUR HOLST, PH.D.
WIDENER UNIVERSITY

Hudson River

"THE RIVER THAT Flows Both Ways" is located between Lake Tear of the Clouds and the southern tip of Manhattan in NEW YORK. It flows 315 mi (507 km) and empties into the ATLANTIC OCEAN. Its low and high tides, along with its salty water, account for the river's bidirectional water flow.

The Lower Hudson River (an estuary) is the dwelling place of a wide variety of plants and animals. Tidal marshes (called Meadowlands) filter the water that passes through them, and over 200 species of fish inhabit the river. The Hudson is located on the Atlantic Flyway and provides food to birds migrating in this north-south corridor. Its base of limestone rock enhances the river's ability to manage common pollution.

In 1609, the Englishman Henry Hudson, captaining a Dutch ship, stumbled upon the river in his search for the legendary Northwest Passage. At the time, native tribes already occupied the 4 million acres of the Hudson Valley, which expands from its narrow watercourse from below Waterford until the approaches to the Palisades Cliffs in NEW JERSEY, across the river from New York City. The indigenous population called the valley's river, Mukheakantuk, which means "Great Water in Constant Motion," or "The River That Flows Two Ways." The Dutch first named it "River of the Prince Mauritius"; it was not until 1664 that the English named it after Henry Hudson.

During the American Revolutionary War, the Hudson River facilitated the transportation of troops and supplies from the port of New York to military campaigns taking place to the north. The completion of the Erie Canal in 1825 enabled water transportation to cross upstate New York, all the way to Lake ERIE, and from there, farther into the Great Lakes region. The canal system made the Hudson River viable for significant trade along the Atlantic and deep into the American heartland.

The Hudson's visual landscape found wide application by artists seeking a characteristic backdrop for their paintings depicting the lush New World. The first major artist known for his sketches inspired by the Hudson River Valley was Thomas Cole. His sketches became an inspiration to other artists, and the Hudson River School of Painting was formed. Today, these works of art are located in the Historical Society of Newburgh Bay and the Highlands in Newburgh, New York.

After the Revolutionary War, ferries became popular for transporting citizens across the river. But river cruises and ferry transportation in general declined with the increase in pollution, which rose to dangerous levels in the mid-20th century. Once the source of drinking water to many cities, the Hudson River experienced a time when companies and residents dumped anything from caustic chemicals to motor oil to dis-

carded food into it, bringing about levels of toxic pollution that rose to a level lethal to humans who swam in it. Over the years, actions from residents and the government have been successful in the cleanup and restoration of the river. One of the most recent actions was an Act of Congress in 1996, which designated the Hudson River Valley National Heritage Area.

Despite the river's ghastly polluted past, the Hudson River is now in better condition than many Atlantic Coast estuaries. Its cruises have become popular again. The George Washington and Tappan Zee bridges span the river and provide a view of the Palisades cliffs for its travelers. The Metro-North Hudson River line railroad serves as a commuter line and as a means of transportation for residents and visitors to visit shops in many valley cities, including New York itself. As a geographic route to New York City, the hijacker piloting one of the planes that struck Manhattan on September 11, 2001, used the river to navigate the plane south from the Boston area.

BIBLIOGRAPHY. Beczak Environmental Education Center, www.beczak.org (October 2004); "The Hudson River," www.hudsonriver.com (October 2004); The Hudson River Valley Institute, www.hudsonrivervalley.net (October 2004); A. Keller, *Life Along the Hudson* (Fordham University Press, 1997); New York State Department of Environmental Conservation, www.dec.state.ny.us (October 2004).

Tonya L. Darby and A. Chiaviello
University of Houston, Downtown

human-environment relationships

THE ENVIRONMENTAL approach in geography, history, anthropology, psychology, and other spheres of humanitarian thought remains one of the most alluring and unequivocal since its origin in ancient times. At the beginning of the 21st century, interdisciplinary studies of cultural history using contemporary methods of instrumental analysis are often bringing scientists to the necessity of taking environmental issues into account when conceptualizing local peculiarities of cultural evolution. As a result, an impressive variety of theories and concepts has been elaborated in the framework of scientific and humanitarian thought during the second half of the 20th century.

Trying to navigate in this boundless theoretical space, an inquisitive researcher inevitably faces the necessity to create some kind of exploratory structure encompassing these approaches and notions. Here, comprehension of human agency in natural environment evolution has been chosen as the main criterion of examination of human-nature interaction. On this base, three main directions of environmental thought development can be distinguished in contemporary humanitarian and scientific thought.

One of them, known as geographical determinism, concentrates on the environmental impact on human history. The second, concerning human agency in nature development, became popular alongside global reconsideration of the human role in the universe, which took place at the time of the scientific revolution. Adepts of a third direction tend to interpret human-environment interaction as an integrated system, all elements of which are of equal importance and are engaged in complicated reciprocal influence.

GEOGRAPHICAL DETERMINISM

Geographical determinism comes from the "man as sufferer" paradigm in the adaptation concept. Interpretation of a human being as a passive sufferer originated in ancient natural philosophy. Since its very beginnings, the human being was regarded as a creature deeply dependent on its natural habitat.

As early as at the middle of 4th century B.C.E., geographic determinism had been designed as a specific direction of philosophic thought with at least two extreme schools: one of climatic psychology and another of climatic ethnology. Later climatic astrology also originated.

The Enlightenment ideology reconsidered these ideas about total human dependence on nature and elaborated the wide circle of geographically and climatically deterministic theories (C. Montesquieu, L. Mechnikov, E. Semple, E. Huntington, E. Reclus).

At the second half of the 20th century, this idea was treated in the framework of the adaptation concept, which is interpreted by representatives of "new" (L. Binford) and behavioral (B. Schiffer) archaeology, environmental psychology (A. Bell), and phenomenological (T. Ingold) and actional (E. Markaryan) approaches to culture studies as phenomena inherent for an active and creative human being.

In Soviet science, most attention was paid to the biological aspects of adaptation, with an emphasis on the human capacity to fit the requirements of natural environment.

At the end of 20th century, adaptive reaction has become the subject of special attention. As a result,

general theory of stresses and ecological stress concept were elaborated (P. Bell). To identify the possible character of human beings and human society in response to natural environmental changes, the concept of social and ecological resilience was introduced (A. Neil).

Thus, the idea of humans as nature modifiers and creators came to be. Roots of the idea of human domination over nature are traced as early as the Enlightenment times, when the human ability to solve rationally all his or her vital tasks was declared for the first time (T. Hobbes, C. Linney).

ECOLOGICAL CRISIS

Along with the beginning of industrial development and the origin of first ecological crisis at the middle of 19th century, scientists study the results of a transforming impact from human activity on the natural environment (J. Marsh, A. Voyeikov). During the 20th century, the idea of humans as nature-creators was conceptualized in the context theories of cultural (O. Schluter, C. Sauer) and anthropogenic (in Soviet science) landscape, and the notion of landscape as series of sequent occupancies (D. Whittlesey).

In frameworks of postmodern methodology, this idea is conceptualized in the idea of landscape as artifact, based on two ideas (T. Darvill, P. Criado Boado). One of them is that landscape should be interpreted as a mental image, which could not exist without human beings who elaborate it. At the same time, humans consciously and purposefully form their geographical environment, and their decisions about living space ordering are deeply motivated by their vital needs. Consequent application of these postulates inevitably results in partial or total negation of the natural landscape existence.

This idea has become the starting point for the theory of human ecodynamics, which concerns the analysis of changes made by humans in the landscape in a long-lasting perspective.

In spite of principal differences in theoretical backgrounds of the "human as nature-creator" concept, they incorrectly tend to date the beginning of the human impact on nature with the origin of agriculture and farming. Thus, the possibility of hunters and gatherers substantially reshaping their landscape is practically excluded or regarded as a minimal and non-permanent one, displayed only in connection with so-called secondary landscape components.

There is also the idea of mutual creativity in human-environment interaction. The process of formation of so-called integral direction of man-environment interpretation is long and ambiguous. These ideas, originating for the first time in ancient natural philosophy, obtained theoretical scientific background during the second half of the 19th century (J. Raskin, K. Ritter).

The fundamental theoretical background for this idea was elaborated during the late 19th and early 20th centuries in the framework of anthropogeography (F. Ratzel) and biosphere theory (V. Vernadsky). The traditional variant of anthropogeography envisages attention to all spheres of human culture and to humanity itself, taken as social and biological creature (F. Ratzel, A. Hettner, A. Grigorev, A. Borzov). At the same time, some researchers proposed to limit their subject field by the phenomena, which directly and materially display themselves in the landscape (O. Schluter, O. Brun) or by the human being as an organic form of life (W. Davis).

AN EMPIRIC BACKGROUND

Intensive deepening of our knowledge about climate, relief, flora, and fauna and their chronological and spatial distribution, which took place at the middle of the 20th century thanks to active development of environmental archaeology, geoarchaeology, and paleogeography, has created an empiric background for theoretical conceptualization of the "man in nature" idea in Western European and American archaeology, prehistory, and paleogeography. In Soviet science, such ideas were reflected in research activity of proponents of the so-called paleoenvironmental approach to prehistoric studies (S. Bibikov).

As a result, at the beginning of the 21st century, one can trace the gradual growth of popularity of the idea of mutual and interdependent evolution of nature and society. A specific form of its interpretation is proposed by representatives of mainly postmodern directions of contemporary geography, scholars who have introduced a wide spectrum of variants of landscape understanding.

BIBLIOGRAPHY. C.J. Glacken, *Traces on the Rhodian Shore: nature and Culture in western Thought from Ancient Times to the End of the 18th Century* (University of California Press, 1967); A. Holt-Jensen, *Geography: History and Concepts* (Paul Chapman Publishing, 1988); L. McDowell, "The Transformation of Cultural Geography," *Human Geography: Society, Space and Social Science*, D. Gregory, R. Martin, S. Smith, eds. (Macmillan Press, 1988); R. Muir *The New Reading the Landscape: Fieldwork in Landscape History* (University of Exeter Press, 2000); Y.G. Saushkin, *Isto-*

ria i metodologia geograficheskoi nauki (Izdatelstvo MGU, 1976); O.V. Smyntyna, "The Environmental Approach to Prehistoric Studies: Concepts and Theories," *History and Theory* (v.42/ 4, 2003).

OLENA V. SMYNTYNA
MECHNIKOV NATIONAL UNIVERSITY, UKRAINE

human geography

HUMAN GEOGRAPHY FOCUSES on interpreting and describing the various ways in which humans in all places and cultures adapt to and possibly modify their natural geographic environments. At the local or national scale, human geographers look at how economic, political, and cultural issues are related to spatial organization in different parts of the planet. How have humans modified topography, changed microclimates developed and changed rivers, lakes, and even coastlines?

Human geography is distinguished from PHYSICAL GEOGRAPHY by its focus on human activities regardless of specific cultures. Changes related to different cultures or social systems is the realm of CULTURAL GEOGRAPHY.

REGIONAL AND GLOBAL PATTERNS

Today, human geography also looks at regional and global patterns. Because of the spread of modern technology, humans today can make changes in the natural environment at a much faster rate and much grander scale than at any other time in human history. In addition, conflicts such as war can cause immediate and widespread environmental damage, such as the oil fires and spills during the various conflicts in IRAQ and KUWAIT or the widespread killing of wildlife in various African conflicts. The extent of environmental degradation and pollution in the former Soviet Union and the demise of the ARAL SEA are other examples of human-created changes in geography. Like physical geography, human geography is divided into a wide range of subtopics. These include economics, transportation, cultural geography, urban geography, and political geography. For example, human geography, when dealing with environmental issues, is not limited to natural dynamics but also takes into account the fact that there are distinct social, economic, and political environments.

The problems dealt with by human geography have varied over the course of time. In addition, new models and technical abilities affect how problems in the human-physical environment are approached. Given the diversity of philosophies and models for the description and analysis of human-environment interaction, it would be more appropriate to speak of a plurality of human geographies rather than one single human geography.

In 1992, David N. Livingstone noted different approaches to the discipline. According to this English geographer, the whole set of problems, subjects, and concepts that have developed over time have come to form part of this tradition and to be called human geography. From the time of the explorers to the drawing of maps, from the days of proposals for the study of industrial locations to the study of the distribution of wealth throughout the world, or the spaces constructed by "gay" communities or "okupas," up to the time of the survey of the surface of the Earth with remote sensors and cartography based on GEOGRAPHICAL INFORMATION SYSTEMS (GIS)—all of this has come to be a part of the human geography tradition and to distinguish it from the type of history and cultural or social and political analyses and description developed by other disciplines.

It is possible to distinguish four significant events that help to understand how different problems and subjects have become defined as human geography. These events were: 1) the formal recognition of modern geography (1870–1920); 2) the development of pragmatic perspectives (1950s); 3) the manifestation of radical and critical views (1970s); and 4) the development of what has been termed a postmodern approach along with more traditional cultural geographies (1980s and 1990s).

FORMAL RECOGNITION (1870–1920)

Human geographies developed in most European countries were influenced by the German, French, or British schools of geography and sometimes by all three together. Some maintain that it was in these three countries that the discipline was first institutionalized as something distinct from history and geology. Between the mid-19th century and the beginning of the 20th century, and in the context of the construction and redefinition of national states and the process of imperialism and colonial expansion, geographical knowledge was clearly related to the extension of political and economic power. As it was considered, geography offered the kind of knowledge that made it

possible to acculturate and integrate or control local populations.

During this period, human geography became a distinct program in the curriculum. For example, the first chair in human geography was set up in 1870 within the context of the unification of Germany, and under the responsibility of the geographer-philosopher, Friedrich Ratzel (1844–1904). France recognized the importance of teaching geography after the loss of the territories of Alsace-Lorraine to the Germans. This was France's context for the creation of the chair of human geography under Paul Vidal de la Blache (1854–1904).

Ratzel developed a project for anthropogeography based on the analysis of the influence of natural conditions on humanity. For Ratzel, the greater the attachment to the earth (as Ratzel called the territory) the greater would be the need for a society to maintain its physical possession. Ratzel believed that it was for this reason that the state was created. The analysis of the relations between state and space was one of the main topics in anthropogeography. The development of any society would imply, as he saw it, the need to increase the size of territory and hence to conquer new areas. One can readily see the seeds of Adolf Hitler's *Lebensraum* (expanding "living space") in this approach to the state.

On the other hand, the Frenchman de la Blache was critical of anthropogeography as an approach to human geography. Rather than being interested in the influence of natural conditions on societies, de la Blache sought to analyze how societies could challenge nature and come to develop an environment suited to their needs. Within this framework he formulated his concept of the *genre de vie* (lifestyle), understood as a historically constructed relationship built up by different human groups with their surroundings, based on the use of available technologies. This was a view that emphasized human abilities and influences in modifying the physical earth itself.

For this geographer, natural human regions and regional study was seen as an expression of lifestyle, which was the whole object of study of human geography. Seen in the light of this approach, for de la Blache, the map of France was the result of the harmonizing of its different regions. In the light of the theories and studies of de la Blache, the concept of a region (the study of the particular relation of a set of diverse elements in a given area) became one of the key concepts in human geography.

In Great Britain, the Royal Geographical Society was responsible for the institutional and financial organization of the two chairs in geography, one at Oxford University and the other at Cambridge University. The first was assumed by Halford J. MACKINDER (1861–1947). The second fell to Francis Guillermard (1852–1933). Mackinder considered that geography could be useful to statesmen since it is an integrating discipline, in which studies can be made of the causal relationships between man and the environment. These studies can be conducted in specific areas and have as their purpose the analysis of these relationships on a global scale.

Toward the end of the 19th century, biology was considered to be the most modern discipline. In the light of such thinking, the concept of geography as a natural science was the guarantee required for it to achieve qualification as a science. Thus, no one hesitated to qualify geography as a natural science, thus placing evolution at the heart of any geographic explanation.

For French geographers, human geography was a discipline that leads to knowledge of the relationship of societies with their environment. For Russian geographers, the purpose of geography in education was to reveal the battle of human beings with nature, thereby leading to a better understanding of the relationship between the two. Some Russians stressed that the study of human diversity implies showing what constitutes families of different peoples, bringing them together regardless of any racial differences, beliefs, or lifestyles. On the other hand, other geographers placed work as the mediator between the physical environment and society.

North American cultural geography presented by Carl SAUER (1889–1975) is one of the few proposals of that time that attempts to rise above a global and North American evolutionary framework (particularly as it was developed in the UNITED STATES by H. Barrows, T.G. Taylor, R.D. Salisbury along with W. DAVIS and E. Semple.). In fact, Sauer makes the whole concept of culture, linked to the anthropology, the key to the transformation of the natural landscape or the visible forms of nature into a cultural landscape.

PRAGMATIC PERSPECTIVES

In the years after World War II and to the beginning of the Cold War, regional analyses of human geographies had become the principal activity conducted in the field.

As financial capital was directed to the reconstruction of postwar Europe, a group of planning organizations were attempting to define the best locations for

productive activities. They did this using a set of geographical engineering models that would define social problems before they became major challenges. The joining of the human geographies of political and economic perspectives led to the birth of a focus on pragmatic versus theoretical or descriptive studies. For most of the geographers following this line, mathematical language and models were considered to be the most appropriate methods for true science.

Human geographies that emphasized descriptive accounts were exchanged for a more statistical approach. Neoclassical models were formulated in the belief that subjects would behave rationally and with a view to seeking a maximization of earnings and opportunities. Now human geography became viewed as a discipline entrusted with spatial analysis. From this time on, spatial organization became the principal object of study of human geographers. The scientific base was to begin with an assumption that geographic space was to begin de novo—that is, without people or prior history. In short, the approach was, "All things being equal, then..." Within these assumptions, the planner could move freely within geographic space, the only variables would be questions of distance, direction, and connection (linkage). The geographic region was understood as the space in which internal differences are minimized and everything beyond its boundaries would be of much greater difference or variance.

Under the idea of pragmatic geography (as another subset of human geography) there were various trends. First, quantitative geographers studied the relations and interrelations of different geographic phenomena, local variations of physical landscape, and the impact of nature on societies and of the latter on the environment. These were all to be numerically expressed and understood. The second trend involved geographers who used systems theory. Hence it is given the name "systematic" or "modeling geography." For example, the geographer Brian Berry defines models as key to the formulation of explanations. A third school of pragmatic geography is represented by a focus on the geography of perception. Drawing on the instruments of behavioral psychology, the followers of this trend try to analyze the subject valuation of space, both in the case of the behavior of the urban dweller and that of native African communities. It is focused on culture as the key to the creation of human geographies.

A part of the legacy of the approach of pragmatic geography has been the evolution of GEOGRAPHIC INFORMATION SYSTEMS (GIS), a field in which the concept of abstract space and the formulations of a mathematical character continue to play an important role. And it is a practice now widely used by many fields of science, both physical and social.

RADICAL, CRITICAL HUMAN GEOGRAPHIES

As part of the sociology of its era, the decades of the 1960s and 1970s witnessed a major political convulsion in the field of human geography. In addition to the controversies surrounding the Vietnam War, it was a period during which many of the long-established European colonies in Africa achieved independence. Various social movements, such as those in defense of human rights, the rights of women, and the protests of an ecological character all appeared on the world scene and were especially intense in Europe and the United States. Many contemporary human geographers not only participated in these movements but also began to question their own practices. The crisis to be identified in geography was a symptom of other crises occurring in capitalism, politics, and science.

Some human geographers now considered that scientific knowledge should not only serve for the understanding of society, but should also help to guide and transform it. Positioning themselves simultaneously against both classical and quantitative geography, they tried to establish the basis of a new science that, as they saw it, would help create the basis of a new society. These geographers called themselves radicals in the UNITED STATES; in other contexts, such as the French, the Italian, the Spanish or the Latin American world, they were referred to as critical geographers.

Both the radical and critical schools of human geography followed the philosophical tenets of Marxism and stressed first and foremost the importance of economics (not the natural environment) when it comes to the interpretation of social dynamics. Second, they stressed the role of ideology in the production of knowledge, opposing the idea that there is any possibility of creating an objective or value-neutral science. Both the radical and the critical geographers reverted to the concept of space already worked upon by the pragmatic geographers in order to provide the social content it had consciously omitted (to be scientifically neutral) when formulated in the 1950s. Radical and critical geography was a shift to emphasizing human economics and philosophies in the creation of human geographies.

The approach known as active geography was opposed to the applied geography promoted over the decade of the 1950s. The first manifestation of this approach was found in *Géographie Active* (1964) the

name of a book written by Pierre George, Yves Lacoste, Bernard Kayser, and R. Guglielmo. This book undertook an analysis designed to reveal the contradictions of capitalism in different regional geography frameworks. Thus, a type of geography was formulated for regional analysis that would reveal inherent social contradictions such as poverty, malnutrition, and precarious housing. The proposal for active human geography also gave a new significance to actual fieldwork in the countryside. Thus, for example, in the Anglo-Saxon context, William Bunge proposed the organization of expeditions to communities living in conditions of poverty in order to help them overcome their situation, establishing a priority for social welfare over academic work.

Following the interdisciplinary exchanges that were opened up by the pragmatists, the radical and critical geographers exchanged ideas with other social sciences. This further removed human geography from its earlier emphasis on the physical environment. This exchange can be seen in the influence of the book by sociologist M. Castells, *La Question Urbaine* (1972), or the philosopher Henri Lefèbvre's *La production de l'espace* (1974) on the urban geography of the period.

Among the radical geographers, one of their most representative works is *Urban Justice and the City* (1973) by David Harvey. Harvey questioned the liberal theories of the city and took on a socialist posture. He adopted the Marxist theory of rent in order to analyze the valuation of urban space. He then studied the use of the land in terms of use/value categories and exchange value. This sort of analysis enabled him to understand the active importance of spatial forms in social processes, an approach that he later developed in works dealing with the role of capital in the generation of unequal space usage and the compression of space/time. In one sense, it was a human geographic view of socialism rather than the traditional emphasis on how humans modify the physical environment.

For critical geographers, particularly in Latin America, one of the most important texts has been *Por uma Geografía Nova* (1978). The author, Milton Santos (1926–2001), showed that it was possible to conceive of personal ideas that could be applied to the interpretation of the third world. In effect, his analysis of the specific nature of urban processes in underdeveloped countries and his theory of banal space (the daily space for solidarity where men, living and feeling, have the opportunity to create a new history) are an example of this. Santos offers multiple ways of perceiving social space. First, space appears as a social product, born of human action. Second, it signifies accumulated work, the incorporation of capital into land surface, which creates lasting forms known as "roughness." These manifestations of "roughness" turn out to be space legacies that end up by influencing the pattern of contemporary action.

In this sense, spatial patterns are the product of past processes that also condition the future. Now there were human geographers who simply followed a modified version of the philosophical view *cogito, ergo sum*—"I think, therefore I am."

POSTMODERNISM

This approach led the human geographer Edward Soja to argue that while modern times have granted primacy in their explanatory role to history and time, postmodernism should open up the way to the social sciences, allowing them to achieve "the spatial turn." Such an approach would be ideal for the analysis of changes and urban dynamics in cities such as Los Angeles. Such a city has particularly diverse geographic landscapes, constantly undergoing a process of change, and continually being reshaped by local and state practice, by internationalization, and by the globalization of work and trade. The city is produced, lived in, and provides for not only a middle class society, but also women, children, old folk, gays, lesbians, a multitude of ethnic minorities, the unemployed and the poor. With the approach prevailing until now, it was difficult to identify any urban fragments associated with the place of residence of certain social classes or the establishment of certain productive activities. In fact, global cities such as NEW YORK or LONDON are composed of a discontinuous collage of partial human geographic landscapes that no longer respond to the old center-periphery pattern. Approaches such as this may also be used for geographies of rural and regional environments.

Human geographers now have conducted a series of studies governed by this perspective. Some of the work along these lines includes: a) gender studies, b) postcolonial studies, and c) new cultural geographies.

Gender studies originated in the feminist movement of the 1960s. One of the basic points of these studies was the recognition that distinct gender geographies are to be found throughout all societies and all of history and that they always play a role in the organization of patterns of human geography. For example, in the geographic patterns resulting from the locations chosen by multinational clothing manufacturers and shoemakers in different parts of the world, there is a pattern that increasingly seeks female workers willing

to work at home on a part-time basis. This means there is a new kind of manufacturing geography, one not focused solely on large plants. Gender studies have also shown the need for a new kind of urban management and planning geography, an approach that looks at the needs for mobility and recreation of women and families.

Postcolonial geographic analysis begins by studying the human organization of space both under the colonial experience and how it exists today. The book *Orientalism,* written by Edward Said (1935–2003), is considered to be the basis for this field. Said maintained that the way in which we view the East from the West is the result of a Western cultural bias or mindset. Such a vision of the East was presented as exotic, sensual, culturally inferior, and backward—all of which supported imperial European expansion between 1870 and 1914.

In this context, colonial space (geographies) became an area of contact where diverse cultures met, collided, and fought. Postcolonial studies have also tried to listen to those who were the object of the colonial experience and had often been silenced by the European discourse.

They attempt to give space to voices that come from countries that went through colonial and live their consequences today (economic dependence, dictatorship, exclusion of women). Postcolonial researchers are also now being called upon to research image construction for the Muslim world and the immigrant Muslim populations, as reflected in political discourse and in the communications media. These findings are proving useful both for the justification of the new imperialism and for the imposition of restrictive migration policies, both aspects are used for understanding the political transformation and the globalization of the world.

CULTURAL GEOGRAPHIES

This section of human geography is the one that has turned out to be the most dynamic, although some authors believe that all human geography is really cultural geography. It was Denis Cosgrove, Chris Philo, and Peter Jackson who, in the decade of the 1980s, introduced politics into the field of culture, aiming at showing its influence on history. This was the so-called cultural turn. Completely new subjects began to be studied in detail, such as daily objects, images of nature in art and the cinema, and even the significance of landscape and how territorial identity is socially constructed.

The discussion of GLOBALIZATION is a very important issue here, especially when it is viewed as a strategy to reduce all cultures to a single model, with the threat that this implies to individual local identity. And other authors think that new subjects should be considered for future studies, such as the interaction between various factors and beliefs, while local and global factors also need to be taken into account.

One element that has always formed a part of human geography and is now the object of renewed interest is "landscape." Cosgrove, a North American geographer, headed this movement in the 1980s. Cosgrove tried to show how economic processes such as capitalism and the relation between capital and labor influence cultural patterns. He undertook a revision of the landscape art of the 18th and 19th centuries and found a link between the different ways in which landscape was represented at that time and such things as different types of land ownership and social relations in rural areas over the centuries. He found that these paintings had an "iconographic" content, transmitting symbols that had a meaning in the places concerned because they reflected the interests of particular social classes. Today, for example, human geography landscapes are being viewed in the light of the postcolonial gender perspective.

Studies like this show how the paintings of the time formulated an exotic world of paradise, associated with tropical environments such as those of the Caribbean. This idealization hid any reference to social injustice, such as the political and economic exploitation that lay behind work on the plantations. Gender studies today clearly show that this type of painting also reflects a purely masculine point of view. They underline the differences between colonial images created by traveling women and those created by men, yet the men led the work of exploration, conquest, and control of the colonies.

Landscape paintings are also being studied with a view to showing their contribution to the concept of a "pure" national identity, a founding stone for nationality. All of this contributed to the construction of "new" geographies. James and Nancy Duncan feel that these paintings can be read as "texts." They say it is possible to recognize the painters and that this is an expression of values, tastes, and aspirations presented in a "codified" way. As is the case with texts, different people can interpret the images in different ways. In a city, architects, dwellers, visitors or spectators each have their own view and thus the geography of landscapes is rewritten.

MODERN GEOGRAPHERS

While the above discussion may give the impression of human geography being a fragmented field, in fact it is the only way in which we can understand the form and complexity of the organization of the modern world.

The practices of human geography today are no longer limited to simple geometric space. Rather, modern human geographers stress how geographic space is organized by and for relationships. This means that geographic space no longer is conceived of as a static container for objects, processes, or flows—Euclidian geometry (with the dimensions *x, y, z*) or simply by the laws of physics and natural processes. There has been increasing importance of the concept of "place" (local human-physical environments) in the work and thinking of human geographers over the last two decades. The visions of place of John Agnew and Doreen Massey are complementary in this connection and allow a better understanding of the concept. For Agnew, the concept of place has three dimension: the first is connected with the idea of location as it refers to the social and economic processes that endow it with a material character; the second—locale—refers to daily social relationships that lead to the creation of an environment (setting); and the third refers to the creation of a subjective feeling as to this environment.

The processes participating in the creation of this place involve actors operating at different levels. This English geographer points out that the relationships places maintain with each other are the product of particular power arrangements, be these of an individual, institutional, material or imaginative character; they are the specific interrelationships that define the particular characteristics of each one. Clearly, for human geographers, the role and impact of nature upon humans and how humans and their various idiosyncratic cultures and politics affect the organization and manifestation of geography remain key topics for analysis and description.

POLITICAL GEOGRAPHIES

The field of human geography made big strides forward between 1910 and 1930. The imperial expansion of the European countries was inspired by the idea that the power of a state was based on its capacity to expand its territory, a concept used to justify German expansion in Europe between 1933 and 1945. After the fall of Hitler, the whole field of political geography suffered from a corresponding loss of prestige that lasted until 1980, when it began to arouse interest once again.

Two factors have played a big role in its rebirth; first the fall of the Berlin Wall in 1989, considered a symptom of the decline of the Soviet bloc, the end of the Cold War, and the birth or rebirth of states that would be incorporated into the market economy; and second, the assault on the World Trade Center towers in New York City and the Pentagon near Washington, D.C., in 2001. On the one hand, the United States appeared as the world's leading military power and adopted an aggressive Middle East foreign policy.

The interest of the United States lies in controlling territories, populations, and natural resources, particularly oil, in that part of the planet; but, on the other hand, it is faced with the organization of radical groups using tactics to destabilize institutionalized methods of warfare. (They do such things as attacking civilians who are then beset by a feeling of insecurity. It is no longer a question of confrontation between states, but now within states; there is no such thing as neutrality or the laws of cease-fire. And the financing providing for the activities of these groups is usually of criminal origin.) Both of these series of events have unleashed processes that form a part of the agenda for political geography today.

The new agenda includes the reformulation of the relationship between territories and power. In this sense, the state is no longer the only basic legal and administrative unit that creates other types of international political relations. Organizations such as the EUROPEAN UNION and other international and transnational agencies have powers that have often been redefined, so that they can now acquire authority over the territorial states themselves (one example would be the World Trade Organization). These new types of organizations have also redefined the sovereign power of individual states, now sometimes reduced to certain limited fields such as the management of labor markets. In fact, functions that before were the exclusive domain of the state, such as capital attraction promotion, are now also encouraged by political groups associated with specific regions or cities. State territorial sovereignty is increasingly questioned in some countries like COLOMBIA, where guerrilla fighters and drug traders control part of the national territory. Frontier building also depends to a great extent on the new vision of power and of territory. Processes designed to stimulate the free circulation of goods and persons exist at the same time as the increase in measures destined to prevent displacement.

Thus, while the European Union extended its eastern frontier in 2004 to include SLOVAKIA, SLOVENIA,

MALTA, POLAND, LATVIA, LITHUANIA, CYPRUS, and the CZECH REPUBLIC, the Mediterranean area is becoming a major monitoring system designed to restrict the access of people from Africa to the New Europe. The wall raised by Israel on the Gaza Strip is also aimed at preventing the Palestinian population from entering the country. And, finally, the direct relation between state-nation-territory is now being challenged. It is true that the so-called equivalence of these terms was fractured when local national claims led to an increase in the autonomy of certain regions of the planet (Cataluña, the Basque Country, Northern Ireland, Quebec and Casamance) or to separatism (with the division of the ex-Yugoslavia and the independence of EAST TIMOR).

The agenda for political geography in the 21st century is engaged in redefining the relationship between state, nation, and territory. It is also incorporating new concerns (such as the environment or HIV/AIDS) and new subjects (guerrilla fighters, mafias, emigrants, and refugees), new conflicts (particularly of an ethnic-religious character such as between INDIA and PAKISTAN or in SUDAN), and studies of future scenarios, such as the political and economic role in this new century of countries like CHINA.

BIBLIOGRAPHY. H. Capel, *Geografía Humana y Ciencias Sociales: Una perspective histórica* (Ed. Montesinos, 1987); David N. Livingstone, *The Geographical Tradition: Episodes in a Contested Enterprise* (Blackwell, 1992); R.J. Johnstone, Gregory D. Pratt, and G. Watts, eds., *The Dictionary of Human Geography* (Blackwell, 2000); D. Hooson, ed., *Geography and National Identity* (Blackwell, 2000); A. Goldweska and N. Smith, *Geography and Empire* (Blackwell, 1994); M. Escolar, *Critica do Discurso Geográfico* (Editora Hucitec, 1993); R.J. Johnston, *Geography and Geographers* (Edwin Arnold, 2000); P.C. Da Costa Gomes, *Geografía e Modernidade* (Editora Bertrand, 1996); P. K. Hubbard, R. Bartley and B. Fuller, eds., *Thinking Geographically* (Continuum, 2002).

PERLA ZUSMAN
UNIVERSITY OF BUENOS AIRES, ARGENTINA

Humboldt, Alexander von (1769–1859)

ALEXANDER VON HUMBOLDT is significant to the study of geography because of the breadth of his scientific inquiries and knowledge and his ability to integrate these studies within larger works that were both scientifically advanced and appealing to a wider nonspecialist public. He was known as a natural scientist of the highest order, working to unify studies of botany, zoology and ecology, but also as a competent writer, whose works provided a wide audience with their only glimpse of South America, a continent previously known to most through myth and speculation.

Born into a relatively minor noble Prussian family, he was equally a product of his class and of his times. Young men of wealth and breeding in mid-18th-century Berlin were influenced by the philosophies and scientific principles of reason inherent in the French Enlightenment, as well as the more adventurous spirit of early German Romanticism. Accordingly, Humboldt went to the leading universities of the day, especially Göttingen, to immerse himself in the natural sciences. A natural polymath, he undertook courses in botany, literature, archaeology, electricity, mineralogy, and so forth.

After a short career working for the state as a mineralogist, he did what so many of his generation and social class did and set out for his grand tour. But Humboldt did not head to ITALY or GREECE as was fashionable. Instead, he set his sights on South America, which at that time was mostly unknown and in fact restricted to non-Spaniards. But the king of SPAIN was attracted to the fact that Humboldt had his own means to support his travels and to his potential as a mineralogist, since the silver mines of Peru were not infinite.

Humboldt spent five years traveling with his companion Aimé Bonpland across VENEZUELA, COLOMBIA, ECUADOR, PERU, MEXICO, and CUBA, returning in 1804 to Paris, FRANCE, where he began to publish accounts of his travels. His descriptions captured the imaginations of both scientists and the general public, with its details of canoe trips up the Orinoco River and horseback rides across passes high in the ANDES, and also with vivid illustrations done by Humboldt himself. Scientifically, his work was useful not simply as a collection of botanic and zoological specimens, but also as a work of analysis on the forces of nature and on how the geographic environment influences plant and animal life of a certain region.

This was an innovation in a scientific community generally more interested in description and classification than in interpretation. Humboldt's writings stress again and again the importance of unity in nature and man's understanding of environmental systems in their entirety. His monographs on Mexico and Cuba set a

standard for the field of geography, being the first to incorporate natural science, politics, and economics all within one study.

By the 1820s, Humboldt was arguably the most famous natural scientist in the world. He was well known in all scientific quarters but was also personal friends with the leading Romantics, such as Johann Wolfgang von Goethe. His prose was admired not just as brilliant science, but as beautiful literature. His major work, *Rélation historique du voyage aux régions équinoxiales du nouveau continent*, was published eventually in 30 volumes, appearing intermittently between 1814 and 1834. One of these volumes was called *Personal Narrative*, and it included Humboldt's thoughts on his own place within the world he was studying. Later travels included a voyage to RUSSIA and Central Asia in 1829, with an account published in 1843. His third major work was a comprehensive survey of all creation, *Kosmos*, published posthumously in 1862.

Humboldt was part of a long tradition of polymath scholars or Renaissance men, whose specializations reached into numerous disciplines and subfields, bringing comparative knowledge to its greatest advantage in whatever arena he explored. And like many of those before him, Humboldt was not merely content to be a scientist. He remained loyal to the calling of his noble origins, serving many years at the court of the Prussian kings in Berlin, first as chamberlain to Friedrich Wilhelm III, then as councilor of state to Friedrich Wilhelm IV. His works served as inspirations for the next generation of scientists, most notably Charles Darwin, who knew parts of *Narrative* by heart.

BIBLIOGRAPHY. Alexander von Humboldt, *Personal Narrative*, abridged and translated by Jason Wilson (Penguin Books, 1995); Wolfgang-Hagen Hein, ed., *Alexander von Humboldt: Life and Work* (Ingelheim am Rein, 1987); Andrew Cunningham and Nicholas Jardine, eds., *Romanticism and the Sciences* (Cambridge University Press, 1990).

JONATHAN SPANGLER
SMITHSONIAN INSTITUTION

Humboldt Current

THE HUMBOLDT CURRENT, also known as the Peru Current, is an ocean current that flows along the western coast of South America, affecting the water and air temperatures of coastal CHILE and PERU. It is one of the largest ocean currents in the world, bringing cold water north from the South PACIFIC for thousands of kilometers (from 40 degrees S to 5 degrees S) before it dissipates in the warmer waters around the equator.

The Humboldt Current, named for Alexander von Humboldt, the German naturalist and geographer who devoted much of his work to South America, is small and sluggish in comparison to the GULF STREAM and works in the opposite fashion, bringing colder water north from ANTARCTICA and lowering temperatures along the Pacific coast by 45 to 47 degress F (7 to 8 degrees C).

Its chilled air carries less moisture, thus helping to create the world's driest desert in Chile's Atacama Desert. The northern end of the current flows past the Galápagos Islands (off the coast of ECUADOR). Cooler water and air temperatures have affected the wildlife of the Galapagos, allowing cold-water species such as penguins and fur seals to thrive as far north as the equator.

The Humboldt Current creates one of the largest and most productive marine ecosystems in the world. Cold waters with low salinity and high levels of nutrients are brought to the surface through upwelling, providing sustenance to fish and marine mammals. The coasts of Peru and Chile are therefore one of the largest fisheries in the world, with approximately 18 to 20 percent of the world's fish catch. Dominant species include sardines, anchovies, and mackerel. Upwelling occurs off the shores of Peru year-round, sometimes as far as 620 mi (1,000 km) into the Pacific Ocean, but only during the summer off the coasts of Chile.

The current is periodically disrupted by climatic changes known as EL NIÑO, referred to by meteorologists and oceanographers as El Niño Southern Oscillation (ENSO) events. El Niño usually consists of winds blowing perpendicular to the current, coming from the west across the Pacific, bringing warmer waters with lower nutrients and unusually heavy rains. The causes of this are uncertain, potentially related to changes in circumpolar currents, which may in turn be caused by effects of global warming on the poles. ENSO events can seriously disrupt the local climate and the productivity of the region's fisheries.

Human activity, however, is also a cause itself in the disruption of the area's rich biodiversity. Overfishing threatens the marine mammals and sea birds that rely on fish for their basic diet. Such species include sea otters, sea lions, large sea birds, and whales. Pollution is also a problem in the region, caused by sewage and

industrial waste, particularly related to petroleum and copper exploitation.

Both Peru and Chile are increasing levels of development in tourism and urbanization along their lengthy coasts, but they have also joined together in regional cooperation for managing the Humboldt Current ecosystem. International groups such as UNIDO (United Nations Industrial Development Organization) and GIWA (Global International Waters Assessment) are working to develop comprehensive sustainable aquaculture management proposals.

BIBLIOGRAPHY. "Humboldt Current," na.nefsc.noaa.gov (August 2004); K. Sherman, L. Alexander, and B. Gold, eds., *Large Marine Ecosystems: Stress, Mitigation, and Sustainability* (American Association for the Advancement of Science, 1993); "Humbodt Current," www.galapagos online.com.

JONATHAN SPANGLER
SMITHSONIAN INSTITUTION

Hungary

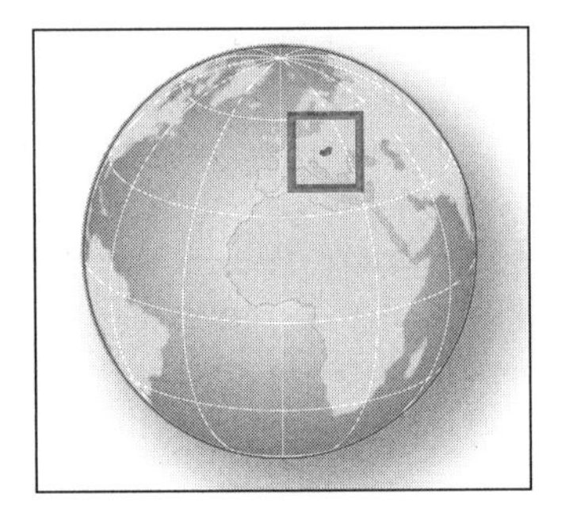

Map Page 1133 **Area** 35,919 square mi (93,030 square km) **Population** 10,045,407 **Capital** Budapest **Highest Point** 3,346 ft (1,014 m) **Lowest Point** 257 ft (78 m) **GDP per capita** $13,300 **Primary Natural Resources** bauxite, coal, natural gas.

HUNGARY IS UNIQUE among the nations of Europe in many ways: a relatively flat country among hillier neighbors, a homogenous population with a language and culture unrelated to the peoples around them, and a long history as a unified nation in the midst of a region known historically for its diversity and political fragmentation. The nomadic Magyar tribes who emigrated from the area around the Urals around the year 900 C.E. founded a state that has dominated this region of the middle Danube for over a thousand years.

Since the collapse of the Soviet Union's dominance in Eastern Europe in the 1990s, Hungary has emerged as the strongest economy once again in the region and led the pack in its reintegration with the Western world, formally joining the EUROPEAN UNION in May 2004.

Hungary is located at a strategic crossroads between Western Europe and the Balkan Peninsula, and between RUSSIA and the MEDITERRANEAN SEA. Here, the DANUBE RIVER flows through a broad plain with some of the most productive agricultural land in Europe. To the east of the river especially, the Great Hungarian Plain, or Alföld, spreads out for miles of almost completely horizontal fields and grasslands, or *puszta*. The plain is surrounded by the arc of the Carpathian Mountains, which form the borders with Hungary's neighbors to the north (SLOVAKIA, UKRAINE) and to the east (ROMANIA). To the south, the plain merges into the more hilly neighbors of the Balkan Peninsula (SERBIA AND MONTENEGRO, CROATIA, SLOVENIA), while the west rises slightly in elevation at the foothills of the ALPS on the frontier with AUSTRIA.

Through much of its history, many of these borders were in fact within the kingdom of Hungary, which once encompassed the entirety of Slovakia and Croatia, much of northern Serbia (the Vojvodina), and over a third of Romania (the historic province of Transylvania). After World War I and the Treaty of Trianon (1920), however, non-Hungarian peoples were granted self-rule, and Hungary lost three-fourths of its territory, a fact that continues to influence politics and relations with Hungary's neighbors today.

Landlocked Hungary is dominated by the river Danube (called here the Duna), with its chief tributary, the Tisza, flowing through the eastern part of the country. Other major rivers include the Dráva on the southern border, the Rába in the northwest, and the Körös, which flows in from the east. In the central western part of the country, Lake Balaton forms the largest lake in Europe, covering 233 square mi (598 square km). Balaton is well-known for its resorts and for its pike, a national dish of Hungary, served with dill and paprika.

Much of Hungary sits atop large reserves of geothermal, mineral, and curative waters, and Hungarian cities are famous for their spas, some of which have been in existence for centuries. Some of the geothermal spas are located in the mountains of the northern part of the country, where Hungary's earliest industries were centered, mining coal, iron, silver, zinc, and gold until these were mostly depleted. The mines were closed in the 1990s to the great detriment of the local economy. This part of Hungary is also known for its Eger and Tokaj wines.

Budapest, until 1873 the twin cities of Buda and Pest, is the center of the country in every way, geographically, historically, and economically. A fifth of the population lives there, and most of Hungary's in-

Hungary's Budapest is the center of the country in every way, geographically, historically, and economically.

dustry is centered in the Budapest region. Hungary's other cities are much smaller but play important roles as regional centers: Debrecen, Miskolc, Szeged, Pécs, and Györ. The eastern plains are sparsely settled, populated instead by herds of horses and cattle and covered with farms raising Hungary's chief products: wheat, corn, sunflower seed, potatoes, and sugar beets. Today, Hungary's economy is the most dynamic in the region, with more foreign investment per capita than any other country in eastern or central Europe.

Tensions remain with some of its neighbors, however, over treatment of their minorities within Hungary's borders, and privileges given to Magyars living outside current borders (especially in Slovakia and Romania). The divisive spirit of nationalism also has called for renewed discrimination against one of the largest populations of Romany in Europe, with numbers running from an official count of 142,683 in 1990, to unofficial estimates of up to 600,000.

BIBLIOGRAPHY. *World Factbook* (CIA, 2004); Wayne C. Thompson, *Nordic, Central and Southeastern Europe 2003*, The World Today Series (Stryker-Post Publications, 2003); Éva Molnár, ed., *Hungary: Essential Facts, Figures and Pictures* (MTI Media Data Bank, Budapest, 1995); "Welcome to Hungary," www.magyarorszag.hu (August 2004).

JONATHAN SPANGLER
SMITHSONIAN INSTITUTION

Huntington, Ellsworth (1876–1947)

CONTEMPORARIES WERE impressed by the clarity of Ellsworth Huntington's style, the intensity of his reasoning, the simplicity of his grand theories, and his unique ability to generalize: "With devastating logic and sound scholarship, Ellsworth Huntington shows how climate, weather, geographical location, diet, health and heredity control the character of a nation—and determine its dominant or defensive position in history and the advance of civilization." Such is how the geographer was described on the book jacket for his *Mainsprings of Civilization* (1947).

Huntington was probably the geographer who defended the cause of environmental determinism with the highest degree of conviction. His work on how climate has influenced the evolution of human societies was informed by the fieldwork he first did in Asia and the MIDDLE EAST, and later in the American southwest. He had traveled widely over North and South America, Europe, Africa, Asia and Australia. Huntington's writings reached an audience well outside the small circle of American geographers.

Geographers did not unanimously approve Huntington's bold generalizations, and some refuted early on the conclusions of Huntington's *The Pulse of Asia*. Historians were annoyed by the way in which Huntington overworked his materials to restate his thesis on the development of high-type civilizations. His racist deductions were not exempt of contradictions. For instance, past migration through the Arctic would explain Native Americans' inability to innovate, but Eskimos would be especially ingenuous because they have remained in the Arctic.

Colleagues called his work an enthusiastic toast to himself because New England was always depicted as the most advanced area on his world maps of the "dis-

tribution of general progress." Most scholars criticized Huntington's speculations on the influence of weather on culture and his measurement of civilization level. Conducted largely by botanists and geologists, later studies on climate change have proved, however, that his speculations on the succession of wet and dry cycles were correct.

Huntington was born in Galesburg, Illinois, went to Beloit College for his B.A. degree, then taught at Euphrates College (Harput, Turkey) from 1897 to 1901. He resumed his studies in the United States and received an M.A. from Harvard University in 1902. As a graduate student, he took part in two scientific expeditions in central Asia, which he described in *Explorations in Turkestan* (1905) and his famous *The Pulse of Asia* (1907). Yale awarded him a Ph.D. in 1909. After graduation, he worked at the Carnegie Foundation for three years. Huntington returned to Yale, where he taught geography until 1913. He became a research associate in 1917, a position that gave him the freedom needed to concentrate on his ambitious research program on climate and civilization. He occasionally taught seminars at Clark and Chicago universities but did not seek contacts with students. Ellsworth Huntington died a professor emeritus of geography at Yale.

Huntington was an unusually prolific geographer. His articles were published in academic as well as nonacademic journals: the *Bulletin of the American Geographical Society, American Historical Review, Harper's Magazine,* and the *Journal of Race Development.* He wrote no fewer than 29 books and co-authored many others, in which he either described the countries he visited or exposed his theories on climate change and eugenics. His bibliography includes major monographs such as *Palestine and Its Transformation* (1911), *The Climatic Factor as Illustrated in Arid America* (1914), *Civilization and Climate* (1915, re-edited twice), *World Power and Evolution* (1919), and *Climatic Changes* (1922).

BIBLIOGRAPHY. Ellsworth Huntington, *Civilization and Climate* (University Press of the Pacific, 2001 reprint); Ellsworth Huntington, "Geography and Natural Selection: A Preliminary Study of the Origin and Development of Racial Character," *Annals of the Association of American Geographers* (v14/1, March 1924); Ellsworth Huntington, *Mainsprings of Civilization* (Mentor Book, 1964 reprint).

PHILIPPE FORÊT, PH.D.
FEDERAL INSTITUTE OF TECHNOLOGY, SWITZERLAND

Hunza

THE HUNZA VALLEY in north PAKISTAN is noted for two features: Burushaski, a language that appears to be unrelated to any other language in the world, and secondly, the myth of the health and longevity of its inhabitants, the Hunzakuts. This valley was the reputed location of novelist James Hilton's Shangri-La.

The valley is located north of Gilgit, the major town in mountainous north Pakistan sandwiched in between the eastern end of the HINDU KUSH mountains, the western end of the HIMALAYA MOUNTAINS, and the Karakorum range to the north. In ancient times, it provided a very difficult footpath for Buddhist pilgrims journeying from the Chinese Pamirs in the 7th century down to the famous Buddhist monasteries in the Swat valley and neighboring Taxila.

The valley, 25 mi (40 km) long, has settlements on both banks of the Hunza river, one properly called Nager populated by "Twelver" (Shia) Muslims on the left bank with paths leading to the Hispar glacier to the east and Braldu in neighboring Baltistan, and the right bank occupied by the dispersed settlement around Karamabad, a name derived from Karim, the son of the 49th imam, the Aga Khan, of the "Sevener" Isma'ili branch of Islam.

In the upper reaches of the Hunza valley, in Gujal, Hunzakuts gained superiority over the Wakhi ethnic group that is indigenous to the high Pamir. The British conquered Hunza in 1891, bringing it under control of the Gilgit Agency, nominally under Kashmiri domain, but the British Resident in Gilgit had suzerainty over all Gilgit. The British supported a local despot, the Thum (later Mir), a hereditary chieftain who, with the aid of a *wazir* (prominent adviser), ruled his subjects from Baltit.

In 1936, a British nutritionist published a book on diet and nutrition and featured Hunzakuts as possessing longevity because of their diet. To date there is no credible evidence that determines that the Hunzakut diet of old, not to mention the current diet of the past four decades, contributes to longevity. Nevertheless, modern-day Western pilgrims converge on Hunza seeking the secrets of longevity through consuming the mythical Hunza diet that is vegetarian in origin.

In the 1930s Nazi era, the Swiss-German vegetarian, Dr. Ralph Bircher, conducted research about the Hunza diet. Adolf Hitler, the Nazi leader, and Heinrich Himmler, the Nazi SS commander, both vegetarians, had sponsored several plant gathering and physiological research expeditions to Hunza seeking "Aryan"

plants suitable for transplanting to Nazi-conquered lands. They promoted the consumption of a diet replicating the Hunza diet, later to be popularized as breakfast muesli.

The Hunzakut vegetarian diet was not by choice; in the steep landlocked mountain valley they lacked access to high-altitude pastures for ruminants. Almost all vegetation in the valley is anthropogenic and field crops were focused on grain because the local chieftain demanded grain as tax revenue. Prior to World War II, the Hunza diet was estimated to be 50 percent derived from apricots. It is true that apricots have substantial antioxidant qualities, but any reason for longevity has to be found elsewhere.

The stunted stature of some of the residents reveal extreme malnutrition when young and recollections of elderly resident tell of spring starvation when food stocks were depleted. Many elderly women tell of children dying during periods of minimal dietary intake. Alexander Leaf, a noted gerentologist who has studied the Hunzakuts, has suggested that many other factors contribute to longevity, such as vigorous physical activity, status granted long-lived persons, and modest caloric intake.

In the early 1950s, German scientists (including a well known Nazi) who made an early foray into Hunza after the jeep track was constructed, recall that the major food product that was conveyed to Hunza at great expense was British canned corned beef produced in Argentina. Goats are kept in Hunza, but their husbandry is traditionally governed by social taboos found throughout that region of northern Pakistan and AFGHANISTAN. The Neolithic goat cult bars women from goat husbandry; it is limited to male work because of the relation of the domestic goat to the wild ibex goat found in the high mountains.

BIBLIOGRAPHY. Nigel J. Allan, "Household Food Supply in Hunza Valley, Pakistan," *Geographical Review* (v. 80, 1990); Hermann Berger, *Die Burushaski-Sprache von Hunza* (Otto Harrassowitz, 1998); Ralph Bircher, *Das Volk: Das keine Krankheiten kannt* (Huber, 1942); Alberto M. Cacopardo and Augusto S. Cacopardo, *Gates of Peristan: History, Religion, and Society in the Hindu Kush* (Instituto Italiano per l'Africa e l'Oriente, 2001); Alexander Leaf, "Long-lived Population: Extreme Old Age," *American Geriatric Society Journal* (v.30); Robert McCarrison, *Nutrition and Health* (Faber and Faber, 1936).

NIGEL J.R. ALLAN
UNIVERSITY OF CALIFORNIA, DAVIS

Huron, Lake

LAKE HURON IS ONE of the five Great Lakes and is named for the Hurons, a Native American people who once lived along its shores. Lake Huron covers 23,000 square mi (59,570 square km) and is 206 mi (332 km) long. At its greatest width, it is 193 mi (295 km) across. It lies at an elevation of 577 ft (176 m) above sea level and is 750 ft (229 m) deep at its deepest point.

Lake Huron is connected to Lake SUPERIOR via the St. Mary's River. The Soo Canal, with four locks on the U.S. side and one on the Canadian side, allows ships to pass the rapids, at which important hydroelectric stations have also been built. Lake Huron is connected to Lake MICHIGAN by the Straits of Mackinac, which are so wide that some hydrologists argue the two lakes are actually two lobes of a single lake and are regarded as separate only because of a historic accident: The first French explorers in the area discovered the lakes' southern ends long before discovering the straits. Lake Huron flows into Lake ERIE via the St. Clair River, Lake St. Clair, and the Detroit River.

MACKINAC ISLAND

Lake Huron contains the largest number of islands of any lake in the world. Manitoulin Island is the largest of them and delineates the North Channel and the west side of Georgian Bay. It is also the largest island in a body of fresh water. Mackinac Island, in the Straits of Mackinac, is a tourist resort well known for its picturesque scenery and complete absence of motor vehicles. During the Gilded Age, the great captains of industry regularly stayed on Mackinac Island to escape the sweltering summer air of their Chicago townhouses, and their wives enjoyed promenading on the Grand Hotel's large veranda.

Manufacturing is the source of 80 percent of the area's income but has resulted in dangerous levels of pollution. Among the notable industrial cities on the shores of Lake Huron are Flint, Midland, Bay City and Saginaw in the U.S. state of MICHIGAN, and Sarna, Ontario in CANADA. The lake itself carries considerable traffic in raw materials and finished goods, with seven deepwater U.S. ports and four Canadian ports.

The Lake Huron area is rich in mineral resources, timber, and mining. Agriculture in the area centers primarily around dairy farming, although there is some row-crop and fruit cultivation. The shores of Lake Huron remain rich with wildlife, and the lake itself boasts a wide variety of fish. However, industrial pollution has made the consumption of lake fish medically

risky, leading to the severe curtailment of commercial fishing in the area. Sports fishermen have continued to pit their skills against the lake's fish but are warned against eating substantial amounts of their catch.

BIBLIOGRAPHY. Ann Armbruster, *Lake Huron* (Children's Press, 1996); Pierre Berton, *The Great Lakes* (Stoddart, 1996); Sara St. Antoine, ed, *Stories from Where We Live: The Great Lakes* (Milkweed Editions, 2000).

LEIGH KIMMEL
INDEPENDENT SCHOLAR

Ibn Battuta (1304–1377?)

BORN IN TANGIER, MOROCCO, Muhammad ibn Abdullah ibn Battuta was a famous Arab traveler and writer who explored in Africa, Europe, and Asia. Ibn Battuta's journey began in North Africa in 1325 with travels that included visits to EGYPT, SYRIA, Mecca on the Arabian Peninsula, northeastern IRAN, southern IRAQ, the RED SEA, YEMEN, East Africa, Asia Minor, AFGHANISTAN, INDIA, Bengal, INDONESIA, CHINA, and SPAIN; his travels ended in 1353 after a journey across the SAHARA and Western Africa. The main motive behind the Moroccan traveler, like other Muslims of his time, was to perform the ceremony of hajj or pilgrimage of the holy places of Mecca in western Arabia or in what is now SAUDI ARABIA. He left Morocco, he tells us, "swayed by an overmastering impulse within me and a desire long-cherished in my bosom to visit these illustrious sanctuaries." For most of his travels, Ibn Battuta was either returning to Mecca or journeying away from the holy city.

During his adventures, he spent months of study in DAMASCUS, Syria, and sought employment and generous rewards in government offices in the Sultanate of Delhi in India under an Islamic kingdom of Turkic origin. As a devoted Sufi inclined in the mystical and ecstatic dimension of Islam, mostly popular in North Africa in the 14th century, Ibn Battuta stayed with other Sufi devotees and frequented places where saintly masters resided, seeking to gain divine grace or baraka under their guidance for a personal communion with God. He also traveled for the sake of curiosity and adventure to some unknown places in the Islamic world, or dar al-Islam. Describing his final travels, Ibn Battuta's rare accounts of East African city-states and the Mali Empire in the 14th century have long been important to historians.

Ibn Battuta came from a family of jurists and judges, with his ethnic ancestors from rural Berbers of northern Morocco. As a legal scholar, he was educated in Islamic sacred law (Sharia) and well versed in classical Arabic writing and poetry. Based on his travels in Asia Minor, India and western Africa, he expanded the genre of travel writing recorded in the book *Rihla,* or *Book of Travels,* published in 1357. Although Muslim writers and travelers from northwestern Africa produced descriptions of the places they traveled to while making the pilgrimage, Ibn Battuta further developed the genre of travel writing by making it more of a detailed account of the traveled lands and encountered people rather than a mere extension of religious preoccupation.

The *Rihla* is mainly significant in its descriptions of the Turkish chiefdoms of Asia Minor, India, and East Africa. It offers the most comprehensive and detailed depiction of the Islamic world after the disintegration

of the Muslim empire in 1258 and the Mongol dominance of the Eurasian landmass in the 14th century.

BIBLIOGRAPHY. Abdullah Yosaf Ali, *Three Travels to India: Being a Simple Account of India as Seen by Yuan Chwang (Hiuen Tsieng), Ibn Batuta, and Bernier* (al-Biruni, 1978). Ibn Battuta, *The Travels of Ibn Battutah* (Picador, 2002); Ibn Battuta, *Travels in Asia and Africa*, H.A.R. Gibb, trans. (Routledge, 1929); Ross E. Dunn, *The Adventures of Ibn Battuta: A Muslim Traveler of the 14th Century* (University Press of California, 1989); Said Hamdun and Noël King, eds., *Ibn Battuta in Black Africa* (Markus Wiener, 1994).

Babak Rahimi
Independent Scholar

Iceland

Map Page 1130 **Area** 39,769 square mi (103,000 square km) **Population** 280,798 **Capital** Reykjavík **Highest Point** 6,950 ft (2,119 m) **Lowest Point** 0 m **GDP per capita** $30,200 **Primary Natural Resources** hydropower, geothermal heat.

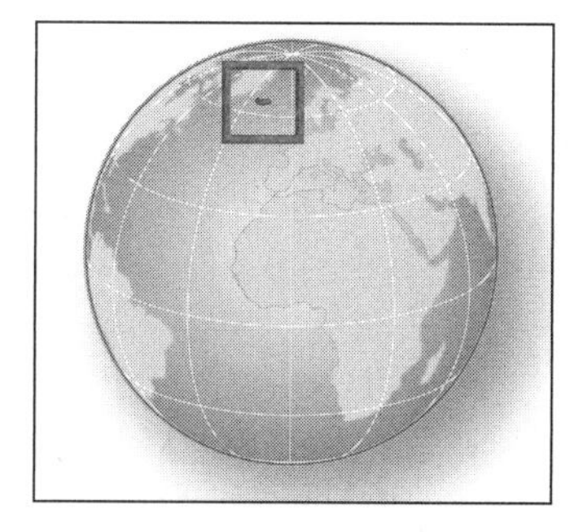

ICELAND IS A volcanic island lying on the Mid-Atlantic Rift 170 mi (280 km) southeast of GREENLAND and 500 mi (800 km) northwest of Scotland. Iceland is a mountainous country with an average elevation of 1,640 ft (500 m). Coastal lowlands (from sea level to 650 ft or 200 m elevation) ring the island and constitute one-quarter of its area. Most of the nation's population lives in the coastal lowlands, especially in the capital region. Glaciers cover 11 percent of Iceland, which includes Europe's largest, Vatnajökull, at 3,240 square mi (8,400 square km). Glacial activity and marine erosion have carved coastal fjords and numerous valleys in the landscape; deep fjords are the hallmark of northwest (West Fjords) and east (East Fjords) Iceland.

Despite its northerly latitude, Iceland has a maritime climate: cool summers and mild winters, owing to the moderating influence of the warm Irminger Current. Mean annual temperature ranges from 36 to 43 degrees F (2 to 6 degrees C) in the coastal lowlands and 37 to 39 degrees F (3 to 4 degrees C) in the mountainous interior. Reykjavík—the most northerly national capital in the world—experiences average temperatures of approximately 32 degrees F (0 degrees C) in January and 51 degrees F (10.6 degrees C) in July. Annual precipitation varies with local topography, particularly from rain shadows created by mountains and glaciers, and is typically greatest in fall and early winter. The highest annual precipitation (156 in or 400 cm) occurs in southeast Iceland on the Vatnajökull and Mýrdalsjökull glaciers; the lowest (16 in or 40 cm) occurs in the volcanic desert north of Vatnajökull's rain shadow.

Rivers are numerous in Iceland because of abundant rainfall and glacial melt. The Þjórsá (143 mi or 230 km) and Jökulsá á Fjöllum (128 mi or 206 km) are Iceland's longest rivers. Many rivers are being dammed for hydropower, which currently supplies 83 percent of Iceland's electricity. Waterfalls are common in Iceland's young volcanic landscapes, including Dettifoss (144 ft or 44 m), Europe's most powerful waterfall. Lakes are also numerous; some of the largest, Þingvallavatn (32 square mi or 84 square km) and Þóisvatn (27 square mi or 70 square km), are of tectonic origin, while others are created by lava or ice dams or volcanic explosions. Volcanic activity has created an abundance of geothermal areas marked by hot springs, mudpots, and geysers. Nonpolluting geothermal energy supplies 89 percent of Iceland's heating needs.

Iceland's isolated locale has greatly influenced its biological diversity; flora is largely northern European in origin and includes 485 species of vascular plants, 560 species of bryophytes, and 550 species of lichens. Seventy-two species of birds are known to nest in Iceland; waterfowl and seabirds are prominent. The arctic fox (*Alopex lagopus*) is Iceland's only native land mammal.

A striking aspect of the Icelandic landscape is the lack of tall trees. Prior to Norse settlement in 874 C.E., dwarf birch woodland covered about 25 percent of Iceland, chiefly in the lowlands. Cutting of trees for fuel and housing and heavy grazing by sheep diminished the woodland area and limited regeneration. Today birch woodland covers only 1 percent of Iceland. Sixty-three percent of Iceland is poorly or nonvegetated and erosion of exposed volcanic soils is a national problem.

BIBLIOGRAPHY. Graeme Cornwallis and Deanna Swaney, Iceland, *Greenland and the Faroe Islands* (Lonely Planet, 2001); Gunnar Karlsson, *The History of Iceland* (University of Minnesota Press, 2000); Terry Lacy, *Ring of Seasons: Iceland—Its Culture and History* (University of Michigan Press, 2000); Ministry for the Environment and Icelandic Institute of Natural History, *Biological Diversity in Iceland:*

National Report to the Convention on Biological Diversity (Reykjavík 2001); Johannes Nordal and Valdimar Kristinsson, eds., *Iceland: The Republic* (Central Bank of Iceland, Reykjavík); D. Roberts, *Iceland: Land of the Sagas* (Villard Books, 1990); K. Scherman, *Daughter of Fire: A Portrait of Iceland* (Little, Brown and Company, 1976).

CHARLES E. WILLIAMS
CLARION UNIVERSITY OF PENNSYLVANIA

Idaho

IDAHO'S VAST areas of unspoiled beauty and natural wilderness give every reason for the 1,293,953 residents (2000) to know their home as the "Gem" state. Idaho is bounded by CANADA in the north, MONTANA and WYOMING in the east, UTAH and NEVADA in the south, and WASHINGTON and OREGON in the west. The capital and largest city is Boise, but Pocatello and Idaho Falls are also important commercial/agricultural centers.

Much of the state's 83,557 square mi (216,413 square km) can be divided among three land forms (ROCKY MOUNTAINS, Columbia Plateau, and Basin and Range) sharing some of the most rugged, unspoiled beauty anywhere in the united states.

In the Rocky Mountain region, the Bitterroot Range dominates the Panhandle region of the north and along the Montana border, where just 45 mi (72 km) separates Idaho from its two neighboring states of Washington and Montana. The mountains then run south and a little east along the border with Wyoming. The names of some other important mountain ranges (Sawtooth, Lost River, Bighorn Crags, Coeur d'Alene, and Seven Devils) bring forth strong images of the natural beauty and ruggedness that sets this landscape apart from that of other states. The state is also home to the deepest gorge in North America. From the summit of the Seven Devils the Hells Canyon gorge drops nearly 7,900 ft (2408 m) to the waters of the Snake River below; deeper than the GRAND CANYON.

The Columbia Plateau region extends east out of Washington into Idaho at the base of the Idaho Panhandle then follows the path taken by the Snake River across southern Idaho. The Snake River Plain is itself a unique geography, built up from ancient lava flows that cover strips of land 20 to 40 mi (32 to 64 km) wide on each side of the Snake River. Shoshone Falls on the Snake River, drops 212 feet (64 m), making it higher than Niagara Falls. Between the Rockies and the Columbia Plateau lies the Basin and Range region, with numerous grassy plateaus, rolling hills, deep valleys, and mountains as you move toward the east.

The state's waters are divided among major rivers including the Coeur d'Alene, Snake, St. Joe, St. Maries, and Kootenai, and Pend Oreille Lake, American Falls Reservoir, Bear Lake, and Coeur d'Alene Lake.

Idaho's climate is diverse but strongly influenced by Pacific weather patterns that serve to moderate weather extremes. Generally, the northern part of the state is cooler than the southern parts with greater precipitation. Typically, summers are hot and dry, particularly in the south, with cold, snowy winters in the north.

Early economic activity in the region was tied closely to the early explorers and fur traders that moved eastward out of the Columbia River region of the Oregon country into what is today Idaho. By the 1840s, however, fur supplies from the region had been severely depleted and it wasn't until the discovery of gold in 1860 that economic activity picked up again. The typical rush of settlers followed, all hoping to strike it rich either from finding gold themselves or from taking it from the miners in one of the many towns that seemed to spring up overnight.

A second rush occurred in 1882 when gold was discovered at Coeur d'Alene. But the boom was short-lived because the results of the strike were disappointing. Fortunately, a major discovery of silver at the same time was able to sustain the mining boom. The late 19th and early 20th centuries witnessed a rise in cattle and sheep ranching as agriculture moved to the front of the state's economic activity. Recently, manufacturing has replaced agriculture as the most important industrial sector. Cattle and dairy goods are among the leading agricultural products in addition to potatoes (for which the state is famous), hay, wheat, peas, beans, and sugar beets. The state's unspoiled natural beauty has also led to a strong year-round tourist industry, with Sun Valley serving as one of the state's leading resorts.

BIBLIOGRAPHY. John Gottberg, *Idaho* (Compass American Guides, 1996); Paul Evan Lehman, *Idaho* (Curley, 1993); Betty Derig, *Roadside History of Idaho* (Mountain Press, 1996); F. Ross Peterson, *Idaho: A History* (Norton, 1976); U.S. Census Bureau, www.census.gov (August 2004).

RICHARD W. DAWSON
CHINA AGRICULTURAL UNIVERSITY

Illinois

ILLINOIS IS A midwestern state located in the north-central part of the UNITED STATES. Lake MICHIGAN and INDIANA border the state in the east, while WISCONSIN borders the state in the north. KENTUCKY lies to the southeast, across the Ohio River, while the MISSISSIPPI RIVER separates the state from MISSOURI and IOWA in the west. The state's 57,918 square mi (150,007 square km) ranks Illinois 25th in size in the United States. However, with 12,419,293 residents, the state ranks fifth in terms of total population and 11th overall in population density. Interestingly, the top-ten largest cities in the state (led by Chicago at 2,806,016) account for about 32 percent of the total population. In terms of per capita income, Illinois ranks 10th at $32,875. The state capital is in Springfield in the west-central part of the state.

The state's lands were fashioned by late Cenozoic glaciation when rugged hills in the north were leveled and the valleys filled in. Elevations across the state range from 279 ft (85 m) above sea level near the Mississippi River to 1,235 ft (376 m) above sea level at Charles Mound in the northwestern corner near Dubuque, Iowa. The result is a rich and interesting landscape that can be grouped into three geographic regions. The Central Plains cover most of the state running from Lake Michigan west and south. The Shawnee Hills is a narrow strip of land that rises as you move south from the Central Plains toward the Ohio River. The region is relatively small, however, ranging from 5 to 40 mi (8 to 64 km) wide and about 70 mi (112 km) long, with numerous rivers, valleys, and woodlands. The Gulf Coastal Region is another small area just to the south of the Shawnee Hills and bounded by the Mississippi River to the west and the Ohio to the east. The area is an extension of the Gulf Coastal Plain that extends up from the Gulf of Mexico. The area is generally hilly, although it flattens out somewhat toward the border with Kentucky.

The prairies of Illinois are tornado country: Above, an F4 strikes Woodford County in central Illinois in 2004.

Enriched by numerous rivers, most of which flow to the Mississippi-Ohio system, the broad, level prairies gave rise to the nickname of Prairie State as the early settlers arrived. These rivers also provided early explorers with convenient access to the interior of the continent and played an important role in the settlement of the prairies further west.

With the completion of the St. Lawrence Seaway, Illinois had access to the Atlantic seaboard via the Great Lakes for oceangoing ships. Illinois has been an important transportation hub since the early 1800s. Today, in addition to important water transportation corridors, the transportation complex also includes railroads, airlines (Chicago's O'Hare airport is one of the busiest in the world), and an extensive highway system.

ILLINOIS HISTORY

The earliest Europeans to visit the area were French explorers and missionaries during the early 1600s who ventured in to lands long occupied by Sac, Fox, and other Native Americans. By 1680, Robert de La Salle had established Fort Crèvecoeur, and although occupation of the area was sparse, the region was highly valued for its fur trade. In 1787, Illinois became part of the Northwest Territories following the signing of the Treaty of Paris at the end of the American Revolution. Illinois became a separate territory in 1809 and a state in 1818.

Industrial development came with the founding of an agricultural implements manufacturing company by

Cyrus McCormick in Chicago in 1847 and the building of railroads during the 1850s. The 1850s also saw the rise of a young lawyer named Abraham Lincoln whose arguments against slavery brought national attention and ultimately the American Presidency in 1861. The rapid rise of industry, large inflows of immigrants from Europe, and key transportation connections provided the state with a strong economic base for success.

Rich lands with adequate rainfall and a long growing season make Illinois an important agricultural state. It consistently ranks among the top states in the production of corn and soybeans. Hogs and cattle are also principal sources of farm income. Other major crops include hay, wheat, and sorghum. Beneath the fertile topsoil lies mineral wealth, including fluorspar, bituminous coal, and oil. Illinois ranks high among the states in the production of coal, and its reserves are greater than any other state east of the ROCKY MOUNTAINS. Its agricultural and mineral resources, along with its excellent lines of communication and transportation, made Illinois industrial; by 1880 income from industry was almost double that from agriculture.

Leading Illinois manufactures include electrical and non-electrical machinery, food products, fabricated and primary metal products, and chemicals; printed and published materials are also important. Metropolitan Chicago, the country's leading rail center, is also a major industrial, commercial, and financial center. The Chicago suburbs have also become important business centers. There are cities scattered across the northern half of the state with specialized industries—Elgin, Peoria, Rock Island, Moline, and Rockford, while the industrially important cities in central Illinois include Springfield and Decatur.

BIBLIOGRAPHY. Ronald E. Nelson, ed., *Illinois: A Geographical Survey* (Kendall/Hunt, 1996); A. Doyne Horsley and Ruquiyah Islam, *Illinois: A Geography* (Westview Press, 1996); Kathy P. Anderson, *Illinois* (Lerner Publications, 2002). Marlene Targ Brill, *Illinois* (Benchmark Books, 1997); Adade Mitchell Wheeler and Marlene Stein Wortman, *The Roads They Made: Women in Illinois History* (Charles H. Kerr, 1977); Robert P. Howard, *Illinois: A History of the Prairie State* (William B. Eerdmans, 1972); Pygman Kilduff, *Illinois: History Government Geography* (Follett, 1962); U.S. Census Bureau, www.census.gov (August 2004).

RICHARD W. DAWSON
CHINA AGRICULTURAL UNIVERSITY

immigration

ALIENS OR NONCITIZENS who reside or seek to reside temporarily or permanently within the borders of a country are generally termed as immigrants. The term *immigrant* refers to someone who enters a country, while the word *emigrant* refers to someone who leaves a country. In the early times, the tendency was to look upon the alien as an enemy and to treat him or her as a criminal and outlaw. Aristotle, probably reflecting the common view in the ancient world, saw non-Greeks as barbarous people who were slaves "by nature." The *jus gentium* of the Roman law applied to both citizens and foreigners and tended to favor the idea that aliens had rights; humanity toward aliens was also fostered, in theory at least, by the Christian idea of the unity of all persons in the church. The legal and ideological expression of humanity toward the alien, however, is a relatively modern development.

As sovereign national states began to take shape, the founders of international law asserted that natural rights were vested in all persons, without regard of citizenship or alienage, rights of which they ought not to be deprived by civilized societies or their governments. There was no general agreement on the content or scope of these natural rights as they affected aliens, but the existence of some minimum standard of civilized treatment was asserted. The minimum standard, it was conceded, did not include the right of aliens to own property or to engage in gainful professions. To meet this situation , states entered into treaties which provided that each of the contracting states would treat the nationals of the other state on an equal footing with its own nationals in the admission into trades and professions, ownership or possession of property, access to courts, enjoyment of liberty of conscience, and freedom of worship. Some treaties do not claim to extend to aliens, however, rights that are by municipal law reserved exclusively to the nationals of the country; thus municipal law, rather than conventional international law, is actually controlling. In particular the desire of nations to protect citizens in their jobs, professions, and businesses against both unemployment and competition is a very strong force restricting the latitude of aliens.

With the discovery of new and unsettled continents, steps were taken by European countries to colonize and populate these lands. The UNITED STATES, CANADA, AUSTRALIA, NEW ZEALAND, ARGENTINA, and BRAZIL have been the principal immigrant-receiving countries. The common economic needs of nations, on

the other hand, have had some liberalizing effects on the immigration process and the treatment offered to immigrants. The treaties constituting the EUROPEAN UNION, for instance, provide that citizens of member states should be free to reside in any signatory country. The United States has a long history of immigration, from the first Spanish and English settlers to arrive on the shores of the country to the waves of immigration from Europe in the 19th century to immigration in the present day. The history of immigration to the United States of America is, in some senses, the history of the United States itself, and the journey from beyond the sea is an essential element of the American myth. From early in the 19th century to 1930, at least 60 percent of the total world immigration was to the United States. Of a total immigration to the United States of 41 million persons admitted from 1820 to 1960, 34 million were from European origin.

The population of the colonies that later became the United States grew from zero Europeans in the mid-1500s to 3.2 million Europeans and 700,000 African slaves in 1790. At that time, it is estimated that three-quarters of the population were of British descent, with Germans forming the second-largest free ethnic group and making up some 7 percent of the population. Between 1629 and 1640, some 20,000 Puritans emigrated from England, most settling in the New England area of North America. From 1609 to 1664, some 8,000 Dutch settlers peopled the New Netherlands, which became New York and New Jersey.

Between 1645 and 1670, some 45,000 Royalists and/or indentured servants left England to work in the Middle Colonies and VIRGINIA. From about 1675 to 1715, the Quakers made their move, leaving the Midlands and North England behind for PENNSYLVANIA, NEW JERSEY, and DELAWARE. The Quaker movement became one of the largest religious presences in early colonial America. Germans migrated early into several colonies but mostly to Pennsylvania, where they made up a third of the population by the time of the Revolution. Between about 1710 and 1775, about 250,000 Scotch-Irish, mostly Presbyterian Protestants of Scottish descent from Northern Ireland, immigrated to and generally settled in western Pennsylvania and in Appalachia and the western frontier, which later would become KENTUCKY and TENNESSEE.

LAWS REGULATING IMMIGRATION

Prior to World War I, the laws of the United States permitted immigration without numerical limitation and were concerned chiefly with barring undesirables. The initial limitations upon immigration prohibited the importation of oriental slave labor, prostitutes, and alien convicts, pursuant to laws enacted in 1862 and 1875. The mentally ill, epileptics, physical defectives, tubercular persons, anarchists, beggars, those likely to become public charges, Chinese laborers, contract laborers, those suffering from loathsome or dangerous diseases, polygamists, paupers, persons whose passages were paid by others, and aliens convicted of crimes involving moral turpitude were added by successive enactments in 1882, 1855, 1891, 1903, and 1907. In 1907 an agreement was concluded with JAPAN limiting the entry of labor from that country. In the same year an immigration commission was appointed; its report in 1911 led to the Immigration Act of 1917, which remained one of the basic immigration statutes until its repeal by the Immigration and Nationality Act of 1952 (the McCarran-Walter Act). To the class of aliens previously inadmissible were added Hindus and other Asians, illiterates, persons of constitutional inferiority, those seeking entry for an immoral purpose, alcoholics, stowaways, and vagrants. The 1917 act was proposed as a restrictive measure to stem the tide of free immigration of the past.

A temporary quota law, restricting the number of admissible aliens, was enacted in 1921. This was followed by first permanent quota law in 1924. Under the 1924 immigration act and presidential proclamation issued thereunder, quotas were distributed on the basis of national origin of the population of the United States in 1920. Aliens seeking entry into the Unites States were divided into three categories: those racially ineligible for citizenship (Asians) were barred from permanent admission; those who were born in the Western Hemisphere countries came in without any quota limitation; all others were subject to a numerical limit assigned on the basis of their country of birth. Thus, immigration from northern and western Europe was encouraged and immigration from the southern and eastern parts discouraged. The 1924 act also initiated the visa requirement; that is, the procuring of a permit from a U.S. consular officer abroad as a condition to immigration to the United States. In 1940 Congress promulgated a law requiring the finger printing and registration of aliens.

In 1952, the U.S. Congress enacted the McCarran-Walter Act, known as the Immigration and Nationality Act of 1952, which retained the nation-origin quota system, though it eliminated race as a complete bar to immigration. In addition to these permanent immigration laws, the United States has enacted temporary

statutes authorizing the admission of refugees. Immigrants who are admitted legally to the United States may be certified and granted "green cards" that entitle them to rights that include employment. But they are still subject to limitations under local laws. The immigrant in the United States is afforded a large measure of economic opportunity; he may invoke the writ of habeas corpus; in criminal proceedings he is entitled to the guarantees of the Bill of Rights; and his property can not be taken without just compensation. But to remain in the country "is not his right" but is a matter of "permission and tolerance." As long as an immigrant is in the United States, the Constitution is his protection; but Congress, not the Constitution, decides whether or not he is to remain.

The laws of the various nations of the world regulating the admission and exclusion of aliens seeking permanent residence differ considerably. Marked contrast also appears in the manner in which immigration laws are administered. In the United States, prior to 1940, the Department of Labor administered and enforced the exclusion and deportation laws, whereas in Australia, the minister of trade and customs is in charge of the issuance of landing permits and the admission of aliens. In New Zealand the immigration laws are administered by the customs department. In Brazil the minister of foreign relations supervises immigration.

Most countries impose no numerical restrictions or quotas on the entry of aliens but enacted preferential systems to maintain designated racial characteristics of the population of the countries employing such policies. The United States, through its quota system, favors immigration from countries of North and South America and northern and western Europe. Australia's preference is for the British stock; Canada leans to persons from the UNITED KINGDOM and FRANCE; Argentina prefers the nationalities of its early settlers, including the Spanish, Italian, Portuguese, German, and Swiss. Brazil gives preferential treatment to Portuguese and to a lesser degree to Italians and Spanish. A number of countries have established educational qualifications for entry; however, England, Brazil, Argentina, and many other countries have no literacy limitations.

Many countries have set forth financial qualifications in their immigration laws. Under the laws of the United States an alien must establish to the satisfaction of U.S. consul abroad and to immigration authorities upon arrival that he will not become a public charge. This is done by proof of his financial resources or by affidavits demonstrating that he will have employment. England, Australia, and Canada have similar provisions excluding those who are likely to become a public charge. New Zealand authorities bar even British subjects by reason of economic conditions. Brazil can exclude immigrants who do not have sufficient funds. Furthermore, physical and in some cases mental conditions are grounds of inadmissibility under the laws of several nations. Also, persons with criminal records and immoral aliens, such as prostitutes and procurers, may be denied entrance to many countries. Very few countries require aliens to serve in their military forces. In the United States, every male alien admitted for permanent residence who is between the age of 18 and a half and 26 is subject to military service. Visitors of this age group who are in the country for more than a year are likewise required to serve but may be exempt from a claim of alienage, which forever debars them from citizenship and permanent residence.

Aliens admitted to the United States for permanent residence are authorized under the immigration laws to engage in any occupation. However, some states laws require U.S. citizenship as a prerequisite to practicing certain professions such as law, medicine, dentistry or engineering. Generally, U.S. citizenship is required for many positions in federal and state governments. Visitors are not permitted to engage in employment in the United States without the permission of the U.S. Citizenship and Immigration Services. Aliens may own real estate and other property and are permitted access to the courts. They do not have the right to vote. In Canada, domiciled aliens are accorded all the civil rights of citizens except the right to vote or hold public office. In New Zealand aliens may acquire both real and personal property and may even vote in municipal elections. Foreigners generally enjoy the same civil rights as citizens in Argentina but are restricted in the practice of certain professions.

ILLEGAL IMMIGRATION

The world has come a long way from the mercantilist days when nations competed for immigrants. Throughout the 20th century, tight restrictions have been placed on entry. Population pressures, intensified nationalism, and increased ease of movement have all contributed to fears of uncontrolled immigration. One consequence of laws restricting the number and ethnicity of persons entering the United States is a phenomenon referred to as illegal immigration, in which persons enter a country and obtain work without legal sanction. In some cases, this is accomplished by entering the country legally with a visa, and then simply choosing not to

leave upon expiration of the visa. In other cases, most notoriously Mexicans in the United States without legal sanction, people enter the country surreptitiously without ever obtaining a visa. Often, people entering in this fashion are economic refugees, a class of refugee not recognized by the U.S. Citizenship and Immigration Services; these persons have left their home country in a desperate bid to provide financial support for themselves and/or their families. This is particularly true in cases where "minimum wage" in the United States is several times what the average laborer earns in a given country; such immigrants often send a substantial part of their income to their countries and families of origin.

IMMIGRATION IDEOLOGY

Much of the controversy today with immigration to the United States involves anti-illegal immigration ideologies. Critics of these ideologies say that those who call for an end to "illegal immigration" really advocate an end to all immigration but do not realize it. This occurs for two reasons: 1) all the problems associated with illegal immigration (race to the bottom in wages, etc.) also apply almost equally to legal immigrants; and 2) anti-immigrant ideologues allegedly misunderstand the immigration process and do not realize that many immigrant workers—who they see as replacing American citizens in jobs they can do—have immigrated completely legally, albeit without citizenship (this number exceeds the number of illegal immigrants on a per-country basis).

At the dawn of the 21st century, the controversy revived when many high-tech and software-engineering workers started to arrive from INDIA on "H1" visas. Critics claimed that these people worked for less money and displaced American citizens. The companies who imported the workers usually argued that the United States lacked enough American citizens to do the work.

A few economists argued that, whatever the truth of that assertion, importing the workers provided more benefits to the United States, and otherwise the recruiting companies would simply offshore the entire operation to India itself. This would likely prove worse for the U.S. economy as a whole, because in the first scenario Indian workers living in the United States would at least spend money in the United States, while the supranational corporations that would purportedly export the jobs to India would probably not pass down as much of the savings to the U.S. consumer who purchased for them.

Thc industrialized nations of the world have adopted policies to defend themselves against influxes from the struggling reaches of the THIRD WORLD. The Federation of American Immigration Reform, which favors reduced immigration, estimates that just to keep pace with the population, the nations of Latin America must create more new jobs each year than the United States has ever succeeded in doing in a single year. Similarly, the International Labor Organization projected that the workforce of the third world countries grew by 600 to 700 million between 1980 and 2000, more than the current total jobs in all industrialized countries combined. Many people in the developed world fear that they are simply losing control. There are serious objections raised on economic grounds. Most visible is the taking of jobs by newcomers, but the prominent question, of course, is the balance between jobs taken and jobs created.

RULES AND REGULATION

The whole debate about immigration led to tighter entry rules and regulations. Thus, when legal immigration from the Western Hemisphere was for the first time subjected to strict limits in 1968, illegal immigration predictably soared. The U.S.-Mexico border is notoriously porous, and there is a suspicion on both sides of the immigration debate that certain economic interests want in that way. Clearly more could be done to curb illegal immigration. Deportation laws contain wide variations. It has been estimated that there are 700 different grounds for deporting or expelling aliens from the United States. Aliens who enter illegally, overstay their visits, become public charges, commit crimes involving moral turpitude, engage in immoral conduct, or are considered subversive may be deported. Except for those who become public charges, an alien may be deported no matter how long he has been a resident of the United States. He secures a hearing and may appeal to the Board of Immigration Appeals in the Department of Justice and to the courts.

In Canada, aliens who enter illegally or overstay a visit, become connected with prostitution; are convicted of criminal offenses; become public charges or inmates of an insane asylum or of a public charitable institution; or are considered subversive are subject to deportation. A hearing is held with an appeal to the minister of mines and resources. Aliens who enter legally as landed immigrants and remain in Canada for five years may not be deported except for subversive grounds. In England, when an alien is tried for a crime for which imprisonment may be imposed, the court

can impose deportation in addition to or in lieu of the criminal sentence. In Australia, an alien may not be deported after he has maintained a residence of three years. In Brazil, aliens who are married to citizens and who are responsible for the support of citizen children may not be deported.

BIBLIOGRAPHY. Maurice R. Davie, *World Immigration* (Taylor & Francis, 1936); Jack Wasserman, *Immigration Law and Practice* (1961); Leon F. Bouvier, *The Impact of Immigration on U.S. Population Size* (Population Reference Bureau, 1981); Mary M. Kritz, ed., *U.S. Immigration and Refugee Policy: Global and Domestic Issues* (Lexington Books, 1993); Simon Kuznets and Ernest Rubin, *Immigration and the Foreign Born* (National Bureau of Economic Research, Occasional Paper No. 46, 1954); W.L. Westermann, "Between Slavery and Freedom," *American Historical Review* (v.50, January 1945); Aga Khan, *Study on Human Rights and Massive Exoduses, United Nations, Economic and Social Council*, Commission on Human Rights (E/CN.4/1503, December 31, 1981).

JITENDRA UTTAM
JAWAHARLAL NEHRU UNIVERSITY, INDIA

imperialism

FEW TOPICS ARE MORE emotion-provoking than imperialism, which many consider to be a product of the 20th century and of Europeans. Nothing is further from the truth. By definition, imperialism is the act of imposing one's will (personal or national) on another culture or state to create an empire, or an empire is the consequence of the political act of imperialism. *Imperare* is Latin and means "to rule or to command."

The geographic aspect of imperialism was for one country (sometimes even a mercantile city) to rule over widely spaced geographic areas. The person who ruled such areas was referred to as emperor or empress. Unfortunately, in recent times it has been common to use the term as a political epitaph, and generally incorrectly. However, any time one country seeks to impose its political authority on another, it is, by definition, an act of imperialism.

The most recent examples of imperialism include the U.S. application of massive military and technology to impose its values on countries around the world, most demonstrably in IRAQ and LIBYA. Others include al Qaeda and even the Wahabbis of Saudi Arabia, who seek to control the values and behavior of Muslims around the world.

History is filled with empires and imperialism. Those created by the Chinese, the Persians, the Romans, the British, the Russians, various Africans, and even the Incas are only a few of the better known examples. In virtually all cases, these "empires" and their "imperialism" affected areas and peoples contiguous to the borders of the ruling country or monarch. But of interesting contrast is that there was no Greek Empire. Greeks had empires created by city-states such as Athens or Corinth, but they were dispersed around the Mediterranean, similar to other commercial maritime empires such as those of the Dutch, Portuguese, Spanish, and British.

It is common to refer to the 15th century's Age of Discovery as the Age of Imperialism. The two processes did go together for Europeans. Certainly the major distinction between this age of imperialism and imperialism of earlier times was its geographic extent. At no other time in history did the process of imperialism reach around the globe and impose political authority on such disparate peoples and cultures, most wholly unrelated to the political center. Thus, we found the small country of Holland creating the Dutch Empire, reaching from the Caribbean to the islands of INDONESIA. It was truly said that the sun never set on the British Empire.

Imperialism is not a European invention or practice, but rather it is a political practice, normally generated by economics found at all time periods and in all geographic areas. From the Scythians to the Mongols to the Chinese, all had empires in their time.

IMPERIAL ECONOMICS

What generates imperialism? Many like to think solely in terms of political power, even conspiracy theories. In fact, history demonstrates that the primary motive has been and continues to be economics. In this context, one can use a modified version of the geographic model that explains trade to also explain empires. That is, there must be complementarity (something one area has that the other wants: raw materials, ports, etc.). This should be unique and not available elsewhere close (no intervening opportunity).

The modern version, with strong historic precedents, are multinational companies that seek dominance (control) of markets or raw material (oil is the most prominent today, but McDonald's, Coca-Cola, and Microsoft are relevant as well). Thus, political imperialism becomes cultural imperialism.

The primary factor in the demise of empires was their geographic extent and their demographic and cultural diversity. There never has been any administrative or social system that could hold the inherent diversity of an empire together. Only when the empire becomes a nation, such as CHINA or the UNITED STATES, and remains geographically contiguous does it seem to last.

The voluntary version of empire, a federal state, in which member parts voluntarily join together for mutual benefit, also reveals strains that so far have prevented its long-term survival. Current examples would include the EUROPEAN UNION, the Russian Federation, and the United States. All have central governments for purposes of national defense and economics, and all face strains of the parts chafing against lost local independence.

No empire can exist without a significant and efficient transportation and communication system. Every early empire had a system of "royal" roads: EGYPT, Mesopotamia, Persia, China, JAPAN, the Mongols, Rome, the Incas, the Ottomans, and Islam. Maritime empires from the Phoenicians and Greeks to the Venetians, Genoese, Spanish, Portuguese, Dutch, and British all fought for dominance and control of key shipping routes, CHOKE POINTs, and harbors. In many instances, use of these "highways" was limited to official couriers.

In addition, every empire required an involved system of administrative centers and officials. Most often, these were existing cities and often local officials were co-opted to serve the new rulers. In addition, empires required records and legal systems as well as warehouses and systems for collecting taxes and sending goods and money to the imperial capital. This in turn created bankers and merchants.

BIBLIOGRAPHY. David B. Abernethy, *The Dynamics of Global Dominance: European Overseas Empires* (Yale University Press, 2002): Susan E. Alcock, ed., *Empires: Perspectives from Archaeology and History* (Cambridge University Press, 2001); Nial Ferguson, *Empire: The Rise and Demise of the British World Order and Its Lessons for Global Power* (Basic Books, 2003); Edward Gibbon, *Decline and Fall of the Roman Empire* (Random House, 2003); Adam Hochschild, *King Leopold's Ghost: A Story of Greed, Terror, and Heroism in Colonial Africa* (Houghton Mifflin, 1999); H. Kamen, *Empire: How Spain Became a World Power* (HarperCollins, 2003).

R.W. McColl, Ph.D.
General Editor

India

Map Page 1123 **Area** 1,284,215 square mi (3,287,590 square km) **Population** 1,049,700,118 **Capital** New Delhi **Highest Point** 28,201 ft (8,598 m) **Lowest Point** 0 m **GDP per capita** $462 **Primary Natural Resources** coal, iron ore, manganese, mica, bauxite.

INDIA IS LOCATED in South Asia bordering the INDIAN OCEAN, ARABIAN SEA, and the BAY OF BENGAL in the south. It has a 4,375-mi- (7,000-km-) long coastline. India, with a rich and long history, is one of the oldest civilizations of the world. The Indus Valley civilization, which flourished in the Indus Valley crescent, is almost 5,000 years old. This agricultural surplus-based civilization had urban and advanced culture. As India was very rich in natural resources, it was continuously invaded by foreign powers. Around 1500 B.C.E., the country was invaded by Aryan tribes from the northwest.

They brought and proliferated their religion, which was later known as Hinduism. The Aryan tribes also introduced the hierarchy of society based on four-caste system: Brahmins, Kshatriyas, Vaishyas and Sudras. The concept of the caste system still persists among the Hindus. Interestingly, as most of the Muslims and Christians in India are converted from Hinduism, many of them have retained their caste distinction. In 711 C.E., India was again invaded by Arabs, followed by Turks in the 12th century. During their long reign, a new feature in the form of Islam religion was introduced in the society.

By 1757, Great Britain had virtually assumed political control over a large part of India. The Indian society yet again was fused with many different elements of the Western world. India finally got its independence from Britain in 1947. However, the Indian subcontinent was divided into two different countries—India and PAKISTAN. India is full of imprints transmitted by diverse racial, religious, and cultural groups, which shaped the Indian culture and society as it exists in the present era.

India is slightly more than one-third the size of UNITED STATES and comprises 28 states and seven Union Territories with diverse physical characteristics. The entire country can be divided into four principal physiographic regions. They are:

1) The Deccan Plateau in the south and center, which is a part of the ancient Gondwanaland. It consists of numerous mountain ranges, mesalike Deccan lava country, and scarplands, and rift valleys.

2) The lofty HIMALAYAS in the north, which were formed in the recent geological era (Tertiary) and support 20 of the highest peaks of the world. The Himalayas rose from the floor of the sea, called Tethys, as a result of pressure from the Indian Plate moving northward and colliding with the Asian Plate. The Himalayas are 1,491 mi (2,400 km) long and are a sequence of parallel or converging ranges intersected by gigantic valleys and widespread plateaus.

3) The great Indo-Gangetic plain, in the north-central part, is a gently sloping land intercepted by the landforms imprinted by rivers. Scholars confirm that the great crescent of alluvium from the delta of the INDUS RIVER to that of the GANGES RIVER represents the infilling of foredeep warped down between Gondwanaland and the Himalayas. The depth of the alluvium at places surpasses 6,000 ft (1,829 m). The plains provide the most fertile land for agricultural use.

4) Coastal plains border the coasts of the plateau in the west and east. These physiographic divisions not only give rise to diverse landforms, but also fabricate assorted human responses to the use of land and resources.

India is drained by many mighty river systems. Rivers, particularly the Ganges (Ganga), are considered sacrosanct in India, and several religious towns have developed at the bank of these rivers. Varanasi, by the side of the Ganges, is considered the most sacred of the Hindu pilgrimage places. The rivers of India originate either in the Himalayas or in the Deccan Plateau. The Himalayan river system of the Ganges and Brahmaputra is younger than its plateau counterpart. The rivers are engaged in swift and extensive downcutting, making a steep V-shaped valley in the mountainous stretch of their courses.

The plateau river systems, Mahanadi, Godavary, Krishna, and Cauvery, are commonly characterized by an older or mature stage, with extensive valleys flowing down the moderate slopes. Narmada and Tapti are the west-flowing plateau rivers, curving through the structural faults. Rivers are incredibly vital for the economic and agricultural development of India. Many of the perennial rivers are utilized for navigation and irrigation. Only some of the rivers are used for hydroelectric power generation.

The climate of India ranges from tropical monsoon in the south and the central part to temperate in the extreme north in the Himalayas. Wet summers and dry winters characterize monsoons. A monsoon climate is actually caused by the differential heating and cooling of land and water in the summer and the seasonal reversal of winds. Indian agriculture is highly dependent on the monsoon rains. As the quantity, timing, and extent of the monsoon rains are extremely erratic, the farmers are fairly undecided about their future. The unpredictable and irregular behavior of monsoons may cause droughts or destructive and extensive flooding, leading to failure of crops. It also results in human and animal deaths. It is from such uncertainty that Indians have developed a belief in fate. Six major climatic regions of India are tropical rainy, humid subtropical, tropical savannas, STEPPE, mountain, and DESERT.

In the tropical rainy region in northeast India and the west coast, the average annual temperature varies from 77 degrees F (25 degrees C) to 80 degrees F (27 degrees C), and rainfall ranges from 78 in (200 cm) to 156 in (400 cm). The region enjoys high rainfall reliability and thus it is less dependent on irrigation. The humid subtropical and tropical regions, covering most of the plateau and Indo-gangetic plain, receive annual precipitation between 39 in (100 cm) and 78 in (200 cm), and average temperatures range from 68 degrees F (20 degrees C) to 77 degrees F (25 degrees C). The region is moderately irrigation dependent. In the tropical savanna region, the annual rainfall varies from 24 in (60 cm) to 32 in (80 cm). Consequently, the region is highly dependent on irrigation for successful agriculture. The mountain region in the north is a narrow strip along the Himalayas and is very cold during the winter and mild in summer. The desert region in the western part of the country has scanty rainfall. No cultivation is possible without irrigation.

India, with a population of some 1.05 billion (2003), is rapidly increasing its number of people. Its total population is second only to that of China. The growth of population was sluggish in the beginning of the 20th century. Nevertheless, since 1921 there has been a large-scale net growth of population. Between 1921 and 1951, the nation's population grew by approximately 1.2 percent annually. However, the stunning growth of population has occurred since the end of World War II. The decade of 1951 to 1961 recorded an annual growth rate of 2.2 percent, increasing to 2.5 percent between 1961 and 1971. Such high growth was part of the post-World War II "population explosion," which was the result of a sharp decline in death rates (8.49 deaths/1,000 population, 2003) and only a gentle decline in birth rates (23.28 births/1,000 popu-

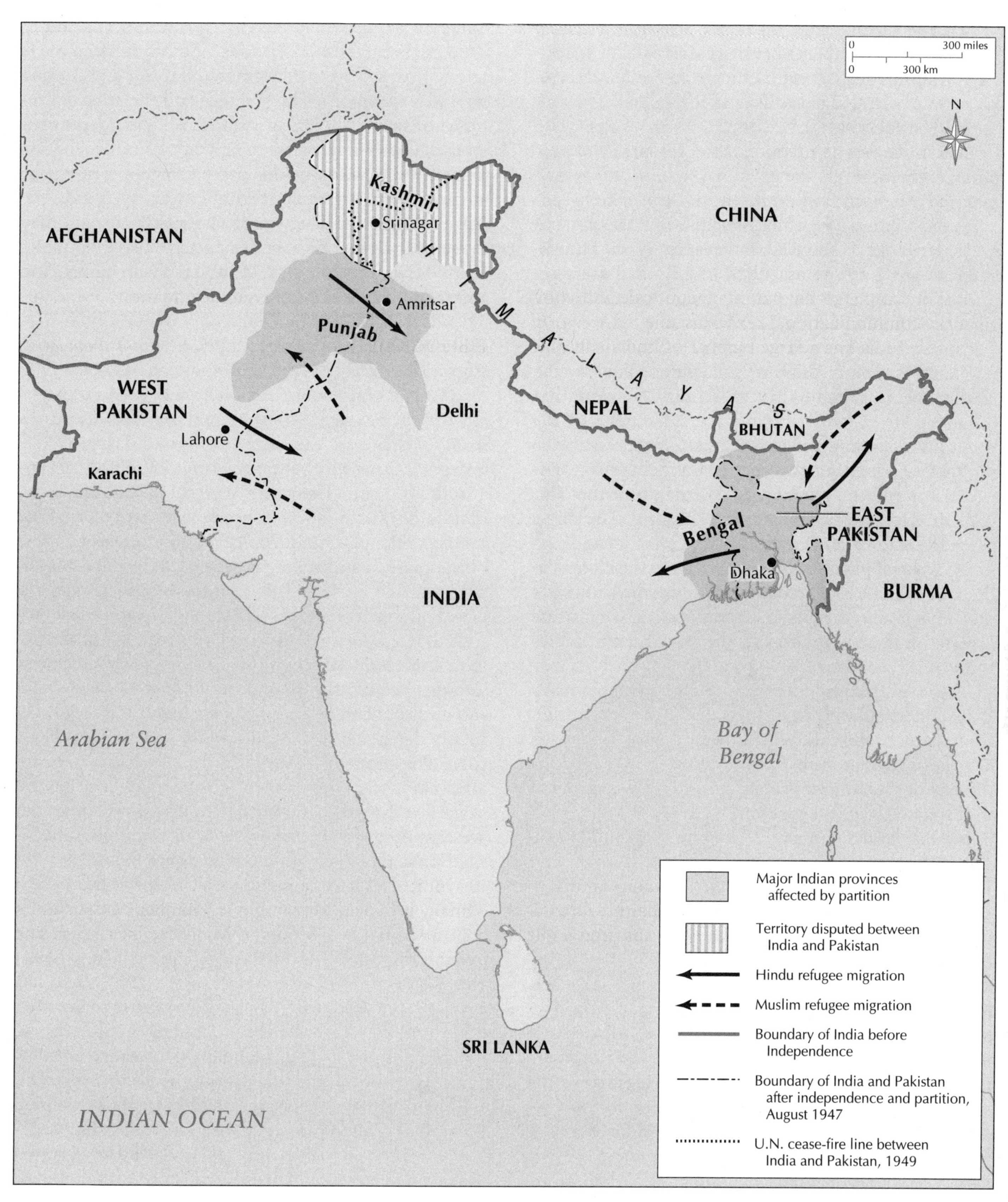

In achieving independence from the British Empire, the geographic area of the Indian subcontinent was divided into the nations of India and Pakistan. Later eastern Pakistan broke away to form Bangladesh.

lation, 2003). Currently, the population growth rate is declining (1.47 percent) because of increasing acceptance of family planning methods. The population is highly concentrated in the fertile plains, irrigated lands, and industrial centers.

Indian society is predominantly Hindu (81.3 percent); other believers are Muslim (12 percent), Christian (2.3 percent), Sikh (1.9 percent) and others (2.5 percent). There are 18 official languages. Hindi, the national language, is spoken by 40 percent of the population. "English enjoys associate status but is the most important language for national, political, and commercial communication," explains the CIA *World Factbook*. India has a large number of knowledgeable people proficient in the English language though only 59.5 percent of the total population of the country is literate.

India, since its independence, has maintained a democratic system of government based on free voting rights for persons over the age of 21, making it the largest functional democracy of the world. The economy of India has been maintaining an outstanding average growth rate of 6 percent since 1990. The economy is based on conventional agriculture, contemporary farming, handicrafts, contemporary small- and large-scale industries, and a large number of support services (especially customer support services for multinational high-tech companies). India supports an agrarian economy where 60 percent of the labor force is engaged in agriculture and about 40 percent of national income is earned from it.

The increased population pressure has resulted in immigration and intense cultivation of the arable land more than once in a year, usually referred as double or triple cropping. This, combined with the development of an irrigation system, has resulted in augmented agricultural production. The Food and Agriculture Organization (FAO) has predicted that India can feed triple the size of its 1985 population by the year 2010.

Industries in India are clustered in the specific areas based on the economies of location, leading to five prominent belts. An array of industries ranging from heavy (chemicals, iron and steel, petroleum, textile) to highly skilled (software) provides a strong base to the country's economy. India maintains trading relations with a number of developed countries. The cities of MUMBAI (BOMBAY) and Bangalore, in particular, attained global importance in the development of information technology.

India is developing at an incredibly fast pace with the onset of modern technology. It provides an enormous market to international and national businesses. This colossal growth has adversely affected and polluted the environment. Deforestation, DESERTIFICATION, air and water pollution, and soil erosion are a few of the intense environmental problems experienced by the Indian population. Constantly mounting population pressure is also overstressing the natural resources. In spite of the remarkable gains in economic investment, India has to go a long way to fully stabilize its economy, settle its international disputes, eradicate poverty, control overpopulation, settle ethnic and religious conflicts, and limit environmental degradation.

BIBLIOGRAPHY. A. Bose, ed., *Patterns of Population Change in India, 1951–61* (Allied Publishers, 1967); A.K. Dutt and M.M. Geib, *Atlas of South Asia* (Oxford University Press and IBH, 1998); J.H.K. Norton, *India and South Asia* (McGraw Hill/ Dushkin Company, 2004); R.L. Singh, ed., *India: A Regional Geography* (National Geographical Society of India, 1971); O.H.K. Spate and A.T.A. Learmonth, *India and Pakistan* (Meuthuen, 1967); *World Factbook* (CIA, 2004); S. Ganguli and N. Devota, eds., *Contemporary India* (Lynne Reiner, 2003).

Ashok K. Dutt
Meera Chatterje
University of Akron

Indian Ocean

MEASURING approximately 26.5 million square mi (68.5 million square km), the Indian Ocean includes the Andaman Sea, ARABIAN SEA, BAY OF BENGAL, Flores Sea, Great Australian Bight, Gulf of Aden, Gulf of Oman, Java Sea, Mozambique Channel, PERSIAN GULF, RED SEA, Savu Sea, Straits of Malacca, Timor Sea, and other tributary water bodies. The ocean is located between Africa, Southern Ocean, Asia, and AUSTRALIA.

The Indian Ocean is the third-largest of the world's five oceans (after the PACIFIC OCEAN and ATLANTIC OCEAN, but larger than the Southern Ocean and Arctic Ocean). The Southern Ocean was delineated in a spring 2000 decision by the International Hydrographic Organization, which consolidated a fifth world ocean from the southern portions of the Atlantic Ocean, Indian Ocean, and Pacific Ocean. This new ocean extends from the coast of ANTARCTICA north to 60 degrees South latitude or the Antarctic Treaty Limit. Five major choke points along commercial sea lanes

provide access to the Indian Ocean. These waterways are the Suez Canal (EGYPT), BAB EL MANDEB (DJIBOUTI-YEMEN), Strait of Hormuz (IRAN-OMAN), and Straits of Malacca (INDONESIA-MALAYSIA), and the Lombok Strait (INDONESIA).

The Indian Ocean has held the historic TRADE ROUTES from Occident to Orient since the dawn of maritime commerce in the ancient world. Spices, slaves, and marvelous though modest handcrafts moved among the many ports the Indian Ocean carried by centuries of monsoon winds and currents. This trade now bears the modern manufactured goods of industrialized nations and the petroleum required to fuel the world's economies. Never an insular sea, the Indian Ocean now serves not only the ports of its own shores but the far reaches of the countries of Europe and the Americas.

Historical accounts include the tales of Egyptian, Greek, and Roman commercial and colonial exploits. The settlement of Madagascar by peoples from Indonesia in ancient times points to the reciprocal flow of culture and commerce from the east. Centuries of trade and exploration were carried out by the Islamic merchants and seaman of the Arabian Peninsula, who took their goods and their faith to the lands of the African coast, India, and Southeast Asia. Before the Europeans made good a sea passage to the waters of the Indian Ocean, Admiral Zheng He of the Ming Dynasty of China led several maritime expeditions across these fated waters from 1405 to 1433.

By 1497, Vasco da Gama navigated around the Cape of Good Hope and began the Portuguese fight for domination of the spice trade across the Indian Ocean. This model of bold commercial venture supported by national navies and prestige was followed by the Dutch, French, and British over the next two centuries. European colonial interests in the Indian Ocean LITTORAL continued until after World War II. The independence of regional states from their colonial masters, growth and development among the newly industrialized states, and the increased flows of manufactured goods and petroleum have made the Indian Ocean truly an international sea.

During the Cold War, the Indian Ocean was an arena of competition between East and West. Regional navies were often overshadowed by the number and modern capabilities of the U.S. and Soviet fleets. Establishment of a permanent support and operational base on the island of Diego Garcia, strategically central in the Indian Ocean, was a clear sign of the American intent to remain active in the region. The United States maintains the most active, modern, and capable military forces in the Indian Ocean.

Throughout the long history of foreign economic exploitation and domination, the Indian Ocean has continuously been a rich resource to the peoples of the region. It constantly supplied transportation routes and rich fisheries along the coasts. The Indian Ocean continues to be a major fishing ground with fleets from many nations vying for the limited resource. Fishing stocks are being depleted by a combination of overfishing and pollution along the Indian Ocean coasts. The overexploitation of fish stocks is mostly due to large fishing vessels operating illegally near the coast. The growth of regional populations, particularly in INDIA and Indonesia, will add pressure on the already challenged marine resources.

This population increase and industrial development creates major pollution problems in the most critical fishing areas. Industrial effluents contain heavy metals and chemical wastes. Pesticides and organic wastes flow untreated into the coastal waters from cities and agricultural land. Oil pollution from accidents, ballast dumping, and offshore oil extraction is on the increase. The nations that share the management and use of the Indian Ocean have no comprehensive plan for conservation and management of the resources or uses of this global asset.

Mineral resources and especially offshore petroleum extraction are continuing to grow as a commercial interest to nations. Large reserves of hydrocarbons are being tapped in the offshore areas of SAUDI ARABIA, IRAN, INDIA, and AUSTRALIA. An estimated 40 percent of the world's offshore oil production comes from the Indian Ocean. Beach sands rich in heavy minerals and offshore placer deposits are actively exploited by bordering countries, particularly India, SOUTH AFRICA, Indonesia, SRI LANKA, and THAILAND. The mining of polymetallic nodules from the seabed remain a tempting, but technologically challenging operation.

Climate and weather patterns are dominated by the annual monsoon. This weather cycle is attributed to low atmospheric pressure over Southwest Asia created by hot, rising summer air, which causes southwest-to-northeast winds and currents during the summer months. A high pressure over North Asia created by cold, falling, winter air results in northeast-to-southwest winds during the winter months.

For half the year (April to October), the winds in this region are from the southwest, reversing in the other half of the year. This monsoon (season) weather pattern dominates the region on land and sea, setting

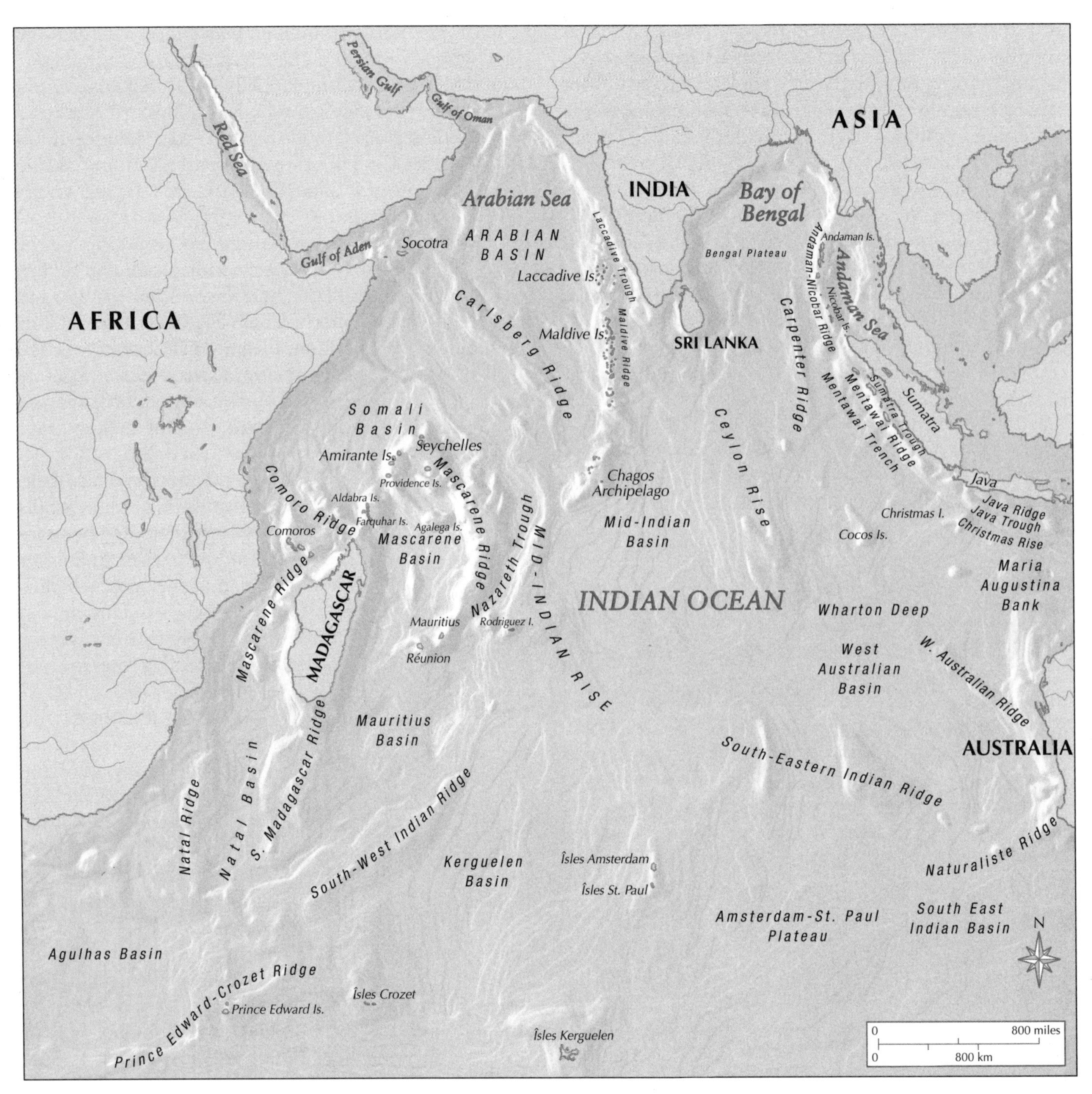

The eastern Indian Ocean, near Sumatra, was the site of a major underwater earthquake, registering 9.0 on the Richter scale, on December 26, 2004. The resulting tsunamis, or tidal waves, killed thousands along the Indian Ocean's land borders, from Indonesia to Africa.

the pace of life onshore and off. Tropical cyclones occur during May/June and October/November in the northern Indian Ocean and January/February in the southern Indian Ocean.

Primary adjoining arms of the Indian Ocean are the Persian Gulf and the Red Sea. The deepest known point is 25,443 ft (7,758 m), off the southern coast of Indonesia in the Java Trench. The Indian Ocean contains numerous islands, the largest of which are MADAGASCAR and SRI LANKA. Smaller islands that constitute independent countries include the MALDIVES, the SEYCHELLES, and MAURITIUS. The major rivers that flow

into the Indian Ocean include the waters of the Limpopo and Zambezi rivers from Africa, as well as the IRRAWADDY, Brahmaputra, GANGES, and INDUS from east to west along Asia. The combined waters of the Tigris and Euphrates rivers mix with the Indian Ocean via the Persian Gulf.

The Indian Ocean is of geostrategic interest as it is a transit route for a major portion of the world's oil supply. In addition the commerce flowing by ship from Asia to Europe also sails this sea. Some major ports and harbors of the Indian Ocean are Chennai (Madras; India), Colombo (Sri Lanka), Durban (SOUTH AFRICA), JAKARTA (Indonesia), KOLKATA (Calcutta; India), Melbourne (Australia), MUMBAI (Bombay; India), and Richards Bay (South Africa).

BIBLIOGRAPHY. *World Factbook* (CIA, 2004); K.N. Chaudhuri, *Trade & Civilisation in the Indian Ocean: An Economic History from the Rise of Islam to 1750* (Cambridge University Press, 1985); Louise Levathes, *When China Ruled the Seas* (Oxford University Press, 1996); H.J. de Blij and Peter O. Muller, *Geography: Realms, Regions, and Concepts* (Wiley, 2002).

IVAN B. WELCH
OMNI INTELLIGENCE, INC.

Indiana

KNOWN AS THE Hoosier State, the north central state of Indiana ("Land of Indians") was formed from the Northwest Territory in 1800 and entered the Union in 1816 as the 19th state. Roughly rectangular in shape, Indiana is approximately 270 mi (434 km) north to south and 140 mi (225 km) east to west. The geography of Indiana encompasses 36,420 square mi (94,327 square km), which makes it the 38th largest state in the UNITED STATES. It is 14th in population. Indiana is bounded on the north by Lake MICHIGAN and MICHIGAN, on the south by KENTUCKY, on the east by OHIO, and on the west by ILLINOIS. In addition to Indianapolis, the capital, Indiana's largest cities are Fort Wayne, Evansville, South Bend, Gary, Hammond, Bloomington, Muncie, Lafayette, and Anderson.

Most of Indiana has a humid continental climate, with four distinct seasons, resulting in long, warm summers and cool winters. The southernmost area of the state has a humid subtropical climate, experiencing more frequent rainfall and less extreme temperatures in the winter. Northern Indiana may experience up to 40 in (101 cm) of snow in the winter, but the southern sections rarely receive more than 10 in (25 cm) a year. Temperatures range from highs of 70 to 80 degrees F (21 to 26 degrees C) to lows of 25 to 35 degrees F (-4 to 2 degrees C). The state experiences an annual precipitation rate of around 40 in (101 cm). Tornadoes are likely in the spring.

Indiana is divided into three geographic areas: the Great Lakes Plains of northern Indiana, the Till Plains found in central Indiana, and the Southern Plains and Lowlands of southern Indiana. The Great Lakes Plains contain large sand dunes along the shores of Lake Michigan. This landscape varies from flat to gently rolling areas, interspersed with a number of lakes and bogs. The soils in this area are generally acidic grays and browns.

The Till Plain is part of the Midwestern Corn Belt. The acidic gray and brown soils found in the low hills and valleys produce extremely fertile land. The more rugged terrain of the southernmost part of Indiana is located in the Southern Plains and Lowlands, which is covered by steep hills known as knobs. This area is home to a number of well-known caves, such as Wyandotte and Marengo Caves, and mineral springs such as those found in West Baden and French Lick.

Elevations in Indiana range from 320 ft (97 m) above sea level at the Ohio River to 1,257 ft (383 m) above sea level at Hoosier Hill. Bordering on Lake Michigan, Indiana lays claim to approximately 230 square mi (595 square km) of this major water source and to a 41-mi (66-mi) shoreline. Other Indiana rivers include the Ohio River, the Wabash River and its tributaries, the White River and the Tippecanoe River, and the Kankakee River. Saint Joseph and Saint Mary's rivers converge at Fort Wayne in northeastern Indiana. The state's major lakes include Michigan Lake, Wawasee Lake, and Monroe Lake.

Approximately 4.5 million acres of Indiana's land area are forested, including the Beech-Maple Forest, the Oak-Hickory Forest, the Southern Floodplain Forest, and the Elm-Ash Forest. Indiana forests provide a home to tulip (the state tree), oak, hickory, maple, walnut, ash, spirea, barberry, and mock orange trees. Wildflowers include peony (the state flower), violet, daisy, columbine, gentian, trillium, and various kinds of orchids. Abundant wildlife in Indiana includes whitetail deer, raccoon, opossum, gray fox, coyote, beaver, rabbit, squirrel, skunk, muskrat, mink, and weasel. Bobcat and badger have been identified as endangered species in Indiana. The state's game birds in-

clude wild turkey, ruffed grouse, quail, and pheasant. Sparrows, blue jays, wrens, and cardinals (the state bird) are found throughout the state. Indiana is also home to 75 species of endangered birds, such as the peregrine falcon, Kirtland's warbler, and the bald eagle.

As a major industrial center, Indiana specializes in the iron, steel, and oil products industries. Bituminous coal, found mostly in southwestern Indiana, is also a major revenue producer for the state. Manufacturing in Indiana also includes the production of transportation equipment, motor vehicles and equipment, industrial machinery and equipment, electric and electronic equipment, mobile homes, farm machinery, wood office furniture, and pharmaceuticals.

The state's major crops are corn, soybeans, wheat, oats, rye, and nursery and greenhouse products. Other successful agricultural activity revolves around the production of tomatoes, onions, popcorn, fruit, hay, tobacco, mint, and livestock. Oak, tulip, beech, and sycamore are the chief products of Indiana's timber/lumber industry.

BIBLIOGRAPHY: "About Indiana," www.ai.org (November 2004); Darrel E. Bigham, *Southern Indiana* (Acardia, 2000); "The Geography of Indiana," www.netstate.com/ (November 2004); Dan Golenpaul, ed., *Information Please Almanac* (McGraw-Hill, 2003); John C. Hudson, *Making the Corn Belt: A Geographic History of Middle-Western Agriculture* (Indiana University Press, 1994); Robert C. Kingsbury, *An Atlas of Indiana* (Indiana University Press, 1970); Ron Leonetti et al., *Unexpected Indiana: A Portfolio of Natural Landscapes* (Indiana University Press, 2004); William E. Wilson, *Indiana: A History* (Indiana University Press, 1996).

ELIZABETH PURDY, PH.D.
INDEPENDENT SCHOLAR

Indonesia

Map Page 1124 **Area** 741,110 square mi (1,919,440 square km) **Population** 238,452,952 **Capital** Jakarta **Highest Point** 16,502 ft (5,030 m) **Lowest Point** 0 m **GDP per capita** $3,200 **Primary Natural Resources** petroleum, natural gas, nickel, copper.

COMPOSED OF 17,000 islands, Indonesia is the world's sixth-largest nation and the world's largest archipelago, stretching over 5,000 mi (8,047 km) along the equator. It lies southeast of the Malay Peninsula and Indochina. AUSTRALIA is due south and the PHILIPPINES lie to the north. About half the population lives on the island of JAVA where the capital, JAKARTA, is located. The island of BORNEO is shared with MALAYSIA, while New Guinea is shared with PAPUA NEW GUINEA. Other larger and populated islands are SUMATRA and Sulawesi.

EARTHQUAKES AND TSUNAMIS

Because of its location on the edges of tectonic plates, Indonesia is prone to earthquakes, which cause TSUNAMIS (as exemplified in the 2004 tsunamis that violently hit the island of Sumatra). Indonesia also has many volcanoes. Terrain of the islands varies and includes mountains, rainforests, and miles of beaches. The country has a hot, humid tropical climate, with a more moderate climate in the highlands.

There are two seasons: the dry season from June to October, and the rainy season from November to March. It is hot all year round, although coastal regions are cooler. Indonesia is home to a wide variety of flora and fauna. Elephants, tigers, leopards, and orangutans live there, as well as sea turtles and Komodo dragons. The world's largest flowers grow in Sumatra. Unfortunately, rainforests are rapidly disappearing from the region.

Indonesia's history has added to its rich diversity of culture. From the 7th to the 14th century, Hindus and Buddhists formed several kingdoms on these islands. Arab traders later introduced Islam. In the early 16th century, Europeans in pursuit of the spice trade arrived. By the 17th century, the Dutch were the dominant colonial force in the region. Indonesia was still under Dutch rule when World War II broke out.

During the war, Japan conquered and occupied the Dutch colony. They found the Indonesian elite to be cooperative and several were decorated by the Japanese emperor in 1943. After the war, a group led by Sukarno declared the country's independence. The Dutch attempted to quell the independence movement and regain control of Indonesia. After several years of bloody fighting, the Dutch accepted Indonesia's independence.

Sukarno was the first president. His government, now called the Old Order, saw military conflict with British Malaya and economic difficulties at home. In 1967, Army General Suharto took over, saying he

Indonesian schooners ply the vast seas that lie between the archipelagoes that comprise the country of Indonesia.

could secure the country against an alleged coup attempt by communists. His 32-year reign is known as the New Order. Suharto's corrupt regime brought wealth to himself and his family, and he was forced to step down in 1998 after massive popular demonstrations against him. During the next three years, the country had three presidents. In 2004, Susilo Bambang Yudhoyono was elected president.

The culture of Indonesia has been influenced by its many immigrants. Most Indonesians are of Malay or Polynesian descent. The population also includes significant numbers from INDIA, CHINA, SAUDI ARABIA, IRAN, and Europe. The official language is Bahasa Indonesia, but English, Dutch, and local dialects are widely spoken as well.

Indonesian art shows influence of other cultures. Javanese and Balinese *wayang kulit* shadow theater shows are popular, as are famous Javanese and Balinese dances. Some islands are famous for their batik and itak cloth. The unique martial art of Silat originated in Indonesia. Indonesia is primarily a Muslim nation, but religious tolerance is widespread. Muslims account for 88 percent of the population, with 8 percent Christian, 2 percent Hindu, and 1 percent Buddhist.

NATURAL RESOURCES

Indonesia is blessed with a number of natural resources. The country is the second-largest exporter of natural gas. Other natural resources include gold, copper, tin, and oil. Indonesia's agricultural products include rice, tea, coffee, spices, and rubber. The UNITED STATES, JAPAN, SINGAPORE, Malaysia, and Australia are Indonesia's major trading partners. Agriculture or farming accounts for 45 percent of the workforce, while 16 percent work in industry, and 39 percent in service industries.

Indonesia faces many problems, both economic and social. The judicial system does not function well and the banking system is weak, making Indonesia a poor climate for foreign investment. There are ongoing threats against Western interests in Jakarta. The 2002 terrorist bombing of the Sari nightclub and a bomb explosion outside the Australian embassy in 2004 add to the unrest. Poverty is a problem, particularly in Jakarta. About 27 percent of the people live below the national poverty level.

BIBLIOGRAPHY. *World Factbook* (CIA, 2004); "Indonesia," www.wikipedia.com (November 2004); "Indonesia," www.geographia.com (November 2004); *South-East Asia* (Time-Life Books, 1987); "Destination Indonesia," Lonely Planet World Guide, www.lonelyplanet.com (November 2004).

PAT MCCARTHY
INDEPENDENT SCHOLAR

Indus River

THE INDUS RIVER in PAKISTAN flows 1,900 mi (3,000 km) from southwestern TIBET to the Arabian Sea. From the dawn of human culture, the Indus River has sustained societies along its banks; the earliest civilization that can be reliably distinguished is called the Indus Valley civilization (or Harappan). This sophisticated society is dated back to 2500 B.C.E. The Harappan culture was a well organized civilization built upon surplus agricultural production and bolstered by commerce that reached as far Mesopotamia. The nascent agricultural predecessors of the Harappan culture may well reach back into the 5th millennium B.C.E.

The name *Indus* comes from its Sanskrit name *sindhu* (meaning "water, flood, ocean"). The drainage basin of this great river has caught a continuous stream of invaders and conquerors of various ethnic, linguistic, and religious origins. Each of these succeeding groups would wind their way through the mountain barriers and flow down toward the Indus River seeking purchase in its fertile valley. The records of these peoples come to us through spoken and literary traditions

that reach back without break to 2000 B.C.E. The Vedas are a uniquely ancient traditional history that tells us of the Aryan tribal groups of pantheistic pastoralists that supplanted the earlier Harappan culture. This pattern of migration, conquest, and decline would repeat itself over the centuries to follow.

Alexander the Great brought his conquering army to the Indus around 326 B.C.E. and left a lasting impression of Hellenist culture in art of the region. This influence survives in the great Buddhist statuary art treasures that grace the Indus Valley. Buddhism gained dominance with the coming of Ashoka (236 B.C.E.) and his turning of Taxila into a major Buddhist center of learning and culture. The ebb and flow of empires continued with rulers from the east and west vying for control. Buddhist, resurgent Hindu, and then Islamic conquerors ruled until the passing of time brought 15th century Europeans to the Indus River.

The British came as all conquerors before them, but were the first to seek to subdue the Indus River as well as its people. They spent 100 years attempting to harness the irrigation power of the waters. This radically changed the landscape and impacted social and political structures, changes that continue today. The dams and barrages built by the British remain the primary infrastructure to fight annual flooding.

The Indus River shares all the challenges that accrue with major man-made improvements. Dams reduce flows in lower portions of the system and limit the transport of fertile sediments downstream into the DELTA. Dams compartmentalize river systems and isolate aquatic life into smaller communities. Extraction of irrigation water, especially during low water periods can threaten fish stocks. Introduced species often compete with indigenous ones.

Density of population along the Indus and its tributaries combined with an almost total lack of conventional public sanitation systems mean many water courses are little more than sewage carriers at low flow periods. This leads directly to the contamination of drinking water and agricultural produce. Low-lying land is commonly used as solid waste dumping sites contributing to illness and mortality. The vaunted "green revolution" has brought increased food production but also introduced large quantities of fertilizers and pesticides into the waters and sediments of the Indus. Growth of the textile industry and other manufacturing has radically increased the flow of toxic industrial wastes into the riparian system. Pakistan is only in the earliest stages of studying and addressing the sustainment needs of the Indus River.

The head of the Indus rises in southwestern Tibet. It then flows northwest through Kashmir before bending to the south and leaving the mountains to become a slow-flowing, highly braided river course. It is dammed near Peshawar to form the Tarbela Reservoir. The catchment area of the Indus is estimated at almost 386,100 square mi (1 million square km), and all of Pakistan's major rivers flow into it. In its upper basin of Punjab (meaning "land of five waters") are the Jhelum, Chenab, Ravi, and Sutlej rivers. Passing by Hyderabad, it ends in a large delta to the southeast of KARACHI.

BIBLIOGRAPHY. "Pakistan," www.loc.gov (Library of Congress, 2004); "The Ganga Basin," State University of New York, www.cs.albany.edu (April 2004); H.J. de Blij and Peter O. Muller, *Geography: Realms, Regions, and Concepts* (John Wiley & Sons, 2002); Robert Eric Mortimer Wheeler, *Early India and Pakistan: To Ashoka* (Textbook Publishers, 1968); Joseph E. Schwartzberg, ed., *A Historical Atlas of South Asia* (Brill Academic, 1992)

IVAN B. WELCH
OMNI INTELLIGENCE, INC.

Industrial Revolution

THE INDUSTRIAL Revolution is one of the most dramatic events in modern world history, an event which has more or less influenced all nations of the world in one form or another. The Industrial Revolution began in the late-18th century in England, UNITED KINGDOM, in a region of the country known as the West Midlands—the largest city today in this region is Birmingham.

The Industrial Revolution is characterized by the rise of industrialization, a process of industrial, social, and economic changes that revolutionized societies from agrarian to industrial. Central to this change is the development of technology that allows for the process of manufacture to occur, that is, the changing of raw materials into finished goods (often in factories) for sale in the marketplace. To simply define what is the Industrial Revolution is problematic, but in simple terms, it may be said to be the application of power-driven machinery to the process of manufacturing products.

The Industrial Revolution must be viewed as a revolution, not in the political sense of the word, but a

revolution in that fundamental changes occurred as a result of its existence, particularly in the social and economic structure of countries it affected. The changes that brought about the Industrial Revolution in Enland occurred in a gradual manner and can be seen in the previous decades and centuries before fundamental change happened. Thus, the Industrial Revolution was the consequence of a long period of change prior to the application of power-driven machinery within the process of manufacture.

CHARACTERISTICS OF THE REVOLUTION

The Industrial Revolution began in the late 18th century in England. As to the precise date of the event, historians have been unable to provide an answer. Regardless of people's perspectives, what has been widely noted is that from the end of the 18th century, fundamental social and economic change occurred in England and subsequently other places, which included a dramatic rise in national population sizes, brought about by changing birth and death rates; a more rapid growth of existing towns and cities, particularly capital cities; the appearance of new social classes related to people's position as owners of industry or as workers in the industrial process; and developments in transportation and networks of communication.

For the Industrial Revolution to have happened, historians have noted a number of significant changes in society. These have included developments in agriculture, such as the adoption of new systems of cultivation and the invention of new machinery which allowed for an increased supply of food, and the development of new machinery in industrial production from increases in knowledge that were mostly the result of practical experiences and informed empiricism.

These new machines that greatly affected industrial behavior included, by way of example, machines like the flying shuttle (John Kay), the water frame (Richard Arkwright) and the spinning jenny (James Hargreaves). Coupled with this particular development were inventions in other areas that were also applied to modern industry.

For instance, Thomas Newcomen's steam engine, first used to help pump water out of mines, and then James Watts's 1763 steam engine opened the door for steam-powered machinery in the workplace. Steam providing the power for machines that previously had used water as a source of energy. Such inventions helped bring about a dramatic rise in output, in turn stimulating further industrial growth and wider social and economic change.

For the Industrial Revolution to have happened, historians such as Michael Zell have noted an important economic/industrial stage immediately prior to the onset of industrialization. This stage of development, a "putting-out" industrial phase commonly referred to as proto-industrialization, is widely accepted to be the phase of modern economic development that preceded the Industrial Revolution proper.

Proto-industrialization can be characterized by two very distinct features: First, the spread of domestic manufacturing in rural places linked previously remote family groupings to not only regional but national and international marketplaces; second, rural manufacture became so widespread in geographical extent, plus so economically and socially powerful, that it helped push economies toward industrial manufacture in the factory situation, a situation associated with capitalist economics and production in urban centers.

PUZZLED HISTORIANS

Historians have also been puzzled as to why England became the first nation to be affected by the Industrial Revolution. Despite much fervent discussion and writing on the subject, no one single answer has been produced; instead, a number of suggested reasons have been proposed.

In a general sense, both the economic and political contexts were suitable for rapid societal transition but also a number of other factors were significant. These include the abundance of natural resources that could be used as raw materials in the industrial process (e.g. coal and iron ore); the availability of capital to invest in modern industry; a growing marketplace in part based on domestic population growth, foreign trade agreements, and the expansion of the British Empire; the availability of people to work in factories, many of whom were to migrate from the countryside into existing towns and cities; and people of sufficient intellectual capacity—that is, as managers of industrial units—but also people of ideas who could create new machines, workplaces, systems of industry, etc.

CRADLE OF INDUSTRY

One of the places known as the "cradle of industry" in the West Midlands region of England is the county of Shropshire. Despite being one of the most rural parts of England, it was in Shropshire that a number of industrial firsts happened, probably as a result of the abundance of raw materials (iron, coal, lead) in the area. For instance, within the area of Coalbrookedale is Ironbridge, where the world's first bridge to be

erected of iron was built. The bridge (built in 1779 by Abraham Darby from a design by Thomas Pritchard) helped bring about improvements in iron smelting that formed an important element in the early phases of the Industrial Revolution. Today, the value of Ironbridge is demonstrated by its World Heritage Site status (given by UNESCO).

Central to the Industrial Revolution was the growth of the factory, a large industrial building within which goods are manufactured. Although the world's first factories were not in England, the establishment of thousands of factories as part of the process of industrialization, some of which employed thousands of people, was central in bringing industrial change. The absorption of large numbers of workers into the factory was not a smooth process, as it involved a new labor routine within the workplace. In addition, some workers in England were so aggrieved by free market principles and the introduction of machinery in the workplace, in that it was a threat to their position of labor, that they destroyed equipment.

The Luddites, as they were known, and their movement gained such notoriety in early-19th century England that machine breaking became a capital crime. Many of the first factories in England were placed in rural locations so as to deter possible Luddite attacks.

LONG HOURS, LOW PAY

Inside factories in the 19th century, laborers worked long hours for low pay, often in dangerous conditions, so as to maintain mass production (the production of large amounts of standardized goods) and industrial output. The development of the factory system was assisted by numerous technological developments at the end of the 18th century and further inventions in the 19th and 20th centuries.

Developments in machinery able to create power as well as the creation of large looms allowed manufacture to be more efficient. In terms of industrial Britain, the cotton factory was the most common factory type, within which cotton was produced. Raw materials from the Americas and the Caribbean, produced under conditions of slavery by workers taken from Africa until the 19th century, helped allow Britain to produce much prosperity, although this wealth was firmly kept within the hands of the factory owners, who would often invest their profits into establishing more factories. As a consequence, huge industrial empires were created in Britain and also in other countries.

The effect of the Industrial Revolution has been felt worldwide. At first the effects of the Industrial Revolution spread from England across Europe and into Northern America. However, even in Europe the dates at which different countries industrialized differed because of the influence of localized circumstances. FRANCE, for example, despite being geographically close to England did not industrialize until maybe 60 to 70 years after England (in the mid-19th century). SPAIN, too, did not start to industrialize until this time. GERMANY by this time already had a highly advanced industrial economy.

In other parts of the world, countries did not begin to industrialize until the 20th century. Many of today's economic powerhouses in Asia, such as JAPAN, CHINA and TAIWAN, did not industrialize as such until the mid-20th century. In the case of China, industrialization and economic development did not start in earnest until the change of economic policies by the communist national government toward the end of the 20th century.

The effect of the Industrial Revolution has been to accelerate existing previous social and economic trends. In the case of England, as the first industrial nation, a number of these trends will now be discussed, and the consequences of industrialization also given some attention. The growth of industrialization in England at the end of the 18th century brought massive societal upheaval in the following decades. For example, the national population began to increase markedly. Existing towns and cities also grew dramatically. In England the demographic change was particularly pronounced, but all countries that have experienced industrialization at some time have also experienced marked population growth and urban growth.

URBANIZATION

With the increase in the national population came a significant increase in the urban population of England and Wales, partly because of the location of factories in urban places so as to utilize economies of scale (land, labor and capital). By the middle of the 19th century (1851), England and Wales already had the majority of their population residing in urban centers.

While the growth in urban populations is an important consequence of industrialization, the changes brought about by the Industrial Revolution established a removal of the medieval or preindustrial urban hierarchy in England and Wales. Places, such as Norwich, Exeter, Shrewsbury, Cambridge, Canterbury, and York, urban places with long established histories closely associated with the church, were pushed down the urban

hierarchy because of the sudden surge of people living in industrial places. Additionally, these places resisted industrial pressure and did not industrialize in the late 18th century. So, although large in size and still of national importance, they were demographically surpassed by other places (that is, those that did industrialize). Whereas prior to industrialization settlements like Norwich and York were at the top of the urban hierarchy in terms of their demographic size, by as early as 1801 their relative importance was declining and by 1851 a new urban hierarchy based on industrial towns and cities at the top was firmly in place. By 1851 industrial places such as Sheffield, Newcastle, Bradford, Hull, and Leeds were positioned at the upper echelons of the urban hierarchy.

CAPITAL GROWTH

A noticeable consequence of the Industrial Revolution was that it allowed already large urban centers to markedly increase their demographic size and urban regions to appear for the first time. Across Western Europe, for instance, capital cities grew dramatically. London increased its population size from 1.1 million to 6.6 million people between 1801 and 1901. By about 1900, there were approximately 150 large cities (of 100,000 of more people) in Europe whereas in 1800 there had been about 20 places of this size.

The marked shift in urban living has been the source of much investigation, and a number of reasons have been given as to how the Industrial Revolution influenced such change. These include the need to have concentrations of labor in areas close to raw materials where factories would be based (e.g., the Ruhr in Germany), the need for marketing finished products at places with access to rail or water (e.g., Liverpool in England, Hamburg in Germany and NEW YORK, UNITED STATES), and finally the tendency for banking and financial institutions to base themselves in existing political and cultural centers (e.g., PARIS and LONDON).

Examples of urban regions that developed in 19th century Europe include the Ruhr (Essen, Dortmund, Bochum, Dusseldorf) in Germany, and the West Midlands (Birmingham, Coventry, Wolverhampton) in England.

To refer back to British industrial society, the early decades of the 19th century witnessed major changes to the economic, political, social, and aesthetic values of Britain as a consequence of industrialization, urbanization, and demographic transition, which were reflected in the changing appearance and form of urban land. Urbanization not only affected the building industry but virtually swamped the administrative practices and building codes that had safeguarded the urban environment, as resultant chronic overcrowding, poverty, inadequate sanitation, dirt, and disease were to testify. These were not unusual occurrences, but now they occurred on a scale never witnessed before. In terms of housing much change occurred as well. In order to capitalize on the rush of migrants into towns and cities to work in factories, speculative builders hastily erected poor quality houses within which worker families would reside, and about 99 percent of houses erected in London in the 19th century were done so speculatively.

Significantly, these houses were packed together, often close to the location of local factories, in high-density fashion, which exacerbated social tensions and problems. Disease, for example, was often rife in working people's districts, and often epidemics of diseases such as cholera would kill thousands. Houses were erected without any sanitation, maybe one outside toilet for an entire street, and without any adequate water supply. Water had to be gathered from local rivers or from water pumps located nearby, but the quality of urban water was far from perfect, as water used for drinking also contained various waste materials.

By the late 1830s and early 1840s, British governments at both the local and the national level were forced into finding solutions for the poor environmental and moral conditions in which people lived. The unfolding of such rational approaches to the urban form marked a fundamental change in the comprehension of the association between social and economic growth and the urban environment, a consequence of the development of the understanding that the urban conditions created under the forces of industrialization and urbanization were not conducive to salubrious living for not just many urban dwellers but a large proportion of the populace.

However, despite such knowledge, the wage structure, and so the working classes' ability to compete in the housing market, had to be strangled in the capitalism system in order to underpin national economic prosperity, which was the basic cultural bequest given to Britain by the Industrial Revolution. Slum housing, which was to be found in every town and city in Britain, was therefore an unfortunate though imperative byproduct of culture and its economy at that time.

The dynamics of any proposed Victorian public involvement thus had to improve the plight of the poor—that is, improve the quality of life for people living in the cheapest and worst housing—without encroaching

upon society's cultural machinery for creating and maintaining its wealth, that is, capitalism. This was achieved through intervening in the housing market by introducing rules to improve the quality of new privately built houses. Therefore, as much as environmental improvement was about improving the plight of individuals, it was also about maintaining the capitalist machine that the Industrial Revolution had helped establish. Such a situation has been highlighted, for example, by government reports from the 1840s, which show the thought that unhealthy workers were not able to fully contribute their economic worth to industrial production. A healthy person could work more and better so it was in the nation's economic interest at least to keep people healthy. For an insight into living conditions in England at the start of the 19th century refer to Friedrich Engels's *Condition of the Working Class* in England in 1844.

CAPITALISM AND MARXISM

The social system produced by the Industrial Revolution can be characterized by a significant number of working people engaging in industrial labor, working in sometimes large-sized industrial units (factories) owned by a single person or a small group of individuals. Philosopher Karl Marx wrote extensively on the development of capitalism and the social tensions it created between the main two social groups that he called the bourgeois, the owners of production, and the proletariat, the industrial workers who were viewed largely as being commodities and not as people by the factory owners.

The relationship between the bourgeois and proletariat, and Marx's views on their association, has been greatly influential on wider social and political thought since Marx and Engels wrote the *Communist Manifesto* in 1848. The ideas of Marx were greatly influential upon political revolutions such as that instigated by the Bolsheviks in early-20th-century RUSSIA and upon governments in countries such as CHINA, CUBA, North KOREA and parts of Eastern Europe in the 20th century. It can argued that Marx's influence upon the Chinese communists in the 20th century was such that it allowed the Industrial Revolution to occur later in the country than would have happened otherwise.

It has been shown that the Industrial Revolution influenced all aspects of society. Developments in other areas of society also affected the industrialization process. For instance, while brief mention has been given to the significance of technological developments, such as the steam engine, these developments also assisted industrial growth by other means. If the steam engine is taken as an example, its development also allowed by the start of the 19th century transport developments via the invention of the steam train. The invention of the train not only allowed greater speed of movement around regions and countries, thus assisting the migration of people from one place to another, but it also helped establish a network of communications between places that were previously isolated from each other. Therefore, industrial markets could open up, raw materials could be speedily brought from one area to another, and finished goods taken to the marker place at a much quicker speed than before.

With regards to developments in rail, Britain again took a leading role. By as early as 1830 Liverpool and Manchester, two of the largest cities at that time in England, were joined by rail line. At this time the first passenger station in Manchester was opened. Steam engines were also developed by the British in boats, which allowed for quicker travel between Britain and the American continent and the quicker import/export of goods and raw materials. Such developments naturally did not hinder industrial progress.

SECOND INDUSTRIAL REVOLUTION

Toward the end of the 19th century, developments and societal change, particularly in Germany, have been called a Second Industrial Revolution. Changes within the chemical, steel, petroleum, and electrical industries from the 1870s mark the start of the second industrial transformation, a new phase of industrial and social development.

Whereas Britain was the cradle of the first revolution, Germany was the cradle of the second. Within Germany from the 1870s, much technological and social change occurred. Market prices, for example, were controlled by cartels, investment was plowed into advancing matters of science, and new technologies utilized in the production process.

While the second phase of the Industrial Revolution had the same social problems that were evident during the first period—for example, low wages, poverty, crime, unemployment, and the major element of workers engaging in industrial production—what may be noted as being wholly different about the two phases is their sources of power. While wood and coal were employed as means to generate steam power at the end of the 18th century and start of the 19th century, by the start of the second industrial period (from the 1870s), power was being generated by electrical motors and combustion engines.

BIBLIOGRAPHY. M.E. Beggs-Humphries, *Industrial Revolution* (Allen and Unwin, 1959); Friedrich Engels, *Condition of the Working Class in England in 1844* (Oxford University Press, 1999); Richard Fleischman, *Conditions of Life among the Cotton Workers of South-eastern Lancashire during the Industrial Revolution* (Garland, 1985); Robert Glen, *Urban Workers in the Early Industrial Revolution* (Palgrave Macmillan, 1985); D.C. Goodman and C. A. Russell, *Science and the Rise of Technology Since 1800* (Open University Press, 1972); S. Lilly, *Men, Machines and History* (International Publishers, 1966); Karl Marx, *The Communist Manifesto* (Signet Classics, 1998); F.L. Mendels, "Proto-Industrialisation: The First Phase of the Industrialisation Process," *Journal of Economic History* (v.32, 1974); Charles More, *Understanding the Industrial Revolution* (Routledge, 2000); E.P. Thomson, *The Making of the English Working Class* (Knopf, 1985). Michael Zell, *Industry in the Countryside* (Cambridge University Press, 1994).

IAN MORLEY
MING CHUAN UNIVERSITY, TAIWAN

insurgent state

INSURGENT STATE is a term that refers to the creation of a geographic area that not only claims independence from the larger state, but whose leaders and policies also seek to replace the existing political order and state, or to become independent. It is not a segment of the country that seeks independence (civil war). It is a state within a state whose leaders seek total control of the existing state. It is not an ENCLAVE, as it is not a base or colony of another external, state or power.

While not a new phenomenon, the conscious use of GUERRILLA BASES to form a proto-state is largely the product of Mao Zedong in his war to become ruler of all CHINA in the 1920s. Since then, the practice of creating distinct geographic areas that are beyond the control of the central government, and which also seek to replace that government or become wholly separate, has become widespread. The Sendero Luminoso in PERU, Che Guevara's *foco* model for BOLIVIA, the current creation and recognition of separate political areas in COLOMBIA, and the numerous rebel-controlled areas in Africa, all provide examples of this process.

It is a case of using geography to change politics (the geography behind politics). It is not the same as creating colonies and thus colonialism. If the base should seek only regional autonomy or a redress of local grievance, then it is not an insurgent state. It would be a guerrilla base or bandit area.

In the 21st century, the al Qaeda movement created by Osama bin Laden has changed the scale for the creation of a politically focused insurgent state from local to global. This movement seeks to establish insurgent states among conservative Muslims, and then to use these bases (insurgent states) to change regional and perhaps global governments. However, this is not the first such movement to seek a global change using small enclaves. The early Christians and later the Roman Catholic Church have used similar techniques in pursuit of their goals.

BIBLIOGRAPHY. R.W. McColl, "The Insurgent State: Territorial Bases of Revolution," *Annals of the Association of American Geographers* (v.59/4, 1969); R.W. McColl and David Newman, "States in Formation: The Ghetto State as Typified in the West Bank and Gaza Strip," *GeoJournal* (v.28/3, 1992); H. J. De Blij and Peter O. Muller, *Geography: Realms, Regions, and Concepts* (Wiley, 2001).

R.W. MCCOLL, PH.D.
GENERAL EDITOR

intercropping

THE CULTIVATION OF TWO or more crops in combination in the same field at the same time is known as intercropping. This is one of two types of multiple cropping, the other being sequential cropping, whereby two or more crops are grown in sequence in the same field. There are four types of intercropping:

1) Mixed intercropping is the cultivation of two or more crops that are randomly distributed rather than grown in rows. This practice is typical of slash-and-burn agriculture, which relies on cutting and firing.

2) Row intercropping involves the cultivation of different crops in adjacent rows. This is typical of agricultural systems with intermediate technology such as plows.

3) Strip intercropping utilizes strips of land rather than narrow rows. Each strip is of sufficient magnitude that it can be cultivated independently, but the width does not preclude interaction between the crops, such as the prevention of disease as individual crops act as barriers. This practice is characteristic of commercial large-scale farming, which is mechanized.

4) Relay intercropping involves the planting of a second crop into a first crop that is partway through its growth but is not yet ready for harvesting. It may incorporate elements of some of the above.

Where a tree crop is present, between the rows of which other crops or grass for fodder are grown, the system is referred to as agroforestry, but it is nevertheless a type of intercropping.

There are several advantages to intercropping that relate to socioeconomic factors and environmental factors. In an economic context, farmers practicing intercropping rarely experience total crop failure and so have a safety net provided by the successful crop(s). Farmers and their families may develop self-sufficiency when they cultivate crops for various purposes such as food, fiber, the feeding of animals and medicines. The practice also facilitates the spread of labor because intercropped species are planted, tended, and harvested at different times. Disadvantages are few but harvesting can be difficult if machinery or special skills are required for the different crops.

The ecological/environmental aspects of intercropping reflect the varied requirements of crop plants for nutrients, shade, light, length of growing season and disease resistance/susceptibility, as well as beneficial relationships with soil flora and fauna. It is, however, essential that an appropriate combination of species is selected. Overall, productivity increases per unit area of land in intercropped systems when compared with monocultural systems.

For example, it is advantageous to grow tall and short crop species—maize (corn) with peanuts or a root crop—especially those with different growing times, in alternate rows, as this reduces competition of light. Or two tall crops may be planted provided they have different growth rates so that they mature at different times. Leguminous crops are also important in intercropping systems because of their association with nitrogen-fixing bacteria that occupy root nodules. These fix nitrogen from the atmosphere and convert it into nitrates. Once in the soil, these salts benefit all crops and reduce or eliminate the need for costly artificial fertilizers.

An example of such an association for agriculture in a temperate environment is maize with oats and soybean, of which the latter is the legume. For a tropical environment, peanut is the legume that is intercropped with sorghum and millet.

Combinations of crops with different nutrient requirements are also desirable in order to utilize the nutrient store in the soil sustainably. In this respect, the crops complement each other and the agricultural system is similar to natural vegetation communities that tend to be diverse with the coexistence of species with complementary environmental requirements. Complementarity replaces competition.

Intercropping generally reduces the outbreak and impact of diseases and pests and so less of the produce is lost in the field. The crop variation appears to favor a wider variety of beneficial insects that prey on pests when compared with monoculture. Rows of crops not preferred by specific insects will act as barriers. Alfalfa, another legume, is especially beneficial in intercropping because it attracts more beneficial insects than most other crops. With careful management, pesticide use can be reduced. The spread of disease, such as fungal and viral pests, can also be limited by such barriers. The maintenance of a crop cover can reduce the incidence of weeds as can the establishment of a good root mat below ground if crops with different root systems are chosen.

BIBLIOGRAPHY. K. E. Giller, *Nitrogen Fixation in Tropical Cropping Systems* (CAB International, 2001); D.Q. Innis, *Intercropping and the Scientific Basis of Traditional Agriculture* (Intermediate Technology Development Group, 1997); P. Sullivan, "Intercropping Principles and Practices, 2001," www.attra.ncat.org (March 2004).

A. M. Mannion
University of Reading, United Kingdom

international date line

THE INTERNATIONAL date line (IDL) is the imaginary line on Earth slicing through the center of the PACIFIC OCEAN that separates two consecutive calendar days. Countries in the Eastern Hemisphere to the left of the line up to the Prime Meridian are always one day ahead of countries to the right of the line and west of the prime meridian. The international date line also marks 180 degrees of longitude. In spite of its name, the IDL has no force in any international treaty, law, or agreement. As a result, there have been numerous changes in the location of the line in the past, and today the line is not straight but jagged, its line in some parts extending as far as the 150 degrees longitude line.

The formation of the line has roots in a problem called the circumnavigator's paradox, in which travelers going westward would find upon returning home

that one more day than they thought had elapsed, and eastward travelers found that they seemed to take one day less than they recorded upon reaching their destination. The first known mention of the circumnavigator's paradox was in the *Taqwin al-Buldan*, written by the Syrian prince-navigator Abu 'l-Fida between 1273 and 1331.

The circumnavigator's paradox was noted by explorers of some famous journeys, including in 1519 during the first circumnavigation of the globe led by Ferdinand Magellan and during Francis Drake's westward circumnavigation of the globe in 1577–80. The problem also fascinated the world of fiction; the idea of gaining time going eastward was crucial for Jules Verne's fictional character Phileas Fogg in the classic *Around the World in Eighty Days*.

The international date line can be placed anywhere on the world, but it is most conveniently located 180 degrees from the prime or zero meridian, found at Greenwich, England. The line is also most convenient passing through only water, as was suggested by one of the early advocates of the new meridian, Erik de Put. It was not until the 1800s that new strides were taken to set a new date line, but even then, authorities were skeptical about the use of such a line, especially since countries for a long time could not even agree on a common prime meridian.

The international date line has not been static over time, its shift in time and place owing to local interests and political affiliations. One notable shift was the Philippine adjustment, where prior to the present position, the Philippine Islands had been to the east of the line. This was because Spanish travelers used to journey to the then-Spanish colony via South America, so it was convenient for them not to have to adjust their dates when they reached the PHILIPPINES. During the early 1840s the commercial interests of the Philippine Islands turned increasingly away from South America (which had become independent from SPAIN) and toward trade with the nearby Malay Peninsula, the Dutch East Indies, AUSTRALIA, and CHINA. In 1844, in order to facilitate the growth of these new trading and communication links, the American day reckoning was abolished in favor of the Asian day reckoning.

Other significant changes include ALASKA, which was to the west of the International Date Line when Alaska belonged to RUSSIA, because most travelers arrived there by way of SIBERIA. When the UNITED STATES bought Alaska in 1867 the line was moved to the west of it. The most recent change was in 1995, when KIRIBATI was moved to the east of the line so that the entire nation would be on the same side of the line and the same day.

BIBLIOGRAPHY. Robert H. van Gent, "A History of the International Date Line," www.phys.uu.nl (September 2004); United States Naval Observatory Astronomical Applications Department, "The International Date Line," www.aa.usno. navy.mil (September 2004)

THERESA WONG
OHIO STATE UNIVERSITY

Intifada

INTIFADA IS THE name given to the popular uprising of the Palestinian population against the continuation of Israeli occupation of the West Bank and Gaza Strip. There have been two Intifadas, the first, essentially spontaneous and indigenous, began in December 1987 and continued sporadically until the early 1990s. The second Intifada began after the collapse of the Camp David peace talks in September 2000. The Intifada was then under the control of various political parties and the Palestinian Authority and was a conscious military tactic, using the name of a popular precursor.

Many political commentators have attributed the holding of the Madrid peace talks in 2001 and the signing of the Oslo Accords in 1993 as resulting, in part, from the events of the first Intifada and the realization that a democratic Israel, even with the strongest military force in the region, could not continue to control a civilian population of 3 million people against their wishes and their demand for sovereignty and independence.

The first Intifada broke out at the end of 1987 initially in the Gaza Strip. It soon spread to all of the occupied territories. It was the first time in 20 years of Israeli occupation that a geographically coordinated mass civilian uprising against Israeli occupation and the occupation military authorities had taken place. It was characterized by widespread acts of civil disobedience and the use of nonlethal weapons. The throwing of stones and the burning of tires and blocking of roads, followed by Israeli military retaliations and the imposition of curfews on Palestinian villages, became an almost common occurrence during this period. The actions united and coordinated once separate and distinct villages and people. It was an organization prima-

rily of youth and not controlled by the Palestine Liberation Organization or Yassir Arafat. For the first time, it became unsafe for Israelis to travel throughout the West Bank. Former places of Israeli-Palestinian interaction, such as the weekend markets in the Palestinian town of Kalkliyah and the commercial nodes of East Jerusalem, ceased to function as a result of the fear on the part of Israelis.

The second Intifada broke out following the breakdown of the Camp David Peace Talks in 2000. Each side blamed the other for the failure to reach a final peace agreement. The frustration among the Palestinian population, who had expected, finally, to achieve their independence, became evident through renewed acts of violence against Israel and the Israeli military authorities. Unlike the first Intifada, the second Intifada was not spontaneous. It also was characterized by a shift to the use of lethal weapons, many of which had been accumulated during the 1990s.

As a result of the Oslo Accords, the Palestinians had achieved limited autonomy in parts of the West Bank and Gaza Strip. The influence and direct intervention of the more radical political organization, the Hamas (which was in its infancy during the period of the first Intifada) also resulted in the introduction of suicide bombers in Israeli civilian centers causing significant carnage and death inside Israel itself. The conflict was now carried outside the occupied territories and into once safe districts of Israel itself.

The demand by Israelis for greater security resulted in the election of a hard-line right-wing government under the leadership of Prime Minister Ariel Sharon, who in turn proceeded to use the full might of the Israeli army to retaliate against the Palestinian population, in some cases almost totally destroying the civilian infrastructure that had been established during the post-Oslo period.

The Israeli government argued that it was the responsibility of the Palestinian Authority, under he leadership of Chairman Arafat, to reign in the violence, despite the fact that he had gradually lost control of the more radical elements within the Palestinian population, and especially throughout the Gaza Strip which had become the headquarters of the Hamas movement.

The Israeli government also targeted leading Hamas leaders, most notably Sheikh Ahmed Yassin, for assassination (in 2004) while claiming that they could no longer do business with Arafat and that, in the absence of an alternative Palestinian leader, there was no "partner" for political negotiations. This position was backed up by the George W. Bush administration, bringing about the almost total isolation of Arafat and the suspension of any peace talks between the two sides. With Arafat's death in 2004, newly elected Palestinian leaders brought a fresh approach to negotiations.

BIBLIOGRAPHY. T. Friedman, *From Beirut to Jerusalem* (Times Books, 1999); Don Peretz, *Intifada: The Palestinian Uprising* (Westview Press, 1990); Jamal R. Nassar and Roger Heacock, *Intifada: Palestine at the Crossroads* (Praeger, 1990); Debeorah Gerner, *One Land, Two Peoples: The Conflict over Palestine* (Westview Press 1994); R.W. McColl and David Newman, "States in Formation: The Ghetto State as Typified in the West Bank and Gaza Strip," *GeoJournal* (1992).

David Newman
Ben Gurion University, Israel

Iowa

IOWA IS A midwestern state in the north-central part of the UNITED STATES bordered on two sides by major rivers. To the east is the MISSISSIPPI that separates Iowa from ILLINOIS and WISCONSIN. In the west, the Missouri River separates it from NEBRASKA, while the Big Sioux River separates it from SOUTH DAKOTA from Sioux City northward. To the north lies MINNESOTA and to the south MISSOURI.

Interestingly, there is a small heel of Iowa that extends south into Missouri in the southeastern corner of the state along the MISSISSIPPI RIVER. Here, Iowa actually has a western boundary with Missouri, since you can leave the city of Keokuk and drive westward into Missouri.

The state's rich, fertile plains are generally low and gently sloping for the eastern three-quarters of the state. Rivers in the area, including the Des Moines River, which flows from Minnesota through the middle of the state, all flow in a southeasterly direction toward the Mississippi River. There is a small region in the extreme northeastern corner that was not glaciated during the last ice age that Iowans often refer to as the Switzerland of Iowa because of the visual effect the sharp cuts in the river bluffs along the Mississippi River give to the landscape.

The western one-quarter of the state is somewhat higher in elevation, with numerous rivers dissecting the hill country along the western border as they flow in a

southwesterly direction toward the Missouri River. Elevations in the state range for a low of 380 ft (116 m) above sea level in the boot heel region in the very southeastern corner to 1,670 ft (509 m) at Hawkeye Point in the northwestern corner of the state. Slightly more than 100 years ago, you would find extensive woodlands of hickory and black walnut in the state as well as extensive prairies with grass higher than the wagon wheels of the settlers' wagons. But rapid settlement during the 19th century brought on rapid deforestation and the plow turned nearly all of the prairie grasslands into farmland for corn and other grains.

Des Moines (pronounced without the "s" sound in either word) is the capital and largest city, with 198,682 residents (2000). Other important cities near or above the 100,000 mark are Cedar Rapids (120,758) and Davenport (98,359). Iowa ranks 30th in total population, with 2,926,324 residents, and 26th in land area at 56,276 square mi (145,756 square km). The state's 10 largest cities account for 29 percent of the total population, enough to drop the density feeling as you drive around the state to 36 persons per square mi (94 per square km) from the official figure of 52.37 per square mi (136 per square km). Iowa ranks 33rd in per capita income ($26,431) and 35th in disposable per capita income ($22,949).

EARLY SETTLERS

Prior to the first Europeans coming to Iowa, the area was home to prehistoric Native Americans mound builders who were primarily farmers. By the time the Europeans arrived in the 17th century, the Iowa, Sac, and Fox roamed over the land, in addition to the Sioux who dominated the area. Fur traders found the land profitable during the late 1600s, establishing a number of small towns along the Mississippi River and Des Moines River junction.

By the late 1700s, the area near present-day Dubuque was leased from the Indians for lead mining. Iowa came into the control of the UNITED STATES in 1803 following the LOUISIANA PURCHASE. Until 1821, Iowa was part of the Missouri Territory. After 1821, Iowa was part of both the Michigan Territory and Wisconsin Territory prior to being organized as the Iowa Territory in 1838. With a reputation for rich soil, Iowa grew rapidly as new immigrants flocked to the state. Many of these came from Europe, notably Germans, Czechs, Dutch, and Scandinavians, bringing their agricultural skills and own customs to enrich Iowa's rural life. One group of German Pietists established the Amana Church Society, a successful attempt at communal organization still recognizable as the Amana Colonies. Iowa became a state in 1846 and in 1857 the capital was moved from Iowa City to Des Moines.

Iowa truly is a breadbasket for the United States with 90 percent of the land devoted to farming. The deep, porous soils help sustain yields in corn and other grains in tremendous quantities. Iowa often leads the nation in the production of corn, soybeans, hogs, and pigs, and is ranked in the top 10 in the raising of cattle. Iowa corn-fed hogs and cattle are nationally known. And Iowa consistently ranks high among farm states in terms of farm income. Abundant and consistent agriculture yields have also benefited the state's food processing industry. Non-electrical machinery, farm machinery, tires, appliances, electronic equipment, and chemicals are among the other manufactured products. Mineral production is small and centered around cement (the most important mineral product) stone, sand, gravel, gypsum, and some coal. Communications, finance, and insurance are especially important in Des Moines; the area is often referred to as the Hartford (from Hartford, CONNECTICUT, insurance companies) of the Midwest.

BIBLIOGRAPHY. Sandra J. Christian, *Iowa* (Capstone Press, 2003); Rita C. LaDoux, *Iowa* (Lerner Publications, 1992); H.L.Nelson, *A Geography of Iowa* (University of Nebraska Press, 1967); Mark R. Doty, *An Introduction to the Geography of Iowa* (Great Raven Press, 1979); Dorothy Schwieder, *Iowa: Past to Present* (Iowa State Press, 1989); U.S. Census Bureau, www.census.gov (August 2004).

RICHARD W. DAWSON
CHINA AGRICULTURAL UNIVERSITY

Iran

Map Page 1122 **Area** 636,296 square mi (1,648,000 square km) **Population** 68,278,826 **Capital** Tehran **Highest Point** 18,605 ft (5,671 m) **Lowest Point** -91 ft (-28 m) **GDP per capita** $6,800 (2002) **Primary Natural Resources** petroleum, natural gas, coal.

IRAN IS A MIDDLE EASTERN country, bordering the Gulf of Oman, the PERSIAN GULF, and the CASPIAN

SEA, between IRAQ and PAKISTAN. Iran is a large country that is dominated by rugged mountains and large deserts. At its center is the Iranian Plateau, with great mountain ranges such as the ZAGROS in the west, the ELBURZ in the north along the Caspian Sea coast, and the mountains of Khurasan to the northeast. These mountains and the high plateau create a bowl-shaped arid expanse broken by oases that mark the ancient caravan routes. Along the Caspian coast in the north a subtropical climate exists, supporting the primary fertile region of Iran.

The Iran of today stands upon an ancient base of civilization and culture. Long a crossroads for conquerors, it has been known in the West as Persia since the beginnings of historical records associated with Cyrus the Great in the 6th century B.C.E. Alexander the Great brought Hellenistic culture to Persia in the 4th century B.C.E. and opened the gate to a long procession of foreign dynasties that would rule there. Greeks, Turks, Arab-Muslims, Mongols, and Tamerlane all made their mark as centuries passed. It was during the Safayid dynasty (1502–1736) that one of the great Persian empires arose. Under Shah Abbas, the Shia branch of Islam was made predominant and the holy city of Esfahan was rebuilt as the capital.

Invasion, succession of dynasties, and finally a nationalist uprising brought Iran into the 20th century. In 1925, a final dynasty was created by an army officer who had seized power and proclaimed himself shah, creating a new Pahlavi dynasty. Under Shah Reza Khan, Iran began to modernize and consolidate power as a nation-state. In 1941, during occupation of western Iran by UNITED KINGDOM and Soviet Union forces, the shah was forced to abdicate and his son Mohammad Reza Pahlavi became the new shah. At the end of World War II, allied forces were slow to withdraw, but with pressure from the UNITED STATES, Soviet forces left and the young shah reestablished his power over all Iran. He became closely aligned with the West. His pro-modernization and pro-Western stance did not hold well with the people of Iran.

Over his three decades of rule, repression and economic difficulties were the norm. The shah fled the country and an exiled religious leader, Ayatollah Ruhollah Khomeini, returned to witness a popular Islamic revolution sweep the country. The United States had long supported the now-shunned Shah Pahlavi and was associated with the feared and hated secret police, the SAVAK. In 1979, the U.S. embassy in Tehran was seized and the embassy personnel were held hostage for over a year.

Conservative clerical forces established a theocratic government within an Islamic republic. Soon after Ayatollah Khomeini's return and rise as Iran's religious leader, neighboring Iraq invaded Iran along its western border and started a bloody eight years of warfare. The West no longer supported the government of Iran and backed Iraq with arms sales and military intelligence during the deadly and futile conflict that raged until 1988.

Relationships with the West remain difficult today as questions of Iran-sponsored terrorism and assassinations are topics of official communications. The United States maintains a trade embargo against Iran. Calls for United Nations scrutiny of Iranian atomic energy programs have increased in light of concerns with proliferation of nuclear weapons technology and pursuit of weapons of mass destruction.

The Persian language, also known as Farsi, is spoken by over half of the population and is written with Arabic script. Turkic dialects make up the next largest language group. The peoples of Persia have long been known for their artistic excellence in handcrafts, architecture, and poetry. Iran has a rich ethnic mix across its landscape. Whereas more than 60 percent of the population is of Aryan origin, large groups of Turkic, Persians, Kurds, Lurs, and Baluchi are found as well. Nearly two-thirds of the population now resides in urban areas where some modernity has taken hold, but the remainder lives in a vast land close to the rhythms of rural life.

Iran's constitution now proclaims that Shia Islam is the official religion. Some 89 percent of the Islamic population of Iran belongs to the Shia branch of ISLAM. The Sunni branch has historically controlled leadership in neighboring Muslim countries. Clerics now dominate politics and almost all aspects of urban and rural life in Iran. Iran is most closely associated with strict Islamic beliefs yet is also the birthplace of the ancient Zoroastrian faith and the more recent Baha'i religion.

Iran's economy is based upon oil production, with over 80 percent of all revenues stemming from this industry. The nation's overall economy requires more diversification and modernization. Iran's economy is a mixture of central planning, state ownership of oil and large scale enterprises, village agriculture, and small-scale private trading and service ventures. Economic reforms continue but at a measured pace and with mixed results.

BIBLIOGRAPHY. "Background Note: Iran," U.S. Department of State, www.state.gov (April 2004); *World Factbook*

(CIA, 2004); *Lonely Planet World Guide*, www.lonelyplanet.com (April 2004); Peter Avery, Gavin Hambly, and Charles Melville, eds., *The Cambridge History of Iran* (Cambridge University Press, 1991); Hanns W. Maull and Otto Pick, eds., *The Gulf War* (Printer Publishers, 1989); H.J. de Blij and Peter O. Muller, *Geography: Realms, Regions, and Concepts* (Wiley, 2002).

IVAN B. WELCH
OMNI INTELLIGENCE, INC.

Iraq

Map Page 1122 **Area** 168,754 square mi (437,072 square km) **Population** 24,683,313 **Capital** Baghdad **Highest Point** 11,847 ft (3,611 m) **Lowest Point** 0 m **GDP per capita** $2,400 **Primary Natural Resources** petroleum, natural gas, phosphates.

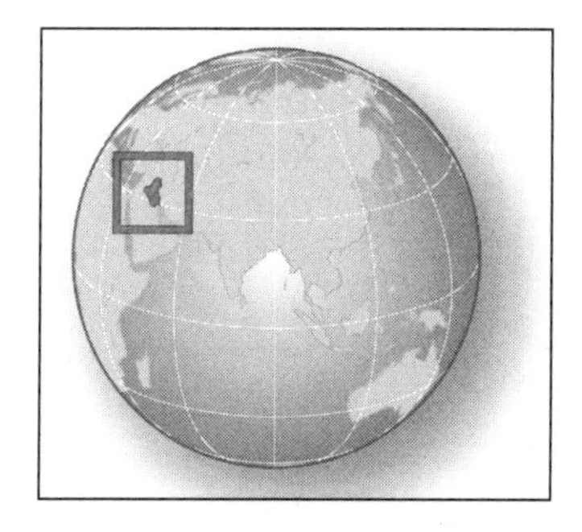

IRAQ IS A MIDDLE EASTERN country, bordering the PERSIAN GULF, between IRAN and KUWAIT. Iraq is also bordered by TURKEY, SYRIA, JORDAN, and SAUDI ARABIA. The country contains most of the large flat basin created by the Tigris and Euphrates rivers. Both great rivers have their origin in high mountains outside of Iraq. The expansive landscape slopes down from these heights descending gradually to the Persian Gulf in the southeast.

Here, the two rivers have created a large DELTA that forms the Shatt al-Arab shared with Iran. To the south, the broad plains mingle with the vast deserts of Saudi Arabia. The mountainous region of northern Iraq receives appreciably more precipitation than the central or southern desert regions, yet extensive irrigation makes much of the country arable along the two great rivers.

From ancient times, the area of Iraq has been known as Mesopotamia. One of the great hearths of human civilization, the Tigris-Euphrates river complex has supported many kingdoms, empires, and dynasties. Sumerian, Babylonian, and Parthian cultures emerged and flourished before the faith of ISLAM made its way to Iraq in the 7th century C.E. A great Islamic empire, the Abassid caliphate, established its capital at Baghdad, which, in turn, eventually became an outpost of the OTTOMAN EMPIRE. Centuries of commercial trade and cultural exchange continued in a stark land as the fortunes of the great Turkic empire of the Ottomans and other Islamic dynasties, such as the Safavids, rose and fell. As the Ottoman Empire expired at the end of World War I, Iraq was occupied by the British and in 1920 became a British-mandated territory under the auspice of the League of Nations. When declared independent in 1932, the Hashemite family, which also ruled Jordan, ruled as a constitutional monarchy.

A republic was proclaimed in 1958, with the killing of the monarch and establishment of power by a military strongman. Continuing political assassinations and overthrows eventually brought the Arab Socialist Renaissance Party (Ba'ath Party) to power in 1963. In 1979, a protégé of the party, Saddam Hussein, was selected by the outgoing ruler and began his role as supreme leader.

Acting in the bellicose style of his predecessors, Hussein used territorial disputes with Iran as cause to launch a bloody and inconclusive war that lasted for eight years (1980–88). The West supported Iraq by sale of arms and provision of military intelligence. This war saw the first large-scale use of chemical weapons since the battlefields of Europe in World War I. The use of these weapons was to haunt Hussein's regime. Facing tremendous debt, degraded commercial access to the gulf, and dismal return on vast military expenditures, Saddam used similar logic and bluster when in August 1990 Iraq invaded and annexed Kuwait. The threat to the free flow of oil to the industrialized West and Asia galvanized a United Nations (UN) coalition, launching a U.S.-led attack that freed Kuwait in 1991 and destroyed much of Iraq's military capability. The UN Security Council required Iraq to end all production and procurement programs related to weapons of mass destruction and long-range missiles. This was to be verified by UN inspectors, but because of noncompliance by Iraq, a decade of sanctions, embargoes, and military actions ensued.

In March 2003, the UNITED STATES led an invasion of Iraq and the overthrow of the Hussein regime. The occupying U.S.-led coalition and recognized Iraqi leaders established advisory and governing councils on the local, regional and national level to ensure that the path into the political future results in elections to establish an internationally recognized representative government in Iraq.

Most of Iraq's population lives on the ALLUVIAL PLAIN stretching from the mountains of the north toward the Persian Gulf off to the southeast. The Tigris

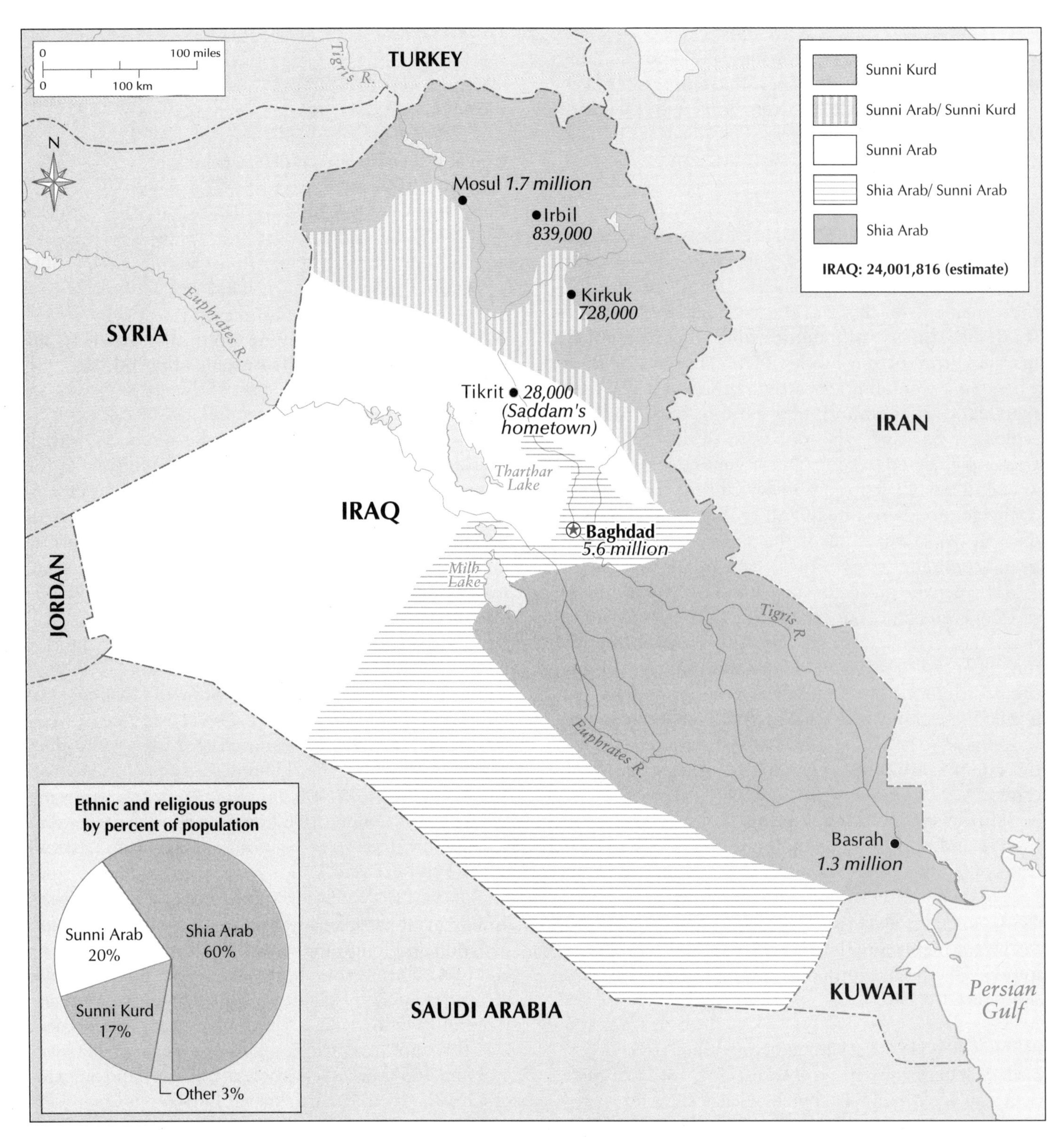

The Arabs of Iraq are predominantly followers of the Shia sect of Islam, but have long been oppressed by political leaders holding to the Sunni sect. The U.S.-sponsored elections of 2005 reflected the Shia majority population base.

and Euphrates rivers continue to be the focus of human settlement as they have been since ancient times. The ruins of great cities of history such as Ur and Babylon share this landscape with the biblical location of the Garden of Eden. Over three-quarters of the Iraqi population is Arab, but the Kurds make up a significant

ethnic minority. The Kurds are most closely associated with the mountainous regions and hold to the Sunni sect of Islam. The Kurds of Iraq have long shared the misfortunes of a stateless nation along with their kinsman in Turkey, Iran, and Syria. The Kurds remain culturally distinct in language, customs, and politics.

MARSH ARABS

The majority of Arabs in Iraq are followers of the Shia sect of Islam but have long been oppressed by political leaders holding to the Sunni sect. One of the more notable incidents of this was the systematic effort to destroy the Marsh Arab culture found in the southern marshes near of the mouth of the Euphrates River. These Shia communities had long thrived in the reedy marshlands that comprised a natural food-producing region. A Ba'ath regime policy to drain the marshes was implemented to control insurgent movement in the area and introduce irrigation agriculture. This massive drainage project resulted in an ecological and cultural disaster, as salinization of the soil spread and traditional communities were displaced.

The petroleum sector has dominated the economy of Iraqi even before nationalization of its oil industry in the 1970s. The disastrous war with Iran followed by the ill-conceived invasion of Kuwait left Iraq with hundreds of billions of dollars of international debt, economic losses, and war reparations. Further economic sanctions by the UN related to disarmament removed the country from any meaningful role in the world economy.

However, with the world's third-largest proven oil reserve and promising unexplored oil-bearing regions, Iraqi oil stands to create significant revenue for some time to come. After the U.S.-led invasion, the UN Security Council lifted all sanctions against Iraq and passed resolutions to ensure that Iraq's oil export earnings are immune from legal proceedings, such as debt collection, until the end of 2007.

BIBLIOGRAPHY. "Background Note: Iraq," U.S. Department of State, www.state.gov (April 2004); *World Factbook* (CIA, 2004); "Iraq," Energy Intelligence Agency, www.eia.doe.gov (April 2003); Ofra Bengio, *Saddam's Word, Political Discourse in Iraq* (Oxford University Press, 1998); Phebe Marr, *The Modern History of Iraq* (Westview Press, 1985); H.J. de Blij and Peter O. Muller, *Geography: Realms, Regions, and Concepts* (Wiley, 2002).

IVAN B. WELCH
OMNI INTELLIGENCE, INC.

Ireland

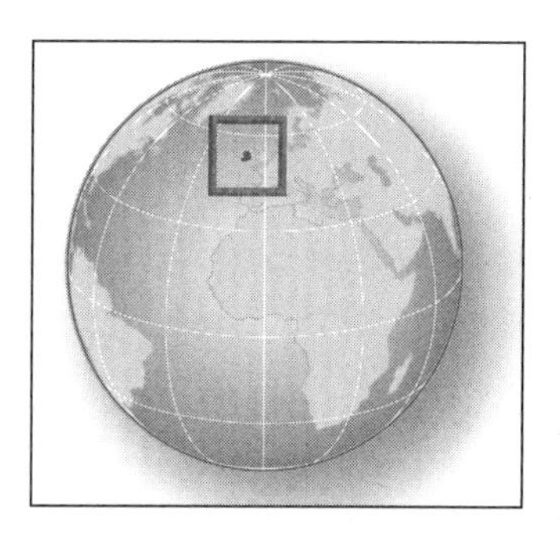

Map Page 1131 **Area** 27,135 square mi (70,280 square km) **Population** 3,924,140 **Capital** Dublin **Highest Point** 3,414 ft (1,041 m) **Lowest Point** 0 m **GDP per capita** $30,500 **Primary Natural Resources** zinc, lead, natural gas, barite, copper, gypsum.

THE REPUBLIC OF Ireland covers five-sixths of the island of Ireland and shares its only border with Northern Ireland, which is part of the UNITED KINGDOM of Great Britain and Northern Ireland, also known by the historic name of Ulster. Ireland is a republic with a president as head of state and a prime minister as the head of government. Its legislative branch is a bicameral parliament. Ireland is divided into four provinces and 26 counties. English is widely used throughout Ireland, but Gaelic is also spoken along the western coast. Ireland's major cities are Dublin, Cork, Limerick, Waterford, Galway, Dundalk, and Kilkenny.

Ireland's terrain is mostly level to rolling interior plain with rugged hills and low mountains. The west coast is studded with sea cliffs, while the east coast of Ireland has few indentations. The central part of Ireland consists of bogs, meadows, and lakes. The chief rivers in Ireland are the Shannon, Boyne, and Blackwater. The temperature ranges from 40 degrees F (4 degrees C) in the winter to 62 degrees F (16 degrees C) in the summer. Ireland receives an average rainfall of 40 in (102 cm) per year.

Human settlement in Ireland goes back more than 10,000 years during the Mesolithic period. Ireland experienced waves of migrations in its early history. The Gaels, who are the ancestors of the modern Irish people, first settled in Ireland in 700 B.C.E. Christianity arrived in Ireland around the third century, although it was through St. Patrick's missionary efforts between 432 and 465 that it took root, making Ireland a center of Christianity in early medieval Europe.

England began its control of Ireland in the 12th century when Dermott MacMurrough of Leinster sought the assistance of Henry II of England in his battles against other Irish chieftains. The conquest of Ireland was complete in 1541 when Henry VIII was declared king of Ireland by the Irish Parliament following a rebellion. Under the Tudors, England began the "plantations," which was systematically settling parts

of Ireland with English settlers. Ireland was also pulled into the English Civil War and suffered under the resulting rule under Oliver Cromwell in the 17th century.

During the Restoration in 1660 under Charles II, the Irish Catholics were relieved of the persecutions imposed upon them. With the succession of James II in 1685, they were hopeful of having a Catholic on the throne again. Even though James II was deposed during the Glorious Revolution of 1689, Irish Catholics continued to recognize him as their king. In 1690, the forces of King William defeated the forces of James II in the Battle of the Boyne, resulting in the Penal Laws, which not only marginalized Irish Catholics from the political and economic life but also witnessed the rise of an Anglo-Irish elite. Economic conditions under absentee landlords led to the Rebellion of 1798.

Through the Act of Union in 1801, Ireland became part of the United Kingdom of Great Britain and Ireland. By 1829, Irish Catholics benefited from the Catholic Emancipation Act. Between 1845 and 1851 the Great Famine ravaged throughout Ireland, which reduced its population from 8 million in 1845 to 2 million in 1851 due to starvation and emigration. The famine exposed the flaws in the Irish tenure system and British governance. The latter half of the nineteenth century was consumed by the question of home rule for Ireland. Home rule remained an intractable question because Ireland was divided between the Irish Catholics who supported an independent parliament and internal autonomy and Protestants in Ulster who wished to remain loyal to Westminster.

In 1920, after the cataclysm of World War I and a guerrilla war against the British army, the Government of Ireland Act divided Ireland into twenty-six southern counties that were represented by its own parliament in Dublin, and the six counties in Ulster that continued to be represented in Westminster. The Anglo-Irish treaty retained allegiance to the British sovereign and naval bases on the Irish coast. In 1937, the Irish Free State became Eire, which repudiated the Anglo-Irish Treaty of 1921.

During World War II, Eire declared its neutrality, depriving Britain use of naval bases in the southern coast. In 1948, the Republic of Ireland Act severed all ties to Britain, the Empire, and the Commonwealth, while Northern Ireland remained loyal to the Crown. In turn the British parliament passed the Ireland Act, which gave rights to all citizens of the Republic of Ireland who traveled to Britain. With the outbreak of violence in Northern Ireland in 1972, the Republic of Ireland and Britain have since cooperated against the Irish Republican Army and other terrorist groups. In 1973, the Republic of Ireland joined the European Economic Community. In 1998, the Republic of Ireland has been a participant in the Good Friday Agreement, in the resolution of "The Troubles" in Northern Ireland, though efforts have stalled in recent years. In 1999, Ireland adopted the euro as its currency. In 2001, a majority of Irish citizens vetoed the Treaty of Nice because of doubts on the expansion of the EUROPEAN UNION (EU). In 2004, the Republic of Ireland assumed the presidency of the EU, tackling issues such as drafting a new constitution for Europe.

Ireland has had a rapidly growing economy in recent decades, growing at an annual rate of 8 percent between 1995 and 2002. Industry makes up for 38 percent of the GDP, while services account for 49 percent of the gross domestic product. Much of its economic growth was due to its exports in technology, followed by consumer spending, construction, and business investment. Ireland's chief trading partners are Britain, the UNITED STATES, GERMANY, FRANCE, JAPAN, and the NETHERLANDS.

BIBLIOGRAPHY. *World Factbook* (CIA, 2004); Mike Cronin, *A History of Ireland* (Palgrave, 2001); "Ireland Country Profile," Economist Intelligence Unit (August 2004); Alvin Jackson, *Home Rule: An Irish History* (Weidenfield and Nicolson, 2003); James Lydon, *The Making of Ireland: From Ancient Times to the Present* (Routledge, 1998).

DINO E. BUENVIAJE
UNIVERSITY OF CALIFORNIA, RIVERSIDE

Irrawaddy River

THE IRRAWADDY is the chief river of MYANMAR, or Burma. It is formed from the confluence of the Mali and N'mai rivers far in the northern highlands on the borders with CHINA, and flows 1,350 mi (2,177 km) before entering the Andaman Sea (a section of the BAY OF BENGAL). The river's extensive DELTA begins about 140 mi (225 km) before it reaches the sea, and splits into nine main channels. It is estimated that the waters of the delta lay down 260 million tons (236 million metric tons) of silt per year.

The capital city of Yangon (Rangoon) is located on one of these delta channels, though not the main river course, located about 70 km (43 mi) to the west. About

a dozen large sea inlets form the mouths of the Irrawaddy, spanning roughly 300 km (180 mi) from west to east. The delta is protected by levees for hundreds of kilometers and is one of the chief rice-exporting areas of Southeast Asia, formerly awarding the area the nickname, the "rice bowl of Southeast Asia." Steamers can transport goods upstream as far as Bhamo (about 1,000 mi or 1,600 km), nearly to the borders with CHINA's Yunnan Province. The country's major oil pipeline follows the course of the river from oilfields near Chauk down to export facilities at Yangon.

The river valley dominates the shape of Myanmar, particularly the central historic province of Burma proper, hemmed in by the parallel chains of north to south mountains, called the Pegu Yoma and Shan Highlands in the east, and the Arakan Yoma and Chin Hills in the west.

The country's other main river, the Salween, also travels through the same sort of north-south valley, though much narrower (and actually for a much greater distance, originating far inside the Chinese border). A third river, the Sittang, while much shorter, is more comparable to the Irrawaddy in its importance as a chief rice-growing area. Most of the Burmese population lives along these valleys, both in the north, around the city of Mandalay, and in the southern delta.

The river, officially spelled Ayeyarwady, can be divided into two main sections, above and below Mandalay. Above this point it is swift and narrow, with several rocky defiles. But just below Mandalay the Irrawaddy joins with its chief confluent, the Chindwin River, and it becomes broad and slow-moving, ranging from 1 to 4 mi (1.5 to 6.5 km) wide. During seasonal floods, however, the currents can be much quicker and hazardous to river traffic. Forests along the river have virtually been eliminated and continue to be cut down at a rate of approximately 2 million acres (800,000 hectares) a year. The Mon River is also a major tributary, with one of the earliest (and largest) dams in the region, built in 1906 and recently renovated, providing extensive irrigation to the upper reaches of the country's western regions.

As the commercial center of the British colony of Burma, the Irrawaddy River valley attracted numerous commercial wet-rice planters from the 1850s, funded by money lenders from Calcutta (KOLKATA), INDIA, a source of later friction between independent India and Burma. From 1855 to 1930, the area cultivated for rice increased from 988,000 acres (400,000 hectares) to 9.8 million acres (4 million hectares), and the population increased from 1.5 million to 8 million. Production dwindled during decades of socialist rule but is starting to pick up again under programs of economic liberalization and remains the regime's most important source of foreign revenue.

Today, the river is seen as a unifier of the nation's diverse ethnic groups who populate its banks—including the Kachin, Shan, and Chin minorities—whose independent spirits frequently threaten to pull the state apart.

BIBLIOGRAPHY. Barbara A. Weightman, ed., *Dragons and Tigers: A Geography of South, East and Southeast Asia* (Wiley, 2002); *Encyclopedia Americana* (Grolier, 1997); "Myanmar," www.myanmar.gov.mm (April 2004).

JONATHAN SPANGLER
SMITHSONIAN INSTITUTION

irrigation

IRRIGATION IS THE application of water to crops in addition to what normal local precipitation supplies. It is primarily used in areas with less than 20 in (51 cm) of rain per year (semiarid or arid climates) or in areas of monsoon rains with long dry periods. Irrigation allows growth of nondryland crops in semi-arid regions or can extend the potential cropping season to allow multiple crops. It is also used in some wealthier areas (for example, CALIFORNIA, MISSOURI, and ALABAMA) to take advantage of special crop subsidies and grow water intensive crops like rice or cotton.

Irrigation has been used for thousands of years. Some of the earliest hydraulic civilizations were the FERTILE CRESCENT societies spanning from EGYPT'S NILE to Mesopotamia's Tigris and Euphrates. Other ancient irrigation cultures have been found in PAKISTAN, CHINA, INDIA, the Andean regions of South America, MEXICO, and the southwestern UNITED STATES.

Various methods of diverting and delivering water produced different levels of efficiency. The first irrigation schemes used simple diversions off of a river to flood entire fields. This method was the easiest to implement but wasted the most water. More specific application diverted water and then let gravity move water toward multiple ditches dug alongside crops. Simple yet ingenious devices were developed to raise river water up on to higher ground. For example, Egyptians first used the shaduf, which was a series of

buckets on counterweighted poles arranged at various levels on the riverbank, to lift water to the level above. Later, they used the Archimedes's screw, which was a large carved screw fitted into a hollow log. Draft animals turned large gears that turned the screw, catching and lifting water through the log to the higher land. Later, treadle pumps were devised in many poorer areas of the world so humans could power the pumps with their feet rather than use fossil fuels.

Sometimes irrigation water was transported many miles to where it was needed. In Pakistan, *qanats* were deep tunnels dug from the distant mountains down to where villages needed water. *Qanats* moved water by gravity and protected the water from evaporation since the water moved underground. Shafts emerging at intervals at the surface allowed the workers to enter the *qanats* for maintenance.

Irrigation generally required the combined effort of many farmers, so a variety of organizations and water laws were created to provide rules for individual use. The *acequias* of NEW MEXICO are an example of old Spanish water organizations created to maintain and oversee the distribution of irrigation water and maintenance of the waterworks. In the early 1900s, the United States created the Bureau of Reclamation to build massive dam projects and manage the diversion of water in the dry West to numerous irrigation districts.

The green revolution of the mid-20th century combined new seed, fertilizers, fossil fuel energy, and irrigation to significantly expand crop output worldwide. Although this was a boon for the rapidly climbing global population, the use of irrigation seems to have only delayed a reckoning between an ever-increasing human population versus a finite supply of fresh water.

Irrigation has continued to expand, and while it helps to produce higher yields, it has significant drawbacks that are becoming more obvious each year. First, irrigation changes the hydrologic regime of a region by reducing the historic flow of streams, dewatering some stream stretches or reducing recharge to connected groundwater supplies. Surface habitat is then degraded along these stream segments.

The ARAL SEA is a good example of a large, once productive sea that is now polluted and shrunk to one-third of its original size because of upstream diversion of water to irrigate cotton. Stranded fishing boats now sit in the sand far outside the current shoreline, and the local economies and human health have never recovered. Many great rivers of the world have also lost so much water to upstream diversions that they rarely reach the sea anymore. The Yellow (HUANG) River in CHINA, and the Colorado River and the RIO GRANDE in the United States, are just a few examples of this.

Groundwater tables have also dropped through over-pumping aquifers with fossil-fueled pumps. Some known examples of pumping these fossil aquifers are the Ogallala in America's high plains, the Libyan desert, YEMEN, and SAUDI ARABIA. Some farm communities over the Ogallala have already disappeared once local irrigators ran out of water. Sanaa, the capital of Yemen has scant years before running out of its water supply.

In areas where diverted water is delivered, groundwater levels can rise and move salts up into the root zones of crops, killing the crops and permanently ruining the soil. This is commonplace in arid areas, where thousands of hectares worldwide have become sterile and been abandoned because of soil salinization. Besides soil degradation, agricultural runoff carries pesticides, fertilizers and salts back into the stream or groundwater, degrading downstream habitat and poisoning groundwater needed for drinking water.

Modern irrigation is also energy-intensive. When fossil fuels were more abundant, it made sense to use them to access as much water as possible. Relatively cheap oil and pumps helped proliferate irrigation around the globe and helped to speed the unsustainable rates of irrigation common today. Now in the face of increasingly limited supplies and even more global demand, the cost of pumping water will become a far more significant portion of the overall costs of irrigation and add to the cost of food production.

Since the beginning of the 20th century, large dam projects have been built by wealthier countries to supply regions with irrigation water as well as flood control, municipal water supply, and hydroelectricity. More dams are now being built in developing countries, with the World Bank and other lending institutions helping to finance construction. Because 90 percent of the world's children between now and 2050 will be born in the developing countries, the need for irrigation water to grow additional food is obvious.

However, these dam projects are not without numerous costs, and their benefits do not flow equally to the inhabitants of the region the project is in. First, the location of the dam project inevitably displaces many villages, with little or no attention or assistance to help relocate those people. Impounding water spreads a variety of waterborne diseases in poor regions with little or no access to medicine or doctors. Aquatic habitats are drastically changed upstream and downstream of

the dams, reducing the types and quantity of fish that were there before. Nutrients in the sediment that once flowed downstream during flood events is now caught behind the dam, leaving downstream valleys in need of alternative fertilizers. Impounding the water also changes the speed and temperature of the water upstream and downstream, again altering aquatic habitat and reducing fish species. Once the dam is filled and irrigation water is delivered, it does not tend to be delivered efficiently, effectively, or equitably. Consequently, vast amounts of money can be spent on irrigation projects that ultimately benefit only a few people relative to the needs of a region.

Almost 90 percent of water consumed by humans is used for irrigation, and demand for water for all uses is increasing as human population increases and as consumption increases in parts of the developing world. Consequently, competition for water is rising everywhere. Irrigation water is applied to over 40 percent of the world's crops, so any gains in efficiency in this sector would supply additional water for other competing uses.

Modern irrigation practices have increased efficiency by using drip hoses laid at the base of plants or have installed low-hanging sprinklers from center pivot systems to apply the water directly to the plants rather than spraying it up in the air and losing more of it to evaporation.

A rising trend in developing countries is to use sewage effluent for irrigation water. Ten percent of the world's irrigated crops are now irrigated with sewage effluent, which is either partially treated or not treated at all. Despite the health threat to farmers and consumers from using sewage effluent, farmers continue to use it since it is more reliable than local rains and so guarantees them a more reliable income. Farmers will often sell this produce at distant markets so no one will know its provenance or the pathogens it may contain.

Irrigation had fed the world's people for millennia but must keep adapting, refining, and reinventing itself to continue as a solution for agricultural production.

BIBLIOGRAPHY. Harm J. de Blij and Peter O. Muller, *Geography: Realms, Regions and Concepts* (Wiley, 2002); Arthur N. Strahler and Arthur H. Strahler, *Physical Geography: Science and Systems of the Human Environment* (Wiley, 2005); John E. Oliver and John J. Hidore, *Climatology: An Atmospheric Science* (Prentice Hall, 2002).

Laurel E. Phoenix
University of Wisconsin, Green Bay

Islam

MOST RELIGIONS, such as Confucianism, Buddhism, Jainism, and Judaism have a geographic point or place of origin and remain largely focused upon that nexus. Some, such as Christianity and Islam, have engaged in specific efforts to proselytize and spread their faith, often as an integral part of trade. Among the significant geographic impacts of Islam are its often highly parochial nature and obsessive control and defense from foreigners and nonbelievers of its holy cities of Mecca and Medina. Islam spread its political, economic, and cultural dominance over North Africa and Spain, and eventually much of East Africa.

This geographic impact shifted trade routes, and thus culture and political and economic importance, away from the once dominant cities of the eastern Mediterranean (such as Ephesus, Antalya, Sidon and Tyre) to the coasts of Arabia, Mesopotamia (modern Iraq), and Central Asia for control of the silk and spice trade. This meant an almost immediate decline in once prosperous cities and a growing demand by Europeans to regain some kind of control or participation in the lucrative trade. One result was the Crusades, in the end more an economic event than a religious one.

Islam's behavioral requirements and origins in the deserts and barren mountains (Al Hijaz) of Arabia make it ideally suited to tribal societies, individuals, and environments where people are dependent upon themselves, the vagaries of nature, and a supernatural being. With its prohibition of usury (interest on loans) and the highly limited agricultural opportunities (including nomadism) in the lands of its origin, the followers of Islam were virtually forced to emphasize trade and the control of richer agricultural lands.

Thus, the geographic dispersal of Islam, from its origins and core in the Mecca and Medina area, clearly followed Arab trade routes, not those of the Greeks and Romans. It was only after Islam had control of the silk and spice trade that it began to spread into the Mediterranean, and then only where there was profitable trade. Islam's key cities and cultural centers remained focused upon the axis from Samarqand to Baghdad. One might say you could follow the camel and dhow to define the geographic spread and influence of Islam.

A WAY OF LIFE

Islam derives its name from the Arabic words *Salm* or *Silm*. In both cases, it stands for submission, surrender, and peace. The two words derived from three Arabic

radicals *s-l-m*, which connotes peace among other meanings. When the letters are pronounced as *salima*, it means "to submit" or "to obey" and may also mean "to propose peace," as in the Muslim greetings: *As-salam 'alaykum*, or "May Allah's peace abide with you."

Succinctly put, Islam is a way of life based on peace and submission. It is based on a peaceful relationship between human beings and their creator, Allah, on the one hand; and a peaceful relationship between fellow human beings and other creatures, on the other hand. Islam therefore claims to have precepts that give sufficient guidelines to its votaries on all aspects of lives, spiritual and ephemeral.

It is also seen as a universal way of life that has evolved since the time of Prophet Adam, and all prophets mentioned in the "revealed scriptures" are regarded as Muslims: Prophets Adam, Ibrahim (Abraham), Nuh (Noah), Musa (Moses), and even cIsa (Jesus Christ) are referred to in various passages of the Qur'an as Muslims. The Prophet Yacqub (Jacob) is reported to have inducted his children into Islam and ensured, even on his deathbed, that they were counseled to remain steadfast as Muslims.

The sources of guidance (and information) in Islam are primarily the Qur'an and the Sunnah. Other sources of secondary importance derive their authorities from the two primary sources. Ijma, for example, as a secondary source, is based on the unanimity of all the learned Muslims in a particular age who have attained the level of Ijthad. Ijthad is the capacity for individual juridical interpretation upon a certain issue after the death of Muhammad, the prophet. The authority for Ijma is based on a prophetic statement that "my Ummah [community, society] will not agree on an error," pointing to the fact that jurists among the Muslims will not be misguided while making decisions on issues of mutual importance to the Ummah.

Muslims see the Qu'ran as the first primary source of guidance and information, as the original source from which all ordinances of Islam are drawn. The Qur'an is variously divided for easy reference. It has 114 *surahs*, each of which is begun with the verse "In The Name Of Allah..." except one, the 9th *surah*. It is also divided into 30 parts in which, each of them is referred to as a *juz'*, or 60 parts, each of which is called a *hizb*.

The Qur'an is seen as the compendium of all divine revelations issued to prophets before Muhammad, including 10 of such scriptures sent to Adam, 50 to Prophet Shith, while Prophet Idris received 30. Zabur, Tawrah, Injil, and Qur'an were revealed to Prophets Daud, Musa, cIsa, and Muhammad. The Qur'an, from the perspective of Islam is the last divine revelation in form of a scripture, which was revealed at a time when human intellectual development attained the level of proper comprehension of the divine message. It thus contains guidance that will be useful for the human race till the end of time.

Islam sees every aspect of life as being noticed by Allah, be it mundane or spiritual, and for each of these aspects, commensurate recompense will be given for every action carried out. In order to prepare its adherents for a purposeful interpersonal relationship, it seeks to train them for this through the fundamental rituals and spiritual activities, which include *As-Salat* (canonical/contact prayer); *Az-Zakat* (compulsory alms); *Sawm* (compulsory fast during the holy month of Ramadan); and hajj (holy pilgrimage to Mecca and Medina).

POLITICS AND ECONOMICS OF ISLAM

The Islamic provisions on most of the concepts of life adopt the medium position in all cases. This is why the Qur'an refers to a nation built on Islamic principles as a justly balanced nation. The political concept of Islam is such that it can operate under various systems if they are inherent with the Islamic polity. This includes the principle of *Tawhid*, in which sovereignty is for Allah and not the state; the principle of *Khilafah*, which states clearly that those in authority are only ruling in trust as vicegerents of Allah the Supreme; and the principle of *Shurrah*, which depicts that in all affairs of the state, the citizenry must be consulted and that the government should be constituted by the majority of the people.

The principle of *Akhirah* promotes accountability, a situation whereby those in authority are made to be conscious of the fact that accountability is twofold. With these principles in place in a federal, unitary, monarchical type of government, such government will be viewed as Islamically compliant.

The economic system in Islam is built on zero interest. Islam thus encourages charity and trade. In fact it incorporates the five pillars of the religion (*Zakaat*) in its economic system. It allows making profits through trade but frowns at multiplying wealth through usury or any form of exploitive tendencies. Instead of encouraging the rich to be idle and multiply their wealth through multiple interest accruing from loans given to the less fortunate, it encourages them to join them as partners in business, sharing both profits and losses.

The Islamic work ethic stipulates that an adherent should strive to earn his means of livelihood or any form of labor embarked upon and it is viewed as honorable even if a man performs menial tasks. Islam frowns at any form of begging, yet it does not close its doors to charity. Whenever the need arises, it urges its votaries to aid those who genuinely need assistance.

The socio-moral teaching in Islam is based on the abilities of the votaries to avoid shameful acts, openly or secretly, and to endeavor to abide by the standard of ethical norms given by the faith. Islam further presses its insistence that adherents should live by the fundamental principles. For examples, the Holy Prophet Muhammad was reported to have said in regard to *As-Salat* (contact prayer) and *As-Sawm* (compulsory fast) that whoever these would not inhibit from lewdness and indecent acts had no reason to embark on any of the principles at all.

ABUSE OF ISLAM

Moreover, on the social level, Islam encourages its votaries to relate to others in the best of manners, irrespective of their creed or color. The Prophet was reported to have vowed that he would stand against an oppressive Muslim and in defense of a *dhimmi* (a non-Muslim under the protection of an Islamic state) on the day of judgment. He also counseled that honoring one's neighbor or guest, irrespective of his or her religious affiliation constitutes an act of faith in Allah and the last day. If all these are part of the basis of Islamic social teachings, then it cannot support aggression or terrorism. Some Muslims, such as Osama bin Ladin and his terrorist organization, have abused Islamic teachings for their own purposes.

In history, Muslims were known to have ruled large parts of the world and their political impact ceased with the abolition of the caliphate in Turkey in 1927, during the reign of Kemal Atatürk. In the contemporary world, a few countries are still referred to as Islamic states using different versions of the interpretation of Islamic principles of government to administer their respective states. IRAN, LIBYA, and SAUDI ARABIA are good examples of these states.

BIBLIOGRAPHY. M. O. Abdul, *The Religions of Islam: Series In The Studies Of Islam* (Islamic Publication Bureau, 1984); L.M. Adetona, *Introduction to the Practice of Islam* (Free Enterprise Publishers, 2001); M.M. Ali, *The Religion of Islam* (Taj Company, 1984); Judith E. Tucker, *Gender and Islamic History* (American Historical Association Publication, 2003); Richard M. Eaton, *Islamic History as Global History* (American Historical Association Publication, 2003); P. K. Hitti, *History of The Arabs* (Macmillan, 1970).

LATEEF M. ADETONA
LAGOS STATE UNIVERSITY, NIGERIA
R.W. MCCOLL, PH.D.
GENERAL EDITOR

island

AN ISLAND IS A LANDMASS smaller than a continent surrounded by water. The largest islands are GREENLAND, NEW GUINEA, BORNEO, MADAGASCAR, Baffin Island, and SUMATRA. The smallest islands are a few square miles or kilometers. An extended line of islands is an island chain. A cluster of islands is an archipelago. Islands are of either the continental, oceanic, or coral type.

A continental island is an exposed part of a continental shelf. The largest islands are of this type. Either tectonic subsidence or melting glaciers, or a combination of both, causes the edge of the ocean to spread over a continent to create islands. Narrow and shallow waters (straits) separate the islands from the larger landmass. During glacial periods, the sea level was much lower and land bridges joined such islands to continents. For this reason, continental islands' plants and wildlife, as well as their geology, are similar to those of the nearby mainland. Entire nations can be continental islands, such as IRELAND and MADAGASCAR. Other such islands can be parts of larger political entities. Vancouver Island, for instance, is part of the Canadian Province of British Columbia. Examples of prominent continental islands belonging to U.S. states are Kodiak Island (ALASKA) and Long Island (NEW YORK).

Continental islands do not have to be solid rock extensions of continents. Some are large deposits left by glaciers on continental shelves. For example, in the United States, part of Long Island, New York, is a glacier's terminal moraine and so are parts of Nantucket Island and Martha's Vineyard, MASSACHUSETTS. Additionally, ocean waves, and longshore currents deposit sand to form small-scale barrier islands that parallel shorelines. The U.S. southeast coast and gulf coast are classic examples of such shorelines.

An oceanic island is generally a single volcano, an assemblage of volcanoes, or a volcanic plateau that grew from the deep ocean floor rather than a continen-

tal shelf. Many island countries are oceanic islands, ranging from relatively large ones, such as ICELAND, NEW ZEALAND and PHILIPPINES, to smaller groupings, such as FIJI and FRENCH POLYNESIA.

It takes an enormous pile of lava to accumulate before a volcano or plateau breaches the surface water. Geographers call an oceanic island a high island because it is mountainous with rugged peaks. When a volcano of an oceanic island erupts, its flanks shudder spasmodically against the water to create seismic sea waves (TSUNAMIS). The largest TSUNAMI can travel across an ocean in a day and cause devastating floods and serious erosion on other islands and on mainland shores that lie in the wave's path.

Island-forming volcanism occurs in association with seafloor spreading, which is the movement of the oceanic lithosphere in opposite directions away from the mid ocean ridges. As a result, oceanic islands occur in three general locations: 1) on the mid-ocean ridge; 2) along edges of oceanic trenches; and 3) above stationary hot spots. The AZORES and ASCENSION are examples of islands sitting on the mid-ocean ridge. JAPAN and the Aleutian and Mariana islands are island arcs occurring on the edges of the oceanic trenches. The Hawaiian Islands are an island chain forming above a hot spot, a location where a plume of magma sits fixed in the mantle, just below a moving lithosphere.

Coral islands are made of former coral reefs. Reefs are ridges of rock or coral that are at or near the water's surface. A coral reef is an accumulation of the skeletal remains of coral polyps (invertebrate animals). Reef-building corals live in tropical waters, so coral islands are in relatively low latitudes. When corals die, their lime skeletons remain behind to build reefs made of limestone. The remains of an immense number of coral polyps make up a single reef. The reefs grow so large and in such abundance that they can become small islands. Geographers call coral islands "low islands" for their lack of mountainous relief.

Coral islands appear on continental shelves and in the deep ocean. Continental coral islands are the high parts of large limestone platforms situated on continental shelves. People living in the Caribbean call these islands cays. In FLORIDA, people call them "keys" as in Florida Keys. The islands associated with the GREAT BARRIER REEF in AUSTRALIA are also on a limestone platform. Other coral islands appear far out at sea as atolls. An atoll consists of a ring-shaped island that sits like a crown atop a submerged volcano. The volcano was once an island with a coral reef growing around its outer edge, but the seafloor slowly sank, taking the volcano below sea level. The reef grew upward as fast as the volcano subsided, and thus became a circular atoll. Virtually all atolls are in the PACIFIC OCEAN, owing to the scarcity of volcanism and seafloor subsidence in the other oceans.

BIBLIOGRAPHY. Patrick Nunn, *Pacific Island Landscapes* (Institute of Pacific Studies, 1998); Harold V. Thurman and Allan P. Trujillo, *The Essentials of Oceanography* (Prentice Hall, 2001); Robert E. Gabler, James F. Peterson, and L. Michael Trapasso, *Essentials of Physical Geography* (Brooks/Cole, 2004).

RICHARD A. CROOKER
KUTZTOWN UNIVERSITY

Isphahan

FOR THE LAST 900 years, Isphahan (Isfahan, Esfahan) has been the capital of the province of the same name in the center of the empire that was known as Persia, now IRAN. The city lies in a basin at an altitude of 5,150 ft (1,570 m) above sea level in the foothills of the ZAGROS MOUNTAINS. The region is desert punctuated by numerous oases that were the source of sustenance for the caravans that once traversed central Asia.

Isphahan is one such oasis that lies on the banks of the Zayandeh River, 272 mi (435 km) from Tehran, Iran's capital. It is the third-largest city of Iran, with a population of about 1.5 million and was acclaimed as a beautiful city in the 16th century by its inhabitants, whose phrase *Esfahan nesf-e Jahan* ("Esfahan is half the world") is frequently repeated today.

Isphahan was founded more than 2,000 years ago and because of its location and resources, it has experienced many invasions and changes of fortune. It was originally known as Aspadana and was an important center in Sassanian times, between 200 and 650 years C.E. It was taken by invading Arabs in the 7th century when Islam was established and when Isphahan became the provincial capital. Four hundred years, later it was annexed by Seljuk Turks when it rose in stature to become the capital of their empire. Like so many cities of central Asia, Isphahan was then captured by the Mongols under Genghis Khan in the 1220s and then by Tamerlane in 1338, when it was reputed that 70,000 people were killed.

Its golden age of artistic and architectural achievement began under Shah Abbas during the period of ap-

proximately 1587 to 1621 of the Safavid dynasty, which had been established in Persia in 1502. Mosques, palaces, gardens, and bridges were constructed, carpet making and artistic endeavors were encouraged, and the city increased its wealth. Its population swelled to about 600,000, and it became one of the great metropolises of the time.

Its heyday was short-lived, as it was taken by the Afghans in 1723 with much bloodshed. It lost its status as the capital, which was bestowed on Shiraz. Following 200 years of relative peace, Russians occupied Isphahan in 1916.

Modern Isphahan is still dominated by the art and architectural heritage of Shah Abbas. Among the most famous world-class sights are Imam Square, with its bazaars, mosques, and flower gardens; the Friday Mosque, Imam Mosque, and Hakim Mosque; Madraseh-ye Emami and Madrasah-ye Mulla Abdollah; as well as numerous minarets, teahouses, mausoleums, palaces, and museums. They attract visitors from throughout world, and its handicrafts, which include carpets, silver- and copperware, and miniature paintings, are much prized. Its industries also reflect the rich agriculture of its oasis HINTERLAND and its location for trade. Iron and steel production, established in 1971, reflects a degree of industrialization in this rapidly expanding city.

BIBLIOGRAPHY. M.T. Faramarzi, *A Travel Guide to Iran* (Yassaman Publications, 1997); M. Hattstein and P. Delius, eds., *Islam Art and Architecture* (Könemann, 2000); "Historical Cities of Iran: Isfahan," www.art-arena.com/esfahan (March 2004); *Essential Geographical Dictionary* (Oxford University Press, 2003).

A.M. MANNION
UNIVERSITY OF READING, UNITED KINGDOM

Israel

Map Page 1121 **Area** 8,019 square mi (20,770 square km) **Population** 6,116,553 **Capital** Jerusalem **Highest Point** 3,974 ft (1,208 m) **Lowest Point** -1,338 ft (-408 m) **GDP per capita** $19,000 **Primary Natural Resources** timber, potash, copper ore.

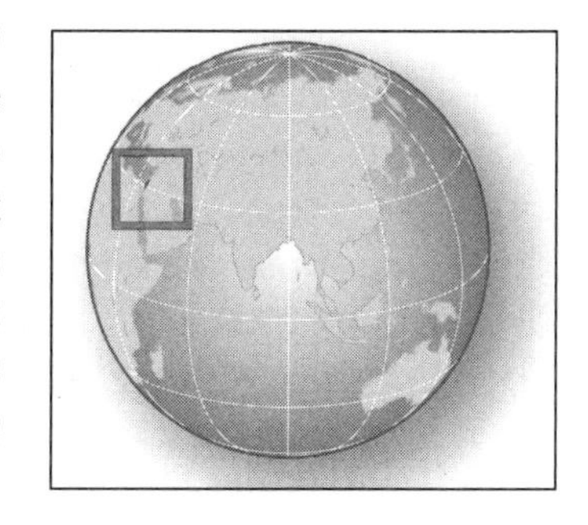

THE STATE OF ISRAEL is located on the eastern shore of the MEDITERRANEAN SEA, bordering the Gaza Strip to the southwest, EGYPT to the southwest, the West Bank to the east, JORDAN to the east and southeast, SYRIA to the northwest, and LEBANON to the north. Israel is a parliamentary democracy with the Knesset as the legislature. The president serves as chief of state, while the prime minister serves as the head of government. Hebrew and Arabic serve as the official languages of Israel, but English is widely used as a foreign language. Jerusalem, Tel-Aviv, and Haifa are the major cities.

The landscape of Israel is varied. The coastal plain stretches from Gaza in the south to Haifa in the north, covering 291 mi (469 km). Mountains traverse north to south in the central part of the country. The Sharon Plain stretches from Haifa to the Yarkon River, from which begins the Shefala Plain, which continues through Gaza. The Jordan River is Israel's main source of water and forms the border with Jordan. The Negev Desert makes up the southern portion of the country. Northern Israel receives average rainfall of 39 in (1,000 mm), while Eilat receives .8 in (20 mm). The country is susceptible to sandstorms during the spring and summer and periodic earthquakes.

Israel was part of the FERTILE CRESCENT that stretched from Mesopotamia. The Hebrew-speaking Semitic people who became the Jews settled in this region 3,500 years ago. A Judean kingdom was founded by King David, which survived until 586 B.C.E. when the Babylonians destroyed the First Temple and Jerusalem and sent part of the population into exile. Thereafter, the region fell under the sway of Persians, Greeks, Romans, Muslims, Crusaders, and the Turks.

In 1897, Theodore Herzl, after witnessing European anti-Semitism, founded a Zionist movement, which called for a Jewish homeland in Palestine. Between 1882 and 1903, Jews from all over Europe settled in Palestine and founded communities. These new settlers faced problems of poor soil, lack of experience, and opposition from Arabs and Turks.

A turning point in the formation of Israel as a modern state came during World War I. When the war broke out in 1914, the Ottoman government began expelling Jewish settlers in PALESTINE, whom it declared "enemy aliens." On November 2, 1917, the Balfour Declaration established official British support of a "national home for the Jewish people." At the end of the war, the OTTOMAN EMPIRE was dismantled, and the League of Nations officially recognized the British Mandate for Palestine. In 1939, with war against Adolf

Hitler's Germany looming, Britain issued the White Paper, which voiced its support for the creation of an Arab state for all of Palestine, barring Jewish emigration to the area.

By February 18, 1947, the British government ended the mandate on Palestine, leaving its fate to be decided by the newly formed United Nations (UN). The new UN Security Council faced the decision of whether to vote on the partition of Palestine into a Jewish and Arab state. As a result of the influence of the United States, the Security Council voted for the partition of Palestine.

On May 14, 1948, David Ben-Gurion declared the independence of the State of Israel. Between 1948 and 1973, Israel fought a series of wars with American support against the Arab states in order to maintain its existence. After the Six Day War in 1967, Israel gained control of the West Bank and East Jerusalem from Jordan; Gaza Strip and the SINAI PENINSULA from Egypt, and the GOLAN HEIGHTS from Syria. After meeting in Camp David in 1977, Prime Minister Menachem Begin of Israel and President Anwar Sadat of Egypt signed the Israeli-Egyptian Peace Treaty in 1979, marking the first time Israel made peace with an Arab state.

Since 1979, there has been a movement toward securing peace between Israel and its Arab neighbors. In 1991, the Madrid Conference called for talks on a final peace settlement. In 1993, through the Oslo accords, Israel and the Palestinians worked toward ending occupation of the West Bank and the Gaza Strip, paving the way for a Palestinian state. In 1994, Israel signed a peace treaty with Jordan. However, by September 2000, renewed hostilities flared between Palestinians and Israel, undermining the gains that had been made toward a permanent settlement.

The Israeli population is 80 percent Jewish. Of that percentage, 32 percent is from Europe, while 15 percent is of Asian descent and 13 percent is of African descent. The remainder of Israel's population is mostly Arab.

Israel has a market economy that includes a significant government role in economic policy. There have been great advances in the increase of agricultural output despite its limited arable land. There has been a significant growth in the technological sector of the economy. The addition of Jews from the former Soviet Union has also revitalized the economy. However, the government has a sizable foreign debt, particularly with the UNITED STATES. Israel's economic prospects continue to be overshadowed by the uncertainty of the Israeli-Palestinian conflict.

BIBLIOGRAPHY. Arnold Blumberg, *The History of Israel* (Greenwood Press, 1998); Ahron Bregman, *A History of Israel* (Macmillan, 2002); Robert O. Freedman, *Israel's First Fifty Years* (University Press of Florida, 2000); *World Factbook* (CIA, 2004).

DINO E. BUENVIAJE
UNIVERSITY OF CALIFORNIA, RIVERSIDE

Istanbul

THE CITY OF ISTANBUL in TURKEY has one of the most interesting and important physical locations of any city in the world, the crossroads of TRADE ROUTES by both water and land. It is the only city to straddle two continents, Europe and Asia, and has been at the center of regional commerce for nearly 3,000 years. As Constantinople, it was the most important city in the Western world after the fall of the Roman Empire in the West and was then transformed into the political center of the OTTOMAN EMPIRE, which dominated the eastern Mediterranean until its demise 1923.

The western bank of the Bosporus, the narrow channel connecting the BLACK SEA and the Sea of Marmara, was first settled about 3,000 B.C.E. At its narrowest, the Bosporus is only 2,640 ft (800 m) wide, an ideal location for a settlement to participate in and control any and all trade passing between the Black Sea and the MEDITERRANEAN. Greek colonists established cities on both sides of this channel in the 7th century B.C.E., Chalcedon on the east side, and a city named for one of their early leaders, Byzas, on the west bank. This city, taking the name Byzantium, was built above the finest natural harbor on the Bosporus, a narrow inlet called the Golden Horn because of its curved shape and the amount of wealth that flowed across its piers. This is the heart of today's Istanbul and forms the northern boundary of the old city. Called Haliç in Turkish, and Keration in Greek, the six-mile-long Golden Horn dominated shipping then, as it does today. The city grew wealthy by charging tolls from any ship passing through the narrows of the Bosporus.

Byzantium remained a fairly minor city until the 4th century C.E., when Roman emperor Constantine the Great chose the city as his new capital. Constantinople reached the height of its intellectual sophistication and architectural grandeur in the 6th century, under Emperor Justinian, who constructed some of the grandest buildings in the world, including the Church

The Blue Mosque, completed in 1616, is one of the prominent landmarks of Istanbul in Turkey.

of Hagia Sophia ("Holy Wisdom"), the largest church in Christendom. After nearly eight centuries of continual attacks, Constantinople fell to the Ottoman Turks in 1453 and became the center of Ottoman power in the eastern Mediterranean. While retaining the name Constantinople officially (Qostantiniyeh in Turkish), gradually the city began to be called Istanbul (or Stambul) locally, a corruption from the Greek words for "to the city." It was also sometimes known to the Turks as Dersaadet, "Abode of Felicity," known for its luxurious palaces and lush gardens. The Turkish sultans ruled a city whose climate was indeed felicitous, warm and not too dry, suitable for the extensive gardens that came to dominate much of the old city within the walls, known as the Surici. The city continued to thrive, with a population of about 500,000 in 1500.

Today, Istanbul is the principal city of Turkey, though it ceased to be the capital after the fall of the OTTOMAN EMPIRE in 1923. With a population exceeding 9 million, Istanbul ranks among the top 10 largest cities in the world. Istanbul is the capital of a *vilayet* (province) of the same name—officially changed from Constantinople only in 1930. The city has three main divisions: Old Istanbul (the city within the ancient and medieval walls), Galata-Beyoglu across the Golden Horn, and the Asian Quarters across the Bosporus. The city is more heterogeneous than the rest of Turkey, with some of its quarters dominated by specific minorities: Greeks, Armenians, and others.

Istanbul proper encompasses a peninsula between the Golden Horn and the Sea of Marmara to the south, from the old wall across the end of the peninsula in the west, to Sarayburnu (Palace Point) in the east. Where once there was a sea wall, Ottoman sultans built several elaborate palaces and gardens, the most famous being the Topkapi Palace, which is today one of the city's major museums and tourist attractions.

FAMOUS MOSQUES

The terrain for most of the city is very hilly, with mosques and funeral monuments built to crown most of the primary hills: the most famous mosques include Mehmet II and Yeni Cami (New Mosque). The main thoroughfare is the Divan Yolu, from the Hagia Sophia to the Bayezid II Mosque. Other tourist sites include the twin fortresses of Anadolu ("Asia") and Rumeli ("Europe") built by the Turks on the shores of the Bosporus just before the conquest and the vast covered markets. The quarter of Eyüp, the supposed site of the tomb of the Prophet Mohammed, was for centuries the site of royal ceremonies and burials.

Galata-Beyoglu, on the northern shore of the Golden Horn, was historically the residence of foreign merchants, and is today the center of modern Istanbul, with its largest shops and hotels. Across the Bosporus lie the Asiatic Quarters connected to the European side by two major suspension bridges. Smaller residential and industrial towns line the Bosporus on both sides, and along the northern edge of the Sea of Marmara, a sector of beach resorts and summer homes.

BIBLIOGRAPHY. *Planet Earth World Atlas* (Macmillan, 1998); *Encyclopedia Americana* (Grolier, 1997); "Istanbul," www.exploreistanbul.com (August 2004); "Turkish History," www.allaboutturkey.com (August 2004).

JONATHAN SPANGLER
SMITHSONIAN INSTITUTION

Italy

Map Page 1131 **Area** 116,306 square mi (301,230 square km) **Population** 57.6 million **Capital** Rome **Highest Point** 15,577 ft (4,748 m) **Lowest Point** 0 m **GDP per capita** $8,914 **Primary Natural Resources** limited natural gas, minerals, beef, arable land.

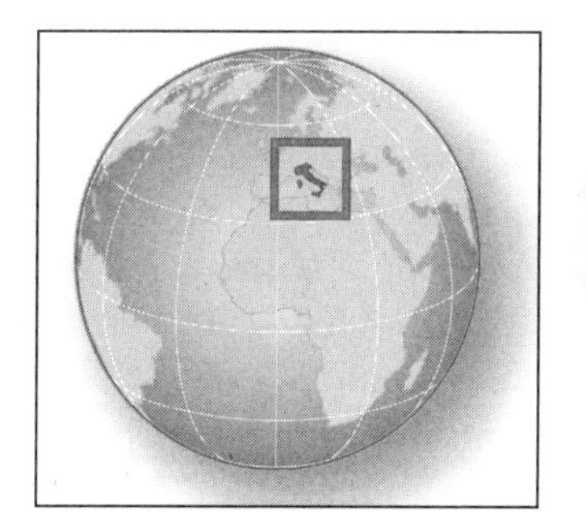

ITALY, IN SOUTHWESTERN Europe, is a peninsula bordered by FRANCE to the northwest, SWITZERLAND to the north, SLOVENIA to the northeast, the ADRIATIC SEA to the east, the Ionian Sea to the south, and the Tyrrhenian Sea to the west; its famous boot shape juts into the MEDITERRANEAN SEA. In both its physical and human geographic expressions, Italy presents a distinct and immediately recognizable character. Italy's landscape has provided the scene for the Roman republic and empire, and its peninsular form has opened it to commerce, culture, and war. The geography of Italy colored the background of Renaissance art and has been the setting for fragmented city-states and a unified state. The geography of Italy today is a rich story of a people and a land that not only coexist but that are strongly tied together by history and opportunity.

Italy occupies the entirety of a peninsula extending southward from the European continent into the Mediterranean Sea, in addition to two large—and many small—islands. The Italian (or Apennine) peninsula is bounded by the highest crest of the ALPS in the north and northwest. These ranges curve to the south and southeast forming the Apennine ranges which serve as the structural framework of the peninsula. Within the curve created by these mountains is the Po River valley, the largest valley on the Mediterranean.

Drainage from the mountains fills several large lakes; among them Lakes Como, Maggiore, and Garda in the north and Lakes Trasimeno, Bracciano, and Bolseno in the central part of the country. Surrounding the peninsula, the Mediterranean Sea is divided into several distinct parts: the Adriatic Sea, with Italy to the west and the former Yugoslavia and ALBANIA to the east; the Ionian Sea, between the southern tip of Italy and Greece; the Tyrrhenian Sea to the west of the peninsula containing the large islands of Sicily and Sardinia; and the Ligurian Sea, between the island of Corsica (French) and the northwestern coastline of Italy.

Italy's climate and weather are typical of Mediterranean climate regimes. The range of temperatures throughout the year is 43 degrees F (24 degrees C) in the north and only 26 degrees F (14 degrees C) in the south. Winter temperatures in the north can average below freezing, while southern low temperatures can be substantially above that mark. The cooling effects of altitude are felt in the Alpine and Apennine highlands. Rainfall is sufficient for agriculture in most of the country with up to 52 in (1270 cm) at some locations in the north, down to 30 in (76 cm) or less in the south. The dry summer season extends over at least June, July, and August in the south, and during these periods irrigation can be necessary for agriculture, and increases in population can strain the limited water resources.

As one of the major water-traffic corridors in the city, the Grand Canal is the largest waterway in Venice.

The Italian landscape is considerably wooded, with 34 percent of the total land area forested; 9.25 percent of the land is engaged in permanent agriculture, while an additional 28 percent of the land is arable. In addition to agricultural potential, Italy's natural resources include limited supplies of mercury, potash, marble, sulfur, natural gas and crude oil reserves, fish stocks, and coal.

Most of the extreme events that occur in Italy are related to its regional physical geographic characteristics. Heavy rains are associated with landslides and mudflows where steep mountainous terrain predominates and with flooding in river valleys and coastal lowlands. Heavy snowfall in the north can generate conditions suitable for avalanches. Active volcanoes are not uncommon in the south; examples include Mount Etna on Sicily, Mount Vesuvius, and Stromboli. Many of the smaller islands have been forged from volcanic activity and such activity continues to the present. Earthquakes can accompany volcanic activity and the associated tectonic movement. Land subsidence is of concern in some coastal areas, most notably in the city of Venice on the Adriatic Sea.

The people of Italy are as distinct as their physical geographic environment. Although some of the prevailing demographic trends in Italy are similar to those of the European continent as a whole, there are many elements of Italy's human geographic character that are wholly unique. In terms of population statistics, Italy

has an aging population with a declining rate of natural increase. That is, their death rate is greater than their birth rate, leading to declining population numbers in the absence of immigration. There is, however, substantial immigration into Italy, particularly from Eastern Europe, Africa, and the MIDDLE EAST.

The Italian language—which does not predominate in any other country of the world—is in the Romance group of the Indo-European language family. There are small areas in northern Italy where French, German, and Slovene are the predominant languages. The population is overwhelmingly Catholic, with the VATICAN CITY in Rome as the administrative center of the Catholic Church. There are small Protestant and Jewish communities and a growing Islamic immigrant community.

Politically, Italy has historically been at one time the heart of an empire, at another a loosely connected set of regional fiefdoms, and at still another the victim of fascist totalitarianism. Today, Italy is a democratic republic. The Italian federal government employs a parliamentary system, with a president and a prime minister. Parliament consists of two houses, the senate and the chamber of deputies. There are several dozen active political parties seeking seats in those houses. There are 20 regional governments with varying amounts of regional autonomy, and a large number of municipal political institutions.

Internationally, Italy is a member of the EUROPEAN UNION and of the NORTH ATLANTIC TREATY ORGANIZATION (NATO). Italy was comparatively late in securing African colonies but did so in the case of ERITREA, and in part Ethiopia and Somalia. These colonial claims were lost at the close of World War II, but even so, there are substantial political and economic ties between Italy and Eritrea to this day.

ECONOMIC GEOGRAPHY

Italy's capitalist economic system is based primarily on a diverse set of industrial activities. Generally speaking, the north of the country is heavily developed and industrialized (particularly in the cities of the Po River valley). Private development of manufacturing and processing dominate the economic activity in this region.

In contrast, the south of Italy is less industrialized and more dependent on agricultural activities. This region receives a greater share of social welfare subsidies to support the larger unemployed population. Some areas, both coastal and mountainous, are sought after as vacation locales by tourist from across Europe and around the world, and these areas depend on tourist spending for their economic base.

With Italy's limited supply of natural resources, most of the raw materials for processing goods—and the energy supplies with which to do so—must be imported. As a member of the European Union, Italy has followed a severe fiscal policy in recent years in order to meet the requirements of that international body. Italy accepted the euro as its sole currency for all transactions on January 1, 2002. Italy places a relatively high tax burden on its citizens in order to allow government support of the labor market and a generous pension system for retirees.

Italy's largest cities are centers of both economic and cultural activities. The capital, Rome, has been a metropolis through millennia. In addition to containing the Vatican City, Rome serves as the center for government and professional services and is an important cultural center. Both Florence in north-central Italy and Venice on the northern end of the Adriatic Sea serve as important centers of culture and history, in addition to supporting diverse economic activities. Turin, in the north, serves as an important center of manufacturing, and Milan, its neighbor to the east, is a center for transportation and business, notably the business of high fashion. Naples serves as the surrogate capital for southern Italy, while the islands of Sicily and Sardinia have concentrated metropolitan areas in Palermo and Cagliari, respectively.

BIBLIOGRAPHY. E. Crouzet-Pavan, *Venice Triumphant* (Johns Hopkins University Press, 2002); A. Carnahan, *The Vatican: Behind the Scenes in the Holy City* (Farrar Straus, 1949); J.A. Gottmann, *Geography of Europe* (Holt, Rinehart and Winston, 1969); H. Hearder, *Italy: A Short History* (Cambridge University Press, 2001); D. Sassoon, *Contemporary Italy: Economy, Society and Politics since 1945* (Longman, 1997); D. Randall-MacIver, *Italy before the Romans* (Cooper Square, 1972); "Italy Profile," www.nationmaster.com (August 2004).

KEVIN M. CURTIN, PH.D.
UNIVERSITY OF TEXAS, DALLAS

Jakarta

JAKARTA, LOCATED on the island of JAVA, is the capital of INDONESIA and serves as a gateway to the country. Java is located in a chain of islands, with SUMATRA to the northwest, Bali to the east, BORNEO to the northeast, and CHRISTMAS ISLAND to the south. It is the world's 13th-largest island.

The huge city of Jakarta covers more than 410 square mi (650 square km) and has a population of over 9 million people. Besides serving as government headquarters, Jakarta is the center of Indonesian business and industry. Jakarta is different from other cities in Indonesia because it has the status of a province and its government is administered by a governor rather than a mayor.

Jakarta has a colorful history. As the port of Sunda Kalapa, it was the last Hindu kingdom in the area when the Portuguese arrived in 1522 to take advantage of the spice trade. Their tenure was short-lived, as they were driven out in 1527 by the Muslim leader Sunan Ganugjati. He named the city Jaykarta, meaning "City of Great Victory." By the early 17th century, both English and Dutch merchants were in the area. When the Dutch took over Indonesia, they changed the name to Batavia. In World War II, the Japanese captured the city and changed its name to Jakarta, mainly to gain the sympathy of the Indonesians. When the war ended and Indonesia gained its freedom, the name *Jakarta* was retained.

The city has a definite cosmopolitan flavor and diverse culture. Jakarta attracts many immigrants whose cultures have contributed to the overall lifestyle of the city. The Taman Mini Indonesia Indah (Beautiful Indonesia in Miniature Park) pays tribute to the cultures of Indonesia's 27 provinces. The 250-acre (100-hectare) park is Jakarta's most visited attraction.

Jakarta's major problems are the result of the rapid growth of the city in the past 40 years. During that time, the population has skyrocketed from 2.7 million to over 9 million. The government has not been able to provide for the basic needs of its residents. Jakarta suffers from floods during the wet season, when sewage pipes and waterways become clogged with debris. The depletion of the RAINFOREST on the hills south of the city has also contributed to flooding.

LIFE IN JAKARTA

About a third of Jakarta's population lives in abject poverty, many in squalid settlements made up of huts with earthen floors. They eke out a meager living by selling cigarettes, shining shoes, and scavenging food. The heat and smog of the city make it a hard existence. Traffic in Jakarta is horrendous, with motorcycles, three-wheeled taxis, dented buses, and pedicabs jockeying for position. Residents and tourists spend count-

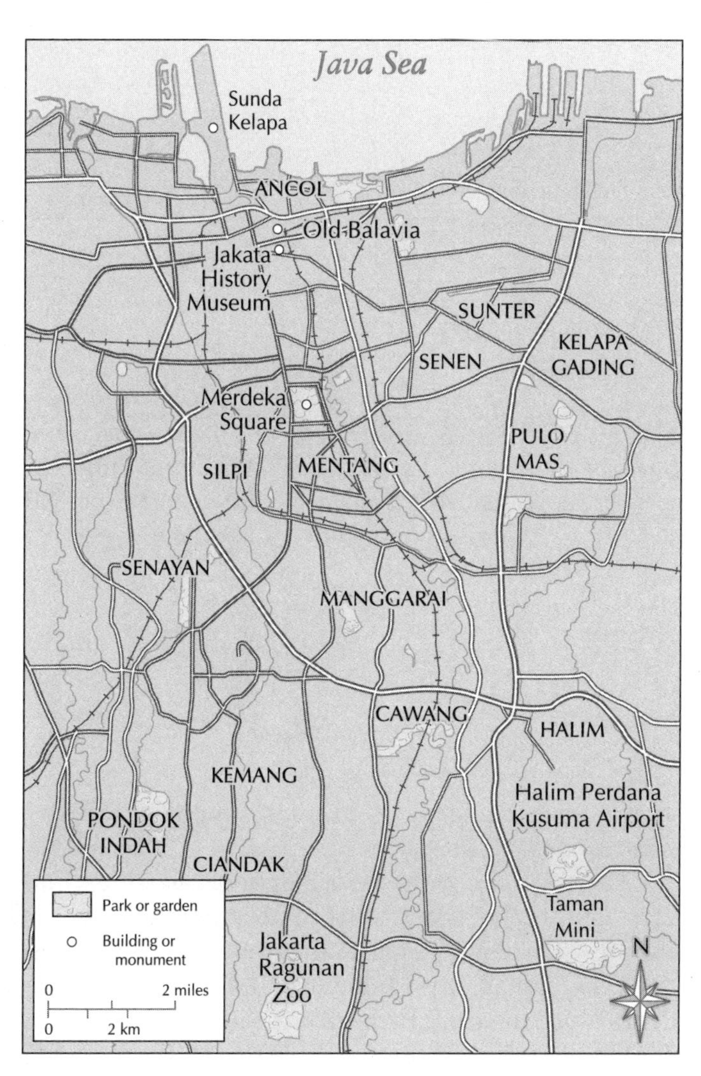

The huge city of Jakarta covers more than 410 square mi (650 square km) and has a population of over 9 million people.

less hours stuck in traffic jams. In an attempt to reduce traffic jams, some major roads now allow only cars with at least three people to be operated during rush hour. Other forms of transportation include railroads. Two monorail systems are being built, and the government is considering a network of water buses along the canals of Jakarta.

BIBLIOGRAPHY. *Southeast Asia* (Time Life Books, 1987); "Jakarta History," www.indonesia-tourism.com (November 2004); *Planet Earth World Atlas* (Macmillan, 1998); "Jakarta," Lonely Planet Guides, www.lonelyplanet.com (November 2004); *Oxford Essential Geographical Dictionary* (Oxford University Press, 2003).

PAT MCCARTHY
INDEPENDENT SCHOLAR

Jakota Triangle

THE JAKOTA TRIANGLE is an East Asian region comprising three countries: JAPAN, SOUTH KOREA, and TAIWAN. The concept originated with Harm de Blij and it has become popularized in the many editions of *Geography: Realms, Regions and Concepts*, first appearing in the 8th edition, published in 1997. The term *triangle* was inspired by the three-sided figure defined by the three capital cities of the region: TOKYO, Seoul and Taipei. But the Jakota triangle is unified by more than its three-cornered geometry. This group of East Asian states shares high population density, a high level of urbanization, rapid growth in manufacturing in spite of its dependence on imported raw materials, and lingering geopolitical problems that may be traced to the end of World War II.

The Jakota Triangle is distinctive for its high average population density. Yet each of the three member countries has a rugged and sparsely settled interior: Japan's 60 active volcanoes are legendary, most notable of which is Mount Fuji, a snow-capped stratovolcano rising to 12,388 ft (3,776 m). The eastern side of the Korean peninsula is dominated by the desolate spine formed by the Taebaek Mountains (Taebaek Sanmaek), just as the most conspicuous physiographic feature of eastern Taiwan is the Chungyang Mountains, which rise to 13,114 ft (3,997 m) at Yu Shan (also known as Mount Morrison). Thus the population distribution of the Jakota countries is uneven with extremely high densities along fertile coastal plains and river valleys and notable concentrations in massive rapidly growing cities.

The Jakota triangle countries have levels of urbanization ranging from 65 percent in Japan to 80 percent in Korea. This feature makes the region distinct from the remainder of East Asia and especially from China, which has less than 40 percent of its population in urban areas.

Tokyo, Seoul, and Taipei are massive primate cities; each is well over twice the size of the second largest urban center in the country and each accounts for about one-quarter of its nation's total population. Tokyo is the largest metropolitan area on the planet with a population of 35 million. Seoul is approaching megacity status with a population of 9.7 million, while Taiwan's capital is smaller with 6.5 million. The primate city dominance of the Jakota countries extends to their pivotal role as centers of political and financial power and leadership in industrial technology. The Jakota triangle countries are also distinguished by the

massive size and rapid growth of their manufacturing sectors, energy-intensive heavy manufacturing such as primary iron and steel and shipbuilding, labor-intensive light manufacturing such as textiles and garments, and traditional handicrafts such as paper, wood, and ceramic products. Just as notable has been the explosive growth in Jakota exports of manufactured goods. Japan is now the world's second-largest manufacturing economy (after the UNITED STATES), while South Korea ranks seventh and Taiwan is in twelfth place. These nations are globally competitive because they have developed leading-edge manufacturing technologies and have highly skilled workers, an extremely low-cost structure, and aggressively entrepreneurial firms that are striving for world dominance in their industrial specialties. But surprisingly the Jakota countries have achieved these industrial wonders with a modest resource base, and they remain highly dependent on imports of raw materials such as energy, minerals, agricultural commodities, and forest products.

BIBLIOGRAPHY. Harm J. de Blij and Peter O. Muller, *Geography: Realms, Regions and Concepts* (Wiley, 2002); Peter Dicken, *Global Shift* (Guilford, 2003); *Oxford Essential Geographical Dictionary* (Oxford University Press).

IAN MACLACHLAN
UNIVERSITY OF LETHBRIDGE, CANADA

Jamaica

Map Page 1137 **Area** 6,829 square mi (10,991 square km) **Population** 2,695,867 **Capital** Kingston **Highest Point** 7,401 ft (2,256 m) **Lowest Point** 0 m **GDP per capita** $3,800 **Primary Natural Resources** bauxite, gypsum, limestone, sugarcane.

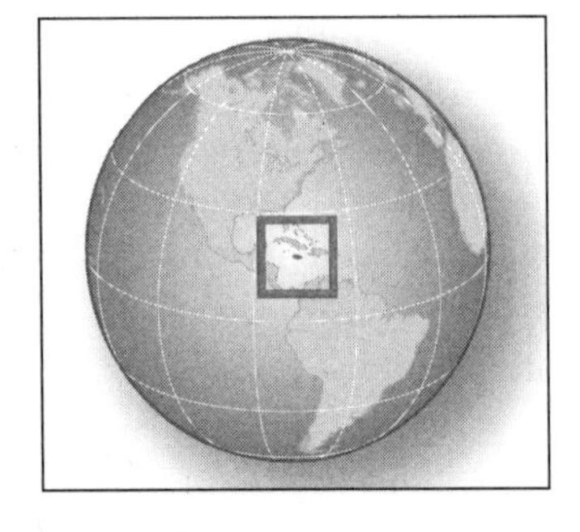

JAMAICA, SOME 62 mi (100 km) south of CUBA, is the third-largest island in the CARIBBEAN SEA and is full of numerous terrain features and vegetation. Surrounding the island is a coastal plain with numerous bays and broad flatlands, especially along the southern coast. Along the north coast, lush vegetation and white sandy beaches attract many visitors. The interior of the country is full of vales and deep ridges. Bush-covered hummocks, sinkholes, and underground caves carve out the limestone-rich region. The Blue Mountains in the east provide a dramatic sight.

Numerous animals and plants are scattered throughout the lush land that supports a tropical moderate climate. The Jamaican hutia is the only native land mammal still alive and 20 species of bats inhabit the country. Crocodiles are found in the swamps, and lizards and frogs are present all across the island. Egrets are commonly seen flying through the air, and John Crows (turkey buzzards) are found in all areas. Woodpeckers, owls, and doves are also some other birds in the island. Throughout the surrounding waters, brain corals, soft-flowering corals, and over 700 species of fish are supported by the reefs in this part of the Caribbean Sea.

Christopher Columbus discovered Jamaica on May 4, 1494, and the Spanish controlled the island until the signing of the 1670 Treaty of Madrid, which handed direct power to the British. The British created a representative system of government, which gave the white settlers power to implement laws. This legislative system lasted until 1866.

Slavery in Jamaica lasted until 1834. By that year, the country consisted of more than 311,000 slaves and only around 16,700 whites. For almost 200 years, slaves were found throughout Jamaican sugar plantations. Estimates state that over 1 million slaves were transported from Africa to Jamaica during this period. Runaway slave communities were created on the island and they even fought successful small-scale battles against British soldiers.

With the decline of plantation life and rise of black revolts, the British government took direct control of the government. They implemented rules where landowners were required to produce titles of ownership. Many of the black peasants did not have the titles and were thus forced off their land. The plantation economy of Jamaica formed once again with the sugar and banana industries, and thousands of the blacks began to migrate to other countries.

In January 1958, Jamaica joined a collective West Indian state of nine British territories, the Federation of the West Indies. However, after a national referendum, Jamaica withdrew from the federation and began to negotiate with Britain for independence, which was ultimately granted in 1962. The Jamaica Labor Party (JLP) won the elections and Alexander Bustamente became the prime minister.

Throughout the 1960s, attempts were made to bring foreign manufacturing companies to the country.

The companies were given tax breaks, but many implemented racist policies, which gave white workers higher positions. In 1972, the JLP was voted out of office, and the progressive party, PNP, returned, led by Michael Manley.

Manley began a program of democratic socialism. Foreign companies were nationalized and employment policies were revised with blacks gaining higher positions in companies. Education was also funded by the government. Middle- and upper-class residents fled the country, and Jamaica fell into an economic crisis. In 1980, political violence swept throughout the country, and the JLP, led by Edward Seaga regained power. The new government abolished many of the social programs and implemented new strategies to bring in foreign assistance.

The PNP and Manley returned to power in 1989. In 1992, Manley retired and Percival James Patterson became the first black to hold the post of prime minister. Patterson currently remains in this position and the Jamaica government continues to attempt to revive the poor economy and end the rising unemployment throughout the country.

BIBLIOGRAPHY. V. Satchell, "Jamaica," www. Africana.com (April 2004); *World Factbook* (CIA, 2004); U.S. Department of State, "Background Note: Jamaica," www.state.gov (April 2004); "Jamaica," Lonely Planet World Guide, www.lonelyplanet.com (May 2004).

GAVIN WILK
INDEPENDENT SCHOLAR

Japan

Map Page 1120 **Area** 145,843 square mi (377,835 square km) **Population** 127,214,499 **Capital** Tokyo **Highest Point** 12,389 ft (3,776 m) **Lowest Point** -13 ft (-4 m) **GDP per capita** $28,700 **Primary Natural Resources** coal, copper, rice, sugar beets.

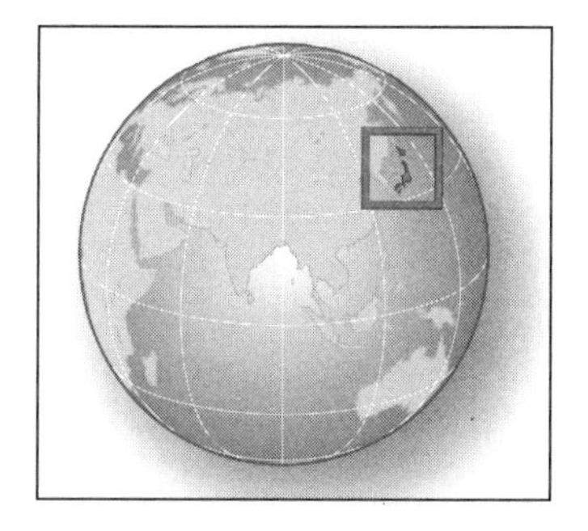

JAPAN IS AN ISLAND nation occupying a long, relatively narrow mountainous archipelago of four large and about 3,000 smaller islands on the northwestern edge of the PACIFIC OCEAN. The main chain of islands hugs the coast of the Asian continent from RUSSIA in the northeast to CHINA in the southwest for about 2,361 mi (3,800 km). From north to south, the four largest islands, constituting about 98 percent of Japanese territory, are HOKKAIDÔ, Honshû (the main island), Shikoku, and Kyûshû.

Separating Japan from the mainland are the Sea of Okhotsk to the northwest, and the Sea of Japan and the East China Sea to the west. The closest point of contact with continental Asia is with the Korean Peninsula, about 124 mi (200 km) across the Korea Strait. Extending 603 mi (970 km) further south from Kyûshû (almost to TAIWAN) are the more than 200 Ryûkyû Islands, administered as Okinawa Prefecture, after the name of the largest island. Far to the southeast, in the Pacific north of the Marianas lie Iwo Jima and the small group of Bonin Islands (known as the Ogasawara Islands in Japan).

Although Japan looks small alongside its huge neighbor of China, it is larger than any of the countries of Europe. Its area is slightly smaller than California.

Japan emerged in the 6th century C.E. as a variant of the great East Asian civilization developed in ancient China. Its relative closeness to the continent of Asia facilitated the borrowing of China's sophisticated language and institutions, but the distances were far enough that in premodern times Japan was also fairly isolated from affairs on the continent. The country was never successfully invaded from the continent (the famous failed attempt of the Mongol Empire to extend its rule over Japan in the 13th century was the only serious attempt to do so).

Thus, Japan was both part of the Asian civilization derived from China but had a distinct insular (island) character. Some observers see Japan as remaining insular, if not provincial, to this day and compare Japan's national character to that of Great Britain (although the distance separating Japan from Asia is five times the distance between England and the European continent).

Japan's natural isolation in the world was magnified by its artificial isolation due to a political decision in the early 17th century to impose the "closed-country" policy, expelling Christians and restricting contact with the West to a single Dutch fleet a year. The closed-country policy lasted until the middle of the 19th century—200 years of willful isolation from outside developments.

The mapping of the land and systematic observation of Japanese geography began quite early, in the 8th century, when the government compiled regional

geographies (Fûdo-ki) for each of Japan's provinces (KUNI). But even though Japanese pirates terrorized the coasts of Asia at times, the country never emerged as a great maritime explorer. Overall, the Japanese have tended to be familiar with nearby waters, not far waters, engaging in fishing rather than exploring or seafaring. For the most part, the seas around Japan have served as boundaries rather than linkages with other lands. It was not until the 1700s that Japan charted its own coasts.

TOPOGRAPHY

Three-quarters of Japan is covered by mountains. This mountainous terrain is the product of the country's geological origins. The Japanese islands were the product of plate tectonics, as the Pacific Plate and the Philippine Plate subducted (slid under) the eastern edge of the Eurasian Plate. After this process was already well under way, about 15 million years ago, the Sea of Japan opened in the gap between the emerging Japanese archipelago and the continent of Asia. The process of subduction is ongoing, driven by relentless if slow forces of PLATE TECTONICS. Japan's geological history also explains the abundant natural hot springs found throughout the country.

Seen from the ocean floor, the Japanese archipelago is the summit of huge underwater mountains, rising from the deepest trenches of the Pacific Ocean, sloping steeply away from Japan to depths of 33,000 ft (10,000 m) or more. From this perspective, the highest mountains in the world could be said to be in Japan. The Pacific Ocean side of Japan is rimmed with mountains ranging from 4,921 to 9,842 ft (1,500 to 3,000 m) high. Most of these mountains are volcanic, such as the country's highest peak, the symmetrical volcanic cone of Mount Fuji (dormant since 1707). The western, Sea of Japan side of the country has lower, more moderate mountains, rising to heights of 1,640 to 4,921 ft (500 to 1,500 m).

Some of the mountains of Japan are dormant volcanic cones, such as the beautiful Mt. Fuji. But dormant volcanoes can become active at any time. The southern island of Kyûshû has several active volcanoes, such as Mt. Aso. In 1990 another of these Kyûshû volcanoes, Mt. Unzen, suddenly became active after 198 years of dormancy. The previous time Mt. Unzen was active, in 1792, it generated an earthquake, landslide, and tidal wave (tsunami) that killed about 15,000. There have been nearly 1,300 recorded volcanic eruptions since the eruption of Mt. Aso in the year 710 C.E. Japan today remains an active locus of volcanic and seismic activity. Japan has more than 40 active volcanoes, and the country experiences an average of more than 1,000 seismic events (earthquakes and tremors) a year. The most disastrous earthquake of modern times hit the Tokyo/Yokohama area in 1923, killing an estimated 140,000. The worst earthquake since the 1923 disaster devastated the Kôbe area in 1995, killing more than 6,000.

The mountainous origins of the archipelago makes flat areas rare. There are only five major plains, and none are very large. The Kantô Plain, by far the nation's largest, with an area of 8,078 square mi (13,000 square km), extends inland from TOKYO. The other plains, all much smaller, include the Nôbi Plain near Nagoya, the Kinki Plain near Osaka and Kyoto, the Sendai Plain in northeast Honshû, and the Ishikari Plain on Hokkaidô. The mountainous topography of the land also means that Japan's rivers are short, rapid, and torrential, typically carrying significant amounts of water only in the spring and summer months. The longest river, the Shinano, is only 228 mi (367 km) in length. None of Japan's rivers is navigable for any significant length. Since Japanese mountains often rise directly from the water's edge, there are only two natural deep-water ports in Japan: Kôbe facing the Inland Sea, and Yokohama, just south of Tokyo.

Japan's rugged mountainous terrain and lack of any major rivers has impeded transportation and communication until modern times. Today's high-speed railway lines and highways are testimony to the extraordinary challenges and achievements of Japanese engineering. Long stretches of highways and rail lines are nearly continuous tunnels and bridges.

Most Japanese people are aware of the impact of geography on their island nation. They will often recite phrases such as "we live on a rocky archipelago with few natural resources." If Japan were a poor country today, people might point to its basic geographic features: a harsh and rocky island country with few natural resources. But while some modern Japanese continually stress how Japan is a harsh environment, stricken by earthquakes and volcanic eruptions and plagued by skimpy natural resources, this is more a problem in the 20th-century context of industrial Japan than in premodern times.

Japanese industry today is highly dependent on imported raw materials, especially petroleum, and the need to pay for those imports helps to explain the urgent drive to export. But Japan was self-sufficient for most of its long history, and even today, Japan, outside urban congestion, is not an especially harsh environ-

ment, but a lovely country. The nurturing and beneficent side of Japan's geography stems from its special climatic situation.

CLIMATE AND POPULATION

Certainly, the high winds and torrential rains accompanying the seasonal typhoons (West Pacific hurricanes) that descend upon Japan can wreak terrible damage and destruction. But in general, climatic conditions have favored Japan. With the exception of parts of the northern island of Hokkaidô, Japan lies in the temperate climate zone. But there is tremendous regional variation, from semitropical regions in the south akin to northern FLORIDA to the cool temperate climate similar to New England in the northern regions. Indeed, as the distance between the northern island of Hokkaidô to the southern island of Kyûshû roughly corresponds to the distance between Maine and Florida, so do the climatic conditions of Japan vary to a corresponding degree.

In general, Japan is part of the monsoon climate region of coastal Asia. The monsoons are seasonal winds created by changes in temperature in the region. In the cold winter months, the air over Central Asia, far from the coastal waters, becomes very cool and dry. As heat rises, so does cold air sink, and this cool dry air sinks down over Asia, bringing cold dry air to the south and east. (The impact of these winter monsoon winds is somewhat different in one region of Japan, discussed below.)

In the warm summer months, the reverse movement takes place. The air over central Asia becomes warmer, rises up, hence pulling moist air from the coasts toward the interior of Asia. This drops often heavy amounts of rain on the eastern regions of Asia (including all of Japan except the very far north of Hokkaidô) in the warm months. In sharp contrast to the precipitation pattern of northwestern Europe, where rains tend to fall in the cold winter months, the monsoon-induced rains of Asia fall exactly in the summer growing season. It is these rains that enable the wet paddy-field rice agriculture that has supported such high populations in Japan and in Asia since premodern times.

Japan's arable land may be quite small—only about 12 percent of the total land area is suited for cultivation—but it has been highly productive. The oceans surrounding Japan have been another great source of foodstuffs, raw and cooked.

These rather fortuitous conditions have permitted a population of 40 or 50 million people in preindustrial times to live in an area about the size of MONTANA or California.

In terms of population, Japan has long been one of the world's biggest countries, dwarfed by China, but for centuries larger than all of the countries of Western Europe combined. Japan's population of about 127 million people today makes it the fifth-largest country in the world. The predominantly mountainous topography concentrates this large population on a very limited land area. The resultant very high population density make Japanese cities and towns among the most crowded in the world. The population is quite homogeneous, with only about 1 percent ethnic Koreans and other minority groups. (An estimated 16,000 Ainu, descendants of the original inhabitants of northern Japan, survive, mostly on the northern island of Hokkaidô.)

JAPAN'S EIGHT REGIONS

The different regions of Japan are strikingly diverse and have distinct climatic and topographical characteristics. The country's 47 prefectures are often grouped into eight separate regions, three for the outlying major islands, and five within the main island of Honshû.

1. The northern island of Hokkaidô has a far less-dense population and a colder climate than the rest of Japan. Some of Hokkaidô's other features stem from its late development. Hokkaidô constitutes nearly 21 percent of Japan's total land area, yet is home to only 5 percent of the Japanese population (5.7 million). The landscape includes soaring volcanic mountains, much hilly lowland, and flat terraces near the coasts, unlike terrain anywhere else in Japan. Agriculture, fishing, and tourism dominate Hokkaidô's economy. Unlike the rest of the country, there is little industry in this northern island. The only major city in Hokkaidô is the prefecture's capital, Sapporo, with a population of about 1.7 million. (Sapporo is the largest Japanese city north of Tokyo.) Japan's northernmost island is separated by the Tsugaru Straits from Honshû.

2. Moving south across the Tsugaru Straits to the main island of Honshû, one enters the region known as Tôhoku (literally "Northeast"). Tôhoku occupies an intermediate position between Hokkaidô and the central region of Honshû in more ways than one. The region constitutes about 18 percent of Japan's land area (slightly less than Hokkaidô) but has about 10 percent of its population (twice Hokkaidô's). Hence, the region's population density of 235 persons per square mi (46 persons per square km) is almost three times that of Hokkaidô's, but substantially less than in other

parts of Japan. Tôhoku includes six prefectures but, like Hokkaidô, is relatively underdeveloped when compared to the rest of Japan, with more of the population engaged in agriculture and less in manufacturing than in more developed regions. Like most of Japan, Tôhoku is hilly and mountainous, with few substantial plains or other flat areas. Like Hokkaidô, Tôhoku produces rice and attracts many tourists.

3. Southwest of Tôhoku is Chûbu (literally "the central part"), the large central region of Japan (and the widest part of the big island of Honshû). In this central zone, the terrain varies tremendously. There are four major subregions within this central zone. The mountainous region of Chûbu is the "rooftop" of Honshû, capped by the massive Hida, Kiso, and Akaishi mountain ranges that constitute the Japanese Alps. Many peaks in the Alps have elevations above 9,843 ft (3,000 m), and are snow-covered for most of the year. There are few flat areas in the mountainous parts of Chûbu, so the population is clustered in six small mountainous basins, where residents traditionally were engaged in silk production. The contrast with the Kantô region of Chûbu could hardly be greater. The Kantô region includes by far the largest flat area of the country, the Kantô Plain, extending inward from Tokyo, as well as many of the major manufacturing and population centers along the Pacific coast. The Kantô region includes seven prefectures and great concentrations of modern industry. Its population density is also by far the highest in the land.

4. Extending south from the Kantô region along the Pacific is the Tôkai region, a quieter and less developed part of central Japan, comprising three prefectures. The most important part of the Tôkai is a narrow band of land between the ocean and the mountains. The old Tôkaidô Highway traversed this strip in the Tokugawa period (1600–1868), and modern highways and railways still follow that route today. Further to the southwest in the Tôkai lie the second-largest flat part of Honshû, the Nôbi Plain, along with Japan's third-largest city, Nagoya.

The final region of the Chûbu, the narrow band of hilly country north of the Alps and adjacent to the Japan Sea, is known as Hokuriku, which encompasses four prefectures. The Tôkai region is much less developed than the superadvanced industrial zones of the Kantô, but the isolated Hokuriku region, on the "back side" of Japan, is even slower and less developed, although somewhat ahead of the northern Tôhoku region. The Hokuriku region is characterized by damp winter air and enormously heavy snowfall, as the winter monsoon winds passing over the Sea of Japan drop their wet cold snow on the region in sometimes astonishing quantities.

A religious shrine near Miyajima Island, Japan, symbolizes the rich and ancient mixture in the country's cultural geography.

5. Moving further west along Honshû brings us to the region of Kansai. (Kansai, especially the part near the old capital of Kyoto is also referred to as Kinki.) From this area south, the subtropical climate of this part of Japan sets it apart from lands to the north and east. The southwestern part of Honshû is divided into the Kansai and Chûgoku regions, but the climatic tendency toward warm sunny weather generally affects the two islands of Shikoku and Kyûshû as well.

The Kansai region is the second most densely populated and developed part of Japan after Kantô. Not only was this region the center of premodern Japan, it was home to the imperial capital of Kyoto for a thousand years. It is also a modern commercial and industrial zone. It encompasses six prefectures and three of the country's largest cities, Kyoto, Osaka, and Kôbe. Kansai also features Japan's largest freshwater body, Lake Biwa, source of water for Kyoto to its south. The ancient capitals of Kyoto and Nara, along with the nearby regions, are among Japan's most-visited tourist destinations, for their temples, historical sites, and long pedigree.

Osaka, merchant city of the Tokugawa age, remains a major industrial and commercial center, second only to metropolitan Tokyo. Though great quantities of goods are transshipped at Osaka, nearby Kôbe is one of Japan's two great natural harbors (along with Yokohama), and serves as the main trading ENTREPOT for western Japan.

6. Continuing to the southwest from Kansai we arrive at the tip of Honshû and the region known as Chûgoku. Usually divided into San'in ("shady side") facing the Sea of Japan to the north, and the San'yô ("sunny side"), facing the warm shallow waters of the Inland Sea to the south, Chûgoku is also home to the major industrial city of Hiroshima. Beyond Chûgogu lie the large islands of Kyûshû to the far west, and Shikoku to the south. Encircled by the three is the Inland Sea, which has served as a protected waterway since prehistoric times.

7. There are four prefectures on Shikoku, the smallest of Japan's four main islands. Shikoku's climate is even more subtropical than the Kinki region, and its agriculture includes mandarin oranges and other warm-weather crops. The island was also site of the traditional pilgrimage around the 88 Buddhist temples and monasteries of the island. Today's pilgrims often make the same voyage not on foot but via tour bus. Shikoku remains less developed than the Chûgoku section of Honshû, which prompted the government to build the spectacular Seto-Ôhashi Bridge linking the island with Honshû via Awaji Island. Whether this bridge, or the proposal for a highway bypass along Shikoku's northern shore, will bring further development is unclear.

8. Finally, at the southwestern tip of the main cluster of Japanese islands, Kyûshû is Japan's third largest island, divided into seven prefectures. As we have seen, Kyûshû has some of Japan's most active volcanoes and other mountains.

In northwest Kyûshû are major coal fields (second only to Hokkaidô in production of coal) that provide much of Japan's energy and are the location of major iron and steel plants. The island is mostly divided between an industrial north and an agricultural south. Kyûshû's major cities include Fukuoka, Kitakyûshû, and the port city of Nagasaki. Between Kyûshû and the largest island of Honshû is the Shimonoseki Strait, spanned in modern times by the Kammon Bridge and three separate tunnels.

BIBLIOGRAPHY. *World Factbook* (CIA, 2004); Martin Collcutt, Marius Jansen, and Isao Kumakura, *Cultural Atlas of Japan* (Facts On File, 1988); John K. Fairbank, Edwin O. Reischauer, and Albert M. Craig, *East Asia: Tradition and Transformation* (Houghton Mifflin, 1989); Isida Ryuziro, *Geography of Japan* (Japan Cultural Society, 1969); *Japanese Geography, 1966: Its Recent Trends* (Association of Japanese Geographers, 1966); *Kodansha Encyclopedia of Japan* (Kodansha Ltd., 1983); Glenn T. Trewartha, *Japan: A Geography* (University of Wisconsin Press, 1965); *Oxford Essential Geographical Dictionary* (Oxford University Press, 2003).

LAWRENCE FOURAKER, PH.D.
ST. JOHN FISHER COLLEGE

Java

WITH A population exceeding 120 million, crowded into a space about the size of NEW YORK state (but with nearly six times the population), the island of Java is one of the most densely populated places on Earth (2,070 inhabitants per square mi or 864 per square km). Java is not the largest of the islands of INDONESIA, but it is certainly the nation's political, historical, and economic core, with 60 percent of Indonesia's total population, and most of its wealth concentrated into less than 7 percent of the nation's total land area. But the future may change this status, as mineral and fossil fuel wealth in the outer islands leads to increased migration away from Java and increased demands for political and economic autonomy.

The 13th-largest island in the world, Java is the fifth-largest island in the Malay Archipelago. It is the smallest of the Greater Sundas (which also include BORNEO, SUMATRA, and Sulawesi), with a total area of 48,830 square mi (125,205 square km). The island is bordered on the south by the INDIAN OCEAN, which drops relatively quickly to great depths, plummeting up to 24,440 ft (7,450 m) in the Java Trench, the deepest point in the Indian Ocean. The rest of the island is surrounded by shallower seas, on the west, the Sunda Strait, with the island of Sumatra only 16 mi (26 km) distant; on the north, the Java Sea; and on the east, the Bali Strait, separating Java from Bali by only 1 mi (1.6 km) at its narrowest.

Part of the Pacific RING OF FIRE, Java was formed from a series of volcanoes running west to east. There are more than 100 volcanoes, 13 active in recent history. Most of the high peaks are concentrated along the southern edge of the island, about 20 of which exceed 8,000 ft (2,424 m). Semeru, in the eastern part of the island, is the highest active volcano on Java (12,131 ft or 3,676 m) and erupted most recently in April 2004. Merapi erupted in 1994, killing 37 and forcing 6,000 to evacuate the area. The most notable, however, is located just offshore to the west: Krakatau, site of the largest volcanic explosion of modern times (in 1883,

killing 36,000 people in western Java and reddening the sky as far as Europe and America).

From the mountainous south, the island's topography then slopes generally downward toward the north shore, where there are more level and fertile plains, followed by mangrove swamps and ample harbors. Most agriculture and population centers are thus concentrated on the north shore. The mountains also contain significant natural resources, including tin, sulfur, asphalt, manganese, limestone and marble. Petroleum deposits are located near Rembang and Surabaya in the north and northeast, and on the island of Madura. Rice is cultivated across Java, along with sugarcane, though much of the island is covered with dense forests, rich in plant and animal wildlife like the Javanese tiger, and teak and mahogany trees. Ornithology is especially diverse, including numerous species of peacock, quail, heron, cuckoo, and hornbill. Temperatures can vary from the mountaintops to the lowlands by up to 20 degrees.

One in three Javanese lives in cities, four of which have over 1 million inhabitants. The largest of these is JAKARTA, the capital of Indonesia, with its port of Tanjung Priok, 6 mi (9 km) to the east. Jakarta has recently become the center of one of the world's densest and fastest growing conurbations, referred to as JABOTABEK (Jakarta, Bogor, Tangerang, and Bekasi), home to nearly 20 million people. Java's other largest cities include Surabaya, the second-largest and center of much of Java's industry and trade; Bandung, on a high plateau in the northwest, formerly the heart of Dutch coffee and tea plantations, and more recently home to high-tech industries such as aeronautics; and Semarang, the island's fastest-growing city, on the northern coast. Jogyakarta, in central Java, is a smaller city but has been the spiritual and cultural heart of Javanese culture for centuries, and its ancient palaces and temples continue to make this Java's main center for tourism.

Today, Java is almost entirely Muslim, but before Islam arrived in the area, the island was ruled by Hindu and Buddhist princes (Bali remains predominantly Hindu today). The name *Java* itself probably derives from a Sanskrit (ancient Hindu) word for the type of grain grown locally.

Numerous temples survive from this period, notably Prambanan (from the 10th century), and Borobudur, the largest Buddhist temple in the world. These princes were at their height in the 1300s, when the Majapahit Empire dominated the entire archipelago, and Malayan became the lingua franca for most of southeast Asia. This empire was dismantled by waves of Arabic traders, until it disappeared altogether in 1518. Bantam and Mataram were the strongest sultanates in the region when the Europeans arrived in the late 16th century. The Dutch established themselves in 1619 at an old Javanese fort called Jakarta and renamed it Batavia.

Total control of Java was achieved in 1756 after the bloody Java War, followed by gradual assertion of Dutch control over the neighboring islands of Sumatra, Borneo, and the rest of what is now Indonesia by the end of the 19th century. External trade was controlled by Europeans, while internal trade was taken over by Chinese immigrants encouraged by the colonial administration. The Javanese economy was transformed into plantations for sugarcane—the largest producer in the world by the end of the nineteenth century. Conditions for the Javanese were rough under Dutch rule, and the Japanese occupation during World War II encouraged them to fight for independence, which was proclaimed in Jakarta in 1945 but not recognized by the Netherlands until 1949.

ETHNIC GROUPS

Three major ethnic groups occupy the island, each with their own language: Sundanese in the west (including Jakarta), Javanese over most of central and eastern Java (with about 75 percent of the population), and Toba Batak in the northeast corner and on Madura. Javanese culture is famous for its shadow puppets and gamelan orchestras. These three groups are by far the largest ethnic groups in all of Indonesia (Javanese 45 percent, Sundanese 14 percent, and Madurese 7.5 percent), and their language formed the basis for the national language, Bahasa Indonesia.

Current issues facing the Javanese include controlling the population expansion; programs for reducing family sizes were begun in the 1960s, accompanied by government resettlement projects, which also has a secondary aim of attempting to "Javanize" the Outer Islands of Indonesia, particularly in sensitive areas like Aceh in Sumatra, West Papua, and Kalimantan, all with strong separatist movements. Farmers are especially encouraged to emigrate to the other islands, but face resistance from locals who resent dominance from Java, politically or culturally. While Java is by no means dependent on the Outer Islands for its existence, the central government in Jakarta is certainly concerned with the loss of potential riches as the mineral wealth of these regions becomes more commercially developed and exploited.

BIBLIOGRAPHY. Barbara A. Weightman, ed., *Dragons and Tigers: A Geography of South, East and Southeast Asia* (John Wiley & Sons, 2002); *Encyclopedia Americana* (Grolier, 1997); Bernard Comrie, Stephen Matthews and Maria Polinsky eds., *The Atlas of Languages* (Quarto Books, 1996); BPS Statistics, bps.go.ind (April 2004).

JONATHAN SPANGLER
SMITHSONIAN INSTITUTION

Jerusalem

THE CAPITAL CITY of the state of ISRAEL, Jerusalem is located in the center of the country. Jerusalem is Israel's largest city, with an area of 48 square mi (126 square km) and a population of 657,500 (2000), of which 68 percent are Jewish and 32 percent mostly Arab. The city is divided into two components: West Jerusalem, or the New City, and East Jerusalem, or the Old City, annexed in 1967. Jerusalem is sacred to Judaism, Christianity, and Islam. The Wailing Wall, the Church of the Holy Sepulcher, and the Dome of the Rock are holy to Jews, Christians, and Muslims. In 1949, Prime Minister David Ben-Gurion declared Jerusalem to be the "eternal capital," and even though Costa Rica and Honduras hold embassies there, the United States and many countries hold their embassies in and around Tel-Aviv, the previous capital.

The site of present-day Jerusalem has been settled by humans since the Stone Age. According to Biblical tradition, King David captured the city from the Jebusites in approximately 1000 B.C.E. and made it the capital of the newly established Kingdom of Israel, until the destruction of the city and the First Temple by the Babylonians in 586 b.c.e. Jerusalem again became the capital of a short-lived Judean kingdom under the

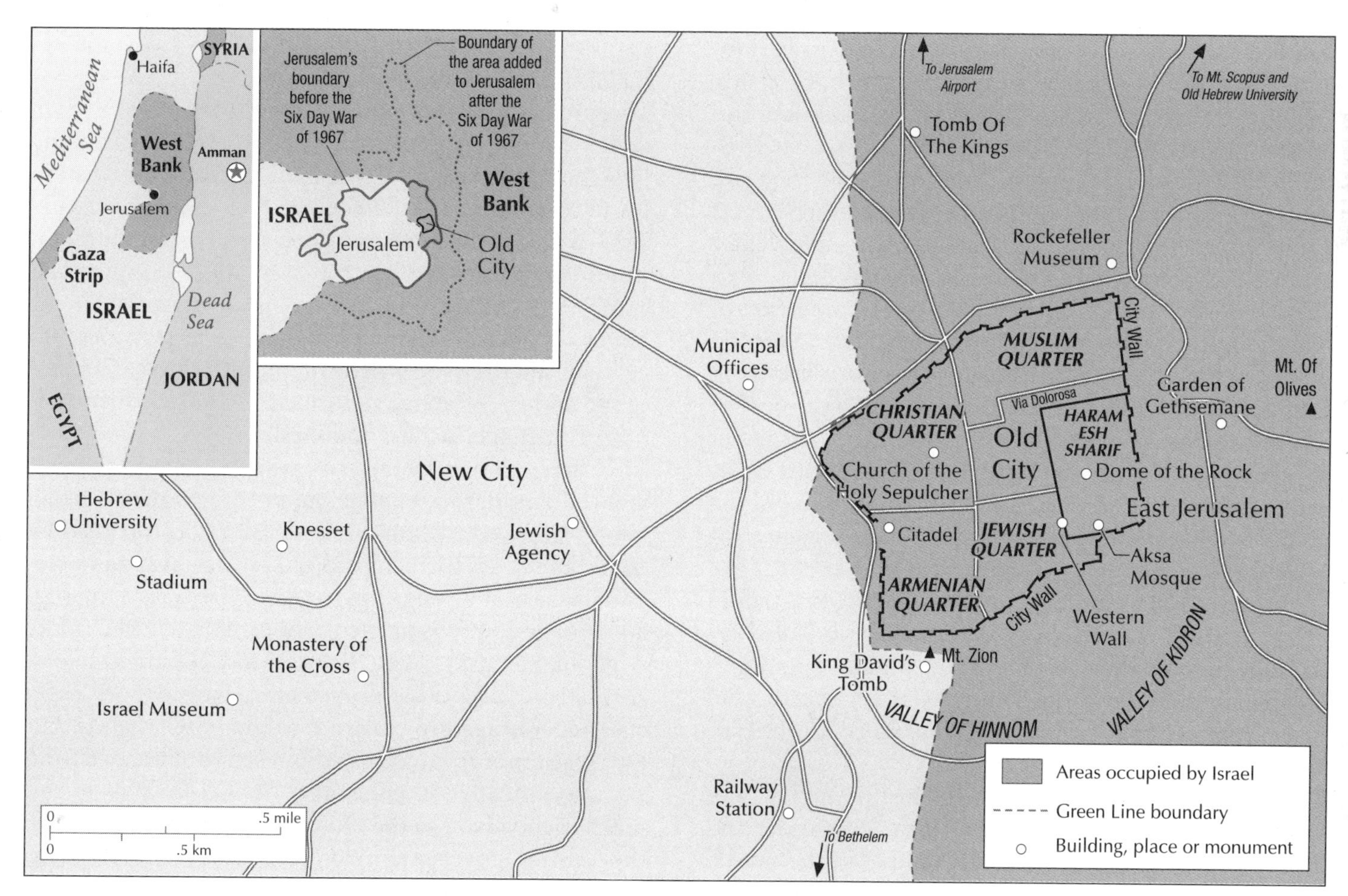

Jerusalem is divided into two components: West Jerusalem, or the New City, and East Jerusalem, or the Old City, annexed in 1967. The city has been a sacred site to Judaism, Christianity, and Islam for centuries.

Maccabees in 142 B.C.E. It fell under Roman influence, and in 70 C.E., the Romans under General Titus destroyed the Second Temple during a rebellion. In 135, after the rebellion led by Simon Bar Kokhba, the Romans destroyed Jerusalem, renamed the city Aelia Capitolina, and expelled the Jews from Judea.

In 638, the Muslims gained control of Jerusalem. After the First Crusade in 1099, Jerusalem served as the capital of the Latin Kingdom of Jerusalem established by the Franks. Muslims regained Jerusalem under the Mamelukes in 1291 and by the Turks in 1517. In 1917, British troops took control of Jerusalem and established the British Mandate in Palestine. In 1949, with the end of the British Mandate, Jerusalem was divided into the New City, the capital of the new state of Israel and the Old City, under Jordanian control. Jerusalem was reunified after the Six Day War in 1967. Palestinians hope to see East Jerusalem as the capital of a Palestinian state. As part of the Oslo Accords in 1993, the fate of East Jerusalem was to be resolved by the Israelis and the Palestinians. However, renewed violence in 2000 has prevented such a settlement.

The city is the country's cultural center and is home to theaters, concert halls, and museums, notably the Yad Vashem, which chronicles the Holocaust, and the Israeli Museum, which holds the Dead Sea Scrolls. The Hebrew University has campuses in the Old City and the New City. Despite its turbulent history, Jerusalem continues to be a vibrant city.

BIBLIOGRAPHY. Kamil J. Asmali, ed., *Jerusalem in History; 3000 B.C.E. to the Present Day* (Kegan Paul International, 1997); Marshall J. Breger and Ora Ahimeir, eds., *Jerusalem: A City and its Future* (Syracuse University Press, 2002); Abraham E. Milgram, *A Short History of Jerusalem* (Jason Aronson, 1998).

DINO E. BUENVIAJE
UNIVERSITY OF CALIFORNIA, RIVERSIDE

jet stream

A JET STREAM IS A relatively narrow band of strong horizontal winds in the atmosphere. This band is 100 to 200 mi (250 to 500 km) wide and 1 to 2 mi (several km) deep. The winds in this band are stronger than those in horizontal or vertical regions adjacent to this band. The boundaries of the jet stream, therefore, contain significant wind shear. Traditionally, the term jet stream has referred to such bands that occur in the upper troposphere, especially those that consist of westerly winds. There are other atmospheric phenomena that consist of bands of strong winds. The tropical easterly jet consists of easterly winds and is associated with the Asian monsoon. The polar night jet occurs during the winter in the stratosphere. Low-level jets form at the top of the planetary boundary layer—usually at a pressure of 900 to 800 mb. These jets have been associated with thunderstorm activity in many parts of the world.

Military pilots discovered the upper tropospheric westerly jet streams during World War II. Aircraft flying in an easterly direction would sometimes encounter winds so strong that they would seriously impede their mission. There are two types of these westerly jets: the subtropical jet and the polar front jet. The jet streams are located 30,000 to 40,000 ft (9 to 12 km) above the Earth's surface. They tend to be higher in summer than in winter and the subtropical jet is usually higher than the polar front jet. Meteorologists generally consider these jets to have wind speeds in excess of 60 mi (97 km) per hour. Average winds are about 135 mi (217 km) per hour in winter. The location of the most intense winds varies from day to day.

The subtropical jet stream forms as a result of the physical principle of the conservation of angular momentum. Figure skaters use this same principle to increase their spinning rate by bringing their arms towards the center of their bodies. Air is heated and rises over the part of the globe receiving the strongest solar radiation. At the top of the troposphere, this rising air moves toward the poles. As it moves toward the poles, the distance between it and the Earth's axis decreases. As this distance decreases, the air accelerates toward the west. The air reaches its maximum westward velocity between 20 and 40 degrees from the equator and then starts sinking. The latitude of maximum winds tends to be more poleward in the summer and more equatorward in the winter. Wintertime jets tend to have faster speeds than summertime jets. The intensity of the jet varies with longitude, but there is a tendency for it to be stronger off the east coast of Asia.

As its name implies, the polar front jet stream forms along the polar front, where warm air from the tropics meets cold air from the poles. The strong temperature contrasts that exist here cause south to north pressure gradient to increase with height. The intensity of the pressure gradient increases up to the troposphere. The Coriolis force then causes winds to go

from west to east. This jet stream is also more intense and located more poleward during the winter than during the summer.

Within both types of jet stream there are regions where winds are stronger than those located both upstream and downstream. Such an area is known as a jet maximum or jet streak. Air flows through the jet maximum accelerating in the entrance region and decelerating in the exit region at an average speed of 125 mi (201 km) per hour. The jet maximum has a distinct structure with areas of convergence in the poleward entrance region and equatorward exit region and areas of divergence in the poleward exit region and the equatorward entrance region. Below areas of divergence, especially the one on the poleward side, are especially favorable areas for midlatitude cyclones to develop.

Jet streams affect our daily lives in other ways as well. Since the polar front jet stream is located where the north-south temperature gradient is strongest, places that are south of the current location of the jet stream tend to have warmer than normal temperatures and places to the north, colder than normal temperatures. The location of jet streams is very important to aviators. Aircraft flying with the jet stream can get a tailwind and aircraft flying against it, a headwind. These winds can significantly influence travel time and fuel consumption. The areas bordering jet streams have high wind shear and are prone to develop turbulence.

BIBLIOGRAPHY. C.D. Ahrens, *Meteorology Today* (Brooks/Cole-Thompson, 2003); T. N. Carlson, *Mid-latitude Weather Systems* (Routledge, 1991); Chester Newton and E.O. Holopainen, *Extratropical Cyclones: The Erik Palmen Memorial Volume* (American Meteorological Society, 1990).

Donna Tucker
University of Kansas

Jordan

Map Page 1121 **Area** 35,536 square mi (92,300 square km) **Capital** Amman **Population** 5,460,265 **Highest Point** 5,689 ft (1,734 m) **Lowest Point** -1,340 ft (-408 m) **GDP per capita** $3,500 **Primary Natural Resources** phosphates, potash, shale.

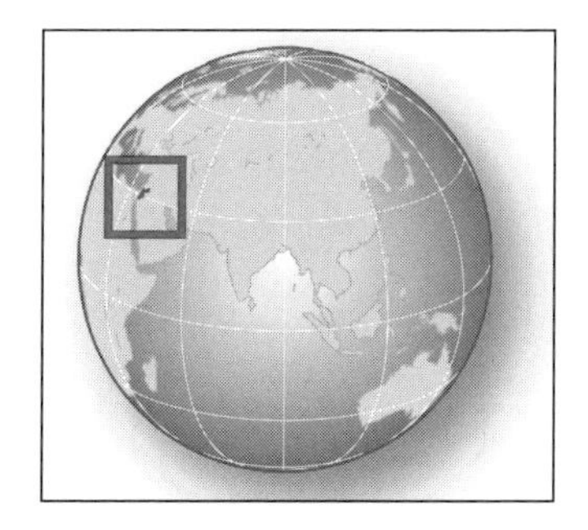

JORDAN IS A MIDDLE Eastern country located in southwest Asia. Its official name is the Hashemite Kingdom of Jordan and it covers an area slightly smaller than INDIANA. SAUDI ARABIA borders it on the south, IRAQ on the east, SYRIA on the north, with ISRAEL and the West Bank territories (Palestinian Hills) on the east. Jordan is almost landlocked, with a small coastline on the Gulf of Aqabah, which is an arm of the RED SEA.

JORDAN TRENCH

Jordan has two major land features—the Jordan Trench, and the Jordanian (Transjordan) Plateau. The Jordan Trench is a part of the Great Rift Valley extending from Syria to MOZAMBIQUE. It holds the Sea of Galilee, the Jordan River, and the DEAD SEA, all of which are below sea level. Shortly after the Jordan River leaves the Sea of Galilee it joins the Yarmuk River. The Yarmuk is Jordan River's principal tributary south of the Sea of Galilee, and the Kingdom's border with Syria. From there, the Jordan River meanders south until it empties into the Dead Sea; below sea level, it is the lowest surface area of the world. The lack of an outlet and the high level of evaporation have made the Dead Sea an area of extreme saltiness. It is mined for chemical resources including phosphates and potash.

South of the Dead Sea is the Dead Sea Valley (Wadi al-'Arabah). It stretches 111 mi (179 km) from the Dead Sea to the port of Aqabah. It is a very dry area with limited agricultural use. The Jordanian Plateau is the second prominent land feature. Most of Jordan lies on a tilted plateau. It rises abruptly at the eastern edge of the JORDAN VALLEY and the Dead Sea. The plateau's greatest heights are at the edge of the Dead Sea Valley. The heights range from 2,000 to 3,000 ft (610 to 915 m).

The Jordanian Plateau borders the Jordan Valley. Forests cover parts of the northwestern part of the plateau, while brush covers the central and southern sections. As it descends gradually to the south, east and north it becomes the Jordanian Desert Plateau. It ends in the Great Syrian Desert with little to identify the boundaries between Iraq, Syria, and Jordan. Grasses grow along the western edge, but much of it is featureless and arid.

The southern portion of the Jordanian Plateau is part of the Arabian Plateau. It has many deep canyons and mountain elevations rising to approximately 4,900 ft (1,500 m). The area holds the Wadi Rum, which has strange rock formations called jebels, and Petra, capi-

tal of the ancient Nabateans. The OTTOMAN EMPIRE built the Damascus-Medina Railroad (Hejaz Railway) in the region, which is still operating as a local line. During World War I, Colonel T.E. Lawrence, British army officer and leader in the Arab revolt, inflicted great damage on the Ottoman army along the rail line. Many scenes in the movie *Lawrence of Arabia* were shot here.

Jordan's climate is modified Mediterranean in the west. It has sharp seasonal variations in both temperature and precipitation. Temperatures below freezing often do occur in January in the plateau. The Jordan Valley has summer temperatures that may reach 120 degrees F (49 degrees C). Amman is usually moderate in temperature averaging 78 degrees F (26 degrees C). Most precipitation is rain (although snow and sleet are not uncommon) and falls in the winter season. It averages about 26 in (66 cm) in the northwest corner of the country to less than 5 in (12 cm) in the extreme east. The Jordan River Valley and the desert in the east receive less than 2 in (5 cm) of rain per year.

BIBLIOGRAPHY. Mohamed Amin and Duncan Willets, *Spectrum Guide to Jordan* (Interlink Books, 1999); Evan Anderson and Ewan W. Anderson, *Middle East* (Routledge, 2000); Susan Arenz, *Landscapes of the Holy Land: Israel, Jordan, Sinai* (Sunflower Books, 1997); Neil Folberg. *In a Desert Land: Photographs of Israel, Egypt, and Jordan* (Abbeville Press, 1998); Aharon Horowitz, *Jordan Rift Valley* (Balkema Publishers, 2001); Tony Howard, *Tracks and Climbs in the Wadi Rum, Jordan* (Cicerone Press, 2001);

ANDREW J. WASKEY
DALTON STATE COLLEGE

Jordan Valley

THE JORDAN VALLEY is part of the Great Rift Valley stretching from East Africa to northern SYRIA. Massive cracking of the Earth's crust caused by tectonic plate movements under the continents of Africa and Asia formed a trench over 4,000 mi (6,438 km) long. The northern section created by this faulting activity is the Jordan Trench.

In Syria, the Great Rift separates the Lebanon Mountains to the west and the Anti-Lebanon Mountains to the east. At Mount Hermon (9,232 ft or 2,814 m) melting snows and springs are the heads of the Jordan River, which traces the course of the Jordan Valley to its mouth at the DEAD SEA. The distance from Mount Hermon to the Dead Sea is only about 120 air mi (193 km), but the surface distance is 223 mi (360 km) owing to the river's meandering.

The word *jordan* means, "that which goes down." The elevation drops from over 1,000 feet (305 m) above sea level at the foot of Mount Hermon to 1,340 ft (408 m) below sea level at the Dead Sea, which is the lowest surface area of the world. From Mount Hermon the Jordan River first descends to the basin of Lake Huleh, a dramatic drop in elevation of 900 ft (275 m) in a distance of about 10 mi (16 km). Lake Huleh was 3 or more mi (4.8 km) wide, but today it is a marshy area because the lake has been drained for agriculture. From Lake Huleh the Jordan River descends nearly 1,000 ft (305 m) through steep, rocky gorges. Just before entering the Sea of Galilee the descent slows and waters a plain near the village of Bet Zayda (Bethsaida). The Sea of Galilee (Lake Tiberias) is a freshwater body 689 ft (209 m) below sea level. When the Jordan River leaves the Sea of Galilee, it flows for about 24 mi (39 km) through a fertile region that supported a variety of agriculture in ancient times as it does today. It is also is joined by the Yarmuk River soon after it leaves the Sea of Galilee.

The Yarmuk arises on the Transjordanian plateau and doubles the volume of the Jordan's waters. The volume of water varies considerably from the wet season to the dry.

South of ancient Pella the Jordan flows through the Ghor Plain, an ancient seabed composed of chalky limestone marls. As the Jordan descends to the Dead Sea, it is joined on its eastern (Jordanian) side by the Wadis Jurm, Kufrinjeh, Rajeb, and finally by the Jabbok River. In wet weather streams in the wadis come tumbling down from the granite Transjordanian Plateau. On the western (Israeli) side, the Jordan is joined as it flows south by the Wadis Bireh, Jalud, Malih, and Farah. The valley has several lateral faults that create the Jezreel Valley and the Saddle of Benjamin.

As the Jordan River flows south it traverses a number of geological and climatic zones. The descent to the Dead Sea is accompanied by an increase in temperature. The Jordan's southern end flows through badlands of soft gray saline marls. The whole rift valley floor is called the Ghor (Plains of Moab). Its narrow floodplain (Zor) contains thickets of thorn scrub and tamarish along its banks ("jungle of the Jordan").

The Dead Sea (Salt Sea or Sea of Arabah) is in the center of the Ghor Plain. At 1,340 ft (408 m) below sea

level it is the lowest surface area of the world. The Dead Sea Valley (Wadi al-'Arabah) is an extension of the Jordan Valley. It stretches 111 mi (179 km) from the Dead Sea to the port of Aqabah. It is a very dry area with little agricultural value. The Jordan River Valley figures frequently in the landscape of the Bible. Numerous passages include the separation of Abraham and Lot, the Hebrew children crossing the Jordan to fight the Battle of Jericho, and the baptism of Jesus in the Jordan River.

BIBLIOGRAPHY. Barbara Ball and Lorraine Kessel, *The River Jordan: An Illustrated Guide from Bible Days to the Present* (Carta, 1998); Brian Bell, *Insight Guide: Jordan* (Insight Guides, 1998); A. Horowitz, *Jordan Rift Valley* (A.A. Balkema Publishers, 2001); Isaac Schattner, *The Lower Jordan Valley: A Study in the Fluviomorphology of an Arid Region* (Magnes Press, 1962); Yehoshua Ben-Arieh, *The Changing Landscape of the Central Jordan Valley* (Magnes Press, 1968); Edward Rizk, *The River Jordan* (Arab Information Center, 1964); Georgiana G. Stevens, *The Jordan River Valley* (Carnegie Endowment for International Peace, 1956).

ANDREW J. WASKEY
DALTON STATE COLLEGE

Junngar Basin

THE JUNNGAR BASIN (also Dzungar) is one of the two major basins that make up much of northwestern CHINA's XINJIANG province (Xinjiang Uygur Autonomous Region). The basin, covering some 69,500 square mi (180,000 square km), is located to the north of the TIAN SHAN MOUNTAINS and has a mixed landscape that includes mountains, deserts, steppes, salt lakes and swamps. The region has much the same pattern as the TARIM BASIN, which is located to the south of the Tian Shan. However, because the Junngar is approximately 1,968 ft (600 meters) lower than the Tarim Basin, it is not quite as arid, receiving between 4 to 12 in (10 to 30 cm) of precipitation per year. In addition, there is also much greater snowfall and snow cover during the winter season, which helps increase the growth of natural vegetation.

The basin is triangular in shape, with the towering Tian Shan Mountains to the south, the Altai Mountains along the border with MONGOLIA to the northeast, and the Ala Tau Mountains along the border with RUSSIA and KAZAKHSTAN in the northwest. The Junngarian Gate, a rift valley in the Ala Tau Mountains, provides an outlet from the Junngar Plain to Lake Balkhash and Kazakhstan to the northwest. The gate is famous for its roaring winds that blow for more than 150 days per year. Between the sand desert at the basin's core and the gravel GOBI is a narrow plain often referred to as the "spring line," where the soil has some clay and silt, allowing limited agriculture and human settlement.

The basin is somewhat geographically cut off from CHINA proper. It is possible to travel over a pass between the Bogdo Ula and the Tian Shan at 3,600 meters high and descend into the Turfan area on the south. Otherwise, the route to China leads through some very desolate valleys parallel to the Altai Mountains in the east to the Mongolian Gobi. Because the slope of the area is to the west, all of the major rivers flow east to west. The upper waters of the Irtysh River, the only river in China that flows to the Arctic Ocean, provides access to Zaisan Nor and Semipalatinsk in Russia. The two main rivers flowing within the basin are the Manas and the Urungu.

These, together with numerous streams descend to the plains before being lost in the reedy marshes and sands of the basin proper. The high mountain ranges surrounding the basin receive as much as 30 in (75 cm) of precipitation a year. The Altai Mountains are notably forested, and the lower slopes covered with willows, poplars, alders, and white birch as you move higher. Above about 6,000 ft (1,800 m), the Siberian larch is dominant, with many of the trees more than 98 ft (30 m) in height. Given this heavy forest cover, the mountain areas bordering the basin are rich in fur-bearing animals such as fox, wolf, sable, ermine, bear, and wolverine.

In contrast to the Tarim Basin, where there are extensive desert lands between oases, the Junngar has a STEPPE and steppe-DESERT landscape with some natural vegetation in the wide valleys that border the central desert lands. Thus, nomadic pastoralism is practiced here as the dominant way of life. Because annual rainfall is typically less than 10 in (25 cm), the pastoralists must be constantly on the move in search of fresh grass. This is much different than in the Tarim Basin to the south of the Tian Shan, where human activity is oasis oriented and people move in trade-oriented patterns from oasis to oasis.

Despite the limited rainfall and desert conditions, major land reclamation projects have been undertaken to introduce arable farming to the region. About half of the reclaimed land is under cotton production, with

the rest in rice, wheat, corn, soybeans, various fruits, and sugar beets. However, more recent economic interest has centered on the discovery of oil at Karamai. Following the discovery of coal and iron deposits on the north side of the Tian Shan, there has been rather rapid industrial development over the past decades.

Urumchi, the capital of Xinjian Province, is located on the southern edge of the basin on a high desert plateau. Before 1949, it was a quiet market town of less than 60,000 people. But with the discovery of oil, coal, iron and significant gold deposits, the city has grown to more than 1.75 million. It is also a center for iron and steel, textiles, and fertilizer production. The other interesting area in Junngaria is the Ili Valley. Lying north of the Tian Shan, but cut off from the rest of the basin, the valley flows westward to Lake BALKHASH in Kazakhstan. Because of the westward opening, the area is open to influence by more western climate characteristics, giving the area a special quality that supports arable farming and permanent settlements. The valley's high production of wheat and other grains has earned it the title of "the granary of Xinjiang."

BIBLIOGRAPHY. John MacKinnon, *Wild China* (MIT Press, 1996); John Man, *Gobi: Tracking the Desert* (Yale University Press, 1997) ; E. Mikhail Murzayev, "The Deserts of Dzungaria and the Tarim Basin," *World Vegetation Types*, S.R. Eyre, ed. (Columbia University Press, 1971); Zhao Ji, Zheng Guangmei, Wang Huadong, Xu Jialin, eds., *The Natural History of China* (McGraw Hill, 1990)

RICHARD W. DAWSON
CHINA AGRICULTURAL UNIVERSITY

Kalahari Desert

THE NAME *KALAHARI* is derived from the Tswana word *Kgalagadi*, meaning "the great thirst." The Kalahari Desert, an arid region on the interior plateau of southern Africa, covers area in central and southwestern BOTSWANA, parts of west-central SOUTH AFRICA and eastern NAMIBIA, encompassing an area of about 100,000 square mi (260,000 square km).

The Kalahari Desert is part of a larger sand basin that extends into ANGOLA and ZAMBIA in the north, through Botswana into ZIMBABWE in the east, to the Orange River in South Africa, and west to the highlands of Namibia. The Kalahari is mostly flat, with an average elevation of about 3,000 ft (1,000 m) above sea level. The sands of the Kalahari are red, brown, or white by region. Parallel dunes run north to south or northwest to southeast, depending on the winds and cover large areas of the Kalahari. The dunes vary in height from about 20 (6 m) to 200 ft (60 m) are up to 50 mi (80 km) long, and are separated by chasms of varying width.

The Kalahari has a semiarid climate with frequent droughts. The region receives about 8 in (20 cm) of precipitation a year, mainly between the months of October and May. The rainfall pattern is unpredictable, and precipitation can vary by more than 100 percent between years. The rains generally come in the summer months, September to March, as thunderstorms. Daytime temperatures range between 95 to 113 degrees F (35 to 45 degrees C) from October to March, which are the hottest months, and can drop below freezing between June and August. Temperatures ranges in the Kalahari Desert are some of the highest and lowest in southern Africa. The daily temperature ranges are also large.

The only permanent surface water in or around the Kalahari is the Boteti River. A swampy region exists on the northern border of the Kalahari in northern Botswana. Heavy rains occasionally flood the delta, and the Boteti carries the overflow east into Lake Xau and the Makgadikgadi Pan on the northeastern fringe of the desert. During the rainy seasons systems of streams and rivers flow temporarily. Water can be found below ground in the temporary waterways that flow into depressions in the desert, known as pans. Pans vary in size from a few meters to tens of kilometers in diameter and provide temporary sources of surface water. After heavy rain, these become pans or lakes, and water is then also found in mud-bottomed pools along the beds of the rivers. Pans also occur in craterlike depressions where rock rises above the desert sands.

Tough sun-bleached grasses grow in patches on the sand dunes. Other plant life consists of scrub bushes, patches of forest with fine leaf trees. The most amazing

plant found in the Kalahari is the watermelon. There are both the bitter and sweet varieties.

Wildlife includes the lion, leopard, hippopotamus, rhinoceros, buffalo, zebra, several kinds of antelope and gazelle, baboon, ostrich, spotted and brown hyena, wildebeest, elephant, giraffe, and eland. The hunting of these three last animals is prohibited, and for all game there are strictly enforced hunting seasons. Most animals are migratory, moving seasonally between sources of available water during the rainy season in summer.

BIBLIOGRAPHY. Michael Mares, *Encyclopedia of Deserts* (University of Oklahoma Press, 1999); David McDonald, "Kalahari Desert," Encarta, http://encarta.msn.com (October 2004); Lauren Van Der Post, *The Lost World of the Kalahari* (William Morrow, 1988).

CLARA HUDSON
UNIVERSITY OF SCRANTON

Kamchatka Peninsula

RUSSIA'S FAR EAST IS MARKED by the large mountainous peninsula known as Kamchatka. Kamchatka peaks form one of the most volcanically active regions on earth, counting over thirty active cones, and at least a hundred more that are now inactive. Few people live in this remote corner of Asiatic Russia, where ties to central government, or any government at all, have traditionally been weak at best.

The Kamchatka Peninsula, 750 mi (1,200 km) in length, is formed of two parallel ridges of mountains, the northern extension of the Pacific RING OF FIRE, a continuation of a line of volcanoes stretching south across the Kuril Islands to the ranges of HOKKAIDÔ, JAPAN. The western range is older, and forms the central spine of the peninsula. The eastern range was formed more recently and contains most of the active volcanoes, notably in the Kluchevskaya Complex, where twelve cones create a plateau of ash and hardened lava that so resembles the landscape of the moon that it has been used as a testing site for Soviet lunar vehicles. Kluchevskaya Sopka, at 15,584 ft (4,750 m), is the tallest volcano in Eurasia.

This intensely active region—bubbling with mud pools, geysers, sulfur springs and fumaroles—is caused by the collision of the Pacific and Eurasian plates of the earth's crust, plus the jointure with another active tectonic fault that runs from here to the east, forming Alaska's ALEUTIAN ISLANDS.

Kamchatka is joined to the mainland of Russia by a narrow isthmus, approximately 60 mi (110 km) wide. To the north lies the Koryak Plateau and the Chukchi Autonomous Republic, the land of Russia's Arctic nomads, similar to Alaskan natives. Kamchatka itself is populated partially by indigenous Koryak tribes, and by Russians. Its population of roughly 250,000 live mostly in the regional capital, Petropavlovsk-Kamchatskiy, or in small coastal settlements. The eastern coastline borders the PACIFIC ocean and its subsidiary, the Bering Sea, and is roughly indented with bays and steep inclines resulting from the presence of the eastern mountain chain and individual volcanoes. The western coast is, by contrast, smooth and low in elevation, with numerous swamps, tallgrass meadows, and swift rivers. The largest river, the Kamchatka, is not on this plain but runs south to north between the western and eastern ranges before turning east to empty into the Pacific at the peninsula's only other city of significance, Ust' Kamchatsk.

Russian expansion to the Pacific was checked in the mid-17th century along the AMUR RIVER by the Chinese, so trappers, merchants and adventurers turned further north and crossed the Sea of Okhotsk to reach Kamchatka in the early years of the 18th century. The city of Saints Peter and Paul, Petropavlovsk, was built in 1740, designed to be a springboard for imperialist ambitions in Japan, CHINA, and even INDIA. Instead, the settlements here became a launch-pad for exploration of Alaska and the northwest coast of North America. Russian occupation was brutal, sparking off a series of mass suicides among the Koryaks and Kamchadals, who preferred death to slavery, until administrators were finally sent from the imperial government to establish some sort of law and order.

By the mid-19th century, Russian interests were again looking south to China and the rich Amur Valley, so Kamchatka faded in importance. American and British whalers, however, continued to trade regularly with Russian Far East settlements. Japanese fishing fleets also came in large numbers, attracted to the vast numbers of salmon, red roe, and other fish in Kamchatka's waters. Trade with Americans and Japanese was encouraged during the 1930s (a formality, since Moscow's central authority was so far away anyway) but was sharply curtailed after World War II, when the area became one of the most completely sealed-off regions in the entire Soviet Union. One of the primary focuses during this time was the creation of geothermal

power stations, several of which are now under sharp scrutiny for environmental violations. Other resources—coal, copper, gold, iron and sulfur—are still mostly unexploited because of great distances from commercial markets, and extreme harshness of the climate. Kamchatka has formed its own separate district since 1956 (178,746 square km or 69,711 square mi), but the Koryak area to the north declared itself independent in 1990.

BIBLIOGRAPHY. Sergei Petrovich Suslov, *Physical Geography of Asiatic Russia*, N .D. Gershevsky, trans. (W.H. Freeman, 1961); John J. Stephan, *The Russian Far East: A History* (Stanford University Press, 1994); John Sparks, *Realms of the Russian Bear: A Natural History of Russia and the Central Asian Republics* (Little, Brown, 1992).

JONATHAN SPANGLER
SMITHSONIAN INSTITUTION

Kansas

LOCATED MIDWAY between the east and west coasts of the UNITED STATES, Kansas has a unique position as a geographic hub. Despite images to the contrary, Kansas has a unique geography as a transition between the historic farmlands of the east and the open range of the west. The land rises gently from 679 ft (207 m) above sea level in the southeast to 4,039 ft (1,232 m) at Mt. Sunflower in the northwest. In between, the land is interspersed with a mixture of large and gently rolling hills and soils that become progressively drier once you pass the middle of the state near Salina on the way west.

The Flint Hill region in the east and west-central part of the state represents one of the last true tall grass GRASSLANDS with its golden prairies in spring and early summer, gentle streams, free-roaming cattle, and a quietness that brings to mind images of cowboys and campfires. In the northeast, the hills are more sharply defined by a rockier terrain and are heavily wooded, a remnant of the last great glacial advance that moved across the North American landscape. From the 100th meridian westward, the land becomes drier and the hills less prominent as you approach the western boarder. Here the Great Plains dominate and a sand hill region is exposed, typifying a PHYSICAL GEOGRAPHY that is prominent from MEXICO far into CANADA. It is also representative of a time when great parts of the continental interior of the United States were covered by ocean waters, which as much as glaciers from the ice ages has helped shape the American landscape. Kansas is bordered by MISSOURI to the east, OKLAHOMA to the south, COLORADO to the west, and NEBRASKA to the north. With a total area of 82,282 square mi (213,064 square km), the state's 2,688,418 residents live in the 15th largest of the 50 states. Kansas was the 34th state to join the United States, doing so on January 29, 1861.

Kansas's waters have played an important role in its development. Early settlers to the region found their way by following the Missouri River upstream from St. Louis to the point where the river is joined by the Kansas or Kaw River and makes a major turn to the north. By the 1820s, this convenient location had become an important warehousing and distribution center as early Kansas City took form in providing support for the Santa Fe and Oregon trails. Other major waterways that continue to serve the state are the Kaw, Republican, and Smoky Hill rivers in the north, and the Arkansas, which makes a somewhat east-west transect of the southern half of the state on its way from the ROCKY MOUNTAINS to the Gulf of Mexico. Kansas is also home to one of the largest underground aquifers (the Ogallala) in the United States. Use of the waters from this aquifer has greatly expanded the agricultural potential of the drier, sandier soils of the western part of the state by allowing a shift from dryland to irrigated production practices. Once known primarily as the "wheat state," Kansas now has significant corn, soybean, and sunflower crops in addition to a thriving cattle industry.

Kansas is historically an agricultural state, ranking third behind TEXAS and MONTANA in total agricultural acreage. While Kansas is the nation's top wheat grower and also the leading producer of grain sorghum and corn, manufacturing and service industries have recently surpassed agriculture as the major income producers. The two leading industries are the manufacture of transportation equipment and industrial and computer machinery. Other important manufactures are petroleum and coal products and non-electrical machinery. In addition, the state is a major producer of crude petroleum and has large reserves of natural gas, helium and salt. Cattle and calves represent the single most valuable agricultural product, and the Kansas City stockyards are among the nation's largest.

Economically, the state can be divided into several distinct production zones. In the wooded northeast, the manufacturing of a variety of products, including auto-

mobiles in the Kansas City region, dominates. The southeast has a strong aircraft manufacturing sector centered in Wichita, as well as railroad shops, flour mills, meatpacking plants, grain elevators, and oil fields. Coal production along the eastern border corridor with Missouri has long helped serve the electric energy needs within the region. The southwestern part of the state around Garden City is home to several major meatpacking firms in addition to feedlot operations and natural gas and helium production facilities. The central, northwest, and north-central portions of the state retain the economic stability of what has come to be called the "heartland," with large-scale scientifically managed farming operations.

The state's main population centers include the Kansas portion of metropolitan Kansas City in the northeast; Wichita, the state's largest city in south central part of the state; Topeka, the state capital; and Lawrence in the northeast.

BIBLIOGRAPHY. Hubert Self, *Environment and Man in Kansas: A Geographical Analysis* (Regents Press of Kansas, 1978); Hubert Self, *Historical Atlas of Kansas* (Regents Press of Kansas, 1989); Ruth Bjorklund, *Kansas* (Benchmark Books, 2000); Allan Carpenter, *Kansas* (Childrens Press, 1979); C.C. Howes, *This Place Called Kansas* (University of Oklahoma Press, 1984); R. Richmond, *Kansas: A Land of Contrasts* (Forum Press, 1989); U.S. Census Bureau, www.census.gov (August 2004).

RICHARD W. DAWSON
CHINA AGRICULTURAL UNIVERSITY

Karachi

AFTER PAKISTAN'S independence, Karachi was the national capital for 13 years (1947–60). In 2004, it was the capital of the Province of Sind. With a population of a little over 11 million, Karachi is the 16th largest metropolis of the world. Karachi has a semi-desert climate with a summer mean temperature of 80 degrees F (26.7 degrees C). Situated adjacent to the INDUS RIVER delta, it opens up to the ARABIAN SEA; it is not only the largest city of the country but also its largest port. During the British colonial times, it was a medium-sized city; the 1941 population was 435,000. After the partition of colonial India into INDIA and PAKISTAN in 1947, there was a huge migration of Urdu and Gujarati speaking Muslims from India to Karachi, and an exodus of Sindhi- and Gujarati-speaking Hindus from Karachi to India. The Indian migrants in Karachi are called Muhajirs. Such population exchanges made Karachi linguistically heterogeneous, as 66 percent of the city dwellers became Urdu speakers and native Sindhi citizens turned into a minority community, and homogeneous in terms of religion, with over 95 percent Islamic believers.

Moreover, the attraction of Karachi as a thriving industrial city pulled many migrants from the Punjab province of Pakistan, who were also Muslims; they were bilingual, as they spoke both Punjabi and Urdu. During the Afghan War of the 1980s, when Soviets virtually occupied Afghanistan, a large number of Pushtu-speaking Pathans (Pashtuns) came as refugees, creating additional diversity. Many of Pushtu-speaking Pathans returned to their country after 2001. No other South Asian metropolis has so large a nonnative population base as Karachi. In the 1960s, only 16 percent of Karachi's population was born in the city.

Karachi's industries, mostly in large, planned industrial estates at the western periphery of the city, include cotton, engineering, leather, and consumer goods. Since the 1980s, development of heavy industries has been prohibited in the city proper because of increased level of air pollution. The center of Karachi has bazaar-type commercial establishments with high density of population. To the east are Civil Lines and the planned military cantonment dating back to 19th-century colonial times. New rich and upper-middle-class housing is located adjacent to the Civil Lines. The poor live in the central area as well as toward the west near their workplaces.

Ethnic and linguistic rivalry, associated with fertile ground for al Qaeda recruitment and availability of U.S. supplied arms during the 1980s Afghan War, have turned Karachi into a playground of unruly militants. Muhajirs are divided into three warring groups: Muttahinda Qaumi Movement, Muhajir Quami Movement, and Basic Association of Citizens of Karachi. They operate as gangs from their own exclusive territorial strongholds. Apart from the law and order situation, Karachi's problems include a severe shortage of drinking water, very high growth rate of population because of migration, and proliferation of slum settlements (*katchi abadi*).

BIBLIOGRAPHY. Ashok K. Dutt and George M. Pomeroy, "Cities of South Asia," *Cities of the World: World Regional Urban Development*, Stanley D. Brunn, Jack F. Williams, and Donald J. Zeigler, eds., (Rowman and Littlefield, 2003);

United Nations, "Population Growth and Policies in Mega-Cities: Karachi," policy paper # 13 (United Nations, 1988).

ASHOK K. DUTT, PH.D.
UNIVERSITY OF AKRON

karst

KARST IS A TERRAIN characterized by sinkholes, caves, and disappearing streams that have been created by chemical weathering in thick carbonate bedrock. Karst is named for the Krs Plateau along the ADRIATIC SEA and comes from an old Slavic word meaning "barren land."

In the karst hydrologic cycle, moving water slowly dissolves away the limestone at and below the surface of the Earth, creating bumpy surface features and subsurface cavern systems. Landscapes generally appear pockmarked and gently rolling. Sinkholes or surface depressions are created when the roof of an underground cavity becomes thin and collapses under the weight of overlying beds. Sinkholes are visible at the surface and usually resemble small lakes and ponds. Some of them are steep-sided and deep, while others are minor dents in the land and relatively shallow. Some sinkholes fill with soil eroded from nearby slopes and are barely distinguishable. As sinkholes grow in size and abundance, cavern ceilings often collapse, thus exposing broad, flat-floored valleys.

Hydrogeologists cannot yet predict the location and timing of sinkhole creation. The dissolution of underground limestone creates honeycombs of caves columned with stalactites and stalagmites, generally circuitous and extending for miles. Streams run underground over long distances and are usually periodic at the surface, gushing up only in the event of heavy rains. Some cave patterns are dendritic or treelike; others have intersecting joints or meander like streams. Springs are a common feature in the karst areas, and because of their cool and constant temperatures, they have unique plant and animal life such as watercress, flatworms, and snails. Karst regions typically have caves in all stages of evolution, from water-filled holes to mature caves with large passageways and sporadic water flow. In order for karst to appear, limestone formations must contain at least 80 percent calcium carbonate, be aerated, and have joints for water to flow,

Karst regions are found throughout the world and cover about 15 percent of the Earth's land area. The most notable examples are the Mammoth Cave region of KENTUCKY; Dalmatia, along the ADRIATIC SEA; the haystack hills of CUBA, PUERTO RICO, southern CHINA; JAPAN; the Yucatan Peninsula of MEXICO; the MISSOURI Ozarks; INDIANA; NEW MEXICO; and northern and central FLORIDA. In Florida, karst is associated with the largest artesian system in North America. Artesian springs abound and are often the source of rivers. Silver Springs, Florida, discharges nearly half a billion gallons of water daily, the largest known artesian outflow in the world. Some parts of the Florida artesian system discharge on the sea floor at considerable distances from shore. With a consistently high water table, the caverns of Florida karst are typically filled with water and the sinkholes are lakes. In fact, there are so many sinkholes in central Florida that the region is called "the lake district."

A coastal karst terrain on Navassa Island exemplifies the alien landscape typical of the landform.

In Missouri, "the cave state," the Division of Geology and Land Survey has counted more than 5,000 caves. Most of them are dolomite or limestone and some of them run more than 15 miles. Mammoth Cave

in Kentucky is the longest cave in the world (350 mi or 560 km); almost 40 percent of Kentucky has karst features. Tower karst in China is composed of resistant blocks of rock that extend up 660 ft (200 m).

Caves have been utilized by human beings since ancient times, some as human burial places, others as hiding places and storage sites for furs, metals, alcohol, and food supplies. More than two dozen caves underlay the city of St. Louis, and historically many of them were used as underground breweries, saloons, beer gardens, and storage areas. Today karst topography poses serious hazards for construction and settlement. Chasms can suddenly appear and swallow entire houses and sections of highway.

Because the bedrock is unstable and cavernous, homes in karst regions are generally built without basements, and highways and cemeteries must be sited with care. Subsidence along a line of sinkholes and cavern collapse can quickly render an area off limits to human development. Sinkholes become repositories for everything from kitchen trash to bulldozers and thus become serious and elusive environmental challenges. Sinkhole contaminants flow with the water underground and travel far afield along the labyrinth of pathways, making cleanup virtually impossible.

BIBLIOGRAPHY. Robert W. Christopherson, *Geosystems: An Introduction to Physical Geography* (Prentice Hall, 2005); Tom L. McKnight and Darrel Hess, *Physical Geography* (Prentice Hall, 2005); A.G. Unklesbay and Jerry D. Vineyard, *Missouri Geology* (University of Missouri Press, 1992).

ANN M. LEGREID
CENTRAL MISSOURI STATE UNIVERSITY

Kashmir

KASHMIR IS THE name given to the Vale of Kashmir, a valley situated on the Jhelum River between the Pir Panjal range and the main range of the HIMALAYAS. By extension, the name is used to refer to the Indian state of Jammu and Kashmir and to a portion of the old preindependence territory that included what is now called Northern Areas and Azad (Free) Kashmir in PAKISTAN. In addition, INDIA claims a 14,500-square-mile (37,554-square-km) section of the Tibetan plateau, "Aksai Chin," as part of the old state of Kashmir. Kashmir was also a state under British soveriegnty until 1947 when the Hindu maharaja of the largely Muslim state attempted to obtain a degree of independence for his land by not joining Hindu India or Muslim Pakistan. The delay led to disputed territory between the two powers. The Indian government claimed that the Hindu ruler of Kashmir state had signed an Instrument of Accession to India, although the original document has never been seen, nor is there any evidence to indicate that the maharaja signed the document on the day cited, as he was being driven from Srinagar to Jammu. The state fell into outside hands, once again having been under the rule of the Moghuls and the Sikhs in previous centuries. Although the United Nations agreed to the suggestion of a plebiscite to which India also agreed, no such vote has ever taken place.

India has claimed territory, the former Gilgit Agency and Baltistan, that is now under Pakistani control and known as the Northern Areas. A small portion of this agency in the extreme north and lying in the watershed draining into CHINA was deeded to China by Pakistan in 1964. At no time did Gilgit Agency come under direct Kashmir rule, because the British kept a civil and military delegation in the agency prior to independence. Because of the continuing disputes of jurisdiction of Kashmir, Pakistan has agitated for a plebiscite, which now seems remote. Armed conflict by India with China and Pakistan has resulted in small wars in 1962 with China and in 1965 and 1971 with Pakistan. A transitory border, "The Line of Control," was established but another dispute broke out in 1984 when Pakistan issued permits to tourists into territory that India claimed as Kashmir India territory. This dispute revolved around the Siachen glacier, the headwaters for the Nubra River flowing into the Shyok River, a tributary of the Indus.

This continued territorial dispute mars the attributes of Kashmir in its several guises today. Geographically, Jammu is in the Punjab lowlands and is Hindu or Sikh. The Vale of Kashmir is almost all Muslim today, the Hindu castes having left. Territorially, the largest area is Ladakh in the trans-Himalaya Tibetan plateau, where local culture is similar to that of Tibet but where there are a substantial number of Muslims, both Sunni and Shia. These people formerly conducted caravan trade across the infamous "Five Passes" route, that is, five passes over 18,000 ft (5,500 m), on the route from Leh in Ladakh to Sainju Bazaar down on the edge of the Takla Makan Desert bowl of XINJIANG province in western China. This lucrative trading route was the reason for the prosperity of old Kashmir. Its

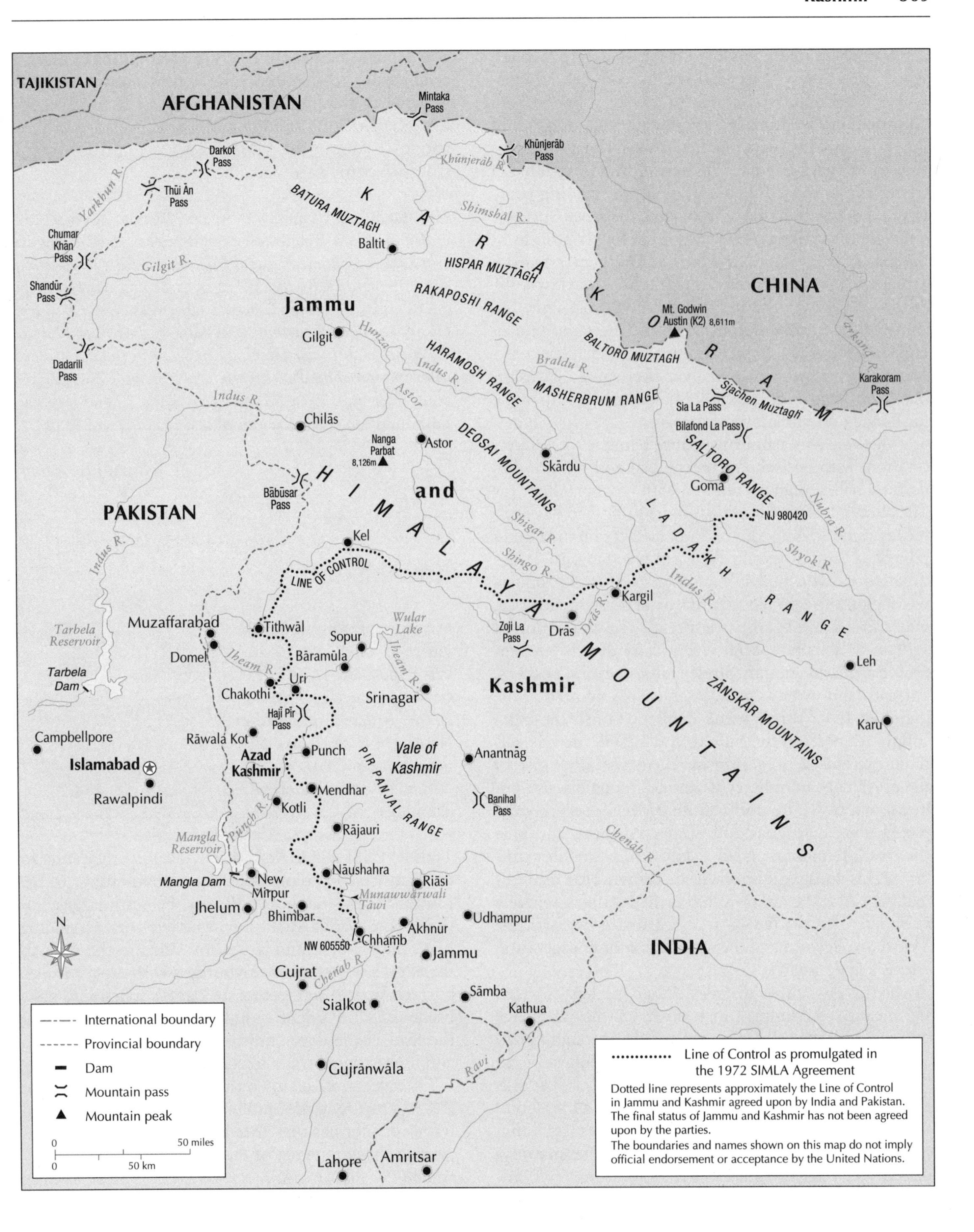

TAJIKISTAN
AFGHANISTAN
Mintaka Pass
Khūnjerāb Pass
Darkot Pass
Thūi Ān Pass
Chumar Khān Pass
Shandūr Pass
Dadarili Pass
KARAKORAM
BATURA MUZTAGH
HISPAR MUZTĀGH
RAKAPOSHI RANGE
HARAMOSH RANGE
BALTORO MUZTAGH
MASHERBRUM RANGE
Siachen Muztagh
SALTORO RANGE
DEOSAI MOUNTAINS
LADAKH RANGE
ZĀNSKĀR MOUNTAINS
PIR PANJAL RANGE
HIMALAYA MOUNTAINS
CHINA
Mt. Godwin Austin (K2) 8,611m
Karakoram Pass
Sia La Pass
Bilafond La Pass
Jammu and Kashmir
Baltit
Gilgit
Chilās
Astor
Nanga Parbat 8,126m
Skārdu
Goma
NJ 980420
Bābūsar Pass
PAKISTAN
Kel
LINE OF CONTROL
Kargil
Drās
Zoji La Pass
Muzaffarabad
Tithwāl
Sopur
Wular Lake
Bāramūla
Domel
Tarbela Reservoir
Tarbela Dam
Uri
Chakothi
Srinagar
Kashmir
Leh
Karu
Hajī Pīr Pass
Campbellpore
Rāwala Kot
Azad Kashmir
Punch
Vale of Kashmir
Anantnāg
Islamabad
Rawalpindi
Mendhar
Kotli
Banihal Pass
Rājauri
Mangla Reservoir
Mangla Dam
New Mirpur
Nāushahra
Riāsi
Munawwarwali Tāwi
Jhelum
Bhimbar
Udhampur
Akhnūr
NW 605550
Chhamb
Jammu
INDIA
Gujrat
Sāmba
Sialkot
Kathua
Gujrānwāla
Lahore
Amritsar
Yarkhun R.
Gilgit R.
Khūnjerāb R.
Shimshāl R.
Hunza
Indus R.
Braldu R.
Astor
Yarkand R.
Nubra R.
Shigar R.
Shingo R.
Shyok R.
Drās R.
Jhelum R.
Punch R.
Chenāb R.
Ravi
N
International boundary
Provincial boundary
Dam
Mountain pass
Mountain peak
0 50 miles
0 50 km
Line of Control as promulgated in the 1972 SIMLA Agreement
Dotted line represents approximately the Line of Control in Jammu and Kashmir agreed upon by India and Pakistan. The final status of Jammu and Kashmir has not been agreed upon by the parties.
The boundaries and names shown on this map do not imply official endorsement or acceptance by the United Nations.

geography created a passland, that is, a state through which passes were essential for its commercial existence. The route was dangerous with robbers from Shimshall in the Hunza area pillaging pack ponies and Bactrian camels who often died from lardug, high-altitude pulmonary edema. The hybrid Buddhist-Muslim trader community in Leh and along the trading route ensured the viability of the caravans. Between Srinagar and Leh, the Zoji pass was snowed in for six months of the year, thereby disrupting traffic. Another route from Srinagar went north over the Burzil pass down the Astor valley, over the Indus, and north to Gilgit. Because of the narrow defiles north of Gilgit and Hunza and over the crest of the Karakorum range little interregional trade was carried over this route. Transhumants of sheep and goats still migrate seasonally in the Pir Panjal and Himalaya mountains.

The Vale of Kashmir was once a major tourist destination, with houseboats on the Dal and Wular lakes nestling below high mountains with local products for sale. Many Indian films customarily showed Kashmir scenes. Trout fishing on the Liddar River on the Hindu pilgrimage route beyond Pahlgam was famous. Traditional woodworking, silk making, and wool products were highly sought. Agriculture is marginal, with rice, maize, peas, and beans widespread. The cultivation of saffron is a local specialty. Basic foodstuffs are imported from the Punjab. Stone fruits, cherries, peaches, apricots, and walnuts and almonds are exported to the plains of India through the Banihal tunnel, the main supply route from the Vale to India. Wildlife has suffered extreme depredation in recent years because of the prevalence of military weapons among divisive political factions. The noble sambar deer is severely depleted; ibex wild goat and urial wild sheep and blue sheep still remain in places in Ladakh, although markhor seems to have disappeared. Brown bear and Himalayan black bear now have their ranges severely curtailed. The elusive snow leopard still exists in the Markha valley in Ladakh, where a major conservation effort is under way.

In the past three decades, language has changed significantly. Kashmiri, an Indo-Aryan hill language cognate with the hill languages spoken through northern Pakistan to Pashai in Kabul Kohestan in Afghanistan, reports a decline in the number of speakers. Bazaar languages, Punjabi in the Vale of Kashmir, like Pashto in northern Pakistan and Farsi in Kabul Kohestan, have smothered regional languages in everyday usage. English schools are ubiquitous. Two villages in Ladakh sitting on the "Line of Control," Dah and Hannu, retain pre-Islamic, pre-Buddhist cultural celebrations to this day. Nineteenth-century linguists referred to these villages, and others, as Dardic, although there is no validity to this term today. Eurocentric myths about the "Dards" being the true Aryans still surface today in Ladakh.

BIBLIOGRAPHY. Andrew L. Adams, *Wanderings of a Naturalist in India* (Edmoston and Douglas, 1867); August Francke, *A History of Ladakh with Critical Introduction and Annotations by S. S. Gergen and Fida M. Hassnain* (Sterling, 1977); Joseph E. Schwartzberg et al., *The Kashmir Dispute at Fifty, Charting New Paths to Peace; Report on the Visit on an Independent Study Team to India and Pakistan Sponsored by the Kashmir Study Group* (The Kashmir Study Group, 1997); J.E. Schwartzberg, "Who Are the Kashmiri People?" *Environment and Planning* (v.29, 1997).

Nigel J.R. Allan
University of California, Davis

Kazakhstan

Map Page 1119 **Area** 1,049,155 square mi (2,717,300 square km) **Population** 16,763,795 **Capital** Astana **Highest Point** 23,084 ft (6,995 m) **Lowest Point** -436 ft (-132 m) **GDP per capita** $6,300 **Primary Natural Resources** petroleum, natural gas, coal.

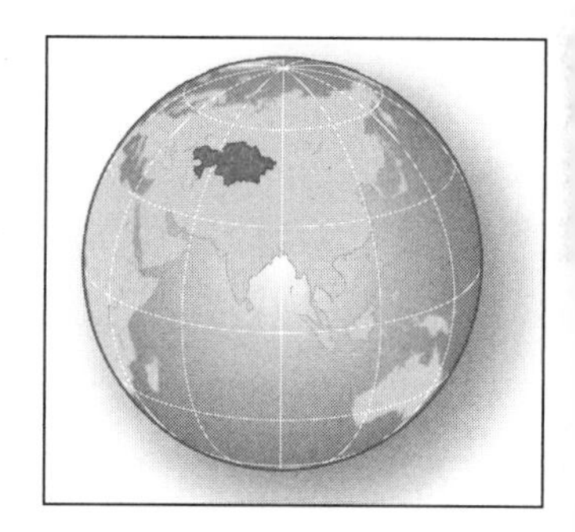

THE REPUBLIC OF Kazakhstan occupies the center of the Eurasian supercontinent, the second largest of the former Soviet republics and today the ninth-largest nation in the world. Most of its vast territory, stretching from the Caspian and Volga lowlands in the west to the ALTAI MOUNTAINS on the borders with CHINA, is flat, covered in arid semidesert or STEPPE. Its people have been nomadic herders for millennia but are now taking the lead in the development of industry, agriculture, and international trade for Central Asia.

Kazakhstan has no coastline, except for 1,174 mi (1,894 km) along the landlocked CASPIAN SEA. Its main trade corridor has therefore been across the land, connecting the commerce of its neighbors (RUSSIA, TURKMENISTAN, UZBEKISTAN, KYRGYZSTAN, and China), starting with the SILK ROAD in the Middle Ages, contin-

uing with the development of railroads in the 20th century, and once again building its transportation links with China today. The Kazakh people are a mixture of Turkic and Mongol nomadic tribes who ruled the region through interlocked, kin-based khanates until gradual annexation by the Russian Empire in the 18th century.

After a brief attempt at autonomy in the 1920s, the region became a Soviet Republic in 1936. The Soviets initiated intense agricultural and industrial development projects in the northern steppes during the 1950s, which brought numerous immigrants—Russians, Ukrainians, Germans, and others—who eventually out-numbered the native peoples. The Kazakhs declared their independence in 1991, and the population has once again shifted, through both mass emigration and a higher birthrate among ethnic Kazakhs.

THREE REGIONS

Geographically, the country is mostly homogeneous, a great flat tableland of lowlands, plains, and plateaus. This can be broken up into three basic regions: the Volga and Caspian lowlands of the west (with the narrow, low Mugodzhar Mountains), the Turgai Plateau, and the Kazakh Steppe. The Turgai Plateau is characterized by a central depression with a chain of small lakes. Most of the area's rivers flow into these lakes, some of which evaporate and disappear completely during the dry season.

The Ulu-Tau mountains divide this area from the eastern third of the country, the Kazakh Steppe, which is mostly flat with a few scattered massifs. The primary of these elevated areas divides the watershed between the upper Irtysh River valley (whose waters flow across the West Siberian Plain all the way to the Arctic Sea) and Lake BALKASH, one of the largest lakes in the world in area, but not in volume, since it is very shallow. Lake Balkash is also very salty and has very little wildlife. South of these broad semi-desert and steppe plains are the northern edges of one of the great deserts of Central Asia, the Kyzyl Kum. Across this desert runs the other major river of Kazakhstan, the Syr Darya, which flows into the ARAL SEA. To the east and south rise the great mountain chains that divide Central Asia from China, the Altai, and Tian Shan.

Along the northern edges of Kazakhstan, several large cities were developed under the Soviet regime, the mining and industrial heart of Central Asia. Kazakhstan is a leading producer of coal, iron ore, manganese, copper, and many other minerals. It is oil and gas, however, that hold the most promise for the Kazakhs, with reserves estimated to put Kazakhstan into the top 10 oil producers in the world by 2015, especially around the Caspian Sea basin—the Tenghiz field ranks as one of the largest deposits in the world. Agriculture is still a major part of Kazakh economy, chiefly in grain exports and livestock.

Actively courting western investment and friendship, the Kazakhs declared their territory nuclear free in 1995, and the new national capital (Astana, moved in 1998, formerly known as Akmola, then as Tselinograd) lies inside a special economic zone with reduced trade barriers and tax incentives for investors. Russia continues to run the space center Baykonur Cosmodrome (formerly called Leninsk), but the Kazakh government is encouraging greater participation by ethnic Kazakhs.

BIBLIOGRAPHY. *World Factbook* (CIA, 2004); M. Wesley Shoemaker, *Russia, Eurasian States and Eastern Europe 1994*, The World Today Series (Stryker-Post Publications, 1994); Paul E. Lydolph, *Geography of the U.S.S.R.* (Misty Valley Publishing, 1990); Sergei Petrovich Suslov, *Physical Geography of Asiatic Russia*, N.D. Gershevsky, trans. (W.H. Freeman, 1961); "Welcome to Kazakhstan," www.president.kz (August 2004).

JONATHAN SPANGLER
SMITHSONIAN INSTITUTION

Kentucky

NICKNAMED FOR the bluegrass that covers much of this U.S. state, and known for its whiskey and for the Kentucky Derby that takes place annually at Churchill Downs in Louisville, the Commonwealth of Kentucky was named for the Kentucky River, which was thought to have been named from the Iroquoian word(s) for either "meadowland" or "land of tomorrow." Kentucky is bounded on the north by the states of ILLINOIS, INDIANA, and OHIO, on the south by TENNESSEE, on the east by VIRGINIA and WEST VIRGINIA, and on the west by MISSOURI.

The Ohio River runs along the entire northern boundary of Kentucky, and the MISSISSIPPI RIVER runs along the western boundary. The total area of Kentucky is 39,728 square mi (102,895 square km). The state ranks 36th in size and 25th in population among the 50 states. Kentucky's largest cities are Lexington, Louisville, Owensboro, Bowling Green, Covington,

Hopkinsville, Frankfurt (the capital), Henderson, Richmond, and Jefferstown.

Approximately 679 square mi (1,758 square km) of Kentucky are covered by water. In addition to the Kentucky, Mississippi, and Ohio, Kentucky's rivers include the Cumberland and the Green. Most of Kentucky's lakes were created by damming river waters, including Barkley Lake, Rough River Lake, Green River Lake, Dewey Lake, and Cumberland Lake. Kentucky has a number of significant waterfalls. The average elevation of Kentucky is 750 ft (228 m) above sea level. The highest point in the state is 4,139 ft (1,261 m) above sea level at Black Mountain, and the lowest point is 261 ft (79 m) above sea level at the Mississippi River. The state is approximately 380 mi (611 km) from east to west and approximately 140 mi (225 km) from north to south.

Kentucky's climate is temperate with annual temperatures ranging from 52 degrees F (11 degrees C) in the northeastern part of the state to 58 degrees F (14 degrees C) in the southwestern section. Temperatures throughout the year range from below freezing in the winter to warm and humid in the summer. Annual precipitation is approximately 45 in (114 cm). Winter snow averages range from 40 in (101 cm) in Kentucky's highest elevations to 25 in (63 cm) in the northeast, to 10 in (25 cm) in the southwest. Kentucky is frequently beset by storms, particularly from March to September.

The geographic area of Kentucky is varied, encompassing five distinct physiological regions: the Bluegrass Region, the Cumberland Plateau, the Western Coal Field, the Pennyroyal Region, and the Jackson Purchase Region. The north central section of Kentucky falls within the Bluegrass Region, sometimes called the Lexington Plain, which extends into the neighboring state of Ohio.

Much of the Bluegrass Region is filled with rolling meadows and low, steep sandstone hills known as knobs, which are found in the Knobs Region that forms the eastern, southern, and western sections of the Bluegrass Region. The Knobs Region separates the Bluegrass Region from the Cumberland or Mississippian Plateau. The Cumberland Plateau is part of the Appalachian Plateau, which extends along much of the eastern part of the UNITED STATES from NEW YORK to ALABAMA. In Kentucky, the area is characterized by mountains, plateaus, and valleys that are underlain with sandstone, shale, and limestone. Kentucky's Cumberland, Pine, and Black mountain ranges are located in the Cumberland Plateau. The section of the plateau known as the Eastern Coal Fields is a mountainous area covered with forests and streams. These mountain ridges are frequently crossed by gaps such as the well-known Cumberland Gap. Much of the Daniel Boone National Forest lies at the western end of the Eastern Coal Fields.

The northwestern section of Kentucky contains the Western Coal Field, which is a hilly area that lies within the Illinois Basin, extending to the Ohio River on the north and to the Pennyroyal Region on the east, west, and south. The Western Coal Field is named for the large deposits of coal that appear throughout the area. The soil on the borders of the Ohio River is highly fertile. The Pennyroyal Region, sometimes referred to as the Pennyrile Region for the small herb that grows in the area or, alternately, as the Highland Rim, covers a stretch of land between the southern border of Kentucky in the Appalachian Plateau to Kentucky Lake. The southern section of the Pennyroyal Region is made up of flat lands interspersed with occasional rolling hills, while the northern section is comprised mostly of rocky ridges containing numerous underground caves and tunnels. The best known of the caves is Mammoth Cave. A treeless area within the center of the Pennyroyal Region is known as The Barrens.

The Jackson Purchase Region is located at Kentucky's western tip and is part of the Gulf Plains Region, which extends from the Gulf of Mexico to Illinois. The Jackson Purchase Region stretches to the Kentucky Lake in the east, to Illinois in the north, and to the Mississippi River in the west. Low hills and flooded plains make up the land in the Jackson Purchase Region. The Madrid Fault is located in this area. When earthquakes hit the area in 1811 and 1812, the Mississippi River began to flow backward, creating Reelfoot Lake.

Approximately 40 percent of Kentucky is forested. The coffee tree is the state tree. Kentucky hardwoods include oak, beech, hickory, maple, and walnut. Softwoods include cypress, hemlock, cedar, and pine. Flowering shrubs include buckeye, dogwood, laurel, azalea, rhododendron, redbud, blueberry, pennyroyal, and goldenrod (the state flower).

Earlier in Kentucky's history, bison and elk roamed the forest, but now the woods are filled with fox, groundhog, muskrat, opossum, rabbit, raccoon, squirrel, and deer. In addition to the cardinal, which is the state bird, Kentucky is home to the eagle, egret, mockingbird, yellow-billed sapsucker, crow, kingfisher, and woodpecker. Migratory birds are also frequently seen

in the area. Kentucky has four state forests and 59 state parks.

Agriculture dominated Kentucky's economy until the middle of the 20th century when services and manufacturing gained prevalence. Tobacco, hay, corn, soybeans, wheat, fruit, dairy, and livestock are the major products produced by Kentucky farmers. Industries include motor vehicles (the state's largest industry), health services, furniture, aluminum ware, brooms, apparel, lumber products, machinery, textiles, and iron and steel products. Kentucky's most important mineral resources are bituminous coal, petroleum, natural gas, stone, sand and gravel, clay, fluorspar, gemstone, limestone, lead, zinc, and fluorite. Kentucky's coal mines provide 85 percent of the state's mineral income.

BIBLIOGRAPHY: Mark Matsson, ed., *Macmillan Color Atlas of the States* (Macmillan, 1997); Dan Golenpaul, ed., *Information Please Almanac* (McGraw-Hill, 2003); "Kentucky" www.netstate. com (April 2004); "Think Kentucky" www.thinkkentucky. com (April 2004).

ELIZABETH PURDY, PH.D.
INDEPENDENT SCHOLAR

Kenya

Map Page 1114 **Area** 224,900 square mi (582,488 square km) **Capital** Nairobi **Population** 31,639,091 **Highest Point** 17,058 ft (5,199 m) **Lowest Point** 0 m **GDP per capita** $1,100 **Primary Natural Resources** coffee, tea, corn, vegetables, livestock.

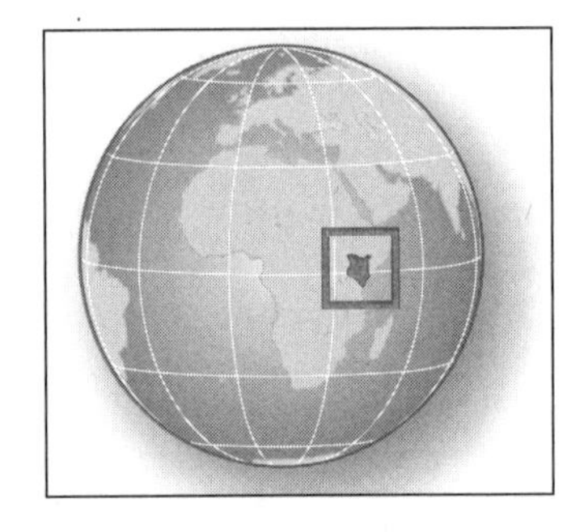

KENYA, WHICH LIES astride the equator on the eastern coast of Africa, is bordered in the north by SUDAN and ETHIOPIA, in the east by SOMALIA, on the southeast by the INDIAN OCEAN, on the southwest by TANZANIA and to the west by Lake Victoria and UGANDA.

The country is notable for its' geographical variety. While most people immediately think of African wildlife and great expanses of GRASSLAND, the land is made up of several distinctive geographical regions. The first is a narrow, low-lying, fertile coastal strip along the shores of the Indian Ocean that is fringed with coral reefs and islands. A series of low hills separates the coastal region from the vast bush-covered plains of the eastern plateau forelands, which sprawl between the central highlands to the west and the coastal strip on the east and slope gently toward the sea.

In the southwestern corner of the country is the Lake Victoria basin, which extends eastward from the lake to the central rift highlands, in which about 85percent of the population and the majority of economic enterprise are concentrated. The central rift highlands run from north to south down the western half of Kenya and are split by the famous Rift Valley into two sections, the Mau Escarpment on the west and the Aberdare Range and accompanying highlands on the east. The Mau Escarpment in the west rises above 9,000 ft (2,740 m) and stretches for more than 200 mi (320 km) northward from the Tanzanian border to the west-central border with Uganda.

The Aberdare Range to the east, which forms the eastern border of the Rift Valley, rises to nearly 10,000 ft (3,050 m). In the high plateau area, known as the Kenya Highlands, lie Mt. Kenya 17,058 ft (5,200 m), Mt. Elgon 14,176 ft (4,322m), and the Aberdare Range rising to over 13,000 ft (3,963 m). The plateau is bisected from north to south by the Rift Valley. The Great Africa Rift Valleys runs from North to South through the whole of Kenya. The Kenyan Rift Valley is a section of a 3,700 mi (6,000 km) rift system that stretches from the Dead Sea in the MIDDLE EAST, south through the RED SEA, Ethiopia, Kenya, Tanzania, MALAWI, and into MOZAMBIQUE. The whole area contains several lakes, extinct volcanoes, and numerous small game parks. The scenery in the Rift Valley is breathtaking, particularly at the viewing points just north of Limuru and Naivasha or from the top of the Mau escarpment. In the south, the valley narrows and deepens with walled escarpments rising 2,000 to 3,050 ft (610 to 930 m) above the valley floor, while in the west, the plateau descends to the plains that border Lake Victoria.

A network of small, seasonal rivers and streams drains most of the Kenyan landscape. Kenya's most important river, the Tana, rises in the central highlands and drains some 16,300 square mi (42,200 square km), roughly 7 percent of the country's total land area, before flowing into the Indian Ocean. Northern Kenya receives little rainfall, but regions in the southern part of the country are plentifully watered. Kenya has two rainy seasons: the long rains from late March to May and the short rains from October to December. Evergreen forests can be found in the highlands along with

wide expanses of bamboo grass. East and west of the highlands, the vegetation gives way to low trees that are casually scattered throughout a predominantly grassy landscape. Semidesert conditions exist below 3,000 ft (915 m) in the north, with thick expanses of thornbush interspersed with massive baobabs trees. In the coastal belt, dense, high bush alternates with limited areas of forest.

Kenya is noted for its wildlife, safaris, and much of our image of Africa. Game and national parks, such as the Masai Mara National Reserve and the Ambosili National Park, are filled with lions, leopards, wild dogs, elephants, buffaloes, rhinoceroses, zebras, antelopes, gazelles, hippopotamuses, and crocodiles. Equally famous, but not a part of Kenya, are the Serengeti National Park and Mt. KILIMANJARO, both of which lie just across the border in Tanzania. Only about 4 percent of the land is arable and nearly all of this is cropped, mainly with corn and other grains.

Kenya became an independent country in 1963. Jomo Kenyatta, a member of the Kenya African Nation Union (KANU), became the country's first president on December 12, 1964. Kenya soon became a one-party state, after the voluntarily dissolving of the Kenya African Democratic Union (KADU).

After Kenyatta's death in 1978, Vice President Daniel arap Moi became the interim president. Under Moi's presidency, political oppression continued with the outlawing of political activities. Thus, many aspiring politicians and political parties went underground for many years. Sensing an opportune time because of resentment of the government in the 1990s, many of the underground parties emerged to challenge KANU. Finally in 2002, a coalition of opposition parties ended the 24-year presidency of arap Moi. Mwai Kibaki became the country's third president.

In 1998, terrorists bombed the American Embassy in Nairobi. Many people, particularly Kenyans, were killed in the horrific violence. The incident further damaged the Kenyan struggling tourism industry. A major challenge to Kenya's stability is population growth. With so many people being added to the total population, there will be severe pressure on land, and subsequent intensity in the rural-urban migration phenomenon.

BIBLIOGRAPHY. David L. Clawson and Merrill L. Johnson, eds., *World Regional Geography: A Development Approach* (Prentice Hall, 2004); Jeffress Ramsay and Wayne Edge, eds., *Global Studies: Africa* (McGraw-Hill, 2004); Marshall S. Clough, *Mau Mau Memoirs: History, Memory and Politics* (Lynne Rienner Publishers, 1998); *World Factbook* (CIA, 2003).

SAMUEL THOMPSON
WESTERN ILLINOIS UNIVERSITY

Khartoum

KHARTOUM IS THE capital city and administrative region of the Republic of the SUDAN, located at the confluence of the Blue and White NILE rivers. Khartoum is the second-largest city in North Africa, with an estimated population of 2.5 million in the city proper, and up to 7 million in the Greater Khartoum district, which includes Khartoum General, Khartoum North, and Omdurman, each city linked by bridges. Khartoum has a thriving market in cotton products, woven textiles, and knitwear and is a railroad hub that facilitates the transportation of a number of goods, including glass, tile, foodstuffs, gum, and oil. Khartoum has an international airport and two major universities.

Khartoum was officially founded in 1821 as an Egyptian army camp and the base of operations for the Ottoman conquest of the Sudan. Muhammad Ali, the Turkish pasha of Egypt, conquered the portion of the Sudan originally controlled by the Funj Empire in order to take advantage of the region's vast human and economic resources and made Khartoum a depot for the Arab slave trade in Africa as well as a major mercantile center. Khartoum remained under Egyptian military occupation until 1881, when Muhammad Ahmad, the self-proclaimed Mahdi (divinely guided seeker of justice) of the Sudanese people, staged a rebellion that resulted in the city's liberation.

Ahmad and his followers, the Madhists, took advantage of Egypt's instability in the wake of the British conquest of Northeast Africa but were themselves soon involved in it directly as a British company under General Charles Gordon, who arrived to take the city in 1884. He arrived at Khartoum on February 18, with orders to evacuate a small force of Egyptian soldiers trapped there, but decided to mount an offensive against the Madhist rebels instead, believing that he had the resources to defend the city. Gordon's tactical error led to a 10-month siege on his company that eventually resulted in his death on January 26, 1885, after the Madhists broke his defenses, destroying much of the city in the process. Khartoum was eventually retaken by the British under Field Marshal Herbert

Kitchener in 1898, which marked the advent of joint British and Egyptian colonial rule that lasted until Sudan gained its independence in 1956.

The greater Khartoum area has experienced its troubles internally and with the world postindependence. It is houses one of the largest refugee populations in the world, largely the result of civil wars in the region. Oil interests have done little to help the city's growing problems, especially escalating poverty resulting from overpopulation and lack of resources. Slums are a major problem in Khartoum. In 1998, a pharmaceutical company in Khartoum was bombed by the UNITED STATES, as it was believed to have been a chemical weapons factory for terrorist groups. Khartoum has been regarded as a center for potential terrorist activity, though no direct evidence has been found directly linking the city's government with any terrorist organization.

BIBLIOGRAPHY. *Oxford Essential Geographical Dictionary* (Oxford University Press, 2003); Thomas Parkenham, *The Scramble for Africa: White Man's Conquest of the Dark Continent 1876–1912* (Avon Books, 1991).

PILAR QUEZZAIRE-BELLE
HARVARD UNIVERSITY

Kilimanjaro, Mount

Tanzania's Mount Kilimanjaro, at 19,340 ft (5,895 m) above sea level, is Africa's highest elevation point.

A MAJESTIC ROCKY GIANT, Mount Kilimanjaro is crowned by an icecap, impressively dominating the scenery. Mount Kilimanjaro towers above the Masai Steppe or the Great Rift Valley, which is believed to be the site of the origin of humankind. Pictures that shaped our imagination of East Africa are firmly connected to Kilimanjaro. The highest mountain of Africa is a volcanic massif situated in the territory of northeastern TANZANIA. The mountain consists of lava-dominated shield volcanoes and has three main volcanic centers, named Shira, Kibo, and Mawenzi. The highest point of the Kilimanjaro at the crater Kibo is called Uhuru Peak. At 19,340 ft (5,895 m) above sea level, it is Africa's highest elevation point. Mawenzi (east of Kibo) rises to 16,896 ft (5,149 m), and Shira (west of Kibo) to 13,000 ft (3,962 m).

Kilimanjaro is a very young volcanic massif: it started to grow less than 1 million years ago and ceased to grow about 450,000 to 300,000 years ago. Volcanic activity subsequently became sporadic, and today the inner crater of the Kibo shows only residual activity. The last blow of ash from the Kibo could be witnessed probably about 200 years ago. The volcanoes of the Kilimanjaro are part of a chain of Cenozoic volcanoes in East Africa. The major factors influencing the volcanic activity of this area are the PLATE TECTONICS of the East African Rift System (EARS), which marks the lines along which the eastern subplate is separating from western subplate.

Kilimanjaro rises from 2,297 ft (700 m) above sea level to its highest peak in a relatively confined space, which stretches from east to west about 53 mi (85 km) and from north to south about 50 mi (80 km). It is thus an ideal example of the geomorphologic and ecological change in different altitudes and is characterized by a distinctive differentiation of the natural area. Five major vegetation zones can be found at Kilimanjaro: the lower slopes; the tropical mist- and mountain-forest zone between 5,905 and 9,187 ft (1,800 to 2,800 m); the low alpine zone with heath and moorland; the highland desert; and the summit. Fertile volcanic soils and good climatic conditions allow a variety of crops to grow in the Kilimanjaro region. This resulted in manifold and diversified pre- and early-colonial agricultural activities of the Chagga, the native people who inhabit the region. The colonial and postcolonial agriculture focused increasingly on the industrialized production of the cash crops banana and coffee. The consequences subsequently led to a fundamental economical and structural change, extensive clearings and destruction of forest and pasture land, increasing erosion, and high population growth.

Another factor with major economic and ecological impact for the Kilimanjaro region is mountain tourism. Hans Meyer, a German colonial geographer and rich heir of a huge Leipzig publishing house, first ascended the Kibo crater in 1889 and called it Kaiser-Wilhelm-Spitze (since 1962, Uhuru Peak). Since this first ascent of Kilimanjaro, more and more mountaineers have been attracted by the mountain.

Each year, almost 20,000 tourists try to ascend the summit. Added to the high number of local guides and porters who are hired to succeed in this endeavor, a yearly total of more than 70,000 people climb up the mountain through ecologically sensitive terrain. Negative side effects of the economic benefits are the inevitable damages to the environment, although since the 1970s the main area of the Kilimanjaro enjoys the status of a national park. Moreover, for reasons of global warming, the impressive ice sheets and glaciers of the mountain (only 10 years ago, glaciers covered most of its summit) have been receding rapidly. More than 80 percent of the icecap that crowned the mountain when it was first thoroughly surveyed in 1912 is now gone. If recession continues at the present rate, it is projected that the glaciers of the Kilimanjaro, the last glaciers of Africa, are in serious threat of vanishing in the next 15 years.

BIBLIOGRAPHY. Thomas Schlüter, *Geology of East Africa* (Gebrüder Borntrӓger, 1997); François Bart, M.J. Mbonile and François Devenne, *Kilimandjaro: Montagne, mémoire, modernité* (Presses Universitaires de Bordeaux, 2003); Imre J. Demhardt, "Hochgebirge: Der Kilimandscharo," *Petermanns Geographische Mitteilungen* (v.146/3, 2002).

BERND ADAMEK-SCHYMA
LEIBNIZ-INSTITUTE OF REGIONAL GEOGRAPHY
GERMANY

Kiribati

Map Page 1128 **Area** 313 square mi (811 square km) **Population** 98,549 (2004) **Capital** Bairiki **Highest Point** Banaba Island 243 ft (81 m) **Lowest Point** 0 m **GDP per capita** $800 **Primary Natural Resources** phosphate (depleted by 1979), copra.

THE REPUBLIC OF Kiribati consists of 33 atolls straddling the equator and the INTERNATIONAL DATE LINE. The country is composed of the Gilbert Islands, the Phoenix Islands, and the Line Islands. The Gilbert Islands are made up of 17 atolls, are home to the majority of the population, include the capital Bairiki, and harbor the once phosphorous-rich island of Banaba. The Phoenix Islands are composed of 8 atolls, none of which has a permanent population. These islands were part of a government relocation program in the 1930s and 1940s as an answer to overcrowding on other islands, yet by 1952 the plan was considered a failure. Resurrection of a similar plan began in 1995 and may lead to future permanent settlements. The Line Islands, of which only three of the eight are inhabited, include the largest of the islands, Kiritimati (CHRISTMAS ISLAND).

Kiritimati is the largest atoll in the world and encompasses approximately half of the Kiribati landmass. A remote country, Kiribati is primarily composed of atolls with little variation in topography. The reefs, flats, and lagoons surrounding the atolls are natural attractions for the growing tourism industry. Global warming issues and the resulting rising sea levels have become a concern for the nation, as most of the country is at sea level. The climate is tropical with the potential for typhoons occurring primarily from November to March.

The majority of the population is of Micronesian ancestry with many of the ancestors originating from Tonga and Fiji. English is the official language, while I-Kiribati is widely spoken. In 1892, the British proclaimed the island group a British protectorate. From 1963 until 1979, the islanders were given an advisory position in the political decisions of the nation and gained final independence from the British on July 12, 1979. With the independence came the name change from the Gilbert Islands to Kiribati, which is the local translation for the Gilberts.

This sovereign democratic republic is resided over by a president, composed of three administrative units that are split into six districts and 21 island councils, and holds a parliament consisting of 42 representatives. Local affairs are handled by local councils. Prior to 1995, the islands were split by the international date line (180 degrees longitude), but in a unilateral move, Kiribati moved the international date line so the entire country could share the same time zone.

The economy is presently based on copra and fish exports, the granting of fishing rights to foreigners, the remittance of income from overseas workers, and in-

ternational aid. Tourism has become a major export earner bringing in approximately one-fifth of the GDP. The local currency is the Australian dollar.

BIBLIOGRAPHY. *World Factbook* (CIA, 2004); Word IQ, www.wordiq.com (September 2004); Pacific Island Travel www.pacificislandtravel.com (September 2004); Lonely Planet, www.lonelyplanet.com (September 2004).

MARK KANNING
WAIARIKI INSTITUTE OF TECHNOLOGY, NEW ZEALAND

Kola Peninsula

THE KOLA PENINSULA IS one of RUSSIA's regions of great contrast: at once an area of stark natural beauty and of severe ecological danger; rich in minerals, but one of Russia's poorest regions; and an area that is both closely linked to the concerns of Western Europe and strictly isolated from the outside world. The peninsula is the westernmost of Russia's numerous peninsulas that jut out into its Arctic seas and has provided strategic sheltered harbors for Russia's northern fleets since the 18th century.

Geographically, the Kola Peninsula forms the heel of the Scandinavian landmass, pointing in the opposite direction from the other three Scandinavian projections that form NORWAY, SWEDEN, and FINLAND. The peninsula covers roughly 50,000 square mi (128,000 square km), separating the White Sea from the much larger Barents Sea to the north. Its coasts on these two seas differ sharply: along the White Sea and its tributary, Kandalaksha Bay, the coast is low and smooth; the coast along the Barents Sea, called the Murman Coast, is mountainous and heavily indented. These coastal mountains form the easternmost extension of the Scandinavian ranges that form the backbone between Norway and Sweden.

The center of the peninsula is a granite and gneiss plateau, with several large lakes, and is covered with tundra in the north and forests in the south. The plateau rises to form some moderate peaks, the highest reaching 3,930 ft (1,191 m). This mountainous region is rich in minerals. Several large rivers drain the peninsula to the north or south, including the Tuloma and the Kola. The Kola River connects the north coast and the region's chief city of Murmansk to the White Sea via Lake Imandra and a short canal. This passage allows ships from Russia's main northern port, Arkhangelsk, on the White Sea, to reach the Murman Coast, one of the few Arctic coasts to remain unfrozen throughout the year.

Because these waters offer Russia's only year-round access to the ATLANTIC OCEAN, it has served as home to Russia's Northern Fleet since the 19th century. Murmansk was the site of a 1918 Allied troop landing and an important base in World War II. It then became a major center for the Soviet Union's nuclear submarine program. The numerous fjords between Murmansk and the Norwegian border (less than 60 mi or 100 km away) are stocked with naval and air bases, and have thus provided a source of heightened tensions with NORTH ATLANTIC TREATY ORGANIZATION (NATO) member Norway. NATO maintains numerous early warning systems on its side of the border, keeping an eye on the stocks of nuclear weapons kept on the Kola Peninsula since the 1950s.

The decay of Soviet power and the collapse of the Soviet Union have left behind about 155 nuclear submarines in the fjords of the Kola Peninsula, about half of which are unfit for use. Most of these are even unfit to be moved or dismantled, creating one of the world's largest nuclear waste problems. It is estimated that roughly two-thirds of the world's nuclear waste has been dumped off the peninsula, not counting that which is sitting in rickety ships near Murmansk. The nearby Kola Nuclear Power Station provides about 60 percent of the region's power, but has been declared one of the world's least safe reactors, having come close to a meltdown as recently as 1993.

The Russian government continues to restrict travel to and from this region, but its heavily impoverished population looks to Norway and Finland for aid rather than to Moscow. The western part of the peninsula forms part of Lapland, the homeland of the seminomadic Saami people that stretches across northern Norway, Sweden, and Finland. The Saami (or Lapps) live in the southeastern parts of the peninsula, while areas in the west are populated by Karelians, close kin to Finns. The languages of both Saami and Karelians are related to other Uralic languages, whose people live elsewhere in northern Russia.

BIBLIOGRAPHY. Wayne C. Thompson, *Western Europe 2003*, The World Today Series (Stryker-Post Publications, 2003); Bernard Comrie, Stephen Matthews, and Maria Polinsky, eds., *The Atlas of Languages* (Quarto, 1996).

JONATHAN SPANGLER
SMITHSONIAN INSTITUTION

Kolkata (Calcutta)

CALCUTTA, RECENTLY renamed Kolkata after its original village site of Kolikata and an on-site temple of goddess Kali, was founded by Job Charnock in 1690 on behalf of the British East India Company. The same location in northwestern INDIA was also a Portuguese and Dutch encampment site during the mid-17th century. It is, however the British encampment that started the impetus to develop the area as a trading center that eventually formed the core of a huge Calcutta metropolis leading to the evolution of a 50-mi- (80-km-) long linear conurbation (continuous urban agglomeration) along the Hughly (Hooghly) river.

Calcutta was the capital of British India (1757–1911) and thereafter it was relegated to the capital of the undivided Bengal Province. It has been the capital of the state of West Bengal since 1947. The Calcutta Metropolitan District was formed in 1961 as a planning unit for the entire conurbation. It consists of over 500 administrative units, including the two largest cities: corporations of Kolkata and its twin city of Haora (Hawrah).

Only 1 degree south of TROPIC OF CANCER, Kolkata falls in the monsoon climatic regime with most of its annual average rainfall of 64 in (162 cm) falling in the four-month period of June through September. Its January temperature average is 67 degrees F (19 degrees C), while July averages 85 degrees F (29 degrees C).

The British established a port, Fort Williams, and an initial water transport network to connect the city with an extensive HINTERLAND that contained the richest agricultural and mineral resource region of India. An extensive railroad network was developed in 1850s. Using the hinterland, Calcutta became the primary manufacturing center of India specializing in jute textiles, paper mills, heavy engineering, and rubber. By 1921, 35 percent of India's industrial workers were based in Calcutta.

Calcutta's economy stagnated in the 1970s because of the antipathy of both domestic and foreign investors in the face of labor union agitation and the Marxist West Bengal state government's anti-capitalist rhetoric. However, Calcutta was the pacesetter for the country for political, educational, and economic leadership. In recent years, Calcutta's importance has been displaced by MUMBAI, the population of which in 2000 was 16 million, in comparison to Calcutta's 13 million.

Nonetheless, Calcutta remains the second-largest metropolis of India; its world rank stands at seventh. The City of Calcutta, the main hub of the metropolis, has one of the highest densities of population, over 62,000 people per square mi (24,000 per square km); one out of every three lives in slums; and the rich/poor contrast in housing is very vivid. Hindus constitute 83 percent of city population and Muslims make up 14 percent. The remaining 3 percent are Christians, Jains, and Sikhs. The city has high linguistic diversity.

The port of Calcutta was the leading export and import center of the country during the 19th and most of the 20th century. It has lost its position to Mumbai. Calcutta is the home of the Calcutta (1857) and Jadavpur (1955) universities. Four Nobel laureates are associated with the city. Rabindranath Tagore had his home here; Mother Teresa did social work and lived here; C.B. Raman worked and researched here; Amartya Sen studied, taught, researched, and lived here, too.

Kolkata, jammed with a combination of vehicles, automobiles, buses, trains, hand-pulled carts, and thousands of pedestrians, introduced a 10.2-mile (16.4-km) long subway system in 1984. The subway, which is being extended farther south and north, carries 25 percent of the commuters.

BIBLIOGRAPHY. Ashok K. Dutt and George Pameroy "South Asian City," *Cities of the World*, Stanley Brunn, Jack Williams and N. Zeigler, eds. (Rowman and Littlefield, 2003); Geoffrey Morehouse, *Calcutta* (Harcourt Brace Jovanavich, 1971); Ashok K. Dutt, "Planning Constraints for Calcutta Metropolis" *Indian Urbanization and Planning*, Allen G. Noble and Ashok K. Dutt, eds., (Tata McGraw Hill, 1978).

Ashok K. Dutt
University of Akron

Kopet Mountains

THE SMALL RANGE of mountains that forms the northern boundary of IRAN with TURKMENISTAN is known as the Kopet Dag, or Kopet Mountains (Köpetdag in Turkmen). The range stretches for 400 mi (645 km), from a point near the CASPIAN SEA to the Harirud River in the east, which flows from northwestern AFGHANISTAN to a desert delta in southeastern Turkmenistan (where it is called the Tejen River).

The highest point in the range is the Kuh-e Quchan (10,466 ft or 3,191 m), in the southern part of the range in Iran. The highest point in the northern range,

which forms the border between the two countries, is Mount Shahshah (9,600 ft or 2,912 m), which looms over the city of Ashgabat to the north.

POPULATION GEOGRAPHY

Most of the population of Turkmenistan is clustered in towns and cities in the foothills of the Kopet Dag, including Ashgabat, its largest city and capital. The northern and southern ranges are divided by the valley of the Atrak River (entirely in Iran, until it forms the border west of the Kopet Dag ranges) and the largest city of the Iranian province of Khorasan, Mashhad. Many of the local inhabitants are semi-nomadic sheepherders, while others cultivate the region's fruit specialties in the rich LOESS foothills and the mountain gorges: pomegranates, plums, figs, almonds, walnuts and pistachios, plus plants known for centuries for their medical properties, aromas, and dyes. This cultivation is aided by the fact that the mountains receive more rain than any other part of Turkmenistan.

The isolation of these mountains by deserts on both sides has produced a large variety of flora and fauna that are found nowhere else, a fact that has led the local Turkmen government to create the Syunt-Hasardag Nature Reserve—over 74,000 acres (30,000 hectares)—in the southwest portion of the range. The reserve is home to the region's most famous wildlife, its hunting birds, including the golden eagle, black griffin, and desert kestrel. Other animals include leopards, boars and desert hyenas.

The mountains are tectonically active and frequently prone to earthquakes—the most devastating in recent history was in 1948, which completely destroyed Ashgabat. The mountains have served as a barrier for various empires centered on the Iranian plateau from invasions from the north but also as a meeting point between Central and South Asian civilizations, as the terminus (end point) of the Great SILK ROAD from CHINA to Persia.

BIBLIOGRAPHY. Sergei Petrovich Suslov, *Physical Geography of Asiatic Russia*, N. D. Gershevsky, trans. (W.H. Freeman, 1961); John Sparks, *Realms of the Russian Bear: A Natural History of Russia and the Central Asian Republics* (Little, Brown, 1992); Central Asian Mountain Information Network, www.camin.org (May 2004); "Turkmenistan Struggling to Preserve Threatened Flora and Fauna," www.newscentralasia.com (May 2004).

JONATHAN SPANGLER
SMITHSONIAN INSTITUTION

Korea, North

Map Page 1120 **Area** 46,541 square mi (120,540 square km) **Population** 22,466,481 **Capital** P'yongyang **Highest Point** 9,055 ft (2,744 m) **Lowest Point** 0 m **GDP per capita** $1,000 **Primary Natural Resources** coal, lead, tungsten, zinc, graphite.

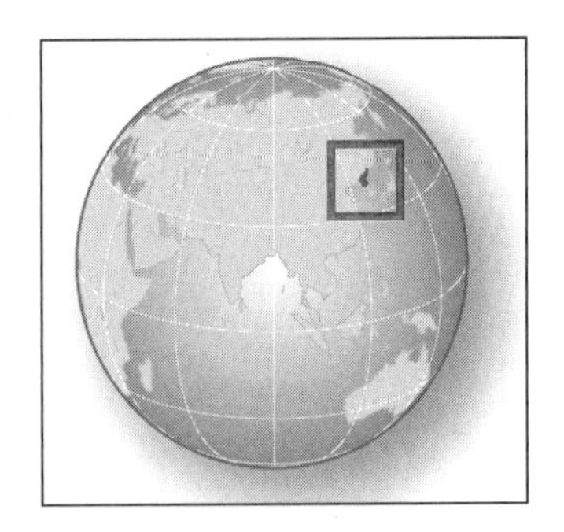

THE DEMOCRATIC People's Republic of Korea, North Korea, is one of the world's most politically isolated countries. Separated from its sister nation (South KOREA) to the south since 1953, the Democratic People's Republic holds on to its Cold War ideologies long after its chief sponsors have either collapsed (the Soviet Union) or turned toward liberalized economy and better relations with the West (such as CHINA). Less populated and less developed economically than the south even before the split, North Korea is almost completely unable to feed its population because of extreme economic centralization and international blockades.

North Korea occupies the northern portion of the Korean peninsula, roughly from the 38th parallel (38 degrees north latitude, the Demarcation Line of July 27, 1953) to the border with Manchuria (China) and SIBERIA (RUSSIA). Its is bordered on the east and west by the Sea of Japan and the Yellow Sea. Most of its northern border is with China, with only 11.8 mi (19 km) bordering Russia, in the far northeastern corner. The southern border with the Republic of Korea (South Korea) is 151 mi (248 km) long, along the demilitarized zone (the DMZ), still the area of the world with the highest concentration of permanent military presence, with North Korea's 1.1 million-person army (the world's fifth largest) facing off against 600,000 South Korean troops and a sizable American military force.

The terrain is mostly hilly, with higher mountains in the interior, separated by deep, narrow valleys. There are wider coastal plains in the west, in which is located the country's only major city, P'yongyang, and its port city, Namp'o.

The mountains of North Korea are generally higher than in the south—as a result, most of the best farmland is in the south (only about one-sixth of the land in the north is suitable), while the north is richer in timber and minerals (coal, iron, copper, zinc, and other ingredients for heavy industry). The mountain ranges that run the length of the Korean peninsula

from north to south are mostly along the eastern edge of the country, with the terrain sloping gradually toward the west. The eastern coast is thus generally more steep, while the west coast has more coastal indentations and small islands (though in no way comparable to the thousands in the south). The central range for the northern end of the peninsula is called the Nangnim Range.

The highest mountains, however, are further to the northeast, including Paektu-san (Korea's highest), on the border with China, from which emanate the two main rivers in the north that form most of the border with China, the Yalu River, flowing west into the Korea Bay and into the Yellow Sea, and the Tumen River, east to the Sea of Japan, its mouth only 80 mi (129 km) from Vladivostok, Russia's major city in the Far East. Both of these rivers are navigable for a considerable distance, and both provide China and North Korea with hydroelectric power. Many of the mountains in this far northeastern corner are extinct volcanoes.

Further south, North Korean territory includes the northernmost end of the Taebaek Range, mountains of medium height that hug the eastern coast, including one of the best known mountains in all of Korea, the Kumgang-san (Diamond Mountain), with its "12,000 peaks," famous scenery, spring foliage, and ancient legends about odd-shaped rocks and ravines. The Taedong River, flowing through the center of P'yongyang, is not the longest in North Korea, but it is the major transport river from the interior, past the capital and into the Korea Bay. The capital is the only city with over a million inhabitants in the north (versus six in South Korea).

Though most of the cultural and political centers in Korean history have been in the southern end of the peninsula, P'yongyang was the center of the Goryeo kingdom (the origin of the name Korea), dominating this region and much of Manchuria before the seventh century. The North prefers to use the name of Korea's former ruling dynasty, Joseon, or Choson, with more ties to these northern kingdoms, rather than the name preferred by the South, Han'guk. The Choson ruled a united peninsula until the Japanese conquest in 1910.

After 1945, Korea was divided into spheres of influence by the UNITED STATES and the Soviet Union at the 38th parallel. The Soviets refused to submit to United Nations elections to decide on the form of government, so a republic was set up in the south in 1948, followed a few weeks later by the north. War from 1950 to 1953 effected little change in the nation's divided status but caused a massive depopulation of the north when nearly 1 million refugees fled south.

Political and economic mismanagement for the past 50 years has driven the country's industries into the ground: official government estimates for famines between 1995 and 1998 claim the loss of over 200,000 lives, while international groups claim much higher figures, around 1.5 million. The population has shifted dramatically from 20 percent urban in 1953 to 60 percent in 1987. Reunification talks have become a possibility only in the last few years, with the first meeting between the two presidents in 2000. Korean athletes marched together for the first time at the Sydney Olympic Games in 2000, and limited border crossings have begun, allowing families to visit relatives they have not seen in over 50 years.

BIBLIOGRAPHY. Barbara A. Weightman, ed., *Dragons and Tigers: A Geography of South, East and Southeast Asia* (Wiley, 2002); *World Factbook* (CIA, 2004); *Encyclopedia Americana* (Grolier, 1997); "DMZ," www.korea-dmz.com (April 2004); "North Korea," www.nationsonline.org (April 2004).

JONATHAN SPANGLER
SMITHSONIAN INSTITUTION

Korea, South

Map Page 1120 **Area** 38,294 square mi (98,190 square km) **Population** 48,289,037 **Capital** Seoul **Highest Point** 6,435 ft (1,950 m) **Lowest Point** 0 m **GDP per capita** $19,400 **Primary Natural Resources** coal, tungsten, lead, hydropower potential.

AN ECONOMIC GIANT of East Asia, South Korea's economy took off after the Korean War of the 1950s. South Koreans now enjoy a per capita income 20 times that of the North but desires for reunification remain strong. Relations began only in 2000, but South Koreans worry about the vast sums of capital that would be required to bring their northern cousins up to similar living standards.

Like its neighbor to the north, North KOREA, most of South Korea (Republic of Korea) is mountainous,

with higher peaks concentrated in the Taebaek Range, the "backbone," running along the eastern coast of the peninsula. The western coast is far more level, with numerous inlets and thousands of islands—South Korea's overall coastline is more than 1,488 mi (2,400 km), mostly because of these features. South Korea shares a border of 151 mi (248 km) with North Korea, and lies across the Korea Strait from Japan, and the Yellow Sea from CHINA. The climate is said to be one of the healthiest in the world, temperate most of the year, with the southernmost regions (notably the island of Cheju) lying within the subtropical zone, influenced by the warm Japan Current from the south.

The southeast also has resorts and hot springs near Pusan. While the south has more agricultural resources than the north, it lacks the north's mineral resources and must import most of the raw materials needed in its factories. The central-west regions are the most fertile, growing rice, root crops, and fruit. Ginseng has been one of the country's signature crops for centuries, centered on the town of Komsan, where richly mulched soil is carefully shaded by thatch or fiber mats, then left to fallow for 10 to 15 years before another planting. Most rivers are relatively short but navigable, like the Han (Han-gang), which flows through Seoul to its port city, Inchon, 24 mi (40 km) away. The many islands that lie off the west and southwest coast of Korea are mostly of volcanic origin and are noted for their beauty, coral beds, diverse marine life, sponges, and pearls. The largest, Cheju, 62 mi (100 km) south of the mainland, is mostly mountainous and famous for its mother-of-pearl. Its largest peak, Halla (6,435 ft or 1,950 m), looks white from a distance because of the large outcroppings of rock.

South Korea has six cities with populations over 1 million (and containing over half the total population): Pusan, Taegu, Taejon, Kwangju, and the conurbation of Seoul and Inchon, with a total of 13 million people. Seoul means "capital" and has been the center of various Korean kingdoms for centuries. The Koreans originally migrated from Manchuria and set up several small principalities, which were gradually united into one state by the end of the 7th century. These earlier tribal groupings were called the Han, from which the preferred modern name for the state is derived: Han'guk, or Tae-han.

The Kingdom of Koryo emerged in the 10th century, providing the country's modern name, but it was transformed in the 14th century and renamed Joseon (or Choson, the name still preferred by North Korea). Influenced by Chinese politics and religion, Korea in turn influenced Japanese society, notably through the introduction of Buddhism. A golden age of Korean history followed, during which the Koreans invented the world's first movable metal type, developed a world-famous silk industry, and invented a new (non-Chinese) alphabet. Weakened by internal factions, Christian ideologies from Western missionaries, and government corruption, the kingdom closed its ports to all foreigners but the Chinese in 1864, until its ports were forced open by the Japanese, who gradually extended their influence over the peninsula from protectorate (1905) to outright annexation in 1910. Brutal imperialist rule by the Japanese over the succeeding decades remains a source of great resentment among Koreans. Liberated from Japanese rule by the UNITED STATES and the Soviet Union in 1945, the two powers' rival occupation led to the Korean War of 1950–53, and the partition of the country into North and South.

The DMZ (demilitarized zone) dividing the two countries lies only 25 mi (40 km) north of Seoul, and the city grew significantly in the 1950s from the nearly 1 million refugees who fled the North. Land reform, education and economic expansion following the war has transformed the country from a poor agrarian society to one of the world's most industrialized. Light industry changed over to heavy industry in the 1970s (steel, iron, chemicals), then to high-tech industries such as automobiles, ships, and electronics. Giant corporate conglomerates (known as *chaebols*, "fortune clusters") are now known all over the world—Hyundai, Daewoo, and Samsung—and have become global in outlook (Daewoo Motors, for example, has factories in POLAND, UZBEKISTAN, and INDIA).

In 1996, the top four *chaebols* accounted for 80 percent of the GDP and 60 percent of the exports. Despite this prosperity, Koreans face increased population problems, especially with urban crowding and severe housing shortages. The government is encouraging the development of new centers of technological research and industry, notably in the cities of the south, especially Pusan, the country's second largest city, and the country's premier harbor.

BIBLIOGRAPHY. Barbara A. Weightman, ed., *Dragons and Tigers: A Geography of South, East and Southeast Asia* (Wiley, 2002); *World Factbook* (CIA, 2004); "DMZ," www.korea-dmz.com (April 2004); "South Korea," www.nationsonline.org (April 2004).

JONATHAN SPANGLER
SMITHSONIAN INSTITUTION

Kosciusko, Mount

MOUNT KOSCIUSKO (also spelled "Kosciuszko") is the highest point (7,310 ft or 2,228 m) in mainland AUSTRALIA. Named after the Polish military leader Thaddeus Kosciuszko (1746–1817), the mountain is a core component of the Kosciuszko National Park in the state of New South Wales. One of the Australian Alps national parks, the park is a UNESCO (United Nations) Biosphere Reserve. It contains six wilderness areas, and its alpine and subalpine areas contain herb fields, bogs, feldmarks and plant and animal species found nowhere else in the world. The park is also home to the headwaters of Australia's biggest river system the Murray-Murrumbidgee, as well as several glacial lakes, including Blue Lake, which is also a wetland reserve.

Kiandra, in the north of the park, was the scene of one of the shortest gold rushes in Australia. The discovery of gold in 1859 led to a sudden influx of gold-seekers, so that by early 1860 there were about 4,000 people in the town, which had 25 stores, 14 hotels, and a jail. After 1861, once the rush was over, the population stabilized between 200 and 300. Kiandra is now a ghost town and a heritage attraction along with a number of other historic huts and homesteads that have become conserved as part of the cultural heritage of the region.

The first proposals to protect Mount Kosciusko and the surrounding regions were developed in the 1930s by the Sydney Bushwalking Club and the National Parks and Primitive Areas Council (NPPAC). In 1935 and 1936 the NPPAC exhibited a plan for the reservation of the region in Sydney. The proposed park was designed for the special purposes of water conservation, wildlife propagation, and public recreation of various kinds, including trail riding, recreational walking and motor camping. Despite the enthusiasm of the NPPAC for the conservation proposals, the main force for its eventual acceptance was more utilitarian conservation notions of the need for irrigation water, the production of hydro-electricity and soil protection.

On April 19, 1944, the Kosciusko State Park Act was assented to by the New South Wales Parliament. Besides the protection that the act gave to the Kosciusko region, it is significant for two reasons. First, it was the first national park in New South Wales that was provided with security of tenure through permanent reservation that could be revoked only by a special act of parliament. Second, it provided for the establishment of a primitive (wilderness) area. Despite its conservation significance, the park served as the core of the Snowy Mountains Hydroelectric Scheme which was completed in 1972. The scheme left an extensive road network and lakes in the park that is now heavily used by tourists.

BIBLIOGRAPHY. Australian Academy of Science, "The Future of the Kosciusko Summit Area: A Report on a Proposed Primitive Area in the Kosciusko State Park," *Australian Journal of Science* (v.23/12, 1961); Colin Michael Hall, *Wasteland to World Heritage: Wilderness Preservation in Australia* (Melbourne University Press, 1992).

MICHAEL HALL
UNIVERSITY OF OTAGO, NEW ZEALAND

Kuroshio Current

THE KUROSHIO IS a warm northeasterly ocean current off the coast of JAPAN. This current is also called the gulf stream of the Pacific or Japan Current. Kuroshio means "the black stream" in Japanese, named after the deep ultramarine color of the high salinity water, which is found flowing north of the current's axis. The system includes the following branches: Kuroshio, up to 35 degrees N; Kuroshio extension, extending eastward into two branches up to 160 degrees E longitude; North Pacific current, a further eastward continuation, which throws branches to the south as far as 150 degrees W; Tsushima current, branches of the main current that run into the Japan Sea, along the west coast of JAPAN; and Kuroshio counter-current, the large swirl or eddy on the east and south east of the Kuroshio.

The Kuroshio originates from the greater part of North Equatorial current, which divides east of the PHILIPPINES. The Kuroshio is the current running from Formosa to about 35 degrees N latitude. It continues directly as a warm current known as the Kuroshio Extension, from there it is continues as the North Pacific current. Water enters the Kuroshio over a broad front, 621 mi (1,000 km) in width, which then accelerates and narrows. A narrow band less than 62 mi (100 km) in width and about .6 mi (1 km) of maximum depth runs for 1,864 mi (3,000 km) along the western edge of the Pacific, between the Philippines and the east coast of Japan. A narrow, intense flow persists for 930 mi to 1,240 mi (1,500 to 2,000 km) after the current leaves Japan's east coast, after which there is a marked drop

in velocity. Here, there is no land boundary on the left-hand side to generate a fractional boundary layer.

Kuroshio is a fast ocean current (2 to 4 knots). Every second, the current carries some 50 million tons of sea water past Japan's southeast coast, a flow equal in volume to about 6,000 rivers the size of the DANUBE or the VOLGA. On the whole, there are two distinct types of water in the current: warm, saline water on the right and cold, dilute water on the left The current undergoes marked changes in speed in the location of its axis, which varies from place to place and with seasons. Apart from changes resulting from tides, short-term changes from a major shift in the axis of the Kuroshio can occur as it flows past southern Japan.

When meanders develop, cold water is brought up toward the surface between the Kuroshio and the coast, and the temperature drops to as much as 50 degrees F (10 degrees C) below normal. This change has a profound effect on coastal and offshore fisheries.

The Kuroshio Current plays a vital role in the circulation of the North PACIFIC OCEAN. The current involves great volumes of water capable of carrying large amounts of heat. The heat, which is carried north by this flow, has an effect on climate of the adjacent land areas. Water temperature offshore strongly influences cloud cover and rainfall. On the southern coast of ALASKA, the effect of the Kuroshio Extension creates a somewhat more temperate climate.

BIBLIOGRAPHY. Philip Lake, *Physical Geography* (Macmillan, 1994); David F. Tver, *Ocean and Marine Dictionary* (Cornell Maritime Press, 1979); Richard Barkley, "The Kuroshio Current," *Oceanography, Contemporary Readings in Social Sciences* (Oxford University Press, 1977).

PRABHA SHASTRI RANADE
JAWAHARLAL NEHRU UNIVERSITY, INDIA

Kuwait

Map Page 1122 **Area** 6,880 square mi (17,280 square km) **Population** 2,183,161 **Capital** Kuwait City **Highest Point** 1,004 ft (306 m) **Lowest Point** 0 m **GDP per capita** $17,500 **Primary Natural Resources** petroleum, fish, shrimp, natural gas.

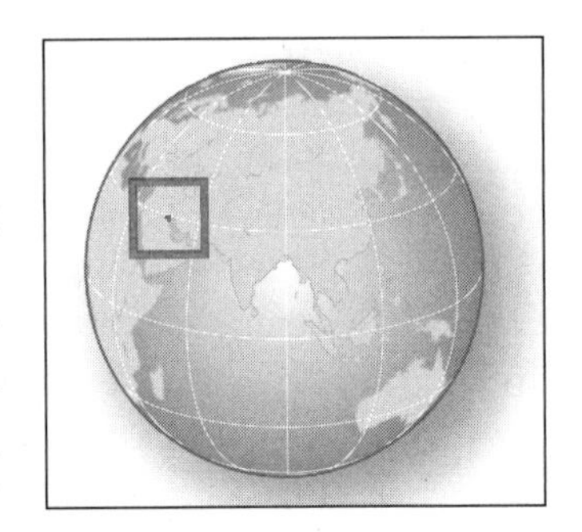

KUWAIT IS A COUNTRY in the MIDDLE EAST, bordering the PERSIAN GULF, between IRAQ and SAUDI ARABIA. The establishment of Kuwait took place in the 18th century by an Arab tribal group that came north from QATAR. They selected a defensible position at a prominent point on the southern edge of a broad bay near the outflow of the Euphrates River, at the head of the PERSIAN GULF.

Kuwait, as its place name implies in Arabic, began as a fortification near the water. The location provided a suitable point for transfer of goods to and from ships and liaison with the camel caravans of Arabia and Mesopotamia. Kuwait lay just on the margins of the OTTOMAN EMPIRE and acted as a trading center for goods entering or leaving the established routes serving that vast domain. In 1751, the Al Sabah family became the principal leaders of the small coastal commercial state. This ruling dynasty continues in a hereditary emirate type of government. As other gulf emirates before him, the Kuwait emir formalized relations with the BRITISH EMPIRE and became a protectorate by treaty in 1899.

UNIMAGINABLE WEALTH

Kuwaitis survived as traders and fishermen with meager wealth rising from the sea in the form of pearls. Oil was discovered in 1937 and once full exploitation began after World War II, unimaginable wealth began to rise from the desert. Kuwait quickly became a super-rich oil country and developed a robust welfare state. On June 19, 1961, Kuwait became fully independent from Britain. The next 30 years saw continued development of the petroleum industry and massive wealth accumulation by the Kuwaiti government. Virtually every aspect of a citizen's life, education, health care, and housing was subsidized or provided for by the state government.

In 1990, neighboring Iraq used alleged theft of oil from a shared oil field as a pretext for invasion and annexation. Several United Nations (UN) resolutions demanded Iraq restore the Kuwait government and withdraw its troops. A UN-sponsored, U.S.-led coalition of nations established a large military force in Saudi Arabia. Once the UN-declared deadline of January 15, 1991, passed, coalition forces conducted an air and ground campaign that resulted in the liberation of Kuwait. Kuwait rapidly rebuilt its ransacked nation and reestablished oil production. Kuwait then became the staging area for U.S. and coalition forces, which maintained UN-mandated economic and military sanctions against Iraq for over a decade. In 2003, a U.S.-led

U.S. Marines train in Kuwait in 1991 in preparation of driving Iraqi forces from the small, oil-rich nation.

coalition force launched a successful invasion of Iraq from Kuwait and other neighboring areas.

Kuwait's economy continues to be centered on the petroleum industry and the government sector. Its proven oil reserves account for about 10 percent of the world total. Kuwait's service and market sectors stand to benefit from the large-scale reconstruction and reorganization of the economy of Iraq. More than 90 percent of the population lives within the environs of Kuwait City and its harbor. An estimated 65 percent of the citizens of Kuwait, along with the royal family, belong to the Sunni branch of Islam.

BIBLIOGRAPHY. "Background Note: Kuwait," U.S. Department of State, www.state.gov (April 2004); *World Factbook* (CIA, 2004); Rick Atkinson, *Crusade, The Untold Story of the Gulf War* (HarperCollins, 1993); Ahmad Mustafa Abu-Hakima, *The Modern History of Kuwait 1750–1965* (Luzac and Company, 1983); H.J. de Blij and Peter O. Muller, *Geography: Realms, Regions, and Concepts* (Wiley, 2002).

IVAN B. WELCH
OMNI INTELLIGENCE, INC.

Kyrghiz Steppes

THE KYRGHIZ STEPPES is a historic name for the region currently forming central and eastern KAZAKHSTAN. It is a broad plain with few to no trees and little moisture. It is a land of horses and cattle and wide-open plains. The name is confusing, and thus used less frequently today, since the actual Republic of KYRGYZSTAN contains no STEPPE at all, while the Republic of Kazakhstan is home to very few Kyrgyz people.

The confusion stems from the 18th- and 19th-century conquest of the region by imperial RUSSIA. Russian authorities were unclear about the differences between the Turkic peoples of the plains and those of the high mountain valleys to the south and east, and for a time, they were both known as Kyrghiz, or Kyrghiz-Kazakh and Kara- (or Black-) Kyrghiz, respectively. The languages of the two groups are nearly the same, and the Russians already used the term Kazakh, or Cossack, to refer to similar nomadic (though Slavic) people who lived in southern Russia. It was not until the 1920s, when the communists began to separate the peoples of Central Asia into ethnically defined autonomous republics, that the name Kazakh was used to distinguish the peoples of the steppes from those of the mountains.

Like the rest of the steppes that cross most of southern Russia, the steppes of northern Kazakhstan are broad flat plains that contain enough moisture to support grasses, but not enough to allow for denser vegetation and forests.

The Kyrghiz Steppes in particular forms the northern third of Kazakhstan and can be divided into two zones. The western zone is in the center of the country, known as the Turgai plateau, starting north and northeast of the ARAL SEA. This plateau is marked by a central depression, with a chain of lakes stretching up to the Russian border. This was once a strait connecting two inland seas, millions of years ago. The Turgai and Irgiz rivers flow into semisalty lakes, which sometimes disappear altogether in especially dry periods. The eastern region is known as the Kazakh folded steppe

and is generally hillier, with scattered higher massifs, including the Ulu-Tau, Karkaral, and Chingiz-Tau mountains. Geologically, these folds are related to the folds in the Altai and TIAN SHAN ranges. Some of these areas are rich in mineral resources, and cities were developed during the Soviet era, such as Karaganda, the fourth-largest coal-producing city in the former Soviet Union. These cities always struggled to provide themselves with enough water, however, both for their growing populations and for industrial needs. Water resources from the Irtysh Valley in the northern edge of this steppe were used for these purposes, as were waters from the Ili River, which flows into Lake BALKHASH, and waters that were diverted from the Syr Darya far to the south, resulting in the serious shrinkage of the Aral Sea.

Other projects initiated during the Soviet era converted large percentages of the formerly open steppes into cultivated agricultural land, again with serious drain on local water resources and a change in the traditional nomadic lifestyle of the local population.

BIBLIOGRAPHY. Paul E. Lydolph, *Geography of the U.S.S.R.* (Misty Valley Publishing, 1990); Sergei Petrovich Suslov, *Physical Geography of Asiatic Russia*, N.D. Gershevsky, trans. (W.H. Freeman, 1961).

JONATHAN SPANGLER
SMITHSONIAN INSTITUTION

Kyrgyzstan

Map Page 1119 **Area** 76,641 square mi (198,500 square km) **Population** 4,892,808 **Capital** Bishkek **Highest Point** 24,407 ft (7,439 m) **Lowest Point** 433 ft (132 m) **GDP per capita** $1,600 **Primary Natural Resources** hydropower, gold, coal, oil.

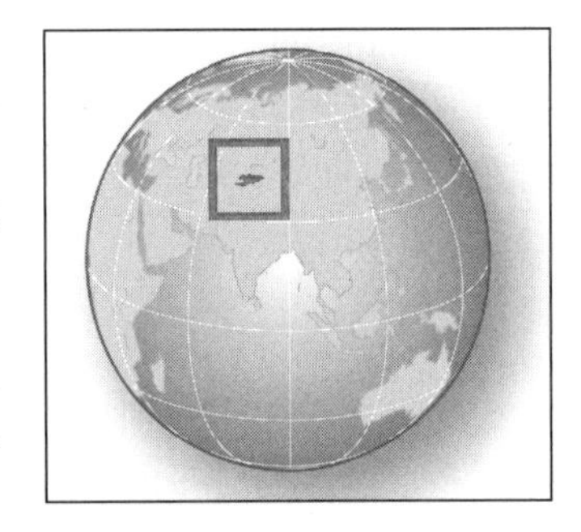

LANDLOCKED AND MOUNTAINOUS, the Kyrgyz Republic achieved its independence in 1991 following the collapse of the Soviet Union. Kyrgyzstan features spectacular mountain vistas and incredible natural beauty reminiscent of SWITZERLAND. Despite its natural beauty and recent attempts to develop a thriving tourist industry, Kyrgyzstan remains mired in poverty. Additional challenges include implementing democracy, combating ethnic tensions, and thwarting terrorism. Central Asia's second-smallest country in terms of area, Kyrgyzstan borders KAZAKHSTAN to the north, CHINA to the east, TAJIKISTAN to the south, and UZBEKISTAN to the west.

Kyrgyzstan is dominated by the TIAN SHAN (primarily) and Pamir (in the south) mountain ranges. The vast majority of the country (roughly 75 percent) is continuously covered by snow and glaciers. Traversing the Tian Shan remains relatively difficult, as a summer trip from the northern capital of Bishkek to the southern second-largest city of Osh (a distance of 186 mi or 300 km) takes more than 10 hours by automobile. Kyrgyzstan is also home to numerous alpine lakes, the largest and deepest of which is Lake Issyk-Kul, located near the Kazakh border in the north. The lake reaches a depth of 2,300 ft (700 m); its clear, sky-blue water and health resorts make the lake a popular tourist destination. For a country its size, Kyrgyzstan has surprising climatic variability, ranging from polar to dry continental through the mountains, to temperate northern foothills, to subtropical in the southwest. Kyrgyzstan's most valuable natural resource may be its gold deposits. The Kyrgyz republic was home to the Soviet Union's largest gold mine (Makmal), which continues to be one of the largest proven gold reserves in the world.

Kyrgyzstan's population is ethnically diverse, including Kyrgyz (64.9 percent), Uzbek (13.8 percent), Russian (12.5 percent), Dungan (1.1 percent), Ukrainian (1 percent), and Uygyr (1 percent) peoples. Population distribution is concentrated in the Fergana, Talas, and Chu valleys and is centered in the cities of Bishkek (the capital, 2004 population 866,300) and Osh (2004 population 229,700). Most citizens are adherents to the religion of Islam (75 percent), although a sizable minority of Russian Orthodox (20 percent) exists. A secular state, Kyrgyzstan has two official languages, Kyrgyz and Russian.

Kyrgyzstan's economy, like that of other poor countries, is dominated by the agricultural sector. A full 55 percent of the labor force is engaged in farming. Nomadic herders raise sheep (for both meat and wool), cattle, and yaks. Other agricultural products include cotton, tobacco, and a variety of vegetables. Industry, which accounts for just 15 percent of the labor force, is limited to gold, small machinery, textiles, and food processing. During its first decade of independence, Kyrgyzstan implemented more market-oriented economic reform but experienced slower economic

growth than the other former Soviet republics of Central Asia.

Perhaps the most pressing geographical/political issue facing Kyrgyzstan is its complex western boundary with Uzbekistan and Tajikistan. Three large Tajik exclaves exist entirely within Kyrgyzstan's borders, and a serious boundary dispute continues with Uzbekistan. Here, seemingly arbitrary boundaries fragment ethnic groups and unite dissimilar peoples. Kyrgyzstan's relative location has also fostered a growing problem of illegal narcotics traffic. The country has become a corridor for the movement of opium and heroin produced in AFGHANISTAN and Tajikistan, bound for the European market. Combating terrorism represents an additional problem confronting Kyrgyzstan. Radical Islam has penetrated the country, and Osh is considered by many to be the Soviet Central Asian headquarters of Wahhabism.

BIBLIOGRAPHY. Mark Elliot and Wil Klass, *Asia Overland* (Trailblazer Publications, 1998); H.J. de Blij and Peter Muller, *Geography: Realms, Regions, and Concepts* (Wiley, 2002); *World Factbook* (CIA, 2004).

KRISTOPHER D. WHITE, PH.D.
KAZAKHSTAN INSTITUTE OF MANAGEMENT

Kyushu Mountains

THE KYUSHU MOUNTAINS form the high, elevated central portion of the Japanese island of Kyushu, the southernmost of JAPAN's four main islands. Running roughly diagonally across the island, northeast to southwest, they cut Kyushu into a northern and southern sector. These sectors differ markedly from each other in many ways, from geology to economics: the north is urban and industrial, while the south is agricultural and poorer. The central part of the range has peaks over 3,300 ft (1,000 m), with the highest elevations at the northern end, overlooking the Aso ash and lava plateau. Mount Aso is the world's largest volcano and last erupted in January 2004, highlighting the range's status as one of the most geologically active places on the planet, with numerous volcanoes and hot springs, such as the famous resort at Beppu.

The island of Kyushu lies at the intersection of three tectonic plates. The core of the island was formed where the Seinan mountain arc (coming south from the island of Honshu) intersects with the mostly submerged arc of the Ryukyu Islands, which penetrates Kyushu from the south. The topography is broken up into narrow valleys cutting through steep slopes. The Kuma River is the chief waterway and flows northward into a gorge famous among trekkers. Restricted lowland area means that there has been a high degree of terracing for rice cultivation, though the population in general is rather sparse compared to the rest of Japan.

Orange groves and forestry dominate the local economy, though there has been recent growth in mineral processing industries (gold, copper, petroleum) on the eastern coast, where the ruggedness of the coast—with mountains descending directly into the sea in places—has created small protected natural harbors with relatively deep waters. The Ono River provides the needed water for these factories, as well as hydroelectric power.

BIBLIOGRAPHY. Glenn T. Trewartha, *Japan: A Geography* (University of Wisconsin Press, 1965); "Japan," www.countrystudies.us (Library of Congress, 2004); Barbara A. Weightman, ed., *Dragons and Tigers: A Geography of South, East and Southeast Asia* (Wiley, 2002).

JONATHAN SPANGLER
SMITHSONIAN INSTITUTION

lacustrine plain

ALSO CALLED A lake plain, a lacustrine plain is an area created out of deposition largely related to the past existence of lakes in the area, although in some cases, the original lakes still exist, having shrunk in size over time. Lacustrine refers to the condition of being affected by a lake or several lakes. Lacustrine plains are some of the flattest of all landform features and have few surface interruptions, although they may contain freshwater marshes, aquatic beds and lakeshore environments. Lacustrine plains are of varying origin, but most are underlain by fine, flat-bedded silt and clay deposited in lakes. The plains are typically related to the impoundment of water by one of the following processes: GLACIATION, differential uplift, and lake creation in now-arid inland basins.

Lacustrine plains that are glacial in origin are known as glaciolacustrine plains, and these are largely created from the trapping and ponding of water on the irregular land surface left by the former continental glaciers. In regions where there were thick masses of stagnating continental ice, steep-sided holes through the dead ice occasionally held lakes. Water was retained in these lakes by the ice walls. Fine-grained sediments (muds) accumulated on the lake bottoms. Once the ice walls melted away, however, the lakes drained, leaving the lake bottoms as plateau-like features underlain by fine-grained sediments. These lakes are largely ephemeral or temporal, eventually draining after the ice is gone. The present Great Lakes region in North America is bordered in many places by extensive lacustrine plains showing the former extent of the lakes. The Lake Agassiz plain is the biggest of these, reflecting the size of the former lake which was bigger than all the present Great Lakes put together.

The lack of topography of glaciolacustrine plains is due to the infilling of deep parts of the lake with clays and silt and wave erosion on the shallower parts of the lake. In North America and some other parts of the world, glaciolacustrine plains are of greatest interest because they often provide suitable land for intensive agriculture, and their flat topography permits mechanized farming. Lake ERIE and Saginaw Bay on Lake HURON were once much higher and extended inland, and the bottom of those bodies of water now makes up the lacustrine plains near Saginaw and Monroe. Large cities such as Chicago, Cleveland, Detroit, and Toledo also originated on the flat plains, where the dry beach ridges of the former lake edges served as roads. The flat lacustrine plains continue to absorb the urban expansion of these cities today. One handicap is that such areas are also poorly drained. Chicago lies on a plain formed when Lake MICHIGAN stood higher, and the city is often beset by the flooding of sewers, basements, and underpasses.

Other lake plains not associated with glaciers include the Congo Lake Plain and the lake plain of south SUDAN in Africa. These broad, flat plains of fine sediment were formed originally as enormous, in-filled basins created through differential uplift during the middle and late Pleistocene. The lakes were drained when the NILE and CONGO rivers eventually eroded their valleys, exposing large expanses of the former lake floors.

Lacustrine plains are also found in inland sites of present arid areas. The lakes associated with these plains were formed in a time of increased rainfall and reduced evaporation. The Chad Basin Plain in Africa, the Lake Eyre Plain in central AUSTRALIA, the plains around the CASPIAN SEA and the plains formed by Lake Bonneville in western UTAH are notable examples.

The size of lacustrine plains varies according to the size of the original lake. The Superior Lake Plain, covering parts of the U.S. states of MICHIGAN, MINNESOTA, and WISCONSIN, is roughly 1,910 square mi (4,950 square km) in size. The Chad Basin Plain extends to about 919,554 square mi (2,381,635 square km) in size and is shared by seven countries: NIGERIA, NIGER, ALGERIA, Sudan, CENTRAL AFRICA REPUBLIC, CHAD, and CAMEROON.

BIBLIOGRAPHY. C.P. Patton et al., *Physical Geography* (Duxbury Press, 1974); William E. Powers, *Physical Geography* (Meredith Books, 1966); Michigan State University Department of Geography, www.geo.msu.edu (September 2004); North Dakota State University, www.ndsu.nodak.edu (September 2004); Food and Agricultural Organization, www.fao.org (September 2004).

THERESA WONG
OHIO STATE UNIVERSITY

land bridge

IN GEOMETRICAL TERMS, a square also qualifies as a rectangle—a four-sided plane figure with four right angles—but a rectangle does not always meet the criteria to be a square—a figure having four equal sides. Such is the case with land bridges and isthmuses. An isthmus qualifies as a land bridge—a strip of land linking two landmasses, allowing free migration in both directions—but a land bridge is not always an isthmus, which is a narrow land bridge; the Isthmus of Panama, measuring 30 mi (48 km) at its narrowest point, for example, is an isthmian land bridge, whereas the Bering land bridge is believed to have been approximately 1,000 mi (1,609 km) wide during the Pleistocene Ice Age and would not have been considered an isthmus.

Land bridges are temporary in nature, and can disappear and reappear when geologic changes occur to the land or when the sea levels rise and submerge them or lower to expose these bridges of land. In addition to the above-mentioned Bering land bridge between Siberian Asia and ALASKA, the SINAI PENINSULA (23,500 square mi or 61,000 square km) is a triangular land bridge, linking northeast Africa with southwest Asia, and is home to over 200,000 people (1986). The Torres Strait waters (90 mi or 145 km wide) between PAPUA NEW GUINEA and AUSTRALIA have contained various land bridges when the sea levels were lower, exposing the continental shelf.

The migration of people and species across land bridges during glacial periods is what interests many scientists. Numerous species of flora, fauna, and animals have extended their ranges to new lands because of the isthmuses and land bridges that have intermittently connected different lands. Today, it is believed that the first humans in North America entered by way of the Bering land bridge, also referred to as Beringia, and although many Native Americans dispute these claims based on spiritual beliefs, archaeologist finds in both SIBERIA and the Bering land bridge region indicate similar tools, dwellings, and practices distinct to the Siberian region, suggesting there was a human migration between the two regions. Beringia was wide grassland and it is highly likely that people made it a home—however brief because of the cold climate.

The population of the Sinai Peninsula is primarily along the coast, and the main industries include fishing, mining, and tourism. Harsh weather makes this land bridge a natural barrier between the competing interests in the surrounding countries. The ongoing disagreements have led to several conflicts, primarily over the Suez Canal along the western perimeter of the peninsula. Active land bridges serve as important trade routes, and the Sinai Peninsula controls much of the trade between Asia and Africa, illustrating the role past land bridges have played.

In the Torres Strait, several islands composed primarily of granite in the western waters are all that remains of the land bridge. These islands include Waiben, Badu, Kiriri, and Gebar, among others. It is possible to see Pleistocene volcanoes in the area, and approximately 17 islands in the Torres Strait are cur-

rently inhabited. The islands act almost like stepping-stones between Australia and Papua New Guinea, with a "trail" of flora, fauna, people, and customs traceable from landmass to landmass.

Geological similarities between Africa, South America, Australia, INDIA, and ANTARCTICA indicate that 150 to 300 million years ago, a supercontinent, referred to as Gondwana or Gondwanaland, made up of these countries could have existed. Recent studies suggest that land bridges might have connected the regions, rather than one large land mass, and new developments in the fields of geology and geography will likely help to answer these questions in the future.

BIBLIOGRAPHY. H.J de Blij and Peter O. Muller, eds., *Geography: Realms, Regions, and Concepts* (Wiley, 1997); "Sinai Peninsula," and "Torres Strait," Microsoft Encarta Encyclopedia (Microsoft Corporation, 2002); National Park Service, "Bering Land Bridge" (U.S. Department of the Interior, 1992); D.H. Tarling, "Continental Drift," AccessScience @McGraw-Hill (McGraw Hill, August 21, 2002).

TARA SCHERNER DE LA FUENTE
UNIVERSITY OF CINCINNATI

landlocked

APPROXIMATELY ONE-FIFTH of the world's countries have no access to the oceans or ocean-connected seas, classifying them as landlocked. Today, there are 42 landlocked countries, including LIECHTENSTEIN and UZBEKISTAN, considered doubly landlocked because they have no access to the oceans and neither does any country that surrounds them. The main issues to consider with a landlocked country include the high transportation costs of trade, coordinating logistics and working trade relationships with neighboring countries, and in some cases, volatile climates.

Fifteen of Africa's 47 continental countries, including BOTSWANA, BURKINA FASO, BURUNDI, CENTRAL AFRICAN REPUBLIC, CHAD, ETHIOPIA, LESOTHO, MALAWI, MALI, NIGER, RWANDA, SWAZILAND, UGANDA, ZAMBIA, and ZIMBABWE, have no access to the ocean.

Most of these countries are among the poorest in the world, and only those rich in gem and mineral resources have escaped extreme poverty. Poor transportation routes have impeded trade and prevented advances in technology from being readily available to many of these countries, facilitating the spread of disease, including HIV/AIDS, in naturally isolated areas. Improved education within the countries only empowers highly trained students to emigrate, and politically, domestic policies do little to combat the geographical barriers between landlocked and neighboring nations.

When the Democratic Republic of the CONGO was created, the country negotiated for a thin strip of land on the north end of ANGOLA, providing the country with just 23 mi (37 km) of access to the ATLANTIC OCEAN—enough to cut transportation costs by half of what they would have been without the ocean access. Botswana, Lesotho, NAMIBIA, and Swaziland have formed a customs union, allowing them greater economic control with South Africa, their main trading partner.

KAZAKHSTAN is the largest landlocked country (1.03 million square mi or 2.67 million square km) and is bordered by CHINA, KYRGYZSTAN, RUSSIA, TURKMENISTAN, Uzbekistan, and the CASPIAN SEA, a landlocked body of water. Kazakhstan is rich in oil and natural gas resources and has become more integrated in the world economy and the development of trade resources among other landlocked countries. In 1994, Kazakhstan joined with the two adjoining landlocked countries—Uzbekistan, doubly landlocked, and the Kyrgyz Republic—to establish a "free-trade zone" among the countries, strengthening their economic standing in Asia. The borders between Kazakhstan and Russia, AZERBAIJAN (across from Kazakhstan on the Caspian Sea), turkmenistan, and the Caspian Sea are currently under negotiation.

Other landlocked countries include the Asian nations of AFGHANISTAN, BHUTAN, LAOS, MONGOLIA, NEPAL, and TAJIKISTAN. The European landlocked nations include ANDORRA, ARMENIA, AUSTRIA, BELARUS, CZECH REPUBLIC, HUNGARY, Liechtenstein, LUXEMBOURG, former Yugoslav republic of MACEDONIA, MOLDOVA, SAN MARINO, SLOVAKIA, SWITZERLAND, and VATICAN CITY, the world's smallest country. BOLIVIA and PARAGUAY are the only landlocked countries on the American continents and are both located in South America, though Paraguay is able to access the ocean via a long series of river connections over 1,000 mi (1,600 km) long.

Historically, countries have made extreme efforts to avoid being landlocked. In the 16th century, RUSSIA was considered landlocked part of the year when the ARCTIC OCEAN froze the country's only ports. A prime motivating factor in the country's expansion was the economic necessity of warmer ports. Ocean access can be an important part of political negotiations because of the economic resources a country can access. Once a

part of the Third Reich during World War II, Gdansk (Danzig) was decreed to POLAND in 1945 as part of the Potsdam Conference, providing the country with its only ocean access.

Several countries, including ERITREA, MONTENEGRO, and the Republic of BOSNIA AND HERZEGOVINA have negotiated independence with access to the ocean as a key element in defining borders. The 12 mi (20 km) of coastline along the ADRIATIC SEA that is part of Bosnia and Herzegovina actually splits the Croatian territories into two segments. But knowing the important economic impact a country's ocean access plays, compromises to get even a small amount of ocean access was crucial to a successful bid for Croatia's national independence.

BIBLIOGRAPHY. *World Fact Book* (CIA, 2004); H.J. de Blij and Peter O. Muller, eds., *Geography: Realms, Regions, and Concepts* (Wiley, 1997); Ira Martin Glassner, *Bibliography of Land-locked States* (Sijthoff & Noordhoff, 1980); R. Hausmann, "Prisoners of Geography," *Foreign Policy* (January-February, 2001); "Map Center," Microsoft Encarta Encyclopedia (Microsoft Corporation, 2002); Dudley Seers, ed. "Development Options," *Dependency Theory* (Frances Pinter, 1981).

TARA SCHERNER DE LA FUENTE
UNIVERSITY OF CINCINNATI

Laos

Map Page 1124 **Area** 91,429 square mi (236,800 square km) **Population** 5,921,545 **Capital** Vientiane **Highest Point** 9,240 ft (2,817 m) **Lowest Point** 230 ft (70 m) **GDP per capita** $326 **Primary Natural Resources** timber, hydropower, gypsum.

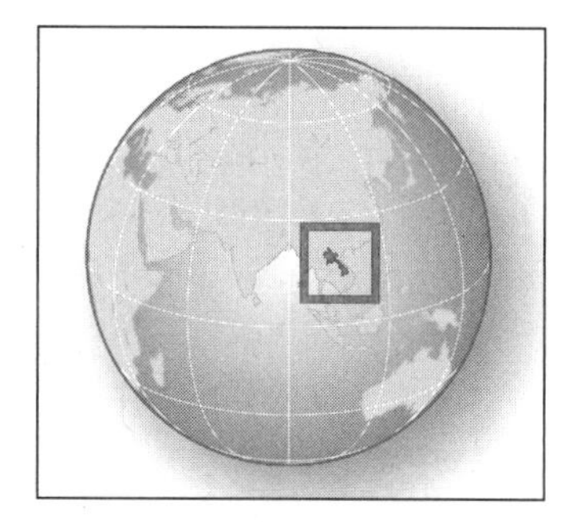

LAOS, THE ONLY landlocked (without any ocean coastline) Southeast Asian country, is one of the poorest of the world, with 40 percent of its population living below the poverty line. Its six-century-old monarchy, which also included French occupation (1893–1953), had dual capitals in Vientiane and Luang Prabang. The monarchy ended in 1975, when the communist Pathet Lao rebel forces, backed by North Vietnam, took control of the government. Laos turned into a communist satellite of the Soviet Union and VIETNAM but maintained a more neutral position than Vietnam and CAMBODIA. After aid from the Soviet Union ceased in 1991, the UNITED STATES, JAPAN, and international agencies provided the aid, without which the country would have been in great difficulty. The post-1975 collectivization of farms and nationalization of a few industries were replaced by a return to market economy and liberal foreign investment laws. Laos remains a communist country.

Western and northern parts of Laos are mountainous; the former includes a part of the Annamite Cordillera, where there are areas that receive 80 to 120 in (203 to 305 cm) of rainfall. Vientiane, the capital receives 68 in (173 cm) annually. Being in a monsoon climatic regime, there is a great deal of uncertainty about rainfall. Only 3.47 percent of the land is arable though 80 percent of the labor force is engaged in agriculture. Rice dominates the food crops and accounts for about three-fourths of the total crops produced.

Laos is self-sufficient in rice, but it needs money to run the government and other activities. Its primary production-related industries (tin, beef, pork, cigarettes, and wood) are in their primitive stages. Upper reaches of the MEKONG RIVER on the western part of the country collect most of the drainage from the rest of the country. The Mekong is suitable for navigation only in sections because of several rapids.

POPULATION GEOGRAPHY

The population of Laos shows characteristics of a less-developed country: more people in the lower age groups (42 percent are in the age group 1 to 14 years); high birth rates (37 births per 100 people); low life expectancy (54 years); high infant mortality rate (89 per 1,000 births); and a high fertility rate (4.94 children born per woman in reproductive age). Sixty percent of Laotians are Buddhist and the remaining 40 percent are animists and others.

There are three strata of people in Laos: 1) Austro-Asiatic group, consisting of 25 percent of the country's population. They were the earliest settlers but were driven into the mountains above 3000 ft (914m) by the Tai and Lao; 2) Lao Loum, who originated from South China and live in the most productive lowland river valleys, growing glutinous rice, accounting for 68 percent of the country's population. They are the most educated and are the major decision makers; 3) Lao Sung, consisting of 9 percent of the population, are the 19th-century migrants from South China, living in the

northern highlands. There is an ongoing government policy of "Laoization," in which efforts are made to acculturate the Lao Theung and Lao Sung minorities in Lao Loum cultural traits. The government encourages ethnic harmony.

BIBLIOGRAPHY. Cecile Cutler and Dean Forbes, "Vietnam, Laos, and Cambodia," T.R. Leinbach and R. Ulack, eds., *South East Asia; Diversity and Development* (Prentice Hall, 2000); A. Dutt, "Laos and Core Areas in the Upper Mekong Valley," A. Dutt, ed., *Southeast Asia: A Ten Nation Region* (Kluwer Academic, 1996); Frank M. LeBa and Adrienne Suddard, eds., *Laos: Its People, Its Society, Its Culture* (Hraf Press, 1963); Grant Evans, "Planning Problems in Peripheral Socialism: The Case of Laos," J.J. Zasloff and L. Unger, eds., *Laos: Beyond the Revolution* (St. Martin's Press, 1991).

ASHOK K. DUTT
UNIVERSITY OF AKRON

latitude and longitude

LATITUDE AND LONGITUDE are points on lines that graph the Earth and allow cartographers and others, by assigning measurements to the lines, to fix the location of any place. Latitude lines run east and west and are also called parallels; longitude lines run north and south and are known as meridians. The measurements for both are given in degrees; more exact locations are expressed by increments of minutes and seconds.

Lines indicating latitude circle the globe in an east-west direction, between the North and South Poles. Latitude lines parallel the equator, itself an imaginary line but one that can be determined with exactness. The equator lies midway between the North and South Poles and is assigned a latitude of 0 degrees. All other points are given in relation to their distance north or south of the equator. The highest latitude possible is 90 degrees, which is the latitude of the North and South Poles: 90 degrees N and 90 degrees S. Each degree of latitude extends 69 mi (111 km).

Ptolemy was the first to use latitude and longitude lines and measure them in degrees in his book, *Geography*, written around 150 B.C.E. Today, well-known lines of latitude enclose the ANTARCTIC and ARCTIC CIRCLES at 66 degrees 33 minutes south or north, which defines the area that experiences at least one full day of darkness in winter. Other named latitude lines are the Tropics of Capricorn and Cancer, which are at 23 degrees 27 minutes south or north. The TROPIC OF CAPRICORN and the TROPIC OF CANCER mark the points furthest south and north, where the sun can be seen directly overhead at least one day during the year.

Longitude lines, which had been envisioned on maps from Ptolemy's time, were far more difficult to fix with accuracy. While the north-south lines can be drawn anywhere on a globe, the questions of where to place 0 degrees and how to measure from it were not resolved until recently. Unlike latitudes, which parallel each other, longitudinal lines are further apart from each other at the equator, and converge at the poles.

Since the measurement of any such line could follow the sun, going from east to west, it can be a measurement of time as well as distance. The Earth is 360 degrees, divisible by 24-hour periods; the Earth rotates 15 degrees each hour. For ships at sea, figuring longitude—and thus, their own location—meant knowing the exact time in their home port, as well as the exact time at sea, and measuring the difference. In 1714, the British Longitude Act offered a prize of 20,000 pounds to anyone who could track longitude with an accuracy of .5 degrees.

In the 1770s, the prize was won by a clockmaker named John Harrison, whose chronometers were proved accurate on voyages with Captain James Cook. In 1884, the British government declared that the meridian running through the Greenwich Observatory near LONDON, England, would be the prime meridian, with a measurement of 0 degrees. Previous to that, and previous to acceptance of that decree by other countries, other meridians had served as 0 degrees. Washington D.C. and PARIS, FRANCE, for example, measured longitude from their own Prime Meridians.

Traveling west from the Greenwich Observatory, any spot between 0 degrees and 180 degrees, is considered west longitude. The line at 180 degrees is exactly opposite the Prime Meridian on the other side of the Earth, and serves as the INTERNATIONAL DATE LINE. Going east, any location between 0 degrees and 180 degrees is designated east longitude.

Fixing latitude and longitude in the 21st century relies more on satellite technology than chronometers. The Global Positioning System (GPS) was developed and deployed by the U.S. Department of Defense to find coordinates on or above the Earth. Electronic receivers decode and triangulate the information from this system to give latitude, longitude, and elevation. GPS receivers have varied uses: scientists and engineers

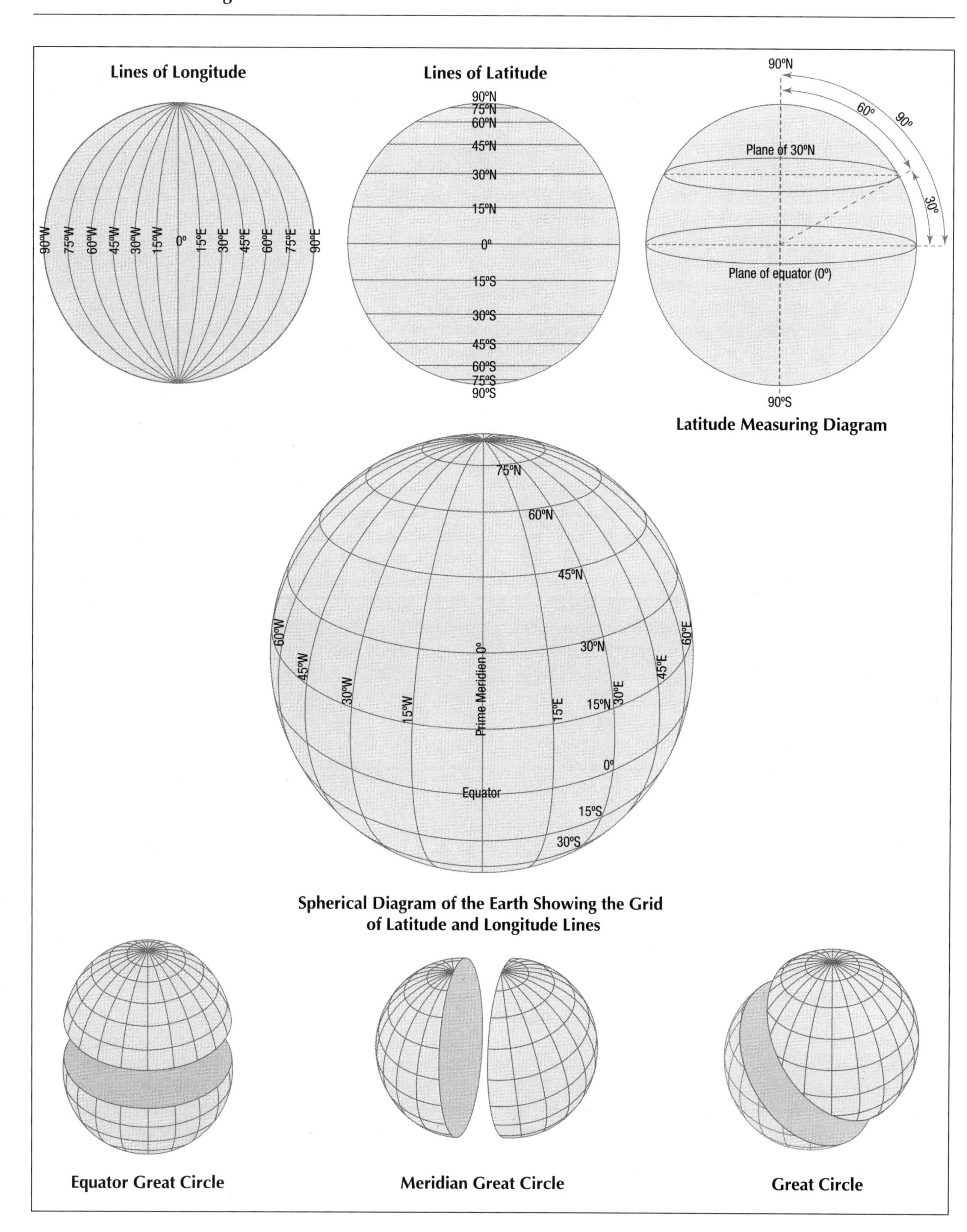
Lines of Longitude
90°W
75°W
60°W
45°W
30°W
15°W
0°
15°E
30°E
45°E
60°E
75°E
90°E
Lines of Latitude
90°N
75°N
60°N
45°N
30°N
15°N
0°
15°S
30°S
45°S
60°S
75°S
90°S
90°N
60°
90°
Plane of 30°N
30°
Plane of equator (0°)
90°S
Latitude Measuring Diagram
75°N
60°N
45°N
30°N
15°N
0°
15°S
30°S
Equator
Prime Meridian 0°
60°W
45°W
30°W
15°W
15°E
30°E
45°E
60°E
Spherical Diagram of the Earth Showing the Grid of Latitude and Longitude Lines
Equator Great Circle
Meridian Great Circle
Great Circle

measure tectonic movement, industries track their vehicles, and ordinary consumers navigate while sailing or hiking. Civilian use is deliberately degraded and limited to 100 meters, while military use of Precision (P) code is accurate to 20 meters.

GPS is called NAVSTAR (an acronym for Navigation System with Timing and Ranging) by the U.S. military, and became fully operational in 1994. Between 1989 and 2004, 50 GPS satellites had been deployed. A minimum of 24 circle the globe on six orbital planes.

Longitude and latitude coordinates may be collected by a GPS receiver and uploaded into a geographic information system (GIS). Data capture (the insertion of information into the system) requires identifying the objects on a map and noting their precise global positions and their spatial relationships. Information from satellite images or aerial photographs may also be extracted with computer software and placed into the database. Existing digital information that is not in map form can be converted by GIS into usable form.

BIBLIOGRAPHY. Lucia C. Harrison, *Sun, Earth, Time and Man* (Rand McNally, 1960); Derek Howse, *Greenwich Time and the Discovery of the Longitude* (Oxford University Press, 1980); Dava Sobel, *Longitude* (Walker Publishing, 1995); "Global Positioning System (GPS)," Jet Propulsion Laboratories, http://leonardo.jpl.nasa.gov (August 2004); Paul Lowman, *Exploring Space, Exploring Earth* (Cambridge University Press, 2002).

VICKEY KALAMBAKAL
INDEPENDENT SCHOLAR

Latvia

Map Page 1130 **Area** 24,937 square mi (64,589 square km) **Population** 2,348,784 **Capital** Riga **Highest Point** Gaizinkalns 1,017 ft (311 m) **Lowest Point** 0 m **GDP per capita** $8,300) **Primary Natural Resources** peat, limestone, dolomite, amber.

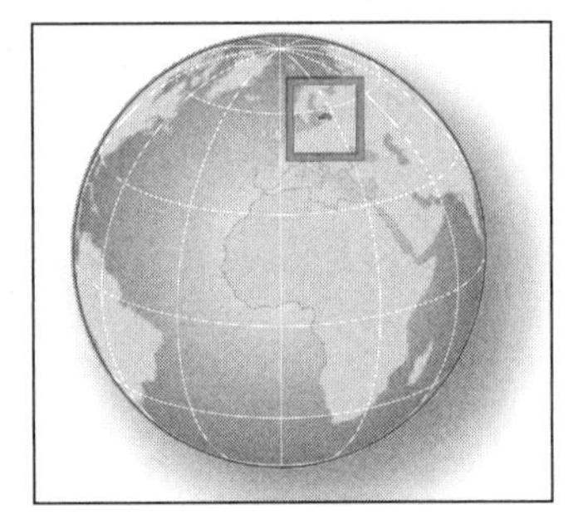

ON THE EASTERN shore of the Baltic Sea in northern Europe, the Republic of Latvia is a flat country that borders ESTONIA to the north, LITHUANIA to the south, BELARUS to the southeast, and RUSSIA to the east. Latvia is a parliamentary democracy with the supreme council, or Saeima, serving as the legislature and a president as head of state.

Most of Latvian topography is a flatland consisting fields, forest, lakes, marshes, and navigable rivers, with the exception of small hills east of Riga and to the southeast. Its chief rivers are the Daugava, Guja, Venta, and Lielupe. The major cities in Latvia are Riga, Daugavpils, Liepaja, and Jurmala. The country is marked by a long coastline indented by the Gulf of Riga to the northwest and some natural harbors. The climate is humid with only 30 to 40 days of sunshine per year. Temperatures range from 28 degrees F (-2 degrees C) in January to 63 degrees F (17 degrees C) in June.

The first permanent human settlements in what is know Latvia date at least to 9000 B.C.E. by migrations from the south and the southwest. From the 12th century onward, Latvia transferred from the rule of the Teutonic Knights, to the Poles, to the Swedes, and to the Russians. A Latvian national consciousness was formed by the early 19th century by its intellectuals, later to be transformed into an independence movement. Russian military weakness in World War I provided the opportunity for Latvia to gain independence in 1920.

The fledgling republic was beset by conflicts between fascists on the right and communists to the left. In June 1940, Latvia was occupied by the Soviet Union and later invaded by Nazi Germany. In 1944, Latvia was reoccupied by the Soviet Union and incorporated as a Soviet republic after World War II. In 1991, Latvia gained its independence from the Soviet Union and, with its neighbors Lithuania and Estonia, did not join the Commonwealth of Independent States, which arose from the demise of the Soviet Union. The last Soviet troops withdrew from Latvian soil in August 1994. Latvia became a member of the EUROPEAN UNION in May 2004.

Latvians or Letts make up 54 percent of the population. The next largest groups include Russians, Belarussians, Ukrainians, and Poles. Incorporating its Russian minority poses a challenge for the new Latvian society, still bitter over Soviet and earlier occupations.

Latvia's fastest-growing exports are in the fields of biotechnology, pharmaceuticals, and timber, the country remains heavily dependent on energy, particularly from Russia. Independence from the Soviet Union also meant the start of a free-market economy. Despite hardships caused by economic reform, Latvia has been

generally successful in emerging as an independent state.

BIBLIOGRAPHY. Andres Plakans, *The Latvians, A Short History* (Hoover Institution Press, 1995); David J. Smith, Aris Pabriks, Aldis Purs, and Thomas Lane, *The Baltic States: Estonia, Latvia, and Lithuania* (Routledge, 2002); *World Factbook* (CIA, 2004).

DINO E. BUENVIAJE
UNIVERSITY OF CALIFORNIA, RIVERSIDE

Law of the Sea

THE LAW OF THE SEA is a compilation of international and national laws regulating the demarcation of areas of maritime jurisdiction appertaining to maritime states. While its origins were military and defensive today it focuses on respective rights of resource exploitation—oil and minerals as well as fisheries. The importance of international innocent passage via geopolitical choke points and along multinational rivers also is relevant.

The oceans had long been subject to the freedom of the seas and innocent passage doctrine, a principle put forth in the 17th century designed essentially to limit national rights and jurisdiction over the oceans to a narrow belt of sea surrounding a nation's coastline. The remainder of the seas was proclaimed to be free to all and belonging to none. While this situation prevailed into the 20th century, by mid-century there was an impetus to extend national claims over offshore resources.

There was growing concern over the toll taken on coastal fish stocks by long-distance fishing fleets and over the threat of pollution and wastes from transport ships and oil tankers carrying noxious cargoes that plied sea routes across the globe. The hazard of pollution was ever present, threatening coastal resorts and all forms of ocean life. The navies of the maritime powers were competing to maintain a presence across the globe on the surface waters and even under the sea.

All maritime countries have claimed some part of the seas beyond their shores as part of their sovereign territory, a zone of protection to be patrolled against smugglers, warships, and other intruders. At its origin, the basis of the claim of coastal states to a belt of the sea was the principle of protection; during the 17th and 18th centuries, another principle gradually evolved: that the extent of this belt should be measured by the power of the littoral sovereign to control the area.

In the 18th century, the so-called cannon-shot rule gained wide acceptance in Europe. Coastal states were to exercise dominion over their territorial seas as far as projectiles could be fired from cannon based on the shore. According to some scholars, in the 18th century the range of land-based cannons was approximately one marine league, or three nautical miles. It is believed that on the basis of this formula developed the traditional 3-mi (4.8-km) territorial sea limit.

By the late 1960s, a trend to a 12-mi (19.3-km) territorial sea had gradually emerged throughout the world, with a great majority of nations claiming sovereignty out to that seaward limit. However, the major maritime and naval powers clung to a 3-mi limit on territorial seas, primarily because a 12-mi limit would effectively close off and place under national sovereignty more than 100 straits used for international navigation.

In 1973, an international conference aimed at reaching an agreement was convened in New York. Nine years later in 1982, it adopted a constitution for the seas: the United Nations Convention on the Law of the Sea. During those nine years, representatives of more than 160 states sat down and discussed the issues and bargained and traded national rights and obligations in the course of the marathon negotiations that produced the convention.

Among the more important aspects of the convention are navigational rights, territorial sea limits, economic jurisdiction, legal status of resources on the seabed beyond the limits of national jurisdiction, passage of ships through narrow straits, conservation and management of living marine resources, protection of the marine environment, a marine research regime, and, a more unique feature, a binding procedure for settlement of disputes between states. In short, the convention is an unprecedented attempt by the international community to regulate all aspects of the resources of the sea and uses of the ocean and thus bring order to one of mankind's very source of life.

TERRITORIAL WATERS

Territorial waters are the belt of sea adjacent to shores that states claim as being under their immediate territorial jurisdiction, subject only to a right of innocent passage by vessels of all nations. As to the breadth of the belt and the exact nature of this right of innocent passage, however, there is still much controversy. The

3-mi limit recognized and practiced by Great Britain, FRANCE, and the UNITED STATES seems to have been derived from the cannon range of the period, when it was adopted between Great Britain and the United States, toward the close of the 18th century.

The doctrine satisfied a requirement of the time and became a maxim of international law throughout northern Europe, both for the protection of shore fisheries and for the assertion of the immunity of adjacent waters of neutral states from acts of war between belligerent states. GERMANY still holds, in principle, to this varying limit of cannon range. NORWAY has never agreed to the 3-mi limit, maintaining that the special configuration of its coastline necessitates the exercise of jurisdiction over a belt of 4 mi (6.4 km). SPAIN lays claim to jurisdiction over 6 mi (9.6 km) from its shores.

Traditionally, smaller states and those not possessing large, ocean-going navies or merchant fleets favored a wide territorial sea in order to protect their coastal waters from infringements by those states that did. Naval and maritime powers, on the other hand, sought to limit the territorial sea as much as possible, in order to protect their fleets' freedom of movement.

As the work of the conference progressed, the move toward a 12-mi territorial sea eventually gained universal acceptance. Within this limit, states are in principle free to enforce any law, regulate any use, and exploit any resource. The convention retains for naval and merchant ships the right of innocent passage through the territorial seas of a coastal state. This means, for example, that a Japanese ship, picking up oil from a gulf state, would not have to make a 3,000-mi (5,000-km) detour in order to avoid the territorial sea of INDONESIA, provided passage is not detrimental to Indonesia and does not threaten its security or violate its laws.

In addition to their right to enforce any law within their territorial seas, coastal states are empowered to implement certain rights in an area beyond the territorial sea, extending for 24 nautical miles from their shores, for the purpose of preventing certain violations and enforcing police powers. This area, known as the contiguous zone, may be used by a coast guard or its naval equivalent to pursue and, if necessary, arrest and detain suspected drug smugglers, illegal immigrants, and customs or tax evaders violating the laws of the coastal state within its territory or the territorial sea.

Largely ignored were the problems such as multiple states claiming the same maritime zones, as in the South China and East China seas, and disputes over control of the sea when two nations, such as CHINA and TAIWAN, claim different governments. In short, the devil is in the geographic and geopolitical details.

EXCLUSIVE ECONOMIC ZONE

The exclusive economic zone (EEZ) is one of the most revolutionary features of the convention, and one that already has had a profound impact on the management and conservation of the resources of the oceans. Simply put, it recognizes the right of coastal states to jurisdiction over the resources of some 38 million square nautical miles of ocean space. To the coastal state falls the right to exploit, develop, manage, and conserve all resources—fish or oil, gas or gravel, nodules or sulphur—to be found in the waters, on the ocean floor, and in the subsoil of an area extending 200 mi (321.8 km) from its shore.

The EEZs are a generous endowment indeed. About 87 percent of all known and estimated hydrocarbon reserves under the sea fall under some national jurisdiction as a result. So too will almost all known and potential offshore mineral resources, excluding the mineral resources (mainly manganese nodules and metallic crusts) of the deep ocean floor beyond national limits. And whatever the value of the nodules, it is the other nonliving resources, such as hydrocarbons, that represent the presently attainable and readily exploitable wealth.

The most lucrative fishing grounds, too, are predominantly the coastal waters. This is because the richest phytoplankton pastures lie within 200 miles of the continental masses. Phytoplankton, the basic food of fish, is brought up from the deep by currents and ocean streams at their strongest near land and by the upwelling of cold waters where there are strong offshore winds. The desire of coastal states to control the fish harvest in adjacent waters was a major driving force behind the creation of the EEZs.

Today, the benefits brought by the EEZs are more clearly evident. Already 86 coastal states have economic jurisdiction up to the 200-mi limit. As a result, almost 99 percent of the world's fisheries now fall under some nation's jurisdiction. Also, a large percentage of world oil and gas production is offshore. Many other marine resources also fall within coastal-state control. This provides a long-needed opportunity for rational, well-managed exploitation under an assured authority

Coastal states also have certain rights in the CONTINENTAL SHELF, comprising the seabed and its subsoil that extend beyond the limits of its territorial sea throughout the natural prolongation of its land terri-

tory to the outer edge of the continental margin, or to a distance of 200 mi from the baselines from which the territorial sea is measured, where the outer edge of the continental margin does not extend up to that distance.

In cases where the continental margin extends further than 200 mi, nations may claim jurisdiction up to 350 mi (482.8 km) from the baseline or 100 mi (161 km) from the 8,202-ft or 2,500-m depth, depending on certain criteria such as the thickness of sedimentary deposits. These rights would not affect the legal status of the waters or that of the airspace above the continental shelf. To counterbalance the continental shelf extensions, coastal states must also contribute to a system of sharing the revenue derived from the exploitation of mineral resources beyond 200 mi. These payments or contributions, from which developing countries that are net importers of the mineral in question are exempt, are to be equitably distributed among state parties to the convention through the International Seabed Authority

The convention also contains a new feature in international law, which is the regime for archipelagic states (such as the PHILIPPINES and INDONESIA, which are made up of a group of closely spaced islands). For those states, the territorial sea is a 12-mi zone extending from a line drawn joining the outermost points of the outermost islands of the group that are in close proximity to each other. The waters between the islands are declared archipelagic waters, where ships of all states enjoy the right of innocent passage. In those waters, States may establish sea lanes and air routes where all ships and aircraft enjoy the right of expeditious and unobstructed passage

The Convention on the Law of the Sea holds out the promise of an orderly and equitable regime or system to govern all uses of the sea. But it is a club that one must join in order to fully share in the benefits. The convention, like other treaties, creates rights only for those who become parties to it and thereby accept its obligations, except for the provisions that apply to all states because they either merely confirm existing customary law or are becoming customary law. The convention was adopted as a "package deal," with one aim above all, namely universal participation in the convention. No state can claim that it has achieved quite all it wanted. Yet every state benefits from the provisions of the convention and from the certainty that it has established in international law in relation to the law of the sea. It has defined rights while underscoring the obligations that must be performed in order to benefit from those rights. Any trend toward exercising those rights without complying with the corresponding obligations or toward exercising rights inconsistent with the convention, must be viewed as damaging to the universal regime that the convention establishes.

In the 21st century, the military aspects again became important. Now it is the right of a country's national defense to board and stop military cargoes to "rogue" nations or for use by terrorists. Coastal zones and CHOKE POINTS were no longer the primary focus of the Law of the Sea.

BIBLIOGRAPHY. Thomas A. Clingan, *Law of the Sea: Ocean Law and Policy* (Austin & Winfield, 1994); Erving Newton, "The United Nations Convention on the Law of the Sea 1982 and the Conservation of Living Marine Resources," www.solent.ac.uk (June 2004).

DAVID NEWMAN
BEN GURION UNIVERSITY, ISRAEL

Lebanon

Map Page 1121 **Area** 4,015 square mi (10,400 square km) **Population** 3,727,703 **Capital** Beirut **Highest Point** 10,131 ft (3,088 m) **Lowest Point** 0 m **GDP per capita** $4,800 **Primary Natural Resources** limestone, iron ore, salt, water, arable land.

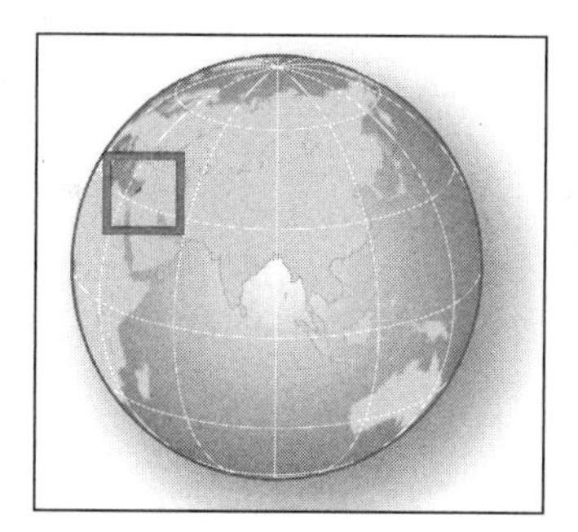

LEBANON, ONE OF the world's smallest countries, is on the eastern shore of the MEDITERRANEAN SEA. SYRIA is its neighbor to the north and east, and ISRAEL is located to the south.

Although a small country, Lebanon has a wide range of geographical regions. All of the major cities are located on the coastal strip. The Mount Lebanon Range located inland, provides majestic peaks and ridges. The Bekaa Valley, located parallel to the coast, is home to a multitude of wine vineyards.

Thousands of years ago, the mountains of Lebanon were covered with great cedar forests. Only a few cedar forests remain, but Lebanon is still recognized as the most densely wooded country in the Middle East. Many pine trees cover the mountain land, and fruit trees are present all across the coastal plain. The mountains are home to many different birds, including ea-

gles, red kites, Sardinian warblers, and Scop's owls. The climate in Lebanon is as diverse as the topography. Along the coast, hot conditions are prevalent during the summer, and cool, moist weather during the winter months. Snow and wind are common in the Bekaa Valley during the winter, and the mountains provide a typical alpine climate year-round.

The French created Lebanon in 1920. Originally, the Maronites, the largest religious community in the country, were placed in control of the government. However, to ensure the Maronites would remain loyal to the French, the French enlarged Lebanon to include mainly Muslim areas in the state. After this inclusion, only 30 percent of the population was Maronite. The tight grip of the French would last for decades, and the religious differences would last for the rest of the century. In 1926, the constitution was passed; a single chamber of deputies was created that could elect a president. The president had limited authority to choose a prime minister and the cabinet. However, the French rule continued to control Lebanese foreign relations and the military. By 1936, French and Lebanese government officials signed a treaty that included a guarantee of fairness to all religious sects in the country. The Maronite Christian Emile Eddé was elected president and he chose a Muslim, Khayr al-Din al-Ahdab as the prime minister. This power sharing formula continued until the late 1980s.

The French left Lebanon after World War II, and for the next two decades, the Lebanese people tried to create a separate identity. In the early 1970s, many Palestinian Liberation Organization (PLO) fighters infiltrated the country. They conducted raids against the Israelis from southern Lebanon and urged many Palestinian refugees in Lebanon to fight for their cause. By April 1975, tensions reached new levels as civil war broke out in Lebanon. The conflict sided the Christian Maronites against the Muslim Lebanese National Movement (LNM). The Palestine Liberation Organization (PLO) soon joined the Muslim forces. One year later, the Syrian army intervened and, with Arab League support, worked out a cease-fire. By October 1976, an Arab Deterrent Force, mainly of 40,000 Syrian soldiers, occupied Lebanon. However, tensions were still high.

On June 6, 1982, after years of PLO attacks, Israel invaded Lebanon with the goal of eliminating the PLO and creating a 25-mile security zone in southern Lebanon. The Israelis bombarded Beirut, hoping the Christian militias would seize control of the government. By August 1982, the PLO evacuated Beirut and took refuge in eight different countries. The Christian Bashir Gemayel was elected Lebanese president but was assassinated a few weeks later. The Israelis immediately returned to Beirut. A multinational peace force returned to the country and Amin Gemayel became president. In June 1985, Israeli forces withdrew from Lebanon, but maintained a presence in the southern Lebanon security zone. The country remained divided for the next six years. On May 22, 1991, the Lebanese government signed a Treaty of Cooperation and Brotherhood. Syria was given control over Lebanon's internal affairs and had around 16,000 troops stationed in the country. In 2005, under international pressure, Syria began to withdraw those troops.

Lebanon has attempted to pick up the pieces from the years of war. Muslims have been given a greater role in the government. Although the militant group, Hezballah still retains its weapons, most of the militias have been weakened or eliminated. The Lebanese armed forces have central government control over the majority of the country and Israel had withdrawn its forces from the southern security zone.

BIBLIOGRAPHY. William L. Cleveland, *A History of the Modern Middle East* (Westview Press, 1994); Ian J. Bickerton and Carla L. Klausner, *A Concise History of the Arab-Israeli Conflict* (Prentice Hall, 1998); *World Factbook* (CIA, 2004); Lonely Planet World Guide, "Lebanon," www.lonelyplanet.com (May 2004).

GAVIN WILK
INDEPENDENT SCHOLAR

leeward and windward

THE TERMS *leeward* and *windward* are used in a number of ways to describe specific places, physical features, and climatic processes. In one sense, windward and leeward generally refer to the location of a place relative to the prevailing wind direction. A windward location is one that is exposed to the prevailing winds. Conversely, a leeward location is protected from the prevailing wind.

For example, Concepcion, CHILE, on the west side of the ANDES would be in a windward location relative to westerlies moving inland from the PACIFIC OCEAN simply because it is exposed to the approaching wind. However, the east side of the Andes would be a leeward location because of the protection afforded by

the intervening mountains. The windward and leeward designations illustrated here are equally applicable to orographic lifting, the process involved when winds strike the front face of a mountain, are forced up the windward face of the mountain and then descend on the leeward side. If the winds are laden with moisture and the mountain is high enough, the moisture carried by the wind may condense and produce precipitation. The resulting precipitation will in the highest volumes on the windward side of the mountain and the leeward side will invariably receive a lesser amount.

There are a number of places in the world where this process is clearly evidenced. Among them are the moisture-laden westerlies reaching the Pacific coast of WASHINGTON, which are forced aloft by the Coastal Ranges. The windward side of this region receives an abundance of rainfall while the leeward side on the eastern slopes receives little or no rainfall. A classic example of orographic lifting is found in the HIMALAYAS when the summer monsoons bring warm, moist winds across the Indian subcontinent and are then forced aloft by the imposing Himalayan barrier. Areas on the windward side on the mountains may receive as much as 100 in (254 cm) of rainfall. However, because of the great heights of the Himalayas, little or no moisture reaches the leeward side. So within a relatively short distance the climatic results vary from near tropical conditions to the true deserts of Central Asia.

The terms are used in a more formal manner to name particular groups of islands. For example, the islands in the Lesser Antilles in the West Indies all lie in the pathway of the northeast trade winds. This wind belt moves from approximately 30 degrees north latitude toward the equator where it meets its counterpart from the Southern Hemisphere, the southeast trade winds. Historically, British sailing ships entered the region with the northeast trade winds at their backs. The first island encountered on these voyages was usually BARBADOS, the island farthest east and most to windward.

The Windward Islands, as they came to be called, include Barbados, the Caribees (a cluster of small islands), DOMINICA, MARTINIQUE, GRENADA, SAINT LUCIA, and SAINT VINCENT AND THE GRENADINES. The Windward Islands, a former British colony, are the southernmost islands in the Lesser Antilles and were once collectively named the Federal Colony of the Windward Islands and later the Territory of the Windward Islands. The northern continuation of the Lesser Antilles includes islands that are farther downwind from the Windward Islands. First discovered by Columbus in 1493, these are the Leeward Islands, which includes ANTIGUA AND BARBUDA, the British Virgin Islands, MONTSERRAT, SAINT KITTS AND NEVIS, and ANGUILLA. A string of leeward islands is also found northwest of the Hawaiian Islands, and this group has become a national bird sanctuary.

In addition, the Society Islands in French Polynesia, a region east of the COOK ISLANDS in the South Pacific, are identified as leeward islands. Reference may be made as well to another use of the word *windward*. The narrow sea-lane separating eastern CUBA and HAITI lies in the path of the northeast trade winds. As such, vessels traveling between the ATLANTIC OCEAN and the CARIBBEAN SEA are using the aptly named Windward Passage. Those traveling through the pass from northeast to southwest have the advantage of the northeast trade winds pushing them along.

BIBLIOGRAPHY. Robert W. Christopherson, *Geosystems: An Introduction to Physical Geography* (Prentice Hall, 1997); Edward J. Tarbuck and Frederick Lutgens, *Earth Science* (Prentice Hall, 1999); Neil Wells, *The Atmosphere and the Ocean: A Physical Introduction* (Wiley, 1997).

GERALD R. PITZL, PH.D.
MACALESTER COLLEGE

Lena River

THE LENA RIVER is one of Russia's great northern rivers, draining an area of 899,641 square mi (2,306,772 square km) and encompassing a region rich in wildlife and natural resources, including one of the world's largest deposits of gold. This is also one of the most inhospitable regions on the planet, with extremes of temperature and vast stretches of northern forests (84 percent). Few people live in the region and there is only one city in the entire Lena basin, Yakutsk, the administrative seat of the former Yakut Autonomous Socialist Republic and now the capital of a semi-independent Yakutia, renamed the Republic of Sakha in 1991.

The river, 2,850 mi (4,597 km) long, flows mostly through Sakha, but its origins are in the Irkutsk District, immediately west of Lake BAIKAL. Its tributaries have their headwaters in the autonomous republic of Buryatiya and the districts of Chita, Amur, and Khabarovsk. The river starts in the Baikal range, only a short distance from the lake itself, though the lake's

waters flow into the Angara River and into the Yenisey, thousands of kilometers from the Lena. It then flows north and east to be joined by its first major tributaries, the Vitim and Olekma, whose headwaters originate in the Yablonov and other parallel mountain ranges east of Lake Baikal (which are also the source of the headwaters of the AMUR RIVER).

The river then turns again north in a large arc, following the contours of the Aldan Plateau to the south, from which emerges the Lena's largest tributary, the Aldan, and the Verkhoyansk Range to the east. This range of mountains forms a steep escarpment for over 600 mi (1,000 km), stretching all the way to the Lena's delta on the Laptev Sea, an arm of the ARCTIC OCEAN. This semicircular area bounded by the Aldan and Verkhoyansk highlands forms a sort of climatic vortex, producing some of the coldest temperatures ever recorded on the planet: -88.9 F (-67 degrees C).

To the west the terrain is much flatter, with larger tributary rivers, notably the Vilyui, which extends far into the Siberian Plateau. From the confluence of the Lena with the Aldan, 800 mi (1,300 km) from the sea, the river becomes very broad, sometimes reaching 5 mi (8 km) across. The delta, covering 12,352 square mi (31,672 square km), is the largest in RUSSIA and third largest in the world. It is formed of numerous islets, marshes and sandbars. The largest islands (hundreds of square kilometers) are covered with damp, mossy tundra and frozen lakes that do not permit the construction of roads, so travel between the eight permanent settlements continues mainly by dogsled.

The river is almost entirely navigable, with an abundance of fish, but is frozen eight months of the year. Ice has been measured at 53 in (136 cm) in the south, and up to 90 in (231 cm) at the delta. Because the Lena is almost entirely fed by mountain snows, spring thaws can bring disastrous floods, followed by equally breakups of river ice, sizable chunks of which can destroy entire sections of the riverbank and any settlements alongside it. The annual flow of the river is very irregular, with 90 to 95 percent of all of its discharge in spring and summer, when its volume increases by as much as 10 times that of the winter months. This irregular flow has limited the development of hydroelectric projects in the Lena basin, though there are two large dam-reservoir complexes on the Vilyuy. The Lena was used as a highway for trappers and traders in Russia's expansion to the Pacific coast, with its main town of Yakutsk founded in 1632.

Russia's relations with the indigenous Yakut and Evenki peoples were not always harmonious, and the region saw a good deal of oppression, lawlessness, and unbridled greed in the race for lucrative furs. Another boom period followed in the 19th century, with the discovery of gold in the Lena valley near the confluence with the Vitim River. Privatization since the 1990s has returned the region to its "Wild West" frontier days, ruled by hustlers, speculators and black marketeers. The 1 million inhabitants who live in the Lena basin are looking to their mineral wealth, still undeveloped because of great distances, difficulties building roads and buildings on permafrost, and difficulty getting water half the year.

BIBLIOGRAPHY. Sergei Petrovich Suslov, *Physical Geography of Asiatic Russia*, N. D. Gershevsky, trans. (W.H. Freeman and Company, 1961); John J. Stephan, *The Russian Far East: A History* (Stanford University Press, 1994); C. Revenga, S. Murray, et al., *Watersheds of the World* (World Resources Institute, 1998).

JONATHAN SPANGLER
SMITHSONIAN INSTITUTION

Lesotho

Map Page 1116 **Area** 11,720 square mi (30,355 square km) **Population** 1,861,959 **Capital** Maseru **Highest Point** 11,425 ft (3,482 m) **Lowest Point** 4,265 ft (1,301 m) **GDP per capita** $550 **Primary Natural Resources** water, diamonds, minerals.

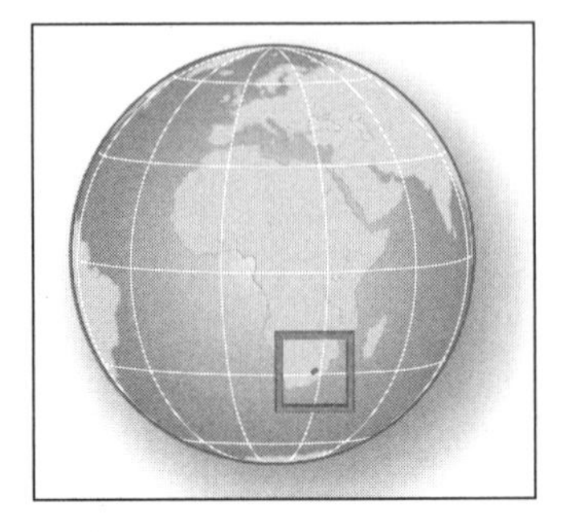

LESOTHO IS LOCATED literally within SOUTH AFRICA: This small southern African country is landlocked and surrounded on all sides by the Republic of South Africa. Lesotho covers an area slightly smaller than MARYLAND.

It is the only country in the world that lies entirely above 3,280 ft (1,000 m) and more than 80 percent of Lesotho is 5,905 ft (1,800 m) above sea level. Lesotho is mainly mountains, hills, and highlands with plateaus. It has a very temperate climate, with cool to cold dry winters and hot, wet summers.

Lesotho has four major mountain ranges within its borders. The DRAKENSBURG MOUNTAIN range, the Central range, and the Thoba Rutsoa range are all in the

central and eastern parts of Lesotho and the Maloti range is in the western region. Thabana-Ntlenyana is the highest peak in all of southern Africa. It is also called "The Kingdom in the Sky," "The Roof of Africa," and "The Switzerland of Africa." Deep river valleys and canyons cut these mountains; the water that flows through these canyons is the major water supply for all of southern Africa. Many of the rivers on the southern part of the African continent start in these hills, including the Orange River and its many tributaries. This water is Lesotho's main natural resource.

Lesotho is exploiting its water through the Lesotho Highlands Water Project. This project is designed to capture, store, and transfer water from the Orange River system and send it to the Republic of South Africa. This project not only moves water but produces hydroelectricity as well. There are going to be approximately seven dams and two tunnels when the project is complete in 2020. With the first phase of the project already completed, Lesotho has become almost completely self-sufficient in the production of electricity. Already it has generated approximately $24 million annually through the sale of electricity and water to South Africa.

The kingdom of Lesotho is heavily populated especially in the western part of Lesotho where there is more arable land. There, the people raise crops of corn, sorghum, wheat, beans, peas, asparagus, tomatoes, and peaches. Most of the land is used for raising animals such as sheep, goats, cattle, pigs, chickens, and horses. These people, called the Basotho, are remnants of various ethnic groups made up mainly of various Bantu-speaking people and some of the original San or Bushmen who were Lesotho's earliest inhabitants. The Basotho managed to fend off both the Zulus and the Boers in the 1800s.

On October 4, 1966, Lesotho gained independence. The Lesotho government is now a constitutional monarchy with a king as the head of state and a prime minister as the head of government. Currently, the king serves primarily as a ceremonial figure with no executive power. However, all land in Lesotho is held by the king and is allocated to the Basotho people through local chiefs. Foreigners in Lesotho are strictly forbidden from owning land. Lesotho has one of the highest literacy rates in Africa; official languages are English and Sesotho, a Bantu language.

BIBLIOGRAPHY. Institut géographique national, *The Atlas of Africa* (Éditions Jeune Afrique, 1973); Kwame Anthony Appiah and Henry Louis Gates, Jr., *Africana* (Basic Civitas Books, 1999); Saul B. Cohen, ed., *The Columbia Gazetteer of the World* (Columbia University Press, 1998); Bureau of African Affairs, "Background Note: Lesotho," (U.S. Department of State, November 2003).

Christy A. Donaldson
Montana State University

Lewis and Clark

IN 1804, MERIWETHER Lewis and William Clark began an expedition across the newly acquired LOUISIANA PURCHASE territory. These two army captains were chosen by the American president, Thomas Jefferson, to explore new land the UNITED STATES had purchased from FRANCE one year earlier, and to find a direct water route across the nation. For a little over two years, Lewis and Clark led their corps through some 8,000 mi (12,800 km) of unexplored lands, acquiring scientific samples, creating maps, and conveying to the local Native Americans that an acquisition of their territory by the United States had just occurred.

The idea of an expedition into the Louisiana Purchase territory began in early 1803 after approval by the U.S. Congress. Lewis, the personal secretary and friend of Jefferson, was chosen to lead the expedition. Lewis entered the army in 1794, where he served in the Frontier Army and rose to the rank of captain. He possessed quality leadership, scientific, and cartography skills, which contributed to the goals of the expedition. In June 1803, Lewis wrote a letter to his friend, Clark, expressing his desire to lead the expedition with him, as well as to recruit volunteers. Lewis had previously served under Clark during the latter half of the 1790s, and he viewed Clark as an able-bodied and highly intelligent leader.

Clark accepted the proposal, and in the middle of October 1803, he met Lewis in Clarksville, in the Indiana territory. The two men began preparations and enlisted men for their new Corps of Discovery. Two months later, the crew was in St. Louis, MISSOURI, making final preparations for the journey. After a winter of training their corps and stocking up on goods and materials, Lewis, Clark, and the Corps of Discovery embarked on their exploration on May 14, 1804. Lewis and Clark proved to be adept at leadership, and each used individual skills to the advantage of the group. Lewis was the planner and scientist. He proved his strength and stamina when he saved himself from a fall

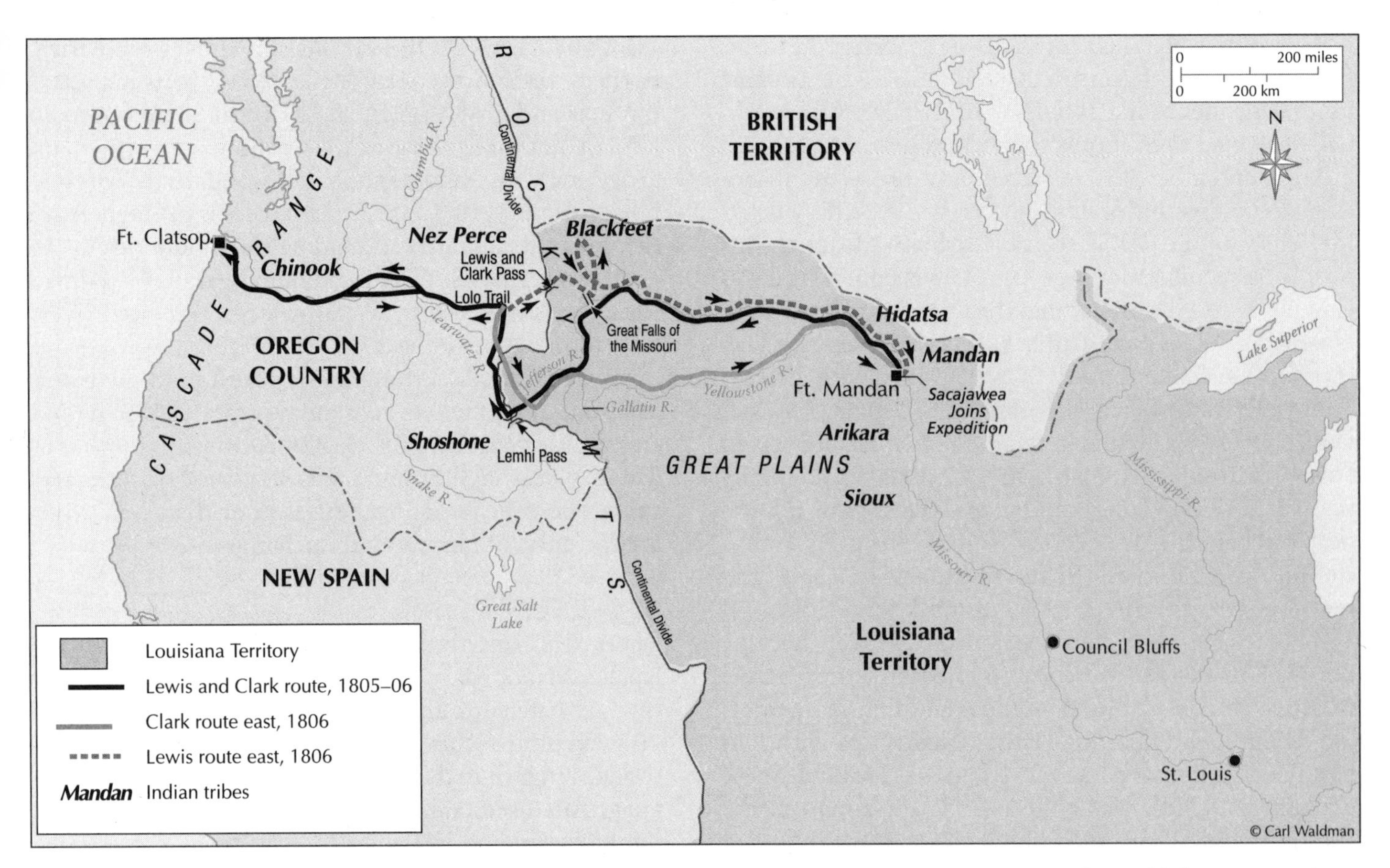

By September 1806, after a successful completion of the expedition, Lewis and Clark returned to St. Louis with careful descriptions of newly discovered animals and plants. Overall, they had accumulated a complete and impressive journal of their experiences.

off a cliff by wedging his knife into a crevice and breaking the fall. He cured a case of accidental poisoning through a homemade remedy. Lewis even survived an accidental gunshot wound from one of the corps members.

Clark proved to be a fine boatman and navigator as well as a skillful cartographer throughout the journey. His first map was made during their 1804–05 winter's stay at Fort Mandan. He used the knowledge obtained from traders and natives to create a map of the upper Missouri and MISSISSIPPI rivers region. He also created a skillful map of the Great Falls of the Missouri with references to botanical features. This map would continue to be used by expeditionary forces years later.

By September 1806, after a successful completion of the expedition, Lewis and Clark returned to St. Louis with careful descriptions of newly discovered animals and plants. Overall, they had accumulated a complete and impressive journal of their experiences. Lewis wrote letters to Jefferson that spoke of a triumphant exploration and of the discovery of a navigable route across the continent, which included stretches of the Missouri and Columbia rivers. His letters spoke of the difficult conditions, but also of the profitable fur trade that was occurring and the abundant animals that occupied the area. Although a direct water route across the continent did not exist, Jefferson was extremely impressed with the exploration.

After returning to their families, Lewis and Clark traveled separately to Washington, D.C., to receive their pay and honors. Each received double pay for service, which amounted to $1,228, and a title to 1,600 acres of land. Lewis was named governor of the Territory of Upper Louisiana, and Clark was made brigadier general of militia and superintendent of Indian Affairs for the Territory of Upper Louisiana.

In 1808, Clark married Julia Hancock of Fincastle, Virginia, and became a business partner of the Missouri Fur Company. This company was to send militia, hunters, and boatsmen into the newly explored territory and to implement the fur trade. Lewis, however, did not thrive with similar success. His effort to get his and Clark's journals published proved unsuccessful; he

had a couple of failed attempts at marriage; and his friendship with Jefferson floundered as he became more dependent on alcohol. By March 1808, Lewis finally traveled to St. Louis to work as governor.

By September 1809, after many problems in administering his duties with his constituents, Lewis left for Washington, D.C., to ask for help. However, his problems seemed too large, and Lewis committed suicide in a roadhouse along the Natchez Trace near Nashville, Tennessee. Upon hearing the news, Clark traveled to Washington, D.C., to grieve with Lewis's family. Afterward, he traveled to Philadelphia, Pennsylvania, where he arranged for the rewriting of his and Lewis' journals. By 1814, the journals were finally published, with remarkable maps, but few of the explorer's scientific discoveries. Remarkably, the American public exhibited little interest in the journals, and few copies were sold. It was not until new editions in 1893 and 1904–05 that the journals finally became popular. In 1813, Clark was named governor of the Missouri Territory. He spent the remaining years of his life meeting and creating lasting friendships with Native Americans, traders, and trappers. Through these encounters, Clark was able to update his map of the West. In 1838, Clark died of natural causes.

BIBLIOGRAPHY. Landon Y. Jones, "Commanding, Cooperative, Confident, Complimentary: Why Lewis and Clark Were Perfectly Cast as co-CEOs," *Time Magazine* (July 8, 2002); PBS, "Lewis and Clark, Inside the Corps: Captain Meriwether Lewis," www.pbs.org (April 2004); PBS, "Lewis and Clark, Inside the Corps: Captain William Clark," www.pbs.org (April 2004); Ron Fisher, "Lewis and Clark, Naturalist Explorers," *National Geographic* (October 1998).

GAVIN WILK
INDEPENDENT SCHOLAR

Liberia

Map Page 1113 **Area** 43,000 square mi (111,370 square km) **Population** 3,317,176 **Capital** Monrovia **Highest Point** 4,540 ft (1,383 m) **Lowest Point** 0 m **GDP per capita** $1,000 **Primary Natural Resources** rubber, coffee, cocoa, timber.

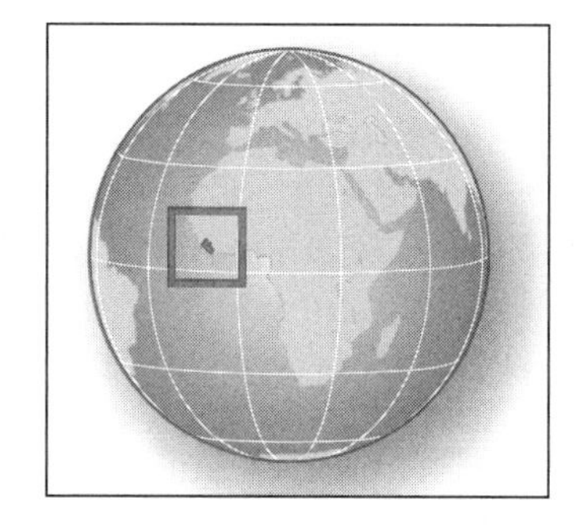

LIBERIA, "LAND OF THE FREE," in western Africa, borders the north ATLANTIC OCEAN, between CÔTE D'IVOIRE and SIERRA LEONE. It has 10 to 50 mi (16 to 80 km) of flat coastal plain that contains creeks, mangrove swamps, and lagoons. Beyond that, forested hills, from 600 to 1,200 ft (180 to 370 m) high cover the rest of the country, excluding the mountains in the northern highlands. The maximum peak in the Nimba Mountains is 4,540 ft (1,383 m). Six principal rivers flow to the ATLANTIC OCEAN.

Vegetation is predominantly forest, and the tropical, humid climate sees rainfall averages of 183 in (465 cm) on the coast and 88 in (224 cm) in the southeast. The dry season (harmattan, December and January) splits two rainy seasons. Cities other than Monrovia are the ports of Harper and Buchanan.

POPULATION GROUPS

There are 16 ethnicities in Liberia, including the Kpelle, Mano, Baso, Grebo, Kru, and Vaj. Seventy percent are native, traditional religion practitioners, while 20 percent are Muslim and 10 percent Christian. Although English is the official language, the native languages are used commonly. The Americo-Liberians, a minority residing in the cities, tend to be Protestants. Other population groups include Lebanese merchants and European and American technicians.

Liberia was founded in 1820 on the Grain Coast; it was the gift of the American Colonization Society, which received it from the Cape Mesurado chiefs. The founders fought bloody battles with the indigenous peoples. Eighty-six freed slaves from the United States established Christopolis, later Monrovia, in February 1820. Approximately 15,000 freed slaves emigrated from the United States until the American Civil War. In 1847, Liberia declared itself an independent republic.

Until 1980, the Americo-Liberian or True Whig Party (TWP) ruled Liberia. The first president was Joseph Jenkins Roberts, American-born. The government and constitution were modeled on that of the UNITED STATES. The republic traded with other parts of West Africa, but it modeled its style of living on those of the United States.

Modernization efforts led to a crisis of foreign debt in 1871. Conflicts with FRANCE and Britain led to losses of territory in 1885, 1892, and 1919. Liberia used the European rivalries and the support of the United States to remain independent. Still its exports declined and its debts rose, leading to foreign interference. Bankruptcy came in 1909. International loans saved Liberia. In 1926, Firestone, the American tire company, leased

large rubber plantations. In 1930, the League of Nations investigated charges that Liberia was exporting labor, that is, engaging in the slave trade. The president resigned.

The new president, William Tubman, opened the country to international investment, allowed indigenes greater participation, and allowed the exploitation of iron and other minerals. Education, roads, infrastructures, and healthcare were improved. Tubman died in 1971. His vice president, W.R. Tolbert took over. A proposed increase in the price of rice in 1979 set off rioting.

The TWP's rule ended on April 12, 1980, when Master Sergeant Samuel K. Doe, a native Krahn, pulled off a successful coup. Doe's forces executed Tolbert and several other Americo-Liberians. Doe's People's Redemption Council promised a return to civilian rule, then began repressing the opposition and abusing human rights. Doe instituted constitutional changes, survived numerous coups, and saw flight of thousands of refugees to the Côte d'Ivoire. The refugees returned in 1989, led by Charles Taylor. The war waxed and waned until finally in 2003 the Taylor regime ended, and the country and government struggled to maintain the uneasy truce.

BIBLIOGRAPHY. *World Factbook* (CIA, 2004); IRIN News, "Liberia: Peace Process Still Has a Long Way to Go," www.irinnews.org (June 2003); Patricia Levy, *Liberia* (M. Cavendish, 1998); Library of Congress, "History of Liberia: A Time Line," www.loc.gov (1998); D. Harold, Gloval Security, "Liberia: A Country Study," www.globalsecurity.org (1985); Paul Rozario, *Liberia* (Gareth Stevens Publishers, 2003).

John Barnhill, Ph.D.
Independent Scholar

Libya

Map Page 1113 **Area** 679,362 square mi (1,759,540 square km) **Population** 5,499,074 **Capital** Tripoli **Highest Point** 7,437 ft (2,267 m) **Lowest Point** -154 ft (-47 m) **GDP per capita** $7,600 **Primary Natural Resources** crude oil, petroleum products.

A RELATIVELY LARGE country, similar in size to the state of ALASKA, Libya largely consists of broad rolling deserts, barren rock inselbergs and immense dune fields or ERGS. It is a landscape of sandstorms; hot dusty wind, or ghibli; an expanding desert; and scarce water. More than 90 percent of the country is considered arid or semiarid. It primary cities are all located on the MEDITERRANEAN SEA coastline, which has facilitated its links across North Africa to Europe and western Asia.

Generally speaking, the Saharan plateau covers most of Libya. The exceptions are in the northwest corner in a region known as Tripolitania and in the northeast in Cyrenaica, Libya's largest region. The Tripolitania region, which runs north to south, is a string of carefully cultivated coastal oases in addition to the triangular Al-Jifarah plain, and the Nafusah Plateau, 200 mi (320 km) of limestone between 2,000 and 3,000 ft (600 to 915 m) in elevation.

Libya has no perennial rivers, but there are extensive underground aquifers that support artesian wells and springs. Libya's arid desert climate is moderated along the coast by the Mediterranean Sea. Precipitation ranges from 16 to 20 in (40 to 50 cm) in the northern hills to less than 5 in (12 cm) throughout most of the south, and to 1 in (2.5 cm) in the Libyan Desert. Droughts are common, meaning natural vegetation is minimal. Libya's principal mineral resource is its reserves of petroleum, Africa's largest and among the world's largest.

Since it earliest days as a major Phoenician and Roman territory on the North African coast of the Mediterranean Sea, Libya has been raided and colonized by Vandals, Arabs, Ottoman Turks, and Italians until its independence in 1951. Only a few years later, the country changed dramatically with the discovery of enormous oil reserves. In 1969, a 27-year-old Muammar Qaddafi led a successful coup to gain control of the nation. Qaddafi has been victorious in removing any imprints of previous cultures to create a landscape from his own vision. Based upon his Third International Theory, he created a political system combining Islam and socialism.

Using petroleum revenues in the 1970s and 1980s to promote political ideologies (including supporting terrorist activities) throughout the region, Libya prompted the United Nations (UN) to impose economic sanctions after the Lockerbie terrorist bombing was suspected to have had Libyan ties. The sanctions were then lifted in April 1999 when Qaddafi handed over Lockerbie bombing suspects.

BIBLIOGRAPHY. J. Azema, *Footprint Libya Handbook* (Footprint Guide Press, 2001); World Factbook (CIA, 2004); K. Park, World Almanac 2004 (World Almanac Publishing, 2003); B. Turner, *The Statesman's Yearbook 2003* (Palgrave Macmillan, 2002).

TOM PARADISE
UNIVERSITY OF ARKANSAS

Liechtenstein

Map Page 1131 **Area** 62.4 square mi (160 square km) **Population** 33,145 **Capital** Vaduz **Highest Point** 8,577 ft (2,599 m) **Lowest Point** 1,419 ft (430 m) **GDP per capita** $25,000 **Primary Natural Resources** hydroelectric potential.

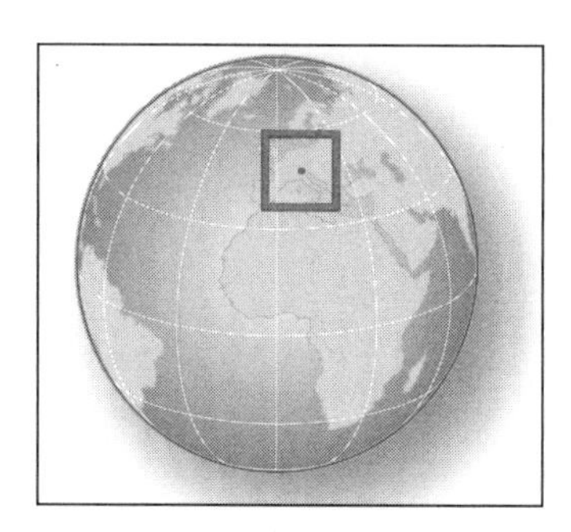

ONE OF WESTERN Europe's five microstates, the story-book principality of Liechtenstein is in reality one of the few "absolutist" states in the world, thanks to a favorable referendum voting extensive powers to the ruling sovereign in March 2003. His Serene Highness, Prince Hans-Adam II von und zu Liechtenstein now has more political prerogatives than any other monarch in Europe.

Sandwiched between the Swiss cantons of St. Gallen to the west and Graubünden to the south and the Austrian province of Vorarlberg to the east, Liechtenstein's history has naturally been closely linked to its larger neighbors. The family had served the Habsburgs in AUSTRIA for centuries, and in fact the family derives its name from its main castle just south of Vienna. Prince Johann Adam bought the county of Vaduz and the adjacent lordship of Schellenberg in 1699 and 1712, and the two territories were united as a fully independent member of the Holy Roman Empire in 1719. But the land itself was of secondary importance to the family: no sovereign prince of Liechtenstein even visited until the middle of the 19th century.

When the Holy Roman Empire fell apart thanks to Napoleon in 1806, Liechtenstein fell between the cracks and was not consolidated into one of the larger German states. Successive princes maintained their independence and forged a beneficial customs union with Austria in 1852. After the collapse of the Austro-Hungarian Empire, however, the princes turned west and formed a similar customs and monetary union with SWITZERLAND in 1923, which is still in effect today. The princely family still owns large estates in Austria and lays claim to numerous others in the CZECH REPUBLIC that were confiscated by the communists (equal to ten times the size of the principality itself). Still, the prince's wealth is estimated at more than $2 billion.

The principality—15 mi (26 km) long, and an average of 4 mi (6 km) wide—lies on the eastern bank of the RHINE RIVER, between its emergence from the high Alpine valleys of Switzerland and Lake Constance to the north. Two canals running on either side of the river through this valley maintain its water levels to reduce risk of spring floods. The eastern third of Liechtenstein is divided from the rest by a range of mountains, forming the upper Samintal, or valley of the Samina River, which runs northward into the Ill in Austria, which in turn joins the RHINE just a few kilometers north of Liechtenstein. To the east and south of this valley rise the much greater Alpine heights dividing the country from Austria, including a number of peaks above 8,250 feet (2,500 m).

The main town, Vaduz, has a population of about 5,000. High above the town, the prince's castle, dating from the 14th century, boasts one of the largest private art collections in the world and is a major tourist attraction. Much of this art collection has recently been transferred to Vienna's newest major art gallery, the restored Liechtenstein Palace, opened to the public in March 2004. Liechtenstein itself has been transformed since World War II from a sleepy agricultural community to a modern industrialized society with one of the world's highest standards of living.

Revenue is generated locally through skiing and the sale of rare stamps, but it is the income from numerous so-called post-office-box companies, attracted by low business taxes, that has boosted the national economy (providing as much as 30 percent of state revenues). Concerns over tax evasion schemes and money laundering, however, have recently caused increased pressure from the EUROPEAN UNION and the principality's authorities.

BIBLIOGRAPHY. Wayne C. Thompson, *Western Europe 2003* (Stryker-Post Publications, 2003); *World Factbook* (CIA, 2004); "Fuerstenhaus," www.fuerstenhaus.li (April 2004); www.liechtenstein.li (April 2004).

JONATHAN SPANGLER
SMITHSONIAN INSTITUTION

Lisbon

WITH A metropolitan-area population of 2,682,687, Lisbon is the capital and largest city in PORTUGAL. It is located on the right bank of the Tagus River, where it forms a large estuary providing a natural safe harbor close to the ATLANTIC OCEAN. The city evolves over a series of hills and enjoys a Mediterranean climate, with mild winters and warm dry summers.

Lisbon was the capital of a worldwide empire for more than 400 years and is the main political, economic, and cultural center of Portugal. The origin of the settlement is remote. Occupied by the Romans in 205 B.C.E., the town was confined to old Castle Hill and the slope leading to the river. It was conquered by the Moors in 714 and recovered by the Crusades and the king of Portugal in 1147. Lisbon was promoted to capital of the kingdom in 1255, and thanks to commerce between the Mediterranean and northern Europe, it became a primary urban center during the 13th and 14th centuries.

Serving as base for the Portuguese expeditionary navies in the age of discoveries, the city benefited the most from this period and became a significant cultural and economic center: infrastructures and especially monuments were built, and the city expanded to the west. In 1527, Lisbon already stood out among Portuguese cities and gained importance at a European level. An earthquake in 1755 destroyed most of the city. The downtown area (Baixa) was rebuilt as a regular grid aligned with the river, replacing tortuous medieval streets and turning it into a hallmark of Lisbon for its unity and architectural value.

Until the development of railroad transportation, the river was the main thoroughfare for the transport of people and goods between the city and more interior areas of the country. During the 19th century industrialization, mass transit was introduced and larger factories were installed along the river, attracting peasants from the countryside and originating typical worker neighborhoods. At the end of that century, the opening of Avenida da Liberdade, a wide boulevard, changed the form of the city by directing growth toward the north. However, by 1940 Lisbon was still concentrated and close to the river. On the hill to the west, the 2,540-acre (1,028-hectare) Monsanto forest park was created on still undeveloped land.

After World War II, a sharp increase in urbanization pushed growth beyond the limits of the municipality, and toward the west and north along main transportation routes (especially railroads). Population of the city stabilized in the 1950s, and by 2000 only 20 percent of the population of the metropolitan area lived within city limits. In the 1950s and 1960s, new industrial areas were located in Cabo Ruivo and on the south bank, connected by a bridge in 1966. More recent transformations include the expanding network of expressways and construction of large peripheral shopping centers, underlining the ongoing strong suburbanization. Part of the east riverfront was rebuilt to replace heavy industry with an area of residential use.

Eduardo VII Park in Lisbon recalls the glory days when Portugal was a major colonial power across the world.

BIBLIOGRAPHY. José Tenedório, ed., *Atlas da AML* (AML, 2003); Teresa B. Salgueiro, *A Cidade em Portugal* (Afrontamento, 1992); *Oxford Essential Geographical Dictionary* (Oxford University Press, 2003).

SERGIO FREIRE
PORTUGUESE GEOGRAPHIC INSTITUTE

Lithuania

Map Page 1130 **Area** 25,173 square mi (65,200 square km) **Population** 3,592,561 (2003) **Capital** Vilnius **Highest Point** Juozapine/Kalnas 954 ft (292 m) **Lowest Point** 0 m **GDP per capita** $8,400 **Primary Natural Resources** peat, arable land.

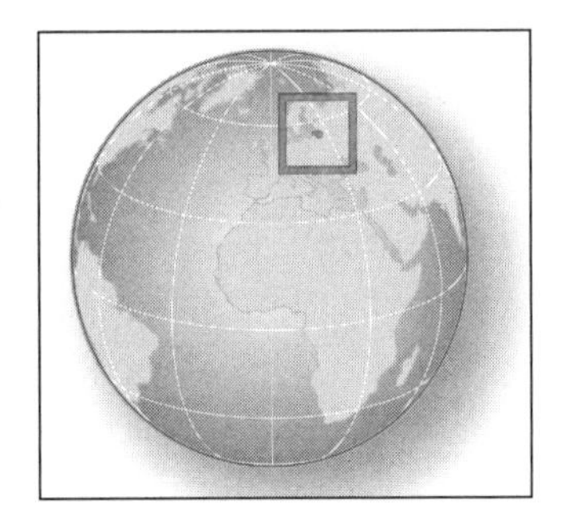

THE REPUBLIC OF Lithuania in northern Europe is a lowland country that borders LATVIA to the north, BELARUS to the east and southeast, POLAND and the Russian enclave of Kaliningrad to the southwest, and the Baltic Sea to the west. Lithuania is a parliamentary democracy with the supreme council or Seimas serving as the legislature and the president serving as head of state. Its major cities are Vilnius, Kaunas, and Klaipedia.

The Lithuanian countryside consists of lowlands and small hills and is dotted by 3,000 lakes. The climate is generally humid, with peak rainfall in August. Temperatures range from 23 degrees F (-5 degrees C) in January to 63 degrees F (17 degrees C) in June. The chief river is the Nemunas, which flows to the Baltic Sea.

Permanent human settlements in what is now Lithuania date to about 8000 B.C.E. In the 13th century, the Teutonic Knights were sent to Lithuania to convert the pagan population to Christianity. In 1385, Grand Duke Jogaila married Queen Jadwiga of POLAND, which began the Christianization of Lithuania and its special relationship with Poland. In 1569, the Treaty of Lublin joined the Grand Duchy of Lithuania and the Kingdom of Poland into a Polish-Lithuanian Commonwealth to balance the growing power of Moscow. Thereafter, Lithuania joined Poland's decline, ultimately culminating in the partitions by AUSTRIA, Prussia, and RUSSIA between 1772 and 1795.

Lithuania was incorporated into the Russian Empire after the final Polish partition in 1795. The population was subjected to harsh policies forbidding use of the Lithuanian language, resulting in revolts in 1830 and 1863. In 1918, Lithuania declared its independence amid the destruction and chaos of World War I and the Russian Revolution. In August 1939, the Molotov-Ribbentrop Pact split Poland and the Baltic States between the Soviet Union and Nazi Germany. In 1941, the Nazis invaded Lithuania, only to be reoccupied by the Soviet Union. After World War II, Lithuania was incorporated into the Union of Soviet Socialist Republics.

In 1991, Lithuania regained its independence with the other Baltic States. The last Soviet troops withdrew from Lithuanian soil in 1993. Like its Baltic neighbors, Lithuania has taken steps to integrate into Europe by joining the NORTH ATLANTIC TREATY ORGANIZATION (NATO) in 2003 and the EUROPEAN UNION in 2004.

Lithuanians make up 80 percent of the population, with the remainder consisting of Russians, Poles, and Belarussians. Roman Catholicism is the religion practiced by the majority of Lithuanians. After the demise of the Soviet Union, the state of Russians residing since the Soviet era poses a challenge for Lithuania.

The Lithuanian economy has had to transform from a command economy under the control of Moscow to a free market economy. This process has been successful in bringing prosperity but has also created inequities in the standard of living. Lithuania's chief exports are meat, milk, dairy products, and television parts, while remaining dependent on oil and natural gas from Russia.

BIBLIOGRAPHY. Thomas Lane, *Lithuania: Stepping Forward* (Routledge, 2001); Kevin O'Connor, *The History of the Baltic States* (Greenwood Press, 2003); *World Factbook* (CIA, 2004).

DINO E. BUENVIAJE
UNIVERSITY OF CALIFORNIA, RIVERSIDE

littoral

THE WORD *LITTORAL* comes from the Latin root *littus*, or "seashore." The littoral zone of a lake or ocean refers to the shallow waters closest to shore. In lakes, this is the zone dropping from the shoreline to roughly 10 ft (3 m) deep where there is enough sunlight for rooted plants to exist. Only in lakes with strong wave action will the littoral zone have algae instead of rooted plants.

Since the littoral zone is the interface between the lake and the surrounding watershed, it receives and accumulates sediment and nutrients that can support a wide variety of plants and animals. Plants include emergent wetland vegetation to submergent plants that may or may not reach the lake surface. This vegetation provides nutrients and habitat for fishes, birds, amphibians, invertebrates, and zooplankton.

Because of tides and wave action, the littoral zone of an ocean is subdivided into many parts. The supralittoral zone (spray zone) is found above the high tide mark up to the point that ocean spray cannot reach. This area usually receives only ocean spray, except for very high tides or storm surges that can inundate it. The intertidal littoral zone is bordered by the high and low tide marks. The sublittoral zone starts at the low tide mark and goes out to roughly 650 ft (200 m) deep, which is the average depth of the edge of the continental shelf. Numerous species live in these zones,

each adapted to the abundance and presence of water they receive each day.

Littoral zones are the interface between land and water, and so they are very productive, meaning they support many plants and animals. However, these same areas are also easily altered by human development or pollution. For example, more people in the UNITED STATES are building homes on the remaining edges of lakes and rivers. Disturbing the soil to grade the lot and build the house causes erosion, which moves sediment into the water. Often, people will dump a load of sand on the littoral zone so that they don't have to walk through plants to go swimming and so it will look neater. Piers built off of these lots can also damage the plants in the littoral zone. As houses age, septic systems start to fail and this nutrient-rich pollution moves laterally into the water, promoting eutrophication. Polluted runoff from driveways and lawns can also impair the waterways.

Coastlines in every country have been feeling additional pressures from growing cities. The more human activity near a coastline results in more damage to the littoral zone. In wealthier areas, beach replenishment (dumping new sand on the beach) buries the current plants and animals, and the building of seawalls and groins to protect one beach will disturb currents to more seriously erode the next unprotected beach downcurrent. The major problem, however, is pollution. Sewer pipes and polluted rivers both dump myriad pollutants into the littoral and sublittoral zones, damaging the life forms there.

Sometimes people hear about this type of pollution when local health departments close down shellfish beds after sewage pollution has been high (because of storm sewer overflows or treatment plant malfunctions). Similarly, each year more beachgoers arrive to find beaches closed where water quality monitoring has detected pathogens, particularly after rain events. Surfers on the west coast of the UNITED STATES have long complained of intestinal sickness and skin sores after spending time in the water near sewage and creek outfalls. There is increasing pressure on coastal cities to repair and upgrade their aging treatment facilities to solve this problem.

Other laws now prevent ocean liners from dumping their sewage (treated or not) near shore, and only to dump treated sewage many miles out to sea. In some countries, laws require that ships not dump their waste oil or oily bilgewater close to shore or that only double-hulled oil tankers may enter port. Degrading littoral zones, particularly along oceans, damages habitat that marine and shorebird animals need, thus reducing this common resource both regionally and globally. The LAW OF THE SEA expects that signatories respect and protect these areas, but oftentimes little or no protection exists.

BIBLIOGRAPHY. Harm J. de Blij and Peter O. Muller, *Geography: Realms, Regions and Concepts* (Wiley, 2002); Arthur N. Strahler and Arthur H. Strahler, *Physical Geography: Science and Systems of the Human Environment* (Wiley, 2005). John E. Oliver and John J. Hidore, *Climatology: An Atmospheric Science* (Prentice Hall, 2002).

LAUREL E. PHOENIX
UNIVERSITY OF WISCONSIN, GREEN BAY

llanos

ALSO KNOWN AS the Orinoco River plains, the llanos of COLOMBIA and VENEZUELA are the northernmost section of the central lowlands of South America. The llanos are formed by a large geologic depression that once was an arm of the sea. This location is an alluvial basin between the northern ANDES MOUNTAINS and the Guyana Highlands. The region is generally very level, with a largely featureless landscape. Most of the llanos is tropical grassland, with some woodlands along the rivers.

The Orinoco River borders the Venezuelan llanos to the south. Shallow depths, rapids, and waterfalls limit travel along the river. In the 1950s, Venezuela dredged the river, opening it to larger ships that could reach Ciudad Guayana. The Guaviare River, a tributary of the Orinoco, marks the southern border of the Colombian llanos.

The lives of inhabitants of the llanos are regulated by rainfall that swell the region's rivers and flood the plains. The rainy season is from April to November, with June through August generally the rainiest months. During this rainy season, the region's grasslands turn green and grow quickly. During the dry season from November to March, rivers and lakes diminish, trees shed their leaves, and vegetation is covered with dust. Early in the dry season, stagnant pools remain, which traditionally have been breeding grounds for disease-carrying mosquitoes. Later, even these pools dry up completely, leaving a hard, sunbaked landscape. Grazing land is often scarce during the dry season and animals are sometimes short of

water. This drought-flood cycle has traditionally limited farming in the llanos. There have been some attempts at flood control and irrigation, such as along the Guarico River in Venezuela.

After the Spanish conquest of the coastal regions of South America in the 16th century, many Native Americans fled to the llanos. Soon, a llanero culture emerged that mixed native, Spanish, and African traditions. The economy came to depend on horses and cattle, as the Spaniards released these Old World animals into the wild. The animals roamed the plains with only limited supervision. Before the 19th century, there was no systematic animal husbandry. The semi-nomadic llaneros drove the animals to upland basins near the coastal and Andean cities for slaughter. Because the llanos were isolated, most of the meat was for local consumption. Hides were for both local consumption and export. Since the livestock was not part of international trade, there was little scientific breeding until the mid-20th century.

The llanos played an important role during the Spanish colonies' wars of independence in the early 19th century. Llaneros made excellent cavalrymen. They generally fought more for concrete material goals rather than abstract political ideals. At first, they fought for the Royalist side from 1813 to 1814. Later, many plainsmen joined with the llanero Patriot José Antonio Páez, who had joined with Simón Bolívar, the most important independence leader.

The llanos continued to be a largely pastoral region after Colombia and Venezuela achieved independence. Low population and limited economic activity predominated until World War II. In the postwar period, oil and gas deposits near El Tigre, Venezuela in the eastern llanos changed the region's economic focus. Also, Ciudad Guayana along the Orinoco River at the southern edge of the llanos grew into an important industrial center of more than 500,000 people. An example of growth pole industrialization popular in Latin America during the mid-20th century, Ciudad Guayana has become one of South America's leading steel, aluminum, and heavy manufacturing centers.

BIBLIOGRAPHY. Brian Blouet and Olwyn Blouet, *Latin America: A Systematic and Regional Survey* (Wiley, 2004); John Lombardi, *Venezuela* (Oxford University Press, 1982); Frank Safford and Marco Palacios, *Colombia: Fragmented Land, Divided Society* (Oxford University Press, 2002).

RONALD YOUNG
GEORGIA SOUTHERN UNIVERSITY

loess

LOESS (PRONOUNCED "LUSS") is a very fine light soil, often buff, yellow, or gray in color. The word stems from the German word *löss* or "loose," and loess is generally easily eroded by water or blown by wind. Major loess deposits are found in Shaanxi, CHINA, in some of the U.S. plains states, notably western IOWA and western NEBRASKa, and some parts of Europe, such as central BELGIUM. Loess can mix with other soil types, yielding categories such as loess, sandy loess, and clayey loess

Recent scientific studies suggest that many of the world's loess deposits previously believed to have been transported by water were actually wind-blown (eolian) in origin. In North America, the process began during the last Ice Age, as glaciers ground rock into a fine, flour-like sediment. As temperatures rose and the glaciers melted, the "flour" was deposited on flatter terrain, where it was later blown to its current locations by the wind.

China's loess deposits are the world's thickest and most ancient. Rather than glacial dust, most of these deposits of loess soil are believed to have originated in the vast deserts of Central Asia, and to have been deposited by the wind on China's Loess Plateau, an area the size of FRANCE centered on Shaanxi Province in north central China. China's Loess Plateau has an average thickness of 492 ft (150 m), and reaching 1,082 ft (330 m) near Lanzhou on the western edge of the plateau. China's second-longest river, the HUANG (Yellow) has its headwaters on the plateau and takes its name from the color of the loess sediment it carries downstream. Some of these deposits have been dated to more than 2 million years old. Loess deposits in the Yellow River valley have historically been easily tilled with simple wooden tools, and archaeologists suggest this relates to why the birthplace of Chinese civilization is found in these areas, centered upon the ancient capitals near Xian and Loyang.

Loess also gives the Yellow River its characteristic color, and its reputation as the "sorrow of China." As much as 1,600 tons of loess a year is carried downstream by the Yellow River, which has silt content measuring as high as 98 times that of the muddy MISSISSIPPI. When the river slows on the floodplains of central and coastal China, the loess is redeposited. Since the riverbank in these heavily populated areas is much higher than the surrounding land, disastrous floods have been a common occurrence throughout Chinese history.

Today, China's government is struggling not only to control floods downstream, but also to deal with some of the world's most severe soil erosion on the Loess Plateau, home to a population of about 40 million.

BIBLIOGRAPHY. D. Derbyshire, ed., *Loess and Palasols: Characteristics, Stratigraphy, Chronology, and Climate* (Pergamon, 2001); D.N. Eden and R.J. Furkert, eds., *Loess: Its Distribution, Geology, and Soils* (A.A. Balkema, 1988); Liu Tungsheng, ed., *Loess, Environment, and Global Change* (Sciences Press, 1991); A.L. Lugn, *The Origin and Sources of Loess* (University of Nebraska Studies, 1962); Cornelia F. Mutel and Mary Swander, eds., *Land of the Fragile Giants: Landscapes, Environments, and Peoples of the Loess Hills* (University of Iowa Press, 1994).

LAWRENCE FOURAKER, PH.D.
ST. JOHN FISHER COLLEGE

Loire River

TO MANY PEOPLE, the Loire River defines FRANCE. It is the country's largest river and draws together its different regions, east and west, north and south. The Renaissance chateaus along its lower course are the second-largest tourist draw after Paris, and the river provides water and transportation for the nation's agricultural center.

From its origins in the remote peaks of the Cévennes in southeastern France to its wide estuary on the Bay of Biscay, the river crosses through, or forms the border of, 13 departments and crosses six historic regions, from Languedoc to Brittany. Seven cities with populations greater than 100,000 are included in its watershed (Orléans, Tours, Angers, and Nantes on the river directly, and Clermont-Ferrand, Limoges, and Le Mans on tributaries). The Loire flows for 627 mi (1,011 km), draining a watershed of 44,956 square mi (115,271 square km), one-fifth of France.

It is the last "wild river" of Europe, with no dams on its main course, and several of those that have been built on tributaries were recently destroyed under a government initiative, Plan Loire Grandeur Nature. This program has set aside funding for 2000 through 2006 to secure the river basin from floods, to regulate the production of nuclear and hydroelectric power, and to restore the region's natural habitat, notably its spawning grounds for Atlantic salmon, whose numbers have decreased over the last century from hundreds of thousands to mere hundreds.

Recognizing the importance of the lower valley's natural and historic beauty to the nation's economy, the government also heavily supports development funds for historic towns such as Blois, Chinon and Saumur and internationally famous castles such as Chambord and Chenonceaux. These efforts are reinforced by UNESCO (United Nations), which declared the Val de Loire a World Heritage Site in December 2000. Other projects are aimed to restore the environment of the Loire estuary between Nantes and its mouth at Saint-Nazaire, home to huge populations of freshwater and saltwater fish and migrating birds, which has been heavily polluted from harbor activities and large oil refineries on the north shore.

The Loire starts as a trickle from a small pipe on a volcanic peak (Gerbier-de-Jonc) in the Ardèche, 4,700 ft (1,425 m) above sea level, only a few meters from streams that find their way east into the RHÔNE, and to the Mediterranean. The swift mountain stream cuts deep gorges through the Massif Central, before passing the iron and coal centers of the 19th century at St.-Étienne ("the French Birmingham") and Roanne. From Roanne, the river opens up onto the plains of the north. Because the river is fed largely from mountain snows, it is liable to swift and disastrous floods in the spring.

Below the town of Nevers, the Loire is joined by its chief tributary, the Allier, and becomes the wide and voluptuously slow river that is famous among painters and poets for its reflective surfaces and rose mists. The river arcs gradually westward, passing the wine region of Sancerre and the first of the grand Renaissance chateaus, Sully-sur-Loire. Other large tributaries enter the river in its lower courses, the Maine from the north and the Vienne, Cher and Indre from the south. From here to the sea, the Loire is the major transport highway for wine, vinegar, grain, salt, timber, stone, coal, and iron, though it has always been hampered by slow currents and a shallow riverbed with frequent and shifting sandbars.

Islands in the lower Loire valley disappear and reappear overnight, causing serious dangers to transport craft. Boats take grain—the region produces 50 percent of France's total—either downriver, or to canals connecting the Loire to the Saône-Rhône and Seine river valleys.

Although the river passes several large and historically potent cities—for example, Orléans, where Joan of Arc first defeated the English in 1429—the country-

side is mostly rural, with numerous small villages. One in three "Ligerians" live in the countryside. The Loire Valley ends at Nantes, before opening up as the Loire estuary, with 35 mi (56 km) of rich wetlands. Nantes was traditionally France's gateway to the world's oceans (before being superseded by the deeper, more predictable waters of the Seine at Le Havre), and many French residents of North America trace their ancestry to emigrants from its ports.

BIBLIOGRAPHY. Piers Paul Read, "The Danube," *Great Rivers of the World*, A. Frater, ed. (Little, Brown, 1984); C. Revenga, S. Murray, et al.,*Watersheds of the World* (World Resources Institute, 1998); "Loire," www.rivernet.org (April 2004); Plan Loire Grandeur Nature, www.eau-loire-bretagne.fr (April 2004).

JONATHAN SPANGLER
SMITHSONIAN INSTITUTION

London

LOCATED ON THE Thames River in southwestern England in the British Isles, London is the capital of England and the UNITED KINGDOM. The history of London can be traced back nearly 2,000 years to its founding as Londinium in 50 C.E. by the Romans. Much debate has occurred as to the exact type of Roman settlement that originated at the site, civilian or military. Archaeological evidence points to the original settlement starting as a civilian effort. For the next 400 years, the Romans controlled the strategic site on the edge of the Thames but eventually abandoned it.

Later, the Saxons established Lundenwic to the west of what would become the walled City of London in the 7th century. The walled City of London came into prominence during the Norman control of the area beginning in the 10th century C.E. It was at this time that present-day London began to take shape. Norman control of London continued for nearly 700 years, during which time a significant landmark, the Tower of London, was constructed. Stuart control of the city and the whole of England saw London's importance continue to grow despite some major disasters such as the Great London Fire of 1666 and the previous year's plague, which wiped out a large portion of the city's population.

The 19th century saw London obtain the major global city status that it still enjoys today. The population of the city rose from 1 million at the turn of the 19th century to over 6 million at the turn of the next. The city was the largest city in the world during this period, the capital of the BRITISH EMPIRE, and the global leader in politics, finance, and business. This period in London's history is also marked by extreme social polarization with millions of the city's inhabitants living in extreme poverty and appalling slums in the inner-city areas. Numerous landmarks were constructed during this century, including Big Ben, the Houses of Parliament, Trafalgar Square, and the Tower Bridge.

One of the biggest changes to London occurred in the 19th century: the introduction of the railroad, with the first line being opened in 1836 connecting Greenwich and London Bridge. Soon after, a large number of rail stations were constructed linking the city to the rest of the British HINTERLAND. In 1850, the London Underground was opened and soon the outflux of those who could afford to move to the open spaces of the periphery of London left the inner city residents to combat extreme poverty and disease.

POPULATION PEAK

The population of London peaked in the 1930s around 8.6 million. The city was heavily damaged by German bombings during World War II. Over 35,000 Londoners were killed during the Blitz and over 10,000 buildings in the city were destroyed. The city was rebuilt after the war and continued to expand, consuming the surrounding landscape. The city added to its reputation as one of the world leaders in finance and banking by becoming a center of Western cultural and fashion change in the 1960s, led by such musical artists as the Rolling Stones and the Beatles. This cultural leadership continued into the 1970s and 1980s as the city was the epicenter for the punk and new wave movements.

While the population of the city itself dropped to around 7.2 million, the larger London metropolitan area has a population estimated at nearly 14 million, making it the largest metro area in Europe. London's diverse ethnic makeup is the result of the city's role as the capital of the former British Empire that spanned the globe.

Seventy-one percent of the population consider themselves white, 10 percent Indian, Bangladeshi, or Pakistani, 5 percent black African and 5 percent black Caribbean. Over 300 languages are spoken, and the 2001 census shows that 29 percent of London's population belonged to a minority ethnic group.

London is a leading world city when it comes to banking, finance, and insurance and one of the leaders

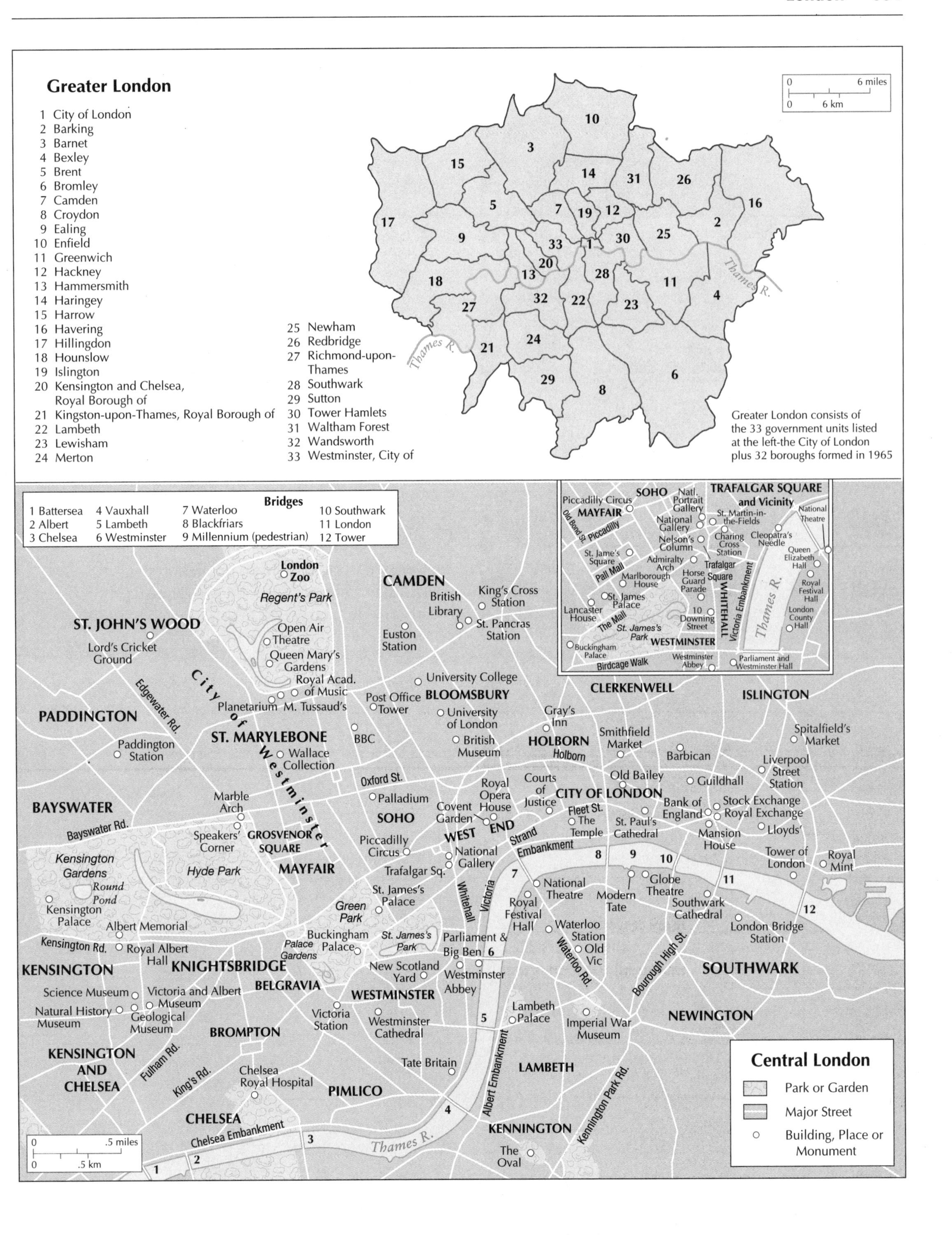

Greater London
1 City of London
2 Barking
3 Barnet
4 Bexley
5 Brent
6 Bromley
7 Camden
8 Croydon
9 Ealing
10 Enfield
11 Greenwich
12 Hackney
13 Hammersmith
14 Haringey
15 Harrow
16 Havering
17 Hillingdon
18 Hounslow
19 Islington
20 Kensington and Chelsea, Royal Borough of
21 Kingston-upon-Thames, Royal Borough of
22 Lambeth
23 Lewisham
24 Merton
25 Newham
26 Redbridge
27 Richmond-upon-Thames
28 Southwark
29 Sutton
30 Tower Hamlets
31 Waltham Forest
32 Wandsworth
33 Westminster, City of
Thames R.
0 6 miles
0 6 km
Greater London consists of the 33 government units listed at the left-the City of London plus 32 boroughs formed in 1965
Bridges
1 Battersea
2 Albert
3 Chelsea
4 Vauxhall
5 Lambeth
6 Westminster
7 Waterloo
8 Blackfriars
9 Millennium (pedestrian)
10 Southwark
11 London
12 Tower
TRAFALGAR SQUARE and Vicinity
SOHO
MAYFAIR
Piccadilly Circus
Natl. Portrait Gallery
National Gallery
St. Martin-in-the-Fields
National Theatre
Nelson's Column
Charing Cross Station
Cleopatra's Needle
Queen Elizabeth Hall
Royal Festival Hall
London County Hall
St. James's Square
Admiralty Arch
Trafalgar Square
Marlborough House
Horse Guard Parade
St. James Palace
10 Downing Street
Lancaster House
The Mall
St. James's Park
WESTMINSTER
WHITEHALL
Victoria Embankment
Buckingham Palace
Birdcage Walk
Westminster Abbey
Parliament and Westminster Hall
London Zoo
Regent's Park
CAMDEN
British Library
King's Cross Station
St. Pancras Station
Euston Station
ST. JOHN'S WOOD
Lord's Cricket Ground
Open Air Theatre
Queen Mary's Gardens
Royal Acad. of Music
Planetarium
M. Tussaud's
City of Westminster
University College
BLOOMSBURY
Post Office Tower
University of London
British Museum
CLERKENWELL
ISLINGTON
PADDINGTON
Edgewater Rd.
ST. MARYLEBONE
BBC
Gray's Inn
HOLBORN
Holborn
Smithfield Market
Barbican
Spitalfield's Market
Paddington Station
Wallace Collection
Liverpool Street Station
Oxford St.
Old Bailey
Guildhall
Royal Opera House
Courts of Justice
CITY OF LONDON
BAYSWATER
Marble Arch
Palladium
SOHO
Covent Garden
Bank of England
Stock Exchange
Royal Exchange
Fleet St.
The Temple
St. Paul's Cathedral
Lloyds'
Mansion House
Bayswater Rd.
Speakers' Corner
GROSVENOR SQUARE
Piccadilly Circus
WEST END
Strand
Embankment
Tower of London
Royal Mint
Kensington Gardens
Hyde Park
MAYFAIR
National Gallery
Trafalgar Sq.
Round Pond
Kensington Palace
St. James's Palace
Whitehall
Victoria
National Theatre
Modern Tate
Globe Theatre
Southwark Cathedral
Green Park
Royal Festival Hall
London Bridge Station
Albert Memorial
Buckingham Palace
Palace Gardens
St. James's Park
Parliament & Big Ben
Waterloo Station
Old Vic
Waterloo Rd.
Borough High St.
Kensington Rd.
Royal Albert Hall
KENSINGTON
KNIGHTSBRIDGE
New Scotland Yard
Westminster Abbey
SOUTHWARK
BELGRAVIA
Science Museum
Victoria and Albert Museum
WESTMINSTER
Natural History Museum
Geological Museum
Victoria Station
Westminster Cathedral
Lambeth Palace
Imperial War Museum
NEWINGTON
BROMPTON
KENSINGTON AND CHELSEA
Fulham Rd.
King's Rd.
Chelsea Royal Hospital
Tate Britain
Albert Embankment
LAMBETH
Kennington Park Rd.
PIMLICO
CHELSEA
Chelsea Embankment
KENNINGTON
Thames R.
The Oval
0 .5 miles
0 .5 km
Central London
Park or Garden
Major Street
Building, Place or Monument

The history of the global city of London can be traced back nearly 2,000 years to its founding as Londinium.

in business. The city is host to 463 foreign banks, 56 percent of the global foreign equity market, and 429 foreign companies listed on the London Stock Exchange. Seventy-five percent of Fortune 500 companies have London offices. The city is the world's leading market for international insurance, with the worldwide premium income reaching over £150 billion in 2001.

Transport plays a vital role in the success of London not only internationally (almost £1 billion in overseas earnings are generated by the maritime industry) but locally as well. Heathrow International Airport is the one of the world's busiest airports serving over 63 million passengers, 90 airlines, and approximately 170 destinations. No fewer than four other international airports also serve the metropolitan area. The London Underground, the world's first underground rail network, provides transportation for a large number of Londoners. The system serves over 3 million daily, and nearly 1 billion annually, on its 253 mi (408 km) of track. Congestion on the city's streets and motorways has become so notorious that in 2003 London implemented a £5 per day fee for driving private automobiles in the central area during weekdays, as a way to reduce traffic congestion.

The city is home to five major symphonies, over a dozen major theaters, and numerous art galleries including the National Galleries and the Tate Galleries. A large number of world-famous museums have their homes in London, including the British Museum, the Science Museum, and the Victoria and Albert Museum. Churches and cathedrals are also part of London's cultural heritage, including Westminster Abbey and St. Paul's Cathedral.

BIBLIOGRAPHY. *Oxford Essential Geographical Dictionary* (Oxford University Press, 2003); "UK Statistics," www.statistics.gov.uk (April 2004); "What Is London?" www.cclondon.com (April 2004).

TIMOTHY M. VOWLES, PH.D.
VICTORIA UNIVERSITY, NEW ZEALAND

Lop Nor

ALSO SPELLED Lob-Nor, Lop Nur, and Lop Nuur, this name invites confusion as it has been applied to a lake (actually two), a village, and a region in eastern XINJIANG province (CHINA). Lop is also the name of a "great desert" and a "large town" that Marco Polo mentioned and Faxian described. The city of Lop was well located at the center of a network of commercial roads leading to China proper, Tibet, and Turkestan. Caravans used to rest for one week in Lop before they undertook a one-month-long crossing of the desert. The SILK ROAD merchants sought to avoid the shorter but hazardous route through Lop. Only the bones of those who died during sandstorms helped travelers orient themselves though the immense plains of the Lop desert.

Some 217 mi (350 km) away from the Lop Nor Lake, Chinese maps show a city named Lop Nur (or Yuli) on the Konqi River that feeds the lake. "Lop"

designates also a village in Kara Buran, the southern basin of Lop Nor Lake. Curiously, the Mongolian-sounding name "Lop Nor" is not in use among the Lopliks, the Turkish people who inhabit the region. They employ the term *Lop* for the area that stretches from the Ughen Daria River to the city of Charkhlik. The Russian geographer Prschevalskij would have applied the name of the region to the lake (then desiccated) by mistake. Transcribed Luobu bo in Chinese, Lop Nor Lake used to be one of the terminal lakes of the Tarim River.

The Swedish explorer Sven Hedin called Lop Nor a "wandering lake" because the fluctuations of the Tarim have created two terminal lakes that are alternatively full and empty. When the river flows east, the Lop Nor fills up; when the river flows south, another terminal lake, the Kara Koshun, receives its waters while the Lop Nor dwindles. Dam and irrigation projects have severely altered the water supply of the Tarim-Konqi system to the entire lake area during the 20th century. Between 1921 and 1952, Lop Nor Lake covered 926 square mi (2,400 square km). The lake has been totally dry since 1964. All the other terminal lakes disappeared after 1972, when the construction of a large reservoir near Tikanlik, China, was completed.

BIBLIOGRAPHY. Christoph Baumer, *Southern Silk Road: In the footsteps of Sir Aurel Stein and Sven Hedin* (White Orchid Books, 2000); Sven Hedin, *The Wandering Lake* (Dutton, 1940); Nils Horner and Parker Chen, "Alternating Lakes," *Geografiska annaler* (Supplement to v.17, 1935); Zhao Songqiao and Xia Xuncheng, "Evolution of the Lop desert and the Lop Nor," *The Geographical Journal* (v.150/3, 1984).

PHILIPPE FORÊT, PH.D.
FEDERAL INSTITUTE OF TECHNOLOGY, SWITZERLAND

Los Angeles

LOS ANGELES, CALIFORNIA, is the second most populous city in the UNITED STATES with an estimated 2002 population of approximately 3.8 million. Los Angeles (also known as L.A.) is the principal city of a metropolitan region with a population of over 12 million.

Located in the Los Angeles basin, the city is 470 square mi (1,215 square km). The extreme north-south distance is 44 mi (71 km), the extreme east-west distance is 29 mi (46.5 km), and the length of the city boundary is 342 mi (550 km).

Los Angeles, America's western cultural geographic capital, is mostly at sea level elevation or a few feet above.

The San Gabriel, Santa Monica, and Santa Ana Mountains surround the basin. The city is bordered to the west and southwest by the PACIFIC OCEAN. Earthquakes are probably the most prevalent natural hazard that the residents of Los Angeles face. It is estimated that there are nearly 30 magnitude 2.0 and higher earthquakes in southern CALIFORNIA everyday. The last significant earthquake to strike the Los Angeles area was the 1994 Northridge quake, measuring 6.7 on the Richter scale. It claimed 61 lives and caused over $20 billion in property damage.

The city's name is attributed to the Spanish California's military governor, Felipe de Neve, who founded a new settlement and named it El Pueblo de la Reyna de los Angeles (the Town of the Queen of the Angels). The name was eventually shortened to Los Angeles. Founded in 1781, Los Angeles was an outpost under a succession of Spanish, Mexican, and American rule. The city became incorporated in 1850 and got a boost following its linkage by rail with San Francisco in 1876. The city was selected as the southern California rail terminus because its natural harbor seemed to offer little challenge to San Francisco, home of the railroad barons. After the introduction of the rail link, the city began to steadily grow in population.

The Mediterranean climate and vast open spaces mixed with agricultural and oil production, the creation of a deep water port, and the completion of the city-financed Owens Valley Aqueduct to provide addi-

tional water, all contributed to Los Angeles's continued growth.

The population swelled to 1.5 million by 1940. Following World War II, Los Angeles became the focus of a new wave of migration, with its population reaching 2.4 million by 1960. The population topped 3 million during the 1980s. After the 2000 census, the foreign-born population was 1.5 million, which was 41 percent of the city's total population. Racially, the city is a melting pot, in the 2000 census, 47 percent of the population identified themselves as white, 11 percent black, and 10 percent Asian. The white population percentage numbers include all those individuals who identified themselves as Hispanic or Latino. One in three residents in the metropolitan area was born outside the UNITED STATES.

ECONOMIC DIVERSITY

Los Angeles's growth continues, anchored by its economic diversity. Services, wholesale and retail trade, manufacturing, government, financial service industries, transportation, utilities, and construction contribute significantly to local employment. Los Angeles County is the top-ranked county in manufacturing in the United States, producing more than 10 percent of the nation's production of numerous items ranging from aircraft, aircraft equipment guided missiles, and space vehicles, to television shows, movies, games, toys, and clothing.

L.A. GATEWAY

The Los Angeles area is also a transportation gateway into the United States for both passengers and cargo. Fueled by trade with the Pacific Rim countries, the ports of Los Angeles and Long Beach combined rank first in the nation in volume. Los Angeles International Airport is served by 68 different passenger carriers and serves over 30 million passengers annually. The airport is the fifth-busiest cargo airport in the world, handling more than 1.96 million tons of origination and destination air cargo in 2002. The private automobile plays an important part of the Los Angles lifestyle. Los Angeles County has 527 mi (848 km) of freeway and 382 mi (614 km) of conventional highway. Altogether, there are over 21,000 mi (33,796 km) of public roads in the county. On an average day, 92 million vehicle miles are driven in Los Angeles County.

As home to the film, television, and recording industries, as well as important cultural facilities, Los Angeles serves as a principal global cultural center. The area boasts 83 radio stations. Twenty of the stations are in Spanish and nine others are in languages other than English including, Chinese, Korean, and Iranian. Twenty-six daily and 86 monthly/weekly newspapers are published within Los Angeles. Among them is the only Danish weekly newspaper published in the United States and a large array of papers published in other foreign languages.

The Los Angeles area is home to 29 television stations. Included among these stations is KSCI Channel 18, which is Los Angeles County's most linguistically diverse television station and the leading Asian language television station in the United States. It provides programming and entertainment in 14 languages, including Arabic, Armenian, Khmer, Cantonese-Chinese, English, Farsi, Hindi, Hebrew, Japanese, Korean, Mandarin-Chinese, Tagalog, Urdu, and Vietnamese. Los Angeles, like its sister city NEW YORK, is a multicultural metropolis.

BIBLIOGRAPHY. *Oxford Essential Geographic Dictionary* (Oxford University Press, 2003); "Los Angeles," www.losangelesalmanac.com (March 2004); "Los Angeles," www.lacity.org (March 2004).

TIMOTHY M. VOWLES, PH.D.
VICTORIA UNIVERSITY, NEW ZEALAND

Louisiana

SITUATED AT THE MOUTH of the MISSISSIPPI RIVER on the coast of the Gulf of Mexico, Louisiana and its people have long been influenced by the intersection of these two major water features. The physical geography of Louisiana can be examined in terms of its five natural regions: the Coastal Marsh; the Mississippi Flood Plain; the Red River Valley; the Terraces; and the Hills. Each has played an important role in the history of the state. The southernmost of these regions is the Coastal Marsh, which serves as a transitional area between land and sea. Characterized by fresh and salt-marsh vegetation as well as peat soils, this natural region serves to provide rich fishing grounds that support the second-largest seafood industry in the UNITED STATES today.

Heading northward away from the coastal areas, rivers dominate not only the physical landscape but also the economic geography of Louisiana. The Red and Mississippi rivers are distinguished as separate natural regions because of soil and drainage differences,

however, they are inherently interconnected and provide an extensive waterway transportation system upon which the Louisiana economy has been built. Combined with five deepwater ports and proximity of the Gulf of Mexico, the river system of Louisiana serves as natural gateway for the exchange of not only goods produced in the state but much of the U.S. midwest as well. While the petrochemical and mineral resource industries are most often associated with the importance of waterway transportation today, the fertile flood plains of the Mississippi River have long been a source of agricultural wealth for the state by producing large quantities of cotton, soybeans, and rice. Rising above the river floodplains into the Terraces and Hills, the forest industry prevails because of the availability over 13 million acres (5.2 million hectares) of hardwood and pine forests.

Natural resources alone do not provide all of the state's economic foundations. Louisiana has also developed a strong tourism industry premised upon its unique cultural heritage. The idea of "cultural gumbo" is often used to describe the people of Louisiana, as they are like the famous gumbo dish created out of many separate ingredients that blend together to create a delightful experience. Throughout its history, the territory that now encompasses Louisiana has been governed under 10 different flags. Although originally claimed by Hernando de Soto for SPAIN in the early 1540s, Louisiana remained largely ignored by Europeans until Robert de La Salle claimed the territory for FRANCE in 1682. The first permanent settlement was finally established in 1714.

Despite the strong French and Spanish presence in the territory, other Europeans, including Germans farmers, began to arrive, each adding their own influences to the Louisiana culture and landscape. With a military victory over France and Spain in 1763, Great Britain also laid claim to the portion of Louisiana east of the Mississippi River basin. While European nations struggled with each other for control of Louisiana, other cultural groups continued to arrive. The rich soils of the river lands fostered the growth of a plantation economy that depended upon the importation of African and Afro-Caribbean slaves who contributed to the formation of the culture in south Louisiana. This area was also settled by Creoles, the French-speaking Acadians who fled British control of Nova Scotia and made their way into south-central Louisiana and are today recognized by the name Cajuns.

Even after the LOUISIANA PURCHASE in 1803 and its incorporation into the United States in 1812, portions of Louisiana would be governed under a foreign flag. In 1810, a controversy arose between the United States and Spain over the control of portions of eastern Louisiana, resulting in the declaration of the short-lived independent Republic of West Florida. Finally, in 1861, Louisiana seceded from the Union and after only a six-week period as an independent republic joined the Confederacy and its efforts in the Civil War. After the conclusion of the war, Louisiana was readmitted to the Union in 1868.

Despite the passage of over 130 years of continual U.S. control, the people of Louisiana have seldom forgotten their past and continue to draw upon it today to create a diverse cultural experience blended from a unique history and environment.

BIBLIOGRAPHY. Gwendolyn Midlo Hall, *Africans in Colonial Louisiana: The Development of Afro-Culture in the Eighteenth Century* (Louisiana State University Press, 1992); Fred B. Kniffen and Sam Bowers Hilliard, *Louisiana: Its Land and Life* (Louisiana State University Press, 1988); Peirce Lewis, *New Orleans: The Making of an Urban Landscape* (Ballinger Publishing Company, 1976); Cecyle Trepanier, "The Cajunization of French Louisiana: Forging a Regional Identity," *Geographical Journal* (v.157, 1991).

TONI ALEXANDER, PH.D.
KANSAS STATE UNIVERSITY

Louisiana Purchase

THE LOUISIANA PURCHASE (1803) included all of the present-day states of ARKANSAS, OKLAHOMA, MISSOURI, KANSAS, IOWA, and NEBRASKA, as well as parts of MINNESOTA, SOUTH DAKOTA, NORTH DAKOTA, MONTANA, WYOMING, COLORADO, NEW MEXICO, TEXAS, and, of course, LOUISIANA. The area is approximately one-third of the continental UNITED STATES. Initially, it included portions of CANADA—southern Manitoba, southern Saskatchewan, and southern Alberta—that drain into the Missouri River.

On April 30, 1803, for 60 million francs (approximately $15 million) under a treaty with FRANCE, the United States received the Louisiana Territory, land in excess of 800,000 square mi (2 million square km). The purchase incorporated territory from the MISSISSIPPI RIVER to the ROCKY MOUNTAINS. The treaty specified that France would receive $11,250,000 in cash, and for the remainder of the price, the United States would

assume French debts to American citizens. The territory had belonged to France until the end of the French and Indian War. In 1762, France ceded it to SPAIN. Spain gave it back in 1800 under the secret Treaty of San Ildefonso. The French ruler, Napoleon Bonaparte, wanted to re-establish French presence in the Americas, and the Mississippi Valley was to serve as the center of trade and food production for Hispaniola, the Caribbean heart of the new empire. Haitian slaves led by Toussaint L'Ouverture upset this dream in 1801 when they rose against their masters and seized control of the country. Napoleon's efforts to suppress the rebellion failed, in part because of yellow fever among the French troops, and Napoleon abandoned his dream of a western empire. Napoleon needed his troops for his anticipated war with Great Britain. He also needed money for his European adventures, so he decided to sell Louisiana.

The U.S. president, Thomas Jefferson, had negotiators in Paris, trying to get a tract on the lower Mississippi or the right of free navigation through New Orleans. American vulnerability in New Orleans had become apparent in October 1802 when the Spanish intendant at New Orleans closed the port by suspending American right of deposit, prohibiting Americans from storing their cargoes in the city. Westerners were concerned, to put it mildly, and the giving of Louisiana to France did not calm them at all.

Jefferson responded by instructing his minister in Paris to buy land on the lower Mississippi to serve as an alternative port. James Monroe went over in early 1803 as minister plenipotentiary to assist in the negotiation. Monroe had authority to offer $10 million for New Orleans and West Florida. Napoleon had already decided to sell before Monroe arrived. When Napoleon offered all of Louisiana, the American negotiators quickly arranged the deal. The purchase more than doubled the size of the United States, provided land for settlement, and guaranteed free navigation on the Mississippi. Jefferson had concerns that the U.S. Constitution didn't authorize the acquisition of new territory by treaty, but he decided that the good to the nation outweighed his philosophical concerns about violating the Constitution. The Senate ratified the treaty on October 20, 1803. On November 30, the Spanish, who had remained in occupancy during the French ownership, yielded Louisiana to France, which in turn ceded it to the United States on December 20.

Louisiana, in 1803, had a population of about 50,000. Ten thousand lived in New Orleans. Over half of the population, 28,000 people, were slaves. The count included residents of British West Florida (given to Spain after the American Revolution) and transplanted Acadians by way of NEW YORK and St. Domingue.

Spanish immigrants from the CANARY ISLANDS reflected the attraction of Spain's liberal land policies in Louisiana. Approximately 10,000 settlers—master and slave—fled the slave revolt in St. Domingue. And some French left France because of the nature of Napoleon's regime. At the same time, Spanish officials and settlers left for Spain. There were free blacks in New Orleans and Natchitoches. Anglo immigration into the area had begun over a decade before the purchase. Spain disapproved of the sale, but Congressional approval forestalled any action. Jefferson appointed a territorial governor, William Charles Cole Claiborne, and sent him along with General James Wilkinson to take possession, which they did on the December date. A formal ceremony on March 10, 1804, completed the transfer from France to the United States.

Not everyone applauded the deal. The Federalists, who wanted the United States to support Britain instead of France, claimed that the purchase was unconstitutional. In their view, the United States had spent a lot of money simply to declare war on Spain. Furthermore, the new land had the potential to shift the balance of political power away from the Atlantic coast to the west. Eastern bankers merchants and western farmers often had clashing interests anyway. Timothy Pickering, senator from Massachusetts, led a Federalist cabal that sought to separate New England into a separate confederacy, with Vice President Aaron Burr as president. Alexander Hamilton intervened, prevented secession, and eventually died at the hand of Burr in an 1804 duel.

LEWIS AND CLARK

Jefferson wasted no time in finding out what he had bought. From the time that Jefferson was secretary of state, he had been interested in the Spanish territory blocking the United States from the Pacific. Even before the finalization of the purchase, when Spain refused his request to explore the territory, he had authorized an expedition by his secretary, Meriwether Lewis, to study the land and find a river route to the PACIFIC OCEAN.

His January 18, 1803, secret request to Congress cited the need to subdue natives and counter infiltration by the French. The next month a "commercial venture" was okayed at a price of $2,500. Lewis began buying supplies and learning scientific observation.

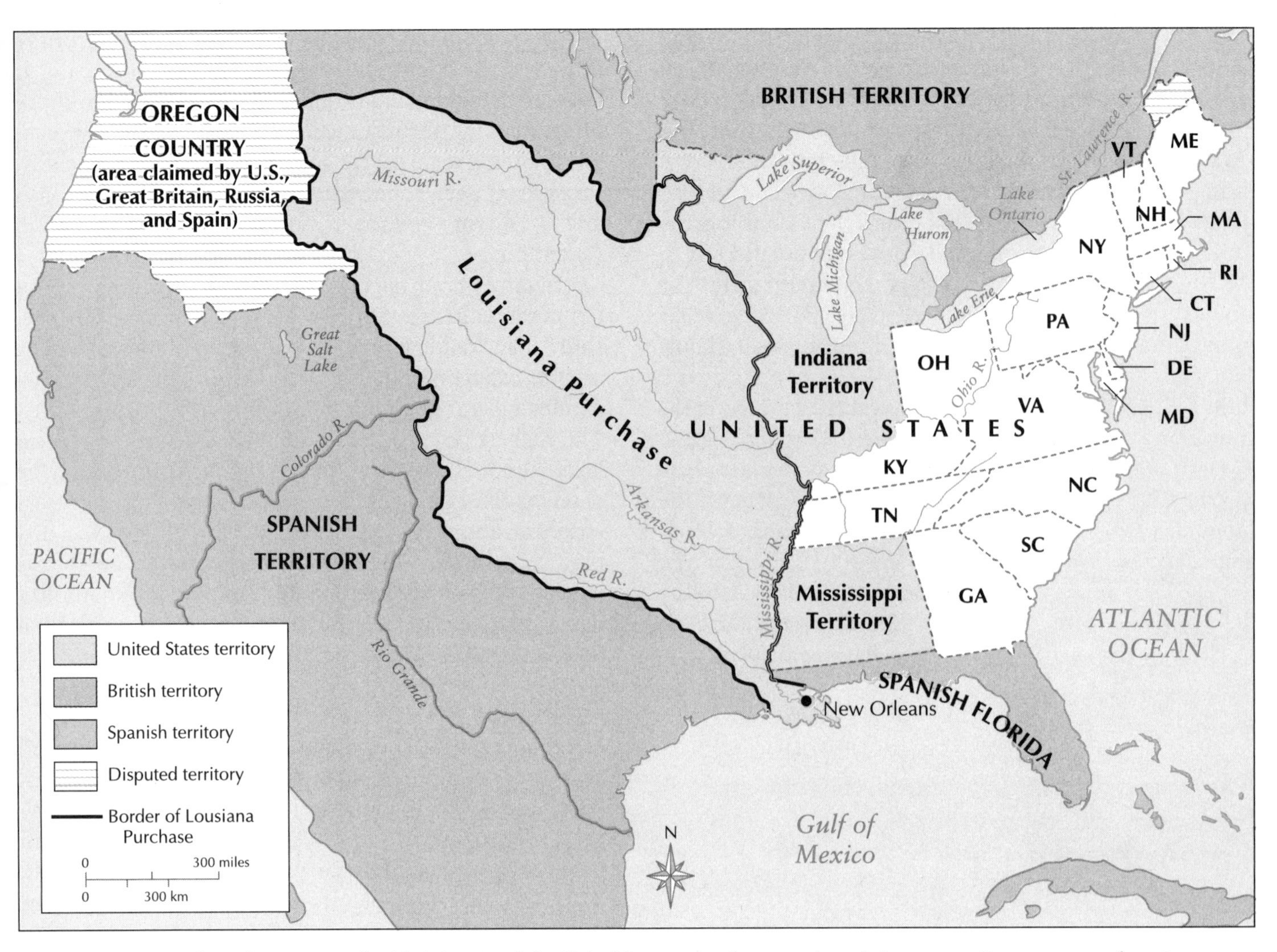

The Louisiana Purchase from France doubled the size of the United States, greatly strengthened the country in resources and territory, and provided a powerful impetus to westward expansion, especially after the Lewis and Clark expedition.

When Napoleon was selling Louisiana to the United States, Lewis was in Pittsburgh, Pennsylvania, buying instruments, guns and ammunition, medicine, trade goods, and a keelboat. In May, with William Clark, Lewis began the three-year exploration of the Louisiana Purchase northern area. When they returned in 1806, they had amassed invaluable information about the plains, the mountains, and the rivers that would later be crossed by the Oregon Trail. They also brought information about the various Indian groups they encountered en route to the Pacific, and they gave the United States a claim to OREGON that was much firmer than that of Captain Robert Gray, who had been in Puget Sound as early as 1789.

Although Lewis and Clark could not find the river route Jefferson wanted, they did end the hope for a northwest passage to CHINA. They reported the presence of obstacles such as the Great Falls of the Missouri and the Dalles on the Columbia that required exhausting portages. With peaks rising 2 mi (3.2 km) into the air, the ROCKY MOUNTAINS themselves were intimidating. Lewis and Clark established that wagon traffic could not exploit their route.

A large part of the purchase was a "wasteland." Between the line of settlement in western Missouri and the ROCKY MOUNTAINS, there was an area where average annual rainfall is less than 20 in (8 cm), insufficient for 19th-century American farmers. Jefferson was aware; he wanted the port of New Orleans, not the wasteland he felt upper Louisiana to be. He got affirmation from Lewis and Clark, whose journals frequently noted treeless and arid land whose rivers were trickles vanishing into the sand. In 1806, Zebulon Pike explored from the Missouri to the Rockies on the lati-

tude of Kansas, and he encountered sand dunes and land too arid for farming or timber. Stephen Long crossed at the latitude of Nebraska in 1819–20 and reported that the drainage basin of the Missouri and Arkansas rivers, as well as a great portion of western Kansas and Nebraska was a great American desert. Even as migrants crossed the land to the valuable areas in Oregon and California, the desert remained. Only in 1870 did the designation change. Immigrants to Kansas from RUSSIA brought dryland farming techniques that made the great desert into the Great Plains.

BIBLIOGRAPHY. J.S. Aber, "Historical Background of the Lewis and Clark Expedition," www.athena.emporia.edu (March 2004); Michael Burgan, *The Louisiana Purchase* (Compass Point Books, 2002); James A. Corrick, *The Louisiana Purchase* (Lucent Books, 2001); Thomas J. Fleming, *The Louisiana Purchase* (Wiley, 2001); Elizabeth D. Jaffe, *The Louisiana Purchase* (Bridgestone Books, 2002); LSU Libraries Special Collections, "The Louisiana Purchase; A Heritage Explored," www.lib.lsu.edu (March 2004).

JOHN BARNHILL, PH.D.
INDEPENDENT SCHOLAR

Luxembourg

Map Page 1131 **Area** 998 square mi (2,586 square km) **Population** 454,157 (2004) **Capital** Luxembourg **Highest Point** 1,834 ft (559 m) **Lowest Point** 436 ft (133 m) **GDP per capita** $55,100 **Primary Natural Resources** arable land.

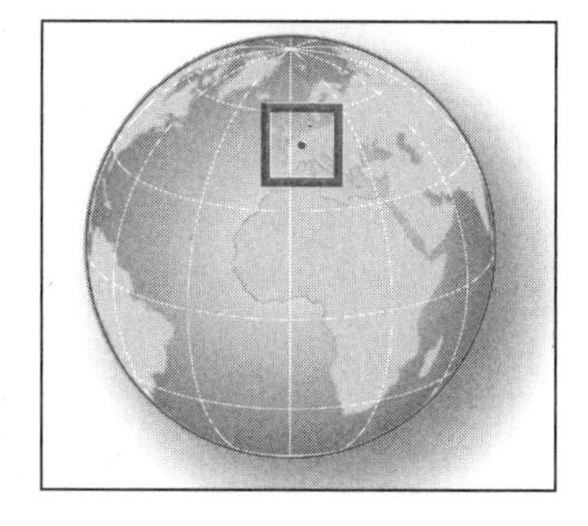

LUXEMBOURG IS a small LANDLOCKED western European country, within the EUROPEAN UNION (EU), and given its small population size, the country does not contain any large urban centers. After Luxembourg City, about 75,000 persons, the next largest settlements are Esch-sur-Alzette (27,000), Differdange (18,000), and Dudelange (17,000). Split into three administrative regions (Diekirch, Grevenmacher, and Luxembourg), Luxembourg is governed by a constitutional monarchy. Bordered by GERMANY to the east (the border being formed by the Our, Sûre and Moselle rivers), FRANCE to the south, and BELGIUM to the west and north, the country has been historically and culturally influenced by countries neighboring it. Linguistically at least, the neighboring influence persists. Legal and political matters, for instance, are still written in French, while police records are noted in German. Officially, three languages (Luxembourgish—a mix of Frankish and old German—French, and German) are spoken in Luxembourg.

In terms of physical geography, Luxembourg has a temperate climate that can be characterized by cool, sometimes cold winters and warm summers. The northern part of the country forms part of the Ardennes hill range and has an undulating topography. The highest point is some 1,834 ft (559 m) above sea level, although the other regions of the country may be described as being somewhat hilly as well. In terms of land use, about 25 percent of land is cultivated, a further 20 percent is used for pasture, and 20 percent is wooded. Given its small spatial extent, Luxembourg has a relatively high population density, even though its urban places are not large demographically.

MODERN ECONOMICS

In economic terms, Luxembourg may be characterized as a low-unemployment, low-inflation, and high-income society. Industry, particularly steel, has historically been significant in Luxembourg but has in recent times been surpassed by the growth of other industries, such as rubber and chemicals, as well as the rise of the tertiary economic sector, especially the banking sector which is largely based in Luxembourg City.

Located at the junction of the Alzette and Pétrusse rivers, Luxembourg City, the capital, is today a settlement of modern international banking repute and also the location of several European Union institutions. These include the European Court of Justice, the European Investment Bank, and the Secretariat of the European Parliament. Because of its exceptional location and the surrounding natural environment, Luxembourg City has historically been a place of military significance.

As an area noted in written history, Luxembourg dates from the 10th century when it was known as Lucilinburhuc ("little fortress"). At this time Siegfried, Count of Ardennes, erected a castle (now in Luxembourg City). From 1506 to 1890, Luxembourg formed part of the territories of numerous European countries, including SPAIN, France, AUSTRIA, and the NETHERLANDS, and it was not until 1815 that the process of independence in Luxembourg began, a consequence of the forming of a Grand Duchy by the Congress of Vienna. The

Congress, held in 1814 and 1815, redrew the continent's political map following the defeat of Napoleonic France, and one of the principle results of the event, aside from the confirmation of France's loss of territories it had annexed between 1795 and 1810, was that the Luxembourg Grand Duchy was formed and the Netherlands became an independent kingdom.

As part of this Europe-wide development, the Luxembourg Duchy was handed to the Dutch monarch, William I. In 1838, political autonomy was granted to by the Netherlands, yet it was not until 1890 that Luxembourg became independent as such from the Dutch, a result of the death of William III. Upon his death William was succeeded by his daughter, but as only men could inherit the title of Grand Duke, it passed to another blood line of the Dutch royal house. The throne thus was offered to William's cousin, Duke Adolf of Nassau, who subsequently became Grand Duke Adolf I of Luxembourg. Today Luxembourg remains the world's only grand duchy.

BIBLIOGRAPHY. Jul Christophory, Emile Thoma, and Carlo Luxembourg Hury, *Luxembourg* (ABC-Clio Press, 1997); Patricia Sheehan, *Luxembourg: Cultures of the World* (Benchmark Books, 1997).

NEIL BIRCH
UNIVERSITY OF ALBERTA, CANADA

Luxor

LUXOR IS LOCATED on the banks of the middle reaches of the NILE RIVER, in Qena governorate, east-central EGYPT. Approximately, 440 mi (700 km) south of Cairo, it is a bustling city with a population of about 150,000. As a modern city, it caters to tourists who are attracted to the famous archaeological sites that attest to Egypt's rich history. These are present in the nearby Valley of the Kings and in Denderah and Abydos just to the north, as well as in Luxor itself. The city comprises three areas: the City of Luxor on the east bank of the Nile, Thebes (also known by the ancients as Waset) on the opposite bank, and Karnak to the north.

Luxor was the capital of Middle Egypt for 1500 years. It became prominent in the 21st century B.C.E.; as Thebes it maintained its status as capital until 661 B.C.E. when it was attacked by the Assyrians. Its significance in the ancient world is recorded by Homer, the famous Greek poet who lived in the 8th century B.C.E., as "the city of a hundred gates." It survived as an influential center until it was destroyed by the Romans in 30 B.C.E. Although it never fully recovered, Luxor has been a place of interest, almost of pilgrimage, for centuries and it ranks as one of the first true tourist destinations.

There is some evidence for settlement in the Luxor area relating to the Old Kingdom of ancient Egypt which had its focus in Upper Egypt, notably in the Nile delta region where the Great Pyramids were constructed and where the capital, Memphis, was located. This period dates from 2686 B.C.E. to 2181 B.C.E. and is represented at Luxor by ancient tombs; one such necropolis is el-Khoka. There may also have been a temple in Karnak, and it seems likely that Luxor (then Waset) was a provincial town.

The Old Kingdom was followed by an Intermediate period from 2180 B.C.E. to 2040 B.C.E., and then the Middle Kingdom from 2040 B.C.E. to 1730 B.C.E. During the Intermediate period, Luxor began to grow in size and significance under the direction of the 7th to 9th dynasties of rulers and achieved capital status during the reign of Mentuhotep of the 11th dynasty, who once again unified Egypt into one kingdom. He also constructed the temple at Deir el-Bahri on the west bank where his mortuary temple is also located. Luxor (then Thebes) benefited from trade. Its location, on the fertile bank of the Nile adjacent to Nubia and the eastern desert, was advantageous for the acquisition and control of resources being traded north to south and east to west. There was a decline in its political status during the 12th dynasty, as the ruler Amenemhat 1 conferred capital status on El-Lisht close to Memphis, but Luxor's religious influence escalated as the local god, Amun, became a principal deity throughout the kingdom. The remains of a temple to Amun, dating from this period, are present at Karnak.

The advent of the New Kingdom, dating from 1552 B.C.E. to 1069 B.C.E., saw Luxor achieve renown as a religious center. It had become a center of life and death. Active worship in its many temples was juxtaposed with the numerous and rich tombs and mortuary temples excavated in the cliffs of the west bank. Included in the latter are the mortuary temples of Rameses II and Seti I as well as the remains of the temple of Amenhotep III. Further west into the desert, so as not to use fertile land, lie the tombs of the Valley of the Kings that have helped to make Egypt so famous. Not least of these is the tomb of the boy king, Tutankhamen of the 18th dynasty, who ruled Egypt from 1361 B.C.E. to 1352 B.C.E. This was discovered in 1922 by Lord

George Carnarvon and fueled much renewed interest in Egypt and its history. On the east bank, Luxor and Karnak became the location of major temple complexes whose architecture and inscriptions are testament to the stonecraft and building technology of the Middle and New Kingdoms. The majesty of the temples is encapsulated in the 134 huge columns of Karnak temple's Hypostyle Hall. The temple of Luxor was built by Amenophis III.

Luxor retained some importance during the Late Period from 664 B.C.E. to 30 B.C.E., but its golden age was over. Much later, Christian churches, Islamic mosques, and a tourist infrastructure added to the heterogeneity of one of the world's oldest cities.

BIBLIOGRAPHY. C. Gates, *Ancient Cities: The Archaeology of Urban Life in the Ancient Near East and Egypt, Greece and Rome* (Routledge, 2003); A. Gill, *Ancient Egyptians: The Kingdom of the Pharaohs Brought to Life* (HarperCollins, 2003); I. Shaw, ed., *The Oxford History of Ancient Egypt* (Oxford University Press, 2000).

A.M. MANNION, PH.D.
UNIVERSITY OF READING, UNITED KINGDOM

Luzon

THE ISLAND OF LUZON is the largest and most populous island in the PHILIPPINE archipelago (17th largest in the world), the center of Philippine political and economic life, and the site of the capital city, MANILA. With roughly 42 million people occupying a total land area of 40,420 square mi (104,688 square km)—roughly the size of VIRGINIA, with nearly 10 times the population—Luzon has one of the highest population densities in the world, in large part because of the fertility of its volcanic soil and abundant rainfall.

Luzon is also one of the most Europeanized areas in Asia, thanks to its convenient location at the western terminus of the Pacific TRADE ROUTES used continuously by the Spanish from the early 16th century. Spanish language and religion dominated the island for the next three centuries, until the islands came under the influence of the UNITED STATES at the end of the 19th century. Luzon is divided into seven regions, each with numerous provinces (30 total on Luzon), and three cities with over 1 million people: Quezon City (2.17 million), Manila (1.6 million), and Caloocan City (1.2 million).

The island is the northernmost in the Philippine group, with the exception of the smaller Batan and Babuyan islands in the Luzon Strait. This strait separates the Philippines from TAIWAN by about 220 mi (370 km). To the east lies the Philippine Sea (part of the PACIFIC OCEAN, bounded by the northern MARIANA ISLANDS, GUAM, Yap, and PALAU), and the South China Sea lies to the west.

The southern coasts of the island border the complex Sibuyan Sea lying between the nearby islands of Mindoro, Masbate, Samar, and Visayan. Formed about 50 million years ago from a combination of volcanoes and the buckling of the Earth's crust, Luzon lies along one of the most active edges of the RING OF FIRE, which extends northward from Sulawesi, INDONESIA, to JAPAN.

The Philippine Trench lies just off the east coast of Luzon, marking the convergence zone of the Philippine and Asian tectonic plates. The trench runs parallel to the island its entire length, plunging to depths of 33,000 ft (10,000 m). A number of Luzon's mountains are volcanoes, including its most active, Mayon (7,943 ft or 2,407 m), which last erupted in January 2004 and rivals Mount Fuji, Japan, in perfection of its symmetry. Mount Pinatubo (4,874 ft or 1,477 m), northwest of Manila, erupted in June 1991 and devastated an area of 154 square mi (400 square km), blanketing much of Southeast Asia with ash. Modern prediction techniques, however, prevented a potentially massive loss of life, and the death toll reached only 250.

The island is a construction of these volcanic chains and owes its erratic shape to this; it is essentially two landmasses: the larger northern portion and the narrower extension to the southeast, the Bicol Peninsula, connected to the rest of Luzon by the Tayabas Isthmus, which is only 8 mi (13 km) wide at points. The Zambales mountain range forms the westernmost part of the island.

Two mountain chains dominate the northern part of the island, the Cordillera Central—including the highest peak, Mount Pulog (9,682 ft or 2,934 m)—and the Sierra Madre along the northeast coast. In these mountains are some of the most inaccessible human populations on Earth, including the Bontoc and other mountain peoples. Another mountain chain forms the Bicol Peninsula. The longest river on Luzon flows between the Cordillera and the Sierra Madre, the Cagayan, which is navigable for a considerable distance and has potential for hydroelectrical power. The most important river for traffic and commerce is, however, the much shorter Pasig, which flows through Manila,

and links Leguna de Bay (the country's largest lake) with Manila Bay.

Luzon has mineral wealth, including two of the principal gold mines in the Philippines, in Mountain Province and Camarines Norte. Other resources include copper, asbestos and chrome. But the island is primarily an agricultural producer, mostly of rice, but also corn, tobacco, sugar, cotton, and several tropical products (coconut, mango, banana, cacao, etc). Nearly all industries are concentrated in Greater Manila. The people are related ethnically and linguistically to Malays; Tagalog, the dialect spoken in the region around Manila, became the official national language (called Filipino).

From conquest by SPAIN in the 1520s, the island was developed by large plantations (mostly in the north of the island, where there is the flattest land); using mostly forced labor, the Spanish produced tobacco at first, followed by sugar, coffee, indigo, and pepper—to compete with exports from the Dutch East India Company. Missionaries naturally followed, and the island was 80 percent Catholic by the time of the U.S. occupation in 1898. The island retains much of its Spanish culture, despite the growing influence of British and American traders from earlier in the century. Status as a U.S. protectorate was generally beneficial to the island in many respects, for example in literacy, which was five times the rate of neighboring INDONESIA by the 1940s. The U.S. naval base at Subic Bay and Clark Air Base provided numerous jobs and economic activity. Hopes for transforming this area into a free trade zone since the bases' closure in 1992 have so far failed to materialize. Christian Luzon also faces increasingly hostile separatist movements from the Muslim-dominated southern islands.

BIBLIOGRAPHY. Barbara A. Weightman, ed., *Dragons and Tigers: A Geography of South, East and Southeast Asia* (Wiley, 2002); Sylvia Mayuga and Alfred Yuson, *Philippines* (Apa Productions, 1980); "Philippines," www.gov.ph (April 2004); "Luzon," www.volcano.und.nodak.edu (April 2004).

JONATHAN SPANGLER
SMITHSONIAN INSTITUTION

Macau

MACAU IS LOCATED in the southern part of CHINA's Guangdong Province, near the tip of the peninsula formed by Pearl River on the east and Xinjiang River on the west. Macau is situated 37 mi (60 km) west of HONG KONG and comprises the Macau Peninsula and the islands of Taipa and Coloane.

Macau has mostly flat terrain resulting from land reclamation done over an extended stretch. However, Macau does have some steep hills as well. The climate of Macau is subtropical and is hot and humid for most of the year. The average year-round temperature is about 77 degrees F (25 degrees C); June to September is the hottest period, with temperatures exceeding 86 degrees F (30 degree C). Macau receives an average of 80 in (203 cm) of rainfall annually; typhoons hit Macau and adjoining areas with a high degree of frequency during the monsoon season.

Macau was colonized by the Portuguese in the 16th century and was the first European settlement in East Asia. China and PORTUGAL signed an agreement on April 13, 1987, which made Macau a Special Administrative Region (SAR) of China, effective December 1999. In modern times, Macau has developed industries such as textiles, electricity, as well as a first-rate tourist industry with a wide assortment of hotels, resorts, and sports facilities. Macau's economy is closely linked to HONG KONG and Guangdong Province in the Delta region. Macau provides banking training, communication and transport facilities in Far East and Northeast Asia.

Chinese and Portuguese are the official languages in Macau. Cantonese, however, is most widely spoken. The official languages are used as link languages in government departments and in all official documents and communication sectors inside Macau. English is generally understood and is used in trade, tourism, and commerce. In terms of ethnic groups, Macau's population comprises Han Chinese at 95 percent; Portuguese, 3 percent; and other, 2 percent. Of those, 44 percent are born in Macau, 47 percent in China, and 9 percent elsewhere. Although the term *Macanese* is used to describe the citizens of Macau, the term also applies to biracial Chinese-Portuguese individuals, persons of exclusive Portuguese descent, and Chinese or mixed Chinese-Portuguese individuals who have been baptized and taken Portuguese names.

Macau has a high percentage of literacy: 90 percent of the population age 15 and above can read. Among males, 93 percent have achieved basic literacy, and among females, 86 percent are literate. Buddhism remains the most dominant religion in Macau, with over 45 percent of Macau's population subscribing to it. Roman Catholics represent 7 percent, Protestants 1 percent, and other religions (Hindu, Muslim, and

other) 1.2 percent, whereas 45.8 percent of population has no official religious preference.

Textile and garment manufacturing are major industries in Macau accounting for 83.8 percent of principal domestic exports. Other major sectors are electricity, footwear and toys. Macau's industries have benefited a great deal from its close geographical proximity with China, which provides Macau basic raw materials. China, Hong Kong, and the EUROPEAN UNION rank among the top three countries in terms of Macau's principal import trade partners.

In the latter half of the 20th century, Macau's economy flourished with increased tourism. The tourism-related service sector has provided an estimated 30 percent of labor force and has contributed approximately 50 percent gross national product to Macau's economy.

BIBLIOGRAPHY. Fei Chengkang, *Macau: 400 Years* (Shanghai Academy of Social Sciences, 1996); Gullen Nunez Cesar, *Macau Streets* (Oxford University Press, 1999); Hing Lo Shiu, *Political Development in Macau* (Chinese University Press, 1995); Porter Jonathan, *Macau: Culture and Society* (Westview Press, 2000); Roberts Elfed Vaughan et al., *Historical Dictionary of Hong Kong and Macau* (Scarecrow Press, 1992).

MOHAMMED BADRUL ALAM
MIYAZAKI INTERNATIONAL COLLEGE, JAPAN

Macedonia (FYROM)

Map Page 1133 **Area** 9,781 square mi (25,333 square km) **Population** 2,063,122 **Capital** Skopje **Highest Point** 9,085 ft (2,753 m) **Lowest Point** 165 ft (50 m) **GDP per capita** $5,000 **Primary Natural Resources** chromium, lead, zinc, manganese.

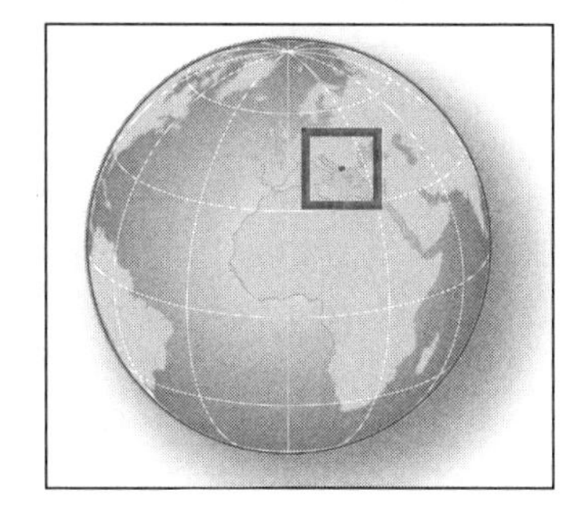

THE REPUBLIC OF Macedonia is a country in the southern Balkans but covers only part of the larger historical and geographical area known as Macedonia. The northern region of GREECE is also known as Macedonia, which has been a source of cultural conflict since the republic declared its independence from Yugoslavia in 1991. Greeks assert that the name "Macedonia" is a specifically Greek term, and not applicable to the Slavic people who today make the region their home. Pending United Nations-moderated negotiations, Greece and most international bodies officially refer to the republic as FYROM, the Former Yugoslav Republic of Macedonia.

Macedonian nationalism, as distinct from other south Slavic peoples is, moreover, a relatively new concept, introduced and encouraged by dictator Josip Tito upon the creation of a separate Macedonian Republic within the Yugoslav Federation in 1946. Prior to this, the area generally known as Vardarska banovina (the district of the Vardar River) was considered simply an extension of its southern Slavic neighbors, either Serbians to the north, or Bulgarians to the east. Slavs arrived in the Balkan Peninsula only in the 6th century, and therefore have nothing to do with the well-known classical kingdom of Macedonia, which dominated the rest of Greece, the Near East, EGYPT, and Persia under Alexander the Great in the 4th century B.C.E. The region was controlled by the Ottoman Turks dating from the 14th century, then contested by Greece, BULGARIA and SERBIA after the First Balkan War of 1912, which immediately led to the Second Balkan War of 1913. The result of this war was a partition of Macedonia between these three powers, the southern half going to Greece, the northern half incorporated into the kingdom of Serbia, and a small eastern portion, called Pirin Macedonia, given to Bulgaria. The Yugoslav Republic of Macedonia was set up after World War II in part to offset the continued claims of Bulgaria on the Vardar valley.

The Vardar River dominates Macedonia, nearly bisecting the country, and providing its chief outlet to the sea. Because Macedonia is LANDLOCKED, its economy relies heavily on transport down the Vardar across northern Greece (where the river is called the Axiós) to the major Aegean port city of Thessaloniki. Greece's trade embargo against FYROM from 1994 to 1995 was thus of great significance. Similar international embargoes against Serbia during its wars of ethnic cleansing in the 1990s cut Macedonia off from its largest trading partner.

The economy was strained even further by the flood of more than 350,000 Albanian refugees in 1999, most of whom have today returned to Kosovo. Macedonia had its own ethnic conflict in 2001, when a group of Albanians began an insurgency in the western part of the country, demanding greater rights for the country's Albanian minority. Ethnic Albanians make up roughly 23 percent of the population, chiefly con-

centrated in the western part of the country, across the borders from the Republic of ALBANIA and the Serbian province of Kosovo.

Macedonia is almost entirely mountainous; 14 peaks exceed 6,600 ft (2,000 m). The highest are in the western part of the country, running roughly from northwest to southeast. The Šar mountains in the northwest, and the Crna Gora (Black Mountains) form the border with Serbia in the north, while the Osogovo Mountains run along the eastern border with Bulgaria. The southern border with Greece is defined by the Nidze mountains. Most settlements are in the larger river valleys, the Crna, Treska, Pcinja, and Bregalnica, all tributaries of the Vardar. Skopje, the country's capital (population about 500,000), is located on the Vardar River. Other major cities are Tetovo, Kumanovo, Gostivar, Bitola, Ohrid, and Titov Veles.

The climate in Macedonia is a mixture of Mediterranean, continental, and mountain. Crops are therefore varied depending on elevation. Agriculture is dominant but there are also significant industries, including chemicals, steel, and textiles. Macedonia was Yugoslavia's poorest republic, but political and cultural disturbances of the past decade have exacerbated its economic struggles—roughly 40 percent are unemployed—and the area is also prone to earthquakes and drought. Large numbers of NORTH ATLANTIC TREATY ORGANIZATION (NATO), United Nations, and EUROPEAN UNION (EU) troops are stationed in Macedonia (nearly 20,000 in 1999), mostly in relation to peacekeeping missions in neighboring Kosovo, which both stimulates the economy and causes further strain on it.

A change in the flag (dropping Alexander the Great's 16-point Vergina Sun for a simpler eight-point star) in 1995 and a change in the wording of the constitution have allayed Greek fears of south Slavic expansion and allowed for greater recognition in the diplomatic world and economic aid from the EU and the rest of the international community. Macedonia in 2004 was a candidate for membership in both NATO and the EU.

BIBLIOGRAPHY. *World Factbook* (CIA, 2004); *Encyclopedia Americana* (Grolier, 1997); Wayne C. Thompson, *Nordic, Central and Southeastern Europe 2003*, The World Today Series (Stryker-Post Publications, 2003); "Balkan Information," www.b-info.com (August 2004); "FYROM," www.macedonia.org (August 2004).

JONATHAN SPANGLER
SMITHSONIAN INSTITUTION

Mackenzie-Peace River

THE MACKENZIE-PEACE River system is made up of the Mackenzie and Peace rivers. Located in western CANADA, this is the 10th-longest river system in the world, at 2,635 mi (4,240 km). Sir Alexander Mackenzie, a Canadian explorer, discovered the river in 1789 and it was named for him. By itself, the Mackenzie River is the longest river in Canada, with a length of about 1,118 mi (1,800 km). At places it is 2 mi (3 km) wide. It starts in the Northwest Territories at Great Slave Lake and flows north into the Beaufort Sea of the ARCTIC OCEAN.

The Mackenzie-Peace River system drains .70 million square mi (1.8 million square km), about one-fifth of Canada, and includes Great Slave Lake, Great Bear Lake, and Athabascan Lake. Small ships can travel about 1,700 mi (2,750 km) along the Mackenzie-Peace Waterway. The river is navigable from about June until October each year. The rest of the year it is frozen over.

The region where the river empties into the Beaufort Sea is known as the Mackenzie Delta. The area was once a large bay. Here land has built up from the sediment carried by the river and dropped as it slows down to enter the sea. The Mackenzie Delta is the largest delta in Canada and the twelfth largest in the world. Unlike most deltas, this one is confined by high landforms on both sides. The Richardson Mountains are to the west and the Caribou Hills are on the east. This keeps the delta from expanding in either direction. About 3,000 Inuvialuit live in the Delta communities of Aklavik, Inuvik, and Tuktoyaktuk. These Native Americans have made a comeback after being nearly obliterated by diseases in the early 1900s.

The most important tributaries are the Liard River, the Peel River, and the Bear River. Most of the land along the waterway is not settled. Much of it is heavily forested. The small communities that do exist along the water are connected only by the river. There are no roads in the area. The most northerly of the Athapaskan-speaking people live in the boreal forests of the Mackenzie Delta and rely, as they always have, on the caribou for food and clothing. Muskrat trapping has also long been a part of the delta economy. Muskrats are plentiful in the many lakes on the delta. Other mammals found on the Mackenzie Delta include black bears, grizzly bears, moose, red fox, Arctic fox, snowshoe hare, Arctic wolf, and musk ox. Beluga whales winter in the Bering Sea, but enter the waterways of the delta in the spring. Tundra swans breed on the tundra, then migrate as far south as central California. Beauti-

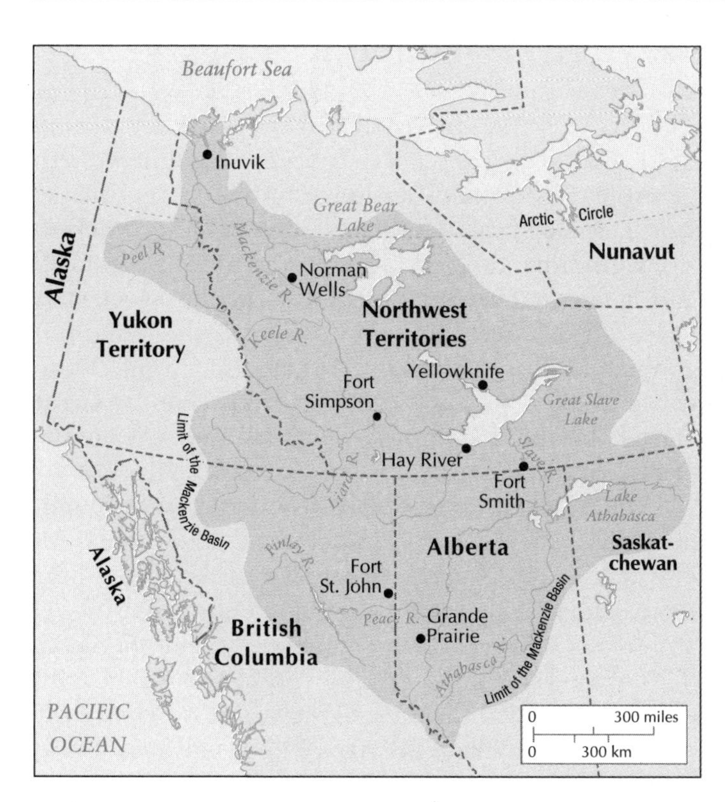

By itself, the Mackenzie River is the longest river in Canada with a length of about 1,118 mi (1,800 km).

ful wildflowers cover the tundra during the brief summer. About 100 species grow here, including purple crocus; white, pink, and purple Indian paintbrush; yellow cinquefoil; blue Arctic lupins; and red, yellow, and purple sweetpeas.

The delta contains deposits of minerals, gas, and oil. The Dene natives knew for a long time that there was oil along the lower Mackenzie River. They called the area "Le Gohini," which means "where the oil is." They finally showed the site to geologists in the early 1900s. The first oil well was drilled there by Imperial Oil in 1919, then a small refinery was built that supplied nearby communities with oil. In the 1940s, the U.S. Army helped with construction of a pipeline from the Norman Wells Oil Fields on the Mackenzie to Whitehorse, Yukon. They were afraid the Japanese would attack their west coast refineries.

However, the war ended just as the pipeline was completed. In the early 1980s, production there reached full capacity. Today Norman Wells averages 10 million barrels of oil per year, Canada's fourth-largest oil-producing field.

BIBLIOGRAPHY. Robert P. Sharp, *Living Ice: Understanding Glaciers and Glaciation* (Cambridge University Press, 1988); Andrew H. Malcolm, *The Land and People of Canada* (Harper Collins, 1991); "The Oil Fields of Norman Wells," www.greatcanadianrivers.com (March 2004); "Mackenzie Delta," http://www.bmmda.nt.ca (March 2004).

PAT MCCARTHY
INDEPENDENT SCHOLAR

Mackinder, Halford J. (1861–1947)

HALFORD JOHN MACKINDER was born on February 15, 1861, in Gainsborough, a small port and market town at the river Trent in England. He was the eldest of six children born to Dr. Draper and Mrs. Fanny Anne Mackinder. His father was well educated and trained as a scientist. It was he who taught Mackinder to look for "interrelationships between factors in the environment." In 1870, he went to Gainsborough Grammar School and from 1874 he was educated at Epsom College. His father always wanted him to be a doctor, and Epsom had a sound reputation for its advanced training in the sciences. At school Mackinder showed strengths in essay writing, languages, public speeches, and environmental sciences. Furthermore, he developed an enthusiasm for historical geology. In 1880 he and his friend Thomas Walker won a five-year Junior Studentship in physical sciences and in October of the same year the two men went up to Oxford and entered Christ Church school. Here, Mackinder specialized in animal morphology, but he also took courses in physics, chemistry, physiology, and botany. During his last two years he attended classes in geology, history, and law.

At the university he was involved in a wide range of school-related activities. He spent a lot of time in the laboratories and the University Museum. He joined the Union and helped found the Junior Scientific Club in 1882. In the same year he enlisted in the Oxford University Army Volunteer Reserve. During the summer vacations, he often took part in exercises and long marches across the countryside. Another interest related to the military was his affinity for war games and his membership in the Kriegsspiel Club. There he met many other people who later would be of importance to him. Among those at Oxford was Herford George, who taught military history. He wrote on the relationship of history and geography and on the historical ge-

ography of the British Empire. And it was George who later proposed Mackinder for membership in the Royal Geographical Society in 1886.

In a paper entitled "The Scope and Methods of Geography" that he wrote for the Royal Geographical Society, Mackinder outlined his ideas of a "New Geography." He defined geography as "the science of distributions" based on a biological tradition in which forces interconnect and play upon each other. As a result of the paper, Mackinder is sometimes labeled as an environmental determinist. It was his conviction that physical and human geography formed one subject, and he consequently drew the conclusion that history and geography can never be studied separately.

In addition to his educational work, Mackinder also planned the first ascent of Mount Kenya. In 1896, the same year the Uganda Railway started, Mackinder received permission for the expedition to Kenya. In Kenya, the group had to face many problems, such as an outbreak of smallpox and a famine. Nevertheless the expedition to East Africa turned out to be a great success.

In 1904, he presented his well-known paper "Geographical Pivot of History" at the Royal Geographical Society. He wrote that "my aim will be to exhibit human history as part of the life of the world organism." Moreover, he argued that sea power was declining relative to land power, and railroads led the way to continental areas. His thesis, widely known as Mackinder's Heartland Theory, suggests that there was a pivotal area "in the closed heartland of Euro-Asia," which was most likely to become the seat of world power because of its inaccessibility. His theory was more or less a model based on world history and geographical facts. Mackinder defined a "world island" that consisted of the two continents Eurasia and Africa. He saw history as a struggle between land-based and sea-based powers. He saw that the world had become a closed system, with no new lands left for the Europeans powers to discover, to conquer, and to fight over without affecting events elsewhere.

Sea- and land-based powers would then struggle for dominance of the world, and the victor would be in a position to set up a world empire. The determining factor in this struggle was geography. The world, he argued, was now a closed political system and a worldview had to be taken.

Mackinder's geographical work was often criticized as being too political, and his scientific reasoning was often accused of being too primitive. In brilliant lectures he expounded the principles of the "new" geography by synthesizing the study of the physical landscape and human activity within a historical context. According to Mackinder there are three correlated arts that he thought were characteristic of geography: observation, cartography, and teaching. Mackinder always pushed for the founding of a geographical institute in London, and in 1893 he became involved in the founding of the Geographical Association to stress the necessity of teaching geography in schools. Mackinder's contribution to the founding of Reading University is considered one of his most important achievements.

Finally, Mackinder was convinced that geography was a distinct discipline with its own methodology, thus deserving its own place within the academic world.

BIBLIOGRAPHY. Halford Mackinder, *Democratic Ideals and Reality* (W.W. Norton, 1962); Charles Clover, "Dreams of the Eurasian Heartland," *Foreign Affairs* (March/April 1999); Jean Gottman, "The Background of Geopolitics," *Military Affairs* (Winter 1942); "Problematizing Geopolitics: Survey, Statesmanship and Strategy," *Transactions of the Institute of British Geographers* (1994); Nicholas J. Spykman, *The Geography of Peace* (Harcourt and Brace, 1944); Richard E. Neustadt and Ernest R. May, *Thinking in Time: The Uses of History for Decision-Makers* (The Free Press, 1986); W.H. Parker, *Mackinder: Geography as an Aid to Statecraft* (Clarendon, 1982)

RICHARD W. DAWSON
CHINA AGRICULTURAL UNIVERSITY

Madagascar

Map Page 1116 **Area** 364,500 square mi (587,040 square km) **Population** 17,501,871 **Capital** Antananarivo **Highest Point** 9,435 ft (2,876 m) **Lowest Point** 0 m **GDP per capita** $800 **Primary Natural Resources** chromite, coal, bauxite, salt, quartz.

MADAGASCAR, THE WORLD'S fourth-largest island, after GREENLAND, New Guinea, and BORNEO, is located 250 mi (400 km) off the southeast coast of Africa. The climate is temperate, with average temper-

atures ranging between 48 degrees F (8 degrees C) and 81 degrees F (27 degrees C). The island coastal region is tropical with a rainy season that extends from November through April.

The island of Madagascar has been described as an "alternate world" or a "world apart" because of the uncommonness of many of its plant and animal species. Most life forms on the island have an African (or South American) origin. Millions of years of isolation have allowed old species, elsewhere extinct, to survive and new species unique to the island to evolve. A great number of plant, insect, reptile, and fish species are found only in Madagascar. Madagascar once was covered almost completely by forests but has been deforested by the practice of burning the woods to clear the land for dry rice cultivation. Wood and charcoal from the forests are used to meet 80 percent of domestic fuel needs. As a result, fuel wood is in short supply. Rainforests survive on the hillsides along a slender north-south border on the east coast. Secondary growth, which has replaced the original forest and consists to a large extent of traveler's trees, raffia, and baobabs, is found in many places along the east coast and in the north.

The landscape of the central highlands and the west coast is savanna or STEPPE, and coarse prairie grasses grow where erosion has not taken over. The remaining rainforest contains a great number of unique plant species. The country has some 900 species of orchids. Banana, mango, coconut, vanilla, and other tropical plants grow on the coasts, and the eucalyptus tree, brought from Australia, is prevalent.

Madagascar was once an independent kingdom but became a French colony in 1896. It regained its independence in 1960. The current system of government is a that of a republic. A constitution was adopted through a referendum vote on August 19, 1992, and free elections were held in 1992 and 1993, ending 17 years of single-party government. Madagascar's independence encouraged the practice of economic privatization and liberalization. This strategy has placed the country on a slow and steady growth path from a 71 percent poverty level.

Agriculture, including fishing and forestry, is the driving economic force of the economy, accounting for more than one-quarter of gross domestic product and employing four-fifths of the population. Other predominant industries on Madagascar are meat processing, soap making, breweries, tanneries, sugar, textiles, glassware, cement, an automobile assembly plant, paper, petroleum, and tourism. The primary agriculture products are coffee, vanilla, sugarcane, cloves, cocoa, rice, cassava (tapioca), beans, bananas, peanuts, and livestock.

The official languages are Malagasy and French although additional regional dialects are spoken. More than half the population follows Animist beliefs (Animism founded upon the belief that the things around us are infused with more than mere existence. Animists believe that the hills, valleys, waterways, and rocks are spiritual beings, as are the plants and animals.) Christianity is practiced by about 43 percent of the population with the remaining being Muslim. There are approximately 20 ethnic groups that make up the population of Madagascar: the central highlanders (Merina and related Betsileo) and côtiers of mixed Arab descent, African, Malayo-Indonesian, Comoran, French, Indo-Pakistani, and Chinese. The life expectancy is 56.54 years of age.

BIBLIOGRAPHY. *World Factbook* (Central Intelligence Agency, 2004); *Europa World Year Book 2004* (Europa Publications, 2004); "A Country Study: Madagascar," http://lcweb2.loc.gov (October 2004).

CLARA HUDSON
UNIVERSITY OF SCRANTON

Madrid

WITH A POPULATION of 3,092,759 and crossed by the Manzanares River, Madrid is the capital and largest city in SPAIN. Located approximately at the center of the Iberian Peninsula, at the foot of the Sierra de Guadarrama, Madrid has a dry continental climate, with hot summers and cool winters.

After enduring conquest by the Moors in 939 and reconquest by the Christian kings in 1083, Madrid was made the capital only in 1561, becoming the center of the largest world empire of the time. At the beginning of the 17th century, the city lost some of its medieval character (irregular blocks and narrow streets), with the new Plaza Mayor (Main Square) becoming the new center. A great number of churches and convents were built in this period.

Madrid attained a new level of development in the second half of the 18th century, with construction of the Paseo (boulevard) del Prado, the Botanical Garden, and the Puerta de Alcalá; elegant palaces were added by the nobility.

By 1750, Madrid had 160,000 inhabitants and its population density increased until 1850 when 280,000 people were living in a cramped city, calling for urban renewal. Response was the tearing down of the city walls in 1857 and the project of the Ensanche (expansion), originating the neighborhoods of Salamanca, Chamberi, and Arguelles to the north and east, known as Plan Castro of 1870. By the end of the century, construction of Arturo Soria's innovative Linear City was initiated to the northeast as an example of liberal urbanism conciliating urban and rural life. Because of its relative isolation, only in the 19th century did Madrid become the largest city in Spain, becoming a place of production.

In 1910, the Gran Via tore through the city to connect the new large neighborhoods. Madrid was bombed during the Spanish Civil War (1936–39), creating an opportunity to better plan a renewed city and to articulate a ring of peripheral settlements and greenbelts. In the 1950s and 1960s, la Castellana was extended to the north and the area became the new financial and commercial center. Because of its growth and importance, in 1984, Madrid became an Autonomous Community.

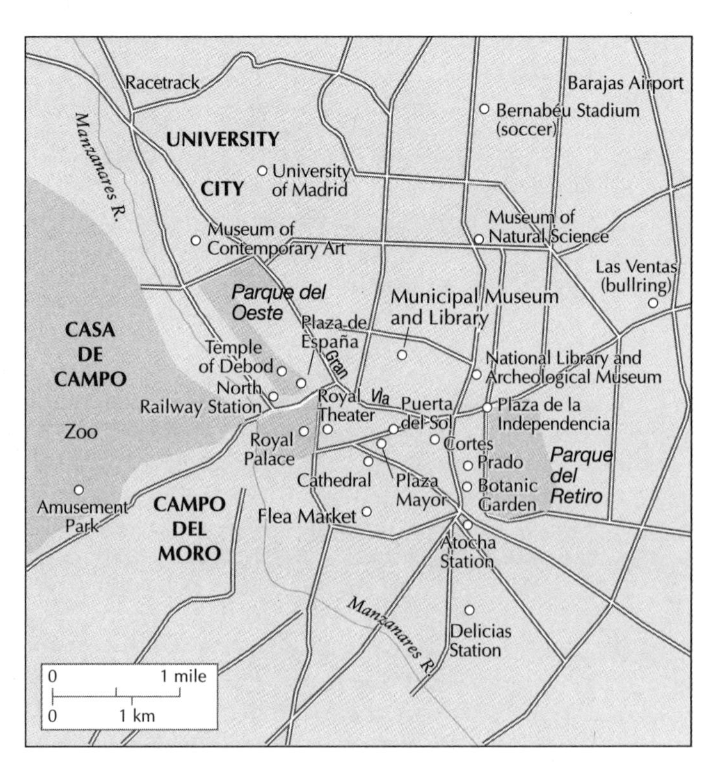

Madrid was made the capital of Spain in 1561, becoming the center of the largest world empire of the time.

MODERN MADRID

Madrid is today a large and dynamic cosmopolitan city with world-class museums and three large urban parks: the Parque del Oeste, the central Parque del Buen Retiro, and the huge Casa de Campo forest park with its 4,448 acres (1,800 hectares) sprawling to the west of the royal palace. The city is the seat of the government and of the royal family, but despite the importance of services, Madrid keeps relevant industrial activity, producing automobiles, electrical equipment, farming machinery, chemicals, and other goods. A web of expressways and railroads radiate from the city in almost every direction, crossing residential suburbs and industrial areas to connect the capital to all regions of Spain.

BIBLIOGRAPHY. *Oxford Essential Geographical Dictionary* (Oxford University Press, 2003); Carlos Sambricio, *De la Ciudad Ilustrada a la Primera Mitad del Siglo XX* (Comunidad de Madrid, 1999); George Kean, *Madrid* (Presença, 1997); Michelin Guides, *Spain* (Michelin, 1995); Manuel Guàrdia, *Atlas Historico de Ciudades Europeas* (Salvat, 1994).

SERGIO FREIRE
PORTUGUESE GEOGRAPHIC INSTITUTE

Magellan, Ferdinand (1480–1521)

FERDINAND MAGELLAN, one of the most distinguished explorers in world history, was born in to a middle-class Portuguese family. In his youth, Magellan was awed by Christopher Columbus's seafaring adventures, and he developed a strong interest in the sea and enlisted in the navy in 1505. While serving as a naval officer, Magellan participated in various victorious battles that gave PORTUGAL immense power in the INDIAN OCEAN. One such victory resulted in the conquering of Malacca, which gave Portugal control of vital trade routes in the region.

Magellan traveled on ships that traded throughout the world, studied charts, and listened to tales of great adventures. Magellan was convinced that a water route existed somewhere between the enormous landmass from the North Pole down to South America. He dreamed of one day finding this route and sailing around the world. In 1494, Pope Alexander VI orchestrated the Treaty of Tordesillas, which divided the unknown world. The eastern part of the world was given to Portugal and the western part to SPAIN. However, the

Though he did not survive the journey, Ferdinand Magellan's expedition set the precedent for success for future explorers.

treaty's boundaries in the east were virtually unknown, for no one had actually sailed around the world.

Throughout Magellan's naval years, he and the Portuguese king, Dom Manuel developed a particularly strained relationship. He once asked the king for a rise in rank and pay and the king refused. The king also refused to allow Magellan to command a royal ship sailing to the Spice Islands.

Thus, Magellan turned for sponsorship for an epic voyage to the Spanish monarch, King Charles I. He asked for approval from the Spanish king to conduct a voyage, which would possibly create permanent boundaries between Spain and Portugal in the PACIFIC OCEAN. He also wanted to determine whether the Spice Islands were actually within Spain's boundaries. King Charles approved of the plans, and on September 20, 1519, five ships, the *Trinidad*, *San Antonio*, *Concepcion*, *Victoria*, and the *Santiago*, along with a crew of 270 men, set sail on a journey full of mutiny, discovery, and death.

The voyage began with numerous problems. Before departure, the king of Portugal hired secret agents who loaded empty water barrels onto the ship, and changed the ships' cargo records. When Magellan left Spain, he had mistakenly believed that his ships were fully equipped for the sailing. The five ships were also not of the best quality, and the Spanish crew made plans to lead a mutiny and kill Magellan at sea. While stopping at the CANARY ISLANDS to resupply, Magellan heard of the plans, and subsequently put the conspirators in chains.

On December 13, the five ships landed in the present-day city of Rio de Janeiro, BRAZIL. There, they met the Guarini natives, who believed the men were gods. After a few weeks in virtual paradise, the sailors stocked up on food and water and sailed further south along the east coast of South America, hoping to discover a route to the Pacific Ocean. However, with winter approaching in the Southern Hemisphere, the seas became increasingly treacherous and the weather turned cold and stormy. Many crew members were calling on Magellan to turn back to Rio de Janeiro, but the leader persisted in his quest. After hugging the South American coast for 60 days, through storms and high seas, Magellan instructed the ships to pull into the present-day bay of Port St. Julian.

From Palm Sunday to late August, Magellan and his crew were anchored in the bay. Throughout this period, mutiny was discussed by crew members. A rebellion took control of three of the ships, and 170 out of the 265 men were involved. Magellan ordered numerous executions and marooned mutineers. With these actions, he regained the control and respect of the sailors, however, he lost the Santiago in May, after it was wrecked in rough seas while on a reconnaissance mission. Although the ship was destroyed, the 37 crew members were saved.

By October, 13 months after first setting sail, Magellan and his crew once again traveled south along the South American coast. On October 20, they found a beautiful bay that led into a strait. For weeks, the ships navigated their way through the treacherous strait with strong currents, gusty winds, and steep cliffs. The crew on the ship *San Antonio*, however, decided that the risks were too high and actually turned around and headed back to Spain. Unfortunately, this ship carried much of the food and supplies.

On November 28, 1520, the *Trinidad, Concepcion*, and *Victoria* entered a vast tranquil sea, which was named the Pacific Ocean. The crew was overjoyed to escape the straits but the excitement was short-lived. Facing nothing but water in front of them and horrid conditions aboard the ship, many of the starving crew

wanted to turn back for Spain. However, Magellan would not turn around and believed the Spice Islands were only a few days away. With the absence of any charts, Magellan's estimates of the proximity of the Spice Islands were not even close, and the three ships continued northward across the endless ocean.

For 96 days, the three ships did not see land as they crossed the Pacific Ocean. They sailed along the peaceful ocean with the aid of trade winds and, unbelievably, missed every single Pacific island. The food supplies and water were dwindling, and the crew was inflicted with scurvy. Many of the crew members were dying, and toward the end of the journey, the men began to eat ox hides and sawdust. Finally, on March 6, 1521, Magellan and his crew landed on present-day GUAM. For three days, the crew made minor repairs and tried to renourish themselves. They set sail again, and on March 16 the crew arrived in the PHILIPPINES. They became the first Europeans to set foot on the islands. For six weeks, the men explored the islands and began to try to convert the islanders to Christianity.

HOSTILITIES

Some of the indigenous people were hostile toward the spread of Christianity and waged a battle against Magellan and 48 of his men on Mactan Island. While dressed in full battle armor, Magellan was struck and killed by an islander. With their leader now dead, the remaining crew of 115 had to somehow return to Spain. The *Concepcion* was burned, and the *Trinidad*, after being blown off course, was captured by a Portuguese battle group off the coast of JAPAN.

The *Victoria* was now the only remaining ship from the expedition. Sebastián del Cano became the leader, and he and a crew of 47 battled through months of monsoons, heat, and frigid cold. On September 6, 1523, the battered *Victoria* returned to Spain. The remaining crew had dwindled to only 18 Europeans. Although Magellan did not survive the expedition, his glory in Portugal has survived through centuries. The leader of the first expedition to circumnavigate the globe traveled to where no European had been before and set a precedent of success for future European explorers.

BIBLIOGRAPHY. Simon Wincehster, "After Dire Straits, An Agonizing Haul Across the Pacific," *Smithsonian* (April 1991); Raymond Schuessler, "Ferdinand Magellan: The Greatest Voyager of Them All," *Sea Frontiers* (September-October 1984); The Mariners' Museum, "Ferdinand Magellan," www.mariner.org (March 2004); BBC History, "Ferdinand Magellan," www.bbc.co.uk (April 2004); World History Sourcebook, "Ferdinand Magellan's Voyage Round the World, 1519–1522," www.fordham.edu (April 2004).

GAVIN WILK
INDEPENDENT SCHOLAR

Maghreb

BROADLY DEFINED, the Maghreb is the triangular region of northwest Africa bounded by the ATLAS MOUNTAIN ranges, the ATLANTIC OCEAn from the border of western SAHARA to the Strait of Gibraltar, and the MEDITERRANEAN SEA from the Strait of Gibraltar to the northeast of LIBYA. Politically, the Maghreb includes the countries of MOROCCO, ALGERIA, TUNISIA and sometimes LIBYA. In the Middle Ages, Moorish SPAIN was included in the Maghreb. The Maghreb (*maghrib*) is Arabic for "the west." The original Arabic name for the northwestern end of the African continent was Maghreb-el-Aqsa, "the land farthest west," or "the place where the sun sets."

The Maghreb is watered by westerly winds from the Atlantic Ocean. Ninety-five percent of the people of the Maghreb countries live between the Atlantic or Mediterranean coasts and the Atlas Mountains.

In the Moroccan part of the Maghreb, the Mediterranean coast is rugged with cliffs and coves. The Rif Mountains rise sharply from the Mediterranean to a height of 7,218 ft (2200 m). The Taza Gap runs east and west between the Rif Mountains and the Middle Atlas ranges to link Algeria with Morocco. Southeast from the Strait of Gibraltar are the Middle Atlas (Moyen Atlas) Mountains. Much of the area is high, partially forested plateaus. Farther south the High Atlas (Haut Atlas) Mountains reach to some of the highest peaks in Africa. Winter snows frequently cover the 400-plus mountaintops that exceed 9,850 ft (3,000 m). The highest peak is Mount Toubkal at 13,665 ft (4,165 m).

The Algerian part of the Maghreb has four east-west zones. The coast stretches from Morocco 600 mi (970 km) to Tunisia. It extends inland from 50 to 120 mi (80 to 190 km). The coastal zone is a lowland strip dotted with mountains. The region is fertile and called the "Tell" in Arabic. The zone of the Tell Atlas Mountains (highest point about 7,570 ft or 2,310 m) is south of the coastal zone. South of the Tell, is a semiarid region of plateaus (average elevation is about 3,500 ft or

1,070 m) containing a number of shallow salt lakes (chotts). These wet-season lakes dry into salt pans during the dry season. A fourth zone holds the semi-arid Saharan Atlas Mountains, a broken series of mountain ranges and massifs.

TUNISIAN MAGHREB

The Tunisian area of the Maghreb contains two branches of the Atlas Mountains that extend eastward across the country from Algeria. Along the coast is a fertile plain. The northern range lies inland a short distance and is called the Atlas Mountains. Most of the mountains are low with peaks under 2,000 ft (610 m) in all but a few locations. The Tabassah Mountains form the southern Atlas branch of the Tunisian Maghreb. The highest is Mount Shanabi at 5,066 ft (1,544 m). The area includes the Grand Dorsal Range and the two-mile-wide Kasserine Pass. South of the Tabassah range the elevation descends across a plateau with salt lakes and date palm oases to the SAHARA DESERT.

The northwestern plain of Libya and the northeastern highland, an extension of the Atlas ranges, across the Gulf of Sidra are sometimes included in the Maghreb, for reasons of culture, if no other.

BIBLIOGRAPHY. Samir Amin, *The Maghreb in the Modern World: Algeria, Tunisia, Morocco* (Penguin, 1970); Marc Coté, *Le Maghreb* (La Documentation française, 1998); Owen Logan and Paul Bowles, *Al Maghrib: Photographs from Morocco* (Edinburgh University Press, 1989); Anthony G. Pazzanita, *The Maghreb* (Clio Press, 1998); Barnaby Rogerson, *A Traveller's History of North Africa* (Interlink Books, 2001); Paula Youngman Skreslet, *North Africa: A Guide to Reference and Information Sources* (Libraries Unlimited, 2000).

ANDREW J. WASKEY
DALTON STATE COLLEGE

Magna Graecia

MAGNA GRAECIA (or "Greater GREECE") was the geographic expression of Greek colonization originating from many different Greek cities. It was a process that began in the 7th century B.C.E., largely because of overpopulation. Competing city-states such as Sparta, Corinth, and Athens began to found new cities (colonies) that, in turn, became centers of an economically thriving and internally competitive expansion of Greek culture. These new colonies were concentrated south from the Bay of Naples to the Gulf of Taranto and along the southern and eastern coasts of Sicily in the MEDITERRANEAN SEA. Because these colonies remained closely linked to their home cities in Greece proper, they together were known as Magna Graecia.

Despite the existence of earlier trading colonies established by the Phoenicians, Greek mythology and folklore eventually asserted the greatest influence on Sicily. In the 7th, 8th, and 9th centuries B.C.E., Sicily and the southern part of the Italian peninsula (today the poorest areas in ITALY) were colonized by Greeks, and it is claimed that the area boasted more Greeks and Greek temples than homeland Greece itself. This process dispersed Greek culture and arts throughout the central and western Mediterranean.

EARLY CONTACTS

Unlike Greek Sicily, Magna Graecia on the Italian peninsula began to decline by 500 B.C.E., probably because of malaria and endless warfare among the colonies, but certainly with the onslaught and emergence of the Roman Empire. Culturally, Magna Graecia also was the center of two philosophical groups in the 6th century B.C.E., Parmenides, who was at Elea, and Pythagoras (originally from Samos), who resided at Croton. And it was through contacts and trade with Magna Graecia that the Etruscans and the Romans first came into early contact with Greek civilization, especially its pottery and the practice of minting coins.

The chief city colonies of Magna Graecia (and their home city) were Tarentum (colonized by Spartans); Heraclea (by people from Tarentum); Metapontum (settled by Achaeans); Sybaris (Achaean immigrants) and then known as Thurii (settled by Athenians who replaced earlier colonists from Sybaris); Paestum, or Posidonia (settled by people from Sybaris); Laos (also settled by people from Sybaris); Siris (migrants from Colophon); Caulonia (people from Crotona); Epizephyrian Locris (settlers from Locris); Hipponium (migrants from Epizephyrian Locris); Cumae (people from Chalcis); Rhegium (now Reggio de Calabria, settled by people from Chalcis); Neapolis (now Naples, earliest people came from Cumae); and Elea (migrants came from Phocaea in Ionia).

It is amazing to historians and geographers that Magna Graecia, once a center of immense wealth and culture geographically, was concentrated in what today many consider to be the poorest and most backward areas of Italy.

BIBLIOGRAPHY. Michael Bennett and Aaron J. Paul, *Magna Graecia: Greek Art from South Italy and Sicily* (Hudson Hills, 2003); G. Carratelli, Greek *World Art and Civilization in Magna Graecia and Sicily* (Rizzoli International, 1996); "Magna Graecia," www.bartleby.com (March 2004): Ancient Coins and Maps, www.bio.vu.nl (March 2004).

ROBERT W. MCCOLL, PH.D.
GENERAL EDITOR

Maine

MAINE IS THE easternmost state in the UNITED STATES and the northernmost of the 48 contiguous states. It is bounded to the east and south by the ATLANTIC OCEAN, to the northeast by the Canadian Province of New Brunswick, and to the northwest by the Canadian Province of Quebec. To the west, Maine is bordered by the U.S. state of NEW HAMPSHIRE.

Of the six New England states, Maine has by far the largest area, consisting of 33,215 square mi (86,026 square km). The state's famous rocky coast was shaped by glacial activity more than 12,000 years ago. The receding glaciers left behind thousands of islands, bays, and coves, resulting in 3,500 mi (5,632 km) of shoreline. For all its size, Maine has a low population density—there are only 1.3 million inhabitants, most of whom are concentrated in the south. Portland is the largest city, with a population of 64,000. Augusta is the capital. Other major urban areas include Bangor and Lewiston.

Maine is called the Pine Tree State. In its interior are 17 million acres (6.8 million hectares) of forest. While much of the northern forest areas are privately owned, there are many protected areas scattered throughout the state. Among them, Acadia National Park is a popular tourist destination. Other major recreational areas include the Allagash Wilderness Waterway and Baxter State Park.

Most visitors to the state enjoy the sandy beaches along the southern coast, between Kittery and Portland. Inland, there are approximately 6,000 lakes and ponds, and there are rugged, mountainous areas to the west. Of note, Moosehead Lake is the largest body of water wholly contained by any state, and Mt. Katahdin (the northern terminus of the Appalachian Trail), at 5,268 ft (1,605 m), is the state's highest peak. Maine typically sees short summers (with warmer weather along the southern beaches and cooler weather inland) and cold winters. The beach season typically starts in July and runs through early September.

The more remote, undeveloped areas of the state include the "Downeast" region along the upper shore in Washington County. The largely rural Aroostook County lies in the far north of the state. Maine was first settled as early as 3000 B.C.E., although little is known about these first inhabitants. Later, Native American tribes settled there, included the Micmacs, Abnakis, Passamaquoddies, and Penobscots. The first Europeans to explore Maine's coast may have been the Norse Vikings, in the 11th century; later, in the 1490s, Englishman John Cabot may have visited, although neither claim has been proven. Throughout the 1500s, English and French ships visited briefly, and in 1607, a group of English tried, and failed, to establish a colony. Permanent settlers did not arrive until the 1620s.

There is no consensus on the origin of the state's name, which first appeared in 1622. It is commonly believed that the name was used to refer to the mainland, distinguishing it from the numerous islands along the coast. The area later became part of the Massachusetts Bay Colony. Throughout the 1700s, area settlers experienced near continuous warfare, with both the native inhabitants and with the French. Because of its long coastline, Maine suffered from British sea attacks during the Revolutionary War. After the war, Mainers began to clamor for statehood, which was not realized until March 15, 1820, when Maine became the 23rd state as part of the Missouri Compromise—a Congressional policy aimed at preserving the balance between slave and free states. As part of that agreement, Maine was admitted to the Union as a free state, and Missouri as a slave state a year later. A strong abolitionist movement sprang up in Maine during the 19th century, and the state sent thousands of young men to fight in the Civil War. The state enacted prohibition laws in 1846 that lasted until the end of national Prohibition in 1934.

Maine's top agricultural commodities are potatoes, dairy products, eggs, seafood, and blueberries. The state is renowned as a source of fresh lobster, yielding 57 million pounds (25 million kg) of these saltwater crustaceans in 2000. In recent years, the state has experienced losses in its traditional base of manufacturing jobs. Tourism is now the primary industry, driven by an increasing number of visitors to the state's quaint coastal villages, rugged coastal headlands, and inland recreational areas. Because of their relative isolation

and distance from the rest of the nation, Mainers (sometimes called "Mainiacs") have staked out a reputation as a self-reliant, traditional people.

BIBLIOGRAPHY. Neal R. Peirce, *The New England States: People, Politics, and Power in the Six New England States* (W.W. Norton, 1976); Paul Karr, *Frommer's Vermont, New Hampshire and Maine* (Wiley, 2004); State of Maine, www.maine.gov (October 2004); E.D. Brechlin, *Adventure Guide to Maine* (Hunter Publishing, 1999).

MANNY GONZALES AND A. CHIAVIELLO
UNIVERSITY OF HOUSTON, DOWNTOWN

Malacca Straits

A PLAQUE IN THE gardens of the sultan's palace in Melaka reads: "Whoever is lord of Malacca has his hand on the throat of Venice." For centuries this passage of water, 620 mi (1,000 km) long, connecting the INDIAN OCEAN and the South China Sea, running generally northwest to southeast, lived up to this reputation as one of the most important TRADE ROUTES in the world, the meeting place of east and west that influenced the economies of states as far away as Venice, ITALY. The straits flow between the Malay Peninsula and the island of SUMATRA, ranging in width from 9.2 to 125 mi (14.8 to 233 km).

The narrowness of these straits, combined with their economic importance as the shortest sea route between three of the world's most populous countries—INDIA, CHINA, and INDONESIA—makes them one of the most strategic economic zones in the world today.

The first major empire in Southeast Asia, the Srivijaya, based its power on control of the straits in the 7th century. Located roughly at the midway point between INDIA and China, the straits drew traders from both of these nations who established settlements along its banks. Attracted mainly to the spices of the east but also to trade in hardwoods, silks, porcelain, slaves, and exotic animals, Arab traders arrived in the area, followed by Portuguese and Dutch traders.

The city of Melaka ultimately became a British possession, along with the important island trading settlements at the northern and southern ends of the straits, Penang and SINGAPORE. In 1824, these settlements were joined together by the British East India Company to form the Straits Settlements Colony (with its capital at Singapore). The Dutch East India Company controlled the southern shores of the straits (on the island of Sumatra) but were never able to consolidate a counterbalance to British control because of the swampiness of this coast and the ongoing difficulties they encountered with the local sultanate of ACEH.

MORE TRAFFIC

The opening of the Suez Canal in 1869 created even more traffic in the straits, and British Singapore developed into one of the world's major trade ENTREPOTS, rivaling Spanish Manila and Dutch Jakarta. Ships from Europe, India, and China had to pass by Singapore, and early on it was a free port, unlike its colonial rivals. The northern coast of the straits was transformed by the British into rubber plantations and settled by thousands of Chinese and Indian workers. This area still has one of the greatest concentrations of Indians outside of South Asia. Another big industry stimulated by the British was tin because of the close proximity of its source in the Main Range to the coast (about 25 to 30 mi or 40 to 50 km).

The northern coast of the straits, hemmed in by this mountain range, has only about a quarter of the land of the Malay Peninsula, but 85 percent of all economic activity, and the 10 largest cities, including the capital of MALAYSIA, Kuala Lumpur, and its port city, Kelang. Singapore is mostly populated by Chinese, and continues to dominate trade in the region. The southern coast, now part of Indonesia, remains largely undeveloped. Both coasts of the Malacca Straits are mostly unbroken mangrove swamps and mudflats, with a few bays that shelter coconut palms.

Today, the Malacca Straits are considered one of the world's oil transit CHOKE POINTS, with an estimated 11 million barrels per day passing between its shores from the MIDDLE EAST to JAPAN, China, and the rest of the Pacific Rim. Some 900 ships per day (or 50,000 per year, carrying nearly $1 trillion in goods) pass through this natural choke point bottleneck, with significant potential for disaster from collisions and spills, piracy, and terrorism. The straits also support large fishing and tourism industries, which are also threatened by shipping accidents and oil spills.

BIBLIOGRAPHY. Donald B. Freeman, *The Straits of Malacca: Gateway or Gauntlet?* (McGill-Queen's University Press, 2003); Barbara A. Weightman, ed., *Dragons and Tigers: A Geography of South, East and Southeast Asia* (Wiley, 2002); "World Oil Transit Chokepoints Country Analysis Brief," www.eia.doe.gov (April 2004); Malacca Straits Research and Development Centre, www.fsas.

upm.edu.my (April 2004); *Oxford Essential Geographical Dictionary* (Oxford University Press, 2003).

JONATHAN SPANGLER
SMITHSONIAN INSTITUTION

Malawi

Map Page 1116 **Area** 45,745 square mi (118,480 square km) **Population** 12,000,000 **Capital** Lilongwe **Highest Point** 9,843 ft (3,002 m) **Lowest Point** 105 ft (37 m) **GDP per capita** $220 **Primary Natural Resources** almost entirely agricultural.

MALAWI IS A SMALL, landlocked country in central southeastern Africa. It is about the size of LOUISIANA or somewhat larger than CUBA. Like Cuba, Malawi is long and narrow. It stretches north to south for about 520 mi (835 km). It is only 50 to 100 mi (80 to 160 km) wide. Malawi's borders MOZAMBIQUE on the east, south and west. Its western border is with ZAMBIA and TANZANIA lies to the north. Most of Malawi's eastern border is Lake Nyasa (called Lake MALAWI in Malawi), opposite Mozambique in the south and Tanzania in the north.

The east African Rift Valley cuts through the central African plateau for the whole length of Malawi from south to north. The Rift Valley creates Malawi's two most prominent features: Lake Nyasa and the adjacent plateaus. Lake Nyasa fills much of Malawi's part of the Rift Valley. The lake's surface is about 1,500 ft (472 m) above sea level. The lake's shoreline is a flat plain with many swamps. The Songwe River feeds Lake Nyasa in the north. The Shire River, the outlet for Lake Nyasa, flows south through Lake Malombe to join the Zambezi River as a tributary.

PLATEAUS AND HIGHLANDS

Plateaus cover about three-fourths of Malawi's land area. The plateaus have a varied terrain that includes plains, rounded mountains, inselbergs ("island mountains"), and forests. West of Lake Nyasa is a broad plain on top of the central plateau region. It averages between 3,000 and 4,000 ft (900 and 1,200 m) in height. The Dedza highlands on the southwestern edge of the central plateau border Mozambique. In the north are the Chimaliro Hills and the Viphya Highlands. The north's unique Nyika plateau section reaches an elevation of about 8,000 ft (2,400 m). South of Lake Nyasa on the eastern side of the Shire River are the Shire Highlands. This plateau region holds Malawi's (and Central Africa's) highest point, Sapitwa (Mount Mlanje).

Malawi's climate is subtropical, with a rainy season (November to May) and a dry season (May to November). The dry season is much cooler than the rainy season. The highlands are much wetter and cooler than the lowlands. Rainfall in the highlands is about 90 in (230 cm) a year. In the Shire River Valley the rain is about 30 in (80 cm) per year. Temperatures in Malawi are affected by the season, altitude, and latitude from the south to north.

Temperatures are hottest in the Shire River Valley just before and after the rains. Cooler temperatures occur in the higher elevations. Malawi's ancient volcanic soils are very fertile. This has encouraged extensive agricultural activity as the basis of the economy. Most of the people of Malawi are Bantu speaking; English and Chichewa are both designated official languages. Since July 6, 1964, Malawi has been an independent republic.

BIBLIOGRAPHY. Samuel Decalo, *Malawi* (Clio Press, 1995); John Douglas and Kelly White, *Spectrum Guide to Malawi* (Interlink Publishing Group, 2002); Owen J.M. Kalinga and Cynthia A. Crosby, *Historical Dictionary of Malawi* (Rowman and Littlefield, 2001); C.G.C. Martin, *Maps and Surveys of Malawi* (A.A. Publishers, 1980); Jocelyn Murray, *Cultural Atlas of Africa* (Checkmark Books, 1998).

ANDREW J. WASKEY
DALTON STATE COLLEGE

Malawi, Lake

LAKE MALAWI IS LOCATED in southeastern Africa lying within the crease where the eastern and western branches of the Great Rift Valley meet. According to the Center of Great Lakes Studies, Lake Malawi is the ninth-largest lake in the world by area and the fifth-largest lake in the world by volume. Malawi is the third-largest lake in Africa, covering 11,600 square mi (30,044 square km), and is 375 mi (603 km) long and

from 15 to 54 mi (24 to 87 km) wide. The lake is at 1,550 ft (471 m) elevation and reaches a maximum depth of 2,487 ft (758 m). It is bordered on the south and west by the country of MALAWI, on the north and east by TANZANIA, and by MOZAMBIQUE to the east. The Malawi-Tanzania border follows the northeastern shoreline of Lake Malawi. The Malawi-Mozambique border runs along the center of the southern part of the lake, with the exception of the islands of Likoma and Chisumulu, which belong to Malawi. Likoma is the largest island on the lake.

The lake is a major tourist destination in this part of Africa, and many resorts have been developed along the southern shores of the lake, where the water is shallower and there are many beautiful white sandy beaches. With the exception of the southern portion of the lake, steep mountain slopes surround Malawi. The Livingstone Mountains rise from the northeastern shore of the lake and are covered by dense forests. Along the north shore of the lake grow thick beds of papyrus reeds, which are the nesting grounds for several species of birds, including kingfishers, cormorants, fish-eagles, and numerous others.

Other inhabitants of the lake include hippopotamuses and crocodiles. Much of the lake's economy is based upon tourists and fishing. The natives that live around the lake bring in over 7,000 tons of fish per year. Most of the fish is dried and used locally, but some of it is exported.

From the north in Tanzania, Lake Malawi is fed by the Ruhuhu River and from the west three small mountain streams, Bua, Dwangwa, and Songwe, feed the lake. The Shire River flows out of Lake Malawi at its southern end and is a tributary of the great Zambezi River. The Shire is greatly affected by the level of the lake, which varies by up to 20 ft (6 m). These changes in water level have been studied and, because they occur every 11years, are believed to be linked to the sunspot cycle.

Lake Malawi could also be known as the lake with many names. Prior to the 18th century, the lake was known as Zaflan, Zambre, Hemozura, and Lake Maravi. In 1859, David Livingstone was the first European to map Lake Malawi. At that time, it was named by Livingstone as Lake Nyasa, which was its native name meaning simply "lake."

In 1964, Nyasaland, formerly a British Commonwealth, became the fully independent country of Malawi, and Lake Nyasa became Lake Malawi to the Malawi people. In Mozambique, they still call the lake Niassa, according to the Portuguese spelling of Nyasa. In literature, it is called Lake of Stars, for the sun glittering off the lake during sunset, and Lake of Storms, for the unpredictable and extremely violent gales that sweep through the area.

BIBLIOGRAPHY. Vera Garland, *Malawi: Lake of Stars* (Central Africana Limited, 1993); R. Kay Gresswell and Anthony Huxley, eds., *Standard Encyclopedia of the World's Rivers and Lakes* (Weidenfeld and Nicolson, 1965); Saul B. Cohen, ed., *The Columbia Gazetteer of the World* (Columbia University Press, 1998).

CHRISTY A. DONALDSON
MONTANA STATE UNIVERSITY

Malaysia

Map Page 1124 **Area** 127,316 square mi (329,750 square km) **Population** 23,522,482 **Capital** Kuala Lumpur **Highest Point** 13,450 ft (4,100 m) **Lowest Point** 0 m **GDP per capita** $4,530 **Primary Natural Resources** tin, petroleum, timber, copper, iron ore.

MALAYSIA IS LOCATED in the heart of Southeast Asia and is divided into two geographical sections: Peninsular Malaysia and the East Malaysian provinces of Sabah and Sarawak, which are located in northern BORNEO, some 403 mi (650 km) across the South China Sea. Malaysia's peninsular neighbors are THAILAND to the north and the island country of SINGAPORE in the south. The Andaman Sea and the MALACCA Straits are on the west coast of Peninsular Malaysia, while the South China Sea borders both the east coast of Peninsular Malaysia and the East Malaysian provinces of Sabah and Sarawak. Sabah and Sarawak border Kalimantan (the Indonesian part of Borneo) and Sarawak surrounds the tiny enclave of BRUNEI. Because of Malaysia's location on a peninsula, it has 2,900 mi (4,675 km) of coastline.

GEOGRAPHY

Peninsular Malaysia's 50,580 square mi (131,000 square km) accounts for 40 percent of the country's landmass. There are several mountain ranges running north to south along the backbone of the peninsula,

and a narrow causeway that connects it with Singapore to the south. A wide, fertile plain extends along the West Coast next to the Malacca Straits, while a narrow coastal plain runs along the east coast next to the South China Sea.

East Malaysia (76,458 square mi or 198,000 square km) lies east of Peninsular Malaysia across the South China Sea and occupies a broad strip running from the westernmost to the northernmost tip of the island of Borneo. Dense jungles and tropical RAINFORESTs cover East Malaysia. Because of the rugged conditions, most of the local natives use the large river networks as the main means of transportation.

Malaysia is hot and humid all year round and over 60 percent of the country is considered to be rainforest. The east coast of Peninsular Malaysia has a distinct rainy season because of the monsoon climate. The wettest season on the west coast of the peninsula is between September and December, while on the east coast and in Sabah and Sarawak, the rainy season comes between October and February. The rains often come in short, strong bursts where the water seems to come straight down as if being poured out of a bucket. When they just as quickly disappear, the humidity level soars and you wish for another rain.

PEOPLE

As the economy develops, Malaysia is becoming more and more urban, with cities accounting for 40 percent of the total population. The principal urban concentration is in the Klang Valley, around the capital city of Kuala Lumpur (1.4 million), where the population has been growing annually at a rate of 7 percent. Peninsular Malaysia is also more heavily populated than East Malaysia with nearly 85 percent of the country's total population. Malaysia is also a very young country, with 38 percent of the total population under the age of 15 and only 4 percent over the age of 65. Malays and other indigenous ethnic groups (together known as Bumiputras) account for 59 percent of the population, the Chinese 24 percent and Indians 8 percent. The official language is Bahasa Malaysia (Malay), but English still predominates in industry and commerce. The official religion of the country is Islam, but there is freedom of worship.

ECONOMY

Malaysia has historically been an exporter of primary products such as rubber, tin, palm oil, and timber. Malaysia has long been an important exporter of palm oil (50 percent of world production), generating ex-

Malaysia's capital city of Kuala Lumpur (1.4 million) has a population that has been growing annually at a rate of 7 percent.

ports of about $2 billion annually. In the mid-1970s export manufacturing, in particular of textiles and electronics, began to develop, based on an accommodative government attitude toward incoming foreign investment and a skilled low-cost labor force. As a result, Malaysia has become one of the world's three largest manufacturers of semiconductors and air conditioners. Malaysia's economy has enjoyed very positive real annual growth during the last decade, with the manufacturing sector contributing more than 40 percent of the total growth, and the finance, agriculture, and transport sectors each contributing about 10 percent. Malaysia is often cited as one of the Asian Tigers because of its relatively high gross domestic product per capita. The key industries in Peninsular Malaysia include rubber and oil palm processing and manufacturing, light manufacturing industry, electronics, tin mining and smelting, and logging and processing timber. Sabah in East Malaysia is known for its logging and petroleum production, while agricultural processing, petroleum production and refining, and logging are done in Sarawak.

HISTORY

Because of the Malacca Straits, which provide an inside passage from the South China Sea to INDIA, Malaysia has been a strategic factor in trade for thousands of years. From the 900s on, there were numerous Chinese and other settlements established to support the great sailing fleets moving goods between Asia, India, and the MIDDLE EAST. With the formation of the Dutch East India Company, the region became a stronghold for

Europeans trying to protect their interests in the highly profitable spice trade and ultimately led to the founding of SINGAPORE. In 1957, the Federation of Malaya, comprising the 11 states of Peninsular Malaysia, gained independence from the UNITED KINGDOM. The federation became known as Malaysia with the accession of three further states, Singapore, Sarawak, and Sabah, in 1963. In August 1965, Singapore withdrew from the federation to become an independent sovereign state. Malaysia is a federal constitutional monarchy. Its head of state is the king, who is elected for a five-year term by the nine hereditary sultans of the eleven states of Peninsular Malaysia.

BIBLIOGRAPHY. Keith Taylor, "The Early Kingdoms," *The Cambridge History of Southeast Asia* (Cambridge University Press, 1999); Anthony Reid, *Southeast Asia in the Age of Commerce: The Land below the Winds* (Yale University Press, 1988); Mary Somers Heidhues, *Southeast Asia: A Concise History* (Thames and Hudson, 2000); Jack-Hinton, Colin, *A Sketch Map History of Malaya, Sarawak, Sabah and Singapore* (Hulton Educational, 1966); Jan Pluvier, *Southeast Asia from Colonialism to Independence* (Oxford University Press, 1974).

RICHARD W. DAWSON
CHINA AGRICULTURAL UNIVERSITY

Maldive Islands

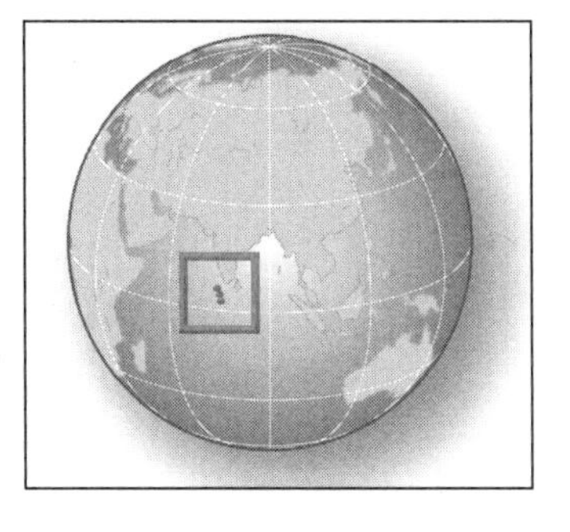

Map Page 1123 **Area** 115 square mi (300 square km) **Population** 329,684 (2003) **Capital** Malé **Highest Point** 7.8 ft (2.4 m) **Lowest Point** 0 m **GDP per capita** $3,900 **Primary Natural Resources** coconuts, corn, sweet potatoes, fish, tourism.

THE COUNTRY OF THE Maldive Islands is located in southern Asia, in the INDIAN OCEAN, as a group of atolls south-southwest of INDIA. This remote island nation is comprised of 1,191 coral islands amid major shipping routes in the Indian Ocean. Archaeological evidence of settlement from South Asia is datable to around 500 B.C.E. The native language of Dhivehi points to Sri Lanka as the source of the Maldives' earliest continuous cultural roots. The details of the conversion of the Buddhist king to Islam is wrapped in legend but venerated as the key point of the islands' history. In 1153, the king adopted Islam and thereupon the title of Sultan Muhammad al Adil. This initiated a series of six dynasties consisting of 84 sultans and sultanas that lasted until 1932, when the sultanate became elective.

For the vast bulk of its sultanate history, the Maldives remained independent. Only in the 16th century did the Portuguese first come upon the Maldives in pursuit of the spice trade. In 1558, the Portuguese invaded Malé only to be ejected 15 years later. The vying colonial powers continued to supplant one another. In the mid-17th century, the Dutch controlled the spice trade but put no colonial government in the Maldives. The British moved the Dutch aside, and by 1887, it officially made Maldives a British protectorate, leaving internal affairs to be managed by the sultan.

During the 1950s, British military activity dominated Maldivian economics and politics. In 1956, the British gained license to reestablish its wartime airfield on Gan Island at very concessionary terms. Newly elected Prime Minister Ibrahim Nasir questioned this agreement and called for official review. Peoples of the southern atolls and leaders who benefited most from the British presence started a secessionist movement to establish an independent state. In 1962, Nasir sent government police to successfully suppress this insurrection. Nasir also renegotiated the terms of the lease of the airfield at Gan.

On July 26, 1965, the Maldives gained independence from Britain. Three years later they voted to abolish the sultanate and established a republic. Ibrahim Nasir dominated politics for the next 10 years until he fled the country for SINGAPORE, taking more than his share of the nation's wealth.

Economic downturns in the fishing industry and the British decision to close the airbase on Gan in 1975 were great a challenges for the government. A former college lecturer and ambassador, Maumoon Abdul Gayoom was elected president in Nasir's place. Gayoom has been in power since 1978, surviving three coup attempts and remaining popular with the Maldive people.

Recent years have seen continuing economic growth and modernization. This growth has been primarily in the tourist industry with profits going to improve other sectors as well. The Maldives have been the recipient of generous and consistent foreign aid. There remains a tension between the progressive nature of the tourist industry and the conservative tone of

a traditional Muslim way of life. Sustained use of the marine resources in the tourist and fishing industry are a continuing challenge.

BIBLIOGRAPHY. "Background Note: Maldives," U.S. Department of State, www.state.gov (April 2004); *World Factbook* (CIA, 2004); H.J. de Blij and Peter O. Muller, *Geography: Realms, Regions, and Concepts* (Wiley, 2002).

IVAN B. WELCH
OMNI INTELLIGENCE, INC.

Mali

Map Page 1113 **Area** 478,640 square mi (1,240,000 square km) **Population** 11,626,219 **Capital** Bamako **Highest Point** 3,789 ft (1,155 m) **Lowest Point** 75 ft (23 m) **GDP per capita** $900 **Primary Natural Resources** gold, phosphates, kaolin, salt, limestone.

A LANDLOCKED REPUBLIC and the largest country in West Africa, Mali is bordered by ALGERIA to the northeast, MAURITANIA to the northwest, SENEGAL and GUINEA to the west, CÔTE D'IVOIRE and BURKINA FASO to the south, and NIGER to the southeast. The capital, Bamako, is by far Mali's largest city, though there are a number of smaller urban areas that include Kayes, Sikasso, Segou, Mopti, Djenné, Timbuktu (Tombouctou), and Gao.

Mali's climate is one of the harshest in Africa, as most of the country is arid or semi-arid: only 4 percent of the country is arable land. Three major rivers run through Mali: the Niger and the Bani meet at Mopti in the center of the country, and the Senegal in the southwest.

Mali's population is among the most ethnically diverse in Africa and is divided geographically north to south. Mande peoples (Bambara, Malinke, Soninke) in the south make up half of Mali's population; the central areas contain largely Peuls/Fulanis (17 percent) Voltaic peoples (12 percent), and the Songhai (6 percent); and the northern desert is largely occupied by the Tuareg and the Moors (10 percent). The remaining 5 percent consist of minuscule ethnic groups scattered throughout the country. Ninety percent of Mali is Muslim, 9 percent practice indigenous religions, and 1 percent is Christian. French is Mali's official language, though Bambara is also widely spoken.

Mali has consistently ranked amongst the poorest countries in the world. The country imports most basic resources because of lack of farmland and consistent water sources. While Mali is mineral-rich, economic woes and political struggles have prevented most of those materials from being fully exploited.

HISTORY OF MALI

Mali's history is among the most ancient in the world, with evidence of civilization from the 11th century B.C.E. Mali was also the seat of the great African trade empires of GHANA, Mali (for which the country is named), and Songhai from the 6th to the 16th century, and its political and economic dominance in West Africa is well documented by Arab geographers, European travelers and local historians.

Mali fell into virtual anarchy after the destruction of the Songhai Empire in the late 16th century. The region was briefly united under the Tukolor Empire in the mid-19th century, then was conquered by the French between 1893 and 1898. Mali became part of French West Africa until 1958, though it gained total independence in 1960.

Mali since independence has experienced considerable political and economic instability. The country became a one-part socialist state and tried to separate economically from France, which caused a financial crisis during the 1960s. During the chaos that resulted, Moussa Traoré installed himself as dictator in 1968. Severe droughts in the 1970s destroyed Mali's delicate economy, and led to unrest and escalating conflicts with Burkina Faso and the Tuareg ethnic minority in the north during the 1980s and 1990s.

Though Traoré was ousted by a coup in 1991, Mali did not establish a true multiparty republic until the elections of 2002. Amadou Touré, a former interim military ruler, became Mali's first unopposed elected president.

BIBLIOGRAPHY. *World Factbook* (CIA, 2004); N. Levtzion, *Ancient Ghana and Mali* (Heinneman Library, 1973); P. J. Imperato, *Historical Dictionary of Mali* (Rowman and Littlefield, 1986); G. Connah, *African Civilizations: A Historical Perspective* (Cambridge University Press, 2001).

PILAR QUEZZAIRE-BELLE
HARVARD UNIVERSITY

Malta

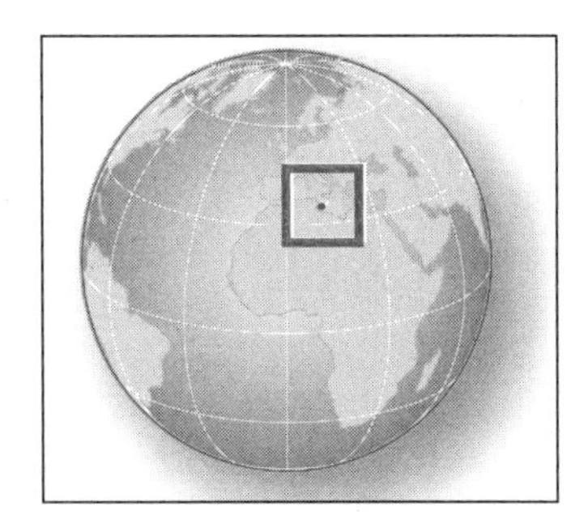

Map Page 1131 **Area** 122 square mi (316 square km) **Population** 400,420 **Capital** Valletta **Highest Point** 830 ft (253 m) **Lowest Point** 0 m **GDP per capita** $10,434 **Primary Natural Resources** globigerina limestone, petroleum, tourism.

THE COUNTRY OF MALTA, which calls itself the Repubblika Ta' Malta, is an archipelago made up of five islands: Malta, Gozo, Comino, Cominotto, and Filfla, as well as two small islets known together as St. Paul's Islands. Only the first three, Malta, Gozo, and Comino, are inhabited. Malta is near the center of the MEDITERRANEAN SEA, 58 mi (93 km) south of Sicily and 179 mi (228 km) north of LIBYA, although its latitude is south of Tunis. It is about midway between GIBRALTAR and Tel Aviv, ISRAEL, at the opposite end of the Mediterranean. The main island of Malta is farthest east, and, in parts, is densely populated. On its southeast is the Grand Harbor, which made it an attractive port for the ancient Phoenicians, Carthiginians, and Romans for centuries. Today it is considered an underdeveloped area of fishing villages.

None of the islands have permanent lakes or streams, and 70 percent of the water used on Malta comes from desalinization plants. Malta has a high population density, particularly in the areas near its largest city, Birkirkara, and its capital, Valletta, both located on the northeast section of Malta Island. Malta's tourism industry is fueled by its beaches and mild climate and by its historical sites. Tourism accounts for one-quarter of Malta's gross domestic product, and the islands host over 1 million visitors per year. Many come to enjoy 85 mi (137 km) of beaches and year-round sunshine; Malta's average monthly temperature ranges from 54 to 88 degrees F (12 to 31 degrees C). Diving, windsurfing, sailing, and other sports attract visitors year round.

Megaliths and religious temples from the 4th millennium B.C.E, predating the rise of Sumer and EGYPT, have been excavated on Malta and Gozo. Roman temples and villages, walled medieval towns, domed Renaissance churches, and forts built by knights of the Crusades are popular tourist sites. Phoenicia and Carthage maintained ports on Malta for trade until the islands were subsumed into the Roman Empire in 218 b.c.e., after the Second Punic War. In Biblical times, St. Paul was shipwrecked in a bay that still bears his name. From the 6th through the 9th century C.E., Malta was part of the Byzantine Empire, then it passed into Arab hands. In 1090, Normans drove out the Arabs and made Malta an appendage of Sicily. As such, it passed through the ownership of several European kingdoms until 1530. At that point, Charles V of the Holy Roman Empire ceded Malta to the Knights Hospitalers of St. John of Jerusalem, who are sometimes known as the Knights of Malta. Many of the spectacular ruins of Malta were built by this group, which successfully defended the island against Suleiman the Magnificent and the Turkish Empire. Napoleon took over Malta in 1798, and the BRITISH EMPIRE was in control by 1814. The scene of much fighting in World War II, Malta gained independence in 1964.

Malta today is a republic, with an elected president and unicameral legislature and an appointed prime minister. It officially joined the EUROPEAN UNION in 2004. The country is 98 percent Roman Catholic, and the official languages spoken are English and Maltese, a language that developed from North African Arabic and Sicilian Italian.

BIBLIOGRAPHY. Brian Blouet, *A Short History of Malta* (Frederick A. Praeger, 1967); Malta Tourism Authority, "Malta, the Heart of the Mediterranean," www.visitmalta.com (March 2004); U.S. Department of State, "Background Note: Malta Profile," www.state.gov (March 2004).

VICKEY KALAMBAKAL
INDEPENDENT SCHOLAR

Manila

CONSIDERED TO BE the best harbor in the Far East, known as the "Pearl of the Orient," Manila has been a commercial and cultural hub of the PHILIPPINES, and of Southeast Asia, for over four centuries. Ferdinand MAGELLAN claimed the island of LUZON for SPAIN in 1521, as a strategic point of access to the Spice Islands of the Far East. Manila, on the northwest coast of the island, was captured from its sultan in 1571 and soon became the main exchange point for goods traveling between Macao and Acapulco—silver from MEXICO was traded here for silk from south CHINA. Manila became a rival port to Batavia, the main port of the Dutch East India Company (today's JAKARTA, INDONE-

SIA). By the 17th and 18th centuries, it was essentially a Chinese city, following the pattern of most commercial cities in Southeast Asia, with manual labor provided by Chinese-Filipino or Mexican-Filipino mestizos.

The revolt against Spain in 1896 started here, and the arrival of the UNITED STATES in Manila Bay in 1898 signaled the end of nearly four centuries of Spanish rule. The United States did not depart as planned, however, and maintained a protectorate over the emerging state until it was granted full independence in 1946. American presence remained heavy after independence, however, particularly in the Manila area, because of its large military presence at Subic Bay naval base and Clark Air Force Base, which were given up only in 1992.

The city, originally called Maynilad ("there is *nilad*," a local white-flowered mangrove), was a Muslim sultanate from the early 16th century. Geographically situated on a plain at the confluence of several mountain ranges, the Laguna de Bay and Manila Bay, it was an obvious location for an urban area that today forms the core of the world's seventh-largest metropolitan area, with about 10.3 million residents in Manila proper and 16 other cities (including Quezon City, which is actually more populous than Manila and was the capital city from 1939 to 1976). It is an amazing mixing bowl of cultures, combining elements of Spanish, Chinese, English, and Tagalog culture.

The city is centered on both sides of the Pasig River, where it enters Manila Bay. The old walled city, or Intramuros, is on the southern, or left, bank of the river. It consists of old Spanish houses, churches and convents, and narrow streets and is surrounded by a defensive wall, built around 1790 to defend against attacks form the Dutch and British. The old town also includes the old city hall (Ayuntamiento) and the University of Santo Tomás, founded by the Dominican Order in 1605. The city's other major university, and the oldest in Asia, is the University of the Philippines, founded by King Philip II of Spain in 1585.

North of the river lies the Binondo section, Manila's Chinese commercial district, famous for its wild bargain basement bazaars in the Divisoria. Nearby is the old governor's palace, Malacañang, now the residence of the Philippine president. North and south of the river the city spreads out in newer neighborhoods, including Ermita and Malate, the center for tourist hotels, restaurants and nightclubs, and Quiapo, Manila's congested downtown. As development continued south along the bay, the large Rizal Park was surrounded by more aristocratic suburbs in the 19th century, and by later American developments, which bear the names of men like President William McKinley and Admiral George Dewey, the victor of the battle of Manila Bay in 1898.

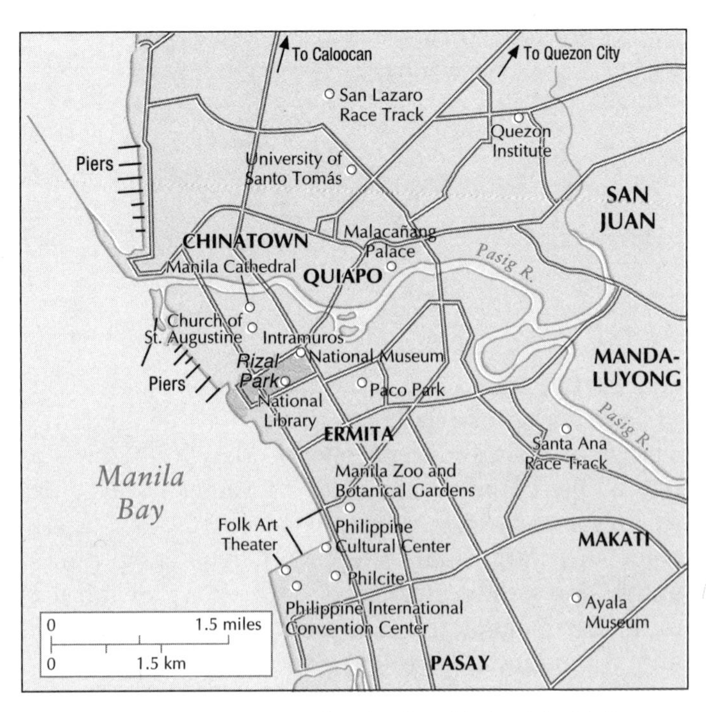

Manila has been a commercial and cultural hub of the Philippines, and of Southeast Asia, for over four centuries.

The city has spread out from the bay to join into an urban conglomeration with Quezon City to the northeast and other smaller towns. The language spoken around the bay, Tagalog, has become the official national language, though English is the lingua franca used in many of the southern islands. Manufacturing is highly concentrated in metro Manila, more than half of the total of the Philippines.

Manila is the center of international trade for the Philippines, plus domestic trade between islands. Raw materials brought in from the interior of Luzon or from other islands is processed and exported, from sugar, rope, and cigarettes/cigars to shoes and woven textiles. The city struggles with air and water pollution. The government has designs to transform the former U.S. military bases into free-trade zones. The search for foreign investment in this project has found some success from Taiwan, but not enough, because of the economic downturn throughout Asia in the 1990s.

BIBLIOGRAPHY. Sylvia Mayuga and Alfred Yuson, *The Philippines* (Apa Productions, 1980); Barbara A. Weightman, ed., *Dragons and Tigers: A Geography of South, East*

and Southeast Asia (Wiley, 2002); "Manila," www.cityofmanila.com.ph (May 2004).

Jonathan Spangler
Smithsonian Institution

maps and globes

A MAP IS AN abstract representation of a selected set of features on or related to the surface of the Earth. The map reduces these selected Earth features to points, lines, and areas, using a number of visual resources such as size, shape, value, texture or pattern, color, orientation, and shape. Whereas aerial photographs and satellite images are realistic representations of the Earth, maps are an abstraction designed to focus and communicate specific information. The map is often drawn to scale, has a coordinate reference system to locate features, and is constructed on a flat medium such as paper or plastic film.

A globe, on the other hand, is a scaled representation of features on or related to the surface of the Earth and constructed on a three-dimensional surface such as a sphere. Globes are both objects of decoration and scientific value. Cartographers refer to a hypothetical "reference globe" as a scaled representation of the Earth that is then transformed point by point on a flat medium in an important process known as map projection. But globes are expensive to make, difficult to reproduce, inconvenient to store, and difficult to make measurements on. Maps created on a flat surface are not affected by these challenges.

OLDEST MAPS

The oldest known maps are preserved on clay tablets from the Babylonian period (2300 B.C.E.). The Greeks also possessed advanced mapping knowledge and the concept of the spherical Earth was well known to Greek philosophers such as Aristotle (350 B.C.E.). During the 12th century onward, maps became influenced by religious views, exploration endeavors, and political ambitions and were highly valued for their economic and military uses. A significant benefit of historic maps is that they provide important clues about the social and cartographic traditions of past societies and civilizations.

Modern maps became increasingly accurate and factual from the 17th century onward. Developments in astronomy and cartography had provided a scientific basis for preparing maps. An appreciation of maps and their uses begins with an understanding about how maps work. All maps are about spaces and places that are represented by shape, area, distance, direction, and location in a graphical medium. The surface of the Earth is not a flat plane, so a modification is required to transform the positions of places on the curved Earth to the flat sheet of the map so that distortions in shape, area, distance, direction, and location are minimized. This process is called map projection, and the transformation is governed by rigorous mathematical rules. The projection process takes the lines of latitude and longitude of the spherical Earth and arranges them on a flat plane as a uniform grid. These grids together with a scale that links the relative linear proportions of the spherical Earth and its representation on the flat plane allow the map space to be structured such that map properties can be determined to a high level of accuracy. Some examples of projections include the Mercator and Robinson projections.

MAP SCALE

The map SCALE determines the level of detail the map can show. Maps of a large scale show more detail with greater accuracy. As the map scale becomes progressively smaller, larger swaths of geographic areas are shown and so features on the map must be generalized to avoid congestion. The generalization procedure involves stages of simplification, selection, enlargement, displacement, and merging. Simplification involves the progressive collapsing of map features from area to line to point representation as scale decreases. As an example, a lake may be represented as an area at one scale but as a point at a smaller scale. Selection attempts to retain features that are important given the goals and uses of the map. However, some of these important features might not be clearly visible at the desired scale and so enlargement artificially distorts their dimensions to enhance visibility. Displacement shifts overlapping features so that they become separate elements and more clearly identifiable. Merging aggregates multiple features into simpler ones to correct map overload that can arise from too much detail.

The mapping process also includes the symbolization of the real world using a standardized graphical language. The symbols used have dimensions—point, lines, areas, volumes, and duration—and can be distributed in a discrete, continuous or sequential manner to communicate feature patterns. Lettering and text labels also form an important part of the feature encoding process. Information about each feature, such as

type of road or population, is encoded using variations in nine graphic variables. The graphic variables are size, shape, orientation, color value, color intensity, level of grayness, texture, focus, and arrangement. It is these graphic variables together with strict rules for their use that allow the cartographer to encode and distinguish the diversity of features from the spherical real world into the flat map.

MAP CATEGORIES

Generally, there are two categories of maps: topographic and thematic maps. The topographic map shows the outlines of selected natural and anthropogenic features, and is used mostly as a reference tool or as a base for integrating other types of existing or new information. The thematic map is used to communicate geographical notions such as population densities and land use. Thematic maps are especially important because they are used extensively in geographical information systems (GIS) as digital mapping outputs. Choropleth thematic maps use units such as counties or census tracts to display aggregate data about income and population for example. The area class thematic map shows units of constant attributes, such as vegetation. The isopleth thematic map shows an imaginary surface constructed by using lines to join points of equal value such as in a temperature and contour map.

Maps provide useful and concise spatial information in a meaningful manner and we use them directly or indirectly in many aspects of our daily lives. Navigating the roadways, finding new places, or simply reflecting on the aesthetics of the map are typical examples of how we use maps. Maps are also used for storing data, as a spatial index for labeling features or integrating multiple map sheets, and as a spatial data analysis tool for planning and decision making. A simple but powerful analysis tool is the map overlay process developed by Ian McHarg in which multiple map layers on transparent film are overlaid to identify regions of interest.

GLOBES

The globe is a three-dimensional representation of the Earth and has an entirely different but familiar construction than a flat map. The globe is composed of longitude lines (meridians), which run north-south, and latitude lines (parallels), which run east-west. The longitude lines converge at the North Pole in the Northern Hemisphere, and at the South Pole in the Southern Hemisphere. The equator divides the Earth into two hemispheres. All meridians and the equator are GREAT CIRCLES since they can form planes that cut the surface and pass through the center of the Earth. Small circles such as latitude lines form a plane that cuts the surface but does not pass through the center of the Earth. In this system of reference, geographic coordinates are measured in units of angular degrees. There are 360 degrees of longitude around the equator, with each meridian numbered from 0 to 180 degrees east and west such that the 180 degree meridian is on the opposite side of the Earth from Greenwich, England. There are 180 degrees of latitude from pole to pole, with the equator being 0 degrees and the north and south poles being 90 degrees. Each degree can be divided into 60 minutes and each minute is divided into 60 seconds. The north-south line is called the prime meridian, which has been set to pass through the Royal Observatory in Greenwich.

The longitude is measured as the angle between the point, the center of the earth and the prime meridian at the same latitude. West is positive and east is negative, meeting at 180 degrees at the international dateline. The east-west line follows the equator and is midway between the north and south poles. Degrees of latitude are measured as the angle between the point on the surface, the center of the earth, and a point on the equator at the same longitude. Here, the size of the cells is largest at the equator and the zones are square. At the poles the zones are smallest and mostly triangular.

BIBLIOGRAPHY. Borden Dent, *Cartography* (McGraw Hill, 1999); Nathaniel Harris, *Mapping the World: Maps and their History* (Thunder Bay Press, 2002); Ian McHarg, *Design with Nature* (Doubleday, 1969); Phillip Muehrcke, *Map Use: Reading, Analysis, Interpretation* (JP Publications, 1986); Arthur Robinson, Joel Morrison, Phillip Muehrcke, Jon Kimerling, and Stephen Guptill, *Elements of Cartography* (Wiley, 1995); Terry Slocum, *Thematic Cartography & Visualization* (Prentice Hall, 1999).

SHIVANAND BALRAM
MCGILL UNIVERSITY, CANADA

market geography

MARKET GEOGRAPHY is a subfield of ECONOMIC GEOGRAPHY that focuses on the spatial nature of market forces. It derives its rationale from the central place theory, first argued in 1933 by German economic geog-

rapher Walter CHRISTALLER in his book on central places in southern Germany. Central place theory is fundamentally concerned with the patterns through which wholesale, retail, service, and administrative functions, plus market oriented manufacturing, are provided to consuming populations.

As such, it complements the theory of agriculture production originally formulated by J.H. von Thunen and the theory of location of industry, which has its roots in the work of Alfred Weber. The increasing dominance of market-based economic functions has established a clear linkage between the workings of market forces and spatial distribution of economic activities. Though location-market linkage is in the agriculture, manufacturing, trade, and transportation sectors, it's more pronounced in the sphere of production and finance.

GEOGRAPHY OF MONEY AND FINANCE

Ron Martin (1999) claims the emergence of an economic geography of money and finance as a subdiscipline, with four identifiable lines of inquiry: 1) Marxist theorization of the geographically uneven and crisis-prone tendencies of the capitalist space economy; 2) specific studies of spatial organization and operation of particular financial institutions, services, and markets; 3) studies of dynamics through which the world's major financial centers are molding global geographies of money (such as LONDON, NEW YORK, and TOKYO); and 4) studies of services to link regional financial flows and regional industrial development. He demonstrates that financial systems are inherently spatial in terms of "location structure" and "institutional geographies" that influence the way money moves between locations and communities as well as "regulatory spaces" and the "public financial space of the state," including the cooperation and competition within the international monetary system. He notes that the money "has a habit to seeking out geographical discontinuities and gaps" to reinforce, rather than reduce, uneven regional pattern.

David J. Porteous (1995) offers a theoretical viewpoint that highlights the significance of "information externalities" and "an information hinterland" to account for urban agglomeration of financial activities. He also explains "spatial switching" as a reversal of the rank order of importance of rival financial centers within national economies, for example, Montreal/Toronto and Melbourne/Sydney. Gordon L. Clark (1993) stresses the importance of U.S. and British geography of pension funds and reveals why this source of capital has such strong influence upon global economy. Clark and other scholars (1997) have attempted an analysis of the "spatial logic" of the financial industry.

David Harvey (1982) stresses the importance of time and geographical space to argue that contradiction in the primary or productive circuit of capital could be dampened down by foreign direct investment to secondary and tertiary circuits. These circuits export wider financial instability in new forms through the creation of fictitious credit money in advance of actual production and consumption—thereby producing the "system's instability." On this platform, Harvey identifies different types of crises, including "sectoral switching crises" (where fixed capital formation is switched to another sphere, such as education), "geographical switching crises" relating to different geographical scales, and "global crises." Harvey considers the further dimension of "GEOPOLITICS" at each scale to suggest that crisis tendencies are not resolved in an orderly manner but involve competitive struggles influenced by discriminatory institutional practices, circuits of money, and cost reduction according to financial, economic, political, and military power.

Stuart Corbridge (1993) suggests limitations of Keynesian measures of financial debt in the consideration of the geography of commercial bank lending in Latin America, the Caribbean, Eastern Europe, and the THIRD WORLD. Others have explored the "world of paper and money" with the following aims: to describe the economy of international money, to link this economy with the distribution of social power, to show some of the ways in which the world of money is discursively constituted through social-cultural practices, and to show how the world of money is constructed out of and through geography. Authors follow neo-Gramscian regulatory theory and operate within four spatial frames: the global monetary economy, the national space of Britain, the regional space of south of England, and the concentrated urban space of the city of London.

The book *Money/Space* is a critique of the rise and fall of "Thatcherism" and "Reaganomics," which marked unprecedented growth of the financial services industry, followed by the rise of disintermediation, a wave of financial innovation, and then more sophisticated forms of risk management. One chapter gives attention to "geographies of financial exclusion" in Britain and the UNITED STATES and suggests that "financial citizenship" is needed as a means of exerting pressure on governments to reform their financial system. The impact of changing geopolitics upon the circula-

tion of capital is enlarged by other scholars. Iain Black (1989) has reconstructed highly fluid spatial mobility of finance within a London-centered political economy during the INDUSTRIAL REVOLUTION and a more contemporary analysis of metropolitan London responses to the financial revolution of the 1980s.

Andrew Leyshon's (1997, 1998) analyses keep abreast of the growth of geographical treatments of money and finance. In one analysis, he links such contributions to the concept of a "geo-political economy," explores the "geo-economics of finance," and reflects upon wide unease over the geographies of financial exclusion. His other analysis draws upon Viviana Zelizer's theory of the social meaning of money, then reviews geographical analysis of the social and cultural construction of financial centers.

GEOGRAPHY OF INDUSTRY

The heart of market geography lies in the spatial patterns and physical landscapes that industry creates. As spatial division of labor disperses industry, such as automobile manufacturing, into a thousand pieces—tire factory here, engine plants there, electronic ignitions and engineering plant somewhere else—it is increasingly difficult to knit together the production function into discrete units called factories. To make matter worse, these webs of production overlap and interconnect in surprising ways that can never be entirely untangled.

This immense geography of production is in constant motion, rendering moot all fixed ideas about industry location patterns. Industrialization drives sectors and places along divergent paths of growth, and disrupts all established geographic habits. That divergence and instability are essential to the uneven development of the industrialized world.

But the movement alone does not capture the creative (and destructive) powers of modern industry. Successive industrial revolutions have built up the great cities, transport systems, and landscapes of production that surround us; industry does not locate in a known world so much as it produces the places it inhabits. This jagged process of industrial development repeatedly outruns prediction and liquidates the geographies of the past; generating endless novelty that makes market geography such a lively and challenging subfield of inquiry.

As the world moves toward the more information-rich forms of production and products, like computer software and video games, there are more products that come in small packages, like CDs, and fewer bulky objects like steel girders. But production is, in all cases, an act of human labor; it involves work, plain and simple. This means that securing a labor force is a prime task of any industrial operation and critical to its locational calculus. Firms must recruit labor either by locating near to where workers live or by attracting them from long distances; this matching of labor demand and supply is the base point for market geography. Different kinds of work demand different kinds of capacities from workers and provide varying levels of wages and other rewards, and here lies an elemental force for spatial differentiation of industrial activities or spatial divisions of labor.

MARKET ECONOMIES

Today's market economies produce millions of different commodities for sale and employ hundreds of millions of people. They are immensely complex systems of production, made up of an extraordinarily large number of pieces. Those bits and pieces constitute "the division of labor" and are the basic building blocks of the industrial system and of market geography. Without the division of labor, there would be no differentiation of economic activities, no factories to site, and no localization of industries. The pied and dappled geography of modern economies comes about precisely because of the wide variety of work being done at different places.

Yet the division of labor is not infinite. Work today is mostly collective labor, where each person is responsible for a part of the whole. These collectivities range from small groups, such as fabric cutter, to whole garment factories, to entire commodity chains. A basic concept of the study of industrial geography is, therefore, a social division of labor.

The term *industrial location* was largely replaced by *spatial division of labor* in the lexicon of economic geographers during the 1980s. The former had come to mean the optimal siting of production units according to their specific needs for inputs, in the tradition of Alfred Weber, or the optimal spatial allocation of sellers according to a highly abstract calculus of access to customers, in Christaller's (1935) central place theory. Industry was assumed to conform to preexisting patterns of people and resources on land.

Doreen Massey (1984) turned this around. For her spatial division of labor signified a view of industrial pattern that recognized powerful forces for spatial differentiation coming out of industry itself and projected onto the landscape. The power of industry and market created a dominant force that effectively shaped the

spatial concerns and in the process created a market geography where market is a senior partner and geography became a junior partner.

A survey of literature in the subfield of market geography indicates that research since 1954 can be divided into three categories, on the basis of approach and methods: qualitative interpretation, usually with substantial numerical evidence and sometimes making use of case-study technique; quantitative classification, in a more or less descriptive sense, with qualitative elaboration and explanation and involving a specific procedure applicable to different areas and time periods; and formulation and testing of specific hypotheses and models. These approaches have been applied, with varying degrees of intensity, to most facets of market geography.

BIBLIOGRAPHY. Chauncy D. Harris, "The Market as a Factor in the Localization of Industry in the United States," *Annals of the Association of American Geographers* (v.44, 1954); Edwin N. Thomas, "Toward an Expanded Central Place Model," *Geographical Review* (v.51, 1961); John Agnew and Stuart Corbridge, *Mastering Space: Hegemony, Territory and International Political Economy* (Routledge, 1995); Leslie Budd and Sam Whimster, eds., *Global Finance and Urban Living: A Study of Metropolitan Change* (Johns Hopkins University Press, 1993); Risto Laulajainan, *Financial Geography: A Banker's View* (School of Economics and Commercial Law, 1998).

JITENDRA UTTAM
JAWAHARLAL NEHRU UNIVERSITY, INDIA

Marshall Islands

Map Page 1125 **Area** 70.7 square mi (181.3 square km) **Capital** Majuro **Population** 56,429 **Highest Point** 33 ft (10 m) **Lowest Point** 0 m **GDP per capita** $1,600 **Primary Natural Resources** coconut products, marine products, seabed minerals.

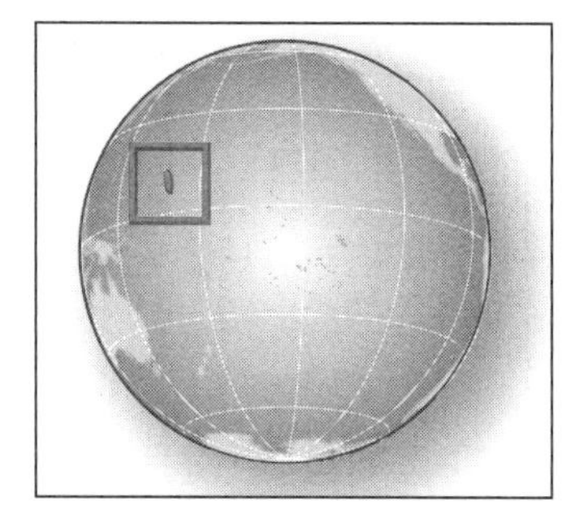

THE REPUBLIC OF THE Marshall Islands, formerly a component of the United Nations Trust Territory of the Pacific Islands (administered by the UNITED STATES since the end of World War II), became an independent state in free association with the United States in 1986. The island country consists of 29 atolls and five islands arranged in two parallel chains, known as Ratak ("sunrise") and Ralik ("sunset"), covering about 750,000 square mi (1,923,077) of open sea. The islands retain a strong U.S. military presence, strategically located halfway between HAWAII and AUSTRALIA, and remain the site of controversial missile testing ranges on a U.S. base at Kwajelein.

The Marshall Islands are the easternmost part of the former Trust Territory, with the Federated States of MICRONESIA (formerly the Caroline Islands) lying to the west and the former British colony of KIRIBATI immediately to the south. The U.S.-held island of Wake is about 500 mi (800 km) to the north, and Hawaii is located 2,300 mi (3,700 km) to the northeast.

Each of the major atolls, crowns of submerged mountains, consists of numerous small islets set in coral reefs surrounding irregularly shaped lagoons—the total number of these islets reaches about 1,150. Kwajalein Atoll has the world's largest lagoon, with 98 associated islets. But only a few of the islets are large enough to support settlements—most of the population lives on Majuro (Mãjro), and the other major islands of Kwajelein (9,311), Jaluit (2,000), and Arno (1,000). Other atolls or islands with populations above 700 include Ebon, Enewetak, Mile, Namorik, and Namu. Residents of Bikini, Enewetak, and Rongelap were relocated to other islands in the 1950s after U.S. nuclear testing made the islands uninhabitable.

The climate is tropical, hot, and humid. The islands are formed of low coral limestone and sand, limiting agricultural output to coconut groves and small plots for melons, taro, and other tropical fruits. The economy is reliant on tourism and fishing, as well as craft items produced from shell, wood, and pearls. Revenue also comes in the form of U.S. financial assistance ($102 million in 1998), both to the central government and to the victims of the nuclear testing of the 1950s, and the usual economic opportunities surrounding a U.S. military installation. The reefs are a major attraction, with several species of coral and five species of marine turtle known only in these islands.

The two chains were settled by Micronesians over 2,000 years ago, who named them Aelon Kein Ad ("our islands"). Encountered by Spanish explorers in 1529, the islands remained off the path of European expansion in the Pacific until they were "rediscovered" and named for the British captain William Marshall in the late 18th century. American whalers and missionaries competed with German entrepreneurs, who set up

the Jaluit Company in the 1870s as a trading post. Formally declared a German Protectorate in 1886, the islands did not remain under German administration, being occupied by the Japanese from 1914 until they were occupied by the United States in 1944. Nuclear testing began on Bikini in 1946, and the first hydrogen bomb was tested on Enewetak in 1952. The islands were self-governing starting in 1979, and the Compact for Free Association with the United States was finally approved and went into effect from 1986. The United States remains in charge of foreign affairs, and provides financial services and certain federal benefits.

Marshall islanders are not U.S. citizens, but are termed "habitual residents," which allows them to enter the United States and work without visas or work permits. The Marshall Islands became a member state of the United Nations, but continue to be reliant on the United States (financial aid is about 55 percent of the national budget). The Free Association Compact was renewed in 2003.

BIBLIOGRAPHY. Ron Crocombe, *The South Pacific* (University of the South Pacific, 2001); Frederica Bunge and Melinda W. Cooke, *Oceania: A Regional Study* (Foreign Area Studies Series, 1985); K.R. Howe, Robert C. Kiste, Brij V. Lal, eds., *Tides of History: The Pacific Islands in the Twentieth Century* (University of Hawaii Press, Honolulu, 1994); Marshall Islands, www.rmiembassyus.org (March 2004).

JONATHAN SPANGLER
SMITHSONIAN INSTITUTION

Martinique

MARTINIQUE, IN the CARIBBEAN SEA, is one of the four French overseas departments and regions. Temperatures range from 25 to 30 degrees C (76 to 86 degrees F) and the climate is tropical and humid. The dry season runs from December to May (*carême*). Drought is relatively common from February to April. During the rainy season (*hivernage*), June to November, temperatures generally rise. Martinique's climate is tempered by trade winds but cyclones or hurricanes can be quite frequent.

The highest point in Martinique is Mount Pelée in the north, a live volcano that last erupted in 1902 and destroyed St. Pierre, which then was a busy port town: 30,000 were killed in the eruption. Mount Pélée is today under constant surveillance. The north is mountainous and precipitation is high. Low hills (mornes) characterize the rest of the island, except for the flat Lamentin area on the west coast where the airport is situated.

Christopher Columbus discovered Martinique in 1502. The island was colonized by the French in 1635 and in spite of spells of British occupation in 1762 and during the Napoleonic Wars (1794–1802), the island remained a French possession. The bulk of today's black and mulatto population is descended from the slaves imported into Martinique to labor on the sugar plantations. East Indians and some Chinese and Lebanese also contribute to the diversity of the population. Slavery was abolished in 1848 and the colony became a French department in 1946 after a campaign led by one of Martinique's most well-known figures, Aimé Césaire. The island became a region in 1982. As such, Martinique is a member of the EUROPEAN UNION. The official language is French, but French Creole is spoken by the mainly black and métis population. Most Martinicans are Roman Catholics.

Martinique enjoys the same advantages as any other mainland French department or region in terms of health, education, and social welfare. The islands also benefit from special measures aimed at encouraging economic development (lower income tax for example). Martinique has a General Council (Conseil Général) and a Regional Council (Conseil Régional). The General Council has 45 elected members, one from each constituency, and the Regional Council has 41 members. Both councils elect their respective presidents. The Martinicans elect three deputies to the French National Assembly and two senators to the French Senate.

The president of FRANCE appoints a prefect (*préfet*), who then becomes the official representative of the French government in Martinique. The prefect is responsible for public order and safety and ensures that civil liberties are properly respected, but he/she does not exercise executive power, which is in the hands of the two councils. The principal export is fruit (mainly bananas). Sugarcane is grown on the island; rum is a derivative and distilled on site. The islands are also popular holiday destinations for the French and tourism represents 7 percent of the gross domestic product.

Nevertheless, government subsidies are essential to Martinique's economic survival. The island imports far more (mainly from France) than it exports (mainly to France). Unemployment is high; in 2002 one in four

persons was out of work compared to approximately one in 10 in metropolitan France.

BIBLIOGRAPHY. Sarah Cameron, ed., *Caribbean Islands Handbook* (Footprint Handbooks, 1998); *World Factbook* (CIA, 2004); Government of France, www.martinique.pref.gouv.fr (April 2004).

SANDHYA PATEL
UNIVERSITÉ PASCAL, FRANCE

Maryland

"THE OLD LINE STATE," covering an area of 12,407 square mi (32,134 square km) is located in the mid-Atlantic UNITED STATES and features a diverse geography, prompting the nickname "a miniature America." Maryland is one of the more irregularly shaped states, with a long straight northern border and an irregular southern border that is delineated by 285 mi (459 km) of the Potomac River. At its most narrow point, Maryland is only a scant few miles from its northern border to its southern border; at its widest point it is about 124 mi (200 km) from northern to southern border. Along the northern border, the state is bounded by PENNSYLVANIA for 190 mi (320 km), and on the western border is WEST VIRGINIA. On the south, Maryland is bordered by West Virginia at the far western reaches of the state and by VIRGINIA. Over the years, border disputes have occurred with Virginia, West Virginia, and Pennsylvania.

The nation's capital, Washington, D.C., is located between Virginia and Maryland about 39 mi (63 km) southwest of the largest city in Maryland, Baltimore. The easternmost portion of Maryland is split into two by the Chesapeake Bay. This part of the state shares what is known as the DELMARVA PENINSULA with DELAWARE to the east and Virginia to the south. The highest point in the state is Backbone Mountain, located in the western reaches, 3,360 ft (1,024 m) above sea level.

The easternmost section of the state is a coastal plain at or about sea level and has considerable wetlands and marshes, especially along the Chesapeake Bay. The central part of the state, known as the Piedmont Plateau, is more hilly, while the westernmost part of Maryland is mountainous, containing a part of the Allegheny Mountains. Farming was historically prevalent in the central part of the state, while coal mining was done in the mountains of the west. The Piedmont area is also known for its racehorse breeding. The southernmost area of Maryland has for hundreds of years been known for its tobacco farming, though this is now in serious decline. Forests are prevalent in western Maryland, with oak, pine and hickory trees. On the coastal plain, cypress and gum trees are common.

The chief city of western Maryland is Cumberland, and Hagerstown is the largest city in central Maryland. Baltimore, the most populous city in the state, is located on the Chesapeake Bay, while the capital, Annapolis, a much smaller city, is located 30 mi (48 km) south of Baltimore, also on the bay. The major tourist attractions are Baltimore, which has revitalized its historic waterfront. This city, the birthplace of Babe Ruth, has a notable maritime museum and is home to the USS *Constitution*, an 18th-century naval warship. On the Atlantic side of the DELMARVA PENINSULA, Ocean City is a popular vacation destination for beachgoers. The forested mountains of the west are a popular destination, and the historic town of Harpers Ferry, West Virginia, is just a couple of minutes from Maryland's border. The northern suburbs of Washington, D.C., are in Maryland, including the populous Silver Spring. Many of these neighborhoods are connected to the nation's capital by the subway system, known as the Washington Metro. Camp David, the presidential retreat, is also in Maryland.

Maryland's roots go back to the early 17th century, when Captain John Smith mapped the area in 1608. The king of England granted Cecil Calvert, the Lord of Baltimore, a land patent in 1632. The first ships bearing European settlers arrived in 1634. They founded what is now St. Mary's City on the bay. A border dispute between the colonies of Pennsylvania and Maryland resulted in two surveyors, Mason and Dixon, being called upon in the 1760s to chart the exact border. Their Mason-Dixon line has since been the unofficial line between north and south. Annapolis was the nation's capital for a few years beginning in 1783. Maryland was in the midst of events during the War of 1812, when the British laid siege upon Washington, D.C. Francis Scott Key wrote the poem that became the national anthem, "The Star Spangled Banner," in 1814 while watching the British bombardment of Fort McHenry in Baltimore Harbor.

Development of the state continued to be rapid. The first railroad in the country was begun in Maryland in the 1820s. In the Civil War, Maryland sided with the north, but thousands of its citizens joined the Confederate Army, believing their tobacco plantations

were at risk if slaves were freed. The bloody Battle of Antietam was fought in Maryland in 1862, resulting in more than 20,000 deaths. Nineteenth-century Baltimore was a major destination for immigrant ships from Europe.

BIBLIOGRAPHY. *America on Wheels: Mid-Atlantic* (Macmillan, 1997); Edward B. Espenshade, Jr., ed., *Goode's World Atlas* (Rand McNally, 1987); James E. DiLisio, *Maryland* (Westview Press, 1983); Leslie Rauth, *Celebrate the States: Maryland* (Benchmark Books, 2000)

RICHARD PANCHYK
INDEPENDENT SCHOLAR

Massachusetts

THE COMMONWEALTH of Massachusetts has a population of 6,379,304 (2001), more than half of whom live in the greater Boston metropolitan area. The largest city in the state is Boston (the capital), with a population of 589,141, and the smallest town is Gosnold, with a population of 86. It borders NEW HAMPSHIRE and VERMONT to the north, NEW YORK to the west, and CONNECTICUT and RHODE ISLAND to the south. Massachusetts has an area of 10, 555 square mi (27,337 square km) and its highest point is Mount Greylock, at 3,487 ft (1,062 m). The lowest point is sea level or 0 ft.

Massachusetts has a widely varying topography, containing beaches, rocky shores and salt marshes along the coast and rolling and wooded hills throughout the central and western parts of the state. The state itself can be divided into six separate land regions: the coastal lowland, the Eastern New England Upland, the Connecticut Valley Lowland, the Western New England Upland, the Berkshire Valley, and the Taconic Mountains.

The coastal lowlands comprise the coastal lands along the ATLANTIC OCEAN, including low hills, swamp land, shallow rivers, and small ponds. Massachusetts also contains several islands, including the two relatively large islands of Nantucket and Martha's Vineyard, as well as the small group of Elizabeth Islands lying between Buzzards Bay and Vineyard Sound. (Boston Harbor itself, part of the Eastern New England Upland, is also home to a handful of small islands, some of which hosted prisoners during the Civil War, and others of which have been used at times for fortification of the city). The Eastern New England Upland contains over half of the state's population, and spreads between the northern border with New Hampshire down to the state's southernmost border with Connecticut and Rhode Island. The state's most fertile region, however, is the Connecticut Valley Lowland, containing most of the state's farmland, lying between the Eastern and Western New England Upland. The only mountainous region is the Taconic Mountains, lying in the far western section of the state, where a narrow swath of mountains is found in the northwest corner.

The original inhabitants of Massachusetts were Native American tribes, including the Nauset and Wampanoag Algonquin Indian tribes settled along the costal section of Massachusetts. European settlement and colonization of Massachusetts began with the arrival of the *Mayflower* in 1620. Europeans established a colony in Plymouth and several smaller fishing villages, including present-day Weymouth, Quincy, and Cape Ann. By 1630, the first large-scale Puritan migration from England occurred, with boats transporting approximately 900 settlers to Boston. In its initial incarnation, the Massachusetts Bay Company was governed as a private company for the first nearly half-century of its existence. The Puritans exerted a tremendous influence over the early development of the commonwealth, with many of their religious and educational influences lasting today. They established Boston Latin School in 1635 and, in the following year, Harvard University, founded with a grant from the General Court of the Massachusetts Bay Colony, both established with the idea of generating well-educated Puritan ministers. During this same time, Boston Common became the first public park in America. The period of 1684 to 1694, however, was a time of upheaval, with the Bay Colony's charter revoked, and the political and religious unrest playing out in dramatic fashion in the Salem witchcraft trials of 1692.

Massachusetts bore witness to a number of key events during the Revolutionary War, including the Boston Massacre (occurring in 1770 at the intersection of what is now State Street and Devonshire Street) and the Boston Tea Party. The famous April 19, 1775, ride of Paul Revere and the firing of the "shot heard round the world" culminated in the Battles of Lexington and Concord, both located only a few miles outside of Boston. In 1780, following the Battle of Bunker Hill and the British evacuation of the commonwealth, the Massachusetts Constitution was ratified, making it the oldest written constitution currently at use anywhere in

Faneuil Hall and Quincy Market in Boston date back to the city's colonial days as a commercial hub, and remain so today.

the world, according to the secretary of the commonwealth's Citizen Information Service.

Massachusetts has again played a significant role in national politics, beginning with the election of former Massachusetts Governor Calvin Coolidge to the presidency in 1923. Former Massachusetts Senator John F. Kennedy followed in his footsteps, winning the 1960 presidential election; 1988 saw Michael Dukakis, another former Massachusetts governor, run unsuccessfully for president against George H.W. Bush. In 2004, another Massachusetts senator, John Kerry, again unsuccessfully sought the presidency.

Massachusetts has also given birth to a number of notable individuals who have had a notable impact on the country in more than the pure political arena. These include not only such historical revolutionaries as Samuel Adams, Benjamin Franklin, Paul Revere, Eli Whitney and Alexander Graham Bell, but intellectuals and writers, including Jack Kerouac, Edgar Allan Poe, Sylvia Plath, Emily Dickinson and Dr. Seuss (Theodore Geisel), as well as Jay Leno, Leonard Nimoy, Conan O'Brien, and Ben Affleck.

Massachusetts is well-known for being an intellectual and cultural hub, home to over 100 colleges and universities, including some of the more well-known: Harvard University (in Cambridge), Amherst College (in Williamstown), Boston College (in Chestnut Hill), Brandeis University (Waltham), Smith College (Northampton), and Wellesley College (Wellesley), to name but a few. The intellectual history of Massachusetts and the influence some of its more prominent citizens throughout American history have contributed to the political and liberal development of the entire United States.

BIBLIOGRAPHY. Richard B. Brown, *Massachusetts, A History* (W.W. Norton, 1972); J. Joseph Huthmacker, *Massachusetts People and Politics* (Atheneum, 1969); Samuel Eliot Morrison, *Maritime History of Massachusetts* (Northeastern University Press, 1997).

AMY WILSON
UNIVERSITY OF WASHINGTON

Mauritania

Map Page 1113 **Area** 397,955 square mi (1,030,700 square km) **Population** 2,912,584 **Capital** Nouakchott **Highest Point** 2,985 ft (910 m) **Lowest Point** 0 m **GDP per capita** $377 **Primary Natural Resources** fish, iron ore, gypsum, copper, phosphates, salt.

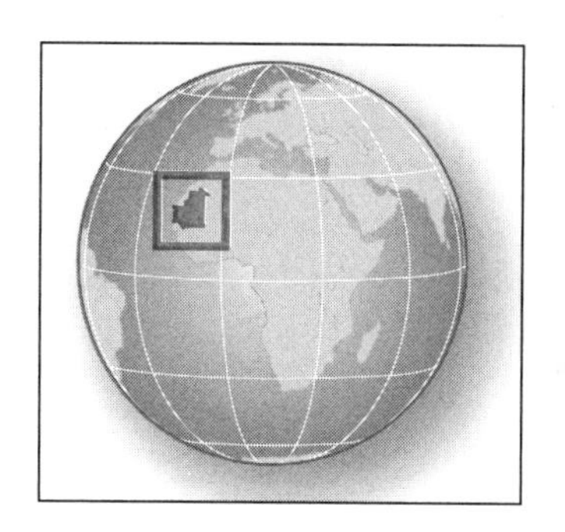

MAURITANIA IS LOCATED in Northwest Africa on the coast of the ATLANTIC OCEAN. It covers an area slightly larger than TEXAS and NEW MEXICO combined. It is bordered on the northwest and north by WESTERN SAHARA, on the northeast by ALGERIA, on the east and southeast by MALI, on the southwest by SENEGAL, and on the west by the Atlantic Ocean. Mauritania has 435 mi (700 km) of ocean coastline, which is extremely rugged with no natural harbors and has waves of ex-

traordinary height. The Senegal River forms the border with Senegal, helps to form an alluvial fan that supports agriculture in the area, and is Mauritania's most densely populated region.

Two-thirds of Mauritania is covered with DESERT. The country lies entirely within the SAHARA DESERT which covers an area of 3,320,450 square mi (8,600,000 square km). The 17th parallel is the dividing line between the true desert to the north and the Sahelian zone made up mostly of savanna to the south. The desert has no vegetation in the eastern parts of Mauritania, but the western section closer to the Atlantic has some temporary pasturage for nomadic camel herders. This western section also has some oases, the largest being the town of Atar. The desert is dominated by wind, sand, and erosion. It has many shifting dunes in immense basins that are grouped into "ranges." The only real elevation is an almost horizontal sandstone plateau rising 1,500 ft (460 m) that runs through the center of the country from north to south. There are also occasional buttes and steep rims covered in sand or sometimes pebbles.

The TROPIC OF CANCER crosses the northern half of the desert in Mauritania. In this area, there is a prevailing continental wind blowing year-round. Rainfall in this area amounts to less than 4 in (10 cm) per year, and when it does occur, rain is usually extremely violent and brief. To the south, in the Senegal valley, the area receives about 26 in (66 cm) of rain per year, mainly during the three or four months of summer. The temperature averages about 100 degrees F (37.8 degrees C) during the day in most of Mauritania and much cooler at night, sometimes down to near 32 degrees F (0 degrees C).

Mauritania gained its independence from FRANCE on November 28, 1960, and since then has been a constitutional republic. Soon after independence, Mauritania developed a modern mining industry. Miners found large deposits of high-grade iron ore in northern Mauritania and have since developed them for exporting. They also mine copper and salt in Mauritania. Most of the population, however, is busy raising crops or tending livestock in southern Mauritania. The ethnic groups that live in this area are mainly the Tukolor, Soninke, Bambara, and Wolof. The nomadic and seminomadic tribes of the Berber, Arab, Tuareg, and Fulani make up the majority of the population and range about Mauritania following their herds.

BIBLIOGRAPHY. Institut géographique national, *The Atlas of Africa* (Éditions Jeune Afrique, 1973); Kwame Anthony Appiah and Henry Louis Gates, Jr., *Africana* (Basic Civitas Books, 1999); Saul B. Cohen, ed., *The Columbia Gazetteer of the World* (Columbia University Press, 1998); Bureau of African Affairs, "Background Note: Mauritania," (U.S. Department of State, 2003).

CHRISTY A. DONALDSON
MONTANA STATE UNIVERSITY

Mauritius

Map Page 1116 **Area** 788 square mi (2,040 square km) **Population** 1,210,447 **Capital** Port Louis **Highest Point** 2,732 ft (828 m) **Lowest Point** 0 m **GDP per capita** $11,000 **Primary Natural Resources** arable land, sugarcane, tea, cattle, goats, fish.

MAURITIUS IS AN island nation in the INDIAN OCEAN, located about 500 mi (800 km) east of MADAGASCAR. Governed by successions of Dutch, French, and British colonial administrations, and settled with workers from East Africa, INDIA, and CHINA, Mauritius today is an interesting hodgepodge of these various elements in everything from language to culture to politics. More economically and politically stable than some of its island neighbors, with one of the highest per capita incomes in the developing world, it is a leader in the Indian Ocean Rim Association, and maintains strong ties with several east African, Arab, and Far Eastern nations.

The country of Mauritius includes the main island, plus Rodrigues Island (372 mi or 600 km to the east) and the much smaller (and much farther away) Agalega Islands and Cargados Carajos Shoals to the north. Together with the island of Réunion (an overseas department of FRANCE), Mauritius and Rodrigues are collectively called the Mascarene Islands. The islands to the north are very small and unpopulated, but they form a key part of Mauritius's claims to an EEZ (exclusive economic zone) covering 468,000 square mi (1.2 million square km) of the Indian Ocean off the coast of Africa.

There are tensions over claims put forward over Tromelin Island to the northwest (held by France) and Diego Garcia far to the northeast (1,197 mi or 1,931

km) in the Chagos Archipelago (held by the UNITED KINGDOM).

The islands are in an area of geologic activity. The Mid-Indian Ocean Ridge lies 200 mi (320 km) to the east, with the Rodrigues fracture zone and the Mauritius Trench as perpendicular offshoots. Mauritius is of relatively recent volcanic origin, about 12 million years, but has been dormant for the last 100,000 years. It is formed of a ring of mountains (about 18 percent of the land area) encircling a central plateau with a smaller coastal plain. Rodrigues is of similar volcanic formation, but in a ridge running west to east, and much younger (about 1.5 million years). Both are ringed by coral reefs. The smaller islands are not volcanic, but coralline, and are home to coconut palms and fishing boats.

The islands' tropical climate and position along southeast trade winds attracted Arab and Malay traders, followed by Portuguese explorers, who named the main island Ilha do Cirne ("island of the swan") early in the 16th century—possibly for the native population of large flightless birds called dodos. The Dutch attempted settlements and plantations twice in the 17th century. Neither attempt took root, but the island's new name did—Mauritius, named for Maurits van Nassau of Holland—as did their environmental impact, that is, the extinction of the dodo, the introduction of rats, and the replacement of much of the native ebony forests with sugarcane. The French East India Company claimed the island in 1715, having already begun settlements on Rodrigues and Bourbon (today's Réunion). Port-Louis was an important center for trade, privateering, and naval operations against the British, and French planters grew very rich on the backs of African slave labor.

The British took the islands in 1810 and retained the colony after the 1814 settlements ended the Napoleonic Wars back in Europe, but most of the French sugar planters stayed on. The small number of English settlers meant that Mauritius retained much of its French culture, and many people today speak a French patois (mixture). With the abolition of slavery in 1834, the British imported indentured workers from India in vast numbers; they made up about two-thirds of the population by 1870.

Today, the descendants of South Asians continue to outnumber the descendants of freed African slaves (Creoles) and the small numbers of people of Chinese and European origin. Tensions between these groups have largely been settled since independence from the United Kingdom in 1968.

The economy was almost entirely sugar-based until recently, but is now being diversified into production of textiles, chemicals, and other light industry, plus the encouragement of tourism. It remains a member of the Commonwealth, but became a republic in 1992. Population density on Mauritius is one of the highest in the world, while only 36,000 people live on Rodrigues.

BIBLIOGRAPHY. Anthony Toth, "Mauritius," *Indian Ocean: Five Island Countries*, Helen Chapin Metz, ed., (Foreign Area Studies Series, 1995); *World Factbook* (CIA, 2004); "Mauritius," www.mauritius-info.com (March 2004).

JONATHAN SPANGLER
SMITHSONIAN INSTITUTION

McKinley, Mount

MOUNT MCKINLEY is the highest peak in North America with a summit at 20,329 ft (6,196 m). Located in the 600-mi- (965-km-) long Alaskan Mountain Range, it is approximately 150 mi (246 km) south of Fairbanks, ALASKA. Mount McKinley is home to more than 650 species of flowering plants, as well as mosses, lichens, fungi, and algae. The growing season is short, and only plants that have adapted to the region's cold winters can survive.

Mount McKinley is located near the center of the 6-million-acre Denali National Park and Wildlife Preserve. The terrain was carved by glacial activity approximately 10,000 to 14,000 years ago. Mount McKinley, or Denali, consists of two peaks that are permanently covered in snow. It is located by a major fault system, known as the Denali Fault, which makes it subject to continual tectonic uplift. A major earthquake in 1912 caused a significant segment of the south face of the mountain to shear off. Many smaller earthquakes occur each year, although they are generally unnoticed. Denali is surrounded by the rugged peaks of other members of the Alaska Range as well as open tundra, glacial rivers, and flood plains. The region is home to a diverse range of wildlife, including 167 species of birds, 10 species of fish, 39 species of mammals, and 1 species of amphibian.

Denali's climate is subarctic. Summers are short, dry, and generally cool. Cloud cover obscures the mountain's peaks much of the time. Winters usually bring heavy snows and extreme cold. The road system through the park is closed in the winter months. The

rangers who are responsible for overseeing the protection and maintenance of the park often use dog sleds, going out on patrols of two or more weeks.

Mount McKinley was originally named Denali ("great white one," or "high one") by the Athapaskan natives, who lived and hunted around the mountain. McKinley was renamed in 1896 by William Dickey. Dickey, a Princeton University graduate who was performing a survey of the mountain following a prospecting trip, renamed it after presidential candidate William McKinley. Although Dickey claimed discovery of McKinley, the first sighting by a nonnative was actually nearly 150 years earlier by Vitus Bering in 1741.

Denali, as it is now called by the Alaskans, has a rich history of exploration. Unpredictable weather and rugged terrain, including swamp and crevasse-laden glaciers, that surround the mountain make climbing it a challenging and dangerous venture.

The first attempt to scale the mountain, which proved unsuccessful, was made around 1900 by Judge James Wickersham. Frederick Cook claimed to have reached its summit in 1907, but his claims were proven untrue three years later. The first successful scaling of the North Peak was accomplished by the so-called Sourdough Expedition Party in 1909. The South Peak was conquered by the Hudson Stock expedition four years later, in 1913.

The introduction of air travel drastically improved access to the region for exploration and surveying. Most of the region's bush pilots, including the legendary pilot-explorer Don Sheldon, fly out of the small town of Talkeetna, which is approximately 121 km (75 mi) south of Denali.

In 1936 and 1937, noted explorer-photographer Brad Washburn embarked on the first of many expeditions to climb and photograph Denali and the surrounding region. Washburn, whose expeditions were later joined by his wife, Barbara, provided an unparalleled glimpse of the mountains for those outside of the exploration community. Hundreds of books on Denali and the region have been produced that utilize Washburn's photographs.

Because of the extremity of McKinley's terrain, the mountain and its surrounding regions were a popular site for the development of search-and-rescue procedures by the U.S. Army's Tenth Mountaineering Core in the middle of the 20th century. Search-and-rescue remains an important part of life in the area. In 1980, under pressure from Alaskans, the UNITED STATES passed a bill to rename the mountain and its surrounding lands the Denali National Park and Preserve. The park is currently administered as a wildlife preserve by the U.S. National Park Service, which is a part of the U.S. Department of the Interior. In the mid-2000s, there have been pressures from the American political right to commence drilling for oil within the boundaries of the preserve.

BIBLIOGRAPHY. William N. Beach, *In the Shadow of Mount McKinley* (Derrydale Press, 2000); William E. Brown, *Denali: Symbol of the Alaskan Wild: An Illustrated History of the Denali-Mount McKinley Region, Alaska* (Alaska Natural History Association, 1993); Denali National Park and Wildlife Preserve, www.nps.gov/dena (March 2004); Bradford Washburn and David Roberts, *Mount McKinley: The Conquest of Denali* (Abrams, 1991).

JESSICA M. PARR
SIMMONS COLLEGE

Mediterranean Sea

THE TRANSLATION of its Latin name ("in the midst of land") indicates that the Mediterranean Sea is located between the landmasses of Europe, Asia, and Africa. Some students of geography may view the Mediterranean as a physical separation between Europe and Africa. Nothing could be further from the truth. The Mediterranean, in fact, has been the great unifier and link for cultures from Africa, the MIDDLE SAST, and Europe as well as Central Asia (via the BLACK SEA passage). Its moderate climate and rich land and sea resources have made the Mediterranean Basin and Sea an historic storehouse of foods as well as a means of commerce. Perhaps a better image is one of a giant mixing bowl that collects and spins out various cultures and economies. Even geologically, the Mediterranean Sea is where the African and European plates meet, their friction creating the rich marble quarries and volcanoes that mark much of ITALY and the Sea of Marmara (marble) in TURKEY.

The sea's total surface area is 967,000 square mi (2.5 million square km). Acting as a major heatsink, it is an important climate modifier (its latitude is the same as that of the much colder areas of Manchuria and north CHINA). It is approximately 2,400 mi (3,900 km) long, with a maximum width of 1,000 mi (1,600 km), and while relatively shallow at GIBRALTAR and the Dardanelles, it is over 16,000 ft (5,400 m) deep near Cape Matapan, GREECE. Its waters have a higher salin-

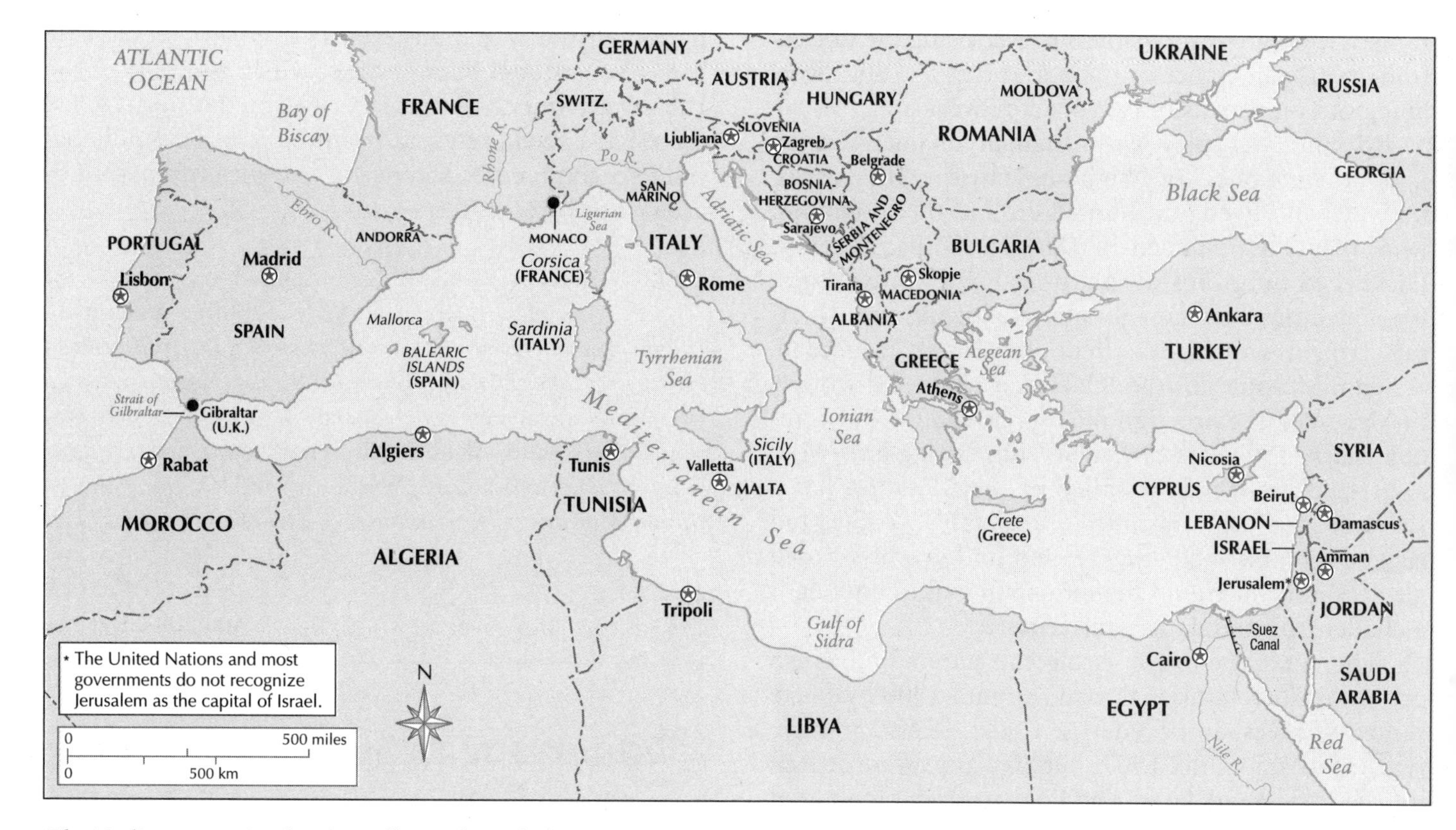

The Mediterranean Sea has for millennia been the great unifier and link for the dozens of countries and cultures from Africa, the Middle East, and Europe as well as Central Asia (via the Black Sea passage).

ity than the ATLANTIC OCEAN in part because there is so little tidal movement. It also has very few rivers that add fresh water, and its high sunshine creates massive rates and volumes of evaporation. Its rich marine resources include over 400 varieties of fish, along with sponges and corals and recently, oil and natural gas also have been found in several sections. Combined with large and often poor populations, there is much overfishing and pollution. But its mild climate has also created rich agricultural resources, and its geology has added important minerals—especially tin and copper so important to the Bronze Age.

Because of its size and complexity, the Mediterranean can be considered a combination of basins or "seas," and not as a monolithic whole. For example, in addition to the shallow sills at Gibraltar and the Dardanelles, there is a shallow sill (the Adventure Bank) between Sicily and TUNISIA that clearly divides the Mediterranean into two primary basins: east and west. In addition, history and geography have created a number of distinct local "seas" essential to understanding the basin's history. These include the Ligurian Sea north of Corsica; the Tyrrhenian Sea enclosed by Sardinia, Italy, and Sicily; the ADRIATIC, which separates Italy and the Dalmatian coast of the former Yugoslavia, the Ionian Sea between Italy and Greece, the AEGEAN between Greece and TURKEY, and the Thracian Sea to the north. And this omits the various seas associated with the Black Sea.

Today, the Mediterranean forms a maritime link for 23 countries (more if we count the countries that share the coast of the Black Sea), including: SPAIN, FRANCE, MONACO, Italy, MOROCCO, ALGERIA, TUNISIA, LIBYA, EGYPT, SLOVENIA, CROATIA, BOSNIA AND HERZEGOVINA, ALBANIA, Greece, Turkey, MALTA, CYPRUS, SYRIA, LEBANON, and ISRAEL. It provides for maritime passage from the heart of RUSSIA and Central Asia via the Black Sea, to the New World via the ATLANTIC OCEAN, and to the INDIAN OCEAN and Pacific Far East via the Suez Canal. Its rivers create land links that tie it to northern Europe as well as to eastern Africa and the Indian Ocean.

The populations with direct access to its shores exceed 425 million (650 million including countries bordering the Black Sea). This makes the Mediterranean Basin potentially one of the world's major economic

zones. It also makes it a focus for political intrigue from Asia and the Middle East to northern and eastern Europe.

In addition to its rich resources and history, it is the sea's strategic geopolitical characteristics—past, present, future—that have dominated its existence. The strategic maritime CHOKE POINTs and history of GIBRALTAR, MALTA, Taormina, and the Dardenelles and Bosporus provide countless examples of the geography behind history.

And the mixing of cultures also has made the Mediterranean Basin a historic center for unique and rich art traditions, from the ancient cave paintings of France and Spain, to the work of modern artists like Marc Chagall, Pablo Picasso, and Paul Cezanne to the mosaics and unique architecture of the Byzantines and Italians. There are naval arts and maps associated with the rise of maritime powers such as the Ottomans, the Venetians, the Genovese, the Spanish, and more.

The LITTORAL (coast) of the Mediterranean was the focus of the Phoenician, Greek, Minoan, Egyptian, Syrian, Macedonian, Roman, Venetian, Genoan and other city-states, empires, and civilizations. Trade, navies, and armies moved over its surface and along the passes to and from its shores.

The shallow and narrow passage linking the Mediterranean to the Atlantic at Gibraltar is only 8 mi (12.9 km) wide, but strong winds and extremely strong tidal flows and currents often forced ancient people to find land routes around the barrier. Similarly, the strong currents over the shallow and narrow sill linking the Mediterranean and Black Seas at the Dardanelles (1 mi or 1.6 km wide) long halted direct shipping between the two basins and was the focus of such epic battles as that of Troy in ancient times and Gallipoli in more modern times.

The Mediterranean is also a center of land trade, both along the shores and at various key land passages, because of its rivers and valleys that link it with the people and resources of northern Europe.

BIBLIOGRAPHY. David Abulafia, ed., *The Mediterranean in History* (Thames and Hudson, 2003); M.M. Philip, *Mediterranean Geography: A Background to Ancient History* (Harrap and Co., 1954); Ellen C. Semple, *The Geography of the Mediterranean Region: Its Relation to Ancient History* (H. Holt, 1931). "Mediterranean Sea," www.infoplease.com (March 2004).

R.W. MCCOLL, PH.D.
GENERAL EDITOR

megaliths

A GREEK-DERIVED word meaning "large stone," *megaliths* refer to huge, uncut, or roughly cut boulders that people have moved to stand upright from the ground. Some megaliths, also called menhirs, have been in place for thousands of years, and the reasons for their positions are not fully known. They are found all over the world.

The most famous megaliths are at Stonehenge, a United Nations World Heritage Site of about 100 acres on the Salisbury Plain in England, UNITED KINGDOM. Stonehenge's construction began around 3150 B.C.E., predating most known cultures of Britain. Simply transporting the stones, which weigh as much as 50 tons (45.3 metric tons), from quarries 17 mi (27 km) and 150 mi (241 km) away, then erecting them on sloping ground in careful circles, involved incredible feats of engineering. The stones are positioned to align with sunrise on the summer solstice, indicating knowledge of astronomy. The organization involved in such an undertaking implies a powerful social administration as well.

Other stone circles have been found in the British Isles and elsewhere. One recently discovered circle dating to 4500 B.C.E., is in Nabta Playa, SUDAN. Research indicates it served as a calendar. The study of archaeoastronomy, started in the 1960s, links many megalithic circles and groups to alignment with astronomical events, such as solstices, eclipses, and the 18.5-year metronomic cycle of the moon's transit.

In addition to circles, megaliths are found in multiple rows. The stones at Carnac, near Morbihan Bay in Brittany, FRANCE, are famous examples of these. Most date from Europe's Iron Age. Megaliths may also be arranged in long avenues, pairs, single rows, triads with a taller center stone, and many other formations. At Callenish on the Isle of Lewis in Scotland, pairs of stones form an avenue that leads to a stone circle surrounding a tomb, with three short rows emanating from it. In 1993, Aubrey Burl catalogued hundreds of megalithic pairs and groupings in Britain, Ireland, and Brittany in his book, *From Carnac to Callenish*.

Some megaliths are formed as dolmens: the large upright stones form a room, with a stone on top as a roof or capstone. These are also called cromlechs, although the term is not used consistently. Dolmens and chamber tombs have been found above- and belowground in the British Isles, France, SPAIN, ITALY, MALTA, Scandinavia, ETHIOPIA, JORDAN, southern INDIA, CHINA, the KOREAS, JAPAN, and COLOMBIA. In some instances,

rough impressions have been carved on the stones. Stones have also been toppled, moved, and reused, and recarved throughout the years.

Standing stones, in any formation or alone, are located all over the world. Granite stones stand in YEMEN, near artifacts dating from 2400 to 1800 B.C.E. In the CAUCASUS MOUNTAINS of RUSSIA, hundreds of dolmans and standing stones, some carved in bas-relief, served as burial chambers. Rows leading to semi–circles of stones have been described in Tibet. Even in the New England states of the UNITED STATES, megaliths have been found standing alone or in groups; the undated complex at Mystery Hill in MASSACHUSETTS is called America's Stonehenge.

Not all megaliths were raised in the distant past. Recently constructed megaliths exist in the Sarawak area of BORNEO, scattered over the foothills near Mount Kinabalu. They were moved and erected in living memory, according to residents, as memorials, sometimes for the dead, and to affirm status or land claims. In MADAGASCAR, megaliths were raised into the 19th century as ancestral tombs.

BIBLIOGRAPHY. Aubrey Burl, *From Carnac to Callenish* (Yale University Press, 1993); English Heritage, "Information on Stonehenge," www.english-heritage.org.uk (March 2004); Roger Joussame, *Dolmens for the Dead* (Cornell University Press, 1985); Caroline Malone and Nancy Stone Bernard, *Stonehenge: Digging for the Past* (Oxford University Press, 2002); R. Schild and F. Wendorf, "The Megaliths of Nabta Playa," www.pan.pl (March 2004); "America's Stonehenge," www.stonehengeusa.com (March 2004).

VICKEY KALAMBAKAL
INDEPENDENT SCHOLAR

Mekong River

THE MEKONG IS the 12th-longest river in the world, traveling nearly 2,600 mi (4,200 km) from its source in the Tibetan plateau to its enormous DELTA in southern VIETNAM. It is the major transportation highway and supplier of life for most of Southeast Asia—with five cities over 100,000 and one-third of the populations of THAILAND, LAOS, CAMBODIA, and Vietnam (60 million people) in the lower Mekong basin—yet much of it is still undeveloped (the first bridge was built in 1993, between Thailand and Laos) and even unexplored (the source was located definitively only in 1994). This is due to the torturous path the Mekong cuts through rough mountain terrain on its journey from the Tanggula Mountains in CHINA to the SOUTH CHINA SEA. It changes names many times along the way, from Lancang Jiang ("turbulent river") in China, to Mae Khong ("mother of waters") in Laos and Thailand, Tonle Thom ("great water") in Cambodia, and Tien Giang or Cuu Long ("river of nine dragons," for the delta's many channels) in Vietnam.

The river is navigable by seagoing vessels as far as Phnom Penh, the capital of Cambodia, a distance of 340 mi (550 km), but they are prevented from going much further upstream by numerous sandbanks and rapids. Communities upriver use smaller vessels on the river, in some places the only route for communications and transport of goods. European powers were excited by the possibility of using the Mekong as a passage to the riches of the interior of China, which never materialized.

The French established a protectorate over the kingdom of Laos in the 1860s (in haste, to beat the British) before fully exploring this option, however, and were disappointed by the river's limited capability for commercial transport in the interior. Laos was therefore mostly ignored as a colony (benign neglect), and remains one of the most underdeveloped nations in the world. The 1990s witnessed a burst of hydroelectric projects (23 are planned in Laos alone), generating a good deal of international protests, because of the enormous impact these projects will have on the ecology of the region and the lives of millions of its inhabitants. The Mekong River Commission, based in Phnom Penh, predicts dire consequences for rice production and the fishing industry in the lower basin, which accounts for about 2 percent of the world's total. Manwan Dam in Yunnan Province (China), for example, will radically alter the course of the river, preventing downriver annual flooding essential to the production of rice. Regional governments ignore these warnings, however, seeing the projects as major symbols of their countries' modernity.

The Mekong originates near the roof of the world, in the Jifu Mountains (about 17,160 ft or 5,200 m), near the town of Zadoi on the borders between the Chinese provinces of Xizang (Tibet) and Qinghai. Here the small mountain river moves swiftly through barren chasms and gorges. The river flows south into Yunnan province through narrow gorges, only a few kilometers from the parallel valleys of the upper Salween and Yangzi (CHANGJIANG) rivers, which ultimately end their courses thousands of kilometers away in the BAY OF

BENGAL and the East China Sea, respectively. The river then flows from China into the valley separating eastern MYANMAR from Laos, which has served as a conduit for Chinese migration into Southeast Asia. Thousands of Chinese emigrated to Mandalay in Myanmar, where they now dominate commercial life. This route is also a passage for illegal arms shipments from China and for drugs (opium and heroin) into China.

The Mekong then forms the main corridor of settlement for the lowland Lao peoples, first in the interior of the country, then along its border with Thailand. One of the oldest cities in the region, Loang Prabang, was established on its banks in the 14th century, followed by a rival Buddhist kingdom centered at Vientiane (Viangchan), about 250 mi (400 km) downriver. Loang Prabang retains much of its beautiful architecture, Buddhist temples, and French colonial buildings and has recently been classified as a World Heritage Site.

Below Pakxé (Laos), the Mekong becomes much wider, and, after the dramatic drop at Khone Falls on the border with Cambodia, enters the wide, flat Cambodian plain. Thousands of years ago, this plain was at the bottom of the sea, but material brought downstream from the mountains gradually filled it in to form a fertile plain, with numerous confluent rivers and large lakes. The largest of these lakes, Tonle Sap, is the remaining evidence of the sea's onetime incursion into this area. The lake is formed by one of the rivers that flow into the Mekong, which reverses course from backflow during the rainy season, expanding the lake up to six times its normal size. This area, with its natural irrigation systems and fertile soils, was home to many early civilizations in Southeast Asia, including the empire of the Khmers with its famous temples at Angkor Wat.

The Mekong delta starts in Cambodia and extends 175 mi (282 km) to the sea. Forming a vast alluvial plain (called the Plain of Reeds), it is the heartland of southern Vietnam, covering 26,000 square mi (67,340 square km). The river splits into two main channels, and numerous smaller passages, forming a broad delta on the South China Sea.

The delta was the site of many powerful civilizations, notably the Cambodian kingdom of Funan, which was a busy ENTREPOT between Chinese and Indian civilizations as early as the 3d century C.E. (even Roman coins have been discovered here). Commercial wet-rice production was established by the French colonial powers from the middle of the 19th century. It was the economic center of the former South Vietnam, and remains one of the most densely populated areas in Southeast Asia.

As independent nations, the countries bordering the Mekong River have struggled to coordinate activities and to limit destructive overdevelopment in the production of rice and logging. Since 1957, regional associations have proposed numerous projects for hydroelectric power, navigation, and flood control, but wars and political turmoil have allowed few to become a reality. Much of the area remains very poor and very rural, relying on rice farming and subsistence fishing.

BIBLIOGRAPHY. Barbara A. Weightman, ed., *Dragons and Tigers: A Geography of South, East and Southeast Asia* (Wiley, 2002); Milton Osborne, *The Mekong: Turbulent Past, Uncertain Future* (Atlantic Monthly Press, 2000); "Dams as Development: Hydropower on the Mekong River," www.volatile.org/research/Mekong (May 2004); "Mekong," www.mekong.net (May 2004).

JONATHAN SPANGLER
SMITHSONIAN INSTITUTION

Melanesia

BEGINNING WITH THE FRENCH explorer Jules d'Urville Dumont in the 1830s, geographers have grouped the far-flung islands of the PACIFIC OCEAN into three great island worlds: Melanesia, MICRONESIA, and POLYNESIA. Melanesia, derived from the Greek words for "black" and "islands," consists of those islands that extend from New Guinea in the northwest to the FIJI Islands in the southeast. This distance is approximately 3,500 mi (5,600 km). Melanesia includes, besides the large island of New Guinea, at least seven major island chains: the Admiralty Islands, the Bismarck Archipelago, the SOLOMON ISLANDS, the Santa Cruz Islands, the New Hebrides (now VANUATU), New Caledonia and nearby islands, and the Fiji Islands.

New Guinea was first settled by sea from Southeast Asia, perhaps as long ago as 40,000 years, and the practice of agriculture was under way in New Guinea by 9,000 years ago, one of the earliest dates for agriculture known in the world. The indigenous Melanesian peoples are all dark-skinned—hence the name "Melanesia"—and all practice agriculture with an emphasis on root and tree crops (taro, yams, sweet potatoes, coconut, and sago palms) and pig husbandry.

Settlements are small, although villages of up to 1,000 people exist in some areas, for example, in New Guinea's Sepik River Valley. Melanesia is still largely dependent on subsistence agriculture and most of the people live in rural areas. The largest city of today's Melanesia is Port Moresby, the capital of PAPUA NEW GUINEA. Port Moresby's population is approximately 300,000.

TRADE ROUTES

Melanesian societies, though small and sometimes widely spaced, were not isolated. TRADE ROUTES, very often emphasizing trade in ceremonial objects, were very common. An example is the famous trade system described by anthropologist Bronislaw Malinowski. This "Kula" trade linked the various settlements of the Trobriand Islands near New Guinea into a single economic unit. Indigenous Melanesian peoples speak languages that can be grouped into two main divisions. Languages of the Austronesian or Malayo-Polynesian language family are spoken along the coast of northern and eastern New Guinea and also in the island groups east to Fiji. So-called Papuan languages (not a language family but a grouping of non-Austronesian languages that may or may not be related) are spoken in interior and southern New Guinea and in the Bismarck Archipelago.

Late in the 19th century, Melanesia was absorbed into the colonial empires of the NETHERLANDS, Great Britain, GERMANY, and FRANCE. In many cases, one European power would establish a colony only to prevent a rival colonial power from doing the same. The Netherlands absorbed the western half of New Guinea as part of the Dutch East Indies, now INDONESIA. Great Britain and later Australia colonized southeastern New Guinea. Great Britain also absorbed the southern Solomon Islands and the Santa Cruz and Fiji Islands. Germany, succeeded by Australia after World War I, colonized northeastern New Guinea, the Bismarck Archipelago, the Admiralty Islands, and the northern Solomon Islands including Bougainville. France took over New Caledonia and nearby islands, and, in a rather strange arrangement, administered the New Hebrides as a joint colonial possession with Great Britain. New Guinea was thus divided among three colonial powers; the Solomon Islands were divided in two.

In 1970, Fiji was the first Melanesian nation to achieve independence. Politically, Melanesia is now divided into the Indonesian province of Papua, formerly called Irian Jaya (population about 1.8 million); independent Papua New Guinea (population about 5.4 million); the independent Solomon Islands (population about 523,000); independent Vanuatu, formerly the New Hebrides (population about 203,000); the independent Republic of Fiji (population about 881,000); and the French Overseas Territory of New Caledonia (population about 214,000). The total population of Melanesia today is thus about 9 million people, approximately 60 percent living in Papua New Guinea.

BIBLIOGRAPHY. H.C. Brookfield and Doreen Hart, *Melanesia: A Geographical Interpretation of an Island World* (Mehuen, 1971); Bronislaw Malinowski, *Argonauts of the Western Pacific* (Routledge, 1922); Paul Sillitoe, *Social Change in Melanesia: Development and History* (Cambridge University Press, 2000).

JAMES A. BALDWIN
INDIANA UNIVERSITY-PURDUE UNIVERSITY

Mercator, Gerardus (1512–1594)

GERARDUS MERCATOR is the Latin name of Gerhard Kremer, one of the greatest geographers and cartographers, who lived in the 16th century. He was born on March 5, 1512, in Rupelmonde in the Flanders (now in BELGIUM), seventh child of Hubert and Emerentia Kremer. His father was a shoemaker. Thanks to his paternal uncle Gisbert Kremer, he went to Bois le Duc for two years to study philosophy and mathematics at Lovanio University, where he graduated as magister.

He studied under the guidance of mathematician Gemma Frisius and instrument maker Gaspar van der Heyden (Gaspar a Myrica). Mercator devoted himself to making mathematical instruments, such as astrolabes and armillary spheres, and to the art of the copperplate engraving, soon becoming so famous he received orders from the emperor, Charles V.

In 1534, he opened a house of cartographic production, and in 1537 he printed his first work, the map of Palestine *Amplissima Terrae Sanctae descriptio ad utriusque testamenti intelligentiam*, in six sheets. In 1538 he printed his map of the world and in 1540 the map of Flanders in four sheets. In 1540, his work *Litterarum latinarum quas italicas cursoriasque vocant scribendi ratio* was printed, written to uniform the toponymy and to use Latin handwriting on maps.

In 1544, in a moment of religious conflict during the Lutheran Reformation, Mercator was imprisoned

under suspicion of heresy, but he was released in a short time thanks to the help of some very eminent people. In 1552, he moved to Duisburg (GERMANY), where in 1554 he printed his map of Europe in 15 sheets that had a second edition in 1572. Mercator was appointed court cartographer by Duke Wilhelm of Cleve in 1564. During this year, the map *Angliae Scotiae et Hiberniae nova descriptio* was printed and in 1569 the great map of the world, *Nova et aucta orbis terrae descriptio*, in 18 sheets, was issued to help in navigation.

Thanks to this work, Mercator is heralded as the founder of modern cartography as a science, founding it on the mathematical method and not on the artistic empiricism prevalent at his time. He, for the first time, used a new way to draft maps with isogonic projection, today called the Mercator projection (1556). For the map to conserve the angles and the representation of the loxodromic curve with a straight line, Mercator's projection is used today also to draft navigation charts. On a globe, the lines of longitude (measuring east-west position) converge at the poles and the lines of latitudes (measuring north-south position) are at equal distance. In a Mercator projection, the lines of longitude are straight vertical lines at equal distance at all latitudes, and horizontal distances are stretched above and below the equator: this alteration is excessive near the poles. The Mercator projection mathematically stretches vertically distances from the same proportion as the horizontal distances so that shape and direction are conserved.

Mercator wished to compile a great encyclopedic work containing the entire geographical knowledge of the universe and the history of geographical science, but so great project was impossible to realize by one person only, and it remained unfinished. He started publishing the *Chronologia* (1569) and finally the *Atlas, sive cosmographicae meditationes de fabrica mundi et fabricate figura* (1585–95), republished complete only posthumously in 1602 by his heirs. This work is composed of a first part, *Galliae, Belgii Inferioris, Germaniae tabulae geographicae*, containing 51 maps, and a second part, with 29 maps, edited posthumously by his son Rumold. The *Atlas of Mercator* had in the following years several editions, the great part abridged. He was also the first to use the term *atlas* for a collection of maps.

Mercator also made an earth globe (1541) and a celestial globe (1551) with great success and built numerous duplicates. He was also the author of the philosophical work *De mundi creatione ac fabrica liber*, written when he was still young but published only in 1592, *Evangelicae historiae quadripartite Monas, sive Harmonia quatuor evangelistarum*, to defend his *Chronologia* from the accusation of being mistaken about the birth year of Christ. On May 5, 1590, Mercator had a massive stroke that left his left side paralyzed. Frustrated since he could no longer work, he slowly recovered but suffered from his inability to continue his map making projects. By 1592, he was able to do a small quantity of work again, but he was almost blind. He had a second stroke at the end of 1593, which took away his ability to speak, and although he tried hard, regaining some speaking capability, a third stroke was fatal to him. Mercator died on December 2, 1594, in Duisburg.

The works of Mercator are regarded as the product of the synthesis of the Renaissance culture, critically reconstructing the text and maps of the past, and imposing a new scientific way to compile maps.

BIBLIOGRAPHY. Jean van Raemdonck, *Gerardus Mercatore, sa vie et ses œuvres* (Dalschaert-Praet, 1869); Arthur Breusing, *Gerhard Kremer, genannt Mercator, der deutsche Geograph* (Raske, 1878); Matteo Fiorini, *Gerardo Mercatore e le sue carte geografiche* (in "Bollettino della Società Geografica Italiana," 1890); Gerhard Mercator und seine Zeit: 7 Kartographiehistorisches Colloquium, 1994 (Braun, 1996); Cornelis Koeman, *The Folio Atlases Published by Gerard Mercator, Jodocus Hondius, Henricus Hondius, Johannes Janssonius and Their Successors* (HES, 1997); Nicholas Crane, *Mercator* (Holt, 2003).

Elvio Ciferri
Leopoldo and Alice Franchetti Institute, Italy

MERCOSUR

ON MARCH 26, 1991, the presidents of ARGENTINA, BRAZIL, PARAGUAY, and URUGUAY signed a common market agreement, the Treaty of Asunción, that set the basis for the creation of the Common Market of South America (MERCOSUR, or Mercado Común del Sur). MERCOSUR was the final step in a long struggle for economic and political cooperation in the region. The common market agreement has its origins in a previous agreement between Argentina and Brazil. The two most powerful countries in South America had signed in 1986 an agreement of economic and political cooperation known as the Argentine-Brazilian Economic

Integration Program (ABEIP). The MERCOSUR agreement called for the creation of a free trade zone and the full integration of the regional economies. Argentina and Brazil joined the common market in December 1994, and Paraguay and Uruguay a year later. Its most important dispositions included a progressive reduction of trade tariffs and the establishment of common external tariffs. The four countries also agreed on coordinating their macroeconomic policies. According to the treaty, trade tariffs would be progressively reduced. By 2000, as Jeffrey Cason explains, "their custom union has made duty-free in approximately 90 percent of all goods, and the four core member nations have agreed to a common external tariff on nearly all goods."

MERCOSUR responded to both political and economic motivations. From a political perspective, it was an effort to guarantee regional political stability and consolidate democratic institutions. In a presidential summit in 1992, the four countries agreed that "an indispensable assumption for the existence and development of MERCOSUR is that democratic institutions are in force," Peter Smith exlains. This compromise was reinforced in 1998 in a meeting in Usuahia, Argentina, where the four MERCOSUR countries and two new associated members (CHILE and BOLIVIA) agreed on the compromise to defend democracy and maintain peace in the region.

From an economic perspective, MERCOSUR brought together the most important and stronger economies of South America in an effort for economic integration and liberalization of international trade. Brazil, however, has always played the leading and stronger role. MERCOSUR had an immediate impact on regional trade. Between 1990 and 1996, internal trade increased from $4.1 billion to $17.0 billion, according to Smith. MERCOSUR has also maintained discussions with the EUROPEAN UNION to create a free trade agreement.

Although MERCOSUR has been an important agent since the mid-1990s, it has also suffered from institutional limitations and faced strong barriers. The creation of a common market and a common custom system has been imperfect, and by 2000 there were still some conflicts and disputes with tariffs. In addition, the internal contradictions of the regional economies have limited the benefits and growth of a common market. The most visible contradictions have been the social and economic structures. During the 1990s, while this subregion concentrated about 200 million people, about 20 percent lived in poverty.

BIBLIOGRAPHY. Jeffrey Cason, "On the Road to Southern Cone Economic Integration" *Journal of Interamerican Studies and World Affairs* (v.42/1, 2000); Luigi Manzetti, "The Political Economy of Mercosur," *Journal of Interamerican Studies and World Affairs* (v.35/4, 1993–94); Peter H. Smith, *Talons of the Eagle: Dynamics of U.S.-Latin American Relations* (Oxford University Press, 2000).

ANGELA VERGARA
UNIVERSITY OF TEXAS PAN AMERICAN

metric system

LEGALLY RECOGNIZED in the UNITED STATES by the Metric Act of 1866 but devised by French scientists in a 1791 report to the French National Assembly, the metric system of measurement is the decimal system of weights and measures based on the meter, liter, and gram with the prefixes deci-, deca-, and kilo-. Originally not universally accepted, the metric system, designed to simplify the traditional system of weights and measurement used in Europe, replaced all the tradition units of measurement except the units of time and angle measure. The establishment of the metric system is widely regarded as the first step in the development of the International System of Units (SI), which links all systems of weight and measures.

The SI, the modern version of the metric system, was established in 1960 by the 11th General Conference on Weights and Measurements. This intergovernmental treaty organization, which itself was created by the Meter Convention in 1875, is the international authority that ensures the dissemination and modification of the metric system to reflect the latest changes in science and technology. Linked to the SI through the Meter Convention is the International Bureau of Weights and Measurements. This organization's mandate is to provide the basis for a single coherent system of measurement, through direct dissemination of units to coordinated international comparisons of national measurements standards, throughout the world.

The essential feature of the metric system of measurement, adopted by nearly every major industrialized country and viewed as a coherent system of units for physical science, is based on the length of a meter in relation to a platinum bar with a rectangular cross section and polished parallel ends. That is, the ideal behind the metric system was the use of only one measure per physically measured quality. Using the Earth as

the measuring stick, the meter, intended to be one ten-millionth part of the quadrant of the Earth, was defined at the Meter Convention as the distance between the polished end faces at a specified temperature. This definition was based on a measurement of a meridian between Dunkirk, FRANCE, and Barcelona, ITALY.

Later used as the prototype for the base units of length and mass, the meter along with the kilogram, were the models for a three-dimensional mechanical unit system of measurement. Endorsed by the American Metric Association (now the United States Metric Association) in 1916, the length of the meter served as the U.S. primary metric system standard until 1960.

The advantages to using the metric system, outside of being the world standard of measurement, are numerous. A few of its greatest advantages are that it has only one unit for each type of measurement that is easy to use and pronounce, it is never necessary to convert from one unit to another within the metric system, and there are no conversion factors to memorize. In addition, the metric system uses decimals instead of fractions or mixed numbers.

In relation to the prefixes, one of the mathematical advantages to using the metric system is the combination metric terminology with its decimal organization. Because there are several prefixes associated with a decimal position, they can be attached to the base metric unit in order to create new metric units.

In 1975, the United States, through the Metric Conversion Act and the United States Metric Board, designated the metric system of measurement as the preferred system of weights and measurements for trade and commerce, and it directed federal agencies, during the construction of federal facilities, to convert, when feasible, to the metric system. The Metric Conversion Act, amended by Executive Order 12770 in 1991 by President George H.W. Bush, directed agencies to convert to the metric system of measurement, designating the Secretary of Commerce to direct and coordinate this effort. The hope is that through uniform use of the metric system there will no longer be misunderstanding, confusion, or error in universal weights and measurements.

BIBLIOGRAPHY. Ken Alder, *The Measure of All Things* (The Free Press, 2002); E.F. Cox, "Metric System: A Quarter-Century of Acceptance," *Osiris* (1959); M. Darton and J. Clark, *The Dictionary of Measurement* (J.M. Dent, 1994).

GLEN ANTHONY HARRIS
UNIVERSITY OF NORTH CAROLINA, WILMINGTON

Mexico

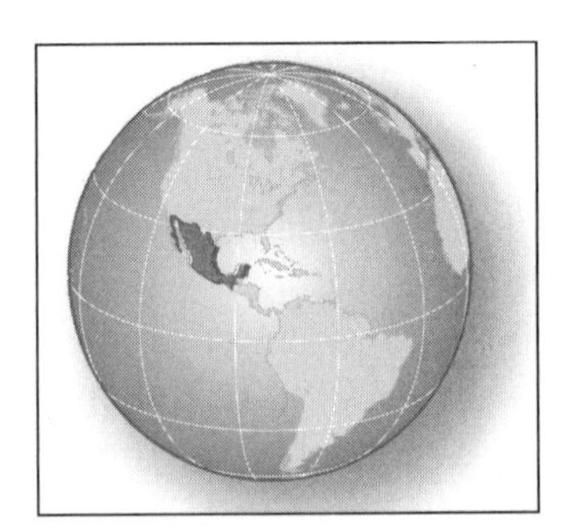

Map Page 1136 **Area** 761,605 square mi (1,972,550 square km) **Population** 104,959,594 **Capital** Mexico City **Highest Point** 18,700 ft (5,700 m) **Lowest Point** -33 ft (-10 m) **GDP per capita** $9,000 **Primary Natural Resources** petroleum, silver, copper, gold.

OFFICIALLY KNOWN AS the Estados Unidos Mexicanos (United Mexican States), Mexico is a country in North America. It is bordered by the UNITED STATES to the north, by the PACIFIC OCEAN to the west and south, by the Gulf of Mexico and CARIBBEAN SEA to the east, and by GUATEMALA and BELIZE to the southeast. Mexico has historically been a country of great mineral resources, ranging from silver to oil. The country has also become largely urban, as some two-thirds of the national population lives in cities, including MEXICO CITY, one of the largest metropolitan areas in the world.

Before the arrival of Europeans in the early 16th century, Mexico was home to several great Native American civilizations, including the Olmecs, Maya, and Aztecs. From the early-1500s until the 1820s, Mexico was part of the Spanish Empire in the New World. After gaining independence from SPAIN, Mexico became a republic in the 19th century. However, the

Miramar Lagoon in Chiapas, Mexico, has the dense jungle terrain similar to the country's Central American neighbors.

country lost much of its territory to the United States, including much of the present-day U.S Southwest from TEXAS to CALIFORNIA.

The altiplano, also called the Mexican Plateau, rises from the border with the United States to the central Mexican highlands. Close to the U.S. border, the plateau reaches heights of about 4,000 ft (1,219 m); south of Mexico City, the altiplano has an altitude of some 8,000 ft (2,438 m). The altiplano is comprised of two parts. The northern section, from the U.S. border to San Luis Potosí, is a largely arid region known as the Mesa del Norte. To the south is the Mesa Central, which is higher, moister, and flatter. The Mesa Central contains a number of intermontane basins, many of which are fertile, and contains the country's largest settlements. The Guanajuato basin has historically been the "bread basket" of Mexico, providing much of the country's food.

VALLEY OF MEXICO

The largest basins are those of Mexico, Puebla, and Guadalajara, all of which contain large cities of the same names. The Valley of Mexico, for example, has been the site of many great ancient Mexican civilizations, including the city of Teotihuacán, famous for its Temples of the Sun and Moon. The Aztec capital of Tenochtitlán was also in this basin. After the Spanish conquest of Mexico, the Spaniards built their colonial capital over the ruins on Tenochtitlán. Today, as part of a Federal District, Mexico City serves as the country's capital. Many of the valleys once had lakes that European settlers later drained and filled. In cities such as Mexico City, this practice has led to weak and unstable soils, often causing buildings to sink.

The altiplano is flanked by mountain ranges on each site. To the west is the Sierra Madre Occidental, which can reach 8,000 to 9,000 ft (2,438 to 2,743 m). River erosion in these volcanic mountains have led to the formation of many canyons, including Copper Canyon, the so-called Grand Canyon of Mexico. To the east is the somewhat lower Sierra Madre Oriental. To the east and west of the two mountain ranges are low plains that extend to the coast.

The altiplano is cut off by a string of volcanoes known as the volcanic axis that stretches from the Pacific to the Gulf coast. A number of the volcanoes are still active. Many of the volcanoes retain their Nahuatl-language names from the pre-Hispanic period. Located near Mexico City, Popocatéptl, or "smoking mountain," and Ixtaccíhuatl, or "white lady," reach over 17,000 ft (5,181 m).

To the east of the Cordillera Oriental is the Gulf Coast Plain. This lowland region stretches 900 mi (1,448 km) from Texas to the YUCATÁN PENINSULA. It consists of many lagoons and swamps to the east of the abrupt escarpment of the Cordillera. On the western side of Mexico is the narrower, less well-defined Pacific Coastal Lowlands. It runs for some 900 miles from the Mexicali Valley to Tuxpan. Despite the region's name, most of it actually faces the Gulf of California, not the Pacific Ocean. This lowland area is largely arid. However, it has also seen a rise in the importance of irrigated agriculture.

Baja California, or lower California, is a long, narrow peninsula that separates the Gulf of California from the Pacific Ocean. Most of the peninsula is mountainous and arid. However, agriculture has developed around Mexicali. There are also important fisheries on the peninsula. Tourism became an important economic activity at the end of the 20th century in areas such as Ensenada and Cabo San Lucas. Near the border with the United States, there are many factories known as *maquiladoras*, which use inexpensive Mexican labor to assemble products manufactured elsewhere.

South of the volcanic axis are the Southern Highlands, an area of old crystalline rock. This region is one of the most rugged areas of Mexico. The highlands are located to the south and east of "Old Antillia," which is believed to have once connected the Caribbean to the mainland. The Southern Highlands consist of steep ranges and valleys. On the southwestern side of the highlands is a series of ranges known as the Sierra Madre del Sur. These relatively low mountains often reach the sea, forming a rugged coastal margin. This areas is sometimes called the Mexican Riviera because of the important tourist destinations such as Acàpulco. The inland basins of the region are much less hospitable. To the northeast is the Mesa del Sur, with numerous valleys between 4,000 and 5,000 ft (1,219 and 1,524 m). The largest and most densely settled is the Oaxaca Valley, with its large native population.

THE YUCATÁN

Located north of "Old Antillia," the Yucatán Peninsula, sometimes called the Antillean Foreland, is a flat area of limestone. The parts of the foreland that are above sea level form the Yucatán Peninsula, the largest lowland plains in Mexico. Because of the limestone's porous nature, there are few surface streams in the Yucatán. Water can be found instead in sinkholes known as cenotes. With the lack of water and a poor-quality soil, this region possesses little fertile land. Historically,

henequen was one of the region's most important crops. In the 20th century, resort areas such as Cancún and Cozumel located here, drawing tourists. The oil industry also developed in the area Yucatán in the 20th century.

The Chiapas Highlands are an extension of mountain ranges in Central America. The region consists of a series of mountains that surround the high rift valley of the Grijalva River. This hot and humid region of southern Mexico is remote and sparsely populated. Chiapas possesses rapidly shrinking forests of dyewoods and hardwoods. The region has also been a retreat for Native American groups such as the Lacondones. In the 1990s, a group known as the Zapatistas began a rebellion in Chiapas to protest the plight of Mexico's Native American poor.

BIBLIOGRAPHY. Peter Bakewell, *A History of Latin America* (Blackwell, 2004); Brian Blouet and Olwyn Blouet, *Latin America: A Systematic and Regional Survey* (Wiley, 2004); Preston E. James, C.W. Minkel, and Eileen W. James, *Latin America* (Wiley, 1986); Harry Robinson, *Latin America: A Geographical Survey* (Praeger, 1967).

RONALD YOUNG
GEORGIA SOUTHERN UNIVERSITY

Mexico City

MEXICO CITY IS THE capital of the country of MEXICO. The city is located in the Valley of Mexico on the central Mexican plateau. The core of the city is comprised of the Federal District. However, the Mexico City metropolitan area goes well beyond the boundaries of the Federal District into the surrounding states. With a population of around 20 million, the Mexico City metropolitan area is one of the largest cities in the world.

The current site of Mexico City is located on top of Tenochtitlán, the capital of the Aztec Empire. The Aztecs founded Tenochtitlán in the 14th century on two small islands in Lake Texcoco. Three causeways connected the city to the mainland. Tenochtitlán grew into the largest city in the Americas before the arrival of Europeans.

When the Spanish arrived in the early 1500s, the city had a population of more than 100,000 people. The city contained a large and impressive marketplace at Tlatelolco. Tenochtitlán also possessed an extensive palace complex for the Aztec emperors and hundreds of religious temples. The Spanish destroyed much of the city between 1519 and 1521.

After conquering the Aztecs, the Spanish constructed their own city over the ruins of the Aztec capital. Constant flooding soon led the Spanish to fill in the lakes in the Valley of Mexico. Mexico City came to be the biggest and most important city in the New World. It served as the capital of the Viceroyalty of New Spain and rivaled European cities in size and wealth. Much of the wealth from Mexico's silver mines made the city prosperous. Already by 1560, Mexico City had a university, the printing press, and large, impressive public buildings and churches. It served as the political, religious, economic, and cultural center of New Spain. Mexico City was also a study in contrasts, as along with the great wealth there was much poverty. There was also much ethnic diversity in the city, as the population consisted of Spaniards, Native Americans, and Africans, along with new racially mixed groups such as mestizos and mulattoes.

After Mexico achieved its independence from SPAIN, Mexico City became part of a Federal District, which the new government created in 1824 to serve as the political capital of the country. The city grew slowly in the first half of the nineteenth century. A number of important events and developments occurred starting around mid-century. During the 1840s, troops from the UNITED STATES occupied Mexico City. In the 1850s, during a movement simply known as La Reforma, the government expropriated Church lands in the city, which in turn led to an opening up of the real estate market beyond the historical central district. This situation led to the city's first significant population shift, as many elite families moved to the west. Mexico City expanded in the 1860s under Maximilian I, who built the well-known avenue Paseo del Emperador, today known as the Paseo de la Reforma. During the dictatorship of Porfirio Díaz, which began in 1876, Mexico City was greatly modernized along the lines of PARIS, showing much French influence. There were many improvements in utilities and services, such as a new drainage system, gas and electrical lighting, and streetcars. Improvement in urban transportation in particular allowed for the continued spatial expansion of the city.

From 1900 to about 1930, Mexico City experienced even more urbanization. During the Mexican Revolution of the 1910s, the population of Mexico City grew as many people fled the countryside. Building in the city resumed in 1924 with the construction

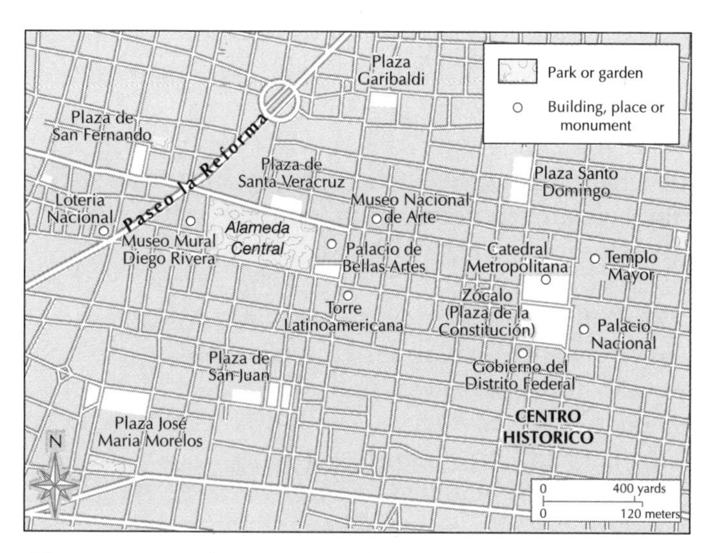

The current site of Mexico City is located on top of Tenochtitlán, the capital of the Aztec Empire.

of the Avendia Insurgentes. In this period, many members of the middle and upper classes moved to the south and west. The historic center of the city took on an increasingly bureaucratic function. Nevertheless, most residents still lived within the city proper. As late as 1930, 98 percent of the population lived within the city limits.

During the 1930s and 1940s, Mexico City experienced even more spatial and demographic growth. The earliest skyscrapers appeared in the 1930s, although the threat of earthquakes kept them small. The city became less concentrated, as residents moved north and south, in large part because of industrial expansion. More of the population lived in areas of the Federal District outside the legal limits of Mexico City.

Beginning about 1950, the city grew significantly because of a prospering economy fueled by oil exports. The population began to spill over the border of the Federal District into the state of Mexico. An example of the population movement was the 1957 creation of Ciudad Satélite, a middle-class suburban development. Population expansion continued in the 1960s with more industrial expansion and the growth of squatter settlements. For example, the squatter settlement of Nezahualcoytl grew from 65,000 to 650,000 in the 1960s, then reached 1.3 million by 1975. By the 1980s, nearly half of the city's population resided in slums and squatter settlements.

Mexico City hosted the Summer Olympics in 1968 and opened its subway the following year. The year 1968 also witnessed a massacre of student demonstrators at the Tlatelolco housing complex. The economic crisis of the 1980s led to decreased spending on urban services even as the city grew in population. A major earthquake struck Mexico City in 1985, destroying many buildings and killing some 7,000 people. This earthquake led to a new construction code for the city. By the end of the 20th century, Mexico City faced a number of critical urban problems, including severe air pollution and major traffic congestion.

BIBLIOGRAPHY. R. Douglas Cope, *The Limits of Racial Domination: Plebian Society in Colonial Mexico City* (University of Wisconsin Press, 1994); Diane Davis, *Urban Leviathan: Mexico City in the Twentieth Century* (Temple University Press, 1994); Michael Johns, *The City of Mexico in the Age of Díaz* (University of Texas Press, 1997); Jonathan Kandell, *La Capital: The Biography of Mexico City* (Random House, 1988).

RONALD YOUNG
GEORGIA SOUTHERN UNIVERSITY

Michigan

IN THE NORTH-CENTRAL UNITED STATES, Michigan is known as the Wolverine State. Michigan entered the Union in 1837 as the 26th state, measuring 490 mi (788 km) north to south and 240 mi (386 km) east to west. The total area of Michigan is 96,716 square mi (250,493 square km), making Michigan the 11th-largest state. Michigan ranks 8th in population. Bordering on four of the Great Lakes, some 38,575 square mi (99,908 square km) of Michigan is made up of water from the Great Lakes. With only 56,809 square mi (147,134 square km) of land, Michigan ranks 22nd in land size. Lansing is the state capital. Other large cities include Detroit, Grand Rapids, Warren, Sterling Heights, Flint, Lansing, Ann Arbor, Livonia, Dearborn, and Westland.

Michigan is divided into the Upper and Lower Peninsulas, which were physically separated from one another until the construction of the Mackinac Bridge in 1957 provided easy access from one section to another. The Upper Peninsula, which is bordered on the north by Lake SUPERIOR, on the south by Lake MICHIGAN and Lake HURON, and on the west by WISCONSIN and MINNESOTA, is filled with low rolling hills and occasional swamps in the east and higher hills with a rugged terrain in the west. The eastern section of the Upper Peninsula is part of the Great Lakes Plain, while

the western section from Lake Superior into the Porcupine Mountains lies within the Superior Upland. The Upper Peninsula contains 34 rapids and waterfalls.

Two major landforms are found in Michigan's Upper Peninsula. The Eastern Upper Plains Lowlands are composed of flat lands interspersed with hills formed from glaciers. Agriculture is limited in this area, but state and national parks abound. Locks were used to eliminate a number of rapids and falls to facilitate travel between Lakes Huron and Superior. The Crystalline Upland, located in the western section of the Upper Peninsula, contains the Porcupine and Huron mountains and the Gogebic and Copper mountain ranges. Forestry is a major activity in this area.

The Lower Peninsula, which is also part of the Great Lakes Plains, is bordered on the west by Lake Michigan, on the east by Lake Huron and Lake ERIE, and on the south by INDIANA and OHIO. The terrain is made up of law rolling hills in the southernmost section and flat lands interspersed with hills in the northernmost section. The Lower Peninsula includes four major landforms. The Hilly Moraines, covering the bottom half of the area, is composed of moraines, or low ridges, occurring at 10- to 25-mi intervals. The Beaches and Dunes section of the Lower Peninsula is comprised of low forest-covered areas alternating with high bare dunes. The High Plains and Moraines section, located north of Muskegon-Saginaw Bay, contains higher ridges. The Eastern Lower Plains Lowlands, extending from the Saginaw Bay area to the tip of the Lower Peninsula, encompasses the most industrialized section of the state, including Detroit.

The humid continental climate of Michigan is tempered by the Great Lakes, which absorb heat in the summer months and cool off slowly during the winter months. The state experiences well defined seasons. Only ALASKA ranks higher than Michigan as the wettest state in the United States. Average temperatures in Michigan range from 83 to 14 degrees F (28 to -10 degrees C). The highest elevation in Michigan is 1,979 ft (603 m) above sea level at Mount Arvon. The lowest elevation is 572 ft (174 m) above sea level where land meets Lake Erie.

Major rivers found in Michigan include the Detroit, the Grand, the St. Clair, and the St. Mary's. Rivers dammed to generate hydroelectricity include Manistee, Père Marquette, Muskegon, Grand, Kalamazoo, Saint Joseph, and Au Sable.

Michigan's flowering plants include arbutus, daisy, goldenrod, iris, lady's slipper, tiger lily, and violet. The apple blossom is the state's flower. Michigan's wildlife includes squirrels, foxes, woodchucks, rabbits, deer, hares, porcupines, black bears, and bobcats. Moose and timber wolves are found on Isle Royale. The state's birds include robin (the state bird), thrush, meadowlark, wren, bluebird, oriole, bobolink, and chickadee. Michigan's game birds are geese, duck, grouse, pheasant, and quail.

Michigan's industries are chiefly concerned with manufacturing, services, tourism, agriculture, forestry and lumber. Home to Detroit, the automobile capital of the United States, automobiles and automobile products are the state's most lucrative industry. Other industries include the production of transportation equipment, machinery, fabricated metal, food products, plastics, cereals, machine tools, airplane parts, refrigerators, hardware, and office furniture.

Soil in the Upper Peninsula and in the top half of the Lower Peninsula are generally acidic grays and browns with limited fertility, while the most fertile section of the state is found in the heavy loams near Saginaw Bay and in the bottom half of the Lower Peninsula. Michigan's major crops include corn, wheat, soybeans, dry beans, hay, potatoes, apples, cherries, sugar beets, blueberries, and cucumber. Maple, oak, and aspen form the basis of Michigan's commercial timber/lumber industry. Minerals found in Michigan include iron, copper, iodine, peat, natural gas, shale, gypsum, bromine, salt, lime, and sand and gravel.

BIBLIOGRAPHY. "About Michigan," www.michigan.gov (November 2004); "The Geography of Michigan," www.netstate.com (November 2004); Dan Golenpaul, ed., *Information Please Almanac* (McGraw-Hill, 2003); Sabine Helling, *Earthbound: Understanding People and the Environment* (Michigan State University Press, 1997); Kenneth E. Lewis, *West to Far Michigan: Settling The Lower Peninsula* (Michigan State University Press, 2002); Bruce Rubenstein et al., *Michigan: A History of A Great Lakes State* (Harlan Davidson, 2000); Lawrence Sommers, *Michigan: A Geography* (Westview, 1984).

ELIZABETH PURDY, PH.D.
INDEPENDENT SCHOLAR

Michigan, Lake

LAKE MICHIGAN IS one of the five original Great Lakes, which are located near or along the border between the UNITED STATES and CANADA; a sixth lake,

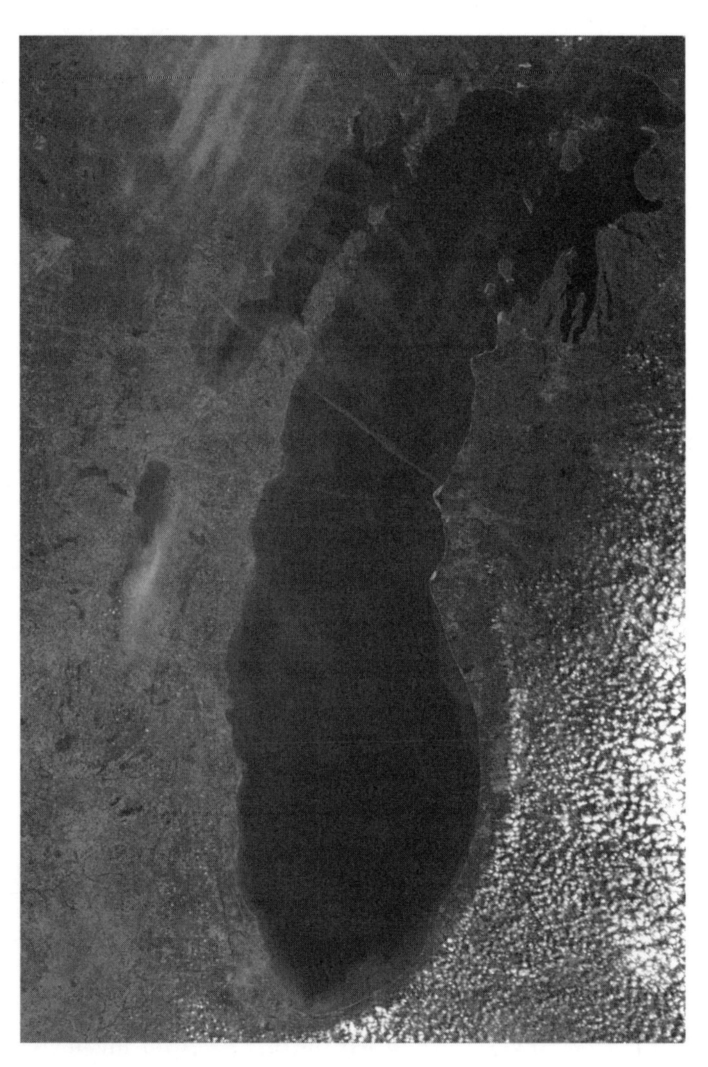

As seen from space, Lake Michigan's tear-drop shape is apparent. Chicago is at the bottom, left (Illinois) side of the lake.

Lake Champlain, was added to the Great Lakes category in 1998. The Great Lakes are the largest group of freshwater lakes in the world, and Lake Michigan is the only one located entirely within the United States; its waters are contained within the states of INDIANA, ILLINOIS, WISCONSIN, and MICHIGAN.

This body of water is 22,300 square mi (57,800 square km). It has the third-largest surface area of the Great Lakes, and is the sixth largest lake in the world. It connects to Lake HURON through the Straits of Mackinac and the two lakes are, hydrologically speaking, inseparable. The lake's drainage basin, the area of land where streams and rivers that drain into the lake exist, nearly doubles the lake's scope at 45,600 square mi (118,000 square km). Rivers such as the Fox-Wolf, Grand, Kalamazoo, Menonimee, Muskegon, and Saint Joseph are among the most notable. Via the Chicago River, Lake Michigan also connects to the MISSISSIPPI RIVER basin and then the Gulf of Mexico.

Lake Michigan is 307 mi (494 km) long and 118 mi (190 km) wide, with an average depth of 279 ft (85 m and a maximum depth of 925 ft (282 m). It contains 1,180 cubic mi (4,920 cubic km) of water, and averages 577 ft (176 m) above sea level. The shoreline is 1,638 mi (2,633 km) long when its island shorelines are considered.

The Great Lakes formed slowly; 600 million years ago, a shallow sea covered the area, and sand and silt deposited there gradually compressed into limestone, sandstone, and shale. The sea eventually dried up and about 1 million years ago, thick glaciers advanced and retreated over the land, carving large holes in soft sandstone and shale. Approximately 10,000 years ago, the last glacier retreated. As the Earth warmed up, the resulting water, called meltwater, filled the holes carved out of sandstone, and the Great Lakes were formed.

Lake Michigan's name also went through a series of changes. Explorer Samuel de Champlain named it Grand Lac in the early 17th century. Later names included Lake of the Stinking Water, and Lake of the Puants, after people living nearby, as well as Lac des Illinois, Lac St. Joseph, and Lac Dauphin. Explorers Louis Jolliet and Père Jacques Marquette gave this body of water the name by which we know it today, which is derived from a Native American designation, Michi-guma, meaning "big water."

The Sleeping Bear Sand Dunes, a spectacular feat of nature located in a national park, hovers 465 ft (142 m) above Lake Michigan and extends 35 mi (60 km) along its eastern shoreline. At this site, visitors can observe windswept dunes and other glacial formations.

A wide variety of fish live in Lake Michigan, including salmon, perch, bass, and walleye. Trout and whitefish are especially prized. By the late 1950s, many fish had been killed by an eellike creature called sea lamprey; although that problem has been alleviated, another greater problem still exists: pollution.

Lake Michigan's waters have been polluted from sewage and industrial waste, in part because its large surface area is a ready target for toxins falling from the atmosphere. At one time, Chicago's waste was dumped into the Chicago River, which flowed directly into Lake Michigan, and many other towns also dumped their raw sewage into the lake. Lake Michigan has a retention time of 99 years, which means that a molecule of water that enters the lake will, on average, take 99 years to leave its confines. Clearly, then, the contaminants entering the lake were accumulating; aggravating

the problem was the fact that Lake Michigan was a closed ecological system. Although many rivers flowed into the lake, none flowed out. To alleviate the pollution problem, engineers reconfigured the course of the Chicago River in 1900 so that it actually flowed away from the lake, thereby reducing the surge of sewage and chemical waste into Lake Michigan. Stricter environmental laws have also reduced the amount of industrial pollutants entering the water. In 1990, Congress passed the Great Lakes Critical Programs Act, which mandated more programs to reduce toxic pollutants so that Lake Michigan and other Great Lakes can be restored to a more healthful and stable condition.

BIBLIOGRAPHY. Ann Armbruster, *Lake Michigan* (Grolier Publishing, 1996); Harry Beckett, *Lake Michigan: Great Lakes of North America* (Rourke Corporation, 1999); Lake Michigan Federation, www.lakemichigan.org (April 2004); Great Lakes Information Network, www.great-lakes.net (April 2004); Lake Michigan Forum, www.lkmichiganforum.org (April 2004).

KELLY BOYER SAGERT
INDEPENDENT SCHOLAR

microclimates

MICROCLIMATES ARE climates of small areas, such as gardens, cities, lakes, valleys, and forests. A microclimate is an expression of the temperature, humidity, and wind within a few feet or meters of the ground. Such expressions exist because surfaces vary in their ability to absorb, store, or reflect the sun's energy, making some areas warmer or colder, wetter or drier, or more or less prone to frosts. Microclimates may be natural or human made. They are important for their effects on comfort, crops, and natural surroundings.

Microclimates can extend for several miles because of the presence of large bodies of water, urban areas, and topography. Large bodies of water, such as the Great Lakes, Chesapeake Bay, and the ATLANTIC OCEAN ,moderate temperatures of adjacent inland areas. Such water bodies store huge amounts of heat during the summer and release it slowly in winter. Consequently, land areas near the water tend to have low temperatures in winter that are not as cold or prone to fall and spring frosts. Small lakes and bays have the same but less extreme effects. Urban areas also have microclimates. In winter, buildings, parking lots, and streets of cities absorb heat during the day, and then radiate it back into the air at night. Temperatures may be moderate enough to lengthen the growing season slightly in urban areas. In summer, the heating affect of concrete walls, metal and tile roofs, and asphalt parking lots can create sweltering temperatures. The excessive heat dries soils, wilts plants, and endangers the health of infants and the elderly. Topography also affects microclimates. In the Northern Hemisphere, south-facing slopes receive direct rays from the sun and are therefore warmer and drier than north-facing slopes. Additionally, cold air, which is heavier than warm air, tends to spill down mountain slopes and pool in valleys at night. Some valleys are 10 degrees F (18 degrees C) cooler than adjoining slopes on winter nights. Such valleys are at risk to frosts in late spring and early fall from the downslope drainage of cold air.

Microclimates can also be much smaller. The backyard has an assortment of possible small-scale microclimates. Surfaces of homes, balconies, rooftops, paved surfaces (such as patios, driveways, and sidewalks), lawns, trees, and soil types have subtle effects on the temperature, humidity, and motion of air. Like urban areas, a home absorbs heat during the day and radiates it back at night. Like mountain slopes, the side of a home facing the sun receives more solar energy and is warmer than the opposite side. Less cold-tolerant trees and shrubs are better suited to the side exposed to direct sunlight; hardier plants that are less prone to spring frosts can survive the cooler, shady side of the house. The eve of a smartly built roof hangs out over the windows just enough to shade the windows from the "high" sun of summer, but the overhang is not too far to block the "low" sun of winter. Fences, walls, and large rocks protect plants from wind and radiate heat. In winter, paved surfaces around the home absorb the day's heat and reradiate it at night, moderating nocturnal temperatures. The same solid surfaces also absorb and reradiate heat to moderate winter temperatures and to raise summer temperatures.

Gardeners, architects, and farmers change the ground surface (such as by changing the reflectivity or heat transmission of the surface or by modifying surface roughness) to create microclimates. Human-made microclimates can be deliberate or unintentional, large or small in scale. For instance, a skillful gardener creates several small-scale microclimates in a single garden to assure a variety of attractive flowers, shrubs, and trees. The gardener knows that certain plants benefit from the coolness of shade trees; other plants require windbreaks for protection from desiccating

winds; and plants that are frost sensitive do better near walls or large rocks that absorb and reradiate heat. Architects design landscapes, homes and office buildings to take advantage of sunlight, solar energy, wind direction, and water drainage.

Farmers utilize microclimates in similar ways in large fields; they select crops according to field exposure to sunlight and wind, as well as moisture retention capacity of the soil. Some of the largest changes that people make to microclimates are unintentional results of widespread suburbanization, forest clearance, and agricultural expansion. Alterations of microclimates will not affect the general climate of a region, but the change can and does result in large local changes in climatic conditions.

BIBLIOGRAPHY. Rudolf Geiger, *The Climate Near the Ground* (Harvard University Press, 1971); T.R. Oke, *Boundary Layer Climates* (Routledge, 1988); Charles W.G. Smith, *Weather-Resilient Garden: A Defensive Approach to Planting and Landscaping* (Workman, 2004).

RICHARD A. CROOKER
KUTZTOWN UNIVERSITY

Micronesia

Map Page 1128 **Area** 271 square mi (702 square km) **Population** 108,143 **Capital** Palikir **Highest Point** 791 m (2,595 ft) **Lowest Point** 0 m **GDP per capita** $1,760 **Primary Natural Resources** forest products, fish, marine products, deep sea minerals.

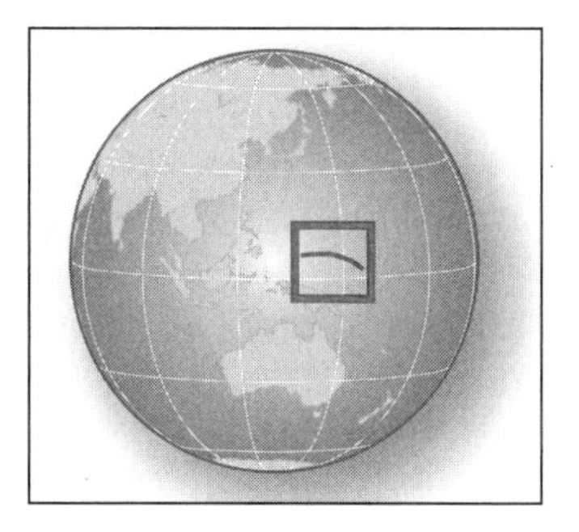

MICRONESIA CONSISTS of four Pacific island groups located between the Marianas and MARSHALL ISLANDS, 1,000 mi (1,600 km) north of PAPUA NEW GUINEA and some 3,000 mi (4,500 km) west of HAWAII. Although they cover an expanse of ocean larger than Western Europe, the total land mass of the 607 islands in the four groups is approximately that of a medium-size city.

A stone fortress built on a reef at Nan Madol gives evidence of advanced human occupation by 500 C.E. Discovered by Spanish explorers in the early 16th century, the islands of Micronesia were originally called New Philippines but were renamed the Caroline Islands around 1700.

Micronesia's islands include the high volcanic groups around Pohnpei, Kosrae, and Chuuk (Truk), as well as those around Yap, which are part of a raised section of the Asian continental shelf. All have tropical climates, abundant rainfall, and sparse populations. Pohnpei is a rocky, circular island with steep cliffs, tropical jungles, and extensive mangrove swamps. Kosrae is a solitary island with few outlying islets, and is encircled by a large coral reef. Chuuk includes 15 large islands, 192 outer islands, and 80 islets and atolls, with a total population of 50,000 people. Yap includes four main islands and 10 islets within the bounds of a large coral reef and 130 outer islands.

European colonial powers became interested in the islands in the early 19th century when whaling vessels began operating in the South Pacific. Westerners brought new languages, new religions, and new illnesses. A smallpox epidemic in 1854 killed half of Pohnpei's native population. In 1899, GERMANY purchased most of the Micronesian islands from SPAIN, but lost them to JAPAN, which seized them in World War I. In December 1941, Japan used its outposts in Micronesia to launch attacks on nearby islands, including Guam. The UNITED STATES oversaw the islands after World War II, and sponsored their transition to independence in 1991.

Micronesia's islands have no native mammals, but they do support a large array of marine and plant life, birds, and insects. The islands have one of the most uniform year-round temperatures in the world, at 81 degrees F (27 degrees C).

BIBLIOGRAPHY. Kenneth Brower, *Micronesia: The Land, the People, and the Sea* (Louisiana State University Press, 1981); Francis X. Hezel, *The New Shape of Old Island Cultures: A Half Century of Social Change in Micronesia* (University of Hawaii Press, 2001); Mark R. Peattie, *Nanyo* (University of Hawaii Press, 1983).

LAURA M. CALKINS, PH.D.
TEXAS TECH UNIVERSITY

Mid-Atlantic Ridge

A MID-OCEAN RIDGE is a system of rifts and parallel mountain ranges or hills found in all major oceans. It is thought to be the site of upwelling new ocean floor

material from Earth's mantle, from which ocean floors are gradually spreading out laterally. The Atlantic ridge is the most striking bottom relief feature of the ATLANTIC OCEAN.

A German oceanographic vessel, *Meteor,* discovered the ridge. The ridge, which is shaped like the letter "S," extends for 9,000 mi (144,000 km) across the ocean from north to south, and seldom falls more than 13,290 ft (4,000 m) below sea level. It is a submarine longitudinal rise traversing the ocean from ICELAND in north through the equator to BOUVET ISLAND in the south. The general course of the ridge is parallel to the coastline of the bordering continent and throws off many branching ridges towards the coast. Following the curves of the coastline and remaining in the center of the ocean, it divides the Atlantic into two broad deeper basins on either side, which are further divided into many subbasins The greater part of the ridge is submerged below sea level and its central backbone rises 5,000 to 10,000 ft (1,430 to 3,000 m) above the sea floor. The width of the ridge also varies from north to south. The Northern Atlantic ridge is known as Dolphin rise and the southern part is named Challenger rise.

The ridge, though under the sea level, has many peaks thrust out above the surface of the ocean. These peaks are the islands of the mid-Atlantic, for example, the AZORES and CAPE VERDE Islands. The sharpest peak of the Atlantic ridge is the cluster of islets known as Rocks of St. Paul near the equator. In the South Atlantic, the ridge produces ASCENSION and ST. HELENA islands.

The Atlantic ridge rises near Iceland. Between Iceland and Scotland, this ridge is known as the Wiley Thompson ridge and forms the boundary between the Atlantic and Arctic oceans. South of the continental shelf of GREENLAND and Iceland, the ridge widens near about 55 degrees N, where the depth of water is between 7,000 and 10,500 ft (2,000 and 3,000 m). This part is known as Telegraph plateau.

To the west of this plateau near 40 degrees N, the ridge bifurcates toward the Newfoundland coast bounded by the contour of 10,500 ft (3000 m). It is known as the Newfoundland rise. Moving southward in the middle of the ocean up to 12 degrees N, the Central Atlantic ridge follows the southwest direction between the coast of Africa and America. Here it is seldom broken and disrupted. South of 40 degrees N, the Atlantic ridge widens toward the coast of Africa, and here the Azores rise bifurcates from the Central ridge. At the equator, the Sierra Leone rise bifurcate toward the northeast coast of Africa, and the Para rise moves to northwest coast of South America.

At 10 degrees S, the Guinea rise, formed by a contour that is 10,500 ft (3,000 m) deep and less, runs northeast toward the Guinea coast. The ridge bifurcates near 40 degrees S, where the Central ridge has the maximum breadth of 600 mi (960 km). Walvis Ridge is bifurcated from here and merges into the African continental shelf. To the west, the Rio Grande rise is between 10,500 and 14,000 ft (3,000 and 4,000 m) and moves toward the coast of South America to 30 degrees S. Here is the Bromley Plateau, at a depth of 2,437 ft (697 m). After 40 degrees S, the central ridge moves toward the southeast and forms Meteor Bank and Cape Swell, near 45 degrees S. It proceeds toward the Cape of Good Hope, south of which it is known as the Mid Atlantic-Antarctic ridge.

The origin of the ridge is attributed to compression and continental drift. The evolution of the mid-Atlantic rise could be dated back to the Pliocene age. The volcanoes are found on the central ridge from Iceland to Bouvet. Some differential horizontal movement is in progress here. Such movements bring strips of crystal rocks near the sea level. The pattern of the Atlantic floor is suggestive of east-west stretching. It is postulated that the horizontal mantle motion between the zone of rising mantle rock (midocean ridge) and sinking rock (beneath the island and mountain arcs) exerts a dragging force on the lithospheric plates.

BIBLIOGRAPHY. R.W. Fairbridge, ed., *The Encyclopedia of Geomorphology* (Dowden, Hutching & Ross, 1968); *Ocean and Marine Dictionary* (Parmer Books, 1979); David F. Tver, et al., *Oceanography for Geographers* (Chaitanya Publishing House, 1980).

PRABHA SHASTRI RANADE
JAWAHARLAL NEHRU UNIVERSITY, INDIA

Middle East

THE TERM *Middle East* came to modern use after World War II, and was applied to the lands around the eastern end of the MEDITERRANEAN SEA including TURKEY and GREECE, together with IRAN and, more recently, the greater part of North Africa. The old Middle East began at the river valleys of the Tigris and the Euphrates rivers or at the western borders of Iran and extended to Burma (MYANMAR) and Ceylon (SRI LANKA).

Some geographers today say the Middle East region stretches 6,000 mi (9,656 km) eastward from the dry Atlantic shores of MAURITANIA to the high mountain core of AFGHANISTAN. Other geographers begin the Middle East with EGYPT. It includes numerous separate political states, most of which were created by colonial government cartographers in the 19th and 20th centuries. A good deal of the Middle East is too dry or rugged to sustain human life, and only 5 to 10 percent of the entire region is cultivated. As a result, a stark contrast exists between core areas of dense human settlement where water is plentiful, and the empty wastes of surrounding deserts and mountains.

Four regions can be identified in this vast, diverse, and distinct area: the Northern Highlands, a 3,000-mi- (4,828-km-) long zone of plateaus and mountains in Turkey, IRAN, and AFGHANISTAN, stretching from the Mediterranean Sea to Central Asia; the Arabian Peninsula, a million-square-mi- (2.6-million-square-km-) desert quadrilateral jutting southward into the INDIAN OCEAN and flanked on either side by the Persian (or Arabian) Gulf and the RED SEA; the Central Middle East, the rich valleys of the NILE in Egypt and of the Tigris and Euphrates in IRAQ, and the intervening fertile crescent countries of ISRAEL, JORDAN, LEBANON, and SYRIA; and North Africa, a band of watered mountains and plains set between the SAHARA DESERT and the Mediterranean Sea. Known by the Arabs as al Maghrib al Aqsa ("Land of the Setting Sun"), it includes the nations of TUNISIA, ALGERIA, MOROCCO, and LIBYA.

THE NORTHERN HIGHLANDS

The Taurus and ZAGROS MOUNTAINS of southern Turkey and western Iran form a physical and cultural divide between Arabic-speaking peoples to the south and the plateau-dwelling Central Asian people of Turkey, Iran, and Afghanistan. Around one-third of the people of the Middle East and North Africa live in the Northern Highlands, on the Anatolian and Iranian plateaus and the flanks of the HINDU KUSH range of Afghanistan.

Turkey is a large, rectangular peninsula plateau bounded on three sides by water—the BLACK SEA on the north, and the AEGEAN SEA to the west, and south. The Turkish coast is rainy, densely settled, and intensively cultivated. About 40 percent of the population is clustered onto the narrow, wet Black Sea coast, on the lowlands around the Sea of Marmara in both European and Asiatic Turkey, along the shores of the Aegean, and on the fertile Adana Plain in the southeast.

By contrast, the center of Turkey—the dry, flat Anatolian Plateau—is sparsely settled. Cut off by the Pontic Mountains to the north and the Taurus to the south, the dead heart of the plateau is too dry to sustain dense agricultural settlement; in the east, the rugged terrain of the Armenian highlands limits agricultural development.

Although the environmental base of Iranian society is similar to Turkey's, the topography is more dramatic, and contrasts are more sharply drawn. High mountains ring the dry Iranian Plateau on all sides except the east.

In the west and south, the folded ranges of the Zagros Mountains curve southeastward for a distance of 1.400 mi (2,253 km) from the northwest Turkish frontier to deserts of Sistan in the southeast. In the north, the steep volcano-studded ELBURZ range sharply divides the wet CASPIAN SEA coast from the dry Iranian interior. The encircled plateau covers over half the area of Iran, with large uninhabited stretches of salt waste in the Dashti Kavir to the north and of sand desert in the Dashti Lut to the south.

Along the Caspian Littoral, which receives up to 60 in (152 cm) of rainfall per year, the intensive cultivation of rice, tea, tobacco, and citrus fruits supports a dense rural population. Similarly, in AZERBAIJAN in the northwest and in the fertile valleys of the northern Zagros, rainfall is sufficient to support grain cultivation without irrigation. But in the rest of Iran, rainfall is inadequate and crops essentially require irrigation. Oasis settlement based on wells, springs, or underground horizontal water channels called qanats is common.

In the small remote country of Afghanistan, the easternmost nation of the Northern Highlands, the processes of population growth, agricultural expansion, and urbanization have barely begun. The country's center is occupied by the ranges of the Hindu Kush; a rugged, snowbound highland that is one of the least penetrable regions in the world. Deserts to the east and south are cut by two major rivers, the Hari Rud and the Helmand, both originating in the central mountains of Afghanistan and disappearing into the deserts of eastern Iran. In the north, the AMU DARYA (Oxus) flows into the Russian STEPPE. Settlements are found in scattered alluvial pockets on the perimeter of the Hindu Kush, where there is level land and reliable water supplies.

Over 70 percent of Afghanistan's population lives in the scattered villages as cultivators of wheat and barley and herders of small flocks of sheep and goats. An additional 15 percent are nomadic tribesmen, whose political power is still felt in this traditional society. In central and eastern Afghanistan, Pathans are domi-

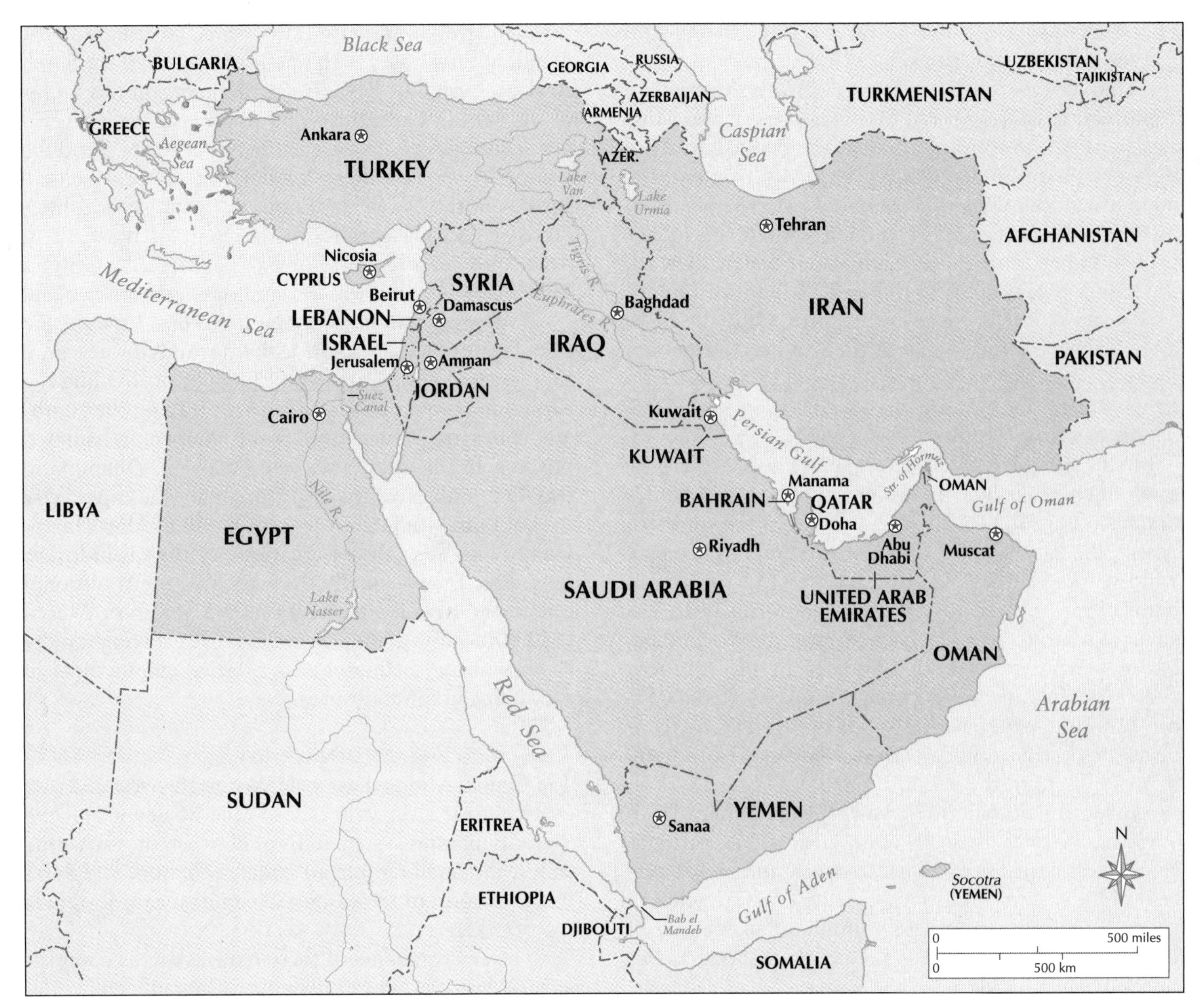

A good deal of the Middle East is too dry or rugged to sustain human life, and only 5 to 10 percent of the entire region is cultivated. A stark contrast exists between core areas of dense human settlement where water is plentiful and the surrounding deserts and mountains.

nant; in the north, the Turkish-speaking Uzbeks and Persian-speaking Tadzhiks predominate.

THE ARABIAN PENINSULA

The Arabian Peninsula may be described as a great plateau sloping gently eastward from a mountain range running along the whole length of its west side. It is a huge desert fault bounded on three sides by water and on the fourth by the deserts of Jordan and Iraq. In the west, the rugged slopes of the Hijaz and the highlands of YEMEN form the topographical spine of this platform. The remainder tilts eastward to the flat coasts of the Persian Gulf, rising only in the extreme southeast to the height of the Jabal al Akhdar (Green Mountains) of Oman.

Although the peninsula is the largest in the world and nearly four times the size of the state of TEXAS, it supports a population of less than 18 million people. The majority of these people live in two nations: SAUDI ARABIA (25.79 million, excluding 5.57 million non-nationals), which governs nine-tenths of this region, and Yemen (20 million), whose highlands trap sufficient moisture to support cultivation without irrigation. Smaller states on the eastern and southern perimeters

of the peninsula include KUWAIT, QATAR, the UNITED ARAB EMIRATES, and OMAN.

Although the TROPIC OF CANCER bisects the Arabian Peninsula, passing south of Medina, Riyadh, and Muscat, most of the southern half of the peninsula is too high or isolated to be characteristically tropical, the main exception being the lowland coast. The principal historical determinant of human settlement in the peninsula has been the availability of water. Overall, the region receives less than 3 in (7.62 cm) of rainfall each year, with a bit more in the north. Only the highlands of Yemen and Oman at the southern corners of the peninsula receive more than 10 in (25.4 cm). Daily temperatures commonly rise above 100 degrees F (37.7 degree C).

Fully one-third of the central plateau is covered by a sea of shifting sand dunes and much of the rest lies under boulder-strewn rock pavement. In the southern desert, the forbidding Rub al Khali (Empty Quarter), wind-worked dunes 500 to 1,000 ft (152 to 305 m) high, cover an area of 250,000 square mi (402,336 square km) to form a bleak, rainless no-man's-land between Saudi Arabia and the states of the southern coast. Arching northward from the Rub al Khali, a 15-mi- (24-km-) wide river of sand, the Ad Dahna, connects the southern sands with the desert of Nafud 800 mi (1,287 km) to the north.

Given this harsh environment, the Arabian landscape has no permanent lakes or streams. Vegetation is sparse. Settlement is confined to oases, and only 1 percent of the region is under cultivation. Vast stretches of the peninsula are completely uninhabited, devoid of human presence except for the occasional passage of Bedouin camel herders.

Within this difficult physical setting, two-thirds of the people of the Arabian Peninsula are rural agriculturalists, seminomads, and nomads. Their lives focus on oasis settlements where wells and springs provide water for the cultivation of dates—the staple food of Arabia—and the maintenance of herds of camels, sheep, and goats. The distribution of these oases is determined by a network of dry river valleys (wadis) carved into the surface of the plateau in earlier and wetter geological periods. These wadis provide the most favored locations for commercial and agricultural settlement and the most convenient routes for caravan traffic.

In the western highlands, where population density is above average, the largest urban centers are Mecca, Medina, and Taif. In central Arabia, underground water percolates down from these uplands and surfaces through artesian (gravity-flow) wells, creating a string of agricultural oases both north and south of the Saudi Arabian capital of Riyadh. Farther east, on the shores of the Persian Gulf, this same water emerges as freshwater springs in Kuwait, eastern Saudi Arabia, and the United Arab Emirates. Similarly in South Yemen, springs in the Wadi Hadramuat, a gash several hundreds miles long parallel to the coast of the Gulf of Aden, provide the basis for oasis settlement. Only in Yemen and Oman is this dryland oasis pattern broken.

Today, oil resources in Saudi Arabia, Kuwait, the island state of BAHRAIN, the United Arab Emirates, and, to a lesser extent, Qatar and Oman are providing the capital for rapid economic growth, leaving the southern states of Yemen and South Yemen in isolated poverty. In the gulf, cities like Dhahran, Dhammam, Ras Tannurah, Kuwait City, Manamah (the capital city of Bahrain), and emirate centers like Abu Dhabi, Dubai, and Sharjah are creations of the oil industry. Less directly but equally dramatically, the traditional centers of Riyadh (population: 3.5 million), Mecca (550,000) and Jeddah (2.8 million) are growing rapidly as farmers and Bedouins seek salaried employment in expanding urban industries.

THE CENTRAL MIDDLE EAST

The Central Middle East is flanked to the west and east by two great river valleys, the Nile of Egypt and the Tigris-Euphrates system in Iraq. Between these riverine states, the small nations of Israel, Lebanon, and Syria line the shores of the eastern Mediterranean; Jordan is LANDLOCKED.

The environments of these nations are as complex as their histories. Four millennia of human civilization have left an essentially denuded landscape—barren hills, steppes overgrazed by sheep and goats, and rivers chocked by the erosional silt of human activity. In Egypt, the Nile Valley, a narrow trough 2 to 10 mi (3.2 to 16 km) wide cuts northward across the dry plateau of northeastern Africa to the Mediterranean Sea. East of the Nile Valley, the heavily dissected Eastern Highlands border the coast of the Red Sea, continuing past the Gulf of Suez into the SINAI PENINSULA. Barren and dry, these highlands are occupied by nomadic herders.

The sources of the Nile River lie 2,000 mi (3,218 km) south of the Mediterranean in the wet plains of the Sudan and the equatorial highlands of East Africa. The Nile's largest tributary, the White Nile, originates in Lake Albert and Lake VICTORIA and flows sluggishly through a vast swamp, the Sudd, in southern Sudan, before entering Egypt. The other major tributaries, the

Blue Nile and the Atbara River, flow out of the Ethiopian highlands, draining the heavy summer rains of this region northward toward the desert. This summer rainfall pours into the Nile system, causing the river to flood regularly from August to December, and raises its level some 21 ft (6.4 m).

For centuries, this flood formed the basis of Egyptian agriculture. Specially prepared earth basins were constructed along the banks of the Nile to trap and hold the floodwaters, providing Egyptian farmers with enough water to irrigate one and, in some areas, two crops of wheat and barley each year.

In the 20th century, British and Egyptian engineers constructed a series of barrages and dams on the Nile to hold and store the floodwater year-round. This transformation of Nile agriculture was largely completed in 1970 with the construction of the Aswan High Dam, a massive earthen barrier more than 2 mi (3.2 km) across, 0.5 mile (0.8 km) wide at the base, and 120 ft (37 m) high.

Behind it, Lake Nasser, the dammed Nile River, stretches 300 mi (483 km) southward to the Sudanese border. The dam added about one-third to the cultivated area of Egypt. Hence, Egyptian population expanded from estimated 10 million at the turn of the century to its current 76 million. During this same period, urban population expanded eight times, and Cairo (7.76 million) and Alexandria (3.9 million) emerged as the two largest cities on the African continent.

In contrast to Egypt, the central problem in the other great river valley of the Middle East, the Tigris-Euphrates of Iraq, is not overpopulation but environmental management. Both these rivers rise in the mountains of eastern Turkey and course southward for more than 1,000 mi (1,609 km) before merging in the marshes of the Shatt al Arab. North of Baghdad, both rivers run swiftly in clearly defined channels. To the south, they meander across the flat alluvial plains of Mesopotamia. East of the valley, the Zagros Mountains rise as a steep rock wall separating Iraq from Iran.

To the west, a rocky desert plain occupied by nomadic herders stretches the borders of Saudi Arabia, Jordan, and Syria. Only in the northeast, in the Kurdish hills, does rainfall sustain non-irrigated cultivation. Elsewhere in Iraq, human existence depends on the water of the Tigris and Euphrates rivers. But unlike Egypt, where every available acre of farmland is intensively utilized, Iraq's agricultural resources are largely wasted.

The Tigris and Euphrates rivers have always proved less manageable than the Nile. Fed by melting snows in Turkish highlands, spring floods 8 to 10 ft (244 to 309 cm) above normal pour down the river channels to Baghdad and then spread out over the vast plains of Mesopotamia, where the land is so flat that elevations change only 4 to 5 ft (122 to 152 cm) over distances of 50 mi (80 km).

In the Fertile Crescent countries of Syria, Lebanon, Jordan, and Israel, which lie along the eastern coast of the Mediterranean between the river valleys of Egypt and Iraq, environmental patterns are extremely complex. The coastal plain, narrow in the north but widening southward, is backed by dissected, rugged highlands that reach elevations of more than 10,000 ft (3,048 m) in Lebanon. Throughout their length, these uplands have been denuded of forests, notably the famous cedars of Lebanon, by centuries of overgrazing and cutting for economic gain. Winter rainfall is plentiful in the north but less in the south. In Syria, the highlands capture this ample rainfall in stream that support life in the oasis cities of Aleppo, Homs, Hama, and Damascus. In Lebanon and northern Israel, runoff from the highlands sustains important commercial and agricultural areas along the coast. Further south, the highlands flatten out into the rainless wastes of the NEGEV DESERT.

Inland, a narrow belt of shallow, flat-bottomed intermontane valleys separates these western highlands from the dry uplands plateaus and mountains of the east. Between Israel and Jordan south of the Galilee, the Jordan River flows along one of these valleys 150 mi (241 km) southward to the Dead Sea, 1300 ft (396 m) below sea level. Farther north, a similar trough in Lebanon, the Beqaa Valley, is drained by the Litani and Orontes rivers. East of these lowlands, rugged highlands grade inland to the grass-covered steppes of Syria in the north and the dry stone pavement of the Jordanian desert in the south. In this varied terrain, the distribution of population is extremely uneven.

NORTH AFRICA

North Africa is the largest subregion of the modern Middle East, covering an area larger than the United States, but inhabited by only 50 million people grouped together on the southern shore of the Mediterranean Sea between water and sand. Much as Egypt is truly the gift of the Nile, cultural North Africa is the result of a physiographic event, the ATLAS MOUNTAINS, which separate the Sahara Desert from the Mediterranean Sea and Europe beyond.

Most of the territory of the modern nations of Maghreb-Morocco, Algeria, Tunisia, and Libya consists of Saharan wastelands that stretch 3,000 mi (4,828 km) across Africa from the Atlantic Ocean to the Red Sea. One-seventh of this area is sand dunes; the remainder is rock-strewn plains and plateaus. Aridity in the Sahara is not interrupted even by the jutting peaks of the AHAGGAR and TIBETSTI massifs at 6,000 ft (1,829 m) which receive as little as 5 in (21.7 cm) of rainfall per year. Here, as well as in other scattered Saharan oasis environments, an estimated 3 million people wrest a living from what is Earth's most difficult cultural environment outside the polar regions. Only in the north, along the mountain-backed coast of the Mediterranean, is rainfall sufficient to sustain substantial concentrations of people. The Atlas Mountains form a diagonal barrier isolating the nomads of the deserts and steppes of the south and east from sedentary agriculturalists in the Mediterranean north.

The Spanish Sahara was the only North African country that was totally desert. Before the recent discovery of extensive phosphate deposits in the north and the possibility of rich iron ore lodes, this territory was of little interests to anyone, but both Morocco and Mauritania have claimed sovereignty. With the departure of SPAIN in 1976, the territory has been divided between its two larger neighbors.

In Morocco, the Atlas Mountains form a succession of four mountain ranges dominating the landscape. In the north, the Rif, which is not geologically associated with the Atlas, is a concave arc of mountains rising steeply along the Mediterranean, reaching elevations of 7,000 ft (2,134 m) and orienting Morocco toward the Atlantic. In the center of Morocco, the limestone plateaus and volcanic craters of the Middle Atlas reach elevations of 10,000 ft (3,048 m); contact with Algeria is channeled through the Taza corridor, and this mountain barrier has isolated the Moroccan Sahara until modern times. Farther south, the snow-capped peaks of the High Atlas attain elevations of 13,400 ft (4,084 m) and separate the watered north from life in the Sahara. Finally, the Anti-Atlas, the lowest and southernmost of the Moroccan ranges, forms topographic barriers to the western Sahara. Historically, the Atlas range has provided a refuge for the original Berber-speaking inhabitants of Morocco, whose descendants today make up half the nation's population.

Throughout mountainous Morocco, Berber populations maintain an agrarian tradition of transhumance of goats and sheep wedded to cultivation of barley, centered around compact mountain fortresses. Density of settlement in the mountains depends on rainfall, which in general diminishes from west to east, and on altitude, which prohibits year-round settlement because of cold winter temperatures in areas much over 6,000 ft (1,829 m). Most of Morocco's 32.2 million people are Arabic-speaking farmers who till the fertile lowlands plains and plateau stretching from the Atlantic to the foothills of the Atlas.

Farther east along the Atlas complex, the primary environmental contrast in Algeria is once again the distinction between the fertile, well-watered, and densely settled coast and mountain ranges of the north and the dry reaches of the Sahara Desert in the interior. The Algerian coast is backed by the Tell Atlas, a string of massifs 3,000 to 7,000 ft (914 to 2,133 m) in elevation, which have formed an important historical refuge for Berber-speaking tribes. In the interior, a parallel mountain range, the Saharan Atlas, reaches comparable elevations in a progressively drier climate. Between these two ranges in western Algeria, the high plateaus of the Shatts, a series of flat interior basins, form an important grazing area. In eastern Algeria, the two ranges of the Atlas merge to form the rugged Aures Mountains. South of these ranges, Algeria extends 900 mi (1,448 km) into the heart of the Sahara.

In Tunisia and Libya, the topography is less dramatic than in Morocco and Algeria, but the same environmental sequence from northern coast to southern desert prevails. Two-thirds of Tunisia's 9.9 million people live in the humid northeast and in the eastern extension of the Arres Mountains. The central highlands and interior steppes, marginally important in the past, have become sites of innovative development projects. Tunisia remains an example of self-motivated and successful state planning. In Libya, the population (5.6 million) is concentrated on the coast in the hilly back country of Tripolitania and Cyrenaica.

BIBLIOGRAPHY. W.B. Fisher, *The Middle East: A Physical, Social and Regional Geography* (Methuen, 1971); R.H. Sanger, *The Arabian Peninsula* (1954); Ibn Haukal, *The Oriental Geography*, Sir W. Ouseley, trans. (1800); James Jankowski and Stephen H. Longrigg, *The Middle East: A Social Geography* (Aldine, 1970); Malcolm Wagstaff, *The Middle East: A Geographical Study* (Oxford University Press, 1976); Daniel Bates and Amal Rassam, *Peoples and Cultures of the Middle East* (Prentice Hall, 1983); Nikshoy Chatterji, *A History of the Modern Middle East* (Envoy Press, 1987); Deborah J. Gerner, ed., *Understanding the Contemporary Middle East* (Lynne Rienner Publishers,

2003); Albert Hourani, *A History of the Arab Peoples* (Harvard University Press, 1991).

JITENDRA UTTAM
JAWAHARLAL NEHRU UNIVERSITY, INDIA

migration

MIGRATION IS referred to as "any residential movement which occurs between administrative units over a given period of time," according to geographers Paul White and Robert Woods (1980). Other scholars have defined migration as the change in the center of gravity of an individual's mobility pattern. The destinations of the mobility flows need not, themselves, change as a result of the change in their center of gravity.

For example, in the local intraurban move, the destinations of journey-to-work, recreation, and shopping may remain the same, while in an interurban move, they are likely to change. The perception of spatial differentiation of opportunities—the idea that different geographical locations offer different levels of potential well-being to various sections of human population—explains why migration occurs. It is these perceived differences between places that are important rather than any simple "push and pull" mechanism.

Hence, migration occurs because migrants believe that they will be more satisfied in their needs and desires in the place that they move to rather than the place from which they come. An importance must be placed on the word *believe*. Migration occurs as a result of decisions made by individuals in the light of what they perceive the objective world to be like: it does not matter if the migrant holds an erroneous view—it is that erroneous view that is acted upon rather than the objective real-world situation. Thus, there may be cases where migration occurs despite the lack of an objective reason for it, and other cases where an objective appraisal of the world, were it possible, might suggest that migration should occur where it is, in fact, absent.

In recent times, international migration is at record levels and is unlikely to slow down in the near future. The number of long-term international migrants (that is, those residing in foreign counties for more than one year), according to the United Nations Population Division, in 1965 was only 75 million, but that number rose to 84 million by 1975 and 105 million by 1985. There were an estimated 120 million migrants in 1990, the last year for which the detailed statistics are available. In the 1990s, migration growth continued with the same pace; hence, in 2000, an estimated 150 million people resided outside their county of birth or nationality. Even with the numbers of international migrants large and growing, it is important to keep in mind that less than 3 percent of the world's population has been living outside their home countries for a year or longer.

International migrants come from all parts of the world and they go to all parts of the world. In fact, only a few countries are unaffected by international migration. Many countries are sources of international flows, while others are net receivers, and still others are transit countries through which migrants pass to reach to receiving countries. Such countries as MEXICO experience all three capacities, as source, receiving, and transit countries.

The noteworthy fact about migration is that it tends to be within regions; migrants often remain within the same continent. More than half of international migrants traditionally have moved from one developing country to another. In recent years, however, migration from poorer to richer countries has increased significantly. While the traditional immigration countries—the UNITED SATES, CANADA, and AUSTRALIA—continue to see large-scale movements as a result of labor recruitment that began in the 1960s and 1970s, Europe, the oil-rich Persian Gulf states, and the "economic tigers" of East and Southeast Asia are now also major destinations for international migrants.

MIGRATION IN PREINDUSTRIAL SOCIETIES

In the history of migration, there is no comprehensive evidence of migration until the 19th century, however, information from census and vital registration provides indirect evidence of the nature of migration. It was once thought that there was little migration in pre-industrial societies, on the grounds that transportation was slow, costly, and often dangerous, and that institutions such as serfdom—or, in England after the 16th century the laws of settlement—hindered mobility. There is evidence supporting this view. In 1841, when four-fifths of Sweden's population lived in rural areas, 92.8 percent of the population was still in the country in which they had been born.

Research on parish registers shows that most women married men from their own or nearby villages; until after 1750, four-fifths of all French women married men from places within a few miles or kilometers of their village. But it does not follow from this

that preindustrial communities were completely isolated and lacked mobility. Studies of English villages in Tudor and Stuart times suggest a remarkable turnover in inhabitants' names that can only be explained by migration, albeit of short distances. A study of 10 parishes near York, England, shows that of the inhabitants at the end of 18th century, only 40 percent had ancestors there in 1700.

Before 1000, villages were the basic unit of rural settlement; in the following centuries people moved into the woodland between the primary settlements and established hamlets and isolated farmhouses; by the middle of the 13th century, there was little good farmland left, and there was a movement on to marginal land. Many of the upland areas of Europe were settled for the first time—in NORWAY, in the French ALPS, and in the Vosges and the PYRENEES.

The long-distance movement was more dramatic between 1050 and 1340; there was a movement of Germans east of the Elbe River into sparsely populated Slav areas; in the 12th and 14th centuries, some 400,000 moved into this area. In southern Europe, the principal movement was by the Spanish southward into the Muslim-occupied areas of Iberia. But while these two movements have received much attention from geographers and historians, they were numerically less important than the innumerable short-distance moves that completed the rural settlement of western Europe. By the beginning of 17th century, the great age of rural-rural migration in the west was ending.

It is difficult to estimate the importance of rural-urban migration in preindustrial Europe; even in the 18th century, the overwhelming majority of Europe's population lived in villages, hamlets, and small market towns, except in the Low Countries and England. Until the 19th century, rural birth rates generally exceeded urban birth rates, and urban death rates exceeded rural death rates; in most towns, and particularly in large ones, death rates exceeded birth rates. Thus, it is generally assumed that before the 19th century, towns could grow only by immigration from the countryside.

In 1000, Europe had no more than 100 places that could be called towns, and half of them were in ITALY; by 1300 there were 4,000 or 5,000 such places. The 12th and 13th centuries were a period of urban growth, standing in marked contrast to the preceding centuries. By 1300, Venice, Milan, and PARIS probably had 100,000 people each; LONDON 50,000. More important was the proliferation of small towns. In the 16th century, there were 500 market towns in England.

The 16th century saw not only a renewed growth of population but also rapid urban growth. The most striking feature of this period was the emergence of a number of very large cities. London grew from some 60,000 in 1520 to 250,000 in 1600 and had exceeded 500,000 by the end of the 17th century. Paris, the largest city in medieval western Europe, grew from 250,000 in 1600 to 500,000 in 1700, while Amsterdam, which had only 31,000 people in 1585, had reached 200,000 by 1650.

Epidemic death rates consistently exceeded birth rates in towns: 30,000 people died in 1580 in Paris from typhus; the plague caused 33,000 deaths in 1603, and 41,000 in 1625. Most of the migrants came to town from comparatively short distances away.

Rural-urban migration thus played an important role in reducing the rate of increase in the rural areas of preindustrial Europe. In most parts of Europe, there was little increase in urbanization before the 18th century, but the difference in the rates of natural increase between the town and country—it was negative in the former—ensured that some of the rural surplus was absorbed. It is likely that the rapid urban growth of this period was partially a function of the "push" from the countryside and not only the "pull" of the towns.

It is true that in the medieval and early modern periods, emigration overseas had no significant impact upon the countries of western Europe, but things drastically changed after the discovery of the Americas when considerable numbers did move. Some 100,000 Spaniards left for the New World in the 16th century. From 1620 to 1640 in England, 80,000 people left for IRELAND, the West Indies, and the North American colonies. By 1700, more than 500,000 people of English extraction were living outside England, compared with just over 5 million in England.

THE ERA OF INDUSTRIALIZATION

After 1750, European countries experienced a pronounced increase in their total population size, so that by 1850 the population of most countries doubled, except in the case of Ireland and France, and increased rapidly for the rest of the century. In England, the century after 1750 saw the beginnings, and indeed the maturity, of industrialization.

But in the rest of Europe, industrialization did not get under way until later, and although there was urban growth, rural population increase was almost as rapid, so by the middle of the 19th century, the portion of population living in the rural areas had declined very little.

As a consequence, the countryside of rural Europe had to absorb a great increase in numbers before industrialization had made any progress. Thus, many parts of western Europe were suffering from rural overpopulation by the 1830s, and this was apparent in the growth of landlessness, the subdivision of farms, underemployment, and falling real wages. It was only in England that migration to the towns helped reduce the problems of the countryside before 1850, for it was only in England that industrial growth was rapid enough to absorb the rural surplus. Many European countries experienced remarkable population growth but without any significant industrialization. Undoubtedly, there was stress in the rural areas of these countries, and while there was some rural migration, the surplus rural population could not be absorbed in the towns. At this juncture, fortunately, there was an alternative open to the population: emigration to North America.

Many scholars have tended to emphasize the "pull" of North America as the major reason for the mass European migration of the 19th century, but until the 1870s the "push" element in the rural areas of western Europe must had been at least as important. In many parts of Europe, emigration was the only solution to rapid population growth. For example, in Norway the population doubled between 1800 and 1865. Nearly all this increase had to be absorbed in the rural sector. Migration was the only viable solution. The first emigrants left in 1825, but they were few until the 1860s, when a crop failure prompted the first mass migration from Norway in 1868 and 1869. By then, crossing the Atlantic was much cheaper, and the early emigrants had sent home good news and money. In SWEDEN, urbanization and industrial growth were more rapid after 1870 than in Norway, and the towns took more of the rural surplus.

Nonetheless, 1,105,000 emigrated between 1840 and 1914, the equivalent of 25 percent of the natural increase of this period. Irish emigration is often thought of as beginning with the famine, but it was already running at a high rate before 1845. The Irish population doubled between 1754 and 1821, and in the next 30 years there was acute subdivision of farms, an increase in landlessness, falling real wages, and an ever-increasing dependence on the potato crop. There was little industrial growth and the country remained overwhelmingly rural. In the absence of any internal outlet for the excess population, emigration was the only solution, and between 1780 and 1845, around 1,700,000 people left Ireland, one-third to Britain, the rest to North America. The famine then merely accelerated the trend; between 1845 and 1851 about 1 million left, and in the following decade another million.

Since 1950, there has been movement out of developing countries, but on a very small scale. Thus, by the 1950s annual migration of Latin Americans into the UNITED STATES exceeded that of Europeans, and there has been a considerable flow of migrants from North Africa but not on a scale sufficient to reduce population pressure at home. In the 19th century, it was possible for some 40 percent of Norway's natural increase to be removed in 50 years, a total of 750,000. A movement of a comparable proportion of INDIA's natural increase between 1950 and 1970 would have involved over 70 million, approximately equal to the total number of emigrants from Europe since 1800.

Although international migration may have helped relieve population problems in some West Indian islands, and in the future may have similar localized effects elsewhere, there seems no prospect of overseas migration affording a solution to the population problem of densely populated developing countries.

BIBLIOGRAPHY. C.H. Alstorm and R. Lindelius, "A Study of Population Movements in Nine Swedish Subpopulations in 1800–49 from Genetico-Statistical Viewpoint," *Acta Genetica et Statistica Medica* (v.16, 1996); F. Braudel, *The Mediterranean and the Mediterranean World in the Age of Phillip II* (Collins, 1972); K.H. Connell, *The Population of Ireland, 1750–1845* (Clarendon Press, 1950); G. Dubey, *Rural Economy and Country Life in the Medieval West* (Arnold, 1962); M. Drake, *Population and Society in Norway, 1735–1865* (Cambridge University Press, 1969); F.V. Emery, "England circa 1600," H.C. Darby, ed., *A New Historical Geography of England* (Cambridge University Press, 1973); T. Hagerstrand, "Migration and Area," D. Hannerberg et al., eds., *Migration in Sweden* (Arnold, 1957); B.A. Holderness, "Personal Mobility in Some Rural Parishes of Yorkshire, 1777–1822," *Yorkshire Archeological Journal* (v.42, 1970); H. Kamen, *The Iron Century: Social Change in Europe, 1550–1660* (Weidenfeld and Nicolson, 1971); C.P. Kindelberger, "Mass Migration: Then and Now," *Foreign Affairs* (v.43, 1965); R. Koebner, "The Settlement and Colonization in Europe," M.M. Postan, ed., *The Cambridge Economic History of Europe* (Cambridge University Press, 1966); E.H. Kossman, "The Low Countries," J.P. Cooper, ed., *The New Cambridge Modern History* (Cambridge University Press, 1970); J.S. Lindberg, *The Background of Swedish Emigration to the United States* (Minnesota University Press, 1930); D.C. North and R.P. Thomas, *The Rise of the Western World: An Economic History* (Cambridge Uni-

versity Press, 1973); N.J.G. Pounds, *An Economic History of Medieval Europe* (Longman, 1974); I. Semmingsen, "Norwegian Emigration in the Nineteenth Century," *Scandinavian Economic History Review* (v.8, 1960); United Nations, *The Determinants and Consequences of Population Trends: New Summary of Findings on Interaction on Demographic, Economic and Social Factors, Population Studies No. 50* (United Nations, Department of Economic and Social Affairs, 1973); A.F. Weber, *The Growth of Cities in the Nineteenth Century: A Study in Statistics* (Cornell University Press, 1963); Paul White and Robert Woods, eds., *The Geographical Impact of Migration* (Longman, 1980).

JITENDRA UTTAM
JAWAHARLAL NEHRU UNIVERSITY, INDIA

military geography

FROM EARLIEST HISTORICAL writings, the nature of warfare is shown to be a struggle for positional advantage at the tactical, operational, and strategic levels. Xenophon writing of the fate of the 10,000 Greeks fighting in Persia in the 4th century B.C.E., tells of the constant consideration commanders gave to all aspects of the terrain and the disposition of resources as they fought their way homeward. River banks, hilltops, and forests gave advantage to groups of soldiers during tactical combat.

Mountains, deserts, and sea coasts gave advantage at the operational level as generals planned campaigns that spanned many miles. Cultural, economic, and political factors, all meshed with the dictates of the physical environment, combined to make up the concerns of military geography. Even earlier (5th century B.C.E.), Thucydides wrote of the consideration of these strategic factors of military geography as Sparta and Athens waged their great Peloponnesian war. It seems as long as mankind has maneuvered for positional advantage in a fight, geography has had military importance.

The two broad categories of consideration for the military geographer are the physical and the cultural aspects. Political leaders and their military commanders must deal with the realities of the physical and human worlds they strive in. To ignore the salient characteristics of the Earth and the people who live on it will imperil their military-political effectiveness, if not their existence.

Relative location and general spatial relationships are of primary concern. Distances dictate modes of transportation, types of weapons, and communications requirements. Sequencing of events and objectives, prioritization of efforts, and assessments of vulnerability will be affected by questions of "how close" and "how big."

The characteristics of the ground topography will determine ease of movement, location of water obstacles, line-of-site for observation and engagement, and general protection from weapons' effects. The underlying geology and soil types will affect all manner of military engineering works dealing with mobility, fortification, and the effects of weather.

WEATHER AND TERRAIN

Weather and climate have a direct effect on the operation of equipment, the level of physical work sustainable by troops, and the amounts of supply and fuel required. Ancient and modern history is replete with ill-advised military operations in the face of predictable weather patterns and known climate regimes. Armadas have been sunk, armies frozen, and air forces negated by the annual weather cycle.

The terrain and climate combine to create vegetation cover that again affects visibility, communication, and mobility. Most of these physical considerations continue across the coastal area into seas and waterways. Underlying geology has an impact according to water depth, while weather and climate remain cogent concerns.

The periods of illumination available from sunrise to sunset, along with moon rise and moon stage, are ever more critical with continuous and worldwide military operations. In the nonvisible areas of the electromagnetic spectrum, magnetic forces and radiations affect navigation and communications.

Human factors are of significant concern to the political and military planner, as they must consider the nature of their chosen enemy as well as the disposition and characteristics of all those who may be effected by the military operation and campaign. Linguistic requirements must be anticipated for the gathering of intelligence, control of refugees, and interface with populations, as well as communication with allies. Knowledge of cultural concerns of religion, holy sites, and historic sites is required for compliance with international law during military operations.

Long-term productive relationships with a liberated or conquered population must be built on a foundation of cultural understanding. Social mores and taboos must be considered during psychological operations. Cultural food habits will determine if rations

provided to affected populations and allies are acceptable or uneatable.

Land use patterns in rural areas, the nature of urbanization, the location and extent of industrialization, and transportation networks of a theater of operations are required knowledge for the military planner. Military forces are limited by these cultural factors as they seek to execute schemes of maneuver, develop logistical support, and maximize weapons' effects against opposing forces while mitigating damage to civilian populations and life support infrastructure.

TACTICAL LEVEL

At the tactical level, military geography translates to the near considerations of terrain and vegetation, weather, and the cultural landscape. These military aspects of terrain are commonly known as observation and fields of fire, cover and concealment, obstacles, key terrain, and lastly, avenues of approach and mobility corridors.

Observation refers to the ability of a ground or near ground actor to see across the battle space unhindered by terrain relief or vegetation. The purpose of this observation is for surveillance and target acquisition. Typically this is done by line-of-sight optics or radar emitters. Observation and fields of fire go hand in hand as direct fire weapons follow line-of-sight trajectories. Intervening terrain or vegetation must be taken into account for the optimum operation of weapons, radios, radars, and lasers. The ability to see and shoot across the landscape is a primary consideration for situating of forces for both offense and defense. Line of sight works both ways.

Cover is protection from the effects of weapons. Masking terrain, defiles, and caves can provide some protection from the impact, blast, and fragmentation of direct and indirect fires. Concealment is protection from observation and can be afforded by terrain or vegetation. Concealment is a minimum requirement for secure military operations. Ideal combat positions are covered and concealed with good fields of fire.

Obstacles are any feature in the battle space that can slow, stop, or canalize military movement and maneuver. The obstacle can be natural, man-made, or a combination of both. The effects of weather can exacerbate the impact of obstacles.

Any point or area in the battle space that would provide a tactical advantage to the force that occupies it or controls it is considered key terrain. Of course the general scheme of maneuver, the precise disposition of forces, the scale of the operation, and enemy capabilities will determine key terrain in each individual situation.

Avenues of approach are routes across the terrain that will allow a force to reach a desired objective. These avenues are evaluated in terms of their width, relative location to adjacent avenues of approach, cover and concealment, observation and fields of fire, and intervening obstacles. Mobility corridors are areas within the avenues of approach that bear specific characteristics best suited for specific types of mobility (for example, mounted, dismounted, or air). They are both further defined by the doctrinal movement rate and maneuver space for the size and type of military force anticipated.

All these aspects of terrain are modified by the weather. Some military aspects of weather are visibility (light data, fog, dust), winds aloft, precipitation, cloud cover, and temperature and humidity. These have the greatest effect on aviation, but also have direct effect on all military forces and their operations. Seasonal rains, temperatures, wind patterns, and general climatic conditions must be factored into every military plan.

OPERATIONS OTHER THAN WAR

With the increased use of military forces in disaster relief, peacekeeping operations, and in many operations other than war, knowledge of the cultural factors of geography have grown in importance. Increased interaction with civilians in the battle space require not only that military strategic planners be cognizant of military geography, but also that individual soldiers and small unit leaders come to grips with cross-cultural issues.

In the broader view, military geography must include considerations of geostrategic issues, areas, and players. Disputes that inevitably involve military forces arise and fester along several boundaries. Man-made virtual boundaries of national frontiers have proven to be very contentious. These border disputes involve economic resources, ethnic populations, and pure territorial claims. Ethnic boundaries are found across the globe and bear the scars of war both ancient and recent. Resource boundaries emerge and fade as oil, fisheries, shipping lanes, and commerce ebb and flow. All these areas of cultural, economic, and political concern can rapidly move into the scope of military geography.

In 1996, the Association of American Geographers acknowledged military geography as a subfield of geography and defined it as the application of geographic information, tools, and techniques to solve military problems in peacetime or war.

The consideration of terrain, culture, politics, and economics in the pursuit of warfare will remain a dynamic field of geographic study and a practical area of military application.

BIBLIOGRAPHY. Xenophon, *Anabasis* (Harvard University Press, 1998); Thucydides, *History of the Peloponnesian War*, Rex Warner and M.I. Finley, trans. (Viking Press, 1954); John M. Collins, *Military Geography for Professionals and the Public* (Brassey's, 1998); Patrick O'Sullivan, *Terrain and Tactics* (Greenwood Press, 1991); C. Peltier and G. Etzel Pearcy, *Military Geography* (D. Van Nostrand, 1966); "Military Aspects of Terrain," Geospatial Terrain Analysis and Representation, www.darpa.mil (June 2004).

IVAN B. WELCH
OMNI INTELLIGENCE, INC.

Mindanao

MINDANAO, THE SECOND-largest island (36,537 square mi or 94,630 square km) in the Republic of the PHILIPPINES, is one of the three groups of over 7,100 islands in the country, the other two being LUZON and Visayas. The Mindanao group to the south consists of some 400 islands. The island Mindanao, about the size of INDIANA, occupies about a third of the Philippines's total area. Its highest point is on Mount Apo, which is 20 mi (32 km) to the west of Davao, often described as the world's largest city in terms of land area (943.5 square mi or 2,443.6 square km).

Mindanao lies in a zone of earthquakes and volcanic activity with an irregular, long winding coastline interspersed with numerous peninsulas, promontories, and bays some of which are very large and picturesque. As a tropical island, Mindanao has a mean annual temperature of 80 degrees F (27 degrees C) and a relative humidity of 77 percent. The rainy season stretches from May to November which is the summer monsoon, while the dry season lasts throughout December to April. Although much of Mindanao is sheltered by highlands, a typhoon is quite common during the peak rainy period (June to October). The average annual rainfall in the lowlands is about 80 in (203 cm).

Mindanao has a long history of political developments beginning with the arrival of Ferdinand MAGELLAN in 1521, which led to colonial rule in 1565. Spanish occupation lasted over three centuries until 1898, when SPAIN signed a treaty passing control of the Philippines to the UNITED STATES. Two years after Japanese occupation (1942–44), the Philippines gained independence but inherited two major security threats from the communist-led Huk rebels in the north of Luzon and the Moro secessionist rebels in southern Mindanao. The Philippine army finally defeated the Huk rebels in 1954, but the Moro insurgency continued into the 1970s when President Ferdinand Marcos declared martial law in 1972, lasting until 1981, in an attempt to destroy the Moro National Liberation Front (MNLF) led by Nur Misuari. With material support from LIBYA the MNLF was able to sustain military campaigns against government troops demanding independence for areas under predominantly Muslim population.

In 1996, the government under President Fidel Ramos initiated a dialogue that led to the signing of a peace agreement providing for self-rule for Muslims in the southern Philippines. While the agreement ended three decades of rebellion that claimed over 120,000 lives, minor skirmishes are still faced by government security forces from two other groups outside the MNLF: the Moro Islamic Liberation Front (MILF) and the Abu Sayyaf. The latter group became notorious after a number of kidnapping incidents involving both local and foreign victims, including the well-publicized kidnapping of 21 foreign tourists on the island of Sipadan on April 23, 2000, and another episode involving six workers on an eco-farm resort in Lahad Datu, Sabah, on October 5, 2003. The long years of political turmoil have left Mindanao, especially the four MNLF-controlled southern provinces, with widespread disruption to the infrastructure, public services and the economy, causing the average income of the local population in the region to drop to less than half of the national average.

Mindanao, by virtue of its location in close proximity to MALAYSIA, BRUNEI, and INDONESIA, has become a natural focus of a growth triangle involving economic cooperation initiatives between the four bordering countries under a multilateral cooperative scheme called Brunei-Indonesia-Malaysia-Philippines East ASEAN Growth Area (BIMP-EAGA). The September 11, 2001, terrorist attacks on the United States brought more condemnation of the MILF, which is alleged to have close links to al Qaeda, thus weakening both domestic and international support for the long-drawn struggle of the Muslim rebel groups. For the non-Muslims and the moderate Muslim nationalists, this may bode well for the future of the island, which can begin to lay arms and start concentrating with greater resolve

on its neglected economic development programs based on the rich natural endowments of the region. Before the Marcos era, the region, with its fertile soils, abundant rainfall and sunshine, was one of the world's leading producers of several cash crops such as abaca (Manila hemp), sugar, and coconut. The region also produces banana, pineapple, coffee, pineapple, palm oil, and cotton for export, besides supplying the locally popular durian and flowers for the domestic market.

BIBLIOGRAPHY. Patricio N. Abinales, *Making Mindanao: Cotabato and Davao in the Formation of the Philippine Nation-State* (Ateneo de Manila University Press, 2000); Moshe Yegar, *Between Integration and Secession: The Muslim Communities of the Southern Philippines, Southern Thailand, and Western Myanmar* (Lexington Books, 2002); David Joel Steinberg, *The Philippines: A Singular and a Plural Place* (Westview Press, 2000).

KADIR H. DIN
OHIO UNIVERSITY, ATHENS

Minnesota

WITH THE EXCEPTION of ALASKA, Minnesota is the northernmost U.S. state, on account of the notch in its northern border, where Lake of the Woods abuts CANADA. Long known as the "Land of 10,000 Lakes," the state is also known for its progressive society, common Scandinavian heritage, sturdy agriculture, advanced technology, and outdoor tourism. Minnesota has a population of 5,019,720 (2002) and covers an area of 86,943 square mi (225,182 square km). Minnesota officially entered the Union in 1858.

The landscape of Minnesota was carved out by the glacial action of the last Ice Age. This glacial erosion left considerable areas of gently rolling hills. It also scooped out lowlands that filled with water, and melted ice, thereby creating the ubiquitous lakes of Minnesota. Depending on how size is used to define "lake," there actually are some 11,000–15,000 lakes in the state. The preponderance of water in the state is striking, especially in the north. When viewed from above in an airplane, the state can appear to be more water than land.

The glacial retreat of the ice age also deposited a thick layer of fertile soil across the lands of Minnesota. Combined with warm summers and typically ample rainfall (a humid continental climate classification), Minnesota is blessed with an excellent agricultural setting. Minnesota is among the country's leading producers of corn, soybeans, livestock, and wheat. Many small towns and villages dot the landscape, acting as agricultural centers of activity amid a mixture of family farms. With a significant share of its land still forested, the state is a major producer of wood and forest products, with about 60,000 people employed in the industry. Historically, the timber industry was epic, featuring the folklore tales of Paul Bunyan and his blue ox Babe, statues of whom can be seen in the northern city of Bemidji.

The land also holds minerals. The Mesabi Range in northern Minnesota has for years produced a wealth of iron ore. Even as some of the richest veins of ore were emptied, extraction of lesser-grade ore (taconite) has continued. Residents of the Iron Range gained income directly and indirectly from the wealth brought in by the ore. Vast quantities of iron ore were shipped out of Duluth harbor, across Lake Superior, and on to steel mills in states such as PENNSYLVANIA. Even wind serves as a resource over the land, with high-tech windmills on wind farms in southwestern Minnesota.

The lakes and forests of Minnesota serve as magnets for tourism. Some 3 million people use Minnesota sites for recreation each year. Lakes provide thousands of settings for swimming, fishing, boating, and water skiing. Even in winter, recreational activity is prominent in the state; snowmobiling, ice fishing, and cross-country skiing are among the activities.

The eastern half of Minnesota's land was acquired from the British upon the American victory in the Revolutionary War. The vast LOUISIANA PURCHASE of 1803 included the western portion, with a small northern strip acquired from Britain in 1818. Numerous place names recall the original native presence in the region. Native American tribes were progressively pushed from their lands, eventually resettling in several reservations, mainly in northern areas. As in several states, some of these tribes built and currently manage gambling casinos.

Although Minnesota features a population of diverse origins, there is a clear imbalance in numbers toward those of Scandinavian origin. Migrants from NORWAY and SWEDEN came to the UNITED STATES in large numbers, particularly in the late 1800s and early 1900s. Their descendants are still there: Several Minnesota counties in the northwest have over 80 percent Norwegian heritage. Approximately half of the state's 5 million people reside in the broad Twin Cities (Minneapolis-St. Paul) metro area. While Minneapolis is the

larger of the paired cities, the state capital is St. Paul. A friendly rivalry exists between the two cities, as well as between "the Cities" and the out-state areas. Politically, the state has long displayed progressive tendencies and has sent forth prominent figures such as U.S. vice presidents Hubert Humphrey and Walter Mondale.

The Twin Cities also serve as the state's economic core. Amid substantial manufacturing and service sector production, several national companies (General Mills, Honeywell, Northwest Airlines, Pillsbury, and 3M) base their headquarters there. The Mall of America in suburban Bloomington is the country's largest indoor shopping center and attracts shoppers from great distances. An hour south of the Twin Cities, Rochester is dominated by the medical industry, featuring the internationally renowned Mayo Clinic.

BIBLIOGRAPHY. Mark Mattson, *Macmillan Color Atlas of the States* (Macmillan 1998); "Minnesota," www.netstate.com (March 2004); "The State of Minnesota," www.state.mn.us (March 2004); "Scandinavian Immigration," Library of Congress, www.memory.loc.gov (March 2004); "Norwegian Ancestry," www.mnplan.state.mn.us (March 2004).

JOEL QUAM
COLLEGE OF DUPAGE

Mississippi

KNOWN AS THE Magnolia State, Mississippi is bordered on the north by TENNESSEE, on the south by LOUISIANA and the Gulf of Mexico, on the east by ALABAMA, and on the west by LOUISIANA and ARKANSAS. The MISSISSIPPI RIVER, which flows along most of the western boundary of the state, is the origin of the state's name. Legend says the Mississippi was named by the Chippewa Indians who called it "Father of Waters." Mississippi is proud of its antebellum history, and the state retains many vestiges of the Old South, including well-preserved plantations and Civil War reenactments.

The total area of Mississippi is 48,434 square mi (1,225,443 square km). The state ranks 32nd in size and 31st in population among the 50 states. Mississippi's largest cities are Jackson (the capital), Gulfport, Biloxi, Hattiesburg, Greenville, Meridian, Tupelo, Southaven, Vicksburg, and Pascagoula.

Approximately 1,520 square mi (3,936 km) of Mississippi are covered by water, and the coastline of the state runs 44 mi (113 square km) along the southern tip at the Gulf of Mexico. In addition to the Mississippi, the state's major rivers are the Big Black, the Pearl, and the Yazoo. Major Mississippi lakes, many of which have been created by damming river waters, include Ross Barnett Reservoir, Arkabutla Lake, Sardis Lake, and Grenada Lake. Over time, the Mississippi River, in its meanderings, has changed course, creating a number of oxbow lakes, such as Eagle Lake on the Mississippi-Louisiana border. The average elevation of Mississippi is 300 ft (91 m) above sea level. The highest point in the state is Woodall Mountain, which is only 800 ft above sea level. The lowest point in Mississippi is at sea level where the land meets the Gulf of Mexico. The state is approximately 340 mi (547 km) from north to south and approximately 170 mi (273 km) east to west.

The climate of Mississippi is moist and semitropical and the state is subject to severe afternoon thunderstorms, particularly from June through November, with most thunderstorms occurring in July. Mississippi has long, hot summers and short, mild winters. The average temperature ranges from the just above freezing in the winter to the mid-90s degrees F (low-30s degrees C) in the summer. Annual precipitation ranges from 50 in (127 cm) in the northwestern section of the state to 65 in (165 cm) in the southeastern region. While northern Mississippi occasionally sees snow, the phenomenon is rare in the rest of the state. Occasional hurricanes assail the state.

The land of Mississippi varies greatly from the deltas along the banks of the Mississippi and Yazoo rivers to coastal terraces, piney woods, and prairies, to the flatlands of Mississippi's highlands. Most of Mississippi falls within the East Gulf Coastal Plain, and the remaining area is encompassed within the Mississippi Alluvial Plain. In the East Gulf Coastal Plain, low hills predominate. The coastline of Mississippi is dotted by a number of large bays, including Bay Saint Louis, Biloxi, and Pascagoula. The relatively shallow Mississippi Sound divides the landmass from the Gulf of Mexico. The Mississippi Alluvial Plan, commonly referred to as the Delta, is relatively narrow but widens as it extends north toward Vicksburg. Much of the soil in the Delta has been enriched by the floodwaters of the Mississippi River.

Approximately 55 percent of Mississippi's land is forested. Northern forests are covered with hardwoods such as elm, hickory, oak, cedar, short leaf pine, and tu-

pelo. Pines, such as loblolly longleaf, and slash, cover Mississippi's southern forests. Live oak, magnolia (the state tree), pecan, and sweet gum are found throughout the state. Flowering plants include magnolia (the state flower), azalea, black-eyed Susan, camellia, dogwood, iris, Cherokee rose, trillium, and violet. Large animals such as the white-tailed deer roam Mississippi's forests, which are also home to beaver, fox, opossum, rabbit, skunk, and squirrel. Mississippi is host to a number of songbirds, including the mockingbird (the state bird). The state has 29 state parks.

Until the beginning of the 20th century, Mississippi's economy was dominated by "King Cotton." After 1907, when the boll weevil destroyed cotton as the mainstay in many states of the Old South, Mississippi farmers turned to other crops such as soybeans. Agriculture continued to support the state's economy until the middle of the 20th century, when manufacturing gained an edge. Mississippi continues to be highly rural, with approximately 40 percent of the state covered by farms. Farmers in the area continue to grow cotton, which is again the state's largest cash crop.

By the beginning of the 21st century, only Texas farmers produced more cotton than those in Mississippi. The state also produces rice, hay, corn, peanuts, sugar cane, sweet potatoes, and wheat. Mississippi leads the United States in the production of upholstered furniture. Other income-producing industries include chemicals, plastics, foods, and wood. The economy of Mississippi is also dependent upon large deposits of petroleum and natural gas. Mississippi is the world's leading producer of pond-raised catfish, and the seafood industry flourishes in coastal areas. Legal gambling in Biloxi also supplements the state's income.

BIBLIOGRAPHY. Dan Golenpaul, ed., *Information Please Almanac* (McGraw-Hill, 2003); "Mississippi" www.netstate.com (March 2004); "Mississippi," www.mississippi.gov (March 2004); "Weather and Climate Data for Mississippi" www.srcc.lsu.edu (March 2004).

ELIZABETH PURDY, PH.D.
INDEPENDENT SCHOLAR

Mississippi River

MARK TWAIN begins the novel *Life on the Mississippi* by writing "The Mississippi is well worth reading about. It is not a commonplace river, but on the contrary is in all ways remarkable." No better words can be used to summarize of one of the world's largest, and North America's greatest, river.

The Mississippi River is born as little more than a trickle of water from Lake Itasca in the northwoods of MINNESOTA. It takes just nine small steps to cross, and is no more than ankle deep. From this beautiful but unimpressive brooklet grows the monstrous stream that drains the heart of a continent. The name *Mississippi* comes from the Chippewa (Ojibwa) word *Misipi*, which translates to "big water" or "great river." From Lake Itasca, derived from Latin meaning "the true head," the Mississippi River travels about 2,300 mi (3,705 km) to the Gulf of Mexico. Because the river's course is constantly changing, its exact length is hard to accurately know. Different sources will inevitably offer many different numbers.

The length has been stated as long as 2,552 mi (4,107 km). When combined with the Missouri River, the longest tributary of the Mississippi, the river system is the fourth longest in the world, behind only the NILE, AMAZON, and CHANGJIANG (YANGZI) rivers. Along the route, the river falls 1,475 ft (450 m), nearly half of which is lost before the river reaches Minneapolis, Minnesota.

Depth also varies along the river. The U.S. Army Corps of Engineers keeps a navigation channel of at least 9 ft (3 m) to the end of navigable water at Minneapolis for cargo vessels. However, the river is much deeper than that in its lower reaches. Near New Orleans, LOUISIANA, the river is not as wide as it is farther upstream but carries more water. This is possible because of the increased depth there. Off Algiers Point in New Orleans, the river reaches 200 ft (61 m) in depth.

Water volume increases steadily as the river flows south. In the upper reaches, tributaries like the Minnesota, Chippewa, and Wisconsin rivers join. The largest jump in volume occurs in the middle reaches, where the two largest tributaries, the Missouri and Ohio rivers, combine with the Mississippi. In the lower reaches, the Mississippi gets a boost from rivers like the Arkansas, Yazoo, and Red. The Mississippi River ranks fifth in the world in average discharge.

The Missouri and Ohio rivers add the most volume, and contribute to flooding more than any other tributaries. The Missouri River drains an enormous area. Its coverage stretches from southern CANADA to the middle of KANSAS, and as far west as IDAHO. Its major tributaries include the Yellowstone, Platte, and Kansas rivers. The Ohio is also impressive; though not

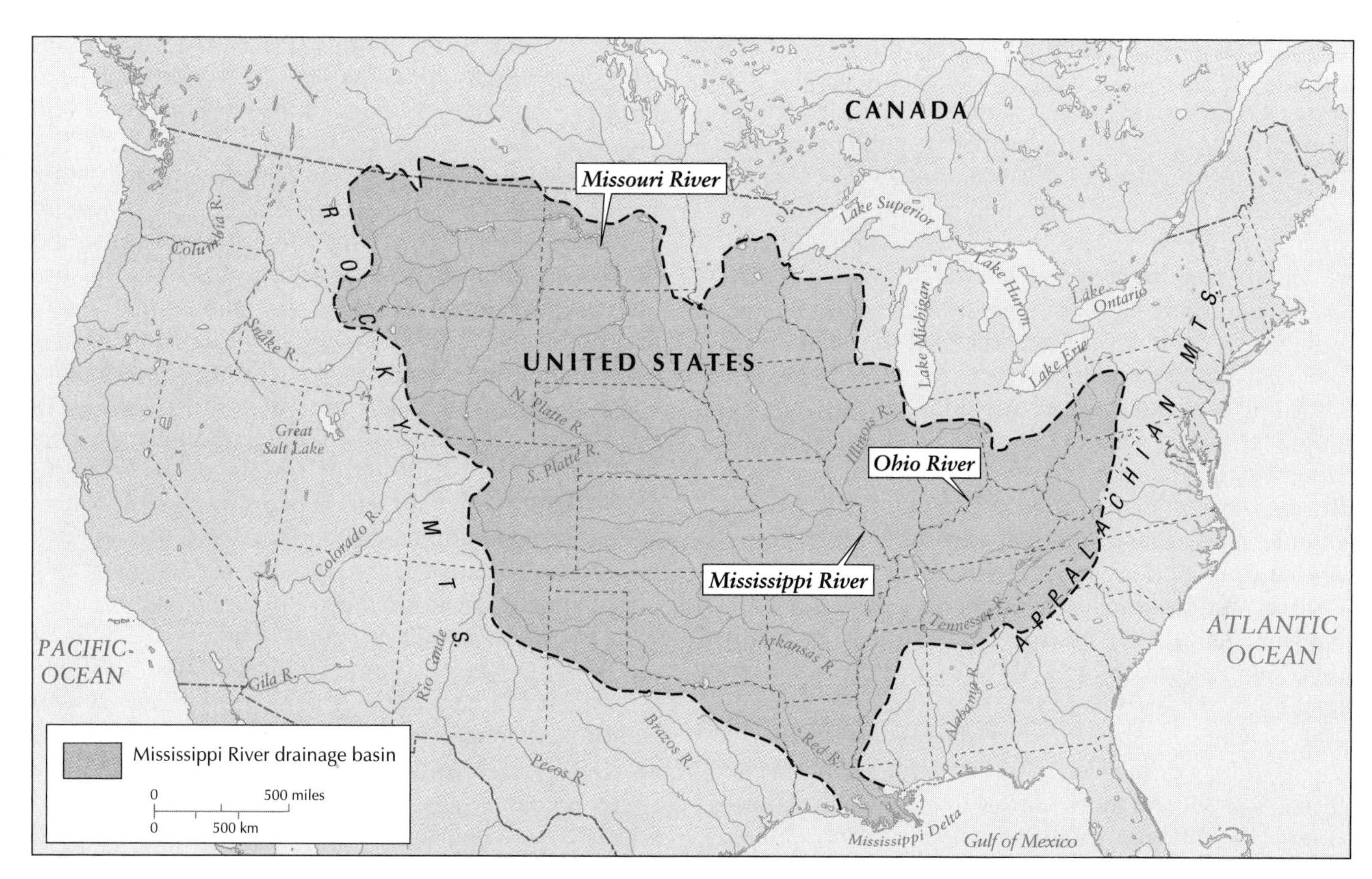

The length of the Mississippi River has been stated as long as 2,552 mi (4,107 km). When combined with the Missouri River, the longest tributary of the Mississippi, the river system is the fourth-longest in the world.

as long as the Missouri, its drainage receives more annual precipitation. Its major tributaries are the Allegheny, Kanawha, Cumberland, and Tennessee rivers.

Water from Lake Itasca will travel through or along 10 U.S. states, reaching the Gulf of Mexico in about 90 days. The average speed of the river near the headwaters is around 1.2 mi (1.9 km) per hour. In the lower reaches the speed can exceed 3 mi per hour (4.8 km per hour). It could be said that the river travels about as fast a person walks.

MISSISSIPPI WATERSHED

The watershed of the Mississippi River is the second largest in the world. It drains an area of over 1.25 million square mi (4.76 million square km). The watershed reaches from the Allegheny Mountains in the east to the ROCKY MOUNTAINS in the west. This area makes up 41 percent of the United States, including all or some of 31 states and two Canadian provinces, Manitoba and Saskatchewan. The river drains one-eighth of North America.

Europeans did not discover the Mississippi River until 1541. In that year, Hernando De Soto came upon the river south of present-day Memphis, TENNESSEE. He arrived there after traveling overland from the Gulf Coast of FLORIDA in search of El Dorado and the Seven Golden Cities of Cibola. De Soto died of fever within a year of the discovery and was buried in the river. Interest in the Mississippi River by Europeans would not rise again for over 100 years. Upon hearing rumors of a great river, and in need of water routes, the French sent two men from the Great Lakes to explore the river in 1673, Louis Jolliet and Jacques Marquette. The men found the river and traveled on it south as far as the confluence of the Arkansas River. Sieur de La Salle later claimed the river for FRANCE in 1682.

Development along the river was slow to come after its discovery. Natchez (in present-day Mississippi) was founded in 1716, and was closely followed by New Orleans, Louisiana, in 1718. The LOUISIANA PURCHASE occurred in 1803, opening western lands and bringing settlers to the river. Fertile agricultural lands

and the discovery of minerals such as lead attracted many. The steamship was invented in 1807, making upstream travel feasible, and commerce increased rapidly. Today, nearly 300 million tons (272 billion kg) are shipped on the river annually.

In 1861, the river gained strategic importance. As the Civil War began, the Mississippi was vital to the Union's war plan. If the river could be controlled, the Confederacy would be split in two. Thomas Jefferson once said, "He who possesses the Mississippi possesses power." The Union also thought so, and General Ulysses S. Grant captured Vicksburg, Mississippi, on July 4, 1863, opening the river and taking a large step toward Union victory.

MISSISSIPPI FLOODS

There are still a few battles being fought on the Mississippi River. One is being fought by the U.S. Army Corps of Engineers against the river itself. Large floods are very destructive to agriculture, property, and human life. The first levee built to hold back floodwaters was constructed in 1717. It was 3 ft (1 m) high and stretched just over 1 mi (1.6 km) to protect New Orleans. The Mississippi River Commission (MRC) was created in 1879 to improve the river for navigation and to control flooding. Engineer James B. Eads designed and constructed jetties at the river's mouth to make the Mississippi deepen its own channel. The levee system was expanded and its height increased. In 1882, the Mississippi flexed its muscle and spilled over the levees and compromised them in 284 places, proving their ineffectiveness. The MRC decided the system needed rapid expansion after a similar flood in 1890.

Flooding continued after expansion. Floods occurred in 1897, 1903, 1912, 1913, 1922, and in 1927 the largest flood to date created 226 levee breaks and inundated an estimated 11 million acres (4.4 million hectares); 246 people died and damages may have been as high as $400 million. It was decided again to increase the size of the levees and create floodways to funnel water. Floods followed in 1937, 1973, and the most destructive Mississippi flood occurred in 1993. Levees have been shown to increase the magnitude of flooding while decreasing their frequency.

Today, locks and dams have been constructed between St. Louis and Minneapolis to aid navigation, 29 in all. In 1963, the Old River structure was completed in Louisiana to keep the Atchafalaya River from capturing the Mississippi and rerouting it away from New Orleans. The Atchafalaya Basin offers a steep, short route to the Gulf of Mexico, which the Mississippi will take if allowed. Flood control is being reevaluated and the river is being allowed to reclaim some of its floodplain to reduce the extent of flooding. Flood control measures have come under heavy fire from the public because of continued flooding, and by environmental lobbies because of the presence of endangered species and the interruption of natural systems.

BIBLIOGRAPHY. "Mississippi River History, People and Places, " Mississippi River Parkway Commission, www.mississippiriverinfo.com (March 2004); "General Information about the Mississippi River," Mississippi National River and Recreation Area, www.nps.gov (March 2004); Theodor Geus, *The Mississippi* (University Press of Kentucky, 1989); USGS, "Mississippi River," biology.usgs.gov (March 2004); John C. Hudson and Edward B. Espenshade, Jr., eds. *Goode's World Atlas* (Rand McNally, 2000); "Mission," Mississippi Valley Division, www.mvd.usace.army.mil (February 2004); Todd Shallat, "Before the Deluge: The Nature of the Mississippi before the Millennial Flood," Works in Progress Essay 2, www.mvd.usace.army.mil (February 2004); Mark Twain, *Life on the Mississippi* (Book-of-the-Month Club, 1992).

DANE BAILEY
UNIVERSITY OF KANSAS

Missouri

IN THE MIDWESTERN UNITED STATES, Missouri is the 19th state in terms of size of all the states. Its name is derived from a Native American word for "town of large canoes." Missouri covers an area of 69,686 square mi (180,487 square km). A LANDLOCKED STATE, inland waters cover an area of 691 square mi (1,790 square km). The largest of these waters is Lake of the Ozarks in central Missouri. The state's north-south extent is 308 mi (496 km) and the east-west extent is 284 mi (457 km). With the MISSISSIPPI RIVER running along its eastern border, Missouri is bounded by eight states: IOWA, ILLINOIS, ARKANSAS, KANSAS, KENTUCKY, TENNESSEE, OKLAHOMA, and NEBRASKA.

Missouri, thanks in part to a lack of major physical features, has a continental climate with hot and humid summers coupled with cold winters. The average July temperature is 78 degrees F (26 degrees C), and the average January temperature is 31 degrees F (-1 degrees C). Missouri sees an average of 40 in (102 cm) of precipitation a year, the bulk of it falling in the south. The

lack of features also creates space for contrasting air masses to meet, causing summer storms or tornadoes.

MISSOURI REGIONS

Missouri has several specific regions, all similar in their moist soil. In the north, the Dissected Till Plains run north of the Missouri River. This area is characterized by rolling hills and fertile plains with red soil. Very similar in the west are the Osage Plains, with slightly lower hills and less fertile soil. The Ozark Plateau covers most of the south. This is the largest physical area in Missouri and contains the state's highest point, Taum Sauk Mountain, at 1,722 ft (518 m).

In the southeast corner of the state are the lowlands, and the most fertile region, as the soil is nourished by the Mississippi River. The Mississippi has been both a blessing, in terms of navigation and agriculture, and a curse, as river flooding can be extensive and economically ruinous (the 1993 flood was particularly disastrous).

Missouri's forests, covering a little over a quarter of the state, are almost all located in the south. In the mid-southwest of the Ozark region are hickory and oak trees, while in the plains bald cypress and sweet gum are prevalent. Most of the wildflowers are found in the alluvial plains of the Mississippi, with milkweed, sweet William, mistletoe, and hawthorn widespread. Missouri is home to many common mammals, such as deer (the most prevalent), bears, rabbits, squirrels, minks, and muskrats. Common songbirds such as blue jays, cardinals, woodpeckers, and finches are all residents of Missouri, along with game birds like turkeys and quail.

Missouri was first explored by Native Americans, as much as 4,000 years ago, but was officially discovered by Jacques Marquette in 1673. Missouri, part of the LOUISIANA PURCHASE, became a U.S. territory in 1803, and a U.S. state 18 years later in 1821. Despite it being a slave state, Missouri was on the Union side during the Civil War. Though Jefferson City is the state capital, Missouri's St. Louis became the gateway to the west during the westward expansion and gold rush eras.

BIBLIOGRAPHY. Paul Nagel, *Missouri: A History* (University Press of Kansas, 1988); Mark Mattson, *Macmillan Color Atlas of the States* (Prentice Hall, 1996); "Missouri," www.wikipedia.com (October 2004).

MARK A. GOLSON
GOLSON BOOKS, LTD.

Moldova

Map Page 1132 **Area** 13,067 square mi (33,843 square km) **Population** 4,439,502 **Capital** Chisinau **Highest Point** 1,419 ft (430 m) **Lowest Point** 6.6 ft (2 m) **GDP per capita** $2,500 **Primary Natural Resources** lignite, phosphorites, gypsum.

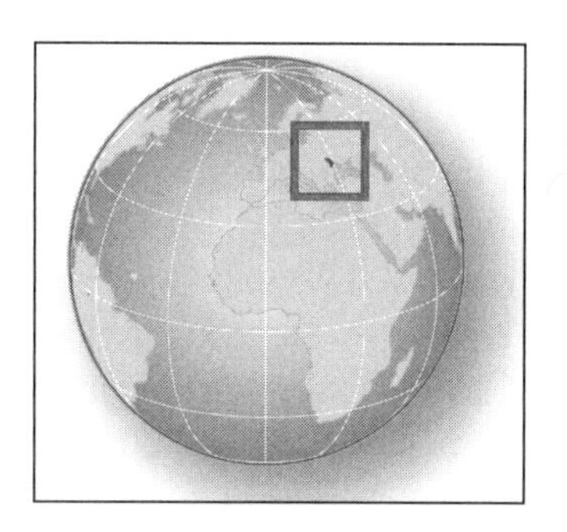

THE REPUBLIC OF MOLDOVA is the second-smallest former Soviet republic and the most densely populated, located in the borderlands between the UKRAINE and ROMANIA, to which it is linked ethnically and linguistically. It is actually only half of the area traditionally known as Moldavia; the western portion has been an integral part of ROMANIA since its independence from the Ottoman Turks in 1878.

The eastern portion, known as Bessarabia, lies between the rivers Prut and Nistra (Dniester in Russian) and has been passed back and forth among the Ottomans, RUSSIA, and Romania several times since the early 19th century. Independent since the breakup of the Soviet Union in 1991, many Moldovans wish for full union with their kindred in Romania, but this move has been blocked by Russians and by Romanians, who do not wish to antagonize their much larger neighbor. One of the reasons for this is the large number of Russians who live in Moldovan territory, on the eastern banks of the Dniester.

This region has proclaimed its independence from Moldova as the "Republic of Trans-Dniestria" (called Stinga Nistrului in Moldovan), and although it is de facto an autonomous state (with its own government and currency), no other nation recognizes its sovereignty, not even Russia, whose troops remain to "protect the peace" (after a short but bloody civil war in 1992). Moldova's boundaries also enclose another region that has been given extensive autonomy (in 1994), albeit much more peaceably, Gagauzia, home to a people who speak a Turkic language but converted to Orthodox Christianity centuries ago.

Moldova is LANDLOCKED, but only by a thin strip of land belonging to Ukraine (south of Odessa) that separates it from the BLACK SEA. Access to the sea is gained through the lower reaches of the Dniester and Moldova's tiny stretch of the DANUBE (0.4 mi or 0.7 km) at its confluence with the river Prut before it enters its broad DELTA. The Dniester River dominates the east-

ern portion of the country, at points forming the boundary with Ukraine. A major feature along this river is the 60-mi or 97-km lake, Dubossary, formed by a large hydroelectric dam built in the 1970s. Other Soviet hydrological projects included a plan to divert much of the water from the Dniester across southern Ukraine to the Dnieper and Bug rivers, but this was never constructed. Two medium-sized tributaries flow into the Dniester, the Reut and the Byk. The Byk flows past the capital, Chisinau (formerly Kishinev), which, although relatively small, was the third-fastest-growing city in the Soviet Union in the 1980s.

One of the poorest states in Europe, Moldova has nevertheless made economic progress through rapid privatization. One of the former Soviet Union's agricultural heartlands, Moldova's broad and fertile plains contain an abundance of rich farmland, but very few mineral resources, so the region remains reliant on Russia for oil, coal, and natural gas. The terrain is slightly hillier in the north, the start of low foothills descending from the Carpathian Mountains to the northwest, where the Prut and Dniester have their sources. The country's topography then declines in a gradual slope toward the Black Sea.

The climate is mostly temperate but becomes warmer and almost Mediterranean in the far south, allowing for cultivation of Moldova's most significant agricultural product, wine, which used to supply as much as one-third of the total wine sales in the Soviet Union. Much of this industry is concentrated in the autonomous Gagauzia region, in its capital, Comrat. Tobacco is also a major export crop, supplying the material for cigarettes for most of Eastern Europe and Russia. The rich black soil of the country's rolling plains (*chernozem* in Russian, among the most fertile soil in the world) are a great resource for potential growth.

The southernmost part of Moldova is the westernmost portion of the STEPPE that extends across southern Ukraine toward Central Asia. The land here is much drier but well-irrigated and easy to cultivate. The richness of this soil and the most favorable climate in the whole of the former Soviet Union (similar to the Crimea), have lured settlers and conquerors throughout history, one of the reasons Moldova is experiencing its independence for the first time.

BIBLIOGRAPHY. *World Factbook* (CIA, 2004); M. Wesley Shoemaker, *Russia, Eurasian States and Eastern Europe 1994*, The World Today Series (Stryker-Post Publications, 1994); Paul E. Lydolph, *Geography of the U.S.S.R.* (Misty Valley Publishing, 1990); "Moldova History," www.moldova.org (August 2004); "Welcome to Gagauzia," www.comrat.iatp.md (August 2004); *Planet Earth World Atlas* (Macmillan, 1998)

JONATHAN SPANGLER
SMITHSONIAN INSTITUTION

Monaco

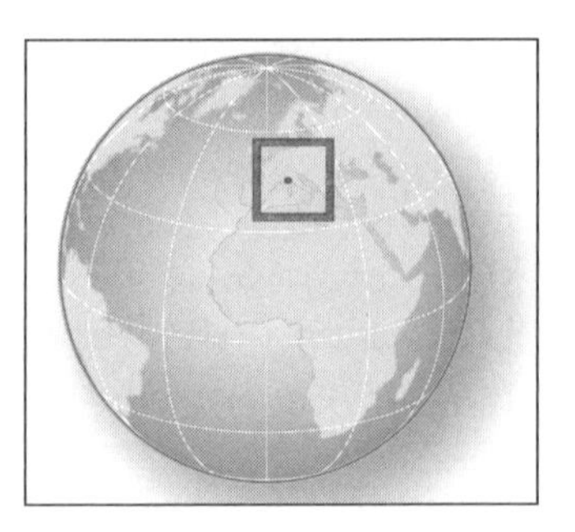

Map Page 1131 **Area** .75 square mi (1.95 square km) **Population** 32,130 **Capital** Monaco-Ville **Highest Point** Mount Agle 460 ft (140 m) **Lowest Point** 0 m **GDP per capita** $27,000 **Primary Natural Resources** climate for tourism.

WHILE EXTREMELY SMALL at only 48 acres (19.4 hectares) in area bordering FRANCE and ITALY, Monaco still has several distinct geographic areas. The official residence of the government and prince of Monaco is a rocky point called Monaco-Ville. The Monte Carlo district, with its famous casino and expensive boutiques, is at the opposite end of the principality. Between the Rock and Monte Carlo is the small harbor and shopping area known as La Condamine. Fontvielle is a new residential area reclaimed from the sea. It has

Monte Carlo Bay in Monaco on the Mediterranean Sea is famous as a haven for the wealthy.

a large marina as well as a sports stadium and some small light industry.

Monaco remains one of the few officially royal countries in the world. Its leaders are hereditary, descending from the founder of Monaco, François Grimaldi, who sneaked into the original castle disguised as a monk on January 8, 1297. Since then the Grimaldi family has been the recognized rulers of Monaco. Should the family die out, Monaco will become an autonomous district of France. It was recognized by the United Nations in 1993.

Monaco's primary economic value today is as a major tourist and banking center as well as a tax haven for the very wealthy. There is no income tax paid in Monaco. Its political neutrality has been recognized virtually since its inception. Often overlooked is Monaco's world-class Oceanographic Museum, founded by Prince Albert I in 1906. It is one of the most prominent centers for marine studies in the world and may be best known for its association with Jacques Cousteau.

BIBLIOGRAPHY. *Oxford Essential Geographic Dictionary* (Oxford University Press, 2003); "Monaco," www.cde.mc/uk (March 2004); *World Factbook* (CIA, 2004); "Monte Carlo," www.monte-carlo.mc (March 2004).

R.W. McColl, Ph.D.
General Editor

Mongolia

Map Page 1120 **Area** 604,250 square mi (1,565,000 square km) **Population** 2,712,315 **Capital** Ulaanbaatar **Highest Point** 14,350 ft (4,374 m) **Lowest Point** 1,699 ft (518 m) **GDP per capita** $1,840 **Primary Natural Resources** oil, coal, copper, tin, nickel.

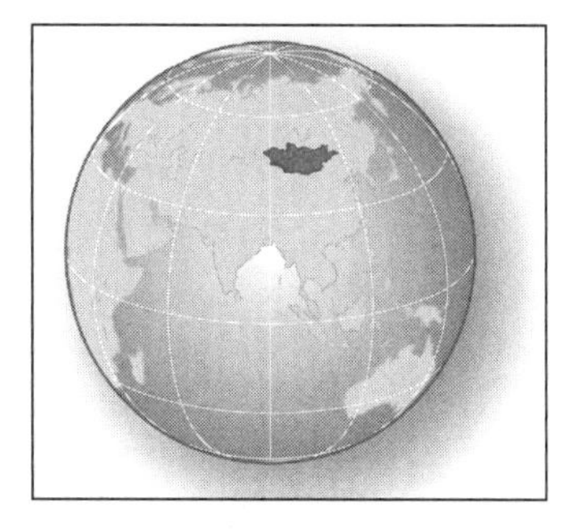

ONCE ONE OF THE world's largest empires (under Genghis Khan), Mongolia eventually became a territory of CHINA, then was dominated by the former Soviet Union. LANDLOCKED (bordered by China and RUSSIA) and often considered to be all DESERT, Mongolia, in fact, contains numerous mountains, forests, and lakes. Only its southern margins bordering on China are drylands and deserts. However, even these support sparse grasses and nomadic herders.

Mongolia can be divided into a series of north-south and east-west grids. Generally, the northern third of the country is mountainous and forest-covered and has several large lakes. Its natural features include the Henti and Hangai mountains and lakes Hovsgol and Ulaangom. The north also is the most economically developed and urbanized segment of the country. The middle section is a STEPPE or GRASSLAND mixed with forests. Desert and grasslands are in the south. Here are concentrated the major herding areas and the nomadic population that emphasizes raising a combination of camels, goats, sheep and horses. It also is the area most susceptible to massive herd losses from winter blizzards known as *dzud*. Fortunately, animals are a renewable resource.

GRASSES AND DESERTS

The southernmost section is dominated by sparse grasses and hard desert surfaces. And, as with most deserts, it also is scoured by often tornadic winds and sandstorms. This is the area where the famous dinosaur remains have been concentrated; the first discovery of dinosaur eggs in 1923 was during an expedition led by the scientist Roy Chapman Andrews.

Along the axis of an east-west grid, eastern Mongolia is generally a moist and rich grassland. It also has recently been the site of oil and natural gas development. Its closest economic link is with northeast China (the former Manchuria). The western third is dominated by the Altai Mountains and its basins. The population here generally has large numbers of nomadic Kazakhs.

There are a number of large inland lakes, such as Ulaangom, that are important wildlife (migratory waterfowl) refuges. The central core area is the most urban. It is dominated by the only rail line in Mongolia. This line links Mongolia with the rest of the world via transit through either China (the shortest route) or Irkutsk in the Russian Federation.

Today, Mongolia is a republic with democratic elections and an elected parliament (Hural). An interesting part of Mongolia's culture is that women have full suffrage, serve in the Hural, and have always had full equality with men, even in the time of Khan.

BIBLIOGRAPHY. Roy Chapman Andrews, *The New Conquest of Central Asia* (The American Museum of Natural History, 1932); R. Grousset, *The Empire of the Steppes* (Rutgers University Press, 1998 reprint); Owen Lattimore,

Inner Asian Frontiers of Asia (Oxford University Press, 1989); David Morgan, *The Mongols* (Blackwell, 1986); American Geographical Society, "Mongolia," *Focus on Geography* (v.47/1, Fall 2002).

R.W. McColl, Ph.D.
General Editor

monsoon

THE HEAVY RAINS of the monsoon appear in June and subside in September every year in the Northern Hemisphere. Meteorologists and scientists are greatly helped by the development of sophisticated weather analysis and forecasting technology in understanding monsoon mechanisms. Numerous people on the earth are affected by this significant climatic system. The monsoon directly governs the destiny of cultivators in South, East and Southeast Asia, as farming is heavily dependent on the seasonal rains. The summer monsoon winds bring heavy showers over most of South Asia.

An improved understanding of monsoons has embraced the majority of the phenomena connected with the weather system in the tropical and subtropical part of the Asian, Australian, and African continents and the neighboring water bodies. ARIZONA, NEW MEXICO, UTAH, and COLORADO in North America are also affected by monsoons in late summer.

A monsoon-type of climate is associated with wet summers and dry winters. The word *monsoon* seems to be derived from the Arabic word *mausim*, meaning "seasons." It is recurrently misused as a synonym for "heavy rainfall," though the misnomer is justified. Ancient merchants used the word to explain a system of seasonal reversal of wind while sailing over the INDIAN OCEAN and adjacent Arabian Sea. Their direction of sailing movement was governed by these seasonal winds. The winds gust steadily from the southwest during one half of the year (summer) and from the northeast during the other (winter). Thus, *monsoon* has been defined by climatologists as a large scale wind system that predominates or strongly influences the climate of large regions, and in which the direction of the wind flow reverses from winter to summer.

Three essential mechanisms are involved in causing monsoons: 1) differential heating and cooling of land and water, as land absorbs/releases heat faster than the water; 2) deflection of wind in response to the Coriolis effect, caused by the rotation of the Earth; and 3) latent heat exchange—the exchange of energy involved while changing the state of water from liquid to vapor and back.

The most renowned of the entire monsoon is the Asian monsoon, which is truly a seasonal, cyclic climate. To understand the causes for the seasonality, certain dynamics need to be analyzed. During the summer, the sun is vertically over the Asian continent. The Central Asian landmass gets intensely heated, resulting in expansion and ascension of air above the surface, creating a low-pressure area over Central Asia. Alongside it, the INDIAN OCEAN watermass remains relatively cooler, making a high-pressure region.

The variation in temperature between landmass and water could be much as 68 degrees F (20 degrees C). Such differential heating and cooling causes the moisture-laden air from the Indian Ocean to move toward South and Southeast Asia. This occurs because wind essentially moves from high pressure to low pressure regions. As the wind is deflected toward the right in the Northern Hemisphere (Coriolis effect), these winds become southwest monsoons. During the fall, landmass and water begin to cool and the land starts losing heat faster than the ocean.

Therefore, in the winter, the pressure system reverses because the cold polar airmass moves southward, and Tibet and the surrounding areas in the north become a high-pressure center. The Indian Ocean, over which the sun rays fall directly, turns into a low-pressure zone. Thus, the wind direction is reversed, becoming the northeast trade winds of the winter.

Another important factor that directs the commencement or disappearance of summer monsoons in Asia is the situation of the upper-level JET STREAMS. These are a "meandering belt of strong winds in the upper atmosphere," as explained in *Introduction to Geography* by Arthur Getis et al.

When a lofty mountain system like the HIMALAYAS lies in its path, the westerly jet streams divide into two branches (south and north of the Himalayas). These two branches prevail from November through April. The disappearance of the jet streams south of the Himalayas in November causes the sudden "retreat" of monsoon rains from Asia. During the months of July and August, there is an easterly jet stream prevailing over the Indian peninsula.

The southwest winds produce most of the rainfall over the entire Indian subcontinent. These winds, however, ignore Rajasthan or the THAR DESERT of INDIA and western PAKISTAN because they are deflected by the

earth's rotation (Coriolis effect) toward the northeast, and, therefore, the northwestern part of the subcontinent remains relatively dry. There are only 20 rainy days in Kutch (India), and some places in the Rajasthan desert have fewer than five rainy days in a year. Nevertheless, a large part of India, BANGLADESH, and Pakistan receive protracted heavy rains and associated flooding.

The monsoon provides valuable and abundant water to Asia. Most of the monsoon climatic region that encompasses South, East, and Southeast Asia supports agrarian economies dependent on rainfall. Consequently, the route of the monsoon winds decides the fate of the region. Deviation of the monsoon from its normal pattern disturbs agriculture operations, which can become disastrous for rain-dependent agrarian economies.

Unfortunately, the pattern, distribution, amounts, and magnitude of rainfall vary considerably each year. The variability can lead to severe drought and floods that can significantly amplify the price or reduce the availability of food. Immense deviation from the normal pattern might ultimately lead to famine. Correct forecasting of the onset of the monsoon has become absolutely indispensable. Computer models of monsoons are intricate and need to be precise; helping to alleviate the implication of a weak monsoon.

BIBLIOGRAPHY. A. Dutt and M.M. Geib, *Atlas of South Asia* (Oxford University Press, 1998); Arthur Getis et al., *Introduction to Geography* (McGraw Hill, 2002); "Monsoon," www.whyfiles.org (September 2004); "Monsoon," www.valuenotes.com (September 2004).

ASHOK K. DUTT
MEERA CHATTERJE
UNIVERSITY OF AKRON

Montana

AS THE TREASURE STATE, Montana is known for its mountains and extensive natural resources, such as copper, silver, gold, coal, lead, zinc, oil, limestone, antimony, phosphates, and gypsum. This mountain state encompasses 147,046 square mi (380,821 square km) of land area; Montana is the fourth largest state in the UNITED STATES, but it is 44th in population, partly because parts on the state are virtually inaccessible. The state is approximately 630 mi (1,014 km) east to west and 280 mi (450 km) north to south. Montana is bounded on the north by CANADA, on the south by IDAHO and WYOMING, on the west by Idaho, and on the east by NORTH DAKOTA and SOUTH DAKOTA. The ROCKY MOUNTAINS cover the western third of the state. The other two-thirds of Montana are covered with gently rolling hills. Montana, which became a territory in 1864, was admitted to the Union in 1889 as the 41st state. Yellowstone National Park, which covers 2,219,791 acres (898,317 hectares), brings approximately 3 million visitors a year to Montana. In addition to the capital city of Helena, Montana's largest cities are Billings, Missoula, Great Falls, Battle, Silver Bow, Bozeman, Kalispell, Havre, Anaconda-Deer Lodge County, and Miles City.

The highest point in the state is 12,799 ft (3,901 m) above sea level at Granite Peak. The lowest point in Montana is 1,800 ft (548 m) above sea level at the Kootenai River. Geographically, the eastern three-fifths of Montana falls within the Great Plains, while the remaining two-fifths of the state are part of the Rocky Mountain region. As part of the Interior Plain of North America that runs from Canada to MEXICO, Montana's Great Plains section is covered with gently rolling land interspersed with hills and river valleys. Bear Paws, Big Snowy, Judith, and Little Rocky Mountains are found within this area. In the southeastern section, Montana's badlands are dominated by red, yellow, brown stone formations.

The western area of Montana is dominated by mountains, flat grass-covered valleys, and fir, spruce, and pine forests. Valleys in this region may spread out for as much as 40 mi (64 km). Some 50 mountain ranges are located in the Rocky Mountain region, including the Absaroka, Beartooth, Beaverhead, Big Belt, Bitterroot, Bridger, Cabinet, Crazy, Flathead, Gallatin, Little Belt, Madison, Mission, Swan, and Tobacco Root. Soils in the mountains tend to be poor and thin, unlike the more fertile areas of the plains. Grama, buffalo, and bluestem grasses are common in Montana.

MONTANA CLIMATE

The Continental Divide separates Montana into two distinct climates. Eastern Montana has a cold, continental climate. The temperature in the west is modified North Pacific maritime. Montana mountains may be covered with snow for as much as eight or ten months a year, and the Great Plains area is beset with freezing arctic air several times each winter. The average temperature in the state ranges from 70 degrees F (21 degrees C) in the summer to 8 degrees F (-13 degrees C)

in the winter. While western Montana may receive as much as 43 in (109 cm) of rain annually, the eastern section receives less than 20 in (51 cm) per year. Earthquakes are not uncommon along the fault lines that lie within the Rocky Mountain region.

Among the 50 states, Montana is the only state that has rivers draining into the Gulf of Mexico, the Hudson Bay, and the PACIFIC OCEAN. Montana's major rivers are the Clark Fork River, the Missouri, and the Yellowstone. The state's major lakes are Lake Flathead and Fort Peck. At the highest levels of the states, glacial drift may still be found.

Agriculturally, grain is Montana's major crop. Wheat, barley, rye, oats, hay, flaxseed, sugar beets, potatoes, and livestock bring in substantial revenue for the state. Other revenue evolves from mining, the tourist trade, dude ranching, hunting, fishing, and skiing. Timber is also a major industry for Montana because two-thirds of the state's forests are suitable for commercial use. The most valuable timbers are western yellow pine, Douglas fir, larch western white pine, and spruce. In addition to timber, Montana's major manufacturing industries deal with food products, wool, and printing and publishing.

Mountain goat, bighorn sheep, elk, moose, mule deer, white-tailed deer, grizzly bear, black bear, mountain lion, and fox frequent Montana's mountains. Other areas provide homes for mule deer, pronghorn antelope, coyote, and various game and waterfowl.

BIBLIOGRAPHY. Phil Condon, *Montana Surround: Land, Water, Nature of Place* (Johnson Books, 2004); "Discovering Montana" www.state.mt.us (November 2004); Dan Golenpaul, ed., *Information Please Almanac* (McGraw-Hill, 2003); National Park Service, "Yellowstone," www.nps.gov (November 2004); Eric Peterson, *Frommer's Montana and Wyoming* (Wiley, 2004); K. Ross Toole, *Montana: An Uncommon Land* (University of Oklahoma Press, 1959).

ELIZABETH PURDY, PH.D.
INDEPENDENT SCHOLAR

Montserrat

THE ISLAND OF Montserrat, an overseas territory of the UNITED KINGDOM, is part of the chain of islands in the CARIBBEAN SEA that were formed millions of years ago by volcanoes, some of which remain active, while others are completely dormant. Montserrat itself is composed of seven active volcanoes of varying ages, which have been mostly inactive since the time of the island's colonization in the 1630s. But in July 1995, the Soufriere Hills began to emit hot ash and gases, forcing over two-thirds of the population from their homes and the abandonment of nearly half the island.

Like its neighbors to the north, ANTIGUA AND BARBUDA, and SAINT KITTS AND NEVIS, Montserrat was colonized by the British mostly for its potential for raising sugarcane. Its neighbor to the south is the French overseas department of GUADELOUPE. These islands also are volcanic in origin, and the Soufriere de Guadeloupe is an active volcano. Nevertheless, there had been little sign of imminent danger on Montserrat, and no emergency plans had been laid. Some evidence indicates that the island saw its last eruption about 18,000 to 19,000 years ago, but other evidence suggests it could have been as recent as the early 17th century, just before colonization. Indeed, the name of the most recently active area is the Soufriere Hills, named for the French term *soufrière*, a vent or fumarole that emits sulfurous gases and vapors (the French word itself is related to the English word *sulfur*).

CONTINUING ERUPTION

Certain recent seismographical readings indicated a rise in activity beneath the island, still the island was unprepared for the slow but steady eruption starting in July 1995, which has continued ever since. By August 1995, the capital, Plymouth, and all of the villages in the southwest and east were evacuated. Volcanic activity began with pyroclastic flows—mixtures of hot ash, boulders, and gases—followed by more serious explosions beginning in September 1996. In June 1998, the British government unveiled an aid package, but about half of the island will be uninhabitable for a decade.

Before the eruption, the island was relatively quiet. With a pleasant tropical climate and only a narrow coastal lowland, tourism was a mainstay, followed by production of rum, textiles, and electronic appliances. The future remains uncertain, with most of the population living in temporary quarters or off the island altogether. Scientists from around the world continue to monitor the situation at the new Montserrat Volcano Observatory, hoping to learn all they can from this occurrence to better protect against similar disasters in the future.

BIBLIOGRAPHY. Howard A. Fergus, ed., *Eruption: Montserrat versus Volcano* (University of the West Indies, School of Continuing Studies, 1996); Jalil Sued-Badillo, ed.,

General History of the Caribbean (Macmillan 2003); *World Factbook* (CIA, 2004).

JONATHAN SPANGLER
SMITHSONIAN INSTITUTION

Morocco

Map Page 1113 **Area** 172,413 square mi (446,550 square km) **Population** 31,167,783 **Capital** Rabat **Highest Point** 13,665 ft (4,165 m) **Lowest Point** -180 ft (-55 m) **GDP per capita** $3,700 **Primary Natural Resources** agriculture, phosphate mining, minerals.

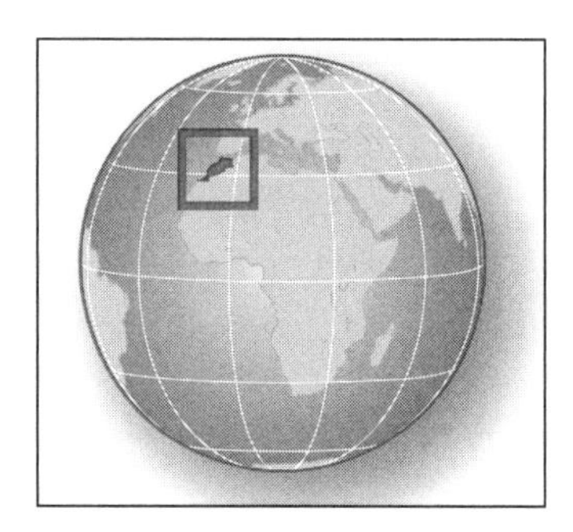

MOROCCO REPRESENTS the finest example of multicultural fusion, combining 2,000 years of languages, rituals and customs, cuisine and foods, and religions into one colorful and productive blend called Al Maghreb. Once hosting vital Roman outposts at the pastoral Volubilus and coastal Lixus, Morocco has seen North African, European, and Asian allies, marauders, and colonizers come and go. However, it has always emerged enriched, and maybe that is why Morocco has developed a culturally embedded sense of acceptance along with its native Berber roots of hospitality.

Located on the western edge of North Africa and situated on both the MEDITERRANEAN SEA and ATLANTIC OCEAN, Morocco shares borders with the countries of ALGERIA and WESTERN SAHARA. Slightly larger than the state of CALIFORNIA, Morocco's long coastlines and mountainous spines afford it varied climatic regimes and agricultural production. The country is dominated by a number of ranges, including the Rif, Middle, High and Anti-ATLAS MOUNTAINS. Towering above the plains of Marrakech lies Mount or Jebel Toubkal with its often cloud-hidden peak of some 13,000 ft (3,000 m) above sea level.

Morocco's change in relief from sea level to its high peaks produces a wide assortment of fruits, nuts, vegetables, and livestock—adding to Morocco's famous cuisine. Integrated Berber, Arab, French, and Spanish residents have all helped create a culture like no other. Crisp and succulent baked pigeon pies, stews of lamb and prunes, and chicken with raisins have elevated Moroccan cuisine to among the most favored and sought in the world.

From the calm blue Mediterranean Sea to the rough, grey Atlantic Ocean, Morocco's coastal plains support thriving farms and orchards, urban and industrial centers, and its primary highways and roads. And although the Atlas Mountains separate the coastal plains from the broad SAHARA DESERT and prevent easy access to either side, these impressive plains, peaks, coasts, and valleys have helped create the unique character of Morocco.

The Kingdom of Morocco was a French protectorate from 1912 to 1956, when Sultan Mohamed became the king. From the house of Alawi and a direct descendant of Muhammad, the Prophet of Islam, the sultan was succeeded in 1961 by his son Hassan, who ruled for 38 years. King (or *Malak*) Hassan, a popular ruler and clever statesman, played a major role in the Palestine-Israel peace process, since many of the earliest immigrants to Israel came from Morocco. In 1976, Hassan inspired thousands of Moroccans, with green-bound Qur'ans in hand, to march into the nation of WESTERN SAHARA to the south, demanding annexation to Morocco. This "Green March" fomented the occupation of Western Sahara, whose status and boundaries are still unresolved.

While the whole of Morocco mourned the death of their leader King Hassan in 1999, his son ascended to the throne to become King Mohammed VI. Following his father's death, King Mohammed declared his support of the constitutional monarchy, a successful political plurality, a moderate economy, and a new thrust to alleviate poverty and unemployment throughout the kingdom. He has proved himself to be opposed to Islamic radicals and has supported private media and the freedom of speech. However, in 2003, the French organization Reporters sans Frontieres condemned Morocco for its "regular interference" in media censorship by the kingdom's intelligence services. Increasingly, self-censorship within media organizations is widespread.

BIBLIOGRAPHY. *World Factbook* (CIA, 2004); M. Ellingham, D. Grisbrook, and S. McVeigh, *Rough Guide to Morocco* (Rough Guide Press, 2002); Ken Park, ed., *World Almanac and Book of Facts, 2004* (World Almanac Publishing, 2003); Barry Turner, ed., *The Statesman's Yearbook, 2003* (Palgrave Macmillan, 2003).

TOM PARADISE
UNIVERSITY OF ARKANSAS

Moscow

MOSCOW IS THE capital city of the Russian Federation, and although superseded as the official capital during the 18th and 19th centuries, it has dominated Russian politics, culture, and economics since the 14th century. Today, it is RUSSIA's largest city and one of the largest urban centers in Europe.

Geographically, Moscow lies at the center of European Russia, the center of the East European Plain. It lies on both sides of the Moscow River, a tributary of the much larger VOLGA, a short distance to the east. The Moscow Region is slightly hilly, wooded STEPPE. The climate is cool: Moderate temperatures in the short summer and bitterly cold in the long, dark winter, when temperatures generally are in the mid-teens F (-8 or -9 degrees C), but occasionally as low as -44 degrees F (-42 degrees C). The city itself is roughly circular, having been built in concentric waves out from the Kremlin, the city's physical and administrative heart. Most major roads in Moscow either circle or radiate from the Kremlin.

The city's boundary corresponds to the outer ring road situated 10 to 13 mi (17 to 21 km) from the city center, encompassing roughly 350 square mi (900 square km). Much of Russia's highway and railroad network radiate from this central point for thousands of miles in every direction. Since the 1930s, Moscow has been a port as well, with the Moscow Canal linking the city to the Volga and its vast internal waterway network connecting the BLACK SEA to the Baltic. Even during the period when Russia's Imperial capital was in SAINT PETERSBURG (1712–1918), Moscow's location at the center of the empire led to its development as Russia's center for industry, as well as a focus for the nation's emerging scientific and artistic communities. Today, Moscow is home to the Academy of Science and numerous colleges and universities, Russia's two main newspapers, and some of the most famous theaters and art galleries in the world, including the Bolshoi Theatre and the Tretiakov Gallery.

Moscow appeared first as a small village in the mid-12th century, strategically located at the center of TRADE ROUTES between the Volga River system and the rivers of the south and west. In 1156, a Russian prince, Yury Dolgoruky, built a wooden fortress on the north bank of the Moscow River, the first Kremlin, which became the center of the medieval city and eventually of an independent principality, called Muscovy. The princes of Muscovy became rich and powerful during Russia's dark period of Mongol occupation, due to the

The Kremlin is not only the political center of Russia, but also the geographic center of Moscow.

fact that they were employed by their Mongol overlords as tax collectors. Eventually, the princes became strong enough to overthrow the Mongols and to unify the other Russian principalities into one state. The Kremlin was rebuilt in the early 15th century, and it remains one of the greatest monuments to native Russian architectural styles, notably in the many churches and monasteries contained within its walls: the churches of the Assumption, Annunciation and the Archangel Michael. Prince Ivan "the Terrible," completed the task of unifying Russia and in 1547 assumed a new title, tzar, reflecting his claims to imperial status and his conceptualization of Moscow as the "Third Rome," spiritual heir to the fallen empire of Constantinople.

The capital was moved from Moscow to Saint Petersburg by Peter the Great in 1712, as part of Peter's efforts to bring Russia closer to the West and wider European culture. Over the next two centuries, Saint Petersburg and Moscow competed for position as Russia's major city, with Moscow frequently advocating conservative, traditional Russian and Orthodox culture, while aristocratic Saint Petersburg focusing on becoming as Western as possible. Saint Petersburg was considered Russia's head, but Moscow remained its

heart. Much of the city was burned in 1812 by its own citizens, successfully forcing Napoleon's French troops to evacuate the region, a moment heroically immortalized in Peter Tchaikovsky's "1812 Overture."

MODERN MOSCOW

Much of the architecture of the city reflects the rebuilding of the city in the following decades. The palaces in the Kremlin were rebuilt, including the Great Palace and the Palace of Congresses which today house the major government organs of the Russian state. Large squares replaced narrow streets and rickety wooden buildings, most notably Red Square along the east side of the Kremlin, the site of parades and ceremony to this day. After the revolutions of 1917, Vladimir Lenin recognized the value of Moscow's central location, as well as its defensibility from foreign invasion, and moved the capital back to Moscow. It became the capital of the Union of Soviet Socialist Republics officially in 1922.

Growth of railway lines and heavy industry was sped up, and millions of peasants, now landless and homeless, flocked to Moscow to find work. During the 1920s, the percentage of working-class residents living with the city's Garden Ring (the central region) increased from 5 percent to 45 percent. Cramped living conditions were finally tackled in massive building campaigns in the 1960s, resulting in today's huge apartment complexes around the city's edges, many built too fast and of inferior quality. Other buildings included monumental Soviet architecture, including Josef Stalin's infamous "seven sisters" skyscrapers.

Today, more than 8 million people live in Moscow, 11 million including its suburbs. In the past decade, large numbers of non-Russians have moved to the city, including Tatars, Chechens, and other peoples from the Caucasus. Moscow's speedy industrialization of the mid-20th century has left behind numerous air- and water-pollution problems. Moscow is a city of great contrasts, between ancient monasteries and ultra-modern office buildings, New Russian millionaires and poverty-stricken communist pensioners.

BIBLIOGRAPHY. *Oxford Essential Geographical Dictionary* (Oxford University Press, 2003); "Moscow," www.moscow-guide.ru (August 2004); "The Moscow Guide," sunsite.cs.msu.su (August 2004); "Moscow," www.lonelyplanet.com (August 2004).

JONATHAN SPANGLER
SMITHSONIAN INSTITUTION

Mosquito Coast

THE MOSQUITO COAST, or Mosquitia, is located on the east coast of NICARAGUA and HONDURAS. The name is derived from the Miskito, the indigenous people of the region. The Miskito are descendants of the Chorotega, an aboriginal people of South America. Because of the absence of historic ruins, little is known of the Chorotega people, except that they were contemporaries of the Maya to the northwest. The word *Miskito* was corrupted into *Mosquito* by European settlers. Although its name sometimes applies to the whole eastern seaboard of Nicaragua and to Mosquitia in Honduras, the Mosquito Coast more accurately consists of a narrow strip of territory, along the CARIBBEAN SEA, stretching inland for about 40 mi (60 km). The area extends from the San Juan River in northeastern Honduras and the Bluefields Lagoon in Nicaragua, centering on Cape Gracias a Dios on the border between Nicaragua and Honduras at the hump of the Central American isthmus. The primary towns in Mosquito include Bluefields, Magdala on Pearl Cay, Prinzapolca, Vounta, and Carata. Bluefields, being the largest town, serves as the unofficial capital.

The Miskito natives, of whom there are several tribes, are short and dark skinned. The expanse of the Mosquito Coast is a combination of coral-lines, low shorelines, reefs, shoals, sandbars, swamps and small islands. It is a desolate region infested with black flies and rampant with yellow fever and malaria. Moving away from the coast, the land rises into savannas and pine woods that feed into the mountains. It is a hot, humid, and swampy region. It has also been historically a political and international hotbed.

Seventeenth-century Spanish settlements were primarily located on the Pacific coast of Central America. The Spanish didn't care for the barren Mosquito region and the hostile native Indians. Pirates and buccaneers during the period were viewed by the Mosquita as allies against the Spanish. The Mosquito Coast historically encompassed the area that is now Nicaragua and was long under control of the British. The first European settlement in the Mosquito region was founded in 1630, when the English-chartered Providence Company occupied two small cays and established relations with the local inhabitants.

From 1655 to 1850, Great Britain claimed a protectorate over the Miskito natives. There was little interest in colonization by British settlers because of the adverse climate. SPAIN and the UNITED STATES opposed British authority out of territorial concerns over a pro-

posed canal and coastal ports. The 1848 uprising by the Mosquito Indians, supported by the British almost led to war. An agreement was reached not to fortify, colonize or exercise dominion over any part of Central America by Britain and the United States. Great Britain relinquished its protectorate of the Miskito native tribes to Honduras in 1859, which resulted in another Indian revolt.

The treaty of Managua in 1860 transferred to Nicaragua domain over the entire Caribbean coast but granted autonomy to the Miskito natives. Nicaragua was limited by native right of self-government. After enjoying independence for almost 14 years, the natives voluntarily surrendered their position and territory in 1894 and the Republic of Nicaragua was formally established. The Mosquito Coast became part of Nicaragua under president José Santos Zelaya. The northern area was awarded to Honduras in 1960 by the International Court of Justice, thus ending a long-standing dispute. The Nicaraguan portion was officially given partial autonomy in 1987, but little real change has resulted and the area remains impoverished. Rubber, lumbering, slash-and-burn cultivation for rice and beans as cash crop farming, mining, banana and plantain plantations are the primary occupations of the indigenous Miskito people. Lobstering has replaced banana cultivation as the major economic activity.

BIBLIOGRAPHY. John Armstrong Crow, *The Epic of Latin America* (University of California Press, 1980); *Encyclopedia of Latin American History and Culture* (Scribner's, 1996); Edwin Williamson, *Penguin History of Latin America* (Penguin Press, 1992).

CLARA HUDSON
UNIVERSITY OF SCRANTON

Mozambique

Map Page 1116 **Area** 309,496 square mi (801,590 square km) **Population** 17.6 million **Capital** Maputo **Highest Point** 7,992 ft (2,436 m) **Lowest Point** 0 m **GDP per capita** $1,100 **Primary Natural Resources** coal, natural gas, titanium ore, iron ore.

MOZAMBIQUE IS located in southeast Africa and is about twice the size of CALIFORNIA. It is bordered on the north by TANZANIA, on the east by the MOZAMBIQUE CHANNEL of the INDIAN OCEAN, on the south and southwest by SOUTH AFRICA and SWAZILAND, and on the west by ZIMBABWE, ZAMBIA, and MALAWI. The only natural borders are Lake MALAWI in the northwest between Malawi and Mozambique and the Rovuma River, which forms part of the northern border with Tanzania. The Mozambique Channel separates Mozambique from the island of MADAGASCAR. Mozambique has 1,600 mi (2,575 km) of coastline that is interrupted by numerous river mouths. The rivers that run through Mozambique to the ocean include the Rovuma, Lurio, Incomati, Lugela, Revue, Save, Limpopo, and the famous Zambezi. South of the Zambezi, the coast is very narrow and the northern coast near Rovuma features rocky cliffs with numerous islets and lagoons. To the far south, just above South Africa, are Maputo Bay and the capital city of Maputo.

The Zambezi River flows through the north-central area of Mozambique and is the most fertile part of the country. It is the only navigable river in the country with a heavy flow of water traffic from its mouth to the city of Tete. Above Tete, about 400 mi (645 km) inland on the Zambezi, is the huge hydroelectric plant Cabora Bassa. Most of the electricity created there is exported to South Africa.

The Zambezi River is the fourth-longest river in Africa, stretching some 2,200 mi (3,540 km) from where it begins, looping through northwestern Zambia to the spectacular cataract of Victoria Falls on the border of Zambia and Zimbabwe, through Lake Kariba and finally entering Mozambique.

SAVANNAS AND LOWLANDS

Most of Mozambique is covered with tropical savanna and coastal lowlands that rise slowly inland to where they form plateaus, which are then broken by isolated mountain peaks. The highest of these lies to the north along the Zimbabwean border east of Lake Malawi and is called Monte Binga. One-third of Lake Malawi lies within the Mozambique border and this area is covered in tea and sisal plantations. Most of Mozambique's income comes from agriculture, making it one of the poorest countries in the world: 75 percent of the country is rural, growing cashews, copra, tea, sisal, and cotton. In the south, they grow rice, sugar-cane, bananas, and citrus fruits. They do raise cattle, sheep, and goats, but their numbers are kept low by the disease-carrying tsetse fly. Some fishing takes place in the

Mozambique Channel and mainly brings in shrimp for export.

Most of the population speaks a Bantu language and comes from 10 major ethnic groups. The Makua-Lomwe, living in northern Mozambique, make up nearly 50 percent of that region, along with the Yao and Makonde. In the center of Mozambique live the Thonga, Chewa, Nyanja, and Sena; in the south live the Shona and Tonga. Mozambique gained its independence on June 25, 1975, from PORTUGAL. The country is structured as a multiparty democracy with a free market economy. Mozambique has suffered through spells of droughts and heavy flooding in the decade that have severely hurt the economy.

BIBLIOGRAPHY. Institut géographique national, *The Atlas of Africa* (Éditions Jeune Afrique, 1973); Kwame Anthony Appiah and Henry Louis Gates, Jr., *Africana* (Basic Civitas Books, 1999); Saul B. Cohen, ed., *The Columbia Gazetteer of the World* (Columbia University Press, 1998); Bureau of African Affairs, "Background Note: Mozambique," (U.S. Department of State, 2003).

CHRISTY A. DONALDSON
MONTANA STATE UNIVERSITY

Mozambique Channel

LOCATED OFF the country of MOZAMBIQUE in Africa, the Mozambique Channel lies within the INDIAN OCEAN, between the African continent and the island of MADAGASCAR. The currents in the Mozambique Channel usually form an anticyclonic system but sometimes they do flow directly south into the Agulhas Current off the coast of SOUTH AFRICA. This area is a seismically active region that is thought to be an offshore continuation of the eastern branch of the East African Rift. On April 29, 1952, a 6.0-magnitude EARTHQUAKE struck within the Mozambique Channel, and since then several earthquakes of this magnitude and higher have been recorded.

The Mozambique Channel is narrowest at 250 mi (400 km), across from the African mainland at Mozambique to Madagascar. Within the channel flows the Mozambique Current, which takes warm water from the South Equatorial Current to the north and moves it south along the coast of Mozambique, then in a counterclockwise circle, back up the channel moving north along the coast of Madagascar. During the monsoon season in October and November, this current has the potential of attaining velocities exceeding 3.7 mi (6 km) per hour. The tidal ranges in the channel can be up to 21 ft (6.4 m), with surface temperatures varying seasonally between 71.6 degrees F (22 degrees C) and 80.6 degrees F (27 degrees C).

The marine life in the Mozambique Channel is extremely diverse thanks to the warm water currents of the area. There have been more than 2,000 species of fish identified in the channel, along with 180 species of birds. Five different types of marine turtles live in the waters of the Mozambique Channel: Pacific green, loggerhead, hawksbill, leatherback, and olive ridley. There are also many types of dolphins, invertebrates, and macroalgae. Southern right whales migrate through the channel, while humpback and minke whales live year-round here. The marine habitats of the Mozambique Channel include coral reef, sandy beach, rocky shore, seagrass beds, and deep water.

It is generally agreed that Madagascar once was linked to the African mainland. The Mozambique Channel is underlain by continental crust, layers of sandstone and limestone, which are thought by many researchers to have once been bridges from Africa to Madagascar. There is some debate as to when exactly Madagascar moved to its present position, with the spreading of the Mozambique Channel. Some researchers say it was during the Cretaceous period and others say it was during the early to middle Cenozoic.

BIBLIOGRAPHY. Michael Myers and Mark Whittington, "Mozambique," *Seas at the Millennium: An Environmental Evaluation* (Pergamon, 2000); J.R. Heirtzler and R.H. Burroughs, "Madagascar's Paleoposition: New Data from the Mozambique Channel," *Science* (v.174/4, 1971); Didier Bertil and Jean Marc Regnoult, "Seismotectonics of Madagascar," *Tectonophysics* (v.294, Elsevier, 1998).

CHRISTY A. DONALDSON
MONTANA STATE UNIVERSITY

Mumbai (Bombay)

BOMBAY, RENAMED Mumbai in 1995, is a city in western INDIA and the capital of the Indian state of Maharashta. With its three totally computerized stock exchanges that handle 70 percent of the country's stock transactions and manufacturing (particularly in textiles, which employs 11 percent of India's factory

workers), Mumbai is the most important economic center of South Asia. It is India's largest port in tonnage handled.

Mumbai is not only the largest city of India, with an estimated population of 16 million (2000), but is also the fifth-largest metropolis in the world. Mumbai is projected by the United Nations (UN) to be the third-largest city by 2015, when its population will reach 22.6 million.

The old city of Bombay, an island, has a lowland terrain and has gone through several stages of reclamation. It forms the southern tip of Greater Bombay. The main part of the metropolis is separated from mainland India by a narrow water body; to the west lies the Western Ghats mountain range. Located south of the TROPIC OF CANCER and falling in the monsoon climate zone, its annual rainfall averages 84 in (213 cm), mostly from June through September. Mumbai's average January temperature is 66 degrees F (19 degrees C) and the May average is 81 degrees F (27 degrees C).

THE CITY'S HISTORY

Historically, Bombay was a fishing village ceded to Portuguese in 1534 by the regional ruler, Bahadur Shah. In 1661, it was transferred to the British as a wedding gift to King Charles II of England, when he married the sister of Portuguese king. The British king then leased the territory to the newly formed East India Company. A fort built in 1717 became the main center of colonial activity. The center of the fort had three radiating roads (still in existence) leading to the city's Apollo, Church, and Bazar gates. When congestion increased considerately within the fort, settlements spread outside the walls. The wall was torn down in 1861.

Mumbai is the most diverse Indian metropolis: 67 percent are Hindu, 14 percent Muslim, 7 percent Sikh, 6 percent Christian, 4 percent Jain, and 4 percent Buddhist. Though English is the language of the elites, 42 percent are Marathi speakers, 18 percent Gujarati, 11 percent Urdu, 10 percent Hindi, 3 percent Tamil, 3 percent Sindhi, and 2 percent Punjabi. The city has several masterpieces of colonial architecture: Victoria (Chatrapati Shivaji) Terminus, the Municipal Building, and the University of Bombay. It is a leading educational center; the University of Mumbai was founded in 1857 and several colleges, including medical colleges, are also prevalent.

Mumbai makes the largest number of films compared to any other city of the world. The metropolis is rightly called "Bollywood, the star machine of India."

BIBLIOGRAPHY. *Planet Earth World Atlas* (Macmillan, 1998); Sujata Patil, Sujata and Alice Thorner, eds., *Bombay: Mosaic of Modern Culture* (Oxford University Press, 1995).

ASHOK K. DUTT, PH.D.
UNIVERSITY OF AKRON

Myanmar (Burma)

Map Page 1123 **Area** 261,970 square mi (678,500 square km) **Population** 42,510,537 **Capital** Rangoon **Highest Point** 9,692 ft (2,954 m) **Lowest Point** 0 m **GDP per capita** $200 **Primary Natural Resources** petroleum, timber, tin, antimony, zinc, copper.

A KITE-SHAPED country with a 1,243-mi (2,000-km) north-south stretch, Myanmar has its apex in the north, bordering INDIA and CHINA, and a tail that extends for 900 km (559 mi) in the south. The British colonized the area known as Burma in 1875 and ruled it mainly as a province of India. Japanese occupation in 1942 lasted for three years, after which the British regained their colony, only to hand over power to the Aung San-led antifascist People's League in 1948. During the British period, most administrative, rail/port operations, postal, banking and trading activities were handled by Indians, who constituted 8 percent of the population.

Moreover, exposure of Burma to the foreign market induced the IRRAWADDY delta-based farmers to produce surplus rice, and the country became the leading rice exporter in the world. In 1940, Burma was exporting 3 million tons of rice annually. The country has not regained that position because of increased demand in the local market and the impractical policy of the "Burmese way to Socialism," which introduced a command economy.

Following independence in 1948, Burma had elected governments, but since 1962, it has been ruled by military dictatorships. After being pressured by the United Nations, the country held a national election in 1990, resulting in 82 percent of the parliamentary seats being won by the Aung San Suu Kyi-led prodemocracy party, the National League for Democracy. Aung San Suu Kyi is the daughter of the "Father of the Nation,"

Aung San. She was awarded the Nobel Peace Prize in 1991. The 1990 election was invalidated by the military junta and Suu Kyi was put under house arrest. Brutal suppression of free expression still continues.

Starting in the 1970s, the agricultural High Yield Variety (HYV)-based green revolution technique has been widely used, resulting in increased rice production. Some analysts believe Myanmar seems to be well on its way to reclaiming its previous position as the world's premier rice exporter.

Myanmar is primarily an agricultural country, with 60 percent of its gross domestic product coming from agriculture, while 78 percent of the labor force is engaged in that occupation. Industries, including agricultural processing, knit and woven apparel, wood and wood products, textiles, and pharmaceuticals employ about 7 percent of the population. Myanmar had been producing petroleum since before World War II. The present production, however, remains stagnant.

Twenty-five percent of people live below the poverty line. High adult mortality from HIV/AIDS is slowing the population growth rate.

BIBLIOGRAPHY. Robert E. Huke, "Myanmar: Promise Unfulfilled," Ashok K. Dutt, ed., *Southeast Asia: A Ten-Nation Region* (Kluwer Academic Publishers, 1996); Robert E. Huke, "Burma," Thomas R. Leinbach and Richard Ulack, eds., *Southeast Asia: Diversity and Development* (Prentice Hall, 2000).

Ashok K. Dutt, Ph.D.
University of Akron

Namibia

Map Page 1116 **Area** 318,696 square mi (825,418 square km) **Population** 1,927,447 **Capital** Windhoek **Highest Point** 8,550 ft (2,606 m) **Lowest Point** 0 m **GDP per capita** $4,500 **Primary Natural Resources** diamonds, gold, tin, copper, lead, zinc.

NAMIBIA, FORMERLY known as South West Africa (SWA), is located in southern Africa. It borders the South ATLANTIC OCEAN, NAMIBIA, ANGOLA, BOTSWANA, SOUTH AFRICA, and ZAMBIA. Namibia can be divided into four major geographical segments. In the west, stretches the great Namib Desert, which extends along the Atlantic coast from the northern part of South Africa to the southern border with Angola. This desert belt varies in width from about 62 mi (100 km) in the south to as much as 680 mi (1,100 km) in the north, and has mighty sand dunes in the central part up to 1,968 ft (600 m) high. The northern and the southern extremes of the desert are predominantly gravel fields.

Moving inland, the desert belt gives way to the escarpment, a mountain wall up to 6,500 ft (2000 m) high. Beyond the escarpment, the land changes into the Central Plateau region, which slowly descends toward the east. The majority of Namibian towns and villages lie on this plateau, including the capital of Windhoek at 5,425 feet (1,654 m). Farther to the east lies the Kalahari Basin and the great KALAHARI DESERT, which is characterized by wide sandy plains and long-dunes with scarce vegetation.

Another unique geographical area is the relatively rainy Kavango and Caprivi region in the extreme northeast. It is flat and covered with dense bushveld. Because the climate is continental, tropical, and very dry, most of Namibia's rivers are dry except during rains. Perennial rivers, such as the Okavango, Kunene, Zambezi, and Orange are confined to Namibia's northern and southern borders areas. The coast is cooled somewhat by the BENGUELA CURRENT. The meager and highly variable precipitation is not very effective in watering the land because of a high rate of evaporation. Average rainfall increases from the southwest to the northeast. The territory suffers from prolonged, periodic droughts. The vegetation is generally sparse except in the far north.

Namibia's widely varied animal life includes the lion, leopard, elephant, rhinoceros, giraffe, zebra, ostrich, and antelope. The country's most renowned game reserve is the Etosha National Park, one of the largest in the world. Namibia has relatively abundant natural resources. Among the most significant are dia-

monds, particularly those that can be found in the sands of the Namib Desert along the Atlantic coast. Like South Africa, there are also important deposits of gold. Namibia is a sparsely populated multiracial country where half of the population is concentrated in the northern region known as Ovamboland. The Ovamboland is home to many ethnic groups, including the Ovambos, who represent 50 percent of the population, and the Kavangos and Caprivians. The white settlers (6 percent) occupy the central and southern highlands along with other smaller, ethnic groups. The white population consists of Afrikaners, British, and Germans. In 1990, the country became an independent nation after a long period of liberation struggle. Since then, the country has been stable politically under Sam Nujoma, who led a prolonged fight against South African rule in Namibia.

BIBLIOGRAPHY. David L. Clawson and Merrill L. Johnson, eds., *World Regional Geography: A Development Approach* (Prentice Hall, 2004); Jeffress Ramsay and Wayne Edge, eds., *Global Studies: Africa* (McGraw-Hill, 2004); *World Factbook* (CIA, 2003).

SAMUEL THOMPSON
WESTERN ILLINOIS UNIVERSITY
RICHARD W. DAWSON
CHINA AGRICULTURAL UNIVERSITY

Nap of the Earth

NAP OF THE EARTH, or NOE, is a Vietnam War-era term for very low-level flight, particularly of helicopters. The concept behind this type of flying is that a pilot guides his or her craft as close to the Earth's surface as terrain, vegetation, and other obstructions will allow and maintains that low altitude while approaching a chosen target or transiting an area. This flight technique becomes a tactic for surprise and survival as the aircraft comes into and passes out of a ground observer's field of vision before an enemy can target and engage the aircraft. It also has the advantage of getting the aircraft below acquisition radar detection envelopes and blending in with the ground clutter provided by vegetation and intervening topography. This type of flight is included in the more inclusive term *terrain flight* or TERF.

An official Department of Defense definition is: "Flight close to the Earth's surface during which airspeed, height, and/or altitude are adapted to the contours and cover of the ground in order to avoid enemy detection and fire."

BIBLIOGRAPHY. "Nap of the Earth Flight," *DOD Dictionary of Military Terms*, www.dtic.mil (June 2004); "Nap of the Earth Flight," www.aviation-terms.com (June 2004); Dennis Dura, "Terrain Flight," International Association of Natural Resource Pilots, www.ianrp.org (June 2004).

IVAN B. WELCH
OMNI INTELLIGENCE, INC.

nationalism

THE MODERN CONCEPT of nationalism was born with the Treaty of Westphalia in 1648. Before then, Europe was a checkerboard of small states, cities, principalities, and alliances united by religions, language, history, and politics. As recently as the 1800s, such nations as CHINA, INDIA, and even ITALY looked nothing like they do today but instead were divided into such multiple states, cities, principalities, and alliances. The concept of nationalism was foreign to much of Africa and Asia as well, which were divided by language, culture, tribal ethnicity, politics, and geography.

The Westphalia peace agreements ended the Eighty Years' War between SPAIN and the NETHERLANDS as well as GERMANY'S Thirty Years' War. SWEDEN'S and FRANCE'S borders were clearly identified. The United Provinces of the Netherlands became a nation. A variety of mountainous city-states calling themselves the Swiss Confederation became an independent republic. Germany's treaty ended a century-long struggle between the Holy Roman Empire and 300 German princes who ruled over a variety of dominions. The Peace of Westphalia recognized the full territorial sovereignty of the member states. The princes were empowered to contract treaties with one another and with foreign powers. They became absolute sovereigns in their own dominions: nations.

The Versailles Treaty ending World War I further recognized the principle of nationalism with Europe and the MIDDLE EAST divided into autonomous entities empowered to take care of their own affairs. A number of such brand-new states were carved out of the defeated Ottoman Republic as British administrators created with the stroke of a pen on a map such countries as JORDAN, SAUDI ARABIA, IRAQ, and IRAN—ignoring ge-

ography as well as historic, ethnic, and religious differences. For example, what might have been a homeland for the Kurds was separated by artificial borders and assigned to the new Iraq, Iran, and TURKEY.

Created by diplomats, Yugoslavia was made up of several intensely rivalrous Balkan states with historical differences and competing interests. It held together until the death of its head of state, Josip Broz Tito, dissolving into such nations as SLOVENIA, BOSNIA, and SERBIA AND MONTENEGRO. One of the former Yugoslav republics goes by the official United Nations-assigned title of the Former Yugoslav Republic of MACEDONIA since adjacent Greece is worried that declaration of an independent Macedonia would prompt a wave of nationalism among Greek Macedonians, who would want to secede from Greece and join the new all-Macedonian nation.

The collapse of Yugoslavia and the dissolution of the Soviet Union at the close of the Cold War resulted in even more independent nations. Nationalism continued to assert itself in such ways as DENMARK refusing to give up its national currency in favor of the EUROPEAN UNION's new currency, the euro. IMMIGRATION became a controversial issue in Britain, with some vocalizing that the British identity was blurring. Nationalistic parties did well in French and Dutch elections. Polls showed that most people continued to have a strong sense of attachment to their nationality. GLOBALIZATION was violently opposed in massive worldwide street demonstrations.

Yet, significant antinationalistic trends also took place. The EUROPEAN UNION transferred significant power from the national level to both local and continental bodies. Historic trade agreements such as the NORTH AMERICAN FREE TRADE AGREEMENT (NAFTA) lowered the economic borders between the UNITED STATES, CANADA, and MEXICO. Such counternationalism increased the internationalization of trade markets while weakening the sovereignty and authority of the nation states. Even so, nationalism has maintained its appeal. Belonging to a culturally, economically, or politically strong nation seemingly makes citizens feel better regardless of whether they have made any contribution to that strength.

Regrettably, nationalism can have extreme negatives. In the 1980s, a very negative nationalism was projected by an Argentine military junta desperate to avert popular attention from inflation and unemployment as well as institutionalized corruption and the outright murder of thousands of political opponents. Amid loud proclamations of national pride and destiny, Argentina invaded the remote FALKLAND ISLANDS, proudly proclaiming that Las Islas Malvinas (as Argentines call the islands) had been "liberated" and restored to the Argentine motherland. The few hundred inhabitants, mostly shepherds, spoke English and traced their roots to England. They appealed to Great Britain for rescue. Argentina was startled when the British mobilized, sinking Argentine navy ships, destroying the Argentine air force, and invading the islands, thereby precipitating the collapse of the Argentine government—deposed in a twist of nationalism as the Argentine people repudiated their actions. Whereas this manifestation of Argentine nationalism was politically motivated and manipulated, nationalism has many forms, which can be positive as well as negative and which include ethnic, religious, historical, linguistic, geographical, and civil nationalism.

ETHNIC AND RELIGIOUS NATIONALISM

Ethnic nationalism exists when the state derives political legitimacy from hereditary groupings and ethnicities. A very negative example would be RWANDA and BURUNDI when, in 1994, frenzied by radio broadcasts, leaders called for a national "cleansing" of both nations. Half a million Tutsi tribespeople in Rwanda and another 300,000 in neighboring Burundi were murdered over a three-month period in which families were hacked to death by machete and refugees huddling in church sanctuaries were burned alive. Perpetrators expressed little remorse, explaining that they were merely ridding the world of Tutsis.

A positive example would be the experience of SWAZILAND. According to Swazi legend, in the late 18th century, Chief Ngwane II led a small band of followers over the Lebombo Mountains, found other African peoples, made peace with them, and together they became what today are the ethnic Swazi. After a difficult British colonial period, Swazi independence was granted in 1968. A new constitution written in 1973 took care to reflect Swazi national traditions, including the rule of the Ngwenyama or king as the country's hereditary head of state, assisted by a council of elders and the Ndlovukazi or mother of the king, who for centuries has been in charge of national rituals. Today, Swaziland is a leader among southern Africa's emerging nations—united by a rich history, proud culture, and unique ethnicity.

Religious nationalism exists when the state derives political legitimacy as a consequence of shared religion, such as Judaism in ISRAEL, ISLAM in PAKISTAN, Catholicism in Italy, or Shintoism in JAPAN. One familiar ex-

ample would be the independence movement on the Indian subcontinent following World War II. Protesting the policies of an occupying nation that insisted that India was British and would always remain so, thousands of indigenous Indians put aside language, geographical, religious, and historical differences to unite, refusing to cooperate with colonial administrators. Mohandas K. Gandhi exhorted the masses to shame the British into leaving the Indian subcontinent through passive revolt, refusing to submit to the authority or rule of the occupiers, and instead asserting that the right to govern belonged to those whose ancestors had been there for thousands of years, and that no government would be better than a government by outsiders. On the eve of independence, however, religious differences between Muslims and Hindus split the proposed new nation apart. Muslim provinces on the east and west sides became Pakistan. What had been scores of small rival states united as the predominantly Hindu nation of India. Although it also had been administered as part of British India, the Buddhist kingdom of Burma had no interest in joining the confederation and instead sought its own independence as MYANMAR.

Another example is IRELAND. The inhabitants of most of the island are Catholic. The northern province of Ulster is Protestant as a result of the politically motivated importation of Scottish Presbyterians centuries ago. These Scots-Irish have remained geographically, religiously, ethnically, and linguistically separate from the rest of Ireland, electing to remain a part of Great Britain. Irish nationalists who have used violent means to seek the unification of the island are nominally motivated by religion, yet one never hears the Irish Republican Army offer to debate the Scots-Irish Protestants of Ulster on celibacy, predestination, or the authority of the papacy. Instead, they fight over nationalistic ideologies, views of Irish culture, and the status of the British army. Nevertheless, the conflict is popularly regarded as religious.

HISTORICAL NATIONALISM

Another kind of nationalism is historical nationalism. In this case, nationalism is projected because of the pride that inhabitants take in the history and heritage of their nation. An example is GREECE. Citizens on the Greek peninsula and islands feel great pride over their nation's rich past as the birthplace of democracy as well as the incubator of Western philosophy and the historic home of the Olympics. During World War II, Greek partisans were fierce defenders of their homeland, driven by a love of their homeland, culture, traditions, and their distinct national identity.

In China during the 1800s, furious that foreign occupiers had subjugated their leaders and introduced nationwide opium abuse, Shaolin monks led their followers in a futile attempt to drive European occupiers out of China. Armed with only bare fists and deep beliefs, these predecessors of today's kung fu and karate practitioners failed to liberate China in what Westerners would dub the Boxer Rebellion.

Instead, China was forced at gunpoint to sign over a number of Chinese coastal cities as foreign enclaves, serving as doorways and giving the Europeans trade access to China. The last of those, Britain's HONG KONG and PORTUGAL's Macao, were ceded back to China only at the turn of the millennium—the result of nationalistic demands on the part of the China, which had never forgotten its historic humiliation. The entire Chinese nation celebrated the return of the two cities with nationwide festivities and proclamations of Chinese national pride.

A common language has been one of the main presuppositions for nationalism. In FRANCE, for example, before the French Revolution, dialects such as Breton and Occitan were spoken in the various regions and were incomprehensible to each other. Following the Revolution, French was imposed as the national language. Nationalism has also prompted the revival of languages, such as Gaelic after Ireland won its independence, and Hebrew upon the founding of the state of Israel.

Civil nationalism exists when the state derives political legitimacy from the active participation of its citizenry, as in the "will of the people." An individual in such a nation must believe that the state's actions somehow reflect his or her will, even when specific actions go against that will. It is the theory behind constitutional democracies such as the United States. A dramatic example of civil nationalism is South KOREA. With their national economy threatened by a 1990s recession, thousands of South Koreans sold family heirlooms and melted down precious jewelry so they could make voluntarily contributions toward paying off the national debt and restoring the prosperity of their industrialized, market economy.

Many citizens remembered well the difficult days of the Korean War and, before that, the Japanese occupation in World War II. Although united by thousands of years of history with the ethnically and linguistically identical inhabitants of the communist northern half of their peninsula, South Koreans remain skeptical of re-

unification with North KOREA, which has a hereditary, non-democratic government and a centralized socialist economy. South Koreans have fought and sacrificed for their nation's security and prosperity—and take strong nationalistic pride in being South Koreans.

BIBLIOGRAPHY. John Hutchinson and Anthony D. Smith eds., *Nationalism* (Oxford University Press, 1995); Paul Treanor, "Structures of Nationalism," Sociological Research Online 1997, www.socresonline.org.uk (February 2004); Malcolm Anderson, *States and Nationalism in Europe Since 1945: The Making of the Contemporary World* (Routledge, 2000).

ROB KERBY
INDEPENDENT SCHOLAR

Nauru

Map Page 1128 **Area** 8 square mi (21 square km) **Population** 12,570 **Capital** None, government offices in Yaren District **Highest Point** 200 ft (61 m) **Lowest Point** 0 m **GDP per capita** $5,000 **Primary Natural Resources** phosphates, fish.

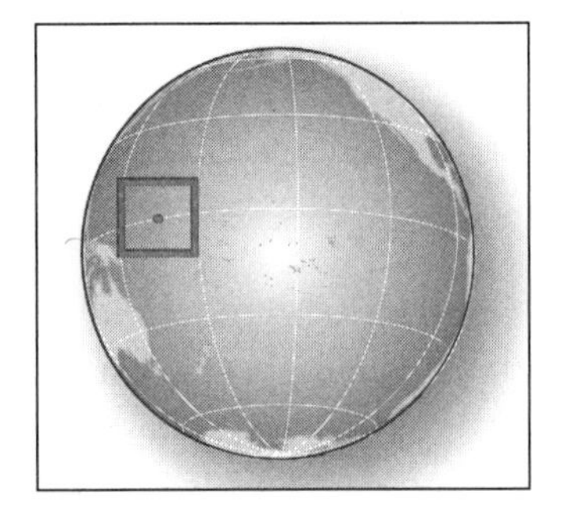

NAURU, LOCATED near the equator south of the MARSHALL ISLANDS in the south PACIFIC OCEAN, is the world's smallest independent republic. The country, officially known as the Republic of Nauru, gained its independence from the AUSTRALIA-, NEW ZEALAND-, and UNITED KINGDOM-administered United Nations trusteeship on January 31, 1968. The exportation of phosphate, principally to Australia, New Zealand, South KOREA, and INDIA, is the primary economic source for the island. However, the phosphate deposits are predicted to become exhausted in the coming decade. To prepare for this, the government has begun to pursue other economic strategies such as the registration of offshore banks and corporations and tourism.

Sea-faring Polynesians and Melanesians first inhabited the island. First contact with Europeans came in the 1830's when whaling ships first encountered the island. During World War II the island came under naval fire from first the Germans and then was later occupied by the Japanese. The Japanese occupied the island and forced over 1,000 Nauruan to work as laborers in the Caroline Islands, where nearly half died. Following the war, the island became a United Nations Trust Territory until its independence in 1968.

The main ethnic group on the island is the Nauruan, comprising 58 percent of the population. Other groups include Pacific Islander, 26 percent; Chinese, 8 percent; and European, 8 percent. The official language of Nauru is Nauruan which is a distinct Pacific Island language. English, though, is widely understood and spoken and used for most government and commercial purposes on the island. A majority of the population is Christian; two-thirds of the population is Protestant and the other third Roman Catholic.

The climate is tropical. The island averages 78 in (200 cm) of precipitation per year, and the monsoon season runs from November to February. Average daily temperatures range between 75 to 93 degrees F (24 to 34 degrees C) with an average humidity around 80 percent. The central plateau is where the phosphate mining takes place, and as a result, four-fifths of the total land area of the island has been lost. Jagged coral pinnacles, up to 49 ft (15 m) high, and coral cliffs dominate the resulting landscape on the plateau. The island is surrounded by a coral reef bordered by deep water on one side and sandy beach on the other. Next to the beach is a wide fertile coastal strip.

BIBLIOGRAPHY. *World Factbook* (CIA, 2004); Northwestern University, www.earth.nwu.edu (April 2004).

TIMOTHY M. VOWLES, PH.D.
VICTORIA UNIVERSITY, NEW ZEALAND

nautical mile

A UNIT OF DISTANCE used primarily at sea and in aviation, the nautical or sea mile is based on the average meter distance on the Earth's surface represented by one minute of latitude. Because it is based on the earth's dimensions, the nautical mile is extremely convenient to use for any type of navigation. Adopted in 1954 by the U.S. Secretary of Commerce and the Secretary of Defense, and used in maritime and aerial navigation, in relation to how boat speeds and wind velocities are measured (one knot is one nautical-mile-per-hour), a nautical mile is approximately one minute of latitude and it is used to express distance. The nautical mile, which is outside of, but accepted by, the In-

ternational System of Units (SI), the modern version of the metric system, was for years established by Britain at 6,080 ft or 1853.18 m. There is no internationally agreed symbol for the nautical mile.

A nautical mile is approximately a minute of arc (a unit of angular measurement equal to one sixtieth of one degree) along a great circle of the Earth. That is, if the Earth was sliced into two equal halves through the center along the equator and then divided along the perimeter into 360 degrees, and each degree into 60 arc minutes, the result (although the Earth is not a perfect sphere) would be close to one nautical mile.

At sea level one minute of angle equals 1.15 miles or one nautical mile. As such, the nautical mile differs slightly from definitions of other miles including the international nautical mile, which was adopted by the International Extraordinary Hydrographic Conference in 1929 at exactly 1,852 meters. Around since the time of the Roman Empire, the term *mile* had many meanings over the centuries. The Roman mile was 5,000 ft in length or 1,479 m in modern dimensions.

Most people are more familiar with the land or international mile. Used in the United States and Britain as part of the imperial system of units, the land mile is defined, by a 1959 international agreement, to be exactly 5,280 international ft or 1,609.344 m. The United States makes use of the statue mile, which was adopted by Congress before the international nautical mile was established.

The statue mile, derived from U.S. geodetic surveys after the adoption of the international mile and used by the U.S. Coast Guard, is one-quarter inch longer than the international nautical mile at 5,280 ft. The geographical mile, a unit of length determined by one minute of arc at 6,087.15 ft, is also closely related to the nautical mile.

Nautical miles are also related to speed in that the nautical mile and the knot, a unit of speed equal to 1 nautical mile per hour (1,852 m), are the bases of sea and aerial navigation. During the days of wooded boats, sailing speed was calculated by unraveling a knotted rope into the water behind the ship. The number of knots that passed over the side of the ship during a given time (most ships during this time used a 28-second sand hourglass) would indicate how fast the ship was traveling. Today, aircraft velocity gauges and flight routes are calibrated and denoted in knots and nautical miles.

BIBLIOGRAPHY. Robert Henderickson, *The Ocean Almanac* (Doubleday, 1984). Peter Kemp, *The Oxford Companion to Ships and Sea* (Oxford Press, 1994); Robert McKenna, *The Dictionary of Nautical Literacy* (International Marine/Ragged Mountain Press, 2003); Helen Gaillet de Neergaard, *Nautical Terms and Abbreviations* (Near Field Press, 1994).

GLEN ANTHONY HARRIS
UNIVERSITY OF NORTH CAROLINA, WILMINGTON

Nebraska

KNOWN AS THE Cornhusker State, the west north-central state of Nebraska was named from a Native American word meaning "flat or broad water." It lies within the Missouri River basin; the western part of the state also lies within the Platte River basin. Nebraska covers an area of 77,354 square mi (200,346 square km), making it the 16th largest state in the UNITED STATES. Nebraska is 38th in population among the 50 states. The state is 430 mi (692 km) east to west and 210 mi (338 km) north to south. Some 95 percent of Nebraska is devoted to agriculture in some way.

Nebraska makes up the westernmost edge of North America's Central Lowlands and is part of the Great Plains. It is bounded on the north by SOUTH DAKOTA, on the south by COLORADO and KANSAS, on the east by IOWA and MISSOURI, and on the west by Colorado and WYOMING. Nebraska entered the Union in 1867 as the 37th state. The state capital is Lincoln. Other large cities include Omaha, Bellevue, Grand Island, Kearney, Fremont, Norfolk, North Platte, Hastings, and Columbus.

The highest point in Nebraska is 5,424 ft (1,653 m) above sea level in Johnson Township; the lowest point is 840 ft (256 m) above sea level at the Missouri River in Richardson County. Much of Nebraska's topography is dominated by till plains, with lowlands evident in the eastern third of the state, which is part of the Dissected Till Plains that extend to the Great Plains in northern Nebraska. The Dissected Till Plains are characterized by rolling hills dotted with numerous streams and rivers, providing the richest farmland in the state.

Nebraska's Great Plains contain the Loess Hills, made up of loamy dust distributed by the wind, varying in color from buff to a yellowish brown. The terrain is rough and hilly. In the Loess Plains that lie in the southeastern part of the region, the land is flatter and more conducive to farming. Because of its frequent

rainfall, this area is sometimes called the Rainwater Basin or the Rainbasin. In LOESS areas of Nebraska, silty soils that are conducive to farming are the most common. In the eastern third of Nebraska, soils are more likely to be low-permeability clays that make irrigation difficult.

A 20,000-square-mi (51,799-square-km) section of Nebraska lying north of the Platte River is known as the Sand Hills, which is the largest area of sand dunes on the North American continent. Abundant grass and readily available water from the hundreds of lakes found throughout the area make this area excellent for grazing.

Nebraska's High Plains lie north and west of the Sand Hills. In this dry area, advances in irrigation technology have allowed some farming, while rougher terrain is used for grazing. Nebraska's badlands are located in the northwestern section of the state. Distinctive formations have been formed in the badlands by wind and water.

Nebraska's climate is continental, resulting in hot summers with frequent thunderstorms and occasional hail and tornadoes. Harsh winters are accompanied by an annual snowfall of 29 in (73.6 cm), with occasional blizzards. The state is somewhat drier than other midwestern states. Average temperatures range from 89.5 degrees F (32 degrees C) in the summer to 8.9 degrees F (-12.8 degrees C) in the winter. Annual precipitation ranges from 33 in (84 cm) in southeast Nebraska to 18 in (45.7 cm) in the west, resulting in a growing season of 170 days in the southeast and 120 days in the northwest section of Nebraska. Draughts are common. Nebraska's largest rivers are the Missouri, the Niobrara, the Platte, and the Republican. The state's largest lakes, including Lewis and Clark Lake, Harlan County Lake, and Lake C.W. McConaughty, are man-made.

Recognized as a major grain producer, Nebraska's approximately 48,500 farms and ranches produce corn, sorghum, soybeans, hay, wheat, dry beans, oats, potatoes, sugar beets, and livestock. Nebraska is also home to the greatest number of forage grasses in the country. Natural resources include oil, natural gas, cement, stone, sand and gravel, and lime. Manufacturing in Nebraska is dominated by processed fuels, industrial machinery, publishing, electric and electronic equipment, metal products, mobile homes, pharmaceuticals, chemicals, and transportation equipment.

Oak, hickory, and elm trees are plentiful in the river valleys of Nebraska's east, and cottonwood and willow, as well as elm, are found in the west. Ponderosa pine predominates in the Great Plains area. Bluestem grasses are most commonly found on Nebraska's prairies, while grama, buffalo, and sagebrush grow in Nebraska's dry Panhandle. Sand sage and grama grow throughout the southwest. Nebraska's wildlife population includes the coyote, antelope, deer, fox, badger, prairie dog, pheasants, and quail. Bison are found in isolated areas of the state.

BIBLIOGRAPHY. Bradley H. Baltensperger, *Nebraska: A Geography* (Westview, 1985); Michael Flocker, *Nebraska: The Cornhusker State* (Gareth Stevens, 2002); "The Geography of Nebraska" www.netstate.com (November 2004); Dan Golenpaul, ed., *Information Please Almanac* (McGraw-Hill, 2003); Charles B. McIntosh, *The Nebraska Sand Hills* (University of Nebraska Press, 1996); "State of Nebraska" www.nebraska.gov (November 2004); William Wyckoff and Lary M. Dilsaver, *The Mountainous West: Explorations in Historical Geography* (University of Nebraska Press, 1995).

ELIZABETH PURDY, PH.D.
INDEPENDENT SCHOLAR

needs and wants

WANTS AND NEEDS are concepts that have come to be developed around three different spheres of discussion, biological, social, and economic. Each is an expansion in size and scale over the one preceding it. In addition, it is important to remember that the subject is biological in character. This means that we do not discuss the wants and needs of inanimate things. The second important point to recognize is that "wants" are subservient to "needs" and, as far as we know, restricted to the realm of humans. From this perspective, we can think of wants as some specific expression of how we choose to meet those needs.

The discussion of wants and needs has an origin in Greek thought more than 2,500 years ago in works by Xenephon, Aristotle, Plato, and Protagoras. But their discussion of the issue was fairly broad and aimed primarily at what constitutes good administration as the driving force in society.

However, Protagoras did make a distinction between human perception and physical phenomena, stressing that hedonic calculus (self-interest) is a vital element in individual decisions. These ideas held for more than 1,200 years until the subject was once again discussed at length during the Scholastic Period beginning in the 1200s. Here, the writings of Albertus Mag-

nus and Thomas Aquinas began to explicitly explore what they called indigentia, or human wants. But the writing was vague in the sense that the distinction between wants and needs was far from clear. At the end of the 18th century, the Classical School of economic writers once again found room for wants and needs in their writings. But the development of economic ideas during this time considered human wants and needs to be insatiable (and generally identical), and as such, the real interest was in finding ways to economically balance unlimited wants with limited resources. The importance here is that the reference to limited resources marks the first time that geographic factors are openly included as an important element in the wants and needs discussion.

During the 19th century all forms of science began to blossom, including psychology. One of the areas of development within psychology was a new examination of human need. Psychologists quickly established certain primal needs such as those involving physiology: the need for food and shelter, where food also includes drink and where shelter can refer to clothing, housing, or other individuals. By the mid-20th century a wide variety of needs were being identified and debated, but they were generally related to biology, achievement, or power. Probably the most well known of these were put forward by Abraham Maslow (1943). Maslow developed a conceptual framework that suggested a hierarchy of human needs addressing issues as either growth needs or deficiency needs and where each lower-level need must be meet before the one above it can be fulfilled.

If you review all that has been discussed, you will find that most of the discussion can be put into one of two boxes, things that are instinctive and things that are learned responses. In doing this, the differences between needs and wants not only becomes much clearer, but it is possible to extend the thinking to areas beyond the biological and make use of the ideas within a geographic context.

HUMAN INSTINCTS

Instincts are hardwired into the brain, the genetic legacy of millions of years of evolution. Instincts make a creature react automatically to stimuli, with the brain carrying out instinctive reactions without conscious control. But we must not think that instinct is fixed. There can, of course, be evolution in instinct. On occasion an aberrant instinct can be more efficient as a survival or reproductive mechanism than the original instinct. Thus, the aberration has a better chance of survival, of reproducing, and of passing on to future generations, the original instinct dying out.

The second element of the subconscious mind is learned response. A learned response is one that mitigates or modifies an instinctive reaction. It is instilled in a person through a series of steps: emotions, belief, attitude, feelings, and behavior. Emotions, loosely defined as feelings, are the result of two things: the instincts discussed earlier and a priori assumptions. A priori assumptions are those conclusions based on theory rather than experience or observation, premises arrived at without examination of evidence. The basis for a priori assumptions is culture and upbringing, what sociologists call socialization or programming. Basic instructions on how to regard the world first come from the family.

HUMAN PROGRAMMING

In human terms, programming tells one how to regard what comes in through the senses. The family is not the only source of programming, of course. As a person gets older, other influences come into play: school, church, peers, teachers, television, books, anyone and anything that provide ways of regarding the world and the people in it. Programming leads to belief: what a person is programmed to believe is what that person does believe. What leads from a person's beliefs is that person's attitude toward the world and the things in it, the next step in learned response. When one believes something is bad, the attitude toward it is one of distaste. The attitude leads to the next step: feelings. Feelings are the actual human emotions that result from the instinctive and learned response reaction to a stimulus. On the basis of feelings, humans respond. Their initial response is to their instinctive reaction to stimuli. However, this is coupled with their learned response and becomes their behavior. Actual behavior is the final level. It is what people do in response to a stimulus, the external manifestation of their attitudes.

Each of the points indicated above suggests the geographic character of the wants/needs issue. We can take needs as given and apply them anywhere we want. Wants, on the other hand, are socially determined and as such reflect the degree to which the geographic environment (our physical and cultural surroundings that provide the basis for our emotions, beliefs, attitudes, feelings, and behavior) gives specific meaning to satisfying needs. While we accept human need as uniform among all people, the world's resources for satisfying them are not equally distributed among regions. This difference is what often leads to stereotyping and

sometimes conflict. It is also a strong driving force behind territoriality, a key factor in GEOPOLITICS and the game of nations.

BIBLIOGRAPHY. Aristotle, *The Works of Aristotle*, 12 vol., W.D. Ross, ed. (Clarendon Press 1908 to 1952); Campbell R. McConnell, *Economics* (McGraw Hill, 1975); S. Todd Lowery, *The Archaeology of Economic Ideas* (Duke University Press, 1987); Xenephon, *Memorabilia and Oeconomicicus*, E.C. Merchant, trans. (G.P. Putnam's Sons, 1923); Alexander Gray, *The Development of Economic Doctrine* (Longman, 1980)

RICHARD W. DAWSON
CHINA AGRICULTURAL UNIVERSITY

Negev Desert

THE NEGEV IS A part of the Saharo-Arabian desert belt running from the ATLANTIC OCEAN across the SAHARA and Arabia to the Sind desert of INDIA's Indus Valley. The Negev, also spelled Negeb, means both "the southland" and "dry." Negev is derived from the Hebrew verbal root n-g-v meaning "to dry." The Ha-Negev (Hebrew) covers about 60 percent of southern ISRAEL. It is a triangular shaped arid region of some 5,019 square mi (13,000 square km). The apex of the triangle is the port city of Elat (Eilat) on the RED SEA. The eastern leg of the triangle is formed by the Syro-African Rift valley from Elat to the dead sea. The western leg of the triangle is bounded by the SINAI PENINSULA (EGYPT). The western boundary extends from the southern end of the Gaza strip coastal plain to Elat. The base of the triangle lies along an east-west line at about 30 degrees 25 minutes N.

The Negev is small in area but has five regions: the Northern Negev, the Western Negev, the Negev Highlands, the Southern Negev, Arabah (Arava). Some scholars also include the Dead Sea as a sixth area. The Northern region is a small ribbon of land running east-west that includes the city of Beersheba (Beer Sheva) which is "the capital of the Negev." The soils are fairly fertile, with enough rain for wheat to grow. The Western Negev stretching from the Gaza strip and the Sinai Peninsula to the Negev highlands, receives about 10 in (25 cm) of rain annually. The LOESS soils in the area are good for growing cereals.

The Negev Highlands is a high, dissected plateau averaging 1,200 to 1,800 ft (400 to 600 m) in elevation. The highest point is 3,395 ft (1,035 m). The area receives only about 4 in (10 cm) of rain per year. In this area are many erosion craters (makhteshim) formed by the erosion of the upward-folded (anticlines) limestone strata. The two largest are the Ramon Crater and the Great Crater.

The Arabah Valley (Wadi al-Jayb) is 111 mi (179 km) long, from Elat to the Dead Sea, and is bounded on the east by JORDAN. It is very arid. While the Negev is a DESERT, it is not as arid as the Sahara to the west or the Jordanian desert to the east. Many desert plants, animals, reptiles and birds flourish in the Negev. Modern irrigation projects have made many areas bloom.

BIBLIOGRAPHY. Michael Evenari and Nephtali Tadmor, *Negev: The Challenge of a Desert* (Harvard University Press, 1982); Nelson Glueck, *Rivers in the Desert: A History of the Negev* (Farrar, Straus, and Cudahy, 1959); Daniel Hillel, *Negev: Land, Water and Life in a Desert Environment* (Praeger, 1982); Lesley Hazelton, *Where Mountains Roar: A Personal Report from the Sinai and Negev Desert* (Henry Holt, 1980).

ANDREW J. WASKEY
DALTON STATE COLLEGE

Nepal

Map Page 1123 **Area** 87,489 square mi (140,800 square km) **Population** 27,070,666 **Capital** Katmandu **Highest Point** 29,035 ft (8,850 m) **Lowest Point** 230 ft (70 m) **GDP per capita** $1,400 **Primary Natural Resources** quartz, water, timber.

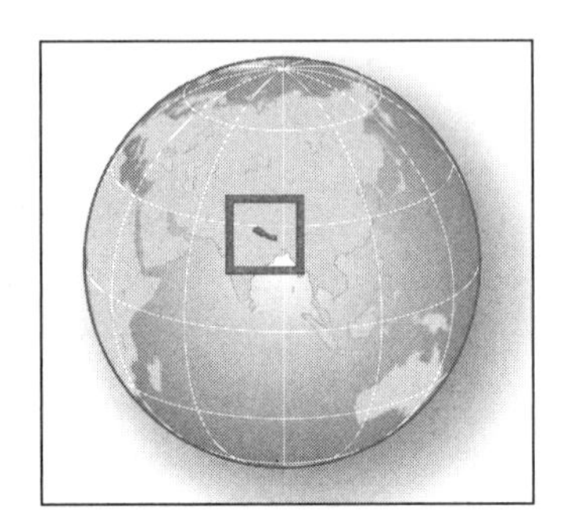

NEPAL, LOCATED between two Asian giants INDIA and CHINA, is a small country of approximately the size of the American state of FLORIDA. It extends some 500 mi (805 km) from east to west in an elongated rectangle along the arc of the HIMALAYAS.

The northeastern section of the country lies in the same latitude as northern Florida; the southeastern extremity, in the latitude of Fort Lauderdale; and Katmandu, the capital, in the latitude of Tampa. Its altitude varies from 197 to 722 ft (60 to 220 m) in the south to the highest mountains in the world in the

north. Nepal is a land of great diversity. There are dense swampy jungles, rich rice-clad valleys, bleak alpine highlands, and towering snow peaks within a comparatively few miles of each other. The northern interior has bitterly cold winters, whereas the southern Terai, less than 100 mi (161 km) away, has a humid, subtropical climate year-around. Into this diverse physical setting many ethnic groups have immigrated over the years to give the nation a racial and cultural pattern as varied as the land itself.

Except for a narrow strip along its southern border, Nepal lies entirely within the great mass of the Himalayas. By altitude, Nepal is divided into three distinct zones: One, the Terai, consists of the low Siwalik or Churia range of Hills, Bhabar, and Terai along the southern border of Nepal; two, the central mountainous belt, varying in altitude from 1,000 to 8,000 ft (305 to 2,438 m); and three, the alpine zone, comprising the higher slopes and valleys of the main Himalayan range and the trans-Himalayan districts of Manangbhot, Mustangbhot, and Charkhabhot.

The backbone of the Nepalese Himalayas contains many of the highest mountains in the world; from east to west there are: Kangchenjunga (28,168 ft or 8,586 m); Makalu (27,790 ft or 8,481 m); Everest (29,035 ft or 8,850 m); Cho Oyu (26,750 ft or 8,153 m); Manaslu (26,658 ft or 8,125 m); Himali Chuli (25,801 ft or 7,864 m); the Annapurna I (26,391 ft or 8,091 m); Dhaulagiri (26,790 ft or 8,174 m); and, in the extreme west, Api (23,399 ft or 7,132 m).

These ranges divide the country into four distinct regions. The western region extends from the Sarda, or Mahakali River, to the Dhaulagiri range. The central region comprises the basin of the Gandak and its tributaries. The great southern offshoot from Gosainthan bifurcates to form the third region, the true Nepal, or valley of Katmandu, lying at an elevation of slightly more than 4,000 ft (1,219 m). To the east lies the fourth region, which is formed by the basin of the Sapt Kosi draining the mountain from Gosainthan to Kangchenjunga.

Rivers in Nepal flow mainly from north to south, which means they originate from the Himalayas and flow into the GANGES RIVER in India. The major rivers in the country of Nepal include the Mahakali, Karnali, Bahai, Rapti, Narayani, Bagmati, Kamala, Sapta Koshi, and Kankai.

In the early history of Nepal it is difficult to distinguish fact from legend. Many successive dynasties followed until King Man Deva, in the 5th century established trading links with India and Tibet and enriched the country. Little is known of the ancient kingdom of Gurkha, but tradition and legends say that the ruling family descended from the Rajput princes of Udaipur, India. After the conclusion of Anglo-Nepali War in 1815, the strong prime minister of Nepal, Bhim Sen, greatly increased the power of prime minister's office and paved the way for the establishment of the Rana line of hereditary prime ministers.

From about 1850 to 1950, the hereditary prime ministers of the Rana family wielded supreme power under the aegis of titular kings. The Ranas were not progressive; they controlled great wealth, and continuance of their position seemed to rest upon an economically depressed Nepal.

As a result, a palace revolt occurred in 1950, when the king regained his position of authority, and in February 1951, King Tribhuvana's proclamation of a constitutional monarchy ended the reign of the 104-year-old Rana oligarchy. With a new constitution, which came into effect from 1959, many encouraging political developments took place in Nepal; however, economic depression still continues and currently Maoist rebels have taken a political path of armed rebellion.

BIBLIOGRAPHY. W.M. Jenkins, *Nepal: A Cultural and Physical Geography* (1960); W.B. Northey, *The Land of the Gurkhas* (1937); D.L. Snellgrove, *Himalayan Pilgrimage* (1961); P.P. Karan and William M. Jenkins, Jr., *The Himalayan Kingdoms: Bhutan, Sikkim, and Nepal* (Van Nostrand, 1963).

JITENDRA UTTAM
JAWAHARLAL NEHRU UNIVERSITY, INDIA

Netherlands

Map Page 1131 **Area** 16,033 square mi (41,526 square km) **Population** 16,150,511 **Capital** Amsterdam **Highest Point** 1,063 ft (322 m) **Lowest Point** -23 ft (-7 m) **GDP per capita** $26,900 **Primary Natural Resources** natural gas, petroleum, arable land.

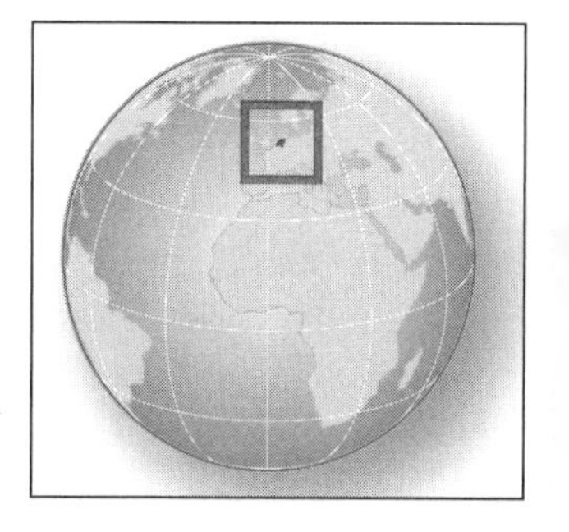

THERE IS A SAYING that God created the world, but the Dutch created Netherlands. Since roughly the 12th

century, the people living at the multiple mouths of the RHINE RIVER have reclaimed over a third of the national territory from the sea, through extensive use of dikes, dams, and polders (drained land). The struggle against the sea concentrated much of the national energy into cooperation and efficiency, resulting in one of the most tolerant and most productive societies on the planet.

The Netherlands—sometimes called Holland, which is actually just one of its 12 provinces—is almost entirely flat. Nearly a third of it is below sea level and requires continual pumping to remove excess water. The region's stereotypical windmills are not merely scenery, but have, since the 15th century, provided the basis for Dutch livelihood and prosperity. Having almost no natural resources, the Dutch instead created one of the largest merchant fleets in Europe, and, at the height of their empire in the mid-17th century, controlled ocean trade from the Caribbean to the East Indies. Raw materials brought back to the Netherlands were processed in ever-growing factories, leading to the development of some of Europe's fastest-growing cities: Amsterdam, Rotterdam, the Hague, Delft, Leiden, and Utrecht.

Today, this cluster of six cities is home to about 7 million people, nearly half the Dutch population, in one of the most densely populated regions of the world. The ports are still among the busiest—Rotterdam is the largest port in the world and has grown in importance through the development of large oil refineries. Canals provide easy transport into the interior of the continent, and it is estimated that 40 percent of all EUROPEAN UNION inland shipping is done by Dutch-owned companies.

Three major European rivers enter the North Sea at the northwest corner of the Great North European Plain: the Rhine, the Maas/Meuse, and the Scheldt. Most of what used to be the Rhine Delta is now canalized and divided into navigable channels. The river's two main channels, the Nederrijn and the Waal flow west into the North Sea, while a significant third branch, the IJssel, flows northward instead, into what used to be the Zuider Zee (South Sea), but since 1932 has been closed off by an immense barrier dam and has been renamed the IJsselmeer (Lake IJssel).

Since then, this large body of water has been slowly drained and reclaimed as massive polders, one of which was recently (1986) named the Netherlands' 12th province, Flevoland. Other polders are being planned for this area. Further to the southwest, the large estuaries of the Scheldt have been secured against flooding by the construction of massive dams, linking together many of the islands of this southwestern province (Zeeland) and shortening the total coastline by 44 mi (710 km). The final dam to be constructed was designed especially to allow the tides to continue to come in but can be lowered in case of flood: a movable dike, considered the world's most expensive insurance policy. It was completed in 1986 and considered a major victory for European environmentalists.

The Netherlands' capital, Amsterdam, has been a commercial center for centuries, dating back to the Dutch colonial empire.

Aside from the historic center of the Netherlands in these western and southern provinces of Holland (North and South), Zeeland, North Brabant, and Utrecht, the Netherlands also consists of its more agricultural eastern and northern provinces. Gelderland,

Overijssel, and Drenthe are more hilly and are home to much of the Netherlands's vast dairy and cheese industries. Most Dutch farms are small, and only 3 percent of the population is engaged in this profession, but the Netherlands is the third-largest agricultural exporter in the world.

The northern provinces of Friesland and Groningen are again very flat and also composed of many polders. The culture is different here from in the industrialized south; it is closely connected to the North Sea region of neighboring GERMANY, and has its own language, Frisian. Frisian is close to Dutch but also shares a strong affinity with Old English. The Frisian Islands—several long, flat, sandy islands, split between the Netherlands and Germany—form a barrier between the North Sea and the mainland. Finally, there is Limburg Province, far to the southeast, with the Netherlands's highest elevations. Limburg has historically had greater ties with Germany and BELGIUM, and was attached to the Kingdom of the Netherlands only in 1839, mostly for strategic purposes, but also for its valuable coal fields.

Administratively, the Kingdom of the Netherlands also consists of a few overseas territories, which enjoy full integrated status as equal partners in the Dutch kingdom: ARUBA and the NETHERLANDS ANTILLES (Bonaire, Curaçao, Sint Eustatius, Saba, and Sint Maarten). These have been autonomous since 1954, Aruba separately since 1986, and, since the mid-1980s have expressed varying degrees of interest in independence. The Dutch government has made clear its goal for independence for these West Indian islands, but at present the status quo continues. Former colonies INDONESIA and SURINAME achieved independence in 1945 and 1975, respectively, but have contributed to the modern cosmopolitan culture of Dutch cities because of large postindependence immigration, especially from Suriname (former Dutch Guiana), where Asians immigrated to Holland rather than face persecution from a black majority.

Aside from coal in Limburg, the Netherlands has few natural resources. There is salt in the far eastern provinces and one of the largest-producing fields of natural gas in Europe, located near Slochteren, in Groningen province. New discoveries of offshore oil have reduced Dutch dependency on Middle East oil but is not enough to supply Holland's numerous large industries. Principal export products include chemicals, plastics, machinery, and electronics, produced by some of the largest global corporations: Phillips, Unilever, and Royal Dutch Shell. For such a small, poorly endowed country, the Dutch economy is one of the strongest—ranked 14th in the world—and Dutch statesmen and businessmen are the leaders of much of today's united Europe.

BIBLIOGRAPHY. *World Factbook* (CIA, 2004); Wayne C. Thompson, *Western Europe 2003*, The World Today Series (Stryker-Post Publications, 2003); Dutch Ministry of Culture, www.minbuza.nl (August 2004); "Netherlands," www.netherlands-embassy.org; "Netherlands History," www.history-netherlands.nl (August 2004).

JONATHAN SPANGLER
SMITHSONIAN INSTITUTION

Netherlands Antilles

AN OVERSEAS TERRITORY of the NETHERLANDS, the Netherlands Antilles consists of four and a half islands, separated by about 545 mi (880 km) of the CARIBBEAN SEA. The southern group, Curaçao and Bonaire, known as the Windward Islands, lie 40 to 50 mi (60 to 80 km) off the northern coast of VENEZUELA. Formerly known as the ABC Islands, the A (Aruba) left the group in 1986 and is administered separately. The northern, or Leeward Islands group are the islands of Sint Eustatius and Saba, plus the southern half of the island of St. Martin (Sint Maarten in Dutch), sandwiched between the British dependency of ANGUILLA to the north and the independent nations of SAINT KITTS AND NEVIS, and ANTIGUA AND BARBUDA to the south and east. The northern part of St. Martin is part of France, as is the nearby island of St.-Barthélemy.

The islands are a constituent part of the Netherlands and have had full autonomy of internal affairs since 1954. The departure of Aruba was mostly due to resentment over dominance in the group by its larger neighbor, Curaçao, but initial plans for independence were scrapped in 1996. Other separatist movements have met with defeat in referenda, most recently for Curaçao in 1993 and for Sint Maarten in 1994. The latter was by a lower percentage than previously (59.8 percent), so there may yet be change to the administrative structure of the five constituent parts of what is known as the Dutch Antilles Federation.

Physically, the two groups are very different. Saba, Sint Eustatius, and Sint Maarten are all considerably more elevated volcanic formations and have about twice the amount of rainfall as Bonaire and Curaçao.

The latter islands are low, barren, and arid (less than 25 in or 63 cm of rain per year) and have more of a desert climate than one would think for the Caribbean—aloe and cacti thrive, plus numerous small herds of goats. But this lack of moisture is also a problem for the southern islands' tourist industry. Large seawater processing stations have been built, a side product of which is salt. Salt was one of the chief attractions for Europeans in the 17th century, and Bonaire is still a major exporter. The tourist industries of both islands benefit from constant sunshine and scant rainfall. Bonaire is also known for its rich underwater life, and its large flocks of flamingos (now protected). Curaçao is known for its characteristic liqueur made from locally grown orange rinds. The southern islands lie outside the hurricane belt, but hurricanes occasionally cause severe damage to the northern group. The most recent case was in 1995, when Hurricane Luis devastated resorts and the island infrastructure on Sint Maarten.

The northern group (sometimes called the 3 S's) are each distinct from each other: Sint Eustatius (locally called Statia) is the poorest, Saba the smallest (only 5 square mi or 13 square km), and Sint Maarten, the most developed. All depend on tourism because they lack any major resources. All were settled by the Dutch in the first half of the 17th century and have been officially administered by the Netherlands since 1816. Yet while Dutch is the official language and the language of schooling, most people on all three islands speak English. Each island has its specialty: Saba's rich underwater life, Statia's historic charm, and Sint Maarten's endless beaches. Saba and Statia are both of fairly recent volcanic origin, while Sint Maarten is part of the older, outer arc of the Antilles (like Antigua) and considerably flatter and drier. Saba has recently created a marine park, circling the entire island, to protect its so-far unspoiled underwater environment from increasing traffic of divers. Efforts to increase tourism on Statia have been less successful.

Because none of these five islands was suitable for the establishment of large-scale plantation economies, they became instead centers for trade in sugar, tobacco, cotton, and especially African slaves. In the 18th century, Sint Eustatius and Curaçao were two of the biggest marketplaces for slaves in the entire Caribbean. Merchants, pirates, and administrators from all over Europe settled in the major Dutch port of Willemstad, on Curaçao, which led to the development of that island's unique trading language, Papiamento, a mixture of Portuguese, English, Dutch, and Spanish, still used by most residents. The abolition of slavery in 1863 in all Dutch colonies hit the islands hard, especially Sint Eustatius, which still today has one-fourth the population it had in the 18th century. Curaçao and Aruba were given a boost with the discovery of oil in nearby Venezuela in the early 20th century and the building of an oil refinery by Royal Dutch Shell on Curaçao in 1917 (one of the largest in the world). Oil formed the backbone of the local economy until the 1980s, when Venezuela opened its own refineries. The Dutch government bought the refinery on Curaçao to prevent it from closing and now leases it to the Venezuelan state oil company. Tourism is most important on Curaçao and Sint Maarten, both of which are lined with resorts for visitors, mostly from America and Venezuela. Duty-free shopping plazas are an additional lure in Willemstad, along with offshore banking, which first developed on the islands in World War II, during the occupation of the Netherlands by Nazi Germany.

TOURISM BOOM

Sint Maarten, the driest and flattest of the northern group, held little interest to European colonial powers, save as a source of salt because of the extensive salt ponds on the southern and western ends of the island. Peaceably divided between the Netherlands and France since 1648, the island was mostly ignored by outsiders until the tourism boom of the latter years of the 20th century. The population of the island has mushroomed, from about 1,500 on both sides of the island in 1950, to over 33,000 today in just the southern half. But duty-free zones and limited customs controls have caused the Dutch government concern over trafficking in drugs, weapons, and even people, as illegal immigrants attempt to take advantage of the numerous cruise ships that stop here (and in Curaçao) daily.

Willemstad is the largest town in the Netherlands Antilles (population 140,000). Other towns are Kralendijk on Bonaire and Philipsburg on Sint Maarten. The main towns of Saba and Sint Eustatius (The Bottom and Oranjestad, respectively) are much smaller. The Bottom has only 350 residents. About half the population is black (except on Saba); the population of Curaçao retains a significant degree of Amerindian (Arawak) blood, more than most other islands in the Caribbean.

BIBLIOGRAPHY. Brian W. Blouet and Olwyn M. Blouet, eds., *Latin America and the Caribbean: A Systematic and Regional Survey* (Wiley, 2002); David L. Clawson, *Latin America and the Caribbean. Lands and Peoples* (Times Mir-

ror Higher Education Group, 2004); Population Statistics, www.world-gazetteer.com (March 2004).

JONATHAN SPANGLER
SMITHSONIAN INSTITUTION

Nevada

KNOWN AS THE Silver State, the Battle-Born State, or the Sagebrush State, the mountain state of Nevada was admitted to the Union in 1864 as the 36th state. Covering an area of 110,540 square mi (286,297 square km), Nevada is 485 mi (780 km) north to south and 315 mi (507 km) east to west, making it seventh in size among the 50 states. Nevada has significant areas of land that remain uninhabited and under federal control.

Nevada, which means "snow-capped," ranks 35th in population The state is bordered on the north by OREGON and IDAHO, on the east by UTAH and ARIZONA, on the southeast by Arizona, and on the southwest and west by CALIFORNIA. Carson City is the capital of Nevada. The largest cities include Las Vegas, Henderson, Reno, North Las Vegas, Sparks, Carson City, Elko, Boulder City, Mesquite, and Fernley.

Nevada's arid/semiarid climate results in many days of sunshine broken by minimal precipitation, making it the driest state in the Union. Summers are long and hot, and winters are short and mild. Average temperatures range from a high of 70 degrees F (21 degrees C) in southern Nevada to 45 degrees F (7.2 degrees C) in the northern part of the state. The state's highest temperature of 125 degrees F (51.6 degrees C) was recorded in 1994 at Laughlin. Annual rainfall is a scant 4 in (10 cm) in drier areas, while other parts of the state may experience up to 40 in (101 cm) each year. Snow may fall at any time of the year in northern Nevada.

VARIED TERRAIN

Covered with more than 300 mountain ranges and a section of the Mohave Desert, Nevada's terrain is varied. Most of the state lies within the Great Basin, which also includes parts of Utah, California, Oregon, Idaho, and Wyoming. Geographically, Nevada is divided into the Columbia Plateau, the Sierra Nevada, and the Basin and Range Region. The Columbia Plateau is located in the northeastern section of Nevada, which is underlain with lava bedrock. Over time, water has shaped this bedrock into the colorful canyons and ridges for which the state is known. As Nevada nears Idaho, the land becomes open prairie.

To the south of Carson City, the Sierra Nevada Region is made up of rugged mountain ranges. Glacier-formed Lake Tahoe is located in one of the valleys of this region. The rest of Nevada falls within the Basin and Range Region, which contains over 150 mountains. Toiyabe and Toquima mountain ranges are located in the central section, while the Snake and Toano are found in the east. Nevada's other mountains include Boundary Peak, Wheeler Peak, Mount Charleston, and North Schell Peak. Single hills known as buttes and flat-topped mountains called mesas, along with geysers and hot springs, dot the landscape. Soil tends to be thin and alkaline. The highest point in the state is 13,140 ft (4,005 m) above sea level at Boundary Peak. The lowest elevation of 479 ft (146 m) above sea level is located at the southern part of the state at the Colorado River.

Nevada's principal rivers are the Humboldt, the Truckee, the Carson, the Walker, and the Colorado. Natural lakes are Lake Tahoe, Pyramid Lake, Walker Lake, Topaz Lake, and Ruby Lake. Nevada also has a number of man-made lakes, including Lake Mead, Lake Mohave, Lake Lahontan, and Rye Patch Reservoir.

A land of abundant natural resources, Nevada produces gold, silver, copper, zinc, brucite, magnesium, magnesite, manganese, tungsten, uranium, mercury, lead, titanium, oil, coal, iron, opal, barite, molybdenum, diatomite, talc, gypsum, dalomite, lime turquoise, fluorspar, antimony, perlite, pumice, salt, and sulfur. The state's major agricultural products include cattle, horses, sheep, hogs, poultry, hay, wheat, corn, potatoes, rye, oats, alfalfa, barley, vegetables, dairy products, and various fruits. Pinon pine, juniper, and fir are the major revenue makers for Nevada's timber industry. The state's chief manufactured products are food products, gaming equipment, monitoring devices, chemicals, aerospace products, lawn and garden irrigation equipment, and seismic and manufacturing equipment.

Nevada's wildlife includes mule deer, pronghorn antelope, bobcat, leghorn sheep, coyote fox, badger, rabbit, porcupine, muskrat, marmot, wild horses, and donkey. Lizards, tortoises, and snakes abound in the desert. Nevada's birds include thrush, horned lark, Nevada creeper, pheasant, partridge, and sage grouse.

Once known as the divorce capital of the UNITED STATES, Nevada has become synonymous with gambling and entertainment. Millions of tourists flock to

Reno and Las Vegas each year, and Lake Tahoe, on the border with California, has become one of Nevada's biggest tourist attractions.

BIBLIOGRAPHY. Gary Bedunnah et al., *Discovering Nevada* (Gibbs Smith, 1998); Larry Ford and Ernie Griffin, *Southern California Extended: Las Vegas to San Diego and Lost Angeles* (Rutgers University Press, 1992); "The Geography of Nevada," www.netstate.com (November 2004); James W. Hulse, *The Silver State: Nevada's Heritage Reinterpreted* (University of Nevada, 2004); "Nevada Facts" http://dmla.clan.lib.nv.us (November 2004); Ann Ronald, *Earthtones: A Nevada Album* (University of Nevada, 1995); Genny Schuma Smith, *Sierra East: Edge of the Great Basin* (University of California Press, 1999); "Welcome to Nevada," www.nv.gov (November 2004).

ELIZABETH PURDY, PH.D.
INDEPENDENT SCHOLAR

New Caledonia

NEW CALEDONIA IS A French protectorate located in southwestern Oceania, approximately equidistant from AUSTRALIA, NEW ZEALAND, and FIJI. It consists of the principal island, Grande Terre, as well as the outlying Loyalty Islands and their nearby atolls and reefs. Archaeological evidence suggests human settlement dates from approximately 4000 B.C.E. Distinct forms of pottery and imported obsidian objects found in New Caledonia indicate that its early human inhabitants had conquered long-distance ocean travel, and the islanders were part of a broad seafaring civilization prominent in MELANESIA and western Polynesia between 3000 and 3500 B.C.E.

The climate is tropical, with heat and humidity modified by southeasterly trade winds and seasonal typhoons. The terrain of Grande Terre is characterized by sandy coastal plains with rugged interior mountains. These heavily forested peaks cover some 40 percent of the country's land area. The country hosts more than 3,000 indigenous plant species, many of which are seen as potential sources for new pharmaceuticals. Its tropical rainforest is one of the most botanically diverse in the world. There are also some 4,300 fauna species, including unique birds and freshwater fish as well as bats, pigs, and lizards.

The population has traditionally been concentrated along the coast. The ethnically Polynesian and Melanesian population first came into sustained contact with Westerners when French Catholic missionaries arrived in 1843. FRANCE established its sovereignty over New Caledonia in 1853. The capital city, Port de France, changed its name to Noumea in 1866. Plans for establishing a French penal colony, similar to Britain's in Australia, were abandoned when it was discovered that the island had substantial deposits of valuable minerals.

The island's hillsides were initially developed as coffee plantations, but mining for nickel soon became the chief industry. The mountainous terrain doomed plantation agriculture, and by the mid-twentieth century animal grazing gained popularity. During the 1950s, labor migrants from Wallis Island came seeking work in the nickel mines and remained as permanent residents.

FRENCH RULE

After World War II, a modest nationalist movement began to develop. In PARIS, France, however, colonial officials reasserted French authority throughout its colonial empire, including Oceania. A referendum on independence held in 1958 produced an overwhelming popular endorsement of continued French rule in New Caledonia. This result was repeated in a similar referendum held in 1998. Nonetheless there were anti-colonial tensions. A lethal confrontation between native nationalists and French police at Uvea in April 1988 helped change the political situation. French and Caledonian delegates meeting in France signed the Matignon Accords, which granted substantial autonomy to New Caledonia. However, French remained the official language and French mining interests remained largely intact.

New Caledonia's international relations are generally harmonious. Nonetheless, its claims to Hunter and Matthew Islands have been contested. These islands are valued because of their nearby undersea oil fields and seabed minerals. New Caledonia has a multiethnic population, with native residents and those of French descent almost evenly divided at 40 percent of the total. Many minority groups, including Wallisians, Indonesians, and Vietnamese, are also resident in New Caledonia.

BIBLIOGRAPHY. Richard Aldrich, *France and the South Pacific since 1940* (University of Hawaii 1993); John Connell, *New Caledonia or Kanaky? The Political History of a French Colony* (National Centre for Development Studies, 1987); Michael Spencer, Alan Ward, and John Connell, eds.,

New Caledonia: Essays in Nationalism and Dependency (University of Queensland Press, 1988).

LAURA M. CALKINS, PH.D.
TEXAS TECH UNIVERSITY

New Delhi

NEW DELHI, by current estimates, has a population of about 300,000 persons, although Greater Delhi's population is estimated to be between 12 and 14 million, making it the third-largest urban settlement in INDIA. Located at the western side of the Gangetic Plain in India, New Delhi is also located next to the city of Delhi, often referred to as Old Delhi. Both of the cities, however, while being in proximity to each other, have very different histories and urban forms: Old Delhi is an urban sprawl filled with narrow, winding streets, forts, bazaars and mosques; the city was the Muslim center of India from the 17th to 19th centuries, although the origin of the present city begins in the 12th century. New Delhi, on the other hand, is greatly different because of its much shorter history and its urban morphology of a monumental context—broad, straight, tree-lined Baroque-style avenues, large-scale governmental buildings, and open spaces. While Old Delhi may be said to be chaotic in form and life, New Delhi is a place of order. Covering an area of almost 579 square mi (1,500 square km), New Delhi is a vast city in spatial extent.

In terms of its history, New Delhi's origins are closely associated with the rise of the BRITISH EMPIRE in India, the declining power of the Mughal Dynasty, and the British government's decision in the early 20th century (1911) to establish a new imperial capital city in India. The previous capital city was Calcutta in the northeast of the country. In January 1931, New Delhi was inaugurated as the new capital of India.

The task of planning and designing the new city was given in 1912 to architects Sir Herbert Baker and Edwin Lutyens (Lutyens was prominent in the development of the English vernacular house design, and Baker was an architect who had designed widely in colonial SOUTH AFRICA). The city plan of New Delhi is similar in form to 19th-century PARIS and Washington, D.C., having a geometric-shaped urban form dominated by long, wide avenues that lead up to nodes created by significant public buildings or spaces, such as the India Gate, the Parliament Building, and Rashtrapati Bhawan, the official residence of the president of India, once the British viceroy's house.

The most important of the city's roads is the Rajpath, a monumental boulevard used for ceremonial parades that provides a huge vista toward the former British buildings on the top of Raisina Hill. Architecturally, the grand edifices of the British were designed to explicitly show imperial power, hence classical forms were employed. Despite India gaining independence from the British in 1947, New Delhi's capital status was confirmed by the Indian government in 1950, and with its increasing size and cultural importance, the city was granted state status in 1992.

Today, Delhi, that is, Greater Delhi, is the richest city in India because of it is the economic, trade, and industrial center of northern India. Despite planning restrictions, the last decades have seen a major increase in industrial development; among the most important local products are chemicals, clothes, and electrical and electronic goods. Engineering and banking are also of growing significance.

BIBLIOGRAPHY. Gavin Stamp, "Lutyens: New Delhi and the Monumental," *New Delhi News* (June 1, 1981); Lawrence James, *Raj: The Making and Unmaking of British India* (St. Martin's Press, 1999).

IAN MORLEY
MING CHUAN UNIVERSITY, TAIWAN

New Hampshire

LOCATED IN THE northeastern UNITED STATES, the state of New Hampshire has a total area of 9,304 square mi (23,380 square km) of land and 277 square mi (23,3380 square km) of inland water, with its geographic center lying in Belknap County, 3 mi (5 km) east of the town of Ashland. New Hampshire is bordered on the north by the Canadian province of Quebec, on the east by state of MAINE, on the south by MASSACHUSETTS, and on the west by VERMONT. New Hampshire is one of six New England states (the others being Maine, Massachusetts, Vermont, RHODE ISLAND, and CONNECTICUT). New Hampshire's highest point is Mount Washington, which rises to 6,288 ft (1,918 m) above sea level, and measures 190 mi long by only 70 mi wide. The state's population is 1,185,000. New Hampshire has a small area along the ATLANTIC OCEAN, with the seaport at Portsmouth.

Five of New England's major streams originate in the hills of New Hampshire, which has resulted in the state's being nicknamed the "Mother of Rivers." The Connecticut River rises in its north; the Merrimack River rises in the Franconia Mountains; the Piscataqua River forms at Dover; and the rivers Androscoggin and Sacco flow east to Maine, becoming two principal rivers of that state. New Hampshire has some 1,300 lakes and ponds and about 40 rivers.

New Hampshire has a dynamic climate. Its proximity to the ocean and its mountains, lakes, and rivers keep its temperature in almost constant flux. The state experiences all four seasons, with wonderful autumn foliage and long, cold winters. Some of the coldest temperatures and strongest winds ever recorded in the continental United States have been observed in New Hampshire. Spring and summer, on the other hand, are short and cool.

New Hampshire contains many forests, abundant with all types of flora, such as elm, maple, beech, oak, pine, hemlock, and fir trees, as well as rare forms such as the balsam fir, willow, dwarf birch, Labrador tea, and Alpine bearberry. The state is also known for its wildlife. Mammals include white-tailed deer, muskrat, beaver, porcupine and snowshoe hare. Some of New Hampshire's wildlife is endangered, including the bald eagle, lynx, and Atlantic salmon.

New Hampshire's nickname is the "Granite State" because of its many granite mountains. Tourist attractions bring 1 million visitors every year, visiting mountain, lake and seashore scenery. Because of its fertile soil, New Hampshire is also famous for its horticulture, such as apples, strawberries, blueberries and peaches. New Hampshire is known as a socially conservative state with an independently minded populace that reflects its motto, "Live Free or Die."

BIBLIOGRAPHY. "The New Hampshire Almanac," www.state.nh.us (November 2004); R. Conrad Stein, *New Hampshire* (Scholastic Library, 2000); "New Hampshire," www.netstate.com (November 2004).

LENA DAHU AND ANTHONY CHIAVIELLO
UNIVERSITY OF HOUSTON, DOWNTOWN

New Jersey

THE STATE OF New Jersey, located on the east coast of the UNITED STATES between NEW YORK, PENNSYLVANIA, and DELAWARE, is home to hundreds of flora and fauna. The landscape of the state is comprised of pine forests and flatlands in the interior, swampy meadowlands in the north and the dry Pinelands in central and southern portions of the state. New Jersey is located in between the northern and southern region of the United States. The state has a moderate climate, hundreds of miles of coastline, and distinct geographic regions, and it is home to over 500 kinds of animals such as frogs, deer, raccoons, and dolphins.

The state is 166 mi (267 km) long from north to south and the greatest width is 65 mi (104 km) east to west. These dimensions result in a wide variety of weather, but overall, New Jersey has a moderate climate, with cold winters and hot summers; temperatures reach at times over 100 degrees F (38 degrees C) in the summer, and drop on occasions to around 0 degrees F (-18 degrees C) in the winter. In 1524, Giovanni da Verrazano was reportedly the first to visit the New Jersey coast. In 1609, Henry Hudson and his crew landed in New Jersey and explored the undiscovered land. The Dutch West India Company was chartered in 1621, and these merchants created settlements in New York and New Jersey.

In 1664, the land in New Jersey was included in a grant Charles II gave to his brother James, the Duke of York. James gave the land to two of his friends, Lord Berkeley and Sir George Carteret. Carteret was formerly the governor of the island of Jersey off the coast of England. Carteret sent his relative Philip Carteret to lead a group of emigrants to his land, and they created a settlement at Elizabethtown.

Settlers began to arrive from New England and they founded Newark. Carteret formed a government in which religious liberty was granted to all Englishman and more began to arrive. In 1676, the province was divided into East Jersey and West Jersey. These two Jerseys ultimately became united and were placed under the governorship of New York and New England. In 1702, New Jersey became a royal province united with New York. This lasted until 1738, when the two colonies were finally separated.

By 1760, 75,000 people occupied the colony of New Jersey. Almost all were farmers from English descent. As New Jersey, along with the 12 other American colonies, began to protest certain British measures, a rebellion against British rule seemed inevitable. The 13 colonies worked together to stage a joint resistance, and by April 1775, the war for independence was under way. For the next eight years, the colonists in New Jersey witnessed a number of battles on their soil,

including fights in Trenton and Monmouth. George Washington's Continental Army camped out in Morristown in the winter of 1779–80 and marched up and down the colony throughout the war. After an astounding victory against the British in 1783, representatives from New Jersey and the newly freed colonies convened four years later in Philadelphia, Pennsylvania, for the Constitutional Convention. On December 18, 1787, New Jersey ratified the constitution and became the third state in the United States.

Between 1790 and 1820, as the United States grew, New Jersey seemed to develop at an extremely slow pace. The population rose from 95,000 to 227,500 between this period, and its agriculturally based economy lagged behind as the United States entered the INDUSTRIAL REVOLUTION. During the Civil War, thousands of troops from New Jersey fought for the Union. After the war, New Jersey grew at an incredible rate: 1,883,669 people inhabited the state in 1900, and factories increased by 230 percent. Railroads were developed, and Jersey City became a major transportation center. Immigrants began to populate the larger cities, and the face of the state underwent a major transition.

SUBURBS

After World War II, the American people began to settle into suburbs of major cities. Towns in New Jersey became major suburbs for New York City and Philadelphia. Between 1950 and 1960, the state's population dramatically rose by 1.2 million. Farm owners were selling their land to developers, and homes were built at a staggering rate. New Jersey also became home to a multitude of research centers.

Ethnic groups in the larger cities such as Newark were also changing. Over 130,000 African Americans moved into Newark between 1950 and 1970. Many of these citizens faced unemployment and extremely difficult living conditions. In July 1967, the city of Newark seemed to explode as rioting and burning broke out: 23 people were killed and $10 million of property was destroyed. Around 23,000 private jobs were lost between 1967 and 1972. Into the 2000s, revitalization projects are occurring in Newark, Jersey City, and Camden, and the shoreline (Atlantic City and its casinos) remains a vacation spot for many.

BIBLIOGRAPHY. Thomas Fleming, *New Jersey: A History* (W.W. Norton, 1977); Richard P. McCormick, *New Jersey from Colony to State* (Rutgers University Press, 1964); Henry William Elson, *History of the United States of America* (Macmillan, 1904); New Jersey Climate Publications, "The Climate of New Jersey," climate.rutgers.edu (May 2004); NJAS: New Jersey's Wildlife, www.njaudubon.org (May 2004).

GAVIN WILK
INDEPENDENT SCHOLAR

New Mexico

NEW MEXICO IS THE "Land of Enchantment." Nowhere else in the UNITED STATES can such a variety of landscapes, histories, and traditions be found. Located in the American southwest, New Mexico is part of the dry and warm Four Corners region. Although primarily arid in the southern deserts to semi-arid in the high deserts, there are also some pine-studded mountain ranges and High Plains GRASSLANDS. New Mexico is bordered to the north by COLORADO, to the west by ARIZONA, to the south by MEXICO and TEXAS, and to the east by OKLAHOMA and Texas.

The RIO GRANDE flows along a rift valley down the state's center, providing a study of contrasts. From Albuquerque's verdant bosque along the river, one can look east to the snow-capped Sandia mountains or west to dry buttes harboring Anasazi petroglyphs. Capulin, a volcano extinct now for 10,000 years, rises from the ancient beds of this former inland sea. The Malpais (badlands) are lava flows found in several parts of New Mexico. They are are difficult to cross because of the rough rock and are often threaded by lava tubes later used by locals for transportation routes or housing. A visitor can climb up the white gypsum sands of White Sands National Monument and gaze at the uniquely blue skies, or descend deep into the Earth to navigate the dark and winding Carlsbad Caverns.

The Rio Grande is not the only wayfinder through this land. Ancient trails of the Anasazi Indians can still be found in Mesa Verde, Chaco Canyon, and the Aztec ruins. Several trails used by Spanish explorers cross several parts of the state, like the old Santa Fe Trail leading from Independence, MISSOURI to the oldest capital in North America (Santa Fe), and the trail leading to Morro Rock where 17th-century explorers carved their names. Along several valleys are old cattle drive trails used to bring Texas and New Mexican cattle first up to the gold miners of the Rocky Mountains and later to the railroads in Colorado for shipment to the east. Interstate 40 parallels the famous Route 66 that once connected Chicago to Los Angeles.

Histories are made of the interplay between cultures and the landscape they occupy. The first inhabitants of New Mexico were the Anasazi, a collective name for the first indigenous peoples who arrived in the area around 200 C.E. By the time the Spanish arrived in the mid-1500s and established Santa Fe as their capital, numerous Indian tribes lived in New Mexico. The Spanish had established a stronghold before the Anglo migrants from the East arrived. Today New Mexico reveals the results of these migrations as the state incorporates the multi-faceted heritage of the many Indian, Latino, and Anglo cultures.

New Mexico proudly celebrates its many traditions. There is Albuquerque's International Balloon Fiesta, the Inter-Tribal Indian Ceremonial near Gallup, Santa Fe Indian Market, Taos Valley Acequia Festival, and Hatch Valley Chili Festival, among others. New Mexico has long been at the forefront of innovation through the Los Alamos and Sandia labs (nuclear research), very large array (VLA) radio telescope (astronomical research), and the White Sands Missile Range (space flight research). The state has a growing information technology industry and is moving to the forefront in wind and solar power research.

BIBLIOGRAPHY. Harm J. de Blij and Peter O. Muller, *Geography: Realms, Regions and Concepts* (Wiley, 2002); *New Mexico* (Lonely Planet, 2002); *Merriam-Webster's Geographical Dictionary* (Merriam-Webster, 2003).

LAUREL E. PHOENIX
UNIVERSITY OF WISCONSIN, GREEN BAY

New York

NEW YORK—OFTEN called New York State to distinguish it from NEW YORK CITY—is a northeastern state of the UNITED STATES, bordering VERMONT, MASSACHUSETTS, and CONNECTICUT to the east; PENNSYLVANIA and NEW JERSEY to the south; and the Canadian provinces of Ontario and Quebec to the north. Lakes ERIE and ONTARIO form the greater part of the boundary between New York and Ontario. Lake Champlain forms more than half of the boundary between New York and Vermont.

New York is roughly triangular in shape, with angles jutting north to Quebec, west to Lake Erie, and southeast toward Long Island. New York's highest point is Mount Marcy, which reaches 5,344 ft (1,629 m) in the Adirondack Mountains of the northeastern part of the state. The total area of New York is 54,556 square mi (141,299 square km), which makes the state 27th in size among thc 50 states.

From 1810 through the U.S. census of 1960, New York was the largest state in population. New York's estimated 2003 population totals 19,190,115, approximately 4 out of 10 of whom live in New York City. If New York City's Long Island and northern suburbs are included, almost two-thirds of the state's population lives in the New York City metropolitan area.

New York's capital is Albany; its largest cities are New York City (with a 2000 census population of 8,008,278), Buffalo (292,648), Rochester (219,773), Yonkers (196,086), and Syracuse (147,306). New York State, and in particular New York City, dominates the nation in finance and in the publishing and fashion industries. Agriculture is important to the upstate New York economy, particularly milk production, the growing of fruits and vegetables, and the production of wine. Tourism is of great importance to New York State, with attractions ranging from Niagara Falls to the Finger Lakes, from the Adirondacks to New York Harbor's Statue of Liberty.

New York is the only state to possess shoreline on both the ATLANTIC OCEAN and the Great Lakes. The state includes four islands just offshore from the mainland. Fishers Island lies off the Connecticut coast. Long Island, which includes parts of New York City, extends eastward from New York Harbor more than 110 mi (177 km) and separates Long Island Sound from the open Atlantic. The island of Manhattan, including the business and tourist sections of New York City, lies across the Hudson River from New Jersey and across the East River from Long Island. Staten Island, which also comprises a portion of New York City, lies off the mainland of New Jersey, across New York Harbor from Long Island. These islands form part of the terminal moraine deposited by the last of the Pleistocene continental glaciers. The valley of the HUDSON RIVER, which flows into New York Harbor at New York City, extends north to the Adirondacks. The Hudson's channel, carved out to form a fjord by the Pleistocene glaciers, is a tidal stream all the way up to Albany, more than 150 mi (241 km.) from the river's mouth.

North and west of the Hudson Valley are New York State's two major mountainous areas, the Adirondacks and the Catskills. The Mohawk Valley, containing the Mohawk River (the major tributary of the Hudson), separates the Adirondacks and Catskills and stretches almost 100 mi (160 km) west from Albany to

the Lake Ontario plain. Southwestern New York is a portion of the Appalachian Plateau, into which the Pleistocene glaciers carved the Finger Lakes, the largest of which are Seneca and Cayuga. Lake Erie empties into Lake Ontario through the Niagara River, which forms part of the Ontario-New York border. The difference between the elevations of these two Great Lakes, together with the large volume of water carried by the river, accounts for a spectacular vertical drop of 182 ft (55 m) at Niagara Falls.

NEW YORK STATE

Two groups of Native Americans inhabited New York State when the state was first visited by Europeans. Algonquin peoples inhabited the southeastern portions of the state. The Iroquois Confederacy (an alliance of the Seneca, Cayuga, Onondaga, Oneida, and Mohawk) dominated central and western New York. The first European settlement was established by the Dutch near Albany in 1624. This settlement soon become part of the Dutch colony of New Netherland, established in 1626 with the founding of the settlement of New Amsterdam at the southern tip of Manhattan Island. The colony of New Netherland capitulated in 1664 to the British, who changed the name of the city and colony to New York.

New York State played a major role in the American Revolution; the American victory at Saratoga is generally considered the decisive battle of the war. New York State grew rapidly throughout the 19th century, especially after the completion of the Erie Canal through the Mohawk Valley and west to Buffalo in 1825. The canal provided the easiest, most level transportation link between the northeastern states and the rapidly developing Midwest, and the commerce and influence channeled by the canal soon justified New York's "Empire State" nickname. The Erie Canal also prompted other advances that helped link the expanding American empire closer together: the New York Central Railroad, the shipping businesses that grew to become Wells-Fargo, and the telegraph industry (the profits of which would eventually endow Ithaca's Cornell University).

In the 20th century, New York State, in spite of increasing competition from rapidly growing CALIFORNIA and TEXAS, was able to maintain a strong influence over the economic and cultural life of the nation.

BIBLIOGRAPHY. David M. Ellis, *A History of New York State* (Cornell University Press, 1967). Milton M. Klein, ed., *The Empire State: A History of New York* (Cornell University Press, 2001); John H. Thompson, ed., *Geography of New York State* (Syracuse University Press, 1977).

JAMES A. BALDWIN
INDIANA UNIVERSITY-PURDUE UNIVERSITY

New York City

"THE BIG APPLE," covering 303 square mi (785 square km), is located at the southernmost point of New York State. It is bounded to the north by Westchester County, to the east by Nassau County, and to the west by the state of NEW JERSEY, across the HUDSON RIVER. The city is comprised of five counties, known as boroughs: Manhattan (New York County), Brooklyn (Kings County), Queens (Queens County), Staten Island (Richmond County), and the Bronx (Bronx County). Each county has a diverse geography and history. All boroughs except for the Bronx are either islands unto themselves (Manhattan and Staten Island) or part of islands (Brooklyn and Queens). Manhattan is 13.4 mi (21.6 km) long and 2.3 mi (3.7 km) wide at its widest point and is located between the East River on the east side and the Hudson River on the west side. The northern reaches of Manhattan are separated from the Bronx by the Harlem River. In total, New York City has 578 mi (930 km) of waterfront.

Most of Manhattan island is near sea level and relatively flat. Uneven terrain used to mark much of the island, but as development continued, the land was flattened. The northernmost tip of the island is hilly and contains Fort Tryon Park and Inwood Hill Park. The bedrock of Manhattan makes it excellent for the deep foundations needed to construct major skyscrapers. This bedrock is closest to the surface in midtown Manhattan and again in downtown Manhattan, explaining the lack of very high skyscrapers between 14th Street and the Wall Street area.

New York City was the site of the terminal moraine during the last Ice Age; the southernmost point to which the glaciers advanced. New York City owes many of its geological features to these advancing and receding glaciers. Though it is the most populous city in the United States, New York has an impressive amount of parkland. The 840-acre (340-hectare) Central Park was created in the mid-19th century in the middle of Manhattan by the architects Fredrick Law Olmsted and Calvert Vaux. Completion of the entirely man-made park involved the planting of 5 million trees

and the creation of numerous lakes and ponds. Other major parks include Prospect Park in Brooklyn, and Forest Park in Queens. An intricate aqueduct system brings potable water to the city from reservoirs in the Catskill Mountains.

The original area of Manhattan Island was smaller, but over the course of the years demand for prime real estate caused landfill to be added to create a wider lower Manhattan. Water Street, on the east side of the island, was originally right on the water but is now two blocks inland. Battery Park City, on the west side of the island, is also on landfill. A 70-acre (28-hectare) freshwater body called the Collect Pond was filled by the early 18th century. The northernmost part of central Queens was also landfilled to create Flushing Meadows Park, site of the 1939–40 and 1964–65 World's fairs. Wildlife is sparse in the city, but the manmade Jamaica Bay Wildlife Refuge in southern Queens harbors many species of birds. The least populated borough, Staten Island, is also the hardest to reach. It is accessible from Manhattan by ferry or from Brooklyn by the Verrazano-Narrows Bridge. One of the largest airports in the country, Kennedy International Airport, is located in southern Queens.

Manhattan is a destination for many tourists from around the country and around the world. Attractions include the Empire State Building, the Theater District (Times Square), historic South Street Seaport, Macy's Herald Square (the world's largest department store), the Statue of Liberty, and numerous museums including the Metropolitan Museum of Art. One of the nation's most prestigious universities, Columbia, is located in northern Manhattan; dozens of other colleges and universities are located throughout the city, including New York University and the exclusive Cooper Union, an engineering school.

Manhattan and the outer boroughs are connected by several bridges and tunnels. Manhattan and New Jersey are connected by the Lincoln Tunnel and the George Washington Bridge. Numerous rapid transit lines (including subways and commuter railroads) also cross under the East and Hudson rivers, transporting commuters to and from the suburbs as far away as southern New Jersey and Connecticut.

Manhattan's rich heritage is evident in its many different neighborhoods with distinct geographies and traditions. Some of the best known and most colorful include Greenwich Village, TriBeCa (triangle below Canal Street), SoHo (South of Houston Street), Chelsea, Hell's Kitchen, Little Italy, and Chinatown. Different neighborhoods have been trendy over the years. For example, Greenwich Village has nurtured numerous artists and writers during the 1950s and 1960s. In the outer boroughs, present-day neighborhoods such as Flushing and Jamaica were actually separate towns until consolidation with Manhattan.

Most of Manhattan's streets were designed in a numbered north-south grid pattern, called the Randall Grid, that was first developed in the early 19th century. South of 14th Street, the street layout is more irregular. In the Wall Street area at the southern tip of the island, the streets are exceptionally narrow and winding.

In 1524, the Italian explorer Giovanni da Verrazzano, sailing from Europe, discovered New York harbor. The explorer Henry Hudson sailed up what is now the river that bears his name in 1609. An explorer named Adrien Block spent about six months on the site of the future city when his ship caught fire and was destroyed, but the first permanent settlers did not arrive from Holland until 1625. By 1628, there were about 270 people living in the settlement, then known as New Amsterdam. The four other boroughs were also

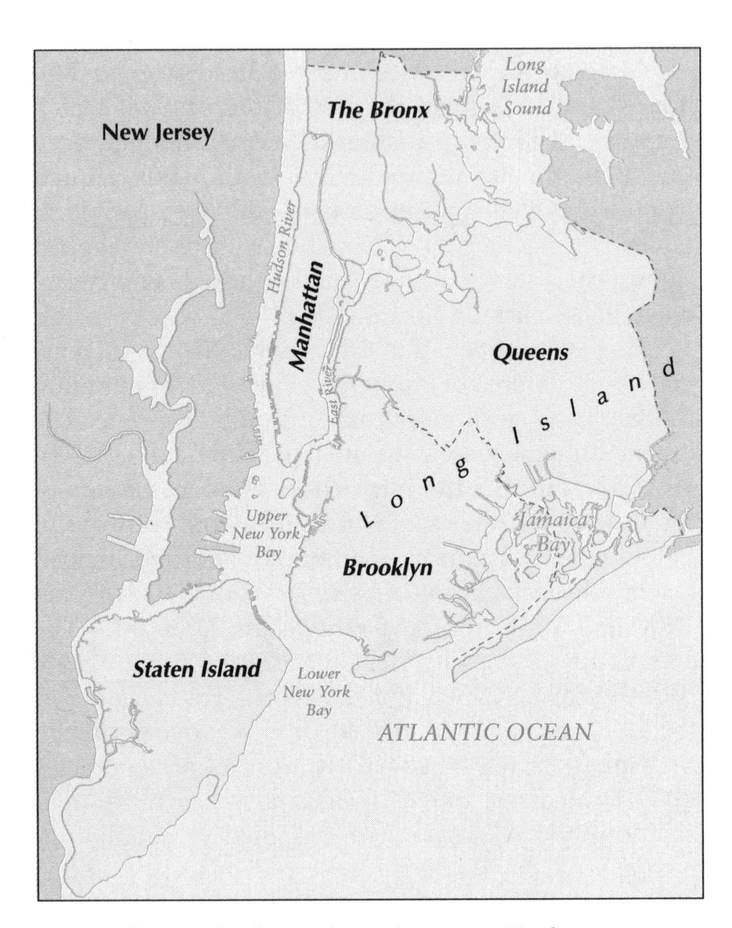

New York City's five boroughs are known as: Manhattan, Queens, Brooklyn, the Bronx, and Staten Island.

settled during the 1600s, by Dutch and English colonists.

Violence broke out between Manhattan settlers and the local Native Americans in 1643. In 1653, the Dutch built a protective wall across the island at the site of what is now Wall Street. Other surviving street names are indicative of the Dutch era; Beaver Street is named after the animal that fur traders sought in the 17th century, and Beekman Street is named after an early Dutch mayor of the city. In 1664, several English warships sailed into New York harbor and peacefully took over the city. The city was briefly restored to Dutch rule from 1673 to 1674.

By 1700, the prosperous city still looked very Dutch, with numerous step-gabled buildings. Settlement of the city had begun at the southern tip of Manhattan Island and proceeded north. With each passing year, the city limits crept further and further north and farmland was turned into homes. By 1760, the population had reached 18,000. The English occupied the city during the American Revolution. After the war, New York City briefly served as the capital of the nation and the first president, George Washington was inaugurated there.

A great fire in December 1835 destroyed 700 houses and with them the last of the original Dutch buildings. Thanks to Irish and German immigration, the population grew rapidly during the 19th century, from 300,000 people in 1840 to 800,000 people in 1860. During the Civil War, the city's poor revolted against the military draft. Several hundred people were killed in the ensuing riots in 1863.

Ellis Island opened in 1892 to handle the increasingly large volume of immigrants; in 1907 immigration reached its peak of more than 1.2 million people. The outer boroughs were consolidated with Manhattan in 1898 to create Greater New York. The construction of the Brooklyn Bridge in 1883 and the first subways in 1904 made transportation among the boroughs much faster. The population of the city doubled between 1900 and 1930, reaching more than 7,000,000. The completion of the Chrysler Building and Empire State Building in the 1930s and the World Trade Center in the 1970s helped create the most impressive skyline in the United States. The terrorist attacks of September 2001 leveled the Twin Towers and killed more than 2,700 people. Planners soon designed an equally imposing Freedom Tower to stand in its place.

BIBLIOGRAPHY. W. Parker Chase, *New York: The Wonder City 1932* (New York Bound, 1983); Susan Elizabeth Lyman, *The Story of New York* (Crown Publishers, 1975); *Let's Cover the Waterfront* (Circle Line Sightseeing Yachts, 1965); Henry Moscow, *The Street Book: An Encyclopedia of Manhattan's Street Names and Their Origins* (Fordham University Press, 1990); Floyd M. Shumway, *Seaport City: New York in 1775* (South Street Seaport Museum, 1975); Frank D. Whalen, Wallace West, and Claudia West, *New York Yesterday* (Noble and Noble, 1949); Gerard R. Wolfe, *New York: A Guide to the Metropolis* (McGraw Hill Book Company, 1988); Richard Saul Wurman, *NYC Access* (Access Press, 1989)

Richard Panchyk
Independent Scholar

Newton, Isaac (1642–1727)

SIR ISAAC NEWTON WAS one of the most famous and influential men in the world of science, in both mathematics and natural philosophy. His laws of gravity and motion formed the basis of classical mechanics, principles that are at the heart of modern engineering, physics, and astronomy. His work provided a mathematical mechanism to prove earlier theories of heliocentrism—the sun-centered system—and allowed later scientists to correctly determine the orbits of planets, comets, and even galaxies. Newton is also known in other areas of science and mathematics for his work on optics and differential calculus.

Newton was born in Woolsthorpe, England, in the rural eastern county of Lincolnshire. He was educated at Trinity College, Cambridge, where he continued as a professor from 1667 until his resignation in 1701. The years just prior to this appointment, however, were among his most productive. During this time he formulated his ideas on calculus, as well as his earliest thoughts on gravity and optics. He analyzed the pull of the planets around the sun, the moon around the Earth, and the pull of the Earth on everyday objects, such as Newton's proverbial apple. From these observations, he was able to calculate the force applied to these bodies or objects, as well as their orbit, velocity, and mass. This eventually led him to conclusions about a universal force applicable to all things, large and small, a force that he named gravity, for the Latin word for weight, *gravitas*.

The relevance of these laws to geographers in particular lies in Newton's theories on the shape of spinning spherical objects (like the Earth), and on the tides.

For the first time, the tides could be mathematically explained through application of Newton's law of universal gravitation.

His first publications were in the area of optics, examining the diffusion of white light into colors of the spectrum through a prism. One of his chief contributions to science was not necessarily in the content of these early publications, but in their presentation, which relied on empirical observations and experimentation alone, rather than mixing them with hypothesis, which had been the accepted practice since the days of Aristotle.

Newton developed a more practical telescope and theorized about the nature of light, considering it to be made up of particles, each affected by gravity. Today, we understand light more as a series of waves, not particles, but the material is again worthy as pure observed data. His system for advanced mathematical calculations, known as the calculus, provided the means for scientists to test and prove ideas that until then existed only as hypotheses. The German mathematician Gottfried Wilhelm Leibniz also created such a system, concurrently and independently, and the credit for the initial idea was heatedly disputed by Newton, and clouded his scientific relationships (and creative output) for the rest of his life.

His work in mathematics and optics quickly became popular among the scientific community and he was named a Fellow of the Royal Society in 1672. It was the publication of his major work in 1687, the *Philosophiae Naturalis Principia Mathematica*, however, that made Newton a household name, not just in England, but across Europe. This was an extension of his first work on the laws of motion, *De Motu Corporem* ("On the Movement of Bodies"), published in 1684.

The *Principia* established the three universal laws of motion that would not be improved upon for the next 300 years. In general, the work is a unification of numerous isolated physical facts developed by previous natural philosophers, but codified by Newton into a satisfying system of laws. The work also presents analysis of the speed of sound in air, and preliminary thoughts on the laws of thermodynamics. There were three reprints of the *Principia* in his lifetime, and numerous others in the centuries to follow. As Newton's fame spread, he was elected president of the Royal Society in 1703, and associate of the French Académie des Sciences. He was knighted by Queen Anne in 1705, and was buried with full pomp and ceremony in Westminster Abbey, England, in 1727.

In addition to his academic career, Newton was also briefly a member of Parliament for Cambridge (1689 and 1701) and served as master of the Royal Mint (1699), in charge of reorganizing the system of British coinage. Newton's work as a scientist cannot be entirely separated from his intense, lifelong passions for both religion and alchemy. He believed firmly that gravity explained the motions of the planets but could not explain who got this motion started. Many of his writings later in life were religious tracts dealing with the literal interpretation of the Bible. Secretly a Unitarian, he disbelieved in the Trinity, so much of this work was published posthumously. As an alchemist, he was very interested in matters of the occult and in ideas of the attraction and repulsion of particles.

The life of Newton overall reflects a fundamental shift in values in Western Europe across the 17th century, the period now known as the Scientific Revolution. Whereas Galileo's work on the movement of celestial bodies had nearly got him burned at the stake in the first third of the 17th century, less than a century later, Newton's work in the same area earned him universal praise.

BIBLIOGRAPHY. Gale Christianson, *In the Presence of the Creator: Isaac Newton and His Times* (Simon & Schuster, 1984); James Glieck, *Isaac Newton* (Knopf, 2003); G. Holton and S. Brush, eds., *Physics, The Human Adventure* (Rutgers University Press, 2001); "Newton Biography," www.newtonproject.ic.ac.uk (August 2004).

JONATHAN SPANGLER
SMITHSONIAN INSTITUTION

New Zealand

Map Page 1127 **Area** 107,737 square mi (268,680 square km) **Population** 3,951,307 **Capital** Wellington **Highest Point** 12,283 ft (3,754 m) **Lowest Point** 0 m **GDP per capita** $20,100 **Primary Natural Resources** natural gas, iron ore, sand, coal.

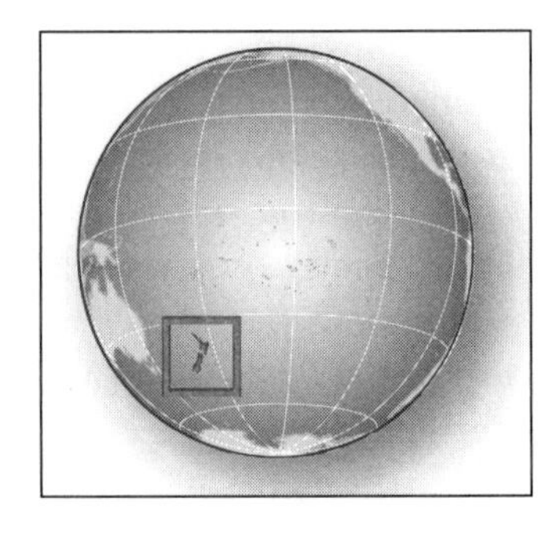

IN THE SOUTH PACIFIC OCEAN, the country of New Zealand lies to the southeast of AUSTRALIA. However, while Australia is continental in size and largely desert

(or outback) in landscape, the land of New Zealand consists almost entirely of two islands (unimaginatively named North Island and South Island) separated by the 20-mi- (32-km-) wide Cook Strait. Quite small compared to Australia, about the size of COLORADO in area, New Zealand displays a wealth of landscapes, usually rich in scenic beauty.

South Island is headed by the Southern Alps, a string of mountain peaks running for about 300 mi (500 km). Much of New Zealand was created from volcanic action. The islands lie along the collision lines of tectonic plates (Pacific and Australian). The resulting geologic activity has caused volcanic eruptions throughout New Zealand's history. The North Island retains most of the country's volcanic activity. There, in 1886, Mt. Tarawera's especially violent eruption was heard hundreds of miles away but locally buried the prosperous neighboring village of Te Wairoa. This general region is now known as Rotorua, where continued geothermal activity creates a wondrous and startling landscape. Similar to the geothermal features found in Yellowstone Park in the UNITED STATES, the elements of Rotorua include geysers, mudpots, hot pools, and more. In contrast, the Southern Alps often feature glaciers, most notably the Tasman Glacier.

On both islands the presence of mountains prompts OROGRAPHIC PRECIPITATION, which, coupled with the moist air inherent to the island setting, produces plentiful rainfall throughout nearly all of the country. Many areas are intensely forested, with the western regions of South Island having RAINFORESTS with 80 to 100 in (203 to 254 cm) of rain annually. Only the Canterbury Plain receives less than 30 in (76 cm) of precipitation yearly. As a group of islands, New Zealand abounds in beaches, with 9,404 mi (15,134 km) of coastline.

Because New Zealand climate is mainly an oceanic temperate climate, the far northern reaches are nearly subtropical and New Zealand features very successful commercial agriculture. The kiwi fruit was introduced in about 1900; New Zealand now harvests the fruit in world-leading numbers. Apples and other fruits also are exported, as well as wine produced from New Zealand grapes. Boosted by the advent of refrigerated ships in the 1880s, New Zealand farmers were able to ship dairy products and meat as far as England. Sheep and cattle grazing expanded rapidly, to the current numbers of about 60 million sheep and 8 million cattle, impressive figures for a country whose human population is less than 4 million. Overall, agriculture yields about 30 percent of the country's export earnings. The industrial and service sectors of the economy have grown rapidly in recent years. An "Australianization" of the economy, via rapid expansion of Australian companies into the New Zealand market, has become a cause of concern for many New Zealanders.

Settled originally by Polynesians, the arrival of Europeans, principally British, prompted accelerated growth of the population. Known as the Maori, the Polynesian sailors reached New Zealand from the north about 1,000 years ago. These early settlers preferred the climate and setting of the North Island, where modern day Maori also tend to live. Official European discovery came in 1642 with the voyage of Dutch captain Abel Tasman. Attempting to reach the great southern land (Australia), Tasman sailed too far south (hitting the island now known as Tasmania) and too far east, reaching New Zealand (FIJI and TONGA too). British settlement was spurred by the voyages of Captain James Cook in the 1760s. British sovereignty was established by the 1840 Treaty of Waitangi, though sporadic conflict with the natives culminated in the British-won Maori Wars of the 1860s. A gold rush during the 1870s brought a surge in the population. Self-rule was granted by the British in 1852, dominion status was given in 1907, and formal independence was established in 1947.

Maori now account for about 10 percent of New Zealand's population. As a modern developed society, New Zealand currently experiences slow population growth of 1 percent annually.

BIBLIOGRAPHY. *Oxford Essential Geographical Dictionary* (Oxford University Press, 2003); *World Factbook* (CIA, 2004); "New Zealand," www.newzealand.com (April 2004); New Zealand Geography," www.innz.co.nz (April 2004); Statistics New Zealand, www.stats.govt.nz (March 2004).

JOEL QUAM
COLLEGE OF DUPAGE

Ngorongoro Crater

LOCATED IN NORTHERN TANZANIA, 80 mi (129 km) west of the city of Arusha is the Ngorongoro Crater. This crater is a caldera formed by the collapse of a large volcano in the Great RIFT VALLEY. It is considered to be the largest subsidence crater in the world with a width of over 12.5 mi (20 km). It is the fifth-largest caldera in the world, and of these, it is the

largest inactive, unbroken, unflooded caldera in the world. The rim varies between 7,480 ft (2,280 m) and 8,005 ft (2,440 m) in elevation, and the floor of the crater has an average depth of 2,000 ft (610 m) below the rim. The floor area of the crater covers 100 square mi (260 square km). The crater hosts several thousand tourists per year but has no actual human inhabitants at this time. No one seems to know the origin of the name Ngorongoro though the Maasia, a nomadic people living in the area, say it means "the great or big hole."

Besides the size of the crater itself, the crater floor has some unique features such as the Ngoitokitok springs that are a year-round hippo bath, two patches of woodland—the Lerai forest and the Laiyanai forest—the Munge River, several fresh and brackish ponds, and just west of the center of the crater, Lake Magadi. This lake is 20 mi (32 km) long and 2 mi (3.2 km) wide and exists thanks to the volcanic springs that feed it. These volcanic springs produce a large amount of carbonate of soda, creating a crust on the lake that is dredged and processed into soda ash to be used in glassmaking. Mainly though the crater floor is a wide grassy savanna.

Ngorongoro is located at the center of the Ngorongoro Conservation Area, making up 3 percent of the area covered by the park. The Ngorongoro Conservation Area was established as a World Heritage site in 1979.

The park is home to approximately 25,000 large mammals, including gazelles, buffaloes, wildebeests, elands, elephants, and the black rhinos. It is also has the densest populations of predators out of any of the African parks, made up of lions, leopards, hyenas, and jackals. Lake Magadi and Munge River attract large numbers of greater and lesser flamingos, pelicans, ostriches, grebes, storks, cranes, and more. The first conservator of Ngorongoro, Henry Fosbrooke, considered the Ngorongoro Crater to be the eighth wonder of the world, and many tourists, after visiting, would agree with him.

BIBLIOGRAPHY. John N. Kundaeli, "Ngorongoro Serengeti: An Irreplaceable Natural Heritage," *World Heritage Review* (UNESCO Publishing, June 1998); Henry Fosbrooke, *Ngorongoro: The Eighth Wonder* (Deutsch, 1972); Saul B. Cohen, ed., *The Columbia Gazetteer of the World* (Columbia University Press, 1998).

CHRISTY A. DONALDSON
MONTANA STATE UNIVERSITY

Nicaragua

Map Page 1136 **Area** 49,998 square mi (129,494 square km) **Population** 5,128,517 **Capital** Managua **Highest Point** 7,998 ft (2,438 m) **Lowest Point** 0 m **GDP per capita** $2,300 **Primary Natural Resources** gold, silver, cooper, tungsten, lead, zinc.

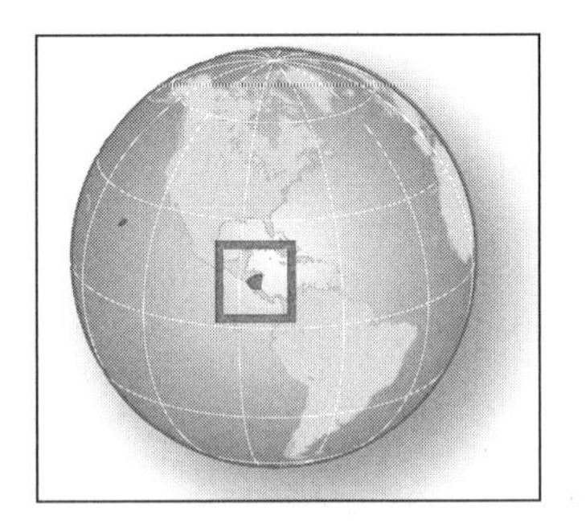

BORDERED by COSTA RICA and HONDURAS, Nicaragua is the largest country in Central America. The country is divided into three distinctive geographical regions: the Pacific Lowlands, the North-Central Mountains and the Caribbean Lowlands (also known as the MOSQUITO COAST). The Pacific Lowlands is a narrow strip of highly fertile land that is composed of approximately 40 volcanoes. North-Central is mountainous, but the northwestern part of the country includes RAINFOREST. Beyond the western mountain range, the western coast is lined with savannas.

Eastern Nicaragua is bordered by rainforest, lagoons, and swamps, and tropical diseases are common. Two large freshwater lakes, Lake Nicaragua (Lago de Nicaragua) and Lake Managua (Lago de Managua), dominate the southwestern landscape. Lake Nicaragua, which is the more southern of the two large lakes, is the largest freshwater body in Central America. Approximately 20 percent of Nicaragua is arable.

Nicaragua is prone to severe earthquakes, volcanic eruptions and particularly hurricanes. Hurricane Mitch, which struck the Atlantic coast in 1998, devastated the country, killing some 10,000 people, causing catastrophic mudslides, and destroying many bridges and roads. The country has also suffered several volcanic eruptions and a major earthquake since the hurricane, which has made the country slow to recover its economy.

Nicaragua's climate is tropical. The western part of the country is hotter and drier and experiences a rainy season between May and November. The Eastern region is rainy nine months of the year and is subject to hurricanes. Nicaragua's key environment concerns are deforestation, soil erosion, and water pollution.

Among Nicaragua's earliest settlers were the Aztec people, who migrated down from the Mexican lowlands during the 10th century. The Aztec culture remains an influence today. Christopher Columbus is believed to be the first European to see Nicaragua,

when he stumbled across the land during an expedition in 1502. The first European settlers were a Spanish exploratory mission that reached the southern shores of Lake Nicaragua around 1522. Nicaragua became a Spanish colony and remained so until gaining its independence from Spain in 1821. Nicaragua officially became an independent republic in 1838. After its independence, the country was headed by several conservative regimes. In 1934, it fell under the power of a repressive military regime, which held power until 1979, when the Sandinista rebel forces overthrew the military. The country held its first democratic elections in 1994.

Local governance of Nicaragua is divided into 15 administrative departments and two autonomous regions. The autonomous regions are Atlantico Norte and Atlantico Sur. The administrative departments are Boaco, Carazo, Chinadega, Chontales, Esteli, Granada, Jinotea, Leon, Madriz, Managua, Masaya, Matagalpa, Nueva Segovia, Rio San Juan, and Rivas.

Nicaragua is a developing nation. The country is a significant producer of cotton and coffee. There are also moderate gold and copper mining industries located in Nicaragua. However, an estimated 50 percent of the Nicaraguan population lives below the poverty level.

BIBLIOGRAPHY. Eduardo Crawley, *Nicaragua in Perspective* (St. Martin's Press, 1984); *World Factbook* (CIA, 2004) Michael K. Steinberg and Paul F. Hudson, eds., *Cultural and Physical Expositions: Geographic Studies in the Southern United States and Latin America* (Geoscience Publications, Louisiana State University, 2002); *Planet Earth World Atlas* (Macmillan, 1998).

JESSICA M. PARR
SIMMONS COLLEGE

Niger

Map Page 1113 **Area** 489,191 square mi (1,267,000 square km) **Population** 11,058,590 **Capital** Niamey **Highest Point** 6,634 ft (2,022 m) **Lowest Point** 656 ft (200 m) **GDP per capita** $800 **Primary Natural Resources** uranium, coal, iron ore, tin.

A LANDLOCKED republic in West Africa, Niger is bordered by ALGERIA to the northwest, MALI and BURKINA FASO to the west, BENIN to the southwest, NIGERIA to the south, CHAD to the east, and LIBYA to the northeast. Its capital, Niamey, is the largest city; other urban centers are Agadez, Maradi, Tahoua, and Zinder. Niger's climate and terrain rank among the hottest and driest in the world, since four-fifths of the country consists of desert; the remainder is dry savannah. Less than 4 percent of the land is arable. Only one major river, the Niger, runs through the southwest.

Niger's population is largely rural and engaged in pastoralism or subsistence agriculture. The majority of the population lives in the south. The Hausa, Niger's largest ethnic group, are 56 percent of the population, followed by the Djerma (Zarma) at 22 percent, and the Beri Beri (Kanouri) at 4.3 percent. In the northern deserts live the Fula/Fulani, who make up 8.5 percent of the country's population, followed by the Tuareg at 8 percent, with 1.2 percent represented by Arabs, Toubou, and Gourmantche.

Niger is ranked amongst the poorest countries in the world, and is heavily dependent upon foreign aid for basic needs. The country's economy has taken a recent downturn because of the decreasing demand for uranium, which was Niger's largest export.

Neolithic remains have been found in Niger, and the Aïr Massif was explored by the Romans in the first century C.E. Major trans-Saharan trading states were established from the 11th to the 16th century by the Tuareg (Agadez), the Kanuri (Bilma in eastern Niger), and the Hausa (Zinder.) Much of Niger fell under the control of the Songhai Empire throughout the 15th and 16th centuries; eastern and southern Niger were annexed by Bornu after the fall of the Songhai Empire, and the Djerma settled in southwest Niger in the 17th century.

The Fulani gained control of southern Niger in the early 19th century during the jihad of Usman dan Fodio. The French made Niger a colonial holding in 1885 then established military posts in south Niger, but because of Tuareg resistance, Agadez along with much of the north did not fall under French control until 1904. Niger stayed a colony until the late 1950s, gaining full independence in 1960.

Since independence, Niger has undergone a series of crises. The 1960s saw relative stability despite rebel insurgencies. Niger's economy then plummeted during the Sahelian drought of 1968–75, which destroyed much of Niger's livestock and agricultural resources and prompted a military coup. During the 1980s, the

uranium boom boosted Niger's economy but led to huge disparities in wealth that caused civil unrest. The 1990s saw armed conflict with the Tuareg in the north, several changes to the government, and the assassination of military dictator Ibrahim Baré Mainassara in 1999. France suspended aid to Niger following the assassination, which prompted Niger to hold free elections and create a civil government.

BIBLIOGRAPHY. *Oxford Essential Geographical Dictionary* (Oxford University Press, 2003); *World Factbook* (CIA, 2004); *Planet Earth World Atlas* (Macmillan, 1998).

PILAR QUEZZAIRE-BELLE
HARVARD UNIVERSITY

Niger River

LOCATED IN West Africa, the Niger River stretches 2,610 mi (4,200 km) in length, making it the third-longest river in Africa, after the NILE and the CONGO. It is ranked 14th among the longest rivers in the world. The drainage basin of the river encompasses 807,000 square mi (2,090,000 square km). The Niger River runs in a long crescent from GUINEA to MALI, right up to the edge of the SAHARA DESERT, before heading south to the Gulf of Guinea.

The unusual course of the river mystified Europeans for many years. The Europeans thought that the section of the river near Timbuktu was part of the Nile River. By the early 17th century, Europeans thought that the river flowed west and joined the Senegal River. Finally, in 1795, Mungo Park became the first European to describe the upper river. While local people knew the actual course of the Niger, Westerners knew about it only through a series of explorations late in the 19th century. The source of the river rises in the Fouta Djallon Highlands at a point near the border between Guinea and SIERRA LEONE, 150 mi (241 km) from the Atlantic coast.

The main source of the Niger River is called the Tembi River. Away from the source but within Guinea, several tributaries join and replenish the Niger. From then on, the Niger traverses the interior plateau in a northeast direction toward the Malian border. As the Niger crosses the border between Guinea and MALI, other rivers such as the Fie join the main channel of water near Kangare, Mali. Thereafter, the Niger flows northeasterly until it reaches the interior delta in Mali. In the interior the Niger is joined by the Bani River, often regarded as among the important tributaries of the Niger. The Bani River is 696 mi (1,120 km) long and has its source in CÔTE D'IVOIRE and BURKINA FASO.

The Niger is precious to life in Mali. It provides fish, drinking water, and water for farming. It also represents a major means of transportation in Mali, particularly in some of the remote areas in the country. From this interior delta, the river flows in a northeast direction before turning to the southeast to form the great bend.

From this point on the Niger slowly meanders in the arid areas in proximity of Gao (a great center of trade and education during the Mali and Songhai ancient civilizations) and enters the country of NIGER.

While flowing through the country, Niger River is joined by several tributaries such as the Faroul, Dargol, Sirba, Garoubi, and Tapoa. The Niger River then continues and forms the boundary between the Republic of Niger and BENIN. This part of the Niger receives the Mekrou, a tributary from Benin. Thereafter, the river enters the Federal Republic of NIGERIA, where the Benue River joins to create an important confluence of the two rivers at Lokoja in Nigeria. From the confluence with the Benue, the Niger heads southward and discharges through a massive delta into the Gulf of Guinea or the ATLANTIC OCEAN.

BIBLIOGRAPHY. David L. Clawson and Merrill L. Johnson, eds., *World Regional Geography* (Prentice Hall, 2004); Samuel Aryeetey-Attoh. *Geography of Sub-Saharan Africa* (Prentice Hall, 2003); Bonaya Adhi Godana, *Africa's Shared Water Resources:* (Lynne Rienner Publishers, 1985); *Atlas of World Geography* (Rand McNally, 2003).

SAMUEL THOMPSON
WESTERN ILLINOIS UNIVERSITY

Nigeria

Map Page 1113 **Area** 356,667 square mi (923,768 square km) **Population** 133,881,703 **Capital** Abuja **Highest Point** 7,936 ft (2,419 m) **Lowest Point** 0 m **GDP per capita** $900 **Primary Natural Resources** oil, coal, tin, palm oil, peanuts, cotton, rubber.

NIGERIA, THE MOST populous nation in Africa, is bordered to the south by the Gulf of Guinea in the ATLANTIC OCEAN, and to the west and east by the Republic of BENIN and CAMEROON, respectively. The Republic of NIGER is across the border in the north. Nigeria's northeast neighbor is CHAD. It is slightly more than twice the size the state of CALIFORNIA.

Nigeria is a tropical country with tropical forests, savannas, mangroves, swamps, and rivers. The Benue and NIGER are the two main rivers in Nigeria. The Niger River flows through the south-central area and forms a large delta to the Atlantic Ocean. There are savannas in central Nigeria, mangrove forests along the coast, and plateaus in the north-central part of the country.

The most important plateau is the Jos Plateau with an elevation of 6,000 ft (1,829 m) above sea level. The far north is characterized by DESERT and some patchy grassland. The climate of Nigeria is hot and humid, though humidity is less than that encountered in the American South.

The Federal Republic of Nigeria is composed of 36 states with a federal capital territory at Abuja. Nigeria was a British colony until October 1, 1960, when it became an independent nation. The people of modern-day Nigeria have a rich history that predates European contacts. They possessed remarkable ancient culture such as the Nok culture, the Yoruba civilization, the Igbo states, and the Ife and Benin civilization. When Portuguese traders and sailors reached the coast of Lagos and the Niger Delta area around 1472, they were surprised to see the people organized, engaging in trade and commerce and displaying their arts and culture. Thus, the first contact was not to trade in humans across the Atlantic; rather, it was for legitimate commerce.

INTERNAL STRIFE

This soon changed, however, because of internal strife and civil wars among the people, as well as European demand for and encouragement of trafficking in human beings. The Atlantic slave trade, like the trans-Saharan slave trade, was successful because African compradors were willing to cooperate with European merchants. To this group, it was beneficial. The toll on African communities was devastating and perilous. British imperial interests along the west coast of Africa began with the need to stop the Atlantic slave trade, partly to ensure the supply of necessary materials for the emergent industrial factories in England. The British squadron ships patrolled to ensure that no slave was transported across the Atlantic. By 1851, a consulate was established in Lagos and the Niger Delta area to encourage trade in legitimate commerce. This was the harbinger of British colonial rule in what later became Nigeria.

In 1861, Lagos became a colony, and by 1885 the activities of the United African Company (Royal Niger Company) had paid dividends with the establishment of Oil Rivers Protectorate for the coastal region. A direct administration was not established until consolidation beyond the coastal areas in 1891. And by 1893, the Niger Coast Protectorate was established, which sealed British imperial goals in what later formed the eastern region. The northern part became a protectorate in 1903, giving Britain a greater claim to what later became northern Nigeria.

FORMING NIGERIA

The name *Nigeria* did not emerge until the amalgamation of 1914 by Frederick Lugard. At the end of World War I, the League of Nations gave German Cameroon to Nigeria as a mandated territory. In 1939, the British divided the Southern Protectorate into eastern and western provinces, with the north remaining intact. Lagos, however, continued to enjoy the benefits of a capital city and administrative nerve center. By 1950, an urban administration was set up to address the growing urban problems in Lagos. With the institution of the mayoral office, Africans were given the opportunity to find solutions to urban problems. Although the mayoral office was canceled in the Lagos colony in 1953, the same office was constituted in Port Harcourt to address urban needs.

Other efforts made were in the realm of constitutional changes and reforms, economic developmental plans, training of junior and senior civil service officials, and other reforms aimed at decolonization. The British did not pursue these efforts as part of a benevolent administrative gesture; rather, they were partly a result of Nigerian anticolonial movements. The leftist groups perhaps played a significant role in British reforms and the pace of decolonization after World War II. With the growing emergence of leftist groups, Marxist literature, and funds from international communist fronts, Nigerian leftist groups were able to make British officials and pro-Western Nigerian nationalist and labor leaders uncomfortable.

Anticommunism thus became an essential part of decolonization, and its success became an ingredient in the eventual transfer of power in 1960. It should not be surprising that postcolonial administration pursued

anti-leftist policies both domestically and internationally. The military interfered in Nigerian politics in 1966 with a coup d'etat that overthrew the civilian administration. The country was soon plunged into a civil war in 1967 that lasted for three years. Since 1966, the military has played a dominant role in governance and civil relations in Nigeria.

Although a transfer of power was smoothly ensured in 1979 by the military, the Second Republic (1979–83) did not survive, as the military saw itself in command of development and reforms. The contrary however is the case. The Nigerian military, like most military in the developing world, is allegedly corrupt and contributed to the underdevelopment of the country. Nigeria is, however, experimenting with a Third Republic, which began in June 1999 when the erstwhile military head of state, Olusegun Obasanjo, was elected in a democratic election.

Nigeria has three main ethnic groups: Hausa/ Fulani, Igbo, and Yoruba. The ethnic groups also have some 250 different subgroups and language dialects. Although English remains the official language, there are three main languages represented by the three main ethnic groups. The Hausa/Fulani dominate the northern part, with the Igbo people are in the east, and the Yoruba are in the west. Christianity, Islam, and traditional religion permeate the lives and homes of the people.

While Nigeria is a rich country in terms of proceeds from oil sales, the people are largely poor and marginalized. The benefits from oil profits are not adequately redistributed among the people, nor is it judiciously used in the development of the country. Oil discovery at Oloibiri in 1958, and subsequently in other parts of the Niger Delta area accounts for 95 percent of Nigeria's exports. Although the economy is supported by abundant hydroelectric power, the country has witnessed more blackouts than any third-world nation. There is a growing effort in textile, cement, and automobile industries.

BIBLIOGRAPHY. Akinjide Osuntokun and Ayodeji Olukoju, eds., *Nigerian Peoples and Cultures* (Davidson Publishers, 1997); Toyin Falola, ed., *Nigeria in the Twentieth Century* (Carolina Press, 2002); Adebayo Oyebade, ed., *The Foundation of Nigeria* (Africa World Press, 2003); "Nigeria: A Country Study," Library of Congress, www.loc.gov (March 2004); *World Factbook* (CIA, 2004).

HAKEEM IBIKUNLE TIJANI
LYNDON B. JOHNSON LIBRARY

Nile River

AT 4,132 mi (6,650 km) long, the Nile River is the longest river in the world. It flows from two principal sources in equatorial Africa that join at KHARTOUM, SUDAN, and continues north through Sudan and EGYPT, emptying into the MEDITERRANEAN SEA. Historically, the Nile has been used in irrigation, farming, and transportation for thousands of years, from the beginning of Egyptian civilization into the 21st century.

The Nile is fed by many tributaries, but ultimately it has two main sources: the White Nile in BURUNDI, and the Blue Nile in ETHIOPIA. Except when swollen by rains in August and September, the Blue Nile contributes less than 20 percent of the water; the White Nile provides the larger share. The ultimate source of the White Nile remained a tantalizing mystery until 1937, when Burkhart Waldecker discovered it while seeking asylum in the Belgian Congo from Nazi persecution in his native GERMANY. Waldecker pinpointed a stream that became the Kasumo River, flowing from Mount Kikizi in Burundi. This river eventually joins the Mukesenyi, the Ruvyironza, the Ruvubu and finally the Nyabarongo River. Also vying for recognition as the source is another small stream from Mount Bigugu that feeds into the Lukarara River and then joins the Nyabarongo.

WHITE NILE

After the Nyabarongo and Ruvubu rivers join to form the Kagera River, the waters, which will become the White Nile, flow east over Rusumu Falls and then into Lake VICTORIA. The river, now called the Victoria Nile, leaves the lake in the north, dropping over Owens Falls Dam into Lake Kyoga. The river then flows east into the Great African Rift Valley by cascading over Kabarega Falls to Lake Albert. It leaves the lake as the Albert Nile, going north through UGANDA. At the border with Ethiopia the river's name changes again, to Bahr al-Jabal, Mountain River. The Mountain River cascades through highlands before reaching the Sudd, a swampland that brings the waters to a halt, where half the river is lost to evaporation. The marshes and lagoons of the Sudd have existed for millions of years, although the expanse of the Sudd varies depending on seasonal rains and the outflow from the lakes and rivers that feed into it.

After hundreds of miles, the Nile waters emerge from the reservoir of Lake No in the north as the White Nile and join with the Sobat River near Malakal in Sudan. Here it finally takes on the color that gives it its

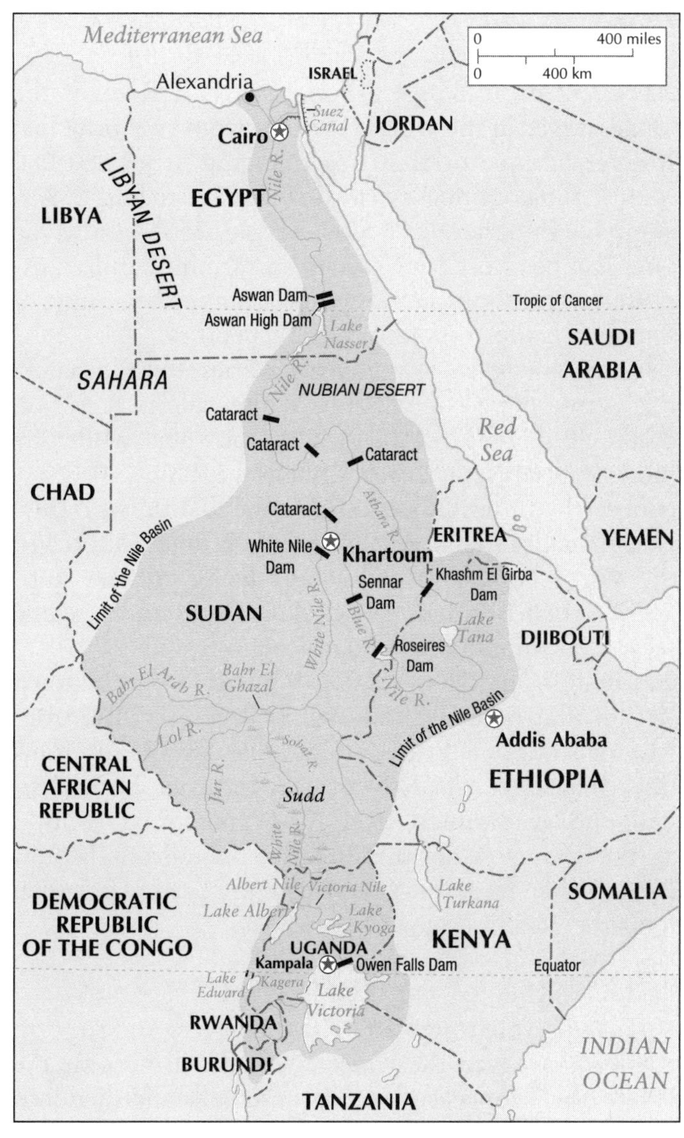

The Nile is fed by many tributaries, but has two main sources: the White Nile in Burundi, and the Blue Nile in Ethiopia.

name, and 500 mi (805 km) later it joins with the Blue Nile. Just 30 mi (48 km) south of Khartoum, the White Nile is dammed into a reservoir by the Jabal Auliya Dam. The original purpose of this dam was to hold back the White Nile during the late summer and fall months when the Blue Nile swelled with rain.

Waldecker's 1937 discovery of the White Nile's source resolved thousands of years of mystery. Greek historian Herodotus had speculated about the Nile's origin in 460 B.C.E. Roman Emperor Nero sought the source during his reign, but the centurions he sent to explore were stopped by the Sudd. Until 1841, the Sudd blocked all attempts to trace the Nile, but in that year an Egyptian expedition did pass through the swampland. Over the next century, many explorers failed and some died attempting to trace the Nile.

BLUE NILE

The Blue Nile's source is a stream issuing from Mount Gish in Ethiopia called Sakala. The stream is considered sacred and develops into a river named Little Abbai. This feeds into the southern end of Lake Tana, and Great Abbai, the Blue Nile, flows south out of this lake to drop 150 ft (46 m) from a naturally occurring lava dam and form the Tisisat Falls (the name means "Smoking Waters"). From this point, the Blue Nile makes a wide semicircle and turns east, then north through Sudan. Many rivers feed into it: the Didessa, Dabus, Beles, Dinder, and Rahad among them; some of the rivers contribute water seasonally.

The Blue Nile swells in late summer from heavy rainfall. Dammed in several places, the Blue Nile provides irrigation for the Gezira Plain in Sudan. After leaving Lake Tana, the Blue Nile meets the White Nile at Khartoum, at a place called the Mogren, meaning "the meeting" in Arabic. James Bruce of Scotland found the source of the Blue Nile in 1770 and was appointed governor of Sakala by Ethiopian Emperor Tekla Haymanot II. The Ethiopians had long been aware that their sacred spring was the source of the Blue Nile; a visiting Spanish priest wrote about being shown the source of the Nile in 1615. Bruce was not believed in Europe when he announced his discovery, but his published works detailing his journey were verified by later explorers.

From Khartoum, the Nile flows north. Six sections of rapids, called cataracts, mark it; the first is at Sabaluqa, 50 mi (80 km) from Khartoum. The northernmost, or first, cataract is at Aswan.

The Atbara River, another major tributary of the Nile, joins the river between the fifth and sixth cataracts. Like most of the other waters contributing to the Nile, the Atbara comes from Ethiopia, where summer monsoonal rains cause the waters to peak in August in September; the waters drop to lower levels from December through April. Coming from the tablelands, the waters that join the Nile carry rich alluvial sediments that for 8,000 years flooded into the farmlands along the banks of the Nile. The original name for Egypt, in fact, was Kemet, the same word used to describe the black silt deposited by the Nile.

ASWAN DAM

The Nile Waters Agreement of 1959 was negotiated between Egypt and Sudan; Ethiopia, RWANDA, and Bu-

rundi were excluded. This agreement divided riparian rights to the Nile between the two negotiating countries, and cleared the way for the building of a new Aswan Dam. From 1959 through 1970, the Soviet Union financed the construction of the hydroelectric Aswan High Dam. To provide power for the construction, the old Aswan Dam was electrified and equipped with turbines to generate up to 1.8 million kilowatts. By the time it officially opened in 1971, the dam had formed Lake Nasser, which is called Lake Nubia in Sudan. The lake, a reservoir holding up to 162 billion cubic meters of Nile water, is 8 mi (13 km) in width and over 300 mi (483 km) long.

The Aswan High Dam controls the annual flooding of the Nile and supplies electricity for Egypt, but it has created problems as well. The new reservoir displaced 50,000 Nubians initially. Its weight creates instability on local fault lines in the Earth. The dam traps sediments that would be delivered to farmlands in Egypt, and silt that would replenish the Nile Delta, which is now eroding. Trapped behind the dam, the sediments produce algae that destroys oxygen in the lake.

The Nile provides waters to over 250 million people; that figure may double by 2025. While the population of countries fed by the Nile grows by about 3 percent a year, the waters do not increase. Evaporation from Lake Nasser claims 10 billion cubic meters of water yearly, roughly 9 percent of the average annual flow of the Nile into Aswan.

Another major and controversial flood control project on the Nile, the Jonglei Canal in Sudan, was begun in 1978 to circumvent the Sudd marshes. This idea had first been proposed by William Garstin, a British hydrologist, in the early 1900s. Construction stopped in 1983 due to civil war and has never been completed.

BIBLIOGRAPHY. Robert O. Collins, *The Nile* (Yale University Press, 2002); Haggai Erlich and Israel Gershoni, eds., *The Nile: Histories, Cultures, Myths* (Lynne Rienner Publishers, 2000); Alan Moorehead, *Blue Nile* (HarperCollins, 2000 reprint); Alan Moorehead, *White Nile* (HarperCollins, 2000 reprint); Nile Basin Initiative Secretariat, "Introduction to the Nile River Basin," www.nilebasin.org (March 2004); Christopher Ondaatje, *Journey to the Source of the Nile* (Firefly, 1999); J.V. Sutcliffe and Y.P. Parks, *The Hydrology of the Nile* (International Association of Hydrological Sciences, 1999).

Vickey Kalambakal
Independent Scholar

Niue

Map Page 1125 **Area** 100 square mi (260 square km) **Population** 2,145 **Capital** Alofi **Highest Point** unnamed hill near the town of Mutalau 223 ft (68 m) **Lowest Point** 0 m **GDP per capita** $3,600 **Primary Natural Resources** agriculture.

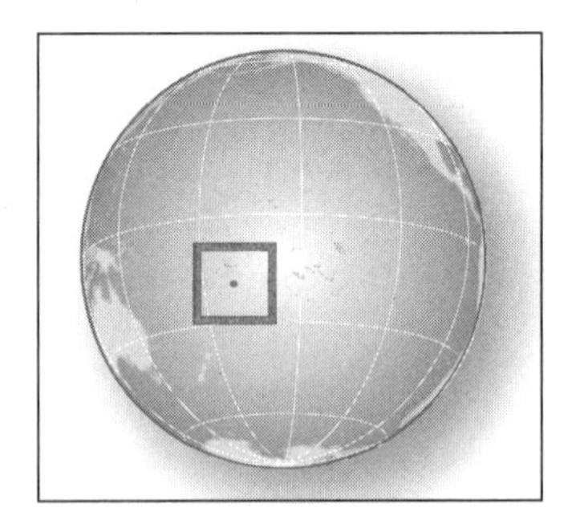

ONE OF THE WORLD'S largest coral islands and smallest countries, Niue is about the size of metropolitan District of Columbia in the UNITED STATES and has a population the size of a small town. It has a rugged coastline with high limestone cliffs and a grassy interior plateau. Niue is located 1,491 mi (2,400 km) northeast of NEW ZEALAND, on the eastern side of the INTERNATIONAL DATE LINE and is 11 hours behind Greenwich Mean Time. Most of the population lives in 13 small settlements found along a 41-mi (67-km) road that circles the island.

The island is dotted with limestone caves and is without streams or rivers. Rain water filters through the porous limestone out into the ocean with little silt run-off. This allows the surrounding ocean to be crystal clear with dive visibility often up to 229 ft (70 m). Niue lies in the zone of the southeast tradewinds, giving it a tropical climate with temperatures averaging 75 to 80 degrees F (24 to 27 degrees C) year-round. Niue's isolated location and cultural and linguistic differences have enabled it to maintain independence from other Polynesian islands despite its minuscule size. However, a lack of employment opportunities has caused its population to continue to drop from a peak of 5,200 in 1966 to about 2,100 in 2002 with substantial emigration to New Zealand.

Niuean folklore tells of the first settlement of Niue by forefathers Huanaki and Fao, together with fire gods from Fonuagalo, or the "Hidden Land." Anthropologists say Polynesians voyaging by outrigger canoe settled on Niue from TONGA, SAMOA, and Pukapuka Island and called the island Motusefua. The English navigator James Cook, sighted Niue in 1774 but was refused landing three times by warriors in canoes. Cook departed but claimed the island for the UNITED KINGDOM and named it Savage Island. Missionaries from the London Missionary Society established Christianity in 1846. Niue chiefs gained British Protectorate status in 1900, but in 1901 Niue was annexed to New

Zealand. In 1974, Niue gained independence in free association with New Zealand. The people of Niue enjoy dual citizenship and are bilingual, speaking both Niuean and English.

Niue's economy reflects its geographic isolation, few natural resources, and its tiny population. Most residents are subsistence farmers with some cash crops grown for export. Employers include several small factories that process passion fruit, lime oil, honey and coconut cream. Sales of stamps and commemorative coins to foreign collectors are important sources of revenue. Government expenditures regularly exceed revenues, and the shortfall is made up by grants from New Zealand to pay public employees. Economic aid from New Zealand in 2002 was about $2.6 million.

The official languages are English and Niuean, a Polynesian language closely related to Tongan and Samoan. The predominant religion is Ekalesia Niue (Niuean Church—a Protestant church closely related to the London Missionary Society).

BIBLIOGRAPHY. *World Factbook* (CIA, 2004); Niue Tourism, www.niueisland.com and www.niueisland.com (April 2004); Niue Government, www.gov.nu (April 2004).

Rob Kerby
Independent Scholar

Norfolk Island

NORFOLK ISLAND, AN AUSTRALIAN territory, is located in the south PACIFIC OCEAN, 1,041 mi (1,676 km) northeast of Sydney, AUSTRALIA. With a total land area of 13 square mi (35 square km), the island is slightly larger than the District of Columbia; it is approximately 5 mi (8 km) in length and 3 mi (5 km) in width. The highest point on the island is Mount Bates at 1,047 ft (319 m) and the lowest is the Pacific Ocean beach at 0 m. Norfolk Island's terrain is rough with volcanic formations and some plains. The climate is subtropical with little seasonal change. The capital is Kingston; the other major towns are Cascade and Burnt Pine. Norfolk Island is an external territory of Australia and includes the small islands of Nepean and Philip Island to its south.

The population of Norfolk Island is estimated to be 1,853 (2003), with almost 64 percent of the people between the ages of 15 and 64. The primary economic activity is tourism, which has increased in recent years to such a point that the small island population enjoys a high level of prosperity. While most finished products must be imported, Norfolk Island has achieved a level of self-sufficiency in some foodstuffs, notably poultry, beef and eggs. Other agricultural products include Norfolk Island pine seed, Kentia palm seed, cereals, vegetables, and fruits. Principal export revenue is acquired through the sale of postage stamps to collectors, and the Norfolk Island and Kentia seeds. Norfolk Island has one airport, no railroads, and a limited network of paved roads; communications with Australia have improved during the last decade as a result of satellite networks.

Captain James Cook is credited with being the European who discovered Norfolk Island in 1774. Named after the Duke of Norfolk, this island became a possession of the Australian colony of New South Wales in 1788. During the late 18th and early years of the 19th century, Norfolk Island served as a penal colony. In the 1850s the descendants of the HMS *Bounty* mutiny were forced to relocate from PITCAIRN ISLAND to Norfolk Island.

In 1914, Norfolk Island became a territory of Australia. The current constitution of Norfolk Island was adopted in 1979 and is based on Australian, Norfolk, and British law.

BIBLIOGRAPHY. Evelyn Colbert, *The Pacific Islands: Paths to the Present* (Westview Press, 1997); Merval Hoare, *Norfolk Island: An Outline of Its History* (University of Queensland Press, 1988); Tom L. McKnight, *Oceania: The Geography of Australia, New Zealand, and the Pacific Islands* (Prentice Hall, 1995); Richard Nile and Christian Clerk, *Cultural Atlas of Australia, New Zealand, and the South Pacific* (Facts On File, 1996).

William T. Walker, Ph.D.
Chestnut Hill College

North American Free Trade Agreement

IN 1993, THE UNITED STATES, CANADA, and MEXICO signed the North American Free Trade Agreement (NAFTA), which created a regional free trade zone that lowered tariffs and trading restrictions and encouraged greater opportunities for cross-border investments and movement of goods and services between the three

countries. The agreement went into effect in 1994. While all three states had well-established trading relations, a new arrangement was deemed necessary, in part, as a reaction to a similar economic initiative proposed by the European Community (EC). It was also part of the liberalizing trend in international trade, which has driven international economics since World War II.

In 1944 and 1945, representatives from the Allied nations met at the NEW HAMPSHIRE resort of Bretton Woods to establish new international organizations that would regulate and systematize the international economic order. The first effort was the International Bank for Reconstruction and Development and the International Monetary Fund to aid in the development and stabilization of international currencies. Part of the proposed economic structure addressed the rampant economic nationalism that had pervaded trade in the 1930s and that was believed to have contributed to the deterioration in international relations that had led to the outbreak of the World War II.

In 1947, representatives from 23 states met in Geneva, SWITZERLAND, to establish the General Agreement on Tariffs and Trade (GATT), which provided a framework to promote multilateral international trade and development, and thereby limit the threat to the international order of economic nationalism. GATT did not establish an organization that centralized trading agreement; rather it specifically encouraged multilateral free trade agreements, and was strongly supported by the United States. Since its signing, GATT has undergone several additions, the latest being the Uruguay Round in 1995 that established the World Trade Organization (WTO), which adjudicates free trade disputes.

Since GATT was signed, several regional (multilateral) free trade zones have been established. One of the first was the European Economic Community (EEC), later known simply as the European Community (EC); it was created by West GERMANY, FRANCE, ITALY, BELGIUM, and LUXEMBOURG in 1957 to facilitate trade among its membership. The EC expanded to 15 nations as the EUROPEAN UNION and in 1993 became a single free trade market with a common currency, eliminating trade and fiscal restrictions and allowing for the internal movement of goods and services. Growing concern among European trading partners, particularly Canada and the United States, and later Mexico, spurred an effort to form a similar agreement between the United States and Canada.

The first step in the formation of NAFTA occurred in the late 1980s, when the United States and Canada signed a comprehensive trade agreement, the Canada-United States Free Trade Agreement (CUSFTA)to compensate Canada and the United States for any potential lost market share from the EC policy. The CUSFTA expanded on an earlier agreement that covered autos. Then in 1990, at the request of Mexico, a larger free trade agreement was proposed among the United States, Canada, and Mexico. NAFTA gradually phased out the bilateral CUSFTA terms, as well as preexisting quotas, tariffs, and investments, enlarging the market and labor available to American, Canadian, and Mexican trade.

NAFTA required the three members to eliminate their respective trade barriers and restrictions to cross-border movement and transportation of goods and services, establish a framework for the extension of NAFTA to other regional members, such as a potential free trade agreement for the hemisphere, and provide for increased investment opportunities, thus combining trade and economic policies. The main benefits to the members are cheaper consumer goods, greater investment opportunities, and increased jobs as a result of trade. Removing tariffs forces industries to become more efficient and productive, thereby lowering costs. NAFTA is also intended to help international development by spurring industrial growth in Mexico and, it is hoped, a stronger middle class, while creating higher-paying, high-technology and white-collar jobs in the United States and Canada.

There are, however, several problems with NAFTA. First, energy remains regulated and not subject to free trade. Labor unions criticized the agreement for outsourcing jobs to Mexico, where there are lower labor and environmental regulations and workers' wages. A larger issue, however, is the misuse of the rules of origins, which allow members to declare certain goods not to be subjected to NAFTA because of the presence of nonmember components. Finally, NAFTA has hurt nonmember economies, especially in the Caribbean and South America.

BIBLIOGRAPHY. Khosrow Fatemi, "New Realities in the Global Trading Arena," *The North American Free Trade Agreement* (Pergamon, 1994); Dominick Salvatore, "NAFTA and the EC: Similarities and Differences," *The North American Free Trade Agreement* (Pergamon, 1994); Gary Hufbauer and Jeffrey Schott, *NAFTA: An Assessment* (Institute for International Economics, 1993); Nora Lustig, Barry Bosworth, and Robert Lawrence, eds., *Assessing the Impact: North American Free Trade* (Brookings Institute, 1992); Jerry Rosenberg, *Encyclopedia of the North Ameri-*

can Free Trade Agreement, the New American Community and Latin-American Trade (Greenwood Press, 1995).

Frederick H. Dotolo III, Ph.D.
St. John Fisher College

North Atlantic Treaty Organization

AT THE END OF World War II (1939–45), Europe lay in wasted ruin with most industrial and transportation infrastructure damaged or destroyed. Some 40 million deaths had occurred in Europe and RUSSIA in the six years of conflict. Approximately 25 million displaced persons were stranded far from original homelands with little resources. The UNITED STATES and its allies were eager to return to a national lifestyle of peace and sought to rapidly demobilize the vast military might they had built up over the years of the war. Men in uniform wanted to return to jobs, education, and opportunities and those at home yearned for the hard won fruits that victory was promised to bring. The scars of war were evident across the landscape of Europe and the human costs of death, disruption, and dislocation cut deep into the heart of every nation. The Soviet Union declined to demobilize and decided to maintain the vast armies and military might that had occupied Eastern Europe and portions of Central Europe.

Efforts by the Allies of WWII to address the rebuilding and reorganization of Europe soon polarized. The Soviets encouraged government changes in areas they occupied to follow socialist or communist patterns. In the west, America, Britain, and FRANCE supported a return to democratic forms of government. Within the areas of Soviet influence, imposition of pro-Soviet undemocratic forms of government became the norm. By the end of 1947, Polish elections had been canceled, the elected prime minister of BULGARIA was forced out of the country, and the People's Republic of ROMANIA was formed. A call to worldwide communist revolution stirred fears of aggression from without and subversion from within.

Soon after the conclusion of the war, as Europe struggled to recover, political events boded ill for a lasting peace. The Soviets made demands of NORWAY, GREECE, and TURKEY that were seen by Western Europe as direct threats to these nations' sovereignty. BELGIUM, France, LUXEMBOURG, the NETHERLANDS, and the UNITED KINGDOM joined in the Brussels Treaty of March 1948 to build a mutual system of defense against ideological, political, and military threats to their national futures. The Soviet Union began a blockade of Allied-occupied Berlin in April 1948. Civil war in Greece and the elected government of Czechoslovakia's overthrow in a communist coup in June 1948 set an ominous backdrop.

Within a span of three decades, the United States had chosen to come to the aid of Europe in two major world conflagrations without any treaty obligations to do so. Now America was financially and politically committed to the rebuilding of Europe through the European Recovery Plan (Marshall Plan). The perceived threat of the Soviet Union moved the Western European nations and the United States to formalize their security relationship in the Treaty of Washington signed on April 4, 1949.

FORMATION OF NATO

This, in effect, was the formation of NATO, the charter members being Belgium, CANADA, DENMARK, France, ICELAND, ITALY, Luxembourg, the Netherlands, Norway, PORTUGAL, the United Kingdom, and the United States. The treaty committed the 12 nations to collective defense, meaning that an attack on one was like an attack on all. The North Atlantic Treaty Organization was seen as assurance against military conflict among the agreeing parties, a check to any aggressive German military resurgence, and a clear counter to the military power of the Soviet Union.

The early years of the NATO alliance were characterized by its growth as an institution, delineation of roles and responsibilities, and integration into the military establishments of the governments of Western Europe. The combined military strengths of the NATO nations were quantitatively inferior to those of the Soviet Union concentrated in Central and Eastern Europe. Consequently, NATO's nuclear weapons capability became a central component of its military and political strategy of deterrence and containment.

NATO expanded in 1955 by allowing the re-armed Federal Republic of GERMANY (West Germany) to join. The Soviet Union saw this as an overt threat to their security and countered by organizing a treaty organization of their own. The Warsaw Pact was signed by ALBANIA, Bulgaria, Czechoslovakia, German Democratic Republic (East Germany), HUNGARY, POLAND, Romania, and the Soviet Union. Now all the actors were assigned for this increasingly dramatic play of international tensions. In 1956, a popular uprising against the

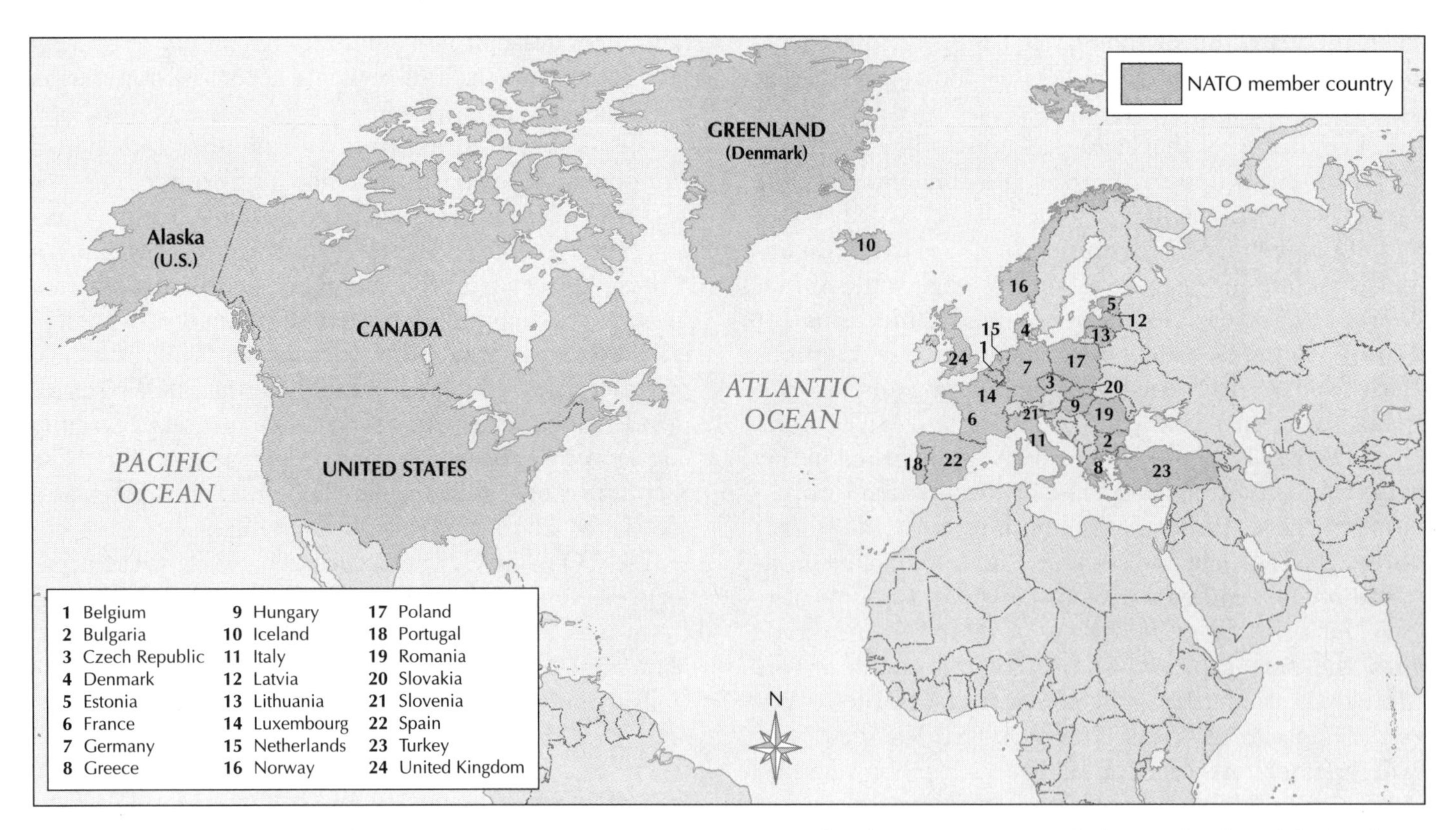

The expansion of the North Atlantic Treaty Organization (NATO) into Central and Eastern Europe with the membership of the Czech Republic, Hungary, and Poland ties the new identity of NATO to the new identity of Europe.

communist government in Hungary was quelled by invading Soviet forces. The United Nations condemned this response, but NATO took no actions.

In the 1960s, tensions grew in Europe with the building of the Berlin Wall and NATO began to publicly debate the requirement of NATO having direct control of nuclear weapons in the European theater. It was felt that this increased capability and responsiveness would add to the deterrence available to check Soviet-sponsored aggression. After the Cuban Missile Crisis between the United States and the Soviet Union, the Americans gave command and control of nuclear capable forces stationed in Europe to NATO. In 1966, France removed its forces from the military structures of NATO, citing the dominance of the United States in the organization. NATO continued to set the pace for Western military strategy and devised a "First Use" policy, which warned that NATO held the right to use nuclear weapons in response to a conventional military attack against any of its member states.

The 1970s saw a time of détente between east and west. The Strategic Arms Limitation Talks (SALT) produced treaties that called for the drastic reduction of nuclear stockpiles. NATO played a critical role in the implementation of agreed-upon conventional forces reductions brought about by the Conference on Security and Co-operation in Europe. The end of that decade saw the Soviet invasion of Afghanistan and a renewed interest by America to match Soviet actions with increases in military readiness, capability, and deployment. NATO sponsored the deployment of nuclear ballistic and cruise missiles into Europe.

During these times of increased military activity and general angst, Ronald Regean and Mikhail Gorbachev began a series of summits that changed superpower relations and opened the door to significant arms reduction talks. NATO again became a major implementing agency of these agreements and other confidence-building measures.

The late 1980s saw the collapse of Eastern European communist regimes as the Soviet Union withdrew military support and Gorbachev encouraged them to seek political and economical reforms. Long-term economic failures and the overall shift in political climate caused change to sweep across Soviet-dominated Europe. NATO takes credit for acting as a continuous deterrent to Soviet military aggression that finally allowed the breakaway of Eastern Europe and perhaps

even the break up of the Soviet Union. The west declared victory in the Cold War as the former Soviet Union dissolved in 1991.

The decade of the 1990s saw vast political change in Central and Eastern Europe and the demise of the Soviet Union. Having watched its rivals dissolve, NATO now faced a redefinition of its role and responsibilities. Interim steps were taken to engage the former Warsaw Pact nations through the Partnerships for Peace program, which developed military relations with former adversaries. As Yugoslavia broke up into smaller republics ethnic violence erupted. NATO took on its first ever combat operations as it worked in line with United Nations (UN) resolutions to create conditions for peace in Bosnia by bombing Serbian military forces. NATO nations provided forces for UN-sponsored peacekeeping forces in the former Yugoslavia.

The defining of new roles for NATO continued into the new century. The terrorist attacks on the United States in 2001 were considered an issue of mutual defense and NATO elected to join America in its war against terrorism embodied in the invasion of Afghanistan. However, as the United States prepared to invade Iraq in 2003, Germany and France blocked all efforts made to gain the political and military support of NATO for the U.S.-led coalition operation. NATO took over command and coordination of the security in and round Kabul, AFGHANISTAN, in August 2003. This is the first mission outside the Euro-Atlantic area in NATO's history, and it raises new questions and concerns as to future of the organization.

TRANSATLANTIC LINK

NATO holds to its central concept that it embodies the transatlantic link of Europe's security and prosperity. It is the concrete expression of the ideal that the security of North America is permanently tied to the security of Europe. The expansion of NATO into Central and Eastern Europe with the new membership of the CZECH REPUBLIC, Hungary, and Poland ties the new identity of NATO to the new identity of Europe. What path that will take is unclear and the outcome remains uncertain.

NATO continues to operate a robust organization that provides an unprecedented forum in which members can raise issues of concern regarding their security. It also maintains an institutional structure that supports consultation and cooperation between members in political, military and economic, as well as scientific and other nonmilitary fields.

For the foreseeable future NATO will continue to act as an intergovernmental agency that seeks to maintain integrated combined military capability to prevent war, manage crisis, and promote security cooperation in Europe and abroad.

NATO's structure consists of a military and civilian component. The civilian side has four primary components. The North Atlantic Council constitutes the executive body of NATO and is formed by permanent representatives from all members. Decisions reached must be unanimous. The Defense Planning Committee is a consultative body that focuses on development of military plans, it includes permanent members from all member nations except France. The Nuclear Planning Group is comprised of national defense ministers. The Secretary General is appointed by member states and acts as the Chair of the senior committees.

NATO's military structure is built on three area-focused commands. Allied Command Atlantic, Allied Command Europe, and Canada-United Sates Regional Planning Group share the military responsibilities. The United States provides the NATO Supreme Commander.

BIBLIOGRAPHY. P. Duignan, *NATO: Its Past, Present and Future* (Hoover Institution Press, 2000); Stephen Smith, "The International Refugee Crisis," Project Safecom, www.members.westnet.com.au (June 2004); "NATO, Origins of the Alliance," www.nato.int (June 2004); "Czechoslovakia Coup," The Cold War Museum, www.coldwar.org (June 2004).

IVAN B. WELCH
OMNI INTELLIGENCE, INC.

North Carolina

KNOWN AS BOTH the Tar Heel State and Old North State, North Carolina was part of colonial America. Like its sister state SOUTH CAROLINA, North Carolina was named for Kings Charles I and Charles II of England. North Carolina is bounded on the north by VIRGINIA, on the south by South Carolina and GEORGIA, on the west by TENNESSEE, and on the east by the ATLANTIC OCEAN. The total area of North Carolina is 52,669 square mi (136,412 square km), making it the 29th state in size. North Carolina ranks 11th in population among the 50 states. North Carolina's largest cities are Charlotte, Raleigh (the capital), Greensboro, Durham, Winston Salem, Fayetteville, Cary, High Point, Wilmington, and Asheville.

North Carolina has 3,826 square (9,909 square km) of inland water, and the North Carolina coast runs for 301 mi (484 km). Elevation in the state ranges from 6,684 ft (2,037 m) above sea level at Mount Mitchell, which is the highest point in the eastern section of North America, to sea level at the Atlantic Ocean. The state is approximately 500 mi (804 km) east to west and 150 mi (241 km) miles north to south. North Carolina's major rivers are the Roanoke, the Tar, the Neuse, Cape Fear, the Catauba, and the Pee Dee, which is also called the Yadkin. The state has a number of natural lakes, such as the Mattamuskeet, the Phelps, and the Waccamaw.

The climate of North Carolina is varied. In the Atlantic Coastal Plain and the Piedmont Plateau, the climate is subtropical, while in the Blue Ridge Region the climate is humid continental. In the higher, mountainous elevations, cold winters are followed by relatively cool summers. North Carolina winters are mild in the Piedmont Plateau and Atlantic Coastal Plains, since mountains protect the area. While rain falls throughout the year, it is more prevalent in the late winter and in summer. Snow is frequent throughout the Blue Ridge Region.

North Carolina's coasts are often subject to hurricanes, but the interior can be struck as well—in 1989 Hurricane Hugo wrecked havoc on the inland city of Charlotte. Tornado season also poses a threat to the state.

Approximately two-fifths of the entire state of North Carolina falls within the Atlantic Coastal Plain made up of coastal lands and tidewater. Soft sedimentary rocks form the foundation of the land in this section. Swamps, marshes, and sounds are also found throughout the area. The best known of the swamps is the Dismal Swamp, which spans parts of both North Carolina and Virginia. Transportation in this area was treacherous until the Dismal Swamp Canal was created, connecting North Carolina's Albemarle Sound with Virginia's Chesapeake Bay. The Dismal Swamp Canal is the oldest man-made canal in continual operation in the United States.

OUTER BANKS

Albemarle Sound as well as Pimlico Sound are located between the mainland of North Carolina and a group of offshore islands known as the Outer Banks, which include the islands of Cape Hatteras, Cape Lookout, and Cape Fear. The Piedmont Plateau, with its moderately fertile soils and low elevations, comprises an additional two-fifths of North Carolina. This area is separated from the Atlantic Coastal Plain by the Fall Line where rivers descend, creating rapids and waterfalls as the two geographical areas converge.

BLUE RIDGE REGION

The remaining fifth of North Carolina makes up part of the Blue Ridge Region of the southeastern United States. The North Carolina section of the APPALACHIAN MOUNTAINS includes both the Blue Ridge Mountains and the Great Smoky Mountains, containing more than 200 mountains with elevations of more than 500 ft (152 m). These mountains tend to be round in shape with steep gorges. The area also has a number of basins such as the one located in Asheville. Roads and tunnels provide access along the Blue Ridge Parkway. The Great Smoky Mountains, part of the southern Appalachian chain, hosts a national park where park officials have identified over 10,000 plants and animals in the area and believe that another 90,000 species have yet to be identified.

Forests of shortleaf, longleaf (the state tree), loblolly, and Virginia pine are prevalent throughout North Carolina's section of the Southeastern Pine Forest. Cypress and various hardwoods are found in wet areas of the state. Oak, hickory, tulip, and poplar are common among the lower slopes of the Blue Ridge Region, while the higher slopes are home to birch, beech, maple, and hemlock. Spruce and balsam are found in the highest regions of the mountains. North Carolina flowers include rhododendron, azalea, camellia, dogwood (the state flower), and orchid. Animals found throughout the state include Virginia deer, opossum, raccoon, squirrel, and fox. Black bear are common in the Coastal Plain forest and in the Blue Ridge Region. The North Carolina coast serves as a home to migrating birds such as ducks, geese, and various waterfowl. The state's numerous songbirds include cardinal (the state bird), blue jay, woodpecker, Carolina wren, mourning dove, purple finch, and towhee.

The economy of North Carolina was heavily dependent on agriculture until the 1920s. At that time, the manufacturing of textiles, furniture, and tobacco products became the mainstays of the economy. Tourism bolsters the North Carolina economy by approximately $1 billion a year. By the 1990s, agriculture and forestry also assumed significance for the state economy. Major farm products for the state include tobacco, corn, cotton, hay, soybeans, peanuts, wheat, sweet potatoes, and various other vegetables. Industries include metallurgy, chemicals, and paper. North Carolina's most important minerals are quartz, lime-

stone, phosphate rock, sand and gravel, granite, feldspar, talc, lithium, mica, asbestos, and gemstone. North Carolina is the country's leading producer of mica and lithium. The quartz found in the Blue Ridge Mountains is used in microprocessors throughout the world because of its unusual purity.

BIBLIOGRAPHY. Dan Golenpaul, ed., *Information Please Almanac* (McGraw-Hill, 2003); "North Carolina," www.netstate.com (March 2004); "North Carolina at Your Service," www.ncgov.com (March 2004); "North Carolina Geography," statelibrary.dcr.state.nc.us (March 2004).

ELIZABETH PURDY, PH.D.
INDEPENDENT SCHOLAR

North Dakota

NORTH DAKOTA, THE Flickertail State, is a land of endless fields of corn, sunflowers, wheat, and buffalo. The state has a small population but is rich in natural resources. Located in the north central UNITED STATES, North Dakota was named after the Sioux Indians who called themselves the Dakota. Rural depopulation is a serious problem in much of the state, but North Dakota also has several growing cities, such as its largest city, Fargo; its capital, Bismarck; Grand Forks; Dickinson; and Pembina. Theodore Roosevelt came here in 1883 and became a famous rancher. Today, there remains a national park named after him in the western part of the state.

North Dakota is the 17th-largest state in land area with a total area of 70,704 square mi (183,123 square km). It is 360 mi (579 km) long from its eastern boundary with MINNESOTA and its western boundary with MONTANA. The state extends for 210 mi (338 km) from the Canadian provinces of Saskatchewan and Manitoba in the north to its sister state of SOUTH DAKOTA in the south. The 2003 estimated population was 633,837. About 95 percent of its residents are white and 4 percent are Native American. Most are of German and Norwegian heritage and subsequently most are Lutheran or Roman Catholic.

North Dakota's physical landscape can be divided into three steps going from east to west. The first, which is the most densely populated, is the Red River Valley. It has fertile soil and productive farming. North Dakota's lowest point is in Pembina County at the Red River at 751 ft (229 m). Going west, one hits the higher Drift Prairie region, made up of dark, fertile soil, rolling hills, lakes and streams, and the Turtle Mountains on the central northern boundary of the state. The Missouri Plateau, which is part of the Great Plains, covers the southwestern part of the state. This sparsely populated area includes the famous Badlands of the Little Missouri River and North Dakota's highest point, White Butte at 3,507 ft (1,069 m).

The state's climate is subhumid continental with short, hot summers and long, very cold winters. Rainfall is sparse to moderate with some drought periods. The precipitation increases as you go farther east. The climate and topography of North Dakota also impact its vegetation. The state is primarily prairie and plains, with only 1 percent of its land area covered by forests.

At the time of European arrival, the native Mandan, Arikara, Hidatsa, Cheyenne, Yankton Sioux, Chippewa, and Dakota tribes were found throughout the state. The area was first explored by the French, was acquired by the United States through the LOUISIANA PURCHASE in 1803, and was later explored by Meriwether Lewis and William Clark. In 1861, North Dakota became a territory with South Dakota, and in 1889, North Dakota became a state. Railroads simultaneously encouraged Norwegian, Swedish, German, and other immigrant farmers to settle the area.

Partially because of its extremely low unemployment rate, North Dakota has a strong economy. It is heavily based on agriculture, and the state is the country's leading producer of barley, oats, flaxseed, and other crops. Other important commodities include wheat, sunflowers, potatoes, corn, and livestock. The state also has generous reserves of oil, gas, and lignite which contribute to the fact that energy production is North Dakota's second most important part of the economy. The service sector contributes the largest part of the state's gross product. Manufacturing, including food processing and machinery, is also important.

BIBLIOGRAPHY. Federal Writers' Project of the Works Progress Administration, *The WPA Guide to 1930s North Dakota* (State Historical Society of North Dakota, 1990); Martin Hinzt, *North Dakota* (Children's Press, 2000); North Dakota Department of Agriculture, www.agdepartment.com (August 2004); United States Census Bureau, quickfacts.census.gov (August 2004); Robert D. Wilking, "North Dakota" *Worldmark Encyclopedia of the States*, Timothy L. Gall, ed. (Gale Research, 1995).

ANTHONY PAUL MANNION
FORT HAYS STATE UNIVERSITY

North Slope

THE NORTH SLOPE of ALASKA stretches from the high mountains of the Brooks Range to the ARCTIC OCEAN on the north. Miles of barren coastal plains and low, rolling hills, caused by freezing and thawing of the ground, make up the region. No trees are in evidence, as the short growing season will support only tundra plants. The word *tundra* comes from the Finnish word *tunturia*, meaning treeless plain. Tundra is characterized by extremely cold climate, limited drainage, a short growing season, and few plants and animals. Soil forms slowly and there is a layer of permanently frozen subsoil called permafrost. There is very little daylight in winter and temperatures are bitterly cold. In Barrow, the northernmost city on the North Slope, the winter temperatures can get down to -50 degrees F (-46 degrees C). Summer temperatures in Barrow average 40 degrees F (4.5 degrees C). Some parts of the North Slope are a bit warmer, but the whole area has a chilly arctic climate.

The North Slope receives only about 5 to 15 in (12.5 to 37.5 cm) of precipitation a year. During warmer weather, water stands on the surface, causing numerous small lakes. Wide shallow rivers flow north to the ocean. Sedges, reindeer mosses, low shrubs and grasses grow on the tundra, as well as 400 varieties of flowers. The plants must adapt to strong winds, low temperatures, and poor soil. Herds of caribou roam the area in summer. Other animals that live on the North Slope at some time of year include wolves, foxes, polar bears, seals, whales, and small mammals such as lemmings, voles, and ground squirrels.

The entire North Slope, which encompasses over 89,000 square mi (230,509 square km), is organized into one borough, making it the largest municipality in the world. The region is about the size of the state of MINNESOTA and lies entirely above the ARCTIC CIRCLE. Many of the people live a subsistence lifestyle and depend on hunting, trapping and whaling for much of their food.

There is no commercial agriculture within the North Slope, because of the short cool growing season and poor soil. Some people have gardens for their own use. The growing season has very long hours of sunlight, but produce has to be limited to plants that will grow and mature in temperatures ranging from 50 to 60 degrees F (10 to 16 degrees C). Besides Barrow, the North Slope includes the villages of Anaktuvuk Pass, Atqasuk, Kaktovik, Nuiqsut, Point Hope, Point Lay, and Wainwright. Most of the people who live here are Inuit. They used to be known as Eskimos, but prefer the term *Inuit*, which means "the People."

Petroleum was discovered at Prudhoe Bay in 1968. An 800-mi (1,287-km) pipeline was built and began carrying oil south to the port of Valdez in 1977. There is still controversy over whether or not oil drilling should be allowed in the ARCTIC NATIONAL WILDLIFE REFUGE in the far northeast part of the North Slope. The refuge includes 10,039 square mi (76,000 square km) of land. Research shows that lasting environmental damage has resulted from oil drilling at Prudhoe Bay.

BIBLIOGRAPHY. Walter R. Borneman, *Alaska: Saga of a Bold Land* (HarperCollins, 2003); "North Slope Borough, Alaska," www.quickfacts.census.gov (March 2004); "North Slope Borough Home Page," www.co.north-slope.askus (March 2004); "The Tundra Biome," www.ucmp.berkeley.edu (March 2004).

PAT MCCARTHY
INDEPENDENT SCHOLAR

Northern Mariana Islands

A COMMONWEALTH IN political union with the UNITED STATES, the Northern Mariana Islands (or CNMI) consists of a string of fourteen volcanic and limestone islands in the far western PACIFIC OCEAN. The islands have closer ties to American and European culture than most others in the region and have therefore opted to retain political and economic integration with the United States rather than seek the free association status of its neighbors.

The Marianas form a segment of a chain of volcanic islands that stretch north from eastern INDONESIA to JAPAN, following alongside a deep ocean trench (the Mariana Trench, which includes the deepest points on the planet) along the edges of the Philippine and Pacific tectonic plates. The three larger islands at the southern end of the chain (Saipan, Tinian and Rota) have 99 percent of the population (86 percent on Saipan), and most of the economic activity. They are volcanic in origin but older than the more recently formed islands to the north and have had longer to form fertile soil through erosion. The small northern islands are thus much taller and steeper (Agrihan is the highest point in all of MICRONESIA) and continue to see volcanic activity: Farallón de Pájaros, Asuncion, Pagan, and Guguan

have all erupted in the 20th century, Pagan most recently in 1981, forcing the evacuation of its small population, and Agrihan continually erupting since May 2003. These northern islands have little soil and insufficient rain so have little population.

The island of GUAM is the southernmost island of the Mariana chain but has been administered separately since it passed from Spanish rule to U.S. jurisdiction in 1898. The Northern Marianas were also governed by the Spanish, starting in the mid-16th century, but were sold to GERMANY in 1898, and annexed by Japan in 1914, before finally joining Guam under U.S. administration as part of the United Nations Trust Territory of the Pacific in 1947. The main island, Saipan, served as headquarters for the entire Trust Territory from 1962, while the neighboring island of Tinian was an important Central Intelligence Agency training base for Nationalist Chinese troops. The third major island is Rota. A covenant establishing the islands as a commonwealth with full internal autonomy plus U.S. citizenship was developed in the 1970s and went into effect in 1986.

The local economy benefits significantly from U.S. financial assistance, but is beginning to establish itself independently through tourism, mostly from JAPAN—Saipan is 5,625 mi (9,073 km) from San Francisco, CALIFORNIA, but only 1,272 mi (2,052 km) from TOKYO, Japan.

The tourist industry now employs 50 percent of the workforce and accounts for about half of the total gross domestic product. This, and the emerging garment industry has attracted heavy immigration from CHINA and the PHILIPPINES. Total population figures have risen from just under 17,000 in 1980 to over 80,000 today. Non-U.S. citizens make up about half the population, with Filipinos alone forming roughly a third of the population, but figures are shaky because of illegal migration, and the local native Chamorro people are increasingly dissatisfied with being a minority in their own territory.

BIBLIOGRAPHY. Ron Crocombe, *The South Pacific* (University of the South Pacific, 2001); Frederica Bunge and Melinda W. Cooke, eds., *Oceania: A Regional Study* (Foreign Area Studies Series, 1985); K.R. Howe, Robert C. Kiste, Brij V. Lal, eds., *Tides of History: The Pacific Islands in the Twentieth Century* (University of Hawaii Press, 1994); U.S. Office of Insular Affairs, www.doi.gov/oia (March 2004).

JONATHAN SPANGLER
SMITHSONIAN INSTITUTION

Norway

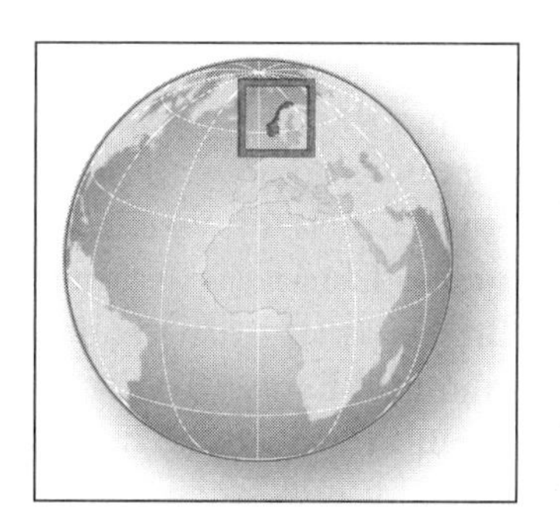

Map Page 1130 **Area** 125,182 square mi (324,220 square km) **Population** 4,546,123 **Capital** Oslo **Highest Point** 8,148 ft (2,469 m) **Lowest Point** 0 m **GDP per capita** $31,800 **Primary Natural Resources** petroleum, copper, natural gas, pyrites, nickel.

NORWAY IS A LONG, narrow country, stretching from the North Sea to the ARCTIC OCEAN, roughly the same length as the eastern coast of the UNITED STATES. Most of its terrain is rocky and mountainous, leaving little room for settlement or agriculture. Norwegians have thus traditionally turned to the sea for their livelihood; and although fishing remains a crucial industry, along with shipping—Norway's fleet is the fourth largest in the world—since the 1970s, it has been oil and natural gas pumped from the sea that has transformed the economy into one of the most dynamic in all of Europe. And because Norway generates nearly all of its electricity from hydroelectric stations on its swift mountain rivers, most of its oil is exported, making Norway the second largest oil exporter and providing a generous trade surplus that has been translated into one of the most comprehensive social welfare systems and the highest standard of living in the world.

Measuring Norway's coast can yield varying results: the direct distance from point to point is about 1,100 mi (1,770 km), while measurements that include all of the thousands of fjords and inlets bring the number closer to 15,592 mi (25,148 km). Including the 50,000 or so islands off Norway's coast, the distance rises again to 36,042 mi (58,133 km), giving Norway the longest coastline in the world. Norway occupies the western part of the Scandinavian Peninsula, shared with SWEDEN. The mountainous backbone of the peninsula, the Kjølen, forms the 972-mi (1,619-km) border with Sweden, except in the south, where Norway's territory includes the relatively flat basin east of the mountains, the location of the city of Oslo and the country's principal river, the Glåma.

Norway also shares a border in the far north with FINLAND, and a small stretch of border with RUSSIA (122 mi or 196 km), along the coast of the Barents Sea. The kingdom also includes several dependencies: BOUVET ISLAND in the South Atlantic, Peter I Island in the South Pacific (both close to ANTARCTICA), Jan Mayen in the

North Atlantic, and Svalbard, an archipelago north of the mainland, bordering on the Arctic Ocean.

Norway's geologic features are very old, dominated by the ancient Fennoscandian (or Baltic) Shield, and the mountainous chain that rises above it—a range that matches in age and structure the highlands of northwest Scotland and the Appalachian mountains in the eastern United States. The rocky surfaces of these land masses are worn down and marked by several waves of glaciation; most of Norway's coastal regions are bare or thinly covered rock.

Glaciation is also responsible for Norway's most characteristic feature, its long, narrow fjords, bordered by rugged peaks. Nearly 2,000 glaciers are left over from the last period of ice cover. The climate is cold, but moderated by warmer air from the Gulf Stream. In the far north, arctic conditions are more common. This area is home to Norway's Lapp, or Sami, population, many of whom continue to lead a traditional lifestyle of reindeer herding, though today this is largely done by snowmobile rather than on foot. In these areas, land of the midnight sun, the sun never drops below the horizon from May to July.

Norway has been fully independent only since 1905. Several centuries before, a Viking maritime empire had dominated the North Sea and North Atlantic, with settlements in ICELAND, GREENLAND, parts of Scotland and IRELAND, and even as far south as northwest FRANCE, an area of which they gave their name, Normandy. From the late 14th century, however, Norway lost its independence to Denmark and remained a satellite kingdom for four centuries before being transferred to the jurisdiction of Sweden during the 19th century. Norway was one of the first nations to join the NORTH ATLANTIC TREATY ORGANIZATION (NATO) after World War II but was less willing to join the common market, rejecting EUROPEAN UNION (EU) membership in referenda in 1972 and 1994. Besides the petroleum and gas industries, Norway's economy is dominated by food processing, shipbuilding, pulp and paper products, machinery, chemicals, and the centuries-old mainstay, fishing.

BIBLIOGRAPHY. *World Factbook* (CIA, 2004); Wayne C. Thompson, *Nordic, Central and Southeastern Europe 2003*, The World Today Series (Stryker-Post Publications, 2003); Clifford Embleton, ed., *Geomorphology of Europe* (Wiley, 1984); "Norway," www.norway.org.uk (July 2004).

JONATHAN SPANGLER
SMITHSONIAN INSTITUTION

Rugged and mountainous, Norway's coastline has been measured as the longest in the world.

Nubia

NUBIA, KNOWN AS THE Gateway to the SUDAN, does not exist as a political entity in the 21st century. Geographically, the section of Northeast Africa that was once Nubia has been encompassed into northern Sudan and southern EGYPT, with cataracts (areas where geological forces have formed outcroppings of rock) along the NILE RIVER determining Nubia's so-called boundaries. The land once known as Nubia is located

between the First Cataract south of Aswan and the Sixth Cataract near KHARTOUM. Contemporary Nubia ranges from the Nubian Desert, which makes up the easternmost section of the Sahara Desert, to the more fertile Nile Valley area.

Since 1970, as a result of the High Dam Project initiated by Egyptian president Gamal Abdul Nasser, most of Nubia has been covered by Lake Nasser, the world's largest lake. Some 120,000 Nubians were relocated to cities such as Kom Ombo, Sukkot, Mahas, and Halfawi in Egypt and New Halfa and Khashm al-Girba in the Sudan. Some Nubians refused to leave the area, preferring to move to higher grounds away from the dam. In new communities, approximately 85 percent of all Nubian males were forced to find work outside the areas. Some Nubians continue to be employed in traditional service jobs. However, as Nubians have become more educated, large numbers have become doctors, lawyers, teachers, and other professionals. Recently, after years of exile, a number of Nubians returned to the banks of the Nile River through the efforts of the High Dam Lake Development Authority with government grants to establish new agricultural communities.

NUBIAN HISTORY

Nubian history is portioned into three major periods: ancient, Christian, and Islamic. Ancient Nubian culture dates from 34th century B.C.E. to the 4th century C.E. Nubia began as a pastoralist area until urban centers emerged around the 26th century B.C.E. These cultures are referred to as A-Group and C-Group respectively. A major trading center, Kerma, flourished in Upper Nubia around 2500 to 1550 B.C.E. Kerma's unique burial mounds and delicate objects suggest a highly advanced civilization, and it is known from Egyptian records that Kermans traded extensively with Egypt and other states in the eastern Mediterranean.

When Egypt lost power in the region during a period of fragmentation about 1700 B.C.E., the Kerman kings also gained control of parts of Lower Nubia, absorbing C-Group cultures. The Egyptians retook most of Nubia, including Kerma, after the latter's resurgence between 1550 and 1450 B.C.E. Lower and Upper Nubia were separated into the Egyptian provinces of Wawat and Kush. During that time, most of the region's polities absorbed Egyptian culture and customs, though they retained a separate identity.

Kush eventually overtook Egypt to form the 25th "Cushite" Dynasty that ruled Egypt and Nubia between 750 and 663 B.C.E. Subsequent Nubian capitals at Napata and Meröe consecutively lasted until the 3rd century C.E., when Meröe was destroyed by the Ethiopians.

The X-Group or Ballana culture marks Nubia's transition to the Christian era. Ballana replaced Meröe as the predominant culture in the region, and was absorbed into the Kingdom of Noabdia, which became fully Christianized under Byzantium influence during the 5th and 6th centuries. Two successive states, Makuria (which absorbed Nobadia) and Alodia emerged in the 7th century. Nubia became an isolated region, cut off from the Byzantine world during Muslim expansion; despite this, the region managed to retain its Christian identity until the 15th century, when it dissolved through conquest and intermarriage. As the Ottomans occupied Nubia proper, the Funj Empire was created in southern Nubia. Both areas were conquered by the Egyptians, then the British, during the 19th century. Nubia became a joint Egyptian and British protectorate in 1899, and was renamed the Anglo-Egyptian Sudan.

NUBIAN GEOGRAPHY

Geographically, Nubian land is made up of arid desert filled with sand and black rocks. The desert climate results in hot, dry summers and moderate winters. The Egyptian section of Nubia rarely sees rain, and the Sudanese section sees only small amounts. Khartoum, for instance, rarely sees more than 7 in (18 cm) annually. In addition to persistent draught, Nubians must also contend with dust storms, earthquakes, flash floods, and landslides that can damage crops and threaten lives and property.

The rock throughout Nubia is either sandstone or granite. In sandstone areas, the Nile Valley is made up of wide alluvial floodplains. This area near Aswan is the most fertile section of Nubia. Nubian land is bordered on the east and west of the Nile by cultivated fields. The western section is dotted with a number of small watering holes. In Upper and Southern Nubia, sections of granite formation have resulted in narrow, sharp cliffs that are generally infertile. This section has been divided into the Batn el Hajar, the Abri-Delgo Reach, the Dongola Reach, the Abu Hamed Reach, and the Shendi Reach, making up five distinct topographic regions along the Nile River.

Nubia was considered a major location along the African trade route and provided essential natural resources such as gold, ivory, copper, ebony, and dolerite. Since ancient times, the palm tree has also provided a major component of the Nubian economy. Palm trees

are frequently given as gifts. They are also used to manufacture a number of products such as furniture and rugs and are used as fuel and animal food. The fruit of the date palm continues to be the most economically viable byproduct of Nubian palms. Other agricultural products of modern Nubia include sorghum and millet.

Transplanted Nubians maintain many of their traditions and cultural characteristics. In addition to Arabic, Nubians have continued to speak dialects such as Sukot, Halfawi, Mahas, and Dongolawi. Understanding the importance of customs and rituals to a displaced people, Nubians form their own closely knit communities in areas of migration. Ancient Nubian rituals are often centered on the all-important Nile River. Newborn infants were immersed in its waters for protection. Newlyweds bathed in the Nile to promote fertility. As part of the mourning ceremony, Nubian women washed mud and blue dye from their bodies. Henna and perfumes were submitted as offerings to the great Nile River. The belief in water angels survived the introduction of both Christianity and Islam into Nubia.

The New Nubia Museum, which opened in November 1997, was erected near the Cataract Hotel in Aswan to bring together over 2,000 Nubian treasures and relics, including jewelry, pottery, and scarabs. The centerpiece of the museum is a gigantic statue of Rameses II. Overall, 23 Nubian temples and shrines were relocated, including the temple of Queen Hatshepsut, which was dismantled and transported in crates to Egypt's National Museum. The temple of Kalabsha, the memorial chapel of Rameses II, and the Kiosk of Kertassi have been reconstructed near the High Dam. While most of the archaeological finds remain in the area, Nubia has become familiar to people in other areas through the relocation of significant relics to museums around the world in response to donations to UNESCO's (United Nations) massive preservation project of the 1960s. Such relics include the Temple of Dendur, which has been relocated to the Metropolitan Museum in New York City, and the Greco-Roman temple of Debod, now located in Madrid.

BIBLIOGRAPHY. Mark S. Copley, et al., "Processing Palm Fruits in the Nile Valley: Biomolecular Evidence from Qasr-Ibriam," *Antiquit,* (September 2001); Walter A. Fairservis, Jr., *The Ancient Kingdoms of the Nile And The Doomed Monuments of Nubia* (Thomas Y. Crowell, 1962); Kathy Hansen, *Egypt Handbook* (Moon Publications, 1993); Jill Kamil, *Aswan and Abu Simbel: History and Guide* (American University of Cairo Press, 1993); Helen Metz, "Sudan: A Country Study" (Library of Congress, 1991); Richard Lobban, *Historical Dictionary of Ancient Civilizations and Historical Eras* (Scarecrow Press, 2004); David O'Connor, *Ancient Nubia: Egypt's Rival in Africa* (University of Pennsylvania Museum Publications, 1994); Think Quest, "Nubia: Geography and Topography," http://library.thinkquest.org (January 2005).

Nathalie Casavin
Waseda University, Japan
Elizabeth Purdy, Ph.D.
Independent Scholar

Ob-Irtysh River

THE OB AND THE IRTYSH rivers together form one of the largest river BASINs in the world, but also drain an area among the least populated and least known to outsiders. The rivers flow from the isolated mountain ranges of Central Asia (the ALTAI and Sayan ranges) across the sparsely populated Western Siberian Lowland to the Kara Sea, a subsidiary of the ARCTIC OCEAN. Population density for the basin as a whole is only nine persons per square km, but there are several large cities clustered around the mineral wealth of the river's southern watershed: Omsk on the Irtysh and Novosibirsk on the Ob are the largest cities in SIBERIA, along with Chelyabinsk, located on a tributary in the western part of the Ob-Irtysh Basin, in the URAL MOUNTAINS, which form the basin's western boundary. The basin covers 1,159,274 square mi (2,972,497 square km)—roughly the same as the MISSISSIPPI basin—and lies mostly within RUSSIA, though the southernmost courses of the Irtysh flow through northern KAZAKHSTAN and small corners of MONGOLIA and XINJIANG, CHINA. Altogether the rivers and their tributaries connect about 17,000 miles (27,400 km) of navigable waterways, though most of these are frozen for much of the year.

The Ob and Irtysh Rivers both have their headwaters in the highlands of the Altai Mountains, on the borders of Mongolia, where peaks reach heights of 13,200 ft (4,000 m) or more. This is one of the most remote spots on Earth, where four countries come together (Russia, Kazakhstan, Mongolia, and China), over 3,000 mi (4,800 km) from the sea. The two rivers start on different sides of this range, however, and do not meet up until both rivers have crossed most of the flat Siberian plains. The Irtysh is longer, but the Ob has more volume, and when they meet, at Khanty Mansiysk (a town named for the two dominant local indigenous groups), their course becomes sluggish and marshy. The far north of Russia is mostly flat and marshy, with little precipitation. Agriculture is severely limited by harsh climate and year-round permafrost. The Ob becomes divided into many ribbons, subject to enormous spring floods and dangerous ice flows during summer thaws. Here the river valley can at times reach 25 mi (40 km) wide. The Ob enters the sea through the 600 mi (375 km) Gulf of Ob, a forked indentation of the Kara Sea.

The region was sparsely populated by nomadic peoples (Mansy, Khanty, Nenets, and Samoyedic peoples) for centuries until Russians became attracted to the area for its "soft gold": furs of numerous squirrels, otters, ermine, mink, and sable. Fortified wooden stockades were built at river junctions as trappers and merchants moved eastward, including such cities as Tobolsk (1587) on the Irtysh, and Salekhard (1595) on the Ob. This latter city is one of the furthest north in all

of Russia, located close to where the Urals meet the Arctic at the Gulf of Ob and where the forests meet the Arctic tundra. Later cities were founded on the upper Ob (Narym and Tomsk) before the push for furs moved on eastward into Siberia. The large industrial cities of the south were built later, with the development of coal and iron ore industries, especially in the Kuznetsk Basin, and more recent pumping of oil near Surgut on the middle Ob. Because of the river's swampiness and lengthy periods of ice cover (generally October to May), it is not used much for navigation. There has been some harnessing of the great volumes of water flowing off of the Altai Mountains, notably at a hydropower station on the Irtysh in northeastern Kazakhstan, at Ust-Kamenogorsk, close to where the river (called the Ertix) flows out of Mongolia and into the large Zaysan lake. This region is one of Kazakhstan's most industrialized and holds a large portion of its population. The lower parts of the rivers are spawning grounds for sturgeon, salmon and whitefish, and the Ob estuary forms one of the largest fishing industries in the Russian Arctic.

BIBLIOGRAPHY. Sergei Petrovich Suslov, *Physical Geography of Asiatic Russia*, N. D. Gershevsky, trans. (W.H. Freeman, 1961); John J. Stephan, *The Russian Far East: A History* (Stanford University Press, 1994); C. Revenga, S. Murray et al., *Watersheds of the World* (World Resources Institute, 1998).

JONATHAN SPANGLER
SMITHSONIAN INSTITUTION

Ohio

OHIO IS A MIDWESTERN state in the Great Lakes region of the UNITED STATES. Interestingly, when the first settlers were arriving, the area was considered to be America's great northwest. In the northeast, Ohio is bordered by PENNSYLVANIA from Lake ERIE southward to the Ohio River near East Liverpool. The Ohio River forms a natural boundary between the states of WEST VIRGINIA (southeast) and KENTUCKY (south), while INDIANA provides a western border, and MICHIGAN and Lake Erie provide borders in the north. With an area of 106,765 square km (41,222 square mi), Ohio ranks 34th in size in the United States. With 11,353,140 residents, the state ranks 7th nationally in terms of total population and 9th in terms of population density. Cleveland is the center of the state's largest metropolitan statistical area (MSA), although Columbus is the largest city and the state capital. Other major cities are Cincinnati, Dayton, Toledo, and Akron. Overall, the ten largest cities in Ohio, six with populations greater than 100,000, account for 22 percent of the state's total population. Ohio was the 17th state to join the United States (March 1803).

From the dunes along the shores of Lake Erie to the gorge cut by the Ohio River in the south, the land is generally flat over the eastern half of the state, with rolling hills in the east and small rugged hills in the southeast. Before the first settlers came to the region, much of the land was covered with miles of virgin forest, including numerous Buckeye trees from which the state derives its nickname. Today, only vestiges of the dense woodlands that helped to build the many cities remain.

Ohio's topography consists of three easily identifiable regions with a general northeast to southwest trend: the Great Lakes Plains, the Central Plains, and the Allegheny Plateau. There are also two smaller notable physiographic zones. One is a small strip in the north bordering the Lake Erie shoreline called the Lake Erie Plains. This region, which varies in width from 50 mi (81 km) at Toledo's Maumee Bay to 10 mi (16 km) at Conneaut near the Pennsylvania border, extends for all 312 mi (503 km) of the Ohio portion of the Lake Erie shoreline. The shoreline itself also has two very different geographies.

Along the eastern half of the shoreline in the northern extreme of the Allegheny Plateau are clay bluffs often 8 to 10 ft (2 to 3 m) high, while the western half of the shoreline in the Central and Great Lakes Plains has beaches of clay and sand. The other is a narrow region extending southeastward along the Miami River from Indian Lake in the east central part of the state to the Ohio River in the very southeastern corner.

The state's highest point (1,549 ft or 472 m)—just southeast of Bellefontaine in the east central part of the state—and the lowest point (455 ft or 139 m), on the Ohio River at the Indiana and Kentucky borders, also lie within this region.

Other important rivers with impact on both the physical and economic geography include the Scioto, which runs from north of Indian Lake south through the middle of the state to the Ohio River at Portsmouth; the Muskingum, which drains a large portion of the southeastern Allegheny Plateau entering the Ohio at Marietta; and the Maumee, which runs northeasterly across the northeastern part of the state to

Maumee Bay at the eastern end of Lake Erie and Toledo.

In prehistoric, times Ohio was home to mound builders, many of whose mounds are preserved in state parks. Prior to the arrival of Europeans, the Native Americans living in the area included the Iroquois, Erie, Miami, Shawnee, and the Ottawa. The French were the first Europeans to claim the area when La Salle was exploring the Ohio Valley in 1669. The region was soon a haven for fur traders and land grabbers. By the 1750s, the last of the French and Indian Wars saw the French losing and control of the area given to the British. In 1763, the British issued a proclamation forbidding settlement west of the Appalachian Mountains, furthering unrest in the region, and in 1774 issued the Quebec Act, putting the region within the boundaries of CANADA. In 1783, the Treaty of Paris ceded the area to the United States, and by 1787, the area became the first region in the Old Northwest to be organized under the Ordinance of 1787. In 1788, Marietta became the first permanent American settlement founded on the Old Northwest. In 1802, a state convention drafted a constitution, and in 1803 Ohio entered the Union, with Chillicothe as its capital. Columbus became the capital in 1816.

Following the War of 1812, Ohio's growth was spurred by the building of the Erie and other canals, and toll roads. The National Road was a vital settlement and commercial link. After the Civil War, increased shipments of ore from the upper Great Lakes and the development of the petroleum industry in northeastern Ohio helped shift the center of economic activity from its fur-trading origins along the banks of the Ohio River to the shores of Lake Erie. Ohio was hit hard by the Great Depression but rebounded during and shortly after World War II. The state economy was particularly depressed during the 1970s and 1980s as the automobile, steel, and coal industries virtually collapsed, with many of the northern industrial centers losing significant portions of their populations. Since the late 1980s, the state has sought to diversify its economy through enlargement of the service sector.

Although highly industrialized, the availability of mineral resources has kept the state among the national leaders in the production of lime, clays, and salt. The state is also a historic center of the nation's ceramic and glass industries. Ohio has extensive farmlands in those areas enriched by limestone during the last ice age, producing large amounts of corn, soybeans, hay, wheat, cattle, hogs, and dairy items. Although agricultural production remains important to the state, the number of family farms is rapidly declining.

Railroads, canals, and highways continue to provide significant transportation linkages for raw materials and manufactures. The state's ports on Lake Erie, especially Toledo and Cleveland, handle iron and copper ore, coal, and oil in addition to finished goods such as steel and automobiles parts. In spite of the general industrial decline since the late 1960s, the state has retained many important manufacturing centers for transportation equipment, primary and fabricated metals, and machinery. Nationally, Ohio ranks 7th in terms of total gross state product ($374 million) and 20th in per capita income ($27,977).

BIBLIOGRAPHY. Carl H. Roberts and Dean W. Moore, *History and Geography of Ohio* (Laidlaw Brothers, 1981); Alfred J.Wright, *Economic Geography of Ohio* (Department of Natural Resources, 1957); Leonard Peacefull, *A Geography of Ohio* (Kent State University Press, 1996); Andrew R. Cayton, *Ohio: the History of a People* (Ohio State University Press, 2002); U.S. Census Bureau, www.census.gov (August 2004).

RICHARD W. DAWSON
CHINA AGRICULTURAL UNIVERSITY

Okavango

COVERING AN AREA roughly the size of CONNECTICUT, Okavango is the largest inland DELTA in the world. It is located in north-central BOTSWANA in the middle of the KALAHARI DESERT and is formed by the Cubango and Cuito rivers flowing out of mountains in ANGOLA and into the Okavango River.

Okavango covers a swampy depression that was once a prehistoric lake. The northern part of the swamp is always wet with lots of papyrus growing in it and covers about 2,320 square mi (6,000 square km) of the Okavango area. A small shelf of sediment about 16 ft (5 m) high from the Kalahari confines the delta here. The southern part of Okavango fills seasonally during the rains and covers another 2,700 square mi (7,000 sq km) and up to 4,630 square mi (12,000 sq km) depending on how much rainfall they get each year. In this area the delta branches out into three main channels: the Thaoge to the west, the Jao/Boro in the center, and the Nqoga/Maunachira to the east. Average water depth in the swamps of Okavango are about 5 ft

(1.6 m). Much of this area is also covered in vegetation that grows profusely due to the rich silt constantly being brought in by the Cubango and Cuito Rivers. These rivers bring a flood of water during the rainy season peaking in March or April. The water is really a blessing to this area since otherwise the delta would only get about 19 in (50 cm) of rain per year. It would have turned into part of the Kalahari Desert long ago were in not for the water flowing into the area. Only 2 percent of water in Okavango actually flows out of the delta area via the Boteti River. This is due to the huge amount of evaporation that occurs in this area. More than 72 in (180 cm) of water evaporate each year because of the average daily temperatures of 86.7 degrees F (30.4 degrees C) in Okavango. The papyrus is well adapted to this type of climate and is one of the fastest growing plants on Earth. It has spread throughout Okavango, sending seeds downstream to become new plants. Papyrus forms beds, which become their own islands in the swamps of Okavango, towering overhead like a jungle.

The upper region of Okavango that is permanent swamp is often called the Panhandle portion of the delta. The region below this is a broad and slightly fan-like area with a gentle slope to it. At the edge of this fan area, there are two major depressions: the Mababe Depression in the east and Lake Ngami, which is currently dry, in the west. In response to the amount of silt in the area, the dense vegetation, shifting of the Earth's crust, and the constantly moving water, over time Okavango's delta has changed shape. Both of these areas contained actual lakes in the past and are now wide grassy areas with rich soil where the Damara people of the region herd sheep and cattle.

Okavango is an extension of the East African Rift System of faults. Two parallel fault lines in the Panhandle confine the upper portion of the delta where the Okavango River enters the Kalahari Basin. The Thamalakane and Kunyere faults define the southeast limit of the delta. At 125 mi (200 km) long, the Thamalakane fault acts like a natural dam from which the Boteti River flows out of the delta. The parallel Gomare fault, a continuation of the Great Rift Valley of eastern Africa, runs southwest to northeast at the southern end of the Panhandle.

After passing through this fault, the swamps then branch out in the alluvial fan formation of the rest of Okavango's delta. The rift between the Gomare and Thamalakane faults is what is believed to have originally formed the geographical basis for the Okavango delta of today.

This whole area supports a wide variety of wildlife. In the swampy Panhandle area live hippopotamuses, crocodiles, and predatory fish along with other smaller species of fish, numerous invertebrates, and thousands of insects. The crocodiles of this area can grow to extraordinary size because they have a tendency to feed on larger mammals that graze along the delta's floodplains. In the Panhandle are many areas of exposed sandbanks that serve as a breeding ground for the Okavango's crocodiles and several African birds including the graceful skimmers and the magnificent Pel's fishing owl.

During the dry time in the delta, huge herds of large mammals migrate in to take advantage of the plentiful grasses. Elephants, buffalo, zebra, wildebeest, impala, bushbuck, warthog, eland, impala, gnu, springbok, and various other antelope come here by the hundreds and thousands, followed by their natural predators, such as lions, leopards, and the nomadic tribal hunters of the area. There are birds of almost every species here, including geese, ducks, teal, bitterns, egrets, ibises, flamingos, and hundreds of other waterfowl. It is said that the game viewing here is better than in the Serengeti of Kenya and Tanzania.

BIBLIOGRAPHY. Karen Ross, *Okavango: Jewel of the Kalahari* (Macmillan, 1987); Frans Lanting, *Okavango: Africa's Last Eden* (Chronicle Books, 1993); Creina Bond, *Okavango: Sea of Land, Land of Water* (St. Martin's Press, 1984); T. S. McCarthy, "The Great Inland Deltas of Africa," *Journal of African Earth Sciences* (v.17/3, Elsevier, 1993).

CHRISTY A. DONALDSON
MONTANA STATE UNIVERSITY

Oklahoma

LOCATED MIDWAY BETWEEN the east and west coasts of the UNITED STATES and just south of the geographic center of the United States in KANSAS, Oklahoma is one of the Great Plains states. The name comes from the Native American Choctaw language as *okla* meaning "people," and *humma* meaning "red." Despite images to the contrary, Oklahoma has a unique geography that serves as a transition zone between the forested mountain woodlands of the east and the deserts and mountains of the west. The land rises gently from 207 ft (87 m) above sea level in the extreme southeast to 4,973 ft (1,516 m) above sea level at Black

Mesa in the northwestern corner in an area known as the Panhandle.

Oklahoma offers a variety of features, from grassland plains in the west to forests and mountains in the east. Most of the state is a great, rolling plain, sloping gently from northwest to southeast. Although the region is considered part of the Great Plains, Oklahoma has four mountain ranges: the Ouachita in the southeast, the Boston in the northeast (part of the Ozark Plateau that runs across northwestern ARKANSAS and MISSOURI), the Arbuckle in the south-central part of the state just north of the TEXAS border, and the isolated Wichita in the southwest. Approximately 24 percent of the state's total area is forested, generally in the mountainous regions along the Missouri and Arkansas border. Throughout the northwest and the Panhandle are sudden outcrops of sandstone and gypsum, sharp ravines, and stark hills. The Panhandle may be one of Oklahoma's most recognized map features; that strip of land in the northwest that extends west from the main body of the state as if it were pointing at something. In the north-central region there are several salt flats near the border with Kansas. The largest of these, the Great Salt Plains, covers about 25 square miles (65 square kilometers) near the city of Cherokee.

Oklahoma is bordered by COLORADO and Kansas in the north, by Texas in the south, by Missouri and Arkansas in the east, and by NEW MEXICO and Texas in the west. With a total area of 69,956 square miles (181,186 sq km) the state ranks 18th nationally in terms of size, including 1,137 square miles of water. Oklahoma's estimated resident population at the end of 2003 was 3,511,532, ranking 27th nationally. Because 41 percent of the state's population lives in the 10 largest cities, the state feels incredibly open and free. Oklahoma City, with a population of 523,303, is the state's largest city and capital, while Tulsa, with 387,807 people, is the state's other major metropolitan area.

The state has a number of important rivers, but the four major rivers are the Arkansas, Red, Canadian, and Grand (or Neosho). The Red River also forms the state's southern boundary with texas. The other significant rivers and streams, all flowing into either the Arkansas or the Red, are the Illinois, Verdigris, Poteau, Canadian, Cimarron, Salt Fork of the Arkansas, and the Washita. Several of the larger rivers in the eastern half of the state have been dammed creating a number of large lakes. Largest of the more than 60 such reservoirs are Eufaula, Texoma, the Grand Lake O' the Cherokees, and Robert Kerr.

Most of Oklahoma has a warm, dry climate. The northwestern part of the state is cooler and drier than the southeast. Precipitation varies greatly throughout the state. Annual average precipitation ranges from 50 in (128 cm) in the southeast to 15 in (38 cm) in the western Panhandle. Snowfall ranges from 2 in (5 cm) a year in the southeast to 25 in (63 cm) in the northwest; the Panhandle receives the most snow.

ECONOMY

Oil made Oklahoma a rich state, but natural-gas production has now surpassed it. Oil refining, meatpacking, food processing, and machinery manufacturing (especially construction and oil equipment) are important industries. Other major commodities include nonelectric machinery, petroleum and coal products, food products, fabricated metal products, glass, rubber and plastic products, and transportation equipment. The state's principal minerals are petroleum and natural gas, followed by helium, gypsum, zinc, cement, coal, copper, and silver. Oklahoma's rich plains produce bumper yields of wheat, as well as large crops of sorghum, hay, cotton, and peanuts. Chief agricultural products are beef cattle, sheep, hogs, poultry, milk, wheat, hay, sorghum and other grains, peanuts, pecans, and cotton.

HISTORY

Native American tribes inhabited Oklahoma when Spanish explorer Francisco Coronado ventured through the area in 1541. The land was included as part of the 1803 LOUISIANA PURCHASE. From the early 1820s until about 1846, the U.S. government waged a continuing effort to move Native Americans from the eastern United States to the Indian Territory (Oklahoma). The Five Civilized Tribes (southeastern United States) were among those tribes forcibly relocated over numerous routes during this time, the most famous being the Cherokee "Trail of Tears."

Immediately after the Civil War, the long cattle drives from Texas to the Kansas railroad centers began, ushering in the age of the cowboy. The cattle were fattened on the rich prairie grasses of Oklahoma as they slowly made their way north to towns such as Dodge City or Abilene. With time, Western folklore became rich with names such as the Chisholm and Cimarron trails. Western expansion of the United States reached the territory in the late 1800s, sparking a major controversy over the fate of the land. The government decided to open the western parts of the territory to settlers by holding a total of six land runs between

1889 and 1895. Included in this land grab was the famous Cherokee Strip or Cherokee Outlet, a narrow piece of land in northern Oklahoma along the Kansas border. Settlers came from across the nation and even other countries to stake their claims. In part because of the discovery of oil, which made Oklahoma the "place to go to strike it rich," statehood had become a sure thing at the turn of the 20th century, and on November 16, 1907 Oklahoma became the 46th state of the United States.

BIBLIOGRAPHY. Victor E. Harlow, *Oklahoma History* (Harlow Publishing, 1967); Edwin C. McReynolds, *Oklahoma: A History of the Sooner State* (University of Oklahoma Press, 1971); Arrell Morgan Gibson, *Oklahoma: A History of Five Centuries* (University of Oklahoma Press, 1973); Edwin C. McReynolds, Alice Marriott, and Estelle Faulconer, *Oklahoma: The Story of Its Past and Present* (University of Oklahoma Press, 1982); James Shannon Buchanan and Edward Everett Dale, *A History of Oklahoma* (Row, Peterson and Company, 1924); Luther B. Hill, *A History of the State of Oklahoma* (Lewis Publishing, 1910).

Richard W. Dawson
China Agricultural University

Olduvai Gorge

OLDUVAI GORGE IS LOCATED in the East African Rift Valley of northeastern TANZANIA, on the eastern edge of the Serengeti Plain. Discoveries of fossils dating back 1.8 million years were made there by Louis and Mary Leakey, and Donald Johansen.

The gorge is now part of the Ngorongoro Conservation Area and lies just to the west of two lakes: Natron in the northeast and Eyasi in the south. Its original name was Oldupai, a Maasai word for the sisal plants that grew there.

The Olduvai Gorge extends for 31 mi (50 km), and its steep sides are up to 295 ft (90 m) high. It forms a Y shape, with a branch called the Side Gorge joining with the Main Gorge. Millions of years ago, volcanic eruptions set down the rocks and layered ash that would become Olduvai Gorge.

By 2 million years ago, the area was a shallow alkaline lake, an attractive, swampy habitat for many animal and plant species. About 1.5 million years ago, the area's climate changed drastically. The lake size lessened, and the fossil records show that species more adapted to a dry savanna moved into the area. Within the last 500,000 years, tectonic activity had created the Olbalbal Depression to the west of the lake, and at some point waters from the lake began to flow into the depression. The Olduvai Gorge was formed by the draining waters, which over millennia carved through the deposits of ash and fossils. Erosion exposed layers of the gorge that date to the lower Pleistocene Age, and have yielded archaeological finds up to 1.8 million years old.

Today, archaeologists divide the stratification into six areas, or beds, with Bed One being the oldest, closest to the basalt bedrock. Although fossils were first found in the area in 1911, nearly 50 years passed before its importance was realized. In 1959, Mary Leakey discovered the skull of a hominid over 1.8 million years old. Originally named Zinjantropus by the Leakeys, it is today called *Australopithecus Boisei*. In the 1960s, the Leakeys also found the first remains of *Homo habilis*, a 1.8-million-year-old hominid, possibly the first hominid to use the tools found at Olduvai Gorge. Both of these were found in Bed One, along with pebble tools.

In Bed Two, which dates to 500,000 years ago, the remains of *Homo erectus* have been found, along with hand axes. Remains of *Homo sapiens* (modern man) have been found in Bed Four.

BIBLIOGRAPHY. Earthwatch Volunteer, "Early Man at Olduvai Gorge," www.episcopalhs.org (March 2004); Donald Johanson and J. Shreeve, *Lucy's Child* (Early Man Publishing, 1989); Phillip Tobias, et al., *Olduvai Gorge, Volumes 1–4* (Cambridge University Press, 1965–1991).

Vickey Kalambakal
Independent Scholar

Oman

Map Page 1122 **Area** 82,031 square mi (212,460 square km) **Population** 2,807,125 **Capital** Muscat **Highest Point** 9,776 ft (2,890 m) **Lowest Point** 0 m **GDP per capita** $8,300 **Primary Natural Resources** petroleum, copper, asbestos, natural gas.

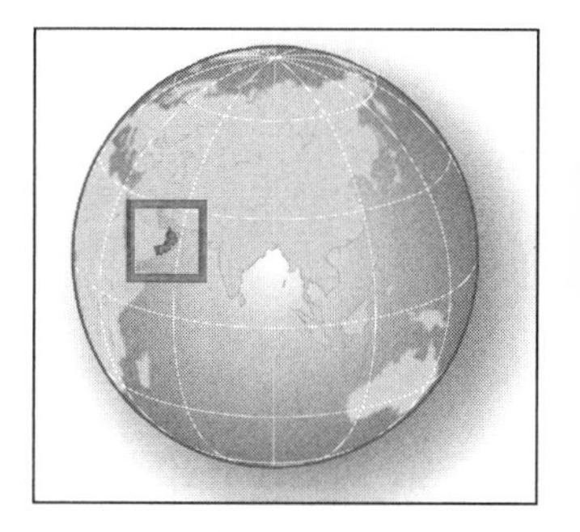

OMAN IS LOCATED in the MIDDLE EAST, bordering the ARABIAN SEA, Gulf of Oman, and PERSIAN GULF, between YEMEN and the UNITED ARAB EMIRATES (UAE). It is a country in the lands associated with earliest civilizations.

By 2000 B.C.E., Oman was known for its production of copper in the north, while the south produced frankincense, which was essential to the social religious life of ancient peoples. Oman adopted ISLAM in the 7th century C.E., during the lifetime of the Prophet Muhammad. The Omanis consolidated their power along the coast of the Arabian Sea, flourishing in maritime trade, but this expansion was checked by a Portuguese invasion in 1506. A century of occupation followed, and afterward, once again the Sultan of Oman spread the faith and Arab culture as he extended his conquests to Mombasa and Zanzibar, other portions of the southern Arabian Peninsula, and the Makran coast (now PAKISTAN). At this height of power in 1856, sons of the sultan fought over succession and the empire was split into Zanzibar and Oman. Decline came swiftly, and soon the sultan lacked the funds to placate the Imams of the interior.

By 1913, there was open rebellion. Oman slipped into medieval repression until Sultan Qaboos overthrew his father in 1970. He launched aggressive reform and modernization. The new sultan was confronted with insurgency in a country plagued by endemic disease, illiteracy, and poverty. He judiciously used foreign military support and progressive measures to defeat the separatist revolt and reintegrated the effected provinces.

With Sultan Qaboos's modern and progressive leadership, considerable change has come to ancient and traditional Oman in the past three decades. Many of his economic, educational, and health care improvements have been the first of their kind. The oil industry is modest in comparison to gulf neighbors but has supplied the revenues for national infrastructure improvement and diversification of the economy.

The government has invested in copper mining and refining, as well as the development of light industry. Adoption of modern techniques and equipment in agriculture and the fishing industry is increasing yields and profitability. There is a concerted effort to decrease the dependency on expatriate labor in the public and private sectors. With improved education and training now available, the "Omanization" of the managerial labor force is progressing. The sultan has opened Oman to tourism (another first) and is making efforts to liberalize foreign investment and joint ventures.

Mutrah Fort in Oman is nestled within the mountains and desert of the Middle Eastern country.

Most Omanis are Ibadi Muslims, belonging to one of Islam's earliest fundamentalist movements. The Ibadi are distinguished by their conservative doctrine and their system of selecting religious leaders by consensus. This makes Oman unique in the gulf and continues to influence the role of the Sultan to the interior communities and their religious leaders. Oman remains a very conservative society, but one that has a history of contact with the wider world. With its strategic location on the Strait of Hormuz and the mouth of the Persian Gulf, Oman will remain of interest to the industrialized world.

BIBLIOGRAPHY. "Background Note: Oman," U.S. Department of State, www.state.gov (April 2004); *World Factbook* (CIA, 2004); "Oman," World Guide, www.lonelyplanet.

com (April 2004); Ian Skeet, *Muscat and Oman* (Faber and Faber, 1974); H.J. de Blij and Peter O. Muller, *Geography: Realms, Regions, and Concepts* (Wiley, 2002).

IVAN B. WELCH
OMNI INTELLIGENCE, INC.

Ontario, Lake

LAKE ONTARIO, BORDERED by the UNITED STATES (NEW YORK state) and CANADA (province of Ontario), is the smallest and easternmost of the five Great Lakes, covering an area of 7,320 square mi (18,960 square km). It is 193 mi (311 km) long and 53 mi (85 km) wide, with a shoreline of approximately 480 mi (772 km). It lies 243 ft (74 m) above sea level, the lowest of any of the Great Lakes, while its deepest point is 802 ft (244 m) below the surface. Because its depth serves as a thermal reservoir, Lake Ontario does not freeze in winter except along the shore where the water is shallow. The surface is cool in the summer and warm in winter, and serves to moderate the climate of the surrounding land.

Water comes into Lake Ontario from Lake ERIE by means of the Niagara River and the Welland Canal, which was built to bypass the famous Niagara Falls. Other rivers flowing into Lake Ontario include the Black, Genesee, Oswego, Trent, and Humber. The Genesee and Oswego rivers, combined with the Erie Canal, allow ship traffic to pass from Lake Ontario to the Hudson River, which leads to NEW YORK CITY. Lake Ontario empties into the ATLANTIC OCEAN via the St. Lawrence River.

The lake has long enjoyed an important status for shipping, and the oldest U.S. lighthouse on the Great Lakes was built at Fort Niagara, just beyond the falls, in 1818 to aid in navigation. Lake Ontario boasts many excellent harbors, and their economic importance has only increased since the completion of the St. Lawrence Seaway made it possible to transport goods by water from any of the Great Lakes to the Atlantic Ocean. On the U.S. side of Lake Ontario, Rochester and Oswego are located in New York state. Canada has Coburg, Toronto, Hamilton, and Kingston. Toronto, capital of the province of Ontario, is the largest city on the shores of Lake Ontario.

Beyond these cities, the shoreline is still primarily rural. Although there are excellent orchards all around the lake, the region is not intensively farmed. However, there is enough runoff from the agricultural regions to exacerbate the pollution load coming from industrial cities upstream. Because Lake Ontario is the last of the Great Lakes, pollutants from all the other four lakes flow through it on their way to the Atlantic Ocean, making it the most polluted of the Great Lakes.

Lake Ontario has seen some commercial fishing, particularly of eels. Large numbers of eels have historically been harvested from Lake Ontario, but eel populations have recently declined to the point that some scientists believe the population may vanish altogether. Ironically, the destruction of eel stocks became a vicious cycle as scarcity drove prices up, making it worthwhile for commercial fishing concerns to go to greater lengths to harvest them.

Excellent state and provincial parks on its shores make Lake Ontario a very popular tourist destination. It boasts excellent sport fishing and hunting, although industrial pollution has made it medically risky for sports fishermen to actually eat their catch. Guidelines generally suggest that an adult should consume wild-caught Lake Ontario fish no more than once a month and that children and pregnant women should not consume any at all.

Geologists believe that Lake Ontario was created as the result of glacial erosion and that it is the remnant of a much larger lake they call Lake Iroquois for the Native American people who lived in the area. The shrinkage of Lake Iroquois into modern Lake Ontario left behind sediments that make for a very rich horticultural region. Lake Ontario was originally named Lake St. Louis in 1632 by the French explorer Samuel de Champlain. This name appeared on maps as late as 1656, but in 1660, Francis Creuxius gave it the name of Ontario, from an Iroquois word meaning "beautiful lake."

BIBLIOGRAPHY. Ann Armbruster, *Lake Ontario* (Children's Press, 1996); Pierre Berton, *The Great Lakes* (Stoddart, 1996); Sara St. Antoine, ed., *Stories from Where We Live: The Great Lakes* (Milkweed Editions, 2000).

LEIGH KIMMEL
INDEPENDENT SCHOLAR

Oregon

OREGON IS A STATE located on the west coast of the UNITED STATES and has a landscape for just about every-

one: mountains; FORESTS; plains; DESERTS; the Columbia River Gorge; waterfalls such as Multnomah Falls (620 ft or 189 m), the tallest in Oregon and the fourth-tallest in the country; Mills End Park, the smallest park in the world at 24 in (61 cm) across; and Hells Canyon, the deepest gorge in North America.

Oregon, bordered by CALIFORNIA and NEVADA to the south, WASHINGTON to the north, and IDAHO to the east, has an area of 97,073 square mi (251,419 square km). Its population is 3,521,515, the capital is Salem, and the highest point is Mount Hood at 11,239 ft (3,428 m). The lowest point is sea level at 0 m.

Eight regions divide Oregon: the Coast Range, Willamette Valley, Cascade Mountains, Klamath Mountains, Great Basin, Blue Mountains, Columbia Plateau, and Snake River.

The PACIFIC OCEAN and Willamette Valley border the Coast Range. The highest point is Mary's Peak, which is still lower than many Cascade Range passes. The rolling hills force the ocean winds through the narrow strip of land between the range and the ocean. The Coast Range also acts as a wind and moisture barrier to regions east of the range. Many coastal lakes dot the landscape, which consists primarily of sandstone and shale.

Oregon's largest cities, including Portland (529,121 people) and Salem (136,924), are located in the Willamette Valley, east of the Coast Range and west of the Cascade Range. The Willamette River travels north through the valley to the Columbia River, where Portland serves as a major ocean port. Westerly winds moderate the climate, and the year-round precipitation is responsible for the lush forests.

The Cascade Plateau is a mountainous strip east of the Willamette Valley, running from the Columbia River to the California border, and is the recreational jewel of the state, including the Columbia River Gorge, waterfalls, much of the Oregon forest, and volcanic formations, including Mt. Hood and the famous Crater Lake.

The Coast and Cascade mountain ranges border the Klamath Mountains in the southwest corner of the state. The Rogue and Klamath rivers are among many that dissect this primarily conserved area, home to the richest mineral deposits in the state. In addition to the regional lumber and dairy industries, the Oregon National Cave Monument makes this region popular among tourists and the nearby residents. The area is generally moderate in climate, except for the Rogue River valley, which varies from -10 to 110 F (-23 to 43 degrees C).

The Great Basin (30,000 square mi or 48,270 square km) covers most of the southeastern corner of Oregon with the topography continuing into California and Nevada. The Cascade Range to the west cuts off this high desert from eastward winds. Although sparsely populated, some small towns persist, mainly by ranching, but manufacturing and telecommunications, as well as new work on the geothermal potential of the area have kept the towns from diminishing.

North of the Great Basin, the Blue Mountain area is primarily comprised of lava plateaus and several higher mountain ranges, which drain rainwater to the Columbia River. Outdoor recreation, along with some cattle grazing, is the main activity of the region.

Wedged in north and east of the Great Basin, and west of the Cascade Range, the Columbia Plateau is what remains after lava flowed from cracks all over the region, and erosion began forming deep canyons and irregular, mountainous "plateaus." Wheat ranches are a main staple of the area, and several cities and towns sprinkle the region.

Finally, the Snake River region occupies a strip of land along the eastern border of Oregon. The Snake River and Hells Canyon, which descends in some places to a depth of 7,900 ft (2,400 m), dominate the region. Hydroelectric power from the three dams on the Snake River is one of the key economic resources of the area. From border to border, the landscape's diversity finally inspired the travelers on the Oregon Trail to make a home.

BIBLIOGRAPHY. "Oregon," *Encyclopædia Britannica* (2004); "Oregon," *The Columbia Encyclopedia* (Columbia University Press, 2003); *Physical and Economic Geography of Oregon* (Oregon State System of Higher Education, 1940); William Loy, *Atlas of Oregon* (University of Oregon Press, 1976).

TARA SCHERNER DE LA FUENTE
UNIVERSITY OF CINCINNATI

orogeny

OROGENY, A GEOLOGIC CONCEPT, comes from the Greek *óros*, meaning "mountain," and *genés*, meaning "stemming from." The etymology associates the term with the concept of building mountains, and in the past, orogeny was associated with mountain formation. Its specific use, today however, relates to the

deformation of rocks that leads to postorogenic mountain building. The thrusting, folding, and faulting of crustal plates form structures within fold belts (referred to as mountain ranges). When one plate moves under another, there is uplift, or stacking, which is termed epeirogeny. Epeirogeny is the actual formation of mountains.

The theory of plate tectonics explains the action of volcanoes, earthquakes, continental drift, and ocean-basin widening. The Earth's lithosphere contains the crust and upper portion of its mantle. Twelve or more plates comprise the principal regions of oceanic and continental crust. Plate tectonics involves the formation, lateral movement, interaction, and destruction of the lithospheric plates. These plates become stressed, and in fact break, causing wrinkles and folds. The process of plate movement relieves the Earth's core buildup of internal heat, thus creating many of the Earth's structural formations.

One theory in support of plate tectonics is that convection in the mantle drives sea-floor spreading and continent formation. Orogenous deformation of rocks occurs because of the crust movement. Pressure within the crust results from upper-layer rock weight as this weight compresses the lower layers. Because of crustal movement, rocks at the surface crack and fragment, referred to as joints, whose fracture lines are called faults. Rock displacement creates horizontal and vertical fault lines. Plates overriding or sliding against each other produce the Earth's crustal instability.

New oceanic lithosphere forms through volcanism in the form of fissures at mid-ocean ridges. Heat escapes the interior as this new lithosphere emerges from below, causing plate divergence or ocean-floor spreading. The lithosphere gradually cools, contracts, and moves away from the ridge, traveling across the seafloor to subduction zones. The existence of younger rock near these ridges is considered proof of continuous seafloor spreading.

Convergent oceanic plates overriding each other become heavier at one end and plunge as much as 62 mi (100 km) into the mantle. The edges of the plates melt from the intense heat and rise to the surface as molten rock. The colliding plates produce deep ocean trenches or subduction zones. Magma rises from the mantle to the ocean floor, creating continental rift valleys and plateaus of basalt. These processes create volcano mountain formations from lava accumulation on the ocean floor.

Transform-fault boundaries become earthquake zones because of plates sliding horizontally against each other. Most of these faults found on the ocean floor are associated with divergent plate boundaries, while other trenches form convergent plate boundaries. Earth instability generates the tremors of plate movement.

The continental lithosphere is about 62 mi (100 km) thick, with a low-density crust and upper mantle. Because of their granite composition, continental plates tend to be more porous and thus float above the oceanic plates. When the heavier (stacked) oceanic plates plunge into the mantle, the continental plates move along with the oceanic plate movement.

Continents move laterally as convection cells move upward, away from hot mantle zones toward cooler ones. This process is known as continental drift. Continental drift was believed to occur when North America and South America split apart from Eurasia and Africa. Because of this split, the once-small ATLANTIC OCEAN is now the second-largest ocean, covering one-fifth of the Earth.

There are active fractures along the Mid-Atlantic Ridge that continually open and release lava flow. These rock-formed ridges continually move the continental plates by broadening the ocean basins. Another form of movement is when one continental plate overrides another. This type of movement, over periods of geologic time, forms high mountains on the Earth's surface.

BIBLIOGRAPHY. J.P. Burg, "Orogeny through Time," *Geological Society* (Dutch, Steven, 1998); R.L. Hamilton, "Earth's Interior and Plate Tectonics," www.solarviews.com (October 2004); Cliff Ollier and Colin Pain, *The Origin of Mountains* (Routledge, 2000); "The Remarkable Ocean World: The Theory of Plate Tectonics," www.oceansonline.com (October 2004); "What is Orogeny?" www.uwgb.edu (October 2004).

P. CLOWE AND A. CHIAVIELLO
UNIVERSITY OF HOUSTON, DOWNTOWN

orographic precipitation

OROGRAPHIC PRECIPITATION is caused or enhanced by one or more of the effects of mountains on the Earth's atmosphere. These effects include the upward or lateral motions of air directly caused by mountains acting as a barrier as well as the thermal effects of the mountains that cause them to be elevated heat or

cold sources. Mountains can generate both stratiform precipitation, which takes place in a statically stable atmosphere, and convective precipitation, which results from the release of static instability.

The most obvious effect of mountains is that they can cause the air encountering them to rise. Rising air cools adiabatically, and if it is sufficiently humid, condensation and perhaps precipitation can occur. Precipitation formed by this mechanism is widely referred to as upslope precipitation. It is widely recognized that the slope on the upwind side of the prevailing wind (the windward side) generally receives more precipitation than the leeward side. In contrast, some of the world's deserts are on the leeward side of mountain ranges. Because of interactions with other processes, however, there is no general rule about the location on the mountain slope where the maximum upslope precipitation will occur.

SEEDER-FEEDER

Upslope precipitation may be enhanced by the seeder-feeder mechanism. A low-level stratus cloud (feeder cloud) forms near the top of a mountain. Its temperature is below freezing but warm enough so that it lacks ice nuclei. Ice crystals fall from a higher (and colder) seeder cloud into the feeder cloud. These ice crystals grow at the expense of the water droplets in the feeder cloud by the Bergeron process. The large ice crystals then precipitate out of the feeder cloud to the Earth's surface. This process increases the precipitation efficiency of the feeder cloud because the moisture from small water droplets, which otherwise would not have reached the Earth's surface, evaporates and is deposited on the ice crystals.

Mountains can generate precipitation through the release of static instability by providing a lifting mechanism for air parcels as well as by orographic processes that destabilize the air column. Precipitation can be generated by three categories of processes: orographic lifting, thermal effects, and obstacle effects. Orographic lifting can bring air to the Level of Free Convection (i.e., the level at which it becomes positively buoyant). If the atmosphere has a sufficiently thick layer above this level that is statically unstable, precipitation may result. Lifting of an air column can cause it to become more statically unstable.

The thermal effects of mountains occur because orographic features can serve as elevated heat or cold sources. During the day, the sun heats the mountain slope. It warms the air next to it, which then becomes warmer than the atmosphere at the same elevation away from the mountain over lower land. Thus, there is low pressure next to the mountain toward which the air moves—producing upslope flow. This heating can be modified by surface conditions. Bare rock will transfer more sensible heat to the atmosphere than snow cover. At night the mountain surface cools, which causes the air above it to cool and results in air that flows down the slopes. Air moving across the mountain top during the day is heated by the mountain. This heating helps to destabilize the air column.

In addition, convergence (which promotes upward air motion) may be produced at the ridge crest as air flows up the slope from opposite directions. The combination of the heating and the heating-induced convergence can generate thunderstorms. At night, downslope winds from mountains on opposite sides of a valley can converge in the valley and encourage the generation of thunderstorms as well.

THERMAL EFFECTS

Mountain ranges and high plateaus can have larger-scale thermal effects on precipitation. During the summer, heating causes low pressure to form near the surface over high elevation areas. Convergence into this low pressure area can encourage the formation of thunderstorms. Since heating is stronger during the day, the location of the lowest pressure varies diurnally. The changes in this location can affect the time of day that is most favorable for precipitation to form for a particular place in or near a mountainous region.

Mountains also act as obstacles to air flow and in so doing can encourage the development of convective precipitation. Convergence may be generated when air moves through a progressively narrower valley or when air moving around the mountain base arrives on the other side. These processes can cause moisture convergence as well and static stability is quite sensitive to atmospheric humidity.

Under stable static stability conditions, mountains may generate buoyancy waves—commonly known as gravity waves. These waves propagate downstream (and sometimes upstream as well) of the mountain. The parts of the wave that contain upward-moving air may aid in the development of convective storms, especially if they interact with other mountain-generated circulations such as the diurnal mountain-valley winds.

Orographic variations can contribute to the development of convective storms in their surrounding areas as well as in the mountainous area itself. As they interact with the large-scale wind field, mountains may generate downstream eddies in the flow under some

conditions. An example of such an eddy is the so-called Denver cyclone. Such a cyclonic circulation can have areas of convergence and vorticity that encourage the development of convective storms. Outflow from thunderstorms generated within the mountains may initiate convective storms in the surrounding lowlands.

BIBLIOGRAPHY. R.M. Banta, "The Role of Mountain Flows in Making Clouds," *Meteorological Monograph* (v.23, 1990); R.G. Barry, *Mountain Weather and Climate* (Routledge, 1992); E.R. Reiter and M.C. Tang, "Plateau Effects on Diurnal Circulation Patterns," *Monthly Weather Review* (v112, 1984); D.F. Tucker, "Diurnal Precipitation Variations in South-Central New Mexico," *Monthly Weather Review* (v.121, 1993); C.D. Whitman, *Mountain Meteorology: Fundamentals and Applications* (Oxford University Press, 2000).

DONNA TUCKER
UNIVERSITY OF KANSAS

Ottoman Empire

THE OTTOMAN DYNASTY created the most enduring empire in human history. The Ottomans originally migrated from Central Asia as nomads and settled in the early 14th century as a military Turkic principality in western Anatolia (present-day TURKEY), between the frontier zone of the Seljuk state and the Byzantine Empire. The Ottomans emerged into a dominant Muslim force in Anatolia and the Balkans and became the most powerful Islamic state since the breakup of the Abbasid caliphate in 1258.

At its height in the 16th and 17th centuries, the empire was the most powerful in the world. Made up of diverse ethnic and religious groups, including Arabs, Armenians, Greeks, Kurds, and Slavs, the empire stretched from Central Europe in the west to Baghdad (IRAQ) in the east, from the Crimean Sea in the north to the Upper NILE in EGYPT and the Arabian Peninsula (SAUDI ARABIA) in the south.

Named after the founder and first sultan (ruler) of the dynasty, the Ottomans came into prominence with their gradual invasion of the Byzantine Empire that had occupied parts of Asia Minor and southeastern Europe for nearly a thousand years. With the conquest of the Byzantine capital, Constantinople, in 1453 under the rule of Muhammad II (1451–81), famously known as "Mehmet the Conqueror," the Ottomans extended their dominance over much of Anatolia and southeastern Europe. Constantinople then became the capital of the Ottoman Empire and was renamed ISTANBUL. After taking Constantinople, the Ottomans, conquered the FERTILE CRESCENT, North Africa from Egypt up to MOROCCO, and the Arabian Peninsula, including the Hijaz, seizing control of the holy cities of Mecca and Medina. Under Suleiman the Magnificent (1520–66), they expanded into the Balkans in 1521, capturing Belgrade (SERBIA AND MONTENEGRO) and even besieging the Habsburg capital of Vienna (AUSTRIA), forming the largest and one the most powerful empires of the 16th-century world.

After the death of Suleiman the Magnificent in 1566, the Ottoman Sultans, also known by the Persian title of Padeshah, became increasingly dependent on the crops of Janissaries, captured Christian slaves trained into elite soldiers, and the clergy class or the ulama, who gradually gained power at the court. The Janissaries were not only a military organization that protected the sultan, but also a warrior, spiritual fraternity, an association inclined in the mystical dimension of ISLAM that upheld a chivalric code of ethics; they were a powerful elite body in the Ottoman Empire. Although the sultans made the important decisions for the empire, including exercising the power to appoint officials to collect taxes and maintain stability within the empire, the grand mufti, the chief religious cleric, legitimized the authority of the sultan as the ruler of the empire. In contrast to their contemporary Muslim states, the Safavids and the Mughals, the Ottoman Sultans shaped the clerics into a state bureaucracy rather than allowing them to evolve into an independent institution.

OTTOMAN HIERARCHY

The Ottoman government was organized in the form of a hierarchy with the sultan in the top and ministers and advisors known as vizirs below him, followed by court officials and military officers. Central to the Ottoman sultanate was an organized bureaucracy drawn from the sultan's court that strictly controlled local provinces of the empire.

The empire was divided into two distinct classes. The ruling elites, primarily the imperial family, landowners, and military and religious leaders, ruled the conquered territories without paying taxes; whereas the ordinary Muslim population, mainly comprised of peasants and craft workers, paid an annual tax to the state in return for protection against invasion and abuse of power. The non-Muslim section of the

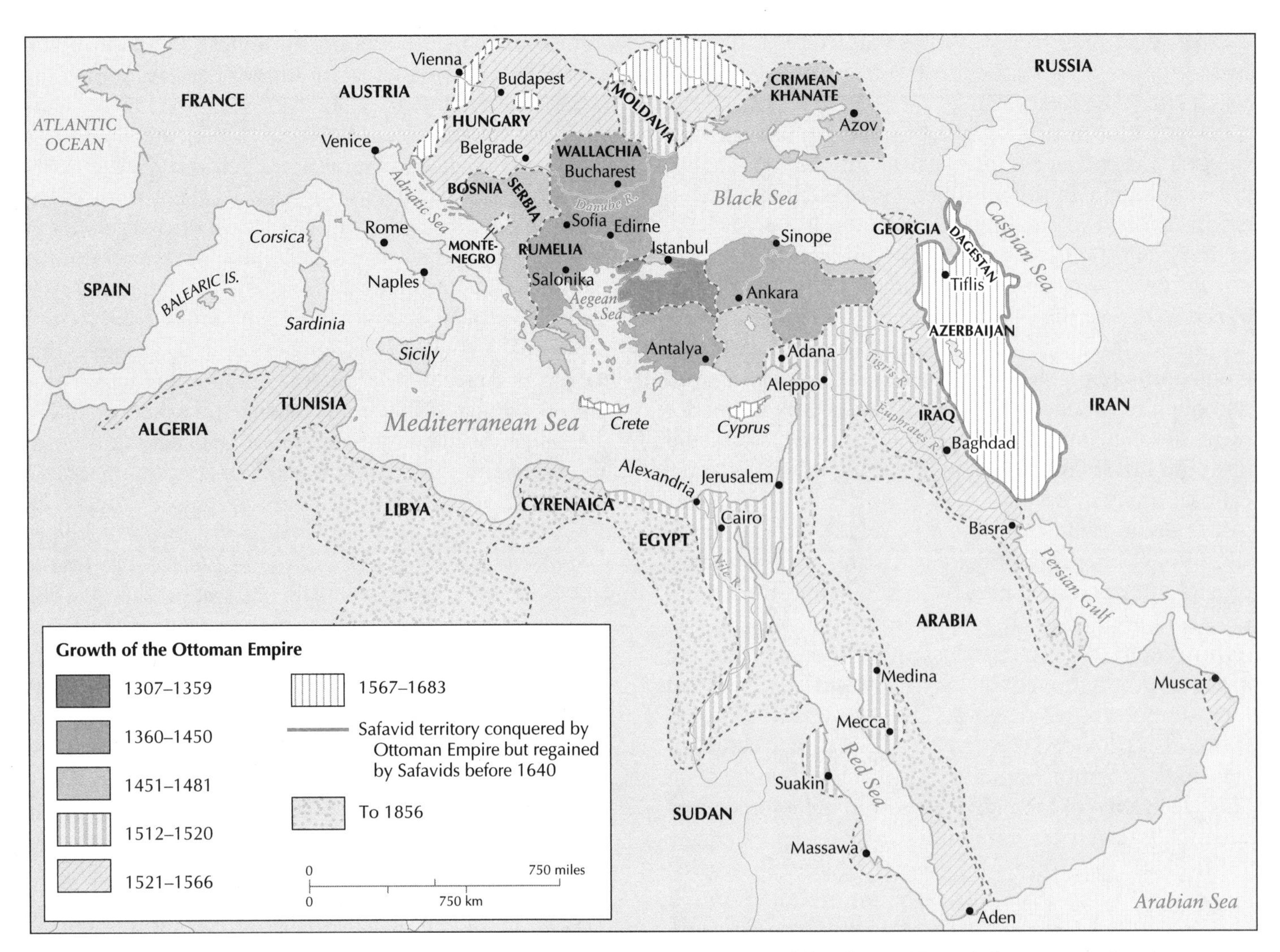

At its height in the 16th and 17th centuries, the Ottoman Empire was the most powerful in the world. With the creation of Turkey in 1923, the oldest imperial power in the world was finally abolished and replaced by a secular republic.

empire was divided into millets, which were administrative units set up on the basis of religious affiliation; they mainly included Armenian, Catholic, Orthodox Christians and Jews.

The Ottomans saw the first serious sign of the supremacy of Europe with the naval defeat of Lepanto in 1571 by the Spanish, the papal states of ITALY, and the Venetians under John of Austria. Although Murad IV restored the Ottoman military in the 1638 victory over Safavid Persia, Vizir Kara Mustafa's army surrendered to Polish forces in 1683. The ensuing years saw the loss of HUNGARY and parts of the Balkan territories, such as the Mora Peninsula and GREECE, through a 1699 treaty, after a series of defeats to various European powers, including the military forces of Charles V of Lorraine and Eugene of Savoy.

In the 18th century, the process of decline began to accelerate. Following the war of 1716–18 against Austria and Venice, the peace of Passarowitz led to the additional loss of the remaining parts of Hungary and Transylvania, including Belgrade.

RISE OF RUSSIA

This marked the end of the Ottoman expansion into European territories—with the exception of the recapture of Belgrade in 1739. With the rise of RUSSIA to power as a formidable opponent, the Ottomans faced further complications in the 18th century. The formation of Crimea (present-day UKRAINE) into an independent region, along with the rise of various Danube principalities under the protection of Russia in the late 18th century, identified a major military weakness of

the Ottoman Empire, a weakness that led to further defeats by Karim Khan Zand of Persia (1776) and the war against the Russians.

The 19th century saw the beginning of reform and the gradual end of the dynasty. In this period, the increasing expansion of European powers into Muslim territories forced the Ottomans to initiate economic, military, and political reform. The tanzimat, the modernization reform movement, inaugurated an era that aimed to review the body of Islamic knowledge and the economic, social, and technological apparatus of the empire and adapt the Ottomans to the modern world. Although reforms were originally under way under the reign Sultan Selim III (1789–1807), in 1839 Sultan Abd al-Majid (1839–61) issued a decree including a significant set of civil reforms.

The reign of Sultan Abd al-Aziz (1861–76) saw the rise of new liberal political parties and the emergence of new elites and state bureaucrats to mange the changing economy and political system of the empire. The reforms marginalized the artisans, the merchants of the bazaar, and the travelodge or caravanserai, which had long supported the ulama and the sufi classes. The reforms also provided the empire with major technological transformations, such as the introduction of steam and electricity power, the telegraph, telephone, and eventually television communications.

In 1908 a group of Ottoman military officers, named the Young Turks, forced Sultan Abdul-Hamid II (1876–1909) to reinstate the imperial constitution, originally drafted in 1876 and long suspended because of the Russian invasion of 1877. The Young Turk movement then successfully managed to limit the authority of the sultan as the supreme sovereign of the empire by expanding the authority of an elected parliamentary government.

With the outbreak of World War I, the Ottomans lined up with the Central Powers and faced a humiliating defeat as a result. After World War I ended in 1918, the empire was under the occupation of several Allied powers, including Britain and Greece. It was not until the Kemalist nationalist movement—named after its leader, Mustafa Kemal, famously known as Kemal Atatürk (1881–1938)—which ended the foreign occupation of Turkey in 1922, that the Ottoman Empire saw its demise. With the creation of Turkey in 1923, the oldest imperial power in the world was finally abolished and replaced by a secular republic.

BIBLIOGRAPHY. Suraiya Faroqhi, *Subjects of the Sultan: Culture and Daily Life in the Ottoman Empire* (I.B. Tauris, 2000); Godfrey Goodwin, *The Janissaries* (I.B. Tauris, 1997); Daniel Goffman, *The Ottoman Empire and Early Modern Europe* (University Press of Cambridge, 2002); Jason Goodwin, *Lords of the Horizons: A History of the Ottoman Empire* (Picador, 2002); Halil Inalcik, *Economic and Social History of the Ottoman Empire* (University Press of Cambridge, 1997); Inalcik, *The Ottoman Empire: The Classical Age* (Phoenix, 2001); Camal Kafadar, *Between Two Worlds: The Construction of the Ottoman State* (University Press of California, 1996); Naim Turfan and M. Naim Turfan, *Rise of the Young Turks: Politics, the Military, and Ottoman Collapse* (I.B. Tauris, 2000).

BABAK RAHIMI
INDEPENDENT SCHOLAR

Pacific Ocean

THE PACIFIC OCEAN HAS an area of approximately 68,767,000 square mi (178,106,000 square km) with its adjacent seas; 65,329,000 square mi (169,202,000 square km) without them. As the world's largest ocean, it has the following characteristics.

Volume: 160,489,000 cubic mi (674,052,000 cubic km) including adjacent areas; 158,065,000 cubic mi (663,871,000 cubic km) if they are not included.

Average depth: with adjacent seas -13,480 ft (-4,110 m); without them -14,038 ft (-4,280 m).

Greatest depth: -35,798 ft (-10,911 m) in the Challenger Deep in the Mariana Trench.

Width: varies from north to south reaching about 12,300 mi (19,800 km) between INDONESIA and the coast of COLOMBIA in South America.

Coastline: 84,300 mi (135,663 km); adjacent areas include the Bering Sea (in the north); the Gulf of California (in the east); Ross Sea (in the south); and the Sea of Okhotsk, the Sea of Japan, and the Yellow, East China, South China, Philippine, Coral, and Tasman seas (in the west).

Because the ATLANTIC OCEAN was home to extensive exploration and commercial use long before the Pacific, research on the Pacific is not as well developed as that on the Atlantic. The Pacific BASIN is almost triangular in shape, narrow in the Arctic north and broad in the Antarctic south. In the west, it touches Asia and AUSTRALIA, while in the east it borders the Americas. Its north to south extent between the Bering Strait in the north and ANTARCTICA in the south is more than 9,300 mi (15,000 km). In the tropics, between the MALACCA STRAITS to PANAMA, the Pacific reaches its greatest width, spanning a distance of nearly 12,427 mi (20,000 km).

If you include all of its adjacent seas, the Pacific covers about one-third of the Earth's surface and 40 percent of the surface area of the world's oceans, an area greater than that of all of the continents combined. The boundary between the Pacific and INDIAN oceans follows a line from the Malay Peninsula through SUMATRA, JAVA, Timor, Australia at Cape Londonderry, and Tasmania. From Tasmania to Antarctica, it follows the meridian from the South East Cape on Tasmania at 147 degrees E.

The dividing line between the Atlantic and the Pacific follows the line of shallowest depth between Cape Horn and the Antarctic Peninsula. In the north, the boundary between the North Pacific and the ARCTIC OCEAN lies along the shallow shelf of the Bering Strait that extends between the Chukchi Peninsula in eastern SIBERIA and the Seward Peninsula in western ALASKA. There is also a boundary between the Pacific's northern and southern zones, formed by the equatorial countercurrents that circulate just north of the equator in an

area known as the Intertropical Convergence Zone (ITCZ).

PHYSICAL FEATURES

Typically, the depth of the Pacific basin is only 1.8 to 2.5 mi (3 to 4 km), so the horizontal dimension of the basin is about 1,000 times greater than its vertical dimension. Another unique feature of the Pacific Ocean is its large number of seamounts, particularly in the Northwest and Central Pacific Basins. Seamounts are found in all oceans, but the volcanism of the northwestern Pacific Ocean produces them in such numbers that in some regions they cover a fair percentage of the ocean floor.

Because the rim of the Pacific Basin is ringed with volcanoes, it is often referred to as the RING OF FIRE. This ring, which extends from Alaska through the UNITED STATES, MEXICO, and South America, then on to NEW ZEALAND and up to JAPAN and RUSSIA includes about 75 percent of all the world's volcanoes.

In great contrast to the Atlantic, the Pacific contains thousands of islands. These are usually one of four basic types: continental islands, high islands, coral reefs, and uplifted coral platforms. In the west, there are a number of large islands such as Japan, TAIWAN, BORNEO, and NEW GUINEA. In the south, there is a cluster of smaller islands included in POLYNESIA that extend east as far as TAHITI. Further away still to the east are several isolated islands such as PITCAIRN, EASTER ISLAND, and the Galapagos, which is just 621 mi (1,000 km) west of ECUADOR in South America.

All of the Pacific's adjacent seas are grouped along its western edge. Some of them are large shelf seas, while others are deep basins. In contrast to the Indian and Atlantic oceans, adjacent seas of the Pacific Ocean have only a small influence on the hydrology of the main ocean basin. The Australasian Mediterranean Sea, the only Mediterranean-type sea in the Pacific Ocean, is a major region of water mass formation and an important element in the mass and heat budgets of the world ocean system. Unlike the CARIBBEAN SEA in the Atlantic, however, its influence on the Pacific's hydrology is of only minor importance.

In the Atlantic and Indian oceans, an interoceanic ridge system divides the basins into compartments of roughly equal size. In the Pacific Ocean, the ridge system runs close to the eastern boundary, producing divisions in the southeastern Pacific Ocean similar in size to those of the Atlantic and Indian basins. The vast expanse of deep water in the central and northern Pacific Ocean, on the other hand, is subdivided more by convention than topography into the Northeast Pacific, Northwest Pacific, Central Pacific, and Southwest Pacific Basins. Further west, New Zealand and the Melanesian islands provide a natural boundary for two adjacent seas of the Pacific Ocean, the Tasman and Coral seas, while in the north the West and East Mariana Ridges and the Sitito-Iozima Ridge offer a natural subdivision.

WATER AND CLIMATE

In the Pacific the systems of winds and currents follow a clockwise movement in the Northern Hemisphere and counterclockwise in the South. Between the two run the Equatorial Currents. The North Equatorial Current runs 9,000 mi (14,416 km) from PANAMA to the PHILIPPINES and is the longest westerly running current in the world. When the North Equatorial Current reaches Japan, it turns north and is known as the Japan Current. The Japan Current is the western Pacific's counterpart of the Gulf Stream in the Atlantic. When it meets the cold waters of the Arctic, fogs and storms are caused, just as they are at the meeting of the Labrador Current and the Gulf Stream. In the southern half of the ocean, the Humboldt Current flows north from Antarctica along the coast of South America.

At regular intervals a vast area of the Pacific's surface waters becomes warmer. This change in temperature and level causes an alteration in the pattern of the winds, a phenomenon known as EL NIÑO. As the warm water travels across the Pacific, it deflects the Humboldt Current near Peru and cuts off the food supply to plankton. The warmer waters along the coast of Peru also cause increased rainstorms and flooding in many parts of South and Central America. The monsoon rains in Southeast Asia are also interrupted, often leading to crop failures in Australia and New Guinea. Every time there is an El Niño, there is even crop damage as far away as Africa, where the maize crop in zimbabwe has been known to decline. On the other hand, in israel, El Niño years bring more rain and the crops are more abundant.

Few rivers shed their waters into the Pacific Ocean, and the few that do have very small catchments. The largest rivers all enter the marginal seas along the western rim of the North Pacific basin, where they have a strong impact on the hydrology. Although the rivers coming down from the mountain ranges are small, they are numerous and their combined freshwater output is comparable to that of the MISSISSIPPI RIVER. This also constitutes about 40 percent of all the freshwater input into the northeast Pacific Ocean. In the Southern

Hemisphere river contributions are negligible because the river catchments are restricted by the ANDES in the east and Australia's Great Dividing Range in the west.

HISTORY AND ECONOMY

The Pacific is the world's oldest ocean. Theories of the Pacific's origin are related to PLATE TECTONICS, where movement of the land masses of the Earth are either drifting apart or sinking and sliding under one another. The role of humans in the Pacific has been the result of migration and exploration. The island peoples of MELANESIA, MICRONESIA, and Polynesia traveled extensively settling in Australia, New Zealand, Hawaii, and thousands of other islands.

People such as Thor Heyerdahl believe the migration might have been in the opposite direction because the sweet potato, a native plant of PERU, grew in Polynesia before the arrival of western explorers. The first European to see the Pacific was the Spanish explorer Balboa in September 1513. Later, Ferdinand MAGELLAN set sight on the Pacific after sailing round Cape Horn at the southern tip of South America in 1520. The first English explorers were Francis Drake in the 16th century and Captain James Cook in the 18th century.

Although there is no written proof, it is believed that ships from the great Chinese voyage of discovery under Zheng He may have sailed around Cape Horn sometime between 1421 and 1423 on the return trip to CHINA. For the remainder of the 16th century, Spanish influence was paramount, with ships sailing from SPAIN to the Philippines, PAPUA NEW GUINEA, and the SOLOMON ISLANDS. During the 17th century, the Dutch, sailing around southern Africa, dominated discovery and trade. The 18th century marked a burst of exploration by the Russians in Alaska and the ALEUTIAN ISLANDS, the French in Polynesia, and the British in the three voyages of James Cook.

The Pacific Ocean is a major contributor to the world economy and particularly to those nations its waters directly touch. It provides low-cost sea transportation between East and West, extensive fishing grounds, offshore oil and gas fields, minerals, and sand and gravel for the construction industry. One of the Pacific Ocean's greatest assets is fish, including herring, salmon, sardines, snapper, swordfish, tuna, and shellfish. Pearls are harvested along Australia, Japan, Papua New Guinea, NICARAGUA, Panama, and Philippine coasts.

ENVIRONMENTAL ISSUES

There are a number of endangered marine species in the Pacific, including the dugong, sea lion, sea otter, seals, turtles, and whales. Current major environmental issues include oil pollution in the Philippine Sea and South China Sea. There is also a zone of violent volcanic and earthquake activity known as the Ring of Fire surrounding it. Southeast and East Asia are subject to typhoons from May to December, while hurricanes may form south of Mexico and strike Central America and Mexico from June to October. The monsoon region lies between Japan and Australia in the far western Pacific. The greatest typhoon frequency exists within the triangle from southern Japan to the central Philippines to eastern Micronesia. The southern shipping lanes are subject to icebergs from Antarctica.

BIBLIOGRAPHY. G. Dietrich, K. Kalle, W. Krauss and G. Siedler, *General Oceanography* (Wiley-Interscience, 1980); G. Neumann and W. J. Pierson, Jr., *Principles of Physical Oceanography* (Prentice-Hall, 1966); P. Tchernia, *Descriptive Regional Oceanography* (Pergamon Press, 1980); Reilly Ridgell, *Pacific Nations and Territories: Islands of Micronesia, Melanesia, and Polynesia* (Bess Press, 1995); Richard Barkley, *Oceanographic Atlas of the Pacific Ocean* (University of Hawaii Press, 1969); Andrew Kippis, *Narrative of the Voyage round the World, performed by Captain James Cook, with an Account of His Life* (Bickers Kamp and Son 1889); Charles Darwin, *The Voyage of the Beagle: A Naturalist's Voyage Round the World* (John Murray, 1913); Gardner Soule, *The Greatest Depths* (MacRae Smith, 1970); John Gilbert, *Charting the Vast Pacific* (Doubleday, 1971); Thor Heyerdahl, *Kon-Tiki: Across the Pacific by Raft* (Rand McNally, 1951).

RICHARD W. DAWSON
CHINA AGRICULTURAL UNIVERSITY

Pakistan

Map Page 1123 **Area** 310,413 square mi (803,940 square km) **Population** 150,000,000 **Capital** Islamabad **Highest Point** 28,251 ft (8,611 m) **Lowest Point** 0 m **GDP per capita** $2,100 **Primary Natural Resources** natural gas, petroleum, coal, iron ore.

PAKISTAN CAME INTO being in 1947 when British colonial India, upon independence, was divided into

Muslim-dominated East Pakistan (BANGLADESH as of 1971) and West Pakistan. Its 97-percent-Muslim population is divided between 75 percent Sunni and 25 percent Ithna'Ashariya (Shia), with a few of Isma'ili and Ahmadis. Parsis (Zoroastrians) and Christians provide the remaining proportion.

The official language is Urdu, a blend of Hindi, Persian, and Turkic languages, but the dominant language is Indo-European Punjabi (65 percent of speakers), followed by Sindhi (15 percent), Baluchi (10 percent), and Pakhto/Pashto (10 percent). A remnant of Dravidian languages, Brahui, is still spoken in the mountains of Baluchistan. Many languages, almost all non-literate, are spoken in the mountainous north. Burushaski, a language heretofore not connected to any other extant language is still spoken in the Hunza and Yasin valleys in the northern areas. English is spoken by 10 percent of the population. The coastal city of Karachi has a population of almost 12 million, while the ancient Mughal city of Lahore has 5 million, the capital conurbation of Islamabad/Rawalpindi 2.5 million, and western cities of Quetta .7 million and Peshawar .25 million.

GEOGRAPHIC CONNECTION

Although there is no direct cultural historical connection between the territory occupied by present day Pakistan with ancient times, there is a geographic connection. The INDUS RIVER civilization founded 4,200 years ago was one of the earliest civilizations, rivaling Mesopotamia and the NILE civilizations in cultural achievements, with archaeological sites such as Mohenjo Daro and Harappa yielding artifacts in bronze, glass, jade and lapis lazuli. The ancient history of the Indus region lasted until 1500 B.C.E. when Aryan invaders from Central Asia poured through the Hindu Kush-Himalayas and overran the remnants of the preceding civilization. These invaders became the dominant cultural feature of today's Pakistan, INDIA and Bangladesh. They composed the Rig Veda, the ancient sacred hymns. Persian intrusions were succeeded by Alexander of Macedon, who penetrated in 326 B.C.E as far as the known world, the Beas River of the Punjab—the land of the five rivers.

Repeated invasions from Central Asia came from the Graeco-Bactrians from their redoubt in Balkh on the northern plains of today's AFGHANISTAN, and the Kushan kingdom based in Bagram north of present day Kabul, Afghanistan. Buddhist archaeological sites of that era include Taxila and Swat near Peshawar and Bamyan in Afghanistan.

The major impact in the first millennium C.E. was the invasion of Arabs and with them, Islam, which replaced Buddhism and Hinduism in many places. Later, in the 12th and 13th centuries, pillaging by invaders from the South Asian frontier, the crest of the Hindu Kush Mountains in present-day Afghanistan, continued to wrack northern regions of cultural India. Another Indo-Islamic period started in the 16th century, when Babur, a Central Asian now buried in Kabul, inaugurated the Mughal civilization, which lasted until it was replaced by the British East India Company in the 18th century. Great advancements were made in literature, the arts, land use, and egalitarian public life during this period.

Periodic battles with Sikhs in the Punjab lasted for several centuries. During the first half of the 20th century, Indian Muslims pressed for a state for Muslims. In 1947 this was realized when Pakistan was created under the premier Mohammad Ali Jinnah, a Bombay Muslim. Many Indian Muslims became émigrés in the new state of Pakistan. Political stability in Pakistan has been uncertain over the years since then because of excessive military and religious interference in civil affairs.

The Pakistani military, almost exclusively drawn from Pushtun and lesser Punjabi populations, repeatedly interfered with popularly elected civil authorities. Armed skirmishes with India in 1965 and 1971, and later in Kargil in the Karakorum mountains on the cease-fire line in 1999, continued the strife with India. In 1971, disaffection with West Pakistan led East Pakistan, composed from the Muslim districts of Bengal province, to secede after a civil war with West Pakistan authorities. The resulting nation-state of Bangladesh was created. Military and civil personnel continued to squabble for the next three decades with the military Inter-Services Intelligence (ISI) interfering in public policy debates. Continued low-level strife with India over the contested erstwhile state of Kashmir has not been resolved because India has refused to honor a United Nations mandate. Since the terrorist attacks of September 2001 in America, Pakistan has been a strong ally of the U.S. War on Terrorism, providing crucial geographic support for the American invasion of Afghanistan in 2002.

BIBLIOGRAPHY. Akbar S. Ahmed, *Jinnah: Pakistan and Islamic Identity* (Routledge, 1997); Akbar S. Ahmed, *Resistance and Control in Pakistan* (Routledge, 2004); Robert Kaplan, *Soldiers of God: With Islamic Warriors in Afghanistan and Pakistan* (Vintage, 2001); Hafeez Malik,

ed., *Founders Aspirations and Islamic Identity* (Oxford University Press, 2001); Ian Talbot, *Pakistan: A Modern History* (C. Hurst, 1998).

NIGEL J.R. ALLAN
UNIVERSITY OF CALIFORNIA, DAVIS

Palau

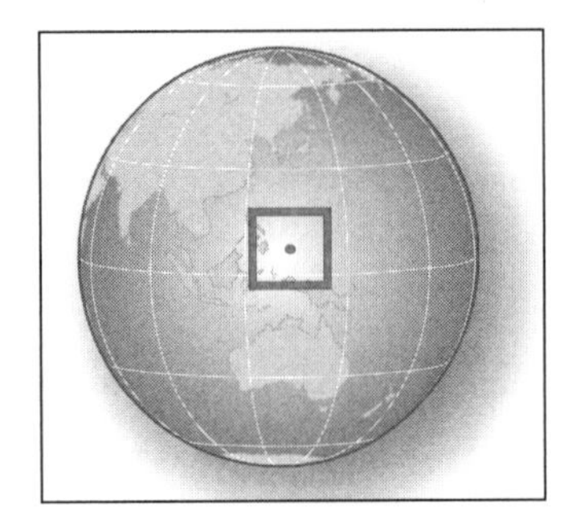

Map Page 1128 **Area** 179 square mi (458 square km) **Population** 19,717 **Capital** Koror **Highest Point** Mount Ngerchel-chuus 798 ft (242 m) **Lowest Point** 0 m **GDP per capita** $9,000 (2001) **Primary Natural Resources** forests, minerals, gold.

PALAU (OR BELAU to its natives) is one of the world's newest nations, finally securing independence (in Free Association status) with the UNITED STATES in 1994 after nearly two decades of struggle. As part of the United Nations Trust Territory of the Pacific, Palau formed the westernmost component of the Caroline Islands but opted for separation in 1978 instead of joining the rest of that group as the independent Federated States of MICRONESIA. Today, known as one of the world's most attractive diving and snorkeling centers, Palau continues to develop its own identity separate from U.S. military requirements yet dependent on U.S. fiscal subsidies.

The Republic of Palau consists of six island groups, made up of a more than 300 islands, with a total coastline of 942 mi (1,519 km). The largest island, Babelthuap, dwarfs the others in size, and is the only one of the group with significant elevation. The rest are low coral islands, fringed by large barrier reefs. The chain stretches across 434 mi (700 km) of the western PACIFIC OCEAN, 496 mi (800 km) east of the PHILIPPINES. Originally two main confederations of chiefdoms, the islands were annexed by SPAIN in 1886, purchased by GERMANY in 1899, annexed by JAPAN in 1914, then occupied by the United States in 1945. Formally put under the jurisdiction of the United States as part of the Trust Territory in 1947, the islands were developed as an important naval station. This was a major cause for delay of ratification of Palau's independence, because of the locals wanted to be nuclear-free. The Free Association Compact that was finally ratified in 1993 provides Palau with up to $700 million in aid over 15 years in return for continued use of military facilities for 50 years. Yet in August 2003, Palau joined with several other countries in signing a comprehensive Nuclear Test Ban Treaty, provoking a statement from the U.S. secretary of state that America would not rule out the possibility of future testing in Palau.

The population (mostly living in the Palau cluster) relies on industries related to tourism (services and craft items from shell, wood, pearls) and developing fisheries and garment industry. One of the largest agricultural products is marijuana. The constitution of 1979 was designed to incorporate both Western ideas of democracy and individual liberty and traditional forms of communal ownership and hereditary political systems. The people thus retain more of their indigenous culture than most postcolonial societies, highlighted by the fact that Palauans have the only active indigenous movement in Micronesia: the United Sect (Ngara Modekngei), practiced by about a third of the population. Palau's strength lies in this cultural heritage and the islands' remarkable beauty.

BIBLIOGRAPHY. Ron Crocombe, *The South Pacific* (University of the South Pacific, 2001); Frederica Bunge and Melinda W. Cooke, eds., *Oceania: A Regional Study* (Foreign Area Studies Series, 1985); K.R. Howe, Robert C. Kiste, Brij V. Lal, eds., *Tides of History: The Pacific Islands in the Twentieth Century* (University of Hawaii Press, 1994); U.S. Office of Insular Affairs, www.doi.gov/oia (March 2004); "Palau," www.motherearth.org/news (March 2004).

JONATHAN SPANGLER
SMITHSONIAN INSTITUTION

Pamir Knot

THE UNIQUE OROGRAPHIC feature known as the Pamir Knot takes its name from the mountains on which it is centered, the Pamir. The "knot" refers to the convergence of some of the world's major mountain ranges, including the TIAN SHAN, Karakorum, Kunlun, HINDU KUSH, and Pamir systems. The origin of the word *pamir* is unclear, although the Tajik name for the region is Bom-i-Dunyo, or "roof of the world." Other sources claim "feet of the sun" and the high altitude grasslands of the region as sources of "pamir." While a number of countries claim to be home to the Pamir

Knot, it is actually centered in the Gorno-Badakhshan autonomous region of eastern TAJIKISTAN. Fringe areas extend into AFGHANISTAN, PAKISTAN, CHINA, and KYRGYZSTAN.

In terms of relative location, the Pamir Mountains region remains one of the least accessible areas in the world. High peaks of the Pamir include Communism Peak (24,590 ft or 7,7495 m), Lenin Peak (23,403 ft or 7,133 m), and Peak Evgenia Korjenevskaya (23,311 ft or 7,105 m). In terms of mountaineering, these towering peaks have been scaled a number of times, although lesser peaks of the Pamir remain as yet unconquered. Considerable temperature variation exists within the Pamir region, with winter daily lows ranging from 1 degree F (-17 degrees C) in eastern areas to 21 degrees F (-6 degrees C) in the west. The Pamir region is heavily glaciated, including Murghab Pass, which stretches for 144 mi (231 km).

GEOPOLITICS

This mountain region is also home to a number of geopolitically important features. The Karakoram highway, linking Gilgit with Kashgar, is the highest international highway in the world. The Wakhan Corridor, a narrow corridor (as narrow as 10 mi or 16 km) running through the Pamir mountains between Tajikistan and Pakistan, can be seen on political maps as the eastern "finger" of Afghanistan. This corridor was an annexation to Afghanistan by Great Britain in the late 18th century. The main purpose of the Wakhan Corridor was to thwart Russian advances toward British India during the Great Game. The Great Game, waged throughout the region, was an 18th century Cold War-style conflict between Britain and Russia for control of Central Asia. The Wakhan Corridor effectively separated British and Russian territory.

The Tian Shan mountain range radiates north from the Pamir Knot through the border region of Kyrgyzstan, Kazakhstan, and the western Xinjiang region of China. The range includes Pik Pobeda, the highest point in Kyrgyzstan, which rises to an elevation of 24,407 ft (7,439 m). The lengthy Kunlun mountains stretch for 1,863 mi (3,000 km) through western China, south to the Pamir Knot, and northeast through northern tibet. The Kunlun range includes Kongur Tagh, rising to an elevation of 25,326 ft (7,719 m). The Hindu Kush mountain system stretches for over 559 mi (900 km) through eastern Afghanistan, forming much of the boundary with Pakistan. Notable peaks include Tirich Mir (25,230 ft or 7,690 m) and Noshaq (24,581 ft or 7,492 m). The Karakoram, stretching for 310 mi (500 km) through the border region of Pakistan, India, and China, is the westernmost range of the himalayas.

BIBLIOGRAPHY. Mark Elliot and Wil Klass, *Asia Overland* (Trailblazer Publications, 1998); Wikipedia Encyclopedia, "Pamir Mountains," www.nationmaster.com (September 2004); *Planet Earth World Atlas* (Macmillan, 1998).

KRISTOPHER D. WHITE, PH.D.
KAZAKHSTAN INSTITUTE OF MANAGEMENT

pampas

THERE ARE MANY types of GRASSLANDS worldwide, especially in the continental interiors of temperate to subtropical regions. That of South America is known as the pampas, a Spanish term. The pampas occupies some 270,270 square mi (700,000 square km) in the countries of ARGENTINA, URUGUAY, and southeast BRAZIL. Here, there are extensive plains with isolated low hills in a region where the major drainage system is the River Plata. Climatically, average temperatures range from 43 to 79 degrees F (6 to 26 degrees C); only mild and short-lived frosts occur during the winter months (July and August), but summer temperatures can reach 104 degrees F (40 degrees C).

The pattern of precipitation mirrors that of other continental grasslands such as the North American prairies with an east-to-west gradient that ranges from 47 in (120 cm) to 17.5 in (45 cm). Although rain occurs year-round in the east, there is a concentration in the winter. Toward the mountains that border the pampas in the west, there is a concentration in the spring that is advantageous for new growth in an areas that is semiarid. The geology comprises LOESS, a wind-blown sediment derived from volcanic ash, that gives rise to gray, brown, or black soils with high organic and nutrient content. They are fine-grained to the east and coarse-grained or sandy to the west, rich in calcium carbonate, and generally neutral or alkaline.

The vegetation communities are dominated by tall grass in the east, which benefits from a higher rainfall than the more arid west, where there are mainly medium- and short-grass communities. Herbs are present throughout, and where there is a local abundance of water, small woodlands occur. There are also wetlands of international importance. Species of the grass

Stipa are common throughout the region and the composition of the grass-herb-shrub communities depends on moisture and nutrient availability and topography. Most of the grasses form tussocks or clumps between which herbs and sedges grow. In the drier, western region, many of the species are xeromorphic (that is, tolerant of drought), and biodiversity is reduced compared with the eastern pampas. Here the grasslands merge with semidesert communities that occupy the rainshadow area created by the Andes. Locally, where soils are salt-rich, there are salt-tolerant species, that is, halophytes.

PLANT SPECIES

There are few plant species that are endemic (native to the pampas) and considerable changes have occurred since the advent of the Europeans in the 1500s and 1600s. First, many introduced species have become naturalized, including clover and numerous grasses from Europe, and these have replaced native pampas species.

Second, there has been much land-cover transformation, especially in the east, as the natural ecosystems have been replaced by agricultural systems. This area is particularly suited to the cultivation of maize (corn) and wheat and to cattle ranching, though resulting environmental problems include soil erosion, which is linked to overgrazing, and pollution from the overuse of ferilizers. The crops and meat produced are very important to Argentina's economy. Apart from the home market, they are major exports and generate significant foreign income.

While much emphasis is usually placed on the plant communities of the pampas, the animal communities can be overlooked despite the fact that there are several endemic animals. These include the pampas deer and various birds such as Olrog's gull, the curve-billed reedhaunter, and the pampas meadowlark, all of which are in danger of extinction. Other animals of the pampas include the guanaco, rhea, puma, geoffroy's cat, and the pampa fox, though many are rare and their populations are diminishing as their natural habitats are reduced.

BIBLIOGRAPHY. Roy Hora, *Landowners of Argentine Pampas* (Oxford University Press, 2001); *Oxford Essential Geographical Dictionary* (Oxford University Press, 2003); *Planet Earth World Atlas* (Macmillan, 1998).

A.M. Mannion
Kansas State University

Panama

Map Page 1136 **Area** 30,193 square mi (78,200 square km) **Capital** Panama City **Population** 2,960,784 **Highest Point** 11,400 ft (3,475 m) **Lowest Point** 0 m **GDP per capita** $6,200 **Primary Natural Resources** bananas, rice, corn, coffee, sugarcane.

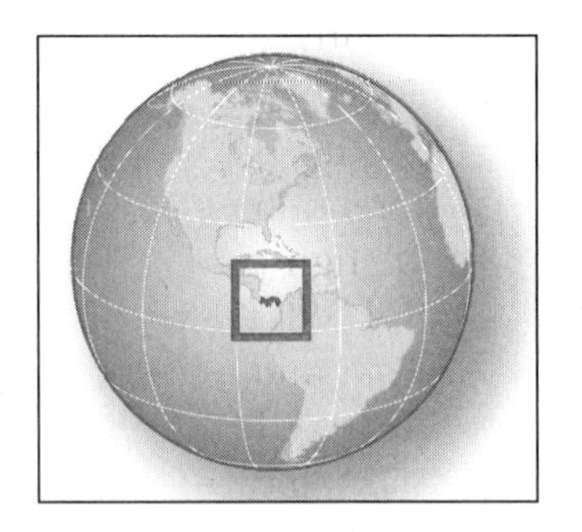

The Republic of Panama is a Central American country bordered both by the CARIBBEAN SEA and the PACIFIC OCEAN, between COLOMBIA and COSTA RICA. Slightly smaller than the state of SOUTH CAROLINA, the most dominant feature of the Panamanian landscape is the highlands, which form the continental divide through the center of the country. The highest constant elevations of the hills and mountains known as the Cordillera de Talamanca and the Serranía de Tabasará that create the continental divide are near Panama's borders, while the lowest elevations are along the Panama Canal, where most of the country's population is concentrated.

The country's Caribbean coastline has an abundance of natural harbors, but the only one that is significantly developed is located at Cristóbal, one of the terminuses of the Panama Canal, connecting the ATLANTIC and Pacific oceans. The port of Balboa is the major trading city on the Pacific coastline, which does not have an abundance of good natural harbors as a result of extraordinarily shallow waters and an extreme tidal range. Panama has almost 500 rivers, some of which provide crucial reservoirs of water for the canal and generate significant quantities of hydroelectric power.

Panama's climate is tropical, and temperatures and humidity remain high throughout the year, with little seasonal variation, usually remaining between 75 and 87 degrees F (24 and 32 degrees C). Temperatures on the Pacific side of the highland chain are typically lower than the Caribbean. In the highlands themselves, some areas experience temperatures that are quite cooler, and frosts occur occasionally in the Cordillera de Talamanca in western Panama.

Unlike temperature, rainfall throughout Panama varies substantially from 39 to 118 in (100 to more than 300 cm) per year. The overwhelming majority of the rains come during a marked rainy season, which usually lasts from April to December. The amount of

rainfall is determined by two factors in general: moisture from the Caribbean and the windshield effect of the continental highlands. Consequently, rainfall is much higher on the Caribbean side. In addition, the country is located outside the hurricane track, but is known to suffer from volatile thunderstorms.

Panama boasts a relatively strong economy, based upon its governmental stability and control over the PANAMA CANAL, which was turned over to Panama by the United States at the end of 1999. A well-developed service sector makes up the majority of its economy, mostly stemming from the Panama Canal, which continues to make its mark upon Panamanian history. Originally, U.S. plans for the Panama Canal provided the impetus for Panama's succession from Columbia in 1903, after which the new country was immediately recognized by the United States and an agreement for the commencement of construction was signed. Now, control over the canal permits Panama to continue to dominate trade in the region, as well as flagship registration and, more recently, tourism.

BIBLIOGRAPHY. *World Factbook* (CIA, 2004); "Panama: Area Handbook Series," Library of Congress, www.loc.gov (March 2004); *Merriam-Webster's Geographical Dictionary* (Merriam-Webster, 2003).

ARTHUR HOLST, PH.D.
WIDENER UNIVERSITY

Panama Canal

A CANAL THAT LINKS the ATLANTIC and PACIFIC oceans through the isthmus of PANAMA from northwest to southeast, the Panama Canal is 51 mi (82 km) long. A trip through the canal takes approximately eight hours. The Panama Canal consists of a system of six locks that help ships negotiate the water levels of the canal. On the Atlantic side are the three locks at Gatun, and on the Pacific side are a lock at Pedro Miguel and two locks at Miraflores. The locks are in six pairs so that ships can travel in opposite directions. The locks are 1,000 ft (300 m) long and 110 ft (33 m) wide. The Panama Canal was formerly part of the Canal Zone, a 10-mi (16-km) strip of land that was under the jurisdiction of the UNITED STATES from 1903 until 1999.

The desire to find a shortcut from the Atlantic to the Pacific stretches back to at least 500 years ago. The reality of construction on a canal began in 1880 when Ferdinand de Lesseps, who oversaw construction of the Suez Canal, gained a concession from the Colombian government, which ruled Panama, to begin work on the canal. However, after eight years, de Lesseps's plan ended in disaster as a result of poor planning, disease, and political upheavals. In 1889, de Lesseps's company declared bankruptcy and operations were turned over to the New Panama Canal Company in 1894. The same problems of disease and worker unrest also plagued the new owners, and after 1898, the New Panama Canal Company sought out potential buyers to take over construction.

The United States had been interested in constructing an isthmian canal since the early 19th century. In 1850, representatives from the United States and Great Britain signed the Clayton-Bulwer Treaty, which stated that neither the country would have sole ownership or defense of any future canal through Panama. By the end of the 19th century, the United States began its imperial expansion. The Clayton-Bulwer Treaty was replaced by the Hay-Pauncefote Treaty in 1901, giving the United States the sole monopoly on an isthmian canal and smoothing relations with Britain. In 1902, the United States began negotiations with COLOMBIA over the purchasing of the rights from the New Panama Canal Company to build the canal.

The Colombian government, however, rejected the Hay-Herrán Treaty because it would lose money in the long term. On November 3, 1903, a revolt erupted in Panama, and it declared its independence from Colombia, with American support. Thus, the United States was in a better position to negotiate for a canal treaty. Under the Hay-Bunau-Varilla Treaty, signed on November 18, 1903, the United States gained the Canal Zone and sole administration and defense of the canal "in perpetuity."

BUILDING THE CANAL

The building of the Panama Canal is considered one of the great engineering feats of history. Construction began in 1904 under the Army Corps of Engineers led by George W. Goethals. In order to compensate for the varied terrain and the unnavigability of Panama's rivers, the Chagres River was dammed, creating a human-made lake 85 ft (278 m) above sea level, later named Lake Gatun. Ships entering from the Atlantic would enter the Gatun Locks to get to the lake. The Gaillard Cut was then constructed for the sharp descent to the Pacific. The number of laborers it took to build the canal totaled some 17,000, mostly from the

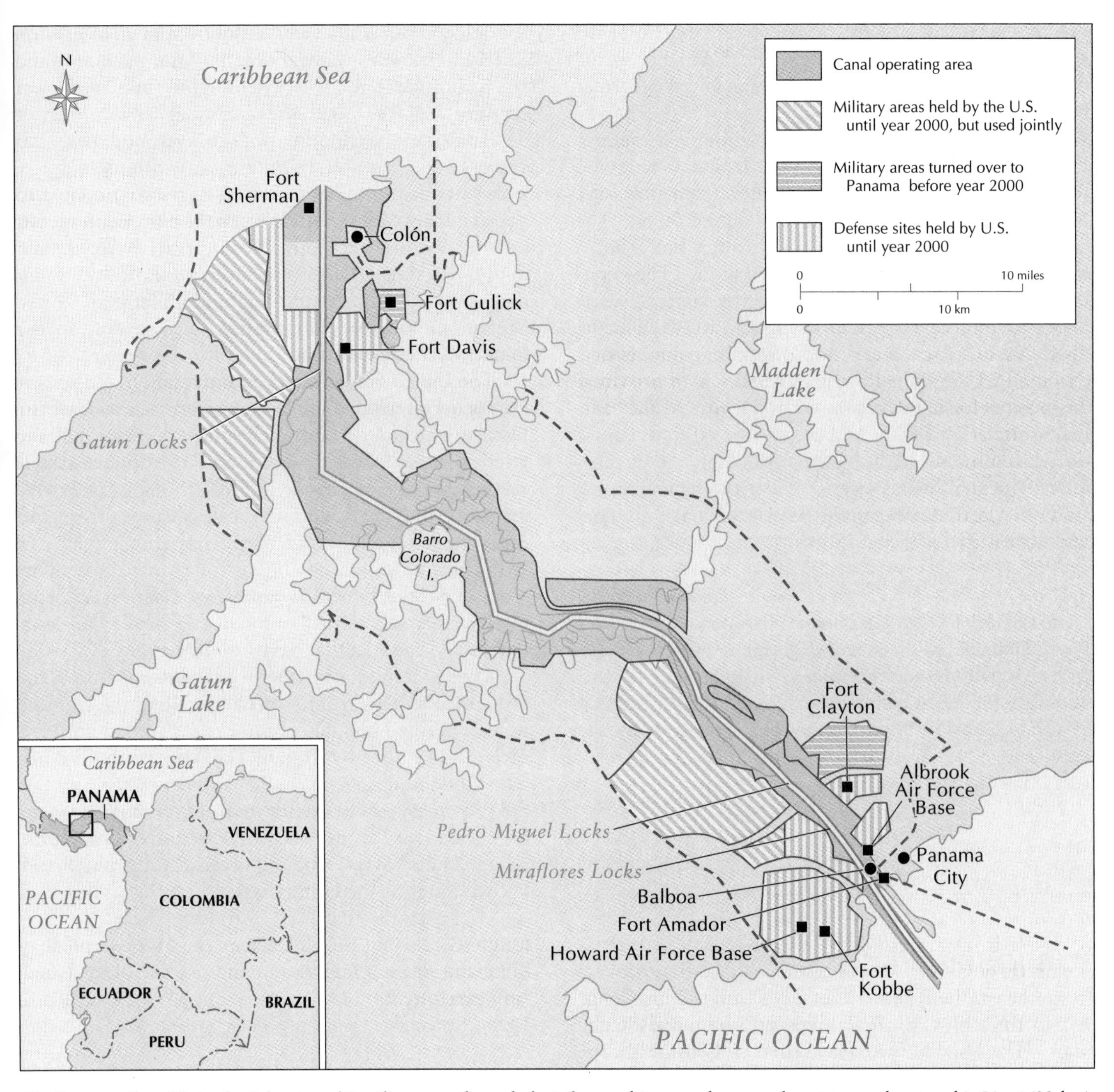

The Panama Canal links the Atlantic and Pacific oceans through the isthmus of Panama from northwest to southeast and is 51 mi (82 km) long. A trip through the canal takes approximately eight hours.

Caribbean. Another obstacle to overcome was eradicating disease, which was the undoing of previous attempts to build the canal. Under William C. Gorgas, the U.S. Army eradicated diseases such as yellow fever and malaria by destroying nesting areas of mosquitoes. The project cost $380 million and lasted about 10 years. The Panama Canal opened in August 1914.

The people of Panama were never satisfied with their relationship with the United States regarding the Panama Canal. After negotiations during the 1960s and 1970s, a new treaty replaced the Hay-Bunau-Varilla Treaty, whereby the United States would gradually hand over control of the canal to Panama. In 2000, Panama gained full control over the canal. Issues from

the American control of the canal remain, however, such as the cleanup of the ecological damage in the Canal Zone and unexploded ordnance from former test sites.

The Panama Canal contributes about 7 percent to Panama's gross domestic product. In 2001, revenues from the canal came to $215.2 million from annuities, tolls, dividends, and aid from the United States. The majority of the traffic that goes through the canal is from East Asia to the eastern United States. The countries that use the canal heavily are the United States, JAPAN, CHINA, CHILE, South KOREA, and PERU. During the 2002–03 fiscal year, 11,725 oceangoing vessels transited through the Panama Canal. Use of the canal has been declining in recent years because of the ever-increasing size of ships being constructed that cannot fit into the locks. To solve this problem, the government of Panama has begun work on widening the Culebra Cut and deepening the Gatun Lake channel and is considering building a third set of locks.

BIBLIOGRAPHY. Michael L. Connif, *Panama and the United States: The Forced Alliance* (University of Georgia Press, 2001); Thomas Leonard, *Panama, the Canal, and the United States: A Guide to Issues and References* (Regina Books, 1993); John Lindsay-Poland, *Emperors in the Jungle: The Hidden History of the U.S. in Panama* (Duke University Press, 2003); "Panama Country Profile," Economist Intelligence Unit (August 2004).

DINO E. BUENVIAJE
UNIVERSITY OF CALIFORNIA, RIVERSIDE

Pannonian Plain

ALSO KNOWN AS the Hungarian Plain, the Pannonian Plain is one of the flattest parts of Central Europe and one of the most agriculturally productive. The plain occupies a wide depression between the Alpine ranges to the west and south and the Carpathian chain to the north and east. At its greatest extent, the plain covers all of HUNGARY, eastern AUSTRIA, northern CROATIA and SERBIA, and southern CZECH REPUBLIC. The central feature, the DANUBE RIVER, divides the plain roughly in half, with the larger portion forming the Great Hungarian Plain to the east (known as the Alföld), and the smaller portion, the Western Lowlands (or the Little Hungarian Plain, the Kisalföld). To the south, the plain merges into the more hilly terrain of the Balkan Peninsula (where the DANUBE merges with its largest tributaries, the Tisza, the Mur, the Drava and the Sava), while the west rises slightly in elevation at the foothills of the ALPS on the frontier with Austria.

The Danube floodplain south of Budapest can reach 12 to 19 mi (20 to 30 km) in width. A high rise between this floodplain and that of the Tisza River to the east is also a division between the eastern and western portions of the plain. This elevated ridge (only about 165 ft or 50 m) is mostly formed of sand dunes and LOESS hills, such as those at Gödöllö or Nagykörös. Several million years ago the area was covered by the Pannonian Sea, which accounts for the richness of its soil, formed from organic sedimentary deposits, at points a kilometer thick. The western segment of the plain still includes a large water-filled depression, Lake Balaton, which, at 233 square mi (598 square km) is the largest lake in Europe in area (though not in volume, since it is very shallow). The eastern plains spread out for miles of almost completely horizontal fields and grasslands, or *puszta*, with rich soil formed from alluvium deposited by the Danube and Tisza rivers. This region is sparsely settled, populated instead by herds of horses and cattle, and covered with farms.

The plain takes its name from the Roman province of Pannonia. This frontier province along the Danube was established in 9 C.E. but not really settled until the early 2nd century by Trajan. The Romans built cities like Vindobona (Vienna) and Aquincum (Buda), but the province was settled not by Romans, but by waves of "barbarians" from the east: Ostrogoths, Lombards, Avars, Huns, Slavs, and finally by Magyars, better known today as the Hungarians.

BIBLIOGRAPHY. Clifford Embleton, ed., *Geomorphology of Europe* (Wiley, 1984); Éva Molnár, ed., *Hungary: Essential Facts, Figures and Pictures* (MTI Media Data Bank, 1995); "Pannonian Plain," www.hungary.com (August 2004).

JONATHAN SPANGLER
SMITHSONIAN INSTITUTION

Pantanal

LOCATED IN SOUTH-CENTRAL BRAZIL and extending into the adjacent parts of BOLIVIA and PARAGUAY, the Pantanal is the world's largest wetland. It can expand during the rainy season to cover about 77,000 square mi (200,000 square km), which is about

one-third the size of FRANCE. Water depths may exceed 18 ft (6 m), with only the highest ground escaping inundation at this time. Water levels recede about 13 ft (4 m) during the dry season.

Rainfall in this part of South America totals 40 to 60 in (100 to 150 cm) per year, most of it concentrated in the hot months of November to April. The climatic impact of the Pantanal extends far beyond its boundaries. It acts as a giant sponge, receiving rainfall during the wet season and then releasing moisture to the atmosphere during dry spells. Climatologists assert that nearby areas would become less habitable if it were not for this store-and-release mechanism.

The Pantanal is an immense reservoir of biological diversity. It contains an astounding variety of plant life. Floating plants cover huge areas. A rich layer of silt supports many aquatic plants. On higher ground, bushy vegetation called cerrado predominates. Palm trees and fig trees punctuate these landscapes. Trees may form galleries along the banks of some rivers (gallery forests).

The aquatic vegetation supports large numbers of animals. Scientists have already identified over 650 species of birds such as the colorful macaw, 250 species of fish, 80 mammals, and 50 reptiles in this ecological paradise. Aquatic birds such as ducks filter small animals and algae from the water. Herons have established large colonies in trees along the riverbanks. They and storks feed primarily on fish, frogs, snails, crabs, and insects. The Pantanal is home to thousands of varieties of butterflies and a myriad of other insects. Larger animals living here include capuchin and howler monkeys, tapirs, capybaras, giant river otters, jaguars, caimans, and anacondas.

Despite the difficulty of access in this part of South America, the region's resources have attracted a great deal of attention, not all of it welcome. For example, poachers unlawfully kill caimans for their skins and capture macaws for export to Northern Hemisphere customers. Although reptile-breeding farms have helped reduce the illegal export of caiman skins, the Brazilian government lacks the resources to effectively combat the poachers. The arrival of commercial fishing boats, supported by refrigerated trucks, has complicated the problem of protecting natural resources here.

In addition, cattle ranchers have recently expanded operations on the margins of the Pantanal. The ecological impact of about 4 million cattle on these ranches has been considerable. For example, ranchers have removed cerrado vegetation from interfluves, leading to increased silting of rivers such as the Taquari. Farmers wishing to cultivate rice, sugarcane, and soybeans have also had a negative impact upon the natural environment. Mercury pollution from gold mining and sewage from nearby cities such as Cuiabá add to the problem. Fish absorb the toxic mercury, which is then passed up the food chain by fishermen and fish-eating birds.

The Pantanal is drained by the Paraguay River to the south and by tributaries of the AMAZON RIVER to the north. Cable News Network reported in April 1997 that engineers have proposed dredging a waterway to connect the two river systems. Proponents say that the proposed Hidrovia project would lower transportation costs and provide a vital transportation link between members of the MERCOSUR trading block.

Opponents assert that the project would be an ecological catastrophe, draining the Pantanal, killing many of the area's plants and animals, and negatively impacting the Native Americans living there. So far, the Brazilian government has yet to approve the Hidrovia project.

BIBLIOGRAPHY. Vic Banks, *The Pantanal: Brazil's Forgotten Wilderness* (University of California Press, 1991); David L. Pearson and Les Beletsky, *Brazil, Amazon and Pantanal* (Academic Press, 2002); Frederick A. Swarts, ed., *The Pantanal: Understanding and Preserving the World's Largest Wetland* (Paragon House Publishers, 2000); Wilderness Research Institute Pantanal, http://www.pantanal.org (September 2004).

JAMES N. SNADEN
CHARTER OAK STATE COLLEGE

Papua New Guinea

Map Page 1128 **Area** 178,707 square mi (462,840 square km) **Population** 5,295,816 **Capital** Port Moresby **Highest Point** 14,793 ft (4,509 m) **Lowest Point** 0 m **GDP per capita** $2,400 **Primary Natural Resources** gold, silver, copper, natural gas, timber.

PAPUA NEW GUINEA IS an archipelago that includes the eastern half of the equatorial island of New Guinea, which lies between the Coral Sea and the South PACIFIC OCEAN, east of INDONESIA and north of

AUSTRALIA. The country also encompasses much of the Bismarck Archipelago, including the islands of New Ireland and New Britain and the northernmost island of the SOLOMONS chain, Bougainville. The region has a monsoonal climate, with the wettest season lasting from December to March. Port Moresby, the capital of southern New Guinea, receives 39 in (1 m) of rainfall annually, while western New Britain regularly sees more than 20 ft (6 m) of rainfall per year.

The island of New Guinea was populated by migrants from Asia some 50,000 years ago. Evidence of complex human societies based on irrigated agriculture dates from 9000 B.C.E. Portuguese and Spanish explorers visited the area during the 16th century, but the island's size, large population, and difficult climate limited Europeans' advances. Dutch officials asserted their control over the western half of the island in the early 1800s, and it eventually became part of modern Indonesia.

GERMANY and the UNITED KINGDOM established claims over the eastern half of New Guinea in the 1880s, and Britain transferred its sector to AUSTRALIA in 1906, renaming it Papua. Germany lost its sector to Australia during World War I, and under an international mandate Australia governed the eastern half of the island. Much of it was seized by the Japanese, who also took control of the outlying islands, in World War II. After the war, New Guinea and many of the associated islands again were overseen by Australia, which guided them to full independence in 1975 as Papua New Guinea.

New Guinea is the world's second largest island, after GREENLAND. It is a geologically complex island dominated by rugged interior mountain ranges, high plateaus, and precipitous valleys, while its coastal areas feature powerful rivers, reedy deltas, and mangrove swamps. On the southwestern coast the country's longest river, the Fly, rises in the Victor Emmanuel mountains and flows through forested gorges before crossing plains and swamps. At is estuary it is 33 mi (53 km) wide. Also in the west is Lake Murray, the country's largest lake, which grows to five times its usual 400 square km (155 square mi) during the rainy season. In the northeast the great Sepik River travels 700 mi (1,126 km). With no natural delta, it deposits silt as far as 31 mi (50 km) out to sea.

The major islands of New Britain, New Ireland, and Bougainville, east of New Guinea, are part of an active volcanic formation known as the RING OF FIRE. The islands are surrounded by some of the largest coral reef complexes in the world. In 1994, 1997, and 2002, New Britain's principal city, Rabaul, was devastated by eruptions of the nearby volcano Tavurvur.

Papua New Guinea hosts 9,000 plant species, 700 species of birds, and some 200 species of mammals. Intensive logging of its rainforests and heavy metals pollution from mining operations, particularly on New Guinea and Bougainville, present difficult environmental challenges.

BIBLIOGRAPHY. J.R. McAlpine, *Climate of Papua New Guinea* (Australian National University Press, 1983); K.J. Pataki-Schweizer, *A New Guinea Landscape: Community, Space, and Time in the Eastern Highlands* (University of Washington Press, 1980); Diane Ranck, *Exploring Geography through Papua New Guinea* (Oxford University Press, 1994).

LAURA M. CALKINS, PH.D.
TEXAS TECH UNIVERSITY

Paraguay

Map Page 1141 **Area** 157,047 square mi (406,750 square km) **Population** 6,036,900 **Capital** Asunción **Highest Point** 2,762 ft (842 m) **Lowest Point** 151 ft (46 m) **GDP per capita** $4,300 **Primary Natural Resources** cotton, sugarcane, soybeans.

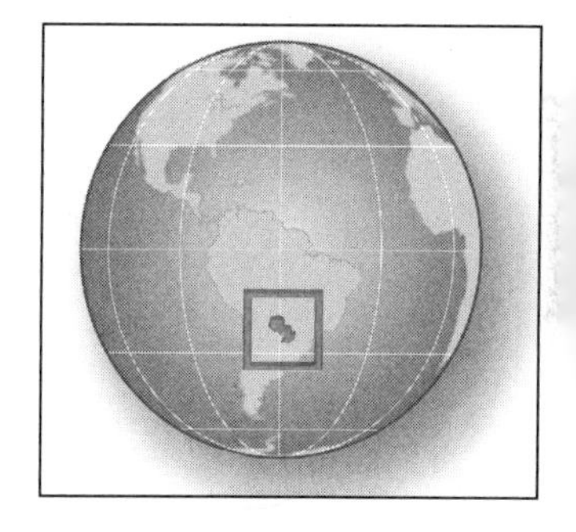

THE REPUBLIC OF PARAGUAY is a LANDLOCKED country located in the central region of South America bordered by BOLIVIA, ARGENTINA, and BRAZIL and is slightly smaller than the state of CALIFORNIA. Paraguay is divided by the Rio (River) Paraguay into the eastern Paranena region and the western Chaco region. In the Paranena region, lands range from low plains to mountains, with the highest elevations occurring near the border with Brazil.

The Chaco region consists of a vast low-lying plain that makes up more than 60 percent of Paraguay's territory. Near the Rio Paraguay, the plains are quite marshy, while at the interior, they are dry and often parched from drought.

As a result of being landlocked, Paraguay depends upon the navigable Rio Paraguay for most of its trade, which flows through to BUENOS AIRES, Argentina.

Paraguayan foreign policy is mostly dominated by Argentine decisions.

The climate is tropical and subtropical but becomes more temperate toward the south of the southern portions of the country. The Paranena region is notably humid and has abundant rainfall, which is distributed in relatively equal amounts throughout the year. In contrast, the Chaco region has a clearly distinguishable wet and dry season. The lack of efficient drainage renders the Chaco region susceptible to flooding. Temperature variations are modest, ranging between hot and humid to mild and damp.

Paraguay's population is mostly concentrated in the southern part of the country. The government is based at the capital of Asunción, which has a special government status, and the remainder of the country is separated and administered into 17 departments. The country is governed by a strong executive who shares authority with a bicameral legislative branch and a supreme court, all of which was established by the most recent national constitution, signed on June 20, 1992.

HISTORY

The acceptance of the constitution marked a return to democratic government, following 35 years of military dictatorship, led by Alfredo Stroessner. Paraguay's history has almost always been marked by conflict, whether it is political infighting or border disputes.

The War of the Triple Alliance, which lasted from 1865 to 1870, proved disastrous for Paraguay, which lost much of its territory and more than two-thirds of its adult males as a result. Following the horrific costs of the war, Paraguay remained stagnant for well over a half-century, until it was able to invade some Bolivian lands and to retain them following the settlement of the Chaco War, which lasted from 1932 to 1935. Even after the conclusion of the major wars in which Paraguay participated, government instability still reigned, allowing Stroessner to seize power for over three decades.

Paraguay's economy is market-based but dominated by a large informal sector, made up of thousands of microenterprises and street vendors. The majority of the population is able to survive as a result of their own family-level agricultural activity that generates enough for subsistence. Even though the formal economy has continued to grow at modest levels, most international observers blame political corruption, uncertainty about reform, and substantial debt for the lack of a stronger and more positive growth.

BIBLIOGRAPHY. *Oxford Essential Geographical Dictionary* (Oxford University Press, 2003); *World Factbook* (CIA, 2004); "Paraguay: Area Handbook Series," Library of Congress, www.loc.gov (March 2004).

Arthur Holst, Ph.D.
Widener University

Paris

PARIS IS NOT ONLY the capital city of FRANCE, it is also the largest city in the country. The census of 1999 showed that although the city itself, which covers 40.5 square mi (105 square km), is relatively demographically small by world standards, having a population of about 2.15 million, the greater Paris metropolitan (*aire urbaine de Paris*) population is significantly larger. In 1999, it was shown to be almost 11.2 million people. In terms of its administration, the city of Paris is broken into 20 districts or *arrondissements* that are each governed by elected councils. Each *arrondissement* has its own mayor, and some local council members are also members of the Council of Paris, which serves as a municipal government and general council of the Paris *département* (Paris political region). Today, Paris is not only at the forefront of the French economy but also that of Europe. With its many skyscrapers in the La Defense area, Paris has a financial core that can rival that of other European powerhouses. Much of the city's industry is based in the service sector and many multinational companies can be found within the city.

Paris has a long history—the name being derived from the Parisis peoples, a Gallic tribe who resided in the area at the time of the Roman conquest in 52 B.C.E. Upon invading France, the Romans named the Parisis settlement Lutetia, "marshy place," although within a handful of decades the settlement had another new name, Paris. Historically, Paris has centered upon a small island within the Seine River that is known as the Île de la Cité, today the site of the Notre Dame Cathedral, while urban development first took place on the left bank (*rive gauche*) of the Seine, which meanders through the settlement.

While the history of the city is long and at times bloody, it should be understood that the settlement's prosperity from the Middle Ages was closely tied to trade and cultural developments. From this age, the city has become known as an intellectual center that was highlighted from as early as 1257 with the found-

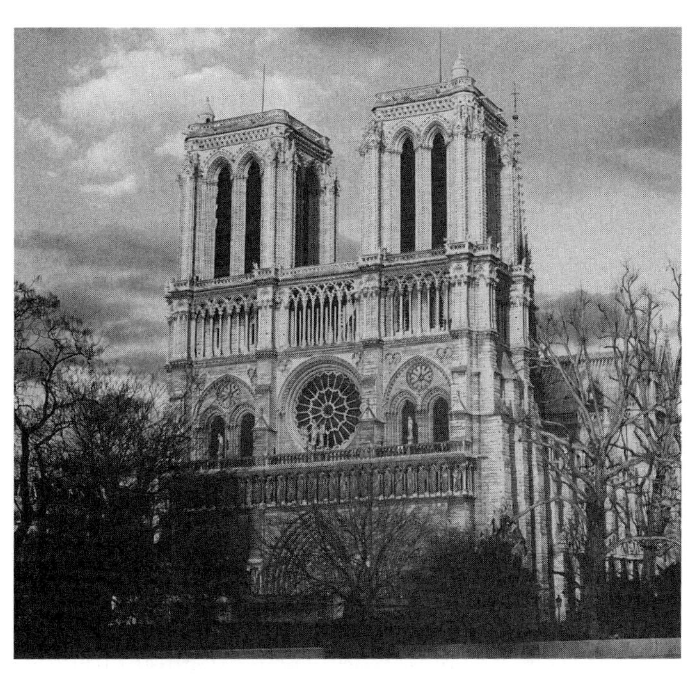

Paris is centered upon a island in the Seine River that is known as the Île de la Cité, today the site of the Notre Dame Cathedral.

ing of Sorbonne University (formerly known as the University of Paris). However, Paris is almost equally famous for its turbulent history, none more so than during the French Revolution (1789), the Franco-Prussian siege (1870), the Paris Commune (1871), and World War II (1939–45).

Topographically, Paris is generally flat because it lies on the former flood plain of the Seine River, although a number of prominent hills exist within its bounds. These include the hills of Belleville and Montmarte, the latter area famous for impressionist painting during the late-19th century and for the Basilica of the Sacré Coeur, which can be seen from almost all areas of the city. In terms of sophistication and art, few cities can match Paris and many of its museums are world famous for their collections. Visually, too, few cities can compare with Paris, with its large classical buildings scatterered throughout the city and the vast sways of buildings erected alongside braod boulevards under the redevelopment project of Baron Georges-Eugène Haussmann during the 1850s and 1860s.

BIBLIOGRAPHY. Alistair Horne, *Seven Ages of Paris* (Vintage Books, 2004); Anthony Sutcliffe, *Paris: An Architectural History* (Yale University Press, 1996).

IAN MORLEY, PH.D.
MING CHUAN UNIVERSITY, TAIWAN

peninsula

A PENINSULA IS AN area of land that projects out into a body of water that surrounds it on three sides. Some peninsulas, like Baja California, are joined to the mainland only by a narrow neck of land called an isthmus. Others, such as INDIA, are connected to the mainland by a wide area of land. A cape is similar to a peninsula, but it is shorter and smaller. Some areas have many peninsulas and others have few. Peninsulas and islands make up more than one-third of Europe's area. Kenai Fjords, in ALASKA, is made up of many jagged peninsulas formed by the ice fields. As the glaciers in the area recede, the fjords deepen. This enlarges and exposes the peninsulas that reach out into the sea.

Peninsulas are formed in other ways, too. In some areas along the shores of the ocean, limestone and clay form the foundation of cliffs. Capes and peninsulas are formed as erosion wears away some of the clay and limestone. The higher the shore and the harder the rock is, the slower the rate of abrasion will be. Some peninsulas are formed when mud and sand are deposited between a series of islets, making a solid land mass. Others are formed by rivers. Erosion sometimes causes a peninsula to break off from the mainland and form an island.

The Arabian peninsula is the world's largest peninsula. It is made of crystalline rock that developed the same time the ALPS did. The Arabian Peninsula used to be joined to North Africa When geologic movements caused a fault in Africa, called the Great Rift, the RED SEA was formed. The Great Rift runs from the Jordan valley to Central Africa. Geologists believe the whole peninsula is rotating slowly counterclockwise and in about 10 million years, the Persian Gulf will be closed off and become a lake. The country of SAUDI ARABIA takes up over half the area of the Arabian peninsula, but it shares the peninsula with other countries. They are JORDAN, ISRAEL, LEBANON, SYRIA, KUWAIT, BAHRAIN, QATAR, UNITED ARAB REPUBLIC, OMAN, and YEMEN.

The FLORIDA Peninsula reaches from the mainland of the UNITED STATES out into the ATLANTIC OCEAN and the Gulf of Mexico. It is a limestone plateau, formed many years ago when the area was covered by a shallow sea. The limestone, which is several thousand feet thick, is made up of the bodies of tiny sea creatures deposited over a period of millions of years. The limestone in Florida is only 50–60 million years old, compared to limestone in Kentucky, which is 430 million years old. The Florida limestone is very soft and very white.

Scientists say that about 200 million years ago, Africa and North America collided, then drifted apart again. A piece of Africa broke off and became the basis for Florida, deep under thc ocean. It was on this piece of land that the limestone built up. About 30 million years ago, enough land had been formed to rise above the level of the ocean. During the past 30 million years, Florida has alternately been covered with salt water and been dry land. Today Florida's position between the Atlantic Ocean and the Gulf of Mexico, combined with its pleasant warm climate, has made it a prime tourist area. Many retirees relocate to Florida every year.

On the opposite side of the United States is a very different peninsula, called Baja California. The Baja Peninsula is part of Mexico. This is a long narrow peninsula extending south from the American state of California. Some scientists think that it used to be part of the mainland of MEXICO. They believe a rift developed and the peninsula moved away from the mainland. This took tens of millions of years. The gap filled with water and became the Gulf of California. Baja California developed its present shape during the last 5–10 million years. The peninsula tilted westward, forming fault block mountains. It is believed that the Baja Peninsula will eventually become an island.

BIBLIOGRAPHY. Lothar Beckel, ed., *The Atlas of Global Change* (Macmillan, 1998); "Saudi Arabia: Topography and Natural Regions," reference.allrefer.com/country-guide-study (March 2004); The Florida Speleological Society Site, "Basic Central Florida Geology," www.caves.com (March 2004).

PAT MCCARTHY
INDEPENDENT SCHOLAR

Pennsylvania

THE COMMONWEALTH OF Pennsylvania, also known as the Keystone State, is located in the northeastern UNITED STATES. With a total area of 46,058 square mi (119,290 square km), Pennsylvania extends 307 mi (494 km) from east to west and 175 mi (282 km) north to south. It is bordered by NEW YORK and Lake ERIE to the north, WEST VIRGINIA and OHIO to the west, NEW JERSEY to the east, and MARYLAND, West Virginia, and DELAWARE to the south. The state, made up of 67 counties, is an industrial leader in the country and is famous for its many historical, cultural, and recreational sites, including Gettysburg, Amish Country, the Poconos, Independence Hall, Fallingwater, and Fort Necessity. As of 2003, Pennsylvania ranked sixth in the nation in population, with an estimated 12,365,500 residents. The most common ancestral heritages reported by Pennsylvanians are German, Irish, Italian, and English. Ten percent of the population is African American. Pennsylvania has a moderate climate with hot and humid summers and cold and snowy winters. Harrisburg is the state's capital.

Pennsylvania's diverse economy is dominated by the service sector, contributing 74 percent of the state's gross product. After the decline of the steel-making industry, the state had to adjust its manufacturing sector which now makes up 20 percent of the gross product. Food processing, chemical production, and electrical equipment dominate this part of the economy. Agriculture—dominated by livestock and dairy—and mining, mainly of petroleum, iron ore, natural gas, and coal, contribute 1 percent each toward Pennsylvania's gross product. One notable agricultural product activity is the harvesting of mushrooms in the southeast part of the state. Pennsylvania is the only producer of anthracite coal in the United States. However, the more common bituminous coal is also extracted.

In the early 1600s, Dutch sailors were the first Europeans to explore Pennsylvania. At the time of European arrival, native tribes in Pennsylvania included the Lenape, Susquehannock, Monongahela, Shawnee, Huron, and Erie. The first European settlement, south of present-day Philadelphia, however, was Swedish. While there were territorial disputes between the Swedes and the Dutch, England eventually took over the territory. King Charles II made a royal charter for Pennsylvania, or "Penn's Woods," in 1681 and named the colony after Admiral Sir William Penn, the father of the founder of the state, William Penn. Penn, a member of the Society of Friends, or Quakers, became the caretaker of the new colony and wanted Pennsylvania to be a refuge for the persecuted Quakers in England. Penn set up Philadelphia as the first permanent English settlement in Pennsylvania.

Soon after this time, the French began to build forts in western Pennsylvania, but they were eventually all taken over by England. During the American revolutionary efforts against Great Britain, Philadelphia became the capital of the new country. Pennsylvania later became a state in 1787. The economy expanded in the 1800s with Pittsburgh and Philadelphia evolving into important shipping hubs. The first oil well in the world

was drilled at Titusville in 1859 by Edwin Drake. Pennsylvania additionally benefited greatly from the Industrial Revolution and the two World Wars. With many coal sources within and near Pennsylvania, the commonwealth, especially the city of Pittsburgh, became the heart of the American steel industry, managed by entrepreneur Andrew Carnegie.

Pennsylvania's natural environment has changed drastically from the time of European arrival. The state has many old, winding rivers, and most of the state is hilly with plateaus, ridges, and valleys. Most of the lakes are man-made, and three-fifths of the state is covered by forests. Pennsylvania can be divided into seven main landform regions. Starting in the southeastern corner of the state, there is the Atlantic Coastal Plain, a small strip of low-lying and level ground that includes the state's largest city, Philadelphia. Immediately to the northwest is the Piedmont Plateau, an area of farmland, low hills, and rolling plains. This area covers most of southeastern Pennsylvania.

To the immediate northeast of the Piedmont, there is a narrow finger-shaped region extending out of New Jersey called the New England Upland. The industrial cities of Allentown, Bethlehem, and Easton are found in this area of low rolling hills and forests. At the southwestern part of the Piedmont, coming out of Maryland, is another narrow region called the Blue Ridge. The Blue Ridge province is hilly and has many productive farms, orchards, and dairies.

The Appalachian Ridge and Valley physiographic region is in the center of Pennsylvania and covers about half of the state, crossing from Maryland and into New Jersey. It is part of the Appalachian Mountains and includes the highest point in Pennsylvania, Mt. Davis at 3,212 ft (979 m). To the west and north is the Allegheny or Appalachian Plateau. This area consists of valleys and divides and covers the majority of western and northern Pennsylvania. Pittsburgh, the state's second largest city, is located here, as well as Allegheny National Forest. The region also once had very productive coal and natural gas mines.

Finally, on the northwest tip of the Pennsylvania panhandle is the Erie Lowland. Found on Lake Erie, the area is flat, has significant vegetable and fruit production, and also includes the city of Erie, a major shipping port for transport through the Great Lakes and the St. Lawrence Seaway.

BIBLIOGRAPHY. Ari Hoogenboom, "Pennsylvania" *Worldmark Encyclopedia of the States*, Timothy L. Gall, ed. (Gale Research, 1995); Pennsylvania Historical and Museum Commission, www.phmc.state.pa.us (July 2004); Wilbur Zelinsky, "Geography," *Pennsylvania: A History of the Commonwealth*, Randall M. Miller and William Dencak, eds. (Pennsylvania State University Press, 2002).

ANTHONY PAUL MANNION
FORT HAYS STATE UNIVERSITY

Persepolis

ONE OF THE MOST magnificent cities of the ancient world, Persepolis was the political, cultural, and religious center of the Achaemenian PERSIAN EMPIRE for over 200 years before its complete destruction at the hands of Alexander the Great in 330 B.C.E.

The ruins of Persepolis lie at the foot of Kuh-i-Rahmat, the "Mountain of Mercy," in the plain of Marv Dasht about 400 mi (650 km) south of the present capital of IRAN, TEHRAN. The site is remote, in a large barren plain surrounded by sharp cliffs, and the ruined city lay hidden for over a thousand years before it was identified in 1620. It was occasionally visited by the curious, but it was not until the 1930s that a scientifically planned expedition was sent to excavate and systematically map and catalog the ruins.

The Persepolis Expedition was sponsored by the University of Chicago's Oriental Institute, later joined by the University Museum in Philadelphia and the Boston Museum of Fine Arts. Its leaders were Ernst Herzfeld and Erich Schmidt, and the project employed up to 500 men recruited from local villages: diggers, draftsmen, recorders, mechanics, and others. World War II put an end to this project, which was taken up again after the war by the Iranian Antiquity Service and the Italian Institute of the Middle and Far East in Rome.

Persepolis was founded sometime in the reign of Darius I, probably around 518 or 516 B.C.E. Inscriptions show that he imagined the city to be a magnificent showpiece for the mighty new empire he and his predecessors, Cyrus and Cambyses, had created. It was called Parsa, City of the Parsis ("Persepolis" in Greek); it is known today in Iranian as Takht-e-Jamshid, "the Throne of Jamshid," a mythical king of Persia. The Achaemenid kings made it an administrative center as well as a stage for receptions and ceremonial festivities. Wealth from all corners of the empire was featured in its decoration to showcase the immensity and diversity of the empire. Darius did not live long enough to finish

his project. In fact, most of Persepolis was built by his son, Xerxes. Still, the complex was not entirely completed until the reign of Artaxerxes I, about 100 years later.

The main feature of the city was the palace complex, a large terrace, about 990 by 1,485 ft (300 by 450 m), rising 33 to 66 ft (10 to 20 m) above the surrounding ground. On this terrace were built the main ceremonial buildings: the Apadana, the Throne Hall, the Gate of Xerxes, the palaces of Darius and Xerxes, the Harem, the Treasury, and the Council Hall. The Audience Hall of the Apadana is the largest and most splendid of these buildings, with 72 columns (of which 13 still stand), two monumental stairways, and rows of reliefs with representations of the 23 subject nations of the empire—each bearing tribute gifts—followed by Persian and Median court notables, soldiers, horses, and royal chariots. A short distance from the city, other grandiose buildings were built to house the tombs of the Achaemenid kings, similar to the Valley of the Kings in EGYPT.

Persepolis was renowned for its beauty and splendor, but because it was such a strong symbol of Persian power and glory, it was an obvious target for Alexander the Great. Alexander's destruction of the city was thus highly symbolic of his complete victory over the Persians. According to Plutarch, Alexander's men burned the entire complex to the ground and carried away its treasures on 20,000 mules and 5,000 camels.

BIBLIOGRAPHY. "Persepolis," Oriental Institute, www.oi.uchicago.edu (August 2004); Erich F. Schmidt, *Persepolis I: Structures, Reliefs, Inscriptions* (University of Chicago Press, 1953); Ann Britt Tilia, *Studies and Restoration at Persepolis and Other Sites in Fars* (Rome, 1972).

JONATHAN SPANGLER
SMITHSONIAN INSTITUTION

Persian Empire

THE PERSIAN EMPIRE was one of the first of the world empires to emerge in the ancient Middle East, the first to unify several different peoples and cultures into one large heterogeneous state. Much of this work was achieved by the emperors Cyrus and Darius, who recognized the strength of diversity, picking the best of a variety of practices and customs and welding them into a system that worked best for the empire as a whole. The original Persian Empire endured only two centuries, but it set a standard to be emulated by successive large multinational states set up by the Greeks and the Romans for centuries to come.

The name *Persia* comes from a specific province within the empire, in what is today southwestern IRAN. Here, in Pars (or Fars), the people spoke a language related to other peoples in INDIA and Europe, but different from the Semitic peoples of Mesopotamia to the west. The Persian tribes (or Parsis) lived in the hilly region between the great empires of the Medes to the east and the Babylonians to the west and were ruled by each of them at various times. Parsis were descended from tribes known as Aryans, which eventually gave its name to modern Iran.

In about 559 B.C.E., a Persian leader, Cyrus, unified the tribes, and led a revolution that overthrew the Medes and took over all of their territory (most of modern-day Iran). Cyrus the Great (559–530) established the first major Persian dynasty, the Achaemenid Dynasty, and took the title shah, or king. Instead of oppressing their former rulers, the Persian rulers united the Parsis and the Medes into one people, incorporating the Medes's strong central government rather than trying to re-create it from scratch. By 539, the combined armies of Persians and Medes had conquered not only neighboring Babylon, but parts of Asia Minor (Anatolia) and Central Asia as well.

CONTINUED EXPANSION

The empire continued to expand under Cyrus the Great's son, Cambyses (530–522 B.C.E.), who conquered Egypt, and his distant cousin, Darius I (522–486), who led his armies as far as the Indus valley in the east and GREECE in the west. Darius was stopped by the Greeks famously after the Battle of Marathon in 490, but his legacy is concerned more with the brilliance of his skills as an administrator of an empire. He organized the empire into a series of regional governments, or satrapies, ruled by men loyal to the emperor alone, responsible for collecting taxes and organizing local militias.

Darius also created public works—irrigation, canals, and public buildings—and built good roads for the improved communication and trade between parts of the empire. He created one single currency and a postal system, and standardized weights and measurements to be used throughout the empire. Darius also built a new capital, on plans laid down by his predecessor, Cyrus, at PERSEPOLIS, one of the largest collections of palaces and buildings the world had ever seen.

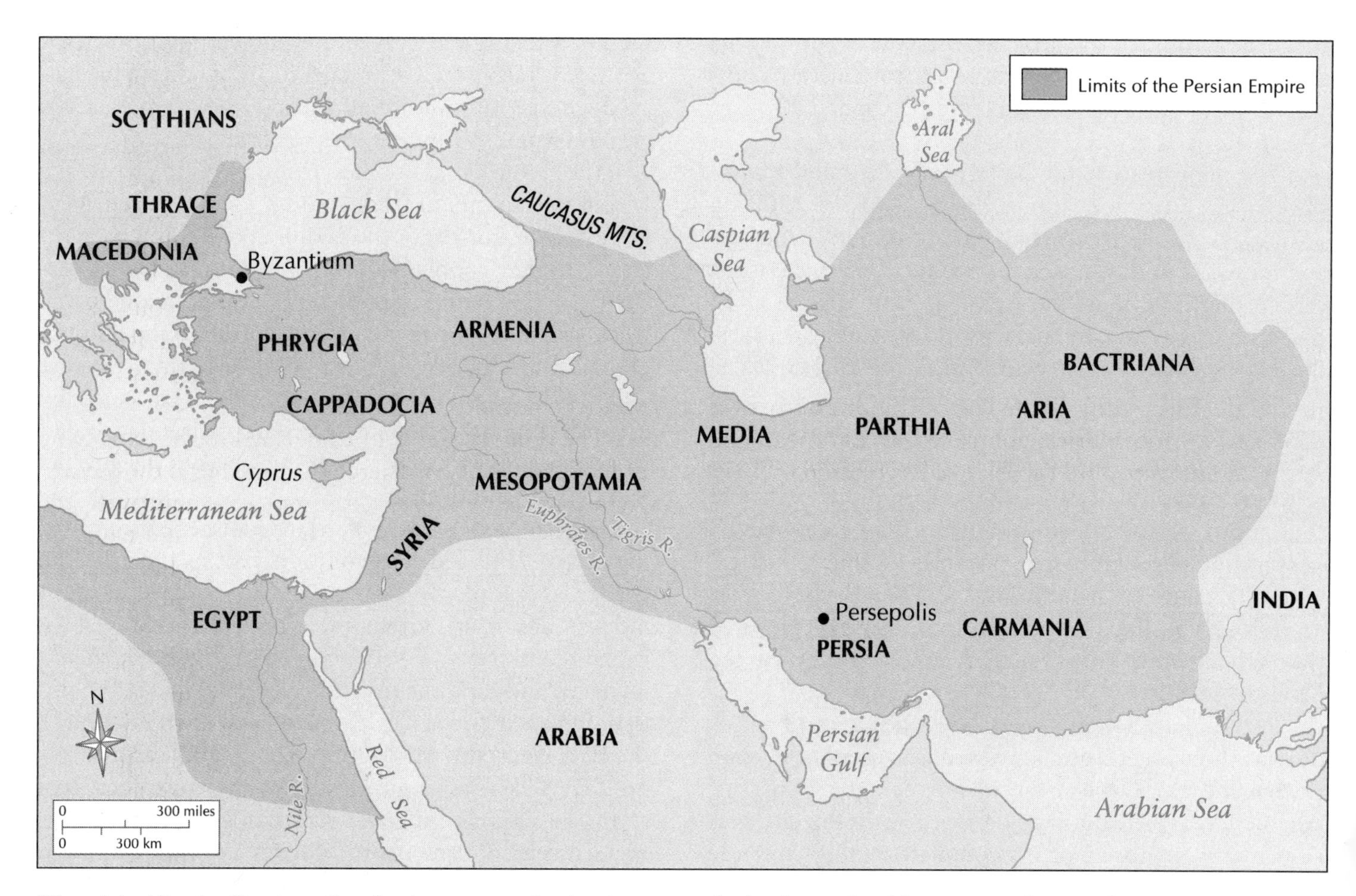

The original Persian Empire endured only two centuries, but it set a standard to be emulated by successive large multinational states set up by the Greeks and the Romans for centuries to come.

Persia at this time became a center for learning, collecting, and synthesizing the technological and intellectual advances of its many subject peoples, including mathematics from India, irrigation from Mesopotamia, and ship-building from the Greeks. Amidst this cultural flowering, a distinctively Persian religion emerged, one of the first world religions centered on one god instead of many.

The teachings of the religious prophet Zoroaster (c. 628–551 B.C.E.) focused on the two fundamental aspects of a supreme being, the aspect of light, truth, health and goodness (Ahura Mazda) and the aspect of darkness, sickness and evil (Ahriman), which eternally strove together for dominance in the universe. Unlike the religions of Babylon and EGYPT, Zoroastrian leaders, known as Magi, taught that there was life after death (for everyone, not just kings) and provided hope for the ultimate victory of Ahura Mazda.

The Achaemenid Persian Empire was the largest and most powerful empire the world had thus far seen, affecting nearly every Eurasian civilization except CHINA. Its greatest impact was in the sharing of intellectual and technological advances from societies as geographically far apart as Egypt and the Indus Valley. But ultimately the size of the empire was too much for one single ruler to handle. The son of Darius the Great, Xerxes, renewed his father's attempts to subdue the Greeks and was again pushed back at Salamis in 480 B.C.E. His defeat in the west encouraged rebellions among the tribes of the east. Political weakness grew through internal and dynastic struggles, inviting Greek armies led by the young King of Macedonia, Alexander the Great, to invade in 334 B.C.E. Within only eight years, Alexander brought the mightiest world empire to its knees and had incorporated it into a new, Helenic (or Greek) empire. The last emperor, Darius III, was defeated at Gaugemela and killed in 331, and his capital at Persepolis was burned to the ground.

For the next three centuries, the main language of imperial rule and culture would be Greek, but the em-

pires of Alexander and his successors did learn a great deal from the Persians about respecting the individuality of their subject peoples, and incorporating and adapting their diverse cultures to benefit the whole. This practice was in turn adopted by the Romans after the first century, and again by the Islamic caliphates established over much of the same area in the 6th century C.E. The territories of the ancient Persian Empire went through many successive rebirths, as the empires of the Parthians, Sassanians, and Safavids rose and fell. The last imperial dynasties, the Qajars and Pahlavis, witnessed the final decay of the Persian Empire across the 19th and 20th centuries, controlled by European and American powers until the Islamic Revolution of 1979 established the theocratic state of Iran.

BIBLIOGRAPHY. "The Persian Empire and the West," *The Cambridge Ancient History* (Cambridge University Press, 1923–39); Pierre Bryant, *From Cyrus to Alexander: A History of the Persian Empire*, P. Daniels, trans. (Winona Lake, 2002); "Persia," www.mnsu.edu (August 2004).

JONATHAN SPANGLER
SMITHSONIAN INSTITUTION

Persian Gulf

THE COUNTRIES BOUNDING the Persian Gulf include OMAN, UNITED ARAB EMIRATES, QATAR (a peninsula off the SAUDI ARABIA coast), BAHRAIN (an island), Saudi Arabia, KUWAIT, IRAQ, and in the north, IRAN. The Persian Gulf is an arm of the ARABIAN SEA extending some 600 mi (970 km) from east to west. It covers an area of approximately 89,000 square mi (230,000 square km) and the greatest depth is 335 ft (102 m). It is connected to the Arabian Sea in the east by the Strait of Hormuz. Its southern coastal area is characterized by low desert plains. The western coast continues as desert plains with a coastal escarpment that leads into a major river delta known as the Shatt al-Arab. The northern coast is more rugged but equally desert. The ancient Greeks named the gulf as "Persian" and so it has appeared in sources from antiquity and is in common use until today. Some Arab states and authors sponsor a modern revision to the "Arab Gulf."

The Persian Gulf has acted as a route for trade since the earliest millennia of human civilization. From 4000 B.C.E. onward, increasing complex trading relationships are supported by archaeological evidence. Trading centers such as Dilham (most likely modern Bahrain) and Madgan (Oman) linked Mesopotamia to the Indus Valley in South Asia. As the dynasties of history grew greater, the more elaborate the trade links. Intervening cities and ports flourished as suppliers and middlemen along this flow of metals, wood, spices, incense, and finished goods. The Persian Gulf proved to be perfect for this gradual growth of trade and enabling technologies. The simplest boats could move down these feeding rivers to the salty shores of the gulf and sea and make their way along coastlines to their partners in trade and commerce.

The domestication of camels led to the expansion of land routes of commerce that enhanced the Persian Gulf trade. Shipbuilding technologies improved and the discovery of the monsoons for seasonal travel to and from India modified routes and trade links. A Persian Gulf coast culture based upon cities of commerce developed and prospered.

The settled people of the Persian Gulf coast differed from the nomadic peoples of the interior of the Arabian Peninsula and the lands along the northern coast of the gulf. The nomads were children of the deserts, and the coastal peoples were dependent upon the sea. One moved across endless deserts to points of water that gave life to their flocks. The other settled on the shores of the endless waters to capture the largess that made their livelihood. If the deserts could not feed their flocks, the nomadic tribes pushed into the settled towns of agriculture and trade.

Cycles of tribal incursion onto the coastal settlements blended Arab culture with that from Mesopotamia and the lands of INDIA, continuing to build a unique Persian Gulf character. Also, these settlements and the tribes that plied the Persian Gulf began to take on an increasingly Arab flavor because of the consistent out migration from the Arab interior. Some Arab tribes from the rocky wastes and scattered oases of the peninsula came to the coasts and adopted the less mobile life of the settlement. Other tribesmen, masters of the sand, became masters of the sea. They took up sailing and began to build the legend of Arab traders and pirates.

In 325 B.C.E., Alexander the Great sent fleets of ships from India to explore the shores and islands of the Persian Gulf. Greek influence was not able to take permanent hold; by 250 B.C.E. the Parthians had brought the gulf under predominantly Persian control for the first time. This control of the Persian Gulf from a northern empire would last until the coming of ISLAM. Islam spread to the gulf during the life of the Prophet

Muhammad and soon reached to all peoples on its shores. The prosperity of the Persian Gulf continued as in 750 C.E. Baghdad became the seat of the caliph and the main center of Islamic civilization and power.

Oman was positioned geographically to take advantage of sea routes in the Persian Gulf and to the RED SEA. The lands around the mouth of the Persian Gulf became the stronghold of those seeking to challenge the powers that controlled the Persian Gulf trade from along the Tigris and Euphrates. Thus began a dance of influence in the Persian Gulf as powers in Mesopotamia vied with those in Muscat. This foretold the long history of conflict between those in the Persian Gulf against outside powers who sought to control this strategic body of water because of the wealth it could command.

By the year 1000, Persian Gulf merchants were traveling regularly to Southeast Asia and beyond to CHINA. Their trading efforts were instrumental in spreading Islam, first to India and then to INDONESIA and MALAYSIA. The tribes of the interior remained culturally distinct from the gulf coastal peoples even under Islam. Empires based in Persian and Iraq depended on customs duties from the East-West trade in the Gulf. Those Arabs along the coasts always had to deal with these external factors. This reality led to political compromises and some cultural concessions. Those in the interior often remained more conservative and traditional. The coastal cities and accompanying wealth of trade passed through the hands of many Islamic rulers over the years be they from Mesopotamia, Persia, or Arabia.

EUROPEAN INTERESTS

European influence in the Persian Gulf dawned with the Portuguese invasion of Oman in 1502. Lasting for only a century, it heralded the coming contentions of colonial powers in the Persian Gulf. Britain gradually took lead as it established protectorates of the Persian Gulf emirates in the 19th century. With the discovery and exploitation of the vast oil potential of the Persian Gulf in the 20th century, the interests of the industrialized world were inextricably tied to these ancient waters of political and economic power. Britain relinquished its role as protector of Persian Gulf commerce to the UNITED STATES soon after World War II.

The United States played a limited part until Iraq invaded Iran in 1980 and the ensuing war threatened the safe passage of oil tankers out of the Persian Gulf. As the land conflict between Iraq and Iran became a stalemate, both sides increasingly moved their attacks into the Persian Gulf. In March 1984, Iraq initiated sustained naval operations in its self-declared maritime exclusion zone that extended from the mouth of the Shatt al Arab to Iran's port of Bushehr. Since the beginning of its invasion of Iran, Iraq had attacked Iranian oil infrastructure as well as neutral tankers and ships trading with Iran.

Iraq expanded the "tanker war" in 1984 by using its French-supplied combat aircraft armed with sophisticated guided missiles. Unprotected neutral merchant ships became favorite targets, and with the very capable Western war technologies, attacks moved farther and farther south. Seventy-one merchant ships were attacked in 1984 alone, compared with 48 in the first three years of the war. Repeated Iraqi attacks on Iran's main oil-exporting terminal at Khark Island failed to stop oil exports but added to the mounting petroleum releases into the Persian Gulf. Iran retaliated by attacking first a Kuwaiti oil tanker near Bahrain and then a Saudi tanker in Saudi waters, making it clear that if Iraq continued to interfere with Iran's shipping, no gulf state would be safe. The entire Persian Gulf was embroiled in the war and awash with the flotsam and jetsam of a focused battle against oil production and export.

Saudi Arabia shot down an Iranian Phantom jet over Saudi territorial waters in 1984, but there was no concerted effort to stop the attacks on shipping. Iraq increased its air raids on tankers that were serving Iran and Iranian oil-exporting facilities in 1986 and 1987. They even began attacking vessels that belonged to the conservative Arab states of the Persian Gulf. Iran responded by escalating its attacks on shipping serving Arab ports in the Gulf. Kuwait became a focus of Iran's attacks. The Kuwaiti government sought protection from the international community in the fall of 1986. The Soviet Union responded first, agreeing to charter several Soviet tankers to Kuwait in early 1987. Washington, which has been approached first by Kuwait and which had postponed its decision, eventually followed Moscow's lead.

On May 17, 1987, Iraqi aircraft launched a missile attack on the USS *Stark*, killing 37 crew members and crippling the ship. Baghdad issued official apologizes and the U.S. chose to blame Iran for escalating the war and launched a full-scale naval campaign to escort Kuwaiti tankers that were "reflagged" as American vessels and manned by American crews. Iranian small boat attacks and mine-laying operations caused damage to U.S. protected shipping. The U.S. navy retaliated by destroying Iranian offshore oil platforms.

The U.S. naval forces skirmished with Iranian forces and limited their ability to interdict Iraqi supplies and equipment arriving through ports in Kuwait. The USS *Vincennes*, a missile cruiser, mistook an Iranian commercial airliner as a combatant and shot it down, killing 290 civilians. The Persian Gulf had become a crowded theater of operations with at least ten Western navies and eight regional navies patrolling the area. Merchant ships continued to be damaged and the shipyards along the Arabian coast were operating at full capacity effecting repairs.

The Western nations generally favored Iraq in the conflict, providing military intelligence and selling large amounts of military arms. Both sides attacked civilian targets with missiles and aircraft, increasing the level of casualties and destruction. Iraq used chemical warfare against Iranian military forces. The war ground to a bloody stalemate and both sides agreed to a UN-sponsored cease-fire in 1988.

In 1990, Iraq chose Kuwait as a target for invasion and annexation. Iraq made several spurious claims about Kuwait regarding oil and finances. Rhetoric about illegally pumping oil from under their border, violating production quotas to drive down prices, and not forgiving previous debts were all proffered as complaints against Kuwait. Other Arab states were assisting with mediation when Iraq invaded on August 2, 1990. A U.S. lead military coalition flooded to the Persian Gulf lodging in Saudi Arabia. In 1991 the UN-sponsored military coalition defeated the Iraqi army and liberated Kuwait on February 28. U.S.-led coalition forces enforced continuing UN sanctions against Iraq by military attacks and maritime interdiction for a decade.

In 2003, U.S. and British forces invaded Iraq from Kuwait. U.S.-led coalition naval forces continue to operate to ensure the free flow of commerce and security along the shipping lanes in the gulf. The Iraqi navy is operating once again in the Persian Gulf. The United States and other Western powers have shown that they will act against any new instability in the gulf that endangers their national interests.

PETROLEUM

In 2002, the Persian Gulf countries (Bahrain, Iran, Iraq, Kuwait, Qatar, Saudi Arabia, and the United Arab Emirates) produced about 25 percent of the world's oil. These countries hold almost 65 percent of the world's proven crude oil reserves. The Persian Gulf region also has some 36 percent of total proven world gas reserves. Another significant factor to world industrial interests is that the Persian Gulf countries maintained about 32 percent of the world total oil production capacity. Most strikingly, the Persian Gulf countries normally maintain around 90 percent of the world's excess oil production capacity. Excess production capacity is important in the event of oil supply disruption in other major oil-producing regions or unexpected depletion of stocks.

With the advent of oil exploration, boundary and territory disputes abounded in the gulf. Most have been resolved and the trend is favorable, as all parties look to increased revenues that only peaceful resolution can bring. The flow of oil from the Persian Gulf continues to be primarily by tanker. In 2002, some 88 percent of oil exported from the Persian Gulf transited by tanker through the Strait of Hormuz. Almost 40 percent of all the oil produced for export and trade in the world passes through this CHOKE POINT. JAPAN receives over 75 percent of its crude oil from the Persian Gulf.

POLLUTION

The primary factor in pollution of the Persian Gulf is the massive extraction and transport of oil and associated petroleum products. There is a continuous discharge of petroleum effluents from offshore wellheads, underwater pipelines, loading terminals, and runoff from onshore facilities. Estimates of amounts are difficult to gather, but enforceable standards and controls are limited or nonexistent in most national jurisdictions.

The movement of 40 percent of the world's total oil trade through these waters creates shipping-associated discharges of marine diesel fuel, waste, and, of most concern, millions of gallons of contaminated ballast water. The oil sludge, released by the aggregate of tankers traversing the Persian Gulf, is estimated to be around 8 million metric tons per year. The sea bed along the primary routes to the regional oil terminals is often covered with oil sludge. No enforceable international requirements exist to address this daily pollution by oil tankers.

The circulation parameters of the waters in the Persian Gulf along with a high rate of water evaporation as opposed to its fresh water supply add to pollution management challenges. The coastal zone, with its intertidal mudflats and near-shore islands, is important for breeding sea birds and other migrating species. The extensive coral reefs and coastal mangrove forests are especially susceptible to damage and destruction caused by petroleum pollution. The long-term effects

on commercial fishing stocks have yet to be determined.

BIBLIOGRAPHY. "The Tanker War, 1984–87," Country Studies, Library of Congress (April 2004); The Persian Gulf Oil and Gas Exports," Energy Information Administration, www.eia.doe.gov (April 2004); World Wildlife Organization, "Persian Gulf Desert and Semi Desert," www.worldwildlife.org (April 2004); "Iran-Iraq War," Jewish Virtual Library, www.us-israel.org (April 2004); Arnold T. Wilson, *The Persian Gulf* (Clarendon Press, 1928); H.J. de Blij and Peter O. Muller, *Geography: Realms, Regions, and Concepts* (Wiley, 2002); Rick Atkinson, *Crusade, The Untold Story of the Gulf War* (HarperCollins, 1993).

IVAN B. WELCH
OMNI INTELLIGENCE, INC.

Peru

Map Page 1139 **Area** 496,226 square mi (1,285,220 square km) **Population** 27,148,000 **Capital** Lima **Highest Point** 22,205 ft (6,768 m) **Lowest Point** 0 m **GDP per capita** $2,126 **Primary Natural Resources** copper, silver, gold, petroleum, timber, fish, iron.

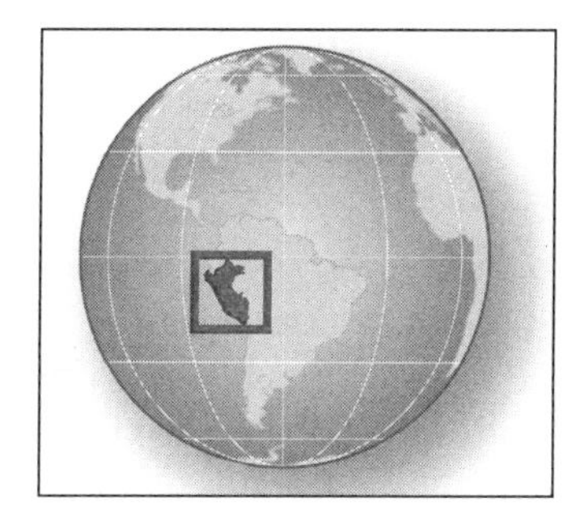

PERU IS A country of contrasts. The fourth-largest country in Latin America, Peru has an amazing geography, a rich and long history, a diverse population, and enormous natural resources. Its geography and environment have shaped Peru's history. Its few rivers, poor and dispersed soils, and extreme altitudes have presented enormous challenges and difficulties to its inhabitants. Peru has three distinctive geographical areas: the Pacific coast, the Andean region, and the Amazon basin.

The Pacific coast has been the focus of Peru's modern political and economic life. The Pacific coast was also the home of the first civilizations in the area, such as the Moche, Nazca, and Chimu cultures. In January 1535, the Spanish conquerors founded the capital city—the City of the Kings, or Lima—on the coast far away from the Inca center of power located in the ANDES. Today, Lima has about 8 million people and is the largest city in the country. A permanent migration from the countryside has created acute social and urban problems in Lima, as most internal migrants have been poor and have settled in shantytowns located in the periphery of the city called Pueblos Jóvenes (young towns).

The first export-oriented economic activity developed along the Pacific coast was the guano industry. Extracted from small islands located along the Peruvian central coast, guano, or dried excrement of seabirds, was exported as a fertilizer to countries such as England, FRANCE, AUSTRALIA, and the southern UNITED STATES as a fertilizer. Between 1845 and 1880, Peru exported 11 million tons of guano. The guano boom was followed by an agricultural boom. In the north coast region, sugar production grew steadily between the 1880s and the 1930s, and although production stagnated in the following decades, it has remained a relatively important economic activity for the country.

Cotton also became an important commodity in the late 19th century, and especially after World War I. In the second half of the 20th century, the fishing industry became one of the fastest-growing economic activities in the country. By the late 1960s, Peru was exporting 12 million tons of anchovetas, and the port of Chimbote became a growing economic and urban center.

The Andean region includes a wide range of climates and microenvironments and has been an important source of mining and agricultural products and livestock. The Puna, the territory located at about 4,000 ft (1,291 m) of altitude, has rich and green grasslands and abundant livestock such as llamas, alpacas, and vicuñas. The lower valleys have been historically devoted to agriculture and have been the home of traditional peasant and Indian communities. The products changed according to the altitude. Products such as wheat, barely, quinoa, and rye can be cultivated at a relatively high altitude; at lower levels, peasants have grown maize, alfalfa, and other vegetables.

Toward the south, the Puna ends in the ALTIPLANO or high plateau, where conditions for agriculture and human settling are extremely difficult. Despite the environmental limitations, in pre-Columbian times, Andean cultures were able to domesticate the potatoes, growing about 200 types of potatoes on the eve of the Spanish conquest. The Andes also has rich mining resources such as copper and silver. The Andean region was also the center of the Inca Empire. From the city of Cuzco, the Incas dominated the Andes until the Spanish conquest in the 1530s.

The jungle region has two distinctive areas: the *montaña*, or eastern slope of the Andes and the lower and flat jungle. The jungle area started growing as a result of the rubber boom in the late 19th century. Throughout the 20th century, the jungle has been important for its production of coca leaves and, lately, the development of a tropical agriculture that includes products such as coffee, sugar, and fruit for the national market.

The population of Peru is diverse. As in other Andean countries, Peruvian society is divided along ethnic-social lines that have their origin in its history of conquest and conflict. The three major groups are Indians, *mestizos*, and whites. Peru has also influential immigrant communities such as Afro-Peruvians, Japanese, who started migrating at the beginning of the 20th century, and Chinese, who migrated to work in the guano and railroad industries in the mid-19th century.

BIBLIOGRAPHY. Peter Flindell Klarén, *Peru: Society and Nationhood in the Andes* (Oxford University Press, 2000); Franklin Pease, *Breve Historia Contemporánea del Perú* (Fondo de Cultura Económica, 1995); Thomas Skidmore and Peter Smith, *Modern Latin America* (Oxford University Press, 1997).

ANGELA VERGARA
UNIVERSITY OF TEXAS PAN AMERICAN

Petra

PETRA HAS BEEN described in *National Geographic* magazine as being "JORDAN's city in the rock." There is no doubt, however, that Petra, located south of Amman on a semiarid site at the edge of the Wadi Araba mountainous desert, is a wonder of the ancient world. Originally developed at the crossroads of overland trading routes, Petra developed as the capital of the Nabatean Empire (c. 400 B.C.E. to 106 C.E.).

Utilizing the local sandstone rock faces—the colors of which vary from white to cream to red and brown—the Nabateans carved tombs and temples into the rocky outcrops. Using their knowledge of hydraulic engineering the Nabateans cut water channels and tunnels into the rock so as to bring drinking water into the developing city from a local dam.

Entering Petra via a narrow channel (the Siq) between two large rock faces, the walkway leads directly to one of the city's most important former edifices, the 131-ft- (40-m-) high Khazneh (treasury), the front of which is dominated by a huge Roman portico. Despite the remarkable exterior, the Khazneh's interior contrasts greatly and is simply formed.

The old heart of Petra lies on open ground around an area known as Wadi Musa. A Roman roadway from the Khazneh leads to the area and is lined with impressive columns. The Roman legacy can be further seen by the layout of a marketplace and amphitheater. However, other prominent structures include the gateway of the temenos and temples such as Qasr-al Bint Firaun (the Castle of the Pharoah's Daughter) and the Temple of the Winged Lions.

Annexed by the Romans in 106 C.E., Petra slowly declined as a commercial center in the following centuries. The arrival of the Byzantines in the 4th century did not greatly change the fortunes of the settlement, partly because of earthquakes and an economic lull. Thus, at the end of the Byzantine Empire (circa 700 C.E.), Petra was in a poor condition, buildings were in near ruin, and the incredible dignity of the city established under the Nabateans was all but lost. In the following centuries, the name of Petra was maintained in local folklore but lost to the West until 1812 when a Swiss explorer, Johann Burchhardt, discovered the ancient settlement. Today, to protect the remaining city, the Jordanian government has moved the local Bedouin population to houses away from the historic core.

BIBLIOGRAPHY. *National Geographic*, "Petra," www.nationalgeographic.com (March 2004); Udi Levy, *The Lost Civilization of Petra* (Floris Books, 1999); Jane Taylor, *Petra* (Aurum Press, 2000).

IAN MORLEY, PH.D.
MING CHUAN UNIVERSITY, TAIWAN

Philippines

Map Page 1124 **Area** 115,831 square mi (300,000 square km) **Population** 84,619,974 **Capital** Manila **Highest Point** 9,691 ft (2,954 m) **Lowest Point** 0 m **GDP per capita** $912 **Primary Natural Resources** timber, petroleum, nickel, cobalt, gold.

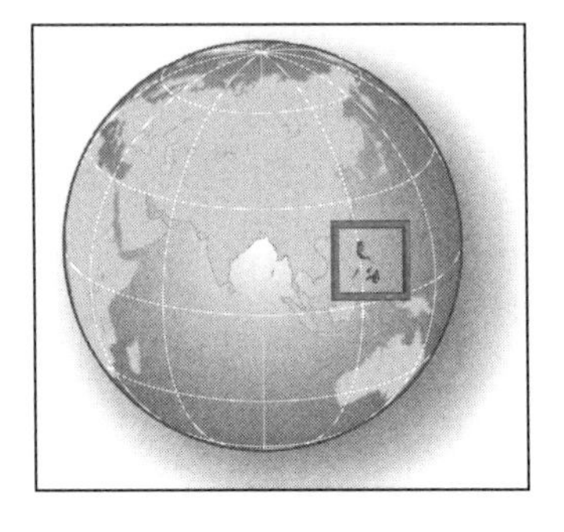

OFF THE SOUTHEAST COAST of mainland Asia lies the Philippines, a southeast Asian country that consists of an archipelago in which there are over 7,100 islands and islets. Most of these islands are of volcanic origin. The latest massive volcanic eruption of Mt. Pinatubo in 1991 caused about 600 deaths, as well as flooding and mudflows.

The north-south extent of the country is 16 degrees of latitude. The two largest islands are LUZON in the north and MINDANAO in the south. They are separated by about 200 mi (322 km). In its tropical marine type of climate regime, temperature varies from 75 degrees F (30.8 degrees C) to 87 degrees F (35.7 degrees C). Average rainfall is 120 in (304 cm), with distinct regional variations. Dry areas are in northern Luzon, Cebu, southern Negros, and interior Mindanao, Sulu and Palawan islands.

Historically, the Philippines was first inhabited by pygmy-like Negroids, followed by Mongoloids and Caucasian groups starting around 11,000 years ago. The largest migration wave was of the Malays (lowland Filipinos) starting from 300 B.C.E. After explorer Ferdinand MAGELLAN's visit in 1521, Spanish colonization in the form of settlements began in 1565. The settlement of MANILA dates back to 1671. Spaniards successfully converted the natives to Catholicism.

The UNITED STATES became the master of the Philippines in 1898 as a result of the Treaty of Paris. Japanese occupation during World War II of the Philippines (1941–45) was brutal, but soon after U.S. reoccupation in 1945, the country became independent in 1946. A democratic form of government was introduced and elections took place every four years. Ferdinand Marcos, first elected president in 1965, put an end to democracy in 1972 and declared martial law in the name of upholding law and order. Democracy came back to the country in 1986 after Marcos was removed from power and a new constitution was adopted. Elections occurred every six years since then. Political instability, combined with Muslim fundamentalist rebellion in Mindanao, which has been recently infested with al Qaeda influence, accentuates centrifugal (disintegrating) forces within the country.

Economic growth since the 1970s has been erratic and lurching. Gross domestic product growth rate in 2002 was 4.6 percent. Forty percent of the country's population lives below the poverty line. Of the total labor force, 40 percent are engaged in agriculture, 27 percent in services, and a mere 10 percent in manufacturing. Rice is the principal crop, followed by corn, sugar, and copra. Industries include pharmaceuticals, food processing, electronic assembly, petroleum refineries, textiles, and wood products.

The Philippines is a Catholic country, a legacy of Spanish colonization; 83 percent of the population is Roman Catholic; 9 percent is Protestant, and 5 percent is Muslims, mostly on Mindanao. Because of strong religious beliefs held by Catholics and Muslims, family planning methods are resisted, which leads to a high fertility rate of 3.29 children born per reproductive-age woman (2003). The two official languages are Filipino and English.

BIBLIOGRAPHY. Amando Doronila, *The State, Economic Transformation, and Political Change in the Philippines* (Oxford University Press, 1992); Alden Cutshall and Anindita Parai, "Philippines," Ashok K. Dutt, ed., *Southeast Asia: A Ten-Nation Region* (Kluwer Academic Publishers, 1996); Richard Ulack, "The Philippines," Thomas R. Leinbach and Richard Ulack, eds., *Southeast Asia: Diversity and Development* (Prentice Hall, 2000).

ASHOK K. DUTT, PH.D.
UNIVERSITY OF AKRON

Phoenicia

ANCIENT PHOENICIA, A LOOSE confederation of city-states existing between 3000 B.C.E. and 146 B.C.E., was located where LEBANON is now. The culture was known for seafaring commerce and trade, and colonizing. The most important legacy left by the Phoenicians, though, was a 22-character alphabet that became the basis of Hebrew, Greek, and Roman script. In 1600 B.C.E., an alphabet called proto-Canaanite was used in the east Mediterranean area. Proto-Canaanite consisted of 28 symbols standing for syllables and was based on Egyptian hieroglyphics. Over a few centuries, the Phoenicians developed their own alphabet from this. Because of their widespread trade, the Phoenicians were in a position to spread the writing they used all over the Mediterranean region. The Greeks adapted the alphabet in the 8th century B.C.E. and added vowel sounds and characters to indicate individual sounds, rather than syllables. This version spread to the Balkans, RUSSIA, and ITALY.

The Phoenicians had arrived in the Levant around 3000 B.C.E.; their original homeland is not known. They settled between the MEDITERRANEAN SEA and the mountains of LEBANON on a narrow strip of land. What

they called themselves is not known; some inscriptions dating back to the 15th century B.C.E. refer to them as Canaanites. Initially, their major city was Sidon, and they were also called Sidonites in the Old Testament of the Bible. By 1200 B.C.E., the Phoenicians had built major port cities along the Mediterranean. Tyre soon surpassed Sidon in influence; other cities were Byblos, Akka, Aradus, and later, Berytus. These port cities were governed as independent city-states and ruled by hereditary kings.

The Phoenicians exploited all available resources in their homeland, including the rich forests. It was as traders, however, that they were best known. They harvested and shipped the cedars, pines, and cypresses of the Lebanon forests. Other products included textiles; Phoenicians became noted for the rich purple color of their cloth, which came from the snail in murex shells. Small hills of these discarded shells, several meters high, have been found while excavating ancient Sidon. Phoenicians were also known for their glassware, which was often clear; they may have invented glass-blowing. Phoenicians may also have trafficked in slaves. Metal working became a Phoenician art. The metals came from as far away as Britain, IRELAND, SPAIN, and Brittany; the raw materials were imported through the network of colonies Phoenicians had established throughout the Mediterranean region.

Phoenician colonies, like their home cities, were often built on rocky promontories and islets with a view over harbors in highly defensible positions. The earliest Phoenician colony was Kition, on CYPRUS, a source of copper. In 1110 B.C.E., the Phoenicians founded Gades (Cádiz). Utica, in North Africa, was built in 1101. Malaca (Málaga), Joppa (Yafo), Leptis Magna (near present-day Tripoli), and many other cities along the coasts of North Africa, southern Iberia, and the islands of the Mediterranean followed. By the 8th century, Phoenicians were trading beyond the Mediterranean, along the Atlantic coasts of Spain and MOROCCO, especially for metals such as copper, tin, and gold. In 814 B.C.E. they founded their most important colony, Carthage, near what is now Tunis. Herodotus claimed that the Phoenicians circumnavigated Africa, which took three years.

By 875 B.C.E., most—but not all—of the cities in Phoenicia were paying tribute to the Assyrian Empire, and within 150 years Phoenicia was annexed to that empire. The Assyrian Empire fell in 612 B.C.E., and in 539 B.C.E. Phoenicia became part of the PERSIAN EMPIRE. Phoenician colonies continued to trade and dominate the Mediterranean, governing themselves autonomously under the leadership of Carthage. Carthage controlled trade from the Straits of Gibralter, founded its own colonies in southern Spain, and continued to trade in metals as the original Phoenician cities had done. The area around Carthage produced rich harvests of grain, and the city controlled the export and sale of that as well. Carthage engaged in long-standing rivalry with Greek trading colonies such as Massalia (Marseilles).

The strong navy of Carthage dominated the Mediterranean until 241 B.C.E., when Rome, having consolidated power in Italy, wrested control of Sicily from Carthage in the First Punic War. During the Second Punic War in 218 B.C.E., the Carthaginian general, Hannibal, crossed the Alps into Italy with his army and elephants and defeated the Romans. But in 202 B.C.E., the Roman general Scipio Africanus beat Hannibal at Zama, and forced the Carthaginians to give up their control of Spain, limit their fleet, and pay heavy taxes. Over the years, Carthage began to return to strength, and in 146 B.C.E., Rome went to war with the city again. After a six-month siege, Rome completely destroyed the city, killed or enslaved all the inhabitants, and sowed the surrounding fields with salt so that nothing would grow there again.

BIBLIOGRAPHY. Nigel Bagnall, *The Punic Wars 264–46 B.C.E.* (Osprey Publishing, 2002); Donald Harden, *The Phoenicians* (Thames and Hudson, 1962); John Haywood, *Historical Atlas of the Ancient World* (MetroBooks, 1998); Gerhard Herm, *The Phoenicians* (William Morrow, 1975); Sabatino Moscati, *The World of the Phoenicians* (Praeger Publishers, 1968).

VICKEY KALAMBAKAL
INDEPENDENT SCHOLAR

physical geography

PHYSICAL GEOGRAPHY IS AN integral part of a much larger area of understanding called geography. Most individuals define geography as a field of study that deals with maps. This definition is only partially correct. A better definition of geography may be the study of natural and human phenomena relative to a spatial dimension. The discipline of geography has a history that stretches over many centuries. Over this time period, the study of geography has evolved and developed into an important form of human scholar-

ship. Examining the historical evolution of geography as a discipline provides some important insights concerning its character and methodology. These insights are also helpful in gaining a better understanding of the nature of physical geography.

Physical geography, a subdiscipline of geography, is a field of knowledge that studies natural features and phenomena on the Earth from a spatial perspective. It primarily focuses on the spatial patterns of weather and climate, soils, vegetation, animals, water in all its forms, and landforms. Physical geography also examines the interrelationships of these phenomena to human activities. This subfield of geography is academically known as the Human-Land Tradition, and has seen very keen interest and growth in the last few decades because of the acceleration of human-induced environmental degradation. Thus, physical geography's scope is much broader than the simple spatial study of nature. It also involves the investigation of how humans are influencing nature. In other words, it focuses on geography as an Earth science, making use of biology to understand global flora and fauna pattern, and mathematics and physics to understand the motion of the Earth and its relationship with other bodies in the solar system. It also includes landscape ecology and environmental geography.

Thus, the discipline, in a sense, is better organized than its human or social counterpart because it rests upon specialist sciences like geology and meteorology which had made great progress. There is no dearth, but rather an embarrassing wealth, of material out of which to construct the subject.

HISTORICAL OVERVIEW

Some of the first truly geographical studies occurred more than 4,000 years ago. The main purpose of these early investigations was to map features and places observed as explorers traveled to new lands. At this time, Chinese, Egyptian, and Phoenician civilizations were beginning to explore the places and spaces within and outside their homelands. The earliest evidence of such explorations comes from the archaeological discovery of a Babylonian clay tablet map that dates back to 2300 B.C.E.

The early Greeks were the first civilization to practice a form of geography that was more than mere mapmaking or cartography. Greek philosophers and scientists were also interested in learning about spatial nature of human and physical features found on the Earth. One of the first Greek geographers was Herodotus (circa 484–425 B.C.E.). Herodotus wrote a number of volumes that described the human and physical geography of the various regions of the PERSIAN EMPIRE.

The ancient Greeks were also interested in the form, size, and geometry of the Earth. Aristotle (c. 384–322 B.C.E.) hypothesized and scientifically demonstrated that the Earth had a spherical shape. Evidence for this idea came from observations of lunar eclipses. Lunar eclipses occur when the Earth casts its circular shadow on to the moon's surface. The first individual to accurately calculate the circumference of the Earth was the Greek geographer Eratosthenes (c. 276–194 B.C.E.). Eratosthenes calculated the equatorial circumference to be 25,000 mi (40,233 km) using simple geometric relationships. This primitive calculation was unusually accurate. Measurements of the Earth using modern satellite technology have computed the circumference to be 24,899.5 mi (40,072 km).

Most of the Greek accomplishments in geography were passed on to the Romans. Roman military commanders and administrators used this information to guide the expansion of their empire. The Romans also made several important additions to geographical knowledge. Strabo (circa 64 B.C.E.–20 C.E.) wrote a 17-volume series called *Geographia*. Strabo claimed to have traveled widely and recorded what he had seen and experienced from a geographical perspective. In his series of books, Strabo describes the cultural geographies of the various societies of people found from Britain to as far east as INDIA, and south to ETHIOPIA and as far north as ICELAND. He also suggested a definition of geography that is quite complementary to the way many human geographers define their discipline today. This definition suggests that the aim of geography was to describe the known parts of the inhabited world and to write the assessment of the countries of the world with clearly highlighting the differences between countries.

During the 2nd century C.E., PTOLEMY (c. 100–178) made a number of important contributions to geography. Ptolemy's publication, *Geographike hyphegesis* (*Guide to Geography*), compiled and summarized much of the Greek and Roman geographic information accumulated at that time. Some of his other important contributions include the creation of three different methods for projecting the Earth's surface on a map, the calculation of coordinate locations for some 8,000 places on the Earth, and development of the concepts of geographical latitude and longitude.

Little academic progress in geography occurred after the Roman period. For the most part, the Middle

Ages (5th to 13th centuries) were a time of intellectual stagnation. In Europe, the Vikings of Scandinavia were the only group of people carrying out active exploration of new lands. In the Middle East, Arab academics began translating the works of Greek and Roman geographers starting in the 8th century and exploring southwestern Asia and Africa. Some of the important intellectuals in Arab geography were Al-Idrisi, IBN BATTUTA, and Ibn Khaldun. Al-Idrisi is best known for his skill at making maps and for his work of descriptive geography. Ibn Battuta and Ibn Khaldun are well known for writing about their extensive travels to North Africa and the MIDDLE EAST.

RENAISSANCE

During the Renaissance (1400 to 1600), numerous journeys of geographical exploration were commissioned by a variety of nation-states in Europe. Most of these voyages were financed because of the potential commercial returns from resource exploitation. The voyages also provided an opportunity for scientific investigation and discovery and added many significant contributions to geographic knowledge. Important explorers of this period include Christopher Columbus, Vasco da Gama, Ferdinand MAGELLAN, Jacques Cartier, Sir Martin Frobisher, Sir Francis Drake, John and Sebastian Cabot, and John Davis. Also during the Renaissance, Martin Behaim created a spherical globe depicting the Earth in its true three-dimensional form in 1492. Prior to Behaim's invention, it was commonly believed in the Middle Ages that the Earth was flat. Behaim's globe probably influenced the beliefs of navigators and explorers of that time because it suggested that one could travel around the world.

In the 17th century, Bernhardus Varenius (1622–50) published an important geographic reference titled *Geographia generalis* (*General Geography*, 1650). During the 18th century, the German philosopher Immanuel Kant (1724–1804) proposed that human knowledge could be organized in three different ways. One way of organizing knowledge was to classify its facts according to the type of objects studied. Accordingly, zoology studies animals, botany examines plants, and geology involves the investigation of rocks. The second way one can study things is according to a temporal dimension. This field of knowledge is of course called history. The last method of organizing knowledge involves understanding facts relative to spatial relationships. This field of knowledge is commonly known as geography. Kant also divided geography into a number of subdisciplines. He recognized the following six branches: physical, mathematical, moral, political, commercial, and theological geography.

Geographic knowledge saw strong growth in Europe and the UNITED STATES in the 1800s. This period also saw the emergence of a number of societies interested in geographic issues. In GERMANY, Alexander von HUMBOLDT, Karl Ritter, and Friedrich Ratzel made substantial contributions to human and physical geography. Humboldt's publication *Kosmos* (1844) examines the geology and physical geography of the Earth. This work is considered by many academics to be a milestone contribution to geographic scholarship.

Late in the 19th century, Ratzel theorized that the distribution and culture of the Earth's various human populations were strongly influenced by the natural environment. The French geographer Paul Vidal de la Blanche opposed this revolutionary idea. Instead, he suggested that human beings were a dominant force shaping the form of the environment. The idea that humans were modifying the physical environment was also prevalent in the United States. In 1847, George Perkins Marsh gave an address to the Agricultural Society of Rutland County, VERMONT. The subject of this speech was that human activity was having a destructive impact on land, especially through deforestation and land conversion. This speech also became the foundation for his book *Man and Nature or The Earth as Modified by Human Action*, first published in 1864. In this publication, Marsh warned of the ecological consequences of the continued development of the American frontier.

Many academics in the field of geography extended the various ideas presented in the previous century to studies of small regions all over the world. Most of these studies used descriptive field methods to test research questions. Starting in about 1950, geographic research experienced a shift in methodology. Geographers began adopting a more scientific approach that relied on quantitative techniques. The quantitative revolution was also associated with a change in the way in which geographers studied the Earth and its phenomena. Researchers now began investigating process rather than mere description of the event of interest. Today, the quantitative approach is becoming even more prevalent because of advances in computer and software technologies.

The history and development of geography, discussed above, suggest a definition that geography, in its simplest form, is the field of knowledge that is concerned with how phenomena are spatially organized. Physical geography attempts to determine why natural

phenomena have particular spatial patterns and orientation.

ELEMENTS AND PHENOMENA

Physical geography and HUMAN GEOGRAPHY are the two major subfields of knowledge emanating from the discipline of geography. It is important to distinguish between these two subfields that use similar methodologies. Knowing what kinds of things are studied by geographers provides us with a better understanding of the differences between physical and human geography.

Phenomena or elements studied in physical geography include rocks and minerals, landforms, soils, animals, plants, water, atmosphere, rivers and other water bodies, environment, climate and weather, and oceans.

Phenomena or elements studied in human geography include population, setllements, economic activities, transportation, recreational activities, religion, political systems, social traditions, human migration, agricultural systems, and urban systems.

Geography is also a discipline that integrates a wide variety of subject matter. Almost any area of human knowledge can be examined from a spatial perspective. Also, the study of geography can involve a holistic synthesis, which connects knowledge from a variety of academic fields in both human and physical geography.

For example, the study of the enhancement of the Earth's greenhouse effect and the resulting global warming requires a multidisciplinary approach for complete understanding. The fields of climatology and meteorology are required to understand the physical effects of adding additional greenhouse gases to the atmosphere's radiation balance. The field of economic geography provides information on how various forms of human economic activity contribute to the emission of greenhouse gases through fossil fuel burning and land-use change. Combining the knowledge of both of these academic areas gives us a more comprehensive understanding of why serious environmental problems occur.

STRENGTH AND WEAKNESS

The holistic nature of geography is a strength and weakness both. Geography's strength comes from its ability to connect functional interrelationships that are not normally noticed in narrowly defined fields of knowledge. The most obvious weakness associated with the geographical approach is related to the fact that holistic understanding is often too simple and misses important details of cause and effect.

Physical geography's primary subdisciplines study the Earth's atmosphere (meteorology and climatology), animal and plant life (biogeography), physical landscape (geomorphology), soils (pedology), and waters (hydrology). Some of the dominant areas of study in human geography include human society and culture (social and cultural geography), behavior (behavioral geography), economics (economic geography), politics (political geography), and urban systems (urban geography).

SPECIALIZATION

Academics studying physical geography and other related earth sciences are rarely generalists. Most are in fact highly specialized in their fields of knowledge and tend to focus themselves in one of the following well defined areas of understanding in physical geography.

The fields of knowledge generally have a primary role in introductory textbooks dealing with physical geography. Introductory textbooks can also contain information from other related disciplines including geology—the study of the form of the Earth's surface and subsurface and the processes that create and modify it; ecology—the scientific study of the interactions between organisms and their environment; oceanography—the science that examines the biology, chemistry, physics, and geology of oceans; cartography—the technique of making maps; and astronomy—the science that examines nature, motion, origin, and constitution celestial bodies and the cosmos.

After 1950, the following two forces largely determined the nature of physical geography:

The Quantitative Revolution. Measurement became the central focus of research in physical geography. It was used primarily for hypothesis testing. With measurement came mapping, models, statistics, mathematics, and hypothesis testing. The quantitative revolution was also associated with a change in the way in which physical geographers studied the Earth and its phenomena. Researchers now began investigating process rather than mere description of the environment.

The Study of Human/Land Relationships. The influence of human activity on the environment was becoming very apparent after 1950. As a result, many researchers in physical geography began studying the influence of humans on the environment. Some of the dominant themes in these studies included environmental degradation and resource use; natural hazards and impact assessment; and the effect of urbanization and land-use change on natural environments.

UNDERSTANDING PHYSICAL GEOGRAPHY

The nature of understanding in physical geography has changed over time. When investigating this change, it becomes apparent that certain universal ideas or forces had very important ramifications to the academic study of physical geography.

During the period from 1850 to 1950, there were five main ideas that had a strong influence on the discipline:

Uniformitarianism. This theory rejected the idea that catastrophic forces were responsible for the current conditions on the Earth. It suggested instead that continuing uniformity of existing processes were responsible for the present and past conditions of this planet.

Evolution. Charles Darwin's *Origin of Species* (1859) suggested that natural selection determined which individuals would pass on their genetic traits to future generations. As a result of this theory, evolutionary explanations for a variety of natural phenomena were postulated by scientists. The theories of uniformitarianism and evolution arose from a fundamental change in the way humans explained the universe and nature.

During the 16th, 17th, and 18th centuries, scholars began refuting belief- or myth-based explanations of the cosmos and instead used science to help explain the mysteries of nature. Belief-based explanations of the cosmos are made consistent with a larger framework of knowledge that focuses on some myth. However, theories based on science questioned the accuracy of these beliefs.

Exploration and Survey. Much of the world had not been explored before 1900. Thus, during this period all of the fields of physical geography were actively involved with basic data collection. This data collection included activities like determining the elevation of land surfaces, classification and description of landforms, the measurement of the volume of flow of rivers, measurement of phenomena associated with weather and climate, and the classification of soils, organisms, biological communities, and ecosystems.

Conservation. Beginning in the 1850s, a concern for the environment began to develop as a result of the human development of once natural areas in the United States and Europe. One of the earliest statements of these ideas came from George Perkins Marsh (1864) in his book *Man in Nature or Physical Geography as Modified by Human Action*. This book is often cited by scholars as the first significant academic contribution to conservation and environmentalism.

Systems Theory. The world of nature is very complex. In order to understand this complexity, humans usually try to visualize the phenomena of nature as a system. A system is a set of interrelated components working together toward some kind of process. One of the simplest forms of a system is a model. Both models and systems are simplified versions of reality. The interaction between perceptible phenomena and theory is accomplished through explanation and validation. This simple model, while an extreme abstraction of reality, illustrates how scientific understanding works. It suggests that in scientific understanding, perceptible phenomena and theory interact through explanation and validation.

In physical geography and many other fields of knowledge, systems and models are used extensively as aids in explaining natural phenomena around us. A system is a group of parts that interact according to some kind of process. Systems are often visualized or modeled as component blocks with some kind of connections drawn. All systems have the same common characteristics. These common characteristics are summarized below:

All systems have some structure.

All systems are generalizations of reality.

They all function in the same way.

There are functional as well as structural relationships between the units of a system.

Function implies the flow and transfer of some material. Systems exchange energy and matter internally and with their surrounding environment through various processes of input and output.

Function requires the presence of some driving force or some source of energy.

All systems show some degree of integration.

Within its defined boundary the system has three kinds of properties: Elements are the kinds of things or substances composing the system. They may be atoms or molecules or larger bodies of matter—sand grains, rain drops, plants, or cows. Attributes are characteristics of the elements that may be perceived; for example: quantity, size, color, volume, temperature, and mass. Relationships are the associations that exist between elements and attributes based on cause and effect.

The state of the system is defined when each of its properties (for example, elements, attributes, and relationships) has a defined value. Scientists have examined and classified many types of systems. These types include the isolated system, a system where there are no interactions outside its boundary layer. Such systems are common in laboratory experiments. A closed

system is closed with respect to matter, but energy may be transferred between the system and its surroundings. Earth is essentially a closed system. An open system is a system where both matter and energy can cross the boundary of the system. Most environmental systems are open.

A morphological system is a system where we understand process relationships or correlations between the elements of the system in terms of measured features. A cascading system concerns the movement of energy and/or matter from one element to another and understands the processes involved. A process-response system involves the movement, storage, and transformation of energy and matter and the relationships between measured features in the various elements of the system. A control system is a system that is intelligently manipulated by humans. An ecosystem is concerned with the biological relationships within the environment and the interactions between organisms and their physical surroundings.

STRUCTURE OF SYSTEMS

Systems exist at every scale of size and are often arranged in some kind of hierarchical fashion. Large systems are often composed of one or more smaller systems working within its various elements. Processes within these smaller systems can often be connected directly or indirectly to processes found in the larger system. A good example of a system within systems is the hierarchy of systems found in the universe.

At the highest level in this hierarchy, we have the system that we call the cosmos or universe. The elements of this system consist of galaxies, quasars, black holes, stars, planets, and other heavenly bodies. The current structure of this system is thought to have come about because of a massive explosion known as the Big Bang and is controlled by gravity, weak and strong atomic forces, and electromagnetic forces.

Around some stars in the universe we have an obvious arrangement of planets, asteroids, comets, and other material. We call these systems solar systems. The elements of this system behave according to set laws of nature and are often found orbiting around a central star because of gravitational attraction. On some planets conditions may exist for the development of dynamic interactions between the hydrosphere, lithosphere, atmosphere, or biosphere.

We can define a planetary system as a celestial body in space that orbits a star and that maintains some level of dynamics between its lithosphere, atmosphere and hydrosphere. Some planetary systems, like the Earth, can also have a biosphere. If a planetary system contains a biosphere, dynamic interactions will develop between this system and the lithosphere, atmosphere, and hydrosphere. These interactions can be called an environmental system. Environmental systems can also exist at smaller scales of size (for example, a single flower growing in a field could be an example of a small-scale environmental system).

The Earth's biosphere is made up of small interacting entities called ecosystems. In an ecosystem, populations of species group together into communities and interact with each other and the abiotic environment. The smallest living entity in an ecosystem is a single organism. An organism is alive and functioning because it is a biological system. The elements of a biological system consist of cells and larger structures known as organs that work together to produce life. The functioning of cells in any biological system is dependent on numerous chemical reactions. Together these chemical reactions make up a chemical system. The types of chemical interactions found in chemical systems are dependent on the atomic structure of the reacting matter. The components of atomic structure can be described as an atomic system.

ENVIRONMENTAL SYSTEMS

An environmental system is a system where life interacts with the various abiotic components found in the atmosphere, hydrosphere, and lithosphere. Environmental systems also involve the capture, movement, storage, and use of energy. Thus, environmental systems are also energy systems. In environmental systems, energy moves from the abiotic environment to life through processes like plant photosynthesis. Photosynthesis packages this energy into simple organic compounds like glucose and starch. Both of these organic molecules can be stored for future use.

The chemical energy of photosynthesis can be passed on to other living or biotic components of an environmental system through biomass consumption or decomposition by consumer organisms. When needed for metabolic processes, the fixed organic energy stored in an organism can be released to do work via respiration or fermentation. Energy also fuels a number of environmental processes that are essentially abiotic: for example, the movement of air by wind, the weathering of rock into soil, the formation of precipitation, and the creation of mountains by tectonic forces. The first three processes derive their energy directly or indirectly from the sun's radiation that is received at the Earth's surface. Mountain building is

fueled by the heat energy that exists within the Earth's interior. Finally, the movement of energy in environmental systems always obeys specific thermodynamic laws that cannot be broken.

It is understood that environment is the complex of physical, chemical, and biotic factors (such as climate, soil, and living things) that act upon an organism or an ecological community and ultimately determines its form and survival. Both human and physical geography provide an important intellectual background for studying the environment. Many environmental studies/science programs offered by universities and colleges around the world rely on the information found in various geography courses to help educate their students about the state of the environment.

FUTURE OF PHYSICAL GEOGRAPHY

The following describes some of the important future trends in physical geography research:

Applied geography. Continued development of applied physical geography will help analyze and correct human-induced environmental problems. A student of applied physical geography uses theoretical information from the field of physical geography to manage and solve problems related to natural phenomena found in the real world.

Remote sensing. Advances in technology have caused the development of many new instruments for the monitoring of the Earth's resources and environment from airborne and space platforms. The most familiar use of remote sensing technology is to monitor the Earth's weather for forecasting.

Geographic Information Systems. A GEOGRAPHIC INFORMATION SYSTEM (GIS) merges information in a computer database with spatial coordinates on a digital map. Geographic Information Systems are becoming increasingly more important for the management of resources.

BIBLIOGRAPHY. Alan H. Strahler and Arthur Strahler, *Physical Geography: Science and Systems of the Human Environment* (Wiley, 2003); R.W. Christopherson, *Geosystems: An Introduction to Physical Geography* (Prentice Hall, 2005); K.J. Gregory, *The Changing Nature of Physical Geography* (Edward Arnold, 2001); J.B. Whitlow, *The Penguin Dictionary of Physical Geography* (Penguin, 2001); A. Allaby and M. Allaby, eds., *Dictionary of Earth Sciences* (Oxford University Press, 1999); R.J. Chorley and B.A. Kennedy, *Physical Geography: A Systems Approach* (Prentice Hall, 1971); A.N. Strahler, "Systems Theory in Physical Geography," *Physical Geography* (v.1/1, 1980); M.E. Harvey and B.P. Holly, eds., *Themes in Geographic Thought* (Croom Helm, 1981); Roger M. Minshull, *The Changing Nature of Geography* (Hutchinson, 1970).

JITENDRA UTTAM
JAWAHARLAL NEHRU UNIVERSITY, INDIA

Pitcairn Island

PITCAIRN (18.1 square mi or 47 square km) is an isolated island located in the east-central PACIFIC OCEAN, approximately 400 mi (644 km) southeast of Mangareva, the closest inhabited island. Pitcairn is the only inhabited island of the British colony of Pitcairn, Henderson, Ducie, and Oeno Islands. The three outer islands of the Pitcairn Islands lie to the north and northeast of Pitcairn. The island of Pitcairn itself is approximately 2 mi (3.2 km) long and 1 mile (1.6 km) wide. Pitcairn has no airstrip and no docking facilities.

The island is fringed with high cliffs, with only one small landing place, Bounty Bay near Adamstown, where longboats can be launched to intercept passing freighters and cruise ships. Communication with the outside world is by radiotelephone and e-mail. Pitcairn's climate is subtropical, tempered by southeast trade winds with occasional typhoons during the November-to-March rainy season. The volcanic soils are fertile, supporting subsistence crops of sweet potatoes, sugarcane, oranges, and bananas. The original forest cover has been much reduced because of the clearing of land for cultivation and the use of wood for handicrafts and construction.

Pitcairn Island was settled by Polynesians sometime before the 15th century. Stone tools and other carvings of Polynesian design have been found on the island, but by the 18th century the Polynesian colony had vanished. Pitcairn was discovered by Europeans in 1767 when British captain Philip Carteret passed by but was unable to land. Carteret named the island after Robert Pitcairn, the midshipman who first sighted the island. Robert Pitcairn was the son of Major John Pitcairn, British commander at the Battle of Lexington.

In 1790, Pitcairn was settled by refugees from the famed mutiny on HMS *Bounty*. Nine mutineers, along with 6 Tahitian men and 12 Tahitian women, settled on the island. Immediately following the mutiny, Fletcher Christian, the leader of the mutineers, had set sail for TAHITI where some of the mutineers elected to stay. Christian, however, rightly reasoned that the

British Navy would send ships to look for the *Bounty*'s crew in Tahiti, where the expedition had been sent to gather breadfruit trees for introduction into the British West Indies as a cheap food for slaves. Christian remembered Pitcairn's description from Carteret's writings and, guessing correctly that Carteret had misjudged the island's longitude, set out to find the island. Christian's small colony soon degenerated into violence, with most of the male settlers, English and Tahitian, being killed in fights over the Tahitian women.

When Pitcairn was rediscovered by an American whaler in 1808, only one of the original mutineers was left alive. This mutineer, John Adams, died in 1829, and in 1831 the colony's entire population was removed to Tahiti. Most soon returned to Pitcairn. In 1856, again because of overpopulation, the entire population was resettled on Norfolk Island, between AUSTRALIA and NEW ZEALAND, but many soon returned to Pitcairn. In 1887 the entire population of Pitcairn joined the Seventh-Day Adventist Church. The peak population of 233 was reached in 1937. Because of out-migration, mostly to New Zealand, the population of Pitcairn has been declining ever since. In 2004, seven Pitcairn men, including the island's mayor, a descendent of Fletcher Christian, were convicted of rape and child sexual abuse charges in a divisive and emotional trial conducted by judges imported from New Zealand. Even though some are once again urging the abandonment of the island, Pitcairn remains Britain's sole remaining colonial possession in the Pacific.

BIBLIOGRAPHY. Ian M. Ball, *Pitcairn: Children of Mutiny* (Little, Brown, 1973); Glynn Christian, *Fragile Paradise: The Discovery of Fletcher Christian* (Little, Brown, 1982); Harry L. Shapiro, *The Pitcairn Islanders* (Simon and Schuster, 1968).

JAMES A. BALDWIN
INDIANA UNIVERSITY-PURDUE UNIVERSITY

place

NOTHING IS MORE important to all geography, descriptive or analytical, natural or human, than place or location. Place can be divided into at least three distinct aspects: geometric location defined by precise latitude and longitude; relative location of where a place is relative to other places, especially in the context of history; and finally, the unique nature of place—a sense of place.

In terms of precise location, geographers use the global grid system of LATITUDE AND LONGITUDE. In addition, the global positioning satellites (GPS) system allows anyone to determine precise global position to within a few meters. GPS is even used by farmers to apply various rates and types of fertilizer and seeds to unique soils and topographic locales on large farms. It also can be (and is) used by insurgents, terrorists, and the military to precisely mark and identify targets.

As the British geographer Gordon East demonstrated in his book *The Geography Behind History*, the importance of place often changes dramatically throughout history. For example, as the focus of world culture, economics and politics shifted from the Mediterranean, many places once of great importance have become historical relics. A city such as Ephesus, once a center of culture, commerce, and even religion, today is nothing more than a tourist attraction. The cause of its decline was a shift of trade from the Mediterranean to Central Asia caused by the rise of ISLAM and a focus that moved to Baghdad, present-day IRAQ). Ephesus's location or place of importance was relative. Rome, once the center of the western world, today is of greatly diminished importance.

A sense of place refers to the psychological and emotional aspects of place. This is the idea that places (or their imagination or romanticization) create unique psychological impacts upon humans. Deserts seem to have consistent impacts in terms of architecture and social values, no matter where they occur—Africa, Asia, North America, South America. The influence of mountains on the imagination, poetry, and art of people around the world is well known. The idea of "Darkest Africa," or "backward areas" or the characteristics attributed to an ethnic community are example of the sense of place, as well as the ideals of "pure or natural environments," "dangerous," "safe," "holy" or "mystic" places.

BIBLIOGRAPHY. Barabra Allen and Thomas Schlereth, *Sense of Place: American Regional Cultures* (University Press of Kentucky, 1991); Denis Cosgrove and Stephen Daniels, eds., *The Iconography of Landscape: Essays on the Symbolic Representation, Design, and Use of Past Environments* (Cambridge University Press, 1988); William Denevan, "The Pristine Myth: The Landscapes of the Americas in 1492," *Annals of the Association of American Geographers* (v.82, 1992); W. Gordon East, *The Geography behind History*

(W.W. Norton, 1965). Edward Relph, *Place and Placelessness* (Pion Press, 1976).

R.W. McColl, Ph.D.
General Editor

planning

WHILE PLANNING IN ITS MOST literal form refers to the creation of a plan in an urban context, in reality the term has a far broader and more complicated definition. For example, planning can refer to matters of safety, that is, the prevention of natural disaster, as well as issues such as aesthetics, the environment, suburbanization and transportation.

The history of planning—that is the act of deliberately arranging the urban form for the sake of beauty and/or convenience—can be said to begin with the history of urban settlements and dates for quite literally thousands of years. Instrumental to its adoption were the Romans, who undertook planning for the purposes of civil convenience and military defense. The basic Roman plan consisted of an outer city wall within which were placed a grid of streets, arranged around an open space known as a plaza. Although city walls are no longer a principal element of city planning, the grid form has persevered and is evident in the morphology of many cities, including BARCELONA (SPAIN), the towns and cities of North and Latin America, and Glasgow (Scotland).

However, it is a Greek architect, Hippodamus, who is widely acknowledged as being the father of urban planning through his plan for the settlement of Miletus, a plan that utilized a grid form. Arguably as a consequence of Hippodamus's interest in city design, the art of planning became closely tied to the ideals of architecture, and at particular times of history, this relationship has been further fortified, for example, in the Roman, Renaissance, and Baroque eras. Urban design, for instance—the practice of smaller scale and more three-dimensional spatial design—still retains a largely architectural outlook and is still closely tied to its sister art of urban or town/city planning.

To define planning is extremely problematic in part because of the broad nature of modern urbanism. However, at its most simple, city, town, or urban planning can be said to be a public-based activity that deals with the large-scale design of the built environment within a municipal or metropolitan context. Regional planning, on the other hand, is very much considered a 20th-century invention, a type of public-based planning that, as its name implies, tends to deal with the distribution of regional activities or infrastructure at, within, and about the largest metropolitan centers. The origins of regional planning center upon the County of London Plan (1943) and Greater London Regional Plan (1944) by Sir Patrick Abercrombie (1879-1957), a former lecturer at the School of Civic Design, Liverpool University. Regional planning is arguably the urban planning of the largest urban places, such is their modern spatial scales.

The term *town planning* was first coined in the early 1900s in Britain as part of the widening professional architectural interest in the built environment, partly a consequence of the Garden City idea by Ebenezer Howard, which was developed in practice by Raymond Unwin and Barry Parker, who were instrumental in the passing of the Housing, Town Planning Etc. Act in 1909—the world's first legislative piece to include town planning in its title. Although early-20th-century town planning was closely associated with Garden City, it has since this time undergone a series of dramatic developments.

By way of example, modernist town planning can be seen to begin after the creation of a European organization, the International Congress of Modern Architecture, in 1928, which not only promoted modernist architectural forms but perceived that urban betterment in the social and economic sense could be attained via the design of buildings and laying out of cities. Charles-Édouard Le Corbusier (1887–1965), for example, was instrumental in the adoption of modernist principles in planning practice.

City planning, on the other hand, is a somewhat different activity from town planning and is generally considered to be the arranging or willful influencing of land-use distribution that can be practiced upon either green-field or brown-field sites (i.e. built or unbuilt urban areas). The Romans are widely noted as being pioneers of this type of planning.

BIBLIOGRAPHY. Patrick Abercrombie, *Greater London Plan* (HM Stationary Office, 1945); John Levy, *Contemporary Urban Planning* (Pearson, 2002); Raymond Unwin, *Town Planning in Practice: An Introduction to the Art of Designing Cities and Suburbs* (Princeton Architectural Press, 1994).

David Newman
Ben Gurion University, Israel

plate tectonics

PLATE TECTONICS IS A geological theory that explains many important features of the Earth's surface through the movement of sections of the crust, known as plates. Among the observed data explained by plate tectonics are the known similarities of features on the east and west sides of the ATLANTIC OCEAN, similarities of plant and animal species in areas now too distant for migrations, the locations of volcanoes and earthquakes, the formation of mountain ranges, and the existence of the MID-ATLANTIC RIDGE.

According to plate tectonic theory, the Earth's crust is made up of approximately 30 rigid sections or plates of various sizes. For instance, the PACIFIC OCEAN is mostly one huge plate, and the major continents each lie on their own plates. Very small plates such as the Gorda Plate in the Pacific Northwest are believed to be the last remnants of ancient plates.

Plates lay over a stratum of hot, elastic rock known as the mantle. Temperatures in the mantle range from 2,400 to 2,600 degrees F (1,300 to 2,000 degrees C), and currents within the mantle drive a slow movement of the plates, at the rate of about 4 in (10 cm) per year. Although this movement is small in human terms, observable only with sophisticated scientific measurements, over millions of years plates can move great distances, even from one pole to the other.

PANGAEA

According to plate tectonic theory, about 200 million years ago all present-day continents were part of a supercontinent known as Pangaea or Pangea (Greek for "all-Earth"). Over millions of years, this supercontinent broke up through the creation of rifts, cracks in the crust that moved apart and allowed magma to rise from lower levels and form new seabed. Water moved into the broken landmass to form enclosed bodies of water such as the Gulf of Mexico and the RED SEA. A sufficiently large rift even formed the present-day Atlantic Ocean.

However, since the Earth has not been growing larger, this expansion must cause collisions between plates elsewhere. One of the colliding plates is driven under the other, forming a deep subocean trench beside the resulting subduction zone. The other plate is crumpled and forced upward, forming tall mountains. The HIMALAYAS, the ALPS, and the CAUCASUS all have been

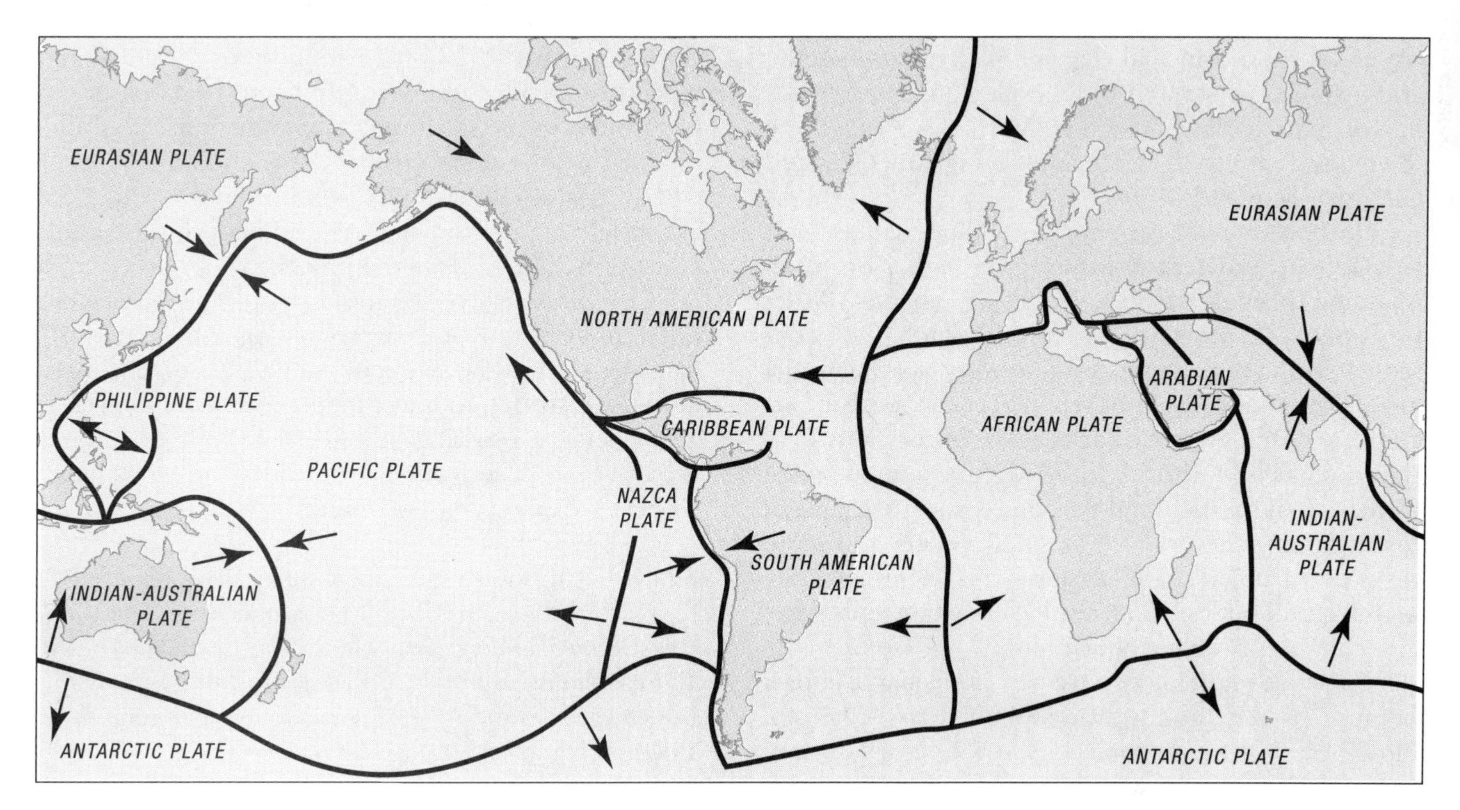

According to plate tectonic theory, the Earth's crust is made up of approximately 30 rigid sections or plates of various sizes. Over millions of years, plates can move great distances, even from one pole to the other.

formed by these geological processes, and in the Himalayas the process will probably continue for the next several million years. Subduction zones form the world's most powerful earthquakes, often near the top of the Richter scale. Seismologists have theorized that deep-focus subduction zone quakes represent the most energy the Earth's crust is capable of releasing. Subduction zones are also responsible for many of the world's volcanoes, particularly the RING OF FIRE that circles the PACIFIC OCEAN and includes the Cascades and Fujiyama.

There are also transform boundaries, at which plates slide against each other. If the two plates become stuck, energy builds up until it releases catastrophically in an earthquake. One of the best-known examples of a transform boundary is the San Andreas Fault in California. Land on the western side of the San Andreas actually belongs to the Pacific Plate, which is slowly sliding north in relation to the North American Plate, but in an irregular way punctuated by destructive earthquakes such as the 1905 San Francisco quake, the 1989 Loma Prieta quake, and the 1994 Northridge quake.

Even volcanoes and earthquakes within plates have been shown to support the theory of plate tectonics. Volcanoes within a plate are created where a rising current known as a mantle plume burns a hot spot in the crust. As the plate moves, it generates a series of volcanic islands, with the older ones eroding away as they are no longer replenished by eruption. The Hawaiian Islands are an example of this process. The big island of Hawaii is still growing as a result of its active volcanoes, while older islands have only extinct volcanoes, such as Punchbowl Crater on Oahu. The westernmost islands, such as Midway Island, have been reduced to tiny spits of land, while the chain continues westward in a series of seamounts that no longer break the surface. Earthquakes within a plate, such as the famous 1811–12 New Madrid quakes, are caused by compressional stresses of plate movement building up within the plate. Such earthquakes can be very destructive because the central part of the plate transmits seismic energy far more efficiently than the broken crust near a plate boundary.

Some geologists have even gone beyond Pangaea to reconstruct the protocontinents that existed before it. Some theories have gone back as far as the Cambrian era, 600 million years ago. In doing so, they are able to explain the existence of such ancient mountains as the Urals and the Ozarks which now lie firmly embedded in the heart of a continental plate. Those mountains are the last bits of evidence of ancient plates otherwise consumed by subduction. The theory of continental drift was originally proposed in 1912 by Alfred Wegener, a German scientist, but he thought in terms of the continents plowing through the seas like ships.

Although he was ridiculed, studies of sea-floor sedimentation in the 1930s and 1940s suggested that the Atlantic was a relatively young ocean, compared to the age of the Earth. Studies of paleomagnetism, in which the magnetic particles in igneous rock reveal the alignment of the Earth's magnetic field at the location where the rock originally solidified, enabled geologists to reconstruct past land movement across millions of years. The 1969 study of the Mid-Atlantic Ridge by the *Glomar Challenger* led to general acceptance of the theory of plate tectonics.

BIBLIOGRAPHY Jon Erickson, *Plate Tectonics: Unraveling the Mysteries of the Earth* (Facts On File, 2001); Roy A. Gallant, *Dance of the Continents* (Benchmark Books, 2000); Linda George, *Plate Tectonics* (Kidhaven Press, 2003); W. Jacqueline Kious, *This Dynamic Earth: The Story of Plate Tectonics* (U.S. Geological Survey, 2002); Brian Knapp, *Plate Tectonics* (Grolier Educational, 2000).

LEIGH KIMMEL
INDEPENDENT SCHOLAR

plateau

A PLATEAU IS A LARGE area of mostly level land elevated high above the surrounding land. Most plateaus have one steeply cliffed side. The rock layers in a plateau are un-deformed and remain flat, unlike the rock layers in mountains, which are tilted. Basaltic plateaus are formed when molten rock forces its way up through a fissure, or crack, in the Earth's surface. The molten material flows out slowly, blanketing a large area with a thick layer of lava. When the lava cools, it forms a dark gray rock called basalt. In order for a plateau to form, the area needs to be uplifted slowly so that all layers are folded to make a flat top.

Plateaus can also be formed when plates of the Earth's crust collide. This causes the crust to buckle. Mountains may be formed, but in the area behind the collision area, a plateau is often created. Erosion can also cause a large flat surface of the Earth to uplift. If rivers cut deeply into the Earth, the area around it becomes raised above the surrounding land. Faults and erosion can divide a plateau into smaller plateaus. Also

a plateau may be eroded into a smaller landform known as a mesa or a butte. Plateaus in humid areas can have many mountains, caused by erosion from rivers. Erosion can form canyons in a plateau when a river eats deeply into the Earth. The edges of plateaus on the shore can be shaped into peninsulas and bays by the action of the ocean waves.

Some well-known plateaus include the TIBETAN PLATEAU in Asia and the Colorado and Columbia plateaus in the UNITED STATES. Scientists believe that for the past 50 million years, since a collision between Asia and INDIA, the Tibetan Plateau has been forming. The plateau is the largest and highest plateau in the world. It measures about 2,175 mi (3,500 km) by 932 mi (1,500 km) and the average elevation is more than 16,400 ft (5,000 m). It is located next to the Himalayan Mountains.

The Columbia River Plateau in the northwestern United States formed when lava flowed from fissures in the earth's surface about 17 million years ago. The lava covered an area approximately 63,321 square mi (164,000 square km) and in some places is about 11,483 ft (3,500 m) thick. The plateau covers parts of the states of WASHINGTON, OREGON, and IDAHO.

The Colorado River Plateau is made up of colorful sedimentary rocks formed millions of years ago when much of the western United States was covered by a vast ocean. The most spectacular feature of the Colorado Plateau is the GRAND CANYON, carved away by the Colorado River. Plateaus are formed in different ways and look different, but they all are relatively flat and are formed of unbroken layers of rock.

BIBLIOGRAPHY. Roger M. Downs, Chief Consultant, *National Geographic Desk Reference* (National Geographic, 1999); W. Kenneth Hamblin and James D. Howard, *Exercises in Physical Geology* (Prentice Hall, 1995); "Creating Flat-Topped Mountains," www.nps.gov (March 2004).

PAT MCCARTHY
INDEPENDENT SCHOLAR

playa

PLAYAS ARE SHALLOW basins that periodically fill with rainwater. They have been described as among the most dynamic geomorphic features, reacting to seasonal and sometimes daily changes in the environment. The term *playa* owes its origin to the Spanish word for "beach" or "shore." A large number of playas began as lakes, drying later as a result of climate change or a reduction in stream inflow. Playas captured the imagination of geologists such as Israel Cook Russell during early (1890) investigations of the western U.S. states. The ubiquitous occurrence of playas throughout the world has led to a variety of regional terms. For example, in CHILE and other Spanish-speaking countries they are called *salar*, in North Africa *sabkha*, and in IRAN *kavir*. It is estimated that there are approximately 50,000 playas on Earth; the majority are less than 1 square mi (2.6 square km) in size.

Playas frequently have annual evaporation to precipitation ratios exceeding 10 to 1. Unlike perennial lakes that are fed by rivers and streams, the water in playas is supplied through rainfall. A nearly impermeable layer of clay on the surface of playas prevents water from percolating downward. In the absence of streamflow or frequent precipitation, the water evaporates quickly, leaving basins dry until rain renews the cycle. Winds remove loose surface particles through deflation.

Climatic conditions favorable to playa formation can be found in tropical and subtropical areas of high pressure with latitudes from 35 to 15 degrees. Playa basins originate in a variety of ways, including downwarping, faulting, and deflation. Subsidence and solution processes can also result in playa formation in areas with underlying limestone or gypsum. Many playas represent the remnants of immense lakes that formed from melting Pleistocene glaciers. The former Lake Manly, the site of present-day Death Valley, extended more than 100 mi (160 km) in length and 600 ft (182 m) in depth and was sustained by three rivers carrying meltwater from Sierra Nevada glaciers. As the climate became drier, the lake disappeared through seepage and evaporation.

The arid appearance of playas may create an impression that they are hydrologically inactive. In reality, playas play an important role in arid-zone hydrology. Where groundwater is found close to the surface, playas discharge large amounts of capillary water into the atmosphere. Playas also play a role in the circulation of groundwater. Occasional flooding of playa surfaces removes surface evaporite accumulations.

Early human use of playas has focused on the harvesting of salt. Playas with large salt accumulations, known as saline playas or salinas, are sources of borax, nitrates, and potash. Playas are also important resources for agricultural operations in arid regions. Al-

though playas themselves are not suitable for growing crops, they are often used to store runoff water for later use. They may also serve as surface focal points for recharging aquifers. In the UNITED STATES, playas have been used for automobile speed trials or as landing sites for experimental air and spacecraft.

The Bonneville Salt Flats in UTAH is the site of land speed records, while the vast expanse of Edwards Air Force Base in CALIFORNIA has been an ideal location for testing military aircraft. In the western United States, playas have been impacted by the widespread lowering of the water table and by the migration of vegetation.

BIBLIOGRAPHY. James T. Neal and S.M. Ward, "Recent Geomorphic Changes in Playas of Western United States," *Journal of Geology* (v.75/5, 1967); C. Reeves, "Pluvial Lake Basins of West Texas," *Journal of Geology* (v.74/3, 1966); David S. Thomas, *Arid-Zone Geomorphology* (Halstead Press, 1989); C. R. Twidale, "Landform Development in the Lake Eyre Region, Australia," *Geographical Review* (v.62, 1972).

THOMAS A. WIKLE
OKLAHOMA STATE UNIVERSITY

Pleistocene geography

THE PLEISTOCENE EPOCH was Earth's most recent Ice Age. The epoch began around 1.8 million years ago (mya) or 2.6 mya. Traditionally, scientists have based the starting date of the epoch on faunal grounds and have estimated it to be about 1.8 mya. A growing number of climate scientists and glaciologists tend to use 2.6 Mya, a time when there was a rapid buildup of ice in the Northern Hemisphere.

There is general agreement that the Pleistocene ended about 10,000 years ago, when temperatures similar to modern times became the norm. The geography of the Pleistocene is of great interest to us for the sheer magnitude of ice coverage, the epoch's wide swings in temperature, and its impacts on the land, people, and animals of the time. Additionally, Earth is presently in a warm, interglacial period, but climate scientists believe that in about 10,000 years or so we could be in the midst of another ice advance; thus, studying the Pleistocene helps us understand humankind's past and contemplate its future.

Climate scientists estimate that the Pleistocene ice sheets covered about 25 to 30 percent of today's land surface. The modern-day ice sheets—the eastern and western Antarctic ice sheets and the Greenland ice sheet—cover only 10 percent of the land. Six or seven ice sheets existed in the Northern Hemisphere 21,000 years ago, when the last glacial maximum occurred. By far the largest ice sheet, the Laurentide, stretched over east-central CANADA and part of the UNITED STATES. Another ice sheet, the Cordilleran, covered the northern Canadian islands, the Rockies of Canada, and parts of MONTANA and IDAHO in the United States. The Greenland ice sheet covered the island of GREENLAND.

The Britain ice sheet spread out over the islands of Britain and IRELAND. The Scandinavian ice sheet spread from NORWAY, SWEDEN, and FINLAND across most of the North European Plain and western RUSSIA. A sixth ice sheet, the Barents-Kara, overlapped north-central Eurasia's northern continental shelf and mainland. A seventh glacier—the East Siberian—might have existed separately from the Barents-Kara sheet. End moraines, glacial lake varves, and carbon-14 dating document where the southern edges of the ice sheets were. In contrast, the positions of the sheets' higher latitude margins are the subject of much scientific debate. The Laurentide and Cordilleran ice sheets may have fused together. The Britain, Scandinavian, and Barents-Kara ice sheets may have also joined.

In addition to Pleistocene ice sheets, mountain glaciers of that time were more numerous at lower latitudes than today. Large highland ice caps draped the summits of the ALPS, South Island of NEW ZEALAND, HIMALAYAS, southern ROCKY MOUNTAINS, and the southern and central ANDES. In addition, small alpine glaciers were more widespread in the tropics and middle latitudes than today.

The Pleistocene epoch's temperatures were cooler than now, but not uniformly so, as repeated episodes of interglacial warmth punctuated its history. Computer models tell us that during glacial advances, global average surface-water temperatures were 9 or more degrees F (5 degrees C) cooler than today. Air over the land was even cooler. For instance, average temperatures of North America were at least 14 degrees F (8 degrees C) cooler. Sea ice was extensive: It was possible to walk from the Brittany coast of FRANCE to IRELAND, for instance. In contrast, the southern limit of sea ice in the present interglacial period is just slightly south of Greenland. Sea ice was farther north in previous interglacials, as temperatures were 2 to 5 degrees F (1 to 3 degrees C) warmer than the present. The warmth of the Pleistocene interglacials resulted in higher evaporation rates and more intense monsoon winds areas than

those of today. For instance, during the last Pleistocene interglacial period, stronger monsoon winds delivered significantly more rainfall to subtropical and middle latitude regions. As a result, large lakes, lush savanna grasses, and riparian woodlands covered areas that are now desert, such the SAHARA DESERT of northern Africa and the Mojave Desert of the U.S. southwest.

ERODED LANDSCAPES

Pleistocene glaciers left behind eroded landscapes in high latitude and alpine landscapes: thin soils, large exposures of ice-scoured rocks, countless lakes, and steep-walled fjords. Additionally, mountain glaciers created signature cirque basins, U-shaped valleys, narrow-crested arêtes and steeple-sharp horns. Rubble of interlobate and end moraines chronicle the size and extent of the glaciers. Moreover, table-flat beds of proglacial lakes and outwash plains, abandoned meltwater channels, sinuous eskers, kames, kettles, and valley trains attest to the voluminous glacial melting that took place after the last glacial maximum.

The rising temperatures that ended the Pleistocene epoch shrank the arctic tundra and cold-tolerant coniferous forests. Simultaneously, there was a dramatic expansion in the boundaries of warmer climes—tropical rainforests, subtropical deserts, and middle latitude forests.

The end of the Pleistocene also witnessed the extinction of many cold-tolerant mammals: the mastodon, woolly mammoth, true horses, saber-toothed tigers, large wolves, giant armadillos, as well as giant ground sloths and ancient bisons, camels, and wild pigs. Biological and cultural evolutions of early human beings also took place during the Pleistocene. Humans' evolving intelligence, as well as hunting and gathering technology, enabled them to adapt to a rapidly changing environment and to spread to all the continents (except ANTARCTICA) before the Pleistocene epoch ended.

BIBLIOGRAPHY. Douglas I. Benin and David J. Evans, *Glaciers and Glaciation* (Oxford University Press, 1998); Martin Williams, David Dunkerley, Patrick De Deckker, Peter Kershaw, and John Chappell, *Quaternary Environments* (Arnold, 1998); R.C.L. Wilson, *The Great Ice Age: Climate Change and Life* (Routledge, 2000); William F. Ruddiman, *Earth's Climate: Past and Future* (William H. Freeman, 2001).

RICHARD A. CROOKER
KUTZTOWN UNIVERSITY

Po Valley

THE PO RIVER VALLEY is the largest and most important economic region in ITALY. It is the center of most Italian industry as well as Italy's agricultural heartland. More than 16 million people—nearly a third of all Italians—live in this fertile basin, in which are located 12 cities with populations surpassing 100,000, including Turin and Milan, with populations over 1 million.

The river itself is not among the longest rivers in Europe, running 405 mi (652 km) from west to east, but together with its 141 tributaries, the Po catchment area stretches across 27,000 square mi (70,000 square km). The river's agricultural and industrial importance has played a primary part in the political and social history of Italy—the basin today accounts for 40 percent of the nation's gross domestic product—but suffers serious environmental consequences through poor water management, industrial and sewage pollution, and agricultural runoff.

The Po River, called the "Padus" in Latin—the origin of the term Val Padana ("Po Valley")—begins as a swift mountain stream in the Cottian Alps on the border with FRANCE, near the peak of Monviso (12,602 ft or 3,841 m). It flows east and north to Turin, a major manufacturing town, then continues east across the Piedmont region, joined by several small rivers flowing down from the ALPS to the west or north. Near Alessandria, it is joined by the river Tamaro, which flows from the south, originating in the Apennines. From this point east, the Po's tributaries continue to be differentiated between those that flow from the Alps to the north and those that rise to the south in the Apennines, with very different characteristics, notably, differences in seasonal flood patterns.

The major Alpine tributaries are (from west to east): the Ticino, Adda, Oglio and Mincio. The headwaters of these rivers generally form the northern boundary of Italy with SWITZERLAND, with the exception of the Ticino, which flows through the southern Swiss canton of the same name. Each of these rivers also flows through a major lake at the point at which the mountains reach the upper Po Valley Plateau (also from west to east): Maggiore, Lugano, Como, Iseo and Garda. These lakes are popular among tourists for their cool climate and rich mountain scenery. The major cities in this region, north of the Po, are located between the major rivers: Novara, Milan, Monza, Bergamo and Brescia, plus the smaller but important historic cities of Cremona and Mantova (Mantua).

The major rivers that flow from the Apennines to the south—usually swifter and with heavier sediment—are (from west to east) the Trebbia, Taro, and Secchia. Major cities on these rivers are Piacenza, Parma, Reggio nell'Emilia, and Modena. The last 75 mi (120 km) of the Po Valley is a narrow corridor to a broad delta on the ADRIATIC, in which the main city is Ferrara. Two other river basins hem in this corridor to the north and south, whose mouths are so close to those of the Po that they can really almost be considered part of the same drainage basin: the Adige, which drains much of the Veneto and northeastern Italy, and the Reno, which collects several of the rivers of Romagna. If these basins are included within the wider Po Valley, even more famous Italian cities can be added to the list, including Bologna and Verona.

IRRIGATION

Parts of the Po River Delta and the Val Padana have been drained for agricultural purposes since Roman times. The Po Valley is Europe's largest rice-growing region, requiring significant amounts of water to be diverted from the river and its tributaries, but even larger amounts of water are diverted for industrial purposes, for example, for the large automotive industry in Turin.

Flooding has always been a major issue in the Po Valley, especially in spring when melted snow and ice from the Alps can double the river's flow within hours. Extensive dykes and levees have been built to curb this problem, along with river straightening projects to aid navigation, but these have mostly served to shift the flooding along to other parts of the river. In 1989, the Italian Parliament created several new bodies to oversee water management throughout the country; the largest of these was created for the Po River Basin.

This agency has the monumental task of restoring the environmental balance of the Po and its tributaries while maintaining the viability of 269 hydroelectric plants, 11 thermal power plants, farming activity that accounts for 35 percent of Italy's gross domestic product, and nearly 40 percent of Italy's industry.

BIBLIOGRAPHY. Erla Zwingle, "Po: River of Pain and Plenty," *National Geographic* (May 2002); C. Revenga, S. Murray, et al., *Watersheds of the World* (World Resources Institute, 1998); The Po River Basin Authority, www.adbpo.it (August 2004).

JONATHAN SPANGLER
SMITHSONIAN INSTITUTION

Poland

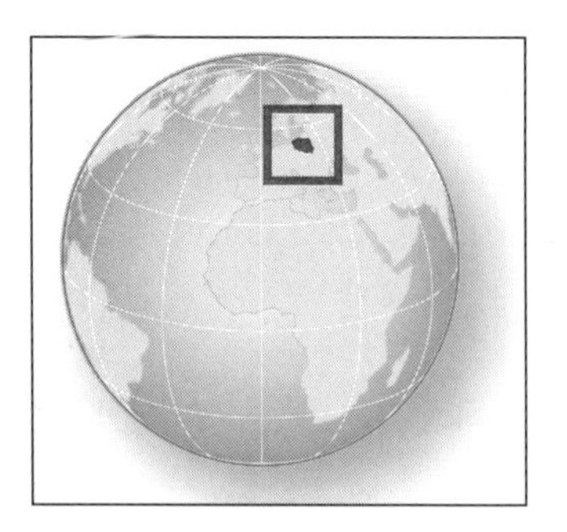

Map Page 1133 **Area** 120,728 square mi (312,685 square km) **Population** 38,622,660 **Capital** Warsaw **Highest Point** 8,199 ft (2,499 m) **Lowest Point** -6.6 ft (-2 m) **GDP per capita** $9,700 **Primary Natural Resources** coal, sulfur, copper, natural gas, silver.

POLAND IS LOCATED in eastern Europe, and is surrounded by the Baltic Sea to the northwest, GERMANY to the west, SLOVAKIA and the CZECH REPUBLIC to the south, and BELARUS, UKRAINE, LITHUANIA, and RUSSIA to the east. The central portion of the country is relatively flat and is an agriculturally driven area. The Vistula River, Poland's longest, flows through this area, and drains into the Baltic Sea. The Sudeten Mountains are located in the west and the Carpathian Mountains run along the south.

About 25 percent of Poland is covered by forests. Wild boar, hare, deer, brown bears, and wildcats can be found among the trees, and numerous storks live in the countryside. The climate is very changeable. A continental climate from the east influences the weather, as well as the maritime climate from the west. Central Poland is the driest area in the country, and the mountain regions receive much rain and snow.

Poland's history can be traced to the reign of Mieszko who converted to Christianity in 966 C.E. During the 14th century, Casimir III the Great built an extensive network of castles and fortifications and one of the first European universities at Krakow. The early Polish state, existed until it was partitioned by Prussia, Russia, and AUSTRIA, in 1795.

After World War I, Poland declared its independence and was under authoritarian rule. On August 23, 1939, the German and Soviet government signed the Ribbentrop-Molotov nonagression pact. This pact secretly separated Poland into Soviet and Nazi controlled zones. On September 1, 1939, the German army marched into the Poland and on September 17, Soviet forces invaded the eastern section of the state. The Soviet troops were eventually overrun by German troops in June 1941.

During the war, exiled Poles created a resistance movement. Some 400,000 Poles fought for the Soviets and 200,000 fought on the western front. However, in 1943, the Soviets broke relations with the Polish gov-

Gdansk, Poland, is the birthplace of the Solidarity movement, which helped overthrow communist control over the country.

ernment and in July 1944, Soviet troops reentered sections of Poland and created the communist-controlled Polish Committee of National Liberation. The Polish citizens who remained in Poland during World War II were brutally treated by the Nazis. About six million Polish Jews were killed and 2.5 million were deported to Germany, where they faced work at forced-labor camps.

After World War II, the Polish Provisional Government of National Unity was formed. Free elections were planned two years later but never occurred. The Communist Party had begun to enforce its rule over the citizens. During 1956, the communist hold over Polish politics seemed to lessen its grip as a period of liberalization occurred under First Secretary Wladyslaw Gomulka. However, this was short-lived, when in 1968, a series of student demonstrations were suppressed. In 1970, many Polish citizens became upset with the living and working conditions in the country, and Gomulka was replaced with Edward Gierek.

During the first half of the 1970s, Poland's economic growth rate was one of the highest in the world. Western credit was infused into the economy, but much of the capital was misspent in the centrally planned economy. As the decade progressed, the country's debt became a major problem and the economic growth halted.

By July 1980, the foreign debt was more than $20 billion, and workers became incensed over the rising prices. Throughout the summer, a series of strikes took place across the nation. On August 31, 1980, workers at a shipyard in Gdansk, led by Lech Walesa, signed an agreement, which guaranteed the right for the workers to create an independent trade union. Other agreements were created as the Solidarity movement grew. Ten million Polish citizens, including 1 million Communist Party members, joined the Solidarity trade union.

As discontent grew with the corruption in the government, Gierek was replaced as first secretary by Stanislaw Kania. The Soviet military also moved along the Polish border, and in December 1981, after the Polish regime declared martial law, the Red Army moved into Poland to eliminate the union. Many of the union leaders, including Walesa, were taken prisoner, and Western countries, including the United States, imposed economic sanctions on the Polish Communist regime and Soviet Union.

One year later, martial law was suspended, and for the next two years, prisoners were slowly released. However, the seeds of the Solidarity movement could not be eliminated, and by 1988, strikes once again occurred throughout the country. As the Soviet and communist influence in Eastern Europe was dwindling with a surge of democracy, a new government led by noncommunists was created in 1989.

The Polish government changed its economy from being centrally planned to free-market-based. In May 1990, the local elections were free, and the Solidarity Citizens' Committee was successful in winning a majority. In December 1990, Lech Walesa became the first democratically elected president in the history of Poland.

Throughout the 1990s, Poland made strides in achieving a democratic government and a market econ-

omy. In 1995, Walesa was defeated by the Democratic Left Alliance candidate, Aleksander Kwasniewski. After a brief departure from the presidency between 1995 and 1997, Kwasniewski regained the presidency and was in power as of 2005. The economy grew rapidly into the mid-1990s with an economic "shock therapy" program but then slowed, which resulted in a rise of unemployment. Throughout its transition process, the United States and other Western nations have reduced Poland's foreign debt and provided economic aid. Trade barriers have also been lowered.

The outlook for Poland seems extremely bright. As the country joins the EUROPEAN UNION, the economy continues to make strides. After hundreds of tumultuous years, Poland seems destined for economic and political success in the 21st century.

BIBLIOGRAPHY. *World Factbook* (CIA, 2004); U.S. Department of State, "Background Note: Poland," www.state.gov (April 2004); Lonely Planet World Guide, "Poland," www.lonelyplanet.com (April 2004).

GAVIN WILK
INDEPENDENT SCHOLAR

political geography

POLITICAL GEOGRAPHY IS the study of the ways geographic space is organized within and by political processes. It focuses on the spatial expression of political behavior. Boundaries on land and on the oceans, the role of capital cities, power relationships among nation-states, administrative systems, voter behavior, conflicts over resources, and even matters involving outer space have politicogeographical dimensions.

Contemplating the state of political geography, Richard Muir observed that "political geography is simultaneously one of the most retarded and most undervalued branches of geography and one that offers the greatest potential for both theoretical and practical advance." Things had not always been so. Many of the early geographers such as Peter Kropotkin, Sir Halford J. MACKINDER, and Isaiah Bowman were explicitly concerned with the relations between politics and geography in both their published work and their public lives.

Mackinder, for example, was a member of parliament, a high commissioner in RUSSIA and chairman of various government committees, and Bowman was an adviser to President Woodrow Wilson at the Versailles Peace Treaty meetings. Sadly, this concern with the very stuff of politics waned after Mackinder and Bowman. Geopolitics became discredited by a Nazi association and political geography became an ossified subdiscipline of a tired subject, often taught, never researched, a prisoner of outdated theories.

From the disciplinary perspective, political geography may be defined as either geography or political science. In the perspective of political science, political geography appears as "the study of political phenomena in their aerial context," as one geographer put it.

FUNCTIONS AND FACTORS

Political geography is the study of relationships among humans, their environment, and their political institutions. The controversy over states' rights, which has been revived again and again in the UNITED STATES, masks geographic problems growing out of the natural environment of the southern states, or out of natural resources of petroleum-producing states, or out of water requirements in dry and semiarid states. All these and more require political accommodation.

The functions of political geography are not confined to one state but embrace the whole globe. It is intriguing to attempt to rank the sovereign states of the world in terms of effective national power, to evaluate the regional importance of one state compared to its neighbors, to range over the world and consider the ever changing power of the (British) Commonwealth of Nations, or the French Community, to analyze the reasons for political tensions between regions in terms of environmental differences—these are the substance of political geography in its broadest terms.

The subject is also dynamic, searching for the effects of change and the rate of such change. Change affects, in every inhabited spot, the elements within the political state that define it, that strengthen or weaken it, that slowly alter the image of a state in the world. The nature of change and its velocity are both little understood, for humans are cursed with a love of the familiar, the usual and ingrained, and their grasp is finite and time bound.

The political requirements of agrarian ARGENTINA altered internally and externally when the Juan Peron government attempted to industrialize the state. The political geography of the (British) Commonwealth of Nations today contrasts in stress and strain with the 19th-century BRITISH EMPIRE. Changing geographic resources and factors buttressed the weapons race between the UNITED STATES and Soviet Union during the Cold War. Political demands for control of French

Louisiana in the early 19th century grew out of the MISSISSIPPI RIVER's role as a transport route for farm products. Today, an interstate superhighway system is of more importance to the agriculture areas of the Midwest than southward river routes. The political importance of the differing economic developments of communist CHINA and democratic INDIA causes scholars to question the suitability of one type of government over another.

Yet scholars admit that the political state is in one sense an abstraction, dependent on written records and some degree of respect for possession or ownership. It could not exist in a world without other political states. As an abstraction, it appears at a certain level of culture, marked by written language, sedentary life, and the need for organization. Today, in certain areas of the world, it appears to be only another stage in the search for unity by groups of people. In newly born states it is national unity. The painfully complex path of Western Europe toward federation is a movement toward regional unity. The latter's course contrasts sharply with the turbulent, uneasy history of newly independent nation-states of the former Soviet Union.

Political geography is functional; it studies the degree of unity reached by the environment and man's political institutions. Laws governing the ownership of water rights that were evolved in moist, cool northwest Europe were unsuccessfully transplanted to the American semiarid southwest. In much of Latin America, most of the land is owned by a small wealthy class. The resultant pressure of population on resources is a continuing specter that threatens to menace the productivity of the environment and to conjure up political revolution.

Subordinate political units in the state also clash with man's use of the environment. It can be witnessed in the U.S. urban trading areas that overlap several states; interstate compacts regulating commerce, navigation, and transportation; and overlapping regional requirements for development of natural resources such as river basins. Above all, there is the increasing role of central political power in the modern industrialized states, which has been forced primarily by the interregional complexities of economic and social problems.

Political geography considers different cultural meanings for similar political and geographic functions. Attitudes, frames of reference, habits, and beliefs—all the rationale of political and cultural action—are explored for their agreement or disagreement with the environment. America, in the colonial period, offered the natives hunting and fishing; to the colonists, it offered farms, lumber, cotton, and tobacco. The prairies of the Midwest or of western CANADA, with their thick, deep-rooted grass, have a different meaning to the settler today than what they had before the invention of the steel plow. These lands were first unsuitable, then invaluable, for profitable settlements. The former accommodates to local tribal government by the patriarch; the latter is the agent of highly centralized, democratic government that is over 1,200 miles distant.

The pace of change today is forcing many peoples to reorient their customs and habits. Unfortunately, cultural inertia often produces only a veneer of change. The oil-rich sheiks of the Arabian peninsula gladly accept the costly consumer goods of the West; many a Cadillac has not been uncrated or may have been run dry of gasoline and abandoned. The distrust of the strange, the foreign, and the unusual continue to haunt most of the world's peoples.

Though moderated after centuries of conflict, religion remains contentious over vast areas. The idea of race, expressed in terms of color of skin and physiognomy causes rioting, murder, economic and social discrimination, and political bias in many countries. Despite the general rise in literacy and the construction of educational systems and rapid communications structures, there probably has not been a corresponding increase in the level of understanding and tolerance. Not less important, and among the slowest to change, are those series of conventions that society impresses upon individuals. These force the control or submission of the instinctive impulses, for the most part, for the general social good.

The political geographer is concerned with the homogeneity and heterogeneity in action within and without the political unit. He or she must attempt to analyze the centrifugal and centripetal forces acting and interacting at different rates.

EVOLUTION AND DEVELOPMENT

Humans remained a pawn of their environment for thousands of years before they became sedentary. Security lay with the tribe and idol, and fears led to primitive worship. The arrival of sedentary agriculture provided the grounding to develop small groupings, implying an intimate association with a single homogeneous landscape. In river valleys such as the NILE, Tigris-Euphrates, INDUS, and HUANG, these collective groupings evolved political forms that were linked to the physical settings of FLOODPLAINS, the presence of

abundant water, the yearly silting of the fields, and the regularity of the sun and the seasons.

The hold of the primeval past remained strong even in the Greek world of Persian invasions, in the lifetimes of Pericles, Plato, and Aristotle. The Greeks were the first known culture to actively explore geography as a science and philosophy, with major contributors including Thales of Miletus, Herodotus, Eratosthenes, Hipparchus, Aristotle, Dicaearchus of Messana, Strabo, and Ptolemy. But Greek scholars began to think logically and abstractly about the meaning of the world around them. Both Plato and Aristotle analyzed the political state, its environmental base, and man's relationships with it. They attempted to clarify cause, space and time.

Although the political world of their day became complex, they agreed to find unity among environment, man, and the state. The polis, the city-state, was their political frame of reference. The influence of topography in fragmenting the Greek peninsula into many small river valleys, separated by hills and mountains but facing the sea, has also been commented on many times. Yet, even for the Greeks it was true, as it is increasingly today, that humans are active, intelligent agents, not the pawns of their environment.

Whenever we study the thought of other people in other cultures and at other times, their frame of reference must be considered to explain their limitations and successes. Greek thinkers were no exception. An early comment on the political environment by Aristotle was both nationalistic and deterministic. He asserted that "the people of cold countries generally, particularly those of Europe, are full of spirit but deficient in skill and intelligence; and this is why they remain free, but show no political development and faculty of governing others. Peoples of Asia are endowed with intelligence but are deficient in skills. This is why they continue to be peoples of subjects and slaves." Aristotle's Europeans were those nomadic tribes in south Russia and the Balkans that periodically raided the Mediterranean world and threatened its colonies. He implied that nomadic tribes are not likely to develop a high degree of political organization. His Asians were the peoples of Asia Minor and the Persian Empire. The first of these were his own people, inhabitants of Greek city-states along the fringe of Anatolia—yet he looked down on them.

Today, we would consider the low level of technology at that time as important, one that produced small surpluses but dense populations. The size of these eastern empires contributed to the necessity for a political organization that depended on a highly centralized monarchy buttressed by military power.

When Aristotle wrote that the Greeks were better than barbarians of the north and the Asians of the east, he emphasized the importance of location; he was writing about the known world. In addition, he believed that climate had a strong influence on qualities such as spirit and intelligence, for in the Greek division of climates, the Greek lived in the temperate zones, the nomad in the cold regions, the Asiatic in the hot areas.

Greek thinkers were cautious of their common nationality, but through most of the ancient period they were more impressed by the consciousness of the value of their individual civic life. The 5th century B.C.E. was the period of highest achievement in Greek political life. Greek unity was imposed by Philip and Alexander of the Macedon. The unity of which Plato wrote was sharply limited in dimension. He was certain that the city-state was the ideal political form for humans. Later, in the age of Caesar and Augustus, the geographer Strabo argued that only with a strong central government with one powerful ruler could a continental empire such as Rome survive and flourish. Through the centuries, humans have altered many times their views on the size, structure, and functions of the political state that they continued to require.

These early observations on the nature of interlinkages between people, the environment, and political institutions could not evolve in a coherent subdiscipline of political geography. The surge of new geography of the 1950s and 1960s bypassed political geography. The new geography, with spatial analysis as its theme, neoclassical economics as its accounting frame, and logical positivism as its methodological underpinning, could not accommodate a political geography. The emphasis of neoclassical economics on the economy as a harmonious, self-regulating system, where each factor of production receives its fair reward, ignored questions of conflict and inequitable distribution, and the focus of logical positivism directed attention to verifiable empirical statements in particular, and data analysis in general, and away from the operation of the more incorporeal power relations within society.

A truly political geography could not flourish in such a climate. Moreover, the explicit analysis of politics was taken over by the last social science discipline, political science. This academic assertion was being conducted by a discipline, which according to some scholars was a device for avoiding politics without achieving science. Ignored by its discipline and lacking

any theoretical substance from political science, it is little wonder that political geography was a moribund subject.

MODERN POLITICAL GEOGRAPHY

The origins of political geography are usually traced to Friedrich Ratzel (1844–1904), who was the brilliant yet ambiguous founder of modern political geography. He was strongly influenced by rapid, vigorous developments in the natural sciences in the 19th century and sought to discover the realities of political society.

Ratzel and Karl Marx both thought that there were natural laws that controlled society. Ratzel's critics have often disregarded his fundamental contributions to the elements of political geography, underestimated the attention he gave to the factors of location and space, and fixed their disliking on his attempt to develop an analogy between political state and living biological organism. In his native GERMANY, the concept of natural selection and survival of the fittest became wedded to a geopolitical jurisdiction of national expansion.

A group of German geopoliticians emerged who gradually discredited his reputation as they abandoned rationale and unbiased geographic thought and turned to justifications of war and conquest.

Ratzel thought that states in all stages of development are considered as organisms that stand in a necessary connection with the ground, and hence must be viewed geographically. He linked the state to a mobile body, to an organism subject to the natural laws of growth and decay. His organism was spiritual and moral. Just as an organism is born, grows, matures, and eventually dies, Ratzel argued, states go through stages of birth (around a culture hearth or core area), expansion (perhaps by colonization), maturity (stability), and eventual collapse.

Strongly influenced by Darwinian thinking, he was interested in the relationships between the state and the Earth, between political institutions and their physical environment. His major contribution came with his representation of the state as a organism needing *Lebensraum* (living space) and the competition between states for that space as a Darwinian contest involving the survival of the fittest. He suggested that only a sporadic absorption of new land and people could stave off the state's decline. In fact, Ratzel proposed a blueprint for imperialism.

He believed the higher the technological and social development of the political state, the farther that state was removed from its organic foundations. In fact, this thought on the analogy of an organism and the state is ambiguous. The true geographic structure of Ratzel's thought suffered because of imperfect distillation from his German and from the rejection by American geographers of any form of determinism. However, the elements of political geography that are thought to be of major importance today were voluminously discussed and analyzed by Ratzel. The state is not an abstraction—it occupies land and water; its size, location, and boundaries are important characteristics.

Ratzel also thought that the surface features of the land together with vegetation and soil were basic to any political analysis. He emphasized the importance of capital in location and function. The beliefs of social groups in the necessity of a political union based on historical, religious, and cultural values; the theory of centrifugal and centripetal forces operative in the state; the idea of the ecumene or heartland; and the vital roles of communication and movement—all these provide the solid substance of his political geography. He lived in an era that saw the growth of colonial empires to their maximum and the partitioning of virtually all the land areas of the world into politically controlled regions. He correctly saw that the increasing ability to overcome space placed a premium on states of large size. Today, we speak of the continental superpowers that have geographic environments that possess varied and immense resources. In addition, Ratzel devoted attention to relationships between states, particularly on the nature and function of boundaries.

Ratzel's ideas were taken up by a number of geographers with political as well as academic interests, notably Rudolf Kjellen, a Swede, and Karl Haushofer, a German who taught and was close to Rudolf Hess, Adolf Hitler's deputy in the 1930s. They developed a school of *geopolitik*, whose writings were used to give an intellectual rationale to 1930s German expansionism—not only the desire to occupy adjacent territories with substantial German populations, such as AUSTRIA and Sudetenland, but also Russian areas further east.

Parallel developments in the UNITED KINGDOM were led by another geographer-politician, Sir Halford Mackinder (1904), whose classic paper related state power to location. In an era when movement of heavy goods and large armies was easier by sea than by land, maritime countries would dominate politically, but as land transport was becoming easier, so "land-based powers" were becoming stronger: he argued that whoever controlled the "world island" (the heartland of Euro-Asia) should be able to control the globe—a geopolitical notion that influenced much strategic

thinking throughout the century, until air power (and then power in space) came to dominate military strategy.

Elsewhere, political geography merged as the study of states and their impact on the landscape, as exemplified by D. Whittelsey (1939) and by R. Hartshorne's (1950) paper on the functional approach: the latter saw the spatial structuring of the state as a resolution of centrifugal and centripetal forces, focused on its core area and capital city. Many of their writings involved identifying typologies of states and dividing the world into geopolitical regions.

Descriptive analyses of the world political map were continued by a number of scholars who at various times posited bipolar, multipolar, center-periphery, and other structures. Other geographers developed interests in boundary demarcation and disputes, on land and at sea. This continuing strand of work on geopolitics had little impact on the wider discipline, despite its links to strategic thinking. It was almost entirely absent from geography in FRANCE, Germany. and Russia from 1945 onward because of the association of political geography with geopolitics and then geopolitik. The Soviet Union, for example, blocked the establishment of commission on political geography within the International Geographical Union until 1984.

Climatic variations have inspired another set of geopolitical hypotheses and critiques. International political patterns have also been linked with the uneven distribution of the various raw-materials requisites of modern industry. There is some disposition to regard areal differentials in technology as the critical variable, a hypothesis that has been linked with demographic distribution to produce a prediction that international political patterns will ultimately be determined by the latter. The prediction is based on the premise that technological primacy will vary with relative numbers of superior scientists and other gifted individuals varying in the long run with the size of population.

A revival of interests in political geography from the 1970s onward was initially linked to the "quantitative revolution," which the wider discipline experienced in the 1960s and 1970s. Work on ELECTORAL GEOGRAPHY started then and geographers increasingly brought their spatial perspective to bear on a range of subjects broadly defined as "political" and relating in some ways to the operation of the state. Location and conflict (over land uses, public goods, and so forth) became topics considered by political geographers. But a political location theory was not as obvious as an economic, or even a social, one.

New texts in the 1970s also stimulated a broadening of the substantive areas of interests within political geography, with chapters on the geography of law, for example, and on spatial variations in the operation of government programs and government spending. Both depth and breadth were brought to the subdiscipline by this concern with theory, which involved moving away from the treatment of space as a given, as the environment within which states operate, toward a perspective that sees space as produced and reproduced by human action—the world political map is a social production. Two "spatial takes" were particularly relevant in this movement. The first was a treatment of scale. Strongly influenced by world-systems analysis, it was argued that world capitalism is organized globally, mediated and regulated regionally by states, and experienced locally.

The second theoretical perspective was introduced to develop the concept of territoriality to show how bounded spaces (including those of nation states) are crucial to the exercise of political, economic, cultural and military powers. The world is a mosaic of nested containers within which power is exercised and people controlled—with the criterion for whether you are subject to a particular rule of law being whether you are within the territory where it is sovereign. The theory is further analyzed in the light of increased globalization and the consequent changing role of the territorially defined nation state.

The most recent area of expansion has been in the study of critical geopolitics, a further outgrowth of the growing theoretical sophistication within human geography. As John O'Loughlin (1994) illustrates, this involves questioning the assumptions upon which geopolitical strategies are based—not so much the "geographical information" employed as the representations of peoples and places (both "selves" and "others"). These are involved in the creation of identities: images of "us" and "them" (as in the 1945 to 1990 Cold War in which the two major powers and their allies each created images of the other on which to base their policies and around which to mobilize popular support). Geopolitical practices are subjected to critical scrutiny as opposing views of the world are deconstructed.

Political geography is now a vibrant component of its parent discipline. Its renaissance and expansion were marked by the launch of the journal *Political Geography Quarterly* in 1983, which is now published eight times a year (as *Political Geography*) and is the major source for tracking developments.

QUANTITATIVE ANALYSIS

Unlike its sister disciplines of economics or political science, political geography has a relatively small amount of published research that contains quantitative analysis. The reasons for the relatively paucity of quantitative work in political geography can be traced to dual trends that have been evident for the past 20 years. First, like the rest of human geography, political geography has seen a rise in interest in poststructuralist and humanistic research methodologies as the 1970s heyday of positivism passed.

Some scholars believe that this trend has acquired because words are more persuasive than numbers, though it seems more likely that political geography is returning to the status quo ante, where quantitative methodology is just one of a plethora of options on the research menu. Second, and connected to the first, quantitative geography (and shortly after, geographic information systems, GIS) was promoted as a response to the challenges of the day, especially the economic stagnation in Western countries. By pursuing spatial analysis and GIS, and later merging these approaches, geography could certify its scientific status and show its various uses.

To paraphrase P. Longley and M. Batty (1996), quantitative political geography now stands at a junction. Either it will be integrated more intensively with the rest of political geography (this has to be a two-way street and will only succeed if non-quantitative political geographers accept quantitative approaches and research results) and more generally with other quantitative social science, or it will become further isolated.

BIBLIOGRAPHY. R.D. Dikshit, ed., *Developments in Political Geography. A Century of Progress* (Sage Publications, 1997); M.I. Glassner, *Political Geography* (Wiley, 1996); M. Jones, R. Jones, and M. Woods, *An Introduction to Political Geography* (Routledge, 2004); J. Agnew, K. Mitchell, and G. Toal, eds., *A Companion to Political Geography* (Blackwell, 2002); M. Blacksell, *Political Geography* (Routledge, 2003); M. Klare, "The New Geography of Conflict," *Foreign Affairs* (v.80/3, 2001); John Agnew, Katharyne Mitchell, and Gerard Toal, eds., *A Companion to Political Geography* (Blackwell, 2003); John Agnew, David Livingstone, and Alisdair Rogers, eds., *Human Geography: An Essential Anthology* (Blackwell, 1996); P. Knox, J. Agnew, and L. McCarthy, *The Geography of the World Economy* (Arnold, 2003).

Jitendra Uttam
Jawaharlal Nehru University, India

Polo, Marco (1254–1324)

MARCO POLO IS THE best known of all Western travelers along the SILK ROAD. He traveled Asia for 24 years, all the way to CHINA, where he became confidant of Kublai Khan (1214–94.) He also wrote of his travels. Marco was born in ITALY, either in Curzola off the Dalmatian coast or in Venice, in 1254. He grew up in Venice, the center of Mediterranean commerce. As befitted the son of nobility, he was educated as a gentleman.

When Marco was six, his father and uncle, merchants of Venice, traveled to Sudak on the Crimea, then to Surai on the Volga River. When civil war broke out, the brothers Polo went to BOKKARA, where they remained for three years. While there, they encountered an ambassador from the Mongol Hulagu Khan. With him they traveled to BEIJING, arriving in 1266. After a year, the brothers returned to Italy with a letter from the Great Khan to Pope Clement IV asking for 100 learned men to teach the Mongols of China about Western ways and religion. Marco accompanied his father and uncle on the next trip to China. Marco was 17. Bearing gifts and letters from Pope Gregory X, the Polos traveled through ARMENIA, Persia, and AFGHANISTAN, over the Pamirs, and all along the Silk Road to China. At Badakhshan, Marco became ill, forcing a year's delay. Once over the Pamirs, the highest place in the world, they crossed the Taklamakan desert. Polo recorded the interesting sights and people. His descriptions were clear and able to generate some sense of the feelings the natural and human wonders he encountered. He noted that although dangerous, the route was well traveled. He also produced a detailed history of the Mongols and the Great Khan. He described life on the STEPPES. He talked of marriage customs and life in general. After the Polos arrived in Shang-tu, the summer capital, Polo began to chronicle China under the Mongol ruler.

As a member of the court and fluent in four languages, Marco Polo traveled to China, Burma, and INDIA, places not seen again by Westerners until the 19th century. He described ceremonies, hunting, and the new capital city. Much of this could be translated into terms easily understood in the West because comparable activities occurred there.

Some things were totally new to him: asbestos, fireproof cloth, paper currency, coal, the imperial post, crocodiles, coconuts. Polo's travels showed him great power and wealth and an extremely complex society. China's economy far surpassed that of all Europe.

China generated 125,000 tons of iron a year, an output not matched until the 18th century in Europe. A canal network linked large cities in a commercial network. Credit was available, as was paper money. The Khan's message system was so efficient that a high-priority message could travel in one day the same distance a lone traveler would cover in 10 days. The level of prosperity was well beyond the wildest European imaginings. Polo even joined the Privy Council in China in 1277. He served as tax inspector in Yanzhou, on the Grand Canal northeast of Nanking.

For 17 years, the Polos remained at the Chinese court. They amassed great stores of gold and jewels. They wanted to be sure they could get their fortunes out of the country before Kublai Khan, then in his late 70s, died. Reluctantly the Khan agreed, and the Polos took two years to get home by sea. In Persia, they learned of the death of Kublai Khan, but his golden tablet of authority still protected its possessors, and they were able to travel through dangerous territory safely. They arrived in Venice in 1295.

In 1298, Polo commanded a galley in the Venetian war against Genoa. Captured, he spent a year in prison. There he met a romance writer, Rustichello of Pisa. Rustichello convinced Marco to dictate *The Description of the World or The Travels of Marco Polo*. Polo's descriptions of China, India, and Africa made his writings one of the most popular books of medieval Europe. Unfortunately, most Europeans referred to it as *Il Milione,* or *The Million Lies*.

With peace in 1299, Polo returned to Venice. He married and had three daughters and remained in Venice until his death at age 70. His deathbed statement was, "I have only told the half of what I saw!" Subsequent scholars wondered about Polo's truthfulness. He didn't mention the Great Wall, nor did he learn Chinese. He ignored common things such as foot binding, calligraphy, and tea. He never became an entry in the Chinese Annals of the Empire, wherein all important visitors were recorded. These discrepancies raise the question of whether or not the Polos actually went to China.

Polo's book retained a strong readership, with hundreds of manuscript editions in the century after his death. It was the most important travelogue of its time, and the most important on the Silk Road ever written in a European language. Eighty manuscript copies of his books in various versions remain extant. Scholars continue to examine and authenticate the travels of Polo. Geographers in the 14th century began using his maps in publications such as the *Catalan World Map of 1375*, used by Christopher Columbus and Prince Henry the Navigator. His distances were accurate, and he was perhaps the precursor of a scientific geographer. Much was validated by 18th- and 19th-century travelers. Even Chinese historians use his work to better understand the events of their 13th century.

BIBLIOGRAPHY. Richard Humble, *Marco Polo* (G.P. Putnam's Sons, 1975); John Hubbard, "Marco Polo's China," www.tk421.net (February 2004); John Larner, *Marco Polo and the Discovery of the World* (Yale University Press, 2001); Silkroad Foundation, "Marco Polo's Travels," www.silk-road.com (February 2004).

John Barnhill, Ph.D.
Independent Scholar

Polynesia

POLYNESIA, MEANING "many islands" (Greek), is one of the three main divisions (the others being MICRONESIA and MELANESIA) of Oceania. This division is based upon the ethnological background of the different islands' inhabitants. Geographically, Polynesia is a triangle area of islands, with the Hawaiian Islands at one corner, NEW ZEALAND at another, and EASTER ISLAND forming the third corner, located in the central and South PACIFIC OCEAN. The larger islands are generally volcanic in origin; the smaller ones are generally coral formations. The principal groups are the Hawaiian Islands, SAMOA, TONGA, New Zealand, KIRIBATI islands, COOK ISLANDS, and the islands of FRENCH POLYNESIA. Malayo-Polynesian languages are spoken in Polynesia.

One of the biggest questions concerning Polynesians is where did they originally come from and how did they get to the distant islands that comprise what today is known as the Polynesian Triangle? Numerous disciplines have become involved in trying to trace Polynesian origins; among them are geographers, anthropologists, archaeologists, and historians.

One way of trying to determine the settlement issue is to study the various languages and oral traditions of discovery by the numerous Pacific peoples. Other avenues of answering the settlement question include archaeological excavations, including those focusing on a distinctively decorated type of pottery called Lapita, and tracing the origins of the various food staples found in Polynesia.

The earliest descendants of the Polynesians were believed to have started their Eastern trek into the Pacific around 1500 B.C.E. Starting from New Guinea and moving east through the SOLOMON ISLANDS then the VANUATU islands. As the early Polynesians moved farther to the east, the distances between the islands increased from tens of miles, to hundreds of miles, to in some cases over a thousand miles. Superior sailing skills is one of the attributes of the Polynesians that allowed them to traverse such great distances and unfavorable conditions to settle the various islands groups that make up Polynesia.

The Polynesia sailing craft was an early catamaran consisting of a double canoe made of two hulls connected by lashed crossbeams. This served two purposes: the first was that the design provided stability and the second was that it increased the capacity to carry large loads. The hulls were large enough to carry all their supplies and equipment. The central platform created over the crossbeams provided working, living, and additional storage space for the seafarers. Sails made of woven matting powered the ships, which were by long paddles enabling them to keep on course.

The canoes were generally 49 to 65 ft (15 to 20 m) in length and could carry 20 people with all of their supplies needed to colonize newly discovered islands. Among the supplies carried on these crafts were domesticated animals (pig, dog, and chicken) and planting materials to begin new cultivations at their homes: sweet potato, taro, bananas, yams, breadfruit, and sugarcane. They developed a portable agricultural system to compensate for the decreased flora and fauna they found on the new islands. The Polynesian navigation system was based on star observations, ocean swells and currents, and the flight patterns of birds along with other natural signs to find their way across the great ocean open distances they traversed.

Polynesia today is a mix of political and economic systems reflecting the large area that it covers. Most of the island groups have their own sovereignty, while others are either under the administration of or have special relations with foreign countries: French Polynesia with France, and Cook Islands, NIUE, and Tokelau with New Zealand. The Hawaiian Islands became the 50th state admitted to the UNITED STATES in 1959. The economies of the various islands are also varied. Tourism plays a very large part in the economies of all the island groups in Polynesia. But because of their small geographic size, natural resource exploitation is generally limited to specialized agriculture and fisheries. A large number of Polynesians make their home in Auckland, New Zealand, which is the largest Polynesian city in the world.

BIBLIOGRAPHY. "Polynesia," Public Broadcasting Service, www.pbs.org/wayfinders (April 2004); "Polynesia," members.optusnet.com.au (April 2004); *Oxford Essential Geographical Dictionary* (Oxford University Press, 2003).

TIMOTHY M. VOWLES, PH.D.
VICTORIA UNIVERSITY, NEW ZEALAND

Portugal

Map Page 1131 **Area** 35,672 square mi (92,391 square km) **Population** 10,102,022 **Capital** Lisbon **Highest Point** Pico 7,713 ft (2,351 m) **Lowest Point** 0 m **GDP per capita** $15,000 **Primary Natural Resources** forests, cork, marble, fisheries.

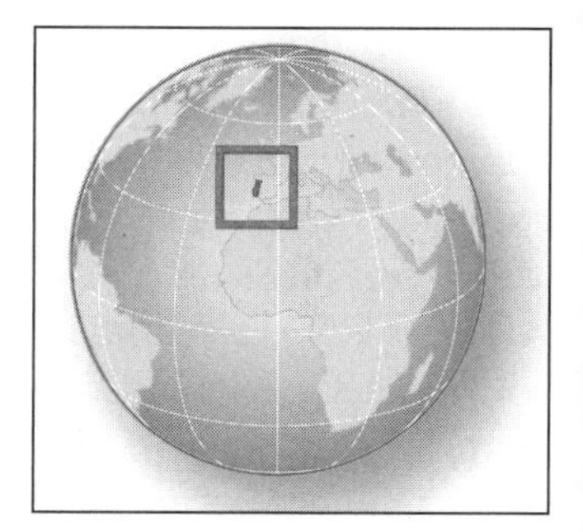

THE PORTUGUESE REPUBLIC is located in southwestern Europe, sharing the Iberian Peninsula with SPAIN, which it borders to the north and east. To the west and south it borders the ATLANTIC OCEAN. Portugal is a parliamentary democracy. The prime minister leads the government and the president is the chief of state. The capital and largest city is LISBON (Lisboa).

The country is organized in 18 districts in the mainland and two Autonomous Regions, the archipelagos of Madeira and AZORES (Açores), located in the North Atlantic Ocean. The state is highly centralized, with only the autonomous regions enjoying a high degree of self-governance. Recently, a more geographic division comprising 5 major regions and 28 minor ones was designed and utilized for statistical purposes.

The area that became the country of Portugal has had a varied human presence since prehistoric times, in part because of its position as the crossroads between the MEDITERRANEAN SEA and northern Europe and the fact that several natural harbors existed along the coast. Settlers included Phoenicians, Carthaginians, Lusitanians, Romans, Suevi, and Visigoths, among others. Roman and Muslim influences endured, especially in the south of Portugal. The country's name seems to have originated from the Roman designation of the area around Porto as Portus Cale (Port of Cale). Mus-

lims from North Africa invaded the peninsula in 711 C.E., sparking a crusade effort by Christians that originated several kingdoms and feudal holdings in the area.

In 1143, Afonso Henriques (Afonso I) obtained independence from the kingdom of Castile and León and became the first king of Portugal, then confined to the area between the Douro and Minho rivers. He proceeded with the reconquest from the Moors southward, seizing Lisbon in 1147, and the coast of Algarve was reached in 1250. The Portuguese borders have remained almost unchanged since 1297.

Cornered in one extreme of Europe and motivated by the will to find a sea route to coveted Asian goods while spreading Christianity, Portugal turned to the sea from the beginning of the 15th century. It first occupied the stronghold of Ceuta (1415), then discovered Madeira (1419), and, supported by the vision of Prince Henry the Navigator, launched the Age of Discoveries that would allow the European expansion into new worlds and shift commerce to the Atlantic.

Portuguese mariners sailed south along the African coast on the way to INDIA and in 1500 claimed the lands that became BRAZIL. The country obtained great economic, political, and cultural influence during the 15th and 16th centuries, culminating its expansionist period with colonies in Africa, Asia, and South America.

In the 1600s and 1700s, Portugal faced increasing competition from other European nations, and lost much of its power with the Napoleonic wars and the independence of Brazil (1822).

After deposing the monarchy in 1910, the country experienced an unstable and short-lived republicanism. In 1926, a military coup initiated a long period of repressive fascist dictatorship, during which the country stagnated and grew more isolated from the world, leading many Portuguese to emigrate. Portugal lost the colony of Goa to India in 1961, and a war for independence erupted in the African province of ANGOLA, soon spreading to the colonies of MOZAMBIQUE and GUINEA-BISSAU.

In 1974, a peaceful military-led revolution reinstated democracy, and the following year Portugal granted independence to its five African possessions, triggering the return of many Portuguese settlers and an influx of immigrants. In 1999, the country's imperial age would finally come to a close, when Macao, the last colonial holding, was returned to China. As if to seal its return to the European boundaries, Portugal became a member of the EUROPEAN UNION in 1986 and was in the first group of countries adopting the euro as official currency in 2002.

LAND AND RESOURCES

Despite its modest dimensions (roughly 342 by 124 mi or 550 by 200 km), mainland Portugal displays surprising physical diversity. The territory is situated on the edge of the Mediterranean basin, bordering the Atlantic Ocean, in the Subtropical Zone of the Northern Hemisphere. This transition area experiences both Atlantic and Mediterranean influences with expression in the climate and in the vegetation. Differences result mainly from the combination of latitude, elevation, and distance to the large Atlantic Ocean, as demonstrated by the complex plant cover. The main limiting factor of the natural vegetation is the irregular precipitation on account of latitude, and botanical domains have organized mostly from north to south.

The climate is maritime temperate (cool, rainy) in the northwest and markedly Mediterranean (warmer, drier) in the south. The northeast suffers similar hot and dry summers but has colder winters. On an average year, Porto receives 50 in (128 cm) of rain, and Lisbon 31 in (78 cm), with some areas in the southeast receiving less than 12 in (30 cm). A few mountainous areas get limited amounts of snow. The main climatic features of Portugal are summer drought and the irregular precipitation, with sequences of dry and wet years alternating frequently. Along the west seaboard, a steady wind blows from the northwest on typical summer afternoons, in response to a pronounced coastal upwelling effect. This process keeps the sea water cool year-round.

In Portugal, the natural conditions also combined with a long human presence to produce diverse landscapes, where native and many introduced plant species coexist. The former provincial division has some geographical meaning and is still used to designate the different regions.

Mainland Portugal sits mostly to the west of the large platform that dominates the center of the Peninsula, called Meseta. The area suffered major tectonic activity during the Miocene era: It was more intense in the north, where the main mountain ranges are found. This activity also originated the Central Range, rising from the southwest to the northeast and culminating in the Serra da Estrela, with the highest elevation of peninsular Portugal (6,539 ft or 1,993 m). The rivers flowing to the west from the interior of the peninsula have cut deep valleys through the western mountains on their way to the ocean. Main rivers, including the

Tagus (Tejo), Douro, Minho, and Guadiana, have their sources in Spain and are heavily dammed, generating some dispute for the water between the two countries.

To the south of this core of older rocks, the basins of the rivers Tagus and Sado filled with sediments during the tertiary Era. Two platforms, starting to form 200 million years ago, spread to the west of the Central Range and along the southern coast of Algarve.

The Portuguese coastline stretches for about 517 mi (832 km). The coast is generally straight and sandy north of Peniche, interrupted by a few rocky capes. In some places it is indented by river estuaries and shallow saltwater lagoons. The broad estuaries of the Tagus and Sado provide good natural harbors, utilized as such since ancient times. The southwest has rocky shores interrupted by sandy beaches.

The Central Range is usually considered the main division between the mountainous north and the south, characterized by rolling plains and a few scattered small ranges.

In the northwest region of Minho, the ranges of granite mountains facing the ocean receive more than 79 in (200 cm) yearly and are the source of the rivers Lima, Cávado, and Ave. The flat coastal areas and river valleys are intensively cultivated, typically with corn fields or pastures surrounded by vines, and dairy farms are common. This green region has the highest population density of Portugal, with numerous villages and houses dispersed along the roads.

RAIN SHADOW

East of the mountains and experiencing a rain shadow effect, the drier northeast is basically a series of plateaus interrupted by valleys. Some of the broader ones are rich agricultural areas, as in Chaves and Vilariça. Isolated villages are surrounded by orchards, vineyards, and grain fields.

In part due to the pronounced summer drought, this region of Trás-os-Montes displays some similarities to the south of the country, with cork oak trees dotting the wide landscapes. Marking the oriental border with Spain, the Douro River runs for 62 mi (100 km) inside a rocky canyon before entering the deep, large valley, where port wine has been made since the 1700s. For thousands of acres on both sides of the river, the slopes were terraced by human hand to sustain the vineyards, in an endeavor of daunting proportions.

South of the Douro, the central region of Portugal is mostly mountainous, with forests covering the lower slopes and sheep and goats grazing at higher elevations. These mountains are the source of important water courses, such as the Zêzere and the Mondego, the latter is the longest river whose source is in Portugal.

People have initially settled along these river valleys, and some towns in the interior regions of Beira Baixa and Beira Alta are still walled or protected by castles, attesting the wars once opposing Iberian kingdoms. In the Beira Litoral, closer to the ocean, crops occupy the flat lower river valleys, and forests of pine trees now cover ancient coastal dunes.

ESTREMADURA

In the hilly region of Estremadura, north of Lisbon, the varied agriculture includes grain, orchards, vineyards, and vegetables to supply the metro area. To the west of the capital city, the humid Serra de Sintra stands out as an island of lush vegetation protected by old palaces and fortresses. South of the 621-mi- (1,000-km-) long Tagus River spreads the savannalike landscape (called *montado*) of the Alentejo, where wheat fields or cattle pastures are punctuated by cork oak trees. The declining population concentrates in settlements of low white houses. Part of the wild coast, with small beaches between rocky cliffs, is protected as a natural park.

The wooded mountains of the Serra de Monchique and Serra do Caldeirão separate the rolling plains from the southernmost region of Algarve. Here, the land slopes gently toward the calm ocean, with orchards and greenhouses for growing vegetables occupying the lower reaches. Along the coast, fishing gave way to tourism as the main activity, with villages becoming summer resorts and golf courses replacing cropland.

AZORES AND MADEIRA

The archipelagos of the Azores and Madeira have a volcanic origin and were both discovered and settled in the 1400s. The nine islands of the Azores occupy 896 square mi (2,322 sq km) and have 241,763 inhabitants (2001). The archipelago straddles the MID-ATLANTIC RIDGE and displays secondary forms of volcanism, such as geysers and fumaroles. The rainy climate and fertile soil allows for agricultural use and presence of lush vegetation cover.

Located 560 mi (900 km) southwest of Lisbon, off the coast of North Africa, the autonomous region of Madeira occupies 303 square mi (785 sq km) and has a population totaling 245,000 (2001). The archipelago includes the namesake island, as well as Porto Santo and two uninhabited group of islets, the Desertas and the Selvagens. The landscape of Madeira, the main is-

land, is heavily eroded, with the mountainous interior revealing deep valleys amid jagged peaks. Fishing villages dot the rugged and rocky coast, almost devoid of sand. A rare and primitive type of forest dating from pre-glacial times occupies the intermediate elevations, giving way to pastures above that. A complex system of canals, called levadas, allows the watering of cereals and vegetables grown at lower elevations and on terraced slopes.

The climate is mild year-round with a mediterranean tone, enabling the presence of semi-tropical vegetation and production of sugarcane and bananas for export. Wine is also a hallmark product of the island, originating from the vineyards that cover the southern slopes. The area around Funchal, the main city, concentrates 70 percent of the population and the chief economic activities. The city is an important port of call for transatlantic cruises, and tourism is a major source of revenue. However, a long history of emigration still causes the population of Madeira to decrease.

Portugal has scarce natural resources, and is heavily dependent on imported fossil fuels. Hydropower accounts for 30 percent of the energetic needs, with the rest coming from coal-based power plants. Use of other renewable sources is slowly increasing. Mining included coal, gold, uranium, wolframite, and copper, but activity has been declining. The south has large reserves of copper and zinc still in exploitation, as well as marble quarries.

AGRICULTURE

The combination of poor soil and Mediterranean climate results unfavorable for intensive agriculture in Portugal. The soils tend to be rocky or sandy, with the exception of fertile alluvial plains, notably along the lower Tagus. Olives, vineyards, and orchards remain important permanent crops, despite the recent increase in irrigated agriculture. Agriculture, forestry and fishing employ 13 percent of the workforce and create 4% of the GDP. Main agricultural products include wine, tomatoes, corn, potatoes, wheat, olives and olive oil, fruits, grapes, beef, and dairy products. Animal production also includes hogs, sheep, and goats. Farms are generally small, but diverse types of land ownership have contributed to differences in farm structure.

Forests cover about one-third of Portugal, but wildfires are a recurrent hazard in the summer, due to drought caused by the mediterranean climate: 1,047,700 acres (424,000 hectares) burned in 2003, the worst year on record. Oak and chestnut trees characterize the north, while pine occupies in the north and center. Introduced for wood pulp production, the eucalyptus has rapidly expanded in the last 20 years. In the south, the cork oak has great economic importance, with Portugal being the leading world producer of this material. Natural shrubs also cover large areas, with mediterranean-type maquis dominating in the drier south.

Portugal has the largest EEZ (exclusive economic zone) of the European Union (617,775 square mi or 1,600,000 square km), but the sea is under-exploited. Commercial fishing is in decline, yet the Portuguese remain important consumers of seafood. Aveiro, Lisbon, Leixões (Porto), and Peniche are major fishing ports.

BIBLIOGRAPHY. Dan Stanislawski, *The Individuality of Portugal: A Study in Historical-Political Geography* (Greenwood, 1969); Mary Vincent, R. A. Stradling, *Cultural Atlas of Spain and Portugal* (Facts On File, 1995); Orlando Ribeiro and Hermann Lautensach, *Geografia de Portugal* (Edições João Sá da Costa, 1987); Orlando Ribeiro, *Portugal, o Mediterrâneo e o Atlântico* (Edições João Sá da Costa, 1967); Suzanne Daveau, *Portugal Geográfico* (Edições João Sá da Costa, 1995); Carlos Medeiros, *Geografia de Portugal* (Editorial Estampa, 1994).

SÉRGIO FREIRE
PORTUGUESE GEOGRAPHIC INSTITUTE, PORTUGAL

prairie

BEFORE NORTH AMERICA was populated by settlers who developed the land, most of the continent was covered by flat grass-covered areas known as prairie, stretching from CANADA to MEXICO. By the beginning of the 21st century, 99 percent of the original prairie had been destroyed. The provinces of Alberta, Saskatchewan, Manitoba, and Ontario are part of Canada's prairie. In the UNITED STATES, prairies are found in NORTH and SOUTH DAKOTA, MINNESOTA, NEBRASKA, IOWA, ILLINOIS, INDIANA, WISCONSIN, KANSAS, MISSOURI, OKLAHOMA, and TEXAS. In South America, prairie land is known as PAMPAS. In Africa, it is called veldt; and in Asia, it is known as STEPPE.

Some 60 different species of grass once could be found among North American prairies. Growth potential seemed almost unlimited because prairie grasses were able to adapt to weather that varied from extreme hot to extreme cold. Prairie grass extended rapidly since it was pollinated by the wind.

Prairie grasses are classified as tall-grass, mixed-grass, short-grass, bunch-grass, mesquite, and tree. The kinds of prairie grass found in an area are dependent on a number of variables, including weather, longitude, latitude, extent of usage, soils, and landforms. Typically, grasses become shorter as the prairie progresses from east to west. Prairie grasses range from eight-foot-tall grasses, such as bluestem, switch grass, and slough grass, to one-half-foot short grasses, such as blue grama, hairy grama, and buffalo grass. Of the six types of prairie grass, tall-grass and mixed-grass regions are the most productive agriculturally. America's wheat belt, which grows abundant amounts of corn, wheat, barley, oats, and rye, is home to such prairie grasses.

The area that extends from southern Manitoba to Missouri to Wisconsin to Illinois to Indiana to Oklahoma and Texas is known as True Prairie. While prairies tend to be relatively flat, knolls, steep bluffs, hills, valleys, and alluvial floodplains exist in this area.

Rainfall varies from one prairie section to another. Southern prairies tend to be dry and hot, with numerous shrubs. In the north, prairies are cool and humid. Some 70 percent of all precipitation in the western prairie falls during the growing season.

In the area that extends from Oklahoma to Illinois, for instance, the prairie receives around 40 in (101 cm) of rain each year. In the northeastern section that reaches from Minnesota to Manitoba, some 23 in (58 cm) fall in a given year. Oklahoma prairies may experience as much as 30 in (76 cm) a year, while Nebraska and North Dakota receive 25 in (63 cm) and 20 in (51 cm) per year, respectively. Droughts may occur in the prairies in the summer and fall, making the area vulnerable to fire. Prairie soil and chernozem soil dominate the plains. Both of these dark soils tend to be rich in humus and conducive to plant production.

One prairie area designated as the High Plains stretches from Alberta, Saskatchewan, in Canada to Montana and North Dakota in the United States. Generally flat in the eastern section, the High Plains rise around 10 feet per mile as they reach the foothills of the ROCKY MOUNTAINS. Since much of this land remains uninhabited, pronghorn antelope and mule deer are still abundant. With average rainfall of less than 20 in (50 cm) per year, the short-grass, which is mostly blue and grama, in the High Plains is too dry to provide irrigation for crops or grazing for livestock. The area's river valleys, coulees, and ravines are too steep for farming and are too bare for logging. Taller and thicker grass populates the eastern section of the High Plains, and forbs, which are broad-leaved, flowering shrubs, are plentiful. Riverbeds and wetlands, with numerous lakes, marshes, sloughs, and potholes provide homes for most of the wildlife remaining in the area.

Originally, large animals found in the prairie included American bison, pronghorn antelope, mule deer, white-tailed deer, elk, lynx, grizzly bear, bobcat, fox, cougar, prairie wolf, and coyote. Jackrabbit, cottontail, mice, weasel, badger, prairie dog, shrew, vole, ground squirrel, gopher, skunk, and raccoon were among the smaller animals found in the area. Bird life on the prairies was abundant, including bobolink, pipits, meadowlark, sparrow, grouse, prairie chicken, hawk, eagle, falcon, and vultures. Pelican, crane, heron, sandpiper, swan, geese, and duck were plentiful in prairie marshes. Only small numbers of the original wildlife remain in the prairie. Buffalo, for instance, are now found only on protected reserves.

Despite their sturdy nature, prairies were vulnerable to land development because most immigrants possessed little knowledge of grasslands. Native Americans also frequently destroyed prairie lands to facilitate travel and provide greater security. Fire has also historically presented a major threat to prairies. In the 1930s, prairie recovery efforts emerged, leading to an increased interest among environmentalists, ecologists, and others interested in preserving the integrity of native prairie grasses.

BIBLIOGRAPHY. Tim Fitzharris, *The Wild Prairie: A Natural History of the Western Plains* (Oxford University Press, 1983); Paul A. Johnsgard, *Prairie Birds: Fragile Splendor in the Great Plains* (University of Kansas Press, 2001); John Madison, *Where The Sky Began: Land of the Tallgrass Prairie* (University of Iowa Press, 1995); Robert F. Sayre, *Recovering The Prairie* (University of Wisconsin Press, 1999); Robert Wardhaugh, *Toward Defining the Prairies* (University of Manitoba Press, 2001); J.E. Weaver, *North American Prairie* (Johnson Publishing, 1954); S. Winckler, *Prairie: A North American Guide* (University of Iowa Press, 2004).

Elizabeth Purdy, Ph.D.
Independent Scholar

precipitation

PRECIPITATION IS WATER falling from clouds. In the tropics, rain is the common form. In colder climates, precipitation also falls as snow, sleet, or hail, if

the air temperature is 32 degrees F (0 degrees C) or less. Water's descent from clouds begins when the size of water droplets or ice crystals becomes large enough for gravity to pull them toward Earth's surface. Precipitation occurs because humid air rises, expands, and cools to form clouds with sufficient moisture to cause a storm. Precipitation will not occur unless there is ample water vapor in surface air, enough air rising from the surface, and sufficient condensation nuclei.

Water vapor is the source of precipitation and bearer of energy that drives storms. The bulk of this vital gas enters the atmosphere when solar energy heats and evaporates surface water and when plants transpire water through leaves. The change of water from a liquid to a gas requires energy that resulting water vapor holds on to, storing it in air as latent heat of vaporization. Latent energy is nonsensible, meaning neither human skin nor a thermometer can detect it. However, this seemingly trivial energy converts to sensible heat when condensation takes place in clouds. As we shall see, this heat (heat of condensation) is the energy that gives birth to storms of all types, including hurricanes and tornadoes.

Condensation occurs because moist air rises, expands, and cools, diminishing the air's capacity to hold water vapor. Condensation starts when air temperature falls below the dew point (the temperature at which air is unable to hold all its water in vapor form). Water vapor collects (condenses) around tiny solid particles (condensation nuclei) to form visible water droplets. The tiny droplets are the beginnings of a cloud.

LIBERATED HEAT

A cloud grows as condensation liberates water vapor's latent energy as heat of condensation. The liberated heat keeps air inside the cloud warmer (and therefore less dense and lighter) than the surrounding air. The warm temperature buoys airlifting inside the cloud and continues condensation and heating by drawing more moisture-laden surface air into the cloud. The circular process of lifting, expansion, cooling, condensation, heat release, and more lifting promotes precipitation. Precipitation stops when moist surface air stops lifting into the cloud.

The principal cause of precipitation is upward movement of moist air resulting from convective, frontal, orographic, and cyclonic (convergent) lifting. Each type of lifting produces a characteristic type of precipitation.

Convective lifting starts when air over a hot surface warms, becomes lighter than the surrounding air, and rises. Cooler surface air then moves in over the hot surface, warms and lifts as well. Airlifting continues as long as surface heating warms air to a temperature that is higher than that of the surrounding air. Convectional precipitation occurs mainly in equatorial and tropical regions and in summer in middle latitudes. The hot summer afternoon thunderstorm, which generates heavy precipitation, thunder, and lightning, is precipitation of this type.

Frontal lifting occurs along boundaries between relatively cold and warm air masses. Meteorologists call these boundaries fronts. Lifting along fronts occurs mainly in the middle latitudes (about 35 degrees to 55 degrees) where polar and subtropical air masses typically meet. The physical basis of frontal lifting is that the cold polar air is denser and of a higher pressure than warm subtropical air, so that the cold air tends to hug Earth's surface and the warm air tends to lift relatively easy. Frontal lifting causes precipitation in two ways: (1) a wedge of cold surface air moves against adjoining warm (lightweight) air to form a cold front; and (2) a warm surface wind overtakes and ascends above the edge of a cold air mass to develop a warm front.

Orographic lifting takes place when a mountain barrier forces a warm, moist wind to rise. Precipitation takes place over the windward (upwind) side of a mountain, as the barrier forces the flowing air to rise, expand, and cool there. Precipitation ceases when the air descends the leeward side, as cooling by expansion ceases. Orographic lifting causes rain shadows (areas of scant precipitation) on leeward (downwind) sides of mountains if moisture-bearing winds come from only one direction during the year. The rain shadow effect is more obvious on western sides of continents in middle latitudes, where mountains block moisture-laden westerlies from oceans from entering leeward areas.

Cyclonic or convergent lifting describes airlifting in two types of storms—extratropical and tropical cyclones. The lifting results from widespread convergence of surface winds toward a low pressure. Moist winds converge in a circular pattern into the center of the low pressure where air is rising in a signature cyclonic spiral. In the Northern Hemisphere, the wind direction is counterclockwise and in the Southern Hemisphere, it is clockwise.

In an extratropical cyclone, the lifting trigger is a small meander in the polar front JET STREAM, a high-speed wind that flows above the front. The jet stream's meander forces upper air to descend and surface air to rise in close proximity. The two resulting surface

fronts—cold front and warm front, respectively—join to form the low-pressure center. The storm grows as cold and warm surface winds converge from all directions toward the center. Meteorologists call this type of storm an extratropical cyclone (or middle latitude cyclone) to distinguish it from a low latitude tropical cyclone, which involves convective (not frontal) lifting.

The tropical cyclone forms over hot tropical oceans as prodigious amounts of heat and moisture transfer into the air. Precipitation begins as a disorganized cluster of relatively small convective thunderstorms. A low surface pressure center develops to centralize the inflow of energy into a rainy tropical depression. Moist air of converging winds lifts, cools, creates dark, moisture-laden clouds, and feeds a massive tropical storm. If wind speeds exceed 73 miles per hour (116 km per hour), the storm is officially a hurricane.

GEOGRAPHICAL DISTRIBUTION

The general occurrence of precipitation in the tropics conforms to latitude. The equatorial zone (roughly latitudes 0 degrees to 10 degrees) receives more precipitation than any other latitudinal zone on Earth; surface heating and convection brought on by the sun's direct rays is the lifting mechanism. The Intertropical Convergence Zone (ITCZ)—an area of convective lifting into which moist trade winds converge—dominates equatorial precipitation. The ITCZ shifts back and forth across the equator as it follows the seasonal path of the sun's most direct rays. The shifts bring precipitation to the equatorial zone all year.

On poleward edges of the ITCZ (about latitudes 10 degrees to 15 degrees), annual precipitation levels drop off rapidly, as convectional showers occur only in the summer. A dry winter is due to the invasion of high-pressure air masses, whose source areas are the subtropics (about latitudes 20 degrees to 35 degrees). The subtropical highs result from descending air, which limits cloud formation. Consequently, the world's greatest deserts are located in subtropical latitudes. Major exceptions to dryness in the subtropics are in Southeast Asia and the Himalayas. In these areas, summer (convective) monsoon showers in combination with orographic lifting bring extremely high levels of precipitation to about 20 degrees N and 30 degrees N, respectively.

The middle latitudes (from about 30 degrees to 60 degrees) have a more complex spatial pattern of precipitation than that of the tropics. Generally, coastal areas are more humid than interior areas, because of onshore flow of marine air. A major exception is severe summer dryness on west coasts between latitudes 30 degrees and 40 degrees. Subtropical high-pressure curbs airlifting there in the summer. In contrast, high amounts of precipitation on west coasts between latitudes 40 degrees to 60 degrees occur because of lifting by of the westerlies by frontal, cyclonic and orographic means. Mountain barriers contribute to rain shadow areas (giving rise to DESERTS) in interiors of North America and Asia. In Europe, open terrain allows westward moving extratropical cyclones greater access to the continental interior. The eastern section of continents in the middle latitudes tends to be humid, as precipitation comes from weather fronts and extratropical cyclones in fall, winter, and spring, and convective showers and weak cold fronts in summer. Tropical cyclones or their remnants add to precipitation totals in the eastern section in the summer and fall.

Cold, dry air masses dominate polar and arctic latitudes (from about 60 degrees to 90 degrees) all year. Precipitation amounts are comparable to those in subtropical deserts. Airlifting is ineffective in generating much precipitation because the region's low temperatures suppress evaporation levels. Cold air subsidence also lessens chances of precipitation occurring. Lower latitudes of this zone receive most of the precipitation, as occluded parts of extratropical cyclones brush the equatorward fringe the year round.

BIBLIOGRAPHY. F.H. Ludlam, *Clouds and Storms* (Pennsylvania State University Press, 1980); Paul E. Lehr and R. Will Barnett, *Weather: Air Masses, Clouds, Rainfall, Storms, Weather Maps, Climate* (Golden Books Adult Publishing, 1987); T.N. Carlson, *Mid-Latitude Weather Systems* (Routledge, Chapman & Hall, 1991); Fredrick K. Lutgens, Edward J. Tarbuck, and Dennis Tasa, *The Atmosphere: An Introduction to Meteorology* (Prentice Hall, 2003); Arthur N. Strahler and Arthur H. Strahler, *Physical Geography: Science and Systems of the Human Environment* (Wiley, 2005).

Richard A. Crooker
Kutztown University

projection, maps

THE EARTH CAN BE assumed as a perfect sphere for purposes of simplicity. However, in reality there is a large difference between the pole-to-pole distance and the equator distance. The Earth is about 1/300th smaller in its pole to pole distance. This unique shape

is called an oblate ellipsoid or spheroid. An estimate of the ellipsoid allows calculation of every point on the Earth, including sea level, and is often called the datum. With time and in different countries many datum have been established.

The current U.S. datum most commonly used is the World Geodetic System 1984 (WGS 84). Map projections are used to transfer the spherical location information on a datum or ellipsoid onto a flat surface. The transformation process always results in distortions in shape, distance, direction, scale, and area. Projections minimize distortions in some of these map properties but at the expense of increasing errors in other properties. Conformal maps preserve shape, as the scale of the map from any point is the same in any direction. Scale is defined as the relation between distances on the map and the same distance on the Earth's surface. Direction is preserved when azimuths (angles from a point on a line to another point) are correct.

In the spherical coordinate system, the Earth is considered to be composed of longitude lines (meridians), which run north-south, and latitude lines (parallels), which run east and west. The longitude lines converge at the North Pole in the Northern Hemisphere and at the South Pole in the Southern Hemisphere. All meridians and the equator are great circles since they can form planes that cut the surface and pass through the center of the Earth. Small circles, such as latitude lines, form a plane that cuts the surface but does not pass through the centre of the Earth.

In this system of reference, geographic coordinates are measured in units of angular degrees. There are 360 degrees of longitude around the equator, with each meridian numbered from 0 to 180 degrees east and west such that the 180 degree meridian is on the opposite side of the Earth from Greenwich. There are 180 degrees of latitude from pole to pole, with the equator being 0 degrees and the north and south poles being 90 degrees.

Each degree can be divided into 60 minutes and each minute is divided into 60 seconds. The north-south line is called the prime meridian, which has been arbitrarily set to pass through the Royal Observatory in Greenwich, England. The longitude is measured as the angle between the point, the center of the Earth, and the prime meridian at the same latitude. West is positive and east is negative, meeting at 180 degrees at the international date line. The east-west line follows the equator and is midway between the north and south poles. Degrees of latitude are measured as the angle between the point on the surface, the center of the Earth, and a point on the equator at the same longitude.

Other coordinate systems are based on a flat surface. The easiest way to try to transfer the information to a flat surface is to convert the curved Earth locations into an X and Y coordinate system, where X corresponds to longitude and Y to latitude. In order to clarify understanding, the transformation process can be considered as a two-stage process. In the first stage, assume that the Earth has been mapped on a reduced size globe that will produce a flat map of a desired size when unfolded.

In the second stage, each point on the surface of the globe is transferred onto a flat surface. The transformation is usually in the form of mathematical functions and the points are "projected" onto one of three flat surfaces, which can then be unfolded into a chart. The three surfaces are a cylinder, cone, and plane. The lines formed by the latitude and longitude on a map are called the graticule. Latitude or longitude lines where the projection surface and location information about the Earth intersect is called the point of contact or standard line. Distortion is minimized nearest to the standard line. A graphical representation called the Tissot Indicatrix is commonly used to visualize the angular and area distortions that occur across the flat map surface because of the transformation process.

A cylindrical projection places the Earth inside a cylinder with the equator touching the inside of the cylinder. Cylindrical projections are mostly used for areas near the equator. If the axis of the cylinder is placed perpendicular to the axis of the Earth, the resulting projection is called a transverse projection. A cone is placed over the Earth to produce a conic projection. A conic projection is best for middle latitudes. An azimuthal projection places a plane over a point on the Earth. The azimuthal projection is used for polar charts because of distortions at other latitudes.

Map projection affect the distance, area and shape, and angle relationships of the Earth features as displayed on the flat map. Equal area projections preserve the property of area. Maps with an equivalent projection show all parts of the Earth's surface with the correct area. However distances along the latitudes are not accurate. Conformal projections preserve the property of shape over small areas. Conformal projections are valuable since they preserve directions around any given point. Moreover, angles relationships within the graticule are shown correctly. Area and shape are normally distorted. Some projections only preserve correct distance relationships along a few places on the map.

There may also be compromise map projections. These may be the average of two or more projections in order to minimize distortion. Some examples of the projections mentioned above are Mercator (conformal projection, preserves shape), Lambert Equal Area (equidistant projection, preserves areas), and Robinson (compromise projection, limits area and angle distortion).

When the three-dimensional Earth is projected onto a flat paper map, there are distortions in distance, area, and direction, and the projection process attempts to minimize these distortions when transferring the curved Earth to the flat paper map. On the flat map there is no natural reference point, but one can be defined using an arbitrary system of coordinates. A grid of intersecting parallel and perpendicular lines is placed over the projected map such that the origin of the grid lines falls on a point of interest on the map. This arrangement is called a grid reference system and every point on the flat map can be located with a unique X and Y coordinate. Usually the X coordinates are referred to as "eastings" and the Y coordinate as "northings."

An example of a grid coordinate system is the Universal Transverse Mercator (UTM) system. The UTM coordinate system is commonly used in geographic information systems because the United States Geologic Survey (USGS) topographic maps use the UTM system. The UTM projection is formed by using a transverse cylindrical projection (the standard line is a line f longitude). The result is to minimize distortion in a narrow strip running pole to pole. UTM divides the Earth into pole-to-pole zones 6 degrees of longitude wide. The first zone starts at the international date line (180 degrees east) and the last zone, 60, starts at 174 degrees east. Northings are determined separately for the areas north and south of the equator. The Southern Hemisphere uses the South Pole, while the Northern Hemisphere uses the equator. Distortion is extreme at high latitudes and so UTM is not normally used above 80 degrees north or south.

Maps provide useful and concise spatial information in a meaningful manner, and we use them directly or indirectly in many aspects of our daily lives. Navigating the roadways, finding new places, or just reflecting on the aesthetics of the map are some examples of how we use maps. Maps are also used as a means for storing data, as a spatial index for labeling features or integrating multiple map sheets, and as a spatial data analysis tool for planning and decision-making. A simple but powerful analysis tool is the map overlay process developed by Ian McHarg in which multiple map layers each on transparent film are overlaid to identify regions of interest.

Understanding projections is especially significant as the concept holds a central position in the design and development of new and emerging spatial technologies such as geographic information systems, remote sensing, and global positioning systems. These technologies provide multiple options to covert among projections. Selecting the correct projection to use must begin with a clearly defined project goal. Thereafter a projection should be selected that has a standard line centered on the geographic region of interest. An assessment of the importance of correct representations of area, angle relations, and distances must also be made. Given these criteria, an appropriate projection for the mapping project can be selected.

BIBLIOGRAPHY. Borden Dent, *Cartography: Thematic Map Design* (WCB/McGraw-Hill, 1999); Daniel Dorling and David Fairbairn, *Mapping: Ways of Representing the World* (Longman, 1997); Allan MacEachren, *How Maps Work* (Guilford Press, 1995); Mark Monmonier, *How to Lie With Maps* (University of Chicago Press, 1996); Arthur Robinson, Joel Morrison, Phillip Muehrcke, Jon Kimerling, and Stephen Guptill, *Elements of Cartography* (Wiley, 1995); Edward Tufte, *The Visual Display of Quantitative Information* (Graphics Press, 1983).

Shivanand Balram
McGill University, Canada

Ptolemy

KLAUDIOS PTOLEMAIOS, known as Ptolemy in English, lived in Alexandria, EGYPT, in the second century C.E. He is considered one of the most important scientists of the later Classical era, with two major treatises in the areas of astronomy and geography. His geocentric model of the universe was the standard accepted view of the cosmos in Europe until the 16th century. He is also credited with the system of assigning systematic coordinates to geographic features on a map using lines of longitude and latitude.

Very little is known about the life of Ptolemy, except that he made astronomical observations in Alexandria between 127 and 141 C.E. Alexandria was the leading city for scientific scholarship in the 2nd century. Ptolemy's works were written in Greek, the

standard scientific language of the day, but were lost to Europeans after the fall of the Roman Empire, and rediscovered in Arabic translation only in the 12th century. For this reason, Ptolemy's most significant work is known by its Arabic name, the *Almagest*, a rough translation of the original Greek title, *Hè Megalè Syntaxis*, or *The Great Treatise*. The *Almagest* is composed of 13 books, compiling the astronomical knowledge of the ancient Greek and Babylonian worlds. The text includes observations of the stars and planetary orbits that are based largely on the work of Hipparchus from three centuries before. Ptolemy was innovative, however, in creating a mathematical system that explained the movement of the sun, moon and planets around the Earth.

This geocentric (Earth-centered) universe had been proposed since the days of Aristotle, but astronomers had been unable to show mathematically how it worked. The Ptolemaic system used three basic geometric constructions—the eccentric, the epicycle, and the equant—to explain the otherwise erratic movement of the celestial bodies. Epicycles, for example, were small circles that revolved around a larger circle, and made sense of the observed orbital peculiarities of the planets Mars and Jupiter, which seemed at times to slow down, stop, and even reverse direction in their course across the sky. This system was adapted by scholars in the Islamic world and added to for centuries before it was reintroduced in the West. It wasn't until the work of Copernicus and Brahe in the 16th century that the fundamental errors of the Ptolemaic system were exposed, and although geocentrism was rejected overwhelmingly in favor of heliocentrism (a sun-centered universe), nevertheless, the mathematics behind Ptolemy's work continued to be praised as a model of sophisticated classical scholarship.

Ptolemy's second major work is the *Geography*, which was also a compilation of the geographical knowledge of the known world of his time. The *Geography* also gave instructions on how to create maps and discussed mathematical projections, or how to plot a spherical object onto a flat plane. For his maps of the known world, Ptolemy assigned coordinates to all features based on lines of LATITUDE AND LONGITUDE. For latitude, he based his calculations on the length of the midsummer day as it increases from south to north. For longitude, he assigned the number zero for the point furthest west that he knew of, the CANARY ISLANDS, in the ATLANTIC OCEAN, then plotted his lines as far east as CHINA, covering 180 degrees. Ptolemy was aware that he knew only about a quarter of the globe, but his calculations for the globe's total circumference were off by a sixth. This mistake and mistakes made in the Almagest regarding the length of the solar year (he was off by 28 hours, a significant error for a mathematician of Ptolemy's stature) have led some recent scholars to question whether Ptolemy was actually the man responsible for the system that bears his name.

Other works demonstrated the breadth of Ptolemy's scholarship. The *Optics* analyzes reflection and refraction of light on flat and spherical mirrors. His innovation here was the first known mathematical relationship between the angles of rays of refracted light and incident light. *Harmonics* is one of the earliest books on music theory, discussing how different notes are produced by lengthening and shortening of a vibrating string. The *Tetrabiblios* (*Four Books*) was meant as an accompaniment to the *Almagest*, to explain how the movement of the planets affected a person's daily life. It was therefore one of the foundations of Western astrology, which, until recently was firmly intertwined with the science of astronomy.

BIBLIOGRAPHY. Owen Gingerich, *The Eye of Heaven: Ptolemy, Copernicus, Kepler* (American Institute of Physics, 1993); "Biography of Claudius Ptolemy," Institute and Museum of the History of Science, www.brunelleschi.imss.fi.it (August 2004); Albert Van Helden, "Ptolemaic System," www.es.rice.edu (August 2004); J.J. O'Connor and E.F. Robertson, "Claudius Ptolemy," School of Mathematics and Statistics, University of St. Andrews, www.gap.dcs.st-and.ac.uk (August 2004).

JONATHAN SPANGLER
SMITHSONIAN INSTITUTION

Puerto Rico

THE ESTADO LIBRE Asociado de Puerto Rico is a densely populated island in the eastern Greater Antilles in the CARIBBEAN SEA, divided into 78 counties. Even though it is a possession of the UNITED STATES, Puerto Rico is an independent commonwealth and has self-governance in all its local affairs. The rectangular-shaped, mostly mountainous volcanic island is approximately 100 mi (160 km) east to west and 35 mi (56 km), north to south and has a total land area of 3,515 square mi (9,103 square km). Four other satellite islands—Mona, Desecheo, Culebra, and Vieques—are also under its control. Puerto Rico's warm and humid

tropical weather varies little throughout the year and is often interrupted by hurricane season from June to November. San Juan, its largest and oldest city, is also its capital. Other large cities include Ponce, Mayaguez, and Caguas. The 2004 estimated population was 3,898,000.

Boriquen was the name given for Puerto Rico by its primary inhabitants, the Taino, at the time Christopher Columbus discovered the island in 1493. The Spanish, in search of gold and resources, eventually wiped out the natives by disease, war, and slavery. In 1511, Puerto Rico became the first Spanish colony in the New World and was attacked periodically by the British and Dutch. Many African slaves later arrived to work on the sugar plantations. The island remained in Spanish hands until after the Spanish American War, when it became part of the United States. Today, there is an ongoing debate between gaining complete independence, keeping its commonwealth status, or becoming the 51st state.

Puerto Rico has a very distinctive and mixed culture, influenced by its Taino, Spanish, African, and American heritages. Approximately 80 percent of the population is white, 8 percent African, and 12 percent mixed or other ethnicities. Both Spanish and English are spoken, and the majority of the population is Roman Catholic.

The island's varied topography can be divided into four main regions. Much of the interior of the island is covered by the Cordillera Central, the central mountain range, which contains the island's highest point, Cerro de Punta at 4,390 ft (1,338 m), and El Yunque rainforest, also known as Caribbean National Forest. On the north part of the island is an area of foothills called the Karst Country, made up of beautiful limestone rock formations and underground rivers. On the southwest corner of Puerto Rico is a region of coastal dry forest, made up of desert trees, cacti, bedrock, and many migratory birds, that is environmentally different from other moister parts of the island.

The fourth geographical zone is the narrow coastal plain around the perimeter of Puerto Rico, which includes beautiful sunny beaches, bioluminescent bays, and cities that help make Puerto Rico a popular tourist destination. There are 45 nonnavigable rivers in total in Puerto Rico which flow into the coastal plain, the longest being Río Grande de Loíza. Puerto Rico has no natural lakes and 12 artificial reservoirs. Common environmental concerns include population growth, urbanization, deforestation, water pollution, and soil erosion.

Puerto Rico's diverse economy is almost equally divided in terms of gross domestic product (GDP) by manufacturing, especially chemical and pharmaceutical production, and service industries, including tourism. Agriculture contributes 2 percent of the nation's GDP, and tourism 6 percent. The island's primary natural resources are clay, limestone, salt, sand, gravel, stone, copper, nickel, and petroleum.

BIBLIOGRAPHY. Barbara Balleto, *Insight Guide Puerto Rico* (Langenscheidt Publishers, 2003); *World Factbook* (CIA, 2004); Romel Hernandez, *Puerto Rico* (Mason Crest Publishers, 2004); Vivo Paquita, "Puerto Rico," *Worldmark Encyclopedia of the States*, Timothy L. Gall, ed. (Gale Research, 1995); Randall Peffer, *Puerto Rico* (Lonely Planet Publications, 2002).

Anthony Paul Mannion
Fort Hays State University

Puncak Jaya

PUNCAK JAYA, THE highest peak of the Sudirman Range of Papua, is also the highest mountain between the ANDES and the HIMALAYAS, rising to 16,503 ft (5030 m) on the tropical island of New Guinea. It was first named Mount Carstenz after the Dutch navigator Jan Carstenz, who first sighted the peak from the coast in 1623. When the Dutch handed western New Guinea to INDONESIA in 1960, the peak was renamed Puncak Jaya, or "Victory Peak." Although Puncak Jaya is located close to the equator, it is high and wet enough to support some small glaciers within approximately 2.8 square mi (7.2 square km) of snow-covered area, stretching from about 1,000 ft (305 m) below the peak. At this height, daytime temperatures vary from 75 to 45 degrees F (7 to 24 degrees C), with frequent rainfall in the afternoons and evenings accompanied by strong winds. The peak was first climbed in 1962 by the renowned Austrian mountaineer, Heinrich Harrer, who lived for a year among the local Dani tribesmen. Since the early 1960s, these tribesmen, who traditionally adorn penis guards (*koteka*), have continued to become a popular subject for photojournalism.

The area is surrounded by unexplored rainforests which makes travel difficult, while it provides effective hideouts for a rebel group called Organisasi Papua Merdeka (OPM), or the Papua Liberation Organization, which is waging a guerrilla campaign for inde-

pendence from Indonesia. OPM's struggle can be traced to the New York agreement signed in August 1962 by representatives of the NETHERLANDS, Indonesia, the United Nations, and Papua. Another agreement signed in Rome, Italy, in May 1963 provided that West Papua would gain independence after 25 years of administrative guidance from Indonesia. While the failure of Indonesia to abide by the terms of the agreements was viewed as imperialistic, Papuan resentment is aggravated by the mining concession given to a U.S. corporation, which operates a large copper mine 18 mi (30 km) to the south of Puncak Jaya. The operation is alleged to have caused serious environmental pollution problems at the expense of the local inhabitants.

Visitors who choose to explore Puncak Jaya can begin their trek uphill from the last airstrip at Illaga village, proceeding on a tough trail for about six days through a cloud-misty rainforest, hilly terrain, muddy trails, and highland swamp and crossing streams with rain showers typically pouring four to seven hours a day. The return journey takes four days before the visitor boards a charter plane that flies directly to Jayapura, the provincial capital of Papua. As one of the "seven summits" of the world, Puncak Jaya attracts intrepid Westerners every year. The unique culture of the isolated hill tribes of central New Guinea is an added attraction to any visit to the area.

BIBLIOGRAPHY. Michael R. Kelsey, *Climbers and Hikers Guide to the Worlds Mountains* (Kelsey, 1990); Greg Child, *Climbing: The Complete Reference* (Facts On File, 1995); G.S. Hope, "Mt. Jaya: The Area and First Exploration," G.S. Hope. J.A. Peterson, U. Radok, and I. Allison, eds., *The Equatorial Glaciers of New Guinea* (A.A. Balkema Press, 1976).

KADIR H. DIN
OHIO UNIVERSITY, ATHENS

Pyrenees

THE PYRENEES MOUNTAINS form the natural border between FRANCE and SPAIN, and have been a more effective barrier between the two nations than other ranges of similar height because of their characteristic ruggedness and lack of usable passes. Even more so than the ALPS, the Pyrenees were impassable to travelers for many centuries. Today, however, they are among the most attractive mountaineering and rock climbing locales in Europe. The chain is made up of two parallel ranges that run for roughly 266 mi (430 km) from the MEDITERRANEAN SEA west to the Bay of Biscay. The chain continues as the Cantabrian Mountains across northern Spain.

The mountains can be divided into three sections—west, central, and east—with the loftiest peaks concentrated in the central section. In total, there are more than 50 peaks over 10,000 ft (3,000 m). The crest of the ridge forms the border between France and Spain, with a few exceptions, such as the Val d'Aran, which is north of the range but within the borders of Spain. The tiny nation of ANDORRA is also sandwiched between France and Spain, at the site of one of the few historic passes across the mountains, reputedly the site of Charlemagne's crossing on one of his crusades against the Saracens in the 8th century. Another Pyrenean site associated with Charlemagne is the gap at Roncesvalles (or Roncevaux), the setting of one of the most famous *chansons de geste*, the Song of Roland, where the Franks' most noble knight, Roland, lost his life in valiant defense of Charlemagne's army in 778. Today, a monastic hospice still aids travelers across the pass, the best in the western Pyrenees, located northeast of Pamplona, Spain; the hospice has been a key stopping point in the pilgrimage route to Santiago de Compostela since the 12th century.

The Pyrenees (Pyrénées in French, Pirineos in Spanish, Pirineus in Catalan) are older than the Alps, but are made of much harder material—mostly granite, some overlain with limestone—which has been resistant to erosion. The mountains have therefore maintained their size and ruggedness across the millennia. A lack of sizeable glaciers and relative dryness has also kept these mountains from appearing too worn. There is more rain in the west than the east, because of the prevailing winds off of the ATLANTIC OCEAN. As a result, the western part of the range is more lush than the east, with its Mediterranean climate.

The Pyrenees are characterized by a lack of any significant lakes and few large rivers. Rainwater flows down the mountains in swift torrents, called "gaves," which frequently tumble great depths over another feature typical to the region, high semicircular rock cliffs at the upper ends of stream valleys, known as cirques. The Gave de Pau contains both the highest waterfall in the range 1,525 ft (462 m) and the most famous cirque formation, the Gavarnie. The Pyrenees are also characterized by numerous caves and underground rivers. Water from the mountains drains south into the Ebro and the Mediterranean and north into either the

Mediterranean via the Aude, the Têt, and the Tech or the Atlantic Ocean via the tributaries of the Adour and the Garonne. The watershed between these northern streams is the small spur of mountains, the Corbières, which also form a sort of climatic barrier between the Atlantic and Mediterranean zones.

The north and south sides of the Pyrenees also differ in their structure. The French side tends to be steeper, with a more united front and broad faces of limestone or granite. The Spanish side is more gradual, includes some minor side ranges, and is also noticeably hotter and drier. The highest peak in the range, Pico d'Aneto at 11,168 ft (3,404 m), lies entirely within Spain. It is the center of the Maladetta Massif, close to the second highest peak, Pico de Posets (11,073 ft or 3,375 m), a few miles to the east.

The third-highest peak, Monte Perdido (or Mont Perdut in Catalan; 11,007 ft or 3,355 m), is considered by many to be the most beautiful in the range, towering over the picturesque valley of the Ordesa, which today forms a Spanish national park. Pic Long (10,472 ft or 3,192 m) is the highest mountain entirely in France and was famously first climbed in 1846 by the Duke of Nemours, son of King Louis-Philippe. Also in France is the Pic du Midi d'Ossau (9,465 ft or 2,885 m), which is an isolated pyramid with two summits and dramatic rock walls, considered to be one of the finest climbing spots in Europe, with more ascent routes than any other peak.

The Pyrenees are not rich in natural resources. There are some iron mines in the Ariège region of France, and some coal and lignite beds on both sides. More significant however, are the hot and mineral springs that are found throughout the range. Bagnières de Luchon and Eaux-Chaudes are both famous sulphurous hot springs. Another attraction to the Pyrenees are the religious shrines, most notably Lourdes. The mountains' flora and fauna are somewhat unique because of their relative geographical isolation. Among the range's unique species are the ibex and water-mole, blind cave insects, and several varieties of mountain rhododendron.

There are few large settlements anywhere in the Pyrenees, with the exceptions of the eastern and western extremities of the range, where the formations are low enough to allow railroads and highways and the development of larger cities, such as Perpignan in France, and San Sebastián in Spain, a major center of Basque culture.

BIBLIOGRAPHY. *Encyclopedia Americana* (Grolier, 1997); "World Mountain Encyclopedia," www.peakware.com (August 2004); "Pyrenees," www.pyrenees-online.fr (August 2004); "Pyrenees," www.parc-pyrenees.com (August 2004); "Pyrenees," www.roncesvalles.es (August 2004).

Jonathan Spangler
Smithsonian Institution

Q-R

Qatar

Map Page 1122 **Area** 4,415 square mi (11,437 square km) **Population** 817,052 **Capital** Doha **Highest Point** Qurayn Abu al Bawl 338 ft (103 m) **Lowest Point** 0 m **GDP per capita** $20,300 **Primary Natural Resources** petroleum, natural gas, fish.

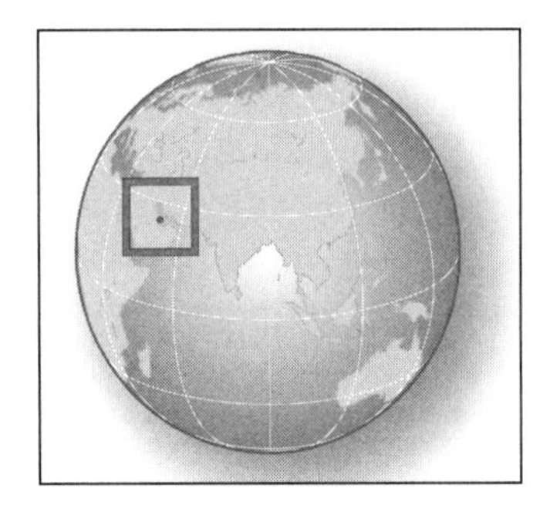

QATAR IS LOCATED in the MIDDLE EAST; it is a peninsula projecting into the PERSIAN GULF and bordered in the south by SAUDI ARABIA. Qatar has been inhabited for millennia but never stood out as a commercial or cultural center in ancient times. The peninsula shared the barren dry nature of the great Arabian deserts to the south and drew the nomads of the interior and the coastal settlers of the Persian Gulf. The Qatar coast held oyster banks that provided the classic gulf item of trade, pearls. This was to be Qatar's main source of wealth for centuries and led to the growth of Doha (Arabic for "port") as the main settlement on the peninsula

Qatar enters historical record in the early 19th century when the Al Khalifa (ruling family of BAHRAIN) domination was challenged by the tribal leaders on the peninsula. Britain was controlling the Persian Gulf at this time as a means of protecting its commercial interests in the INDIAN OCEAN. Attack of the Al Khalifa on Doha and the Qatari counterattack brought the British into the fray which was settled by treaty in 1868. This marked the first time Qatar was widely recognized as separate from Bahrain and the Al Khalifa family. The Al Thani family emerged as the ruling group in Qatar and learned to play the British and the encroaching Ottoman Turks off of one another.

In 1916, the Al Thani Sheikh signed a treaty with the British similar to those entered into by other Persian Gulf emirates and became a British protectorate. A 1934 treaty granted more extensive British protection and opened the way for exploration and exploitation of Qatar's petroleum potential. Across the Persian Gulf, World War II created a hiatus on oil development. In 1949 the beginning flow of oil wealth marked a turning point for Qatar. Slowly, development spread, often impeded by the age-old problems of strife within the Al Thani family and among the leading tribal groups of Qatar.

As Britain prepared to withdraw their protection form the Gulf states in a general reduction of military commitments east of Suez, Qatar considered joining into a coalition of emirate states. When their old nemesis Bahrain and the Al Kalifa family chose independence, so too did Qatar on September 3, 1971. Qatar

participated in the liberation of Kuwait and was supportive of the United Nations sanctions against Iraq in the 1990s. The current (2005) emir deposed his father in 1995 and rapidly increased the pace of national development, as well as opened Qatar to the buildup of U.S. military forces.

The economy centers upon the petroleum sector where oil reserves have diminished. However, Qatar has the third-largest proven natural gas reserves in the world and is rapidly developing and improving extraction and exporting capabilities. Qatar is seeking to replace expatriate workers with nationals by an aggressive program of education and training. Most Qartaris, including the ruling family are followers of the Sunni branch of ISLAM. Islam is the official religion of the state and the people continue to be traditional in dress and custom.

BIBLIOGRAPHY. "Background Note: Qatar," U.S. Department of State, www.state.gov (April 2004); *World Factbook* (CIA, 2004); "Qatar," World Guide, www.lonelyplanet.com (April 2004); Ian Skeet, *Muscat and Oman* (Farber and Faber, 1974); H.J. de Blij and Peter O. Muller, *Geography: Realms, Regions, and Concepts* (John Wiley & Sons, 2002).

IVAN B. WELCH
OMNI INTELLIGENCE, INC.

Quaternary geography

THE QUATERNARY PERIOD is the time in which people became fully human and the dominant animal species on earth. The Quaternary is Earth's most recent geological period and includes the Pleistocene and Holocene epochs. (Eras, periods, and epochs comprise the geological calendar. An era is the longest unit of geologic time. Traditionally, geologists identify periods as beginning and ending with the appearance or extinction of certain life forms. Geologists divide the periods into epochs.) The Pleistocene epoch began about 1.8 to 2.6 million years ago (mya) and ended about 10,000 years ago. (There has been a long debate over when the Quaternary period began. On faunal grounds, it started about 1.8 mya. On climatic grounds, it started 2.58 mya with a rapid build-up of ice in the Northern Hemisphere.)

The epoch was subject to broad swings in temperature, swings that gave rise to growth and decay of glaciers. During the epoch, humans were hunters and gathers. They spread out and established tentative footholds in far corners of Earth. The booming migrations were due to growth in human intelligence, which led to a multiplicity of advances in technical skills, especially in usage of fire, language, cooperative hunting methods, and the making of weapons for hunting. The Holocene epoch began only 10,000 years ago and continues today. It is an interglacial period, a time in which glaciers have largely disappeared. Holocene geography is most notable for the domestication of plants and animals by humans. Agriculture led to a dramatic rise and spread in human populations, tool making, food production, and human influence on the environment.

Quaternary geography is the study of the changing relationships between people and the environment during the Quaternary period. Geographers and other scientists—particularly anthropologists—use evidence from fossils, sediments, ice cores, tree rings, peat bogs, weather records, vegetation, and artifact dating to reconstruct changes in relations between humans and the land during the Quaternary. Quaternary geography has taken two tracks.

Research on the Pleistocene focuses on impacts of climate change on the environment and on early human evolution and migrations. Holocene research focuses on the rapid rise of humans as superior agents of environmental change. The following is a summary of changing human-land relationships during the Quaternary period.

PLEISTOCENE ENVIRONMENT

Temperatures of the Pleistocene fluctuated between greenhouse and icehouse conditions. Swings in temperature varied in magnitude, but in total, there were more than 20 glacial advances (glaciations) separated by intervening periods of warming. Nearly all the glacial ice was in the form of enormous continental ice sheets. Ice sheets spread out over CANADA, GREENLAND, Scandinavia, and eastern SIBERIA in the Northern Hemisphere and ANTARCTICA in the Southern Hemisphere. During the last Pleistocene advance, ice sheets were up to 2.5 mi (4 km) thick. During times when the glaciers were at their maximum size, roughly 5.5 percent of the world's water was in the form of ice (the corresponding value today is 1.7 percent).

When global temperatures plunged to icehouse conditions, the ice sheets grew at the expense of oceans, causing sea levels to fall by up to 500 ft (150 m). During such glacial advances, shallow areas between continents and adjoining islands became land bridges. These land connections allowed animals and

human ancestors to migrate great distances from their areas of origin.

Pre-Pleistocene geography is notable for its legacy of early hominids (human-like apes). Hominids were the first upright-walking animals. They originated on the edges of forests in a volcanic region of East Africa about 4.0 mya. Pollen records in the region suggest that pre-Pleistocene glaciations created cycles of cooling and drying in Africa. Anthropologists believe that climate change, in turn, caused forest habitat to break up into grassland and forced biological advances in our early ancestors, including bipedal walking, which freed up the hands for complex tasks, such as gathering and carrying seeds and fruits and making and using tools. Although anthropologists are still seeking more fossil links, they have accumulated enough artifacts and other evidence to demonstrate an evolutionary line from the hominid *Australopithecus africanus* to us. The lineage includes two intermediary hominids—*Homo habilis* followed by *Homo erectus*—before *Homo sapiens*, or true humans, appeared.

The earliest evidence of *Homo habilis* comes from East Africa and dates to between 2.4 and 2.0 mya. This hominid was the first toolmaker, although the tools were primitive stone scrapers. No evidence suggests that *Homo habilis* migrated out of Africa. By mid-Pleistocene, about 1.0 mya, *Homo erectus* had replaced *Homo habilis*. The use of fire (probably from volcanic origins) enabled this human ancestor to spread out of Africa and into latitudes that are more northerly. *Homo erectus* also migrated east, across Asia as far as Java.

Homo erectus evolved into *Homo sapiens*, but the timing of the latter's appearance is still being debated. The earliest unquestioned evidence of human remains stems from Africa and Europe, where skulls and crude stone tools date to 0.5 mya. Around this time, the population of *Homo sapiens* may have included Neanderthals, but anthropologists sometimes classify them as a separate species (*Homo neanderthalensis*).

Skeletal, artistic, tool, and tool-making evidence proves that the Cro-Magnons of Europe were the first fully modern humans. The Cro-Magnon population grew and spread out rapidly from Europe beginning about 50,000 years ago. The earliest undisputed proof that humans reached far-flung New Guinea and AUSTRALIA dates to about 30,000 years ago. The least contested evidence indicates that humans reached South America, the farthest destination from Africa by land, roughly 11,000 years ago. Genetic and linguistics studies suggest that, as early migrations took place, geographical isolation gave rise to racial and language groupings.

MAMMALIAN EXTINCTIONS

A glut of large mammals became extinct as the Pleistocene epoch ended. Their disappearance may be the first serious human impact on the environment. The list of mammals lost includes nearly 60 species, among them were mammoths and mastodons (larger than modern elephants), horses the size of Clydesdales, camels, giant ground sloths, saber-toothed tigers, and beavers as large as modern bears. Before their disappearance, humans fell prey to some of these mammals. There is no clear explanation for these particular mammalian extinctions, but many scientists argue that was due to a convergence of climatic change and improvements in human weaponry for hunting purposes.

Evidence for human weaponry causing the extinctions is clearest in America. The cold of the glacial maximum was so harsh for some warm-blooded mammals that they had to concentrate in whatever warm habitats remained. Simultaneously, humans began making Clovis points for spears. (The points are so-named because archaeologists discovered the points first near Clovis, New Mexico.) Assiduously fashioned from stone, this double-edged point—when lashed skillfully to a throwing spear—gave humans the advantage they needed to kill dangerous prey at a safe distance. Many archaeological sites in America have yielded evidence of massive overkills by spear-wielding Clovis hunters, but there is some evidence in America and elsewhere to suggest that climate change may have been the main cause of such extinctions. There is an ongoing debate as to the ultimate cause of mass extinctions in the late Pleistocene—overkill or climate change.

The great northern ice sheets were virtually gone by the beginning of the Holocene (10,000 years ago), only a withering GREENLAND ice sheet remains today. (The only other existing ice sheets cover Antarctica in the Southern Hemisphere.) In response to interglacial warming, mixed woodland replaced barren tundra over large areas of Europe and North America and global sea levels rose significantly. Moreover, a combination of climatic and vegetation changes affected stream activity, weathering rates, and soil formation. The same warming extended the geographical range of large grazing mammals, edible plants, seeds, and fruits. The geographical extension of animals and plants aided the spread of humans living in sedentary farming communities.

The adoption of agriculture by people took place first in the FERTILE CRESCENT about 11,000 years ago. Today, tributaries of the Tigris and Euphrates rivers in borderlands separating IRAQ, TURKEY, and IRAN drain this lush arc of foothills and valleys. Main diffusions of food production were from the Fertile Crescent region to Europe, EGYPT, ETHIOPIA, Central Asia, and the Indus Valley. Later independent discoveries and diffusions of agriculture took place from the SAHEL and West Africa to East and southern Africa; from CHINA to tropical Southeast Asia, the PHILIPPINES, INDONESIA, KOREA, and JAPAN; and from Mesoamerica to North America.

These dispersals forever changed the face of Earth by bringing new plant and animal species to new areas. Moreover, in the process, agriculturalists cleared vast areas of natural vegetation to make way for the new crops and livestock. Agriculture made possible the feeding of much larger human populations in many regions. It also opened the way for complex societies in which individuals could specialize in activities other than food production.

Specialization led to the INDUSTRIAL REVOLUTION of the 18th century. Industrialization resulted in an explosion in human population. The present list of environmental problems brought about by industry and population growth is daunting: disappearing forests, plant and animal extinctions, soil degradation, overgrazing, acid rain, water and air pollution, and ozone depletion. Moreover, a large body of scientific evidence suggests that burning fossil fuels and forests has led to excess accumulations of carbon dioxide and other greenhouse gases. The excess gases, in turn, are raising global air temperatures, causing remaining glaciers to melt, sea levels to rise, and ecosystems to change.

BIBLIOGRAPHY. Carl O. Sauer, *Agricultural Origins and Dispersals* (American Geographical Society, 1952); William L. Thomas, Jr., ed., *Man's Role in Changing the Face of the Earth* (University of Chicago Press, 1956); Karl W. Butzer, *Environment and Archeology: An Introduction to Pleistocene Geography* (Aldine, 1964); Martin Bell and Michael J. Walker, *Late Quaternary Environmental Change* (Longman, 1992); Martin Williams, David Dunkerley, Patrick De Deckker, Peter Kershaw, and John Chappell, *Quaternary Environments* (Arnold, 1998); Jared Diamond, *Guns, Germs, and Steel: The Fates of Human Societies* (Norton, 1999); R.C.L. Wilson, *The Great Ice Age* (Routledge, 2000); William F. Ruddiman, *Earth's Climate* (Freeman, 2001).

RICHARD A. CROOKER
KUTZTOWN UNIVERSITY

rainforests

THE WORLD'S RAINFORESTS are often considered the world's "hot spots" since they are often found in tropical climates and they are home to the majority of the world's species. As of 2000, studies and surveillance indicated that rainforests may have from one-half to two-thirds of the world's species, despite only covering 5–7 percent of the world's surface. Biodiversity is just one of the many advantages that rainforests provide for the planet and that make it worthwhile to protect them. Other contributions from rainforests include carbon sequestration, controlling global warming, preventing soil erosion, minimizing carbon emissions through the photosynthetic process, and preventing DESERTIFICATION. Despite these positive aspects of rainforests, they continue to be threatened by market forces, increased demand for agriculture and a lack of a policy framework in most of the developing countries in which these rainforests are found.

Within the tropics, where the temperature lingers at 70 degrees F (20 degrees C) and higher, a rainforest is a common BIOME to encounter. Tropical zones occur primarily within the equatorial zone, 10 degrees within the equator both north and south. The largest areas of tropical rainforest are in the Amazon Basin of South America and in western Africa and INDONESIA. There are many other smaller areas where rainforests abound throughout Central America and parts of the Pacific. With annual rainfall of over 79 in (200 cm), rainforests are rich with flora and fauna. There may be a dry season that lasts for a month or two, but this is substantially milder than the winter and summer contrast experienced in the temperate zones. Seasonal variation in the rainforests is very slight.

The soils found in rainforests are typically old, of decomposed organic matter, and not very fertile to support an abundance of agriculture. Because of the heavy rainfall, leaching of all soluble constituents of the original rock layer occurs, leaving behind a latosol, or a red or yellow soil composed mainly of aluminum and iron oxide. Rainforest soils are not rich in nutrients; the nutrition of the rainforest is lost when logging occurs, since the trees are the main source of nutrients for the ECOSYSTEM.

The components of a rainforest are numerous and diverse. The upper layer or "story" consists of trees 147 to 180 ft (45 to 55 m) tall with round or umbrella-shaped crowns, and they are referred to as "emergents." These trees do not necessarily form the canopy characteristic of rainforests; they are tall and small in

diameter. The second layer of trees is about 98 to 131 ft (30 to 40 m) tall. The third layer of vegetation rises about 33 to 82 ft (10 to 25 m), and this story contains most of the flowering and fruit produce of plants. The shrub story, or the "understory," includes dwarf palms, bananalike leaves, and two flora especially characteristic of rainforests: lianas, or vines rooted in soil that climb to the canopy; and epiphytes, plants that grow on other plants.

COLDER RAINFORESTS

Despite the vast majority of the world's rainforests being tropical and found in the world's warmer regions, not all are located in the warm belt of the globe. The world's oldest forest spans along the western coast of North America, from the panhandle of ALASKA to northern CALIFORNIA. Canopy heights of 197 to 230 ft (60 to 70 m) are usual here, and in the coastal California redwoods, the canopy top regularly reaches 328 ft (100 m). The record height for a redwood is about 367 ft (112 m), the same height as a 28-floor building.

The Tongass Forest, located on the panhandle of Alaska as part of this Pacific Northwest spread, is part of the last remaining temperate rainforest in the world. Tongass covers more than 600 mi (965 km) and is home to the largest collection of bald eagles and grizzly bears.

The other, the Russian Taiga, located in the western region of RUSSIA, covers more than 2 million miles and could cover the whole of the continental UNITED STATES. Temperate forests are often called evergreen forests and are often considered ancient because of their ancient tree populations. They typically consist of coniferous trees and/or evergreen broadleaf trees. Other typical members include Douglas firs, western hemlocks, and western cedar trees. These temperate forests are unrecognized gems in the Northern Hemisphere that may actually host more species than their tropical counterparts because they possess the oldest and tallest trees in the world, and maintain more plant matter as well. The temperate forests are from where hundreds of substances have been extracted for pharmaceutical use, including the famous taxol, a drug used as an anticancer substance.

Tropical or temperate, rainforests are the sources of all kinds of environmental services and commercial commodities to which we are acquainted through the force of our global economy today. Some of the foods that were originally from rainforests around the world include cashew nuts, Brazil nuts, macadamia nuts, bananas, plantains, pineapples, cucumbers, cocoa (chocolate), coffee, tea, avocados, papayas, guavas, mangoes, cassava (a starchy root), tapioca, yams, sweet potatoes, okra, cinnamon, vanilla, nutmeg, mace, ginger, cayenne pepper, cloves, oranges, grapefruit, lemons, limes, passion fruit, peanuts, rice, sugarcane, and coconuts (mostly from coastal areas).

In addition to environmental services and commodities, the world's rainforests are also home to thousands of the world's indigenous peoples. Many of these groups, like the Yanomamo tribe of the Amazon rainforests of BRAZIL and southern VENEZUELA, have lived in scattered villages in the rainforests for hundreds or thousands of years. These tribes get their food, clothing, and housing mainly from materials they obtain in the forests. Indigenous peoples are mostly hunter-gatherers; they get their food by hunting for meat, fishing for fish, and gathering edible plants like starchy roots and fruit. Many also have small gardens in cleared areas of the forest. Since the soil in the rainforest is rather poor and infertile, the garden areas must be moved after just a few years, and another part of the forest is cleared.

THREATS

The world's rainforests and their peoples are being threatened by various forces and global needs. Most indigenous populations are declining from diseases like smallpox and measles, which were inadvertently introduced by Europeans, and governmental land seizure, which has monopolized the rainforests, their homes, in ways that are incomprehensible to their ways of life. Depletion is on the rise in southern Asia, INDONESIA, the PHILIPPINES, Central America, and Brazil. Destruction is somewhat slower in other areas including GUYANA, SURINAME, FRENCH GUYANA, the western part of the Brazilian Amazon, and in Africa's Congo Basin Forest.

Illegal logging occurs virtually everywhere, as certain woods, like mahogany, become more scarce. Attempts to slow its extinction are poorly regulated. Precious woods that are supposedly under the protection of a government are by and large unmonitored since it takes resources and transparency to adequately lead a monitoring process and enforce legislation. Many developing countries in which the rainforests are situated simply do not have the funding to maintain proper monitoring of parks and lands classified as reserved. As illegal logging has skyrocketed in the Central and South America, where forests like the Brazilian Amazon, CHILE's Patagonia, and even forests that have officially been labeled as national parks lie in harm's

way as the global demand for rare, tropical woods increases.

Rainforests are also the target of another global market force: agricultural expansion. In the 1990s, demand for beef in the United States, Europe and Latin America greatly increased as fast-food restaurants proliferated all over the world. The multiplication of fast-food restaurant road signs literally spelled out the imminent global change in rising demand of beef. And as demand for beef rises, so does demand for space where cattle can graze. Unfortunately, even though rainforests do not maintain the most fertile of soils and are not equipped to nurture agriculture, cattle grazing is even harder on the land because of the water requirements, the use of monocultural grasses for feed, and lack of enrichment of the soil.

Agriculture of crops has evolved and been revolutionized with the proliferation of genetically modified organisms (GMO) and has moved offshore to developing countries as developed countries diminish their agricultural sectors. Soy, a very lucrative crop, has been increasingly farmed and harvested in Brazil, leading to rapid deforestation that occurs at an exponential rate since the crop requires unharvested soil every few years. Moreover, while soy is lucrative as an export for the Brazilian economy, less than 20 percent of the Amazonian soil is suitable for growing it. The problem lies in the global market demand for soy and other crops like it, which may satisfy a consumer demand but impairs a developing country's economy as well as its natural resources. The solution to sustainable harvesting of the resources of rainforests is to enable the domestic economy that usually depends on its natural resources—often to the extent of endangering species, rapid deforestation, and extensive environmental threat—to profit from conservation. This may occur through developing sustainable tourism.

Other reasons why rainforests are being threatened and deforestation is escalating to about a loss of 40 million acres (16 million hecatres) a year, the size of the state of WASHINGTON, is due to the needs of the THIRD WORLD poor. When a country suffers from poverty, environmental degradation does not become a priority, and so unsustainable agricultural practices are extremely common in the parts of the world where the world's rainforests are often located. Deforestation also creates a new problem of refugees. Thousands of indigenous tribes that have historically lived in the rainforests of the world have been literally run out by deforestation and the expansion of agriculture. Despite the morose picture of the loss of rainforest around the world, the global economy has become porous enough to distinguish a new market for what are called "environmental services," such as carbon sequestration and preservation of biodiversity. A carbon sequestration program would assign a dollar value to a hectare of forest, essentially assigning a dollar value to the amount of carbon being sequestered in that hectare, and therefore making a hectare of forest more valuable than a hectare of pasture land. The key element in the success of this kind of program is the participation of governments. Because these kinds of relationships usually occur between developed and developing nations, it is imperative that the developing countries play a greater role in negotiations and have the resources to monitor rainforests to ensure their maintenance.

REASONS FOR HOPE

There are also reasons for hope because of the public awareness made possible by civil society organizations advocating on behalf of the rainforests and their indigenous communities. In addition, several intergovernmental institutions, such as the European Forest Institute, represent emerging economies and developed countries working jointly to preserve forests that share political boundaries and that provide a variety of services for their citizens. Other institutions serving the same purpose include the World Commission on Forests and Sustainable Development, a massive joint government effort to create awareness of the services provided by rainforests that the world's population cannot survive without and to formulate sound policy practices that address poverty and trade so that deforestation can be slowed to a sustainable rate.

The ways of business and corporate responsibility are also changing, as are the patterns of consumer behavior as consumers become better informed about the status of the world's rainforests and what remains of them. Companies like Home Depot have been forced to stop selling old growth wood products because their customers do not want to buy products from endangered forests.

Scientists are increasingly vocal about the need to protect what remains, politicians are careful not to be branded as antienvironmental, and every year, many countries add new protected areas. Such public actions and efforts have been evident in the last 20 years. Today's social trends present positive forecasts for protecting the planet's remaining old-growth forests. However, it remains to be seen whether society will change quickly enough to stop logging companies and others who have a head start in destroying rainforests.

BIBLIOGRAPHY. Amazon Alliance, www.amazonalliance.org (September 2004); Marci Bortman, et al., *Environmental Encyclopedia* (Gale, 2003); Richard Brewer, *The Science of Ecology* (Saunders, 1988); European Forest Institute, www.efi.org (September 2004); "Saving the Rainforest," *The Economist* (July 24 2004); World Commission on Forests and Sustainable Development, International Institute for Sustainable Development, www.iisd.org (September 2004).

LINDSAY HOWER JORDAN
AMERICAN UNIVERSITY

Red Sea

THE RED SEA CONNECTS the MEDITERRANEAN SEA to the INDIAN OCEAN via the Suez Canal. It lies in the Great Rift Valley between Africa and the Arabian Peninsula. EGYPT, SUDAN, ERITREA, and DJIBOUTI border it to the west, while ISRAEL is to the north and JORDAN, YEMEN, and SAUDI ARABIA are the to the east. Scientists believe it was formed 20 million years ago when the Earth's crust weakened and was torn apart, creating a jagged rift across Africa. Volcanoes erupted on either side, creating volcanic mountains. Water filled this part of the fault, creating the Red Sea. The sea is still widening by about .5 in (12.7 cm) each year.

The sea has an area of about 170,000 square mi (440,300 square km). The long, narrow body of water is 1,240 mi (2,000 km) long and measures 185 mi (300 km) at its widest point. The Red Sea is a very deep sea, reaching depths of almost 9,840 ft (3,000 m) in the center.

The Red Sea is saltier than any ocean. This is partly due to a high rate of evaporation because of the heat and partly due to low annual rainfall. The water is composed of 3.9 to 4.1 percent salt. At its northern edge, the Red Sea splits into two gulfs: the GULF OF AQABA to the west and the Gulf of Suez to the east, flanking the Sinai Peninsula.

The area has historically served as a departure point for pilgrims traveling to Mecca. Its strategic location made the Red Sea an important trade route in ancient times. During the time of the Roman Empire, trade in spices and other exotic goods flourished between Egypt and INDIA. Other goods included cloth, aromatic woods, incense, coffee, and tea. There was often conflict in the area, as different groups wanted to control the lucrative trade route. The Red Sea's importance declined when an all-water route around Africa to Europe was discovered in 1498.

The Red Sea rose to importance again when the Suez Canal opened in 1869. It provided a more direct connection between Europe and East Asia and AUSTRALIA. The canal was closed for several years after the Arab-Israeli War in 1967. It was reopened and enlarged in 1975 and traffic again increased.

Important ports include Jedda, Mukalla, and Suez. Jedda, located in Saudi Arabia, is just 31 mi (50 km) from Mecca and serves as the main port for pilgrims coming by air or by sea. It's also a commercial center and a center for importing livestock. Jedda has iron and steel plants and oil refineries as well. Mukalla, in Yemen, is located on the Gulf of Aden. It's the area's most important port, but monsoons make it impossible to use between June and September. Mukalla's economy is based on the fishing industry, but boatbuilding is also important. Suez lies at the southern entrance to the canal and has been a commercial port since the seventh century. It was devastated during the Arab-Israeli War, but has managed to recover. Industry includes petrochemical plants as well as cement and fertilizer plants. Fishing is also important to the economy.

The Red Sea area has a diversity of flora and fauna. Marine turtles, including green, loggerhead, leatherback, olive ridley, and hawksbill flourish there. A major bird migration route makes the Red Sea a favorite destination for bird-watchers. Here you might see a pink-backed pelican, a brown booby, a white-eyed gull, or a white-cheeked tern.

The Red Sea has become a popular diving and snorkeling destination, because of the variety of marine life beneath the sea. Seventeen species of fish are endemic to the area, meaning they are not found anywhere else. Dugong, beaked whales, white-tip reef sharks, butterfly fish, giant clams, and several species of dolphins can be seen around the coral reefs.

A few of the sea creatures are dangerous to divers. Shark attacks in the Red Sea are rare, but other species pose a threat to swimmers. The main danger comes from animals with poisonous stings rather than those that bite. The stonefish and scorpionfish lie on the bottom of the sea, perfectly camouflaged. When stepped on or touched, they sting. Their poison causes swelling and unbearable shooting pains. People have died from the stings.

The Red Sea features prominently in the Bible in the book of Exodus. According to the Bible, Moses was leading thousands of formerly enslaved Jews out of Egypt. Pharaoh sent men in 600 chariots to stop

them When the Jews came to the Red Sea, they thought they would be recaptured, but Moses stretched out his hand over the sea and the Lord drove the waters back, making a path of dry land where the sea was divided. The Jews escaped across the dry path, but when the Egyptian chariots followed them, the wall of water engulfed them, drowning them all.

Scientists have advanced several theories explaining the parting of the sea. One says the coral reef beneath the sea used to be closer to the surface. It hypothesizes that if the wind blew all night, it could drive the water back for half an hour, creating a dry path for the Israelis to cross.

BIBLIOGRAPHY. Don Groves, *The Oceans* (Wiley, 1989); "The Red Sea," www.cyberegypt.com (March 2004); "About the Red Sea," www.seaqueens.com (March 2004).

PAT MCCARTHY
INDEPENDENT SCHOLAR

redistricting

THE PROCESS OF redistricting means dividing anew into districts or, in particular, to revise the legislative districts of a certain area, typically of a city or state. In the UNITED STATES, the process of redistricting is also referred to as "legislative reapportionment," which the U.S. Constitution requires so that Congressional representatives are elected according to population. The process of redistricting redraws the legislative district boundaries to reflect population changes and is usually undertaken every 10 years, when the results of the decennial census have been calculated. Redistricting reapportions the population of a state's Congressional, state representative, senatorial, and other legislative districts, ostensibly to ensure the existence of an appropriate number of districts of approximately equal inhabitants.

Although state legislatures are not required to abide by federal constitutional statutes, they are obligated to apportion according to population, geographic size, special interests, and territorial divisions like counties and towns. The legislature is responsible for appropriately redrawing district lines in order to conduct elections for representatives to Congress. Unfortunately, like-minded politicians often create districts that give disproportionate power to small, partisan, unrepresentative minority groups, who sometimes abuse their control. The process of corrupt redistricting is known as gerrymandering, a word whose source lies in the shape that tendentious redistricting often takes: that of a salamander, with extensions into population areas resembling on a map the legs and tail of a salamander.

Historically, some states legislatures avoided redistricting, despite population shifts, for almost 60 years. It was not until 1962 that the U.S. Supreme Court, in *Baker v. Carr*, ruled that a voter could challenge redistricting if it seemed to violate the equal protection clause of the Fourteenth Amendment to the Constitution. Within a year, lawsuits calling for redistricting had been filed in at least 34 states. In 1964, in *Reynolds v. Sims*, the Supreme Court determined that population, according to the one-person, one-vote rule, must be the foremost consideration in all redistricting plans for both upper and lower houses of state legislatures.

More recently, in TEXAS, redistricting caused considerable friction between Republicans—who controlled the redistricting process—and Democrats and minorities. In addition to carrying out the redistricting mandated after the 2000 census, Texas legislators, led by U.S. Congressional majority leader Tom Delay, decided to redistrict the state again in 2003, to create a more "Republican-friendly" Congressional map of the state. While many Democrats held that the new redistricting plan violated the Voting Rights Act, three federal judges upheld the process. Democrats said they intended to appeal the ruling to the U.S. Supreme Court.

In the new plan, the number of Republicans representing Texas in Congress increased. Republicans maintained that the new districts better reflect a growing trend of Republican voters, while Democrats hold that it is intended to weaken minority-voting strength in key districts. The Fort Worth district, for example, stretches from that city to the OKLAHOMA border to include a majority of wealthy white voters who counterbalance the population of African Americans in the inner city area, who traditionally held voting sway in the district. Republicans hoped the redistricting would increase their party's Congressional representation from western Texas and result in the election of a member of Congress who would support oil and gas rather than farming interests.

A similar situation obtained in Austin, in which the city was divided into three districts that extend far into distant, rural ranching areas. For example, the city's southern portion was gerrymandered into a district

that extended 300 miles south, to the Mexican border. While Republicans claim such new redistricting will better represent rural south Texans, Democrats object that the new districts will leave Austin—the capital and most liberal-minded city in Texas—without representation in Congress. As a result of the 2003 redistricting, Republicans said in 2004 that they hoped to win at least 22 of the 32 seats Texas holds in Congress, seats divided evenly between the two parties before the redistricting.

While the Texas case is only one example, it is clear that the redistricting process can be distorted to support a majority party's continued strength, or to deny Congressional representation to certain minorities. Originally thought by founding fathers, such as Thomas Jefferson, to be a process that would ensure continued fair representation and substantial turnover among members of the House of Representatives, the redistricting process has turned out to be the single most effective method for political parties, once in power, to solidify their control of the political process that will determine how a particular seat in Congress is filled. As a result, the House of Representatives has become the more stolid and unchanging house of Congress, where initial election can mean lifetime membership, ensured via the redistricting process.

BIBLIOGRAPHY. Paul Krugman, "Texas Redistricting," *New York Times* (June 13, 2003); New Jersey Legislature, New Jersey State Government, www.njleg.state.nj.us/legislativepub/glossary (April, 2004); North Star Network, www.thenorthstarnetwork.com (April 2004); Texas Legislature Online, Texas State Government, www.capitol.state.tx.us (April 2004); Wyoming State Legislature, http://legisweb.state.wy.us/leg2/redistrict (April 2004).

A. CHIAVIELLO AND RYAN JOHNSON
UNIVERSITY OF HOUSTON, DOWNTOWN

region

THE NOTION OF REGIONS has been central to geographic thinking, first because of the importance of the observation and its corollary, and second because of the classification in classic geography emphasized by the empiricism and primacy of the field survey. The term *region* originates from Old French and stems from the Latin *regio*, meaning "direction and district," and from *regere*, meaning "to rule," or direct and thus give the dimension of political space. Contemporary geography considers the region as a spatial entity of a middle size (between the local and the national level), which may be different from one country to another and one period to another. Regions can be a political entity used, for example, by the United Nations, which divides the world in a certain number of regions with variations in size. It is under this term that it is possible to designate a group of states, or a continental group such as the MIDDLE EAST or Southeast Asia, and also to refer to regional powers or regional conflicts. In addition, the term has changed from the traditional definition that includes a political or scale dimension to include urban, natural, desert, or jungle areas, among many others.

The concept of region has been at the core of the French school of geography. Paul Vidal de la Blache (1845–1918), a French geographer, was the first to formalize the concept of region. It was for him a natural area that is translated in the landscape as the result of an inter-relationship between the natural environment and human activities. In the UNITED STATES, in the beginning of the 20th century, the evolution of the concept of regions changed to be recognized and used in practice. The administration of President Woodrow Wilson (1856–1924) chose Isaiah Bowman, director of the American Geographical Society and region expert, as territorial adviser at the Versailles Peace Conference concluding World War I. Geographers in Europe were consulted for the internal political reorganization in several countries.

Traditional geography, that is to say, the descriptive taxonomy of regions, was criticized in the 1960s and led to new classification, such as the structural taxonomy making a differentiation between formal, functional, nodal, and equitable regions. The study of regions has been associated in geography with area-differentiation and chorology. The regions concerned here have distinguishable cores; however, the regional characteristics are losing their peculiarities with increasing distance from the core. David B. Grigg (1965) has examined the similarities between regionalization and classification and supported the regional taxonomy as the basis for formal region-building algorithms.

On the contrary, the functional regions are concerned with the human organization of space. They are defined as areas in which a higher degree of mutual socioeconomic interactions exist within them than with outside areas. This functional geography joins the space of regional science. The nodal region merged in the 1950s, when Derwent Whittlesey (1954) developed

the concept of cores and regional dominance in networks. Thus, the development of the different regions concepts from the 1950s has led to the importance of planning and the preparation and implementation of regional policy by practitioners and service providers.

In the 1980s, nonfunctionalist formulations emerged and were linked to the renewed interest of the regions in the area of regional geography. Torsten Hagerstrand developed in spatial analysis the concept of time-geography to show the importance of the temporal variable in the constitution of space at different scales in the regions. Continuing on the ideas, Nigel Thrift has conceived the region as a "meeting place of human agency and social structure." He is also inspired by Antony Giddens's work (1984) on structuring theory. Allan Pred (1984), besides his schematic model of the region as process, has opened an approach to the Marxist geography particularly regarding the division of labor, layers of investment and the uneven development all these concepts accentuating the social practices developed above.

It is interesting to note how the idea of regions can survive with the digital technology of information and communication and with the growing role of cities that offer opportunities for convenient meetings to the actors of the new economy in a global environment.

BIBLIOGRAPHY. H. J. de Blij and Peter O. Muller, *Concepts & Regions in Geography* (Wiley, 2002); Paul Claval, *Introduction to Regional Geography* (Blackwell Publishers, 1998); Antony Giddens, *The Constitution of Society* (Polity Press, 1984); David B. Grigg, "The Logic of Regional Systems," *Annals of the Association of American Geographers* (v55, 1965); David B. Grigg, "Regions, Models and Classes," P. Haggett and R. J. Chorley, eds., *Models in Geography* (Methuen, 1967); Nigel Thrift, "On the Determination of Social Action in Space and Time," *Environment and Planning* (v.1, 1983); Alen Pred, "Places as Historically Contingent Process: Structuration and the Time-Geography of Becoming Places," *Annals of the Association of American Geographers* (v.74, 1984).

NATHALIE CAVASIN
WASEDA UNIVERSITY

regionalism

REGIONALISM IS A COMPLEX and contested concept. As such, there is no straight or simple answer to what regionalism is. One thing for sure is that regionalism is closely related to REGION. Since a region may connote geographical contiguity ranging from a small neighborhood to a few cities right up to several states and continents, regionalism thus can exist within a state, as parts of states, among states, and between groups of states. But contiguity is not the only variable in delineating regions, which means that regionalism may well occur irrespective of spatial boundaries. It is therefore safe to say that regionalism entails an intricate set of ideas, behaviors, and allegiance in the conscious minds of individuals as to how they perceive their region, hence giving it a distinct physical and cultural feature. The same group of people could be engaged in more than one form of regionalism as regions overlap and change over time. This happens because of the nature of regions as entities that are socially constructed but can also be predefined. Depending on the objective and purpose of the regionalism pursued, some forms would be more elaborated and focused compare to others.

Within a state, regionalism is both positively and negatively correlated with the idea of region. Pessimistically, regionalism can be one manifestation of ethnic nationalism. This could happen in countries where ethnic groups are identified via regions. These groups are normally the minorities and most often than not unfavorably treated or forgotten mainly because there is a lack of integration between the core and the periphery. The problem of assimilation and stark cultural differences are some factors that give rise to the pursuit of regionalist discourses to secure and preserve their beliefs and rights should they perceive the actions of the state as detrimental to their own. Those with higher aspirations and means may set political agendas for separatism and independence. Such activities no doubt challenge the legitimacy and authority of the state. States with many capabilities will try to suppress those aspirations through carrots and sticks, those with fewer capabilities might designate trouble areas as autonomous regions, and those that are incapable may see their territorial boundaries redrawn.

This mostly affects large states, young independent states, politically turmoiled states, and failed states. Some known examples with varying degrees of regionalism are the Chechens in RUSSIA, the Abkhazian and South Ossetian in GEORGIA, the Uyghurs and Tibetans in CHINA, and the Acehian and previous East Timorans in INDONESIA. While ethnicity plays a central role, other factors such as ideology can be a powerful tool for regionalism. Religion and communism, as separate ele-

ments or in combination with ethnicity like the predominantly ethnic-Chinese supported communist insurgency in Malaya, could push for the same kind of regionalism.

POSITIVE SIDE

On the positive side, regionalism takes a different form. It may simply reflect the desire of communities that are interested in increasing the efficiencies of their respective towns, counties, or cities through better management and administration. This involves building coalitions that are tailored to specific projects such as land use, housing needs, environmental control, health care, job creation, bioterrorism, traffic improvement, poverty eradication, and others.

Many of these problems cannot be easily handled or solved within the existing political boundaries because of changes in demography and cross-border human activities. As they spill over and become regional issues, interconnectedness then requires local administrations to forge strategic alliances so as to effectively address those concerns. Such regionalism can lead to accountability and better cost utility by leveraging on the competitive advantages of those in alliance. States may also decide to merge towns or cities for similar effects.

The same dynamics could exist for regionalism that transcends parts of bordering states. Mostly centered on economic cooperation, three or more countries may choose to pursue regionalism by earmarking a regional space for development. Geographically adjacent areas are linked to form a distinct space in which differences in the factor endowments (land, labor, capital, etc.) and levels of development are exploited for the purpose of promoting external trade and direct investment. This imaginative space has been given various unofficial terms such as growth triangle, growth quadrangle, transnational economic zone, natural economic territory, circle of growth, extended metropolitan region, or, simply, growth area. Regionalism of this sort is attractive because participating countries are able to gain from the differences in comparative advantage which serve to complement rather than compete with one another. Economic complementarity is not the only motivating factor, as it can also include cooperation on natural resources, infrastructure development, and even tourism. Successful projects can be replicated elsewhere with different modalities of cooperation to meet the local needs and conditions of the region in focus.

Furthermore, a country can participate in several projects at the same time. Such undertakings have the ability to prop up unproductive peripheries by overcoming rigid territorial barriers and utilizing shared experiences.

Some examples are the Indonesia-MALAYSIA-THAILAND growth triangle, the ZAMBIA-MALAWI-MOZAMBIQUE growth triangle, the Tumen River growth area, the Greater Mekong subregion, and the Gulf of Finland growth triangle.

On the other hand, cross-border regionalism could have a negative impact on states. Rather similar to regionalism within states, this may involve the manifestation of ethnic nationalism that goes beyond states' borders. One example is the Kurdish minority located in a region that spans across four countries, including parts of IRAQ, TURKEY, IRAN, and SYRIA. The Kurdish people, while lacking in political unity, have a strong belief and aspiration for independence, but infighting often results in their suppression and repression by the countries they reside in.

Between states, regionalism generally aims at finding regional solutions to regional problems through regional cooperation. A group of neighboring countries with common concerns may decide to gather and commit themselves to dialogue on certain areas of cooperation such as economy, finance, security, health, welfare, cultural, environmental, human rights, and crime. The amount and degree of cooperation varies depending on the problems faced. Security issues could cover terrorism, piracy, and nuclear weapons, while economic issues may focus on trade and investment, and environmental issues would concentrate on pollution, smog, and illegal logging.

INTERDEPENDENCE AND GLOBALIZATION

All these issues cannot be tackled alone, as they are the manifestations of interdependence and globalization. This is where regionalism can be effectively employed to deal with the challenges of globalization. Regionalism allows smaller and weaker states to bind their strengths for better results. States may further decide to establish regional groupings or organizations to institutionalize their cause on matters of importance. Some of the typical regional arrangements are alliances, ententes, free trade areas, and custom unions.

The NORTH AMERICAN FREE TRADE AGREEMENT (NAFTA), for example, was established to enhance regional economic cooperation. So were the Economic Community of West African States and the Mercado Común del Sur. The Association of Southeast Asian Nations (ASEAN) was originally an organization concerned with security matters but has since moved to

focus on economic regionalism. Groupings with higher ambitions can take regionalism beyond simple cooperation toward integration, as the EUROPEAN UNION did. But regional integration may imply the pooling of sovereignty and interference in state affairs and so directly challenges the state system. Therefore, regionalism between states can have both positive and negative implications and can be seen in stages with political union as the ultimate goal.

The concept of regionalism can be broadened to include cooperation between groups of states. An established regional organization may want to expand its cooperation with countries or regions outside of its own. In East Asia, for example, the Asian financial crisis resulted in a closer relationship between the Southeast Asian region and the Northeast Asian region. Economic and financial regionalism is being pursued under the framework of ASEAN Plus Three. Interregionalism such as the Asia-Europe Meeting, which links East Asia to Europe, is another case in point. What's more is the notion that regionalism as a social construction can occur out of shared common political, economic, or cultural objectives with no bearing to geographical proximity. The Group of Eight and the NORTH ATLANTIC TREATY ORGANIZATION (NATO) are two examples that bring to the table different sets of countries with different purposes but yet are free from any geographic outlay.

REGIONALISM AND REGIONALIZATION

The concepts of regionalism and regionalization are most prevalent in the study of international relations and international political economy. More often than not, these two concepts are used interchangeably to mean the same thing. Differences, however, do exist between them. There are a few interpretations in distinguishing the two concepts.

Regionalism, as a general phenomenon, may refer to a formal project, policy, or scheme promoted by regional states. As a political project, it contains a certain set of ideas, norms, values, principles, and identity that is shared by the participating members. Hence, the characteristics of one regionalist project would obviously be different from another. The aim of such projects varies ranging from promoting a sense of regional awareness to forming supranational institutions.

The process needed to achieve the aim of the project is what can be termed as regionalization. It is an empirical process with an activist element that harmonizes states policies by changing heterogeneous factors defined as obstacles to closer cooperation toward increased coherence and convergence within the given geographical area. Regionalization can also take a different meaning, one that is not tied to a regionalist project. Here, regionalization takes place as the result of spontaneous forces. It depicts a multidimensional and undirected natural process of social and economic interaction driven by the people as nonstate actors that could plausibly contribute to the growth of societal integration and transnational civil society within a regional space. Such vigor may instead give rise to regionalism with the emergence of regional groups and organizations.

Regionalism and regionalization anchored in the economic domain may have a slightly different interpretation. Regionalism can be understood as a political process whereby states cooperate and coordinate their economic policies across regions. One method pursued by states is to form regional trading arrangements. These arrangements furnish states with preferential access to members' markets. Free trade agreement, as it is often called, dates back to the early 1960s but has become an increasing trend in the post-Cold War period marked by a shift from the old protectionist regionalism to the open and flexible new regionalism. But there is still a constant debate as to whether regionalism through trade agreements complements or contradicts the world trading system. Regionalization, on the other hand, is understood as the concentration of regional economic activities and trade flows of market actors. This may involve transnational corporations, entrepreneurs, consumers, investors, and capitalists that have a regional interest.

All these suggest that not only are there differences but the relationship between regionalism and regionalization is progressive and robust. In short, regionalization can both precede and flow from regionalism. One interesting example is the East Asian region, where decades of economic regionalization with the absence of regionalism created an intricate web of interdependence and then saw a postcrisis emergence of regionalism geared toward facilitating further regionalization.

BIBLIOGRAPHY. Karsten Fledelius, "What is a Region? What Is Regionalism?," *Regional Contact* (v.10, 1997); Louise Fawcett, "Exploring Regional Domains: A Comparative History of Regionalism," *International Affairs* (v.80/3, 2004); Louise Fawcett and Andrew Hurrell, eds., *Regionalism in World Politics: Regional Organization and International Order* (Oxford University Press, 1995); Graham Evans and Jeffrey Newnham, *The Penguin Dictionary of International Relations* (Penguin Books, 1998); Bruce Katz,

ed., *Reflections on Regionalism* (Brookings Institution Press, 2000); Yoshinobu Yamamoto, ed., *Globalism, Regionalism and Nationalism: Asia in Search of Its Role in the 21st Century* (Blackwell Publishers, 1999); Takatoshi Ito and Anne Krueger, eds., *Regionalism versus Multilateral Trade Arrangements* (University of Chicago Press, 1997); Michael Schulz, Fredrik Soderbaum and Joakim Ojendal, eds., *Regionalization in a Globalizing World: A Comparative Perspective on Forms, Actors and Processes* (Zed Books, 2001).

BENNY TEH CHENG GUAN
KANAZAWA UNIVERSITY, JAPAN

religion

RELIGION CAN BE DEFINED as a unified set of beliefs, values, and practices of an individual or a group of people that is based on the teachings of a spiritual leader. It includes codes of behavior, faith in and devotion toward a supernatural power or powers, and a framework for understanding the universe. Religion is an important component of many individuals' lives and identities, and it is deeply embedded into most if not all societies. From a geographer's standpoint, however, religion is meaningful for the following reasons: It has a varying presence in every inhabited locale on Earth; religion displays clear geographical patterns; many of the characteristics of religions and their development and their impact are rooted in geographic factors; in order to better understand a place, we must consider its religious character; religion is one of the most or the most important components of cultural groups; and religious issues are often the root cause of many geopolitical problems.

The geography of religion examines the way in which religion is expressed on the Earth and its social, cultural, and environmental impacts. More specifically, it looks at, maps out, interprets, compares, and analyzes various religions' origins, diffusion, subsequent distributions and effects, and the religious landscapes they create. The geography of religion additionally investigates how religion impacts lifestyles, commerce, demography, gender issues, political geography, the environment, and other components of society and culture. With the growing issues of militant ISLAM, religious fundamentalism, and cultural politics, there is an increasing awareness of the field.

Geographers primarily categorize religion into two types: universal and ethnic. Universal or proselytic religions actively solicit new converts and have widely and quickly diffused across the globe. Christianity, Islam, and some forms are Buddhism fall into this category. Ethnic religions, on the other hand, are associated with a particular ethnic group and customarily do not proselytize. They often dominate a certain cultural group, are usually confined to a particular country, and spread spatially at a slow rate. Common examples are Judaism, Hinduism, and Shintoism. Many individuals also add a third type of religious faith, those that are tribal or traditional. These animistic faiths usually believe in spirits inhabiting inanimate objects such as plants, animals, and other natural features. A religion can be can be either monotheistic, meaning believing in only one god, or polytheistic, which is having belief in multiple supernatural entities. Atheists, in turn, do not believe in an existence of a god or gods.

The geographic study of religion can be divided into five main themes: religious regions, religious ecology, religion and society, religious landscapes, and religious diffusion. Religious culture regions are areas that are based on religious characteristics. They can be formal regions, such as the Muslim dominated northern SUDAN, where there is a particular religion practiced. They can also be functional in nature or serve a specific administrative purpose, such as a Roman Catholic Church diocese or parish. Religious regions can be as small as a city block or can exist worldwide. Their characteristics are a result of diffusion and the interplay of religion with components of culture, society, and the environment.

RELIGIOUS REGIONS

The most commonly described religious regions are those where a religion is practiced. Christianity is the world's largest religion in area and in the number of adherents. The faith, an offshoot of Judaism that is based on the teachings of Jesus Christ, is fragmented into numerous different denominations. The most common division is between Western and Eastern Christianity. Western Christianity is made up of the unified Roman Catholic Church and the highly divided Protestant denominations. Eastern Christianity includes groups such as Coptics, Maronites, Nestorians, and various Eastern Orthodox churches.

Christianity is almost a completely global religion but is most common in the Americas, Europe, RUSSIA, AUSTRALIA, and southern Africa. The Eastern Orthodox Church is strong in the former Soviet Union, parts of Europe, and North Africa. Roman Catholicism is concentrated in Europe and South and Middle America.

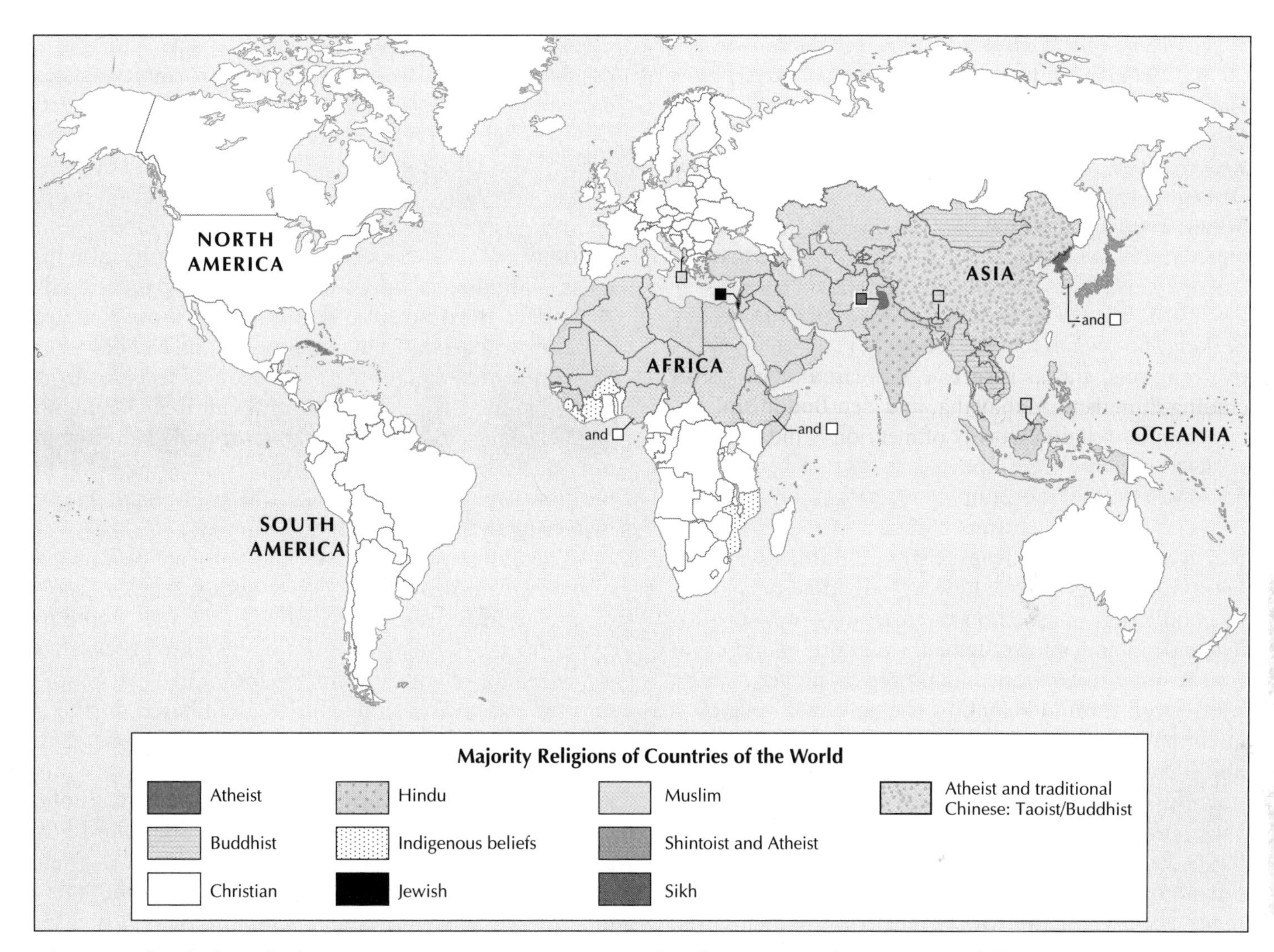

The geography of religion looks at, maps out, interprets, compares, and analyzes various religions' origins, diffusion, subsequent distributions and effects, and the religious landscapes they create.

Protestantism dominates much of North America but is also found in Europe.

U.S. GEOGRAPHY OF RELIGION

The UNITED STATES has a distinctive geography of religion because of its unmatched religious diversity. Baptists and other conservative fundamentalist groups are common in the Bible Belt region in the South. Lutherans dominate most of the Upper Midwest, influenced by Scandinavian settlement of the region.

Roman Catholics are strong in the northeast United States and southern LOUISIANA but are also found in large numbers in the southwest part of the country, which has a strong Mexican influence. The Church of Jesus Christ of Latter-day Saints, or the Mormon Church, is concentrated in UTAH and other parts of the Mountain West. Other smaller denominational religions, often at the county or local level, also exist. These "religious islands" are influenced by the presence of small ethnic groups, religious universities, and other religious group concentrations.

Islam, like Christianity, is proselytic and monotheistic. It is common in Southwest Asia, northern Africa, and parts of Southeast Asia. Its prophet, Muhammad (570?–632), gained many converts during his lifetime, but the faith has spread worldwide since his death. Today there are over 1 billion Muslims worldwide, distributed among two main divisions. Shiites constitute about 10 to 15 percent of all Muslims and are mainly found in IRAN and IRAQ. Sunni Muslims, the largest group, are common in other Muslim-dominated areas, such as North Africa.

Hinduism, which has no specific founder, is a polytheistic faith common in India. It evolved over a 4,000-year period as indigenous faiths and cultural practices merged with a religion brought to India by Aryans tribes. Today, Hinduism is made up of numerous sects and splinter groups such as Sikhism and Jainism. Perhaps the most prominent offshoot of Hinduism is Buddhism. This widespread religious faith is especially common in East and Southeast Asia. It was founded in India around 525 by Siddhartha Gautama, called the Buddha. Buddhism had the tendency to merge with native religions, and as it spread, it created divisions including Lamaism, Mahayaha, and Zen Buddhism.

Another religion worthy of mention is Judaism. Associated with the Jewish people, this monotheistic, ethnic faith of many subgroups is widely dispersed across the globe in small clusters. Many Jews are found in North America and Europe. Eighty percent of the population of the country of ISRAEL is Jewish.

There are also several parts of the globe where animistic or traditional religions are found. Animism is common in less developed parts of Africa, South America, Southeast Asia, New Guinea, and Australia. What makes animism especially interesting is that each tribe has its own unique belief system that creates distinctive practices and geographies.

A Buddha in Sri Lanka dominates the local geographic landscape; Buddhism is diffused throughout the country's culture.

RELIGIOUS ECOLOGY

The second theme of the geography of religion, religious ecology, deals with how different religions perceive and interact with their natural environments. For example, geographers analyze environmental attitudes of various faiths, because some religions appear to promote preservation and others seem to encourage environmental exploitation. Many faiths also view natural features as sacred, including rivers, forests, springs, and mountains.

For example, the GANGES and Bagmati rivers in South Asia are important to Hindus. The Jordan River around Israel, PALESTINE, and JORDAN, for example, is holy to Christians. Mt. Fuji in JAPAN is sacred in Japanese Shintoism.

Religion additionally plays a vital role in environmental perception and whether or not people view the environment as an ally or something to fear. The environment may also influence the characteristics of religious faith. While the influence of the environment on most major universal religions is less significant, it is especially visible in animism and also in terms of the philosophy of feng-shui in Chinese and Korean Buddhism. The geography of religion also concerns itself with how religious belief aids in appeasing the forces of nature.

There is additionally a high level of integration between religion and society. This third theme analyzes how religious faith interacts with other components of culture. Because it is a strong human motivator at the personal and group level, religion consistently impacts other human traits, cultural group history, lifestyles, economic systems, political geography, and demography. Religion is also an important, if not the primary, component of ethnicity. For example, most Arabs are Muslim, most Mexicans are Roman Catholic, and the majority of Norwegians are Lutheran. Their religious faiths play a strong role in group identity.

Religion can also influence economic geography. For example, many religious groups have overt taboos, or the prohibition against certain items or activities.

The Wailing Wall in Israel is a unique feature among many in the country's religious landscape.

Caffeine is forbidden to members of the Church of Jesus Christ of Latter-day Saints and, subsequently, influences the sale of some beverages. Alcohol is forbidden to Muslims and some Christian denominations, and patterns reflecting these taboos are visible on the landscape. Religion can also influence agriculture in the absence or occurrence of certain crops and farming activities. Wine is important to Holy Communion among many Christian groups, and this has demanded a need for vineyards and grape cultivation in Christian regions. Since Jews and Muslims forbid the consumption of pork, pigs are rare in areas dominated by these two groups.

Religious faith also influences birth rates and other demographic characteristics. The Roman Catholic Church and other groups, for example, prohibit most forms of contraception and encourage large families. Such practices often increase the number of adherents and, subsequently, the spreading of the religion.

There is an increasing awareness today that religion can be frequently integrated with politics and government. Many countries are divided by religious faith, and religion is often a rallying point for political action. Religion played an important role in the partition of India in 1947 and also shaped the 20th-century geopolitics of IRELAND and its relationship with the UNITED KINGDOM. Many countries, such as Iran, are theocracies, or governments guided by a church or a religion. The Taliban, a fundamentalist Islamic group, ran AFGHANISTAN from 1996 to 2001. Some countries, including the Lutheran Norway, have an official state church. There is increasing controversy worldwide, including in the United States, about the role of religion in government.

PILGRIMAGES

Pilgrimages provide a unique example of how religion impacts culture and society. This act of religious devotion usually involves large numbers of people traveling in various ways to places that are often the setting of miracles, sacred physical features, or the geographic origin of a faith. A sense of duty or hope of receiving healing or a special blessing may be the motivation behind most pilgrimages. Pilgrimages are especially common among Muslim, Hindu, Shinto, and Roman Catholic adherents. They also have a significant impact on local economies and environments.

Often, religious tourism is created because so many people visit specific places. Many Roman Catholics, for example, travel to Rome, the French town of Lourdes, or the Lady of Guadalupe Shrine in MEXICO. Hindus make pilgrimages to Varanasi on the Ganges River in India. Japanese Shintos visit Ise. Perhaps the most famous pilgrimage is the hajj to Mecca in SAUDI ARABIA. The hajj is an important pillar of the Muslim faith, and it is the duty of all Muslims to travel to the city.

RELIGIOUS LANDSCAPE

The most studied component of the geography of religion is the religious landscape, or the religious imprint on the material cultural landscape. Because there are differences in the characteristics, history, and interaction with the environment and society, religions also have distinctive landscapes. Variations in religious landscapes also give particular regions a unique character and can be seen in the form, orientation, density, and architecture of structures. The most prominent religious landscape feature is the house of worship. These church buildings, temples, synagogues, or mosques frequently have distinctive architecture based on religion. For example, it is usually easy to differentiate between an Islamic mosque and a Hindu or Buddhist temple.

Various Christian church buildings also have unique attributes. Roman Catholics usually perceive the house of worship as being the house of God, and subsequently the structures are very ornate. Most Protestant denominations view a church building as simply a place to worship and congregate. Because of this, the majority of Protestant structures are modest, smaller, and more functional than a typical Catholic church.

Some church buildings are also unique because of the congregation's ethnic background. For example,

there are often visible differences, both exterior and interior, between a German Lutheran church and an Icelandic Lutheran building.

Every religion also has its own practice for burial of the dead. Some religions, such as Hinduism and Buddhism, use cremation, so cemeteries dedicated to these faiths do not exist. Most other religions, such as Christianity and Islam, do bury their dead, but there are often differences in the architecture and orientation of grave sites. Other common man-made religious landscape features include yard and roadside shrines, religious murals and statues, and parochial schools.

It is also important to consider the symbolic and sacred qualities of the religious landscape. Why are some places special and others not? Religions often designate sacred space or features because they are worthy of devotion, loyalty, or fear, or because they have supernatural or mystical qualities. All man-made religious features can be considered sacred, but sacred space also extends to physical features such as Mt. Sinai for Judaism or Ayers Rock for Aboriginal animism in Australia.

RELIGIOUS DIFFUSION

The fifth and final theme is religious diffusion, or the spreading of religion across space. Religious diffusion is important because the landscapes and regions that religions create are all products of spatial expansion. Geographers are interested in how religions diffuse, how they change over time, and what processes encourage these changes. Most religions have spread through contagious expansion diffusion, meaning increasing numerically through direct contact of individuals. This occurs through various conversion methods. Barriers to religious diffusion exist as well.

All religions have a source area or a hearth where their diffusion started. The most widespread religions today originate out of two primary religious hearths. The Semitic Hearth, found in Southwest Asia, is the source area of Christianity, Islam, and Judaism. Judaism started about 4,000 years ago, and Christianity evolved out of Judaism about 2,000 years ago. Islam, with its origin in the Arabian Peninsula, began approximately 1,300 years ago. Hinduism and Buddhism both originated in the Indo-Gangetic Hearth at the northern edge of the Indian subcontinent of Asia.

BIBLIOGRAPHY. Terry G. Jordan-Bychkov and Mona Domosh, *The Human Mosaic* (Longman, 1999); Lily Kong, "Mapping 'New' Geographies of Religion: Politics and Poetics in Modernity," *Progress in Human Geography* (v.25/2); David S. Noss, *A History of the World's Religions* (Prentice Hall, 2002); Christopher Park, "Religion and Geography" *The Routledge Companion to the Study of Religion*, John Hinnells, ed. (Routledge, 2005); Chris C. Park, *Sacred Worlds* (Routledge, 1994); David E. Sopher, *Geography of Religions* (Prentice Hall, 1967); Roger Stump, "Introduction," *Journal of Cultural Geography* (v.7/1); Wilbur Zelinsky, "The Uniqueness of the American Religious Landscape," *Geographical Review* (v.91/3).

Anthony Paul Mannion
Fort Hays State University

resource

THE WORD *resource* is derived from the Latin *resurgere*, meaning to "rise again." Therefore, a resource can be part of a cyclical process that can change and return over time and space. Resources are those elements that humans have the knowledge and technology to utilize to provide desired goods and services. Resources are subjective, functional, and dynamic. They can be tangible things that create, support, and supply material wealth. Resources, such as beauty, peacefulness, or diversity are considered intangible. Both tangible and intangible resources reflect variations in knowledge, technology, social structures, economic conditions, and political systems.

Humans create particular types of resources: labor, entrepreneurial skills, investment funds, capital assets, cultural adaptations, and technology. Natural resources are substances, organisms, and properties of the physical environment. These resources provide the material wealth that humans are dependent upon. Humans use many natural resources that are also important for other species.

Resources can also be considered renewable or nonrenewable. Renewable resources or flow resources naturally regenerate to provide new supplies within a human life span. These resources can be replenished or replaced continuously and sustainably into the future and faster than they can be used. Renewable resources include sunlight, biological organisms, and biogeochemical cycles that provide essential ecological services and generally will not run out.

For example, biological organisms replace themselves by reproduction; ecological processes are self-renewing. Water is considered a renewable resource. It is part of the hydrologic cycle the moves water in differ-

ent physical forms and allows for it replenishment over time and space. Humans can harvest the surplus and use it without diminishing future availability.

Nonrenewable resources or stock resources are those things that take millions of years to form and are in a fixed supply in relation to human terms. These resources may be renewed or recycled by geological or ecological processes, but the time scales are so long by human standards that the resource will be gone once present supplies are exhausted. The use of nonrenewable resources is not dictated so much by the absolute amount available but rather is due to the economic and environmental costs required to extract them. The use leads to adaptation through more efficient use, recycling, substitution of one material for another, and better extraction from other sources. Nonrenewable sources include minerals, fossil fuels, or other materials present in fixed amounts in the environment that will eventually run out.

Energy is one resource that is commonly divided into renewable and nonrenewable categories. Renewable energy includes solar, wind, tidal/wave, biomass, hydroelectricity, and geothermal energy. Solar energy is continually supplied to the Earth by the sun. Geothermal energy is continuously created beneath the Earth's surface from the extreme heat contained in liquid rock—magma within the Earth's core. Biomass describes many different fuel types form sources such as trees, agricultural wastes, fuel crops, sewage sludge and manure. Wind is created when the sun heats the Earth's surface unevenly, because of the seasons and cloud cover, causing warmer air to move toward cooler air.

Nonrenewable energy include such things as coal, petroleum, gas, and nuclear energy. Coal is a rock that is a fossil fuel formed over millions of years from decomposing plants. Another fossil fuel is petroleum, or crude oil. It is formed in a similar fashion as coal but is a liquid that became trapped between layers of rocks. Gas is trapped between layers of rock. Nuclear energy is the energy released when atoms are either split or joined together. A mineral called uranium is needed for this process. At each stage of the process, various types of radioactive wastes are produced.

Another way of understanding resource use is by viewing resources as part of a use-renewability continuum. At one extreme are naturally determined, infinitely renewable resources. The amount of these resources is unrelated to current usage levels. Examples include: solar energy, tidal and wind power, and water resources. The other extreme is where utilization exceeds regeneration. The use of the resource is consumptive, and byproducts result in unusable forms of matter and energy. Examples are fossil fuels, plants, animals, fish, forests, and soils. In between these extremes is where resource renewability is dependent on human decisions, where future supply availability is determined by usage rates and investment in artificial regeneration to ensure supply and quality (e.g., air and water quality, or minerals).

BIBLIOGRAPHY. Willima P. Cunningham and Barbara Woodworth Saigo, *Environmental Science* (McGraw Hill, 2001); Jesse Ausubel and Hedy Sladovich, eds., *Technology and Environment* (National Academy Press, 1989); James Rees, "Natural Resources, Economy and Society," D. Gregory and R. Walford, eds., *New Horizons in Human Geography* (Macmillan, 1986); Education Queensland, "Power for a Sustainable Future," www.sustainableenergy.qld.edu.au (Queensland Government, 2000); EPA, "Electricity from Non-Hydroelectric Renewable Energy Sources," www.epa.gov (September 2004); UNEP, *World Resources 2000–2001* (World Resources Institute, 2000).

Melinda J. Laituri, Ph.D.
Colorado State University

Rhine River

WITH A TOTAL LENGTH of 820 mi (1,320 km) flowing through SWITZERLAND, GERMANY, and the NETHERLANDS, the Rhine is the longest river in Germany and probably the most important waterway of western Europe. The Vorderhein, Hinterrhein, and Alpenrhein are the sources of the river within Switzerland. The Rhine then enters the Bodensee (Lake Constance) and reemerges as the Hochrhein up to Basel on the Swiss-German border. From Basel, it passes as the Oberrhein on the French-German border via Strasbourg/Kehl into Germany to the cities of Karlsruhe, Mannheim/Ludwigshafen, and Mainz. After Mainz, it becomes the Mittelrhein and Niederrhein and goes on via Bonn, Cologne, Duesseldorf, and Duisburg to the German-Dutch border. It terminates its route via Nijmegen at Rotterdam, the Netherlands, on the North Sea.

Flowing roughly northwest, the Rhine's annual flow at its mouth is 16.6 cubic mi (69.3 cubic km). The river is of enormous economic, cultural, and historical importance for all of western Europe. The Romans

came to the Rhine and tried to colonize parts of Germania by crossing the river, but it was in the early Middle Ages, when the new empire of Charles the Great consisted of large territories on both sides of the Rhine. These territories now form—at least in parts—FRANCE, Germany, Switzerland, AUSTRIA, BELGIUM, LUXEMBOURG, and the Netherlands. When the empire broke up after the death of Charles, two succession regions (what is today France and Germany) engaged in longstanding disputes, whereby the Rhine was a symbol as well as a catalyst.

The concept of natural borders developed under the French King Louis XIV embraced all the left (western) side of the Rhine as the French border. This strategic goal was not reached until the military successes of Napoleon Bonaparte. The French always hoped for a confederation of smaller German states on the right (eastern) side of the Rhine allied with France, which they tried to install in the early 1800s and again after 1918. These attempts failed, and since the early 19th century, German nationalism claimed the Rhine not as a border, but as an integral part of Germany on both sides of the river. German nationalists tried to establish this concept by force and with catastrophic consequences in 1870–71, 1914–18, and 1939–45. After more than 200 years of wars concerning the rule over the Rhine, this dispute was finally settled by the outcome of World War II and by the French-German friendship treaty signed by French President Charles de Gaulle and German Chancellor Konrad Adenauer in 1963.

The economic importance of the Rhine is mostly due to its role as a waterway crossing large parts of western Europe (with canals to French rivers and to the German Main and DANUBE). Due to the engineering work started by Johann Gottfried Tulla in the Grand-Duchy of Baden in the early 19th century, the Rhine is today accessible for ships from Basel via Strasbourg through all of Germany to Rotterdam and the North Sea. The economic importance led to international treaties, which were all signed in Mannheim (especially in 1831 and 1868). This waterway helped the development of all sorts of industry along the Rhine (for example: chemical industry in Basel, Strasbourg, Mannheim and Ludwigshafen; steel and coal in the parts of the Rhine near the Ruhr area). The ports of Strasbourg, Mannheim, and Duisburg are important river ports within western Europe, and the port of Rotterdam is the busiest harbor worldwide. During the last decades, environmental protection of the Rhine both for animals (especially birds and fish) and for the prevention

The Rhine is the longest river in Germany and probably the most important waterway of western Europe.

of floods (Rheinauen) has become a subject of international efforts and treaties.

As a historically disputed frontier, the banks of the Rhine have a great number of fortresses, especially between Mainz and Bonn in Germany. Though mostly ruins now, these structures witnessed the importance of the Rhine as a melting pot for cultural and economic traditions for more than 1,000 years. Today, the ruins, along with notable vineyards, contribute to the tourism industry. The battles of the Allied forces during World War II against Nazi Germany in the Rhine region (especially Arnheim) have contributed to the unfortunately bloody heritage of the Rhine.

BIBLIOGRAPHY. Karl Baedeker, *The Rhine: From Rotterdam to Constance* (Baedeker, 1903); Hans Christian Hoffmann, *The Rhine: Our World Cultural Heritage* (DuMont, 2003); Roland Recht, *The Rhine: Culture and Landscape at the Heart of Europe* (Thames & Hudson, 2001).

OLIVER BENJAMIN HEMMERLE
UNIVERSITY OF MANNHEIM, GERMANY

Rhode Island

RHODE ISLAND, SMALLEST of the 50 United States at 1,214 square mi (3,144 square km), is located in New England on the ATLANTIC seaboard. It is bounded

by MASSACHUSETTS on the north and east, the Atlantic Ocean to the south, and CONNECTICUT in the west. While it is typically referred to as Rhode Island, the official name is actually the State of Rhode Island and Providence Plantations. Interestingly, nearly one-third of the state's total area, 500 square mi (1,295 square km), is water.

In 2000, the state had a populatioN of 1,048,319, an increase of 4.5 percent since the last census in 1990. Although more than half of Rhode Island is covered with forests, it is highly urbanized. Providence is the capital and the largest city; other important cities are Warwick, Cranston, Pawtucket, and Newport. Though the population numbers may not seem big, when coupled with the state's small area, Rhode Island is the second most densely populated state with 864 people per square mi (2,238 per square km). Rhode Island was the 13th of the original 13 colonies to ratify the Constitution. It became a state on May 29, 1790.

Most of the state in the south and east can be described as part of the Coastal Lowland Region while the lands in the northwest are part of the Eastern New England Upland. The highest point is Jerimoth Hill at 812 ft (248 m) above sea level. The dominant physiographic feature of the state is the Narragansett basin, a shallow lowland area of carboniferous sediments that extends southeastward into Massachusetts.

In Rhode Island, the sediments are partly submerged as Narragansett Bay, which cuts inland for about 30 mi (50 km) to Providence. The bay contains several interesting islands, including Rhode Island (or Aquidneck), the largest and the site of historic Newport, Conanicut Island with the Jamestown Resort, and Prudence Island. Numerous sand spits and barrier beaches, in addition to small sheltering lagoons and salt marshes, mark the coastline. The general countryside contains many small lakes and a rolling hilly surface that is punctuated by short, swift streams and numerous waterfalls, all remnant of the last glaciation.

The Massachusetts Bay colony established the first settlement in the area at Providence on land purchased from Native Americans in 1636. In 1638, Puritan exiles bought the island of Aquidneck (now Rhode Island) from the Narragansetts established the settlement of Portsmouth (1638). In order to thwart claims made to the area by rivaling colonies (Massachusetts Bay and Plymouth), Roger Williams secured a parliamentary patent in 1644 and by 1647 had organized a government.

The early settlers were mostly English, with many drawn to the colony by the guarantee of religious freedom. The early settlers were allowed to own land that was bought from the Native Americans. Fishing and trade flourished in addition to a sound livestock industry and the more traditional agricultural products. Until the American Revolution, Newport was the commercial center of the colony, thriving on the triangular trade in rum, slaves, and molasses.

Rhode Island's traditional manufacturing economy has diversified to include important services, trade (retail and wholesale), and finance sectors. In addition, many of the traditional Rhode Island products (jewelry, silverware, textiles, primary and fabricated metals, machinery, electrical equipment, and rubber and plastic) are still being manufactured. While recent events have seen a growth in tourism agriculture has become relatively unimportant. The coastal areas are lined with resorts for swimming and boating, and windswept Block Island is a favorite vacation spot. Narragansett Bay remains famous for its sailing and yachting events, including the America's Cup yacht race that has been held in Newport several times, beginning in 1930 and most recently in 1983.

Most of the state's farmland is now used for dairy and poultry production, with Rhode Island Reds a nationally recognized brand of chickens. Commercial fishing remains important, but the industry is on the decline. U.S. naval facilities at Newport also contribute to the state's income.

BIBLIOGRAPHY. Patrick T. Conley, *A Rhode Island Profile* (Rhode Island Publications Society, 1982); David W. Hoyt, *The Influence of Physical Features upon the History of Rhode Island* (Department of Education, State of Rhode Island, 1910); George H. Kellner and J. Stanley Lemons, *Rhode Island, the Independent State* (Windsor Publications, 1982); William Gerald McLoughlin, *Rhode Island: A Bicentennial History* (Norton, 1978); Ted Klein, *Rhode Island* (Benchmark Books, 1999); U.S. Census Bureau, www.census.gov (August 2004).

RICHARD W. DAWSON
CHINA AGRICULTURAL UNIVERSITY

Rhône River

THE RHÔNE IS THE chief river of southeastern FRANCE, draining much of the western ALPS and connecting the regions of the interior to the MEDITERRANEAN coast. From its rise in the Swiss Alps to its

wide DELTA, the Camargue, the river travels 500 mi (800 km), at first through narrow and twisting mountain valleys, but from Lyon to the sea (170 mi or 275 km), the river follows a nearly direct north-south line through a wide valley separating the alps from the mountains of the Massif Central. Although the river itself is too turbulent for commercial river traffic, its valley has been a thoroughfare from the MEDITERRANEAN SEA to the interior of France for centuries.

The Rhône, together with its major tributaries to the north, the Saône and the Doubs, drains a basin of 39,207 square mi (100,531 square km). It begins on the Rhône glacier (altitude 6,000 ft or 1,743 m) on the Saint Gotthard massif in southern SWITZERLAND, then travels about 90 mi (150 km) through the canton of the Valais before it enters the Lake of Geneva (Lac Léman). It exits the lake at its opposite end, passing through the center of Geneva, then winds its way through the mountainous province of Savoy, taking several sharp turns before it reaches the broad valley of the lower Rhône, where it doubles in size with the confluence of the Saône at Lyon.

Twelve cities with populations greater than 100,000 are located in this basin: Lausanne, Geneva, Lyon, Villeurbanne, Valence, and Avignon on the Rhône itself; Belfort, Besançon, and Dijon to the north in the Saône-Doubs basin; Grenoble and Annecy to the east on Alpine tributaries (Isère and Fier); and St-Étienne, which lies to the west, on the divide between the watershed of the Rhône and that of the Loire. Other major tributaries include the Arve in Switzerland, and the Ain, Ardèche, Gard, and Durance in France.

All of these rivers begin as mountain-fed streams, making the currents of the Rhône after spring melts among of the most powerful in Europe. The upper reaches of the river have therefore been harnessed for hydroelectric power—one of the largest, Génissiat, where the Rhône divides the French Alps from the Jura range (southwest of Geneva, between Savoy and southern Burgundy), was built in 1937–49, creates a reservoir 14 mi (23 km) in length, and generates about 2 million kilowatt hours annually. Other dams and barriers are important for flood control on the river's lower courses, lined with prosperous farms, orchards and vineyards.

But enough water is let through to maintain the river's other notable feature, its heavy levels of silt and mud, carried down from the mountains to the great delta of the Camargue. At 360 square mi (923 square km), the Camargue is the largest delta in western Europe. Roughly 700 million cubic ft (20 million cubic m) of mud are deposited each year, slowly moving the coast further out into the Mediterranean (for example, the town of Aigues-Mortes was originally built on the sea, but is now 3 mi or 5 km inland). The delta is a broad plain of reed marshes and lakes, with brine lagoons cut off from the sea by sandbars, watered by two main channels, the Grand Rhône (with 85 percent of the river's flow) and the Petit Rhône. The delta has been protected as a wildlife refuge since the 1920s and is a haven for over 400 species of wild bird, notably the pink flamingo. Canals divert some of the river traffic to the larger cities of France's Mediterranean coast (Montpellier and Martigues). Other canals have been proposed to link this waterway, via the Saône, to the Rhine in northeastern France, but this has been temporarily shelved since 1997 because of environmental pressure groups. Tourism is also important to the economy of the lower Rhône basin, as visitors to France are drawn in great numbers to the Roman ruins at Orange and the Pont-du-Gard, the papal palaces of Avignon, and river sports in the gorges of the Ardèche.

BIBLIOGRAPHY. C. Revenga, S. Murray, et al., *Watersheds of the World* (World Resources Institute, 1998); *Encyclopedia Americana* (Grolier, Inc., 1997); "Rhone," www.rivernet.org (April 2004).

JONATHAN SPANGLER
SMITHSONIAN INSTITUTION

rift valley

A RIFT VALLEY IS A trenchlike basin with steep parallel sides. The valley is essentially a down-faulted crustal block (grabben) between two parallel faults. A rift valley is different from an ordinary grabben for its remarkable length and depth. Such extraordinary basins typically occur where tectonic plates diverge. The release of Earth's interior energy creates most rift valleys in the processes of seafloor spreading or continental rifting. Igneous intrusions and volcanism accompany the rifting process.

According to the theory of plate tectonics, convective movement of material in the asthenosphere transfers Earth's interior energy to the lithosphere, causing sections of the lithosphere (plates) to rift (tear) and slide laterally. Most rifting takes place over the rising limbs of convection cells in the asthenosphere where low-density material upwells beneath the lithosphere.

Some of the rising material is molten and pushes upward into the lithosphere to form new crustal rock. The rest of the material remains in convective flow; sliding laterally until gravity pulls it into the cell's descending limb.

Rift valleys are caused by rifting continents. Continental rifting results as Earth's interior heat collects beneath a thick continental crust. Thermal expansion or swelling of the crust causes a complex system of high-elevation normal fault (grabben) valleys. The basin-and-range topography of the North American Great Basin region might represent thermal rifting of a sort that has since dissipated. As rifting proceeds, Earth's crust stretches and thins and large blocks displace along tensional fractures paralleling the rift margins. A modest amount of volcanism and igneous intrusions accompany the rifting, as magma pushes into the fractures and onto the surface. These land-based rift valleys or basins represent the first stage in the origin of an ocean basin. The best example of this initial stage is the East African rift system.

Rift valleys on the seafloor. As Earth's lithosphere continues to respond to driving motions in the asthenosphere, copious amounts of basaltic magma injects into the parallel tensional fissures and cools to form dense, heavy rock. The additional weight of the new rock deepens the floors of the rift valleys so that the sea inundates them. The Red Sea is a long, linear sea and is an example of this stage of rifting. The narrow basins of the Dead Sea and Sea of Galilee are continental extensions of RED SEA rifting. The broader present-day oceans are results of long-term rifting and seafloor spreading. Enormously deep rift valleys slice along the axes of huge midocean ridges where the spreading takes place. ICELAND, which is a volcanic exposure of the Mid-Atlantic Ridge, has a readily observable rift valley of this kind.

Ancient rift valleys on passive continental margins. Some rift valleys that formed in the initial stage of rifting remain on the margins of separating continents. Typically, these basins fill in with sediment and become parts of submerged continental shelves. However, local geological situations sometimes preserve fragments of these basins on land. The Triassic Lowland of MARYLAND, PENNSYLVANIA, and NEW JERSEY is a classic example. The Lowlands are an area of rolling land and low ridges. They are remnants of down-faulted basins produced as the supercontinent of Pangaea rifted open during the Triassic and Jurassic periods, about 200 million years ago. Similar Pangaea rift basins occur along the east coast of North America from Labrador to GEORGIA. In their day, these basins probably resembled today's rift valleys of eastern Africa.

Rift valleys caused by continental collisions. Although compressional folding and thrust (reverse) faulting dominate continental plate collisions, rifting associated with normal faulting may also take place. A bulge on the edge of one of the converging continents induces rifting as it pushes into the other continent. In a complex manner, rifting on the continent with the bulge takes place, but it occurs at some distance beyond the axis of collision. In this way, relatively isolated rift valleys can form, such as the RHINE grabben in the HINTERLAND of the ALPS. The Lake BAIKAL grabben, although a great distance from the Himalayan mountain system, probably formed in the same manner.

BIBLIOGRAPHY. Philip Kearey and Frederick J. Vine, *Plate Tectonics* (Blackwell Science, 1996); Cindy Lee Van Dover, *The Ecology of Deep-Sea Hydrothermal Vents* (Princeton University Press, 2000); Webster Mohriak and Manik Talwani, *Atlantic Rifts and Continental Margins, Geophysical Monograph, Vol. 115* (American Geophysical Union, 2000); Ben A. Van Der Pluijm and Stephen Marshak, *Earth Structure: A Introduction to Structural Geology and Tectonics* (W.W. Norton, 2003).

RICHARD A. CROOKER
KUTZTOWN UNIVERSITY

Ring of Fire

The Pacific Ring of Fire is an arc of intense EARTHQUAKE (seismic) and volcanic activity stretching from NEW ZEALAND, along the eastern edge of Asia, north across the ALEUTIAN ISLANDS of ALASKA, and south along the coast of North and South America. It is made up of over 75 percent of the world's active and dormant volcanoes. The Ring of Fire is located along the borders of the Pacific Plate and other tectonic plates. It was recognized and described before the development of the relatively new and generally accepted science of PLATE TECTONICS theory.

Volcanoes are not randomly distributed over the Earth's surface, most are concentrated on the edges of CONTINENTS, along island chains, or beneath the sea, forming long mountain ranges. The peripheral areas of the Pacific Ocean Basin, containing the boundaries of several plates are dotted by many active volcanoes. More than half of the world's active volcanoes that are

above sea level encircle the PACIFIC OCEAN to form the Pacific Ring of Fire. The ring is an example of plate-boundary volcanoes. According to the theory of plate tectonics, scientists believe that the Earth's surface is broken into a number of shifting slabs or plates, which average about 50 mi (80 km) in thickness. These plates move relative to one another above a hotter, deeper, more mobile layer that moves at rates as great as a few inches per year. Most of the world's active volcanoes are located along or near the boundaries between shifting plates and are called plate-boundary volcanoes.

Some active volcanoes are not associated with plate boundaries and are called intraplate volcanoes, which form roughly linear chains in the interior of some oceanic plates. The Hawaiian Islands provide an example of an intraplate volcanic chain, developed by the northwest-moving Pacific Plate passing over an inferred hot spot that initiates magma generation and the volcano formation process.

The Ring of Fire around the Pacific represents one type of this volcanism. The chains of volcanoes in the island arcs (such as the Aleutian Islands) and continental margins (such as the ANDES) around much of the ocean form above moving oceanic plates. Plates are like giant floats on the Earth's surface, which slide next to, collide with, and are forced underneath other plates. Around the Ring of Fire, the Pacific Plate is colliding with and sliding underneath other plates.

This is called subduction and the volcanically and seismically active area is known as the subduction zone. There is a great amount of energy created by these plates; it melts rock into magma that rises to the surface as lava and forms volcanoes. Volcanoes are temporary features on the Earth's surface. There are currently about 1,500 active volcanoes in the world of which 10 percent are located in the UNITED STATES. Volcanic areas in the Ring of Fire include:

South America—the Nazca plate colliding with the South American plate has created the Andes and volcanoes such as Cotopaxi and Azul.

Central America—the small Cocos plate is moving into the North American plate, forming the Mexican volcanoes of Popocatepetl and Paricutun (which rose up from a cornfield in 1943 and instantly became mountains).

Northern CALIFORNIA and British Columbia, CANADA—the Pacific, Juan de Fuca, and Gorda plates created the Cascades and Mount Saint Helens (in washington state), which erupted in 1980.

ALASKA—the Aleutian Islands are growing as the Pacific plate hits the North American plate. The deep Aleutian Trench has been created at the subduction zone with a maximum depth of 25,194 ft (7,679 m).

RUSSIA's Kamchatka Peninsula to JAPAN—the subduction of the Pacific plate under the Eurasian plate formed the Japanese islands and volcanoes (such as Mt. Fuji).

The final section of the Ring of Fire exists where the Indo-Australian plate subducts under the Pacific plate and has created volcanoes in the New Guinea and MICRONESIA areas.

BIBLIOGRAPHY. Robert Decker and Barbara Decker, *Volcanoes* (W.H. Freeman, 1989); Jon Erickson, *Volcanoes and Earthquakes* (Tab Books, 1988); NOAA Ocean Explorer, *Submarine Ring of Fire 2003* (U.S. Dept. of Commerce, National Oceanic and Atmospheric Administration, Office of Ocean Exploration, 2003); "Ring of Fire," Internet Resource Computer File Serial, http://oceanexplorer.noaa.gov (October 2004).

CLARA HUDSON
UNIVERSITY OF SCRANTON

Rio Grande

THE RIO GRANDE was once known as "Rio del Norte," and the first to describe it to Europeans was Captain General Juan de Oñate, whose party of exploration first visited the river on April 20, 1598. The river was then called Corre del Norte, meaning that its current ran from the north. It is the second-largest river in North America, running more than 1,800 mi (2,897 km) from its source in the southern ROCKY MOUNTAINS in COLORADO, to Brownsville, TEXAS, where it drains into the Gulf of Mexico. Its size is exceeded only by the MISSISSIPPI-Missouri river system in the central UNITED STATES, and the Rio Grande is the predominant drainage for the state of NEW MEXICO. The river delineates the southern border of the United States from El Paso, Texas, to the Gulf of Mexico.

A systematic and careful monitoring of the movement of the Rio Grande shows that it enters New Mexico via a spectacular gorge and flows south through a series of alluvium-filled basins produced over millions of years by the rifting process. Rising at an elevation of 9,842 ft (3,000 m), the Rio Grande flows southward about 746 mi (1,200 km) to the United States-MEXICO border, separating El Paso from Cuidad Juarez, Chihuahua. From there, the it flows in a more southeast

direction for approximately 1,243 mi (2,000 km), until it reaches its mouth at the Gulf of Mexico.

The Rio Grande was once useful as a major supplier of surface and ground water for residential, commercial, and agricultural use in the Albuquerque, New Mexico area and in other villages, towns, and cities of the state through which the river flows. (Its international frontier in Texas is largely devoid of urban settlement from El Paso to the more southern reaches of the valley at Del Rio and Laredo.) Due to the extensive 20th-century degradation in water quality and natural habitat—a result of its extreme subjugation to agricultural exploitation—the Rio Grande is considered one of North America's most endangered river systems.

DEGRADATION

Being one of the longest rivers in the United States, and as an international border outside the exclusive control of any nation, perhaps it was inevitable that the Rio Grande would experience serious and uncontrolled degradation of water quality, quantity, and associated habitat. Despite the ability of the river to move about and change flow and location in its river course over most of the Rio Grande valley, it has experienced more frequent periods of stability and equilibrium as hydrological engineers have attempted to control its floods with dams and artificial lakes. Prior to this period of river engineering, riparian (riverine) ecosystems had more opportunities to become established on riverbanks and islands, and deposition and erosion of sediments allowed colonization of riparian vegetation across wide areas of the valley as it proceeded southward.

Over the course of the 20th century, the patterns of change in the Rio Grande have been radically diminished or stopped completely. It has, within the last 100 years, gone from a braided and sinuous set of channels to the flow of a single channel, in many places dredged and engineered so that it actually dries up for whole seasons, as its river waters are diverted to irrigation canals.

The Rio Grande's variegated flow over the last century has caused havoc in the Albuquerque area, as the city and its environs have suffered from major flooding. The record cites severe floods occurring on a fairly regular cycle, with major events in 1874, 1884, 1909, and 1920, to mention a few. In the 1874 flood, it was estimated that water covered 24 square mi (62 square km), filling the area between Bernalillo and Albuquerque for more than three months. While the river's ecology has been seriously compromised, it continues to provide services to towns and farms along its course, supplying irrigation water and drainage, and recharging the Ogallala aquifer that lies beneath its route. It continues to support natural habitats, particularly *cienega* (swamp) structures, which provide avian life with important stopovers in north-south migrations.

BIBLIOGRAPHY. D. Earick, "History and Development of the Rio Grande River," www.unm.edu (October 2004); P. Hogan, *The Rio Grande in North American History* (Wesleyan University Press, 1991; A.R. Jose, "Acequia Culture: Water, Land & Community," www.cerc.usgs.gov (October 2004).

WALTER TETE AND A. CHIAVIELLO
UNIVERSITY OF HOUSTON, DOWNTOWN

riparian

RIPARIAN IS DERIVED from the Latin *riparius,* meaning "of the river bank." The term was historically used to describe the area of land lying adjacent to a body of water: primarily streams or rivers but also lakes. Riparian areas are ecotones—transition areas or interfaces between terrestrial and aquatic ECOSYSTEMS. As such, riparian areas possess features and processes influenced by adjacent ecosystems as well as those unique to riparian habitats.

Historically, different disciplines and professions (fisheries biologists, forest managers, plant ecologists, soil scientists) have defined riparian areas in slightly different ways, emphasizing certain soil properties, plant species or communities or management goals as key characteristics. Common to most definitions, however, is that presence of water and its flow pattern or regimen—the hydrology of the system—is the key distinguishing characteristic of riparian areas. Water delivery, routing, and persistence in the soil help define the boundaries of riparian areas. Because of their position in the landscape, typically in valley bottoms adjacent to streams, riparian areas receive surface water from upland runoff and stream flooding and from subsurface groundwater flows or saturated soil horizons. Riparian areas typically have soils that are saturated by water for at least part of the growing season.

Soils in riparian areas can vary over short distances and often form a moisture gradient consisting of saturated hydric soils along streams to moderately well to well-drained soils above stream level. Decomposition

rates of organic materials (leaves, branches) vary with soil moisture in riparian areas. Decomposition is generally slower in saturated soils than in drier soils owing to cooler temperatures and reduced oxygen availability. Riparian soils are often relatively fertile due to nutrient runoff from adjacent uplands and deposition of nutrient-rich alluvial sediments by flooding.

Riparian definitions have evolved over recent years to include multidimensional considerations of ecological structure and function—an ecosystem or landscape perspective. In this regard, riparian areas are defined as three-dimensional, linear ecotones that extend longitudinally along streams and rivers, vertically from the groundwater zone below the stream channel to the vegetation canopy above, and laterally from the stream bank through the FLOODPLAIN to adjacent uplands. The width and boundaries of riparian areas are variable, influenced by the size of the water body they flank and the geomorphology of the landscape in which they are embedded. In general, riparian areas increase in width with increasing stream size and valley width and are constrained by steep slopes in mountainous areas. Thus, riparian areas along geologically constrained headwater streams in a watershed tend to be narrower in width than those along lower-elevation, low-gradient, meandering reaches at the watershed's base.

Riparian areas are among the most diverse, dynamic, and complex ecological systems in the world. The functional processes and linkages that forge riparian dynamism are a function of the interaction of aquatic and terrestrial ecosystems. Vegetation best illustrates the ecological linkage of aquatic and terrestrial ecosystems in riparian areas. Plant species diversity is often high in riparian areas because the mosaic of soil types and stream flooding disturbance (particularly scour) creates a range of soil and site conditions that favor coexistence of many different plant species in a small area. Riparian vegetation in turn influences aquatic ecosystems by shading stream channels and regulating water temperatures, by providing leaves which form the base of the stream food web, and by the input of large woody debris (logs), that affects water and sediment movement and molds in-stream habitat for aquatic organisms.

Riparian areas have important ecological functions in the landscape and are valued as buffers that protect and enhance water resources. Riparian areas may act as filters of sediments and nutrients from uplands, transformers of toxins, nutrients, and microclimate, sources of species and energy, sinks for excess nutrients, and as habitat and movement corridors for organisms. Worldwide, riparian areas have been degraded by deforestation, grazing and urban development. Recognition of the important ecological functions of riparian areas has stimulated efforts to restore and enhance these diverse and dynamic systems in many areas of the world.

BIBLIOGRAPHY. John F. Kundt, Tim Hall, V. Daniel Stiles, Steve Funderburk, and Duncan McDonald, *Streamside Forests: The Vital, Beneficial Resource* (University of Maryland Cooperative Extension and United States Fish and Wildlife Service, 1988); William J. Mitsch and James G. Gosselink, *Wetlands* (Van Nostrand Reinhold, 1993); Robert J. Naiman and Henri DeCamps, "The Ecology of Interfaces: Riparian Zones," *Annual Review of Ecology and Systematics* (v.28, 1997); Robert J. Naiman, Henri DeCamps, and Michael Pollock, "The Role of Riparian Corridors in Maintaining Regional Biodiversity," *Ecological Applications* (v.3, 1993); David J. Welsch, *Riparian Forest Buffers: Function and Design for Protection and Enhancement of Water Resources* (USDA Forest Service, 1991).

CHARLES E. WILLIAMS
CLARION UNIVERSITY OF PENNSYLVANIA

river

A RIVER IS A LARGE STREAM OF WATER flowing in a bed or channel and emptying into the ocean, a sea, a lake, or another stream. Rivers have a starting point called a source, and a mouth, where they empty into a larger body of water. Its source may be a spring, lake, or mountain snowmelt, and its ultimate destination is generally an ocean, but its mouth may empty into a lake or another river along the way. Small tributaries combine with others to create larger streams; this is repeated many times to form the large main-branch rivers that reach the sea.

Rivers are the sculptors of the landscape, shaping the Earth as they continually transport material from the land to the sea like conveyors. Some of the most permanent geographical features, they can outlive the largest mountains, watching them rise and helping to tear them down. Mountains are symbols of strength and durability, yet rivers cut through them like temporary annoyances, finding and exploiting their weak spots to plow water gaps and gouge canyons.

As the demolition of mountain ranges is taking place, so is the creation of other landforms. Sediments

carried from upstream are deposited at the river's mouth, forming a DELTA. Deltas are low, flat, and wet areas that provide much needed wetland habitat. As the river approaches the sea, it will sometimes separate into many channels forming a bird's foot delta.

The deltas of the MISSISSIPPI and NILE Rivers are good examples. Floodplains are created beside rivers when sediments are deposited by floodwater. These plains are broad, flat valleys composed of organic rich sediments, making them ideal lands for cultivation. However, repeated inundation makes them risky places for human settlement.

CRADLES OF HUMANITY

Rivers have played an important role as cradles of human civilization. The Mesopotamians flourished along the banks of the Tigris-Euphrates. The rivers provided them with a food source, transportation, and most important, fresh water.

The Euphrates is an exotic river, or a river that flows through an arid region. In the dry climate of what is now IRAQ, the Euphrates valley was a lush oasis. Its water was not only used for drinking, cooking, and bathing, but also for the irrigation of crops and a source of fish. Canals, early pumps, and augers flooded fields with river water, enabling the desert floor and early human civilization to flourish. This happened along many other rivers as well, such as the INDUS, Nile, and the CHANGJIANG (Yangzi) and HUANG (Yellow) rivers in CHINA.

Humans have not distanced themselves from rivers in modern times. River valleys are still favored locations for population centers. America's early urban areas developed around mills and factories that harnessed the power of rivers cascading over the Piedmont. These rivers provided water, energy, and transportation to the sea. Rivers powered the INDUSTRIAL REVOLUTION in the UNITED STATES, helping to strengthen the new nation. Rivers farther west would be the arteries that helped expand it.

Transportation has always been an important use of rivers. Before jumbo jets and freeways, rivers were the world's highways. They have provided paths of exploration and trade routes of commerce. Movements of people and goods have followed river routes since the ancient Mesopotamians transported grain on reed rafts. In more recent times, rivers served as paths for exploration and expansion of the western United States. LEWIS AND CLARK used the Missouri River to travel and survey the vast lands of the LOUISIANA PURCHASE in 1804.

Robert Fulton's invention of the steamship made river travel viable upstream as well as downstream. Waterways like the Mississippi and Ohio rivers opened up world markets to formerly LANDLOCKED producers. Steamships dominated transportation for a short time in the 1800's, but were replaced by more versatile railroads.

Today, barges still carry large freight loads, but in many places rivers are now considered barriers to transportation rather than the source of it. Networks of highways and railroads can move cargo to more places in less time. After centuries of serving as essential travel routes, the river's role in transportation is quickly fading in many parts of the world. Rivers have not been treated well for their help in advancing civilization. Instead, they have been treated like prisoners and sewers. Rivers today have been polluted by industry and incarcerated by engineers. Very few remain untouched.

Pollution may be the worst illness rivers face. Chemicals from industry, waste from human settlement, and agricultural runoff have all acted to degrade the health of rivers and their ecosystems. Fish, plants and animals are not the only victims. By poisoning our fresh water supply we poison ourselves. Ignorance is the excuse of the past, and with gained knowledge and awareness, river conservation is the responsibility of all humans.

Dams and levees constrict the flow of rivers and confine them to narrow channels. Flood control structures prevent rivers from spreading over the floodplains they took several millennia to create. Natural fertilization of the plains has been restricted, forcing farmers to apply chemical fertilizers. While the frequency of floods has been reduced, the magnitude has increased. Rivers have often been compared to the human life cycle. Upper reaches of rivers are usually small, fast-moving streams that are seen as adolescents with youthful energy. In their lower reaches they become wide and slow in appearance, carrying with them the sediments and memories of the miles they've traversed. Rivers have provided us with water, power, and transportation. But those who live on their banks may argue that their most precious gift is a sense of place.

BIBLIOGRAPHY. John Bardach, *Downstream: A Natural History of the River* (Harper and Row, 1964); Frank Bergon, ed., *The Journals of Lewis and Clark* (Penguin, 1989); John M. Kauffmann, *Flow East: A Look at Our North Atlantic Rivers* (McGraw-Hill, 1973); Andy Russell, *The Life of a River* (McClelland and Stewart, 1989); Ludwik

A. Teclaff, *The River Basin in History and Law* (Martinus Nijhoff, 1967).

DANE BAILEY
UNIVERSITY OF KANSAS

Rocky Mountains

THE ROCKY MOUNTAINS are a chain of mountain ranges 3,000 mi (4,800 km) long and as wide as 350 mi (563 km) running predominantly north to south in the western part of the North American continent. The Rockies run through the U.S. states of NEW MEXICO, COLORADO, UTAH, WYOMING, IDAHO, MONTANA and ALASKA, as well as the Canadian provinces of Alberta and British Columbia and the Yukon and Northwest Territories. They have been described poetically as the continent's spine.

The Rockies were originally formed approximately 100 million years ago, during the Cretaceous period, when dinosaurs still walked the Earth. There was a second period of uplift within the past 25 million years, and Pleistocene glaciers further reshaped many of the major valleys, widening and rounding their bottoms. Some small glaciers have survived high in the northern Rockies, remnants of the great Ice Age. The best known may be found in Glacier National Park in Montana on the Canadian border. Recently there has been concern that global warming is causing these glaciers to shrink and some of the smaller mountain glaciers may disappear altogether.

The Rocky Mountains are rich in wildlife to match the rugged scenery. Piñon pine and juniper are common in the southern Rockies, while they give way to firs, pines, and spruces further north. Beyond the timber line (the altitude above which trees cannot grow), mountain goats and bighorn sheep browse in alpine meadows. The forests downslope are home to bears, deer, elk, hares, minks, cougars, porcupines, squirrels, and other woodland creatures. The rivers of the region are rich with fish such as rainbow trout and grayling.

Within the Rocky Mountains lies the Continental Divide, which separates the waters that flow to the PACIFIC OCEAN from those which will flow into the ATLANTIC. The Rocky Mountains contain the headwaters of such rivers as the Arkansas and Missouri (tributaries of the MISSISSIPPI), the RIO GRANDE (which forms the border between TEXAS and MEXICO), the Colorado (which formed the GRAND CANYON and ultimately

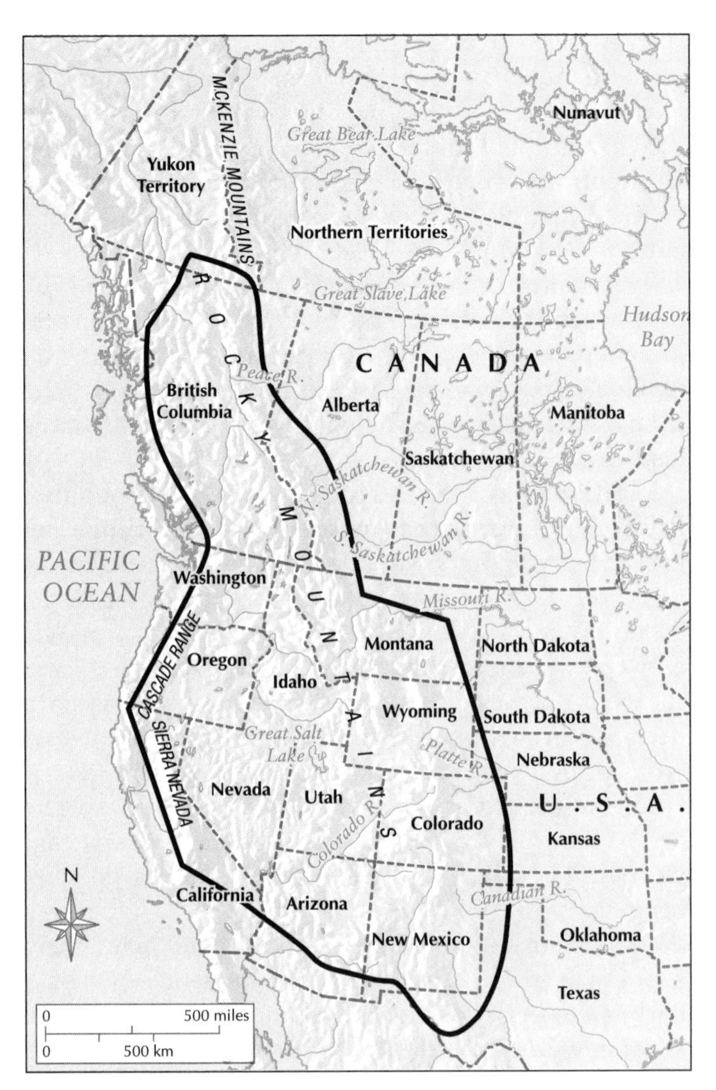

Many of the ranges within the Rockies are fold and fault-block mountains, but there are also extinct and dormant volcanoes.

flows into the Gulf of Cortez) and the Columbia (principal river of the U.S. northwest).

The most famous mountain of the Rockies is Pikes Peak in Colorado, which is 14,110 ft (4,301 m) tall. Another notable mountain in the range is Cheyenne Mountain, which houses the underground command center for NORAD (the North American Aerospace Defense Command). That complex was constructed by blasting an enormous artificial cave out of the rock of the mountain and was intended to survive a direct attack by Soviet nuclear missiles.

However, most activity in the Rocky Mountains is of a more peaceful nature. Historically, mining has been the chief industry in the Rockies. The mountains contain a wealth of minerals, including gold, lead, molybdenum, silver, zinc and soda ash. Fossil fuels

such as coal, petroleum, and natural gas to fuel hungry lowland industries have also been a valuable source of wealth. Lumbering has been an important economic activity, although timber companies have moved from clear-cutting of natural growth to the managed harvesting of planted stands, making it more akin to agriculture than mining.

Agriculture in the Rockies is primarily regulated by the availability of water. Although the use of irrigation has changed the established patterns of agriculture by making water available where it was previously scarce, it has brought other problems, including water table depletion and soil salination.

Herding has been the primary form of agriculture in the high mountains, with cattle and sheep being moved between upper and lower pastures according to the seasons. Because sheep can graze far closer to the ground than cattle, there have been frequent conflicts between shepherds and cattle ranchers, who accuse sheep of destroying the range. These problems have been exacerbated in the UNITED STATES by federal government policies that permitted herders to rent federal lands for grazing at prices far below the market value of the land. In the foothills farmers raise crops that work well in small plots, especially vegetables and chili peppers.

The picturesque scenery of the Rockies make them a popular tourist destination. Yellowstone National Park, an enormous caldera volcano slumbering in the Wyoming Rockies, is famous for its hot springs and geysers as well as its abundant wildlife. The rugged mountains also make for excellent ski slopes. Well-known ski resorts in the Rockies include Vail and Aspen, Colorado; Taos, New Mexico; and Jackson Hole, Wyoming. The abundance of wildlife draws hunters and sports fishermen.

However, the growth of the tourist industry has not been without its problems. In addition to the obvious problems of environmental degradation, developments adjacent to wilderness have brought increasing conflicts with humans and wildlife. Almost invariably wildlife is the loser in the encounter, particularly when the association has led to an erosion of wild creatures' fear of humanity. Bears, cougars, and other predators come wandering through suburban developments, attacking pets and even their masters, leading to the destruction of the predator as a threat. Deer and elk crossing the highways become increasingly likely to be struck and killed by traffic, although passengers in smaller cars are also apt to suffer injuries or even death.

BIBLIOGRAPHY. J. A. Kraulis, *The Rocky Mountains: Crest of a Continent* (Facts on File, 1987); James V. Murfin, *The National Parks of the Rockies* (Crescent, 1988); Jeremy Schmidt, *The Rockies: Backbone of a Continent* (Thunder Bay Press, 1990); Bob Young, *The Story of the Rocky Mountains* (Hawthorn Books, 1969).

LEIGH KIMMEL
INDEPENDENT SCHOLAR

Romania

Map Page 1133 **Area** 91,000 square mi (237,500 square km) **Population** 22,355,551 **Capital** Bucharest **Highest Point** 8,346 ft (2,544 m) **Lowest Point** 0 m **GDP per capita** $2,310 **Primary Natural Resources** oil, natural gas, timber, coal, iron ore.

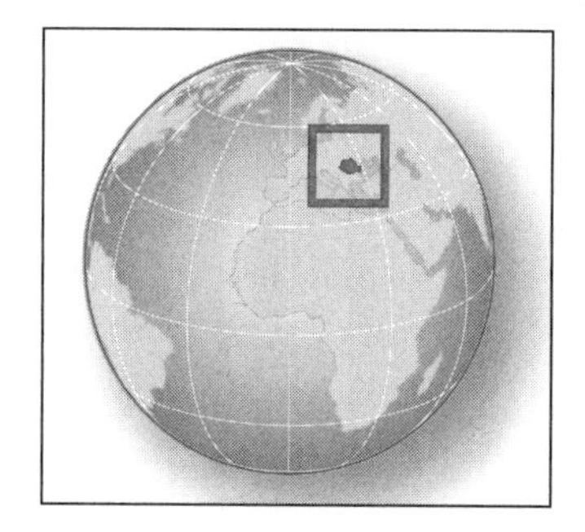

ROMANIA IS AN OVAL-shaped country in southeastern Europe. Situated in the northeastern part of the Balkan Peninsula, it is halfway between the ATLANTIC OCEAN and the URAL MOUNTAINS and also halfway between the North Pole and the equator. It is slightly smaller than OREGON or about half the size of FRANCE.

The eastern border of Romania is the BLACK SEA. Moving in a counterclockwise direction, Romania is bordered by UKRAINE and MOLDOVa in the northeast, Ukraine in the north, HUNGARY in the west, SERBIA AND MONTENEGRO in the southwest, and BULGARIA in the south. The DANUBE river flows for nearly 900 mi (1,400 km) mostly through featureless plains on the southern and eastern border with Serbia, Bulgaria, Moldova, and Ukraine. Romania has six land regions: Dobruja, Moldavia, Bukovinia, Transylvania, Walachia, and Banat.

Dobruja is the driest region of Romania. Its largest city is Constansa, located on the Black Sea. Dobruja is formed by the change in direction from east to north by the Danube River, which bends east again at its confluence with the Siretul and Prut rivers. The northern third of Dobruja is the Danube's DELTA, which borders the Ukraine. The delta is one of Europe's great marsh regions. It is now a protected BIOSPHERE.

North of Dobruja is the Moldavia region. It lies between the eastern Carpathians and the Prut River,

which borders the country of Moldova (Bessarabia). It is a plain with small hills and farmlands centered on both sides of the Siretul River and its tributaries.

Bukovinia is in the highest part of the Carpathian Mountains. It is thickly forested and dotted with small villages and ski resorts. Transylvania is the largest and most varied region in Romania. It lies in the central and northwestern part of the country. The Transylvanian Plateau is a basin bordered on the east by the eastern (Moldovian) Carpathians; on the west, by the Bihor Mountains; and by the Southern Carpathians (Transylvanian Alps) on the south.

Banat lies in southwestern Romania. Bordering Hungary in the northwest, Serbia in the west and Bulgaria in the southwest, it is the region with the greatest concentration of ethnic minorities—Hungarians, Germans, and Serbs. Its major city is Timisoara. Walachia is in the south. It rises in elevation northward from its Danube River border with Bulgaria to the Transylvanian Alps. The Olt River divides Walachia into Oltenia on the west and Muntenia on the east. The Danube forms a natural boundary between Walachia and Dobruja. Bucharest is located in the center of the Muntenia on the Buflea River.

Romania's Transylvanian Alps region is famous for its stories of vampires. The Dracula castle depicted in Bram Stoker's novel *Dracula* is located at Brasov. The historical Count Dracula was Vlad Tepes ("the Impaler"), whose castle was at Targoviste.

BIBLIOGRAPHY. Lucian Boia, *Romania* (Reaktion, 2001); Ronald D. Buckman, ed., "Romania: A Country Study," (Library of Congress, 1991); Julian Hale, *The Land and People of Romania* (J.B. Lippincott, 1972); Nicolae Keppler, *Romania: An Illustrated History* (Hippocrene Books, 2002); Tiberiu Morariu, Vasile Cucu, and Ion Velcea, *The Geography of Romania* (Meridiane Publishing, 1962).

Andrew J. Waskey
Dalton State College

rotation, Earth axis

EARTH HAS TWO primary motions: revolution and rotation. The first refers to the earth's annual orbit of the sun, which takes a bit more than 365 days per year (hence, a leap year with 366 days every four years in order to "catch up"). As it revolves around the sun, the Earth rotates on its axis once every 24 hours, a period of time referred to as a mean solar day or sol. The axis of rotation is an imaginary line that passes through both the geographical North and South Poles.

It is important to note that the Earth's axis is tilted approximately 23.5 degrees from the ecliptic, defined as an imaginary plane described by the revolution of the earth around the sun. The direction of the Earth's rotation may be determined by viewing the Earth from a point in space above either pole.

An observer above the North Pole would note a counterclockwise motion of the Earth to the east. An observer would see a similar eastward motion over the South Pole, but from this perspective the motion would be clockwise. The eastward rotation of the Earth accounts for the apparent westward motion of the celestial bodies (sun, moon, stars) throughout the 24-hour period.

An observer in the Northern Hemisphere on a clear night can gain evidence of rotation by noting the apparent east to west movement of stars in the vicinity of Polaris, which aligns well with the Earth's axis and remains in a stationary location. The speed of the Earth's rotation varies depending on the latitudinal position of the observer. The speed is greatest at the equator where the circumference of the Earth is at a maximum. The general rule for determining the speed of the Earth at any latitude is straightforward:

$$\text{speed} = \text{circumference of latitude}/24$$

The circumference of the earth at the equator is approximately 25,000 mi (40,000 km). Applying the expression using these inputs results in a speed of 1,150 mi per hour (1,700 km per hour). At the latitudes of 90 degrees north and 90 degrees south (the poles) the speed of rotation is zero. Latitudes between the poles and the equator will have rotational speeds more than zero and less than that of the equator. Rotation should not be interpreted as a spinning motion. The Earth rotates, and even though this motion has similarities to the spinning motion of a top or a figure skater, it is not the same.

Spinning implies a rapid whirling motion not linked to a specified axis. Both the top and the figure skater spin in association with a wavering axis of motion. The Earth's rotation, on the other hand, is regular and invariably related to its clearly specified, observable, and measurable axis. The motions of revolution and rotation, both universally accepted now, were not understood until well into the 16th century. For centuries, the Ptolemaic system held that the Earth was the

center of the universe and all celestial bodies revolved around it.

COPERNICAN THEORY

This theory held sway in the scientific word from the time of PTOLEMY throughout the Middle Ages until it was proven false in the 16th and 17th centuries. The first to proclaim that the Earth and the other planets revolved around the sun was Nicolas Copernicus, a Polish astronomer who published his theory in 1543, the year of his death. Copernicus also claimed that the Earth rotated on its axis. Additional support for Copernicus came from Johannes Kepler, a German astronomer who rejected Ptolemy's concept of circular revolution and proposed the idea of the elliptical motion of the planets. Finally, it was Galileo who demonstrated the accuracy of the Copernican theory and developed a comprehensive mathematical proof of the heliocentric system.

The Earth's rotation produces a constantly changing diurnal (daily) system of daylight and darkness, which is sensed and responded to by plants and animals alike. Rotation also produces changes in the amount of heat accumulated and lost during the 24-hour diurnal cycle. Another interesting result of rotation is the diversion of air masses in the atmosphere in predictable directions, phenomena known as Coriolis force.

Due to the Earth's rotation, high-pressure air masses in the Northern Hemisphere will be diverted in a clockwise direction, whereas low-pressure air will divert in a counterclockwise direction. These directional diversions are reversed in the Southern Hemisphere.

BIBLIOGRAPHY. George J. Demko, *Why in the World: Adventures in Geography* (Anchor Books, 1992); William D. Pattison, "The Four Traditions of Geography," *Journal of Geography* (v.63, 1964); Alisdair Rogers, Heather Viles, and Andrew Goudie, *The Student's Companion to Geography* (Basil Blackwood, 1992).

GERALD R. PITZL, PH.D.
MACALESTER COLLEGE

Rub' al-Khali

THE RUB' AL-KHALI (Arabic for "Empty Quarter") is located in the southern part of the Arabian Desert. The Al-Murrah Bedouin, who roam its southern edges call it the ar-Ramlah, the "sand." It covers an area of about 250,000 square mi (647,500 square km). It is somewhat smaller than TEXAS or about as large as FRANCE together with BELGIUM and Holland.

Most of the Rub' al-Khali is in southern and southeastern SAUDI ARABIA. It covers about a fourth of that country. It is about 700 mi (1,127 km) long east to west and about 400 mi (644 km) wide north to south. The northern boundary of the Rub' al-Khali is the central plateau (Nedj) of Saudi Arabia. While most of it is in Saudi Arabia, the southern boundary overlaps the borders of YEMEN and OMAN. In the east, the Rub' al-Khali overlaps the boundaries of the UNITED ARAB EMIRATES.

The Rub' al-Khali slopes from an altitude of about 3,300 ft (1,006 m) in the west to close to sea level in the east. The Rub' al-Khali is connected to the Nafud sand desert in northern Saudi Arabia by the Dahna belt of sand dunes. A northwesterly wind, called the shamal, shapes the dunes making them into an active sea of shifting sand. The shamal grows in force each day with the heating of the air. It also causes sandtorms in the eastern part of Saudi Arabia. During the February to March monsoon season, the wind blows mainly from the south. The wind forms the sand dunes into many shapes.

Aerial and space photography have shown that the dunes are arranged in belts, but take many shapes. In the western area of the Rub' al-Khali are many linear dunes that run for many miles in a northeast-southwest direction. In the north-central area of the Rub' al-Khali is the great Wabar impact crater created by a large meteorite. A number of meteors have been found in the vast sand dunes. The dunes in this area are also often crescent-shaped. Some of the crescent dunes have fish-hook shapes at their end or are like scimitars. Some dunes are red sand.

The eastern part of the Rub' al-Khali fills a broad, shallow basin that slopes toward the southern shores of the PERSIAN GULF. The eastern area is relatively level but covered with salt flats (*sabkhas*) in many areas. The *sabkhas* can be quagmires.

In the south, the water from the wadis flowing off the coastal plateau that borders the entire southern end of the Arabian peninsula disappears in the sands of the Rub' al-Khali. In Oman, this drainage creates a very dangerous region of quicksands and poisonous bogs called the Umm al Samim.

Most of the Rub' al-Khali is uninhabited and much of it is some of the driest land on Earth. Temperatures in the Rub' al-Khali can reach 130 degrees F (50 de-

grees C) or higher. The humidity is usually quite low. Very little fauna and flora are found in the Empty Quarter.

The rains that do fall are either in the northern area from winter rains or from monsoon rains off the ARABIAN SEA. These rains can stimulate the growth of vegetation lasting up to three years. The Rub' al-Khali was much wetter in prehistoric times. Bones of animals and flint arrowheads have been found in a number of locations. Many desert plants do grow when the rare rains come. There are also numerous insects, snakes, and other animals.

In 1992, the "Atlantis of the Rub' al Khali" was uncovered from a layer of sand. The ancient fort is believed to be the legendary lost city of Ubar. American archaeologists located the ruins with radar images taken by the ill-fated space shuttle *Challenger*. Ubar was famous in ancient times as a city of great wealth that produced frankincense.

The east area has many oil and gas fields. The al-Ghawar oil field (discovered in 1948) is one of the largest in the world. The Al-Murrah, the Rashidi, and other Bedouins have wandered the eastern region, but the advent of oil has forced the settlement of most of the Bedouin.

BIBLIOGRAPHY. G, Edgell, "Evolution of the Rub' al Khali," *Journal of King Abdul Aziz University, Earth Science* (v.3, 1989); Edward Elberg, Ruel D. Geirhart, and Leon F. Ramirez, *Geology of the Eastern Rub al Khali Quadrangle, Kingdom of Saudi Arabia* (U.S. Geologic Survey, 1963); Bruce Kirkby, *Sand Dance: By Camel across Arabia's Great Southern Desert* (McClelland & Steward/Tundra Books, 2002); H. Saint John B. Philby, ed., *The Empty Quarter, Being a Description of the Great South Desert of Arabia Known as the Rub' Al Khali* (Henry Holt, 1933); Donald Powell, *Nomads of the Nomads* (Aldine, 1975).

ANDREW J. WASKEY
DALTON STATE COLLEGE

Ruhr Valley

IN TERMS OF LENGTH and volume, the Ruhr River is not one of GERMANY's major rivers. But considered as a center of economic activity, the Ruhr is among the most prominent rivers in Europe. Starting in the early 19th century, the coal that was mined in this region was processed in factories all along the river's course, creating one of the world's largest industrial centers, notably for steel production, until coal ceased to be a dominant energy source in the second half of the 20th century.

The Ruhr Basin covers roughly 1,200 square mi (3,000 square km), from Arnsberg in the east to the river's confluence with the RHINE at Hamborn and Duisberg, overlying one of the largest coal deposits in the world. Today, the Ruhr area, or Ruhrgebeit (Ruhrpott in local slang), forms the largest conurbation in the state of Northrhine-Westphalia, consisting of ten major cities: Duisberg, Oberhausen, Bottrop, Mülheim, Essen, Gelsenkirchen, Bochum, Herne, Hagen, and Dortmund. Altogether, these cities contain about nine million people, the third-largest urban area in Europe. It is Germany's most densely populated region, stretching from the Lippe River in the north to the hills of the Bergisches Land to the south, and from the city of Hamm in the east to the Rhine River in the west.

The river itself begins in the uplands of the Sauerland, near Winterberg, and flows about 100 mi (160 km) to the Rhine. Formerly an area of small villages and pastures, far from the political and urban centers of Germany, the Industrial Revolution brought enormous changes to the area and large cities grew up nearly overnight. The area was a principal target for Allied troops in both world wars as the primary arsenal for German war machinery, the center of the "steel empires" of Krupp and Thyssen. Since the 1960s, the area has had to diversify, first toward manufacturing, then to the service and high-technology industries.

The Ruhr has been a model for eradicating air and water pollution and is today the site of numerous artificial lakes, parks, and bike trails, integrated within the highly populated residential areas. The Zollverein Coal Mine Industrial Complex in Essen, for example, once the largest and most modern colliery in Europe, was converted from a decaying facility into a modern technology museum and a monument to industrial art. A side valley in the Ruhr region has also given its name to a major chapter in human prehistory, the Neanderthal.

BIBLIOGRAPHY. *Oxford Essential Geographical Dictionary* (Oxford University Press, 2003); "Ruhr Valley," www.germany-tourism.de (August 2004); "Ruhrpott," www.about-germany.org (August 2004); "Ruhr Valley," www.dw-world.de (August 2004).

JONATHAN SPANGLER
SMITHSONIAN INSTITUTION

Russian Federation

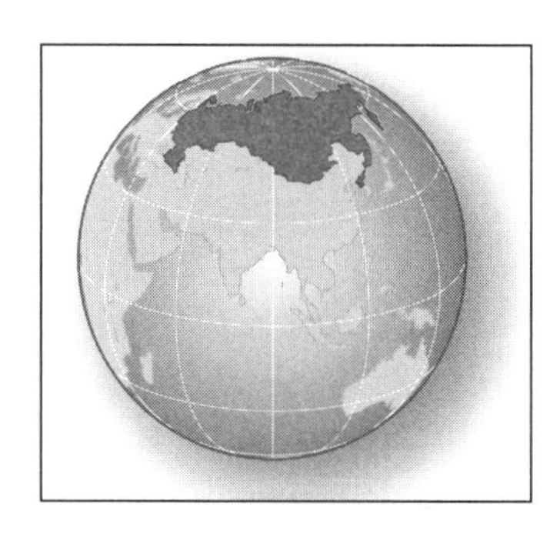

Map Page 1118 **Area** 6.6 million square mi (17 million square km) **Population** 144 million **Capital** Moscow **Highest Point** 18,476 ft (5,633 m) **Lowest Point** -91 ft (-28 m) **GDP per capita** $9,300 **Primary Natural Resources** oil, natural gas, coal.

STRETCHING IN A GIGANTIC arc around the ARCTIC OCEAN and North Pole, the Russian Federation spans 11 time zones, nearly half the globe from east to west. Russia is by far the world's largest country, occupying much of Eastern Europe and northern Asia. The country includes one-eighth of the Earth's inhabitable land area. The smaller European portion is home to most of Russia's industrial and agricultural activity. The URAL MOUNTAINS, which divide Europe and Asia, also separates the Great Russian Plain in the east from the West Siberian Plain.

From east to west, the land gradually rises to form the Central Siberian Plateau. In the south, the Caucasus region separates the BLACK SEA from the CASPIAN SEA. The tundra of northern Russia has scant vegetation of mostly scrub plants and lichen. South of this high Arctic zone is a great forested zone known as the TAIGA, beyond which lie the great STEPPES, or GRASSLANDS of Central Asia.

The Russian Far East is mountainous, and the KAMCHATKA PENINSULA contains active volcanoes and hot springs. Asian Russia is about as large as CHINA and INDIA combined, occupying roughly three-quarters of the nation's territory. But it is the European western quarter that is home to more than 75 percent of Russia's people.

This acutely uneven distribution of human and natural resources is one of the striking features of Russian geography. The country's terrain is diverse, with extensive stands of forest, numerous mountain ranges, and vast plains. On and below the surface the land has extensive reserves of natural resources that provide the nation with enormous potential wealth. Russia ranks sixth in the world in population, trailing China, India, the UNITED STATES, INDONESIA, and BRAZIL. The population is as varied as the terrain. Slavs (Russians, Ukrainians, and Belarussians) are the most numerous of the more than 100 European and Asiatic nationalities. Russia has borders with NORWAY, FINLAND, ESTONIA, LATVIA, LITHUANIA, BELARUS, and the UKRAINE in the west; GEORGIA, AZERBAIJAN, KAZAKHSTAN, MONGOLIA, and China in the south; the PACIFIC OCEAN in the east, and the Arctic Ocean in the north.

Politically, the country is organized as a federation that is divided into 89 regions. The president serves as the head of state, while a prime minister serves as the head of government. In addition to the president and prime minister, leadership is managed through a bicameral legislature consisting of a Federal Assembly that represents the 89 regions and a state Duma that provides popular representation. The major cities are MOSCOW, SAINT PETERSBURG, Kiev, Minsk, Novgorod, Volgograd, Kaliningrad, Murmansk, Novosibirsk, Irkutsk, Pskov, and Vladivostok.

GEOGRAPHIC OVERVIEW

Geographically, it has been traditional to divide Russia's vast territory into five natural zones: the tundra; the taiga, or forest; the steppe, or plains; an arid zone; and a mountain zone. In broad geographic terms, most of the Russian landscape consists of two plains (the East European Plain and the West Siberian Plain), two lowlands (the North Siberian and the Kolyma), two plateaus (the Central Siberian Plateau and the Lena Plateau), and a series of mountainous areas in the extreme northeast or intermittent scattered in pockets along the southern border. The East European Plain encompasses most of European Russia, while the West Siberian Plain (the world's largest) extends east from the Urals to the Yenisey River. Because the terrain and vegetation are relatively uniform in each of the natural zones, the Russian landscape appears to be uniform. Despite this illusion, however, Russia contains all of the major vegetation zones with the exception of a tropical rain forest.

About 10 percent of Russia's land is a treeless, marshy plain or tundra located above the ARCTIC CIRCLE. This northernmost zone stretches from the Finnish border in the west to the Bering Strait in the east before running south along the Pacific coast to the northern end of the Kamchatka Peninsula. It is an area known for its vast herds of wild reindeer, for the so-called white nights of summer (dusk at midnight, dawn shortly thereafter), and for seemingly endless days of total darkness in winter. The long, harsh winters and lack of sunshine allow only mosses, lichens, and dwarf willows and shrubs to sprout in a very narrow zone just above the barren permafrost. Although several major Siberian rivers traverse this region, their partial and intermittent thawing hampers drainage of the nu-

merous lakes, ponds, and swamps of the tundra. This is a landscape that was severely modified by glaciation in the last ice age. Less than 1 percent of Russia's population lives in this zone above the Arctic Circle. Among the regions major employers are fishing, the port industries of the northwestern KOLA PENINSULA and the huge oil and gas fields of northwestern Siberia.

The taiga, the world's largest forest region and the largest natural zone in the Russian Federation, is about equal in size to the United States. The taiga zone extends in a broad band across the country's middle latitudes, stretching from the Finnish border in the west to the Verkhoyansk Range in northeastern Siberia in the east and as far south as the southern shores of Lake BAIKAL near the Mongolian border.

Because much of the Taiga is above 60 degrees north latitude, the forest contains mostly coniferous spruce, fir, cedar, and larch, species well adapted to the long winter conditions that frequently bring the world's coldest temperatures. Isolated sections of taiga also exist along the southern part of the Urals and in the AMUR RIVER valley bordering China in the Far East. About one-third of Russia's population lives in this zone.

The steppe has long been depicted as the typical Russian landscape, although most of the former Soviet Union's steppe zone was located in what are now the Ukrainian and Kazakh republics. The much smaller Russian portion of that steppe is located mostly between those nations then extends southward between the Black and Caspian seas before blending into the increasingly desiccated territory of the Republic of Kalmykia. The steppe itself is a broad band of treeless, grassy plains that extend from HUNGARY in the west across the Ukraine, through southern Russia, and into Kazakhstan and Mongolia before ending in northeast China. Within the vast Russian landscape, the steppe provides the most favorable conditions for human settlement and agriculture because of relatively moderate temperatures and normally adequate levels of sunshine and moisture.

Russia has nine major mountain ranges, with the eastern half of the country being more mountainous than the western half. Russia's mountain ranges can be found along its continental divide (the Urals), in the Caucasus region along the southwestern border, along the border with Mongolia, and in eastern Siberia. The Urals are the most famous of the country's mountain ranges, containing quite large and valuable mineral deposits. As the natural boundary between Europe and Asia, the range extends about 1,304 mi (2,100 km) from the ARCTIC OCEAN to the northern border of KAZAKHSTAN. From Kazakhstan, the divide continues another 854 mi (1,375 km) from the southern end of the Ural Mountains through the Caspian Sea to the CAUCASUS MOUNTAINS. In terms of elevation and vegetation, however, the Urals are far from impressive, nor do they represent any formidable natural barrier. The highest peak, Mount Narodnaya, is 6,212 ft (1,894 m), lower than the highest of the APPALACHIAN MOUNTAINS. In addition, there are several low passes that provide major transportation routes through the Urals eastward into SIBERIA.

East of the Urals is the West Siberian Plain, extending about 1,180 mi (1,900 km) from east to west and about 1,490 mi (2,400 km) from north to south. With more than half its territory below 1,640 ft (500 m) in elevation, the plain contains some of the world's largest swamps and floodplains. Most of the plain's population lives in the drier section, which is generally south of 55 degrees north latitude.

The region directly east of the West Siberian Plain is the Central Siberian Plateau, which extends eastward from the Yenisey River valley to the Lena River valley. The Yenisey valley, which delineates the western edge of the Central Siberian Plateau from the West Siberian Plain, runs from near the Mongolian border northward to the Arctic Ocean. It is also the traditional dividing line between what the Russians think of as eastern and the western Russia. The region is divided into several plateaus, with elevations ranging between 1,050 to 2,400 feet (320 and 740 m) and the highest elevation of about 5,900 ft (1,800 m) in the northern Putoran Mountains.

Truly alpine terrain can be found in the southern mountain ranges between the Black and Caspian seas. The Caucasus Mountains rise to impressive heights, forming a boundary between Europe and Asia. They also create an imposing natural barrier between Russia and its neighbors to the southwest, Georgia and Azerbaijan. One of the peaks in the range, Mount ELBRUS, is the highest point in Europe, at 18,505 feet (5,642 m), and a popular mountaineering climb. The geological structure of the Caucasus extends to the northwest as the Crimean and Carpathian mountains and southeastward into Central Asia as the TIAN SHAN and Pamirs.

The mountain systems west of Lake Baikal in south-central Siberia contain a number of ranges, with peak elevations ranging from the 10,500 ft (3,200 m) in the Eastern Sayan to 14,760 ft (4,500 m) at Mount Belukha in the Altay Range. The Eastern Sayan reach nearly to the southern shore of Lake Baikal, where they

rise impressively some 9,315 ft (2,840 m) above the water. The mountain systems east of Lake Baikal are lower, forming a complex of minor ranges and valleys that extend eastward to the Pacific coast.

Northeastern Siberia, north of the Stanovoy Range, is an extremely mountainous region. The long Kamchatka Peninsula, which juts southward into the Sea of Okhotsk, includes many volcanic peaks, more than 20 of which are still active. The highest of these is the 15,580-ft (4,750-m) Klyuchevskaya Volcano, which is also the highest point in the Russian Far East. The Kamchatka region is also one of Russia's two main centers of seismic activity (the other is the Caucasus), and earthquakes are common. In 1994, a major earthquake largely destroyed the oil-processing city of Neftegorsk.

DRAINAGE BASINS

Russia has thousands of rivers and inland bodies of water, providing it with one of the world's largest surface-water resources. However, most of Russia's rivers and streams are part of the Arctic drainage system, extending across sparsely populated Siberia. Of the Russian rivers longer than 620 mi (1,000 km), 40 are east of the Urals, including the three major rivers that drain Siberia as they flow northward to the Arctic Ocean: the Irtysh-Ob' system, the Yenisey, and the Lena. The basins of these river systems cover about 3 million square mi (8 million square km) and discharge nearly 1.7 million cubic ft (50,000 cubic m) of water per second into the Arctic Ocean.

The northward flow of these rivers, however, means that their source waters come from areas that thaw before the areas downstream. This buildup of water each spring has created vast swamps, such as the Vasyugane Swamp, in the center of the West Siberian Plain. The same is true of other river systems, including the Pechora and the North Dvina in Europe and the Kolyma and Indigirka in Siberia. The result of all this is that approximately 10 percent of Russia's territory is classified as swampland.

A number of other rivers drain from Siberia's eastern and southeastern mountain ranges into the Pacific Ocean. The AMUR RIVER, which forms a long winding boundary between Russia and China, together with its main tributary, the Ussuri, drains most of southeastern Siberia.

Altogether, 84 percent of Russia's surface water is located east of the Urals, where the rivers flow through sparsely populated territory and empty into the Arctic or Pacific oceans. By contrast, the highest concentrations of population, and therefore the highest demand for water supplies, tend to have climates much warmer than those of Siberia and thus higher rates of evaporation. As a result, densely populated areas such as the Don and Kuban river basins north of the Caucasus have barely adequate water resources.

Three basins drain European Russia. The Dnepr, which flows mainly through BELARUS and Ukraine, has its headwaters in the hills west of Moscow. The 1,155-mi- (860-km-) long Don originates in the Central Russian Upland south of Moscow and then flows into the SEA OF AZOV and the Black Sea at Rostov-na-Donu. The VOLGA is the third and by far the largest of Russia's European systems, rising in the Valday Hills west of Moscow and meandering southeastward for 2,180 mi (3,510 km) before emptying into the Caspian Sea. With the addition of several canals, European Russia's rivers have long been linked together as part of a vital transportation system. The Volga system still carries two-thirds of Russia's inland water traffic.

Russia's other inland bodies of water are chiefly a legacy of extensive glaciation during the last several glacial periods. The most prominent of these bodies of fresh water is Lake Baikal, the world's oldest and deepest freshwater lake. Numerous smaller lakes dot the northern regions of the European and Siberian plains. The largest of these lakes are Beloye, Topozero, Vyg, and Il'men' in the European northwest and Lake Chany in southwestern Siberia. In European Russia, the largest lakes are Ladoga and Onega, both of which are northeast of St. Petersburg. A number of other smaller man-made reservoirs have been created on the Don, the Kama, and the Volga rivers to increase the water resources in those areas where population demands exceed natural capacity. There have also been many large reservoirs constructed on some of Siberia's rivers; the Bratsk Reservoir, for example, northwest of Lake Baikal is one of the world's largest.

CLIMATE

Because climate has played such a critical role in Russia's history and development, let alone the mental image one might have, it is important to include some of its major influences. Russia has a largely continental climate because of its sheer size and compact configuration. But weather in the Northern Hemisphere generally moves from west to east.

This means that European Russia and northern Siberia lack any topographic protection from the wintertime extremes of cold air that build in the Arctic and North Atlantic oceans. On the other hand, Russia's

mountain ranges are predominantly to the south and the east, thus blocking any moderating temperatures that might move north from the INDIAN OCEAN or onshore monsoonal flows moving inland from the Pacific Ocean. Because only small parts of Russia are south of 50 degrees north latitude and more than half of the country is north of 60 degrees north latitude, extensive regions experience six months of snow cover over subsoil that is permanently frozen to depths of several hundred meters (hundreds of feet). The average yearly temperature of nearly all of European Russia is below freezing, and the average for most of Siberia is far below freezing. The result is that most of Russia has only two seasons, summer and winter, with very short intervals of moderation between them.

The long, cold winter has a profound impact on almost every aspect of life. It affects where and how long people live and work, what kinds of crops are grown, and where they are grown (no part of the country has a year-round growing season). The length and severity of the winter, together with the sharp fluctuations in the mean summer and winter temperatures imposes special requirements on many branches of the economy.

In regions of permafrost (ground frozen throughout the year), buildings must be constructed on pilings, machinery must be made of specially tempered steel, and transportation systems must be engineered to perform reliably in extremely low and extremely high temperatures. In addition, during extended periods of darkness and cold, there are increased demands for energy, health care, and textiles. Transportation routes, including entire railroad lines, are redirected in winter to traverse rock-solid waterways and lakes.

There are some areas that represent important exceptions to this description, however. The moderate maritime climate of the Kaliningrad Oblast on the Baltic Sea has a climate similar to that of the American Northwest. And the Russian Far East, under the influence of the Pacific Ocean, has a monsoonal climate that reverses the direction of wind in summer and winter, creating sharply differentiating temperatures and extremes. There is even a narrow, subtropical band of territory on the Black Sea that is Russia's most popular summer resort area.

Because Russia has little exposure to ocean influences, most of the country receives low to moderate amounts of precipitation, a critical factor necessary for consistent agricultural production. The highest amounts of precipitation fall in the northwest, with amounts decreasing as one moves to the southeast across European Russia. The wettest areas exist as two small pockets, a lush subtropical region adjacent to the Caucasus in southern Russia and another in the Kamchatka region along the Pacific coast.

Along the Baltic coast, average annual precipitation averages around 24 in (60 cm), while in Moscow the average is about 20 in (52.5 cm). In contrast, the area near the Russian-Kazakh border in Russian Central Asia has an average of only less than 1 in (2 cm) and there are similar measurements along Siberia's Arctic coastline. Another important indicator is the average number of days of snow cover, a critical factor for agriculture. While the actual figure depends on both latitude and altitude, it generally varies from 40 to 200 days in European Russia to 120 to 250 days in Siberia.

NATURAL RESOURCES

Russia is one of the world's richest countries in raw materials, many of which are significant inputs for an industrial economy. Russia accounts for around 20 percent of the world's production of oil and natural gas and possesses large reserves of both fuels. This abundance has made Russia virtually self-sufficient in energy and a large-scale exporter of fuels. Oil and gas were primary hard-currency earners for the Soviet Union, and they remain so for the Russian Federation. Russia also is self-sufficient in nearly all-major industrial raw materials and has at least some reserves of every industrially valuable nonfuel mineral. Tin, tungsten, bauxite, and mercury were among the few natural materials that were imported during the Soviet period. Russia possesses rich reserves of iron ore, manganese, chromium, nickel, platinum, titanium, copper, tin, lead, tungsten, diamonds, phosphates, and gold, and the forests of Siberia contain an estimated one-fifth of the world's timber reserves.

The iron ore deposits close to the Ukrainian border in the southwest are believed to contain one-sixth of the world's total reserves. Intensive exploitation began there in the 1950s. Other large iron ore deposits are located in the Kola Peninsula, Karelia, south-central Siberia, and the Far East. The largest copper deposits are located in the Kola Peninsula and the Urals, and lead and zinc are found in North Ossetia.

ECONOMY

For much of the 20th century, Russia had a command economy in which the government controlled every facet of economic activity. Soviet communism forbade any private property and placed farmers in collec-

tivized farms. Since the disintegration of the Soviet Union in 1991, the Russian economy has been in a difficult transition to a more free market form.

The liberalization of the economy has produced gaping inequalities between the rich and powerful few who control Russia's industries and the rest of the population who barely make enough money for subsistence. Another limiting factor is Russian infrastructure, which dates back to the Soviet era and is well behind Western standards. Russia is heavily dependent on its exports of petroleum, natural gas, timber, and metals, a condition that leads to extreme vulnerability to dramatic market changes.

Foreign investment has been difficult to attract in Russia because of uncertainties in its banking system, because of business laws that have not kept up with Western standards, and because of excesses in government corruption. Although there is still a long way to go, 2002's growth rate of 4 percent was encouraging. However, terrorism and political uncertainties continue to raise doubts over Russia's transition to a free market system.

ECONOMIC REGIONS

The Russian Federation may be conveniently divided into 9 major economic regions: the Central European, the North and Northwest European, the Volga, the North Caucasus, the Ural, the Western Siberia, the Eastern Siberia, the Northern and Northeastern Siberia, and the Russian Far East.

The Central European area is flat, rolling country, with Moscow as its center. It forms a major industrial region for the production of trucks, ships, railway rolling stock, machine tools, electronic equipment, cotton and woolen textiles, and chemicals. The Volga and Oka rivers serve as major water routes, and the Moscow-Volga and Don-Volga canals link Moscow with the Caspian and Baltic seas. Many rail lines serve the area.

The North and Northwest European area is centered on Saint Petersburg. Here the focus is on the production of machine tools, electronic equipment, chemicals, ships, and precision instruments. The hills, marshy plains, lakes, and desolate plateaus contain rich deposits of coal, oil, iron ore, and bauxite, and the area is a prime source of lumber. The chief water routes are the Baltic-Belomor Canal and the Volga-Baltic Waterway.

The Volga region has highly developed hydroelectric power installations, including major dams at Volgograd, Kazan, Samara, and Balakovo. Farm machinery, ships, chemicals, and textiles are all manufactured here. In addition, there are extensive oil and gas fields producing in the region. Agricultural products include wheat, vegetables, cotton, hemp, oilseeds, and fruit. Livestock raising and fishing are also important.

The North Caucasus area, descending northward from the principal chain of the Caucasus Mountains, has rich deposits of oil, natural gas, and coal. The region is an important production source for farm machinery, coal, petroleum, and natural gas. The Kuban River region is one of Russia's chief granaries. Wheat, sugar beets, tobacco, rice, and sunflower seeds are grown, and cattle are also raised. Major rivers include the Don, the Kuma, and the Terek, and the Volga-Don Canal is a major transportation route.

The Ural area, the southern half of the Ural region, has been a major center of iron and steel production in addition to producing a substantial share of Russia's oil. The region also has important deposits of iron ore, manganese, and aluminum ore.

The Western Siberian region is of growing economic importance. At Novosibirsk and Kamen-na-Obi are large hydroelectric stations. The Kuznetsk Basin in the southwest is a center of coal mining, oil refining, and the production of iron, steel, machinery, and chemicals. The area is also served by the Trans-Siberian and South Siberian rail lines, with Barnaul is a major rail junction. Agricultural products include wheat, rice, oats, and sugar beets, and livestock is raised.

Eastern Siberia, with its plateaus, mountains, and river basins, is a major source of coal, gold, graphite, iron ore, aluminum ore, zinc, and lead. There is also livestock industry, but mostly of reindeer. The regions major cities (Krasnoyarsk, Irkutsk, Ulan-Ude, and Chita) are located along the Trans-Siberian Railroad. There are also hydroelectric stations at Bratsk, Krasnoyarsk, and Irkutsk.

Northern and Northeastern Siberia covers nearly half of Russian territory. This is the least populated and least developed area. The Ob, Yenisei, and Lena rivers flow to the Arctic Ocean, but because they are frozen throughout much of the year, they provide little in the way of hydropower. Through the use of atomic-powered icebreakers, the Northern Sea Route has gained increasing economic importance. The Kolyma gold fields are the principal source of Russian gold, and industrial diamonds are mined in the Sakha Republic, notably at Mirny. Fur trapping and hunting are the chief activities in the taiga and tundra regions.

The Russian Far East, which borders on the Pacific Ocean, has the major cities of Komsomolsk, Kha-

barovsk, Yakutsk, and Vladivostok. Machinery is produced, and lumbering, fishing, hunting, and fur trapping are important. The Trans-Siberian Railroad follows the Amur and Ussuri rivers and terminates at the port of Vladivostok.

BIBLIOGRAPHY. Paul Bushkovitch, *Peter the Great: Struggle for Power* (Cambridge University Press, 2001); *World Factbook* (CIA, 2004); Brian Crozier, *The Rise and Fall of the Soviet Empire* (Prima, 1999); Ronald H. Donaldson, Joseph L. Nogee, *The Foreign Policy of Russia: Changing Systems, Enduring Interests* (M.E. Sharpe, 2002); Sheila Fitzpatrick, *The Russian Revolution* (Oxford University Press, 1994); Vera Tolz, *Russia* (Arnold, 2001); Andreas Kappeler, *The Russian Empire: A Multiethnic History* (Longman, 2001); Charles E. Ziegler, *The History of Russia* (Greenwood Press: 1999); "Russia," Committee on Land Resources and Utilization (Moscow, 1996); Organization for Economic Co-operation and Development, "OECD Economic Surveys: The Russian Federation 1995," (Paris, 1995).

RICHARD W. DAWSON
CHINA AGRICULTURAL UNIVERSITY

Ruwenzori Mountains

SHROUDED IN MISTS sitting under the hot sun of the equator is a mountain range believed to be the mysterious "Mountains of the Moon" spoken of by the ancient Greeks. Their discovery is actually credited to Henry Morton Stanley of Britain, who in 1888, while on expedition, glanced up one day and saw what the native porters at the time believed to be a mountain covered in salt.

The Ruwenzori Mountains are located on the border of UGANDA and the Democratic Republic of CONGO. Sitting between lakes Albert and Edward, they are the highest mountain range in Africa. Ten of the Ruwenzori summits are over 15,750 ft (4,800 m); the highest are Mt. Margherita, at 16,763 ft (5,109 m), and Mt. Alexandra, at 16,750 ft (5,105 m).

These mountains are a fault-block range composed of ancient crystalline rock. The Kilembe copper mine is located in the eastern foothills of the Ruwenzoris in Uganda. The vegetation of this area is profuse and prolific because of the ever-pervading cloud cover, which harbors drenching mists and rain. This Afro-alpine foliage is some of the most highly studied botany on the continent of Africa because of its varying species and vegetation belts as one climbs higher up the peaks.

ALPS OF AFRICA

Many of these mountains, called "the Alps of Africa," contain high mountain glacial lakes, and several have glaciers at their peaks. These glaciers have recently been the center of several studies because of their rapid rate of decline and concerns over the loss of snow on the mountain peaks. These studies have shown that some of the glaciers have receded over 1,000 ft (300 m) in the past decade alone. They also forecasted that at the present rate of global warming, all of the glaciers of the Ruwenzoris could be melted completely by 2025.

These mountains were formed in the western region of the Great Rift Valley between 40 and 10 million years ago, when the crust of the east African plateau started on a faulting process. This has taken place very slowly, only millimeters a year, to form these snow-capped equatorial mountains. Ruwenzori, unlike other surrounding areas, is nonvolcanic. These mountains are tilt-block, with steep faces on the western rift and gentler slopes running east. When these mountains tilted up out of the plains of Africa, they carried with them the metamorphosed volcanic rock with which they were formed.

BIBLIOGRAPHY. Guy Yeoman, *Africa's Mountains of the Moon: Journeys to the Snowy Source of the Nile* (Universe Books, 1989); Curtis Abraham, "Going, Going, Gone," *New Scientist* (v.176/2367); Christian Amodeo, "African Glaciers in Retreat," *Geographical* (v.75/12); Saul B. Cohen, ed., *The Columbia Gazetteer of the World* (Columbia University Press, 1998).

CHRISTY A. DONALDSON
MONTANA STATE UNIVERSITY

Rwanda

Map Page 1114 **Area** 16,365 square mi (26,338 square km) **Population** 7,810,056 (2004) **Capital** Kigali **Highest Point** 14,826 ft (4,519 m) **Lowest Point** 3,100 ft (950 m) **GDP per capita** $1,300 **Primary Natural Resources** gold, cassiterite (tin ore).

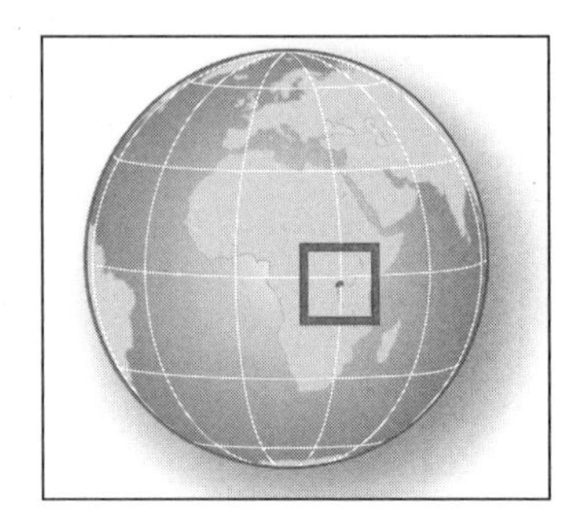

RWANDA, LOCATED in Central Africa, is a landlocked country of savanna GRASSLAND with a population that is predominantly rural. It is bordered by BURUNDI, the Democratic Republic of the CONGO, TANZANIA, and UGANDA.

The terrain is generally grassy uplands and hills with mountains extending southeast from a chain of volcanoes in the northwest. The climate is temperate with two rainy seasons from February to April and November to January. Frost and snow can occur in the mountainous regions. The country is troubled by the prospect of deforestation that is occurring as a result of unimpeded cutting of trees for fuel, overgrazing by livestock, overuse of farm land, and erosion. Additional environmental and climate difficulties include poaching, droughts, and volcanic activity in the Virunga Mountains.

The ethnic and cultural makeup of the Rwandan people is: Hutu, 84 percent; Tutsi, 15 percent; and Twa Pygmoid 1 percent. Rwanda is the most densely populated country in Africa. Sixty percent of the population lives below the poverty line. Estimates for life expectancy in Rwanda take into account the effects of excess mortality rates from an uncontrolled AIDS epidemic, which significantly lowers the life expectancy, causes higher infant mortality and death rates, and lowers the population and growth rates.

This has changed the distribution of population by age and sex. The current life expectancy for the total population is 39 years. Roman Catholicism is the primary religion, practiced by 56 percent of the population; Protestant, 26 percent; Adventist, 11 percent; Muslim, 4.6 percent; indigenous beliefs 0.1 percent; and 1.7 percent claim no religious beliefs. Rwanda is a poor rural country with about 90 percent of the population engaged in subsistence agriculture. The main industries include cement production, agricultural products, small-scale beverages, soap, furniture, shoes, plastic goods, textiles, and cigarettes. Agriculture products that are produced include coffee, tea, pyrethrum (insecticide made from chrysanthemums), bananas, beans, sorghum, potatoes, and livestock. Products that are used for foreign trade are coffee and tea. The 1994 war and genocide destroyed Rwanda's already delicate economy and even further served to impoverish the population. Rwanda has made little progress in economic recovery. Despite Rwanda's fertile land, food production often does not meet the needs of population growth.

Three years before its independence from BELGIUM, in 1959, the Hutus, who are the majority ethnic group, overthrew the ruling Tutsi king. Over the next few years, thousands of Tutsis were killed, and 150,000 were driven into exile and escaped to neighboring countries. The children of these exiles later formed a rebel group, the Rwandan Patriotic Front, and started a civil war in 1990. This war, along with political and economic instability, aggravated ethnic tensions, resulting in the April 1994 genocide of roughly 800,000 Tutsis and Hutus. The Tutsi rebels defeated the Hutu regime and ended the killing in July 1994. This time, 2 million Hutus, many fearing Tutsi retribution, fled to neighboring Burundi, Tanzania, Uganda, and other countries. Subsequently, many of the refugees have returned to their homes.

Despite substantial international assistance and political reforms, including Rwanda's first local elections in 1999 and presidential and legislative elections in 2003, the country continues to be divided. Tutsi, Hutu, Hema, Lendu, and other conflicting ethnic groups, political rebels, armed gangs, and various governmental forces continue to fight, crossing into the borders of Burundi, the Democratic Republic of Congo, and Uganda to gain control over populated areas and natural resources. In spite of government and United Nations efforts to end the conflicts, localized violence continues.

BIBLIOGRAPHY. *World Factbook* (CIA, 2004); *Countries of the World and Their Leaders Yearbook* (Gale, 2005); *Europa World Year Book* (Europa Publications, 2004).

CLARA HUDSON
UNIVERSITY OF SCRANTON

Made in the USA
San Bernardino, CA
29 March 2017